34th ICAF Conference and 28th ICAF Symposium (ICAF 2015)

Helsinki, Finland
1 – 5 June 2015

Volume 1 of 2

Editor:

Aslak Siljander

ISBN: 978-1-5108-5774-2

Printed from e-media with permission by:

Curran Associates, Inc.
57 Morehouse Lane
Red Hook, NY 12571

Some format issues inherent in the e-media version may also appear in this print version.

Printed by Curran Associates, Inc. (2018)

For permission requests, please contact VTT Information Service
at the address below.

VTT Information Service
PO Box 1000
FI-02044 VTT
Finland

Phone: +358 20 722 111
Fax: +358 20 722 7001

info@vtt.fi

Additional copies of this publication are available from:

Curran Associates, Inc.
57 Morehouse Lane
Red Hook, NY 12571 USA
Phone: 845-758-0400
Fax: 845-758-2633
Email: curran@proceedings.com
Web: www.proceedings.com

ICAF 2015

Aslak Siljander *Editor*

ICAF 2015 Structural Integrity: Embracing the future – respecting the past; supporting aging fleets with new technologies

Proceedings of the 34th Conference and the 28th Symposium of the International Committee on Aeronautical Fatigue and Structural Integrity, 1–5 June 2015, Helsinki, Finland

ICAF
International Committee on Aeronautical Fatigue and Structural Integrity

VTT

Editor

Aslak Siljander
VTT Technical Research Centre of Finland Ltd
B.O.Box 1000
FI-02044 VTT, Finland
Telephone
Fax
Email: aslak.siljander@vtt.fi

Espoo, May 2015

General Information

ICAF General Secretary

Prof. Anders Blom
Defence & Security, Systems and Technology
Swedish Defence Research Agency (FOI)
SE-164 90 Stockholm, Sweden

ICAF National Delegates

AUSTRALIA
Phil Jackson
Airworthiness and Life Evaluation Group
Defence Science and Technology Organisation
506 Lorimer Street
Fishermans Bend VIC 3207, Australia

CANADA
Nick Bellinger
National Research Council Canada
1200 Montreal Road, Bldg. M-14
Ottawa, Ontario K1A 0R6, Canada

FINLAND
Dr. Aslak Siljander
VTT Technical Research Centre of Finland Ltd
Machines and Vehicles
P. O. Box 1000
FI-02044 VTT, Finland

FRANCE
Dr. Thierry Ansart
Structures Division
DGA Aeronautical Systems
BP 93123, 47 rue Saint Jean
31131 Balma, France

GERMANY
Dr. Claudio Dalle Donne
EADS Innovation Works
81663 München, Germany

ISRAEL
Dr. Yuval Freed
Head of F&DT Dept., Engineering & Development Group
Israel Aerospace Industries
Tel Aviv, Israel

ITALY
Prof. Luigi Lazzeri
Universita di Pisa
Department of Aerospace Engineering
Via G. Caruso, 8
56122 Pisa, Italy

JAPAN
Prof. Nobuo Takeda
University of Tokyo
Dept. of Aeronautics and Astronautics
7-3-1 Hongo, Bunkyo-ku
Tokyo 113-8656, Japan

POLAND
Dr. Antoni Niepokólczycki
Director, Materials and Structures Research Center (MSRC)
Institute of Aviation
Al. Krakowska 110/114
02-256 Warsaw, Poland

SWEDEN
Hans Ansell
Saab Aerosystems
Saab AB
SE-581 88 Linköping, Sweden

SWITZERLAND
Dr. Michel Guillaume
ZHAW Zurich University of Applied Sciences
Head of the Center for Aviation
Technikumstrasse 9
8400 Winterthur, Switzerland

THE NETHERLANDS
Marcel J. Bos
National Aerospace Laboratory (NLR)
P.O. Box 153
NL-8300 AD Emmeloord, The Netherlands

UNITED KINGSOM
Dr. Steve Reed
Technical Advisor Structural Integrity
Physical Sciences Department
Defence Science and Technology Laboratory
Desk 236, Room 102, Bldg 5 (Porton Down)
Salibury, Wilts SP4 0JQ, UK

UNITED STATE OF AMERICA
Dr. Ravinder Chona
Structural Sciences Center
US Air Force Research Laboratory (AFRL / RBSM)
Wright-Patterson Air Force Base
Ohio, USA

Organisation

Aslak Siljander
Conference & Symposium Chair
VTT

Management Team, ICAF 2015 Organizing committee

Tuomo Hokkanen, VTT Technical Research Centre of Finland Ltd
Paula Kauppinen, VTT Technical Research Centre of Finland Ltd
Satu Koskela, VTT Technical Research Centre of Finland Ltd
Maria Rajasalmi, VTT Technical Research Centre of Finland Ltd
Aslak Siljander, VTT Technical Research Centre of Finland Ltd
Tomi Viitanen, VTT Technical Research Centre of Finland Ltd
Christina Vähävaara, VTT Technical Research Centre of Finland Ltd

Technical Program Committee

Captain (eng) Kalle Vaaraniemi, Finnish Air Force Command (AFCOMFIN)
Mr. Ari Kivistö, Finnish Defence Forces Logistics Command, Joint Systems Centre, Aeronautical Systems Division
Mr. Jukka Raunio, M.Sc., Vice President, Engineering, Patria Aviation
Mr. Mika Mustakangas, M.Sc., R&D Project Manager, Patria Aerostructures
Mr. Sakari Pasanen, Director, Business Development, Patria Aerostructures
Mr. Jaakko Sotkasiira, M.Sc., Finflo Ltd.
Dr. Timo Siikonen, Professor, Finflo Ltd.
Dr. Olli Saarela, Professor, Aalto University
Dr. Ari Visa, Professor, Tampere University of Technology
Mrs. Liisa Poussa, M.Sc., Research Team Leader, VTT Technical Research Centre of Finland Ltd
Mrs. Eetta Saarimäki, M.Sc., Senior Scientist, VTT Technical Research Centre of Finland Ltd
Mr. Tomi Viitanen, M.Sc., Senior Scientist, VTT Technical Research Centre of Finland Ltd
Mr. Risto Laakso, M.Sc., Senior Scientist, VTT Technical Research Centre of Finland Ltd
Mr. Keijo Koski, M.Sc., Principal Scientist, VTT Technical Research Centre of Finland Ltd
Mr. Kari Lahdenperä, Lic.Tech., Principal Scientist, VTT Technical Research Centre of Finland Ltd
Dr. Aslak Siljander, Principal Scientist, VTT Technical Research Centre of Finland Ltd

Preface

ICAF is coming to Helsinki, Finland, for the first time in the history of ICAF. The 34th Conference and the 28th Symposium of the International Committee on Aeronautical Fatigue and Structural Integrity is taking place in Helsinki from the 1st to the 5th of June 2015. ICAF 2015 is hosted and organized by VTT Technical Research Centre of Finland Ltd and its venue is Marina Congress Center in downtown Helsinki.

The International Committee on Aeronautical Fatigue and Structural Integrity (ICAF) was formed in 1951 under the name "The International Committee on Aeronautical Fatigue" in response to growing concerns regarding fatigue problems in metal aircraft structures. In 2010 the name was changed to the present one in order to clarify that the scope of the committee had broadened over the years and now also includes topics such as damage formation and growth in composite structures, structural health and loads monitoring, probabilistic modelling of structural integrity, corrosion control, engine integrity etc. The acronym ICAF was maintained. The stated aims of ICAF are to encourage contacts between people actively engaged in aircraft structural integrity problems and to exchange information, experience, opinions and ideas concerning aeronautical fatigue and fatigue-related subjects. To this end, a Conference and, since 1959, a Symposium are organized every two years for attendance by representatives of industry, universities and institutes, military specialists, regulatory agencies and aircraft operators throughout the world.

ICAF 2015 begins with a two-day Conference, June 1–2, consisting of presentations of the National Reviews given by the National Delegates from the fourteen member states of ICAF. The fourteen national reviews will describe research, development and airworthiness activities of international interest which has taken place in each country during the two years since the previous ICAF Conference in Jerusalem. In addition, I am indeed pleased to welcome the national reviews from three important new countries: Brazil, China and Russia.

The ICAF 2015 Conference will be followed by a three-day Symposium, June 3–5. The Symposium's main theme – *"Embracing the future – respecting the past: supporting aging fleets with new technologies"* – highlights that the global air travel is increasing, yet the structural safety must be guaranteed, new materials are increasingly applied (e.g. composites), new manufacturing methods are introduced (e.g. Additive Manufacturing), many experts retire, young researchers both scientists and engineers graduate and enter the industry at large. Furthermore, nations and end-users may not have the funding to replace their fleets as planned, meaning that life extension programs with associated risk analyses may need to be considered. In any case, the gap between theory and practice must be bridged also in the future of civil and military aviation.

Each ICAF Symposium since 1951 begins with a lecture honoring the memory of Dr. Frederick J. Plantema, the founder of ICAF. At the 28th Symposium, the Plantema Lecture will be delivered by Jerzy Komorowski, whose lecture is entitled *"Structural Integrity – the unfinished business"*. Mr. Komorowski, previously the Canadian National Delegate to ICAF (2001–2011), is a member of the Canadian Aeronautics and Space Institute; the Canadian member of the NATO Science and Technology Organization Advanced Vehicle Technology Panel (AVT); and Chair of the AVT Support Committee. He serves as the Canadian National Resource Specialist at the Aerospace Engineering Group of The Technical Cooperation Program.

The Jaap Schijve Award and Lecture will be awarded to a young researcher, selected by a special committee, for his or her outstanding performance. A total of 42 lectures including the Plantema Lecture and the Jaap Schijve Award Lecture and 57 posters are presented in three days. All the papers are of high technical quality and only the limited time prevents all papers from being orally presented.

All the oral and poster presentations will be delivered on a USB stick only. The ISBN-endowed USB stick is the ICAF 2015 Proceedings for the first time. As all the papers are of high quality, the reader, intentionally, cannot distinguish whether the paper was an oral or a poster presentation.

The year 2015 marks the 20th anniversary at VTT of aircraft structural integrity program (ASIP) related applied research activities aimed at serving the needs of the Finnish Air Force (FINAF) and the national research network described in more detail in the Finnish National Reviews. Finland started seeking the ICAF membership in 2001 and became the 13th full member in 2005. Without the in-country ASIP research activities, solely orchestrated and funded by the FINAF during the past 20 years, Finland would not be an ICAF member. On behalf of the in-country ASIP network and VTT in particular, I sincerely thank the FINAF for their vision, initiative, enthusiasm, diligence and trust over these years.

Without good teamwork the organization of a major international conference would not have been possible. The ICAF 2015 in-country Technical Program Committee, responsible for reviewing and ranking the abstracts, consisted of the following individuals: Captain (eng) Kalle Vaaraniemi (Finnish Air Force Command), Mr. Ari Kivistö (Finnish Defence Forces Logistics Command, Joint Systems Centre, Air Systems Division), Mr. Jukka Raunio (Patria Aviation),

Mr. Mika Mustakangas and Mr. Sakari Pasanen (Patria Aerostructures), Dr. Timo Siikonen and Mr. Jaakko Sotkasiira (Finflo Ltd.), Dr. Olli Saarela (Aalto University), Dr. Ari Visa (Tampere University of Technology), Mrs. Liisa Poussa, Mrs. Eetta Saarimäki, Mr. Tomi Viitanen, Mr. Risto Laakso, Mr. Keijo Koski and Mr. Kari Lahdenperä (VTT Technical Research Centre of Finland Ltd).

I am greatly indebted to the Organizing Committee (Ms. Maria Rajasalmi, Ms. Christina Vähävaara, Ms. Paula Kauppinen, Ms. Satu Koskela, Mr. Tuomo Hokkanen and Mr. Tomi Viitanen, all with VTT Technical Research Centre of Finland Ltd) for their invaluable assistance in shaping the Conference and Symposium program.

Financial support for ICAF 2015 has been received from various organizations. Lockheed Martin (Gold Partner), Patria (Bronze Partner), Aerospace Testing International (Media Partner), Fatigue Technology (Exhibitor), nCode (Exhibitor), HBM (Exhibitor), Curtiss-Wright (Exhibitor), SwRI and NLR have stepped forward as Conference and Symposium Partners. The contribution of the city of Helsinki for hosting the Reception is also greatly appreciated.

Finally, I would like to thank all the authors and their organizations, ICAF's General Secretary and my fellow ICAF National Delegates for their valuable contributions to the success of ICAF 2015. Without their contributions the proceedings would have never been published.

Aslak Siljander
ICAF National Delegate Finland
ICAF 2015 Chairman

Table of Content

Advanced analytical, numerical and experimental methods

Fatigue prediction of additive manufactured Ti-6Al-4V for aerospace: Effect of defects, surface roughness

Daniel Greitemeier[1], Vitus Holzinger[1], Claudio Dalle Donne[1], Jens Eufinger[2], Tobias Melz[2]

[1] *Airbus Group Innovations, 81663 Munich, Germany*

[2] *Fraunhofer Institute for Structural Durability and System Reliability, Bartningstr. 47, 64289 Darmstadt, Germany*

Abstract: Ti-6Al-4V is the most common used titanium alloy in aerospace. Additive manufacturing (AM) allows to build these Ti-6Al-4V parts, rapidly and directly from computer-aided design information, with better material utilisation and reduced lead time, compared to wrought or cast material. However the understanding of reliability in terms of mechanical properties is still at an early stage. Therefore, this study should give first insights on fatigue properties of direct metal laser sintered (DMLS) material, with special focus on defects and surface roughness influences. Fracture mechanical approaches are used to estimate the influence of defects based on data obtained from computer tomography (CT) scans, as well as the influence of process inherent surface roughness by the use of an equivalent initial flaw size concept. It is shown, that the prediction is conservative for defects, while the estimated results of rough surfaces is in good agreement with the experimental determined values.

INTRODUCTION

Additive manufacturing (AM), often referred as additive layer manufacturing or 3D printing, is a process of building parts layer by layer from computer-aided design information. In this study a powder bed processes is investigated: Ti-6Al-4V powder is selectively melted by a laser (DMLS). The main advantages of this AM approach for the aerospace industry are better material utilisation, lead time reduction and weight saving due to topology optimisation. Despite the high potential for aerospace applications, the current uncertainty of the mechanical properties has not been investigated thoroughly yet.

Fatigue performance of surface milled AM specimens, show to some extend, large scatter within the S-N curves [1-3]. Beside various factors, process inherent defects seems to be the main reason [4]. Hot isostatic pressing (HIP) therefore is applied to increase the fatigue performance and decrease the scatter due to a reduced defect size [4, 5]. The values are reported to be comparable to wrought material. A recent study displays the influence of process inherent surface roughness on the fatigue performance. A significant lower endurance limit compared to polished specimens was determined [6] because of cracks nucleating from the surface asperities.

Mastering the influence of defects and surface roughness from a damage tolerance assessment point of view, will contribute to understand the mechanisms leading to low fatigue performance and scatter of AM Ti-6Al-4V.

In the framework of a lager effort undertaken in Airbus Group to qualify AM for aerospace parts, the fatigue performance of Ti-6Al-4V samples, produced by DMLS, with milled and rough surfaces (as-built), were investigated. The aim is to offer a better experimental documentation, as well as to elaborate a practical engineering approach to estimate the fatigue performance of defect afflicted and as-built specimens, based on linear elastic fracture mechanics (LEFM).

EXPERIMENTAL METHODS

Manufacturing process

The builds are carried out by direct metal laser sintering (DMLS) on an EOSINT M 280 (EOS) platform, with 370 W, a scan speed of $\leq$ 7000 mm/s and a layer thickness of 60 µm. A shell and core concept is used, rotating each layer by an angle of 30°. The builds take place under argon atmosphere to avoid e.g. oxygen contamination of the melt pool. The average chamber temperature is ~35°C.

Post treatments

All specimens are stress relieved under vacuum at 710°C for 2 h, followed by furnace cooling under argon atmosphere. Fatigue specimens are tested with process inherent surface roughness or machined surfaces. Final machining takes place after the heat treatment.

Mechanical characterisation

For all specimens, the load is applied perpendicular to the layer deposition direction. The tests take place at ambient temperature.

Axial fatigue tests are performed in accordance to EN 6072 on a resonance test machine, type Microtron (Rumul), under constant load (R: 0.1). Figure 1 shows the used specimen geometry.

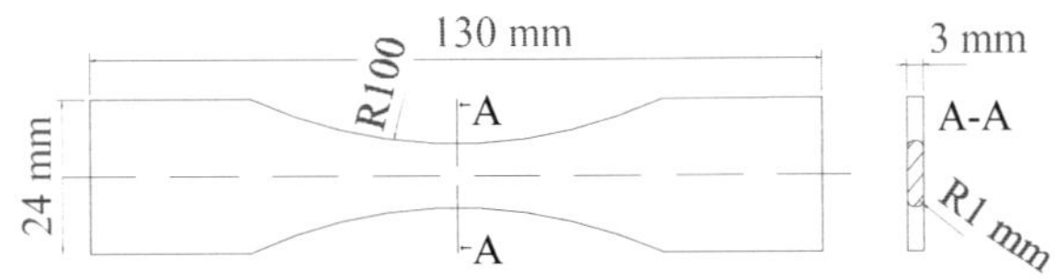

Figure 1. Axial fatigue specimen.

Fatigue crack growth tests are performed on a servo hydraulic test device, type PC 160N (Schenck), in accordance to ASTM E 647. C (T) 40 specimens (Figure 2) are used under a constant load ratio of 0.1. The threshold is determined according to ASTM E 647 at 10^{-10} m/cycle.

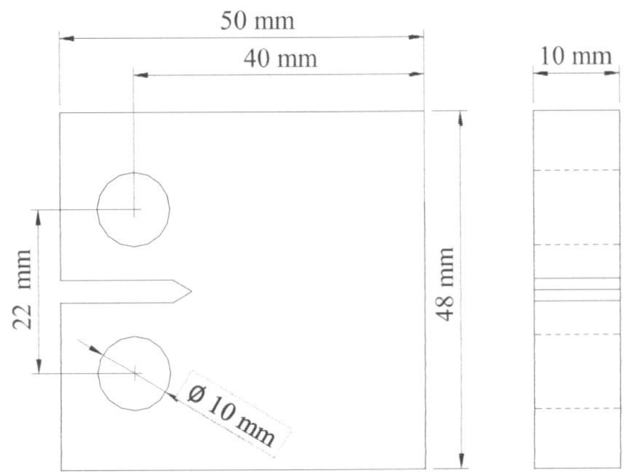

Figure 2. Fatigue crack growth specimen.

Non destructive testing

Defects are determined by CT-scans with the use of a RayScan 150 with Perkin Elmer detector (1024 x 1024 à 200 µm DRZ, 16 Bit). The scans are analysed with the software VG Studio max (Volume Graphics). The surface roughness is determined with a tactile measuring device, type DektakXT (Bruker), according to EN ISO 4288. The average value of three measurements, describes the maximum height of the profile (R_t).

ESTIMATION MODELS

Two different models are proposed to estimate the influence of defects (e.g. pores or lack of fusion defects) and process inherent surface roughness. The goal is to create a simple engineering approach to estimate the fatigue performance of AM Ti-6Al-4V. To keep both models as simple as possible, the following assumptions are made:

- Predictions are based on linear elastic fracture mechanics.
- Defects and surface roughness are described as a crack.
- The crack nucleation phase is neglected.

Influence of defects

The aim of the model is to estimate the influence of defects on the fatigue properties based on CT-scans. Each step, starting from the calibration of the model followed by CT-investigations and defect analysis for prediction, as described below.

Model calibration. The fracture mechanical model is calibrated based on experimentally determined S-N curves. The initial defect area is measured at the fracture surface and transferred into a 2D-model (Figure 3). This post-diction allows to establish the accuracy of the approach.

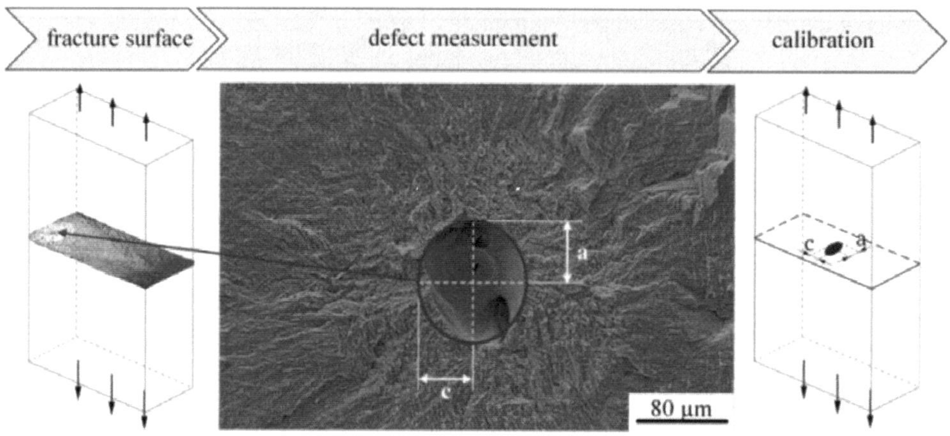

Figure 3. Defect measurement based on experimental determined results and transfer into a 2D defect.

The AFGROW fatigue crack growth analysis software tool [7] is used for the predictions. In this tool three defect locations can be selected as shown in Figure 4. The parameters a and c therefore describe the size and geometry of the defect.

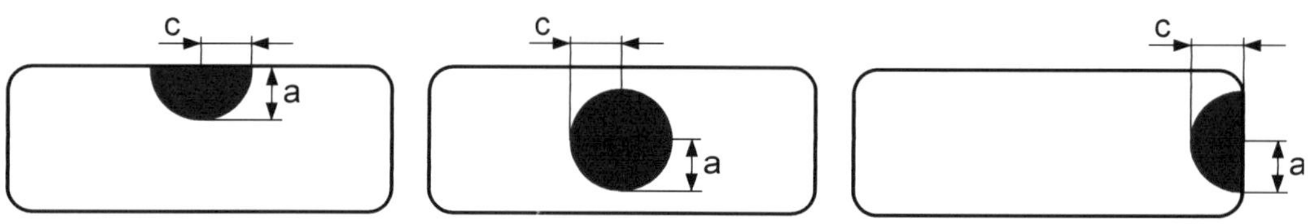

Figure 4. Possible defect locations within the model.

Non destructive testing. The CT scans serve as a basis for the model. Therefore, the defect geometry, the defect area and the amount of defects are analysed. For simplicity reasons the defect geometry is approximated as an ellipitical projected area perpendicular to the applied load. The maximum projected area of a defect is regarded to be the critical 2D-crack. A Monte-Carlo simulation is used to account for the defect distribution in the volume of the specimen.

Defect analysis. The above mentioned Monte-Carlo simulation is used to vary the defect distribution within the investigated volume. In this way, 1000 specimens with different defect distributions are generated. For each defect, a stress intensity factor (SIF) is then calculated based on an approach from Murakami [8]. It accounts for the different SIF depending on the defect location (surface or internal defect).The dominant features are then used for the prediction. Defect interactions are not taken into consideration.

Prediction. The prediction is based on the dominating defects from the Monte-Carlo simulation and a scatter of the da/dN-ΔK curve of ~20%. All defects are described by a semi-elliptical surface flaw. The determined S-N curve therefore describes the probability of failure for AM Ti-6Al-4V based on the CT-scans.

Influence of process inherent surface roughness

The model is based on an equivalent initial flaw size (EIFS) approach, to estimate the total fatigue life. In short, this is a method where found cracks are analytically grown backwards to time equal zero (time or cycles) to determine an initial flaw, referred to as an EIFS. If a number of cracks are grown backwards a distribution of the EIFS can be determined. In this work a deterministic approach is used for simplicity. The goal is to link the experimentally determined surface roughness to an EIFS, which can be used for fatigue life estimations of AM parts with as-build surfaces.

RESULTS

Non destructive testing

Defects as well as surface roughness can influence the mechanical properties. For a fracture mechanical evaluation it is essential to describe the influential factors as exact as possible. The results of the CT-scan and roughness measurements are therefore summarised below. Figure 5 shows the typical inherent surface roughness of a DMLS specimen. A maximum profile height (R_t) of ~109,8 µm ± 0.5 % is determined perpendicular to the layer direction [9]. In comparison the milled specimens showed a maximum profile height of ~1.6 µm ± 13%.

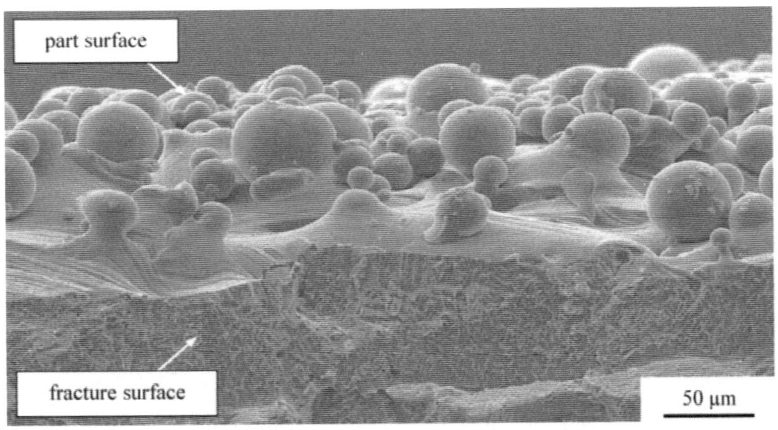

Figure 6. Surface roughness of DMLS Ti-6Al-4V.

Figure 6 displays the maximum defect diameter with regards to the total detected number. Within the volume of 323 mm³, 133 defects are detected. More than 90% of the defects show a diameter less than 150 µm. The maximum determined diameter is 240 µm.

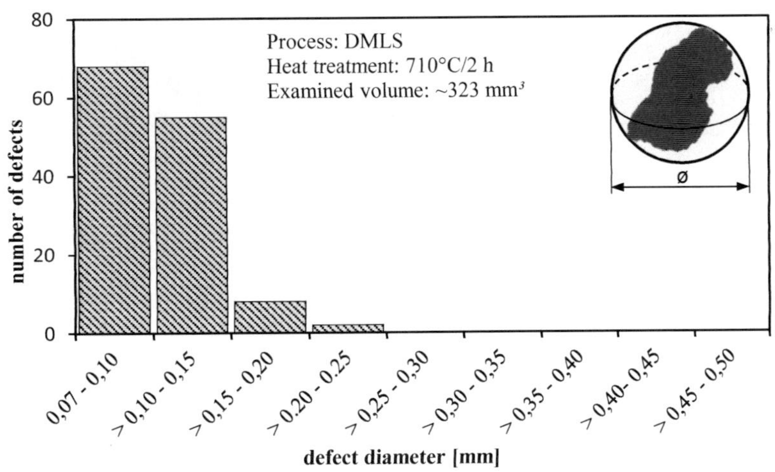

Figure 6. Defect diameter distribution based on CT-scans of the investigated DMLS material.

Additionally, data evaluation of the aspect ratio of defects can be seen in Figure 7. The aspect ratio is calculated based on the quotient of the longest and shortest defect diameter in XY, XZ or YZ plane. The maximum aspect ratio is 2,4:1.

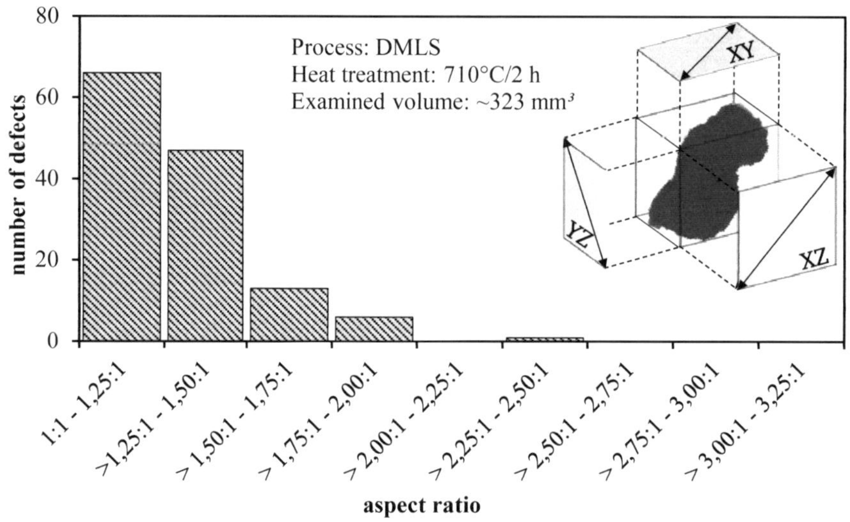

Figure 7. Aspect ratio of defects based on CT-scans from the investigated DMLS material.

Destructive testing

Fatigue crack growth of the heat treated DMLS specimens are measured and plotted logarithmically as a function of the stress intensity range (Figure 8). For comparison reasons a heat treated (730°C/2 h) Ti-6Al-4V design curve (plate material) is plotted in Figure 8 as well. Differences can be seen between plate [10] and DMLS Ti-6Al-4V [9], in the threshold range (ΔK_{th}). With an average value of ΔK_{th}=3.2 MPa√m, the DMLS specimens showed higher resistance against crack growth than the reference plate material (ΔK_{th}=1.8 MPa√m).

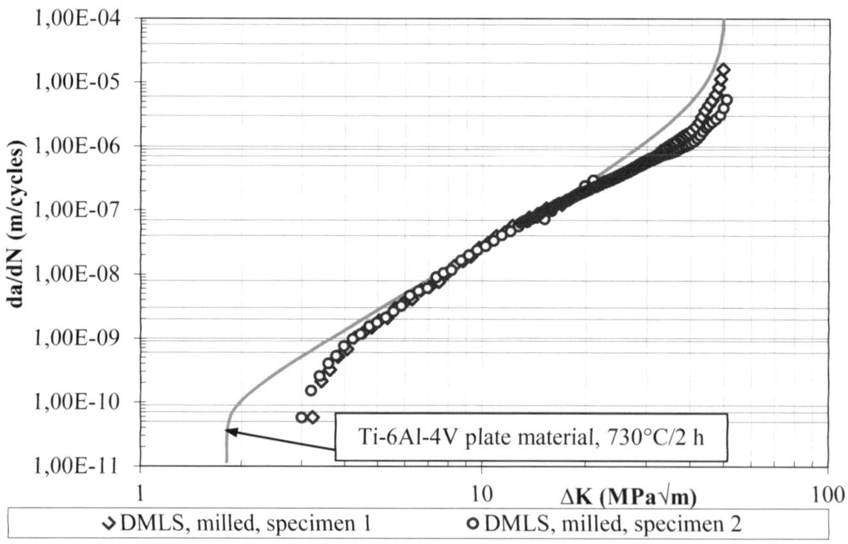

Figure 8. Fatigue crack growth properties of DMLS Ti-6Al-4V [9], specimen geometry: C (T) 40, build direction: XZ, heat treatment: 710°C/2 h, surface: milled, r-ratio: 0.1, compared to plate material [10].

Besides the fatigue crack growth measurements, S-N data are determined to validate the models. Specimens with milled and as build surfaces are tested with a r-ratio of 0.1 (Figure 9). Run outs are marked by an arrow. The S-N curves are determined according to Weibull with a failure probability of 50%. For comparison reasons scatter bands of cast and wrought material are plotted.

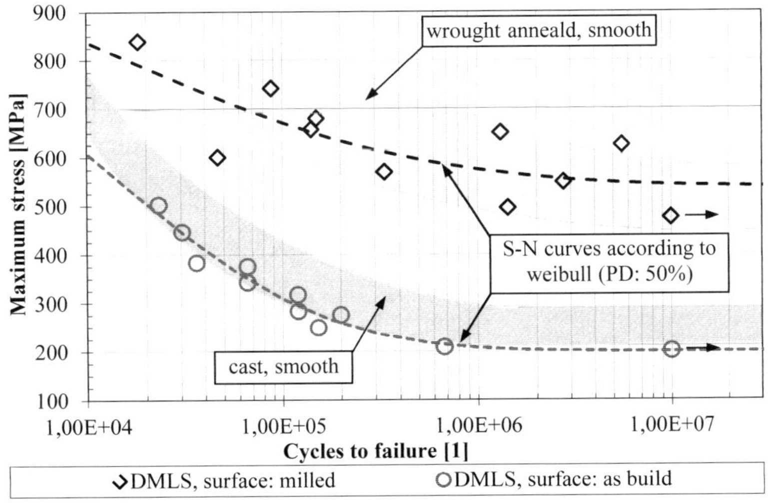

Figure 9. Fatigue properties of DMLS Ti-6Al-4V with milled and as build surfaces [9], S-N curves according to weibull (PD: 50%), build direction: ZX, HT: 710°C/2 h, r-ratio: 0.1, compared to wrought and cast material [11].

Influence of defects. The deterministic calculations are based on the crack nucleating defect size measured on the fracture surface, its geometry as well as on the determined crack propagation curves. By comparison of the experimental and estimated fatigue performance, a linear regression of 0.9943 is achieved up to cycles to failure of $1.5*10^6$ (Figure 10).

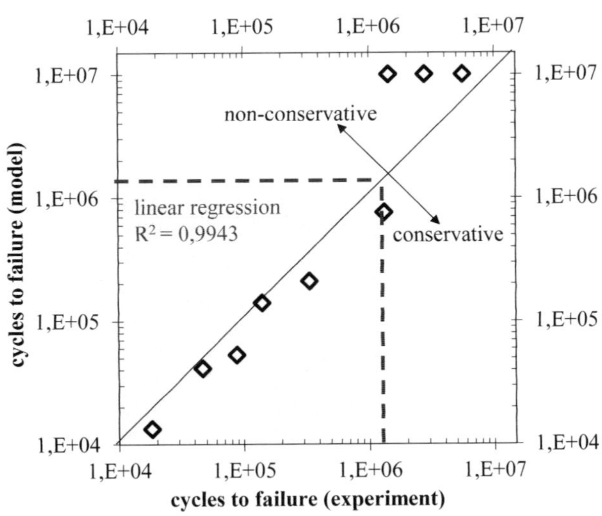

Figure 10. Calibration results for DMLS Ti-6Al-4V based on experimental examined fatigue properties in comparison to the predicted results, linear regression (R^2) without estimated run outs.

The estimated values are conservative compared to the experimental results, up to a limit of $1.5*10^6$ cycles. Additional, numerical tests are performed to determine the sensitivity of the model, based on S-N curves with 50% probability to failure. The defect diameter is varied in each direction by ±2 µm and the da/dN-ΔK curve by ±20%. By superimposing both factors of uncertainty a range of scatter, within the fatigue limit of 48 MPa, is observed. It is therefore assumed that the prediction is robust enough for estimations based on CT-scans. However, two simplifications are made:

1. The aspect ratio of defects is set to 1:1. Numerical tests with the largest defect detected (240 µm) and the maximum aspect ratio (2.5:1) are performed to verify this simplification. The fatigue limit is calculated to be 138 MPa for an aspect ratio of 1:1. By applying the extremes, the fatigue limits vary between 130–170 MPa. Therefore, the aspect ratio of 1:1 delivers conservative values in the majority of cases.

2. The defect area is based on the maximum projected diameter, perpendicular to the load direction. Due to this circumstance an overestimation of the defect sizes takes place.

The prediction results depending on the CT-scans, are shown in Figure 11.

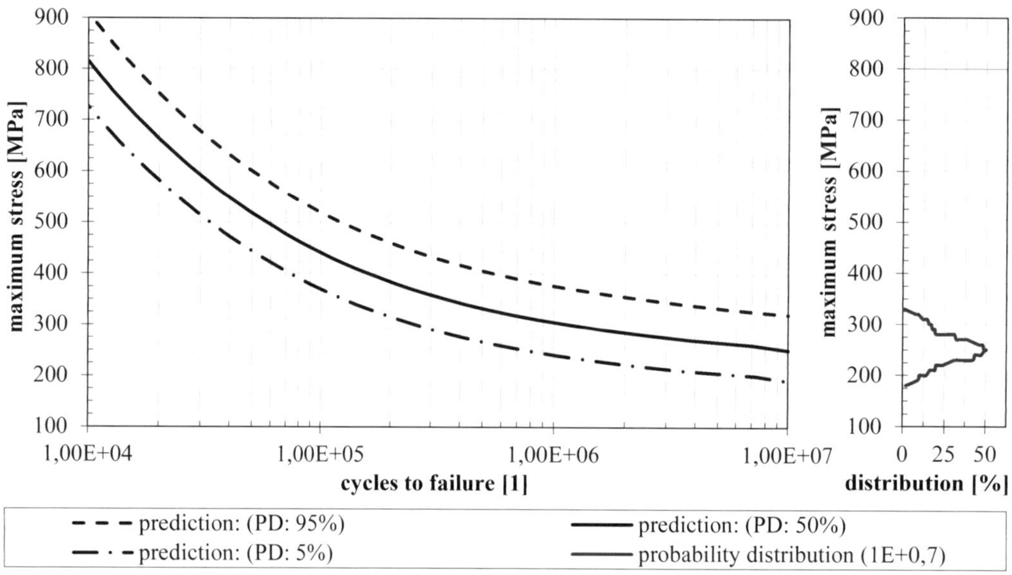

Figure 11. Prediction of the cycles to failure (R=0.1) of DMLS Ti-6Al-4V specimens, based on computertomography investigations with the distribution of the probability of failure (1.00E+7).

As expected due to the simplifications, the prediction is conservative compared to the calibration results based on defects measured at the fracture surface (Figure 10).

Influence of surface roughness. The fatigue performance of specimens with surface roughness can completely be described by a semi elliptical surface flaw (EIFS) with a radius of 251 ±111 μm (Figure 12). In comparison data from Wycisk et al. [6, 12] are shown, for DMLS Ti-6Al-4V with slightly lower surface roughness (R_t: 90 μm).

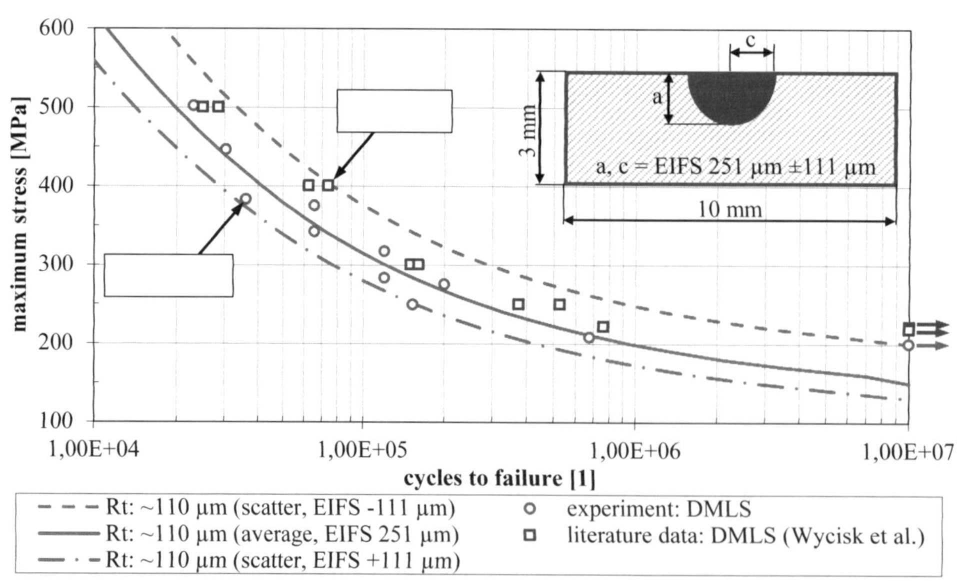

Figure 12. Description of the process inherent surface roughness of DMLS Ti-6Al-4V by an EIFS compared to literature data [6, 9, 12].

DISCUSSION

Fatigue performance

The fatigue performance of DMLS Ti-6Al-4V is influenced to a large extend, by stress raisers like notches, surface roughness or defects. The CT-scans reveals defects with a maximum size of 240 µm whereas the EIFS of the specimens with process inherent surface roughness has an average value of 251 µm. An influence of defects on the fatigue performance can therefore not be excluded completely for rough specimens. However, the probability of such cases is assumed to be small, since such large pores are scarce. Moreover, stress raiser close to the surface are more critical than internal cracks, in fact all cracks nucleated from the surface asperities in the as-build specimens.

In the absence of process inherent surface roughness, for example due to machining, internal defects dominate the fatigue performance and scatter.

Estimations based on defects

By considering the limits of LEFM, the predicted values of the calibration are in good agreement with the experimentally determined fatigue performance. However, the prediction based on CT-scans leads to more conservative values, as a result of the over-estimated defect sizes. The variations of the defect distribution, based on the Monte-Carlo simulation, enable to estimate the scatter of fatigue life. The reliability of the predicted scatter could be further improved by additional CT-scans from different build jobs.

Estimations based on surface roughness

The estimated fatigue performance is based on an EIFS approach. The results indicate, that an estimation of the fatigue performance of rough surfaces is possible by an EIFS and the use of fracture mechanics in accordance to Takahashi and Murakami [13], who related surface roughness values to an EIFS. It should be mentioned however, that the EIFS was calculated for rough specimens under constant amplitude loading and low variability of surface roughness. Further studies with different surface roughness would be needed to validate the estimation model. Also, it is expected that more realistic specimens shapes (for example with notches) as well as variable amplitude loading will lead to additional effects on the EIFS.

CONCLUSIONS

This paper covers first insights on the prediction of the influence of process inherent defects and surface roughness on the fatigue performance of DMLS Ti-6Al-4V. Experimentally determined S-N curves as well as crack propagation curves are presented. Non destructive testing methods were used to determine the distribution and size of internal defects by CT-scans and surface roughness by tactile measurements. The following conclusions can therefore be made:

- Process inherent surface roughness tend to significantly lower fatigue properties and scatter, compared to machined DMLS specimens.
- The maximum projected defect diameter obtained from CT-scans leads to conservative fatigue estimations.
- Scatter due to the distribution of defects is accounted by a Monte-Carlo simulation.
- The influence of surface roughness on the fatigue performance can be estimated by an EIFS for the investigated test specimens.

REFERENCE

[1] A. A. Antonysamy. (2012). In: *Microstructure, texture and mechanical property evolution during additive manufacturing of Ti6Al4V alloy for aerospace applications*, Faculty of Engineering and Physical Sciences.

[2] G. Kasperovich, J. Hausmann. (2015). *Journal of Materials Processing Technology*, vol. 220, p. 202-214.

[3] P. Edwards, M. Ramulu. (2014). *Materials Science and Engineering: A*, vol. 598, p. 327-337.

[4] D. Greitemeier, K. Schmidtke, V. Holzinger, C. Dalle Donne. (2013). *27th ICAF Symposium.*

[5] E. Brandl. (2010). In: *Microstructural and mechanical properties of additive manufactured titanium (Ti-6Al-4V) using wire, evaluation with respect to additive processes using powder and aerospace material specifications*, Faculty of Mechanical, Electrical and Industrial Engineering.

[6] E. Wycisk, C. Emmelmann, S. Siddique, F. Walther. (2013). *Advanced Materials Research*, vol. 816, p. 134-139.

[7] J. A. Harter. (2008). Lex Tech Inc., Ohio.

[8] Y. Murakami. (2002). In: *Metal fatigue: effects of small defects and nonmetallic inclusions*, Elsevier, Amsterdam.

[9] D. Greitemeier, C. Dalle Donne, F. Syassen, J. Eufinger, T. Melz. (2015). *Materials Science and Technology.*

[10] ESTEC, ESA. (1995). In: *ESACRACK user's manual*, ESA Publications Division, ESTEC Netherlands.

[11] U. S. D. o. Defense. (1974). In: *Military handbook: titanium and titanium alloys*.
[12] E. Wycisk, J. Kranz, C. Emmelmann. (2012). *DDMC Direct Digital Manufacturing Fraunhofer Conference*.
[13] K. Takahashi, Y. Murakami. (1999). *Engineering Against Fatigue*.

Validation Of The USAF Characteristic K Approach For Long Cracks Under Variable Amplitude Loading And Naturally Occurring Small Cracks Under A Combat Aircraft Load Spectrum

Dinaz Tamboli, Pu Huang, Rhys Jones

Centre of Expertise Structural Mechanics, Department of Mechanical and Aerospace Engineering, Monash University, Wellington Rd, Clayton, Australia 3800.

Abstract: Damage tolerance analysis of structures requires reliable predictions of fatigue crack growth rates for both microscopically small and long cracks. Due to the largely empirical nature of crack growth under cyclic loading, several models have been developed to capture the growth rate of cracks. The development of a model that can be used to predict the growth of both long and short cracks under a random loading spectrum, however, has continued to challenge researchers. The present paper seeks to address this issue by validating the USAF characteristic K approach using the Hartman-Schijve variant of the NASGRO equation. In order to achieve this, two sets of data were analysed; the first set uses the long crack growth data for Al 7075-T6, Al 2024-T3 and Mil-Annealed Ti-6Al-4V under variable amplitude loading and the second set uses the test results for small cracks growing from naturally occurring material discontinuities under a representative combat aircraft load spectrum. In each case it was shown that by defining a value of the characteristic stress intensity factor, K that best captures the load history, the Hartman-Schijve variant of the NASGRO equation was able to predict the growth of cracks reasonably well.

INTRODUCTION

Accurately predicting the fatigue life of aerospace alloys in challenging conditions is critical for continued aircraft airworthiness. More specifically, military aircraft are expected to achieve long lives under demanding operational conditions which consist of complex and variable spectra. Fatigue crack growth behaviour of structures subjected to variable amplitude loading is very complex. This is due to load interaction effects like retardation and crack acceleration when tensile or compressive overloads are interspersed with low amplitude cycles. A real-life service load history contains several high and low stress amplitudes with positive and negative overloads in random order. Accounting for these retardation and reduction in retardation effects can become quite complicated, but qualitatively the loading produces a behaviour that is similar to constant amplitude history with incidental overloads. The USAF Damage Tolerant Design Handbook [1] states that a characteristic K approach can often be used for several practical engineering problems where the load spectra can be approximated by a number of repeating load blocks and each block consists of a spectrum with *n* cycles that have peak stresses of σ_i, $i = 1\ldots n$, with the associated cyclic ranges being $\Delta\sigma_i$, $i = 1\ldots n$. It also requires that: a) the slope of the crack length versus block curve has a minimal number of discontinuities and b) there are a large number of blocks before failure.

In such cases, crack growth can often be approximated by

$$\frac{da}{dB} = D(K_{char})^m \tag{1}$$

where da/dB is the increment in the crack length per load block, m and D are constants and K_{char} is the characteristic stress intensity factor in the load block. Here the term K_{char} is defined as:

$$K_{char} = \beta\sigma_{char}\sqrt{\pi a} \tag{2}$$

where β is the geometric correction factor for the mean crack length associated with a given load block, a is the crack length and σ_{char} is the stress in the spectrum, typically σ_{max} the maximum value of stress in the block or $\Delta\sigma_{rms}$ the root mean square value of the stress intensity factor range in the block. The corresponding value of K_{char} then becomes K_{max} or ΔK_{rms}. Gallagher [2] used a particular version of equation (1), viz:

$$\frac{da}{dB} = D(K_{char})^m/(1 - K_{max}/A) \tag{3}$$

For long cracks subjected to a randomised spectrum, it has been reported in literature [5-10] that the crack growth rate can be characterised reasonably well by using ΔK_{rms}. Barsom and Rolfe [7] reported that the da/dN vs ΔK_{rms} curve for bridge steels under a typical bridge load spectrum coincides with the constant amplitude *da/dN* vs ΔK curve. The same phenomenon has been reported by Kim et al. [9] for crack growth under a randomised sill longeron spectrum of an aircraft. In contrast to these findings, Wanhill [16] has stated in his review paper that "Good characteristic K correlations of long/large fatigue crack growth were obtainable except in the near threshold crack growth region". However, as explained in [5] this was due to an error in the tests used in [16] to determine the constant amplitude *da/dN* vs ΔK relationship and when using the *da/dN* vs ΔK relationship determined by tests at NASA the poor correlation mentioned in [16] vanishes. Reference [5] presented a characteristic K variant of the the Hartman-Schijve equation [3-4], viz:

$$\frac{da}{dN} = D\left(\Delta K_{char} - \Delta K_{char,thr}\right)^p/(1 - K_{max}/A)^{p/2} \tag{4}$$

where $\Delta K_{char,thr}$ and A can be best interpreted as parameters chosen so as to fit the measured data, see [3, 5]. As explained in [5] equation (4) is a special case of the NASGRO equation [5, 11], used in both AFGROW [12] and NASGRO [13], viz:

$$\frac{da}{dN} = D[(1-f)/(1-R)]^m \Delta K^{(m-p)}(\Delta K - \Delta K_{thr})^p/(1 - K_{max}/A)^q \tag{5}$$

where $f = K_{op}/K_{max}$, m, p and q are additional experimentally determined constants and ΔK_{thr} is the associated effective threshold value of ΔK. Indeed, equation (4) follows from equation (5) by setting $m = p$ and $q = p/2$.

Whilst this approach was first developed for long cracks the recent review paper [5] has shown that it also holds for cracks that arise from small naturally occurring material discontinuities. It was also shown [5] that, for small near micron initial cracks in Al 7050-T7451 tested under combat aircraft spectra, when taking the value of the characteristic K to be ΔK_{rms} the short crack data and long crack constant amplitude data could all be represented by the same characteristic K NASGRO equation. This paper builds on the results presented in [5] and presents a number of problems whereby a characteristic K variant of the Hartman-Schijve equation is used to compute crack growth. The first data set used to illustrate the applicability of this approach involves the long crack data presented in the paper by Porter [14] which was obtained from a series of fatigue tests performed at Boeing on Al 7075-T6, Ti-6Al-4V and Al 2024-T3 panels. For this data set, ΔK_{char} in equation (4) is set to be proportional to the maximum stress intensity factor in the spectrum, i.e. K_{max}. The second set involves the growth of cracks that initiate from small naturally occurring material discontinuities at a fastener hole [15] in Al 7050-T7451 subjected to a combat aircraft spectrum. In this instance it is shown that both the crack growth histories and the scatter in the measured crack growth histories can be captured reasonably well by setting ΔK_{char} in equation (4) equal to ΔK_{rms} and using the values of D and p determined from tests on long cracks growing under constant amplitude loading.

GROWTH OF LARGE CRACKS UNDER VARIABLE AMPLITUDE LOADING

Porter [14] conducted a series of tests at Boeing to demonstrate the effect of load interaction in large cracks under variable amplitude loading. The tests were performed on transverse grain panels that were 304.8 mm wide by 914.4 mm long and 4.06 mm thick, as shown in Figure 1.

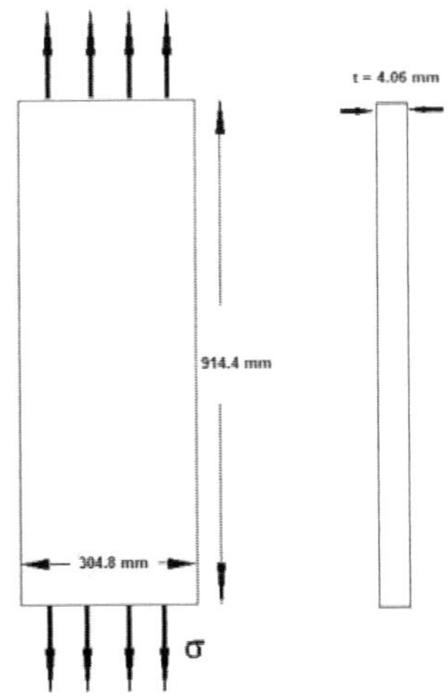

Figure 1. Schematic of the panels made of Al 7075-T6, Al 2024-T3 and Mil-Annealed Ti-6Al-4V used by Porter [14]

The materials used were Al 7075-T6, Al 2024-T3 and Mil-Annealed Ti-6Al-4V. It was observed, as seen in Figure 2, that for Al 7075-T6 varying the number of overloads has a significant effect on the overall fatigue life of a structure. For instance, for the same baseline stress levels, coupon 2-A, which had only 1 overload and 50 baseline cycles, showed nearly 50% increase in life with a slower growth rate as opposed to coupon 1-A, which was tested at a basic 50-cycle spectrum with no overloads. Subsequently increasing the number of applied overloads to 50 in coupon 7-A dramatically reduced the fatigue life of the coupon.

The results of tests with variable spectrum size in different materials are shown in Figures 3, 6 and 8. In all cases there was only 1 overload cycle per spectrum. These results indicated that the largest increase in life was realised only when minimum number of overload cycles was applied per spectrum and this trend held for all the three tested materials. Similar behaviour was observed for repeated spectrum loading. Figure 4 shows that when a high peak load is applied once in a 30-cycle spectrum, the specimen life was increased by the application of overload cycles and the longest life corresponded to the highest level of applied overload, except in one case (coupon 2-D) where there was only a slight increase in life.

The results in Figure 5 show the test data that included minimum peak stresses. The influence of load sequence (hi-lo or lo-hi) was also investigated and the results indicated that the minimum stress tends to cancel the effect of the peak overload and this effect is particularly significant if the minimum immediately follows the peak overload. Figures 7 and 9 represent data for 2024-T3 and Ti-6Al-4V that also illustrate the load order effects. For all the coupons tested, the final crack length reported before failure was greater than 50 mm.

All of these tests gave results that are consistent with fracture mechanics theory and provided a good range of data under a variety of loading conditions that could be used to test the predictive ability of equation (4) using the USAF characteristic K approach where the crack length is portrayed as a function of the number of load blocks applied.

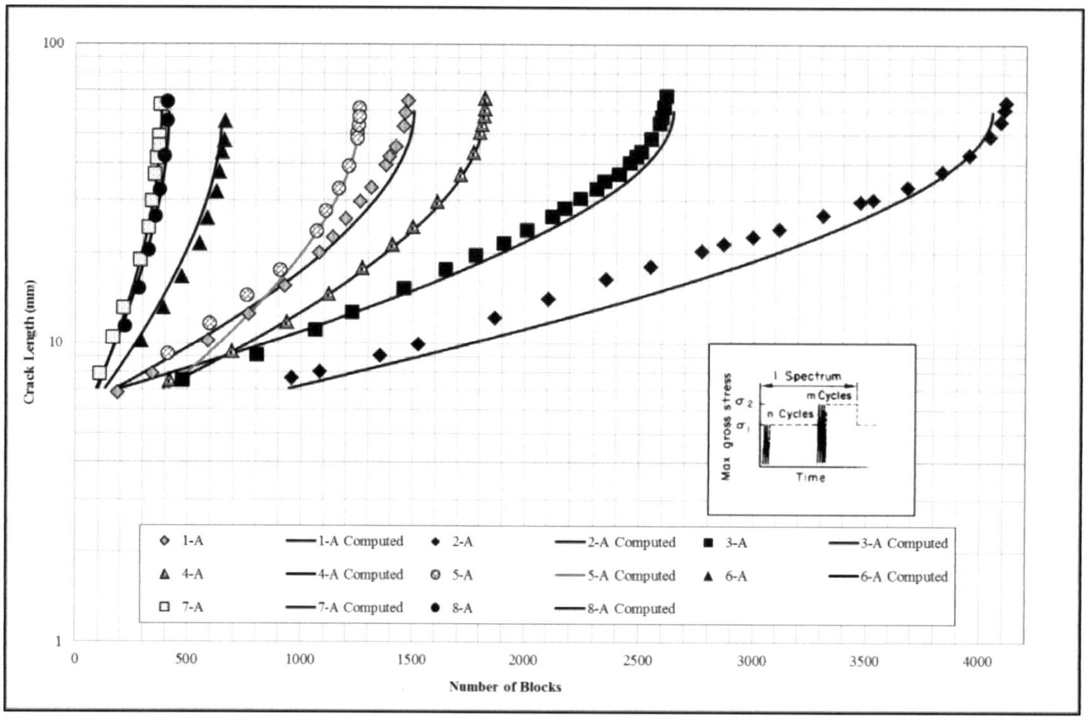

Figure 2. Measured vs computed data showing the log of crack length vs cycles behaviour of 7075-T6 panels with variable overloads.

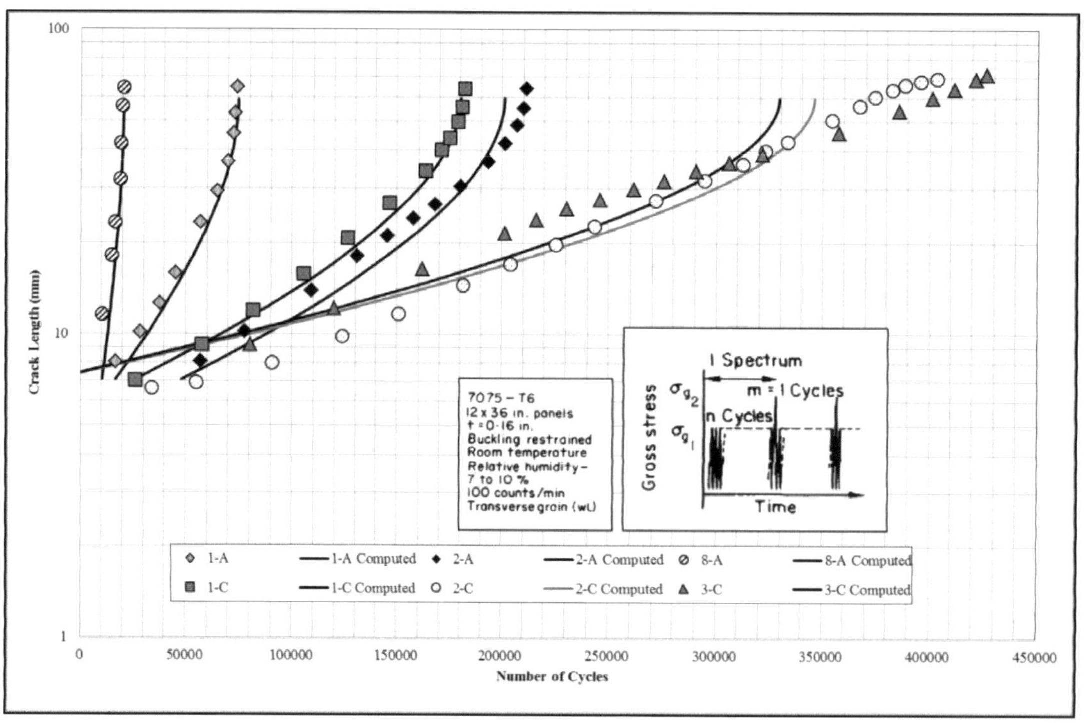

Figure 3. Measured vs computed data showing the log of crack length vs cycles behaviour of 7075-T6 panels with variable block size.

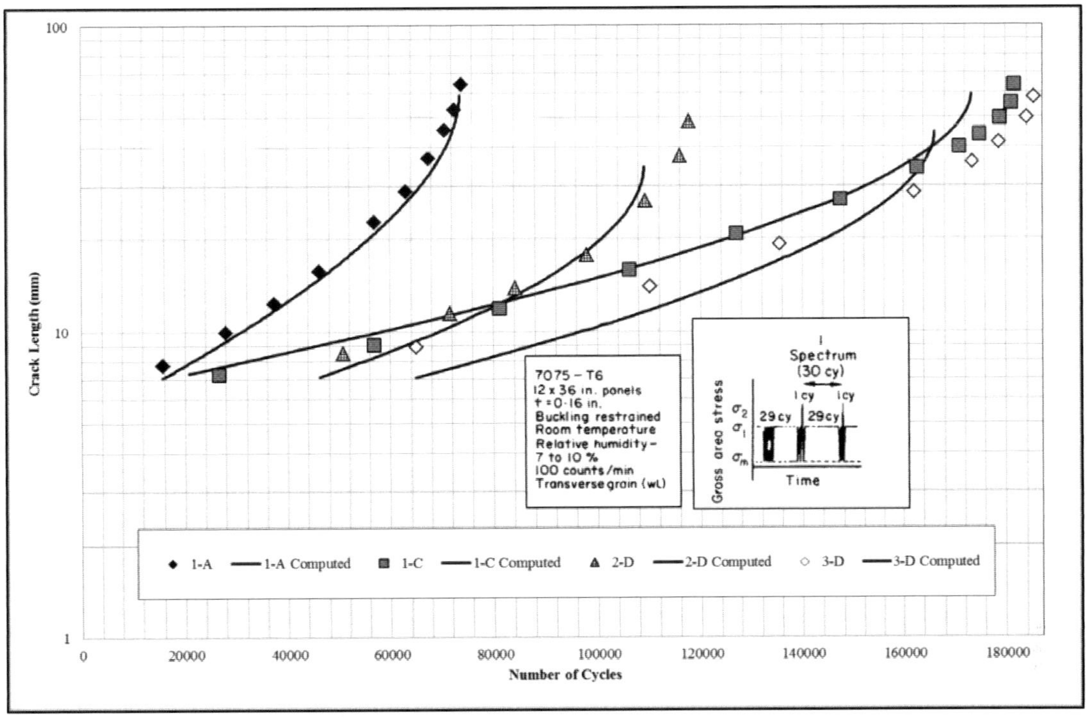

Figure 4. Measured vs computed data showing the log of crack length vs cycles behaviour of 7075-T6 panels with variable overload stress.

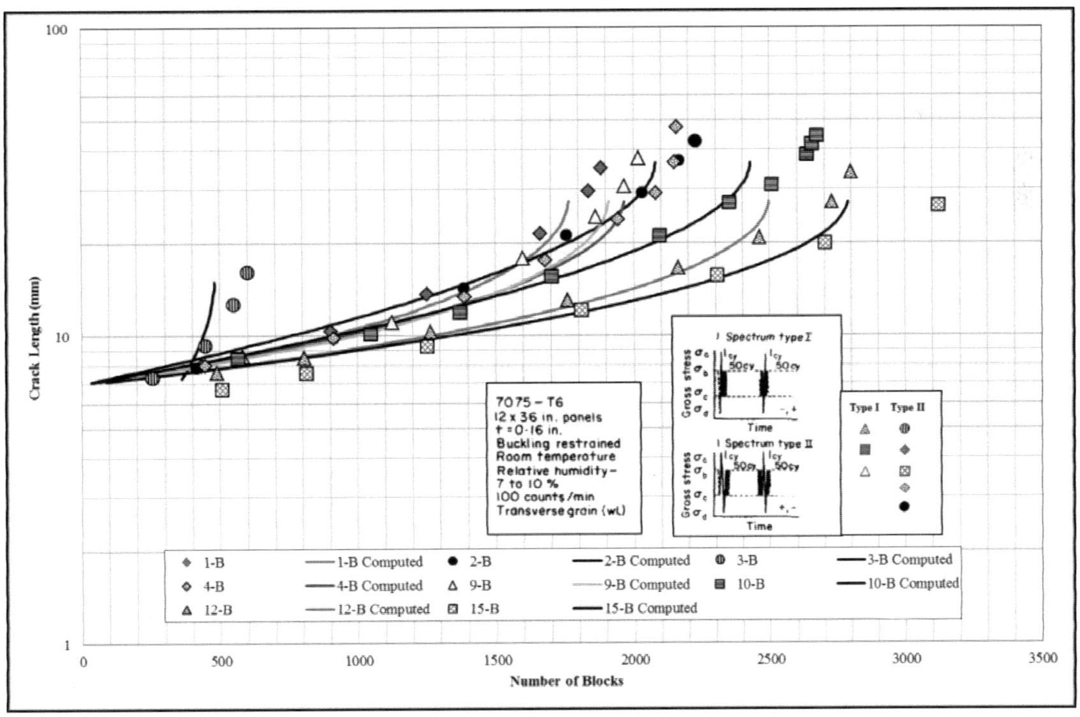

Figure 5. Measured vs computed data showing the influence of peak cycle sequence on crack growth in 7075-T6.

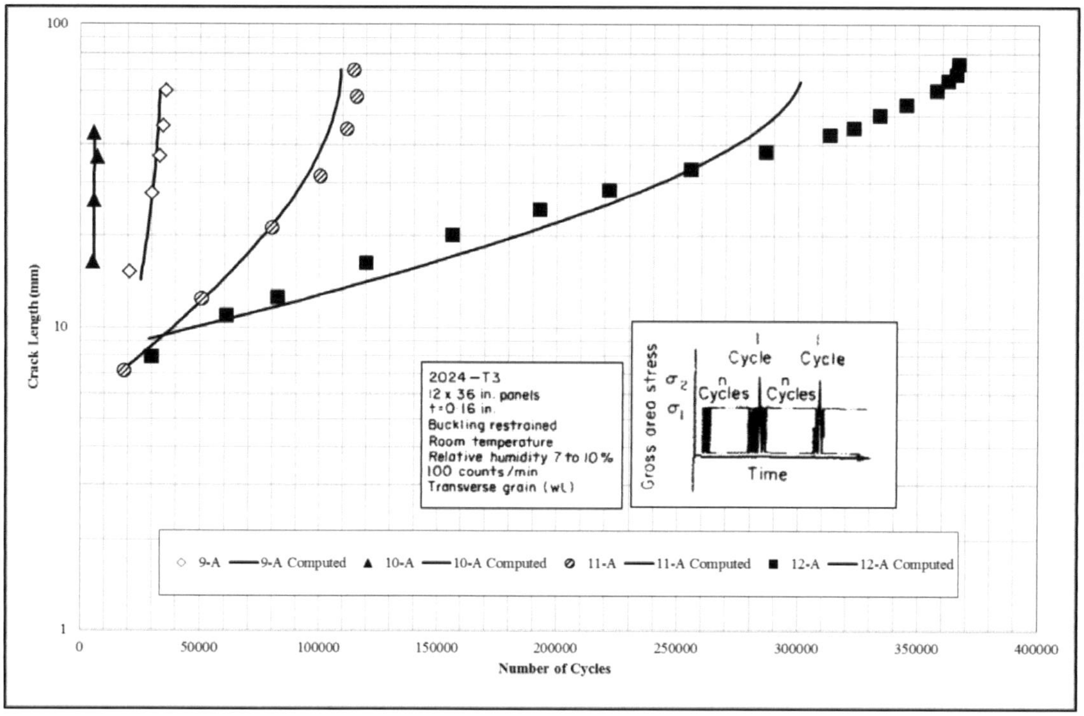

Figure 6. Measured vs computed data showing the behaviour of 2024-T3 panels with variable block size.

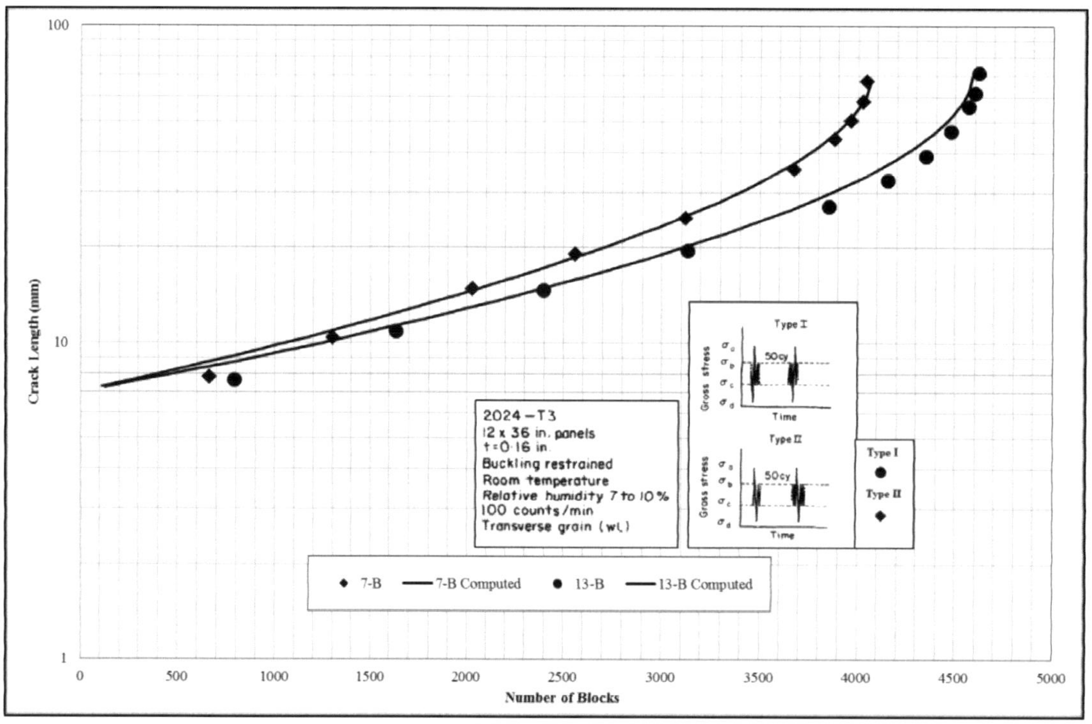

Figure 7. Measured vs computed data showing the influe4nce of peak cycle sequence on crack growth in 2024-T3.

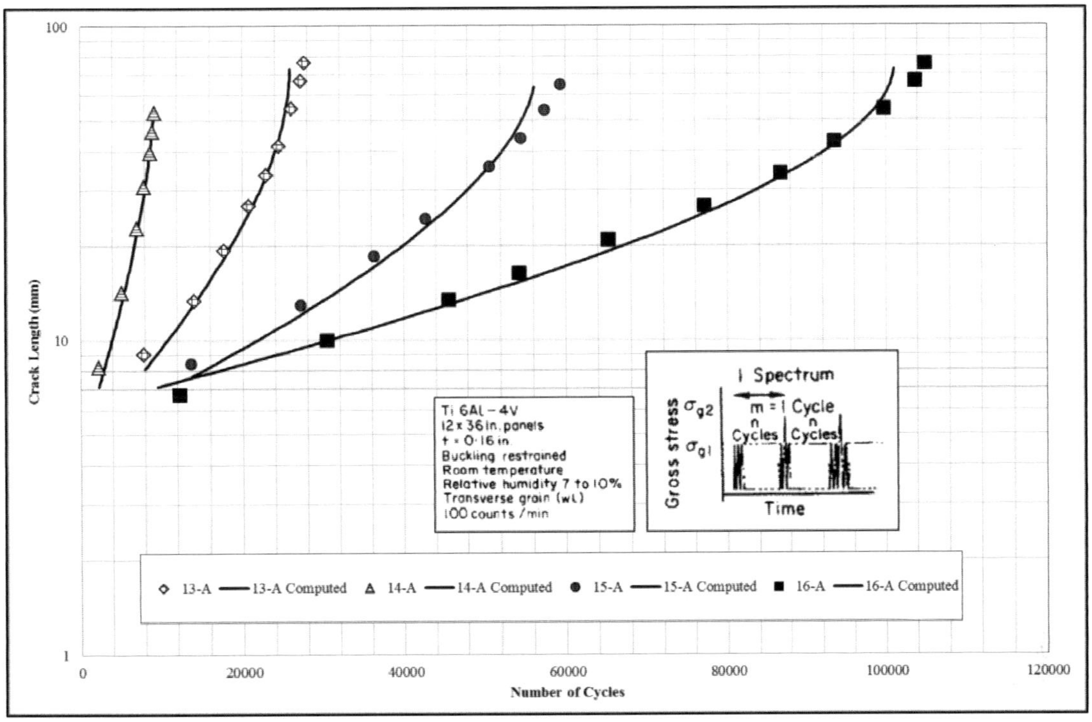

Figure 8. Measured vs computed data showing the log of crack length vs cycles behaviour of Mil-Annealed Ti-6Al-4V panels with variable block size.

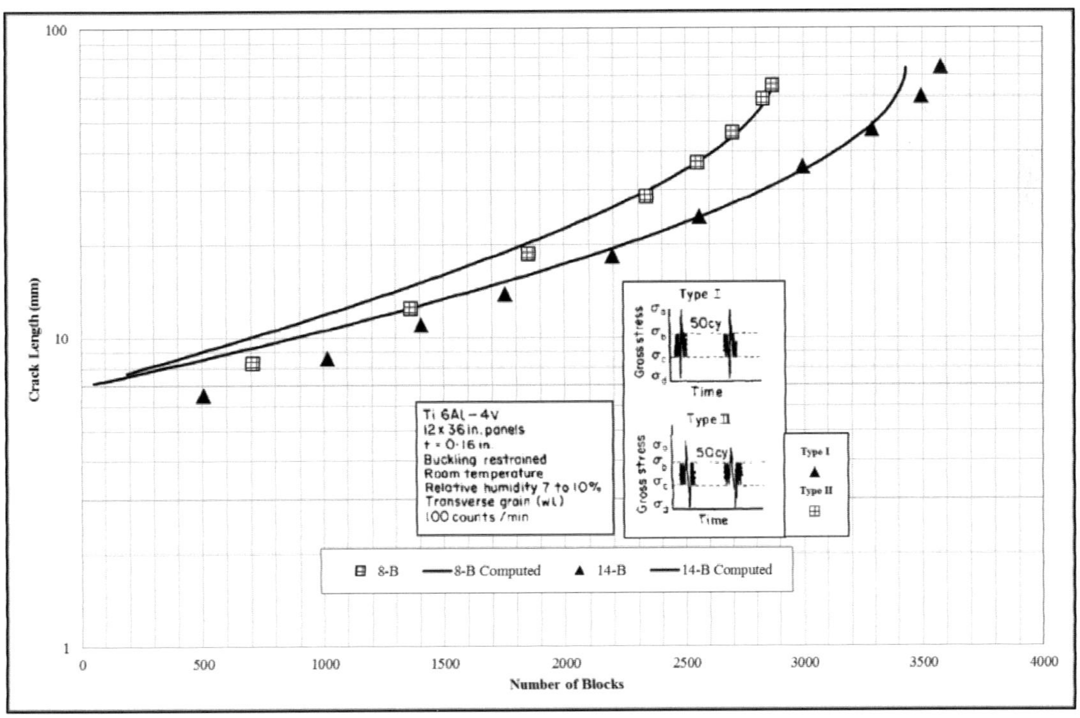

Figure 9. Measured vs computed data showing the influence of peak cycle sequence on crack growth in Mil-Annealed Ti-6Al-4V.

Discussion of Results

Barsom and Rolfe [7] state that using ΔK_{rms} as the characteristic stress parameter in equation (1) is subject to certain limitations. While it is applicable to relatively smooth and continuous crack length vs cycles curves, which can be obtained from several variable amplitude loadings or frequently applied overloads, it does not accurately represent data which has single or infrequently applied overloads. It is also noted in the USAF Damage Tolerant Design Handbook [1] that retardation effects associated with single or infrequently applied overloads are ignored by ΔK_{rms}. It is for this reason that K_{max} was chosen as the characteristic stress parameter in equation (4). The equation used to model the crack growth behaviour then became

$$\frac{da}{dN} = D\left(K_{max} - K_{max,thr}\right)^{p} / (1 - K_{max}/A)^{p/2} \qquad (6)$$

The value of p in the above equation was chosen to be equal to 2 because previous studies by Jones et al. [4] have shown that a Paris exponent of 2 in equation (6) accurately represents crack growth rate at various stress ratios for wide range of aerospace aluminium and titanium alloys. For each material, the value of A was chosen so that the data fit equation (6) at one stress ratio and it was then kept constant for all the other tests. Since there is a large variation in the spectra used in these tests, different values of D were needed for each spectrum to best represent the crack growth curves.

To capture the scatter in tests which used the same spectrum, viz: specimen 2-A in Figure 3 and all the specimens in Figure 5, the value of D kept constant and, as explained in [5], the value of K_{thr} was varied to get the curves that best represented the crack growth data for each particular test. The same approach was applied to other test specimens viz: 1-C, 2-D and 3–D in Figure 4 for 7075-T6 and all the 2024-T3 and Ti-6Al-4V specimens in Figures 7 and 9. The full list of all the constants used in the tests is tabulated in Table 1 below. The table also shows how, for a given spectrum, the variability in the crack growth histories is captured well by allowing for changes in the threshold term, $K_{max,thr}$. For the same material, there is only a slight variation in the D values for most cases except the fast growing cracks subjected to a large number of overloads. It can be seen in the graphs above that by setting the characteristic K value to be equal to K_{max}, the Hartman-Schijve variant of the NASGRO equation given above represents the crack growth data reasonably well.

Table I. List of variables used in the Characteristic K_{max} test.

Figure	Material	Coupon	Baseline Cycles	Overloads	σ_{max} MPa	σ_{min} MPa	A MPa√m	D	$K_{max,thr}$ MPa√m
2	7075-T6	1-A	50	0	103.4	3.448	46	4.20E-10	0.3
		2-A	50	1	103.4	3.448	46	1.70E-10	0
		3-A	50	3	103.4	3.448	46	2.10E-10	0.1
		4-A	50	6	103.4	3.448	46	3.40E-10	0.1
		5-A	50	10	103.4	3.448	46	5.40E-10	0.2
		6-A	50	25	103.4	3.448	46	7.10E-10	0.2
		7-A	50	50	103.4	3.448	46	9.80E-10	0.3
		8-A	0	50	103.4	3.448	46	1.70E-09	0.2
3	7075-T6	1-A	∞		103.4	3.448	46	4.20E-10	0.3
		2-A	50	1	103.4	3.448	46	1.70E-10	0.3
		8-A	0	1	103.4	3.448	46	2.80E-09	0.1
		1-C	29	1	103.4	3.448	46	1.80E-10	0.3
		2-C	300	1	103.4	3.448	46	7.50E-11	0
		3-C	1000	1	103.4	3.448	46	7.85E-11	0
6	2024-T3	9-A	50	0	155.1	5.171	76	9.30E-10	0.3
		10-A	0	50	155.1	5.171	76	1.60E-08	0.1
		11-A	29	1	155.1	5.171	76	1.50E-10	0.1
		12-A	30	1	155.1	5.171	76	4.60E-11	0.1
8	Ti-6Al-4V	13-A	50	0	258.5	8.619	130	2.60E-10	0.2
		14-A	0	50	258.5	8.619	130	7.20E-10	0.2
		15-A	29	1	258.5	8.619	130	1.14E-10	0.3
		16-A	30	1	258.5	8.619	130	5.58E-11	0.1
4	7075-T6	1-A	50	0	103.4	3.448	46	4.20E-10	0
		1-C	29	1	103.4	3.448	46	1.80E-10	0.5
		2-D	29	1	137.9	3.448	46	1.80E-10	3.1
		3-D	29	1	120.6	3.448	46	1.80E-10	2.1
5	7075-T6	1-B	50	1	155.1	5.171	46	1.70E-10	11
		2-B	50	1	134.4	31.02	46	1.70E-10	1.2
		3-B	50	1	206.8	5.171	46	1.70E-10	0
		4-B	50	1	155.1	31.02	46	1.70E-10	7.2
		9-B	50	1	155.1	5.171	46	1.70E-10	11.6
		10-B	50	1	134.4	31.02	46	1.70E-10	2.6
		12-B	50	1	155.1	31.02	46	1.70E-10	8.9
		15-B	50	1	155.1	51.71	46	1.70E-10	5.9
7	2024-T3	7-B	50	1	155.1	5.171	76	7.90E-11	2.6
		13-B	50	1	155.1	5.171	76	7.90E-11	4.3
9	Ti-6Al-4V	8-B	50	1	258.5	8.619	130	3.50E-11	0
		14-B	50	1	258.5	8.619	130	3.50E-11	4.5

GROWTH OF CRACKS FROM SMALL NATURALLY OCCURRING DISCONTINUITIES

A study [15] was carried out at the Defence Science and Technology Organisation (DSTO) using a dog-bone type specimen made of Al 7050-T7451 that was 10 mm thick and 32 mm wide with a 6.35 mm diameter hole in the centre, see Figure 10. The two sets of specimens that were investigated had different surface finishes to simulate some areas of the F/A-18; one was termed "as-machined" which was intended to simulate the reamed holes present in the aircraft and the other was termed "etched" and was intended to simulate the surface that is found under the Ion Vapour Deposited (IVD) aluminium coating of the Al 7050-T7451 F/A-18 components. The etch solution was 50% (volume) nitric acid (HNO_3, 70%) and 1% volume percent Hydrofluoric acid (HF, 70% Technical grade), remainder tap water, with an etching time of 5 min at room temperature, see [17] for more details. A minimum of five coupons were tested for each surface finish to obtain statistical confidence. The specimens were tested under a typical RAAF F/A-18 Hornet spectrum with a peak net section stress of 155MPa and 325 simulated flight hours per block.

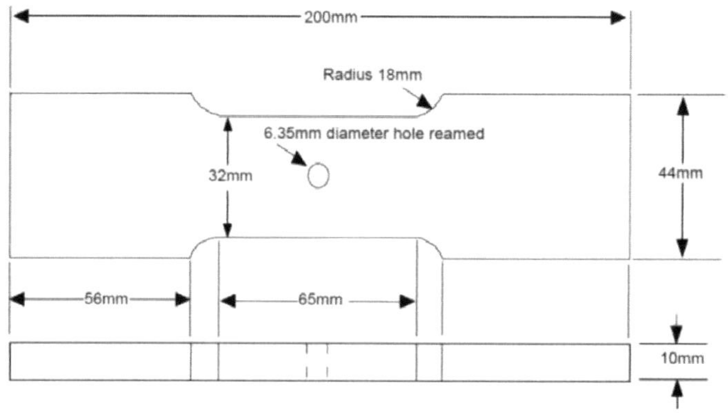

Figure 10. Schematic of the specimen used in the DSTO test program [15].

The resultant crack growth curves are shown in Figures 11 and 12. Quantitative fractography data found that the cracks observed in etched specimens KK1H353, KK1H348 KK1H339 and KK1H193 originated at etch pits whilst those from specimens KK1H83 and KK1H375 initiated at small machining defects.

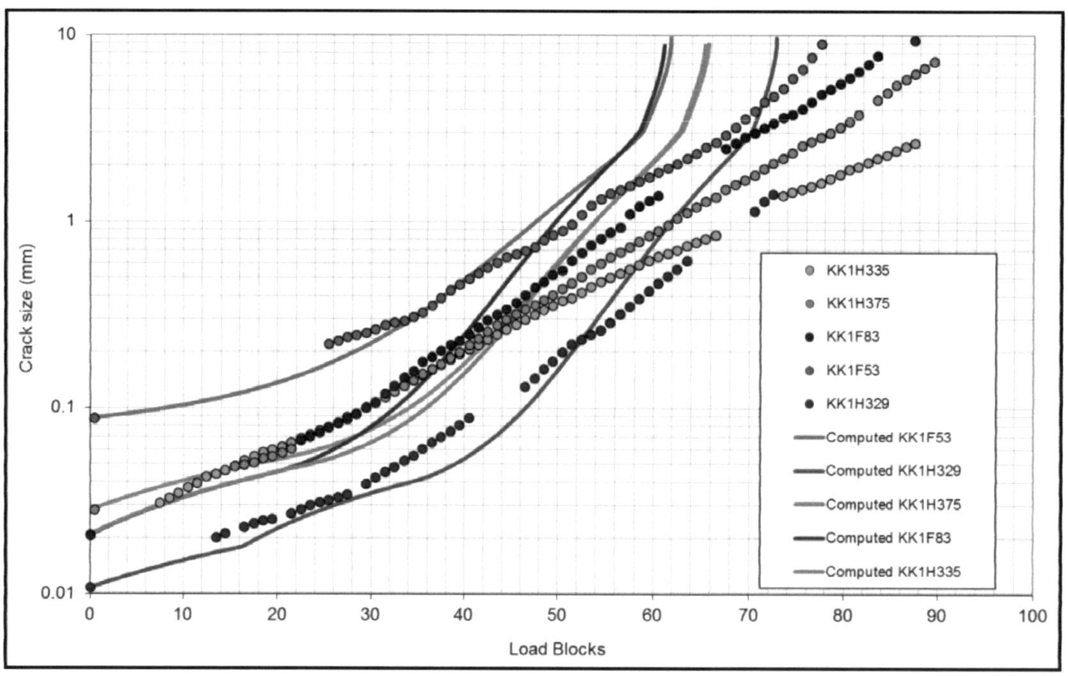

Figure 11. Measured vs computed crack growth histories for "as-machined" coupons under a representative F/A-18 Spectra.

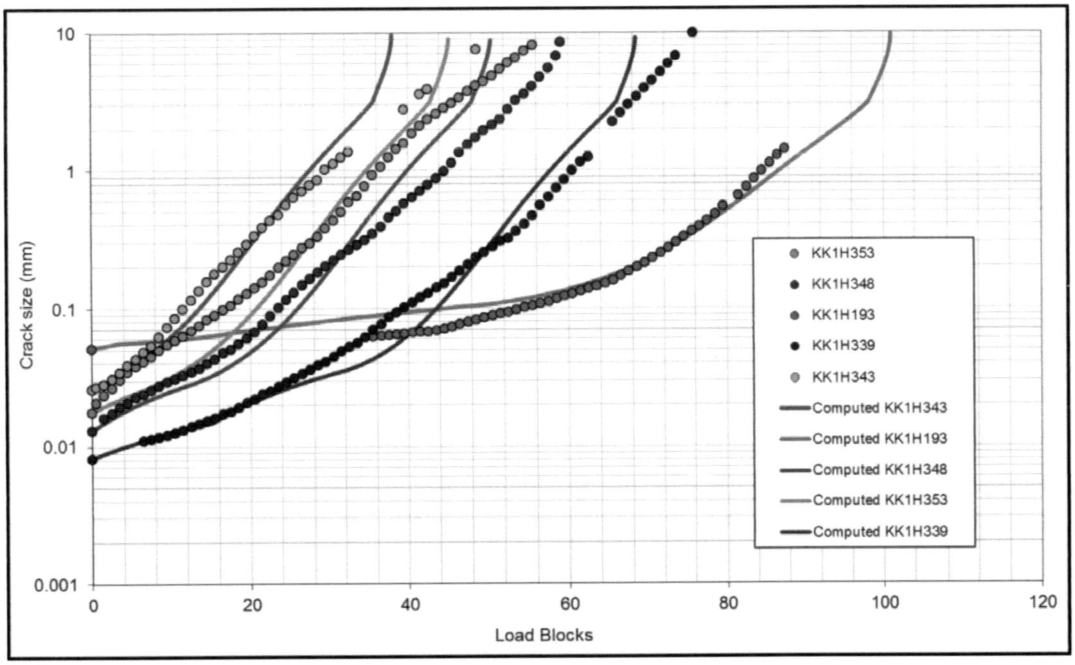

Figure 12. Measured vs computed crack growth histories for "etched" coupons under a representative F/A-18 Spectra.

Discussion of Results

For this study, the root mean square value of stress intensity factor, ΔK_{rms}, was chosen to represent the crack growth data given in Figures 11 and 12 above. Equation (4) thus became

$$\frac{da}{dN} = D\left(\Delta K_{rms} - \Delta K_{rms,thr}\right)^p / (1 - K_{max}/A)^{p/2} \tag{7}$$

The value of ΔK_{rms} was calculated using the formula

$$\Delta K_{rms} = \sqrt{\frac{\sum_{i=1}^{N} (\Delta K_i)^2}{N}} \tag{8}$$

The values of $D = 7.0 \times 10^{-10}$, $A = 47$ MPa$\sqrt{\text{m}}$ and $p = 2$ obtained in [5] from constant amplitude tests for Al 7050-T7451 were used. Equation (7) was then used to model the crack growth behaviour of both, the machined and etched coupons by varying the $\Delta K_{rms,thr}$ value in small increments and the resultant predictions were found to be of reasonably good agreement to the measured crack growth curves. A large scatter was observed in the as-machined test as opposed to the etched test, which could be because etching increases the density of discontinuities from where multiple cracks could grow. The scatter associated with the growth of these cracks from small material discontinuities was captured well by the variations in the threshold value.

Table II. List of variables used in the Characteristic K_{rms} test.

Figure	Surface Finish	Coupon	A MPa√m	D	K_{thr} MPa√m
11	As-Machined	KK1H329	47	7.0E-10	0.61
		KK1H375	47	7.0E-10	0.85
		KK1H335	47	7.0E-10	0.76
		KK1F83	47	7.0E-10	0.72
		KK1F53	47	7.0E-10	1.30
12	Etched	KK1H343	47	7.0E-10	0.51
		KK1H193	47	7.0E-10	1.28
		KK1H348	47	7.0E-10	0.51
		KK1H353	47	7.0E-10	0.51
		KK1H339	47	7.0E-10	0.53

CONCLUSION

The present paper validated the USAF characteristic K approach using the Hartman-Schijve variant of the NASGRO equation to model the growth of long cracks under variable amplitude loading and cracks growing from small naturally occurring material discontinuities under a representative fighter spectrum. In the first study K_{max} was chosen as characteristic stress parameter and it was shown that by varying the values of the constant D in the Hartman-Schijve equation, the crack growth under variable amplitude loading with various overloads for three different materials, Al 7075-T6, Al 2024-T3 and Mil-Annealed Ti-6Al-4V were captured well by the model. For tests which used the same spectrum, changes in the threshold value of the maximum stress intensity factor accounted for the scatter associated with the spectrum fairly well.

The second study involved the growth of cracks from small material discontinuities under a representative fighter spectrum at a fastener hole of coupons with two different surface finishes viz: "as machined" and "etched". The crack growth curves obtained from tests done on these coupons displayed a larger scatter in cracks growing from as-machined coupons as opposed to etched coupons. It was shown that by using the root mean square value of the stresses in the spectrum to characterise the stress intensity factor, changes in the threshold term of the Hartman-Schijve equation were able to capture the scatter associated with cracks growing from small material discontinuities at a fastener hole in Al 7050-T7451 reasonably well.

REFERENCES

[1] Gallagher JP, Giessler FJ, Berens AP and Engle RM. (1985) USAF Damage Tolerance Design Handbook: Guidelines For The Analysis And Design Of Damage Tolerant Aircraft, AFWAL-TR-82-3073, Air Force Research Laboratory, Wright-Patterson Air Force Base, USA.

[2] Gallagher JP and Stalnaker HD (1978). Developing normalised crack growth curves for tracking damage in aircraft, American Institute of Aeronautics and Astronautics, Journal of Aircraft, vol 15, pp 114-120.

[3] Jones R and Tamboli D (2013). Implications of the lead crack philosophy and the role of short cracks in combat aircraft, Engineering Failure Analysis, vol 29, pp 149-166.

[4] Jones R, Molent L and Walker K (2012). Fatigue crack growth in a diverse range of materials, International Journal of Fatigue, vol 40, pp 43-50.

[5] Jones R., (2014). Fatigue crack growth and damage tolerance, Invited Review Paper, Fat Fract Eng Mat and Struct, 37, 5, pp. 463-483.

[6] Gallagher JP, Miedlar PC, Juarez V (1990). Rapid repair DTA technology for F-16 aircraft, Proceedings of The 1989 Structural Integrity Program Conference, Edited by TD Cooper and JW Lincoln, WRDC-TR-90-4051, Wright-Patterson AFB, pp 497-518.

[7] Barsom JM and Rolfe ST (1999). Fracture and Fatigue Control in Structures: Applications of Fracture Mechanics, Butterwoth-Heinemann Press: Woburn, Ma, USA.

[8] Miedlar PC, Berens AP, Gunderson A, Gallagher JP (2003). Analysis and support initiative for structural technology (ASIST), AFRL-VA-WP-TR-2003-3002.

[9] Kim SA, Tadjiev D and Yang HT (2006). Fatigue life prediction under random loading conditions in 7475-T7351 aluminium alloy using the RMS model, International Journal of Damage Mechanics, vol 15, pp 89-102.

[10] Zhuang W, Boykett R, Phillips M and Molent L (2008). Effective block approach for damage tolerance analyses of F-111 D/F model wing structures, DSTO-TR-2124.

[11] Forman, RG, and Mettu, SR (1992) Behavior of Surface and Corner Cracks Subjected to Tensile and Bending Loads in Ti-6Al-4V Alloy, Fracture Mechanics 22nd Symposium, Vol. 1, ASTM STP 1131, H.A. Ernst, A. Saxena and D.L. McDowell, eds., American Society for Testing and Materials, Philadelphia, 1992.

[12] Harter JA. AFGROW users guide and technical manual. AFRL-VA-Wright-Patterson-TR-1999-3016, 1999.

[13] Fatigue Crack Growth Computer Program "NASGRO" Version 3.0: Reference Manual, JSC-22267b, March 2002.

[14] Porter TR (1972). Method of analysis and prediction for variable amplitude fatigue crack growth, Engineering Fracture Mechanics, vol 4, pp 717-736.

[15] Jones R, Molent L and Barter S., (2013). Calculating crack growth from small discontinuities in 7050-T7451 under combat aircraft spectra, International Journal of Fatigue, vol 55, pp. 178-182.

[16] Wanhill RJH (2009). Characteristic stress intensity factor correlations of fatigue crack growth in high strength alloys: review and completion of NLR investigations 1985-1990, NLR-TP-2009-256.

[17] Huynh J, Molent L, Barter S (2008). Experimentally derived crack growth models for different stress concentration factors, International Journal of Fatigue, vol 30, pp 1766-1786.

Is The World Ready for HOLSIP?

Paul N. Clark[1], Nicholas C. Bellinger[2], David W. Hoeppner[3]

[1]*Southwest Research Institute, United States of America*
[2]*National Research Council Canada*
[3]*University of Utah, United States of America*

Abstract: HOLSIP – What does this mean? HOLSIP is an acronym for Holistic Structural Integrity Process. This paper will review the development of the HOLISTIC LIFE ASSESSMENT APPROACH FOR AIRCRAFT STRUCTURES. It will describe HOLSIP as a philosophical approach as well as comparing and contrasting the HOLISTIC approach to the safe life approach and the damage tolerant approach. The emphasis will focus on modeling the physical characteristics of a system which is an approach inherent to the newer HOLISTIC methodology. The paper will cover some of the basics of the approach and the history of its evolution.

Starting in the 1960s a systems oriented approach to fatigue and damage tolerance began to emerge. This was in part based on many of the idealized assumptions of mechanics of solids and elasticity. However, as early as the 1950s some began to notice that fatigue cracks had their origin at various particles and discontinuities in aircraft structural materials. Indeed this followed the pioneering work of Dr. Alan Arnold Griffith. In additional extensive work in the 1960s and 1970s it was clearly established that other "extraneous" influences also acted as the origin of fatigue cracks. These include fretting, corrosion of various mechanisms, wear, creep and others. By the late 1970s a major review written by one of the authors termed the origin of fatigue cracks at either intrinsic or created discontinuities an IDS which is a now accepted term for either "Initial Discontinuity State" or "Initial discontinuity state". This is now part of the building block of the HOLISTIC lifing procedure. Moreover, the quantification of evolution of the degradation mechanisms both over the entire range of fatigue crack propagation and the incorporation of corrosion and corrosion fatigue consideration into the HOLISTIC framework has provided much optimism for enhanced lifing of aircraft structures in the future. The basis of this framework will be discussed in the paper related to several models already incorporated into the HOLISTIC framework.

Concepts for design and management of aircraft have matured dramatically over the past half century. However, the world continues to fly aircraft that are matured, and maturing as well. Much of the world's aircraft continue to age, with their usage continually being stretched to meet longer service lives. Many of the aircraft flying today were designed under either the safe life or the damage tolerant design paradigm; these design philosophies are powerful but make assumptions that are not necessarily applicable for sustainment of aircraft beyond their original design lives. Moreover the safe life paradigm does not deal with the physical characteristics of damage per se and also does not couple NDI/NDT with the design. HOLSIP presents a methodology and framework that aids in bringing multiple solutions to the design engineer and life management team. Modeling physical characteristics of a system, recognizing and modeling the Initial Discontinuity State, Evolving Discontinuity State (EDS) and Modified Discontinuity State (MDS) are all part of the HOLSIP process. Examples of modeling the physical characteristics of a system as well as in-depth discussions regarding the various discontinuity states and their relevance will be presented.

HOLSIP is about total life of components and subsystems and related issues. Wholeness or consideration of all aspects affecting life is one of the primary objects based on a physics of behavior approach. This is not a trivial task and much remains to be considered, captured, understood and developed. This paper will present several topics to be considered as part of the HOSLIP approach and demonstrate their utility for the design and management of structural integrity of both new and maturing aircraft.

BACKGROUND

In order to answer the question, "Is the world ready for HOLSIP?" a brief review of aircraft design philosophies and methodologies is in order.

Early aircraft designs were concerned with one goal, flying. Since the accomplishment of Sir George Cayley (the "father of the aeroplane") to the Wright brothers among many others, the focus of aircraft design has changed to efficiency, purpose, comfort and reliability, including structural integrity. For more than half a century after the history making success of the Wright Flyer engineering designs would operate in what the authors refer to as a "No Life" paradigm. Fundamentally, this type of design, from a structural integrity perspective was concerned with static strength.

There was little consideration for failure modes such as fatigue, creep, corrosion or wear and combinations of these beyond an aesthetically driven layer of paint. This type of design experienced surprising success with models such as the DC-3 primarily due to a "conservative design". However, the push for more efficient designs toward increasing performance would challenge this design approach. The late 1940s and early 1950s would usher in an era of aluminum alloy development for airframes that focused on increasing strength without regard for parameters that would ultimately affect the life of the aircraft designs, namely, fatigue. Perhaps the most significant lessons learned with respect to aircraft and fatigue came from the events surrounding the Comet.

The de Havilland Comet (DH-106) was a significant milestone for aviation; it was the world's first jet propulsion commercial aircraft. At the time, the overarching philosophy of design was that the Comet was utilizing state-of-the-art principles, and that issues would be worked out along the way. "Lifing" or planning for time based failures was not considered as the recent history had shown aircraft to become obsolete well before reaching their economic service life. An inadequate full-scale fatigue test was performed, but proved fruitless and did not prevent fatal inflight structural failures. The events that unfolded surrounding the de Havilland Comet revealed significant problems with the engineering design and testing approaches used and much was learned as a result.

On 10 January 1954, British Overseas Airways Corporation Flight 781, G-ALYP "Yoke Peter" crashed off of the coast of Elba into the Mediterranean Sea, fatal for all on board. Comet aircraft, G-ALYY "Yoke Yoke" decompressed in-flight causing it to crash fatally in route from Ciampino Airport to Cairo, Egypt in the Mediterranean Sea off of the coast of Naples, Italy. As determined by the Royal Aircraft Establishment the probable cause of these accidents was rapid decompression due to the fatigue cracking that nucleated from fastener holes near windows and other openings in the fuselage [1]. A previous accident was then called into question: Comet aircraft G-ALYV "Yoke Victor" crashed on a flight out of Calcutta, India. The Indian officials reported the cause as either weather (severe thunderstorms) or over controlling (pilot error). The investigating technical expert reported that the wings had failed symmetrically from a wing down bending event. He suggested that this occurred as a secondary event [2].

The loss of life from these events was catastrophic; however, lessons learned from de Havilland's Comet were extraordinary and contributed dramatically to the body of knowledge utilized by the aerospace community. These tragic events marshalled the adoption of safe-life and fail safe designs.

By the mid-1960s, the world was experiencing dramatic advances in aircraft, particularly for the military. 1967 introduced the General Dynamics F-111 "Aardvark" to the USAF and later in 1973 to the Royal Australian Air Force (RAAF). The F-111 sported a pivoting wing, a pioneering technology for the age. The wings were swept forward for landing or low speed flying and swept back to achieve supersonic speeds reaching above Mach 2. The F-111 entered service with a demonstrated service life of 6000 flight hours (full-scale fatigue testing had been performed to 16,000 simulated flight hours). On December 22, 1969 an F-111 experienced a structural failure at the left wing pivot fitting. This failure occurred during an estimated 4g service load (the F-111 was designed for and statically tested to 11g). The accident aircraft had logged 105 total service hours a scant 1.75% of its estimated SAFE LIFE. The root cause of this accident was concluded as a significant and undetected crack-like manufacturing discontinuity, shown in Figure 1.

Another example of when safe life design and management practices failed aerospace was in 1970 when a F-5 wing root fatigue fractured also from a fatigue crack nucleating at a manufacturing flaw. These events, among others, prompted the adoption of damage tolerance over safe-life design. The engine community was slower to implement damage tolerance. Two Lockheed L-1011 aircraft with Rolls-Royce turbofan engines experienced first stage fan disc bursts (two different fatality free events in December 1972 and January 1973). Another event in 1989, United Airlines flight 232 was forced to make an emergency landing (113 fatalities, 171 survived) at Sioux City, Iowa, when a Douglas DC-10, General Electric CF6-6 turbofan engine experienced an uncontained engine separation rendered the control systems inoperable. This event was attributed to a hard alpha defect (discontinuity) in a Ti-6Al-4V compressor fan disc that served as a nucleation site for fatigue which ultimately failed the component. These engine components were designed and tested to a safe-life or "initiation" based analyses. The inspections of opportunity had missed the fatigue crack seven times during inspection due to lack of empowering the inspectors with the proper approaches to have a higher probability of detecting the fatigue cracks prior to disc burst.

All of these "safe-life failures" are excellent examples of a lack of characterization of the material initial discontinuity state, a concept that is important to Holistic Structural Integrity.

Figure 1. F-111 Manufacturing Induced Discontinuity.

The era of damage tolerance began in the USAF in 1974 with significant scientific and engineering advance. Linear Elastic Fracture Mechanics (LEFM) is the foundation of crack growth analysis. The basis for LEFM is attributed to Dr. A. A. Griffith and his early work using glass as a model material circa 1920 [3,4]. Dr. Griffith postulated that the energy lost was equal to the increase in surface energy created by the newly exposed surfaces due to the crack propagation. The Griffith criterion associates change in elastic energy due to crack growth as equal to the change in the energy required to cause the crack to grow. The Griffith energy balance is shown in Eqn. 1. While ground-breaking, Dr. Griffith's work was complicated by the challenges of measuring surface energy, but this soon proved manageable. Circa 1940, Dr. George Irwin built upon the revelations of Griffith, taking the Griffith energy criterion, rearranging terms and establishing the stress intensity factor K. He set the right side of Griffith's equation equal to the stress intensity factor, K, resulting in the equation shown in Eqn. 2, providing tremendous flexibility and utility [5].

$$\sigma\sqrt{\pi a} = \sqrt{2E\gamma} \qquad (1)\ [6]$$

Where: σ = stress (ksi or MPa)
a = half crack length (inch or meter)
E = Young's Modulus
γ = surface energy of the material

$$K = \sigma\sqrt{\pi a} \qquad (2)$$

Where: K = stress intensity factor (ksi$\sqrt{\text{in}}$ or MPa$\sqrt{\text{m}}$)

The pioneering work by Griffith, Irwin and others (e.g.: Inglis, Orowan, Williams) represented an important paradigm shift; moving toward *the recognition* of discontinuities within materials; acknowledging that nearly all materials contain discontinuities and *are not* continuous, isotropic and homogenous. This was a critical change for the future of aircraft design and management. The assumption that a crack existed in the most critical location at the onset of service, forced designers to develop and use materials with slow crack growth characteristics as well as high fracture toughness. This powerful idea became known as Damage Tolerant Analysis (DTA). Damage tolerance involves the integration of structural analysis, codes, methods, and training. It involves determination, storing and provision of material characteristics such as K_{Ic}, K_c, residual strength, fatigue crack growth behavior, sustained load crack growth behavior, influence of thermal and chemical environments on subcritical crack growth behavior, and related issues. This also ushered in a time for non-destructive inspection (NDI) as it involves assessment of inspection capabilities and the quantification of NDI. It involves integration of all of the above in a consistent framework. To date, some effort has been expended on the physics of failure (nucleation, corrosion, fretting, etc.), and related issues, but much more must be achieved.

Although damage tolerance proved to be a formidable instrument for design, maintenance and management of aircraft, it was not without its challenges in application. On April 28, 1988, at 1346, a Boeing 737-200, N73711, operated by Aloha Airlines Inc., as flight 243, experienced an explosive decompression and structural failure at 24,000 feet, while en-route from Hilo, to Honolulu, Hawaii. Approximately 18 feet of the cabin skin and structure aft of the cabin entrance door and above the passenger floor-line separated from the airplane during flight. There were 89 passengers and 6 crewmembers on board. One flight attendant was swept overboard during the decompression and is presumed to have been fatally injured; 7 passengers and 1 flight attendant received serious injuries. The flight crew performed an emergency descent and landing at Kahului Airport on the Island of Maui [7]." The primary culprit behind this incident was corrosion. The environment of the Hawaiian Islands was able to infiltrate a crevice created by a failed cold-bond; this created conditions for corrosion to pervade the lap joints, providing an uncountable number of nucleation sites and material residual strength degradation. This produced widespread fatigue damage (WFD) in the forms of multisite damage (MSD) and multi-element damage (MED). This single event called attention to corrosion as a failure mechanism for aircraft.

In cases where safety of flight is threatened by corrosion, NDI techniques have been developed and successfully implemented in commercial operation. In other cases where imminent danger is deemed unlikely, such as splice joints and wing planks, NDI is in general restricted to visual inspection only. For commercial operators who achieve design service objectives in a much shorter time frame than military operators, this mode of operation has been somewhat successful from a safety point of view [8]. Notable exceptions are the decompression accidents of a Boeing 737 operated by Far Eastern Air Transport in 1981, and another Boeing 737 operated by Aloha Airlines in 1988. Other incidents are documented in a report by Hoeppner et al. [9]; a key conclusion of this report was that the lack of consistent and stringent reporting requirements makes it difficult to determine when corrosion and/or fretting were important factors in the cause of incidents.

Does the world need HOLSIP? These history making incidents as well as many others of lesser notoriety have answered this question with an unequivocal yes. What follows is an introduction to the HOLSIP methodology, some examples of recent and current efforts to support HOLSIP and finally an attempt to answer the questions posed by this article; "Is the world ready for HOLSIP?"

INTRODUCTION

HOLSIP, an acronym for Holistic Structural Integrity Process provides a framework for the interconnected nature of the various failure mechanisms. HOLSIP attempts to model the physics of failure, recognizing one failure mechanisms may be overtaken by another mechanism and then transition to another or working synergistically to hasten failure. Along with modelling failure mechanisms, HOLSIP recognizes the intrinsic nature of materials to contain discontinuities, as promoted by Griffith, Irwin, Inglis, Hoeppner, and others. Recognition of an Initial Discontinuity State or IDS is key to understanding HOLSIP. Further, HOLSIP embraces the knowledge that the IDS of a component will continually change as it encounters operational loads and various environments. This is referred to as the Evolving Discontinuity State or EDS. HOLSIP also attempts to account for Non-Destructive Inspection (NDI) issues related to maintenance, inspectability, accessibility, and detectability. The extraordinary value provided from failure evaluation to provide feedback toward continuous process improvement is also incorporated into HOLSIP.

The following is one definition of HOLISM:

> "Holism (from ὅλος holos, a Greek word meaning all, whole, entire, total), is the idea that natural systems (physical, biological, chemical, social, economic, mental, linguistic, etc.) and their properties, should be viewed as wholes, not as collections of parts. This often includes the view that systems somehow function as wholes and that their functioning cannot be fully understood solely in terms of their component parts." [10]

Griffith attempted to see things in a more complete or "whole" sense through the recognition of discontinuities. The following excerpts are from Griffith's work [11]:

> "Hence all the material in the immediate neighborhood of the critical value a crack will form."...
> "Further alterations of stress will cause this crack to spread until complete fracture occurs. This theory makes the limiting safe range of stress equal to that which just fails to maintain repeated sliding in the most favorably disposed crystals."..."The safe limit of alternating stress will usually be less than the apparent stress necessary to initiate yield in a static test, on account of initial stresses, including those due to unequal contraction of the crystals."
>
> "Suppose, in accordance with the foregoing theory of fatigue, that one crystal has been fractured, then the general criterion of rupture shows that the crack cannot spread unless the material is subjected to a certain minimum stress, which is greater the smaller the crack."

Griffith began one of the concepts of HOLSIP circa 1920; nearly a century later much work remains to be accomplished in order to embrace and utilzed this methodology.

Holistic Life Prediction Methodology

Fundamentally, engineering is applied science. In recent years, there have been significant advances in the sciences used for aircraft design and management making the concepts of HOLSIP, or highlighting the whole and the synergistic effects of the mechanisms and processes, much more within reach [12].

Circa 1971, the concepts of holistic life prediction methodologies were developed by Hoeppner [13]. He and others had not yet envisioned this phraseology but the use of systems based approach to structural integrity had been proposed. In 1994 Hoeppner offered a presentation relating inherent discontinuities and heterogeneities as they relate to safe life, damage tolerance, and Holistic Structural Integrity Design [14].

HOLSIP is concerned with many aspects of engineering, some of the more prudent include, design as "closed loop" and concepts of failure mechanisms and "pathology" of structures pervade all phases of design and operation. Materials are characterized for an IDS capturing details inherent to the material as well as those introduce by the manufacturing and assembly processes. The process attempts to adequately characterize the EDS. All aspects of NDI are considered from the design process through field access and all stages in between. Fatigue is modeled in phases with other failure mechanisms contributing to the process. Eqn. 3, Figure 2, and Table I describe some of the details of the HOLSIP modeling methodology. The overarching effort is to capture all contributing factors and model the physics of failure, attempting to capture the causes for and potential solutions toward curtailing the degradation and failure processes.

$$\text{Total Life (L)} = L_1 + L_2 + L_3 + L_4 \tag{3}$$

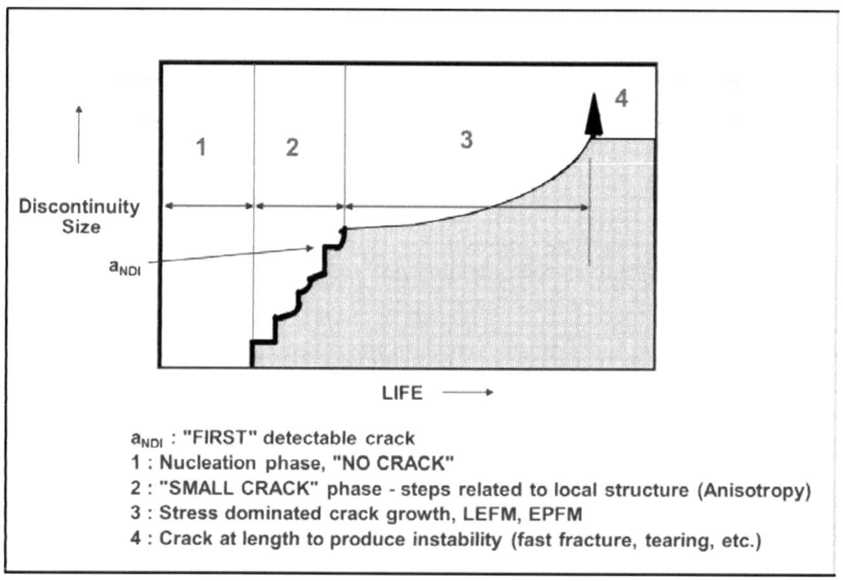

Figure 2. Schematic Representation of the HOLSIP Failure Process [15].

Table I. Phases of Fatigue Life.

L_1	L_2	L_3	L_4
NUCLEATION	"SMALL CRACK" GROWTH	STRESS DOMINATED CRACK GROWTH	FAILURE (FRACTURE)
Material failure mechanism with appropriate stress/strain life data Nucleated discontinuity (not inherent) type, size, location Presence of malignant D*, H* Possibility of extraneous effects: • Corrosion • Fretting • Creep • Mechanical Damage	Crack Propagation Threshold related to structure (micro) Structure dominated crack growth Mechanisms, rate, onset of stress dominated crack growth Effects of: • R (stress ratio) • Stress state • Environment *(time, chemical, Temp)* • Spectrum ⇒ waveform	Fracture mechanics: • similitude • boundary condition (LEFM – EPFM) Database** Appropriate stress intensity factor Initial D*, H* size, location, type Effects of: • R (stress ratio) • Stress state • Environment *(time, chemical, Temp)* • Spectrum ⇒ waveform	K_c, K_{Ic}, K_{Iscc} etc. C.O.D. Cracking mode Tensile/compressive buckling

*Discontinuity, Heterogeneity, **i.e. Mil. Handbook 5 [16]

Capturing all of these variabilities is virtually impossible in a determinstic framework. Probabilistic approaches have been implemented to attempt to capture the HOLSIP modeling methodology. In 1999 it was shown that HOLSIP can be utilized in a manner in which to insert the effects of corrosion fatigue as part of the analytical process [17]. Integrating corrosion into the USAF Aircraft Structural Integrity Program (ASIP) using USAF ASIP tasks per MIL-STD-1530C can be used to illustrate how a revised program could be structured [18].

THE NATURE OF HOLSIP

As fleets of aircraft around the world continue to age, the need for HOLSIP becomes more apparent. By incorporating scheduled inspections as part of the damage tolerance philosophy, the structural integrity community has achieved great success in extending the service life of aircraft. However, damage tolerance still treats failure mechanisms as extremely fragmented. That is, there is not a strong sense of wholeness in the approach to managing aircraft. This can lead to unnecessary inspections, early retirements, over conservative assumptions or worse, catastrophic accidents due to synergistic effects as a result of corrosion and/or fatigue and perhaps additional material degradation mechanisms. Fretting fatigue is a perfect example of a fragmented approach to a naturally holistic failure mechanism. Currently, there exists little to no guidance for incorporating fretting fatigue into safe life and damage tolerant design approaches; HOLSIP blends these effects and accounts for the initial discontinuity state (IDS) as well as for the evolving discontinuity state (EDS) which is hypercritical to capturing the nucleation phase of fatigue cracking. Profound elucidation on these concepts exists in a 2011 article by Hoeppner *"Fretting fatigue considerations in holistic structural integrity base design processes (HOLSIP) – A continuing evolution."* [19] The title of the article is revealing on the nature of HOLSIP as it is indeed an evolutionary process. As has become more evident, HOLSIP is an extraordinarily powerful concept and there is a significant amount of work to be done. The HOLSIP methodology attempts to capture and model failure mechanisms in a synergistic and logical stepwise process; it introduces a state of mind that encourages modeling the physics of failure versus "one-size-fits-all" and overtly conservative modeling assumptions.

HOLSIP TERMINOLOGY

At the 26th ICAF Symposium, Steve Swift presented a paper titled "Sticks and Stones (Could the Words of Aeronautical Fatigue Hurt Us?)." The paper discusses the need for care with the language of aeronautical fatigue. It discusses how communication is as important as calculation and the potential for poorly chosen words to overcome sound engineering [20]. A list of potential definitions for *some* of the critical terms related to fatigue, crack nucleation and some of their contributors is provided here. These proposed definitions are taken from a list of draft definitions as developed from the HOLSIP workshop committee [21].

CLD-Crack like discontinuity. A discontinuity that meets the criteria for a crack. A stress singularity exists near or at the tip of the discontinuity; no "traction forces" exist on the surfaces of the discontinuity.

Crack detection- The detection or observation of a crack (or cracks) by a stated detection or NDI procedure with reference to a probability of detection to a stated confidence level. The NDI may be either directed or random on a given structure.

Crack Formation-The process by which a crack forms in a solid body from deformation, fatigue, corrosion, creep, wear, or synergisms or combinations of these. It is not recommended the term crack initiation be used as it has led to much confusion about what it is in reality. Either crack formation or crack nucleation emphasize the PROCESS nature of mechanisms that form/nucleate a crack given it is not there initially.

Crack Nucleation-The process by which a crack nucleates in a solid body from deformation, fatigue, corrosion, creep, wear, or synergisms or combinations of these.

Defect-(Various definitions exist.) The most common definition is any feature that is outside the boundary conditions of a given component/product design that will make the component/product incapable of meeting its requirements when it is needed. Defects are also defined related to product manufacturing and also related to representation of the product.

Examples are: Manufacturing Defects – when the product departs from its intended design, even if all possible care has been exercised, and,

Design Defects – when the foreseeable risks of harm posed by the product could have been reduced or avoided by the adoption of a reasonable alternative design, and failure to use the alternative design renders the product not reasonably safe.

Inadequate Instructions or Warnings Defects – when the foreseeable risks of harm posed by the product could have been reduced or avoided by reasonable instructions or warnings, and their omission renders the product not reasonably safe.

DS-Discontinuity state. See IDS, EDS, and MDS below.

EDS-Evolving discontinuity state. The description of the evolution of the discontinuity and the progression of changes to the discontinuity or population of discontinuities over time and cyclic load exposures. (Subsequent to either the nucleation of a discontinuity or the activation of an IDS by a specific physical or chemical process acting alone or conjointly the resultant discontinuity or population of them may evolve in state with time or cyclic load exposure. Various metrics are used to describe the EDS).

IDS- Initial discontinuity state. The initial (intrinsic) population of discontinuities that are in a structure made of a given material as it was manufactured in a given geometric form. The IDS is a geometric and material characteristic that is a function of composition, microstructure, phases and phase morphology, and the manufacturing process used to process the material. The geometric and material discontinuities can be modeled separately.

Examples of material IDS types include constitutive particles, inclusions, grain boundaries, segregated phases, phase boundaries, voids (vacancies, micro-porosity, and porosity), intrinsic cracks, etc. Manufacturing processes such as machining and assembly can introduce additional discontinuities at fasteners, fillets, etc., that extend the tail (larger discontinuity sizes) of the IDS distribution.

IDS_{mat} Initial discontinuity state associated with a material. The initial population of intrinsic material discontinuities. See IDS.

IDS_{mfg}-Initial discontinuity state associated with manufacturing of a given type. The resultant effect on the population of discontinuities from a given manufacturing process or sequence of manufacturing processes including joining of the three major types (viz. mechanical joining, thermal joining, adhesive bonding).

IDS_{geom} - Initial geometric discontinuity. The initial geometric discontinuities in a product. These often are generally referred to as a "notch."

MDS-Modified discontinuity state. The physical state of a discontinuity or damage state at any given time in its evolution. Various metrics may be used to describe the state. Example-a crack has grown to a given size and it is an MDS at a specific time and thus size. Example 2-A corrosion pit has grown to a state at some point in time. The IDS may progress (EDS) to various MDS values through the mechanisms of corrosion, creep, fatigue, wear or combinations of these over time.

The definitions provided here are critical to the HOLSIP mindset.

RECENT WORK IN HOLSIP

At the USAF Aircraft Structural Integrity Conference, Dr. Calcaterra discussed how "Not one Air Force (USAF) system operating today is made completely to blueprint" and highlighted the need for validated analysis tools that can handle blueprint deviations such as residual stresses and corrosion couples [22]. The following outlines several steps taken by the A-10 and T-38 ASIP teams in conjunction with Southwest Research Institute towards understanding residual stresses as well as efforts to incorporate these into analysis tools.

Basic USAF damage tolerance analysis guidance assumption is a 0.050 inch "initial flaw size" or IFS when analyzing circular holes in primary structure. When cold expansion is applied the IFS assumption is revised to 0.005 inches. This guidance is provided by JSSG 2006 and reads as follows:

> "The limits of the beneficial effects to be used in design should be no greater than the benefit derived by assuming a 0.005 inch radius corner flaw at one side of an as-manufactured, non-expanded hole containing a neat fit fastener in a non-clamped-up joint" [23].

This assumption has been shown experimentally to generally be conservative. Pilarczyk and Carlson experimentally demonstrated that the 0.005" assumption is dramatically conservative [24]. However, as shown by Warner [25] and also by Andrew [26], this assumption is not always true. Warner and Andrew in separate studies showed this assumption to be less than conservative for some loading conditions (σ_{max}=43.25ksi, R=0.1, constant amplitude) and for edge margins of 1.2 (e/D=1.2). Additionally, as demonstrated in Figure 3 and explained by Thomsen, Clark and Babish, field results can prove less than conservative for this 0.005 inch assumption.

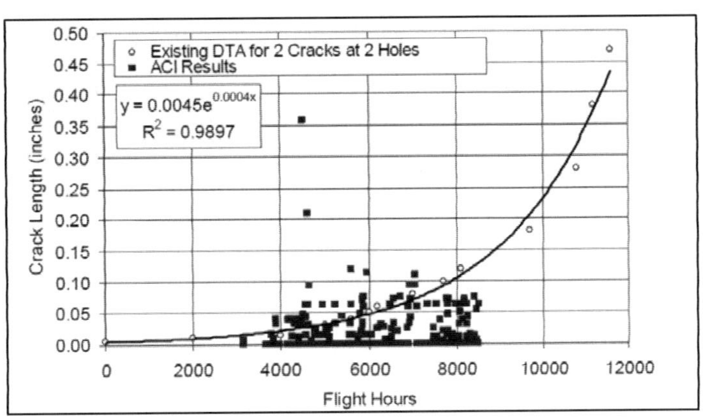

Figure 3. Inspection Results vs. Damage Tolerant Analysis [27].

These examples begin to show the inconsistencies associated with the generic assumption guidelines provided by JSSG. Evidence for modelling the physics of failure and improving engineering analysis tools becomes apparent as does the need for further research. Circa 2006, in tandem studies, Carlson and Pilarczyk modeled the crack growth behavior of open (not filled) cold expanded holes by developing beta solutions and applying those through the crack growth software AFGROW®. Additionally, they captured and modeled the unique crack shape, as shown in Figure 4, witnessed in many cold expanded holes. They were successful in capturing the effect of cold expansion, which has served as a significant step towards understanding the physics of fatigue crack propagation under these conditions.

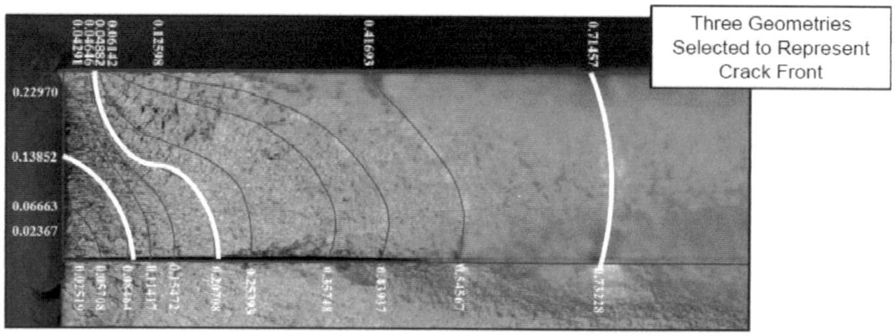

Figure 4. Modelling the Physics of Crack Progression in Cold-Expanded Holes [28].

Another challenging area for sustainment of aging aircraft is the recognition that non-destructive inspection (NDI) does not capture existing cracks 100% of the time. Frequently in sustainment engineering support, the engineer has been asked to provide a solution to repair the existence of a crack. The crack is sometimes visible but more often it is represented by a signal from some form of NDI (e.g.: Eddy Current, Ultrasound, etc.). If the crack is believed to be small, the crack can be removed by the process of reaming or oversizing the hole. Most commonly, bolt-hole eddy-current NDI is relied upon to determine if the crack has been successfully reamed out. However, the threshold of detection at the $N_{90/95}$ level in aluminum structure is 0.050 inch. This is an order of magnitude larger than the 0.005 inch IFS assumption as directed by JSSG. The question was raised: "What if a 0.050" crack is missed in a fastener hole that is then cold expanded? Will the "missed crack" propagate? Are the inspection intervals adequate to ensure structural integrity?" These questions were researched by Warner [29], and Andrew [30] investigating the "missed crack" considering the effects of loading and edge margin. Warner and Andrew each showed that for certain circumstances the 0.005 inch assumption was less than conservative and a threat to structural integrity.

In 2008 the A-10 fleet experienced a threat to the lower wing skin in the form of corrosion fatigue cracking that could have potentially grounded the entire fleet [31]. The following year (2009) another corrosion fatigue cracking scenario, this time threatening the structural integrity of the fuselage was encountered; this threat was associated with a maintenance induced discontinuity, an example of part of a Modified Discontinuity State (MDS) [32]. At the time, analysts had to make assumptions in order to develop stress intensity solutions required to manage the needs of the fleet. Crack shapes were assumed to be either straight or elliptical (once transitioned through thickness) and did not capture realistic crack propagation or crack link-up. Stress intensity solutions at that time were resolved at the surface and along the bore for corner cracks. For through cracks, a stress intensity solution was selected as an average of multiple points within the center of the elliptical through crack. A stepwise process of modeling was developed in order to produce confident load redistribution and establish a more physics based approach to establishing fatigue crack growth curves to be used for fleet management. This stepwise process used various modeling tools that involved manually transferring the results from one tool into another, which was time consuming. This was surmounted by coupling StressCheck®, a finite element software used to develop stress intensities with the crack growth software AFGROW®. A schematic representing the flow of modelling and predicting crack propagation is shown in Figure 5. The result of this coupling was a tool called Broad Application for Modelling Failure (BAMF). BAMF allowed for crack propagation estimation or prediction in a manner that more closely represents structural crack growth [33]. This development was a significant step towards a more holistic approach toward modelling the physics of failure.

The efforts to more closely represent the structural crack growth provided USAF sustainment groups with fleet management options that prevented fleet groundings, which would previously have been the only viable option. The methodology incorporated into BAMF takes advantage of load redistribution and results in a significantly larger critical cracking scenario which presents many more options for strategic decision making purposes.

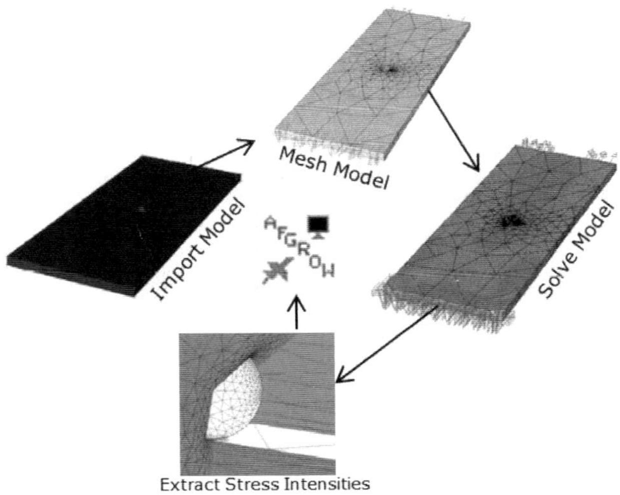

Figure 5. Scheme for Broad Application for Modeling Failure [34].

Failure is a term that must be characterized. Accounting for the state of mind with regard to HOLSIP it is proper at this time to discuss failure. Failure is difficult to define as it means different things to different people and different applications. Is failure the presence of a crack? Deformation? Fracture into two or more pieces? Etc. MSG-3 provides a sound definition of failure as: "The inability of an item to perform within previously specified limits" [35,36].

As part of a Rapid Innovation Funding program, the A-10 and T-38 ASIP team has worked with Analytical Processes/Engineered Solutions, Inc. (AP/ES) and Hill Engineering to attempt to characterize the residual stress state of a fastener hole with high fidelity results. The contour method was employed to take measurements of coupons with cold expanded holes. Preliminary results from these experiments are illustrated in a 2014 NATO publication [37]. These measured and characterized residual stresses will then be coupled with BAMF modelling software providing another step toward understanding and predicting the physics of failure.

Capturing the physical phenomenon present within a structure is key to holistic modeling. Generally, analysts make assumptions due to not knowing and not understanding the physical phenomenon of the structures for which they work. These assumptions tend to be overly conservative but can be tragically under conservative as has so often been the case as highlighted in the Comet, F-111, and other catastrophes previously within this article.

HOLSIP WORKSHOPS AND EDUCATION ACTIVITIES

In 2002, the first workshop dedicated to Holism and Holistic Structural Integrity Processes was held in Park City, Utah. The HOLSIP workshop, as it is now referred to, was started by a small group of engineers associated with work in aircraft structural integrity and has continued to meet annually. The workshop has an open invitation and is regularly attended by members of academia, industry and military. Every year the workshop has been well attended by many international participants including Australia, The Netherlands, Canada, Japan, Poland, and England.

> VISION: By 2025, HOLSIP is the state of mind driving the framework for design and support of engineering structures. The framework will significantly enhance safety, reliability and reduce cost of design, manufacturing and operation.
>
> MISSION: Working collaboratively with industry, government agencies and academia, we will develop and deliver a physics-based structural integrity framework that takes into account both intrinsic and extrinsic factors for existing and next generation aerospace materials.

Recently, the workshop gathered for the 14th meeting with 53 participants, a record number of attendees for HOLSIP; there were many international attendees and multiple representatives from the United States Air Force.

Beginning in 1971, Dr. Hoeppner introduced concepts which were later to become known as HOLSIP through University of California Los Angeles-Extension and shortly thereafter at the University of Missouri, University of Toronto and University of Utah; these courses offered unique approaches to management of fatigue, corrosion, creep and wear as well as concepts regarding reliability and quality, all from an *holistic* perspective. These innovative concepts evolved to the state of HOLSIP as is it exists today. Dr. Clark continues instructing HOLSIP philosophies through University of California Los Angeles-Extension. Currently at Delft University of Technology (TU Delft), there is a dedicated effort toward implementing a Holistic Structural Integrity Process (HolSIP) in aerospace and wind energy infrastructure. The Faculty of Aerospace Engineering and DUWIND (TU Delft Wind Energy Institute) are working on maximizing operational component life of wind turbine and aerospace infrastructure through a physics based program entitled "Holistic Structural Integrity Process" [38]. Dr. Martinez and Dr. Rans co-offer a course emphasizing: "The course discusses a Holistic Structural Integrity Process (HolSIP) founded upon the primary idea that all failure mechanisms involved in the degradation of the structure are interconnected and should not be analyzed as merely the sum of individual mechanisms" [39].

Structural integrity of aging and modern aircraft is a careful balance of multiple failure mechanisms as well as design and management philosophies. Holistic Structural Integrity Processes requires understanding degradation and failure mechanisms, characterizing initial and constantly evolving discontinuity states, and appropriately modeling the physics of failure of these items to compliment the tools available to analysis, research, design, and sustainment engineer.

SUMMARY

Nearly a century ago, Alan Arnold Griffith documented the existence of discontinuities in materials. This fact has since been challenging for the engineering community to manage. In order to advance structural integrity into a holistic mindset, engineering must embrace the presence of and demand the characterization of discontinuities. Inadequate characterization of discontinuities and their continual evolution can lead to inefficient use of components (early retirement) or far worse, catastrophic consequences. Engineering is a discipline rooted in science, it is not an exact science; history has shown time and again how often engineering assumptions have proved deadly. HOLSIP presents a framework as an attempt toward rooting engineering in physics based modeling and to remove as much of the guesswork and assumptions associated with unknowns as possible. The various HOLSIP related activities going on around the globe highlight that The World is Ready for HOLSIP! And the time is now!

REFERENCES:

[1] Ministry of Transport and Civil Aviation. (1955). Civil Aircraft Accident Report of the Court of Inquiry into the Accidents to Comet G-ALYP on 10th January, 1954 and Comet G-ALYY on 8th April, 1954. London: Her Majesty's Stationery Office.

[2] Ministry of Civil Aviation. (1953). Civil Aircraft Accident Report of the Court Investigation of the Accident to COMET G-ALYV on 2nd May, 1953. London: Her Majesty's Stationery Office.

[3] Griffith, A. A. (1921). In: *"The Phenomena of Rupture and Flow in Solids",* Philosophical Transactions, Royal Society of London, Series A, Volume 221.

[4] Griffith, A. A., (1924). In: *"The Theory of Rupture",* Proceedings of the First International Congress of Applied Mechanics.

[5] Irwin, G. (1948). *"Fracture Dynamics"*, Fracturing of Metals, American Society for Metals.

[6] Broek, D. (1986). In: *Elementary Engineering Fracture Mechanics, 4th Edition,* Martinus Nijhoff Publishers.

[7] NTSB Report No. NTSB/AAR-89/03 (1988). In: "Aircraft Accident Report—Aloha Airlines, Flight 243, Boeing 737-200, N73711, Near Maui, Hawaii, April 28, 1988.

[8] Komorowski, J. P., Forsyth, D. S., Bellinger, N. C., Hoeppner, D. W. (2001). In: *"Life and Damage Monitoring-Using NDI Data Interpretation for Corrosion Damage and Remaining Life Assessments"*, Published in the Proceedings of the RTO Specialist's Meeting on Life management for aging air Vehicles, Paper No. 13, Manchester, UK.

[9] Hoeppner, Grimes, Hoeppner, Ledesma, Mills, Shah (1995). In: *"Corrosion and Fretting as Critical Aviation Safety Issues,"* Proceedings of the 18th ICAF Symposium, vol. I, pp. 87-106, Grandage, J.M. and Jost, G.S. (Eds.), Melbourne, Australia .

[10] http://en.wikipedia.org/wiki/Holism, February 16, 2015 at 4:47pm MST.

[11] Griffith, A. A. (1921). In: *"The Phenomena of Rupture and Flow in Solids",* Philosophical Transactions, Royal Society of London, Series A, Volume 221, pp. 191, 194, 195.

[12] Komorowski, J. P., Forsyth, D. S., Bellinger, N. C., and Hoeppner, D. W. (2001). In: *"Life and Damage Monitoring-Using NDI Data Interpretation for Corrosion Damage and Remaining Life Assessments",* Published

in the Proceedings of the RTO Specialist's Meeting on Life management for aging air Vehicles, Paper No. 13, Manchester, UK.

[13] Hoeppner, D. W. (1972). In: *"Corrosion Fatigue Considerations in Materials Selections and Engineering* Design *"*, Corrosion Fatigue: Chemistry, Mechanics, and Microstructure, NACE, pp. 3-11.

[14] Hoeppner, D. W. (1994). In: *"The Role of Inherent Discontinuities (D) and Heterogeneities (H) in the Process of Nucleation of Fatigue* Cracks in Aircraft Structural Materials", presented at the Third DND-Canada Conference on Structural Integrity, Ottawa, Canada.

[15] Hoeppner, D. W. (1985). In: *"Parameters that Input to Application of Damage Tolerance Concepts to Critical Engine Components,"* AGARD Conference, San Antonio, Texas, Published in Conference Proceedings AGARD-CP 393, Damage Tolerance Concepts for Critical Engine Components, pp. 4-1 to 4-16, NATO-AGARD, France.

[16] Hoeppner, D. W. (1985). In: *"Parameters that Input to Application of Damage Tolerance Concepts to Critical Engine Components,"* AGARD Conference, San Antonio, Texas, Published in Conference Proceedings AGARD-CP 393, Damage Tolerance Concepts for Critical Engine Components, pp. 4-1 to 4-16, NATO-AGARD, France.

[17] Brooks, C. L., Prost-Domasky, S., and Honeycutt, K. (1999). In: *"Corrosion is a Structural and Economic Problem: Transforming Metrics to a Life Prediction Method"*, Fatigue in the Presence of Corrosion, Report AC/323(AVT)TP/8; RTO-MP-18, North Atlantic Treaty Organization, Brussels, Belgium.

[18] Brooks, C. L., and Simpson, D. L. (1999). In: *"Integrating Real Time Age Degradation Into the Structural Integrity Process",* Fatigue in the Presence of Corrosion, Report AC/323(AVT)TP/8; RTO-MP-18, North Atlantic Treaty Organization, Brussels, Belgium.

[19] Hoeppner, D. W. (2011). *Tribology International,* vol. 44, 1364-1370.

[20] Swift., S. (2011). In: *Structural Integrity: Influence of Efficiency and Green Imperatives*," Proceedings of the 26st ICAF Symposium, pp. 27-37, Komorowski, J. (Ed.), Springer, Montreal, Cananda.

[21] Hoeppner, D. W. (2010). *HOLISTIC Related Definitions of Terms*, 1998-current, copies available from dhoeppner@eng.utah.edu, Last draft 2010.

[22] Calcaterra, J. (2008). In: *"The Effect of Manufacturing Variability on Air Force Systems,"* ASIP Conference, San Antonio, Texas, Published in Conference Proceedings.

[23] Department of Defense, United States of America (1998). Joint Service Specification Guide, Aircraft Structures, JSSG-2006, Washington D.C., USA.

[24] Carlson, S., Pilarczyk, R. (2007). *"Experimentally Derived Beta (β) Corrections to Predict/Model the Fatigue Crack Growth Behavior at Cold Expanded Holes – A Bridge to the Future,"* ASIP Conference, San Antonio, Texas, Published in Conference Proceedings.

[25] Warner, J. J., Clark, P. N., Hoeppner, D. W. (2014). *"Cold Expansion Effects On Cracked Fastener Holes Under Constant Amplitude and Spectrum Loading In The 2024-T351 Aluminum Alloy,"* International Journal of Fatigue, doi: http://dx.doi.org/10.1016/j.ijfatigue.2014.05.002.

[26] Andrew , D., Clark, P. N. , Hoeppner, D. W. (2014). *"Investigation of cold expansion of short edge margin holes with pre-existing crack in 2024-T351 aluminum alloy,"* Fatigue and Fracture of Engineering Materials & Structures, Vol. 37, pp. 406-416, Wiley Publishing Ltd.

[27] Thomsen, M. L., Clark, P. N., Babish, C. (2003). In: *"Re-Establishing an Aircraft Structural Integrity Program for the A/OA-10A,"* ASIP Conference, San Antonio, Texas, Published in Conference Proceedings.

[28] Carlson, S., Pilarczyk, R. (2007). *"Experimentally Derived Beta (β) Corrections to Predict/Model the Fatigue Crack Growth Behavior at Cold Expanded Holes – A Bridge to the Future,"* ASIP Conference, San Antonio, Texas, Published in Conference Proceedings.

[29] Warner, J. J., Clark, P. N., Hoeppner, D. W. (2014). In: *"Cold Expansion Effects On Cracked Fastener Holes Under Constant Amplitude and Spectrum Loading In The 2024-T351 Aluminum Alloy,"* International Journal of Fatigue, doi: http://dx.doi.org/10.1016/j.ijfatigue.2014.05.002.

[30] Andrew , D., Clark, P. N. , Hoeppner, D. W. (2014). *"Investigation of cold expansion of short edge margin holes with pre-existing crack in 2024-T351 aluminum alloy,"* Fatigue and Fracture of Engineering Materials & Structures, Vol. 37, pp. 406-416, Wiley Publishing Ltd.

[31] Pilarczyk, R., Carlson, S., Stowe, G. (2009). In: *"Is ASIP Still Alive? The A-10 Lower Wing Cracking Issue of 2008-2009,"* Proceedings from USAF ASIP Conference, Jacksonville, FL.

[32] Sedgwick, H., Clark, P. N., Pilarczyk, R., Stowe, G. (2010). In: *"FEA & DTA Development for A-10 Fuselage Longeron Cracking,"* Proceedings from USAF ASIP Conference, San Antonio, TX.

[33] Hodges, J., (2014). *"BAMF! Look How Far We've Come,"* Proceedings of the 2014 AFGROW Workshop, Layton, UT, USA.

[34] Hodges, J. (2012). *Applications of Advanced Fracture Mechanics Utilizing StressCheck and AFGROW,"* USAF, T-38 Structural Integrity and Analysis Group, 11-11-22_GH_035.

[35] ATA, MSG-3 (2007). Operator/Manufacturer Scheduled Maintenance Development, Air Transport Association of America, Washington DC, USA.

[36] Swift., S. (2011). In: *Structural Integrity: Influence of Efficiency and Green Imperatives*," Proceedings of the 26st ICAF Symposium, pp. 27-37, Komorowski, J. (Ed.), Springer, Montreal, Cananda.

[37] Carlson, S. C., Hodges, J. L. , Pilarczyk, R. T., Clark, P. N. (2014). In: *"21st Century Crack Growth Analysis Methods & Tools – Building New Damage Tolerance Capabilities in the United States Air Force (USAF) Aircraft Structural Integrity Program (ASIP) Environment,"* Proceeding from NATO STP-MP-AVT-222 Continuing Airworthiness of Ageing Aircraft Systems, Brussels, Belgium.

[38] DUWING, TU Delft Wind Energy Institute, R&D Programme 2015-2020, TU Delft, The Netherlands.

[39] Martinez, M., Rans, C. (2015). Course syllabus, "Holistic Structural Integrity Process (HoLSIP), Course No. AE4770, TU Delft, The Netherlands.

How proper similitude principles could have improved our understanding about fatigue damage growth

R.C. Alderliesten

Structural Integrity & Composites, Faculty of Aerospace Engineering, TU Delft

Abstract: This paper discusses similitude parameters for predicting fatigue damage growth, and demonstrates that proper principles exploiting the strain energy release could have improved our knowledge on fatigue damage growth. The paper demonstrates how the original Stress Intensity Factor concept introduced by Irwin has been introduced inconsistently for describing fatigue crack growth. Subsequently, it illustrates how this has led to misinterpretation of certain phenomena, like for example plasticity induced crack closure.

Using proper similitude in agreement with the physics of observed fatigue damage phenomena, it is demonstrated how the average strain energy release over a single load cycle should relate to the crack surface extension in that same load cycle. The paper concludes with illustrating how phenomena like for example plasticity and fibre bridging in composites can be understood and quantified.

INTRODUCTION

To describe fatigue in metallic structures and materials, different approaches are adopted. Fatigue life, or fatigue initiation life, is generally described based on stress amplitudes and stress concentration factors, whereas for crack propagation, methodologies adopting Linear Elastic Fracture Mechanics (LEFM) have been developed. Reviewing the literature illustrates that most methodologies have been developed in a rather empirical fashion. A parameter was taken to describe similitude and using this parameter, observed data was correlated and formulated as material characteristic. For stress based methods, these material characteristics are generally provided as S-N curves, while for the LEFM methods, these characteristics may be presented in the form of Paris-type relationships [1].

Despite the differences between these two categories, they both seem to have in common that various additional corrections or factors are required. Some corrections are attributed to geometry or finite width, others to the mean stress or stress ratio, and some corrections are specifically attributed to phenomena like crack closure or plasticity [2,3]. This paper discusses the meaning of the necessity of additional corrections in light of the similitude taken in these methodologies. The hypothesis here is that some corrections have been introduced with an explanation that considering the similitude principle adopted, is incorrect.

To illustrate these discrepancies, the present paper first discusses the origin of in particular the methodology using Stress Intensity Factors (SIF). The objective is to demonstrate how the original SIF concept introduced by Irwin [4] has been introduced inconsistently for describing fatigue crack growth. Then, an alternative methodology is proposed that utilizes strain energy release over individual load cycles as similitude parameter to describe fatigue damage growth.

PROBLEM STATEMENT

The correction factor that has received most attention within the literature on fatigue crack growth is the stress ratio correction. This correction has been explained by numerous authors to be caused by plasticity induced crack closure. In an earlier paper [5], the author has demonstrated that this correction to great extent can be described by considering cyclic energy rather than cyclic stress. A more thorough discussion on the nature of the stress ratio correction is given in [6], where the following observations are reported:

- Stress ratio effect is present in any material system, even brittle composites, not only ductile alloys
- Plasticity induced crack closure correction is similar to common mean stress corrections for the fatigue limit in S-N fatigue analyses
- Paris curves for different stress ratios seem to collapse when dividing the SIF by the Young's modulus
- Stress ratio dependency on threshold SIF vanishes when recalculated to threshold Strain Energy Release Rates (SERRs).

A straightforward demonstration that one can make is the second observation reported in [6]. Figure 1 (a) presents the correction proposed by Schijve for aluminium 2024-T3 [7,8]. This correction is generally plotted in terms of S_{op}/S_{max} versus stress ratio R. With the mechanical properties for this aluminium known, the correction factor can be recalculated to an amplitude stress S_a versus mean stress S_m, assuming that the effective stress cycle is between S_{op} and S_{max}. The result of this recalculation is presented in Figure 1 (b) where the curve is compared with the Gerber mean stress correction for the fatigue limit.

The correlation between the crack closure correction and the mean stress correction is striking. In particular when one considers that one correction aims to correct for plasticity induced closure in the wake of macroscopically long cracks, while the other correction corrects for the fatigue limit, i.e. where the contribution of crack growth to the life is nil [8].

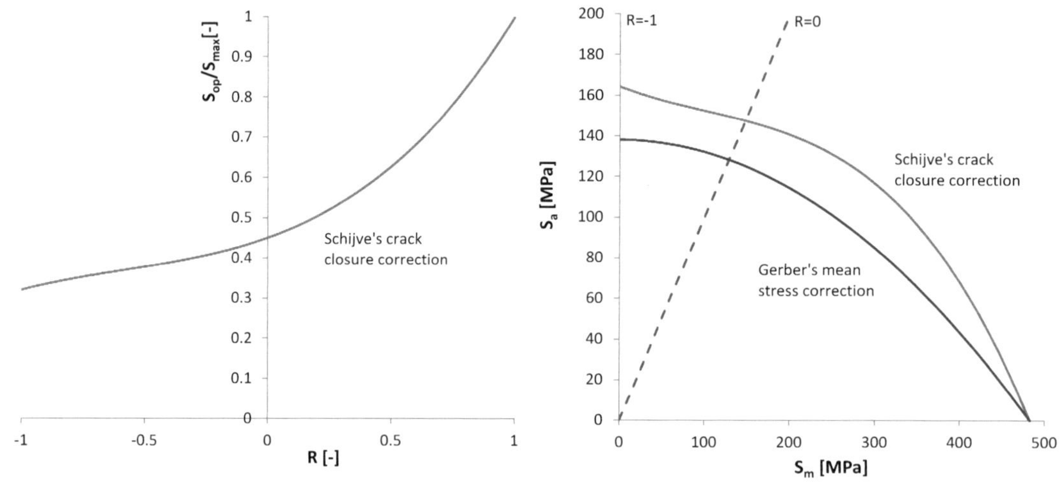

Figure 1. Correlation between Schijve's crack closure correction for aluminium 2024-T3 (a) and the mean stress correction for the fatigue limit for that alloy (b).

The problem here is that an apparent effect has been observed: the Paris relationship between da/dN and ΔK depends on the stress ratio R. But the physics of that effect is not well understood. Hence, hypotheses have been proposed that could have been falsified if data had been considered beyond the limited scope of fatigue crack growth in ductile alloys.

To fully understand the mechanism of fatigue crack propagation in metallic alloys, one has to understand the physics of fatigue damage growth in engineering materials in general. Because all mechanisms in these materials are subject to the first and second law of thermodynamics, this implies understanding the nature of energy dissipation in the process of fatigue. Any intrinsic or extrinsic shielding mechanism, like plasticity, crack closure, fibre bridging (in fatigue delamination), etc. should then be described with similar energy balance considerations. However, this requires the establishment of proper principles of similitude that reflect both the physics and mechanics of fatigue damage growth. It seems that despite the claims in the literature, such principles have not yet been well established.

STRESS INTENSITY FACTOR FOR CRACK GROWTH

In 1921, Griffith [9] proposed considering strain energy, rather than using stress based methods, as was common in those days. In his hypothesis, the prediction of crack growth is based on the balance of energy required to break atomic bonds, and the energy that is available. This concept was further developed by Irwin [4] who quantified the available energy, defined as the SERR. For fixed grip conditions in which a crack increment does not impose application of extra work, the release in strain energy as result of the crack increment can be compared with the energy required to further increment the crack. Hence, the SERR could be used as a quasi-static failure (or stability) criterion, in which the SERR constitutes a 'driving force' for crack growth.

Furthermore, Irwin related the SERR to the SIF describing the singular stress field at a crack tip. This relationship is generally accepted to be valid, in particular for the quasi-static fixed grip conditions that Irwin studied. The release of elastic strain energy as result of an instantaneous infinitesimal increment of crack growth is thus considered to be equivalent to the elastic stress field at the crack tip for the applied load prior to that crack increment.

This relationship was used by Paris [1] who demonstrated that the crack growth rate, da/dN, plotted against the SIF range, $\Delta K = K_{max} - K_{min}$, could be described by a power law

$$\frac{da}{dN} = C\Delta K^{n} \tag{1}$$

In this equation, C and n are curve-fitting parameters. The argument of Paris for using K was that the crack tip stress field controls the rate of crack extension, and that in K the effects of external load and configuration were reflected. At present, this 'Paris relationship' is the basis for nearly all crack growth prediction models. Later, modifications to equation (1) have been proposed to include a critical K_c to capture the asymptotic behaviour at high ΔK [10], and/or a threshold ΔK to capture the lower asymptote [11,12,13].

STRAIN ENERGY VERSUS STRESS INTENSITY

For the growth of disbonds in adhesive joints and for delaminations in composites, the 'Paris relationship' is also commonly used. Because these cracks are often at an interface between two dissimilar materials, the SERR is easier to compute than the SIF. Hence, rather than ΔK, G_{max} or ΔG are used in equation (1), but as parameter to represent the stress state at the crack tip, rather than to represent strain energy release as proposed by Griffith.

Within LEFM this relationship between K and G is taken generally for plane stress conditions as

$$G = \frac{K^2}{E} \tag{2}$$

It seems that most people who apply LEFM to fatigue crack propagation consider G equivalent to K in that it is a driving force, rather than the consequence of crack growth. However, equation (2) was derived for fixed-grip conditions where an instantaneous and infinitesimal crack increment releases strain energy that can be related to the stress field in the crack tip vicinity. That does not necessarily imply that the release in strain energy by crack extension during a single fatigue cycle relates to ΔK with equation (2)!

This misunderstanding is illustrated by Paris [14] when he refers to earlier observations by Anderson, and Harris and himself [15], who reported that measured fatigue crack growth rates (in inert environments) were identical for various base materials and their alloys, if K was normalized with the elastic modulus E. The implication of this observation, i.e. fatigue crack growth is governed by energy rather than stress, was not considered.

SHORTCOMING OF SIF RANGE ΔK

In the quasi-static analysis of Irwin, the correlation between the SIF and SERR according to equation (2) is valid. However, due to the cyclic nature of the fatigue crack extension process, this relation requires some elaboration. Either the process is governed by stress, and then $\Delta S = S_{max} - S_{min}$ suffices, or the process is governed by energy and then, to describe the crack driving force one should use

$$\Delta U = \frac{P_{\max}^2}{2E} - \frac{P_{\min}^2}{2E} = \frac{V}{2E}\left(S_{\max}^2 - S_{\min}^2\right) \tag{3}$$

In this equation V is the volume of the sample or system. The correlation between strain energy and load (and thus stress) is illustrated in Figure 2 (a). In this illustration it is evident that in case the mean stress increases, substantially more work is applied to the specimen, which is not considered with ΔS. Hence, a stress ratio effect is to be expected [5].

Another aspect that is apparent from Figure 2 (a) is that the minimum stress represents part of the total applied work to the specimen following $U_{tot} = U_{mon} + U_{cyc}$. Hence, deducting the minimum stress (or minimum strain energy) from the maximum stress (or total strain energy) may be deemed inappropriate as it constitutes a contribution to the crack driving force. In particular when high stress ratios are considered, this contribution may be significant.

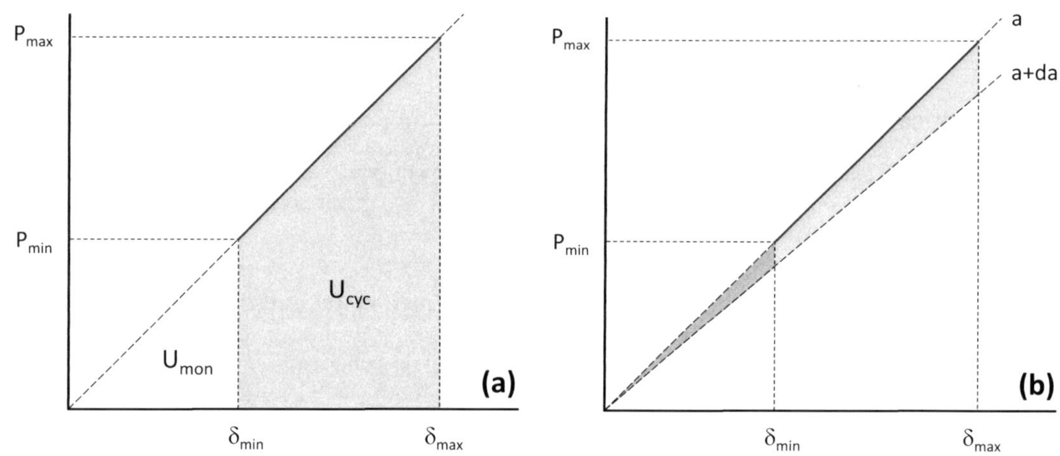

Figure 2. Illustration of linear elastic strain energy applied when loading from P_{min} to P_{max} (a) and the strain energy release due to crack extension *da* (b).

Another reason for considering the strain energy below the minimum load P_{min} is that in case of a crack increment *da*, strain energy is released even below P_{min}, see Figure 2 (b). In that respect it indeed seems more appropriate to consider G_{max} to describe fatigue delamination in composites, rather than ΔG [16]. However, when taking data from the literature on aluminium 7075-T6 [17] it appears that when plotting *da/dN* against ΔG rather than G_{max} a good correlation is obtained. In particular for higher crack growth rates, there is substantially less difference between the curves compared to the conventional Paris relationship. This is illustrated in Figure 3.

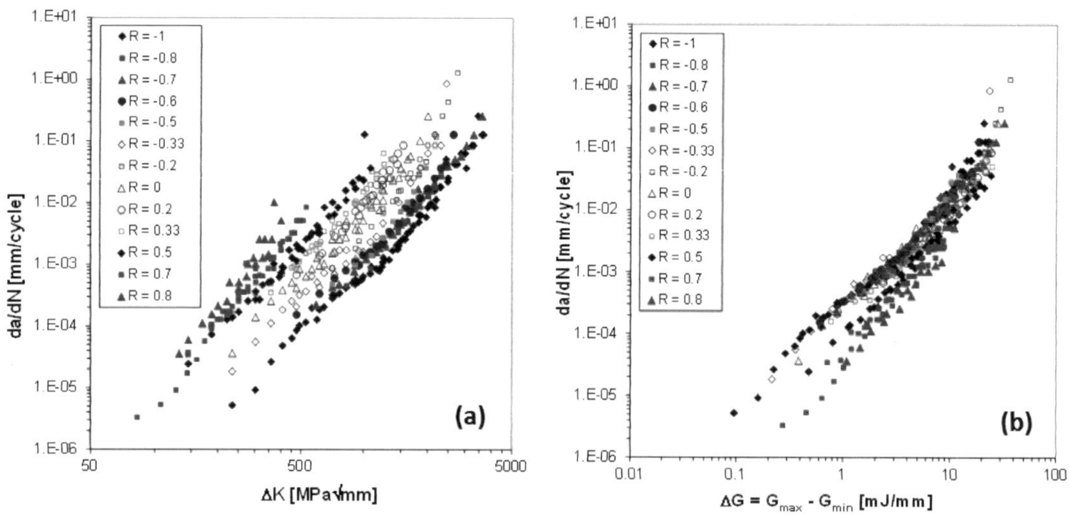

Figure 3. *da/dN* versus ΔK for different stress ratio for aluminium 7075-T6 (a), the same data plotted as *da/dN* versus ΔG (b); data from [17].

Evidence as provided with Figure 3, and observations reported in [6] indicate that the SIF range ΔK does not represent a proper parameter to describe similitude in fatigue. The arithmetic difference between K_{max} and K_{min} does not reflect the cyclic nature of the energy dissipation process that takes place during fatigue crack propagation. Hence, a better concept is needed to describe fatigue damage growth in line with the energy balance originally proposed by Griffith, but based on energy physically released rather than artificial SERR values.

THEORY VERSUS PREDICTION MODEL

In the following discussion, one has to acknowledge the fundamental difference between a theory describing the physics of fatigue, and a prediction model. A scientific theory constitutes the body of knowledge on how processes are understood and described consistent with certain laws. A prediction model is an engineering tool that basically relates input parameters that an engineer can work with to the consequence, like failure, or in this particular case: crack growth.

This distinction is important. When characterizing materials for their response to cyclic loading, one must be careful that the format of representation comprises material behaviour only. In particular this aspect is generally neglected when interpreting the Paris relationship. In that relationship, material response (da/dN) is plotted against a model (ΔK). Hence, any trend observed in that representation could either be attributed to some physical mechanism, or to the adopted model. With the example of the stress ratio effect, generally plasticity induced crack closure was considered as a physical mechanism to explain the trend, where it would have been more appropriate to reconsider the use of ΔK for describing similitude.

PHYSICS OF STRAIN ENERGY RELEASE IN FATIGUE

If one considers the concept of strain energy balance to the problem of fatigue crack growth, several distinct differences with the quasi-static approach of Griffith and Irwin can be identified. First, the crack does not increment instantaneously under a quasi-static load, but it progresses incrementally with every load cycle. In fact, with the increase in load during the load cycle, the crack gradually develops, and is generally considered to have reached its maximum extension at the maximum load of the cycle.

This implies that strain energy release dU should be related to the crack increment da generated in a particular cycle N. Thus dU/dN should physically correlate to da/dN in the form

$$\frac{dU}{dN} = \frac{dU}{da}\frac{da}{dN} \tag{4}$$

where dU/da is the effective or average strain energy release rate G_{eff} per unit width for that particular load cycle. If crack growth is the only strain energy dissipating mechanism present, then one would expect this G_{eff} to be constant and independent of either a or da/dN.

The second observation relates to the applied work W. Instead of linear-elastically increasing the load to a maximum value and subsequent instantaneous load reduction induced by crack extension, OAB in Figure 4, the force displacement curve gradually develops from O to B, because the crack gradually develops.

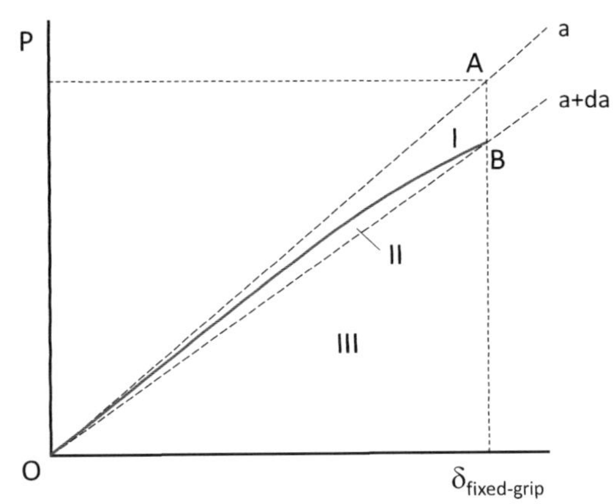

Figure 4. Illustration of energy partitioning of a load-displacement curve.

In general the SERR for this fixed grip condition is taken equal to the area I+II, while the applied work to the specimen W is characterised by the area I+II+III. However, during fatigue crack growth, the effective SERR is equal to area II, while the effective work applied is equal to area II+III. Area I represents strain energy that never has been applied to, nor released from the specimen. Note that this effectively changes the ratio of energy dissipated (dU) over the work applied (W).

Another observation is that SERR in equation (3) includes all dissipating mechanisms that may occur during that load cycle. In particular plasticity has been observed to provide a major contribution to the strain energy dissipation [18]. In fact, strain energy may be released by plasticity even before a crack may increment. This implies that equation (4) is part of

$$\frac{dU}{dN} = \frac{dU_a}{dN} + \frac{dU_{pl}}{dN} \tag{5}$$

where dU_{pl}/dN is the energy released by plasticity in a particular cycle, and dU_a/dN the energy released by crack extension. Note that dU_{pl}/dN may be substantially greater than dU_a/dN [18]. In addition, dU_a/dN mostly relates to the decohesion of material and formation of crack surfaces. Thus only an increase in effective crack surface per projected crack area, equivalent to for example an increase in roughness, may alter the magnitude of dU_a/dN. If the roughness would not change, the value for dU_a/dN would remain constant, independent of the crack length. This is not the case for dU_{pl}/dN, which seems to be linearly related to the crack length in case of force controlled tests [6]. Hence, the ratio of energy dissipated by crack extension to the energy dissipated by plasticity is not constant along the crack length, which was the implicit assumption in the application of LEFM to fatigue crack growth.

INTERPRETING STRAIN ENERGY RELEASE RATES IN FATIGUE

To fully understand the concept of strain energy release rates for the problem of fatigue damage growth, and to properly interpret the experimental data, one has to first look at what equation (4) represents. This equation can be illustrated with Figure 5 (a). In that illustration, the slope of the curve represents the magnitude of the effective SERR. It implies that indifferent of the rate with which the crack extends under the applied loading, the amount of energy dissipated by per unit of crack extension is the same. Obviously, this could only be true if the physical fracture surface at small *da/dN* and at large *da/dN* would be identical, which is not the case as numerous papers in the literature report.

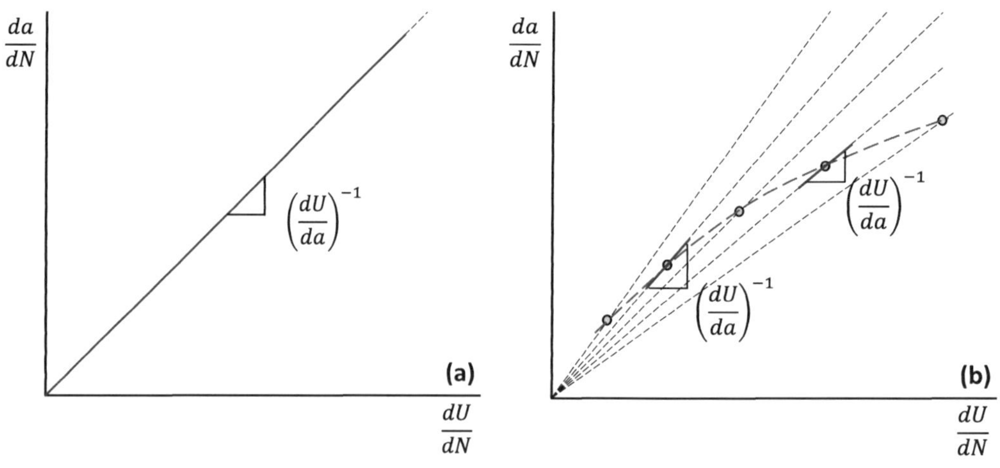

Figure 5. Theoretical constant SERR (a) and the definition of SERR in case the effective fracture surface increases with *da/dN* (b); note that both graphs contain linear scales.

Thus the fracture surface develops with crack extension, for example, by an increase in roughness or a transition from a mode I crack to a mixed mode crack by shear lip formation. The crack length a generally is a planar projection of the fracture surface. This increase in roughness in that case should be interpreted as an increase in effective fracture surface per crack area da (unit width). An increase in fracture surface constitutes an increase in decohesion of atomic bonds, and an increase in strain energy released. This is illustrated in Figure 5 (b), where at higher crack growth rates effectively more strain energy is released. Hence, dU/da is not constant and dependent on the crack growth rate da/dN.

But one has to be careful on the definition of the slope of the curve. Figure 5 (b) illustrates with the blue dashed line the curve that one may obtain with a single crack growth tests. With an increase in da/dN relatively more strain energy is released per cycle. But the slope representing the SERR is not the tangent of the dotted line, but as illustrated in Figure 5 (b), it is the slope between the origin and the point on the curve. A decrease in that slope, is equivalent to an increase in SERR.

In the illustrated cases in Figure 5, the assumption is that all strain energy released is related to crack propagation and not to other dissipating mechanisms. With equation (5) it is stated that the strain energy that is released in an experiment comprises more energy dissipating mechanisms. A major contribution to energy dissipation for example is provided by plasticity. In fact, as mentioned before, strain energy may dissipate by plasticity, prior to any crack increment. The consequence of this is illustrated inFigure 6 where $dU/dN \neq 0$ at $da/dN = 0$, which constitutes a physical threshold.

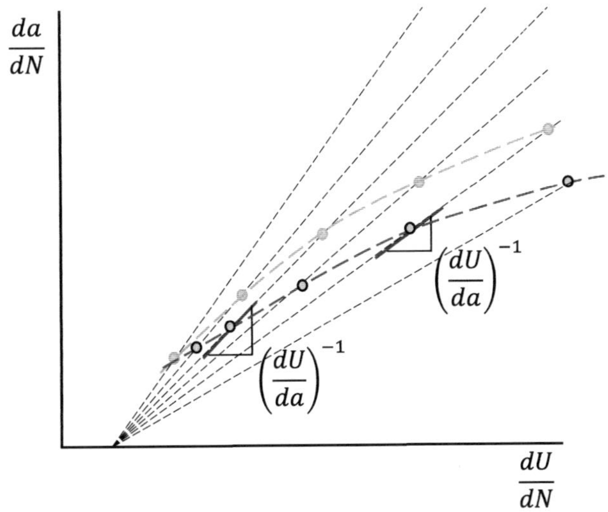

Figure 6. Definition of SERR in presence of plasticity: $dU/dN \neq 0$ at $da/dN = 0$ (a) and further shift in the curve with increasing plasticity at higher da/dN (b); note that both graphs contain linear scales.

In [6] is is explained that the energy dissipation due to plasticity increases approximately with crack length, whereas the energy dissipation due to crack formation remains approximately constant. Hence, the dissipation dU_{pl}/dN is greater for the case more plastic deformation occurs, than for the case little or no plastic deformation occurs. Thus, the amount of plastic deformation for small cracks is substantially less than the plastic deformation at large cracks. This explains the observation reported by various authors that for small cracks very small or even zero thresholds have been observed [20,21].

SHIELDING MECHANISMS

Various shielding mechanisms have been reported that improve the crack growth resistance. These mechanisms may be classified as intrinsic and extrinsic shielding mechanisms [19]. A very distinct extrinsic shielding mechanism in interlaminar delamination growth in composites is fibre bridging. The above described procedure applied to the problem of fatigue delamination has revealed that fibre bridging as shielding mechanism does not contribute to the position of the *da/dN* versus *dU/dN* curve. The elastic strain energy that these fibres carry is released upon unloading and is therefore not dissipated. Only when these fibres are pulled out of the matrix, or when they fail, is strain energy permanently dissipated [22].

Therefore, the hypothesis currently is that intrinsic (shielding) mechanisms directly influence the *da/dN* versus *dU/dN* curves, whereas extrinsic shielding mechanisms will only do that once they fail in the course of fatigue damage propagation. An example of intrinsic mechanisms [19] is for example plasticity in ductile alloys. Strain energy dissipates in each load cycle due to additional plastic deformation, in addition to dissipation due to the crack incrementing. However, one has to keep in mind that the intrinsic mechanism does not necessarily relate to crack propagation linearly, as illustrated by the difference between small cracks and large crack. Theoretically, both a small crack and large crack could experience the same SIF, but due to the different levels of plasticity yield different values for *dU/da*.

RELATION BETWEEN QUASI-STATIC AND FATIGUE CRACK GROWTH

With the introduction of the Paris relationship, the observed asymptotic behaviour of the relationship at high ΔK has been related to a quasi-static SIF K_c. Here, the idea is that once the maximum SIF is equal to the monotonic value of K_c, failure occurs in a single cycle. With modifications to equation (1), often forms have been proposed that normalise by this quasi-static K_c [10]. Similarly, researchers proposed to normalise G_{max} or ΔG by a critical G_c to capture the asymptotic behaviour [23].

The appropriate method to relate quasi-static SERRs with fatigue SERRs is demonstrated by Amaral [24] for delamination growth in composite laminates. The concept is illustrated in Figure 7 (a). Amaral shows how quasi-static crack growth data can be discretised and analysed as if it constitutes low cycle fatigue data. Depending on the level of discretization, multiple points can be derived from the same data set, that linearly line up as indicated with the solid line in Figure 7 (b).

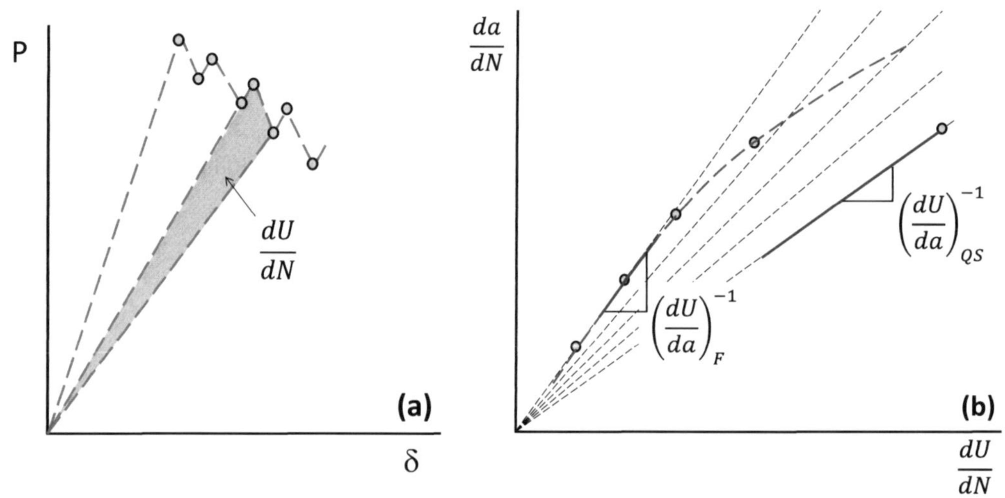

Figure 7. Discretization of the quasi-static force-displacement data equivalent to low-cycle fatigue (a), relation between fatigue SERRs and quasi-static SERRs (b); after [24].

Hence, the SERR obtained from fatigue tests can be correlated to the quasi-static SERR, in which the quasi-static curve represents the highest strain energy release per crack increment. Physically, quasi-static crack growth may be considered the least efficient form of crack propagation, i.e. per crack increment it dissipates the most strain energy.

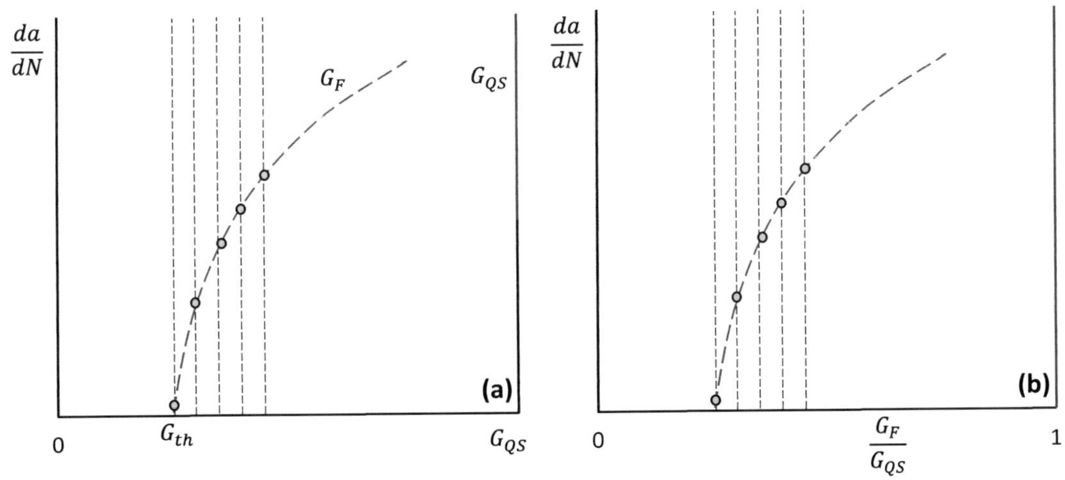

Figure 8. Correlation between the fatigue SERR G_F and the quasi-static SERR G_{QS} (a) and the same trend normalised with G_{QS} (b).

The graphs of *da/dN* versus *dU/dN* can be recalculated to graphs plotting *da/dN* versus $G_F = dU/da$, as illustrated in Figure 8 (a). In this illustration the SERR G_F increases from a threshold value for increasing *da/dN*, which could either relate to changes in effective fracture surface (roughness and changes in fracture mechanisms), or to additional strain energy dissipation mechanisms (plasticity). Note that the representation of Figure 8 (a) appears to be similar to common Paris relationships utilizing the SERR. However, instead of plotting *da/dN* against artificial ΔG or G_{max} values, the physical crack growth rate is plotted against the physical average strain energy released per load cycle. Consequently, the representation of Figure 8 (a) can be normalised by the quasi-static SERR G_{QS} as illustrated in in Figure 8 (b).

DISCUSSION

The representation of the crack growth resistance according to equation (4) relates the physical crack extension in a fatigue load cycle, to the physical strain energy released in that same load cycle. Hence, instead of plotting a measurement (*da/dN*) against a model (ΔK), a measured material characteristic is obtained. This characteristic can be interpreted physically for trends observed, because any trend in this representation must relate to a physical mechanism.

However, one has to keep in mind that both *da/dN* and *dU/dN* are consequences of the application of a load cycle. In fact, *dU/dN* could be the consequence of *da/dN*, if no other energy dissipating mechanisms are involved. Hence, in order to predict damage growth, a 'driving force' must be related to the consequence. Theoretically, the fatigue damage process can be described by

$$U_{\uparrow} = U_{\downarrow} + \frac{dU}{dN} \tag{6}$$

where $U_{\uparrow}$ is the work applied to the specimen or structure during loading, $U_{\downarrow}$ the work recovered during unloading, and *dU/dN* the strain energy dissipated in that load cycle according to equation (4).

Thus, the theory states that the strain energy dissipated should relate to the intrinsic fracture mechanisms involved, but it does not constitute a prediction model as it does not link a 'driving force' to damage growth. The appropriate parameters that represent the 'driving force', correlating to the physical material characteristic, still must be established.

Nevertheless, the benefit of plotting *da/dN* against either *dU/dN* or G_F is that the material characteristic is not polluted by modelling aspects, which currently is the case when *da/dN* is plotted against for example ΔK. Hence, any trend observed in such a representation can be related to physical mechanisms, and therefore contribute to understanding the problem of fatigue damage growth.

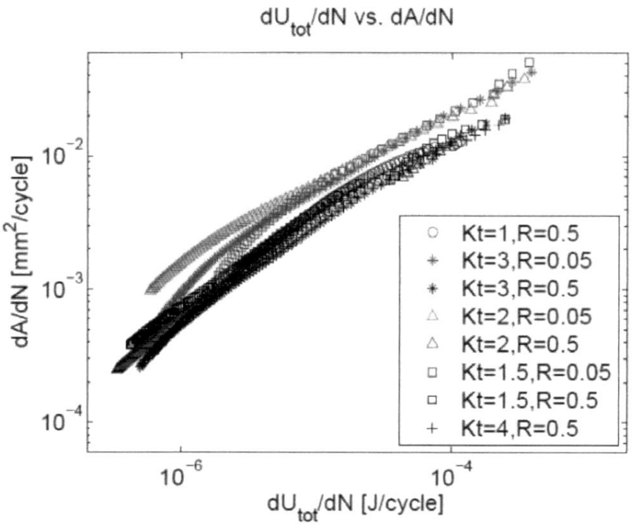

Figure 9. The total strain energy dissipated per load cycle versus the crack growth rate in a solid round metallic bars containing shoulder fillets [26].

For example, although the stress ratio effects greatly vanish when plotting data in the form of *da/dN* against *dU/dN*, still a stress ratio effect remains. This is illustrated by Pascoe for disbond growth in adhesively bonded joints [25], but recently also by Pasman [26] for fatigue crack growth in metallic round bars containing shoulder fillets. Because the material characteristic *da/dN* against *dU/dN* does not contain any implicit assumption related to a model, such stress ratio effect must relate to physical mechanisms.

Because extrinsic shielding mechanisms like crack closure (or fibre bridging in composites) are not present in the *da/dN* against *dU/dN*, the explanation should relate to intrinsic damage mechanisms. To obtain the same *da/dN* for both low and high stress ratios, different levels of maximum loading are applied yielding different fracture mechanisms and surfaces, and different levels of plasticity ahead of the crack tip. Hence, the strain energy dissipation at different stress ratios may yield different curves as illustrated in Figure 9.

CONCLUSIONS

This paper discusses similitude parameters currently adopted to describe fatigue damage growth in various material systems. Where fatigue crack growth in metallic materials generally is described using the SIF, damage growth in adhesively bonded structures and composites often adopt the SERR. Either way, the method of characterization is highly empirical, and constitutes the correlation between a measurement (*da/dN*) and a model (ΔK, ΔG, G_{max}, etc).

In conjunction with earlier work, it is demonstrated that the current approaches lead to misinterpretation of phenomena. For example, attributing the stress ratio effect in the Paris relationship predominantly to plasticity induced crack closure is invalid. This has been demonstrated by correlating the corrections for plasticity induced crack closure to mean stress corrections in fatigue life analyses, and to the fact that the effect is also present in brittle material systems that do not exhibit closure in fatigue.

With the introduction of the average strain energy released over a full fatigue load cycle, *dU/dN* or G_F, different phenomena have been explained, like for example the contribution of plasticity to fatigue thresholds. The average or effective SERR can be correlated to the physical intrinsic damage mechanisms that contribute to damage growth in individual load cycles. In addition, a physical correlation has been established between fatigue damage growth and damage growth under quasi-static load conditions.

The author believes that developing prediction models that utilize physical material characteristics in the form of *da/dN* against G_F eventually leads to more understanding about fatigue phenomena, and to more efficient use of fatigue data.

REFERENCES

[1] Paris, P., Erdogan, F. (1963). *Journal of Basic Engineering*, vol. 85(4), p. 528-533.

[2] Elber, W. (1970), *Engineering Fracture Mechanics*, vol. 2, p. 37-45.

[3] Elber, W. (1971), *Damage tolerance in aircraft structures.* ASTM STP 486, p. 230–242.

[4] Irwin, G. R. (1957). *ASME Journal of Applied Mechanics*, vol. 24, p. 361-364.

[5] Alderliesten, R.C. (2014). *Advanced Materials Research,* vols. 891-892, p. 289-294.

[6] Alderliesten, R.C. (2015). *International Journal of Fatigue,* revision submitted.

[7] Schijve, J. (1981), Engineering Fracture Mechanics, vol. 14, p. 461-465.

[8] Schijve, J. (2009). *Fatigue of Structures and Materials,* 2nd ed., Springer Science+Business Media B.V.

[9] Griffith, A. A. (1921). *Philosophical Transactions of the Royal Society of London. Series A*, Containing Papers of a Mathematical or Physical Character, vol. 221, p. 163-198.

[10] Forman, R.G., Kearney, V.E., Engle, R.M. (1967), *Journal of Basic Engineering*, p. 459-463.

[11] Priddle, E.K. (1976), *Int. J. Pressure Vessels & Piping,* vol. 4, p. 89.

[12] Forman, R. G., Mettu, S. R. (1992). In: *Fracture Mechanics: Twenty-second Symposium*, vol. 1, ASTM STP 1131, p. 519-546, Ernst, H. A., Saxena, A., McDowell, D. L. (Eds), American Society for Testing and Materials, Philadelphia.

[13] Jones, R., Pitt, S., Brunner, A.J., Hui D. (2012). *Composite Structures*, vol. 94, p. 1343-1351.

[14] Paris, P.C. (1998), *Fatigue & Fracture of Engineering Materials & Structures*, vol. 21, p. 535-540.

[15] Paris, P.C., Harris, D.O. (1970). In: *Proceedings of the Beryllium Conference: A Report of the National Materials Advisory Board,* Vol. 1, p. 537–568, NMAB-272, NAS, Washington, D.C.

[16] Martin, R.H., Murri, B. (1990). In: *Composite Materials: Testing and Design (Ninth volume)*, ASTM STP 1059, pp. 251–270, Garbo, S.P. (Ed.), ASTM Philadelphia.

[17] Hudson, C.M., Scardina, J.T. (1969). *Engineering Fracture Mechanics,* vol 1, p. 429-446.

[18] Orowan, E. (1950), *Fundamentals of Brittle Behavior in Metals', Fatigue and Fracture of Metals*, Wiley, New York.

[19] Ritchie, R.O. (1990), *International Journal of Fracture*, vol. 100, p. 55-83.

[20] Wanhill, R.J.H. (2009), Characteristic stress intensity factor correlations of fatigue crack growth in high strength alloys: reviews and completion of NLR investigations 1985–1990. NLR-TP-2009-256.

[21] Jones, R., Molent. L., Walker, K. (2012), *International Journal of Fatigue*, vol. 40, p. 43-50.

[22] Yao, L., Alderliesten, R.C., Zhao, M., Benedictus, R. (2014), *Composites: Part A*, vol. 66, p. 65–72.

[23] Pascoe, J.A.P., Alderliesten, R.C., Benedictus, R. (2013), *Engineering Fracture Mechanics,* vol. 112, p. 72-96.

[24] Amaral. L., Alderliesten, R.C., Benedictus, R. (2015), In: Proceedings of the 28th ICAF Symposium, Siljander, A. (Ed.), International Committee on Aeronautical fatigue, Helsinki.

[25] Pascoe, J.A., Alderliesten, R.C., Benedictus, R. (2013), *Engineering Fracture Mechanics,* vol. 133, p. 1-13.

[26] Pasman, D.J. (2015), *Fatigue crack growth in solid round metallic bars with a shoulder fillet - An experimental research project*, MSc Thesis, Delft University of Technology.

An Algorithm for the Prediction of Very High Cycle Fatigue Life

RJ Amos, A Whitelaw, MA Hussain

MBDA – UK Limited, Six Hills Way, Stevenage, Hertfordshire, SG1 2DA, United Kingdom

Abstract: Weapon systems can have long service lifetime requirements in which the number of vibration cycles can be very high – over 100 million cycles. Conventional fatigue testing to such high numbers of cycles would lead to excessively long test durations. Recent development of piezoelectric fatigue machines has drastically reduced test times and in 1999 Claude Bathias, of the University of Paris, reported that steels can fail at stresses below their endurance limit at very high numbers of cycles contrary to popular belief.

Since then many experimental results, observations and theories have been reported in the literature. However, these remain both academic in the sense that no practical method to include very high cycle fatigue (VHCF) into engineering design has been proposed, and piecemeal in the sense that effects are considered in isolation rather than in combination.

This paper reports on work done to address these limitations. Work from the open literature has been reviewed and extended, and an algorithm has been developed to predict the effects of vibration on the metallic components of a weapon system. The algorithm combines the effects of VHCF ($>10^8$ cycles), fretting at contacting surfaces and corrosive media, with classic fatigue using cumulative damage sums. The algorithm has been implemented as an excel spreadsheet. This paper describes the methodology on which the computer algorithm is based. The work provides a new capability for designers to estimate fatigue life over worst case combinations of vibration environments including VHCF. No such capability was found in the open literature.

INTRODUCTION

Weapon systems see vibration loading throughout their service life: for example in road, sea and air transport, and in operation. With long lifetime requirements the number of vibration cycles can be very high – over 100 million cycles. In the past fatigue life testing, analysis and prediction have generally only considered load cycles up to around 10^7. Conventional fatigue testing above 10^7 load cycles can lead to excessive test durations and costs: conventional electro-hydraulic pistons cannot provide fatigue loads to specimens at frequencies above around 100 Hz. By 10^7 load cycles some engineering materials, in particular many conventional steels and titanium alloys, appear to exhibit a fatigue endurance limit (i.e. a maximum stress at which the material can sustain an unlimited number of cycles). Above 10^7 load cycles, therefore, it has been usual to keep the applied stresses well below this endurance limit arguing that the loads are then small enough not to cause fatigue failure.

In recent years piezoelectric fatigue machines have evolved, which have enabled efficient and reliable testing at very high cycles. They are capable of producing 10^{10} cycles, for example, in less than one week (at 20 kHz) while it would take more than three years for the same experiment using conventional equipment. Early experiments using this equipment have found that for most materials, including steels and titanium alloys which may show an apparent fatigue endurance limit at 10^6 to 10^7 cycles, their fatigue strength decreases, and failures still occur, as fatigue life reaches 10^8–10^{11} load cycles. In 1999 Claude Bathias, of the University of Paris, reported in the literature that steels can fail at stresses below their endurance limit at very high numbers of cycles contrary to popular belief [1]. A typical plot obtained is shown in Figure 1 obtained from [2].

Since then many experimental results, observations and theories have been reported in the literature. However, these remain both academic in the sense that no practical method to include very high cycle fatigue into engineering design has been proposed, and piecemeal in the sense that effects are considered in isolation rather than in combination. To address these limitations work has been carried out with the objective of providing an algorithm that would predict the effects of vibration on the metallic structural components of a weapon system. The algorithm would address the combined effects of very high cycle fatigue (VHCF) $>10^8$ cycles, together with fretting at contacting surfaces, and corrosive media both found to have a greater effect in VHCF than in low or high cycle fatigue, and with classic fatigue cumulative damage sums to combine multiple stress levels in low, high and very high cycle fatigue. The algorithm has been implemented as an excel spreadsheet. This paper describes the methodology on which the computer algorithm is based. The use of this algorithm will allow weapons designers to better estimate service fatigue life over worst case combinations of vibration environments including VHCF. No equivalent capability was found in the open literature.

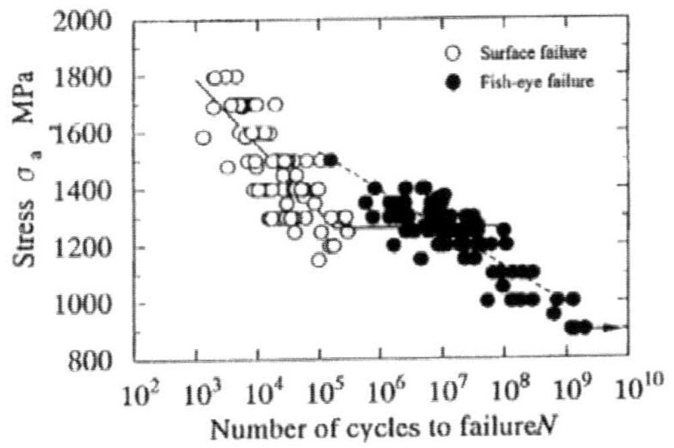

Figure 1. Typical S-N curve for a high strength steel showing drop in fatigue strength above 10^7 cycles [2].

FATIGUE LIFE

The algorithm is based on the concept of ***fatigue life***. Fatigue life is an important characteristic of an engineering component and is measured by the number of load cycles it can withstand before fatigue failure takes place. Based on the fatigue life concept, mechanical fatigue can be divided into three regimes: ***low cycle fatigue*** **(LCF)** – up to around 10^3 cycles to failure; ***high cycle fatigue*** **(HCF)** – between around 10^4 and 10^7 cycles to failure and ***very high cycle fatigue*** **(VHCF)** – over 10^7 cycles to failure. VHCF is also known as ***ultra-high cycle*** or ***gigacycle*** fatigue. High cycle fatigue is defined as when the applied stresses are fully within the elastic regime. Low cycle fatigue is defined as when the applied stresses are sufficiently high to induce plasticity. HCF and VHCF are best described by stress-based parameters. LCF is best described by strain-based parameters.

The fatigue life of an engineering component cycled at a single stress amplitude can be predicted from a material ***S-N curve*** defined as a plot of the amplitude S of a cyclic stress against the number of cycles N to failure at that stress. For load inputs comprising multiple stress amplitudes the S-N curve is used with a ***cumulative damage law***. The simplest approach would be to use the classic ***Palmgren-Miner*** rule. This assumes a linear accumulation of damage with load cycles. In a more complex non-linear rule the damage accumulation with load cycles can be non-linear and dependent on stress level applied. In this case the load order becomes significant. This is the case in low cycle fatigue where the Palmgren-Miner rule is generally found to give conservative results. For load inputs defined as a general vibration signal then this can be converted to the above form by standard procedures such as ***rainflow analysis*** in the time domain and ***Dirlik's method*** in the frequency domain.

LITERATURE REVIEW

A review of the open literature was carried out and the major findings are summarised below. Most references on VHCF found were from Japan, China and France. Most of these were published since 2000. There are two major textbooks on VHCF due to Murakami [3] and Bathias and Paris [4]. These provide a detailed introduction to the subject. Results and findings from over 70 references were used to develop and support the methodology used in the algorithm.

Most materials, including steels and titanium alloys, do not have a fatigue endurance limit at 10^6–10^7 cycles, but instead their fatigue strength decreases, and failures still occur, as fatigue life reaches 10^8–10^{11} load cycles. More precisely, the general S-N curve takes the form shown in Figure 2 taken from [5]. Here, region I is classic high cycle fatigue dropping to an apparent endurance limit at around 10^6–10^7 cycles; shown as region II. Then, at around 10^7–10^8 cycles the curve drops again; shown as region III. Finally, the curve may level off at a second endurance limit; shown as region IV.

Below 10^7 cycles fatigue failures generally initiate at the positions of worst stress which are generally at the material surface. Above 10^8 cycles fatigue failures are generally found to initiate from material micro-structural defects. These defects are often subsurface or internal due to the higher probability of finding critical defects in the bulk of the material than at the surface. Hence, in Figure 2 regions III and IV are referred to as ***internal fatigue***. However, it is found in a number of papers that stress concentrators at the surface such as notches, and damage due to fretting and/or corrosion, may lead to crack initiation at the surface. In Figure 1 the solid circles are subsurface or internal failures and the hollow circles are surface failures.

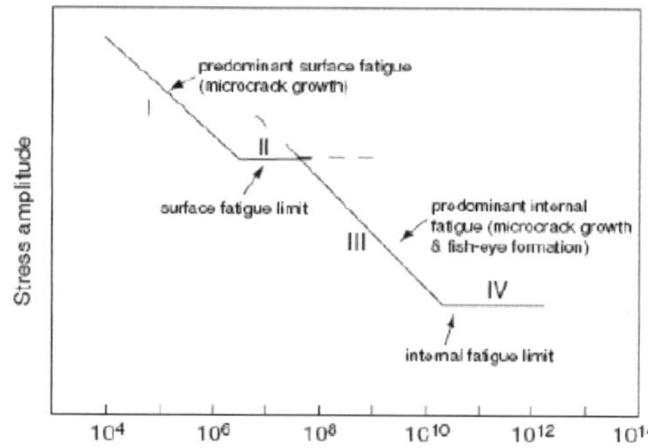

Figure 2. General Form of Material S-N Curve in HCF and VHCF [5].

Most of the material data and virtually all the predictive models presented in the literature are for high strength steels. On this basis the algorithm has been developed only for high strength steels. Other materials remain an area for further study. For high strength steels the main types of defect causing fatigue failure in VHCF are non-metallic inclusions. In modern material specifications the micro cleanness of a material, i.e. the maximum allowed size and number of inclusions, is generally defined by the material specification. Where fatigue test data is presented in the literature it is common that the actual inclusion sizes within the test specimens have been determined by inspection.

Figures 1 and 2 both refer to ***fish-eye*** failures. This refers to the form of the failure at an inclusion which looks like a fish-eye; see the example in Figure 3 obtained from [6]. According to Murakami [3] the fatigue crack growth proceeds as follows: first, the crack initiates from a material defect; second, it grows within the fish-eye, intermittently and accordingly at a very slow rate with the assistance of hydrogen trapped by the inclusion (hydrogen embrittlement); third, when the size of the inclusion plus the fish-eye reaches a critical value then crack growth starts outside the fish-eye without the assistance of hydrogen; and fourth, the crack rapidly grows to failure.

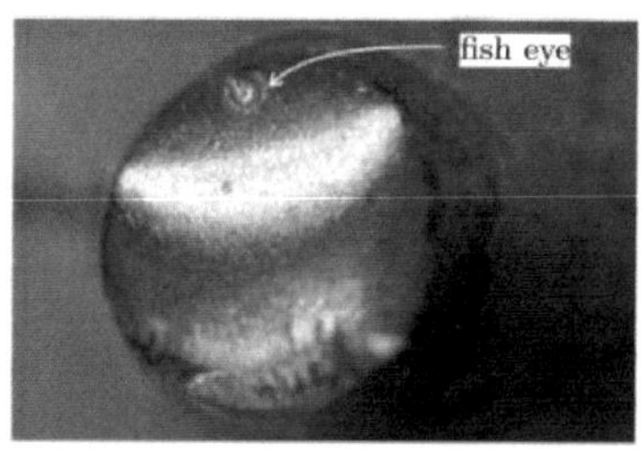

Figure 3. Fracture Surface in VHCF showing Fish-Eye [6].

It is suggested in [6] that fatigue fracture does not initiate from inclusions below a critical size. A method of estimating the ***critical inclusion size*** (CIS) of high strength steels is presented in [7].

In VHCF it is found that the number of cycles to crack initiation is very large compared to the number of cycles to propagate that crack to failure. This is different from HCF and LCF. In VHCF the number of cycles to crack initiation is of the order of 99% of the full lifetime to failure whereas in HCF it is of the order of 90% and in LCF it is of the order of 50%, [8].

Both Fretting and Corrosion fatigue can have significant effects on fatigue life and fatigue strength. These are found to be worse in VHCF than in LCF or HCF and are therefore important considerations in assessing fatigue life in service.

Very little work was found on a cumulative damage law in VHCF. Only two papers were found giving data which could be used for evaluation and these appeared to find contradictory effects; [9] and [10]. There is some evidence in the papers found that the Palmgren-Miner rule gives a good estimate erring on the conservative side. On this basis it is difficult to justify using anything other than the basic Palmgren-Miner rule in the algorithm. There is insufficient understanding and data to fit a better model. This lack of study into a cumulative damage law is a major hole in current knowledge and more experimental investigations need to be carried out.

A number of methods and models were found to predict fatigue behaviour in the 10^8–10^{11} load cycles regime for, at least, high strength steels and including the effects of fretting and corrosion fatigue. These will be discussed further below.

THE ALGORITHM

The algorithm is divided into two main parts. The first part performs calculations to predict the S-N curve for the material and conditions selected based on a limited set of material mechanical properties. Conditions include the effects of different mean stresses and stress ratios (i.e. the ratio between the minimum and maximum stress within a cycle), corrosive media and fretting.

The second part performs the calculations to predict the fatigue life based on the selected cumulative damage law against an input stress spectrum specified as a number of cycles against each of a number of cyclic stresses defined by their amplitude and mean values. The stress input must be determined by stress analysis prior to running the algorithm. The predicted life is then given in terms of the number of repeats of the spectrum to failure.

PREDICTION OF THE S-N CURVE IN VHCF

This paper is concerned with the prediction of the S-N curve in VHCF, i.e. regions II, III and IV in Figure 2. Region I is the classic S-N curve in high cycle fatigue. Experimental data is widely available for many materials in HCF and tests are relatively quick to carry out. Predictive methods are available, but because the mode of failure is different in HCF they are not applicable to VHCF. Although equations and results will be presented here for a stress ratio of -1, i.e. fully reversed stress cycles, other stress ratios can be accounted for by using standard methods such as described in section 6.5 of [3].

To obtain the apparent high cycle fatigue endurance limit, σ_{w0}, shown as region II in Figure 2, there is a well-known estimate in terms of the Vickers Hardness, Hv, of the material. If the hardness is not known there is also a well-known estimate of this in terms of the ultimate tensile strength, UTS. These estimates are given in [3], for a stress ratio of R = -1 (i.e fully reversed cycles), as

$$\sigma_{w0} = 1.6\ \mathrm{Hv}\ \underline{\textit{and}}\ \mathrm{Hv} = \mathrm{UTS} / 3.2 \qquad (1)$$

The very high cycle fatigue endurance limit, σ_w, shown as region IV in Figure 2, can be obtained using a method due to Murakami [3]. The method determines the maximum stress at which the stress intensity factor at the defect lies below the threshold value for crack initiation. It depends on only two basic quantities, namely the Vickers hardness Hv (kgf/mm^2) as the representative material parameter, and $\sqrt{area}$ (μm) as the representative geometrical parameter for defects and cracks. The parameter $\sqrt{area}$ is defined as the square root of the area obtained by projecting a small defect or crack onto a plane perpendicular to the maximum principal stress, and $\sqrt{area}_{max}$ is the maximum possible value of $\sqrt{area}$ for any inclusion in the material. Then, for a stress ratio of -1, σ_w is given by

$$\sigma_w = 1.41\ (\mathrm{Hv} + 120) / (\sqrt{area}_{max})^{1/6} \qquad (2)$$

The S-N curve in VHCF lies between these two limit stresses, σ_{w0} and σ_w estimated by equations (1) and (2).

Many authors are investigating predictive models for specific materials and claiming success in that predictions compare reasonably well with experimental results; see for example [8]. However, in some of the papers the arguments are circular in that the same data has been used to fit the model as to validate it. These models have been reviewed and two selected for investigation:

The first of these is based on Murakami's method of predicting the fatigue limit stress described above. It assumes a similar form to equation (2), i.e. the number of cycles to failure N_f at the applied stress level σ_f is defined by

$$\sigma_f = (A - B \log N_f)\ (\mathrm{Hv} + 120) / (\sqrt{area}_{max})^{1/6} \qquad (3)$$

where the parameters A and B are fitted from empirical data.

The second method is a fracture mechanics approach based on a dislocation model of Tanaka and Mura [11] for predicting crack initiation life. Since, in VHCF, the number of cycles to crack initiation is virtually the full fatigue life to failure then the crack initiation life is a good approximation to the full fatigue life in VHCF.

Following the derivation described in [8] it is possible to obtain the following estimate for the number of cycles to failure N_f at the applied stress level σ_f:

$$N_f = 9(K_{th})^2 / 2(1+\nu)\, D\, (\sigma_f - \sigma_w)^2 (\sqrt{area}_{\,max}) \qquad (4)$$

where K_{th} is is the threshold stress intensity factor for inclusions, ν is the Poisson's ratio of the material, D is a damage accumulation factor, and σ_w can be obtained from equation (2).

It is noted that the parameter D is left out of the model presented in [8]. It is explained in [12] that because the model in [11] was derived using complete slip irreversibility a damage accumulation factor D should be introduced representing the fraction of plastic energy accumulated at each load cycle. It is introduced to correct the discrepancies between the assumptions of the model and experimental results. Without the factor D included it is not possible to reproduce the results in [8].

Following Murakami's analysis [3] the value of K_{th} can be estimated from

$$K_{th} = 2.77 \times 0.5 \times 10^{-3} (Hv + 120) (\sqrt{area}_{\,max})^{1/3} \qquad (5)$$

Parameters in both models must be derived from experimental data: A and B for Murakami's method and D for the Tanaka-Mura method. To avoid a circular argument the widest possible set of data was obtained from the literature on VHCF and the curve fit based on this. These selected materials covered a wide range of UTS and Hv values. The values obtained were then used in the models to predict the S-N curve for any high strength steel. Figure 4 shows a comparison between the predicted S-N curves obtained using each method and the mean test data for the high strength steel shown in Figure 1. The test data is shown in red, the prediction from the Murakami model in blue, and from the Tanaka-Mura model in green. The two models give very similar predictions very close to the test data. Similar results were obtained for other steels. A further example is given in Figure 5 below.

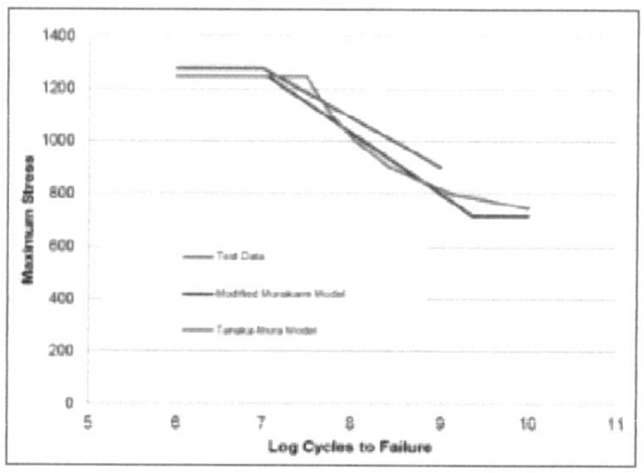

Figure 4. Comparison of Different Models for Predicting the S-N Curve of the High Strength Bearing Steel of Figure 1-

It is interesting to note that the estimates of the parameters A and B given in [13] were found clearly not to give the best curve fit when viewed by eye. The curve fit was repeated using Excel on a wider set of data, and essentially the same result was obtained as in [13]. The mathematical reason for this is unclear, but the curve fit by eye was accepted as the best approach in this case.

EFFECT OF CORROSIVE MEDIA

The effects of corrosive media are considered in [14] and [15]. The model used there is essentially the same as Tanaka-Mura except that fracture toughness is used rather than threshold stress intensity factor. It is shown that for the same number of cycles N_f, then $(\sigma_f - \sigma_w)$ is proportional to fracture toughness.

This result still contains some uncertainty in that the S-N curve in the corrosive medium could lie anywhere between the case where both σ_f and σ_w are proportional to fracture toughness and the case where $\sigma_w = 0$ and σ_f alone is proportional to fracture toughness. This is illustrated in Figure 5 for a non-corrosion resisting steel. The test results are close to the first case between 10^6–10^7 cycles and closer to the second case above 10^7 cycles.

The steel used in this example is designated 40Cr in China. This is a frequently quoted steel in the literature on VHCF for which there is a good amount of data published. It has therefore been used as a test example throughout this study. For corrosion resisting steels, where the effect of corrosion is much less, it is also found that the first case where both σ_f and σ_w are proportional to fracture toughness gives predictions close to test results in the 10^6–10^7 cycles region.

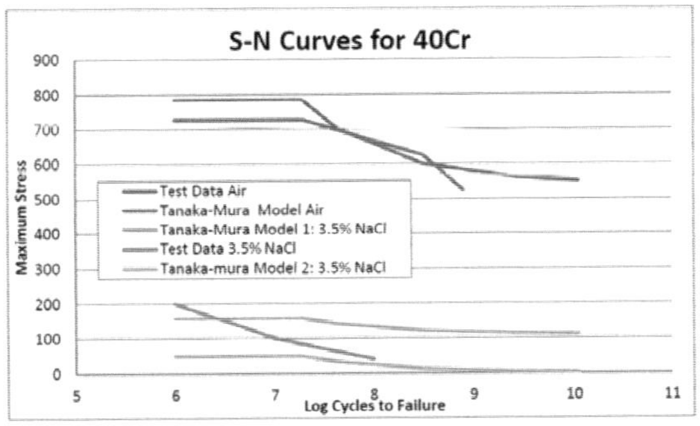

Figure 5. Results from the Tanaka-Mura Model for Predicting the S-N Curve in Air and in a 3.5% NaCl Corrosive Medium.

EFFECT OF FRETTING

The most promising approach found by the literature review for predicting fretting fatigue is due to Toshio Hattori and various co-workers. This was concluded on the basis that their approach is fracture mechanics based and therefore fits the general approach adopted here, covers LCF, HCF and VHCF regimes, and has had some success in correlation against experimental results. The work is described in a large number of papers; with [16] being a good introduction. However, even though the literature, such as [4] and [16], refers to fretting fatigue in VHCF no previous study has included the effects of crack initiation at inclusions or defects with fretting.

Fretting fatigue results when two structural elements are in contact under a contact pressure, and cyclic stress and relative displacement occur along their contact surfaces. Under these fretting conditions the fatigue strength is found to decrease significantly compared to that under non fretting conditions. The strength is reduced because of the concentration of contact stresses such as contact pressure and tangential stress at the contact edge, where the fretting fatigue crack forms and propagates. This is illustrated in Figure 6 taken from Hattori's publications.

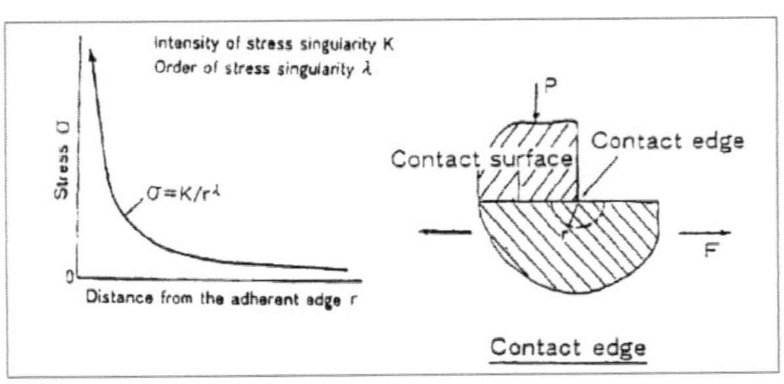

Figure 6. Stress Distribution near a Contact Edge in Fretting (Hattori).

Finite element or other stress analysis must be carried out to predict the stress singularity at the edge of the contact surface. The fretting fatigue process model developed by Hattori et al. then consists of three mechanisms.

1. **Crack initiation:** Cracking due to fretting fatigue starts very early in the fretting fatigue life. The stress singularity parameters at the contact edge are used to estimate the initiation of these cracks.
2. **Wear extension:** The effect of wear on the contact surface then reduces the contact pressure near the contact edge, and cracks gradually start to propagate. Initially, fretting fatigue cracks tend to close and propagate very slowly, especially in the low stress amplitude range, due to the high contact pressure acting near the contact edge. As wear increases and a crack grows then the threshold stress intensity factor increases from the short crack to the long crack value. The crack may arrest if the operating stress intensity factor range falls below the threshold value for the current crack length. Wear analysis can be carried out using the classic Archard's equation.
3. **Crack propagation:** Finally, if the applied stresses are high enough, wear increases further and the operating stress intensity factor exceeds the threshold value for long cracks and the crack propagates to failure. The propagation life in the long crack length region can be estimated using ordinary fracture mechanics, i.e. the Paris Law.

It was remarked above that for prediction of VHCF a good estimate is obtained by considering crack initiation only. It is found that crack propagation amounts to less than 1% of the total life in VHCF. For VHCF therefore only the first two mechanisms significantly affect fatigue life. On the other hand, in the case where the stress range is high i.e. in LCF, then wear is not found to be significant and only the first and third mechanisms will significantly affect fatigue life. For HCF all three mechanisms will be significant.

In [16] it is stated that fretting fatigue life in the low stress amplitude range will be dominated by the propagation of the small cracks initiated at the contact edge. So to estimate the fretting fatigue strength or life in the low stress amplitude range, the precise estimation of the fretting wear progress is important. However, the stress singularity parameters at the contact edge will depend on the precise geometry. As wear occurs due to the repeated loading then this geometry changes which, in turn, changes the stress distribution. Modelling the fretting wear progress would result in a highly complex iterative process where the stress analysis must be continuely updated to reflect the changing geometry and stress distribution. Sufficient iterations would be required to adequately model the wear progression.

For simplicity the first version of the algorithm has not included modelling of the fretting wear process, i.e. the geometry and stress distributions are assumed not to change from their initial values. It would be expected that this would give a conservative prediction of fretting fatigue life.

In [17] Hattori et al define their crack initiation criterion in VHCF using two strength properties of the material: firstly, σ_{w0}, the endurance limit in HCF from the S-N curve; and secondly, K_{th}, the threshold stress intensity factor for crack initiation. They assume the endurance limit in HCF extends into the VHCF regime. This is clearly not correct from the studies reported here. To model fretting crack initiation in VHCF the method of [17] must therefore be modified to take account of inclusions and defects. The criterion developed by Hattori et al is illustrated by the blue curve in Figure 7. This defines the threshold intensity value of the stress singularity against the order of the stress singularity. The curve intersects the y axis, i.e. where the order is zero and there is no stress singularity, at the value of σ_{w0}.

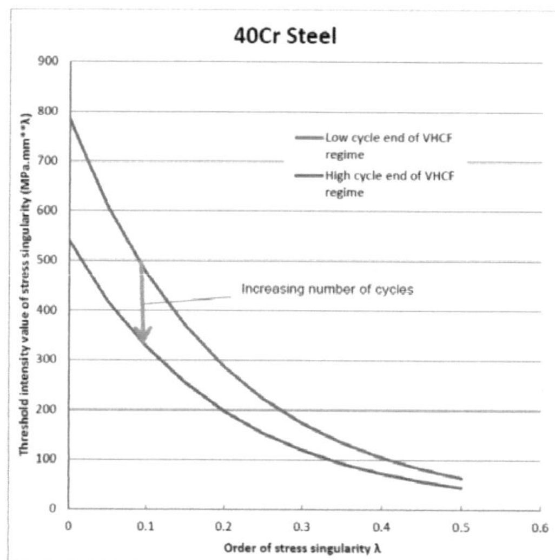

Figure 7. Crack Initiation Criterion at a Stress Singularity in VHCF.

The S-N curve in VHCF with no singularity has already been determined. This drops from σ_{w0} at the low cycle end (10^6–10^7 cycles) to σ_w at the very high cycle end. For crack initiation at inclusions it is proposed that the whole crack initiation criterion curve, the blue line in Figure 7, is reduced by the same proportion as at zero order. This defines a family of curves depending on number of cycles ranging from the blue line down to the red line which is the criterion at very high cycles and intersects the y axis at a value of σ_w.

Using this crack initiation criterion it is then possible to derive the S-N curve under the assumed fretting conditions. An example is shown in Figure 8. This shows a significant reduction due to the fretting, but the fretting conditions assumed may have been excessive and unrealistic. The S-N curve obtained will depend on geometry at the contact edge, on material properties of both sides of the contact, and on the applied contact pressure. No experimental data has been found to verify the predictions.

Finally, the effects of a corrosive medium can easily be combined with fretting by using the material properties and S-N curve for the corrosive medium in the fretting calculations.

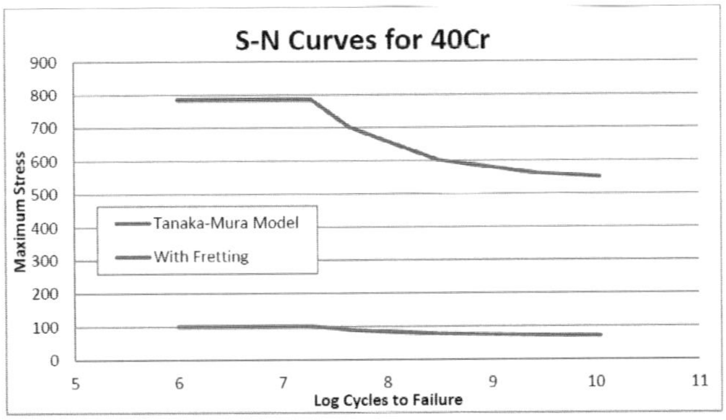

Figure 8. Predicted S-N Curve with Fretting.

CONCLUDING REMARKS

1. The study of VHCF has only become possible in recent years due to the development of new experimental procedures to obtain data.
2. An algorithm is described which can predict the effects of vibration on the metallic structural components of a weapon system under the combined effects of very high cycle fatigue, fretting at contacting surfaces, and corrosive media, and provides an output which can be used with classic fatigue cumulative damage sums.
3. Verification of the algorithm from experimental data has been achieved for the S-N curve for VHCF in air and in a corrosive medium. Insufficient data is available to verify the algorithm for fretting fatigue.
4. There is some evidence that the Palmgren-Miner law for cumulative damage provides a good estimate.
5. A model of the effect of inclusions on fretting fatigue in VHCF has been proposed. No previous study was found in the literature to address this.
6. The major benefits from the work are an improved understanding and a new capability to carry out fatigue life predictions for very high numbers of vibration load cycles over worst case combinations of vibration environments including corrosive media and fretting.

ACKNOWLEDGEMENT

This work was funded by MoD through the Weapons Technology Centre.

REFERENCES

[1] Bathias C (1999), *There is no infinite fatigue life in metallic materials*, *Fatigue & Fracture of Engineering Materials & Structures* 1999 07;22(7):559-65.

[2] Sakai T, Harada H and Oguma N (2006), *Crack Initiation Mechanism of Bearing Steel in Very High Cycle Fatigue*, In *Fracture of Nano and Engineering Materials and Structures 2006*, pp 1129-1130.

[3] Murakami Y (2002), *Metal Fatigue: Effects of Small Defects and Non-metallic Inclusions*, Elsevier, 2002.

[4] Bathias C and Paris PC (2005), *Gigacycle fatigue in mechanical practice*, NY 10016, USA: Marcel Dekker; 2005.

[5] Kazymyrovych V (2009), *Very High Cycle Fatigue of Engineering Materials*, Karlstad University Studies, 2009-22.

[6] Wang Q, Khan MK and Bathias C (2012), *Current Understanding of Ultra-High Cycle Fatigue*, *Theoretical and Applied Mechanics Letters*, 2, 031002 (2012).

[7] Yang ZG, Zhang JM, Li SX, Li GY, Wang QY, Hui WJ, and Weng YQ (2006), *On the critical inclusion size of high strength steels under ultra-high cycle fatigue*, *Materials Science and Engineering* A 427 (2006) 167–174.

[8] Wang QY, Bathias C, Kawagoishi N and Chen Q (2002), *Effect of inclusion on subsurface crack initiation and gigacycle fatigue strength*, *International Journal of Fatigue* 24 (2002) 1269–1274.

[9] Huang Z, Wang Q, Wagner D and Bathias C (2012), *Effect of Low Cycle Fatigue Pre-Damage on Very High Cycle Fatigue*, *Theoretical and Applied Mechanics Letters*, 2, 031007 (2012).

[10] Nakajima M, Tokaji K, Itoga H, and Shimizu T (2002), *Fatigue Life under Variable Amplitude Loading in a High Strength Steel Showing Step-Wise S-N Curve, European conference on fracture, ECF 14, fracture mechanics beyond 2000*, 525-532.

[11] Tanaka K and Mura T (1981), *A dislocation model for fatigue crack initiation*, *Trans ASME J. of Applied Mechanics,* Vol.48, pp97-103, 1981.

[12] Zhou RS, Cheng HS, and Mura T. (1989), *Micropitting in rolling and sliding contact under mixed lubrication, Trans ASME J Tribol* 1989;111:605.

[13] Wang Q, Berard JY, Dubarre A, Baudry G, Rathery S and Bathias C (1999), *Gigacycle fatigue of ferrous alloys, Blackwell Science Ltd. Fatigue Fract Engng Mater Struct* 22,1999, 667–672.

[14] Qian G, Zhou C, and Hong Y (2011), *Experimental and theoretical investigation of environmental media on very-high-cycle fatigue behavior for a structural steel, Acta Materialia* 59 (2011) 1321–1327.

[15] Qian G, Zhou C, and Hong Y (2013), *Crack propagation mechanism and life prediction for very-high cycle fatigue of a structural steel in different environmental medias, Frattura ed Integrità Strutturale*, 25 (2013) 7-14; DOI: 10.3221/IGF-ESIS.25.02.

[16] Hattori T, Yamashita M and Nishimura N (2005), *Fretting fatigue strength and life estimation in ultra high cycle region considering the fretting wear process, JSME International Journal*, Series A, Vol. 48, No. 4, 2005.

[17] Hattori T, Nishimura N and Yamashita M (2008), *Standardization of strength evaluation methods using critical distance stress, Engineering Fracture Mechanics*, 2008 | 75 | 7 | 1725-1735.

Composite Door Pressure Fatigue Test

Martin Semsch, Frank Wießner

IMA Materialforschung und Anwendungstechnik GmbH Dresden
Wilhelmine-Reichard-Ring 4, 01109 Dresden, Germany

Abstract: A pressure fatigue test of an airplane composite door in the framework of the European research program MAAXIMUS was performed in the test lab of IMA Dresden. The purpose of this test was to validate a new door design under cabin pressurization taking into consideration the deformations of the door surround structure. The composite door consisted of a CRFP structure including new joining technologies like stitching for further aircraft application. It was designed and manufactured by Latecoere.

The target for the validation process was to demonstrate the resistance to low-cycle fatigue of the composite structure with respect to 3 DSG and to show the damage tolerance behaviour of the design. In total 180.000 flight cycles were simulated during the fatigue test. In regular intervals inspections of the door structure were performed. The tested composite door withstands the simulated fatigue loading without any findings.

Furthermore, a damage growth campaign of 30.000 flight cycles with single element failure and a fatigue test with an included repair solution of additional 20.000 flight cycles were realized. At the end of the test campaign, the residual strength behaviour of the door structure was investigated.

INTRODUCTION

Composite materials are more and more used in modern airframes. Significant improvements are still achievable. The aim of the European research project MAAXIMUS (More Affordable Aircraft structure through eXtended, Integrated, & Mature nUmerical Sizing) is to demonstrate the fast development and right-first-time validation of a highly-optimised composite airframe. The objectives of the project are a reduction of the assembly time of large composite parts and a reduction of the structural weight. To achieve these targets, a complete fuselage section including windows, the floor structure and doors was investigated and optimized. At the end of the development process, a validation of the achieved results was performed by structural testing and a comparison of the test results with the predictions based on the virtual simulation.

One of main component, which was investigated and optimized in the frame of the project, was a composite passenger door. To validate the design and to demonstrate the behaviour of the composite door structure, a test campaign including static, fatigue and damage tolerance tests was performed. Therefore, a special test method for door testing was developed and a new test rig with respect to this method was designed, manufactured and assembled. Basis of the test method are the loading conditions influenced by the internal pressurization of the airframe and the deformation of the door surround structure. The deformation of the door surround structure has a significant influence on the stress distribution and has to be considered in door structure testing.

TEST OBJECT

The test object was a new designed composite door structure (Figure 1). The door was designed by Latercoere and consists of a RTM structure (skin, beams, frames), stop fittings, guide rollers and the complete sealing on the edge of the passenger door. Furthermore, a new joining technology was applied during the assembly process of the door structure. The stitching technology was used to join beams and frames with the outer skin. The overall main dimensions of the door are 1254 mm (width) x 2097 mm (high). The radius of the door is not constant and varies between 2040 and 2400 mm.

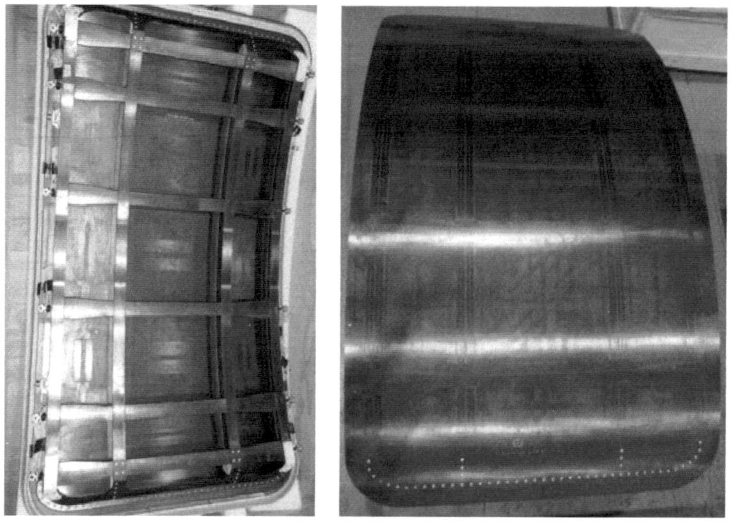

Figure 1. Test object: composite door.

TEST SET-UP

The test set-up for the door test campaign (Figure 2) has to be able to simulate the internal pressurization of the airframe and the deflection of the door surround structure. A suitable test concept consists of a pressure chamber with included controlled hydraulic jacks at each door stop fitting position (Figure 3). The pressure chamber has to be able to integrate the door structure and has to represent a closed and airproof pressure system together with the door.

The interface between pressure chamber and the passenger door were represented by the original door stop fittings of the aircraft system to guarantee the realistic boundary conditions. To simulate the deflection of the door surround structure each stop fitting was linked to a controlled hydraulic jack. The control system was based on the displacements of the stop fittings. Furthermore, the reaction forces at each door stop position were monitored during the test run. By this method realistic boundary conditions for door structure in flight condition can be simulated. The advantage of this method is the possibility to adapt the reaction forces at the stop fittings and an improved quality of the test results.

Figure 2. Test set-up

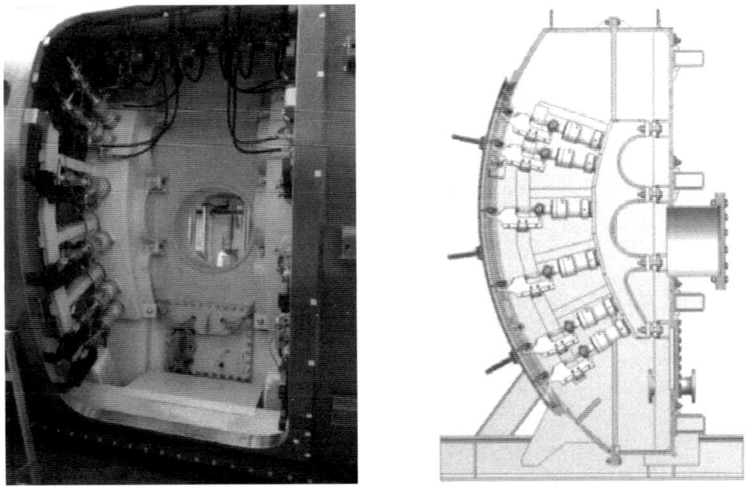

Figure 3. Interior of the pressure box.

TEST PERFORMANCE

The test performance consisted of a static, fatigue and damage tolerance test. The damage tolerance test included a damage growth test campaign with single element failure, a fatigue test with an included repair solution and a residual strength test at the end.

Static test

Static tests before fatigue test were performed to demonstrate the correct stress and strain distribution at the door. A measurement system with load cells, pressure transducers, displacement transducers and strain gauges were used to monitor the test results. These results were used for a comparison with the predictions of the FE- simulation. Furthermore, a measurement campaign with the optical measurement system ARAMIS (Figure 4) was carried and out and an evaluation of the displacements and strain on the basis of these results was performed. The results of the static test demonstrate a high quality of the test and a good correlation with the predicted values.

Figure 4. Measurement campaign wit the optical measurement system ARAMIS.

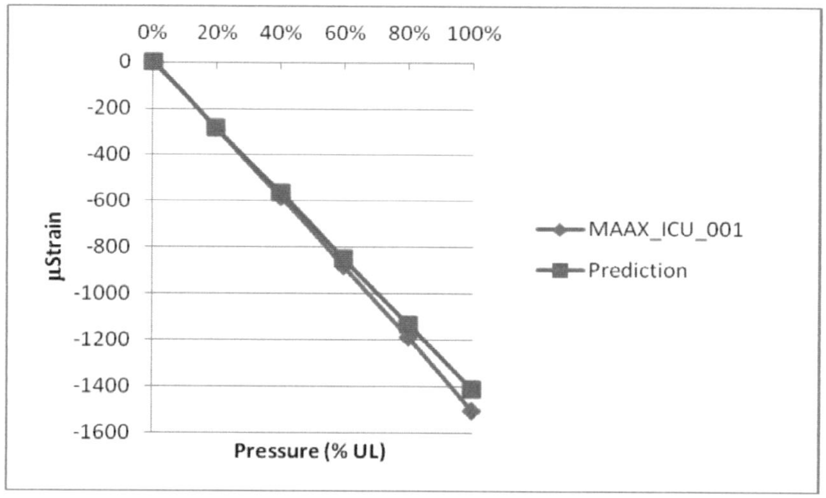

Figure 5. Comparison of the measurement results with predictions.

Fatigue test

A fatigue test with 180.00 simulated flight cycles was carried out. A flight cycle consisted of the simulation of one differential pressure cycle. The pressure cycle was applied sinusoidal with a maximum level of 794 mbar, a load rate of 0,1 and frequency of 0.05 Hz. The stroke of the hydraulic jacks at the door stops were adjusted based on the theoretical deformation of the door surround structure and controlled in phase with the differential pressurization. The theoretical deformation of the door surround structure has to be estimated before testing, usually by FE-calculation.

In regular intervals inspections of composite structure and static tests including a measurement campaign were carried out. No significant changes of the strain distribution were and no indications were found during the inspections. The door structure passed the simulated fatigue test with 3 DSG and 180.000 pressure cycles successful.

Damage tolerance test – damage growth test and fatigue test

The door test specimen was subjected to barely visible impact damage (BVID) and large visible impact damage (LVID). In total 6 impact damage were applied. Therefore, a gas gun and a drop tower were use and different impact energy levels between 35 J and 200 J were applied. Two examples of the impact damage are shown in Figure 6.

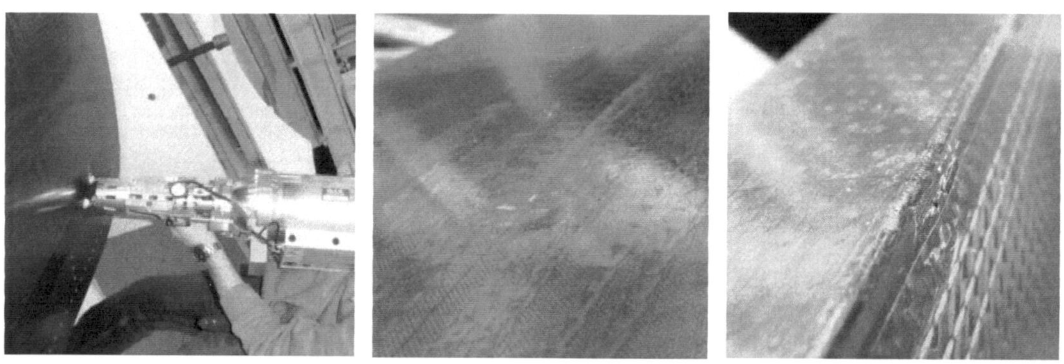

Figure 6. Impact damage.

Furthermore, one single element failure was included in the damage growth campaign. Therefore, one stop fitting was removed. Afterwards, 30.000 pressure cycles were simulated and inspections in regular intervals were carried out. The focus of the inspections was the locations of the damage. No significant growth of the damage could be identified.

After the damage growth test, the LVID was repaired, the removed stop fitting was re-installed and additional 20.000 cycles were accomplished to show the behaviour of a repair solution of a composite structure. No finding on the repair solution could be identified during this test phase.

Damage tolerance test - residual strength test

At the end, a residual strength test up to failure was performed. The pressure was increased continuously with a dwell time at different load levels. Details are shown in Figure 7. The composite door withstands the test up to a pressure of 1.505 mbar. Afterwards, the sealing between door and test rig failed.

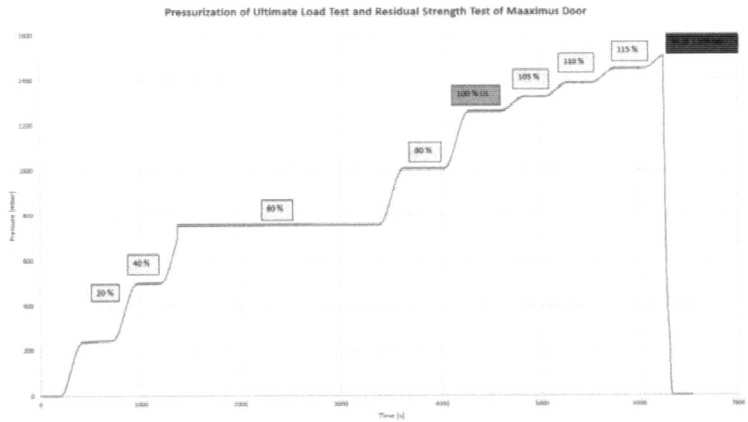

Figure 7. Residual strength test – load sequence.

CONCLUSION

A static, fatigue and damage tolerance test including damage growth test, singe element failure test and residual strength test were carried out on a new designed composite door. Therefore, a special designed test rig and a new developed test method were used. In total, 230.000 pressure cycles were simulated. The door withstands all test successful.

[1] Wießner, F. (2014). *Test Report: Door pressure test*, MAAXIMUS-WP6.7-IMA-DEL-D6.7.7-V1.0

Stress Analyses and Stress Intensities for Part-Through Cracks in Bolted Joints Exhibiting Secondary Bending

Zlatan Kapidžić, Hans Ansell, Lars Djärv

Saab AB, SE-581 88 Linköping, Sweden

Abstract: Crack growth was investigated in single strap, shear loaded bolted joints exhibiting different amount of secondary bending. Detailed stress analyses were performed in order to examine the stress field in the crack plane of the joint plates and the secondary bending along the joints. The analyses showed a constant bending moment along the joint plates, except in the area of bolt holes where high gradients were observed, especially in the crack planes. The stress field in the joint crack plane was compared to the combined stress field of the three unit cases: bearing, by-pass and bending, which are used to build the stress intensity factor solution for bolted joints in the crack growth software AFGROW. It was found that the normal stress distributions were equal but the bending stress had a higher concentration in the joint than in the unit case. Crack growth predictions were performed in AFGROW using the superposition of the three unit cases with different amounts of bending and using the stress intensity factor solution computed by a detailed finite element model. The predictions were compared to results from the experiments with constant amplitude cycling. Using the bending unit case with the moment calculated in the crack plane under predicted the crack growth compared to the experiments. Using the constant moment outside the bolt hole area is concluded to be conservative.

INTRODUCTION

Aircraft structure consists of large numbers of riveted and bolted shear loaded joints. The main problems associated with fatigue and damage tolerance in such joints are the crack nucleation and the crack growth at fastener holes. Performing fatigue and crack growth analyses requires a thorough understanding of the stresses in the vicinity of the fastener and in the crack plane. The fatigue and crack growth life are affected by several load parameters: the load transferred through the fasteners, the by-passed load and the secondary bending. The secondary bending is a result of the joint eccentricity and it can have a considerable effect on the fatigue and crack growth life, which is why it is important to properly account for. The superposition technique to obtain the combined stress intensity factors for cracks at fastener holes is made available in software such as AFGROW [1], but it requires the proportion of each unit load case to be known. The methods for determining the transferred and the by-passed load are rather well established, while the secondary bending is more difficult to quantify. One reason for this is that the secondary bending varies along the joint. Another one is that the bending stress distribution in the unit case might be incapable of representing the bending stress distribution in the actual joint.

One-dimensional models, such as the Neutral Line Model [2, 3], can be used to calculate the tensile and the bending stresses along the joint. In this model, the joint is idealised as a single structural element, in which the bending is determined from the out-of-plane displacement of the neutral line. The Neutral Line Model is suited for calculation of the bending stress variation along the joint and can be used to indicate the most critical fastener row in the joint. However, due to its one-dimensionality, the model gives no information about the bending stress variation in the width direction of the joint. Finite element analyses of a three rivet asymmetric lap-splice joint, Fawaz [4], showed that the bending stress varies in the width direction, being maximum at the fastener plane and decreasing to a minimum at half of the fastener pitch. It is reasonable to expect that such stress variation over a crack plane would have an effect on the crack growth.

In the present paper, the tensile and the bending stress variations along and in the width direction of a shear loaded, bolted single strap joint are studied by detailed finite element (FE) analyses. The results are used to estimate the unit load case proportions for crack growth calculations with the superposition technique in AFGROW. Two different joint configurations are considered, one with an aluminium splice plate and one with a steel splice plate, in order to study different amounts of the secondary bending. The detailed FE-modelling, with extended finite element method (XFEM) [5], and the superposition technique in AFGROW are both used to compute the stress intensity factors for part-through cracks at the outmost fastener holes in the joint members. Constant amplitude crack growth is then calculated by AFGROW and compared to the experimental results from the tests performed earlier.

EXPERMENTAL PROCEDURE

Two bolted single strap joint specimens, see dimensions in Fig. 1, were cyclically loaded at the load ratio R = 0.1 and a maximum force of 25.2 kN. In one of the specimens, the splice plate, marked 2 in Fig. 1, was made of aluminium and of steel in the other one. We refer to the aluminium splice specimen as Al-specimen and to the steel splice specimen as St-specimen. In both specimens, both of the joint member plates, marked 1 in Fig. 1, were made of aluminium alloy AA 7010-T73651. Single corner cracks were initiated from saw-cut notches at the outer bolt holes in both joint member plates at the faying surface side. The crack plane is denoted A-A in Fig. 1. We refer to the cracks in the specimen with the aluminium splice as Al-1 and Al-2, and in the specimen with the steel splice as, St-1 and St-2. Titanium bolts with protruding heads were used and a low pretension torque was applied.

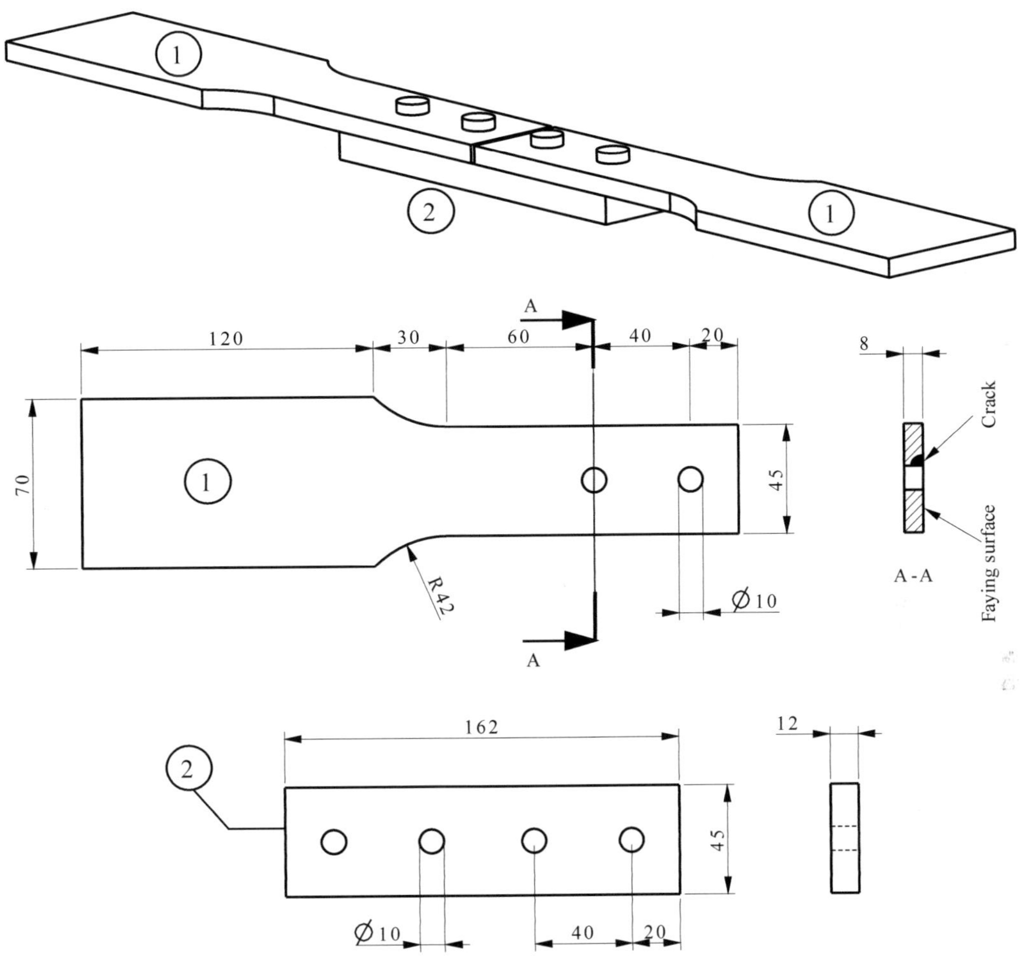

Figure 1. Test specimen, bolted single strap joint.

During the testing, the crack surface lengths were measured every 1000 cycles by a microscope, except in the aluminium splice specimen, at crack lengths longer than half of the thickness, where the lengths were measured every 500 cycles. Each time, the joint was dismantled and the outer bolts and the splice plate were removed. A shorter splice plate, see Fig. 2, which connected only the two inner bolts, was then assembled with the joint member plates and a constant load was applied to the joint. This way, the cracks in the outer bolt holes opened up and became accessible for measurement by the microscope. The cyclic tests were stopped before the cracks went through the thickness.

Figure 2. Test setup with the short splice plate.

An additional test was performed on an open hole specimen with the same geometry as the joint member plates and under the same loading conditions. The purpose of this test was to serve as a validation case of the material crack growth rate data used for the joint calculations.

FE-MODELLING

Two types of models were used in the FE-analyses: models of the joint and models of the unit load cases that build up the AFGROW solution. The analyses were performed using Abaqus/Standard [5].

Joint models

One fourth of the joint was modelled, see Fig. 3, in order to take advantage of the joint symmetry in the planes perpendicular to the x- and y-axis. Boundary conditions simulating the symmetry conditions were applied to the joint members in these planes. The load, corresponding to the maximum applied force during cycling, was applied at the free end of the joint member plate, where also the rotation of the cross-section in the y-direction was constrained. Solid, eight-node brick elements, with linear elastic material behaviour were used in all joint parts.

The bolt geometry was based on the hexagon bolt standard LN29943K and the bolt-hole clearance was assumed to be 49.5 μm, according to the mean value stated by ISO f7/H10 fitting standard. Frictionless contact conditions were assigned between all free surfaces in the joint. The contact enforcement was penalty-based with surface-to-surface formulation.

A finer mesh was generally used in the area of contact, especially in the joint plate and particularly in the area around the outer hole and in the crack plane. This finer mesh area was connected to the coarser mesh in the joint member plate by a tie constraint. Two different mesh configurations were used in the refined area of the joint member plate around the outer bolt: one with a uniform, un-cracked, element distribution through the thickness and another one with a fan-shaped mesh pattern containing a crack, see Fig. 3. The first one was used for stress distribution studies of the un-cracked plate, and the other one was used for stress intensity factor evaluation along the crack tip. The crack was modelled with the mesh-independent technique, XFEM, available in Abaqus and the stress intensity factors were extracted by evaluation of J-integral.

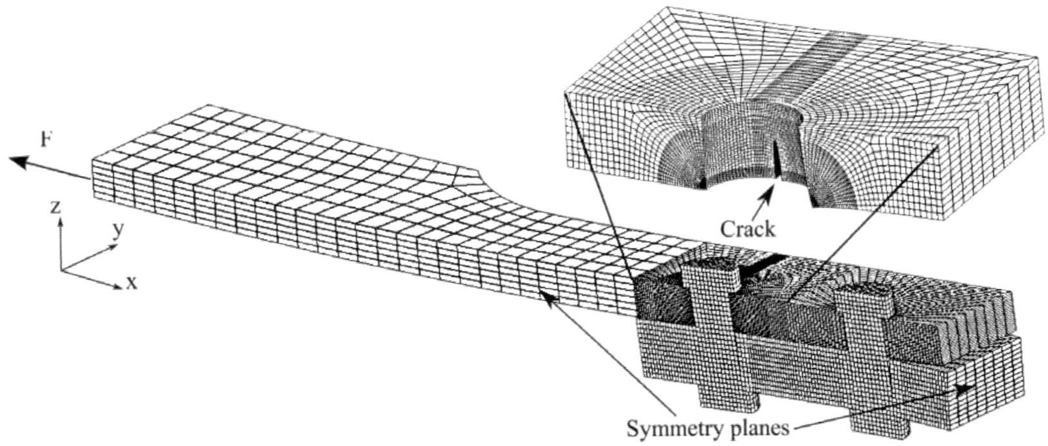

Figure 3. FE-model of the joint.

Unit case models

The unit case models were created for the purpose of studying the stress fields of the unit cases that build up the superposed AFGROW solution, and comparing them to the stress fields in the joint models. The unit cases are referred to as axial (by-pass), bending (secondary bending) and bearing (transferred bolt load). The same element type and material behaviour as in the joint models were used. In the axial and the bending case, the loads were simply applied as a far-filed force and moment, respectively, to a plate with an open hole and the same thickness and the width as the joint plate. In the bearing case, the fastener load was applied to the hole edge through a contact with a rigid cylindrical surface of the same diameter as the nominal bolt diameter.

CRACK GROWTH CALCULATIONS

All crack growth calculations are performed in AFGROW. The crack growth in the joint is calculated in two different ways:

- by extracting the stress intensity factors from the XFEM solution of the joint and inserting those as user defined entities into AFGROW, which then calculates the crack growth. Assuming the symmetry in the FE-model actually gives the solution for two symmetrically placed cracks at the hole. The results are corrected to the one-crack solution using a conversion factor as described in [6].
- by extracting the proportion of axial, bending and bearing load cases from the un-cracked FE-model of the joint and inserting those into AFGROW, which then constructs the superposed stress intensity factors from the built-in solutions and calculates the crack growth.

AFGROW offers two built-in options for calculation of stress intensity factors in bolted joints: Classic (AC) and Advanced (AA). Both options are considered in this paper. The stress intensity solutions in AC are based on the superposition of classic cases for axial and bending loading in [6] with a bearing solution [5]. The AA solution is based on the work in [7]. Generally, the stress intensity factors calculated by the two options differ somewhat for all three load cases. Also, the pin load in the bearing solution is applied in a different manner. In AC option, the pin load is applied through springs attached to the hole edge, which constrains the deformation of the hole. The load case is intended to model the neat fit pin. In AA option, on the other hand, the pin load is applied as a cosine-squared distributed pressure, which allows the hole edge deformation.

The crack growth in the reference specimen is calculated by the built-in solution for an open hole specimen with both AC and AA options, see Fig. 5, where c denotes the surface crack length and a is the through-the-thickness crack length. The crack growth rate data used in all predictions is shown in Fig. 4. Figure 5 indicates that the AA option gives closer correspondence to the test results than the AC option, especially for the through-the-thickness crack length, a.

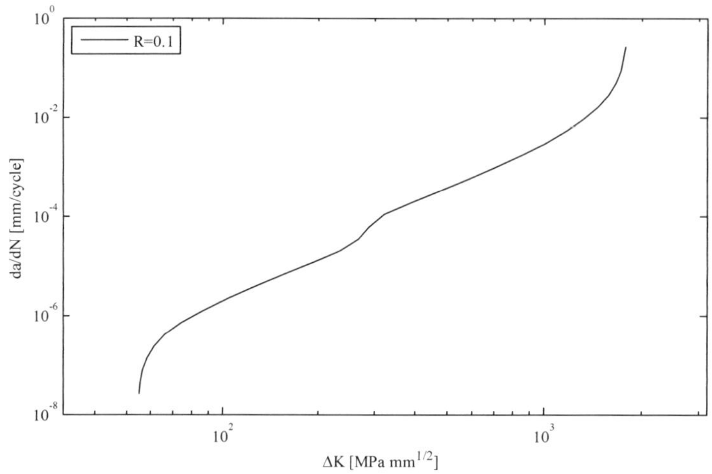

Figure 4. Crack growth rate for AA 7010-T73651 at R = 0.1.

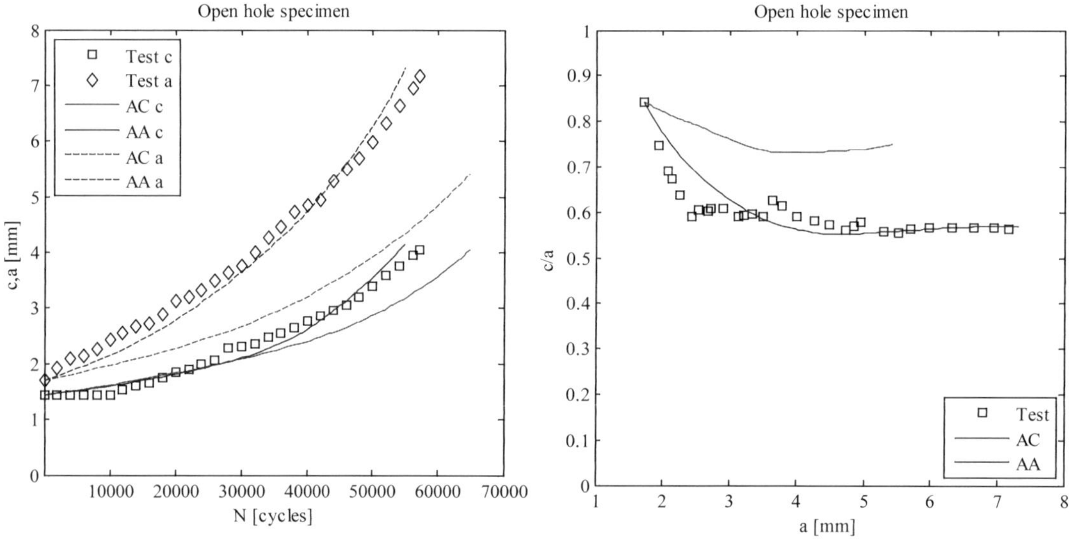

Figure 5. Crack growth in the reference specimen.

RESULTS AND DISCUSSION

First, we consider the FE-results for the joints without cracks. We refer to the coordinate system in Fig. 3 and set x = 0 at the crack plane. The variation of the cross-sectional normal force and bending moment along the joint members of the two specimens in the x-direction are shown in Figs. 6 and 7. The normal force is F = 25.2 kN for x < 0 and decreases between the bolts to 14.7 kN in Al-specimen and to 12.9 kN in the St-specimen. Since no load in the model is transferred by friction, the amount of the axial and the bearing load can be determined. The bending moment is constant for x < -5 mm, M_b = -18.5 kNmm in the Al-specimen and M_b = -8.1 kNmm in the St-specimen. It then varies with a large gradient and changes sign in the area -5 mm < x < 5 mm, which contains the crack plane. At the crack plane, i.e. x = 0, the bending moment is M_b = -8.2 kNmm in the Al-specimen and M_b = -2.4 kNmm in the St-specimen. Assuming a homogenous bending state, the average secondary bending, SB, in the crack plane is defined as the ratio of the bending and normal stress

$$SB = \frac{6|M_b|/(ht^2)}{F/(ht)} = \frac{6|M_b|}{Ft} = \begin{cases} 0.24 & Al-specimen \\ 0.07 & St-specimen \end{cases} \qquad (1)$$

where t is the plate thickness and h is the net-section width. As expected, the effect of secondary bending is less pronounced in the St-specimen. In the area of constant moment, for x < -5 mm, the average secondary bending is SB = 0.55 in the Al-specimen and SB = 0.24 in the St-specimen.

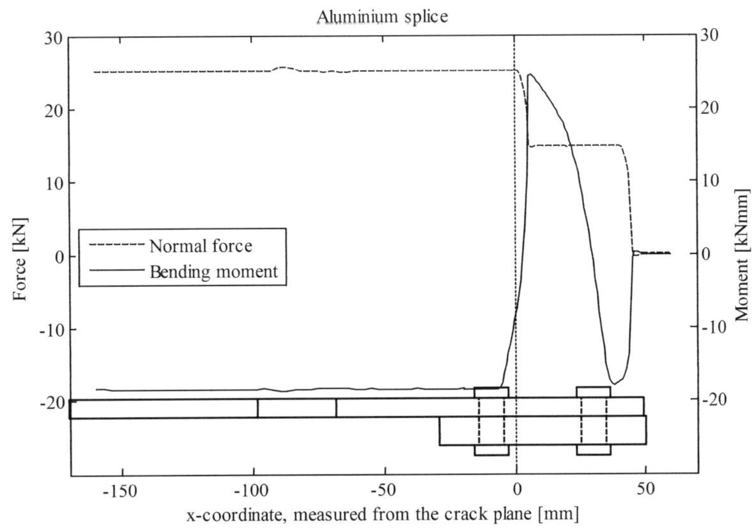

Figure 6. Normal force and bending moment along the joint member in the Al-specimen.

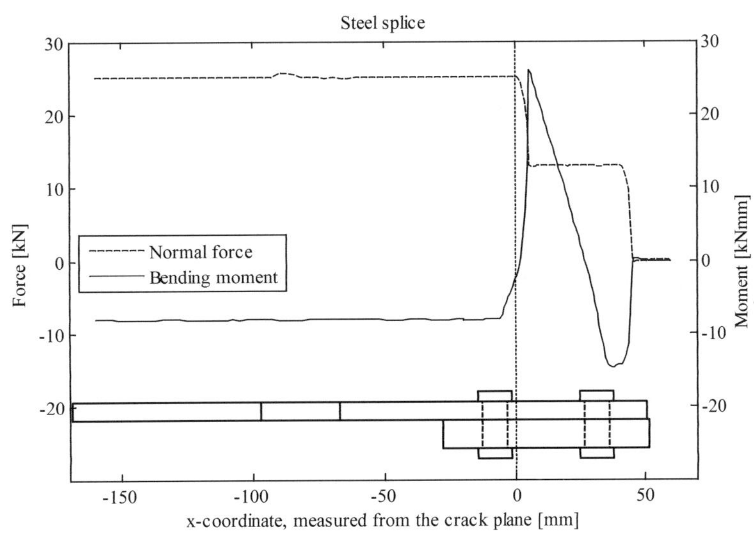

Figure 7. Normal force and bending moment along the joint member in the St-specimen.

Let us now consider the stress distribution in the crack plane of the joint and the unit case models. The same axial, bearing and bending (Eq. (1)) loads, as computed by the joint model, were applied to the unit case models. The normal and bending stress, σ_n and σ_b, are computed from the stresses at the lower and upper side of the joint member plate, σ_L and σ_U, along the path from the hole edge to the free edge in y-direction as

$$\sigma_n = \frac{\sigma_L + \sigma_U}{2}$$
$$\sigma_b = \frac{\sigma_L - \sigma_U}{2} \qquad (2)$$

For the unit cases, the normal stress is plotted as the sum of contributions from the axial and bearing case. In both the Al-specimen and the St-specimen, the normal stress complies well with the normal stress from the unit cases, see Fig. 8. However, the bending stress concentration in the joint models is much higher than in the unit case models. The difference becomes even more apparent, see Fig. 9, expressed in terms of the secondary bending defined as

$$sb = \frac{\sigma_b}{\sigma_n} \tag{3}$$

It is obvious that the bending stress distribution in the unit case is not able to represent the distribution in the joint. An explanation is that the tilting of the bolt shifts the contact point between the bolt shaft and the hole edge towards the faying surface, thereby inducing a local bending moment which influences the bending stress.

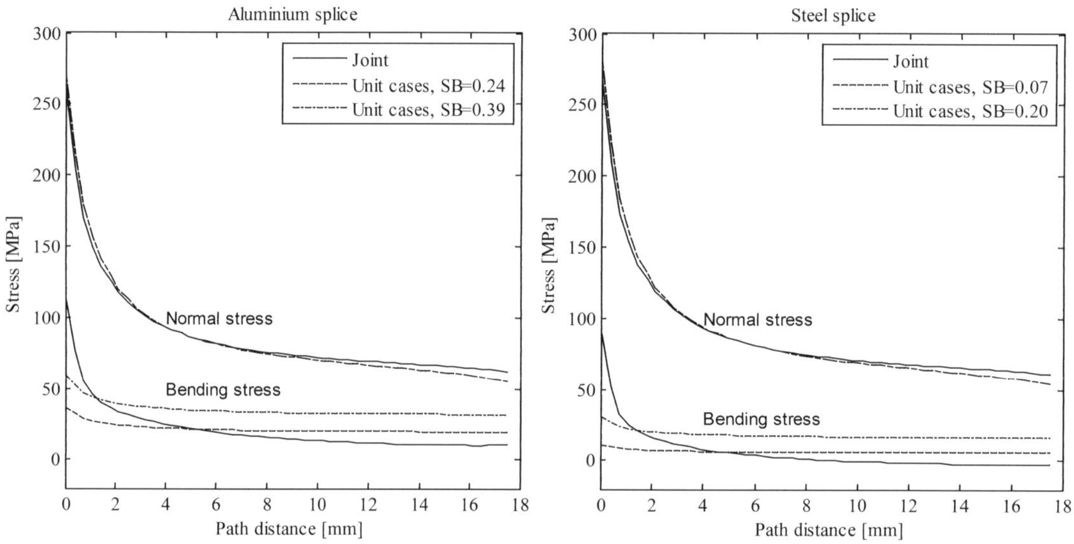

Figure 8. Normal and bending stress along the crack plane.

Because of the high bending stress concentration in the joint, it is reasonable to believe that the unit case with *SB* according to Eq. (1) would under predict the crack growth of part-through cracks. The question is how large *SB* would have to be applied to the unit case in order to get better predictions. In this paper, we try to estimate that value by requiring that the integral of the bending stress, in the crack growth area, is the same in the unit case as it is in the joint. The resulting values are

$$SB = \begin{cases} 0.39 & Al-specimen \\ 0.20 & St-specimen \end{cases} \tag{4}$$

The bending stress and the secondary bending, *sb*, for values in Eq. (4) are also shown in Fig. 8 and 9.

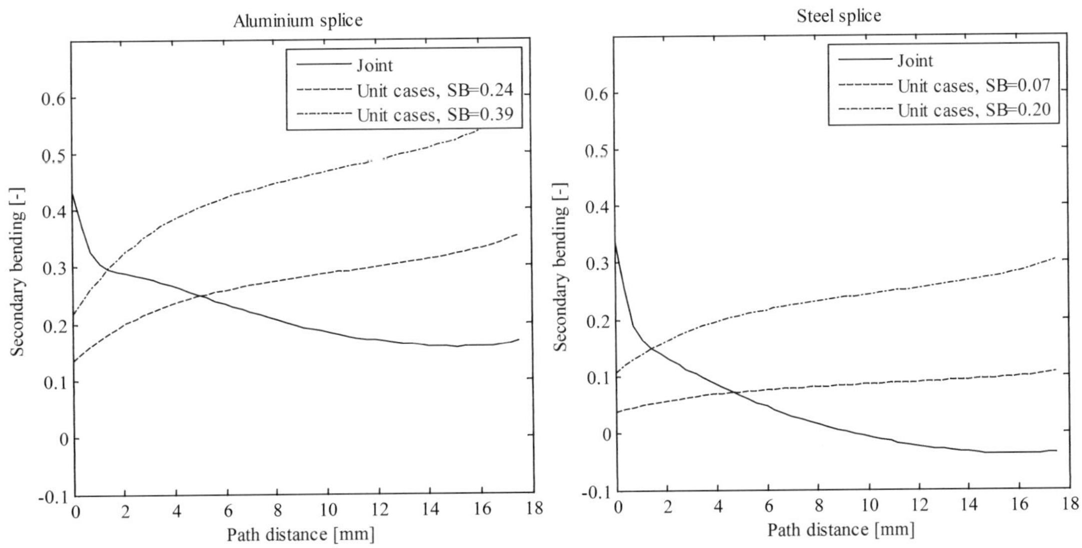

Figure 9. Secondary bending, *sb*, along the crack plane.

In the following, we will compare the crack growth measured from the tests with the prediction results from the XFEM solutions and from the unit cases with *SB* values in Eq. (1) and (4). Figure 10 shows the growth of the surface crack length, *c*, during the cycling in both joint specimens. The XFEM solution results seem to be closest to the experimental results, except for the St-2 crack. Also, the predictions with *SB* from Eq. (4) comply better with the tests than the predictions using Eq. (1), as anticipated. It is also obvious that AC gives better results than AA, which may be attributed to the difference in the stress intensity solutions of the two methods.

Similarly, the through-the-thickness crack growth complies best with XFEM, and *SB* from Eq. (4) gives a slightly better correspondence than *SB* from Eq. (1), see Fig. 11. On the other hand, AA option gives results that are somewhat closer to the test results than the AC results. The crack length ratios in the different analyses are shown in Fig. 12.

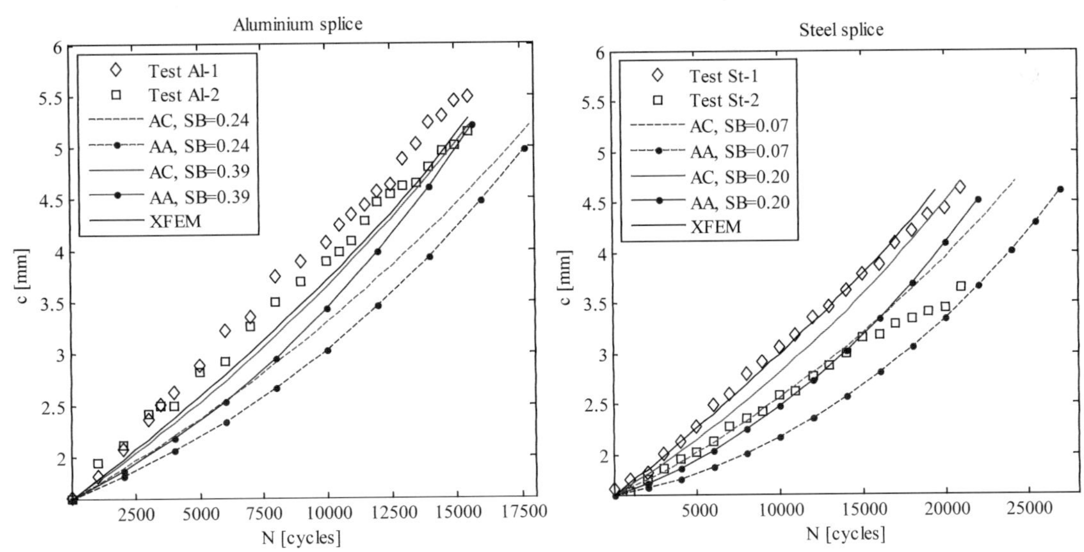

Figure 10. Surface crack length, *c*.

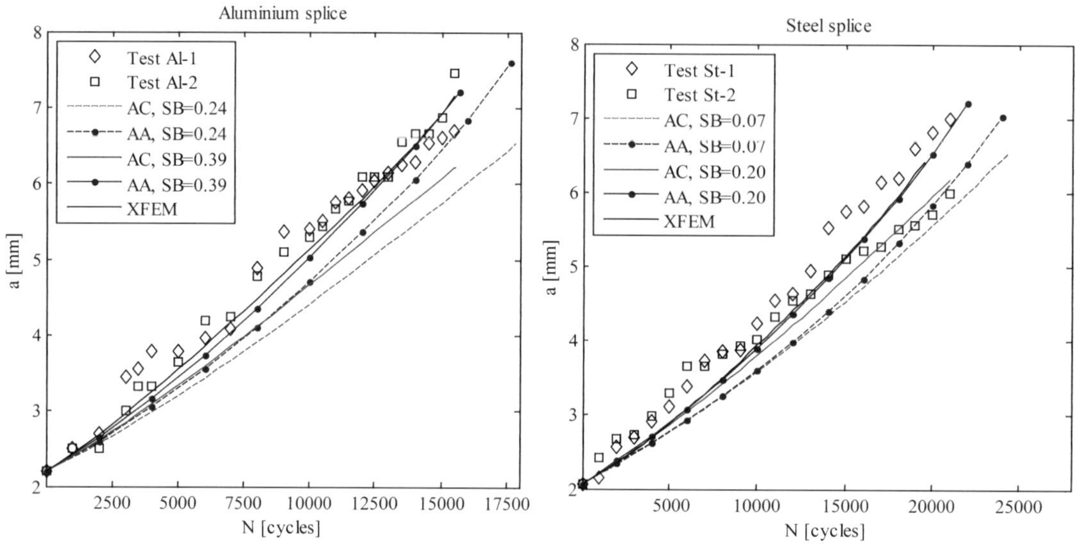

Figure 11. Through-the-thickness crack length, a.

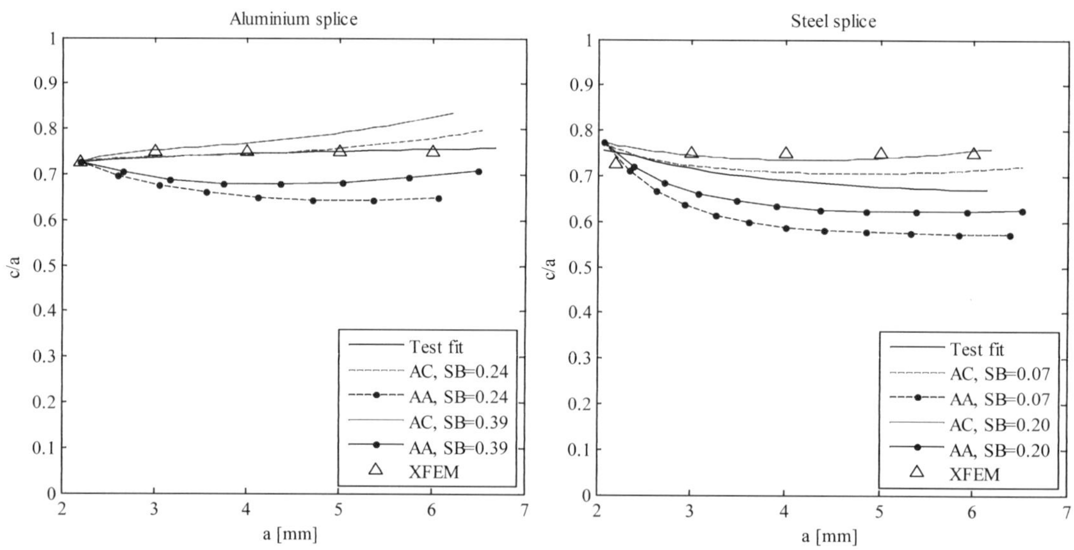

Figure 12. Surface to through-the-thickness crack ratio.

Finally, we compare the stress intensity factor ranges at the surface crack tip, see Fig. 13, and at the through-the-thickness crack tip, see Fig. 14. The figures show that the XFEM solution, which gave the closest predictions to the crack growth in the test, roughly follows the AC solution with SB according to Eq. (4) for the c-tip and AA solution with SB according to Eq. (4) for the a-tip.

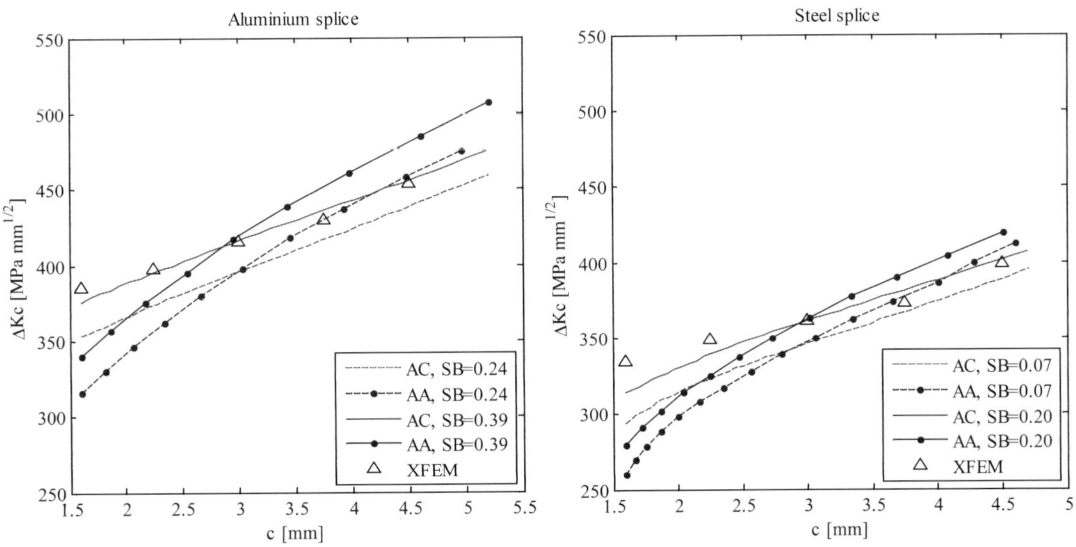

Figure 13. Stress intensity factor range at the surface crack tip.

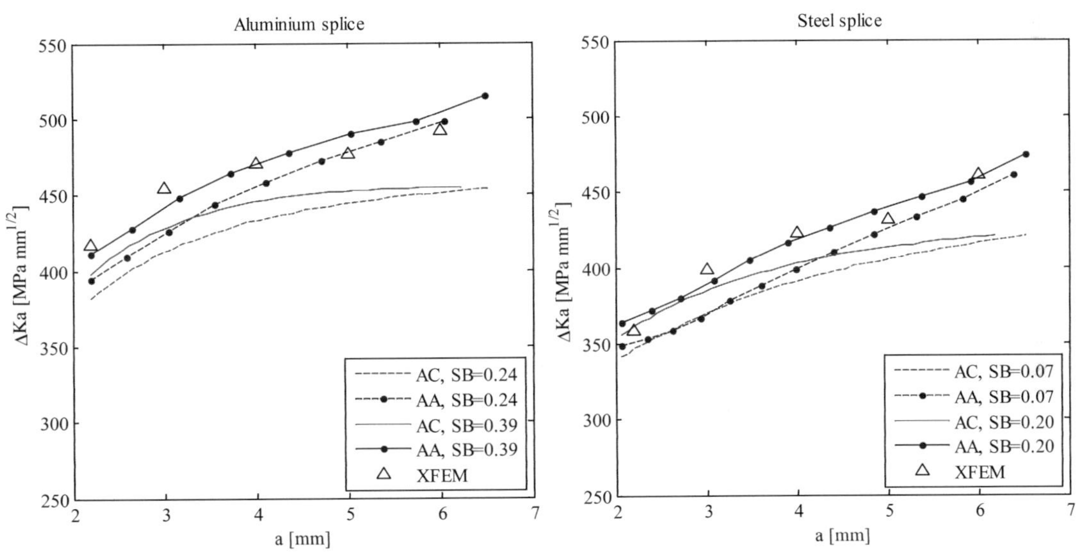

Figure 14. Stress intensity factor range at the through-the-thickness crack tip.

The performed comparisons indicate that the amount of secondary bending, assigned to the bending unit case in AFGROW, according to Eq. (4) gives closer predictions then according to Eq. (1), for part-through cracks. For longer through cracks Eq. (4) might over predict the crack growth and Eq. (1) might give better predictions, considering that the *sb* tends to be lower in the joint model than in the unit case at larger distances from the hole, see Fig. 9. When simplified models are used, for instance the Neutral Line Model or shell element FE-models where the fastener is represented by a line element, only the moment outside the bolt area, i.e. $x < -5$ mm, can be calculated. This moment gives an even higher *SB* than Eq. (4), and leads to conservative crack growth predictions for both part-through and through cracks.

CONCLUSIONS

Crack growth was investigated in single strap bolted joint specimens exhibiting different amounts of secondary bending. The stress fields in the crack plane of the joints were studied by detailed FE-analyses. Also, stress fields in the unit cases, which built up the superimposed solution in AFGROW, were examined and predictions of crack growth were performed using different amounts of secondary bending. The predictions were compared to test data. The stress intensity factor ranges calculated by AFGROW were compared to the XFEM solutions. The following conclusions were drawn:

- Bending moment varies along the joint length, with a large gradient in the vicinity of the crack plane.
- The normal stress distribution in the joint crack plane is correctly represented by the axial and bearing unit cases.
- The bending stress distribution in the joint crack plane cannot be represented by a bending unit case.
- The bending stress distribution in the joint has a larger stress concentration at the hole than the bending unit case.
- Using the bending unit case with the bending stress determined from the moment at the crack plane under predicts the crack growth for part through cracks.
- Using the bending case with the bending stress determined from the moment in the constant moment area, outside the bolt area, over predicts the crack growth. This option results in a conservative approach and can be used in conjunction with simplified models which do not account for the bending stress variations in the vicinity of the crack plane.

The growth of longer through cracks was outside of the scope of this paper. However, in order to understand the effect of secondary bending field on the total crack growth life, from a small corner crack to a long through crack, further studies need to be performed.

REFERENCES

[1] Harter J. (2008). *AFGROW users guide and technical manual. AFRL-VA-WP-TR-2008-XXXX*, Air Force Research Laboratory.

[2] Schijve, J. (1972). *Some elementary calculations on secondary bending in simple lap joints.* NLR-TR-72036. Amsterdam, National Aerospace Laboratory.

[3] De Rijck, J.J.M., Fawaz, S.A., Schijve, J., Benedictus R. and Homan, J.J. (2001). In: *Durability and damage tolerance of aircraft structures: metals vs. composites*, Proceedings of the 24st ICAF Symposium, vol. I, pp. 172–187, Lazzeri, L. and Salvetti, A. (Eds.), Pacini Editore, Pisa.

[4] Fawaz, S. (1997). *Fatigue crack growth in riveted joints.* Dis. Delft University of Technology, Delft University Press.

[5] Abaqus-Inc. Abaqus User Manual, Version 6.13. Dassault Systèmes Simulia Corp Providence, RI, USA, 2013.

[6] Newman, J.C. and Raju, I.S. (1986). Stress intensity factor equations for cracks in three-dimensional bodies subjected to tension and bending loads. Chapter 9, Computational methods in the mechanics of fracture, Elsevier Science Publishers, B.V.

[7] Fawaz, S.A. and Andersson, B. (2004). *Eng. Fract. Mech.*, vol. 71, pp. 1235-1254.

Improving the prediction of small crack growth in 7XXX aluminium alloys

S Barter, M Burchill, M Jones

Aerospace Division, Defence Science and Technology Organisation, 506 Lorimer Street Fishermans Bend, Vic. Australia, 3207

Abstract: This paper presents crack growth rate curves for 7XXX series alloys that can be used to robustly predict the crack growth for typical service cracks in aircraft structures for a range of spectra. Using a standard crack growth rate algorithm two sets of crack growth rate data are examined; the first used reverse-engineered crack growth rate curves based on a limited set of small crack growth data from fighter spectrum loading. The second used crack growth rates curves based only on measured small crack growth rate data collected from constant amplitude loading bands applied within other constant or variable amplitude loading. The two models were then used, along with handbook long crack growth data curves, to predict the short crack growth and total fatigue lives of coupons tested with a range of complex variable amplitude loading spectra typical of those that occur in service aircraft. These predictions were compared to the experimentally measured total lives of the coupons and also to the crack growth of each lead crack in the coupons tested. All predictions used a cycle-by-cycle damage summing method and were carried out both with and without a simple closure/retardation correction. Of the two crack growth rate models developed, the reverse-engineered crack growth rate curve appeared to provide the better overall predictions for a wide variety of spectra as compared to the rate curve models developed based solely on constant amplitude crack growth rate data.

INTRODUCTION

In conventional total crack life predictions, based on linear elastic fracture mechanics (LEFM) principles, standard material crack growth rate data collected from long cracks, typically via the ASTM method outlined in E647-13a [1], is often used in fatigue life prediction codes. However, it is regularly observed that for the natural cracks that may develop in an aircraft structure, the use of these *long crack* based growth data can result in a significant under prediction of the rates of growth while cracks are small (< 1 mm) as well as when they are mechanically small (≈1–5 mm)[1], as defined in [2], [3]. In typical aircraft aluminium alloys, these predictions may be particularly poor. A number of reasons have been offered in the literature [1], [4]–[10] that identify various crack closure mechanisms as the key and that by accounting for these by diverse corrections, the standard ASTM data may be used in predicting small crack growth. Due to the difficulty in obtaining small crack growth data the validity of these various corrections has not been extensively investigated. It should be noted that, this *small crack issue* is avoided to some extent by limiting crack growth prediction to crack sizes, typically greater than 1.27 mm (0.05") deep (USAF damage tolerance [10] for a corner crack), that are less influenced by near threshold effects. Nevertheless, even at these larger crack sizes the cracks remain mechanically small and with growth rates that may be important to the overall accuracy of crack growth prediction [1], [12], since many of the typical load cycles may only produce cyclic stress intensity ranges (ΔKs), at any crack tip that are near the threshold for crack growth.

In making predictions using an LEFM approach, the fundamental inputs are the crack growth rate data for several different stress ratios, R ($\sigma_{min}/\sigma_{max}$), starting from the threshold value (ΔK_{th}) in the material in the orientation of interest. This is typically given as crack growth per load cycle, da/dN as a function of the applied ΔK, which is a function of the geometry, applied stress and the depth of the crack: $\Delta K = \Delta\sigma\beta\sqrt{(\pi a)}$, where β is the finite size correction factor and a is the crack depth. The cyclic change stress intensity threshold for no growth occurs when $\Delta K = \Delta K_{th}$ for the material of interest at the R being considered. Gathering these data, particularly the threshold and growth rate for small cracks is difficult, and while various methods are suggested in E647-13a, these tend to produce data with a high degree of scatter (fanning) at the lower threshold ΔK values [13][2] and leaves significant questions about what the true value for ΔK_{th} is for many materials, [14]–[16]. In this paper a brief discussion of work conducted at the Defence Science and Technology Organisation (DSTO) in the Thresholds and Short Cracks (TASC) program is presented where growth rates in this regime are examined and measured for two common aluminium alloys (AA) used in fighter and transport aircraft. This work is aimed at measuring crack growth rates for small cracks near the fatigue crack growth threshold,

[1] Definition from [1] **Error! Reference source not found.**where mechanically small cracks are those for which near-tip plasticity is comparable to the crack size.

[2] This reference discusses an alternative to the method used in E647-13a and to that used here.

and to increase the understanding of the crack growth behaviour in this regime. The rate data collected for this material was used to develop two material rate curves that were further modelled to produce a family of rate curves for different R values. These were than used for crack growth predictions for a range of different types of spectra (16 types in all, several of which are reported on here) applied at different stress levels typical of those that may be found in highly stressed regions of aircraft, for cracks that that have initiated from small discontinuities (of approximately 0.01 mm in size).

The main material of interest here is thick plate AA7050-T7451 (AA7050) used extensively in the Royal Australian Air Force (RAAF) Hornet F/A-18A/B, Super Hornet F/A-18F and JSF F-35 airframes. Also, although only preliminary crack growth rate results are available, material data from the newer AA7085-T7452 (AA7085) alloy are presented for comparison with the AA7050 results. The small crack AA7050 data was compared to two sets of data that are available from in-built material libraries (NASGRO [17] and Harter-T [18]) within AFGROW that are typical of *long crack* data obtained using the method outlined in ASTM E647-13a. The crack data presented here was collected via quantitative fractography (QF) measurements of fatigue crack growth in small coupons tested to failure at DSTO. The test loading used small bands or sub-blocks of constant amplitude (CA) loading with various different stress ratios and/or variable amplitude (VA) loading.

Also presented and discussed here are the test results and predictions for fatigue crack growth under a range of VA loading. Results of which raised several questions regarding the premise behind many closure/retardation predictive methods such as the ΔK effective (ΔK_{eff}) damage summing approach based on the notion of plasticity-induced closure. Quantitative measurements of crack growth, along with qualitative observations of the fracture surface morphology, provided some evidence that these closure/retardation models are not robust for predicting crack growth in the threshold region when cracks are small and are strongly influenced by local material properties, particularly crystallographic orientations with respect to the growth path [19], [20].

Finally, this paper briefly discusses how results from this work suggest improvements in the understanding of crack growth for the small crack growth region in the main material examined. The target of this work remains to produce a robust set of crack growth data that can be used in LEFM models to accurately predict typical service cracks for a range of different types of spectra, stress levels and geometries typical of critical locations in RAAF (and other) airframes.

USING QUANTITATIVE FRACTOGRAPHY TO DETERMINE SMALL CRACK GROWTH RATES

The small crack growth rate measuring method used to collect the data here is described in [21] and [22]. This method uses test load spectra selected to produce local crack path and topological changes [23]–[25] to make observable features on the fracture surface that can be related to the loads applied. An example of a set of loads that has produced such features is shown in Figure 1a; with sub-blocks that have the same σ_{max} but differing R. Load sequences such as this, when tested on AA7050 (and AA7085) coupons, result in a contrasting banded pattern forming on the fracture surfaces due to the crack path and topographic changes. Crack growth per cycle data can be determined from inspection and measurement of the width of these growth bands during QF and dividing by the number of cycles in each band, resulting in an average crack growth rate rather than the measurement of individual striations. These bands can be found optically, as in Figure 1b, or by scanning electron microscope (SEM) as in Figure 1c, where the SEM was used for very small cracks with very low ΔKs where individual CA sub-blocks crack growth is as low as approximately $2x10^{-8}$ m [24].

As an alternative to varying the R for alternate sub-blocks of CA load cycles, small sequences of VA can also be used to mark the fracture surfaces between the CA load sub-blocks. An additional advantage of using a small sub-block of VA loading is that possible retardation effects on small crack growth rates, observed previously [26], from accumulating CA cycles are expected to be lessened. The VA sub-block was a small segment of a larger spectrum that is representative of wing root bending loading from a fighter aircraft full-scale fatigue test. To prevent potential retardation from the VA sub-block loads affecting the growth from the CA loads, the peak of the CA sub-block was set to equal the peak load in the VA sub-block as shown in the example spectrum in Figure 2a. In this work the sub-blocks of CA loading, as shown in Figure 2a, were typically either R = 0.8 or R = 0.1 referred to as *high R* and *low R* respectively.

a) b) c)

Figure 1. Observed effect of a load spectrum of a) sub-blocks of CA on fatigue crack path/fracture in AA7050 visible via b) optical microscope at crack depth 1.47 mm and c) SEM at high tilt at crack depth 1.28 mm.

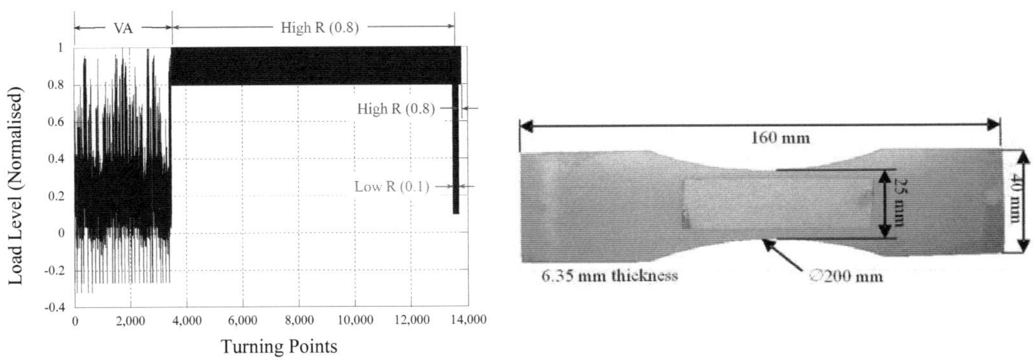

a) b)

Figure 2. a) An example of one of the simple spectrum used to generate the small crack growth rate data. b) Coupon configuration; coupons were etched to produce small surface pits as crack starters (~0.01 mm) and then shot-peened (except where masked as shown). Material: specification AMS 4050, AA7050-T7451 (6" plate), yield strength = 418 MPa.

The tests were carried out on coupons of the form shown in Figure 2b; an hourglass shaped coupon with a constant radius in the net-section. These coupons were tested to failure in a 100 kN, computer-controlled, servo hydraulic fatigue test machine at ambient room temperature and humidity. The peak spectrum stress applied was either 200 or 220 MPa, at the net test section. The coupons were etched to produce small surface pits as crack starters[3] (~0.01 mm) and then shot-peened in the areas that were undesirable for crack growth, i.e. the edges and corners, this forced initiation on the faces away from the edges of the coupons, allowing a simple geometry correction factor for calculations of K. The peening had no effect on the crack growth over the crack depths measured (as determined by separate tests). Cracks that entered the peened areas were eliminated from the data set.

Crack growth rate results

Shown in Figure 3 is an example of a crack surface in an AA7050 coupon, produced with the spectrum shown in Figure 2a. Using the combination of low R CA, high R CA and VA load sequences in sub-blocks, the *crack growth due to each sub-block of loading was clearly visible as alternating bands on the fracture surfaces*. During QF, these individual bands were measureable down to ~7 μm in thickness via optical microscopy, using the variation in contrast produced by the crack path angles and roughness differences. With the SEM (electro-optical) examination, using a field emission gun instrument, the texture differences at very short crack depths (~ 0.2 mm) allowed the measurement of individual bands as short as about 20 nm wide. The measurements of growth band widths under CA loading were then divided by the number of cycles applied to calculate a crack growth per cycle. The results were then plotted against ΔK, where the geometry factor (β) was calculated from a standard solution for a centre semi-elliptical surface crack (allowing for

[3] The etching process used here results in a surface condition similar to that observed in RAAF airframe components that have corrosion prevention treatments applied during manufacturing. Fatigue cracks have been observed, in both fleet and test articles, to consistently initiate from the ~0.01 mm etch pits [27].

angular correction from the origin [28]). Growth at *ΔK*s down to less than 0.6 MPa√m were measured (a long crack threshold is typically quoted as ~ 1.5 MPa√m for AA7050) with average crack growth per cycle values approaching 1×10^{-11} m/cycle. The $R = 0.8$ data is presented in Figure 3b, upon which the two *long crack* $R = 0.8$ curves from the in-built libraries[4] in AFGROW are overlayed. These data indicate higher growth rates at much lower *ΔK* values than the long crack data below about 2 MPa√m.

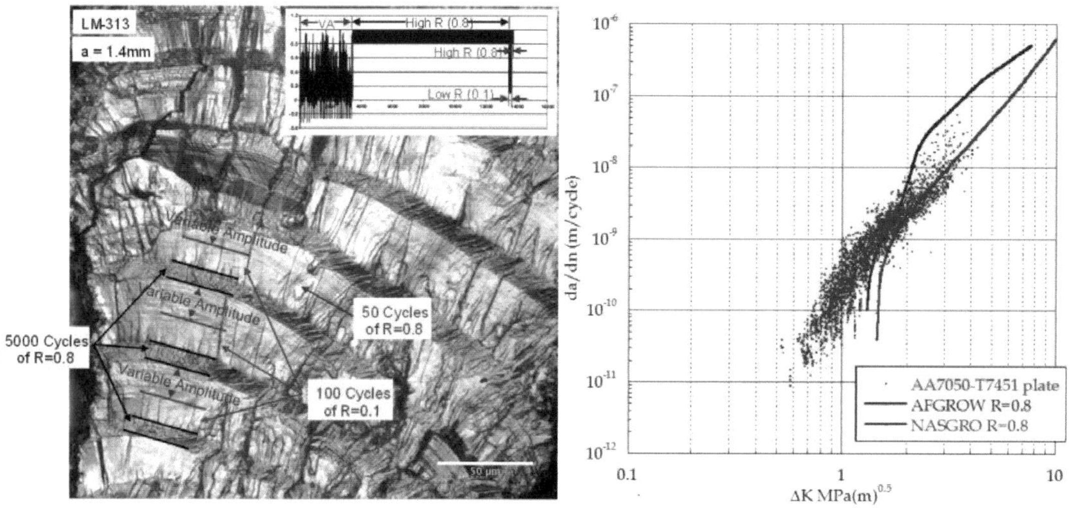

a) b)

Figure 3. Results from testing: a) example of a fracture surface from loading with a spectrum of alternating load sequences of High R CA, Low R CA and VA; and b) crack growth: da/dN v ΔK data, calculated from QF measurements of individual CA crack growth bands for the R = 0.8 bands in AA7050.

While the above data provided evidence of a much lower crack growth threshold and much faster growth in the threshold region than standard test methods suggest, it is also notable that the increase in scatter at lower ΔKs is less evident than other methods used to collect small crack growth rate data, for example [29]. By way of comparison, a similar exercise was carried out using material cut from a 6" (152.4 mm) thick hand forging of AA7085. This material gave very similar results as the AA7050, with the exception that the growth rates below about 1.0–1.5 MPa√m diverged progressively from the AA7050 data to give lower growth rates and an apparent higher threshold for crack growth as seen in Figure 4.

[4] Within the in-built material libraries in AFGROW (Version 5.2.1.18), two different data sets were chosen to represent long crack data, typical of that which is obtained from ASTM E647-13a testing. The first from the Harter-T library for AA7050-T74 plate, labelled hereafter as Harter-T. The second was from the NASGRO material library for AA7050-T7451, labelled hereafter as NASGRO.

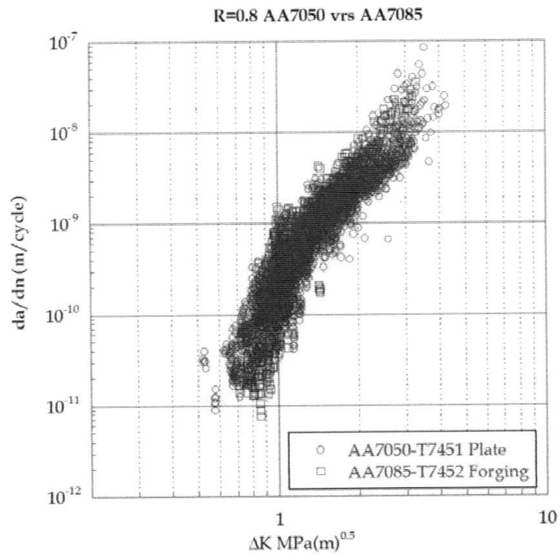

Figure 4. The results of the testing of AA7085 – green points. Material specification AMS 4414, AA7085-T7452 (hand forging), yield strength = 463 MPa. Only the R = 0.8 da/dN v ΔK data is shown compared to the AA7050 data – red points.

MODELLING THE CRACK GROWTH RATE CURVES

As noted in the introduction, two crack growth rate curves sets were developed; the first based on the data collected here, and the second reverse–engineered to produce good predictions for a limited set of VA tested coupons.

Reverse-engineered crack growth rate curve: Prior to the completion of the collection of the small crack growth data for AA7050, predictions of the likely form of this curve were developed by reverse-engineering from a limited CA and VA data set. This method was reported in [30], and is briefly described below.

- Known growth data, for cracks that were produced from VA loading based on either a fighter or a transport aircraft wing root bending moment spectra, were predicted via the use of two crack growth rate curves for two stress ratios (R = 0.5 and 0) inserted into the crack growth code AFGROW. The Harter T method, also known as the point-by-point Walker Shift method, was used to develop models of the growth rate curves for other stress ratios for these simulations.
- The primary curves, which were simple in form, were made up of six constant power curves each. These were adjusted so as to give what was assessed, after running them in AFGROW, as a reasonable prediction of the VA crack growth observed in the tests.

Two parameters were used to assess the success of the predictions made using the curves; the shape of the total predicted crack growth curves needed to match the shape of the measured crack growth curves and the total lives needed to be similar. Crack growth measurements and predictions were compared for growth from small initiating discontinuities (where possible) to failure, i.e. from 0.1 mm to approximately 5 mm in the case of the cracks that were used. The resultant curves are shown in Figure 5a, hereafter called the DSTO12 curves. Guidance for the initial shape of the crack growth rate curves was suggested by some small crack growth data that had already been collected from other coupon tests as reported in [30]. Although some small crack data existed, the process of reverse-engineering was required to aid development of the complete DSTO12 crack growth rate curves.

Crack growth rate curves from the small crack growth rate testing: With the completion of further testing, sufficient small crack growth rate data of R = 0.8 (shown in Figure 3b) and R = 0.1, R = 0.5 and R = -0.5 down to low ΔK values was then available for the development of equations to represent this data. The equation was created by using a non-linear least squares optimisation technique to fit cubic curves to the data with modifications that allow for the R effect and for the increase in rate of growth due to approaching fracture toughness values, [31]. This produced a second set of crack growth rate curves as shown in Figure 5b (only two of the R curves are shown along with their measured data), hereafter referred to as the DSTO14 curves. The DSTO12 and DSTO14 curves are compared with the long crack curves for two stress ratios in Figure 6. It is notable that the DSTO curves have generally faster growth and lower thresholds compared to the long crack curves as would be expected.

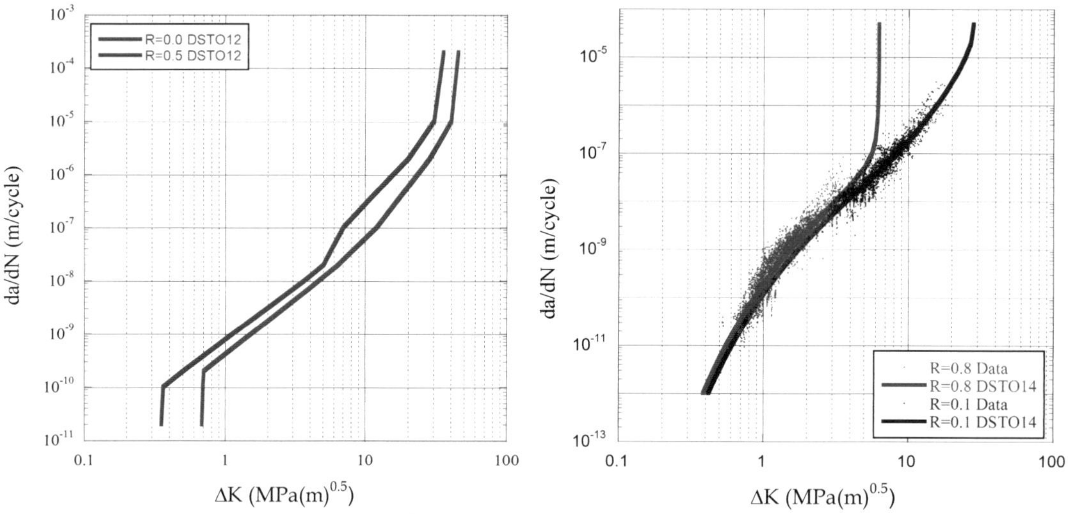

Figure 5. The two DSTO da/dN v ΔK material curves are shown for two R values: a) the DSTO12 curves; and b) DSTO14 model along with the measured data of growth rates from small sub-blocks of CA loading.

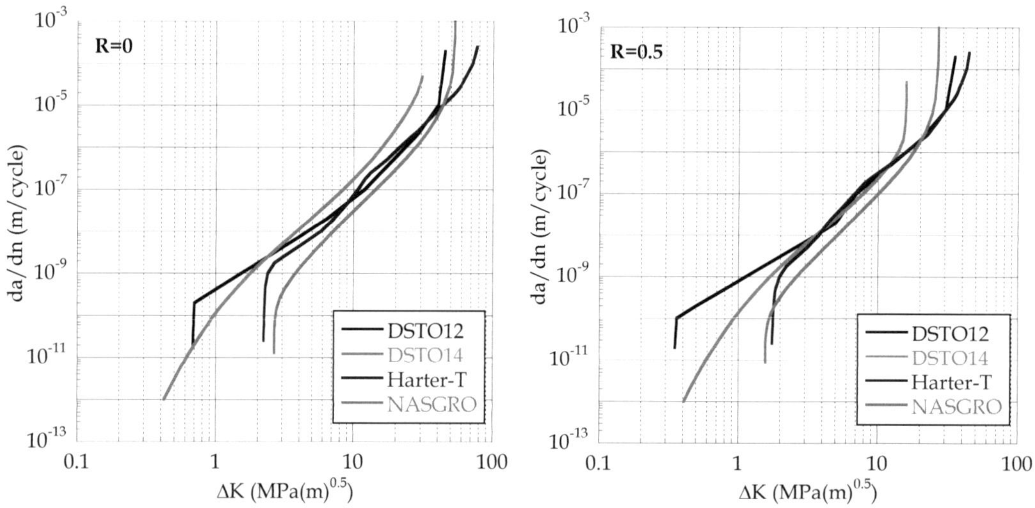

Figure 6. The DSTO12 (black) curves and the DSTO14 (orange) curves compared to AFGROW HARTER-T and NASGRO curves for $R = 0$ and $R = 0.5$.

PREDICTIONS USING THE SMALL AND LONG CRACK GROWTH RATE CURVES OF AA7050 CRACK GROWTH DATA

The DSTO12 and DSTO14 growth rate curves were used, along with the long crack growth rate curves, to predict the growth of a number of cracks grown in coupons of the same type as shown in Figure 2b by use of the AFGROW program. These cracks were grown with a wide variety (16 types) of VA spectra that included combat aircraft wing root bending moment, transport wing root bending moment, helicopter and buffet dominated spectra for cracks growing from small typical surface discontinuities to failure. The first two examples of the results obtained for two standard spectra; Mini-TWIST and Mini-FALSTAFF, generated using GENISIS4fatigue[5] [32], are shown in Figure 7. (Note that neither of these examples was used in the reverse-engineering of the DSTO12 rate curves.)

[5] The mini or shortened versions of the two standard spectra were used. These are for a transport aircraft wing and a fighter aircraft wing respectively.

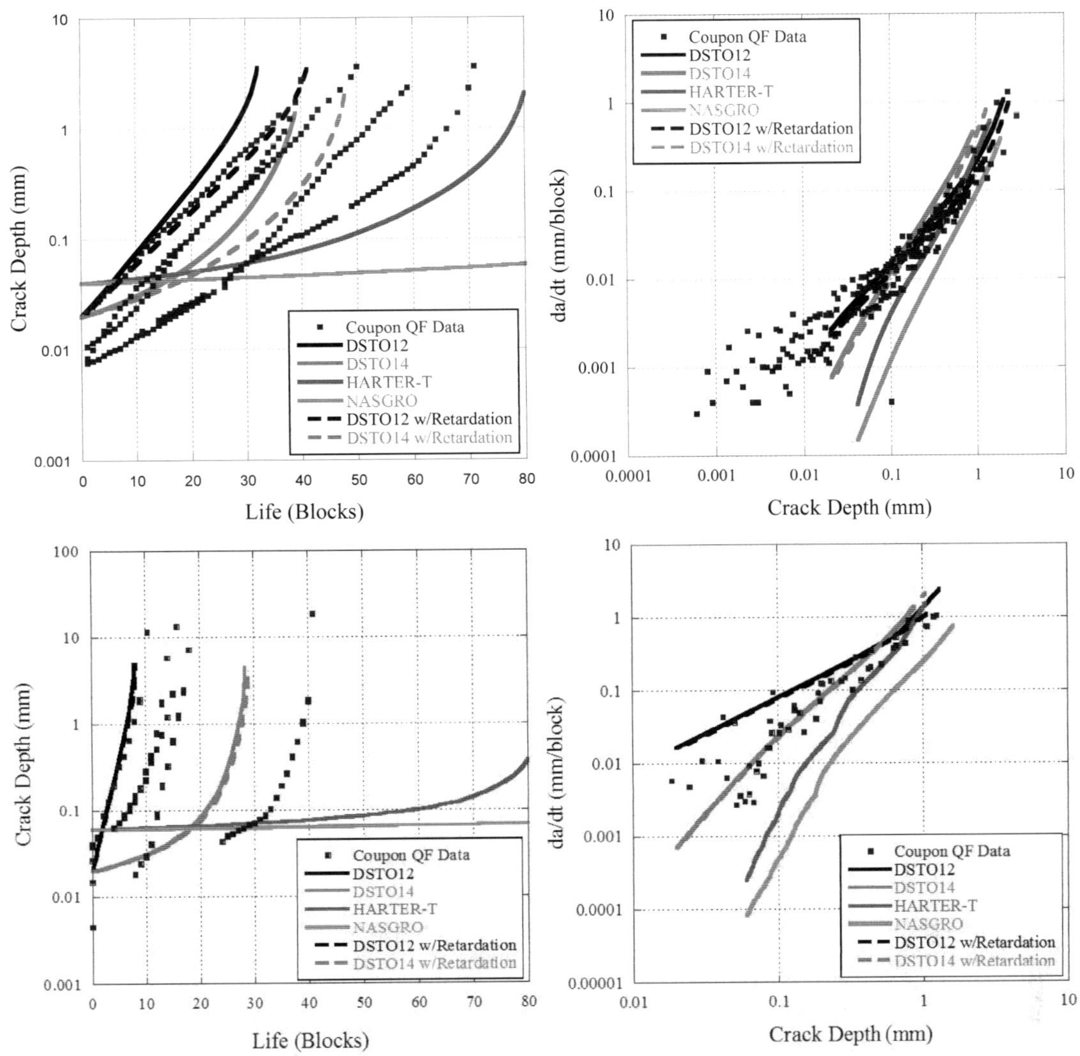

Figure 7. Experimental data and predictions of crack growth in simple coupons produced by the Mini-FALSTAFF (top) (peak stress = 390 MPa) and the Mini-TWIST (bottom) (peak stress = 300 MPa) spectra. Both the crack growth (left) and the crack growth rates (right) are presented.

As can be seen in Figure 7, the predictions were compared to the crack growth of the coupons over the entire lives of the test coupons, as this could be accurately measured since each block of the test spectrum included unique marker band load sequences [33]. These created visible bands on the fatigue crack surfaces in a manner similar to the banding that was used to generate the crack growth data in the development of DSTO12 and DSTO14. The presence of such marker bands greatly aid the accurate measurement of crack depths down to tens of microns of growth.

Included in Figure 7 are the DSTO12 and DSTO14 predictions of crack growth from an assumed an initial crack depth of 0.02 mm, which is approximately the average depth of all the discontinuities that were observed to initiate fatigue cracks in these coupons. The Harter-T and NASGRO results, based on the long crack growth curves required larger initial crack depths since no growth would occur at the depths that were measured in the test coupons. Note that the crack growth predictions presented here, based on DSTO12 and DSTO14 material curves were re-run with the Willenborg [34] retardation correction included in the AFGROW software, with a *Shutoff OverLoad Ratio* (SOLR) parameter set to 2.8.

These results indicated that there was an improvement in the prediction of the test cracks using both the DSTO12 and DSTO14 data curves compared to the long crack growth curves. What is also clear in from comparison of the curves to the crack growth data presented in Figure 7 is that the crack growth per block of loading is more accurately predicted with the DSTO12 data, in particular for the Mini-FALSTAFF spectrum, while it was on the conservative side (predicted more rapid growth) for the Mini-TWIST spectrum.

To further demonstrate the improvement in predicting fatigue crack growth rates and total fatigue life using the DSTO derived material rate curves, results are presented below from another similar coupon program using representative wing root bending loading spectra, Figure 8. The test spectra are based on actual service data from two individual combat aircraft [35], noting that the results from the more severe spectrum are presented in Figure 8 upper, however both spectra have the same peak spectrum stress. Figure 8 presents both the crack growth data from a series of simple coupon tests along with the predictions using both the DSTO and long crack material data curves. Predictions of the crack growth per block of spectrum loading (Figure 8 right) show the DSTO12 material data curves predict the fastest short crack growth well, and result in conservative results for total life.

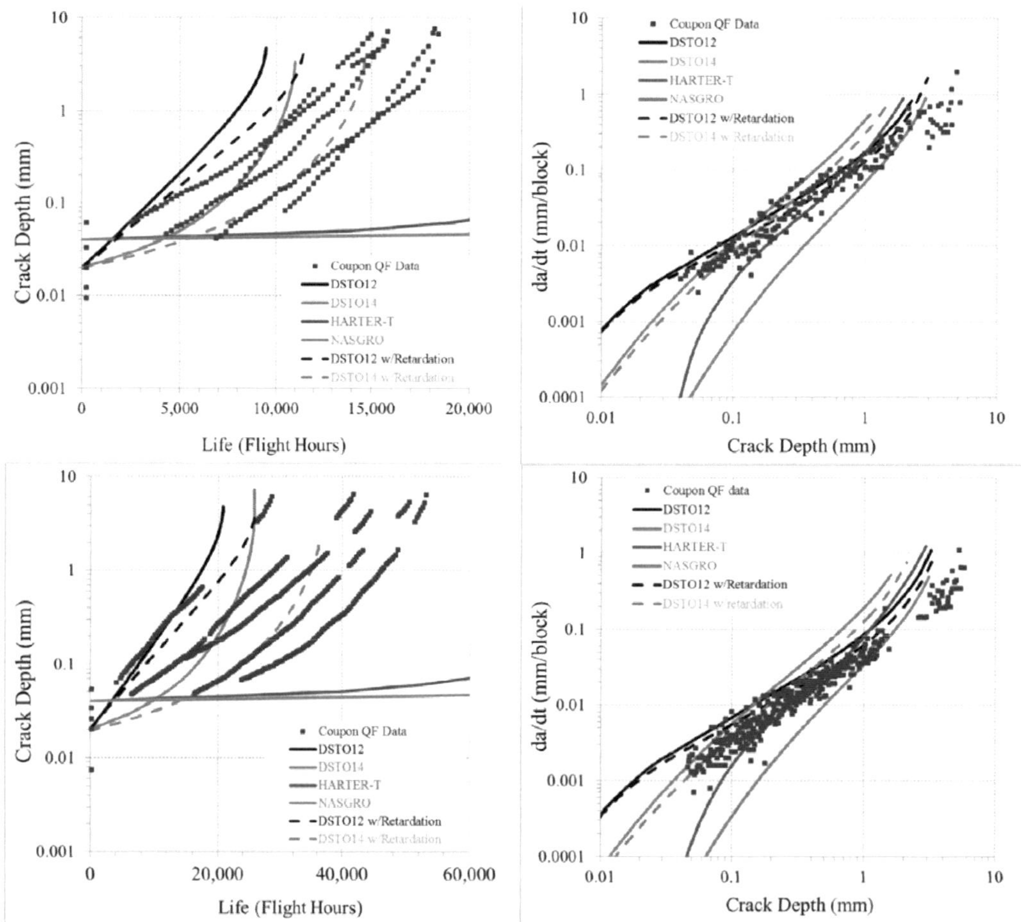

Figure 8. Experimental data and predictions of crack growth in simple coupons produced by a severe (top) and benign (bottom) combat wing root bending moment (peak stress = 324.1 MPa) spectra. Both the crack growth (left) and the crack growth rates (right) are presented.

Fatigue crack growth predictions using the DSTO12 material curves produce results that appeared to more realistic, for the combat spectra, over much of the growth of these cracks; particularly the growth up to about 1mm. Generally all of the rate curves appeared to over predict the rate of growth for cracking beyond about 1mm in depth – with DSTO12 being the best match again. The inclusion of a retardation algorithm is shown in the results presented above, with the slowing of the predicted growth as the crack depth increases, although this effect is less pronounced for the mini-TWIST predictions. Although not presented here, the DSTO12 material curves were also very effective in predicting the growth of cracks in other combat and helicopter aircraft spectra, over a wide range of peak stress levels. However, the predictions tended to become conservative (predicted shorter lives) as the peak stress level dropped or the number of small cycles increased relative to the large cycles (as can be seen in the Mini-TWIST results).

While it was expected that the crack growth rate curves engineered in DSTO12 curves would fare well in the predictions, since they had been developed in part using similar test data, is was surprising how much better it predicted small crack growth in all the spectra examined. This suggests that the measured rate data that was used to develop the DSTO14 growth rate models (based as they were on the mean crack growth rates) were not capturing the faster rates at very small ΔKs that the DSTO12 model suggests should exist. To investigate this further, coupon tests with reducing numbers of CA cycles in each of the bands were tested and some qualitative results are presented in the next Section.

OBSERVATIONS OF SMALL CRACK GROWTH BEHAVIOUR IN AA7050

In many experiments on growing small fatigue cracks, it is often seen, as demonstrated in the preceding Section, that they grow at lower ΔKs and faster than long crack growth rates suggest. It has also been noted that VA loading tends to grow cracks at faster rates than would be predicted from similar cracks that have been grown with CA loading only [35], [37]. While the first of these anomalies is explained by showing that small cracks can be initiated and grown at ΔKs well below the long crack data thresholds, this does not fully explain why VA loading may appear to have a magnifying effect of this growth rate. From the results presented in the preceding Sections, where DSTO12 has faster rates than the measured small crack growth rate data in DSTO14 at the lower ΔKs, and that DSTO12 better predicts the VA data below about 1mm of crack depth in the test conducted, it can be seen that the DSTO14 growth rates are still not fast enough to explain the fast small crack growth found in the VA coupon tests.

It is postulated that one cause of this discrepancy is the *breakup* of the crack front throughout the growth of the CA sub-block. An example of this crack front breakup is shown in Figure 9, where the crack growth from two sub-blocks of high *R* CA are shown for a sub-block of 2000 cycles in Figure 9a and for a sub-block of 10,000 cycles in Figure 9b. Comparing these two figures shows that the crack has become *rougher* and the crack front more disjointed by the end of the CA sub-block (top of the figures). Individual striations are faintly observable either side of the high R CA load sub-blocks where no striations can be seen. The high R CA load sub-blocks show *chevrons/slip plane* effects and high relative roughness that includes path divergence in the form of tilting and local rotations about the original path in the 10,000 cycle example. It is hypothesised that with accumulating high R load cycles the crack front becomes fragmented and hence less efficient. On a micro-scale, the localised crack growth appears to follow varying paths through the material dictated by the local planes and directions that are the easiest for the crack to propagate on. The breakup of the crack front on a micro-scale has resulted in micro-river patterns that form as the crack front separates and re-joins during its progress. In contrast, as soon as the loading changes significantly i.e. to low R CA cycles, the crack reforms as a consistent crack front and the micro-river patterns disappear. Since the crack front is breaking-up and becoming more uneven with the application of more cycles of the same type (both ΔK *and* K_{max}), the growth appears to be progressively retarded as a result of this crack path inefficiency.

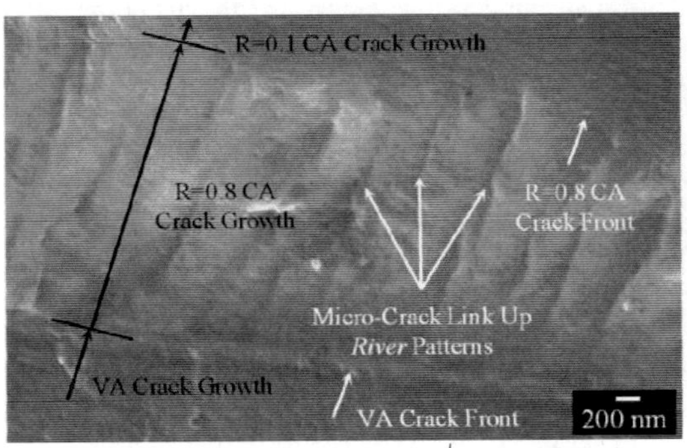

a) a ~ 0.93 mm; ΔK~1.66 MPa√m; 2,000 cycles

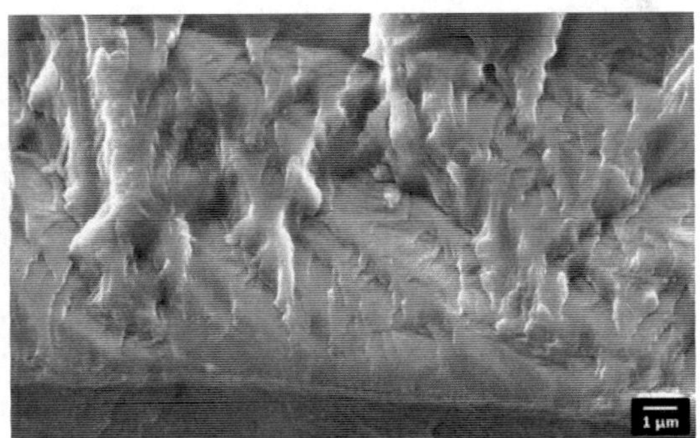

b) a ~ 0.89 mm; ΔK ~1.62 MPa√m; 10,000 cycles.

Figure 9. SEM images taken at approximately same crack depth showing details of the fatigue fracture surface of CA bands and the changing appearance of the crack front at the end of these bands (note scale varies).

While the development of local roughness and crack front breakup is one observable possibility for the development of progressive local retardation of the growth under CA loading, other causes are possible and this is an area that still requires further investigation. Examinations of the growth of cracks using the simple two R loading spectrum, in the regime where striations are just visible, suggests that CA loading at these stress intensities will also lead to retardation even when there is little crack surface roughness developing as the crack grows i.e. see the change in striation width demonstrated in Figure 10 for a band of R = 0.1 CA loads. The cause of this retardation is less clear and is most likely the result of the residual damage state that is forming ahead of the crack in the form of a cyclic plastic zone. In such a zone, residual stresses remain after unloading as per the Willenborg concept or alternatively, the mean stress may decay within this zone since the crack extension is less than the cyclic plastic zone size. Notably at these short crack depths the cyclic plastic zone size [38] is considerably larger than the incremental growth for a single cycle (Figure 10), and suggests that the material at the crack tip may not be governed by continuum mechanics of LEFM but may be strongly influenced by the crystal orientations on the crack path. These explanations remain incomplete and need further examination, nevertheless they do indicate that the growth for a single cycle of a particular *R* embedded in cycles of different *R* and σ_{max} can be faster than that which has been measured in the small crack growth rate data presented earlier.

Figure 10. SEM image of the fatigue fracture surface of a CA R = 0.1 band having 100 cycles showing that the first few cycles are larger than the latter cycles, cf. box A with box B. ΔK = 3.9 MPa√m.

SUMMARY AND CONCLUSIONS

In the Thresholds and Short Cracks Program at DSTO, a series of test spectra were applied to simple coupons to produce cracks that grew from small discontinuities. The tests were aimed at:

- Generating high fidelity material crack growth rate data from fracture surface measurements.
- Producing CA small crack growth rate data, that in addition to standard long crack growth data could be used to predict crack growth under VA loading.
- Examining the fracture surfaces of fatigue cracks to identify causes for retardation/~~closure~~ effects.

The general objectives of this work are to improve the accuracy of crack growth predictions for typical fatigue critical locations in metallic airframe components. The following can be concluded from the present study:

- The reverse-engineered DSTO12 material crack growth rate curves performed better than the DSTO14 material curves developed using small crack growth data under CA loading at predicting total life crack growth under fighter spectra. It was observed that the predictions tended to become noticeably conservative when there were many small cycles such as may be the case for transport (or helicopter) spectra.
- Although the CA developed DSTO14 material curves overall produced reasonable results, the predictions tended to under predict growth for small cracks and over predict at longer crack depths i.e. the predicted crack growth curves were initially slower than that observed experimentally and faster for larger cracks. The under prediction appears to be the result of a subtle CA retardation effect that is occurring while cracks are small, while the over prediction at larger depths appears to require corrections for retardation/closure.
- The cause of the *subtle* retardation effect due to using only CA loading to measure crack growth rates, even for relatively short periods needs further investigation. One possibility appears to be the breakup of the crack front's uniformity that was observed to correlate with increasing numbers of CA loading cycles in any one band, and in so doing possibly increasing crack closure due to roughness. In addition, from crack front features identified during the examination, it was postulated that a variation or deviation from an efficient crack path was occurring. This is similar to roughness closure but different in that it is the path of the crack that is being fragmented. This may be best termed path closure due to local crack front tilts, rotations and breakup, although other influences are clearly at play and demonstrated.
- Retardation was demonstrated in the results of this investigation, but the predictions carried out with the analytical crack growth models and settings described here showed little effect in the growth of small cracks.

Further investigations including the testing of different simple spectra and improving the analytical models to account for closure are on-going. This includes the measurement of small crack growth in materials other than AA7050-T7451 and AA7085-T7452. Current plans are to examine small crack growth in additional aluminium alloys, such as AA7075 and AA2124, as well as a common titanium alloy used in RAAF airframes, Ti-6Al-4V.

REFERENCES

[1] ASTM E647-13a - Standard Test Methods for Measurement of Fatigue Crack Growth Rates, ed.: American Society for Testing and Materials, 2014.

[2] Suresh, S. and Ritchie, R. O., Propagation of short fatigue cracks. International Metals Reviews, Elsevier, 29, pp. 445-76. 1984.

[3] McClung, R. C., Chan, K. S., Hudak, S. J. and Davidson, D. L., Behaviour of small fatigue cracks, in ASM Handbook, Vol. 19: Fatigue and Fracture, ASM International, Materials Park, Ohio, pp. 153-158. 1996.

[4] Blom, A. F., Relevance of short crack growth data for durability and damage tolerance analysis of aircraft. Proceedings of the Second Engineering Foundation International Conference/Workshop on Small Fatigue Cracks, Metallurgical Society Inc, pp. 623-638. 1996.

[5] Forth, S. C., Newman Jr, J. C., and Forman, R. G., On generating fatigue crack growth thresholds, International Journal of Fatigue, vol. 25, pp. 9-15, 2003.

[6] Liao, M., Dislocation theory based short crack model and it's application for aircraft aluminium alloys, Engineering Fracture Mechanics, vol. 77, pp. 22-36, 2010.

[7] Newman Jr, J. C., Phillips, E. P., and Swain, M. H., Fatigue-life prediction methodology using small-crack theory, International Journal of Fatigue, vol. 21, pp. 109-119, 1999.

[8] Shyam, A., Allison, J. E., Szczepanski, C. J., Pollock, T. M., and Jones, J. W., Small fatigue crack growth in metallic materials: A model and it's application to engineering alloys, Acta Materialia, vol. 55, pp. 6606-6616, 2007.

[9] El Haddad, H., Topper, T. H., and Smith, K. N., Prediction of non propagating cracks, Engineering Fracture Mechanics, vol. 11, pp. 573-584, 1979.

[10] Elber, W., The significance of fatigue crack closure. ASTM STP486 American Society for Testing and Materials, West Conshohocken, Pennsylvania, USA, pp. 230-242. 1971.

[11] Military Specification Airplane Damage Tolerance Requirements, MIL-A-83444, United States Air Force, The Pentagon, Virginia. 1974.

[12] Wallbrink, C., Jackson, P. and Hu, W. Crack growth rate curves: which part dominates life prediction and when? In: International Committee on Aeronautical Fatigue, Montreal, Canada: May 29th - June 3rd. 2011.

[13] Newman, J. C. and Yamada, Y., Compression precracking methods to generate near-threshold fatigue-crack-growth-rate data, International Journal of Fatigue, vol. 32, pp. 879-885, 2010.

[14] Murakami, Y. & Endo, M. Effects of hardness and crack geometries on ΔK_{th} of small cracks emanating from small defects. The behaviour of short fatigue cracks, EGF1, (Edited by K. J. Miller, and E. R. de los Rios), Mechanical Engineering Publications. pp. 275-293. 1986.

[15] Lewis, A, C. and Geltmacher, A. B., Image-based modelling of the response of experimental 3D microstructures to mechanical loading. Scripta Materialia, Elsevier, 55, 1, pp. 81-85. 2006.

[16] Ritchie, R. O. & Yu, W., Short crack effects in fatigue: a consequence of crack tip shielding. Small fatigue cracks, proceedings of the Second Engineering Foundation International Conference/Workshop. Editors Ritchie and Lankford. pp. 167-185. California University Berkeley. 1986.

[17] Forman, R., Shivakumar, V., Cardinal, J., Williams, L., and McKeighan, P. Fatigue crack growth Database for Damage Tolerance Analysis, US Department of Transportation, DOT/FAA/AR-05/15, August 2005.

[18] Harter, J. A., AFGROW Users Guide and Technical Manual, Air Vehicles Directorate, Air Force Research Laboratory, WPAFB OH USA AFRL-VA-WP-TR-206, 2006.

[19] Suresh, S., Crack deflection: implication for the growth of long and short fatigue cracks, Met. Trans., 14A, p. 2375, 1983.

[20] Ritchie R.O., Yu W., Blom A.F., Holm D.K. An analysis of crack tip shielding in aluminium alloy 2124: A comparison of large, small through-crack and surface fatigue cracks, Fatigue and Fracture of Engng Materials and Structures, 10, 5, pp. 343-36, 1987.

[21] Barter, S and Burchill, M. AA7050-T7451 thick plate short crack growth data. Defence Science and Technology Organisation Minute 2012/11109727/1, 2012.

[22] Burchill, M., Barter, S., and Amsterdam, E., Improved fatigue life predictions for combat aircraft fatigue life from a novel testing program, presented at the AIAC, Melbourne, Australia, 2013.

[23] White, P., Barter, S., and Molent, L., Observations of crack path changes caused by periodic underlaods in AA7050-T7451, International Journal of Fatigue, vol. 30, pp. 1267-1278, 2007.

[24] Barter, S. A., Wanhill, R.J.H., Marker loads for Quantitative Fractography (QF) of fatigue in Aerospace Alloys, National Aerospace Laboratory NLR, the Netherlands. NLR-TR-2008-644, 2008.

[25] White, P., Barter, S. A., and Wright, C., Small crack growth rates from simple sequences containing underloads in AA7050-T7451, International Journal of Fatigue, vol. 31, pp. 1865-1874, 2009.

[26] Burchill, M., Barter, S. A., and Jones, M. The effect of crack growth retardation when comparing constant amplitude to variable amplitude loading in an aluminium alloy, 11th International Fatigue Congress. Advanced Materials Research, 891-892:948-954, 2014.

[27] Barter, S.A., Molent, L. and Wanhill, R.J.H. Typical fatigue-initiating discontinuities in metallic aircraft structures, International Journal of Fatigue, vol. 41, pp. 11-22, 2012.

[28] Murakami, Y. Editor in-chief, *Stress intensity factors handbook*, Pergamon Press. Solution from Page 709, 1986.

[29] Edwards, P., and Newman, J. Short-crack behaviour in Various Aircraft Materials, AGARD-R-767, North Atlantic Treaty Organization, 1990.

[30] Walker, K. F., and Barter, S. A., The critical importance of Correctly Characterising Fatigue Crack Growth Rates in the Threshold Regime, 26th ICAF Symposium, pp. 249-263, 1-3 June 2011.

[31] White, P. and Mongru, D. MSMP3 A proposed fatigue damage tracking algorithm for RAAF F/A-18 A/B aircraft. DSTO Minute BM2/121, 08 July 2014.

[32] GENErate Standardised Sequences for fatigue analysis and testing of structural components and assemblies: GENISIS4fatigue, NRL, 2001.

[33] Burchill, M., Barter, S., McDonald, M., and Jones, M. The development of bar-coded marker band load sequences to enhance fatigue test outcomes presented at the AIAC, Melbourne, Australia, 2015.

[34] Willenborg, J., Engle, R. M. and Wood, H. A., A crack growth retardation model using an Effective Stress Concept, Technical repot AFFDL-TR-71-1, Air Force Flight Dynamics Laboratory, Wright-Patterson AFB, Ohio, Jan. 1971.

[35] Pell, R. A., Molent, L., and Green, A. J. The fractographical comparison of F/A-18 Aluminium Alloy 7050-T7451 Bulkhead Representative Coupons Tested Under Two Fatigue Load Spectra at Several Stress Levels, DSTO-TR-1547, Defence Science and Technology Organisation. 2004.

[36] McMillan, J. C. Pelloux, R. M. N. Fatigue crack propagation under program and random loads, Fatigue crack propagation, ASTM STP 415, Am. Soc. Testing Mats., pp. 505-535. 1967.

[37] Gassner, E. and Jacoby, G. Experimentelle und rechnerische lebensdauerbeurteilung von bauteilen mit start-lastwechsel, Raumfahrttechuik 11, pp. 138-148. 1965.

[38] Rice, J. R., Mechanics of crack tip deformation and extension by fatigue. Fatigue crack propagation, ASTM STP 415, Am. Soc. Testing Mats., pp. 247-309. 1967.

Experimental determination of the crack tip stress intensity factor in integrally stiffened panels

Caio Magno de Assunção, Carlos Eduardo Chaves

EMBRAER S/A

Abstract: Integral structures offer large benefits in terms of manufacturing cost, but suffer from a lower degree of fail safety when compared to built-up structures. In order to achieve an improvement on the fatigue crack propagation (FCP), crack containment features (also known as crenellations) have been used on these structures. The source of the FCP improvement is the stress intensity factor (K) modification due to the geometry change. In the current study, an analysis about means of estimating K from the experimental information, and also to verify the K behavior while the crack propagates was performed. The study tested two AA 7475 panels, one with crenellations and another without. As the crack propagates, the K values were estimated in two forms, based on the crack propagation rate and by using a digital image correlation (DIC) system, coupled with strain gages. Based on DIC system, it was possible to evaluate the K estimation, the singularity dominated zone size and the K increase, as long as the crack propagated, for both test specimens. A comparison between the two methods was also made, and finally the use of a DIC system as a tool for estimating the K parameter was discussed.

INTRODUCTION

Integral structures offer large benefits in terms of manufacturing cost, but suffer from a lower degree of fail-safety when compared to built-up structures. The concern is that the degree of fail-safety is reduced when altering from a multiple load path component – in this case, riveted stringers to skin – to a single load path component – i.e., an integrally stiffened panel – whose scenario of a completely cracked panel should be considered during design for damage tolerance.

The benefits, in terms of fatigue crack growth, of implementing a crack containment feature (also known as crenellation) were already shown on previous studies [1][2]. These benefits are caused mainly by the stress intensity factor (K) modification due to the geometry change. Prior to the crenellation, the growing crack faces a stiffer region ahead, where part of the remote loads to open the crack are transferred from the crack front to the crenellation, alleviating the stress condition immediately ahead of it.

This paper presents the main results for experimental K determination obtained from crack propagation tests of integrally stiffened lower skin panel representative of an executive jet wing, using two different methods for K determination – indirectly, from crack growth rate and directly, using a digital image correlation (DIC) system and measuring the stress fields on the crack tip.

The aim of the tests was to evaluate the damage tolerance characteristics and the stress intensity factor variation during the crack propagation in an integrally stiffened panel with and without the crack containment feature.

The tests comprised two specimen types, one with crenellations and the other without. Both have the same cross section area, resulting in an equivalent weight. Two specimens for each type were manufactured and tested. The specimens were paired into two sets. Each set was composed of one specimen with crenellation and one without. Different complex load spectra were applied to each set. The main objectives were to evaluate the crenellated specimen improvement and the stress intensity factor behavior at different load level settings. From these two sets, the results from the one with lower stress level are presented in this paper.

The stress and strain determination through Digital Image Correlation (DIC) was performed by using a stochastic pattern in a small area around the crack and a set of cameras in order to detect changes on the pattern and translate these changes into strain values. The main advantages of this method are (a) the possibility of measuring directly the strain in any direction, including out-of-the-plane deformation (in order to detect any panel bending), giving a better view of the crack tip stress distribution; (b) the possibility of detecting the strain peaks and (c) the possibility of measuring the strain distribution in an entire area, not only a hot spot. The main disadvantages are (a) accuracy, which depends of the measured area size, and (b) need of a standard for calibration.

Based on the results of this test, an assessment was performed on the geometric correction factors and crack propagation rates for integrally stiffened panels. In the following chapters the test specimen description, the types of instrumentation and data acquisition performed, the stress intensity factor determined based on the crack growth rate and on the crack tip stress field and finally the comments and conclusions will be presented. Also, a more detailed description about the Digital image Correlation principle will be presented in Annex A.

TEST SPECIMENS

The tests were conducted with two test specimen types, one with crack containment features known as crenellations, and one without, where both specimens have the same cross section areas.

Both test specimens consist of an integrally stiffened 2 meters long and 714 mm wide panel. Four longitudinal stringers with 2.7 x 34 mm cross section area are laid symmetrically, the panel edges remaining free.

The test specimen with crenellations is shown in Figure 1, and a sketch of the cross section is shown in Figure 2. The test specimen without crenellation is shown in Figure 3, and a sketch of the cross section is shown in Figure 4.

The test specimens were fabricated from a 63.5 mm (3.5 in) thick plate made of aluminum alloy 7475-T7351.

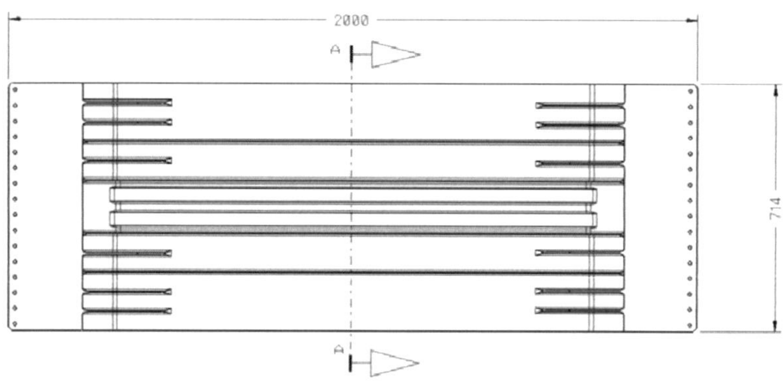

(dimensions in mm)

Figure 1. Test specimen with crenellation.

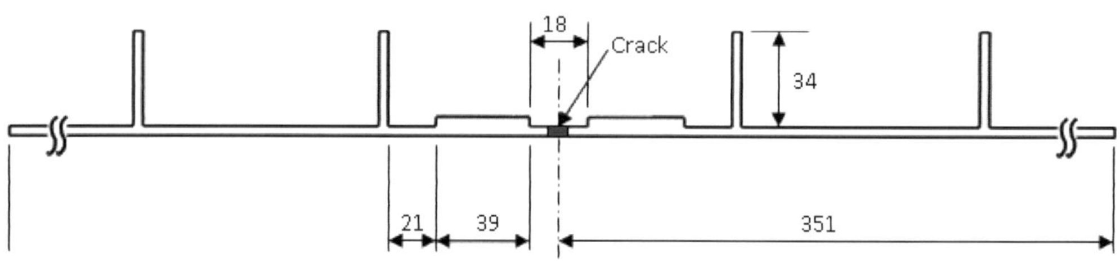

(dimensions in mm)

Figure 2. Test specimen with crenellation – sketch of cross section.

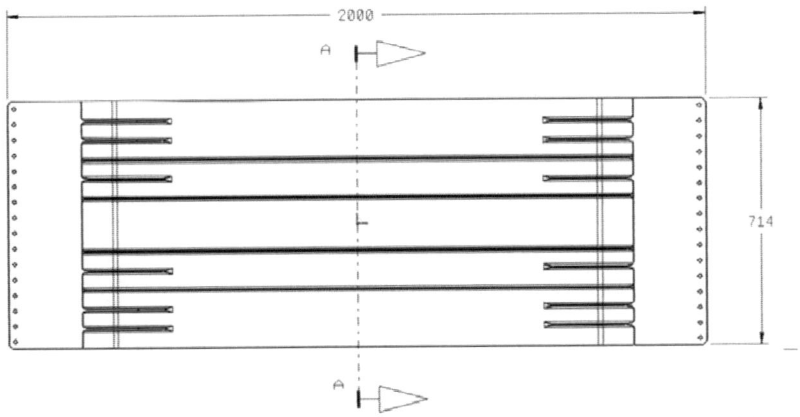

(dimensions in mm)

Figure 3. Test specimen without crenellation.

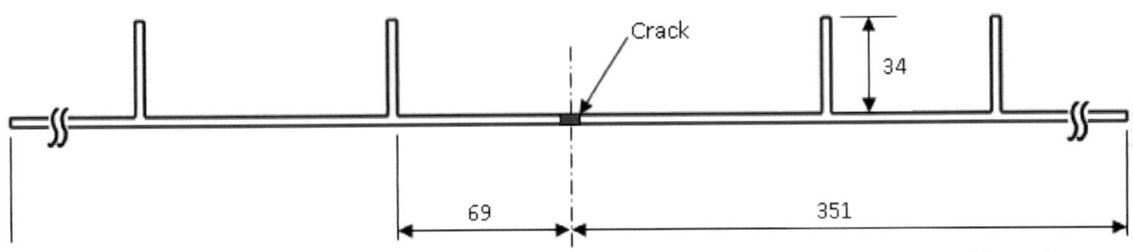

(dimensions in mm)

Figure 4. Test specimen without crenellation – sketch of cross section.

Instrumentation

Digital Image Correlation Data Acquisition. The test specimen external surface was used for digital image correlation data acquisitions, using the DANTEC® Q400 system.

The region around the crack tip was prepared by using a grey pattern and a set of cameras in order to detect changes on the pattern and translate these changes into strain values.

The DIC acquisition was performed in an area of 12,000 mm^2 (200 x 60 mm), on upper side of skin, with reference to the central crack in the test specimen, as presented in Figure 5.

Annex A of this paper brings a more detailed description about the Digital image Correlation principles and means for application of this technology.

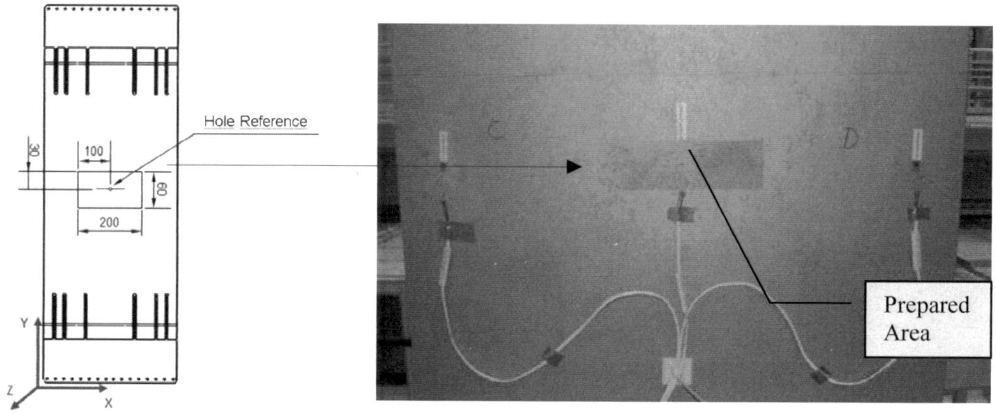

Figure 5. Area for DIC acquisition and instrumentation (external view).

Strain gages. Some strain gages were installed, in order to validate the DIC acquisition, as shown in Figure 6. Additional strain gages (D5 to D8 in Figure 6) were placed in crack plane, close to panel edges, for specimen alignment verification.

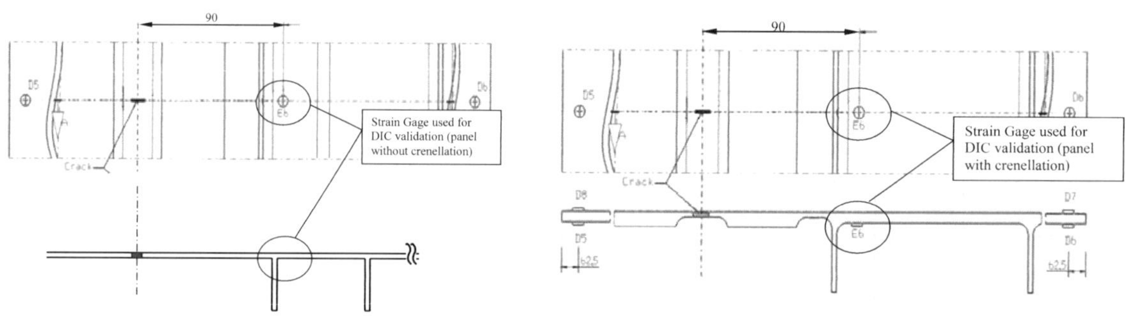

Figure 6. Test specimen instrumentation (strain gages).

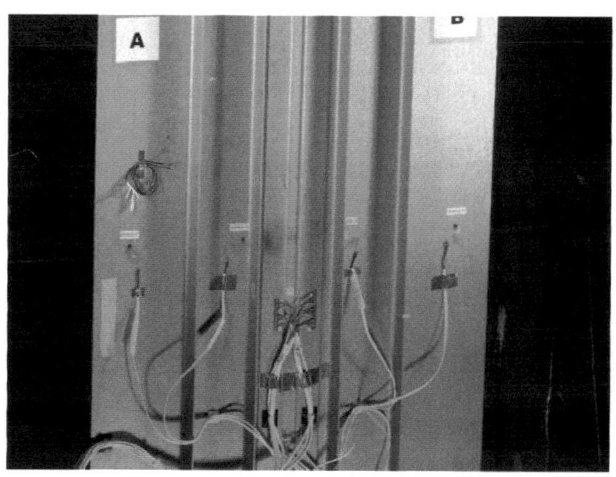

Figure 7. Test specimen instrumentation (internal view).

Test Devices. The test device was composed of two rigid columns, supported by a set of beams as indicated on Figure 8. Whiffletrees were used on test specimen extremities in order to distribute the loading. The upper beams supported test specimen to react the loads whilst the lower beams supported the hydraulic actuator to apply the loads.

The whiffletree was designed in such way that the loading did not induce any bending on the panel.

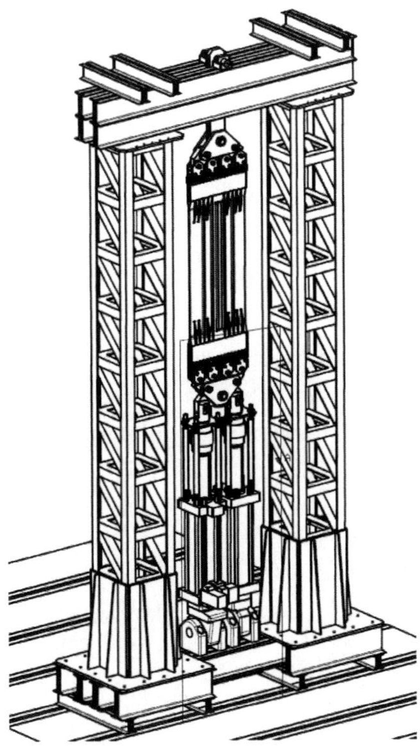

Figure 8. Test Device - General View.

THE STRESS INTENSITY FACTOR DETERMINATION

In this work, the stress intensity factors were determined by two ways: based on the crack propagation rate (da/dN x ΔK), and based on the crack tip stress fields, measured through the DIC data acquisition coupled with a set of strain gages.

Stress intensity factor determination based on crack growth rate

When the crack size is measured as function of the number of flights or cycles, it becomes possible to estimate the stress intensity factor for the increasing crack size from the *a vs. N* curve. For such, one needs to derivate this curve, then obtaining *da/dN* as function of the crack size or as function of the number of cycles. One example of experimental determination of K through the *a vs. N* curve is presented in Ref. [3].

If the principle of similitude is valid for the different configurations and for the load conditions analyzed, then the *da/dN vs. ΔK* relationship at intermediate values of ΔK is the same for these configurations, and such principle allows that K_{max} (assumed as ΔK when R = 0) is estimated for the configurations under analysis [4].

Figure 9 shows ΔK as function of the crack growth rate for Al 7475-T7351, for R = 0. The fitting equation showed in the figure was used for estimation of ΔK, once da/dN is known for a certain crack size or number of cycles.

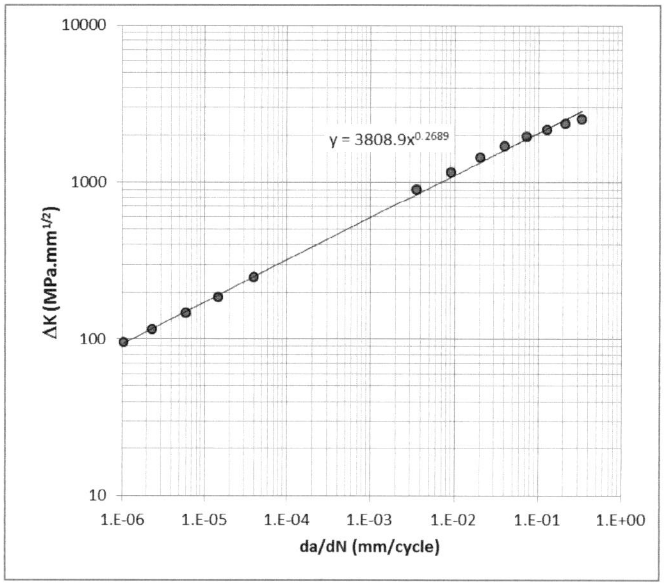

Figure 9. Al 7475-T7351 – Estimation of ΔK for R = 0 (fitting equation included).

For both configurations evaluated, Figure 10 presents the crack size as function of the number of flights. The advantage of designing with crenellations is clearly seen in this figure, where the time to reach a condition close to instability is about three times slower for the configuration with crenellations.

Figure 11 shows the crack growth curve for the specimen with crenellations. The last events prior to the residual strength evaluation, when the crack length ranges from 62 mm to 67 mm, were not included in this study, and are omitted from this figure. Figure 11 includes a schematic representation of the crack growing along the cross section, which was conveniently divided in three phases in order to provide a better understanding about the changes in crack propagation rate as long as the crack advances. The following is a description of each phase:

- Phase I represents the period when the crack grows with a constant thickness, before it faces the crenellation. The presence of the crenellation relieves the stress condition in the crack location, and consequently the crack grows slower, as it is observed in the plot.
- Phase II corresponds to the period when the crack is facing an increase of stiffness in front of it, and the crack growth rate initially increases and then tends to decrease.
- Phase III corresponds to the period when the crack front already reached the crenellation thickness. From this point, it grows steadily and even increases its growth rate as long as it approaches the end of the crenellation, prior to the thickness reduction.

The *a vs. N* curves were fitted to high degree polynomial equations to avoid excessive deviations from the average trends at each crack growth phase. Then the fitting equations were derived in order to obtain da/dN, and then ΔK was estimated from the plot in Figure 9. However, while the crack propagation curve for the geometry without crenellations shows a smooth trend, allowing a good fitting from the minimum to the maximum crack length evaluated, the curve for the structure with crenellations required three distinct adjusts, each one corresponding to each of the phases above described, and showed a "sudden deviation of trend" in the derivatives, as it will be noticed later in Figure 18, when the stress intensity factor results are presented for this configuration.

Despite being possible to estimate the stress intensity factors through this experimental procedure, this approach seems not to be robust, because small changes in trends of *a vs. N* result in somehow large deviations in ΔK. For the consolidated stress intensity factor results presented later in Figure 17 and Figure 18, scatter bands of ±10% from the average K values will be assumed, in order to accommodate eventual deviations due to adjustments of experimental trends.

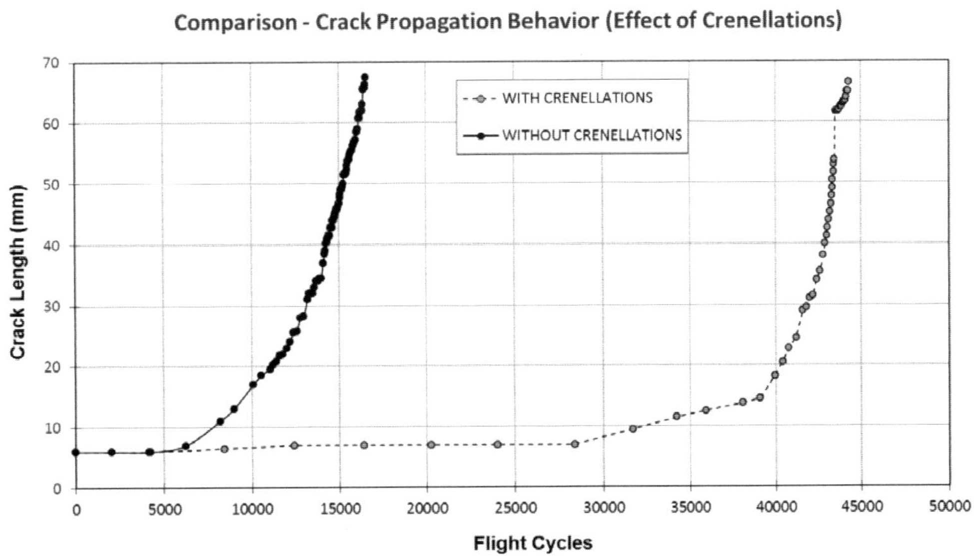

Figure 10. Crack growth (a vs. N) curves for both configurations.

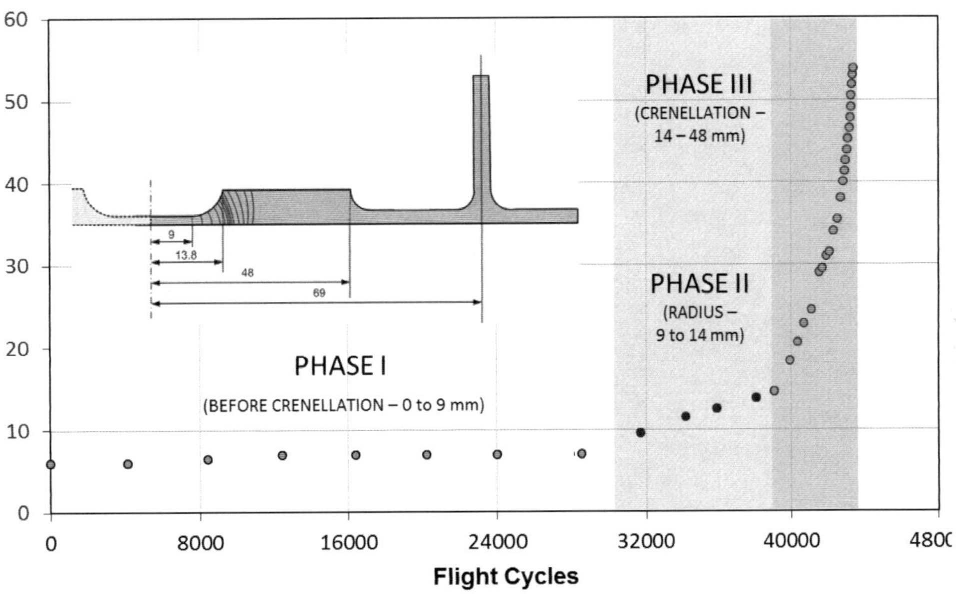

Figure 11. Crack growth phases for the configuration including crenellations.

DIC acquisition validation

In order to validate the DIC acquisition, a comparison was made between the data obtained from the strain gages and the data obtained from DIC. This validation was made by comparing the strain values measured from the strain gages with the strain values obtained from the DIC acquisition, with the same 300.000 N reference load applied.

As the strain from DIC is a full field data, the strain from the strain gage was compared with the mean strain of a rectangle, with the same area of the strain gage grid and on the same position.

Many measurements were made for increasing crack sizes. Figure 12 and Figure 13 show the results of this comparison.

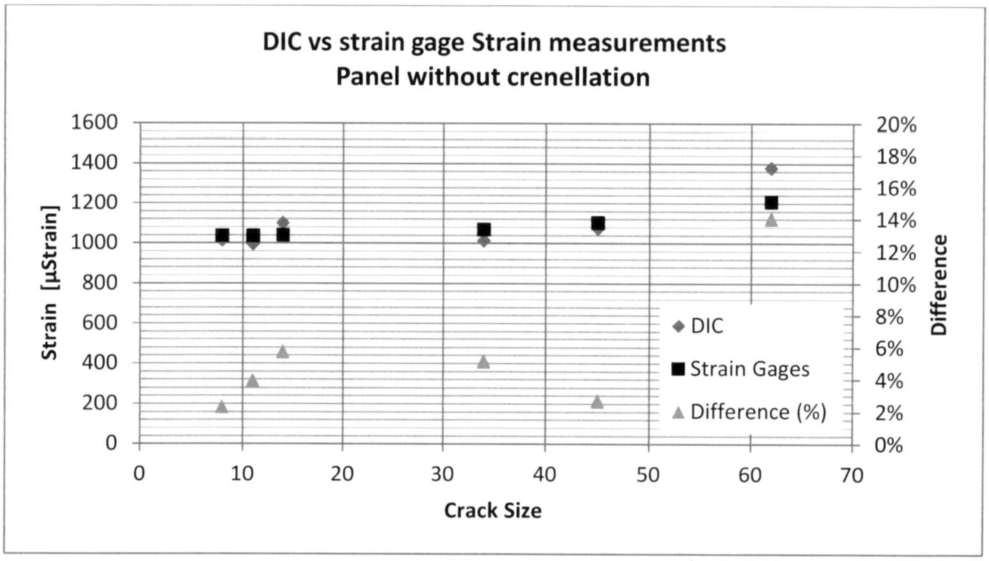

Figure 12. Comparison between DIC and Strain gages (panel without crenellation).

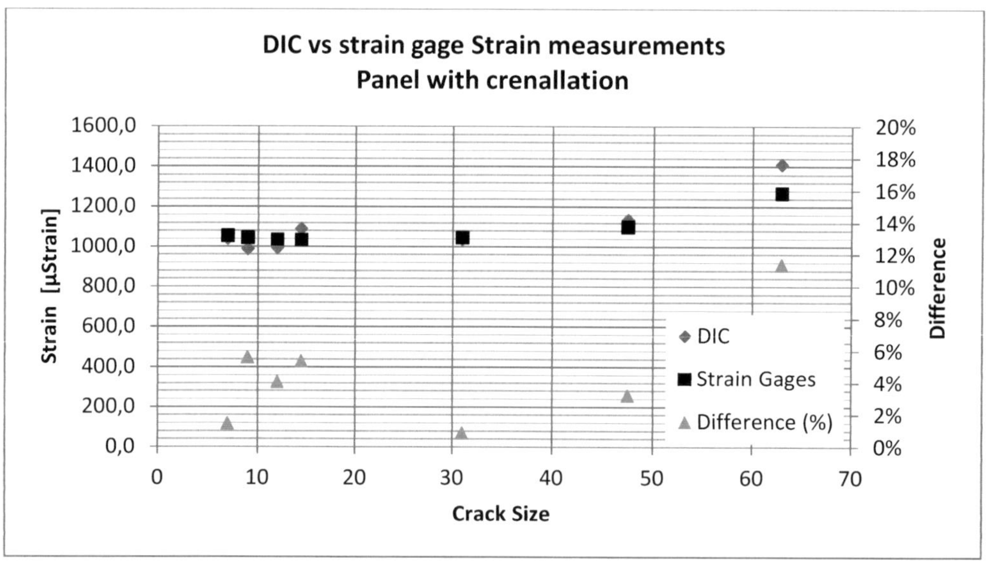

Figure 13. Comparison between DIC and Strain gages (panel with crenellation).

As the crack size increased, it was noted an increase on the difference between the strain gage value and the value obtained from the DIC. The maximum difference observed between the systems was 13% for the panel without crenellations and 11% for the panel with crenellations.

One cause that could have contributed for this difference is the fact that the gage was placed relatively close to the adjacent stringer and, as the crack approaches the stringer, the load that has to be resisted by the stringer increases. This fact, combined with the stress intensity factor due to the stringer radius could have influenced the strain gage measurement, less sensitive to a large stress gradient inside its grid.

Stress intensity factor determination based on the crack tip stress field (DIC measurement)
This section shows the methodology used to determine the stress intensity factor based on the DIC data acquisition.

DIC acquisitions were performed by forming an image with no load applied on test specimen and another with 300.000 N of applied load. The raw data from DIC system was presented in form of a full field strain image and the data corresponding to the values for principal strain in crack plane were exported to an ASCII file.

Figure 14 shows a typical data obtained from DIC system. It can be seen that as the system is based on the deformation of the grey pattern, for coordinates inside crack size it reads the crack opening as a deformation.

Assuming isotropic linear elastic material behavior, it can be shown [4] that, for pure mode I loading, the stress on the crack plane contains a leading term that is proportional to $1/\sqrt{(r)}$, in the singularity dominated zone. Based on this equation, the data plotted from the crack tip was fitted by a curve proportional to $1/\sqrt{(r)}$; were r represents a polar coordinate, with the origin on the crack tip. The value, or the scale parameter, that best fits the curve $1/\sqrt{(r)}$ to the acquired data was assumed as the value for K.

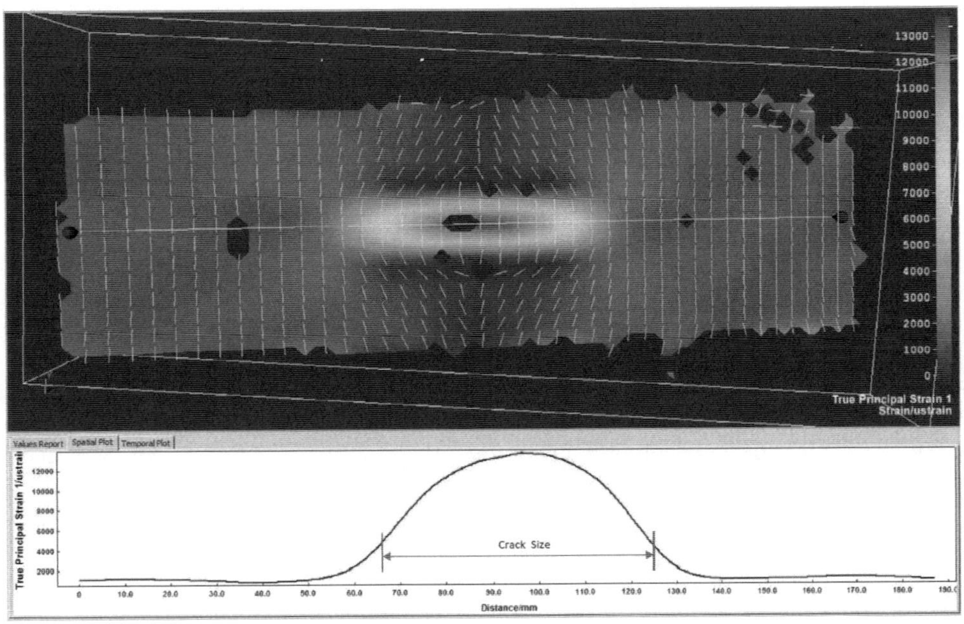

Figure 14. DIC Acquisition Raw data – crack size 62 mm (tip to tip).

Figure 15 and Figure 16 show a data acquisition file, and its correlation with the curve used for the K determination.

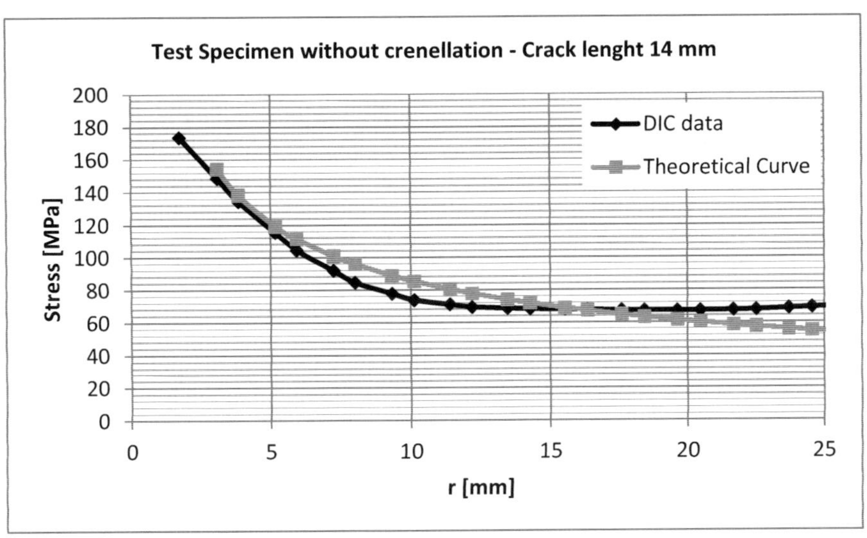

Figure 15. DIC versus Theoretical curve.

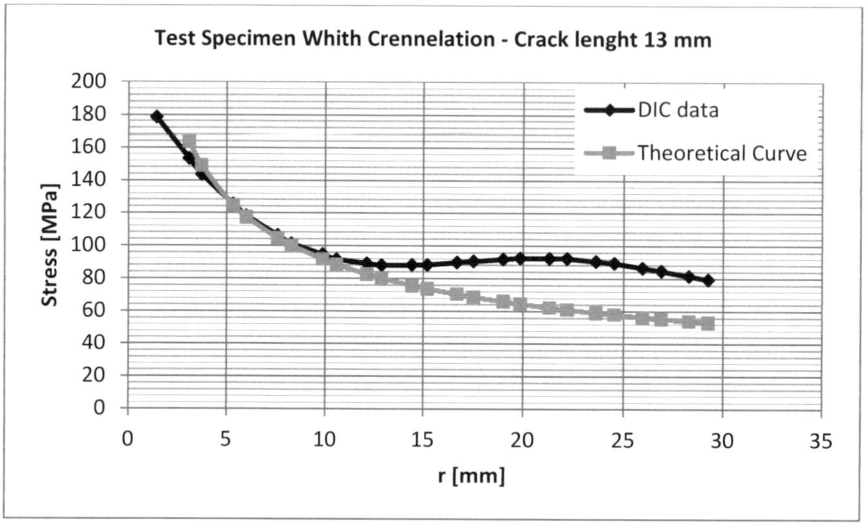

Figure 16. DIC versus Theoretical curve.

During the data acquisition and treatment, for both configurations, it was noted that the $1/\sqrt{(r)}$ curve was not representative for distances from the crack tip smaller than 1 mm, for which the linear elastic stress fields are no longer valid. It was also noted that the singularity dominated zone size was around 10 mm for any crack size evaluated, including when the crack tip approached the crenellation.

Table II presents the K values determined from the DIC acquisition, for selected crack length.

Table II. K values from DIC acquisitions.

Test specimen without crenellation	
Crack Length [mm]	K [MPa mm$^{1/2}$]
7	520
10	650
11	680
32	1000
61	1400

Test specimen with crenellation	
Crack Length [mm]	K [MPa mm$^{1/2}$]
7.5	510
9	500
11	520
12	520
14.5	770
31	950
47.5	1050
63	1200

Comparison between DIC and Crack propagation rate methods

As the final task of the study, the values for K determined on the previous sections were compared, in order to evaluate the panel behavior and the accuracy of each method.

The results are presented on the Figure 17 and Figure 18.

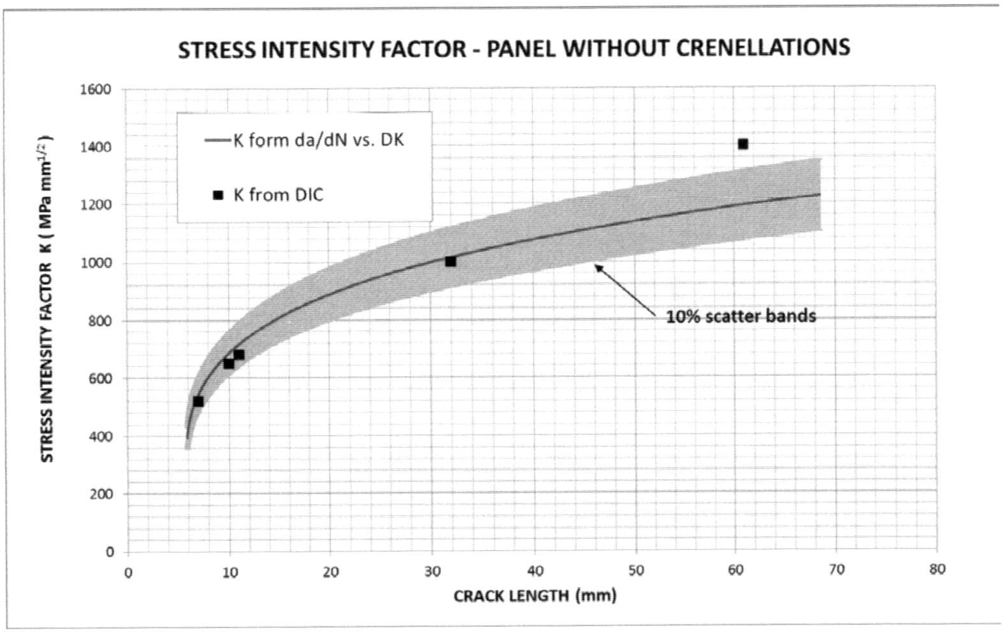

Figure 17. Values for Stress Intensity Factor – comparison between the methods.

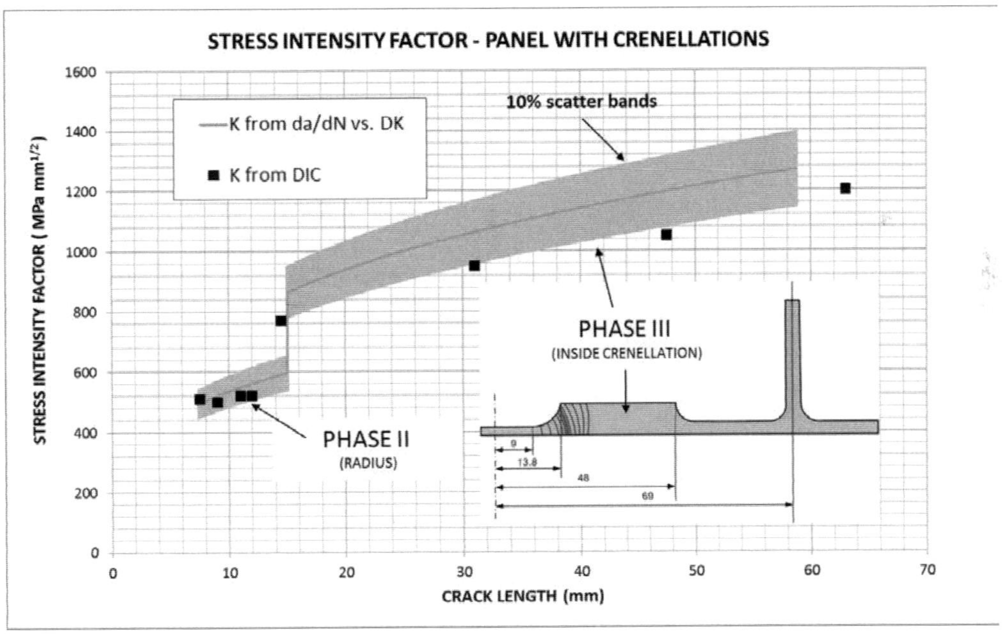

Figure 18. Values for Stress Intensity Factor – comparison between the methods.

As it can be seen on the figure, the data from the DIC measurements and the data from da/dN have a reasonable correlation, showing that the full field strain data acquisition can be used as a fast and precise method for an experimental K determination.

In the case of the panel with crenellation, the DIC measurements were particularly useful for assessing the behavior for the transition region.

COMMENTS AND CONCLUSIONS

This work described experimental procedures for obtaining the crack tip stress intensity factor (K). Two test specimen configurations were evaluated, for which the main difference was the difference in their cross section geometry due to the presence of crenellations.

As expected, the data obtained from the tests showed a significant increase on the FCP life on the test specimen with crenellations.

As the crack propagates, the stress intensity factor was determined conventionally, by the crack propagation rate (da/dN) and also using a Digital Image Correlation system, coupled with a set of strain gages. The stress intensity factor determined by the two techniques was presented for several crack tips and for both test specimens.

The DIC system allowed to determine, with reasonable accuracy, the crack tip stress and strain gradient, and based on this data allowed the crack tip stress intensity factor determination.

The authors recommend the use of Digital Image Correlation as a tool for crack tip stress intensity factor determination, for panels configuration in general.

Further, the following are some recommendations for DIC utilization:

- The DIC technique should be used in one crack tip at time, for better accuracy.
- A reference for the crack tip position should be created previous to the DIC use, applying another technique, as for example dye penetrant.
- The use of some strain gages are recommended, in order to validate the acquired data.

Finally, a comparison with numerical analysis based on 3D crack propagation and evaluation of the stress intensity factors in the crack front for the growing crack is also advised.

REFERENCES

[1] Uz, M.V. and Chen, Y.J. and Huber, N. In: *Fatigue Life Improvement of Metallic Aerospace Structures via Crenellations*, Proceedings of the 26th ICAF Symposium, Komorowski, J. (Ed.), Montreal.

[2] Ehrström, J.C. and Veen, S.V.D. and Arsène, S. and Muzzolini, R. In: *Improving Damage Tolerance of Integrally Machined Panels*, Proceedings of the 23rd ICAF Symposium, Donne, C. D. (Ed.), Hamburg.

[3] Poe, C.C., Jr., (1971). *Damage Tolerance in Aircraft Structures*, ASTM STP 486, American Society for Testing and Materials, pp. 79-97.

[4] Anderson, T.L., (1995). Fracture Mechanics Fundamentals and Applications, CRC Press, pp. 25-101.

[5] Sutton, M.A. and McNeil, S.R. and Helm, J.D. and Chao, Y.J, (2000). *Photomechanics*, P.K. Rastogi, New York pp. 323-372.

[6] Sutton, M.A. and Orteu, J.J. and Schreier H.W., (2009). *Image Correlation for Shape, Motion and Deformation Measurements*, Springer Science+Business Media.

[7] Siebert, T. and Crompton, M. J. In: *Application of High Speed Digital Image Correlation for Vibration Mode Shape Analysis*, Proceedings of the SEM annual Conference 2010, Indianapolis, USA.

[8] Jr, J.M.F. In: *Medição de Deformações em Ensaios Estruturais e Materiais Através de Sistemas Ópticos*, Proceedings of the 2nd EMBRAER Technology and Innovation Seminar, Brazil.

ANNEX A – THE DIGITAL IMAGE CORRELATION (DIC) SYSTEM

In recent years optical full-field measuring techniques are increasingly being used in research and industry as development and design tools for improved characterization of materials and components. The digital image correlation method is currently widely used for full-field displacement measurements [5][6]. Digital Image Correlation (DIC) is a full-field image analysis method, based on grey value digital images, that allows determination of the contour and surface displacements of an object under load in three dimensions.

Since the system determines the absolute position and displacement of the object in space, deformation measurements with very high resolution are possible even under the presence of large deformation amplitudes and macroscopic rigid body movements.

The displacement resolution depends on the field of view (typically down to 10-5 of the field of view) and therefore scalable, e.g. a few μm for an A4 paper size field of view depending on the number of pixels used. The measurable displacement range for this configuration is from μm up to several centimeters. Out of the contour and displacement measurement, the image correlation technique provides material parameters far into the range of plastic deformation. Further data analysis tools allow the determination of the location and the amplitude of the maximum strain, the global strain distribution and behavior of crack growth [7].

The Principle of Digital 3D Correlation

In classical image correlation the deformation of an object is determined by observation with a CCD camera. Then a digital image correlation process determines the shift and/or rotation and distortion of little facet elements determined in the reference image. Such correlation algorithms can determine the maximum of the displacement with an accuracy of up to 1/100 pixel. This procedure allows the determination of the object deformation in a plane parallel to the image plane of the camera.

For three-dimensional measurement, two cameras are used. If the object is observed by two cameras from different directions, the position of each object point is focused on a specific pixel in the camera plane. If the positions of the two cameras relatively to each other, the magnifications of the lenses and all imaging parameters are known, the absolute 3-dimensional coordinates of any surface point in space can be calculated, (See Figure 19). If this calculation is done for every point of the object surface, the 3D surface contour of the object can be determined in all areas, which are observed by both cameras. However, it is important, that the object surface shows enough structure to allow the algorithms to correlate identical points from both cameras.

The correlation algorithm based on the tracking of grey value pattern in small local neighborhoods. If $G(x, y)$ is the grey value of a pixel with the coordinates x and y inside of the subset or facet the correlation algorithm minimized the sum:

$$\sum_{x,y}\left(G_t(x_t, y_t) - G(x, y)\right)^2$$

where $G_t(x_t, y_t) = g_0 + g_1 G(x_t, y_t)$ and

$$x_1 = a_0 + a_1 x + a_2 y + a_3 xy$$

$$y_1 = a_4 + a_5 x + a_6 y + a_7 xy$$

By the variation of the illumination parameters (g_0, g_1) and the parameters of the affine transformation (a_0 ...a_7) an accuracy for the matching of better than 0.01 pixel can be achieved [5-8]. Once the 3D contour has been determined, the second step in digital 3D correlation is the measurement and determination of the three-dimensional deformation of the object surface. This process is carried out by correlation of the images, taken by both cameras with their original reference images [5-8].

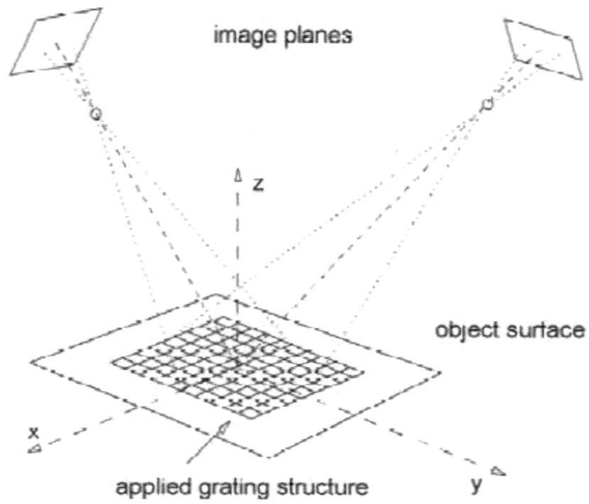

Figure 19. Principle of 3D image correlation with 2 cameras.

In Figure 20, the displacement vector of a surface element is shown. The center point P has been displaced from the reference state *u* to the deformed state *v*. additionally, the surface element has been rotated, tilted and distorted.

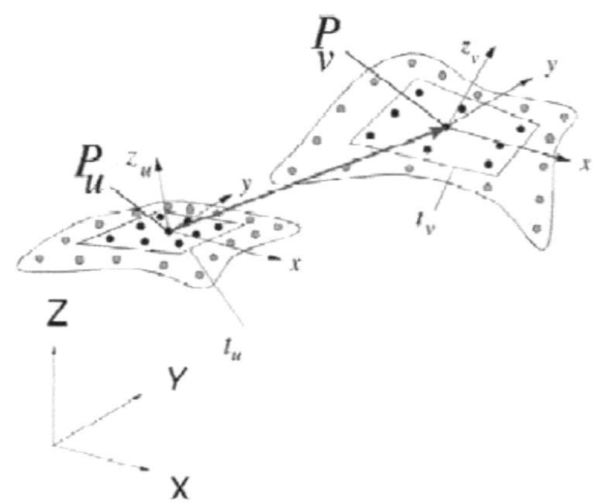

Figure 20. Determination of the three-dimensional displacement vector.

With the known displacement vectors of each surface point and the reference contour, the strains can be calculated. They can be derived either directly by the differentiation of the displacements of adjacent surface points or by the analysis of the distortion of each local facet, which has been used for correlation.

Summarizing, the following steps must be done, in order to perform a DIC measurement:

- Preparing the surface (Black and white paint and grey spray);
- Assembly and calibration of the cameras;
- Acquisition of a reference image (unloaded structure);
- Loading and acquisition of images
- Information post-processing.

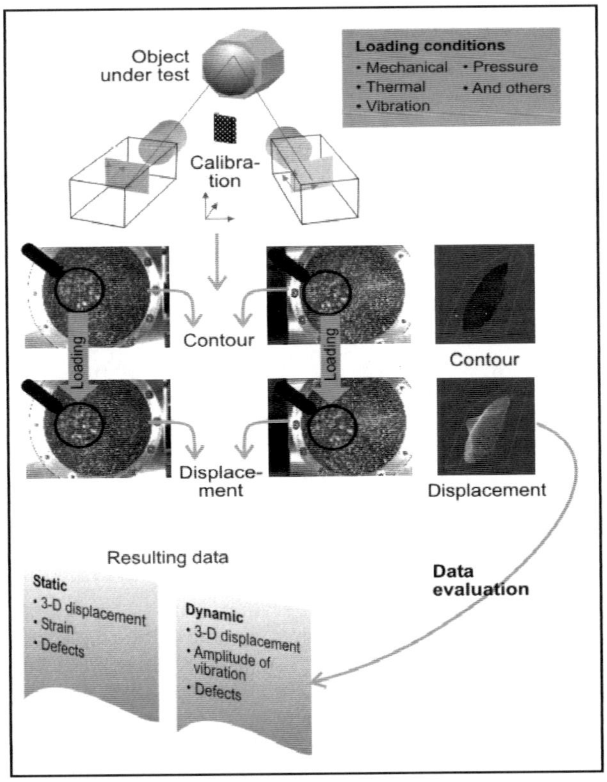

Figure 21. DIC measurement sequence.

Figure 22 shows a typical DIC assembly, including the test specimen with the prepared surface installed in a universal testing machine and the cameras installed in a calibrated bar. The distance between the cameras and the test specimen affects directly the system accuracy. The tests presented were performed by using a DANTEC® Q400 Equipment.

Figure 22. DIC typical assembly.

Challenges and progress in modelling fatigue crack growth under transport aircraft and helicopter type spectra

Weiping Hu, Beau Krieg, Phil Jackson, Matt McCoy, Kai Maxfield, Rob Ogden

Defence Science and Technology Organisation, Australia
506 Lorimer Street, Fishermans Bend, Victoria, Australia 3207

Abstract: Fatigue crack growth analysis plays a critical role in the management of aircraft structural integrity, for ensuring safety, maximising availability and minimising cost of ownership. For load histories experienced by transport aircraft and helicopter airframes, the accurate prediction of crack growth is still a challenging task, despite the great achievements made during the last fifty years in fatigue and fracture mechanics. These load spectra are characterised by a large number of load cycles with very small amplitudes (referred to as "small cycles" herein), often at high stress ratios, interspersed with relatively severe ground-air-ground cycles representing take-off and landing events of individual flights. The total number of stress cycles can run into tens, if not hundreds of millions. To conduct full scale structural fatigue test under these service loads, it is necessary to truncate the spectrum by removing a large number of small cycles, and relying on the analytical method to interpret and evaluate the test results, so that they can be reliably applied on the actual structure that experience the non-truncated spectrum.

This paper reports progress made in quantifying fatigue damage caused by small cycles removed through the truncation processes. Experimental crack growth data were obtained from two coupon test programs, one for C-130J and the other for a helicopter test demonstrator. The numerical tool, FASTRAN, was used for analysis. Three sub-models within FASTRAN were explored: (1) the frequency at which the crack opening stress is calculated; (2) the threshold region crack growth models; and (3) the variable constraint regimes, using the most recent threshold region crack growth rates. Detailed numerical analyses of these sub-models were carried out, to illustrate their effects on crack growth at different stages. Based on the analyses, numerical calibrations were performed, to obtained optimal sets of parameters that best correlate the numerical results to test data. Significant improvement has been achieved for both the C-130J case and the helicopter case. For the cases studied, it has been found that the variable constraint regime played a major role in improving the quality of the numerical results.

INTRODUCTION

Fatigue crack growth analysis plays a critical role in the management of aircraft structural integrity, for the determination of inspection thresholds and inspection intervals, as well as for interpretation and evaluation of full-scale fatigue test results. It ultimately helps the aircraft operators to ensure safety, maximise availability and minimise the cost of ownership. Great achievements have been made during the last fifty years in fatigue crack growth predictions, underpinned by advancement in fracture mechanics, experimental techniques and computational mechanics. However, for load histories experienced by transport aircraft and helicopter airframes it is still a challenging task. These load spectra are characterised by a large number of load cycles with very small amplitudes (referred to as "small cycles" herein), often at high stress ratio, interspersed with relatively severe ground-air-ground cycles representing take-off and landing events of individual flights. The total number of stress cycles may run into tens, if not hundreds, of millions. This is particularly so for helicopters which are subject to cyclic loads related to the main and tail rotor frequencies. A long spectrum with a large number of small cycles makes it difficult to accurately predict the crack growth, as any error in fatigue damage is accumulated and compounded as the crack grows.

This may become a challenging issue during the interpretation and evaluation of full-scale fatigue test results. The load spectrum applied during the test is necessarily a truncated version of the target service load, with a large proportion of the small cycles removed, in order for the test to be completed within a reasonable timeframe. An important task during test result interpretation and evaluation is, then, to account for the damage caused by those small cycles to a reasonable degree. To gain confidence and to identify shortfalls in the numerical methods used for test interpretation and evaluation, coupon tests are typically conducted using the loads derived by applying different levels of truncation, and the numerical methods are checked for their capability in quantifying the fatigue damage caused by the small cycles.

The Defence Science and Technology Organisation (DSTO) has been making collective efforts in developing methods and tools for improving the accuracy of crack growth analysis under service loads. For instance, Wallbrink, Jackson and Hu [1] investigated the importance of the crack growth rates in the threshold region in crack growth analysis involving service loads. Walker and Barter [2] developed improved techniques of determining crack growth threshold rates which led to improvements in crack growth predictions for both long and short cracks.

Recently, two coupon test programs have been conducted at DSTO. The first program was conducted with the objective of supporting the test interpretation and evaluation of the results of a C-130J full-scale wing fatigue test [3], which is a collaborative program conducted by Australia and the UK. The objective of the second coupon program, known as HAFT-TD [4], was to examine the capabilities and to identify shortfalls of the current methods and tools in predicting fatigue damages under helicopter load spectrum. Both of these programs involved very long load spectra with a large number of small cycles.

This paper reports progress made in quantifying fatigue damage caused by small cycles removed through the truncation processes, for both test programs, using the numerical tool FASTRAN [5]. In FASTRAN, the crack growth model is essentially two-dimensional, with the out-of-plane constraint on in-plane deformation at the crack tip represented by a constraint factor. A number of sub-models in FASTRAN may affect the prediction we are interested in, and our investigation focused on three areas: (1) the frequency at which the crack opening stress is calculated; (2) the threshold region crack growth models; and (3) the variable constraint regimes, using the most recent threshold region crack growth rates. Based on the analyses of these sub-models, numerical calibrations were carried out for both test programs, to determine the optimal parameters. Using these optimal parameters, FASTRAN was able predict the crack growth behaviour of the un-calibrated load spectra, achieving very good correlation to the test data.

EXPERIMENTAL DETAILS

C-130J Coupon Test Program

The objective of the C-130J coupon test program is to verify and calibrate the analytical procedures in interpreting and evaluating the test results obtained from the full-scale wing fatigue test, for application on un-tested service spectra. Currently, a collaborative C-130J wing fatigue test is being conducted jointly by the Royal Australian Air Force and the UK Ministry of Defence. In order for the test results to be representative of the service load of both parties, a compromised load spectrum was developed. Each party is then required to interpret the test results for its own service loads and to evaluate its effect on the subsequent life assessment of its own airframes. A key part of this interpretation process is to develop, calibrate and verify the numerical capabilities for durability and damage tolerance analysis, including accounting for the damage caused by the small cycles which are not present in the test spectrum. A coupon test program was conducted at DSTO for this purpose. A brief summary of the test program is given below, and more details are available in a separate document [3].

Geometry. The specimens were designed in accordance with ASTM standard E647-05 [6]. They are of rectangular plates with an overall dimension of 380×100×6.35 mm. The length was in the rolling direction. A hole of 2 mm was drilled at the centre of the plate, and two slits with a depth of 0.5 mm were cut at the edge of the hole in the direction perpendicular to the load, as illustrated in Figure 1.

Material. The materials used for the coupons were aluminium alloy 7075-T7351. The static mechanical properties in the length-traverse (LT) direction, including the fracture toughness, are listed in Table I. The crack growth properties are given later in the paper.

Load. The load spectra used in the calibration activity were developed by Marshall Aerospace and are representative of the full scale wing fatigue test stress sequences. The load spectra for six wing critical structural locations, including central wing and outer wing locations, were used for coupon tests and analyses. These loads are very different in both the maximum and minimum stresses, as well as in the spectrum variations.

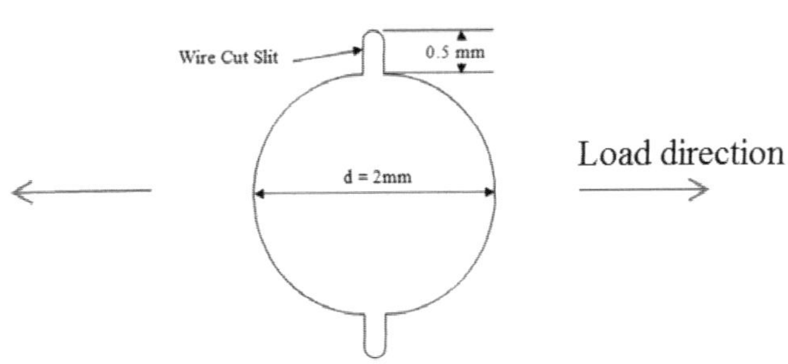

Figure 1. Notch geometry for the C-130 test specimens.

Table I: Material Properties from MMPDS [7] for aluminium alloy 7075-7351 (MPa, m).

Yield Stress	Ultimate Stress (LT)	Young's Modulus	Fracture Toughness
393.0	475.7	71016.0	413.7

Helicopter Test Program for HAFT-TD

HAFT-TD stands for the helicopter advanced fatigue test – technology demonstrator. It is a collaborative program between the US Naval Air System Command (NAVAIR) and DSTO. The objective of the program is to evaluate the feasibility of simultaneously applying flight representative, quasi-static and vibratory loads to a military helicopter airframe in a laboratory environment, and one aspect of the program is to test the capabilities and to identify the shortfalls of the current analytical crack growth methods and tools in interpreting and evaluating fatigue test results.

Geometry. The specimens used for the HAFT-TD test are similar to those for the C-130J program, cf. Figure 1. Rectangular specimens were machined in the L-T direction, with a centre open hole of 2 mm in diameter. Two slits were cut, using electrical discharge machining technique, at the edge of the hole, perpendicular to the loading direction, and the length of the slits are 0.5 mm. The overall dimension of the specimen is 380×100×4 mm.

Material. The material for the specimens is 4 mm thick aluminium alloy 7075-T7351 rectangular plates.

Loads. The load spectrum used in the HAFT-TD was constructed based on actual usage data from a strain gauge on a Black Hawk helicopter, on which cracking was observed in the main transmission support beam near the intersection of the frame and the butt line. Different levels of truncation were applied to the raw load spectrum to obtain the test spectra, using DSTO developed index rainflow filtering method:

1. The spectrum was Rainflow cycle counted [8], whilst retaining the original index of each turning point;
2. A mean stress correction was applied to each half cycle in the raw spectrum, and the number of repetitions of this cycle to cause failure (N_{ff}) was determined using published stress-life curves (for AL7075-T7351, Mil Handbook 5 was used [9]). Where necessary, the published data was extrapolated to higher value of N via curve fitting);
3. All half cycles for which $N_{ff} > N_{cut}$, where N_{cut} is a user defined value, were removed;
4. All remaining half cycles were sorted into their original order;
5. All intermediate points were removed, leaving only a sequence of peak/valley turning points.

The truncated spectra are designated by the magnitude of the number of cycles removed during the truncation process, e.g., the spectrum E15 corresponding to $N_{cut} = 10^{15}$. It should be noted that larger N_{cut} corresponds to a lower truncation level, *i.e.*, the removal of a smaller number load cycles, as indicated in Table II.

Table II. Spectrum truncation levels (note that the "RAW" spectrum was not tested).

1	Full Spectrum Load Turning Points	Fraction of Raw Spectrum (%)
E13	26,038	0.13
E15	582,682	2.94
E16	2,605,358	13.17
E17	6,030,946	30.48
Raw	19,788,760	100.00

NUMERICAL MODELS

The numerical tool used to model the crack growth is FASTRAN5D, which is the version of FASTRAN 5.42 as implemented in DSTO-developed CGAP 2.03 [10]. In this paper, FASTRAN and FASTRAN5D are used interchangeably. FASTRAN implements a crack closure model, which simulates the observation made in [11] that a crack tip may remain closed under a tensile load, if previous load history has developed residual plastic deformation at the crack tip. Crack growth does not occur if the crack tip remains closed. Hence, under the application of a load cycle, the portion of the stress intensity factor which is effective for crack growth is given by $\Delta K_{eff} = K_{max} - K_{open}$, where K_{open} is calculated using the crack opening stress. The crack closure model is considered to be more capable in dealing with growth retardation and acceleration caused by over-load and under-load, as well as the sequence effect [5].

In addition to this central model, FASTRAN also provides a number of sub-models to simulate the crack growth behaviour in the threshold region, the fast fracture region, as well as the better-defined Paris region. A sub-model is also used to simulate the effect of three-dimensional deformation at the crack tip, using a parameter that scales up the yield stress. When FASTRAN is used for complicated analysis such as the cases discussed here, the parameters that define the sub-models need to be calibrated using a set of test data. In essence, the calibration process attempts to improve the predictive capability of FASTRAN by optimising the parameters for a fixed set of sub-models. The details of these sub-models are discussed below, which are used to guide the calibration process.

Rate Modifiers in the Transition Regions

First, we examine the modifiers that were introduced in FASTRAN5D to modify the crack growth rate behaviour in the threshold region and the near-fracture region. These modifiers are applied to crack growth rates given in either equation or tabular format. For example, to calculate the crack growth rate from a given ΔK_{eff}, the tabular input is treated as piecewise Paris laws, defined by C_k and m_k, $k = 1, \ldots, N$, where N is one less the number of data sets in the tabular input. For a ΔK_{eff} falling on the kth Paris law, the crack growth rate is calculated as

$$\frac{da}{dN} \equiv \dot{a} = C_k \Delta K_{eff}{}^{m_k} \delta. \tag{1}$$

In equation format the crack growth rate is simply defined as

$$\frac{da}{dN} = C \Delta K_{eff}{}^{m} \delta. \tag{2}$$

In the above equations δ is the rate modifier. The specific modifiers implemented in FASTRAN5D are defined below. In this paper, $\dot{a}$ and $\mathrm{d}a/\mathrm{d}N$ are used interchangeably.

Modifier 1 – the crack growth rate is set to zero if the applied stress intensity factor is smaller than a threshold.

$$\delta_1 = \begin{cases} 1, & if\ \Delta K_{eff} \geq \Delta K_o \\ 0, & otherwise \end{cases}. \tag{3}$$

Obviously, this is a simple cut-off of the crack growth rate curve, without a smooth transition. Any ΔK_{eff} that is below the threshold will not cause any crack growth. Therefore, if the threshold ΔK_{o} is set too low, there will be barely noticeable effect on crack growth, as the second condition in Eqn (3) may never be triggered. If the threshold is set too high, then suddenly the cut-off condition is triggered frequently and there will be a significant effect on crack growth. This model does not affect crack growth in the near-fracture region.

Modifier 2 – the crack growth rate as determined by the piecewise Paris law is modified by the following function:

$$\delta_2 = \frac{1 - \left({}^{\Delta K_o}/_{\Delta K_{eff}} \right)^{C_7}}{1 - \left({}^{K_{max}}/_{K_C} \right)^{C_6}} \tag{4}$$

The function attempts to modify the crack growth rate behaviour in the threshold as well as the near-fracture region. For a given threshold ΔK_{o} the threshold behaviour is fine-tuned by C_7, and for a given cyclic fracture toughness K_C, the near-fracture behaviour is adjusted by C_6. Unlike model 1, this model simulates a smooth threshold region crack growth.

Modifier 3 – the crack growth rate as determined by the piecewise Paris law is modified by the following function:

$$\delta_3 = \frac{\left(1 - {}^{\Delta K_o}/_{\Delta K_{eff}} \right)^{C_7}}{\left(1 - {}^{K_{max}}/_{K_C} \right)^{C_6}} \tag{5}$$

Next, we examine the behaviour of these modifiers in order to shed light on the numerical calibration process. The behaviour may be examined by assuming a set of baseline parameters and then varying the individual parameters. For this purpose, the following values are considered to be the baseline: $C_6 = 2$, $C_7 = 2$, $R = 0.5$. We further assume $\Delta K_o = 2$ MPa. Figure 2 shows the behaviour of the second modifier, δ_2, in comparison with that of the first modifier. The solid red lines represent the baseline behaviour in each plot, and the variations are given by the dashed black lines and the dash-dotted blue lines. The response of the first modifier, Eqn (3), is given by the olive horizontal lines. Figure 2a illustrates the change in the threshold region behaviour, when C_7 takes the value of 1, 2 and 3. Clearly, larger C_7 would make the transition steeper, approaching the behaviour of modifier 1, δ_1, while smaller C_7 makes the transition smoother, as shown by the dashed line. Figure 2b shows the variation in the modifier behaviour in the near-fracture region. Here, larger C_6 delays the start of the transition, but makes the transition steeper once it starts, as shown by the blue dash-dotted line. In both plots, the mid-section of modifier 2 is close to modifier 1, which is desirable, as the crack growth rates in this region is generally well represented by the test data.

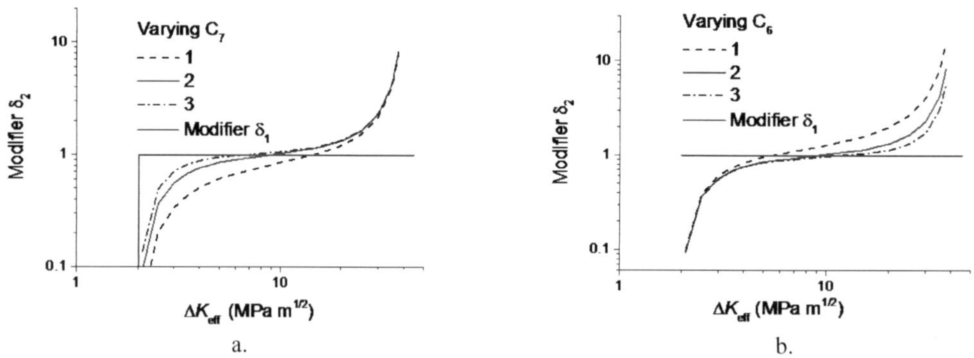

Figure 2. Illustration of the behaviour of transition modifier 1 and 2.

Figure 3 shows similar behaviours for transition modifier 3. It is interesting to note that slopes of the trends are reversed, in comparison to modifier 2. For example, in Figure 3a larger C_7 would lead to a steeper transition. More importantly, this modifier significantly deviates from modifier 1 in the mid-section, which will materially change the crack growth rates in this section. This is an undesirable feature.

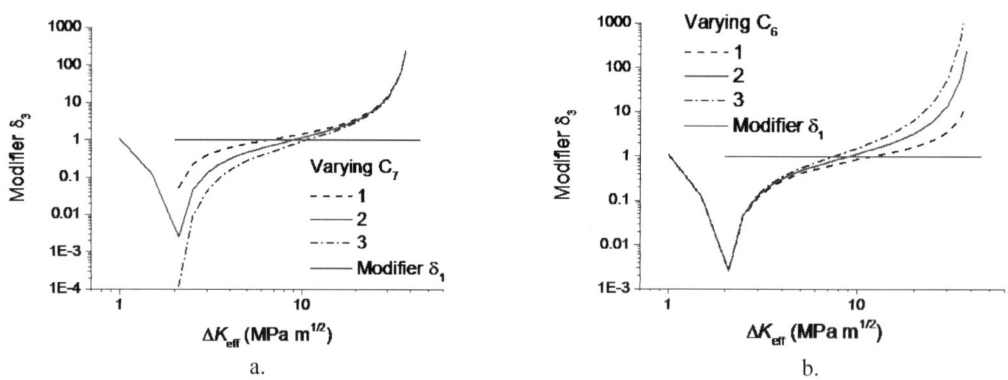

Figure 3. Illustration of the behaviour of transition models 1 and 3.

Threshold Models

In the above transition models, the threshold ΔK_o is assumed to be known. In FASTRAN, ΔK_o itself is defined by two parameters C_3 and C_4, to model the effect of stress ratio on the threshold ΔK_o. Experimentally, it has been observed that larger stress ratio would lead to higher threshold stress intensity. Two models are available in FASTRAN5D,

$$\Delta K_o = \begin{cases} C_3(1-R)^{C_4} & for\ R > 0 \\ C_3, & for\ R \le 0 \end{cases}, \quad (6)$$

or

$$\Delta K_o = \begin{cases} C_3(1 - C_4 R) & for\ R > 0 \\ C_3, & for\ R \leq 0 \end{cases} \quad . \tag{7}$$

The above functions are plotted in Figure 4 to illustrate the effect of parameter variation on the threshold ΔK_o. From Figure 4a it can be seen that varying C_3 leads to a linear change in ΔK_o for negative stress ratios and a nonlinear change in ΔK_o for positive stress ratios. In contrast, varying C_4 has no effect on ΔK_o for negative stress ratios, and it has a more subtle effect for positive stress ratios, with larger C_4 corresponding to lower ΔK_o. Ideally, the parameters C_3 and C_4 should be obtained by fitting the equations to experimental data, but such data are not always available, hence they are determined by correlating the numerical data to test data through a trial-and-error process.

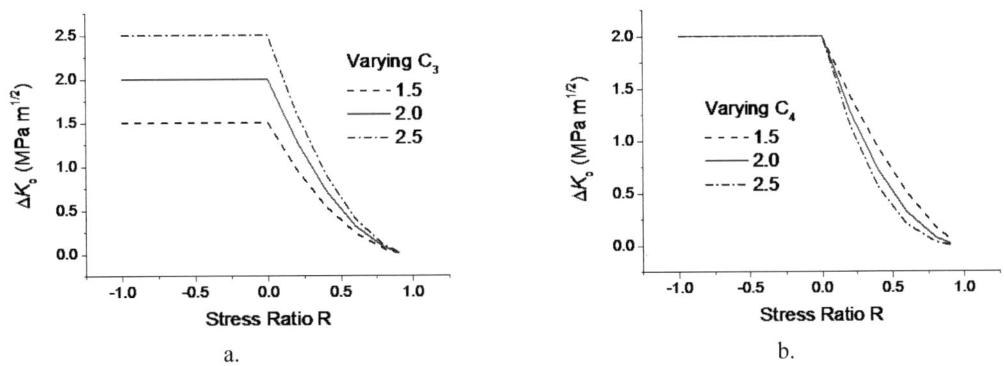

Figure 4. Illustration of the behaviour of the threshold models. The baseline values are $C_3 = 2$ and $C_4 = 2$.

Crack Tip Constraint Loss Models

FASTRAN5D is in essence a two-dimensional crack growth model. To account for the effect of out-of-plane constraint on plastic deformation at the crack tip, a scalar parameter, known as α, is introduced. Its effect is to raise the flow stress for the material at the crack tip. This is known as the fixed α model. By varying α between 1 and 3, the flow behaviour of plane stress to plane strain and the intermediate stress state may be simulated. This technique is used when the crack growth rate data are developed from specimen tests, and it is also used when that data is used for crack growth analysis.

To better correlate the crack growth rates, a more elaborate constraint model was introduced [5], referred to as variable α model. This is based on the observation [12] of crack surface, which showed that when the growth rate is low, the crack front tends to be straight, indicating the plane strain condition; and when the growth rate is high, the crack front tends to be curved, with a tunnelling effect, indicating the plane stress condition. From the plane strain to plane stress state, there is a transition region. In FASTRAN, two constraint factors α_1 and α_2, and the corresponding crack growth rates at which these constraint factors begin to take effect, $\dot{a}_1$ and $\dot{a}_2$, are introduced to simulate the transition. Before the start of the transition, the constraint factor is α_1, and at the end of the transition period, the constraint factor is α_2. During the transition, the constraint factor varies linearly on a log-log scale, as shown in Figure 5. Experience shows that the constraint regime has a significant effect on the quality of FASTRAN predictive capability.

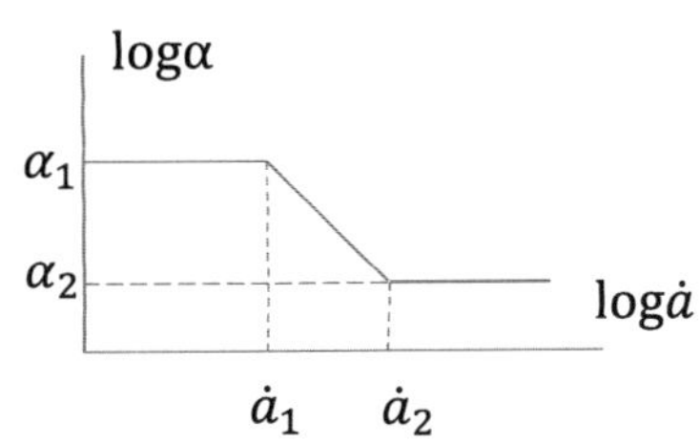

Figure 5. Variable constraint model.

applied. The predetermined crack length is programmatically set as 0.05 times the reverse plastic zone size under the maximum load, and the number of load cycles, denoted N_{max}, is set to 1000 by default, but it may be changed by the user. For example, the crack opening stress may be calculated at every load cycle.

NUMERICAL CALIBRATIONS AND RESULTS COMPARISON

Based on the above discussion of the sub-models in FASTRAN that affect the crack growth behaviour, numerical calibrations were conducted to obtain the set of model parameters that best correlate the numerical results to the test data. The calibration process was guided by the analyses of the sub-models presented above and previous experience.

Crack Growth Rate Data

The baseline crack growth rate curve for 7075-T7351 aluminium alloy is shown in Figure 6. These data were sourced from references [12-13]. The baseline rates are subsequently modified during the calibration process to obtain the best correlation between the numerical results and the test data.

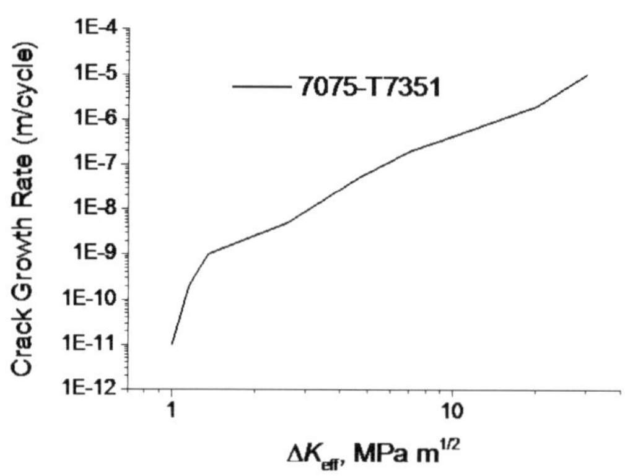

Figure 6. Baseline crack growth rates for aluminium 7075-T7351[13].

C-130J Test Program

The calibration of the numerical model was conducted by a trial-and-error process, guided by the understanding of the model presented in the previous section. For each load spectrum representative of a structural location, the numerical model was adjusted to match the mean crack growth curve of the test data, and the set of parameters that best predicts all the test results are chosen as the calibrated. The calibrated parameters are listed in Table III.

Table III. Calibrated FASTRAN5D parameters for C-130J test program.

Threshold and near-fracture region transition				Start transition		End transition		N_{max}
C_3	C_4	C_6	C_7	α_1	$\dot{a}_1$	α_2	$\dot{a}_2$	
0	0	2	2	1.9	1.08×10^{-6}	1.1	4.83×10^{-6}	1000

Using the calibrated parameters, crack growth analyses were conducted for each spectrum representative of a structural location. A comparison of the numerical and test results is shown in Figure 7. Figure 7a shows the results for location CW-1, for two stress levels, where the symbols represent the test data and the lines the numerical predictions. It can be seen that for both stress levels, very good correlation was achieved. Figure 7b shows a similar comparison for OW-54. Generally, this level of correlation has been achieved for all the spectra representative of the service loads at six wing structural locations on C-130J.

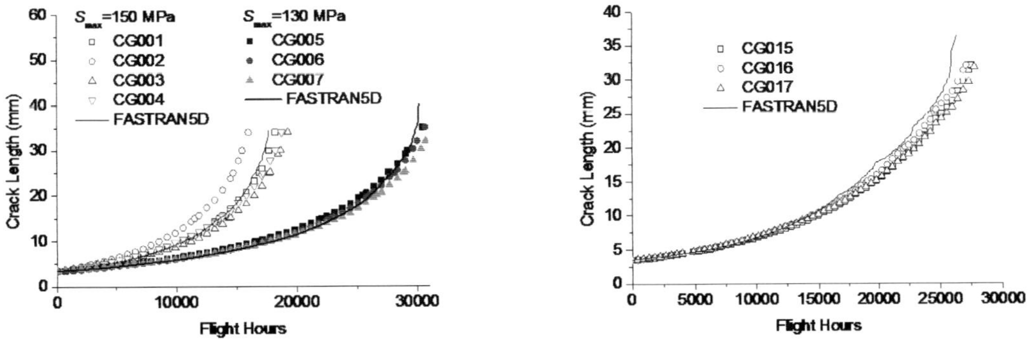

a. b.

Figure 7. Comparison of numerical and test crack growth data for (a) CW-1 and (b) OW-54 for C-130J coupon test program.

HAFT-TD

In the case of HAFT-TD, the test data for spectrum E16 was used for calibration. An initial set of values for the parameters were obtained by taking values based on experience to obtain a benchmark case. The parameters were then further fine-tuned by treating it as an optimisation problem, and the R scripts [14] were used to automate the process. The optimal solution was sought by varying three parameters, C_3, $\dot{\alpha}_1$ and $\dot{\alpha}_2$. Considering that $\dot{\alpha}_1$ should be greater than the threshold growth rate (~10^{-9} m/cycle), it was considered that the optimal values should lie in these ranges: $\dot{\alpha}_1 \in [10^{-9}, 10^{-7}]$, and $\dot{\alpha}_2 \in [2 \times 10^{-7}, 2 \times 10^{-6}]$. This allowable range is illustrated in Figure 8. Figure 9 shows the sensitivity of C_3 to crack growth under spectrum E16. Figure 10 plots sample crack growth curves during the optimisation process, which show that the selected constraints for the parameters do ensure that the numerical results are generally on the conservative side. The optimal parameters were obtained when the error between the numerical and the test values were minimised. The optimised parameters are listed in Table IV.

Table IV. Calibrated FASTRAN5D parameters for HAFT-TD.

Threshold and near-fracture region transition				Start transition		End transition		N_{max}
C_3	C_4	C_6	C_7	α_1	$\dot{\alpha}_1$	α_2	$\dot{\alpha}_2$	
0	0	2	2	1.7	4.0×10^{-9}	1.1	2.0×10^{-7}	1

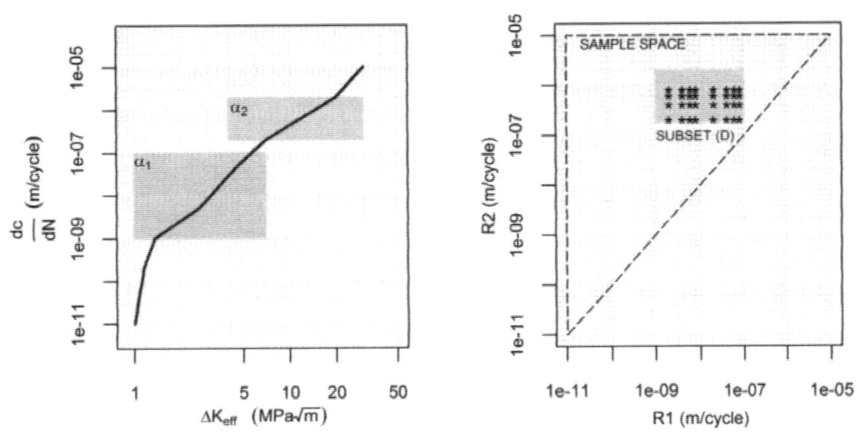

a. b.

Figure 8. Constraint factor rate search range (left) and sample space (right), with test points shown as "*", illustrating the allowable subset of ($\dot{a}_1$, $\dot{a}_2$) values (after taking into account informed constraints).

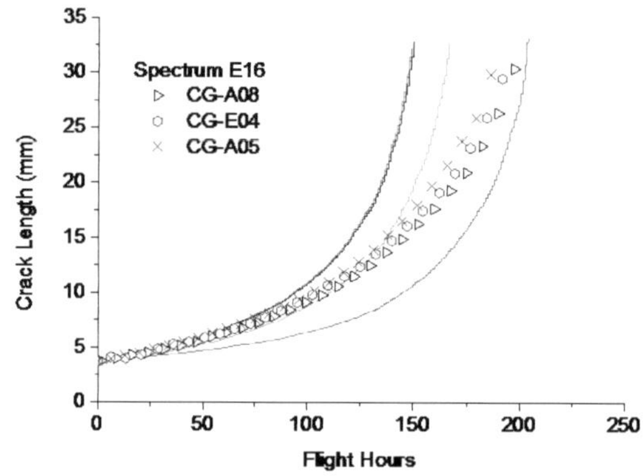

Figure 9. Sensitivity of C_3 to crack growth under spectrum E16 ($C_4 = 0$).

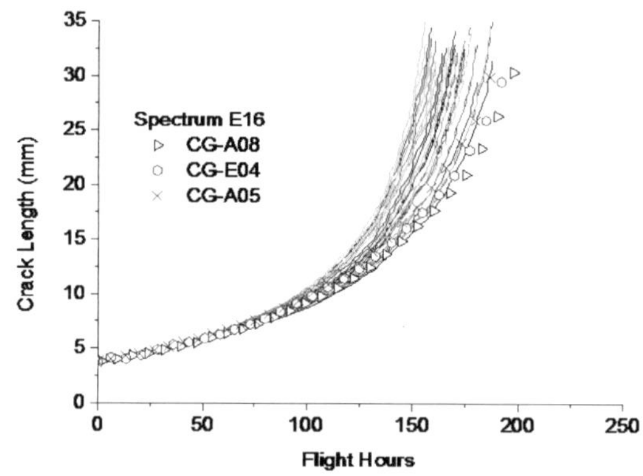

Figure 10. Simulations (lines) for selected subset D (Figure 8b) of rates ($\dot{a}_1$, $\dot{a}_2$), with (α_1, α_2) set to (1.7, 1.1).

Crack growth analyses were then conducted for load cases E13, E15 and E17, and the results are summarised in Figure 11, together with the calibrated E16 data. Here the symbols represent the test data, and they are grouped according to the load spectra under which there were acquired. The black dashed line represents the baseline result for E13, and the black solid line represents the calibrated numerical result for the same spectrum. The difference between them indicates the level of improvement achieved through the calibration process.

In Figure 11 the test data for load spectrum E13, E15 and E17 were not used during the calibration process, and the contents of these load spectra are substantially different from that of E16 due to the removal of different amount of small cycles. Therefore, the numerical results obtained for these spectra may be considered as "blind predictions". In can be seen in the figure that for all the spectra considered, very good correlation has been achieved between the numerical and the test data. It should be noted that for the HAFT-TD calculations, the crack opening calculation was done for every load cycle, with $N_{\max} = 1$.

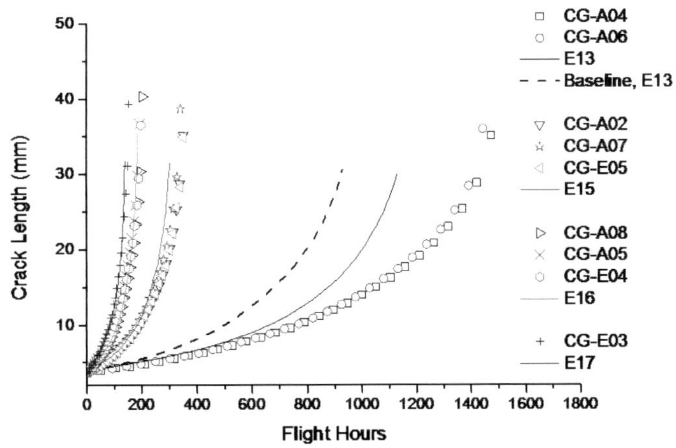

Figure 11. Experimental crack growth data (symbols) compared with the numerical results from FASTRAN5D. The dashed black line shows the un-calibrated numerical results under E13. The solid lines represent the results obtained using the calibrated parameters.

DISCUSSION

The approach taken in the work reported here is to numerically calibrate FASTRAN crack growth model to obtain a set of optimal parameters that could be used for un-calibrated load cases for better predictions. This has been achieved by analysing the sub-models that affect the crack growth rate behaviour in the threshold region and the near-fracture region. Guided by this analysis, calibration was carried out by minimising the error between the numerical results and the test data selection for calibration. The actually process used in the case of C-130J test program was trial-and-error, while on HAFT-TD was an automated optimisation process.

In HAFT-TD, the FASTRAN5D calibration using E16 test data has generated a set of parameters which are able to accurately predict the crack growth for the other un-calibrated load spectra. It is worth noting that the contents of the un-calibrated spectra E13, E15 and E17 are substantially different from that of E16, as different amounts of small cycles were removed from each spectrum. The level of correlation achieved between the test data and numerical results then demonstrate the predictive capability of the software. The calibration process indicated that the set of parameters are robust, in the sense that any small change in the parameters only leads to small change in crack growth prediction, as shown in Figure 10.

The models that describe the crack growth rates in the threshold and the near-fracture regions are more effective when the crack growth rates are available for the Paris region only. For tabulated rates that cover the full range from threshold to near-fracture regions, these models are less effective. This has been observed in both the C-130J and the HAFT-TD cases. In both cases, the main optimisation was achieved by modifying the constraint regime.

The current calibration was completed by varying only three parameters C_3, $\dot{\alpha}_1$ and $\dot{\alpha}_2$, with the others kept constant, including C_4, α_1 and α_2. The optimal set of parameters were obtain by searching part of the solution space, although trial-and-error has been used to fix the partial space. Therefore, it is possible that another set of parameters may be obtained that could outperform the set derived here. This is more so for the case of C-130J where the optimal parameters were essentially obtained by trial-and-error. Given that FASTRAN5D is a deterministic approach to crack growth, calibration could be performed by numerically optimising all six calibration variables, over an appropriate error function for multiple test data sets (i.e. multiple load levels/spectra/truncation levels). This would ensure the suitability of a single calibration to a wide range of applications and may increase the generalizability of the model. Future work could use additional coupon testing to validate this approach. Two separate sets of crack growth curves could be generated using different load levels/spectra/truncation combinations. One set would be used as a training set, to optimise the six calibration parameters in the model, which would then be used to predict the second set.

CONCLUSION

The crack growth analysis software CGAP/FASTRAN5D has been used to predict the crack growth for the C-130J and HAFT-TD test program. Optimal parameters were obtained for the sub-models for crack growth rates in the threshold region and the near fracture region, as well as the model for threshold stress intensity factor range, by carefully calibrating against selected test data. Using these optimal parameters, numerical crack growth results were obtained for the un-calibrated load spectra, and the results correlate very well with the test data. It has been observed in the calibration process that the constraint-loss model plays a more significant role in correlating the numerical and test data for the cases considered here.

ACKNOWLEDGEMENT

The authors wish to acknowledge Dr Manfred Heller of DSTO for carefully reviewing the paper and providing constructive feedback, which significantly improved clarity of the paper.

REFERENCES

[1] C. Wallbrink, P. Jackson, W. Hu (2011). In *Influence of Efficiency and Green Imperatives - Proceedings of the 26th Symposium of the International Committee on Aeronautical Fatigue* (Ed.: J. Komorowski), Springer, Montreal, Canada, pp. 303-314.

[2] K. F. Walker, S. A. Barter (2011). In *Influence of Efficiency and Green Imperatives - Proceedings of the 26th Symposium of the International Committee on Aeronautical Fatigue* (Ed.: J. Komorowski), Springer, Montreal, Canada, pp. 249-263.

[3] M. McCoy, K. Maxfield, R. Ogden (2015), Royal Australian Air Force C-130J Wing Fatigue Test Analytical Tool Calibration and Verification (FASTRAN 5D & FAMSH 1.6). DSTO-TR-3099. Defence Science and Technology Organisation.

[4] B. Krieg, P. Jackson (2015), HAFT-TD: Calibration and verificantion of analtyical crack growth and fatigue life tools for use with heliocpter spectra. DSTO-TR-XXXX. Defence Science and Technology Organisation.

[5] J. C. Newman, Jr. (1992), FASTRAN II - A fatigue crack growth structural analysis program. NASA TM-104159. NASA.

[6] ASTM International (2005), Standard Test Method for Measurement of Fatigue Crack Growth Rates. Document E647-05.

[7] Batelle Memorial Institute (2010), Matallic material properties development and standardization Document MMPDS-05.

[8] (2004), Standard Practices for Cycle Counting in Fatiuge Analysis. ASTM E 1049-85. American Society For Testing and Materials.

[9] (2010), Metallic Materials Properties Development and Standardization (MMPDS-05) Handbook. Batelle Memorial Institute.

[10] W. Hu, C. Wallbrink (2013). In *15th Australian International Aerospace Congress (AIAC15)*, Melbourne.

[11] W. Elber (1971), in *ASTM STP 486*, pp. 230-242.

[12] J. C. Newman, Jr., X. R. Wu, S. L. Venneri, C. G. Li (1994), Small-crack effects in high-strength aluminum alloys. NASA Reference Publication 1309. NASA.

[13] J. Newman (2007), Analyses of Fatigue Crack Growth Databases for use in Damage Tolerance Approach for Aircraft Propellers and Rotorcraft. DOT/FAA/AR-07/49.

[14] R Development Core Team (2010). R: A language and environment for statistical computing. Vienna, Austria. http://www.R-project.org.

Closed Form Equations of Stress Intensity Factors for Through Crack in Angle Shaped Cross-section Parts

Toshimi Taki

Kawasaki Heavy Industries, Ltd., Japan

Abstract: Structural parts with angle shaped cross-section are commonly used in aircraft primary structures. The examples of such parts are spar chords of wing, stringers of wing skin, chords of fuselage frame, etc. Calculation of stress intensity factors (SIF's) is necessary in the damage tolerance analysis assuming initial cracks in the parts. For angle section parts, initial cracks are usually assumed at the edge of the fastener holes in the parts. Unfortunately, SIF's of such crack configurations are not published in the handbooks of SIF's or research papers. SIF's of initial cracks at the edge of fastener holes in angle shaped parts can be calculated by "similarity method" or "compounding method", if SIF solutions of various configurations of angle section parts with through cracks are available. The purpose of this paper is to develop closed form equations of SIF's for angle section parts under uniform tension load.

INTRODUCTION

Angle shaped cross-section parts are very common in aircraft structures, such as spar chords of wing, stringers of wing skin, chords of fuselage frame, etc. One of such examples, spar chord of wing is shown in Figure 1. Damage tolerance analysis is required for principal structural elements and angle section parts are often classified as principal structural elements.

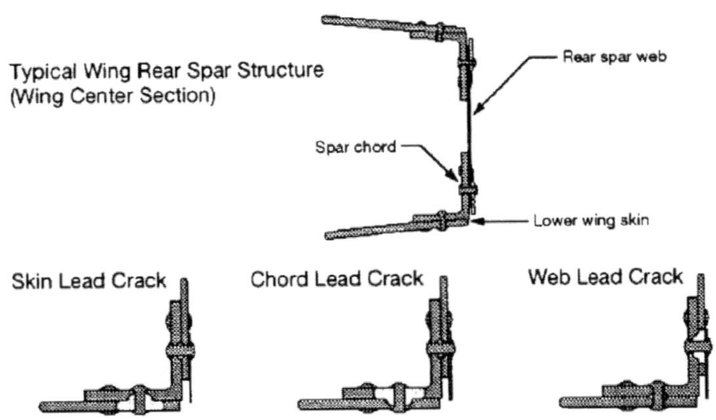

Figure 1. Example of Wing Spar ([3]).

Calculation of SIF's is necessary in the damage tolerance analysis assuming initial cracks in the parts. For angle section parts, initial cracks are usually assumed at the edge of the fastener holes in the parts. A simplified approach for the SIF calculation, which assumes a strip with equivalent width to the angle section, has been commonly used in the aircraft industries. Usually, the width of the equivalent strip is set to the sum of the widths of the both legs (see Figure 2([1])). But the accuracy of the approach is not evaluated. In reference [4], another simple model using similarity method is proposed, but the accuracy is evaluated for only limited examples. This means that finite element analysis is the only method to calculate accurate SIF's of angle section parts.

From practical standpoint, closed form equations are convenient for some purposes such as parametric study of the crack growth analysis. Unfortunately, SIF's for angle section parts are not published in the handbooks of SIF's or research papers. Closed form equations for angle section parts are expected to be developed and the accuracy of the equations should be evaluated. This paper presents closed form equations of SIF's of through cracks in angle section parts under uniform tension load. In this study, two crack configurations, (1) internal through crack in one leg and (2) through crack from free edge, were considered. Various angle section configurations were considered in the study, equal leg widths with equal or different thicknesses, different leg widths with equal or different thicknesses.

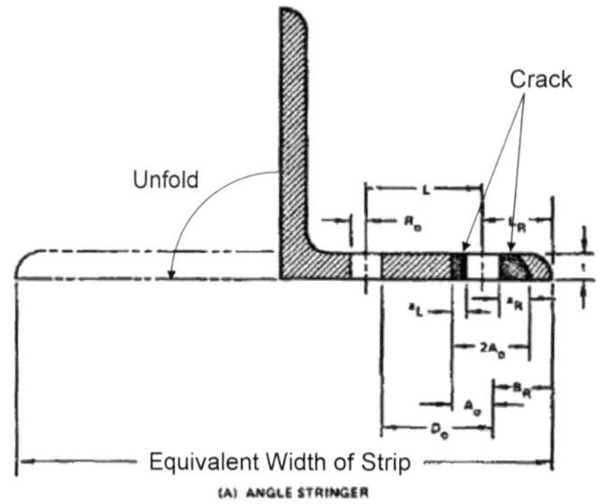

Figure 2. A Simplified Model of SIF Calculation for Angle Section ([1]).

CRACKS IN ANGLE SHAPED CROSS-SECTION PARTS

Cracks in Angle Shaped Cross-Section Parts

Figure 3 shows a cross section of a wing structure which has angle shaped cross-section spar chords. The lower skin is attached to the lower chords with two rows of fasteners. Tensile stress is produced in the chords and the lower skin by wing up-bending load during flight. Initial cracks to be considered in the damage tolerance analysis are shown in the figure. Corner cracks are assumed to be at the edge of the fastener holes. Various cracking sequences from the initial cracks should be considered. Typical cracking patterns appeared in the cracking sequences are shown in Figure 4.

Similarity Method and Compounding Method

In order to calculate the SIF's for the cracking patterns, similarity method ([2]) or compounding method ([5]) can be applied as shown in Figure 5. Internal crack from a hole in an angle is modelled as a combination of an internal through crack in the angle and a crack at the edge of the hole in infinite plate. Edge crack from a hole in an angle is modelled as a combination of an edge crack in the angle and a crack at a notch root in semi-finite plate. Solutions of the SIF's for a crack at the edge of a hole in infinite plate and a crack at a notch root in semi-finite plate are available in the handbooks of SIF's ([6], [7]). The solutions are expressed in closed form equations. Because solutions of the SIF's for cracks in angle cross-section are not available in handbooks or papers, this paper presents the solutions in closed form equations.

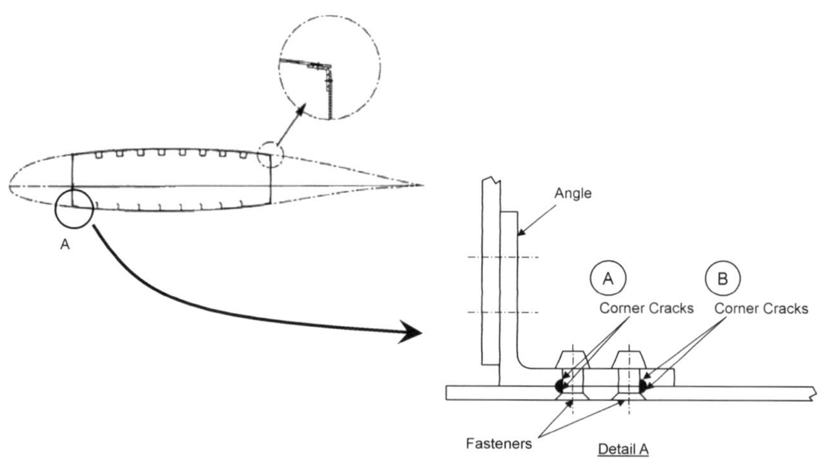

Figure 3. Examples of Initial Cracks in Angle Shaped Cross Section.

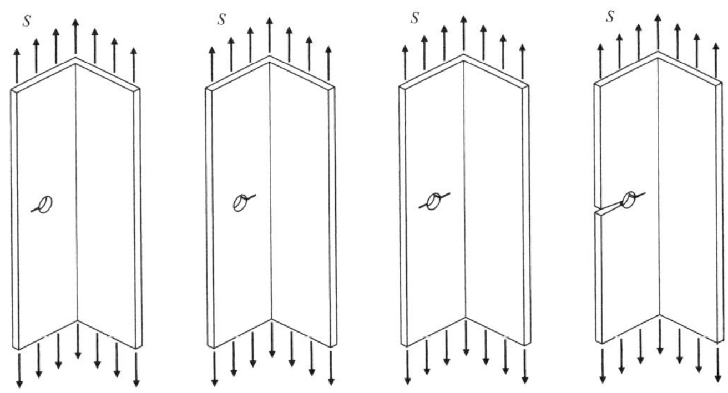

Figure 4. Typical Cracking Patterns.

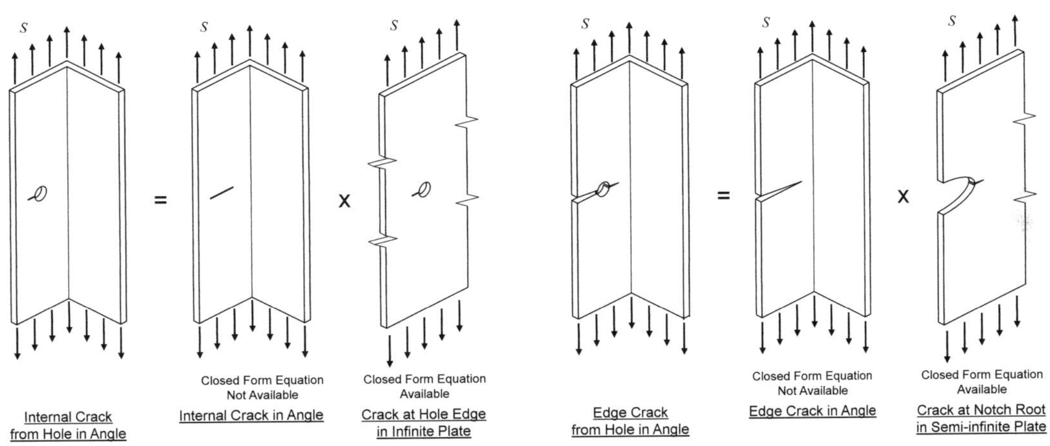

Figure 5. Similarity Method or Compounding Method for Crack from Hole.

CALCULATION OF STRESS INTENSITY FACTORS FOR THROUGH CRACKS IN ANGLE SHAPED CROSS-SECTION

Definition of Stress Intensity Factor for Through Crack in Angle Shaped Cross-Section

Two configurations of crack patterns in angle shaped cross-section are considered as shown in Figure 6. One is internal through crack and another is edge crack. Dimensions of the angle and cracks are indicated in the figure. Uniform tensile stress S is applied on the both ends of the angle. The SIF is expressed in the following equation:

$$K = \beta\sqrt{\pi a}$$

where K : stress intensity factor

β : geometric correction factor

a : crack half length of internal crack, or crack length for edge crack

Finite Element Analysis

Finite element analysis is used to calculate the SIF's for the two configurations shown in Figure 6. Definition of the dimensions is also shown in the figure. Upper half of the angle is modelled and the finite element analysis model is shown in Figures 7 and 8. MSC-NASTARN is used for the analysis and plate elements, CQUAD4 and CTRIA3 are used. Examples of the deformation are shown in Figure 9. Configurations of angles for edge crack analysis are summarized in Figure 10.

Stress Intensity Factor

The total strain energy stored in the angle, U is calculated from the grid point deflection and applied grid point force at the upper end of the angle for crack length a:

$$U(a) = 2U_{half\ model}(a) = \sum_{i=1}^{n} P_{y,i}(a) v_i(a)$$

where $U_{upper\ half}$: Total strain energy stored in the upper half of the angle
U : Total strain energy stored in the angle
$P_{y,i}$: Applied grid point force at the upper end of the angle
v_i : grid point deflection (y-direction) at the upper end of the angle

Strain energy release rate, G is expressed as follows:

$$G = \frac{dU}{tda}$$

Because the bending deformation is very small around the crack tip, stress intensity factor for mode I is expressed as follows:

$$K = \sqrt{EG} = \sqrt{E\frac{dU}{d\partial a}}$$

where E : Young's modulus of the angle

Strain energy release rate is calculated from the total strain energy by the following equation:

$$G(a) = \frac{dU}{tda} = \frac{1}{2}\left[\frac{U(a+\Delta a) - U(a)}{t\Delta a} + \frac{U(a) - U(a-\Delta a)}{t\Delta a}\right] = \frac{U(a+\Delta a) - U(a-\Delta a)}{2t\Delta a}$$

where Δa : Small crack increment

Calculated SIF's for internal crack in equal-leg angle with thickness/width ratio (t/b), 0.06, 0.10 and 0.15 are shown in Figures 11 to 13. It is found that the SIF's are not sensitive for thickness/width ratio. Closed form equation of SIF for internal crack in finite strip (reference [2]) is also plotted in the figures. The closed form equation is conservatively applicable to the internal crack in angle. Width of the equivalent strip should be b_1 for SIF of crack tip near to free edge (Point A) and $1.2b_1$ for SIF of crack tip far from free edge (Point B). This means that the simple model shown in Figure 2 is not conservative.

Calculated SIF's for edge crack in various angle dimensions are shown in Figures 14 to 16. The effects of thickness/width ratio (t_1/b_1, t_1/t_2) and different leg width are evaluated in the figures. Closed form equations developed in this paper are also plotted in the figures. It is found that the simple model shown in Figure 2 is not applicable.

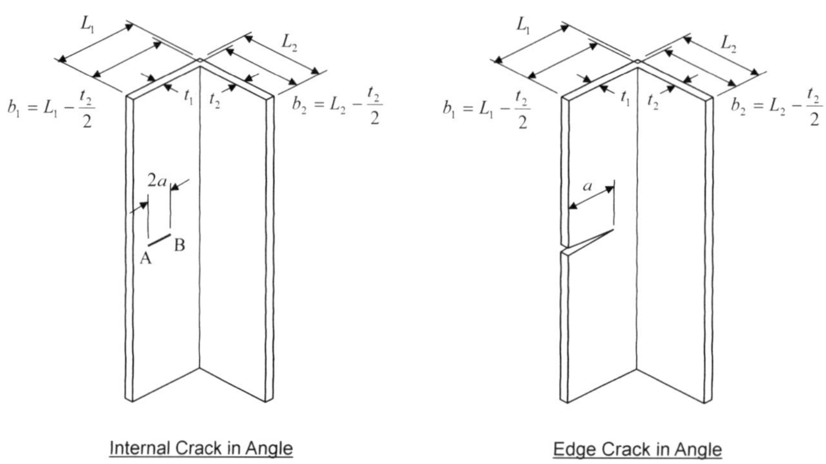

Figure 6. Dimension of Through Crack in Angle Shaped Cross-section.

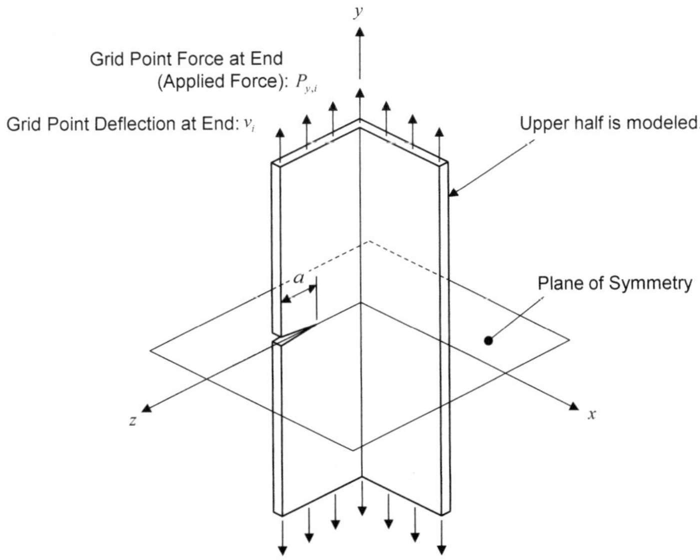

Figure 7. Analysis Model.

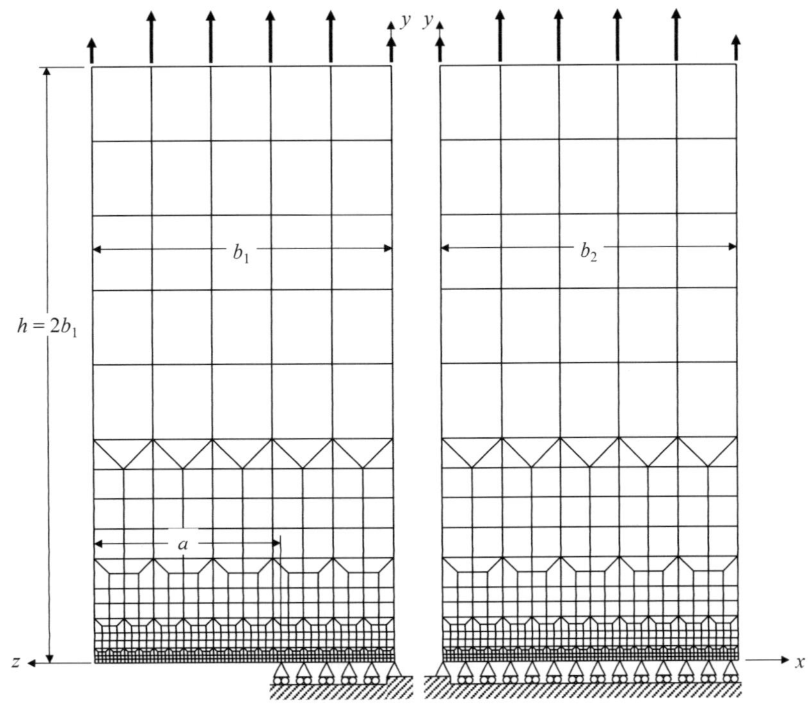

Figure 8. Finite Element Analysis Model – Model for Edge Crack is shown.

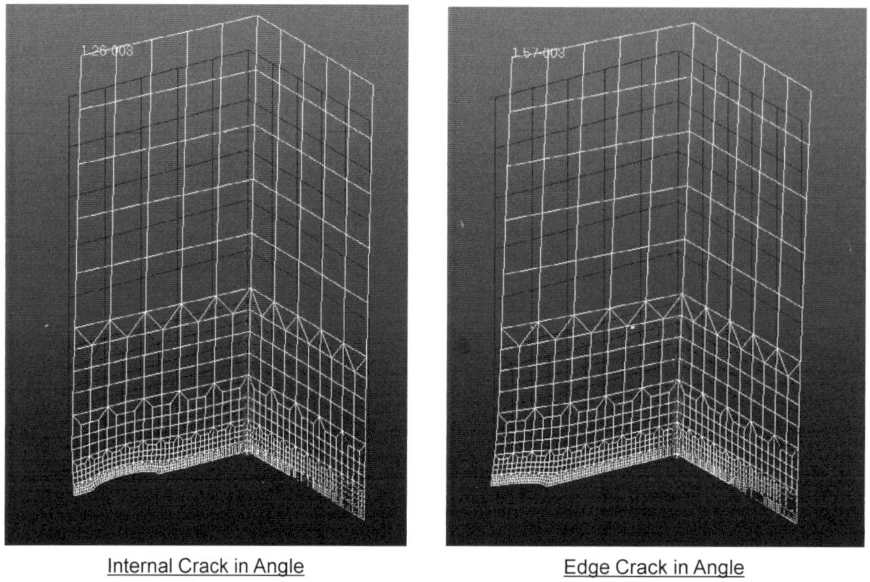

Figure 9. Examples of Deformation of Cracked Angle.

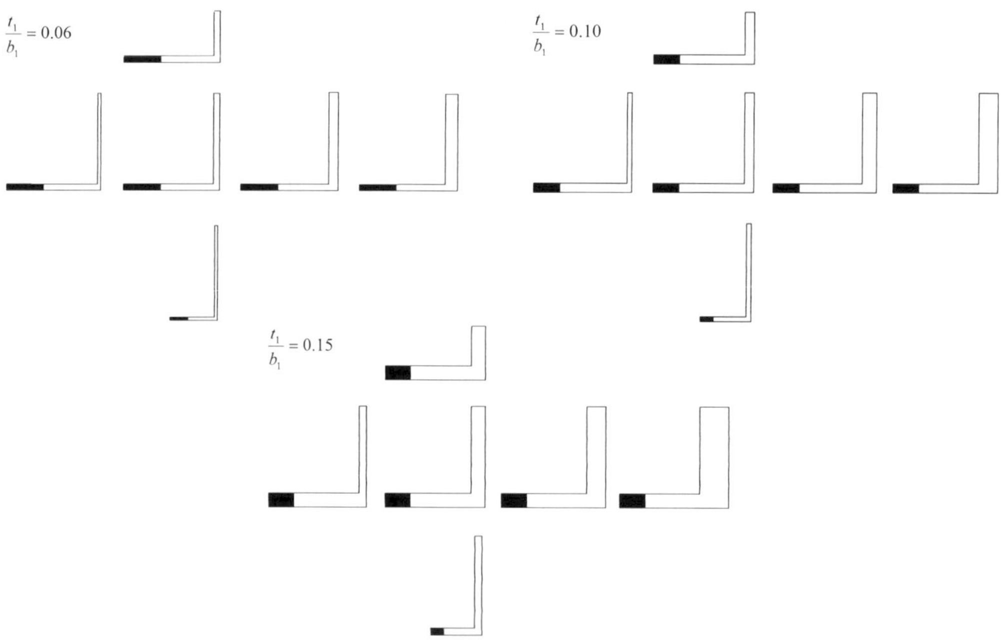

Figure 10. Configurations of Angles for Edge Crack Analysis.

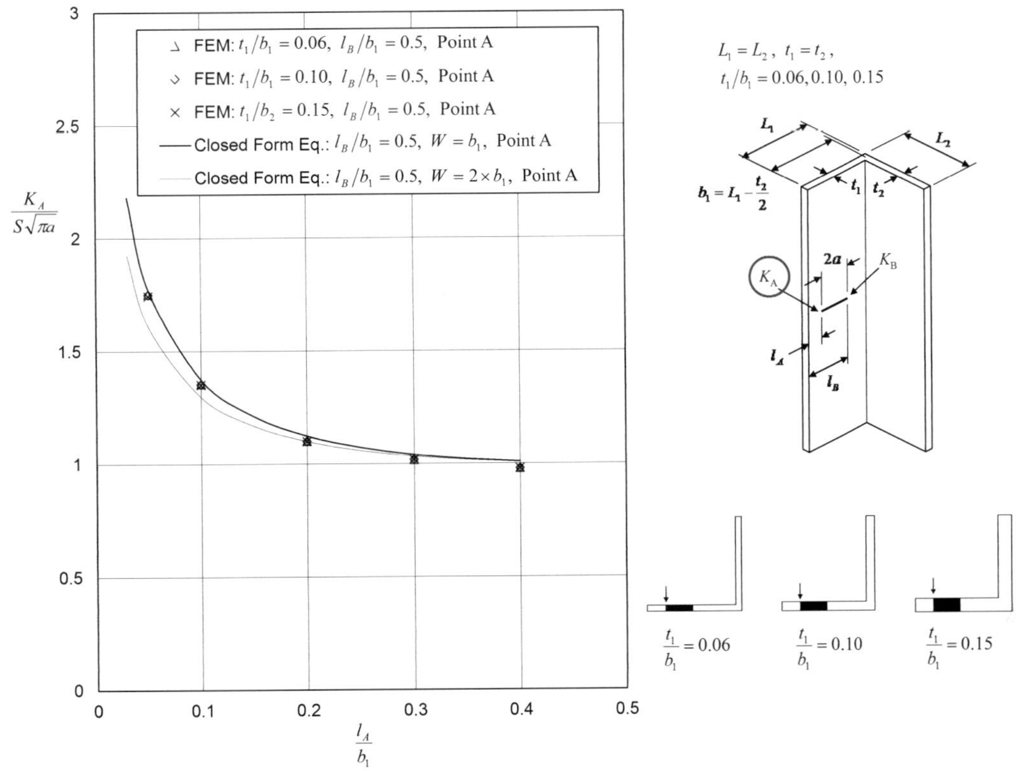

Figure 11. SIF at Point A for Internal Crack, Angle with Equal Legs.

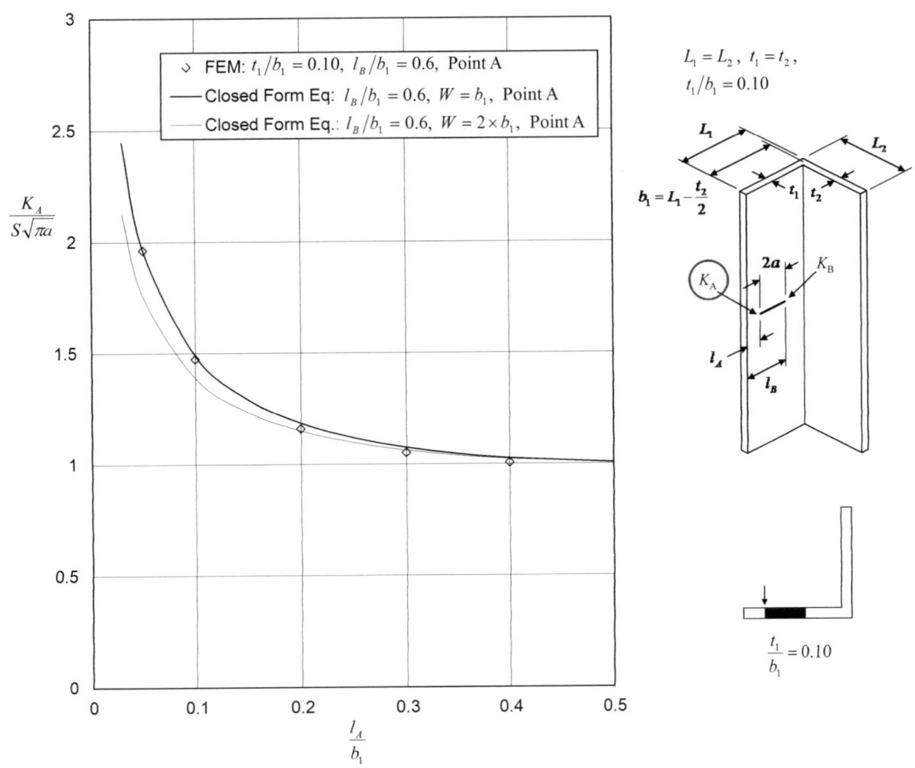

Figure 12 (1/3). SIF at Point A for Internal Crack, Angle with Equal Legs.

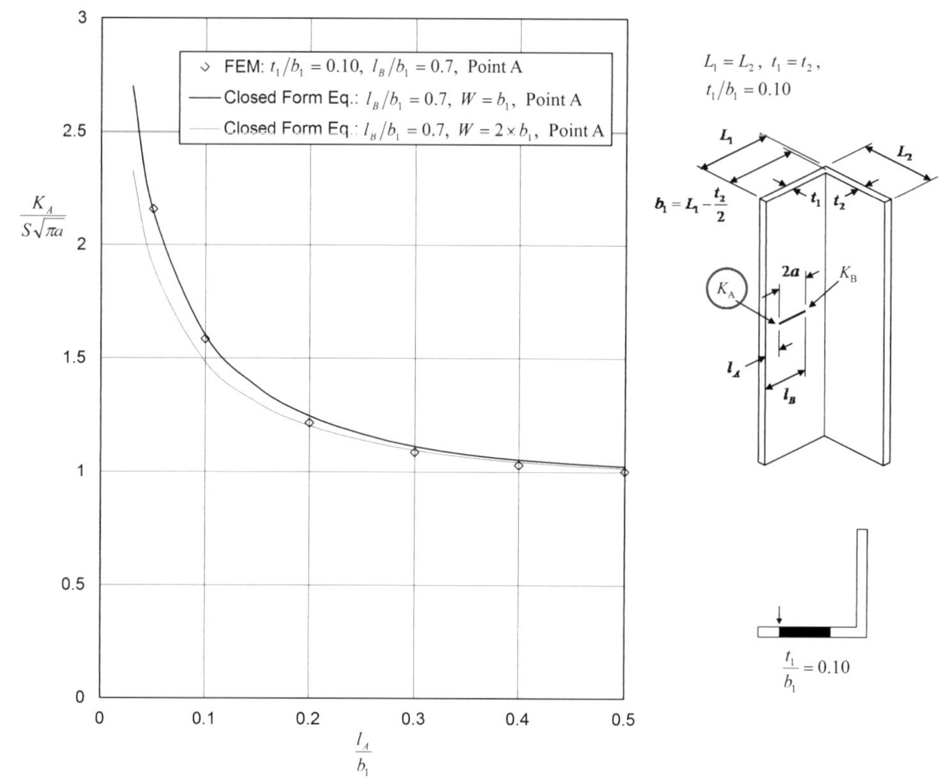

Figure 12 (2/3). SIF at Point A for Internal Crack, Angle with Equal Legs.

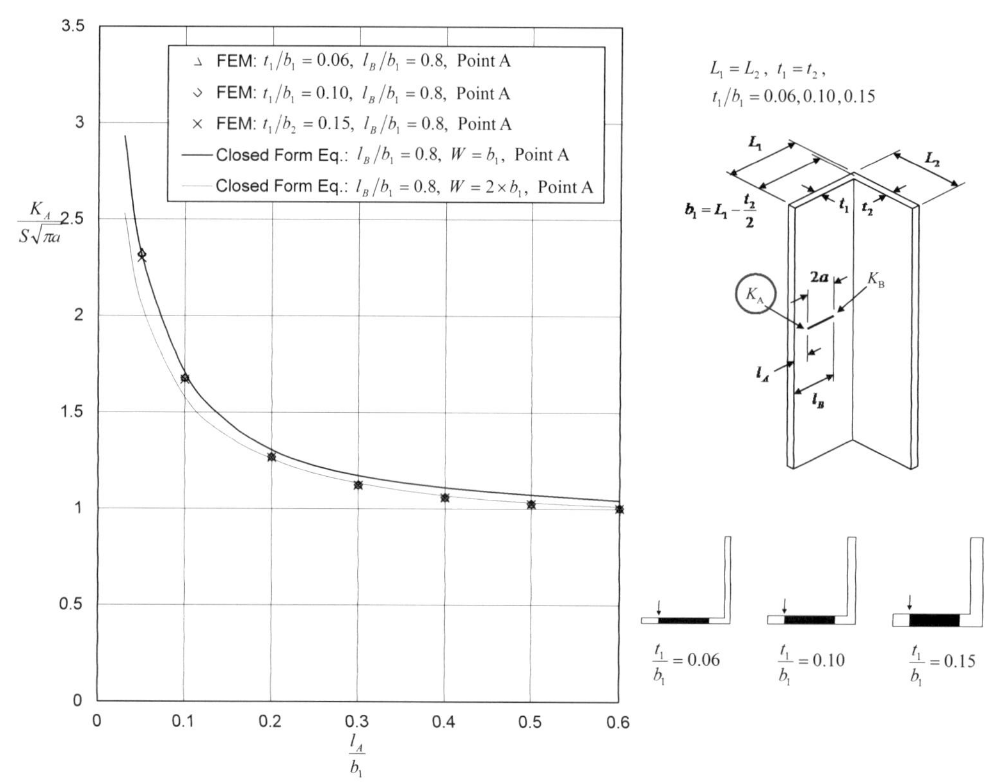

Figure 12 (3/3). SIF at Point A for Internal Crack, Angle with Equal Legs.

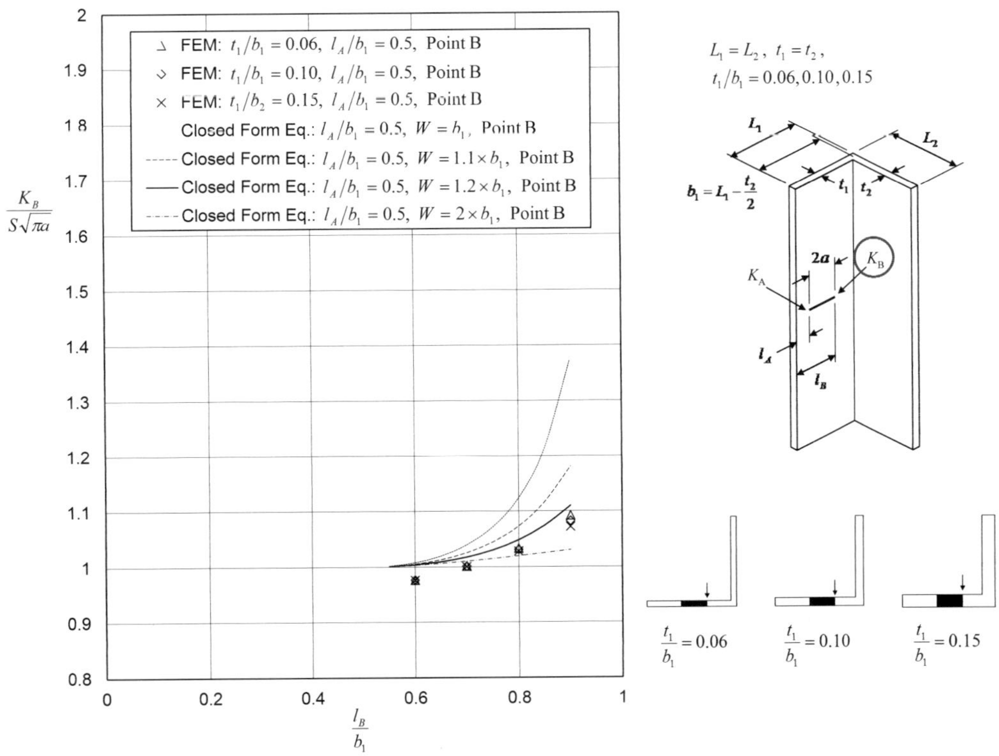

Figure 13 (1/2). SIF at Point B for Internal Crack, Angle with Equal Legs.

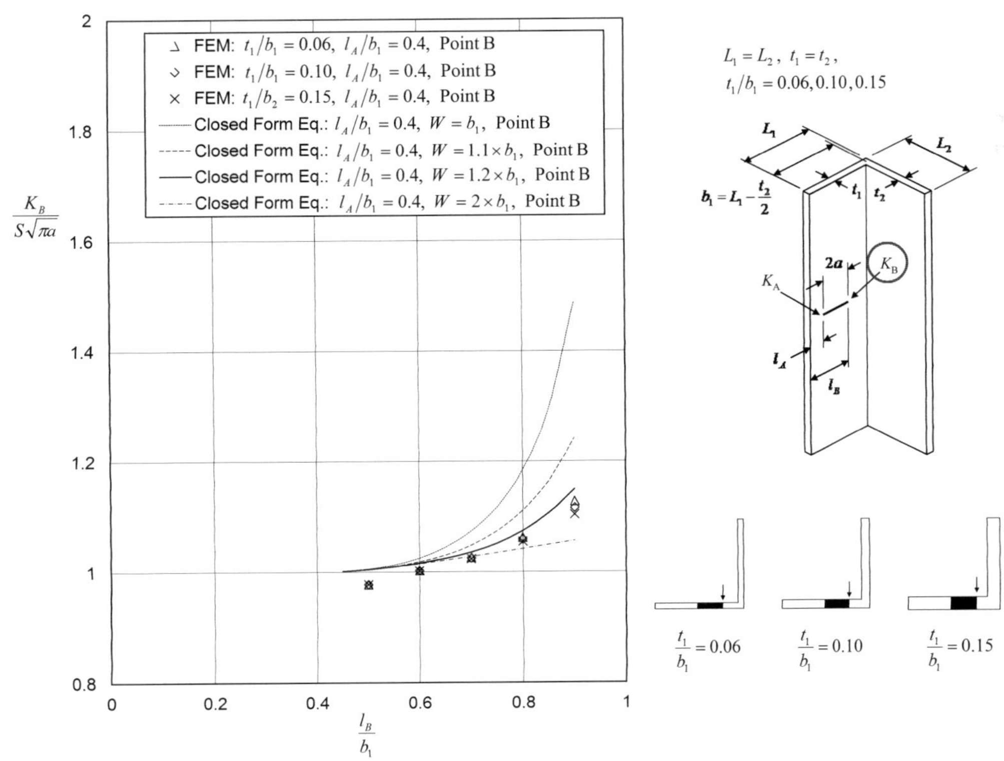

Figure 13 (2/2). SIF at Point B for Internal Crack, Angle with Equal Legs.

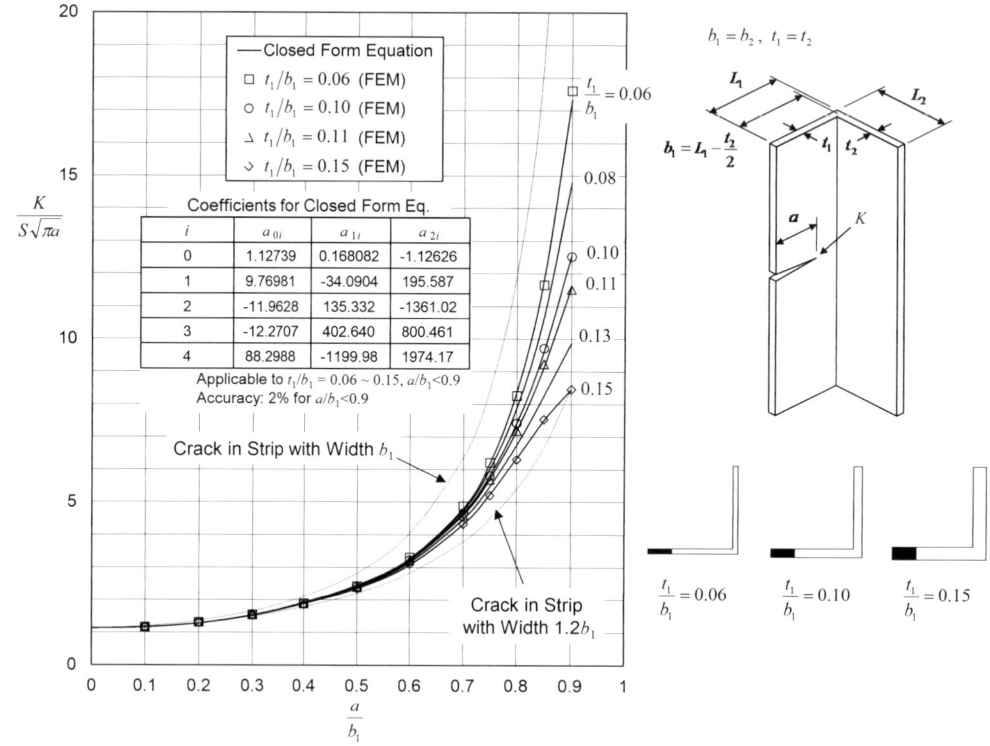

i	a_{0i}	a_{1i}	a_{2i}
0	1.12739	0.168082	-1.12626
1	9.76981	-34.0904	195.587
2	-11.9628	135.332	-1361.02
3	-12.2707	402.640	800.461
4	88.2988	-1199.98	1974.17

Figure 14. SIF for Edge Crack, Angle with Equal Legs.

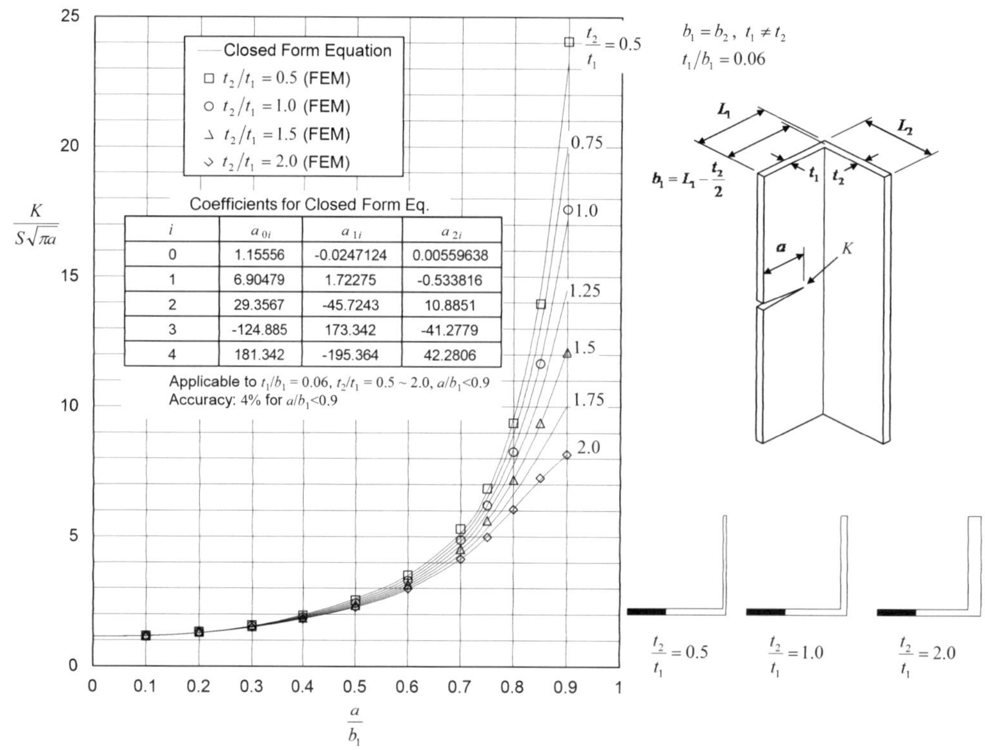

i	a_{0i}	a_{1i}	a_{2i}
0	1.15556	-0.0247124	0.00559638
1	6.90479	1.72275	-0.533816
2	29.3567	-45.7243	10.8851
3	-124.885	173.342	-41.2779
4	181.342	-195.364	42.2806

Figure 15 (1/3). SIF for Edge Crack, Equal Leg Width, Effect of Leg Thickness.

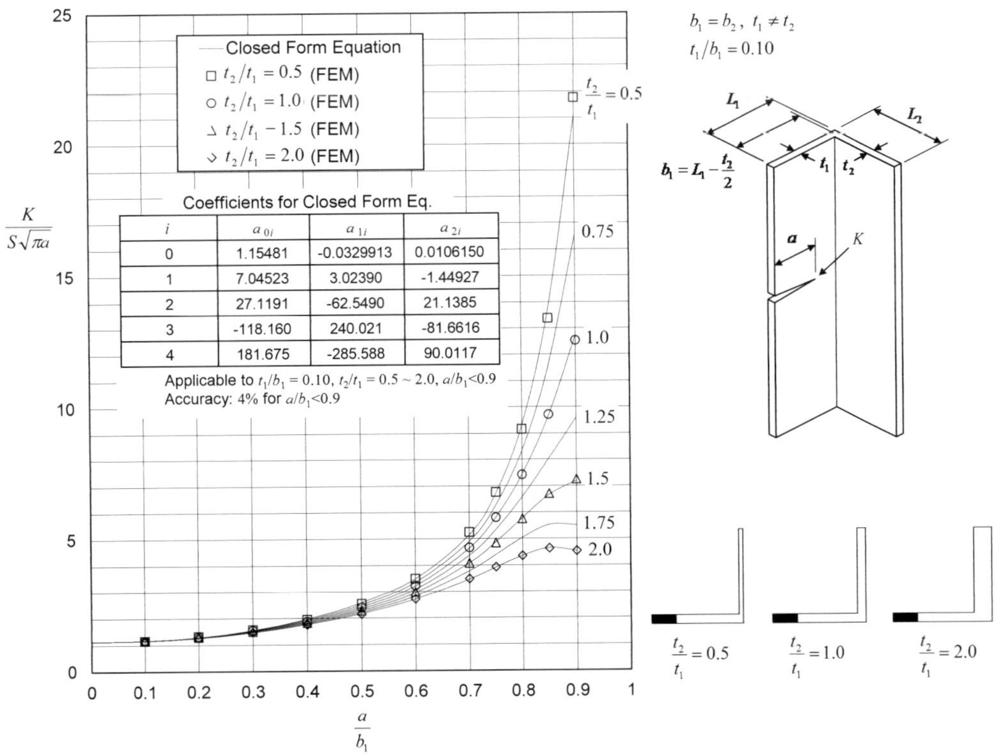

Figure 15 (2/3). SIF for Edge Crack, Equal Leg Width, Effect of Leg Thickness.

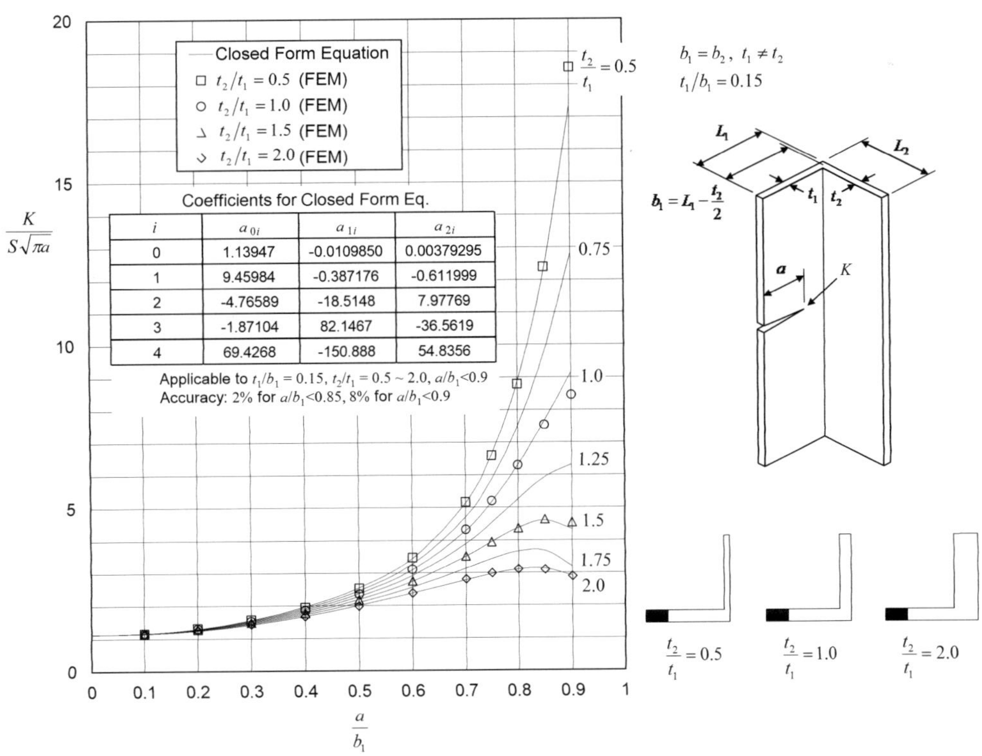

Figure 15 (3/3). SIF for Edge Crack, Equal Leg Width, Effect of Leg Thickness.

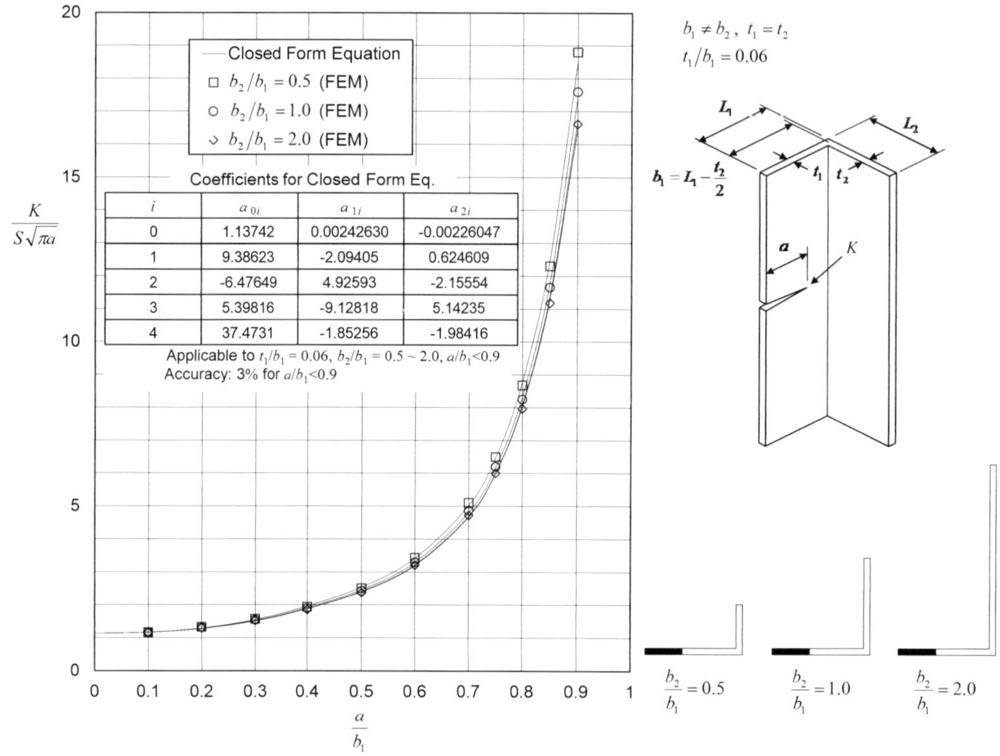

Coefficients for Closed Form Eq.

i	a_{0i}	a_{1i}	a_{2i}
0	1.13742	0.00242630	-0.00226047
1	9.38623	-2.09405	0.624609
2	-6.47649	4.92593	-2.15554
3	5.39816	-9.12818	5.14235
4	37.4731	-1.85256	-1.98416

Applicable to $t_1/b_1 = 0.06$, $b_2/b_1 = 0.5 \sim 2.0$, $a/b_1 < 0.9$
Accuracy: 3% for $a/b_1 < 0.9$

Figure 16 (1/3). SIF for Edge Crack, Equal Leg Thickness, Effect of Leg Width.

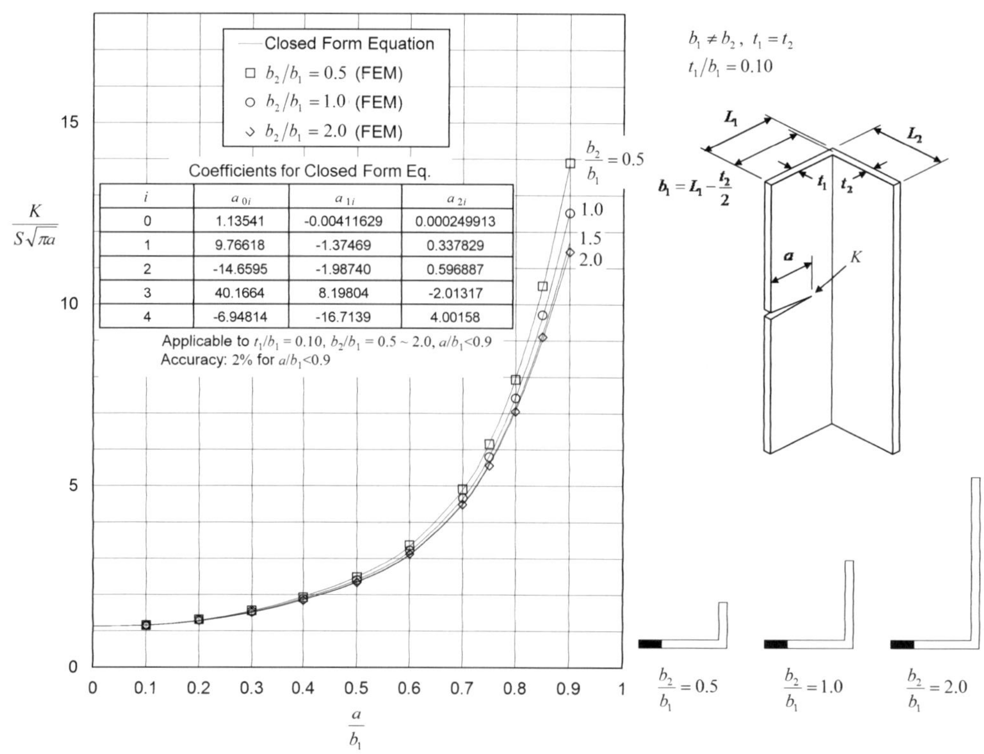

Coefficients for Closed Form Eq.

i	a_{0i}	a_{1i}	a_{2i}
0	1.13541	-0.00411629	0.000249913
1	9.76618	-1.37469	0.337829
2	-14.6595	-1.98740	0.596887
3	40.1664	8.19804	-2.01317
4	-6.94814	-16.7139	4.00158

Applicable to $t_1/b_1 = 0.10$, $b_2/b_1 = 0.5 \sim 2.0$, $a/b_1 < 0.9$
Accuracy: 2% for $a/b_1 < 0.9$

Figure 16 (2/3). SIF for Edge Crack, Equal Leg Thickness, Effect of Leg Width.

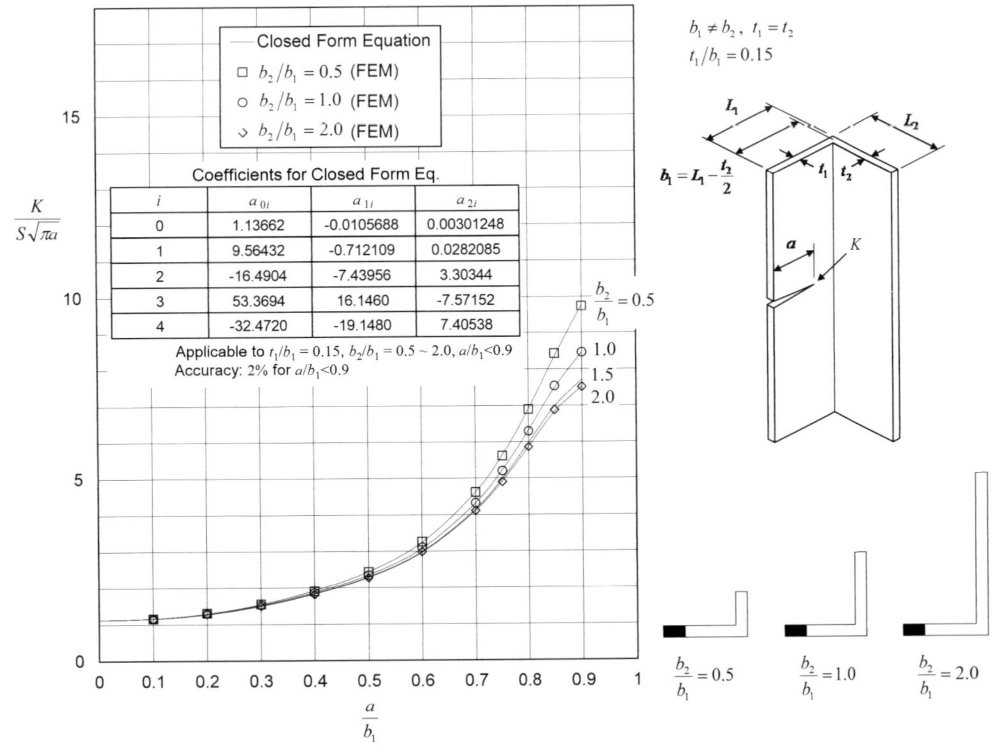

Figure 16 (3/3). SIF for Edge Crack, Equal Leg Thickness, Effect of Leg Width.

CLOSED FORM EQUATIONS FOR STRESS INTENSITY FACTORS FOR THROUGH CRACKS IN ANGLE SHAPED CROSS-SECTION

Internal Crack

For the internal crack, SIF is not sensitive to thickness/width ratio as shown in Figures 11 to 13. The closed form equations for internal crack in finite strip (proposed in reference [2]) is applicable for the SIF of the internal crack in angle. Equivalent width, W_e is introduced. Different equivalent widths should be used for Point A and Point B as follows.

✧ For Point A (near to free edge):

$$\text{Applicable to : } 0.06 \le \frac{t_1}{b_1} \le 0.15,\ t_1 = t_2,\ b_1 = b_2$$

$$\text{Put } W_e = b_1 = L_1 - \frac{t_2}{2}$$

$$\beta_A = \frac{K_A}{\sqrt{\pi a}} = \left[\gamma + \left(\frac{1-\gamma}{4}\right)\left\{1 + (\cos\lambda_1)^{0.25}\right\}^2\right]\sqrt{\sec\lambda_1} \ :\ \text{when } l_A \le W_e - l_B$$

$$\text{where } \gamma = \sin\left(\frac{\pi\lambda_2}{\lambda_1 + \lambda_2}\right),\ \lambda_1 = \frac{\pi a}{2(l_A + a)},\ \lambda_2 = \frac{\pi a}{2(W_e - l_B + a)},\ \lambda_{12} = \frac{4}{7}\lambda_1 + \frac{3}{7}\lambda_2$$

$$\beta_A = \frac{K_A}{\sqrt{\pi a}} = 1 + \frac{\sqrt{\sec\lambda_{12}} - 1}{1 + 0.21\sin\left[8\tan^{-1}\left\{\left(\frac{\lambda_1 - \lambda_2}{\lambda_1 + \lambda_2}\right)^{0.9}\right\}\right]} \ :\ \text{when } l_A > W_e - l_B$$

where $\gamma = \sin\left(\frac{\pi\lambda_2}{\lambda_1+\lambda_2}\right)$, $\lambda_1 = \frac{\pi a}{2(W_e - l_B + a)}$, $\lambda_2 = \frac{\pi a}{2(l_A + a)}$, $\lambda_{12} = \frac{4}{7}\lambda_1 + \frac{3}{7}\lambda_2$

Accuracy: 8% for $\frac{a}{l_A + a} \le 0.9$, 3% for $\frac{a}{l_A + a} \le 0.8$

✧ For Point B (far from free edge):

Applicable to : $0.06 \le \frac{t_1}{b_1} \le 0.15$, $t_1 = t_2$, $b_1 = b_2$

Put $W_e = 1.2b_1 = 1.2\left(L_1 - \frac{t_2}{2}\right)$

$$\beta_B = \frac{K_B}{\sqrt{\pi a}} = \left[\gamma + \left(\frac{1-\gamma}{4}\right)\left\{1 + (\cos\lambda_1)^{0.25}\right\}^2\right]\sqrt{\sec\lambda_1} : \text{ when } l_A > W_e - l_B$$

where $\gamma = \sin\left(\frac{\pi\lambda_2}{\lambda_1+\lambda_2}\right)$, $\lambda_1 = \frac{\pi a}{2(W_e - l_B + a)}$, $\lambda_2 = \frac{\pi a}{2(l_A + a)}$, $\lambda_{12} = \frac{4}{7}\lambda_1 + \frac{3}{7}\lambda_2$

$$\beta_B = \frac{K_B}{\sqrt{\pi a}} = 1 + \frac{\sqrt{\sec\lambda_{12}} - 1}{1 + 0.21\sin\left[8\tan^{-1}\left\{\left(\frac{\lambda_1 - \lambda_2}{\lambda_1 + \lambda_2}\right)^{0.9}\right\}\right]} : \text{ when } l_A \le W_e - l_B$$

where $\gamma = \sin\left(\frac{\pi\lambda_2}{\lambda_1+\lambda_2}\right)$, $\lambda_1 = \frac{\pi a}{2(l_A + a)}$, $\lambda_2 = \frac{\pi a}{2(W_e - l_B + a)}$, $\lambda_{12} = \frac{4}{7}\lambda_1 + \frac{3}{7}\lambda_2$

Accuracy: 8% for $\frac{a}{l_A + a} \le 0.9$, 3% for $\frac{a}{l_A + a} \le 0.8$

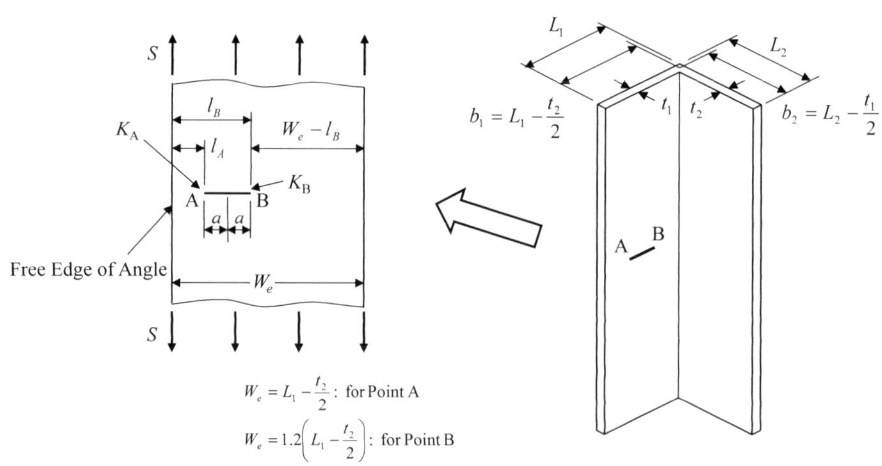

Figure 16. Simplified Model for Internal Crack.

Edge Crack

For the edge crack, new closed form equations are developed. The coefficients of the closed form equations are optimized to fit the curves to the SIF values calculated by finite element analysis. The accuracy of the equations is shown in Figures 14 to 16.

Closed for equation:

$$\beta = \frac{K}{S\sqrt{\pi a}} = c_0 + c_1\left(\frac{a}{b_1}\right)^{2.5} + c_2\left(\frac{a}{b_1}\right)^{5} + c_3\left(\frac{a}{b_1}\right)^{7.5} + c_4\left(\frac{a}{b_1}\right)^{10}$$

where $c_i = a_{0i} + a_{1i}\left(\frac{t}{b_1}\right) + a_{2i}\left(\frac{a}{b_1}\right)^2$

See Figures 14 to 16 for coefficients, a_{0i} and applicability.

CONCLUDING REMARKS

SIF's for internal crack and edge crack in angle shaped section were calculated by finite element analysis. Closed form equations of the SIF's were developed from the SIF values obtained by finite element analysis. It is found that the simplified model of SIF calculation which has been commonly used in the aircraft industries is not conservative. The closed form equations developed in this paper are more accurate than the simplified model.

The closed form equations can be used for damage tolerance analysis of aircraft structures by applying compounding method or similarity method.

REFERENCES

[1] Brussat, T. R., Chiu, S. T. and Creager, M. (1977). *Flaw Growth in Complex Structure, Volume I – Technical Discussion*, AFFDL-TR-77-79, Vol. I.

[2] Kathiresan, K., Hsu, T. M. and Brussat, T. R. (1984). *Advanced Life Analysis Methods – Crack Growth Analysis Methods for Attachment Lugs*, AFWAL-TR-84-3080, Volume II.

[3] Goranson, U. G. (1997). *Int. J. Fatigue*, vol. 20, n. 6, p. 413.

[4] Kuo,A., Yasgur, D., and Levy, M. (1986). *Assessment of Damage Tolerance Requirements and Analysis – Task I Report, Volume II – Analytical Methods*, AFWAL-TR-86-3003, Volume II.

[5] Rooke, D. P. (1986). *Compounding Stress Intensity Factors: Applications to Engineering Structures*, The Parthenon Press, Carnforth, Lancashire, England, U.K.

[6] Tada, H., Paris, P. C. and Irwin, G. R.(1973), *The Stress Analysis of Cracks Handbook*, Del Research Corporation.

[7] Miedler, P. C., Berens, A. P., Gunderson, A. and Gallagher, J. P. (2002), *Analysis and Support Initiative for Structural Technology (ASIST), Delivery Order 0016: USAF Damage Tolerant Design Handbook: Guidelines for the Analysis and Design of Damage Tolerant Aircraft Structures*, AFRL-VA-WP-TR-2003-3002.

Effective Structural Health Monitoring System for 3D Printed Components

M. F. Hinderdael[1], D. De Baere[1], M. Strantza[2], R. Vafadari[3], W. Devesse[1], P. Guillaume[1]

[1]*Department of Mechanical Engineering, Vrije Universiteit Brussel, Brussels, Belgium*
[2]*Department of Mechanics of Materials and Constructions, Vrije Universiteit Brussel, Brussels, Belgium*
[3]*Department of Materials Science and Engineering, Ghent University, Ghent, Belgium*

Abstract: 3D printing techniques have shown high potential for aeronautical applications. However, due to fatigue behavior, the structural integrity of 3D printed components is still an important challenge nowadays. In this paper, a Structural Health Monitoring system is discussed that can be integrated into 3D printed structures. Embedded capillaries are initially pressurized and the pressure is continuously monitored. The presence of a crack is signaled when the pressure increases due to the air leak at the crack location. Besides the detection of fatigue cracks, useful localization information can also be obtained from the pressure waves that are generated inside the embedded capillaries. Initial test results and the localization feasibility of the system will be discussed in this paper. Information about the crack location is crucial for further, more detailed inspections of the crack zone and/or remaining life-time prediction. The localization capability further increases the benefits for the end users.

INTRODUCTION

Additive manufacturing, or 3D printing, is a relative new production technique whereby physical objects are built by means of sequential addition of material on basis of a 3D virtual model. A 3D printer adds material, layer by layer, in order to build the component as in the 3D model. Additive manufacturing techniques have shown high potential for a variety of applications. The ability of manufacturing very complex shaped components as light weight as possible is a very appealing benefit of 3D printed components certainly for aeronautical applications. In addition, there is a clear economic benefit to optimize structures with respect to weight, because the lighter the component is made, the cheaper it becomes to produce as less material is needed. 3D printed components also possess the property of embedding smart technologies during the production process, generating overall smart solutions.

As additive manufactured components are constructed layer by layer, their metallurgic structure is different than conventionally produced materials. Fatigue and crack propagation properties of additive manufactured materials are not very well known yet [1], but multiple sources report a reduced fatigue lifetime. In a work of Chan [2], fatigue lifetime of titanium Ti6Al-4V alloys fabricated by means of additive manufacturing were compared to conventionally produced Ti6Al-4V alloys (rolled or cast). Conventionally rolled Ti6Al-4V alloys showed approx. twice the fatigue lifetime as the best additive manufactured Ti6Al-4V alloys. Gong [3] stated that as-built surfaces of Selective Laser Melted (SLM) Ti-6Al-4V samples become crack initiation sites, resulting in a reduced fatigue lifetime when the stress level exceeds 500MPa. Similar results were also recorded in a paper of Strantza [4], where the SLM Ti6Al-4V samples indicated a stress level at failure under fatigue loading of 590MPa and 570MPa.

Despite the numerous benefits, the reduced fatigue behaviour is currently jeopardizing the introduction of additive manufactured primary components in industries where safety and product availability are very important. Inspection intervals must then be shortened, increasing inspection costs and product unavailability which inevitably induce large economic losses. Including an effective Structural Health Monitoring system that monitors the fatigue behaviour of 3D printed components can largely eliminate this major drawback. Structural Health Monitoring (SHM) is the integration of sensing and possibly also actuation devices to allow the damaging conditions of a structure to be recorded, analysed, localized and predicted in a way that non-destructive testing becomes an integral part of the structure [5].

Because of the continuous health indication of the component, safety levels are enhanced (see the curve shift on Fig. 1). As a result of the introduction of a SHM system, the inspection intervals can be prolonged beyond the conventional inspections times if the loading remains constant. The same reasoning also allows stating that when one keeps the maintenance intervals fixed, the loading on the component can be increased when introducing a SHM system. Practically, this could also mean that components, for which the design process was fatigue driven, can be slimmed to a lower weight. This clearly indicates the huge economical savings that can result from the introduction of a SHM system.

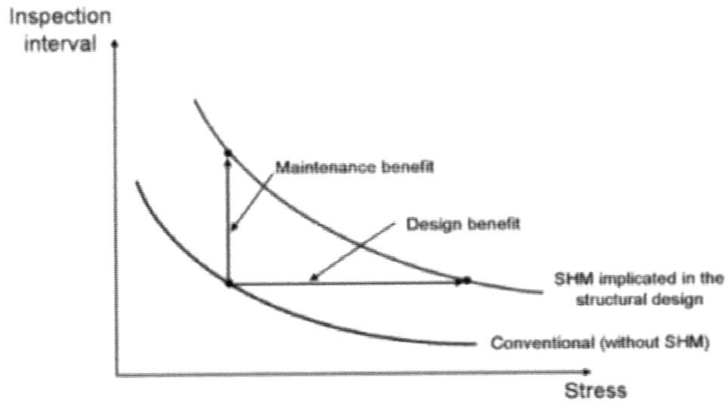

Figure 1. Benefits of the use of a Structural Health Monitoring system [6].

Besides the economic impact, the enhanced safety as such might even be more important when human lives are at risk. In May 1974 a Sikorsky S-61N helicopter crashed in the North Sea with the loss of six lives. After a profound investigation, it was found that a growing crack had caused one of the main rotor blades to break during flight. Yet, the rotor blades were pressurised (with Nitrogen) and the pressure level was checked (on the ground) every time before flight. In case a crack would have been present in the rotor blade, the nitrogen would have leaked causing the internal pressure to drop. Notwithstanding the rotor blade didn't show the presence of any cracks before the flight (rotor blade was still pressurised), it crashed. The investigators found that cracks could grow from an undetectable size to a detrimental size within the time of a normal helicopter flight (2.1-4.3hrs). The crash investigation report finally concluded on a remedial action by introducing blade pressure indicators in the cockpit, i.e. the installation of a SHM system [7].

Recently, a new effective Structural Health Monitoring (eSHM) system for 3D printed components has been invented. Close to the component's surface where fatigue cracks are expected to grow, small capillaries are embedded in the structure by means of additive manufacturing techniques. After pressurization, this capillary pressure becomes very sensitive for leaks. A crack reaching the capillary will result in a leak flow, equalling the internal capillary pressure to the outside atmospheric pressure. On basis of the pressure measurement inside the capillary, it is thus possible to derive the presence of a fatigue crack. In this paper, the crack detection and crack localization ability of the presented eSHM system are studied.

EFFECTIVE STRUCTURAL HEALTH MONITORING

A SHM system must satisfy a lot of requirements and be optimized to obtain the highest added value. As stated by C.R. Farrar, one of the most fundamental challenges of a SHM system is the ability to detect very local phenomenon, such as cracks, when considering the global response of a structure. This response may not be significantly being changed under the local effect of a crack compared to other influencing factors such as environmental conditions [8].

It is also important to reduce the number of parameters to be monitored in order to derive the global health of the component. The amount of sensors that have to be installed and maintained, directly impacts the cost of the system. The amount of data that has to be transmitted and stored increases for each required sensor. The less sensors and variables needed for the SHM system, the easier its implementation for real-life applications and the more straightforward the post-processing.

Even more important is the requirement that the SHM system must at least be as robust as the structure. It is of no added value if the SHM system is prone to false detections of damage of the structure. A self-test functionality is therefore also of high added value as this quickly allows checking the SHM system's performance.

As a last important property, the SHM system must be cheap and easy to install on the structure. A Structural Health Monitoring system that satisfies the above mentioned necessities is hereafter referred to as an effective Structural Health Monitoring system, abbreviated as an eSHM system.

PRESSURE MONITORING OF EMBEDDED CAPILLARIES

Capillaries are embedded in the 3D printed structure by means of additive manufacturing techniques there where fatigue cracks are expected to pass through during the crack growth phase. The capillaries are then pressurized, making them very sensitive for leaks. The capillary pressure is continuously monitored and a deviation w.r.t. the initial capillary pressure indicates the presence of a crack. A graphical representation of a capillary structure inside a cogwheel can be retrieved in Fig. 2.

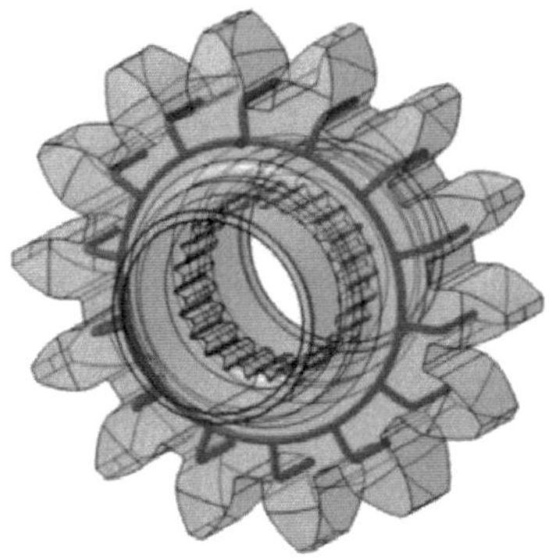

Figure 2. Embedded capilalries of the eSHM system shown inside a cogwheel.

The presented system for 3D printed components meets a lot of the posed requirements of an effective Structural Health Monitoring system. Embedding the capillary in a way that it connects all locations prone to fatigue cracks allows monitoring the critical areas without the need of multiple sensing devices. It is thus able to monitor the structure globally, by evaluating the individual regions prone to fatigue.

Only one pressure measurement is required in order to detect the presence of a fatigue crack, which largely facilitates the installation, data storage and post-processing of the data of the eSHM system. A second backup sensor can be installed, and used for localization purposes as discussed later (see "Crack localization").

Being fully embedded in the structure, this eSHM is not prone to external accidental damage. Only the pressure sensor may be protruding, but this component is easily replaced if damaged. A self-test functionality results from the comparison of the pressure read-out with a second sensor installed on the same capillary. The system has the advantage that the absolute pressure signal is always present in contrast to other techniques that are based on rarely emitted signals (e.g. Acoustic Emission).

Embedding the capillaries in the structure during the 3D print process doesn't induce any additional cost for the process and a cheap solution can be found for the installation of the pressure sensor. Despite the fact that the benefits of the eSHM system are the largest for purely additive manufactured components, it is not a necessity that the whole component is produced using additive manufacturing techniques. Given a conventionally produced component, the capillary network can be added on top of the component's surface (an example is given in Fig. 3). When the final dimensions of the conventionally produced component must be respected, part of the material can be removed and replaced by additive manufacturing techniques (an example is given in Fig. 4).

In comparison to other SHM systems, this system is not surface bonded, allowing contact surfaces to be monitored where other SHM (e.g. CVM) cannot be applied. In Fig. 1, an example of such an application is presented where the teeth of a cogwheel are monitored.

Figure 3. Capillary network added on top of existing, conventionally produced, component.

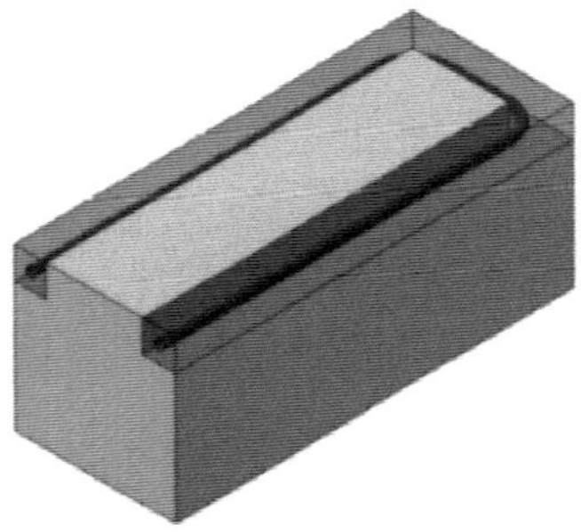

Figure 4. Capillary network added on top of existing, conventionally produced, component keeping the final dimensions intact.

Current additive manufacturing techniques allow accurate embedding capillaries of a minimal diameter of 0.5mm and 1mm using respectively Layerwise Laser Melting (LLM or SLM) and Laser Metal Deposition (LMD) techniques. Multiple capillaries can be embedded at different depths below the component's surface. Doing so, the fatigue crack growth can be followed as deeper and deeper located capillaries also lose their pressure. For fatigue applications, the first capillary network can be located very close to the surface, corresponding to the detectable limit of fatigue cracks of a non-destructive inspection technique. A second, independent, capillary network located much deeper in the structure can function as an indication of the critical crack length.

Crack initiation

Stress computations indicate the high stress zones and serve as a first indication for the regions where fatigue cracks are expected to occur. These regions will have to be equipped with capillaries that are defined such a manner that growing cracks starting from the prone location would pass through the capillaries without significantly influencing the stress distribution. The stress and crack growth analysis will therefore have to take into account the presence of the capillaries. However, as fatigue cracks are surface initiated, the presence of internal capillaries should not influence the fatigue initiation behaviour.

In the work of Strantza [4], tests with sinusoidal capillaries were carried out to investigate the effect of the capillary depth on the crack initiation location (see Fig. 5). If the majority of the test samples showed crack initiation where the capillary is closest to the specimen edge, it must be concluded that the presence of the capillary influences the crack initiation behaviour. In the same work, it was concluded that the integrated capillaries of the eSHM system did not influence the crack initiation location, as crack initiation was recorded where the distance between the capillary and the specimen edge was the largest.

Crack detection

In the event of a fatigue crack reaching the capillaries, a leak flow will tend to equilibrate the internal and external pressure. A pressure sensor continuously measures the internal capillary pressure. A deviation w.r.t. the initial capillary pressure indicates the existence of a crack. A test campaign was launched to prove the crack detection ability of the system. Four point bending tests were conducted on additive manufactured test samples with embedded capillaries of 3mm diameter.

During the four point bending tests, the specimen is loaded with a constant moment in between the inner rollers. As the investigated test samples were unnotched, the crack initiation location was unknown but expected in between these two inner rollers. Because the fatigue limit of the material was unknown and only a limited number of samples were available, the step method was used to initiate a fatigue crack at the lowest possible stress level but within a certain, prescribed, number of cycles. In our tests, the limited number of cycles was set to 100.000 cycles. Every 100.000 cycles, the sample was inspected for crack initiation using liquid penetrant inspection and the loading on the specimen was increased if no crack initiation was found. When a crack initiation was recorded, the loading was kept constant. The loading frequency was 5Hz and the R ratio of the loading was 0.1.

Before the fatigue tests started, the pressure level inside the capillaries was lowered to 0.57bar. When finally the crack initiated and has grown till the depth of the capillary, the pressure starts increasing till the ambient atmospheric pressure. When a pre-set pressure limit was reached (0.83bar in Fig. 6), the loading on the specimen and the test was stopped. The last seconds of the pressure measurement are shown in Fig. 6.

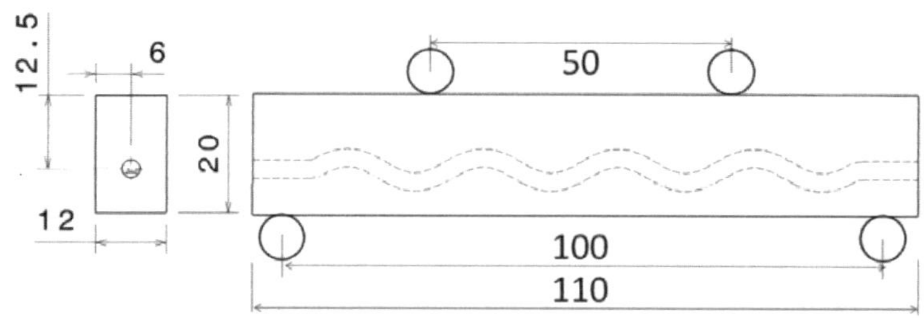

Figure 5. Design of the test specimens for 4 point bending tests (dimensions in mm).

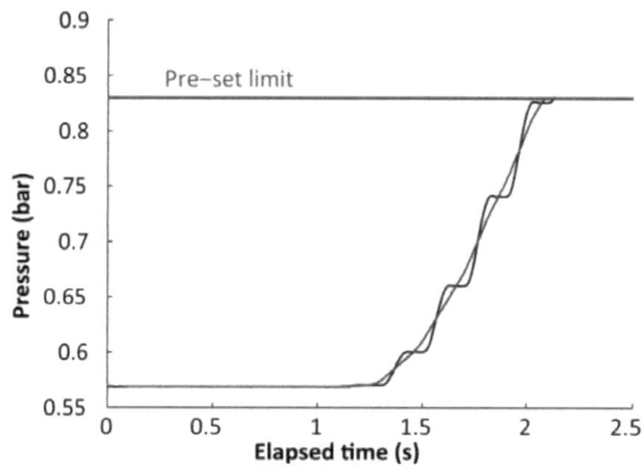

Figure 6. Internal pressure rise due to crack presence.

The pressure rise shows the existence of multiple pressure steps. These are corresponding to the loading frequency of 5Hz whereby during the half cycle of reduced loading, the crack again closes and the pressure stops increasing. Half a period later, the pressure rises again as a consequence of the higher loading that opened the crack.

The work of Strantza [4] reported that all tests were successfully stopped after the occurrence of a crack, without missing any fatigue crack or any false alarm. It was concluded that the experiments constitute the "proof-of-concept" of the presented eSHM system using embedded, pressurized capillaries.

Crack localization

When a crack is detected, further inspections with more elaborated techniques will be needed to evaluate the extent of the damage. The analysis of the phenomena that occur at the moment of the opening of the crack allows retrieving an indication of the crack location. The event of a crack opening the capillary results in a perturbation of the fluid at the crack location and propagates as a wave, the so-called Negative Pressure Waves (NPW), away from the leak to both ends of the capillary. Installing two pressure sensors at both ends of the capillary allows sensing the arrival of these waves at the ends of the capillary. The difference in time-of-arrival of these waves serves as a basis for the localization of the crack.

A test setup was built using cylindrical tubes representing the capillaries inside the 3D printed component in order to investigate the crack localization feasibility of the SHM system. The leak, representing the crack location, could be located at 5 different positions along the tube, as indicated in Fig. 7. A venturi was used to lower the initial pressure level below atmospheric values to approximately 0.2bar. A check valve and shut-off valve separated the tube from the vacuum-equipment when performing the tests.

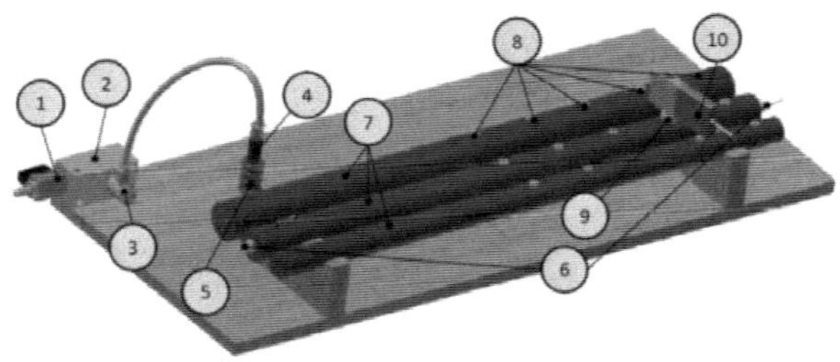

①	Shut-off valve
②	Venturi
③	Flow controller
④	Check valve
⑤	Shut-off valve
⑥	Pressure sensors
⑦	Tubes
⑧	Leak positions
⑨	Leak size
⑩	Solenoid valve

Figure 7: Test setup representing the embedded capillaries to investigate the crack localization capability of the new eSHM system

In Fig. 8, the principle of the leak localization is highlighted. The formula for the crack location can easily be retrieved by expressing the times needed for the pressure waves to travel the distances x_L and L-x_L at the speed of sound c. The difference in this time-of-arrival is a measure for the crack location. The arrival of the pressure waves at the location of the pressure sensors is seen as the first pressure rise, measured by that sensor (see black squares in Fig. 9).

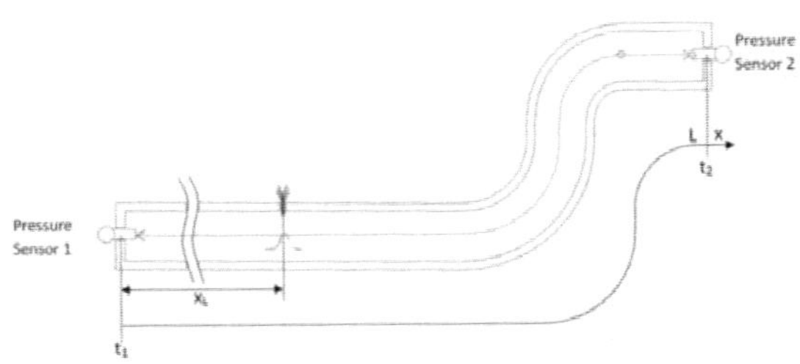

$$x_L = \frac{L - c(t_2 - t_1)}{2}$$

Figure 8. Time of flight principle for crack localization.

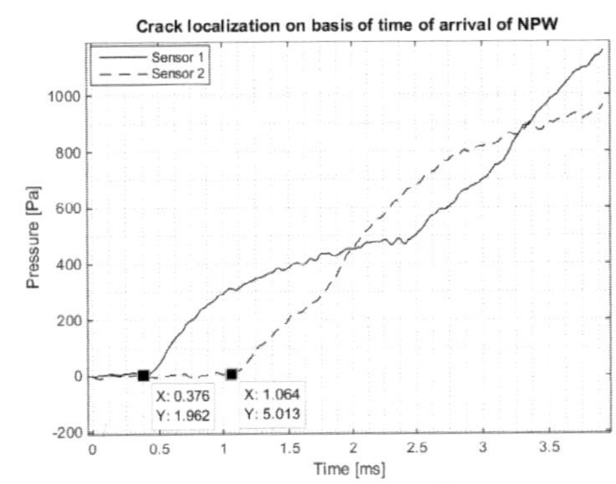

Figure 9. The difference in time of arrival of the Negative Pressure Waves (NPW) at the pressure sensor locations is an indication of the crack location.

Knowing that these pressure waves travel at the speed of sound of the fluid inside the capillary, which was air at room temperature (T = 298K), and knowing the overall length of the capillary (L=0.462m), it is possible to estimate the leak location on basis of the measurement of the difference in time-of-arrival of the pressure waves indicated in Fig. 9.

$$x_L = \frac{L - c(t_2 - t_1)}{2} = \frac{0.462 - 343(0.001064 - 0.000376)}{2} m = 11.3cm$$

The real leak location was 11.4cm. It must be remarked that the obtained accuracy largely depends on the sampling frequency of the pressure measurement. The data shown in Fig. 9 was acquired at a sampling rate ($f_s = \frac{1}{\Delta T_s}$) of 204.8kHz, leaving a theoretical uncertainty on the localization of

$$\Delta x = c\Delta T_s = \frac{c}{f_s} = 1.7mm$$

Together with more measurements, it could be concluded that this method is able to retrieve the crack location with high precision, but depending on the sampling frequency.

CONCLUSION

The design freedom and resulting weight gains of additive manufactured components are very appealing for aeronautical applications, but their weak structural integrity is still jeopardizing their introduction in such safety critical applications. An effective Structural Health Monitoring system embedded in the 3D printed component allows monitoring the health of the component continuously and enables the introduction of additive manufactured components in safety critical applications. The combination of an eSHM system installed on 3D printed components enlarges the benefits that come along with these new technologies. Safer operation, economic savings due to reduced weight and increased component availability, more complex designs, etc. are just some of them. The crack detection and crack localization feasibility of the presented eSHM system were investigated. All cracks were detected by the system without any false alarm and the cracks could be localized with high accuracy if an appropriate sampling rate is chosen.

AKNOWLEDGEMENT

The research work financed by the Agency for Innovation by Science and Technology in Flanders (IWT).

REFERENCES

[8] Wycisk, E., Solbach, A., Siddique, S., Herzog, D., Walther, F. and Emmelmann, C., (2014), *Effects of Defects in Laser Additive Manufactured Ti-6Al-4V on Fatigue Properties*, In: Physics Procedia, 45, pp. 371-378, Schmidt M., Vollertsen F. and Merklein M. (Ed.), Elsevier B.V., Fürth, Germany

[9] Chan, K.. S., Koike, M., Mason, R. L., Okabe, T., (2013), *Fatigue Life of Titanium Alloys Fabricated by Additive Layer Manufacturing Techniques for Dental Implants*, In: Metallurgic and Materials Transactions A, 44, 2,pp. 1010-1022, Laughlin, D. E., et. al. (Ed.), Springer US

[10] Gong, H., Rafi, K., Starr, T. and Stucker, B., (2012), *Effect of defects on fatigue tests of as-built Ti-6Al-4V parts fabricated by selective laser melting.*, Solid Freeform Fabrication Symposium

[11] Strantza, M., De Baere, D., Rombouts, M., Clijsters, S., Vandendael, I., Terryn H., Guillaume P. and Van Hemelrijck D., (2014), *3D Printing for Intelligent Metallic Structures*, Proceedings of the EWSHM 7th European Workshop on Structural Health Monitoring, Nantes, France

[12] Boller, C., and Meyendorf, N., 2004, *State-of-the-art in Structural Health Monitoring of Aerospace Structures*, Proc. Of Internat. Symposium on NDT in Aerospace, Fürth/Bavaria, Germany, pp.5-6.

[13] Calomtirescu, M., (2008), *Lamb Waves for Structural Health Monitoring in Viscoelastic Composite Materials*, Science-report aus dem Faserinstitut Bremen (Band 3), Bremen, pp.2.

[14] Wanhill, R.J.H., (2009), *Some notable aircraft service failures investigated by the National Aerospace Laboratory NLR*, In: Structural Integrity and Life, Vol 9., No 2, pp.71-87.

[15] De Baere, D., Strantza, M., Hinderdael, M., Devesse., W. and Guillaume, P., (2014), *Effective Structural Health Monitoring with Additive Manufacturing,* Proceedings of the EWSHM 7th European Workshop on Structural Health Monitoring, Nantes, France

PGD-Virtual Charts for structural design: application to geometrical parameters

Amaury Courard[1,2], David Neron[1], Pierre Ladeveze[1], Alain Bergerot[2], Ludovic Ballere[2]

[1]LMT Cachan (ENS Cachan/CNRS/Université Paris-Saclay), 61 avenue du Président Wilson, 94235 Cachan Cedex, France

[2]AIRBUS Defence and Space B.P. 20011, 33165 St Médard en Jalles Cedex, France

Abstract: The paper presents the concept of "Virtual Charts" applied to computational structural mechanics. Based on the Proper Generalized Decomposition (PGD), a model reduction method, the idea is to create a numerical decision support tool to help engineers to perform shape optimization. Quantities of interest, computed for the full parametric problem, will be stored in Virtual Charts for a subsequent use. Due to the peculiar nature of geometric parameters, PGD had to be adapted in order to take them into account. In partnership with AIRBUS Defence and Space, particular attention has been paid to integrating the methodology developed into a current industrial design process. Promising results on simplified industrial structure were obtained and PGD-Virtual Charts were constructed.

INTRODUCTION

Experimental testing is still essential during the design stage of industrial structures because numerical resolutions of high sized complex models often remains out of reach, despite constant progress in computing power over the past years. And even when simulations are possible, the fully "virtual" optimization of the structures with respect to the design variables (material coefficients, geometry...) leads to so many computations that the structures tend to be oversized during the design stage in order to maintain the conception time reasonable. The reduction of the computational costs keeping a good reliability is then a crucial industrial issue.

In the framework of this study in partnership with AIRBUS Defence and Space, the design variables correspond to geometrical variations, which are key points of the design process. In the current approach used in design offices, each new structure (corresponding to new design variables) is considered as a new problem, and treated independently of cases previously studied leading to a very large number of simulations. The idea, which is followed in this work, is to create a numerical decision support tool, called "Virtual Charts", for structural design. For this purpose, similar structures, which differ from each other in the value given to a set of parameters, are put together into "families" and the full parameterised solution is computed offline for each family. The quantities of interest are, then, stored in "Virtual Charts" which will be used in real-time by the engineer during the design stage to optimize the conception.

The construction of these charts is based on the model reduction method called Proper Generalized Decomposition (PGD) introduced for the first time in 1985 by P. Ladevèze under the name "Radial Approximation" [1] in order to deal with time-dependent non-linear problems. It allows generating the solution of a problem for a whole set of parameters and its efficiency was proven for time-space [2], multiscale [3], multiphysics [4] and parameterised problems [5]. The PGD method is based on the idea that the displacement u can be approximated with a very good precision in a separated variable way:

$$\boldsymbol{u}(\boldsymbol{X},\alpha_1,\cdots,\alpha_k) \approx \sum_{i=1}^{n} \boldsymbol{\Lambda}_i(\boldsymbol{X}) f_i^1(\alpha_1)\cdots f_i^k(\alpha_k)$$

where $\alpha_1,\ldots,\alpha_n$ are parameters. In practical cases, n happens to be very small so the computation and storage costs are highly reduced. Furthermore, the functions f_i and Λ_i are not given *a priori*. They will be computed through an iterative process. The Virtual Charts are then constructed for each quantity of interest studied during design (stress in a certain area, size of plastic zones...). All that is remaining for the end-user is to look over the Virtual Charts to select the set of parameters meeting the mechanical requirements, as he would have done with a handbook.

PROPER GENERALIZED DECOMPOSITION

Proper generalized decomposition

An extended presentation of the PGD method can be read in [6] and a detailed bibliography in [7]. The PGD method is based on the idea that the solution can be approximated with a good precision thanks to a reduced number of modes.

One can notice the analogy with dynamics where the major part of the information is contained in the first modes and while new modes are added, the solution is enriched.

For this purpose, a separated variables representation of an unknown field f is sought. An example of a PGD approximation is given below for a time-space problem:

$$\forall(\underline{M},t) \in \Omega \times [0,T], f(\underline{M},t) \approx \hat{f}_n(\underline{M},t) = \sum_{i=1}^{n} \lambda_i(t)\Lambda_i(\underline{M}) \tag{1}$$

where Λ_i are the spatial modes and λ_i are the time functions.

In practical cases, m happens to be very small so the computation and storage costs are highly reduced. Furthermore, the functions Λ_i and λ_i are not given a priori. They will be computed through an iterative process. This method can be easily extended to parametric and multidimensional problems as shown in Eqn. 2.

$$f(\alpha_1,\ldots,\alpha_k,t,\underline{M}) \approx \sum_{i=1}^{n} f_i^1(\alpha_1)\ldots f_i^k(\alpha_k)\lambda_i(t)\Lambda_i(\underline{M}) \tag{2}$$

where $\alpha_1, \ldots, \alpha_k$ are the parameters which can be of different natures (material parameters, geometric parameters, loads…).

Geometry variations

We will focus the rest of our study on geometric parameters, which add a difficulty in the PGD framework with respect to other parameters, material parameters for instance [8]. Let us consider a structure Ω and a set of parameters $\boldsymbol{\alpha} = \alpha_1, \ldots, \alpha_k$ defining the geometry. The problem considered is linear elastic. Let us write the variational formulation of the problem:

$$\forall \delta\underline{u} \in \mathcal{U}^\star, \quad \int_{\Omega_{\boldsymbol{\alpha}}} \boldsymbol{\varepsilon}\left(\underline{u}(\underline{M},\boldsymbol{\alpha})\right) : \boldsymbol{\kappa}\boldsymbol{\varepsilon}\left(\delta\underline{u}(\underline{M},\boldsymbol{\alpha})\right) \mathrm{d}\Omega = \int_{\partial_f\Omega_{\boldsymbol{\alpha}}} \underline{f}^T \delta\underline{u}(\underline{M},\boldsymbol{\alpha})\,\mathrm{d}\Gamma \tag{3}$$

- $\boldsymbol{\varepsilon}$: strain tensor
- $\boldsymbol{\kappa}$: Hooke's tensor
- $\mathcal{A}$: space where $\boldsymbol{\alpha}$ belongs to
- $\underline{u}$: displacement
- $\mathcal{U}^\star = \left\{\boldsymbol{v} \in H^1 \left| \boldsymbol{v}_{|\partial_u\Omega} = \boldsymbol{0}\right.\right\}$

The classic approach would be to solve the problem defined in Eqn. 3 for all the values of $\boldsymbol{\alpha}$. However this can be out-of-reach due to the numerous heavy simulations to be computed. In the PGD framework, the solution is sought in a larger space. Indeed we, not only, integrate the problem over space but also over the parameters. Hence, one writes the full variational formulation:

$$\forall \delta\underline{u} \in \mathcal{U}^\star, \quad \int_{\mathcal{A}}\int_{\Omega_{\boldsymbol{\alpha}}} \boldsymbol{\varepsilon}\left(\underline{u}(\underline{M},\boldsymbol{\alpha})\right) : \boldsymbol{\kappa}\boldsymbol{\varepsilon}\left(\delta\underline{u}(\underline{M},\boldsymbol{\alpha})\right) \mathrm{d}\boldsymbol{\alpha}\mathrm{d}\Omega = \int_{\mathcal{A}}\int_{\partial_f\Omega_{\boldsymbol{\alpha}}} \underline{f}^T \delta\underline{u}(\underline{M},\boldsymbol{\alpha})\,\mathrm{d}\boldsymbol{\alpha}\mathrm{d}\Gamma \tag{4}$$

A difficulty now appears. The integration domain Ω_α explicitly depends on the geometric parameters, which makes impossible the separation of the different integrals and their computations, therefore, remain very time consuming. The idea, to overcome this, is simply to define a reference structure Ω_0 and thanks to a change of variables, the solution is sought for that structure. Eqn. 4 becomes:

$$\forall \delta\underline{u} \in \mathcal{U}^\star, \quad \int_{\mathcal{A}}\int_{\Omega_0} \boldsymbol{\varepsilon}(\underline{u}(\underline{M}_0,\boldsymbol{\alpha})) : \boldsymbol{\kappa}\boldsymbol{\varepsilon}(\delta\underline{u}(\underline{M}_0,\boldsymbol{\alpha}))|\det \underline{\underline{J}}_{\boldsymbol{\alpha}}|\,\mathrm{d}\boldsymbol{\alpha}\mathrm{d}\Omega = \int_{\mathcal{A}}\int_{\partial_f\Omega_0} \underline{f}^T \delta\underline{u}(\underline{M}_0,\boldsymbol{\alpha})|\det \underline{\underline{J}}_{\boldsymbol{\alpha}}|\,\mathrm{d}\boldsymbol{\alpha}\mathrm{d} \tag{5}$$

It can be seen that the dependency on the geometric parameters have been transferred from the integration domain to the integrand through the Jacobian. In other words, from the current geometric configuration, it is possible to go back to the reference structure via a geometric transformation.

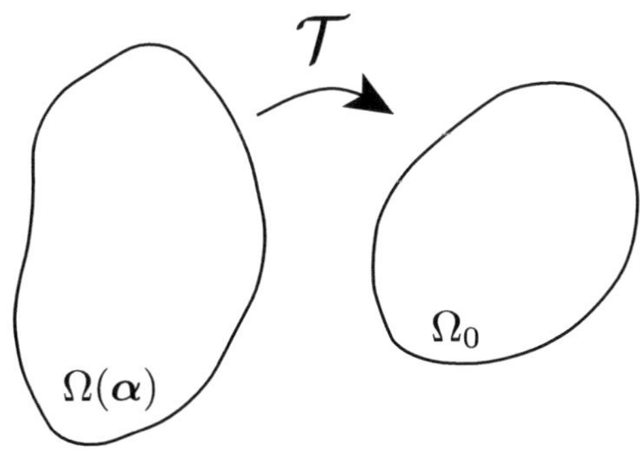

Figure 1. the solution is sought for a reference structure thanks to a geometric transformation.

INDUSTRIAL PROBLEM

The structure considered is shown in Fig. 2. This is an axisymmetric linear elastic problem and the geometry is parameterized by the positions of points 1 and 2, which gives in total four geometric parameters.

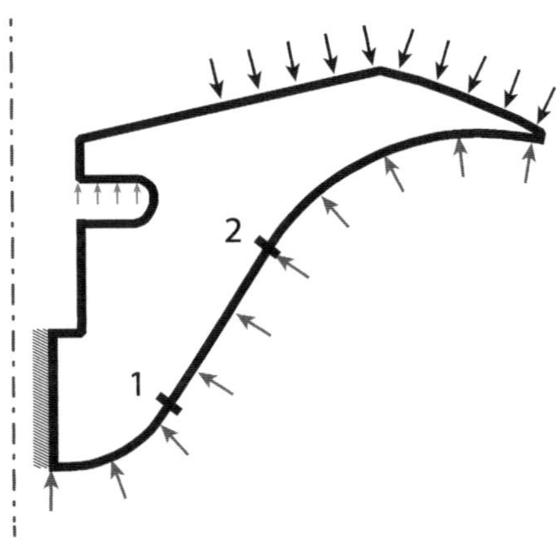

Figure 2. Industrial structure considered. The geometry is parameterised by the positions of points 1 and 2.

The structure is partitioned. Thus, a simple geometric transformation is associated to each sub-domain.

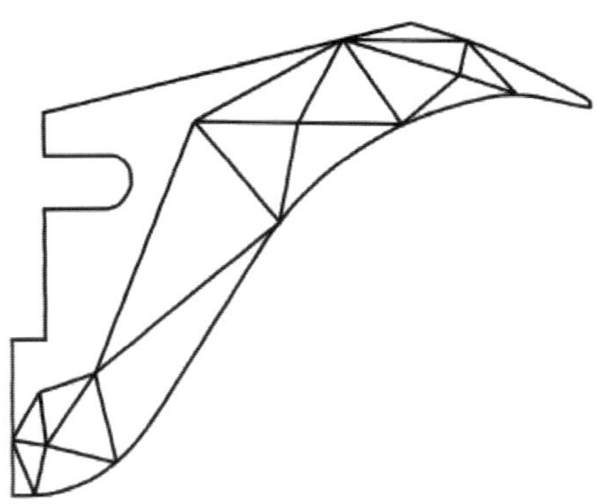

Figure 3. Partitioning of the structure.

In order to be used in a real industrial design process, our methodology combines a MATLAB code, which computes the specific PGD operators and the commercial Finite Elements software SAMCEF. SAMCEF utilities are used as much as possible. The geometry, partitioning, loads, materials, and mesh are defined in SAMCEF and SAMCEF solver accomplishes the resolution of the space functions. Ad hoc Fortran functions were furnished by LMS SAMTECH allowing the communication between SAMCEF and our MATLAB code.

RESULTS

A PGD approximation of the displacement was successfully computed with 17 modes. The space basis, i.e. the functions of the space variables was orthonormalized and the amplitude was transferred to the functions depending on α_1. In Fig. 4 are plotted the first functions associated to the geometric parameters. It can be seen that the first mode is preponderant on the others. Furthermore, parameters α_1 and α_2 have little influence on the solution (low slopes of the functions associated).

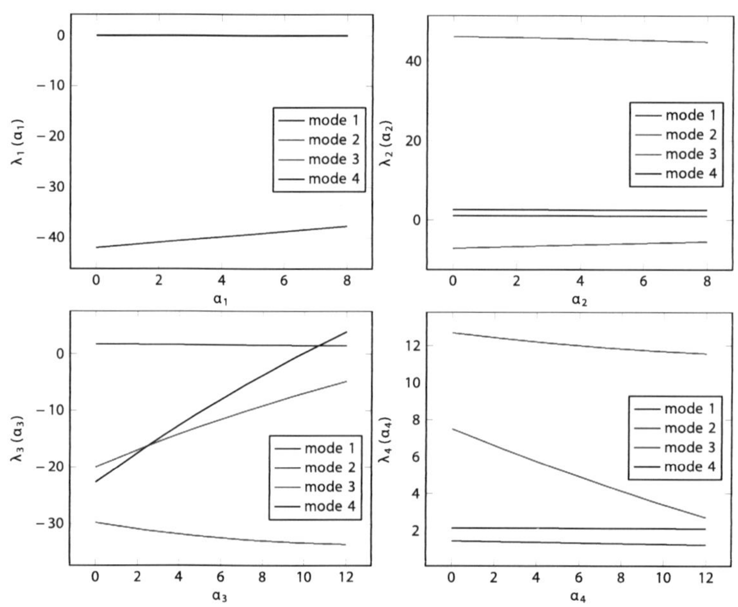

Figure 4. First functions associated to each geometric parameter. α_1 and α_2 are respectively the displacements in x and y of point 1. α_3 and α_4 are respectively the displacements in x and y of point 2.

In Fig. 5, a comparison between the PGD approximation computed and FE simulations for two different configurations was done. It can be noticed that the PGD approximation fits accurately the FE simulations for these configurations.

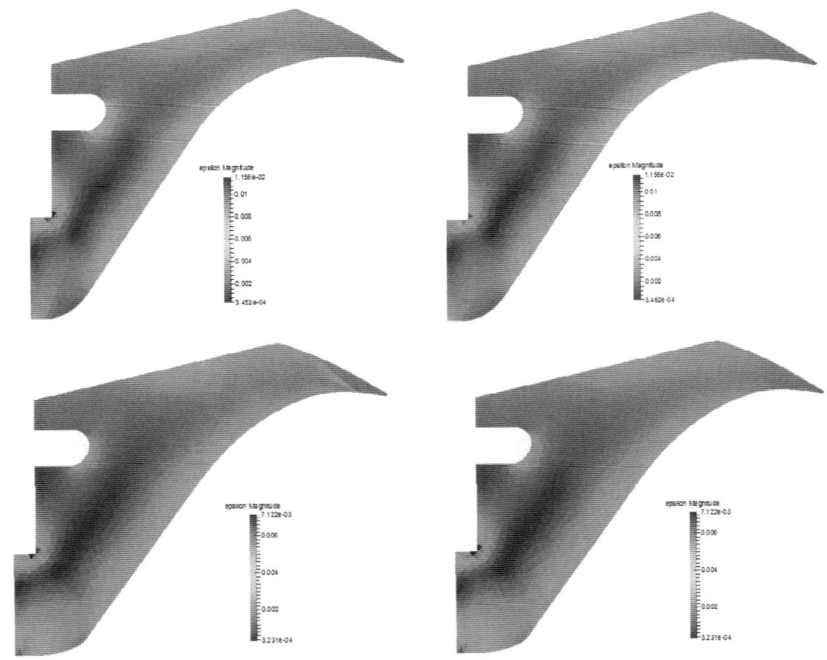

Figure 5. Comparison between PGD/FE for two configurations. On the left is plotted the strain for the PGD approximation and on the right the strain for the corresponding FE simulation for the same configuration.

It was decided to construct a Virtual Chart representing the variation of the plastic zones. Taking advantage of the plugin PXDMF of PARAVIEW developed at Ecole Centrale de Nantes allowing the visualization of separated variables fields, a threshold was defined and each element, which strain exceeds this threshold, is enlightened. Thus it gives an estimation of the size of the plastic zones. The Virtual Chart constructed can be used as a decision support tool for design. Indeed the final-user can visualize in real time the influence of the variations of the different parameters on the behaviour of the structure (here the size of the plastic zones). In Fig. 6, three configurations of the Virtual Chart are plotted highlighting the influence of the parameters on the strain.

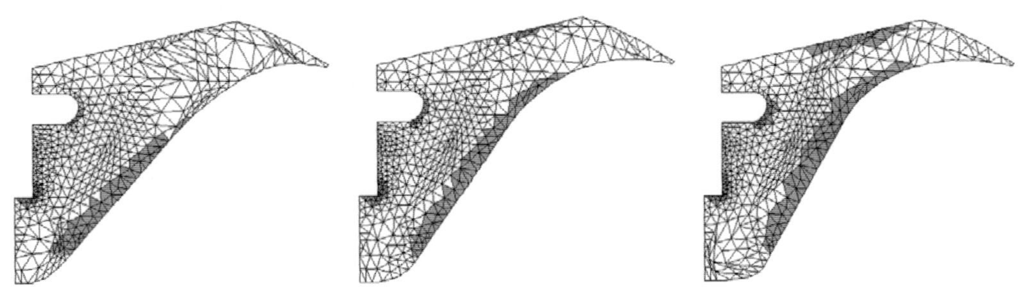

Figure 6. Example of Virtual Chart. An element is enlightened when its strain exceeds a defined threshold, giving a representation of the plastic zones.

CONCLUSIONS

Geometric parameters were introduced with success into the PGD. Furthermore, particular attention has been paid to integrating the methodology developed into a real industrial design process. A PGD-Virtual Chart was constructed given an estimation of the size of the plastic zones taking advantage of the PARAVIEW plugin developed at Ecole Centrale de Nantes for the visualization of separated variables fields. It was shown how it could be used as a decision support tool for design. The next step will be to take into account large displacements and strain in the simulation.

REFERENCES

[1] Ladevèze, P. (1989). The large time increment method for the analysis of structures with nonlinear constitutive relation described by internal variables. *Compte rendu de l'académie des Sciences*, vol. 309, pp. 1095-1099.

[2] Ladevèze, P. (1999). Nonlinear Computational Structural Mechanics – New Approaches and Non-Incremental Methods of Calculation, Springer Verlag,

[3] Ladevèze, P., Passieux, J.-C., Néron, D. (2010). The LATIN multiscale computational method and the Proper Generalized Decomposition. *Computer Methods in Applied Mechanics and Engineering*, vol. 199, pp. 1287-1296.

[4] Néron, D., Ladevèze, P. (2010). Proper Generalized Decomposition for Multiscale and Multiphysics Problems. *Archives of Computational Methods in Engineering*, vol. 17, pp. 351-372.

[5] Chinesta, F., Ammar, A., Keunings, R. (2011). An overview of the proper generalized decomposition with applications in computational rheology. *Journal of Non-Newtonian Fluid Mechanics*, vol. 166, pp. 578-592.

[6] Nouy, A. (2010). A priori model reduction through Proper Generalized Decomposition. *Computer Methods in Applied Mechanics and Engineering*, vol. 199, pp. 1603-1626.

[7] Chinesta, F., Ladevèze, P., Cueto, E. (2011). A short review on model order reduction based on Proper Generalized Decomposition. *Archives of Computational Methods in Engineering*, vol. 18, pp. 395-404.

[8] Ammar, A., Huerta, A., Chinesta, F., Cueto, E., Leygue, A. (2014). Parametric solutions involving geometry: a step towards efficient shape optimization. *Computer Methods in Applied Mechanics and Engineering*, vol. 268, pp. 178-193.

Reducing Weight by Bionic Shape Optimisation of Triangular Levers

Dort Daandels[1], Oliver Seack[1], Sarah Payros[2], Alexander Bruns[3], Mark Knuebel[4]

[1]*Airframe Research & Technology, Airbus Operations GmbH*
[2]*Fatigue and Damage Tolerance, Airbus Operations GmbH*
[3]*Performance and Improvement, Airbus Operations GmbH*
[4]*Design, Airbus Operations GmbH*

Abstract: Nowadays Airbus is gradually improving several aircraft models. Furthermore Extended Service Goals and usage of new mission mixes are projects to extend the usage of existing aircrafts. Because of these changes some components need to be reassessed for fatigue and damage tolerance. A more severe load spectrum applied on an existing structure is a challenge for redesign. The already existing space envelope restricts the possible design solutions and additional weight impacts should be minimized. For a set of triangular levers a redesign was performed. Inspired by bionic engineering some of the locations on the links were selected for redesign by optimising the contour to reduce the stress concentration. The optimisation was started with a finite element model and finalised by normal design drawings. The selection of the locations was supported by manufacturing, ensuring the design could be manufactured within the current production process.

INCREMENTAL DEVELOPMENT – DESIGNING IN AN EXISTING SPACE ENVELOPE

Nowadays Airbus is gradually improving the performance of several aircraft models. Projects like A320 Sharklet or New Engine Option (NEO) for Single Aisle and A330 are well known examples. For the Sharklet and NEO modifications the loads (lift, drag, engine trust and weight) on the wing can change compared to the previous modification standard. Therefore both the fixed wing structure and the movables (e.g. flaps and slats) need to be reassessed for the adopted load set. Depending on the outcome of the structural assessment, the structure might need to be redesigned for static or fatigue reasons.

Apart from aircraft performance improvements, there are other projects that target improvements by assessing options for "new" usage of the aircraft. That is usage which was shown not during the original type certification. In case of Extended Service Goal, the possible amount of flight cycles and / or flight hours is improved. Further investigations aim at fine-tuned usage of the aircraft for certain regions, for example city hopping with an aircraft originally designed for long range missions. For this the influence of new mission profiles or mission mixes are reviewed in the fatigue departments. The goal of these projects is to either extend the life of existing aircrafts or optimise their maintenance program with respect to their actual usage of the aircraft by the airliner. For the latter two scenarios the static load envelope does not necessary change, however the load spectra for the fatigue assessment are significantly impacted. Depending on the outcome of the fatigue assessment parts might need to be redesigned or alternatively the maintenance program might need to be adjusted.

Typical for all the previously described scenarios is the restriction of the space envelope, which needs to be considered for the redesign. The luxury of a relative large freedom during the design of a new aircraft is not applicable. The already existing space envelope can severely restrict the possible redesign solutions. Once the space envelope cannot be maintained a snow ball effect might be triggered and much of the surrounding structure and systems might need to be redesigned as well, potentially affecting their adjacent structure again and again. Furthermore additional weight impacts shall be minimized; otherwise the benefit of new engines or improved aerodynamics might diminish.

Taking the restricted space envelope into account and dealing with a more severe load spectrum applied on an existing assembly, is a challenge for redesigning only a part of the structure of the assembly. However it is also a trigger for innovation, because standard design solutions might not fulfil the requirements, whereas new methods may predict promising new designs. In the next sections of this paper an example of an optimisation, of a triangular lever, support by bionic engineering, in an existing design is presented.

TRIANGULAR LINKS WITH FATIGUE SENSITIVE STRESS CONCENTRATIONS

Identification of typical locations with stress concentration

During the redesign of twenty similar triangular levers several typical locations were identified, which showed a clear stress concentration. These typical locations were the inner corners of the triangular links, as shown in Figure 1.

Changing the material from aluminium into steel or titanium would drive up the weight and cost and is therefore less desirable. The web thickness could be increased, but this would also impact the weight of the design. The inner radii were already relative large, but still showed significant stress concentrations. Consequently further increasing the radius did not seem very promising. However once these stress concentrations were identified, the interest was raised to reduce the stress concentration instead of adding or changing material. "Designers should always try to reduce stress concentrations as much as possible in order to avoid fatigue problems" [1]. Filling up the triangle, while reducing the web thickness, would remove the stress concentrations and might have been weight balanced. However the framework design seemed to be more adequate for the loading and closing the inner section pocket might lead to undesirable pockets where water could be contained. Finally it was decided to try and remove the stress concentrations by locally optimising the structure. Because of the small web thickness a complex shape in the corner was possible without changing the manufacturing process. An optimisation process – called bionic engineering, because of the use of lessons learned from nature – was started and supported with modern 3D Finite Element (FE) solving capabilities.

Figure 1. Typical example of a triangular link, with radial corners in the inner pocket.

Structural test support

In the initial 3D FE models of the triangular levers the inner corners were shown to be sensitive locations to the fatigue spectrum. To confirm the sensitivity of these locations a development test was set-up for two triangular links. The load spectrum for one typical flight was rather long. Therefore the spectrum was reduced to speed up the testing. The reduction was performed by removing the second part of the spectrum, which after a visual check of the spectrum did not seem to apply significant load cycles onto the corner locations. A cross check confirmed the total damage was changed by less than 1% for the locations of interest. The run time of the test was effectively reduced by a factor of >5.

The test was split into a fatigue phase and a damage tolerance phase with initial flaws. The process of spark erosion was selected to create the initial flaws as to get relative sharp crack tips. But the initial flaws were not pre-cracked with constant amplitude loading. Instead in case the initial flaws would not start in the first part of the damage tolerance phase, they could be extended into through the thickness cracks to trigger the crack growth at the start of the second part of the damage tolerance phase. Digital Image Correlation was selected to visualize the stress concentration in the corners. At the start of the fatigue phase of the test one of the specimens was loaded at a constant load level and picture for the Digital Image Correlation were taken, as is shown in Figure 2. Once the fatigue phase of the test was finished and no initiations were detected, initial flaws were introduced at several locations. During the damage tolerance phase a crack initiated from the applied initial flaw in the corner, as shown in Figure 3. This finding supported the detailed FE model assessment. In the crack propagation calculations most of the crack growth life is reached in the vicinity of the stress concentration. A reduction of the stress in the inner corners will be the driver for a successful redesign of the lever.

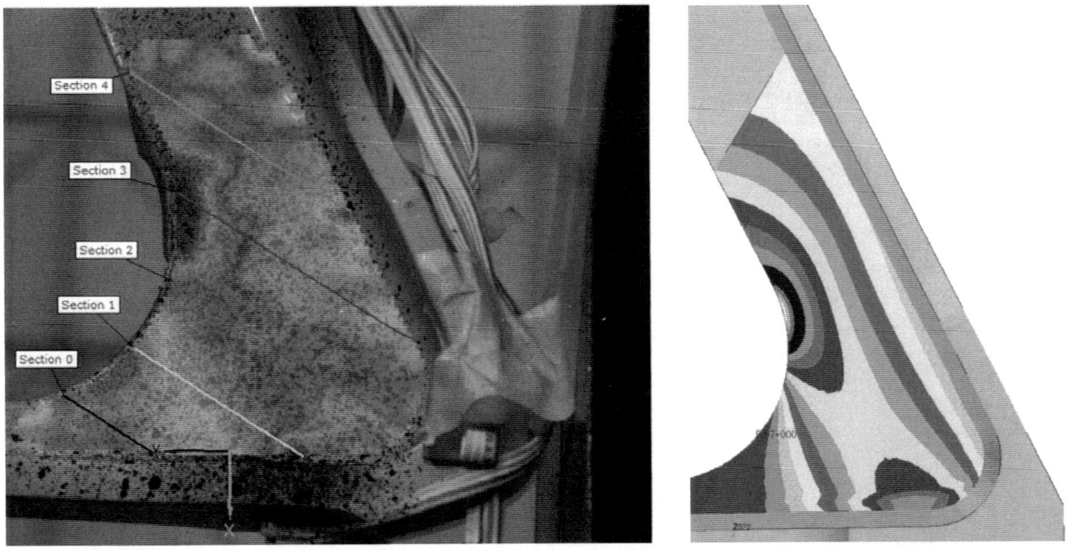

Figure 2. Stress concentration visualised by Digital Image Correlation and in a FE Model.

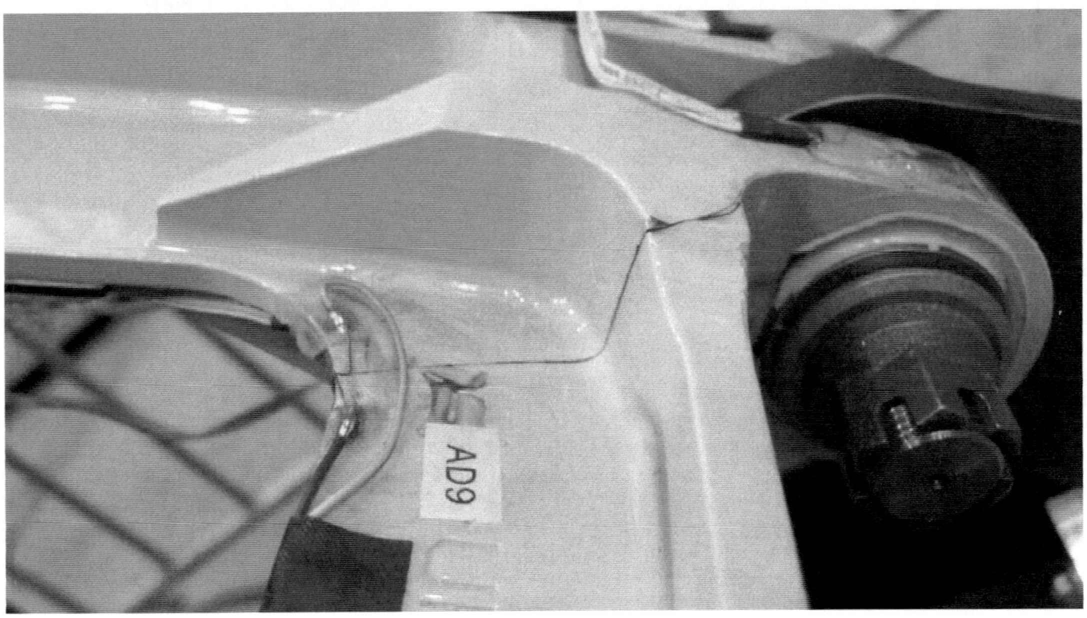

Figure 3. Crack in one of the test specimen.

BIONIC ENGINEERING APPROACH ON THE TYPICAL STRESS CONCENTRATIONS

Bionic engineering – natural design of Y-branches

The locations on the inside of the triangular levers show some similarity with a Y shape of two branches of a tree. For comparison see Figure 1 and Figure 4. References [2, 3] provide examples how trees minimize stress concentrations. Trees use the principle of load adaptive growth to form the outer contour of stems, branches and specifically the transition zone between those, to achieve a condition of constant stress at the outer contour. In areas where the stress is higher than the desired value, a tree generates locally more material; the annual growth ring becomes locally slightly thicker. In areas, where the wood is not loaded up to a critical level, the local growth is slower; the annual growth ring remains thinner. As a result the outer contour of the tree adapts an optimized contour. For our typical location trees feature an optimized use of material by gradually changing the radius in the corner to achieve a uniform stress-level on the surface. At the start of the corner the radius is relatively big. But once the curvature has started, the radius, which is used, becomes sequentially smaller and smaller. The idea behind this approach is that the stress concentration is not driven by the radius itself but rather by the change of the radius, from flat shape (infinite radius) to a finite radius. By using a bigger radius in the beginning and reducing it to the end, the stress concentration can remain very low, while the change in radius is kept small all the time. An approach to determine an optimised contour the thickness jump in a tension loaded structure is described in reference [3]. The basic principle, shown on the right side in Figure 5, is to start with a triangle with two angles of 45° in the corner. Then two more triangles are added with the same basis as the first triangle, which are fitted to the side of the structure and to the triangle already in place. With the guidance of the 4 points an optimised curve can relatively easy be predicted without the use of a computer.

Figure 4. Examples of Y-shape of branches in trees, (photos D. Daandels).

The shape obtained from this method, shown in Figure 5, looks a bit similar like an elliptical curve. According to literature, e.g. ref [1] significant benefits in stress concentrations can be obtained. Reductions of stress concentration of 30 to 40% are possible by changing a circular shoulder fillet into an elliptical one. Of course reduction in stress concentration depends on the detailed design. Because of the many possible elliptical shapes, these features are not very much presented in design hand books. However with the help of detailed FE modelling, the stress concentration can be well determined nowadays. Therefore the qualitative advantage of elliptical or even more complex shapes can be quantified.

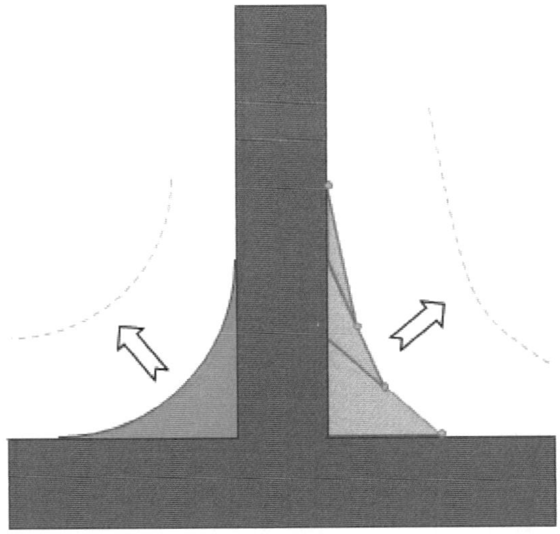

Figure 5. Left side standard radius, right side optimised curve based on principle from reference [3].

Transfer of natural Y-branches to triangular links

To put this idea of bionic engineering into practice the original design was transferred into a 3D FE model. This FE model was run in an optimizer tool, which was allowed to alter a given set of parameters in the model. In this case the optimisation was focused on reducing the stress by changing the web width and the contour of the radii, while maintaining the same web thickness. The load introduction lugs were excluded from the optimisation in order to keep the same interface points and interface design as before. This ensured that the bolt connection with the surrounding structure could be continued to be used in its existing format. Further the flanges to which the web was connected where also excluded from the optimisation. Finally the part has to remain symmetrical so the same part can be built into the left and right hand side wing.

The optimisation tool does not have a routine for fatigue and damage tolerance calculations. Therefore a static allowable stress level was determined. To obtain this stress level fatigue and crack propagation models were run with a detailed spectrum. The existing stress spectrum for the old design was scaled such that the desired fatigue and crack growth lives were exactly reached. From the new spectrum a typical load case was selected and used as a reference load case. This reference load case can now provide the optimisation tool with an allowable stress level for the structural optimisation.

With the defined free design boundaries and the reference load case the optimisation was started. After multiple iteration steps the web and corners were optimized by creating a smooth contour instead of a constant radius. The final result of the FE model was send back to the design department, where the contour was cleaned and adapted to part symmetry. A smooth and constant curvature has been created, instead of tangential transitions with constant radii. Minimum milling tool diameters have been taken into account and applied as well. Nevertheless, the post modification contour in the final drawing was still close to that of the FE model. The finalized design was transferred back into the FE tool. For two typical locations of interest the stress was reduction was reviewed. For the first location the stress was reduced by approximately 25%, while actually removing a bit of material, see Figure 6. This result is actually quite close to that predicted in ref [1] for a quarter elliptical shapes. For the second location a stress reduction factor ≥ 2 was reached, however the web height was also increased therefore it is assumed that the stress reduction is caused by a combination of web height and contour optimisation.

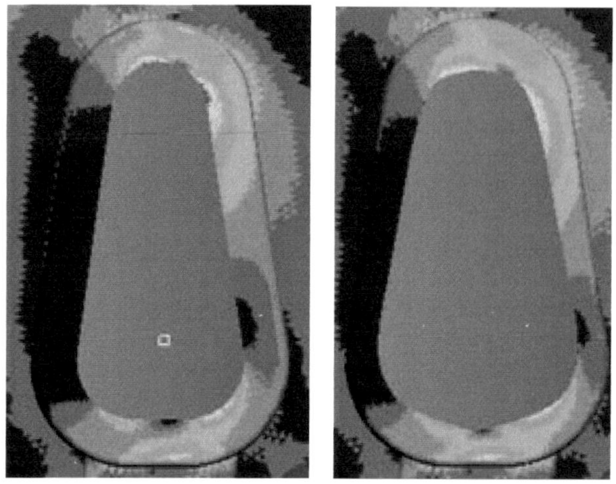

Figure 6. Left design with standard radii, right side optimised contour with FE model.

CONCLUSION

By learning from nature a smart solution for a local redesign of a fatigue sensitive design detail was identified. A combination of iterative 3D FE optimisation supported by analytical optimisation resulted in a new design reflecting the reduced stress concentration by gradually changing the radius. This resulted in a weight optimized redesign without going outside the design space envelope. The stresses for the final structural assessment were obtained from a detailed FE model, which allows taking full benefit from the reduced stress concentration of the non-standard geometry, as opposed to using approximations with standard shapes from handbook solutions. The optimized shape of the part is not only theoretically feasible. By pre-selecting promising locations together with manufacturing, the redesign was implemented successfully into the existing production process.

The selected optimisation method to reduce the stress concentration was applied successful for this case. Following this case it is now also applied in a few in other design areas. Therefore it has been established as a feasible method within the regular design process and it is intended to apply this method in future re-design activities.

REFERENCES

[1] Schijve, J. (2009). *Fatigue of structures and materials, 2nd Edition*, Springer Science+Business Media B.V.

[2] Mattheck, C, (1992). *Design in der Natur – Der Baum als Lehrmeister, 1st ed.*, Rombach Verlag, Freiburg.

[3] Mattheck, C. (2010). *Denkwerkzeuge nach der Natur, 1st ed.*, KS Druck GmbH, Kronau.

An operational load related method for fatigue tests on aircraft components

Martin Semsch, Andreas Schelenz, Alexander Hamann, Silvio Nebel

IMA Materialforschung und Anwendungstechnik GmbH Dresden
Wilhelmine-Reichard-Ring 4, 01109 Dresden, Germany

Abstract: A fatigue life assessment is necessary for the certification of components for aerospace application. Therefore, a quite large number of test campaigns are used to demonstrate the fatigue behaviour of the components or to validate the calculation methods.

IMA Dresden has applied a special method for fatigue tests on aircraft components based on real-time loading information. The real operational loads have to be estimated before testing and will be used to derivate the test loads and to analyse the test results. For these tasks the loading conditions will be analysed with respect to the load level, the frequency range and the influence on the fatigue life of the components.

The controller and special software are able to simulate these complex loading conditions and to monitor the test run. Therefore, an iteration process is necessary to find the best compliance with the loading requirements.

The test method is an efficient tool for the fatigue life assessment of aircraft components. In a first step it was applied on a complex test setup for a high-lift structure. The test set-up included 10 controlled channels to simulate different vibration modes. Hereafter, it was used and evaluated for bearing application. In future, an application with aerospace components on a multi-axial shaker table will be examined.

INTRODUCTION

A fatigue life assessment is necessary for the development and certification of components for aerospace application. Therefore, a quite large number of test campaigns are used to demonstrate the fatigue behaviour of the components or to validate the calculation methods. To reduce the effort for these tasks and time for the development phase, robust methods for calculation and testing are necessary. Thereby, a combination of computed based simulation and experimental investigation provides an optimum for the achievable results.

On the other side, the requirements concerning realistic experimental simulation are increasing and are in line with improved technical possibilities. Especially the industrial progress of equipment and software solutions for control und measurement systems allows the application of new test methods and opens new path in evaluation and analysis of the test results.

To fulfil the requirements of the customers and industrial partners, IMA Dresden has applied an operational load related test method. Concepts for testing and controller solutions, which are used for the automotive industry, were adapted and applied for special needs. The method can be used for component testing with high frequency vibrations, for structural testing with multi-axial loading conditions including non-linear effects an for vibration testing on a multi-axial shaker table.

METHOD

Basis of the test method are real-time data, which represents the demand for test performance and are used for the evaluation and analysis of the test results. The principal of the test method consists of 4 subtasks:

- Load measurement
- Test load determination
- Component testing
- Test result analysis

Figure 1 shows an overview about the subtasks of the test method and interactions between the different tasks.

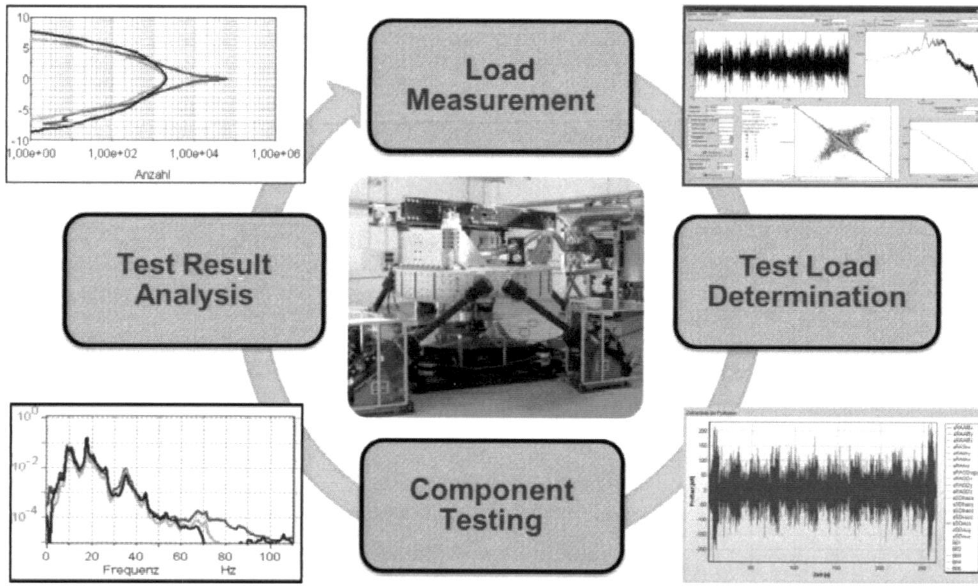

Figure 1. Method for operational load related fatigue and vibration test.

LOAD MEASUREMENT

The basic input for the experimental simulation is the availability of representative real-time loading data. The real-time loading information have to be estimated before testing and can be derived from wind tunnel investigations, full scale test activities or flight tests. The most representative option is to use real-time measurements of in-flight loads [1]. Alternatively, it is possible to derivate time-history signals from dynamic or transient FE-calculation. Depending on the complexity of the test object and the test set-up, different physical data can be used. The most common data are forces, accelerations, displacements and strains.

To determine these measurement data, extensive data acquisition campaigns have to be carried out. It is useful to acquire time stamped recordings of all relevant flight parameters that should be taken into account for the later tests. To get sufficient signal resolutions, the measurement rate should be in the kHz-range. The applied measurement rate depends on the specific application and the expected response signal. To achieve an appropriate signal resolution for testing, a factor 10 for the measurement rate in relation to the maximum of the expected response signal frequency should be used.

However, the measured data are use to determine the test loads and represent the basis for the evaluation and analysis of the test results.

TEST LOAD DETERMINATION

Based on the measured data, the test loads have to be determined. Therefore, the measured data have to be filtered and reduced. A filter technique has to be used to remove non-relevant frequency ranges for the test run. Furthermore, all time intervals of the load-time-history with low amplitudes and no relevant influence of the life assessment should be cut out. This approach has to be done with respect to the loading direction, the maximum values of the signals, the frequency range of the signals and the influence on the fatigue live of the components.

IMA Dresden has developed a special tool for this approach. The algorithm is used to determine test loads on the basis of measured values including a fatigue assessment and a comparison of both signals. As an example figure 2 shows a measured load-time-history and the corresponding test load signal. With this method it is possible to reduce the test duration by a significant factor. Different parameter (safety factor, load enhancement factor) and the choice of the fatigue life assessment method help to find optimum results for the test load determination.

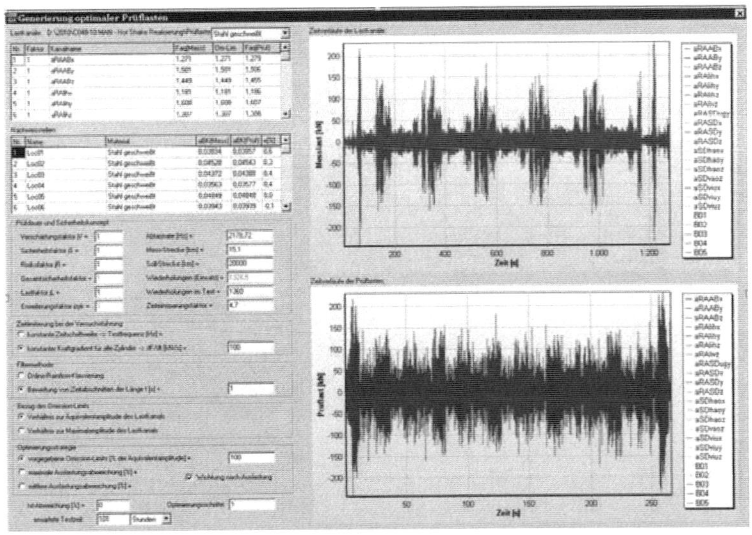

Figure 2. Test load determination by reducing measured values.

COMPONENT TESTING

The operational load related test method can be used for a large range of different applications. Typical applications are

- Component tests with high frequencies
- Component test with vibration modes
- Structural tests with non-linear effects
- Vibration tests on a multi-axial shaker table

For that matter, the test set-up can have a different dimension and complexity to meet the special requirements. Test scenarios with uni-axial or multi-axial loading conditions and a forced controlled test run are possible. For vibrational testing a test run with controlled acceleration is a preferred procedure. IMA Dresden has used and evaluated the method on a

- Ten-channel component test on a high lift structure
- Two-channel bearing test
- Vibration test on a multi-axial shaker table

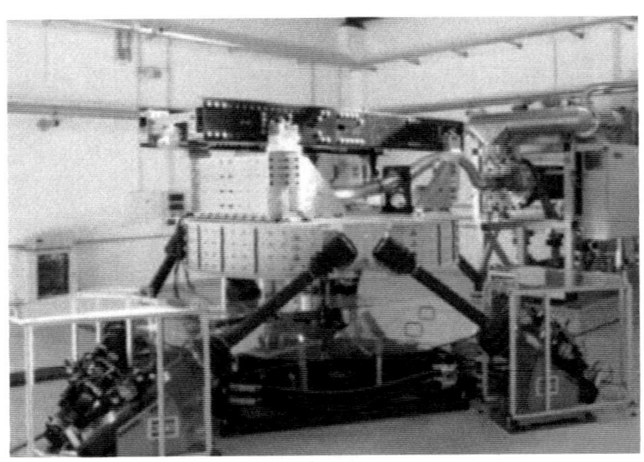

Figure 3. Multi-axial shaker.

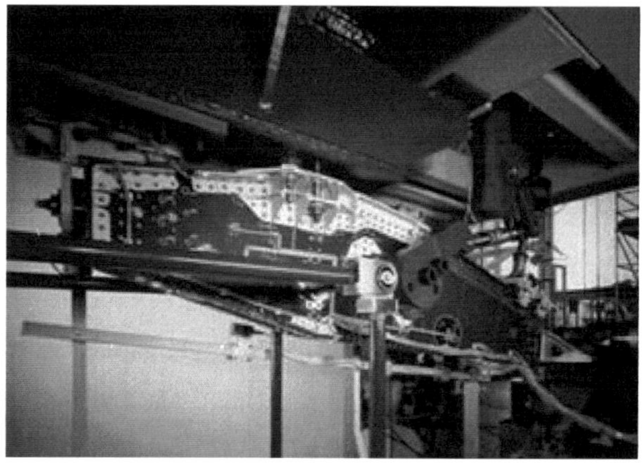

Figure 4. Component test including a high frequency range.

The challenge for the design of the test configuration is linked to the quality of the mechanical test set-up (bearings, interface and fitting parts) and the controller. The mechanical test set-up has to be designed with respect to the natural frequencies of the system. Therefore, all fittings and interface parts must have an appropriate stiffness and the lowest natural frequency of the parts should be higher then the maximum frequency of the test loads. Furthermore, it is important that all joints and bearing should have no interference.

For the test run a special controller software has to be used. The controller is using software concepts that are commonly applied for the automotive industry. It includes the identification of the dynamic transfer function of the complete test set-up [1]. The system response has to be optimized in a multistage iterative process. The aim of test optimisation is to receive a good correlation between the determined test loads (command signal) and the simulated loads (response signal).

TEST RESULT ANALYSIS

All test results will be analysed during and after testing. An approved method is a monitoring of the frequency content and the load spectrum. A change of the frequency content is an indication of a failure or changed loading conditions. The monitoring of the loading spectrum is useful for analysing the different signals and to analyse the fatigue behaviour of the specimen after testing. As an example, figure 5 shows a load spectrum chart and PSD chart. The different graphs represent the measured values (red), the determined lest load (light blue) and the simulated load (dark blue) [2, 3]. A good correlation between the different signals can be identified in the displayed charts.

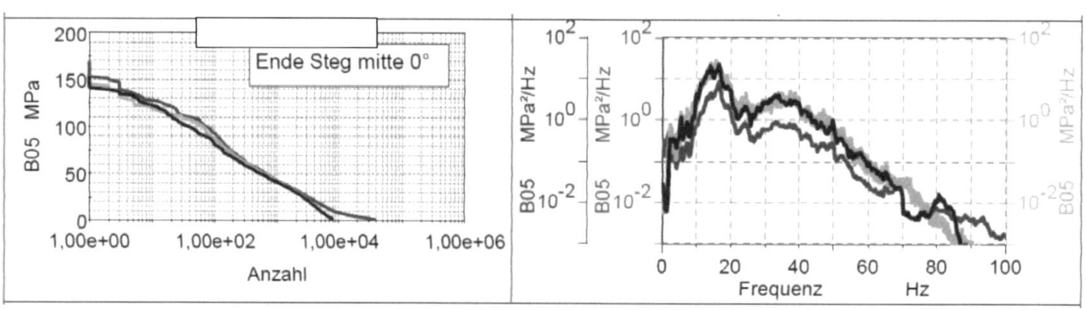

Figure 5. Load spectrum and PSD chart.

CONCLUSION

An operational load related method for aerospace components was applied by IMA Dresden. This method can be used for component testing with high frequency vibrations, for structural testing with multi-axial loading conditions including non-linear effects and for vibrational testing on a multi-axial shaker table. It is an efficient tool for the fatigue life assessment on aerospace components, which are influenced by induced vibration modes or multi-axial loading conditions.

REFERENCES

[1] Nebel, S., Semsch M. and Fleischer, Th. (2013). *Advanced Concepts for Fatigue Testing of Aircraft Structures using Real-time Load Simulation* Proceedings of the 27th ICAF Symposium 2013, Jerusalem, Israel

[2] Schönfeld, F., Streicher, M., Grosskopf, R., Semsch, M., Schelenz, A., and Hamann, A. (2013). *Hot-Shakerprüfung für den betriebsbelastungsgerechten Lebensdauernachweis an EURO VI Abgasanlagen für Nutzfahrzeuganwendungen* 40th Conference of the DVM Working Group Structural Durability, Paderborn, Germany

[3] Schönfeld, F., Streicher, M., Grosskopf, R., Semsch, M., Schelenz, A. and Hamann, A. (2014). *Hot-Shakerprüfung für den betriebsbelastungsgerechten Lebensdauernachweis an EURO VI Abgasanlagen für Nutzfahrzeuganwendungen* Materials Testing, Edition 7-8/2014, 567-574

Development of Analysis and Test Methods for Lightning Current Flow on Composite Fuel Tank Structure

Yasunori Sato, Megumi Hiraki, Hiroyuki Tsubata, Takayuki Nishi

Fuji Heavy Industries LTD.

Abstract: Composite materials are becoming as primary structures of aircrafts. However, since the electric current concentration of composite material is completely different from that of conventional metallic structures, a numerical simulation is getting more important to reduce the practical test for development cost reduction. We have developed the accurate and rapid lighting current simulation by a zooming method. In addition, the fuel tank model include wing to fuselage connection, were analyzed and verified by the practical test. On the other hand, novel optical measurement system was developed as a quantitative sensor for spark detection. Applying these analytical and test methods make the verification of structural integrity for composite structure more efficient, which contributes for cost reduction, productivity improvement and saving weight of aircraft in a future.

CCURRENT DENSITY ANALYSIS

We developed a high-speed calculation method to call a "zooming method". The zooming method is precisely calculates by partial structure with fine mesh model and the boundary condition which are the interpolation of the results from a coarse model. Appropriate electric conductivities, those are measured by a lot of experiments, is the key to accurate simulation.We made the database of conductivities of each part of real structures, and we succeeded to verify the most complicated current density simulation of a fuel tank that includes wing to fuselage connection by using the data.

Figure 1. Current density analysis on the aircraft and the fuel tank.

SPARK DETECTION TECHNOLOGY

We developed a novel optical measurement system with a photo-multiplier tube (PMT) which detect and estimate spark threshold as standard efficient and accurate testing method. The system allows us to simultaneously measure the spark light emission still image, the spark light emission intensity waveform from the fastener, and the emitted electromagnetic (EM) wave during the test together with the applied lightning strike test current waveform. To estimate the system and understand the relationship between the spark energy and the corresponding spark light emission images, universal and accurate standard method could be completed.

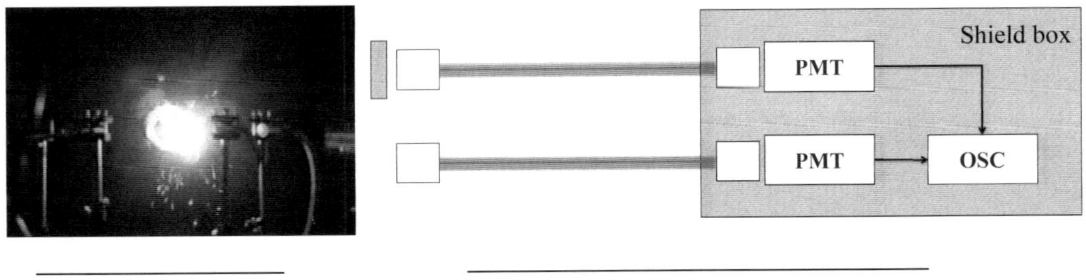

Figure 2. Overview of Spark Detection System

Non-stationary models for predicting strain on aircraft landing gear from flight data measurements

Geoffrey Holmes[1], Valerijan Cokonaj[2], Paul Southern[2], Keith Worden[1], Elizabeth Cross[1]

[1]*The University of Sheffield, United Kingdom*
[2]*Messier-Bugatti-Dowty, United Kingdom*

Abstract: This paper explores the use of two classes of non-stationary frameworks for modelling different physical regimes within datasets collected from production aircraft. The aim is to predict hard-to-measure strains on specific components of the aircraft landing gear from in-flight measurements collected at other locations on the aircraft, including the standard measurements taken by the flight data recorder. Treed Gaussian Processes and Bayesian Mixtures of Experts represent two ways in which the input space of a modelled system can be partitioned. In conjunction with this partitioning, appropriate models are estimated and allocated to the distinct regions. This allows the use of relatively straightforward, stationary, component models but with the benefit of global non-stationarity in the resulting composite model. These two approaches are compared with a single stationary Gaussian Process model. The modelling fidelity, quantification of uncertainty and computational efficiency of the three methods are evaluated. It is found that the two input partitioning methods reduce both the modelling error and associated prediction uncertainty compared to the standard Gaussian Process. Computational efficiency is more variable. The Mixture of Experts, with linear component models has a closed form iterative solution and completes its training in a reasonable time compared to the standard Gaussian Process. The Treed Gaussian Process, however, requires Markov Chain Monte Carlo sampling and thus needs significantly more computation time. Both partitioning methods find divisions of the input space that are reasonably consistent with each another and which suggest changes in the physics of the system that can be explored further.

INTRODUCTION

The landing gear of an aircraft is subject to significant variations of loading regime, where such loads are costly to directly measure in real time. Access to loading history would be beneficial in a number of areas, opening up, for instance, the possibility of a switch to condition based maintenance and automatic hard landing detection systems. In many different contexts, machine learning techniques have achieved considerable success in using existing available measurements to accurately predict hidden or costly-to-measure quantities. Due to physical properties of the system in question, such as shock absorber dynamics, the system is highly non-linear. Previously, stationary Gaussian Process models have been applied to model these non-linear relationships [1] and achieved a good degree of success in predicting strain at different locations on the gear, especially during the latter stages of landing events. However, whilst a Gaussian Process is known to be a powerful and flexible regression tool for non-linear systems, it was found that the stationary Gaussian Process lacked the full flexibility to model the different physical regimes which are assumed to be present within these aircraft landing datasets. In particular, if there is an initial bounce event, when the landing gear wheels leave the runway briefly subsequent to initial contact, the predictions of the stationary Gaussian Process models tend to have a large structural error, when compared to the observed strain outputs, during this early phase.

In this paper, the Treed Gaussian Process and a Bayesian Mixtures of Linear Experts are applied to the aircraft landing data modelling problem and compared to each other, with reference also to the use of a single stationary Gaussian Process. This paper will give a brief introduction to the modelling frameworks used. These methods will then be applied to predict a load on the main landing gear from other available flight data measurements. This is done using flight test data and the results are presented with conclusions drawn for the relative modelling fidelity, computational burden and future avenues to explore. It will be concluded that these approaches provide promising ways forward for modelling aspects of the data that have been problematic in the past. As such they will be useful for accessing important factors for structural health of aircraft which are not otherwise easily measured.

MODELLING APPROACHES

A full introduction to the three sophisticated modelling techniques employed is beyond the scope of this paper but interested readers are referred to [2] for Gaussian Processes (GPs), [3, 4] for Treed Gaussian Processes (TGPs) and to [5, 6] for Mixtures of Experts (MoEs).

Gaussian processes for regression

Gaussian Processes (GPs) are widely used for regression and classification tasks. They have previously been developed for the context of aircraft landing gear for acoustic source location [7]. A GP can be thought of as a distribution over functions. The nature of the functions within this distribution is determined by the specification of a mean function and a covariance structure. Loosely speaking these influence respectively the trend of the functions and qualities of how they changes over the input space such as the smoothness and variability.

GP regression is described as non-parametric because the functions drawn from a GP distribution are not limited to those of a particular parametric form (e.g. polynomial). However, the mean function and covariance structure are themselves controlled by a set of hyperparameters. For instance, a noise scale hyperparameter describes how much measured outputs differ from their true values and lengthscale hyperparameters describe the extent to which these true values are spatially correlated. Determining the hyperparameters which are most appropriate for a given dataset is known as training the GP. Here a zero mean function is specified along with a squared-exponential covariance function. The hyperparameters of this covariance function are determined during model training by minimising the negative log marginal likelihood of the training data as described in [2]. In other words, the chosen hyperparameters are those which make the realisation of the training data more likely than they would be under any other hyperparameter set. The chosen hyperparameters define a particular distribution over functions. Every function from this distribution is then also assigned a probability weighting determined by how much it agrees with the training data. This is achieved via the multivariate Gaussian property of a GP [2]. The result is a GP which is conditioned on the training data and which for any new choice of input data can make a prediction of what the output quantity should be.

The GP approach to regression has a number of advantages: they are analytically simple to define and compact in the equations for their implementation; and despite this compactness they are very flexible in the functional forms they can produce; the GP automatically provides a distributional form for its predictions rather than simply a point value. This enables the uncertainty of a prediction to be quantified and incorporated into any decision subsequently taken.

However, there are two particular disadvantages of GPs that can make their use problematic. The first is that for large training datasets GP training and prediction is computationally very intensive. This is due to the need to invert a $N \times N$ covariance matrix; the computation time needed for this is $O(N^3)$ (though a number of efforts have been made to mitigate this, e.g. [8]). Secondly, a standard GP has a stationary structure. This means that it behaves in a similar fashion at every point in its input space and cannot therefore cope very well with switching of physical regimes within a system.

Treed Gaussian processes

The treed Gaussian process (TGP) model provides a conceptually simple way to address the stationarity issue of standard GPs. It is an example of the more general Classification and Regression Tree (CART) approach [9] which divides an input space into distinct regions. The most simple tree consists of a single leaf, which represents the whole of the input space. A tree can be extended by considering any particular leaf and proposing to divide it into two leaves about a particular value on one of the input dimensions. This is a GROW operation. There are corresponding PRUNE operations along with others of a more complex nature [9, 3] which vary the structure of the tree in order to search over all possible divisions of the input space. For a given tree, the leaves represent separate sub-domains of the input space and each sub-domain is allocated its own model which is trained to describe and predict the data falling within that region. A simple example of a tree partitioning a two dimensional input space is shown in Figure 1.

The sub-models can take any form that the modeller wishes to propose. Original applications of the approach used linear models. More recently the method was extended to allow use of the more powerful and flexible GP models at the leaves [3]. This extension also more firmly embedded the method in a Bayesian framework which amongst other things allows the incorporation of prior beliefs about the model and its parameters in the estimation process. This can be illustrated in particular with reference to the tree structure. An unregularised CART estimation procedure would risk dividing the input domain into a vast number of leaves due to the fact that a more complex model can always give a better fit to data than a simple one. It does this, however, by 'overfitting' the noise in the data, yielding a model that does not generalise well. Specifying a Bayesian prior on tree complexity creates a trade-off between modelling precision and tree complexity which yields a more sensible and generalisable model.

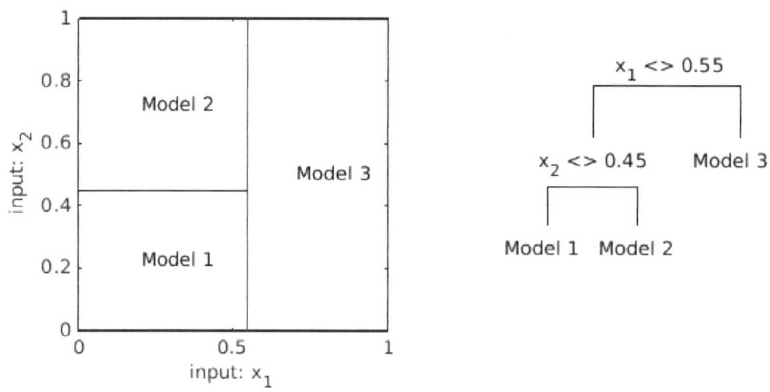

Figure 1. An example of how a TGP tree structure might partition a two dimensional input space. On the left is the partitioned input space and on the right is the corresponding tree structure. The divisions are necessarily parallel to a coordinate axis. The data in each partition can be used to train a separate sub-model.

It can easily be seen that this setup allows for non-uniformity in the overall composite model. The different sub-models can capture quite different behaviours relating to different physical regimes in a system. It thus tackles one of the main disadvantages of the GP. By splitting, it also reduces the amount of data used in the training of each individual GP model which in itself should reduce the computation time. For example two GPs trained with $N/2$ training datapoints, even without any computational parallelisation, will take a quarter of the time needed to train a single GP with N training data points. This is due to training time being $O(N^3)$. However, the incorporation of the tree structure within a full Bayesian framework produces analytically intractable integrals that can only be numerically approximated using reversible jump Markov Chain Monte Carlo methods. The result is a process with computational overheads that are hard to parallelise with the result that computation time actually increases as will be seen in the comparative results.

A further possible disadvantage of TGPs is that the splitting process produces sub-regions that are hypercuboidal, i.e. all their boundaries lie parallel to a coordinate axis. For instance division along a diagonal, if required, can only be approximated by a rough tessellation of the space resulting in far more leaves than actually required and a conflict with the tree prior calling for simplicity. Despite these limitations we will see that the TGP provides a powerful tool for the flight test data problem addressed here.

Mixtures of Experts

To provide a further comparison against the performance of the GP and TGP, models for the flight test data are also demonstrated using the Bayesian Mixtures of Experts (MoE) framework. The approach of [6] is followed where the mixture of experts is set within a variational Bayesian paradigm. The MoE model uses a probabilistic division to divide the input space in a more flexible way than is allowed by the CART approach (see Figure 2). The division is achieved via a gating function which assigns a probabilistic weighting to each expert at each point in the input domain. The sub-region designated to each expert consists of all the points where its weighting is greater than that of any of the other experts. This allows for much more generality of division than is possible with the axis-parallel division of the TGP framework. As in the TGP approach the division of input space is a purely data-driven process and indeed, if the initially specified number of experts to search for is too large, the algorithm will prune it back to only that number required to give an adequate model of the data, subject to the Bayesian priors specified.

The expert models within the mixture can again take any form specified by the user. The current study followed [6] in using linear experts for this study as this results in analytical equation which can be quickly iterated through variational expectation and maximisation steps to reach a solution. For a more complete comparative study it will be necessary to also implement a mixture of GP experts. That is beyond the scope of this investigation and is noted for future research.

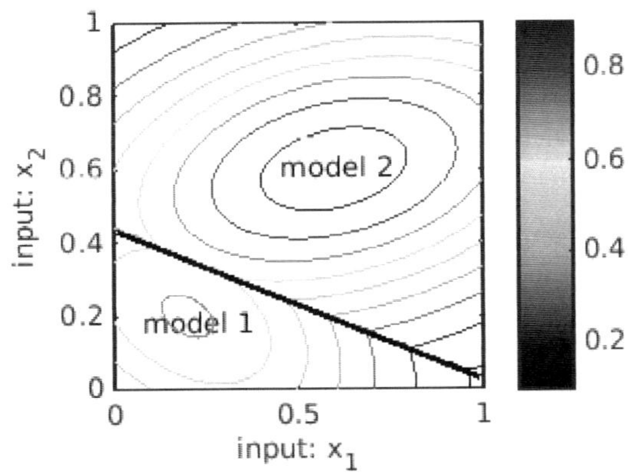

Figure 2. Example of how a MoE model might divide a two dimensional input space. The contours in conjunction with the colourbar represent the probabilities of the experts at each point in the domain. A model applies anywhere where its probability exceeds that of all other models. In contrast to the TGP, divisions need not be parallel to a coordinate axis. Here the division is shown by the black diagonal.

RESULTS

Presented here are the results of two tests carried out on flight test data.

- In Test I the three different modelling frameworks described earlier were applied to flight test data in order to evaluate their relative performance in terms of prediction fidelity, quantification of uncertainty and computational expense.
- In Test II the mixture of linear experts model was applied intensively to flight test data to investigate how the different expert models were allocated by the estimation algorithm to the differing phases of the landing data.

The datasets include a number of candidate measurements which could be used as inputs for the models. Previous work [1] has identified certain of these inputs as useful for modelling of the chosen ouput. Some of these inputs are standard ones logged by the flight data recorder, others come from additional instrumentation on the gear itself. In both tests the output being predicted is a vertical load on the main landing gear.

Test I – Comparison of the three model types

In this test the GP, TGP and MoE models were all trained using the same training data which was subsampled from a quarter of the available landing datasets to give a total of 990 training points. The trained model was tested on approximately 700 datapoints from each available flight, with the training data removed from the candidate testing set for the corresponding flights. Each of the datapoints comprises nine input measurements. The GP models made use of code in MATLAB [10] as did the MoE model [6]. The TGP model made use of R code [9]. The PC used had an Intel® i7-CPU @ 3.40GHz with 16GB RAM. In MATLAB parfor loops, which allocate repeated tasks to run in parallel on available CPU cores, were implemented where possible and in R OpenBLAS was installed to utilise all the processor cores.

Table I shows the training time for each of the three models. It is clear that the two splitting models required significantly greater computing resources, even for the modestly sized training set. The MoE having an iterative analytic solution algorithm is nevertheless much faster than the TGP which needs reversible jump MCMC to sample from the posterior predictive distribution. The mean squared error (MSE) was used to quantify predictive performance on the test data. It was used in its normalised form,

$$\mathrm{MSE} = \frac{100}{n\sigma^2}\ \sum(\mathrm{prediction} - \mathrm{observation})^2$$

where n is the number of test data points and σ is the standard deviation of the test observations. The third column of Table I shows the average MSE where the average was taken across all test flights. The prediction error for the MoE was slightly worse than for the GP. This was probably because the flexibility of having multiple sub-models was undermined by the fact that the sub-models were linear. Hence the composite model did not match the overall flexibility of the GP. On the other hand, the TGP provided a significant reduction in the prediction error. More detail on the distribution of prediction errors over the testing flights is shown in Figure 3. Overall, the errors for the GP and MoE models were very comparable with no significant difference in their distribution. The TGP predictions had a consistently lower MSE.

Table I. Model training times and average prediction errors for Test I.

Model framework	Training time (mm:ss)	Mean test prediction error
Gaussian process	00:08	5.84
Treed Gaussian process	16:60	2.34
Mixture of linear experts	01:07	6.25

Further insight into the comparative results can be seen in Figures 5 & 6 which respectively show portions of the data towards the start and towards the end of a representative flight dataset. The TGP model divided the data into two sub-regions each with its own GP sub-model. It divided the data according to the tree shown in Figure 4. The data in Figure 5 is representative of a time where the landing gear has briefly touched the runway before the main landing event. This initial contact is referred to here as the 'bump'. All of the data modelled in Figure 5 fell within the left leaf of the tree and thus was described by a single model. This model was able to focus on the initial part of the landing and as a result it did a much better job of predicting the bump than either of the other models. For this reason it was also able to be much more confident as shown by the narrower shaded region within which the true observation is expected to lie 95% of the time.

The MoE utilised two sub-models within this portion of the data. This is shown in Figure 5C, where the green and purple sections of the horizontal bar represent the portions of the data where each sub-model applies. One of them appears to be a very good model of the non-contact dynamics (the flat portion immediately before and after the bump) with an appropriately tight confidence region. The second model, however, is quite erratic in its prediction of the bump.

During the later portion of the data, shown in Figure 6, all three models made good predictions. The MoE model is dominated by a third expert sub-model in this portion which provided a very good prediction with an appropriately tight confidence region showing that a single linear model is a very acceptable approximation for this part of the data. The TGP model switched into the right hand leaf of its tree and a GP sub-model which has the most accurate predictive capacity. The confidence region is very tight and yet appropriate, as the observed value is always found within it. Interestingly, although the TGP sub-models have the capacity to switch to a limiting linear model in certain dimensions when appropriate [6], it did not do so in this case. So whilst the linear model of the MoE was acceptable the fully non-linear model was still worth the extra computational expense for the extra fidelity it provides.

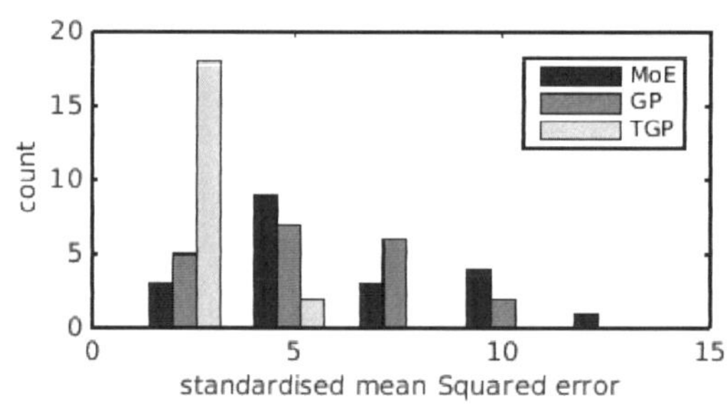

Figure 3. Distribution of standardised mean squared errors of prediction over the flight datasets used for model testing. The actual predictions for an example landing are shown in Figures 5 & 6.

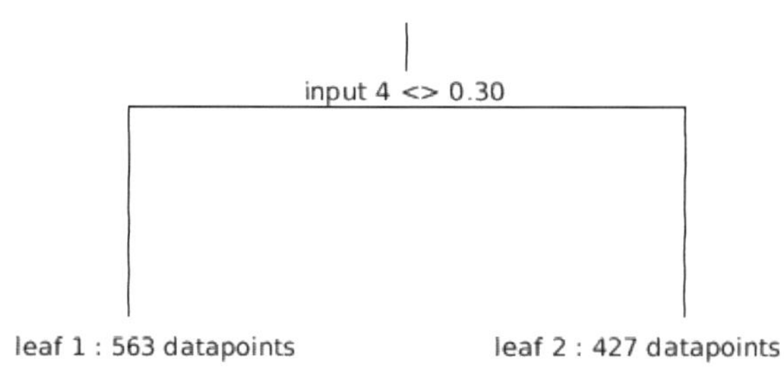

Figure 4. The model tree structure identified by the treed Gaussian process algorithm when applied to the flight test data. The tree only has one split which occurrs at the normalised shock absorber closure value of 0.30.

Test II – Investigating different landing phases

This test investigated how sub-models were allocated when as much training data as possible was given to the model training process. In this case the ability of the model to generalise to unseen data wasn't under scrutiny so no data was held back for validation purposes and all available data was used for training. 31,500 training datapoints were used in all, each comprising ten input measurements. For reasons of computational load already discussed for Gaussian Processes, neither the GP nor TGP model in their current implementations was able to train on this amount of data with the computing facilities available to this research, so the MoE framework was applied to this test. Having linear sub-models and an iterative analytic algorithm, this was able to complete model training in under an hour.

Figure 7 shows the prediction and uncertainty at the training points and the allocation of the expert sub-models for four representative flights. Figure 9 shows the model parameters for the various models. One can see that with the information contained in the comprehensive training dataset the models were able to tailor themselves well to different phases of the datasets. There is now a very low variance model (corresponding to the light blue model) that describes the non-contact periods and two models (dark blue and orange) that are used in different landings to describe the initial contact period. Of particular interest, however, are the model allocations in the later part of the data as shown in Figure 8. There are two dominant models (yellow and green) when the aircraft is settling into its taxiing phase. A major difference between them is the noise scaling which is approximately six times greater in the yellow than the green model. One or other of these two models ends up dominating the taxiing phase of each landing apart from in one case (Figure 8C) where a clear transition is made from the less noisy to the more noisy model. A possible hypothesis based on these results is that the sub-models adapted themselves to different levels of roughness in the runway surfaces. The datasets are collected at several different runways with some runways being used more than once and the correspondences are consistent with this hypothesis.

CONCLUSIONS

This paper has explored the use of two classes of non-stationary modelling frameworks to model different physical regimes within aircraft landing datasets. The Mixture of Experts (MoE) models produced some interesting results which suggested different expert models were being employed for different physical regimes of the modelled system. This observation was especially true during the latter stages of landing with the suggestion that there were two different models corresponding to different levels of runway roughness. During the earlier, more volatile, stage of landing, the MoE had shortcomings in its predictive accuracy and was not able to improve in this regard on the benchmark GP model. This is likely to be because the linear component models are not flexible enough to model what are expected to be highly non-linear processes. The MoE was rather more computationally expensive than the GP model for small datasets. However, for larger datasets the increased computational overhead is far less than for the GP model.

The Treed Gaussian Process Model produced very nice predictive results with a division of the input space into two parts that corresponded approximately to the initial and later phases of landing. The predictions were a significant improvement on those of the benchmark GP model. The TGP framework, in its current form, was nevertheless very computationally expensive which would make it hard to scale up to the larger training sets that were possible with MoEs.

These results suggest that further investigation and development of these modelling frameworks will be beneficial to the goal of predicting hard-to-measure loads on aircraft landing gear. Faster implementations of the TGP framework can be sought, perhaps incorporating computation saving approximations and the MoE framework can be extended to included more flexible component models such as the GP.

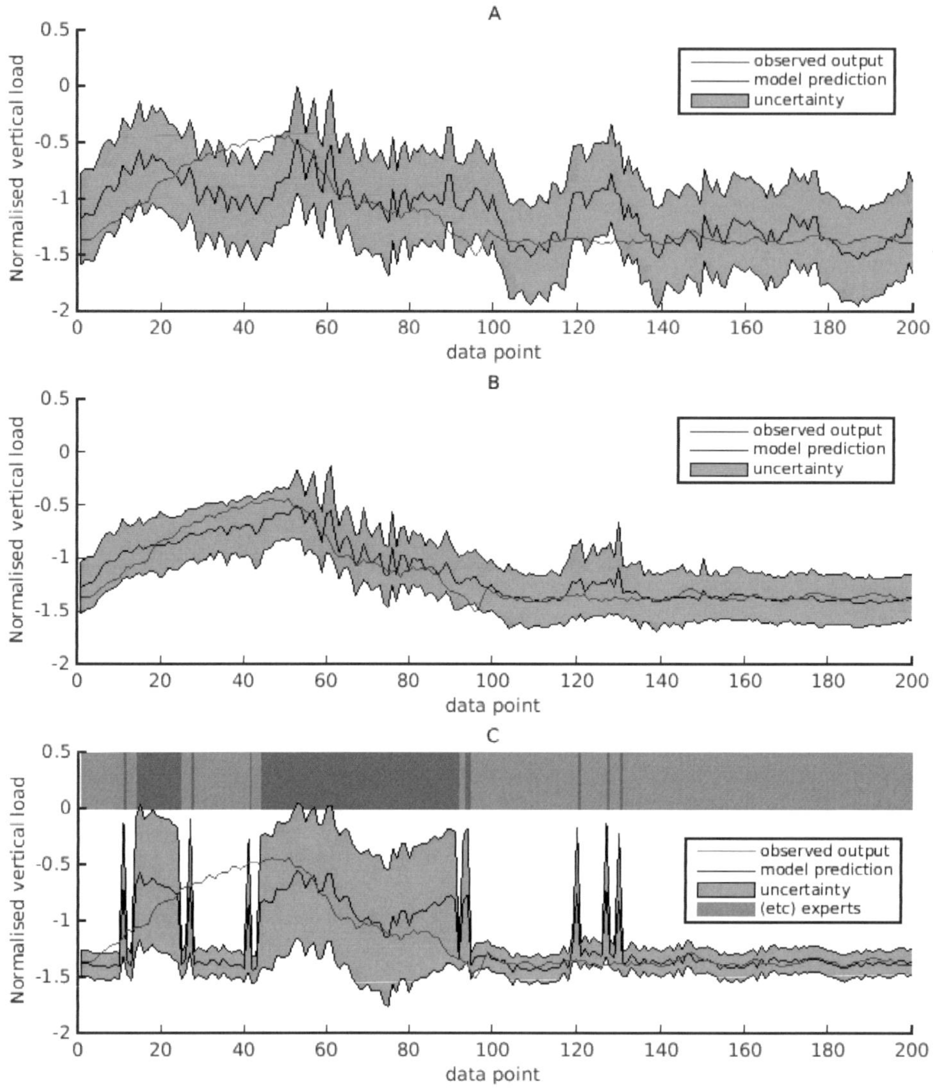

Figure 5. Comparing Gaussian process (A), treed Gaussian process (B) and mixture of linear experts (C) model prediction results during the initial phase of landing. The observed output (red lines) being predicted by the models (central black lines) is the vertical load on the gear which has been normalised to zero mean and unit standard deviation. The confidence intervals (grey shading) represent the region within which the observed output is expected to fall with a probability of 95%. In C, the horizontal colourbar shows that this part of the test data is divided between two different experts, one represented by the purple sections and the other by the green.

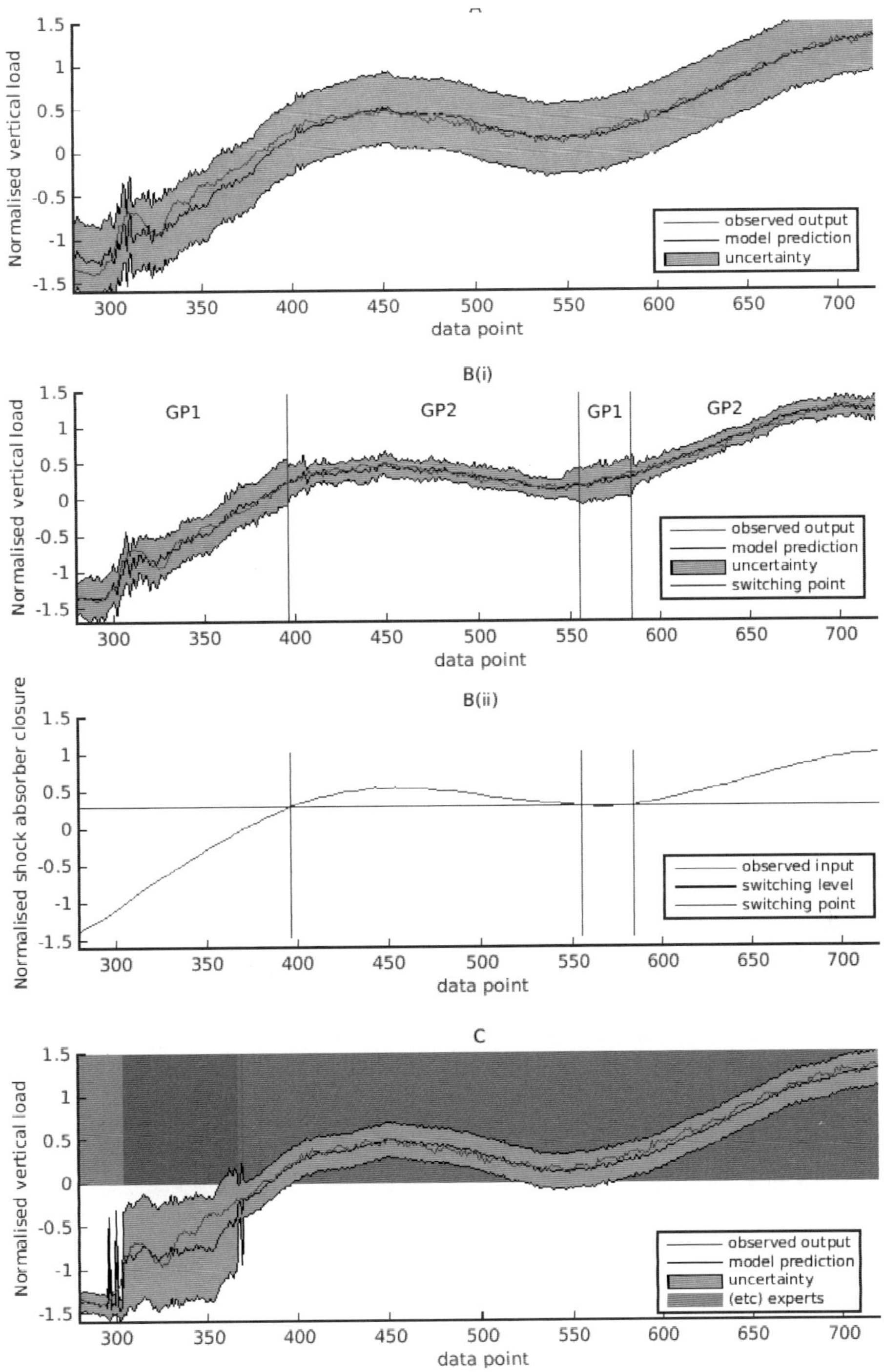

Figure 6. Comparing Gaussian process (A), treed Gaussian process (B(i)) and mixture of linear experts (C) model prediction results during the later phase of landing. The observed output (red lines) being predicted by the models (central black lines) is the vertical load on the gear which has been normalised to zero mean and unit standard deviation. The confidence intervals (grey shading) represent the region within which the observed output is expected to fall with a probability of 95%. B(ii) shows where the model switching occurs in the TGP model as determined by input 4 crossing the switching threshold. The structure underlying the switching is also shown in the tree representation in Figure 4. Thus the input data is divided between two leaves each of which has a GP trained to model the input-output data it contains. Following the data through time, it begins in leaf 1 then moves to leaf 2, briefly returning to leaf 1 before finishing in leaf 2. In D, the horizontal colourbar shows that this part of the test data is divided between three different experts, represented by green and two shades of purple.

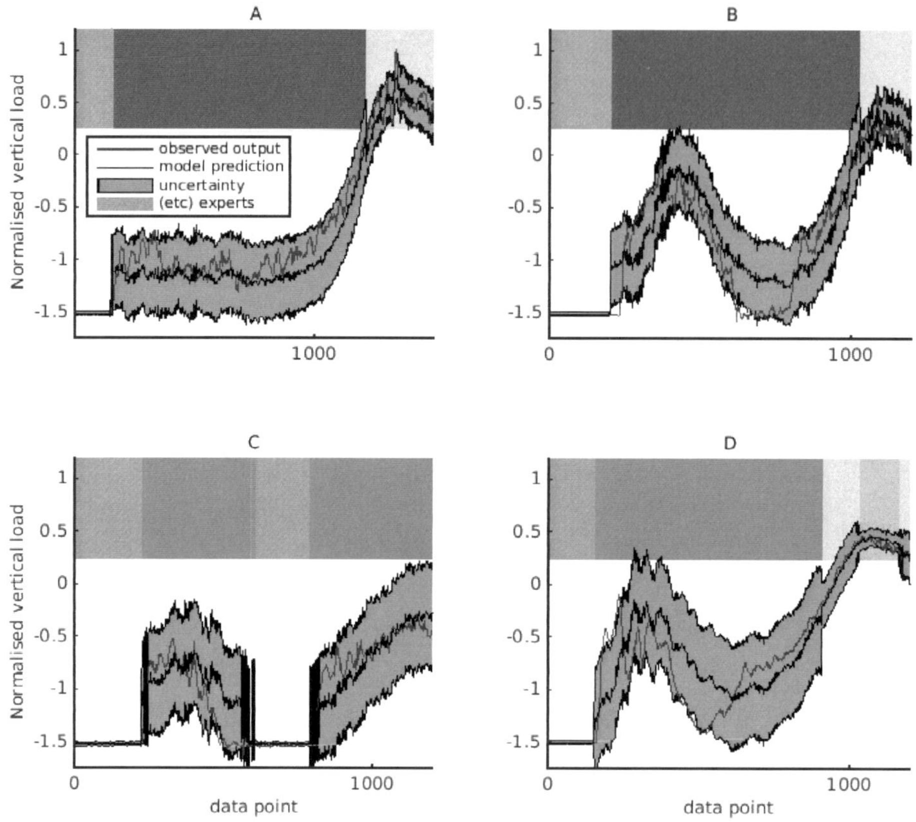

Figure 7. Mixture of linear experts model prediction during the initial phase of landing for four comparative data sets. Five different colour coded 'expert' sub-models are employed. The colourbar in each axes demonstrates the portions of the data where each expert dominates. The expert represented by light blue is a low variance model representating data where the landing gear has no contact with the runway. The dark blue and orange experts predominate during the inital phase of landing. The ratio of flights in which the dark-blue expert appeared to those in which the orange expert appeared was 3:4. They never both appeared in the same landing. The model parameters of the experts are shown in Figure 9.

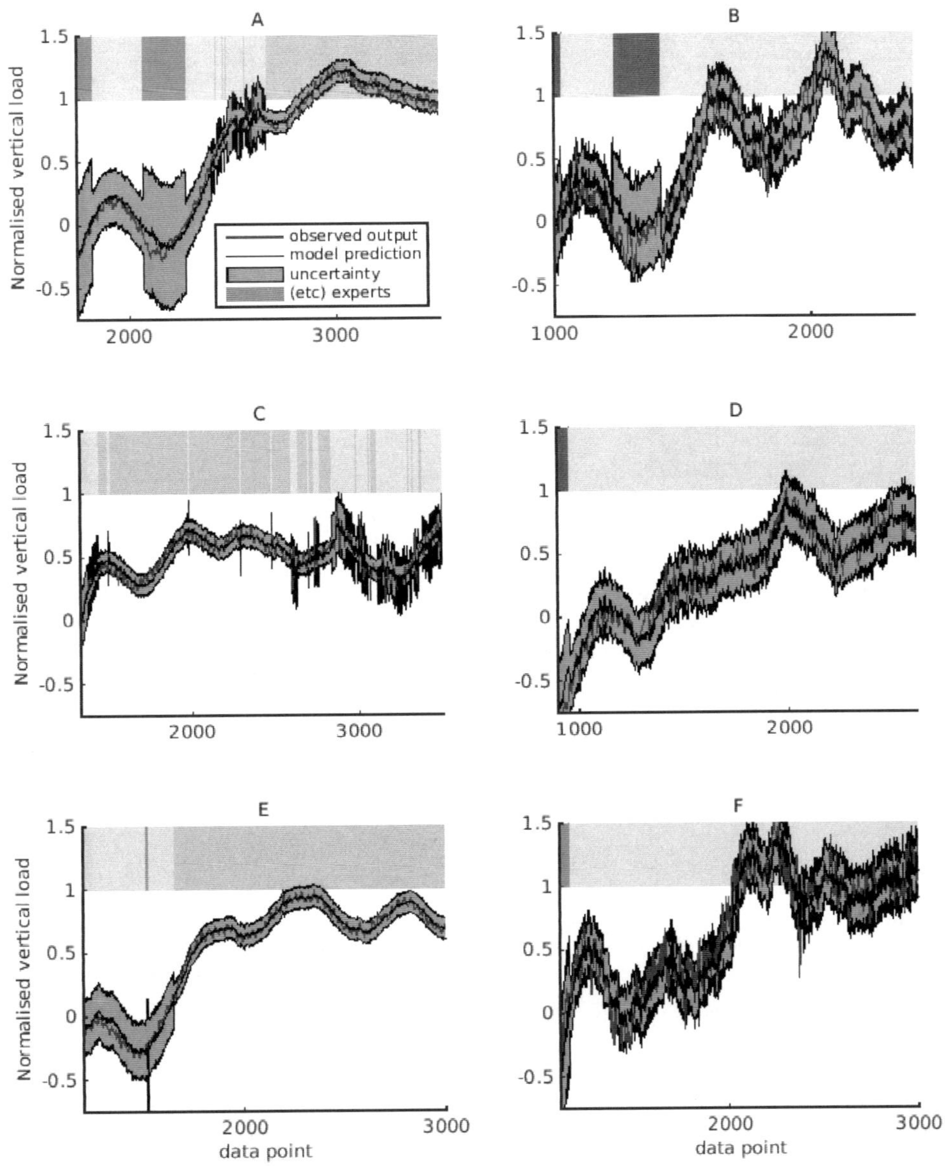

Figure 8. Mixture of linear experts model prediction during the later phase of landing for six comparative data sets. Five different colour coded 'expert' sub-models are employed. The colourbar in each axes demonstrates the portions of the data where each expert dominates. Two experts, represented by green and yellow, eventually dominate during this phase of landing as the aircraft settles into its runway taxiing. The yellow expert eventually dominates exclusively or almost exclusively in approximately three quarters landings considered, including two shown here (B,D&F). The green expert does similarly in the remaining landings, including (A&E). One interesting exception (C) shows a gradual transition between the two models during the latter phase. The model parameters of the experts are shown in Figure 9.

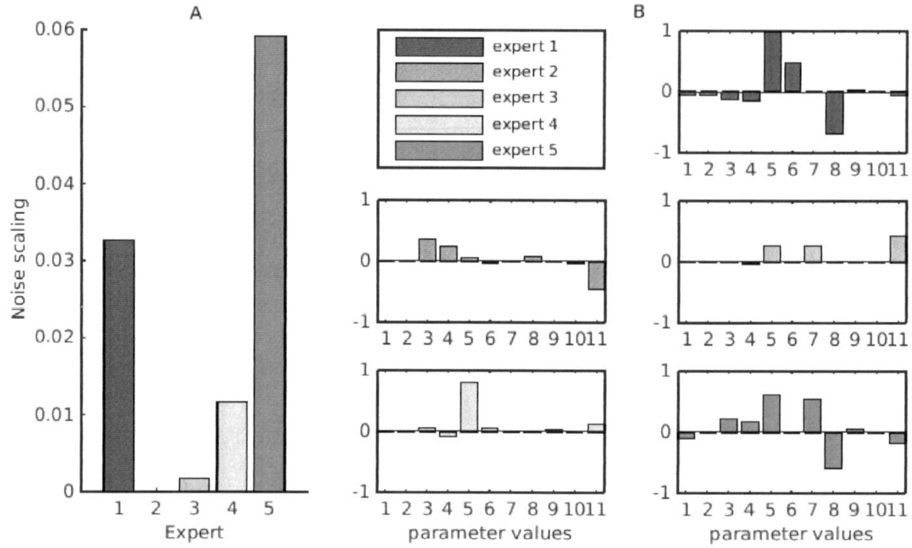

Figure 9. This figure shows the parameters of the five expert models which were identified in Test II. For each expert model (colour-coded to match Figures 7 & 8) there is a noise scale parameter (A) which represents variance of the structure in the data that cannot be captured by the linear model. If the model is correct this represents the process and measurement noise of the system for the output being predicted. Each model also has a parameter set (B). There is one parameter for each of the ten input dimensions. The eleventh value is an intercept parameter. Because each model is linear and the inputs have been standardised the absolute magnitude of the parameter signifies how important the corresponding input is in the model. So for example, in expert 4 (yellow), the fifth input, left main landing gear primary gas pressure, is dominant. In expert 3, the seventh input, left main landing gear secondary gas pressure, is of roughly equal importance to the fifth input.

BIBLIOGRAPHY

[1] E. J. Cross, P. Sartor, K. Worden and P. Southern (2013) "Prediction of Landing Gear Loads from Flight Test Data Using Gaussian Process Regression," in Proceedings of the 6th European Workshop - Structural Health Monitoring, pp. 1056-1063

[2] C. E. Rasmussen and C. K. I. Williams (2006) Gaussian Processes for Machine Learning, MIT Press, Cambridge MA.

[3] R. B. Gramacy and H. K. H. Lee (2008) "Bayesian Treed Gaussian process models with an application to computer modellling," *Journal of the American Statistical Association,* vol. 103, issue 483, pp. 1119-1130.

[4] R. B. Gramacy (2007) "tgp: An R package for Bayesian nonstationary, semiparametric nonlinear regression and design by treed Gaussian process models," *Journal of Statistical Software,* vol. 19, issue 9, .

[5] R. A. Jacobs, M. I. Jordan, S. J. Nowlan and G. E. Hinton (1991) "Adaptive mixtures of local experts," *Neural Computation,* vol. 3, issue 1, pp. 79-87.

[6] T. Baldacchino, E. J. Cross, K. Worden and J. Rowson, "Variational Bayesian Mixture of Experts Models and Sensitivity Analysis for Nonlinear Dynamical Systems," *Accepted for publication in Mechanical Systems and Signal Processing.*

[7] J. Hensman, R. Mills, S. G. Pierce, K. Worden and M. Eaton (2010) "Locating acoustic emission sources in complex structures using Gaussian processes," *Mechanical Systems and Signal Processing,* vol. 24, issue 1, pp. 211-223.

[8] A. Naish-Guzman and S. Holden (2008) "The Generalized FITC Approximation," in *Advances in Neural Information Processing Systems*, vol. 20, pp.1057-1064, Platt J, Koller D, Singer Y, Roweis S (Eds), MIT Press, Cambridge MA.

[9] H. Chipman, E. George and R. McCulloch (1998) "Bayesian CART model search," *Journal of the American Statistical Association,* vol. 93, issue 443, pp. 935-960.

[10] C. E. Rasmussen and H. Nickisch (2010) "Gaussian Processes for Machine Learning (GPML) Toolbox," *Journal of Machine Learning Research,* vol. 11, pp. 3011-3015.

Effect of Periodicity and Reverse Yielding on Cyclic Loading of Notched Specimens

Floyd Caiado, S. Keshavanarayana

Wichita State University, USA

Abstract: This study examines the effect of reverse yielding and its periodicity on cyclic loading on sheet specimens that have a center notch. An overload which causes the stress in the vicinity of a notch to exceed the yield strength induces plasticity at the notch. On removal of this overload, if cyclic loading is applied where the notch stresses are lower than the stress caused by the overload, a comparatively greater number of cycles would be required to cause further damage than would if the overload was not initially applied. An application of an under load, which causes reverse yielding (or compressive yielding), after the overload reverses the beneficial effect. The effect becomes detrimental. Notched specimens of Al 2024 T3 clad were used for the study. The experimental work clearly shows a decrease in life with the application of a tensile notch residual stress and a further decrease in life if the load that causes the tensile notch residual stress is applied more frequently. Reverse yielding negates the effect of the compressive residual stress established after the first half cycle of loading. Test results are compared with estimates using Miner's rule, which over estimates the cycles to failure.

BACKGROUND AND LITERATURE REVIEW

The beneficial and detrimental effects of residual stresses are clearly provided by J. Schijve [1]. In a chapter on residual stresses, the author shows a centrally notched part subjected to tensile loading. A far-field or remote tensile load that causes material in the vicinity of a notch to yield, results in creating a plastic zone at the notch. The removal of this remote tensile load results in a compressive residual stress to exist in the plasticized region. The existence of notch compressive residual stresses in a part or specimen is beneficial to its fatigue life. This is because a compressive residual stress reduces the cyclic mean stress for loading that occurs after the plastic zone has been created in the notch vicinity. A tensile residual stress has the opposite effect. Conversion of a notch compressive residual stress to a notch tensile residual stress can occur through reverse yielding. N. E. Dowling [2] explains the effect of reverse yielding in a chapter on notched members. The experimental work and the analysis in this study shows the effect of a tensile residual stress, which is created as a result of reverse yielding at the notch at the end of the first cycle of loading.

EXPERIMENTAL WORK

Specimens used in this study are of Al 2024 T3 clad sheet with a thickness of 0.09 in. The specimens were machined to have a width of 0.75 in and a centered hole with a diameter of 0.161 in. All the constant amplitude and variable amplitude cyclic tests were conducted on servo hydraulic controlled MTS load frames. The specimens were tested at a frequency of 20 hertz. The specimen dimensions are shown in Figure 1. The unit for stress used in the study is ksi – kilopound per square inch. Failure of a specimen is said to occur when the specimen breaks. Cycles to failure of a specimen are the number of cycles of loading that caused the specimen to break.

An anti-buckling fixture [3] was used to prevent buckling of specimens under compressive loading. The anti-buckling fixture comprised of two HDPE (High Density Polyethylene) plates held together with 6 bolts and nuts tightened to a torque of 10 in-lbs. The surface of the HDPE which contacts the specimen was coated with a thin layer of grease and grooves were made in each of the plates to prevent contact of plates with the notch of the specimen during testing.

In addition to the cyclic tests, an unnotched specimen was subjected to tensile loading till failure in order to determine the material properties (yield strength, ultimate strength, stress-strain curve, true stress-true strain curve, Modulus of Elasticity) of the material used in the study.

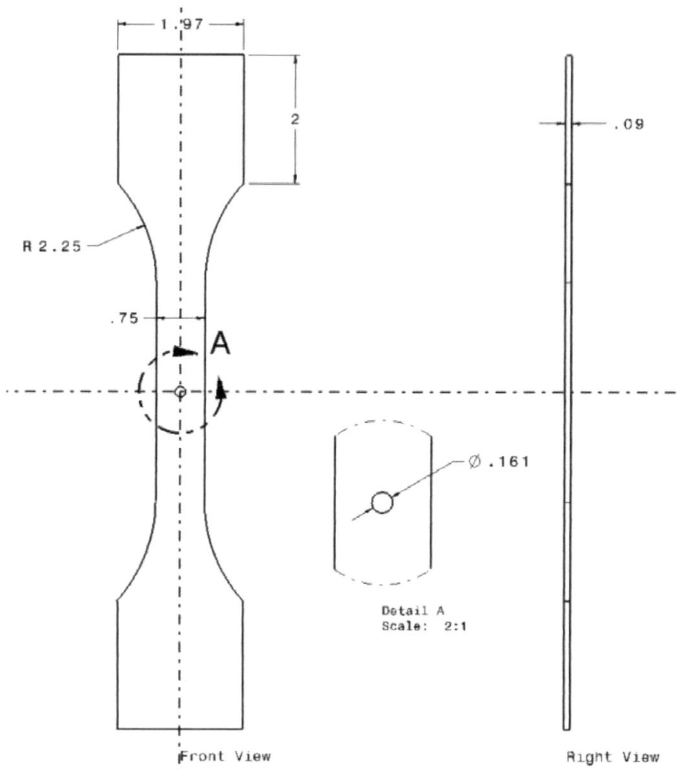

Figure 1. Specimen geometry (all dimensions are in inches).

Table I and Table II provide the test matrices for the constant amplitude cyclic tests and variable amplitude cyclic tests respectively. Figure 2 shows a sample loading sequence and the associated terms from the Table II.

Table I. Test matrix for constant amplitude cyclic tests.

Load set	Number of Specimens	Maximum Gross Stress (ksi)	Minimum Gross Stress (ksi)	Mean Stress (ksi)
A	6	30	-18	6
B	3	35	-18	8.5
C	3	44	-18	13

Table II. Test matrix for variable amplitude cyclic tests.

Loading Sequence	Number of Specimens	Overload: Maximum Gross Stress ($\sigma_{max\ overload}$) (ksi)	Overload: Minimum Gross Stress (σ_{min}) (ksi)	Constant Amplitude: Maximum Gross Stress (σ_{max}) (ksi)	Constant Amplitude: Minimum Gross Stress (σ_{min}) (ksi)	Number of cycles between overload events or periodicity (η)
S1	1	44	-18	30	-18	not applicable
S2	1	44	-18	30	-18	1,000
S3	1	44	-18	30	-18	100
S4	1	35	-18	30	-18	not applicable
S5	1	35	-18	30	-18	1,000
S6	1	35	-18	30	-18	100

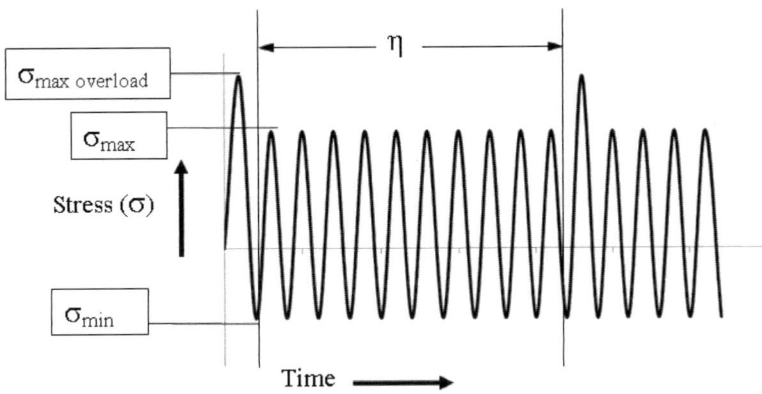

Figure 2. Sample loading condition

The sequence of loading is repeated till failure. If the loading sequence did not have a repetition of the overload then the last column in Table II indicates that as 'not applicable'. In such cases, the overload was only applied at the start of loading. The number of cycles between overloads, or periodicity, was arbitrarily chosen to study the effect.

EXPERIMENTAL RESULTS AND ANALYSIS

Experimental Results

The tensile test of the unnotched specimen provided the following material properties. Modulus of Elasticity of 10500.0 ksi, ultimate tensile strength of 62.8 ksi and yield stress of 48.5 ksi.

The results for the constant amplitude tests and the variable amplitude tests are shown in Table III and Figure 3 respectively. Figure 3 also shows comparison to estimates of cycles to failure using Miner's rule [4].

Table III. Test results for constant amplitude cyclic tests.

Load sets	Average cycles to failure
A	2,416
B	965
C	410

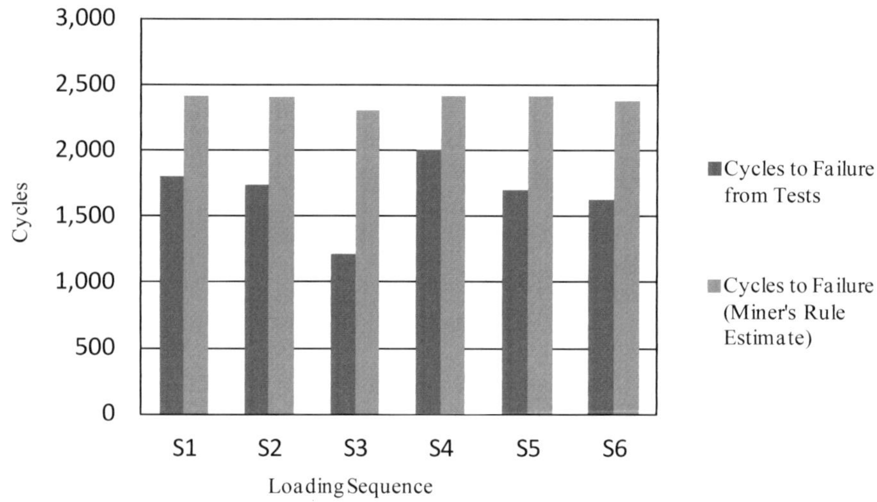

Figure 3. Variable amplitude cyclic test results and comparison of cycles to failure.

Analysis

The experimental results in Table III and Figure 3 show that irrespective of the overload applied (35ksi or 44ksi) there is a reduction in life compared to the case where only the constant amplitude maximum load of 35ksi and minimum of -18ksi is applied (2,416 cycles).

Considering that the material experienced kinematic hardening, the notch residual stresses were determined after the completion of the first overload cycle. The notch residual stresses for the two cases are shown in Table IV.

Table IV. Notch residual stresses.

	Overload Stress (Far-Field)		
	Stress Level I (tensile)	Stress Level 2 (compressive)	Residual notch stress (after completion of the overload cycle or Far-field stress is 0ksi)
	ksi	ksi	ksi
Case 1	44	-18	16.07
Case 2	35	-18	8.41

The notch residual stresses were determined using True Stress-True Strain curve of the material (which was experimentally determined), Neuber's rule [5], Stress concentration factor of 3.16 [6] and considering kinematic hardening behaviour for one complete cycle of loading. From Figure 3 and Table IV, it can be seen that for a given constant amplitude loading, a greater tensile notch residual stress prior to the constant amplitude loading results in lower cycles to failure. Without an overload, the average cycles to failure was 2,416 cycles (Table III), with a tensile notch residual stress of 8.41 ksi prior to the constant amplitude loading resulted in 2,007 cycles to failure and with a tensile notch residual stress of 16.07 ksi prior to the constant amplitude loading resulted in 1,809 cycles to failure. As part of the analysis, it was determined that at the end of first half cycle (in Case 1 and Case 2), the loading resulted in a compressive notch residual stress. However, the completion of the second half cycle caused reverse yielding at the notch, thereby converting the compressive notch residual stress into a tensile notch residual stress.

The increase in frequency of application of the overloads shows a decrease in life. Loading Sequences 2 and 3 in Figure 3 show a decreasing trend in lives (1,744 cycles to 1,222 cycles) as periodicity changed from 1,000 cycles to 100 cycles or in other words, the frequency of the overloads increased. A more frequent tensile notch residual stress caused an earlier failure. Loading Sequences 5 and 6 in Table IV also show a similar trend for the overload case of 35 ksi. A decreasing trend in lives (1,707 cycles to 1,629 cycles) as periodicity changed from 1,000 cycles to 100 cycles. The above shows if the overload applied is more frequent, this would imply a greater and more frequent reverse yielding, which in turn would result in more frequent notch tensile residual stresses, thereby causing an earlier failure. Figure 3 shows that Miner's rule overestimates the cycles to failure.

CONCLUSIONS AND RECOMMENDATIONS

The reverse yielding reverses the beneficial effect of the compressive residual stress caused by the far-field tensile load. A tensile residual stress results in reduction of cycles to failure. If the overloads that cause the tensile residual stress are applied more frequently, then the life or cycles to failure is further reduced. Estimation of cycles to failure using Miner's rule is non-conservative for the loading cases in this study. The current study includes two maximum overload stress levels which are 35 ksi and 44 ksi. Further study could include another load level or a different constant amplitude loading stress (other than 30 ksi). Also, an interactive damage study involving a combination of 35 ksi and 44 ksi loading sequence will also provide an interesting comparison.

REFERENCES

[1] Schijve, J., (2009), *Fatigue of Structures and Materials*, Springer Science+Business Media, Netherlands, p. 623.

[2] Dowling, N. E., (2013), *Mechanical Behavior of Materials*, Pearson Education Limited, England, p. 960.

[3] Keshavanarayana, S., Smith, B. L., Gomez, C., Caiado, F., (2010), *Journal of Aircraft,* vol. 47, n. 1, pp. 181-191.

[4] Miner, M. A., (1945), *Trans ASME. J. Appl. Mech.,* vol. 12, n. 3, pp. A159-A164.

[5] Neuber, H., (1961), *Trans ASME. J. Appl. Mech.*, vol. 28, n. 4, pp. 544-550.

[6] Young, W. C., (2001), *Roark's formulas for stress and strain*, McGraw-Hill Professional, USA, p. 832.

Cold Spot Welding in Riveting Joint

Wojciech Wronicz[1], Jerzy Kaniowski[1], Krystyna Pietrzak[2]

[1]*Institute of Aviation, Poland*
[2]*Motor Transport Institute, Poland*

Abstract: The riveting still remains the most popular and basic method of joining parts in airframes. At the same time, the rivets are critical from the fatigue point of view. This is especially important in the case of ageing aircrafts. The paper presents the selected investigations of local phenomena during the riveting process. At the beginning, some results from the literature were shown. The analysis of these results leads to the conclusion that the increase in fatigue life caused by higher riveting force is so high that the more beneficial residual stresses cannot be the only explanation, but the new phenomena in riveting process occurs. Another premise supporting the conjecture that there is some new phenomena in riveting joints were the 'S'–shaped signals recorded by strain gauges during the riveting process.

The working hypothesis has been assumed that during the riveting process adhesive joints (called cold welding) were formed between the rivet shank and the sheets. The FE analysis indicated that conditions necessary for adhesive joints formulation are met during the riveting process.

Several microstructure tests were conducted in order to verify experimentally the hypothesis of cold welding during the riveting process. Observations of microsections with the use of optical and scanning electron microscope were performed. In the case of several specimens the punctual and linear welding between rivet shank and sheets were observed.

INTRODUCTION

Despite the development of new materials (e.g. composites) as well as new joining techniques (e.g. Friction Stir Welding) riveting remains the most commonly used technique of joining metal parts in airframes. Because of that, the riveting is a subject of interest for designers, materials and other technology scientists. This fact was reflected among others in the IMPERJA project [1], which has been executed under the Eureka Initiative. This project includes extended research programme of phenomena related to riveting joints in airframes.

The influence of riveting process parameters on the fatigue life of joints has been previously widely described by Müller [2]. Fig. 1 presents results from this work obtained for the three-row riveted lap joints consist of 2024-T3 aluminium alloy sheets with thickness of 1.27 mm and NAS1097 rivets with diameter of 4.0 mm made of 2117-T3 aluminium alloy. These results entitle the author of [2] to draw the following conclusions. Fatigue life of riveted lap joints made of 2024-T3 alloy sheets has been substantially increased by increasing the riveting (squeezing) force. The fatigue life can be 10-times higher if the driven head diameter increases from 1.2 to 1.75 of rivet shank diameter. At the same time, a location of crack nucleation has been shifted away from the rivet hole and eventually cracks grow at some distance from the hole edge. Thanks to the better clamping and residual stresses the rivet hole is no longer critical from the fatigue point of view. An increase in fatigue life has been obtained for a few stress levels what means that for the higher riveting forces the S-N curves move upward. An improvement of the fatigue life usually means an increase in time to the crack initiation, what means time to the moment when the crack reaches 1 to 2 mm length. The crack initiation time covers majority of the total fatigue life, i.e. about 90%.

Müller emphasized that increase in riveting force, what means increase in rivet shank diameter to driven head diameter ratio (D/D_0), caused increase in hole expansion. Based on FE analyses Müller has showed that, if the riveting force is high enough then the residual tangential stresses became compressive (Fig. 2.).

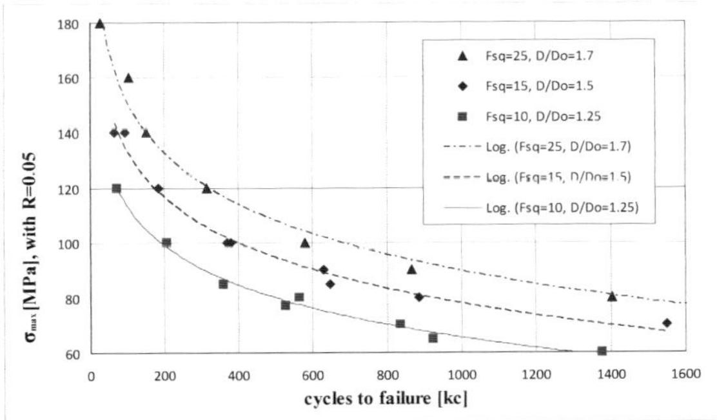

Figure 1. Fatigue life of 3-row riveted lap joint made of 2024-T3 alloy as a function of nominal stresses σ_{max} for various riveting force (sheet thickness t=1.6 mm, rivet pitch s = pitch between the row p = 20 mm, specimen width W=140mm, stress ratio R=0.05, rivet squeeze force F_{sq}) [2, Fig. 11.16].

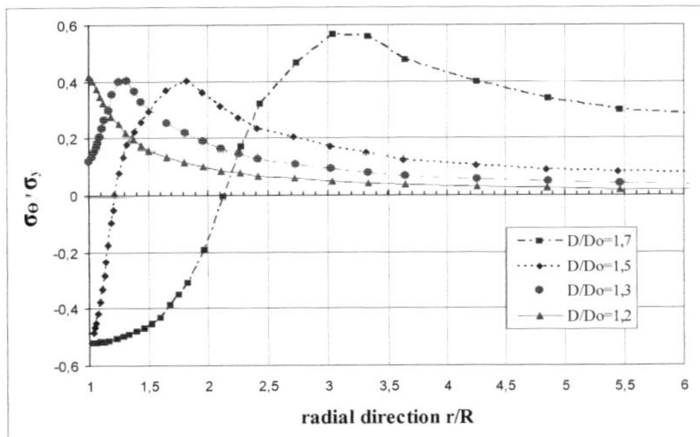

Figure 2. Tangential residual stresses to yield stress ratio σ_Θ/σ_y along the mating surface as a function of radial coordinate to rivet radius ratio r/R. FE symmetrical model (sheet alloy 2024-T3, yield stress σ_y=324MPa, sheet thickness t=0.85mm, shank diameter D_0=3.2mm, shank diameter to driven head diameter ratio D/D_0) [2, Fig. 6.6].

During the realisation of the Eureka-IMPERJA project, the influence of a riveting technology on the fatigue properties of joints was investigated. The researches include among others measurements of hole expansions, fatigue tests and FE analyses. The measurements of hole expansion shows that a higher squeezing force increase an expansion and that rivets with compensator (marked PA24c) increase strongly the expansion on the manufactured head side (Fig. 3.) [4].

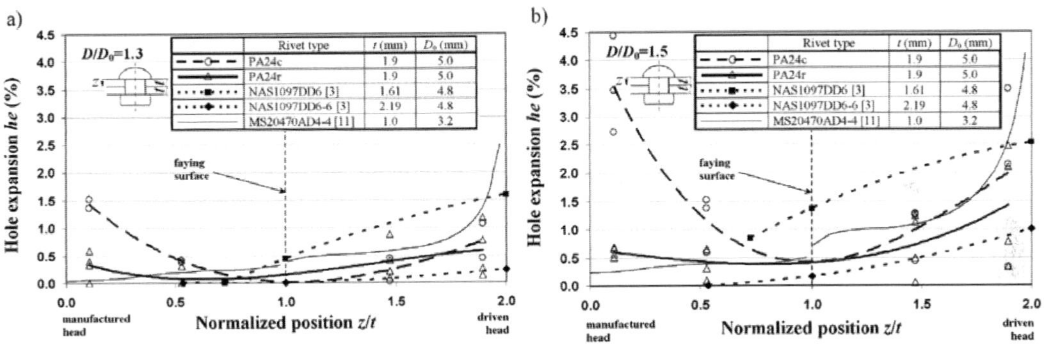

Figure 3. Hole expansion measurements, a) D/D_0=1.3, b)D/D_0=1.3 [4].

The FE analysis of squeezing force influence shows that in the case of tangential stresses on the mating surface, higher squeezing force causes compressive stresses near a rivet and lower maximum peak of tensile stresses which is shifted outside the rivet (Fig. 4. [5]. Fatigue test of joints confirmed that higher squeezing force and rivets with compensator enhance fatigue life of joints (Fig. 5. [4])

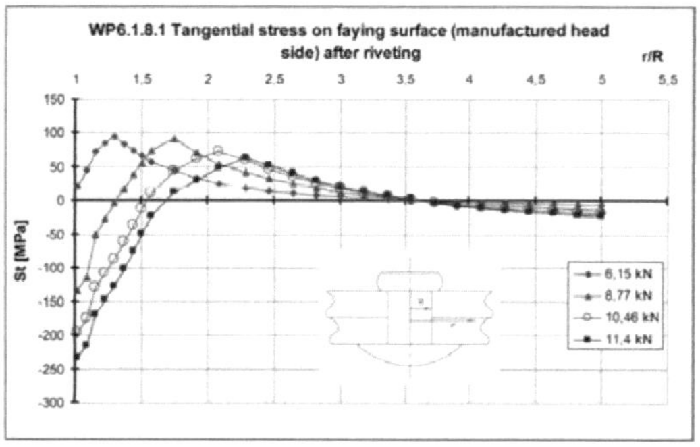

Figure 4. Tangential stresses on mating surface for various squeezing forces [5].

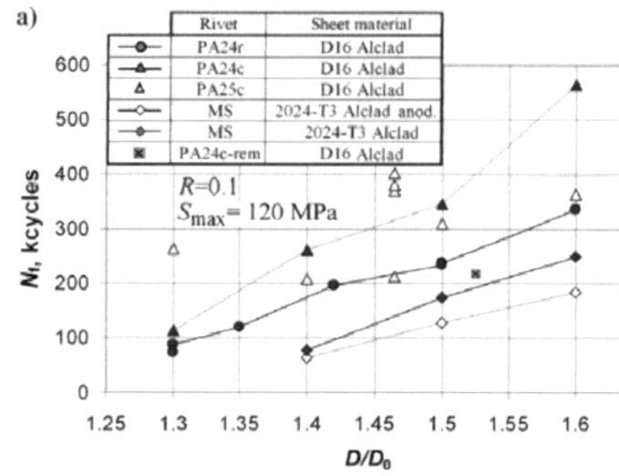

Figure 5. Fatigue life of joint for various squeezing forces and rivet types [4].

In his Ph.D. thesis [Rans] Rans described one more mechanism which explains increase in fatigue life for higher squeezing forces, in the case of countersunk rivets. When a squeezing force exceeds some level, the 'wedge expansion' mechanism takes place (Fig. 6.). FE simulation performed by this author shows that this phenomena do not influence significantly tangential compressive stresses near a rivet hole but at distance equal to about 2 rivet radius differences in stresses are significant. This is an area where fatigue cracks nucleate in the case of high riveting force. At this distance, a local minimum of stresses appears. Similar effect (local minimum) has been obtained in FE riveting simulation for a rivet with a compensator (Fig. 7 [5]).

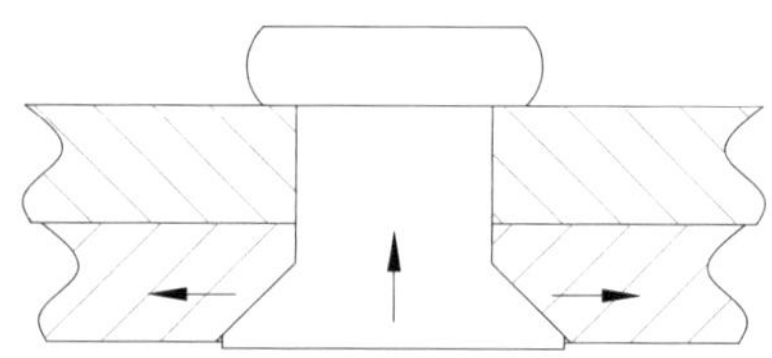

Figure 6. 'Wedge expansion' mechanism described in [6].

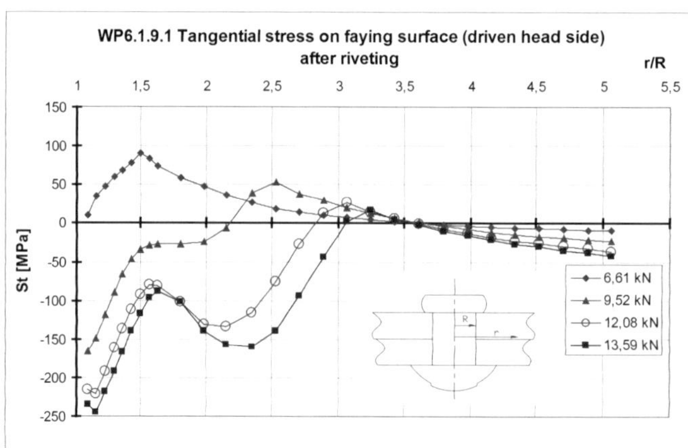

Figure 7. Stresses on mating surface for rivet with compensator and for various squeezing forces [5].

The riveting force influence has been also presented by Müller and Hart-Smith in [3]. In this case the subject of researches were riveted lap joints consist of 2024-T3 alloy sheets (as in [2]) but with thickness 1.6 mm and NAS1097 rivets with diameter 4.8 mm made of 2117-T3 alloy. Table I presents selected results from this paper. The analysis of rows 1–3 (Fig. 8.) shows that the change in riveting force level, from low (12 kN/ D/D_0=1.2) through medium (1.42-fold increase, D/D_0=1.4) to high (3-fold increase, D/D_0=1.7) causes 2.4 and 11.3 respectively increase in fatigue life of joints (see column 7). Further increase in fatigue life has been obtained by application of the NACA riveting method (fig. 9.). In this method, the countersunk hole is prepared and protruding rivet is inserted from the non-countersunk side. During the riveting, the rivet shank is compressed, fill countersunk and form the driven head. At the end of the process, the part of the driven head protruding above the sheet surface is milled. The application of the NACA riveting method in the top row of the rivet joint caused 25.5-fold increase in fatigue life. When this technology has been used in both external rows and with antisymmetric configuration (manufactured heads of rivet in top and bottom rows were on the opposite joint sides) the increase in fatigue life was equal to 34.2 (see column 7 and Fig. 8).

Table I. Selected results of 3-row riveted lap joint fatigue tests [3].

	1	*2*		*3*	*4*	*5*	*6*	*7*	*8*
	Joint diagram	Rivet type		Riveting force [kN]		D/D_0	Cycles to failure [cycles]	Relative increase in fatigue life with respect to the minimal fatigue life	Relative increase in fatigue life with respect to the nominal fatigue life
1		100° flat countersunk NAS1097[1)]		Minimal according to TS[2)]	12	1.2	41 552 37 226 40 112	1.0	0,4
2	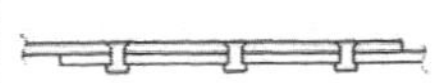			Nominal according to TS[2)]	17	1.4	97 535 99 624 88 442	2.4	1,0
3				High[2)]	36	1.7	451 270 441 556	11.3	4,7
4	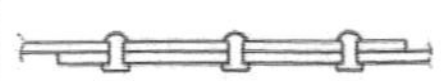	Protruding[1)]			36	1.7	483 714 491 998	12.3	5,1
5	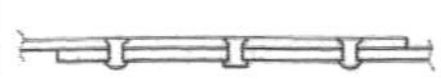	Round rivet installed in countersink[3)]	In outer rows	High	36	–	983 764 1 035 000	25.5	10,6
6	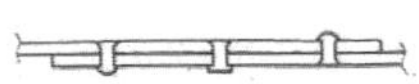		In outer rows, antisymmetric configuration		36	–	1 209 000 1 383 000 1 479 000	34.2	14,2

[1)] Standard riveting method [2)] TS – Fokker technical specification [4] [3)] NACA riveting method.

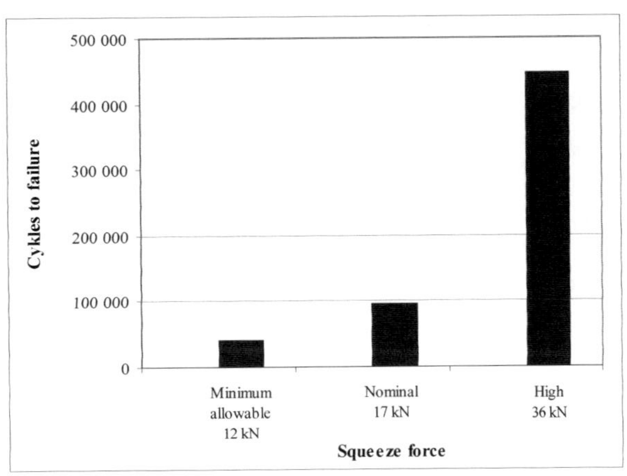

Figure 8. Influence of squeezing force on fatigue life of 3-row riveted lap joint [3, Fig. 4].

The minimum allowable ratio of a driven head diameter divided by a rivet diameter D/D_0 has been increased in currently used riveting instruction (process specifications) to the value of 1.4 [7]. This value is equal to the nominal value used in Müller and Hart-Smith researches. When their results are referred to this value, the increase in fatigue life is still high and its equal to 4.7, 10.6 and 11.4 respectively (see column 8).

The authors of [3] explained this very high increase in fatigue life as a result of high compressive tangential residual stresses in the sheets around the rivet induced by high riveting force. Hart-Smith in his presentation [8] showed that in the case of traditional riveting, compressive tangential stresses are induced near a driven head . In the case of NACA riveting, tangential compressive stresses are induced near the mating surface (fig. 9), where fatigue cracks nucleate in the case of lap joints and unsymmetrical but joints.

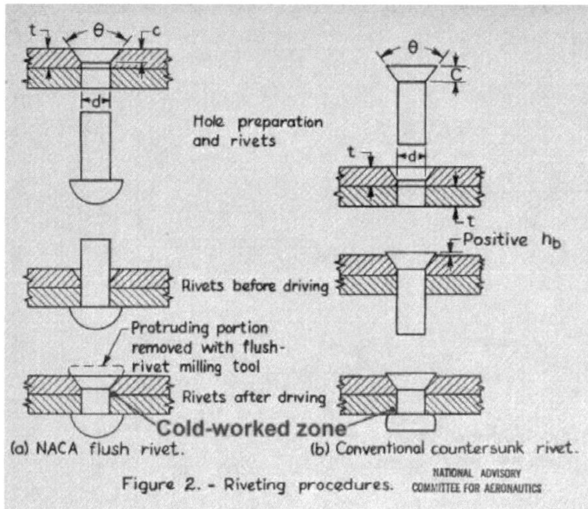

Figure 9. Conventional and NACA riveting [9].

HYPHOTESIS

W. Karliński, after analysis of the Müller and Hart-Smith results [3], has claimed in [10] that this increase cannot be only quantitative but this is a qualitative change. This means that the mechanism of joint formulation has changed. In the case of nominal riveting forces the mechanical joint is formed because the protruding parts of rivet and hole geometry get caught mutually. Between these contact zones there are non-contact areas, so the real contact surface is much smaller than a theoretical one. Increase in riveting force caused only a limited grow of real contact surface. Based on these observations W. Karliński has proposed the hypothesis that in the case of significant increase in riveting force, the plastic strain reaches the level which is necessary for adhesive joint formulation [10].

Another premise supporting the conjecture that there is some new phenomenon in riveting joints were the results of strain measurements during riveting. The specimens with three rivets were used (Fig.10). At the beginning, the outer rivets were installed, afterwards the strain gauges were applied on the sheet surface, near the driven head and the central rivet was installed. During this process, the strain progress was recorded. The experiment has been performed for four types of rivets [11]. In the case of normal (without a compensator) 90° countersunk rivet, the very clear 'S'–shaped strain signals was recorded (Fig.11). In the case of brazier rivet with a compensator, the reversal of strain signal was much less distinct and appear only in the case of radial strains. Similar shapes of strain plots can be found in [12]. Li et all explained this phenomenon by clearance between the sheet and rivet (distance of rivet head above sheet surface). Numerical calculations do not confirm this hypothesis since such effect was no recorded, moreover, this effect has been observed for brazier rivet also. The fact that this shape of strain progress has not been obtained in any FE calculations (axisymmetric and solid [11] suggest, that the reason of it is some new phenomenon which is not modelled.

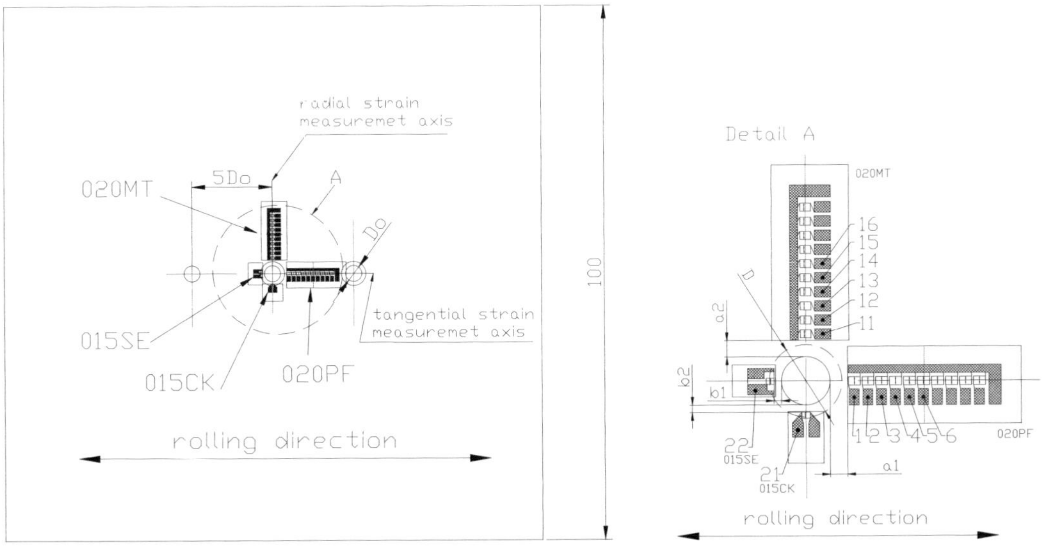

Figure 10. Geometry of specimens used in strain gauge measurements [13].

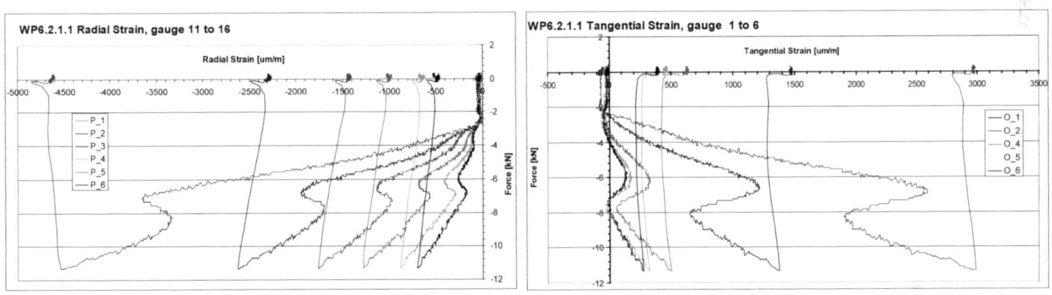

Figure 11. Radial and tangential strains recorded during the riveting process of a 90° countersunk rivet [13].

In the case of aluminium and its alloys, the minimal strain of 60 to 70 % and 75 to 95% respectively is necessary to formulate adhesive joint called cold-welding [14, 15]. This strain levels correspond with stress at least 380 MPa. Because the fact that the real contact surfaces between singular peaks on the rivet-hole interface are very small, the contact stresses are considerably higher than this value and formulation of microwelds seems to be very probably in the case of riveting force equal to 36 kN (high riveting force in the Table I). Significant higher force is required to destruction of such adhesive joint than the mechanical joint. Perhaps, the exceptional increase in fatigue life of joints riveted with high force is a result of micro-welds formulation on the rivet-hole interface. If this is a true, than the strength of joint between rivet and sheets must considerably increase.

Another condition required to formulate adhesive joint is a relative displacements between two surfaces [16]

The axisymmetric FE model was used to verified if conditions necessary to formulate adhesive joint, mentioned above, are fulfilled in riveting process. Nonlinear implicite algorithm were used to simulate quasi-static riveting process (MSC MARC). The model consists of deformable (sheets and a rivet) and rigid (tools) contact bodies. There are 2132 linear elements and 2433 nodes in the model. Material models were developed based on monotonic tests. More information of the analysis can be found in [13]. Figure 12 presents the model and a graph of contact normal stresses between a rivet and a sheets along the path which was marked with the red line. The 380 MPa stress level was marked with a dot line. The series name indicates temporary squeezing force value during the process. The value 0 kN correspond to the end of the riveting. Values of force for which the curves are presented correspond to moments of riveting shown on Figure 13.

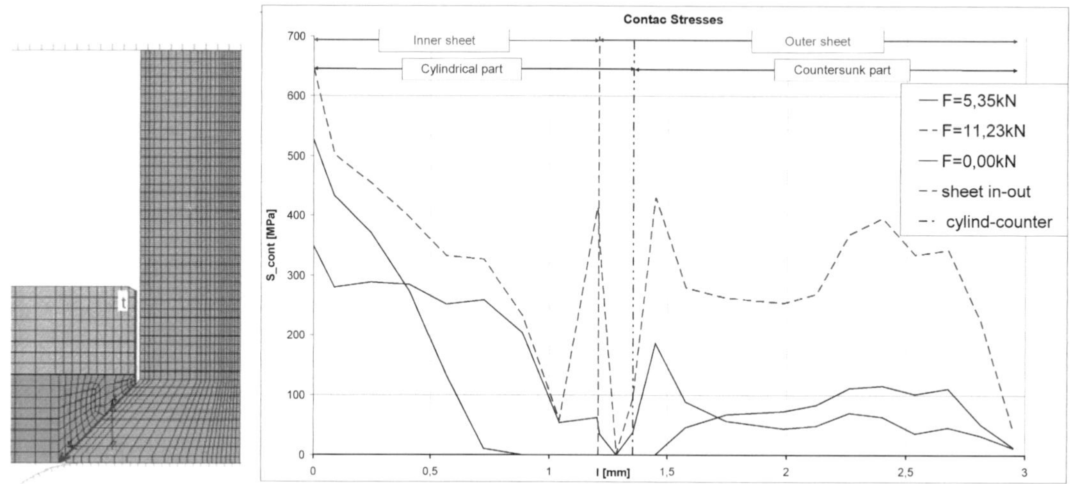

a) b)

Figure 12. a) model, b)contact normal stresses between rivet and sheet along path marked with red line on the model during riveting process.

On the Figure 13, the graph of relative displacements of the rivet and sheet is presented. Each curve shows displacement of singular rivet node in relation to the closest (after contact) sheet node. Numbers of nodes are indicated in series names and are shown on the picture on the left side, which presents a fragment of the model. Small pictures under the graph shows shape of the model in the particular moment of the process. Yellow and red colour indicates contact between the rivet and the sheet. Vertical lines indicate moment of first contact between particular rivet node and sheet element. These lines are marked with 'c' letter and the node number.

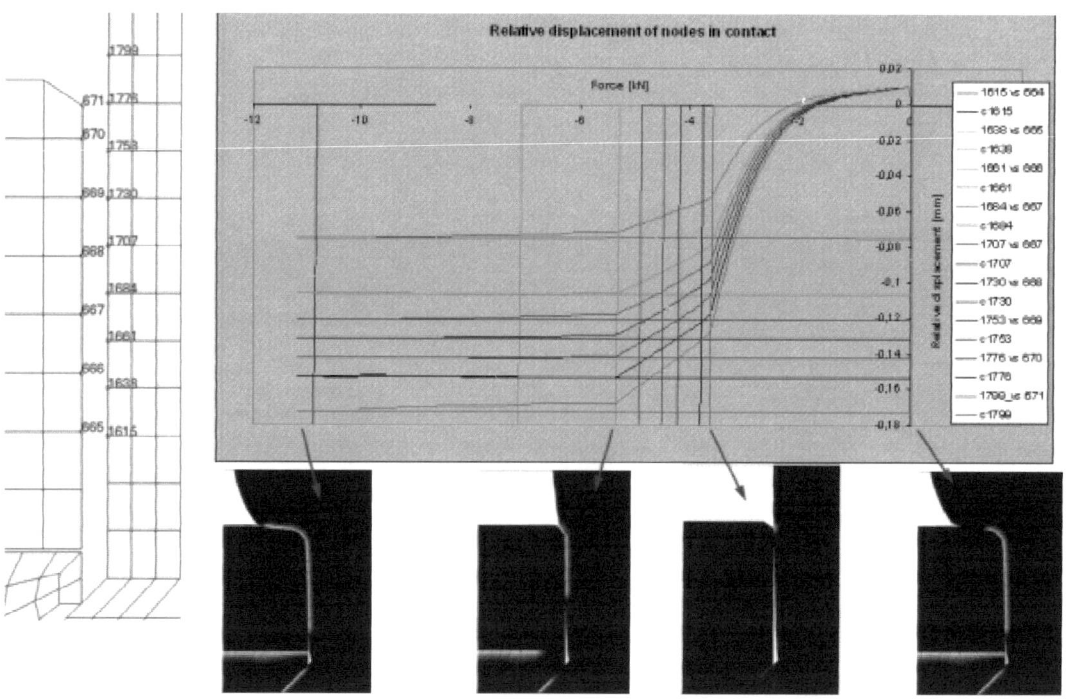

Figure 13. Relative displacements of rivet and sheets nodes during riveting simulation.

At the begging of the process there are small positive displacements which are caused by clamping of the sheets before squeezing the rivet. Then, displacements becomes negative and change quite quickly until the rivet shank touch the sheet (force equal to 3.63 kN). Between this moment and the moment when the rivet shank fill a hole chamfer (force equal to 5.35 kN), the speed of relative displacements is smaller than at the beginning but noticeable. Then, nodes of the rivet almost do not move in relation to nodes of sheets.

The FE analysis indicated that conditions necessary for adhesive joints formulation are met during the riveting process near the driven head.

MICROSTRUCTURE TESTS

A series of microstructure tests have been perform in order to verify the hypothesis formulated in [10]. The first two series have been executed during the last months of the IMPERJA project realization. The specimens were made of 2024-T3 SHT clad sheet with a thickness of 2.5 mm. 90° countersunk and round head rivets with diameter of 5 mm were used. There were anodized and not anodized rivets. The special riveting set were used in order to ensure the perpendicularity of the rivet axis and the sheet surface (Fig. 14a). The riveting force levels have been selected based on calibration tests. During the tests riveting force levels were assumed in such a way as to obtained a relative size of driven head D/D_0 in the range of 1.4 to 2.2. The specimens configurations are presented in Table II. Real force levels as well as dimensions of rivets, holes and driven heads have been measured and recorded. The first series (labelled as I) have been manufactured in the conditions similar to industrial practice. A few rivets with D/D_0 parameter = 2.2 from this series have been examined. The specimens were carefully cut near the rivet axis and microsections were prepared. For any of the studied rivets the microwelds were found between a rivet and sheets. For this reason the tests have been repeated on the second series (labelled as II) which was made in 'clean' conditions. Rivets and sheets has been cleaned with acetone in ultrasonic cleaner and during riveting process clean cotton gloves were used (Fig. 14a). Only in the case of rivet number 3 (D/D_0 = 1.7) local punctual and linear microwelds have been observed in a few location on the rivet-hole interface (Fig. 15b).

Table II. Specimens for the Microstructure tests manufactured under the IMPERJA project.

Item number	Specimens number	Rivet type	Rivet standard	Surface treatment – rivet:	Riveting method	Manufacturing conditions
1	1/I	90° countersunk	BN-70/1121-04	anodised	conventional	'industrial'
2	1/II	90° countersunk	BN-70/1121-04	anodised	conventional	'clean'
3	2/II	90° countersunk	BN-70/1121-04	not anodised	conventional	'clean'
4	3/I	90° countersunk	BN-70/1121-04	anodised	NACA	'industrial'
5	3/II	90° countersunk	BN-70/1121-04	anodised	NACA	'clean'
6	4/II	90° countersunk	BN-70/1121-04	not anodised	NACA	'clean'
7	5/I	round	BN-70/1121-03	anodised	NACA	'industrial'
8	7/II	90° countersunk, round brazier with a compensator	BN-70/1121-03 BN-70/1121-03 OST 1 34040-79	anodised and not anodised	conventional and NACA	'clean'
9	8/II	90° countersunk, brazier with a compensator	BN-70/1121-04 OST 1 34040-79	not anodised anodised	conventional and NACA conventional	'clean'

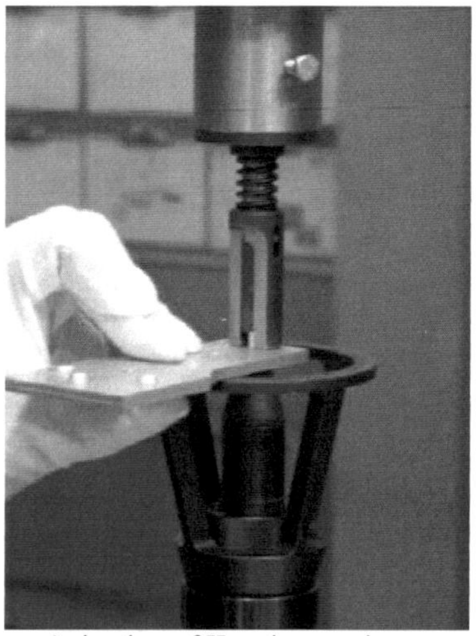

a) riveting of II series specimens

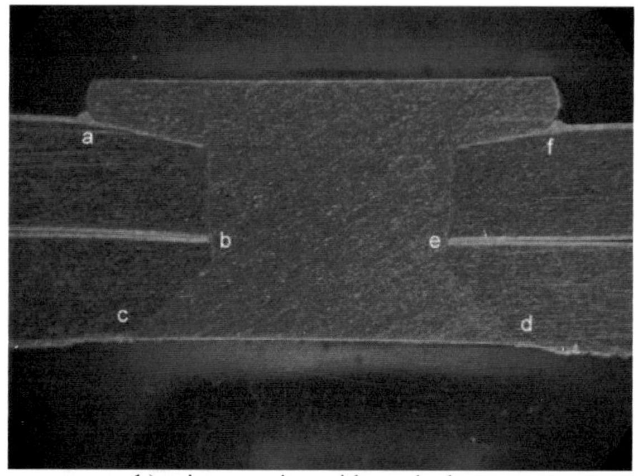

b) miscrosection with marked area

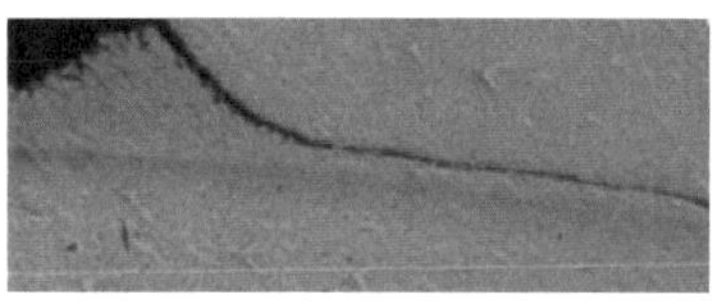

c) area a

d) area b

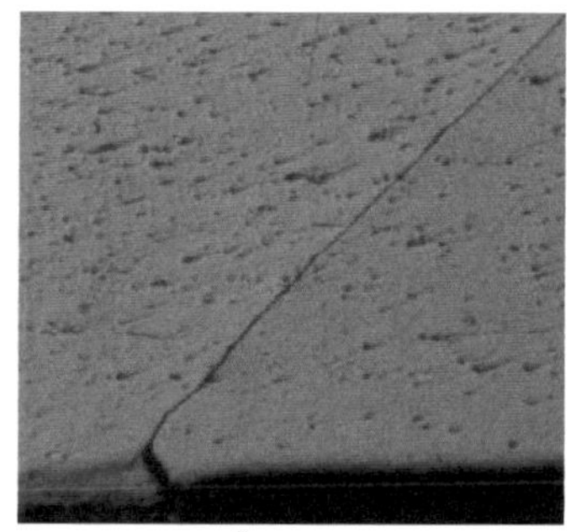

e) area c

Figure 14. Microsection no. 8 (riveting force 93,5 kN) and selected fragment of joint in specimen no. 2/II.

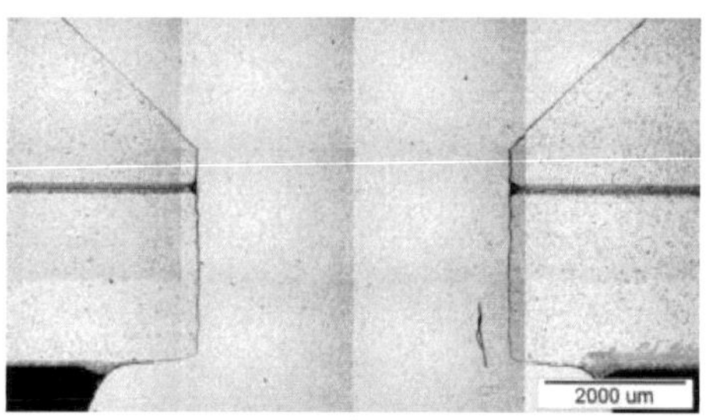

a)

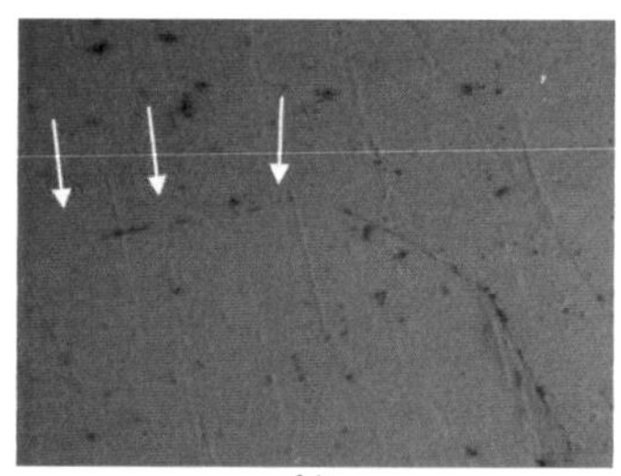

b)

Figure 15. Specimen no. 2/II – 'clean', a) microsection of rivet, b) area with local punctual and linear joints (marked with arrows).

After completion of the IMPERJA project the research has been continued. Another specimens, labelled ILOT.4.1 and ILOT.4.2 (Fig. 16, Table III) has been manufactured from clad D16CzATV (aluminium alloy) sheet with thickness 4 mm. The rivets with round, countersunk and brazier with compensator heads as well as universal rivet were used. All rivets except universal has diameter 5 mm. The universal rivet diameter was 4.8 mm. After riveting the specimens have been cut and mounted in resin. Then the specimens have been polished parallel to the sheet surface. The surface after first polishing was about 0.5 mm under the sheet surface, the distance between surfaces after successive polishing (polishing depth) was about 1 mm. Between polishing observations on a microscope have been performed.

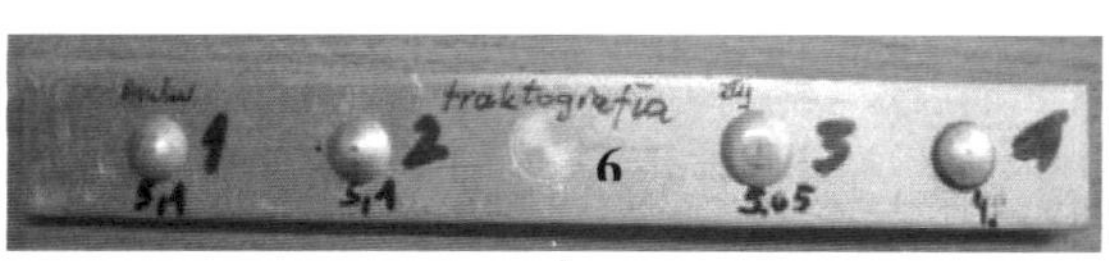

a)

b)

Figure 16. Specimens for the microstructure tests, a) specimen No ILOT.4.1, b) specimen No ILOT.4.2 [17].

Table III. Specimens for the microstructure tests, series ILOT4.

Specimens label	Rivet No.	Rivet head	Rivet material	Surface treatment – rivet:	Riveting method	Remarks	Real D/Do
ILOT.4.1	1	round	PA24	anodised	conventional	brushed	1.72
	2	round	PA24	not anodised	conventional	—	1.72
	3	brazier with a compensator	PA25	not anodised	conventional	hold-on r=14	skew of driven head
	4	universal	2117	alodined	conventional	—	1.71
ILOT.4.2	5	brazier with a compensator	PA25	not anodised	conventional	hold on r=14	1.69
ILOT.4.1	6	90° countersunk	PA24	not anodised	conventional	brushed	1.72

The microsections of specimens after riveting process have been studied with the use of optical and scanning electron microscopes. Local punctual and linear welds have been observed on the microsections of brazier rivets with a compensator (Fig. 17) and round rivets (Fig. 18, 19).

Because of a speculation, that grinding products could had been forced into a gap between rivet and hole edge, after identification areas where cold-welding could exist in the microsection of the countersunk rivet, this miscrosection has been etched. Pictures presented in Figures 18–19 prove that in the investigated specimens a cold-welding has occurred indeed.

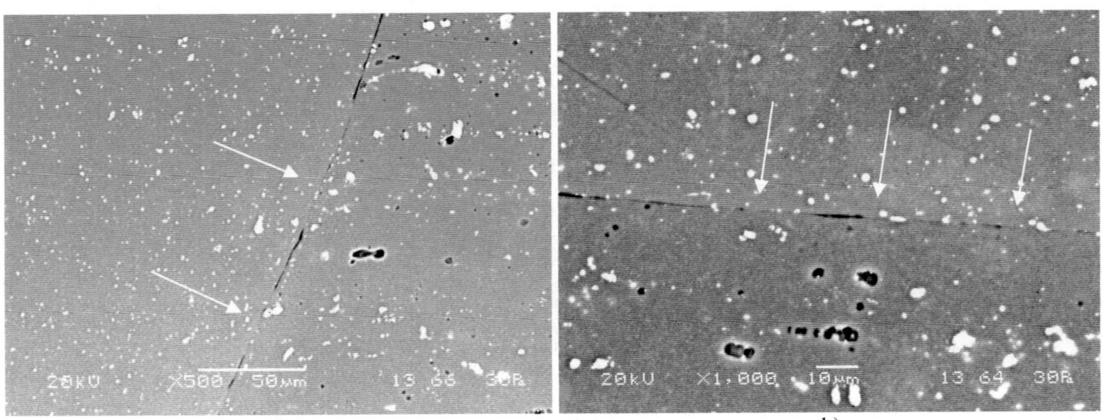

a) b)

Figure 17. Microsections of brazier rivet with compensator, a) rivet No. 3, b) rivet No. 5 (areas of practically ideal joint were marked with arrows) [17].

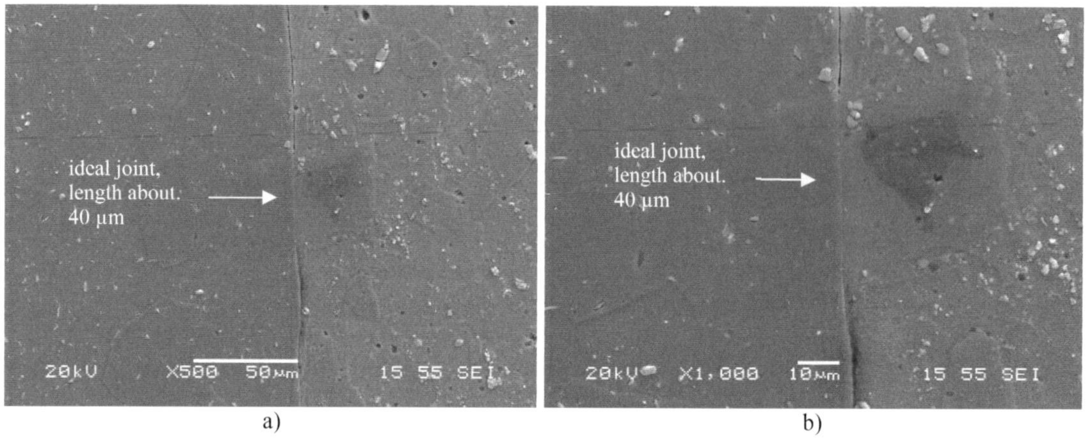

Figure 18. Microsections of round rivet No. 2, pictures obtained with secondary electrons technique, after etching, various magnifications [17].

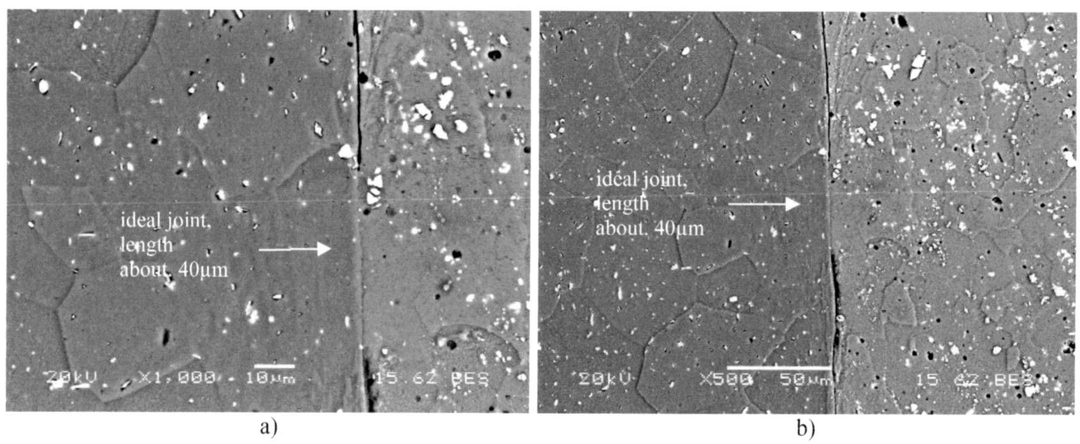

Figure 19. Microsections of round rivet No. 2, pictures obtained with backscattered electrons technique, after etching, various magnifications [17].

SUMMARY

Squeezing force level influence strongly fatigue life of riveted joints. Many authors explained this as a results of more beneficial residual stress system around rivets. After analysis of Müller and Hart-Smith results [3], the working hypothesis has been assumed that increase in fatigue life caused by higher riveting force is so high that the more beneficial residual stresses cannot be the only explanation, but a new phenomenon in riveting process occurs. The hypothesis assumed that, in the case of high riveting forces, a cold-welding on the rivet/hole interface can occur.

Another premise supporting the conjecture that there is some new phenomena in riveting joints was the 'S'-shaped signals recorded by strain gauges during the riveting process. FE analysis indicated that conditions necessary for adhesive joint formation are met during a riveting process.

Several microstructure tests were conducted in order to verify experimentally the hypothesis of cold welding during the riveting process. Observations of microsections with the use of optical and scanning electron microscope were performed. In the case of several specimens the punctual and linear welding between rivet shank and sheets were observed.

The results of preliminary investigation presented in this paper are a premise to recognized as a true the hypothesis formulated at the beginning. The microstructure tests proof that cold-welding during riveting can occur, however this is not a conclusive evidence that cold welding is a reason of huge increase in fatigue life connected with high riveting force and a reversal in strain signal ('S' shaped graph) during riveting.

In order to precisely investigate the phenomenon of cold-welding in riveting joints in airframes the authors are intended to continue presented researches. The objective of this researches will be defining necessary conditions for occurrence presented phenomenon between elements made of aluminium alloy used in aerospace industry and determination possibility of utilize it in riveted joints. Probably it is possible to optimize the riveting process (e.g. parameters, coatings, materials, surface, thermal treatment and geometry of a rivet) in order to maximise the surface of joints between a rivet shank and sheets. This would lead to a tremendous increase in fatigue life of riveted joints.

ACKNOWLEDGEMENTS

The financial support from the governmental research funds within the years 2006–2012 via the Eureka-IMPERJA project No. E!3496, as well as own financial resources of Institute of Aviation and Motor Transport Institute is acknowledged.

REFERENCES

[1] Kaniowski J.(2013) *The Synthetic Description of The Results, Scientific Achievements and Practical Application of the Eureka-IMPERJA Project, E3496!, titled "Improving the Fatigue Performance of Riveted Joints in Airframes"*. In: *Fatigue of Aircraft Structures*, pp.38-57, Niepokólczycki A. (Ed.), Monographic Series, Institute of Aviation Scientific Publication, Warsaw, Poland. http://www.degruyter.com/view/j/fas.2012.1.issue-4/v10164-012-0055-4/v10164-012-0055-4.xml?format=INT, DOI: 10.2478/v10164-012-0055-4.

[2] Muller R.P.G. (1995), *An Experimental and Analytical Investigation on the Fatigue Behaviour of Fuselage Riveted Lap Joint*. Ph.D. thesis, TU Delft, Netherlands.

[3] *Müller* R. P.G., Hart-Smith L.J. (1997), *Making fuselage riveting lap splices with 200-year crack-free fatigue life*. In: *Fatigue in new and ageing aircraft*. Proceedings of the I9-th ICAF Symposium of the International Committee of Aeronautical Fatigue1997, pp. 499-522, Cook R., Poole P. (Eds.) Edinburgh, EMAS Publishing.

[4] Skorupa A., Skorupa M. (2012), *Riveted Lap Joints in Aircraft Fuselage. Design, Analysis and Properties. Solid Mechanics and Application*, vol. 189. Springer, Nederlands.

[5] Wronicz W., Kaniowski J. (2011), *Experimental and Numerical Study of Strain Progress During and After Riveting Process for Brazier Rivet and Rivet with Compensator - Squeezing Force and Rivet Type Effect*. In: *Fatigue of Aircraft Structures*, pp. 165-189, Niepokólczycki A. (Ed.), Institute of Aviation Scientific Publication, Monographic Series. ISBN 2081-7738, http://www.degruyter.com/view/j/fas.2011.1.issue-3/v10164-010-0047-1/v10164-010-0047-1.xml?format=INT, DOI: 10.2478/v10164-010-0047-1, August 2011.

[6] Rans C. D. (2007), *The Role of Rivet Installation on the Fatigue Performance of Riveted Lap* Joints. PhD thesis, Carleton University Ottawa, Ontario, Canada.

[7] Kaniowski J. (2015), *Comparison of Selected Rivet And Riveting Instructions*. In: *Fatigue of Aircraft Structures*, A. Niepokólczycki (Ed.), Monographic Series, Institute of Aviation Scientific Publication (in printing).

[8] Hart-Smith L. J. (2010) *Lessons learned by one aerospace structures engineer in a 40-year career*, 6th Australasian Congress on Applied Mathematic, ACAM 6, 12th - 15th December 2010, Perth, Australia. https://www.engineersaustralia.org.au/sites/default/files/shado/Events/Eminent%20Speaker%20Series/2010%20 ACAM%206%20LESSONS%20LEARNED%20SLIDE%20SHOW,%20Full%20Length%20Show.pdf

[9] Bartone, L. M., Mandel, M., W. (1944), *Tensile Tests of NACA and Conventional Machine- Countersunk Flush Rivets*. In: *NACA Wartime Report L-176*. United States, UNT Digital Library. http://digital.library.unt.edu/ark:/67531/metadc61751/.

[10] Kaniowski J., Gadalińska E., Wronicz W., Szałkowski S., Kowalski W., Karliński W., Korzeniewski B. (2011), Improving *the Fatigue Performance of Riveted Joints in Airframes The final report of the Research Project No. EUREKA/59/2006*, Institute of Aviation internal report, Warsaw, p. 229.

[11] Kaniowski J., Wronicz W. (2013), *Local phenomena during riveting process*. In: *Fatigue of Aircraft Structures*, pp. 66-78, Niepokólczycki A. (Ed.), Monographic Series, Institute of Aviation Scientific Publication.

[12] Li G., Shi G., Berlinger N.C. (2006), *Study of the Residual Strain in Lap Joint*. In: Journal of Aircraft, Vol. 43, No. 4, pp. 1145-1151.

[13] Kaniowski J., Wronicz W. (2012), *Analysis of the Quasi-Static Riveting Process for 90° Countersunk Rivet*. In: *Fatigue of Aircraft Structures*, pp. 141-156. Niepokólczycki A. (Ed.), Monographic Series, Institute of Aviation Scientific Publication, Warsaw, Poland,. ISBN 2081-7738, http://www.degruyter.com/view/j/fas.2012.1.issue-4/v10164-012-0065-2/v10164-012-0065-2.xml?format=INT, DOI: 10.2478/v10164-012-0065-2, April 2013.

[14] Stojman I.M. (1985) Cold Welding of Metals, Mashinostroenie, Leningrad, p.38 and 51. (in Russian)

[15] Sachackij Grigorij P.(1979), The Technology of Cold Welding of Metals, Naukowa Dumka, Kiev 1979, p. 111. (in Russian).

[16] Niewiadomski C., Olszewski M. (1961), *Cold Welding of Metals*, PWT, Warsaw. (in Polish).

[17] Kaniowski J., Wronicz W, Pietrzak K. (2015), *Analysis of the local phenomena during riveting process*. In: *Fatigue and Fracture 2014*, , Solid State Phenomena Vol.224, Skibicki D. (Ed.), Trans Tech Publication Switzerland, 2015, pp. 255-260. DOI. http://www.scientific.net/SSP.224.255.

Strength, Fatigue and Crack Growth Resistance of Al-Alloys that are applied in Current Transport Airplanes Wing and Fuselage Skin

Boris G. Nesterenko[1], Grigory I. Nesterenko[2], Alexander V. Kulemin[2], Valentin N. Bassov[2]

[1]*United Aircraft Company (UAC)*
[2]*Central Aerohydrodynamic Institute (TsAGI)*

Abstract: The results are presented of TsAGI's experimental investigations of strength, fatigue and crack growth resistance of the state-of-the-art enhanced Al-alloys developed at VIAM (Russia) and ALCOA (USA). The experimental data were obtained in the course of testing the standard specimens with use of electro-hydraulic machines of MTS firm (USA), Instron (Great Britain) and Schenk (FRG). The characteristics of fatigue cracks growth rate and the R-curves were estimated based on ASTM standards (USA). The tested specimens were cut out of semi-products manufactured by series technologies.

INTRODUCTION

Currently the aviation alloys still remain the principle structural material for transport aircraft. Nowadays the content of alloys is 70% of total structural materials amount in airframe [1]. To provide the high weight efficiency in combination with high life time and high operational durability characteristics of aircraft structures the Al-alloy are needed to have the following characteristics: high resistance against variable loads, low fatigue cracks growth rate, specified residual strength and high corrosion resistance. The mentioned above characteristics of Russian and foreign enhanced Al-alloys that are recently created and used in existing and being designed aircraft are considered in this paper (excluded the corrosion resistance). Only those characteristics of materials that are testes in TsAGI by one and the same technique are shown. The results of these tests are summarized in Table I and given in Figures 1-12.

Table I. Al-alloys properties.

Airplane	Material	F_{tu}, MPa	F_{ty}, MPa	δ_5, %	*Fe*, %	*Si*, %	*Zr*, %	N_{133}, cycle	*m*	$(da/dN)_{31}$, mm/kcycle	K_{app}, MPa√m	K_c, MPa√m
Wing upper surface												
B777, A380	7055-T7751 plate	620	595	7	0.13	0.12	0.11	300 000	5.9	3.1	90	
	V96ts-3pchT12 plate	635	595	10	0.12	0.03	0.12	320 000	7.4	4.1	70	
IL-96-300	V95ochT2 plate	540	460	10	0.12	0.07	—	170 000	4.7	2.4	160	198
AN-124	1973T2 plate	531	484	14	0.15	0.1	0.12	110 000	4.5	2.6	111	
Wing lower surface												
TU-204	1163T plate	460	340	20	0.07	0.04	—	205 000	4.8	2.6	158	212
IL-96-300	1163T7 plate	500	390	14	0.12	0.06	—	200 000	5.0	2.6	163	
AN-124	1161T extruded panel	474	324	16	0.10	0.03	0.11	220 000	5.3	1.4	155	
A380	2324-T39 plate	500	460	12	0.08	0.04	—	275 000	5.9	2.5	148	
A340	C433-T351 plate	456	320	15	—	—	—	260 000	5.3	1.1		
Fuselage												
IL-96-300	1163ATV sheet	442	315	24	0.12	0.05	—	90 000	4	1.8	156	239
TU-204	1163RDTV sheet	456	350	23	0.14	0.03	—	115 000	4.0	1.7	171	245
A380	2524-T3 sheet	450	363	20	0.07	0.03	—	95 000	3.4	1.8	181	264
	1441RT1 sheet	443	364	13	0.05	0.03	—	100 000	4.34	3	115	141
	6013-T6 HDT sheet	365	333	—	—	—	—	85 000	3.6	1.9	175	252

The characteristics of materials that are developed in Russia (VIAM) and in USA (ALCOA) are as follows:

- 2524-T3 (ALCOA), 1163ATV (VIAM) and 1163RDTV enhanced alloys for fuselage skin;
- 6013-T6 HDT (ALCOA), 144RT1 (VIAM) stain-proof alloys for fuselage skin;
- 2324-T39, C433-T351 (ALCOA), 1163T, 1163T7 and 1161T (VIAM) enhanced crack growth resistance alloys for wing lower surface;
- 7055-T7751 (ALCOA), V95ochT2, V96ts-3pchT12 and 1973T2 (VIAM) enhanced static strength alloys for wing upper surface.

It is to be noted that the development of structural material is carried out differentially for the wing upper surface, the wing lower surface, for the fuselage upper part (above-the-floor line part) and for the fuselage lower part. The level of rated stresses in wing upper surface skin is specified as conditioned by wing upper surface static strength under compressive load. The stresses in the wing lower surface and in the fuselage skin are specified, mainly, as conditioned by skin fatigue resistance and durability.

The properties of a number of alloys mentioned are considered partially in [2-5]. For the listed above alloys the percentage of silicium (Si), iron (Fe) and zirconium (Zr) is shown. The content of these elements was estimated based on results of special chemical analysis by OJSC VILS and VIAM institutes. The specimens analyzed were tested in TsAGI. The difference is observed of destruction regularities in Al-alloys and composite materials. The models of aircraft where these alloys are used are given in this paper.

STRENGTH

The strength characteristics are given in Table I as follows: the F_{tu} damage threshold, the F_{ty} yield strength, the δ_5 (El) percentage extension. These characteristics were evaluated based on testing the specimens by Russian standards.

The table data make it possible to conclude as follows: the strength and the yield strength values of 7055-T7751 and V96ts-3pchT12 alloys that have Zr additives are significantly higher that the corresponding characteristics of Al alloys, which are used for wing upper surface skin of current aircraft.

The δ_5 percentage extensions in Al alloys, which are used for the wing lower surface and the fuselage significantly exceed those of alloys, which are used for the wing upper surface.

The 1441RT1 and 6013-T6 HDT welded stain-proof alloys are developed for fuselage skin. The 1441RT1 Al-Li alloy is developed based on the Al-Cu-Mg-Li complex, the 6013-T6 HDT alloy is developed based on Al-Mg-Si-Cu complex. The sheets of 1441RT1 Al-Li alloy have an advantage in terms of lower density of approximately 7% vs. the other alloys [1].

FATIGUE

The fatigue tests results are given in Table I and in Figures 1–3. The plane specimens that were strips with 6 mm diameter hole in center were tested (the stress concentration factor is $K_t^{gross} = 3.1$). The specimen width was 36 mm. they were tested by Instron, MTS и Schenk electro-hydraulic test rigs. The loading frequency was 3…5 Hz. The specimens were tested under zero-to-tension cycle stresses (asymmetry ratio is $R = 0$) in $\sigma_{max}^{gross} = 80\ldots220$ MPa maximal stresses range. The $\sigma_{max}^{gross} = 133$ MPa zero-to-tension cycle stresses are the basic type of tests commonly used in Russia to compare the properties of aviation Al alloys materials. Table I gives the magnitudes of m exponent quantity of S-N diagram of tested alloys:

$$\sigma^m N = 10^C, \tag{1}$$

where m and C are constants.

The Table I gives also the N_{133} durability magnitudes that present the alloys averaged durability magnitudes under the $\sigma_{max}^{gross} = 133$ MPa maximal gross stresses and the cycle asymmetry ratio $R = 0$.

Based on the experimental data given it follows that the m exponents of S-N diagram are equal:

- 4.5…7.4 for wing skin alloys;
- 3.4…4.3 for fuselage skin alloys.

In logarithmic coordinates the straight lines present the above mentioned σ-N relations in tested durabilities range for sheet specimens that are used for fuselage skin, plates and extruded panels for wing skin.

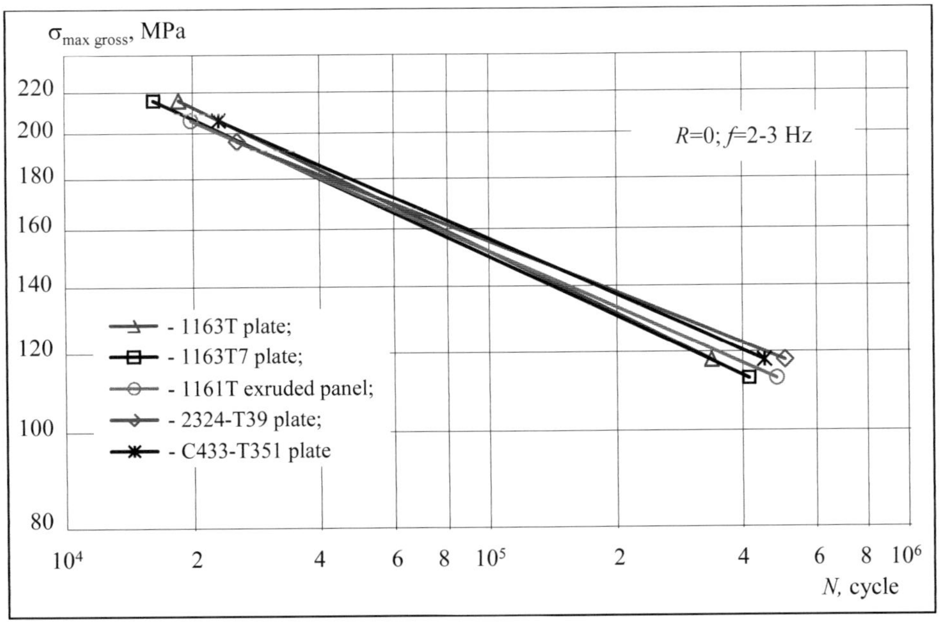

Figure 1. Fatigue of wing lower surface skin material specimens.

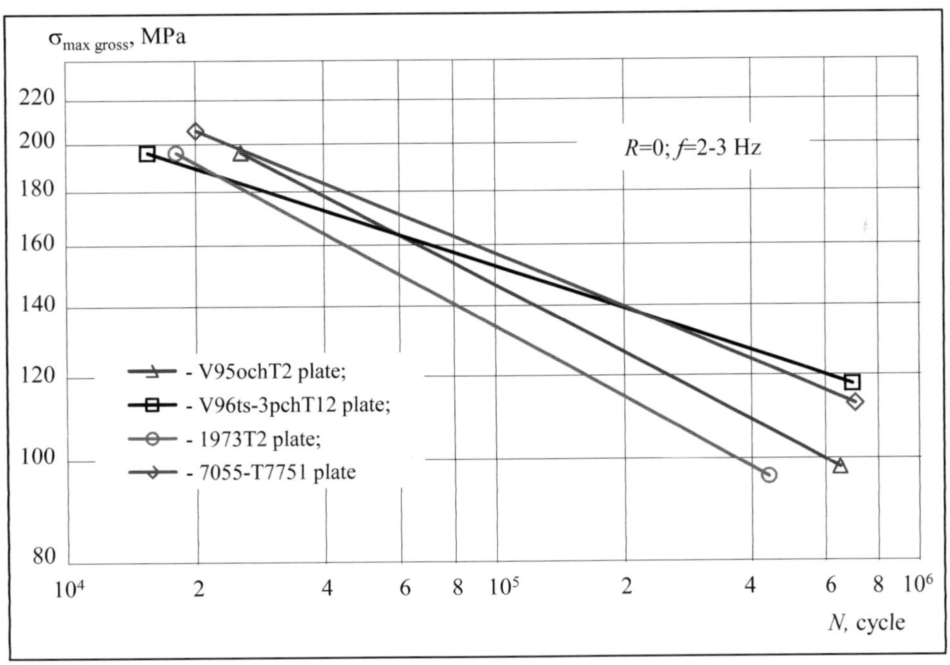

Figure 2. Fatigue of wing upper surface skin material specimens.

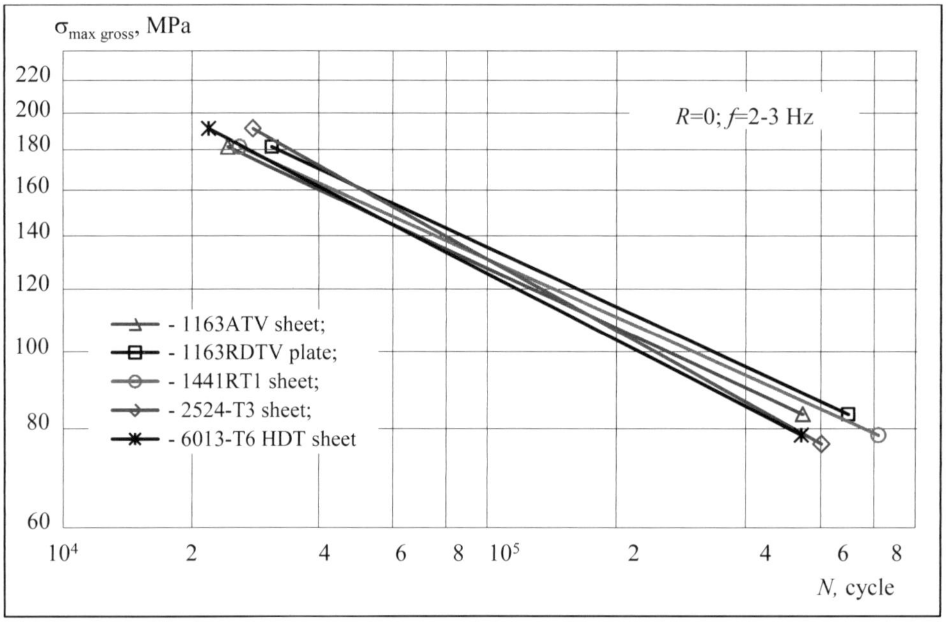

Figure 3. Fatigue of fuselage skin material specimens.

FATIGUE CRACKS GROWTH RATE AND DURATION

The cyclic fracture strength tests results are given in Table I and in Figures 4–9. The flat specimens with central crack were tested. The specimens width was 200 mm for the wing lower surface skin and the fuselage. The specimens width was 100 mm for the wing upper surface skin. The initial through cracks of $2a_0$=6.4 mm length were created in the specimens' center by electrolytic erosion technique. The Instron, MTS и Schenk electro-hydraulic test rigs were used in testing campaign. The specimens were tested under near to zero-to-tension cycle stresses ($R \approx 0$), the maximal σ_{max}=105…130 MPa stresses and loading frequency of 1…3 Hz. The tests were performed without elimination of specimens buckling in crack zone. The crack growth duration was estimated visually by optical microscope.

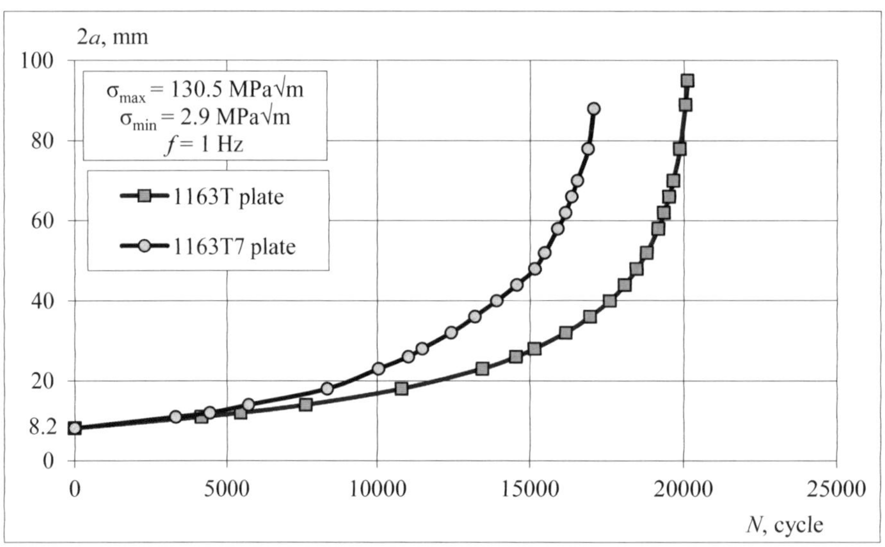

Figure 4. Cracks growth duration in material specimens for wing lower surface skin.

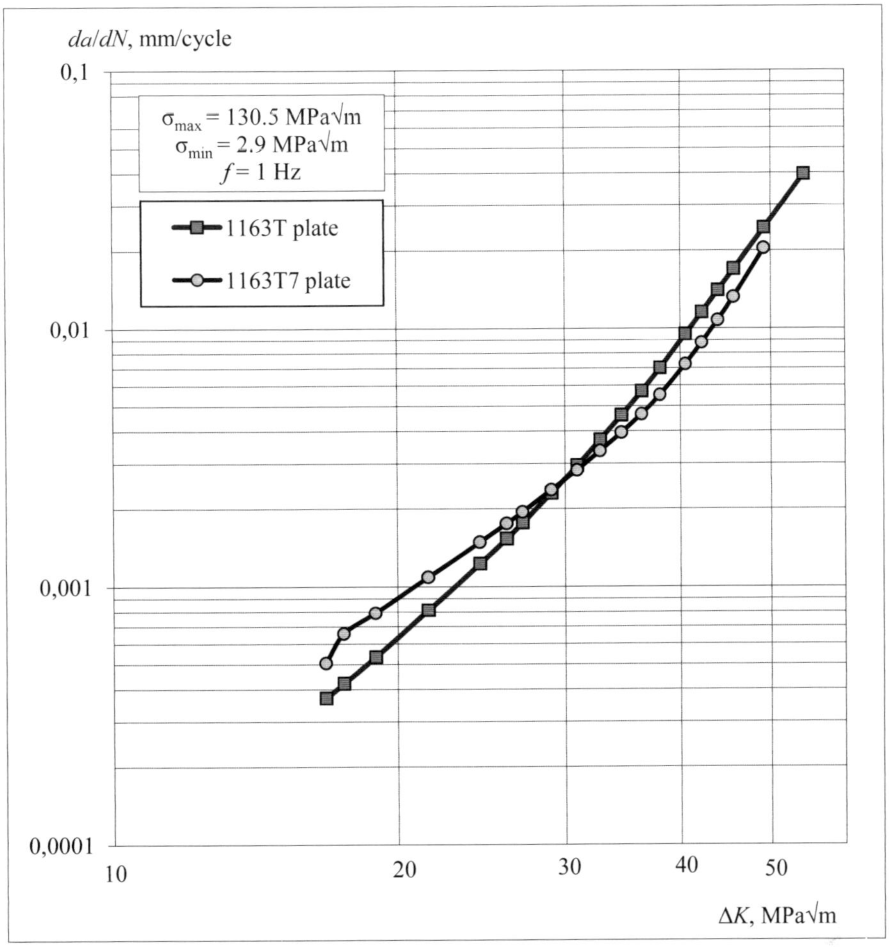

Figure 5. Cracks growth rate in material specimens for wing lower surface skin.

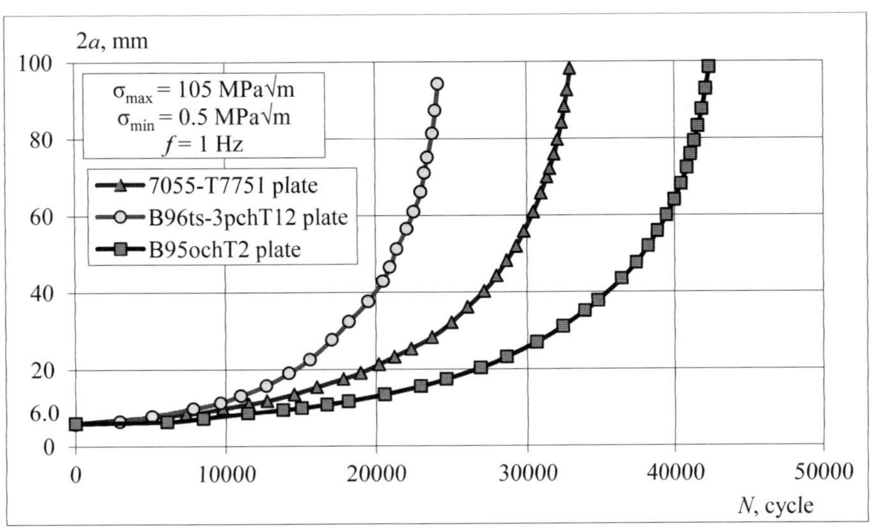

Figure 6. Cracks growth duration in material specimens for wing upper surface skin.

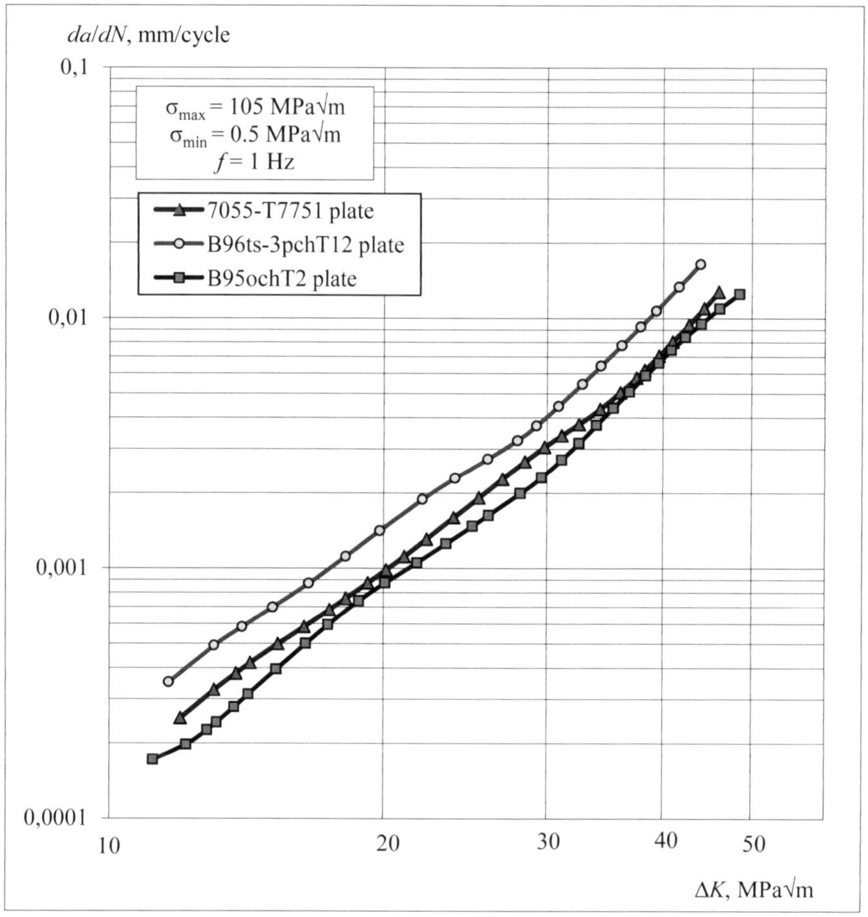

Figure 7. Cracks growth rate in material specimens for wing upper surface skin.

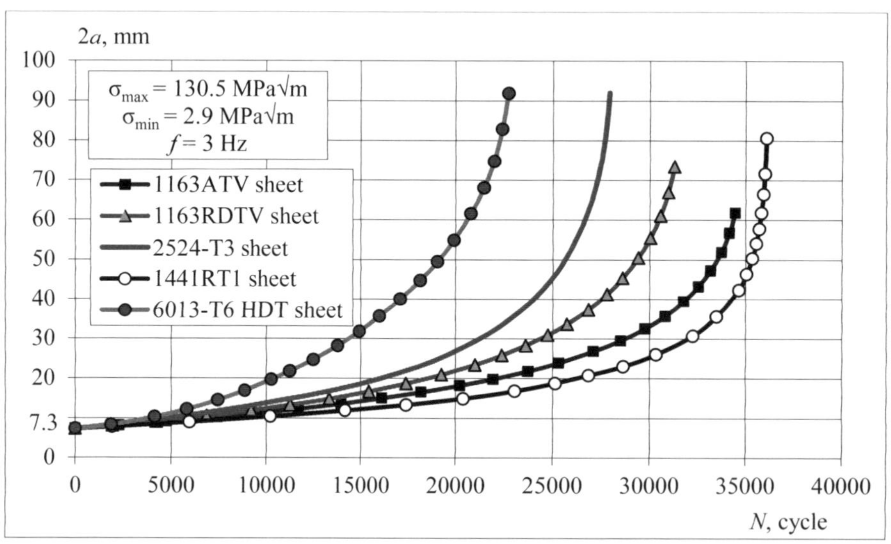

Figure 8. Cracks growth duration in material specimens for fuselage skin.

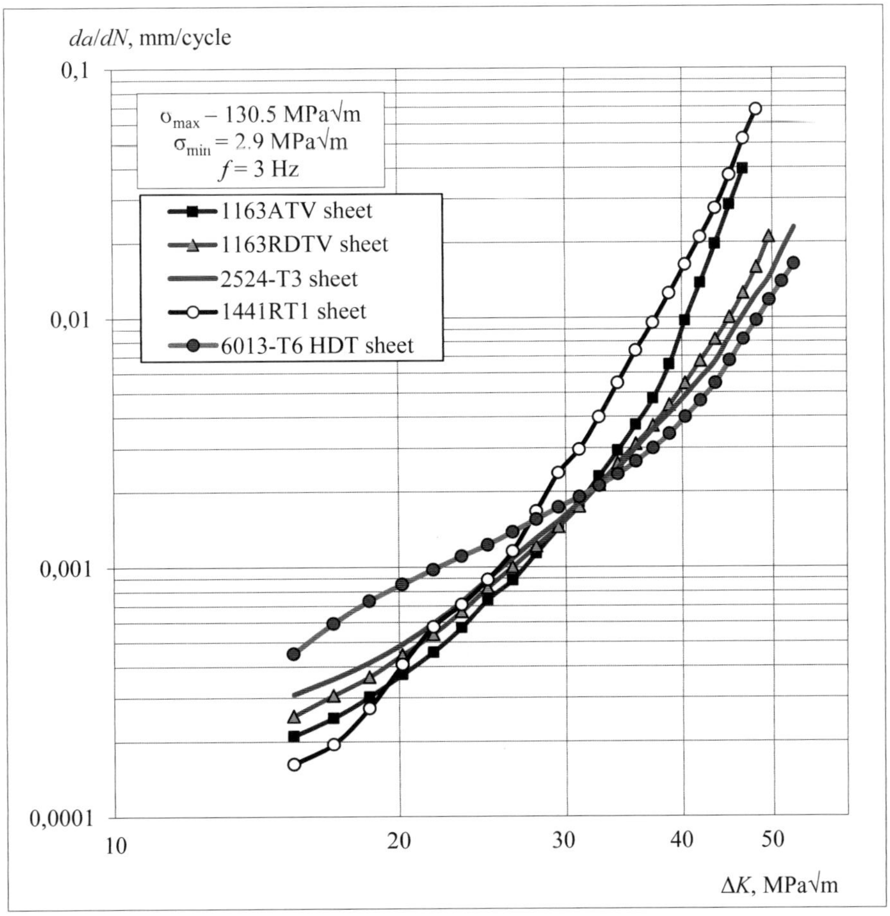

Figure 9. Cracks growth rate in material specimens for fuselage skin.

The cracks growth rate test data were processed by the linear fracture mechanics.

Table I gives the cracks growth rate magnitudes $(da/dN)_{31}$ in mm per kcycle. In this case, *a* denotes a half-length of crack. The values $(da/dN)_{31}$ are equal to cracks growth rate under ΔK =31 MPa√m stress intensity factor amplitude. The ΔK = 31 MPa√m amplitude is used in Russia to compare the cracks growth rates for different aviation Al alloys materials.

Based on the shown experimental data it follows that the *da/dN* fatigue cracks growth rates in wing and fuselage skins materials under ΔK = 31 MPa√m stress intensity factor amplitude are in (1.1 … 4.1) mm/kcycle range.

The diagrams of $(da/dN) = f(\Delta K)$ fuselage skin sheets fatigue cracks growth rates intersect approximately under the ΔK =31 MPa√m stress intensity factor amplitude.

RESIDUAL STRENGTH

The residual strength tests results in form of K_{app} and K_c critical stress intensity factors are given in Table I and Figures 10–12.

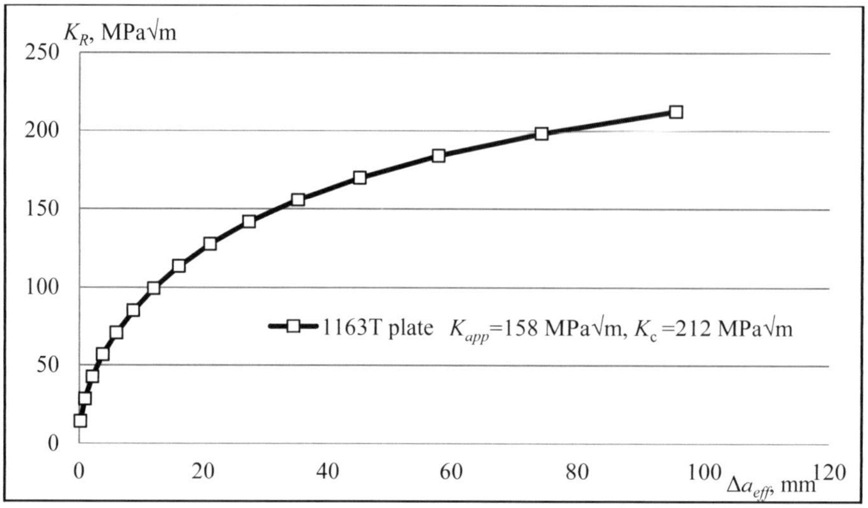

Figure 10. R-curve of wing lower surface skin material.

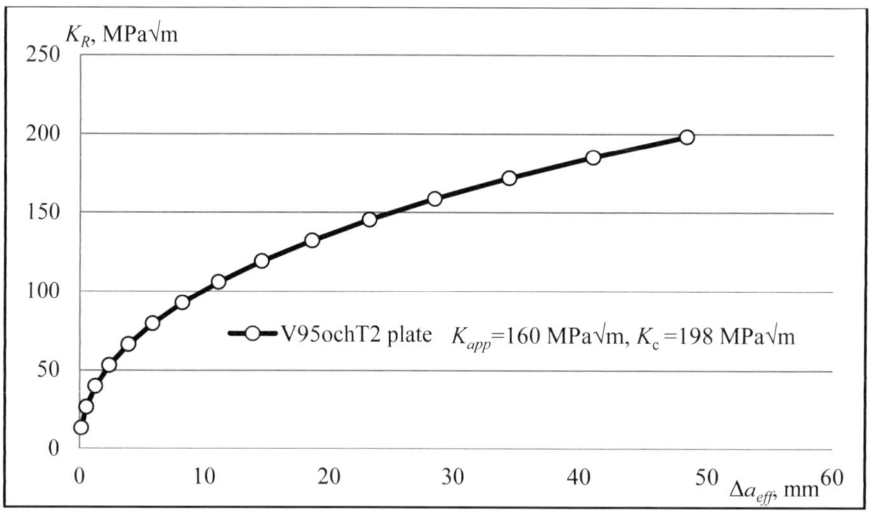

Figure 11. R-curve of wing upper surface skin material.

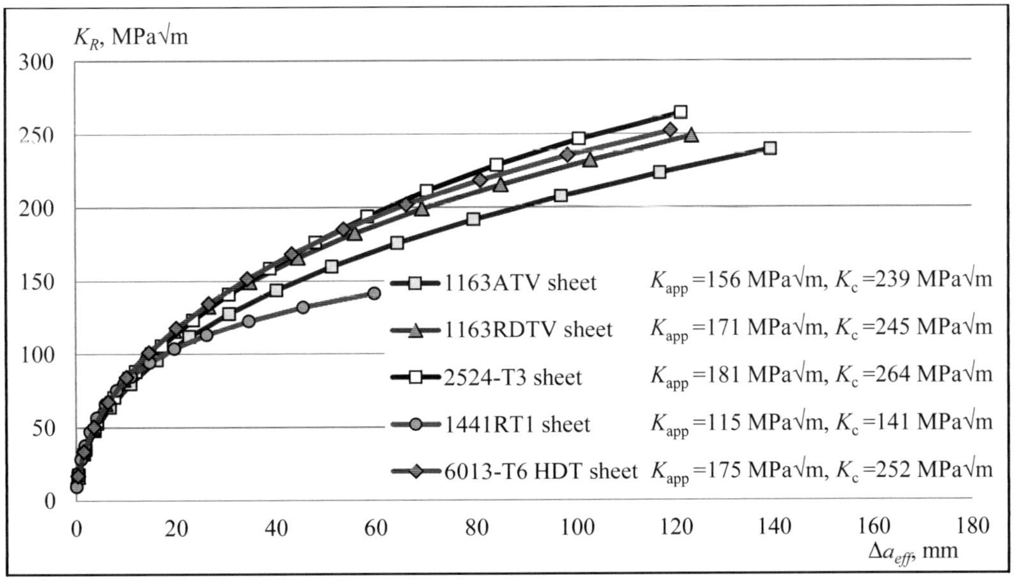

Figure 12. R-curve of fuselage skin material.

The flat specimens with central crack were tested. The W specimens width was 1200 mm. The initial cracks of $2a_0 = 0.33W$ length were created in the specimens' center. The MTS electro-hydraulic test rig was used in testing campaign. The R-curves were constructed based on tests results. These diagrams were used to evaluate the K_{app} and K_c magnitudes. The tests were performed with elimination of specimens buckling (buckling failure) in crack zone by virtue of special means. The tests followed the ASTM standard.

The main tests results are as follows:

The magnitudes of critical stress intensity factors of ductile skin materials for wing lower surface and fuselage are up to twice higher vs. the critical stress intensity factors of high-strength skin materials for wing upper surface.

The specimens were tested under near to zero-to-tension cycle stresses ($R \approx 0$), the maximal $\sigma_{max} = 105\ldots130$ MPa stresses and loading frequency of 1…3 Hz. The crack growth duration was estimated visually by optical microscope.

For the wing lower surface and fuselage ductile skin materials the K_c value is approximately 40…60% higher than K_{app} has.

ON MAIN DIFFERENCES IN DESTRUCTION REGULARITIES IN AL-ALLOYS AND COMPOSITE MATERIALS

The results of investigating the composite materials destruction regularities in Russia and abroad [6] found out the main differences between these two types of materials.

The slow damage growth rate concept is inappropriate for composites. Therefore the damages non-propagation principle must be the primary requirement for them.

Unlike in metal where the single crack develops across the maximal tensile force, in composite material a very complicated phenomenon develops that covers the multiple interlaminar cracks as well as results into fibers destruction.

The fatigue damages of composites represent, mainly, the delaminations, which are oriented in parallel to longitudinal forces.

This are the compressive stresses that are more critical for composites and that cause the buckling loss than the tensile stresses that cause the metal structures fracture.

CONCLUSION

- The following parameters were estimated experimentally: the (F_{tu}, F_{ty}, δ_5) static strength characteristics, the (σ-N) fatigue diagrams, the ($2a$-N) cracks growth duration, the (da/dN-ΔK) fatigue cracks growth rates kinetic diagrams, the (K_R-Δa_{eff}) R-curves for enhanced Al alloys that are applied in current national and foreign airplanes wing and fuselage skin.
- The analysis of these strength, fatigue and crack growth resistance characteristics was carried out.

REFERENCES

[1] Friedlyander I.N. (2002). Al alloys in aircraft in 1970-2000 and 2001-2015. *Light Alloys Technology*, № 4, pp. 12-17.

[2] Nesterenko B.G. and Nesterenko G.I. (2011). *Analysis of requirements on fatigue and damage tolerance for civil transport airplanes*, Proceedings of the 26th ICAF Symposium, vol. I, pp. 39–59, Montreal, Canada.

[3] Bassov V.N., Nesterenko B.G., Nesterenko G.I. (2008). Destruction of high-strength Al alloys. *Polet. 90-th anniversary of TsAGI*, pp. 87-92.

[4] Nesterenko B.G. (2010). Fracture strength of civil aircraft structures skin materials. *The Moscow State Technical University of Civil Aviation (MSTUCA) Scientific Bulletin*, № 153, pp. 7-14.

[5] Bassov V.N., Nesterenko G.I (2010). Strength and fatigue of civil airplanes structures skin materials. *The Moscow State Technical University of Civil Aviation (MSTUCA) Scientific Bulletin*, № 153, pp. 15-23.

[6] Rouchon, J. (2009). *Fatigue and damage tolerance evaluation of structures. The composite materials response*, Proceedings of the 22-d ICAF Symposium, Rotterdam, the Netherlands.

Development of a Fatigue Crack Growth Rate Material Model for 7249-T76511 Aluminium Alloy

Y. Bombardier, M. Liao

National Research Council Canada (NRC) – Aerospace
1200 Montreal Rd, Ottawa, Ontario, Canada, K1A 0R6

Abstract: Fatigue crack growth rate (FCGR) tests were conducted for 7249-T76511 in laboratory air and in a corrosive environment using compact tension (C(T)) and single edge notch tension (SENT) specimens. Overall, the variability of the FCGR test results obtained from the C(T) specimens in laboratory air was small. The tests conducted in the corrosive environment revealed that the cells designed to contain the corrosive solution in the notch may have affected the FCGR in the threshold regime.

The FCGR data obtained from laboratory air was used to develop material models. The accuracy of these models was assessed by comparing the calculated crack growth life against experimental data obtained from C(T) and modified middle tension (M(T)) specimen results tested using three different loading spectra. The first type of material model was based on a tabular look-up approach coupled with a load interaction model. For this type of model, the load interaction model had to be calibrated using variable amplitude loading spectrum data, such that the selected model parameters may not be suitable for other loading spectra. The second type of material model was developed using FASTRAN by tuning the parameters to fit the steady-state FCGR test results only. In summary, the fatigue crack growth life calculated using the material models presented in this paper agree with the experimental results within 33%. However, further testing is required to gain confidence on the accuracy of the proposed material models for different thicknesses and spectra, and to identify their limits.

INTRODUCTION

The 7249–T76511 aluminium alloy is a relatively new alloy which is being used for new transport aircraft wings as a 'drop-in' replacement for existing 7075-T6511 aluminium alloy components. This alloy was optimized to maintain a high level of strength and fracture toughness, while improving its corrosion resistance over 7075-T6511. Despite the numerous advantages of using 7249-T76511 over 7075-T6511, no fatigue crack growth rate (FCGR) data could be found in publically available literature, which limited the ability to conduct fatigue life assessment of structures made of this material. As it is expected that 7249-T76511 will be more commonly used for airframe components in the future, there was a need to obtain FCGR data and develop a model for this material to support current and future platforms.

This paper presents FCGR tests conducted on 7249-T76511 under constant and variable amplitude loadings. The constant amplitude tests were conducted to establish the baseline FCGR material properties at various stress ratios (R=0.05 to 0.85) for crack growth modelling. The variable amplitude test results were used to verify the developed FCGR material models.

MATERIAL

The tested 7249-T76511 aluminium alloy material was made by Universal Alloy Corporation (UAC) in the form of extrusions 0.25 inch thick, 3 inches wide, and 144 inches long. According to the tensile tests conducted by UAC for the longitudinal-transverse direction, the yield and ultimate strengths of this alloy are 76.3 ksi and 83.9 ksi, respectively. Compared to 7075-T6511 extrusions, this alloy has higher strengths and better corrosion resistance.

The National Research Council Canada (NRC) conducted residual stress measurements using two techniques: the X-ray diffraction (XRD) and the slitting techniques. Residual stress measurements obtained using the slitting method, which provides average through-the-thickness residual stress values, indicated that there were insignificant average residual stresses (less than 0.1 ksi) in the extrusion. XRD measurements, however, indicated that low compressive residual stresses (-5 to -10 ksi) may be present at the surface of the extrusion (measurement depth smaller than 0.001 inch).

TEST PROCEDURES

Overview

FCGR tests were conducted according to ASTM E647-08 [1] using compact tension (C(T)) and single edge notch tension (SENT) specimens. To obtained long data, steady-state fatigue tests were conducted on C(T) specimens under constant stress ratios, R, of 0.05, 0.60, and 0.85. The tests were conducted in laboratory air as well as in a 3.5% sodium chloride (NaCl) solution in order to quantify the effect of the corrosive environment on the FCGR. SENT specimens were used to obtain small crack data from naturally nucleated fatigue cracks. Variable amplitude FCGR tests were conducted on C(T) and modified middle tension (M(T)) specimens to acquire experimental data under aircraft spectrum loadings for the validation of the developed material models. As opposed to the standard M(T) specimens with a centrally located crack, the modified M(T) specimens only have a radial crack propagating from a hole.

The geometries of the C(T), SENT, and modified M(T) specimens are shown in Figure 1. The holes of the C(T) specimens were bevelled to maintain load straightness and concentricity to minimize undesired out-of-plane bending moments that could promote unequal crack growth [2]. The surfaces of the C(T) and modified M(T) specimens were polished using progressively finer grits to achieve a mirror like finish to enhance visual crack measurements using optical travelling microscopes.

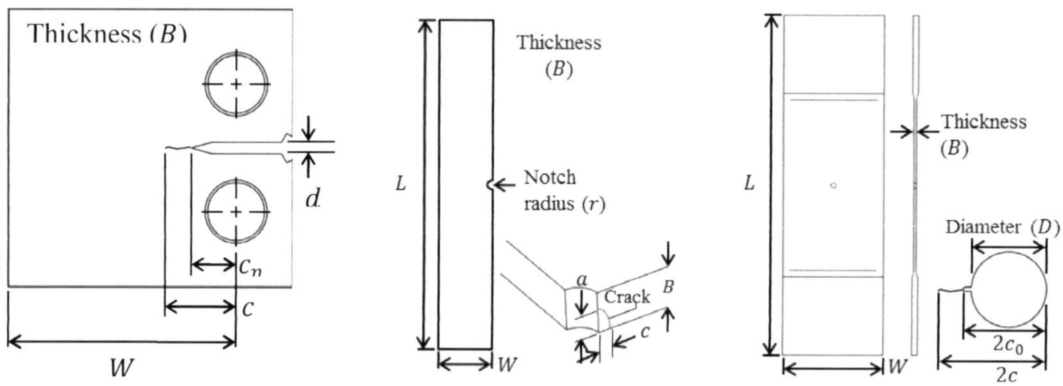

a) C(T) specimen

$\boldsymbol{W}$=2.0 inches; $\boldsymbol{B}$=0.25 inch; $\boldsymbol{c_n}$=0.40 inch; $\boldsymbol{d}$=0.1 inch

b) SENT specimen

$\boldsymbol{W}$=2.0 inch; $\boldsymbol{B}$=0.25 inch; $\boldsymbol{L}$=12.0 inches; $\boldsymbol{r}$= 0.125 inch

Modified M(T) specimen

=3.8 inches; $\boldsymbol{B}$=0.1 inch; $\boldsymbol{L}$=13.0 inches; =0.188 inch; $\boldsymbol{2c_0}$=0.208 inch

Figure 1. FCGR test specimens.

Steady-state FCGR tests

Compact tension specimens. Steady-state FCGR tests were performed according to the guidelines provided in ASTM E647-08 and using the FCGR test system developed by Fracture Technology Associates (FTA) [3]. The compliance method described in ASTM E647-08 was used to monitor crack growth of the C(T) specimens using a clip-on extensometer. The test procedure consisted of four stages: the pre-cracking stage, the K-decreasing stage, the K-increasing stage, and the constant load stage, where K refers to the crack tip stress intensity factor (SIF). All specimens were pre-cracked in K-control mode using a constant maximum SIF, K_{max}, and a constant stress ratio, R. Based on ASTM E 647-08, the specimens were pre-cracked up to a crack length of 0.1 inch. The K_{max} values for the pre-cracking was selected such that the initial FCGR for the K-decreasing stage would be between 2×10^{-7} and 4×10^{-7} inch/cycle. The K-decreasing stage initial K_{max} values were 4 ksi$\sqrt{\text{inch}}$ for R=0.05, 7 ksi$\sqrt{\text{inch}}$ for R=0.60, and 15 ksi$\sqrt{\text{inch}}$ for R=0.85. The K-decreasing stage was conducted with a K-gradient of -6.0 inch^{-1} and was terminated when the FCGR reached 4×10^{-9} inch/cycle or lower. The K-increasing stage was initiated at 2/3 of the initial K_{max} used for the K-decreasing stage, such that the FCGR results from the K-decreasing and K-increasing stages overlapped. The K-increasing stage was ended when the FCGR reached 1×10^{-6} inch/cycle. The terminal K-increasing load was decreased by 2% and was used for the constant load stage until failure. The steady-state FCGR tests were conducted at frequencies between 10 and 12 Hz. Crack size measurements obtained using the compliance method were corrected for crack curvature and visual measurements taken at the front and back surfaces of the specimens. All FCGR tests in laboratory air were conducted using the clevis and pin design described in ASTM E647-08 for a clevis opening equals to 1.05B. In this paper, this clevis and pin assembly is defined as Fixture A. During the tests, the temperature and relative humidity varied from 18°C to 28°C and 12% to 66% relative humidity.

The steady-state FCGR tests were also conducted in a corrosive environment by immersing the C(T) specimens in a 3.5% NaCl solution. The cells, shown in Figure 2, were designed and fabricated to contain the corrosive solution around the crack tip during the fatigue tests. Two cells were bonded to the front and back faces of each specimen using silicone. The silicon plug was moulded in the notch area to prevent the corrosive solution from leaking. The corrosion cells were fabricated using transparent acrylic to allow the test operator to perform visual crack size measurements during the tests. Rigid tubes, illustrated by the cell connectors in Figure 2, were integrated into the corrosion cell design to connect flexible hoses for circulating the corrosive solution through the cells. The corrosive solution was pumped from a tank to the front cell, flowed through the notch, and returned to the tank through the tube connected to the back cell. The FCGR tests were conducted with the pump constantly running. To accommodate the presence of the cells, clevises with an opening width of 0.82 inch were manufactured according to the clevis design suggested in ASTM E 647 for the clevis opening exceeding $1.05B$. In this paper, the corrosion test clevis and pin assembly is defined as Fixture B.

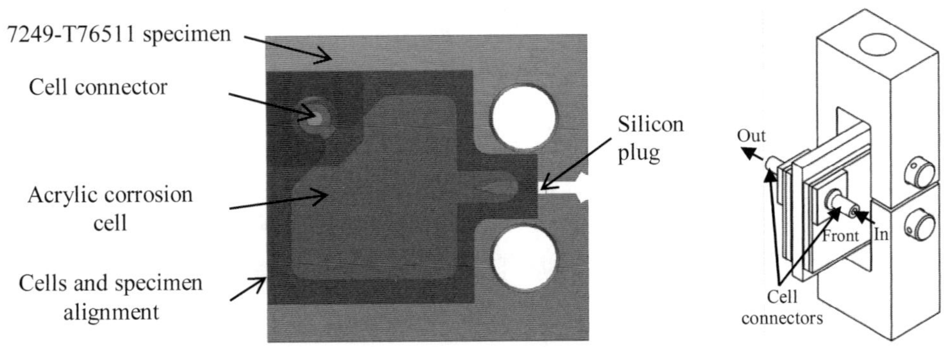

Figure 2. Corrosion cell design and specimen installation.

During these FCGR tests, data were acquired using the FTA FCGR Testing Software and then post-processed using the FTA FCGR Analysis Software [3]. Crack size corrections were applied to all specimens using the methodology outlined in ASTM E 647. FCGR was calculated using the 7-point incremental polynomial technique and the modified secant method [3]. The FCGR results for all test phases were post-processed together by concatenating the data of all test stages. The SIF ranges were also calculated using the ASTM 2% opening load method (ΔK_{OP}) [4], the $2/\pi$ correction to opening load method ($\Delta K_{2/\pi}$) [4], and the adjusted compliance ratio (ACR) method (ΔK_{ACR}) [4]. Unless otherwise specified, the FCGR results provided in this paper are expressed as a function of the applied SIF range, referred to as ΔK.

Single edge notch tension specimens. SENT specimens were tested at a frequency of 10 Hz using a constant amplitude spectrum with built-in marker band (MB) patterns. The spectrum, shown in Figure 3a, was composed of 3 blocks of cyclic loads at a constant stress ratio of R=0.05, separated by marker band loading blocks, shown by 10x MB, 4x MB, and 6x MB. Each marker band block contained a sequence of sub-blocks consisting of 100 cycles at R=0.6 followed by 10 cycles at R=0.05. The number of sub-blocks per marker band block varied (10 repeats for 10x MB, 4 repeats for 4x MB, and 6 repeats for 6x MB), such that it was easier to correlate the observed marker bands with the number of cycles. Examples of the marker bands observed during the fractographic analysis with the corresponding marker band block are shown in Figure 3b. The maximum spectrum load was selected such that a crack would nucleate within 1,000,000 cycles. The maximum load was kept constant throughout the test duration and only the minimum load was changed to modify the stress ratio within the marker band blocks.

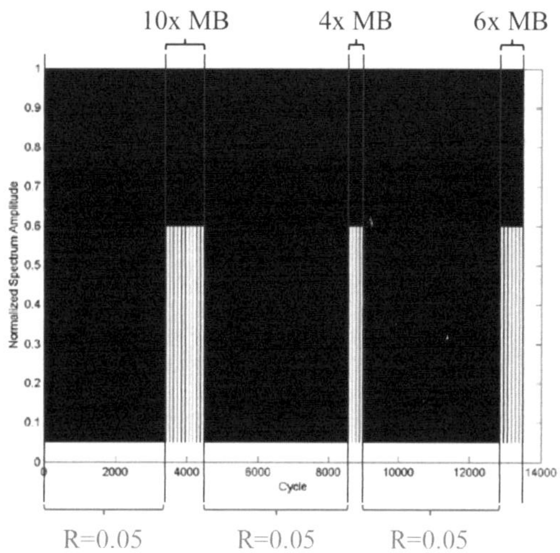

a) Spectrum sequence used for the SENT specimens

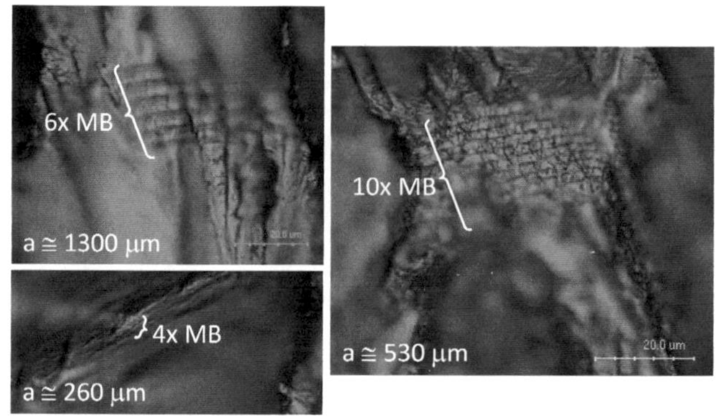

b) Fractographic analysis of the SENT specimens with the resulting marker bands

Figure 3. Spectrum sequence and the resulting marker bands on the SENT specimens.

After failure, the size of the crack as a function of cycles was measured from fractographic analyses. For each identified crack front, the corresponding SIF was calculated using the solution for a corner crack at a semi-circular notch [5] by assuming the crack to have a quarter-elliptical shape defined by the through-the-thickness crack length (a) and the surface crack length (c), as illustrated in Figure 4. However, these dimensions were not provided directly from fractography. Crack size measurements were typically taken at 45 degrees (c_{45}-direction in Figure 4) from the surface using an optical microscope and a scanning electron microscope (SEM). When possible, additional crack size measurements were included for specimens that provided clear marker bands to estimate the crack shape aspect ratio (a/c). For other specimens, a circular corner crack shape was assumed. This simplifying assumption was required due to the limited amount of data that could be collected from fractography. However, it was observed that the aspect ratios of the corner cracks were very close to 1.0 for small crack sizes.

The FCGR between two identified crack fronts was calculated using the secant method. The two crack size measurements used for the secant method typically represented either the crack growth between the end and the beginning of two consecutive marker band blocks (crack growth corresponding to test stress ratio) or between the beginning and the end of a marker band block (crack growth corresponding to marker band stress ratio). As such, only consecutive identified features delimited by known load history and time intervals were used. It is understood that the FCGR calculated within the same marker band block may have been influenced by the lower stress ratios (R=0.05) used to make the bands that could be counted during the examination of the fracture surfaces. However, it was assumed that this had no significant effect on the calculated FCGR due to the small number of lower stress ratio cycles applied during the marker band blocks (10 cycles at R=0.05 per 100 cycles at R=0.6). Also note that since the maximum load remained the same for R=0.05 and R=0.60, there was no additional plasticity induced crack closure caused by changing the stress ratio.

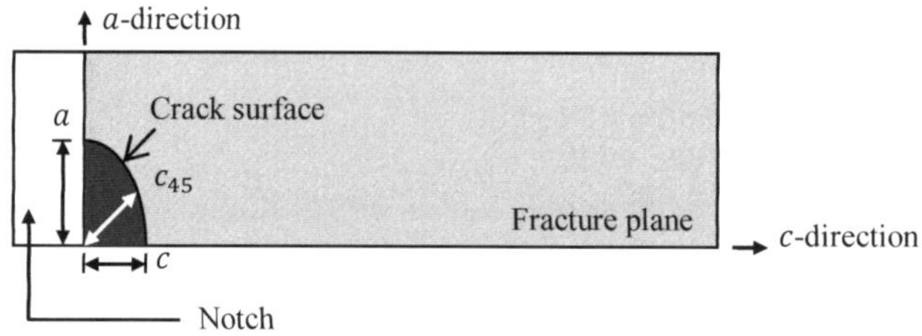

Figure 4. Definition of crack dimensions (cross-section of the SENT specimen shown).

Variable amplitude FCGR tests

Compact tension specimens. Compact tension specimens were tested under variable amplitude loading in laboratory air. The test procedure consisted of a pre-cracking stage under a constant maximum load smaller or equal to the first spectrum cycles, followed by the variable amplitude loading stage. Two spectra were tested: a clipped Mini-TWIST transport aircraft spectrum and a clipped CC-130 military transport aircraft spectrum. These spectra originally included compressive loads, which had to be removed due to the limitation imposed by the C(T) specimens. For the clipped Mini-TWIST spectrum, shown in Figure 5a, the majority (>72%) of the load pairs (min-max) occurred at stress ratios between 0.4 and 0.7 and at stress ranges between 15% and 30% of the maximum applied stress. For the clipped CC-130 military transport aircraft spectrum, shown in Figure 5b, the majority (>81%) of the load pairs occurred at stress ratios above 0.8 and at stress ranges lower than 10% of the maximum applied stress. The average test frequency was 10 Hz for the clipped Mini-TWIST spectrum and 17 Hz for the clipped CC-130 aircraft spectrum.

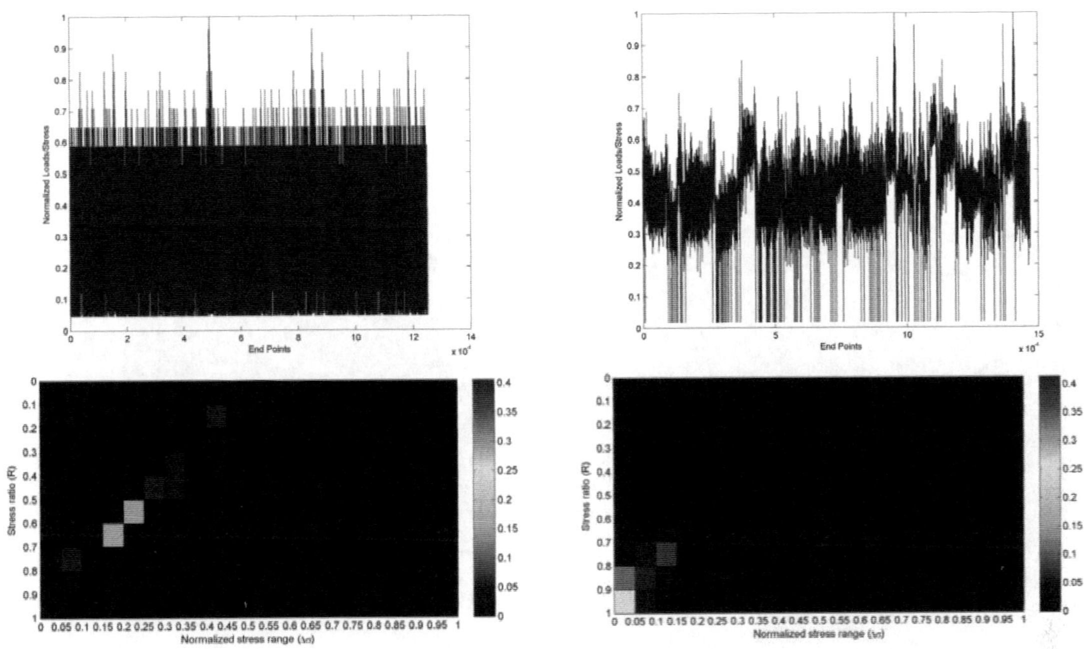

a) Clipped Mini-TWIST spectrum

b) Clipped CC-130 spectrum

Figure 5. Load spectrum and normalized occurrences as a function of stress ratio and stress range.

The crack size was monitored using the crack compliance method. Visual measurements were also taken during the tests to correct the compliance crack size measurements.

Modified middle tension specimens. Modified M(T) specimens were manufactured from three extruded wing panels. The test procedure consisted of a pre-cracking stage under a constant maximum load smaller or equal to the first spectrum cycles, followed by the variable amplitude spectrum loading stage. The spectrum was developed for the CP-140 aircraft wing and includes tensile and compressive loads. The normalized spectrum load history is shown in Figure 6, where the normalized maximum and minimum load values are 1.0 and -0.8454, respectively. Compression-compression load pairs accounted for 21% of the all the load pairs in the spectrum. The majority (>63%) of the load pairs (min-max) occurred at stress ratios between 0.3 and 0.6 and at stress ranges between 15% and 40% of the maximum applied stress.

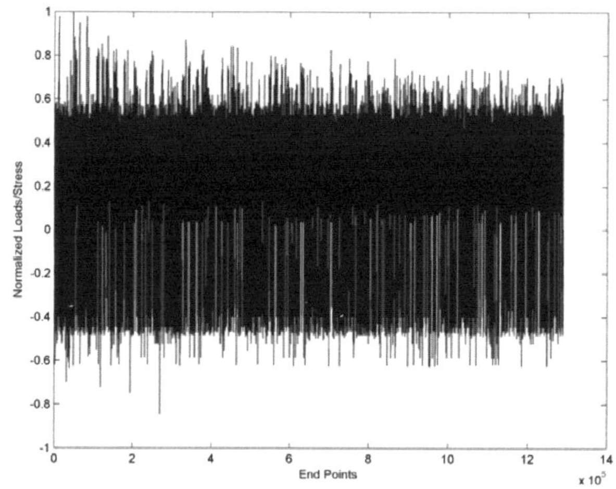

Figure 6. Normalized CP-140 load spectrum.

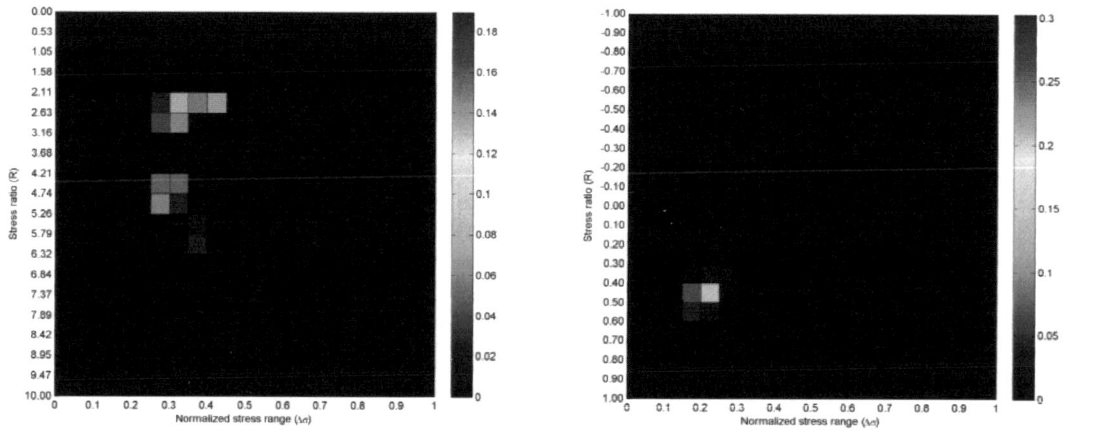

a) Compression-compression load pairs

b) Tension-tension and tension-compression load pairs

Figure 7. Normalized occurrences as a function of stress ratio and stress range for the CP-140 spectrum.

The DCPD method was used to measure the crack size for the modified M(T) specimen tests. Initially, the ASTM E 647 procedure for standard M(T) specimens was used. This procedure is based on Johnson's formula, which was developed for a centre crack symmetrically located about the probe location. However, this formula was found to be inaccurate for the modified M(T) specimens, which have a single radial crack growing asymmetrically from the centreline of the specimen. A brief literature survey indicated that there was no available formula to calculate the size of a radial crack in a M(T) specimen. Therefore, the following mathematical model was developed to fit potential drop (PD) values to visual measurements to accurately calculate the crack size as a function of the normalized PD (PD_n).

$$\left(\frac{c}{c_0}\right)^2 = A_1 + A_2(PD_n)^2 + \frac{A_3}{(PD_n)^2} + \frac{A_4}{(PD_n)^4} \tag{1}$$

$$PD = PD_c\left(\frac{PD_{r_0}}{PD_r}\right) \tag{2}$$

$$PD_n = \frac{PD}{PD_{c_0}} = \left(\frac{PD_c}{PD_{c_0}}\right)\left(\frac{PD_{r_0}}{PD_r}\right) \quad (3)$$

where

PD	Potential drop (corrected)
PD_n	Normalized potential drop (corrected)
PD_c	Potential drop measured from the active probes
PD_{c_0}	Initial potential drop measured from the active probes
PD_r	Potential drop measured from the reference probes
PD_{r_0}	Initial potential drop measured from the reference probes
c	Half crack size as defined in Figure 1c
c_0	Half initial crack size (notch size) as defined in Figure 1c
A_1, A_2, A_3, A_4	Regression parameters calculated by fitting the equation to visual measurements

The procedure for fitting the model to experimental data was integrated into an Excel spreadsheet to facilitate data processing and the fitting parameters were calculated using the Excel built-in optimization solver.

A constant direct current (DC) was supplied by a programmable linear DC power supply. For the modified M(T) specimen a current of 20 amperes was selected to reach sufficient potential drop between the probes installed at the crack (active DCPD signal). Reference probes (reference DCPD signal) were also used to compensate for temperature fluctuations. The two DCPD signals (active and reference) were conditioned, amplified, and filtered using two-channel differential amplifiers. These signals were then recorded by the FTA test system and post-processed using an Excel spreadsheet developed for this purpose.

STEADY-STATE FATIGUE CRACK GROWTH RATE RESULTS AND DISCUSSION

Steady-state FCGR results in laboratory air

Steady-state FCGR tests were conducted on 11 C(T) specimens in laboratory air for three stress ratios (R=0.05, 0.6, and 0.85). The FCGR results for the C(T) specimens are presented in Figure 8a. These FCGR results were calculated based on the applied SIF range (ΔK). Overall, the variability of the C(T) FCGR results is small, although larger data scatter is observed for R=0.05 compared to higher stress ratio data. It is interesting to note that the FCGR for R=0.85 are lower than the ones obtained at R=0.60 for ΔK lower than 2 ksi$\sqrt{\text{inch}}$. Further investigation will be required to explain this behaviour. However, it is understood that the SIF or loads applied to specimens tested at R=0.85 had to be significantly higher than the ones used for lower stress ratios to reach the target ΔK ($K_{max} = \Delta K/(1-R)$). It is therefore possible that the size of the plastic zone at R=0.85 may have made the stress state at the crack tip plane-stress, as opposed to plane-strain for the lower R tests, which may have influenced the FCGR results.

Eight SENT specimens were tested to obtain small crack data from naturally nucleated fatigue cracks, but only five specimens generated reliable FCGR data. The other three either did not crack, failed under the wedge grips, or included multiple nucleation sites that could not be accurately analysed using the available analytical SIF closed-form solution [5]. The time required to nucleate a crack for a given maximum applied stress varied significantly among the SENT specimens. Because of this large variability, silicon-based replica could not be used for this test due to time constraints and the high cost associated with the technique. Consequently, crack sizes were only measured from fractographic analyses and the smallest detected crack size was 45 μm. The FCGR results obtained using the SENT specimens are shown in Figure 8b. Under the maximum applied cyclic loads (18 to 20 ksi), the smallest detected crack was too large to obtain FCGR data in the ΔK threshold regime. Despite the fact that the small crack data could not be obtained, the SENT FCGR results were found to be very similar to the ones obtained using the C(T) specimens for both tested stress ratios.

As it was suspected that the machining process used for the notch may have induced surface residual stresses, the notch of some specimens was chemically etched to reduce the effect of surface residual stresses and the crack nucleation life. Based on the observation of the crack nucleation sites using a scanning electron microscope, the cracks of the etched specimens appeared to have nucleated at pits created by the etching process.

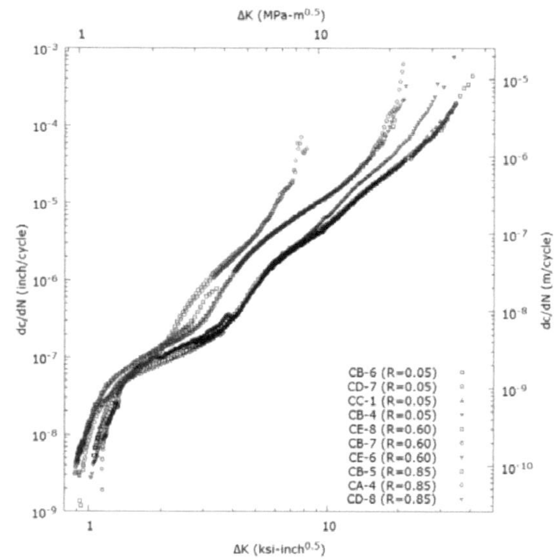

a) C(T) FCGR results
(R=0.05, 0.60, and 0.85)

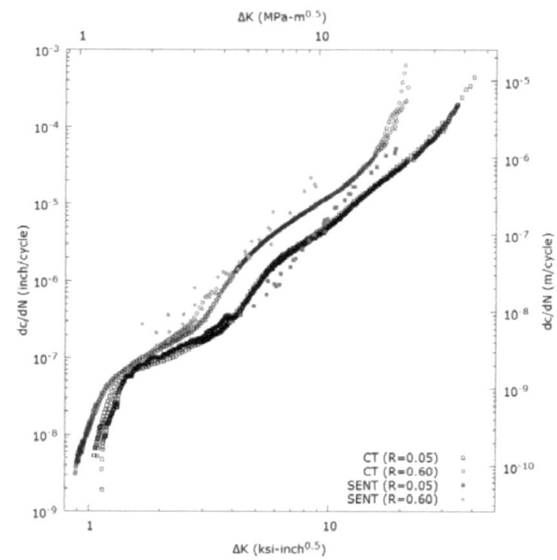

b) Comparison between C(T) and SENT FCGR results
(R=0.05 and 0.60)

Figure 8. FCGR results for laboratory air.

For most C(T) specimens, there was no significant difference between the applied ΔK and the ΔK calculated using the ASTM 2% opening load method (ΔK_{OP}), the $2/\pi$ correction to opening load method ($\Delta K_{2/\pi}$), and the adjusted compliance ratio method (ACR) (ΔK_{ACR}) [3]. However, significant differences were observed for the first test conducted at R=0.05, for which the initial K_{max} exceeded the initial FCGR value recommended in ASTM E 647. Figure 9 presents the effect of the initial K_{max} on the FCGR results by comparing the FCGR results obtained using a K_{max} of 4 ksi$\sqrt{\text{inch}}$ and 7 ksi$\sqrt{\text{inch}}$. As shown in Figure 9a, the differences between ΔK, ΔK_{OP}, and $\Delta K_{2/\pi}$ are very small for the initial K_{max} of 4 ksi$\sqrt{\text{inch}}$. However, for higher initial K_{max} (e.g. 7 ksi$\sqrt{\text{inch}}$), the ΔK, ΔK_{OP}, and $\Delta K_{2/\pi}$ values diverge in the threshold regime. Moreover, the resulting ΔK values are different from the average results (spline fitted results) obtained using lower initial K_{max}. This indicates that the initial K_{max} influences the near-threshold FCGR results. It is also interesting to note that the ΔK_{ACR} results obtained using the higher K_{max} are very similar to the average FCGR results obtained using the lower K_{max}. This observation suggests that the ACR method may reduce the effect of load history induced by high initial K_{max} values, whereas ΔK_{OP} and $\Delta K_{2/\pi}$ may not accurately reduce this effect.

For higher stress ratios, the difference between ΔK and ΔK_{ACR} is typically very small, which suggests that crack closure effects are less significant for higher stress ratios compared to lower stress ratios and/or the that ACR method may not be suitable to capture the closure effects for higher stress ratios.

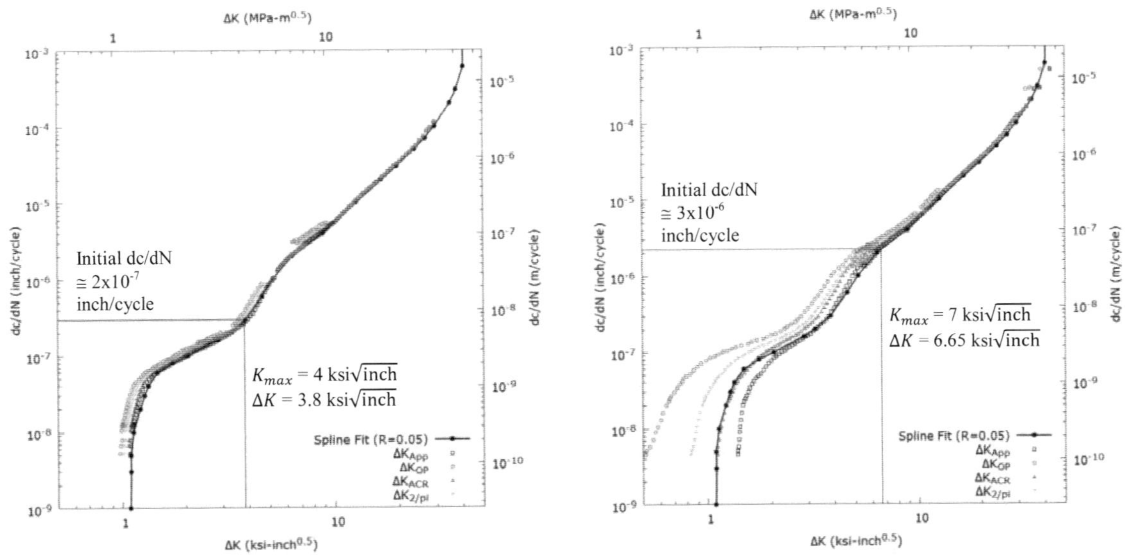

a) Initial K_{max} of 4 ksi$\sqrt{inch}$ b) Initial K_{max} of 7 ksi$\sqrt{inch}$

Figure 9. Effect of initial K_{max} on FCGR results for R=0.05.

The C(T) FCGR results obtained for 7249-T76511 are compared against two 7075-T6511 FCGR material models included in AFGROW [6] (NASGRO and Harter-T material models) for two stress ratios (R=0.05 and 0.60). The comparison with 7075-T6511, shown in Figure 10, reveals that 7249-T76511 has higher FCGR in the threshold regime and has a lower threshold SIF. Above 1x10^{-7} inch/cycle, 7249-T76511 and 7075-T6511 have similar FCGR behaviours. Depending on the loading spectrum, the difference in the threshold regime may have significant impact on the fatigue life, especially if crack closure is significant.

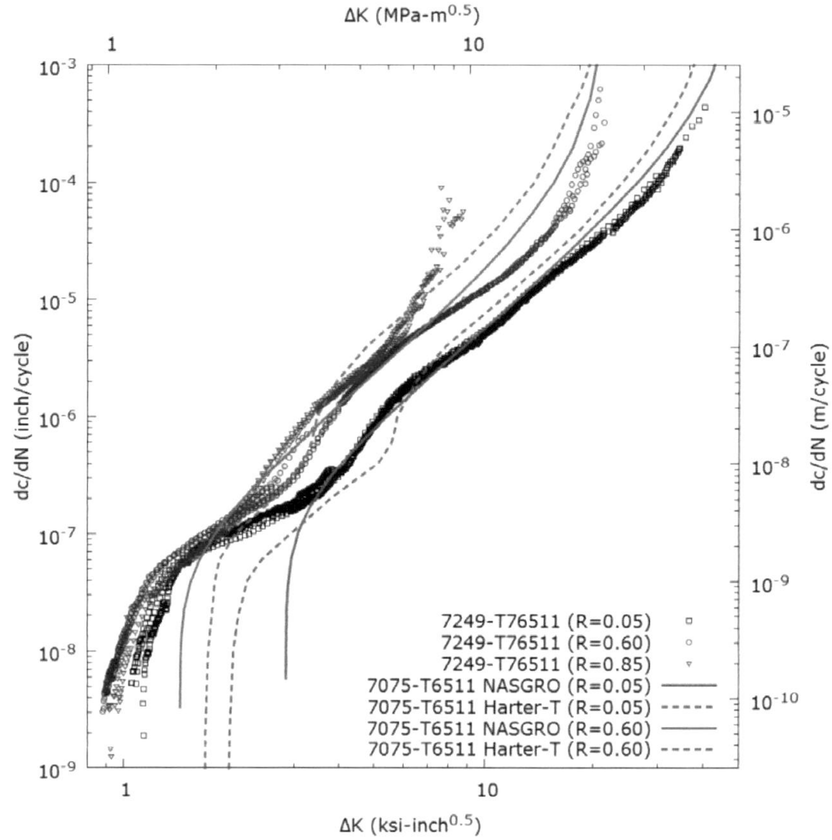

Figure 10. FCGR comparison between 7249-T76511 and 7075-T6511.

Steady-state FCGR results in corrosive environment

The effects of the fixture and the cells on the measured FCGR were quantified prior to testing the C(T) specimens with the corrosion cells filled with the 3.5% NaCl solution. First, a specimen was tested with Fixture B (Figure 2) without installing the corrosion cells. Due to the flat bottom holes and the wider opening, it was found more difficult to position the specimen in order to obtain a linear load vs. displacement signal. A comparison between the FCGR results obtained using fixture A and B is presented in Figure 11. Although the FCGR results obtained using Fixture B are very similar to the ones obtained using Fixture A, the variability of the FCGR results obtained using Fixture B is higher than that of Fixture A. Furthermore, the crack size measurements obtained using the compliance method could not be obtained with Fixture B for very low ΔK due to a degradation of the linearity of the load vs. displacement relationship as the applied load decreased.

Two specimens were tested with empty corrosion cells installed. As shown in Figure 11 (violet symbols), the corrosion cells had a significant effect on the FCGR results and increased the measured ΔK threshold from 1.09 to 1.66 ksi$\sqrt{\text{inch}}$. It is suspected that load transfer occurred between the specimen and the bonded cells and through the silicon plug used to seal the notch. The load transfer reduced the load seen by the specimen and consequently, underestimated the FCGR in the near threshold regime. Due to this behaviour, it was decided to compare the FCGR obtained in the corrosive environment against the average FCGR curve obtained using Fixture B with dry corrosion cells. Based on those results, the FCGR was reduced by at least one order of magnitude for ΔK values lower than 2.25 ksi$\sqrt{\text{inch}}$ and increased by up to 4.6 times for ΔK values larger than 8 ksi$\sqrt{\text{inch}}$. These results are specific to the testing conditions (3.5% NaCl solution at room temperature and a test frequency of 10 Hz).

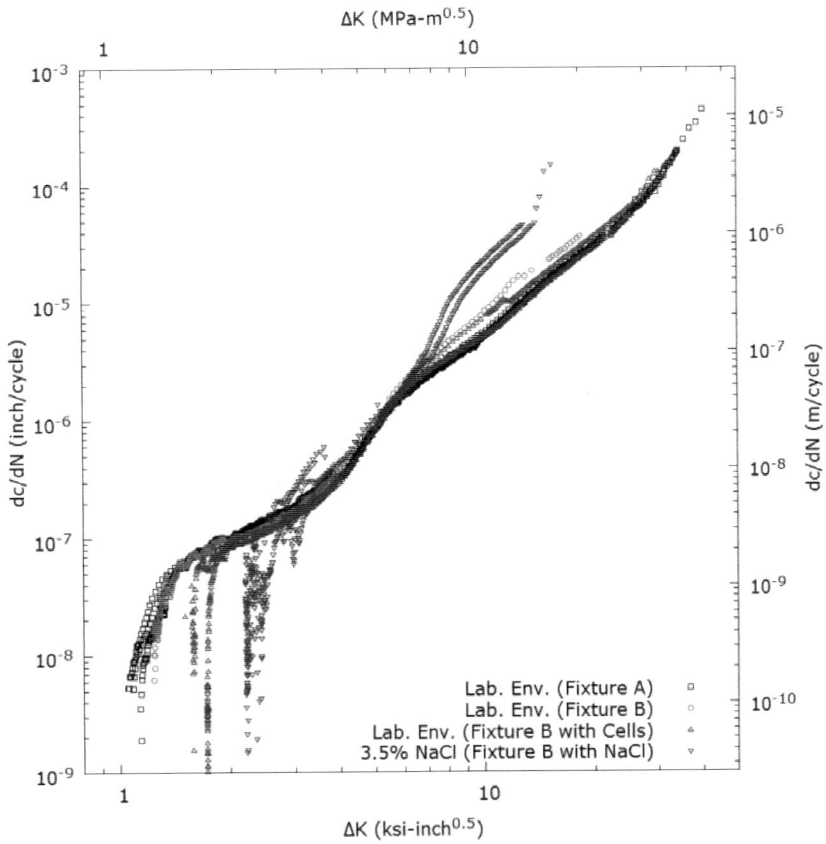

Figure 11. Effects of fixture, corrosion cells, and corrosive solution on the FCGR results.

MATERIAL MODELLING

Two FCGR material models were developed using the steady-state FCGR results. The first model was based on a tabular look-up FCGR material model combined with a load interaction model such as Hsu, Willenborg, or Wheeler. The second model was developed using FASTRAN 5.42 [7].

The developed tabular look-up model uses Walker's equation on a point-by-point basis to interpolate and extrapolate FCGR data at any given stress ratios. As the Walker equation assumes that the FCGR increases with the stress ratio for a given SIF range, the FCGR results obtained for R=0.85 could not be used as it was crossing the results obtained with lower stress ratios. As such, the tabular look-up model was developed using the FCGR obtained using the stress ratios of 0.05 and 0.60 only. The lower and upper stress ratio limits were set to 0.0 and 0.60. The model was implemented in AFGROW version 5.1.6.17 [6] in order to perform crack growth simulations. Three load interaction models, namely Hsu, Willenborg, and Wheeler, were calibrated and evaluated based on their ability to calculate the fatigue lives for the tested spectra. During this process, it was found difficult to select a single set of load interaction model parameters that could provide accurate fatigue life predictions for all tested spectra. The load interaction model parameters were first selected using a trial-and-error process to fit the crack-length-against-cycle behaviour on the Mini-TWIST spectrum test results. The selected parameters were then used to simulate crack growth in the specimens tested using the CC-130 and CP-140 spectra.

The FASTRAN material model was developed using the C(T) FCGR data. The relationship between the effective SIF, ΔK_{eff}, and the FCGR was estimated using the high stress ratio data and the FASTRAN model parameters were adjusted to fit the steady-state FCGR data for all tested stress ratios (0.05, 0.60, and 0.85). The threshold behaviour was modelled using multiple ΔK_{eff} versus FCGR segments and the threshold parameter, ΔK_0, was set to zero (C3=C4=0.0). To fit high FCGR data, the power term on fracture (C6) was set to 4.0 and the limiting value of maximum elastic SIF at failure was analytically adjusted to 54.6 ksi$\sqrt{\text{inch}}$ (60 MPa$\sqrt{\text{m}}$). To fit the FCGR data, a variable constraint factor was used. For FCGR lower than $1.97\text{x}10^{-7}$ inch/cycle ($5\text{x}10^{-9}$ m/cycle), a very high constraint factor of 3.0 was used to fit the low stress ratio FCGR data. This constraint factor corresponds to the theoretical plane strain state. It is understood that plain strain does not practically exist on the specimen surface and the selected constraint factor is significantly higher than the values typically used in FASTRAN, but this constraint factor was required to fit the FCGR data for a stress ratio of 0.05. For FCGR higher than $1.97x10^{-4}$ inch/cycle ($5x10^{-6}$ m/cycle), the constraint factor was reduced to 1.15, which corresponds to the plane stress behaviour. Using the DKEFF13 program provided with FASTRAN, ΔK against FCGR was generated from the developed FASTRAN material model and compared against experimental results obtained from steady-state FCGR tests. This comparison is shown in Figure 12 with the ΔK_{eff} curve.

Comparisons between the fatigue crack growth simulation results obtained using the tabular look-up model (AFGROW), the FASTRAN material model, and the experimental results obtained from the variable amplitude FCGR tests is presented in Figures 12 to 16.

Despite that the tabular look-up model, coupled with one of the tested load retardation models, provided crack growth life results within 33% of the experimental results, it is important to note that the load interaction model parameters had to be tuned using spectrum results, such that it was not a pure prediction. For the tested cases, the selected load interaction parameters provided acceptable results, but further testing would be required to ensure that the proposed material model provides reliable fatigue life predictions for other material thicknesses and loading spectra. Also, none of the tested load interaction models used with the tabular look-up model could capture the kink effects of the CP-140 spectrum spike overloads indicated in Figure 16. FASTRAN was the only model able to predict the sudden FCGR drops caused by these overloads.

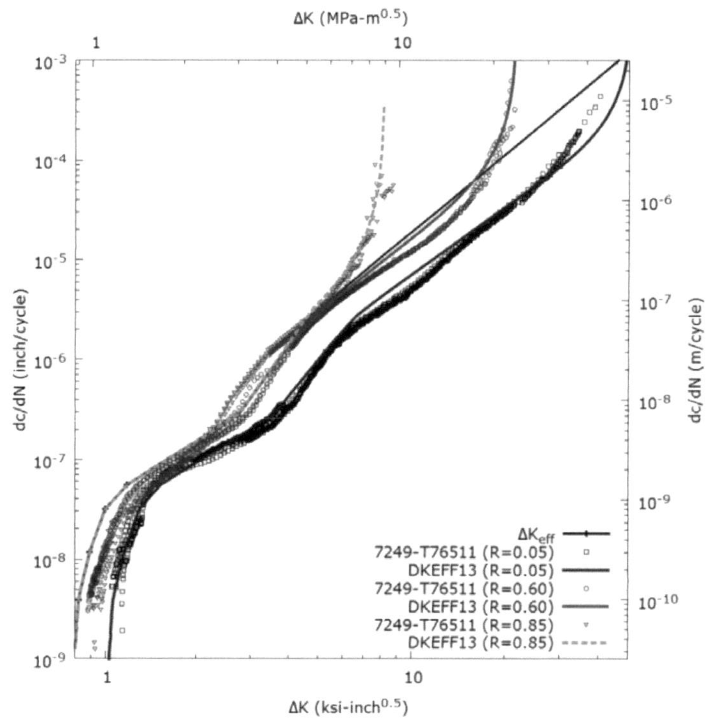

Figure 12. Comparison between steady-state FCGR results and material properties generated using the FASTRAN model.

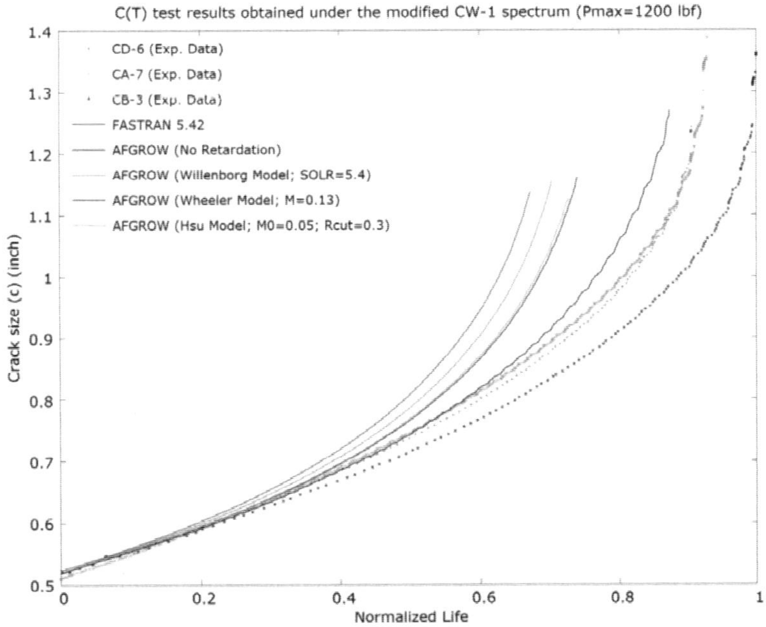

Figure 13. Fatigue life prediction for the CC-130 test spectrum (maximum load of 1200 lb_f).

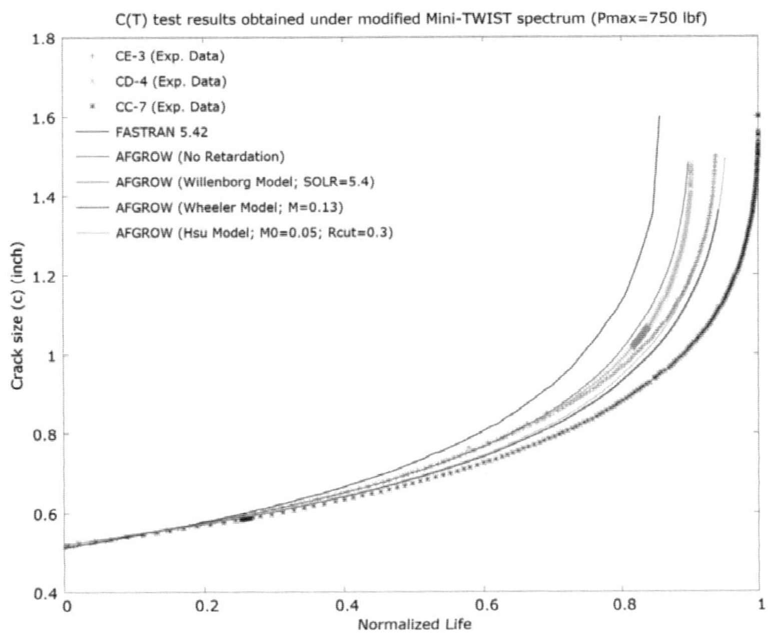

Figure 14. Fatigue life prediction for the modified Mini-TWIST spectrum (maximum load of 750 lb_f).

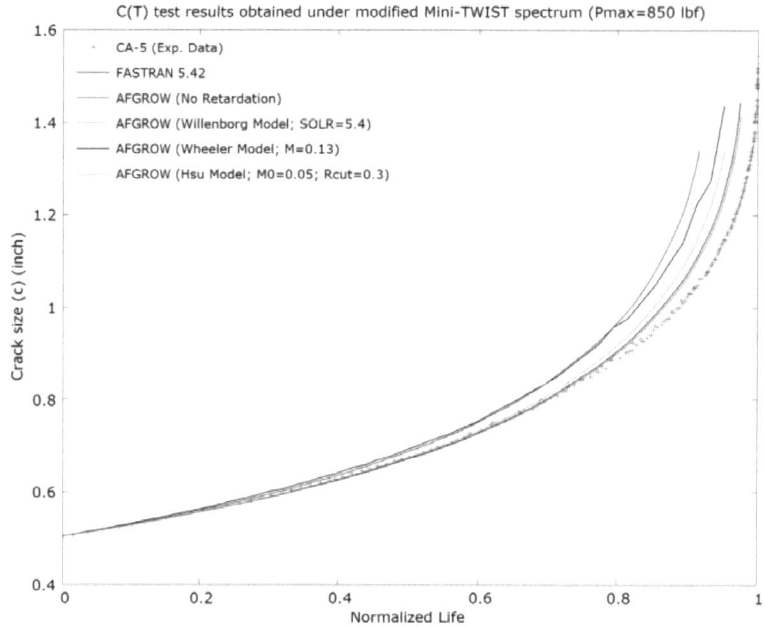

Figure 15. Fatigue life prediction for the modified Mini-TWIST spectrum (maximum load of 850 lb_f).

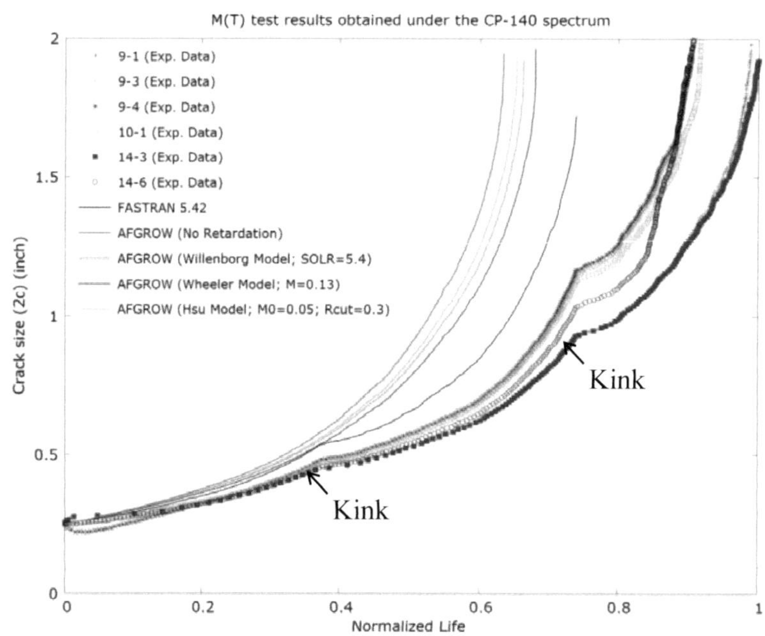

Figure 16. Fatigue life prediction for the CP-140 test spectrum.

CONCLUDING REMARKS

FCGR tests were conducted for 7249-T76511 using C(T) and SENT specimens to characterize the fatigue crack growth properties of the material. C(T) test specimens were tested in laboratory air according to ASTM E 647 under three stress ratios: 0.05, 0.60, and 0.85. Overall, the variability of the FCGR test results obtained from the C(T) test specimens was small. The ACR method, used for the C(T) specimens, was found to reduce the effect of load history in the threshold regime when the initial K_{max} exceed the initial FCGR recommended in ASTM E 647. C(T) FCGR tests were also conducted in a 3.5% sodium chloride (NaCl) solution to quantify the effect of a corrosive environment on the FCGR.

However, it was found that the cells used to contain the corrosive solution around the notch may have influenced the FCGR results and a new test method would be required to provide accurate FCGR results in the threshold regime. SENT tests were performed to determine the FCGR of small cracks naturally nucleating from a notch. The FCGR obtained from these tests were in good agreement with the long-crack FCGR results obtained using the C(T) specimens. However, no FCGR results could be obtained in the stress intensity factor threshold regime using the SENT specimens as the marker bands corresponding to the smallest crack sizes could not be found from fractography.

Using the results from the steady-state FCGR tests, two FCGR material models were developed. A tabular look-up model was first developed to interpolate and extrapolate the FCGR data for different stress ratios. This model was combined with a load interaction model to consider crack retardation effects for crack growth simulations under variable amplitude loadings. The load interaction models were calibrated using a subset of variable amplitude loading test data and verified with other test data. Overall, the selected load interaction parameters provided acceptable results, but there is a strong possibility that these parameters may not be suitable for other loading spectra. Unlike the tabular look-up model with load interaction models, the FASTRAN model presented in this paper was developed solely based on the steady-state FCGR data. Although not finalized yet, the current FASTRAN material model developed by NRC seems promising due to the facts that it was not developed by tuning its parameters to fit variable amplitude spectrum results and that it could simulate the effects of spike overloads on the FCGR.

The FCGR material models developed using the steady-state FCGR results need to be improved and fine-tuned using additional steady state FCGR data (e.g. negative stress ratio, effect of thickness, other coupon geometries) to improve the accuracy of the fatigue life calculations. In addition, additional variable amplitude loading fatigue test data is also required to gain confidence on the accuracy of the proposed material model and identify its limits.

ACKNOWLEDGEMENTS

This work was performed with financial support from Defence Research and Development Canada (DRDC) and the National Research Council Canada (NRC) through the following projects: *Integrated Structural Life Assessment for RCAF Air Fleets* and *Aircraft Structural Life Assessment and Prognosis*.

REFERENCES

[1] ASTM. E647-08e1: Standard Test Method for Measurement of Fatigue Crack Growth Rates. Annual Book of ASTM Standards. 2008. West Conshohocken, PA, ASTM International.

[2] Yamada, Y. and Newman, J., "Crack closure under high load-ratio conditions for Inconel-718 near threshold behavior," *Engineering Fracture Mechanics*, Vol. 76, No. 2, 2009, pp. 209-220.

[3] Donald, K. and Blair, A., "Automated Fatigue Crack Growth Testing and Analysis - Series 2001, Version 3.09 - Users' Steady State Reference Manual," 2009.

[4] Donald, K. and Paris, P. C., "An evaluation of DKeff estimation procedures on 6061-T6 and 2024-T3 aluminum alloys," *International Journal of Fatigue*, Vol. 21, Supplement 1, No. 0, 1999, pp. S47-S57.

[5] Newman, J. C. Jr., "Small-Crack Test Methods," American Society for Testing and Materials, ASTM STP 1149, Philadelphia, 1992.

[6] Harter, J. A., "AFGROW users guide and technical manual (Version 5.01.xx.16)," 2010.

[7] Fatigue & Fracture Associates, L., "FASTRAN - A Fatigue Crack Growth Life-Prediction Code Based on the Crack-Closure Concept - Version 5.4 - User Guide," 2013.

Material Characterization of Aluminum Lithium Alloys Used in Aerospace Applications

Kevin Stonaker[1], John G. Bakuckas, Jr.[1], Ian Y. Won[2], Mark Freisthler[2]

[1] *FAA William J. Hughes Technical Center, Atlantic City International Airport, NJ 08405*
[2] *FAA Transport Airplane Directorate, 1601 Lind Ave SW, Renton, WA 98057*

Abstract: This study was undertaken to assess the material properties and mechanical behavior of the next generation aluminum lithium (Al-Li) alloys being used in aerospace structures through comparisons made to conventional aerospace aluminum alloys. The latest generation of Al-Li alloys purports to offer a significant weight savings over conventional aerospace aluminums resulting in significant use in recent aircraft and aerospace applications. The current public data provided for these alloys is very limited and does not provide a comprehensive understanding of the strengths and weaknesses of these materials. Because previous generations of Al-Li alloys displayed material behaviors that limited their use for aerospace applications, it is necessary to understand the properties of these new alloys and identify if any unique behaviors exist.

Two Al-Li alloys were considered as a case study, namely 2198-T8 and 2196-T8511 alloys used for skin and extrusion applications, respectively. Several properties were assessed and compared with baseline 2024-T3 and 7075-T6 alloys, including static properties; fatigue life and fatigue crack growth behavior; and supplemental properties. Each segment of testing was comprised of multiple test plans designed to assess individual aspects of the overall material behavior. Resources and expertise were leveraged with several organizations to achieve the goals of this program. In general, test results revealed that the mechanical properties of typical third-generation Al-Li alloys were comparable to those of baseline aluminum alloys. However, there were fatigue damage modes unique to the Al-Li materials tested, which included interlaminar and eyebrow cracking in lap joint specimens.

INTRODUCTION AND BACKGROUND

Over the past decade, significant investments and advances have been made by industry in the development of new metallic alloys with optimized material properties for specific structural applications. The latest (third) generation of advanced aluminum-lithium (Al-Li) alloys offers weight savings compared to conventional aluminum alloys, while maintaining mechanical performance. Al-Li alloys have a lower density, higher modulus, and improved resistance against corrosion over the widely used 2xxx and 7xxx series alloys [1]. Previous generations of Al-Li exhibited material behavior, such as low elongation and ductility, low transverse toughness, inadequate thermal stability and fatigue properties, and highly anisotropic behavior with respect to the rolling grain orientation which restricted their use for many aerospace applications [2–4]. However, the latest generations of Al-Li alloys are reported to minimize these effects, while also providing significant weight savings [5].

With these improvements, there has been increased use of Al-Li in recent aircraft and aerospace applications. Bombardier is constructing the entire fuselage of its new C-Series airplane, a Part 25 airplane with 150-seat capacity using Al-Li [6]. Extruded floor beams for the Airbus A380 are now being made of Al-Li. In addition, Al-Li plate applications are being used for the internal structure of the fuselage in the Boeing 787. Although these manufacturers have generated substantial proprietary data to substantiate structural applications, there is a lack of data available in the public domain to thoroughly document the mechanical performance of these latest generation Al-Li alloys.

The objective of this test program was to gain a better understanding of the material properties, mechanical behavior, and unique characteristics of the next generation Al-Li being used in airframe structures through comparison with traditional aluminum alloys. As a case study, two Al-Li alloys were considered; namely, 2198-T8 and 2196-T8511 alloys used for skin and extrusion applications, respectively. Several properties were assessed and compared with baseline aluminum alloys 2024-T3 and 7075-T6, including static properties; fatigue life and fatigue crack growth (FCG) behavior; and supplemental properties. Resources and expertise were leveraged from several organizations, including Constellium, Bombardier, National Aeronautics and Space Administration-Johnson Space Center (NASA-JSC), Naval Air Systems Command (NAVAIR), University of Dayton Research Institute (UDRI), National Institute for Aviation Research (NIAR), Drexel University and the Federal Aviation Administration Technical Center (FAA-TC).

Testing was conducted in three major grain directions (L, LT, and 45°), plus two midpoints (22.5° and 67.5°), to assess the extent of anisotropy. In general, test results revealed that the static; fatigue and FCG; and several supplemental properties of typical third-generation Al-Li alloys were comparable to the baseline aluminum alloys. Anisotropic behavior was observed in the Al-Li, where properties at the 45° grain direction differed from other grain directions. In addition, fatigue damage modes unique to the Al-Li include interlaminar and eyebrow cracking in lap joint specimen configurations. Results from this study contribute to the availability of information in the public domain to gain the knowledge for proper assessments of Al-Li for aerospace applications. The information contained in this paper represents a broad overview of the whole program. Detailed test information and results can be found in [7].

TEST PROCEDURES

This material characterization study considered two Al-Li alloys; 2198-T8 in the sheet form and 2196-T8511 in the extruded form. For baseline comparison purposes two industry standard aluminum alloys were also tested for each Al-Li alloy; 2024-T3 bare (except where noted as clad) and 7075-T6, respectively. The 2196-T8511 alloy was supplied by Bombardier in the form of I- and Z-shaped extrusions and was tested at thicknesses of 1.5 mm, 3.0 mm, and 3.7 mm. The 2198-T8 alloy was supplied by Constellium in the form of sheet material and was tested at thicknesses of 1.8 mm, 3.2 mm, and 6.4 mm. Testing was conducted in three major grain directions (L, LT, and 45°), plus two midpoints (22.5° and 67.5°), to assess the extent of anisotropic behavior in the materials.

As shown in Table I, the overall test program was grouped into three main types: static properties, fatigue and FCG properties, and supplemental properties. In general, for each property, tests were repeated three times and results reported as an average value. Each test type was comprised of numerous individual test plans carried out by one of the partners with expertise in that field. The targeted results will provide a general overview of the basic material properties.

Static Properties

The static properties considered in this study included tension, compression, and shear. Tests to generate these properties were governed by an associated ASTM standard and the results are compared with published values in the Metallic Materials Properties Development and Standardization (MMDPS) Handbook, version 8. Three repeat tests were conducted for each of the three material thicknesses and at five grain directions to assess the anisotropic behavior of the materials.

Tension. The tension properties were generated by three laboratories, namely the FAA-TC, UDRI [8], and NIAR, using ASTM E8 standard for tension testing and ASTM E21 standard for elevated temperatures where applicable. Flat dog-bone specimen geometry was used for each test with modifications as needed. Specimens tested at an elevated temperature had an extended grip section to allow clearance for the oven. For the 2196 T8511 alloys, micro-sized dog-bone specimens were used because of the restrictive size of the extrusions. For this same material, a set of full-size specimens were also tested at the L grain direction to evaluate if there were any effects of specimen geometry. Each test was conducted under displacement controlled conditions with load and strain recorded throughout. The data collected was used to calculate tensile modulus, E_t, yield strength, F_{ty}, ultimate strength, F_{tu}, and percent elongation, *e%*. The elevated temperature tests were run at 93°C and 177°C. The specimens were held at temperature for 20 minutes prior to testing.

Table I. Al-Li Material Characterization Test Matrix.

Test Type	Test Name and Standard	Materials Tested	Variables*	Property Measured	Performing Organization
Static Properties	Tension Properties, ASTM E8	2198-T8, 2196-T8511, 2024-T3 Bare, 7075-T6	t_s (mm): 1.8, 3.2, 6.4 t_e (mm): 1.5, 3.0, 3.7 θ: L, 67.5, 45, 22.5, LT T (°C) = RT, 93, 177	F_{ty}, F_{tu}, E_t, $e\%$	FAA-TC, UDRI, NIAR
	Compression Properties ASTM E9	2198-T8, 2196-T8511, 2024-T3 Bare, 7075-T6	t_s (mm): 1.8, 3.2, 6.4 t_e (mm): 1.5, 3.0, 3.7 θ: L, 67.5, 45, 22.5, LT	F_{cy}, F_{cu}, E_c	FAA-TC, NIAR
	Shear Properties, ASTM B831	2198-T8, 2196-T8511, 2024-T3 Bare, 7075-T6	t_s (mm): 1.8, 3.2, 6.4 t_e (mm): 1.5, 3.0, 3.7 θ: L, 67.5, 45, 22.5, LT	F_{sy}, F_{st}, G	NIAR
Fatigue and Fatigue Crack Growth Properties	Fatigue Crack Growth ASTM E647	2198-T8, 2024-T3 Bare	t_s (mm): 1.8, 6.4 θ: L, 45, LT T (°C) = RT, 93, 177 R = 0.1, 0.4, 0.7, 0.8	da/dN vs. ΔK	NASA-JSC, UDRI
	Spectrum Fatigue Crack Growth ASTM E647	2198-T8, 2024-T3 Bare	t_s (mm): 1.8 θ: L, 45, LT	Crack length during spectrum fatigue loading	UDRI
	Lap Joint Fatigue, ASTM G85	2198-T8, 2024-T3 Bare, 2024-T3 Clad	t_s (mm): 3.2 θ: L, 45, LT R = 0.1	Fatigue life, cycles to first crack detection	NAVAIR, FAA-TC, Drexel U.
Supplemental Properties	Environmentally Assisted Cracking ASTM G129	2198-T8, 2024-T3 Clad	t_s (mm): 3.2 θ: 45 V (rate): 10^{-5}, 10^{-6} 10^{-7}	Time to fracture and strength	NAVAIR
	Fastener Hole Process: Cold Work & Interference-Fit	2198-T8, 2196-T8511, 2024-T3 Bare, 7075-T6	t_s (mm): 1.8, 3.2, 6.4 t_e (mm): 1.5, 3.0, 3.7 θ: L, 45, LT	Cracks and damage	FAA-TC, NIAR, FTI
	Flammability and Burn Resistance, AC 25.856-2A	2198-T8, 2024-T3 Bare	t_s (mm): 1.8, 3.2, 6.4	Burn-through Time	FAA-TC

* Variable definitions
t_s = Typical sheet thickness for the 2198-T8 and 2024-T3, mm
t_e = Typical extrusion thickness for the 2196-T8511 and 7075-T6, mm
θ = Grain orientation, degree from L (rolling) direction
R = Load ratio (applied minimum to maximum)
T = Temperature, °C
V = Applied strain rate per second

Compression. The compression property tests are being performed by the FAA-TC and NIAR using ASTM E9 standard where applicable while using a support jig and specimen geometry based on the ASTM D695 standard. Standard specimen geometries are being used where applicable, whereas micro-sized dog-bone specimens were used for the 2196-T8511 extrusion because of the limited cross-sectional size. Each test was conducted under displacement controlled conditions with load and strain recorded throughout. The data collected was used to calculate compressive modulus, E_c, yield strength, F_{cy}, and ultimate strength, F_{cu}.

Shear. The shear properties were determined by NIAR following ASTM B831 standards where applicable. Because of the limiting size of the extrusions, modified micro shear specimen geometry was used for both materials. The geometry, shown in Figure 1, was determined experimentally through a study plan of various configurations. Each test was conducted under displacement controlled conditions with load and strain recorded throughout. The data collected was used to calculate shear modulus, G, yield strength, F_{sy}, and ultimate strength, F_{st}.

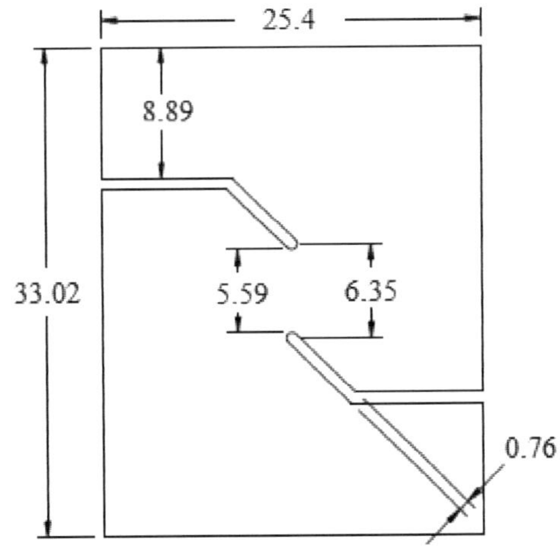

Figure 1. Shear properties test conducted by NIAR: Specimen geometry (all dimensions in mm).

Fatigue and Fatigue Crack Growth

Three separate test programs were used to assess the fatigue life and FCG behavior of the subject materials when subjected to various loading and environmental conditions. Typically, each test was run under load controlled conditions while different specimen configurations were used to achieve the desired results. Three repeat tests were typically run for each test condition. Additionally, the bare and clad forms of 2024-T3 were used for the lap joint fatigue portion due to conditioning in a corrosive environment.

Fatigue Crack Growth. FCG properties at room and elevated temperatures were generated by NASA-JSC and UDRI using ASTM E647 where applicable. The test covered the uniaxial fatigue testing of Al-Li 2198-T8 and 2024-T3 sheet material to determine the crack growth rate, *da/dN*, as a function of stress-intensity factor range, ΔK, for a given load ratio, R. The tests were conducted at both room and elevated temperatures (93°C and 177°C) for two thicknesses (1.8 mm and 6.4 mm) and three grain directions (L, 45°, and LT) using eccentrically loaded single-edge crack tension specimens (ESE(T)) by NASA-JSC. To check for lab-to-lab variation, FCG data were generated by UDRI using middle tension (M(T)) specimens in the same three grain orientations having a thickness of 1.8 mm.

The tests conducted by NASA-JSC utilized MTS machines controlled by Fatigue Technology Associates software for constant *R-K* control test method. This method maintains a constant load ratio while varying the loads throughout the test, allowing FCG data to be generated for the full *da/dN* vs. ΔK range for each specimen. The load ratios used were 0.1, 0.4, 0.7, and 0.8. Crack growth data were recorded using DC potential drop KRAK gages bonded to each specimen. The crack growth data was reduced using the Adjusted Compliance Ratio method to derive *da/dN* and ΔK data for each test. The tests by UDRI were conducted under constant amplitude loading with crack progression tracked using a traveling microscope having a 20X magnification. In general, three repeat tests were conducted for each test condition by both labs.

Spectrum Fatigue Crack Growth. The effects of spectrum loading on FCG properties of Al-Li 2198-T8 and 2024-T3 sheet materials were studied by UDRI [9]. Specimens were machined at one thickness (1.8 mm) and three grain directions (L, 45°, and LT) in a middle tension specimen M(T) configuration. Each test was monitored using a traveling optical microscope and crack length data was recorded throughout the test.

Fatigue tests were performed using a uniaxial tension test machine under load controlled mode using two loading profiles; constant amplitude loading (R = 0.1) and comprehensive flight spectrum typical of the crown section of a fuselage forward of the wing ($R \sim 0$). The comprehensive spectrum was developed as part of a previous FAA program by UDRI using incremental vertical load factor exceedances from the FAA Operation Loads Monitoring Program (OLMP) for the upper crown of the fuselage above the wing front spar for a B737 aircraft [10]. Data from the OLMP was used to create a representative flight-by-flight sequence, as shown in Figure 2 which included stresses from maneuvers and gusts from a typical flight. It should be noted that tests were run at $R \sim 0$ in the previous UDRI program [10], whereas constant amplitude tests were run at $R = 0.1$ in this study, to allow for comparison with the data generated by NASA-JSC.

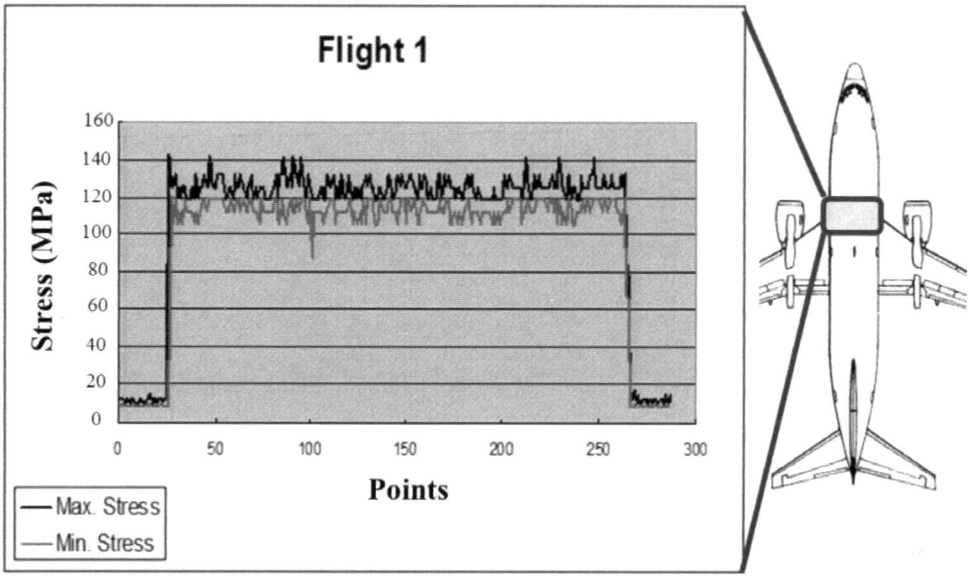

Figure 2. Spectrum FCG test by UDRI: Single flight cycle for comprehensive spectrum for B737.

Lap Joint Fatigue. Uniaxial fatigue testing using a lap joint configuration was performed as a collaborative effort by NAVAIR, FAA-TC, and Drexel University. Lap joint specimens were constructed from Al-Li 2198-T8 sheet material and the bare and clad forms of 2024-T3 material. Each specimen, configured as shown below in Figure 3, consisted of two main sheets machined out of 3.2 mm material at three directions (L, 45°, and LT), separated by a spacing shim machined from 1.6 mm 2024-T3 at the L direction. The joints were constructed using two fastener rows consisting of three fasteners in each row. The fasteners used were A286 Hi-Lok® with a 100° countersunk head and held by a shear collar. The material was used as supplied and was not coated or treated. Additionally, no adhesive or sealant was used in constructing the lap joint specimens.

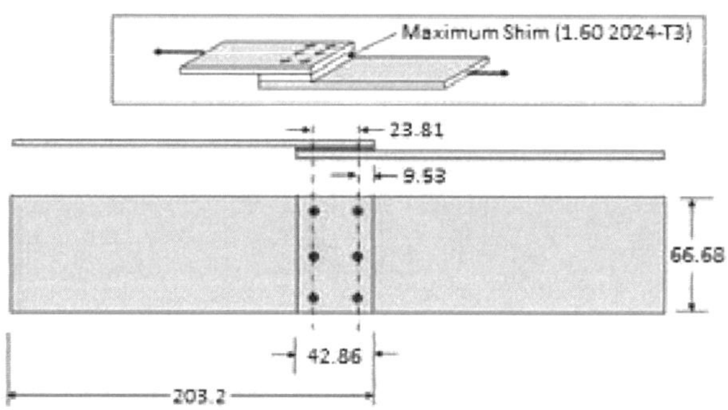

Figure 3. Lap joint fatigue test by FAA-TC: Specimen configuration.

The design of the lap joint was chosen to induce a large amount of secondary bending under load. A number of preliminary static tensile tests were conducted using 2024-T3 specimens in the L direction to determine the amount of bending induced in the configuration shown in Figure 3. For this, back-to-back strain gages were applied to both the outer and inner faying surface (the shim was slotted to allow clearance) of the main plate along the centerline of the first critical fastener row. The tension and bending strain components were calculated and compared to similar data generated at the FAA Full Scale Aircraft Structure Test Evaluation and Research Lab for a full-sized curved fuselage panel [11], as shown in Figure 4 below. Bending in the lap joint specimens used in this study was approximately four times greater than that in the full scale panel having curvature and lap joint configuration representative of a narrow body aircraft.

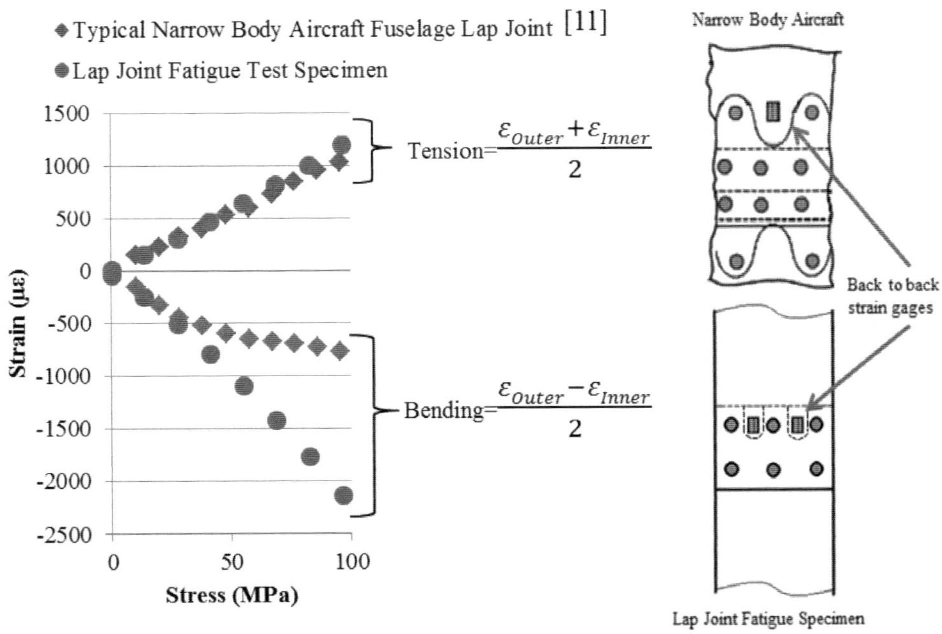

Figure 4. Lap joint fatigue test conducted by the FAA-TC: Bending strain in lap-joint specimens four times larger than that in a typical fuselage lap-joint configuration.

Once assembled, a subset of the lap joints was conditioned by NAVAIR per ASTM G85 Annex A4 using a SO_2/Salt Fog + 5% NaCl environment. The joints were conditioned for 30 days and then stored in a desiccator until tested.

The fatigue tests were then conducted by the FAA-TC using a uniaxial MTS test frame under load control. The initial tests on the 2198-T8 and 2024-T3 bare materials were conducted as simple constant amplitude tests with an $R = 0.1$. The subsequent tests on the 2024-T3 clad material were conducted while implementing marker band spectrums into the test sequence to assist the posttest fractography. The marker bands were run at $R = 0.6$ while maintaining the maximum stress in a similar process detailed in [11]. The reported fatigue life for specimens tested under the marker band sequence was adjusted by subtracting the number of cycles run at $R = 0.6$ from the overall cycle count. All tests were cycled at a maximum initial net stress of 105.5 MPa as determined by a load verification program. Tests were additionally paused at 10,000 cycle intervals (9850 cycles for the marker band specimens) to allow for inspections for the onset of cracking and the subsequent progression throughout the test. Because of the high secondary bending illustrated in Figure 4, cracks always initiated at the inner faying surface near the fastener hole and progressed in a tunneling fashion, rarely breaking the surface of the material before final fracture. As a result, traditional surface inspection techniques were ineffective, so an angled phased array ultrasonic system was used. The system (an Olympus OmniScan MX Flaw Detector phased array ultrasonic system with a 64-element probe attached to a wedge that created a 70° shear wave in aluminum) allowed a cross-section of the critical fastener rows to be monitored, as shown in Figure 5. Here, the first inspection conducted at 0 cycles serves as a baseline for a clean (undamaged) fastener hole and subsequent changes in the signal response indicate the first detection (at 60,000 cycles for this specimen highlighted by the yellow circles) and subsequent progression of a crack.

Figure 5. Lap joint fatigue test conducted by FAA-TC: Ultrasonic inspection results from a 2198-T8 unconditioned specimen. Inspections were conducted at 10,000 cycle intervals.

Post-test fractographic examinations of the fracture surfaces were then conducted by Drexel University using a Zeiss Supra 50VP scanning electron microscope (SEM) as reported in reference 12. This effort focused on reconstructing the crack growth history; that is, determining crack front shapes, crack initiation sites, crack growth rate (*da/dN*), relationship between crack area and fatigue life, as well as any other general observation in the surface topography. Fractographic results are reported for unconditioned 2198-T8 and 2024-T3 bare specimens only.

Supplemental Property Tests

The supplemental property tests were designed to evaluate the material response of the Al-Li alloys in comparison to the baseline aluminums when subjected to external conditions as well as machining processes as described below.

Environmentally Assisted Cracking. Environmentally assisted cracking (EAC) using slow strain rate testing (SSRT) was studied by NAVAIR per ASTM G129 standard to determine the corrosion resistance of the material under a gradually increasing tensile strain. The specimens were machined at one thickness (3.2 mm) and one grain direction (45°) for both the 2198-T8 and 2024-T3 clad alloys. The specimen geometry used was the subsized tension dog bone per ASTM E8 standard.

The test was run at three strain rates (10^{-5}/s, 10^{-6}/s, and 10^{-7}/s) to determine if rate has an effect on the material properties. An aqueous 5% NaCl solution was used to create the corrosive environment during the test while duplicates were also conducted at lab air conditions as a baseline. The tests were run until specimen failure with information on load, extension, and duration recorded.

Fastener Hole Process. The effects of the cold expansion process on open holes and the installation of interference fit fasteners was studied by the FAA-TC. Both the 2198-T8 and 2196-T8511 materials were tested at three thicknesses (1.8 mm, 3.2 mm, and 6.4 mm for 2198-T8; 1.5 mm, 3.0 mm, and 3.7 mm for 2196-T8511). Comparisons were made with baseline 2024-T3 and 7075-T6 alloys. The specimen processing was completed by Fatigue Technology Incorporated (FTI). A split sleeve method was used to create the cold work holes with the sleeve opening oriented at three different angles (L, 45°, and LT). Through this method, a maximum of 5% cold working was accomplished. The interference fit fasteners were selected for a 0.003" interference. Post-processing inspections were then made using surface eddy current nondestructive inspections (NDI) and X-Ray computer tomography. Finally, specimens were sectioned and subjected to high magnification examinations to determine if cracking occurred due to typical fastener hole processes. Tests are currently underway and will be reported when results become available.

Flammability and Burn Resistance. Testing was conducted by the FAA Fire Safety Branch at the FAA-TC to determine flame penetration, flammability, and burn resistance of the 2198-T8 material. The specimens were machined at three thicknesses (1.8 mm, 3.2 mm, and 6.4 mm) while grain directions were inconsequential to this test.

The test was performed using a NexGen burner specified by AC 25.856-2A [13], which created a flame temperature of 982°C. Each specimen was mounted on a test frame and moved into direct contact with the open flame. The tests were run until the flame was visually seen to penetrate the specimen with test duration and general observations being recorded.

TEST RESULTS

Test results are presented for the material properties and mechanical behavior of two Al-Li alloys, namely 2198-T8 and 2196-T8511 used for skin and extrusion applications, respectively. Several properties were assessed and compared with baseline 2024-T3 and 7075-T6 alloys, including static properties; fatigue life and FCG behavior; and supplemental properties per Table I. Full details are provided in [7]. Representative results are presented herein.

Static Properties

Tensile, compression, and shear static properties were measured for 2198-T8 and 2196-T8511 Al-Li and compared to baseline 2024-T3 and 7075-T6 alloys, respectively, as discussed in the subsequent sections.

Tension. Tensile properties were measured by the FAA-TC for the 2198-T8 and the baseline 2024-T3 sheet materials. Overall the 2198-T8 material exhibited higher modulus, higher strength, and lower elongation compared to the baseline 2024-T3 material, with the exception of a few specific points. Results presented are the average of three tests. The modulus measured for the 2198-T8 and 2024-T3 was 74.5 GPa and 71.7 GPa, respectively. The elongation for both alloys is shown in Figure 6 as a function grain orientation and thickness. As shown, the 2198-T8 (solid columns) exhibited lower elongation for all thicknesses and grain orientations compared to the baseline 2024-T3 (hashed columns).

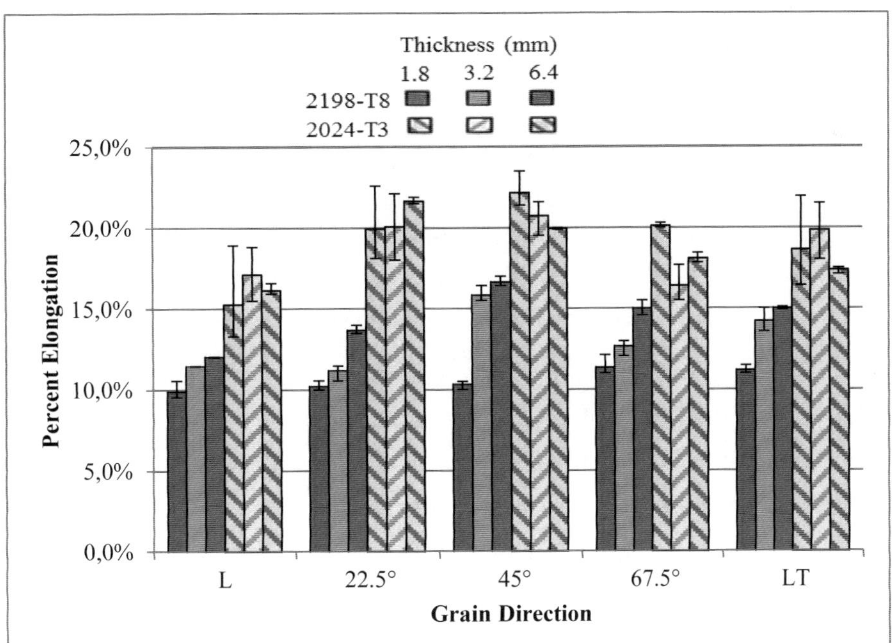

Figure 6. Tension property tests conducted by the FAA-TC: 2198 vs. 2024 Percent Elongation Comparison. Results are an average from three tests.

A comparison of the yield and ultimate strengths are shown in Figure 7 and Figure 8, respectively. In these figures, the variation of yield and ultimate strength as a function of grain orientation and thickness are displayed for both the 2198-T8 and 2024-T3. While the yield strength was always higher for the 2198-T8 material compared to the baseline 2024-T3, the ultimate strength was lower for the 3.2 mm thick material at the intermediate grain directions (22.5°, 45°, and

67.5°) as well as the 45° direction for the thickness of 6.4 mm. For the 45° direction, the ultimate strength was lower by 10.1% and 7.1% for the respective thicknesses.

In general, the 2198-T8 exhibited both grain orientation and thickness effects. For a thickness of 3.2 mm, the 2198-T8 material showed the greatest amount of strength variation with grain orientation having approximately a 17% reduction in both yield strength (Figure 7) and ultimate strength (Figure 8) from the L direction to the 45° direction. In addition, the 3.2-mm-thick material generally recorded the lowest yield and ultimate strengths for each test with the only exceptions being the L and LT direction for yield strength and L direction for ultimate strength, each time only surpassing the 1.8-mm-thick results. On the other hand, for a thickness of 1.8 mm, the 2198-T8 results were consistent across all grain directions. Expectedly, all of the measured values exceeded the applicable MMPDS A and B basis values including the 45° direction. In addition, test results were highly repeatable for both materials. The coefficients of variation (CV), as shown in Table II, for both materials are considerably lower for the strength values when compared to the values listed in ASTM E8 for 2024-T351. And while the variation in elongation was higher for the 2024-T3 results generated in this program, the results for 2198-T8 were comparable with the published value.

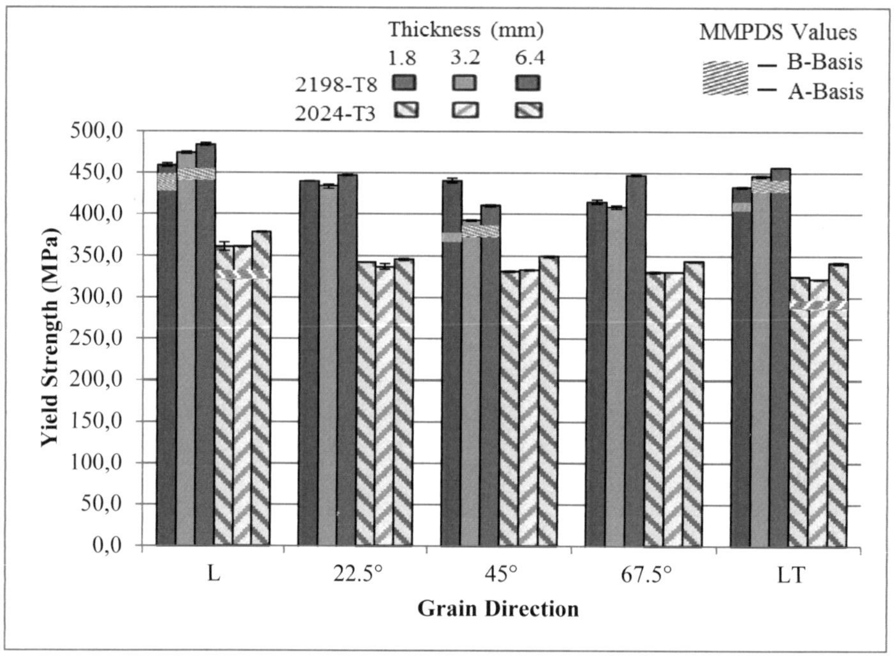

Figure 7. Tension property tests conducted by FAA-TC: 2198 vs. 2024 Yield Strength Comparison. Results are an average from three tests.

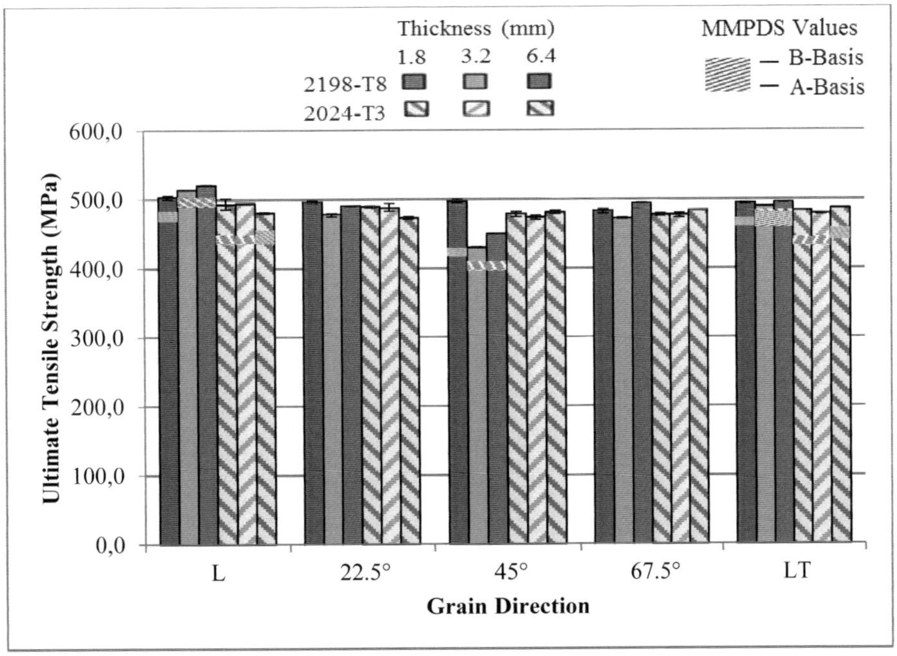

Figure 8. Tension properties tests conducted by the FAA-TC: 2198 vs. 2024 Ultimate Strength Comparison. Results are an average from three tests.

Table II. Tension properties coefficients of variation for tests conducted by the FAA-TC: Comparison of variation to published ASTM E8 value for 2024-T351. Results are an average of 15 variations values for each material.

	Coefficient of Variation		
Material	Yield Str	Ult Str	Elongation
2024-T351 (ASTM E8 Value)	1.41%	1.24%	2.94%
2024-T3	0.29%	0.46%	6.22%
2198-T8	0.29%	0.24%	3.02%

Tensile properties at elevated temperatures of 93°C and 177°C were measured by UDRI [8] and are summarized in Figure 9 for both 2198-T8 and 2024-T3 in terms of percent reduction in ultimate strength from ambient conditions. As expected, there was a general decrease in strength as the temperature increased for both the Al-Li and baseline materials.

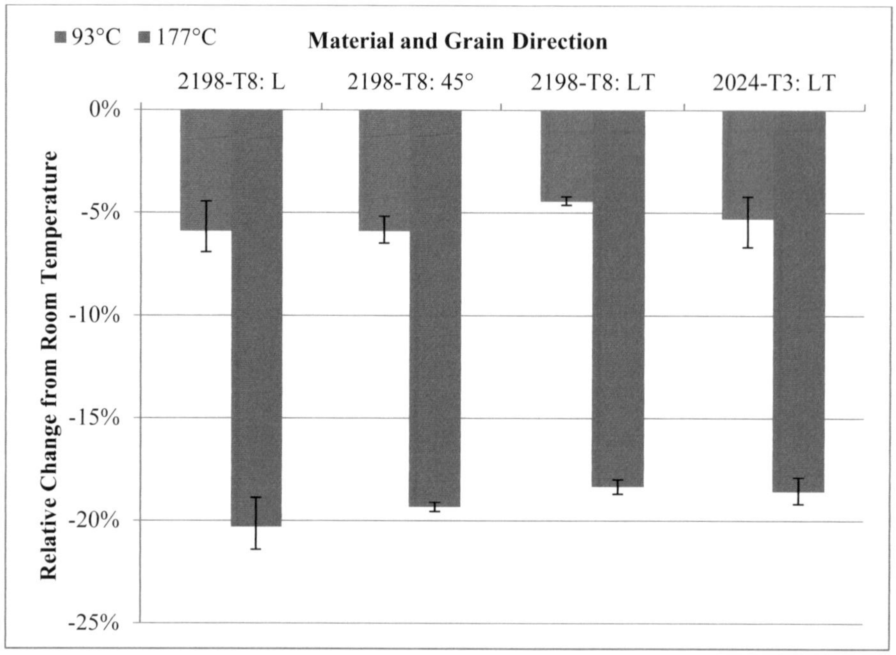

Figure 9. Tension property tests conducted by UDRI [8]: 2198-T8 vs. 2024-T3; 1.8-mm-thick; elevated-temperature effect. Results are an average from three tests.

The results for the 2196-T8511 extrusions generated by NIAR showed the Al-Li to have a higher modulus compared to the baseline 7075-T6 aluminum with average values of 81.2 GPa and 70.6 GPa, respectively. In comparison to the baseline, the ultimate strength values of the 2196-T8511 were generally lower across all grain directions and thicknesses, as shown in Figure 10, with the exception of the L direction. Although the L direction values for the 2196-T8511 extrusions were relatively constant and similar to the baseline, there was a strength reduction in the 45° direction of approximately 10% for all thicknesses tested compared to the 7075-T6 material. Similar to the 2198-T8 tensile strength results (Figures 7 and 8), the 2196-T8511 exhibited anisotropic behavior with lower strengths recorded at the 45° grain direction. The relative reductions in strength compared to the L direction results are approximately 14% for all thicknesses. The ultimate strength values measured were quite repeatable (CV values of 0.42% for 7075-T6 and 0.72% for 2196-T8511 were comparable with the values in Table II above) and exceeded the published MMPDS A and B basis values.

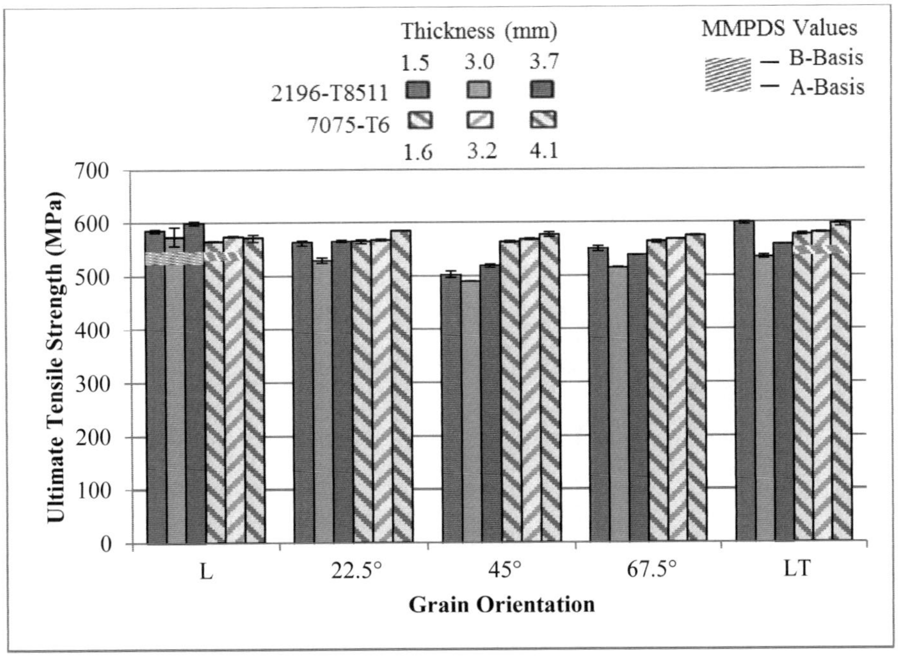

Figure 10. Tension property tests conducted by NIAR: 2196 vs. 7075 Ultimate Strength Comparison. Results are an average from three tests.

Compression. The results from compression tests conducted by NIAR reveal higher modulus values measured in the 2196-T8511 Al-Li at all thicknesses and grain directions compared to the baseline 7075-T6 aluminum with average values of 74.5 GPa and 71.7 GPa, respectively. Compressive yield strength as a function of gain orientation and thickness is shown in Figure 11 for both materials. As indicated in this figure, anisotropic behavior was observed in the Al-Li where the yield strengths across all thicknesses were lowest at the 45° grain direction. The intermediate thickness (3.0 mm) again showed the most dramatic change with reductions of approximately 18% from the L to 45° direction. The yield strengths measured in both materials exceeded the published MMPDS A and B basis values. In addition, there is less scatter in the test results for the 2196-T8511 compared to the 7075-T6 where the CV was approximately 1.9% and 2.9% for each material, respectively.

Shear. The shear properties measured by NIAR reveal that the shear modulus for the 2198-T8 material was higher than the 2024-T3 baseline with values of 29.4 GPa and 24.4 GPa respectively. The ultimate strength values for the two materials were similar and relatively consistent for all thicknesses and grain orientations, as shown in Figure 12. There was little scatter in the ultimate strength test results, with CV of 0.56% and 1.01% for 2024-T3 and 2198-T8, respectively, and all measured values in both materials exceeded the published MMPDS A and B basis values.

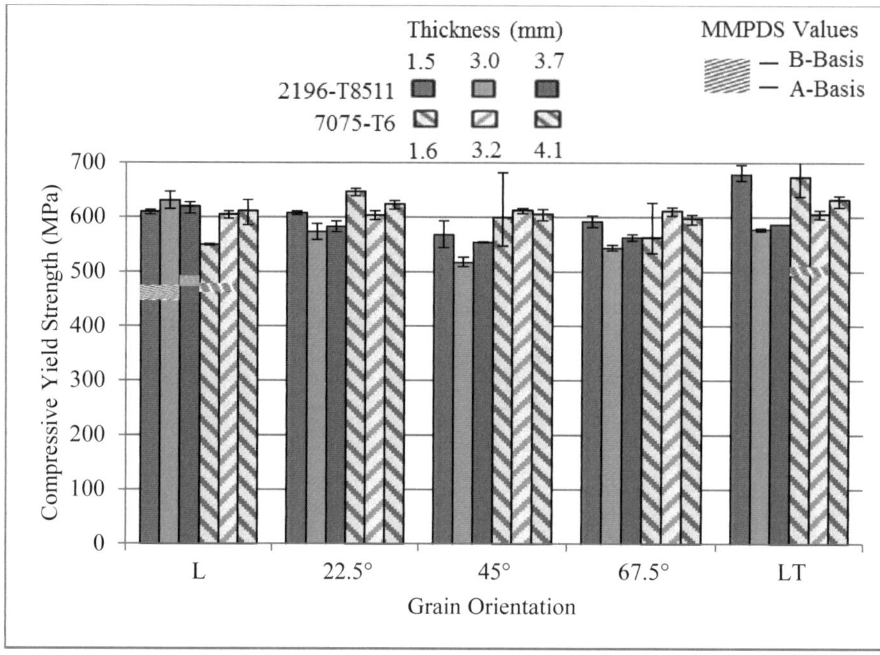

Figure 11. Compression property tests conducted by NIAR: 2196 vs. 7075 Yield Strength Comparison. Results are an average from three tests.

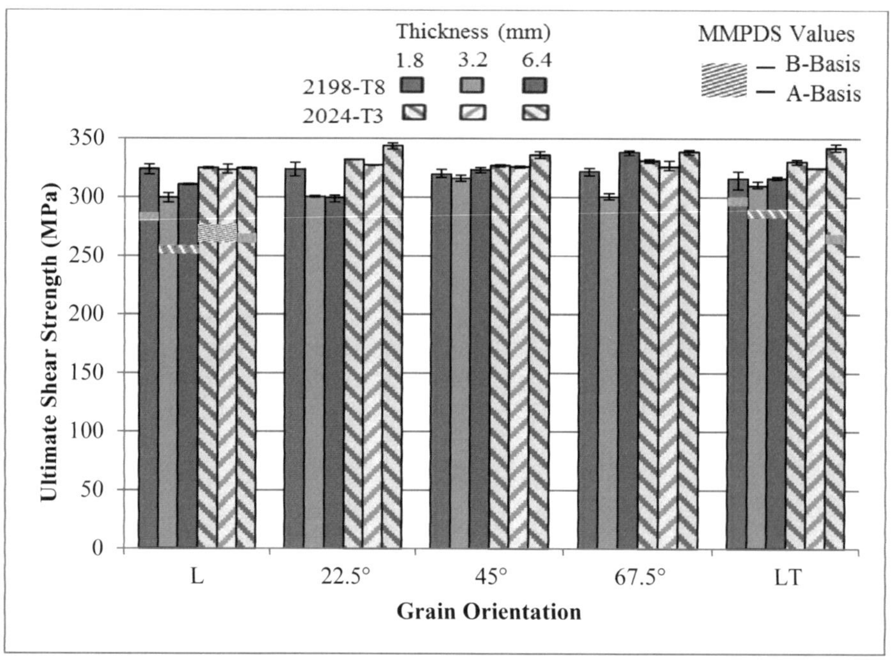

Figure 12. Shear property test conducted by NIAR: 2198 vs. 2024 Ultimate Strength Comparison. Results are an average from three tests.

The results for the 2196-T8511 shear tests were similar to those of the 2198-T8 as shown in Figure 13. Again, the shear modulus was higher for the Al-Li compared to the baseline 7075-T6 at 31.4 GPa and 29.5 GPa, respectively. Also, the material did show signs of anisotropy with the intermediate grain directions having the highest strength. Note that the shear strength, on average, tended to be highest in the 45° direction. In comparison to the baseline 7075-T6 material, ultimate strength values measured in the 2196-T8511 were lower for all thicknesses and grain orientations. Results were repeatable for both materials (CV of 0.63% and 1.10% for 7075-T6 and 2196-T8511, respectively) and exceeded the published MMPDS A and B basis values.

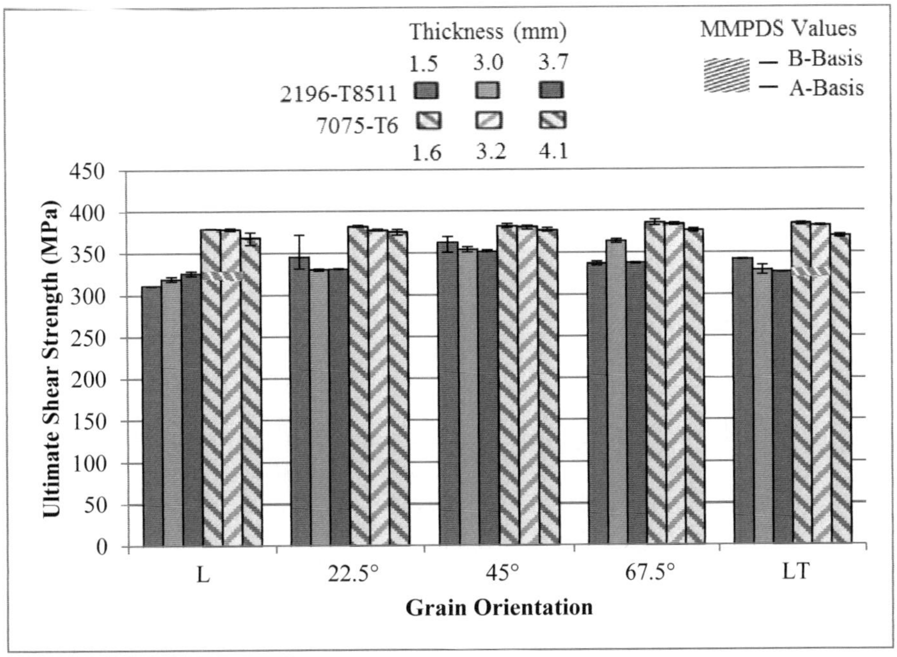

Figure 13. Shear property tests conducted by NIAR: 2196 vs. 7075 Ultimate Strength Comparison. Results are an average from three tests.

Fatigue and Fatigue Crack Growth Property Tests

Fatigue and FCG properties were measured for 2198-T8 Al-Li and compared to baseline 2024-T3 alloy as discussed in the subsequent sections; 2024-T3 clad was additionally tested for the lap joint fatigue test because of the use of environmental conditioning.

Fatigue Crack Growth. The results from the FCG generated by NASA-JSC for the 2198-T8 Al-Li material show no significant anisotropic or thickness effects on the stable crack growth region of the *da/dN* vs. ΔK curves. As shown in Figure 14, there was limited variation in the FCG data for the 1.8-mm-thick Al-Li material tested at room temperature at $R = 0.1$ for the three major grain orientations. In addition, FCG results were similar for the two thicknesses (1.8 mm and 6.4 mm) considered in this study. These results were quite surprising considering the anisotropic behavior observed in static properties for the Al-Li material

Comparing the FCG measured in the 2198-T8 and baseline 2024-T3, the two materials behaved very similarly in the near threshold and stable crack growth regions of the *da/dN* vs. ΔK curves, as shown in Figure 15. However, the curves tend to diverge in the upper region with 2198-T8 showing a higher crack growth resistance compared to the baseline. This trend is further supported by the results from the constant amplitude tests conducted by UDRI [9].

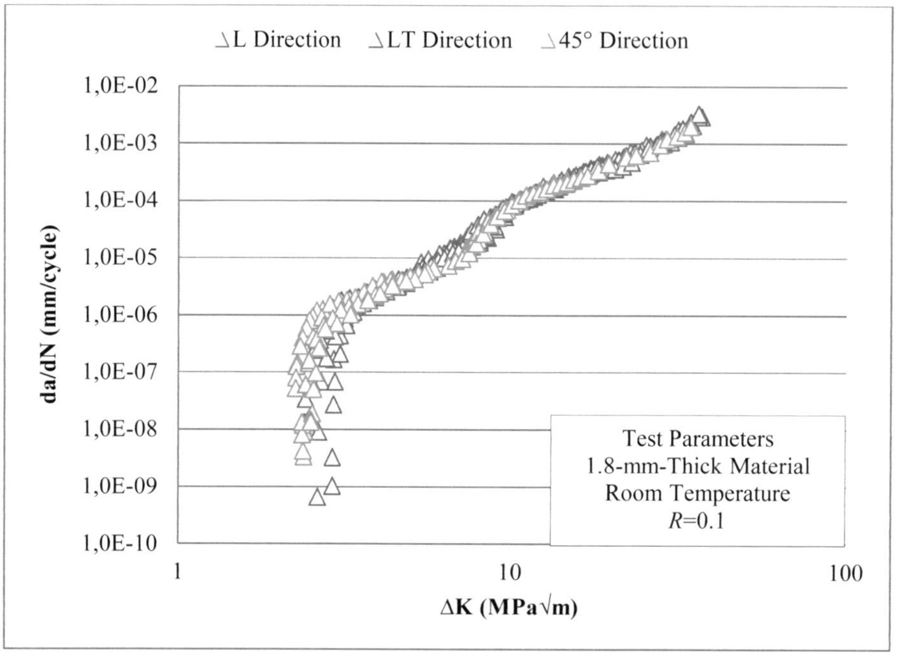

Figure 14. FCG properties generated by NASA-JSC: Effect of grain orientation on 2198-T8 having a thickness of 1.8mm. Three specimens used to generate data for each of the three orientations.

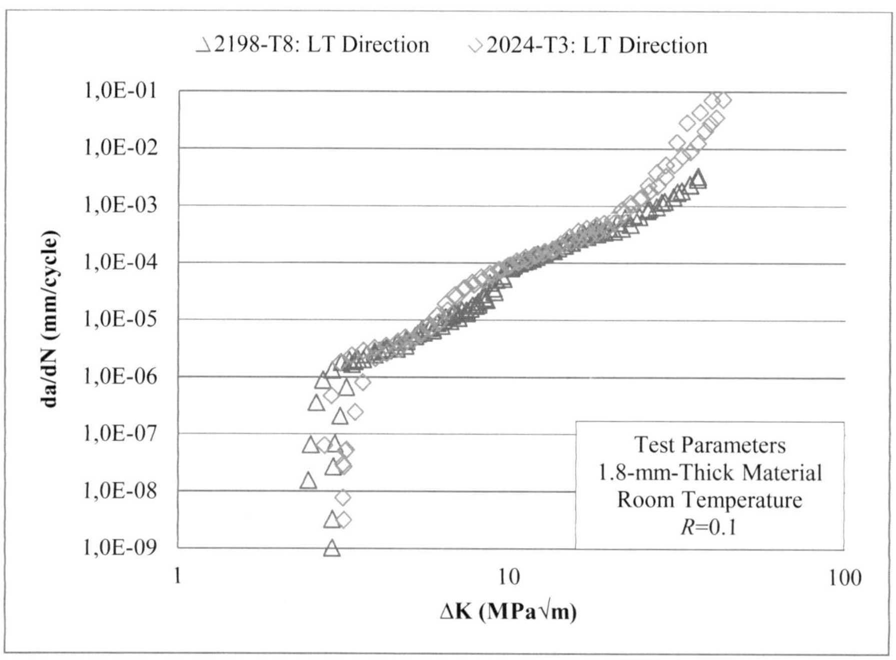

Figure 15. FCG properties generated by NASA-JSC: 2198-T8 vs. 2024-T3; 1.8-mm-thick, LT Grain Orientation. Three specimens used to generate data for each material.

A comparison of the constant amplitude data sets from both labs, shown in Figure 16 below, reveals two important observations: first, there was good repeatability and limited lab-to-lab variation where the FCG data sets overlap closely; and second, the data generated by UDRI, reported in the following section, represents the upper region of the *da/dN* vs. *ΔK* curve.

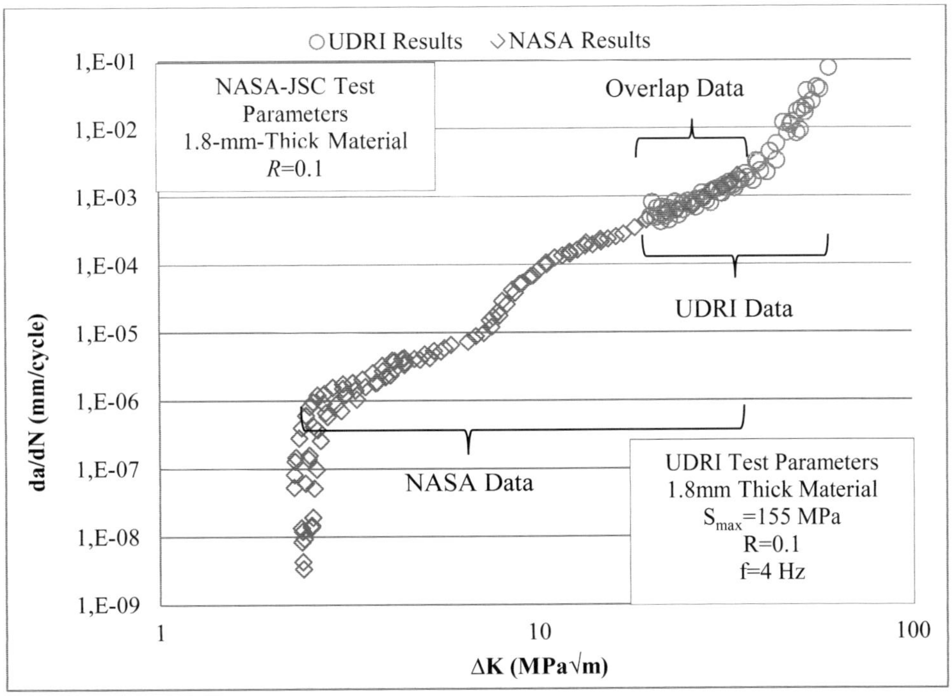

Figure 16. Comparison of FCG data generated by NASA-JSC and UDRI [8]; 2198-T8 material, 1.8-mm-thick, 45° grain direction.

Spectrum Fatigue Crack Growth. The effects of spectrum loading on FCG behavior were investigated by UDRI [9]. In general, the spectrum stress sequence used representing the crown of the B737 just forward of the wing resulted in FCG in the upper region of the da/dN vs. ΔK curve, as shown in Figure 16. The comprehensive spectrum results in Figure 17 show, similar to the NASA-JSC data, that the 2198-T8 material at all three grain directions has a higher crack growth resistance as crack size approaches instability compared to the baseline 2024-T3 L direction. Comparing just the 2198-T8 results reveals that the 45° grain orientation displayed the longest fatigue life despite this orientation having the lowest strength in the static tensile tests. These results were consistent with the constant amplitude tests, which also showed higher crack growth resistance for 2198-T8 material, especially at the 45° orientation. Additionally shown in Figure 17 is the FCG data generated exhibited good repeatability and limited scatter where the individual specimen results (dashed lines) are grouped closely with the average (solid line).

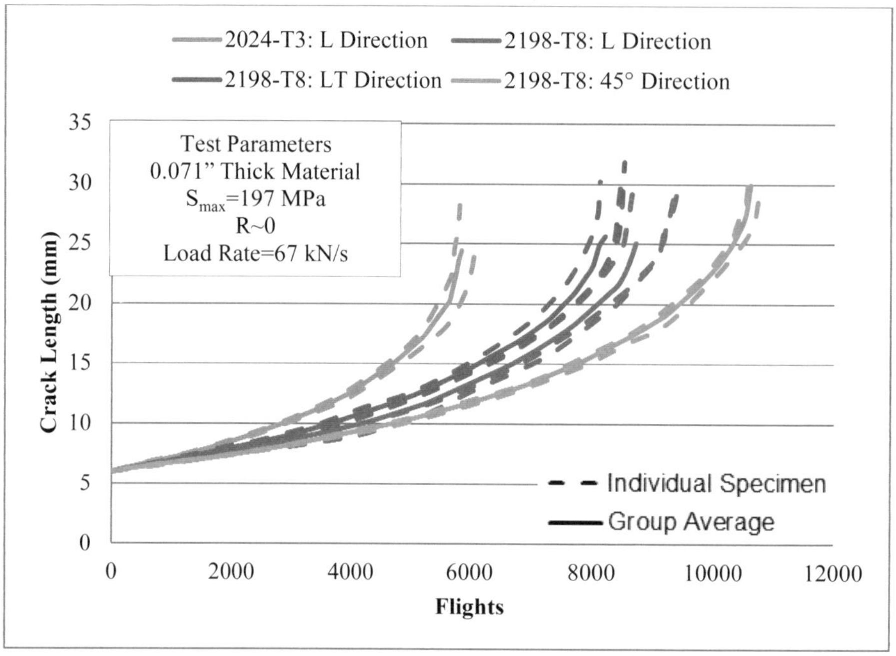

Figure 17. Spectrum FCG tests conducted by UDRI [9]; Flights vs. crack length for comprehensive spectrum tests.

Lap Joint Fatigue. Lap-joint fatigue tests were conducted by the FAA-TC for 2198-T8 sheet and 2024-T3 sheet in both bare and clad forms. The effect of exposure to a highly corrosive environment on the fatigue life was studied for both materials. For this, NAVAIR conditioned a subset of lap joint specimens per ASTM G85 Annex A4 using a SO_2/Salt Fog + 5% NaCl solution for 30 days. This conditioning corroded the surfaces of the specimens to various degrees, as shown in Figure 18 (the images on the right show the material after sanding was done to remove the corrosion byproducts and, in the case of the 2198-T8 material, remove the blistering). The 2024-T3 bare surfaces exhibited signs of local pitting, as shown in Figure 18a, while the 2024-T3 clad surfaces were relatively clean, as shown in Figure 18b. The 2198-T8 displayed signs consistent with superficial exfoliation as per ASTM G34 (Rating EA: tiny blisters, thin slivers, flakes, or powder, with only slight separation of metal) and shown in Figure 18c.

Subsequent fatigue tests were then performed in lab ambient conditions under constant amplitude loading to determine the onset of cracking and the fatigue life for the three materials. The onset of crack detection and subsequent progression were determined through inspections at 10,000 cycle intervals using an OmniScan ultrasonic phased array system, as shown in Figure 5. Figure 19 shows the average number of cycles until first crack indication, as the horizontal bars overlaid on the appropriate cycles to failure column. In general, cracks were detected earliest in the 2024-T3 clad material, followed by the 2198-T8 and 2024-T3 bare materials, respectively. The corresponding fatigue life for the unconditioned specimens, indicated by the solid columns in Figure 19, reflected the first crack indication order as the 2024-T3 bare material had the longest fatigue life across all grain orientations, followed by the 2198-T8 material. As expected, the 2024-T3 clad had a significant reduction compared to the bare and had the lowest fatigue life of the three materials.

The results from the conditioned specimens, shown as the hatched columns in Figure 19, indicate that the 2024-T3 clad had the longest fatigue life, followed by the 2198-T8. The cladding was quite effective in protecting the baseline material from an extremely corrosive environment. The 2024-T3 clad showed no reduction in fatigue life because of the environmental conditioning compared to reductions of approximately 30% and 60% seen for 2198-T8 and 2024-T3 bare, respectively.

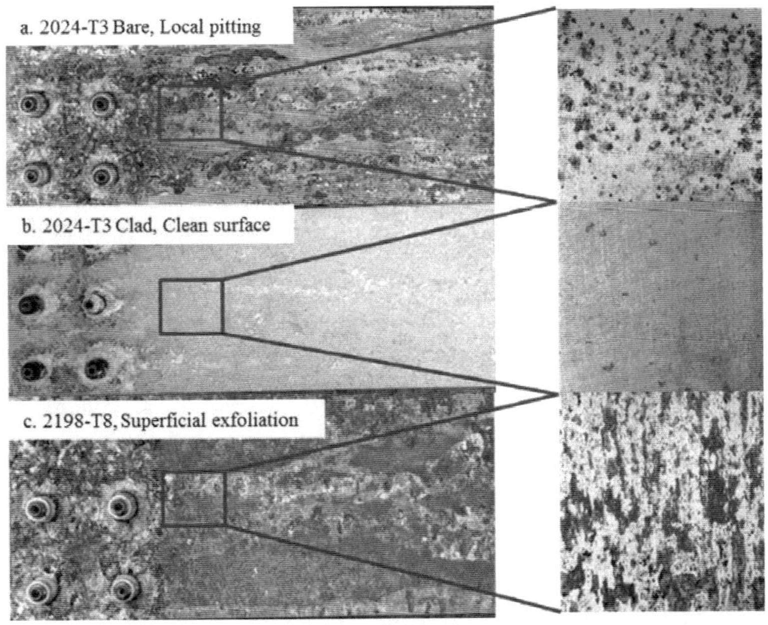

Figure 18. Condition of lap joint specimen surfaces after 30 days exposure to ASTM G85 Annex A4 using a SO_2/Salt Fog + 5% NaCl solution by NAVAIR.

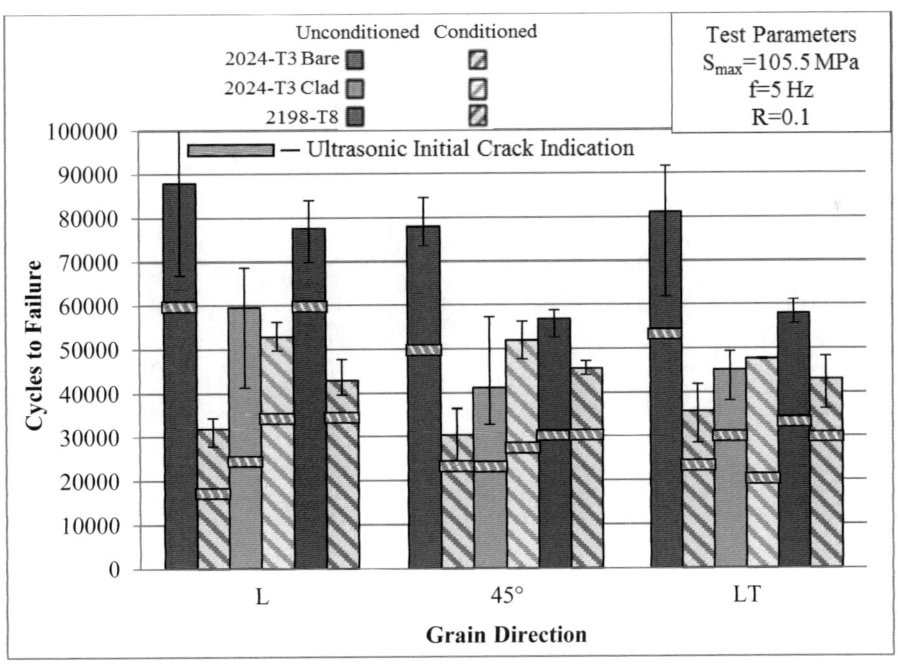

Figure 19. Comparison of average fatigue life from lap joint fatigue tests conducted by the FAA-TC.

Post-test fractographic examinations were then conducted to determine differences in failure modes in the 2198-T8 sheet and 2024-T3 bare specimens. Initial general inspections for the unconditioned specimens reveal failure modes unique to the 2198-T8 sheet material, namely interlaminar cracks, as indicated in Figure 20. In addition, eyebrow cracking was highly prevalent in the Al-Li lap joints, where the major crack forming the final failure surface followed a path that circumvented the fastener holes.

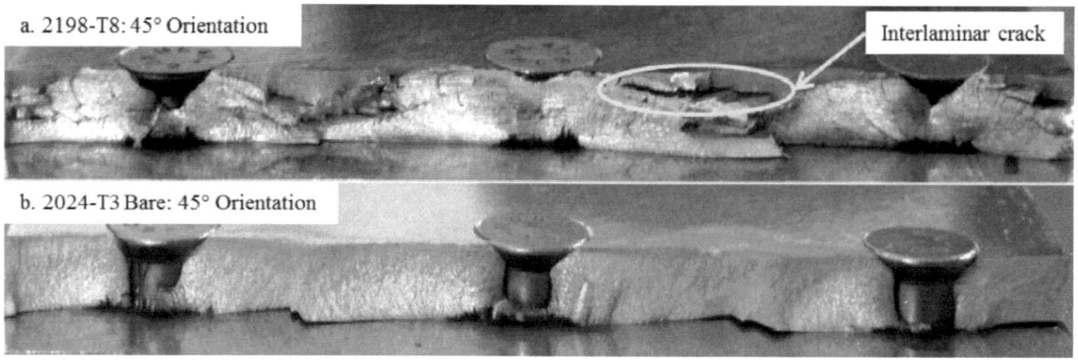

Figure 20. General observations of failure surfaces for the 2198-T8 and 2024-T3 bare sheet lap joint specimens, FAA-TC.

Detailed fractographic examinations using a scanning electron microscope were made by Drexel University for the 2198-T8 sheet and 2024-T3 bare specimens. Details are provided in reference 12 with representative results shown herein. In general, high magnification inspections revealed very different topographies for the two materials, as seen in Figure 21 through Figure 23. The fracture surfaces in the fatigue fracture zone (FFZ) are shown Figure 21. Here the 2024-T3 material displays a very rugged fatigue surface resulting from large constituents in the material, which affects the progression of the crack front, as shown in Figure 21a. The constituents appear to arrest local crack growth and form local initiation sites. In comparison, the FFZ in the 2198-T8 material was much smoother with shallow striations and constituents clustered along the grain boundaries leading to the formation of a single, dominant crack front, as shown in Figure 21b. Crack initiation for both materials occurred at the faying surface slightly removed from the fastener hole, although the actual crack initiation sites were generally obscured because of a large debris field caused by fretting.

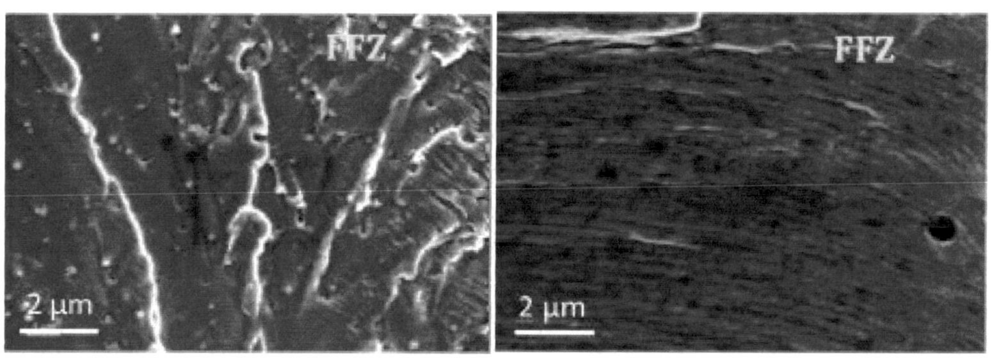

(a) (b)

Figure 21. Lap joint fractography conducted by Drexel University [12]: Fatigue surface topography comparing (a) 2024-T3 bare and (b) 2198-T8.

Also of note was the difference in grain shape and structure where the 2024-T3 material was equiaxed while the 2198-T8 sheet was highly elongated with a pancake-like appearance resulting in differences in the ductile fracture zone (DFZ) of the two materials. The 2024-T3 bare material shows typical dimpled deformation and void coalescence in the vicinity of large constituents (as shown in Figure 22a), while the 2198-T8 material is more lamellar in appearance, (as shown in Figure 22b). The elongated grain structure and presence of constituents along the grain boundaries in the 2198-T8 material coupled with the large secondary bending apparently contribute to the occurrence of interlaminar cracks as mentioned previously and shown again in Figure 23 under higher magnification. While some interlaminar cracks were large enough to be detected through general observations, the SEM study revealed they are prevalent throughout the FFZ for all grain orientation of the 2198-T8.

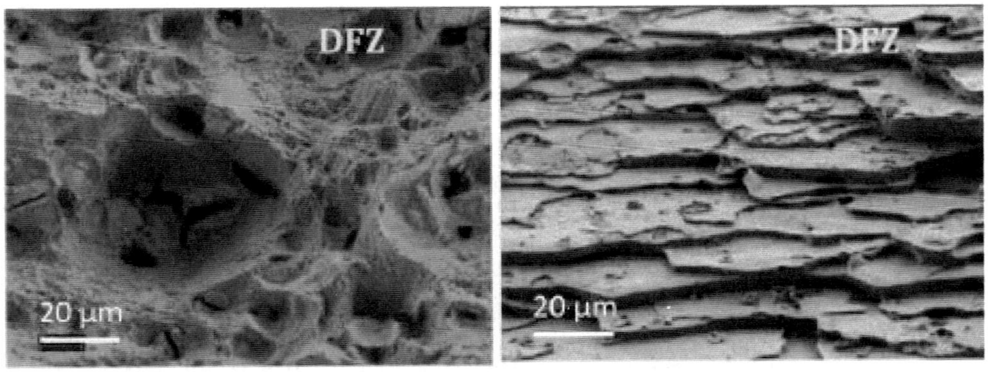

(a) (b)

Figure 22. Lap joint fractography conducted by Drexel University [12]: Ductile fracture surface topography comparing (a) 2024-T3 bare and (b) 2198-T8.

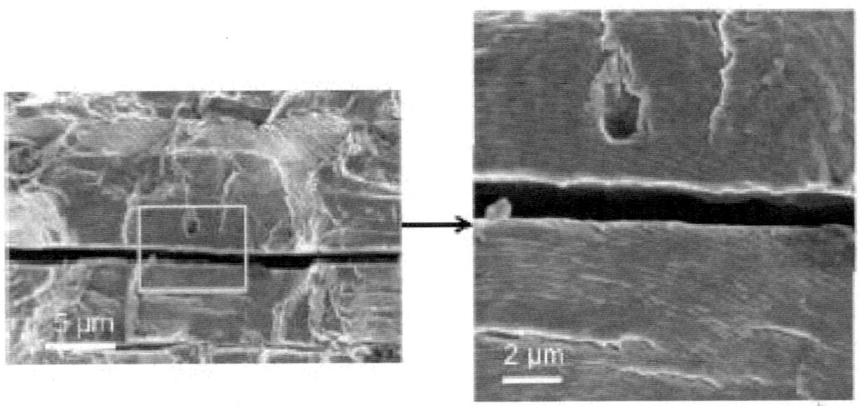

Figure 23. Lap joint fractography conducted by Drexel University [12]: Interlaminar cracking observed within the FFZ in 2198-T8.

Crack growth reconstruction was also conducted by Drexel University to estimate the number of cycles to crack initiation and the shape and size of the FFZ for both unconditioned 2198-T8 and 2024-T3 bare materials at three major orientations. A detailed explanation of the process used and data generated can be found in reference 12. In general, striation counting was conducted to determine the FCG rate at distances from the approximate initiation sites. The crack growth was then back calculated to approximate the cycles to initiation, which correlated well with the first crack indication results from NDI, shown in Figure 19, where fatigue cracks generally initiate earlier in the 2198-T8 compared to the baseline. Subsequent crack growth rates measured in the 2198-T8 were typically lower than that in the 2024-T3 bare material. In addition, the overall size of the FFZ was larger in the 2198-T8, as shown in Figure 24, suggesting that the Al-Li was able to withstand larger crack growth prior to failure.

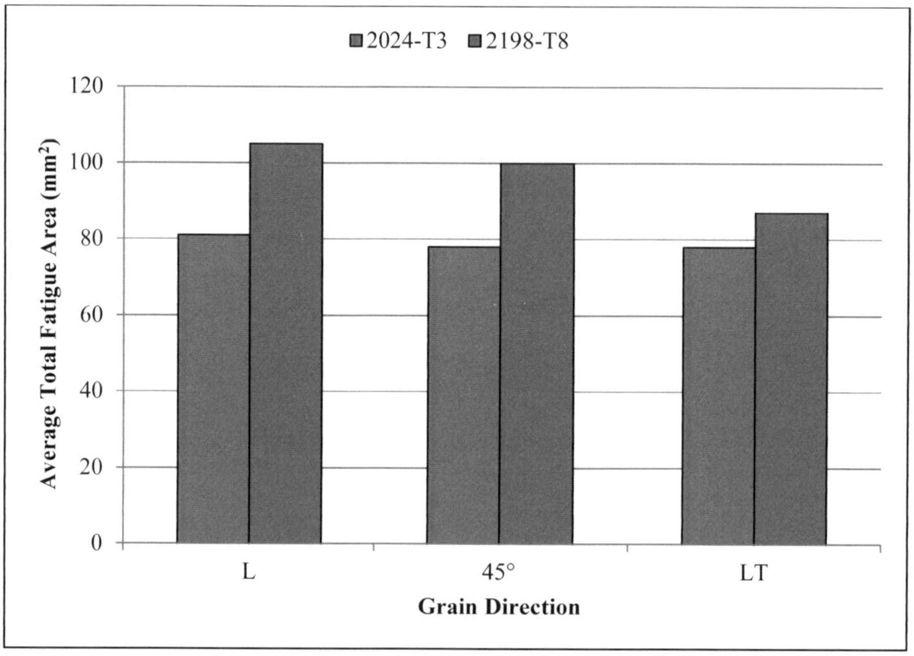

Figure 24. Comparison of the size of total FFZ in the lap joint specimens measured by Drexel University.

Supplemental Property Tests

Results assessing the robustness of the materials subjected to a corrosive environment and to a high temperature fire are summarized below.

Environmentally Assisted Cracking. EAC using SSRT is being conducted by NAVAIR per ASTM G129 standard. Tensile tests are being run at three strain rates, namely 10^{-5}/s, 10^{-6}/s, and 10^{-7}/s up to failure in both an aqueous 5% NaCl solution and in ambient lab conditions for the 2198-T8 and 2024-T3 clad materials in the 45° grain orientation. Of note, the 2024-T3 clad specimens were machined from sheet material and no additional coatings were applied to the machined surfaces. Initial results presented in Figure 25 for two strain rates, namely 10^{-5}/s and 10^{-6}/s, indicate there was no effect of strain rate or test environment on the two material systems. However, tests still need to be conducted for the slowest strain rate of 10^{-7}/s to draw additional conclusions.

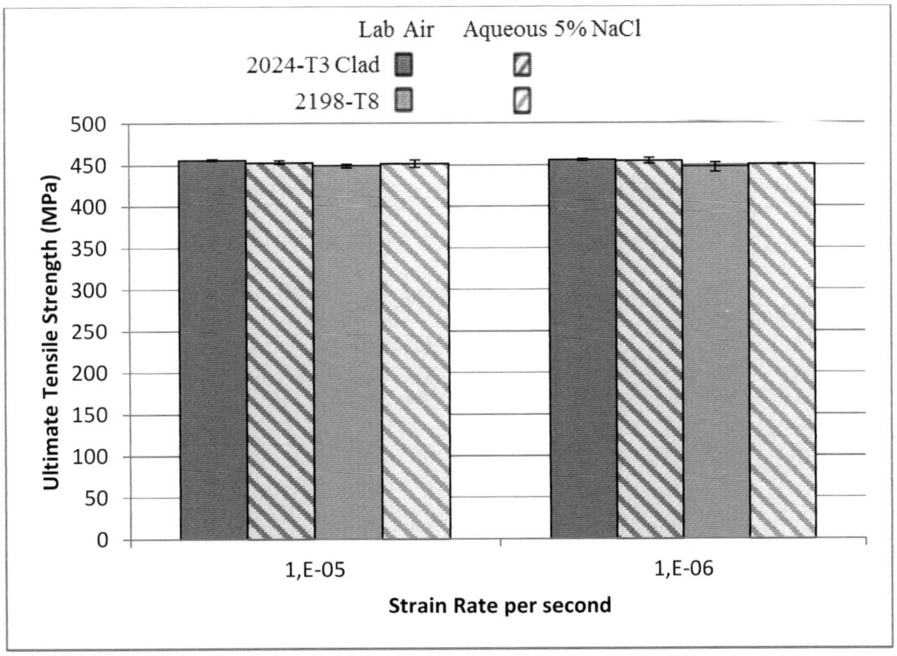

Figure 25. Slow strain rate test conducted by NAVAIR: Ultimate tensile strength comparison. Results represent an average of three test values.

Flammability and Burn Resistance. Testing was conducted by the FAA Fire Safety Branch at the FAA-TC to characterize flame penetration and burn resistance of the 2198-T8 material compared with 2024-T3. For the 2198-T8, three thickness were considered; namely, 1.8 mm, 3.2 mm, and 6.4 mm, while a single thickness of 3.2 mm was considered for the baseline. For both materials, a flame having a consistent temperature of 982°C was applied and the burn though time (time it takes the flame to penetrate though the thickness) was measured.

Overall, as shown in Table III below, the 2198-T8 material had a higher resistance to burn through compared to the 2024-T3 baseline test while neither material showed any sign of ignition or combustion once removed from the flame source. When comparing the results for the two specimens tested at the 3.2-mm thickness, the 2198-T8 material remained intact for approximately one minute longer than the baseline 2024-T3.

Table III. Fire Test: Melt and Burn Through Times.

Material	Thickness (mm)	Burn Through Time
2024-T3	3.2	2:08
2198-T8	1.8	1:52
2198-T8	3.2	3:10
2198-T8	6.4	5:45

SUMMARY

In general, the mechanical properties and behaviors of 2198-T8 and 2196-T8511 materials were in accordance with expectations for the tests completed and comparisons made to the baseline 2024-T3 and 7075-T6 materials, respectively. The static properties for both the Al-Li and baseline materials tested were comparable and the values exceed all MMPDS A and B basis values. Anisotropic behavior was evident in the Al-Li at the intermediate grain directions, especially 45°, at which the tensile and compressive strengths were lowest compared to the other orientations, which may need to be investigated further.

There was a general improvement in the FCG properties in the 2198-T8 compared to the baseline 2024-T3 bare material. Under constant amplitude loading, FCG behavior was similar for both materials in the near-threshold and linear stable crack growth regions. In the upper instability region of the FCG curves, the 2198-T8 displayed a higher crack growth resistance compared to the baseline. Likewise, under spectrum loading, the 2198-T8 exhibited better FCG compared to 2024-T3; interestingly, the 2198-T8 tested in the 45° grain orientation had the longest fatigue life despite this orientation having the lowest tensile and compressive strengths.

Lap-joint fatigue tests conducted on the 2198-T8 and 2024-T3 in the bare and clad forms using specimens configured to induce high secondary bending yielded varying results. Tests on unconditioned specimens demonstrated a shorter fatigue life in the 2198-T8 compared to the 2024-T3 bare sheet, while having a longer fatigue life compared to 2024-T3 clad. Inspections using phased array ultrasonics detected cracks earlier during fatigue for the 2198-T8 than in the 2024-T3 bare due to fretting. For tests conducted on conditioned specimens (30 days of conditioning to ASTM G85 Annex A4), the fatigue lives for the 2198-T8 and 2024-T3 clad were similar, whereas that in 2024-T3 bare was much reduced. Examinations of the fracture surfaces of the 2198-T8 and 2024-T3 bare sheet revealed marked differences. The fracture surface for the 2198-T8 was much smoother with shallower striations compared to the baseline material. Additionally the 2198-T8 grain structure appeared elongated with constituents congregating along the grain boundaries. This, coupled with the high secondary bending, contributes to development of interlaminar cracks not observed in the 2024-T3. Striation measurements and crack area at fracture suggest a slower crack growth rate and larger crack progression, respectively, for the 2198-T8 material compared to the baseline 2024-T3 bare.

The supplemental properties tests also showed the Al-Li compared favorably to the baseline material. For the environmentally assisted cracking using slow strain rate testing, both the 2198-T8 and 2024-T3 clad suffered no reduction in strength at 10^{-5}/s and 10^{-6}/s strain rates when tested in aqueous 5% NaCl solution. Additionally, the 2198-T8 had a longer burn through time compared to the baseline and showed no signs of reigniting when removed from direct contact with the flame.

FUTURE WORK

The results contained in this paper are reflective of the work completed and available at the time of writing. Further work is currently ongoing for the 2198-T8 compression properties, 2198-T8 environmentally assisted cracking at a strain rate of 10^{-7}/s, and 2196-T8511 and 2198-T8 cold work specimen fractography. Additionally, some unique material behaviors were observed in the 2196-T8511 and 2198-T8 alloys that need to be addressed; namely, anisotropic behavior at the 45° grain orientation and unique cracking (eyebrow and interlaminar) in a lap joint configuration subject to high secondary bending. A follow-on test plan is being developed to further evaluate these behaviors and determine their effect on overall material performance.

ACKNOWLEDGEMENTS

The authors would like to acknowledge the integral work done by the numerous partners who have been involved with this program. Their contributions make up a sizeable portion of the efforts detailed above and are due their just credit. A program of this scope would not be possible without their efforts.

Bruce Thomas – Bombardier
Michael Niedzinski – Constellium
Frank Eberl – Constellium
Jonathan Awerbuch – Drexel University
Tein Min Tan – Drexel University
Babak Anasori – Drexel University
Franklin Saillot – Drexel University
David Stanley – Drexel University
Tim Marker – FAA
Yongzhe Tian – FAA
Jeff Panco – FAA
Doug Ostgaard– FAA (retired)
Joy Ransom – FTI
Royce Foreman – NASA
Eun Lee – NAVAIR
Paul Jonas – NIAR
Royal Lovingfoss – NIAR
Mike Bode – Sandia
Ciji Nelson – Sandia
Todd Jones – UDRI

REFERENCES

[1] Prasad, N.E., Gokhale, A.A., and Wanhill, R.J.H. (Eds.), *Aluminum-Lithium Alloys; Processing, Properties, and Applications,* Butterworth-Heinemann, Waltham, Massachusetts, 2014.

[2] National Research Council, "New Materials for Next-Generation Commercial Transports," The National Academies Press, Washington, DC., 1996.

[3] Campbell, F.C., "Elements of Metallurgy and Engineering Alloys". ASM International, 2008

[4] Zink, W., Weike, J., Schwarmann, L., and Rendigs, K.H., "Investigation of Sheet Material of 8090 and 2091 Aluminum-Lithium Alloys". Defense Technical Information Center, 1989

[5] Guimmarra, C., Thomas, B., and Rioja, R.J., "New Aluminum Lithium Alloys for Aerospace Applications" *Proceedings of the Light Metals Technology Conference*, 2007

[6] Alcan Global Aerospace Transportation & Industry, "Bombardier chooses Alcan's Airware Technology for its *CSeries* aircraft program in the framework of a long-term agreement" [Press Release], July 20, 2010.

[7] Stonaker, K., Bakuckas, J., Won, I., and Freisthler, M., "Material Characterization of Aluminum Lithium Alloys Used in Aerospace Applications", FAA Report DOT/FAA/TC-15/12, 2015.

[8] Jones, T., "Assessment of Aluminum-Lithium for Primary Structure: Effect of Temperature on Tensile Properties", FAA/UDRI Report DOT/FAA/TC-14/13, 2015.

[9] Jones, T., "Assessment of Aluminum-Lithium for Primary Structure: Effect of Stress History on the Fatigue Crack Growth Behavior", FAA/UDRI Report DOT/FAA/TC-14/13, 2015.

[10] Jones, T., Rustenburg, J.W., Skinn, D.A., and Tipps, D.O., "Development and Assessment of Simplified Stress Sequences for Fuselage Structures", FAA/UDRI Report DOT/FAA/TC-12/17, 2014

[11] Ahmed, A., and Bakuckas, B., "Development of Multiple-Site Damage in Fuselage Structure", FAA/Drexel University Report DOT/FAA/AR-05/38, 2005.

[12] Stanley, D., Saillot, F., Anasori, B., Stonaker, K., et al., (2015). In: "Reconstruction of Fatigue Crack Growth in AA2024-T3 and AA2198-T8 Fastened Lap Joint," *Proceedings of the XVII International Colloquium on Mechanical Fatigue of Metals.*

[13] Federal Aviation Administration, Advisory Circular (AC 25.856-2A), "Installation of Thermal/Acoustic Insulation for Burnthrough Protection," July 29, 2008.

Energy based study on quasi-static delamination growth treated as a low cycle fatigue process

Lucas Amaral, Liaojun Yao, René Alderliesten, Rinze Benedictus

Faculty of Aerospace Engineering, Structural Integrity & Composites Group, Delft University of Technology, The Netherlands

Abstract: This work proposes to treat quasi-static mode I delamination growth of CFRP as a low-cycle fatigue process. To this end, mode I quasi-static and fatigue delamination tests were performed. An average physical Strain Energy Release Rate (SERR), derived from an energy balance, is used to characterize the energy released in crack extension. The physical SERR is then compared to the SERRs calculated by the ASTM standard, and the physical background of the standard is questioned. Furthermore, when normalizing the physical SERR for fatigue delamination by the one obtained for quasi-static crack extension, a stress ratio dependence is observed. This stress ratio effect is physically explained to be present due to the relation between quasi-static and fatigue loading at high stress ratios.

INTRODUCTION

Carbon Fibre Reinforced Polymers have become, due to their high specific strength and stiffness, a suitable option to build lighter and more damage tolerant aerospace structures [1]. However, their range of applications is limited by the poor interlaminar strength [2], which causes delamination to be the most frequently observed damage mode in CFRP structures [3]. Delaminations represent a crack-like discontinuity between the plies, and they propagate during the application of mechanical or thermal loads, or both [4], causing a reduction in the load bearing capacity of the structure. Therefore, several studies have been conducted to assess delamination growth in composites [1–3, 5–9].

The appropriate similitude parameter to be used to assess fatigue delamination is still under discussion [10]. While some authors use the maximum Strain Energy Release Rate (SERR) G_{max} [11], others prefer the SERR range $\Delta G = G_{max} - G_{min}$ [12] or even $\Delta\sqrt{G} = (\sqrt{G_{max}} - \sqrt{G_{min}})^2$ [13].

Amongst different propositions, some authors propose obtaining an actual physical SERR from measured data. The procedure is based on measuring, during a fatigue test, the crack length a, the crosshead displacement δ, the force P and the number of cycles N. With these data it is possible to obtain a graph plotting of da/dN versus dU/dN. This presentation of the data allows to obtain an average SERR over the cycle, G*, from the inverse of the slope of the curve, as defined in Eqn. 1, where b is the width of the specimen. This procedure is based on an energy balance analysis of the crack growth, and it accounts for the stress ratio in its definition, often collapsing fatigue curves to a narrow band for different stress ratios [14–17].

$$G^* = \frac{1}{b}\frac{dU/dN}{da/dN} = \frac{dU}{dA} \tag{1}$$

Although an ASTM standard has been published [18] to aid the assessment of quasi-static delamination growth, questions related to its physical background, to be discussed in the next section, still remain unanswered. Thus, given the successful use of a real physical SERR G* to characterize crack extensions under fatigue loading, how could this same similitude principle be used to assess quasi-static crack growth?

To answer this question this work proposes to analyse quasi-static data with the same procedure proposed in [15], described by Equation 1. A comparison between G* and the SERR calculated according to the ASTM standard for quasi-static delamination is presented, and the differences are highlighted. Normalization of the data is proposed, relating fatigue and quasi-static delamination. Furthermore, the observation of a stress ratio effect in the physical SERR is discussed.

HYPOTHESES

Quasi-static delamination growth by ASTM D5528-01

The ASTM D5528-01 standard describes the determination of the opening mode I interlaminar fracture toughness of continuous fibre-reinforced composite materials. The Double Cantilever Beam (DCB) specimen is used and the test is limited to carbon fibre and glass fibre tape laminates with brittle and tough single-phase polymer matrices [18]. Three different values of SERR can be used as fracture toughness, according to this standard. The first is the fracture toughness at the point of deviation from linearity in the load-displacement curve, which will be referred to as G_{NL}. The second is the SERR at the point at which delamination is visually observed on the edge of the specimen, referred to in this work as G_{VIS}. Finally, the third value is the SERR at the point at which the compliance has increased 5% or the load has reached a maximum value, referred to as G_{CRIT}. Typically, G_{NL} is the smallest of these three values and it is used as a conservative approach, while G_{CRIT} is the largest value of SERR obtained through this standard.

It is notable the fact that none of these three definitions are correlated to the growth of the crack. The SERR is the strain energy released when a crack extends. However, these three parameters (G_{NL}, G_{VIS} and G_{CRIT}) are not calculated based on measured growth. They are artificial SERR values calculated when no crack extension has occurred yet. Thus, G_{NL}, G_{VIS} and G_{CRIT} have never been correlated to the creation of any crack surface or to the rate at which the crack extended.

Quasi-static data treated as low-cycle fatigue

An illustration of a load-displacement graph plotting for a typical mode I quasi-static test performed on a CFRP DCB specimen in displacement controlled conditions is shown in Figure 1. Point 1 illustrates the conditions just before the test starts. When the measured force *P* increases and reaches a critical value, Point 2, crack extension occurs, which under displacement control conditions causes a decrease in the measured force. The system is thus taken to Point 3, due to the displacement rate imposed by the machine upon the test specimen. Because of the external work applied by the testing machine, the measured force increases once more. This is represented in the illustration by moving from Point 3 to Point 4. This incremental decrease in the measured force by crack extension and subsequent increase by the application of additional displacement to the specimen is repeated continuously in a gradual negative slope of the curve.

Therefore, the quasi static mode I delamination of CFRP DCB specimens seems to be well represented by a low-cycle fatigue process. For example, each drop and increase in force can be considered a cycle *N*, and a strain energy *U* can be associated with each *N*, as illustrated in Figure 1.

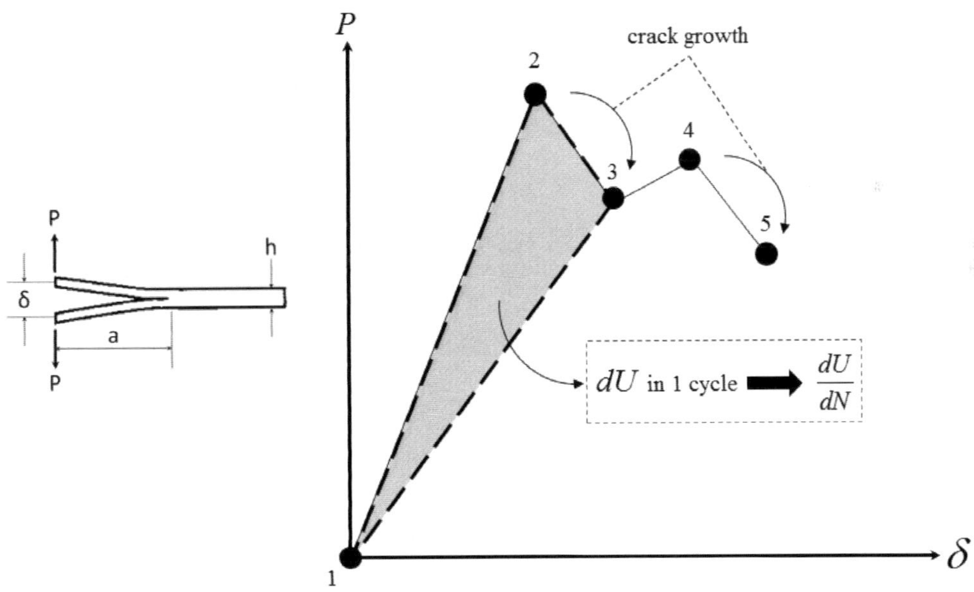

Figure 1. Illustration of a quasi-static force-displacement curve for a DCB CFRP specimen.

The physical SERR: G*

Such as stated in the Introduction section, measuring the crack length *a*, the displacement δ, the force *P* and the number of cycles *N* allows to graph a plot between the crack growth rate and the change in strain energy with the number of cycles: *da/dN* versus *dU/dN*. Fatigue data has been shown to align linearly for CFRP mode I delamination in this type of presentation of data, under different stress ratios [14, 15].

If the fatigue data can be fit by a straight line in a linear scale, such as illustrated in Figure 2, the strain energy released by crack extension is given by Eqn. 1. Therefore, the energy released in crack extension can be characterized by a single physical parameter, the average SERR over the cycle, G*.

It is important to note the difference between G* and G_{NL}, G_{VIS} and G_{CRIT}. The parameters defined by the ASTM standard describe a theoretical and artificial SERR that is calculated when there is no crack extension. However, if no crack extension occurred, no energy was actually released. Thus, what do these values mean?

Meanwhile, G* is a physical parameter calculated from the crack extension itself. It is the average of the actual physical SERR.

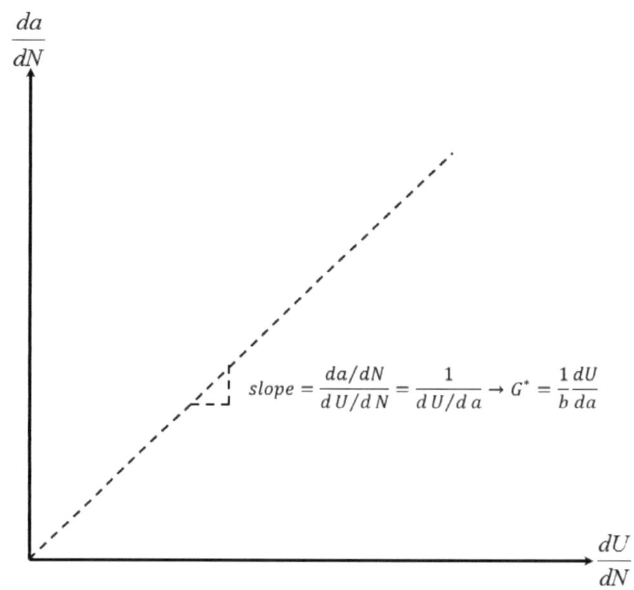

Figure 2. Obtaining a physical SERR from a *da/dN* versus *dU/dN* plot.

When data are aligned along the same slope in the *da/dN* versus *dU/dN* plot, it means they have the same release of strain energy per crack increment, *dU/da*. Therefore, the same amount of energy dissipation corresponds to the same amount of crack surface created. In addition, since the composite material used in this study presents neither relevant plasticity nor other significant dissipation mechanisms, any curve fit to the data in the *da/dN* versus *dU/dN* plot must go through the origin, because it is deemed impossible to dissipate energy without extending a crack [19].

DATA INTEGRATION

Five CFRP DCB specimens were tested quasi-statically. All test specimens were manufactures from the same material batch (M30SC-150-DT 120-34 F) and the tests were reported in [14, 20]. Although the integration of the quasi-static data was thoroughly explained in [19], it is repeated here for convenience for the reader.

"Each of the five sets of quasi-static data was discretized in order to calculate *dU/dN*. This discretization was performed as explained in Figure 3 (a), in which each shaded area is a *dU* in one cycle. As the quasi-static data is being treated as low cycle fatigue, each shaded area in Figure 3 (a) can thus be considered a *dU/dN*.

As an outcome of this discretization, a cloud of points was obtained when plotting the results in terms of *da/dN* versus *dU/dN*. This is shown by the blue markers in Figure 4. The average of these points, in *da/dN* and *dU/dN*, resulted in one point in the graph plotting *da/dN* versus *dU/dN*, shown by the red marker in Figure 4. However, this result is not enough to enable full understanding of the trend of *dU/da* with respect to an increasing *da/dN*, because a single data point contains no information on the slope. In which fashion would this point move if *da/dN* increased or decreased? Thus, new ways of analysing the same data with different discretization methods were necessary. Therefore, in order to analyse the trend of the data properly, each of the five sets of data was discretized in four different ways. The data was discretized to different levels, considering different number of cycles for the same crack growth, such as illustrated in Figure 3 (a) to (d). In Figure 3 (a), for instance, 3 cycles were considered, and thus 3 values of *dU/dN* and *da/dN* were calculated. Meanwhile, the exact same crack was considered to be grown in 6 cycles in Figure 3 (b), resulting in 6 values of *dU/dN* and *da/dN*. In each of these integrations what differs is the step *da* in crack growth that is used to calculate *dU/dN*. In this illustration, each shaded area limited by dashed lines represents a *dU/dN*. In Figure 3 (b) and (c), for example, the steps *da* are smaller than in Figure 3 (a). Meanwhile, in Figure 3 (d) the step *da* is bigger than in Figure 3 (a).

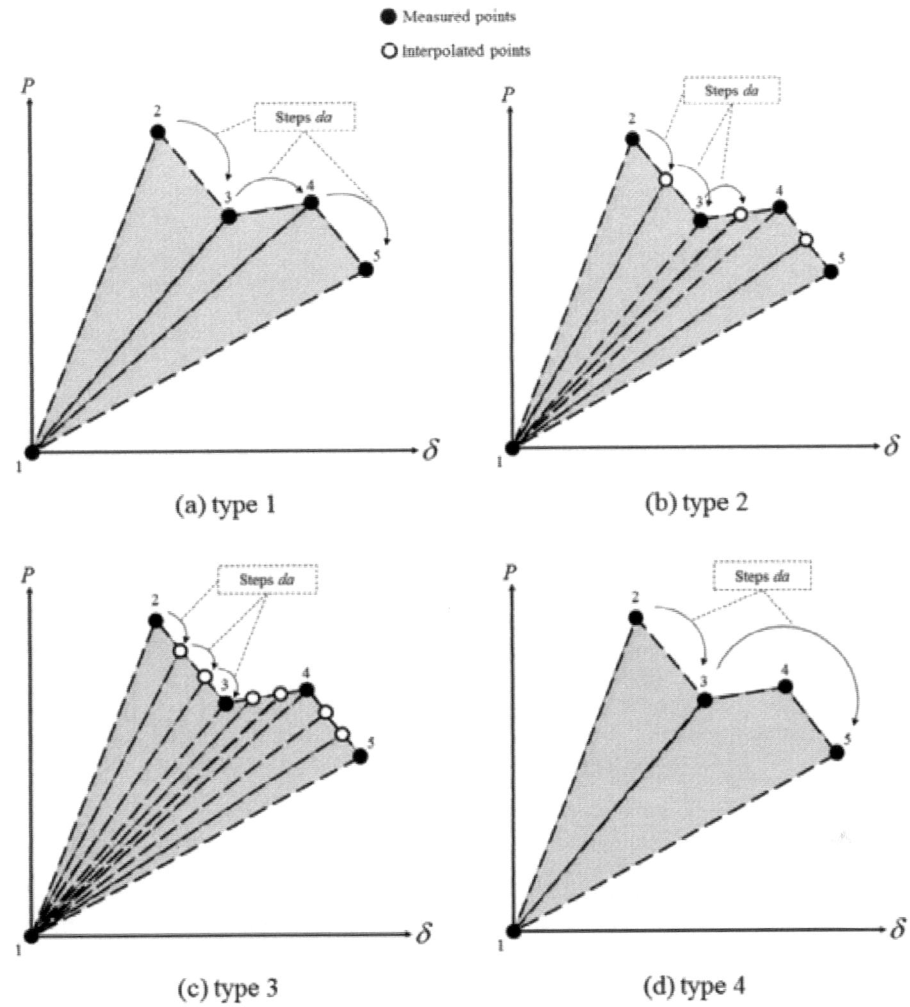

Figure 3. Discretization of the quasi-static data as low-cycle fatigue [19].

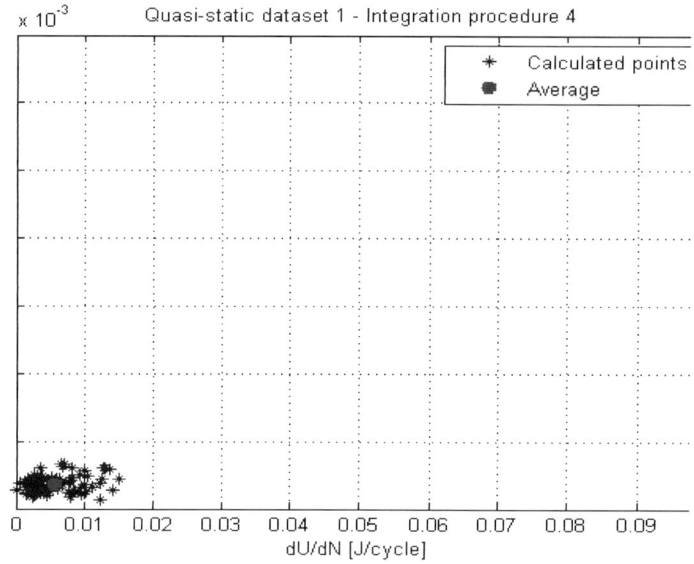

Figure 4. Quasi-static dataset 1: the blue markers show the points that resulted from the integration procedure 4, and the red marker shows the average [19].

It is important to note that, although each set of data is discretized with four different procedures, each set still represents the same crack extension. In other words, the fracture surface and the energy spent in creating this fracture surface quasi-statically is the same, independently of the way *dU/dN* is calculated. This is observed by the fact that all integration procedures, for a given test specimen, yield the same value for the sum of individuals *dU/dN*. This can be easily observed in Figure 3: the sum of the shaded areas always results the same total area. This shows that the procedure is consistent, because the total energy spent in extending the crack is the same for a given dataset.

Therefore, the average values of *dU/dN* and *da/dN* were calculated with these four discretization procedures and plotted together. Data obtained from five specimens were discretized in four different increments, yielding 20 points in the *da/dN* versus *dU/dN* plot. These points were plotted together, and a linear regression was used to produce the best linear fit by the minimization of the sum of the square of the error. The result, presented in Figure 5, shows a good correlation (i.e. coefficient of determination $R^2 = 0.9894$). This graph shows that a quasi-static test can be analysed as low-cycle fatigue in a consistent way, and the SERR can be easily calculated from the slope of the linear fit, i.e. $1/(dU/da)$. Thus, quasi-static crack extension can consistently be considered as a special case of fatigue crack growth, is given by Eqn. 2" [19].

$$G^*_{quasi-static} = 611.6\, J/m^2 \tag{2}$$

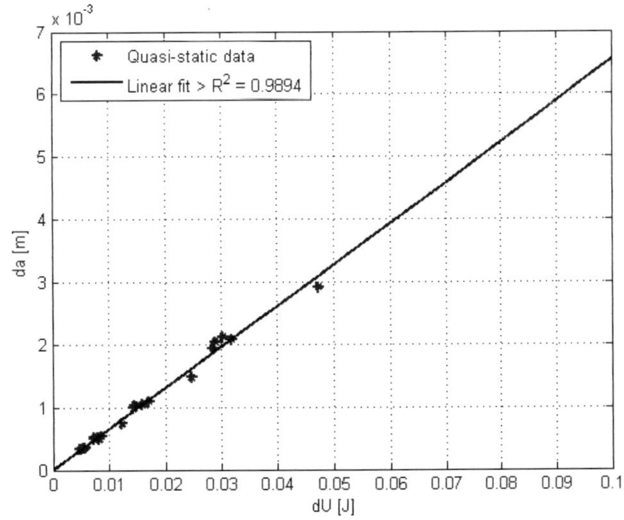

Figure 5. Linear fit through the average values of *da/dN* and *dU/dN* obtained from 5 quasi-static tests discretized in 4 different levels [19].

The values of G_{NL} and G_{CRIT} were calculated for each of the five test specimens according to the ASTM D5528-01 standard [18]. The average value was then considered for comparison, and the results are presented in Table I. One should note the large difference between the values calculated through the ASTM standard and the physical SERR G*. This difference is due to the fact that G_{NL} and G_{CRIT} are artificial values for the SERR that are not calculated from the measured crack growth, as discussed in the "Quasi-static delamination growth by ASTM D5528-01" section. Meanwhile, G* is related to the crack extension that occurred during the quasi-static fracture, representing the average of the actual strain energy released during crack extension.

If G* is the real physical SERR, as it is calculated directly from measured crack extension, then G_{NL} and G_{CRIT} have no physical meaning, as they are not based on physical principles of crack extension and energy balance. Furthermore, using such artificial parameters to assess delamination growth can be misleading and it does not lead to a physical understanding of the phenomenon.

Table I. Comparison between the physical SERR and the SERRs obtained by the ASTM standard

G* [J/m²]	G_{NL} [J/m²]	G_{CRIT} [J/m²]	G_{NL}/G*	G_{CRIT}/G*
611.60	244.02	423.03	0.40	0.71

FATIGUE AND QUASI-STATIC DELAMINATION: NORMALIZATION PROCESS

Fatigue tests were performed at three stress ratios (0.1, 0.5 and 0.7) in DCB specimens made from the same material and with the same dimensions as the ones used in quasi-static tests. These are reported in [14], and the relation between the actual strain energy released in fatigue and in quasi-static delamination is thoroughly discussed in [19].

The fatigue data obtained at three different stress ratios align linearly in a *da/dN* versus *dU/dN* plot along a narrow band [14]. A small stress ratio is still distinguishable, but it is possible to fit the fatigue data with a straight line and a good coefficient of determination ($R^2 = 0.8756$), shown in Figure 6. At low values of *da/dN*, however, this linear fit correlates to a lesser extent with the measured data points. This occurs due to the fact that, at this crack growth rate, the dissipation mechanisms observed in the fracture surfaces consum less energy, as explained in [19].

From the slope of this linear curve fit it is possible to calculate the average physical strain energy released during fatigue delamination growth, which is $G^*_{Fatigue} = 289.70$ J/m². As mentioned before, this is the average value for all the data points obtained in fatigue tests. However, it is observed in the stress ratio effect in Figure 6 that the SERR changes with the crack growth rate, and the use of an average SERR per test is incorrect. In fact, the physical SERR should be calculated at any instant during the test when the crack extends. This is performed calculating the value of $G^*_{Fatigue}$ for each data point obtained during the fatigue test. The procedure consists in plotting a straight line from the origin, because it is deemed impossible to dissipate energy without extending a crack for this material system, to each measured point. From the slope of this straight line the physical SERR can be obtained for that particular measured data.

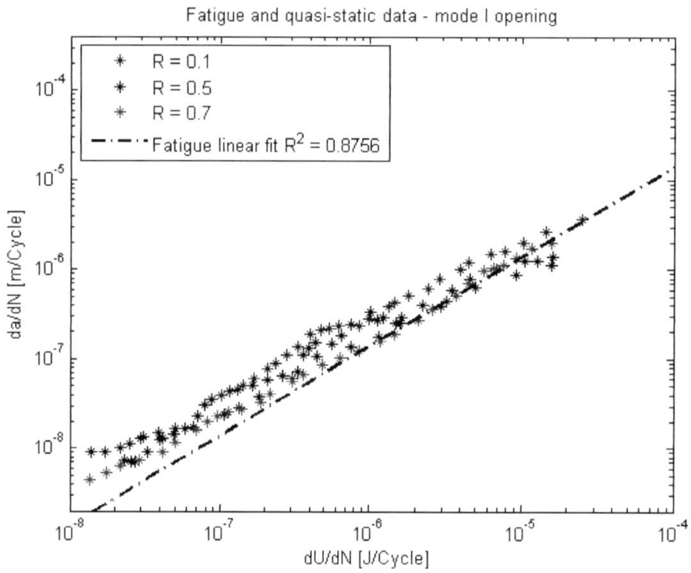

Figure 6. Fatigue data with linear fit.

Normalizing the value of $G^*_{Fatigue}$ for each fatigue data point by the quasi-static physical SERR is then possible, and the result is presented in Figure 7. This graph allows the better observation of what appears to be a stress ratio effect. Figure 7 shows that the extension of a crack by the same rate *da/dN* requires higher energy for higher stress ratios. In fact, the presence of a stress ratio effect can be explained by the relation between fatigue and quasi-static fracture. Growing a crack in fatigue is more efficient than growing it quasi-statically, because requires less energy to extend the same unit surface *dA*. The higher the stress ratio, the closer to the quasi-static condition the load is, and therefore more energy is required for crack extension.

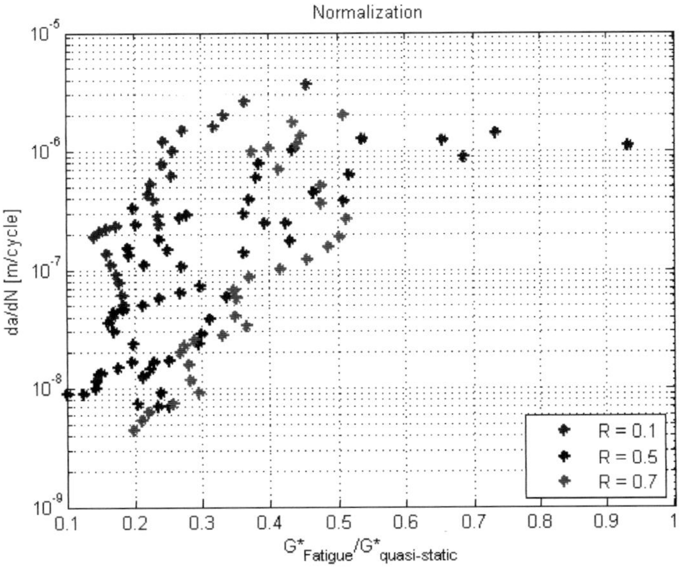

Figure 7. Normalization by the quasi-static SERR.

Moreover, a reasonable scatter can be observed in Figure 7, mainly for $R = 0.5$. This is due to the limited dataset and to the fact that prior to the test the parameters that influence the results were not completely understood. New tests with more accurate measurement systems and better calibrated load cells are recommended.

CONCLUSIONS

Quasi-static mode I delamination data were considered as low-cycle fatigue data, discretized to different levels and characterized by a real physical SERR. The average SERR over the cycle, G*, can be used as the parameter to characterize crack extensions in both fatigue and quasi-static loading conditions, as it is obtained from a material characterization.

The experimentally obtained value of G* was compared to G_{NL} and G_{CRIT}, both obtained according to an ASTM standard. The difference between the results showed that the strain energy values calculated through the ASTM standard are artificial and over conservative. In addition, G_{NL} and G_{CRIT} do not equate the physical condition of crack extension. As G* is a physical value obtained from measured data, further research in this topic is recommended to solidify the use of this parameter to assess delamination growth.

When the physical SERR G* for the fatigue data was normalized by the quasi-static value of G*, a stress ratio dependence is clearly observed. This can be explained by the fact that, at higher stress ratios, the load condition is more similar to quasi-static loading, which requires more energy for crack extension.

REFERENCES

[1] Khan, R. (2013) *Mode I Fatigue delamination growth in composites*, Delft University of Technology: Delft.

[2] Hojo, M., Ando, T., Tanaka, M., Adachi, T., Ochiai, S., Endo, Y. (2006). *International Journal of Fatigue*, 28(10): p. 1154-1165.

[3] Hojo, M., Ochiai, S., Gustafson, C-G., Tanaka, K. (1994). *Engineering Fracture Mechanics*, 49(1): p. 35-47.

[4] Donald F. Adams, L.A.C., Byron Pipes R. (2003) *Experimental Characterization of Advanced Composite Materials*. Third ed. CRC Press. 258.

[5] Stelzer, S., Brunner, A.J., Arguelles, A., Murphy, N., Pinter, G. (2012). *Composites Science and Technology* 72(10): p. 1102-1107.

[6] Jones, R., Stelzer, S., Brunner, A.J. (2014). *Composite Structures*. 110: p. 317-324.

[7] Murri, G.B. (2012). *Evaluation of delamination onset and growth characterization methods under mode I fatigue loading*. in American Society for Composites, 27th Technical Conference. DEStech Publications, Inc.

[8] Gustafson, C.-G.o., Hojo, M. (1987). *Journal of Reinforced Plastics and Composites*. 6(1): p. 36-52.

[9] Alderliesten, R.C., Schijve, J., van der Zwaag, S. (2006). *Engineering Fracture Mechanics*, 73(6): p. 697-709.

[10] Pascoe, J.A., Alderliesten, R.C., Benedictus, R. (2013). *Engineering Fracture Mechanics*, 112-113: p. 72-96.

[11] Roderick, G.L., Everett, R.A., Crews Jr, J.H. (1974). *Debond Propagation in Composite Reinforced Metals*, NASA: Hampton, VA.

[12] Mostovoy, S., Ripling, E.J. (1975). *Adhesion Science and Technology*, L.-H. Lee, Editor. Plenum Press: New York. p. 513-562.

[13] Rans, C., Alderliesten, R.C., Benedictus, R. (2011) *Composites Science and Technology*. 71: p. 230-238.

[14] Yao, L., Alderliesten, R.C., Zhao, M., Benedictus, R. (2014). *Composites Part A: Applied Science and Manufacturing*. 66(0): p. 65-72.

[15] Alderliesten, R. (2015) Submitted to the *International Journal of Fatigue*.

[16] Pascoe, J.A., Alderliesten, R.C., Benedictus, R. (2014). *Procedia Materials Science* 3 (ECF-20): p. 610-615.

[17] Pascoe, J.A., Alderliesten, R.C., Benedictus, R. (2015). *Engineering Fracture Mechanics*. 133(0): p. 1-13.

[18] ASTM Standard D5528-01. (2007). *Standard Test Method for Mode I Interlaminar Fracture Toughness of Unidirectional Fiber-Reinforced Polymer Matrix Composites*. ASTM International: US.

[19] Amaral, L., Yao, L., Alderliesten, R., Benedictus, R. (2015). Submitted to *Engineering Fracture Mechanics*.

[20] Yao, L., Alderliesten, R.C., Zhao, M., Benedictus, R. (2014). *Composites Part A: Applied Science and Manufacturing*. 63(0): p. 103-109.

Widespread Fatigue Damage

WFD Evaluation of Riveted Lap Joint

Takao Okada[1], Min Liao[2], Shigeru Machida[1], Gang Li[2], Guillaume Renaud[2]

[1]*Japan Aerospace Exploration Agency (JAXA)*
[2]*National Research Council Canada (NRC)*

Abstract: A riveted flat panel lap joint was used for evaluating the early stage of widespread fatigue damage (WFD) development, i.e., WFD onset. Constant amplitude fatigue loading was applied to the lap joint panel, and strain data were obtained in order to evaluate the strain distribution around the fasteners. In most cases, the locations of the cracks were above the horizontal centerline of the rivet holes, and apparent fretting damage was observed near the crack nucleation area on the faying surface of the joint and around the fastener holes. Numerical analysis of the riveted lap joint was also conducted using 3D global-local finite element (FE) modeling to predict the stress/strain distribution induced by the riveting process and then during the fatigue loading. Strain data measured from the test was used for validating the FE analysis. The effects of the riveting squeezing force and clearance fit on the joint stress/strain condition and then on the fatigue life assessment were studied. The cycles for a crack to 0.5 mm were predicted using the Smith-Watson-Topper (SWT) equation, and then the life distribution to the first linkup were analyzed using the NRC in-house tool, CanGROW. The analytical models and results were compared and validated with the test data. Both testing and modeling results are expected to improve the knowledge of the WFD feature and to improve research efficiency by reducing test quantity and cost.

INTRODUCTION

The limit of validity (LOV) [1] is a new airworthiness requirement (14 CFR Parts 25, 26, 121) for some airplanes, which includes both safe life and damage tolerant philosophies. It is not only a requirement for the certification for new aircraft but also an operational requirement for existing aircraft, irrespective to the period when they were developed. The damage tolerant philosophy prevents abrupt structural failure by a maintenance program based on engineering data using initial cracks and crack growth behavior during operation. The fatigue life between an initial crack and a crack of a certain length needs to be predicted reasonably well. Many research activities for damage tolerant evaluation have been conducted for many years and there are many tools to evaluate crack growth behavior [2-7]. The range of the research is from coupons to full scale components. On the other hand, the LOV is the period of time (in load cycles, flight hours, or both), up to which it has been demonstrated that widespread fatigue damage (WFD) is unlikely to occur in an airplane structure by virtue of its inherent design characteristics and maintenance actions. The WFD candidates are all fatigue critical locations and all possible locations need to be evaluated. WFD involves the simultaneous presence of multiple cracks, at multiple locations, that are of sufficient sizes and density such that the structure will no longer meet the residual strength requirement. Some WFD cracks are hidden in the structure and very difficult to detect. As such the WFD evaluation usually does not assume the same initial crack size as the damage tolerant evaluation. The fatigue life from the first operation day to a crack growing to a certain length is required for the WFD evaluation. Comparing to the damage tolerant analysis, the WFD evaluation method is not well known and not well established yet. Because of this situation, research on WFD assessment for gaining knowledge of WFD, and research on WFD evaluation by experiments and analysis are imperative for aircraft OEM and some operators in order to comply with the new airworthiness requirement. The followings are examples of items for the WFD assessment: Initial and final cracking scenario, WFD average behavior and LOV, Inspection Start Point (ISP), Structural Modification Point (SMP), etc. Since 2009, the National Research Council Canada (NRC) has been carrying out research on MSD/MED evaluations, such as for the center wing of transport aircraft CC-130 [8].

In this work, the Japan Aerospace Exploration Agency (JAXA) and NRC jointly conducted research work for the WFD evaluation of a typical riveted lap joint. The conventional single lap joint was used for this evaluation because it is a typical fuselage joint and it is known as one of the imperative candidates for WFD evaluation. Constant amplitude fatigue loading was applied to the lap joint panel, and strain data were obtained in order to evaluate the strain distribution around the fasteners. Fatigue crack locations and lengths were also measured for evaluation of fatigue crack growth behavior. Numerical analysis of the riveted lap joint was conducted using 3D global-local finite element (FE) modeling to predict the stress/strain distribution induced by the riveting process and then during the fatigue loading. The effects of the riveting squeezing force and clearance fit on the joint stress/strain condition were also included in the analysis. The life to a 0.5 mm crack was predicted using the Smith-Watson-Topper (SWT) equation [9], and then the life to the first linkup was calculated using a NRC in-house tool, CanGROW [7].

EXPERIMENTAL PROCEDURE

Fig. 1 shows the geometry of the specimen used for the tests. It features a typical longitudinal fuselage joint riveted by 3 rows of 20 fasteners and a narrow plate attached to the middle row to simulate a fuselage panel stringer. Nine button head rivets were used on both sides of the panel while countersunk rivets were used at the other locations. The area fastened by the countersunk rivets was the location to be evaluated. Strain gauges were attached to the joint in order to evaluate the stress/strain distribution around the joint. Locations of the strain gauges are also shown in Fig. 1. Ten rosette gauges and twenty four uniaxial gauges were attached to the specimen. Constant amplitude fatigue load was applied to the lap joint. The maximum stress and stress ratios were 65 MPa and 0.069, respectively. Visual inspection around each fastener using a CCD camera with magnification up to 50 times was conducted during the fatigue test to detect the locations and lengths of the cracks formed and propagating from the rivet holes. Fig. 2 shows the setup of the fatigue test. The crack lengths were measured every 10,000 cycles, after the first fatigue crack was detected.

NUMERICAL PROCEDURE

Finite Element Model

A 3D global-local finite element model was developed in order to calculate the stress/strain distribution in the riveted lap joint. The effects of rivet squeezing and remote loading were simulated in the numerical analysis. The MSC/Patran (pre- and post-processor) and MSC/Marc (solver) FE package were used for the analysis. An isotropic hardening behaviour was assumed for both the rivet and sheet materials. The material parameters of the 2117-T4 Al alloy MS20426AD5-5 rivet [10] are presented in Table I. The material constants *C* and *m* were calculated by substituting uni-axial tensile test data into Eq. (1).

$$\sigma_{true} = C\,(\,\varepsilon_{true}\,)^{m} \tag{1}$$

A cyclic stress-strain curve of 2024-T3 in the transverse direction [11] was assumed for the skins during the application of three load cycles in the FE analysis. During the riveting process simulation, a monotonic stress-strain curve in the transverse direction was assumed by using the monotonic-to-cyclic ratio calculated in the longitudinal direction [12]. The sheet material's Young's modulus, $\boldsymbol{E}$= 70.3 GPa, and Poisson's ratio, $\boldsymbol{\nu}$= 0.33, were used in the elastic deformation stage. The transverse stress-strain curves are presented in Fig. 3. The material was defined as temperature dependent in order to switch from the monotonic properties to the cyclic properties, with a yield strength of 430 MPa, after the riveting process. A dummy temperature change was applied to the model to force an update of the material properties. This temperature change had no other impact in the model (i.e. no thermal expansion).

A global shell model of half the panel was created to take advantage of the symmetrical joint configuration. The displacements (u_x, u_y, and u_z) under the maximum and minimum tensile loading levels were extracted from the shell model. The global FE model is shown in Fig. 4. The panel holes were not introduced in the global FE model. The shell elements were created at the mid-planes of the panels and strap. The fasteners were modelled using rigid link MPCs (multi-point constrains). Nonlinearity in material behaviour was not considered in the global model but geometric nonlinearity, introduced by large displacement, was activated during the remote tensile loading stage. For the global FE model, only one tensile loading step was conducted. MPCs assigning the same u_x displacement were set for all the right-side pins shown in Fig. 4.

A local finite element model of the central region of the joint structure that covers the overlap section and some of the neighboring panel areas was also built. The local model is presented in Fig. 5, with 216 mm length in the loading direction, including the 60 mm overlap section, and 60 mm width in the transverse direction. The local FE model consisted of five contact bodies: four deformable bodies (outer panel, inner panel, strap panel and fastener) and two rigid bodies (fastener pusher and set). To avoid edge effect on the stress at the top-row rivet in the central local model region, the overlap section in the local model covered nine rivets. For simplicity, eight of the nine rivets were modelled as "welded" to the panels in the local FE model. The riveting simulation was only performed for the central top-row rivet. The strap, used to simulate a stringer structure, was assumed welded to the middle row rivets in the local model. The rivet shank diameter, *D*, was 3.969 mm (5/32"), e.g., the nominal diameter of the rivet (MS20426AD5-5) used in the JAXA lap joint. The finished countersunk holes show no knife edges (Fig. 8). Since no actual measurements were available before the riveting, the countersunk depth in the 1.27 mm thick outer panel was assumed to be 0.945 mm (leaving a 0.325 mm thick edge for the hole protruding part) which is close to the dimension after testing (Fig. 8). A regular fine mesh was applied to the rivet, overlap section and nearby region, while a coarser mesh was used for the panel areas that were away from the overlap ends. FE strain data were extracted from a tensile loading stage following the riveting process to be compared with the strain gauge static load readings.

Smith-Watson-Topper equation

The strain-life method was used to estimate the fatigue life to a crack size of 0.5 mm. This calculation was performed using the Smith-Watson-Topper (SWT) equation [9], which takes into account mean stress effects:

$$\sigma_{max}\,(\Delta\varepsilon/2) = (\sigma'_f)^2/E\,(2N_f)^{2b} + \sigma'_f\epsilon'_f\,(2N_f)^{(b+c)} \tag{2}$$

where: σ' : the tensile fatigue strength coefficient
ε' : the tensile fatigue ductility coefficient
c : the fatigue ductility exponent
b : the fatigue strength exponent
N_f : the number of cycles to get, in this case, a 5% decrease in tension load. This was used to calculate to the depth of an elliptical crack in the tested cylinder specimen [11], which is approximately 0.5 mm.
$\Delta\varepsilon$ $= \varepsilon_{max} - \varepsilon_{min}$: the strain range

The SWT parameter, $(\Delta\varepsilon/2)$, was calculated at every node of the model in all directions. For the 2024-T3 aluminum alloy sheets, the transverse-direction (TL) strain-life constants were taken from reference [11] as: 835 MPa, 0.174, −0.644, −0.096, and 70.28 GPa. Using the SWT equation, the cycles (N_f) for a crack to 0.5 mm can be calculated.

Crack Growth Analysis

Fatigue crack growth from a 0.5 mm crack to the link-up of two adjacent cracks in two holes was evaluated using the NRC in-house tool, CanGROW, which includes capabilities to perform Monte-Carlo simulations for MSD problems. However, the current version of CanGROW does not include a model to analyze a part-through crack in a countersunk hole with residual stresses. Therefore, correction curves, expressed as functions of crack length, were developed to consider the effects of the riveting residual stresses, obtained from the FEA, and the effects of the countersunk geometry. These correction curves were calculated using AFGROW 5.2, which includes a countersunk geometry stress intensity factor solution. Using the AFGROW COM interface, the applied remote load (spectrum) was automatically adjusted, based on the crack tip position, to consider the stress fields under the maximum and minimum cyclic loads, including the post-riveting residual stresses. This strategy was used because the residual stress option is not available for the countersunk geometry in the current version of AFGROW. The stress intensity factor curves obtained from AFGROW were then converted to correction curves applicable to straight MSD cracks located at straight holes in a CanGROW model, which could then be used to estimate the life to the first link-up of the two adjacent cracks.

RESULT AND DISCUSSION

During the JAXA fatigue test, using a maximum stress of 65 MPa and a 0.069 stress ratio, the first crack was observed at about 1,950K cycles at the right side of rivet no. 4, and the second crack was found at about 1,955K cycles at the left side of rivet no.12. The first link-up of two fatigue cracks was at about 2,110K cycles between rivet no. 4 and 5 and the structure finally fractured at 2,133,534 cycles. The numbering of the rivet hole starts on the left side of the panel. A picture of the specimen after fatigue failure is shown in Fig. 6 and a-N curves for observed cracks are shown in Fig. 7. As shown in Fig. 7 fatigue cracks were observed until the final failure. After the fatigue test was finished, faying and fracture surface measurements were also conducted to identify the origins and shapes of the cracks. Figure 8 shows the faying and fracture surface for rivet no. 4. As shown in the figure, a fretted area was observed on the faying surface on both sides and fretted products were also found on the fracture surface. These observations indicate that the crack nucleated from the fretted area. Fracture surface observation identified 10 through cracks and 19 part-through cracks before final failure.

The strain values obtained by the FEA were compared with the test data in order to validate the FE models. Fig. 9 shows the comparison of the strain at A-2 and B-2 obtained by the experiment and global FE model, while Fig. 10 indicates the strain obtained by experiment and local FE model. As shown in these figures, the difference between experiment and FE model was mostly found at lower and middle load levels, whereas at higher load both of them agreed very well. Similar good agreement was also confirmed for most of the other locations.

Fig. 11 shows typical stress distributions at the fastener hole obtained by the local FE model for different combinations of hole diameters and rivet squeeze deformations, expressed in terms fastener head deformation, D_{max}/D. It is seen that the residual stresses are lower at the hole bore for smaller holes and larger fastener head deformation. It is also seen for these cases that without post-riveting residual stresses the peak stress is located at edge of the rivet hole, with the riveting-induced residual stresses the peak stress location is away from the hole edge. Also, as seen in Fig. 11b) the stresses at the bore are not always local minimum values; lower values may exist at a small distance away from the hole edge.

The fatigue life to a 0.5 mm crack was predicted using the SWT equation for a different combinations of hole diameters and rivet squeeze deformations, and for several meshing schemes. The rivet driven head deformation range, D_{max}/D, was measured on the test article to be between 1.50 to 1.52 for most of the rivet holes, with an average of 1.516. Rivet driven head deformations of 1.50, 1.51 and 1.52 were assumed in local 3D models with very fine meshes. Based on a JAXA panel joint drawing provided to NRC, the rivet hole diameter was 4.039±0.127 mm (max: 4.166 mm, min:3.912 mm but this size is not likely used in the joint as it would have caused an interference fit). The predicted crack nucleation points, located at the outer panel faying surface, are shown in Figs. 12-14. Table II lists the predicted lives to 0.5 mm crack, obtained for two hole sizes selected within the drawing specifications, and the three driven head deformations. Under the applied test cyclic loading, the predicted lives ranged between 2.15 and 4.59 million cycles. The predicted lives obtained using the selected two hole diameters were slightly higher but relatively close to the lives observed in the test.

It should be noted that, the model predicted a wide range of lives due to different hole sizes and deformation ratios, which were found to have a significant effect on the fatigue lives. The SWT equation used was seen to be very sensitive to the local stress and strain states that are affected by the riveting and assembly process. For instance, the analysis of some smaller holes, although still within the specifications, resulted in lives exceeding 100 million cycles. The material scatter was not considered in the modeling work as no specific data is available at present, however the material scatter could be overshadowed by the riveting process. If all the rivet and hole sizes before the riveting, and the rivet head deformations became available, for example through a qualification process to qualify/certify the riveting process, these data could be used to determine a distribution of fatigue life to the 0.5mm crack. Then this distribution could be used to continue crack growth analysis as described in the next paragraph.

A preliminary assessment of life to first link-up was carried out using the Size 2 diameter and the 1.517 driven head deformation, which values are assumed to be, amongst the cases presented in Table II, the closest to the test nominal dimensions. The life predicted by the SWT equation, 3.4020E+6 cycles, was assumed to be the median life to a 0.5 mm crack. Also, a standard deviation of 0.1296 for Log10 (life), used in DEF STAN 00-970 (PART 1/5, SECTION 3, LEAFLET 35) recommended for metallic airframe full scale tests, was assumed in order to generate a time to a 0.5 mm crack size (TTCS) distribution. The stress distributions calculated by the FE model, shown in Fig. 15a), were used in AFGROW to generate the correction curve that includes the effect of the stresses and countersunk geometry, presented in Fig. 15b). This correction curve was then used in CanGROW to generate an equivalent initial flaw size (EIFS) distribution by regressing the TTCS distribution. That EIFS distribution ensured that the nucleation and short crack phases (to 0.5 mm) would be properly captured in the CanGROW crack growth model. A MSD Monte Carlo simulation of 45,000 trials, each consisting of different random initial cracks sampled from the EIFS distribution and positioned at two adjacent rivet holes, was then performed to estimate the distribution of life to crack link-up between the two holes. When considering the 13 potential link-ups between the 14 adjacent holes, the median of the combined probability of first link-up was calculated to be approximately 2,500K cycles, as shown in Fig. 16, along with the time to first link-up observed in the test. It should be noted that so far, only one lap joint test was completed and used for comparison. The comparison may be changed with more test results available.

CONCLUSION AND DISCUSSIONS

The WFD evaluation of riveted lap joint was conducted experimentally and numerically in this research. Experimental results show that WFD features were found during the test and then one of the WFD scenarios (fatigue lives and crack distributions) was obtained in this test. From the numerical point of view, global/local FEM analysis was conducted to obtain stress/strain distribution around the joint and excellent correlation was achieved between experimental and FE results. Strain life analyses, using the Smith-Watson-Topper equation, predicted reasonable fatigue lives to 0.5 mm cracks. AFGROW was used to calculate correction factors to consider the countersunk geometry and the residual stress obtained by the FE model. CanGROW was used for preliminary Monte Carlo simulation from an EIFS distribution, and then produced a fatigue life distribution to the first link-up.

The modeling work showed that the rivet hole size and rivet head deformation have significant effects on the local stress and strain states of the lap joint, and then can greatly affect the fatigue lives. These parameters should be carefully measured to quantify a riveting process; and then strictly controlled during the manufacturing. Other manufacturing induced facts, such as assembly induced stress, manufacturing induced initial damage, may also affect the fatigue performance of the lap joint, which are not modeled in this study.

The followings are the possible future actions. In order to obtain test data comparable to the numerical result, the life up to 0.5 mm cracks and further fatigue crack growth are necessary. SEM observation on the fracture surface and additional fatigue test with marker band loading will be necessary for this issue. From the numerical view point, StressCheck can be used to develop a Beta factor that includes the countersunk geometry and FE stresses. To further validate the developed WFD methodology for the riveted lap joint, carefully measured rivet and countersunk hole dimensions before and after the riveting process are needed from a number of lap joints and rivet holes ensuring a certain statistical significance. These well measured lap joints could be fatigue tested to provide more data to determine a standard deviation of the fatigue lives. Appropriate probabilistic distributions could be determined for the key geometrical and manufacturing parameters, and used in CanGROW to predict a realistic fatigue life distribution of the lap joints. The life distributions are expected to be used for evaluating the average WFD behavior and the probability of 'unlikely' WFD occurrence.

ACKNOWLEDGEMENTS

The work was performed with financial support from JAXA and NRC collaborative projects. A part of NRC's effort was also supported by the Department of National Defence (DND) Canada.

REFERENCES

[1] Sippel, W.,(2010). In: *Federal Register*, vol.75, n. 219, pp.69746-69789, FAA.

[2] Swift, T., (1992). In: Fatigue of Aircraft Materials, pp. 351-387, Beukers, A., de Jong, Th., Sinke, J., Vlot, A. and Vogelesang, L.B. (Eds.), Delft University Press, Delft.

[3] Bakuckas. Jr., J.G., McIver, K. and Hsu, C. (2009). In *Bridging the Gap between Theory and Operational Practice*, Proceedings of the 25th ICAF Symposium, pp. 407-426, Bos, M. (Ed.), Springer, Dordrecht.

[4] Hu, W. and Wallbrink, C. (2013). In: Proceedings of the 27h ICAF Symposium, pp. 509-524, Blot, A. (Ed.)

[5] NASGRO Ver. 7.1, http://www.swri.org/ 4org/d18/mateng/matint/nasgro/

[6] Harter, J. A., AFRL-VA-WP-TR-2008-XXXX, 2008.

[7] Bombardier, Y., Liao, M., and Renaud, G., "A New Crack Growth Analysis Tool for the Assessment of Multiple Site Fatigue Damage", Aircraft Airworthiness & Sustainment Conference 2010 (AA&S2010) Austin, TX, USA.

[8] Liao, M., Bombardier, Y., Renaud, G., Bellinger, N.C. and Cheung, T. (2009). In: *Bridging the Gap between Theory and Operational Practice*, Proceedings of the 25th ICAF Symposium, pp. 811-837, Bos, M. (Ed.), Springer, Dordrecht.

[9] Smith, K.N.; Watson, P.; Topper, T.H. (1970). *J. Mater*, 1970, vol. 5, p. 767-778

[10] Li, G., Shi, G., Bellinger, N.C. (2011). *Journal of Aircraft*, vol. 48, n. 2, p. 527-539.

[11] Boller, C. and Seeger, T. (1987). In: *Materials Data for Cyclic loading, Part D: Aluminium and Titanium Alloys*. Elsevier Science Publishers, Amsterdam.

[12] Renaud, G., Li, G., Liao, M. (2014). NRC report LTR-SMM-2014-0091.

Table I. Elastic and plastic property for rivet material.

E (GPa)	71.7GPa
ν	0.33
σ_y (MPa)	172 MPa
Hardening parameter $0.02 < \varepsilon_{true} < 0.1$	C=544, m=0.23
Hardening parameter $0.1 < \varepsilon_{true} < 1.0$	C=551, m=0.15

Table II. Predicted lives to 0.5mm cracks.

Hole diameter	D_{max}/D for the countersunk MS20426AD5-5 rivets			65 MPa max stress $\boldsymbol{R} = 0.069$	
	FE	Experimental	Dif.(%)	$\boldsymbol{\sigma}_{\mathbf{max}}$ $(\Delta\varepsilon/\mathbf{2})$	N_f (cycles)
Size 2: 4.115 mm	1.507	1.50	0.47	0.5176	2.4337E+6
	1.517	1.51	0.46	0.4851	3.4020E+6
	1.527	1.52	0.46	0.4578	4.5895E+6
Size 3: 4.166 mm	1.501	1.50	0.07	0.5300	2.1526E+6
	1.509	1.51	−0.07	0.5170	2.4471E+6
	1.519	1.52	−0.07	0.5147	2.5031E+6

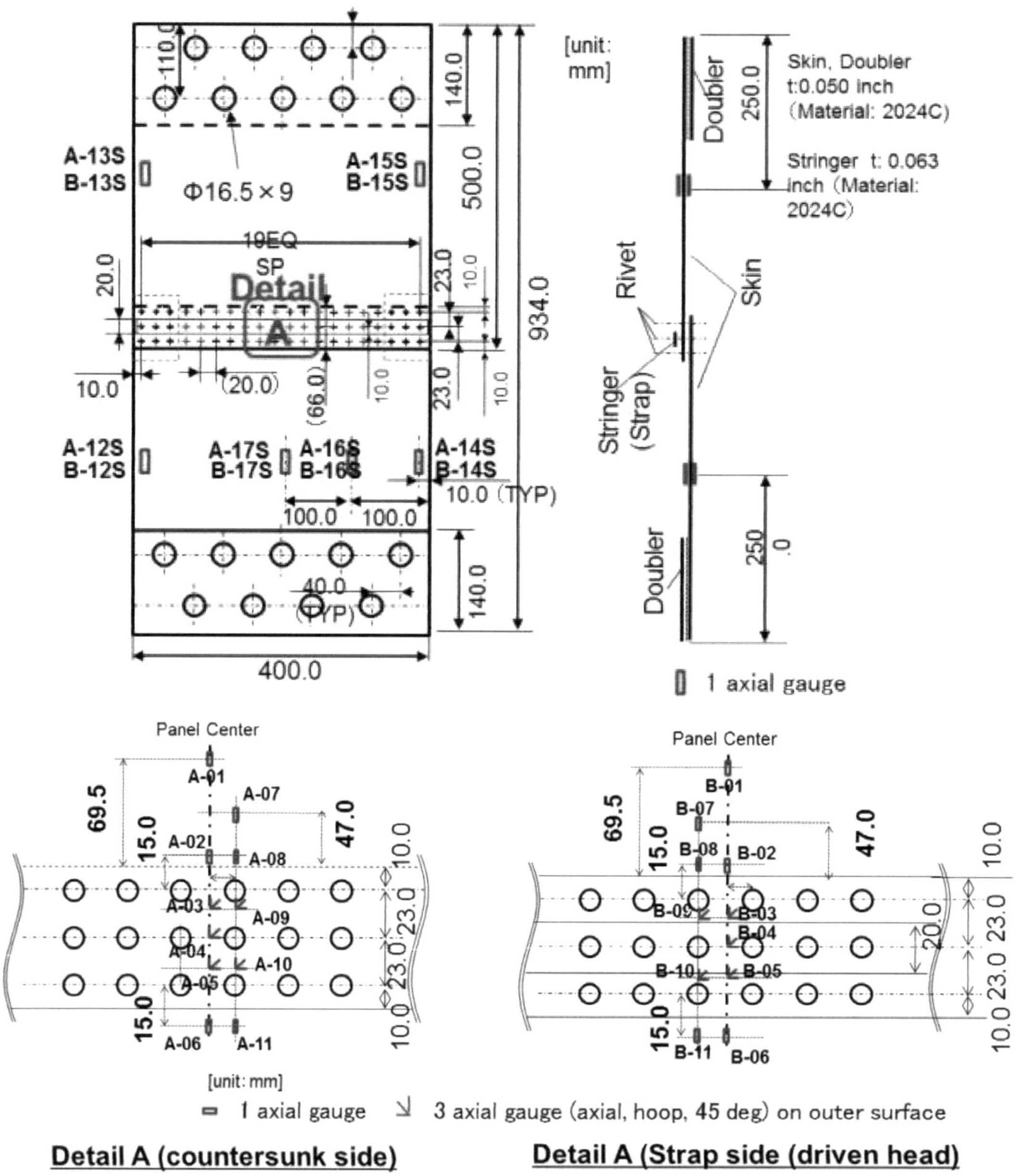

Figure 1. Specimen configuration and strain gauge locations.

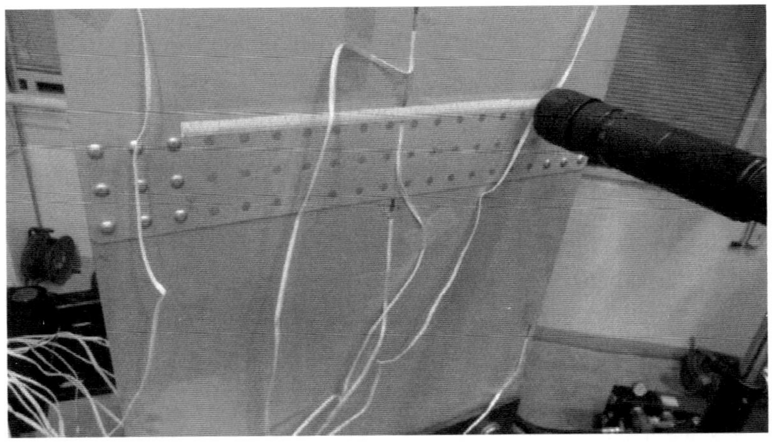

Figure 2. Fatigue test set up.

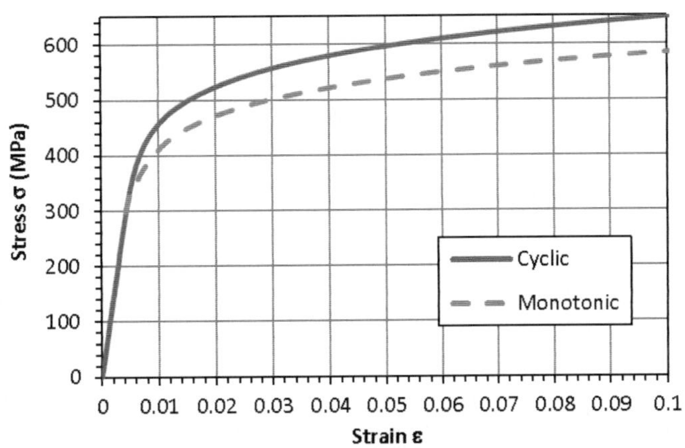

Figure 3. Stress strain relationship for 2024-T3 aluminium alloy.

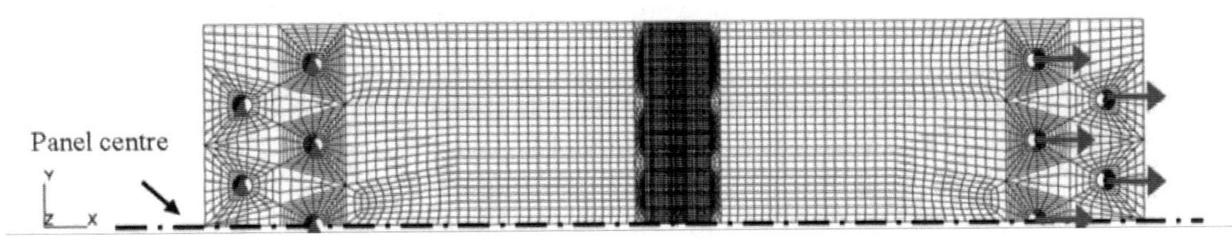

Figure 4. Global finite element model for flat panel joint structure.

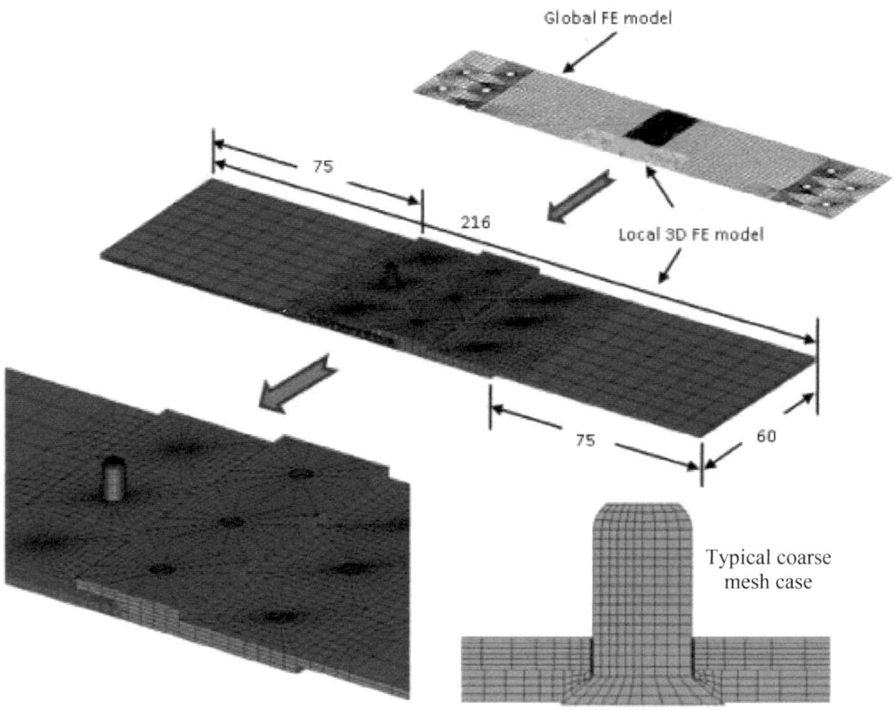

Figure 5. Local finite element model for flat panel joint structure.

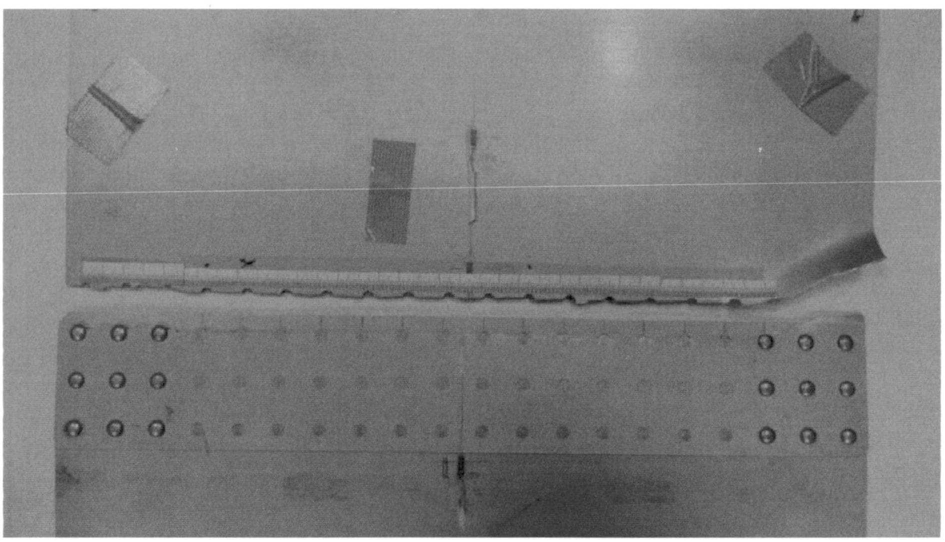

Figure 6. Flat panel joint specimen after fatigue test.

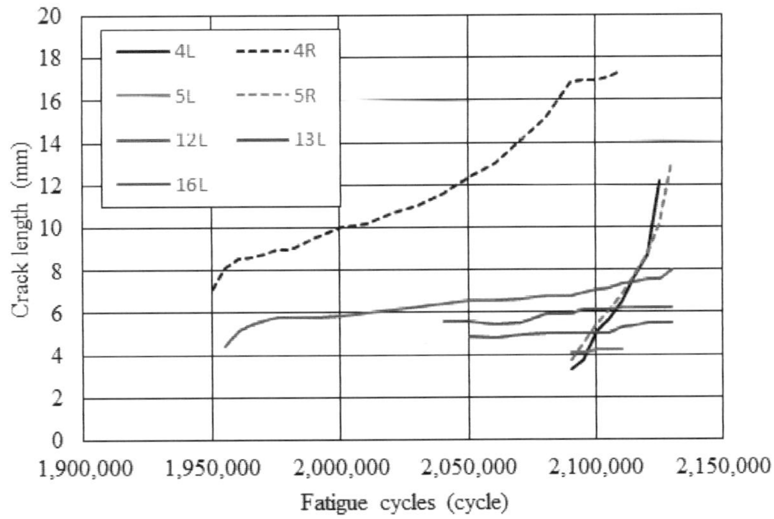

Figure 7. Fatigue crack growth data for flat panel joint specimen.

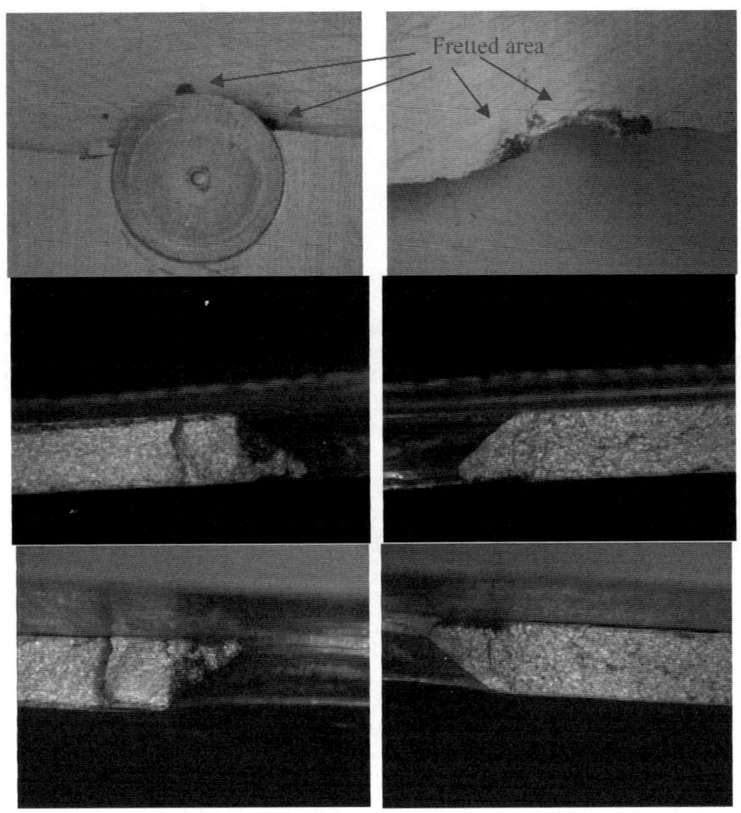

Figure 8. Faying and fracture surface observation.

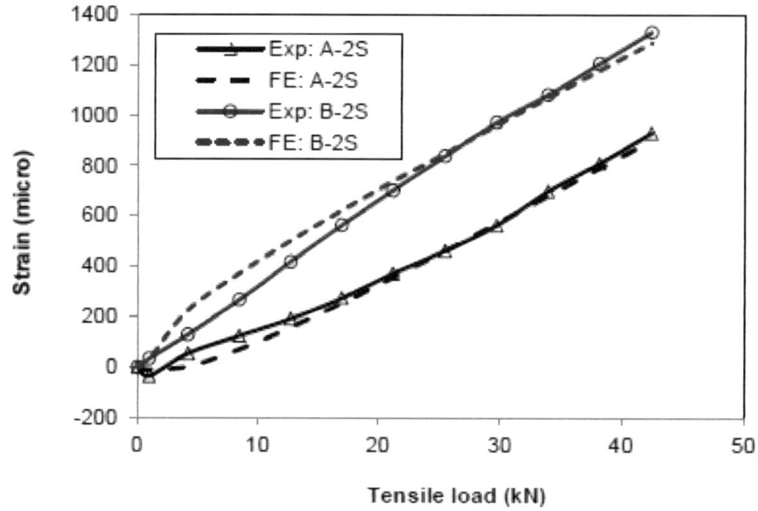

Figure 9. Comparison of strain at strain gauge A-2 and B-2 obtained from experimental and global FE model.

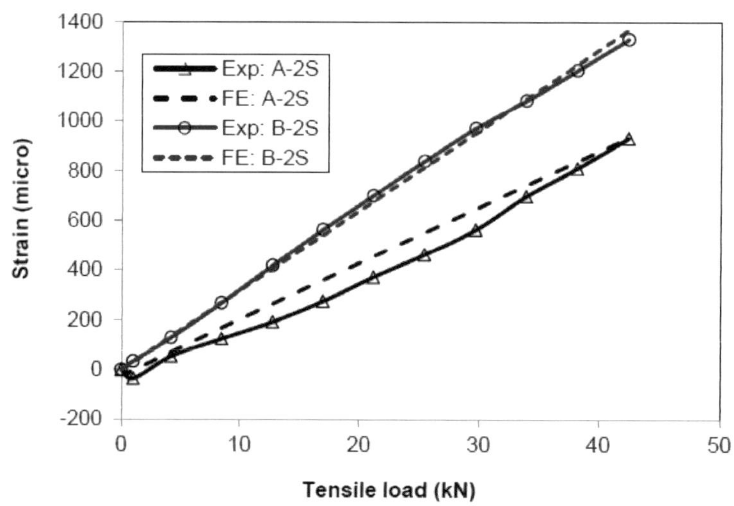

Figure 10. Comparison of strain at strain gauge A-2 and B-2 obtained from experimental and local FE model.

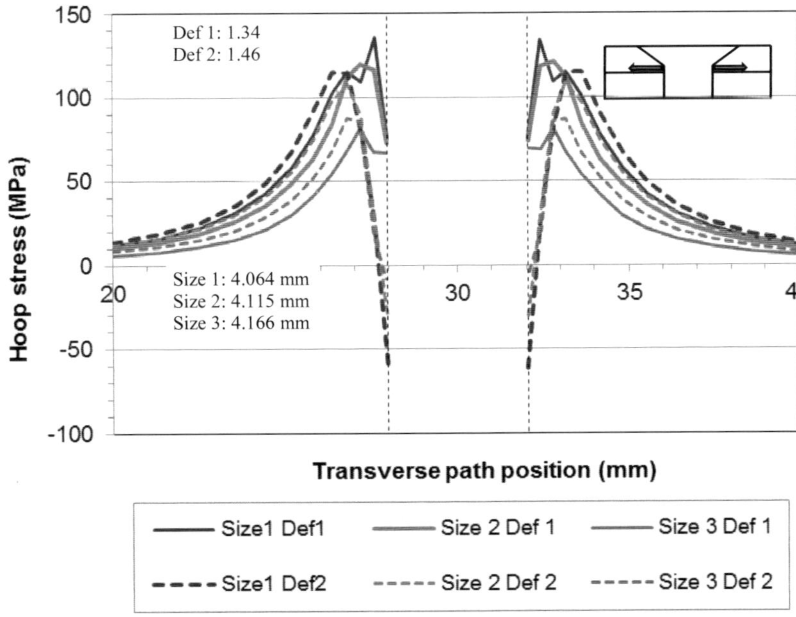

a) Post-riveting residual stresses

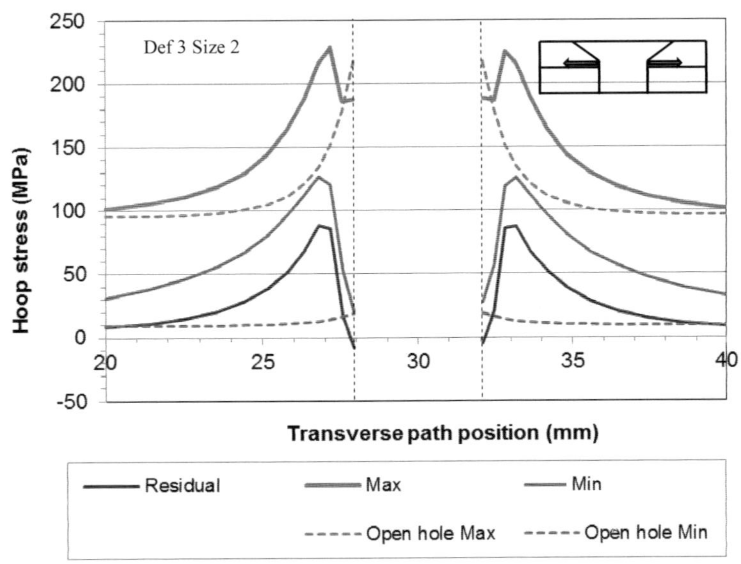

b) Stress distributions under cyclic loading

Figure 11. Typical stress distributions at the rivet hole.

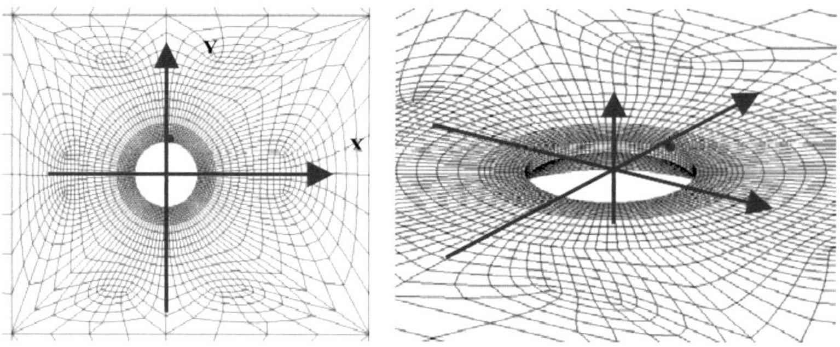

Figure 12. The size 2 FE model and predicted crack nucleation position (in red) on the outer panel faying surface for driven head deformations of 1.507 and 1.517. The X-axis is in the global planar remote tensile direction.

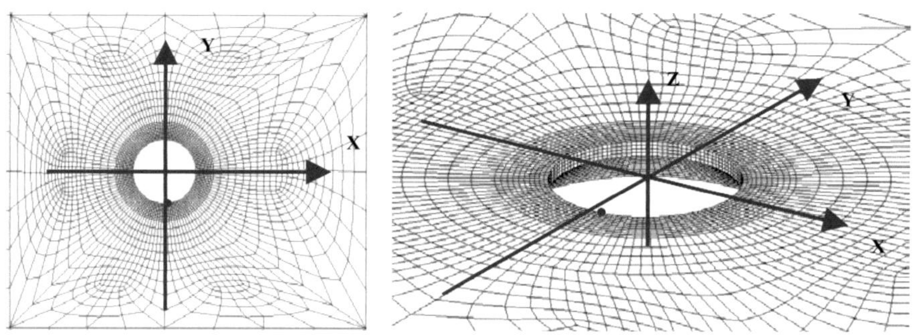

Figure 13. The size 2 FE model and predicted crack nucleation position (in red) on the outer panel faying surface for a driven head deformation of 1.527. The X-axis is in the global planar remote tensile direction.

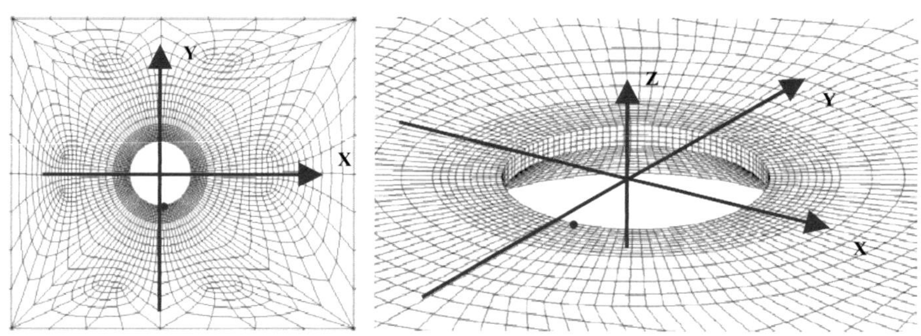

Figure 14. The size 3 FE and predicted crack nucleation position (in red) on the outer panel faying surface for driven head deformations of 1.501, 1.509, and 1.519. The X-axis is in the global planar remote tensile direction.

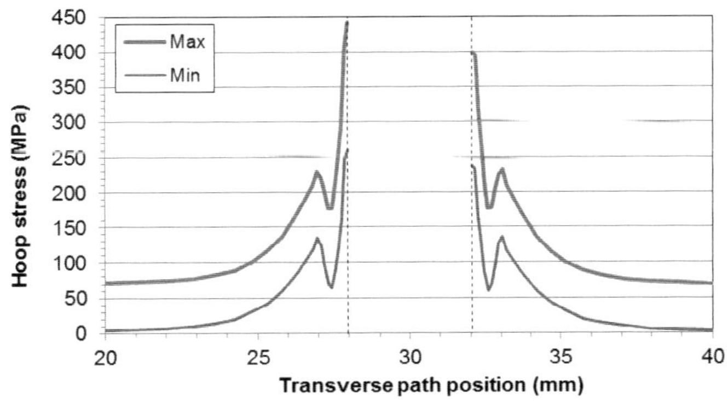

a) Stress distributions under cyclic loading

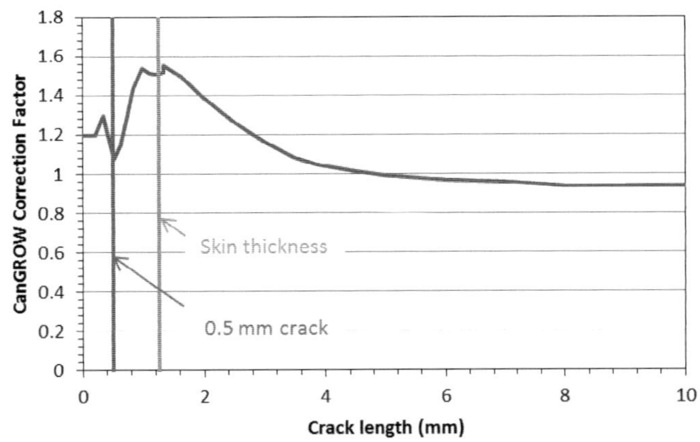

b) Correction factor for countersunk geometry and residual stresses

Figure 15. Stress distributions and CanGROW correction factor for Size 2 hole diameter (4.115 mm) and driven head deformation D_{max}/D = 1.517 under a maximum remote stress of 65 MPa and a stress ratio 0.069.

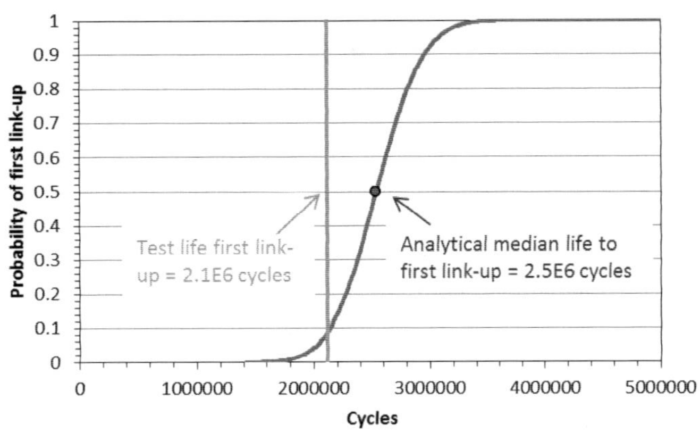

Figure 16. Probability of first link-up between 14 holes.

A Risk Analysis Based Methodology for Widespread Fatigue Damage Evaluation

Yuval Freed, Dvir Elmalich, Abraham Brot, Yael Buimovich, Carmel Matias

Engineering & Development Group, Israel Aerospace Industries, Ben Gurion International Airport, 70100, Israel

Abstract: The Widespread Fatigue Damage (WFD) phenomenon is recognized by the aviation industry as one of the biggest threats on structural integrity, especially for ageing aircraft. This type of damage involves simultaneous evolution of cracks at multiple structural locations that are of sufficient size and density such that the structure might no longer meet its required residual strength. This paper proposes a methodology for WFD assessment, based on probabilistic analysis. Our approach is flexible in choosing Inspection Starting Point (ISP), Structural Modification Point (SMP) and inspection intervals, as long as a desired overall probability of failure is maintained. Based on our proposed approach, the operator can choose, for each structural detail which is prone to WFD, whether to adjust the inspection intervals, inspection threshold or the structural detail Limit Of Validity (LOV), all while maintaining an equivalent probability of failure. Special attention is given to a lap-joint analysis, including validation to test results published in the literature.

INTRODUCTION

Widespread Fatigue Damage (WFD) is the simultaneous presence of cracks at multiple structural locations that are of sufficient size and density such that the structure will no longer meet its required residual strength. It is common to divide the WFD phenomenon into two. *Multiple Site Damage* (MSD) is a source of widespread fatigue damage characterized by the simultaneous presence of fatigue cracks in the same structural elements, whereas *Multiple Element Damage* (MED) is a source of widespread fatigue damage characterized by the simultaneous presence of fatigue cracks in similar adjacent structural elements. Illustrations of MSD and MED are provided in Figure 1. Application of traditional damage tolerance analysis concepts is usually not sufficient to preclude WFD, since cracks related to WFD can grow quickly and interact in such a way that an operator cannot detect them in time before they lead to structural failure. A separate WFD assessment and determination of specific maintenance actions is necessary to adequately address WFD.

The Widespread Fatigue Damage phenomenon was recognized by the aviation industry as one of the biggest threats on structural integrity, especially for ageing aircraft. The dramatic effect of WFD on aircraft structural integrity was demonstrated in April 1988 on Boeing 737 airplane operated by Aloha Airlines in Hawaii, when an 18-foot section of the aircraft fuselage flew off the airplane during flight. This accident drove the certification authorities to revise the relevant Federal Aviation Regulations paragraphs to ensure that the manufacturers account for the effect of WFD during the development and substantiation processes of transport category aircraft. During amendment 96 of FAR 25, paragraph 571(b) was revised to ensure, by "sufficient full-scale fatigue test evident", that the designer shows that WFD will not occur within the service goal of the airplane. Since WFD will happen to every airplane eventually, it was decided in 2011 to set a *limit of validity* (LOV) of the engineering data that supports the structural maintenance program for each airplane. The LOV requirement is limited to large aircraft, in which its maximum takeoff weight is greater than 75,000 lbs. Once the LOV is established, airplanes must not fly beyond that point, unless additional engineering data are developed to support an extended LOV and any associated maintenance actions. This limitation ensures that WFD will not occur within the operational life of the aircraft.

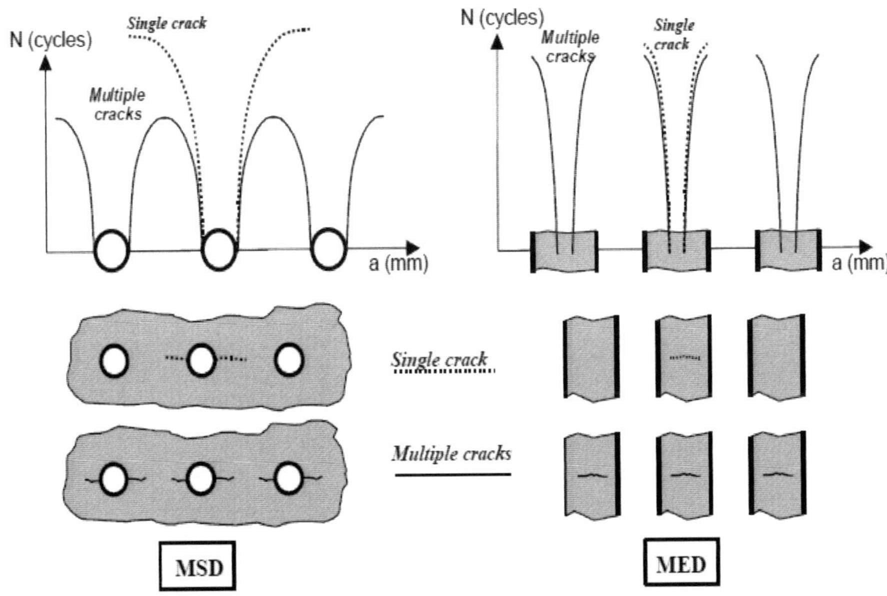

Figure 1. Multiple Site Damage (MSD) and Multiple Element Damage (MED) interaction effects (taken from Ref. [7]).

Recently, a Notice of Proposed Amendment (NPA, Ref. [1]) of ageing aircraft structures was published by the European Aviation Safety Agency (EASA). This NPA, focusing on large aircraft, requires WFD assessment even for local repairs or alterations conducted on aging aircraft. This requirement is unique to the proposed EASA regulations, and is not included at the moment in the equivalent FAR 26 regulations, but it highlights the importance of the WFD assessment for aging aircraft.

Several approaches were suggested in the past for WFD assessment and to establish maintenance program, if applicable, to preclude WFD in fleet [[2]–[8]]. All of them start by reviewing and identifying the structural elements which are suspected to WFD. The corresponding maintenance program is determined as follows. First, an average widespread fatigue lifetime $WFD_{average}$ (i.e., lifetime in which widespread fatigue damage will initiate in 50% of the fleet) is obtained. This can be based on analysis supported by test evidence. Then an Inspection Starting Point (ISP) is defined. The ISP is the point in time when special inspections of the fleet are initiated due to specific probability of having an MSD/MED. The Structural Modification Point (SMP) is defined as well, as the point in time when a structural area must be modified to preclude WFD. Special inspection intervals are determined between the ISP and SMP. An example of the inspection policy drawn from these approaches is shown in Figure 2.

Brot [[8]] presented a methodology to examine the admissibility of the inspection intervals in fatigue critical structures that is based on the risk analysis concept. In his analytical work, the inspection intervals for a typical lap-joint splice, which is prone to MSD, were determined using various NDI methods. The structure was then analyzed for the overall probability of failure (OPOF) using INSIM software [[9]]. The study presented in [[8]] serves as a starting point for the current study.

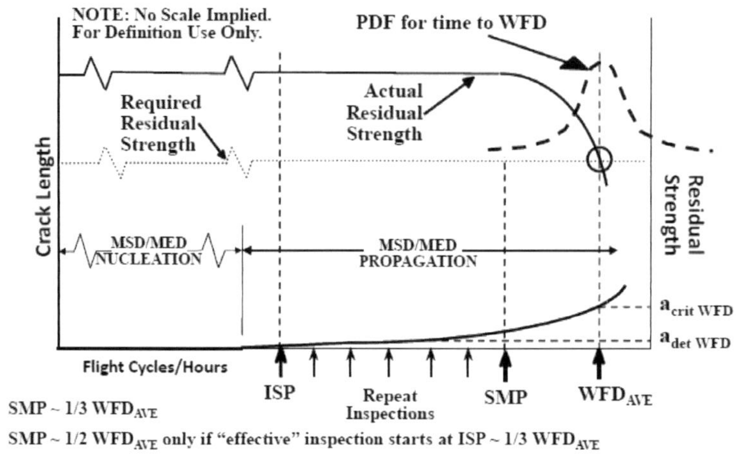

Figure 2. Inspection program to preclude WFD (taken from Ref. [7]).

In this study, a new approach to set the ISP, SMP and the corresponding inspection intervals is presented. Our approach is flexible in choosing ISP, SMP and inspection intervals, as long as the desired overall probability of failure is maintained. Based on our proposed approach, the operator can choose, for each structural detail which is prone to WFD, whether to adjust the inspection intervals, inspection threshold or the structural detail LOV, all while maintaining an equivalent probability of failure. Special attention is given to a lap-joint analysis, including validation to test results [[10]].

Section 2 of this paper provides some useful guidelines for identification of structural elements which are prone to MSD or MED. Section 3 describes the proposed methodology for determination of ISP, SMP and periodical inspection intervals, all with the same overall probability of failure. Section 4 provides an example of a typical lap joint configuration, which is validated by test results. Discussion and concluding remarks are provided in Sections 5 and 6.

IDENTIFICATION OF STRUCTURE SUSCEPTIBLE TO WFD

The identification of structural elements which are susceptible to WFD is the first stage in the WFD assessment of the aircraft. An excellent source for guidelines on which structural elements are candidates to WFD assessment is Appendix 5 of Ref. [7]. Service experience and teardown inspection findings are mostly welcomed, as well. Possible candidates for WFD are longitudinal skin joints, fuselage frames, tear straps, circumferential joints and stringers, shear clips and shear ties, Aft pressure dome elements, window surround structure, over-wing fuselage attachment, door stoppers, latches and hinges, rib to skin attachments, skin runouts and more. Many of these structural elements are illustrated and discussed in Appendix 5 of Ref. [7].

Special attention should be given to pressure loaded structure, which often comprises similar details operating at similar stress levels. Fuselage lap joints are an example of such structure, in which many small cracks nucleate at adjacent fastener holes, none of which of detectable size, and can finally link up to endanger the structural integrity of the aircraft.

As mentioned in Ref. [7], the designer should develop rationale for including or excluding structure from the WFD susceptible structure list. This should be based on criteria developed and tailored for the specific products, related to its design, usage and service objective goal. The flow chart shown in Figure 3 describes the decision making process in the identification of the WFD susceptible structural components. If all questions in the flow chart are answered "YES", the structural element should be added to the MSD/MED candidates list.

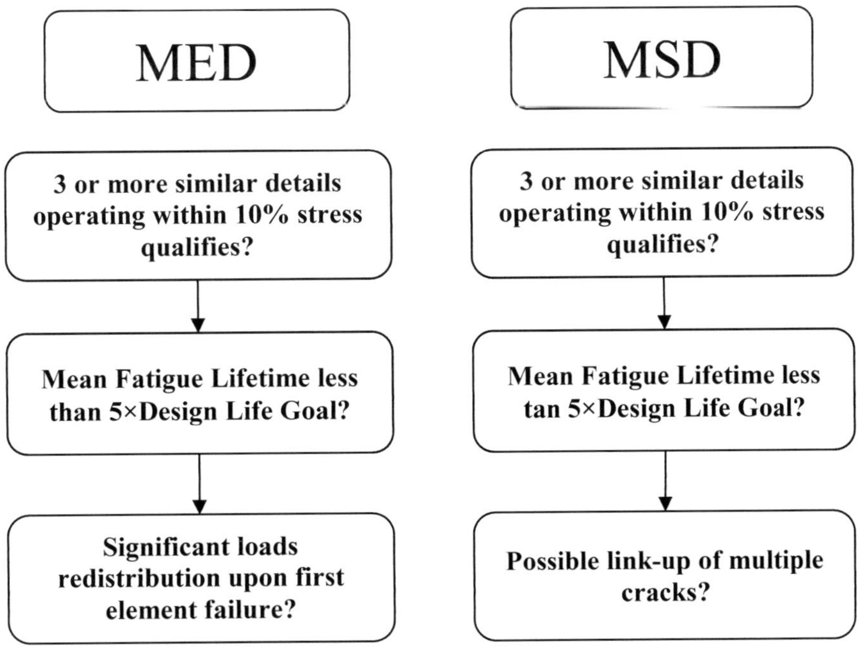

Figure 3. Chart of filtering questions for the decision making of MSD/MED prone structural elements.

In many cases primary structural elements can sustain the required fail-safe loads even in the presence of extensive damage. If such damage is readily evident (due to malfunction or obvious during walk-around inspection), it can be omitted from the WFD candidates list, and no special inspections are needed to be determined for it. If the damage is not considered as obvious or readily inspected, an inspection program that covers WFD aspects should be determined to assure its continuing structural integrity.

Finally, structural details made of composite materials are not included in the WFD candidates list. This is since the conservative design of the composite structural details is based on the "no growth" concept and reduced stress level due to employing damage tolerance allowables.

PROPOSED METHODOLOGY

Ref. [7] proposes a simple method to determine the ISP and SMP, as well as the reduced repeated intervals. Setting the SMP to $WFD_{average}/2$ and the ISP to $WFD_{average}/3$ corresponds to a certain level of safety, which is equivalent to 99% reliability and 95% reliability with a confidence level of 95% for the ISP and SMP, respectively, assuming Weibull distribution. The reduced repeated inspections intervals between the ISP and the SMP can increase the reliability once the crack is of detectable size. Although this approach is systematic and easy to implement, one cannot guarantee that the level of safety of all structural details is equivalent.

The concept suggested in this study assuring a desired level of safety for the structural component that is under consideration. In order to meet the demand for a certain level of safety, combinations of the ISP, the inspections intervals and the SMP are examined by means of risk analysis software. Naturally, an infinite number of options of the trio (ISP, SMP and Inspections interval) can be obtained. However, some combinations are more acceptable from an engineering point of view. These trios along with structural characteristics of the examined component are introduced into risk analysis software which produces the Cumulative Probability Of Failure (CPOF) for the given component. Repeating the risk analysis computation of certain values of the trio allows mapping the domain of the ISP, the SMP and the inspections interval and introducing curves of allowable values.

The first step of the proposed methodology is to determine the $WFD_{average}$ behavior, either experimentally or numerically (based on MMPDS S-N curves, Ref. [11], or by means of fatigue analysis software). It may be noted that the $WFD_{average}$ denotes the median of the population, rather than the average. In other words, it denotes the fatigue lifetime that 50% of the test specimen will survive upon cyclic loading. Another approach is to determine the $WFD_{average}$ by means of risk analysis, assuming that no inspections are performed. In that case the $WFD_{average}$ is the fatigue life that corresponds to 50% CPOF. For that purpose (and for the following risk analysis), a crack growth analysis is carried out to predict the crack growth curve.

Next, a risk analysis is applied. The risk analysis simulations utilize the computer program INSIM (INspections SIMulation, [[9]]). This computer program was developed in order to simulate the entire fatigue environment that an aircraft structure must withstand in a probabilistic manner. The risk analysis accounts for the service life variation, service load severity, crack nucleation period, crack growth history and NDI detection capabilities. Crack-initiation, crack growth and crack detection characteristics of a specific structural location are input into INSIM, which performs the simulation of the fatigue process for the entire virtual fleet. Cracks initiate at various times and grow at variable rates in each aircraft. Inspections are performed according to a predetermined schedule. Cracks are detected during these inspections according to the statistical probabilities of detection. As the simulation proceeds from aircraft to aircraft, cracks are detected and repaired, aircraft retire from service or failures occur. Based on these simulations, INSIM calculates the probability that failure has occurred. INSIM uses several statistical models for the various parameters in the analysis as follows: a two-parameter Weibull Distribution for the crack-initiation life variation, a Normal Distribution with a 5% standard deviation on crack growth rate for the crack growth curve, Normal Distribution with a 5% standard deviation taken on the spectrum loading, a Normal Distribution with the mean service life and the high-time life (mean + 3 standard deviations) for the mean service life to aircraft retirement, a three-parameter Weibull Distribution for the crack detection capability.

Based on the statistical distributions of the various parameters outlined above, the probabilities for aircraft retirement, cracks detection prior to maintenance operation, and failure due to undetected cracks are calculated. Out of the three computed probabilities, the later one, in conjunction with admissible CPOF defines the risk analysis. It is emphasized that in order to provide statistically significant results, the population size must comply with the admissible probability of failure.

The next step in the proposed methodology is to define an admissible CPOF. In Ref. [7] the ISP and the SMP are determined based on predefined factors without direct correlation to a certain CPOF. That is, lap joints that sum up to hundreds of structural details in each individual aircraft are treated as the pressure dome, which consists of only dozens of structural details which are prone to WFD. In this study we propose to define the CPOF as the governing parameter for the WFD analysis, and the ISP, SMP and the inspection intervals are adjusted accordingly. The CPOF accounts for a desired level of safety (corresponds to the factors defined in Ref. [7]), as well as the number of similar structural details in an individual aircraft and the fleet size. For instance, the CPOF for a lap joint is taken as 0.001% for a large fleet (>1,000 aircraft), and can be increased to 0.005% for a small fleet (~200 aircraft).

Another approach is to define an adequate CPOF to be equivalent to a certain mean hazard rate. CPOF goal of 0.001% (0.00001) is equivalent to a mean hazard rate of $0.00001/75{,}000 = 1.33\times10^{-10}$ per flight, for a life goal of 75,000 flights. This is considered to be extremely improbable.

In order to meet the demand for a certain level of safety, the third step in the methodology examines combinations of ISP, inspections intervals and SMP. Naturally, an infinite number of options of the trio (ISP, SMP and Inspections interval) can be obtained with an equivalent level of safety. However, some combinations are more acceptable from an engineering and operational point of view. The three parameters in the maintenance program along with structural characteristics of the examined component are introduced into INSIM (INspection SIMulation [9]). Using the parametric module of the risk analysis software the domain of the ISP, the SMP and the inspections interval is spanned.

In the fourth step of the methodology, the spanned domain of ISP, SMP and Inspection interval in conjunction with the desirable level of safety are utilized to determine the maintenance operations. At this point, economical, practical and operational aspects should be considered in the planning and optimization of the maintenance program. For instance, a relatively late ISP combined with short inspection intervals can be profitable to certain operator, whereas other operators cannot maintain such short inspection intervals in its fleet.

EXAMPLE OF A LAP JOINT ANALYSIS

The proposed methodology is examined for a typical case of lap-joint element. Such elements are prone to MSD in aging aircraft, as discussed in Ref. [[7]]. Test results of 2024-T3 aluminum lap-joints under unidirectional loading studied experimentally in Ref. [10] are used in this analysis. Description of the test articles is given in Figure 4. The load is applied at R ratio of 0.1. The characteristic fatigue life for that lap-joint is extracted from the corresponding MMPDS S-N curves [[11]] taking into account stress concentrations due to countersink, bearing, open-hole, and pin bending. The characteristic fatigue life equals 269,000 flights. In order to take into account the issue of representativeness of the MMPDS test articles, the characteristic fatigue life obtained from the MMPDS S-N curves is scaled down by a factor of 2.

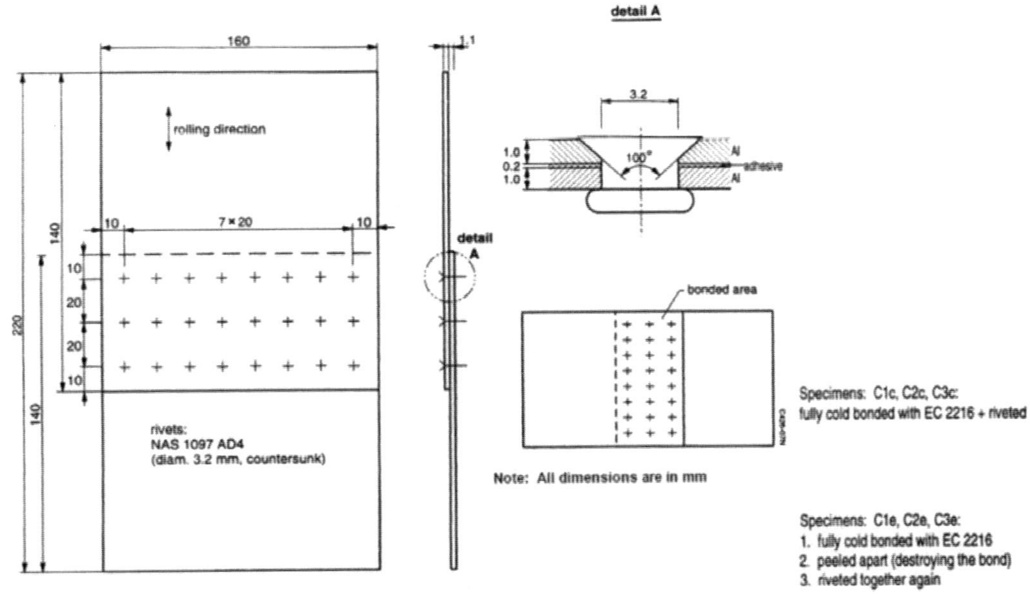

Figure 4. Example of a lap-joint detail under unidirectional tension (taken from Ref. [10]).

It should be noted that the test specimens in Ref. [10] represent two configurations: two specimens had a width of 120 mm while the other two had a width of 160 mm. The test loads were adjusted to yield the identical stress levels on all four specimens.

In order to check the statistical validity of the results, a Weibull analysis was performed using the rank-regression method contained in SuperSmith software (Ref. [12]). The results shown in Table I and Figure 5 show reasonable correlation to a Weibull distribution, with all the results contained well within a 90% confidence band (± 5% from the Weibull result).

Table I. Summary of experimental results per Ref. [10] and the corresponding numerical predictions.

Specimen	*Width [mm]*	*First crack (cycles × 1,000)*	*First linkup (cycles × 1,000)*	*Failure (cycles × 1,000)*
C1e	160	275	311	311
C2e	160	200	201	202
C4e	120	175	218	218
C4f	120	150	182	182
Test average		200	228	228
MMPDS+crack growth		269		282
MMPDS/2+crack growth		134		147
INSIM				134

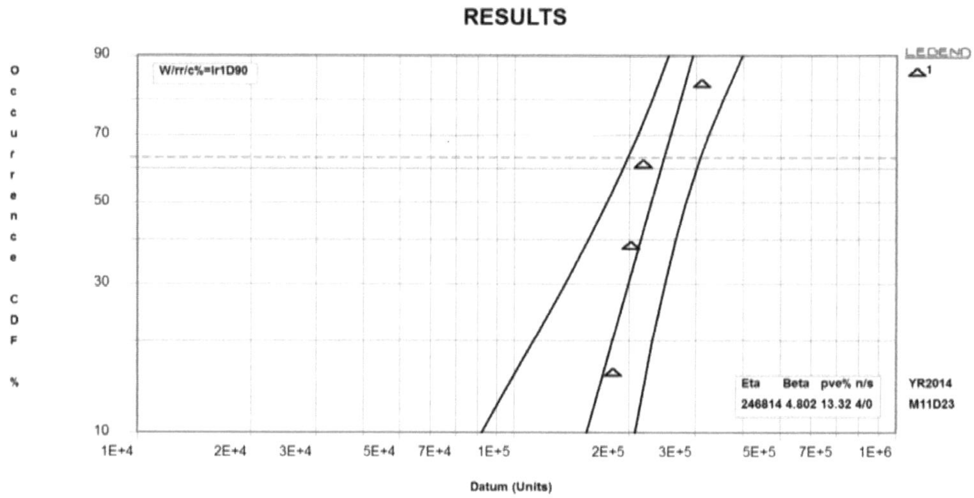

Figure 5. Determination of Weibull shape factor for the lap-joint.

The characteristic life (Eta) was calculated as 246,800 cycles, while the Weibull shape-factor (Beta) was found to be 4.8, as is shown in Figure 5. Since a typical Weibull shape-factor for aluminum structures is generally taken as 4.0 (Ref. [13]), the results clearly show that the four test failures are very likely to have come from the same population. This justifies obtaining the experimental average WFD life as the mean value of the four test failures (228,250 cycles).

The crack growth analysis is carried using NASGRO 7.1 software [14]. A built-in module of through cracks at hole in plate with row of holes (Module TC05) is employed to represent an MSD scenario. An illustration of this scenario is shown in Figure 6.

Cracking scenario of several holes in a row with two initial cracks of 0.01 inches at each hole each is considered. In addition, the crack growth analysis accounts for the countersink using the relevant stress intensity factor compounding. The crack growth curve is presented in Figure 7. The crack growth life equals 13,400 flights, with a period of 4,000 flights of detectable crack size.

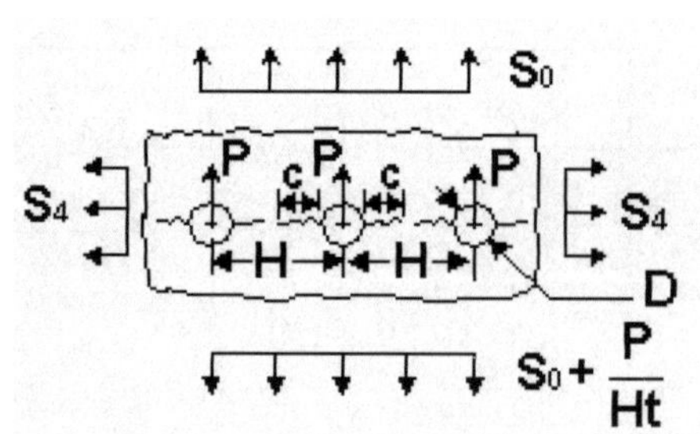

Figure 6. Through cracks at hole in plate with row of holes.

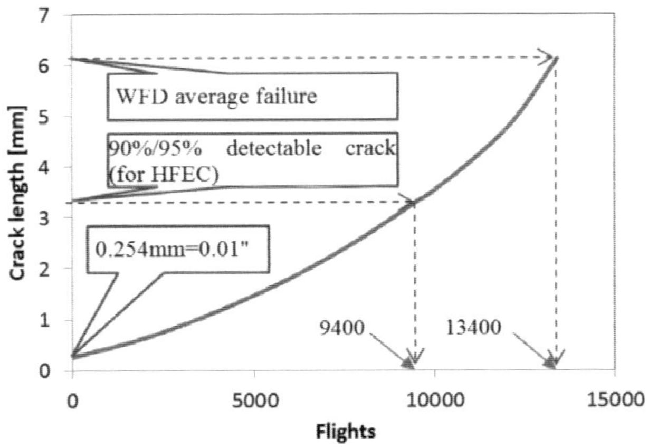

Figure 7. Crack growth curve of a single crack in a two cracks per hole scenario.

Next, the $WFD_{average}$ for the concerned problem is evaluated numerically adopting the risk analysis concept. For that purpose, the crack growth curve and calculated fatigue lifetime are introduced into the risk analysis software along with the assumption that no inspections take place. The $WFD_{average}$ is represented by the expected life that yields a 50% CPOF (see Figure 8). In other words, assuming that the cracks are nucleating and growing with no preventive or corrective actions taking place, 50% of the lap joints in fleet will fail after 134,000 flights. The proposed methodology is validated by comparison to test results provided by Ref. [10]. Table I presents a summary of the corresponding results from Ref. [10], The MMPDS based results and the risk analysis results for the $WFD_{average}$. The deviation from the risk analysis based value and the experimental result is attributed to the representativeness factor introduced to the MMPDS results. In the present case, the MMPDS graphs refer to simple coupon testing, and the contribution of the lap joint configuration is accounted for in the calculation of the stress concentration factor K_{tn}. In addition, the MMPDS specimens are manufactured and assembled very carefully, while maintaining proper manufacture, surface treatment, countersunk drilling, etc. Thus, a factor of 2.0 was applied to the MMPDS results to account for the representation of the MMPDS specimens compared to the actual lap joints test specimens.

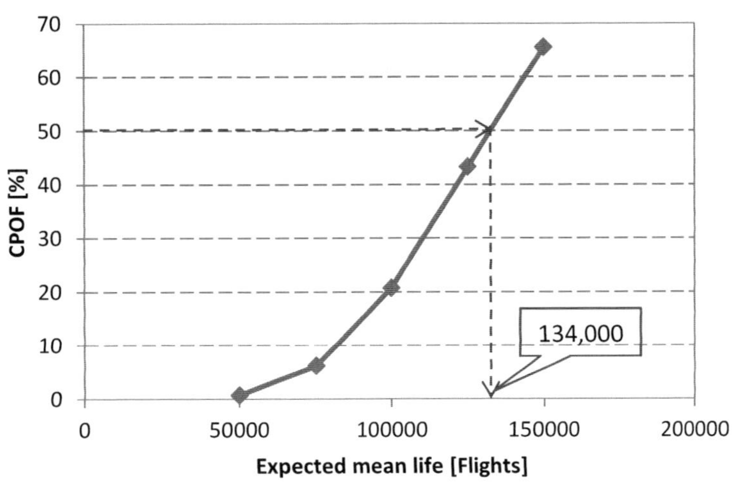

Figure 8. CPOF vs. fatigue life assuming no inspections throughout life time.

Once the $WFD_{average}$ is determined, a WFD analysis can be made. In order to optimize the maintenance program for precluding WFD, series of risk analyses are conducted for various combinations of ISP, SMP and Inspections intervals. In the current example, High Frequency Eddy Current (HFEC) inspections are assumed to take place. As discussed earlier, a CPOF value of 0.001% is considered for the WFD analysis.

The results are presented in two forms. In the first, the results are presented in the form of a two-color map. The x and y axes map the combination of two of the parameters where the colors distinguish between admissible CPOF and inadmissible CPOF. In the following analysis the SMP is chosen based on the Advisory Circular [[7]] recommendation (i.e. SMP= $WFD_{average}/2$ = 134,000/2 = 67,000 flights). The maximum inspection interval is bounded by the number of flights that from detectable crack sizes up to failure (4,000 flights). The latter boundaries limit the WFD analysis to a specific admissible and practical domain.

Figure 9, Figure 10 and Figure 11 study the effect of the ISP threshold and repeated inspection intervals on the cumulative probability of failure. Based on Figure 9, the operator can choose between different combinations of ISP (varying from 6,000 flights to ~20,000 flights) and inspection intervals (varying between 1,000 flights to ~3,000 flights), all of them with admissible cumulative probability of failure (less than 0.001%). Figure 11 reveals that setting the ISP to less than the crack growth period (13,400 flights for the case studied here) yields practically the same CPOF values for all the selected inspection intervals (the curves for ISPs 6,000 to 13,000 practically coincide). It is also observed that, for all of the selected ISPs, inspections intervals of 2,000 flights or higher (about half the detectable crack to failure period) approach or exceed the admissible CPOF. As expected, if the inspection intervals are close to the time between detectable crack length and critical crack length (4,000 flights in the case studied here), the required CPOF cannot be reached. Based on Figure 10 and Figure 11, it can be concluded that the inspection intervals has a major role in reducing the CPOF value to an admissible one.

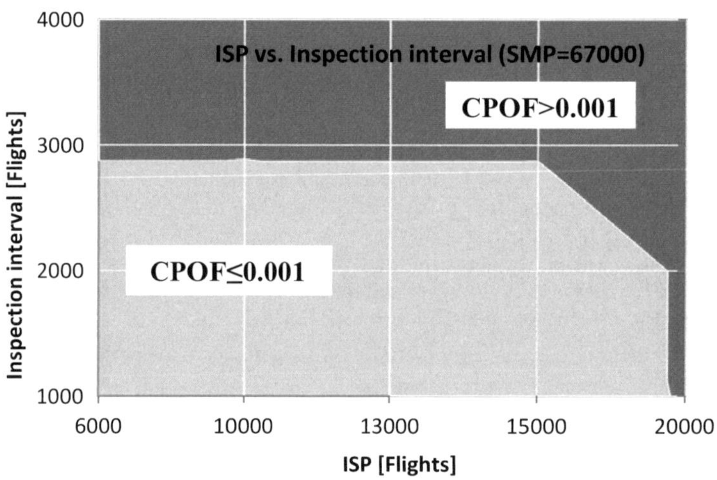

Figure 9. 2D map of admissible combinations of ISP and inspections interval for SMP=67,000.

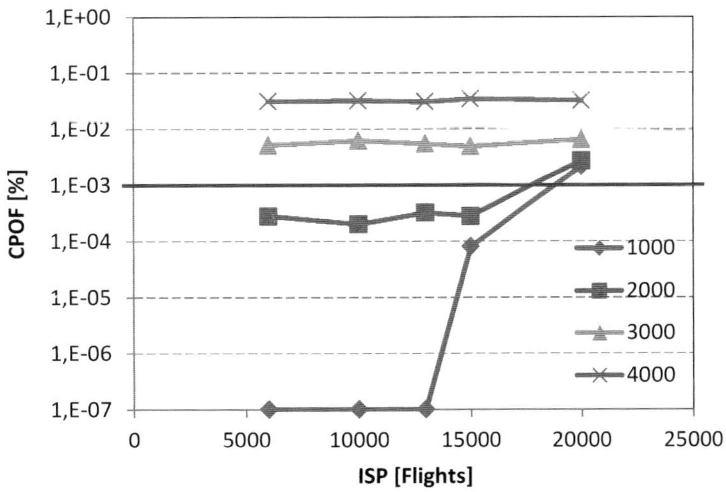

Figure 10. CPOF vs. ISP curves of four levels of inspections interval and SMP=67,000.

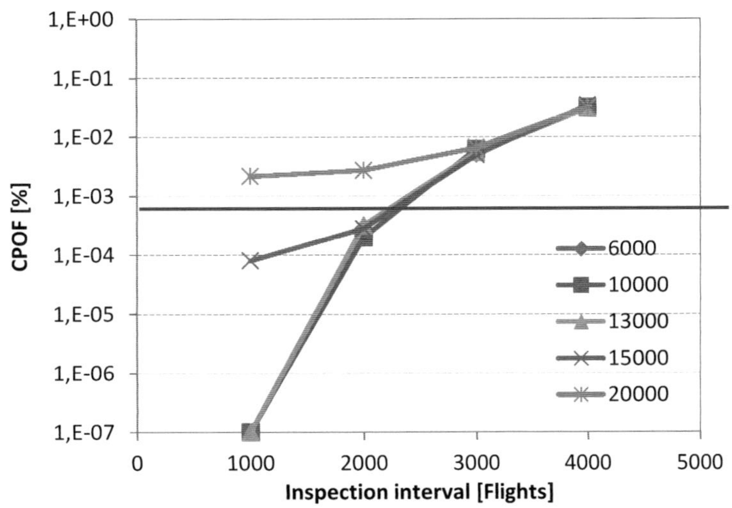

Figure 11. CPOF vs. Inspections interval curves under various levels of ISPs and SMP=67,000.

The color-map in Figure 9 provides a powerful tool for examination of the alternatives for maintenance program. To keep Figure 9 as simple as possible, only two colors which stand for admissible and inadmissible CPOF are shown. However, in order to better estimate the transition from admissible to inadmissible CPOF, it is recommended to use Figure 9 together with Figure 10 and Figure 11. For example, inspection to preclude WFD can start at 10,000 flights and based on Figure 9, in conjunction with Figure 11, the allowed inspections interval can go up to 2,500 flights. Alternatively, starting the inspections at 17,000 flights will require shorter inspections interval.

EXTENDING THE ANALYSIS TO AN ACTUAL FLEET

As discussed in Section 3, the determination of admissible CPOF is a key parameter in the determination of inspection policy to preclude WFD threats. As opposed to the guidelines provided in Ref. [7], in which the values for ISP and SMP points are strictly defined as 1/3 and 1/2 of the $WFD_{average}$, in this study we propose to define the CPOF as the governing parameter for the WFD analysis, and the ISP, SMP and the inspection intervals are adjusted accordingly. It was recommended to correlate between the admissible CPOF and the number of similar details in each individual aircraft and the fleet size.

It can be readily stated that as the test specimen that is used for determination of $WFD_{average}$ is more representative of the actual aircraft, and more accurate inspection instructions can be defined. Full Scale Fatigue Test (FSFT), which is required by the Federal Aviation Regulations (FAR) to demonstrate that WFD will not occur within the service goal of the airplane, is the most recommended tool to be used for this purpose. However, since the FSFT is usually completed few years after the aircraft enters to service, it is recommended to use coupon or elements tests at early stage of the development program to ensure that the design is robust enough in the sense that it is not prone to WFD. It is understood that WFD findings in late stage of the FSFT can result in major engineering effort, as well as possible retrofits in fleet or determination of special inspection program for these details. This can be costly for the OEM as well as the operators.

Special caution should be taken here. Element tests as described in Figure 4 are not fully representative of the actual structural detail in the aircraft and fleet. Taking the lap-joint test specimen described in Figure 4 for example, only 24 fasteners, arranged in three rows of 8 fasteners, are used in these test coupons. Let us assume that our fleet has six lap-joints at different fuselage stringers (3 left and 3 right) and 70 bays along the length of the fuselage. We further define, somewhat conservatively, a catastrophic failure as the failure of a single bay. This is based on the observation that WFD could occur very soon after a single bay failure of a single lap-joint. A common approach that can be used to deal with this issue is the determination of a 'testing factor' and 'scaling factor' to account for the representativeness of the test specimens versus the actual aircraft.

Alternatively, we proposed to decrease the CPOF for WFD assessment which is based on coupon testing at early stage of the program. Figures 9 – 11 show an allowable cumulative probability of failure (CPOF) of 0.001% for the one-bay NLR test specimen. Each aircraft in the fleet has 420 bays, (70 x 6) each of which can be considered to be a prelude to WFD. Therefore the allowable CPOF should be reduced to 2.4×10^{-6} % (0.001%/420). Figure 11 shows that the required inspection interval will need to be reduced from about 2,400 flights to about 1,300 flights in order to achieve the allowable CPOF. Alternately, the SMP could be reduced from 67,000 flights to 50,000 flights with repeated inspection intervals of 1,800 flights. These conservative inspection guidelines should be documented in the aircraft airworthiness limitation section until completion of FSFT and adjustment of the WFD inspection policy based on FSFT findings.

CONCLUSIONS

A new approach for analysis and quantification of widespread fatigue damage is presented in this study. Special attention is given to multiple site damage scenario. The proposed approach is based on probabilistic methods to determine the probability of failure assuming a variety of parameters. The capabilities of the new approach have been studied through WFD analysis of a lap-joint structural member, validated by test results published in the literature. The study has shown that the new approach can be a good engineering tool in the selection of alternatives for establishment of maintenance programs to preclude WFD treats.

REFERENCES

[1] Ageing Aircraft Structures, Notice of Proposed Amendment 2013-07, 23.4.2013.

[2] J. McGuire, J. Foucault, Recommendations for Regulatory Action to Prevent Widespread Fatigue Damage in Commercial Airplane Fleet, Report of the Airworthiness Assurance Working Group of the FAA, 11.3.1999.

[3] H. Trey and H.J. Schmidt, Programme for Life Extension and Widespread Fatigue Damage Evaluation to Ensure Continued Structural Integrity of Airbus Large Transport Category Airplanes, RTO AVT Specialists' meeting on "Life Management Techniques for Ageing Air Vehicles", Manchester, UK, RTO-MP-079(II), 8-11 October 2001, p. 29-1 – 29-15.

[4] M.L. Gruber, B.E. Johnson and I.C. Whittaker, Widespread Fatigue Damage Assessment of Jet Transport Structures, 21th ICAF Symposium, Toulouse, June 25-29, 2001.

[5] I. Buchweitz, A. Santgerma and N. Turrel, , Airbus on the way to show compliance with WFD regulation, 27th ICAF Symposium, Jerusalem, June 5-7, 2013.

[6] P. Safarian, Widespread Fatigue Damage Evaluation Method, 27th ICAF Symposium, Jerusalem, June 5-7, 2013.

[7] Advisory Circular 120-104, Establishing and Implementing Limit of Validity to Prevent Widespread Fatigue Damage, 1.10.2011.

[8] A. Brot, Optimizing the Inspection Intervals for Aircraft Structures Prone to Multi-Site Damage, 53rd Israel Annual Conference on Aerospace Sciences, 2013.

[9] INSIM, INspection SIMulation software, ver. 3.0.

[10] H. Vlieger and H.H. Ottens, Uniaxial and Biaxial Tests on Riveted Fuselage Lap Joint Specimens, Report No. DOT/FAA/AR-98/33, 1998

[11] Metallic Materials Properties Development and Standardization (MMPDS).

[12] SuperSmith Software version 5.0, Fulton Findings, San Pedro CA, USA.

[13] U.G. Goranson, Fatigue Issues in Aircraft Maintenance and Repairs, International Journal of Fatigue, 20, 1997.

[14] NASGRO ver. 7.10, SWRI, 2014.

Curved Panel Fatigue Test for MRJ-200 Pressurized Cabin Structure

Kaoru Tsukigase[1], Toshiyasu Fukuoka[1], Keisuke Kumagai[1], Toshio Nakamura[2], Shunsuke Taba[2]

[1]*Mitsubishi Aircraft Corporation, Japan*
[2]*Mitsubishi Heavy Industries, LTD., Japan*

Abstract: To prevent widespread fatigue damage, FAA has issued 25.571 amendment 25-132 and Part 26 subpart C. To comply with the new regulatory requirement, a full-scale airplane fatigue test of MRJ-200 will be conducted mainly to show free from widespread fatigue damage in operational life. In addition, to find out the fatigue characteristics of the fatigue critical structure in early stage of MRJ-200 substantiation phase, the sub-component level fatigue tests are planned. The fuselage cabin pressurization area was selected and tested as typical widespread fatigue damage susceptible structures, in early substantiation activity. For damage tolerance evaluation of MRJ-200, large numbers of coupon/element/sub-component tests are conducted to support analysis. Since the full-scale airplane fatigue test will be utilized for widespread fatigue damage evaluation, the crack growth evaluation of fuselage structure will be substantiated by sub-component test simulating cabin pressure in conjunction with axial loading due to fuselage bending.

For widespread fatigue damage and damage tolerance evaluation of MRJ-200, a new test fixture for sub-component level tests has been developed to demonstrate fatigue cyclic loading under cabin pressure and axial load due to fuselage bending. As main focus of this test fixture concept, negative pressure loading is used to simulate differential pressure load. Using this test fixture, several curved panel fatigue tests for fuselage critical area are planned as early substantiation activities for widespread fatigue damage and damage tolerance evaluation.

This paper provides overview of the test fixture, some trial test results and usage plan for widespread fatigue damage and damage tolerance evaluation.

BACKGROUND

Widespread fatigue damage (WFD) is a well-known complex fatigue phenomenon. To prevent catastrophic failure due to WFD, FAA has issued 25.571 amendment 25-132 and Part 26 subpart C in late 2010, and they became effective in January 2011. To comply with the new regulatory requirement, a full-scale airplane fatigue test of MRJ-200 will be conducted by simulating actual airplane structural configuration with a typical loads expected in service to show free from WFD in operational life.

For damage tolerance evaluation, some tests are also required to support structural analysis to comply with 25.571.

MRJ-200 TEST PLAN – FATIGUE AND DAMAGE TOLERANCE EVALUATION

The full-scale airplane fatigue test to show free from WFD in operational life requires a great investment of time to apply the spectrum load. Therefore, in order to find out the fatigue characteristics of the fatigue critical structure in early stage of MRJ-200 substantiation phase, the sub-component level fatigue tests to evaluate WFD are planned and conducted. Structural details and elements, which are repeated over wide areas and subject to similar stress levels, are considered the candidate portions as susceptible structures to WFD. In these WFD susceptible structures of MRJ-200, fuselage cabin pressurization area with typical skin splices, typical skin pockets, typical stringers and typical frames are selected to evaluate in early substantiation phase, and a number of fuselage curved panel fatigue tests are conducted.

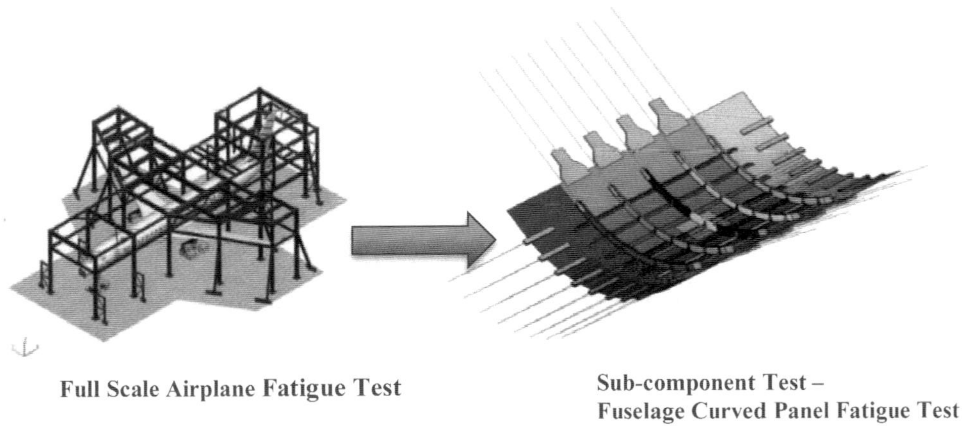

Figure 1. Fatigue test to evaluate WFD for MRJ-200.

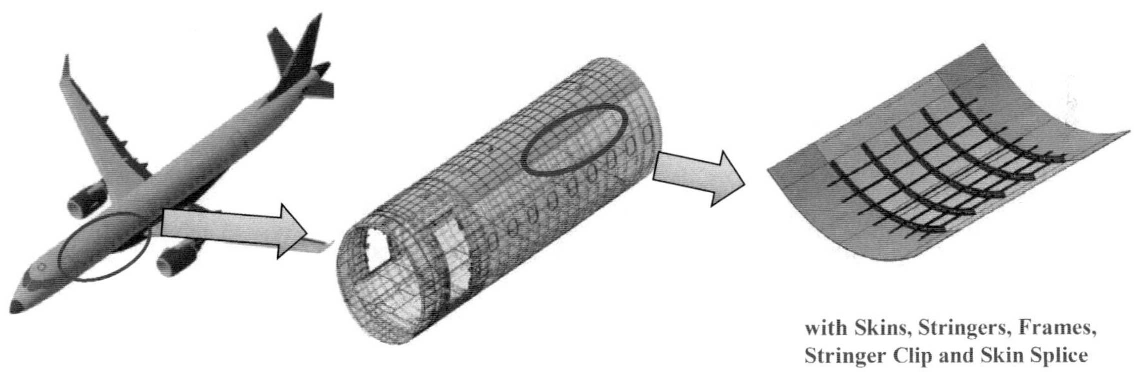

Figure 2. Selection of WFD susceptible structure.

Since the full-scale airplane fatigue test will be mainly utilized for WFD evaluation and any saw cuts will not be induced. For damage tolerance evaluation for MRJ-200, large numbers of coupon/element/component tests are conducted to support structural analyses. Coupon tests are intended to validate material data, large numbers of element tests are to validate stress concentration of some structure details and pin loads, and sub-component tests are to validate stiffness effect under loading, bulge effect under pressure respectively. For fatigue critical portion of complex structure with high pin loading, high stress concentration under complex loading, some component tests are conducted.

The crack growth characteristics for pressurized fuselage structure with bulge effect, stiffness effect, pin loading and some stress concentration is evaluated by sub-component tests of fuselage curved panel with skin splice, stringers, frames and stringer clips, simulating the load due to the combination of cabin pressure and axial loading in the case of fuselage bending.

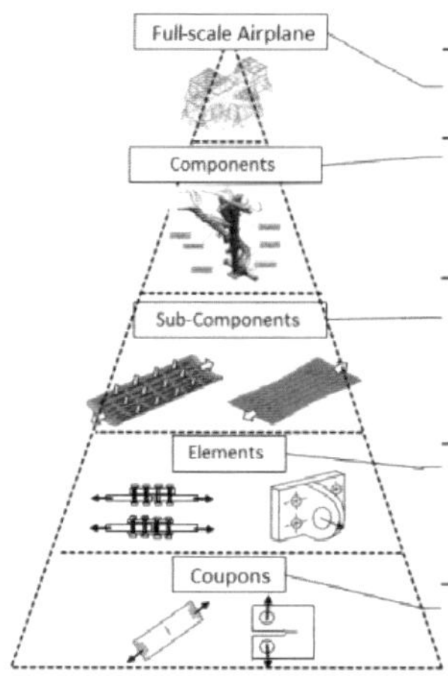

	Name	Number of specimens
Full-scale Airplane	- Full-scale Airplane Fatigue (Mainly for WFD)	1
Components	- Landing Gear (Safe Life) - Engine Mount - Flap Track etc.	2 1 1
Sub-Components	- Fuselage Curved Panel - Fuselage Door Surround - Wing Panel and Spar - Wing Side of Body Joint etc.	Dozens
Elements	- Splice Joint - Lug - Pin and Bolt etc.	Dozens
Coupons	- Crack Growth - Fracture Toughness	Hundreds

Figure 3. MRJ-200 Fatigue test building block.

CURVED PANEL TEST FIXTURE DEVELOPMENT

For the WFD and damage tolerance evaluation for MRJ-200 fuselage, the test fixture for sub-component level tests has been developed to demonstrate fatigue cyclic loading due to the cabin pressure and axial load in fuselage bending. (Figure 4)

To simulate hoop load due to the cabin pressure, one typical testing way is a full scale barrel test. Instead of a full-scale barrel structure, some companies and laboratories had developed other test fixtures to use a curved panel specimen to reduce the cost of test specimen. And in many cases the normal cabin pressure loading is applied using positive (bursting) pressure. (Example Ref. [1], [2])

For MRJ-200 structural test program, a new test fixture has been developed to conduct many test menus of fuselage fatigue and damage tolerance with reducing the cost and schedule time.

Figure 4. Curved panel test fixture for MRJ-200.

As main focus of this test fixture concept, negative pressure loading is used to simulate differential pressure load between inside and outside of cabin compartment. The test article of curved panel is put on pressure box, and the inside of the box is depressurized by pump in a vacuum. At the edge of the pressure box, rubber seal is attached to keep airproof of pressure box. (Figure 5) The fuselage typical structure consists of skin with skin pockets, stringers, stringer clips and frames attached by fasteners. The test specimen of fuselage curved panel is put on pressure box with inside on upper side. The skin outer surface of fuselage curved panel contact to rubber seal. Therefore, rubber seal can contact skin outer surface without any interference with inside structures. And pressure can be applied on the panel without termination of inside frames or stringers, which may induce offset moment of the evaluated skin panel portions. And larger area of curved panel can be evaluated under hoop stress.

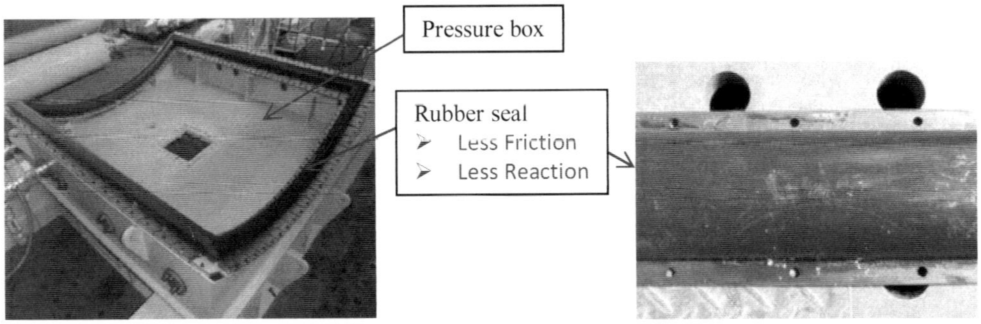

Figure 5. Rubber seal on pressure box.

Negative pressure system of test fixture is shown in Figure 6. Hoop load due to pressure is reacted by circumferential supporting rod attached to frames. To avoid reaction by rubber seal, supporting rod connected to frames will be adjusted. To avoid friction on rubber seal due to panel deformation, lubricating grease is applied. In addition, to avoid moment by fixed at rubber seal attached, seal stiffness is enough low.

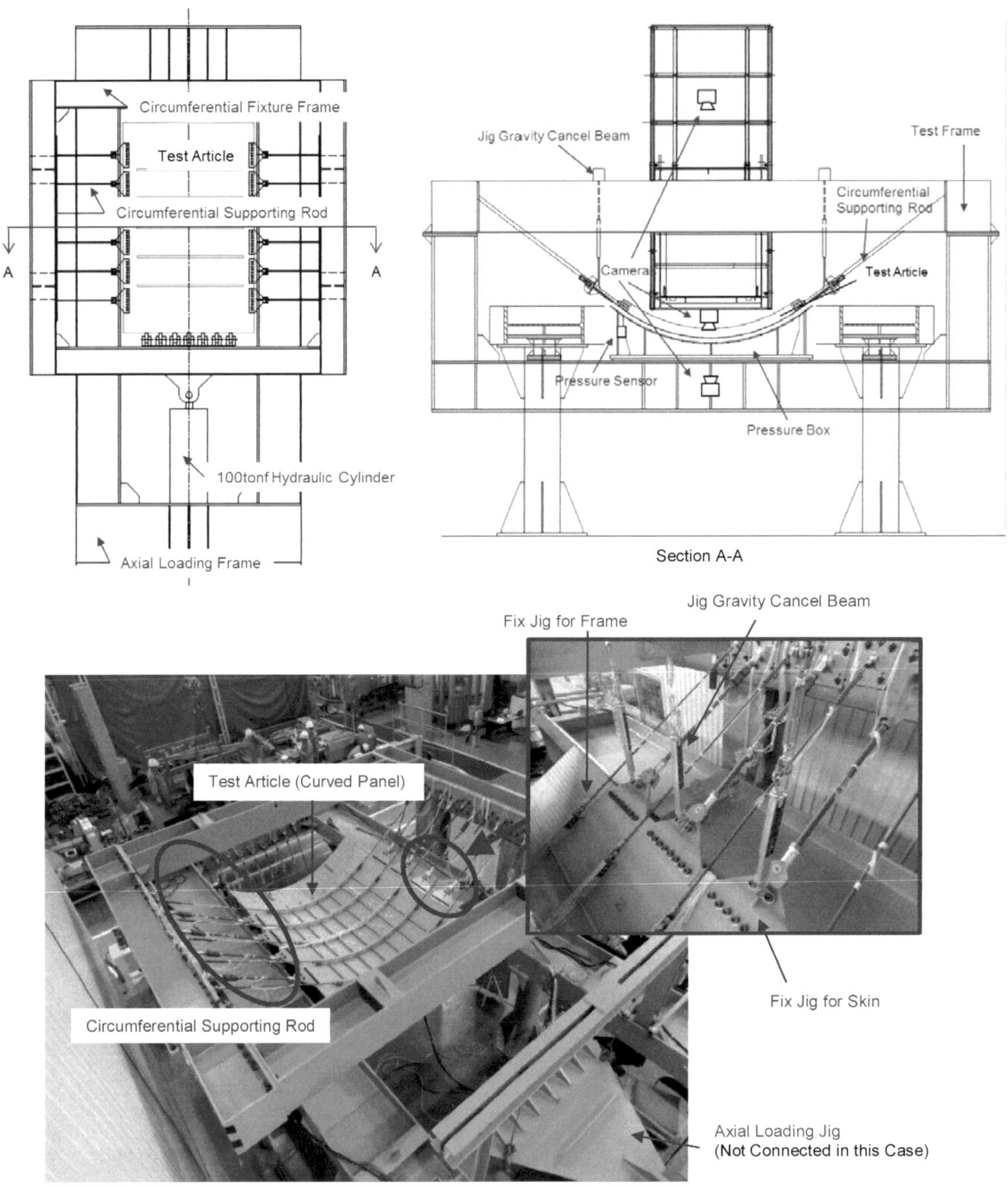

Figure 6. Negative pressure system of test fixture.

This fuselage curved panel fatigue test fixture using negative pressure takes some benefits.

1) First one is test loading speed by optimized pressure box (minimized volume).

The test specimen of fuselage curved panel put on pressure box with inside on upper side. There is no structure part in the pressure box. And inside pressure box is filled with styrofoam to minimize the air volume as shown in Figure 8. To depressurize from pressure box by pump in a vaccume, the large volume of vaccum air in reserver tank is used instead of atomasphere to reduce time. The depressurization system is showin in Figure 9.

For operation pressure level of typical regional jet, one negative pressurization can be applied per only six seconds. (1/6 Hz)

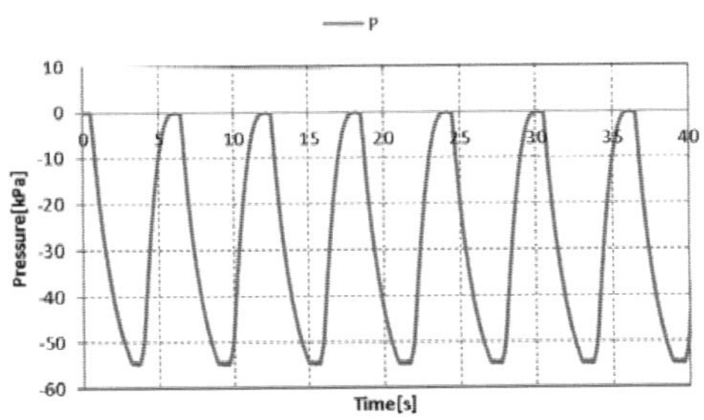

Figure 7. Pressure time history.

Figure 8. Inside box is filled with Styrofoam.

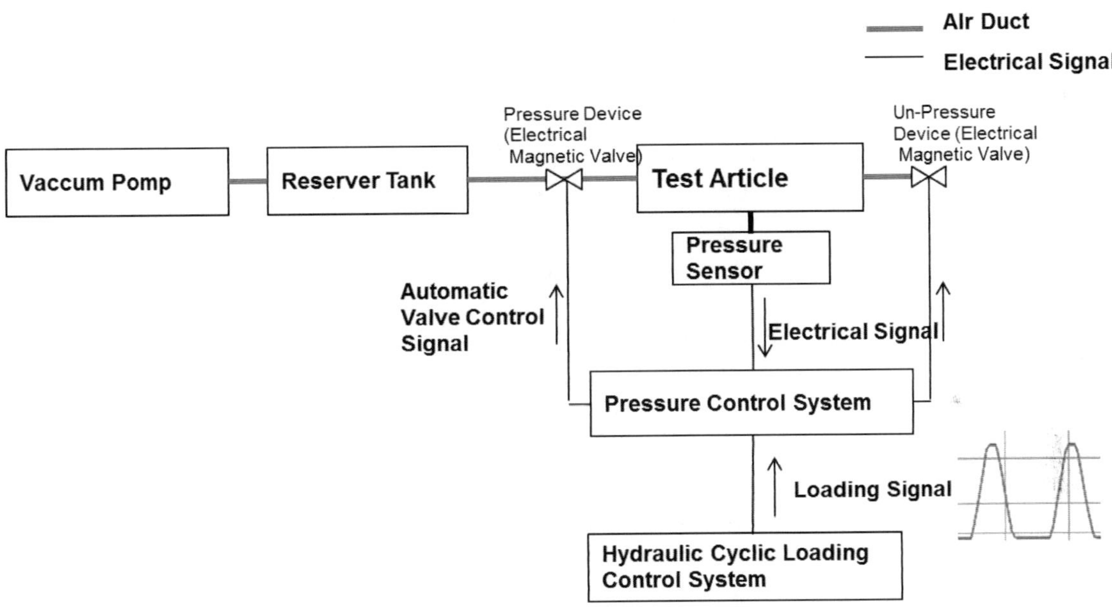

Figure 9. Loading system.

2) Second one is safety.

This test fixture applies negative pressure system. In principle, the pressure higher than atmospheric pressure can't be applied on test panel, therefore test panel has less risk for over pressure and dangerous bursting.

In the fatigue/damage tolerance test, crack can be safely observed in close proximity under pressurized condition. Crack during loading is open and detectable easily. Therefore, fatigue natural crack can be found easily by detail visual inspection. In addition, test does not need downtime for inspection or measurement of crack propagation.

3) Monitoring System

To monitor crack initiation/propagation easily, monitoring provision is prepared on this test fixture. To access test article easily, the monitor bridge with some access panel is installed above the test article. To monitor test article of pressure box inside area, access window and panel is installed on the pressure box. Through these access panel and window, test article can be observed from both side at the extremely close distance.

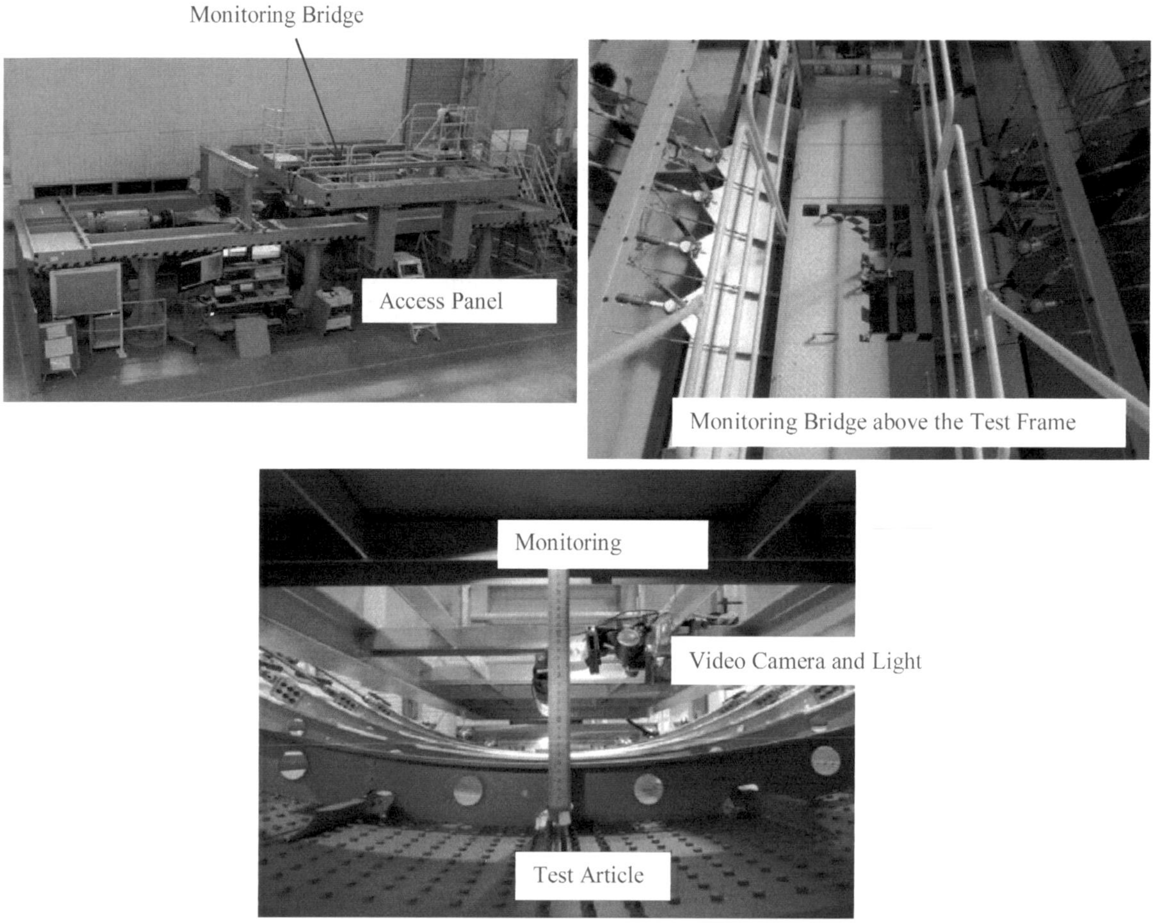

Figure 10. Access to test article from upper side.

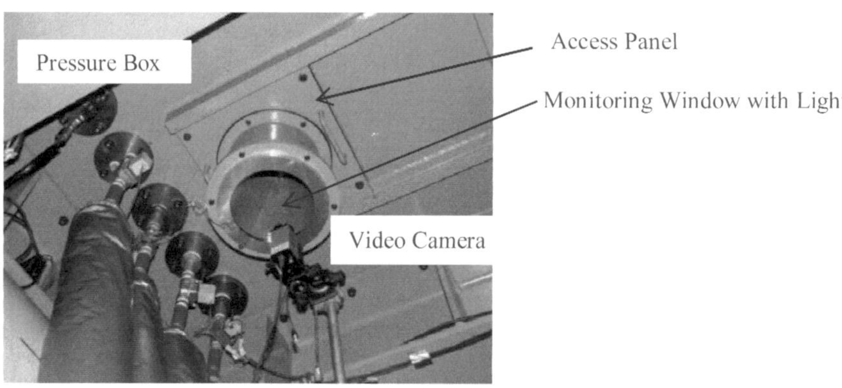

Figure 11. Access to test article from bottom side.

To obtain crack propagation data, video camera is set on monitoring bridge. This data can be obtained with high degree of accuracy without downtime. This monitoring system gives us the benefit of crack observation could be significantly shortened.

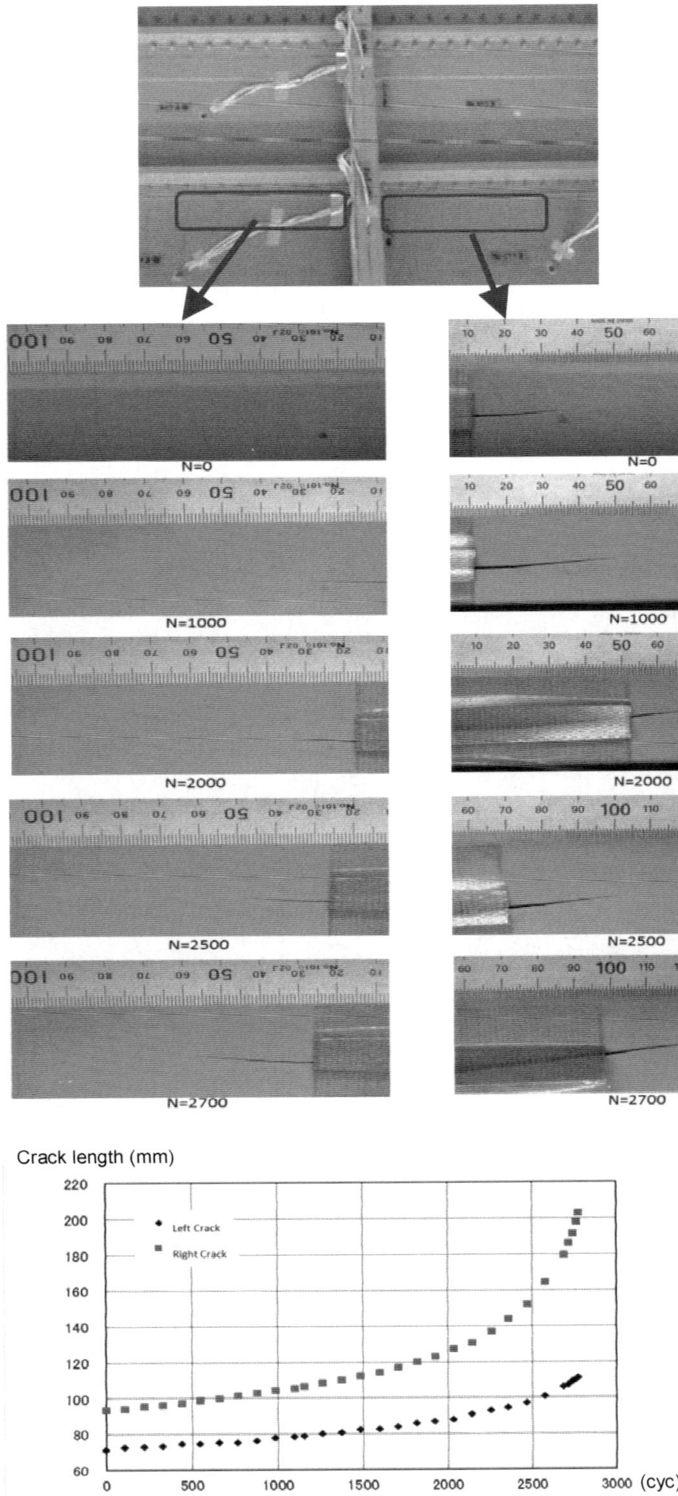

Figure 12. Crack propagation Data obtained from monitoring system.

TRIAL AND TEST RESULT – MRJ-200 FUSELAGE FATIGUE TEST

In early activity, a fuselage fatigue skin panel test for MRJ-200 was completed.

In the test fixture development phase, the stress level and deformation on the curved panel is compared between barrel model simulating actual fuselage and curved panel test model by FEM analysis. (Figure 13) In the curved model, circumferential supporting rod, skin fixed jig, rubber seal are also simulated in this model. As a result in comparison, almost same stress level and same deformation on the skin panel can be obtained. In the strain survey of the fatigue test, circumferential supporting rod, rubber seal stiffness are coordinated to obtain similar stress level with FEM analysis. After some trial, similar stress level can be simulated within small difference. This curved test fixture with negative pressure system can simulate hoop stress and bending stress on fuselage panel under cabin pressure very well.

To evaluate WFD on fuselage panel under cabin pressure in early stage activity, cabin pressure loading corresponding to 3lives (240,000 cycles = 80,000 cycles x 3.0) is applied. This fatigue test was completed in one month.

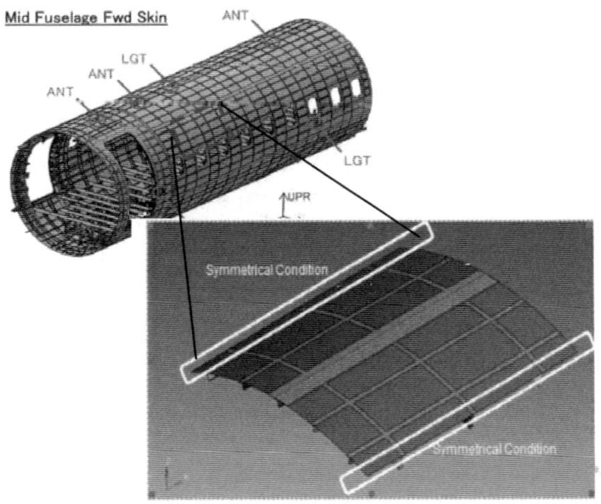

Barrel Model

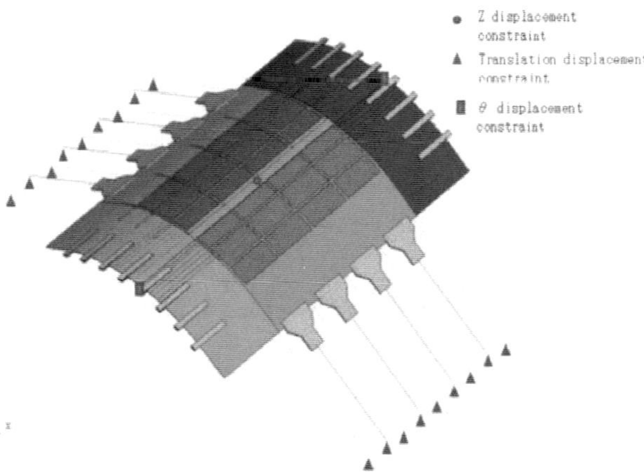

Curved Panel Test Model

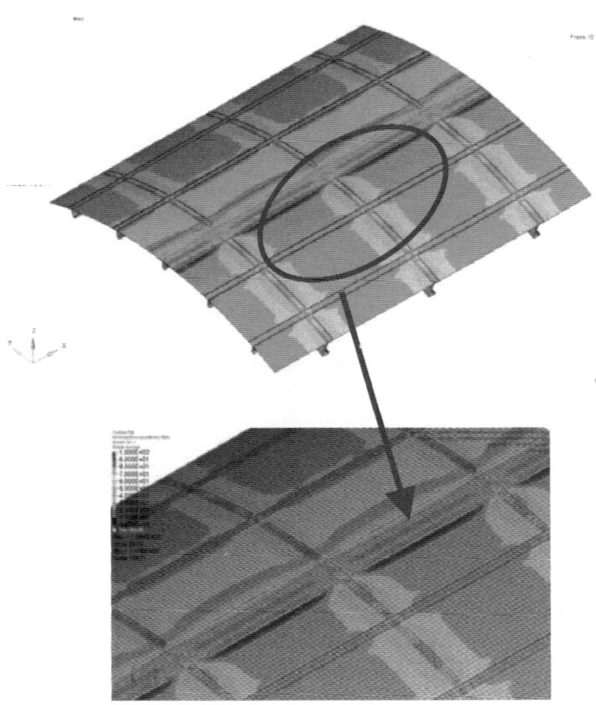

Stress Contour Plot - Barrel Model

Stress Contour Plot -Curved Panel Test Model

Figure 13. FEM analysis stress level comparison – Barrel model vs. Curved panel model.

Figure 14. Test article – MRJ200 fatigue test.

FATIGUE AND DAMAGE TOLERANCE TEST PLAN FOR MRJ-200 USING CURVED PANEL TEST FIXTURE

Using this test fixture, several curved panel of fuselage fatigue critical area (skin splice, stringer joint, cutout surrounding, window surrounding etc.) and repair configurations are planned to test as early substantiation activity for WFD and damage tolerance evaluation.

CONCLUDING REMARKS

In MRJ-200 development phase, an experimental and analytical investigation has been conducted and test fixture for curved panel has been developed successfully. Using this test fixture, large numbers of tests are continued to evaluate damage tolerance in MRJ-200 certification activities. To evaluate more complex configurations with more complex loading, test fixture will be modified and improved accordingly.

REFERENCES

[1] Samavedam G. and Hoadley D., "Fracture and Fatigue Strength Evaluation of Multiple Site Damaged Aircraft Fuselages - Curved Panel Testing and Analysis", DOT-VNTSC-FAA-93-8/ DOT/FAA/CT-94/10, January 1994

[2] Bakuckas J., "Full-Scale Testing and Analysis of Fuselage Structure Containing Multiple Cracks", DOT/FAA/AR-01/46, July 2002

Analytical Modelling of Mechanical Joint for New Monitoring Method

Shigeru Machida, Takao Okada

Japan Aerospace Exploration Agency

Abstract: In order to reduce structural risk, a proactive, rather than reactive, approach should be adopted such as structural monitoring during all life of aircraft. A new method for characterizing degradation and damages inside a mechanical joint from the information of structural damping, a non-linear response against cyclic loading, was presented as an unique monitoring concept for structural joints. Addition to that, the new analytical model which can express structural damping characteristics of riveted lap joints was proposed. The cyclic load tests with single lap joint specimens which has typical three rivet rows were conducted under load control using a constant amplitude sinusoidal waveform. The test results showed the unique relationship between energy dissipation per cycle and number of cycles qualitatively. The new equations for the displacement as the function of tensile load were established as a new analytical model for rivet joints. Using the results from finite element analyses and the load-deflection hysteresis loops from cyclic loading tests with additional specimens, it was shown that the coefficients of the new analysis model can be used for characterizing degradation inside mechanical joints. By using the proposed mathematical joint model and improved analysis method, it is possible to quantify the non-linearity of force/displacement hysteresis loop and to identify the coefficients which can explain condition on the faying surface of shear lap joint. Experimental results of join specimens using round-head rivets indicate that the unique relationship between D and number of cycles N and the trend can be explained in four phases as the previous investigation with countersunk rivet specimens. In comparison with the results of countersunk rivet specimens, the range of changing of energy dissipation value is relatively small.

INTRODUCTION

Structural joins of aircraft structures have a significant role to characterize static and fatigue strength, damage tolerance, stiffness, vibration and structural damping. From a structural dynamics analysis point of view, it is needed to determine structural damping characteristics of airframe carefully for evaluating structural dynamics behaviours. Historically, structural damping values are determined based on data of experiments or existing aircrafts which have similar structural layout and same materials. For a new aircraft which has new materials and unique structural layout, it is necessary to establish a method to estimate structural damping characteristics. Regarding a fatigue evaluation of aircraft structures, "mechanical joint" is one of the susceptible areas to be designed carefully for preventing catastrophic failure, such as widespread fatigue damage. However, an inspection method for widespread fatigue damage is not established yet. In order to reduce structural risk, a proactive, rather than reactive, approach should be adopted such as structural monitoring during all life of aircraft. In a previous paper [1], a new method for characterizing degradation and damages inside a mechanical joint from the information of structural damping, a non-linear response against cyclic loading, was presented as an unique monitoring concept for structural joints. Addition to that, the new analytical model which can express structural damping characteristics of riveted lap joints was proposed.

The term of "Structural Damping" means the energy dissipation properties of structure and it is well known that interfacial slip phenomena at structural joints are the primary source of "Structural Damping", which may be affected by types of joint and conditions of faying surfaces. Therefore, it was strongly expected that structural damping properties can show an important information of conditions on faying surfaces such as fretting wear, corrosion and cracking. Several riveted lap joint specimens were tested under constant cyclic loading. Specimens can represent the part of circumferential single lap joint which have typical three rivet rows. The cyclic load tests were conducted under load control using a constant amplitude sinusoidal waveform. Stopping cyclic loading at specific cycles, an extensometer was installed at the side edge of specimen to measure deflection between two points over the rivet joint under loading. The test results showed the unique relationship between energy dissipation per cycle and number of cycles qualitatively.

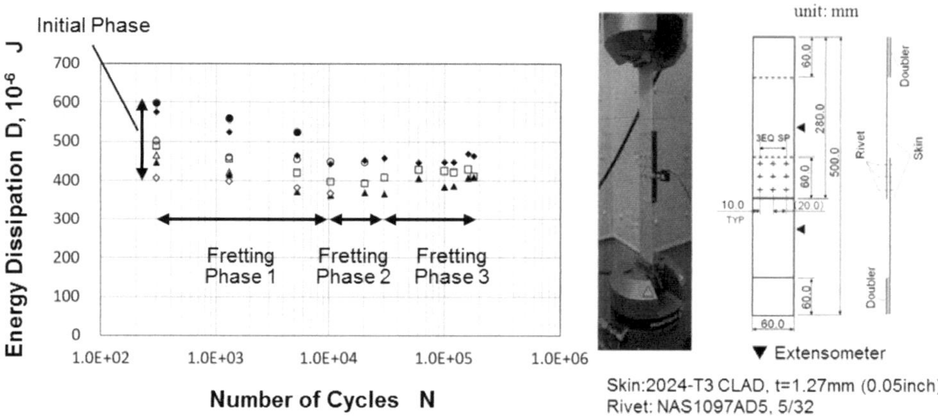

Figure 1. Experimental Results of Energy Dissipation per Cycle.

An analytical joint model was studied by Metherell and Diller[2] who investigated an energy dissipation of single lap joint. Their works gave equations which can express a shape of load-deflection hysteresis loop and energy dissipation per cycle under uniform clamping pressure. However the equations could not consider joint stiffness and a load transfer of stick portion. In order to establish governing equations for single lap joint with rivets which has stick part and slip part to transfer loads, the stiffness of stick portion, K_{ST}, was introduced. The new equations for the displacement as the function of tensile load were established as a new analytical model for rivet joints. From experimental results, it is found that the analytical model can quantify the non-linearity of force/displacement hysteresis loop and its coefficients can explain condition on the faying surfaces of shear lap joint.

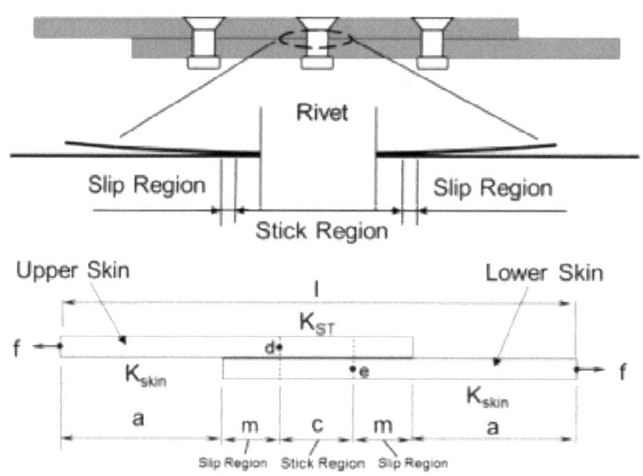

Figure 2. Simplified Analytical Joint Model.

In our previous works, it was assumed that the stiffness K_{ST} was tow times as large as that of skin for simplification. The basis of this assumption is that the stick portion was modelled as doubled thickness of skin in Figure 2. However actual rivet joint has a rivet stiffness and rotation caused by eccentricities in the neutral line of skins. In this paper, improvements of calculation method to analyse coefficients of the proposed analytical model is presented. Using the improved method and the load-deflection hysteresis loops from cyclic loading tests, unknown coefficients of the proposed analysis model are discussed. The experimental investigation with a joint specimen using round-head rivets is also presented.

ANALYTICAL MECHANICAL JOINT MODEL

Analytical Joint Model

One of the final goals of this study is to establish a method to estimate structural damping of full-scale aircraft before conducting its ground vibration tests. Because of the complexity and huge number of joint elements of aircraft structure, a finite element analysis is practical to determine structural damping of aircraft. In order to accomplish that goal, the analytical joint model which can describe distributed joint properties, such as properties per unit length rather than individual small joint, is needed from calculation cost point of view and it is prefer to identify coefficients of the analytical model from experimental data using small structural element specimens.

Metherell and Diller[2] investigated an energy dissipation of single lap joint. Their work gave equations which can express a shape of force/displacement hysteresis loop and energy dissipation per cycle under uniform clamping pressure in a shear lap joint. Their joint model is simplified and practical from engineering point of view. However, in their simplification process, a load transfer at the stick region, which includes rivet and narrow area around rivet, was ignored. It is not realistic because stick region including rivet itself can carry a part of shear load. Here, new mathematical joint model is developed considering the shear load at stick region of a shear lap joint.

In order to establish governing equations for single shear lap joint with rivets which has stick part and slip part to carry loads, the joint is simplified as upper skin and lower skin with the joint region with slip and stick area under constant pressure which represents the clamping force by rivets (Figure 3). The stiffness of stick region including rivet, K_{ST}, is introduced. It is assumed that the length of slip region is proportional to axial force in the skins and shear force per unit length on slip region is express as the product of width of joint, friction coefficient and uniform clamping pressure in the same manner as the Metherell and Diller model.

The displacement of the stick region u_{stick} is given by

$$u_{stick} = \frac{f_{stick}}{\frac{K_{ST}}{c}} \tag{1}$$

where f_{stick} and c are axial force and length in stick region respectively. The length of slip region m and the shear force per unit length in slip region q are given by

$$m = \alpha f \tag{2}$$

$$q = w\mu p \tag{3}$$

where α is the proportionality constant in equation (2) and w, μ, p are width of joint, friction coefficient and uniform clamping pressure of lap joint respectively.

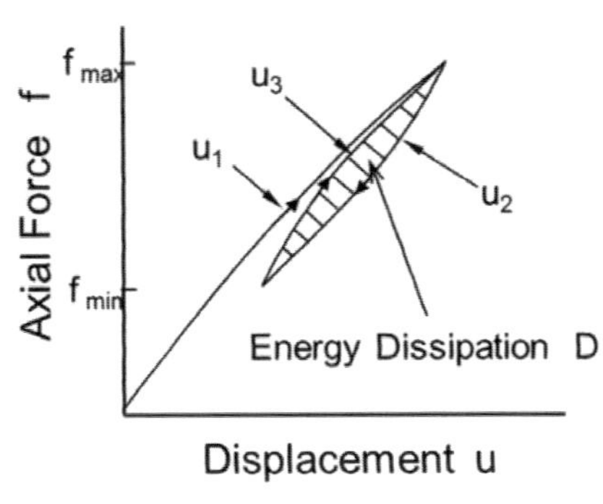

Figure 3. Joint Force/Displacement Hysteresis Loop.

In general, the force/displacement relation in a shear lap joint subjected to axial loading is described as a hysteresis loop in Figure. 3. The two loading and one unloading phases are considered. In the initial loading phase, axial force is increased from zero to maximum force f_{max}. The second and third phases are the unloading condition to minimum force f_{min} and the loading condition toward f_{max} respectively. The total displacement between both loaded ends of a lap joint, u_i, is formulated by summing up the displacement of each segment. The total displacements in three phases are given by

In the initial loading phase;

$$u_1 = C11f^2 + C12f + C13 \quad (4)$$

In the unloading phase;

$$u_2 = C21f^2 + C22f + C23 \quad (5)$$

In the loading phase;

$$u_3 = C31f^2 + C32f + C33 \quad (6)$$

The coefficients in the above equations, which are functions of α_i and q_i for u_i (i=1, 2, 3), express the characteristics of the force/displacement hysteresis loop as follows.

C11, C21and C31: Curvature of the loop or the non-linearity due to energy dissipation.

C12, C22, C32, C13, C23 and C33: Linear part of the loop or joint stiffness at zero axial force.

Identification of the Analytical Joint Model

The coefficients, C11~C33, can be identified by using a least-squares fit of the hysteresis curves from coupon tests. Energy dissipation D can be calculated by the equation (5)-(7) with the identified coefficients, C21~C33.

$$D = \int_{f_{min}}^{f_{max}} (u_2 - u_3) df \quad (7)$$

In the case of $\alpha_2=\alpha_3=\alpha$, $q_2=q_3=q$ and $K_{skin}=K_{upper\ skin}=K_{lower\ skin}$, D is given by Eq. (5)-(7) as equation (8), which shows that the energy dissipation D is proportional to the cube of the range of joint load.

$$D = \frac{2\alpha q}{3K_{skin}} (f_{max} - f_{min})^3 \quad (8)$$

In order to determine condition on the faying surface quantitatively, the coefficients of equations, α_i and q_i, are identified by using experimental data of force/displacement relation. As described above, C11~C33 are identified by curve fitting to the force/displacement data. The coefficients for u2 and u3 are given by

$$C21 = \frac{-2\alpha_2{}^2 q_2}{K_{skin}} \quad (9)$$

$$C22 = \frac{2}{K_{skin}} (a + \alpha_2 f_{max}) + \frac{1}{K_{ST}} (1 - \alpha_2 q_2)(l - 2\alpha_2 f_{max}) \quad (10)$$

$$C31 = \frac{2\alpha_3{}^2 q_3}{K_{skin}} \quad (11)$$

$$C32 = \frac{2}{K_{skin}} \{a - \alpha_3{}^2 q_3 (f_{max} + f_{min}) + \alpha_3 f_{max}\} + \frac{1}{K_{ST}} (1 - \alpha_3 q_3)(l - 2\alpha_3 f_{max}) \quad (12)$$

The coefficients, α_i and q_i, can be derived fromC21, C22, C31and C32. In the previous study, α_i is a coefficient of equation (2) and assumed constant during one cyclic loading. The coefficient q_2 and q_3 defined by equation (3) are used for u_2 and u_3 respectively. Since the skin thickness of stick region is doubled by that of single skin, K_{ST} is given by

$$K_{ST} = 2K_{skin} \quad (13)$$

From equation (9), (10) and (11), α, q_2 and q_3 are determined mathematically.

For the discussion on the results of experiments and calculations, more interpretations to coefficients are described as follows. The proportionality constant in equation. (2), α, means elongation of the slip region per unit axial force (i.e., growth rate of slip region according to axial force). qi are friction forces per unit length of slip region in loading and unloading phase and affected by condition of faying surface including micro-contact area and fretting wear. αq_i is friction force per unit axial force or in other words, what percentage of axial force is carried by friction force at slip region. $\alpha^2 q_i$ expresses the rate of energy dissipation due to a given axial force.

Improvements of calculation method

It was assumed that the stiffness K_{ST} was tow times as large as that of skin for simplification as equation. (13). The basis of this assumption is that the stick portion was modelled as doubled thickness of skin in Figure 2. Since actual rivet joint has a rivet stiffness and rotation caused by eccentricities in the neutral line of skins, it is better to determined K_{ST} from experimental data. Therefore experimental stiffness data of u_3 curve at f_{min}, where there is no slip area, is used. From equation. (6), tangential line at f_{min} is given by

$$\frac{du_3}{df} = C31 \times f_{min} + C32 \tag{14}$$

From equation. (6), (11) and (12), u_3 with no slip area, where α is zero, can be given by

$$u_3 = \left(\frac{2a}{K_{skin}} + \frac{l}{K_{ST}}\right) \times f \tag{15}$$

C_{31}, C_{32} and K_{skin} are known and K_{ST} is given by equaton. (16).

$$K_{ST} = \frac{l}{(C31 f_{min} + C32) - \frac{2a}{K_{skin}}} \tag{16}$$

From equation. (9), (10), (11) and (12), α_2, α_3, q_2 and q_3 are determined mathematically.

ANALYSIS AND DISCUSSION

The analyses to determine those coefficients are conducted to evaluate the condition on the faying surfaces using the proposed mathematical joint model. Figure 4 (a)–(d) show the typical results of changes in α_i, $\alpha_i q_i$ and $\alpha_i^2 q_i$, where i is 2 as loading or 3 as unloading, with number of loading cycles N respectively along with the energy dissipation D.

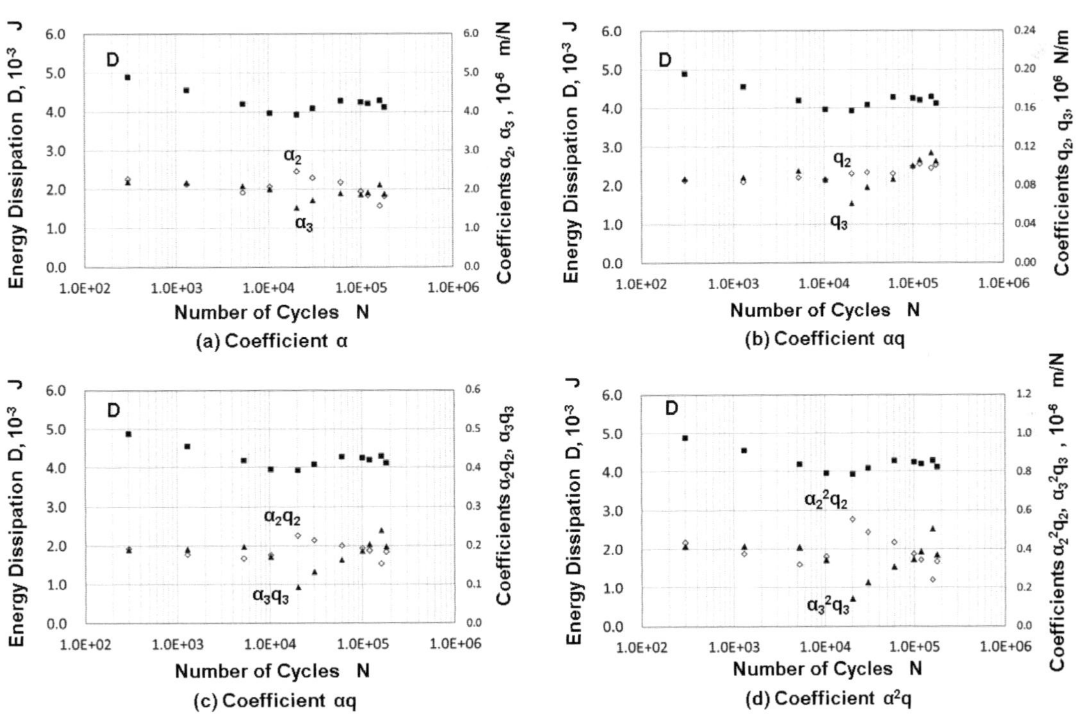

Figure 4. Identified Coefficients of Mathematical Joint Model.

In the fretting phase 1, from 0 to 10,000 cycles, α_i and$\alpha_i q_i$ are decreasing. In contrast, the value of q_i shows few changes. Those behaviours explain the decrease in friction force with N below 10,000 and support the results of the experimental investigation for the fretting phase 1, "In such metal-to-metal contacts, the area of micro-contact on faying surface increases and area of slip region decreases coincidentally."

As the previous investigation, there is an important finding that the values of $\alpha_2 q_2$ are different from that of $\alpha_3 q_3$ with N beyond 10,000 cycles. That difference is caused by the asymmetry of the force/displacement hysteresis loop. The analytical results indicate that friction force during loading shows the different trend from that during unloading. Since the increase of $\alpha_3 q_3$ is predominant over the decrease of $\alpha_2 q_2$, D shows an increase in the fretting phase 3, beyond 30,000 cycles.

By using the proposed mathematical joint model and the improved analysis method, the coefficients of α_i, , q_i, $\alpha_i q_i$, $\alpha_i^2 q_i$ can be determined and it is possible to quantify the non-linearity of force/displacement hysteresis loop.

EXPERIMENTAL INVESTIGATION

In the results of previous investigation [3], rivet joint specimens using countersunk rivets show more complex structural damping response relatively rather than round-head rivets. In order to eliminate such complexity using countersunk rivets, cyclic loading tests were conducted using three-row riveted specimens with round-head rivets (MS2047AD5) which has the same dimension as Figure 1.

The tests were conducted under load control using a constant amplitude sinusoidal waveform at a frequency of 3Hz. The stress applied to the specimens was 83.5Mpa at R=0.1 assumed to be the same level as hoop stress of fuselage skin under cabin pressure condition. The total number of cycles on each specimen was varied from 1,000 to 180,000 to investigate a change of energy dissipation. The test environment was laboratory air at ambient room temperature. Stopping cyclic loading at specific cycles, an extensometer, gage length 140mm, was installed at an edge of each specimen to measure displacement between two points over the rivet joint and cyclic loading was applied at a frequency of 1Hz for measuring force and displacement.

The results, Figure 5, indicate that the unique relationship between D and number of cycles N and the trend can be explained in three fretting phases as the previous investigation with countersunk rivet specimens as shown in Figure. 1. In comparison with the results of countersunk rivet specimens, Figure 1, the range of changing of D value is relatively small.

Figure 6 shows the difference of slipping mechanism between countersunk rivet and round-head rivet. From force balance point of view, countersunk rivet has the possible slip area under head of countersunk rivet due to bending moment caused by eccentricities in the neutral line of skins, but round-head rivet. That is the main reason why energy dissipation of countersunk rivet specimens has lager range of changing rather than round-head rivet specimens.

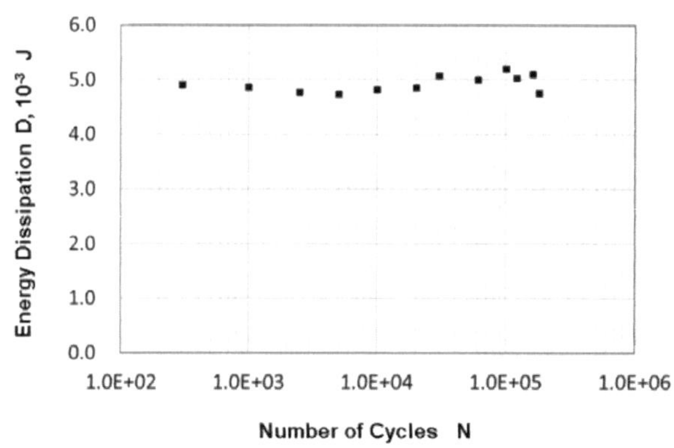

Figure 5. Energy Dissipation per Cycle with Round-head Rivet Specimens.

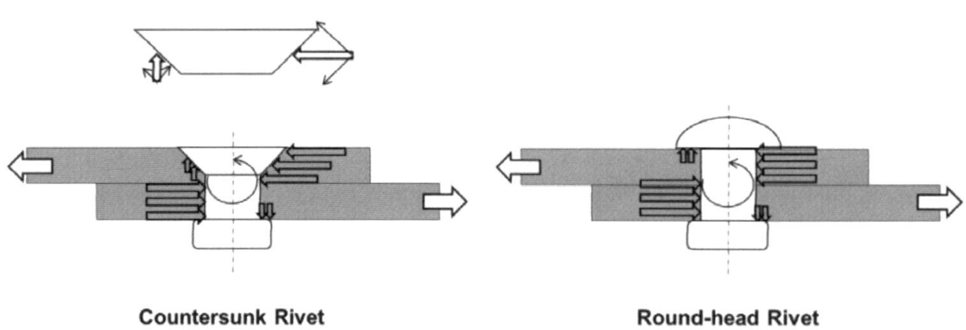

Figure 6. Force Balance Concept of Countersunk Rivet and Round-head Rivet.

CONCLUSION

In order to investigate a new method for characterizing degradation and damages inside a shear lap joint, attention will be given to the structural damping property. The following conclusions can be drawn from present study.

By using the proposed mathematical joint model and improved analysis method, it is possible to quantify the non-linearity of force/displacement hysteresis loop and to identify the coefficients which can explain condition on the faying surface of shear lap joint.

Experimental results of joint specimens using round-head rivets indicate that the unique relationship between D and number of cycles N and the trend can be explained in three phases as the previous investigation with countersunk rivet specimens.

In comparison with the results of countersunk rivet specimens, the range of changing of energy dissipation value is relatively small.

REFERENCES

[1] Machida, S. Kishishita. S. Ito, Y. Okada, T. and Asakawa, M. (2001). In: *ICAF2013 SYMPOSIUM*, Proceedings of the 27th ICAF Symposium, vol. II, pp. 797–807, Brot, A. (Ed.), Jerusalem, Israel.

[2] Metherell, A. and Diller, S. (1968). *Journal of Applied Mechanics*, vol. 35, pp. 123-128.

[3] Machida, S. (2015). In: Proceedings of 56th AIAA/ASCE/AHS/ASC Structures, Structural Dynamics, and Materials Conference, AIAA 2015-1637, Kissimmee, Florida, U. S. A.

Fatigue Life Assessment of PZL-130 Orlik Structure – Final Analysis and Results

Andrzej Leski, Marcin Kurdelski, Piotr Reymer, Krzysztof Dragan, Michał Sałaciński

Air Force Institute of Technology, Poland

Abstract: The following article summarizes the SEWST research program, which aim was to establish new fatigue life limit as well as to introduce damage tolerance based maintenance system for the modernized PZL-130 Orlik TC-II trainer aircraft structure. The whole program consisted of many complex tasks e.g. Operational Load Monitoring, Fatigue Load Spectrum Definition, Full Scale Fatigue Test, Teardown Inspection, Crack Propagation Calculations and definition of the Individual Aircraft Tracking Program.

Within the article authors briefly bring back the previously described tasks [1, 2] and focus on the recently finalized program stages.

Firstly the results of FSFT are presented showing the test progress in time as well as NDI findings and structural repairs which were carried out during the test course. Moreover a specially created database for managing the test results is presented.

Secondly the crack propagation estimations for defined Critical Points (CP) are presented. The methodology was based on the definition of geometry correction factor for defined CP using the Virtual Crack Closure Technique.

Thirdly the teardown Inspection results are presented. The aircrafts structure after the test was shipped back to Poland, where it was carefully disassembled in order to verify structural integrity of the components inaccessible during the test.

Finally the whole program is summarized and the goals completion discussed. After finalization of the planned tasks the AIT program was implemented and the fleet operation modified according to program results.

INTRODUCTION

Air Force Institute of Technology (AFIT) has been involved in the SEWST research program for the last five years (2010-2014). The program was contracted by the Polish MoD and led by EADS PZL-Okęcie, the manufacturer of the aircraft. AFIT was the main subcontractor therefore was responsible for most research activities. The total cost of the program was 7 mln USD whereas the estimated savings after introduction of the program are around 70 mln USD.

The main goals of the program were: reduction of operating costs, increase of accessibility of the PZL-130 TC-II aircrafts, cancellation of major overhauls every 1000 Flight Hours (FH), introduction of periodic inspections depending on actual usage of individual aircrafts.

The main task of the structural part of the program was to carry out the Full Scale Fatigue Test (FSFT). In order to achieve this goal a set of subtask had to be performed. Beginning from definition of the flight profile of PZL-130 aircrafts operated in the Polish Air Force [1], followed by the Operational Load Monitoring Program the load spectrum for the FSFT was defined [2]. During the OLM as well as the FSFT two aircrafts were instrumented with sets of over 100 strain gauges. The aircraft used in FSFT was a withdrawn from service TC-I version overhauled to TC-II version by reinforcing the fuselage and installing a brand new pair of modernized wings. The TC-II version wings have larger span as well as wing area. The overhaul which was carried out to the FSFT test specimen is identical to the modernization which all the TC-II aircrafts undergo.

The FSFT was carried out in Výzkumný a Zkušební Letecký Ústav (VZLU) in Czech Republic. The aircraft was loaded by means of 20 hydraulic actuators [3, 4] which were exerting forces on the main structural elements: 4 actuators for each wing (3 for bending along wing span one for wing torque), 1 actuator for each side of the horizontal tail plane, 1 for vertical tail plane, 3 actuator for each of the main landing gears, 1 for each of the two pilot seats and 1 final for the engine dummy mount.

FULL SCALE FATIGUE TEST SUMMARY

General results

The FSFT started in autumn 2011 and was contracted for 36000 Simulated Flight Hours (SFH). The load spectrum used in the test represented 200 SFH including landing and taxing loads. The loading frequency, achieved after initial tuning of the system, was around 0.5 Hz which for load spectrum consisting of 24 000 load lines allowed to accomplish 1000 SFH every 15 work days excluding NDI test and holidays. Figure 1 shows the general progress of the test throughout time.

Blue diamonds on the graph represent NDI inspections (both more often visual as well as planned inspections using more sophisticated equipment) whereas the thick red line defines the contracted test limit. Figure 1 shows that despite the initial tuning problems and test stops due to occurrence of fatigue damage around 10 000 and 19000 SFH, which necessitated introduction of time consuming repairs, the test course was rather stable which allowed to finalize it close to the estimated deadline. The test ended at 35056 SFH due to critical lower wing spar damage, which repair was both hard to carry out and economically unjustified due to close FSFT limit.

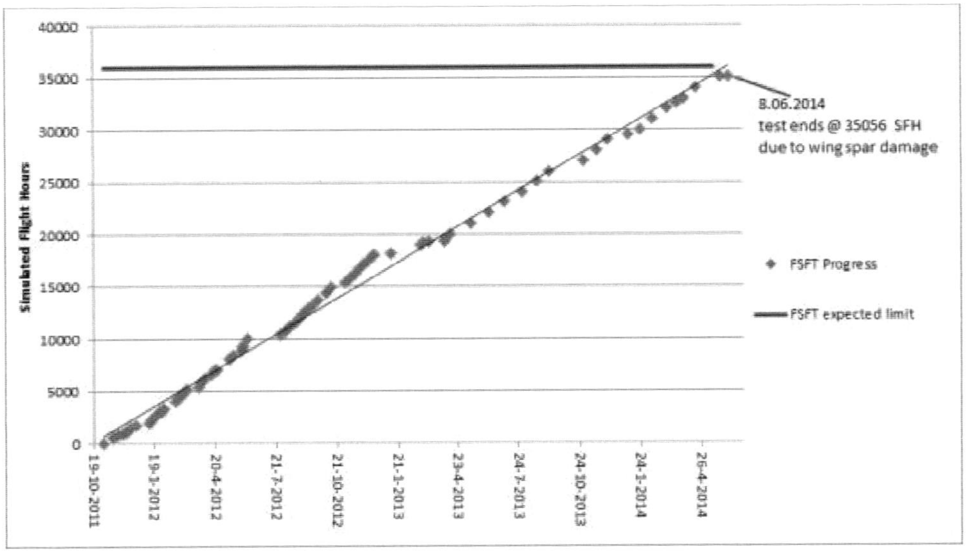

Figure 1. Full Scale Fatigue Test progress chart.

Fatigue damages

During the Test a series of Non Destructive Inspections (NDI) were carried out in order to find fatigue damages before they reached critical sizes. This allowed to check crack propagation rates as well as to undertake necessary mitigation actions in order to proceed with the test. According to the NDI program presented in [2], three types of inspections were carried out:

- Level 1 – every 1000 SFH – basic visual inspection (limited access),
- Level 2 – every 5000 SFH – eddy current and ultrasonic inspection (wing clamps detached),
- Level 3 – every 10000 SFH – whole structure inspection (separation of wing and fuselage).

The general course of inspection program is shown in Figure 2.

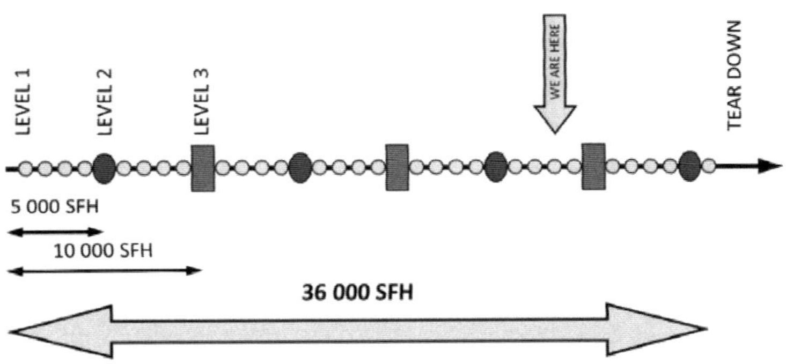

Figure 2. Scheme of NDI planned throughout the Fatigue Test.

Inspections carried out throughout the test allowed finding numerous damages, both cracks propagating in primary structure which introduced risk for the integrity of the aircraft's structure as well as small damages in non critical locations. Figure 3 shows NDI findings from the beginning of the test up to 27 000 SFH. Right Y axis corresponds with the red squares which indicate number of new damages found during consequent NDI. Left Y axis corresponds with the blue line showing the cumulated total number of damages found during the test. From the Figure 3 it can be clearly seen how significant the level of inspection was for the number of damages found. Although Level 1 visual inspections allowed finding most critical damages during the test Level 2, and even more Level 3 inspections, allowed to find many more damages which were undetectable during Level 1 inspections.

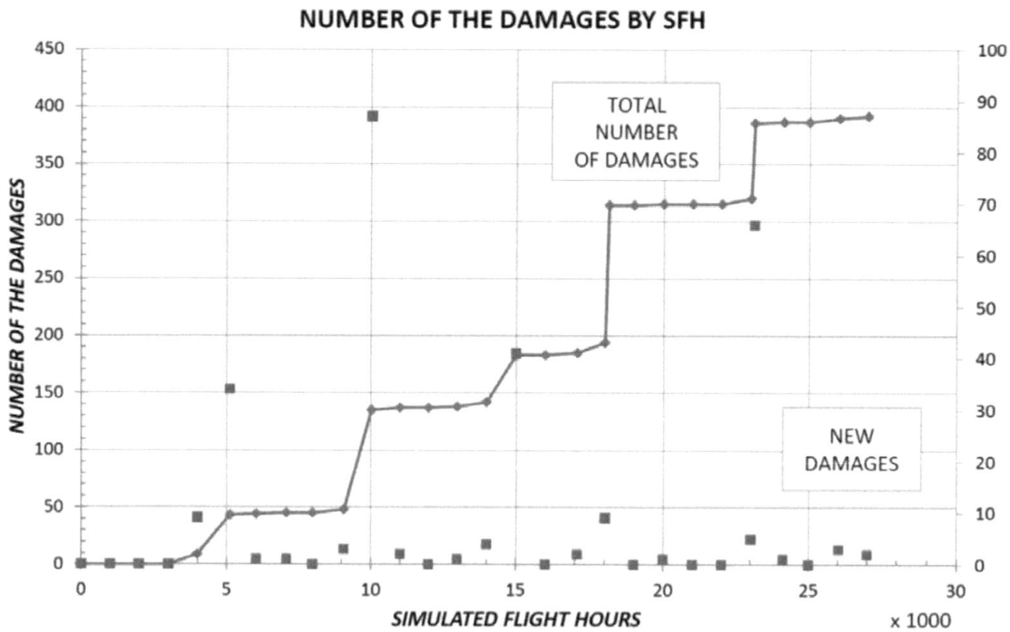

Figure 3. Chart showing damage findings till 27 000 SFH.

Damages found during inspections were located in different structural parts of the aircraft (Fig. 4). Most of the damages were found in the wing area both in the right (137 findings, 35% of total findings) and left (124 findings, 32% of total findings) wing parts. Total of 16 damages which make 27% of the total findings were detected in the fuselage area. Total of 21 damages were found in the horizontal stabilizer area and only 4 damages in the landing gears.

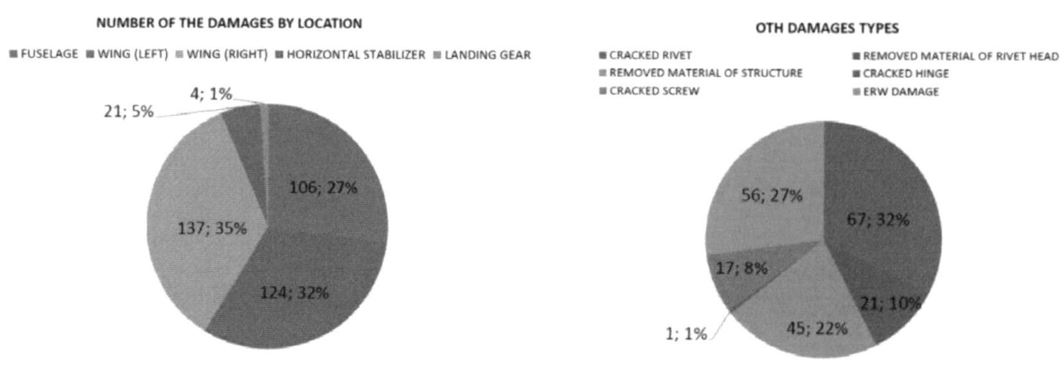

Figure 4. Distribution of found damages throughout the structure and by type of damage.

The damages also differed by the type, e.g. cracked rivets (32%), ERW (27%) removed material or structure (22%), removed material or rivet head (10%), cracked screw (8%).

Several types of damages are presented in the following figures. Figure 5 on the left show a set of damages found around 20 000 SFH inside the fuselage cutout on the rear right side. Damage number 237 was a through crack which probably had originated from the screw socket rivet propagating both to the flanged edge of the cutout strenghthening plateas well as toward the outer edge. Damages number 238 and 239 (found during the same inspection) are failed electrical welded joints. Right side of the Figure 5 shows crack which occurred in the element joining lower spars of the wing. The crack originated on the edge of the elements lower flange and propagated toward the upper flange. This damage was found during 19013 SFH Level 1 inspection and was replaced. Same damage was found on the right side of the replaced element during the 35 000SFH inspection.

Figure 5. Example damages found during the test. ERW and corner crack.

Rivets were one of the more often failing elements during the test. During inspections some damages were found in the early stage, when a crack was propagating within a rivet, some were found only when the rivet had fell of due to fast damage development. Left side of the Figure 6 shows a group of rivets located on the fuselage left front side. Damaged rivets were found during 29 500 SFH Level 2 inspection. Adequate rivets were defined as damaged on the right side of the fuselage during 33 000 SFH Level 1 inspection. Right side of Figure 6 shows the final damage number 818 that caused the test to terminate. The critical damage occurred at 35056 SFH although several smaller damages (failed rivets and small cracks) were detected in this region from 33 000 SFH.

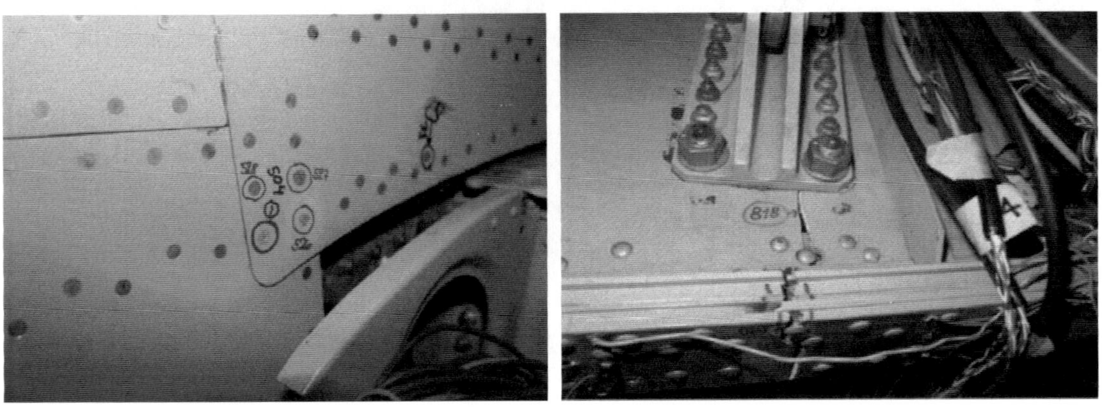

Figure 6. Example damages found during the test. Failed rivets and final damage.

Structural repairs during the test

Several repairs were introduced to the structure during the test. In general they can be divided into two categories: metal patch repairs and composite patch repairs.

The first type of repair was performed by the aircraft manufacturer EADS PZL-Okęcie. In general this type of repair was based on replacing or overlaying the damaged structural element with a specially prepared (bended and rivet holes drilled) metallic patch which was then riveted to the original structure. Such repairs were introduced for an example to the fuselage sides as well as wing upper skin, where due to friction of the wings upper surface on the fuselage wing cutout abrasive damage occurred. Examples of metallic patch repairs are presented in Figure 7.

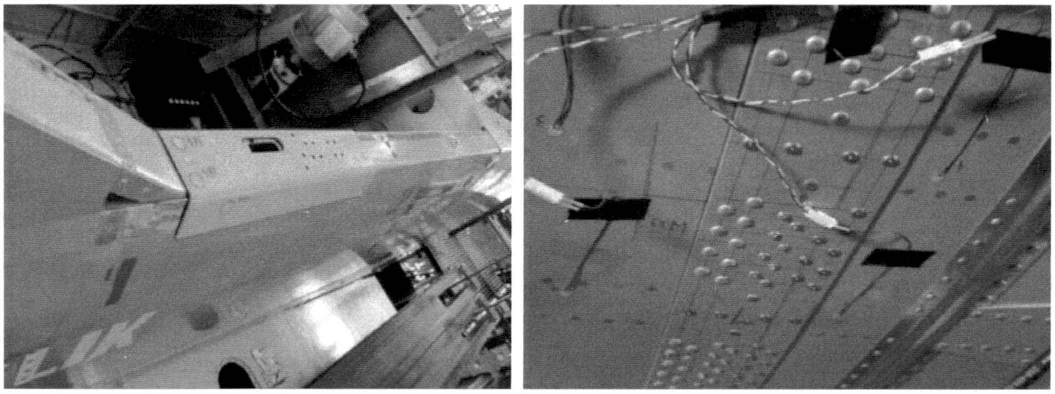

Figure 7. Example of metal overlay repairs performed by EADS – fuselage side and wing surface.

The other type of repair used composite prepreg materials. Two types of composite materials were used: Boron Fiber Reinforced Plastic (BFRP) and Carbon Fiber Reinforced Plastic (CFRP). Two patches, each composed of one type of fibers, were introduced to the left and right wing lower surfaces as shown in Figure 8, where 20 mm cracks were discovered propagating from rivet holes.

Figure 8. Composite bonded repairs. Boron fiber patch on the left and carbon fiber patch on the right [5].

The pre-preg patches were designed and prepared prior to the installation. Each patch composed of set of layers. Before installation the aircrafts surface had to be adequately prepared by means of mechanical and chemical treatments. Finally the patches were installed using Heatcon heat blanket which guaranteed even heat and underpressure distribution throughout the surface. NDI tests performed on coupon tests [5] allowed evaluate the reliability of available NDI techniques in monitoring crack propagation underneath the patches throughout the rest of the test. Test results allowed determining that the thermography method would be the most suitable for detecting disbands both for CFRP and BFRP patches for the thickness used. Moreover the automated eddy current inspections allowed determining the actual crack size under the patch.

Teardown inspection

After finalization of the test the final NDI was performed by VZL and the final report issued after which the test rig was disassembled and the test specimen taken out from the test rig and shipped back to Poland (Fig. 9).

Figure 9. Disassembly and removal of the test article from the test rig.

Teardown Inspection of the PZL-130 TC-II after FSFT was carried out by disassembling the aircrafts structure down to separate parts [6] and performing versatile NDI in order to both examine the elements which were inaccessible during the test as well as to verify damages which were reported during the test. Figure 10 shows the process of disassembling the wings. After additional analysis of the wing's outer surface using MAUS V automated system, the sheeting was carefully removed by drilling out the rivets thus allowing access to the inner wing structure.

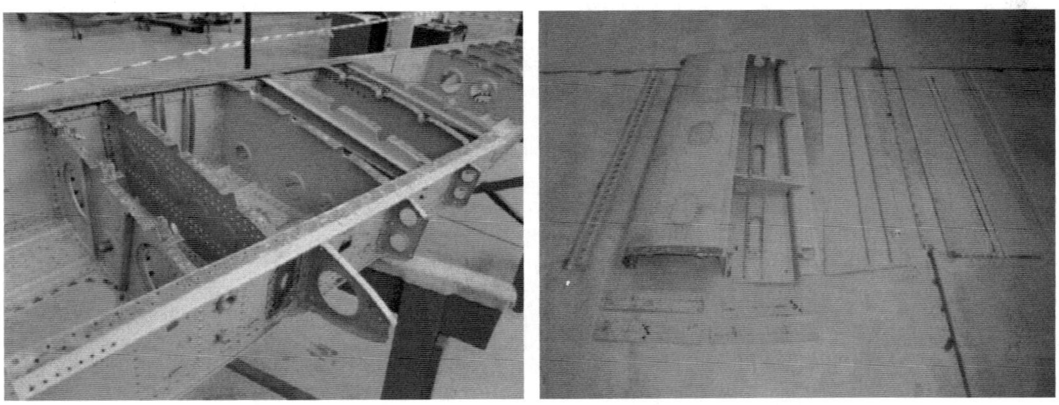

Figure 10. Teardown Inspection of the PZL-130 TC-II wing.

During TI many new damages were found among which majority was located in the wings since the wings inner structure was inaccesible during the test. Similarly to the results of NDI during FSFT the damages were of different types, from failed rivets through partial cracks up to elements cracked through.

After the TI selected cracked elements (both with through and closed cracks) were examined with Philips XL30 Scanning Electron Microscope in order to identify fatigue markers which were introduced in the modified load spectrum [2, 7, 8]. Total of 23 samples were examined from various damages found in the wing-fuselage connection area. During the study magnification range of 10-500000x was used. Although nucleation, fatigue fracture and final overload areas were clearly visible on the samples crack surfacs and fatigue fracture areas of coupons testred before FSFT [2,4] and FSFT aircraft samples showed similarities it was unable to define the fatigue markers caused by the reordering of load spectrum.

Failure database for PZL-130 Orlik

In the course of FSFT numerous damages were found during NDI inspections, both during the FSFT as well as TI, and several repairs were carried out. The team of people taking part in this project consisted of specialists of many fields which were using the FSFT information for purpose of different subtasks and it was essential to allow easy access to the obtained results. Therefore in order to manage all the FSFT data a special database was created. This dedicated software allowed for definition of NDI findings by defining type of damage e.g. Hole Through Crack, Hole Quadrant Crack, Plate Through Thickness Crack etc. as well as estimated size of the damage. The database also holds information about the NDI technique used, personnel which was performing the inspection and inspection parameters which were used e.g. frequency, phase angle, gains etc. The database also allows for uploading pictures of visual damages and moreover to position each finding or repair using included technical drawing of the aircraft.

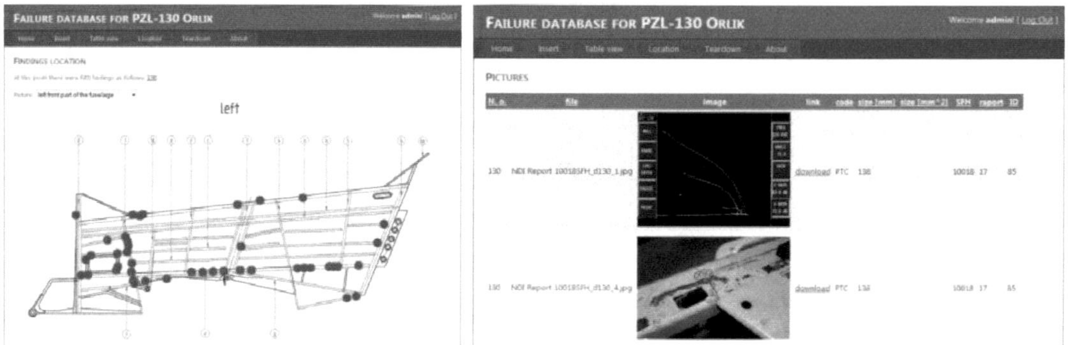

Figure 11. FSFT damage database: damage location, NDI results and photo documentation.

CRACK PROPAGATION CURVES DEFINITION

Critical Points definition

Among all the damages that were found during the NDI program, both during FSFT and TI, 10 locations were selected as Critical Points (CP). The selection was done in agreement with EADS PZL-Okęcie based on the structural integrity risk the occurred damages introduced to the PZL-130 aircraft structure. Due to planned change in operation routine it was essential to define crack propagation curves for selected CP's which would allow scheduling structural inspections based on the actual usage.

Methodology for Crack Propagation calculations

Crack propagation curves were defined using numerical calculations based on the Virtual Crack Closure Technique (VCCT). This method is based on the assumption that energy released during infinitesimally small crack increase Δa is equal to the energy needed to close the crack over Δa counteracting forces that cause the crack to propagate [9-12]. The calculated energy release rate can be then used to calculate the boundary correction factor β (1) (where K is the stress intensity factor, G energy release rateand a is the crack length) which for defined geometry and load type defines the crack propagation rate. Therefore once calculated β can be used for crack propagation calculations for different spectra [13].

$$\beta = \frac{K}{\sigma\sqrt{\pi a}} \qquad (1)$$

In order to perform numerical calculations using mentioned method the following data had to be prepared:

- boundary correction factor versus crack length for the specific CP geometry,
- load spectrum characteristic for the CP damage,
- material properties for the of the damaged structure.

Boundary correction factors definition

The boundary correction factors were defined using detailed FEM models of the damaged structure. Each model consisted of CP including the crack and significant adjacent structure and was prepared basing on the geometrical models delivered by EADS PZL-Okęcie. The detailed models were loaded using displacement fields determined using global FEM model of the aircraft under level flight load conditions.

Example of local FEM model of CP is shown in Figure 12 on the left side. The CP was located on the aircrafts hull left side near limousine lock. In this location damage propagated from both front corners of the rectangular cutout shown on the right side of Figure 12 as sections "a" and "b". Cracks propagated around the whole profile until it was completely damaged.

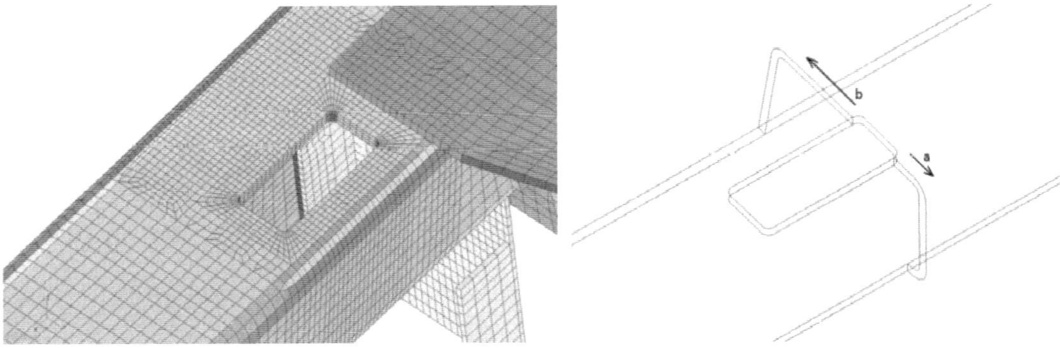

Figure 12. Local FEM model of described CP and visualization of introduced fatigue damage.

Cracks were introduced in the FEM model as discontinuities between elements and the calculations were carried out for consequent crack lengths in order to determine the change in energy release rate versus crack range. Figure 13 shows stress field in the CP due to introduced displacement field from the global model. Graph shown in Figure 13 shows the calculated value of boundary correction factor for "a" crack versus crack length.

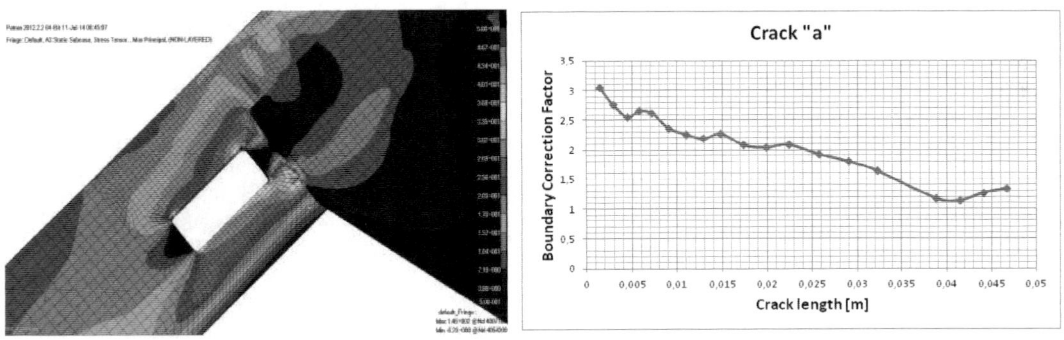

Figure 13. Stress field in the CP region and boundary correction factor for crack "a" versus damage length.

Load spectrum definition

Load spectra for crack propagation calculations were prepared basing on strain measurements carried out during the Full Scale Fatigue Test. Data from the array of over 100 strain gauges as well as additionally installed strain gauges in the CP's were used to define local load spectra for crack propagation calculations. Crack propagation calculations load spectra consisted of 800 SFH in order to include three standard FSFT load spectra and one reordered spectrum [2], each consisting of 200 SFH.

Material properties definition

Most of the PLZ-130 primary structure is made of 2024-T351 aluminum and in general material properties of this alloy from AFGROW material library were used during calculations. Moreover values of K_{IC} and E were determined via material tests carried out in Laboratory for Material Strength Testing at AFIT. The tests were carried out on coupons delivered by the aircrafts manufacturer. The samples were prepared according to adequate standards (PN-EN ISO 12737:2011 and PN-EN ISO 6892-1:2010) and moreover the surface and heat treatment were carried out according to manufacturer standards.

Crack propagation calculations

Final calculations were performed using AFGROW software. Boundary correction factors defined from FEM calculations were used in the analyses as well as material data from laboratory coupon tests and load spectra defined using strain gauge measurements. As a result a set of crack propagation curves was obtained which corresponded well with the NDI findings during the Test.

Crack propagation Curves Summary

Using all the prepared data as described in the previous chapters the crack propagation calculations were performed. Crack propagation curves for sections "a" and "b" are presented in Figure 14.

Estimated durability of the structure sections "a" and "b" were 7 300 13 900 SFH respectively, which is consistent with the time at which the damage was discovered during NDI tests (10 018 SFH). For both curves the detectable (dashed line) and critical sizes (continuous red line) of cracks were marked and the time interval between them determined. This allowed defining safe inspection intervals for which at least two inspections can be carried out before crack propagates from detectable to critical length.

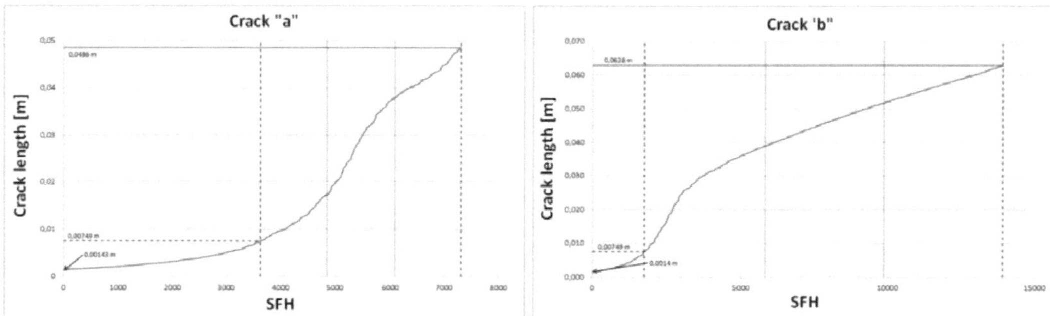

Figure 14. Estimated crack propagation curves for cracks "a" and "b".

SUMMARY

The SEWST program was finalized at the end of 2014. All the goals have been reached and the scheduled tasks fulfilled. New operation procedure has been developed and introduced basing on the results achieved in SEWST program. Obtained results allowed moreover modifying the aircraft's structure in order to increase its reliability.

The full scale Fatigue test allowed determining Critical Points of the PZL-130 TC-II structure. Numerical crack propagation calculations confirmed the NDI findings during the test and moreover allowed defining safe inspection intervals for detection of fatigue damages during operation.

During the program several tools were developed like the investigation/damage/repair database (which will be continuously supplemented with findings which will occur during regular operation), crack propagation methodology along with the global and local FEM model for further analysis as well as the Individual Aircraft tracking program, during which flight severity of each aircraft is analyzed and reported basing on quarterly gathered flight data and compared with the load spectrum used in FSFT in order to define the actual usage and adjust the inspection schedule. On the basis of the tested TC-II version of PZL-130 trainer aircraft EADS PZL-Okęcie developed the glass cockpit version of the aircraft which is now offered for the Polish Air Force.

REFERENCES

[1] Leski A, Reymer P, Kurdelski M. D*evelopment of load spectrum for full scale fatigue test of a trainer aircraft.* ICAF 2011 Structural Integrity: Influence of Efficiency and Green Imperatives: Proceedings of the 26th Symposium of the International Committee on Aeronautical Fatigue (edited by J. Komorowski), Montreal, Canada, 1-3 June 2011 pp.573-584.

[2] Leski A., Dragan K., Reymer P.: *Full-Scale Fatigue Test of PZL-130 Orlik Structure*. Proceedings of the 27th Symposium of the International Committee on Aeronautical Fatigue and Structural Integrity (edited by A. Brot), Jerusalem, Israel, 5-7 June 2013. pp. 199-208.

[3] *Anderson, A., Parker, R. GR.,* Full-scale fatigue test of the Pilatus PC9/A trainer aircraft , *20th ICAF Symposium: Structural Integrity for the next Millennium, Bellevue, Waszyngton USA, (1999).*

[4] *Molent L., Barter S.A., White P., Dixon B.:* Damage tolerance demonstration testing for the Australian F/A-18. *International Journal of Fatigue 31, 1031-1038 (2009)*

[5] Sałaciński M., Leski A., Synaszko P., Stefaniuk M., *Determination of composite patch thickness on the accuracy of nondestructive testing of the repaired structure of PZL-130 Orlik aircraft*, Proceedings of the 16th Australian International Aerospace Congess.

[6] *Handbook for best practice in teardown of aircraft structures*, Technical paper Nr TR-AER-1-2005, The Technical Co-operation Program, April (2005).

[7] *Barter, A.A., Molent, L., Wanhill, R.J.H.,* Marker loads for quantitative fractography of fatigue cracks in aerospace alloys, *25th ICAF Symposium:, Rotterdam, Netherlands, (2009)*

[8] Leski A., Kłysz S., Lisiecki J., Gmurczyk G., Reymer P., Bochenek D., Zasada D., *Introduction of fatigue markers in Full Scale Fatigue Test of an aircraft structure*, Fatigue of Aircraft Structures, 2011

[9] S. N. Atluri, T. Nishioka, *Computational Methods in the Mechanics of Fracture*, Elseviere, 1986, pp. 229-287, 1986.

[10] K.N. Shivakumar, I.S. Raju, *Engineering Fracture Mechanics*, Vol. 42, No. 6, pp. 935-959, 1992.

[11] E.F. Rybicki, M. F. Kanninen, *Engineering Fracture Mechanics,* Vol. 9, pp. 931-938, 1977.

[12] A. Leski, Finite Elements in Analysis and Design 43, (2007), 261-268

[13] Y. Murakami, *Stress Intensity Factors Handbook*,.Volume 1, Pergamon Press, Great Britain, 1987.

Peridynamics for Crack Initiation and Propagation under Cyclic Loading

E. Madenci[1], A. Zamani[2], I. Guven[3], N. Phan[4]

[1]*University of Arizona, Tucson, AZ 85721*
[2]*Global Engineering Research and Technologies, Tucson, AZ 85712*
[3]*Virginia Commonwealth University, Richmond, VA 23284*
[4]*Naval Air Systems Command, Patuxent River, MD 20670*

Abstract: This study concerns the prediction of crack propagation topography in isotropic materials under cyclic loading by using peridynamics. Crack growth and its topography are governed by a single failure parameter referred to as critical stretch through an unguided process. The critical stretch can be expressed in terms of the fracture toughness of the material. Three dimensional simulations concern the propagation of a quarter corner crack and a circular surface crack in a beam subjected to cyclic four-point bending loading. The simulations capture experimentally observed non-self-similar crack propagation topography.

INTRODUCTION

The prediction of crack initiation and propagation through computational models is of great importance for many engineering applications. Critical engineering components are usually designed based on the concepts of fracture mechanics to avoid premature failure under the expected operating conditions. The search for better and accurate predictive capability for crack propagation path especially under cyclic loading continues unabatedly despite exhaustive amount of past research.

Although the existing techniques are applied to many problems with acceptable success, there still exist challenges when predicting the evolution of an arbitrary crack shape that may be non-planar. Often there are multiple cracks exhibiting complex patterns forming non-planar 3-D surfaces. The presence of manufacturing or service related residual stresses and sequence of load interactions introduces additional challenges. Also, the existing numerical methods for calculating fracture parameters encounter challenges due to this topological evolution.

For decades, finite element analysis (FEA) has been the tool of choice for stress analysis which is still widely practiced. Currently, most failure simulations involve FEA. Although mature, powerful and versatile, the FEA simply fails to predict failure initiation and complex crack growth because the FEA formulation is not mathematically suitable for simulation of failure. The governing equations of motion include spatial derivatives of the displacement components which are not valid in the presence of displacement discontinuities such as cracks. Therefore, it requires special treatment of mathematical singularities which are not physical. Furthermore, these equations lack an internal length parameter which appears naturally in many physical problems. These shortcomings present limitations when addressing crack initiation and growth in materials. Damage initiation is not intrinsic, and crack growth prediction is commonly achieved by implementing either an external local or global criteria for guiding the crack path subsequent to numerical discretization and analysis. Even mesh-free codes based on the conventional theories require such criteria. The same is true of cohesive elements and similar methods. Methods to predict crack initiation and its growth remain active areas of research.

Peridynamics converts the existing governing field equations from their local to nonlocal form through a peridynamic differential operator while introducing an internal length parameter. The main difference between the PD theory and classical continuum mechanics is that the former is formulated using integral equations as opposed to differential form of the classical equations of motions. This feature allows crack initiation and propagation at multiple sites, with arbitrary paths inside the material, without resorting to external crack growth criteria. In peridynamic analysis, damage in the material is simulated in a much more realistic manner compared to the classical continuum-based methods. The broken interactions between the material points may align themselves along surfaces that form cracks and the deformation is discontinuous across such as a crack, yet the equations of motion continue to remain valid. Because of the natural treatment of cracks in peridynamics, cracks may appear spontaneously in a body, and they may grow and arrest according to whatever conditions exist in the deformation. They can grow at whatever velocity is determined by the interaction between a crack tip and its surroundings. Similarly, the direction and curvature of crack growth emerges from the peridynamic equations of motion and constitutive model. All of these features are predicted without the need for any special techniques or supplemental equations. Crack growth in the peridynamic theory is "autonomous" or "unguided."

PERIDYNAMIC THEORY

According to the peridynamic (PD) theory introduced by Silling [1], the motion of a body is analyzed by considering the interaction of a material point, $\mathbf{x}_{(k)}$, with the other, possibly infinitely many, material points, $\mathbf{x}_{(j)}$, with $(j = 1,2,..,\infty)$, in the body (Fig. 1). Therefore, an infinite number of interactions may exist between the material point at location $\mathbf{x}_{(k)}$ and other material points. However, the influence of the material points interacting with $\mathbf{x}_{(k)}$ is assumed to vanish beyond a local region (horizon), denoted by $H_{\mathbf{x}_{(k)}}$. Similarly, material point $\mathbf{x}_{(j)}$ interacts with material points in its own family, $H_{\mathbf{x}_{(j)}}$.

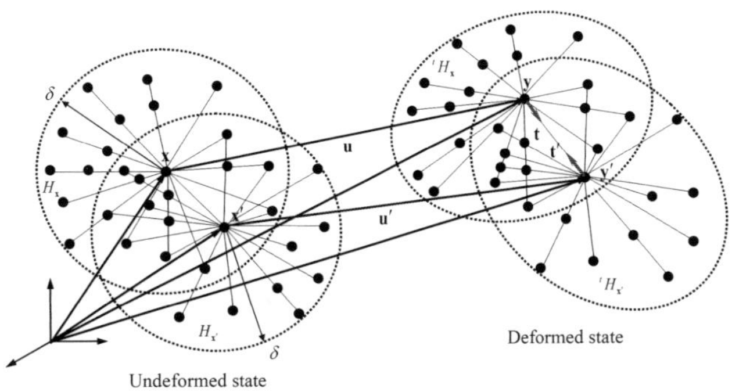

Figure 1. Interaction of peridynamic material points $\mathbf{x}$ and $\mathbf{x'}$ influenced by their own family of material points.

In other words, the PD theory is concerned with the physics of a material body at a point that interacts with all points within its range. The range of material point $\mathbf{x}_{(k)}$ is defined by δ, referred to as the "horizon." Also, the material points within a distance δ of $\mathbf{x}_{(k)}$ are called the family of $\mathbf{x}_{(k)}$, $H_{\mathbf{x}_{(k)}}$. The interaction of material points is prescribed through a micropotential that depends on the deformation and constitutive properties of the material. The locality of interactions depends on the horizon, and the interactions become more local with a decreasing horizon, δ. Hence, the classical theory of elasticity can be considered a limiting case of the peridynamic theory as the horizon approaches zero [2].

The peridynamic equation of motion originally formulated by Silling [1] can be expressed in the form [2]

$$\rho(\mathbf{x})\ddot{\mathbf{u}}(\mathbf{x},t) = \int_H \left(\mathbf{t}(\mathbf{u}'-\mathbf{u},\mathbf{x}'-\mathbf{x},t) - \mathbf{t}'(\mathbf{u}-\mathbf{u}',\mathbf{x}-\mathbf{x}',t)\right) dH + \mathbf{b}(\mathbf{x},t) \tag{1}$$

In an undeformed state of the body, each material point is identified by its coordinates, $\mathbf{x}_{(k)}$ with $(k = 1,2,...,\infty)$, and is associated with an incremental volume, $V_{(k)}$, and a mass density of $\rho(\mathbf{x}_{(k)})$. Each material point can be subjected to prescribed body loads, displacement, or velocity, resulting in motion and deformation. With respect to a Cartesian coordinate system, the material point $\mathbf{x}_{(k)}$ experiences displacement, $\mathbf{u}_{(k)}$, and its location is described by the position vector $\mathbf{y}_{(k)}$ in the deformed state. The displacement and body load vectors at material point $\mathbf{x}_{(k)}$ are represented by $\mathbf{u}_{(k)}(\mathbf{x}_{(k)},t)$ and $\mathbf{b}_{(k)}(\mathbf{x}_{(k)},t)$, respectively. The motion of a material point conforms to the Lagrangian description. The stretch between material points $\mathbf{x}_{(k)}$ and $\mathbf{x}_{(j)}$ is defined as

$$s_{(k)(j)} = \frac{\left(\left|\mathbf{y}_{(j)} - \mathbf{y}_{(k)}\right| - \left|\mathbf{x}_{(j)} - \mathbf{x}_{(k)}\right|\right)}{\left|\mathbf{x}_{(j)} - \mathbf{x}_{(k)}\right|} \tag{2}$$

For an isotropic and elastic, the force density vector $\mathbf{t}_{(k)(j)}\left(\mathbf{u}_{(j)} - \mathbf{u}_{(k)}, \mathbf{x}_{(j)} - \mathbf{x}_{(k)}, t\right)$ can be rewritten in terms of peridynamic material parameters as

$$\mathbf{t}_{(k)(j)}\left(\mathbf{u}_{(j)}-\mathbf{u}_{(k)},\mathbf{x}_{(j)}-\mathbf{x}_{(k)},t\right)=\frac{1}{2}A\frac{\mathbf{y}_{(j)}-\mathbf{y}_{(k)}}{\left|\mathbf{y}_{(j)}-\mathbf{y}_{(k)}\right|},\tag{3}$$

with

$$A=4w_{(k)(j)}\left\{ad\,\Lambda_{(k)(j)}\theta_{(k)}+b\left(\left|\mathbf{y}_{(j)}-\mathbf{y}_{(k)}\right|-\left|\mathbf{x}_{(j)}-\mathbf{x}_{(k)}\right|\right)\right\}\tag{4}$$

Similarly, the force density vector $\mathbf{t}_{(j)(k)}\left(\mathbf{u}_{(k)}-\mathbf{u}_{(j)},\mathbf{x}_{(k)}-\mathbf{x}_{(j)},t\right)$ can be expressed as

$$\mathbf{t}_{(j)(k)}\left(\mathbf{u}_{(k)}-\mathbf{u}_{(j)},\mathbf{x}_{(k)}-\mathbf{x}_{(j)},t\right)=-\frac{1}{2}B\frac{\mathbf{y}_{(j)}-\mathbf{y}_{(k)}}{\left|\mathbf{y}_{(j)}-\mathbf{y}_{(k)}\right|}\quad,\tag{5}$$

with

$$B=4w_{(j)(k)}\left\{ad\,\Lambda_{(k)(j)}\theta_{(j)}+b\left(\left|\mathbf{y}_{(k)}-\mathbf{y}_{(j)}\right|-\left|\mathbf{x}_{(k)}-\mathbf{x}_{(j)}\right|\right)\right\}.\tag{6}$$

in which

$$\Lambda_{(k)(j)}=\left(\frac{\mathbf{y}_{(j)}-\mathbf{y}_{(k)}}{\left|\mathbf{y}_{(j)}-\mathbf{y}_{(k)}\right|}\right)\cdot\left(\frac{\mathbf{x}_{(j)}-\mathbf{x}_{(k)}}{\left|\mathbf{x}_{(j)}-\mathbf{x}_{(k)}\right|}\right)\quad.\tag{7}$$

The parameter, $\theta_{(k)}$ representing dilatation can be rewritten in the form

$$\theta_{(k)}=d\delta\sum_{j=1}^{N}s_{(k)(j)}\Lambda_{(k)(j)}V_{(j)}\tag{8}$$

where the PD parameter d ensures that $\theta_{(k)}$ remains non-dimensional. As derived by Oterkus and Madenci [3], the peridynamic parameters for a three-dimensional analysis can be obtained as

$$a=\frac{1}{2}\left(\kappa-\frac{5\mu}{3}\right),\qquad b=\frac{15\mu}{2\pi\delta^{5}},\quad\text{and}\quad d=\frac{9}{4\pi\delta^{4}}\tag{9}$$

Material damage in peridynamics is introduced through elimination of interactions among the material points. A new crack surface forms when all the interactions between the material points across a plane are eliminated. For a linear elastic material response, the force density vector $\mathbf{t}_{(k)(j)}$ in PD equation of motion, Eq. (1), depends linearly on stretch between material points, $\mathbf{x}_{(k)}$ and $\mathbf{x}_{(j)}$ while including the constitutive properties of these material points. As shown in Fig. 2, it is assumed that when the stretch, $s_{(k)(j)}$ between these material points exceeds its critical value, $s^{c}_{(k)(j)}$ the onset of damage occurs, and the force density vectors, $\mathbf{t}_{(k)(j)}$ between these material points no longer exist. Therefore, damage is reflected in the equations of motion by removing the force density vectors between the material points in an irreversible manner. As a result, the load is redistributed among the material points in the body, leading to a progressive damage growth in an autonomous fashion.

In order to include damage initiation in the material response, the force density vector can be modified through a history-dependent scalar-valued function μ as

$$\mathbf{t}_{(k)(j)}\left(\mathbf{u}_{(j)}-\mathbf{u}_{(k)},\mathbf{x}_{(j)}-\mathbf{x}_{(k)},t\right)=\frac{1}{2}A\mu\left(\mathbf{x}_{(j)}-\mathbf{x}_{(k)},t\right)\frac{\mathbf{y}_{(j)}-\mathbf{y}_{(k)}}{\left|\mathbf{y}_{(j)}-\mathbf{y}_{(k)}\right|}\tag{10}$$

and

$$\mathbf{t}_{(j)(k)}\left(\mathbf{u}_{(k)}-\mathbf{u}_{(j)},\mathbf{x}_{(k)}-\mathbf{x}_{(j)},t\right)=-\frac{1}{2}B\mu\left(\mathbf{x}_{(k)}-\mathbf{x}_{(j)},t\right)\frac{\mathbf{y}_{(j)}-\mathbf{y}_{(k)}}{\left|\mathbf{y}_{(j)}-\mathbf{y}_{(k)}\right|} \tag{11}$$

where μ can be written as

$$\mu\left(\mathbf{x}_{(j)}-\mathbf{x}_{(k)},t\right)=\begin{cases}1 & \text{if } s_{(k)(j)}\left(\mathbf{x}_{(j)}-\mathbf{x}_{(k)},t'\right)<s_c \quad \text{for all } 0<t'<t\\ 0 & \text{otherwise}\end{cases} \tag{12}$$

in which s_c is the critical stretch for failure to occur (Fig. 2) between material points $\mathbf{x}_{(k)}$ and $\mathbf{x}_{(j)}$. In the case of isotropic materials, the value of critical stretch, s_c can be related to the critical energy release rate [4] as

$$s_c = K_{IC}\sqrt{\frac{5}{9E\kappa\delta}} \tag{13}$$

where κ is the bulk modulus of the material. The critical stretch value may also be extracted from the experimental measurements and peridynamic simulations in an inverse manner.

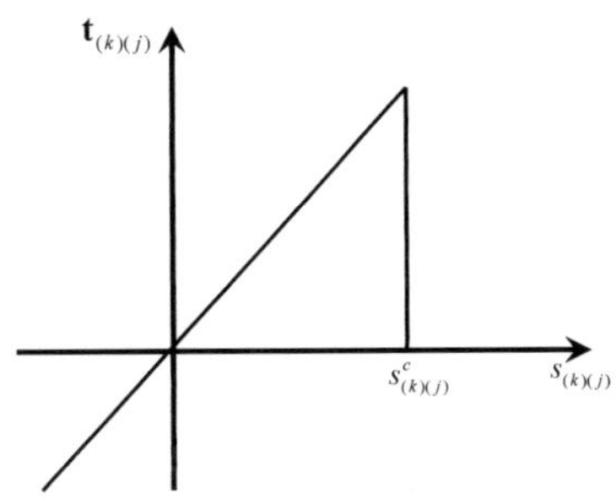

Figure 2. Constitutive relation for elastic stretch between material points $\mathbf{x}_{(k)}$ and $\mathbf{x}_{(j)}$.

During the solution process, the displacements of each material point, as well as the stretch, $s_{(k)(j)}$ between pairs of material points, are computed and monitored. When the stretch between two material points exceeds its critical stretch, failure occurs, and these two points cease to interact with each other; thus, the history-dependent scalar-valued function, μ is zero rendering the force density vector to be zero.

Local damage at a point is defined as the weighted ratio of the number of eliminated interactions to the total number of initial interactions of a material point with its family members. The local damage at a point can be quantified as

$$\varphi(\mathbf{x},t)=1-\frac{\int_H \mu(t,\xi)dV'}{\int_H dV'} \tag{14}$$

The local damage ranges from zero to one. When the local damage is one all the interactions initially associated with the point are eliminated, while a local damage of zero means that all interactions are intact. A damage value of 0.5 and above indicates the creation of new surface in the form of a crack.

Accurate prediction of fatigue life is a challenging task largely due to the long duration of the load spectrum; accounting for the contribution of each load cycle of the load spectrum on the fatigue life of the structure is computationally prohibitive. Thus, simplifying assumptions are unavoidable. However, when the assumptions depart from the underlying physics and rely more on curve-fitting to empirical data, the predictions tend to become inconsistent and inaccurate, which is the case in currently used prediction methods in industry. The PD approach allows "unguided" crack growth modeling of quasi-static and dynamic fracture retaining a close connection to the underlying physics. This property equally well applies to the modeling of fatigue crack growth in engineering materials.

This model introduced by Silling and Askari [5] assumes degradation of critical stretch over time (remaining life); the rate of degradation increases with local deformation and damage. The general form of the relationship between critical stretch s^* and number of cycles N is given by

$$\frac{ds^*}{dN} = -C\left(s - s_o^*\right) \tag{14}$$

This relationship leads to a recursive relationship for critical stretch at a material point in the form:

$$s_{new}^* = s_{old}^* - \left[\frac{\left(s_{old}^* - s_o^*\right)}{s_i^*}\left(s_{\max} - s_f\right)\lambda\, A\, \Delta N\right] \tag{15}$$

in which

s_{new}^* : current critical stretch
s_{old}^* : previous critical stretch
s_o^* : asymptotic critical stretch limit
s_i^* : initial critical stretch
$s_{\max}$: maximum current stretch associated with the material point
s_f : if current stretch is below this value, fatigue is not activated
λ : decay constant
A : load cycle amplitude

In order to solve Eq. (1), a collocation method is adopted and the numerical treatment involves the discretization of the domain of interest into subdomains. The domain can be discretized into subdomains by employing line, square, and cubic subdomains for one-, two-, and three dimensional regions, respectively.

NUMERICAL RESULTS

The robustness of peridynamics for predicting crack propagation topography is demonstrated by simulating the experiments conducted by Riddell et al. [6]. The experiments considered by Riddell et al. involve a beam subjected to cyclic four-point bending loading. The geometry and boundary conditions are illustrated in Fig. 3. The beam has either a circular initial crack at mid-span on one corner (i.e., a quarter circle corner crack) or semi-circular surface crack located off-centered depthwise; initially it is not visible from sides.

Damage contours based on PD simulations along the mid-surface (plane of the initial quarter-circle crack) as the number of load cycles increases are shown from front view in Fig. 4. Same results are presented in Fig. 5 from an oblique view. Note that the initial quarter-circle crack is located at top-left corner of the midplane of the PD model (as opposed to the bottom-left corner depicted in Fig. 3a). The PD simulations predict a slower growth of the vertical edge of the crack (a_1 in Fig. 3a) compared to the horizontal edge growth (a_2 in Fig. 3a). This is consistent with the experimental observations. The PD predictions of crack growth are compared with the experimentally measured values in Figs. 6 and 7 for a_1 and a_2, respectively. The experimental measurements are given in three distinct sets accounting for lower and upper bounds, and the median. The lower and upper bound observations define an envelope for the crack growth. In both a_1 and a_2 results, the PD predictions for the crack growth fall within these envelopes. Finally, the crack front observed experimentally is compared with the PD prediction of crack front in Fig. 8. The peridynamic simulations closely predict the crack front shape and proportions observed experimentally.

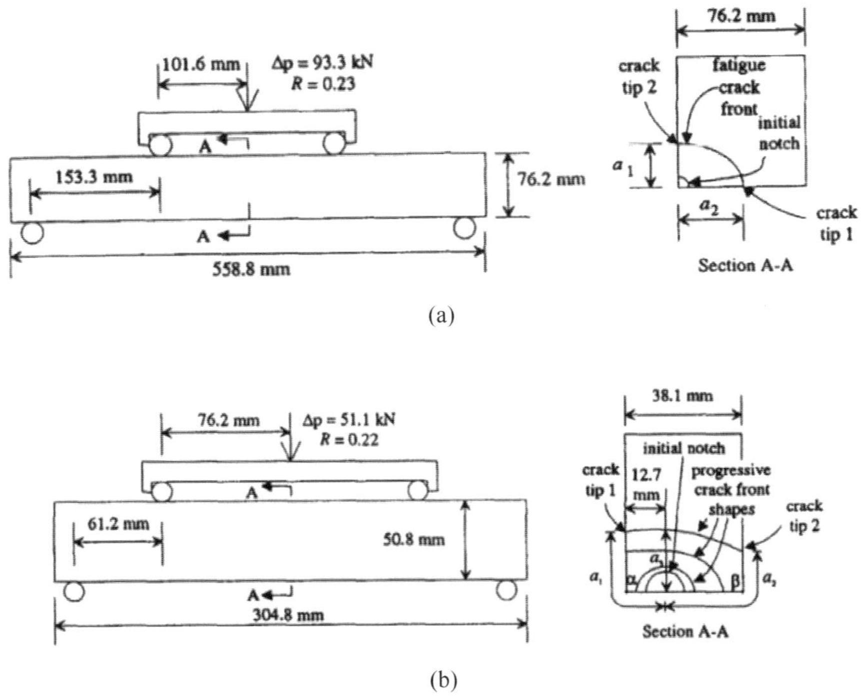

Figure 3. Geometry of the specimens: a) corner crack, and b) surface crack.

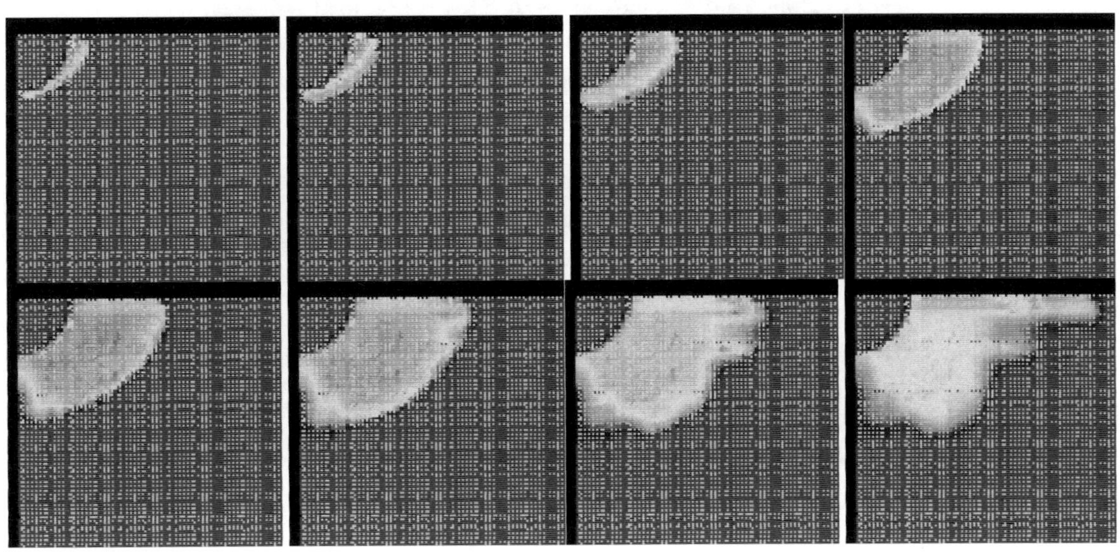

Figure 4. Front view of PD damage contours indicating the crack front at different load cycles.

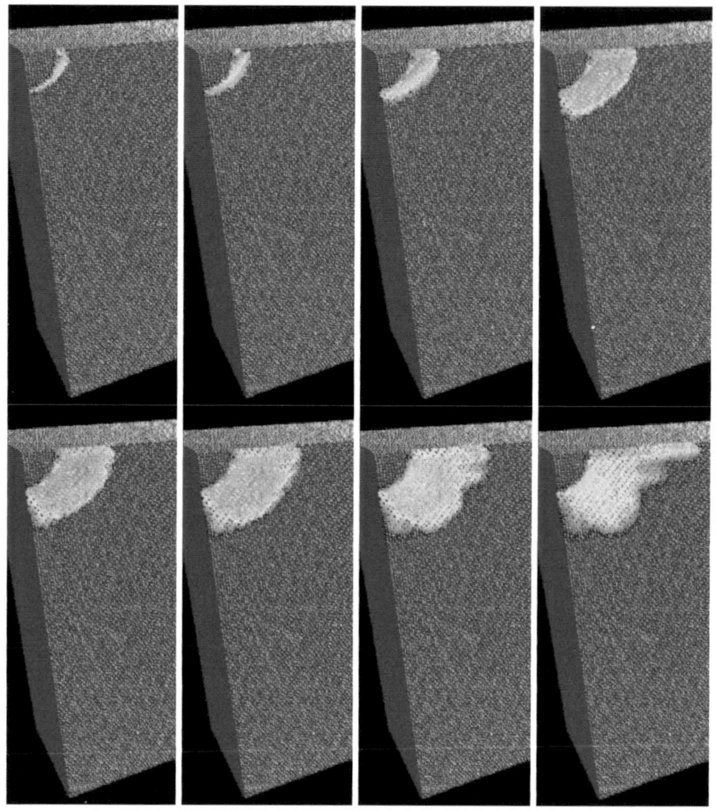

Figure 5. Oblique view of PD damage contours indicating the crack front at different load cycles.

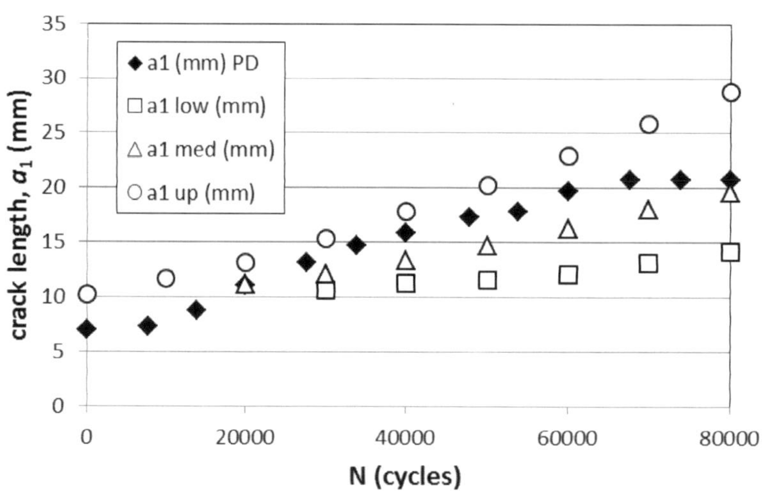

Figure 6. Comparison of PD predictions and experimental measurements of crack growth (a_1).

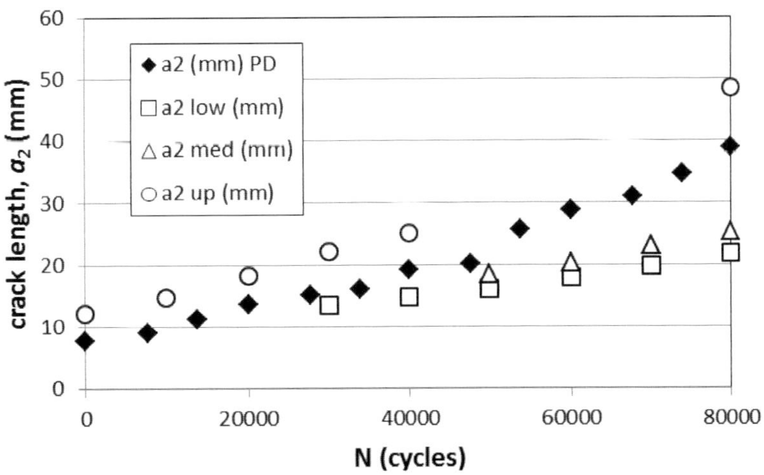

Figure 7. Comparison of PD predictions and experimental measurements of crack growth (a_2).

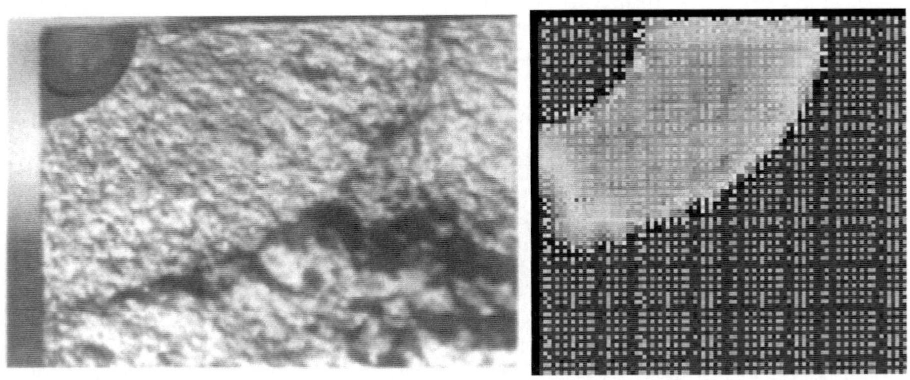

Figure 8. Comparison of the experimentally observed (left) and PD simulated (right) 3-D crack front.

Damage contours based on PD simulations along the mid-surface (plane of the initial semi-circular surface crack) as the load cycles increases are shown from front view in Fig. 9. Same results are presented in Fig. 10 from an oblique view. The PD predictions of crack growth values are compared with the experimentally measured values in Figs. 11, 12 and 13 for a_1, a_2 and a_3, respectively. Similar to the previous problem, the experimental values are given in three distinct sets accounting for lower and upper bounds, and the median. PD predictions of a_1 growth (Fig. 11) are initially consistent with the upper bound experimental observations while falling within the bounds later. In the case of the a_2 growth, the PD predictions are between the upper and lower bounds of the experimental data (Fig. 12). The a_3 growth cannot be observed experimentally as it is completely embedded in the bulk; the PD simulations provide insight in this crack growth as shown in Fig. 13; consistent with the lower bound observations. Finally, comparison of the crack front observed experimentally is made with the PD prediction of crack front in Fig. 14, which indicates a close agreement.

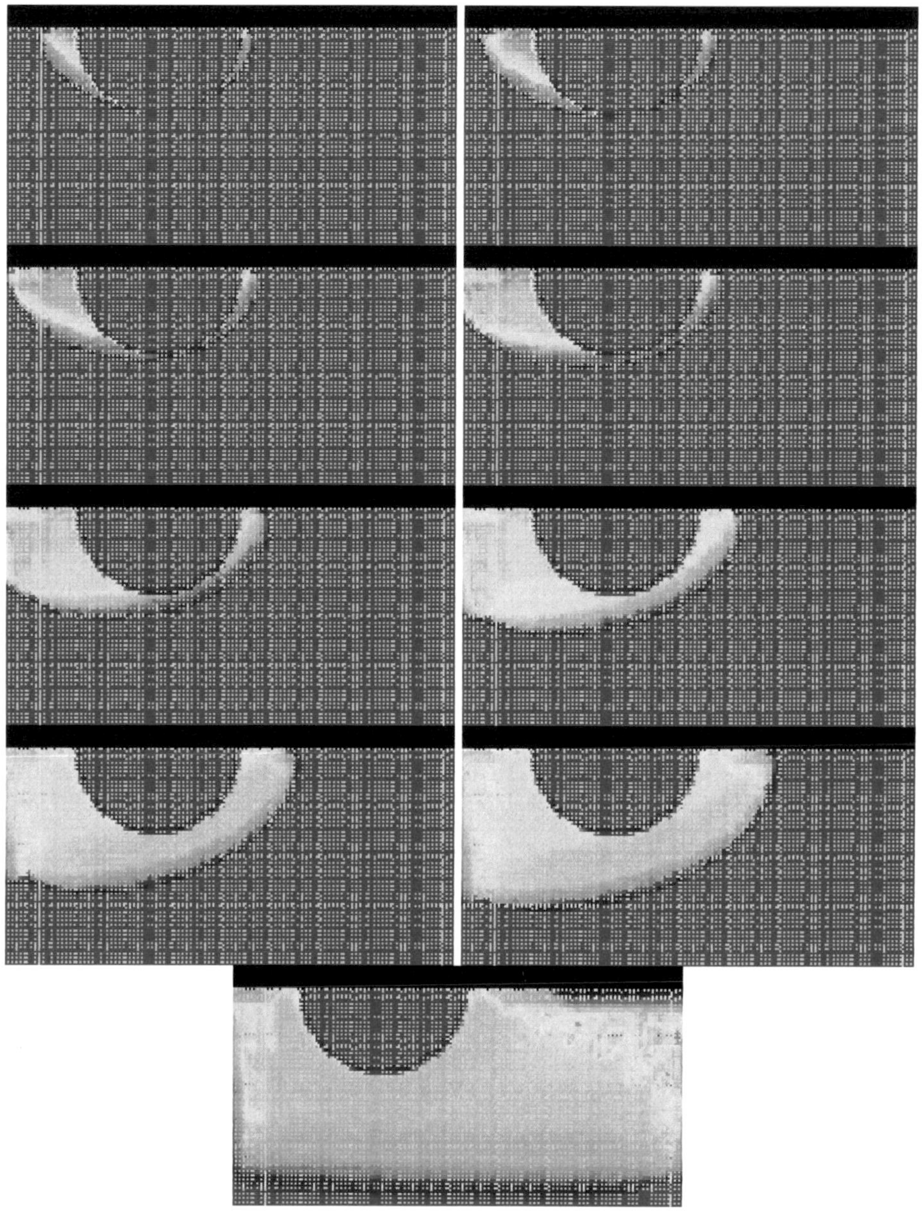

Figure 9. Front view of PD damage contours indicating the crack front at different times.

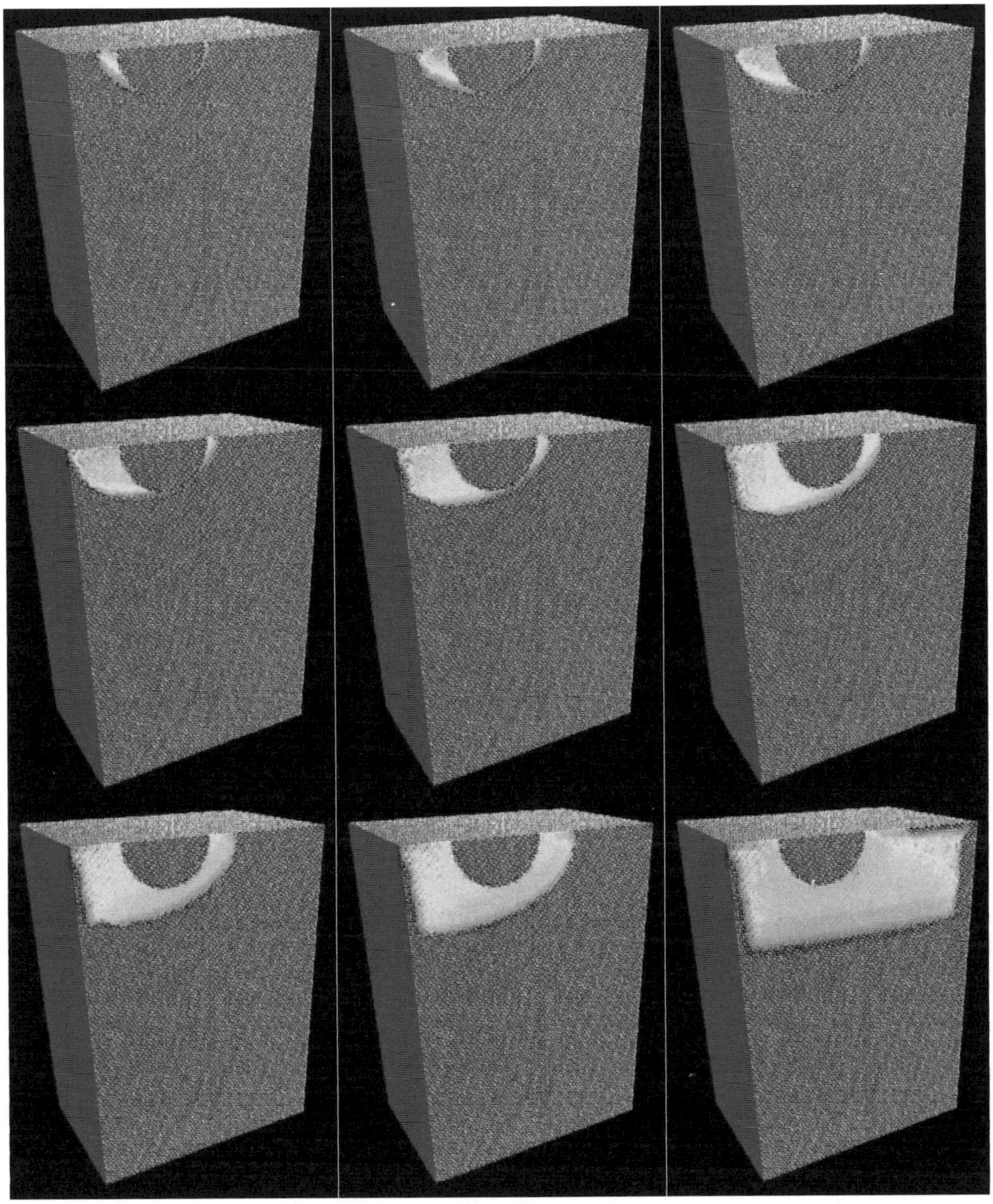

Figure 10. Oblique view of PD damage contours indicating the crack front at different times.

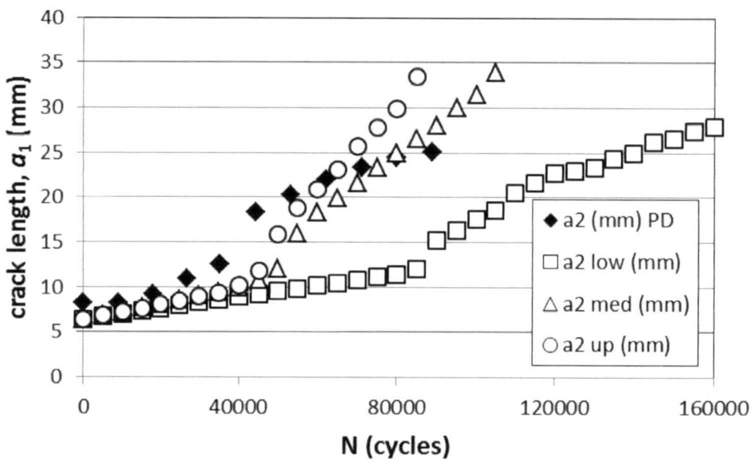

Figure 11. Comparison of PD calculated and experimental measurement of crack growth (a_1).

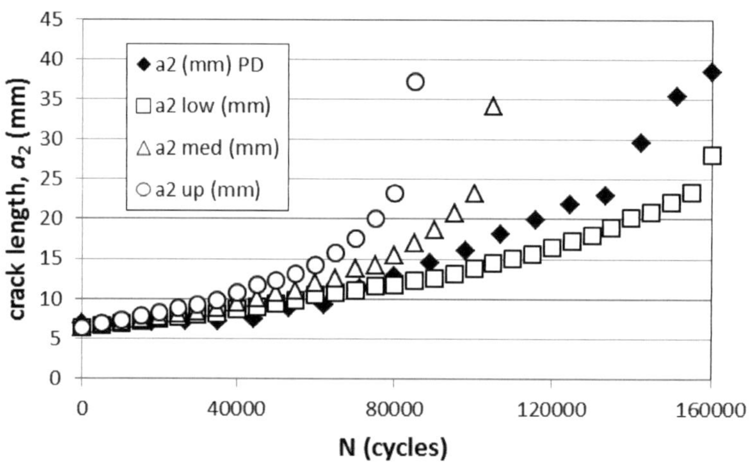

Figure 12. Comparison of PD calculated and experimental measurement of crack growth (a_2).

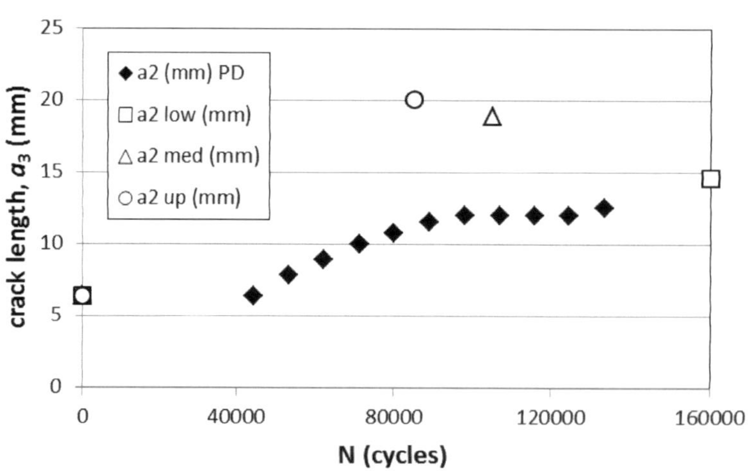

Figure 13. Comparison of PD calculated and experimental measurement of crack growth (a_3).

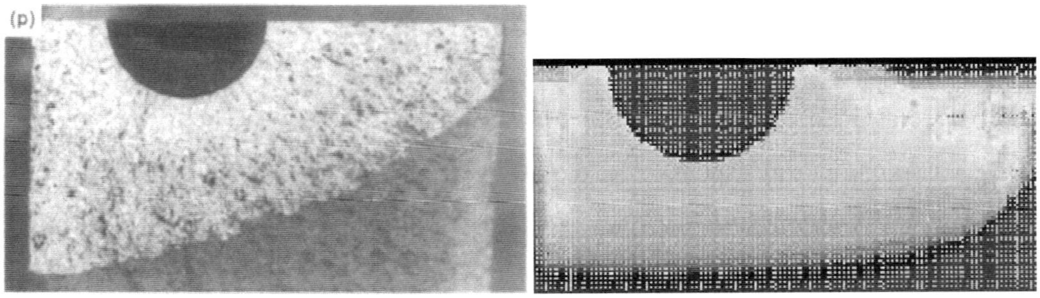

Figure 14. Comparison of the experimentally observed (left) and PD simulated (right) 3-D crack front.

CONCLUSIONS

This study demonstrates the capability of peridynamics for simulation of crack growth under cyclic loading conditions for non-self similar crack propagation topography. Crack growth is not guided through an external criteria. In addition, failure is predicted by using only a single critical failure parameter even under mixed mode loading conditions. This parameter (critical stretch) is related to the fracture toughness under pure opening mode. The approach treats the critical stretch parameter as dependent on number of cycles and loading amplitude; critical stretch decreases (decay) with increasing number of cycles. The simulations agree extremely well with the experimental observations of non-self-similar crack propagation paths.

REFERENCES

[1] Silling, S. A., (2000) Reformulation of Elasticity Theory for Discontinuities and Long-range Forces, Journal of the Mechanics and Physics of Solids, 48, pp. 175-209.

[2] Silling, S. A., Epton, M., Weckner, O., Xu, J., Askari, A., (2007) Peridynamics States and Constitutive Modeling, Journal of Elasticity, 88, pp. 151-184.

[3] Madenci, E., Oterkus, E., (2014) Peridynamic Theory and Its Applications, Springer, New York, (2014)

[4] Silling, S. A., Askari, E., (2005) A Meshfree Method Based on the Peridynamic Model of Solid Mechanics, Computers & Structures, 83, pp. 1526-1535.

[5] Silling, S. A., Askari, E., (2014), Peridynamic Model for Fatigue Cracking, SANDIA REPORT SAND2014-18590

[6] Riddell, W. T., Ingraffea, A. R. and Wawrzynek, P. A. (1997) Experimental Observations and Numerical Predictions of Three-dimensional Fatigue Crack Propagation, Engineering Fracture Mechanics, 58, pp. 293-310.

Engine integrity

Effect of Shot Peening on Low Cycle Fatigue Behavior of Ti-6Al-4V

Hiroshi, Nakamura[1*], Tatsuhito, Honda[1], Joris, Prou[2], Mitsuyoshi, Tsunori[2]

[1]*IHI Corporation, Aero-Engine & Space Operations, Japan*
[2]*IHI Corporation, Research laboratory, Japan*

Abstract: This study investigates shot peening effect on low cycle fatigue (LCF) behavior of Ti-6Al-4V. Two types of specimens were used in this study. One was polished plate specimen. Another was plate with hole which was simulated actual features such as bolt hole of compressor disk. Shot peening was conducted to both types of specimens after machined. Peened plate surface indicated higher and deeper compressive residual stress than that of as machined. Average LCF life of peened plate specimen was longer than that of as machined. This result suggests that LCF capability of peened plate specimen was enhanced due to higher compressive residual stress. Peened hole specimen indicated almost same average LCF life as that of as machined hole, whereas smaller variability of LCF life were observed. This may be attributed to more stable compressive residual stress caused by shot peening than that of as machined hole in actual features. Crack propagation analyses were conducted to assess LCF life of shot peened specimen. Residual stress relaxation was considered if the applied stress to the LCF specimen is larger than material yield stress. As a result, LCF life can be predicted within a factor of 4 scatter band. This scatter agrees with variability of LCF test results.

INTRODUCTION

Ti-6Al-4V has been used for rotating components of aero engine low temperature sections such as disks and blades because of its excellent specific strength and fatigue capability. Rotating turbine components are routinely subjected to cyclic thermal stress, body force and pressure. This situation causes low cycle fatigue (LCF) damage during operation. It is important to improve LCF capability because structural safety and integrity of turbine components are dominated by fatigue failure. Shot peening is a surface enhancement technique that causes compressive residual stress layer on the material surface and improve fatigue life [1-4]. Several studies have reported shot peening effect on high cycle fatigue properties, however, only a few studies deal with LCF capability [5-7]. This paper describes shot peening effect on LCF behavior of Ti-6Al-4V. Two types of specimens were used in this study. One was polished plate specimen. Another was plate with hole which was simulated actual features such as bolt hole of compressor disk. Shot peening was conducted to both types of specimens after machined. An X-ray diffraction method was used to measure residual stresses on the specimen surfaces. Fatigue tests were carried out using as machined specimen and shot peened specimen. Crack propagation analyses were conducted to predict LCF life of shot peened specimen.

MATERIALS AND EXPERIMENTAL PROCEDURE

Material and specimen

Table I tabulates the chemical compositions of Ti-6Al-4V used in this study. The material was heat treated at 1239K for 1 h. water cooled, and 977K for 2hs. air cooled. Figure 1 shows the microstructure after heat treatment. Table II lists the tensile properties of the material at room temperature. The shapes of specimens are shown in Figure 2. Two types of specimens were used in this study. One was polished plate specimen. Another was plate with hole which was simulated actual features such as bolt hole of compressor disk. Table III shows finishing conditions for specimen surfaces. PP is plate specimen that is polished using #400 emery paper. Shot peening was applied to PP, the specimen is PS. HM is hole specimen that was used hole making process and conditions in actual compressor disk. HS is hole specimen subjected to shot peening. Hole making conditions is same as HM. Table IV indicates shot peening conditions. Shot peening was carried out according to AMS2430 using ASR70 cast steel with an intensity of 13N and 100% coverage.

Experimental procedure and conditions

Residual stress was measured at the surface of specimen by using X-ray diffraction method. Directions of residual stress in PP and PS were orientation of the long axis of the specimen. Residual stresses in HM and HS were measured in the hoop direction of the hole. LCF tests in PP and PS were conducted under strain controlled. The strain ratio R_ε was 0. Stress controlled tests were carried out in HM and HS. The stress ratio R_σ was 0. Test environment was ambient air and room temperature.

Table I. Chemical compositions of Ti-6Al-4V(wt%).

C	V	Al	N	Fe	O	Ti
0.008	4.2	6.56	0.003	0.2	0.19	Bal.

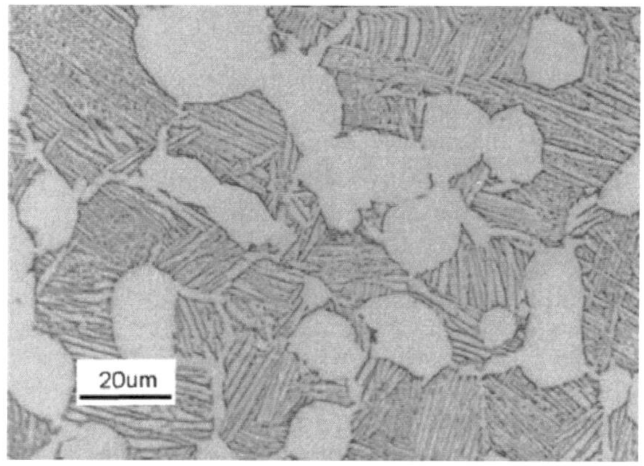

Figure 1. Microstructure of Ti-6Al-4V.

Table II. Tensile properties of Ti-6Al-4V.

0.2% Proof stress (MPa)	Tensile strength (MPa)	Elongation (%)
900	992	16.3

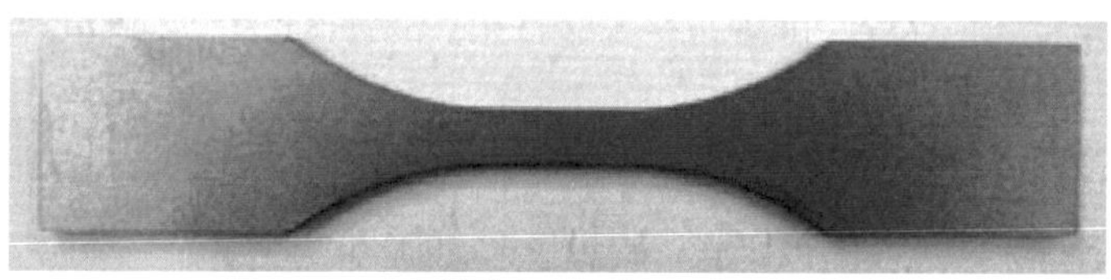

(a) Plate specimen

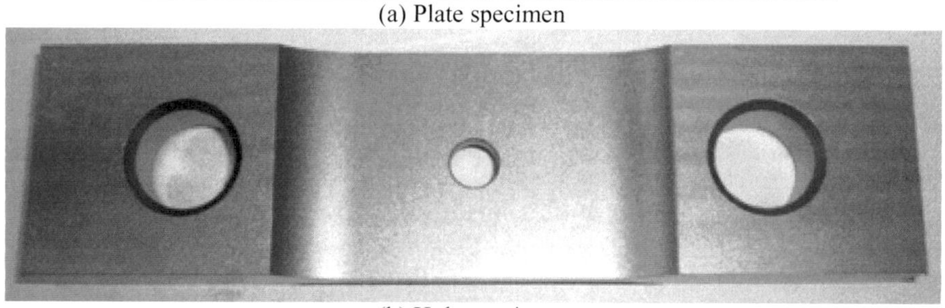

(b) Hole specimen

Figure 2. Shape of the specimens.

Table III. Specimen surface conditions.

ID	Specimen type	Surface finishing conditions
PP	Plate specimen	Polishing # 400
PS		Polishing # 400 → shot peening
HM	Hole specimen	Hole making
HS		Hole making → shot peening

Table IV. Shot peening conditions.

	Conditions
Media	ASR70
Intensity	13N
Coverage	100%

EXPERIMENTAL RESULTS

Residual stress distribution

Figure 3 shows residual stress distributions of as machined and peened specimens. Compressive residual stress layers are observed at the surfaces of PP and PS(Figure 3(a)). The maximum compressive stress of PP is caused at the surface, and the value is -340MPa. The depth of residual stress layer of PP is 30μm. PS shows the maximum compressive stress of -550MPa at the depth of 10μm. Residual stress layer of PS is until the depth of 100μm. Therefore, PS has higher and deeper compressive residual stress than that of PP. HM and HS indicate similar compressive residual stress distribution in Figure 3(b). The maximum compressive stress of HM is induced at the surface, and the value is -540MPa. HS shows the maximum compressive stress of -750MPa at the depth of 10μm. Residual stress layer of HS is until the depth of 200μm.

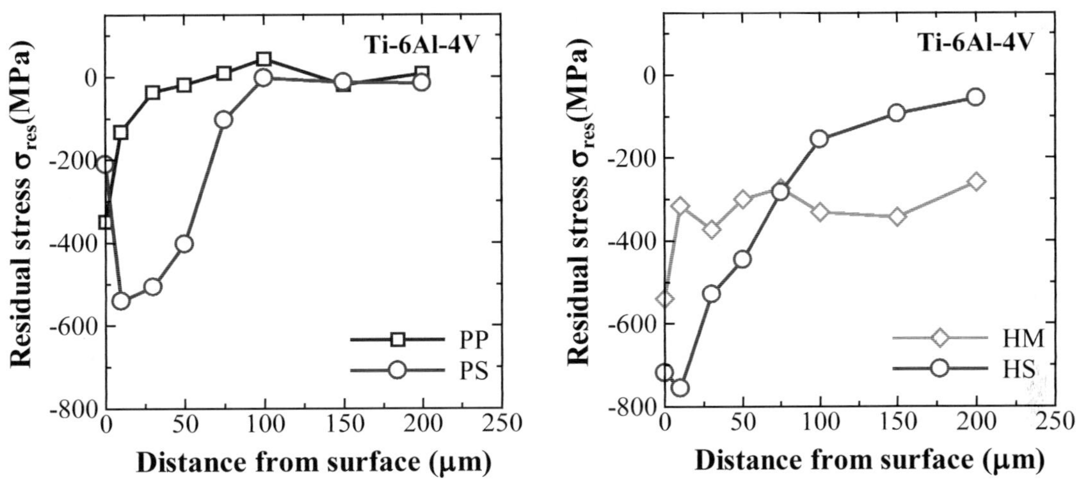

(a) Polished and peened plate.

(b) As machined and peened hole.

Figure 3. Residual stress distributions of as machined and peened specimens.

LCF test results

The relationships between normalized stress and the number of cycles to failure N_f are shown in Figure 4. Solid line is linear regression line, and the line indicates average LCF life. Dot line is S-basis line where at least 99% of population values is expected to equal or exceed the minimum value with a confidence of 95%. S-basis line can be treated as predicted minimum LCF life. Distance between average line and minimum line relates to variability of the LCF test results. Detail definition of S-basis is described in the reference [8].

Figure 4(a) shows that average LCF life of PS is longer than that of PP. Comparison of the distance between average line and minimum line indicates that the variability of LCF life of PS is similar to that of PP. These results suggest that average LCF life of plate specimen is enhanced due to shot peening. Average LCF life of HS is almost same as that of HM in Figure 4(b). However, variability of LCF life in HS is smaller than that of HM. This result indicates that shot peening decreases the scatter of LCF life of hole specimen.

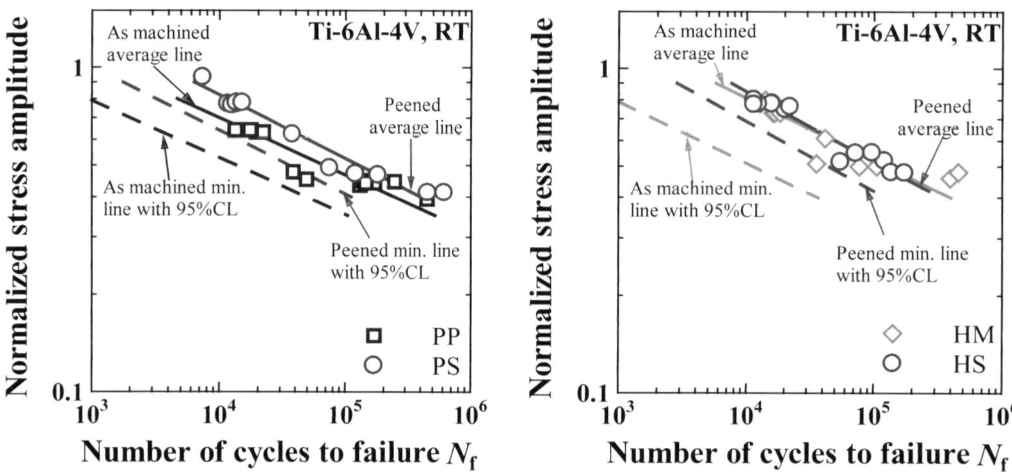

(a) Polished and peened plate.

(b) As machined and peened hole.

Figure 4. LCF test results of as machined and peened specimens.

DISCUSSION

Shot peening effect on LCF life

Shot peening caused higher and deeper compressive residual stress than that of polished plate specimen. Average LCF life of plate specimen increased by shot peening. These results suggest that LCF life of peened plate specimen was enhanced because larger compressive stress layer was induced by shot peening. Peened hole specimen was almost same average LCF life as that of as machined hole, whereas smaller variability of LCF life was observed in peened hole specimen. This may be attributed to more stable compressive stress layer caused by shot peening than that of as machined hole in actual features.

LCF life prediction based on fatigue crack growth analysis

Average LCF life of plate specimen was enhanced by shot peening. This effect was assessed in relation to fatigue crack propagation analysis including to the compressive residual stress. Figure 5 shows flowchart of LCF life prediction. Number of cycles to failure is given by Eqn 1.

$$N_f = N_i + N_c \tag{1}$$

Here, N_i is the fatigue crack initiation life and N_c is the fatigue crack propagation life. To simplify the fatigue life prediction, it is assumed that shot peened specimen and as machined specimen have same fatigue crack initiation lives. Fatigue crack initiation lives were estimated based on the crack propagation analysis using LCF test results of as machined specimen. NASGRO 7.1 was used for fatigue crack growth analysis [9]. Crack growth rate calculation in NASGRO is given by Eqn 2.

$$\frac{da}{dN}=C\left[\left(\frac{1-f}{1-R}\right)\Delta K\right]^n \frac{\left(1-\frac{\Delta K_{th}}{\Delta K}\right)^p}{\left(1-\frac{K_{max}}{K_c}\right)^q} \tag{2}$$

Where N is the number of applied fatigue cycles, a is the crack length, R is the stress ratio, ΔK is the stress intensity factor range, and C, n, p, and q are material constants. f is the crack opening function. Table V lists crack growth properties used in this assessment. The material properties of Ti-6Al-4V were used from NASGRO material data base in which heat treatment is similar to the material in this study.

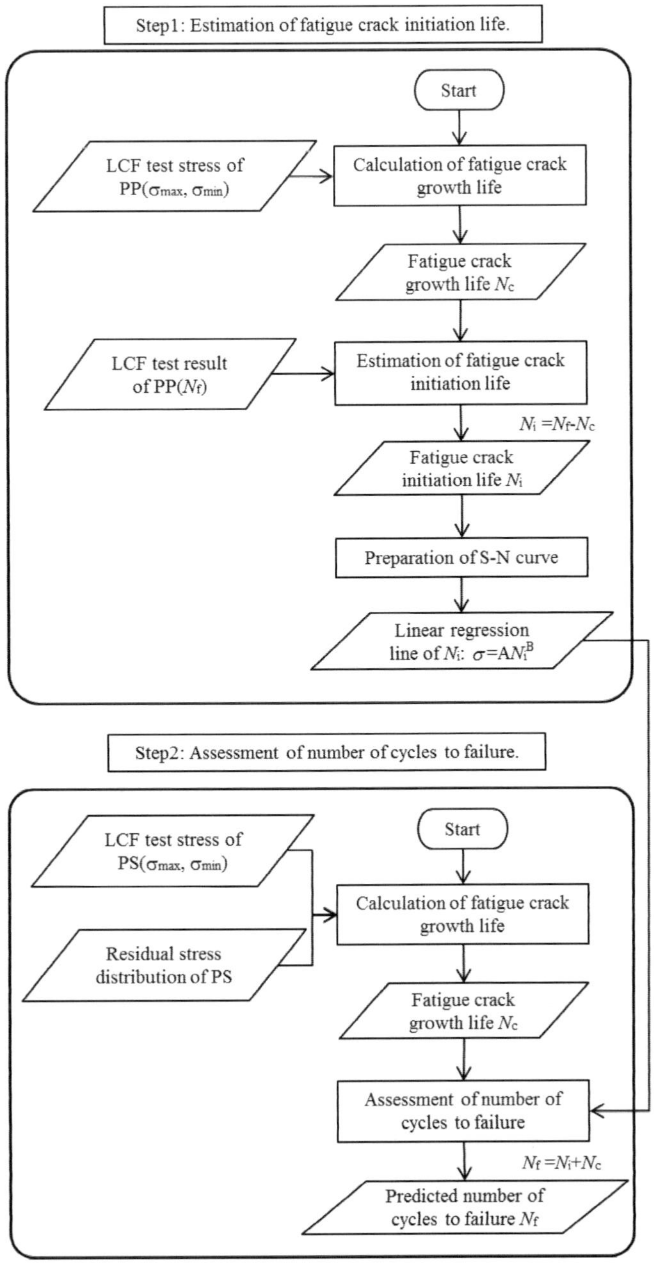

Figure 5. Flowchart of LCF life prediction based on fatigue crack growth analysis.

Table V. Crack growth properties of Ti-6Al-4V.

Material	Material ID	C	n	p	q
Ti-6Al-4V	P3ED20AB1	4.47E-13	3.2	0.25	0.75

Figure 6 shows results of fatigue life prediction based on the crack propagation analysis including residual stress. A factor of 4 scatter band, which agrees with variability of LCF test results, is indicated in this figure. Predicted lives are longer than that of experimental lives at the fatigue life shorter than 2×10^4 cycles. Previous study showed that residual stress relaxation was observed during LCF tests[10]. Thus, longer life prediction may be attributed to the overestimation of residual stress. Figure 7 indicates fatigue life prediction using 50% of the measured residual distribution if the applied stress to the LCF specimen is higher than material yield stress. All predicted lives are within a factor of 4 scatter band. Therefore, life prediction including to the residual stress relaxation agrees with variability of the test results satisfactory.

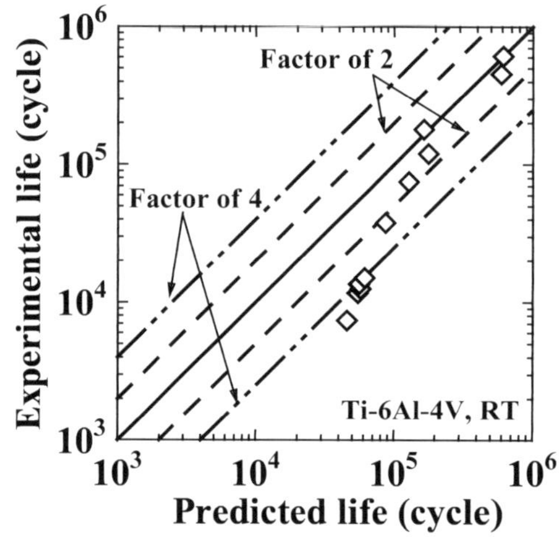

Figure 6. Life prediction based on crack growth analysis with residual stress induced by shot peening.

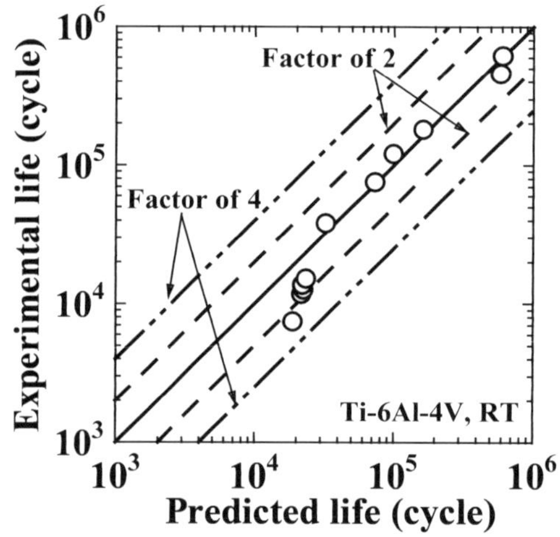

Figure7. Life prediction using 50% of the initial residual stress distribution.

CONCLUSION

This paper dealt with shot peening influence on LCF behavior of Ti-6Al-4V. Two types of specimens that were polished plate and plate with hole were used. Shot peening was carried out after machining. Residual stress measurement and LCF tests were conducted. Peened plate surface had higher and deeper compressive residual stress than that of as machined. Average LCF life of peened plate was longer than that of as machined. This result suggests that shot peening enhanced LCF capability of peened plate because of larger compressive residual stress layer. Peened hole had almost same average LCF life as that of as machined hole, whereas smaller variability of LCF life was shown in peened hole specimen. This may be attributed to more stable compressive residual stress induced by shot peening. Crack growth analyses were carried out to assess LCF life of shot peened specimen. Residual stress relaxation was considered if the applied stress to the LCF specimen was larger than material yield stress. As a result, LCF life can be predicted within a factor of 4 scatter band that agrees with variability of LCF test results.

ACKNOWLEDGMENT

This study was conducted under the collaborative investigation with Japan Aerospace Exploration Agency (JAXA) as a part of "Advanced Fan Jet Research (aFJR)" project.

REFERENCES

[1] Gerhard WK. (2003). In: *Shot Peening*, pp.13-22., Wiley-VCH Verlag GmbH & Co. KGaA.

[2] Kevin K. Liu and Michael R. Hill (2009). *Tribology International*, Vol.42, Issue9, pp.1250-1262.

[3] Gallagher JP. (2001). No. UDR-TR-1999-000079.

[4] Wagner L., Gerdes C. and Lutjering G. (1984). *Proceedings of the fifth international conference on Titanium*, Vol.4, 2147-2154.

[5] Khadhraoui M., Cao W., Castec L. and Guedou JY. (1997). *Materials science and technology*, Vol13, 360-367.

[6] Cao W., Khadhraoui M., Brenier B., Guedou JY. and Castex L. (1994). *Materials science and technology*, vol.10, no.11, 947-954.

[7] Prevey P., Shepard MJ. and Smith R. (2001). *Proceedings of the 6th national turbine engine high cycle fatigue conference*. 1-10.

[8] BATTELLE - BMI - Battelle Memorial Institute (2014). *Metallic Materials Properties Development and Standardization* (MMPDS-09).

[9] National Aeronautics and Space Administration (NASA) and Southwest Research Institute (SwRI) (2014). Fatigue crack growth computer program "*NASGRO*" version 7.1.

[10] Nakamura H., Takanashi M., Itabashi Y., Kuroki H. and Ueda Y. (2011). *Proceedings of ASME Turbo Expo 2011*, GT2011-46847.

Stress gradient effect in fretting-fatigue: a nonlocal approach by exploiting the crack analogue

Claudio Montebello[1,2], Sylvie Pommier[2], Karim Demmou[1], Julien Leroux[1], Jean Meriaux[1]

[1]*Snecma Villaroche, Rond-Point René Ravaud, Réau, 77550, Moissy-Cramayel, France*
[2]*LMT-Cachan (ENS Cachan/CNRS), 61 avenue du Président Wilson, 94230, Cachan, France*

Abstract: This paper describes an interesting method to model the stress gradient effect in fretting-fatigue. The analysis of the mechanical fields in the proximity of the contact edges allows to extract nonlocal intensity factors that take into account the stress gradient evolution. For this purpose, the kinetic field around the contact ends is partitioned into a summation of multiple terms, each one expressed as the product between nonlocal intensity factors, I^s, I^a, I^c, depending on the macroscopic loads applied to the mechanical assembly, and spatial reference fields, $\mathbf{d}^s$, $\mathbf{d}^a$, $\mathbf{d}^c$, depending on the local geometry of the part. This description is obtained through nonintrusive post-processing of FE computation and is conceived in order to be easily implementable in the industrial context.

By using as input the macroscopic load, the procedure consists in computing a set of nonlocal stress intensity factors, which are an index of the severity of the stress field in the proximity of the contact edges.

This description has two main advantages. First, the nonlocal stress intensity factors are independent from the geometry used. Secondly, the procedure is easily applicable to industrial scale FE model.

INTRODUCTION

Computational models for fatigue damage analysis in presence of a steep stress gradient still remain poorly effective. This is the case of fretting-fatigue where a local stress concentration is introduced by contacting components experiencing small amplitude relative motion.

This phenomenon is a major concern for the aerospace industry, in particular for aircraft engine applications where there are several components that may suffer fretting-fatigue failure like the dovetail blade-root connections.

In the last years several fatigue criteria have been developed aiming at improving the accuracy of the fatigue life prediction in the presence of a severe stress gradient. Almost all of them can be divided into two major families. On the one hand, the nonlocal stress fatigue models [1, 2] employ an averaged quantity as input for the prediction criterion in order to mitigate the effect of the strong stress gradient. On the other hand, a different approach consists in computing the evolution of the stress intensity factors, ΔK, as a function of the crack length to analyse whether the crack will stop or not [3, 4]. Even though both the approaches show good results in terms of life prediction accuracy some important limitations make them difficult to implement in the industrial context. For instance, the analysis of the ΔK evolution at the crack tip implies to set up a computational strategy to determine the crack path resulting from the macroscopic load application. This operation is heavily time-consuming. On the other hand, the methods based on an averaged quantity over a *critical distance* require an accurate description of the stress gradient evolution that is obtainable only by using a really fine mesh (micron size range), condition difficult to reproduce in industrial FEM models.

MODEL

Background

Fretting-fatigue is a special wear process that occurs at the contact area between two materials under load and subject to minute relative motion by vibration or some other force. One of the main peculiarities of this phenomenon is the fact that the contact introduces a severe and extremely localized stress gradient in the proximity of the contact edge.

The feature presented above is the same that characterizes cracks or notches where a steep stress gradient is present as well. As a consequence the mechanical fields generated close to the contact edge and the ones arising at the crack tip are comparable.

This analogy can be exploited to apply the mathematical tools already developed for fracture mechanics to fretting fatigue. Giannakopoulos has been the first author to quantify from an analytical point of view this analogy [5, 6]. Two different contact configuration are studied: (i) flat punch over a planar surface which creates a stress singularity at the contact tip (crack analogue) and (ii) round punch over a planar surface characterized by a smooth transition to zero pressure at the contact area edges (notch analogue).

Among the different approaches available in fracture mechanics one in particular can be really useful to describe the stress gradient effect in fretting-fatigue. An original model has been proposed by Pommier [7, 8] aiming at describing mixed-mode cyclic elastic-plastic behavior of the crack tip at the global scale. The purpose was to establish a reasonably precise model but condensed into a set of partial differential equations so as to avoid huge elastic-plastic FE computations.

The model proposed hereunder exploits the results obtained by Giannakopoulos [5, 6] and Pommier [7, 8]. A description of fretting-fatigue using nonlocal quantities has been developed. The tools employed to describe the elastic-plastic behavior of the crack tip at the global scale are redefine and adapted to fretting fatigue contact problem via the crack analogue approach.

Field partitioning

The model presented here, is based on the post-treatment of the velocity field generated by fretting-fatigue close to the contact edges. The velocity field is partitioned into a summation of multiple terms, each one expressed as the product between nonlocal intensity factors, I^s, I^a, I^c, depending on the macroscopic loads applied to the mechanical assembly, and spatial reference fields, $\mathbf{d}^s$, $\mathbf{d}^a$, $\mathbf{d}^c$, depending on the local geometry of the part,

$$\bar{v}(\bar{x},t) \cong \dot{I}^s(t)\bar{d}^s(\bar{x}) + \dot{I}^a(t)\bar{d}^a(\bar{x}) + \dot{I}^c(t)\bar{d}^c(\bar{x}). \tag{1}$$

From the practical point of view, the velocity field is obtained through a finite element computation. In Figure 1 an example of a FE model used in the analysis, is presented. A cylinder-plain contact configuration is employed. This choice is driven by the fact that the numerical procedure is validated by comparing it with experimental tests [9, 10, 11], carried out with a cylinder-plain apparatus. In other word the FE model needs to be as close as possible to the experimental setting.

Concerning the main parameters used in the FE model definition, it consists in an elastic quasi-static computation and plane strain linear elastic quadrilateral elements are employed. To handle the contact at the interface, the technique employed is the Lagrange multipliers, which assures the best accuracy. To modelling the friction, Coulomb's friction law is employed. With regard to the mesh used, it is a structured mesh characterised by an average length of 5–10 microns. The loads applied to the pad to reproduce fretting-fatigue consist in a constant normal load (P) and a cyclic tangential load (Q). A cyclic bulk load (σ_{fa}) is applied to the specimen to reproduce the fatigue effect.

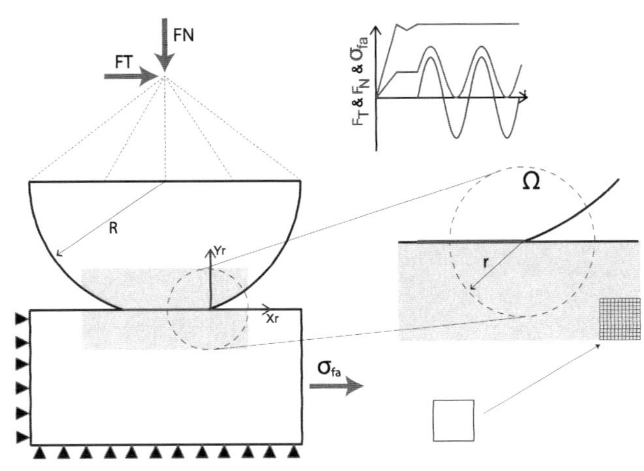

Figure 1. FE model and fretting-fatigue loading variation.

Once the velocity field is computed, the following step is to partition it in order to extract the nonlocal intensity factors. Since the most critical zone in fretting-fatigue is situated close to the contact border, only the value of the velocity field inside a circular region Ω, of radius r, centred at the contact tip is retained.

To obtain the partition presented in equation 1, the first step is the determination of the reference fields ($\mathbf{d}^a$ and $\mathbf{d}^s$). $\mathbf{d}^s$ is selected as the field generated by a ΔP, while $\mathbf{d}^a$ as the one produced by a ΔQ in a situation where all the contact surface is in a stick condition. Further details of the procedure followed can be found in [9].

By expressing the reference fields in polar coordinates, their radial and tangential evolution with respect to the crack tip can be obtained.

$$\overline{\boldsymbol{d}}^s(\overline{\boldsymbol{x}}) \rightarrow \overline{\boldsymbol{d}}^s(\boldsymbol{r},\boldsymbol{\vartheta}) \cong \boldsymbol{f}^s(\boldsymbol{r})\, \overline{\boldsymbol{g}}^s(\boldsymbol{\vartheta}), \tag{2}$$

$$\overline{\boldsymbol{d}}^a(\overline{\boldsymbol{x}}) \rightarrow \overline{\boldsymbol{d}}^a(\boldsymbol{r},\boldsymbol{\vartheta}) \cong \boldsymbol{f}^a(\boldsymbol{r})\, \overline{\boldsymbol{g}}^a(\boldsymbol{\vartheta}). \tag{3}$$

To express the reference fields as a product between two functions depending separately on r and θ, the Karhunen-Loeve decomposition [10] is employed.

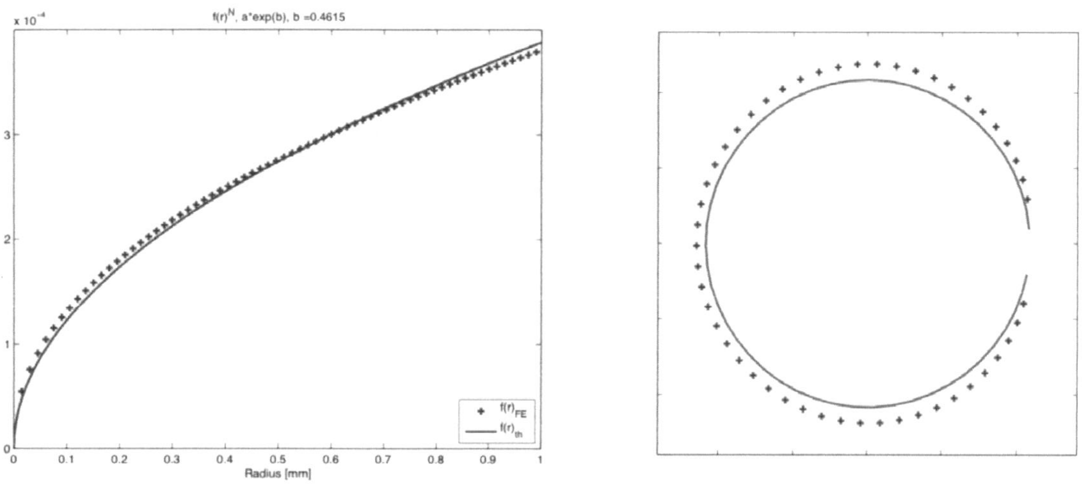

Figure 2. Radial and tangential evolution of d^s.

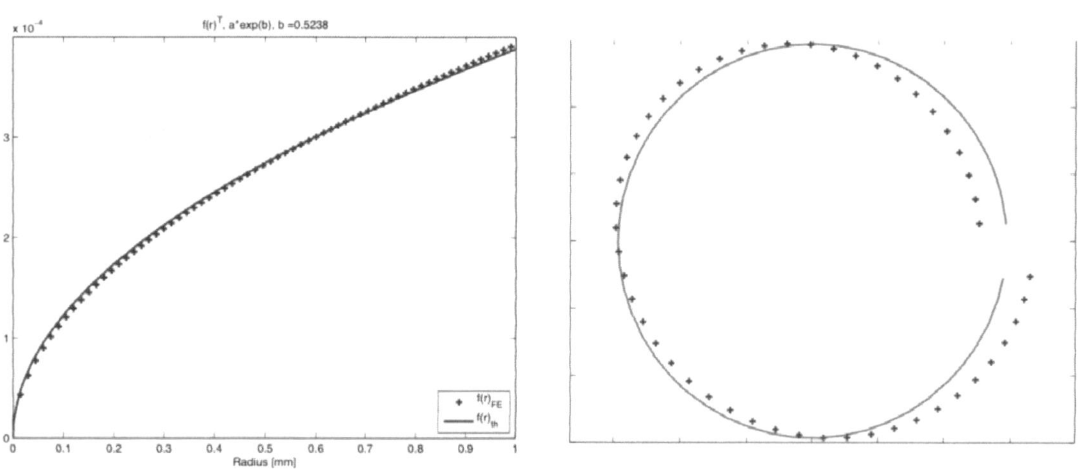

Figure 3. Radial and tangential evolution of d^a.

In Figures 2 and 3 the radial and tangential evolution of $\mathbf{d}^s$ and $\mathbf{d}^a$ are presented. Clearly the analogy between crack and contact problems is justified. Furthermore, both reference fields are normalized in order to correspond to the displacement field obtained at the crack tip during an elastic loading phase with either ΔK_I or ΔK_{II} equal to 1MPa.

The projection of the reference fields on the velocity field allows us to extract I^s and I^a as follows:

$$\dot{\boldsymbol{i}}^s(\boldsymbol{t}) = \frac{\int_\Omega \overline{\boldsymbol{v}} \cdot \overline{\boldsymbol{d}}^s}{\int_\Omega \overline{\boldsymbol{d}}^s \cdot \overline{\boldsymbol{d}}^s} \rightarrow \dot{\boldsymbol{i}}_t^{\,s} = \frac{\sum_i^N \overline{\boldsymbol{v}}_{i,t} \cdot \overline{\boldsymbol{d}}_i^{\,s}}{\sum_i^N \overline{\boldsymbol{d}}_i^{\,s} \cdot \overline{\boldsymbol{d}}_i^{\,s}}, \tag{4}$$

$$\dot{\boldsymbol{i}}^a(\boldsymbol{t}) = \frac{\int_\Omega \overline{\boldsymbol{v}} \cdot \overline{\boldsymbol{d}}^a}{\int_\Omega \overline{\boldsymbol{d}}^a \cdot \overline{\boldsymbol{d}}^a} \rightarrow \dot{\boldsymbol{i}}_t^{\,a} = \frac{\sum_i^N \overline{\boldsymbol{v}}_{i,t} \cdot \overline{\boldsymbol{d}}_i^{\,a}}{\sum_i^N \overline{\boldsymbol{d}}_i^{\,a} \cdot \overline{\boldsymbol{d}}_i^{\,a}}. \tag{5}$$

A first approximation of the velocity field is therefore obtained:

$$\overline{\boldsymbol{v}}^e(\overline{\boldsymbol{x}}, \boldsymbol{t}) \cong \dot{\boldsymbol{i}}^s(\boldsymbol{t})\overline{\boldsymbol{d}}^s(\overline{\boldsymbol{x}}) + \dot{\boldsymbol{i}}^a(\boldsymbol{t})\overline{\boldsymbol{d}}^a(\overline{\boldsymbol{x}}). \tag{6}$$

$\mathbf{v}^e$ is the elastic approximation of the velocity field. Since the friction between the contacting bodies introduces a nonlinearity in the system a third term, $\mathbf{v}^c$, has to be added to the approximation presented in equation 6,

$$\overline{\boldsymbol{v}}^c(\overline{\boldsymbol{x}}, \boldsymbol{t}) = \overline{\boldsymbol{v}}(\overline{\boldsymbol{x}}, \boldsymbol{t}) - \overline{\boldsymbol{v}}^e(\overline{\boldsymbol{x}}, \boldsymbol{t}). \tag{7}$$

Here, the Karhumen-Loeve decomposition is used twice: the first time to partition the residual velocity field in a product between two functions depending separately on time and space,

$$\overline{\boldsymbol{v}}^c(\overline{\boldsymbol{x}}, \boldsymbol{t}) \cong \dot{\boldsymbol{i}}^c(\boldsymbol{t})\overline{\boldsymbol{d}}^c(\overline{\boldsymbol{x}}) \cong \dot{\boldsymbol{i}}^c(\boldsymbol{t})\boldsymbol{f}^c(\boldsymbol{r})\,\overline{\boldsymbol{g}}^c(\boldsymbol{\vartheta}), \tag{7}$$

and the second time to analyse the evolution of $\mathbf{d}^c$ with respect to r and θ, as presented in Figure 4. Also in this case the field is normalized in order to assure that $f^c(r=0)$ is equal to 1. By contrast with the elastic reference fields, here the intensity decreases quickly far from the contact confirming the extremely localized effect of friction.

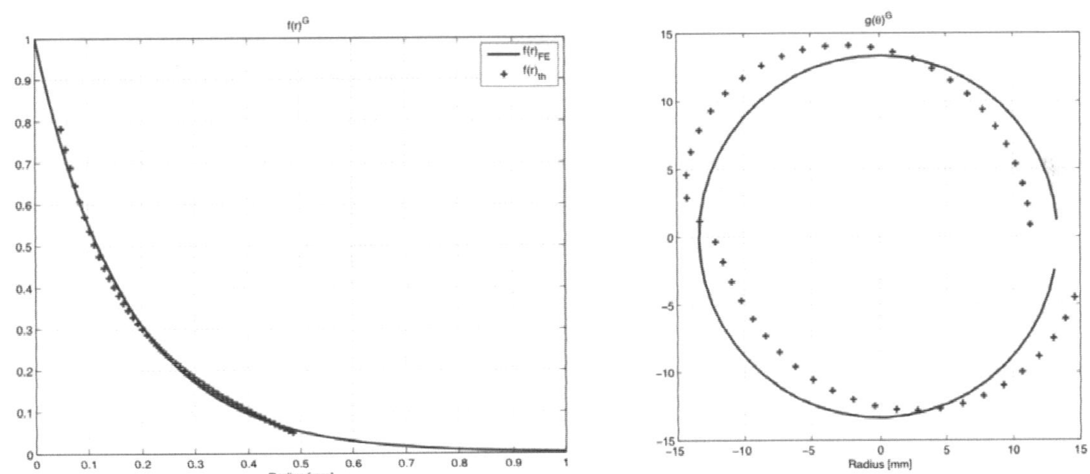

Figure 4. Radial and tangential evolution of d^c.

The error introduced by the approximation can be defined as follows:

$$\xi_e(\boldsymbol{t}) = \sqrt{\frac{\int_\Omega [\overline{\boldsymbol{v}}(\overline{\boldsymbol{x}}, \boldsymbol{t}) - \overline{\boldsymbol{v}}^e(\overline{\boldsymbol{x}}, \boldsymbol{t})]^2}{\int_\Omega \overline{\boldsymbol{v}}(\overline{\boldsymbol{x}}, \boldsymbol{t})^2}}, \tag{8}$$

$$\xi_t(t) = \sqrt{\frac{\int_\Omega [\bar{v}(\bar{x},t) - \bar{v}^e(\bar{x},t) - \bar{v}^c(\bar{x},t)]^2}{\int_\Omega \bar{v}(\bar{x},t)^2}}, \tag{9}$$

where ξ_e is the error introduced considering just the elastic approximation while in ξ_t all the terms presented in equation 1 are taken into account.

In Figure 5 the evolution of the error during a fretting-simulation is shown. It is worth noting that the total error falls below 5 %.

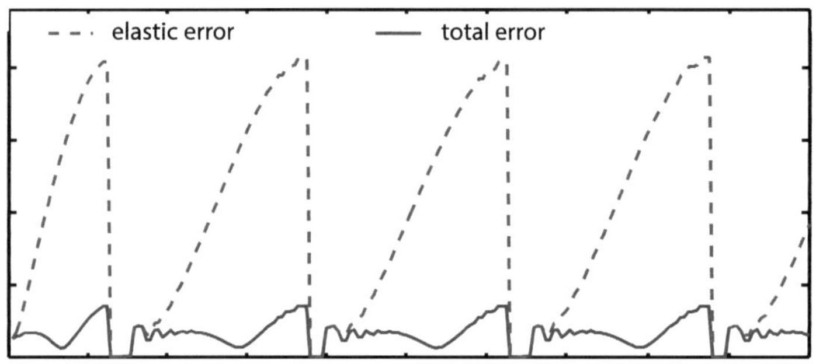

Figure 5. Error evolution during fretting-fatigue simulation.

The maximum of the cumulative integral of the intensity factors rate is computed in order to obtain, I^a, I^s, I^c.

To resume for any given macroscopic load combination (P, Q, σ_{fa}), a univocal set of nonlocal intensity factors (I^a, I^s, I^c) is found which characterizes completely the stress field close to the contact edge.

EXPERIMENTAL VALIDATION

The procedure described in the previous section has been validated by comparing it to the experimental results coming from the work of Fouvry [11]. The author performs several plain fretting tests using a cylinder-plain apparatus in order to determine different crack initiation frontiers as a function of the pad radius used. The material employed in the experimental campaign is 35 Ni Cr Mo 16 low-alloyed steel with the following properties:

- Young's modulus = 205000 MPa
- Poisson's ratio = 0.3
- Yeld stress (0.2%) = 950 MPa
- Ultimate stress= 1130 MPa
- Traction compression fatigue limit (R=-1, 10^7 cycles) = 575 MPa
- Shear fatigue limit (R=-1, 10^7 cycles) = 386 MPa
- Long crack threshold, $\Delta K_0 = 3.2$

Concerning the test apparatus, the plain fretting tests are carried out using a tension-compression hydraulic system as presented in Figure 6. The normal force (P) is kept constant, while a purely alternating cyclic displacement (δ) is imposed. As a consequence, a cyclic tangential load (Q) is generated at the contact surface. After a certain number of cycles, usually 10^5 or 10^6, the test is stopped and the specimen is cut along the median axis of the fretting scar in order to verify whether the crack is initiated or not. This procedure is repeated several times to find the tangential load amplitudes (Q$*$) that provokes crack initiation for a given normal load. Here, the crack is considered "initiated" if its projected length is higher than 10 μm.

To highlight the stress gradient effect, this process is reiterated for different cylinder radii. Varying the normal force, the normal pressure (p0) at the contact surface is kept constant and, therefore, the eventual difference in the initiation threshold needs to be caused by different stress gradients generated by the change in geometry.

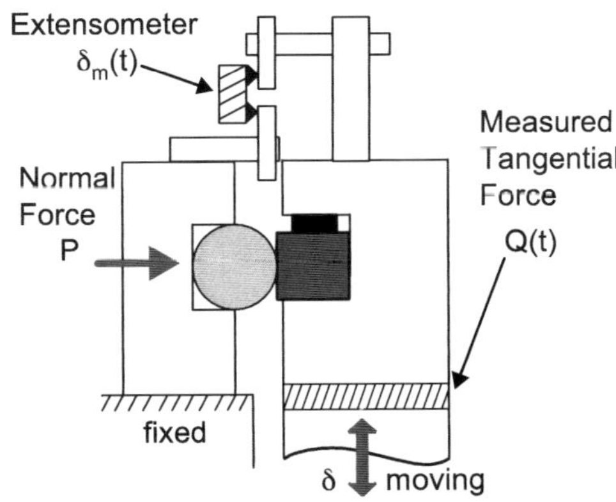

Figure 6. Schematic representation of a plain fretting apparatus.

In Figure 7, the experimental crack initiation frontiers are presented. It is clear that the gradient effect plays an important role. In particular it is evident that the biggest is the pad radius the worst is the effect in term of crack initiation threshold.

To interpret this tendency, it is important to notice that two factors influence crack nucleation; (i) the maximum stress at the hot spot and (ii) the material process volume over which the maximum stress operates. The increase in pad dimension extends the influence of contact stress below the surface, and therefore increases the process volume.

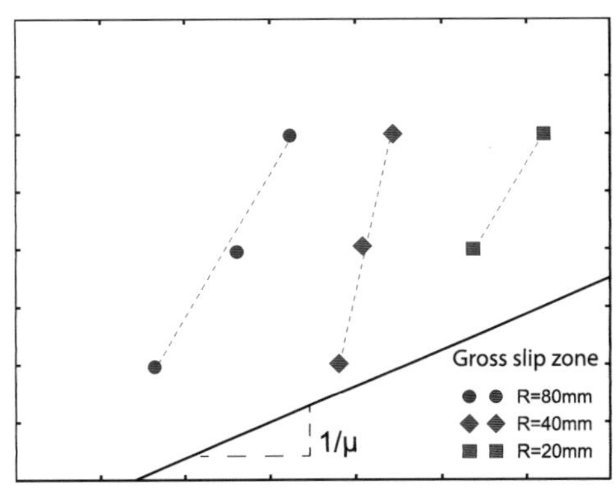

Figure 7. Crack initiation frontier for 35NiCrMo16 low-alloyed steel.

Application of the methodology.

The steps followed in the application of the numerical algorithm presented above are the followings (Figure 8):

1. The experimental data highlight different crack initiation frontiers expressed in (q_{max}-p_0) coordinates. For each point belonging to these frontiers the macroscopic load to apply in the FE model are computed. For cylinder-plain contact configuration, analytical expressions linking q_{max}-p_0 to P and Q are available.
2. A FE computation is run and the velocity field is computed. The forces used as input to load the structure are the ones found in the previous step.
3. The velocity field is partitioned as presented in equation 1.
4. The nonlocal stress intensity factors are extracted and a nonlocal crack initiation map is defined.

Before to comment the results shown in Figure 8, it is worth noting that only the elastic nonlocal stress intensity factors (I^a-I^s) are used in the description of the crack initiation frontiers for two reasons. The third term, I^c, describe the nonlinear effect of friction, which is extremely localized as confirmed by Figure 4. As a consequence the elastic nonlocal intensity factors, I^s and I^a, are sufficient to univocally characterize the contact edge stress condition. This approach is analogue to the one followed in fracture mechanics where the hypothesis of small scale yielding permits to use an elastic stress intensity factor, K, to describe the mechanical field close to the crack tip. On the other hand, all the test data are obtained assuring that the contact is in a partial slip regime. As explained by Vinsdbo and Sodemberg in [12], in this condition the wear rate due to friction is negligible.

The outcome of the application of the numerical algorithm to extract nonlocal intensity factors (Figure 8), shows clearly how these quantities are objective quantities through which is possible to take into account the gradient effect efficiently. The different crack initiation frontiers displayed in a classical q_{max}-p_0 crack initiation map, merge into a single one if the nonlocal intensity factors are used.

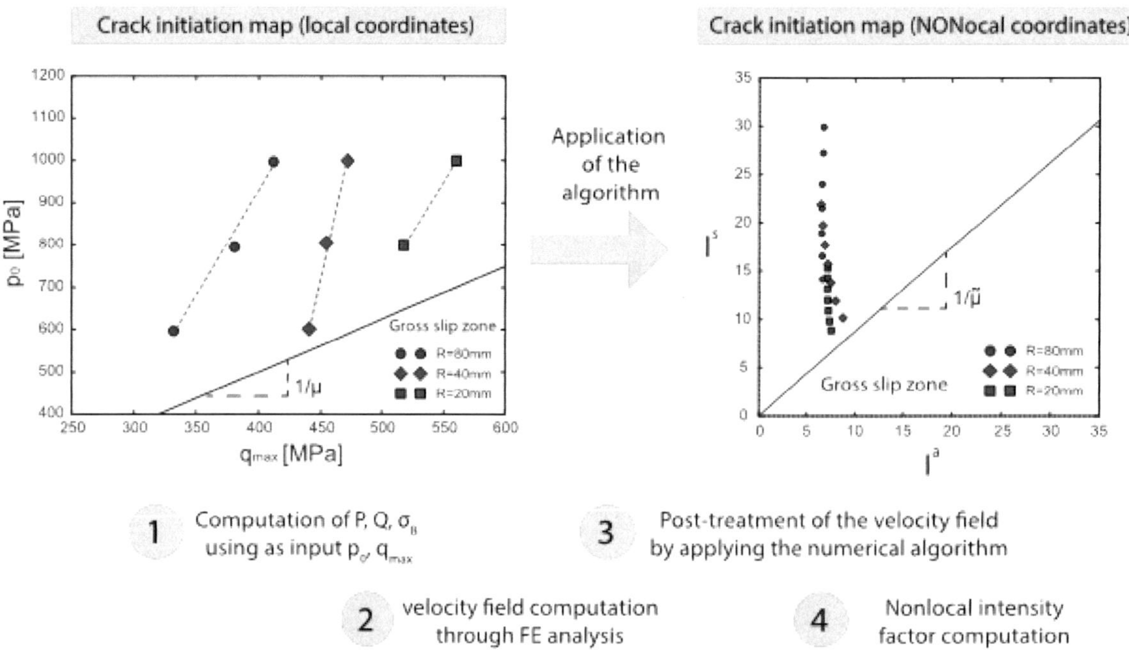

Figure 8. Different steps in the application of the numerical algorithm.

DISCUSSION OF THE RESULTS

In order to understand the importance of the results obtained above, it is useful to remember the main challenges than the industrial sector has to handle when the structures are interested by fretting fatigue problems.

The first main problem is related to the fact that fretting-fatigue introduces a big stress gradient at the contact interface that depends on the local geometry of the part. The fatigue life of the part depends on the stress gradient, which in turn depends on the local geometry. As a consequence, if the example of an aircraft engine is considered, the manufacturer has to certify experimentally several parts to show that no catastrophic failures occur. The limitation here is that the certification performed for a given part cannot be used for another part sharing the same properties except for the geometry. This is a big limitation because it forces the manufacturer to multiply the experimental tests to certify all the different components.

With the methodology introduced here, we propose a possible solution to this problem. If the nonlocal stress intensity factors are used to describe the mechanical field arising close to the contact edges, the gradient effect is taken into account and the change in geometry is no more a problem.

A second important complication is due to the fact that the few multi-axial fatigue criteria suited to be used for fretting fatigue [1, 2, 3, 4], need a really precise description of the stress gradient evolution. The only way to obtain it is by using FE computations characterized by extremely fine meshes. This requirement is not compatible with industrial constraints.

The methodology presented here is well-suited for a multi-scale approach. The nonlocal stress intensity factors can be extracted from rough meshes. In addition, since the procedure is not intrusive, it can be easily implemented in industrial procedures.

PERSPECTIVES

The future works will focus on the development of a fatigue criterion that take as input the nonlocal stress intensity factors extracted from the post-treatment of the velocity field. In this way it should be possible to place the position of the crack initiation frontier easily.

The second main objective is to apply the procedure to an industrial part. In particular the disc-blade root attachment will be analysed.

The author is grateful to Snecma (Safran group) for its financial support, within the IRG COGNAC project framework, and thanks all the co-authors.

BIBLIOGRAPHY

[1] J. Araújo, L. Susmel, D. Taylor, J. Ferro, E. Mamiya, On the use of the theory of critical distances and the modified wöhler curve method to estimate fretting fatigue strength of cylindrical contacts, International Journal of Fatigue 29 (1) (2007) 95 – 107.

[2] From uni- to multi-axial fretting-fatigue crack nucleation: Development of a stress-gradient-dependent critical distance approach, International Journal of Fatigue 62 (0) (2014) 194 – 209, 9th Fatigue Damage of Structural Materials Conference.

[3] S. Fouvry, D. Nowell, K. Kubiak, D. Hills, Prediction of fretting crack propagation based on a short crack methodology, Engineering Fracture Mechanics 75 (6) (2008) 1605 – 1622.

[4] D. Dini, D. Nowell, I. N. Dyson, The use of notch and short crack approaches to fretting fatigue threshold prediction: Theory and experimental validation, Tribology International 39 (10) (2006) 1158 – 1165, the Fourth International Symposium on Fretting Fatigue.

[5] A. Giannakopoulos, T. Lindley, S. Suresh. Aspects of equivalence between contact mechanics and fracture mechanics: theoretical connections and a life-prediction methodology for fretting-fatigue. Acta Mater 1998;46(9):2955-68.

[6] A. Giannakopoulos, S. Suresh, C. Chenut. Similirarities of stress concentrations in contact at round punches and fatigue at notches: implication to fretting fatigue crack initiation. Fatigue Fract Eng Mater Struct 2000;23(7):561-71.

[7] S. Pommier, R. Hamam. Incremental model for fatigue crack growth based on a displacement partitioning hypothesis of mode I elastic-plastic displacement fields. Fatigue & Fracture of Engineering Materials & Structures 30 (7) (2007): 582 – 598.

[8] S. Pommier, P. Lopez-Crespo, P. Decreuse. A multi-scale approach to condense the cyclic elastic-plastic behaviour of the crack tip region into an extended constitutive model. Fatigue & Fracture of Engineering Materials & Structures 32 (11) (2009): 899 – 915.

[9] C. Montebello, S. Pommier, K. Demmou, J. Leroux, J. Meriaux. Analysis of the stress gradient effect in fretting-fatigue through nonlocal intensity factors. Internation Journal of Fatigue, 2015.

[10] M. Loéve, Probability Theory, Van Nostrand, New York, 1955.

[11] S. Fouvry, H. Gallien, B. Berthel, From uni to multiaxial fretting-fatigue crack nucleation: Development of a stress gradient dependent critical distance approach, Internation Journal of Fatigue 62 (0) (2014) 194-209, 9th Fatigue Damage of Structural Materials Conference.

[12] O. Vingsbo, S. Söderberg. On fretting maps, Wear 126 (2) (1988) 131-147.

Shot Peening Residual Stress Evaluation from 3D Finite Element Simulation

Joris Prou[1], Mitsuyoshi Tsunori[1], Hiroshi Nakamura[2]

[1]*IHI Corporation, Research Laboratory, Japan*
[2]*IHI Corporation, Aero-Engine & Space Operations, Japan*

Abstract: Shot peening is a mechanical treatment that induces a surface compressive residual stress layer by impacting hard spheres on a component surface in order to increase its lifetime. In preparation for improving fatigue life through the shot peening conditions, 3D dynamic finite simulations of the peening process for Ti-6Al-4V target material were carried out to assess the effect of peening process parameters and workpiece geometry on the residual stress. Simulation showed that while impact velocity and impact angle have a significant effect on both magnitude and depth of the compressive stress layer, increasing shot size lead to deeper compressive layer without influencing the residual stress magnitude. Moreover, computations performed considering different workpiece features suggest that a decrease in residual stress magnitude is to be observed at the vicinity of edge or hole features over a region of characteristic size similar to the shot diameter.

INTRODUCTION

Shot peening is a surface enhancement process used to improve component fatigue life by impacting hard shots at high velocity on the target material. This process induces a surface compressive residual stress layer that is recognized to improve the fatigue resistance [1-2]. By including the shot peening benefits in the design phase, it is expected that jet engine components such as disks can be lighten without impairing their fatigue resistance. However, to take full advantage of the shot peening process, the effect of the peening conditions and the work piece geometry on the residual stress distribution need to be quantitatively assessed. Performed on a purely experimental basis, such investigations turn out to be costly and time consuming. The use of finite element simulation is thus a promising alternative to reach optimal peening conditions. Finite element simulations of the shot peening process have been developed over 30 years to discuss residual stress formation mechanism, influence of peening parameters and predict residual stress distribution [3-8]. Shot peening simulation models have evolved from single impact axisymmetric models [3,5] to the more realistic multiple impact three dimensional models [6-7], where effects due to inclined shots and peening coverage (percentage of work piece surface impacted) can be taken into account. In the latter case, simulated residual stress distributions are usually found to be in good agreement with residual stress measurement. In this paper, as a first step toward the optimization of the peening process, we focused on 3D dynamic finite element simulations of the shot peening on Ti-6Al-4V. Ti-6Al-4V was selected as the target material for this study since its strength-to-weight ratio and good corrosion resistance make it one of the major alloys used in aerospace engine components. Single shot simulations were conducted to assess the influence of peening parameters and validate the computational approach. Further, single and multiple impact computations were performed to discuss the effect of the target component geometry on the residual stress distribution.

SIMULATION OF THE SHOT PEENING PROCESS

Numerical simulations of the shot peening process have been performed using dynamic finite element analysis to model single and multiple shot impacts on the target material. In this study, an elastoplastic behavior with isotropic hardening was assumed for the Ti-6Al-4V and the material properties specified in [8] were used. Possible strain-rate sensitivity of the target material was omitted in this study. While deformable shot model should be considered when shot and target material have similar stiffness and hardness [7], shots were modeled as rigid spheres since the stiffness of Ti-6Al-4V is small compared to the one of cast steel shots. This assumption was validated by preliminary simulations using elastoplastic and rigid shot models showing that the shot model had little influence on the residual stress distribution. Concerning the boundary conditions, normal displacement constraints have imposed at the bottom and lateral surfaces of the model. It should be noted that care should be taken for the boundary conditions when dealing with thin specimen. As underlined in [9], thin specimen are prone to show bending deformations when subjected to peening process and boundary conditions should be chosen accordingly to allow this type of deformation. In this study, only contact conditions between the shot and workpiece were considered, neglecting possible collisions between shots. The contact was assumed to follow Coulomb friction law and the friction coefficient chosen as 0.3. The input corresponding to the shot peening process parameters are the shot velocity V_0 and the impact angle α which are specified by mean of initial velocity conditions (Fig. **1**).

In single shot simulation, the impact location was taken as the center of the top surface, and the residual stress distribution was evaluated below the impact location as a function of the depth (Fig. 2 (a)). For the multiple shots case, the shots were distributed randomly over an impact region similarly to the procedure defined in [6] and illustrated in Fig. 2 (b). The residual stresses were then estimated by averaging stresses at over an evaluation area for different distance to the surface. Such simulation approach was applied to assess the influence of shot peening process parameters and peened part geometry in order to gain insights into peening process optimization.

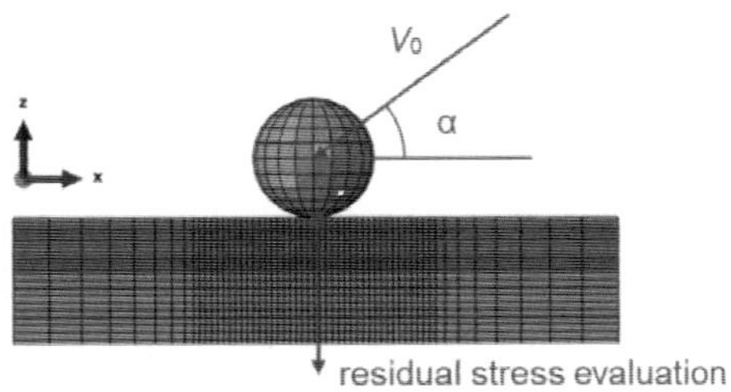

Figure 1. Finite element model showing the shot velocity V_0 and the impact angle α initial conditions.

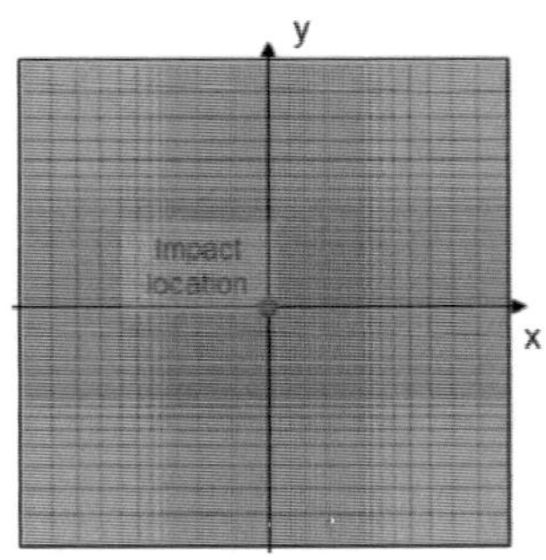

(a) Single impact

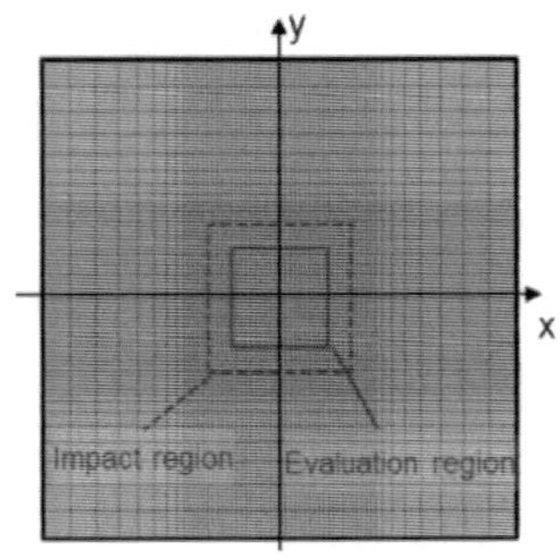

(b) Multiple impacts

Figure 2. Impact location and stress evaluation location for and (b) mutiple impact simulations.

INFLUENCE OF PEENING PROCESS PARAMETERS

The effects of several peening parameters were studied from single shot simulations. Shot diameter, impact velocity and impact angle were considered here in view of their supposedly high influence on peening residual stress. Each parameter influence was then characterized using the following representative values: compressive layer thickness and maximum compressive stress. These values are obtained from the residual stress distribution as indicated in Fig. 3, where the residual stress correspond normal stress along the x axis. When not specified otherwise shot velocity of 50 m/s, shot size of 0.325 mm, and normal impact (α=90°) have been used.

Influence of shot size

Single shot simulation using five shot diameters in the range of 150 to 500 μm have been performed. Simulation results showed that increasing shot size results to an increase of compressive layer thickness, while having almost no influence on the residual stress values (Fig. 4). The compressive layer thickness is found to depend nearly linearly on the shot size. These results are in agreement with the analytical model of Al-Hassani [10] where the plastic layer depth is proportional to the shot radius.

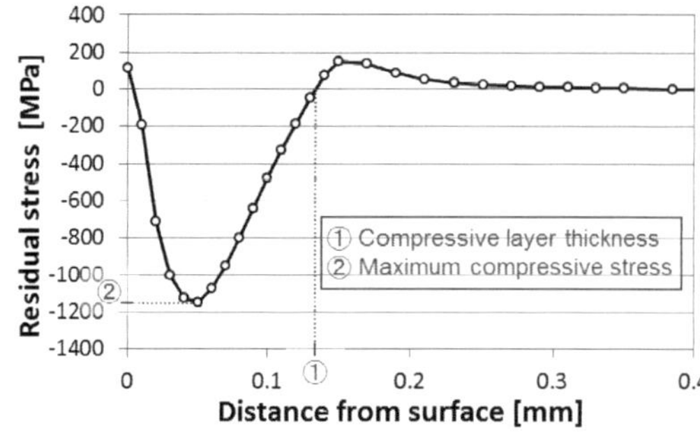

Figure 3. Residual stress distribution and its representative values.

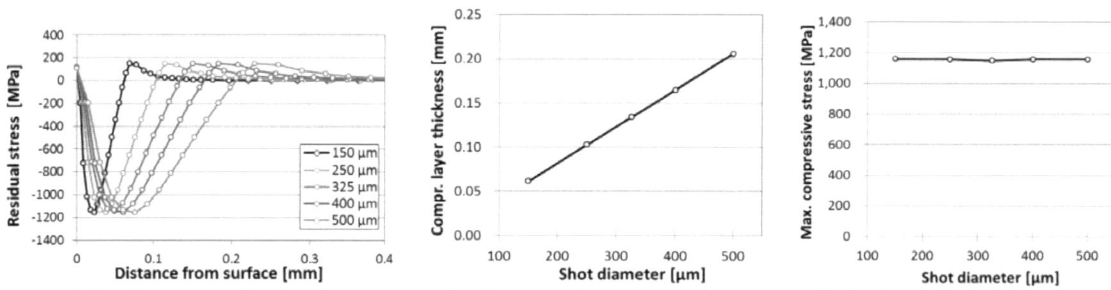

(a) Residual stress distribution (b) Compressive layer thickness (c) Maximum compressive stress

Figure 4. Influence of the shot size on the residual stress.

Influence of impact velocity

Five different impact velocities ranging from 10 to 90 m/s have been used in this study. These values are typical values of shot velocities in the shot peening process. Residual stress results as a function of the velocity are presented in Fig. 5. These indicate that an increase in the impact velocity results in an increase in both compressive layer thickness and compressive stress. Figure 6 presents the results with modified horizontal axes to highlight the dependency of the compressive layer thickness and maximum compressive stress to the velocity. Simulation results show an almost linear dependence of the compressive layer thickness with the root of the impact velocity. This is in agreement with the analytical analysis of Al-Hassani showing that the plastic zone below the impact is proportional to $v^{1/2}$. The maximum compressive stress is found to follow an approximate linear dependence to $\ln(V_0)$.

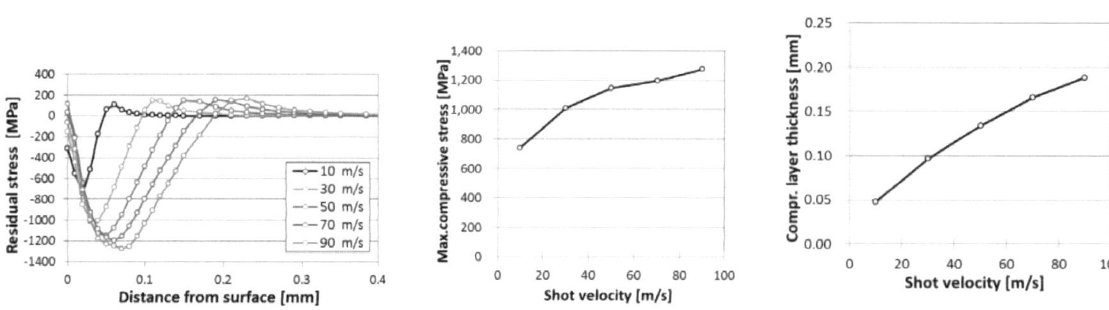

(a) Residual stress distribution (b) Compressive layer thickness (c) Maximum compressive stress

Figure 5. Influence of the impact velocity on the residual stress.

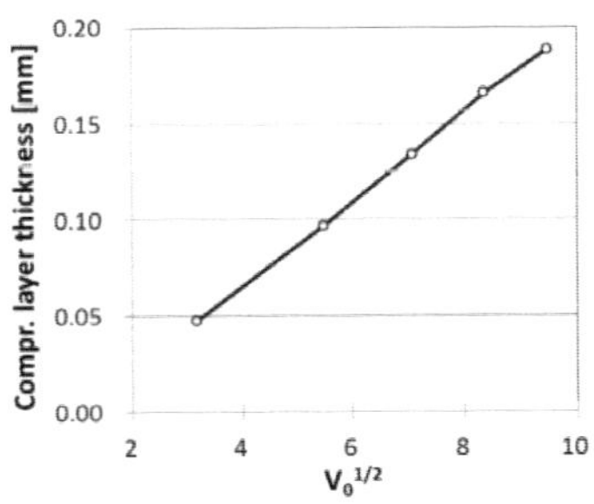

(a) Compressive layer thickness

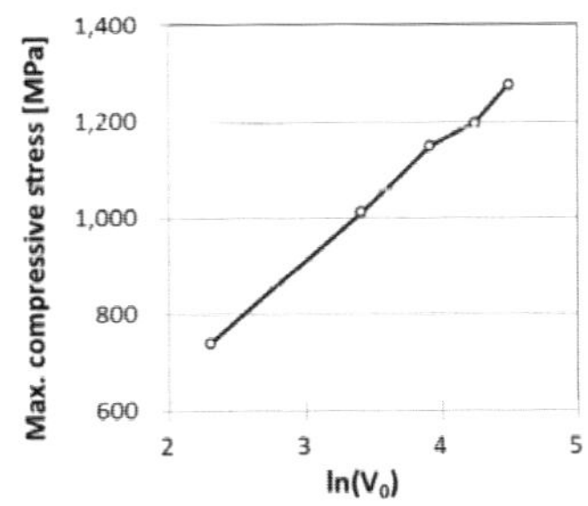

(b) Maximum compressive stress

Figure 6. Dependence of the residual stress representative values on the impact velocity V_0.

Influence of impact angle

The influence of impact angle have been studied using impact angle from 10 to 90°, where the angle of 90° is corresponding to an normal impact. The residual stress distribution results are presented in Fig. 7 (a). As for normal impacts, residual stress distributions were evaluated below the impact location. As can be seen in the residual stress contour in Fig. 7 (b), using such an approach for residual stress evaluation allows a good estimation of the compressive layer thickness but not of the maximum compressive stress. The maximum compressive stresses for oblique impact in Fig. 8 (a) were thus estimated directly from the contour map, while compressive layer thickness results presented in Fig. 8 (b) were estimated directly from the residual stress distributions in Fig. 7 (a). These simulation results indicate that increasing impact angle lead to an increase of maximum compressive stress and compressive layer thickness, the influence of the angle is getting weaker for high impact angles.

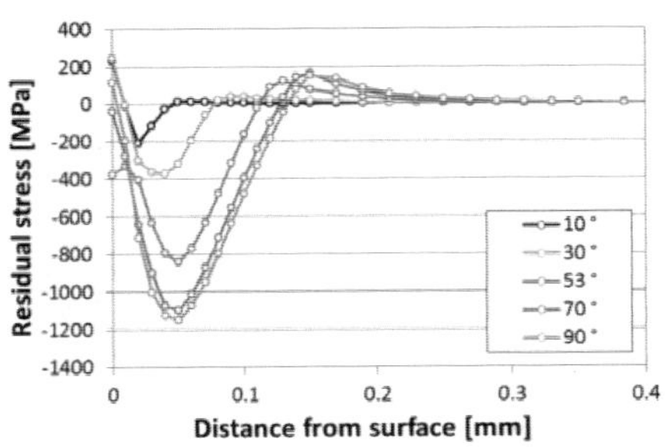

(a) Residual stress distribution

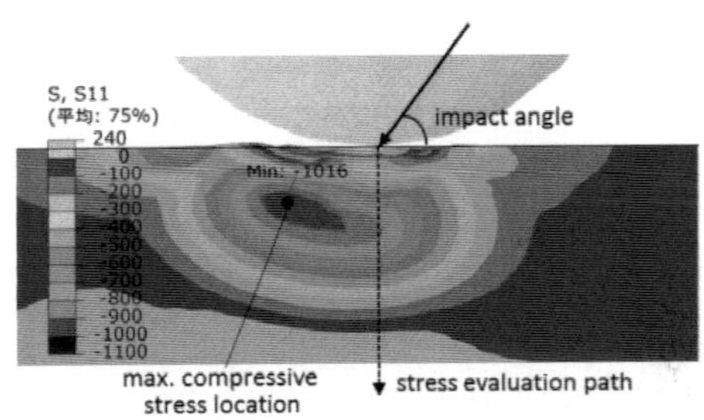

(b) Residual stress contour near impact location (impact angle:53°)

Figure 7. Influence of the impact angle on the residual stress.

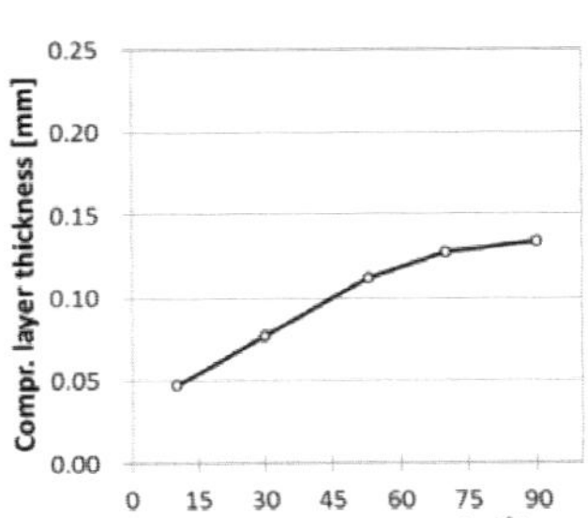

(a) Compressive layer thickness

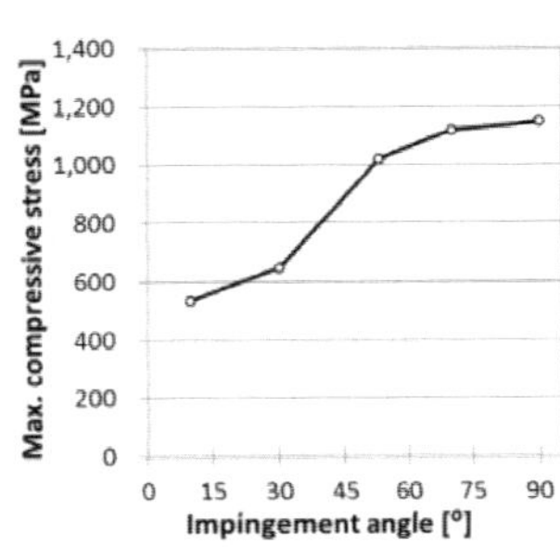

(b) Maximum compressive stress

Figure 8. Influence of the impact angle residual stress representative values on the impact angle.

Assuming that most of the deformation stems from the component of the shot velocity normal to the surface, these results have been plotted in Fig. 9 as a function of the normal velocity V_{0Z} and compared with the velocity results for normal impact displayed in Fig. 6. For normal impact the normal velocity V_{0Z} is simply V_0 while for oblique impact it is given by $V_0 \sin(\alpha)$. Figure 9 reveals that the relation between compressive layer thickness and normal velocity for oblique impact is similar to the one for normal impact. However, at same normal velocity level, the maximum compressive stress is found to be lower for oblique impact than in normal impacts. This can be attributed to the direction of the maximum compressive stress in oblique impacts. Contrary to normal impacts the direction of maximum stress for oblique impacts does not coincide with the x axis used in the stress evaluation. Moreover, Fig. 9 (b) shows a relatively high value of compressive stress for low impact angle (10°), which may be related to friction. Indeed, only normal velocity has been considered, however due to friction part of the kinetic energy due to the horizontal motion of the shot is transferred to the workpiece.

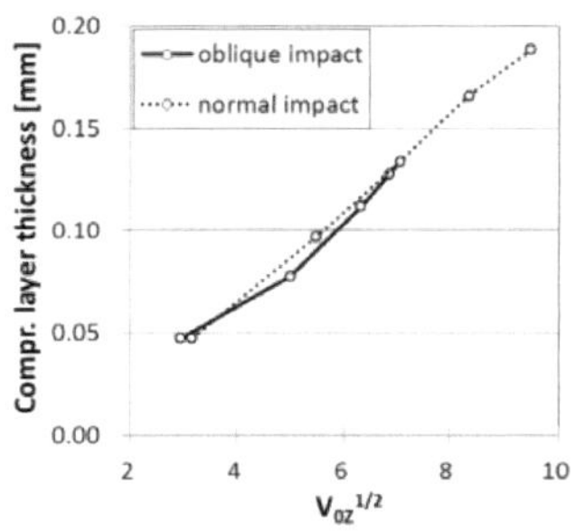

(a) Compressive layer thickness

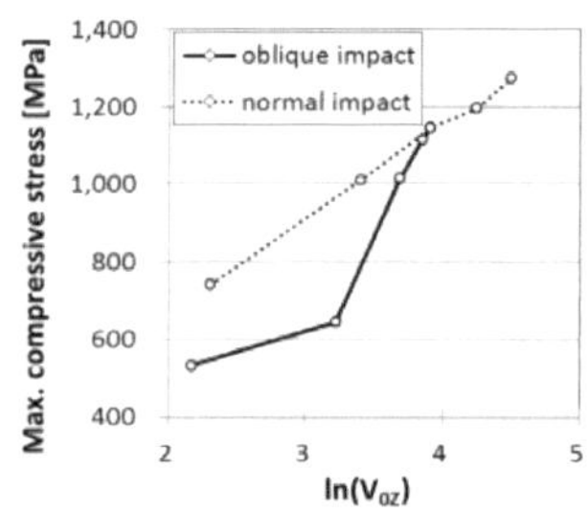

(b) Maximum compressive stress

Figure 9. Comparison of the influence residual stress representative values dependence to the normal impact velocity V_{0Z} for normal and oblique impacts.

INFLUENCE OF WORK PIECE GEOMETRY

The influence of the workpiece geometry on the residual stress distribution was discussed based on single and multiple shot simulations. The chosen workpiece is a fatigue test specimen with hole represented in Fig. 10. The influence of the specimen geometry was discussed based on computations for flat surface, edge and hole features as indicated in the same figure. The effect of the free edge was assessed by comparing residual stress distribution obtained for the edge and hole model to the one of the reference case represented by the flat surface. The three corresponding finite element models are presented in Fig. 11. For the edge and hole feature cases, shot were impacted on two adjacent surfaces to simulate the full peening of the work piece. Simulations were performed assuming normal impact with an initial velocity of 50m/s. The shot diameter was taken as 0.2375 mm and the friction coefficient was assumed to be 0.3. For multiple impacts simulation, the shot were randomly distributed over an impacted region of size 1.5D×1.5D and residual stress averaged over a region of size D×D, where D corresponds to the shot diameter (Fig. 2 (b)). The shot number was chosen to achieve full coverage of the evaluation area. To this purpose, coverage was evaluated for simulation with different shot numbers based on the equivalent plastic strain values of the elements at the surface. The element were counted as impacted when the plastic strain value was greater than a threshold value representative of the plastic strain state within the dimple obtained from single shot simulation. Fig. 12 (a) shows the surface vertical displacement and plastic strain simulation results after single impact using the flat surface model. It can be observed that the lowest plastic strain within the dimple is found at the dimple boundary. This value was therefore chosen as the plastic strain threshold value for coverage estimation. The coverage curve as a function of the shot number is displayed in Fig. 12 (b). Based on the coverage curve a value of 48 shots was chosen to carry on multiple shot simulations on the flat surface model. This shot number correspond to a 98% coverage which is the value usually accepted to define full coverage.

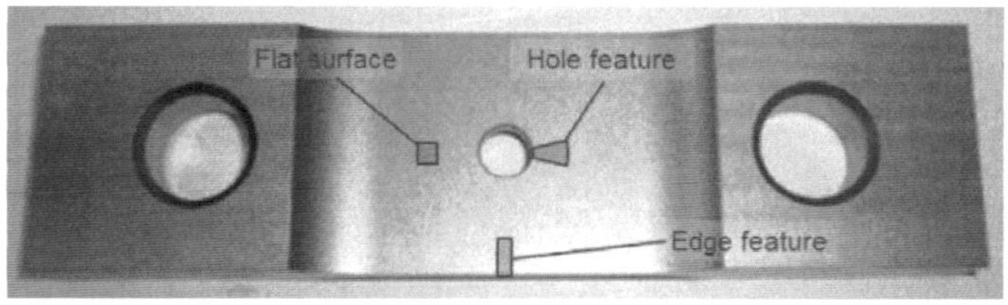

Figure 10. Plate specimen with holes indicating the target features.

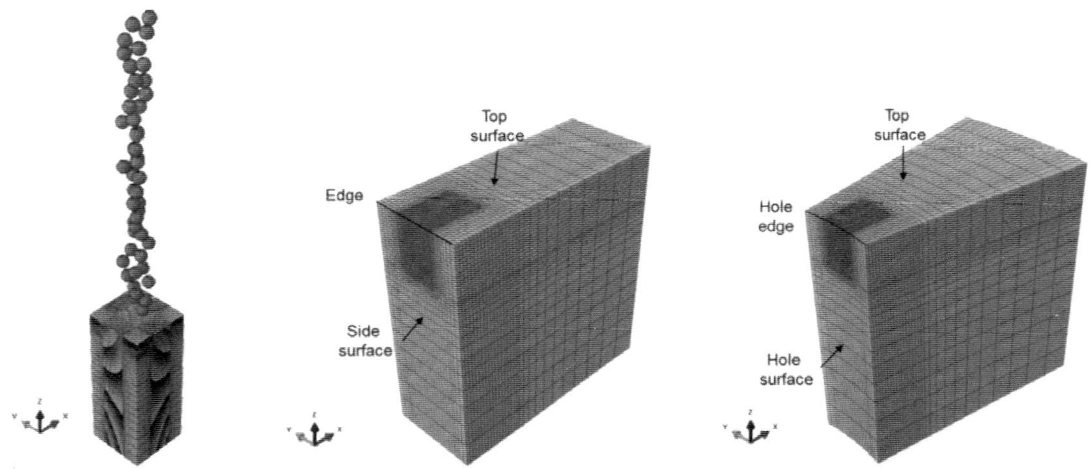

(a) Flat surface (b) Edge feature (c Hole feature

Figure 11. Finite simulation models for different target geometries.

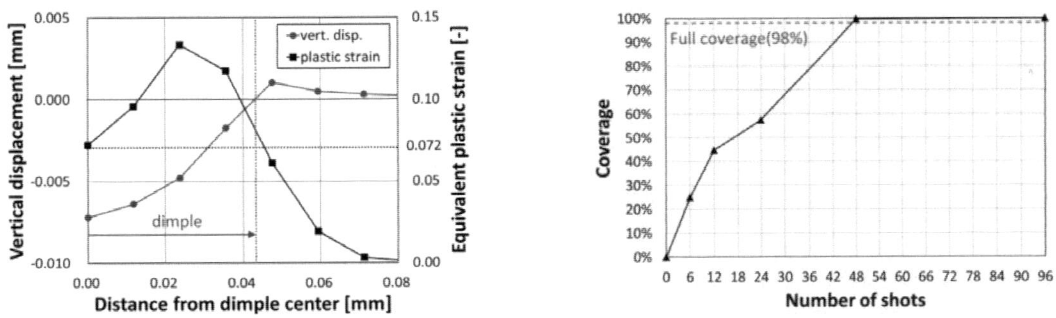

(a) Equivalent plastic strain at the dimple vicinity from single shot simulation

(b) Simulated coverage curve

Figure 12. Determination of the shot number for multiple impact simulation based on coverage estimated from the equivalent plastic strain.

The residual stress results for single and multiple impacts on the flat surface model are shown in Fig. 13. The maximum compressive stress is substantially higher in the single impact case, since stress are evaluated just below the impact where stress are maximum, while in the case of multiple shots stress represent averaged values. On the contrary, compressive residual stresses at the surface are higher for multiple impacts. These residual stress distributions were used as a reference to discuss the influence of edge and hole features

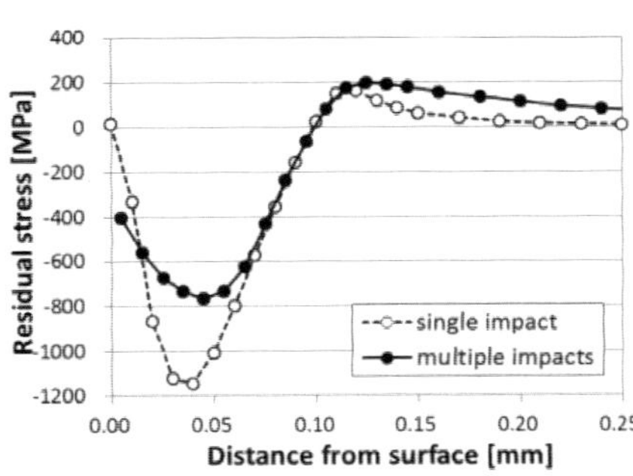

Figure 13. Residual stress distribution for the flat surface model

For the edge and hole cases, single shot simulations were conducted for six different impact locations by increasing the distance to the edge by 0.2 D as indicated in Fig. 14 (a), (b). As previously the residual stress were estimated below impact locations. When considering peening effect on edge crack growth, residual stress acting along the edge directions are thought to a have a predominant effect. Consequently, normal stresses along y axis were evaluated. For multiple impacts simulation, the impacted region was increased along the direction normal to the edge to reach 2.5 D as shown in Figs. 14 (c), (d). The residual stresses were then evaluated as for the flat surface case, by averaging stress over a region of approximate size D×D. To estimate edge effect, the averaging region was then shifted away from the edge by increment of 0.25 D. The required shot number was calculated using the shot number obtained for the flat surface which was then scaled by the impact area. For the edge model 80 shots were required to peen each of the top and side surfaces, while for the hole model the shot number was determined to be 80 for the hole surface and 86 for the top surface.

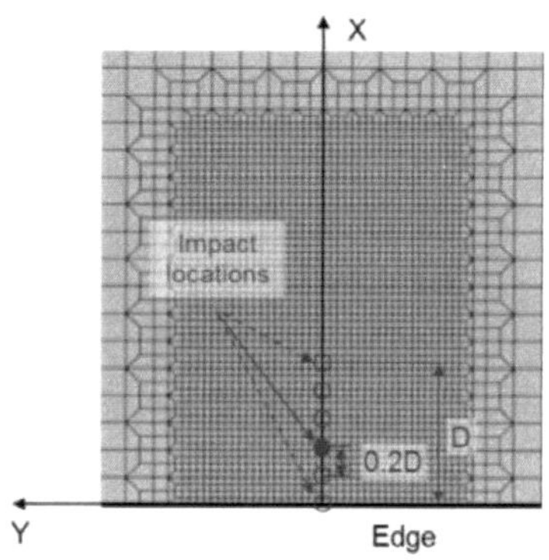

(a) Single impact edge model

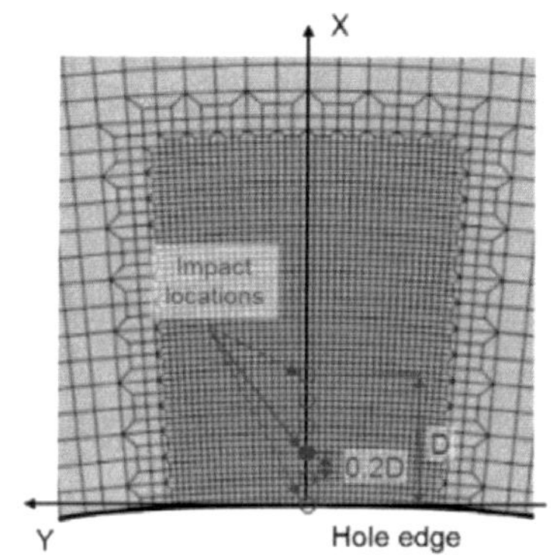

(b) Single impact hole model

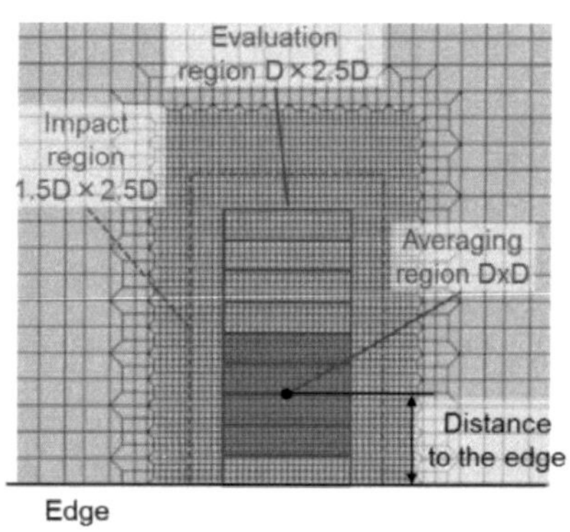

(c) Multiple impacts edge model

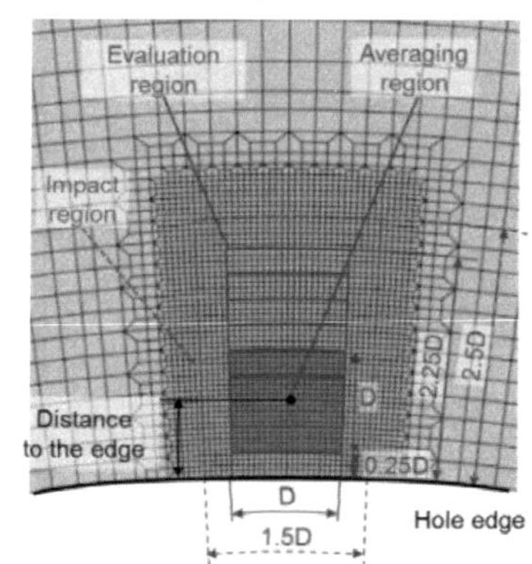

(d) Multple impacts hole model

Figure 14. Edge and hole models near the impact showing the impact location and stress evaluation region.

Figure 15 (a) shows the residual stress distributions results in the single shot case. For impact locations close to the edge, a decrease in the compressive stress as well as partial tensile residual stress were observed. For impact to edge distance over 0.4 D, the residual stress converged to similar distribution obtained for the flat surface. The multiple shots results, which are more representative of the peening process, are presented in Fig. 15 (b). As for the single shot, similar edge effect could be observed, however the effect on the residual stress was found to be comparatively smaller. In comparison with the flat surface case, the maximum compressive stress close to the edge was decreased by approximately 20%. Comparable results were obtained from the simulation of the hole feature as shown in Fig. 16. Such simulation results suggests that the benefit of shot peening may be reduced near edge or hole features, usually subjected to higher working stress.

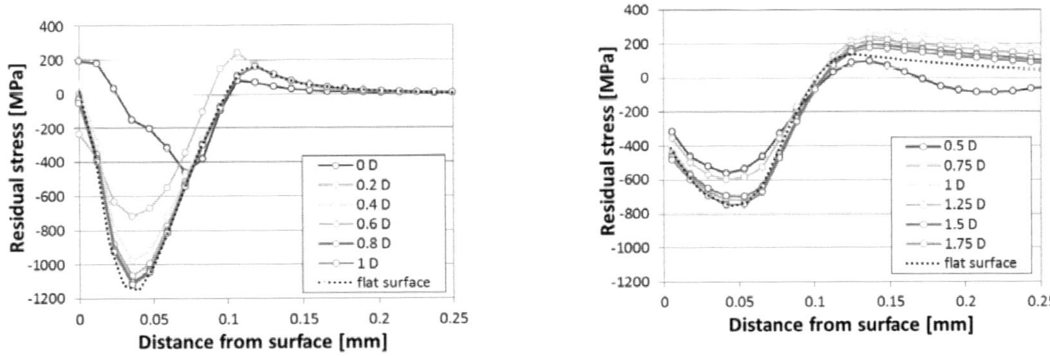

(a) Single impact (b) Multiple impacts

Figure 15. Residual stress profiles obtained from simulation of the edge model for different distance to the edge. Distance are expressed relatively to shot diameter D and represent impact location for single impact and center of the stress averaging region for mutiple impacts.

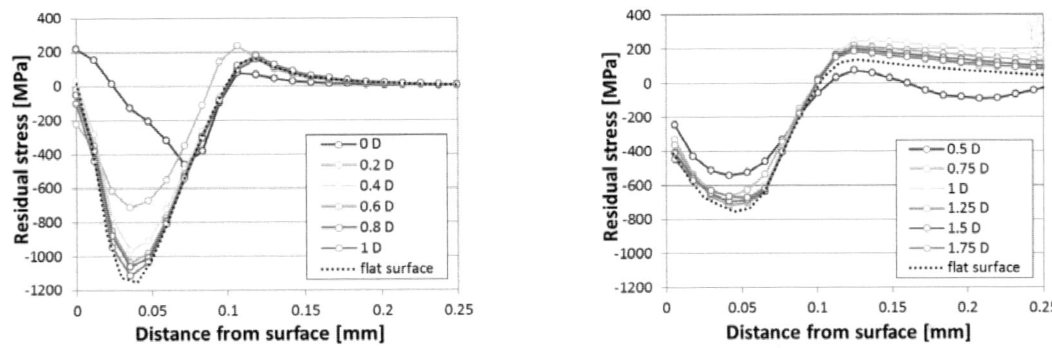

(a) Single impact (b) Multiple impacts

Figure 16. Residual stress profiles obtained from simulation of the hole model for different distance to the edge.

CONCLUSION

In this paper, three dimensional finite element computations of shot peening for Ti-6Al-4V have been performed to assess the effect of process parameter and workpiece geometry on the residual stress. From single impact simulation, the effect of impact velocity, impact angle and shot size were investigated. Simulation results indicated that while shot size affects primarily the depth of the residual stress compressive layer, impact velocity and angle have a large effect on both magnitude and depth of the residual stress. Moreover, simulation conducted on model representing edge and hole features suggested that at the vicinity of free edge there exists a region of size comparable to the shot diameter where the magnitude of the residual is decreased, which may reduce the efficiency of the peening effect process.

ACKNOWLEDGGEMENTS

This study was conducted under collaborative investigation with the Japan Aerospace Exploration Agency (JAXA) as part of the “Advanced Fan Jet Research (aFJR)” project.

REFERENCES

[1] Rodopoulos, C.A., Curtis, S.A. de los Rios, E.R. and SolisRomero, J. (2004), *Int. J. Fatigue*, vol. 26, n. 8, pp. 849-856.

[2] Miao, H.Y. and Lévesque, M. (2011), *Proceeding of the 11th International Conference on Shot Peening*, pp. 7-11

[3] Sinclair, G.B., Follansbee, P.S. and Johnson, K.L. (1985), *Int. J. Solid Struct.*, vol. 21, n. 8, pp. 289-310.

[4] Zimmermann, M. (2010), *Int. J. Comput. Mater. Sci. Surf. Eng.*, vol. 3, n. 4, pp. 865-888.

[5] Rouhaud, E., Oakka, A., Ould, C., Chaboche, J.L. and Francois, M. (2005), *Proceedings of the 9th International Conference on Shot Peening*, pp. 107-112.

[6] Miao, H.Y., Larose, S., Perron, C. and Lévesque, M. (2003), *Adv. Eng. Soft.*, vol. 40, n. 10, pp. 1023-1038.

[7] Kim, T., Lee, H., Hyun, H. C. and Jung. S. (2013), *Mater. Des.*, vol. 46, pp. 26-37.

[8] Meguid, S.A., Shagal, G. and Stranart, J.C. (2006), *J. Eng. Mater. Technol.*, vol. 129, n. 2, pp. 271-283.

[9] Zimmermann, M., Klemenz, M., Schulze, V. and Lohe, D. (2009), In: *High Performance Computing in Science and Engineering '08*, pp. 481-492, Nagel, W.E., Kröner, D.B., Resch, M.M. (Eds), Springer-Verlag, Berlin Heidelberg.

[10] Al-Hassani, S.T.S. (1981), *Proceeding of the 1st International Conference on Shot Peening*, pp. 583-602.

Development of Composite Structural Guide Vane for the New Turbo Fan Engine

Takaomi Inada[1], Hiroyuki Yagi[1], Hideo Morita[1], Rintarou Kajiwara[1], Takehisa Yamada[1], Tsutomu Murakami[1], Koji Miyazawa[1], Takehiko Uchiyama[1], Katsuyoshi, Moriya[1], Shinichi Tanaka[2]

[1]*IHI Corporation, Japan*
[2]*IHI Aerospace Co., Ltd., Japan*

Abstract: Recently, fan diameter of turbo fan engines is tend to increase to enlarge bypass ratio for improvement in fuel efficiency since the bypassed air could generate thrust without direct fuel burn. Larger bypass ratios, however, results in weight increase in fan components, such as fan blades and fan exit guide vanes. In recent high bypass ratio engines, the fan exit guide vanes have a role to remove the swirl from the fan blades as well as to support other fan component, such as fan containment case, and called as structural guide vanes (SGVs). This study describes a development activity of composite SGV for the new turbo fan engine. So called "a building block approach" was applied for the development and several sub-component tests were conducted to validate the design as well as material system for the SGV. The tests were successfully completed and capability as the structural component was validated.

INTRODUCTION

Weight reduction is always a key requirement in development of next generation turbofan engines. To improve fuel efficiency of turbofan engines, fan diameter is tend to increase recently to obtain large bypass ratio. Larger bypass ratios, however, results in weight increase in fan components and composite fan blade and fan containment case have been developed [1]. In recent high bypass ratio engines, fan exit guide vanes (FEGVs), that are located just after the fan blade, are designed to fulfil aerodynamic requirements as well as structural requirement to withstand the engine loads and such FEGVs are called as structural guide vanes (SGVs). Since the size of a SGV increases as the fan diameter increase, weight reduction needs to be considered in the development. Several approaches have been developed and adopted for fan blades design [2] and employing composite materials is one of the best solutions for weight reduction. The authors have been developed lightweight composite fan containment case for high-bypass ratio engines with carbon fiber/thermosetting resin composite [3]. However, neither of the same composite material system nor the same manufacturing system can be applied for composite SGVs since the number of parts per engine is quite different and higher production rate than that for fan containment case needs to be achieved. This means high cycle manufacturing process, such as die mold process, has to be applied for composite SGV manufacturing and carbon fiber/thermoplastic resin system needs to be developed [4]. This paper describes development activity of the composite SGV which has been adopted for PW1100G-JM (Fig. 1) for the Airbus A320neo. So called "a building block test approach" was applied through the development. The in-house prepreg was developed to enhance impact resistance capability against foreign object and validated through coupon tests. A sub-component test was developed for design validation and structural capability assessment of the joint structure between composite vane and metal attachments. A full-size single vane sub-component test was conducted as well to validate structural capability of the SGV under the engine load conditions and to validate impact resistance capability against bird and hail ball. Finally a full-ring structure, in which SGVs were assembled with a dummy fan containment case and a dummy core structure, was tested under static FBO (Fan Blade Off) load and it was confirmed that the SGVs had sufficient load carrying capability.

OVERVIEW OF THE DEVELOPMENT

In order to develop the composite SGV which has high impact resistance capability against foreign object impact, such as bird strike, as well as sufficient structural capability as a primary structure, a building block test approach was employed as shown in Fig. 2.

Firstly, a series of soft body impact tests were conducted to select composite material system for the SGV since higher impact resistance capability is one of the key requirements. Commercially available composite system as well as in-house composite system had been tested. The tests were conducted in the form of laminates and the in-house composite system was selected. After establishment of manufacturing process for the in-house composite system, mechanical properties for the SGV design were obtained by conventional material tests.

In the next step, a special composite laminate which had joint structure to metal attachments were fabricated by using the selected material, and subjected to mechanical loading tests to validate the joint design and structural capability. In addition to the tests, sub-component tests using the full-size prototype SGVs were conducted. In order to assess impact resistance capability against foreign object, a soft body ingestion test and a hail ball ingestion test were employed. For structural capability assessment, a single vane loading tests were conducted by applying simulated engine loads by the hydraulic actuators that were connected to top of the SGV.

As the final step of the development, a full-ring hardware composed of composite SGVs, a dummy fan containment case and a dummy core structure made with carbon steel was prepared and static FBO load test was conducted. The test was successfully completed and the structural capability of the SGV was demonstrated.

(Image credit : Pratt & Whitney)

Figure 1. PW1100G-JM cutaway view [5]

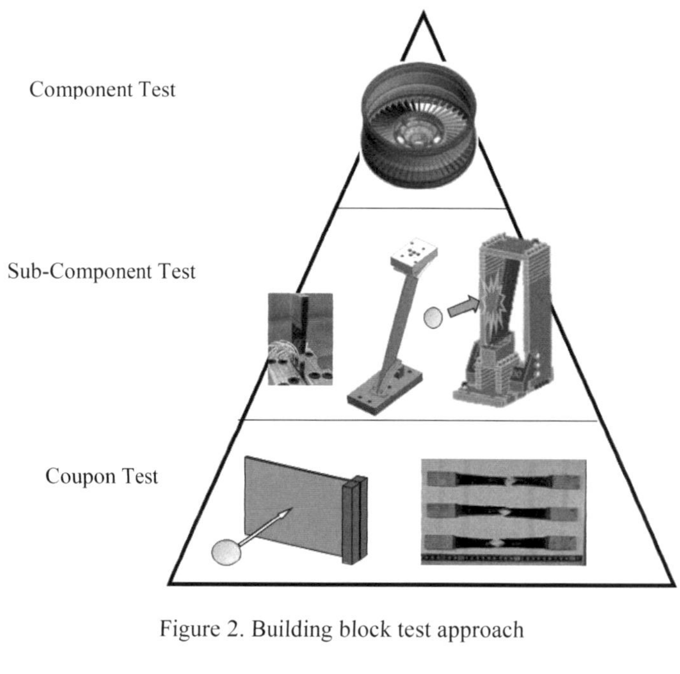

Figure 2. Building block test approach

COUPON TEST

Impact resistance capability of several composite material systems, including commercially available materials and in-house materials, was evaluated for material screening for the composite SGV. A soft body (gelatin ball) impact test was employed as a material screening test. As for commercially available materials, both of tough epoxy prepreg system and thermoplastic prepreg system were evaluated.

Figure 3 shows the test apparatus. The specimen, a flat panel made with composite laminate, was clamped at one end like a cantilever beam. A gelatin ball simulating a bird was ingested to the center of the specimen by using gas gun and impact energy of the ball was calculated from its weight and impact velocity. The behaviour of the ball and the specimen during impact event was observed by a high speed camera. Main deformation mode was bending and waving deformation at the edge of specimen named "edge wave" was observed as well. The major damage was delamination due to the edge wave that was confirmed by high speed camera observation. Each specimen was inspected by ultrasonic and damage area was obtained for evaluation of impact resistance.

The relationship between damaged area and impact energy of each specimen was obtained and summarized as shown in Fig. 4. Material "C", in-house thermoplastic composite developed by modifying the interface property between carbon fiber and resin, shows higher impact resistance capability than tough epoxy composites (material "E") and proven thermoplastic composites (material "F"). Namely, it is evident that damaged area of the material "C" at a certain level of impact energy is much smaller than compared to other two materials. As a result of the test, the material "C" was selected as the prime candidate for the SGV.

On material "C", a series of material tests to obtain material properties for the SGV design were conducted and chemical resistance capability of the material system was also evaluated in parallel.

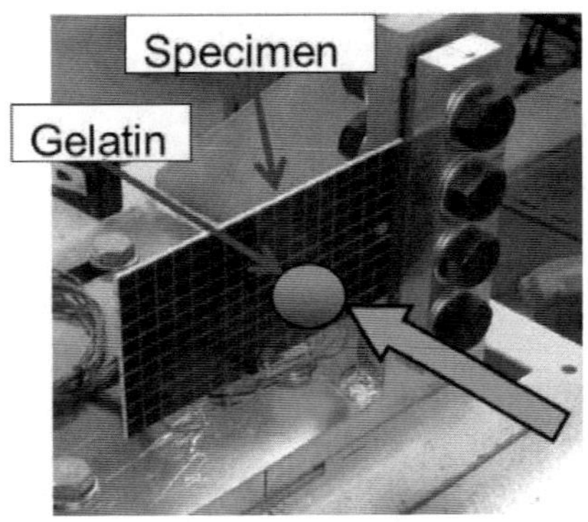

Figure 3. Test apparatus (soft body impact).

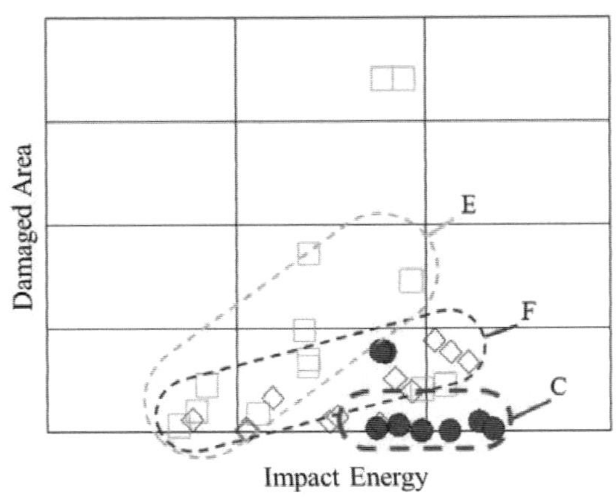

Figure 4. Comparison between damaged area and impact energy.

SUB-COMPONENT TEST

Flat panel loading test

A thick flat composite panel which had special feature for joining to metal attachments was fabricated with selected material, material "C", and subjected to static loading and fatigue loading tests to assess structural capability. Figure 5 shows photo of a flat panel. The composite to metal joint is a key feature in development of the composite SGV since the SGVs are assembled between a core structure and a fan containment case and the joint feature needs to be incorporated at both ends of the SGV. Two kinds of joint feature were tested and the one which had higher bending stiffness compared to another one was finally selected.

Figure 6(a) shows the relationship between moment and displacement during the test. The initial slope of the curve changed slightly at point (A) due to disbond between the composite panel and the metal attachments. Initial AE (Acoustic Emission) response was observed as well at point (A) as shown in Fig. 6(b), time history of moment and hits of AE response. Minor delamination observed at point (B) and the curve in Fig. 6(a) showed clear non-linearity after the point since multiple damages were accumulated as the load increase as observed as AE hits shown in Fig. 6(b). The failure mode at ultimate load was evaluated as well and the joint feature was actually employed to the SGV.

Figure 5. Flat Panel.

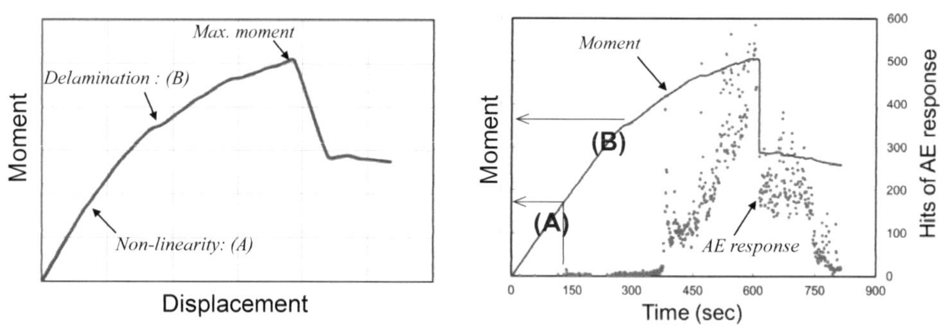

(a) Moment vs displacement curve

(b) Time history of moment and AE

Figure 6. Flat panel test result.

Single Vane Loading Test

Several full-size prototypes of composite SGVs were prepared and subjected to mechanical loading tests. Figure 7 shows a schematic of testing configuration to measure single vane stiffness. Uniaxial load was applied to top of the SGV to measure arbitrary stiffness component of it. Stiffness was simply calculated from applied load and measured displacement in loading direction. Measured stiffness data was employed for analysis model correlation from stiffness viewpoint.

Static loading tests were conducted under simulated engine loads, such as limit load and FBO load, by applying 3-axial loads to top of the SGV simultaneously in the same way as stiffness test though additional fixtures were needed. Since composite material is usually tend to loose strength at elevated temperature, the FBO test was conducted in the temperature controlled atmosphere to simulate the highest temperature condition under engine operation. The measured displacement at maximum load was well agreed to pre-test prediction by FE analysis and confirmed that the composite SGV still had sufficient load carrying capability under FBO load.

Cyclic loading tests were also conducted to confirm low cycle fatigue capability of the SGV under normal engine operation. The tests were done at appropriate temperature levels to assess temperature effect on the composite SGV. 3-axial loads were applied simultaneously to top of the SGV in the same way as static loading test and the cyclic loading test up to required design cycles was completed under loading control condition. It was confirmed that the SGV still had sufficient stiffness after design life and cracks were not observed on the composite vane.

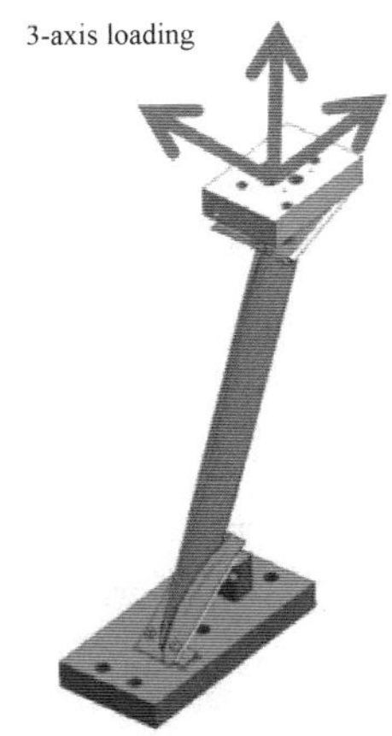

Figure 7. Schematic of testing configuration for stiffness measurement.

Single Vane Impact Test

In order to assess impact resistance capability against foreign objects, two kinds of impact tests were conducted. The first one was the soft body (gelatin ball) ingestion test and the other one was the 2 inch hail ball ingestion test.

Test apparatus and a specimen for gelatin ball ingestion test are shown in Fig. 8. Since the SGVs are located just after the fan blades, birds would not hit the SGVs directly but sliced birds would hit. Size of the gelatin ball and ingestion velocity was determined by the impact analysis using LS-DYNA, a commercially available finite element code, simulating bird strike event of 2.5lb mid-size bird. In order to validate the impact analysis, high speed cameras were employed to capture motions of the SGV during and after impact and strain data was measured as well. Measured strain time history on the vane during gelatin ball impact event was well agreed to pre-test prediction by LS-DYNA analysis, although the figure is not shown. The damage due to gelatin ball impact was localized and well agreed to pre-teste prediction as well.

2 inch hail ball ingestion test was conducted using the same apparatus with that for gelatin ball test. Since a hail ball hit by fan blade breaks into small pieces and impact energy applied to SGVs would be negligibly small and not cause any damage. Therefore, only "pass though" condition was considered to decide the hail ball velocity and span wise location on the vane. High speed camera images and strain time history data were measured in the same way as gelatin ball ingestion test. No visible damage observed after the hail ball ingestion test.

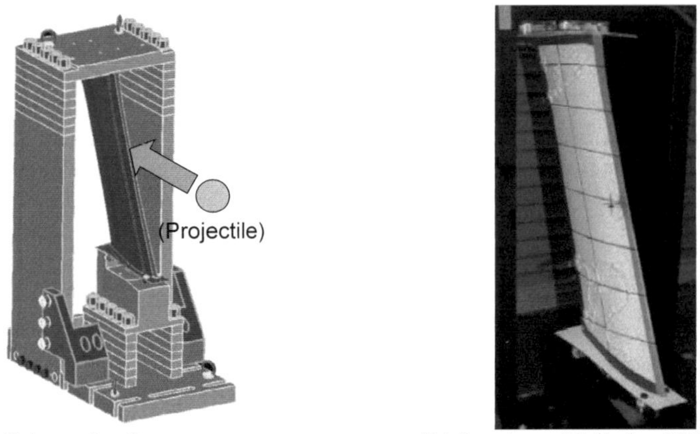

(a) Schematic of test (b) Specimen (Full-size SGV)

Figure 8. Schematic of test apparatus for impact tests.

COMPONENT TEST

Full ring assembly of the SGVs with a dummy fan containment case and a dummy core structure (frame) was prepared and tested. The test loads were applied to the dummy fan containment case by using hydraulic actuators (Fig. 9).

As the first step of the full ring test, the overall system stiffness of the SGV's radial cascade was measured by applying loads for structural analysis model correlation. The measured system level stiffness showed good agreement with the value obtained by structural analysis.

As the next step, FBO loading test was conducted to confirm system level load carrying capability. Since it was impossible to apply dynamic loads by the actuators, simulated static FBO load was applied. The test was successfully completed with no composite failure on the SGVs that was similar to the result observed in the single vane static loading test. The overall system stiffness reduction due to FBO load was also small and within acceptable level.

As the final step of the full ring test, cyclic loading test was conducted to assess fly back capability after the FBO event. A simulated windmilling load, vibration load that occurs due to unbalance of fan blades, was applied using the actuators. The test was also successfully completed without additional composite vane damage and the resultant stiffness change was kept within acceptable level.

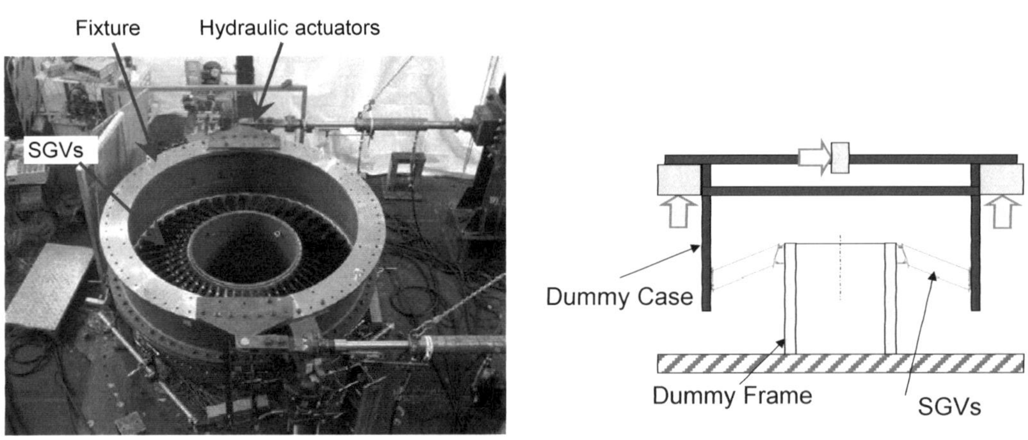

Figure 9. Full ring test fixture.

CONCLUDING SUMMARY

A building block test approach was employed and the composite SGV was successfully developed and demonstrated. The composite SGV is made with in-house carbon fiber/thermoplastic resin prepreg and has high impact resistance capability against foreign object impact. The SGV also has sufficient stiffness and strength to be applied to primary structure. The SGV has been adopted for PW1100G-JM engine, a next-generation civil aircraft engine.

ACKNOWLEDGEMENT

This research was conducted under the contract with Japanese Aero Engines Corporation, partially funded by Ministry of Economy, Trade and Industry of Japan, as a part of the project, "Advanced Materials & Process Development for Next Generation Aircraft Engines".

REFERENCES

[1] http://www.compositeworld.com/articles/composites-in-commercial-aircraft-engines-2014-2023

[2] Leye M. Amoo. (2013). *Progress in Aerospace Sciences,* vol. 60, pp. 1-11

[3] Morita H. et al. (2011). In: *Development of Light Weight Composite Fan Case for Turbo Fan Engine,* Proceedings of the 2nd Joint US-Canada Conference on Composite, pp. 1189-1197

[4] Murakami T. et al. (2013). In: *Development of Composite Parts for Turbo Fan Engine - Fan Case and Structural Guide Vane-*, Proceedings of 13th Japan International SAMPE Symposium and Exhibition

[5] PurePower® PW1000G Engine - Product Cards: http://www.purepowerengine.com/pcards.htm (accessed 2015-4-8)

Adhesively bonded repairs / reinforcements

Evaluation of Adhesively Bonded Repairs to Fuselage Structure Subjected to Environmental Conditions

John G. Bakuckas, Jr.[1], Bin Lei[2], Yongzhe Tian[3], Ian Y. Won[4], Mark Freisthler[4], Kelly Greene[5], Carlyn Brewer[5], Jonathan Awerbuch[6], Tein Min Tan[6]

[1]*FAA William J. Hughes Technical Center, Atlantic City Int'l Airport, NJ 08405, USA*
[2]*FAA-Drexel Fellow, FAA Technical Center, Atlantic City Int'l Airport, NJ 08405, USA*
[3]*Diakon Inc., FAA Technical Center, Atlantic City Int'l Airport, NJ 08405, USA*
[4]*FAA Transport Airplane Directorate, 1601 Lind Ave SW, Renton, WA 98057, USA*
[5]*Boeing Research & Technology, 9725 East Marginal Way, Seattle, WA 98108, USA*
[6]*Drexel University, 3141 Chestnut St, Philadelphia, PA 19104, USA*

Abstract: In this ongoing study, the effects of environment on the durability and fatigue performance of bonded repairs to fuselage panels under simulated service load conditions are being investigated using the Federal Aviation Administration's Full-Scale Aircraft Structural Test Evaluation and Research lab. For this study, modifications were made to the fixture to apply synchronous mechanical, temperature, and humidity loading profiles. A hot-wet environmental condition with a temperature of 73.9°C and a humidity of 85% were applied to both boron/epoxy and aluminum bonded repair patches installed on a B727 fuselage panel subjected to mechanical fatigue loads to at least one design service goal. Bonded repair patches were intentionally made deficient and contained defects to encourage damage growth. During all test phases, damage formation and growth of cracks and disbonds were monitored and recorded using a prototype structural health monitoring system and several nondestructive inspection methods, including visual inspections, eddy current, flash thermography, and resonance ultrasonic. In addition, strains in the vicinity of the repair patches were continuously recording, using strain gages and the digital image correlation method. To date, 80,000 fatigue cycles have been completed for which representative results are summarized herein. In general, the overall fatigue performance of the repairs were reduced as the quality and condition of the repairs were degraded. In addition, environmental effects are a contributing factor to fatigue crack growth measured in the boron/epoxy repair patches where the rates were typically slower under hot-wet conditions compared to ambient conditions. Elevated temperatures of the hot-wet condition apparently reduced the residual strains induced during the original curing process of the boron/epoxy patch installation. Data from this program will be used to assess tools and methods for evaluating and monitoring the repair integrity over the life of the part.

INTRODUCTION

In a joint effort, the Federal Aviation Administration (FAA) and The Boeing Company are investigating the structural robustness and damage-resistance capabilities of adhesively bonded repair technology through the testing and analysis of metallic B727 fuselage panels at the FAA Full-Scale Aircraft Structural Test Evaluation and Research (FASTER) lab. The program objectives are to characterize the durability and fatigue performance of boron/epoxy (B/Ep) and aluminum bonded repairs under simulated service load (SL) conditions over the typical design service goal of an airplane (60,000 cycles for B727) and to investigate tools for evaluating and monitoring the repair integrity over the life of the part.

A phased approach is being taken involving the testing and analysis of several panels containing B/Ep and aluminum bonded patches. A variety of loading and environmental conditions, damage scenarios, and repair conditions are being considered. The first panel tested provided baseline data and was completed in 2010 [1-4]. The results demonstrated that properly designed and installed bonded repairs can be durable under fatigue to at least one design service goal under SL conditions and to at least 20,000 cycles at elevated loads of 115% and 133% of the SL conditions. In addition, they are capable of effectively containing significant damage (e.g., a 660.4 mm, two-bay notch in a lap-joint with central stiffener severed) under severe static loads in excess of ultimate load requirements, 150% limit load.

The second follow-on test was completed in 2012, in which repair patches were intentionally made deficient and contained defects to permit damage growth [5-6]. In general, the capability to detect growing flaws under bonded patches and to monitor the effectiveness of the repairs was demonstrated using several nondestructive inspection (NDI) methods and a prototype piezoelectric-based structural health monitoring (SHM) system. Model predictions were in good agreement with crack growth test results for the majority of repair patch configurations. However, the test and analysis correlations revealed that further research is needed to better account for thermal residual stresses and thick patch configurations.

The two panels previously discussed were tested at laboratory temperature ambient conditions. In this current phase of the program, a third B727 panel is being used to assess the effects of environment on the durability and fatigue performance of B/Ep and aluminum bonded repairs and to continue investigating tools to evaluate and monitor the repair bond integrity. Major modifications were made to fully integrate an environmental system with the FASTER fixture to apply synchronous mechanical, temperature, and humidity load profiles. With this new enhancement, fuselage panels can now be tested under a variety of operating environments, ranging from hot-wet (73.9°C and 85% to 95% humidity) to extremely cold (-45.6°C) conditions.

The primary focus of this study was to determine the effect of environment on the fatigue performance of adhesively bonded repairs to metallic fuselage, including repairs made intentionally deficient to promote damage growth. This paper summarizes the fixture modification for environmental fatigue testing, provides an overview of the test plan, and presents the results up to the 80,000 fatigue cycles for this third panel test. Repair configurations from the first two panel tests were generally retained for comparison purposes to assess the effects of hot-wet (73.9°C and 85% humidity) environmental conditions on repair performance. The effectiveness of an updated SHM system and several NDI methods, including visual inspections, eddy current, flash thermography, and resonance ultrasonic, are being studied throughout the test to monitor the growth of the cracks and disbonds.

In general, the overall fatigue performance of the repairs was correlated to the relative condition and quality of the repairs. Fatigue crack growth measured in the B/Ep composite repair patches was slower under hot-wet conditions as compared to the ambient conditions. Mismatches in the thermal coefficient of expansion of the composite patch and aluminum fuselage result in the development of tensile residual stresses ahead of the notch during the initial patch-installation curing process. At elevated temperatures, tensile residual stresses are relaxed, contributing to slower crack growth. Results for the aluminum repair indicated limited effect of environment on the fatigue crack growth because both the repair and the fuselage skin are made of aluminum. This effort provides the data to calibrate and verify methodologies to assess the integrity of bonded repairs.

FIXTURE MODIFICATIONS

To apply synchronous mechanical and environmental loadings to the test panel, two components, an environmental chamber and a remote conditioner, were integrated with the existing mechanical loading fixture. Figure 1 shows the final configurations of the fixture modifications.

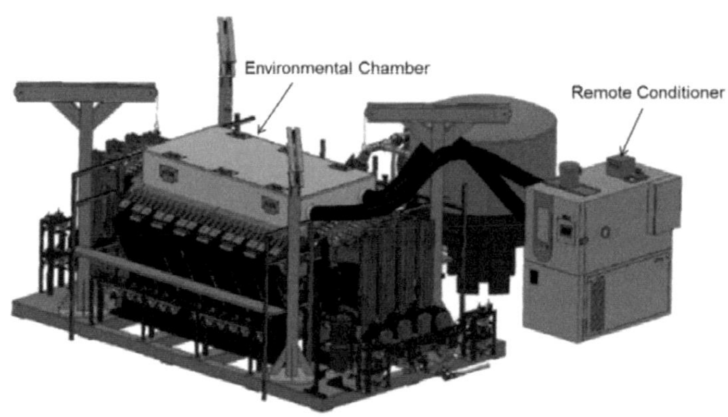

Figure 1. FASTER fixture integrated with environmental chamber and remote conditioner for synchronous mechanical, temperature, and humidity loading profiles to fuselage panels.

Environmental Chamber

An environmental chamber was designed to contain the temperature and humidity conditions outlined in Table I to a 1.2m by 2.7m by 0.3m volume covering the outer surface of the fuselage panel, as shown in Figure 2. The chamber was built using prefabricated insulated panels constructed with 101.6 mm-thick expanded polystyrene foam core insulation material having a thermal resistance rating of 105.9 per mm thickness, sandwiched between 26 gage galvanized steel sheets. An airtight interface between the chamber and fuselage panel was accomplished using a double layer of rubber bulb seals and 25.4 mm-thick weather-resistant neoprene foam. The chamber is secured to the fuselage panel with mechanical straps.

Table I. Temperature-humidity requirements.

Condition	Temperature, °C	Relative Humidity, %
Operational Conditions:		
Hot-Dry	35.6	3–8
Hot-Wet	35.6	85–95
Cold	-31.7	Tending toward saturation
Test Capability:		
Hot-Dry	73.9	3–8
Hot-Wet	73.9	85–95
Cold	-45.6	Tending toward saturation

Figure 2. Environmental chamber with airtight seal. Insert shows the bottom seal consisting of rubber bulb seal and weather-resistant neoprene foam.

Remote Conditioner

A remote conditioner generates the temperature and humidity conditions in the environmental chamber to the specifications in Table I. For this, a commercially available unit was used, namely a Cincinnati Sub-Zero (CSZ) Z-plus remote conditioner (P/N Model ZPRCHS-816-6-6-SC/AC), which can be operated both locally and remotely. Locally, the unit has a built-in 762 mm^3 volume chamber used to condition coupons and small components. Remotely, the unit is interfaced with the environmental chamber using a pair of 203.2 mm inside diameter insulated ducts, one inlet and one outlet, for a closed-loop environmental conditioned system. A control interface was added to the existing FASTER graphic user interface to allow for full operation of the remote conditioner, including system start-up, defining temperature and humidity profiles, adjusting set point values, and setting tolerances while running a fatigue test.

An acceptance test was conducted to verify the functionality of the integrated system. An array of sensors was used throughout the environmental chamber to measure the temperature and humidity distribution. The fully integrated system was run to verify that all conditions in Table I were met (i.e., at steady-state, both temperature and humidity were evenly distributed throughout the environmental chamber within ±1.7°C and ±5% RH respectively).

EXPERIMENTAL PROCEDURE

Panel Description

The fuselage panel used in this study was removed from a retired passenger service Boeing 727-225 aircraft with tail number N675MG and serial number 22553, Figure 3. During service, the aircraft accumulated 48,579 flight hours and 29,821 flight cycles. A large section of the fuselage was removed from the crown of the aircraft from frame stations, STA560 to 660, and stringer stations, S-3R to S-6L, intact with skin, frames, shear clips, stringers, a longitudinal skin splice joint, and a cutout and doubler pad up for an upper rotating beacon.

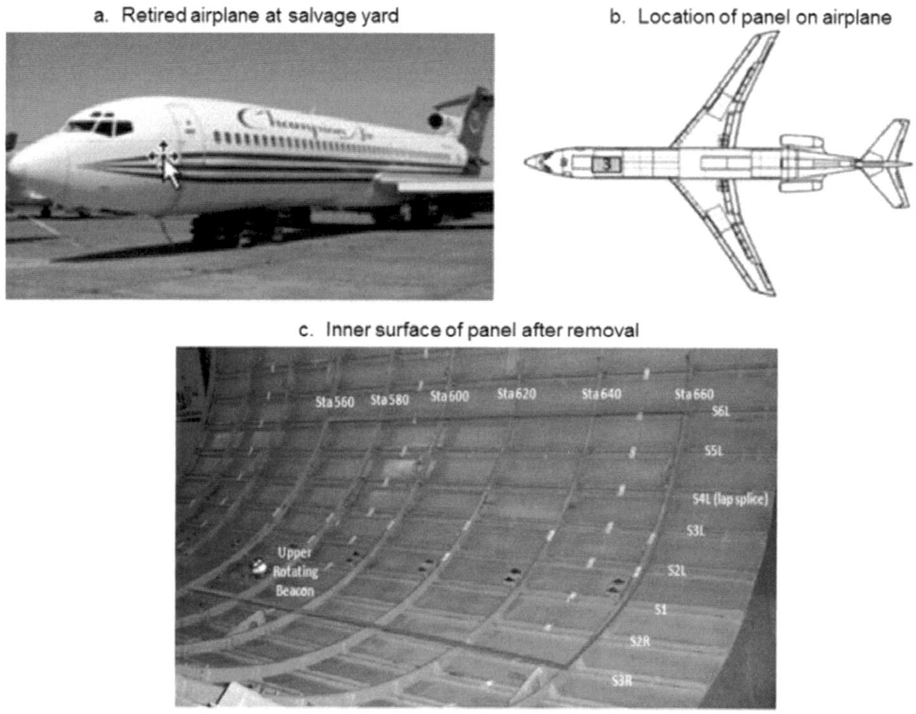

Figure 3. Highlights of B727 airplane and panel, (a) retired airplane in salvage yard; (b) location of panel section on airplane, and (c) internal view of removed panel section.

After the panel was removed, detailed inspections were made to determine the condition of the panel test section and to repair damaged structure. The panel was then cut to size to fit into the FASTER fixture. Reinforcing doublers were installed along the outer perimeter of the skin and to the frame ends for load attachment points of the fixture. Holes of 12.7 mm diameter are drilled along the reinforced doubler edge of the panel at a specific spacing for load introductions in the axial and hoop directions.

The final panel dimensions were 3175 mm by 1854 mm with a radius of 1880 mm, as shown in Figure 4. The skin thickness was 1 mm throughout the test section. The substructure included six stringers in the axial direction with 241-mm spacing and six frames in the hoop direction with 508-mm spacing. A longitudinal lap joint was located along stringer S-4L.

Initial Damage Scenario

The initial damage scenario consisted of mid-bay through cracks for each repair location. Mid-bay through-the-thickness notches of 71 mm in length and 1.5 mm in width (at the notch-tip) were created using a jeweler's saw. Cyclic loads were then applied, which were 75% of the test fatigue loads, to generate natural cracks with sharp crack tips. Figure 5 shows the symmetric and co-linear crack extension from the notch tips. The pre-cracking was continued until the natural cracks generated extended from the notches for a total length of 76 mm.

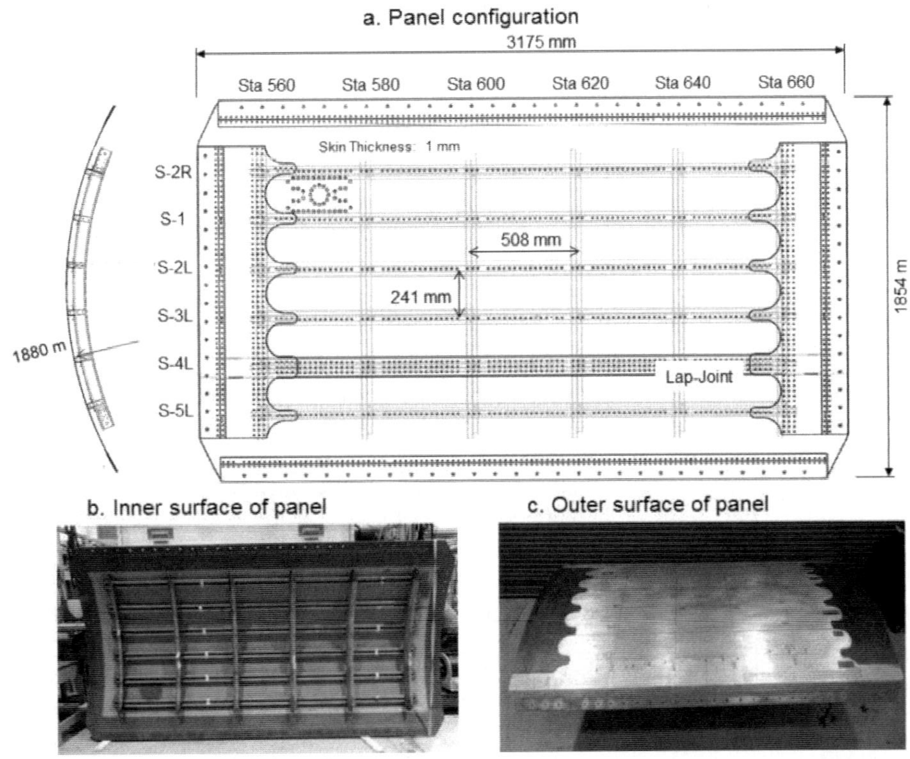

Figure 4. Details of panel 3: (a) configuration, (b) internal surface, and (c) external surface.

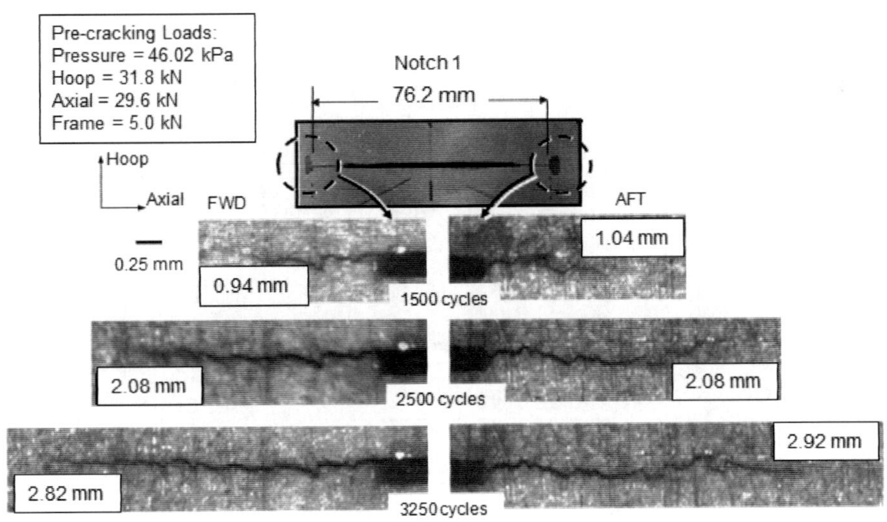

Figure 5. Photographs of crack growth from mid-bay notch during fatigue pre-cracking.

Repair Types

Several types of boron/epoxy (B/Ep) composite and aluminum bonded repairs were considered in this study and are categorized as Type 1 (under-designed), Type 2 (under-designed and partial disbond), Type 3 (weak bond), and reference repairs, as summarized in Table II. Note: Type 1, Type 2, and the reference repairs are identical to those used in the previous second panel test [5, 6]. These repair configurations were retained in this study for comparison purposes to assess the effects of hot-wet environmental conditions (73.9°C and 85% humidity) on repair performance. The Type 3 weak bond repairs were introduced in this study and tested in both ambient and hot-wet conditions. The location of the repair patches in the test panel is shown in Figure 6.

Table II. Summary of Repair Type.

Repair Types	Designation	Description
Type 1: Under-Designed Repairs	UB	Under-designed B/Ep repair with five plies subjected to hot-wet conditions
	UA	Under-designed aluminum repair with two doublers subjected to hot-wet conditions
Type 2: Partial Disbond Repairs	UDB	Partially disbonded and under-designed B/Ep with five plies repair subjected to hot-wet conditions
	UDA	Partially disbonded and under-designed aluminun repair with two doublers subjected to hot-wet conditions
Type 3: Weak Bond Repairs	W1A	Under-designed aluminum repair with two doublers and weak bond using improper primer subjected to ambient conditions
	W2A	Under-designed aluminum repair with two doublers and weak bond using thick adhesive layer subjected to ambient conditions
	W1C	Under-designed aluminum repair with two doublers and weak bond using improper primer subjected to hot-wet conditions
	W2C	Under-designed aluminum repair with two doublers and weak bond using thick adhesive layer subjected to hot-wet conditions
Reference Repairs	RB	Reference B/Ep repair with nine plies subjected to hot-wet conditions
	RA	Reference aluminum repair with three doublers subjected to hot-wet conditions

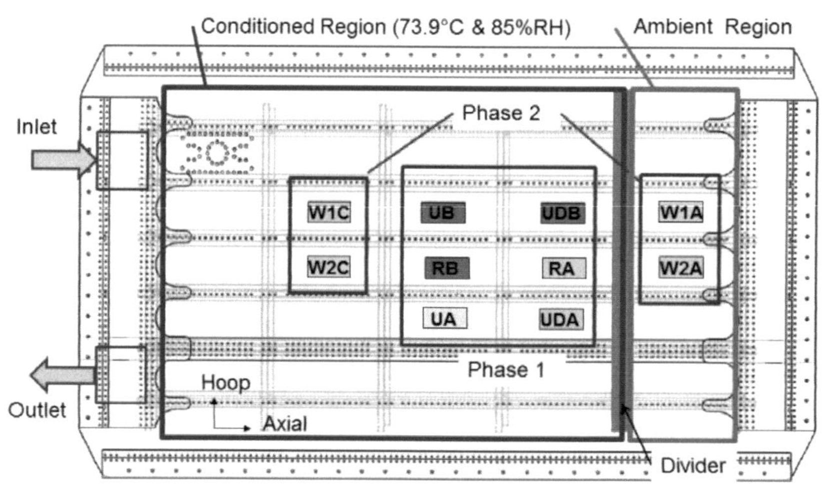

Figure 6. Location and designation of bonded repairs.

Type 1: Under-Designed Repairs. Type 1 repair patches include a B/Ep patch and a two-layer aluminum patch over 76.2 mm mid-bay crack, Figure 7. These repairs were intentionally under-designed by reducing the number of reinforcing B/Ep plies and aluminum doublers and reducing the patch footprint in the hoop direction. The under-designed B/Ep repair had an octagon-shaped footprint approximately 203.2 mm in the axial direction by 76.2 mm in the hoop direction, and consisted of 5 plies, [20°/-20°/0/-20°/20°] with a taper ratio of 20:1. The approximate thickness of each B/Ep ply was 0.145 mm where the 0° fiber direction is perpendicular to the crack, Figure 7a. The under-designed aluminum patch had the same footprint (203.2 mm. by 76.2 mm) consisting of two 0.5-mm-thick aluminum doublers with a taper ratio of 20:1, Figure 7b.

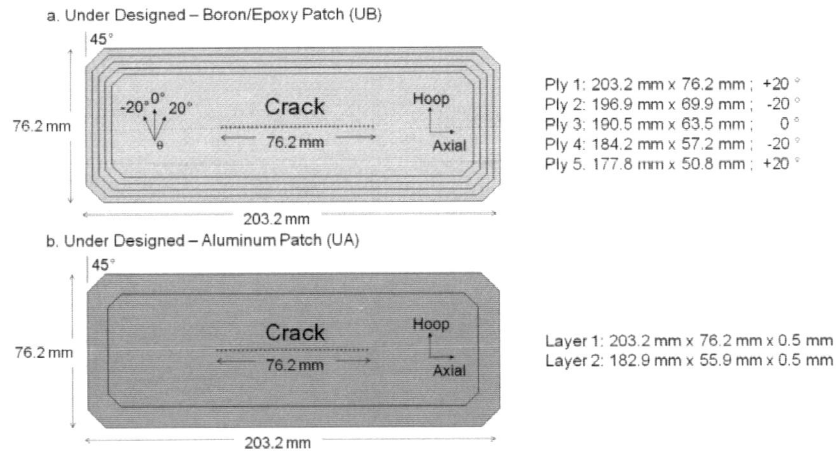

Figure 7. Type-1 repair configuration: (a) B/Ep composite repair, UB, and (b) aluminum patch repair, UA.

Type 2: Partial Disbond Repairs. Type 2 repair patches include a B/Ep patch and a two-layer aluminum patch over a 76.2-mm mid-bay crack, Figure 8. The repairs included engineered disbonds, which were introduced using fluorinated ethylene propylene (FEP) tabs and aluminum shims (coated with a mold release agent). During the curing process, FEP tabs were installed around the crack region between fuselage skin and adhesive, and the aluminum shims were located at the edge of the patch. After curing, the aluminum shims were removed. The patch dimensions and designs for both the B/Ep and aluminum patches are identical to the Type 1 repair configurations. Figure 8 shows the disbond size, shapes, and locations in the repair patches.

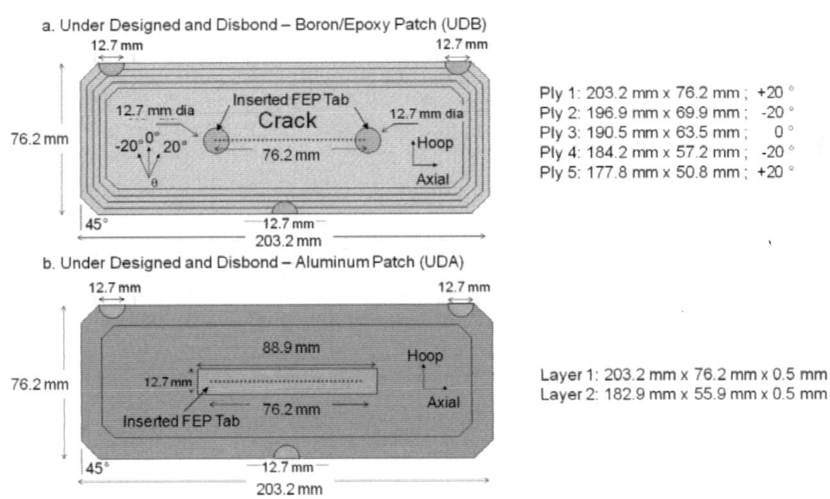

Figure 8. Type 2 repair configuration: (a) B/Ep composite repair, UDB, and (b) aluminum patch repair, UDA.

Type 3: Weak Bond Repairs. For the Type 3 weak bond repair patches, Boeing conducted a comprehensive coupon test program varying key processing parameters for metallic bonded structure. Processing parameters included two major categories: improper surface preparation techniques and issues with adhesive materials and cure. Climbing drum peel, wedge crack, and lap shear tests were conducted to evaluate the improper material and processing parameters. Results that provided indications of consistent weakness were deemed good "weak bond" candidates for application to the aluminum patch repairs.

From the Boeing coupon test results, two types of weak bond repair configurations were selected for application: the use of an incorrect surface primer and the use of an excessively thick adhesive bondline. For both types of weak bond repairs, Type 1 aluminum patches were used consisting of two 0.5 mm aluminum doublers, as shown in Figure 7b. One set of repair patches was treated with the incorrect surface primer, designated W1, and the other set applied multiple layers of adhesive to create an excessively thick bondline, designated W2. Both sets of weak bond repairs were installed and tested in the hot-wet (73.9°C and 85% humidity) conditioned region and the ambient region for comparison purposes.

Reference Repairs. The reference repair patches included a 9-ply B/Ep patch and a three-layer aluminum patch over a 76.2-mm mid-bay crack, Figure 9. These repairs provide a baseline performance for all the other bonded repairs and were designed to contain damage and withstand one DSG (60,000 cycles). The reference B/Ep repair patch had an octagon-shaped footprint approximately 203.2 mm in the axial direction by 127 mm in the hoop direction and consisted of 9 plies, [20°/-20°/20°/-20°/ 0/-20°/20°/-20°/20°/] with a taper ratio of 20:1 where 0° fiber direction is perpendicular to the crack, Figure 9a. The reference aluminum patch also had an octagon-shaped footprint approximately 203.2 mm in the axial direction by 76.2 mm in the hoop, consisting of three 0.5-mm-thick aluminum doublers with a taper ratio of 20:1, Figure 9b.

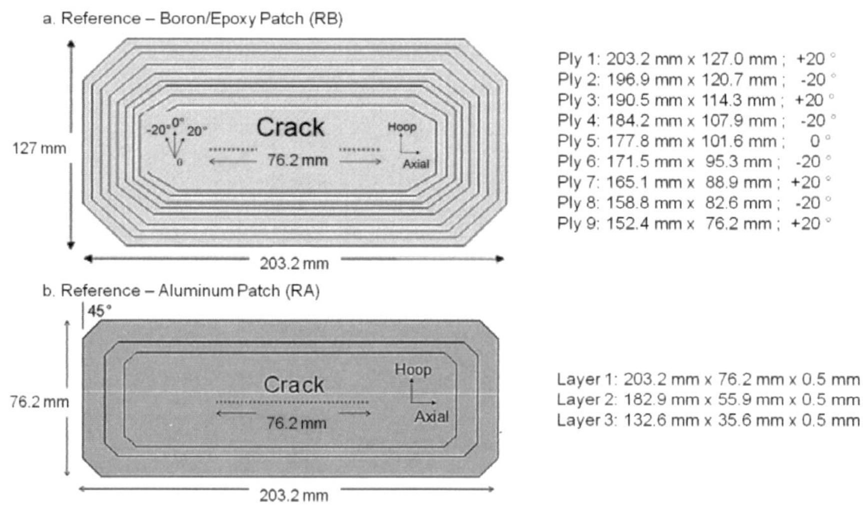

Figure 9. Reference repair configuration: (a) B/EP patch, RB, and (b) aluminum patch, RA.

Test Matrix

The test sequence for each type of repair configuration consisted of a baseline strain survey, fatigue pre-cracking, and fatigue cycling. The details of the applied loads are provided in Table II. The loads used for the fatigue test simulated the service load (SL) conditions, including cabin pressurization (61.36 kPa) and fuselage vertical bending, and were represented by an equivalent constant-amplitude spectrum. The magnitude of the applied loads used in the strain survey and pre-cracking was 75% of the service load conditions used for the fatigue test. For the baseline strain survey tests, quasi-static loadings were applied in ten increments up to the maximum loads listed in Table II. For the pre-cracking and fatigue test, constant-amplitude loading was applied at a frequency of 0.033 Hz (resulting in two pressurization cycles per minute) with an R ratio (minimum to maximum load) of 0.1. The fatigue load history for each repair to date is shown in Table III. As indicated, the Type 1, Type 2, and reference repairs were all subjected to 80,000 fatigue cycles under a hot-wet environment (73.9°C and 85% humidity). The Type 3 repairs were added in a later test phase and were subjected to only 60,000 fatigue cycles under both ambient and hot-wet conditions.

Table II. Maximum Applied Loads.

Test Phase	Load Type	Maximum Load			
		Pressure (*kPa*)	Hoop (*kN*)	Axial (*kN*)	Frame (*kN*)
Strain Survey (75% SL Condition)	Quasi-Static	46.19	31.76	29.69	4.95
Fatigue Pre-Crack (75% SL Condition)	Cyclic (R=0.1)	46.19	31.76	29.69	4.95
Fatigue at SL Condition	Cyclic (R=0.1)	61.36	42.35	39.59	6.72

Table III. Fatigue Load History.

Repair Types	Description	Designations
Type 1: Under-Designed Repairs	Fatigue at SL conditions and hot-wet environment for 80,000 cycles	UB, UA
Type 2: Partial Disbond Repairs	Fatigue at SL under hot-wet conditions for 80,000 cycles	UDA, UDB
Type 3: Weak Bond Repairs	Fatigue at SL under hot-wet conditions for 60,000 cycles	W1C, W2C
	Fatigue at SL ambient conditions for 60,000 cycles	W1A, W2A
Reference Repairs	Fatigue at SL under hot-wet conditions for 80,000 cycles	RA, RB

Inspection Methods

Periodic inspections were made every 5000 cycles. Several NDI methods were used to monitor and record damage formation, growth of cracks, and disbonds. Two methods were used for disbond and delamination detection; namely flash thermography and resonance ultrasonics. The flash thermography system sends a pulse of light to heat the structure and measures the surface temperature as the sample cools. Defects, such as disbonds, trap the heat flow, resulting in a temperature gradient on the surface, which outlines the disbond region. In addition, the resonance ultrasonic technique using Olympus BondMaster is being used to detect disbonds.

For crack detection, visual inspections and eddy current techniques were used. Visual inspections were made on the inner and outer surfaces of the skin using high-magnification cameras, which could be remotely controlled during the test. Low-frequency eddy current (LFEC) was used on the outer surface to inspect for cracks in the skin underneath the B/Ep composite repair patches. High-frequency eddy current (HFEC) was used on the inner surface of the skin. Along with these inspection methods, the strain gages and digital image correlation (DIC) method were used to monitor strains throughout the tests. The details of the inspection methods and strain monitoring methods are provided in references 4–6.

RESULTS AND DISCUSSION

Tests were performed to study fatigue performance of bonded repairs subjected to environmental conditions. The overall loading profile included mechanical fatigue loading simulated service loads synchronized with temperature and humidity profiles. Representative results of strain survey, fatigue crack growth, and NDI are described in the subsequent sections.

Strain Gage Surveys

Strain surveys were taken at 75% of the simulated service load conditions under ambient conditions periodically throughout the fatigue test. Figure 10a shows the typical strain gage pattern and locations instrumenting each repair patch. For this example, Type 1 aluminum patch designated UA is considered. Along cross-section A, gages S19 and S21 were located on the outer surface of the skin, 19.05 mm away from the patch boundary. Gages IS19 and IS21 were corresponding back-to-back gages installed on the same location on the interior surface. Gage S20 was located at the center of the patch on the outer surface. Representative results for the UA patch are shown in Figure 10b. The results indicate bending along the patch boundary where the outer skin surface is under higher tension than the inner skin surface. In addition, high compressive strains were measured in the center of the patch above the notch. The strain measurement history during fatigue is shown in Figure 10c. In general, strains remained relatively constant throughout the fatigue test.

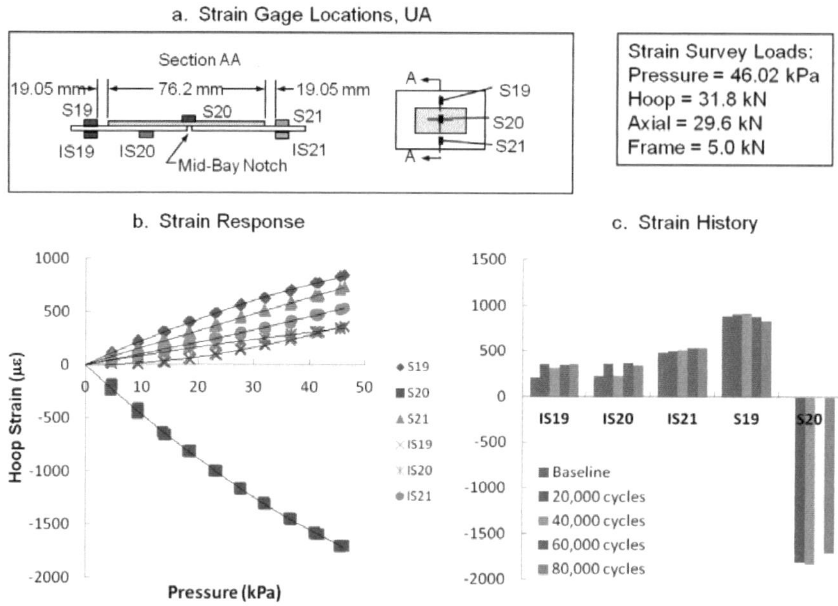

Figure 10. Strain gage measurements during fatigue, Type-1 aluminum patch, UA: (a) strain gage locations, (b) strain survey as a function of pressure, and (c) strain history.

Similar results for Type 1 B/Ep patch, designated UB, are shown in Figure 11. The strain gages instrumenting the patch are indicated in Figure 11a. Along cross-section A, gages S13 and S15 were located on the outer surface of the skin, 19.05 mm away from the patch boundary. Gages IS13 and IS15 were corresponding back-to-back gages installed on the same location on the interior surface. Gage S14 was located at the center of the patch on the outer surface. As shown in Figure 11b, there is a large amount of bending along the patch boundary where the outer skin surface is in tension and the inner skin surface is slightly in compression. High compressive strains were measured on the outer surface of the repair patch over the notch centerline. The variation of strain during fatigue is shown in Figure 11c, which indicates a slight reduction in strains in the early stages of fatigue resulting from the relaxation of residual strains induced during the curing process of the initial patch installation. After 10,000 cycles, the results appeared to be relatively constant throughout the fatigue test up to 80,000 cycles.

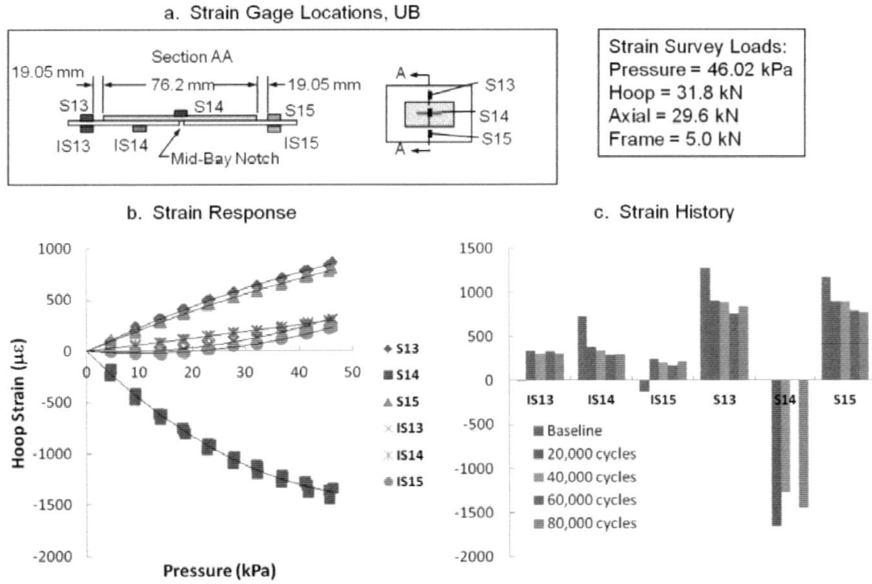

Figure 11. Strain gage measurements during fatigue, Type-1 B/EP composite repair, UB: (a) strain gage locations, (b) strain survey as function of pressure, and (c) strain history.

Full Field Strain Distributions

Global-field-of-view (GFOV) images of the repair patch and surrounding area were taken using the DIC method to monitor for strain redistribution and load transfer. Full-field images of strain were taken periodically under laboratory ambient conditions at 75% service load conditions. Representative results of the full-field hoop strain are shown in Figure 12 for Type 1 B/Ep repair, UB at several fatigue cycles. As shown, tensile strains (red regions) were measured in the vicinity of the patch boundary in the skin, while compressive strains (blue regions) were measured in the center region of the patch along the notch centerline. Similar trends in the results were observed from strain gage measurements, as shown in Figure 11. In general, the global strains on the external surface remained relatively constant throughout the fatigue test up to 80,000 cycles. Strain redistribution is evident from the increasing zone of compressive strains (blue region) in the center of the patches as the fatigue cycle increased, indicating crack growth.

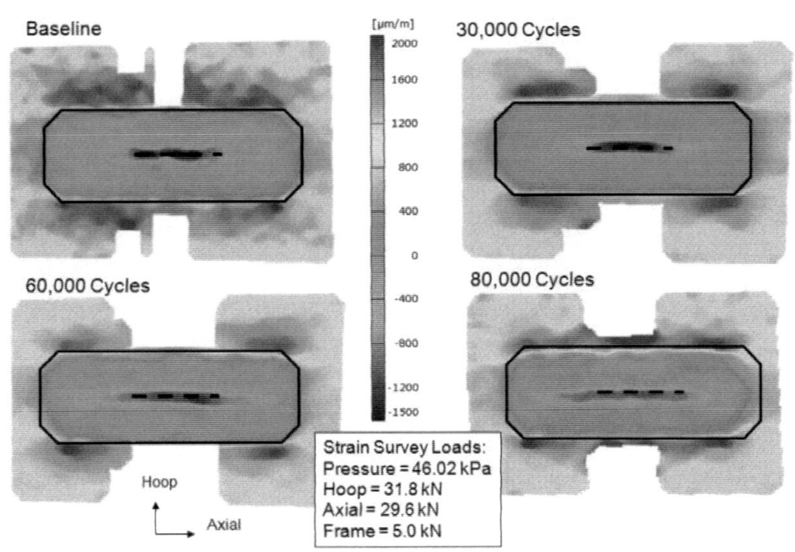

Figure12. Wide field-of-view full-field hoop strain during fatigue, UDB.

Further details from higher resolution full-field strain measurements near the notch region are provided in the narrow-field-of-view (NFOV) images using the DIC method in Figure 13. Shown are strain measurements made at three stages of the fatigue test: baseline (zero cycles), Figure 13a, midway through the fatigue test at 40,000 cycles, Figure 13b, and at the end of the fatigue test at 80,000 cycles, Figure 13c. The hoop strain variation across the patch is provided along three vertical sections spaced at 19.05 mm: 0 centerline of the patch, 1 within the original notch-tip region, and 2 intact regions ahead of the original notch-tip. In general, the hoop strain varied most along the patch boundaries and toward the center of the patch where the notches were located. Strain redistribution occurs with crack extension as seen by the growth of the compressive zone (blue region) in the center of the patch.

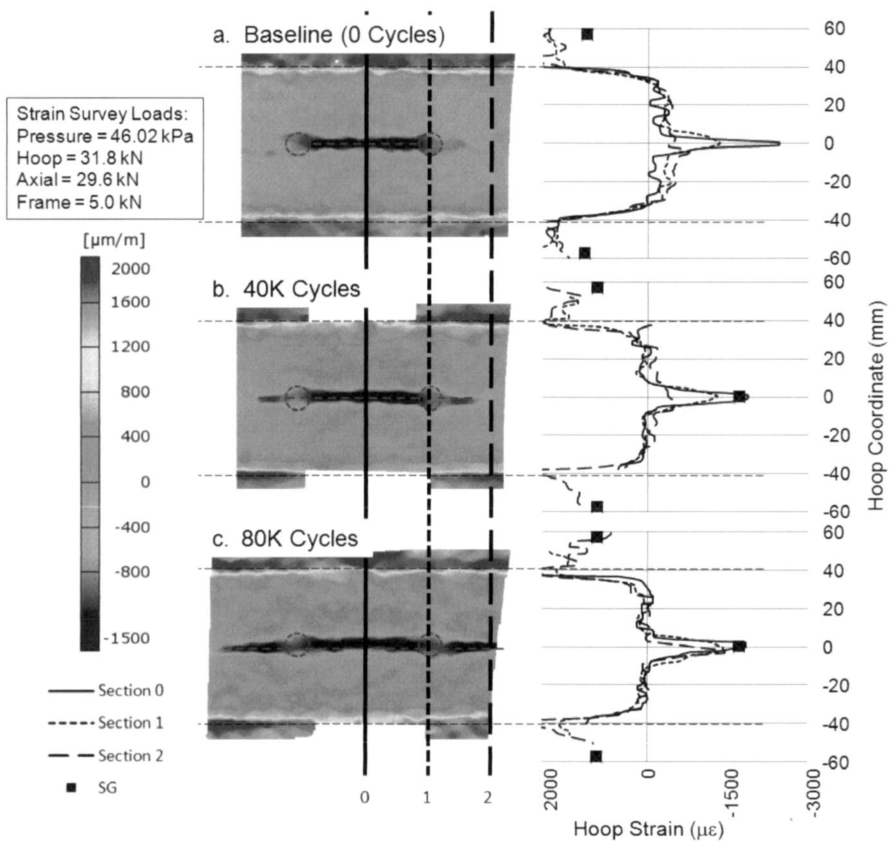

Figure 13. NFOV hoop strain variation in Type 1, B/Ep composite repair UA; comparison between DIC and strain gage measurements.

Along Section 0, the hoop strain values ranged from approximately 2000 με in the skin outside the patch to approximately -2500 με in the center of the patch for the baseline measurement, Figure 13a. High strain gradients were evident over short distances at the patch boundary and the notch region. Across the patch boundary, the strain was reduced from 2000 με to zero over a distance of approximately 5.0 mm. The strain gradient was also quite high in a narrow band of approximately ±12.7 mm around the notch. Over this short distance, the strain ranged from 0 to -2500 με. The magnitude of the compressive strain in the center of the patch was reduced to approximately -2000 με as the number of cycles increased, as shown in Figures 13b and 13c for 40,000 and 80,000 cycles, respectively. As shown in the strain variation plots, data from the DIC were in good agreement with data measured using strain gages that were also located along Section 0.

Similar trends in strain distributions were displayed along Section 1, where high strain gradients occurred over short distances at the patch boundary and the notch region. The values ranged from approximately 2000 με in the skin outside the patch to approximately -1500 με in the center throughout the fatigue test.

Along Section 2, high-strain gradients were initially measured over short distances only at the patch boundary. Hoop strain values ranged from approximately 2000 με in the skin outside the patch to approximately -500 με in the center, as shown in Figures 13a and 13b, at the baseline (zero cycles) and 40,000 cycles, respectively. As seen in Figure 13c, strain redistribution occurs as the crack extends beyond Section 2, resulting in a substantial increase in the compressive hoop strain to -1500 με in the center of the patch.

Fatigue Crack Growth

Fatigue crack growth in the fuselage skin under each repair was continuously monitored visually using high-magnification cameras. Both LFEC and HFEC inspections for fatigue cracking were made in 5000 cycle intervals. In general, limited crack growth was observed in the Type 1 aluminum repair, UA, and in the Type 3 weak bond aluminum repairs, W1A, W1C, W2A, and W2C, as shown in Figure 14. Apparently, the under-designed configuration was conservative for these aluminium patch repairs. In addition, the weak bond was strong and durable enough to maintain proper load transfer throughout the applied fatigue.

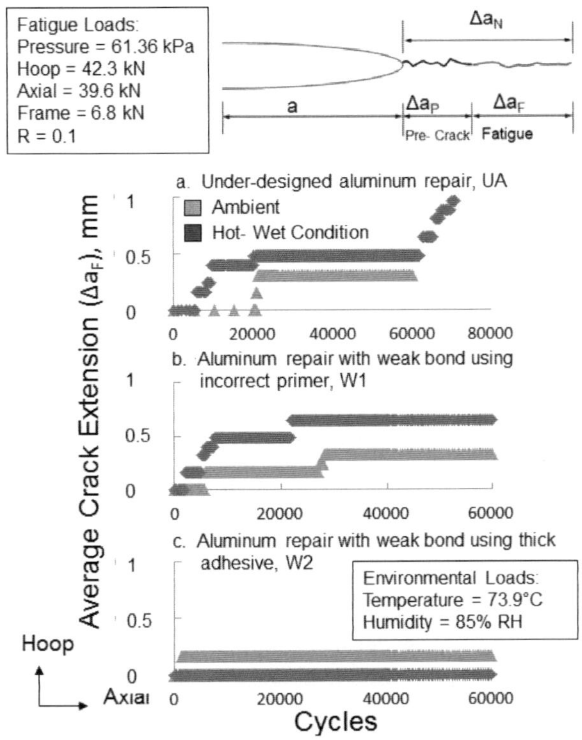

Figure 14. Fatigue crack growth measurements: (a) under-designed aluminum repair, UA; (b) aluminum repair with weak bond using incorrect primer, W1, and; (c) aluminum repair with weak bond using thick adhesive, W2.

For the remaining repairs, namely the Type 1 B/Ep repair UB and Type 2 B/Ep composite and aluminum repairs, UDB and UDA, respectively, considerable fatigue crack growth was observed. Representative results are shown in Figure 15 for the UDB and UDA repairs. As shown, slow and stable crack growth was observed in both repair patches. In addition, the crack growth was symmetric and co-linear and the measure rate was nearly constant after the initial 5000 fatigue cycles. Results from visual and HFEC inspections were in good agreement.

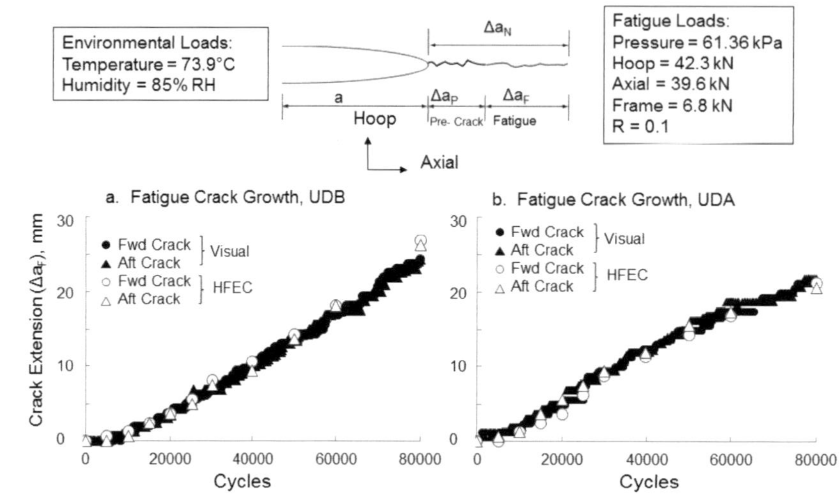

Figure 15. Fatigue crack growth measurements: (a) B/EP repair, UDB, and (b) aluminum repair, UDA.

The effect of repair quality and condition on fatigue performance is summarized in Figure 16 for the B/Ep and aluminum repair patches. As indicated, the fatigue crack growth increased as the repair quality reduced. For both the B/Ep and aluminum repairs, the reference repair, which was properly designed, yielded the least amount of crack growth and the repair configuration with the worst quality, intentionally under designed with disbonds, resulted in the most crack growth.

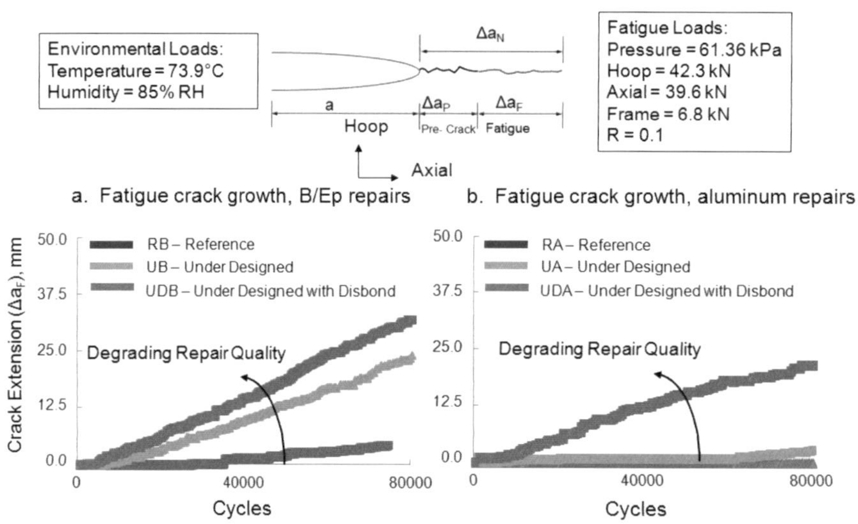

Figure 16. Effect of repair quality on fatigue crack growth in B/Ep and aluminum bonded repairs.

A comparison of the results from Panel 2 tested under ambient conditions [5-6] with those from the current Panel 3 results tested under hot-wet (73.9°C and 85% humidity) conditions reveal differences in which the environment may be a contributing factor on fatigue performance, Figure 17. Although difference in results from the two panel tests is certainly a result of other factors, including inherent scatter from applied loads and the location, geometry, and installation of the repairs, temperature and humidity play a role. Results for the B/Ep composite repairs shown in Figure 17a reveal slower crack growth under hot-wet conditions compared to those under ambient lab conditions for which the average measured rates were 4.35×10^{-4} mm/cycle and 5.01×10^{-4} mm/cycle, respectively. In addition, stable crack growth did not occur until after 5000 cycles under hot-wet conditions. Mismatches in the thermal coefficient of expansion of the composite patch and aluminum fuselage result in thermal residual stresses ahead of the notch from the initial patch installation curing process. At elevated temperatures, these residual stresses are relaxed, as illustrated in

Figure 18a. As shown, the hoop strains measured from two notch gages located the same distance from the crack-tip were approximately 10% lower under hot-wet conditions as compared to ambient conditions after 20,000 fatigue cycles.

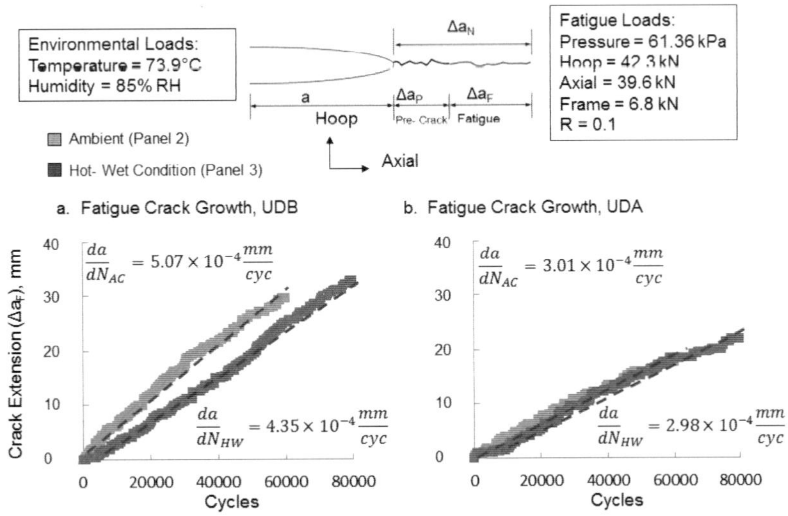

Figure 17. Comparing fatigue crack growth results from Panel 2 under ambient conditions and Panel 3 under hot-wet conditions for Type-2 repairs: (a) B/Ep, UDB and (b) aluminum, UDA.

Results for the aluminum repair indicated limited effect of environment on the fatigue crack growth, as shown in Figure 17b. As shown, the fatigue crack growth behavior was similar for hot-wet and ambient conditions for which the average measured rates were 2.98 $\times 10^{-4}$ mm/cycle and 3.01 $\times 10^{-4}$ mm/cycle, respectively. Because both the repair and the fuselage skin are made of aluminum, the coefficients of thermal expansion were comparable, resulting in limited thermal residual stresses after the initial patch installation curing process. As shown in Figure 18b, the hoop strains measured from two notch gages ahead of the crack-tip were similar for both conditions.

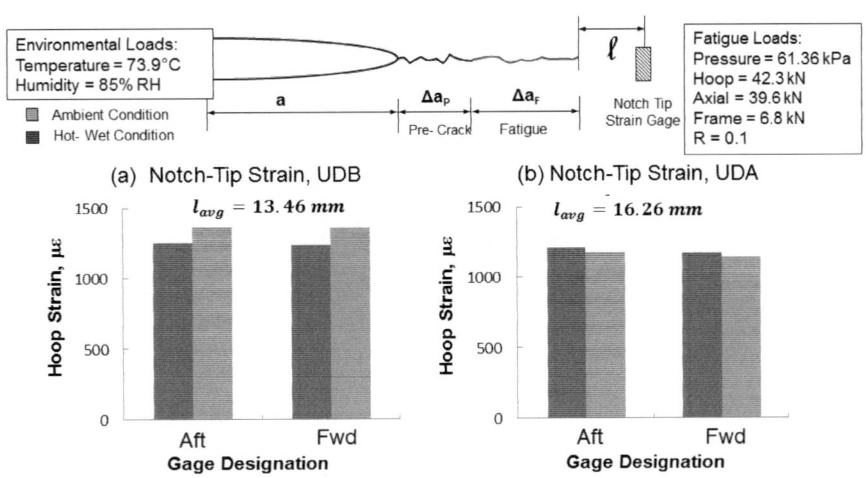

Figure 18. Effect of environment on the notch-tip region strains after 20,000 cycles: (a) B/Ep repair, UDB, and (b) aluminum repair, UDA-

Disbond Inspections

Flash thermography and resonance ultrasonics were used to monitor for disbonds in the repair patches in 5000 cycle intervals. Correlations of disbond detection were made with full-field strains measured using the DIC. Representative results are discussed in the following sections.

Flash Thermography. The flash thermography system worked effectively on all B/Ep composite repairs. Figure 19 shows representative results from Type 2 B/Ep repair, UDB with intentional disbond inserts at the patch boundary and at the notch-tip regions. Photographs in the figure show the results after an exposure time of 2.5 seconds, taken at the beginning of the fatigue and later at 60,000 and 80,000 cycles. The flash thermography clearly indicated the disbond inserts, the notch location, and crack growth from the notch-tips. In general, there was no indication of disbond growth in all B/Ep composite repairs throughout fatigue up to 80,000 cycles.

Correlations between disbond detection and the full-field strains measured using the DIC is shown in Figure 20 for the Type 2 B/Ep repair, UDB. The full-field strains and strain variation across the repair measured using DIC revealed a dispersed compression region outlining the width of the disbond area at the notch region. The width of the compression region measured was 12.7 mm, which was the same as the intentional disbond insert.

Resonance Ultrasonic. The resonance ultrasonic system was used for all repairs. Calibration standards with known disbond defects were designed and built for both B/Ep and aluminum repairs to adjust the settings of the resonance ultrasonic system for flaw detection. Figure 21 shows representative results for Type 2 aluminum repair, UDA, in which the central disbond region was readily detected. For this repair, an 88.9 mm by 12.7 mm FEP tab was inserted over the notch location to create disbond between the repair patch and the fuselage panel skin. In general, there were no indications of disbond growth for all Type 1, Type 2, and reference repairs using resonance ultrasonics.

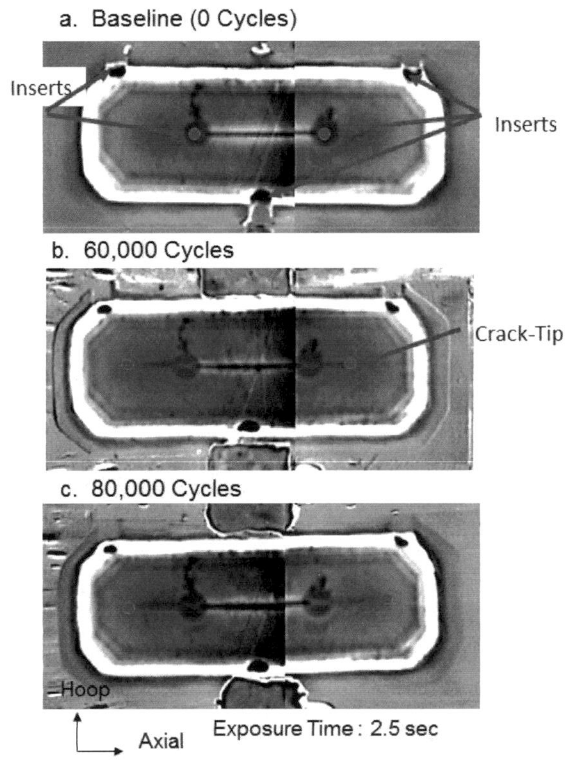

Figure 19. Flash thermography of Type 2 B/Ep repair; UDB illustrating disbond inserts; growth of disbonds not observed; crack growth was detected.

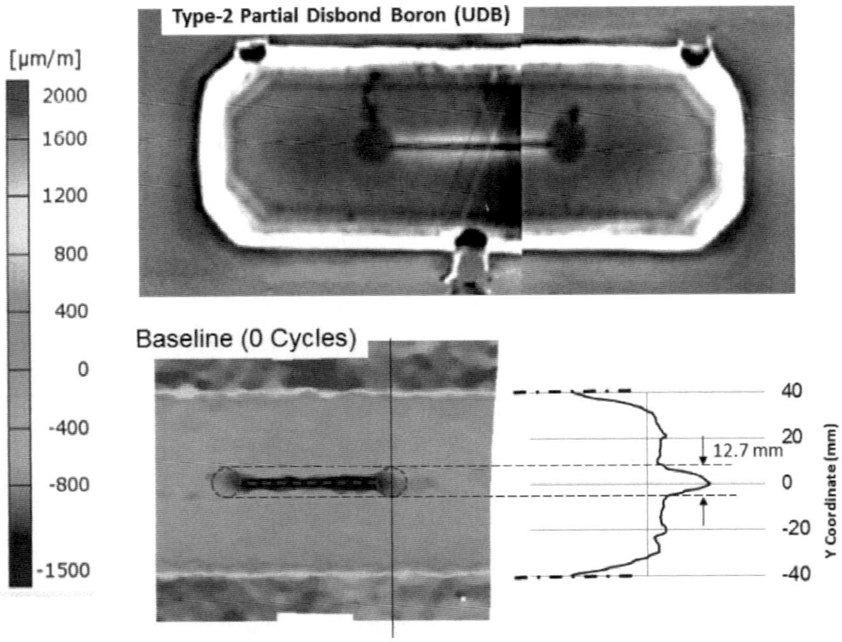

Figure 20. NDI-DIC correlation, Type 2 B/Ep repair with known disbond inserts.

Correlations between disbond detection using resonance ultrasonics and and the full-field strains measured using the DIC is also shown in Figure 21. The outline of the disbond is indicated from the resonance ultrasonics. The corresponding full-field strains and strain variation across the repair measured using DIC revealed a dispersed compression region outlining the width of the disbond area. The width of the compression region measured was 12.7 mm, which was the same as the intentional disbond insert. Thus, a good correlation was found between the strain distribution and the disbond indication for this particular repair.

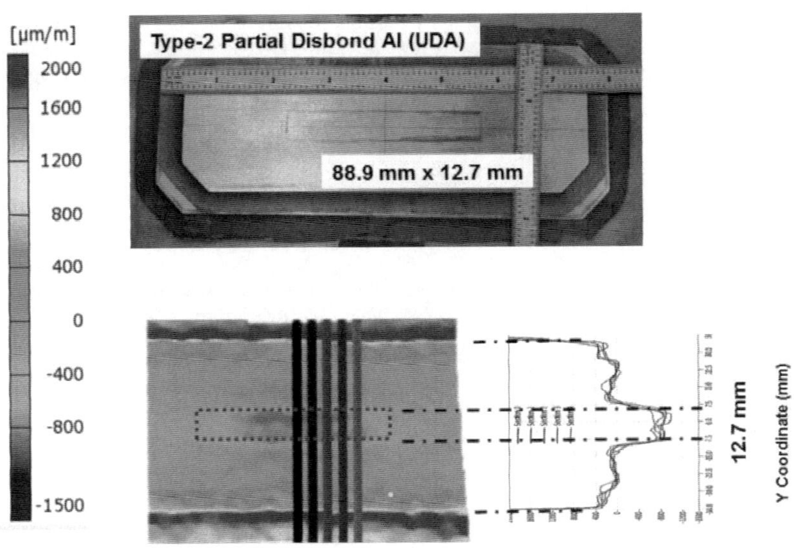

Figure 21. NDI-DIC correlation; Type 2 aluminum repair with known disbond inserts.

SUMMARY

In this study, the effect of environment on the fatigue behavior of bonded repairs to metallic fuselage was investigated. The goal was to collect experimental data to characterize the durability and fatigue performance of bonded repairs under simulated service load conditions over the typical design service goal of an airplane. Both boron/epoxy (B/Ep) and aluminum patch repairs were intentionally made deficient to allow damage growth in the form of crack propagation and disbonding during fatigue cycling. Full-scale fatigue tests were performed using a Boeing 727 fuselage panel. For all phases, a digital image correlation method was used to obtain full-field displacement and strain measurements in the patch and surrounding regions. Several nondestructive inspection (NDI) methods were used to monitor crack and disbond growth, including flash thermography, resonance ultrasonic, and eddy current.

In general, the fatigue crack growth increased as the repair quality reduced for both the B/Ep and aluminum repair patches. Globally, strains remained relatively the same throughout the testing up to 80,000 cycles while local strain redistribution occurred with crack growth. Representative results revealed that hot-wet environmental conditions (73.9°C and 85% humidity) may affect fatigue performance. In the B/Ep composite repair patches, crack growth was slower under hot-wet conditions as compared to the ambient conditions. Mismatches in the thermal coefficient of expansion of the composite patch and aluminum fuselage result in the development of thermal residual stresses ahead of the notch during the initial patch-installation curing process. At elevated temperatures, these residual stresses are relaxed, which may contribute to slower crack growth. Results for the aluminum repair indicated a much lesser effect of environment on the fatigue crack growth because both the repair and the fuselage skin are made of aluminum. The output from this program are verification and calibration data for tools to assess bonded repairs for commercial aircraft applications.

FUTURE WORK

Future work will consider the effects of cold temperatures on durability and fatigue performance of bonded repairs. Data collected will be used to explore emerging NDI methods using existing repairs as a test bed for various inspection methods. In addition, the teardown and destructive evaluation of tested repairs will be conducted. Also, data will be used to further explore repair material and installation process robustness.

ACKNOWLEDGEMENTS

This paper summarizes a team effort consisting of representatives from the Federal Aviation Administration (FAA), The Boeing Company, and Drexel University. From the FAA, the authors wish to acknowledge Jeff Panco and Patrick Sheehan for their diligent efforts running the tests using the FASTER lab. From Boeing, thanks are due to Bud Westerman, Russell Keller, Chris Davis, Nick Room, and Phil Johnson for managing the activities of Boeing; to Ching Hsu, Cong Duong, and Trica Carr for their technical contributions in the design and analysis of the repairs; to Mike Evens and Allan (Josh) Slocum for installing the repairs; and to Paul Rutherford, Tom Williams, and Eric Powell for their efforts in providing NDI support.

REFERENCES

[1] Bakuckas, J. G., McIver, K., and Hsu, C., "Durability and Damage Tolerance of Bonded Repairs to Metallic Fuselage Structure," *Proceedings of the 25th ICAF Symposium*, Rotterdam, May 27–29, 2009.

[2] Chadha, R., Bakuckas, J.G., Won, I., Westerman, E.A, Keller, R., McIver, K., et al. "Characterization of Adhesive-Bonded Repairs to Fuselage Structure," *Proceedings of the 2010 Aircraft Airworthiness & Sustainment Conference*, Austin, Texas, May 10–13, 2010.

[3] Bakuckas, J. G., and Westerman. B., "Fatigue and Residual Strength Performance Bonded Repairs to Metallic Fuselage," *Proceedings of the 26th ICAF Symposium,* Montreal, Canada, June 1–3, 2011.

[4] Chadha, R. and Bakuckas, J.G, "Adhesively Bonded Repairs to Metallic Fuselage Structure: Test 1, Fatigue and Residual Strength Performance," Final Report, DOT/FAA/AR-11/4.

[5] Chadha, R., Bakuckas, J.G., Won, I., Tian, Y., Hunziker, K., Greene, K., et al. "Monitoring Damage Growth in Adhesively Bonded Repairs," *Proceedings of the 2011 Aircraft Airworthiness & Sustainment Conference*, San Diego, California, April 18–21, 2011

[6] Chadha, R. and Bakuckas, J.G, Cong, D, and Hunziker, K, "Adhesively Bonded Repairs to Metallic Fuselage Structure: Test 2, Monitoring Damage Growth," Final Report, DOT/FAA/TC-15/3.

Fatigue Strength of Adhesively Bonded Joints of CFRP with Laser Surface Treatment

Tomohiro Yokozeki[1], Keigo Sato[1], Masaru Ishibashi[2], Hideyasu Shamoto[3], Yutaka Iwahori[4]

[1]University of Tokyo, [2]GSI Creos Corporation, [3]Laserx Corporation, [4]Japan Aerospace Expploration Agency

Abstract: Pulsed laser surface treatment was applied to carbon fiber reinforced plastics (CFRPs) to enhance the work efficiency as well as to establish the stable clean process of surface pre-treatment during the bonding process of CFRP structures. Surface analyses were conducted to CFRP treated by various laser conditions in order to find the optimal laser process. Adhesively bonded specimens were fabricated using CFRPs subjected to laser treatment, and adhesive lap shear strength of CFRPs bonded with adhesive films was evaluated compared with the abrasive paper treatment. Surface conditions, adhesive strength, and adhesive fracture mode depended on the treatment parameters, and adhesive strength similar to abrasive paper treatment could be obtained in the specimens with appropriate laser treatment. Debond fracture toughness and fatigue lap shear strength were also evaluated to confirm the long-term durability and reliability of laser treated CFRP bonded joints. It was concluded that laser surface treatment can be applied to the surface treatment process of CFRP bonded structures.

INTRODUCTION

Light-weight and high-performance structures are required for transportation system (e.g. aerospace and automobile vehicles), and application of carbon fiber-reinforced plastics (CFRP) to the primary structures is expanding. As the large scale structures needs assembly of some structural members, highly reliable technology for the joining of CFRP/CFRP and CFRP/metal parts should be established to achieve the light-weight structures. When CFRPs are used for light-weight structural members, adhesive bonding is one of the effective joining methods by consideration of mechanical characteristics and manufacturing process of CFRPs. It is necessary to develop a powerful method to realize both the high reliability and the efficient process of adhesive bonding [1,2].

Adhesive bonding process of CFRPs generally comprises surface polishing (i.e. manual sanding using abrasive papers or treatment by sand blasting), surface cleaning (chemical coating is applied, if necessary), application of adhesives or adhesive films, and curing of adhesives. During this adhesive bonding process, quality stability and rapid process are required for the surface pre-treatment. Development of a reliable and efficient surface pre-treatment process of CFRPs will result in realization of light-weight structures.

Various types of surface treatment are now available: mechanical treatment by abrasive papers and sand blasting, use of peel ply, plasma treatment, various chemical treatments etc.[3,4]. This study investigates the applicability of laser treatment to surface pre-treatment of CFRPs prior to adhesive bonding. Laser treatment is also applied for surface cleaning (removal of surface contamination) and removal of coating. Laser ablation, which aims to make the surface rough by laser treatment, is focused on in this study. Previous researches (e.g. [5]) investigated applicability of ultraviolet (UV) treatment by excimer laser. Recent development of solid-state lasers activates the researches on laser treatment. Belcher et al. [6] investigated application of UV laser (wavelength: 355nm) to surface pre-treatment of CFRPs. Fisher et.al. [7] evaluated the adhesive strength of CFRPs treated by UV laser (wavelength: 308nm) and near-infrared laser (wavelength: 1064nm).

This study considers the applicability of pulsed CO_2 laser treatment to the surface pre-treatment of CFRPs for adhesively bonded structures. The authors consider that this technology makes it possible to treat large-area surfaces with constant quality rapidly. Pulsed laser treatment is applied to CFRP surfaces and easy of automation, and thus, is expected to contribute to the development of useful surface pre-treatment process of CFRPs by making the process more stable, faster, and cleaner. In the present study, several laser treatments are applied to CFRPs, and the surface conditions of treated CFRPs are analyzed. Adhesively bonded specimens are prepared using adhesive films and treated CFRPs, and subjected to evaluation concerning on adhesive mechanical properties. Applicability of laser pre-treatment to CFRP bonded structures is investigated by comparison with conventional mechanical surface pre-treatment.

EXPERIMENTAL

Laser treatment

The present study applied pulsed CO_2 laser treatment using SLCR (Selective Laser Coating Removal) machine (ML105E, wavelength: 10.6μm, maximum energy: 6.5 J /pulse, pulse frequency: 50 Hz, maximum power: 250 W).

Organic substances exhibit good absorption of laser with this wavelength, and thus, it is expected that only polymers on the surface of CFRP are processed and removed, which results in removal of contaminated release agent and formation of asperity. Laser treatment conditions can be controlled by selecting irradiation energy, laser spot area, and sweeping speed. Although it is possible to apply nitrogen gas during laser treatment, atmospheric laser treatment is selected in the present study. An apparatus of pulsed laser treatment is shown in Figure 1. In the present study, surface cleaning by chemical solvents (e.g. ethanol) was avoided after the laser treatment. The present laser treatment is completely non-contact process.

The CFRP samples were fabricated using unidirectional T800S/2592 prepregs (supplied by Toray). Cross-plied layups ($[0_2/90_2/0_2/90_2/0]s$, 3mm thickness) were utilized and the stacked materials were cured at 130 °C in the autoclave. It is noted that release agent and release films were used during the manufacturing of CFRPs. Several laser parameters were preliminarily investigated, and two laser conditions (A and B), which have same laser power and sweeping speed, but different spot area, were selected for evaluation. The CFRPs treated by two laser conditions are presented in Figure 2. Two treated surfaces exhibit different appearance.

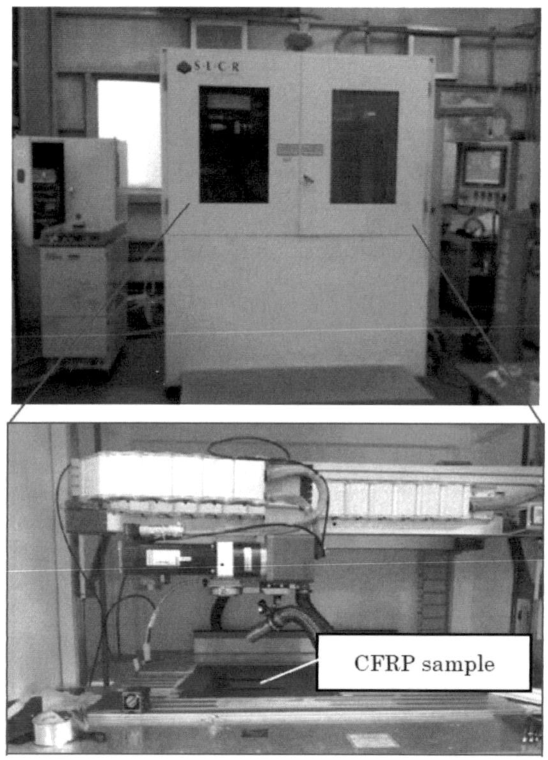

Figure 1. Apparatus of pulsed laser processing.

Surface analysis

Surfaces of CFRP treated by two laser conditions were analyzed by the following methods. For comparison, untreated surfaces and mechanically treated surfaces (mechanical treatment using abrasive papers with #240 and #400 roughness followed by surface cleaning using ethanol) were also analyzed.

Laser microscopy (OLS4000, Shimadzu)
FT-IR (ATR method, Diamond prism, IRAffinity-1, Shimadzu)
XPS (EACS-3400, Shimadzu)

Surface microscopic roughness, functional group on the surfaces, and elements on the surfaces were investigated.

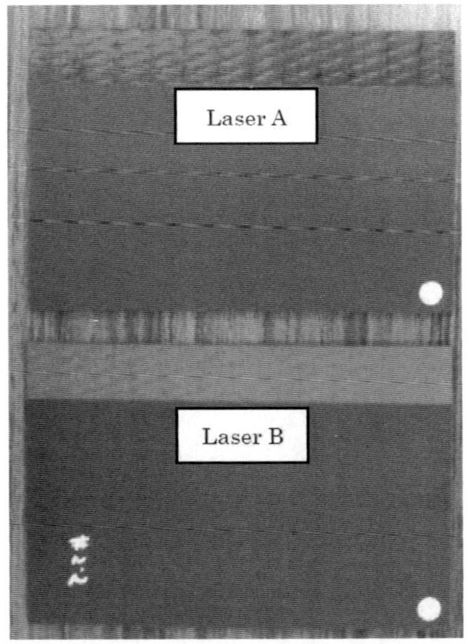

Figure 2. Appearance of CFRP surfaces subjected to laser processing (CFRP panel size: 150mm × 100mm).

Adhesive lap shear strength test

CFRPs with four kinds of surface pre-treatment (i.e. untreated, abrasive paper, Laser A, and Laser B) were prepared, and adhesive lap shear strength tests were performed. Adhesive films (AF-163, 3M) were used as adhesives. Curing temperature was 110°C, curing duration was 90 minutes, and applied pressure was about 100kPa during the bonding process. Adhesive thickness was about 0.2mm. Lap-shear strength tests were conducted in reference to JIS K6850 standard. Five specimens were subjected to tests for all CFRPs with four kinds of surface pre-treatment.

Previous researches indicated that adhesive thickness, residual thermal stress, water uptake, etc. influence adhesive strength [8]. Therefore, CFRPs were stored in the low humidity chamber (relative humidity is less than 25%) before and after the surface pre-treatment. Surface treatment process, bonding process, and strength test were performed at the same timing (without time interval between the processes) for all specimens. The effect of surface pre-treatment on static lap shear strength was evaluated in this section for comparative study.

EXPERIMENTAL RESULTS

Comparison of surface conditions

Microscopic surface views (scale: 256μm×256μm) by laser microscopy are presented in Figure 3. In the case of Laser A sample, resin is almost removed from the surface, and only partial removal of surface resin is recognized in the sample of Laser B. The untreated surface is completely covered by resin. The condition of the surface treated by abrasive paper is apparently similar to that treated by Laser A. It should be noted that, in the laser treatment, treated time duration depends on the in-plane position of the sample, as the sweeping motion is V-shaped. Periodically dappled patterns are induced on the surface by laser treatment. It is confirmed that laser pre-treatment removed only surface resin (no fiber removal).

FT-IR spectrums of the four samples are shown in Figure 4. The vertical axis is the relative intensity of spectrum, and the spectrum curves are vertically shifted to some amount in order to make the comparison clear. The untreated surface reflects the surface resin characteristic, and the Laser B surface exhibits the spectrum similar to the unteated surface. This trend coincides with the microscopic observation (i.e. Laser B surface is mainly covered by resin). Surfaces treated by Laser A and abrasive paper exhibit small peaks corresponding to C=O and C-O groups compared to untreated and Laser B surfaces. It is concluded that surfaces treated by Laser A and abrasive paper are similar in terms of FT-IR spectrums.

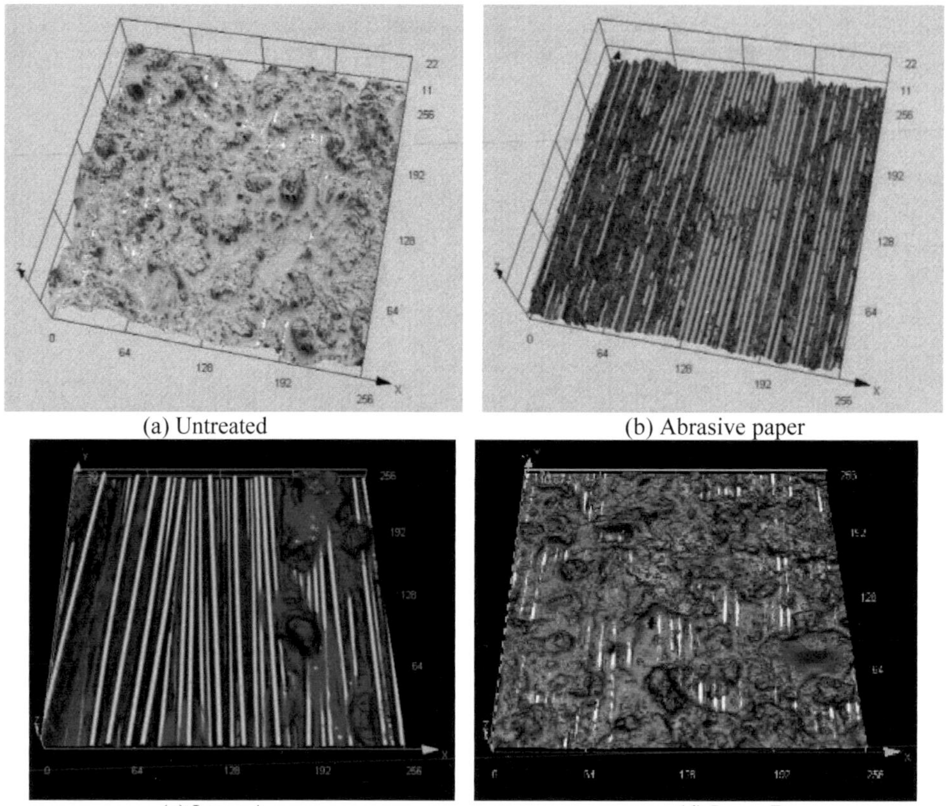

(a) Untreated (b) Abrasive paper

(c) Laser A (d) Laser B

Figure 3. Microscopic surface views of treated CFRP samples.

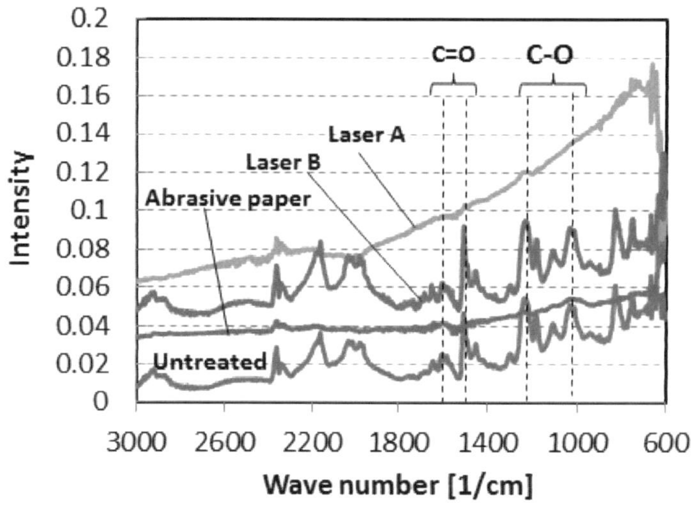

Figure 4. Comparison of FT-IR spectrum of CFRP surfaces.

The XPS wide spectrum of untreated surface is presented in Figure 5. Fluorine and silicone based peaks were recognized in the case of untreated surface. Elemental components of the release film and the release agent remained on the CFRP surfaces. The quantitative analysis of elemental components of the four samples was performed using the XPS narrow spectrums, and the elemental ratios were summarized in Table I. Fluorine and silicone remained on the untreated surface, while they were completely removed in the case of the surface treated by the abrasive paper. Table I also indicates that laser treatment removed fluorine and silicone, specifically in the case of Laser A.

Based on laser microscopy, FT-IR, and XPS, it is concluded that Laser A treatment removes the remained fluerine and silicon, and makes the surface condition of CFRP similar to that by abrasive paper.

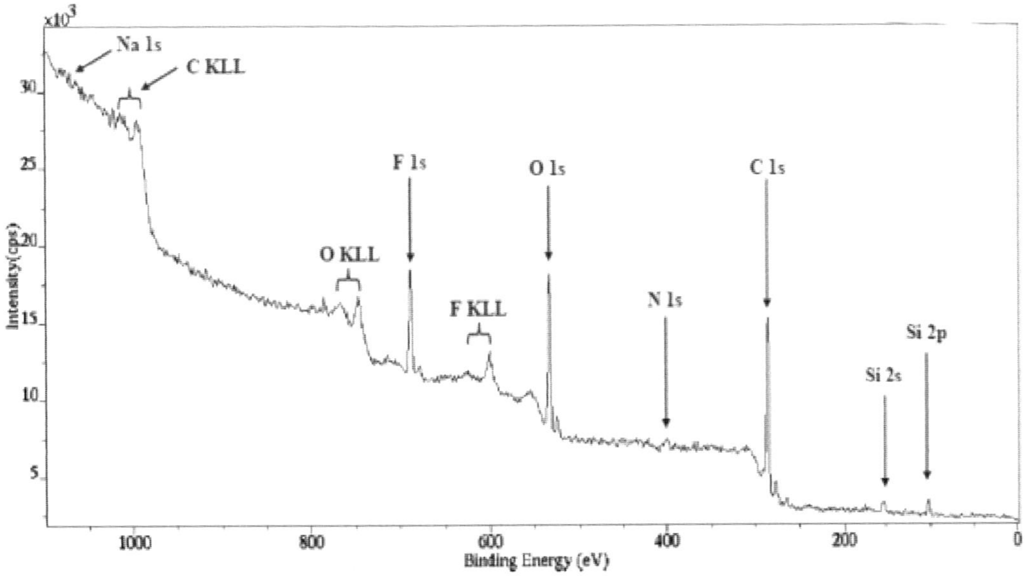

Figure 5. XPS wide spectrum of untreated CFRP surface.

Table I. Comparison of atomic concentration based on XPS analysis.

	F 1s [%]	Si 2p [%]	Others [%] (C 1s, O 1s, etc.)
Untreated	13.1	3.8	83.1
Abrasive paper	0.0	0.0	100
Laser A	1.4	1.2	97.4
Laser B	4.7	2.7	92.6

Adhesive lap shear strength

Adhesive lap shear strength of CFRPs (five specimens) with four kinds of surface pre-treatment was measured at room temperature. Comparison of lap shear strength is summarized in Figure 6. Compared to untreated CFRPs, specimens treated by abrasive paper exhibited high strength. Average strength and data variation of specimens treated by Laser A were similar to those of specimens treated by abrasive paper. Laser B treatment resulted in slightly lower shear strength and larger variation compared to abrasive paper or Laser A.

Fracture surfaces of the adhesive lap joints are shown in Figure 7 for specimens treated by abrasive paper, Laser A, and Laser B. Cohesive failure is recognized in the specimens treated by abrasive paper and Laser A, while interfacial failure is partially observed in the specimens treated by Laser B. These results indicate that pre-treatment by Laser A gives surface conditions similar to abrasive paper in terms of lap shear strength and failure mode. It is concluded that pulsed laser treatment is applicable to the surface pre-treatment of CFRPs.

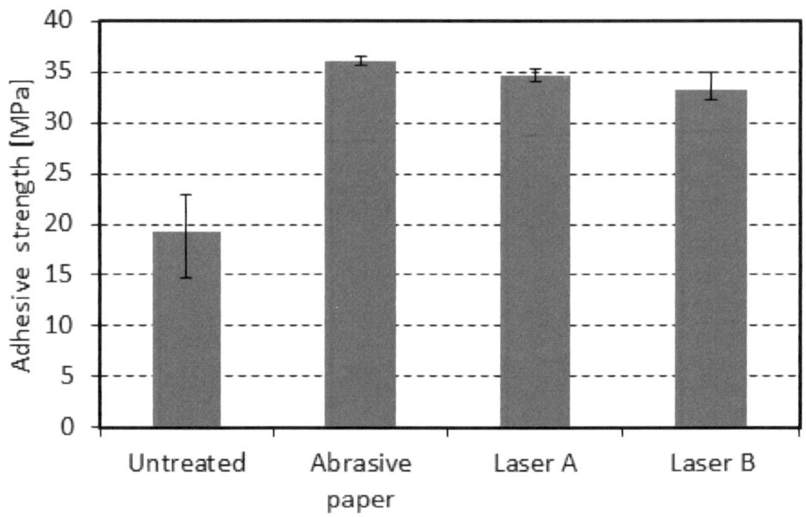

Figure 6. Comparison of adhesive lap shear strength.

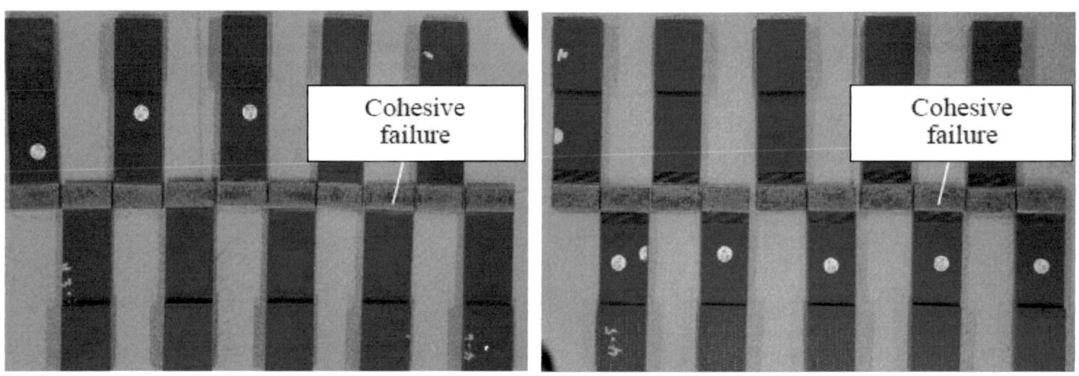

Abrasive paper (b) Laser A

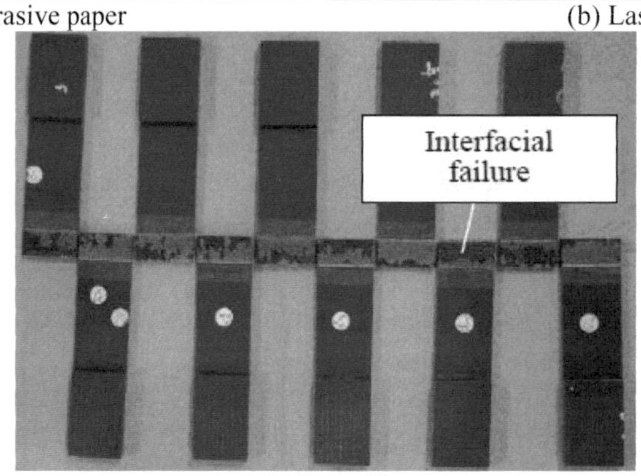

(c) Laser B

Figure 7. Comparison of adhesive failure mode.

DURABILITY EVALUATION

Pulsed CO_2 laser treatment has been applied to surface pre-treatment of bonding process of CFRPs, and it was demonstrated that suitable laser condition gave surface conditions and adhesive strength similar to the conventional treatment by abrasive paper. The following experiments were performed to investigate further the durability and long-term reliability of adhesive bonded joints of CFRP with laser treatment.

i) Debond fracture toughness
ii) Fatigue lap shear strength

CFRP bonded specimens treated by abrasive paper and the optimal laser process (Laser A) were prepared in this section.

Debond fracture toughness

Two kinds of surface pre-treatment (i.e. abrasive paper and Laser A) were applied to CFRPs. Double cantilever beam (DCB) specimens were fabricated using CFRPs (same as the above experiment) and the adhesive film (AF-163). Initial cracks were introduced by partially placing thin release film along the bonding line between the adhesive film and the upper CFRP with surface pre-treatment. Crack pre-extension was processed prior to DCB tests to make natural and sharp crack tips. DCB tests (five specimens for each pre-treatment) were conducted in reference to JIS K 7086 standard. Modified compliance method was applied to evaluate mode-I fracture toughness of adhesive bonded joints as a function of crack growth.

Mode-I debond fracture toughness was summarized in Figure 8 as a function of crack growth. Specimens treated by Laser A exhibited almost same fracture toughness as those treated by abrasive paper. The average toughness and coefficient of variance (COV) are 2.35 [kJ/m^2] and 8.8 % for abrasive paper and 2.32 [kJ/m^2] and 11 % for Laser A, respectively, which were evaluated using the data in the range of 0-60mm crack growth. It is concluded that adhesive bonded joints using abrasive paper and Laser A process exhibit same performance in terms of debond fracture toughness.

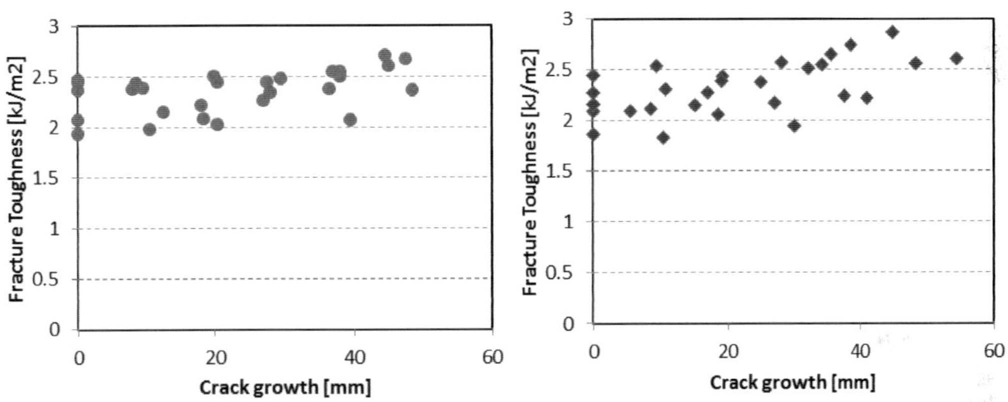

Figure 8. Comparison of mode-I debond fracture toughness (left: abrasive paper, right: Laser A).

Fatigue lap shear strength

Lap shear specimens were prepared using the same CFRP processed by two kinds of surface pre-treatment (i.e. abrasive paper and Laser A) and adhesive film. Tension-tension fatigue tests were conducted with stress ratio of 0.1, varied maximum stress, and frequency of 5 Hz to evaluate fatigue life of bonded joints. Comparison of fatigue life of two kinds of bonded joints is presented in Figure 9. The ordinate is applied maximum stress normalized by static lap shear strength. It is confirmed that fatigue strength is same for abrasive paper treatment and Laser A treatment.

CONCLUDING REMARKS

The present paper investigated the applicability of pulsed CO_2 laser treatment to the surface pre-treatment of CFRPs for adhesively bonded structures. It was demonstrated that suitable laser condition gave surface conditions and adhesive strength similar to the conventional treatment by abrasive paper. The experimental results suggested the usefulness of laser treatment for the bonding process. Durability experiments also indicate the reliability of laser treatment if the process parameter is optimized.

Laser surface treatment holds several process parameters and requires some preliminary tests or some screening tests to find the optimal condition for the target CFRP sample. The present study performed three surface analyses (laser microscopy, FT-IR, XPS) and found out the optimal condition by comparing the results with those by abrasive paper. Similarity of surface conditions between samples treated by the abrasive paper and the optimal laser parameters could be recognized, and the optimal laser condition gave the adhesive strength similar to the abrasive paper. This indicates that, for the laser surface pre-treatment, surface analyses are effective as the preliminary tests. Then, the optimal laser condition can be found out by conducting the adhesive strength tests using several specimens treated by preliminarily selected laser conditions.

Laser treatment is a clean process, and possibly makes the pre-treatment process more stable and faster. Further studies on the applicability of this process to other combinations of CFRP and adhesives and dissimilar bonded joints are encouraged.

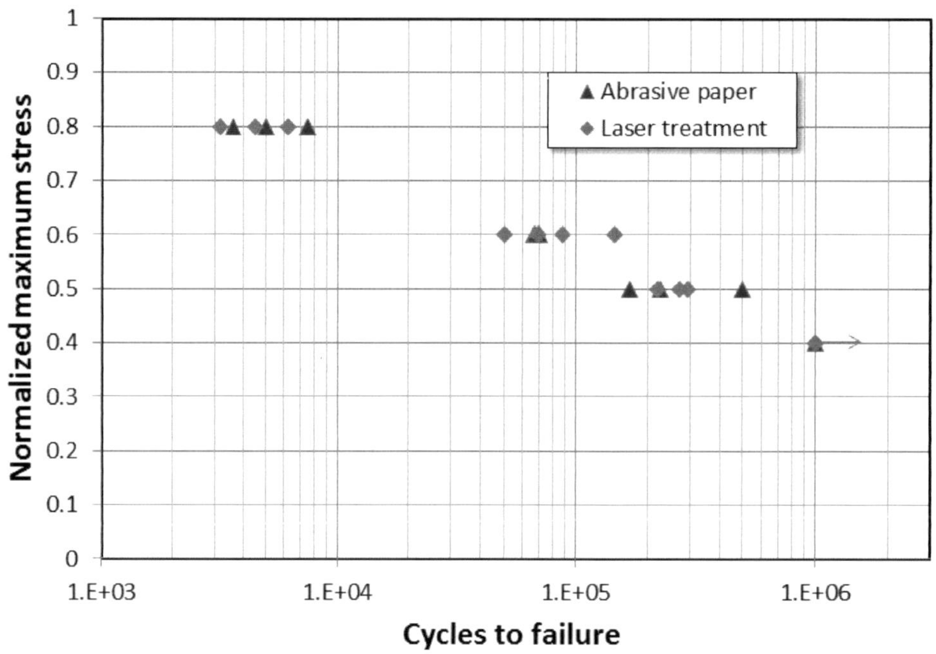

Figure 9. Fatigue strength of lap shear bonded joints.

REFERENCES

[1] Adams, R.D., Comyn, J. and Wake, W.C. (1997). *Structural adhesive joints in engineering. 2nd ed.*, Chapman and Hall, London.

[2] Wachinger, G., Thum, C., Llopart, L., Maier, A., Wehlan, H. and Stöven T. (2013). In: *New trends in CFRP treatment and surface monitoring for automated structural adhesive bonding*, Proceedings of 19th International Conference on Composite Materials, Montreal.

[3] Bénard, Q., Fois, M. and Grisel, M. (2005). *Compos. Part A*,vol.36, pp.1562-1568.

[4] Awaja, F., Gilber, M., Kelly, G., Fox, B. and Pigram, P. (2009). *Progress in Polymer Science*. vol.34, pp.948-968.

[5] Rotel, M., Zahavi, J., Tamir, S., Buchman, A. and Dodiuk H. (2000). *Appl. Surf. Sci*. vol.154-155, pp.610–616.

[6] Belcher, M.A., Wohl, C.J., Hopkins, J.W. and Connell, J.W. (2010). In: *Laser surface preparation for adhesive bonding of aerospace structural composites*, Proceedings of 14th European Conference on Composite Materials, Budapest.

[7] Fischer, F., Kreling, S.K. and Dilger, K. (2013). In: *Surface pre-treatment of CFRP by using laser radiation*, Proceedings of 19th International Conference on Composite Materials, Montreal.

[8] Blackman, B.R.K., Johnsen, B.B., Kinloch, A.J. and Teo, W.S. (2008). *The Journal of Adhesion*. vol.84, pp.256–276.

Full-Scale Fatigue Testing of a Boron-Epoxy Bonded Doubler for the Finnish Air Force F/A-18 Hornet Centre Fuselage

Geoff Swanton[1], Mika Keinonen[2], Janne Linna[2]

[1]Defence Science and Technology Organisation, Australia, [2]Patria Aviation Oy, Finland

Abstract: The effective management of structural integrity matters is an essential element in the safe and cost-effective operation of any aircraft fleet. In today's Defence environment, many military fleets are being operated well beyond their original service life goals, or conversely, the fatigue life consumption trends from some operators suggest that their service life goals may not be met. In both cases, the implementation of structural modifications is often required to achieve the necessary life extensions. The Finnish Air Force (FINAF) have one such example where an indigenous research, development and manufacturing capability was used to design a bonded doubler reinforcement to address a potential fatigue life deficiency on the Y488 bulkhead in their fleet of 'Classic' F/A-18C/D Hornet aircraft.

During the 2012 F/A-18 International Structural Integrity Forum meeting held by the global F/A-18 structures community, a proposal was put forward to consider fatigue testing of the FINAF doubler on a full-scale centre fuselage test article as part of Australia's Defence Science and Technology Organisation's (DSTO's) current well-established testing program for such test articles, also known as 'centre barrels'. The proposal was agreed to, and in 2013 personnel from Patria Aviation visited DSTO to install the doubler on a retired United States Navy F/A-18D centre barrel.

This paper presents a summary of the design, development and initial coupon test aspects of the doubler. It then discusses the installation of a FINAF fleet doubler on a centre barrel test article, followed by a description of the subsequent instrumentation of the specimen, the testing regime, interpretation of data with respect to doubler performance, inspection activities, and examination of the eventual bulkhead failure. The implications of this test program are significant in that it provided realistic full-scale geometric and loading conditions for verification of an in-service modification to primary structure. It revealed information about failure mechanisms of the surrounding structure and a potential deficiency with current fleet inspection practices with respect to disbond detection. Overall, it is anticipated that this collaborative activity will benefit the FINAF's structural integrity management program by providing timely and relevant information associated with the doubler modification.

INTRODUCTION

The core load bearing structure in the F/A-18A-D Classic Hornet is located in the aircraft's centre fuselage and is comprised of three wing attachment bulkheads, each machined from single billets of aluminium alloy (AA) 7050-T7451[6]. These bulkheads are classified as fracture critical primary structure, and along with the connecting longerons, skins, panels and formers, are colloquially known as centre barrels (CBs), see Figure 1. The aft-most of the CB bulkheads is designated the Y488 bulkhead (as it is positioned 488 inches aft of the aircraft datum point). The lower part of the bulkhead is a location prone to fatigue cracking, as evidenced by the failures suffered in the original equipment manufacturer's full-scale fatigue test (FSFT), subsequent FSFTs and numerous analytical studies [1][2][3]. The local names of the details of interest here are marked in Figure 1(d).

[6] With the exception of the Swiss fleet which have titanium bulkheads.

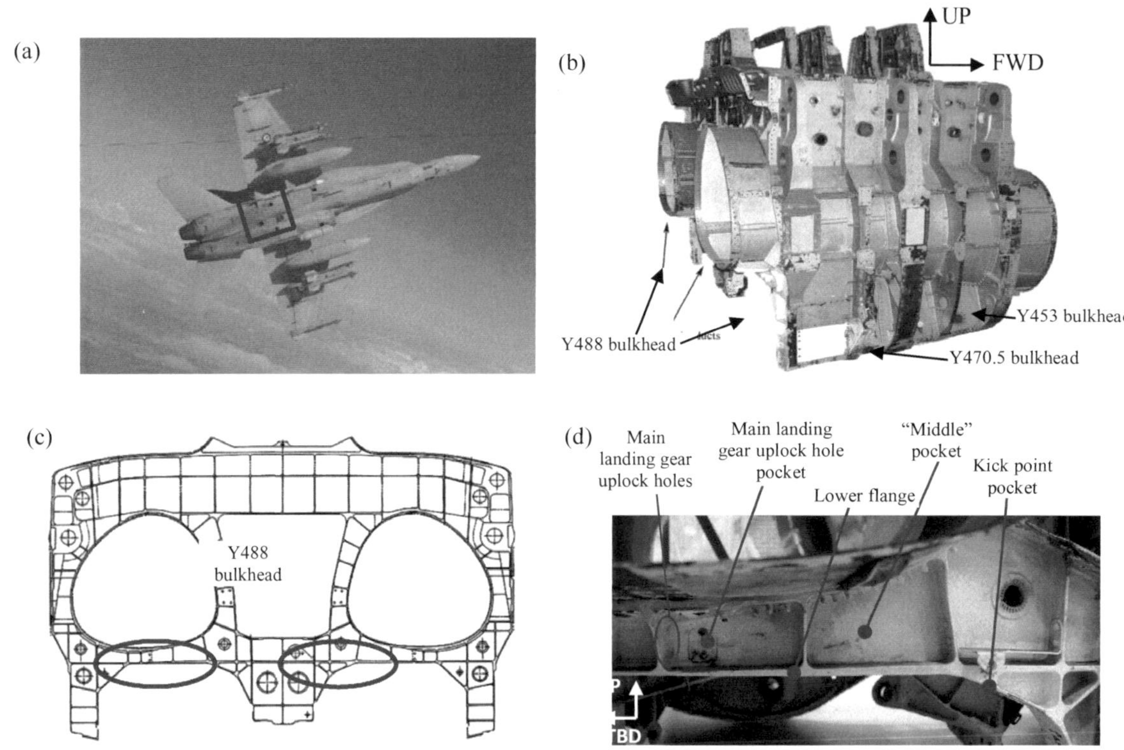

Figure 1. (a) The location of the CB in the F/A-18A-D aircraft. (b) CB structure showing the location of the Y488 bulkhead. (c) Y488 bulkhead with fatigue critical region of interest circled. (d) Nomenclature and structural details in the Y488 bulkhead fatigue critical region (left hand side, view looking forward).

For the Finnish Air Force (FINAF) fleet, the safe life estimation for this critical location, based on their Hornet Operational Loads Monitoring (HOLM) system, suggested that their desired service life goal of 4,500 airframe hours (AFHRS) might be unachievable with a real risk of that some aircraft could face premature retirement. Therefore, the capacity to reduce the stresses in this location was considered beneficial in order to increase the likelihood of the entire fleet achieving the desired service life goal. To address this potential shortfall in life, a bonded boron-epoxy doubler reinforcement was designed, developed and is currently being implemented for this location [4][5]. It is interesting to note that DSTO had also been involved in a conceptual bonded doubler reinforcement for the Y470.5 bulkhead, which was successfully demonstrated on the International Follow-On Structural Test Project (IFOSTP)[7] [6] FT55 FSFT article in the 1990s [7] although this never came to fruition on the Royal Australian Air Force (RAAF) fleet.

The original proof-of-concept testing of the doubler design was conducted using component-level specimens, approximating only part of the lower Y488 bulkhead and omitting some other structural details in the region. Whilst the tests achieved the design objectives, there was not an opportunity to assess the effect of the doubler in a fully representative test scenario in Finland, i.e. on a FSFT article. Meanwhile, at Australia's Defence Science and Technology Organisation (DSTO), the Flaw Identification through the Application of Loads (FINAL) program had been fatigue testing and tearing down ex-service CBs since 2003, in support of the Royal Australian Air Force's (RAAF's) own F/A-18 aircraft structural integrity programs [8][9][10][11][12]. At the 2012 F/A-18 International Structural Integrity Forum (FISIF) personnel representing the FINAF and DSTO agreed to collaborate on a project which would involve the installation of the FINAF doubler on a FINAL CB for subsequent fatigue testing and any potential post-failure analysis.

In the context of the FINAF fleet, the first aircraft to be modified with the doublers was in 2011, and as of November 2014, 35 of 62 aircraft had doublers installed. The entire fleet is expected to be modified by the end of 2017. Once installed, the doublers and bond line are subjected to an ultrasonic non-destructive inspection (NDI). To pass the inspection, no disbonds are allowed to extend to the edges of the doubler, and any disbonds that may be present must

[7] IFOSTP was a series of FSFTs for certifying the F/A-18 for the RAAF and Royal Canadian Air Force fleets. FT55 was the centre fuselage test article, FT46 was the aft fuselage and empennage test article and FT245 was the wing test article.

not exceed a size of 100 mm^2 [13]. Thereafter, once a modified aircraft accrues 2,200 AFHRS or a wing root fatigue life expended (WR FLE) index of 0.44[8], the doublers are reinspected at recurring intervals of 150 AFHRS. The same acceptance criteria remain in effect for these recurring inspections. As of February 2015, there had been no reports of doublers failing the acceptance criteria in the fleet. It is noted however, that after the initial post-installation NDI, the inboard end of the doubler cannot be inspected at the recurring intervals due to the presence of some metal tubing that prevents access to this area.

This paper presents a brief description of the doubler design before discussing the program of work which was conducted at DSTO in support of the collaboration. This includes the preparation of the test article, doubler installation, instrumentation and testing phase through to the catastrophic failure of the bulkhead. The test results were then examined and analysed to inform an assessment of the doubler performance, whilst parallel NDI activities were conducted which supported the assessment. These NDI activities also proved to be an informative study in their own right, where the effectiveness of different inspection techniques for this particular test case was compared. A quantitative fractography (QF) analysis was carried out on the bulkhead failure and the results are presented here. Further activities yet to be completed have been identified and these are presented after the concluding remarks at the end of the paper.

DOUBLER DESIGN

Material Selection and Properties

The design and development of the doubler was carried out by Patria Aviation in Finland. High stiffness combined with a favourable coefficient of thermal expansion (CTE) were the prime reasons for selecting a boron-epoxy 5521 unidirectional (UD) laminate for the doubler. Boron-epoxy also exhibits other favourable material properties such as non-conductivity that enables eddy current inspections of the bulkhead through the doubler, as well as a low risk of galvanic corrosion. Further detail about the various candidate options and the selection criteria are discussed in [14]. The final doubler design consisted of an 8-ply UD laminate with a nominal thickness of 1.05 mm. A summary of the doubler properties is presented in Table I and further background regarding the analysis is in [15].

Table I. Boron-epoxy doubler material and analytical properties.

Lay-Up	Number of Plies	Elastic Modulus, E_x (GPa)	Thickness (mm)	CTE α (10^{-6}/°K)	Predicted Stress Decrease (%)	Margin of Safety (Limit)
UD (0°)	8	186	1.05	4.5	-23	+1.13

For the bonding process, FM300-2 film adhesive was selected due to its compatibility with the bulkhead material during the cure cycle. The temperature distribution survey for the cure cycle was conducted on an actual fuselage bulkhead, and it was found that 100°C could be achieved, albeit with special heater blocks [16].

Finite Element Strength Analysis

The efficiency of the doubler was first analysed using a finite element model (FEM) of the bulkhead and the component-level specimen that was to be used in the testing phase, see Figure 2. The predicted stress decrease with an 8-ply UD boron-epoxy doubler was 23% (although a 28% stress decrease was measured in the specimen lower flange during component testing). Due to the different CTE of the doubler and bulkhead, permanent residual stresses were created during the bonding process. Taking these into account, the overall stress reduction in the bulkhead lower flange area was predicted to be 18% at room temperature, which satisfied the design criteria of a 10%–20% stress reduction. This level of stress reduction was projected to increase the fatigue life in the areas targeted by a factor of 1.5 to 2.5.

[8] A WR FLE index of 1.0 represents the end of the design fatigue life.

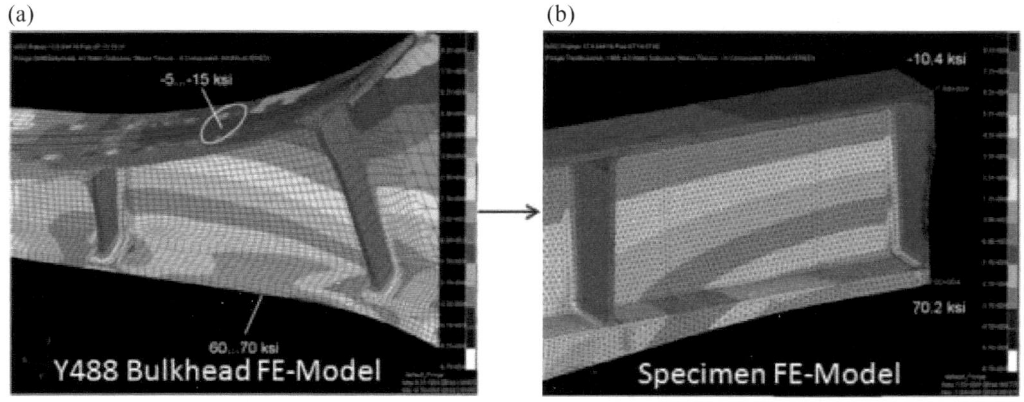

Figure 2. (a) FEM of the F/A-18 Y488 bulkhead lower flange. (b) FEM of the component-level specimen which approximates the same area ("middle" pocket and flange).

The area that was analysed as the next critical location in the lower Y488 bulkhead region, due to the effect of the doubler, was the main landing gear (MLG) uplock hole pocket. Specifically, the doubler was predicted to increase the stress level at the upper uplock hole by 4.6%. As a result, the FINAF fleet will have interference fit bushes installed at the MLG uplock holes, after they have been cold worked, to ensure adequate fatigue life. Further detail regarding the finite element analyses can be found in [15][17][18].

Component-Level Test Program

The Finnish fatigue test program was conducted by the Technical Research Centre of Finland (VTT). The program consisted of three series of component-level tests with specimens manufactured from the same material as the bulkhead, and with a geometry that approximated only part of the lower Y488 bulkhead, specifically the middle pocket[9] (between the MLG uplock hole pocket and the kick point pocket), see Figure 3. The specimens were also etched to replicate the surface condition which forms part of the ion vapour deposition (IVD) corrosion protection process; these etch pits tend to be initiation points for fatigue cracks [19]. Note that the dimensions for the test doublers were different to those on the fleet and they also did not have the same amount of curvature as the fleet doublers.

The three test series were:

- Doublers installed at the beginning of the fatigue tests (three specimens)
- Pre-fatiguing before doubler installation at approximately half life (three specimens)
- Fatigue testing without the doubler for life comparison purposes (three specimens)

The variable amplitude fatigue spectrum that was applied to the specimens approximated average FINAF F/A-18C usage, and was constructed using data from a HOLM system strain gauge. The final test spectrum contained 3,691 turning points, which were applied to produce a maximum stress of 308 MPa (44.6 ksi) in the specimen. Compressive marker band loads were also included in the spectrum to aid QF.

The component-level fatigue tests demonstrated an increase in life by a factor of 1.5, based on the specimens that were pre-fatigued and had the doubler installed at a time equivalent to approximately 3,000 AFHRS of FINAF usage. For the cases where the doubler was installed at the beginning of the test, a factor of 2.0 on life was achieved. The test program and results are presented in detail in [20][21][22].

[9] The doublers were originally designed and tested to reduce the stresses at the middle pocket, which was identified as a problem area before the adjacent geometric details were also known in Finland to be fatigue hot-spots. Therefore the coupons were not designed to be fully representative of the lower Y488 bulkhead.

(a)

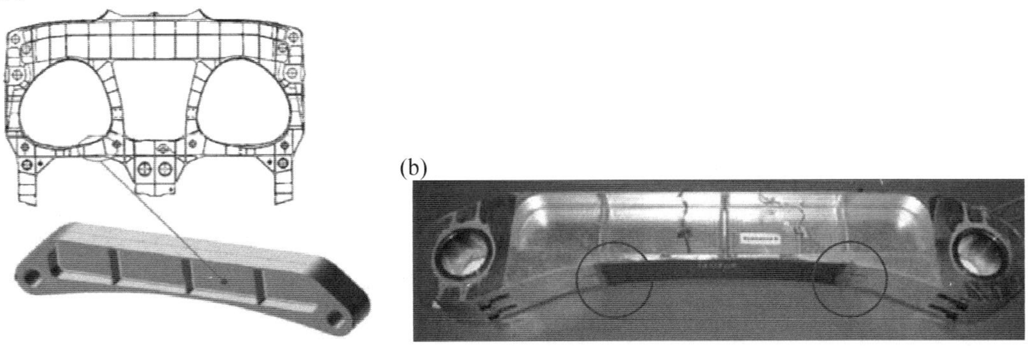

Figure 3. (a) 3D computer model of the component-level test specimen. (b) Photo of one of the test specimens. Note the minimal curvature of the lower flange at each end of the doubler.

DSTO TEST ARTICLE HISTORY, CONFIGURATION AND CONDITION

DSTO was responsible for testing the FINAF fleet doubler on a FSFT article. The CB selected for this activity was a retired United States Navy (USN) F/A-18D Hornet (Buno 164049). At its retirement it had accrued 7,749 AFHRS and a WR FLE[10] index of 0.734. The test article was designated "FINAL CB18" when it was delivered to DSTO since it was part of a consignment containing the eighteenth and nineteenth CBs to be acquired by the FINAL program. Of the 18 FINAL CBs tested to date, it had the second highest number of AFHRS and the third highest US WR FLE index.

The Y488 bulkhead (Part No. 74A324206) on FINAL CB18 had a "-2075" dash number configuration. For most[11] of the area affected by the doubler coverage, this was the same as the Y488 bulkheads in the FINAF fleet which are a newer build and have a "-2081" dash number. It was decided that the LHS of the bulkhead would be used for the doubler installation. Unlike any of the previous CB test articles, FINAL CB18's Y488 bulkhead had six areas where integral stiffeners had been cut out as well as mechanical damage (gouging) at one of the fatigue critical areas on the lower flange. See Figure 4. These were repaired by blending and the addition of reinforcements as required.

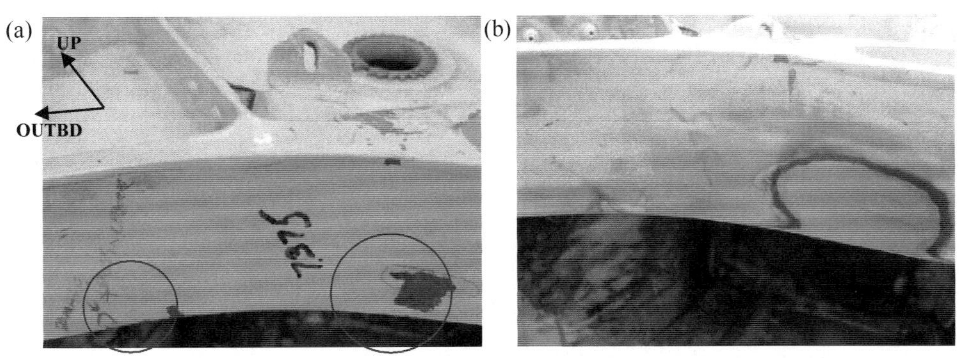

Figure 4. (a) Gouging on the LHS lower flange of the Y488 bulkhead on FINAL CB18 (circled), in the area where the doubler was to be installed. (b) The gouges were removed by blend repairs.

PRE-TEST INSPECTIONS AND STANDARD MODIFICATIONS

A conventional eddy current inspection (ECI) was performed on the LHS and right hand side (RHS) Y488 bulkhead lower flanges and pocket areas to check for cracking. No cracks were found in the flanges, the kick point pocket or the middle pocket. However, the MLG uplock hole pocket was found to contain cracking on either side of the upper uplock hole, see Figure 5(a). Consequently, RAAF modification 7213.006-312 [23] was incorporated (similar to FINAF

[10] USN usage.

[11] In the region where FINAL CB18 eventually failed (MLG uplock hole pocket), the -2081 configuration has a thicker web.

procedures), which involved oversizing the hole diameter to remove all cracking, cold expansion and a finishing ream[12] to 9.017 mm (0.355 inch), Figure 5(b). The lower hole, although not cracked, was subjected to the same modification. The RHS of FINAL CB18 was also subjected to the same procedure. The final part of the modification involves the installation of a bush in the hole but this was omitted in this case as an uplock pin (and hence the correct corresponding diameter) was not used in this CB test. After the modification, another ECI was carried out on the uplock holes and no evidence of cracking was detected.

(a) (b)

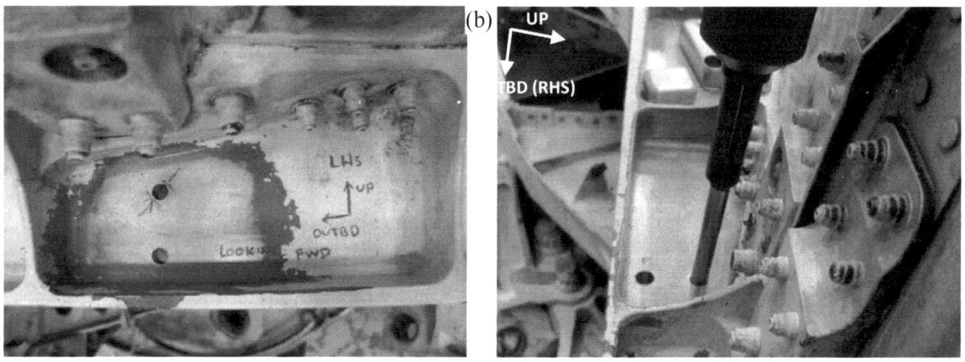

Figure 5. (a) Cracking was detected in the upper MLG uplock hole on the LHS and RHS (LHS shown).
(b) After oversizing the holes to remove the cracks, the holes were cold expanded (RHS shown).

DOUBLER INSTALLATION

A team of four Patria Aviation staff travelled to DSTO to perform the doubler installation. They were responsible for every process of the activity, from the surface preparation (see Table II) to the curing cycle, in order to replicate as closely as possible the same steps that would be taken for a fleet installation in Finland. Some of these processes are depicted in Figure 6. This included shipping out their specialist equipment and tooling. The only things that were supplied by DSTO were the adhesive and other consumable products, as this simplified the work health and safety aspects and logistics of the activity.

The doubler was transported from Patria to DSTO in its final pre-formed state, that is, a pre-cured UD 8-ply (0°) prepreg 5521 boron fibre laminate. The dimensions of the doubler were approximately 403 mm x 58 mm, with 4 mm steps between plies at either end. DSTO supplied the FM300-2 film adhesive. Special heating elements and vacuum bagging were put in place before the doubler was affixed to the bulkhead, and afterwards, this was followed by the installation of a heater blanket, thermocouples and the completion of the vacuum bagging. The curing cycle was performed under vacuum pressure at 95°C +/-5 °C, for approximately 9 hours. These steps were performed in accordance with FINAF F/A-18 Hornet standard repair manual (SRM) processes.

[12] This removed any trace of crack growth generated during the service life of the aircraft at the holes.

Table II. Surface treatment details.

For Bulkhead		For Doubler
(a) Degrease with MEK solvent	(b) Distilled water break test	(a) Degrease with MEK solvent
(c) Grit blast	(d) Wipe with MEK solvent	(b) Oven dry cycle
(e) Silane surface preparation		(c) Grit blast
		(d) Wipe with MEK solvent

(a) (b) (c) (d)

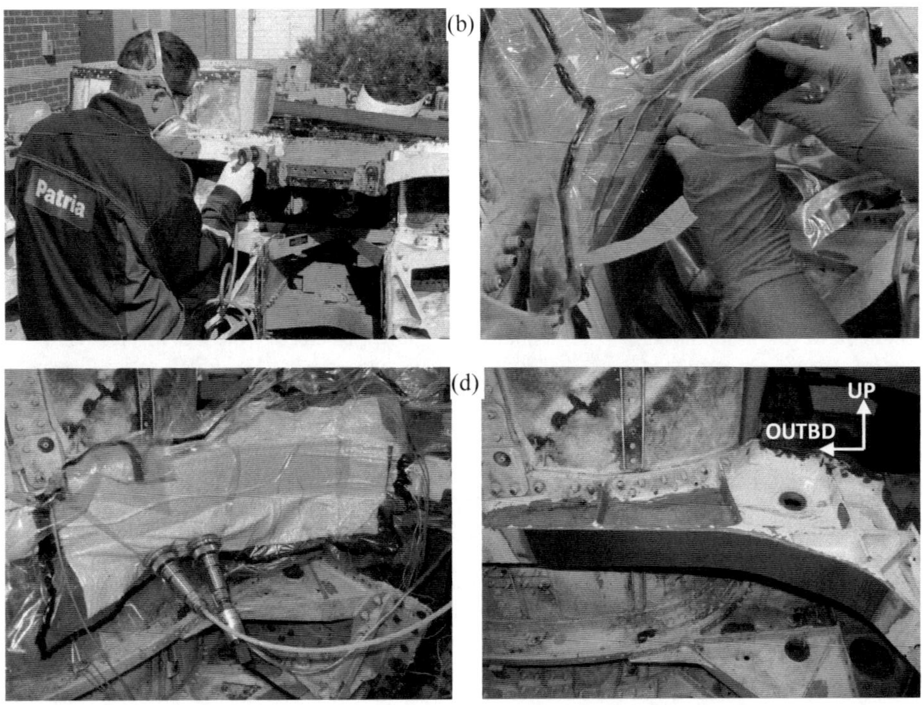

Figure 6. (a) Surface preparation of the LHS Y488 bulkhead lower flange area. (b) Installing the doubler to the bulkhead. (c) Heater blanket and vacuum bagging surrounding the doubler during the cure cycle. (d) The finished product, view looking forward and up.

STRAIN GAUGE INSTALLATION

Eight uniaxial strain gauges were installed on the doubler, see Figure 7. Four of these were gauge pairs (forward and aft edges) whilst one gauge was located at the centre of the inboard and outboard ends of the doubler, and the remaining two were in the centre of the doubler. Four of these gauges were mirrored on the RHS of the bulkhead so that direct comparisons between the reinforced side and unreinforced side could be made. There was also a T-rosette gauge (labelled "PATCH T") installed close to the centre of the doubler for the purposes of the thermoelastic stress analysis (TSA) scanning.

Numerous other gauges were installed in the vicinity of the doubler, including on either side of the doubler on the bulkhead itself (uniaxial), on the upper surfaces of the lower flange (uniaxial) and on the adjacent pocket webs (rosettes) [24]. The strain output files were only analysed at the completion of testing (refer next Section) and the results from some of the doubler, upper surface flange and web gauges, and their RHS equivalents, are presented and discussed later in this paper.

Figure 7. Strain gauge locations on the boron doubler and adjacent bulkhead lower flange. View looking up.

TEST PHASE

Loading

All loads were applied to FINAL CB18 as pure wing root bending moment (WRBM) loads. The test rig is designed such each bulkhead is loaded by an opposing set of hydraulic jacks connected to beams which in turn are connected directly to the wing attachment lugs.

Before any cyclic fatigue loading was applied, the test article was subjected to a 100% static strain survey[13] to confirm the functionality of the strain gauges and check for any non-linear structural response.

Thereafter, a block of constant amplitude (CA) loading was applied to create a distinct marker feature on any fracture surface to aid fractographic investigations post-testing. The CA block consisted of 1,201 cycles alternating between 0% and 80% of the peak FT55 load and generated approximately half the amount of crack growth as a block of the FT55 spectrum.

The variable amplitude FT55 spectrum was adapted from the one originally applied to the joint RAAF and Royal Canadian Air Force (RCAF) centre fuselage FSFT article. One spectrum block is comprised of 15,328 turning points in a peak-valley sequence and represents 324.9 spectrum flight hours (SFH) of average fleet usage for these two operators. The peak load in the FT55 WRBM spectrum is 730 kN-m (6,464 in-kips) and occurs once in each block.

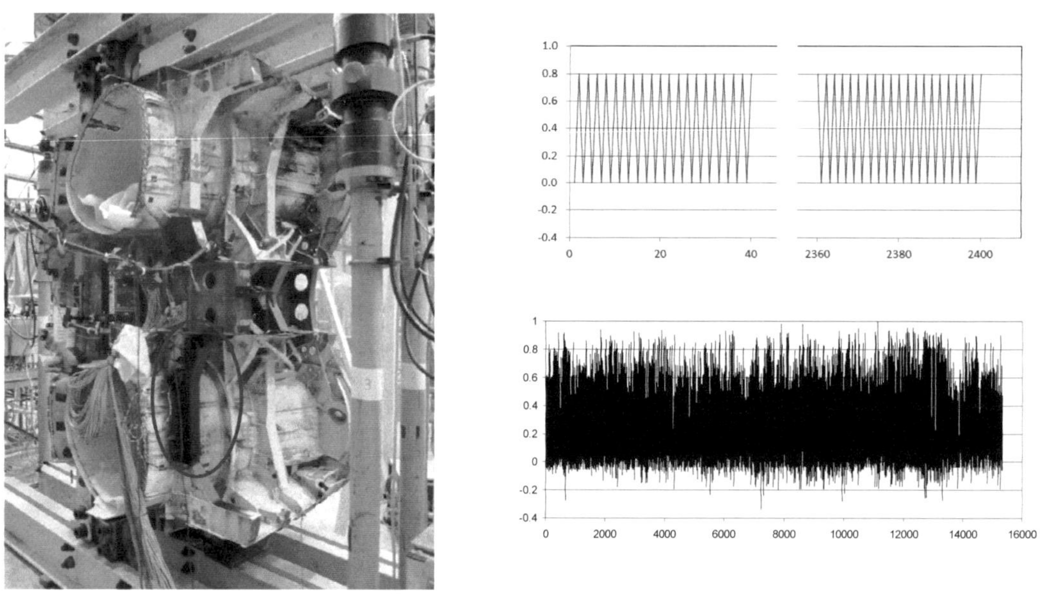

Figure 8. (a) FINAL CB18 in the test rig. The boron doubler and surrounding area (circled) were spray painted in matt black to aid the TSA scans. (b) The CA spectrum block. (c) The FT55 spectrum block.

[13] That is, 100% of the peak FT55 spectrum load.

Thermoelastic Stress Analysis Scans and Additional Bulkhead Modifications

After the completion of the first CA block, the standard process for FINAL CB tests is to apply 10 FT55 blocks. This pattern of one CA bock followed by 10 FT55 blocks is then repeated for the remainder of the test. However, in this case, testing was halted after the completion of the first FT55 block for several reasons. One was to conduct a TSA scan [25][26] of the doubler planform, whereby a microbolometer infrared (IR) camera was used to generate a real time image of the stress gradients in the structure[14], see Figure 9(a). Another was to conduct a separate research project on other parts of the CB, namely supersonic particle deposition (SPD), which is an additive manufacturing process being investigated as a possible repair method (beyond the scope of this paper). Lastly, the missing integral stiffeners mentioned earlier were repaired by removing sharp corners and installing mechanical straps with fasteners.

Test Article Failure

FT55 loading was resumed until the completion of block 10, after which time a second TSA scan was performed on a profile view of the doubler region, see Figure 9(b). Following the scan, the second CA block was applied and then FT55 loading was continued with the next schedule of blocks (11 to 20). At 41% into the third CA block, FINAL CB18 suffered a catastrophic failure in the Y488 bulkhead, at the LHS MLG uplock hole pocket, just outboard of the doubler, see Figure 10. The doubler itself appeared undamaged.

Therefore, the doubler experienced 20 FT55 blocks (6,498 SFH) and 2.41 CA blocks of fatigue loading. Once the test article had failed, the entire lower LHS section of the bulkhead was removed to preserve the fracture surface for fractographic analysis and to facilitate various NDI techniques and doubler removal.

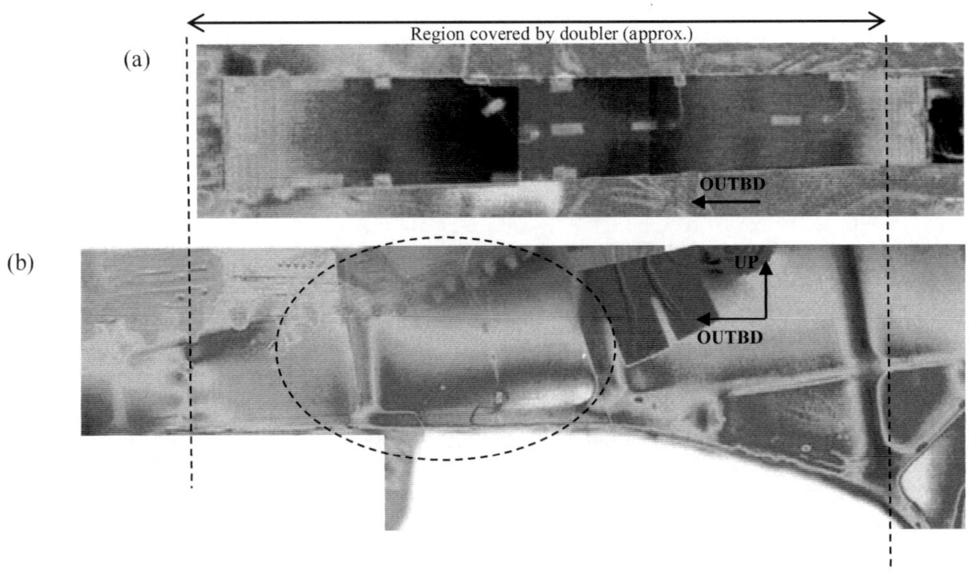

Figure 9. The colours in the TSA scans above denote relative changes in stress, not absolute values, as the scans are performed at low load levels. In this case, red signifies positive stress (tension) and the blue signifies negative stress (compression). (a) TSA scan (composite of multiple images) of the doubler planform conducted after the first FT55 block. View looking up. (b) TSA scan (composite of multiple images) of the doubler and bulkhead profile conducted after the tenth FT55 loading block. View looking forward. Note the similarity of the stress gradients in the middle pocket (circled) compared with the FEM of the same area in Figure 2(a).

[14] The TSA scans are performed using low magnitude CA loading between 0% and 8% of the peak FT55 load. The loading is also applied at a higher frequency than the FT55 block spectrum.

Figure 10. Failure of the LHS Y488 lower bulkhead in the MLG uplock hole pocket. View looking forward.

DOUBLER PERFORMANCE

Static Strain Survey Comparison of Reinforced versus Unreinforced Sides

The 100% static strain survey comparisons of the uniaxial gauges between the LHS (with doubler) and RHS (without doubler) indicated that the LHS strains were reduced by anywhere from approximately 19-35% on the underside of the lower flange and approximately 14-27% on the upper surfaces of the lower flange. Strain data from the rosette gauges on the web of the MLG uplock hole pocket were also examined. The LHS rosette above the doubler indicated a slight reduction in both the maximum principal strain (~ 9%) and maximum shear strain (~ 7%) when compared to the RHS. The LHS rosette outboard of the doubler and MLG uplock holes indicated a very small increase of the maximum principal and shear strains compared to the RHS (both less than 4%), but this could have been due to the accuracy of the gauge placement. The strain survey results and comparative information are presented in Table III.

Table III. Strain data comparisons for LHS and RHS gauges at 100% load.

LHS gauge name (doubler side)	Location	Strain (με)	Stress (MPa)	Strain comparison with RHS	RHS gauge name (mirror of LHS, no doubler)	Strain (με)	Stress (MPa)
40229L	Doubler	3,909	280	-18.8%	40229R	4,813	345
488005	Doubler	2,122	152	-34.2%	488006	3,225	231
488007	Doubler	2,125	152	-34.7%	488008	3,253	233
48800373	Doubler	2,642	189	-26.4%	48800374	3,591	257
48800375	Upper surface*	3,660	262	-14.3%	48800376	4,273	306
48801099	Upper surface*	2,585	185	-26.7%	48801100	3,529	253
488001	Upper surface*	3,075	220	-16.2%	488002	3,668	263
48800377-379-381♦	MLG pocket web	2,465♣	177	+3.5%	48800378-380-382	2,381	171
48800409-411-413♦	MLG pocket web	3,474♣	249	-9.2%	48800410-412-414	3,825	274

** On the upper surface of the lower flange. ♦ 3-element rosette gauge. ♣ Maximum principal strain.*

Fatigue Loading Strain Response of Doubler Gauges and Disbonding Indications

The next analysis involved an examination of the strain data recorded from the CA and FT55 fatigue loading blocks. (This activity had parallels with earlier work done at DSTO whereby strain gauges were used as a structural health monitoring demonstrator for a composite doubler repair on an F-111 wing test [27] in order to detect disbonds). The data from the all the doubler gauges, with the exception of the two at either end, demonstrated linear, repeatable responses throughout the duration of testing. However, the two gauges located at either end of the doubler revealed the likelihood that disbonding at both ends had occurred, although at very different times during the test. This likelihood was to be later supported by NDI, and confirmed by removing the doubler, and is discussed further on in this paper. For the gauge at the inboard end (488LDI), the maximum strain response remained constant during the first CA block (CA1). However, this response started to decrease within the first block of FT55 loading, and then steadily decreased as more blocks were applied, see Figure 11(a). This behaviour suggested the formation of a disbond early on as the doubler was losing its capacity to react load at this location. Around the time the second CA block was applied, it appeared that significant positive gauge drift started to occur as the responses were increasing relative to the preceding block responses, see Figure 11(b).

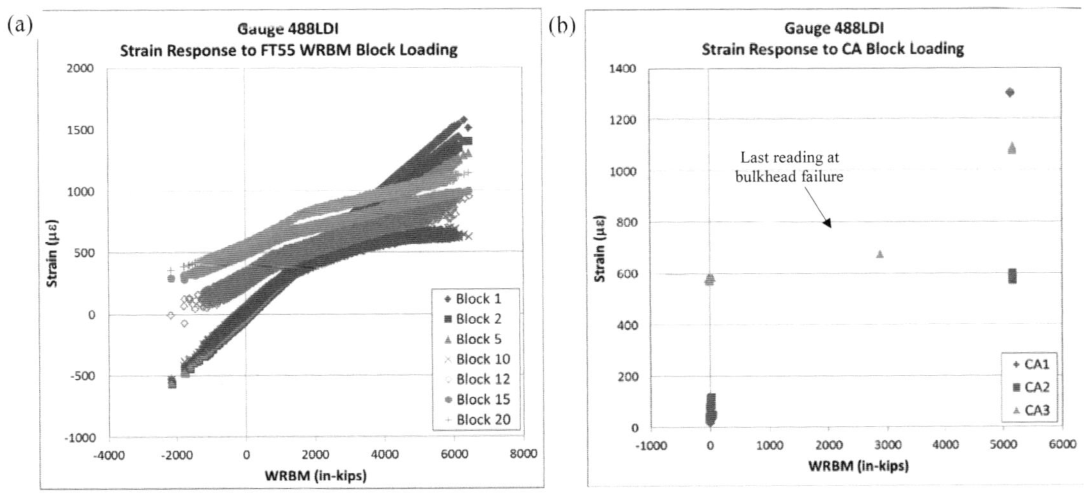

Figure 11. (a) Strain gauge 488LDI reponse to FT55 block loading (inboard end of doubler, refer Figure 7).
(b) Response of same strain gauge to the CA block loading.

The gauge data from the outboard end of the doubler (488LDO) proved harder to interpret in light of the much larger disbond that was subsequently discovered at the end of the test. The strain response remained linear and consistent for the first two CA blocks and throughout all of the FT55 blocks (Figure 12(a)), although some positive gauge drift had begun to occur. In any case, it appeared that the outboard end of the doubler was still effectively transferring load. However, an examination of the third block of CA strains indicated that the doubler, or the structure it was attached to, began to experience a rapid change in structural response during this period, see Figure 12(b). The positive drift offset was readily apparent, but the peak strains started to increase significantly until the bulkhead failed. This suggested that load shedding was starting to occur into the lower flange as the cracks in the MLG uplock hole pocket grew in size. To verify that this gauge behaviour was not in fact anomalous, the responses of other nearby gauges on the lower flange (488LBOD and 488LBODS.2) were also investigated. These were plotted along with 488LDO and are also presented in Figure 12(b). The behaviour of all these gauges was consistent, whether they were on the alloy bulkhead or the boron doubler, in that more load was being shed into the lower flange as the load-carrying capacity of the pocket diminished. Therefore, it is likely that the disbond at the outboard end of the doubler grew very rapidly in overload just before the bulkhead failed.

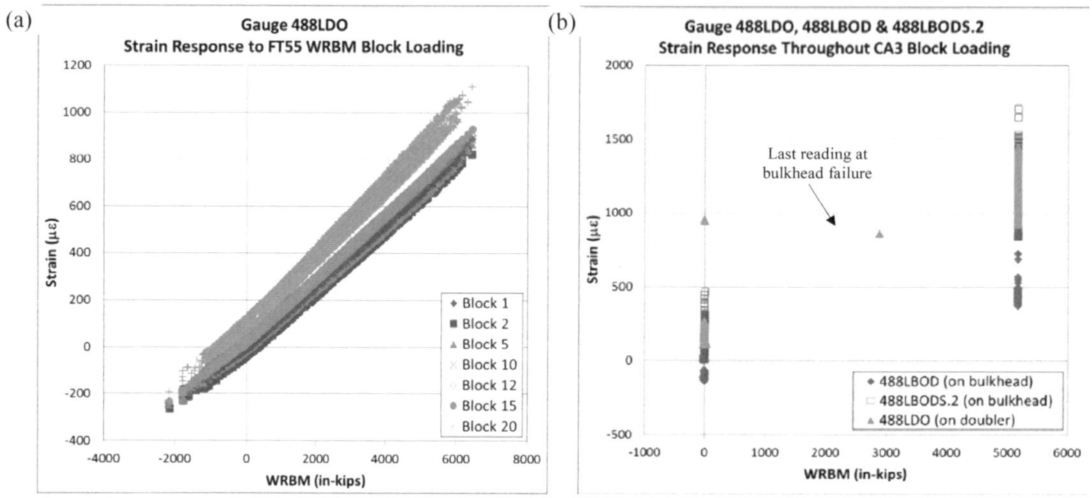

Figure 12. (a) Strain gauge 488LDO reponse to FT55 block loading (outboard end of doubler, refer Figure 7). (b) Response of same strain gauge and two nearby bulkhead gauges to the third block of CA loading.

Disbond Identification using Thermoelastic Stress Analysis

As mentioned earlier, a TSA scan [25][26] of the planform was conducted after the first CA and FT55 blocks had been applied. Analysis of this scan data supported the strain gauge responses at either ends of the doubler. The response of the X-component (proportional to stress) was analysed by drawing a trace across the width of the doubler at the inboard end (Figure 13(a)), which showed that either side of the doubler was carrying load (negative value equates to a positive stress) that meant the bond was effective (Figure 13(b)). However, in the centre region, the X-component changed to a positive value, indicating that this region was not carrying as much load and had therefore possibly disbonded (Figure 13(b)). A trace in the longitudinal direction closer to the edge of the doubler (Figure 14(a)) showed that load was being carried along this path (negative value, Figure 14(b)). At the outboard end, a trace drawn in the longitudinal direction, from the constant thickness part of the doubler to the bulkhead flange, showed that it was carrying load (negative value) and that there was no sign of any disbonding at the time in this region.

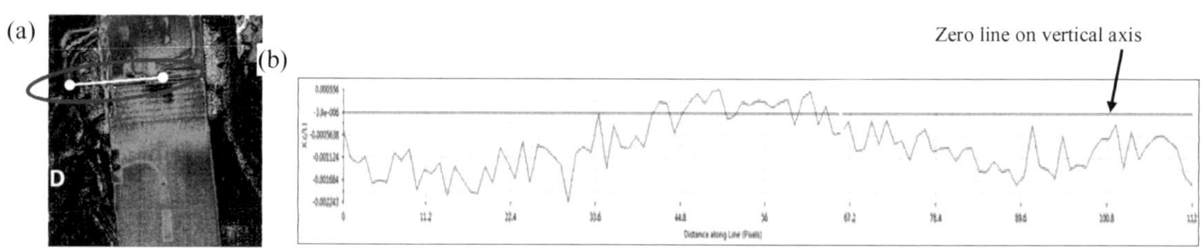

Figure 13. (a) A trace denoted by the white line with round ends, spanning the width of the doubler in the X-component scan of the inboard end. (b) Screen shot of the X-component profile is negative at both ends suggesting tensile stress (i.e. good bond) but near and exceeding zero in the centre region suggesting no load transfer (i.e. possible disbond). The operator qualified this assessment noting the presence of a strain gauge at this location and some excess adhesive which might account for some of the signal variation.

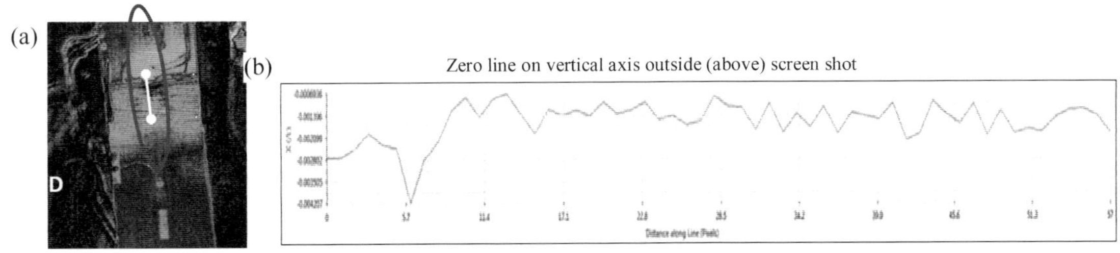

Figure 14. (a) A trace denoted by the white line with round ends, in the longitudinal direction across the tapered end of the doubler in the X-component scan of the inboard end. (b) Screen shot of the X-component profile shows a largely uniform negative value suggesting tensile stress (i.e. good bond), and therefore no obvious sign of disbonding.

PATTI Testing

After the bulkhead had failed and the section containing the doubler removed from the CB, a final validation of the bond integrity in the constant thickness sections of the doubler was demonstrated via two Pneumatic Adhesion Tensile Test Instrument (PATTI) tests [28], see Figure 15. Both cases failed at the boron and not the adhesive indicating a good bond. The first specimen (closer to the outboard end) failed at 24.2 MPa (3,507 psi) and the second one (closer to the inboard end) at 18 MPa (2,609 psi)[15]. The desire to preserve the underlying bulkhead at known cracking locations precluded the conduct of further PATTI tests along the specimen.

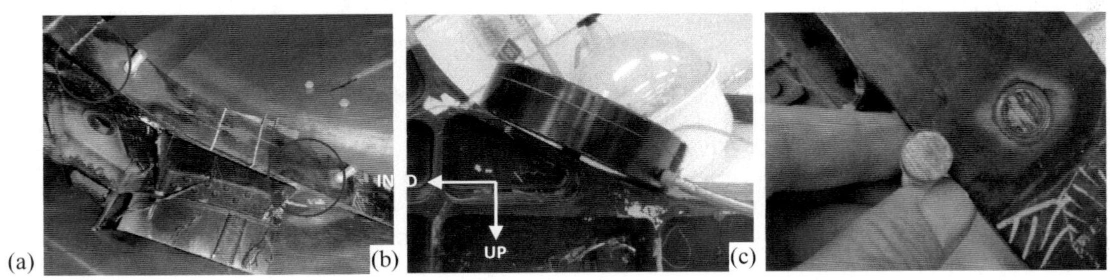

Figure 15. (a) Location of pull stubs for PATTI testing of the doubler. (b) PATTI test unit at the inboard location. Note the curvature of the specimen at this location prevented the loading fixture to sit flush with the surface. (b) PATTI test pull stub from the outboard location after removal.

POST-TEST INSPECTION OF DOUBLER

The section of bulkhead containing the doubler that was removed from the CB was subjected to a variety of different NDI techniques to check for disbonds. Several non-contact techniques were trialled which included shearography, lock-in thermography, flash thermography, and ultrasonic immersion tank C-scan. In addition, several direct techniques, requiring physical contact with the doubler, were also used; the tap test, resonance test, sonic thermography and the prescribed pulse-echo ultrasonic inspection (as interpreted) used on the FINAF fleet. DSTO has extensive experience with most of these techniques and some of these are discussed in [29]. The sonic thermography required an excitation source to contact the bulkhead at a short distance away from the doubler. Obviously, some of these do not have practical applications for employment in fleet aircraft inspections; however, they do highlight their respective relative strengths and/or limitations and provide a reference point for future deliberations.

Shearography

Based on the principle of laser interferometry, this technique recorded images of the specimen in an unloaded (baseline) state and a stressed state (induced by thermal loading with a heat source) [30]. Any defects (i.e. disbonds) respond differently to the thermal loading compared with the surrounding structure and these are revealed with the post-processing of the images to give the relative differences of the responses. Although the shearography camera unit that was employed was ideally suited to larger, flatter composite surfaces, it was still able to yield some positive results despite its limitations. Furthermore, the geometry of the specimen was not optimal with respect to maximising the

[15] The curvature of the bulkhead near this location prevented the PATTI test loading fixture from being operated on a flat surface which was a sub-optimal testing configuration, see Figure 15(b).

potential of the unit, especially at the inboard end of the doubler where there was more curvature and the unit could not sit flush with the structure. The best results were obtained when the specimen was located on the flat part of the flange and the lamp unit was covered to minimise heat diffusion. The inspection detected a suspected disbonded region at the outboard end of the doubler, shown in Figure 16.

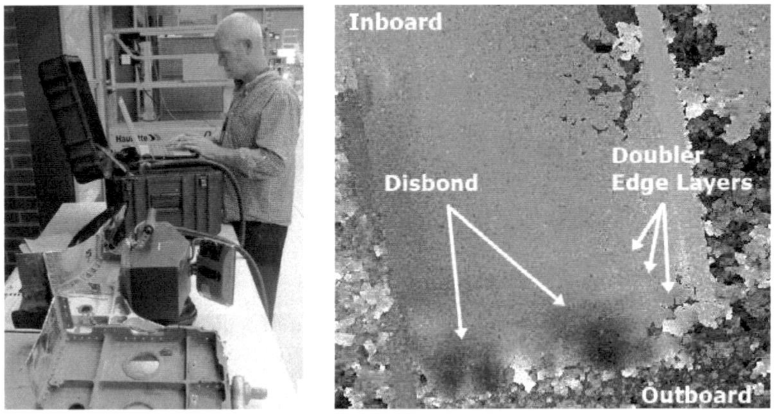

Figure 16. Shearography system used for NDI of the doubler. (b) Scan results of the outboard end of the doubler showing the suspected disbonded region.

Lock-in Thermography

In this scenario the doubler was thermally excited, at periodic frequencies, which caused a thermal wave to propagate from the surface of the doubler into the boron and to partially reflect at interfaces of different effusivity, such as at a disbond. The reflected wave interferes with the incoming wave and causes an interference pattern in the local surface temperature that oscillates at the same frequency as the thermal wave. An IR camera monitored the local surface temperature which was then used to evaluate the amplitude and phase to reveal the sub-surface information about the specimen [31][32]. Initially two heater lamps were used to energise the entire length of the doubler at various frequencies (0.2, 0.5, 1.0 and 5.0 Hz). Both the amplitude and phase of the resultant thermal images were analysed and the results obtained at each of these test frequencies were very similar. A suspected disbond at the outboard end of the doubler was clearly evident; however, no evidence of any suspected disbonding elsewhere in the doubler could be identified. Figure 17 shows the set-up used for the NDI tests as well as an example of a phase image result.

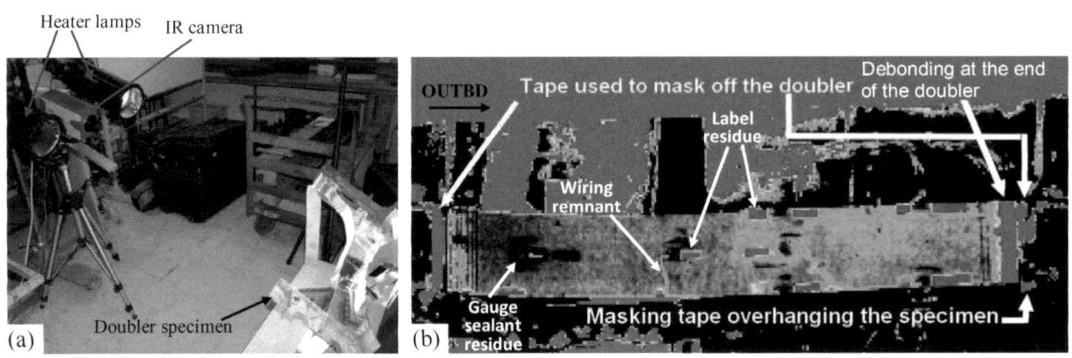

Figure 17. (a) Lock-in thermography set-up at Monash University showing the heater lamps, IR camera and doubler specimen. (b) Phase image of the doubler showing the surface irregularities, strain gauge, wiring and labelling marks, as well as the suspected disbonded region at the outboard end of the doubler.

Flash Thermography

With this technique, the specimen was thermally loaded by a brief pulse of light (as opposed to the periodic excitation used by lock-in thermography) and an IR camera was used to monitor the time-dependent response of the surface temperature [33]. As the heat propagated through the thickness, any defects encountered will cause a local temperature variation at the surface; in this case the areas that were thought to be disbonded retained more heat. The imagery that was generated from the raw data (at different times after the heating) and phase analysis (over different frequency ranges) revealed areas of suspected disbonding at either end of the doubler. Some examples are shown in Figure 18.

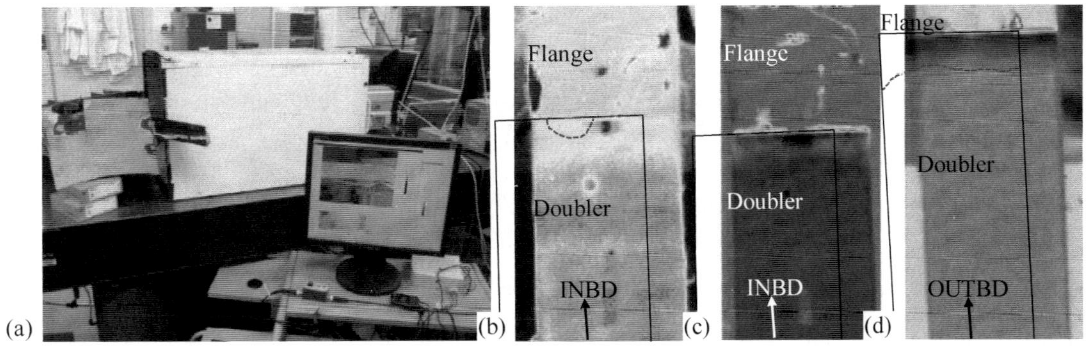

Figure 18. (a) Flash thermography set-up at DSTO showing computer, specimen in the flash box, and IR camera. (b) Raw data image of inboard end of doubler (at approx. 5 seconds). The outline of the doubler and suspected disbonds has been drawn for reference. (c) Phase analysis image of inboard end of doubler (frames 10 to 300, frequency = 156 MHz). (d) Phase analysis image of outboard end of doubler (frames 10 to 300, frequency = 156 MHz).

Ultrasonic Immersion Water Tank C-Scan

This technique used DSTO's ultrasonic water tank facility to produce a C-scan of the doubler. The principle and set up of this technique are described in [34]. The system had limitations with respect to being able to scan the entire doubler, as the probe was not able to follow the curvature of the specimen. Consequently a scan of the inboard end of the doubler could not be performed satisfactorily; however, the amplitude and phase signal responses for the outboard end were successful in detecting the suspected disbond as indicated in Figure 19.

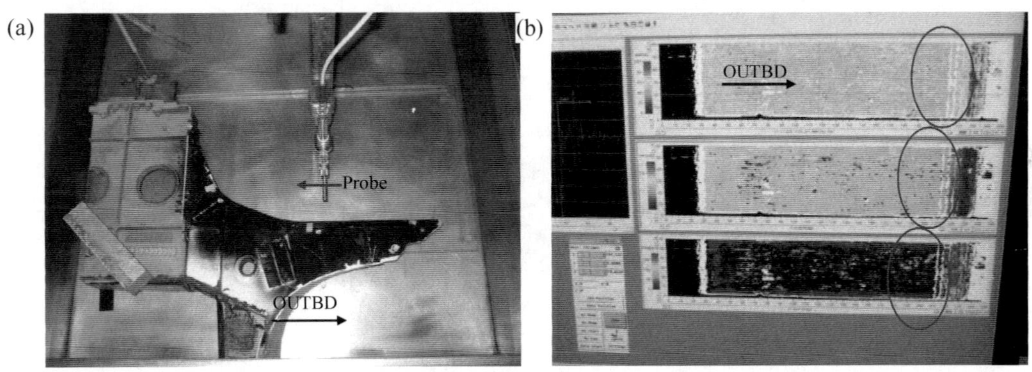

Figure 19. (a) View looking down at the specimen in the DSTO immersion water tank with probe scanning the flat part of the doubler. (b) C-scan results showing from top to bottom: (i) bond line interface amplitude response, (ii) time of flight response (which indicates the appropriate thickness of the doubler) and (iii) bond line interface phase signal response.

Tap Test

One of the cheapest, quickest and simplest of the NDI technologies would have to be the tap test, and consequently it has been used in multiple aerospace applications at DSTO such as in [35]. If performed by an experienced operator, this form of testing can produce very good results. The tap test was performed just after the flash thermography to confirm the results of the flash thermography. The areas of the probable disbonds, as indicated by the change in acoustic tone, were marked up on the specimen as shown in Figure 20. As can be seen, this technique proved to be very accurate and consistent with the flash thermography, albeit with the caveat that prior information from the flash thermography was available.

(a) (b) (c)

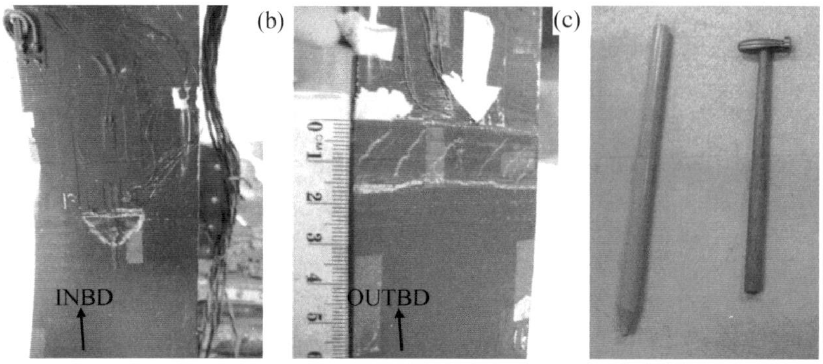

Figure 20. (a) Area of disbond at the inboard end of the doubler identified by tap test at DSTO indicated within the yellow lines. (b) Area of disbond at outboard end of the doubler identified by tap test within the yellow hashed area. (c) Chinagraph pencil and tap test hammer.

Resonance Test

The set up for the resonance test was not unlike the conventional fleet ultrasonic test (see next paragraph), in that a hand held probe, couplant and a portable display unit were required. The resonance test operates at a lower frequency than the conventional ultrasonic test and monitors signal amplitude response and phase changes [29] [36]. It functions as more of a "go/no-go" test. In lieu of a reference standard for the specimen geometry containing known disbonds, the calibration of the system could only be estimated, and therefore the results were considered as suitable for guidance only. The "flying dot" signal on the display indicated a suspected disbonded area when it moved away from the centre of the display; see red arrows in Figure 21.

(a) (b)

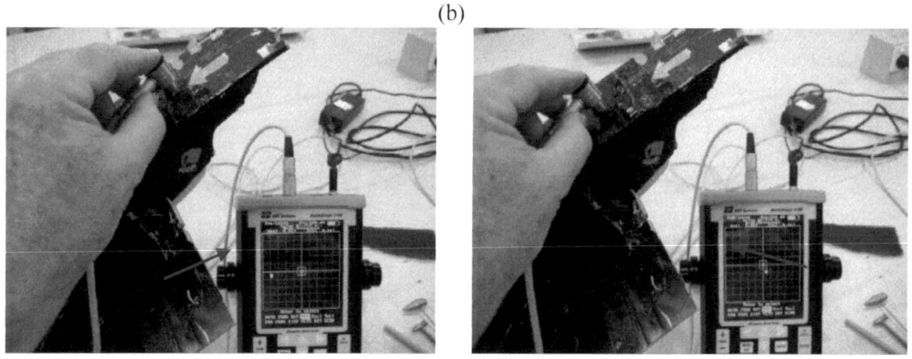

Figure 21. (a) Resonance testing of the outboard end of the doubler using a Bondascope 3100 unit at DSTO. (b) Probe on constant thickness area of the doubler with a good bond as indicated by the proximity of the signal to the centre of the display.

(Fleet) Ultrasonic Inspection

A copy of the NDI procedure [13], as performed on the doublers in the FINAF fleet, was provided to DSTO to simulate an on-aircraft inspection of the doubler bond integrity, noting that there is no reference standard used for the fleet inspections. This inspection technique sends an ultrasonic pulse through the probe shoe and monitors the echo responses from the surface of the doubler, the bond line interface and the back wall of the flange [29]. A loss of the back wall response signifies suspected disbonding as there is no contact between the doubler and flange to transmit the pulse and resultant echo. This behaviour was supported by using the immersion tank C-scan with similar ultrasonic settings to the fleet inspection. Figure 22 shows the ultrasonic probe indicating areas exhibiting a good bond and a suspected disbond at the inboard end of the doubler. Figure 23 shows the C-scan of the flange back wall amplitude response at the mid-to-outboard region of the doubler.

Of particular interest is the acceptance criteria for disbonds in the fleet, which only allows the existence of voids (disbonds) up to 100 mm^2. Discounting the probable outboard end disbond (likely caused by local overload when the bulkhead failed), the suspected disbond at the inboard end was at least 150 mm^2 (see Figure 25(b) further on for measurements), and hence would not have passed the inspection. However, it was also pointed out that after the initial post doubler installation inspection, the presence of a tube assembly in the vicinity of the inboard end of the doubler prevents follow-up inspections at this location in fleet aircraft. This may prove a matter for further consideration for fleet managers.

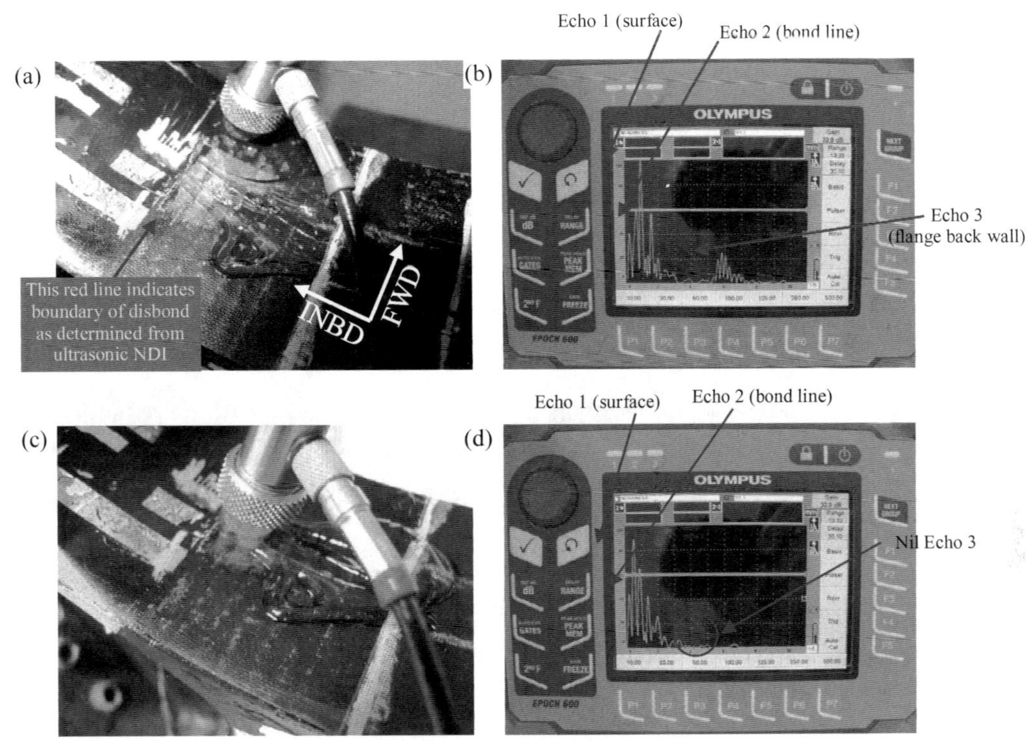

Figure 22. The images illustrate a representative fleet ultrasonic inspection, conducted at DSTO, at the inboard end of the doubler (an area which is not inspected at recurring intervals in the FINAF fleet due to access limitations). (a) Probe on region of good bond integrity. (b) Corresponding A-scan display indicating echo from flange back wall. (c) Probe on suspected disbonded region. (d) Corresponding A-scan display indicating nil echo from flange back wall.

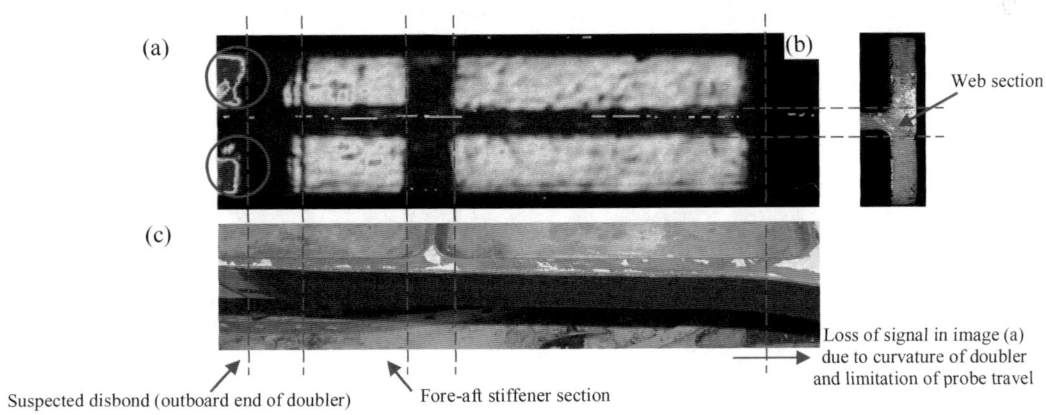

Figure 23. (a) Plan form view of C-scan flange back wall amplitude response display for the mid-to-outboard region of the doubler (using the DSTO immersion tank; refer Figure 19). The horizontal dark blue band indicates there is no back wall response from the web. The vertical dark blue band towards the middle indicates another region where there is no back wall response due to the fore-aft stiffener. The vertical dark blue band at the left indicates where the suspected disbond is. The circled areas would be transitioning in colour from yellow to red if the bond line was intact. This finding suggests that there is the possibility of missing a disbond with the fleet ultrasonic inspection, should it occur in line with the web and/or stiffener. (b) Section view of bulkhead showing flange and web. (c) Profile view of bulkhead and doubler showing disbond and fore-aft stiffener.

Sonic Thermography

As the name suggests, this technique applies an acoustic excitation to the specimen to generate a thermal response which is captured by an IR camera [37]. In this case, the specimen was subjected to a pulse of approximately one second duration with an excitation frequency of 20 kHz. The pulse was applied at a short standoff distance from either end of the doubler, depending on which end was being inspected. The results were processed using principal component thermography (PCT) second mode, to generate the images shown in Figure 24. The areas of suspected disbonding can be plainly seen. The large areas of red identified on the sides of the doubler in Figure 24(b) were thought to be a modal artefact, the likelihood of which was increased by processing using PCT third mode [38]. Further investigation by the operator was suggested, although the other NDI results would seem to support this suspicion.

(a) (b) (c)

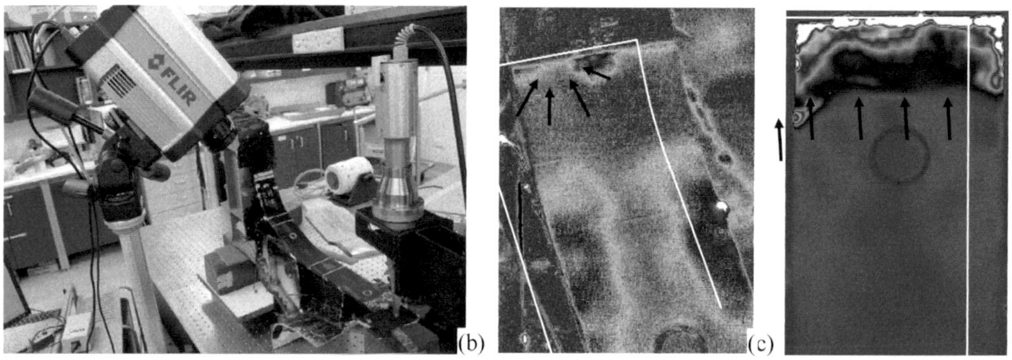

Figure 24. (a) Sonic thermography inspection at DSTO showing the set-up with the specimen, acoustic horn excitation system and IR camera. (b) Image of inboard end of doubler processed using PCT second mode. The large areas of red on either side of the doubler were thought to be modal artefacts. The edges of the doubler have been drawn for reference and the suspected disbond area is indicated by arrows. (c) Image of the outboard end of the doubler processed using PCT second mode. The suspected disbond area is indicated by arrows.

DESTRUCTIVE EXAMINATION FOR DISBOND CONFIRMATION: DOUBLER REMOVAL

The ultimate proof of the extent of the suspected disbonding was made evident when the doubler was separated from the parent bulkhead. The removal process involved heating the specimen in an oven at 190°C for approximately six hours in order to thermally degrade the adhesive. A cutting blade was then used to separate the doubler from the bulkhead. The regions where the disbonds occurred left a distinct mark on the bulkhead as shown in Figure 25. It was evident that the disbonds were cohesive in nature. The ability to remove the doubler in such a manner was a luxury that would not be possible on a fleet aircraft, but in this case it eliminated any doubt about the extent of the disbonding. It also allowed the accuracy of the different NDI techniques to be compared against a known quantity, as highlighted in the previous Sections.

(a) (b) (c)

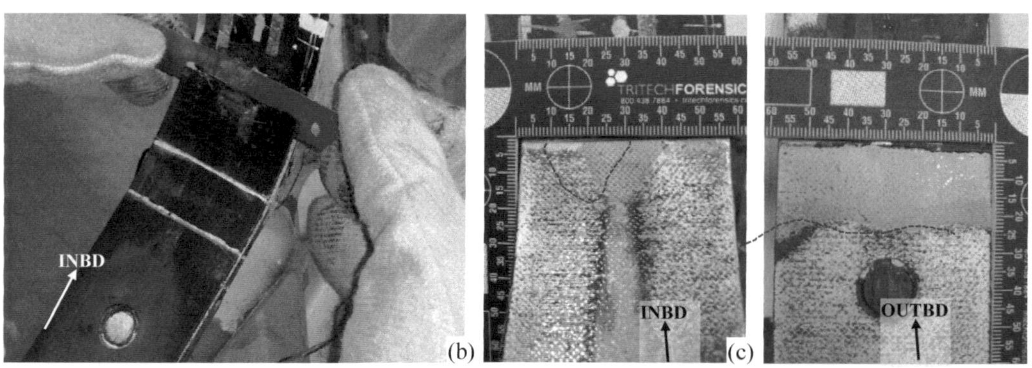

Figure 25. (a) A metal blade was used to separate the doubler from the bulkhead after a long heating period. (b) Witness mark of a disbond at the inboard end of the doubler as indicated by red dashed line. (c) Witness mark of a disbond at the outboard end of the doubler as indicated by red dashed line.

FAILURE ANALYSIS OF THE BULKHEAD FRACTURE

A QF analysis of the fracture surface was performed using high powered optical and scanning electron microscopes [39]. This confirmed the presence of three distinct cracks. The crack that caused the failure (labelled C1 in Figure 26) originated from a machining groove at a sharp radius at the upper forward face of the MLG uplock hole pocket. It grew through to the full thickness of the web and in the inboard direction progressed along the top of the pocket until it severed the upper flange. In the outboard direction the crack progressed at approximately 45 degrees downwards before taking a horizontal turn towards the upper MLG uplock hole. Two other cracks originated from the upper MLG uplock hole: the larger one (labelled C2) originated from an inclusion particle and grew approximately 45 degrees downwards and outboard (7.30 o'clock if looking forward) before severing the lower flange. The smaller crack (labelled C3) originated from a lap created during the rework of the fastener hole and grew approximately 45 degrees upwards and inboard before changing direction towards back towards the main fracture. An overview of the failure and a selection of QF imagery are presented in Figure 26.

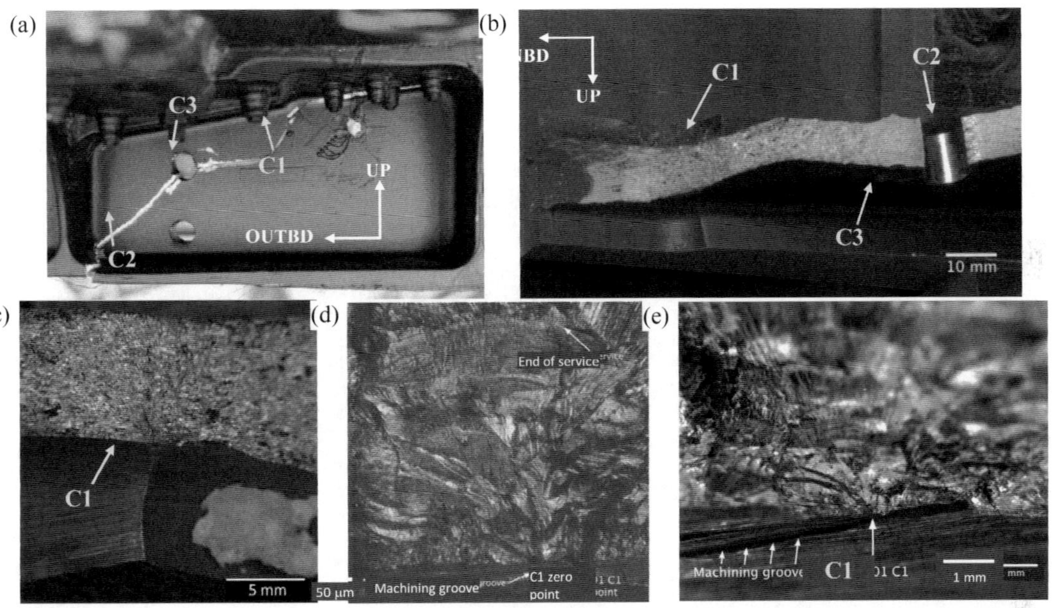

Figure 26. (a) Overview of the LHS MLG uplock hole pocket showing the three cracks discovered at failure. The failure crack is labelled C1. (b) View of one side of the fracture surface after separation of the two pieces of bulkhead. (c) Macro view of the origin region for crack C1. (d) High powered optical image of the C1 crack face indicating the end of service growth. (e) High powered optical image of the C1 crack origin.

Failures from other FINAL Tests

This same failure has occurred on four other FINAL tests which were also loaded with the FT55 spectrum but had no reinforcement in the area [40][41][42][43]. The crack growth curves for the lead cracks in all of these cases, including FINAL CB18 (after its growth measurement), are shown in Figure 27(a). As can be seen for FINAL CB18, the initial crack growth rate was less severe than the other examples, noting that the initial crack depth (i.e. end of service measurement) for this test was very large (0.5655 mm) and which may have also been influenced by the presence of the doubler. Nevertheless, FINAL CB18 also failed at approximately the same time as the failures from FINAL CB8 and FINAL CB17 (all between 22 and 24 blocks).

Cracking from the Radius versus Uplock Holes

Examination of the three FINAL CB18 cracks in isolation reveals the tendency for the upper MLG uplock hole to crack. This is supported by several other FINAL tests whereby the fracture has come from a crack originating at this hole [44][45][46]. Figure 27(b) shows that even with a reworked hole, cracks start growing at the commencement of block loading, and in this case, at a faster initial rate than the lead crack which led to failure. The subsequent decrease in slope of these hole cracks denotes their growth into the zone of compressive residual stress caused by the cold expansion of the hole.

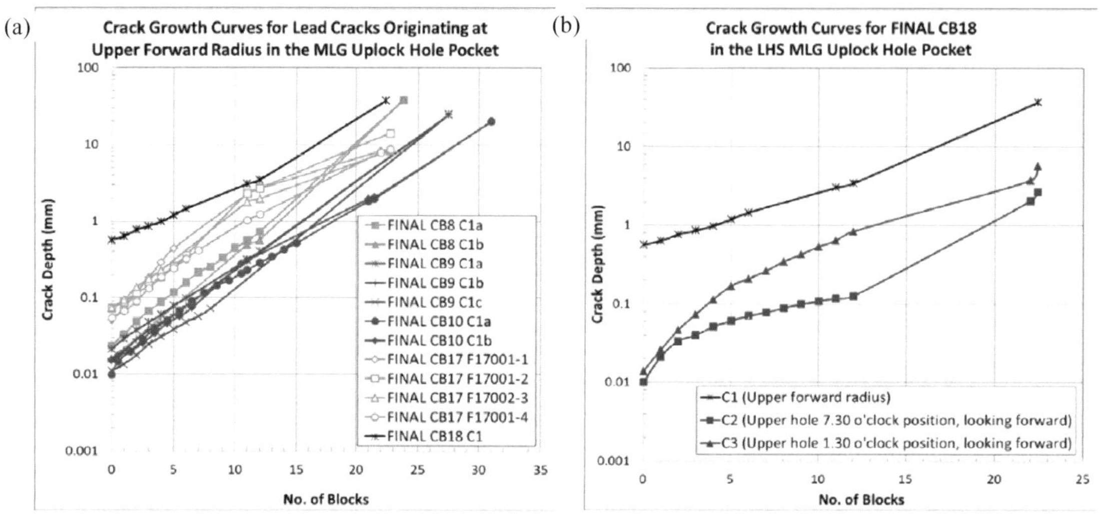

Figure 27. (a) Crack growth curve for the FINAL CB18 failure (black), compared with other FINAL tests which also failed from the same origin. The crack depth at 0 blocks is the initial flaw and service growth. Each marker represents one FT55 spectrum block. (b) Crack growth curves for the FINAL CB18 failure crack (C1) and the two cracks growing from the upper MLG uplock hole (C2 and C3).

CONCLUDING REMARKS

The opportunity to conduct the FINAF doubler fatigue test on a FSFT article was a natural progression from the component-level testing that was previously completed in Finland. By using a CB, the effect of the reinforcement on the surrounding structure could be assessed (noting that the component-level tests used a different sized doubler, with minimal curvature, on a specimen which was only partly representative).

As a result of the testing the following conclusions were made:

- The doubler itself did not fail under the (RAAF) representative F/A-18 fatigue loading for the 6,498 spectrum flight hours that had been applied. This was considered to be a good result, since failure of the doubler may have invalidated its design.
- The doubler did, however, develop a disbond at the inboard end, which appeared to start from the first FT55 block of loading. The size and location of this disbond failed the acceptance criteria specified by the fleet NDI procedure for the doubler. This is a point of potential concern given that this part of the doubler is not inspected at the recurring intervals due to access restrictions caused by tubing in the area in fleet aircraft. The disbond at the outboard end of the doubler seemed to have been generated when the bulkhead failed.
- Despite the bulkhead failing near the outboard end of the doubler and the suspicion that was henceforth placed on it, the failure occurred at a known critical location, as evidenced by the failure of four other FINAL tests at the same location. A comparison of the QF results for this failure indicated that the crack growth did not appear to be more severe when compared with the other failures that had lower flanges without the doubler fitted.

  In addition, there have been other FINAL tests that had failed from the middle pocket and kick point pocket from cracking in the lower flange; two areas that were also covered by the doubler. Therefore, based on the FINAL CB18 results, the doubler was successful in preventing failure from these two locations, while apparently leaving the probability of failure the same at the third (i.e. MLG uplock hole pocket) location. The Patria analysis predicted the stress increase at the upper uplock hole (seemingly supported by the QF results for the hole cracks), which led to a bushing modification for these holes in the FINAF fleet. While these bushed holes require a recurring inspection to check for re-cracking, special attention should also be paid to the sharp radius in the upper part of the pocket, from where FINAL CB18 failed, due to its demonstrated potential for generating catastrophic cracking.

The post-failure application of various NDI techniques to the reinforced section of the bulkhead found that, aside from the FINAF ultrasonic fleet procedure, the other seven techniques trialled all met with varying degrees of success at indicating the disbonded regions. (Noting that some of the equipment was not optimal for this specimen, e.g. the shearography unit, or would not be feasible for a fleet inspection, e.g. water tank C-scan). The various techniques did however demonstrate their relative ease of application and their fidelity, along with their limitations. An outcome from this NDI study might be that consideration be given to complementing the current fleet ultrasonic inspection with another technique to give greater certainty of the precise nature of any indications of disbonding.

Another visual "inspection" technique that was successfully demonstrated during the actual testing phase was the TSA scanning. The post-processing of the results was successful in identifying the disbond at the inboard end of the doubler, showing that this is a powerful tool for testing situations.

Lastly, DSTO's unique FINAL CB test program has been instrumental in supporting significant elements of the RAAF F/A-18 aircraft structural integrity programs, such as major structural upgrade projects and life extension work. The CB test program has also proven to be a useful asset for concept demonstrators and collaboration with international F/A-18 partners, with the FINAF doubler testing being one of several successful examples.

FURTHER WORK

At the time of writing, there were several aspects of this study still to be completed.

- One is a comparison of the relative fatigue life under the FINAF spectrum (HOLM) used for the original design analysis of the doubler, and the FINAL test spectrum (FT55) used for the testing at DSTO. This is required for the FINAF to determine if their desired service life goals have been demonstrated by the test. It is proposed that this will be done using a strain life and/or crack growth analysis, as well as a small coupon test program to compare the cracking from both spectra.
- Another aspect is to assess the effectiveness of the doubler in retarding crack growth in the lower flange. However, if any cracks are present, they are likely to be very small and unable to be detected via NDI. This may involve destructive inspection of areas prone to cracking, based on the evidence from previous (unreinforced) FINAL test articles, in order to find the cracks. Conversely, there is likelihood that cracking may be detected at the RHS upper MLG uplock hole. If this is the case, then a comparison of the QF results from the RHS cracking can be made with those presented here from the LHS to make a further assessment about the influence of the doubler on these cracks.
- An investigation of the chord-wise (fore-aft) stress distribution at the inboard end of the doubler may reveal why the disbond grew in the shape and position that it did. In lieu of being able to instrument another FSFT with an array of strain gauges, a finite element analysis would be required.

ACKNOWLEDGEMENTS

The authors would like to acknowledge the valuable contributions by the following people without whose efforts this paper would not have been possible: Mr. Yrjo Laatikainen, Mr. Kari Hirvonen, Mr. Iiro Ojala and Mr. Jarmo Silvan (Patria Aviation) who visited DSTO to install the doubler. Mr. Mark Fitzgerald (Fortburn) who installed the strain gauges and Dr. Leigh Robertson (Fortburn) for processing and making available the strain data. Mr. Peter Smith (DSTO) for configuring the data acquisition system and running the test. Mr. Dario Endrigo (Fortburn) for preparing the test article and rig. Mr. Richard Mazeika (Fortburn) for the QF of the bulkhead failure. Mr. Jorma Patronen (Patria Aviation) for providing the FINAF NDI procedure. Dr. Nik Rajic (DSTO), Mr. Greg Surtees (DSTO), Professor Rhys Jones (Monash University) and Mr. Shane Wilson (Southern NDT) for their application of the various NDI methods used on the test article. Mr. Ivan Stoyanovski (DSTO) for his assistance with the Patria visit, PATTI testing and doubler removal. Captain Kalle Vaaraniemi (FINAF) and Mr. Lorrie Molent (DSTO) for their backing of the project. Mr. Robert Boykett, Dr. Alan Baker, Dr. John Wang and Dr. Simon Barter for their critique and suggestions for the paper. The staff from NAVAIR (USN) involved in making the test article available.

REFERENCES

[1] MDC (1992). *F/A-18 Teardown and Inspection of Center Fuselage Fatigue Test Article (FT01/ST16/FT24)*, B2378 Vol. 1 Final Report, McDonnell Douglas Corporation, USA.

[2] Bombardier (2001). *Notice of Structural Deficiency*, No. 10180, Iss. 1, International Follow-On Structural Test Project, Bombardier Aerospace Defence Services, Canada.

[3] Molent, L., and Swanton, G. (2014). *The DSTO Contribution to the Fatigue Life Reassessment of the RAAF F/A-18 Hornet Centre Barrel Structure*, DSTO-TR-3062, Aerospace Division, Defence Science and Technology Organisation, Australia.

[4] Siljander, A. (2009). In: *A Review of Aeronautical Fatigue Investigations in Finland May 2007 – April 2009*, Proceedings of the 25th ICAF Symposium, Bos, M. J., (Ed.), Rotterdam, The Netherlands.

[5] Peltoniemi, E., and Siljander, A. (2011). In: *A Review of Aeronautical Fatigue Investigations in Finland May 2009 – March 2011*, Proceedings of the 26th ICAF Symposium, Komorowski, J., (Ed.), Montreal, Canada.

[6] Simpson, D. L., Molent, L., Graham, A. D., Landry, N., Roussel, J., and Schmidt, N. (2002). In: *The Canadian and Australian F/A-18 International Follow-On Structural Test Project*, Proceedings of the 23rd International Council of the Aeronautical Sciences, Grant, I., (Ed.), Toronto, Canada.

[7] Bartholomeusz, R. A., Baker, A. A., Chester, R., and Searl, A. (1999). In: *International Journal of Adhesion and Adhesives,* Vol. 19, pp 173-180.

[8] Dixon, B., and Molent, L. (2003). *Ex-Service F/A-18 Centre Barrel Fatigue Flaw Identification Test Plan*, DSTO-TR-1426, Air Vehicles Division, Defence Science and Technology Organisation, Australia.

[9] Molent, L. et al. (2009). In: *The F/A-18 FINAL Fatigue Testing Program*, Proceedings of the 13th Australian International Aerospace Conference, Melbourne, Australia.

[10] Swanton, G., and Robertson, L. (2011). In: *Developments with the F/A-18 FINAL Centre Barrel Test Program*, Proceedings of the 14th Australian International Aerospace Conference, Melbourne, Australia.

[11] Barter, S., Molent, L., and Robertson, L. (2009). In: *Using In-Service F/A-18 A/B Aircraft Fatigue Cracking as Disclosed by Teardown to Refine Fleet Life Limits*, Proceedings of the USAF Aircraft Structural Integrity Program (ASIP) Conference, Jacksonville, Florida, USA.

[12] Swanton, G., and Louli, Z. (2012). *Flaw Identification through the Application of Loads: Test and Teardown of the Centre Barrel from USN F/A-18C BuNo 164006 (FINAL CB5)*, DSTO-TR-2532, Defence Science and Technology Organisation, Australia.

[13] FINAF, extract from HN4-110-02S1 *(Structural Inspections Manual)*, Section 5 Ultrasonic Inspection.

[14] Aakkula, J. (2008). *Komposiittivahvikkeiden materiaalien ja prosessien esiselvitys*, KRT-T323, Helsinki University of Technology, Laboratory of Lightweight Structures, Espoo, Finland, (in Finnish only).

[15] Kettunen, J., and Linna, J. (2008). *Kaaren Y488 koesauvan ja vahvikkeen suunnittelu*, HN-L-0142, Patria, Finland, (confidential, in Finnish only).

[16] Linna, J., and Laatikainen Y. (2009). *HN-Runkokaaren Y488 lämmityskokeet boorikuituvahvistusta varten*, HN-L-0174, Patria, Finland, (confidential, in Finnish only).

[17] Linna, J. (2009). *Boorikuidulla vahvistettujen kaaren Y488 koesauvojen väsytyskokeiden loppuraportti*, HN-L-0167, Patria, Finland, (confidential, in Finnish only).

[18] Kunnari, A. (2014) *Kaaren Y488 MLG Uplock reikien väsymisanalyysi*, HN-L-0244, Patria, Finland, (confidential, in Finnish only).

[19] Molent, L., Dixon, B., and Barter, S. (2004). In: *Flaw Identification Through the Application of Loading (FINAL)*, Proceedings of the International Conference on Structural Integrity and Fracture, Brisbane, Australia.

[20] Koski et. al. (2009). *Alumiinisten työstettyjen koekappaleiden väsytyskokeet*, VTT-R-01979-09, VTT, Espoo, Finland (confidential, in Finnish only).

[21] Koski et. al. (2011). *Alumiiniseoksisten koekappaleiden väsytyskokeet*, VTT-R-02004-11, VTT, Espoo, Finland, (confidential, in Finnish only).

[22] Anderson, B., et al. (2009). In: *A Case Study of Multiple-Site Fatigue Crack Growth in the F-18 Hornet Bulkhead*, Proceedings of the 25th ICAF Symposium, Bos, M. J., (Ed.), Rotterdam, The Netherlands.

[23] RAAF (2003). *Rework of Y488 Bulkhead and Y497 Former Main Landing Gear Uplock Assembly Attachment Holes*, Modification No. 7213.006-312, AAP 7213.006-100B5, Royal Australian Air Force, Australia.

[24] Swanton, G., Sykiotis, G., and Robertson, L. (2014). *Strain Survey Results for the Finnish Air Force Boron-Epoxy Doubler Installation on the DSTO FINAL CB18 Test Article*, Minute, DSTO internal file BM2/121, Aerospace Division, Defence Science and Technology Organisation, Australia.

[25] Rajic, N., and Galea, S. (2014). In: *Journal of Structural Health Monitoring,* DOI: 10.1177/1475921714548936.

[26] Rajic, N., Street, N., Brooks, C., and Galea, S. (2014). In: *Full Field Stress Measurement for In Situ Structural Health Monitoring of Airframe Components and Repairs*, Proceedings of 7th European Workshop on Structural Health Monitoring, Nantes, France.

[27] Baker, A. (2008). *Structural Health Monitoring of a Bonded Composite Patch Repair on a Fatigue-Cracked F-111C Wing*, DSTO-RR-0335, Air Vehicles Division, Defence Science and Technology Organisation, Australia.

[28] Lee, M., Wang, C., and Yeo, E. (2013) In: *International Journal of Adhesion & Adhesives,* Vol. 44, pp 259-268.

[29] Roach, D. P., and Scala, C. M., (2002). In: *Advances in the Bonded Composite Repair of Metallic Aircraft Structure,* Vol. 2, pp 659-726, Baker, A. A., Rose, L. R. F., and Jones, R., (Eds.), Elsevier, United Kingdom.

[30] LTI (2007). *LTI-6200S Portable Thermal Shearography Inspection System Operators Manual*, Laser Technology Inc., Pennsylvania, USA.

[31] Rahammer, M., Menner, P., and Busse, G. (2012). In: *Thermal Waves for NDE of Aircraft: Comparison of Lockin Thermography and Lockin Interferometry*, Proceedings of the 11th International Conference on Quantitative InfraRed Thermography, Cardone, G., (Ed.), Naples, Italy.

[32] Jones, R., Krishnapillai, M., Cairns, K., and Matthews, N. (2010). In: *Journal of Fatigue and Fracture of Engineering Materials and Structures,* Vol. 33, Iss. 12, pp 871–884.

[33] Krishnapillai, M., et al. (2005). In: *Journal of Composite Structures,* Vol. 67, pp 149-155.

[34] Clark, G., and Bishop, B. (1991). In: *Acoustics Australia,* Vol. 19, No. 1, pp 17-21, Australia.

[35] Callus, P., et al. (2011). *F-111 Adhesive Bonded Repair Assessment Program (FABRAP) – Phase 1 Testing, Preliminary Results*, DSTO-TN-1024, Air Vehicles Division, Defence Science and Technology Organisation, Australia.

[36] Crompton, M. (2014). In: *Ultrasonic Bond Testing as a Quality Tool, Quality Magazine – NDT in Aerospace*, (http://www.qualitymag.com/articles/91999-ultrasonic-bond-testing-as-a-quality-tool).

[37] Tsoi, K., and Rajic, N., (2011). *Non-destructive Evaluation of Aircraft Structural Components and Composite Materials at DSTO Using Sonic Thermography*, DSTO-TN-0986, Air Vehicles Division, Defence Science and Technology Organisation, Australia.

[38] Rajic, N. (2002). *Principal Component Thermography*, DSTO-TR-1298, Airframes and Engines Division, Defence Science and Technology Organisation, Australia.

[39] DSTO (2015). *FINAL – Used F/A-18 Centre Barrel Teardown Fractographic Analysis Report*, QFFINAL 189 CB18, DSTO file 2014/1158793/1, Defence Science and Technology Organisation, Australia.

[40] DSTO (2012). *FINAL – Used F/A-18 Centre Barrel Teardown Fractographic Analysis Report*, QFFINAL 065rev1 CB8 A21-016, DSTO file 2008/1008220/1, Defence Science and Technology Organisation, Australia.

[41] DSTO (2010). *FINAL – Used F/A-18 Centre Barrel Teardown Fractographic Analysis Report*, QFFINAL 081 CB9, DSTO file 2008/1008224/1, Defence Science and Technology Organisation, Australia.

[42] DSTO (2010). *FINAL – Used F/A-18 Centre Barrel Teardown Fractographic Analysis Report*, QFFINAL 112 CB10, DSTO file 2010/1045229/1, Defence Science and Technology Organisation, Australia.

[43] DSTO (2011). *FINAL – Used F/A-18 Centre Barrel Teardown Fractographic Analysis Report*, QFFINAL 131 CB17, DSTO file 2011/1160253/1, Defence Science and Technology Organisation, Australia.

[44] DSTO (2006). *FINAL – Used F/A-18 Centre Barrel Teardown Fractographic Analysis Report*, QFFINAL 018 USN buno 163510, DSTO file 2006/1060212/1, Defence Science and Technology Organisation, Australia.

[45] DSTO (2007*). FINAL – Used F/A-18 Centre Barrel Teardown Fractographic Analysis Report*, QFFINAL 061 CB7, DSTO file 2006/1157226/1, Defence Science and Technology Organisation, Australia.

[46] DSTO (2012). *FINAL – Used F/A-18 Centre Barrel Teardown Fractographic Analysis Report*, QFFINAL 143 CB12, DSTO file 2009/1148141/1, Defence Science and Technology Organisation, Australia.

Evaluation of a UAV Composite Wing Spar Repair Using an Embedded Optical Fiber Rayleigh Back-Scattering Distributed Strain Sensing

A. Bergman[1], U. Ben-Simon[2], A. Schwartzberg[2], N.Y. Shemesh[3], B. Glam[3], J. Burvin[2], I. Kressel[2], T. Yehoshua[1], M. Tur[1]

[1]School of Electrical Engineering, Tel-Aviv University, Tel-Aviv Israel
[2]Engineering & Development Group, Israel Aerospace Industries, Ben Gurion International Airport, 70100, Israel
[3]IAF Aircraft Structure Branch, Tel-Aviv Israel

Abstract: A representing damaged UAV wing spar cap was repaired using bonded unidirectional composite materials. For such a repair concept, its strength and long-term durability depend on optimizing the geometry of the repair in order to minimize stress concentration in the adhesive. In order to assess the stress distribution along the repair bond-line, an optical fiber was embedded during the repair application for distributed Rayleigh back-scattering strain measurement. The fiber was placed inside the structure, therefore enabling taking strain measurements, at a high spatial resolution, under the wing skin directly on the spar repair. This type of sensing concept can also be used to monitor this critical repair over time. It also coincides with the recently introduced airworthiness requirements for UAVs, where the substantiation of structural bonded joint can be based on: "repeatable and reliable non-destructive inspection".

INTRODUCTION

The increasing use of high modulus composite materials for UAVs main structure imposes a challenge when repair of such component is needed, since both stiffness and strength must be restored. Two basic repair concepts can be considered for such structures: a bolted repair, where a patch is attached by fasteners over the damaged area or a more advanced concept, where the patch is bonded to the part [1]. For UAV structures, where high or medium modulus composite material is used at high percentage in the layup, the fastened concept, in most cases, cannot restore the original structure strength and stiffness. Bonded repairs are known for their ability to restore both strength and stiffness [1,2] but their design and application are challenging, especially where high stiffness has to be addressed. Major challenges in applying bonded repair are its sensitivity to adhesive thickness and surface preparation for bonding. It should also be noted that no commercially available non-destructive inspection is capable of evaluating the strength of such a patch.

In this work a representative damaged UAV wing spar cap was repaired using bonded unidirectional composite materials. Two types of bonded repair concepts were evaluated in order to select the best option for application, in terms of both design and quality of application. In order to assess the repair performance, an optical fiber was embedded during the repair application to be later used for distributed Rayleigh back-scattering strain measurement with high spatial resolution [3,4,5].

Three specimens were tested statically and under fatigue spectrum, all were monitored using the embedded optical fiber. The first two specimens, tested up to failure, were used to optimize the repair concept and the third one was a full scale wing spar portion, representing the actual repaired front spar and leading edge. This last specimen was tested up to ultimate load, followed by cyclic loading spectra, representing the full lifetimes of the structure. During this entire test, strain measurements were taken periodically with a spatial resolution of 1cm. The resulting strain distributions and their dependence on the applied load were found to predict the quality of the bonded repair. Thus, it appears that this sensing concept enables direct monitoring of bonded repair durability during both application and service.

THE BONDED REPAIR CONCEPT

The basic analytical model for stress distribution in bonded repairs was presented by Hart-Smith in the 1970s [6]. Typical bonded patch repair is shown in Figure 1. For this configuration where t is the thickness of both the patch and the substrate, E is the elastic modulus, ν is the Poisson's ratio and G_a is the adhesive shear modulus. For the configuration of Figure 1 this model predicts that the maximum shear stress occurs at the patch ends.

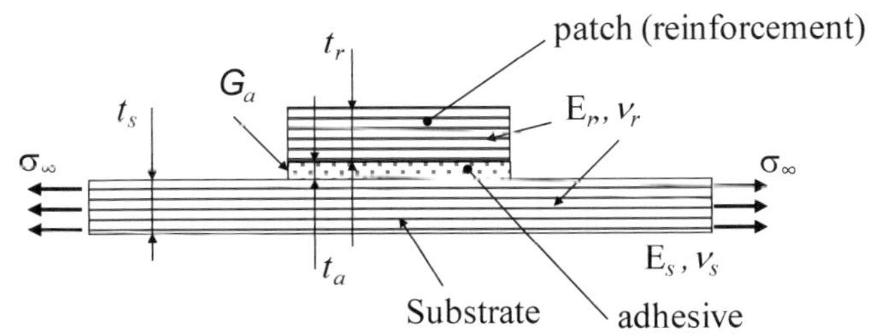

Figure 1. Bonded patch basic concept.

For a simplified arrangement where the thickness and material properties are the same, for both the patch and substrate, the maximum shear stress is equal to:

$$\tau_{a_{\max}} = \frac{1}{2}\beta\sigma_\infty t_s \quad , \tag{1}$$

where β is defined by: Israel Aerospace Industries

$$\beta^2 = \frac{G_a}{t_a}\left[\frac{1-\nu_s^2}{E_s t_s} + \frac{1-\nu_r^2}{E_r t_r}\right] \tag{2}$$

By evaluating equations (1) and (2) it is clearly seen that the stress concentration in the adhesive is proportional to the remote stress and is not a function of the bond line length. Since we consider here unidirectional, intermediate modulus, high strength, carbon fiber composite, this issue of adhesive stress concentration is of great importance and should be minimized by proper design.

The common way to reduce the stress associated with the ends of the bonded patch repair is by introducing a scarf or stepped patch as described in Figure 2. In the two concepts the patch is bonded to the substrate with 9394 paste adhesive at room temperature under vacuum.

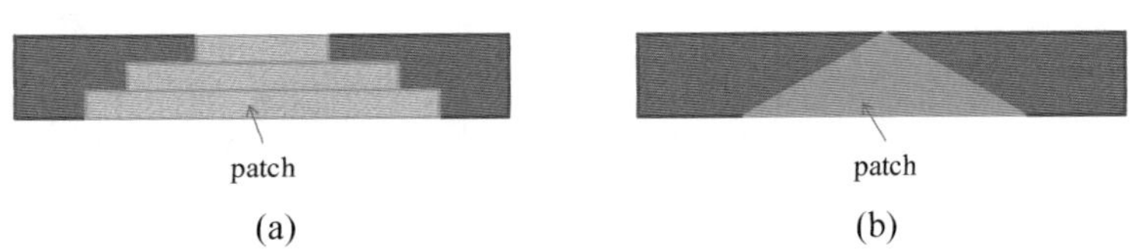

Figure 2. Bonded patch basic concepts: (a) Stepped lap joints. (b) Scarf lap joint

EXPERIMENTAL WORK

In order to assess the stress distribution along the repair bond-line, an optical fiber was embedded during the repair application for distributed Rayleigh back-scattering strain measurement with high spatial resolution. This type of distributed sensing is an excellent alternative to the conventional electrical strain gauges that are discrete and can only be placed on the outer surface where adhesive stress concentrations are masked by the skin and can be missed.

The Rayleigh 'signature' of the fiber is recorded at a reference state and then when the structure with the embedded fiber is strained, the modified new reading is correlated with the reference one to produce a measure of the strain change the fiber had experienced (at a constant temperature). This is a high resolution technique (~1cm), requiring no preparation on the part of the fiber.

Three specimens were tested statically and under fatigue spectrum, all monitored using the embedded optical fiber. The first two specimens one with a stepped lap joints and the other with a scarf lap joint (Figure 3), were tested up to failure.

These two specimens were used to optimize the repair concept. The third one was a full scale portion of a UAV wing front spar and leading edge, including the repair. This last specimen was tested up to ultimate load, followed by cyclic loading spectra, representing many lifetimes of the structure. During this entire test, strain measurements were taken periodically.

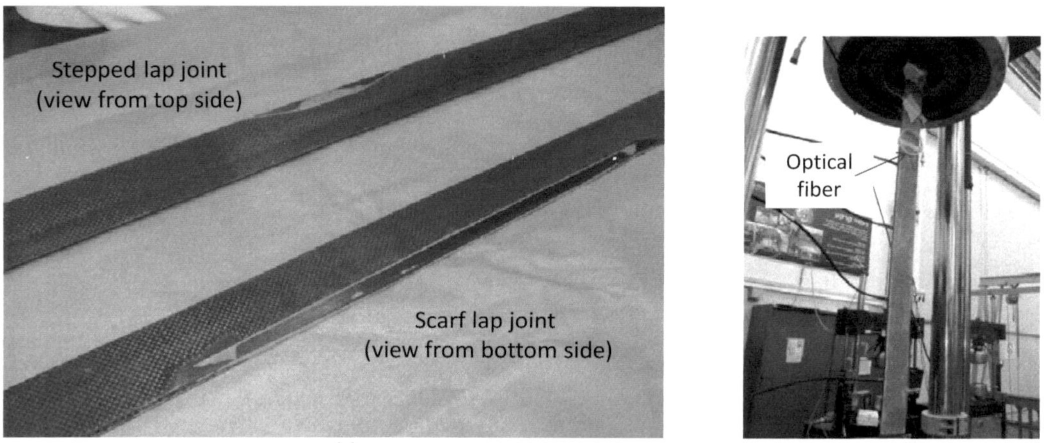

(a) (b)

Figure 3. Bonded repairs specimens. (a) The scarf and stepped joints specimens, for orientation see fig. 2. (b) Joint specimen on the test machine.

The two repair concepts strain measurements, taken during ultimate loading tests are shown in Figure 4. It is clearly seen that the observed dependency of strain to load is highly nonlinear for stepped joint repair. This nonlinearity corresponds to sharp changes in the repair stiffness, stressing the adhesive beyond its linear limit at two locations in the repair. For the second specimen, where the repair concept was designed to avoid stress concentration, it is seen that the stress concentration at the joint critical location is very small and no nonlinearity is observed.

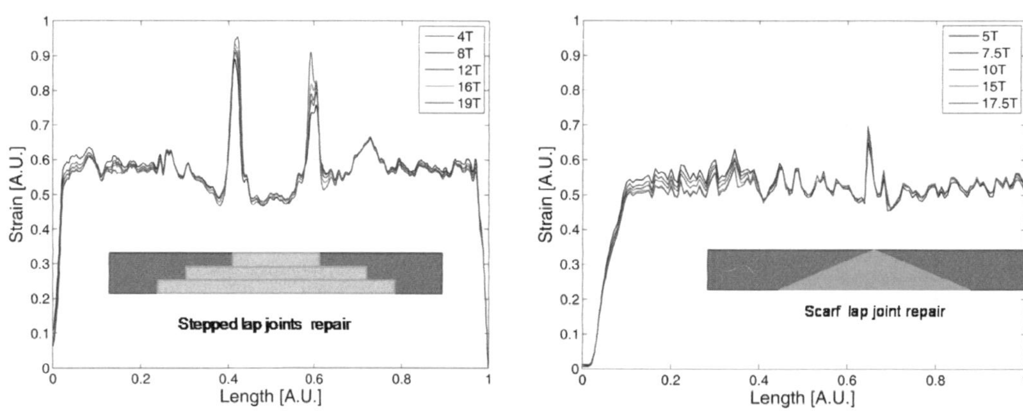

(*a*) (*b*)

Figure 4. Normalized strain distribution in the repair specimens; (*a*) First specimen: stepped lap joints repair concept; (*b*) Second specimen: scarf repair with gradual stiffness change.

As a result of this adhesive overstressing, the stepped lap joints specimen with sharp stiffness changes failed at approximately 66% of the ultimate failure load of the second scarf repair specimen.

Based on these results, the 3rd specimen, a 6m long UAV wing spar repaired by the scarf lap concept, was manufactured and successfully tested statically up to the ultimate design load, followed by fatigue and residual limit loading. The specimen on the test rig and strain measurements are presented in Figure 5 (a) and (b), respectively. Here, the repair strain distribution remained the same during the entire duration of the test, verifying that the repair airworthiness is maintained.

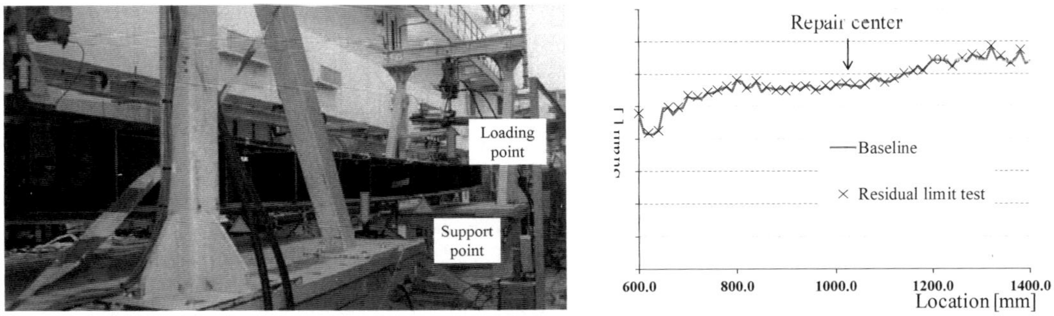

(*a*) (*b*)

Figure 5. UAV wing spar test article. (*a*) The specimen on the test rig. (*b*): Strain measurements at the repair area, no change was observed after static and fatigue tests.

CONCLUSION

Rayleigh back-scattering strain measurement was used for quantitative evaluation of bonded repair concepts designed for high strength and stiffness members, as required in modern UAVs. The repair concept selected withstood static loading up to ultimate, followed by fatigue and residual limit loading, without any change in strain distribution. Since this sensing approach can track such critical bonded repairs also during service, eventually also in real time, it makes it possible to monitor directly any change in stress distribution that may lead to failure. This also coincides with the recently introduced airworthiness requirements for UAVs, like the STANAG 4671 [7] where the substantiation of structural bonded joint requires: "repeatable and reliable non-destructive inspection".

REFERENCES

[1] Baker, Alan A., LR Francis Rose, and Rhys Jones. Advances in the bonded composite repair of metallic aircraft structure. Vol. 1. Elsevier, 2003.

[2] Kressel I., Ghilai G., Gali S., Hackman N., Leibovich H., Peled D., Brot A., Sol U., Grosu E., Cohen I., Ben-Noon M., Jiny S., Meir Y., Halevi R., Alon L. and Atzmoni A., "Design substantiation and application of boron-epoxy patch repair on an F-16 fuselage skin", in: Proceedings of the 22nd ICAF Symposium, EMAS Publishing, 2003.

[3] Froggatt, Mark, and Jason Moore. "High-spatial-resolution distributed strain measurement in optical fiber with Rayleigh scatter." Applied Optics 37.10 (1998): 1735-1740.

[4] Soller, Brian, et al. "High resolution optical frequency domain reflectometry for characterization of components and assemblies." Optics Express 13.2 (2005): 666-674.

[5] Güemes, Alfredo, Antonio Fernández-López, and Brian Soller. "Optical fiber distributed sensing-physical principles and applications." Structural Health Monitoring 9.3 (2010): 233-245.

[6] Hart-Smith, Leonard John. Adhesive-bonded single-lap joints. Hampton, VA: Langley Research Center, 1973.

[7] NORTH ATLANTIC TREATY ORGANIZATION (NATO). STANAG 4671 "UAV Systems Airworthiness Requirements (USAR) for North Atlantic Treaty Organization (NATO) Military UAV Systems", Edition 1, 2007.

Application of Distributed Fiber Optic Strain Sensors for Local Indentation/Impact Damage Detection in Foam Core Composite Sandwich Structures

J. Siivola, S. Minakuchi, N. Takeda

The University of Tokyo, Graduate School of Frontier Sciences, Department of Advanced Energy, Kashiwa-no-ha 5-1-5, Kashiwa-shi, Japan

Abstract: Local indentation or low-velocity impact of foam core sandwich structures can leave only barely visible damage in the face sheet while the core can be notably damaged leading to degraded strength of the structure. In the present study a high spatial resolution Rayleigh scattering based distributed fiber optic strain sensors were applied to foam core sandwich structures to detect the barely visible indentation and impact induced damages. Monitoring ability of the system was studied experimentally by conducting indentation and low-velocity impact tests with sandwich beam and panel structures. Strain distribution in the sandwich structures was monitored using the embedded optical fiber sensors. The measurement results were verified against predictions by finite element analysis and a simple way to estimate the damaged core area was defined. Location and size of damage in the structures was estimated with high precision from the strain distribution data. The used system was seen to provide a fast and easy way to monitor and assess barely visible damage in sandwich structures.

INTRODUCTION

Composite foam-core sandwich structures have high specific strength, stiffness, and bending stiffness properties while being extremely lightweight [1]. They consist of thin composite face sheet from high performance material such as carbon fiber reinforced plastic (CFRP), and lightweight foam core separating the face sheets and can provide the high strength and stiffness properties of the face sheets with enhanced bending stiffness properties without added stringer components. This can lead to significant decrease in the part-count of complex structures thus allowing for easier and faster manufacturing. Due to their attractive properties, sandwich structures have been widely used in various applications ranging from automobile structures to wind energy applications. Increasing interest has also been shown in using foam core sandwiches in the primary structures of aircraft structures [2-5].

Thin faces and the relatively weak core of foam core sandwich structures however make them somewhat vulnerable against transverse loadings. For example, during indentation or low-velocity impact of foam core sandwich structures, the face sheet bends locally under the transverse loading causing a compressive loading in the core material as shown in Figure 1. The foam core then starts to crush and deform plastically as the cell walls buckle and collapse. Under low indentation or impact energies the face sheet remains mostly undamaged and thus, after the loading is removed, tries to return to its initial state while pulling the crushed core. However, as the core deformed plastically during indentation, it resists the pull from the face sheet up to some degree leading to a barely visible residual dent to remain in the face sheet [6].

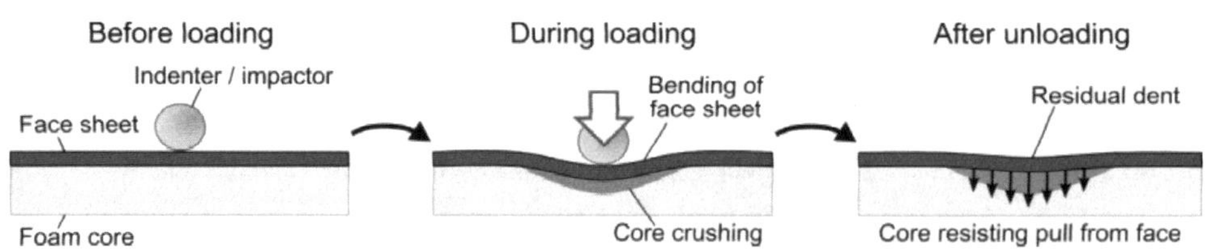

Figure 1. Damage formation in foam core sandwich structure during indentation or low velocity impact.

These kinds of indentation loadings or low-velocity impacts are especially problematic as they can leave only barely visible impact damage (BVID) in the face sheet while the core and adhesive layer can be significantly damaged (Figure 2) thus possibly having notable degrading effect on the strength and stiffness properties of the structure [7-9]. The remaining barely visible residual dent also makes detection and inspection of the damages difficult and time consuming by current visual and other non destructive inspection (NDI) methods. Therefore a way to accurately detect these kinds of damages and to assess their severity is needed to evaluate the integrity and residual strength of the structure and to ensure their safe usage.

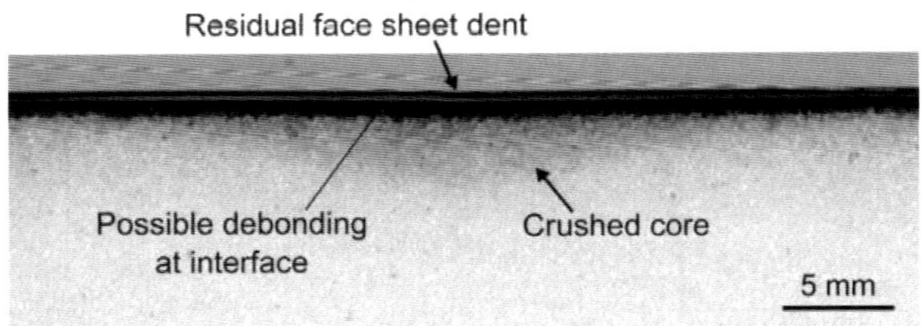

Figure 2. Residual deformation and core damage in foam core sandwich structure after indentation loading.

In this study, optical fiber sensors are applied to foam core sandwich structures to monitor the strain distribution during and after indentation and low-velocity impact events. To study the monitoring ability of the system, indentation loading tests are conducted with sandwich beam and panel specimens and the results are compared with finite element analysis predictions. Based on the experimental and analysis results the strain information is used to identify the damages and to estimate their location and size. Finally usability of the distributed fiber optic strain sensor system is verified by monitoring and assessing low-velocity impact induced damages in a large scale sandwich panel.

DISTRIBUTED FIBER OPTIC STRAIN MONITORING

Strain monitoring with the embedded optical fibers was done using an optical backscatter reflectometry (OBR) based distributed strain measurement system (OBR 4600, LUNA Technologies Inc.). The system uses swept frequency interferometry (SWI) and monitors the Rayleigh backscattered light over an optical fiber sensor [10-13]. The Rayleigh scattering is caused by random but static fluctuations in the index profile along the optical fiber, thus making the amplitude of the scattered signal also a random but static property that is unique for each fiber.

In a damaged sandwich structures with a dent present, bending of the face sheet causes highest strains in the surfaces of the face sheet. It is therefore most efficient to embed the optical fiber into the adhesive layer in the core-face sheet interface where the bending strains of the face sheet are the largest and where the strain distribution can be affected by damages in the interface and the adjacent foam core. In the case of BVID the deformations can however be very small and local, which requires high resolution and accuracy from the monitoring system to be able to detect the damage. The OBR strain monitoring system is thus used to detect these small deformations as it can provide strain data with spatial resolution down to few millimeters. The strain measurement principle of the system is next explained briefly.

During the analysis of the measurement data, the fiber is divided into fiber segments based on the gage length and sensor spacing settings as depcited in Figure 3. Rayleigh scattering spectrum at each segment along the fiber is then obtained from the measured data. Changes in strain or temperature along the fiber change the local period of the Rayleigh scatter which can be seen as shift of the backscattered Rayleigh spectrum. Strain distribution along the optical fiber is therefore calculated by looking at the Rayleigh scattering spectrum at each fiber segment and by comparing it to a reference data. By cross-correlating the measured spectrum with reference spectrum, a shift in the spectrum ($\Delta\lambda$ for wavelength or Δv for frequency) is obtained. This shift depends on the change in the temperature ΔT and strain $\Delta\varepsilon$ of the fiber at each measurement point and can be defined as

$$\frac{\Delta\lambda}{\lambda} = -\frac{\Delta v}{v} = K_\varepsilon \Delta\varepsilon + K_T \Delta T \qquad (1)$$

where λ is the mean optical wavelength, v the mean optical frequency, and K_T and K_ε are the temperature and strain calibration constants which depend on the used fiber. If the reference state is defined as having zero strain and assuming there is no change in temperature, the strain can be calculated as

$$\varepsilon = -\frac{\bar{\lambda}}{cK_\varepsilon}\Delta v = \bar{K}_\varepsilon \Delta v \qquad (2)$$

where $\bar{\lambda}$ is the center wavelength of the measurement and c the speed of light. These form the conversion factor $\bar{K}_\varepsilon$ which can be obtained experimentally by relating applied strain and frequency shift of the Rayleigh scatter spectrum in the used optical fiber.

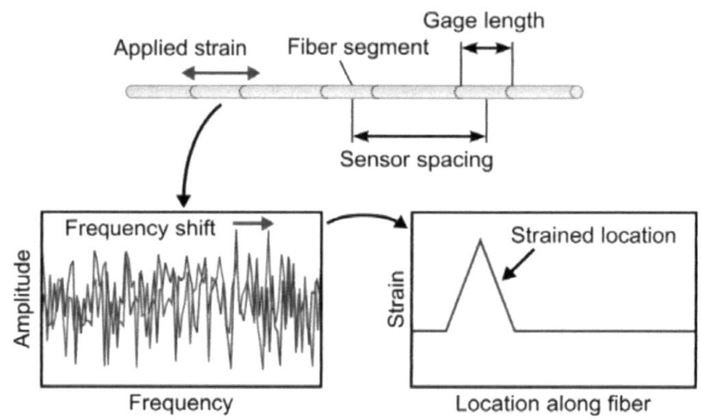

Figure 3. Illustration of the distributed strain measurement along an optical fiber.

INDENTATION TESTS

Indentation test specimens and setup

Performance of the strain monitoring system was first studied by conducting indentation tests with foam core sandwich specimens. In the indentation tests, a sandwich beam and a sandwich panel specimens as depicted in Figure 4 were used. The specimens consisted of T700S/2592 (Toray Co.) CFRP face sheets prepared in advance and Rohacell 51 WF (Evonik Rohm GmbH.) PMI foam core bonded together using AF 163-2K (3M Co.) adhesive film. For strain monitoring a low-bending-loss optical fiber (Heatop 300, Totoku Electric Co., Ltd., outer diameter: 150 μm, cladding diameter: 125 μm) was embedded between the face sheet and adhesive layer during adhesion. The specimens were adhered to a steel or aluminum jig to prevent global bending deformation during the tests.

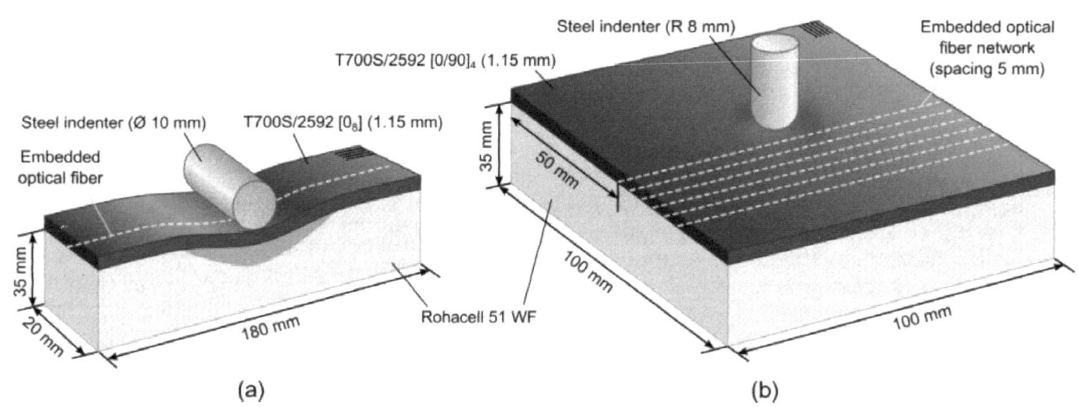

Figure 4. Schematic of the (a) sandwich beam and (b) sandwich panel specimens used in the indentation tests.

In the indentation tests, quasi-static indentation loading was applied at the middle of the test specimens using a 10 mm diameter cylindrical (for beam) or semispherical (for panel) steel indenter. Constant indentation rate of 5 mm/min was used throughout the tests. Loading was applied until certain level of indentation displacement or loading was achieved, after which the loading was removed at the same rate. Strain distribution data along the embedded optical fiber sensor were measured continuously during the tests. Strain distributions were calculated using 1 mm gage length and 0.6 mm sensor spacing.

Sandwich beam indentation test results

The monitoring system was first studied during indentation loading cycle of the sandwich beam specimen. The strain distributions were measured along the embedded optical fiber during an indentation loading-unloading cycle as shown in Figure 5. Each strain distribution data over the length of the sandwich beam specimen in Figure 5 (a) now corresponds with a point selected along the loading cycle as indicated in Figure 5 (b).

Detailed observation of the strain distribution in the structure is now possible due to the high resolution of the monitoring system. During loading a sharp tensile strain peak can be observed at the loading point with areas of tensile strain on both sides. These strains are caused by the bending of the face sheet and increase accordingly as the loading increases and the dent on the face sheet becomes deeper and wider. During unloading peak gradually flattens while the width of the strained area stays nearly the same. Finally after the loading has been removed a flat tensile strain peak remains visible. This indicates that a residual dent is remaining in the face sheet due to compressive plastic deformation in the core resisting the pulling force from the face sheet.

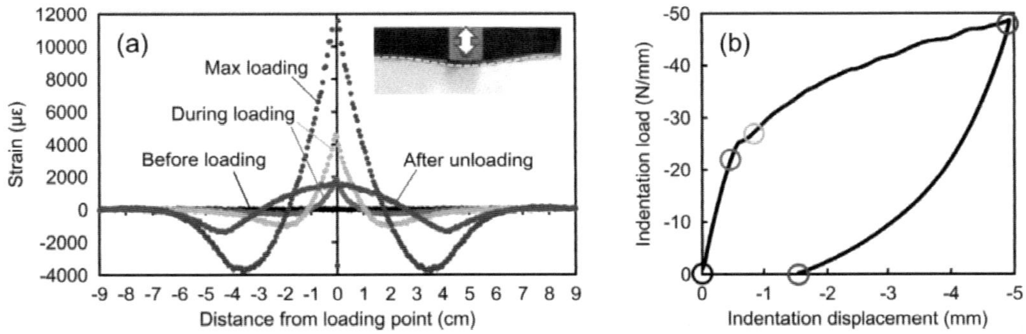

Figure 5. (a) Measured strain distributions at (b) selected locations during indentation loading cycle of sandwich beam specimen.

Sandwich panel indentation test results

Similar indentation tests were also conducted with sandwich panel specimen. Indentation loading was applied to the middle of the specimen in loading-unloading cycles while the maximum indentation loading was increased gradually on each cycle up to 2000 N.

Strain distributions measured during maximum indentation and after unloading, shown in Figure 6, are similar to those of the beam specimen. In the panel case the shape of the measured strain distribution now changes according to the dent profile at the locations of the embedded optical fibers. Therefore further away from the loading point where the dent becomes shallower, smaller strains are observed. Furthermore for tha panel specimen the point loading causes highly local deformation which results in local high strains and possible damage of the adhesive layer and face sheet at the loading point. This causes correlation problems in the strain measurements and can be seen as noise in the strain response at the loading point and its vicinity.

The residual strain distributions measured at the embedded fiber locations after the indetation test are shown in Figure 7 (a). While the visible dent on the face sheet is only few millimeters in diameter, the strained area in the residual strain data is notably larger and therefore it is easy to locate the damage. Looking at the cut cross section of the panel specimen in Figure 7 (b) it can be seen that the crushed core area extends approximately 30 mm from the loading point. This corresponds quite well with the residual strain data in Figure 7 (a), which shows a compressive strain peak near the edge of the crushed core area.

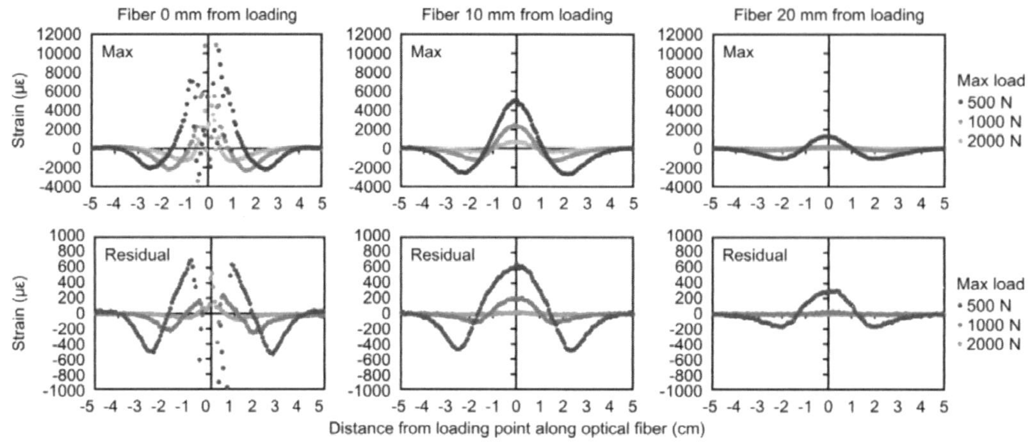

Figure 6. Maximum and residual strain distributions at three fiber locations (0, 10 and 20 mm from loading point) for three maximum indentation loading cases.

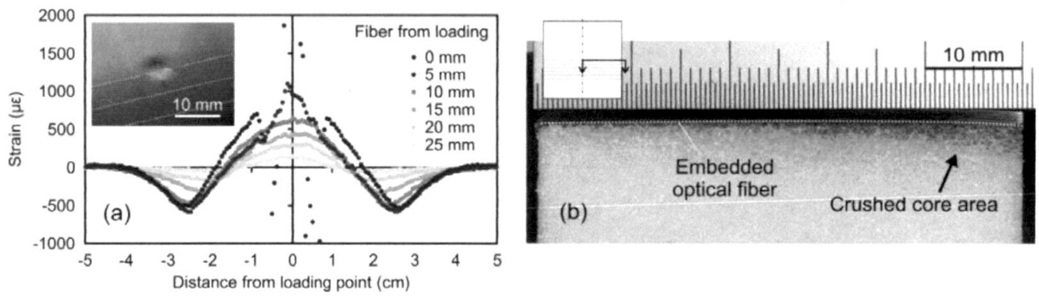

Figure 7. (a) Residual strains and (b) cross section cut of the sandwich panel specimen after indentation test.

Comparison with other strain measurement systems

To better understand the capability of the measurement system, the strain distributions measured in the sandwich beam indetation tests were compared with those measured using a Brillouin scattering based Pulse Pre-Pump Brillouin Optical Time Domain Analysis (PPP-BOTDA) system (Neubrescope NBX-7020, Neubrex Co. Ltd.) and strain gages embedded between the adhesive layer and the face sheet. Comparable settings for gage length and spatial resolution of the BOTDA system were 2 cm and 1 cm respectively. The maximum and residual strain distributions measured by each method are now compared in Figure 8.

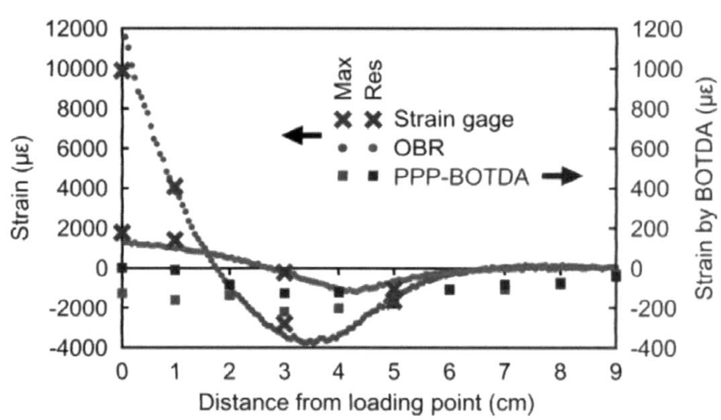

Figure 8. Comparison of strain distributions in indentation loaded sandwich beam specimen measured by the OBR system, strain gages and the PPP-BOTDA system.

The strains measured using the embedded strain gages are close with the strain distribution measured using the current OBR system thus suggesting that the system can measure the strains in the adhesive-face sheet interface correctly. Tho BOTDA system on the other hand shows very small and mostly compressive strains. Due to the low spatial resolution and large gage length of the BOTDA system, it cannot accurately detect the local and high strain peak at the loading point and gives only an average strain over the deformed region. The OBR system provides much better strain measurement for the current application with high resolution and accuracy and is therefore suitable for detection of highly local and barely visible damages in foam core sandwich structures.

FINITE ELEMENT ANALYSIS

To verify the strain measurement results and to understand the relation between the residual strains and damage in the core of the sandwich specimens, finite element analysis was conducted. For finite element analysis, a two dimensional model for the sandwich beam specimen, and a three dimensional model for the sandwich panel specimen were constructed using the ABAQUS 6.11 software. The model configurations are depicted in Figure 9. Due to symmetry only half of the beam and one quarter of the panel was modeled with appropriate symmetry boundary conditions. Boundary conditions were set to be similar with the indentation experiments. Indentation loading was applied similar to the experiments by controlling the displacement of the rigid indenter.

The face sheet and adhesive layer were modeled as elastic using the material properties shown in Table I and Table II. The core was modelled as elasto-plastic using the Crushable foam material model available in ABAQUS using the material properties and model parameters shown in Table III. The foam material model was calibrated using experimental uniaxial compression-tension data for the Rohacell 51 WF foam material.

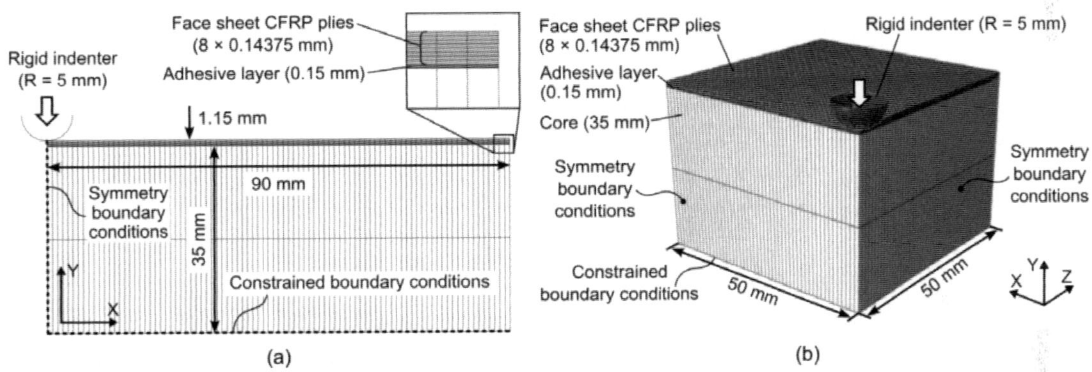

Figure 9. Finite element models of the (a) sandwich beam and (b) sandwich panel specimens.

Table I. Elastic material properties for one ply of T700S/2952 CFRP [14].

Young's modulus (GPa)	E_1 E_2, E_3	135 8.5
Poisson's ratio	ν_{12} ν_{13}, ν_{23}	0.34 0.49
Shear modulus (GPa)	G_{12} G_{13}, G_{23}	4.8 2.7
1: Fiber direction, 2: Fiber transverse direction (in-plane), 3: Through thickness direction		

Table II. Elastic material properties of AF163-2K adhesive layer [15].

Young's modulus (GPa)	E	1.1
Poisson's ratio	ν	0.33

Table III. Foam material properties and parameters used for the Crushable foam material.

Young's modulus (MPa)	E	46.44
Poisson's ratio	ν	0
Compressive yield stress (MPa)	σ_y	0.912
Yield stress ratio for compressive loading	k	1.03659
Yield stress ratio for hydrostatic loading	k_t	0.028

Indentation simulation

The indentation of sandwich beam specimen was simulated by applying displacement controlled indentation loading until 5 mm indentation displacement after which the loading was removed. Predicted strain distributions in the adhesive layer at the location of the embedded optical fiber were compared with the measured strains during maximum indentation loading and immediately after unloading and are shown in Figure 10. Overall the predicted strains correspond well with the measured data although some differences are seen in the residual strains due to the simplified response of the core material model during unloading. The predicted indentation response should however be sufficiently accurate to be usable to assess the formation and magnitude of damage in the sandwich structures.

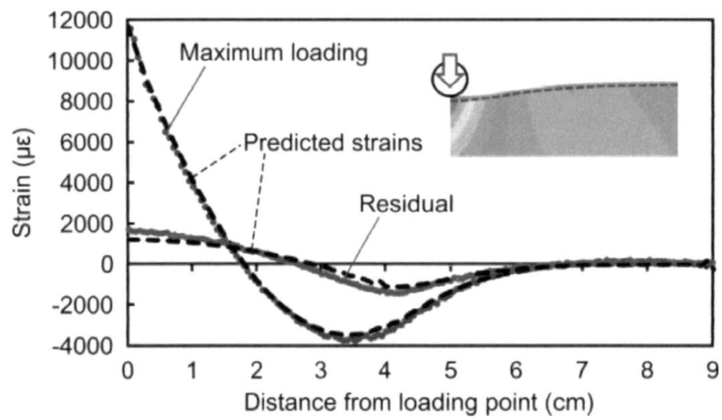

Figure 10. Comparison of measured and predicted strain distributions during maximum indentation loading and after unloading for the sandwich beam specimen.

Estimation of core damage size

In sandwich structures the core is not visible from the outside and thus it is difficult to estimate the damage in the core. As the finite element model could quite accurately simulate the indentation loading response and the residual state of foam core sandwich structures, the model can also be used to find a correlation between damage in the core and strains that can be measured using the fiber optic sensors.

As was shown in the measured strain data in Figure 5, a compressive peak is observed at the outer edge of the dented area as the face sheet bends over the crushed core while being supported by the adjacent undamaged core. Again, during unloading the compressive peak moves further away from the loading point when the dent becomes shallower. The strain distributions and the size of the plastically deformed area in the core of a sandwich beam specimen obtained by finite element analysis are compared in Figure 11 (a). It can be seen that after the loading is removed the compressive strain peak in the residual strain distribution seems to settle near the edge of the damaged core area. Similar behaviour can be seen for the sandwich panel model in Figure 11 (b) where the compressive strain area in the residual strain distribution also settles near the edge of the damaged core area. Approximate size of the damage in the core can therefore be simply estimated from the strain distribution data by observing the location of the compressive strain peaks. Residual strain distributions and cross section of the indentation loaded sandwich panel specimen in Figure 7 also seem to verify this observation.

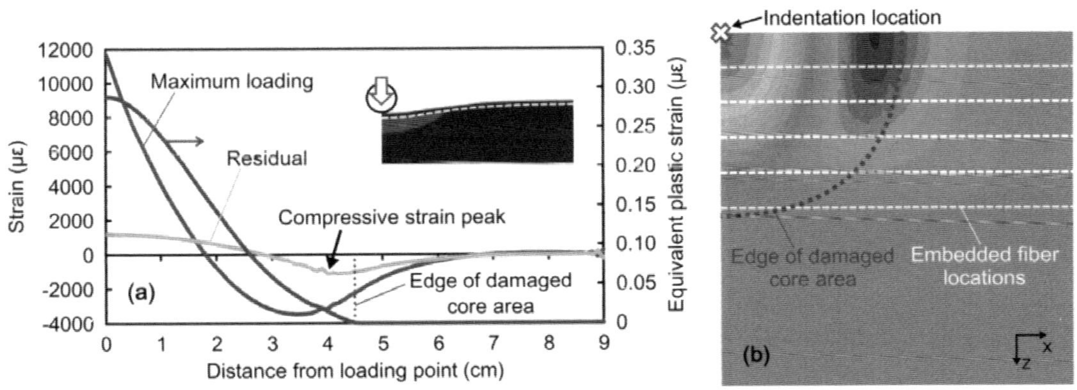

Figure 11. Comparison of predicted damaged core area and residual strain distributions for (a) sandwich beam and (b) panel specimens. Residual strains in x-direction are shown for the panel.

LOW-VELOCITY IMPACT TESTS – LARGE SCALE SPECIMEN

Low-velocity impact test specimen and setup

Monitoring ability of the system under conditions closer to real life applications was demonstrated by low-velocity impact tests using a large scale (700 mm × 700 mm) sandwich panel as depicted in Figure 12. The sandwich panel consisted of faces made of Kawasaki's proprietary high performance and low cost KMS-6115 carbon fiber fabric composite and Rohacell 51 WF foam core. The core thickness is 35 mm and the face sheet consists of 2 fabric plies in the mid area of the panel and 6 fabric plies at the taper regions. An optical fiber network with parallel fiber spacing of 20 mm was embedded between the core and the face sheet.

Impact test were conducted with a 12.5 mm diameter hemispherical impactor at the selected locations in the mid and taper region while the panel was clamped to a supporting structure at the edges. Only small impact energies (2.75, 5.49, 8.24 and 10.98 J) were used so that the damages at the tested regions of the panel were barely visible. If clearly visible damage formed in the face sheet after impact, test using higher impact energies were not conducted at that region.

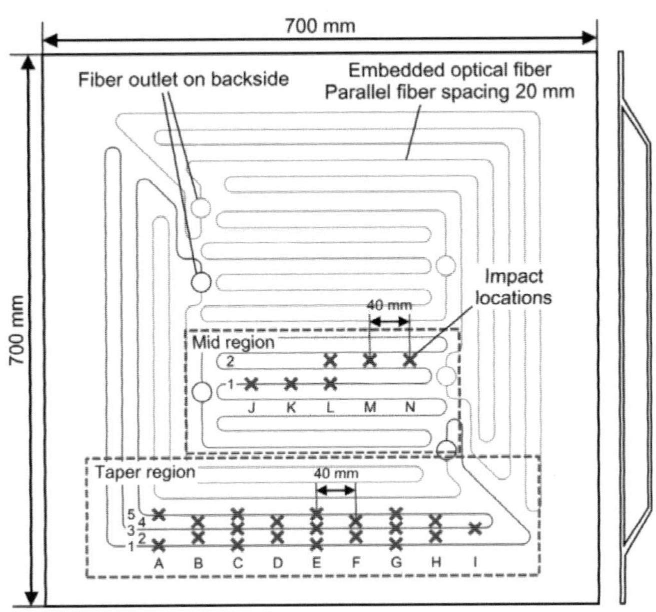

Figure 12. Schematic of the large scale sandwich panel specimen. The tested impact locations are shown in the mid and taper regions.

Test results

In the low-velocity impact tests, the residual strain distributions were calculated by comparing data obtained immediately after the impact with data before the impact. Figure 13 shows a typical strain measurement results. Even though low impact energy of 2.75 J is used, strain peaks giving clear indication of the damage are seen in the residual strain data. The location of the damage can be estimated by interpolating the strain data based on the fiber locations and finding the highest strain peak. In the shown case the damage is on the taper region where the thickness of the core and the face sheet change and therefore slightly higher strain peak is observed near the center of the panel where the core is thicker.

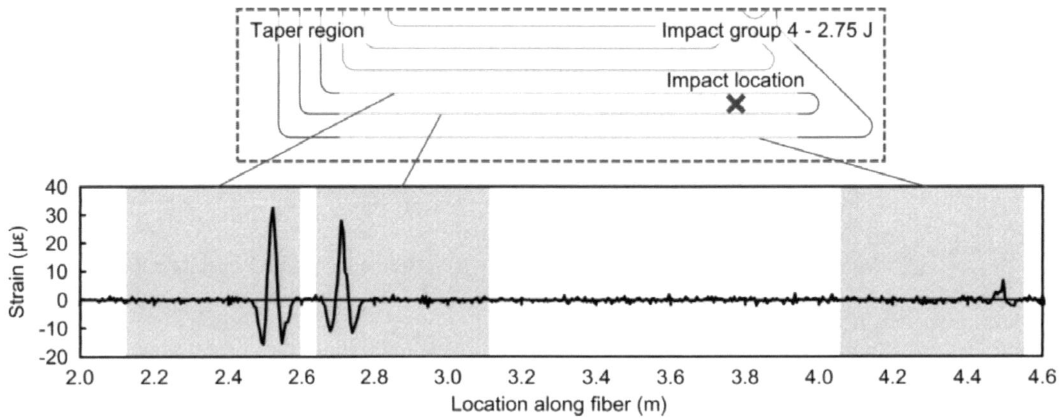

Figure 13. Measured residual strain distributions in adjacent fibers used to estimate the damage location.

Figure 14 shows typical measured strain distributions at the taper region after multiple consecutive impact events. Already at the lowest used impact energies, a clear tensile peak is observed in the strain distribution data at the impact location. Due to the small impact energies and small damage area, the previous impact damages also remained detectable even after multiple impacts. Maximum strains are lower in the taper region where the face sheet is thicker and the core thinner than in the mid region. This however did not affect the detectability of the impact damages at the tested impact energy levels.

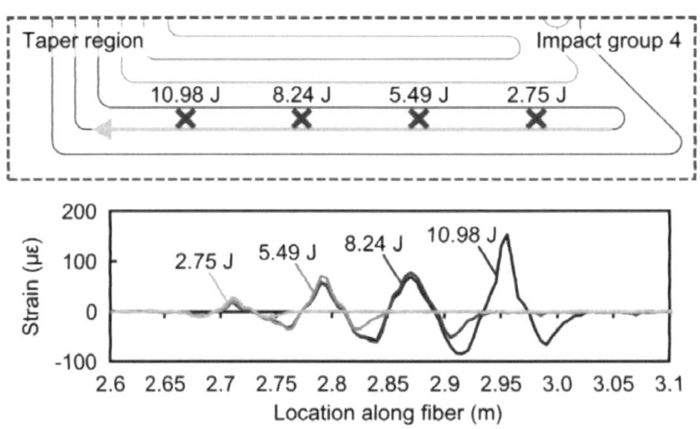

Figure 14. Measured residual strains distributions after multiple impacts.

Comparing the visible damage in the face sheet with the measured strain distribution at the damage location in Figure 15 it can be seen how the strain data shows a clear indication of impact damage while the deformation in the face sheet is barely visible. Detection of BVID is thus possible with the used system and the location and size of the damage can be estimated rather easily from the strain data by using the principles explained previously.

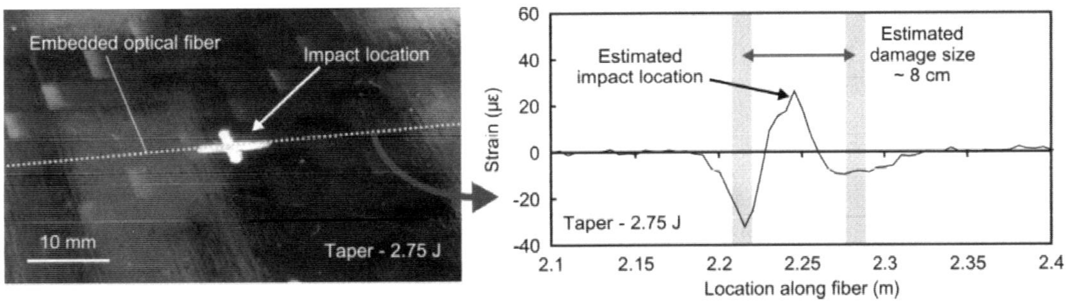

Figure 15. Barely visible residual dent and measured residual strain distribution at corresponding location.

CONCLUSION

A fiber-optic Rayleigh scattering based monitoring system was applied to detect barely visible indentation and impact damages in foam core sandwich structures. Indentation tests were conducted with sandwich beam and panel specimens with an optical fiber sensor embedded in the adhesive layer to study the damage detection ability of the used system. Finite element analysis was also conducted to verify the measurements and to find correlation between the predicted core damage area and residual strains to estimate the size of the damaged core area from the strain data. Finally low-velocity impact tests were conducted using a large scale panel to demonstrate the performance of system. With the used monitoring system the location and size of even very low energy impact induced damages with only barely visible dent remaining in the face sheet could be estimated rather well from the measured strain distribution data. Based on the experiments and finite element analysis, it was seen that the used system can accurately measure the strain distribution in the sandwich structures and detect changes in residual strains that can be used to estimate the location and size of even barely visible damage.

ACKNOWLEDGEMENTS

This work was partially conducted as a part of the project, "Aerospace Industry Innovation Program-Advanced Materials & Process Development for Next-Generation Aircraft Structures" under the contract with The Materials Process Technology Center, founded by Ministry of Economy, Trade and Industry (METI) of Japan. We appreciate Kawasaki Heavy Industries, Ltd. for providing the sandwich panel specimen.

REFERENCES

[1] Zenkert, D. (1997). *The Handbook of Sandwich Construction*, EMAS Publishing.

[2] Hirose, Y., Kosugi, K., Nishitani, M., Sashikuma, H., Imuta, M., Fukugawa, H. and Kikukawa, H. (2000) In: *Proc. of 23rd Congress of the ICAS*, p. 343.1.

[3] Herbeck, L., Wilmes, H., Kolesnikov, B. and Kleineberg, M. (2003) In: *Proc. SAMPE Europe*, March, Paris, p. 1.

[4] Zuardy, M. and Herrmann, A. (2011) *CEAS Aeronautical Journal*, vol. 2, p. 253.

[5] Maia, L. and de Oliveira, P. (2005) *SAE Technical Paper 2005-01-4011.*

[6] Zenkert, D., Shipsha, A. and Persson, K. (2004) *Composites Part B*, vol. 35, p.511.

[7] Akay, M. and Hanna, R. (1990) *Composites*, vol. 21, p. 325.

[8] Tomblin, J., Raju, K., Liew, J. and Smith, B. (2001) *DOT/FAA/AR-00/49.*

[9] Zenkert, D., Shipsha, A., Bull, P. and Hayman, B. (2005) *Compos. Sci. Technol.*, vol. 65, p. 2597.

[10] Soller, B., Gifford, D., Wolfe, M. and Froggatt, M. (2005) *Optics Express*, vol. 13, p. 666.

[11] Soller, B., Gifford, D., Wolfe, M., Froggatt, M., Yu, M. and Wysocki, P. (2006) In: *Proc. OFC 2006*, p. OFN3.

[12] Duncan, R., Soller, B., Gifford, D., Kreger, S., Seeley, R., Sang, A., Wolfe, M. and Froggatt, M. (2007) In: *Proc. 10th Joint DOD/NASA/FAA Conference on Aging Aircraft*, p. 10.

[13] *LUNA Optical Backscatter Reflectometer Model 4600, User Guide 5* (2012) Luna Technologies.

[14] *Torayca Technical Reference*: http://www.torayca.com/en/techref/index.html, Toray Industries Inc., Obtained: March, 2014.

[15] *AF163-2 Data Sheet* (2009) 3M Co.

Residual stress engineering

Predicting fatigue crack growth emanating from cold-expanded holes

Luigi Lazzeri[1], Domenico Furfari[2], Qiang Fu[3], Anja Wetzel[2], Mauro Bruno Biasizzo[2], David Osman Busse[2]

[1]*Pisa University, Italy*
[2]*Airbus Operations GmbH, Germany*
[3]*Airbus Operations Ltd, United Kingdom*

Abstract: Split sleeve Cold eXpansion (CX) is a common technique to improve the fatigue behaviour of open holes as well as fastener holes in fatigue critical joints in aeronautical metallic structures. For several decades it is applied in high loaded aeronautical components to extend their fatigue lives, taking into account the beneficial effect of the compressive residual stress induced by the CX technology on fatigue initiation. In order to predict the crack growth in the presence of residual stress (e.g. for fatigue life prediction in structural components containing an initial flaw) it is very important to calculate accurately the Stress Intensity Factor (SIF) related to a crack growing in 3D residual stress field. For this reason, a dedicated test campaign has been launched with the aim to collect crack growth data associated to a crack front emanating from a quarter circle artificial notch placed at the edge of an open hole, following the development from corner crack to complete through thickness crack.

INTRODUCTION

Split sleeve cold expansion technology is a common technique that has been used in aeronautical industry for over 35 years. For ageing aircraft, fatigue performance can be of significant concern for both military and civil operators. Technical solutions that have no weight penalty and result in extension of fatigue life of otherwise critical components, avoiding premature replacement of the parts or reducing the repetitive maintenance inspection burdens, are always welcome by military and civil operators. Compressive residual stresses induced in the area surrounding holes (e.g. fastener holes in structural joints) by means of split sleeve cold expansion technique is one of the solution commonly adopted.

Amongst other, structural fastened joints are one of the most prone locations where fatigue cracks may occur during the aircraft operational life. According to service findings, up to 70% of the total fatigue failures may occur at fasteners holes [1]. The determination of the residual stress distribution around a cold expanded hole is of primary importance to allow to quantify the fatigue and crack growth benefit (both experimentally and by numerical and analytical means). Fatigue life prediction in structural components containing an initial flaw requires accurate Stress Intensity Factor (SIF) evaluation related to cracks growing in three-dimensional geometrical constraints (e.g. corner crack) and residual stress field. An accurate prediction of crack growth in residual stress field (e.g. after cold expansion) is essential to define efficient damage tolerance programs and appropriate maintenance tasks for the continuous airworthiness in military and civil aviation.

Many test evidence in the past decades demonstrated the capability of the cold expansion process to provide fatigue life enhancement, such as longer inspection intervals or a general increased operational lifetime of the airframe as reported in [2, 3, 4, 5, 6, 7 and 8]. Nevertheless, residual stresses are still not commonly used in the design phase of an aircraft mainly because of the complexity to develop predicting capability to model crack growth in presence of residual stresses.

The residual stress field introduced by the cold expansion process is not straight forward to predict or to measure. Many previous models available in literature assumed that such residual stress field was constant through the thickness; the stress state in reality varies significantly through thickness as reported by the authors in [4, 5, 9 and 10]. Most existing approaches to predict a residual stress field, either determined by closed-form analytical methods or by Finite Element Analysis (FEA), are two-dimensional (2D) plane models, that are based on uniform radial expansion of the hole, which ignores the through-the-thickness variation of stresses.

Other parameters, such as different materials, plate thicknesses or Degree of Cold Expansion (DCE), which is the ratio between the initial hole diameter and the diameter after the cold expansion has been applied, have an important role in determining the residual stress field and they need a deeper understanding.

Test evidence, included in this paper, demonstrated that fatigue crack, emanating from an artificial notch placed as corner flaw, propagated with different crack growth rates and crack front shape if under the influence of the residual stress field induced by cold working technique. Significant portion of the crack growth life was consumed by fatigue cracks from cold expanded holes before a crack became through thickness crack. This behaviour confirms how important the prediction of the crack growth is while the crack front is still confined into the thickness. R. Actis, M. Watkins and others in a recent publication proposed a simplified approach to compute SIFs for cracks in Cold-Worked holes using the contour integral method for loaded cracks (CIM-LC) by means of StressCheck® V10, which includes a module commercially available to calculate the SIF on prediction capability of crack emanating from cold worked holes when only 1 component of residual stress is known [11]. T. Mills, S. Prost-Domasky and others investigated the effect of split sleeve cold expansion in reduced edge margin coupons made of Al2024-T351 alloy, 0.2 inch thick, with 0.2 inch diameter hole [12]. The authors concluded that fatigue analysis based on crack growth prediction modelling in residual stress fields is a highly sensitive process and a more accurate modelling of crack shape is necessary to further enhance the prediction capability.

EXPERIMENTAL APPROACH

Experimental results of a test campaign to collect crack growth data in open hole specimens made of Al 2024-T351 under constant amplitude tensile loading are included in this paper. Three specimens' thicknesses (2, 4 and 6 mm) and two levels of cold expansions (3% and 4%) have been evaluated. Crack growth behaviour in cold expanded hole has been compared with the one emanating from simply drilled hole, referred in this paper as baseline test. Each test condition (i.e. thickness and level of expansion) has been tested at stress ratio of 0.1; additional tests have been carried out also under a 0.7 stress ratio, only for 2 mm thick specimens (baseline and 3% expansion). A fatigue crack initiation notch, as quarter of circle crack at the edge of the hole, was induced by Electric Discharge Machining process with initial length of about 0.25mm in each direction (i.e. front surface and through the thickness). A drawing of the specimen including the artificial notch is shown in Figure 1.

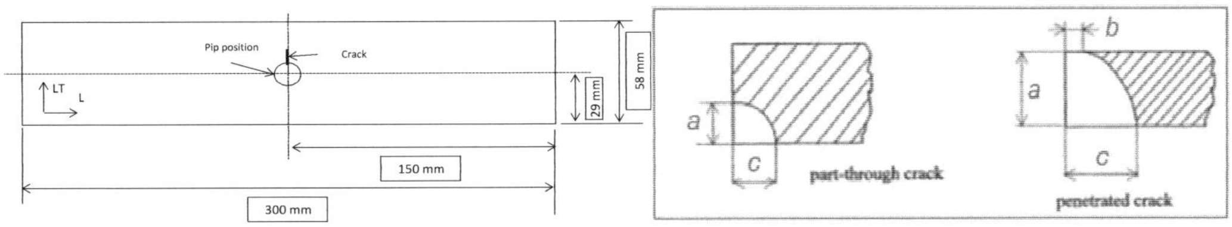

Figure 1. Open Hole specimen geometry (left); details of artificial notch and crack tip nomenclature (right).

The naming of the specimen follows the naming convention shown in Figure 2.

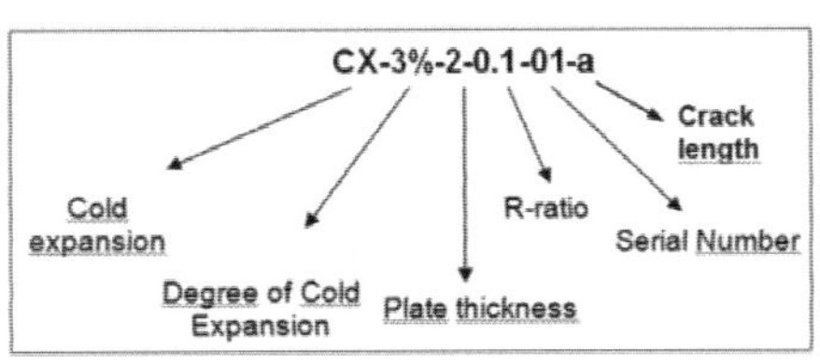

Figure 2. Naming convention for OH test specimens

Tables I report the complete test matrix, which comprised a number of groups of specimens (typically 3 samples) that differed for the thickness or for the cold expansion (three options were available: not present in the basic material batch, 3 % expansion and 4 % expansion). The various groups listed in Table Ia had been tested under a R=0.1 stress ratio; the investigation was complemented by two groups of specimens that were tested using a R=0.7 stress ratio (limited on 2 mm thick specimens, without CX process and with 3% expansion). In total, 39 specimens have been tested.

Table Ia. Tests performed under a 0.1 stress ratio.

Specimen nr.	Thickness	CW	Smax (MPa)	R
NON-CX-2-0.1-01	2.0	No	100	0.1
NON-CX-2-0.1-02	2.0	No	90	0.1
NON-CX-2-0.1-03	2.0	No	90	0.1
				0.1
NON-CX-4-0.1-07	4.0	No	100	0.1
NON-CX-4-0.1-08	4.0	No	90	0.1
NON-CX-4-0.1-09	4.0	No	90	0.1
NON-CX-6-0.1-10	6.0	No	98	0.1
NON-CX-6-0.1-11	6.0	No	90	0.1
NON-CX-6-0.1-12	6.0	No	100	0.1
CX-3%-2-0.1-13	2.0	3%	150	0.1
CX-3%-2-0.1-14	2.0	3%	160	0.1
CX-3%-2-0.1-15	2.0	3%	150	0.1
CX-3%-2-0.1-15b	2.0	3%	160	0.1
CX-3%-4-0.1-19	4.0	3%	180	0.1
CX-3%-4-0.1-20	4.0	3%	167	0.1
CX-3%-4-0.1-21	4.0	3%	150	0.1
CX-3%-4-0.1-21b	4.0	3%	160	0.1
CX-3%-6-0.1-22	6.0	3%	158	0.1
CX-3%-6-0.1-23	6.0	3%	157	0.1
CX-3%-6-0.1-24	6.0	3%	153	0.1
CX-3%-6-0.1-24b	6.0	3%	150	0.1
CX-4%-2-0.1-25	2.0	4%	160	0.1
CX-4%-2-0.1-26	2.0	4%	160	0.1
CX-4%-2-0.1-27*	2.0	4%	150	0.1
CX-4%-2-0.1-27b	2.0	4%	150	0.1
CX-4%-4-0.1-28*	4.0	4%	140	0.1
CX-4%-4-0.1-29	4.0	4%	166	0.1
CX-4%-4-0.1-30	4.0	4%	160	0.1
CX-4%-4-0.1-30b*	4.0	4%	150	0.1
CX-4%-6-0.1-31	6.0	4%	160	0.1
CX-4%-6-0.1-32*	6.0	4%	150	0.1
CX-4%-6-0.1-33*	6.0	4%	150	0.1

*All these specimens have failed by fretting fatigue at the clamping area.

Table Ib. Tests performed under a 0.7 stress ratio load.

Specimen nr.	Thickness	CW	Smax (MPa)	R
NON-CX-2-0.7-04	2.0	No	180	0.7
NON-CX-2-0.7-05	2.0	No	190	0.7
NON-CX-2-0.7-06	2.0	No	200	0.7
CX-3%-2-0.7-16	2.0	3%	250	0.7
CX-3%-2-0.7-17	2.0	3%	250	0.7
CX-3%-2-0.7-18	2.0	3%	240	0.7
CX-3%-2-0.7-18b	2.0	3%	240	0.7

Test Setup

The crack length was measured always from the edge of the hole: front surface, through thickness and back surface. The crack front, initially developed in the front surface and through the thickness of the hole, turned into a through crack after reaching the back surface. Particular attention was paid to monitor the crack front development in the transition phase between 3D crack advancing (as corner crack in front surface and through thickness direction) and 2D crack (as through crack when the crack front reached the back surface).

To monitor the crack growth in the scenario above described, a properly designed test set up has been developed. The servohydraulic fatigue test machine of 100KN capacity was equipped with three high resolution digital cameras. Two cameras were fixed in a stand clamped with the test machine frame columns (and were focusing on the front side of the specimen, i.e. the one with the artificial notch) while the third camera was focusing on the rear side of the sample. One of the cameras on the front side was inclined at 45 degree to follow the crack propagation in the internal surface of the hole, aiming at the through-the-thickness direction crack. A full automatic system, able to control the fatigue loading of the test machine as well as the image acquisitions, was programmed in LabView® code giving the possibility to run the test without requiring the presence of an operator to measure the crack growth. Figure 3 shows the test setup at the servohydraulic fatigue test machine.

Figure 3. Test setup with the three cameras used for taking images of the crack (seen from rear).

The test procedure was the following: after preparation of the surfaces, the specimen was mounted in displacement control, gripped by means of bolts tightened to a prescribed torque and then the actuator control passed in force. The surface preparation was essential and revealed to be, together with the proper light disposition, the most critical event to enhance the readability of the crack tip position. The surfaces where propagation was expected were painted in red, so that the internal surface of the crack reflected the light, appearing brilliant silver. As the number of tests performed increased, the experience showed clearly that a diffused light gave better results than the use of a pair of concentrated spots. A surface free from scratches was highly desirable, as in some circumstances the presence of scratches interfered with the crack path, as will be shown in the paragraph dedicated to the presentation of the test results. In the CX specimens, the lateral surface of the specimen had to be flattened in order to remove the so-called volcano effect. Indeed, the split sleeve expands in the radial direction, forcing the material to plastic deformation not only radially, but also axially, so producing an out-of-flatness situation. This operation was required for defining the hole border, where the notch had to be introduced. It is also necessary in the real industrial world, if a fastener has to be introduced into the hole, so that the head matches a flat surface and not the top of the volcano.

Finally the specimen was slowly loaded up to the minimum load in the cycle, and then the test started: the repetitive load story written in the command file consisted of cycles that were summed to the minimum already applied, i.e. went from zero to the range (difference Maximum – minimum); in this way, the desired cyclic load was applied. The first cycle was applied slowly (25 seconds), while the subsequent 999 cycles were applied at 4 hertz. The story was repeated indefinitely, up to specimen failure. During the application of the cycle at very low frequency, the photos of the crack tips were captured and stored for the post-mortem analysis to measure the crack lengths.

When the workstation cycle counter passed from one thousand, it automatically sent a triggering signal on the serial exit to a second PC that was fully dedicated to the acquisition, elaboration and storage of the images. Three programs, written in the National Instruments LabView® environment, interfaced with each camera and, after waiting for half-cycle, commanded a snap from each camera, i.e. when the sinusoidal load was at the maximum, the most suitable moment for having a good indication of the tip position.

Three groups of pictures, taken by the Front_Camera, by the Rear_Camera and by the Inclined_Camera, were stored in separate folders and were analysed at the end of each test, in a backwards processing; the crack lengths were measured taking advantage of the knowledge of the final crack path. The standard image resolution (i.e. 5 mm field view over 620 pixels) allows easily obtaining an accuracy of 0.05mm, for a well identified crack tip position. For those images where the crack tips were not clearly positioned, a further image analysis was carried out to monitor the advancing crack front. Figure 4 shows the general view of the machine and test set-up.

Figure 4. General view of the machine and test set-up.

To help in fixing a scale on the pictures, before the beginning of the test two scratches were introduced in the axial direction (i.e. parallel to the load application) on the lateral surfaces, on the expected crack trajectory, by means of a sharp point, 5 mm apart. Figure 5 shows an example of the two scratches 5 mm apart to determine the scale of the pictures for the crack growth measurements.

The "*a*" dimension, i.e. through the thickness, was the most difficult to assess, because of the reduced field depth that was available. For the 2 mm thick specimens, the view was quite clear, even when the tip approached the rear surface (and so was far in distance from the notch). For the thicker specimens, the picture was quite clear for the first 2–3 mm, but became foggy for longer cracks. It was decided to privilege the initial part of the growth, with a proper choice of the focus.

In the case of baseline tests on simple open hole specimen, i.e. without cold working, the maximum cyclic stress was selected in such a way that the expected duration of the test was of about 150000 cycles.

Figure 5. The two lines, 5 mm apart, used for determining the scale.

Test Results and Observations

The final results for each test were compiled in three tables, where crack length vs. number of cycles were presented. The three crack lengths have been named "*a*" through the thickness, "*b*" on the back surface, "*c*" on the front surface (see Figure 1). The cracks that start from the notch (*a* and *c*) include the notch in their length, i.e. their length has been measured starting from an ideal corner on the hole border.

In all the tests, it was observed that the crack propagation "*a*" in the axial direction was faster than "*c*" on the front face – more details are presented in the next section. A picture of the initial notch is shown in Figure 6; the red paint applied in preparation of the test has also flooded the notch, which contrasts sharply with the fracture surface, brilliant silver.

On the cold expanded specimens, the nucleation occurred earlier than in the non cold expanded specimens, as if the Electro Discharge Machining operation, by locally inducing high temperatures, could release some residual stresses. After the quick nucleation, the growth indeed proceeds at a very slow rate for a long period, after which it increases considerably, becoming unstable in the final part of the test.

In specimen NON-CX-6-0.1-10 the "*a*" crack changes trajectory, deviating considerably from the straight path as shown in Figure 7. Figure 8 shows an example of tortuous crack path in the front surface propagation (i.e. "*c*") of specimen NON-CX-6-0.1-12; the surface was rich in scratches, and the crack path often followed such scratches.

Figure 6. Detail of the initial notch, observed at optical microscope.

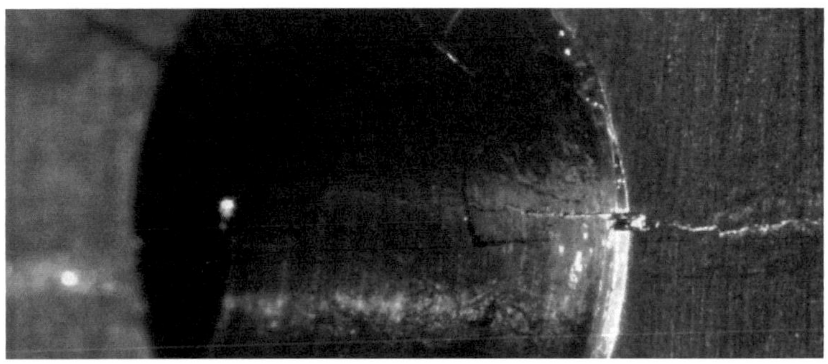

Figure 7. The "*a*" crack changes trajectory, deviating considerably from the straight path. in specimen NON-CX-6-0.1-10.

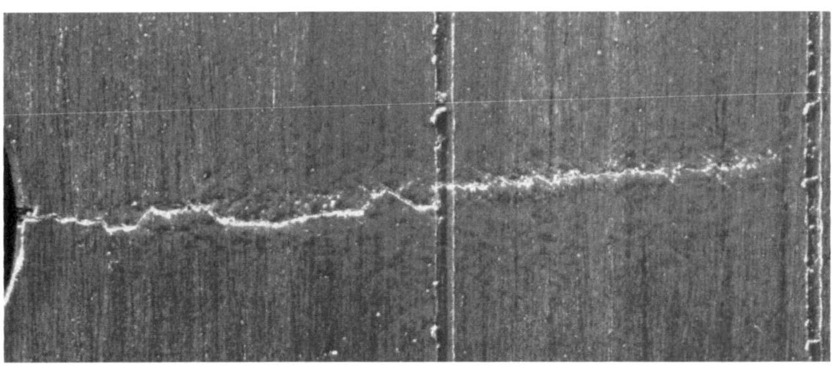

Figure 8. Tortuous crack path in the front surface propagation of specimen NON-CX-6-0.1-12; crack path "*c*" is partly influenced by scratches.

Crack branching was observed during crack propagation test of specimen CX-3%-4-0.1-21b. The same specimen also experienced two crack initiations from the notch, at upper and lower corner of the notch, with the upper one becoming main crack until the final failure of the specimen. Figure 9 shows the multi cracks initiation from the notch at the front surface and the crack branching which occurred after several millimetres of propagation.

Other examples of secondary crack initiation ahead the main crack was also observed in other specimens, such as during the test of specimen CX-3%-6-0.1-23, as shown in Figure 10. In this case a surface crack, nucleated inside the hole surface, joined the main "*a*" crack in less than 10000 cycles.

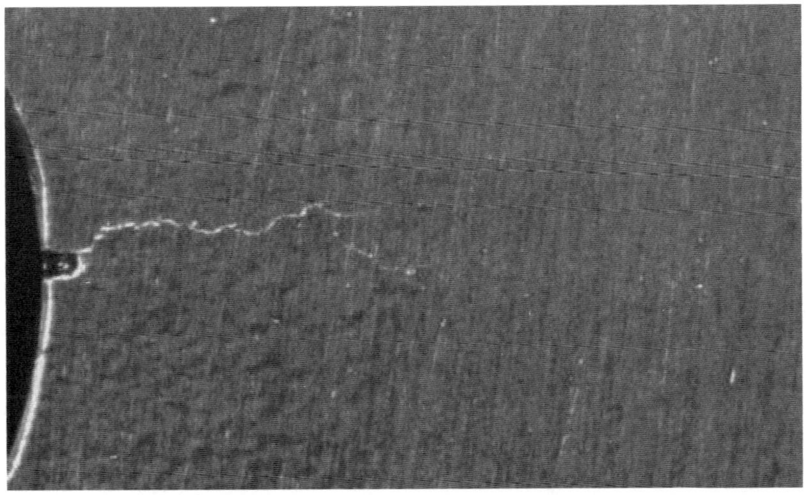

Figure 9. Crack branching during crack propagation test of specimen CX-3%-4-0.1-21b.

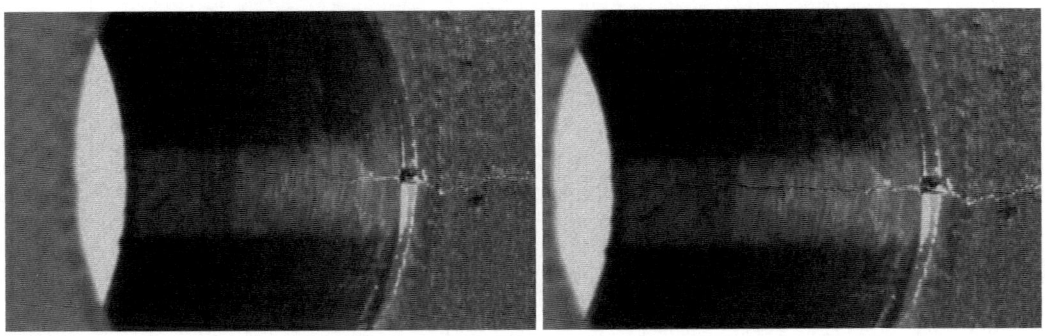

Figure 10. Second "*a*" crack initiated ahead the main crack at specimen CX-3%-6-0.1-23 (left); the crack nucleated inside the hole surface joined the main "*a*" crack in less than 10000 cycles.

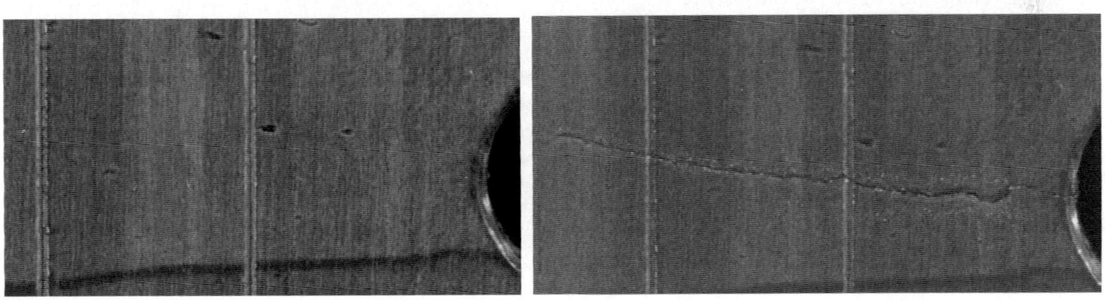

Figure 11. Rear surface view of specimen CX-3%-6-0.1-24b after 279000 cycles (left); crack propagation from rear surface after 285000 cycles (right).

A faster crack growth was observed when the crack front reached the rear surface of the specimens. Figure 11 shows the example of crack propagation of specimen CX-3%-6-0.1-24b, where the crack emanating from the rear surface propagated for several millimetres in only 5000 cycles.

Specimen CX-4%-2-0.1-26 had a second crack initiated as corner crack diametrically opposite to the notch as shown in Figure 12.

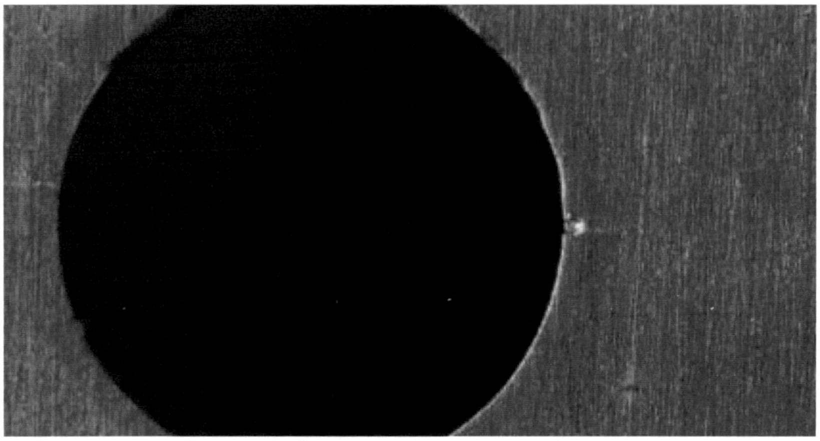

Figure 12. Second corner crack nucleated diametrically opposite to the notch of specimen CX-4%-2-0.1-26.

TEST RESULTS ANALYSIS

This section presents test results and their correlation with analytical approaches. Since the analytical NASGRO approaches [13] do not account for the cold working residual stress field, Finite Element Analyses (FEA) are applied to estimate the residual stress field and to simulate the crack growth afterwards.

Analysis of Test Results

Figure 13 shows the crack growth results for 2, 4 and 6 mm thick Non-CX and CX-3% specimens. All specimens have shown the following behaviour: for the corner crack, the through-thickness crack a grew faster than the front-surface crack length c (for crack length definition see Figure 1). The crack b started slightly after the crack a have reached the back of the specimens. After that, crack b grew faster than the front-surface crack length c to catch up: after a short number of cycles, the front-surface crack length c and back-surface crack length b have shown identical values. Exceptions to this behaviour were mentioned in the previous section.

In order to get similar numbers of load cycles in the tests, the maximum stress was increased from 100 MPa to 160 MPa for the CX-3% specimens (right-hand side of Figure 13). As expected the test results showed that the CX specimens can sustain a higher maximum stress for a similar number of load cycles. It is important to mention that the CX led to a much slower crack growth at the early stage of propagation and a much faster crack growth approaching the final failure.

For specimens of 2 and 4 mm thick, the effect of 3% Cx increases in approximately 50% the number of cycles crack a needs to reach the back of the specimen. On the other hand, the crack growth rate of crack b is much faster in the cold worked specimens. The growth rate of crack c presents also a reduction when compared with non-cold expanded holes. These observations support the known 3D effect caused by the Cx process on the crack growth rate of corner cracks, suggesting also a zone of influence where these "retardation" effects take place. This can be observed in Figure 13 and Figure 15.

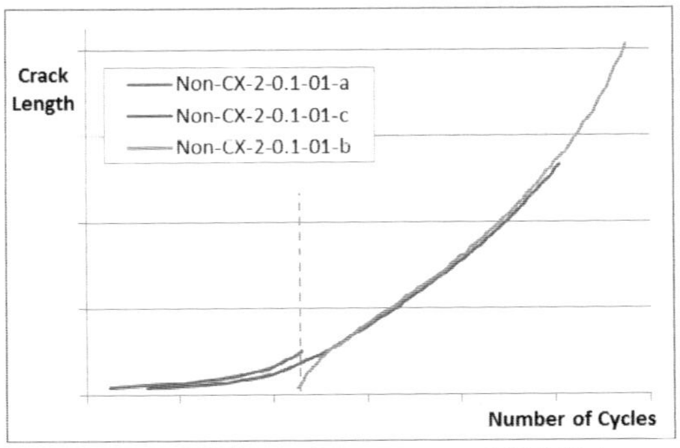

(a) Non-CX, 2 mm (max stress = 100 MPa)

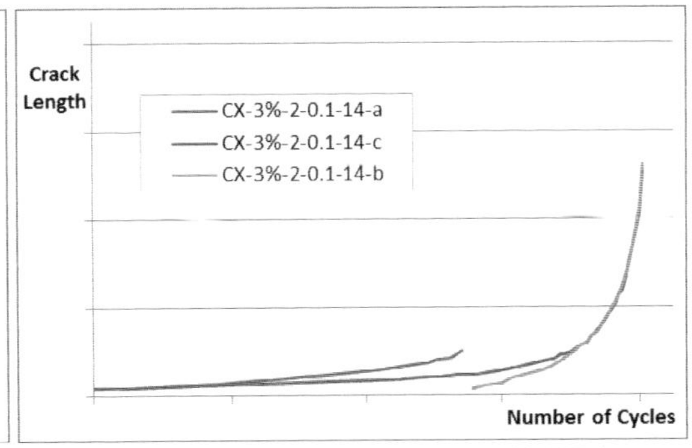

(b) CX-3%, 2 mm (max stress = 160 MPa)

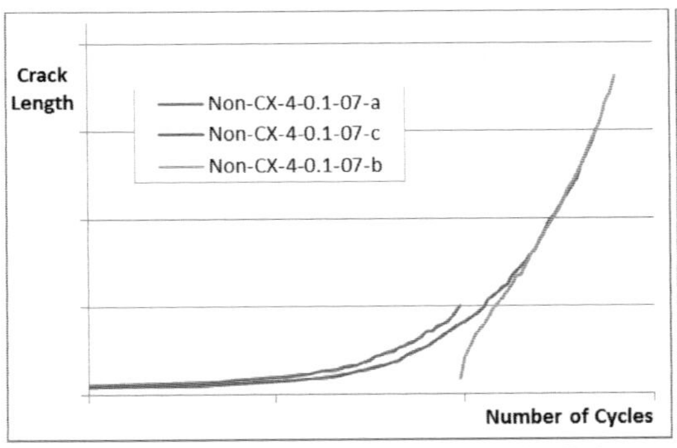

(c) Non-CX, 4 mm (max stress = 100 MPa)

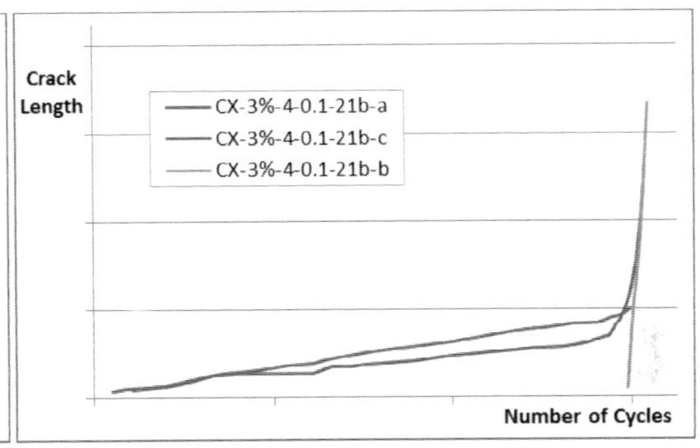

(d) CX-3%, 4 mm (max stress = 160 MPa)

Figure 13. Crack Length vs. Number of Cycles (a, b, c vs. N) for Non-CX & CX-3%.

Figure 14 presents four comparisons between two different levels of cold expansion and three different thicknesses. When comparing the curves in Figure 14 (a) and (b), the effect of the applied stress can be noticed, where the differences between 3% and 4% cold expansion is more significant for the lower applied stress, always considering the scatter inherent to this kind of coupons. For the specimens of 4 and 6 mm, both tested under 160 MPa – Figure 14 (c) and (d) – a factor of approximately 2 in life (number of cycles) can be seen when comparing 3% and 4% cold expansion. A more detailed investigation must be done in order to exactly determine which crack (*a*, *b* or *c*) is driving these effects and how they are affected by the different applied stresses and cold expansion levels.

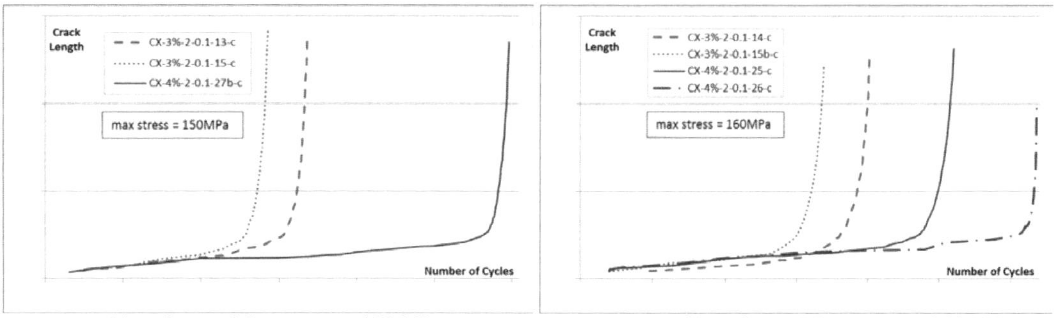

(a) t= 2mm , Max stress = 150 MPa (b) t= 2mm, Max stress = 160 MPa

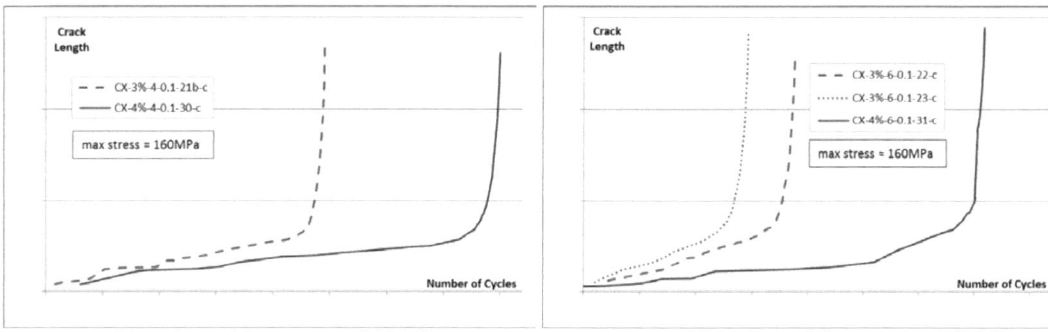

(c) t= 4mm, Max stress = 160 MPa (d) t= 6mm, Max stress = 160 MPa

Figure 14. Crack Length vs. Number of Cycles (*a*, *b*, *c* vs. N) for CX-3% & CX-4%, 2, 4 and 6mm thick.

In order to extend the comparison among specimens tested at different load levels, crack growth rate vs. stress intensity factor range (da/dN vs. ΔK) curves were plotted for the comparison of Non-CX, CX-3% and CX-4% specimens (see Figure 15). At a first glance, the curves for the cold expanded holes present a more complex shape than the ones for non-cold expanded holes. A closer look at these curves allows the identification of three different zones where: the first one presents a low crack growth rate, followed by a zone of very high rate and finally the third zone where the crack growth rate is similar to the one of the non-cold expanded holes.

By comparing the curves for the 3% and 4% cold expanded holes, it can be noticed that the main difference occurs in first zone, while in the second and third there are almost no differences.

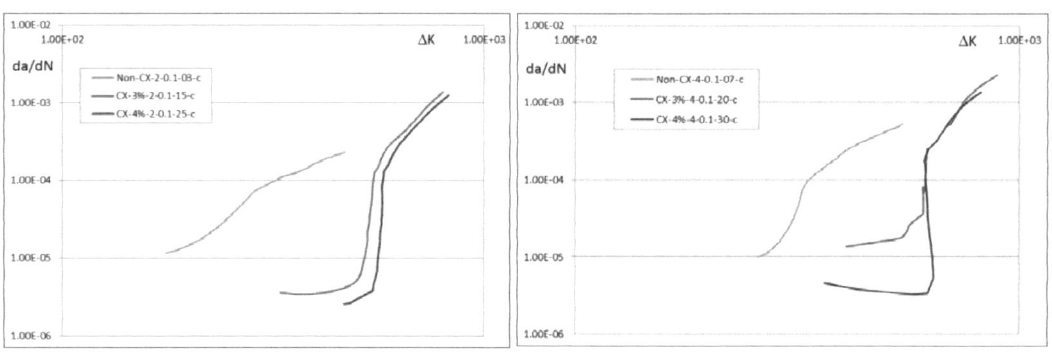

(a) Thickness 2 mm (b) Thickness 4 mm

Figure 15. Crack Growth Rate vs. Stress Intensity Factor Range (da/dN vs. ΔK) for Non-CX, CX-3% & CX-4%.

TEST CORRELATION BY XFEM ANALYSIS WITH RESIDUAL STRESS FIELD

In order to properly capture the 3D effect the cold expansion process has on crack growth a FEM (finite element model) was setup.

To simulate the corner crack propagation after cold-expansion, two essential elements are important: an accurate assessment of the corner crack propagation without residual stress and the determination of the residual stress field introduced by the cold-expansion. This section is focused on these two simulations and a simulation strategy for corner crack propagation from a cold-expanded hole is suggested in the conclusion section.

Commercial FEA software package ABAQUS/Standard was used for corner crack propagation simulation. Corner crack propagations from non-cold-expanded holes were simulated and compared with corresponding test results (2 mm and 4 mm thick specimens). ABAQUS XFEM was applied to model these corner cracks and a quasi-static crack propagation simulation methodology was selected in order to have a better control of the SIF (Stress Intensity Factor). The simulation results were compared with NASGRO CC02/CC16 modules [13] and test results. For the cold-expansion process simulation, ABAQUS/Explicit was used to deal with complex contact convergence issues. The results were compared with reference tests and further simulations were focused on predicting the residual stress fields of the test specimens with different thickness and cold-expansion degree.

Non-Cold-Expanded Corner Crack Propagation

A quasi-static crack propagation simulation strategy was selected and at each step a stationary XFEM crack was inserted into the existing mesh. The XFEM crack front is presented by an elliptical function and the SIFs for determining the crack propagation rate were evaluated. To get accurate SIFs at the crack front, in-house SIFs calculation tool and adaptive mesh technology were applied. Once stabilized, the SIFs were obtained and the propagation of the crack front could be calculated (the Paris law with Elber curve was used) and a new crack was created.

The geometry of a central holed specimen is shown in Figure 16 and the dimensions are given in table II.

Table II. Geometry and material properties of the Specimen and the Hole.

Plate Width mm	Plate Length mm	Plate Thickness mm	Hole Diameter mm (before reaming)	Hole Diameter mm (after reaming)	E MPa	Poisson's Ratio
58	300	2/4	5.766	6.35	6.895e4	0.3

The FE model was created with C3D10 elements and is shown in Figure 16.

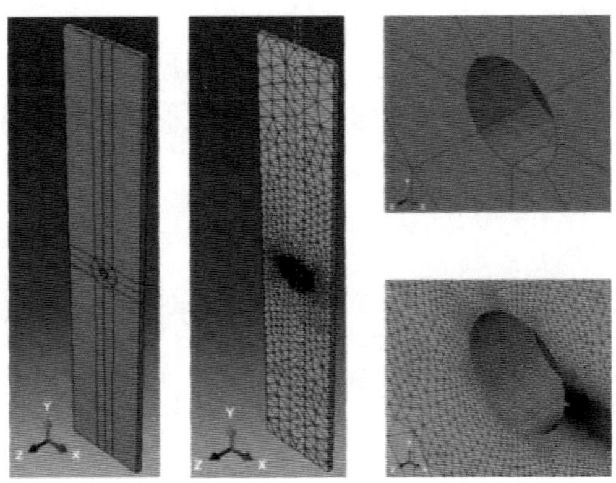

Figure 16. Geometry and mesh of the central holed specimen.

Figure 17 shows the adaptive mesh at the crack location and the finer mesh that follows the crack path of the crack front.

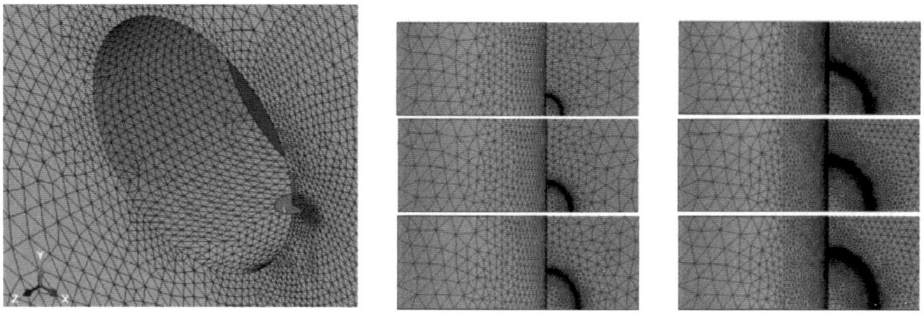

Figure 17. Adaptive mesh results with quasi-static crack propagation.

The crack growth of the simulation for specimen NON-CX-2-0.1-01 (2 mm thickness), specimen NON-CX-4-0.1-01 and NON-CX-4-0.1-02 (4 mm thickness) are shown in Figure 18.

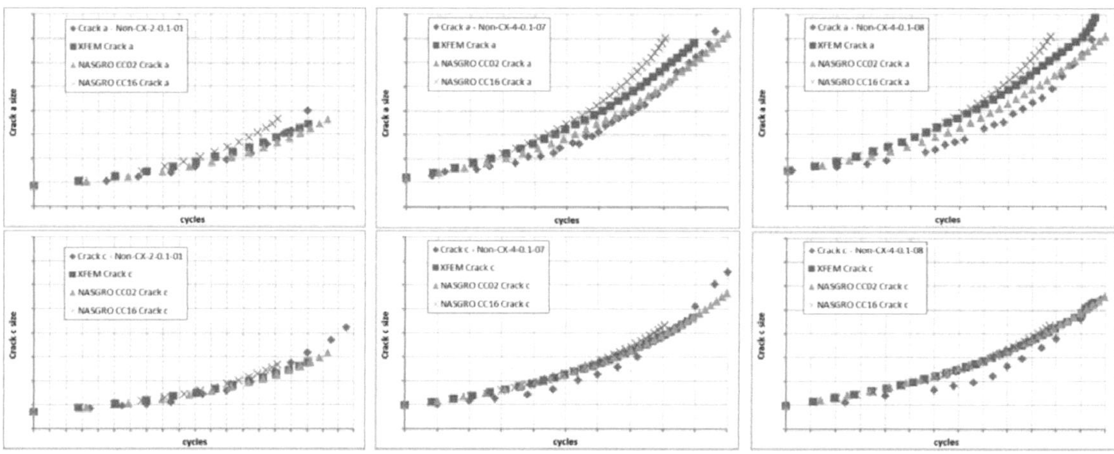

Figure 18. Crack growth results for specimen NON-CX-2-0.1-01 (left column), NON-CX-4-0.1-07 (central column) and NON-CX-4-0.1-08 (right column).

The results show that the corner crack propagation simulations achieve a good correlation with test results.

Residual Stress Prediction

The cold-expansion process simulation was done with ABAQUS/Explicit. The dimensions of the split sleeve tools follow a standard Cold Expansion Process and the geometry is given in Figure 19. Contact pairs were generated among these parts and the assembled models along with mesh are given in Figure 19. A three-back-stresses combined hardening material model was used in the simulation to represent Al 2024-T351 material behaviour. The cyclic stress strain curve for the three-back-stresses material model is shown in Figure 20. The results of these simulations are summarized in the matrix shown in Table III.

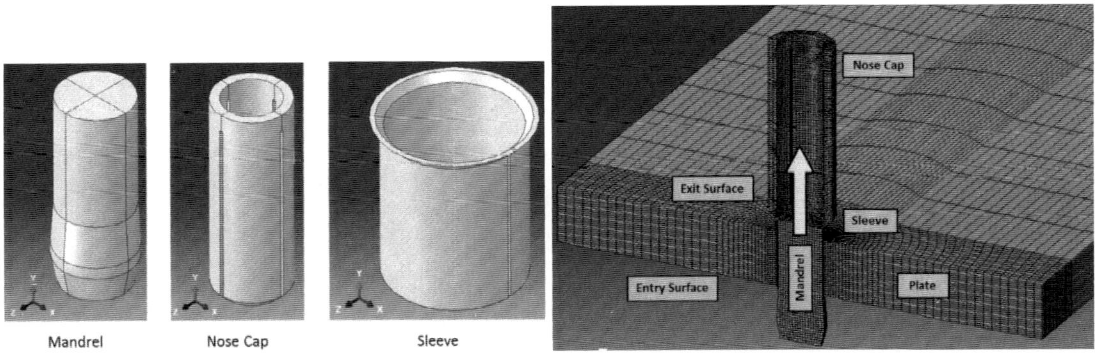

Figure 19. Geometry dimensions of the tools and assembly of the model.

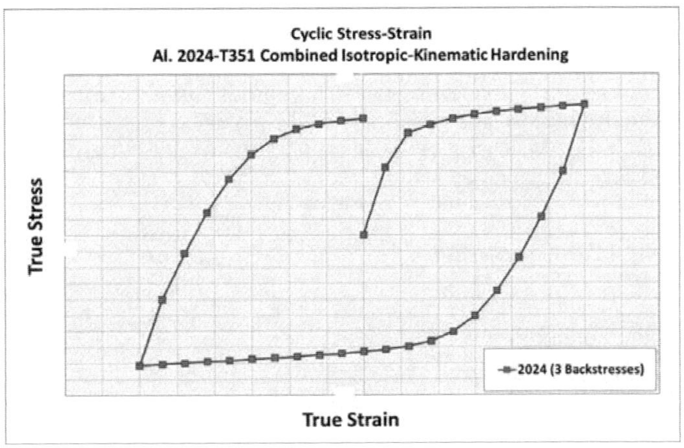

Figure 20. Combined Isotropic-Kinematic Hardening model for Al. 2024-T351.

Table III. Simulation Matrix.

Plate thickness mm	2	2	4	4	6	6	6.35
Hole Diameter mm	5.766	5.72	5.766	5.72	5.766	5.72	5.72
CX Degree	3%	4%	3%	4%	3%	4%	4%
Model naming	CX-3%-2-0.1	CX-4%-2-0.1	CX-3%-4-0.1	CX-4%-4-0.1	CX-3%-6-0.1	CX-4%-6-0.1	CX-4%-6.35-0.1

To validate the simulation and material parameters, a reference model was created with parameters according to the reference tests reported in [10], as shown in Table III. The hoop residual stresses along the radius direction extracted on the entry and exit surfaces from the FE simulation are compared with the reference test [10] in Figure 21.

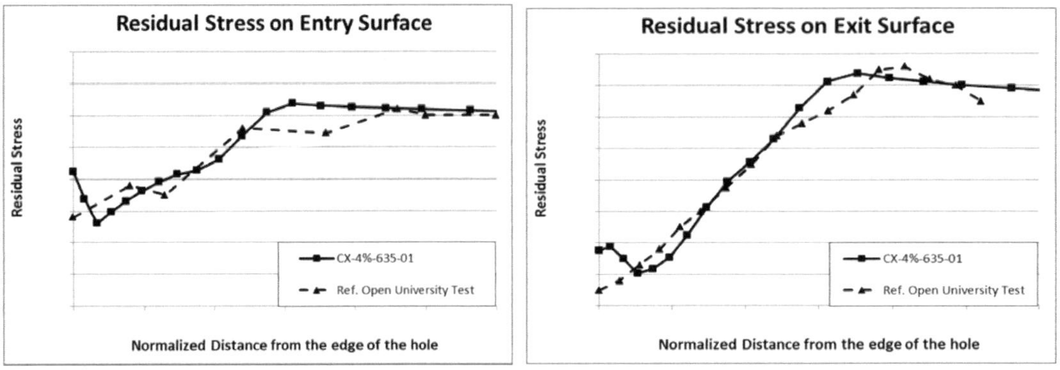

Figure 21. Simulation results against test results from Open University [10].

It has been demonstrated that the current material model and simulation gives a good correlation against the experimental reference stress distribution and are then applied to the simulation cases. In Figure 22, hoop residual stress along the radius direction extracted from different thickness models with 3% cold expansion are compared. It was found that on the entry surface, with the increase of the thickness of the plate, the magnitudes of the residual stress slightly decreases while on the exit surface the residual stress is almost the same for all cases. Similar conclusions can be drawn also for 4% cold-expansion degree simulation results.

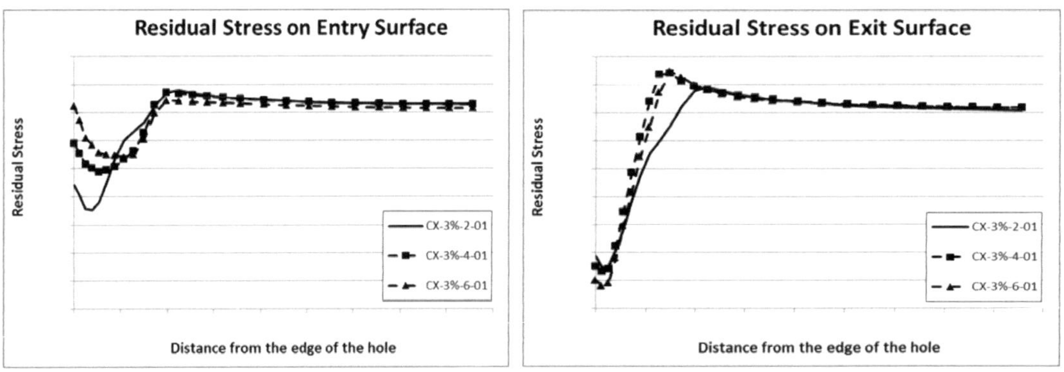

Figure 22. Hoop residual stress on the entry/exit surfaces of 3% cold expanded specimens.

The residual stresses on the entry and exit surfaces with different cold expansion ratios are given in Figure 23.

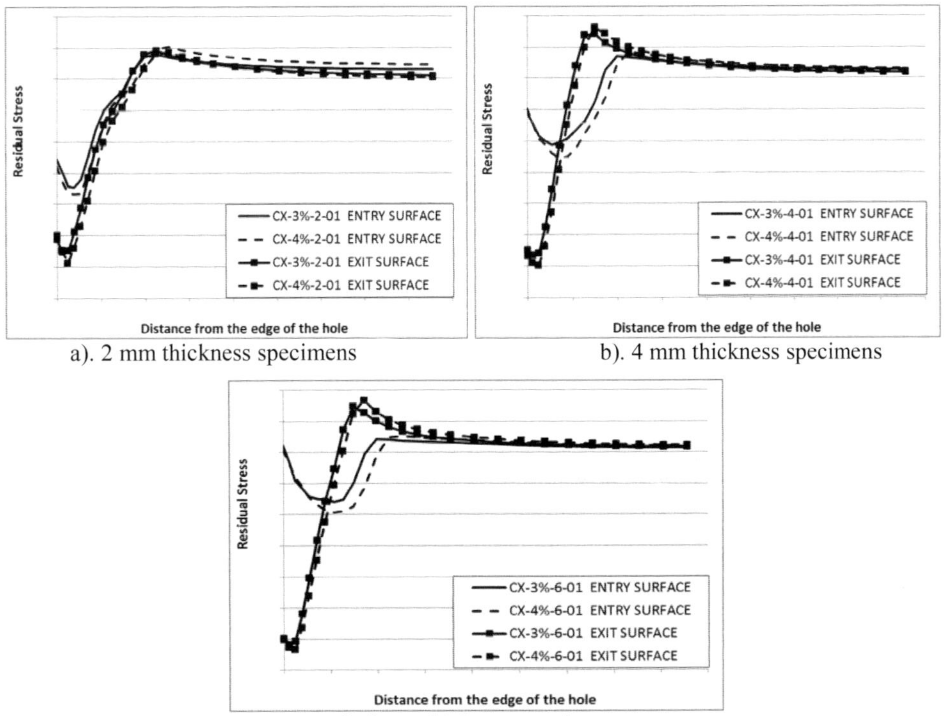

a). 2 mm thickness specimens

b). 4 mm thickness specimens

c). 6 mm thickness specimens

Figure 23. Residual stress on the entry/exit surfaces of different thickness cold expanded specimens.

As expected Figure 23 shows that 4% cold expansion degree gives a higher residual stress magnitude than 3% cold expanded specimens. When compared the magnitudes of residual stresses on entry and exit surfaces, higher residual stress was obtained on the exit surfaces than on the entry surfaces in all thickness specimens. This is due to the free entry face effect and the local equilibrium effect during the mandrel pulling out . The variation of the hoop stresses can be split into four phases along the cold-expansion process as shown in Figure 24. When the mandrel contacts the sleeve, an axial force generates a compressive stress field through the thickness of the hole. Along with the moving of the mandrel, in the region right before the mandrel, a tensile stress field is locally created due to the equilibrium of the compressive stress field generated by the mandrel, and then it turns to a compressive stress field due to the expansion. In the region after the mandrel, a re-equilibrium of the local stress field releases some of the compressive residual stresses. Since the exit surface does not have this re-equilibrium phase, it has a higher compressive residual hoop stress field.

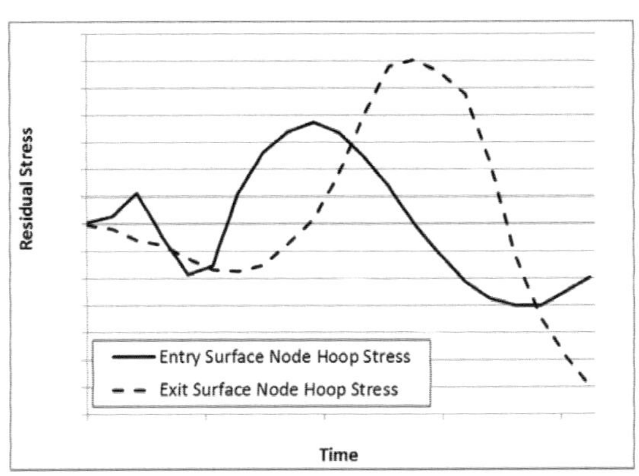

Figure 24. Hoop residual stresses extracted on the entry and exit surfaces.

FUTURE WORKS AND CONCLUSIONS

An experimental program has been carried out at Pisa University to collect test data on fatigue crack propagation of 3D defect (namely corner cracks) in open hole specimens made of 2024-T351 aluminium alloy, in presence of the residual stress field introduced by the split sleeve cold expansion process. The data collected has been used to refine the development of an analysis tool, for the assessment of the fatigue crack growth life enhancement obtainable by means of such a process; this tool together with the valuable tests results should support the better understanding of crack propagation in residual stress fields. For the simulation of the Constant Amplitude fatigue propagation of a corner crack originating from a cold expanded hole, two fundamental types of simulations were carried out, one related to a non-cold-expanded corner crack propagation and the other relevant to the cold-expansion residual stress prediction, respectively.

For what concerns the simulation of the crack growth, a quasi-static crack propagation simulation strategy with ABAQUS XFEM was tested and a very good correlation was achieved between the numerical results and the experimental results.

In order to validate the cold expansion process simulation and the evaluation of the residual stress field, a correlation with tests results was carried out. After the successful conclusion of this material model and process validation activity, the analyses have been performed simulating the various situations of the different specimens tested in the experimental activity, which were characterized by different thickness and cold-expansion degrees. The thickness effect, cold-expansion rate effect, and differences of residual stress on entry and exit surfaces were investigated and compared, showing consistency with similar test results available in the literature.

Further development work is planned, to evaluate improvements of the analysis tool; in particular, the material model that could include consideration of large cyclic stress and strains. Moreover, the ABAQUS Co-Simulation technology will be assessed, to simulate the corner crack propagation from a cold-expanded hole.

Further activities are planned for what concerns the use of residual stress measurements in cracked specimens (available after this experimental activity) to assess residual stress field modifications during the fatigue crack growth phenomenon. Extension to other materials and alloys is part of a future work.

ACKNOWLEDGEMENTS

The authors would like to express their gratitude to Dominique Deloison (Airbus Innovations) and Sjoerd Van der Veen (Airbus SAS) for providing FEM, material models and advices on the simulation of the Cold eXpansion process.

REFERENCES

[1] Buxbaum, O. and Huth, H. (1987), *Expansion of cracked fastener holes as a measure for extension of lifetime to repair*, Eng. Fracture Mechanics, Vol. 28, 689-698.

[2] A. Amrouche, G. Mesmacque, S. Garcia and A. Talha, (2003) *Cold expansion effect on the initiation and the propagation of the fatigue crack*, International Journal of Fatigue, pp. 949-954.

[3] D. L. Ball and M. T. Doerfler, (2003), *Experimental and Analytical Studies of Residual Stress Field Evolution and fatigue Crack Growth at Cold Expanded Holes*, The Aircraft Structural Integrity Program Conference, Savannah (GA), USA.

[4] S. S. Carlson, (2008), *Experimentally Derived Beta Corrections to Accurately Model the Fatigue Crack Behaviour at Cold Expanded Holes in 2024-T351 Aluminium Alloys*, University of Utah, Utah/U.S.A.

[5] M. T. Kokaly, J. S. Ransom, J. H. Restis and L. Reid, (2005), *Observation and Analysis of Fatigue Crack Growth from Cold Expanded Holes*, 8th Ageing Aircraft Conference, Palm Springs (CA), USA.

[6] V. D. Lacarac, D. J. Smith, M. J. Pavier and M. Priest, (2000), *Fatigue crack growth from plain and cold expanded holes in aluminium alloys*, International Journal of Fatigue, pp. 189 - 203.

[7] V. D. Lacarac, A. A. Garcia-Granada, D. J. Smith and M. J. Pavier, (2004), *Prediction of the growth rate for fatigue cracks emanating from cold expanded holes*, International Journal of Fatigue, pp. 585-595.

[8] K. W. Jones and L. M. Dunn, (2009), *Predicting corner crack fatigue propagation from cold worked holes*, Engineering Fracture Mechanics, pp. 2074-2090.

[9] J. M. Tan, S. Pratihar, M. E. Fitzpatrick and L. Edwards, (2005), *Residual Stress Redistribution at Fatigue-Aged StressWave Cold-Worked Holes*, in The 3rd Conference on Stress Evaluation with Neutrons and Synchroton Radiation (Meca Sens 3), Santa Fe, New Mexico, USA.

[10] B. Nadri, L. Edwards, M. Fitzpatrick and A. Lodini, (2004), *Analysis of Residual stresses following overloading of cold expanded holes using the X-ray diffraction and finite element method*, Journal of Neutron Research 12 pp. 219-224.

[11] R. Actis, W. Watkins, S. Prost-Domasky, J. Hodges, (2013). In: *Computation of SIFs for Cracks in Cold-Worked Holes*, The Aircraft Structural Integrity Program Conference, December 3-5, Bonita Springs, Florida, USA.

[12] T. Mills, S. Prost-Domasky, R. Pilarczyk, D. Andrew, J. Hodges, (2013). In: *Modeling Fatigue Performance at Cold Worked Fastener Holes*, The Aircraft Structural Integrity Program Conference, December 3-5, Bonita Springs, Florida, USA.

[13] Anon. (2001). Fatigue crack growth computer program 'NASGRO' version 3.0 – Reference Manual. NASA Technical Report JSC-22267B.

Validation of new experimental and analytical tools to address residual stress effects accurately in the design cycle: Material characterization and design analysis validation

Mark A. James[1], R. Craig McClung[2]

[1]Alcoa Technical Center, USA; [2]Southwest Research Institute, USA

Validation of New Experimental and Analytical Tools to Address Residual Stress Effects Accurately in the Design Cycle: Part 1 – Material Characterization

Mark James[1], Dale Ball[2], Don Millard[1], Mark Newborn[1], John Dalton[1], Robert Bucci[3]

[1] Alcoa Technical Center, USA
[2] Lockheed Martin Aeronautics Corporation, USA
[3] Alcoa Retired, USA

Abstract: Residual stress remains a major obstacle to capturing full benefits and broadening adoption of unitized structure solutions derived from a variety of thick and near net shape aluminum alloy product forms. For instance, the bulk residual stress created during forging thermo-mechanical processing can sometimes introduce bias and scatter into coupon-based fatigue crack growth rate (FCGR) and fracture toughness property determinations, which can lead to uncertainty in the data and its translation to design. The proper accounting of residual stress effects requires that their influence be understood, quantified, controlled, and accounted for throughout the entire product life cycle. Residual stress effects must be excluded from the material property database so their consequence can be managed in the appropriate design and sustainment phases of product life. Proper accounting for residual stress and its management throughout the product life-cycle should be determined based on a range of factors including the amount of conservatism necessary for the given application. The Air Force Research Laboratory sponsored Metals Affordability Initiative (MAI) consortium has conducted a detailed experimental and analytical study of bulk residual stress, fatigue crack initiation, and fatigue crack growth in aluminum alloy coupons extracted from material with intentionally induced residual stress. This Part 1 paper and a companion Part 2 paper summarize a portion of this work directed to validating a next-generation design capability to account for residual stress at the beginning of the design cycle. The main focus of this paper (Part 1) is an improved material FCGR characterization process that partitions residual stress effects from the material "true" FCGR behavior, leading to FCGR design curves that are free of residual stress bias. The Part 2 companion paper describes the fatigue crack growth (FCG) validation test program using design feature (DF) coupons, the analysis methods used to predict the FCG lifetime while addressing residual stress effects in the DF coupons, including variability in the residual stress, and the comparisons between experimental and model results. The advancements described in these papers not only enable quantification and management of residual stress during FCGR property determination, but when integrated and deployed in concert with new design methodologies described elsewhere, these technologies will enable new opportunities for optimization of final part design with respect to material properties, structural design, life management, etc. with the presence of residual stress.

INTRODUCTION

Residual stress has long posed significant challenges for the aerospace damage tolerance community, both in terms of material characterization and subsequent structural design. In forged aluminum alloy parts for instance, bulk residual stress is created during forging thermo-mechanical processing. While high quality cold work has demonstrated ability to dramatically reduce bulk residual stress in forged components, the cold work neither completely eliminates residual stress in the forged part, nor eliminates the possible confounding effects of bulk residual stress on test data derived from the extracted test coupons [1, 2]. When present, residual stress can introduce appreciable scatter and bias into material characterization for fatigue crack growth rate (FCGR) and fracture toughness property data, which leads to interpretive uncertainties in both the data and the subsequent translation to design, with potential consequences ranging from large weight penalties to premature product failure. In short, residual stress continues to impede introduction of promising new technologies suited to growing and capturing the full benefit of large, single-piece aluminum alloy product solutions, e.g., refs 1–7.

Ultimately, the proper accounting of residual stress requires that their influence be understood, quantified, controlled and accounted for throughout the entire design/build/sustain product life cycle for aerostructure. Residual stress effects must be excluded from the material property database so their consequence can be accounted for and managed as a body load in the appropriate design and sustainment phases of product life. Management throughout the product life-cycle should be determined based on a range of factors including the amount of conservatism necessary for the given application. The advancements described in this paper not only enable quantification and management of residual stress during FCGR property determination, but when integrated and deployed in concert with new design methodologies described elsewhere [7-11], these technologies will enable new opportunities for optimization of final part design with respect to material properties, structural design, life management, etc. with the presence of residual stress.

The Air Force Research Laboratory sponsored Metals Affordability Initiative (MAI) consortium has conducted a detailed experimental and analytical study of bulk residual stress, fatigue crack initiation, and fatigue crack growth in aluminum alloy coupons extracted from material with intentionally induced residual stress [12]. The overarching focus of the MAI program is to develop a next generation design capability to assess and account for residual stress at the beginning of the design cycle. This Part 1 paper and a companion Part 2 paper summarize a portion of the MAI work directed to validating a next-generation design capability to account for residual stress at the beginning of the design cycle. The main focus of this paper (Part 1) is an improved material FCGR characterization process that partitions residual stress effects from the material "true" FCGR behavior, leading to FCGR design curves that are free of residual stress bias. The Part 2 companion paper describes the fatigue crack growth (FCG) validation test program using design feature (DF) coupons, the analysis methods used to predict the FCG lifetime while addressing residual stress effects in the DF coupons, including variability in the residual stress, and the comparisons between experimental and model results [11].

BACKGROUND

Residual stress poses a variety of challenges for the fatigue crack growth rate (FCGR) characterization process [2]. Experience has shown that significant caution is warranted any time thick and/or shaped product forms are involved because complete stress relief is difficult to achieve, and as a result, test coupons removed from such products are also likely to contain residual stress. Coupon isolation relieves and redistributes the residual stress, leading to a residual stress state in the coupon that resembles neither the original residual stress in the host material, nor the final residual stress in the machined end-component. Further compounding the challenge is sensitivity to coupon size and spatial position within the original host material envelope and sensitivity to the characterization coupon type (e.g., edge crack coupon versus center crack coupon). While the coupon isolation and type sensitivities are often idealized as linear elastic phenomenon and are realtively easy to model and understand, residual stress redistribution due to nonlinear material behavior such as shakedown during cyclic loading can be more elusive. Among the many references in the literature addressing residual stress, three have some key interesting points to highlight here [1, 14, 15]:

- The Bucci reference [1] outlines a set of guidelines for identifying cases where residual stress may be biasing FCGR data.
- The Parker reference [14] provides a good introduction to many of the key issues central to an application of linear elastic fracture mechanics to fatigue crack growth with residual stress fields, including a salient discussion about the validity of superposition and residual stress redistribution with crack growth, as well as several particularly challenging cases that must be handled carefully, including several examples of crack face contact.
- McClung provides a broad survey of residual stress literature covering a variety of topics [15].

Coupon type sensitivity is central to the results reported later in this paper. As a simple example of coupon-type sensitivity, consider material that has a simple parabolic residual stress distribution representative of a high strength 7xxx aluminum alloy that has been solution heat treated, quenched and artificially aged, but not cold-work stress-relieved. Figure 1 is a schematic with such a parabolic residual stress distribution superimposed on two coupon configurations – one middle crack and one edge crack. In this example the residual stress is assumed in equilibrium, and the notch has not yet been introduced. For the middle crack case, the notch aligns with the highest tensile residual stress, and as a result, a positive stress-intensity factor due to the residual stress (K_{res}) is expected when the notch is introduced. Significant crack growth would be required before the residual stress field transitions from tensile to negative for the middle crack case. In contrast, for the edge crack case the notch is initially in a compressive residual stress field which would be expected to produce a negative K_{res} value. The physical interpretation of a negative stress-intensity factor is usually that crack closure would exist and an

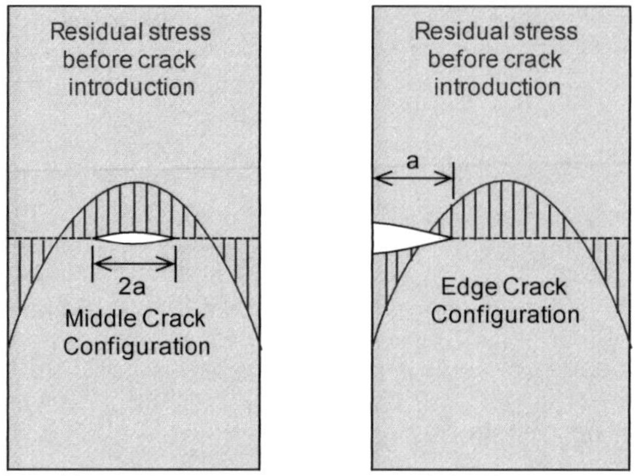

Figure 1. Schematic showing how middle crack and shallow edge crack configurations are affected differently for the same (unflawed) residual stress distribution.

external applied force in proportion to the negative K_{res} would be needed to overcome the closure. Parker [14] discussed several nonlinear contact cases associated with crack closure due to compressive residual stress acting on a cracked configuration. For the edge crack, as the crack grows across the coupon, the compressive residual stress will transition to tensile residual stress. Schematic examples such as these should be approached with a certain amount of caution, and in both cases, equilibrium would need to be established in an analysis to evaluate the actual K_{res} stress-intensity factor. For example, care is especially warranted for deep notch coupons, where far field stress parallel to the crack can create bending moments that unexpectedly influence the stress-intensity factor [1]. Nonetheless, for this discussion the salient point is that for the same residual stress distribution, significantly different stress-intensity factors would result for the middle crack versus the edge crack. Bucci [1, 2] contains example FCGR data for such an example, where the middle crack case has significantly higher FCG rates compared to the edge crack case at equal stress-intensity factor. In Figure 1, the expectation is that the tensile residual stress in the middle crack case would open the crack, decreasing crack closure and leading to increased FCG rates compared with a residual stress free coupon, and the compressive residual stress in the edge crack case would close the crack, increasing crack closure and decreasing FCG rates. These are the trends observed by Bucci [1, 2] both for FCG rates and for closure measurements based on the ASTM opening load compliance method [16].

Two key developments in recent years have enabled significant opportunity to partition residual stress effects from FCGR data. The first of these was the development and standardization [16] of the Adjusted Compliance Ratio (ACR) approach to characterize the effects of crack closure remote from the crack tip [17]. The second of these was the development of the online crack compliance method to measure the stress-intensity factor due to residual stress in a coupon in real time during an FCGR test [18]. The essence of the online crack compliance method is to perform the equivalent of a slitting method residual stress measurement [19] on a coupon using only the usual data collected during a compliance-based FCGR test. Donald and Lados [20] outlined a methodology to integrate the two developments into one post-test data analysis procedure to "collapse" the data into a single normalized stress-intensity factor (K_{norm}) based "master" FCGR curve that is free of residual stress bias and normalized for closure effects. This K_{norm} based approach can then be used with a variety of mechanics-based approaches to reintroduce the stress ratio effect and transform the K_{norm} master FCGR curve to the familiar family format of multiple-stress-ratio FCGR design curves common to today's damage tolerance design handbooks. One mechanics-based approach developed and used in trade studies as well as recent MAI program activities [8, 9, 10] combines the Donald and Lados approach with a crack closure mechanics model [21]. This mechanics-based approach is discussed and used below to interpret the FCGR data. A second approach to collapsing the data and reconstructing FCGR design curves will also be presented that gives essentially the same result as the ACR/K_{norm} method, giving further confidence that closure-based concepts can be used to partition and remove residual stress effects from description of material FCGR behavior by either of two independent approaches. A principal advantage for both approaches described here is that the newly generated FCGR curves will be free of the confounding residual stress bias that has long encumbered design allowable development and life prediction processes for residual stress containing parts. As a result, the design process must be augmented to include residual stress as an additional load case. This premise was also tested in the above referenced MAI program and is documented in the Part 2 companion paper [11], and for which the background has been reported elsewhere [8, 9, 10].

MAI TEST PROGRAM

As part of the MAI program [12], a designed experiment culminated in the testing of various material property coupons containing known bulk residual stress. The host material used in the program was processed intentionally to introduce significant residual stress that could be used to produce coupons having known residual stress, but also that would enable systematic evaluation of the sampling effect sensitivities to within-material spatial residual stress distribution. Alcoa commercial aluminum alloy 7085-T7651 stress-relieved plate served as the starting stock material for this study. The plate product form offers a more consistent and well behaved grain flow microstructure in contrast to typical forging, and hence, for the intended purpose of this experiment, plate was selected over forging to mitigate potential interpretive risks associated with forging grain flow complexity. Figure 2 is a schematic of the manufacturing sequence and related nomenclature to define the sampling locations of the individual coupons. With commercial thick aluminum plate serving as the starting material stock, the coupon fabrication sequence began by dividing the host plate into a number of equally dimensioned logs that afterward were thermal treated in batch to re-introduce residual stress. The individual post-treated logs were next sectioned into multiple machined panels (6 panels per log) and smaller sub-panel blanks from which the various types of test coupons were removed, Figure 2.

Residual stress was intentionally re-introduced using immersion quench at Alcoa Forging and Extrusion's Cleveland Works facility. The logs were solution heat treated, immersion quenched and aged, all in batch, to a T74 temper condition with no stress relief step. The final T74 temper condition was chosen to achieve representative strength level in the material consistent with a typical die forging for the 7085 alloy. For this program, the immersion quench was done in one batch as an experimental control so that consistent and relatively high residual stress would be introduced into all of the log material at the same time.

As a risk mitigation strategy and prior to any thermal treatment, a finite element model of the quench-induced residual stress analysis was done on several configurations to provide an initial estimate of the log post-quench residual stress magnitude and profile. The initial model estimates provided enough information to select a final quench configuration, as well as to better understand sensitivities and make final adjustments to process parameters involved in the residual stress introduction. Because the log is of significant length, a generalized plane strain 2D finite-element formulation was used to predict the as-quenched residual stress in the square-section log. In the simulation the logs were solutionized, quenched and artificial aged consistent with Alcoa Cleveland Works standard practice (as described above). The model assumed uniform heat transfer on the outside of a log and used the heat transfer model of the Alcoa Cleveland Works immersion quench along with a detailed strain and strain rate sensitive material flow law for the material used in the study.

Figure 3 illustrates the predicted residual stress in a log (Figure 3 far left) followed by simulated machining and residual stress measurement experiments for each of the layers excised from a log (Figure 3 center and right). Only half of each coupon is shown to enable viewing of the out-of-plane residual stress distribution (stress along the length of the log LT direction) on each coupon symmetry plane before and after stress rebalance. In the full log cross section on the far left of Figure 3, the color contour shows the out-of-plane component of residual stress (stress along the length of the log LT direction). The expected parabolic residual stress field for the square cross section is apparent (high tensile residual stress in the center of the cross section and compressive stress on the surface of the cross section). To simulate the machining process, the stress tensor from the log cross section is mapped onto the coupon machined envelopes in the center of Figure 3. Then the stress is relaxed to equilibrium in the right of Figure 3 as the final step in simulating the process. The analysis prediction shows that significant relaxation occurs when the coupons are excised from the log and that each of the three panels has a different residual stress distribution. The characteristic parabolic residual stress distribution across the coupon width (L direction) was transferred into the coupons, and the coupon thickness (ST direction) was small enough to minimize the through-thickness gradient in the residual stress distribution. These results provided confidence that coupons excised from different locations within the log would provide opportunity to investigate the sampling effects of residual stress on property determination.

A total of eleven logs of material with approximate dimensions 105 x 105 x 1000 mm were available for the designed experiment, and six locations through the thickness and three locations along the length of each log were identified for coupons. The test matrix included test coupons for three purposes: residual stress characterization, material property determination, and design feature fatigue testing. Each of the coupon types were carefully distributed across the logs to enable sampling of the various locations through-thickness, along the length, and in multiple logs.

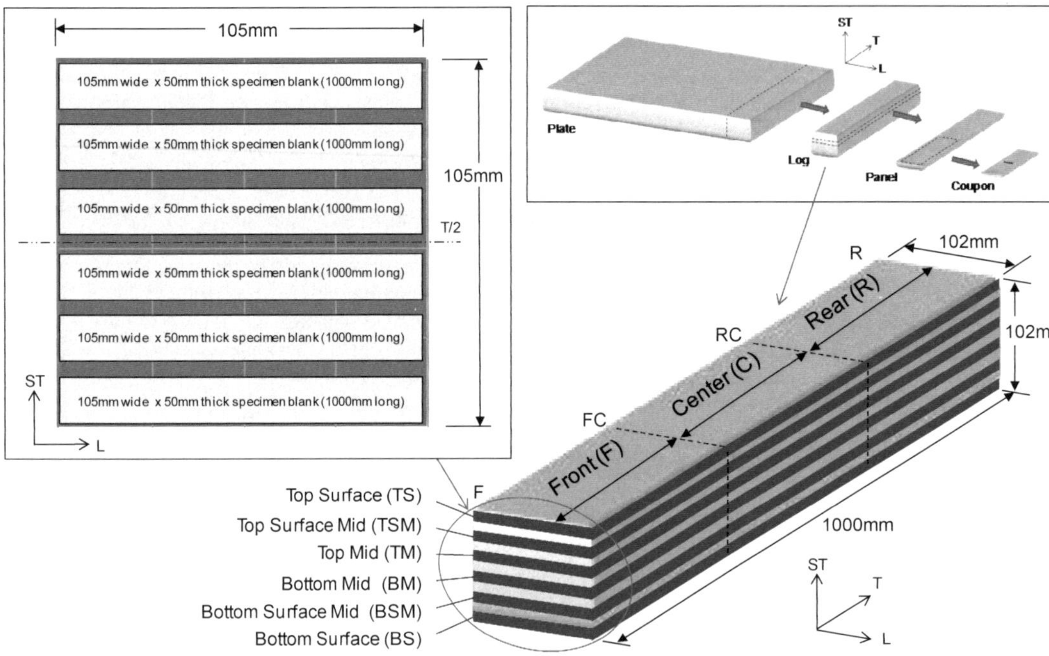

Figure 2. High Level Overview of Log/Panel/Coupon Fabrication Sequence and Nomenclature.

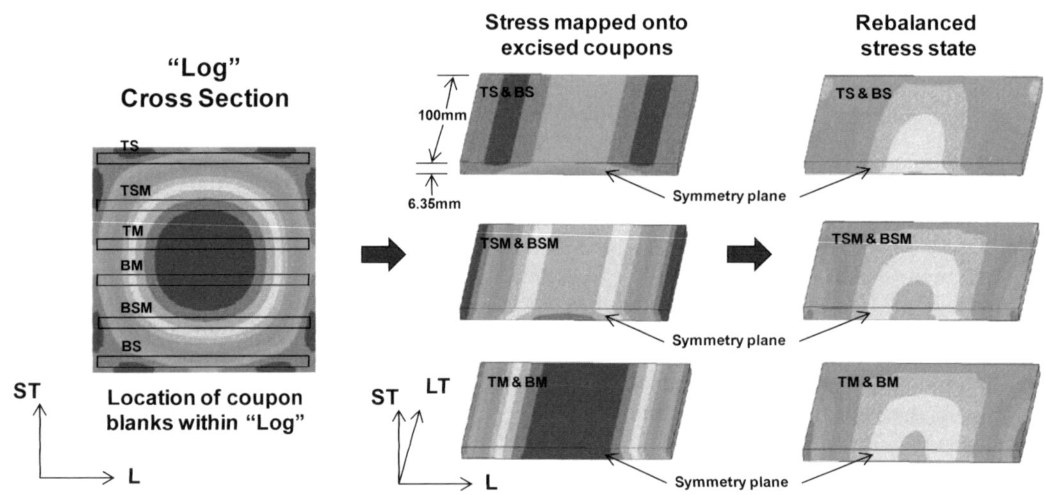

Figure 3. Idealized log finite element quench analysis using uniform heat transfer.

CHARACTERIZATION APPROACH

The primary characterization process used the new Adjusted Compliance Ratio (ACR) method of remote closure correction recently standardized by ASTM International in the E647 test method [16]. In addition, a new test method (not yet standardized) was used to measure the effect of residual stress on each coupon during each FCGR test [18]. In essence, this new method allows the measurement of the stress-intensity factor due to residual stress (K_{res}) that is acting on a coupon, and to measure this in real time for the coupon as a function of crack length. Next the ACR data and K_{res} data are combined in a stress-intensity factor normalization process to collapse the closure- and residual stress-corrected test data into a unique master curve that is free of both closure and residual stress effects [20]. Finally, an analytical reconstruction process is used to reintroduce stress ratio effects for a family of FCGR design curves that have the expected stress ratio relationships, but that are free of confounding residual stress effects. A second approach to collapsing the data and reconstructing design curves will also be presented based on linear superposition of the measured K_{res} for a given coupon test result onto the external applied $K_{applied}$ crack tip stress-intensity factor and combined with ΔK_{eff} closure mechanics [21]. This second approach will be compared with results from the ACR/K_{norm} method, giving confidence that closure-based concepts can be used for partitioning out residual stress effects by either of two independent approaches.

FATIGUE CRACK GROWTH RATE TESTING

This section describes the fatigue crack growth testing for FCG coupons excised from the logs. A subset of these coupons were designated for FCGR testing. The test matrix is outlined in Table I. All tests were conducted in high humidity air (RH > 90%) under constant amplitude loading. All coupons were precracked using constant ΔK at the starting condition for the respective test. Coupon geometries used in this study were the middle crack coupon (M(T)) with width W = 101.6 mm and thickness B = 6.35 mm, and the compact coupon (C(T)) with width W = 76.2 mm and thickness B = 6.35mm. The test procedures nominally followed the ASTM E647 standard. In addition, the new online crack compliance method (not yet standardized) was used to measure the effect of residual stress on stress-intensity factor (K_{res}) in real time during an FCG test [18]. The compliance method was used to monitor crack length to facilitate K_{res} and ACR-method closure data collection during each test. Testing and data reduction was performed using Fracture Technology Associates software [22].

Table I. Test matrix for fatigue crack growth rate testing

Coupon	**Stress Ratio**						**Total no. tests**
	0.7	**0.5**	**0.3**	**0.1**	**- 0.5**	**- 1**	
M(T)				1	1	1	3
C(T)	1	1	1	1	-	-	4

Residual stress modeling at the outset of the program was used to estimate the influence that the expected residual stress distribution would have for the M(T) and C(T) coupons for various applied stress ratios and sampling locations. The simulations indicated that the C(T) would have significant clamping residual stress (negative K_{res}) at the start of a test, and the M(T) would have significant opening residual stress (positive K_{res}) at the start of a test. This observation was true to varying degrees for all of the test locations sampled, and because of this, it was decided that the C(T) coupon would be used for positive stress ratios and the M(T) coupon would be used for negative stress ratios, with both coupon types tested at stress ratio R = 0.1.

RESULTS

Figure 4 contains the FCGR data from the M(T) and C(T) coupon tests on a ΔK_{app} basis. There are a number of indications that these test data are confounded with residual stress effects. In simple terms, the curves do not layer as expected based on applied stress ratios. For instance, the M(T) R = -1 and -0.5 test results cross over to the left and have about the same FCG rates as the C(T) R = 0.7 data. The C(T) R = 0.1 curve has lowest FCGR rates at the outset of the test of all of the coupons, and the C(T) R = 0.1 and M(T) R = 0.1 results differ significantly. For traditional aluminum alloys with minimal residual stress, one would expect that C(T) and M(T) tests would give comparable FCG rates for rates in the range from slightly above 1E-6 mm/cycle to slightly above 1E-4 mm/cycle (called Region-II crack growth, also referred to the as the Paris Power Law region).

The confounded FCGR data in Figure 4 can be rationalized in the context of differing residual stress states within the two coupon types. In this case, the C(T) tests have a strong negative K_{res} at the start of the tests, and the M(T) tests have a strong positive K_{res} at the start of the tests. The results were fully anticipated and are consistent with the interaction of external and internal stress fields for coupons excised from material with the particular quench-induced residual stress field in the logs (e.g., recall discussion relating to Figure 1). Residual stress measurements made from separate slitting tests confirmed that the residual stress was significant [23].

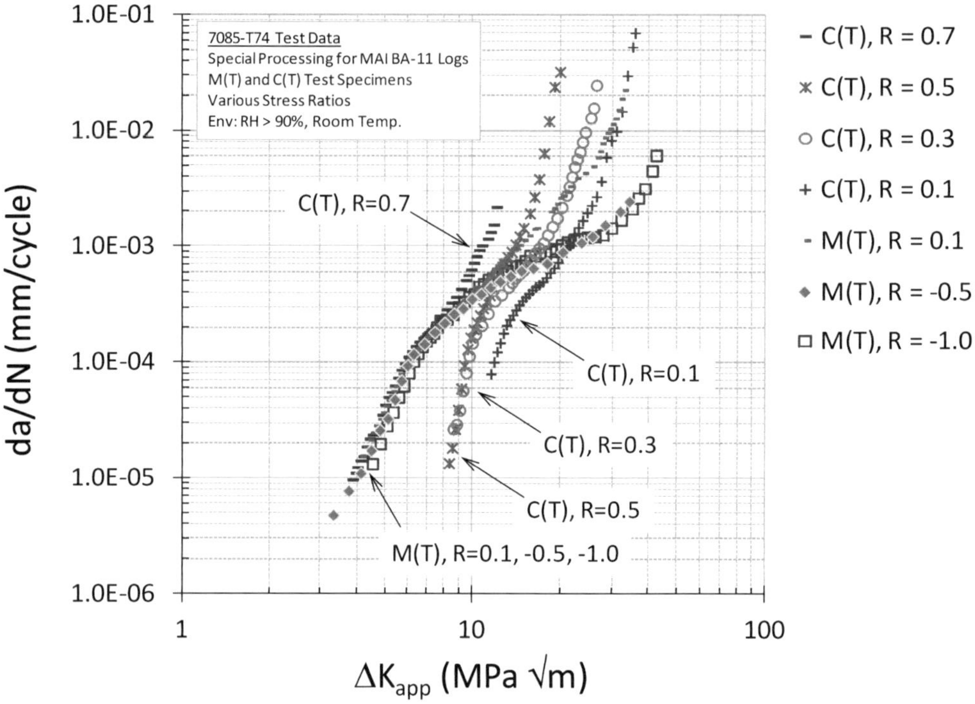

Figure 4. Fatigue crack growth rate data on ΔK applied basis.

Figure 5 is the FCGR data from the M(T) and C(T) tests interpreted on a ΔK_{ACR} basis. As discussed above, the ACR method provides an alternate closure correction to the traditional ASTM opening load method. The ACR method has been interpreted as removing closure remote from the crack tip [24]. The method is most sensitive to remote closure when the crack length is relatively short and is much less sensitive to crack tip local closure. The method also is not particularly sensitive to closure for long crack lengths (e.g., when respective test crack growth rates approach fast fracture), likely due to inherent limitations in the method [24]. Comparing Figures 4 and 5, the ACR data is nominally collapsed excluding the rates for each test approaching fracture. The only three curves that changed significantly from the closure correction are the C(T) coupon results for R = 0.1, 0.3, and 0.5. All three of these tests have FCG rates (for ΔK applied) that appear lower than expected for a 7000 series aluminum alloy. The M(T) results have minimal closure correction because these coupons are already essentially closure free due to the tensile K_{res} acting to open the crack. These results are understandable in the context of tensile residual stress for the M(T) and compressive residual stress for the C(T) (recall for instance the discussion above regarding Figure 1).

Figure 6 is the FCGR data from the M(T) and C(T) tests on a K_{norm} basis. The equation used to calculate K_{norm} is taken from Reference 20 based on the work of Broek and Schjive [25] as well as Walker [26]:

$$K_{norm} = (\Delta K_{ACR})^{1-n} * (K_{max\text{-}applied} + K_{res})^{n}, \qquad (1)$$

where the exponent $\boldsymbol{n}$ is the K_{max} sensitivity and Eqn. 1 is referred to as K_{max} normalization for the ΔK_{ACR} data. The value of $n = 0.25$ was selected for a visual best fit of the data in Figure 6. Also included in Figure 6 is the red diamond symbol labeled "K_{norm} master curve." This K_{norm} master curve was selected to approximate the average of the underlying data and will be used as the input of the reconstruction process. At this point the da/dN data reduction process is complete, and the reconstruction process can begin to restore the FCGR relationship to the customary multiple stress-ratio FCGR curve format (da/dN = f(ΔK_{app}, R)) familiar to most production damage tolerance design analyses.

The reconstruction process is based on mechanics-based plasticity-induced closure equations [21]. Although the referenced closure equations ignore oxide and roughness induced crack closure, the mechanics based approach

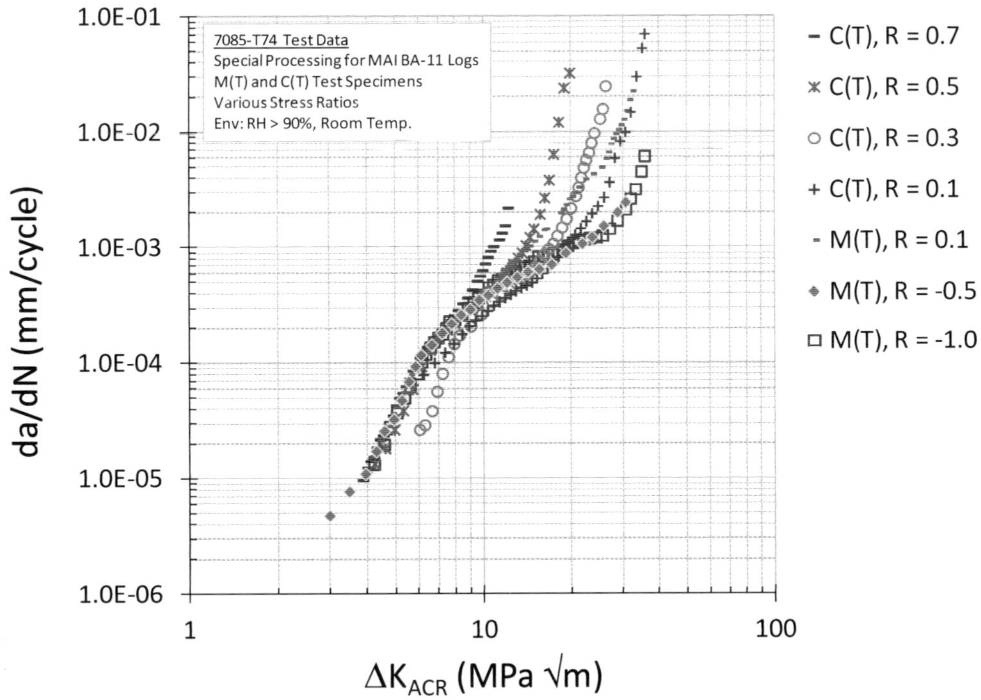

Figure 5. Fatigue crack growth rate data on ΔK_{ACR} basis.

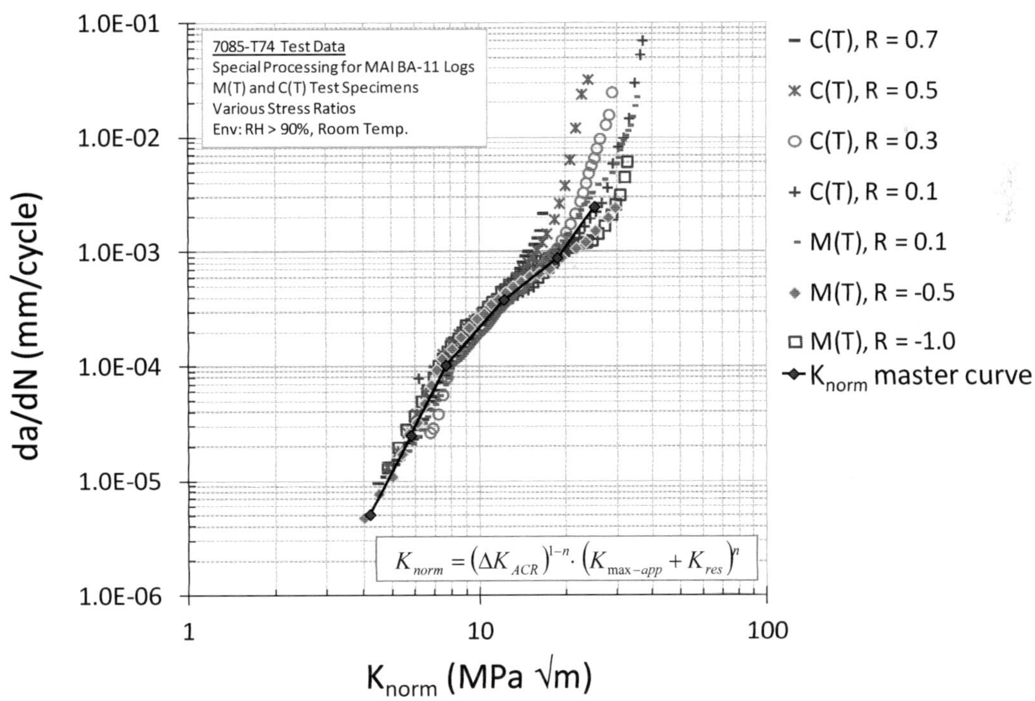

Figure 6. Fatigue crack growth rate data on K_{norm} basis.

has proven both useful and accurate for many cases documented in the literature [27]. The process is first to transform the K_{norm} master FCGR curve into a closure free ΔK_{eff} curve, then to use the closure equations to reintroduce stress ratio effects, but without reintroducing any residual stress bias. The reconstruction is based on the observation that the high-R

ΔK_{ACR} data was nominally closure free to begin with, and that for closure free data the K_{norm} equation is fully reversible via the Walker equation (i.e., for closure free data $\Delta K_{ACR} = \Delta K_{app}$). Thus, the first step to transforming the K_{norm} master curve into a closure free ΔK_{eff} curve is to use the Walker equation to calculate a predicted R = 0.7 $\Delta K_{applied}$ curve from the K_{norm} master curve as:

$$\Delta K = K_{norm} * (1\text{-}R)^{n}, \tag{2}$$

with R = 0.7 and n = 0.25, for each of the K_{norm} master curve data points in Figure 6. Next the predicted R = 0.7 curve is transformed to ΔK_{eff} using the inverse closure equations (the R=0.7 curve is nearly closure free, but does move ever so slightly to the left; the step is included in the analysis for completeness). Finally, the closure equations are used to reintroduce the stress ratio effects to transform the ΔK_{eff} mean curve into the more familiar multiple stress-ratio ΔK_{app} FCGR family of curves, where each of the individual FCGR curves are taken to be representative of the true material behavior free from residual stress induced bias. The closure equations have two parameters that have physical interpretations, but that are nonetheless tuning parameters. These are the ratio S_{max}/σ_o (ratio of maximum applied stress during a fatigue test to the average of yield stress and tensile stress) and the constraint factor α, which ranges from α = 1.0 for pure plane stress to α = 3.0 for pure plane strain. Nominal values used here are S_{max}/σ_o = 0.3 based on nominal maximum applied and residual stress for the M(T) coupon and α = 1.8 as a nominal constraint factor for Region-II crack growth (and because additional information such as spike overload spectrum data useful for tuning the constraint model are not available).

Figure 7 is a comparison of fatigue crack growth rate curves (represented by the solid symbols and lines) reconstructed using the K_{norm} methodology and compared with the ΔK applied data from Figure 4. For clarity, only three stress ratios are included in the reconstructed curves. As expected, the reconstructed curves do *not* agree with the original curves. Comparisons with the original data are difficult because the residual stress effects remain in the data, while the curves represent the material behavior *without* residual stress bias.

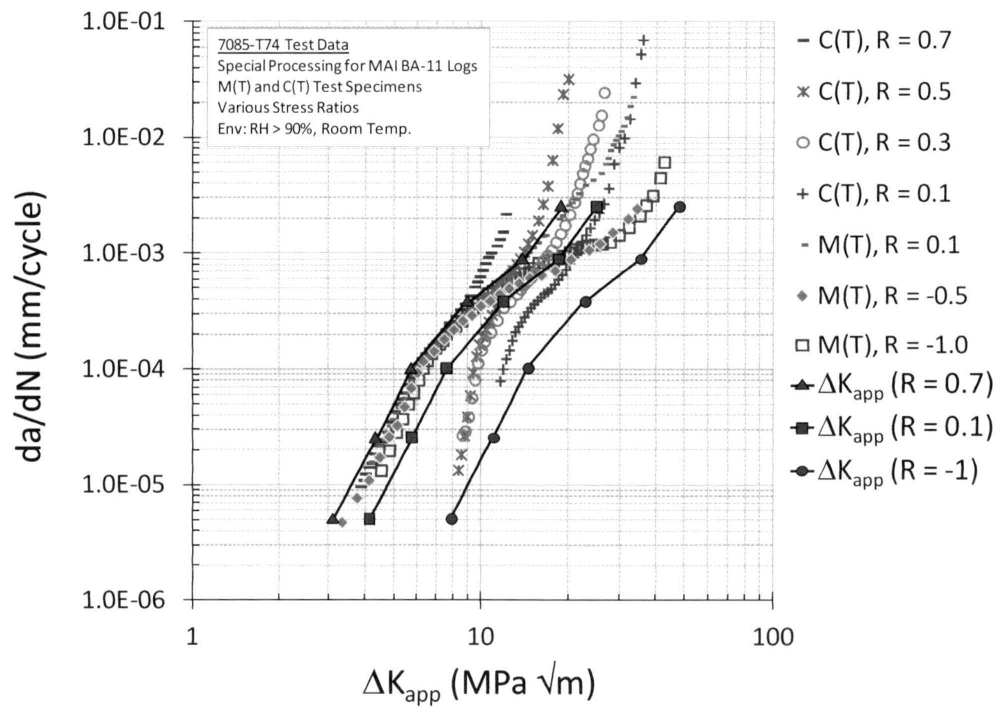

Figure 7. Fatigue crack growth rate curves reconstructed (only at stress ratios R = 0.7, 0.1, and -1) using the K_{norm} methodology and compared with the ΔK applied data.

VALIDATION & DISCUSSION

This section focuses on three approaches used to validate the methodology for generating residual stress free FCGR design curves. In the first example, we compare the previous set of predicted curves derived from MAI log program material to a second test data set generated from a similar alloy (and that for the purpose of this illustration can be regarded as virtually residual stress free through a combination of stress relief processing and specialized coupon selection). In the second example, we reduce the MAI Log program raw ΔK_{app} data using linear superposition and the ΔK_{eff} approach and compare this to results of the ACR/K_{norm} method. In the third example, and for completeness, we show a brief summary of fatigue crack growth life predictions [9]. For a more complete presentation of the fatigue spectrum testing and associated fatigue life predictions, the authors refer the reader to Part 2 of the current work [11].

No "residual stress free" data exists for coupons excised from the log material. As a result, for the first validation example, a second pre-existing data set was used. Data was generated on stress-relieved 7085-T7452 die forging material in the 2006 time frame under a different program where significant effort was used to test relatively small cross-section edge crack EC(T) type test coupons (dimensions W = 20.3mm, B = 5.1mm) to minimize the coupon residual stress and influence of residual stress on generated FCGR data. These data are shown in Figure 8 along with a set of MAI log program reconstructed curves. The comparison is quite favorable. Note that although the microstructure is somewhat different for the MAI log material and die forging for the EC(T) coupons, the expectation is that if residual stress effects are mitigated, the two materials would have similar da/dN-ΔK response for Region-II (Paris region) crack growth because the typical microstructure differences for 7xxx are not thought to have significant influence on the intrinsic FCG rates in Region-II. The ACR/K_{norm} data reduction process and mechanics-based reconstruction based on the K_{norm} master curve has done a remarkable job of removing the residual stress effects and generating curves that represent quite reasonable behavior.

The second validation example is based on the original $\Delta K_{applied}$ data of Figure 4 and uses the ΔK_{eff} approach to collapse the data directly to a closure-free and residual-stress-free data set. Rather than use ASTM opening load closure measurements (which are often problematic due to inability to distinguish crack tip closure from remote closure), in this example plasticity induced closure equations [21] are used to collapse the data to ΔK_{eff}. The key

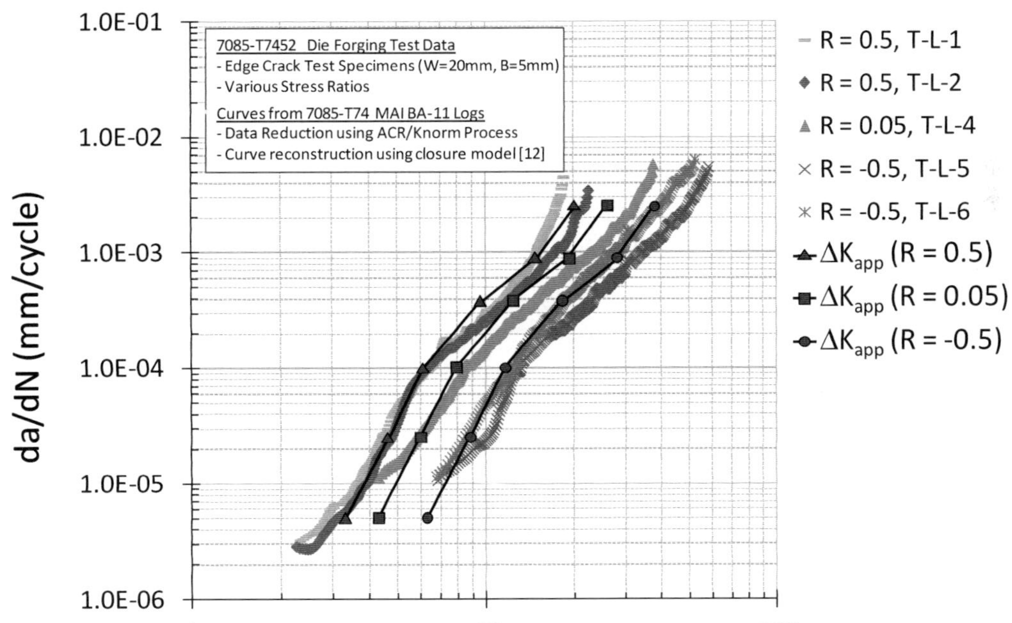

Figure 8. Fatigue crack growth rate curves reconstructed using the K_{norm} methodology and compared with additional residual stress free FCGR data from 7085-T7452 die forged material.

new information that enables this approach is the K_{res} data measured during the FCG test for each coupon. The K_{res} data is introduced using a simple linear superposition approach, where the residual stress-intensity factor is superimposed onto the applied stress-intensity factor, $K_{applied}$, due to the external applied forces for the test (e.g., as measured by the load cell). Thus, the total crack-tip stress-intensity factor can be expressed as:

$$K_{total} = K_{applied} + K_{res}, \tag{3}$$

and then the total stress ratio can be expressed as:

$$R_{total} = \left(K_{min-applied} + K_{res}\right) / \left(K_{max-applied} + K_{res}\right), \tag{4}$$

where $K_{min-applied}$ and $K_{max-applied}$ are the applied stress-intensity factor at minimum and maximum load, respectively. For each data point in Figure 4, the total stress ratio can be calculated using Eqn. 4 and the measured K_{res} for that data point (note that for linear superposition the applied ΔK remains the same; only the stress ratio changes due to the residual stress). Then, the closure equations are used to calculate ΔK_{eff} for each data set of Figure 4 with the total stress ratio (rather than the applied stress ratio, as would be usually done). Figure 9 is a comparison of the ΔK_{eff} curve predicted using the ACR/K_{norm} approach above with the ΔK_{eff} data calculated using the total stress ratio. The data from Figure 4 has largely collapsed to a unique curve, and overall, the agreement between the collapsed data and the ΔK_{eff} curve from the ACR/K_{norm} method is remarkably good.

The third validation example is also one that is likely the most relevant to the aerospace design analysis community and draws on Reference 9 to compare fatigue crack growth life predictions with test data. The data set is documented in detail in the companion Part 2 paper and others [9, 11, 23, 28], and only a very brief summary is provided here. This validation example compares the predicted fatigue crack growth life with constant amplitude and spectrum test results for a variety of coupon configurations containing "design features" (DF) machined into the coupons (e.g., through hole in the tensile field, through hole in the compressive field, pockets with representative radii, etc.). The fatigue crack growth life predictions used da/dN relationship curves

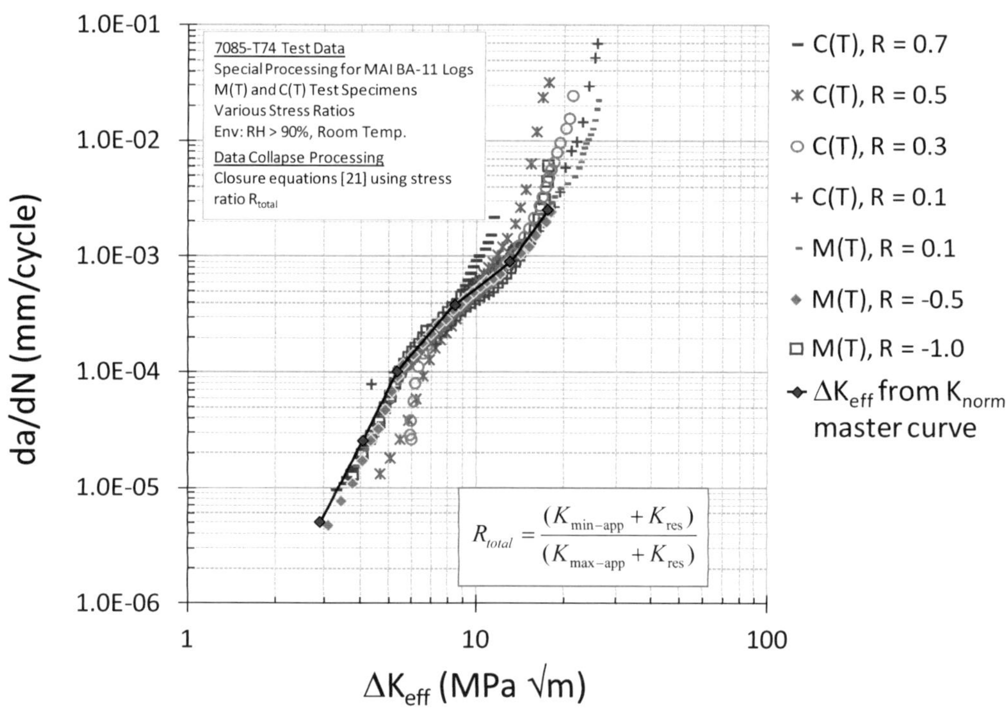

Figure 9. Second approach to removing residual stress effects. In this case the data was processed using total stress ratio (R_{total}) and a closure model [21] and compared to results from the master curve approach.

developed from the above methodology and also used measured / modeled residual stress to reintroduce the residual stress effect as an internal body stress for each coupon configuration. Figure 10 contains example residual stress measurement data and model results for DF1 coupons (which contained a center hole in a simple panel) [9]. The color contours are measurement data using the contour residual stress measurement method [29] and the figure also contains a line-plot comparison of measured and modeled results. The residual stress results were introduced into fatigue crack growth life predictions of the coupons using the weight function approach to calculate stress-intensity factors for the residual stress fields [30].

A large matrix of coupons were tested under either constant amplitude or variable amplitude loading for fatigue crack growth, and the test results have been reported elsewhere [9, 10, 11, 28]. In addition, the fatigue crack growth lives were predicted using two slightly different approaches. The ACR/K_{norm} FCGR relationship outlined above was used as the basis for both the previous publications and for the companion Part 2 paper. The first approach was published recently [9, 10] and used the traditional family of da/dN-ΔK curves as a function of stress ratio with a conservative linear extrapolation of the data in the threshold region for each curve. The second approach is included in the companion Part 2 paper [11] and used the ACR/K_{norm}-predicted ΔK_{eff} curve as input to a strip yield model [31]. In addition, the companion Part 2 paper used an estimate of threshold material behavior from a similar 7xxx alloy to augment the ΔK_{eff} curve to better estimate threshold retardation effects.

Figure 11 is an aggregate comparison of predicted versus experimental fatigue life for the constant amplitude and spectrum cases across all four design feature coupon configurations and summarizes selected results from Reference 9. Each data point represents the life prediction for one test. The analyses were run either with the modeled residual stress or with no residual stress. Reference 23 describes a probabilistic model of the residual stress that led to the deterministic residual stress model for the fatigue life analyses. The results in Figure 11 show that there is significant sensitivity to presence of residual stress and that when residual stress was excluded from the analysis (open symbols in Figure 11) the scatter in predicted life increased substantially and agreement between predicted and actual life decreased substantially, sometimes by as much as an order of magnitude. When residual stress was included in the analysis, the fatigue life was predicted within the standard scatter factor of 2x (or the analyses were conservative), building confidence that when the distribution/magnitude of residual stress is known, the approach to residual stress management advocated here is viable in a standard damage tolerance approach to structural integrity.

This paper validates the concept that a material's true FCGR property characterization can be discerned from test records confounded with residual stress effects, even in cases where the residual-stress-induced bias on the measured test result is substantial. It is also demonstrated that the methodology solves one of the long-standing barriers to developing clean FCGR data and design curve information essential to efficient design of aerospace structure in the presence of residual stress. The approach is believed to be relatively general and thus applicable to FCGR characterization for other materials and for other sources of residual stress such as parts that have been welded, shot-peened, or had other process-induced or engineered residual stress introduced. The second validation example based on superposition also worked well for the material cases investigated in this paper;

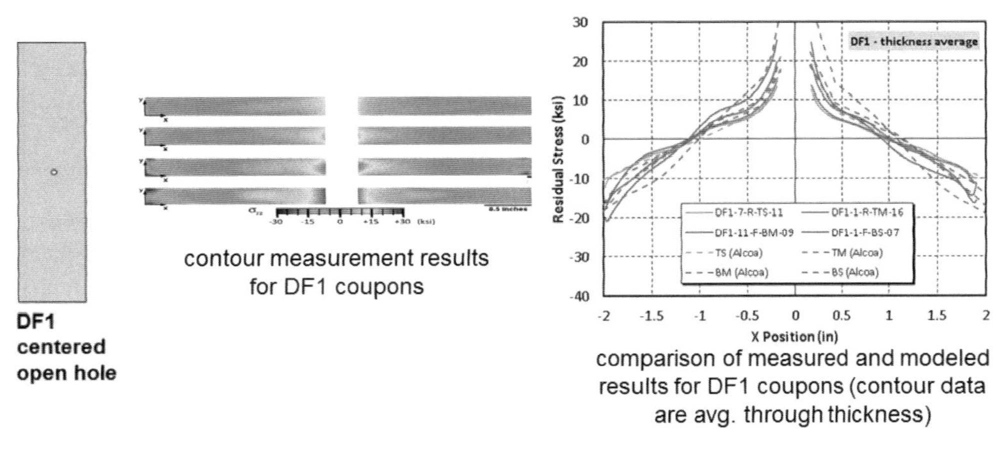

Figure 10. Contour method residual stress measurement data for DF1 coupons (left) and line-plot comparison of measured and modeled results (right) [9].

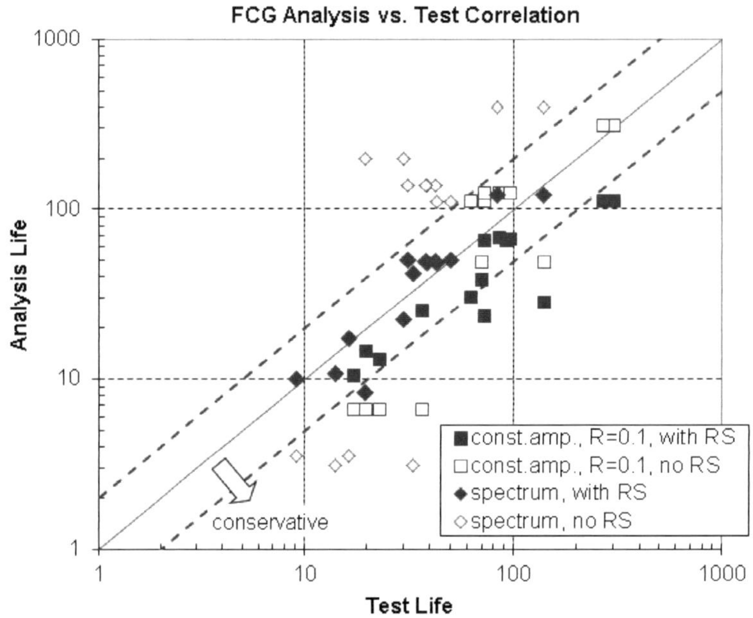

(a) comparison of analysis life versus test life (a straight line is perfect match).

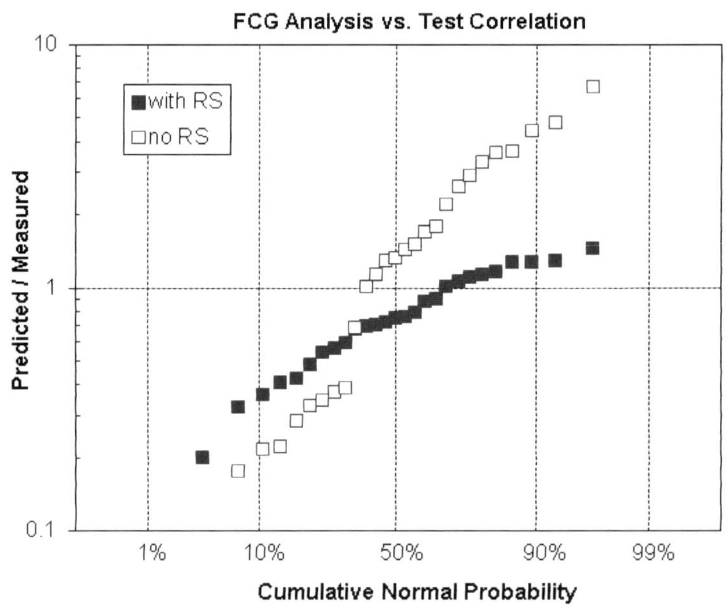

(b) comparison of analysis predicted life versus measured life as a function of cumulative probablility.

Figure 11. Example validation results – FCG analysis vs. test correlation summaries [9].

however, there are ample examples in the literature where superposition did not work well (e.g., see Reference 15 for a literature survey), demonstrating need for new approaches such as the ACR/K_{norm} method. It is important to note that the ACR/K_{norm} method advocated here is intended to be an engineering one, and while it worked well for this particular case, additional work is recommended to demonstrate that the approach has broader applicability for real world cases. Finally, the characterization process is but one of several necessary steps along the path to fully accounting for residual stress in the design/build/sustain process for aerostructure. The Part 2 paper [11] provides detailed validation of the process used to reintroduce residual stress into the design process.

SUMMARY AND CONCLUSIONS

The examples shown here provide compelling evidence that the ACR/K_{norm} approach to partitioning residual stress effects from fatigue crack growth rate data is producing a set of FCGR design curves that are residual stress free and that do represent the "true" material FCGR behavior. The results in this paper provide validation for the vision outlined in the past that when residual stress is known and correctly included in the fracture mechanics analysis, the residual stress effects can quite reasonably be accounted for in an engineering sense, both numerically and trend-wise, and thereby enable full benefit and broadened application of unitized structure solutions that contain residual stress.

Some specific observations include: Results from this study and cited references confirm that the C(T) and M(T) coupon types respond differently but predictably to the residual stress field introduced in the log material and confirm that sampling effects should be considered as part of any test program where residual stress is a concern. The new online crack compliance test method provides measured data for the actual residual stress state in a coupon, and this data has proven essential for the FCGR data reduction process, both in terms of interpreting the raw ΔK_{app} data in general terms before residual stress correction and in terms of partitioning the true, unbiased, material behavior from the raw test data. Two independent data reduction approaches were used to interpret the data, and both approaches produced approximately the same end result, giving confidence that closure-based concepts in combination with the measured K_{res} data can be used for partitioning out residual stress effects and demonstrating that either of the two approaches is acceptable. The results confirm that the net effect of residual stress is simply to shift the mean stress, as indicated by the ΔK_{eff} results calculated using R_{total}. While this simple superposition approach worked well for this data set, there are ample examples in the literature where superposition did not work well, demonstrating need for new approaches such as the ACR/K_{norm} approach outlined here. Additional work is recommended to demonstrate that both the ACR/K_{norm} characterization approach and the broader design approach have applicability for real world cases.

ACKNOWLEDGEMENTS

This work was funded by the Air Force Research Laboratory (AFRL) Materials Affordability Initiative (MAI) Consortium, Technology Investment Agreement FA8650-06-2-5211. The specific project was "Incorporation of Residual Stress Modeling into Next Generation Design, Manufacturing and Sustainment for Advanced Airframe Structures (BA-11)." Other BA-11 team members include Boeing (James Castle), Northrop Grumman (Miles Honkawa) and Rolls Royce (Michael Glavicic). The authors would like to express their gratitude to the BA-11 program manager, Dr. James B. Castle, and the Air Force Research Laboratory program technical monitor, Dr. Pamela Kobryn, for their support. Alcoa acknowledges John Watton for many contributions for residual stress modeling.

REFERENCES

[1] Bucci, R.J. (1981) "Effect of Residual Stress on Fatigue Crack Growth Rate Measurement," *ASTM STP 743*, American Society for Testing and Materials, pp 28-47.

[2] Bucci, R., James, M., Sklyut, H., Heinimann, M. Ball, D. L., Donald, J. K. (2006) "Advances in Testing and Analytical Simulation Methodologies to Support Design and Structural Integrity Assessment of Large Monolithic Parts," SAE Technical Paper 2006-01-3179, 2006.

[3] R. J. Bucci, R. W. Bush, and G. W. Kuhlman. (1997) "Damage Tolerance Characterization of Thick, Wrought Aluminum Products with and without Stress Relief: Focus on Toughness and Crack Growth Characteristics to Capture Advances in Forging Stress Relief Technology," *Proceedings, 1997 USAF Aircraft Structural Integrity Program Conference*, San Antonio, TX, December 2-4.

[4] R. J. Bucci (2002) "Lessons Learned - Effect of Residual Stress on Material Durability/Damage Tolerance Capability Assessment; Coupon-to-Component Issues, Problems and Cures," ASTM-E08 Workshop on Residual Stress Effects on Fatigue and Fracture Testing and Incorporating Results into Design, Pittsburgh, PA, May 8.

[5] R. E. Sanders, W. L. Otto, W. L., and R. J. Bucci (1979) "Fatigue Resistant Aluminum P/M Alloy Development," AFML-TR-79-4131, Final Contract Report, U. S. Air Force Materials Laboratory, Wright Patterson Air Force Base.

[6] L. N. Mueller, R. J. Bucci, D. Selfridge, R. R., Sawtell, E. Balmuth, D. Chellman, J. T. Amin (2005) "Aerospace Aluminum Forgings – A New Perspective," *16th Annual AeroMat Conference & Exposition*, Orlando, Florida, June 6-9.

[7] D. L. Ball. (2005) "A Proposed Analytical Approach for the Incorporation of Residual Stresses in the Design of Large Forged Components," *Proceedings Boeing 7th Industry Wide Unitized Structure Technical Interchange*, Dallas, TX, November 10, 2005.

[8] D. L. Ball (2008) "The Influence of Residual Stress on the Design of Aircraft Primary Structure," Journal of ASTM International, Vol. 5, No. 4 (Paper ID JAI101612).

[9] Ball, Dale L., et al. "The Impact of Forging Residual Stress on Fatigue in Aluminum," AIAA SciTech 2015, Kissimmee FL January 2015.

[10] Ball, D. L., James, M. A., Bucci, R. J., Watton, J. D., DeWald, A. T., Hill, M. R., Popelar, C. F., and McClung, R. C., "A Detailed Evaluation of the Effects of Bulk Residual Stress on Fatigue in Aluminum," *Fatigue 2014, Proceedings 11th International Fatigue Congress*, Vols. 891-892, Advanced Materials Research, Melbourne, Australia, March 2014, pp. 1205-1211.

[11] McClung et al. "Validation of New Experimental and Analytical Tools to Address Residual Stress Effects Accurately in the Design Cycle: Part 2 – Design Analysis Validation," In: ICAF 2015 Proceedings, Helsinki, Finland, June 2015.

[12] Dutton, R., Kobryn, P., Ball, D., Castle, J., James, M., Yavari, P., "From Integrated Computational Materials Engineering to Integrated Computational Structural Engineering," *Proceedings of the 3rd World Congress on Integrated Computational Materials Engineering (ICME 2015)*, May 31 - June 4, 2015, Colorado Springs, Colorado, USA.

[13] Nelson, D. V., "Effects of Residual Stress on Fatigue Crack Propagation," *Residual Stress Effects in Fatigue, ASTM STP 776*, American Society for Testing and Materials, 1982, pp. 172-194.

[14] Parker, A. P., "Stress Intensity Factors, Crack Profiles, and Fatigue Crack Growth Rates in Residual Stress Fields," *Residual Stress Effects in Fatigue, ASTM STP 776*, American Society for Testing and Materials, 1982, pp. 13-31.

[15] McClung, R. C., "A literature survey on the stability and significance of residual stresses during fatigue," *Fatigue Fract Engng Mater Struct* 30, 173–205, 2007.

[16] ASTM E647-13e01: Standard Test Method of Fatigue Crack Growth Rates, Annual Book of ASTM Standards E03.01, ASTM International, West Conshohocken, PA, 2013.

[17] G. H. Bray, J. K. Donald, "Separating the Influence of Kmax from Closure-Related Stress Ratio Effects Using the Adjusted Compliance Ratio Technique," *Advances in Fatigue Crack Closure Measurement and Analysis: Second Volume, ASTM STP 1343,* R. C. McClung, J. C. Newman, Jr., Eds., American Society for Testing and Materials, 1999, pp. 57-78.

[18] Lados, D. A., Apelian, D. and Donald, J. K. (2006) "Fracture mechanics analysis for residual stress and crack closure corrections," *International Journal of Fatigue*, 29, 687-694.

[19] Schindler, H. J., Cheng, W. and Finnie, I. (1997) "Experimental determination of stress intensity factors due to residual stresses," Experimental Mechanics, 37, 272-277.

[20] Donald, J. K. and Lados, D. A., "An integrated methodology for separating closure and residual stress effects from fatigue crack growth rate data," *Fatigue Fract Engng Mater Struct* 30, 2006, 223–230.

[21] J. C. Newman, Jr., "A Crack Opening Stress Equation for Fatigue Crack Growth," *Int. J. Fracture* 24 (1984): pp. R131 – R135.

[22] J. K. Donald, "Automated Fatigue Crack Growth Testing and Analysis Software, Series 2001, Version 3.12, Fracture Technology Associates, Bethlehem, PA.

[23] McFarland, John M., et al. "Probabilistic Modeling of Bulk Residual Stresses," AIAA SciTech 2015, Paper Id AIAA 2015-0385, Kissimmee FL January 2015.

[24] Donald, J.K., Connelly, G. M., Paris, P. C., Tada, H., "Crack Wake Influence Theory and Elastic Crack Closure Measurement," *Fatigue and Fracture Mechanics: 30th Volume, ASTM STP 1360*, P. C. Paris and K. L. Jerina, Eds., American Society for Testing and Materials, West Conshohocken, PA 2000, pp. 185-200.

[25] Broek, D. and Schijve, J., "The Influence of the Mean Stress on the Propagation of Fatigue Cracks in Aluminum Alloy Sheet," NLR-TR M.21ll, Nationaal Lucht-en Ruimtevaart/Aboratorium, January, 1963.

[26] Walker, K., "The Effect of Stress Ratio During Crack Propagation and Fatigue for 2024-T3 and 7075-T6," ASTM STP 462, American Society of Testing and Materials, 1970, pp. 1-14.

[27] J. C. Newman, Jr., "The merging of fatigue and fracture mechanics concepts: a historical perspective," *Progress in Aerospace Sciences*, 34, 1998, pp. 347-390.

[28] McClung, R.C., Popelar, C.F., McFarland, J., Bhamidipati, V., James, M.A., Watton, J.D., Hill, M.R., DeWald, A.T., Ball, D.L., "Predicting Fatigue Crack Growth in Forgings with Bulk Residual Stress," 2014 USAF ASIP Conference, San Antonio TX, Dec. 2014.

[29] M. B. Prime, A. T. DeWald, "The Contour Method," *Practical Residual Stress Measurement Methods*, G. S. Schajer, Ed. John Wiley and Sons, Ltd., 2013, pp. 109-138.

[30] Wu, X-R and Carlsson, A.J., *Weight Functions and Stress Intensity Factor Solutions*, Pergamon Press, Oxford, 1991.

[31] NASGRO Fracture Mechanics and Fatigue Crack Growth Analysis Software, Southwest Research Institute.

Validation of New Experimental and Analytical Tools to Address Residual Stress Effects Accurately in the Design Cycle: Part 2 – Design Analysis Validation

R. Craig McClung[1], Carl F. Popelar[1], John McFarland[1], Vikram Bhamidipati[1], Mark A. James[2], John D. Watton[2], Michael R. Hill[3], Adrian T. DeWald[3], Dale L. Ball[4]

[1]*Southwest Research Institute, USA*
[2]*Alcoa Technical Center, USA*
[3]*Hill Engineering LLC, USA*
[4]*Lockheed Martin Aeronautics Co., USA*

Abstract: Large aluminum forgings offer some manufacturing advantages for aerospace structures, but can also contain complex residual stress (RS). The effects of this RS on component lifetime are also complex, and this complexity has hindered the effective utilization of forgings. However, significant maturation of analytical and experimental tools has now made it possible to characterize and address RS effects accurately in the design cycle. This effort demonstrated and evaluated the current state of the art of these tools. Thirty-two fatigue crack growth (FCG) tests were performed on five different "design feature" (DF) coupon geometries with constant amplitude or spectrum loading. The coupons had all been machined from "logs" of Al 7085 plate that had been heat-treated in a manner to deliberately induce significant levels of bulk residual stress. Probabilistic models of the variability in the RS were derived using Principal Components Analysis from replicate slitting measurements using the original specimen blank geometry and finite element analyses of the DF coupons indicating how bulk RS would be redistributed. FCG life analyses were performed with the NASGRO® software. The analyses of the constant amplitude tests employed a simple non-interaction crack growth model, while the analyses of the spectrum tests employed the NASGRO STRIPY strip-yield (crack closure) load interaction model. Both used tabular representations of FCG properties with RS effects removed. Weight function stress intensity factor solutions were used to address the combined effects of applied and residual stresses. FCG analyses were performed for each test using the mean RS gradient and the ±2 sigma (95%) RS gradients, and another analysis was performed neglecting RS. Nearly all FCG test lifetimes were predicted within ±2x using the mean RS gradient, and most predictions were conservative. Predictions where the fatigue crack was growing primarily in tensile RS fields tended to be relatively insensitive to variability in the RS gradient, while predictions where the fatigue crack was growing primarily in compressive RS fields tended to be very sensitive to variability in the RS gradient. Predictions neglecting RS were nearly always inaccurate, sometimes extremely so.

INTRODUCTION

Large aluminum forgings offer some manufacturing advantages for aerospace structures, but can also contain complex residual stress (RS). The effects of this RS on component lifetime are also complex, and this complexity has hindered the effective utilization of forgings. However, significant maturation of analytical and experimental tools has now made it possible to characterize and address RS effects accurately in the design cycle. This effort demonstrated and evaluated the current state of the art of these tools.

The effort was part of a large research and development program funded by the Air Force Research Laboratory through the Metals Affordability Initiative (MAI) consortium [1]. Selected portions of the overall effort are summarized here in two companion papers. The Part 1 paper [2] provides general background and addresses the material characterization issues, including the material processing used to deliberately induce significant levels of RS in test coupons, the generation of baseline fatigue crack growth (FCG) rate property data, and the use of unique analysis methods to remove confounding RS effects from the resulting FCG properties. This companion Part 2 paper describes the FCG validation test program using design feature (DF) coupons, the analysis methods used to predict the FCG lifetime while addressing RS effects in the DF coupons, including variability in the RS, and the comparisons between experimental and model results. Further background and additional information about other portions of the larger research effort are available elsewhere [3].

EXPERIMENTAL PROCEDURES

All test coupons were excised from a series of "logs" (105 mm x 105 mm x 1000 mm) of commercial aluminum alloy 7085-T7651 plate that had been heat-treated in a manner to deliberately induce significant levels of bulk RS that varied through both the height and width of the logs. The individual coupons were longitudinal slices from the logs taken at different heights from top to bottom. The RS in each coupon was compressive near the edges and tensile near the centerline. The RS also varied systematically as a function of the log height from which the coupon was taken. Further details about the fabrication and heat treatment of the logs and the resulting RS are provided in the Part 1 paper [2].

Residual stresses were destructively measured in selected plain coupons using the slitting method, sometimes referred to as the cut compliance method [4]. The slitting coupons were nominally machined to 203.2 mm long x 101.6 mm wide x 6.35 mm thick blanks, and the slot was introduced by electro-discharge machining (EDM) wire cut. Using back face strain gages, strain release with incremental extension of the wire cut was measured and converted by analysis to a one-dimensional profile of residual stresses acting normal to the slitting plane. The strain data were converted to stress using a series expansion approach with Legendre polynomial fitting method, and polynomial order was selected using a fitting metric to minimize residual error. A total of twenty-three slitting coupons from different heights and lengths within the log were measured.

Five different design feature (DF) coupons were developed to investigate how fatigue cracks growing from representative structural details would interact with complex residual stress gradients. These coupons are shown schematically in Figure 1. The simplest coupon (DF0) was a plain rectangular plate with either an edge through crack growing in a compressive RS field or a center through crack growing in a tensile RS field. Coupons DF1 and DF2 both involved an initial corner crack at the edge of a circular hole (7.87 mm diameter) located either in the center of the coupon (tensile RS field) or near the edge of the coupon (compressive RS field, with the crack on the short ligament side of the hole). Coupons DF3 and DF4 both involved symmetric milled pockets on the front and back of the specimen, with a single centered pair of pockets in DF3 and two pairs of offset pockets in DF4. A single initial surface crack was centered in one of the front milled pockets at the base of the transition radius at the top of the pocket. This crack was in a complex bivariant stress field with predominantly tensile RS in DF3 and mixed RS in DF4. All DF coupons were 101.6 mm wide and 381 mm long. DF0, DF1, and DF2 coupons were 6.35 mm thick, while DF3 and DF4 were 12.7 mm thick. The coupons were classified into four categories (TS, TM, BM, and BS), according to the fixed locations in the rectangular stock (the "logs") from which they were cut. Here T denotes top and B denotes bottom; the M coupons are closer to the middle of the original log, and the S coupons are closer to the top or bottom surface.

FCG tests were performed in a Southwest Research Institute laboratory. Testing was performed using servohydraulic test frames with hydraulic grips and closed-loop feedback control to ensure desired loadings were accurately applied. Cycling frequencies were in the range of 5-10 Hz. Visual crack length measurements on the specimen surface were performed using traveling microscopes with digital scales mounted on mechanical slides. Bore crack length measurements were made using a drill rod positioned in the axis of the hole with a 45° polished surface acting as the mirror. The specimen surface was sometimes lightly polished with diamond paste and sometimes darkened with a black permanent marker along the anticipated crack path to enhance identification/location of the crack tip. The experimental apparatus is shown in Figure 2 for a DF3 coupon test.

Both constant amplitude (CA) and variable amplitude (VA) loading tests were performed. All CA tests were performed at a load ratio (the ratio of the minimum-to-maximum load) of $R = 0.1$. The VA spectrum was based on a tension-dominated design wing bending moment history that had been edited sufficiently to reduce the load sequence to 22059 cycles. The CA and VA loading histories were scaled differently for each DF geometry to ensure practical test lifetimes, and the scale factors were kept constant for all tests with a given DF geometry to permit a direct comparison of RS effects at different locations within the log (TS, TM, BM, BS).

A starter notch was introduced using EDM to facilitate crack nucleation at the desired location. Pre-cracking was performed to nucleate a fatigue crack at the EDM notch prior to beginning the actual FCG testing. Once the crack was initiated, pre-cracking was continued to allow the crack to grow nominally 0.5 mm beyond the EDM notch. Pre-cracking was performed by applying a CA fatigue load with $R = 0.1$. For VA testing, the maximum pre-crack loading was typically less than 65% of the maximum VA load to minimize the potential for pre-crack load history to influence the initial crack growth behavior during VA cycling. For CA testing, pre-cracking was performed at the same load amplitude and load ratio as the regular test conditions. A representative fracture surface for a DF3 coupon is shown in Figure 3, indicating both the EDM notch and the final crack front profile at incipient failure.

ANALYSIS PROCEDURES

Due to the nature of the slitting method, the RS data measured from the slitting coupons did not include RS values at the left and right ends along the width direction. An approximate method based on equilibrium considerations was used to estimate the RS values at the ends of each of the twenty-three coupons. The RS in the slitting coupons was assumed to be constant through the thickness. Although independent contour measurements of RS showed a small variation through the thickness, the slitting data provided no basis for estimating this variation, and the error in ignoring the variation was assumed to be insignificant.

High-resolution three-dimensional finite element (FE) models were generated for each of the DF coupon geometries. These models were used to calculate the applied stress gradients due to the fatigue loading as well as the RS gradients resulting after the design features (holes or milled pockets) were machined from the original specimen blanks. This DF coupon machining process caused significant local changes in the original RS distributions in the coupon blanks.

These changes in RS were calculated by modeling both the original plain plate geometry and the post-machining DF geometry in each FE model using the "partition" feature in the Abaqus FE software and attaching the cut features using the "tied constraint" contact feature. Symmetry conditions were used across the thickness, noting that the RS gradient was not symmetric acros the width of the coupon. The initial RS distribution was input and the model solved for equilibrium. Then the tie constraints were removed and the model was re-solved for equilibrium to determine the RS in the DF coupon. This process was repeated independently for each measured RS profile and each DF geometry.

Representative RS profiles in the original plain coupons and in the corresponding DF coupons are shown in Figures 4, 5, 6, and 7 for DF geometries DF1, DF2, DF3, and DF4. The examples all use a representative RS profile from one of the TS slitting coupons. The before-and-after profiles show the significant changes in the RS profiles, especially at the locations of the initial fatigue cracks. Note also the significant through-thickness variations in RS for the DF3 and DF4 geometries, due to the milled pockets.

The same FE models were also used to calculate stress gradients in the DF coupons due to fatigue cycling. Figure 8 shows the gradients in DF1 and DF2. Figure 9 shows contour plots of the bivariant gradients in DF3 and DF4 (here only the values on a rectangular grid with thickness equal to the ligament between the back-to-back pockets are shown).

The minor through-thickness variations in the design feature RS and fatigue stress were neglected in DF0, DF1, and DF2 for the purposes of FCG analysis, but the significant through-thickness variations in both RS and fatigue stress were included in the FCG analysis of DF3 and DF4.

Probabilistic models to quantify the variability in the RS were developed using Principal Components Analysis (PCA) based on the slitting measurements of the RS in the replicate coupons. Each of these replicate measurements was used to calculate a corresponding RS field in each of the five DF coupon geometries at each of the four different locations in the original logs. Independent PCA models were then constructed for each DF coupon type and log height. Comparisons of the resulting PCA models with a limited number of independent experimental measurements of residual stress in actual DF coupons using the contour method indicated good agreement. More details of this probabilistic analysis process and its results are provided elsewhere [5].

All FCG analysis in this study was performed with the NASGRO computer code, using a special research derivative of version 7.0 (the standard production version when the work was performed) that contained a few special features not released in production versions until v7.1 or v8.0. The CA predictions used the standard non-interaction (NI) mode with a tabular representation of the da/dN versus ΔK relationship at several discrete R values. The VA predictions used the STRIPY strip-yield (SY) crack closure load interaction model based on a tabular representation of the da/dN versus ΔK_{eff} relationship. An alternative formulation of the STRIPY implementation based on a K-analogy theory was used to incorporate the effects of RS on crack closure properly. A few CA tests were also analysed with the STRIPY model to confirm that there were no significant differences between the results from the NI and SY analysis modes.

The material FCG properties in the NASGRO analysis employed the baseline FCG data described in the Part 1 paper [2]. That paper documents a unique method for generating baseline properties that are entirely free of confounding RS effects, even if the baseline test coupons contain significant levels of bulk RS. That approach facilitates generation of da/dN vs. ΔK tables at any desired R value, as well as generation of a da/dN vs. ΔK_{eff} table suitable for use at any R value, both of which were needed for these NASGRO analyses. The baseline data reduction performed in the Part 1 effort was focused in the Paris regime and did not include near-threshold or near-instability data. The tabular property data used in NASGRO were augmented with additional points in the near-threshold regime based on previous Alcoa testing of similar alloys, and with additional points in the near-instability regime based on inspection of the original Alcoa baseline FCG test data, which involved failure of through-cracks in coupons of the same thickness as the DF coupons (similar constraint conditions).

The free (calibration) parameter in the NASGRO STRIPY load interaction model is the constraint factor, α. Alcoa assumed an nominal value of $\alpha = 1.8$ in their data reduction process for the Part 1 effort, which employed a set of equations derived from the similar FASTRAN strip-yield model of crack closure. This same value was used without adjustment in the NASGRO predictions in the current effort, preventing any ability to "calibrate" the predictions to the test results. Therefore, the NASGRO predictions are entirely independent of the test results they are attempting to predict.

The stress intensity factor (SIF) solutions used in the NASGRO analyses were weight function (WF) solutions for corner, surface, or through cracks. This WF formulation facilitates FCG analysis based on the arbitrary stress gradients in the uncracked body along the crack plane. NASGRO permits independent fatigue stress and RS gradients to be used. Most analyses employed SIF solutions based on univariant gradients, since through-thickness RS variations were usually minor. Some analyses of DF3 and DF4, where the stress gradient in the vicinity of the initial crack is bivariant, employed bivariant SIF solutions.

The analyses of RS effects on FCG rates employed simple superposition principles. SIF values were calculated independently from the cyclic stress gradients and RS gradients and then summed. This resulted in no change in the ΔK values for any given cycle but a distinct change in the R value (where $R = K_{min}/K_{max}$) for every cycle. Since FCG rates are functions of R, this resulted directly in a change in predicted FCG rate. The RS gradients were assumed to remain invariant as the test progressed, neglecting any potential relaxation or redistribution due to cycling or crack growth [6]. The applied cyclic loads had been chosen previously to avoid further yielding at stress concentrations due to the superposition of cyclic and residual stresses.

The cross-sections of DF3 and DF4 have variable thickness, which cannot be addressed by any current NASGRO crack models. Therefore, these coupons were approximated by a simple rectangular cross-section corresponding to the minimum thickness (the ligament between the back-to-back milled pockets), using the proper stresses from the actual three-dimensional geometry. The crack remained in the small ligament for most of the life of these coupons.

Several different predictions were generated for each test coupon. One prediction employed the average RS gradient for that particular log level, while another prediction neglected RS altogether. In addition, a first-order estimate of the effect of RS variability on life was obtained by performing life predictions with approximate mean ±2 sigma (standard deviation) RS profiles. For each case, the ±2 sigma RS profiles were obtained by taking the mean profile plus and minus two standard deviations in terms of the first mode shape only. The ±2 sigma level corresponds to a probability of 95%, under the assumption that the principal component scores follow a normal distribution. Since the first mode shape was found to capture most of the RS variation in every case [5], these curves are representative of overall variation.

RESULTS: COMPARISONS OF EXPERIMENTS AND ANALYTICAL PREDICTIONS

The FCG test matrix comprised thirty-two unique combinations of five DF types (actually six, counting edge crack and center crack variants of DF0), four log levels, and two loading history types. There were no replicate tests, and some possible combinations were not tested. Each individual result and its corresponding prediction were interesting and instructive, although occasionally vague due to the possible effects of test scatter. However, it is not practical to review all of the individual results in detail. A few selected tests of particular significance will be highlighted, and then all results will be presented in summary form.

Constant Amplitude Tests

The basic effects of RS and RS scatter on FCG behavior are seen most clearly in the simple CA, through-crack tests (DF0 edge and center cracks). Figure 10 compares the test results with various predictions for a DF0 center crack coupon from the TS position. Here, as in the other figures that follow, the experimental measurements are denoted with symbols, and the predictions are denoted with lines. The solid line is the prediction using mean values of the RS profile. The dashed line in the same color is the prediction neglecting RS altogether. The thin dot-dashed black lines are the predictions using ±2 sigma RS profiles. Since no RS measurements were performed on the actual FCG test coupons, it is not known where each FCG coupon would fall in the general scatter band of RS profiles.

For this particular specimen, where the crack was growing in a tensile RS field, the mean value RS predictions agreed very well with the measurements, while predictions neglecting RS altogether were very non-conservative. The ±2 sigma RS predictions showed very little deviation from the mean value prediction (only ±20% in life), so in this case the observed RS variability from coupon to coupon had little impact on the life prediction.

Figure 11 shows results for a DF0 edge crack coupon (also from the TS position), where the fatigue crack was growing in a compressive RS field. The most striking result here was the dramatic effect of RS variability on the predicted crack growth rate (even though the nominal RS variability was exactly the same as in the previous DF0 TS center crack test). Here the ±2 sigma variability in RS led to a 6x variation in predicted life. The 2x difference between the test result and the mean value RS prediction could have been due to the variation in the actual RS in this particular coupon. However, note that the prediction neglecting RS altogether gave a predicted life much too conservative.

Representative results for a crack at a hole in a highly tensile RS field (DF1) are shown in Figure 12. This graph is more complicated because measured and predicted results are shown both for the initial corner crack, tracking both tips (with smaller symbols and thinner lines), and the eventual through crack (larger symbols and thicker lines). The results are simple, however: the mean value RS prediction did not agree well with the test results, mostly due to significant errors in the corner crack predictions. The effect of RS variability on life was very small, however, and thus not likely the source of the poor prediction. The observed behavior was actually closer to the predictions that neglected RS altogether.

The predictions for all DF1 CA tests were consistently poor, generally worse than any other combination of test conditions (including much worse than the DF1 VA tests, as will be seen later). The reasons for this disagreement are not known, but the DF1 CA tests did exhibit one unique characteristic. The combination of the fatigue stress and the RS at the edge of the hole was extremely high—approaching the yield stress—and the specimen experienced a very large number of cycles, including a long pre-cracking phase (in this case, 80,000 cycles were required to initiate the growing fatigue crack). It is known that residual stresses can significantly relax when cycled mechanically [6], especially when the cyclic stresses are large and total stresses approach yield, and so it is possible that the actual RS at the edge of the hole at the beginning of the crack growth was significantly lower than expected. This would be consistent with the longer lifetimes observed. Further investigations would be required to confirm or deny this hypothesis.

Representative results with a hole in a compressive RS field (DF2) are shown in Figure 13. Here the final life prediction was conservative by about 30%, but the effects of ±2 sigma RS variability are about 2x in life. The mean RS predictions for the early corner crack behavior were quite good. The predictions neglecting RS altogether were excessively conservative. In this case, the combined fatigue + residual stresses at the edge of the hole are relatively low, since the fatigue stresses are tensile and the RS is compressive, so residual stress relaxation would be much less likely.

Representative CA results for the DF3 coupon with a front-and-back pair of centered milled pockets are shown in Figure 14. In this geometry, the initial surface crack is growing in a peak tensile RS field with lower magnitude (and lower peak fatigue stress) than DF1. The FCG rates were predicted accurately early in the life, and then the predictions gradually become conservative as the surface crack grows through the ligament and transitions to a through crack. The NASGRO model makes no attempt to capture the crack growth retardation exhibited very late in the test as the through crack grows out of the ligament and into the full coupon thickness before final fracture. The results were qualitatively similar for two CA tests on the DF4 coupon (see Figure 15), with two pairs of offset pockets. In this case the initial surface crack was growing in a region of mildly tensile RS.

All of the CA predictions of total FCG life based on the mean value RS profile are compared with the corresponding test data in Figure 16. Note that for DF3 and DF4, the total life for this comparison was taken to be the number of cycles until the through crack reached the pronounced change in section thickness, since the model made no attempt to address the complex retardation phenomenon that occurred at this point. There was a consistently conservative bias to all of the predictions, but over 80% of the tests were predicted within 2x, and many were predicted much more closely.

Figure 17 shows the same data but adds the predictions based on the ±2 sigma RS profiles ("+" symbols) and the prediction neglecting RS altogether (open symbols). Note that RS variability consistently has little effect on life for the coupons with cracks in tensile RS fields (DF1, DF3, DF4, and DF0 center crack), but a pronounced effect on life for the coupons with cracks in compressive RS fields (DF2 and DF0 edge crack). The two DF0 predictions that lie outside the ±2x scatter bands are both edge crack coupons, and so the apparent conservatism in those two predictions may actually be due to RS scatter. The predictions neglecting RS altogether are almost all either nonconservative or excessively conservative.

Variable Amplitude Tests

The comparisons of predictions and test results for the DF0 coupons under VA (spectrum) loading (not shown here) looked very much like the DF0 CA comparisons, with one intriguing difference. The DF0 center crack results (crack in tensile RS field) showed excellent agreement between test and prediction, and RS variability had little effect on the predicted FCG rates.

The DF0 edge crack results (crack in compressive RS field) showed a very large impact of RS scatter on predicted lifetime (10x). The individual predictions using mean RS profiles for the two VA edge crack DF0 coupons were 2x nonconservative in one case and 3x conservative in the other case (although both results fell inside the bounds arising from general RS variability). However, these two coupons were extracted from adjacent locations (BS and BM) in the same log, so it seems unlikely that the RS profiles would be extreme variants in opposite directions. Another theoretical possibility (not confirmable) is that the two coupons were inadvertently switched/mislabelled at manufacture; alternative predictions of the BS test using the BM mean RS profile and the BM test using the BS mean RS profile were both highly accurate. Further work would be needed to investigate the actual RS profiles in these two coupons.

The VA predictions for all of the other DF geometries using mean RS profiles generally agreed very well with the test results. Representative comparisons for individual tests are shown in Figures 18, 19, 20, and 21 for DF1, DF2, DF3, and DF4, respectively. As for the CA results, the coupons with cracks in tensile RS fields (DF1, DF3, DF4) exhibited little impact of RS variability on predicted life, while the DF2 coupons with cracks in compressive RS fields exhibited a large impact of RS variability on life. Predictions neglecting RS altogether always exhibited extreme errors that were non-conservative for tensile RS and conservative for compressive RS.

It was previously speculated that the strong conservatism in the DF1 CA predictions might be due to cyclic relaxation of the residual stresses. This possible explanation is not inconsistent with the improved accuracy of the DF1 VA predictions, because the VA loading involved a much smaller total number of large stress cycles (the cycles that might lead to relaxation) and a much larger total number of small stress cycles (which would be inconsequential for relaxation).

These trends are further documented in Figures 22 and 23, which summarize total FCG life predictions and test results for all VA coupons. Figure 22 shows only the mean RS predictions, while Figure 23 also shows all of the ±2 sigma RS predictions and the predictions neglecting RS altogether. Here the pink symbols denote the DF0 edge crack predictions with the hypothetical swap of the RS profiles mentioned earlier. Note that all of the VA tests are predicted within ±2x (with the possible exception of one problematic DF0 edge crack test) using the mean RS profiles, with a mild conservative bias, and most of these predictions were much closer to the test results than 2x.

DISCUSSION

The most significant trends emerging from this study have already been noted repetitively in the presentation of the results, but it is useful to review them here and provide additional comments.

Predictions of nearly all thirty-two CA and VA tests based on mean RS profiles agreed with the test results within ±2x, with some conservative bias. This result indicates that the current state of the art to address RS effects by removing confounding RS effects from baseline data (the Part 1 paper [2]) and re-introducing RS effects explicitly in the assessment of design features (this paper) appears to be robust.

It is even more heartening to note that different investigators, using similar but independent methods, have demonstrated very similar success. In particular, another principal investigator [3] performed and previously published parallel predictions in the same research program [1] in the same time frame using the same FCG test data. The parallel effort employed the same baseline FCG data described in the Part 1 paper, but performed completely independent calculations of the mean RS in the DF coupons, and performed completely independent calculations of the FCG life for this mean RS using independent weight function SIF solutions, an independent load interaction model for the VA spectrum loading, and independent (proprietary) FCG analysis software. While predictions of individual tests sometimes differed from the results documented in this paper, the overall trends of the results (published previously [3] and reprinted for convenience in the Part 1 paper [2]) were strikingly similar: Most predictions fell within ±2x of the test results, those few predictions outside these scatter-bands were conservative, and CA predictions exhibited a conservative bias.

That previous parallel effort did not address the effects of RS variability on predicted life. The current effort showed repeatedly that FCG life predictions for configurations with fatigue cracks in tensile RS fields exhibited very little sensitivity to RS, while life predictions for configurations with fatigue cracks in compressive RS fields exhibited strong sensitivity to RS. These results are not surprising, based on simple arguments about the effects of stress ratio on FCG rates. When cyclic stresses are tension-dominated and RS is also tensile, the resulting stress ratios will generally be strongly positive (especially for the spectrum loading used in this study). Variations in RS will result in some variations in *R*, but FCG rates are usually only a weak function of *R* for high *R* values (in the absence of time-dependent effects such as elevated temperature). From a crack closure standpoint, little closure occurs at high positive *R* values, and small changes in negligible closure will have little impact on FCG rates. On the other hand, when local RS are predominantly compressive, the resulting stress ratios from cyclic + residual stress superposition will tend to be low, perhaps even negative. FCG rates are generally a much stronger function of *R* for negative and weakly positive *R* values, where crack closure effects are much more pronounced.

These effects of RS variability will be especially important in the real world of engineering design and fleet management. In the current study, a large number of replicate coupons were manufactured under closely controlled conditions, and a large number of replicate RS measurements were performed and analysed to generate a rigorous characterization of RS variability. In the real world, a large number of replicate components will be manufactured over a long period of time using production methods that may exhibit some variability. Furthermore, relatively few destructive measurements of RS variability will be performed. At most fatigue-critical locations, there may be no available experimental RS data. Predictive models of bulk residual stress do exist [7], but these are less precise and less mature at the present time, especially for complex production forgings. Therefore, some ambiguity may remain about the exact RS profiles in actual components.

The present study has indicated that if fatigue cracks are forming at locations of tensile RS, the resulting life predictions will not be particularly sensitive to the exact magnitude of the RS, as long as the RS profile used in the calculation is nominally correct. Ignoring the RS altogether in this case will almost always give a non-conservative result. The present study has also indicated that if fatigue cracks are forming at locations of compressive RS, the uncertainty in RS can have a dramatic effect on predicted FCG life. This is, of course, a result of some concern. However, remembering that tensile RS profiles are generally detrimental (resulting in decreased FCG life) and compressive RS profiles are generally beneficial (resulting in increased FCG life), the situation as a whole is encouraging. In short, it appears that when detrimental tensile RS is present, it is not so important to know the exact values of the RS, as long as the RS trends are captured. When RS variability has a large impact on FCG life, because RS is predominantly compressive at a given location, the total FCG lifetimes are likely to be longer and (hopefully) will not have a significant impact on overall structural reliability.

Of course, it must be immediately noted that the trends observed in this particular study should not be overgeneralized. The numbers will be different in different applications. If the fatigue spectrum is fully-reversed or compression-dominated, then variability in tensile RS will likely have a larger impact on variability in FCG life. However, it should also be noted that the test coupons in the present study were deliberately manipulated to produce significant (and unusually high) RS values for the purposes of the study. It should not be construed that the RS levels in this study are representative of bulk RS in actual production components.

However, the most encouraging result of this study is that the current tools for addressing RS effects appear to work relatively well. While further validation is undoubtedly needed (and further validation work is in fact underway), the tools appear able to characterize the general effects of complex bulk RS gradients on FCG behavior in a variety of representative design features to an acceptable level of accuracy.

CONCLUSIONS

1. Engineering models for FCG rates and lifetimes were successfully applied to predict the results of thirty-two tests using six different design feature coupon geometries and several variations in complex bulk residual stress gradients, including fatigue cracks in both tensile and compressive residual stress fields. Nearly all FCG test lifetimes were predicted, without calibration, within ±2x using the mean RS gradient, and most predictions were conservative.
2. Predictions where the fatigue crack was growing primarily in tensile RS fields tended to be relatively insensitive to variability in the RS gradient, while predictions where the fatigue crack was growing primarily in compressive RS fields tended to be very sensitive to variability in the RS gradient. Predictions neglecting RS were nearly always inaccurate, sometimes extremely so.
3. This study has indicated that the necessary experimental and analytical tools are maturing adequately to support a new paradigm for addressing bulk residual stresses in the design of aerospace structure. The central concept of this new paradigm is that any residual stress effects in the baseline test coupons must first be removed from the baseline material properties, and then the effects of the residual stresses in the actual structure must be reintroduced in the design analysis so that they can be explicitly addressed when determining fatigue lifetimes.

ACKNOWLEDGEMENTS

This work was funded by the Air Force Research Laboratory (AFRL) Materials Affordability Initiative (MAI) Consortium, Technology Investment Agreement FA8650-06-2-5211. The specific project was "Incorporation of Residual Stress Modeling into Next Generation Design, Manufacturing and Sustainment for Advanced Airframe Structures (BA-11)". Other BA-11 team members include Boeing (James Castle), Northrop Grumman (Miles Honkawa) and Rolls Royce (Michael Glavicic). The authors would like to express their gratitude to the BA-11 program manager, Dr. James B. Castle, and the Air Force Research Laboratory program technical monitor, Dr. Pamela Kobryn, for their support. Distribution A: Approved for public release; distribution unlimited (Case Number 88ABW-2015-1235).

REFERENCES

[1] Dutton, R., Kobryn, P., Ball, D., Castle, J., James, M., and Yavari, P. (2015). "From Integrated Computational Materials Engineering to Integrated Computational Structural Engineering," In: Proceedings of the 3rd World Congress on Integrated Computational Materials Engineering (ICME 2015), Colorado Springs, Colorado, USA.

[2] James, M., Ball, D., Millard, D.,.0 Newborn, M., Dalton, J. and Bucci, R. (2015). "Validation of New Experimental and Analytical Tools to Address Residual Stress Effects Accurately in the Design Cycle: Part 1 – Material Characterization," In: Proceedings of the 28th Symposium ICAF Symposium, Helsinki, Finland.

[3] Ball, D. L., James, M. A., Bucci, R. J., Watton, J. D., DeWald, A. T., Hill, M. R., Popelar, C. F., and McClung, R. C., (2014). "A Detailed Evaluation of the Effects of Bulk Residual Stress on Fatigue in Aluminum," In: *Fatigue 2014, Proceedings 11th International Fatigue Congress*, Vols. 891-892, Advanced Materials Research, Melbourne, Australia, pp. 1205-1211.

[4] Prime, M. B. (1999). "Residual Stress Measurement by Successive Extension of a Slot: The Crack Compliance Method," *Appl. Mech. Rev.*, Vol. 52, No. 2, pp 75-96.

[5] McFarland, J. M., Bhamidipati, V., McClung, R. C., James, M. A., Watton, J. D., Hill, M. R., and Ball, D.L. (2015). "Probabilistic Modeling of Bulk Residual Stresses," Paper AIAA 2015-0385, 56th AIAA/ASCE/AHS/ASC Structures, Structural Dynamics, and Materials Conference, Kissimmee, Florida.

[6] McClung, R. C. (2007). "A Literature Survey on the Stability and Significance of Residual Stresses During Fatigue," *Fatigue Fract. Eng Mater. Struct.*, Vol. 30, No. 3, pp. 173-205.

[7] Watton, J. D. (2010). "Computational Modeling and Optimization of Bulk Residual Stress in Monolithic Aluminum Die Forgings," In: 4th Residual Stress Summit, Society for Experimental Mechanics, Bethel, Connecticut, USA, pp. 228-258.

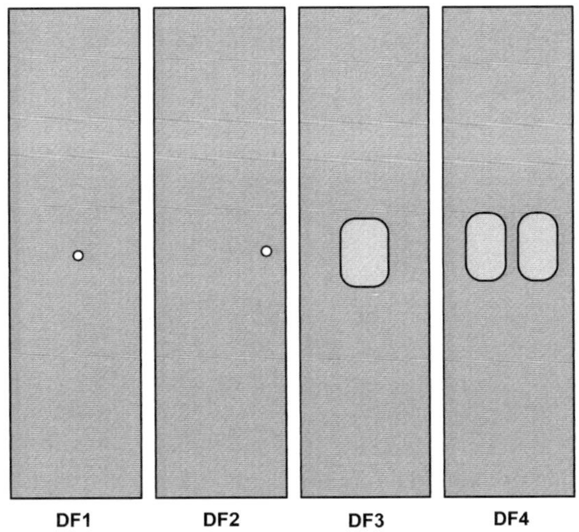

Figure 1. Schematic drawings of design feature coupons.

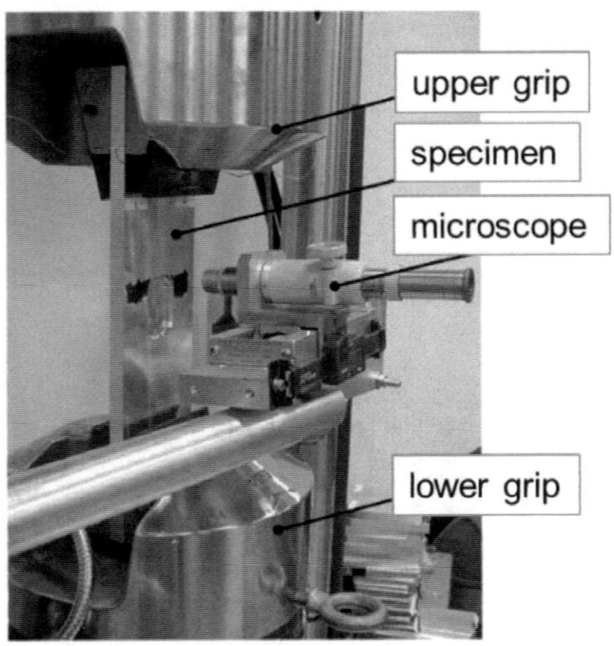

Figure 2. Experimental apparatus showing DF3 specimen and crack measurement system.

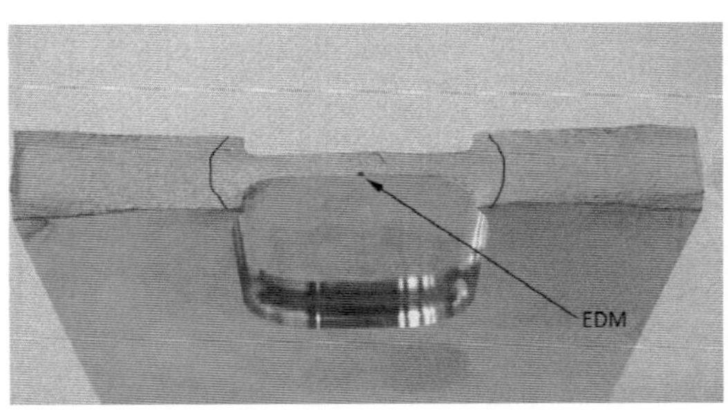

Figure 3. Representative fracture surface for selected DF3 test coupon.

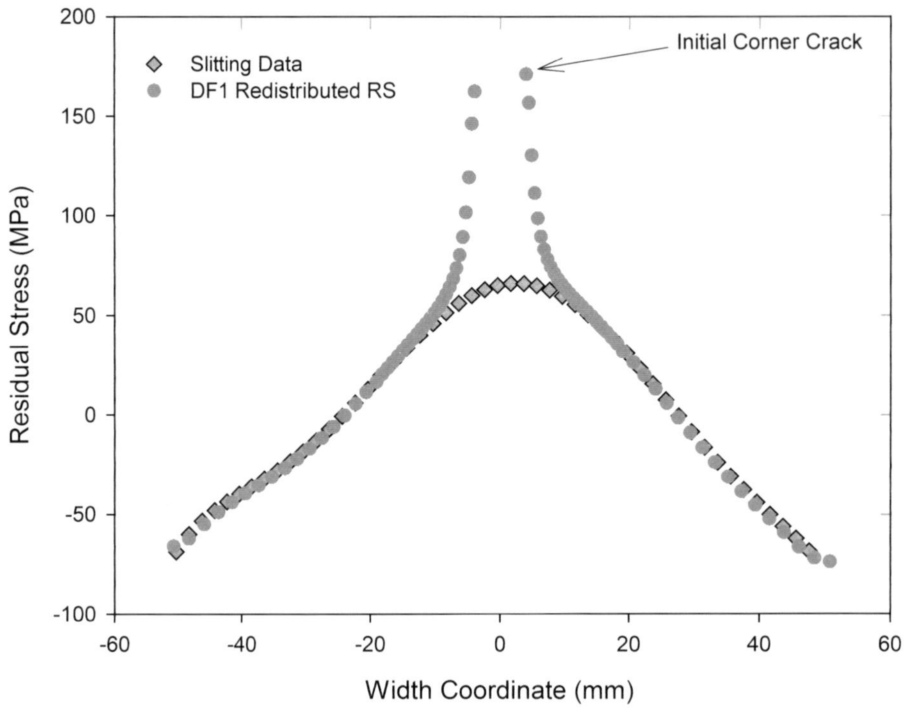

Figure 4. Calculated residual stresses in a representative DF1 coupon.

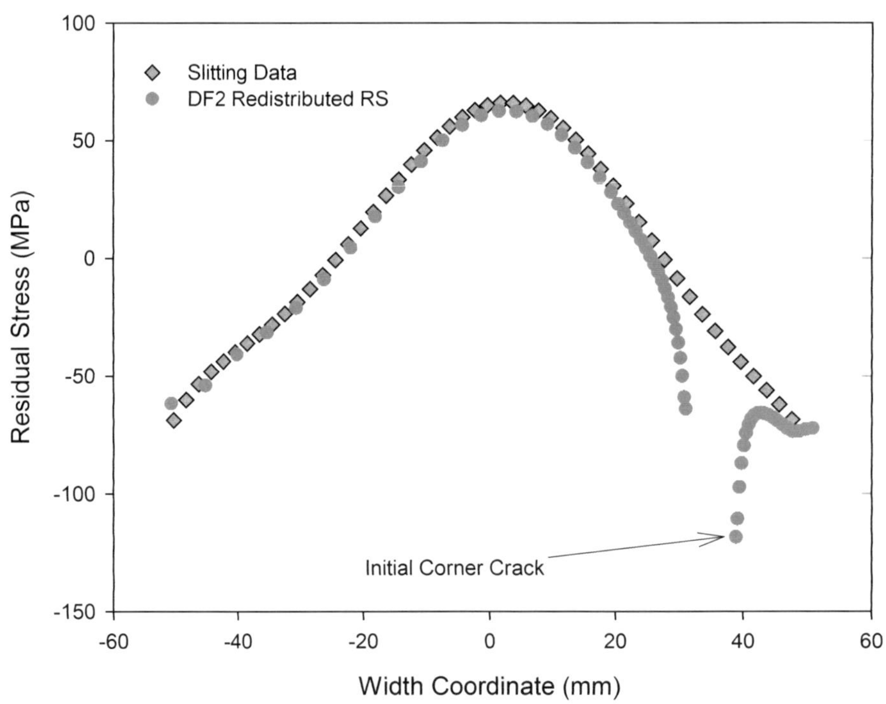

Figure 5. Calculated residual stresses in representative DF2 coupon.

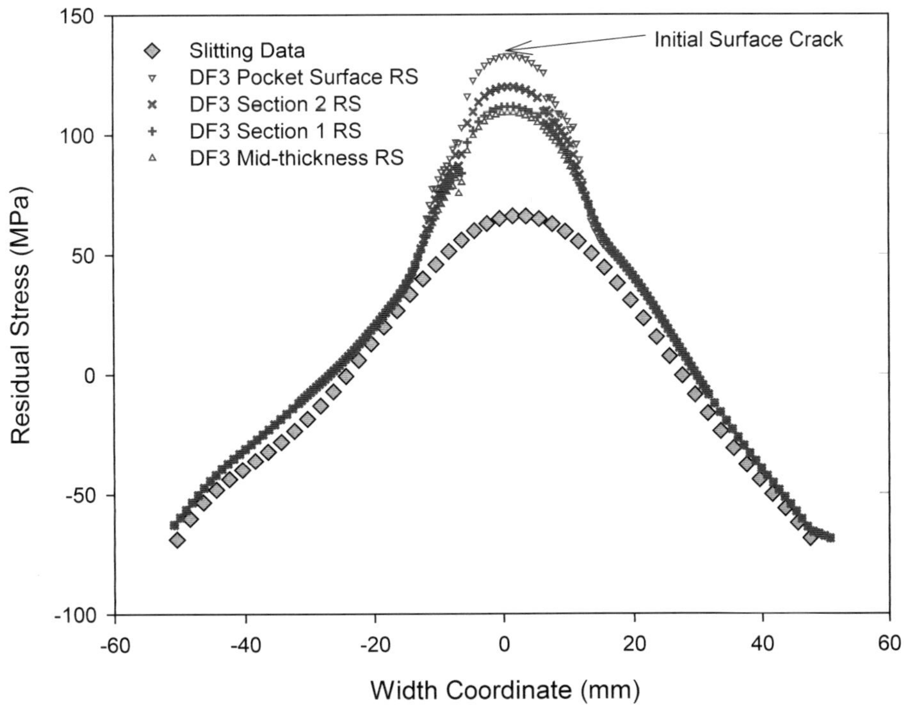

Figure 6. Calculated residual stresses in representative DF3 coupon.

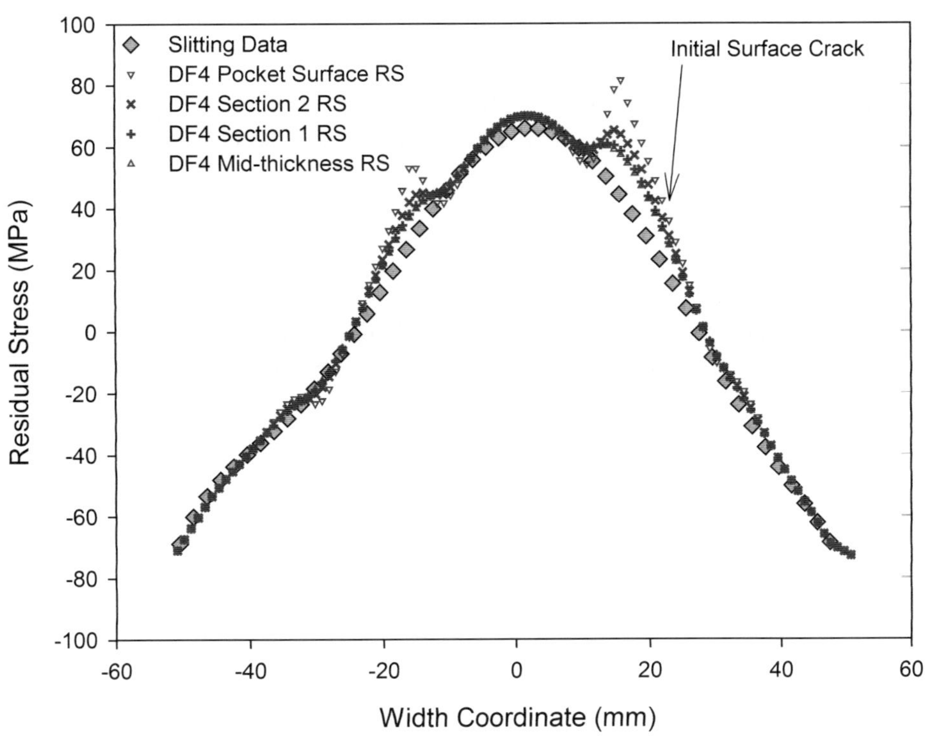

Figure 7. Calculated residual stresses in representative DF4 coupon.

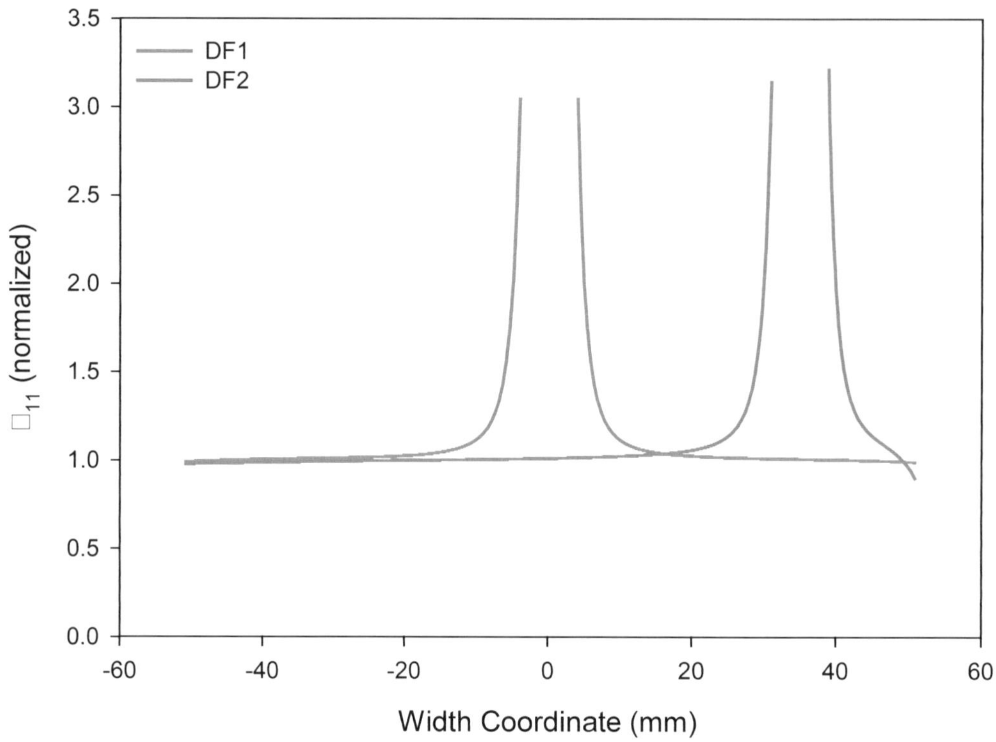

Figure 8. Univariant (thickness-averaged) normalized fatigue stress gradients in DF1 and DF2 coupons.

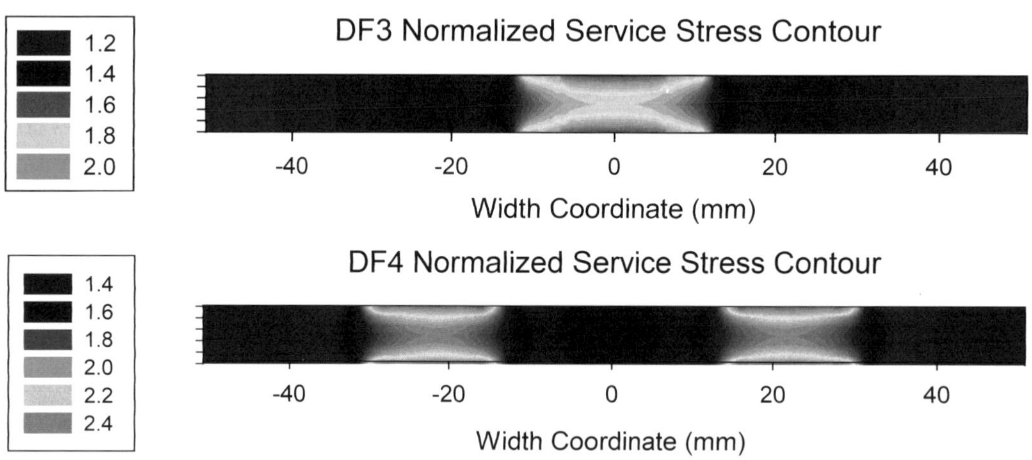

Figure 9. Bivariant normalized fatigue stress contour plots for DF3 and DF4 coupons.

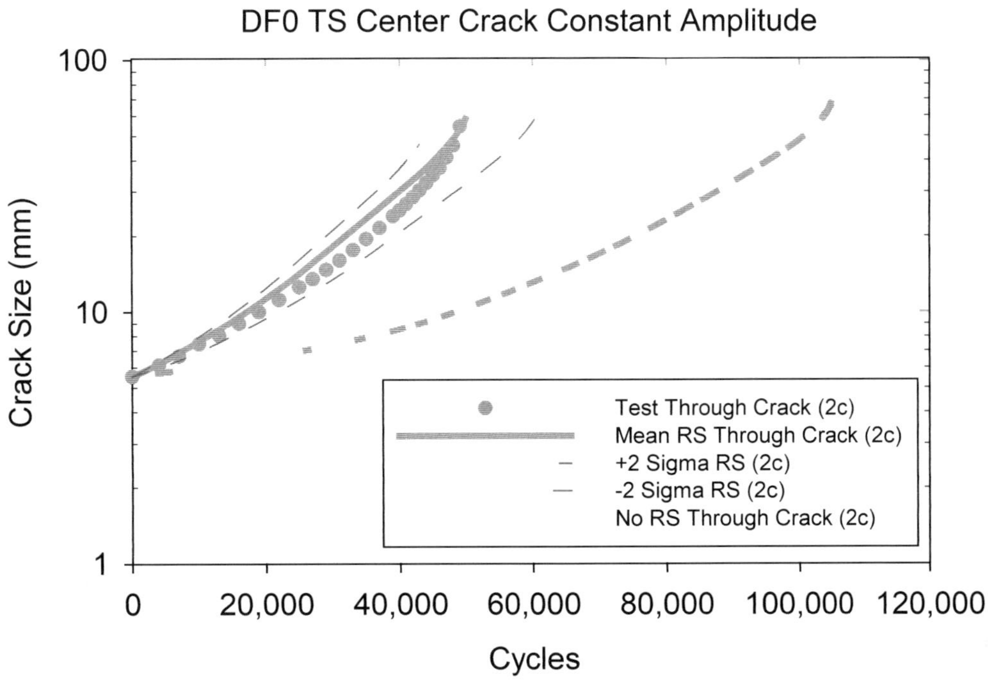

Figure 10. Test Data and Predictions for Representative DF0 Center Crack Coupon Under CA Loading.

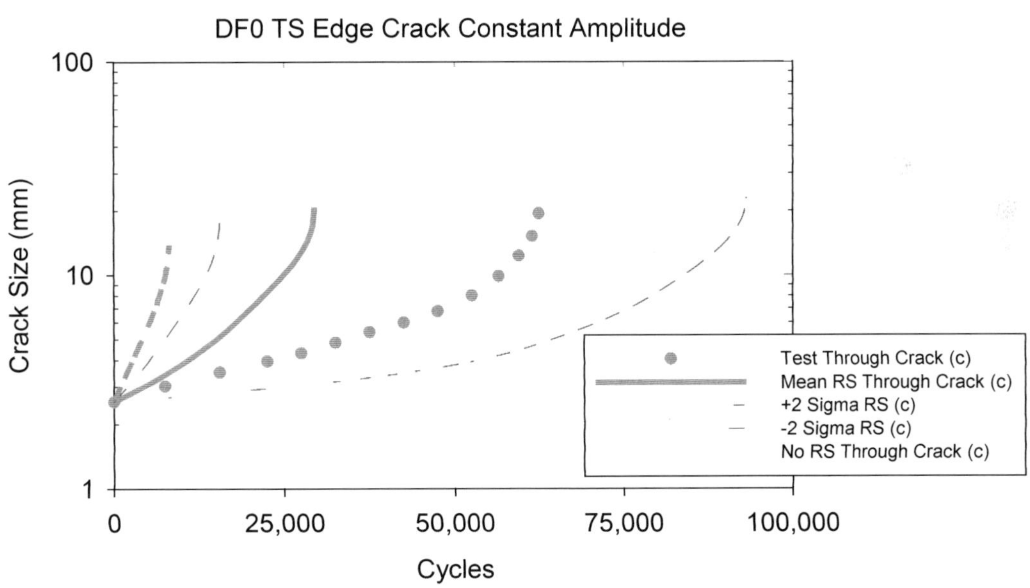

Figure 11. Test Data and Predictions for Representative DF0 Edge Crack Coupon Under CA Loading.

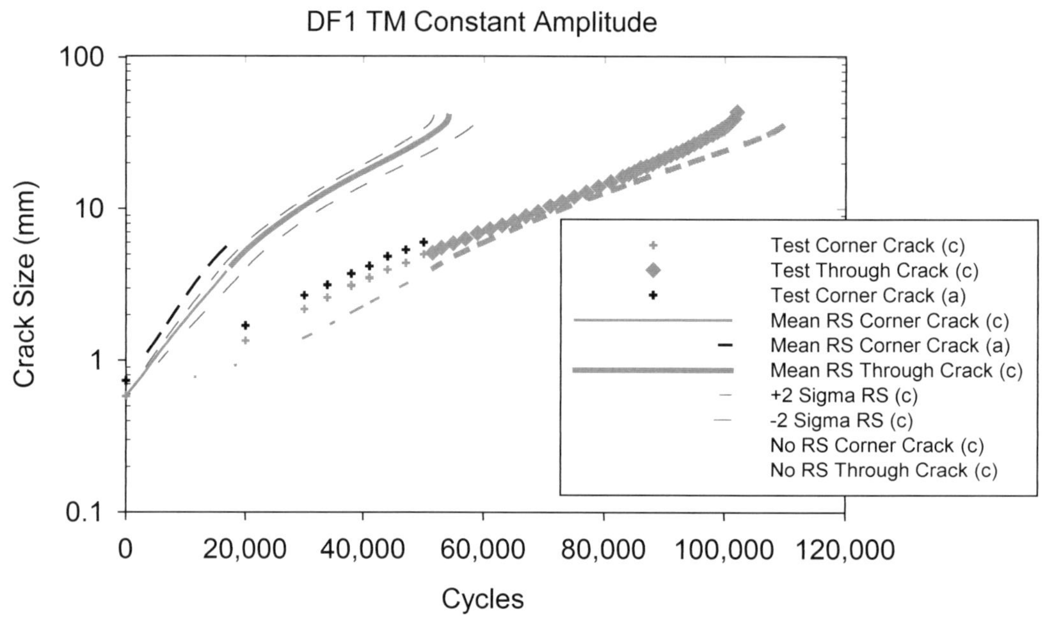

Figure 12. Test Data and Predictions for Representative DF1 Coupon Under CA Loading.

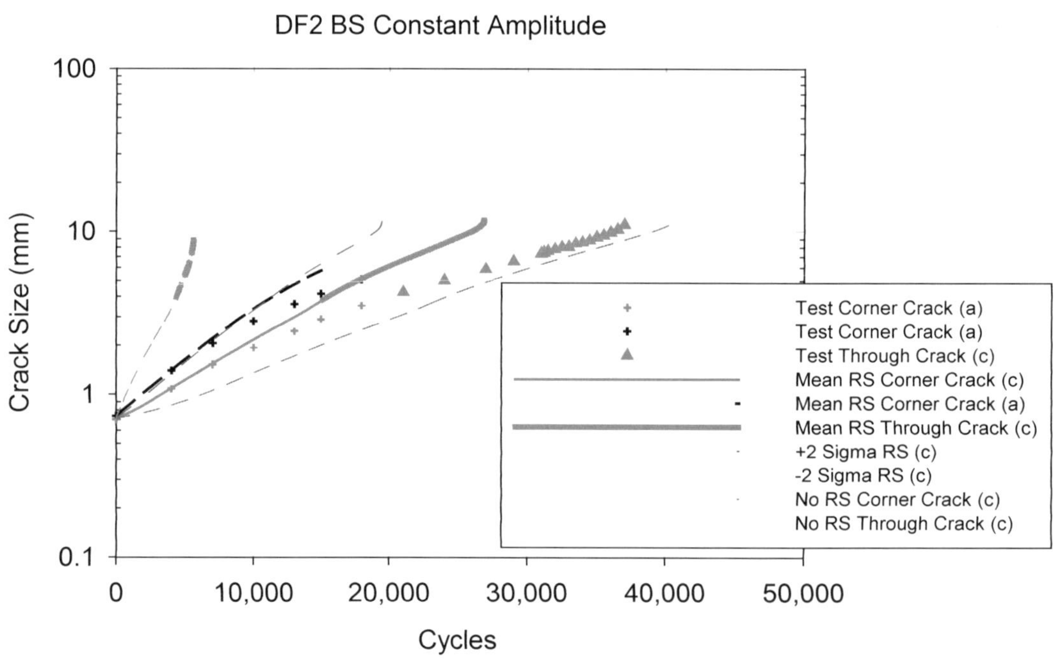

Figure 13. Test Data and Predictions for Representative DF2 Coupon Under CA Loading.

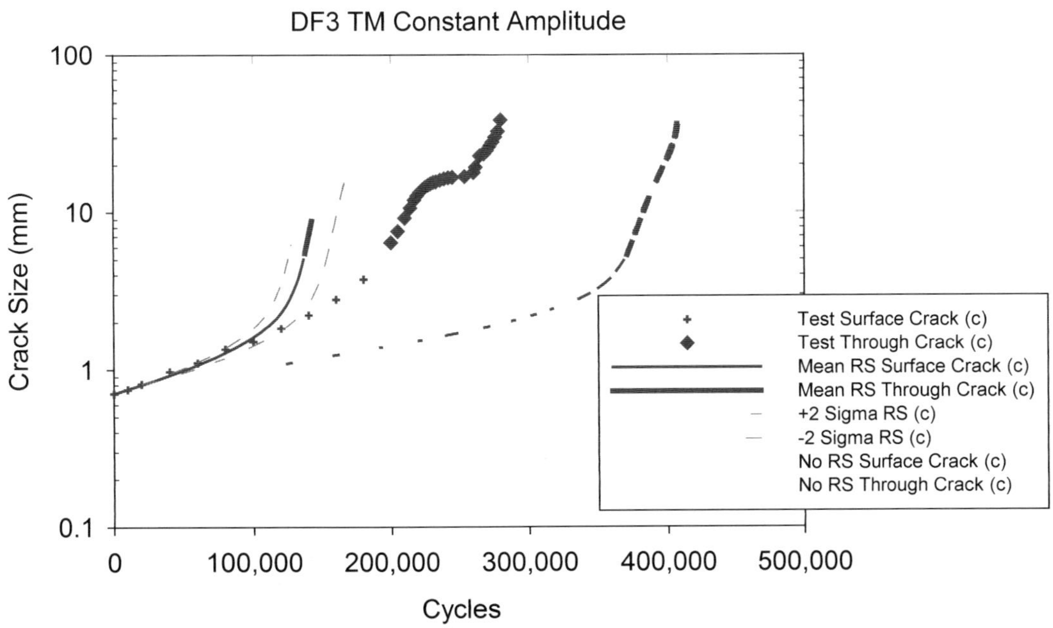

Figure 14. Test Data and Predictions for Representative DF3 Coupon Under CA Loading.

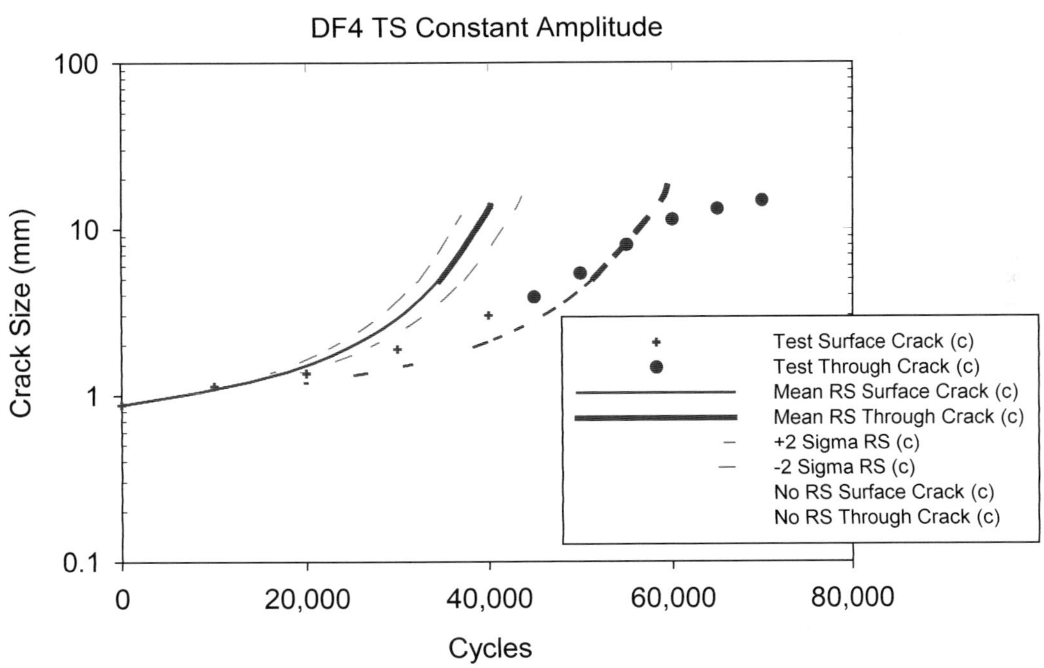

Figure 15. Test Data and Predictions for Representative DF4 Coupon Under CA Loading.

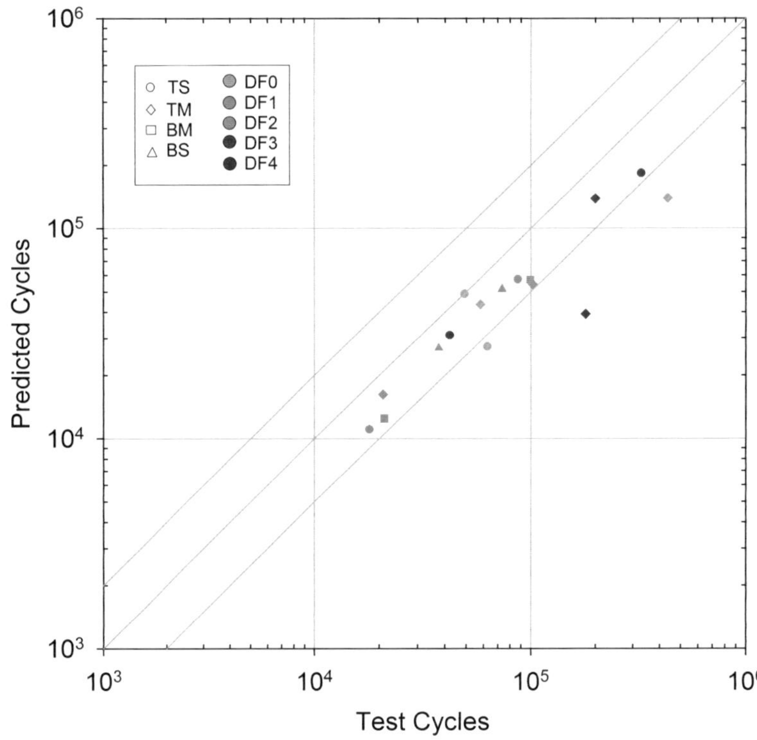

Figure 16. Comparison of Predicted (Mean Residual Stress) and Test Cycles for all CA Tests.

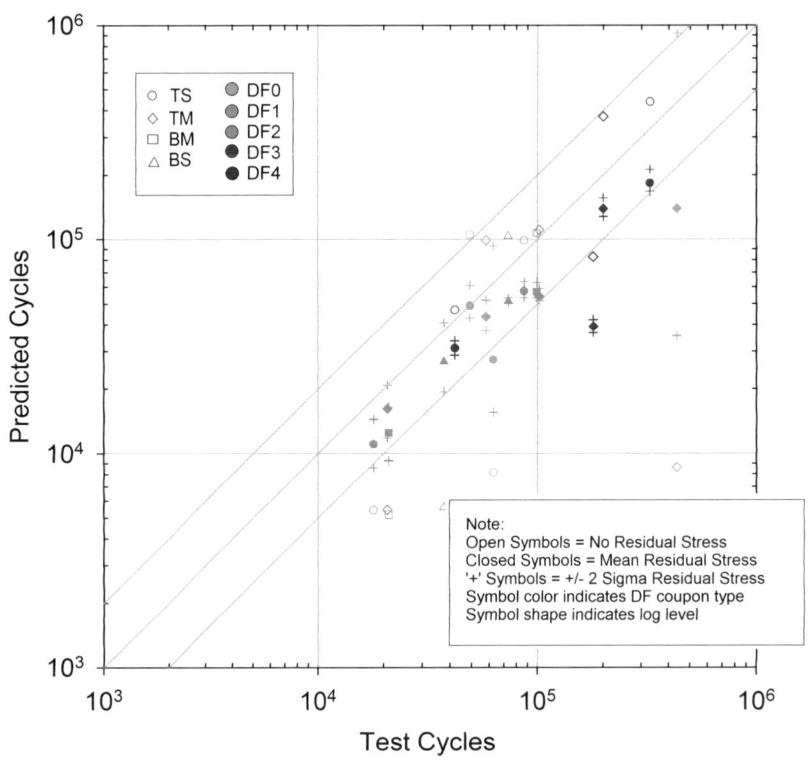

Figure 17. Comparison of Predicted (Considering RS Variability) and Test Cycles for all CA Tests.

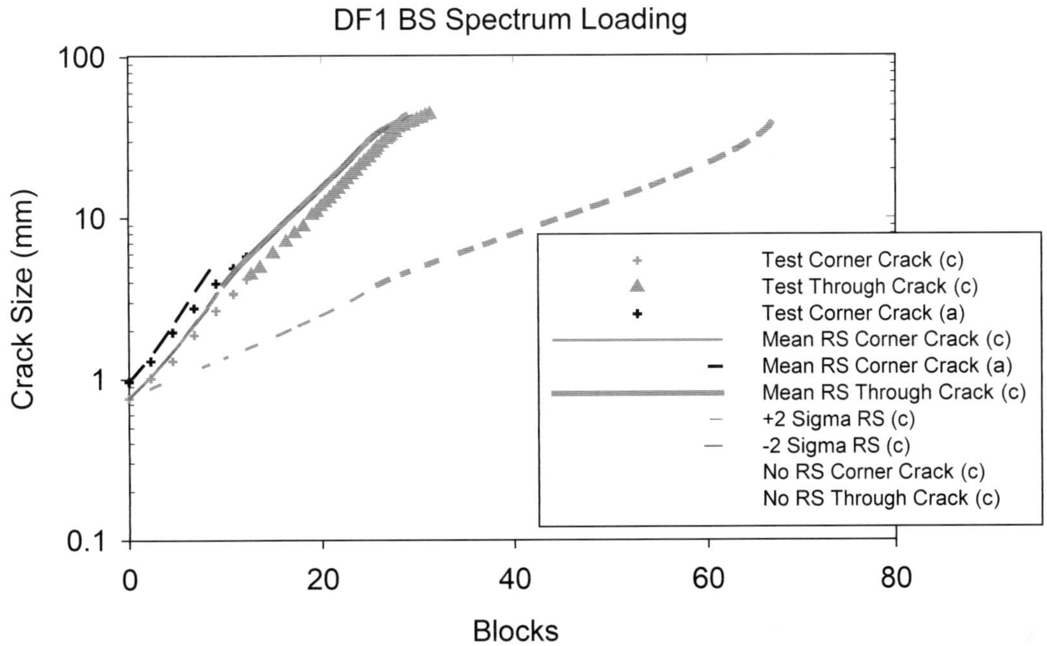

Figure 18. Test Data and Predictions for Representative DF1 Coupon Under VA Loading.

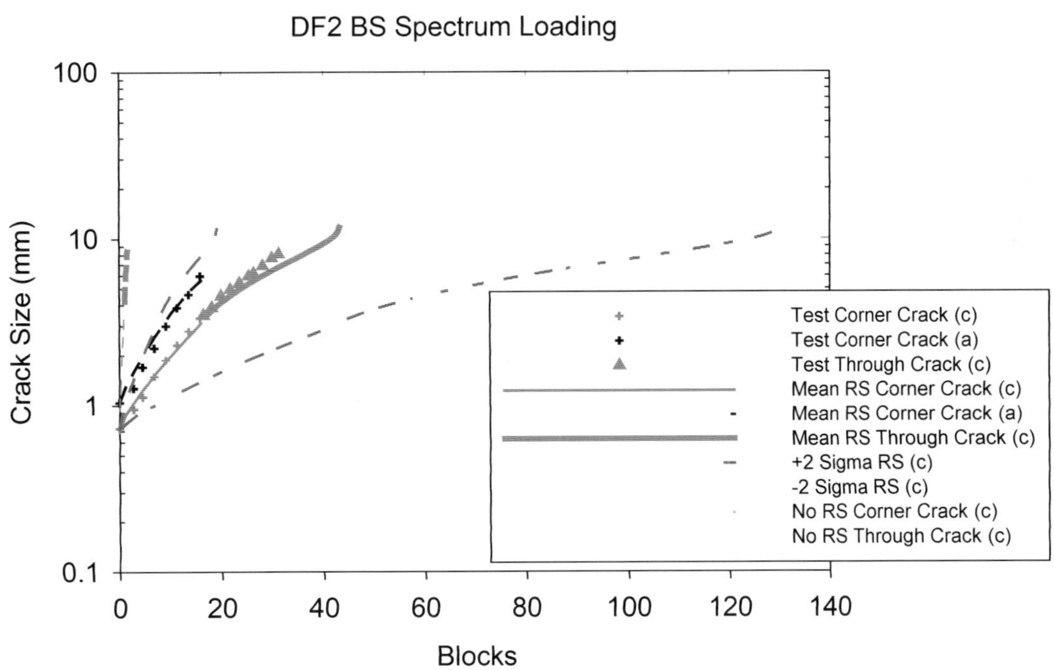

Figure 19. Test Data and Predictions for Representative DF2 Coupon Under VA Loading.

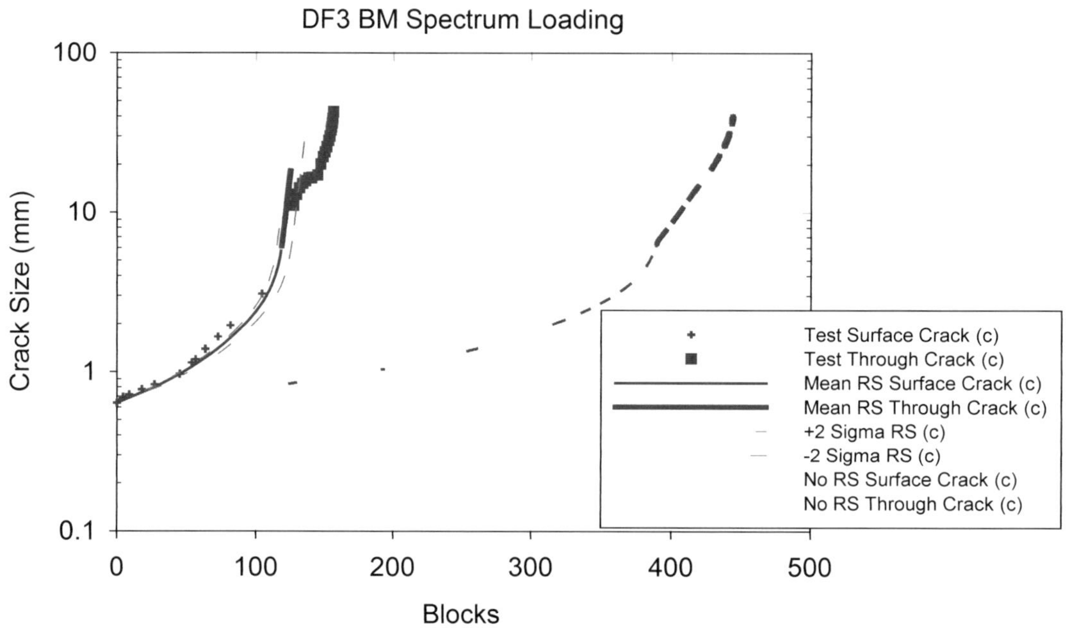

Figure 20. Test Data and Predictions for Representative DF3 Coupon Under VA Loading.

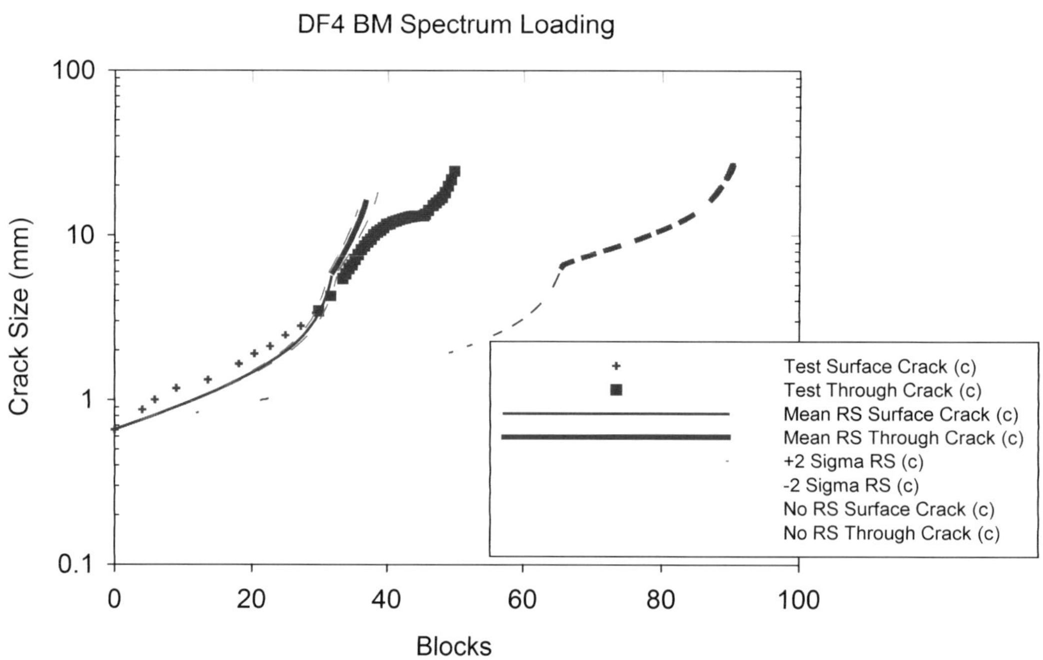

Figure 21. Test Data and Predictions for Representative DF4 Coupon Under VA Loading.

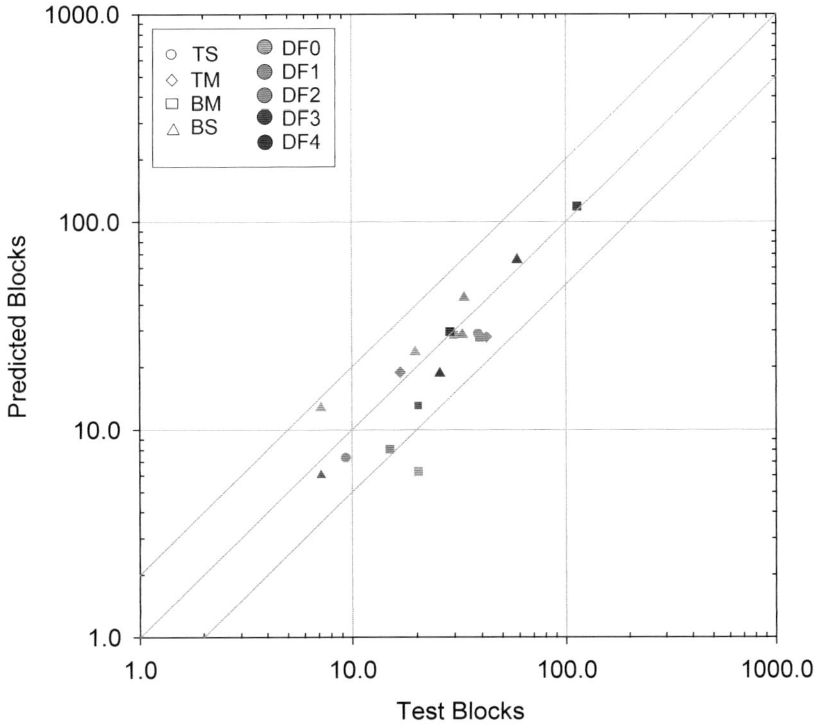

Figure 22. Comparison of Predicted (Mean Residual Stress) and Test Cycles for all VA Tests.

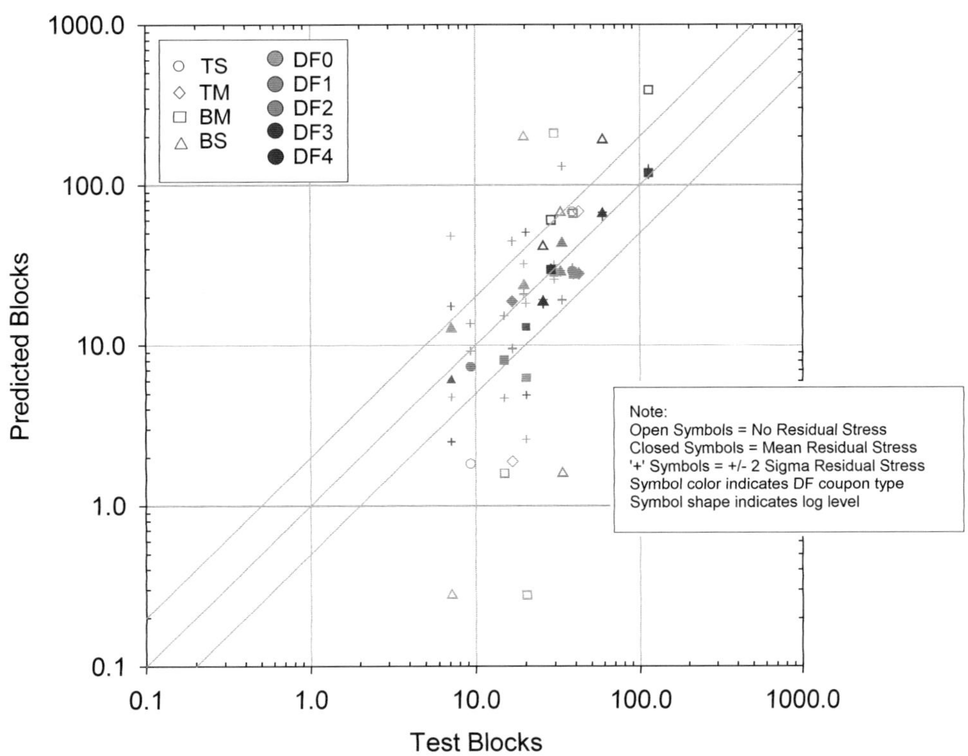

Figure 23. Comparison of Predicted (Considering RS Variability) and Test Cycles for all VA Tests.

Effect of LSP Generated Residual Stresses on Fatigue Crack Propagation in Thin Panels

Enrico Troiani[1], Sara Taddia[1], Ivan Meneghin[2], Gianluca Molinari[2], Claudia Polese[3]

[1]*University of Bologna, Forlì, Italy*
[2]*Sii Deutschland GmbH, Germany*
[3]*University of the Witwatersrand, Johannesburg, South Africa*

Abstract: This work deals with the prediction of the residual stress field after Laser Shock Peening process on aluminum coupons by means of Finite Element (FE) modelling. The commercial software Abaqus Explicit has been used. The work includes investigation on mesh sensitivity, tuning and optimization of a simplified model by using physical (Kinematic Hardening material model) parameters. The models have been validated by comparison with experimental results on simple specimen with multiple peening. Moreover, the effect of the LSP treatment on thin aluminum panels has been evaluated. Starting from the numerical assessment of the self-balancing residual stress distribution along the entire panel width, the fatigue crack growth through the panels has been analytically estimated and compared with experimental results. The comparison highlights the sensitivity of the fatigue crack propagation life to the selected LSP pattern configuration which has to be accurately setup in order to exploit the full potentiality of the LSP process in increasing the fatigue life and avoid undesired reduction of the component performances.

INTRODUCTION

Laser Shock Peening

Laser Shock Peening (LSP) is a technological process used to improve mechanical properties in metallic components. When a short and intense laser pulse irradiates a metallic surface, high pressure plasma is generated on the treated surface; elasto-plastic waves, then, propagate inside the target and create plastic strain, as shown in Figure 1. This surface treatment induces a deep compressive residual stress (RS) field on the treated area and through the thickness; such compressive RS is expected to increase the fatigue resistance, and reduce the detrimental effects of corrosion and stress corrosion cracking. LSP has applications in numerous fields such as turbo-machinery and engines, airframe and automotive structures, nuclear power plants and medical appliances. A detailed outline of the process and its historical development can be found in references [1,2].

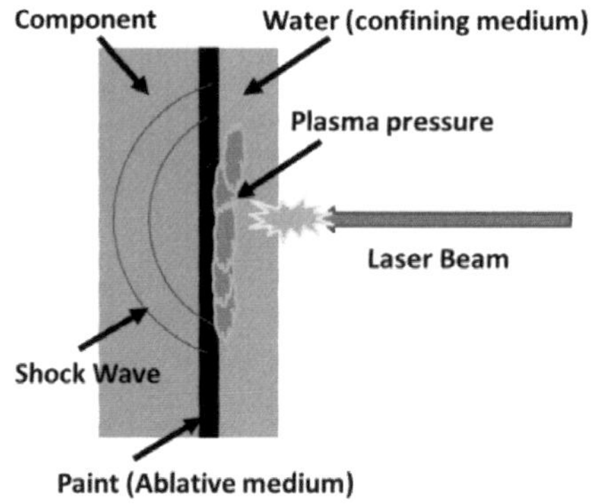

Figure 1. LSP process.

The LSP surface treatment is very effective in bulk structures [3]. Moreover, limited studies have been presented on the fatigue crack growth in thin components [4,5] where, as a consequence of the self-balancing RS field induced by the LSP in the thin-gauge panel, the chosen peening pattern configuration (distance of the crack tip to the laser shot, width of the laser pattern) can affect significantly the fatigue crack propagation performances [6,7].

Research motivation

Through the years, most investigations have dealt with the physical aspects of LSP, focusing the attention to its mechanical effects in order to apprehend the nature of the process, how to control it and determine optimized setups for obtaining the best RS distribution and, hence, performances of the treated metallic material. In this regard, several experimental studies have been carried out to investigate the role of laser parameters and, at the same time, optimize the process. However, dynamic responses related to the application of LSP are very complex and need sophisticated instruments (X-Ray Diffraction and Incremental Hole Drilling methods, for instance) to measure the effects in terms of RS distribution. Furthermore, the involved high strain rates (about 106s-1), combined to the transient nature of LSP, determine real time monitoring of the process very challenging. For these reasons, reliable simulation techniques are widely used in achieving a good knowledge of LSP applications and phenomena.

The advantages in using computational models in order to simulate LSP concern the possibility to analyze different peening conditions and materials by varying laser setups and geometry, providing several configurations and results. Furthermore, LSP simulations allow to reduce time and costs related to the development of laser peening facilities by computational modeling instead of building them up or buying the technology.

Background

The application of LSP can be modeled following two different modeling strategies: by simulating the entire physical process, from the application of the laser pulse and the generation of plasma, to the introduction of induced pressure pulse and shock wave propagation; by simulating only the second phase of the process, that is, the application of a pressure load at the treated area and the propagation of shock waves within the material. In this regard, an analytical model for predicting the influence of laser parameters on pulse pressure has been shown by Fabbro et al. [8], with the possibility to introduce a reliable estimation of the pressure without the need to simulate the expanding plasma, characterizing LSP as a pressure load at the surface and obtaining a more simple modeling of the process.

Concerning LSP modeling, typical simulation methods point out the attention to the influence on RS distribution by varying the following factors:

- Laser parameters: Power density, Pulse duration, number of layers, Overlapping;
- Geometry of the target: thickness, peened area, effects of defects and cracks;
- Peening condition: Specimen's boundary conditions, peening sequence and strategy (e.g. single/double sided peening).

Through the years, researchers have developed several numerical modeling methods in order to get an understanding of the LSP process. Braisted and Brockman [9] defined a model based on the commercial finite element code ABAQUS to determine both the short duration shock wave response and the resulting RS distribution. They performed a two-step analysis using explicit and implicit codes. ABAQUS/Explicit was used to model the application of load and the propagation of shock waves within the material and it is performed until final plastic strain is reached; on the other hand, ABAQUS/Standard was used to determine the RS distribution at static equilibrium. Ocaña [10] presented a theoretical and computational method called Shocklas, in order to simulate each phase of the LSP in a self-consistent way from the physical point of view. The referred phases are the following: analysis of the plasma dynamics, including consideration of breakdown phenomenology in the dielectric media; simulation of the hydrodynamic phenomenology arising from plasma expansion between the confinement layer and the base material; analysis of the propagation and induction of permanent structural changes by shock wave evolution in bulk material. Korsunsky [11] determined RS distribution by means of residual elastic strain using the eigenstrains method; such method defines RS field as a function of permanent inelastic deformation and, once the distribution of eigenstrains is known, explicit analytical integral expressions define residual elastic strain and, hence, RS distribution. Another example of two-steps analysis was developed by Hu et al. [12]. They proposed a method to simulate shock waves propagation and RS distribution within the treated material using LS-DYNA and ANSYS for three-dimensional modeling. Both the software determined full FEM simulation of LSP; in particular, LS-DYNA provided explicit solution for the dynamic analysis while ANSYS code read the stress field defined by LS-DYNA, performed the static analysis defining the stable stress and strain deformation output and, in the end, post-processing tools determined RS field and surface strain.

Work purpose

The aim of this paper is to develop an approach for the prediction of the RS field in a material subjected to LSP, optimizing, at the same time, the amount of work performed by the user and the needed computational time for the simulations. The objective is to realize a user friendly tool: FE analyses have been carried out by using the commercial code ABAQUS/Explicit only and providing an insight of the effects of the process and results in a single step. A huge time and work saving is achieved, in particular for multiple shock peens and multiple layers configurations, where the two-phases analysis procedures would have requested a very high computational time to perform explicit analysis and define implicit inputs in order to obtain the results for each single laser shock peen. The Explicit code will, at the same time, guarantee the convergence even for the first "fast impact" phase.

Furthermore, the effect of the LSP-induced RS distribution on the crack growth in thin panels will be investigated. These stress distributions are assessed by means of experimental tests and finite element simulations, together with an analytical model for the prediction of fatigue crack propagation in thin panels. A sensitivity study in order to determine which LSP parameters can influence the crack growth rate has been carried out.

FEM ANALYSIS OF LASER SHOCK PEENING PROCESS

During the application of LSP, the treated material is subjected to very high value of strain rates (about 10^6s-1) due to the dynamic nature of LSP process, based on irradiation of a high intensity and short duration laser pulse and following shock waves propagation. As already mentioned in the previous section, LSP analysis procedures has been performed with commercial finite code ABAQUS/Explicit, specifically designed for transient and short duration analysis. More efficient implicit codes (e.g. ABAQUS/Standard) could be also employed to simulate LSP and the transient nature of shock waves propagation; however, time requested to perform several simulations and computational expenses needed to determine the initial shock conditions are prohibitive. For these reasons, ABAQUS/Explicit code has demonstrated to be more suitable for getting an investigation of the material response during and after the application of LSP by using a dynamic analysis.

Indeed, a standard explicit/implicit process was developed by some of the authors to simulate an open hole specimen [13,14], representative of a section of an aircraft fuselage lap joint, typically prone to fatigue crack nucleation at the rivet holes. The results were compared in [15] with different solutions for the insertion of compressive RS around fastener holes in thin-walled structures, showing good agreement at the expense of huge computational cost. Therefore, an improved full-explicit model of a single laser shot has been developed and calibrated in terms of RS and displacement distribution with experimental measurements carried out at EADS-IW OTN [16].

Johnson Cook model

Due to the high strain rate loading, the Johnson Cook's (JC) empirical material model for 7050-T7451 aluminum alloy has been used [17]. The JC model describes the flow stress (σ) of the material as a product of three terms: strain hardening, strain rate and thermal dependence as described by eqn.(1).

$$\sigma = [A + B\varepsilon^n]\left[1 + C\ \ln\frac{\dot{\varepsilon}}{\dot{\varepsilon_0}}\right]\ [1 - T'^m] \tag{1}$$

where :

$T'^m = \frac{T - T_r}{T - T_m}$ (*Tr* room temperature and *Tm* melting temperature);
$A,\ B,\ n,\ C,\ \dot{\varepsilon_0}$ are experimentally determined parameters: *A* is the initial yield strength at room temperature, *n* takes into account strain hardening, while parameter *m* models the thermal softening and *C* represents the strain rate sensitivity.

The high plasma pressure resulting from the high-energy laser pulse impact can be defined by a spatial and temporal profile. The pressure peak is dependent on the individual LSP conditions and can be evaluated as function of the laser intensity (GW/cm2) [18]:

$$P(GPa) = 0.01\sqrt{\frac{\alpha}{2\alpha+3}}\sqrt{Z(g\ cm\ s^{-1})}\sqrt{I_0(GW\ cm^{-2})} \tag{2}$$

where I_0 is the incident laser power density, P is the pressure, Z is the reduced shock impedance between the target and the confining medium and α is the efficiency of the interaction.

Since the stress and displacement values must be zero at the end of the analysis, bulk viscosity coefficients (BV) were introduced to limit the numerical oscillations of high speed and high strain processes like LSP. he coefficients are merely numerical parameters available in ABAQUS/Explicit FE software, providing an artificial damping associated with the volumetric straining. The BV effect is inserted only in the unload step in order to avoid their influence on the LSP induced plastic deformations.

The ABAQUS model was calibrated by means of experimental results and proved to be reliable only after an optimization of the bulk viscosity setting. Moreover, the study about the material properties highlighted a gap of the standard models about the simulation of the surface conditions. The solution was in the modelling of the ablative layer employed during the real process, usually a super-thin sheet of pure aluminum stuck on the masterpiece. In the simulation, it has been reproduced as a 100μm layer made by a material with a yield point of 10MPa. All those settings have been applied to different geometry models, with multiple overlapped layers of LSP, in order to verify the robustness of the model [16]. As shown in Figure 2, numerical results in terms of RS have been compared with experimental measurements carried out in EADS-IW for a 4-18-2 laser configurations. Two different reference paths have been taken into account. The comparison in figure 2 has been done with two peened specimens (1 and 2) with the same laser settings.

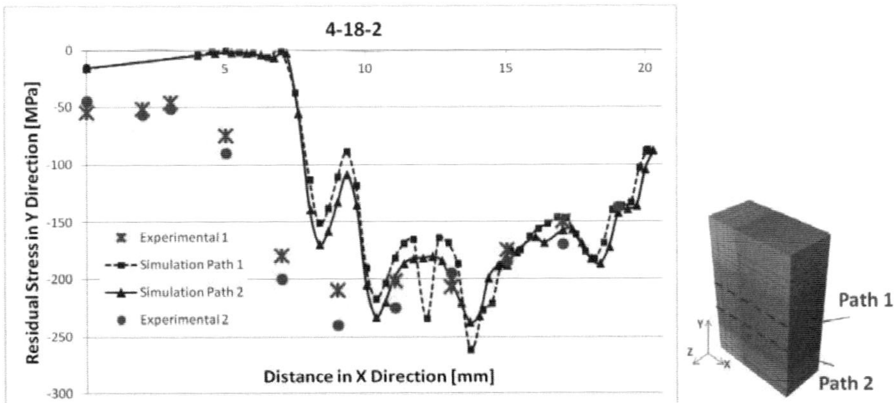

Figure 2. Numerical – experimental comparison of stresses for 4-18-2 laser configuration.

The good correlation between the simulation and experimental tests results proved this model to be reliable for prediction of detailed effects of LSP, but only with a complex calibration of various parameters, such as the bulk viscosity or the properties of the ablative layer.

Kinematic hardening model

In order to reduce the number of parameters, a new approach for the material model simulation has been performed. The effects of material due to the back and forth propagation of shock waves following LSP can be thought as a condition of high strain rate cyclic load conditions. In this regard, a nonlinear isotropic/kinematic hardening (KH) plastic model can be used to simulate the behavior of materials that are subjected to such conditions. Nonlinear isotropic/kinematic hardening model provides an accurate prediction of the so called Bauschinger effect, a change of property and stress/strain behavior of materials as a result of microscopic stress distribution of the material.

The nonlinear/kinematic hardening model consists of two components: a nonlinear kinematic hardening component and an isotropic hardening one. The kinematic hardening components are called Backstresses that are defined according to the following relations:

$$\dot{\alpha}_k = C_k \frac{1}{\sigma}[\sigma - \alpha]\dot{\bar{\varepsilon}}^{pl} - \gamma_k \alpha_k \dot{\bar{\varepsilon}}^{pl} \quad (3)$$

$$\alpha = \sum_{n=1}^{N} \alpha_k \quad (4)$$

where k is an arbitrary backstress, N the total number of defined Backstresses, $\sigma 0$ the value of yield surface size, $\dot{\bar{\varepsilon}}_{pl}$ equivalent plastic strain rate, Ck and γk, respectively, the initial kinematic hardening moduli and the rate of kinematic hardening decrease with plastic deformation. For this work, two different isotropic/kinematic hardening plastic models have been considered, assuming, respectively, one and three backstresses. The characteristics of the two models have been defined by the introduction of the material parameters, without temperature dependence.

Numerical simulations

The simulations performed have dealt with the followings topics:

- simulate LSP for single and multiple peens configurations, investigating the effects on the specimens and RS fields introduced;
- define material selection (JC and KH models);
- determine the influence of the position of the peened area in terms of RS field;
- determine the influence of number of layers.

In the first part of LSP FE analysis, the process has been simulated by means of the application of a single laser spot at the center of the specimen, in order to focus the attention on material response following the application of LSP in regards to the type of plastic model selected. As already mentioned, two Kinematic Hardening plastic models, respectively, with one (KH1) and three (KH3) Backstresses, have been used for the simulations. The results have been compared with the Johnson- Cook all explicit model with ablative layer model.

Figure 3 and Figure 4 show the results in terms of RS distributions in y direction (S22), for all the defined material, respectively, at the surface and in-depth into the material at the center of the laser spot.

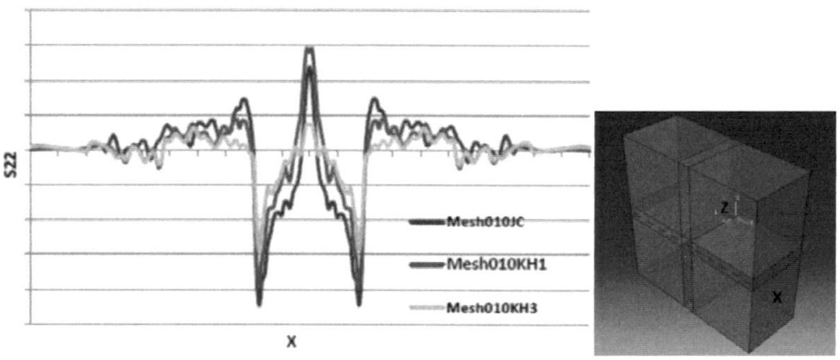

Figure 3. Single Peen Analysis: RS field at the surface along the x path.

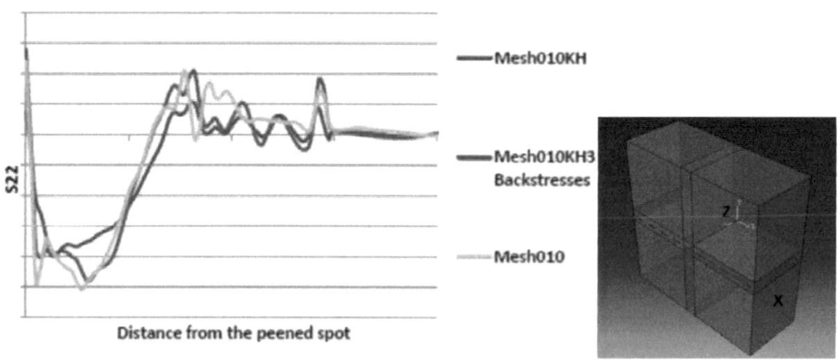

Figure 4. Single Peen Analysis: RS field in-depth along the z path.

The results show the introduction of compressive RS field except at the centre of the spot where a dip of tensile RS has been determined. Such phenomenon is coherent with experimental measurement of a single peen effect and it can be explained by the shockwaves propagation at the surface, interacting each other with a subsequent high increase of plastic strain and a tensile stress field.

Concerning the trend of RS distribution, material response shows little changes by changing selected material modelling. In this regard, KH3 model shows lower value of magnitudes for compressive and tensile peaks whether compared with JC and KH1 models. The reason may be found in the role of hardening that attenuates the values of tensile and compressive RS peaks.

However, a multi-spots LSP strategy is commonly used to cover larger regions of the materials. In this regard, particular attention has been focused on the effects of laser peening sequence, overlapping and frequency of peening in determining uniform distributions of RS and improving performance in terms of fatigue, wear and corrosion resistance of treated materials.

For this kind of analysis, a multiple overlapped layers strategy has been defined, providing also a description of the influence of material thickness on RS field, using single or double sided peening in agreement with material characteristics and geometry. Overlapping has been defined between successive layers and its value has been set according to the number of layers defined.

A parametric study has been performed for evaluating the influence of material thickness on RS distribution; five material thicknesses have been investigated: 3.8mm, 6mm, 9mm, 12.5mm and 20mm. A 4-18-3 peening setting has been selected for such analyses, performing a single sided peening for the 20mm-thick specimen, while double sided peening has been used for the other four models.

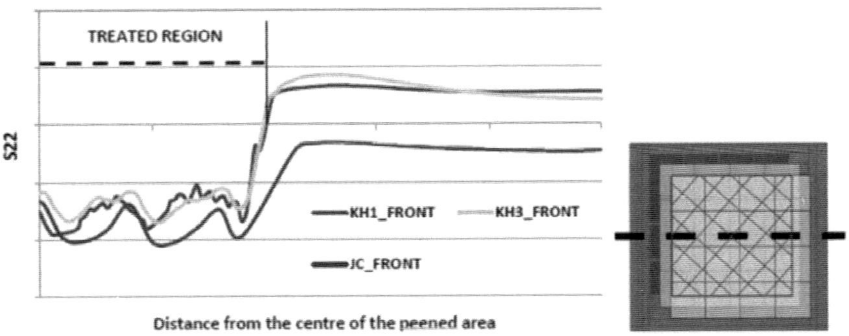

Figure 5. 3.8mm-thick model: RS at the front surface along the dotted path.

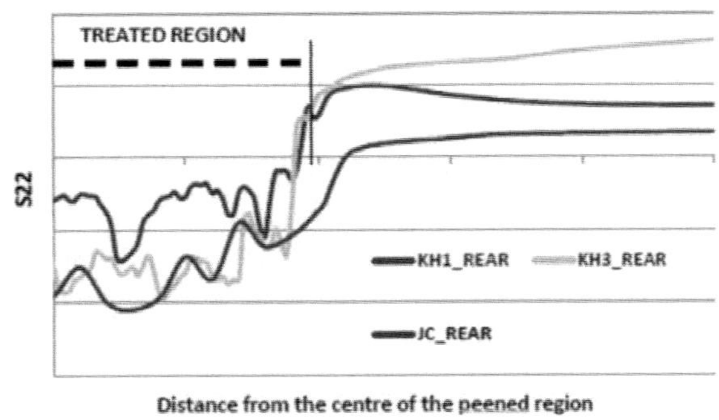

Figure 6. 3.8mm-thick model: RS at the back surface along the dotted path in Figure 5.

For both surfaces, results show a relatively uniform distribution of RS, confirming multiple layers as a good option to prevent the drop of compressive RS at the centre of the spot, as already seen for single spot LSP analysis. By comparing the RS distribution at front and rear surfaces, it can be seen how peening sequence plays an important role in defining RS profiles.

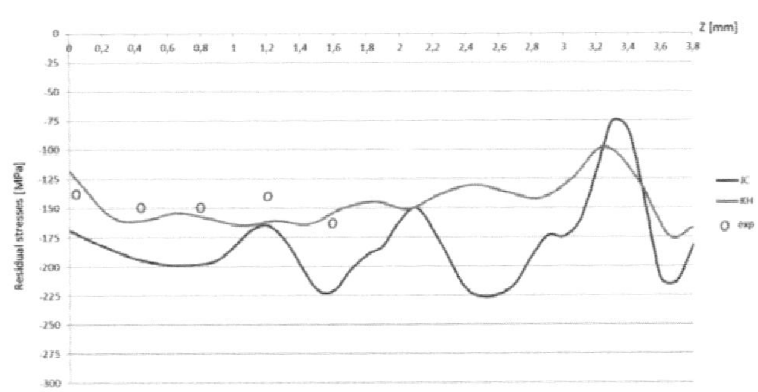

Figure 7. 3.8mm-thick model: RS through-the-thickness.

Figure 7 shows in-depth RS distribution. The plot shows full compressive RS profile, confirming double sided peening as a good option for imparting a uniform RS distribution through the thickness, improving material performance. However, the non-symmetric distribution of RS at the two sides of the specimen also confirms the role of peening sequence in final RS distribution. The results have been compared with experimental test, performed with X-ray diffraction and incremental hole drilling techniques, showing the good performances of the KH 3-backstresses model.

In order to get a better understanding of the potential of LSP in imparting full compressive in-depth RS, a 6mm-thick model has been defined. In this regard, Figure 8 shows the through-the-thickness RS distribution. In this case, the RS profile is not uniform along the thickness, while the core of the material does not show any tensile stress field. Furthermore, results show a weaker but still present non-symmetric profile of RS peaks close to of the surfaces, confirming the role of peening sequence in influencing RS distribution for thin specimen.

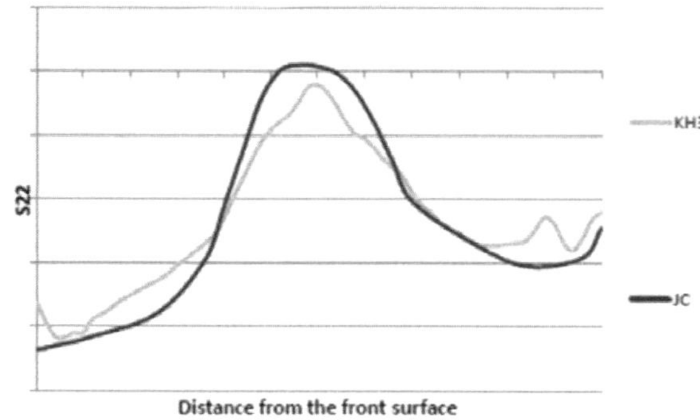

Figure 8. 6mm-thick model: RS through-the-thickness.

Such difference in terms of RS peaks is highlighted in results inherent to double sided peening configuration, as shown in Figure 9. It is quite clear how the rear surface of the model shows a lower in magnitude compressive RS profile, independently from the plastic model considered, whether compared with the front surface. Indeed, as it has been shown previously concerning the temporal evolution of the process for double sided peening, by peening one side of the specimen, the effects in terms of RS field affect the peened area as well as the opposite surface, so that, when the peening is applied at the other surface, a RS field is still present at the first shot surface and through the thickness, influencing the following RS fields.

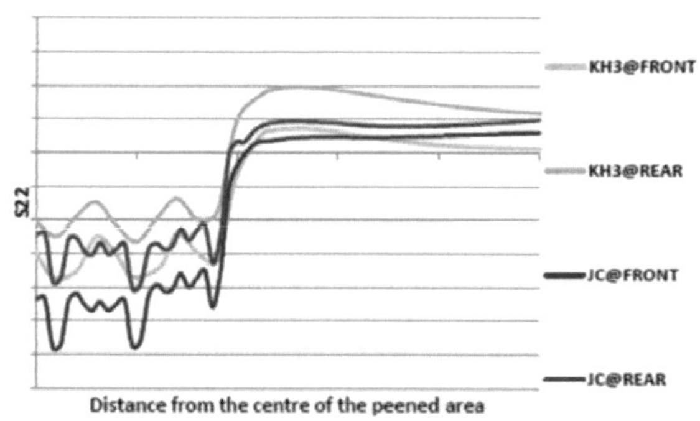

Figure 9. 6 mm-thick model: RS at the front and back surfaces along the dotted path in Figure 5.

The previous two models have shown the capability of double sided peening in determining full compressive RS field at the treated surface and through the thickness in thin specimens. Fig 10 shows the compressive RS field achieved on both shot surfaces of 12.5 mm thick treated specimen. In particular, it is confirmed that the propagating shockwaves are able to plastic strain the material up to a certain value of depth, while the core of the material maintains an elastic behaviour, so that, once the pressure plasma is completely dissipated at the treated surface, equilibrium is not reached only by means of tensile RS in the surroundings of the peened area, but also beneath the surface of the material. Concerning the effects of peening sequence in determining in-depth RS, a good grade of symmetry has been reached for the 12.5 mm-thick model, confirming that, by increasing material thickness, the effects of peening sequence tends to fade away, with a trend already seen in the previous investigated thicknesses.

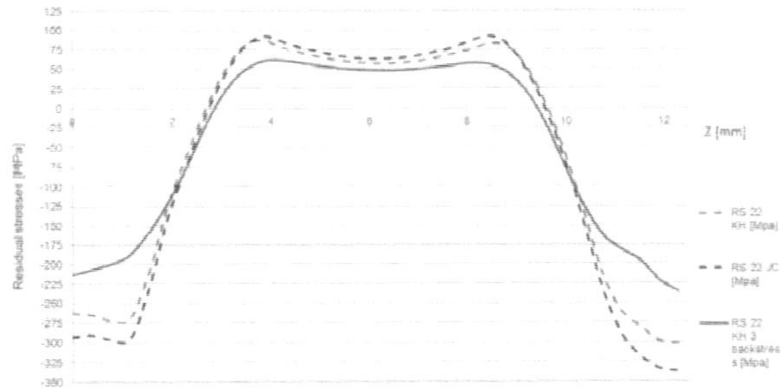

Figure 9. 12-5 mm-thick model: RS through-the-thickness.

The numerical results herein reported show how the Kinematic Hardening material model seems to be the best compromise to predict the residual stress field achieved by laser shock peening process. Indeed, the Johnson-Cook material parameters are obtained by the very complex split Hopkinson bar test, which goes up to strain rate to 10^4 sec^{-1}, that means two order of magnitude slower than the actual laser peening process. Moreover, it has been shown how the JC parameters from literature need a tuning and the overall results are strongly influenced by the essential insertion of an artificial damping (bulk viscosity) and the ablative layer in the simulation. These last two features are not needed in the KH approach, since the high accuracy of the KH parameters in literature.

FATIGUE CRACK PROPAGATION

Starting from the FE numerical evaluation of the through-the-thickness RS in thin specimens with KH material model, the effect of the residual stress distribution introduced by the LSP on the fatigue crack propagation in thin aluminium panels has been evaluated.

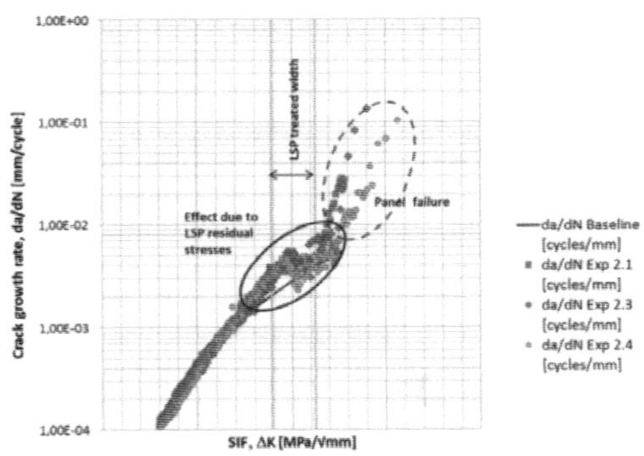

Figure 10. Paris' Curves of 3 LSP coupons vs Baseline.

Experimental results of fatigue crack propagation tests on some central cracked aluminium panels 2mm thick [19] showed no improvement achieved with the selected configuration of the LSP treated pattern. The LSP influence has been evaluated via comparison with a baseline, relative to the non-treated material [6], as shown in Figure 10. Indeed, in the middle of the laser treated pattern, a crack growth slowdown has been highlighted in the crack propagation rate experimental data. Unfortunately, this improvement in the fatigue properties is nullified by a steep increase in the crack growth rate, noticeable in the areas before and after the shot pattern, mostly due to the self-equilibrating tensional residual stress accumulated in this area, as shown in Figures 5 and 6 for a similar thickness specimen. An overall negative effect of the laser shock peening process is therefore noticeable in comparison to the non-treated solution. The laser configuration, in terms of distance of the pattern from the crack origin and the shot width, is unable to exploit the benefit of the compressive RS induced by the LSP, with an overall detrimental effect on the fatigue crack propagation performance of the cracked panel.

In order to predict how the laser pattern dimension and location affect the fatigue crack propagation performances of the panels, an accurate description of the self-balancing residual stress distribution along the entire panel width has been obtained by the all explicit finite element model (Abaqus/Explicit) with KH material model, as shown in Figure 11.

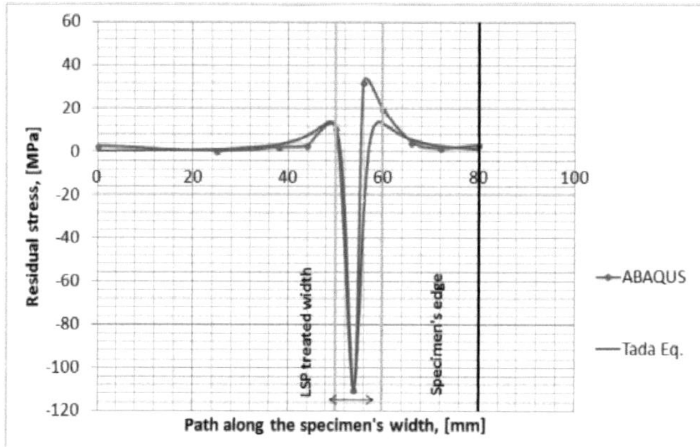

Figure 11. FEM (KH material model) and analytical evaluation of Residual Stress.

Tensile RS are established in both the sides of the LSP pattern in order to restore the global stress equilibrium of the panel. The tensile stresses predicted at the inlet to the shot pattern are lower than that at the exit, explaining the moderate acceleration in crack propagation shown by the tests' results, in respect to the baseline, before reaching the pinned line, and subsequently slowing down in the central area, before the final sharp increase of the FCP. The asymmetry in the RS distribution to the respect of the peened area is related to the finite dimension of the specimen, with the black vertical line indicating the specimen's edge. This asymmetry is neglected by analytical evaluation performed by the authors by means of a modified Tada model [20, 21].

The effectiveness of the predictive model allows therefore a parametric analysis on the effect of the LSP pattern position and width on the fatigue life of thin specimens. Combinations varying the pattern width, the peak value of compressive RS and the distance between the crack origin and the peened area have been examined.

The results summarized in Table I show an increase in FCP life only when the peening pattern is closer to the initial crack origin, at 15 mm. Similarly, a larger laser shot width and higher compressive residual stresses were producing the best results in terms of FCP.

Table I. Percentage variation of the fatigue life referred to the baseline.

Crack origin – LSP distance [mm]	15	26.5	50	61.5
RS = -100 MPa Shot pattern = 10 mm	23.7	0.04	-1.77	
RS = -100 MPa Shot pattern = 33 mm		47.5		-11.9
RS = -150 MPa Shot pattern = 10 mm	74.98	7.9	-2.43	
RS = -150 MPa Shot pattern = 33 mm		291.18		-15.52

CONCLUSIONS

From the results of the present work, LSP is confirmed as an effective method to introduce significant compressive residual stresses in fatigue sensitive areas of metallic structures.

The numerical FE simulation to LSP showed to be extremely difficult as a consequence of the parameters involved in the process. Starting from previous analyses, the JC material model has been compared with the KH model, demonstrating the latter better suitability for the evaluation of the RS field. The numerical investigations led to a reliable and optimized ABAQUS finite element model, calibrated by means of experimental results and valid even for complex geometries.

A good correspondence with real applications of LSP concerning material response in terms of RS and surface displacement is noticeable. The influence of material thickness shows how RS profiles can change in terms of compressive RS peaks – higher in magnitude for thicker material – and in-depth distribution – with the possibility to achieve full compressive RS field for thicknesses up to 6-7mm. Results also show a higher distortion effect for thinner specimens, demonstrating the role of material thickness and the peening sequence in preventing such detrimental effect of the process.

Using the validated FEM approach to predict the residual stress profile along the entire width of LSP treated specimen, together with experimental XRD measurements of the actual residual stresses and an analytical crack propagation model, a comparison of the fatigue crack propagation predicted performances has been done with the experimental data available, showing a good agreement. The model was then used to make predictions about the best peening configuration to achieve benefit in terms of fatigue crack propagation life after LSP treatment. This lead to the conclusion that, in order to achieve an increase in fatigue crack propagation life, the laser shot has to be placed close to the crack starting tip and a larger shot pattern has to be used.

According to the achieved results, LSP results to be a very effective and interesting method for the generation of beneficial RS fields for improving mechanical properties and resistance to fatigue, wear and corrosion for metallic alloys.

REFERENCES

[1] Kruusing, A., (2007), *Handbook of Liquids-Assisted Laser Processing*. Elsevier Science & Technology, Amsterdam

[2] Ding, K. and Ye, L., (2006), *Laser shock peening: Performance and process simulation*, Woodhead Publishing Ltd., Cambridge.

[3] Heckenberger, U., Hombergsmeier, E., Holzinger, V. and von Bestenbostel, W.,(2011), in G. Ivetic (Ed.) Proceedings of the 2nd International Conference on Laser Peening, Emerald Group Publishing Ltd., Bradford.

[4] Ivetic, G, Meneghin, I., Troiani, E. et al., (2012), *Materials Science and Engineering A*, vol.534, p. 573

[5] Ivetic, G., Troiani, E., Meneghin, I. et al., (2011), in Proceedings of ICAF 2011 Symposium, Montreal, Canada.

[6] Troiani, E., Taddia, S., Meneghin, I. and Molinari, G., (2014), *Advanced Materials Research*, vol. 996, p. 775

[7] Glaser D., Polese C., Bedekar R.D., Plaisier J., Pityana S., Masina B., Mathebula T., Troiani E. (2014) *Advanced Materials Research*, vol. 891-892, p. 974

[8] Fabbro, R., Fournier, J., Ballard, P., Devaux, D. and Virmont, J., (1990), *J. Appl. Phys*., vol. 68, p. 775.

[9] Braisted, W. and Brockman., R., (1999), *Int. J. of Fatigue*, vol. 21, p. 719.

[10] Ocaña, J.L., Morales, M., Molpeceres, C. and Torres, J., (2004), *Appl. Surf. Sci.*, vol. 238, p. 242-248.

[11] Korsunsky, A.M. (2005), *J. Strain Analysis*, vol. 41, n. 3, p. 195.

[12] Hu, Y., Yao, Z. and Hu, J., (2006), *Surf. Coat. Technology*, vol. 201, n. 3/4, p. 1426

[13] Ivetic G., Meneghin, I. and Troiani, E., (2011), *Materials Science Forum*, Vol. 681, p. 267

[14] Ivetic G., Meneghin, I., Troiani, E., Molinari, G., et al., (2012), *Materials Science and Engineering A*, Vol. 534, p. 573

[15] Ivetic G., Troiani, E., Meneghin, I., Molinari, G., Lanciotti, A., Ristori, V., Ocaña, J.L., Morales, J., Polese, C. and Venter, A., (2011). In: Proceedings of ICAF 2011 Symposium, Montreal, Canada,

[16] Troiani, E. Crudo, C., Molinari, G. and Taddia, S., (2013), in A. Brot (Ed.), Proceedings of ICAF 2013 Symposium, Jerusalem, Israel.

[17] Liu, Z., Shi, Z. and Guo, Y. (2006) In: Proceedings of ISMNM 2010 – Guilin and Han (Ed.)

[18] Peyre, P., Fabbro, R,, Merrien, P. and Lieurade, H.P., (1996), *Material Science and Engineering A,* vol. 210, p. 102

[19] Furfari, D., Ohrloff, N., Hombergsmeier, E., Heckenberger, U., Holzinger, V., (2013), in A. Brot (Ed.), Proceedings of ICAF 2013 Symposium, Jerusalem, Israel.

[20] Terada, H., (2011), *Fatigue of aircraft structures*, vol.1, p. 5.

[21] Tada, H. and Paris, P.C., (1983), *Journal of Fracture*, vol. 21, p. 279

Retardation of fatigue crack growth in thin AA2024 sheets by laser heating

Anne Groth, Manfred Horstmann, Nikolai Kashaev, Norbert Huber

Institute of Materials Research, Materials Mechanics / ACE-Centre, Helmholtz-Zentrum Geesthacht, 21502 Geesthacht, Germany.

Abstract: Local heating was applied by means of a defocussed laser beam with the aim of generating residual stress profiles with high compressive stresses perpendicular to the crack growth direction. The retardation effect on the fatigue crack growth behaviour was investigated for different heating line configurations. The corresponding experiments were carried out using M(T)200 specimens, fabricated out of sheets of the aluminium alloy 2024 T3 with a thickness of 2 mm. Constant stress amplitude fatigue crack growth tests at a *R*-ratio 0.1 were performed. One laser heating line on each side of the specimen results in an increase in fatigue lifetime of 10% in comparison to the base material. Each additional line further increases the number of load cycles until failure. Finally, with four lines on each side, a gain of approximately 425% in load cycles was reached.

INTRODUCTION

Thermal or mechanical treatment by welding processes has the potential to save weight and therefore costs in structural components. During production of these components thermal and mechanical processes introduce residual stresses. These residual stresses can considerably influence the fatigue lifetime [1].

One method to effectively retard fatigue crack growth is local heating, e.g. by applying spots in front of the crack tip as described by Ray et al. [2]. The effectiveness of this procedure is the result of the non-uniform heating in front of the crack. In detail, Ray et al. determined the optimized position for a C(T) specimen manufactured out of a carbon steel sheet with a thickness of 3.4 mm as 5 mm ahead of the crack tip. According to the overload principle [3], spot heating induces compressive stresses in front of the crack tip, which reduce the available crack tip driving force. The measured fatigue crack growth retardation increased with increase of the spot heating temperature. Another possibility is a surface annealing treatment as described for example in Tsay et al. [4]. The authors of this study used a CO_2 laser with a rectangular beam spot to anneal the surface of a C(T) specimen with a thickness of 7 mm made from AISI 316 austenitic stainless steel. A single line and a double line in a distance of 7.5 and 19.5 mm from the crack tip were analysed. Only slight changes in microstructure and no changes in hardness were observed. But a clearly higher resistance to crack growth was described.

The introduction of residual stresses by local heating also was investigated by finite element approaches. Yilbas et al. [5] predict the residual stress field using a finite element method which couples thermal heat conduction and elongation with mechanical stress development and corresponding local deformation. In this study, a laser surface treatment of aluminium was investigated. More concrete, the residual stresses resulting from a surface treatment of two laser pulses with different pulse lengths were determined. Yilbas et al. also studied the metallurgical changes in the irradiated region. The maximum of von Mises stresses were found in the vicinity of the surface.

A similar combination of experimental and numerical methods was applied by Schnubel et al. [6,7]. In order to retard the fatigue crack growth by residual stresses, the specimen was treated by local heating with a defocused laser beam perpendicular to the crack growth direction. In the direction of the heating line tensile residual stresses were generated which, in the vicinity of the heating line, were compensated by corresponding compressive residual stresses. These compressive stresses significantly retarded the fatigue crack growth and increased the fatigue lifetime. For a C(T)100 specimen fabricated out of a AA2198 T8 sheet with a thickness of 5 mm, an increase of the fatigue lifetime of 100% was achieved.

In this work, the approach proposed by Schnubel et al. is transferred to a M(T)200 specimen made of AA2024 T3 sheets with a thickness of 2 mm. The M(T)200 geometry was chosen because it more realistically mimics the load conditions of an aircraft. The machining of the starter notch for the fatigue crack growth measurements took place after the laser treatment process. In this way the influence of the local residual stress field on the fatigue lifetime behaviour was analysed for a crack-free sample in order to confirm the possible usefulness of this heat treatment already during the construction of an airplane. Additionally, the impact of the application of further heating on the fatigue lifetime was investigated. A considerable improvement of the fatigue properties is expected. The induced compressive stresses produced by each new heating line adds to that induced by previous heating lines and a M(T)200 specimen offers enough surface area to position multiple lines in the path of crack propagation.

EXPERIMENTAL

Sample preparation and laser heating treatment

The experiments were carried out using M(T)200 specimens, fabricated out of AA2024 T3 sheets with a thickness of 2 mm. A sketch, shown in figure 1a, shows the sample geometry including the holes for the clamping system and the position and orientation of the starter notch. Laser heating (LH) was performed using a 2.2-kW Nd:YAG laser with fibre optics (300-µm core diameter) and a 250-mm focal length. The operating power was 600 W and the laser movement velocity was 1.6 m/min. A defocused beam setup was used in order to produce a spot diameter on the specimen surface of 3.5 mm. A black graphite layer was applied to the aluminium surface to reduce its reflectivity. The position of the heating lines is illustrated in figure 1b-e. The lines were positioned alternately one on each side of the crack from inside to outside. For symmetry reasons every specimen has the same number of laser heating lines on each side. The orientation of the heating line was in rolling direction. To further continuously increase the fatigue lifetime one to four heating lines as shown in figure 1b-e were applied. They were added in a distance of 10 mm to the previous ones from inside to outside.

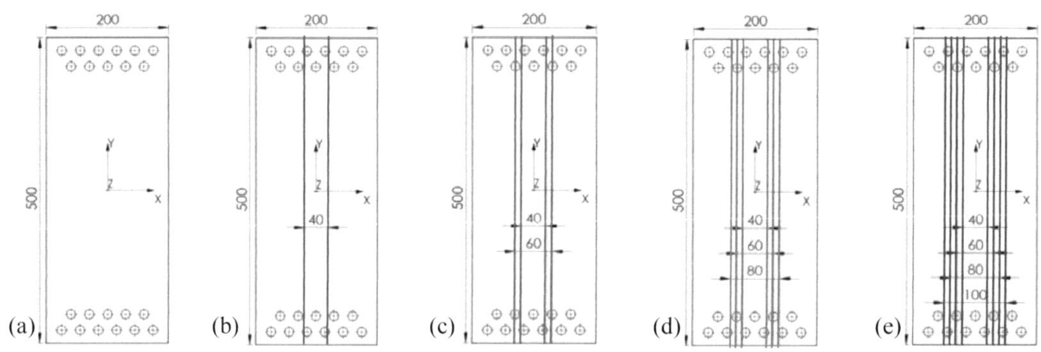

Figure 1. Sketch of the investigated M(T)200 specimen (a). The position of the starter notch is in centre of the coordinate system, it's aligned along the x-axis and has a total length of 7 mm. Marked in colour are the positions of the laser heating lines in the case of two (b), four (c), six (d), and eight heating lines (e).

Sample characterisation

Micro hardness tests were performed to measure changes of local mechanical properties which resulted from laser heating. A Shimadzu HMV-2000 equipped with an automatically x-,y- movable stage was used to measure the micro hardness on the sample surface with a grid spacing of 0.2 mm. The device was equipped with a Vickers indenter tip. The applied load for each indent was 17.6 N with a holding time of 20 s resulting in indentation depths low enough to ensure that neighbouring indentations are not influenced. The measurements were performed in series over cross-sectional polished cuts perpendicular to the heating line. The minimum distance to the edge of the samples also was 0.2 mm. Because of the symmetry it was sufficient to restrict the investigated area to one half of the cross section, setting the zero value for x into the centre of the sample.

Fatigue crack growth tests

Fatigue crack growth tests were performed for the investigation of the damage tolerance behaviour. The base material behaviour was measured as a reference. The tests were carried out as constant amplitude tests with a R-ratio of 0.1 and a frequency of 5 Hz in a Schenck-Instron, 400 kN frame. The starter notch was placed in the middle of the specimen between the heating lines. The crack direction was perpendicular to the rolling direction. The starter notch was 7 mm and the data acquisition started after the crack reached a total length of 10 mm. The crack length was measured optically using a travelling microscope.

Fatigue tests

Dog bone shaped specimens were cut out by spark erosion. The narrow-gauged section was oriented along the heating lines. The exact shape of the samples as well as the position of the heating lines on the sample surface is shown in figure 2. In order to further reduce the influence of surface and edge condition on the test results, all samples were handled with great care and the edges additionally polished prior to the fatigue tests. The device in use was a Schenck

H100 servohydraulic horizontal testing station. The tests were carried out in the tensile regime with a frequency of 15 Hz. The *R*-ratio was kept constant at 0.1 and the maximum stress ranged from 450 N to 150 N.

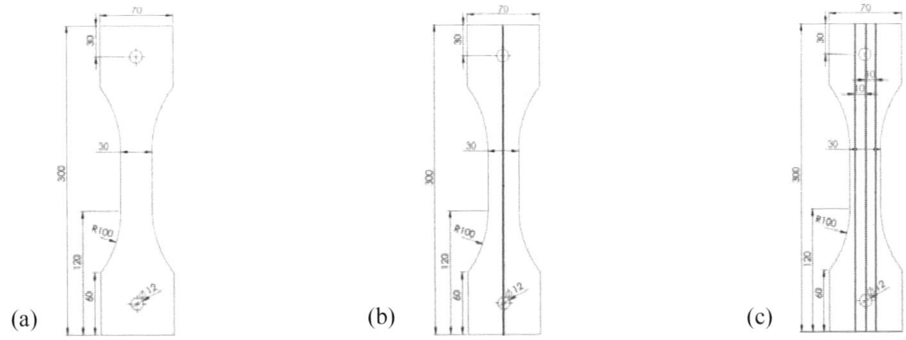

Figure 2. Sketch of the S-N specimen in untreated condition (a), cut-out of material with one heating line (b), and three heating lines (c).

RESULTS AND DISCUSSION

Local modifications of the material due to laser heating

Figure 3 shows the results of the hardness measurements with one, two, and three heating lines on one side of the M(T)200 specimen. For all samples, there is a tendency to observe a lower hardness in the interior of the sheets in comparison to the surface near regions what probably is caused by the rolling production process of the base materials. More interesting, in the vicinity of every heating line a considerably lower hardness, i.e. a reduction of

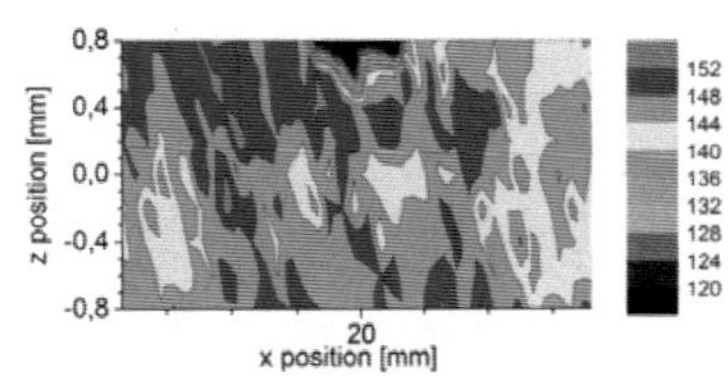

Figure 3. Colour coded plot of the measured micro hardness scan for the specimen with one, two and three heating lines on each side.

up to 20%, can be observed. The lateral size of the area of reduced hardness corresponds well with the size of the spot diameter of 3.5 mm. Additionally, the values are reduced down to a depth of about 0.4 mm from the heated surface.

Obviously, the laser heating treatment only influences a small part of the sample volume despite the fact that the surface becomes slightly molten as can be seen in figure 4. Variations in hardness outside the heat affected zone are small and in the case of multiple lines on each side even negligible. In summary, an unwanted modification of the overall bulk material properties can be effectively prevented when using this laser heating technique. Only local ageing on the welded surface can be observed.

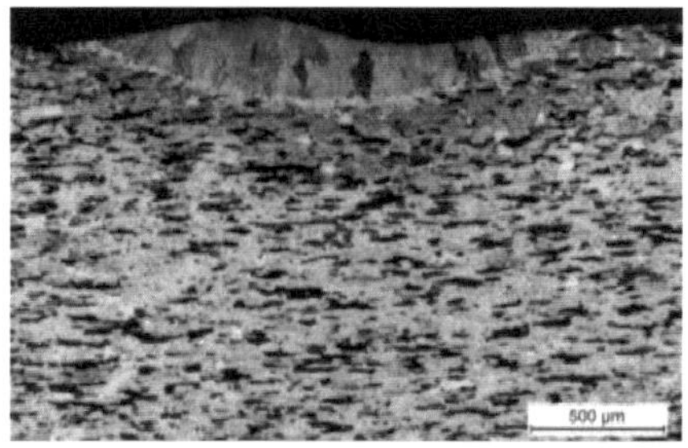

Figure 4. Optical micrograph of a polished cross section of a M(T)200 sample treated with one laser heating line on each side. Shown is the part of the surface which was partially molten by the laser heating treatment.

Effect of laser heating induced residual stresses on fatigue crack growth behaviour

Despite the fact that the main volume of the material seems unaffected by the laser heating treatment, a considerable effect in the fatigue crack growth behaviour is observed, see figure 5. The number of cycles is increased by more than 10% for a single heating line applied to each side of the specimen. Additional heating lines significantly increase the fatigue lifetime. Two lines on each side increase the fatigue lifetime by 20%. With three lines on each side an increase of 70% is achieved and with four lines on each side a 425% increase results.

As in a preceding set of experiments [8], the general trends of fatigue crack growth rate da/dN vs. the load cycles N (see figure 6) are closely related to the development of the crack length vs. the load cycles N. When the propagating crack reaches the area of a heating line, a sudden retardation of the crack growth can be observed. After passing the area of the heating line the crack growth rate reaches its former value again. This trend becomes less distinct if the number of heating lines increases.

This kind of crack growth retardation evidently works very well as long as the length of the crack is low in comparison to the width of the M(T)200 specimen, i.e. at low stress intensity factor values. After reaching a crack length of ca. 30 mm the propagation rate becomes independent of the laser heating treatment. In case of a single heating line on each side the induced tensile residual stresses reach values of approximately 225 MPa directly at the position of the heating line[8]. Compressive stresses of about -40 MPa neighbouring the tensile region were measured [8]. By adding further heating lines the compressive stresses sum up corresponding to the superpositional principle. These higher compressive stresses cause retardation of the fatigue crack growth. Obviously, it's not the change of local material properties but the evolution of a residual stress field which causes this extraordinary improvement of the damage tolerance behaviour.

Figure 5. Fatigue crack growth behaviour of the laser heated M(T)200 specimens with one, two, three and four heating lines on each side at R-ratio 0.1 and the base material.

Figure 6. Fatigue crack growth rate vs. the number of applied load cycles of the laser heated M(T)200 specimens at R-ratio 0.1 and the base material. The dashed lines mark the position of the heating lines. (a) Detail of the lower load cycles for up to 3x2 heating lines; (b) Overview over the whole range of load cycles for all configurations under investigation.

Investigation of fatigue crack initiation

The S-N curves shown in figure 7 testify that the applied laser heating process has only a minor effect on the crack initiation behaviour. The fatigue threshold of the heat treated material turned out to be about 20 MPa lower than that of the base material, independent of the number of heating lines. The effect is relevant for high cycle fatigue. If the applied maximum stress is increased the curves of the heated specimens converge to the curve of the base material. At a maximum stress value (scale of the ordinate on the right-hand side of figure 7a) of about 350 MPa all curves are congruent in the range of experimental error. These results indicate that the influence of the local modification and of the induced residual stresses becomes negligible with respect to the overall fatigue behaviour if the applied loads cause stresses in the material which equal or exceed the magnitude of the induced ones,in this case values more than about 225 MPa.

(a)

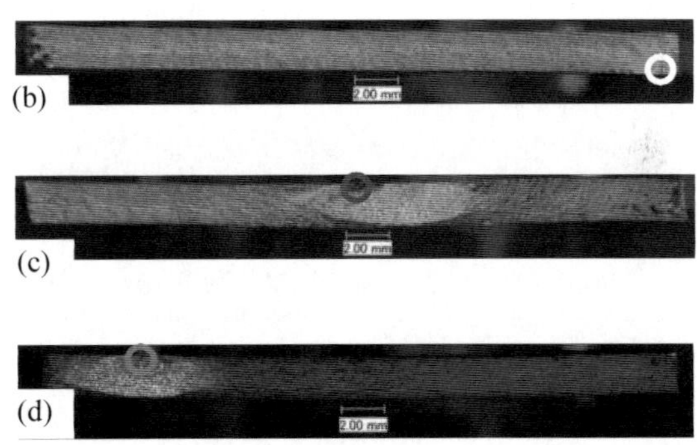

Figure 7. S-N curves with one and three heating lines in comparison to the base material (a), picture of crack surface after the crack initiation tests of the base material (b), treated with one heating line (c), and treated with three heating lines (d).

Figure 7 also shows the typical crack surfaces after failure of the differently treated specimens. As expected for an untreated material, the crack initiation in the base material takes place at varying locations at the surface depending on its local quality, i.e. the cracking starts where the largest local notches are located. One such example is shown in figure 7b. There, the crack started on the lower right-hand side. Totally different conclusions concerning the crack initiation can be drawn from the crack surfaces of the laser heated samples. In practically all cases, the crack started in direct vicinity of the heating line. More precisely, its initiation took place in the region between the small melted part on the surface and the unchanged bulk volume. This means, the crack initiates in the region with high tensile residual stresses at positions where inhomogeneities in the material are present.

In conclusion, the reduction of the fatigue threshold in case of loads lower than 350 MPa as well as the changed crack initiation behaviour as described above have to be taken into account when using laser heating as a processing tool to modify structural elements.

CONCLUSIONS

It is shown, that the application of laser heating lines considerably increases the fatigue lifetime of AA2024 T3 sheets. With each added line more load cycles were accepted by the M(T)200 specimen until failure. One heating line on each side resulted in an increase of 10%, two heating lines increased the fatigue lifetime by 20%, and three heating lines by 70% in comparison to the base material. With four lines on each side the fatigue life increases by 425%. These changes were achieved while the hardness of the material showed only a small decline directly at the position of the heating line and its near vicinity. There are no changes in the hardness visible between the heating lines. The induced residual stress field is primarily responsible for such an improvement in the fatigue life. The influence of the laser heating treatment on the crack initiation was only moderately unfavourable. In the regime of high loads, no influence of the induced residual stresses was detected. In the high cycle regime a drop of 20 MPa in the fatigue threshold was detected.

ACKNOWLEDGEMENTS

This work was carried out within the frame of the LISA project, which was supported by the Light-weight materials Assessment, Computing and Engineering Centre (ACE) of the division Materials Mechanics of the Helmholtz-Zentrum Geeesthacht. The authors also thank Falk Dorn for the sample surface preparation prior to the hardness tests as well as Hamdi Tek for his assistance during the mechanical testing.

REFERENCES

[1] Masubuchi, K. (1980). Analysis of welded structures: Residual stresses, distortion, and their consequences", Pergamon Press.

[2] Ray, P.K., Verma, B.B. and Mohanthy, P.K. (2002). Spot heating induced fatigue crack growth retardation. *International Journal of Pressure Vessels and Piping* vol. 79, p. 373-376.

[3] Darvish, M. and Johansson, S., Fatigue crack growth studies under combination of single overload and cyclic condensation environment. *Engineering Fracture Mechanics* 1995;vol. 52, issue 2, p. 295-319.

[4] Tsay, L.W., Liu, Y.C., Lin, D.-Y., and Young, M.C. (2004). The use of laser surface-anneaaled treatment to retard fatigue crack growth of austenitic stainless steel. *Materials Science and Engineering* A384, p. 177-183.

[5] Yilbas, B.S., Arif, A.F.M., Karatas, C., and Raza, K. (2009). Laser treatment of aluminium surface: Analysis of thermal stress field in the irradiated region. *Journal of material processing technology*, vol. 209, p. 77-88.

[6] Schnubel, D., Horstmann, M., Ventzke, V., Riekehr, S., Staron, P., Fischer, T., and Huber, N. (2012). Retardation of fatigue crack growth in aircraft aluminium alloys via laser heating – Experimental proof of concept. *Materials Science and Engineering* A 546(2012), p. 8-14.

[7] Schnubel, D., Horstmann, M. and Huber, N. (2013). Retardation of fatigue crack growth in aircraft aluminium alloys via laser heating – Simulation-based design optimization. *International Journal of Structural Integrity* vol. 4, No. 4, pp. 429-445.

[8] Groth, A., Kashaev, N. and Huber, N. (2014) Introducing residual stresses in thin aluminium structures by laser generated temperature fields, Proceedings of 29th Congress of the International Council of the Aeronautical Sciences, St. Petersburg 2014, ICAS2014_0244.

Influence of Materials on the Effectiveness of Crenellations in the Damage Tolerant Design

Jin Lu, Nikolai Kashaev, Volker Ventzke, Norbert Huber

Institute of Materials Research, Materials Mechanics, Helmholtz-Zentrum Geesthacht, Max-Planck-Str. 1, 21502 Geesthacht, Germany. Contact: jin.lu@hzg.de

Abstract: Crenelllation is a novel concept to heighten the fatigue resistance of the airframe structures without increasing weight. In this work the impact of materials on the effectiveness of crenellations was investigated. Two candidate alloys for future fuselage skin: AA2139 and AA2198 were selected for this task. Firstly the microstructure and texture of those materials were investigated. Then flat and crenellated panels made of the two alloys were tested under service-related biaxial loading conditions. A removable δ_5 clip gauge was applied at the crack tip to monitor the crack closure behavior during the fatigue tests. After tests the fracture surfaces were examined under SEM and laser-scanning microscope to interpret the respective fatigue behavior. It was found that sharp textured AA2198 alloy, which also showed tortuous shear lip morphology, has higher fatigue resistance and larger fatigue life improvement of crenellations compared with the randomly textured AA2139 alloy. The correlation between the sharp texture and tortuous shear lip morphology was discussed. The source of the additional fatigue resistance and the increased crenellation efficiency in AA2198 was also analyzed.

INTRODUCTION

Fatigue is one of the utmost important concerns in the design of airframe structures. Statistics show that about 60% of failures in aircraft components are caused by fatigue damage [1]. To safeguard the structural integrity of aircraft from fatigue problems, the damage tolerance approach is usually adopted. This approach requires periodic inspections on airframe structures to detect and repair small cracks before they become a threat. The growth speed of these cracks from a detectable size up to an allowable size determines the necessary inspection frequency and the operational cost of the aircraft [2].

To retard the growth of fatigue crack in the airframe structures as much as possible, efforts are made generally in two different approaches. The first approach is to toughen the material against fatigue crack propagation. The toughening mechanism can have either intrinsic or extrinsic sources according to Ritchie [3]. The intrinsic toughening is achieved by enhancing the inherent microstructural resistance to cracking (e.g. modifying the precipitates, changing the bond strength). A good example of this is the optimized damage tolerance properties of the AA2139 and AA2198 alloys developed in recent years [4,5]. On the other hand metallic materials can also be extrinsically toughened by so-called crack tip shielding mechanism. In such a case premature crack closure can occur at a certain load level above the minimum load during the cyclic loading. This can significantly reduce the effective driving force of crack growth [6]. The crack closure is mostly due to the residual plastic deformation at the crack tip and the misfit of crack surface at asperities, which is even pronounced when large relative sliding of the two crack surfaces occurs such as under mixed mode loads [7–10].

In aluminum alloys, the primary structural material for airframes, the crack closure behavior is frequently reported to be related with shear lips on the fracture surface [6]. Shear lips are the slant regions of crack surface (mostly 45° inclined) where the tensile mode decohesion is replaced by the shear mode (Figure 1). Either single or double shear lips can occur at the surface, where the plane-stress state exists. After initiation they are broadened towards the center until a stable shear lip width t_s is achieved (if t_s is larger than half of the panel thickness, the whole fracture surface becomes slant) [11]. Since shear lips are usually the highest points of the fracture surface and the residual plastic deformation is also the largest at the surface where the shear lips normally form [12], crack closure is frequently observed to occur at shear lips. The morphology of shear lips is also reported to have influences on the crack tip shielding effects [13].

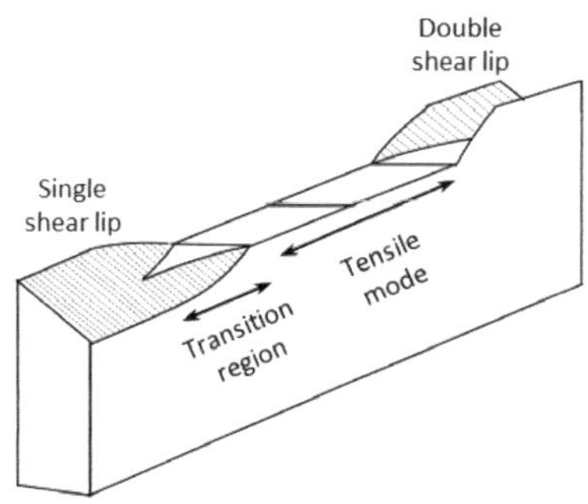

4

Figure 1. Schematic sketch of shear lips form on the fracture surface (after [11]).

Another approach to retard the fatigue crack growth is to modify the driving force via optimization of structural design. In contrast to other approaches, the concept of crenellations has the potential of enhancing the fatigue performance of the structure without introducing additional weight [14-17]. By a systematic thickness variation of the fuselage skin, the fatigue crack is retarded and accelerated in specific regions. In a well-designed crenellation, the fatigue life gain in the retardation region is much larger than the fatigue life loss in the acceleration region. This leads to an overall fatigue life improvement. It was demonstrated that crenellations lead to a 9% fatigue life improvement in unstiffened panels and a 65% improvement in stringer-stiffened panels [18].

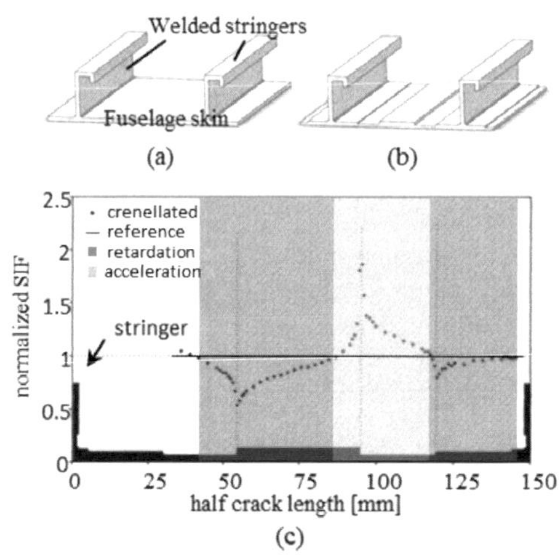

Figure 2. (a) Flat and (b) crenellated structure with the same weight, (c) stress intensity factor (SIF) profile of a center through crack in crenellated stiffened panel normalized to the SIF values of a reference panel (continuous line) with the same structural weight (after [14]).

So far the two aforementioned approaches were only studied separately. However, it is quite possible that the fatigue life improvement via crenellation depends also on the material used. The efficiency of crenellation could be further enhanced if the interaction between the material and geometrical factors is well understood. Thus the aim of this study is to investigate the possible influence of materials on the effectiveness of crenellations. Two candidate alloys for future airframe structures: AA2139 and AA2198 are used in this study for their significant inherent difference. Firstly the grain structure and texture of the two materials were characterized. Then, the fatigue behaviors of crenellated panels made of each material were examined. Possible clues to understand the observed fatigue behavior were obtained through the subsequent fracture surface examination.

MATERIALS AND METHODS

Materials

Metal sheets of the materials AA2139 (4.5 mm thick, T351 heat treatment condition) and AA2198 (5 mm thick, T3 heat treatment condition) were investigated in this study. The nominal compositions of both materials are given in Table I:

Table I. Chemical composition of AA2139 and AA2198 (wt.%).

AA2139 [19]

Cu	Mg	Ag	Mn	Zn	Fe	Ti	Si
4.5-5.5	0.2-0.8	0.15 – 0.6	0.2-0.6	≤0.25	≤0.15	≤0.15	≤0.1

AA2198 [20]

Cu	Li	Mg	Ag	Mn	Zn	Zr	Si	Fe
2.9 - 3.5	0.8 - 1.1	0.25 - 0.8	0.1 – 0.5	≤0.5	≤0.35	0.04–0.18	≤0.08	≤0.01

Microstructure and texture characterization

The grain structure and texture of both materials were investigated using optical microscope and EBSD analysis. Samples for metallographic observation were sectioned parallel to the longitudinal/transverse (L-T) plane, the normal/longitudinal (N-L) plane and the normal/transvers (N-T) plane respectively. The EBSD analysis was performed on N-L sectioned specimens so that the possible variation of texture in through-thickness direction resulted from rolling process can be examined. The Kikuchi patterns were measured using an EDAX Pegasus XM4 system incorporated in a JEOL JSM-6490LV SEM. The magnification at SEM during the EBSD analysis is 500x for AA2198 and 100x for AA2139. The collected data was processed using the software package TSL OIM Data Collection 5.31 and OIM Analysis 5.31.

Biaxial fatigue tests and the corresponding FEM modeling

To investigate the variation of the effectiveness of crenellations with different materials, fatigue tests were carried out on flat and crenelated panels made of the two materials respectively. Square-shaped specimens (560 mm x 560 mm, Figure 3b) representing parts of fuselage panels were cyclically loaded in two perpendicular directions, which is aimed at imitating the periodic fuselage load due to the repeated cabin pressurization and depressurization among flights. The specimens were cut from the metal sheets of both materials. The outer part of the specimens has the original thickness of the sheet, while the inner region (400 mm x 400 mm) was machined to be either flat (thickness: 2.9 mm, machined symmetrically from both surfaces) or crenellated (equivalent thickness: 2.9 mm, machined only on one side of surface). The geometry of the crenellation pattern is depicted in Figure 4, which is in accordance with previous experiments [21]. The red lines in Figure 3 (a) indicate the position of stringers, which will be added to the panel in subsequent studies. The specimens were horizontally fixed in a biaxial servo-hydraulic testing machine with specially designed clamping device to ensure an even distribution of the exerted load. Constant amplitude cyclic loading was applied in both axes of the machine in phase ($F_{major, max}$ = 112.4 kN, $F_{minor, max}$ = 56.2 kN) with R ratio of 0.1 and frequency of 3Hz. The fatigue crack was initiated at both tips of a 12 mm centre notch (tip radius: 0.15 mm), which align perpendicular to the major loading direction. Every time the crack had an approximately 1.5 mm extension at both crack tips, the cyclic loading was halted and the positions of crack tips were recorded using a traveling camera under the specimen.

FEM models for the experimental setup of the fatigue tests were established using the commercial code ABAQUS (version 6.12) (Figure 3c). Quadrilateral 4-node shell elements with reduced integration were applied to model the panels. The clamping device was modeled as a series of beams using quadratic 3-node beam elements. Along the crack path the mesh was refined to 1mm. The stress intensity factors (SIF) during the fatigue crack growth were obtained through the calculation of the strain energy releasing rate (SERR) using a two-step crack closure technique as described by Kruegel [22]. All the FEM models used have been validated in the previous work of the author [23].

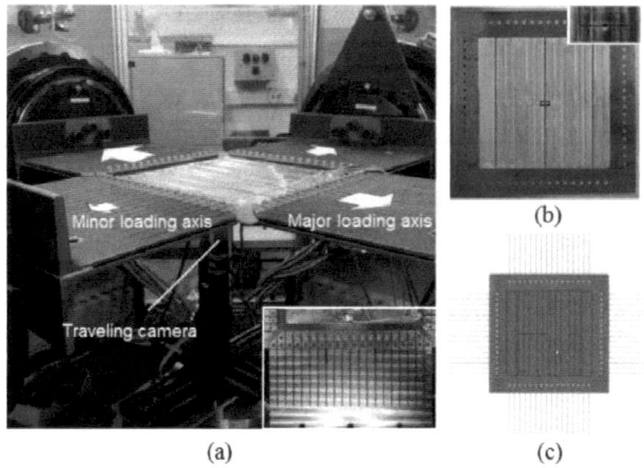

Figure 3. (a) Experimental setup, (b) specimen and (c) corresponding FEM model.

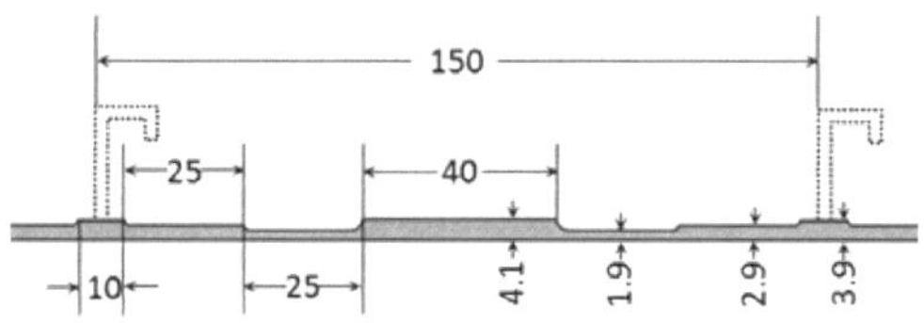

Figure 4. Geometries of crenellation patterns used in this study.

Crack closure measurement technique

The crack opening load was measured by a δ_5 clip gauge mounted at one crack tip, which is the same as used by Heerens et al. [24]. However, in contrast with the stationary clip gauge in their work, the clip gauge in this study was remounted to the new crack tip, every time when the crack extended 10 mm further (Figure 5). During the remounting of clip gauge, two indents with 5 mm distance were firstly made above and below the crack tip. After fitting the two tips of the clip gauge into the indents, fast glue Z70 from the HBM Company was applied to both tips to ensure a good attachment to specimen surface. The glue was afterwards dissolved using a commercial acetone EMSURE® before the next remounting. During the fatigue test, in each load cycle 100 δ_5 values and the corresponding instantaneous load were recorded with equal time interval. The data in uploading phases was used to extract the crack opening load using the compliance method [25]

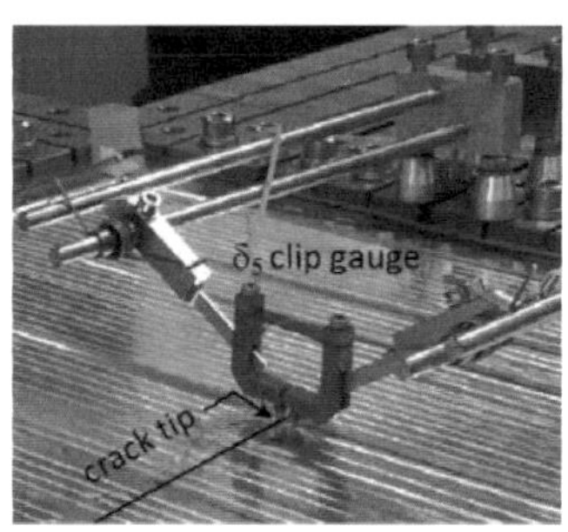

Figure 5. Mounting of δ_5 clip gauge at the crack tip.

Fracture surface observation

The 3D morphology of the fracture surface was measured using a VK-9700 Keyence color 3D laser microscope. The angle between the shear lip plane and the fracture plane was measured on the reconstructed 3D morphology model using the software VK Analyzer 2.5.0.1. A close examination on the fracture surface was carried out using a JEOL JSM-6490LV SEM.

RESULTS

Microstructure and texture

The grain structures of the two materials are compared in Figure 6. The material AA2139 has slightly elongated grains in the rolling direction. The averaged grain sizes in the rolling direction and in the normal direction are 45.5μm and 24.3 μm respectively. In contrast, AA2198 has highly elongated grains, which form a pancake-like structure. The averaged grain size in the normal direction is 3.7 μm whereas in the rolling direction the grain size can exceed 1 mm. The aspect ratio is more than 400:1.

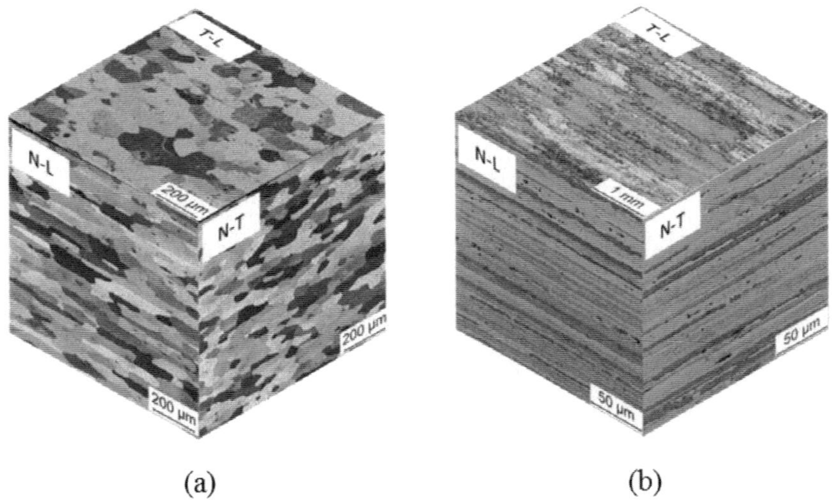

Figure 6. Microstructures of (a) AA2139 and (b) AA2198.

The two materials also show distinct textures as shown in Figure 7. AA2139 has nearly random orientation of grains with very weak cubic texture component. No variation in terms of texture in the thickness direction was found. AA2198 shows very sharp Brass texture in the center of the panel, whereas near the the sheet surface strong S texture components appear.

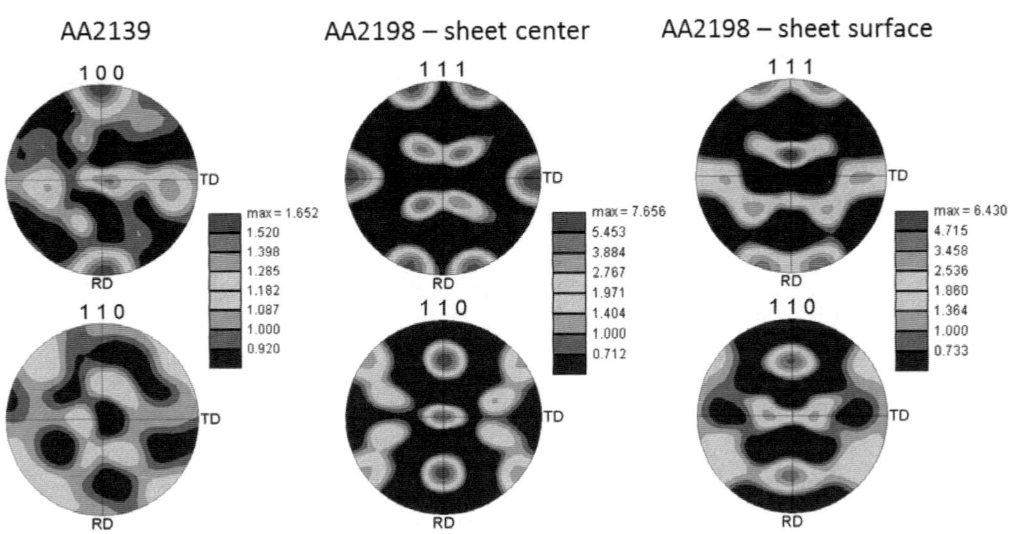

Figure 7. Variation of texture in different materials and at different positions of the panel.

Fatigue tests

The results of the fatigue tests (Figure 8) show that, under the same loading condition, panels made of AA2198 generally have higher fatigue resistance than the panels of the material AA2139. The fatigue life improvement by crenellations in the material AA2198 is two times of that in the material AA2139.

The crack propagation rates calculated based on Figure 8 with a 5-points-polynominal fitting method [25] were plotted against the corresponding stress intensity factor range from the FEM simulation under double logarithm (Figure 9). The data points of both flat and crenellated panels of AA2139 fall into the same linear band as characterized by Paris Law whereas the data points for panels of AA2198 are located in a lower and broader scatter band. The crenellated panel of AA2198 generally has lower crack propagation rates (the da/dN values) at the same ΔK compared to the flat panel. At low ΔK range ($<15 MPa\cdot m^{1/2}$), the da/dN values of AA2198 panel tend to be the same as those of the panels of AA2139. The same trend was also found at high ΔK range ($>35 MPa\cdot m^{1/2}$) where the da/dN values of all the panels tend to be the same.

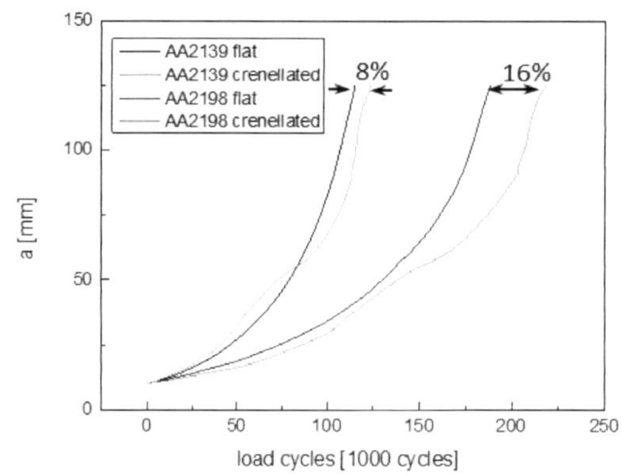

Figure 8. The fatigue performance of flat and crenellated panels with different materials.

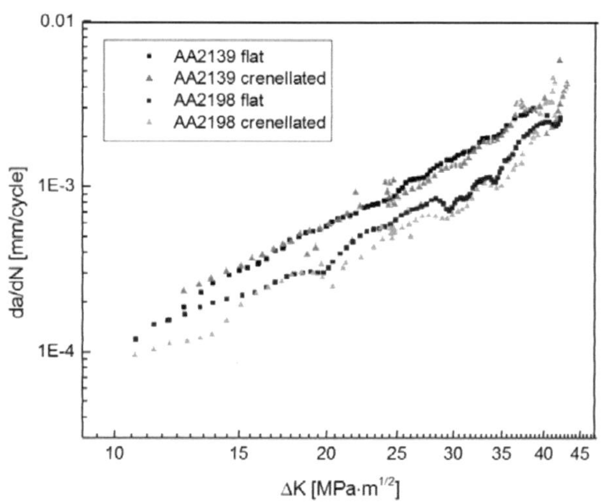

Figure 9. da/dN-ΔK plot of the four tested specimens.

During the fatigue tests, the crack opening load was measured every time when the crack extended by another 2mm. Table II summarizes the averaged crack opening load measured in the flat and crenellated panels of both materials. Both the flat panel and crenellated panels of AA2139 and the flat panel of AA2198 have very close crack opening load S_{op} whereas S_{op} is significantly higher in the crenellated panel of AA2198.

Table II. Average crack opening load (in major loading axis) of the four tested specimens.

	flat	crenellated
AA2139	51 kN	52 kN
AA2198	49 kN	57 kN

Fracture surface observation

Figure 10 compares the difference of fracture surfaces of AA2139 and AA2198 panels at macroscopic scale. Both of the two AA2139 panels show flat crack surfaces with smooth single shear lips. The shear lips start at positions marked by the white arrows and are fully developed at positions marked by the black arrows, where the whole section of the fracture surface is 45° inclined.

In the panels of AA2198, the fracture surface is much more tortuous due to the development of complex shear lip morphology. In the flat panel double shear lips are sometimes observed which are accompanied with deviations of crack path around 10°. In contrast with the gradual rotation of the crack plane during the shear lip development, abrupt changes of the shear lip plane are usually observed as marked by the squares in Figure 10. In the crenellated panel, those abrupt changes are usually found associated with the thickness steps in the crenellations. In the crenellated panels very sharp shear lips are found along the crenellated side since the beginning of the fatigue tests. Those very sharp shear lips are less than 1 mm wide with the local crack surface about 70° inclined to the original crack plane. At large crack length (a>95 mm) stable single shear lips are observed in both flat and crenellated panels as those observed in the AA2139 panels.

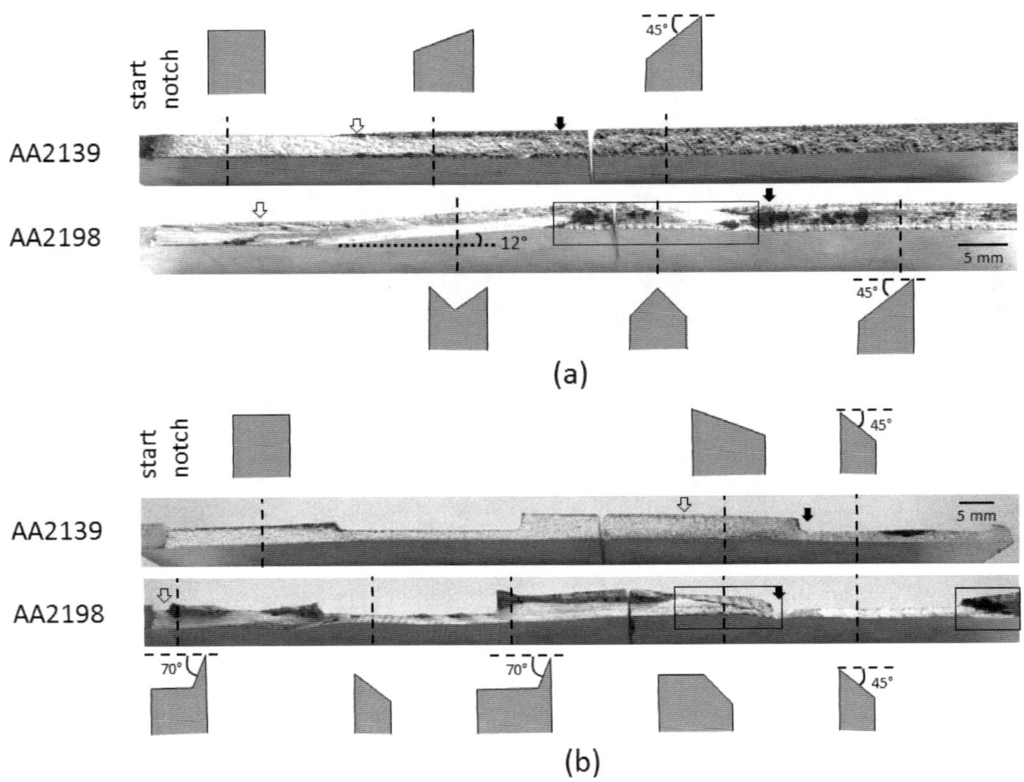

Figure 10. Fracture surface of (a) flat panels and (b) crenellated panels and the different shear lip shapes developed at different positions.

Close examination of the crack surface in the tensile mode region under SEM shows a flat crack surface at microscopic scale in both materials. The irregularities on the surface are found mostly correlated with the positions of grain boundaries as indicated by the optical micrograph. Within the grains the crack surface shows smooth appearance with visible striations. The normal of the crack surface is well aligned with the major loading direction.

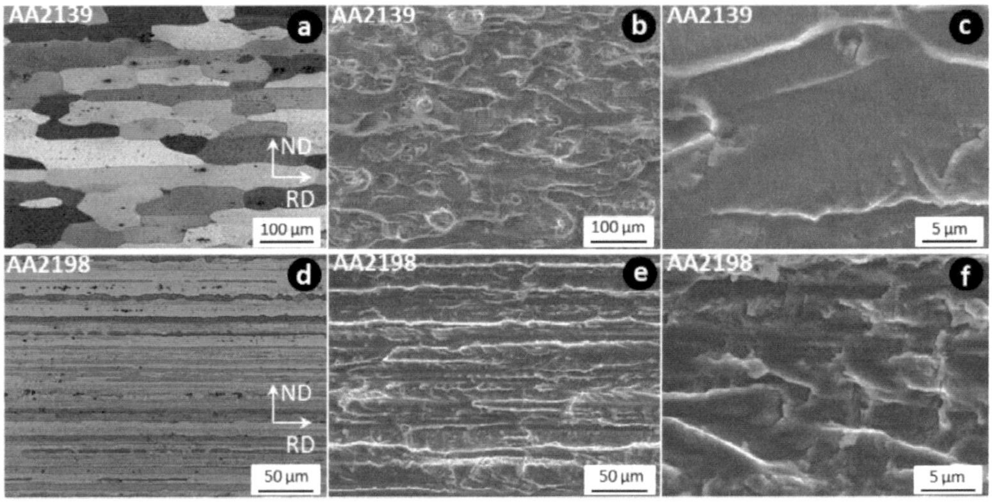

Figure 11. SEM micrographs of crack surfaces (b, c, e, f) of the two materials where the fatigue cracks grow in tensile mode and the comparison of crack surface morphology with the corresponding base material microstructure (a, d).

In the shear mode region (the shear lips), the crack surface is amorphous (Figure 12). Abrasion regions near the sharp shear lips were usually observed in the AA2198 crenellated specimen, where straight scratches are found in the direction perpendicular to the panel. However such abrasion zone is hardly seen in other specimens.

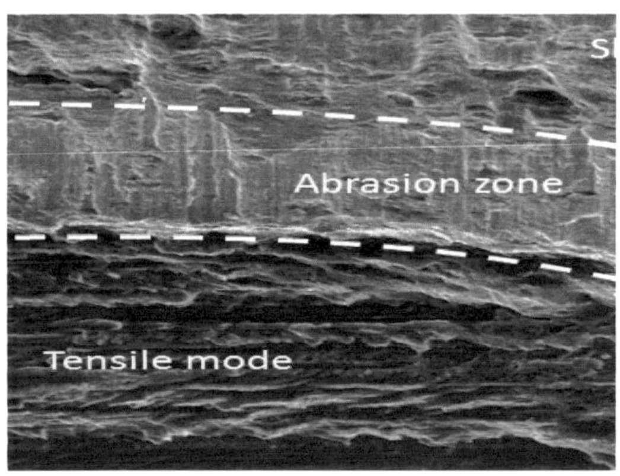

Figure 12. Abrasion zone observed on the fracture surface of the AA2198 crenellated panel.

Observations on the deformation state of the panels under biaxial load

Out of plane bending movement is observed in the FEM simulation of the biaxially loaded crenellated panels which is, however, absent in flat panels. This out of plane bending is found strongly influenced by the magnitude of load perpendicular to the crenellations. Although cyclic out of plane movement is also observed during the fatigue tests of flat panels, its magnitude is much smaller than that observed in the fatigue tests of crenellated panels.

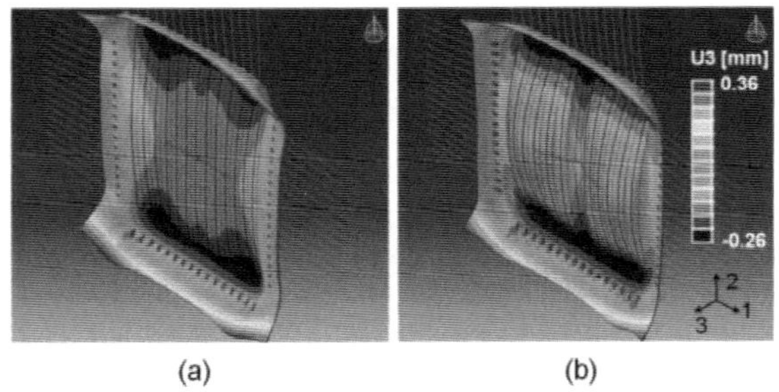

Figure 13. Out of plane bending of crenellated specimens under biaxial loads (a) biaxial load ratio 0.1 (b) biaxial load ratio 0.5.

DISCUSSION

Texture and the shear lip formation

The distinct textures of the two materials are related with the different heat treatment conditions. The T351 heat treated AA2139 is only slightly stretched (1-2%) after the solution treatment whereas the T3 heat treatment of AA2198 involves large amount of cold working [26]. The very sharp Brass texture observed in AA2198 is very typical for cold rolled FCC metals [27]. The strong S texture components near the surface of the plate are due to the increasing degree of deformation caused by the direct contact with rollers [28]. The large amount of cold working in AA2198 is also reflected by the pancake grain structure shown in Figure 6.

The distinct textures of the materials can significantly influence the morphology of shear lips that are developed. This has been reported by several previous researchers. Zuidema et al. suggested the formation of shear lips is related with the type and amount of texture in the material [11]. Tchorzewski and Hutchinson also found the width of shear lips is strongly influenced by the orientation of the fracture plane in the textured Ti-6Al-4V alloy [29].

The influence of texture on the shear lip morphology can be explained in terms of slip characteristic of the material. According to Schijve [30], in the region where shear lips form, the initial tensile mode decohesion of the material is overruled by the shear mode decohesion along the plane of maximum shear stress, which is 45° inclined to the initial crack plane. Thus the formation of shear lips involve large amount of dislocation slip along the plane of shear decohesion or along planes with close orientations. In the strong textured material AA2198, the easy sliding planes – the {111} planes of the grains are not randomly orientated but concentrate in specific orientations. If they have a good alignment with maximum shear stress plane, they can contribute a lot to the slip that is necessary for the shear lip formation. In such a case the morphology of shear lips developed should be quite different from the material of no texture (AA2139), in which the {111} easy sliding planes are randomly orientated.

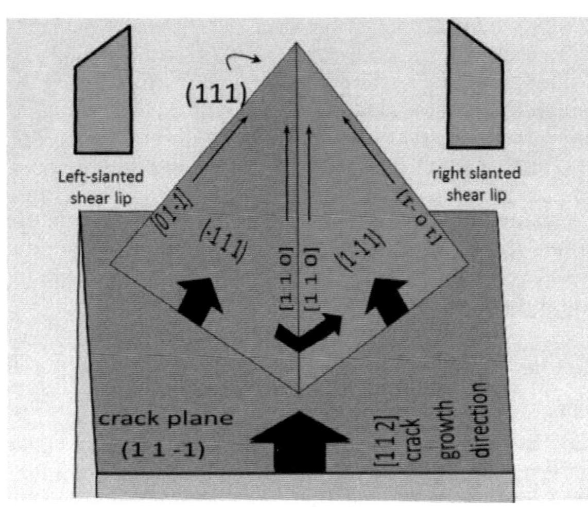

Figure 14. Orientation of {111} planes in the fatigue cracked AA2198 panels and the activated slip systems involved in the shear lip formation.

Table III. Schmid-Factors of different slip systems in the strong textured AA2198 panels subjected to tensile stress field that is perpendicular to the crack line.

Slip System	Schmid-Factor	Slip System	Schmid-Factor
(111) [-110]	0	(1-11) [110]	0
(111) [-101]	0.272	(1-11) [-101]	0.272
(111) [01-1]	0.272	(1-11) [011]	0.272
(-111) [110]	0.272	(111) [-110]	0
(-111) [101]	0	(111) [101]	0
(-111) [01-1]	0.272	(111) [011]	0

Based on the measured sharp Brass texture in AA2198 panels and the alignment of fatigue crack growth direction with the rolling direction in the tests, the orientations of the {111} planes with respect to the sample frame can be determined, which are schematically sketched in Figure 14. The (11-1) plane coincidences with the initial crack plane in tensile mode while the other three {111} planes form a regular tetrahedron above the crack plane with one corner pointing to the opposite of the crack growth direction. The Schmid-Factors of the 12 slip systems on the 4 {111} planes were calculated (Table III) presuming a tensile stress field perpendicular to the original crack plane. It can be seen that 6 slip systems of them are more advantageous to be activated. Since (-111) and (1-11) planes have the best alignment with the maximum shear stress planes, which are 45° inclined to the side surfaces of the panels, the activation of slip systems (-111) [01-1], (-111)[110], (1-11)[10-1] and (1-11)[110] will probably have the greatest contribution to the shear lip formation. If so, the co-activation of slip systems (-111)[01-1] and (-111)[110] will favor shear decohesions along the (-111) plane and will finally generate a left-slanted shear lip; the co-activation of slip systems (1-11) [10-1] and (1-11)[110] will favor shear decohesions along the (1-11) plane and will finally generate a right-slanted shear lip. As shown in Figure 14 the (-111) and (1-11) planes are symmetric and equally advantageous in initiating shear lips. If shear lips are initiated along both planes but on different sides of panel surface then a double shear lip forms as observed in the AA2198 flat panel. This equality of (-111) and (1-11) planes also implies there can be a competition of activated slips along those planes. Local inhomogeneties (e.g. the thickness steps of crenellations) can probably lead to alternate activation of slip along those planes, during which the abrupt change of shear lip plane orientations can occur as observed in experiments (Figure 10, marked by squares).

The texture of AA2198 also plays an important role in the formation of sharp shear lips along the crenellated side in the crenellated specimen. Normally the formation of shear lips in crenellated panel should have been delayed compared to the flat panel due to the reduced crack propagation rate [11] in the retardation region of crenellations. However, the first sharp shear lip observed in the crenellated panel was only 1-2 millimeters from the starting notch. This unexpected early formation of shear lips is probably attributed to the enhanced stress level at the crenellated side due to the bending of the panel (Figure 14). The increased tensile stress finally can activate sufficient slip along either (-111) or (1-11) plane to initiate the shear lip. Under this near threshold condition, since the slips are dominantly provided by the slip systems belongs to either (-111) or (1-11) plane, the orientation of shear lip plane should show strong dependence on the orientation of those planes. In experiment, the angle between those sharp shear lip plane and initial crack plane is around 70°, which is nearly the same as the angle between the different {111} planes (70.72°). When the fatigue crack grows longer (>80mm in our case) and the stress concentration near the crack tip becomes high enough, this tendency of shear lip formation along (-111) or (1-11) plane disappears. 45° inclined shear lips become dominant in the end.

In summary the tortuous shear lip morphology in AA2198 panels either the due to the abrupt change of shear lip plane or due to the formation of sharp shear lips is closely related with sharp Brass texture of the material.

Shear lip morphology and the fatigue behavior

The tortuous shear lip morphology in the AA2198 panels can influence the fatigue behavior in two different aspects. Firstly the sudden change of shear lip plane and the occurrence of double shear lips in the AA2198 panels can lead to crack deflection. In such situation the mode I crack driving force is reduced due to the deviation from maximum tensile stress plane as described by Ritchie [3]. This can be the major additional toughening mechanism compared with the material AA2139. Without considering the regions of shear lips, the mechanism of fatigue cracking in both materials is the same, which results in similar smooth and flat crack surface perpendicular to external loading (Figure 11). As a result before the development of tortuous shear lips and after formation stable single shear lips in AA2198, the fatigue crack in AA2198 develops at a similar speed to that in AA2139 as shown in Figure 9. This agrees with the observation of Bergner et al. [31], that many aluminium alloys show similar da/dN at $\Delta K_0 = 10\text{MPa} \cdot \sqrt{\text{m}}$, and the additional fatigue resistance shown in some alloys is due to the increased fracture surface roughness.

The second influence is the heightened crack closure level in the crenellated AA2198 panel. The out of plane bending increases the chance of surface interference. This plus the tortuous crack surface morphology provides additional sliding crack closure. However, a rough fracture surface alone without relative sliding between the surfaces will not contribute to an increased crack closure level as measured in the AA2198 flat panel. It is the same situation as the roughness induced crack closure observed in low ΔK_0 range, which requires both a rough surface and relative displacement between the two crack surfaces [32]. As a result in all tested four panels only the combination of tortuous shear lip morphology of the AA2198 alloy and the out of plane bending behaviour of the crenelated panel can further enhance the crack closure level (Table IV).

Table IV. Schmid-Factors of different slip systems in the strong textured AA2198 panels subjected to tensile stress field that is perpendicular to the crack line.

	flat	**crenellated**
AA2139	Smooth shear lip; No out of plane bending => No sliding crack closure	Smooth shear lip; Out of plane bending => No sliding crack closure
AA2198	Tortuous shear lip; No out of plane bending => No sliding crack closure	Tortuous shear lip; Out of plane bending => Sliding crack closure

CONCLUSION

The materials used in the crenellated structure can significantly influence the effectiveness of crenellations. Fatigue life improvement via crenellations doubles in the strong textured material AA2198 compared to the random textured material AA2139. The strong texture leads to tortuous shear lip morphology. The out of plane movement of crenellated structure under service-related loading conditions (biaxial loads in our case) favors the formation of sharp shear lips along the crenellated side and leads to an enhanced crack surface interference at those shear lips. The resulted sliding crack closure contributes to the additional fatigue life improvement observed in the AA2198 crenellated panel.

ACKNOWLEDGEMENT

The authos would like to thank Mr. Jürgen Knaack, Mr. Karl-Heinz Balzereit and Mr. Falk Dorn from the department Joining and Assessment of the Institute of Materials Research, Materials Mechanics of the Helmholtz-Zentrum Geesthacht for their support in the experimental work.

REFERENCE

[1] Bhaumik, S.K., Sujata, M., Venkataswamy, M.A. (2007). *Eng. Fail Anal.*, vol. 15, n.6 p. 675-694.

[2] Goranson, U.G. (2007). In *Damage Tolerance – Facts and Fiction*, 1st International Conference on Damage Tolerance of Aircraft Structures, keynote presentation, Delft.

[3] Ritchie, R.O. (1988). *Mater. Sci. Eng. A*, vol. 103, p. 15-28.

[4] Morgeneyer, T.F. (2008). *Micromechanical Studies and Modelling of Toughness in High Strength Almuinium Alloys*, doctor thesis, University of Southampton, Southampton.

[5] Tsivoulas, D. (2010). *Effects of Combined Zr and Mn Additions on the Microstructure and Properties of AA2198 Sheet*, doctor thesis, University of Manchester, Manchester.

[6] Schijve, J. (1986). Keynote paper at the International Symposium on Fatigue Crack, Charleston

[7] Seo, K.J., Lee, J.M., Choi, B.H. (2012). *Mater. Trans.*, vol. 53, n.2 p. 4493-4502.

[8] Freitas, M., Reis, L., Fonte, M., Li, B., (2011). *Eng. Fract. Mech.*, vol. 78, n.5 p. 826-835.

[9] Fonte, M., Reis, L., Romeiro, F., Li, B., Freitas, M. (2012). *Int. J. Fatigue*, vol. 28, p. 609-617.

[10] Akhurst, K.N., Lindley, T.C., Romeiro, Nix, K.J. (1983). *Fatigue Eng. Mat. Str.*, vol. 6, n.4 p. 345-348.

[11] Zuidema, J., Veer, F. and Kranenburg, C.V. (2005) *Fatigue Fract. Eng. Mater. Struct.*, vol. 28, p. 159-167.

[12] Materna, A. and Oliva, V. (2005). In Proceedings of 11th International Conference on Fracture, Turin.

[13] Zuidema, J. and Krabbe, J.P. (1997). *Fatigue Fract. Eng. Mater. Struct.*, vol. 20, p. 1413–1422.

[14] Uz, M.-V., Koçak, M., Lemaitre, F., Ehrström, J.-C., Kempa, S., Bron, F. (2009). *Int. J. Fatigue*, vol. 31, p. 916-926.

[15] Muzzolini, R., Ehrstroem, J.-C. (2004). In Proceedings of the 15th Advanced Aerospace Materials & Processes Conference and Exposition AeroMat, Seattle

[16] Bucci, R. J. (2006). In Proceedings of Aircraft Structrual Integrity Program Conference, San Antonio.

[17] Ehrstroem, J.-C., Muzzolini, R., Arsene, S., Van der Veen, S. (2005) In Proceedings of the 23rd Symposium of the International Committee on Aeronautical Fatigue, Hamburg.

[18] Uz, M.V. (2010). *Improvement of Damage Tolerance of Laser Beam Welded Aerospace Structures via Local Engineering*, doctor thesis, Techischen Universität Hamburg-Harburg, Hamburg.

[19] Morgeneyer, T.F., Starink, M.J., Sinclair, I. (2008). *Acta Mater.*, vol. 56, p. 1671-1679.

[20] Alexopoulos, N.D., Migklis, E., Stylianos, A., Myriounis, P. (2013). *Int. J. Fatigue*, vol. 56, p. 95-105.

[21] Lu, J., Kashaev, N., Huber, N. (2014) In Proceedings of the 29rd Congress of the International Council of the Aeronautical Science, St. Petersburg.

[22] Krueger, R.(2004). *Appl. Mech. Rev.*, vol. 57, n.2, p. 109-143.

[23] Lu, J., Huber, N., Kashaev, N., (2014) In Proceedings of the 2nd International Conference of IJSI, Funchal.

[24] Heerens, J., Schödel, M. (2009). *Eng. Fract. Mech.*, vol. 76, p. 101-113.

[25] ASTM E647, *Standard test method for measurement of fatigue crack growth rates.*

[26] ANSI H35.1/H 35.1M-2009, *Alloy and Temper Designation Systems for Aluminium.*

[27] Suwas, S., Kumar Ray R., (2014). *Crystallographic Texture of Materials*, Springer, London.

[28] Cai, C., Ji, Z., Zhang, H., Wang, G. (2009). *J. Metallurgy*, vol. 2009, p. 1-6.

[29] Tchorzewski, R.M. and Hutchinson, W.B. (1978). *Metall. Trans.*, vol. 9A, p. 1113-1124.

[30] Schijve, J. (1986). Keynote paper at the International Symposium on Fatigue Crack, Charleston

[31] Bergner, F., Zouhar, G., Tempus, G. (2001). *Int. J. Fatigue*, vol. 23, p. 383-394.

[32] Suresh, S. and Ritchie, R.O. (1982). *Metall. Trans. A*, vol. 13A, p. 1627-1631.

Composite materials development of innovative structural concepts

Fatigue and Damage Tolerance Assessment of 3-D reinforced CFRP Bonded Joints

Chiara Bisagni[1], Domenico Furfari[2], Marco Pacchione[2]

[1]*Politecnico di Milano, Italy, currently at TU Delft, Netherlands*
[2]*Airbus Operations GmbH, Germany*

Abstract: An experimental campaign has been carried out to investigate the behavior of co-bonded CFRP joints with a novel design incorporating a through the thickness local reinforcement. Different specimens were manufactured to investigate static and fatigue behavior, as well as delamination size after impact and damage tolerance characteristics. Mechanical performance of specimens with a reinforcement based on the insertion of spiked thin metal sheets between the co-bonded laminates were compared with purely co-bonded joints. This novel design demonstrated by tests a more predictable failure with respect to the standard design based solely on adhesive structural bonding. The tests allowed obtaining a significant number of experimental results that can be used to define preliminary design guidelines.

INTRODUCTION

Failure predictions for bonded fiber reinforced plastics (FRP) composites have been researched extensively, but the mechanisms are not yet well-understood. Bonded FRP shows different failure modes, and ASTM D5573 identifies 7 classes of failure modes [1]:

- Adhesive failure, sometimes referred to as interfacial failure: the separation appears to be at the adhesive-adherent interface;
- Cohesive failure: the separation is within adhesive;
- Thin-layer cohesive failure, sometimes referred to as interphase failure: the separation is very close to the adhesive-substrate interface;
- Fiber-tear failure: failure occurs exclusively within the FRP matrix;
- Light-fiber-tear failure: failure occurs within the FRP substrate, near the surface, characterized by a thin layer of the FRP resin matrix visible on the adhesive;
- Stock-break failure: break of the FRP substrate outside the adhesively bonded-joint region, often occurring near it;
- Mixed modes failure: any combination of two or more of the six classes of failure modes defined above.

The well-known joint configurations regularly studied in literature are the single lap joints, double-lap joints, scarf joints and stepped-lap joints [2].

The failure mode and the ultimate strength load depend strongly on several factors, such as the mechanical properties of adhesives and adherents; the eventual residual stresses; the surface preparation and interaction; the loading condition; and the joint configuration.

Consequently, failure predictions have to consider all these factors, but also different parameters (e.g. surface ply angle, stacking sequence, bond-line thickness), so it becomes difficult to predict failure strength because of its dependence on different bonding methods and parameters.

Single-lap, double-lap and scarf joints configurations strength and in particular the orientation of the composite layer at the adhesive adherent interface was analyzed, among others, by Meneghetti et al. [3]. Single lap joint specimens were tested under tension-tension fatigue. Furthermore, different overlap lengths and corner geometries of the bonded area were investigated to identify initiation and crack growth.

Tong [4] focused on the interlaminar delamination with embedded cracks near the overlap ends, in single lap joints subjected to axial loads. The presented approach can be compared to a damage tolerance design. The presence of cracks in the adherents can change the stresses distributions and influence the failure strength and the crack modes, which depend on the location of the crack initiation. Tong recorded a reduction in the failure load, unless the crack was placed completely outside the overlap area, but in that case it might not affect the joint strength.

Some work related to the properties of the adhesive, include Tomblin, Yang and Harter [5] who investigated the influence of thickness on the shear strength of adhesive joints for both metal-to-metal interface (aluminum alloy 2024) and composite adherents (carbon/epoxy and fiberglass/epoxy). They tested the strength properties of a single-lap joint and its failure modes, with particular attention to the specimen bonding procedure and surface preparation in accordance with the ASTM standards D3165 [6] and D1002 [7]. They observed interlaminar failure in the first ply of the substrate in the composite specimens. Furthermore, the strength of the joint decreased with the increasing thickness of the bond-line. Thus, it was concluded that a ruling factor was the load path eccentricity, which increased the peel stresses in the joints.

Different 3D reinforcement methods have been studied over the past years to improve the performance and in particular the damage tolerance of bonded joints of carbon fiber reinforced plastics (CFRP). The 3D reinforcements establish a mechanical link between two adhesive-bonded CFRP structures. A general overview over reinforced and trans-laminar composites can be found in [8].

Different techniques have been studied and considered, e.g. stitching, tufting and Z-pinning. In stitching and tufting, a needle perforate a fabric preform, inserting a high-tensile-strength yarn made of glass, carbon or aramid. In particular, tufting requires access only on one side of the fabric, since the needle penetrates and retracts through the same pathway. Both the techniques can only be applied to joints made from resin-infused fabric. On the contrary, Z-pins are suitable for pre-preg composite laminates. They can be both thin metallic or composite rods inserted in the through-thickness direction of a laminate during manufacturing.

Tong et al. [9] describe the variety of processes used to manufacture 3D reinforced composites as well as the models for predicting failure strength. Great importance is given to the results in terms of performances, such as in-plane mechanical properties, failure mechanism under tension, compression, bending and fatigue loads.

Failure mechanism and evaluation of the energy release rate have been studied for single lap joints stitched and unstitched specimens in [10], observing that the stitches delayed the debonding phenomenon and increased the failure load of two and a half times. An improvement in the failure load and in the damage tolerance have also been noticed with stitched T-joint specimens in [11] tested under flexure and tension loads.

Freitas et al. [12] investigated the tensile pull-off strength of a carbon-epoxy T-shaped joints with z-pin reinforcement made of titanium alloy. The tests results showed that the z-pins were not effective in suppressing the onset of tensile damage in stiffened panels. However, the reinforced joints continued to support an increasing load 2.3 or 2.6 times higher than the failure load for unreinforced joints.

Detailed studies of z-pinned composite bonded joints and the influence of different parameters like z-pins size, volume content and type of loading can be found also in [13-14], and in [15] where Chang et al. underlined some of the problems that 3D reinforced structures can face, like broken fibers, fibers kinking or misalignment and resin pockets. All these factors can affect the static and fatigue behavior of the joints.

This paper reports the results of an experimental campaign carried out to investigate the behavior of co-bonded joints on CFRP specimens with a novel design incorporating a through the thickness local reinforcement. Different specimens are manufactured to investigate static and fatigue behavior, as well as delamination size after impact and damage tolerance characteristics.

DESCRIPTION OF THE 3-D REINFORCED CFRP BONDED JOINTS

The laminates are manufactured from M21/T800S UD carbon/epoxy prepreg tape from Hexcel, with a nominal ply thickness of 0.193 mm and a fiber mass of 198 g/m2 as well as 0.12mm and fiber mass of 134g/m2. The laminates consist of 10 plies with stacking sequence: [45/-45/0/90/0]S when thicker ply thickness are used and 16 plies with stacking sequence: [45/-45/0/90/-45/45/0/90]S when lower thickness ply are used, both resulting in a thickness of 2 mm.

For the bonding of the laminates a high shear strength modified epoxy adhesive from Cytec is used.

The 3D reinforcements consist of thin spiked metal sheets between co-bonded structures, so that the resulting combination of adhesive and mechanical load transfers allows more progressive and predictable failure mechanisms.

This new joint technology is called RHEA (Redundant High-Efficiency Assembly), and is briefly presented by Nogueira et al. [16-17]. The spiked metal sheet reinforcement, sketched in Figure 1, can be produced out of cold formable Titanium alloys (Ti 3-2.5 or Ti 15-3-3-3) or Stainless steel (304/ DIN1.4301) by metal stamping process in thickness ranging from 0.2 to 0.4mm. Prototypes reinforcements used for the tested specimens are obtained from SAE304 featuring a thickness of 0.2 mm by laser cutting and subsequent bending of the spiked geometry with a special tooling. The optimization of the spike geometry and array is currently under investigation.

The manufacturing process involves the introduction of the spiked metal sheet reinforcement into a CFRP prepreg laminate during its curing cycle, as shown schematically in Figure 2. During the curing process, the autoclave pressure and temperature lead to the insertion of the spikes into the prepreg. In this way, fibers are not cut but simply realigned. Once this first curing cycle is completed, the 3D reinforced cured part is joined with another uncured part or with another 3D reinforced cured part through the use of an additional uncured doubler.

Figure 1. Spiked metallic reinforcement (RHEA).

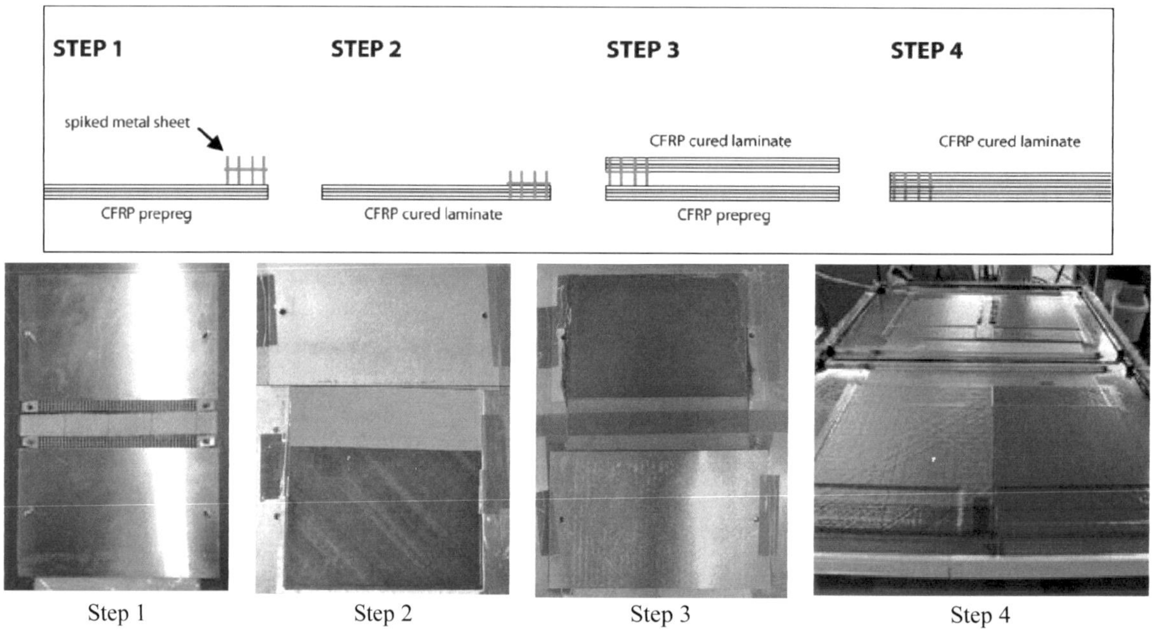

Figure 2. Co-bonding manufacturing process for RHEA joints [16].

Reference specimens without metallic reinforcement are manufactured at the Department of Aerospace Engineering of Politecnico di Milano, using the CFRP material and the adhesive provided by Airbus. The specimens with the metallic reinforcement are manufactured by Airbus Group Innovations (Ottobrunn).

An X-ray scan of a specimen is reported in Figure 3, showing a top view of the spiked metal sheet reinforcement. Figure 4 shows some details of the grid through the thickness taken by photos at the microscope. Depending on where the metallic reinforcement is cut, different aspects can be seen. In photo A the pins are visible through the entire thickness with the typical chessboard scheme. Photo B, instead, highlights the constant cross-section in the thickness, while photo C refers to the cross-section near the pins, although they are not completely visible.

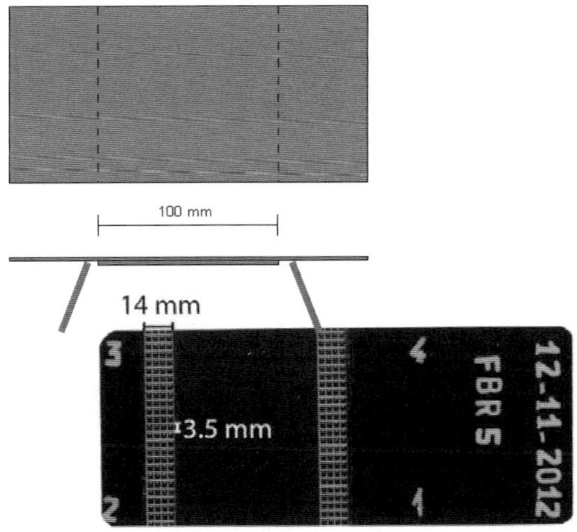

Figure 3. X-ray scan of a RHEA specimen.

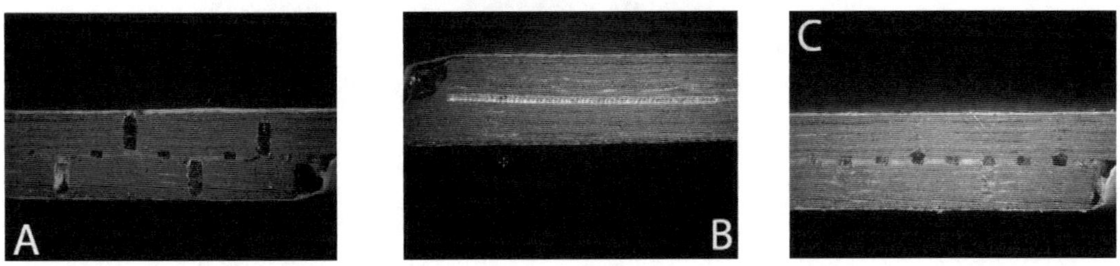

Figure 4. Photos taken at the microscope through the thickness showing different views of RHEA.

SINGLE LAP SHEAR JOINTS

The first type of specimens that is tested consists of unsupported single lap shear joints with overlap distance of 55 mm to represent a load transfer conditions similar to the one at typical high load in aircraft structural components. Both specimens without any metallic reinforcements and with RHEA are tested. These types of specimens allow to assess the static and fatigue behavior of the new joint design under tensile loading with high peel stresses at bonded line run out.

The dimensions of the single lap shear joints specimens are shown in Figure 5, while Figure 6 shows the metallic spiked reinforcement that is present for a width of about 14 mm on each side and extends outside the overlap area for about 2.5 mm.

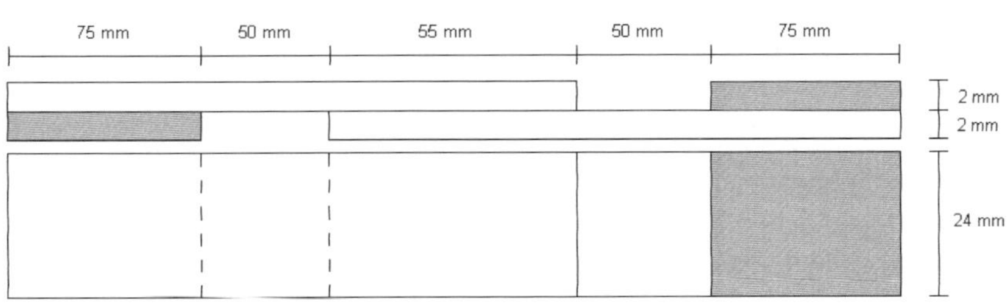

Figure 5. Geometry of single lap shear joint specimens.

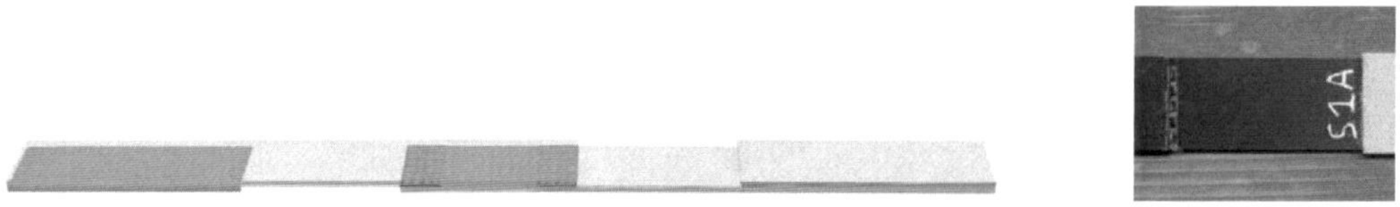

Figure 6. Single lap shear joint specimens with RHEA.

All the tests are performed at Politecnico di Milano using a MTS testing system 810 TestStar IIs with maximum vertical load equal to 250 kN. Figure 7 reports a photo of the MTS equipment with a specimen during a test.

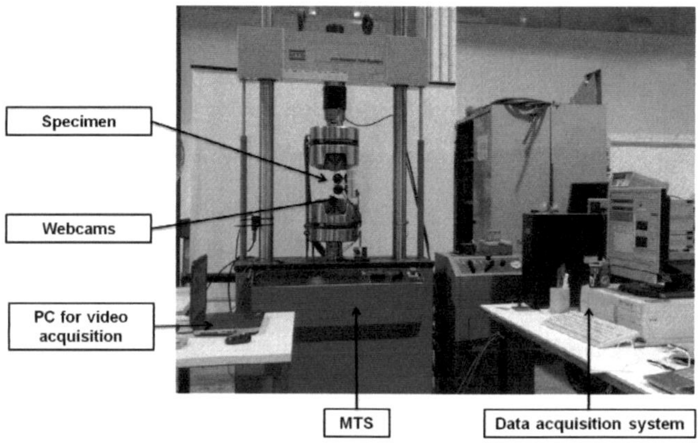

Figure 7. MTS testing equipment during a test on a single lap shear joints.

Six specimens are statically tested without reinforcements and one is tested with reinforcements. The loads and the shear strengths obtained in each test are summarized in Table I. It is possible to note that the collapse load measured on the specimen with RHEA is 53% higher than the average collapse loads of the specimens without RHEA.

Table I. Collapse loads and shear strengths (max load divided by the overlapping surface) of single lap shear joint specimens without RHEA and with RHEA.

Specimen	Collapse load [N]	Shear strength [MPa]
S1 no RHEA	16731	12.6
S2 no RHEA	16345	12.5
S3 no RHEA	15759	12.0
S4 no RHEA	16410	12.4
S5 no RHEA	15404	11.6
S6 no RHEA	14524	10.9
S1 with RHEA	23873	17.6

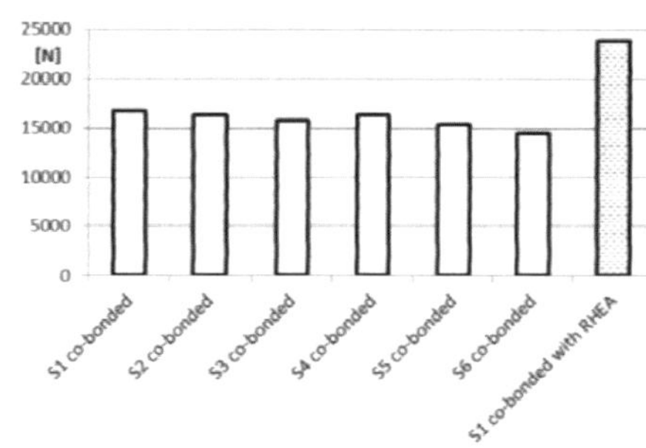

Then, six specimens without reinforcements and three specimens with reinforcements are tested in fatigue. The tests are performed at different level of loads with frequency of 10 Hz.

A summary of all the performed fatigue tests on the single lap shear joints specimens is reported in Table II.

Table II. Summary of fatigue tests on single lap shear joint specimens.

	Specimen	Fatigue min load [N]	Fatigue max load [N]	Max number of cycles
Co-bonded	F1 no RHEA	500	5000	Runout (3 million)
	F2 no RHEA	700	7000	96079
	F3 no RHEA	650	6500	385178
	F4 no RHEA	650	6500	469308
	F5 no RHEA	650	6500	361903
	F6 no RHEA	600	6000	1475961
Co-bonded with Rhea	S2 with RHEA	650	6500	1905775
	S3 with RHEA	700	7000	1283002
	S4 with RHEA	700	7000	3076115

The experimental results show that the static strength as well as the fatigue strength of the reinforced joint is significantly higher than the co-bonded baseline solution.

The crack growth is monitored during the tests. The growth measured during the fatigue test on specimen S4 with RHEA is reported in Figure 8 in function of the number of cycles, and some photos taken on the specimen are shown in Figure 9.

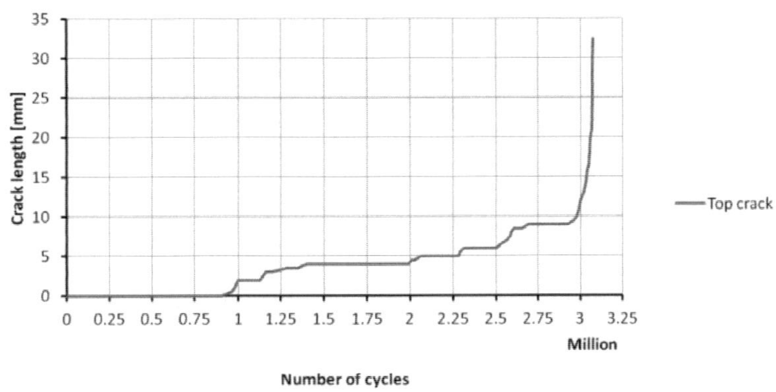

Figure 8. Crack growth measured during the fatigue test on specimen S4 with RHEA.

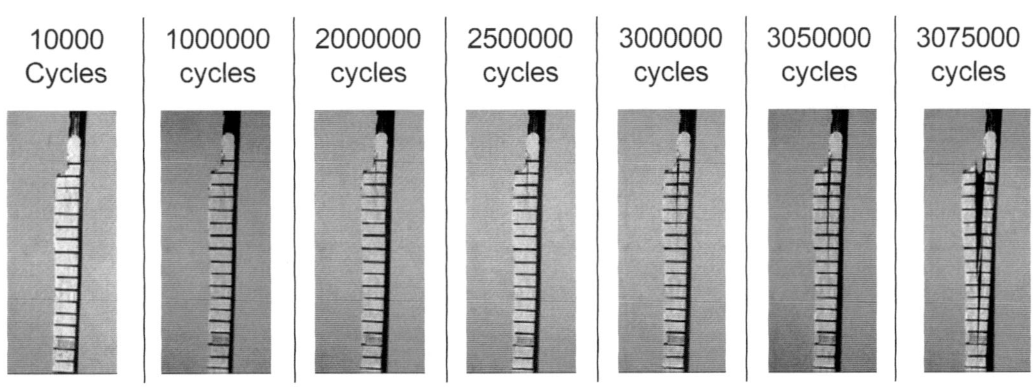

Figure 9. Photos taken during the fatigue test on specimen S4 with RHEA.

The failure happens at different interfaces as shown in Figure 10, where photos of the two parts of specimen S2 with RHEA after the test are reported. It is also possible to see that the metal pins pulled out from the laminate.

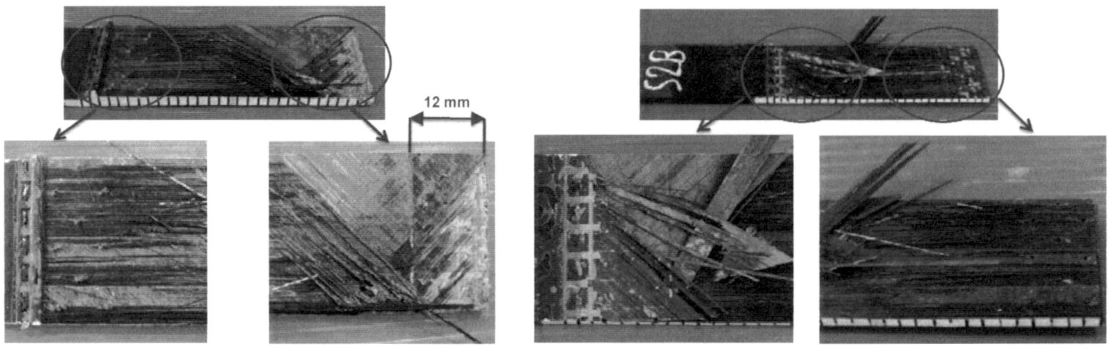

Figure 10. Photos of the two parts of specimen S2 with RHEA after the fatigue test.

BONDED FLANGE FOR 4-POINT BENDING TESTS

The second type of specimens that are manufactured consist of a bonded flange on a rectangular skin and are tested in a 4-point bending setup under static and fatigue loading. In this case the compressive load is applied to assess the static and fatigue behavior of the joints measuring the delamination growth of the reinforced joints compared to the purely co-bonded ones.

Basically, each specimen is constituted by two laminates. The first rectangular laminate is 2 mm thick, 100 mm wide and 200 mm long. In the central part a second laminate is co-bonded using adhesive. The second laminate is 2 mm thick, and both width and length are equal to 100 mm. A sketch of the specimens is shown in Figure 11, together with the set-up of the tests [18].

In order to study the influence of an embedded defect at the bonding layer, specimens with a 7 mm Teflon insert at the adhesive level in one side of the specimens were also manufactured.

The terms B and T were chosen to describe co-bonded specimens without and with Teflon insert, respectively. Six B specimens and six T specimens were manufactured and tested.

In the specimens with RHEA, the reinforcements are placed for about 14 mm along the edges of the flange, where the highest shear stresses occur and the damage is more likely to develop. Fifteen RHEA specimens were manufactured and tested. Twelve specimens are provided without Teflon insert, and they are called FBR specimens, while three specimens are provided with 7 mm Teflon insert, and they are called FBRK specimens.

Different load levels up to 50% of the ultimate static load are considered.

The influence of the type of specimen and load level is investigated, observing a correlation between the damage growth rate and the load level that depends on the different configurations.

The main results of all the static and fatigue 4-point bending tests performed on the co-bonded specimens and on the RHEA specimens are summarized in Table III.

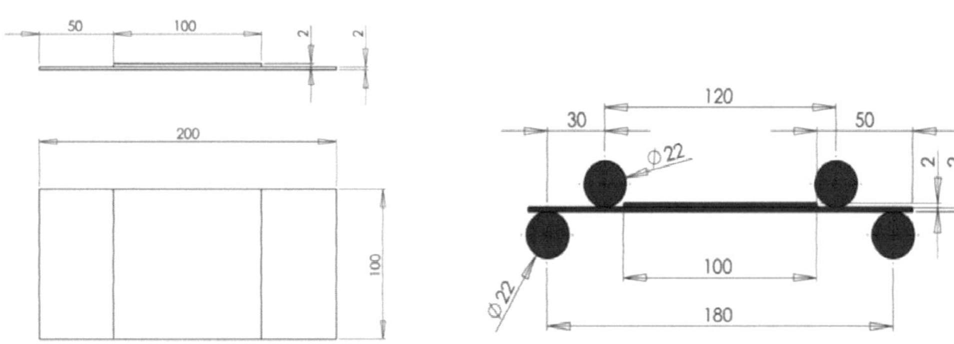

Figure 11. Geometry of bonded flange specimens and set-up for 4-point bending tests.

Table III. Summary of static and fatigue 4-point bending tests on co-bonded and RHEA specimens.

	Specimen	Test	Min load [N]	Max load [N]	Max. N° cycles	Damage onset [n°cycles]	Max damage [mm]	Type of damage
Co-bonded	B1	Static	-	5070	-	-	13	Debonding
	B2	Fatigue	250	2500	216000	~4900	35	Debonding
	B3	Fatigue	250	2500	250000	~8000	46	Debonding
	B4	Fatigue	225	2250	293000	~9000	23	Debonding
	B5	Fatigue	225	2250	301000	~7500	28	Debonding
	B6	Fatigue	200	2000	530000	~10000	21	Debonding
Co-bonded with Teflon	T1	Static	-	4790	-	-	11	Debonding
	T2	Fatigue	250	2500	17000	~3000	14	Debonding
	T3	Static	-	9500	-	-	45	Delamination
	T4	Fatigue	200	2000	520000	~4000	20	Debonding
	T5	Fatigue	200	2000	615000	~2800	28	Debonding
	T6	Fatigue	200	2000	350000	~2000	24	Debonding
Co-bonded with Rhea	FBR1	Fatigue	200	2000	630300	~50000	11	Delamination
	FBR2	Fatigue	200	2000	497500	~60000	8	Delamination
	FBR3	Fatigue	175	1750	975000	~200000	4	Delamination pin area
	FBR4	Fatigue	175	1750	623000	~60000	4	Delamination
	FBR5	Fatigue	175	1750	825000	~23000	6	Delamination pin area
	FBR6	Fatigue	200	2000	650000	~28000	6	Delamination pin area
	FBR7	Fatigue	250	2500	8990	~6000	8	Delamination pin area
	FBR8	Fatigue	250	2500	109620	~10000	11	Delamination pin area
	FBR9	Fatigue	250	2500	110958	~24000	12	Delamination pin area
Co-bonded with Rhea and Teflon	FBRK1	Fatigue	250	2500	3169	~2600	11	Debonding
	FBRK2	Fatigue	200	2000	501700	~4000	14	Debonding
	FBRK3	Fatigue	200	2000	485600	~25000	12	Debonding

The debonding growth is shown, as an example, in Figure 12 for specimen B6. Specimen B6 is tested in fatigue with maximum load equal to 2000 N. In specimen B6 a pure debonding is observed, without any delamination between the plies.

A different behavior is observed, for example, for specimen FBRK3, where both debonding and delamination are present, as reported in Figure 13, where the damage growth is shown.

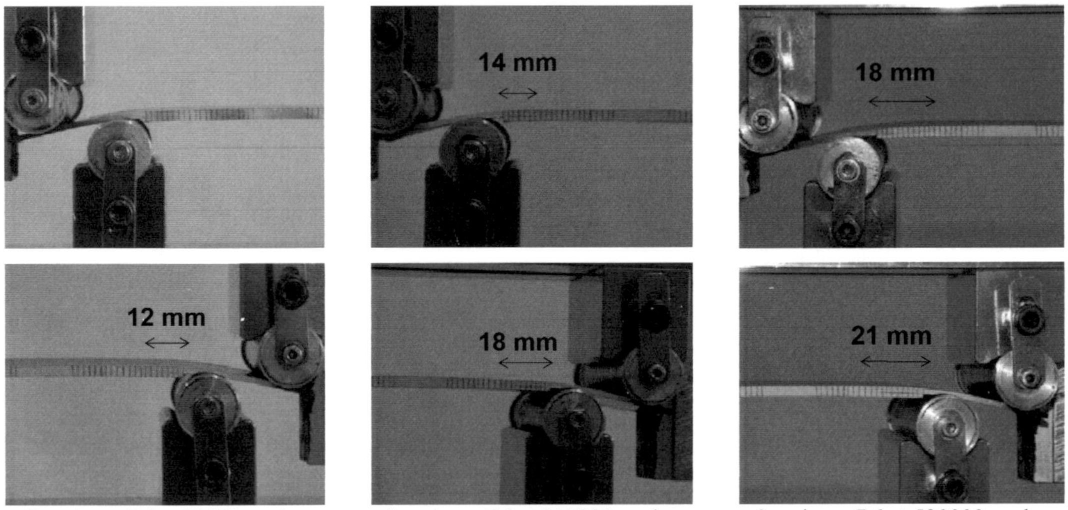

Specimen B6 at 30000 cycles | Specimen B6 at 117000 cycles | Specimen B6 at 530000 cycles

Figure 12. Debonding growth on co-bonded specimen B6.

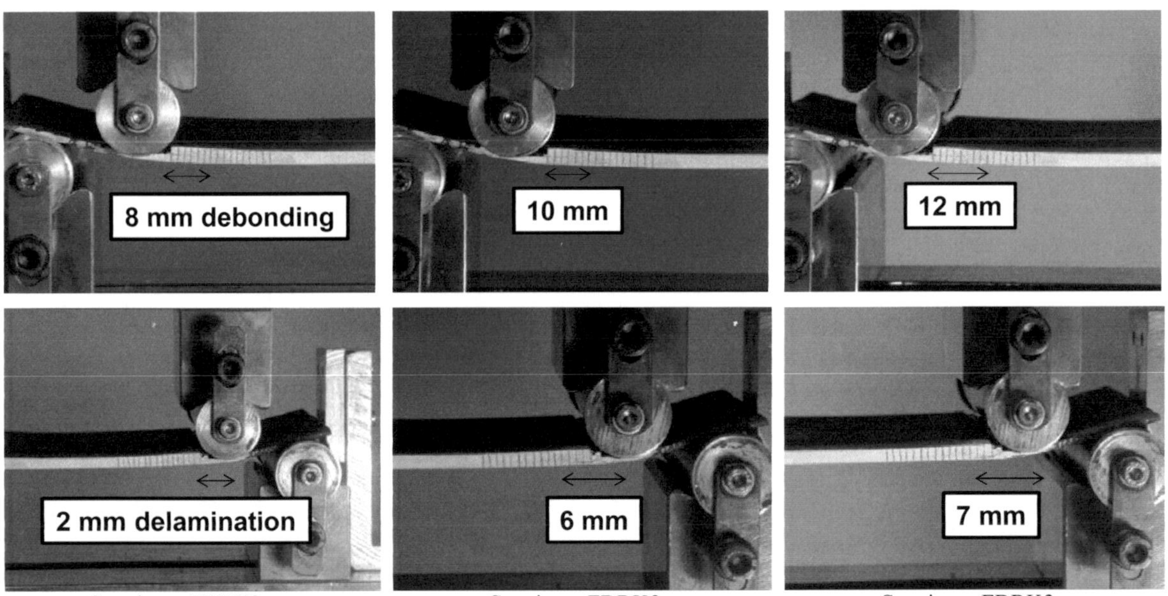

Specimen FBRK3 at 30000 cycles | Specimen FBRK3 at 265000 cycles | Specimen FBRK3 at 485600 cycles

Figure 13. Debonding and delamination growth on reinforced specimen FBRK3.

A comparison among the main results for the FBRK specimens to those ones of the T specimens is here reported. In both the sets of specimens the damage started to propagate at the Teflon side, but different damage modes followed for the two sets. Debonding on the opposite side is observed for T specimens, while it is not noticed for the FBRK set, which is characterized by delaminations at a higher number of cycles.

The debonding growth in the FBRK specimens is more controlled and the maximum debonding lengths is equal to 12-14 mm. This aspect is highlighted in Figure 14, which presents the critical lengths for T and FBRK specimens tested between 200 and 2000 N. The debonding length measured for both FBRK specimens is always shorter than the one of T specimens at the same number of cycles. Besides, the debonding rate does not increase rapidly. In particular, it is interesting to notice that, at the end of the test, the debonding reaches about the length of the RHEA insert at the interface and does not show the steep behavior typical of the co-bonded set.

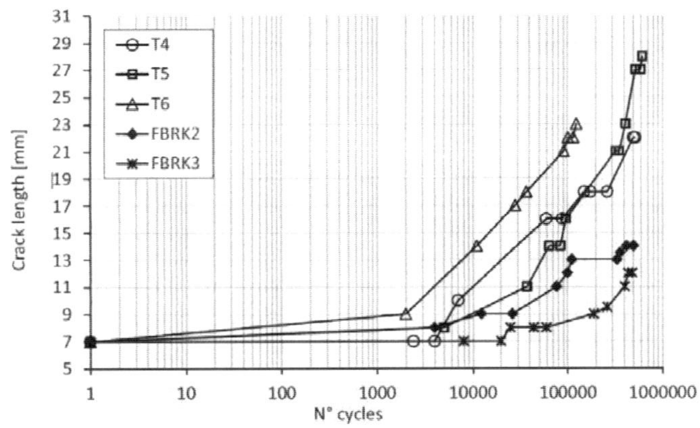

Figure 14. Crack length versus number of cycles for FBRK specimens and T specimens.

It is possible to observe that the damage progression under cycling load is significantly delayed by the RHEA reinforcement enhancing the damage tolerance of the joint by means of bridging mechanisms of the load (i.e. through thickness metallic pins provide a secondary load path) resulting in crack growth rates lower than co-bonded solutions by an order of magnitude.

LOW-ENERGY IMPACT TESTS

The test campaign is completed with low-energy impact tests. In this case the bonded flange specimens are used comparing the purely co-bonded joints with the reinforced ones. Impact damages are introduced by means of drop-weight test machine at bond line joint run out impacting the skin side. In particular, two symmetrical locations on the top of the skin laminate are chosen to be impacted for each specimen.

The impactor has a tip diameter of 25 mm and a mass of 1.022 kg. Rebound hints are prevented by sliding a rigid plate between the impactor and the specimen, after the impactor hit the specimen surface.

Three different boundary configurations are tested: simply supported, fully supported and fully supported with a hole concentric to the impact location [18] as shown in Figure 15. The impact locations are also shown in Figure 15 (left).

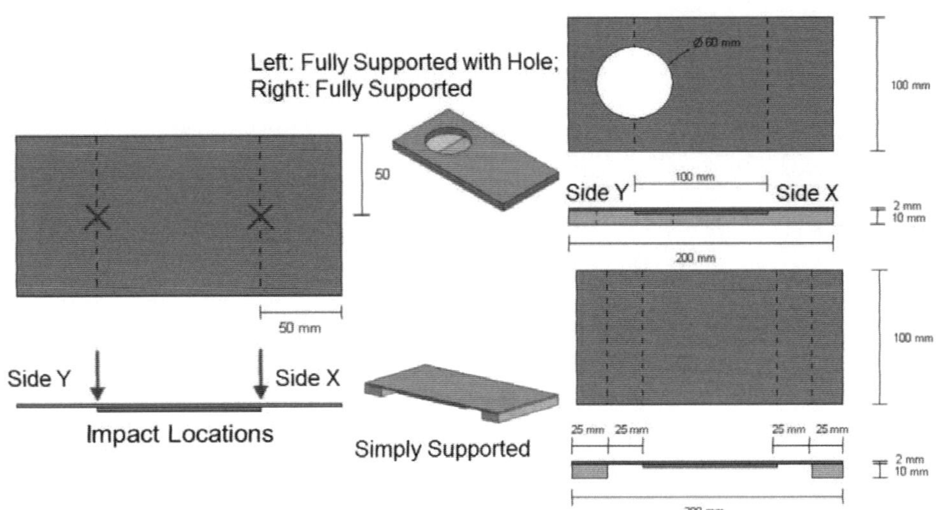

Figure 15. Three boundary conditions used for impact tests: simply supported, fully supported and fully supported with a hole. Impact locations also shown on left.

The extent of the damage caused by the impact event is evaluated by visual inspection and by X-ray techniques. An example of X-ray scan is reported in Figure 16.

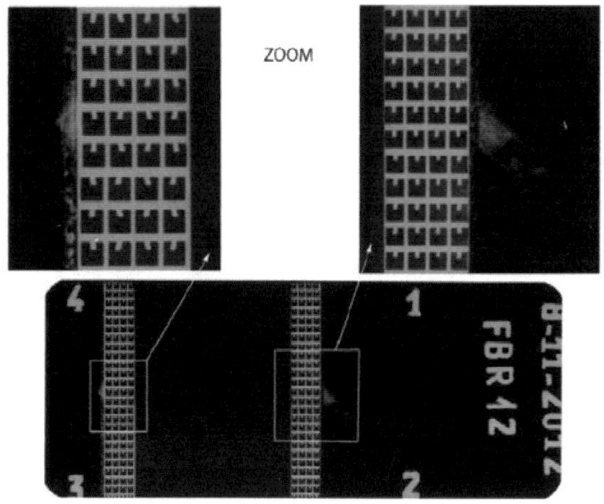

Figure 16. X-ray scan of specimen FBR12 after impacts at 20 J.

The impact behavior of the joints reinforced by the metallic spiked sheets is compared with the same impact energy levels demonstrating the capability of the novel design to contain the damage size. Depending on the energy levels, different failure modes are detected at reinforced joints: delamination, fibers breaking and split at the impact area indicating that the metallic reinforcement is inducing different damage patterns due to the local change of the stiffness driving the different failure modes. This suggests an optimum design of the metallic reinforcement to further reduce the damage size or address the most appropriate failure modes depending on the load to be transferred.

The main results of the impact tests on the co-bonded and RHEA specimens are summarized in Table IV. The terms P and FBR are chosen to describe co-bonded specimens and RHEA specimens, respectively. Six P specimens and six FBR specimens are tested.

Table IV. Summary of impact tests performed on co-bonded and RHEA specimens.

	Specimen	Boundary conditions	Energy level [J]	Max damage [mm]
Co-bonded	P1	Simply supported	10	20
		Simply supported	10	20
	P2	Simply supported	10	22
		Simply supported	10	22
	P3	Fully supported	20	30
		F.S. with a hole	20	23
	P4	Fully supported	15	0
		F.S. with a hole	15	23
	P5	Fully supported	20	18
		F.S. with a hole	17.5	23
	P6	Fully supported	20	24
		F.S. with a hole	17.5	29
Co-bonded with Rhea	FBR10	Simply supported	10	9
		Simply supported	10	8
	FBR11	Simply supported	10	6.4
		Simply supported	10	8.8
	FBR12	Fully supported	20	2.6
		F.S. with a hole	20	6.5
	FBR13	Fully supported	15	-
		F.S. with a hole	15	9
	FBR14	Fully supported	20	-
		F.S. with a hole	17.5	10
	FBR15	Fully supported	20	-
		F.S. with a hole	17.5	9.6

Figure 17 summarizes the results of the impact tests in graphs accounting damage size only for disbonding at bonded flange run out.

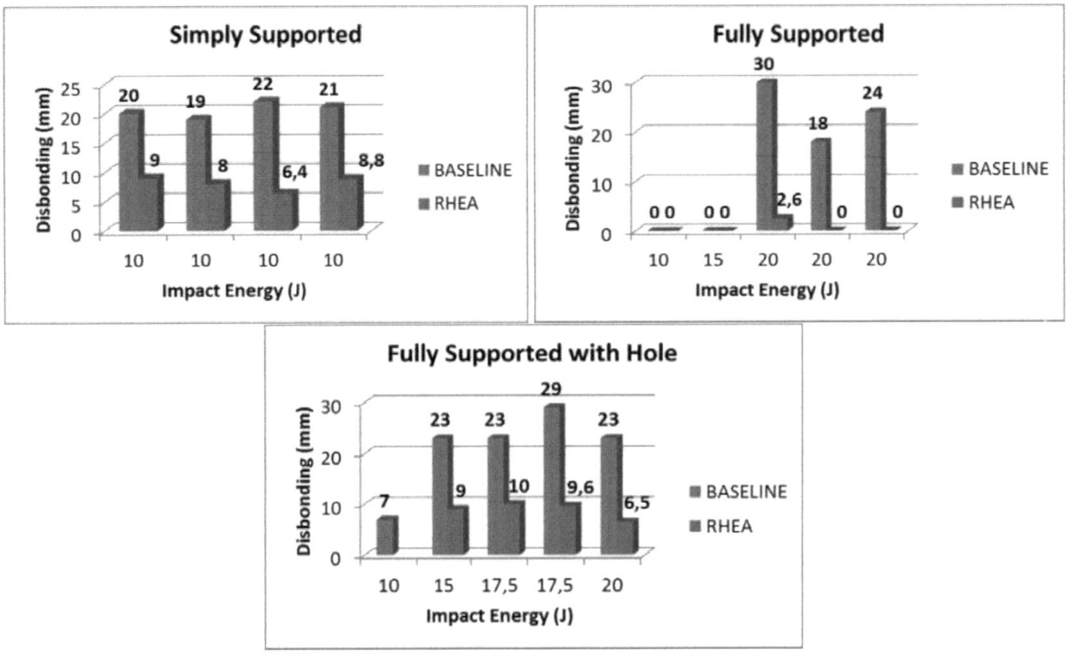

Figure 17. Summary graphs account damage size only for disbonding at bonded flange run out.

CONCLUSIONS

The results obtained during an experimental campaign carried out to investigate the behavior of co-bonded joints on CFRP specimens with a novel design incorporating a through the thickness local reinforcement were presented. This new joint technology is called RHEA (Redundant High-Efficiency Assembly), and is based on the introduction of spiked metal sheet reinforcement into a CFRP prepreg laminate during its curing cycle.

Different specimens were manufactured to investigate static and fatigue behavior, as well as delamination size after impact and damage tolerance characteristics.

The experimental results show that the static strength as well as the fatigue strength of the reinforced joint is significantly higher than the co-bonded baseline solution. Moreover, the damage progression under cycling load was delayed by the novel through-thickness reinforcement enhancing the damage tolerance of the joint by means of bridging mechanisms of the load (i.e. through thickness metallic pins provide a secondary load path) resulting in crack growth rates lower than co-bonded solutions by an order of magnitude for both tensile and compressive loading fatigue.

The tests campaign gathered a significant number of experimental results that have demonstrated a more predictable failure with respect to the solely adhesive structural bonding, and that can be used to define preliminary design guidelines. From the tests results it can be concluded that the joints reinforced by this inter-laminar metallic spiked sheet at the bond line provide a low cost innovative 3D reinforcement to guarantee improved damage tolerance characteristics when compared to a co-bonded joint.

ACKNOWLEDGEMENTS

The research was possible thanks to the financial support of Airbus Operations GmbH, Germany. The authors are grateful to Airbus Group Innovations, and in particular to Elke Hombergsmeier, Ana Carolina Nogueira and Michael Juergens, for having provided the specimens with the RHEA reinforcements. The authors would also like to thank Valentina Buratti, Alberto Caruso and Paolo Rubini, who helped in performing the tests at Politecnico di Milano.

REFERENCES

[1] A.A.V.V. (2012). *Standard practice for classifying failure modes in FRP joints*. Technical report, ASTM International.

[2] Banea, M.D. and da Silva, L.F.M. (2009). Adhesively bonded joints in composite materials: an overview. *Journal of Materials: Design and Applications*, vol. 223, n. 1, p. 1-18.

[3] Meneghetti, G., Quaresimin, M. and Ricotta, M. (2010). Influence of the interface ply orientation on the fatigue behaviour of bonded joints in composite materials. *International Journal of Fatigue*, vol. 32, n. 1, p. 82-93.

[4] Tong, L. (1998). Failure of adhesive-bonded composite single lap joints with embedded cracks. *AIAA Journal*, vol. 36, n. 3, p. 448-456.

[5] Tomblin, J.S., Yang, C. and Harter, P. (2002). Investigation of thick bondline adhesive joints. *Journal of Composites Technology and Research*, vol. 24, n. 2, p. 332-344.

[6] A.A.V.V. (2007). *Standard test method for strength properties of adhesives in shear by tension loading of single-lap-joint laminated assemblies*. Technical report, ASTM International.

[7] A.A.V.V. (2010). *Standard method of test for strength properties of adhesives in shear by tension loadings (metal-to-metal)*. Technical report, ASTM International.

[8] Dickinson, L.C., Farley, G.L. and Hinders, M.K. (1999). Translaminar reinforced composites: a review. *Journal of Composites Technology and Research*, vol. 21, n. 1, p. 3-15.

[9] Tong, L., Mouritz, A.P. and Bannister, M. (2002). *3D fibre reinforced polymer composites*. Elsevier.

[10] Glaessgen, E.H., Raju, I.S. and Poe, C.C. Jr. (2002). Analytical and experimental studies of the debonding of stitched and unstitched composite joints. *Journal of Composite Materials*, vol. 36, n. 23, p. 2599-2622.

[11] Stickler, P.B. and Ramulu, M. (2001). Investigation of mechanical behavior of transverse stitched T-joints with PR520 resin in flexure and tension. *Composite Structures*, vol. 52, n. 3, p. 307-314.

[12] Freitas G., Frusco T., Campbell, T., Harris, J. and Rosenberg, S. (1996). Z-fiber technology and products for enhancing composite design. In *Proc. of the 83rd Meeting of the AGARD SMP on "Bolted/Bonded Joints in Polymeric Composites"*.

[13] Tze Min Koh. (2012). Improving the mechanical properties of aerospace carbon epoxy joints by Z-pinning. PhD thesis, RMIT University.

[14] O'Brien, T.K. and Krueger, R. (2006). Influence of compression and shear on the strength of composite laminates with Z-pinned reinforcement. *Applied Composite Materials*, vol. 13, n. 3, p. 173-189.

[15] Chang P.; Mouritz A.P. and Cox B.N. (2007). Flexural properties of zpinned laminates. *Composites Part A: Applied Science and Manufacturing*, vol. 38, n. 2, p. 244-251.

[16] Nogueira, A.C., Drechsler, K., Hombergsmeier, E. and Pacchione M. (2011). Properties and failure mechanism of a 3D-reinforced joint. *JEC Composite Magazine*, p. 39-44.

[17] Nogueira, A.C., Drechsler, K., Hombergsmeier, E., Furfari D. and Pacchione M. (2011). Investigation of a hybrid 3D-reinforced joining technology for lightweight structures. In *Proc. of 16th International Conference on Composite Structures*.

[18] Buratti, V. (2012). Damage Tolerance of Bonded Composite Joints. Master Thesis, Politecnico di Milano.

Philosophy of Multiple-site Damage Analysis for Fibre Metal Laminate Structures

Wandong Wang, Calvin Rans, René Alderliesten, Rinze Benedictus

Faculty of Aerospace Engineering, Delft University of Technology, the Netherlands

Abstract: This paper outlines the proposed approach and current implementation progress for a multiple site damage model for Fibre Metal Laminate structures. The current state of the art in FML crack growth prediction utilizes linear elastic fracture mechanics, displacement compatibility, and superposition in order to characterize the coupled crack/delamination growth behaviour of a fatigue damaged FML. Directly extending this approach to a MSD scenario introduces several challenges which will be outlined and discussed. A computational simplification of idealizing cracks as "negative stiffeners" in order to simplify the estimation of stress intensity factors at single crack locations will be substantiated and discussed.

NOMENCLATURE

a	Crack length
a_s	Saw-cut length
K_∞	Stress intensity factor due to far-field applied load
K_{br}	Stress intensity factor due to bridging load
K_{total}	Total stress intensity factor at the crack tip
K_{red}	Stress intensity factor due to applied load and redistributed load
n_f	Number of fibre layers
n_m	Number of metal layers
G	Strain energy release rate
E_f	Young's modulus of fibre layer
E_m	Young's modulus of metal layer
t_f	Thickness of fibre layer
t_m	Thickness metal layer
S_f	Stress in the fibre layers
S_{br}	Bridging stress
j	Number of interfaces
v_∞	Crack opening displacement due to far-field load
v_{br}	Crack closing component due to bridging load
δ_f	Elongation of prepreg layer
δ_{pp}	Shear deformation of prepreg layer
da / dN	Crack growth rate
db / dN	Delamination growth rate
Δa	Crack extension
Δb	Delamination bar length extension
ΔN	Applied fatigue cycles

INTORDUCTION

Fibre Metal Laminates (FMLs) are a class of hybrid material and comprise metal layers bonded with composite layers. Their well-known superior fatigue crack resistance and longer critical crack length, which are desirable in the context of damage tolerance design philosophy utilized in the aerospace sector, are attributed to the bridging mechanism offered by intact fatigue resistant fibres in the wake of a fatigue crack in FMLs. See Figure 1. The philosophy of damage tolerance assessment for fatigue crack growth in FMLs has been well established. These damage tolerance analyses for

FMLs mainly focus on predicting the evolution of an isolated crack. As a result, they may be invalid due to the presence of Multiple-site Damage (MSD) cracks in fibre metal laminate structures.

An identified deficiency of damage tolerance design philosophy is its compatibility with an indefinite structural life. The use of this philosophy to ensure structural integrity becomes problematic in the case of MSD scenario within an aging structure. In order to combat the possibility of the failure caused by occurrence of MSD scenario, the airworthiness regulations has been revised in 2010 to include the definition of a Limit of Validity (LOV) placing limits on damage tolerance analyses. It is therefore crucial to examine MSD crack growth behaviour of fibre metal laminates even they possess superior damage tolerant behaviour.

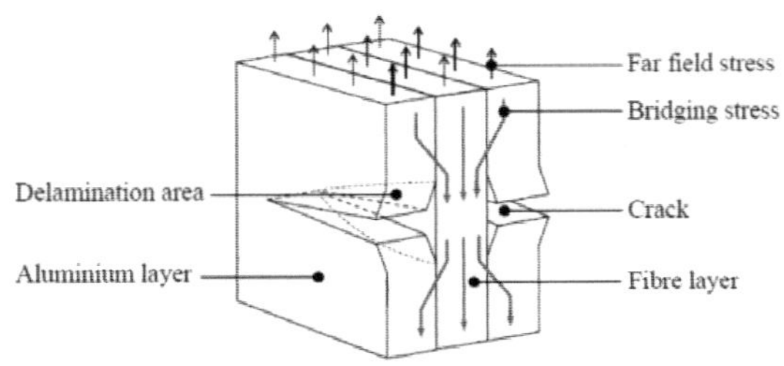

Figure 1. Bridging mechanism.

The models for damage tolerance analyses of FMLs in open literature predict crack propagation in metal layers and extension of delamination at the metal-composites interfaces using classical Linear Elastic Fracture Mechanics (LEFM). Calculating the bridging stress distribution in the intact fibres bridging the fatigue crack is essential for predicting the coupled failure mechanisms in FMLs in these models. However, directly extending these models for simultaneously predicting the crack growth behaviour of MSD cracks in FMLs is cumbersome and computationally inefficient due to the complicated calculation of load redistribution among multiple cracks and corresponding bridging fibres.

This paper describes the philosophy with the aim of proposing a simplified prediction methodology for analysing MSD behaviour in FMLs. This paper explains the challenges of directly applying the models of damage tolerance analysis for FMLs to examine MSD behaviour in FMLs. The flow diagram of the presented philosophy for MSD analysis of FMLs is shown in this paper. The general utilized approach in this proposed philosophy is classical LEFM, in conjunction with the principle of superposition and displacement compatibility. The idealization of crack interaction in terms of load redistribution in FMLs containing MSD scenario will be illustrated.

Additionally, this paper will highlight the importance of predicting behavior of an eccentric crack with two different delamination shapes on both sides of its saw-cut in a FML panel, and the compatibility of this eccentric crack prediction with the MSD growth prediction in FMLs.

NATURE OF FATIGUE IN FMLS

As a result of their hybrid nature, FMLs have different fatigue behaviour compared to monolithic metal. The fatigue resistant fibres which remain intact in the wake of fatigue cracks in metal layers in FMLs act as a second load path and restrain the opening of the cracks. Due to this bridging mechanism, the stress intensity factors are considerably reduced and the crack growth in FMLs is quite slow in comparison with monolithic metal. The bridging mechanism, however, is not effective in the crack initiation phase for FMLs [1]. In addition, the cyclic stress in the metal layer of FMLs is a superposition of the stress induced by the applied load and the curing stress [1, 2]. These two factors result in a quite short initiation life for FMLs.

Figure 2 shows the contributions of the initiation life and crack growth life to the total fatigue life for FMLs and monolithic aluminium. The differences in contributions are typical for FMLs and monolithic metals [1]. As a result of the very fast crack growth in monolithic metal, the crack growth life only accounts for a very small part of its fatigue life. On the contrary, the superior fatigue crack growth resistance of FMLs results in a considerably long crack growth life which covers the main part of the total fatigue life.

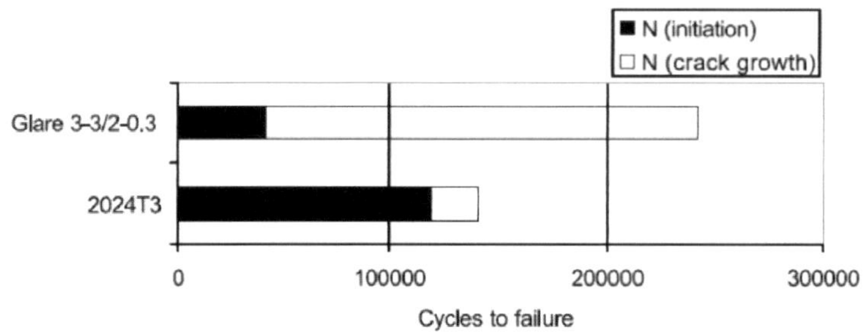

Figure 2. Fatigue life comparison for FMLs and monolithic aluminium [1].

The difference in ratio of crack growth life to overall fatigue life for metals and FMLs has implications on the fatigue behaviour of such materials in an MSD scenario. In order to discuss these differences, the influence of MSD cracks will be divided into two categories: load redistribution effects and crack-tip interaction effects. Crack tip interaction effects occur when the proximity of multiple crack tips results in an interaction of the stress singularities in front of the crack tips. For two cracks approaching each other, this tends to result in rapid fatigue propagation and ultimately linkup of the cracks. Load redistribution effects occur due to a loss in component stiffness due to the presence of multiple damages. These effects will be present over the entire range of damage sizes and proximities (i.e.: they are also present when crack-tip interaction effects are present), although they are easier to describe for cracks sufficiently far that crack-tip interaction effects are negligible. In such a case, the crack propagation of one crack will be increased due to the increase in stress in the undamaged region of the component resulting from other damaged regions.

The relative crack growth behaviour of metals and FMLs, with and without MSD, is shown schematically in Figure 3. The relatively rapid crack growth behaviour of the metal implies that cracks in an MSD scenario will grow towards each other relatively quickly resulting in early linkup. Crack linkup will rapidly increase the size of the primary crack resulting in a reduction of fatigue life. The reduction in crack growth life is thus heavily influenced by crack-tip interaction effects. Conversely, the slower fatigue crack growth rate in FMLs combined with the shorter crack initiation life means that a significant portion of the MSD crack growth life in FMLs can occur well before the presence of crap-tip interaction effects. During this time, the cumulative effects of load redistribution resulting from the presence of multiple damages can result in a significant reduction in fatigue life (ΔN).

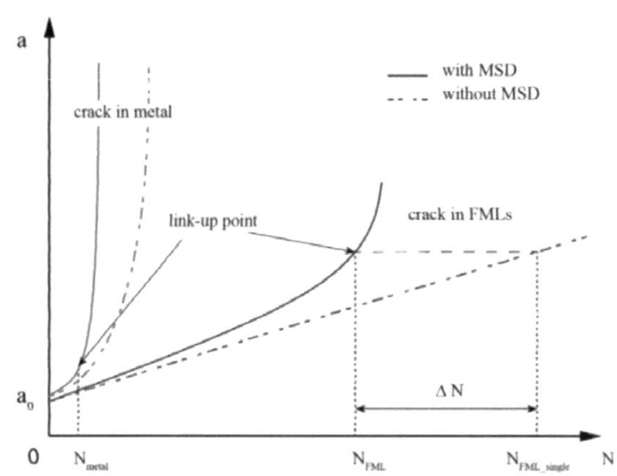

Figure 3. Schematic comparison of crack growth with MSD and without MSD for FMLs and metal.

DAMAGE TOLERANCE ANALYSIS FOR FMLS

Since the proposed philosophy is based on the methodologies used in damage tolerance analysis for FMLs, a brief review of the well-established FML crack growth model developed by Alderliesten [7, 8] is given before the proposed philosophy is detailed.

Using the Linear Elastic Fracture Mechanics (LEFM) together with principles of superposition and displacement compatibility, Alderliesten has successfully made the prediction of coupled propagation of cracksin metal layers and delaminations at metal/composites interfaces [7, 8]. Both crack extension and delamination extension are determined with LEFM method, while stress intensity factor, K, is used to characterize crack extension in metal layers (Eqn. 1) and strain energy release rate, G, is applied to estimate the extension of delamination (Eqn. 2). For both crack growth resistance and delamination extension resistance of FMLs, experimentally determined Paris relations are applied [7].

$$K_{total} = K_{\infty} + K_{br} \quad (1)$$

$$G = \frac{n_f t_f}{2jE_f}\left(\frac{n_m t_m E_m}{n_m t_m E_m + n_f t_f E_f}\right)\left(S_f + S_{br}\right) \quad (2)$$

The principle of superposition is applied in Eqn.1, i.e. the total stress intensity factor, K_{total}, is decomposed into K_{∞} due to the external applied load plus internal residual stress, and K_{br} due to the stress reduction caused by the bridging fibres in the wake of a fatigue crack. The strain energy release rate, G, is also a function of bridging stress (Eqn. 2). The bridging stress distribution therefore needs to first be determined in order to resolve the crack extension and delamination growth.

The principle of displacement compatibility has been applied by Alderliesten to resolve the bridging stress distribution over the cracked region [7]. Since the fibre layers are bonded to the metal layers at the delamination front, the displacement of the fatigue cracked metal layer should be identical to the deformation of the bridging fibres over the delaminated length. See Eqn. 3. The displacement of cracked metal layer, which is given on the left side of Eqn. 3, is decomposed into opening displacement , v_{∞}, due to the external applied load and residual stress and crack closing item, v_{br}, due to the bridging stresses in bridging fibres. The deformation of the bridging fibres over the delaminated length includes elongation of the prepreg layers, δ_f, and shear deformation of the prepreg layers, δ_{pp}.

$$v_{\infty}(x) - v_{br}(x) = \delta_f(x) + \delta_{pp}(x) \quad (3)$$

As the displacement of the metal layer and the deformation of the bridging fibres vary along the crack flank, the delamination shape is evenly divided into bar elements to implement Eqn. 3 at each bar location for solving the bridging stress. Figure 4 schematically illustrates the division of the delamination shape. Moreover, the crack closing component, v_{br}, at one bar location is not only a function of the bridging stress at the bar element location, but affected by the bridging stress at each bar element. As a result, the displacement compatibility has to be implemented simultaneously for all the bar elements, forming a system of linear equations which can be expressed with matrix notation containing a square matrix whose size is equal to the number of bar elements. Once the bridging stress distribution is calculated, the crack growth rate and delamination growth rate can be determined by using Eqn. 1 and Eqn. 2 together with their Paris relations.

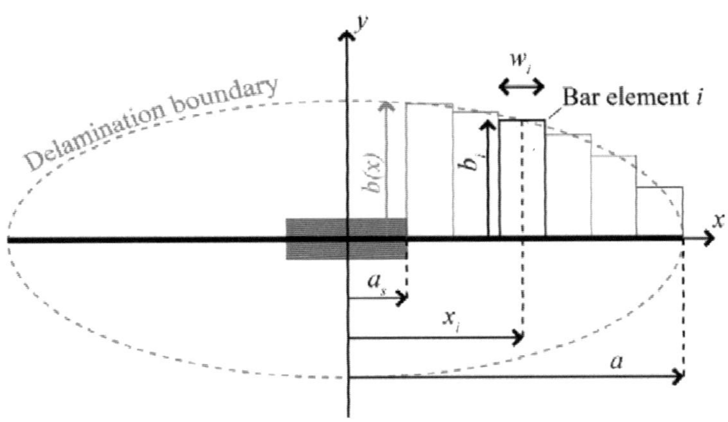

Figure 4. Bar element division for solving bridging stress distribution.

ISSUES OF SIMULTANEOUS PREDICTION OF MULTIPLE COLLINEAR CRACKS

Based on the description in the preceding section, it is essential to calculate the bridging stress distribution by implementing displacement compatibility for determining the crack extension and delamination extension in FMLs.

Extending this approach to simultaneous prediction of crack propagation and delamination growth of multiple collinear cracks (MSD scenario) in FMLs seems to be a natural choice. However, simultaneously predicting all the crack states of the collinear cracks in FMLs has two major issues.

The first issue is to calculate the crack opening displacement (COD) for a row of cracks depending on the configuration [9].This will be explained by an example depicted in Figure 5 which schematically illustrates a row of 3 cracks in FMLs with the delamination zones divided into bar elements. In order to implement the principle of displacement compatibility at the location of bar element i , the crack opening displacement, v_{∞} , of Crack 2 is required as an input in Eqn. 3. However, the COD of Crack 2 is difficult to be calculated due to the presence of other two collinear cracks: Crack 1 and Crack 3. The same issue is valid when the displacement compatibility is implemented at one of the bar elements of Crack 1 or Crack 3. Although the COD solutions for some specific MSD configurations are available in literature [10-12], it is a challenging issue to calculate COD for generic MSD configuration.

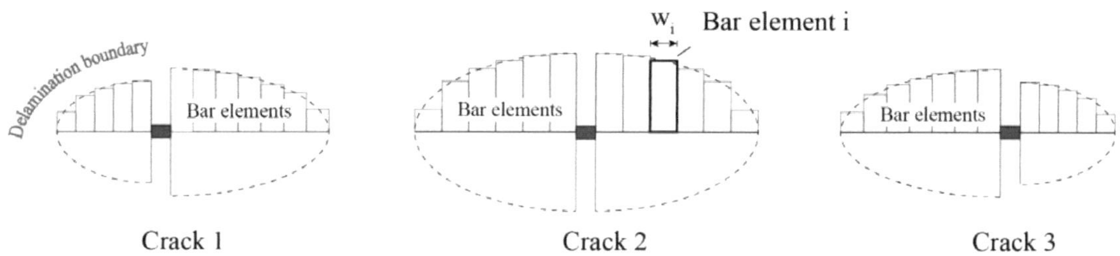

Figure 5. Illustration of bar element division for all crack.

The second issue is the computational efficiency of simultaneous prediction of the bridging stress at all bar elements over multiple crack spans. As an example illustrated in Figure 5, the closing displacement, v_{br} , of bar element i is a function of the bridging stresses at all bar elements of Crack 1, Crack 2 and Crack 3. Thus a matrix whose size could be around 3 times larger than that for one crack configuration needs to be constructed for the same reason described in the preceding section. Depending on the number of cracks and the length of each crack of generic MSD configuration in FMLs, the much larger matrix constructed to predict bridging stress distribution results in computational inefficiency.

ANALYSIS METHODOLOGY FOR FML STRUCTURES CONTAINING MSD

Effect of stiffness variation

The effective stiffness of a structure is related to the material stiffness (Young's modulus) and geometric stiffness of the structure [13]. The material stiffness of a panel is an inherent property which cannot be changed merely by the presence of a crack. However, the presence of a crack can decrease the geometric stiffness of a structure by reducing the net sectional area, resulting in larger deformation under same loading condition as the remaining material has to carry the entire applied load. Then the crack growth rate increases with the crack length [13]. In case the cracked panel is reinforced by stiffeners which increase the effective stiffness of the cracked panel, the deformation is restrained. The crack growth rate decreases when the crack grow towards an adjacent stiffener as a result of the load redistribution from the cracked skin to the stiffeners [13]. Another example is that the crack growth in FMLs is much slower in the comparison of the fast crack growth of one crack in a monolithic metal panel because the intact bridging fibres partially compensate the geometric stiffness reduction in the metal layers [14].

The reverse case of a cracked panel reinforced by stiffeners is the presence of MSD cracks in a panel. In addition to one single crack in a monolithic metal panel, the presence of collinear adjacent cracks prohibitively decreases the effective stiffness of the panel as no other material could compensate the stiffness reductions. As a result, the crack growth rate of the single metal crack prematurely soars up. By contrast, the effect of the presence of adjacent MSD cracks on the growth behavior of the single crack in FMLs is not as pronounced as in metal. See Figure 3. This is attributed to the compensation made by the intact bridging fibres in the wake of MSD cracks to the stiffness reductions in metal layers. The presence of MSD cracks in FMLs still results in stiffness reductions which lead to load redistribution from the locations of MSD cracks to the single crack. Figure 6 demonstrates the load redistribution from an adjacent crack to the centre crack in FMLs.

Properly estimating the effect of stiffness reduction caused by the presence of MSD cracks on the single crack in FMLs and applying the methods used in damage tolerance analysis for FMLs to resolve the single crack state comprise the simplified analysis philosophy for FMLs with MSD problem.

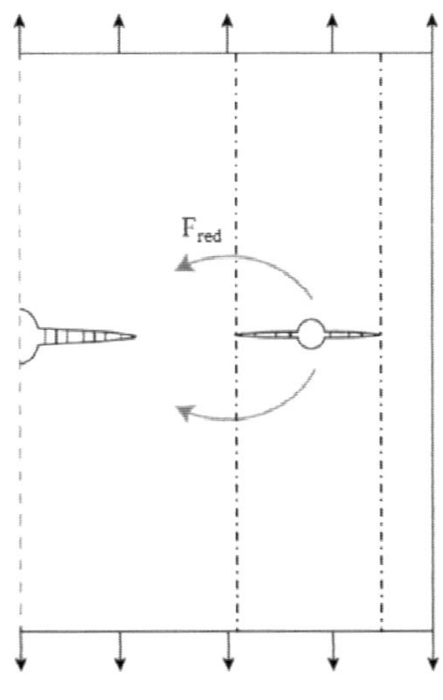

Figure 6. Illustration of load redistribution.

A precursor model has been developed to predict fatigue crack growth in FMLs panels containing discretely notched layers. The objective of the precursor model is to verify the concept of load redistribution caused by stiffness reduction which is achieved by discretely notched layers [9]. The difference between discretely notched metal layers and fatigue cracked metal layers is the severity of stress concentration/plasticity at the notch edge/ crack tip in the crack section. For the development of the model and the test procedure, please refer to [9]. Only Figure 7 from [9] is given to illustrate the effects of different stiffness reductions on the crack growth behaviour in FMLs.

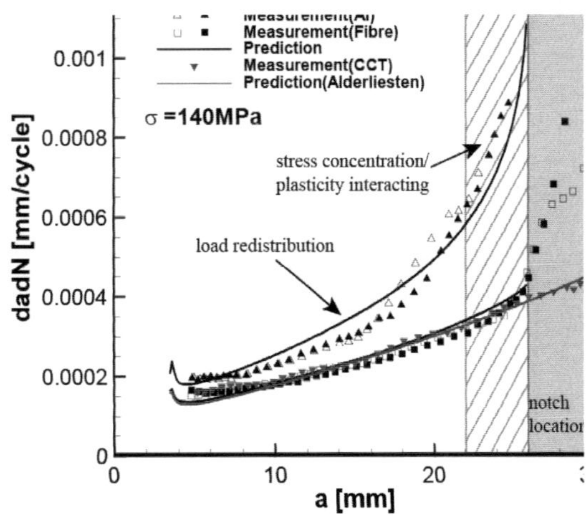

Figure 7. Comparison of effects of different stiffness reductions on the crack growth behaviour in FMLs.

As can be seen in Figure 7, the crack growth rate with discretely notched Al layers is higher than that with notched fibre layers and M (T) specimen. Because the stiffness reduction resulted from notched Al layers is much higher than notched fibre layers, more load transferred from notched area to the crack tip in FMLs. The little stiffness reduction resulted from notched fibre layers leads to limited load redistribution and increase in crack growth rate compared to M (T) specimen [9].

Another feature in Figure 7 is that the possible stress concentration/ plasticity interacting zone covers a small portion of total crack length. Most part of the crack growth is affect by the load redistribution mechanism. In addition, the relatively slow crack growth rate resulted from load redistribution and much higher crack growth rate in stress concentration /plasticity interacting zone result in that neglecting the stress concentration/ plasticity interacting does not affect the total crack growth life prediction too much. The influence should be acceptable then in terms of total fatigue life.

Analysing eccentric crack growth in FMLs

It is notable that most of MSD cracks are not symmetric, each asymmetric crack has different crack lengths with different delamination shapes on both sides of its saw-cut, as shown in Figure 8. The delamination shapes in Figure 8 can be examined by etching away the outer metal layers of FML specimen after fatigue testing [7, 15]. This eccentric crack phenomenon raises new challenges for the analysis of MSD crack growth in FMLs: asymmetric COD for an eccentric crack and asymmetric bridging stress distribution with respect to the crack centre. Consequently, the solutions applied by Alderliesten [7] to resolve bridging stress distribution for a central crack with symmetric delamination shapes and crack lengths in FMLs are no longer applicable to an eccentric crack in FMLs. New analysis methodology for an eccentric crack in a finite FML panel is needed to be extended to fit in with the MSD growth analysis for FMLs.

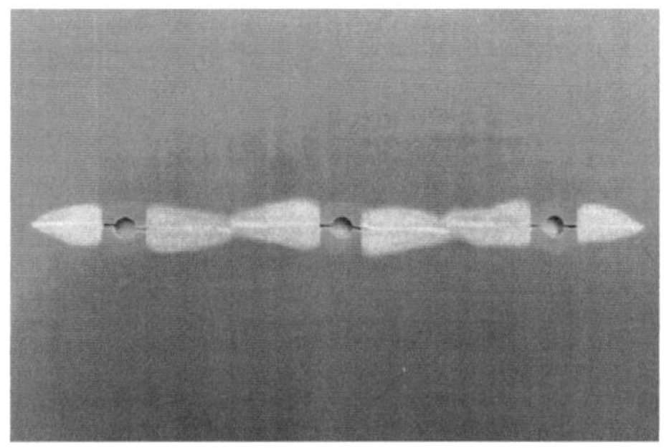

Figure 8. MSD cracks in a FML panel-

The symmetric condition of a finite panel is damaged by the presence of an eccentric crack, the COD for the eccentric crack therefore is not symmetric and the stress distributions in front of the crack tips are not the same either. See Figure 9. In open literature [11, 16, 17], only different stress intensity factors for two crack tips of an eccentric crack in a finite metal panel are given. However, the asymmetric COD of an eccentric crack is yet needed to implement the principle of displacement compatibility to calculate the bridging stress distribution in FMLs. The asymmetric COD of an eccentric crack in metal layer result in asymmetric bridging stress distribution and thus different delamination shapes, as schematically illustrated in Figure 9.

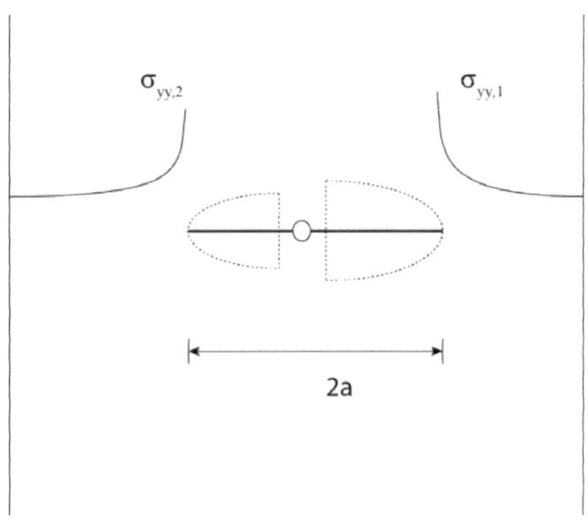

Figure 9. Illustration of an eccentric crack in a FML panel.

Two different Westergaard stress distributions, illustrated in Figure 9, can be assumed in front of the two crack tips of an eccentric crack in metal layers in FMLs. If these two Westergaard stress distributions can be calculated, then the asymmetric crack opening displacement for the eccentric crack can be obtained [18]. This method also permits the possibility of extending the model for eccentric crack analysis in FMLs to estimate the load redistribution caused by the presence of MSD cracks, in the same way as described in [9].

The method used by Alderliesten [7] to calculate the displacement closing component , v_{br} in Eqn 3, for symmetric delamination scenario cannot be used to calculate the bridging stress for asymmetric delamination scenario. Then the Westergaard stress method for two points loading on Page 5.6 in crack analysis Handbook [10, 11] has to be applied to calculate the corresponding closing component in order to implement the principle of displacement compatibility.

Investigating the crack growth behaviour of an eccentric crack in FMLs is a necessary step for analyzing MSD crack growth in FMLs. The analytical model for analyzing eccentric crack growth in FMLs should have the capability to be extended to calculate the effect of load redistribution in the presence of MSD cracks on the crack growth of the eccentric crack itself.

Implementation

The prediction of crack and delamination growth rates of an isolated crack in a FML is a simultaneous analysis of metal crack growth, interface delamination extension and their interaction. After each crack growth step, the growth rates need to be recalculated and the damage configuration must be updated. Therefore, the modelling of the fatigue damage evolution in FMLs is a iterative process [19].

For the analysis of crack growth in a FML panel with MSD cracks, the crack state of each crack has to be calculated. The effect of the presence of other adjacent cracks on each crack needs to be analysed by means of load redistribution mechanism at first, then the LEFM method together with principle of displacement compatibility is used to calculate the crack state. This process has to be repeated until all the crack states have been analyzed. Then the crack lengths and delamination shapes of all cracks are updated by using the parameters of each crack state. Figure 10 illustrates the overview of these two iterative processes and the essential parameters to be calculated.

The layup and the material properties of the FML to be investigated together with loading and temperature parameters are given as inputs. With these inputs, the stresses and corresponding strains in different layers in FMLs due to applied far-field load can be calculated by applying the Classical Laminate Theory (CLT) [2].

The MSD configuration in the FML can be initialized by specifying the number of cracks, k , locations of the two crack tips and two saw-cut tips for each crack with respect to a specified coordinate in the panel, and the size and shape of each delamination. A crack can be referred to as Crack j , where index j ranges from 1 to k . In order to carry out the calculation of bridging stress distribution, the crack lengths and delamination shapes are discretized, specifying the locations where the displacement compatibility principle is implemented.

To aid in the calculation of the state of Crack j, the far applied load is decomposed into two parts to perform the calculation of the load redistribution mechanism and the bridging mechanism separately: the applied load carried by the intact bridging fibres over Crack j, and load in the remaining material. This decomposition follows the same principle in [9].

The load redistribution calculation is performed to evaluate the effect of other cracks on Crack j. The stress intensity factors K_{red} for two crack tips of Crack j due to the applied load and transferred load is determined. The load redistribution also affects the COD of Crack j, which is incorporated in bridging calculation. The bridging calculation is solving a system of equations representing the displacement compatibility principle at discretized nodes / bar elements for Crack j only. With the bridging loads known, K_{br} and G for Crack j can be determined. Then the crack growth and delamination extension rate is determined based upon the experimentally determined Paris relation. All these calculations need to be repeated for every crack until all the crack states are known.

After all the states of cracks in the FML panel have been analyzed, the length of each crack, Δa, and height of each delamination bar element, Δb, are updated based on the growth rates and applied cycles which is determined by the limited crack extension and the maximum crack growth rate. The increments of cracks and delamination shapes are updated iteratively.

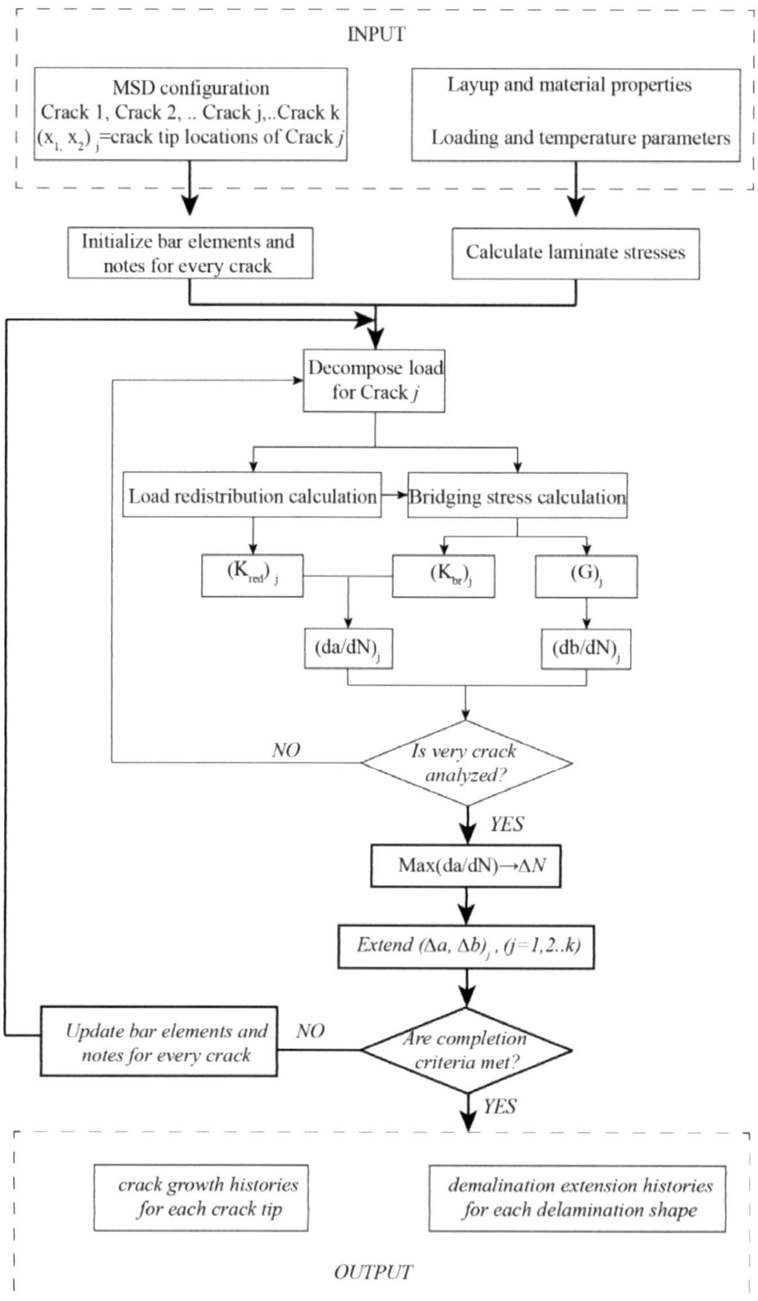

Figure 10. Prediction process of crack growth and delamination extension of MSD in FMLs

The completion criteria can be link-up of two crack tips, the maximum crack length, maximum number of iterations, or maximum life cycles, N. If one of the completion criteria is met, the data can be output and post processed for analysing.

PIN-LOADING EFFECT ON CRACK GROWTH BEHAVIOUR IN FMLS

The mechanically fastened joints in fuselage structures are susceptible to MSD problem [6, 20, 21]. The dominant load for fuselage structures is introduced by the Ground-Air-Ground (GAG) pressurization cycles which create roughly constant amplitude fatigue loading. The load is transferred from one fuselage skin panel to another via the fasteners in the fastened joints. Normally the load path in fuselage joints is eccentric, which introduces secondary bending. The total stress in joints therefore comprises membrane stress, the secondary bending stress and the bearing stress resulted from pin-loading on the holes [22]. The stress concentrations caused by the presence of the fastener holes, pin-loading and secondary bending in fuselage joints affect the fatigue life of the joints [23]. Due to the fatigue loading and stress concentration contributors in fuselage joints, cracks can initiate at edges of multiple holes in the joints forming MSD problem.

The primary objective of investigating MSD is to combat MSD problem in fuselage joints to ensure the flight safety. Addressing MSD issue in fuselage joints without understanding the effect of pin loading on the crack growth behaviour is impossible. Since the proposed philosophy of MSD analysis for FML panels mainly utilizes the superposition principle, the effect of pin loading on the crack growth behaviour in FMLs can be superimposed onto the methodology for predicting MSD to tackle the MSD analysis in FMLs joints. The secondary bending effect can be incorporated by applying the methodology in [19]. A model for analysing pin loading effect on crack growth behaviour in FMLs therefore needs to be developed in order to build up an integral model for combating MSD in FML joints.

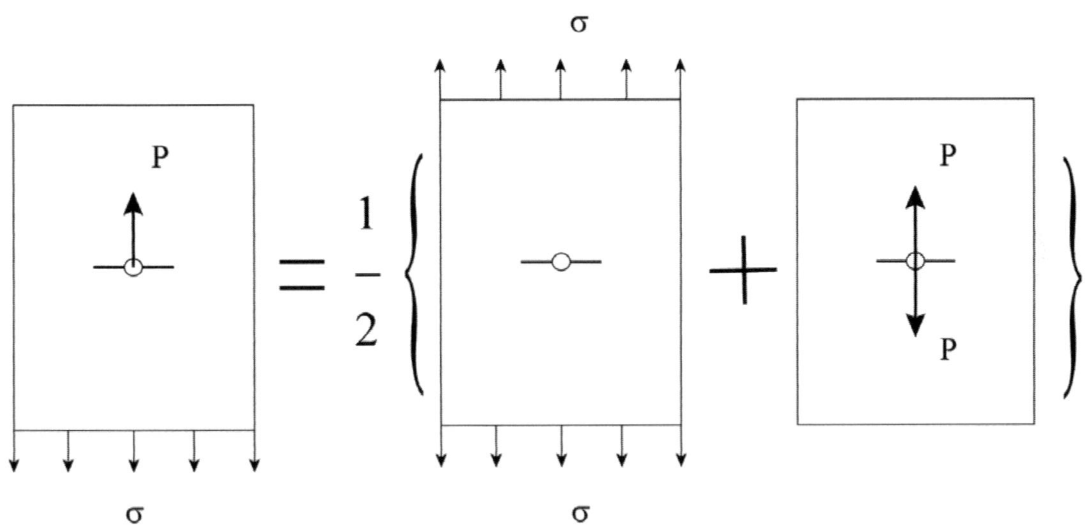

Figure 11. Decomposition of pin loading case.

A simple way of analysing the pin loading effect on fatigue crack growth in FMLs is using the configuration illustrated on the left side of Figure 11, i.e. a FML panel is remotely loaded by uniform tensile stress and pin-loading at the crack centre. The loading case can be decomposed into two symmetric loading cases by superimposing, as illustrated in Figure 11. The same methodology applied by Alderliesten can also be utilized to resolve the crack propagation and delamination growth of the symmetric loading cases on right side of Figure 11 respectively.

LIMITATION OF THE PROPOSED METHODOLOGY

In reality, the stress concentration is severe in front of the crack tips of MSD cracks. Plastic zones are also present in corresponding to the extremely high stress concentration. Neglecting the stress concentrations and plastic zones in front of adjacent cracks in idealization of crack interaction may lead to some error. The growth rate of a single crack before link up can be over predicted provided that the stress concentrations at adjacent crack tips are neglected. The over predicted growth rates lead to conservative results. When the crack tips get closer enough and the plastic zones start to interact, neglecting the influence of stress concentration and plastic zones will result in under prediction of crack growth rate. However, the error is acceptable according to the nature of fatigue in FMLs.

CONCLUSION

The nature of crack growth in FMLs results in different factors to be considered in MSD analysis. Larger possible damage sizes before link-up in FMLs result in the significance of load redistribution in predicting metal crack and interface delamination growth. The interaction between cracks caused by plasticity or stress concentration can be neglected without leading to much error in terms of total fatigue life.

Damage tolerance analysis for FMLs is based on simultaneous prediction of metal crack and interface delamination growth and the compatibility of metal crack and delamination. Extending this approach to simultaneous prediction of all crack states of multiple cracks and their compatibilities in FMLs is computationally inefficient and problematic. As a result, a new analysis methodology for FMLs containing MSD scenario is proposed in this paper based on idealization of the crack interaction in FMLs as load redistribution mechanism. The idealization is consistent with the nature of crack growth in FMLs.

This paper proposes the philosophy of MSD analysis for FMLs in order to develop a simplified methodology for predicting MSD growth in FML structures. Load redistribution mechanism for this simplified methodology has been developed. However, the analysis of eccentric crack growth in FMLs still needs to be developed. The model for eccentric crack in FMLs should be able to be extended to fit in with the MSD analysis. The methodology can be implemented according to Figure 10. In order to predict MSD growth in FML joint, the effect of pin loading effect on crack growth in FML panels needs to be studied. As the methodology is based on principle of superposition and LEFM, the investigation of pin loading effect can be integrated.

REFERENCE

[1] Homan JJ. Fatigue initiation in fibre metal laminates. International Journal of Fatigue. 2006;28:366-74.

[2] Spronk SWF, Şen I, Alderliesten RC. Predicting fatigue crack initiation in fibre metal laminates based on metal fatigue test data. International Journal of Fatigue. 2015;70:428-39.

[3] Pitt S, Jones R. Multiple-site and widespread fatigue damage in aging aircraft. Engineering Failure Analysis. 1997;4:237-57.

[4] Pártl O, Schijve J. Multiple-site damate in 2024-T3 alloy sheet. International Journal of Fatigue. 1993;15:293-9.

[5] Singh R, Park JH, Atluri SN. Growth of multiple cracks and their linkup in a fuselage lap joint. AIAA Journal. 1994;32:2260-8.

[6] Schijve J. MULTIPLE-SITE DAMAGE IN AIRCRAFT FUSELAGE STRUCTURES. Fatigue & Fracture of Engineering Materials & Structures. 1995;18:329-44.

[7] Alderliesten RC. Fatigue crack propagation and delamination growth in GLARE [dissertation]. Delft, the Netherlands: Delft University of Technology; 2005.

[8] Alderliesten RC. Analytical prediction model for fatigue crack propagation and delamination growth in Glare. International Journal of Fatigue. 2007;29:628-46.

[9] W. Wang CR, R.C. Alderliesten, R. Benedictus. Predicting the Infuence of Discretely Notched Layers on Fatigue Crack Growth in Fibre Metal Laminates. Engineering Fracture Mechanics. 2015;(Accepted with minor revision).

[10] Tada H, Paris PC, Irwin GR. The stress analysis of cracks handbook. Hellertown: Del Research; 1973.

[11] Tada H, Paris PC, Irwin GR. The stress analysis of cracks handbook. New York: ASME; 2000.

[12] Wu XR, Xu W. Strip yield crack analysis for multiple site damage in infinite and finite panels – A weight function approach. Engineering Fracture Mechanics. 2011;78:2585-96.

[13] Rans C, Rodi R, Alderliesten R. Analytical prediction of Mode I stress intensity factors for cracked panels containing bonded stiffeners. Engineering Fracture Mechanics. 2013;97:12-29.

[14] Vlot A, Gunnink JW. Fibre metal laminates- an introduction. Dordrecht: Kluwer Academic; 2001.

[15] Chang PY, Yeh PC, Yang JM. Fatigue crack growth in fibre metal laminates with multiple open holes. Fatigue & Fracture of Engineering Materials & Structures. 2012;35:93-107.

[16] Isida M. Stress-Intensity Factors for the Tension of an Eccentrically Cracked Strip. Journal of Applied Mechanics. 1966;33:674-5.

[17] Chen DL, Weiss B, Stickler R. A new approach for the determination of stress intensity factors for finite width plate. Engineering Fracture Mechanics. 1994;48:561-71.

[18] Westergaard HM. Bearing pressure and cracks. Journal of Applied Mechanics. 1939;6:49-53.

[19] Wilson GS. Fatigue Crack Growth Prediction for generalized fiber metal laminates and hybrid materials [PhD dissertation]. Delft: Delft University of Technology; 2013.

[20] Hendricks WR. The Aloha Airlines Accident — A New Era for Aging Aircraft. In: Atluri SN, Sampath SG, Tong P, editors. Structural Integrity of Aging Airplanes: Springer Berlin Heidelberg; 1991. p. 153-65.

[21] Silva LFM, Gonçalves JPM, Oliveira FMF, de Castro PMST. Multiple-site damage in riveted lap-joints: experimental simulation and finite element prediction. International Journal of Fatigue. 2000;22:319-38.

[22] De Rijck JJM, Vogelesang LB, Delft TU, Aerospace E, Tu Delft DUoT. Stress analysis of fatigue cracks in mechanically fastened joints: an analytical and experimental investigation: Delft University Press; 2005.

[23] De Vries TJ, Vogelesang LB, De Borst R, Delft TU, Aerospace E, Tu Delft DUoT. Blunt and Sharp Notch Behaviour of Glare Laminates: Delft University Press; 2001.

Experimental Study on Partial Composite Wing Structure

Hikaru Hoshi[1], Takao Okada[1], Yutaka Iwahori[1], Satoshi Morooka[2], Naoyuki Watanabe[2]

[1]Institute of Aeronautical Technology, Japan Aerospace Exploration Agency, Japan
[2] Department of Aerospace Engineering, Tokyo Metropolitan University, Japan

Abstract: Tokyo Metropolitan University (TMU) and Japan Aerospace Exploration Agency (JAXA) are investigating advanced composite structures and smart technology for the next generation aircraft. This project aims to educate the Asian students, and research on advanced aircraft technology. Manufacturing and structural test of the partial composite wing program is major topic of our research project. This partial composite wing program includes demonstration of materials, manufacturing technologies, and measurement techniques.

A half wing span composite wing structure was designed to evaluate the novel technology for the next generation single aisle transport aircraft. Then 5m long Partial Composite Wing Structure (PCWS) was designed to demonstrate the full scale fabrication, assembly and structural test.

INTRODUCTION

A composite wing structure was designed to demonstrate the novel technology for the next generation single aisle transport aircraft. Our research target is assumed 150 seat class airliner which is same as the Boeing B737, and the Airbus A320. Two conceptual aircraft were planned for this project to conduct the design and manufacturing and test which are the 150 seat class of 6 abreast fuselage airliner. Figure 1 gives an image of 150 seat class airliner on this project. This conceptual aircraft was assumed that is widely used composite materials for wing, fuselage, and tail plane structure.

State of art materials and low cost / high speed manufacturing process are key technology to develop the next generation single aisle aircraft. New stitch composite with Vacuum assisted Resin Transfer Molding (VaRTM) Process, was prepared to meet this demand. Stitch composite has high impact resistance characteristics which derive from binding effect of thickness direction stitch thread. Therefore, VaRTM stitch composite was applied to the wing spar to eliminate the in-flight impact problem, complex shape drape problem, and also cost. Spar splice is necessary for easy handling and manufacturing flexibility in wing assembly process. Electrical Heating Press Forming (EHPF) method was applied to titanium fitting of spar splice . This EHPF process is able to reduce the machining time and material cost due to near net shape forming by press machine process.

Structural test program of PCWS is configured strain survey, impact test, static test fatigue test, non-destructive inspection, and repair. Structural Health Monitoring (SHM) by optical fiber sensor will applied to PCWS test program. This test program is scheduled to reach the final design ultimate load test in 2 years. This program aims to demonstrate the new technologies and to extract the technical problems of practical use. Now, final assembly and test set up of the Partial Composite Wing Structure is underway.

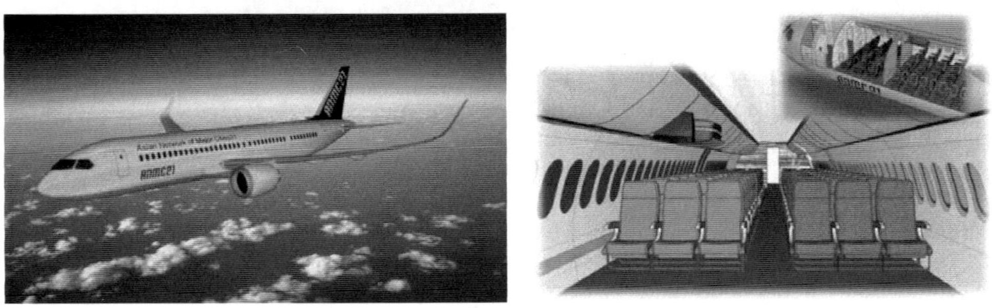

(a) Conceptual aircraft. (b) Cabin image.

Figure 1. Conceptual image of 150 seat class airliner.

DESIGN AND ANALYSIS OF WING STRUCTURE

Summary of conceptual aircraft

Conceptual aircraft' specification was set almost same as current single aisle aircraft to simplify the configuration design process. Summary of conceptual aircraft of this project is shown in Figure 2. Carbon fibre Reinforced Plastics (CFRP) is widely used for principal structures to reduce the weight and maintenance cost on fatigue and corrosion problems. Wing is configured upper and lower covers, front and rear spars, in spar ribs, and winglets which are fabricated by CFRP solid laminates. Aerodynamic design was also conducted to apply to the next generation aircraft at aerodynamic section of this project. Riblet and laminar flow control design are applied for the each fuselage and wing to reduce the fuel burn. Finally, this configuration of conceptual aircraft was used for structural design, analysis and manufacturing on partial composite structure.

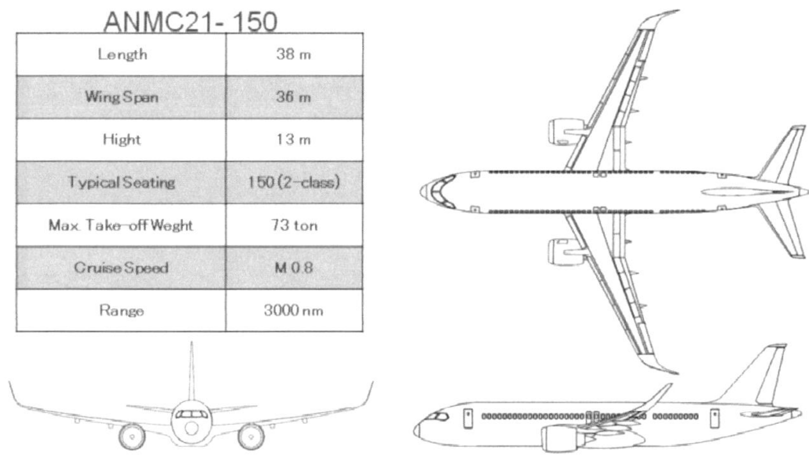

Length	38 m
Wing Span	36 m
Hight	13 m
Typical Seating	150 (2-class)
Max. Take-off Weght	73 ton
Cruise Speed	M 0.8
Range	3000 nm

Figure 2. Summary of conceptual aircraft.

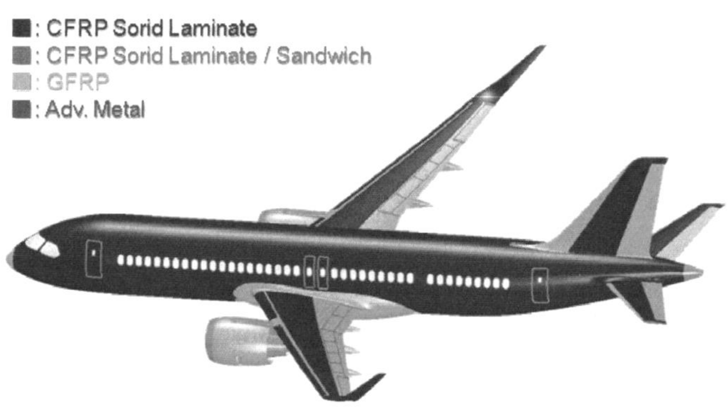

Figure 2. Material usage of airframe.

Aerodynamic design

Simple aerodynamic design was conducted to generate the wing shape, then computational fluid dynamics simulation was conducted to evaluate the feasibility. Wing geometry data of final design given in Table I. Wing shape optimization analysis was also calculated to reduce the drag as shown in Figure 3. But, this optimization process is required long time cycles. Therefor, structural design and external load estimation were hardly related to the aerodynamic design to minimize the design cycle.

Table I. Wing geometry data of final design.

Wing span	[m]	36	Taper ratio	–	0.32
Area	[m2]	120	Thickness ratio (Root)	–	0.15
MAC	[m]	6.6	Thickness ratio (Tip)	–	0.11
Aspect ratio	–	12	Swept back angle 25% chord	[deg]	25

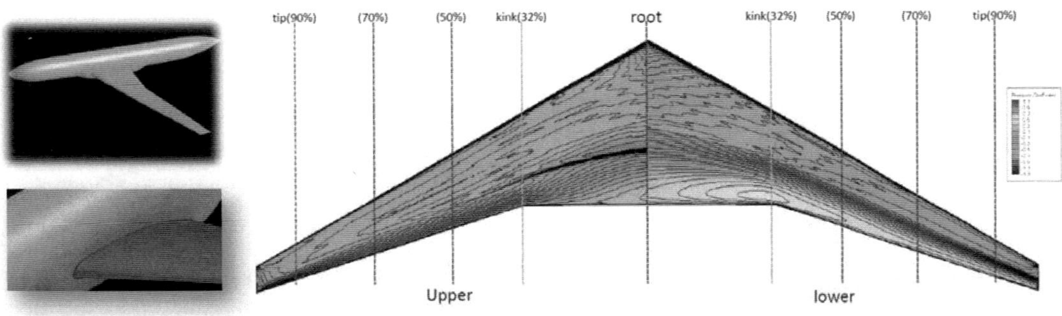

Figure 3. Pressure distribution of optimized wing shape.

Wing structure and material concept

Initial structural lay out was conducted to set the basic configuration of wing structure. Each parameter were decided from structural configuration survey of current aircraft. This wing structure is composed by conventional wing configuration which include multi stringer wing covers, C shaped spar, and stiffened web type rib. The planning time was shortened by use of above conventional structural configuration without the new technology risk. Schematic of wing structure is shown in Figure 4.

A wing structure is fabricated by CFRP. Conventional prepreg system is used for upper and lower wing covers and rib webs, stitch composite by VaRTM process is use for front and rear spars. Here, it is thought that conventional prepreg system is more suitable for the automatic fabrication system by robotics compared with VaRTM process. But, current automation system of the prepreg is expected that the possibility of future weight reduction is little. If VaRTM processed stitch composite is able to use for each spars, low fibre wrinkling at the joggle area is possible to expect by flexibility of the dry preform before the resin injection. Therefor, more lighter and shorter ply drop-off technique is able to apply to the wing covers to prevent the extra thickness of wing skin in the joggled spar region. The structural configuration and material usage were decided by the above reasons.

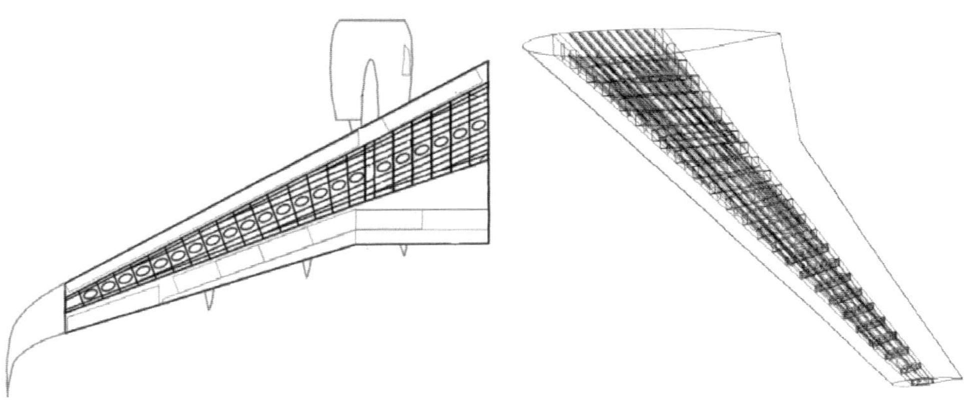

Figure 4. Structural arrangement of wing.

External load

The critical case is selected to the 2.5G pull up maneuvering condition at zero fuel weight. The design external load was estimated by simple wing span loading which is assumed form pressure distribution of aerofoil and wing shape. The external loading conditions without nacelle load at critical case are given as Figure 5. Nacelle and main landing gear inertia loads were also considered to design the inboard wing structure.

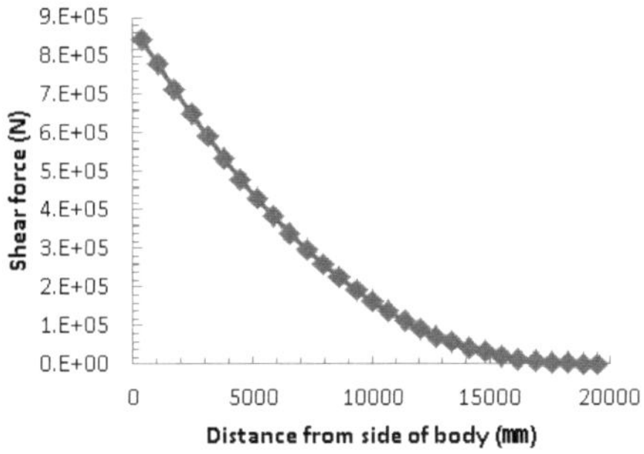

(a) Shear force distribution without nacelle.

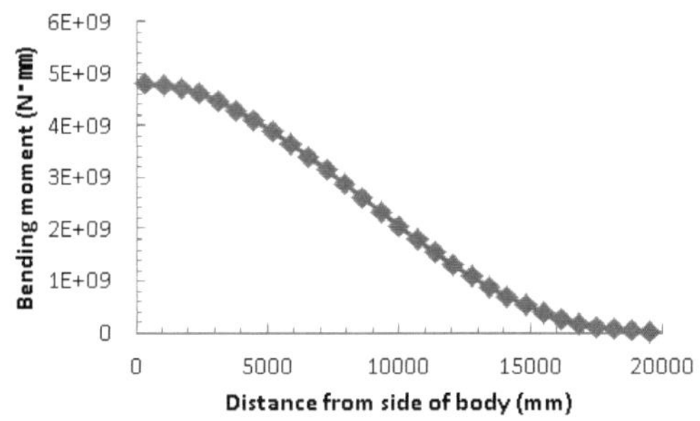

(b) Bending moment distribution without nacelle.

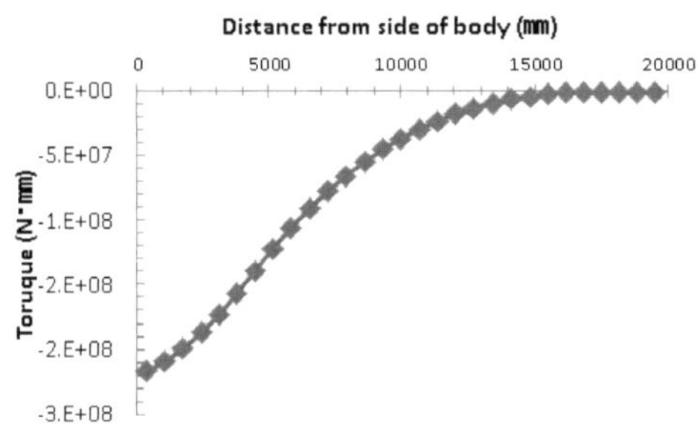

(c) Torque distribution without nacelle

Figure 5. External loading conditions of half wing.

Internal load and structure design

Internal load and structure design were designed by commercial software of ProMECHANICA and NASTRAN. One degital model concept between geometrical design and finite element analysis was adopted to minimize the design cycle time. Digital data transferring and re-modelling was avoided by ProMECHANICA's dual function of geometrical modelling and structural analysis. As a result, design cycle that start from geometrical modelling to internal load analysis was merged within more shorter time.

A structural element design was conducted by Excel based conventional hand calculation. Design allowable was decided from JAXA's advanced composites data base (JAXA-ACDB) and other original coupon tests. To reduce the risk of un-matured failure, conservative design allowable was used for highly loaded composite parts. A example of design allowable for this project is shown in Table II.

Table II. Design allowable.

Test	Prepreg	Stitched composite
Non Hole Tension	11100 μ	7800 μ
Non Hole Compression	9000 μ	8500 μ
Open Hole Tension	5500 μ	5200 μ
Open Hole Compression	3200 μ	3500 μ
Filled Hole Tension	4900 μ	6800 μ
Filled Hole Compression	4400 μ	4200 μ
Compression After Impact	3400 μ	3800 μ
Bearing	500 MPa	300 MPa

After hand calculation, wing digital model was re-fined by geometrical data and ply orientation data of each parts. Then, FEA analysis was conducted to confirm the actual internal load and the minimum margin of safety.

Finally, lay-up design was conducted by CATIA V5 composite function to make manufacturing drawings. Ply drop-off design and fibre drape condition were also decided at this manufacturing drawing phase while keeping the structural and the manufacturing requirements. Digital model is shown in Figure 7.

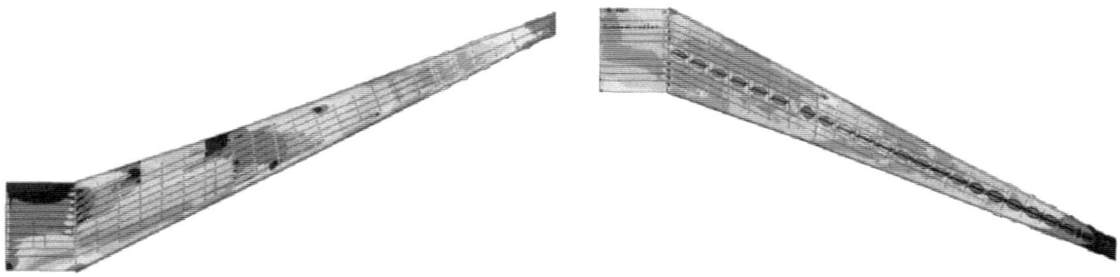

(a) Minimum principal strain distribution of upper cover (b) Maximum principal strain distribution of lower cover

Figure 6. Principal strain distribution at critical up bending condition.

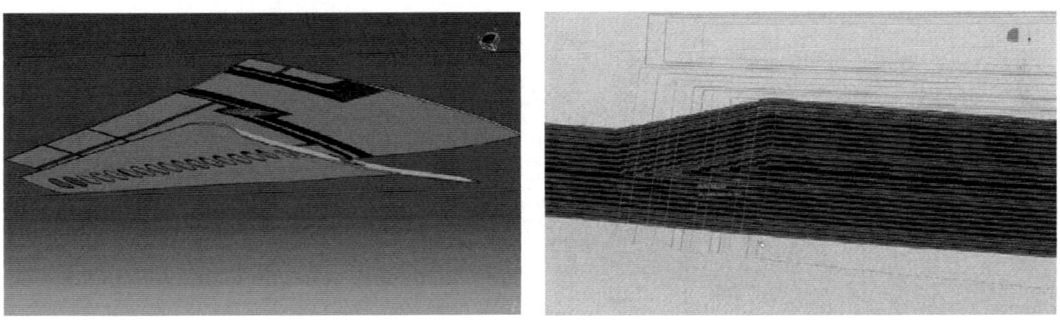

(a) Wing cover digital model. (b) Ply drop-off pattern of skin.

Figure 7. Digital model of wing cover.

DESIGN AND MANUFACTURING OF PARTIAL COMPOSITE WING STRUCTURE

Summary of partial composite wing structure

To reduce the risk and cost, full span wing structure was shrinked to the Partial Composite Wing Structure (PCWS) of wing tip portion. Section of the PCWS is defined between from the root of winglet to the end of outboard flap tip which is almost 5 m long. Upper and Lower covers were not changed material from full span wing, but front and rear spar were changed material to the conventional prepreg from the stitched composite. The reason of this material changing is manufacturing capability problem of stitched composite machine and work area. As this reason , a stitched composite spar was fabricated only 2 m span at the wing tip portion of the front spar. Schematic of the PCWS is given in Figure 8.

Upper and lower covers. Upper and lower cover is consisted from wing skins and stringers which are made by conventional prepreg. Wing skin and stringers were bonded with co-bonding process that was splited two elements of the pre-cured skin and partially cured stringers. This bonding process is able to reduce the low stringer molding accuracy risk, but co-bonding surface is increased to two layers from conventional one layer. In this project, this bonding process was selected to avoided the molding risk. Wing skin and stringer are given in Figure 9.

Stitched composite front spar. A stitched composite spar is consisted from three different materials which are made by carbon fibre, epoxy resin and Vectran fibre of stitch thread. Dry pre-form was fabricated by special weaving machine for the 2 m long spar. C shape formed dry pre-form was stitched by modified rock stitch technique to bind the 13 mm thickness carbon fibre layers. Finally, epoxy resin was infused to the stitched dry pre-form by VaRTM process, then resin was cured at 180 °C. A Stitched composite spar is shown in Figure 11.

Prepreg front and rear spars. The inboard front spar and the rear spar were fabricated by autoclave molding with conventional prepreg to reduce the risk and cost problems. A key technology of spar fabrication is high accuracy molding of small corner region without the fibre wrinkling. To get the approved quality, many molding tries were conducted for spars. Prepreg spar is shown in Figure 12.

Ribs. The rib structure was composed two different material parts which are assembled by the epoxy resin prepreg rib web and the thermo plastic resin L shaped stiffener. The C shaped prepreg rib web is used for out-board region, H shaped prepreg rib web is used for in-board region which are reinforced by L shaped thermo plastic CFRP. Rib webs were fabricated by autoclave molding with conventional prepreg, L shaped stiffeners were fabricated by press forming process with thermo plastic laminate. The rib web and the stiffener are given in Figure 13.

Titanium spar splices and rib posts. The titanium spar splices and rib posts were designed and manufactured by the electrical heating press forming (EHPF) process. This EHPF process can be reduced titanium machining process due to use of sheet metal material. EHPF process is simple. Initially, the titanium sheet is set to the die position, then electrical wire is connected to the titanium sheet, finally material is heated at 800 °C, and it is formed by the press machine. In EHPF process, final trimming process is required, but material cost and machining time can be reduced. The titanium spar splices and rib posts are given in Figure 14.

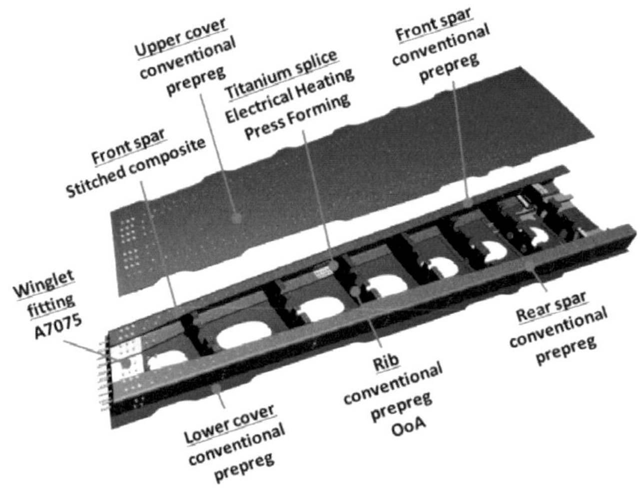

Figure 9. Schematic of the partial composite wing structure.

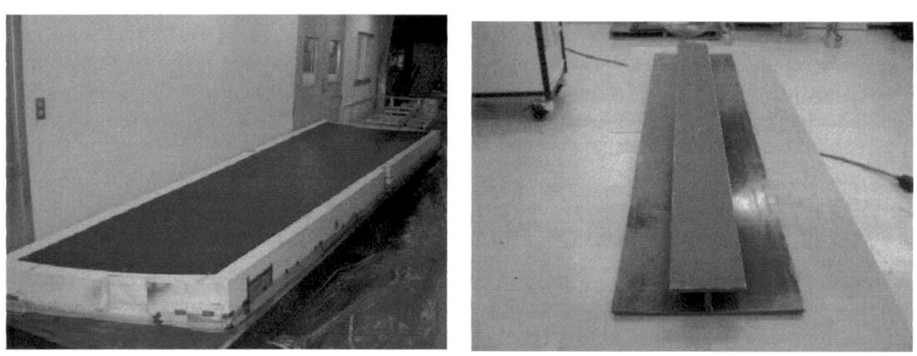

(a) Wing skin of before trimming process. (b) Stringer of before trimming process.

Figure 10. Wing skin and stringer of before trimming process.

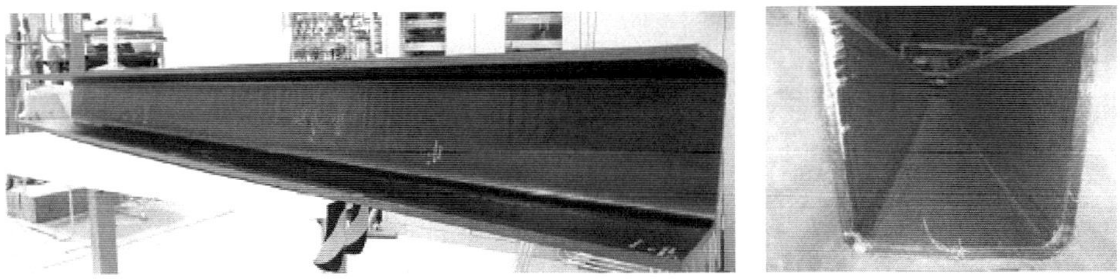

(a) Stitched composite spar (b) Just before the de-molding

Figure 11. Stitched composite spar.

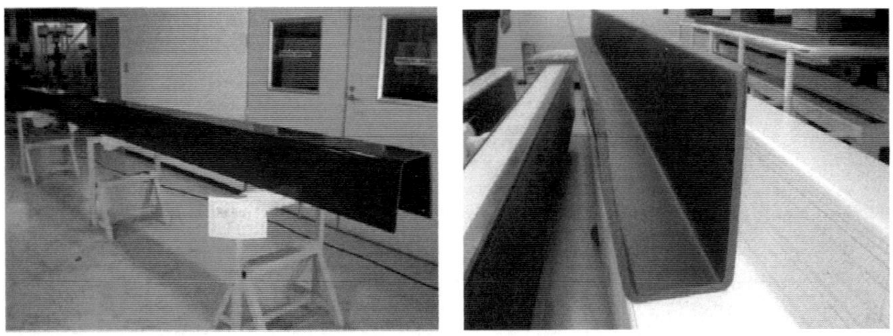

(a) Rear spar of before the trimming process. (b) Rear spar at just after the de-molding.

Figure 12. Rear spar.

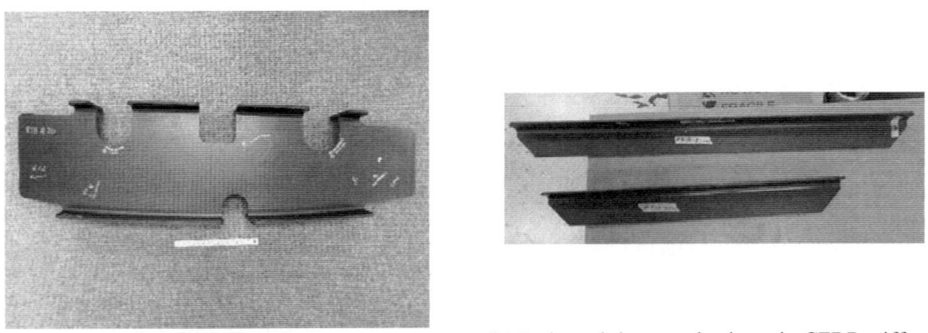

(a) H shaped rib web. (b) L shaped thermo plastic resin CFRP stiffeners.

Figure 13. Rib web and stiffener.

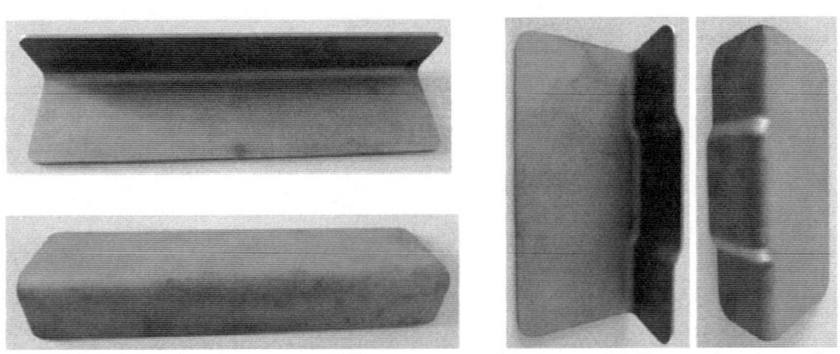

(a) 4 mm thickness titanium spar splice. (b) 2 mm thickness titanium rib post.

Figure 14. EHPF titanium spar splice and rib post.

Final assembly of wing structure
Now, final assembly of the partial composite wing structure is underway. Wing covers, spars, ribs are assembled at simple final assembly jig within titanium fasteners. Final assembly jig concept is given in Figure 15.

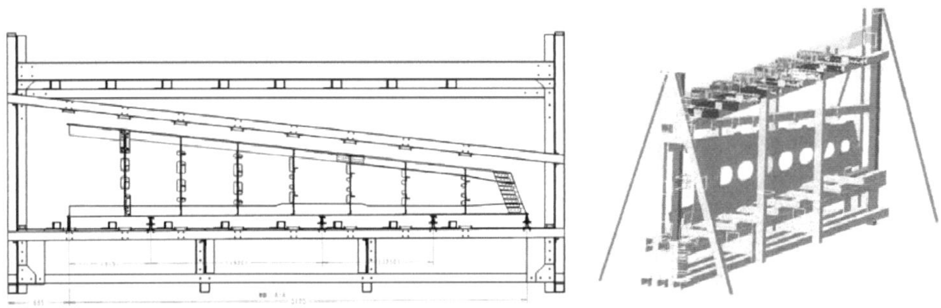

Figure 15. Final assembly jig.

FULL SCALE TEST OF PARTIAL COMPOSITE WING STRUCTURE

Summary of full scale test
PCWS is preparing the full scale bending test to evaluate the design strength and the deformation characteristics. The root end of PCWS is installed tension fitting to transfer the bending load between wing structure and loading jig. The region of tension pads is reinforced by CFRP pad-up plate to prevent the un-matured failure before the design ultimate load. The design air load distribution is simplified by many whiffle tree fixtures with 8points hydraulic actuator's. The loading method was selected the single side tension pad type to remove the upper loading side fixture. This single side loading method can be acquire the clear surface image to measure the skin deformation without obstruction of loading fixtures. Therefor, buckling deformations and strain distributions of the upper cover of PCWS can be measured by Digital Image Correlation (DIC) method. Many Strain gages and optical fibre sensors are installed to the PCWS which can be compare the design strain level between analysis and test.

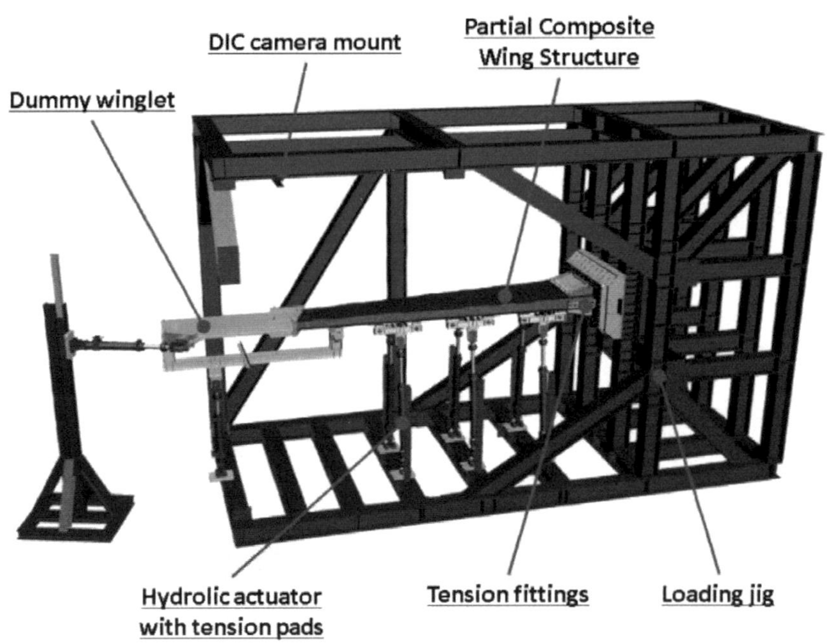

Figure 16. Digital image of test set-up.

Tension pad

The end of loading point is consisted from tension pads and the support plate to alleviate the artificial concentrated load. The tension pad is made by the rubber tip and the aluminium base plate. The bonding between the rubber tip and aluminium base plate is selected the silicon paste adhesive to keep the flexibility and toughness. Tension pads are bonded to the PCWS at just under the cross point of spar and rib, then tension pads are connected to the whiffle tree fixture and the actuator. The photograph of tension pad strength test is given in Figure 17.

(a) Photograph of tension pad test.

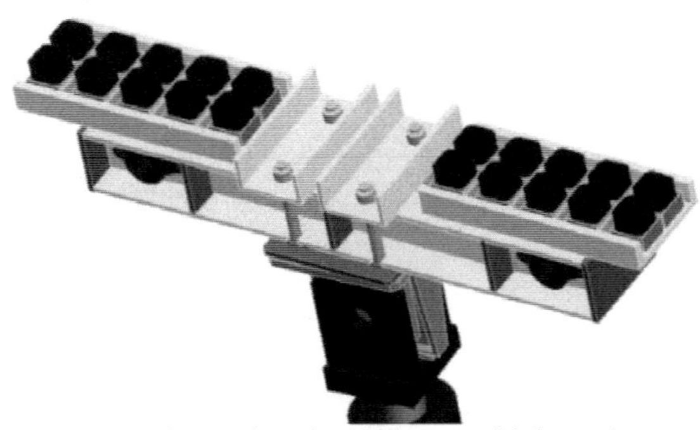
(b) Tension pads and whiffle tree with base plates.

Figure 17. Tension pad test.

Optical fibre sensor

The Optical fibre sensor of Fibre Bragg Grating (FBG) type is installed to the PCWS to measure the strain. Generally, connection of the optical fibre sensor cable is necessary for the special splicer equipment to minimize the attenuation of optical power. On the other hand, this fibre welding splicing equipment is too large to put into the inside of wing structure at the splicing work of the fibre. And, if fibre is completely spliced, flexible routing of the sensor cable is very difficult. Hence, the special mechanical connector has been developed to keep the flexibility of routing of optical fibre sensor cable. The FBG sensor and the connector are shown in Figure 18.

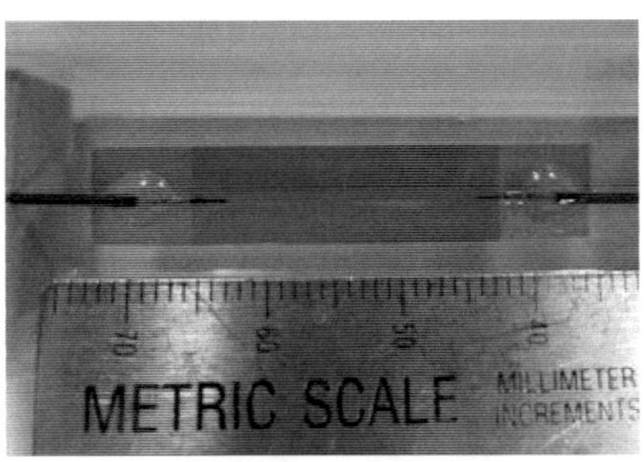

(a) FBG sensor.

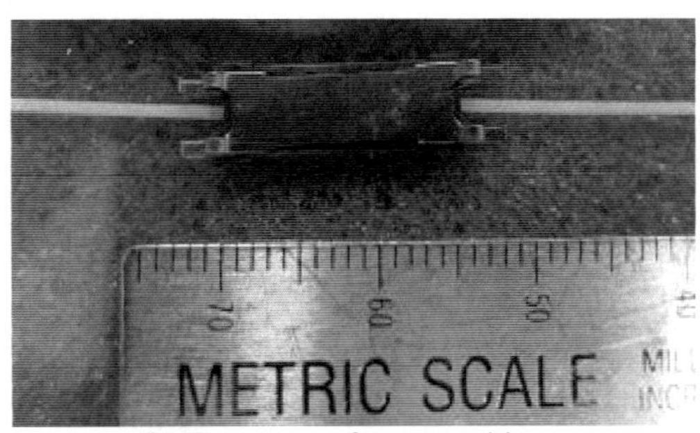

(b) Connector of sensor cable.

Figure 18. Optical fibre sensor and connector.

Loading case

Three major loading cases are prepared to conduct the static test for PCWS to evaluate the strength characteristics. The first loading case is strain survey of PCWS to check the loading and measuring systems at 20% Design Limit Load (DLL). The second loading case is DLL test to evaluate the characteristics of the detrimental permanent deformation. Final case is Design Ultimate Load (DUL). The damage tolerance test is also scheduled at latter phase of the static test. The Damage Tolerance Test (DTT) is also scheduled at latter phase of the static test. The DTT is consisted from impact test, saw-cut test, and repair test. Fatigue test is also scheduled to measure the characteristics of the flight load cycle after the static test program.

CONCLUSION

A 5m long Partial Composite Wing Structure (PCWS) was designed to demonstrate the full scale fabrication, assembly and structural test. VaRTM stitch composite was applied to the wing spar to eliminate the in-flight impact problem, complex shape drape problem, and also cost. Electrical Heating Press Forming (EHPF) method was applied to titanium fitting of spar splice. Assumed air load is loaded by 8 hydraulic actuators with whiffle tree and rubber pads. Rubber pads are bonded to lower panel of just under the spar and near the rib to apply the test load. Tens of optical fiber sensors are prepared to measure the strain without electrical noise and a lot of sensor cables. Digital Image Correlation (DIC) technique is used to capture the post buckling shape of upper panel and entire wing deformation. Conventional strain gages and displacement sensors are also used to validate the design and analysis. Structural test program of PCWS is configured strain survey, impact test, static test fatigue test, non-destructive inspection, and repair. Structural Health Monitoring (SHM) by optical fiber sensor will applied to PCWS test program. This test program is scheduled to reach the final design ultimate load test in 2 years. This program aims to demonstrate the new technologies and to extract the technical problems of practical use. Now, final assembly and test set up of the Partial Composite Wing Structure is underway.

ACKNOWLEDGEMENTS

All the partners involved in this research program. These partners are consisted from TMG, TMU, JAXA, N.A.C.T., Sentan digital, TCM, TOYOTA INDUSTRIES, W2 Design, NTT-AT, Yoshimasu Seisakusho, Kyowa, HBM Japan, Spic, MSYS, Japan Technology, Takeuchi, and IHI Jet Service.

Fatigue Bearing Failure of CFRP Composite in Biaxially Loaded Bolted Joints at Elevated Temperature

Zlatan Kapidžić, Hans Ansell

Saab AB, SE-581 88 Linköping, Sweden

Abstract: Structures that contain composites-aluminium interfaces tend to develop internal loads at elevated temperatures. In long bolted joints, the thermally induced bearing load is superimposed onto the mechanically applied load and can induce a biaxial bearing load state. The paper presents an experimental and numerical investigation of the bearing failure in cyclically loaded carbon-epoxy laminate specimens at 90 °C, in a uniaxial and a biaxial loading state. For the experimental part, a specifically designed rig was used, where the thermally induced load was applied by mechanical actuators. The failure damage was studied in a microscope and matrix cracking was pointed out as the main fatigue driving mechanism. A fatigue prediction model, based on the kinetic theory of fracture for polymer matrices, was implemented in a finite element program. Satisfactory correlations to the experimental results were obtained and the model exhibited reasonably correct behaviour for varying temperature, frequency and load ratios. Both the experimental and numerical results suggested that a longer fatigue life is obtained in biaxial loading than in uniaxial loading, for the same applied maximum resultant bolt load. A smaller matrix stress range in the biaxial load case was offered as an explanation. The found result indicates that dimensioning of biaxially loaded joints, using uniaxial fatigue data and in terms of maximum resultant bolt load, results in a conservative design.

INTRODUCTION

Aircraft structures that contain hybrid composite-aluminium bolted joints develop internal loads when exposed to elevated temperatures. The stresses arise due to the different thermal expansion properties of the two material types. In fuselage and wing structures, which contain long bolted joints, the thermally induced load could be oriented in different direction than the applied mechanical load, creating a biaxial bearing load state at the bolts. Depending on the joint length, the thermally induced bearing load could be of significant magnitude and should then be accounted for. The aim of this study is to characterise and model the fatigue bearing failure of the composite laminate in a hybrid joint and to investigate the influence of the biaxial loading on the fatigue life. Both experimental methods and numerical analyses are employed.

The experiments were conducted on simple two-bolt, double-lap joints with a carbon fibre reinforced polymer (CFRP) laminate specimens, in uniaxial and biaxial cyclic loading at elevated temperature. An experimental setup was specifically designed for the purpose, where a constant thermally induced bolt load is simulated by means of mechanical load actuators and the mechanical bolt load is applied cyclically in the perpendicular direction.

The numerical analysis of the joint fatigue failure is performed using the finite element method (FEM) and includes a fatigue failure model which is a modified version of the work presented in [1-3]. The model includes a failure criterion based on the kinetic theory of fracture for polymers [4-6] and a stress calculation procedure based on the multi-continuum theory [7, 8]. The underlying assumption is that the main driving mechanism of the composite fatigue is matrix cracking. Following this assumption, the multi-continuum theory is employed to obtain the stresses in the matrix which are then passed on as the fatigue driving stresses into the failure criterion. The model describes the appearance of matrix cracks as a thermally activated process promoted by cyclic loading. Consequently, the effects of the temperature, the loading frequency and the load ratio on the crack onset are included in the model. This is an advantage, since the need for extensive experimental characterisation of these effects is omitted.

EXPERIMENTAL PROCEDURE

The experimental work included testing of 18 specimens, manufactured from 48 plies of HTA/6376C prepreg stacked in a quasi-isotropic lay-up sequence, see Fig. 1. The specimens were held by four L-shaped steel plates and two titanium bolts, in a double-lap joint arrangement. The contact area between the steel plates and the composite plate was machined in a shape of a washer. A low pretension force was applied to each bolt. In the vertical direction, the mechanical force, F, was applied cyclically to the free end of the specimen and to the L-shaped plates, see Fig. 2. The load was cycled with a constant amplitude at a load ratio R = 0. A constant thermally induced load, F_{th}, was applied by load actuators connecting the steel plates in the horizontal direction to a surrounding steel frame, which was suspended by chains and week springs, see Fig. 3. During the testing, the specimens were placed inside a furnace where the air was heated to 90 °C. Of totally 18 specimens tested to failure, 8 were loaded uniaxially, i.e. $F_{th} = 0$ kN, and 10 were loaded biaxially with $F_{th} = 25$ kN. Two additional specimens were tested to approximately half of the fatigue life, whereupon they were cut along the bearing planes and studied in a microscope.

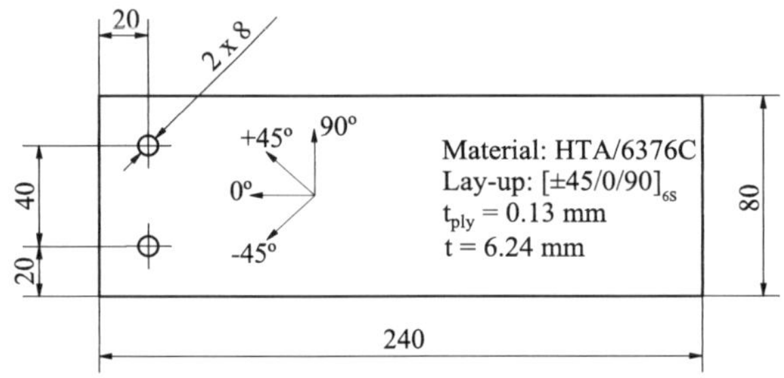

Figure 1. Test specimen.

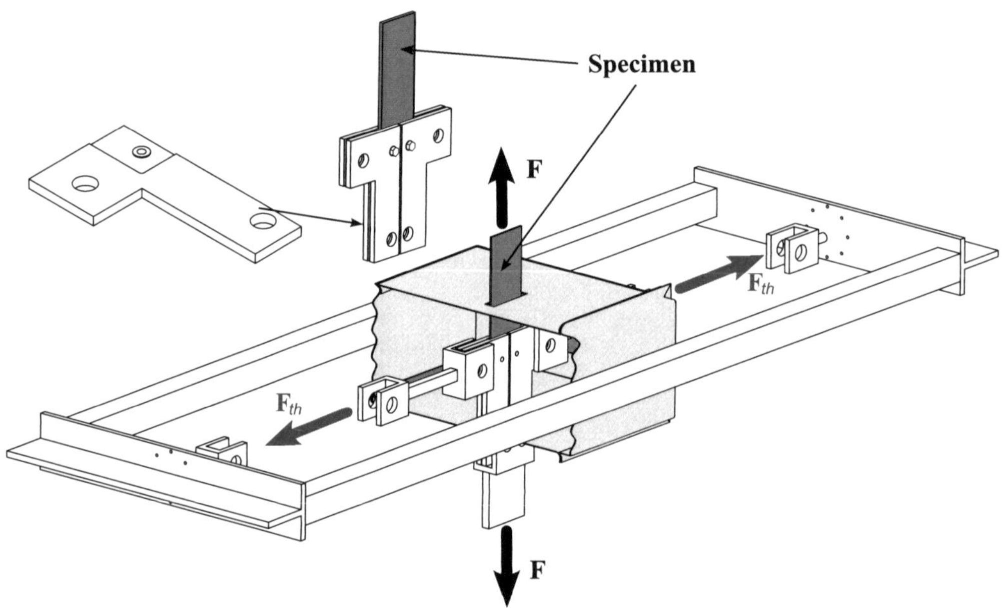

Figure 2. Experimental setup and load application.

Figure 3. Test rig.

MODELLING

The numerical analysis procedure is broken down in several blocks, as shown schematically in Fig. 4. First, finite element analyses of the specimen are performed, where the load cases with the maximum applied mechanical cyclic load, F_{max}, and the thermally induced load, F_{th} are studied. From the resulting displacement field vector, $\mathbf{u}$, the homogenised ply stress vector, $\boldsymbol{\sigma}_c$, in each layer in the specimen is obtained with the classical laminate theory. In the next step, the multi-continuum theory is applied to the homogenised ply stress vector to compute the homogenised matrix stress vector, $\boldsymbol{\sigma}_m$, from which an effective matrix stress, σ, is constructed. The last step contains the failure criterion which uses the effective matrix stress, considers the temperature and the loading frequency, and calculates the fatigue life N_f in each material point. The whole procedure is programmed in MATLAB [9]. In the following, each of the steps is described briefly.

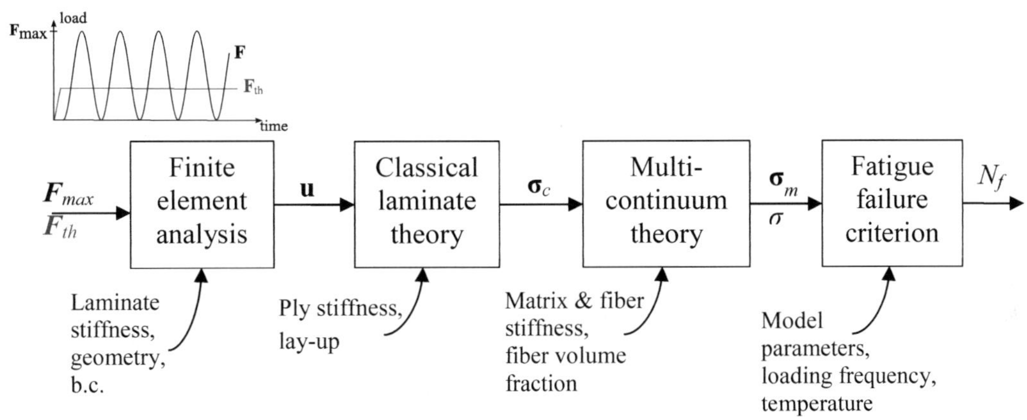

Figure 4. Flowchart of the computation procedure.

Finite element modelling

The FE-model of the specimen consists of iso-parametric quadrilateral, eight-node plain stress elements, integrated with a full Gauss quadrature integration rule. A linear elastic orthotropic model is used to describe the material behaviour. The integration of the stiffness matrix considers the orientation of the principal directions of each ply in the laminate. Only half of the specimen in included into the model, due to the geometrical and load symmetry. The bolt load is applied by enforcing a constraint relation on the nodes on the hole edge, along a semi-circle in the direction of the resultant bolt load. These nodes are only allowed to move on the edge of a circle, representing the bolt, whose centre is assigned a prescribed displacement. A more detailed description of the model is found in [10]. In the uniaxial load case, since the minimum applied bolt load is zero, the cyclic stress range is calculated by applying F_{max} only. In the biaxial case, the difference is taken between the states ($F_{max} \neq 0$, $F_{th} \neq 0$) and ($F_{max} = 0$, $F_{th} \neq 0$) to obtain the stress range.

Stress calculation

The stress calculation of the homogenised ply stress vector, $\boldsymbol{\sigma}_c$, and the homogenised matrix stress vector, $\boldsymbol{\sigma}_m$, is performed at the Gauss points. Based on the nodal displacements, $\boldsymbol{\sigma}_c$ is computed in the principal material direction of each ply using the classical laminate theory. The matrix stress vector is then obtained by the multi-continuum theory [7, 8] as

$$\boldsymbol{\sigma}_m = \mathbf{Q}_m \boldsymbol{\sigma}_c \tag{1}$$

$$\mathbf{Q}_m = \mathbf{C}_m \left(\mathbf{C}_c (\phi_m \mathbf{I} + \phi_f \mathbf{A}) \right)^{-1} \tag{2}$$

$$\mathbf{A} = -\frac{\phi_m}{\phi_f} (\mathbf{C}_c - \mathbf{C}_f)^{-1} (\mathbf{C}_c - \mathbf{C}_m) \tag{3}$$

where $\mathbf{C}_c$, $\mathbf{C}_m$ and $\mathbf{C}_f$ are the composite, matrix and fibre stiffness matrices respectively, ϕ_f and ϕ_m are the fibre and matrix volume fractions, and $\mathbf{I}$ is the identity matrix. The matrix stresses are then extrapolated to the element corner nodes, where the effective matrix stress, σ, is computed as

$$\sigma = \sqrt{\sigma_{m,22}^2 + \sigma_{m,12}^2} \tag{4}$$

where $\sigma_{\mathrm{m},22}$ and $\sigma_{\mathrm{m},12}$ are the transverse normal and shear matrix stress components.

Fatigue failure criterion

The concept of rate dependent kinetic theory of fracture for polymers [4–6] describes the formation submicrocracks as a time-dependent, thermally activated process of inter-atomic bond breakages. It postulates the formation of submicrocracks, as a function of time t, according to

$$\frac{dn}{dt} = (n_T - n)K_b \tag{5}$$

where $n(t)$ is the concentration of submicrocracks normalised by the submicrocrack concentration at rupture, n_T is the normalised concentration of potentially weak points and K_b is the bond rupture rate. Based on the above definition, $n(t)$ works as a damage state variable which becomes 0 for no damage and 1 for a damage state corresponding to formation of a macro crack. Solving Eq. (5) for constant applied stress, and enforcing that $n = 1$ at the time of failure, gives that $n_T = \mathrm{e}/(\mathrm{e}-1)$.

he bond rupture rate of the thermally activated process is obtained as the frequency of thermal oscillations of the molecules, kT/h, times the likelihood of the occurrence of the process, $\exp(-U/(kT))$, where h is Planck's constant, k is Boltzmann constant and T is the absolute temperature. The process is activated when the thermal energy, kT, exceeds the energy barrier U, also referred to as the activation energy. Assuming that the applies stress works in favour of the bond breaking process and thus lowering the effective energy barrier, the bond rupture rate can be expressed as

$$K_b = \frac{kT}{h}\exp\left(-\frac{U - \gamma\sigma(t)}{kT}\right) \tag{6}$$

where γ is the activation volume associated with the stress. The parameters U and γ are assumed to be temperature dependent. Notice that the stress quantity in this context is the effective matrix stress. Inserting Eq. (6) into Eq. (5) and integrating for a variable time history of stress yields

$$n(t) = n_T\left(1 - \exp\left(-\frac{kT}{h}\exp\left(-\frac{U}{kT}\right)\int_0^t \exp\left(\frac{\gamma\sigma(\tau)}{kT}\right)d\tau\right)\right) \tag{7}$$

Inserting the condition that $n = 1$ at the time of rupture t_f gives

$$1 = \frac{kT}{h}\exp\left(-\frac{U}{kT}\right)\int_0^{t_f} \exp\left(\frac{\gamma\sigma(\tau)}{kT}\right)d\tau \tag{8}$$

Considering a constant amplitude cyclic stress variation with frequency f, the number of cycles to failure, N_f, is expressed as

$$N_f = \frac{\frac{h}{kT}\exp\left(\frac{U}{kT}\right)}{\int_0^{1/f} \exp\left(\frac{\gamma\sigma(\tau)}{kT}\right)d\tau} \tag{8}$$

The integral in Eq. (8) can be solved analytically for the saw-tooth shaped stress time history [3], but not for the sine shaped, which is the one considered in this study. In this work, the integration is therefore performed using numerical methods.

Evaluating Eq. (8) for constant maximum stress and temperature and an increasing stress ratio, results in a decreasing number of cycles to failure. Intuitively, the opposite is expected and is usually seen in experiments. This behaviour was corrected in [10] by considering the increase of temperature due to the cycling. The following expression was proposed

$$T = T_0 + \psi f N^q \Delta\sigma^2 \tag{9}$$

where T_0 is the ambient temperature, $\Delta\sigma = \sigma_{max} - \sigma_{min}$ is the stress range, and ψ and q are material parameters. Equation (9) in then inserted into Eq. (8) with $N = N_f$. Although this correction fixes the problem of decreasing N_f with increasing stress ratio, it complicates the model, since Eq. (8) now has to be solved numerically for N_f.

Model parameters

Table I shows the elastic stiffness properties of the unidirectional laminate, matrix and fibre at room temperature and at 90 °C. The elevated temperature properties are used in the stress calculations in the FE-model and in Eqs. (1-3), in the stiffness matrices. The model parameters that need to be determined, by fitting Eq. (8) to fatigue data, are the activation energy U, the activation volume γ and the constants ψ and q. For this purpose, constant amplitude fatigue data from tests performed at room temperature on un-notched, quasi-isotropic specimens was used. Figure 5 shows the fitted curves and the test data in terms of the maximum absolute applied strain at different load ratios and the resulting model parameters. Failure of a specimen is assumed to occur at first failure of the off-axis plies. In this case the ±45° plies fail first.

The activation energy, U, and the activation volume, γ, determined in Fig. 5 are valid for RT, whereas the bolted joint tests were performed at 90 °C. In absence of test data from un-notched specimen tests at elevated temperature to fit the parameters to, we assume an increase of U by 26% and of γ by 300% at 90 °C based on the results on other studies in the literature as described in [10].

Table I. Elastic properties of HTA/6376C at room temperature (RT) and at 90 °C.

	Ply at RT	Ply at 90 °C	Matrix at RT	Matrix at 90 °C	Fibre
E_{11} (MPa)	137 000	134 600	4 670	2 853	238 000
E_{22} (MPa)	10 400	8 700	5 263	2 931	12 800
G_{12} (MPa)	5 190	3 900	2 264	1 233	15 000
ν_{12}	0.3	0.27	0.43	0.41	0.27
ϕ_f	0.65	0.65	-	-	-

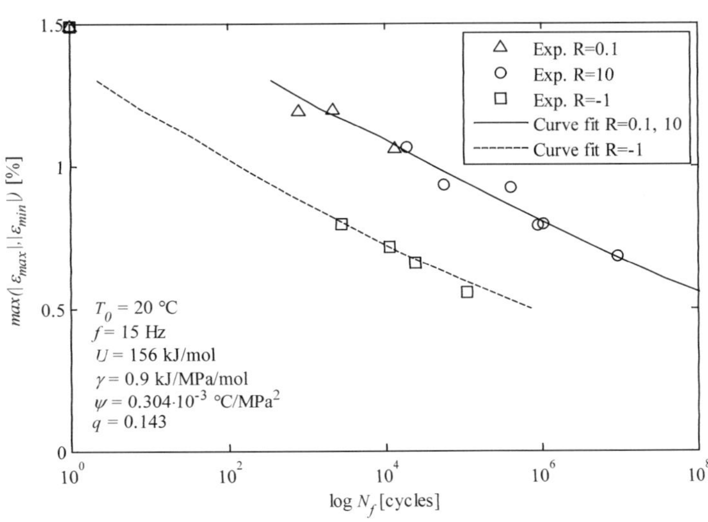

Figure 5. Curve fitting to the experimental fatigue data (Exp.) from un-notched specimen tests.

Equation (8) can also be solved for different pairs of σ_{max} and σ_{min} at constant N_f and a constant life diagram, usually expressed in terms of mean stress and stress amplitude, can be constructed. Figure 6 shows a constant life diagram for an un-notched quasi-isotropic laminate in terms of the applied mean strain, ε_m, and strain amplitude, ε_a, at RT and constant frequency. Notice that N_f increases with a decreasing strain amplitude at constant mean strain, as a result of the correction in Eq. (9). The diagram is symmetric with respect to ε_m = 0, which is a consequence of the choice of the definition of effective matrix stress in Eq. (4).

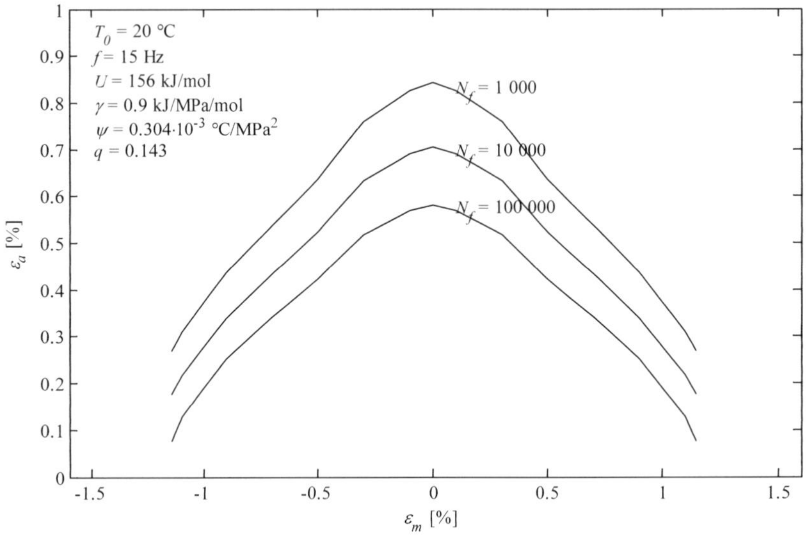

Figure 6. Constant life diagram at RT.

Figure 7 shows that both an increasing cyclic frequency, f, and an increasing ambient temperature, T_0, have a detrimental effect on the constant life diagram. The effect of frequency seems to be lower at elevated temperature. It is also evident that the choice of U and γ at elevated temperature significantly influences the fatigue life. It is therefore desirable to base that choice on actual fatigue data, such as those shown in Fig. 5, at the relevant temperature.

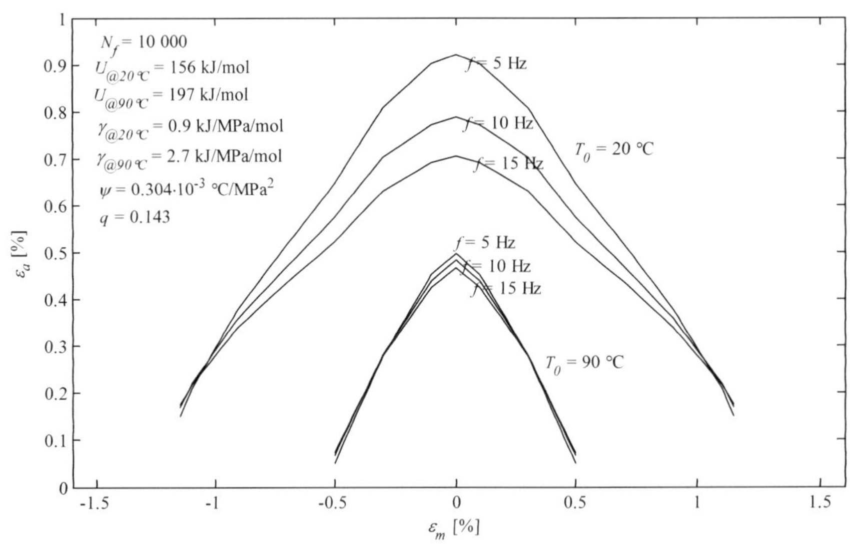

Figure 7. Constant life diagrams at different frequencies for N_f = 10 000 cycles at RT and at 90 °C.

RESULTS AND DISCUSSION

In all tested specimens, successive bolt hole elongation was observed during the cyclic loading. The holes were elongated in the direction of the maximum resultant bolt load. Macroscopically, the failure appeared as crushing of the hole edge in the contact area between the plate and the bolt. The microscopy images of the bearing planes of the two specimens, which were tested to approximately half of the fatigue life, are shown in Fig. 8. The lowest magnification level shows the damage in the bearing plane as a permanent deformation and fracture of the plies near the hole edge. Delamination of the layers is also observed. The plies seem to have bent out of the plane, towards the free surfaces.

Zooming in at the periphery of the damaged area shows the bending deformation of the layers, and in the biaxial case even tensile fracture of the fibres. In contrast to static bearing failure, where fibre kinking bands were observed in the bearing plane [11], the deformation seems to be more continuous, as if the failure is caused by gradual material softening rather than by sudden instability of the on-axis layers. Further magnification on the root of the bent area shows the presence of matrix cracks and fibre-matrix debonding damage in the off-axis plies. Beyond this area, no matrix cracks or fibre-matrix debonding damage is present. These observations support the assumption on which the prediction model is based, i.e. that the driving mechanism of composite fatigue is matrix micro-cracking.

The damage progress can be described as follows. The damage initiates at the hole edge and spreads then inwards the material, in the direction of the bolt load. Matrix cracking gradually lowers the laminate stiffness and reduces the support of the matrix to the fibres, which causes increasing displacement and eventually delaminations, fibre failure and ply fracture. The bending of the plies is, to some extent, restricted by the confinement of the composite between the steel plates. This restriction inhibits the cracking and delays the structural failure of the joint, until the damage has spread outside the washer-shaped area. We therefore define the number of cycles to failure of the joint as the number of cycles it takes to initiate a micro-crack outside the washer area.

Figure 9 shows the FE-predictions of the number of cycles to failure around the bolt hole for the uniaxial and the biaxial load case. The number of cycles to failure is evaluated as the smallest N_f found on the node path along the washer area perimeter. In the uniaxial case this value is found in the ±45° plies and in the 0° and 90° plies in the biaxial case. The predictions of N_f versus the maximum mechanical bearing stress $\sigma^b_{max} = F_{max} /(2tD)$, where t is the laminate thickness and D is the bolt diameter, show good correspondence with the experimental results, see Fig. 10. The biaxial loading gives a shorter fatigue life then the uniaxial loading at the same level of σ^b_{max}, which is due to the presence of the thermally induced load, F_{th}. However, if the same diagram is plotted in terms of the maximum resultant bearing stress $\sigma^b_{res,max} = ((F_{max}/2)^2+(F_{th})^2)^{1/2}/(tD)$ see Fig. 11, the biaxial loading results in a longer fatigue life.

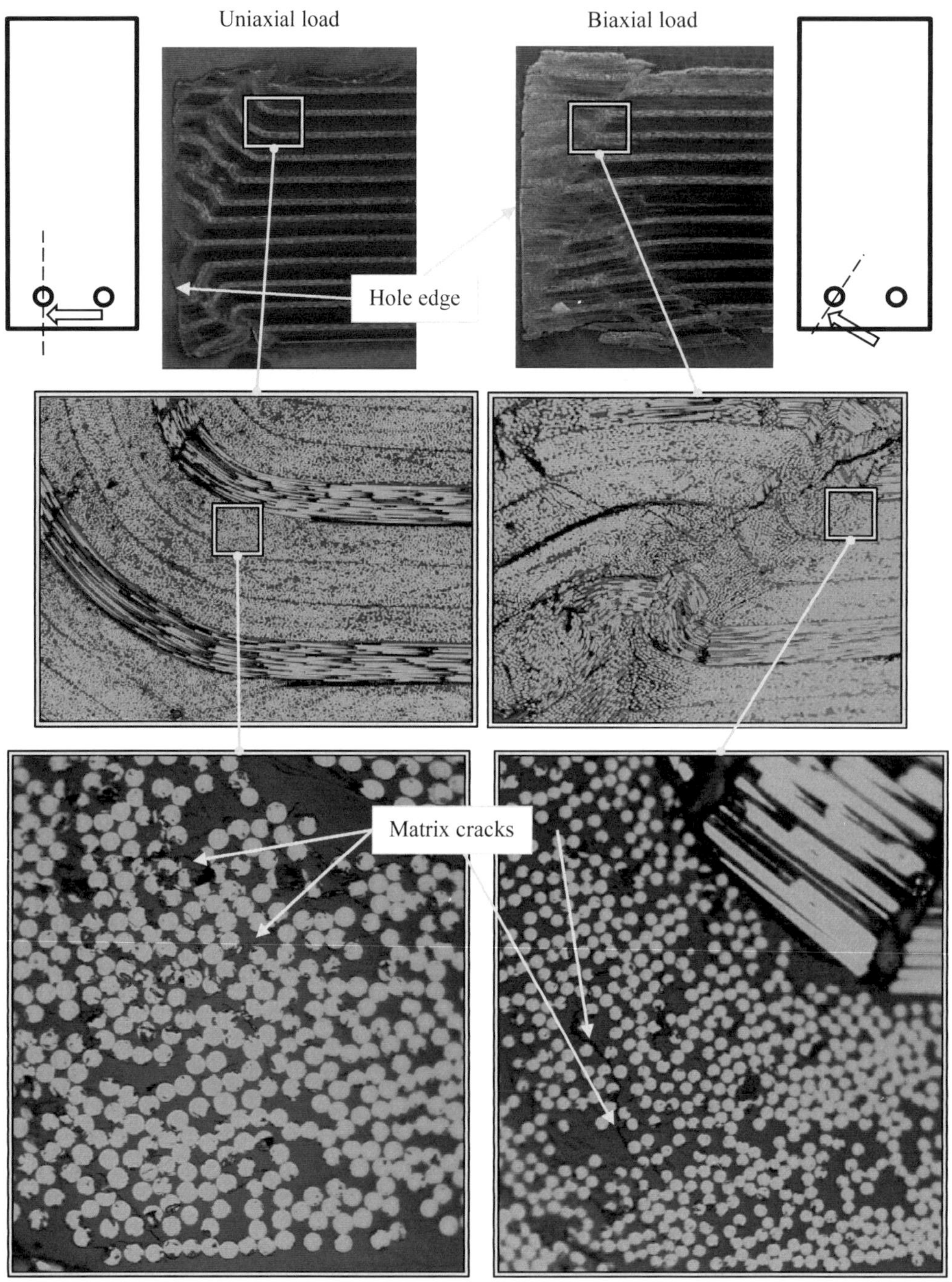

Figure 8. Microscopy images of the bearing plane of the uniaxially loaded (left) and the biaxially loaded (right) specimen, after half of fatigue life, at three magnification levels.

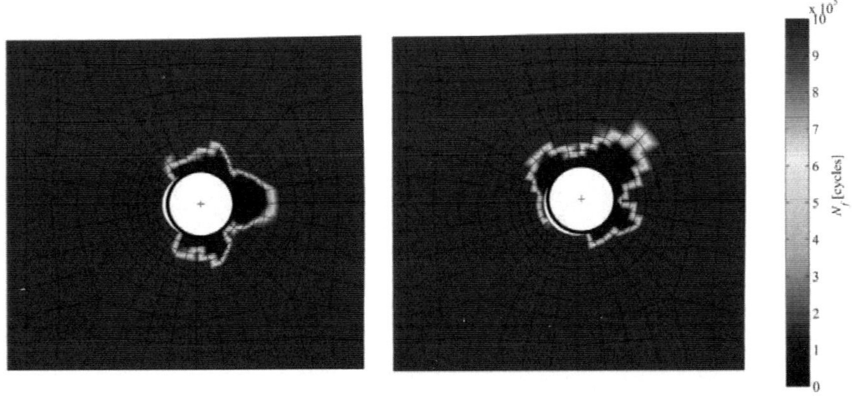

Figure 9. Number of cycles to failure in the uniaxial case (left) and biaxial case (right).

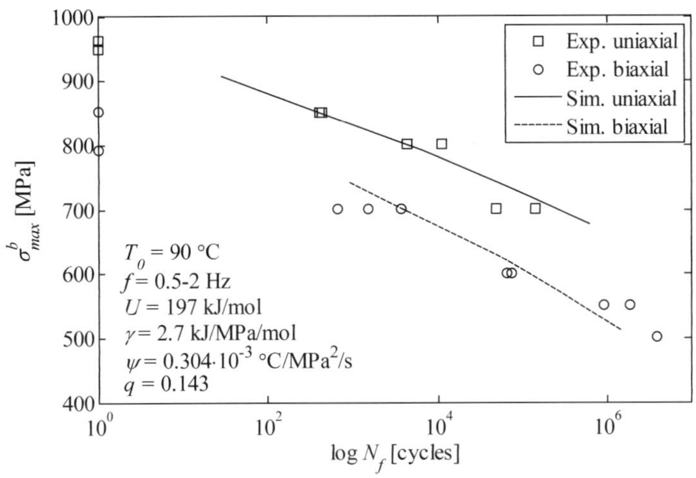

Figure 10. Experimental results (Exp.) from uniaxial and biaxial bearing tests compared to the simulation results (Sim.) in terms of maximum mechanical bearing stress.

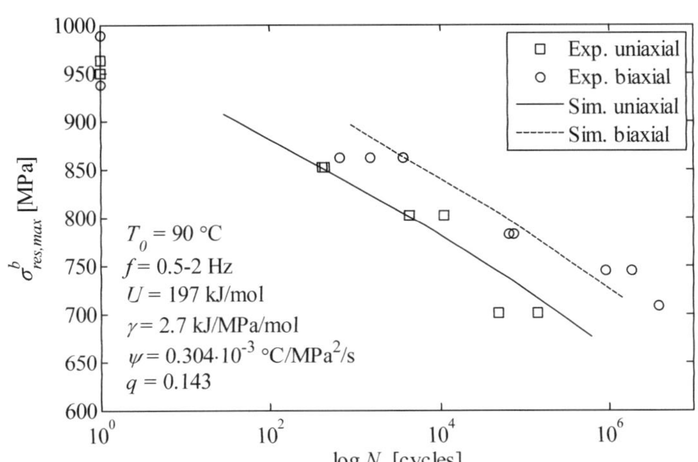

Figure 11. Experimental results (Exp.) from uniaxial and biaxial bearing tests compared to the simulation results (Sim.) in terms of maximum resultant bearing stress.

An explanation for this is that in the biaxial loading, the minimum resultant bearing stress is higher than in the uniaxial loading, thus resulting in a smaller resultant stress range and therefore a longer fatigue life. This can also be seen but looking at the maximum and minimum effective matrix stresses σ_{max} and σ_{min} on the path where the fatigue life is evaluated, see Fig. 12, for the same maximum resultant bearing stress. It appears that the effective matrix stress range, σ_{max} - σ_{min}, is larger in the uniaxial case. The effect of the thermally induced load is, in this view, somewhat similar to that of a pretension load in tensile loaded joints.

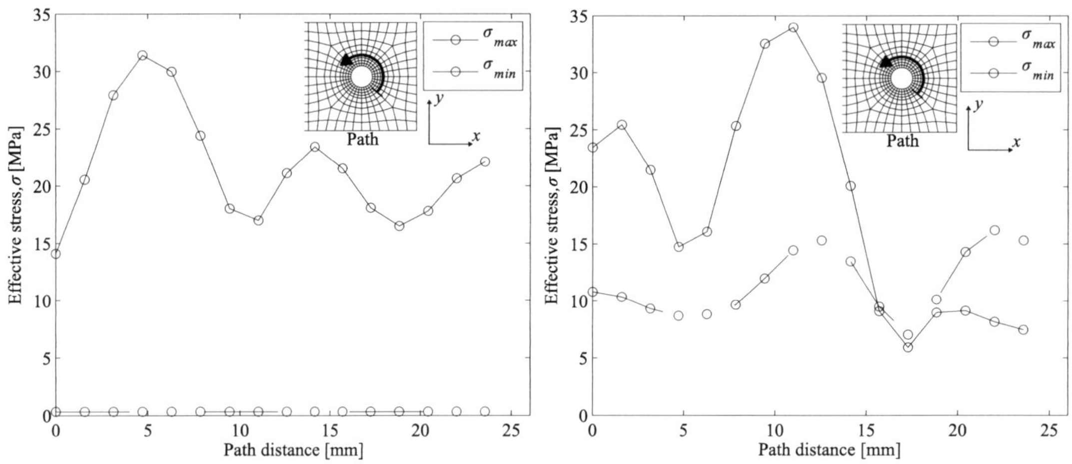

Figure 12. Effective matrix stress along the path around the bolt hole, in uniaxial case (left) and biaxial case (right) at the same maximum resultant bearing stress, $\sigma^{b}_{res,max}$.

Referring to Fig. 11, a useful conclusion can be drawn: using uniaxial data for dimensioning of biaxially loaded joints in constant amplitude, in terms of the maximum resultant bearing stress, leads to a conservative design.

CONCLUSIONS

An experimental and numerical investigation of uniaxially and biaxially cyclically loaded bolted composite specimens was done. A specifically designed rig was used to assess the bearing failure behaviour under thermally induced loads coupled to mechanical loads, at elevated temperature. A microscopy study of the damage development in the bearing plane was conducted. A fatigue life prediction model was implemented to simulate the tests. The following conclusions were drawn:

- Matrix cracking and fibre-matrix debonding were identified as the driving mechanisms behind the fatigue damage initiation. The damage development leads to a successive compliance increase during the cyclic loading.
- The fatigue prediction model based on the mentioned mechanisms gave satisfactory results compared to tests. The model requires a very low material characterization effort and works for fatigue prediction at different temperatures, frequencies and load histories. However, more validation is necessary in order to establish the predictive capability of the model for different geometries, lay-ups, materials, etc.
- Application of thermally induced load had a detrimental effect on the fatigue life of the bolted joint. However, compared in terms of the maximum applied resultant load, biaxial loading is more favourable than the uniaxial loading. Using uniaxial data for dimensioning of the biaxially loaded joints, in terms of the maximum resultant bearing stress, leads to a conservative design.

REFERENCES

[19] Fertig, R.S. and Kenik, D.J. (2011). In: *NAFEMS World Congress*, Physics-based fatigue life prediction of composite structures, Boston, MA.

[20] Fertig, R.S. and Kenik, D.J. (2011). In: *52nd AIAA/ASME/ASCE/AHS/ASC Structures, Structural Dynamics and Materials Conference*, Predicting composite fatigue life using constituent-level physics, Denver, CO.

[21] Fertig, R.S. (2009). In: *SAMPE Fall Technical Conference Proceedings: Global Material Technology: Soaring to New Horizons*, Bridging the gap between physics and large-scale structural analysis: A novel method for fatigue life prediction of composites, Wichita, KS.

[22] Zhurkov, S.N. and Kuksenko, V.S. (1975). *Int. J. Fracture*, vol. 11, n. 4, pp. 629-639.

[23] Zhurkov, S.N. (1984). *Int. J. Fracture*, vol. 26, n. 4, pp. 295-307.

[24] Hansen, A.C. and Baker-Jarvis, J. (1990). *Int. J. Fracture*, vol. 44, n. 3, pp. 221-231.

[25] Hansen, A.C. and Garnich, M.R. (1995). *Compos. Eng.*, vol. 5, n. 9, pp. 1091-1103.

[26] Mayes, J.S. and Hansen, A.C. (2004). *Compos. Sci. Technol.*, vol. 64, n. 3-4, pp. 379-394.

[27] MATLAB. Version 8.3.0. MathWorks Inc., Natick Massachusetts, United States; 2014.

[28] Kapidžić, Z., Ansell, H., Schön, J. and Simonsson, K. (2015). *Compos. Struct.*, vol. 127, pp. 298-307.

[29] Kapidžić, Z., Ansell, H., Schön, J. and Simonsson, K. (2015). *Compos. Struct.*, vol. 125, pp. 60-71.

AUTHOR INDEX

AUTHOR INDEX

AUTHOR INDEX

VTT Information Service
PO Box 1000
FI-02044 VTT
Finland

CURRAN ASSOCIATES INC.
proceedings.com

ISBN 978-1-5108-5774-2

34th ICAF Conference and 28th ICAF Symposium (ICAF 2015)

Helsinki, Finland
1 – 5 June 2015

Volume 2 of 2

Editor:

Aslak Siljander

ISBN: 978-1-5108-5774-2

Printed from e-media with permission by:

Curran Associates, Inc.
57 Morehouse Lane
Red Hook, NY 12571

CURRAN ASSOCIATES INC.
proceedings.com

Some format issues inherent in the e-media version may also appear in this print version.

Printed by Curran Associates, Inc. (2018)

For permission requests, please contact VTT Information Service
at the address below.

VTT Information Service
PO Box 1000
FI-02044 VTT
Finland

Phone: +358 20 722 111
Fax: +358 20 722 7001

info@vtt.fi

Additional copies of this publication are available from:

Curran Associates, Inc.
57 Morehouse Lane
Red Hook, NY 12571 USA
Phone: 845-758-0400
Fax: 845-758-2633
Email: curran@proceedings.com
Web: www.proceedings.com

Aslak Siljander *Editor*

ICAF 2015

ICAF 2015 Structural Integrity: Embracing the future – respecting the past; supporting aging fleets with new technologies

Proceedings of the 34th Conference and the 28th Symposium of the International Committee on Aeronautical Fatigue and Structural Integrity, 1–5 June 2015, Helsinki, Finland

ICAF
International Committee on Aeronautical Fatigue and Structural Integrity

VTT

Editor

Aslak Siljander
VTT Technical Research Centre of Finland Ltd
B.O.Box 1000
FI-02044 VTT, Finland
Telephone
Fax
Email: aslak.siljander@vtt.fi

Espoo, May 2015

General Information

ICAF General Secretary

Prof. Anders Blom
Defence & Security, Systems and Technology
Swedish Defence Research Agency (FOI)
SE-164 90 Stockholm, Sweden

ICAF National Delegates

AUSTRALIA
Phil Jackson
Airworthiness and Life Evaluation Group
Defence Science and Technology Organisation
506 Lorimer Street
Fishermans Bend VIC 3207, Australia

CANADA
Nick Bellinger
National Research Council Canada
1200 Montreal Road, Bldg. M-14
Ottawa, Ontario K1A 0R6, Canada

FINLAND
Dr. Aslak Siljander
VTT Technical Research Centre of Finland Ltd
Machines and Vehicles
P. O. Box 1000
FI-02044 VTT, Finland

FRANCE
Dr. Thierry Ansart
Structures Division
DGA Aeronautical Systems
BP 93123, 47 rue Saint Jean
31131 Balma, France

GERMANY
Dr. Claudio Dalle Donne
EADS Innovation Works
81663 München, Germany

ISRAEL
Dr. Yuval Freed
Head of F&DT Dept., Engineering & Development Group
Israel Aerospace Industries
Tel Aviv, Israel

ITALY
Prof. Luigi Lazzeri
Universita di Pisa
Department of Aerospace Engineering
Via G. Caruso, 8
56122 Pisa, Italy

JAPAN
Prof. Nobuo Takeda
University of Tokyo
Dept. of Aeronautics and Astronautics
7-3-1 Hongo, Bunkyo-ku
Tokyo 113-8656, Japan

POLAND
Dr. Antoni Niepokólczycki
Director, Materials and Structures Research Center (MSRC)
Institute of Aviation
Al. Krakowska 110/114
02-256 Warsaw, Poland

SWEDEN
Hans Ansell
Saab Aerosystems
Saab AB
SE-581 88 Linköping, Sweden

SWITZERLAND
Dr. Michel Guillaume
ZHAW Zurich University of Applied Sciences
Head of the Center for Aviation
Technikumstrasse 9
8400 Winterthur, Switzerland

THE NETHERLANDS
Marcel J. Bos
National Aerospace Laboratory (NLR)
P.O. Box 153
NL-8300 AD Emmeloord, The Netherlands

UNITED KINGSOM
Dr. Steve Reed
Technical Advisor Structural Integrity
Physical Sciences Department
Defence Science and Technology Laboratory
Desk 236, Room 102, Bldg 5 (Porton Down)
Salibury, Wilts SP4 0JQ, UK

UNITED STATE OF AMERICA
Dr. Ravinder Chona
Structural Sciences Center
US Air Force Research Laboratory (AFRL / RBSM)
Wright-Patterson Air Force Base
Ohio, USA

Organisation

Aslak Siljander
Conference & Symposium Chair
VTT

Management Team, ICAF 2015 Organizing committee

Tuomo Hokkanen, VTT Technical Research Centre of Finland Ltd
Paula Kauppinen, VTT Technical Research Centre of Finland Ltd
Satu Koskela, VTT Technical Research Centre of Finland Ltd
Maria Rajasalmi, VTT Technical Research Centre of Finland Ltd
Aslak Siljander, VTT Technical Research Centre of Finland Ltd
Tomi Viitanen, VTT Technical Research Centre of Finland Ltd
Christina Vähävaara, VTT Technical Research Centre of Finland Ltd

Technical Program Committee

Captain (eng) Kalle Vaaraniemi, Finnish Air Force Command (AFCOMFIN)
Mr. Ari Kivistö, Finnish Defence Forces Logistics Command, Joint Systems Centre, Aeronautical Systems Division
Mr. Jukka Raunio, M.Sc., Vice President, Engineering, Patria Aviation
Mr. Mika Mustakangas, M.Sc., R&D Project Manager, Patria Aerostructures
Mr. Sakari Pasanen, Director, Business Development, Patria Aerostructures
Mr. Jaakko Sotkasiira, M.Sc., Finflo Ltd.
Dr. Timo Siikonen, Professor, Finflo Ltd.
Dr. Olli Saarela, Professor, Aalto University
Dr. Ari Visa, Professor, Tampere University of Technology
Mrs. Liisa Poussa, M.Sc., Research Team Leader, VTT Technical Research Centre of Finland Ltd
Mrs. Eetta Saarimäki, M.Sc., Senior Scientist, VTT Technical Research Centre of Finland Ltd
Mr. Tomi Viitanen, M.Sc., Senior Scientist, VTT Technical Research Centre of Finland Ltd
Mr. Risto Laakso, M.Sc., Senior Scientist, VTT Technical Research Centre of Finland Ltd
Mr. Keijo Koski, M.Sc., Principal Scientist, VTT Technical Research Centre of Finland Ltd
Mr. Kari Lahdenperä, Lic.Tech., Principal Scientist, VTT Technical Research Centre of Finland Ltd
Dr. Aslak Siljander, Principal Scientist, VTT Technical Research Centre of Finland Ltd

Preface

ICAF is coming to Helsinki, Finland, for the first time in the history of ICAF. The 34th Conference and the 28th Symposium of the International Committee on Aeronautical Fatigue and Structural Integrity is taking place in Helsinki from the 1st to the 5th of June 2015. ICAF 2015 is hosted and organized by VTT Technical Research Centre of Finland Ltd and its venue is Marina Congress Center in downtown Helsinki.

The International Committee on Aeronautical Fatigue and Structural Integrity (ICAF) was formed in 1951 under the name "The International Committee on Aeronautical Fatigue" in response to growing concerns regarding fatigue problems in metal aircraft structures. In 2010 the name was changed to the present one in order to clarify that the scope of the committee had broadened over the years and now also includes topics such as damage formation and growth in composite structures, structural health and loads monitoring, probabilistic modelling of structural integrity, corrosion control, engine integrity etc. The acronym ICAF was maintained. The stated aims of ICAF are to encourage contacts between people actively engaged in aircraft structural integrity problems and to exchange information, experience, opinions and ideas concerning aeronautical fatigue and fatigue-related subjects. To this end, a Conference and, since 1959, a Symposium are organized every two years for attendance by representatives of industry, universities and institutes, military specialists, regulatory agencies and aircraft operators throughout the world.

ICAF 2015 begins with a two-day Conference, June 1–2, consisting of presentations of the National Reviews given by the National Delegates from the fourteen member states of ICAF. The fourteen national reviews will describe research, development and airworthiness activities of international interest which has taken place in each country during the two years since the previous ICAF Conference in Jerusalem. In addition, I am indeed pleased to welcome the national reviews from three important new countries: Brazil, China and Russia.

The ICAF 2015 Conference will be followed by a three-day Symposium, June 3–5. The Symposium's main theme – *"Embracing the future – respecting the past: supporting aging fleets with new technologies"* – highlights that the global air travel is increasing, yet the structural safety must be guaranteed, new materials are increasingly applied (e.g. composites), new manufacturing methods are introduced (e.g. Additive Manufacturing), many experts retire, young researchers both scientists and engineers graduate and enter the industry at large. Furthermore, nations and end-users may not have the funding to replace their fleets as planned, meaning that life extension programs with associated risk analyses may need to be considered. In any case, the gap between theory and practice must be bridged also in the future of civil and military aviation.

Each ICAF Symposium since 1951 begins with a lecture honoring the memory of Dr. Frederick J. Plantema, the founder of ICAF. At the 28th Symposium, the Plantema Lecture will be delivered by Jerzy Komorowski, whose lecture is entitled *"Structural Integrity – the unfinished business"*. Mr. Komorowski, previously the Canadian National Delegate to ICAF (2001–2011), is a member of the Canadian Aeronautics and Space Institute; the Canadian member of the NATO Science and Technology Organization Advanced Vehicle Technology Panel (AVT); and Chair of the AVT Support Committee. He serves as the Canadian National Resource Specialist at the Aerospace Engineering Group of The Technical Cooperation Program.

The Jaap Schijve Award and Lecture will be awarded to a young researcher, selected by a special committee, for his or her outstanding performance. A total of 42 lectures including the Plantema Lecture and the Jaap Schijve Award Lecture and 57 posters are presented in three days. All the papers are of high technical quality and only the limited time prevents all papers from being orally presented.

All the oral and poster presentations will be delivered on a USB stick only. The ISBN-endowed USB stick is the ICAF 2015 Proceedings for the first time. As all the papers are of high quality, the reader, intentionally, cannot distinguish whether the paper was an oral or a poster presentation.

The year 2015 marks the 20th anniversary at VTT of aircraft structural integrity program (ASIP) related applied research activities aimed at serving the needs of the Finnish Air Force (FINAF) and the national research network described in more detail in the Finnish National Reviews. Finland started seeking the ICAF membership in 2001 and became the 13th full member in 2005. Without the in-country ASIP research activities, solely orchestrated and funded by the FINAF during the past 20 years, Finland would not be an ICAF member. On behalf of the in-country ASIP network and VTT in particular, I sincerely thank the FINAF for their vision, initiative, enthusiasm, diligence and trust over these years.

Without good teamwork the organization of a major international conference would not have been possible. The ICAF 2015 in-country Technical Program Committee, responsible for reviewing and ranking the abstracts, consisted of the following individuals: Captain (eng) Kalle Vaaraniemi (Finnish Air Force Command), Mr. Ari Kivistö (Finnish Defence Forces Logistics Command, Joint Systems Centre, Air Systems Division), Mr. Jukka Raunio (Patria Aviation),

Mr. Mika Mustakangas and Mr. Sakari Pasanen (Patria Aerostructures), Dr. Timo Siikonen and Mr. Jaakko Sotkasiira (Finflo Ltd.), Dr. Olli Saarela (Aalto University), Dr. Ari Visa (Tampere University of Technology), Mrs. Liisa Poussa, Mrs. Eetta Saarimäki, Mr. Tomi Viitanen, Mr. Risto Laakso, Mr. Keijo Koski and Mr. Kari Lahdenperä (VTT Technical Research Centre of Finland Ltd).

I am greatly indebted to the Organizing Committee (Ms. Maria Rajasalmi, Ms. Christina Vähävaara, Ms. Paula Kauppinen, Ms. Satu Koskela, Mr. Tuomo Hokkanen and Mr. Tomi Viitanen, all with VTT Technical Research Centre of Finland Ltd) for their invaluable assistance in shaping the Conference and Symposium program.

Financial support for ICAF 2015 has been received from various organizations. Lockheed Martin (Gold Partner), Patria (Bronze Partner), Aerospace Testing International (Media Partner), Fatigue Technology (Exhibitor), nCode (Exhibitor), HBM (Exhibitor), Curtiss-Wright (Exhibitor), SwRI and NLR have stepped forward as Conference and Symposium Partners. The contribution of the city of Helsinki for hosting the Reception is also greatly appreciated.

Finally, I would like to thank all the authors and their organizations, ICAF's General Secretary and my fellow ICAF National Delegates for their valuable contributions to the success of ICAF 2015. Without their contributions the proceedings would have never been published.

Aslak Siljander
ICAF National Delegate Finland
ICAF 2015 Chairman

Table of Content

Damage Tolerance Assessment of Near Edge Impacts in CFRP Structures

Vjola Ristori[1,3], Maria Pia Falaschetti[1], Enrico Troiani[1], Marco Montemurro[1], André Baeten[2], Goran Ivetic[2], Gianluca Molinari[3]

[1] *University of Bologna, Forlì, Italy*
[2] *Augsburg University of Applied Sciences, Germany*
[3] *Sii Deutschland GmbH, Germany*

Abstract: This work deals with the experimental evaluation of near-edge impacts on the damage tolerant properties of CFRP structures. Though many examples of central impacts exist in literature, very few can be found regarding edge impacts. This experimental activity had the objective of residual strength characterization of the damaged coupons in comparison to the pristine ones. Different types of damages were inserted in the specimens (transverse and near-edge impact) for two different impact energy levels (3J and 5J). Subsequently, compression after impact tests were performed. A total of 25 specimens were tested.

The results of this campaign indicate the higher criticality of near-edge impacts in respect to central impacts and they contribute to better understanding of the behaviour of CFRP materials subject to impact loads.

INTRODUCTION

Literature overview

The challenging requirements set on new full composite aeronautical structures are mostly related to the demonstration of damage tolerance capability of their primary structures, required by the airworthiness bodies [30]. And while composite-made structures inherently demonstrate exceptional fatigue properties, when put in real life working conditions, a number of external factors can lead to impact damages thus reducing drastically their fatigue resistance due to fibre delamination, disbonding or breaking.

Though many examples of transverse impacts exist in literature very few can be found regarding edge impacts [31]-[34]. In [31], it is shown that edge impacts can indeed be more critical in respect to impacts in lamination plane and therefore deserve further investigation. Reference [32] investigates the on-edge impacts in CFRP specimens and presents an analytical model for evaluation of these impacts, comparing it with experimental tests. In [33], it is dealt with edge impacts in glass fibre composites, presenting results for experimental and FEM evaluations of these impacts, while [34] deals with on-edge impacts in CFRP structures. Figure 1 illustrates the three common types of edge impacts, with the near-edge impacts being of interest in the current study.

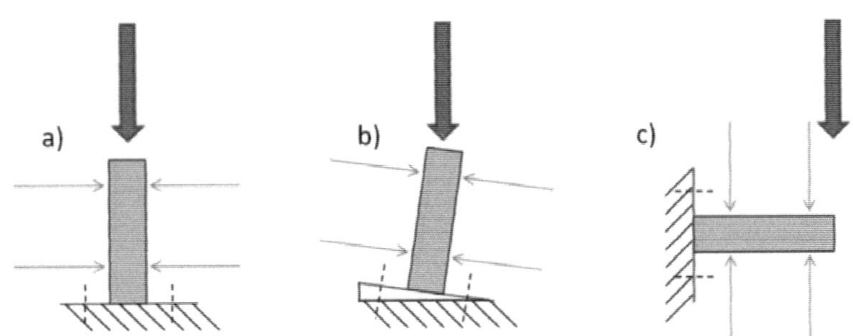

Figure 1. Three types of edge impacts, a) on edge impact, b) oblique impact, c) near-edge impact; adapted from [33]. Arrows indicate the direction of the clamping force on the specimen.

Airworthiness requirements

Federal Aviation Administration (USA) and the European Aviation Safety Agency (EU) prescribe in the Part 25, section 25.571 of their airworthiness regulations, requirements relative to fatigue and damage tolerance in aeronautical structures, whatever their material may be.

CS 25.571, paragraph a) states that [30]:

"An evaluation of the strength, detail design, and fabrication must show that catastrophic failure due to fatigue, corrosion, manufacturing defects, or accidental damage, will be avoided throughout the operational life of the airplane."

Furthermore, additional documents exist that deal explicitly with damage tolerance requirements in composite aeronautical structures [35], [36]. The requirements presented in these documents state:

"Damage tolerance evaluation starts with identification of structure whose failure would reduce the structural integrity of the aircraft. A damage threat assessment must be performed for the structure to determine possible locations, types, and sizes of damage considering fatigue, environmental effects, intrinsic flaws, and foreign object impact or other accidental damage (including discrete source) that may occur during manufacture, operation or maintenance."

In order to meet these requirements, several factors need to be taken into consideration. These factors include damage threat assessment, inspection program definition, damage detectability, fatigue tests and analyses [37].

Damage Tolerance in composite aeronautical structures

Current damage tolerance approach to composite aeronautical structures, the "no-growth" concept of Barely Visible Impact Damage (BVID), needs to be used together with previously mentioned additional factors in order to be able to meet the airworthiness requirements on composite structures.

A damage threat assessment is crucial in order to understand in which area of an aircraft the probability of an impact is higher and of which magnitude this impact can occur. According to [38], more than two thirds of impacts are registered in the door or door surrounding areas, and this information helps greatly in the definition of an inspection program.

An appropriate inspection program is also necessary due to the fact that composite materials show different behaviour when subjected to fatigue loads, compared to metals. And while metals show gradual degradation of their residual strength when subjected to fatigue loads, this degradation in composite materials is usually caused by an external factor, such as an impact, and it brings to step-wise degradation of residual strength, Figure 2, [39]. As a result, this may cause the composite structure to lose its ultimate load carrying capability for a longer period of time, compared to metals, before the damage could be detected and repaired.

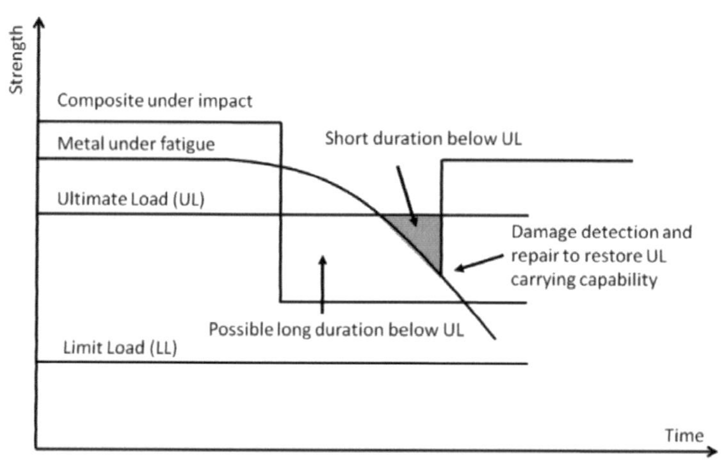

Figure 2. Residual strength of composite vs. metallic material under fatigue or impact loading.

The damage detectability depends on the type of the inspection that is being carried out and it is represented by the BVID. The fatigue tests and analyses need to show the "no-growth" concept, which requires no initiation of new damage and no growth of existing damages (BVID). Impact tests are carried out in order to establish the behaviour of a composite structure when subjected to these impact loads and to determine the severity of the impact, based on different impact positions and energies. The results presented in this work aim to contribute to better understanding of CFRP structures when subjected to near-edge impacts.

EXPERIMENTAL SETUP

CFRP Specimens

An experimental campaign was defined in order to characterize the Compression After Impact (CAI) behaviour of near-edge impacted CFRP specimens, comparing it with the compression strength of the pristine specimens. The specimens used for the experiments are obtained from a rectangular CFRP plate, with the dimensions of 490x420 mm.A total of 25 specimens were produced and used for the experiments, Figure 3.

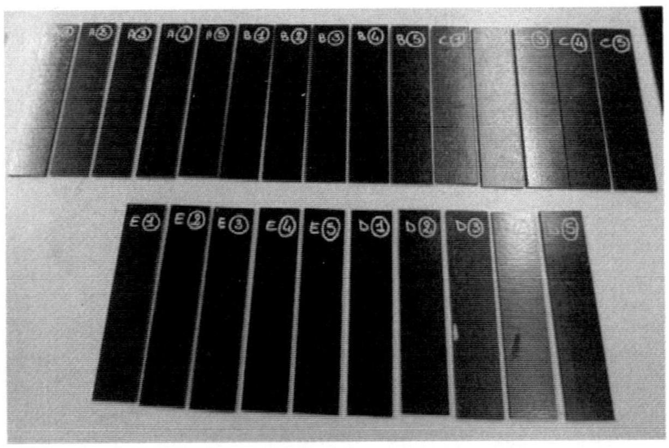

Figure 3. CFRP specimens cut from the laminated plate.

To the best of authors' knowledge there is no prescribed standard for compression after edge impact tests. The specimen size was 140x30 mm, with the length of the specimens taken from ASTM D6641/D6641M-14 standard [11], relative to compression testing of composites, and the width chosen based on the maximum size that could fit into test equipment prescribed by the same standard. The specimens were produced out of 9 plies of cross-ply prepreg fibres, produced by DeltaPreg, with ply thickness of 0.32 mm. The symmetric and balanced stacking used was $[90/0_2/90/\underline{90}]_s$ with the total nominal thickness of the specimens of 2.76 mm.

The specimens were divided in 5 groups, as described in Table I:

Table I. Specimens division.

Group	Impact Position	Impact Energy [J]
A	Not impacted	-
B	Near edge	3
C	Near edge	5
D	Central	3
E	Central	5

Impacting equipment

A Charpy pendulum was used in order to introduce low energy impacts in the specimens. The impacting element of the pendulum was a steel cylinder, having a sphere of 7 mm in diameter at its end, Figure 4. The impacts were performed by appropriately inclining the pendulum, introducing impact energies of 3 J and 5 J.

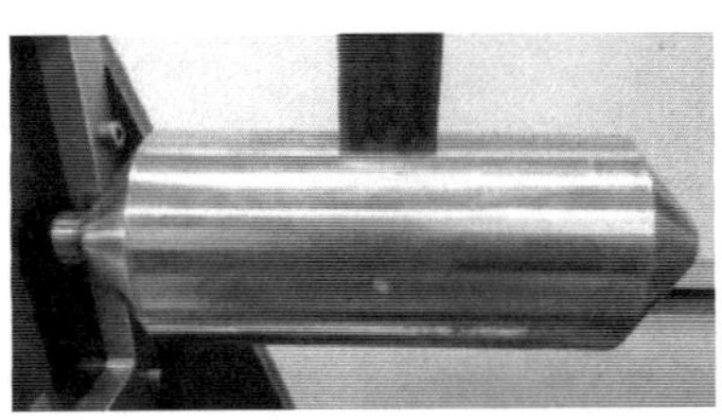

Figure 4. Impacting element.

A fixing mechanism, Figure 5, was realized in order to be able to maintain in an appropriate position the specimen in respect to the impacting element and to be able to contain the specimen itself.

Figure 5. Fixing mechanism (left) and with the specimen mounted (right).

Combined loading compression (CLC) tests

The compression tests were carried out according the ASTM D6641 standard [11]. The CLC fixing mechanism is composed of four steel blocks that allow fixing of standard size specimens of 140 mm length and of a maximum width of 30 mm, used for this experimental campaign, Figure 6. The CLC tests allow introducing both normal and shear loads in the structure, with the possibility of changing appropriately their ratio, by changing the fixing force of the mechanism.

Figure 6. Specimen mounted in the CLC fixing.

An MTS hydraulic press machine, Figure 7, was used for the CLC tests [41], with the maximum nominal loading of 100 kN. In the same figure, a detail showing the mounted CLC fixing at the beginning of the compression test can be seen.

Figure 7. Hydraulic press machine, left and with the CLC fixing mounted, right.

EXPERIMENTAL RESULTS

The impacts were observed via a camera and the recordings were inspected after the tests. This was necessary in order to establish the angle of the spring-back of the pendulum and to determine the amount of energy absorbed by the specimens, by knowing the initial releasing angle of the pendulum. Table II gives an overview of applied energies on different specimen groups.

Table II. Overview of applied energies on different specimen groups.

Specimen	Effective enery [J]	Residual energy [J]	Absorbed energy [J]
B1	3.342	0.907	2.435
B2	3.065	0.831	2.234
B3	3.065	1.244	1.821
B4	3.342	1.244	2.098
B5	3.202	1.336	1.866
C1	4.704	0.907	3.797
C2	4.866	1.244	3.622
C3	4.866	1.069	3.797
C4	4.704	0.986	3.724
C5	4.543	0.831	3.712
D1	3.342	1.155	2.187
D2	3.342	1.069	2.273
D3	3.484	1.069	2.415
D4	3.202	1.069	2.133
D5	3.202	0.986	2.216
E1	5.029	1.431	3.598
E2	5.196	1.529	3.667
E3	4.866	1.336	3.53
E4	4.543	1.244	3.299
E5	4.866	1.431	3.435

The typical surface condition of the specimens, after the impact, is illustrated in Figure 8. The images a) and b) show the impacted area for near edge impacts, with the energy levels of 3 J and 5 J, while images c) and d) show central impact for the matching energy level, 3 J and 5 J respectively.

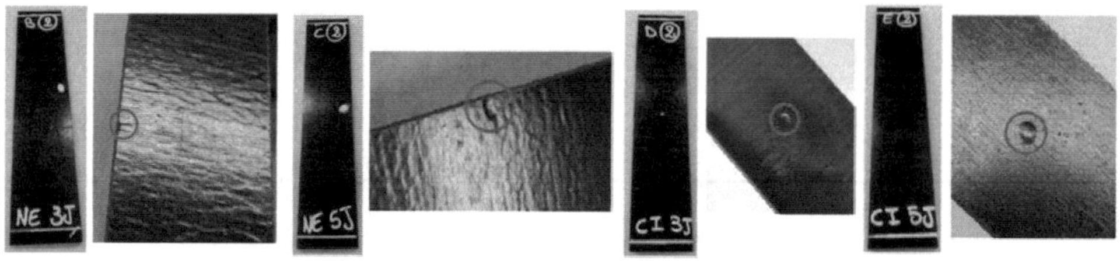

a) b) c) d)

Figure 8. Visual inspection of the specimens after impact.

Subsequently, the impacted specimens, together with the baseline, not impacted specimens, were subjected to CLC tests. The readings of the hydraulic press, relative to the applied force at the compression failure of the specimens, for each of the specimen group, are given in Table III.

Table III. Maximum applied force at failure for different specimen groups.

	Specimen	Nominal cross section [mm^2]	Max applied force [N]
Baseline	A1	82.03	-34172
	A2	87.94	-32845
	A3	87.31	-34216
	A4	85.87	-32593
	A5	84	-31091
Nearedge, 3J	B1	69.76	-23913
	B2	81.87	-20118
	B3	84.52	-29537
	B4	89.13	-38055
	B5	88.12	-30601
Nearedge, 5J	C1	84.11	-20473
	C2	84.94	-20684
	C3	84.36	-29630
	C4	86.78	-26366
	C5	87.14	-16114
Central, 3J	D1	68.67	-21924
	D2	81.16	-30936
	D3	72.44	-22220
	D4	81.67	-30050
	D5	71.01	-25746
Central, 5J	E1	81.42	-26725
	E2	84.8	-24547
	E3	83.58	-31050
	E4	84.23	-35774
	E5	82.11	-19922

After the failure of the specimens following the CLC tests, the failure mode of each of the specimen was examined and classified according to the specifications presented in [39]. The specifications from the standard give a list of acceptable and non-acceptable failure modes, helping in verifying the validity of the test. All of the performed compression tests entered in the category of acceptable failure modes.

Figure 9 shows typical condition of the specimens after the failure following the CLC test.

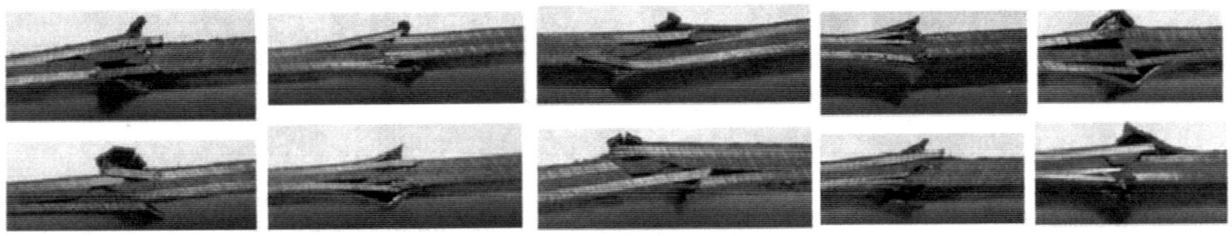

Figure 9. Typical condition of the specimens following the compression failure.

DISCUSSION AND CONCLUSIONS

Based on the results obtained during this research activity, the following conclusions can be drawn:

- The effect of the near-edge impacts is more critical than of central impacts, also reported in the literature for thick CFRP components [31]. This conclusion is drawn based on lower residual strength after impact analysis, as well as after comparing the amount of absorbed energy after the impact. For near-edge impacts, this absorbed energy is higher than for central impacts, especially for the 5 J impact energy case.
- With lower energy levels (3 J), a significant difference in residual strength between central and near-edge impacts was not found. Relative to this result, similar level of absorbed energy for the 3 J impact energy case was registered both for central and near-edge impacts.
- It is assumed that there is a certain threshold energy level below which the reduction of residual strength for the given material, thickness and ply orientation can be disregarded. This reduction should be correlated to the extent of the delamination introduced in the structure by the impact, and this should be done via an ultrasound inspection (e.g. C-scan)

It is recommended that additional analysis should be performed in the future:

- An ABAQUS® FEM model is currently being developed to predict the extension of the damage after a near-edge impact and the consequent reduction of the material static residual strength. This FEM, based on continuum shell and cohesive elements, will be compared against the test results.
- C-scan analysis is recommended, in order to correlate the CFRP thickness, impact type, energy level and the extent of delamination. This information should give important input for FEM model fine tuning
- Moisture and temperature effects, which according to the airworthiness bodies must be taken into account when certifying a composite structure, must be considered. These environmental effects on the reduction of residual strength of the CFRP structure, both on pristine and impacted structures should be assessed
- CFRP structures with higher thicknesses are also of interest and their behaviour after impact should be assessed. This is due to the fact that a full composite aircraft fuselage has a thickened-up skin in the door surrounding area in order to compensate for higher loads present in these areas, and at the same time, the door surrounding areas are the ones in which the probability of an impact is the highest one.

REFERENCES

[30] EASA (2015). CS-25, Book 1, Airworthiness Code, Amendment 16

[31] Breen, C., Guild, F.J. and Pavier,M.J. (2006). *J. Mater Sci.*, vol. 41, n. 20, p. 6718–6725.

[32] Rhead, A.T., Marchant, D. and Butler, R. (2010). *Composites: Part A*, vol. 41, n. 9 p. 1056–1065

[33] Malhotra, A., Guild, F.J. and Pavier, M.J. (2008). *J. Mater Sci.*, vol. 43, n. 20, p. 6661–6667.

[34] Ostré, B., Bouvet, C., Lachaud, F., Minot, C. and Aboissière, J. (2014). *J. Comp. Mater.* 0021998314537325, DOI: 10.1177/0021998314537325

[35] EASA (2010). AMC 20-29.

[36] U.S. Department of Transportation, FAA (2009). Advisory Circular 20-107B

[37] Li, D.S. (2009). In: Proceedings of the 17th International Conference on Composite Materials, Edinburgh, United Kingdom. http://www.iccm-central.org/Proceedings/ICCM17proceedings/Themes/Industry/AEROSPACE%20APPLICATIONS/A2.3%20Li.pdf

[38] Thévenin, R. (2011). In: *Aircraft Composite Repair Management Forum,* Madrid, Spain. http://events.aviationweek.com/html/compos11/4-Roland%20Thevenin.pdf

[39] SAE International (2012). Composite Materials Handbook 17, vol 3.

[40] ASTM International (2014). ASTM D6641/D6641M-14. DOI: 10.1520/D6641_D6641M-14

[41] Scafè M., Raiteri G., Brentari A., Dlacic R., Troiani E., Falaschetti M.P. and Besseghini, E. (2014) *Frattura ed Integrità Strutturale*, vol. 29, p. 399-409, DOI: 10.3221/IGF-ESIS.29.35

Evaluation of Transverse Crack Initiation in Cross-Ply and Quasi-Isotropic CFRP Laminates under Fatigue Loading

Atsushi Hosoi[1], Yuzo Fujita[2], Hiroyuki Kawada[1]

[1]*Departement of applied mechanics and aerospace engineering, Waseda University, Japan*
[2]*Composite Materials Research Laboratories, Toray Industries, Inc., Japan*

Abstract: A method has been proposed to predict the fatigue life to the transverse crack initiation in cross-ply and quasi-isotropic carbon fiber reinforced plastic (CFRP) laminates. The six kinds of the laminates were formed of the two kinds of the prepregs, T800S/3900-2B with toughened interlaminar layers and T800H/3631 without toughened interlaminar layers. The unidirectional and cross-ply laminates of the stacking sequence of $[90]_{12}$, $[0/90_4]_S$ and $[0/90_6]_S$ were formed of the T800S/3900-2B prepreg, and the cross-ply and quasi-isotropic laminates of the stacking sequence of $[0/90_2]_S$, $[0/90_6]_S$ and $[45/0/-45/90]_S$ were formed of T800H/3631 prepreg. Tensile fatigue tests were performed under load control using a hydraulic fatigue testing machine. The transverse crack initiation in the laminates under fatigue loading was evaluated by modifying the Smith-Watson-Topper (SWT) model, which can evaluate the mean stress effect on the fatigue life. Finally, it was found that the initiation of transverse crack in the laminates with the various laminate configuration can be predicted using the modified SWT model, and that means that it is possible to predict the initiation of transverse crack in the laminates with the various laminate configuration using S-N curve of unidirectional laminates in 90° direction.

INTRODUCTION

Carbon fiber reinforced plastics (CFRPs) have been used as the main structural material of the latest commercial airplane. The ultimate design strain level in aircraft is kept low, approximately 4000 με considered compression after impact of CFRP laminates [1]. At this strain level, the composite materials can withstand large numbers of fatigue cycles without failing, and damage growth is therefore not seen to be a major problem. However, to use composite structures to their full potential and to reduce aircraft weight, design strain levels will need to increase and a partial growth criterion needs to be adopted. When the CFRP laminates are subjected to tensile fatigue loading, transverse cracks are generally first damage in the CFRP laminates and then induce more serious damage, such as delamination or fiber breakage. Thus, it is important to evaluate crack initiation to establish and improve the long-term durability and reliability of CFRP structures.

Authors have been proposed a method to predict the initiation of a transverse crack in cross-ply CFRP laminates under fatigue loading using the power law between transverse crack growth rate and energy release rate range associated with transverse crack formation [2, 3]. We have evaluated the effects of laminate thickness [4], fiber volume fraction [5], and toughened interlaminar layers [6] on the transverse crack initiation. Besides, we have proposed a new method to predict the initiation of a transverse crack in arbitrary cross-ply CFRP laminates using the fatigue properties obtained from an S-N diagram of unidirectional laminates in 90° direction [7]. Furthermore, it was cleared that the method can be applied for the initiation of a transverse crack in cross-ply CFRTP laminates, which cause short edge cracks [8]. These studies were focused on predicting the initiation of a transverse crack in cross-ply laminates. However, the initiation of transverse crack in quasi-isotropic laminates in which the edge cracks occur and multiply has not been evaluated. Therefore, the initiation of the transverse crack in both cross-ply and quasi-isotropic CFRP laminates under fatigue loading was evaluated in this study.

EXPERIMENTS

Specimen

Six kinds of the laminates were formed of the two kinds of the prepregs, T800S/3900-2B with toughened interlaminar layers and T800H/3631 without toughened interlaminar layers. The fiber volume fraction of the prepregs is 56% and 57%, respectively. The cure temperature is 453 K for both prepregs. The unidirectional and the cross-ply laminates of the stacking sequence of $[90]_{12}$, $[0/90_4]_S$ and $[0/90_6]_S$ were formed of the T800S/3900-2B prepreg, and the cross-ply and quasi-isotropic laminates of the stacking sequence of $[0/90_2]_S$, $[0/90_6]_S$ and $[45/0/-45/90]_S$ were formed of the T800H/3631 prepreg. Figure 1 and Table I show the specimen geometry and dimensions of each laminate. The edge surfaces of specimens were polished with emery papers and were finished by buffing with diamond powder of 6 μm.

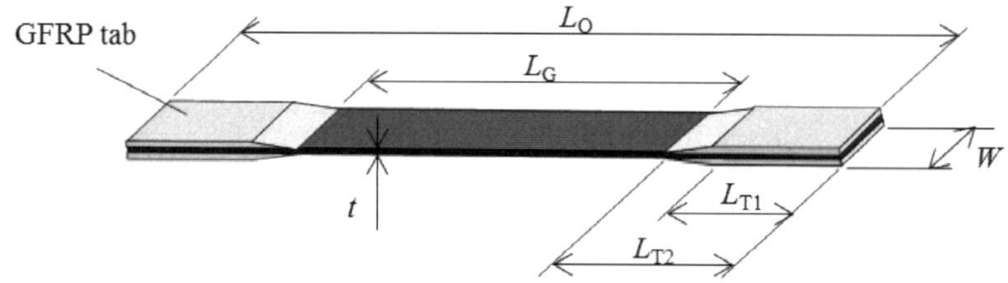

Figure 1. Specimen geometry.

Table I. Dimensions of each laminate.

Prepreg	V_f	Laminate Configuration	Cure Temp.	L_O	L_G	W	t	L_{T1}	L_{T2}
T800S/3900-2B	56	$[90]_{12}$	180	200	100	25	2.3	50*	50*
T800S/3900-2B	56	$[0/90_4]_S$	180	230	120	30	1.9	40	55
T800S/3900-2B	56	$[0/90_6]_S$	180	230	120	30	2.7	40	55
T800H/3631	57	$[0/90_2]_S$	180	210	120	30	0.86	30	45
T800H/3631	57	$[0/90_6]_S$	180	210	120	30	2.0	30	45
T800H/3631	57	$[45/0/-45/90]_S$	180	210	100	30	1.1	40	55

* Emery paper tabs were used for unidirectional laminates in 90° direction.

Tensile fatigue tests

Tensile fatigue tests were performed under load control using a hydraulic fatigue testing machine. All tests were run at a stress ratio of $R = 0.1$ and the frequency of $f = 5$ or 100 Hz. The fatigue test conditions are shown in Table II. Where, σ_{max} and σ_{ti} are maximum stress of the applied loading and transverse crack initiation stress under tensile loading, respectively. In the previous study [9], the effect of the frequency on the damage growth was small within the test conditions. Fatigue tests were interrupted at arbitrary loading cycles to observe damage. The transverse cracks in the laminates were observed by optical microscopy and soft X-ray photography.

Table II. Fatigue test conditions.

Prepreg	V_f	Laminate Configuration	σ_{max}/σ_{ti}	f [Hz]	R	σ_{ti} [MPa]
T800S/3900-2B	56	$[90]_{12}$	0.60-1.00*	5	0.1	68.3*
T800S/3900-2B	56	$[0/90_4]_S$	0.70-1.40	5	0.1	200
T800S/3900-2B	56	$[0/90_6]_S$	0.80-1.30	5	0.1	158
T800H/3631	57	$[0/90_2]_S$	0.60-0.90	5	0.1	871
			0.40-0.50	100		
T800H/3631	57	$[0/90_6]_S$	0.40-0.95	5	0.1	344
T800H/3631	57	$[45/0/-45/90]_S$	0.55-1.09	5	0.1	426
			0.36-0.64	100		

* The stress of crack initiation for unidirectional $[90]_{12}$ laminates was used as not σ_{ti} but σ_b, tensile strength.

ANALYSIS

The thermal residual stress applied in 90° layers of the laminates is different depending on the laminate configuration. Therefore, the stress ratio applied in 90° layers differs according to the laminate configuration and the applied stress level even if the mechanical loading of the constant stress ratio, $R = 0.1$, is applied in the laminates. In addition, the free-edge effect, which is the stress singularity on free edges of the laminates, should be considered. Furthermore, the stress at transverse crack initiation differs according to the laminate configuration because of the size effect. To evaluate the transverse crack initiation, mainly these three factors should be considered.

Some methods have been proposed to evaluate the mean stress effect, such as the methods by Gerber [10], Goodman [11], Soderberg [12], Morrow [13], Walker [14] and Smith-Watson-Topper (SWT) [15]. In this study, STW model was used as given by

$$\sigma_{ar} = \sqrt{\sigma_{max}\sigma_{a}}\,, \tag{1}$$

where σ_{max} is the applied maximum stress, σ_a is the stress amplitude, σ_{ar} is the equivalent fully reversed stress amplitude. The relationship between the fatigue life and the equivalent fully reversed stress amplitude based on Basquin's equation is given as

$$\sigma_{ar} = \sigma'_f (2N_f)^b\,, \tag{2}$$

where σ'_f and b are the fatigue strength coefficient and exponent constant, respectively.

It was cleared that the free-edge effect and the size effect can be cancelled by normalizing the applied stress with the stress at the transverse crack initiation under static tensile loading from previous studies [7, 8]. Considering these effects and using Eqns. (1) and (2), the fatigue life to transverse crack initiation can be expressed as

$$N_i = \alpha \left(\frac{\sqrt{\sigma_{max}^{(90)}\sigma_a^{(90)}}}{\sigma_{ti}^{(90)}} \right)^{\beta}, \tag{3}$$

where superscript (90) shows the 90° layers in the laminates. α and β are constants. The stress applied in 90° layers of cross-ply and quasi-isotropic laminates can be express as the following equation [16].

$$\sigma^{(90)} = \frac{E_A^{(90)}}{E_0}\sigma_0 - \frac{bE_A^{(S)}E_A^{(90)}}{hE_0}\Delta\alpha T = E_A^{(90)}\varepsilon_0 - \frac{bE_A^{(S)}E_A^{(90)}}{hE_0}\Delta\alpha T\,, \tag{4}$$

where superscript (S) is sublaminate adjacent to 90° layers. E_0 is the composite Young's modulus. E_A is Young's modulus of the 90° layers or sublaminate in the axial direction. σ_0 and ε_0 are the stress and the strain applied in composites, respectively. b and h are the one-sided thickness of sublaminate and half the thickness of the cross-ply or quasi-isotropic laminates, respectively. $\Delta\alpha$ is the difference between the thermal expansion coefficients of 90° layers and sublaminate. T is the difference between the test temperature and the stress free temperature.

RESULTS & DISCUSSIONS

Damage observation

Figures 2 and 3 show the damage growth behavior observed by soft X-ray photography in cross-ply $[0/90_4]_S$ laminate formed of T800S/3900-2B prepreg and quasi-isotropic $[45/0/-45/90]_S$ laminates formed of T800H/3631 prepreg, respectively. In cross-ply laminates as shown in Fig. 2, transverse cracks that occurred at specimen edges grew in the width direction of the specimen soon and were multiplied in the specimen with an increase in the number of loading cycles. On the other hand, in quasi-isotropic laminates as shown in Fig. 3, short transverse cracks are initiated at specimen edge, and the transverse cracks propagate to width direction at the same time as increasing at specimen edges. Afterwards, delamination was initiated at the interlaminar of 90°/-45° plies of specimen edges due to the stress concentration of a transverse crack tip. After delamination grew to longitudinal direction at specimen edges, delamination propagates to width direction of the specimen.

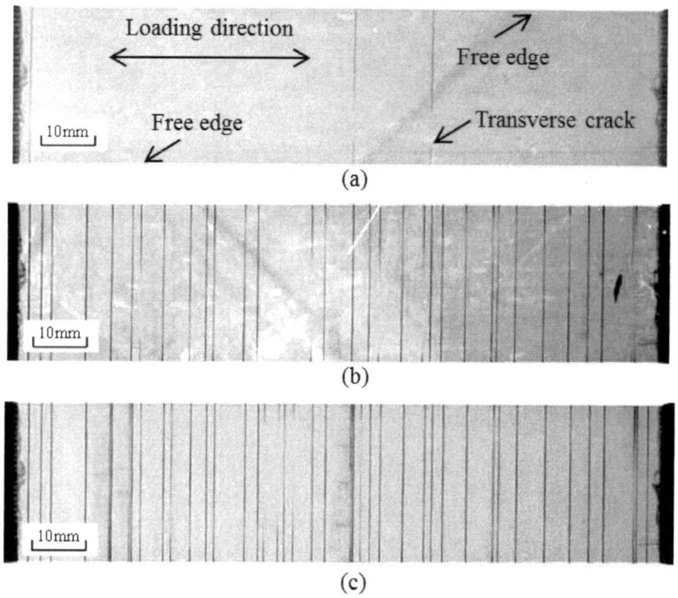

Figure 2. Observation of internal damage growth of $[0/90_4]_S$ laminate under applied stress level of $\sigma_{max}/\sigma_{ti} = 1.2$ with soft X-ray photography; (a) $N = 1.0 \times 10^3$ cycles, (b) $N = 1.0 \times 10^5$ cycles, and (c) $N = 2.0 \times 10^6$ cycles [7].

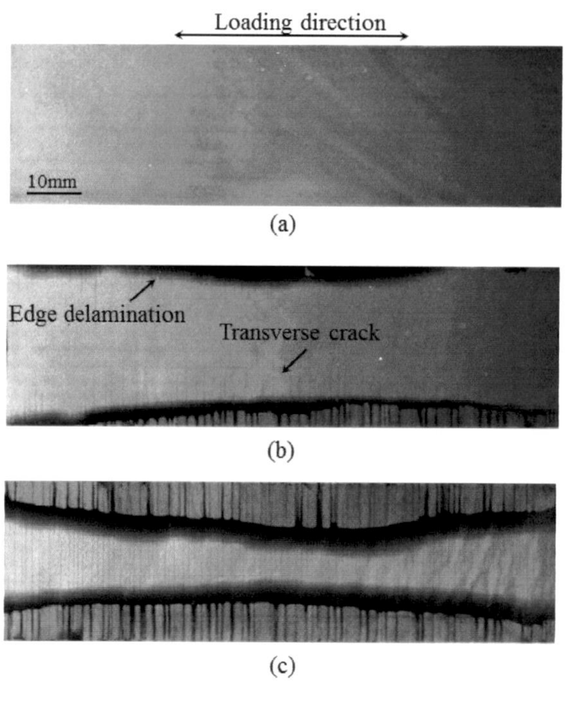

Figure 3. Observation of internal damage growth of $[45/0/\text{-}45/90]_S$ laminate under applied stress level of $\sigma_{max}/\sigma_{ti} = 1.09$ with soft X-ray photography; (a) $N = 5.0 \times 10^0$ cycles, (b) $N = 5.0 \times 10^2$ cycles, and (c) $N = 5.0 \times 10^3$ cycles [17].

Prediction of transverse crack initiation in CFRP laminates

Figures 4 and 5 show the evaluation results of the modified SWT model for the fatigue life prediction to the transverse crack initiation of the laminates formed of T800S/3900-2B and T800H/3631 prepregs, respectively. The prediction results show good agreement with experimental results, and the fatigue life to the transverse crack initiation can be evaluated with a curve regardless of the laminate configuration. From the results shown in Fig. 4, it is found that the initiation of the transverse crack in cross-ply laminates can be predicted from the S-N curve of the unidirectional laminates in 90° direction. On the other hand, from the results shown in Fig. 5, it is found that the initiation of the transverse crack can be predicted for the quasi-isotropic laminates that cause the edge cracks. The transverse crack initiation in $[45/0/\text{-}45/90]_S$ laminates was not observed up to 2×10^8 cycles under the lowest loading condition within this study. The free-edge effect of the $[45/0/\text{-}45/90]_S$ laminates is greater than that of cross-ply laminates. However, it was thought that the free-edge effect can be cancelled by normalizing the applied stress with the transverse crack initiation stress applied in 90° plies as described in our previous study [8].

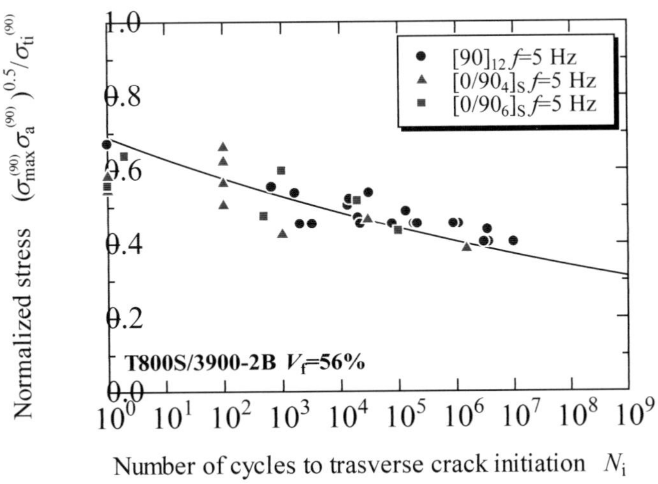

Figure. 4. Evaluation of the fatigue life to the transverse crack initiation of the laminates formed of T800S/3900-2B prepreg using a modified SWT model.

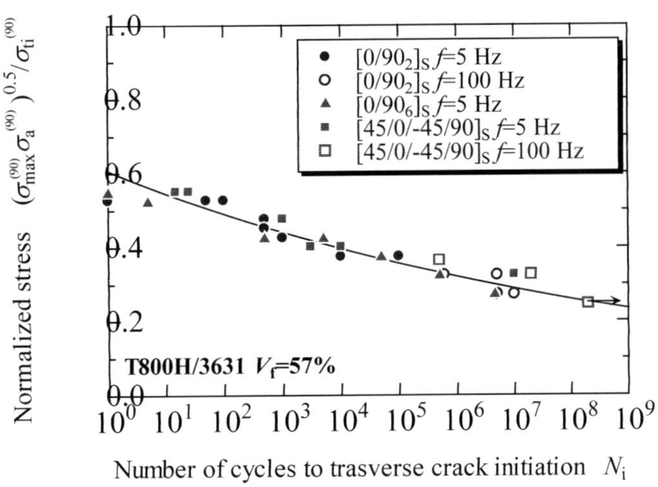

Figure. 5. Evaluation of the fatigue life to the transverse crack initiation of the laminates formed of T800H/3631 prepreg using a modified SWT model.

CONCLUSIONS

In this study, a method was proposed to predict the fatigue life to the transverse crack initiation in cross-ply and quasi-isotropic CFRP laminates using a modified SWT model. The analytical results showed good agreement with the experimental results. It was found that the proposed model can be useful for predicting the fatigue life to transverse crack initiation in not only cross-ply CFRP laminates but also quasi-isotropic CFRP laminates. In addition, it is thought that the initiation of transverse crack in the laminates with the various laminate configuration can be predicted using S-N curve of unidirectional laminates in 90° direction.

REFERENCES

[42] Davies, A. J. and Curtis, P. T., (2003). In: *Fatigue in composites*, Sec. 26, pp. 686-708, Haris, B. (Ed.), Woodhead Publishing Limited, Cambridge.

[43] Takamura, K., Hosoi, A., Sato, N. and Kawada, H. (2011). *Trans. Jpn. Soc. Mech. Eng. Ser. A*, vol. 77, n. 779, p.1123. (In Japanese)

[44] Hosoi, A., Takamura, K., Sato, N. and Kawada, H. (2012). *Trans. Jpn. Soc. Mech. Eng. Ser. A*, vol. 78, n. 791, p. 1000. (In Japanese)

[45] Kurihara, K., Hosoi, A., Sato, N., and Kawada, H. (2013). *Trans. Jpn. Soc. Mech. Eng. Ser. A*, vol. 79, n. 799, p. 249. (In Japanese)

[46] Hosoi, A., Shigemori, K., Fujita, Y. and Kawada, K. (2013). *Proceedings of the ICAF-2013 Symposium*, vol. 2, pp. 925-935.

[47] Kawada, H., Soga, M., Shigemori, K., Hosoi, A. and Fujita, Y. (2014). *Trans. Jpn. Soc. Mech. Eng. Ser. A,* vol. 80, n. 812, SMM0079. (In Japanese)

[48] Hosoi, A., Sakuma, S., Fujita, Y. and Kawada H. (2015). *Compos. Part A-Appl. S.,* vol. 68, p. 398.

[49] Hosoi, A., Fukui, T., Kobiki, A., Kotani, M. and Kawada, H. (2015). *Proceedings of ICFC6*, (in press).

[50] Hosoi, A., Sato, N., Kusumoto, Y., Fujiwara. K. and Kawada, H. (2010). *Int. J. Fatigue*, vol. 32, n. 1, p. 29.

[51] Gerber, W. (1874). *Z. Bayer. Arch. Ing.*, vol. 6, p.101.

[52] Goodman, J. (1899). In: *Mechanics Applied to Engineering*, Longmans Green, London.

[53] Soderberg, C. R. (1939). *Trans. Am. Soc. Mech. Eng.*, vol. 52, p.13.

[54] Morrow, J. D. (1968). In: *Fatigue Design Handbook-Advances in Engineering*, vol. 4, Sec. 3.2, p. 21.

[55] Walker, K. (1970). In: *Effects of Environment and Complex Load History for Fatigue Life,* Special Technical Publication 462, p.1.

[56] Smith, R. N., Watson, P. and Topper, T. H., (1970). *J. Mater.*, vol, 5, p.767.

[57] Nairn, J. A. and Hu, S. (1992). *Int. J. Fracture*, vol. 57, n. 1, p.1.

[58] Hosoi, A., Takamura, K., Sato, N. and Kawada, H. (2011). *Int. J. Fatigue*, vol. 33, no. 6, p.781.

Airworthiness regulations and substantiation methodology development

Damage tolerance approach on recurrent damage of Lynx helicopters

Chloé Kinzelin, Etienne Deshaies, Pierre Madelpech

DGA Aeronautical Systems, Toulouse, France

Abstract: The Lynx fleet of French Navy is ageing and shows poor availability. This is partly due to the findings of recurrent damage on some structural parts of the airframe. Two different zones can especially be pointed out: the first one is the frame which is at the junction between tail boom and fuselage. The existing inspection program is not associated with an authorized delay before application of the repair. The second zone is the lateral panels of helicopter. Currently, Lynx helicopters are not allowed to flight with any damage on panels, and repair solutions are heavy.

Two flight tests campaigns were conducted to assess flight and ground loads of affected zones. Then, limit loads for each studied area are derived from those flight tests. Finite Elements Models were also used in order to extend the results of the measurement to damaged zones. Finally, the establishment of a damage tolerance approach on those specific areas will extend the fleet availability. Once accepted by airworthiness authorities, it will enable consequent improve in fleet availability.

INTRODUCTION

The French Army have some ageing fleets, with life extension in perspective. That is reason why recurrent damage appear on loaded areas; longer and more expensive maintenance operations should be done. Cost and availability of the fleet are crucial arguments to optimize the inspection of risked areas. Damage tolerance approach should contain the determination of location and mode of damage due to fatigue or corrosion. The residual strength must also be evaluated and the remaining structure must be able to withstand loads corresponding to the following conditions [1]. Follow a local damage tolerance approach in area prone to exhibit crack is commonly used to adapt the inspection interval on military fleets.

The Lynx is a Westland helicopter of the French Navy. The fleet contains 22 airframes. The main task of those helicopters is the anti-submarine warfare. The Lynx fleet shows a poor availability. Indeed, recurrent damage appear on some structural parts of the airframe. Two different areas have been specifically studied: the first one is a frame of tail boom and the second zone corresponds to lateral panels of the helicopter. A damage tolerance approach could be applied on the both parts, which is detailed in the following paragraphs of this article.

FRAME 3353A: JUNCTION BETWEEN TAIL BOOM AND CENTER FUSELAGE [2]

Background and objectives

The frame 3353A of a Lynx helicopter is at the junction between tail boom and fuselage. Number of damage appear in the most loaded zones of the frame and adjoining parts: spars, beams and skin. The studied parts are all Aluminium structures. The objective is to define an acceptable size of damage for each considered area in order to keep the aircraft in service without prejudice the flight safety. The current maintenance program is not based on tolerance damage approach. In other zones, some tolerance exists (few flight hour allowed before repair). Thanks to this study, the inspection interval could be adapted to crack propagation rate and residual strength. To identify the critical area, the model but also experiences of maintenance units were used. Following pictures of the Figure 1 show an example of damage observed on the frame 3353A during maintenance operations.

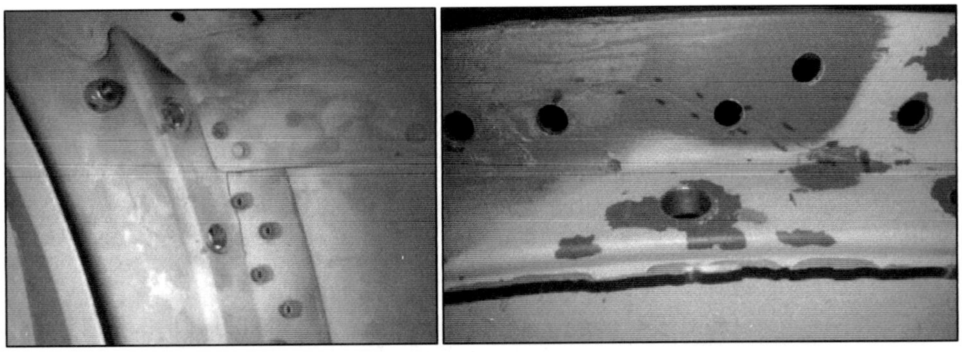

Figure 1. Observable damages on parts of the frame 3353A zone (crack along the angle of frame, crack along the splicing of the frame).

Flight tests campaign

The objective of flight tests was to define loads supported by the airframe. This is complex area in terms of geometry and of loads; the frame has to be equipped precisely in order to determine the contribution of solicitations and to be able to extrapolate to the other areas. Then, thirty-six strain gauges were placed all around the frame.

Different flight and ground tests were performed. The first flight contained a lot of loading manoeuvers for the structure (turns, pitch-up, yawing, landing…) in order to go all over the operating range envelope. The second one was a succession of deck-landings, more or less severe. Indeed, deck-landing is an operation particularly soliciting for the tail boom. Calibrations were performed through ground tests: test with ballasts introduced at diverse sections of the tail boom and tail boom folding test. Results of those ground tests were used to calibrate the strain gauges. The maximal loads of the structure could be estimate thanks to the results of the different flight tests. They were also used to adjust the numerical model. The following figure shows the equipped area with different strain gauges.

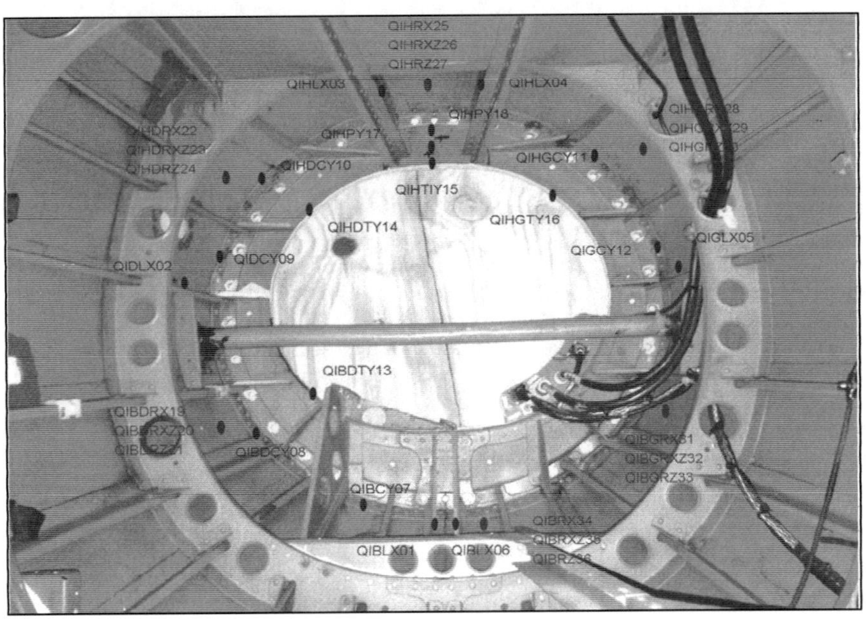

Figure 2. Frame equipped with 36 strain gauges for flight tests campaign

Finite elements modelling

Thanks to results of the strain gauges, loads are known in some zones of the studied structure. However, the most loading areas are often small and inaccessible to place a gauge (for instance, angles radius, fastening zones…). Results on these areas have to be extrapolated. A finite elements model of the structure could enable to obtain loads on the most critical areas. Then, the numerical model has to be adjusted with the tests results (thanks to the measures of gauges) and confirm the most loading manoeuvres.

For these reasons, frame, spars, beams and skin had to be designed precisely. Model was realized with the finite elements analysis software, Samcef Field. It was design from paper plans from the Westland Company. Different parts have been outlined thanks to shell elements: frame 3353A and adjoining skin, spars and beams close to frame, and some doublers. Fasteners have not been represented because of their huge numbers. The different designed parts have been assembled between them thanks to numerical sticking function. Loads were introduced by a node fixed to the structure by nodes representing the attachment of the frame to remains of tail boom (in total, 34 fasteners gather the 3353A frame with the tail boom). As shown in the following figure, the other extremity was clamped.

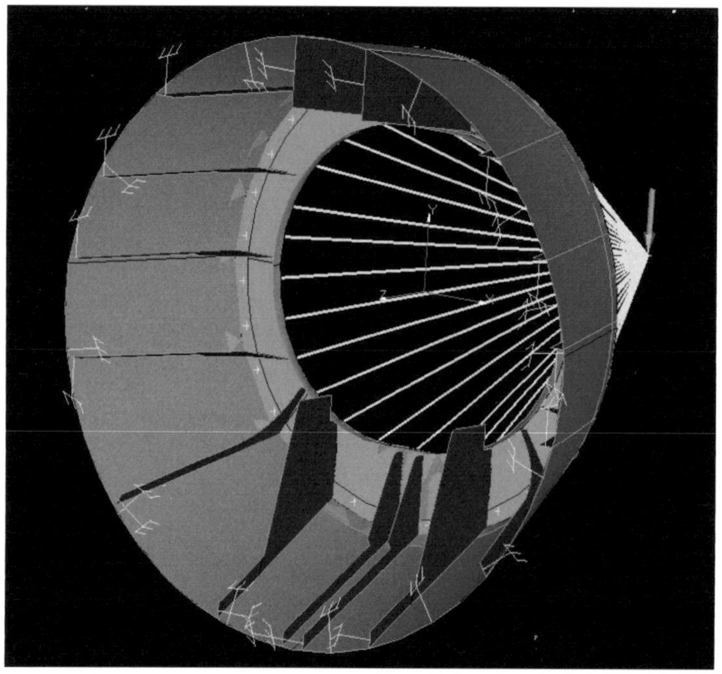

Figure 3. Finite elements model of the studied structure with constraints (clamp and loads)

This model was meshed with 2D elements (15958 linear quadrangles and 9422 linear triangles). Gauges were also included thanks to beam elements (1D element). Then, comparisons between numerical and flight results are immediate. Concerning loads applied on the model, several unit calculations were performed. Indeed, the objective was to determine the macroscopic loads enunciated to test manoeuvres. For this step, a total of six gauges were wisely chosen to characterize the unit loads: stress and bending moment according to the three directions (Tx, Ty, Tz, Mx, My, Mz). Numerical values of the six chosen gauges were extracted for the six computation. A matrix is then obtained (6*6) and could be reversed to obtain an expression of the six "unit" loads from the gauge values. Gauges data is edited during all the flights. Then, each unit load could be plotted for each flight (in this case, unit loads are plotted with the DynaWorks Software). Finally the manoeuvres could be compared in terms of. The results will be detailed in the next part. The following figure is a screen of the unit load curves (for the three bending moment).

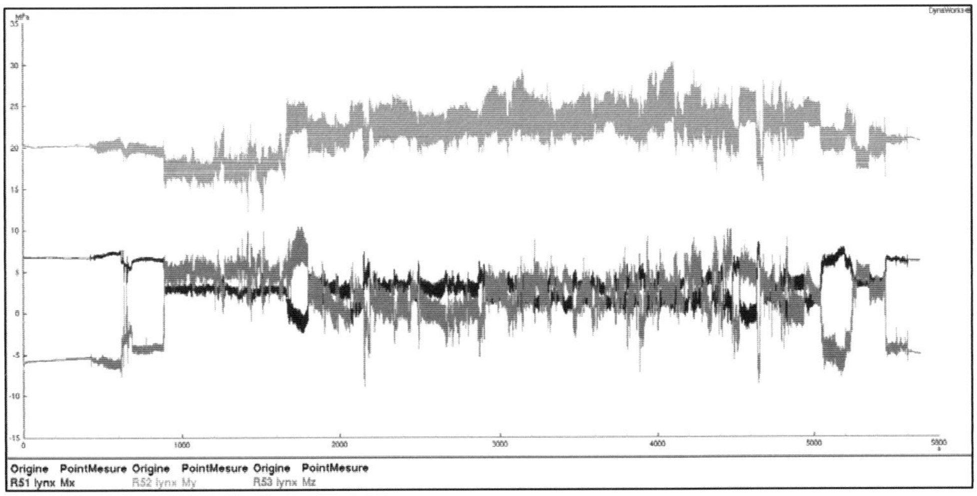

Figure 4. Unit bending moment (in the three direction) plotted for the 'manoeuvres' flight

Important results of the study

This part presents noteworthy results of the study. First of all, the most loading manoeuvres are determined by zone depending on the applied loadings. Those results will be used to define the most critical areas on the frame and associated loading conditions. Then, the fracture mechanics analysis will give the critical size of a potential crack and the crack propagation calculation will be developed for the studied areas. Conclusions give, for each zone, advised interval inspection depending on size crack. Only representative results will be detailed in the following.

Defining the most loading flight manoeuvres. Thanks to those different curves, most loading manoeuvres could be detailed for each unit load. Indeed, maximum, minimum and average of manoeuvres are noticed in a graph (see an example on figure 5) to take into account scattering. Some results concerning the most loading flight manoeuvres:

- Ty : turns and deck-landings are the most loading
- Tz : landings and deck-landings
- Mx : VNE and deck-landings
- My : turns and deck-landings
- Mz : turns and take-off

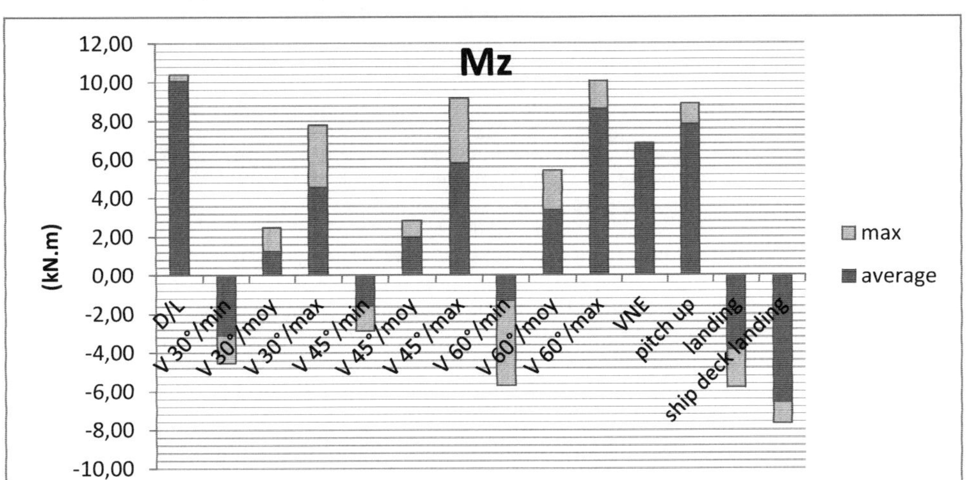

Figure 5. Average and maximal values of the unit bending moment Mz by tested manoeuvre.

Defining the critical zones. The structure is divided in several areas, with the following conditions:

- Same type of loading for the whole zone (same unit load)
- Only one damage maximum reported from feedback of the maintenance units by zone
- Only one hot spot in the area from a stress and fracture mechanics point of view,

Frome those considerations, the frame were divided in seven areas. As the upper part is the most loaded and experience most of in service damage, it is the most refined areas:

- Area 1: on the frame, below the upper spar (red)
- Area 2: on the frame, above the upper spar (green)
- Area 3: on the skin, near the upper spar (yellow)
- Area 4: on the frame, near the upper beam junction (white)
- Area 5: on the upper beam, near the frame junction (blue)
- Area 6: on the frame, near the skin junction, on the left hand side (purple)
- Area 7: on the frame, folding zone (angles) (black)

In order to visualize those areas, the following figure details the locations.

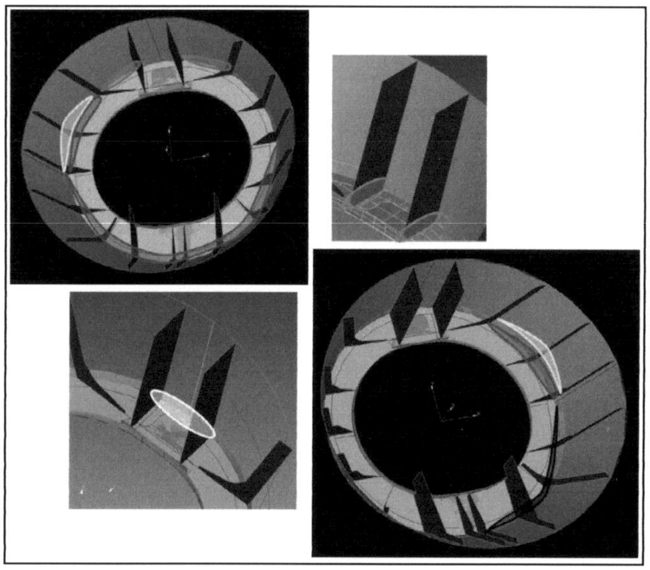

Figure 6. The seven study areas.

The critical areas 1, 2, 4, 6 and 7 are placed on the frame and precisely on the flange, part the most sensitive of the frame, more precisely at locations of fasteners, which hold spars, beams and skin. The area 1 is the most stressed (feedback during maintenance), the other will cover other potential damage in the rest of the frame. For each zone and for each unit loading, principal stress is measured on the finite element model. The values are then combined and post-processed with DynaWorks. Finally, for each critical area, the stress is obtained versus the time all along the flight. The maximal stress value could thus be picked. Every time, it corresponds to 60° turn (equivalent to a 2g turn). This value is then extrapolated in order to obtain the stress value at the envelope of the flight domain, which will be assumed to be the limit loads (the flight envelope is -1g/3g in terms of vertical load factors). The values will be used to identify critical size of crack at limit load (a_{LL}) and ultimate load (a_{UL}) thanks to the Afgrow software.

Residual strength. With the Afgrow Software, it is possible to determine a critical size of crack depending to design and loads of the area. A simplified models library is available within the software. Then, an equivalent geometry was chosen for studied areas: Double through crack at hole. Design analysis and available non-destructive inspection technics give a detectable size of 15 mm (that value will be used as the initial size for crack propagation). The following scheme explains the simplification hypothesis for calculation in critical zones. The hot spot were determined according model results. Different hypothesis were done during propagation: non-redistribution of bearing and immediate re-starting of crack from other edge of a hole when the crack reaches it (as seen on the picture 7).

Different calculations give critical sizes of failure, for each concerning zone. Those values are defined for limit loads and ultimate loads of the studied area (values determined thanks to the finite elements model and results of flight tests campaign). Critical sizes of cracks are exposed in the table 1 for aera 1.

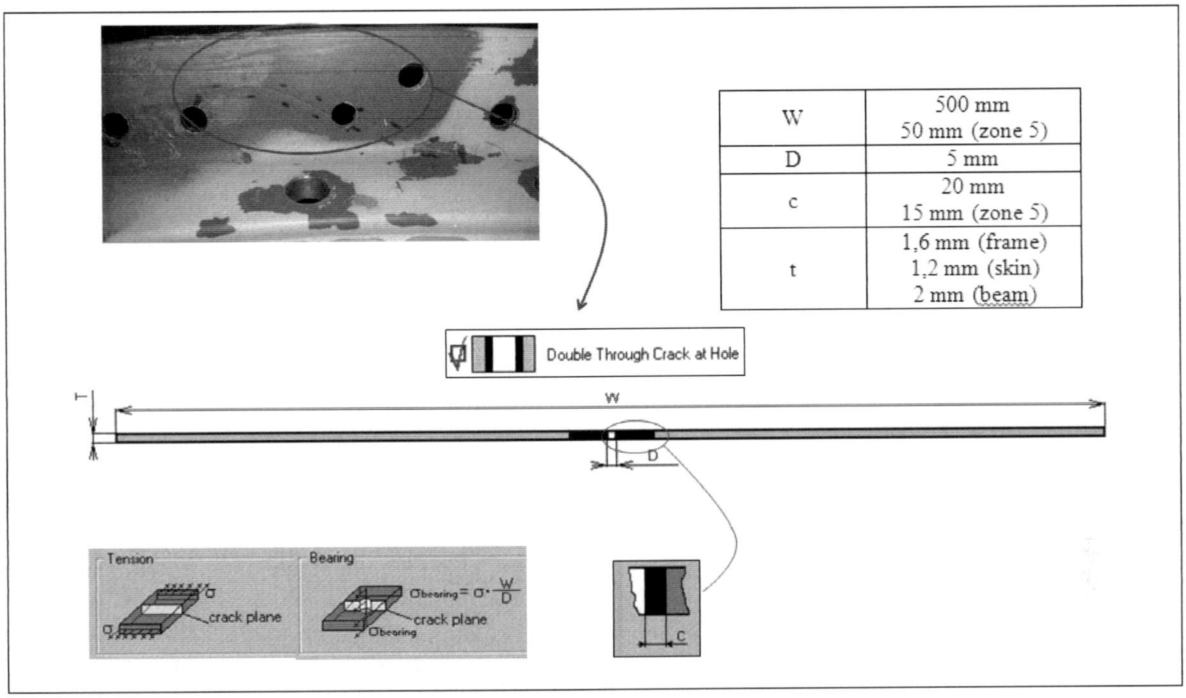

W	500 mm 50 mm (zone 5)
D	5 mm
c	20 mm 15 mm (zone 5)
t	1,6 mm (frame) 1,2 mm (skin) 2 mm (beam)

Figure 7. Simplification hypothesis of geometry of critical areas.

Table I. Critical size of crack for zone 1.

Critical area	$\mathbf{a_{LIMIT\ LOAD}}$ **($\mathbf{a_{LL}}$) (mm)**	$\mathbf{a_{ULTIMATE\ LOAD}}$ **($\mathbf{a_{UL}}$) (mm)**
1	187,95	140,51

Crack propagation calculation. Same software, Afgrow, was used to realize the crack propagation calculation. A simplified load triangular spectrum is applied (R = 0,1) for propagation calculation. Its amplitude is chosen equal to 0.8*limit load, with the hypothesis of 1 cycle = 1 flight. This was chosen because flight spectra of Lynx were unknown and because vibration levels were low. The conservatism were demonstrated by comparing this loads to flight test spectrum Paris law was the equation chosen for the propagation calculation (no threshold). The objective was to find the number of cycles between detectable and critical size of crack. Then, graph of crack size depending on number of cycles could be drawn for each studied area. The following graph gives example of the crack evolution in the critical area 1. For the other zones, results are similar. On the graph, gaps are noticeable. They correspond to the hypothesis of immediate re-starting of crack from hole when crack reaches it. On that graph, critical size of crack at limit load and ultimate load is also drawn. Then, number of cycles at those critical sizes could be directly picked up. That step is done for all studied areas. Results are summarised in the following table.

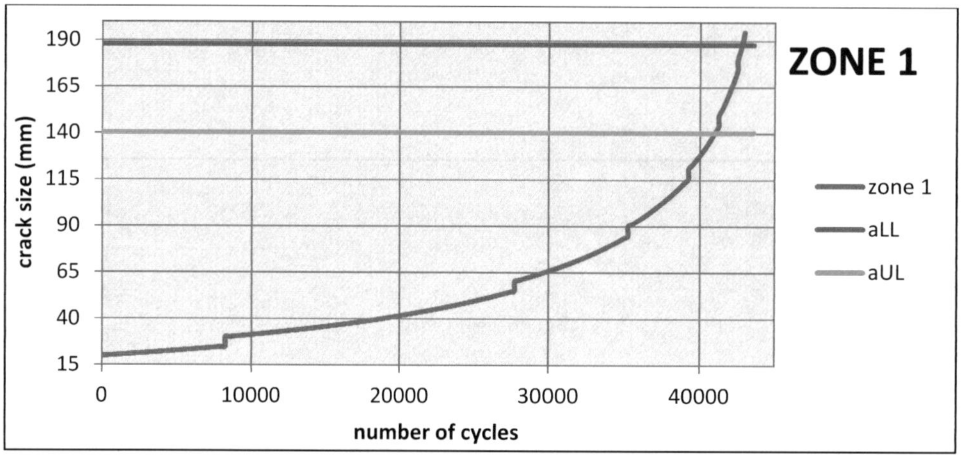

Figure 8. Evolution of crack size depending on number of cycles for critical area 1.

Table II. Crack propagation: number of cycles corresponding to LL and UL.

Critical area	Number of cycles corresponding to a_{LL}	Number of cycles corresponding to a_{UL}
1	42750	41000

Thanks to data at limit load, inspection interval could be defined, with a chosen security factor of 3. They could be used in case of a virtual crack is on the considered area. Results are done on the following table. From the results, it was concluded than no additional maintenance inspection that those existing were necessary.

Table III. Inspection interval for studied zone 1.

Zone	1
Inspection interval (cycles)	14250

To maintain a cracked structure in flight the crack has to sustain ultimate load during the inspection interval The other criterion was that as long as detected damage is fewer than critical size at ultimate load, a ferry flight (single flight with low manoeuvres) is possible. Then, thanks to different results collected for each area, the recap table, presented below, could be given. For each zone, in case of detected damage, advised inspection interval is defined, in order to sustain ultimate loads between to inspection of the rotorcraft.

Table IV. Results of crack propagation calculation.

Zone	Crack size (mm)	Inspection
1	130 mm	330 flight
	140 mm	1 ferry flight

Conclusions

A previous maintenance data edited by manufacturer, Westland, exists but it was revoked. A new document based on damage tolerance was required by maintenance teams. The objective was only to postponed repair to next scheduled maintenance, so inspection intervals must not exceed 50 flight hours in general. Then the following table presents flight recovery conditions depending on damage size, for each studied area. In case of area 1 for instance, until 75 mm for crack size, control each 50 flight hour is permitted. Between 75 and 140 mm, a ferry flight is permitted and for a crack higher than 140 mm, no flight is authorized, immediate repair must occur. Those results were accepted by airworthiness authorities. Fleet availability could then be improved.

Table V. Summary of results of the frame 3353A study

Zone	Crack size (mm)	Inspection
1	75 mm	50 flying hours
	140 mm	1 ferry flight

LATERAL PANELS

Background and objectives

The other main reason for which Lynx fleet presents a poor availability partly is recurrent damage observed on lateral panels. Indeed, those panels are sandwich with very thin sheets, very exposed to accidental damages and no criterion actually allows helicopters to flight with any size of defects. Lateral panels are made of sandwich structure constituted of Aluminium skins and Nomex honeycomb. Two kinds of damage are generally observed: disbonds and indents damages (indent damages can be visually detected). The maintenance feedback does not precisely explain the reason of the defects presence, except safety shoes impacts then mechanics control the MGB before flight on ship deck. However, the non-destructive control, tap-test control, spots very easily damage and helicopter has to be repaired. Change an entire panel is long and expensive but maintenance operators do not have an efficient and fast method to repair those panels. The objective of that study is then to determine the authorized maximum size of damage, depending on the type of damage, without altering the flight safety. That study consist in a flight tests campaign, mechanical tests on damaged panels and finite elements models.

Flight tests campaign [3]

Limits loads had another time be determined from flight tests. In that purpose, a tested airframe was equipped with some strain gauges: they were glued inner and outer the lateral panels. Four different flights were performed: cargo, sling, deck-landing. During all those flights, strain gauges recorded information from the instrumented lateral panels. The objective was to analyse and extrapolate to the flight envelope (-1g/3g) strain gauges data. After that, limit load in compression and shear stress could be estimated. It is important to notice that tension loads are not altered by the existence of damage. Tension measured will therefore not been analized in the frame of this study.

The following figure details the equipped lateral panels with different strain gauges. The equipped zone is in reality consisting of different panels, they are arbitrarily named: panel 1 to panel 5 (cf. figure 7)

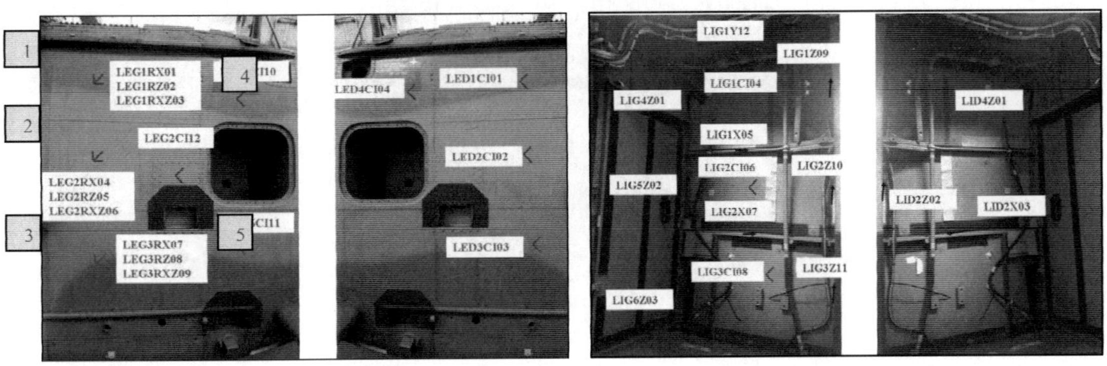

Figure 9. Lateral panels equipped with strain gauges to realize flight tests campaign.

Remarkable results of the flight tests campaign are the determination of limit loads in compression and shear stress, for each panel. Indeed, panels 1 to 5 are not loaded identically. The following table presents those results.

Table VI. Limit loads defined thanks to the flight test campaign

Limit loads	Panel 1	Panel 2	Panel 3	Panel 4	Panel 5
Shear	27 MPa	32 MPa	53 MPa	16 MPa	10 MPa
Compression	-13 MPa	-10 MPa	-7 MPa	Information not available	

Those results show that shear stress is more important than compression stress in amplitude, and that the stress levels are low. Thereafter, a specific study will be done in shear stress. A good approximation of the result could be approached thanks to that analysis.

Mechanical tests

No reliable numerical method exists to determine residual stress with those kinds of damaged structure. Then, a mechanical tests campaign was launch to assess allowable values. Geometry specimens and mechanical tests are depending on service capacity of tensile-testing machine.

Different specimens were tested: compression or shear stress, with indents or disbonds damage at different diameters or healthy specimens. A set of tests were performed.

Shear stress tests. A total of twenty specimens were tested in shear stress. The dimensions of the coupon are 700*700 mm. Different kinds of circular damage (assumed worst geometry at given diameter) were introduced, from 50 to 150 mm diameter at the centre of the specimen (one side). To represent disbonds damage, a sized Teflon disc was introduced during coupon manufacturing. To represent indent damage, a mechanical indentation (2 mm depth) with a good sized punched is introduced.

Static and fatigue tests were realized (10 specimens of each). All the specimens were tested in a tensile machine thanks to a shear frame (capacity of the tensile machine 25T). Stress was applied on specimens thanks to a ramp of 1mm/min. Each specimen was instrumented with 4 strain gauges on each side, all around the defect. The following figure shows an example of a shear instrumented specimen placed in the frame.

Figure 10. Shear test rig (with shear frame).

Results of the fatigue tests allowed to validate the non-growth of the considered damage. Concerning static tests, specimen is loaded to limit load (here, 53 MPa) during 20 seconds. Then, a non-destructive control by ultra-sound. If no evolution of damage is detected, the specimen is loaded to ultimate load (1.5 * limit load) during 3 seconds. Finally specimen is loaded until failure to estimate residual strength. In case of fatigue tests, the first up-loading at limit load is also done. Then, 3000 cycles of fatigue at fatigue load, corresponding to 0.8 * limit load, is launched. At the end of that step, a NDI controls a potential evolution of the damage. Finally, specimen is loaded to ultimate load during 3 seconds then until failure. Before and after mechanical test, non-destructive inspections were realized. Thanks to that, non evolution of damage could be validated and if another defect appears, it could be also seen. For each kind of damage, two coupons were manufactured, for static and fatigue tests..

In case of shear stress tests, all of them have their failure after ultimate load. Residual strength was then estimated. The failure mode analysed is a sandwich buckling mode: crimpling mode. It consists in general buckling of honeycomb due to its poor properties. A global buckling of the specimen then appears and a plastic deformation is observed on thin skins (no more stress is sustained by skins). Figure 9 illustrate the crimping mode.

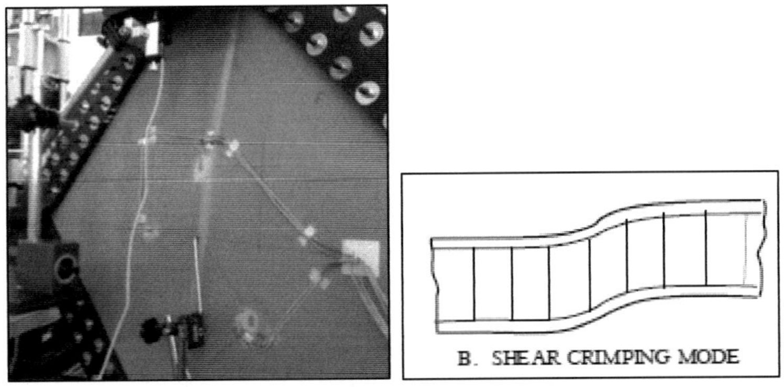

Figure 11. Failure mode of shear specimens: crimping.

Another result shows that fatigue solicitation in shear has no influence on residual strength. Moreover, no propagation was observed whatever damage (indent or disbonds). The following graph gathers failure results according to size of damage together. Disparity is observed on results but all of them are above ultimate load (79.5 MPa).

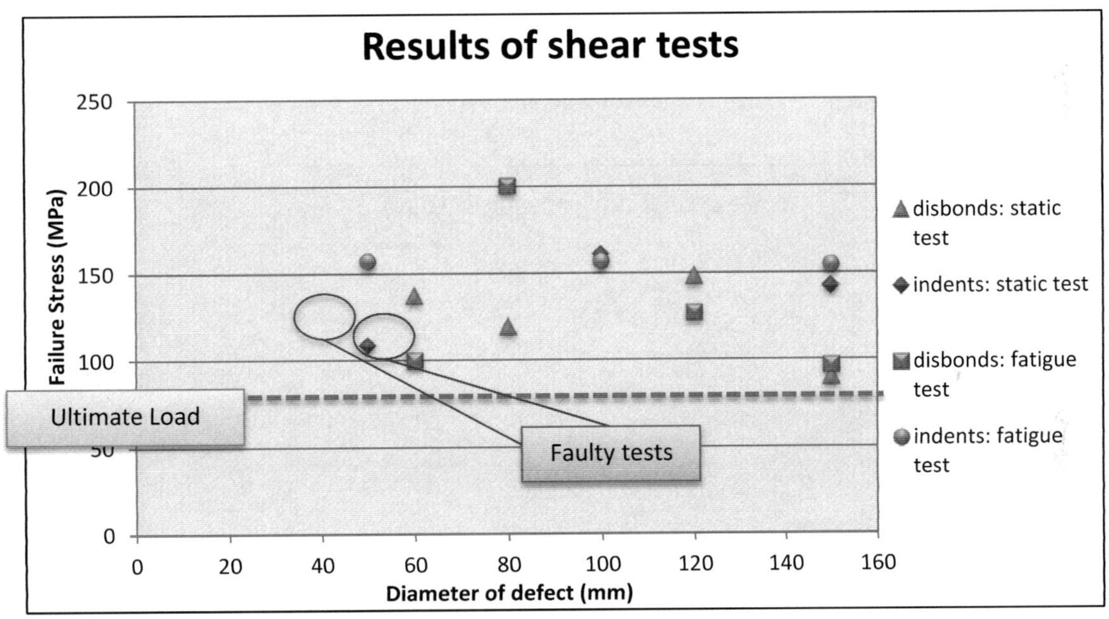

Figure 12. Results of shear tests depending on diameter of defect.

Compression stress tests. Only eight specimens were tested in shear stress. The dimensions are 400*400 mm. As shear stress tests, kinds of circular damage (from 50 to 150 mm) were introduced during manufacturing. Static and fatigue tests were performed on a compression machine (same machine used to shear stress tests, capacity of 25T) with a non-buckling device, as show in the following figure. A total of 4 strain gauges per side were fixed on each specimen all around damaged zone.

Figure 13. Compression test fixture (with non-buckling device).

Same load diagram is applied to specimens with compression loads (limit load during 20 seconds then ultimate load during 3 seconds, and finally loading until failure). Non-destructive inspections were done before, during and after test. All of specimens have their failure well after the ultimate load. Compression stresses are lower than shear stresses and lateral panels have good behaviour in compression stress, so high margin of safety. Priority will be put in this study to shear failure and a correction will be added to cover compression behaviour.

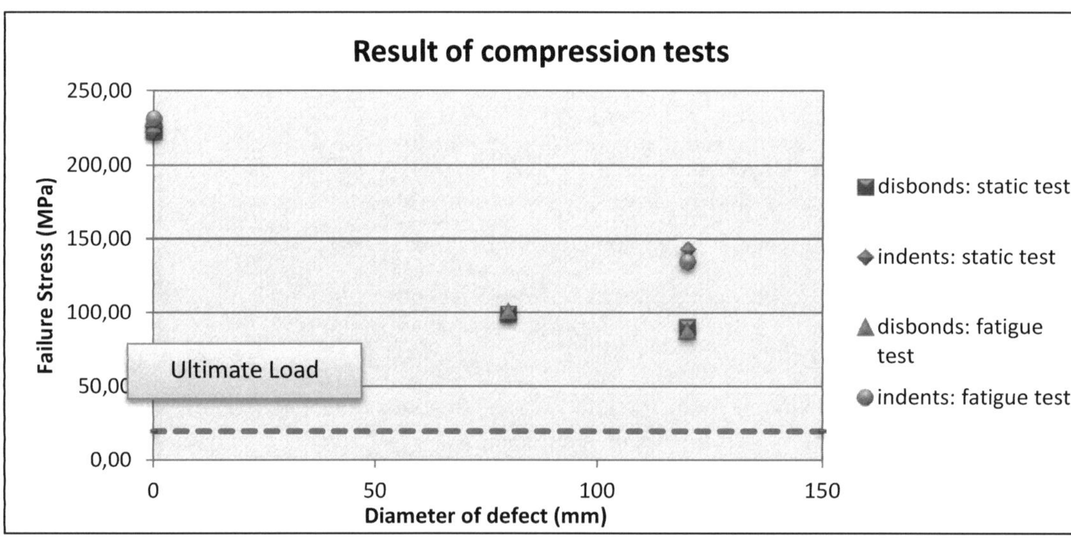

Figure 14. Results of compression tests depending on diameter of defect.

Finite elements modelling

During mechanical tests, it was impossible to test all sizes of defect and define operationally acceptable damage size. Modelling could extrapolate panel behaviour to other sizes. Models were realized thanks to the finite elements software Samcef.

Shear test modelling. Models were consistent with mechanical tests. Indeed, artificial damage (disbonds and indents) were introduced in order to be representative. 3D simulations were done: specimen and shear frame were modelled and computation of the buckling loads and modes (of a linear elastic structure subjected to external loads) was enabled thanks to the Stabi module of Samcef. Thin Aluminium skins and honeycomb were assembled thanks to a glue layer of elements. To model disbonds damage, the glue was suppressed in the defect area. Indent damage were artificially drawn on specimen, because the damaged honeycomb behaviour is unknown. The following figure presents a description of a shear test model.

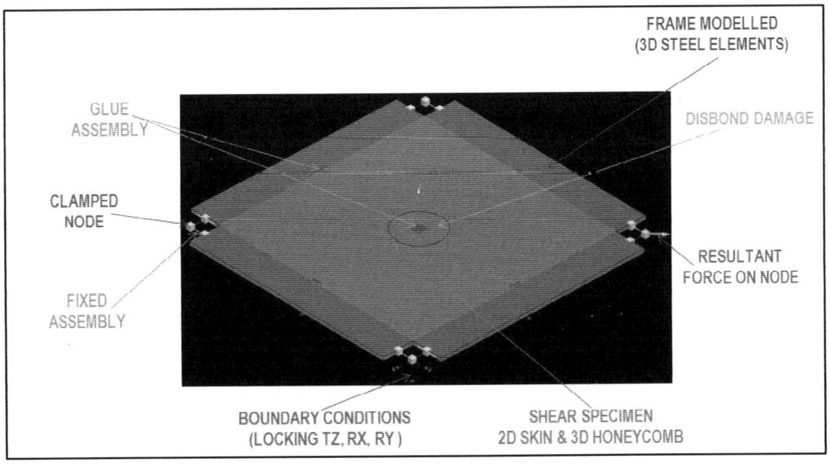

Figure 15. finite elements model: shear specimen with disbonds damage.

Indented and undamaged specimens were also modelled. Damage sizes were consistent with the tested coupons. Numerical models boundary conditions could then be adjusted in order to well represent tests.

Tests/simulations correlations (shear tests). During this phase between mechanical tests and numerical simulations, different points were compared: failure stress level and its location, failure mode and no-growth of damaged zone. Modelled defects were the same as tested panels. Buckling loads of shear stress simulations are comparable with mechanical test failure level according to damage size.

- Healthy specimen : only 1% discrepancy (shear test)
- Disbonds damaged specimens : 8% discrepancy (shear test)
- Indent damaged specimens : 9% discrepancy (shear test)

Simulation results are close to mechanical shear tests results. Then, defect sizes could be extrapolated thanks to model in order to define an acceptable damage size (depending on the kind of damage) to improve the Lynx fleet availability. As expected, linear correlation could be noticed between simulation results with same kind of damage. As figured on the following graph, linear combinations were defined, depending on size of damage. Thanks to those relations buckling values could be extrapolated to other diameters of defect. Moreover, results give a very poor influence of indents size on buckling stress (as observed on mechanical tests).

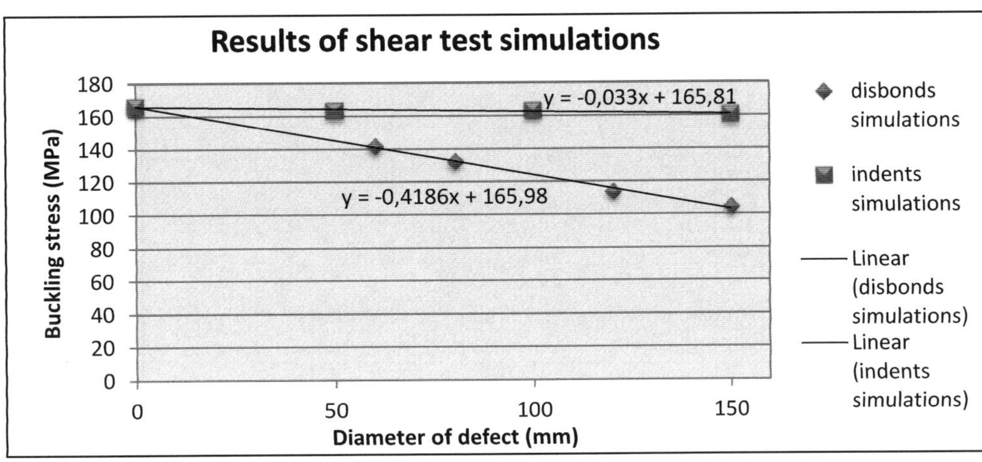

Figure 16. Results of shear tests simulations and linear correlations.

Compression test modelling. As shear test modelling, numerical models were consistent with mechanical tests: 400-400 mm damaged specimens. Artificial damage were introduced on 3D simulations. Then, computation of the buckling loads was realized thanks to the Stabi module of Samcef. Specimen only was modelled (no frame or non-buckling device designed). Constraints were applied on the specimen model boundary conditions to represent the non-buckling device.

Tests/simulations correlations (compression tests). As in case of shear tests, failure stress level, failure mode, and no-growth of damaged zone were compared with mechanical test results. A gap between results of simulations and mechanical test is noticed: whatever kinds of damage (disbonds, indents and even healthy specimen), results of mechanical testing are all higher than simulation. A discrepancy of around 15% is observed for damaged specimens and a bigger relative error of 60% is observed for healthy specimens. Compression simulations underestimate failure loads compared to mechanical tests. That conclusion confirms good behaviour of panel in compression. Fewer compression tests were done than shear tests. However, linear combinations could be defined for disbonds simulations. Then, results could also be extrapolated on other diameters of defect. Concerning indents results, damage have no influence on panel behaviour. On the following picture are detailed results of compression simulations.

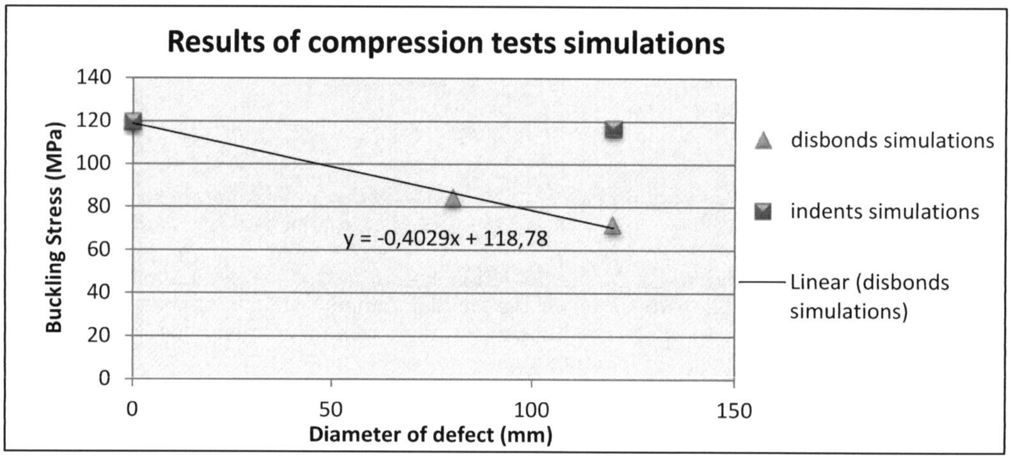

Figure 17. Results of compression tests simulations and linear correlations.

First results

In contrast, a critical size of disbonds has to be estimated (according to the damaged panel). A study of combined shear / compression load is carried by the formula (1) from reference [4] in order to take into account different contribution on the panel. f_S and f_C of shear ultimate load and compression ultimate load. F_{CR} corresponds to failure load. That equation is applied for each size of damage. If that result is higher than 1, damaged panel does not sustain those combined loads.

$$\left(\frac{f_S}{F_{CR}^S}\right)^2 + \frac{f_C}{F_{CR}^C} = 1 \qquad (1)$$

Table VII. Criteria values for disbonds damage for different diameters

Disbonds damage	f_S = 79.5 MPa - f_C = 19.5 MPa				f_S = 48 MPa - f_C = 19.5 MPa			
	D100mm	**D120mm**	**D150mm**	**D180mm**	**D100mm**	**D120mm**	**D150mm**	**D180mm**
Criteria	0.65	0.75	0.91	1.19	0.40	0.45	0.54	0.71

The determined critical size of disbonds damage are diameter of 150 mm in case of the most loaded panel (reminder, the most loaded panel is number 3 with shear ultimate load = 79.5 MPa). Concerning the other panels, shear ultimate load was taken equal to 48 MPa (corresponding to the second most loaded panel, cf. table VI). For those panels, first conclusions give a critical size of disbonds damage as at least 180mm.

The indents have no influence of compression.

Those results have to be corrected, for the following reasons:

- The results are based on mean values of allowable, but airworthiness requires B-values. On that kind of test supported by analysis, adapted knock-down factors have to be agreed with the authorities. They are still under discussion.
- The acceptable diameters are too large to neglect boundary effects, but they are in general beneficial
- The critical sizes are larger than the ones with are very common and are a problem for fleet availability.

For all those reasons, reduced the critical size is necessary but will not be an operational problem, for the study.

For damage tolerance evaluation, mechanical tests show no-growth on damage (whatever kind of damage). Thanks to feedback of maintenance teams concerning probability of occurrence of different defect on lateral panels, operationally acceptable inspection interval will be determined. Same as for frame, a maximal size sustaining ultimate load will be determined and another one for a ferry flight.

CONCLUSION

Damage tolerance study of 3353A frame of Lynx airframe was completed and airworthiness authorities accepted that approach. Maintenance units could have better flexibility and frame is not necessary changed if crack is observed. Fleet shown its availability improved.

Few discussions are still necessary to conclude on effective critical sizes on lateral panels to propose a final solution to airworthiness authorities. Good mechanical behaviour of damaged panels will surely enable to improve availability of Lynx fleet too.

REFERENCES

[1] FAA, Part 29-22, Airworthiness standards: transport category rotorcrafts, chap. 571.b

[2] C. Kinzelin (2015). In: Détermination des critères de tolérance aux dommages sur le cadre 3353A du Lynx.

[3] E. Jacquet (2013). In: Détermination des charges en vol vues par les panneaux latéraux, les cadres 3353A et 406F du Lynx.

[4] M. Niu (1999). In: Airframe stress analysis and sizing, pp 468-469.

Structural Durability and Damage Tolerance in the Next Century of Commercial Aviation

Steven Chisholm, Antonio Rufin, Brandon Chapman, Quentin Benson

The Boeing Company, United States of America

Abstract: As The Boeing Company prepares to enter its second century in commercial aviation (1916-2016), this paper reviews the progress made over the past forty years in the development and application of durability and damage tolerance methods across its commercial airplane product line, and ventures forward to look at future challenges and opportunities. The company's pioneering efforts, which saw significant advances in the 1970's and 1980's with the development of comprehensive internal technology standards have been evolving, driven by shifts in the regulatory and competitive environments and significant technological developments. Successes are evident in terms of significant safety improvements and considerable reductions in service actions on airplanes designed since that time. Managing an aging fleet and the adoption of new regulations affecting maintenance planning have been, and remain some of the greatest challenges. New materials (composite and metallic) and assembly methods, and the ever-constant quest for productivity gains in the factory and in design and analysis processes are all further shaping the way durability and damage tolerance are being assessed and implemented.

PREFACE: A HISTORICAL PERSPECTIVE

In his 1993 ICAF Plantema Memorial lecture [1], Dr. Ulf Goranson described damage tolerance (and indirectly, durability) as the result of an evolutionary process built on decades of experience and lessons learned, made possible only through diligent attention to detail design, manufacturing, and maintenance and inspection procedures. The lecture touched on basic aspects of the damage tolerance philosophy in its application at The Boeing Company, and almost presciently, on the many challenges that the industry was beginning to face, particularly as it addressed widespread fatigue damage (WFD) and the application of damage tolerance to major structural repairs and alterations. Both of these topics have since become the subject of major regulatory actions and significant industry focus.

During these years, technology has evolved. Although the basic analysis principles employed today are not substantively different from the methods described in Ref. [1–5], the tools and processes used have been refined over these past 20 years, extended to broader applications, seeking more efficient use of engineering resources, greater accuracy, and incorporating service experience acquired from an ever-expanding fleet. In that time, durability and damage tolerance (DaDT), moreover, have had to keep pace with major developments in materials and manufacturing processes. Composites, for example, saw their first primary structure application at Boeing in the early 1990's on the 777 empennage. Today, they are used on the new Boeing 787 and 777X in areas of the airframe that previously were the exclusive domain of metals.

The most important outcome from the early developments and their evolution in the more recent past is the unprecedented, steady advance in commercial aviation safety that has been achieved since the advent of commercial jet transports in the 1950's, despite a significant growth in the size of the worldwide fleet (Fig. 1). Although the number of hull loss accidents associated with structural issues in these statistics comprises only a small percentage of the totals depicted in Fig. 1, design for DaDT has certainly been a fundamental contributor to this remarkable safety record. Progress in commercial aviation safety has always been entwined with the evolution of design requirements that are mandated by the regulatory agencies (Fig. 2). The introduction of fail-safe design on the Boeing 707, design and testing for durability starting with the model 727, and incorporation of DaDT methods and more active corrosion prevention and control measures starting with the 757 and 767 airplanes, have all played a role in bringing about a significantly safer and more durable and therefore economically viable fleet.

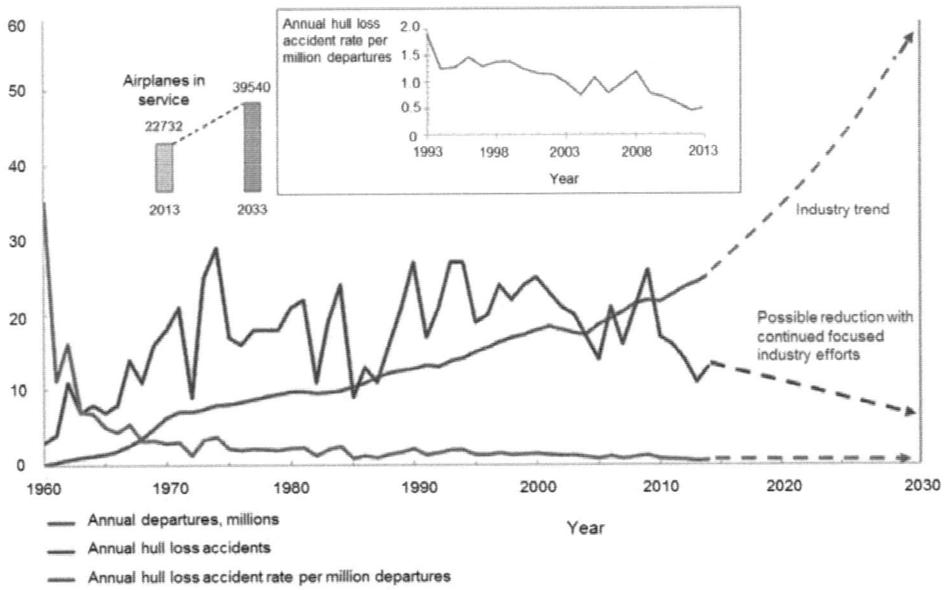

Figure 1. Annual hull loss accident totals and rates for the worldwide commercial jet fleet [6].

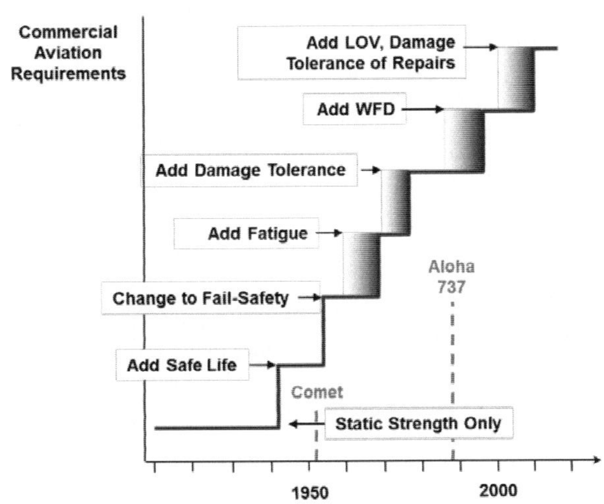

Figure 2. Evolution of design requirements for commercial transports, and some of the events that shaped them.

In another major accomplishment, and supporting these developments, Boeing has created industry-leading technology methods and standards that provide a Boeing commercial airplane product-wide uniform approach to design and analysis for fatigue, damage tolerance and corrosion prevention and control (Fig. 3). The standards capture lessons learned and are periodically updated to incorporate new methods and to gather in further testing and fleet experience, which can then be adopted in new designs or as design improvements. Furthermore, since their inception, these standards have been built for ease of use both in terms of prerequisite analysis skills and tools, and reliance on a manageable number of variables. This paper reviews the origins and evolution of durability and damage tolerance of commercial airplane structures, focusing on basic principles and their current application at Boeing, and their validation. Recent developments on composite damage tolerance are discussed next. The paper concludes with a discussion of new technologies and challenges, and how they are being addressed.

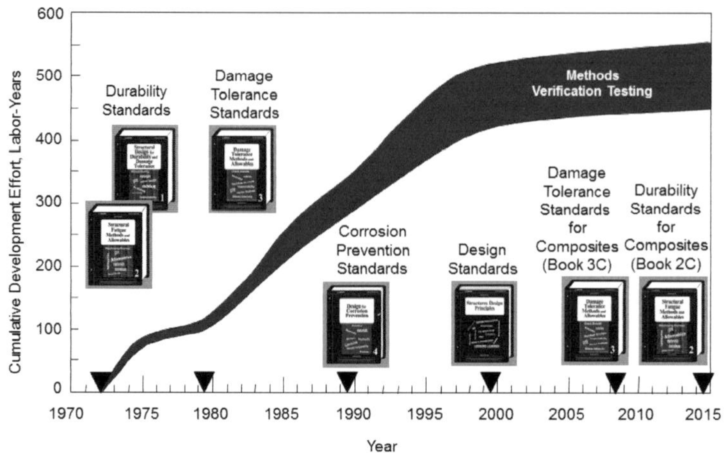

Figure 3. Evolution timeline and resources required for Boeing structural DaDT technology standards.

THE BOEING DURABILITY METHOD

Initially developed in the early 1970's and first implemented with the 757 and 767 models, the Boeing durability analysis standard, commonly referred to as "Book 2", along with its design companion, "Book 1", were first introduced with the overall objective of assuring a competitive economic life for the structure. This goal was specifically defined as an absence of significant fatigue cracking in the first 20 years of service (defined as a probability of cracking of less than 1 percent with 95 percent confidence) and at least 30 years of service before fatigue related maintenance begins to measurably escalate (defined as a nominal probability of cracking of 5 percent with 95 percent confidence), meaning that the airframe truly remains economically viable for a minimum of 30 years. This objective, and the means to attain it, has since been validated by 737 Next Generation (NG) /757/767/777/787 full-scale fatigue test performance and by direct comparisons of service data with earlier airplane experience. In more recent times, the method has been effectively extended from its original economic life focus to also support damage tolerance threshold calculations and WFD analyses.

Durability method fundamentals

For metallic structure, the Book 2 method at its core balances a structural capability – characterized by the Detail Fatigue Rating (DFR) – against a set of requirements (stresses, minimum design service objective) to arrive at a Fatigue Margin analogous to a static strength Margin of Safety, in that it is likewise defined by a ratio of stresses or a ratio of parameters in units of stress (Fig. 4). Statistical factors are used to attain prescribed reliability levels for each component [1–3].

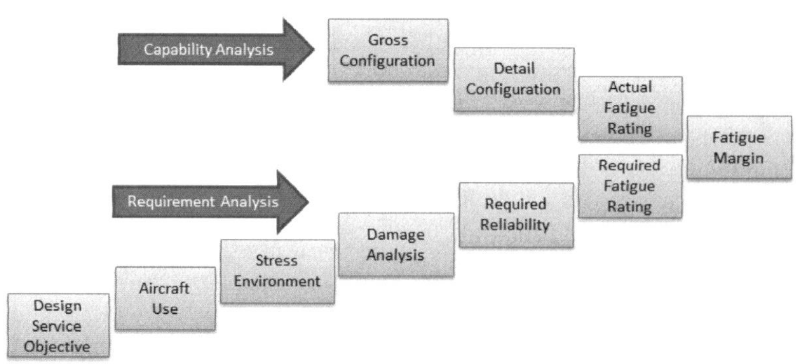

Figure 4. The Book 2 fatigue analysis process.

DFR is a component-level fatigue quality index expressed in units of stress. With a set of standardized stress-life (S-N) curves, it can be used to determine 95 percent reliability, 95 percent confidence fatigue lives at any given stress or allowable fatigue stresses for a particular target reliable life. DFR values can be applied to a wide range of structural features, from the simplest of stress concentrations to complex joints. For joints, the method can capture load transfer effects, material, surface finish, hole fill, hole treatments such as coldworking, and clamp-up. DFR values are determined based on combinations of empirical data and analysis. They can be drawn from test as well as service data, as Figure 5 shows. Service data can be especially useful in developing or improving fleet maintenance planning, or to handle unexpected service findings; for example, as a basis for fatigue-related service bulletin thresholds. Book 2 provides a comprehensive summary of DFR values as well as the procedures and constituent factors used to determine them. Complementing Book 2, Book 1 provides specific design and analysis guidance collected from service and full-scale fatigue test data to aid designers and analysts avoid known issues and adopt best practices.

The requirements aspect (lower branch in the Fig. 4 diagram) of the analysis begins with the definition of design service objectives based on a range of anticipated airplane operational usage rates. Loads are structured as standard mission profile segments, covering all applicable conditions encountered in a typical flight.

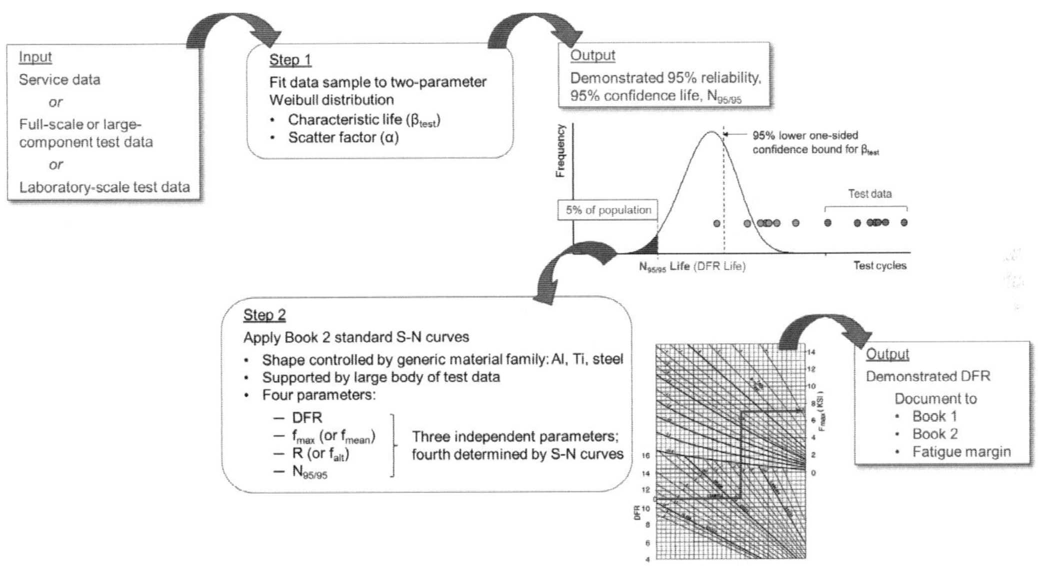

Figure 5. Determining DFR values from empirical data.

The method can accommodate conditions represented by step, constant-amplitude (discrete-cycle), and spectrum loading. The latter rely on standard exceedance curves substantiated by flight test or other operational data. Variable-amplitude profiles are rendered as equivalent once-per-flight constant-amplitude "GAG" (ground-air-ground) loads or stresses, the relationship between flights and the GAG cycle being handled by a proportionality constant termed "GAG Damage Ratio". The GAG Damage Ratio is determined by a complex procedure combining cycle counting and Palmgren-Miner's Rule over large numbers of flights to arrive at a cumulative fatigue damage that is then compared to the GAG cycle damage. This enables transforming design service objectives, which are expressed in numbers of flights into equivalent constant-amplitude GAG cycles that can then be used in conjunction with constant-amplitude S-N curves to calculate stress-based fatigue margins for any given structural component. The entire process can be accomplished seamlessly using specialized software.

A legacy of durability

Recent indications from the worldwide commercial airplane fleet are that average airplane ages may be peaking, but much of the current pool remains quite advanced in age. The size and longevity of the Boeing active commercial fleet, as illustrated by Table I, is a testament to a quality achieved only as a result of decades of development and experience. These statistics are also a reminder of the continuing challenges associated with aging structures and the value of design for durability.

Table I. Commercial airplane usage statistics, Boeing 727, 737, 747, 757, 767, and 777, as of September 2014.

Model	EIS* Date	Airplanes†		Average age, yrs.	High time airplanes	
		Delivered	In service		Flights‡	Hours‡
727	1964	1827	66	35.7	63.2	82.4
737-100/200/300/400/500	1968	3098	902	21.6	75.4	94.3
737-600/700/800/900	1993	4830	4653	7.0	42.8	60.3
747	1970	1472	566	15.1	24.6	125.2
757	1983	1043	798	19.7	44.5	100.0
767	1982	1047	802	17.2	50.0	118.1
777	1995	1227	1205	8.3	35.9	84.4

* Entry into service
† Commercial fleet only
‡ In thousands

The introduction of the durability method as well as active corrosion prevention programs has had a dramatic and almost immediate effect on fleet maintenance, as Fig. 6 shows. The 757 and 767 were the first models to take advantage of these efforts, resulting in a drastic reduction of maintenance service hours, a trend that has continued on newer airplanes, the 737 NG and 777, further enhanced by improvements to the methods and the continuous learning process afforded by fleet observations. This trend is also clearly evidenced in the many full-scale fatigue tests successfully completed or underway since then, which are discussed later in this paper.

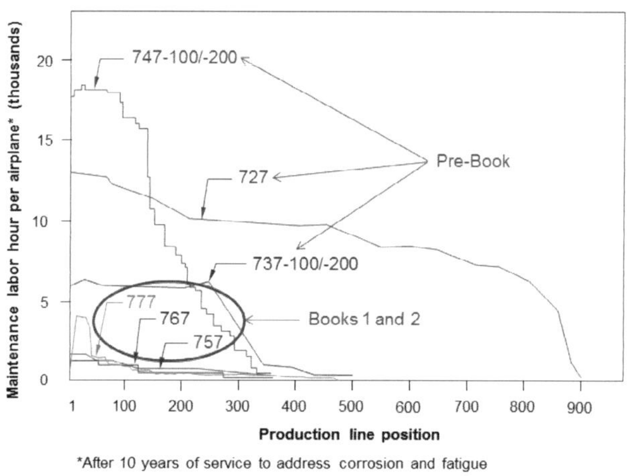

Figure 6. Impact of new design and analysis practices on fleet maintenance hours.

Despite the successes, there have been a few durability challenges as well, none perhaps as great for Boeing as the 737 fuselage skin longitudinal lap splices. Design, processing, and a complex stress environment required a number of iterations to achieve a lap splice design with a high level of durability. The evolution of some of the key lap splice design characteristics is illustrated in Fig. 7. A cold bonding process, which proved to lack the necessary robustness, was eliminated early on. Further tests led to the addition of hot bonded doublers and changes to the tear strap design and rivet diameters on later 737 Classic variants, significantly improving the durability and damage tolerance properties of the laps. The lap splices on the 737 NG airplanes represent a further improvement over earlier designs. Changes include increased skin gages (lower basic stress), widened tear straps (stiffening ratio increased), reduced eccentricity between upper and lower skin (lower local bending stresses), and machine-installed rivets (consistently better hole fill). The net results of these changes are illustrated in Fig. 8.

737 NG fatigue performance was successfully demonstrated on a 737-800 full-scale fatigue test to 225000 simulated flights, or three times the 75000-flight design service objective. High-frequency eddy current inspections of the rivet holes during teardown revealed very few crack indications, except at the manually assembled Stringer 14 lap splices (just below the window belt), where some small incipient cracks were found. Upper and lower row fatigue cracking was otherwise eliminated.

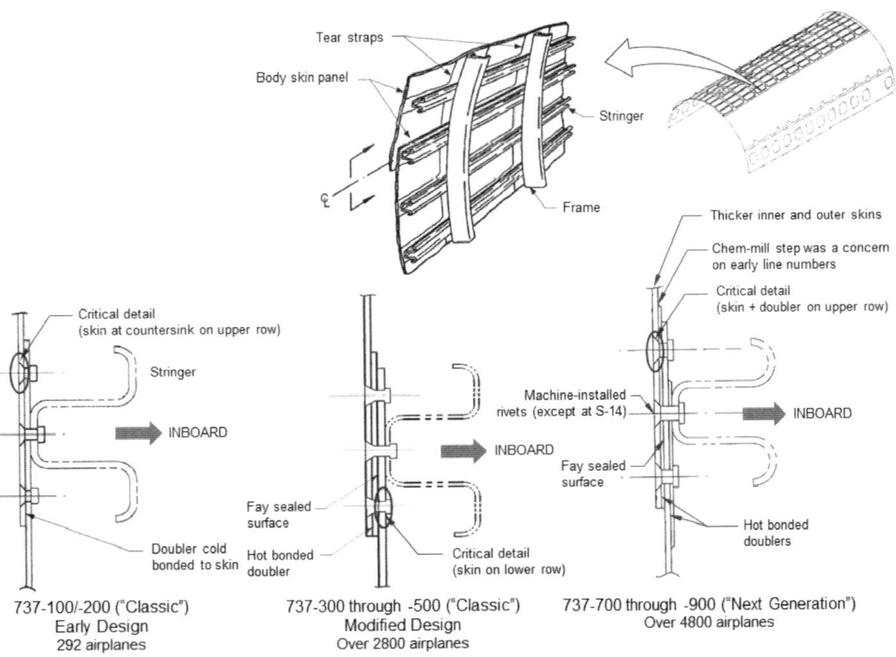

Figure 7. Boeing 737 Fuselage skin lap splice design evolution.

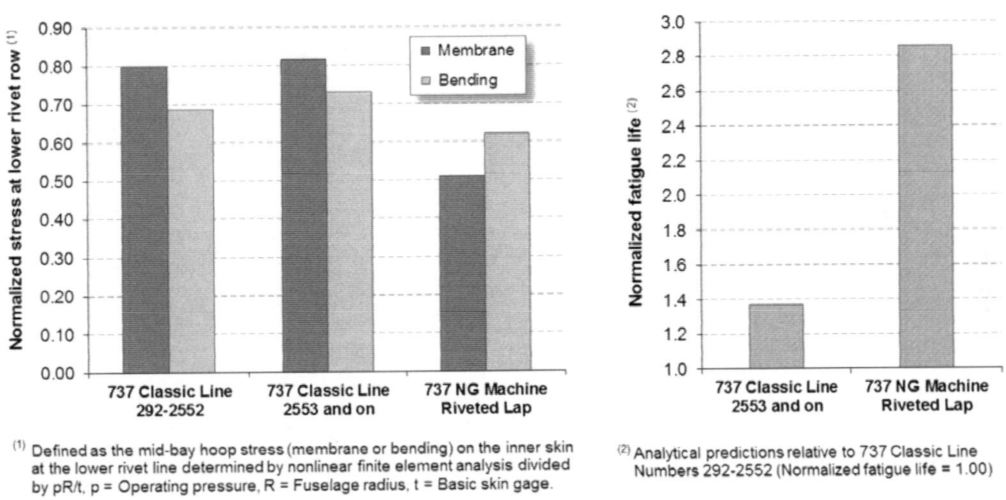

Figure 8. Fuselage crown longitudinal lap splice performance, Boeing 737 Classic vs. 737 NG (NOTE: For the 737 NG, the reference operating pressure is 7 percent higher than for the 737 Classics).

THE BOEING DAMAGE TOLERANCE METHOD

The accident at Lusaka in 1977 involving a Boeing 707 helped stimulate a 1978 revision of the regulatory requirements (14 CFR 25.571) [7]. That airplane, which had been designed and certified as fail-safe, was lost due to a horizontal stabilizer failure caused by fatigue cracking. The event demonstrated the limiting value of maintenance programs based solely on fail-safe design principles, which do not specify *where* to look and do not quantify *how often* to look for cracking. The accident occurred at a time when the United States Air Force (USAF) had already implemented damage tolerance requirements [8], and the commercial airplane industry was beginning to recognize that safe-life, fail-safe, and damage tolerance principles each have some inadequacies and that a combination of all three philosophies is needed in some cases. The subsequent changes in the civil regulations mandated the use of damage tolerant principles in all instances unless it imposes an unreasonable penalty. The Boeing damage tolerance analysis standard, commonly referred to as "Book 3," was first introduced in 1979 in response to the new requirements. Book 3 provides a method and design data for damage tolerance analysis of metallic structure. Similarly to the Boeing durability method, the Boeing damage tolerance approach was first implemented for new design with the 757 and 767 models. Book 3 was also used concurrently for supplemental structural inspection programs on the 727, 737 and 747 models.

Boeing damage tolerance method fundamentals

Damage tolerance is the attribute of the structure that permits it to retain its required residual strength for a period of use after the structure has sustained a given level of fatigue, corrosion, or accidental or discrete source damage. The basic requirements for damage tolerance are set by 14 CFR 25.571, which mandates damage tolerant design for structurally significant items (SSIs) or principal structural elements (PSEs). To clarify what is required, Boeing classifies aircraft structure into one of four categories, summarized in Table II. Any structural detail element or assembly that contributes significantly to carry flight, ground, pressure or control loads, and whose failure could affect the structural integrity necessary for the safety of the airplane, is characterized as a PSE. The changes in the regulations also mandated the development of inspection programs for new airplanes, the assessment of structural maintenance programs for existing airplanes, and initiated the development of supplemental structural inspection documents (SSID) in the late 1970's and early 1980's. Boeing retained the fail-safe requirement even though it is no longer required for certification.

The key objective for airplane structure designed to be damage tolerant has always been to carry minimum regulatory loads until detection and repair of any fatigue cracks, corrosion, or accidental or discrete source damage occurring in service (Fig. 9). The best design has an allowable damage size that is obvious or malfunction evident to flight or ground crew personnel during routine activities around the airplane (Category 2). When this is not possible, some level of inspection is required to assure timely damage detection. Category 3 is primary structure that requires a planned inspection program to maintain structural integrity. Large allowable damage and/or slow crack growth, coupled with an easy inspection access, minimize operator inspection cost.

Table II. Boeing structural classification for damage tolerance.

Structural Category			Technique of Ensuring Safety	Technology Control Method	Structural Classification Examples
Other Structure	① Secondary Structure		Design for safe separation or loss of function	Continued safe flight	Wing spoiler segment (safe separation or safe loss of function)
Primary Structure (Structurally Significant Items or Principal Structural Elements)	Damage Tolerant Design*	② Damage Obvious or Malfunction Evident	Adequate residual strength with extensive damage-obvious during walk around or indicated by malfunction	Residual strength	Typical wing skin/stringer surface (fuel leak)
		③ Damage Detection by Planned Inspection	Inspection program matched to structural characteristics	Residual strength crack growth inspection program	All primary structure not included in two categories above
④	Safe-Life Design	Safe-Life	Conservative fatigue life	Fatigue	Landing gear structure

*Fail-safety required for Cat. 2 and 3. Additional WFD analyses for certain Cat. 3 structure

The fundamental elements of a classic damage tolerance analysis are residual strength, crack growth, and damage detection. An analytical approach to damage tolerance is complex even in its simplest form. In Book 3, Boeing has developed a method for damage tolerance analysis of metallic structure that is suitable for use by analysts having varying levels of familiarity with fracture mechanics concepts. Book 3 provides a frequently updated database of empirically-derived properties that determine the relationship between crack growth rate and stress intensity for various materials and environments [4, 5]. This document also provides a library of factors for a large number of frequently encountered structural configurations. Each of these factors is obtained by compounding individual component effects (e.g., fastener hole, free edge, fuselage curvature). Other factors are used to account for the effects of internal load redistribution caused by cracking of adjacent members as well as spectrum effects on specific configurations [5].

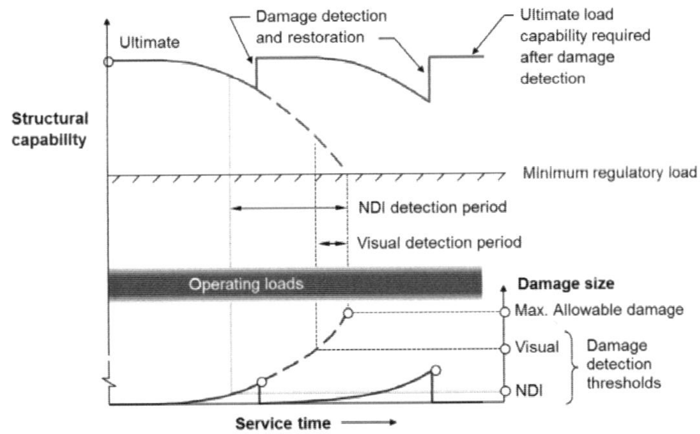

Figure 9. Damage evolution, detection, and repair as a function of service time in damage tolerant structure.

For the purpose of defining and managing the supplemental inspection program, Boeing uses a system referred to as Damage Tolerance Rating (DTR), which is built into the Book 3 method (Fig. 10). The DTR system is used to determine the adequacy of the baseline structural inspection program and if needed, to support the development of a supplemental program for detecting fatigue damage in all Category 3 SSIs/PSEs. This system is used to assess the probability of detecting fatigue damage assumed to have occurred in the fleet before the residual strength of the first cracked airplane falls below allowable limits. The adequacy of the structural inspection program for each individual component is determined by comparing the probability of detection ("Total DTR") to a predetermined acceptable level ("Required DTR"). By establishing damage detection probabilities as a function of inspection methods and crack size, it is possible to significantly increase the damage detection period [9, 10]. In contrast, the commonly used practice of setting inspection intervals to one-half of the damage detection period fails to provide quantitative damage detection reliability and does not capture the combined benefits of visual inspections performed during normal maintenance programs with targeted non-destructive inspection (NDI) methods [9, 10].

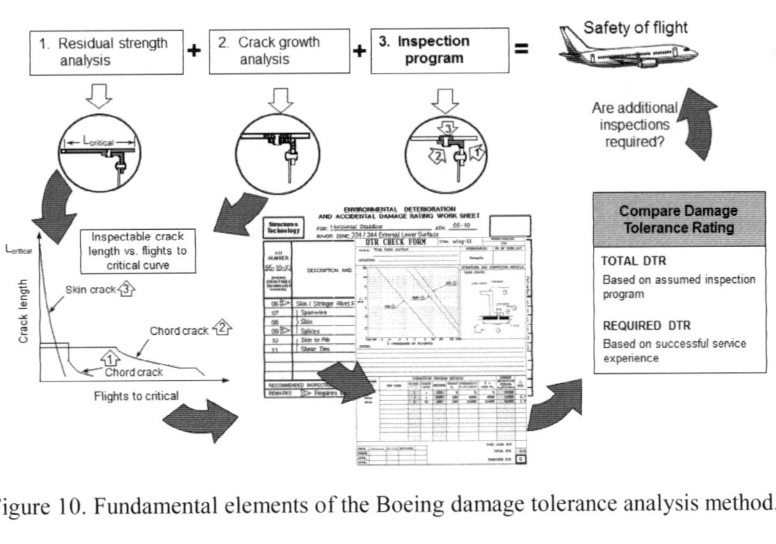

Figure 10. Fundamental elements of the Boeing damage tolerance analysis method.

All damage-tolerant structure requires an inspection plan where inspection methods, thresholds, and frequencies are defined consistent with the stress levels and crack growth propagation characteristics. Normal maintenance inspections should allow the damage (developing at single or multiple sites) to be found before the crack reduces the residual strength capability of the structure below regulatory requirements. If the normal (or baseline) maintenance inspection program is not sufficient to meet the required DTR, supplemental inspections and repeat intervals are established for the airframe structure by the airlines and approved by the regulatory agency. For this purpose, Boeing provides the DTR forms to the operators for each PSE. Operators use these DTR forms to develop their own repeat inspection interval for each PSE with the goal to achieve or exceed the required DTR established by the regulatory authorities. A DTR Check

Form allows operators the flexibility to customize their supplemental inspection programs. Also, Boeing establishes inspection thresholds by crack growth analysis, assuming there is an appropriate initial manufacturing flaw, or by fatigue analysis, depending on the compliance approach set by the applicable airworthiness standard. Traditionally, a single threshold for the supplemental inspections is used for each group of PSEs of each model.

Maintenance programs at Boeing are based on product testing, long fleet experience, and standard industry practices. All known forms of structural degradation are considered, providing the necessary inspections for damage tolerance, allowing for a timely detection of critical forms of damage including environmental (e.g., corrosion and stress corrosion) as well as accidental damage. Comparison with past successful practice is the primary means by which maintenance inspections or other procedures for accidental and environmental damage are substantiated. The process for establishing the baseline program takes into consideration an environmental deterioration rating (EDR) and accidental damage rating (ADR). The EDR/ADR rating systems are used to develop the initial structural inspection program for new Boeing airplanes.

Widespread fatigue damage

The Boeing 737 Aloha accident and a number of findings on other aircraft show that under certain conditions and in certain types of structure, multiple adjacent cracks can nucleate independently and eventually coalesce, resulting in faster crack propagation and a loss of residual strength that would not be accounted for in a classic damage tolerance analysis. This could in turn potentially invalidate a conventionally developed airframe inspection program. The limiting factors in the classic damage tolerant design philosophy are the initial assumptions concerning crack sizes and locations; the analyst needs to assume some form of dependent damage in the structure and in effect construct a potential failure scenario. With WFD, the analysis has to account for possible crack coalescence or other forms of interacting damage nucleating from multiple independent sources (Fig. 11).

Federal Aviation Regulation 14 CFR 25.571 was amended (25-96) in 1998 to include a requirement that sufficient full scale test evidence exists demonstrating that WFD will not occur within the design service life of the structure. In addition, Advisory Circular AC 91-56A was released at the time to provide guidance to airplane manufacturers and operators in the development of a WFD program that would preclude operation in the presence of WFD. The Boeing method for WFD evaluation, accepted by the FAA, makes full use of current industry practices and related FAA Advisory Circular material. The approach utilizes a comprehensive, multistage analysis procedure that combines statistics, fatigue, crack growth, residual strength, and test/service data to provide a simple analysis tool aimed at precluding the occurrence of WFD in the fleet.

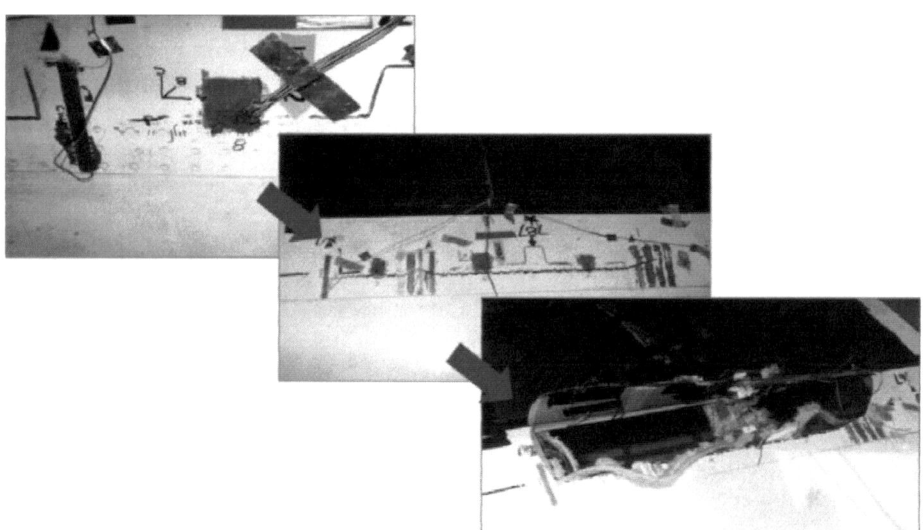

Figure 11. The classic WFD stages on a fuselage skin lap splice: Crack nucleation → Coalescence → Flapping.

In the Boeing process, the Inspection Start Point (ISP) is set at a level that will provide damage detection prior to a certain subset (typically 1%) of the airplanes in the fleet reaching a state where a particular amount of similar structural details are cracked on a number (typically the first) of similar component on the airplane. The Structural Modification Point (SMP) represents a conservative, lower bound estimate of possible WFD occurrence and is the point by which susceptible structure should be modified. The SMP is meant to provide an equivalent reliability level of a two-lifetime full-scale fatigue test, and provide enough opportunities to find damage on the lower bound airplane, typically the worst 5% of the airplanes in the fleet. Collectively, ISP and SMP define the beginning and end points of the damage detection period, sometimes referred to as the monitoring period for multi-site damage/multi-element damage (MSD/MED) detection (Fig. 12). Once the ISP and SMP are determined, the full inspection plan for the WFD program can be developed. The appropriate inspection methods and intervals are tailored to ensure timely crack detection with high confidence during the monitoring period, relying on the use of the Boeing DTR system. The Boeing WFD approach has been discussed previously in some depth [11], so no further elaboration is necessary here.

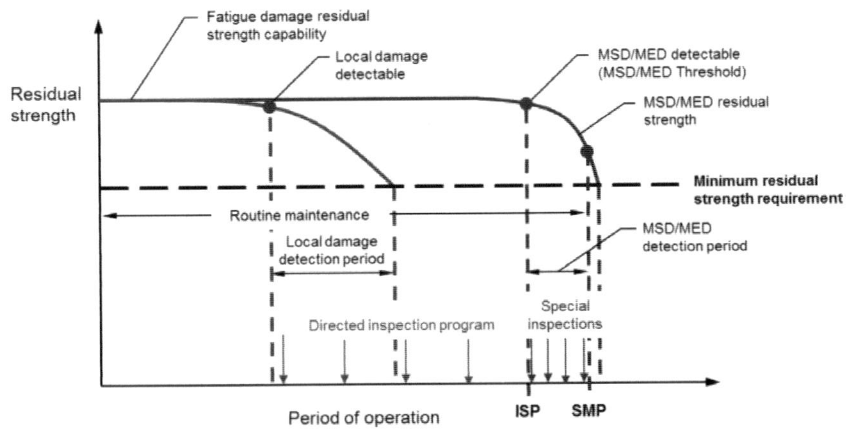

Figure 12. WFD typical implementation schematic.

Limit of validity

In a continuing effort to address aging aircraft issues involving WFD, the FAA issued a new rule, finalized in January, 2011 designed to protect most of today's commercial planes and those designed in the future from WFD as they age [12]. Since then, all airlines that operate airplanes under 14 CFR Parts 121 or 129 have been required to take action to revise their U.S. FAA-approved structural maintenance program to bring it into compliance with the new rule. This final rule amends FAA regulations pertaining to certification and operation of transport category airplanes to prevent WFD in those airplanes.

The limit of validity (LOV) is a point (measured in flight cycles or flight hours) in the structural life of an airplane beyond which there is significantly increased risk of uncertainties in structural performance and the probable development of WFD. It represents an operational limit based on engineering data that supports the maintenance program. Therefore, all identified service actions are required for operation up to LOV. Any LOV extension requires additional fatigue test or service evidence and validation of the maintenance program for efficacy against WFD. The FAA defines the LOV as an airplane-level number, not a limit applicable to any particular part or component. When an airplane reaches its LOV, it must be retired from 14 CFR 121/129 operation; however, serviceable parts and components may be transferred to other airplanes provided the operator has complied with all existing continuing airworthiness requirements. With LOV now added as an absolute operational life limit, safety by retirement is reintroduced as a fundamental element of the safety process.

For certain existing airplanes, the new rule requires design approval holders to evaluate their airplanes to establish a LOV of the engineering data that supports the structural maintenance program. For future airplanes, the rule requires all applicants for type certificates, after the effective date of the rule, to establish a LOV. With the new rule, design approval holders and applicants must demonstrate that the airplane will be free from WFD up to the LOV. The rule furthermore requires that operators of any affected airplane incorporate the LOV into the maintenance program for that airplane and prohibits airplane operation beyond its LOV unless an extended LOV is approved. These requirements can drive a variety of maintenance program outcomes, as Fig. 13 shows.

The rule has been incorporated into 14 CFR regulation amendments 25-132, 26-5, 121-351, and 129-48. Additionally, the FAA has issued AC 120-104, which offers guidance on compliance with the new rule. Guidelines are provided for (1) design approval holders on establishing a LOV and how to address maintenance actions that have been determined necessary to support it, (2) operators on how to incorporate the LOV into their maintenance programs, and (3) anyone considering a LOV extension.

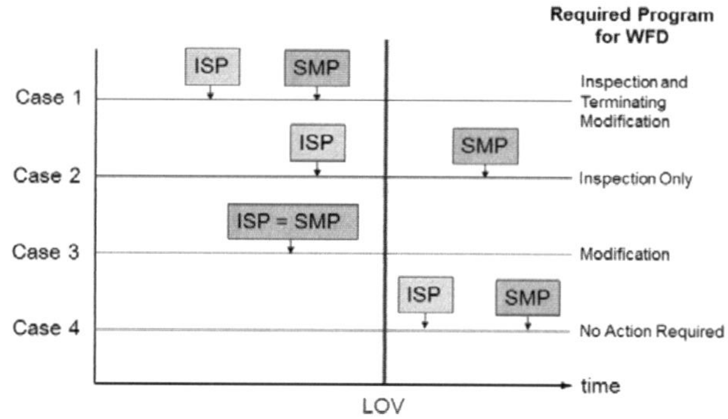

Figure 13. WFD Assessment outcomes as a function of LOV.

Nominally the LOV is assessed in terms of fatigue test evidence divided by a factor. FAA AC 120-104 describes the LOV as one-half of the fatigue test life representing the WFD average behavior. Boeing's approach to determine this value varies based on the amount of fatigue testing conducted for development of the final design requirements. Even though fatigue testing for certification purposes was not required for airplanes certified prior to 14 CFR 25 Amendment 96 (see Table III), in many cases such fatigue testing was actually accomplished. For those models certified prior to Amendment 96, fatigue test evidence is based not only on fatigue tests conducted at time of design, but also relies on post certification testing and in-service experience.

LOV rule implementation is taking place in stages that are a function of the airframe certification basis. The target compliance dates for Boeing are in 2015 for the 737 NG, 747-400, 757, 767, 777-200/300, MD-11, MD-90 and 717 models and in 2016 for the 777-200LR/-300ER/F, 747-8 and 787 models. The LOV values for the Boeing airplanes listed in Table IV (pre-Amendment 45, collectively known as Group 1) were submitted to the FAA in July 2012, and were approved by the FAA at various dates in August 2012. These LOVs have been published in Airworthiness Limitations (AWLs) and Certification Maintenance Requirements (CMRs) documents.

Table III. Part 25 Certification basis.

Model Definitions	
Pre Amendment 45 Airplanes	
Boeing Heritage Models	727, 737CL, 747CL, 747-400
Douglas Heritage Models	DC-8, DC-9, MD-80, DC-10/MD-10
Amendment 25-45 to 25-95 Airplanes	
Boeing Heritage Models	737 NG, 757, 767, 777-200/-300, 747-8
Douglas Heritage Models	MD-11, MD-90, 717
Amendment 25-96+ Airplanes	
Boeing Heritage Models	777-200LR/-300ER/F, 787

737CL: 737 Classic models are 737-100, -200, -300, -400, -500

737 NG: 737 Next Gen models are 737-600, -700, -700C, -800, -900, -900ER

747CL: 747 Classic models are 747-100, -100B, -100B SUD, -200B, -200C, -200F, -300, 747SP, 747SR

Table IV. Currently approved limits of validity for Boeing airplanes.

Airplane Model	FAA Approved LOV	
	Cycles	Hours
727-100 L/N 1-47	50000	50000
727-100/100C/200/200F L/N 48 and on	85000	95000
737-100/200/200C L/N 1-291	34000	34000
737-200/200C L/N 292-1585	75000	100000
737-300/400/500 L/N 1001-2565	75000	100000
737-300/400/500 L/N 2566-3132	85000	100000
747-100/100B/100B SUD/ 200B/200C/200F/300, 747SP, 747SR	35000	135000
747-400/400D/400F, 747-400BCF(SF), 747-400LCF	35000	165000
DC-8 ALL	56000	125000
DC-9 ALL	110000	110000
MD-80 ALL	110000	150000
DC-10 ALL	60000	160000

Damage tolerance of repairs and alterations

On December 12, 2007, the FAA issued 14 CFR 26 Subpart E (26.41–26.49) entitled "Aging Airplane Safety – Damage Tolerance Data for Repairs and Alterations." Under the provisions of 14 CFR 26, Subpart E, any repair, master change service bulletin, or Supplemental Type Certificate (STC) affecting fatigue-critical structure approved by Boeing after January 11, 2008 is required to have a damage tolerance evaluation. Operators are also required to have adopted the damage tolerance provisions provided in the approval documentation for these repairs and alterations. After December 20, 2010, it became the operators' responsibility to ensure that all new repairs to fatigue-critical structure receive a damage tolerance evaluation and are properly documented in the damage tolerance-based maintenance program.

The objective of the rule is to support operator compliance with the damage tolerance requirements of the Aging Airplane Safety Rule (AASR), with respect to repairs and alterations. The rule requires identification of Fatigue Critical Baseline Structure (FCBS) and Fatigue Critical Alteration Structure (FCAS) on all Boeing models. Fatigue-critical structure is a new FAA concept introduced with this rule, defined as airplane structure that is susceptible to fatigue cracking that could contribute to a catastrophic failure, as determined in accordance with 14 CFR 25.571. A damage tolerance evaluation should be performed on repairs or alterations to FCBS or FCAS to determine if supplemental inspections are required. Existing repairs and alterations made before December 20, 2010 that affect fatigue-critical structure and that do not have approvals indicating a damage tolerance evaluation require further action.

The new rule had a significant impact on structural maintenance programs for all Boeing 7-series, as well as heritage Douglas transport-category airplanes. The requirements created a need for new criteria and methods for damage tolerance analysis of new structural configurations. Boeing has since published new and updated material, including lists of FCBS and FCAS for all affected models, making these materials available to assist airlines operating these models in complying with the new AASR [13].

Boeing has also developed a "Low Stress Criterion" for structural repairs that define stress levels below which no damage tolerance analyses and inspections need to be conducted on repairs on FCBS [13]. Often, in certain fatigue prone areas, stresses are so low that the fatigue lives could be several multiples of the Design Service Objective (DSO). Figure 14 shows an example set of criteria taken from Ref. [14]. Boeing has been able to develop these criteria, with a focus on reducing the scope of the effort, without compromising the structural safety of the fleet. This has been of considerable benefit to the industry, enabling operators to focus their attention and resources on the more critical structures that are above the stress levels defined by this criterion.

Figure 15 summarizes the inspection and maintenance philosophy discussed in this section for ensuring safety of the fleet in the presence of fatigue, environmental deterioration and accident damage. Boeing works with customers and regulators to understand any safety issue, then takes appropriate action to mitigate the risk and communicate changes to the fleet. New findings are documented and evaluated across all Boeing airplanes. Continued airworthiness is therefore assured by: (i) development of a baseline maintenance program, (ii) continued support of the baseline maintenance program, (iii) development of a supplemental maintenance program, and (iv) timely addressing of in-service findings.

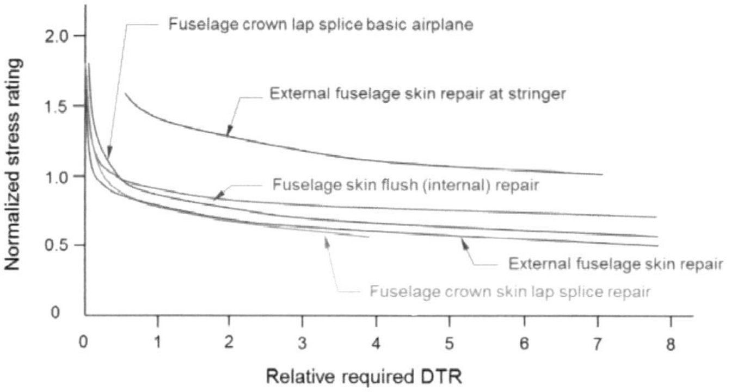

Figure 14. Normalized Stress Rating versus the multiple of required DTR for D/4C surveillance inspection for Pressure Critical Structure (reproduced from Ref. [14]).

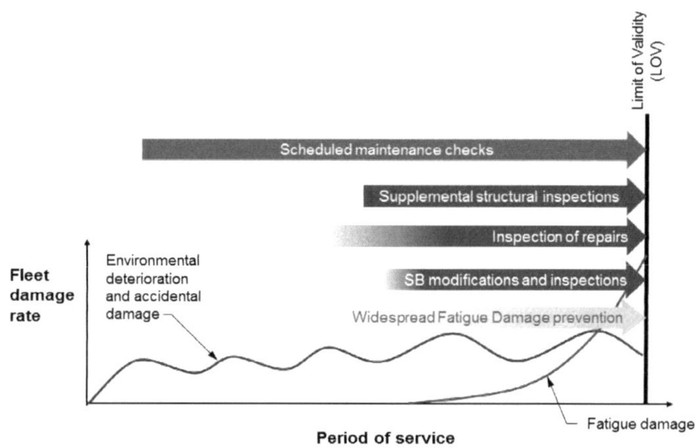

Figure 15. Strategies for ensuring fleet safety in the presence of fatigue, environmental deterioration, and accidental damage.

DURABILITY AND DAMAGE TOLERANCE METHODS VALIDATION

The accumulation of test and service data over past decades has repeatedly shown that the effectiveness of Boeing DaDT methods is reflected in the quality and safety of Boeing airplanes. In this section, some of these trends and findings are highlighted, drawing specifically upon full-scale fatigue test evidence and teardown inspections. For brevity, the Boeing aging fleet survey program (initiated in 1986) and service demonstrated reliability methods based on fleet utilization/sampling are not reviewed in this paper. The interested reader may explore these topics in Ref. 1.

Full-scale fatigue test evidence

A full-scale fatigue test exposes a structurally complete airframe to the typical operating loads experienced by an airplane model fleet. Full-scale fatigue testing has long been a major part of Boeing structural performance data development, for both new models and airplanes retired from service. Though not the only evidence, such testing is primarily used to confirm the DaDT characteristics of the primary airframe structure, supporting a verification of the DSO and the proposed inspection and maintenance program. Additionally, it provides full-scale test evidence demonstrating the effectiveness of the design to preclude the possibility of WFD occurring within the DSO of the airplane, as required by the FAA since 1998 and by other regulatory agencies worldwide. Any fatigue cracking found in a full-scale fatigue test is assessed by the airplane program to determine whether design or manufacturing changes are needed. Internal guidelines are in place at Boeing to help make this determination. Figure 16 summarizes the current fleet data relative to DSO and historical full-scale fatigue test data for all Boeing twin and single aisle-category airplanes.

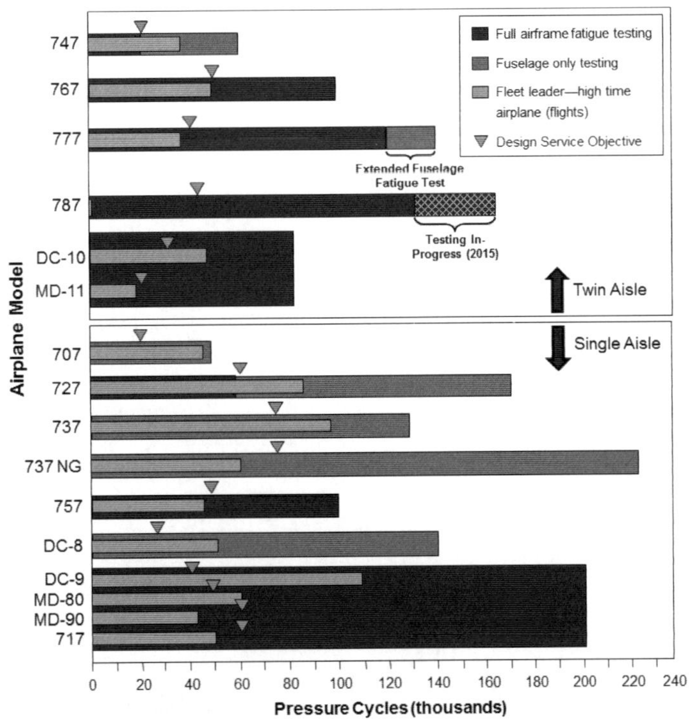

Figure 16. Fleet data relative to DSO and full-scale fatigue test for twin aisle and single aisle-category airplanes.

The effectiveness of the Boeing durability system, as introduced with Book 2 in the 1970's, was first evident from the 757 and 767 major airframe fatigue test findings. These airframes were fatigue tested to 2×DSO, with crack-findings leading to no more than a few dozen major design changes. This can be compared to the previous 747 major airframe fatigue test, which had roughly four times more design changes implemented after being tested to just one DSO. Later, full-scale fatigue testing of the 777 demonstrated further improvements relative to the 767 high-performance testing a decade prior with far fewer damage findings. This positive trend can be explained in part by the continued evolution, development, and consistent application of the Boeing DaDT technology standards.

The success thus far on the Boeing 787 full-scale fatigue test further demonstrates the effectiveness and continued maturation of Boeing durability methods. For the 787 airframe, which recently reached three DSOs, Boeing has set an objective to test to 165000 flight cycles (3.75 × Short Mission DSO) in an equivalent 5×5 fatigue spectrum. As of December 2014, 132000 equivalent flights (i.e., three DSO, equivalent to 60 years of service) have been completed. The 132000 flight inspections were recently performed with no substantial findings. Teardown of the 787 full-scale fatigue test airplane is scheduled be completed in January 2016.

The 787 test data compiled thus far has demonstrated quite clearly that the 787 is outperforming the 767 and 777 fatigue performance. Figure 17 shows a comparison of fatigue damage findings from the 767, 777 and 787 full-scale fatigue tests, up to two DSO. The relatively low number of 787 fatigue damage findings might be expected, to some extent, due to the fact that there is less metallic structure on the 787 airframe. However, one could argue that much of the critical, difficult-to-analyze primary structure remains metallic across these three airplane models. Although there can obviously be no contributions from crack findings in lap joints on the 787, many of the crack findings revealed in the 787 test have occurred in metallic structures within the fuselage.

Figures 18 and 19 show examples of two such structural findings, which are characteristic of the type of findings observed that have been observed beyond the first DSO. Another notable example, discovered early in the testing after 24000 test cycles, was the failure of pins on the bearing pads of the main landing gear trunnion upper housing. A service bulletin was subsequently released in June 2013, and the bearing pad design was changed.

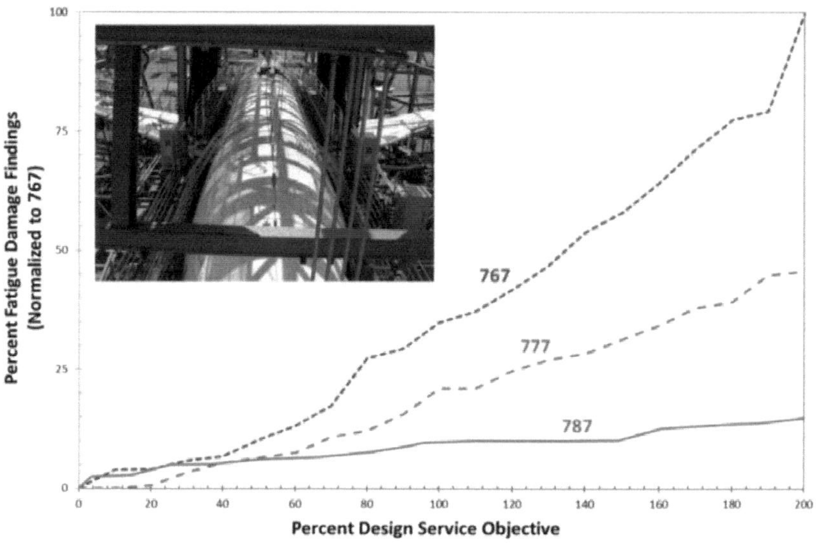

Figure 17.Comparison of findings from the 767, 777 and 787 major airframe fatigue tests.

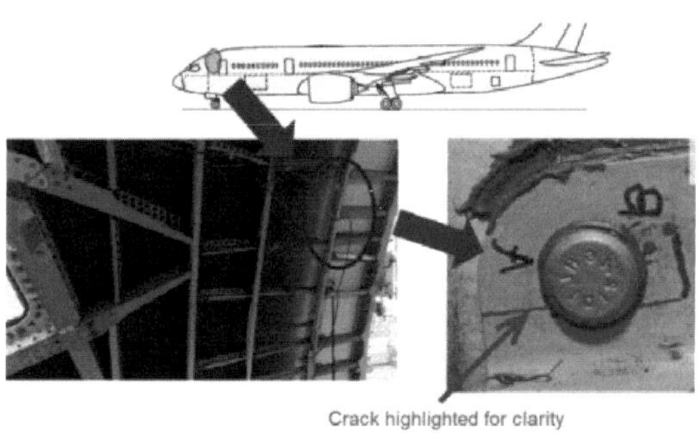

Figure 18. Example of a 787 full-scale fatigue test crack finding. Cracked fuselage door surround intercostal clips, found at the 72000-cycle inspection. In this case, the corrective action was to remove and replace the cracked parts.

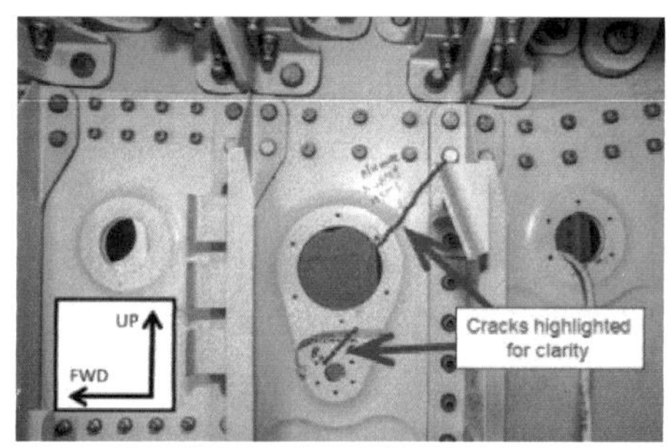

Figure 19. Example of a 787 full-scale fatigue test crack finding. Cracked satellite holes in the left and right (depicted here) side-of-body ribs immediately forward of the rear spar, found at the 132000-cycle inspection. This cracking was deemed acceptable for continuing cycling with increased surveillance.

Teardown inspections

Full-scale airframe testing is followed by extensive teardown inspections to locate any obvious problem areas. Since the introduction of the 707, Boeing has conducted several teardown inspections and evaluation of high-time airplanes. Major teardown efforts supplementing the 707, 727, 737 and 747 airplanes have been described elsewhere in much detail [15]. In this paper, we focus on more recent developments with the Boeing 777 teardown activities.

The 777 airframe was originally tested to 120000 cycles, or 3×DSO, demonstrating excellent durability performance. Later, an extension of the full-scale fatigue testing was conducted on the 777 fuselage structure only, reaching as high as 3.5×DSO [16]. The primary objective of this extended test was to obtain additional crack growth data in the fuselage to support the structural maintenance plans for future aging fleet programs. Another objective was to develop and validate analytical procedures for calculating parameters that characterize WFD. The test demonstrated outstanding fatigue performance with the successful test completion to 140000 cycles, simulating 70 years of service, well above regulatory requirements. The test showed excellent correlation of crack growth data with analysis, validating analysis methods, and verified the damage tolerance capability of the 777 using intentional damage (saw cuts) introduced earlier at 100000 cycles.

To assist in establishing an airplane LOV, the new rule recommends the Type Certificate holder conduct effective teardown inspections of fleet or fatigue test articles to detect and characterize any MSD or MED that could result in a WFD condition. At the conclusion of the 777 full-scale airframe fatigue test, limited (sampling) teardown inspections were conducted for several WFD-susceptible items for evaluation. More recently, extensive teardown inspections were conducted for several WFD-susceptible items including the aft pressure dome, circumferential body joints, chemically milled steps in skin panels, fuselage skin to stringer attachments, and the outboard wing lower surface. Fasteners were removed from the outboard wing lower surface for open-hole high-frequency eddy current (HFEC) inspections (Fig. 20). Fastener locations inspected included skin attachments to splice stringers, shear ties, and the rear spar. No detectable cracking was found among the nearly 1500 fastener locations inspected, demonstrating excellent fatigue performance and supporting the development of the 777 LOV.

Using the limited teardown inspections of the 777 aft pressure dome completed at the conclusion of testing, a preliminary WFD evaluation was conducted to conservatively estimate the required ISP and SMP based on the data from the cracked samples excised from the dome. This analysis demonstrated that the airplane would be free from WFD over its 20-year DSO. To better understand the WFD performance of the 777 pressure dome relative to the proposed 777 LOV, more extensive teardown inspections were recently conducted by removing nearly 2000 fasteners for open hole HFEC inspection. Based on the inspection results, several segments of the radial lap splices were excised and disassembled for laboratory evaluations of crack indications. Using these observations, the WFD behavior of the aft pressure dome lap splices has been re-evaluated to fully characterize the structure in order to meet the certification requirements and ensure safe operation up to the LOV.

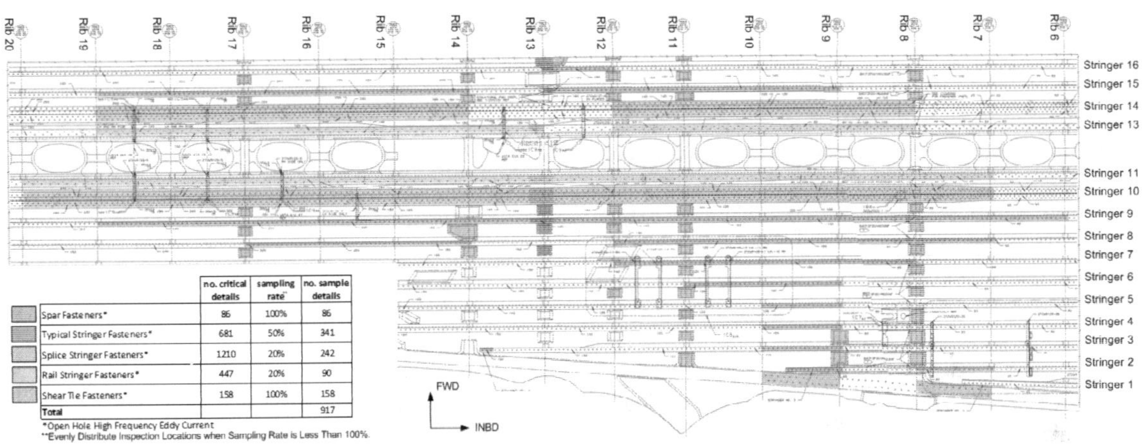

	no. critical details	sampling rate**	no. sample details
Spar Fasteners*	86	100%	86
Typical Stringer Fasteners*	681	50%	341
Splice Stringer Fasteners*	1210	20%	242
Rail Stringer Fasteners*	447	20%	90
Shear Tie Fasteners*	158	100%	158
Total			917

*Open Hole High Frequency Eddy Current
**Evenly Distribute Inspection Locations when Sampling Rate is Less Than 100%.

Figure 20. Boeing 777 Full-scale fatigue test teardown activities on the outboard wing lower surface.

In addition, several fuselage lap splice repairs installed during the 777 full-scale fatigue test were also recently excised in an effort to determine whether repairs are potentially susceptible to WFD. Over fourteen longitudinal bays of fuselage skin repairs accumulated between 40000 and 58000 cycles of fatigue testing. The majority of the repairs performed well, with no visible cracking observed during the test. Partial teardown showed that some small cracks had nucleated at a small percentage of the inspected fastener holes. The results were compared to Book 2 predictions, which matched quite well. The findings were also used to develop large (greater than five-bay) lap splice repairs.

Similar teardown efforts were conducted in other areas of the 777 as well to support the WFD and LOV assessments. Supplemental teardowns over and above the original teardowns were also recently conducted for the 737NG, 757, 767 and 777.

DURABILITY AND DAMAGE TOLERANCE OF COMPOSITES

After many years of limited application on commercial airplane structures, composite materials, especially carbon fiber-reinforced polymer composites (CFRP) are now being used in broad areas of the airframe. The Boeing 787 airplane is a prime example of this technological shift, with composites adding up to roughly one-half of the structure by weight (Fig. 21). Among the benefits composites realize are (1) the ability to more easily integrate large components in production, (2) lighter structure, and (3) greater resistance to fatigue and corrosion. For the 787, the expectation is that the latter attribute in particular will enable a doubling of the standard maintenance intervals, which for a heavy maintenance check will now become 12 years rather than the current 6 used on legacy metallic airplanes, translating into a substantial reduction in operating costs.

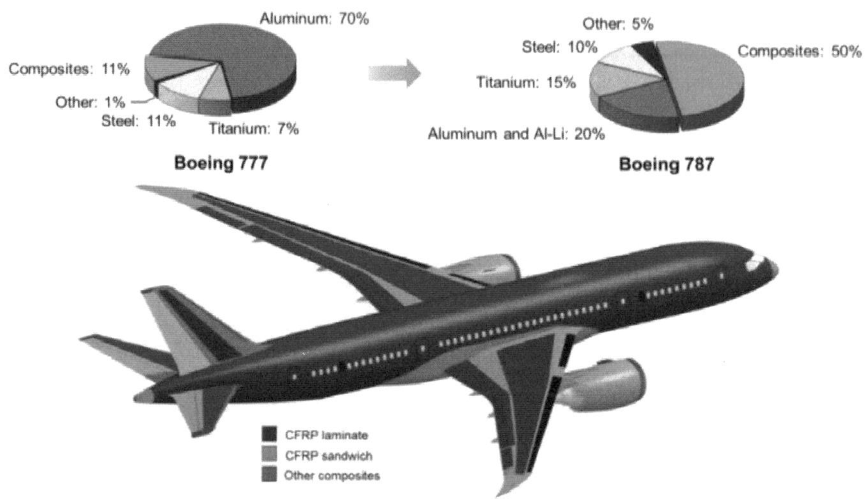

Figure 21. Boeing 787 Composite materials usage.

As Fig. 22 illustrates, expansion of composites technology has been made possible to a significant extent by the introduction of large-scale automated manufacturing and assembly processes [17], allowing—

- Construction of larger assemblies
- Significant reductions in tooling
- 30-40 percent shorter assembly flow
- Repeatable processes that create consistent first-time quality
- Sizable reductions in hazardous chemicals and waste

Figure 22. Large-scale composite structure manufacturing development for the Boeing 787 program. Numerous fuselage barrels were constructed at the Boeing Developmental Center in Seattle. The first composite fuselage section (right) was unveiled in January 2005.

Incorporation of composites into primary structure on Boeing airplanes to the extent that they are used today on the 787 has been the culmination of a process that began in earnest in the late 1970's and saw its first major production application on the 777 empennage. As part of the NASA Aircraft Energy Efficiency (ACEE) program, five 737-200 CFRP stabilizer shipsets were designed in 1978 and placed into service in 1984 (Fig. 23). After several years of service, two of the stabilizers underwent partial teardowns and a third stabilizer had a more thorough teardown inspection, which included mechanical property testing and nondestructive inspections. Inspections found little deterioration due to wear, fatigue, or environmental factors. Test specimens excised from the stabilizers had residual strengths comparable to values measured more than 20 years earlier [18, 19].

Figure 23. Boeing 737 composite horizontal stabilizer technology demonstrator.

Other prototyping and technology development efforts followed, including a full-scale fatigue test of a prototype composite horizontal stabilizer based on the Boeing 767 planform, and the Boeing ATCAS (Advanced Technology Composite Aircraft Structures) program, whose goal was to support development of the technology required for cost- and weight-efficient use of composite materials in transport fuselage structure. This latter work was initiated in 1989 under a NASA contract as part of the Advanced Composites Technology (ACT) initiative [20].

The next major milestone was the successful development of the 777 composite empennage (Fig. 24), which has been in production since the early 1990's. Certification of the empennage was supported by full-scale fatigue testing to three times the design life, following countless other mechanical tests ranging from small coupons to configured elements covering basic material properties, environmental effects, joints, durability, effects of defects, and repairs. In addition to satisfactory fatigue performance, the full-scale tests demonstrated greater than 150 percent of design limit load capability with barely visible impact damage (BVID) after fatigue loading as well as good correlation with analytical predictions [21].

Figure 24. Boeing 777-200 composite horizontal stabilizer and vertical fin full-scale fatigue tests (1993-94).

First flight of the Boeing 787-8 was on December 15, 2009, followed by certification in August 2011 and first delivery to customer All Nippon Airways (ANA) on Sept. 25, 2011. Today, progress continues with the 787-9 and 787-10, the newest members of the 787 family. The 787-9 took flight on Sept. 17, 2013, launching a comprehensive flight-test program leading to certification and first delivery to launch customer Air New Zealand in June 2014. The third and longest 787, the 787-10, achieved firm configuration in April 2014 and is on track for delivery in 2018. Test and service experience on the composite structure has thus far been very positive.

Durability and damage tolerance of composites

The damage tolerance philosophy and regulatory requirements discussed earlier are applicable to all primary structure for the most part without differentiation between metals and composites. However, composite materials have some unique features that affect how damage tolerance principles are implemented.

Relative to metals, composites offer significant benefits in terms of specific strength and stiffness, tailorability, and resistance to fatigue and corrosion. Composites, however, do not possess the ability to blunt large stress concentrations or absorb damage (e.g., impact) by plastic deformation as metals do, and when built into laminates, they contain inherently weak interlaminar planes that can facilitate damage progression. Secondary loads, which are almost impossible to eliminate from complex built-up structure, often produce out-of-plane loading, resulting in interlaminar stresses that can cause delaminations. The diversification of fatigue damage (e.g., fiber breakage, matrix cracking, matrix crazing, fiber buckling, fiber-matrix interface failure and delamination), damage mode interactions, non-uniform damage development, and the inelastic behavior of composites during cyclic loading make analytical modeling difficult. Furthermore, linear damage accumulation hypotheses, such as the Palmgren-Miner Rule, which is often used for fatigue (crack nucleation) evaluations in metals, may not be adequate or even relevant for composites.

In assessing the types of damage posing the greatest risk to the integrity of composite structure, foreign object impact is usually among the chief concerns. Electrical discharge (including lightning strike) and major discrete source damage events such as ground collisions, engine rotor burst, bird strike, and fire/thermal are also considered. Figure 25 shows how design load levels vary in general terms as a function of damage severity, based on the guidance in [22]. Figure 26 takes this concept further using residual strength, illustrating how the guidance is typically applied, starting with barely visible impact damage (BVID, Fig. 27), which requires a demonstrated ultimate design strength capability and no detrimental damage growth during the design service life of the airframe with an appropriate factor on load, and at the opposite end, discrete source damage (DSD), which requires continued safe flight and landing loads. In between, maximum design damage (MDD) is established relative to a limit load requirement, which is also known as visible impact damage (VID, Fig. 27). Note that in general a strength evaluation for the structure in a pristine condition is of relatively little value as a result of this approach. This also means that any composite structure strength evaluation is in effect a damage tolerance assessment, because the analysis must consider the potential presence of undetectable damage such as manufacturing flaws and BVID.

As with other forms of discrete source damage, composite structures must meet the same lightning strike regulatory requirements as their aluminium counterparts. At Boeing, composite structures are designed to withstand significant strikes with only superficial damage, considered to be approximately 80-90th percentile strike energy levels. This damage will be generally expected to be within the Allowable Damage Limit (ADL), and the airplane can as a result be dispatched with a deferred structural repair. In similar aluminum structure, the strike would likely puncture the panel, requiring an immediate structural repair and thus putting the airplane temporarily out of operation.

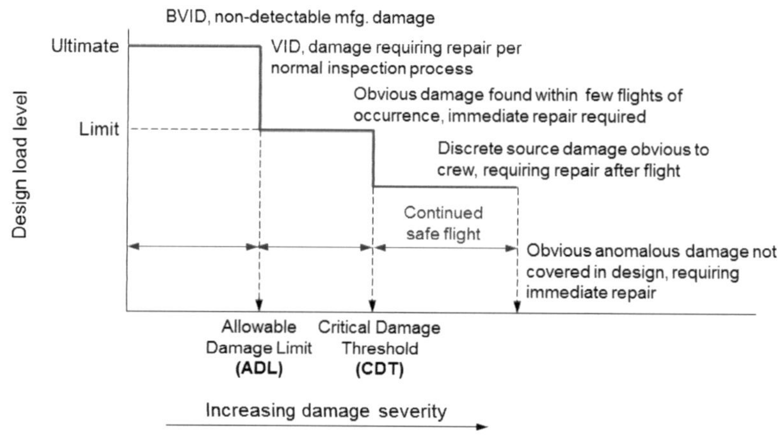

Figure 25. Design load dependence on damage severity (adapted from [22]).

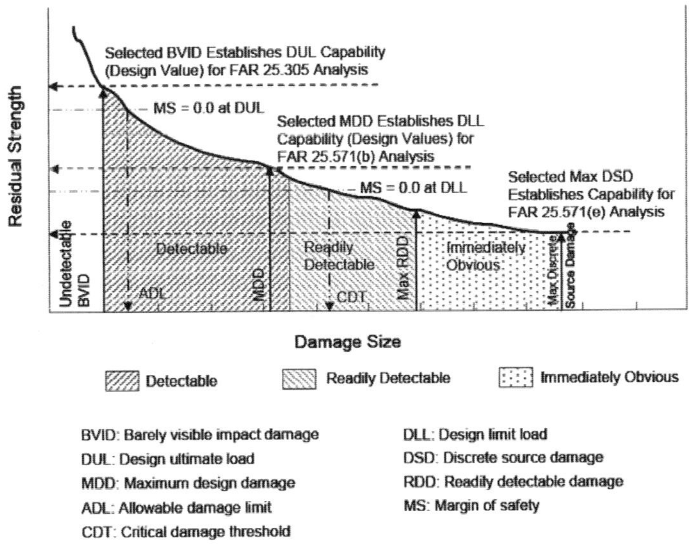

Figure 26. Design damage limits for composite structure.

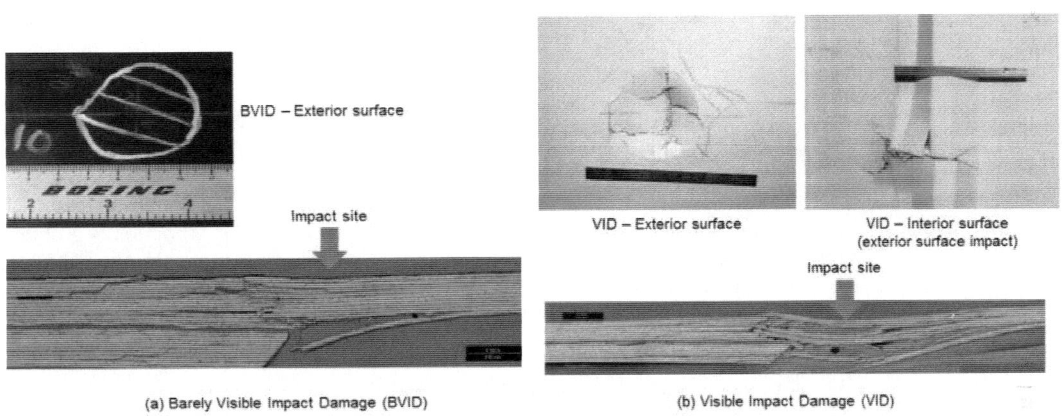

Figure 27. Typical damage: BVID vs. VID (from [23]).

Many of the small defects and impact damage composite structure can be sensitive to cannot be reliably detected by visual inspections in the field. There are effective NDI methods that may detect many of these types of defects or damage; however, the implementation has proven difficult in practice. Furthermore, damage progression analysis methods for composite structure are not mature enough today to enable development of a reliable inspection program for composite primary structure similarly to the way metallic structure is handled. For these reasons, evaluation for composite structure (according to [22]) used at Boeing since the NASA ACEE 737 program, and now including the 787, have thus far been based on a principle of no detrimental damage growth (NDDG). This approach assures no detrimental growth from undetectable defects and damage during the service life of the aircraft with a high reliability and confidence level as substantiated by fatigue test evidence.

Guidelines for demonstrating NDDG can be found in [22]. In keeping with that approach, structural details, elements, and subcomponents representing critical areas of the structure are tested in a manner that replicates anticipated service use and takes into account the effect of environment. Test articles feature damage representative of the types of conditions that may arise during fabrication, assembly, and in service consistent with the inspection techniques used. In these tests, residual strength is the primary test variable. The test goals are a function of whether the damage is categorized as undetectable (e.g., a small embedded manufacturing flaw, BVID) or detectable (from ADL to CDT). For the former, the main objective is usually to demonstrate a static ultimate load capability after cycling to a sufficient number of design lifetimes, taking into account reliability targets and application of an appropriate Load Enhancement Factor (LEF). For detectable damage, a successful demonstration of NDDG will show that following cyclic loading for at least two inspection intervals the damaged structure can sustain DLL with adequate reliability (refer to Fig. 26).

Although Boeing damage tolerance analysis methods cover a wide range of large damage types, large-notch analyses tend to be the most commonly performed evaluations. Experience gained in the design of the 787 shows that residual strength of fuselage structure with VID is enveloped by large-notch analysis. These analyses are usually based on parameters developed using various modeling strategies, including power-law (Mar-Lin [24]) fitting of test data on flat, featureless panels with center notches as the basic structural element. Damage nucleation from fatigue is not usually the norm for composite primary structure that is sized to meet the NDDG requirement. Instead, the environmental deterioration and accidental damage rating (EDR/ADR) system, which is also used for metallic structure, serves as a basis for a baseline structural inspection program for timely detection of environmental deterioration and accidental damage. Accordingly, there are generally no published fatigue thresholds for composite structures, and the baseline structural inspection program will carry on past the DSO and up to the airplane LOV.

Boeing has created companion technology standards for Books 2 and 3 specifically for DaDT of composites. Book 2C (Fatigue) and Book 3C (Damage Tolerance) for composites were released in 2014 and 2008, respectively. These documents are based primarily on experience gained in designing 777 empennage and 787 primary structures, with emphasis on methods and design values for CFRP solid laminates. Book 2C provides standard S-N curves for various structural details, treatment of thermal effects on fatigue, and statistical factors to account for the higher data scatter exhibited by composites relative to conventional metallic materials, including LEF determination criteria. Book 3C provides residual strength methods for composites, encompassing methods and allowables for structures with VID or large-scale damage.

Composite-metal structural interaction

While much attention in an airplane using composites to the extent that the Boeing 787 does is naturally focused on the composite structure, metallic elements cannot be ignored. Not only do they amount to a significant fraction of the structure (about 50 percent on the 787 –see Fig. 21), but they are also used in some of the most critical areas and experience some of the highest loads. One of the significant findings in hybrid structure is that when mated to metal parts, composites can create new conditions or exacerbate damage mechanisms in the metallic elements of the joint that can impact their durability (Table V). Other than perhaps corrosion and thermally induced stresses, this kind of adverse interaction is not always obvious or well understood [25]. The first two items in Table V have been the subject of significant work recently at Boeing and were deemed to be especially relevant to this paper, and are discussed below.

In mechanical joints, frictional load transfer has long been known to result in a lower rate of fatigue damage accumulation than when loads are predominantly transmitted by the fasteners in bearing [26]. Friction is promoted by joint clamp-up, which is driven mainly by fastener preload. When composite materials are present in the joint, some of the clamp-up can be lost due to the low transverse elastic properties of the material, thermal expansion differences, and creep. The latter can cause an irreversible loss in joint load transfer capability.

Table V. Composite effect on the durability of joint metallic elements.

Composite attribute/feature	*Possible impact to metal elements in hybrid joint*
Short-transverse creep	Long-term reduction in joint clamp-up
Reduced electrical conductivity	Arcing damage induced by high electrical currents
Low in-plane thermal expansion	Thermal stresses
Low in-plane strain capability	Limits on hole treatments and fastener installation methods
Electrochemically passive carbon fibers	Susceptibility to corrosion when in contact with anodic materials (e.g., aluminum)

Fig. 28 shows some of the results of an internal Boeing study that collected bolt preload measurements over time in all-metal and CFRP-metal joints assembled with instrumented bolts. Comparing the two plots in the Figure, it is evident that the addition of a composite part causes the preload to drop by at least 10 percent in the first few hours following bolt installation, and that the loss is non-recoverable, contrasting with the stable behavior of the all-aluminum joint. In the Fig. 28 curves, the fluctuations between room temperature (RT) and cold values over time are the result of the differences in thermal expansion between the bolt and the composite, the latter usually having a much higher short-transverse expansion rate than the bolt material. A similar effect has been observed in joints cycled at high temperature, except the cyclic preload now peaks at temperature (rather than at ambient sea-level temperature) and the long-term loss of preload is magnified, the latter attributed to further softening of the composite. In airplane design, the loss of clamp-up can be addressed by either taking a reduction in joint fatigue capability (determined experimentally) of the metal details, by re-application of torque, or offset by specifying more capable fasteners [25].

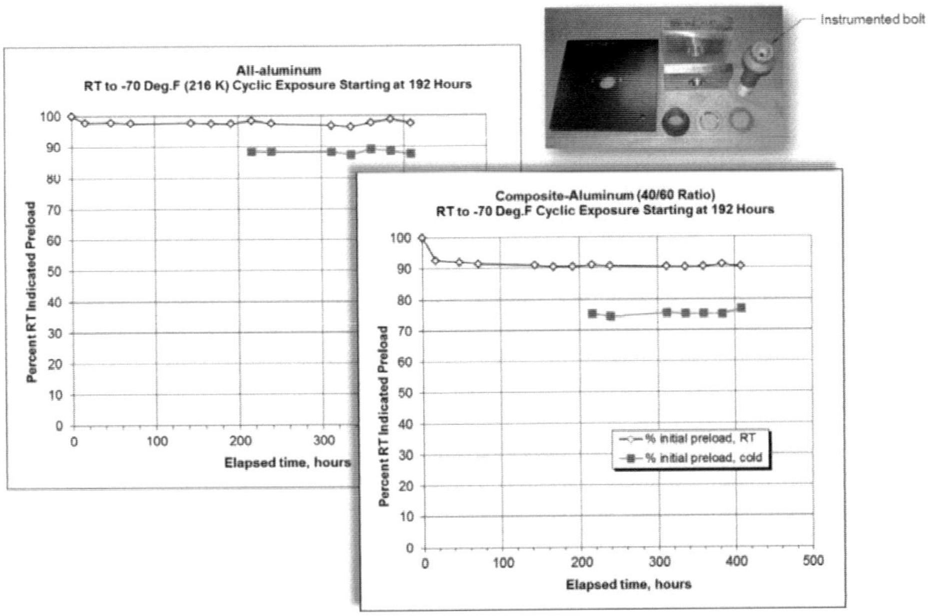

Figure 28. Half-inch (12.7 mm) nickel alloy 718 bolts in faying surface-sealed aluminum and hybrid joints aged at ambient temperature, then cycled cold [25].

Another item of particular relevance to this discussion is electrical discharge. Aircraft structures are exposed to a variety of electromagnetic effects threats: Lightning strike, high-intensity radiated fields/electromagnetic interference (HIRF/EMI), static discharge, and systems ground faults. With regard to lightning specifically, the design challenge is to provide current flow management to (1) protect passengers, crew, and systems, (2) prevent fuel ignition, and (3) minimize operator impact. New requirements, including some of the regulatory safety provisions in 14 CFR 25.981 concerning fuel ignition, make all joints and installations in fuel areas now potentially critical for lightning, and are driving significant changes to materials, finishes, and assembly processes. Additionally, greater composite usage results in a higher degree of sensitivity compared to conventional aluminum structure. Fatigue testing performed at Boeing on CFRP-metal joint specimens exposed to a range of electrical current levels has shown that under certain conditions, arcing can occur between the bolt and the hole as the current is transmitted across the joint, as evidenced by pitting and discoloration indicative of thermal damage in both the fastener and the hole on the metal part. These pits act as fatigue crack nucleation sites (Fig. 29, inset). The tests indicate that the severity of these conditions measured in terms of fatigue life degradation seems to depend on joint configuration, fastener finish (coated vs. bare), fastener fit, and current levels. An example is provided in the plot, illustrating the sensitivity to applied current of standard clearance-fit lockbolts installed in CFRP-aluminum joints.

New requirements and the potential for fatigue life reduction, along with the relatively common occurrence of lightning attachment on aircraft in service, make consideration of the phenomenon in the design process necessary. At joints, typical strategies involve reducing the current in the joint, minimizing the amount of energy discharged, and containing that energy. Energy discharge in particular appears to be best controlled with the help of fastener interference, usually by means of sleeve fasteners. For a given joint, the choice of fasteners and the level of shielding or containment required is usually a function of the location of the joint and the estimated magnitude of the threat.

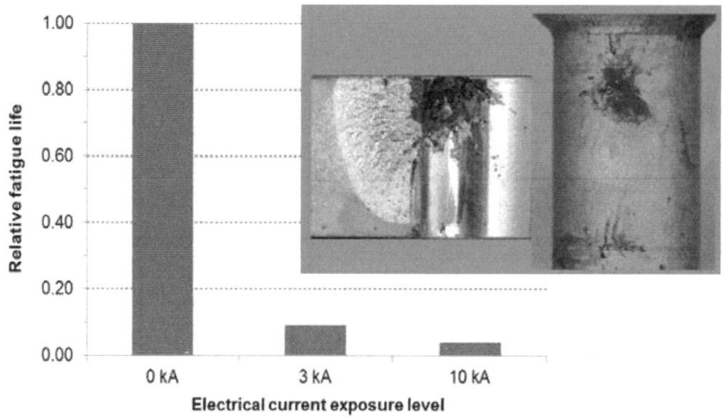

Figure 29. (Inset) Typical pitting and resulting fatigue crack resulting from a 10 kA per fastener step current through a hybrid CFRP-aluminum joint, standard titanium ¼-inch (6.4 mm) lockbolt installed in clearance and matching pin location. (Plot) CFRP-aluminum joint tests, standard titanium lockbolts installed in clearance [25].

Composite testing for the Boeing 787

Development of the 787 program has been made possible by significant advances in composite technology and data. For the 787, in addition to the above requirements, economic operation and maintenance of the airframe are being assured by application of the following criteria [23]:

- Allowable damage limits are based on damage that can be physically measured.
- Visual inspection techniques are the same as for current aluminum airplanes.
- Instrumented NDI will not be required for damage levels within published allowable damage limits.
- No new NDI techniques or equipment; planned inspections based on current Boeing 777 techniques and equipment modified to account for 787 structural configurations.
- Instrumented NDI may be required for damages that exceed published allowable damage limits.
- Methods validated by probability of detection studies and application to test articles.

Successful validation was accomplished thanks to a well-planned progression from laboratory-scale tests, configured elements, large panels (Fig. 30), subcomponents (Fig. 31), to full-scale fatigue testing (Fig. 32), culminating in a full-airframe fatigue test discussed earlier in this paper.

Repairs have also been the subject of a substantial amount of effort on this program. Numerous test articles ranging from coupons to components have featured repairs of the types planned for the Structural Repair Manuals, including bolted and bonded repairs. Tests have included static and fatigue with and without BVID, and including environmental tests.

Figure 30. Example of configured element testing of composites (In the photo: Residual strength testing of stiffened panel with BVID).

Figure 31. Subcomponent testing example: Co-bonded Boeing 787 composite "mini box" test article with VID.

Figure 32. Boeing 787 full-scale fatigue test examples: Left: Section 41 (forward fuselage). Right: Center wing and outboard wing stub box.

LOOKING AHEAD

Materials and basic part fabrication

Although much of the current focus in structural materials development is on composites, metals will continue to play a significant role in many areas of the airframe, and for the foreseeable future, will likely remain the dominant choice in short-haul, high-production rate airplanes. The industry's decades of familiarity with design, analysis, fabrication, and operation of metallic structures, and the considerable metals production and maintenance infrastructure are some of the reasons for that choice. In the meantime, metals technology has continued to evolve, taking advantage of new alloys, highly efficient basic fabrication techniques, and increased assembly process automation.

On the materials front, some of the best examples of this continued evolution are aluminum-lithium alloys. The so-called "third generation" aluminum-lithium alloys currently under development are characterized by reduced amounts of lithium (less than 2 percent by weight), but still provide 2 to 8 percent lower density than their conventional aluminum alloy counterparts [27], and slightly better or at least competitive mechanical properties (Figs. 33 and 34). Parts produced from 2098 and 2099 sheet, plate, and extrusion are already in production in areas of the Boeing 787 fuselage, and newer alloys such as 2060 are being traded against other metals and composites on other models.

Additional efficiencies can be gained by implementing advanced fabrication techniques. For metals, this is especially true for titanium, an impact that is compounded by the rising use of titanium alloys on new airplanes such as the 787. The high buy-to-fly ratios (the ratio of the weight of the raw material used for a component and the weight of the component itself) that are typical of machined titanium detailed parts (fittings, precision parts) are a measure of the significant recurring costs for these parts and high waste rates in the current fabrication process. Because of the intrinsic cost of titanium, even a moderate reduction of the buy-to-fly ratio can lead to significant cost savings. Welding, powder metallurgy and additive manufacturing are among the technologies being explored in pursuit of this goal (Fig. 35). Challenges from a DaDT perspective include development of design values in these uniquely configured, and often complex parts, non-destructive inspection, and characterization of the effects of highly process-dependent defects.

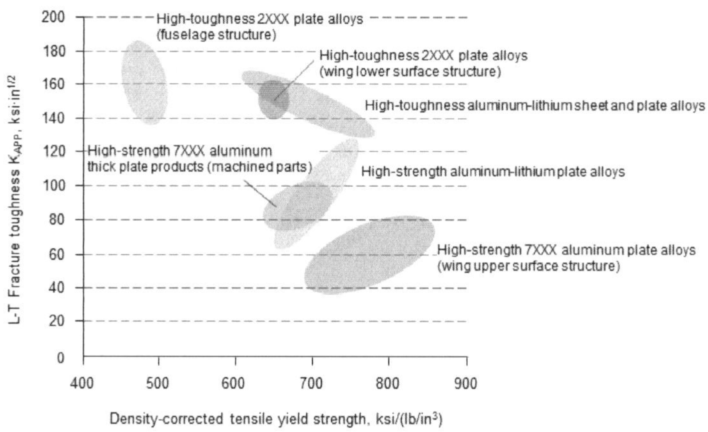

Figure 33. Fracture and strength properties of aerospace aluminum and third-generation aluminum-lithium alloys (Boeing data).

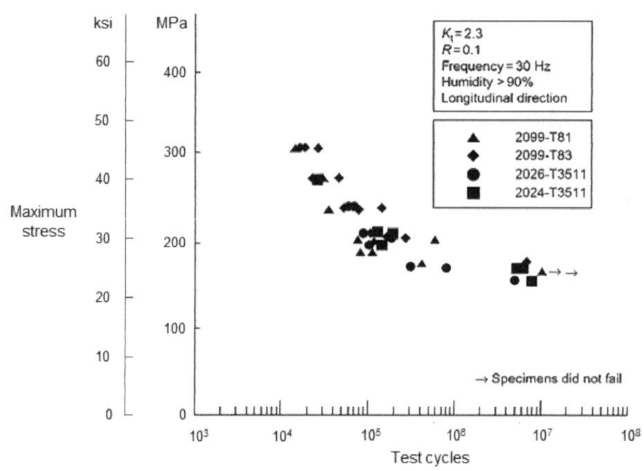

Figure 34. Typical constant-stress amplitude fatigue test data for notched aluminum (2024-T3511) and aluminum-lithium extrusions (reproduced from [27]).

Rough-machined titanium linear friction welded 787 floor beam fitting prototype

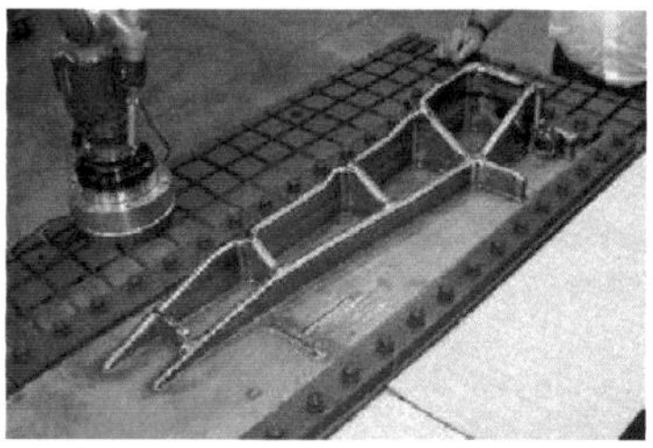

Titanium additive manufacturing 787 side frame demonstration article

Figure 35. Sample advanced titanium fabrication processes.

Lastly, the DaDT community is branching out to applications outside the traditional structures domain, into propulsion, systems, aircraft interiors, and non-traditional designs, where some of the biggest challenges are unique materials, designs, and construction methods (e.g., welding) that are distinct enough from primary structure that they require development of new or application-specific methods. Validation, and requirements that are often not as specific or as well controlled as in primary structure, add to the complexity of the task.

Assembly processes

With fuel costs having doubled in the past decade as a percent of total operating costs and the new generation airplanes being able to deliver much improved fuel consumption, lower emissions, and lower environmental noise, orders for new and replacement aircraft will continue to grow at an extraordinary pace. The result is that a commercial airplane worldwide fleet of just under 21000 airplanes at the end of 2013 is expected to increase to more than 42000 aircraft by the end of 2033, of which roughly 37 percent will likely be replacements and 50 percent will represent new growth [28]. To meet that level of demand, production rates will need to grow to unprecedented levels. For example, the Boeing 737 model will see the current 42 airplane-per-month rate climb to as many as 52 airplanes per month by 2018, including both the Next Generation and MAX variants (Fig. 36).

Figure 36. Production rate forecast through 2018 (left) for the Boeing 737 NG and MAX (right).

On the 737 airplane family, the most significant production changes are impacting the wing line. The existing wing panel assembly process, which rivets wing skins and stringers while the wing is held in a stationary horizontal position will soon be replaced with a new high-precision automated system termed (vertical) PAL (Panel Assembly Line). PAL will assemble the panels on a moving line at higher rates and in a vertical attitude, eliminating flow-intensive temporary fastening and overhead crane movement of parts (Fig. 37).

Qualification of the new PAL assembly system includes a comprehensive fatigue test program the goal of which is to show that for fatigue-critical areas of the wing, the new system meets or exceeds the fatigue performance of the existing processes. The laboratory-scale tests cover the range of fastener types and diameters and material stacks to be implemented on PAL for the 737 Next Generation and MAX models, using test specimens assembled on actual production equipment (Fig. 38).

Figure 37. The 737 PAL wing assembly system. Automating the process of assembling the 737 wing panels is expected to reduce flow time and improve safety and quality.

Figure 38. Fatigue test specimen being assembled in a production 737 PAL gantry system.

On the Boeing 777 program, the focus is on a radically new fuselage one-up assembly process, FAUB, or Fuselage Automated Upright Build (Fig. 39). With FAUB, aluminum major fuselage sections will be built using pre-programmed, guided robot pairs that will fasten the panels of the fuselage together, drilling and filling the more than approximately 60000 fasteners that are today installed by hand, all of these operations with only minimal fixed tooling. In addition to benefiting production rates, FAUB will improve workplace safety and increase product quality. Prototype parts have been built and fatigue qualification testing has begun. By the time the program is completed, hundreds of specimens representative of various joints including riveted and bolted lap splices will have been fatigue tested. Additionally, a 777 Freighter center-forward fuselage barrel will be subjected to the equivalent of three times the design pressure cycles. The 380-inch (9.65 m) long barrel, currently being built, will feature both the machine assembly process as well as manually assembled lap splices, to allow a direct comparison of the current and new assembly methods.

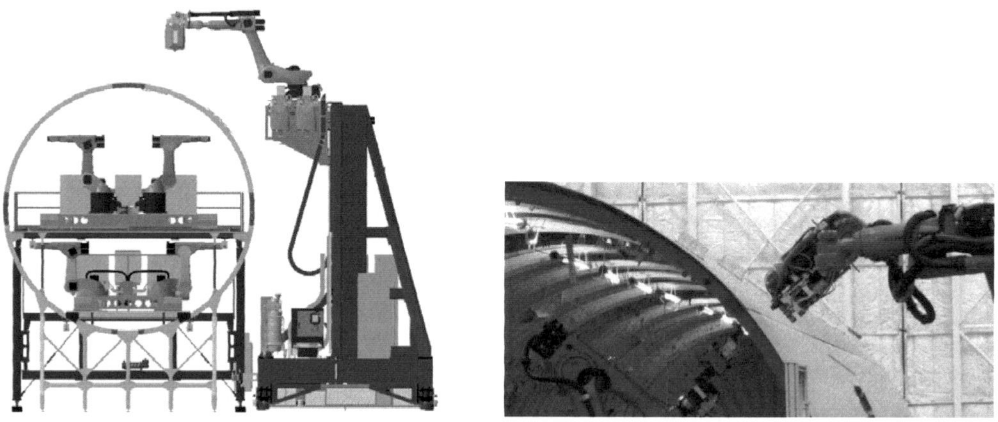

Figure 39. The Boeing 777 FAUB fuselage assembly process.

Analysis methods

The Boeing DaDT methods, as embodied in Book 1 through 3 (and their composite counterparts) have also undergone many changes over the years. These compendia are considered "living documents" that are periodically revised to adopt the latest improvements in analysis, loads, materials, and structural concepts, as well capturing new test and service data as it becomes available. As originally conceived, the methods were intended for a simpler, more conventional analytical treatment of the structure, loads, and stresses. As understanding of the operating loads environment and design data has expanded, methods have become more complex and a more efficient seamless integration of loads and stress data with the DaDT methods is becoming increasingly important.

One area that has been the object of considerable attention over the past decade has been the development of finite element analysis best practices in their application to both fatigue and damage tolerance assessments. Experience shows that the value and versatility of finite element analysis as an adjunct to DaDT evaluations can only be truly realized when the proper modeling strategy (both the type of analysis and physical model) is tailored to the specifics of the problem and evidence from either test or other previously accepted analyses can be used to validate the results. Figures

40 and 41 showcase elements of a particular initiative currently underway to improve the accuracy and consistency of fuselage lap splice fatigue life predictions, by developing and documenting a set of validated nonlinear finite element analysis standards and modeling techniques that can be easily used by analysts across airplane models as a complement to other existing methods [29]. Validation is being accomplished based on experimental data ranging from laboratory techniques such as optical digital image correlation (DIC) to airframe ground pressurization tests.

Similar efforts have been taking place on the damage tolerance side, the emphasis in most of these analyses being on development of stress intensity factors for complex assemblies. The example shown on Figure 42 corresponds to a Boeing 747 passenger door cutout analysis, the goal of which was to generate Book 3 crack growth factors for the cracking scenarios depicted in the Figure. The model in this case represented a multi-bay section around the upper corner of the door cutout. Structural details included in the model were the bear strap reinforcement, edge frame outer chord, and the fasteners connecting the skin, bear strap, and outer chord. The analysis was handled as geometrically nonlinear problem. Stress intensity factors were calculated from the results of the finite element analysis using the Virtual Crack Closure Technique (VCCT) [30]. Validation of the modeling strategy was accomplished by comparing strains drawn from an 'intact' version of the model with strain gage data taken from ground pressure test of an actual airplane.

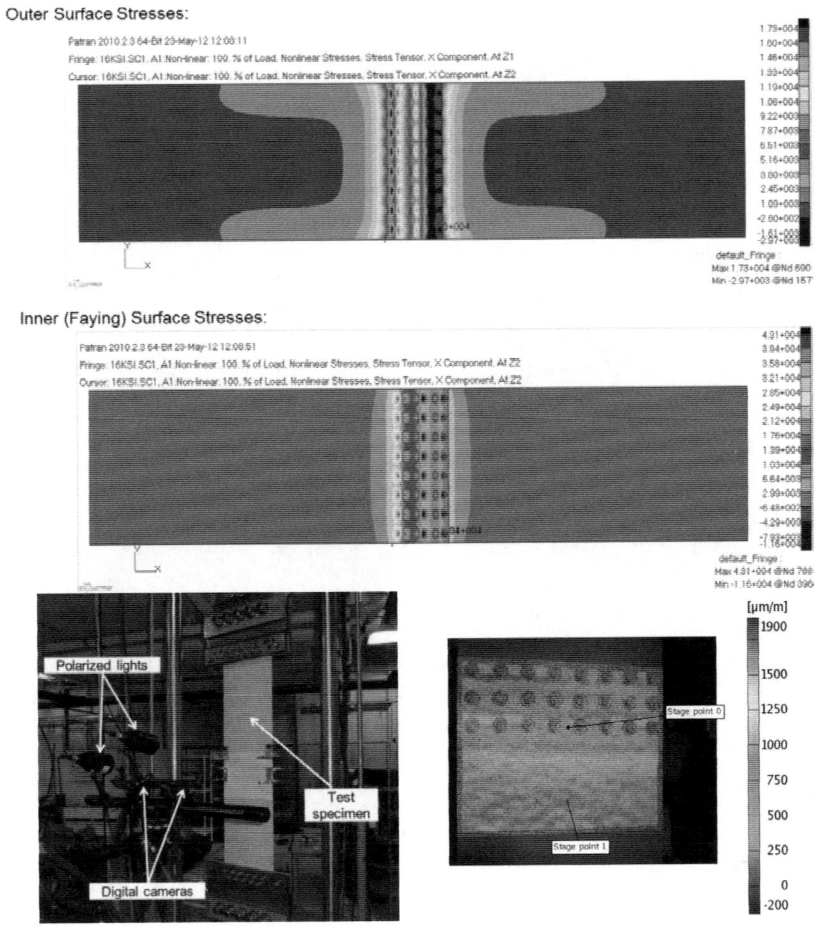

Figure 40. Fuselage lap splice analysis and laboratory-scale validation using optical DIC.

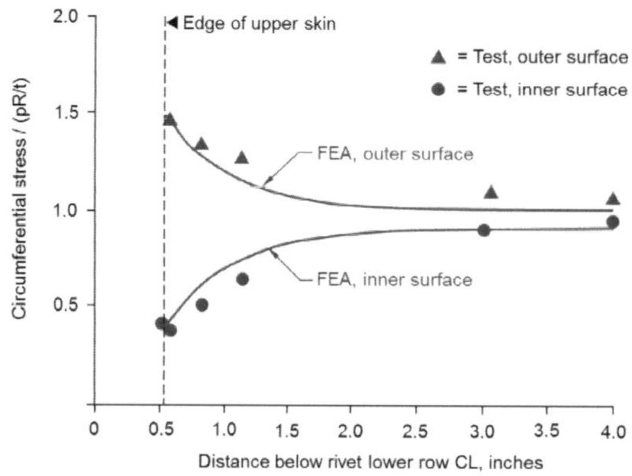

Figure 41. Finite element model correlation with ground pressurization test measurements on a Boeing 747 lower fuselage skin, below a longitudinal lap splice at mid-bay.

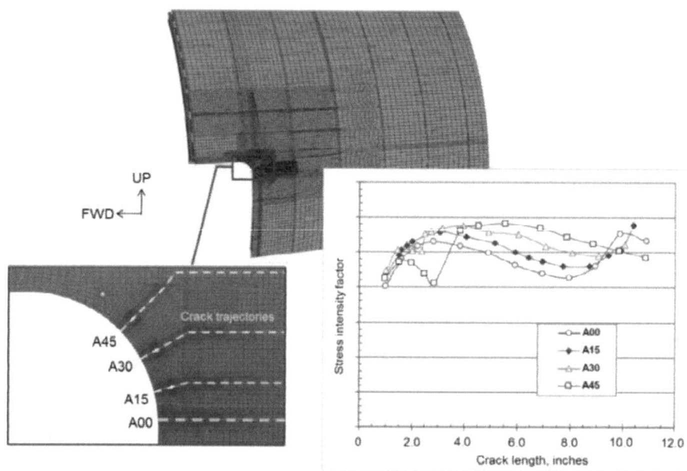

Figure 42. Boeing 747 Passenger door cutout crack growth analysis.

For composites, some of the greatest current challenges are in NDI, particularly in the application of the technology to large composite surfaces and in its ability to reliably detect degraded bondlines or interlaminar disbonds. There are NDI techniques that are capable of finding small delaminations, but they tend to also be time-consuming and require relatively high skill compared to other traditional inspection techniques. Boeing has been developing NDI methods that put energy into the structure transverse to the critical interfaces, scanning the structure as the energy pulse is applied. This technique has been shown to be capable of finding "weak bonds" or delaminations that are on the verge of damage nucleation and growth, but it is not yet ready for deployment in the field. The next issue is how to make use of that type of information once it becomes available in a damage progression or residual strength analysis. The analytical methods need to be quicker than existing VCCT-based damage propagation tools in order to support production-related rework and field repairs, and must be nimble and efficient enough to enable their use by non-experts.

Another area likely to receive some attention in the future is the evolution in composites from NDDG as an underpinning philosophy to a "predictable damage growth" approach. As composite usage is expanded, the desire to operate at higher strain levels in order to reduce weight will become an increasingly stronger motivation to steer away from the current application of AC20-107B toward the kind of safety-by-inspection approach that is commonplace with metallic structure. This will require methods and data capable of predicting the onset of delamination under combined cyclic loading and environment, which are currently very limited or unreliable.

The Human Connection

One challenge that is easy to overlook is the industry's ability to maintain skill continuity in this highly specialized (and sometimes subjective) technical discipline. Although U.S. Government projections indicate relatively low (less than 1 percent) annual employment growth rates in aerospace engineering over the next decade [31], the reality is that the industry is faced with significant turnover due to age demographics, globalization, shifting skill needs, and competition from other industries, at a time when commercial airplane demand is expected to surge. Boeing is projecting that about 50 percent of its top engineers and mechanics will be eligible to retire over roughly the next five years [32]. These are expected to become particularly acute issues in the fields of composite and stress engineering (including fatigue and damage tolerance), systems integration, and manufacturing/production engineering.

Boeing is actively engaged in efforts to maintain an experienced workforce through efforts in training and mentoring. In the DaDT field, training curricula have been developed and are being used across the company's design centers worldwide. Between 2006 and 2014, nearly 85000 hours of specialized DaDT training were imparted to engineers supporting Boeing commercial airplane projects. Graduate certificate classes in partnership with local universities such as the University of Washington are being regularly offered. A further creative illustration of this commitment was the acquisition by Boeing of a retired 737 Classic (-300 L/N 1231) fuselage with slightly over 73000 flights as a live training exhibit for structural engineers (see Figure 43). Since acquiring the retired fuselage in late 2012, over 1500 Boeing engineers have visited the hull. In addition, approximately 30 usage visits have been logged where the fuselage was used to either support a fleet investigation or testing of NDI equipment.

Figure 43. Retired Boeing 737 fuselage training aid.

CLOSING

The DaDT methods whose foundations were laid down at Boeing decades ago have continued to prove themselves through service and test experience. As the regulatory environment has evolved and technology has progressed, these methods have had to keep pace or expand into new areas. Composites are a prime exponent of the latter. One other important success story has to be how Boeing developed supplemental structural inspections and integrated them into operator-applied maintenance programs, essentially converting fail-safe certified airplanes into damage tolerant ones. The success of these programs is evidenced in the ever-improving fleet safety record of the commercial airplane fleet, the majority of which consists of Boeing airplanes. Boeing has a monitored fleet in which events are evaluated using a common damage tolerance methodology, and maintenance programs are adjusted as necessary, while the airplanes are operated out to their 30-year life expectations and beyond.

As a community devoted to the advancement of structural integrity since 1951, ICAF can rightfully take credit too for many similar developments across the industry, and for serving as a forum where regulators, industry, airplane operators, academia, and others have had the opportunity to interact and thus cooperatively advance the state of DaDT, with an obvious benefit to flight safety. The role of ICAF can moreover be considered as relevant today as it was nearly 65 years ago.

Many challenges remain, not all of which are purely technical –for example, how experience and past lessons learned are preserved and built upon. It has been pointed out that with the introduction with LOV of a set of definitive structural life limits, DaDT regulatory requirements have come full circle back to the safe life philosophy of 60 years ago, and though the reality is of course more complicated, the industry would not have been able to progress without learning from experience that spans these many years. Maintaining that continuity in the midst of a major engineering generational transition will be one of our greatest future challenges yet.

ACKNOWLEDGMENTS

This paper is the result of a collaborative effort involving numerous individuals, past and present Boeing employees who generously gave of their invaluable time and experience. The authors are especially indebted to Todd Accetturo, Girindra Das, Kevin Davis, Allen Fawcett, Ulf Goranson, Sean Harper, Matthew Miller, Gary Oakes, Shane Shaffner, Julie Smart, Alyssa Voightmann, and Zachary Whitman.

REFERENCES

[1] Goranson, U.G. (1993). *Damage Tolerance Facts and Fiction*, 14th Plantema Memorial Lecture, In: Proceedings of the 17th ICAF Symposium (Stockholm, Sweden), vol. I, pp. 3–105, Blom, A.F. (Ed.), Engineering Materials Advisory Services Ltd., Warley, UK.

[2] Craig, L.E. and Goranson, U.G. (1979). *Airworthiness Assessment of Boeing Jet Transport Structures*, In: Proceedings of the 10th ICAF Symposium (Brussels, Belgium), Maenhaut, A. (Ed.), Publ. Aeronautics Administration.

[3] Goranson, U.G. and Hall, J. (1980). *Airworthiness of Long-Life Jet Transport Structures*, Royal Aeronautical Society Spring Convention on Long-Life Aircraft Structures, London, England.

[4] Gunther, C.K. and Goranson, U.G. (1983). *Spectrum Loading Effects on Crack Growth in Major Components of Commercial Transport Aircraft*, In: Proc. International Conference on Application of Fracture Mechanics to Materials and Structures.

[5] Miller, M., Luthra, V.K. and Goranson, U.G. (1987), *Fatigue Crack Growth Characterization of Jet Transport Structures*, In: Proceedings of the 14th ICAF Symposium, Simpson, D.L. (Ed.), New Materials and Fatigue Resistant Design, Engineering Materials Advisory Services Ltd., Warley, UK.

[6] Published by BCA Aviation Safety (August 2014), www.boeing.com/news/techissues/pdf/statsum.pdf.

[7] Eastin, R.G. and Bristow, J.W. (2003). *Looking at Lusaka's Lessons*, Aircraft Structural Integrity Program (ASIP) Conference, San Antonio, Texas.

[8] Tiffany, C.F. (1977). *Durability and Damage Tolerance Assessments of United States Air Force Aircraft*, In: Proceedings of the 9th ICAF Symposium (Darmstadt, Germany), Buxbaum, O. and Schütz, D. (Ed.), Publ. LBF (TR 136), pp. 4.4/1-4.4/31.

[9] Hall, J. and Goranson, U.G., *Principles of Achieving Damage Tolerance with Flexible Maintenance Programs for New and Aging Aircraft*, 13th Congress of the International Council of the Aeronautical Sciences/AIAA Aircraft Systems Technology Conference, Seattle, Washington, August 1982.

[10] Goranson, U.G. and Rogers, J.T. (1983). *Elements of Damage Tolerance Verification*, presented at the 12th Symposium of ICAF (Toulouse, France), Labourdette, R. and Deviller, D. (Ed.) Publ. CEAT.

[11] Gruber, M.L., Johnson, B.E. and Whittaker, I.C. (2001), *Widespread Fatigue Damage Assessment of Jet Transport Structures*. In: Proceedings of the 21st ICAF Symposium (Toulouse, France), vol. I, pp. 99–115, Rouchon, J. (Ed.), Cépaduès Editions, Toulouse.

[12] Safarian, P. (2003). *Widespread Fatigue Damage Evaluation Method*, In: Proceedings of the 27th ICAF Symposium (Jerusalem, Israel), vol. I, Brot, A. (Ed.).

[13] Pillo, R.M. and Hoggard, A. *Complying with the Aging Airplane Safety Rule*, in Boeing Aero Magazine, QTR 02-2010, pg. 15 (2010).

[14] Das, G.K. and Miller, M. (2009). *Development of a Low Stress Criterion that Eliminates a Large Portion of the Aircraft from Damage Tolerance Based Maintenance Program for Structural Repairs Required by the New Part 26 Federal Regulation*, In: Proceedings of the 25th ICAF Symposium, Rouchon, J. (Ed.), Springer, Rotterdam, the Netherlands.

[15] Goranson, U.G. and Miller, M. (1989). *Aging Jet Transport Structural Evaluation Programs*, In: Proceedings of the 15th ICAF Symposium (Jerusalem, Israel), Berkovits, A. (Ed.), Engineering Materials Advisory Services Ltd., Warley, UK.

[16] Das, G.K., Lusk, J.W. and Stariha, M.S. (1993). *Extended Full-Scale Fatigue Testing and Tear-down of the 777 Fuselage*, In: Proceedings of the 20th ICAF Symposium (Seattle, U.S.A.), vol. I, pp. 603–625, Rudd, J.L. and Bader, R.M. (Ed.), Engineering Materials Advisory Services Ltd., Warley, UK.

[17] Stickler, P.B. (2004). *7E7 Composites Airframe Structures Development*, SAE 2004 Aerospace Manufacturing & Automated Fastening Conference, Sept. 20-23, 2004, St. Louis, MO.

[18] Aniversario, R.B., Harvey, S.T., McCarty, J.E., Parsons, J.T., Peterson, D.C., Pritchett, L.D., Wilson, D.R. and Wogulis, E.R. (1983). *Design, Ancillary Testing, Analysis, and Fabrication Data for the Advanced Composite Stabilizer for Boeing 737 Aircraft, Vol. I – Technical Summary*, NASA CR 3648.

[19] Boeing internal data (2004).

[20] Scholz, D.B., Dost, E.F., Flynn, B.W., Ilcewicz, L.B., Nelson, K.M., Sawicki, A.J., Walker, T.H. and Lakes, R.S. (1997). *Advanced Technology Composite Fuselage – Materials and Processes*, NASA CR 4731.

[21] Fawcett, A.J., Trostle, J. and Ward, S. (1997). *777 Empennage Certification Approach*, Proc. 11th International Conference on Composite Materials (ICCM-11), July 14-18, 1997, Gold Coast, Australia.

[22] FAA Advisory Circular AC 20-107B, Composite Aircraft Structure (Rev. 2010).

[23] Fawcett, A.J. and Oakes, G.D. (2006). *Boeing Transport Experience with Composite Damage Tolerance and Maintenance*, presented at the FAA/Wichita State University Composite Damage Tolerance and Maintenance Workshop, July 19-21, 2006, Chicago, IL.

[24] Mar, J.W. and Lin, K.Y. (Oct. 1977). *Fracture of Boron/Aluminum Composites with Discontinuities*, Journal of Composite Materials, vol. 11, pp. 405–421.

[25] Rufin, A.C. and Jochum, R.L. (2014). *Fatigue Analysis and Design of Composite-Metal Joints in Primary Aircraft Structures*, Proc. 11th International Fatigue Congress 2014, Melbourne, Australia.

[26] Heywood, R. B. (1962). *Designing Against Fatigue of Metals*, Reinhold, New York.

[27] Wanhill, R.J.H. and Bray, G.H. (2014). In: Aluminum-Lithium Alloys: Processing, Properties, and Applications, Prasad, N.E., Gokhale, A.A. and Wanhill, R.J.H. (Ed.), Butterworth-Heinemann.

[28] *Current Market Outlook* (2014), http://www.boeing.com/boeing/commercial/cmo.

[29] Das, G.K., Sovar, T. and Miller, M. (2001). *Durability Assessment of Fuselage Single Shear Lap Joint with Pads*, In: Proceedings of the 21st ICAF Symposium, Rouchon, J. (Ed.), Cépaduès Editions, Toulouse.

[30] Krueger, R. (2002). *The Virtual Crack Closure Technique: History, Approach, and Applications*, NASA CR-2002-211628.

[31] Aviation Workforce: Current and Future Availability of Aviation Engineering and Maintenance Professionals (2014). United States Government Accountability Office Report GAO-14-237. http://gao.gov/assets/670/661239.pdf.

[32] Nawaguna, E. (2014). U.S. aerospace, aviation industries see threats in aging workforce, Reuters. http://www.reuters.com/article/2014/03/13/us-usa-aerospace-workforce-idUSBREA2C2DR20140313.

F-35 Full Scale Durability Modeling and Test

Dale L. Ball[1,a *], Philip C. Gross[1,b], Robert J. Burt[1,c]

[1]Lockheed Martin Aeronautics Co., P.O. Box 748, Fort Worth, TX 76101, USA
[a]Dale.L.Ball@lmco.com, [b]Philip.C.Gross@lmco.com, [c]Bob.J.Burt@lmco.com

Abstract. The F-35 Joint Strike Fighter program includes three aircraft variants, one of which has been designed and built according to US Air Force requirements, and the other two of which have been designed and built according to US Navy requirements. For all three variants, a system design and development (SDD) configuration aircraft is being subjected to a full-scale durability (FSD) test. In each case, the complete airframe is being subjected to two lifetimes of 'severe' design spectrum loading, with maneuver, catapults/arrestments (carrier variant only) and buffet loads applied as separate, alternating 1000 flight hour blocks during the major test sequence. For the airframe tests, the buffet loads are applied quasi-statically; for the separate vertical tail component tests, they are applied dynamically. In addition, tests of doors and attachments (local tests) are conducted when the full airframe test is down for inspections (as required, for example, between the first and second lifetimes). In this paper, we describe the manner in which the airframe tests were designed, including fatigue spectrum development and test adequacy analyses. In addition, we provide a summary of the test findings to date, along with a description of the analytical simulation for a typical finding. The paper includes an analysis vs test correlation summary that provides an indication of the validity of the fatigue crack initiation (FCI) and fatigue crack growth (FCG) analysis methods used to design the aircraft.

Keywords: full-scale durability test, fatigue crack initiation, fatigue crack growth.

INTRODUCTION

The F-35 Joint Strike Fighter (JSF) is a stealthy, single engine combat aircraft that is currently under development as the next generation strike fighter. The F-35 fighter program is unique in aviation history as three different design variants are under development, with each having unique capabilities, yet with each being derived from a common airframe design. The F-35A variant is a Conventional Takeoff and Landing (CTOL) aircraft that will serve as a multi-role fighter replacement for the US Air Force F-16 Falcon. The F-35B variant is designed for Short Takeoff and Vertical Landing (STOVL) capability and will serve as a replacement for the well known AV-8B II Harrier "Jump Jet." The F-35C Carrier Variant (CV) is designed as the first ever stealthy Naval fighter capable of aircraft carrier operations at sea. See Fig. 1. The aircraft is being jointly developed by an international contractor team comprised of Lockheed Martin Corporation (prime), Northrop Grumman Corporation, British Aerospace Systems, and other partners. The F-35 Program is currently in the System Design and Development (SDD) Phase which began at contract award in October 2001.

Certification of the airworthiness and structural integrity of the F-35 is achieved through the demonstration by analysis and testing that a number of structural capability requirements have been met or exceeded. These structural requirements include static strength, durability, and damage tolerance, among others; they are derived (tailored) from JSSG-2006 [1], the Joint Services Specification Guide for Aircraft Structures. These specific requirements are delineated in the F-35 structural design criteria and they are the product of considerable negotiation between the aircraft manufacturer and the procuring agencies. While the path to certification for each of the F-35 variants includes a long and complex sequence of building block test programs designed to characterize materials, demonstrate structural concepts and validate design methods [2], in each case it culminates with full scale ground and flight tests which demonstrate the capability of the vehicle.

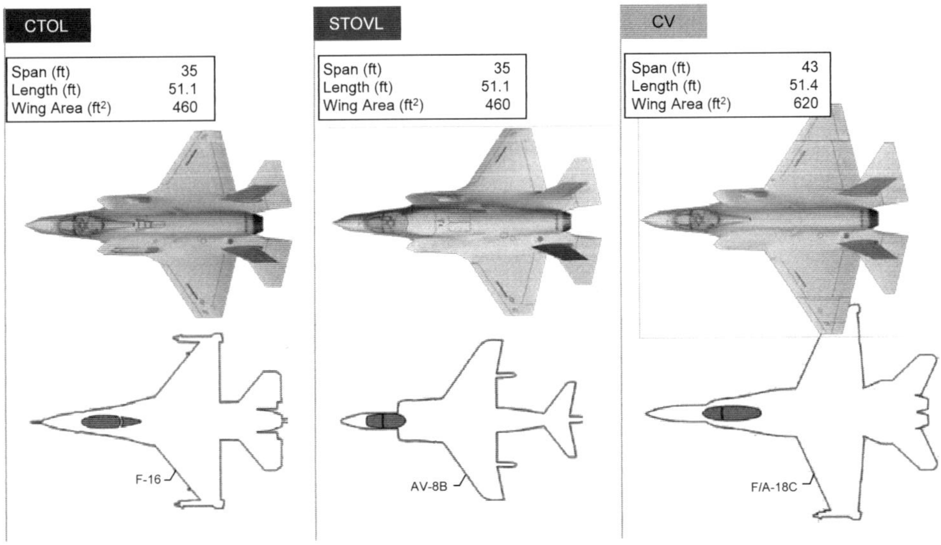

Figure 1. F-35 variant comparison.

DURABILITY AND DAMAGE TOLERANCE ANALYSIS

All three of the F-35 variants have been designed using proven durability and damage tolerance (DaDT) design principles [3]. For the CTOL aircraft, the durability requirement calls for two lifetimes of stable crack growth from an 0.01 inch flaw based on a severe spectrum. While the damage tolerance (DT) requirement calls for two lifetimes of stable crack growth from a DT initial flaw based on a mean spectrum (where the DT initial flaw size is structure and geometry dependent). Both of these requirements are imposed during design using linear elastic fracture mechanics based fatigue crack growth (FCG) analysis. For the STOVL and CV aircraft, the durability requirement states that it must take at least 2.67 lifetimes of severe spectrum loading to form an 0.01 inch flaw and the damage tolerance requirement calls for one lifetime of stable crack growth from an 0.01 inch flaw, based on severe spectrum loading. For these variants, the durability requirement is imposed during design using strain-based fatigue crack initiation (FCI) analysis, and the DT requirement is imposed using FCG analysis. For all three variants, the DT requirement also states that the structure must be able to carry limit load at all times during the growth from the initial to final crack sizes, i.e. it must retain residual strength throughout its design life. These requirements are summarized in [4].

On the F-35 program, an advanced software system for the conduct, archival, retrieval, interrogation and documentation of DaDT analyses was (and continues to be) used [5]. This system uses air vehicle level finite element models, along with design master event sequences for the generation of fatigue spectra. The system was used throughout SDD to impose DaDT requirements through the generation of FCI- and FCG-analysis-based design allowable stresses. It continues to be used for the analysis of both ground and flight test structures, and will be used throughout the life of the program for all aircraft structural integrity program (ASIP) related DaDT analyses.

Fatigue Crack Initiation Analysis. The FCI analysis tool used for F-35 durability analysis is based on the computer program LOOPIN which was originally developed for the U.S. Navy [6]. This algorithm is essentially an implementation of the standard hysteresis loop tracking, strain-based damage accumulation methodology developed by Dowling [7] and others. In this analysis, compound stress concentration factors, also known as stress severity factors (SSF), are computed for the structural detail in question. When multiple load types are applied simultaneously, superposition is used to calculate total SSF (in-phase variation of load types is assumed). The applied stresses for each cycle in the fatigue load spectrum are scaled by the SSF to determine the local or notch stress. Neuber's rule [8] is then used in conjunction with the material cyclic stress-strain curve to determine the cyclic stress-strain response at the notch or stress concentration. With the local response known, the mean stress and stress amplitude and the mean strain and strain amplitude are defined for each closed hysteresis loop. Finally, the equivalent, fully-reversed strain amplitude is calculated using one of several mean stress and strain corrections [9] and a damage increment is calculated. Damage increments are calculated for each cycle in the applied fatigue spectrum as the ratio of the number of applied cycles to the allowable number of cycles at that strain level. The total damage for the spectrum is found by summing the damage increments and the predicted life is found as the reciprocal of the total damage. As a result, the calculated crack initiation life is the number of cycles required for the accumulated damage value to reach 1.0.

Fatigue Crack Growth Analysis. F-35 FCG and residual strength analyses are based on the standard LEFM approach [10]. This analysis uses the crack tip stress intensity factor (SIF) as the single fracture parameter: subcritical crack growth rates are based on SIF range, and failure is based on the fracture toughness, the critical value of SIF. The FCG analysis tool will generate SIFs for the selected geometry, crack type, and loading. The SIF is written in terms of the reference stress and one or more geometric boundary/loading correction terms. As with the FCI analysis, when multiple load types are applied simultaneously, superposition is used to calculate a total SIF, and as in that case, all load types are assumed to cycle in phase with one another. In the crack growth analysis algorithm, for each cycle the SIF is calculated for the cycle minimum and the cycle maximum stresses, as well as for the maximum spectrum stress (MSS) and the residual strength stress (RSS). If the calculated SIF due to either RSS or MSS exceeds the fracture toughness of the material, then failure is indicated. If it does not, then the applied SIF range is found based on the applied cycle minimum and maximum values, and the effective SIF range is determined based on load interaction effects [11] (if any). The fatigue crack growth rate (FCGR) is found using the effective SIF range in one of several empirical crack growth rate equations or, as is typically the case, by 2D table look up in a FCGR data table. From the FCGR, the crack growth increment is found based on the number of applied cycles and the crack size is increased accordingly. This process is repeated on a cycle-by-cycle basis until failure is predicted or the crack reaches a geometric boundary. The calculated crack growth life is found as the number of cycles required to grow the crack from a specified initial size to the final size.

FCI and FCG Analysis Verification and Calibration. During the course of the F-35 building block test programs, the FCI analysis method implementation was repeatedly verified, and the model was calibrated (primarily via adjustment of the equivalent strain amplitude equation coefficients) so as to capture the mean spectrum crack initiation behavior observed in test. In the case of FCG analysis, both legacy and the F-35 building block coupon tests were used to verify and calibrate the model. Calibration was achieved via appropriate selection of load interaction parameters for F-35 materials and spectra [2]. As was the case for FCI, the FCG analysis algorithm was calibrated so as to capture the mean spectrum crack growth behavior observed in test. Once this process was complete, the equivalent strain amplitude equation coefficients and load interaction parameters were fixed at their calibrated, 'design' values, and the calibrated models were used throughout BTP to support structural sizing via the generation of FCI- and FCG-based design allowable stresses.

FULL SCALE DURABILITY TESTING

In order to conclude SDD, both ground and flight testing of full scale aircraft must be performed to demonstrate the structural integrity of the airframe design. The ground tests are laboratory tests in which fully instrumented aircraft are subjected to simulated ground handling and operational flight loads. The F-35 ground test program involves seven aircraft: a static and durability test aircraft for each variant and a third CV aircraft for simulated carrier landing or "drop" tests. All of the static tests, as well as the CV drop test, have been completed, while the durability tests are currently underway [12], [13], [14]. Each durability test is a fatigue spectrum loading test that simulates the mechanical loading aspects of the operational environment. This includes flight maneuver, buffet and both cockpit and fuel tank pressurization loading, as well as take-off, landing and ground loadings. In addition, for the CV variant, both catapult and arrest loadings are included.

Test Development. One of the greatest challenges in the ground test of a complete airframe is the simulation of actual flight external pressure (aerodynamic), inertial, and internal pressure loadings using a system of discrete loading rams or jacks. For the F-35 FSD tests, the loads are applied using a system of approximately 150 load jacks which are tied directly to structural hard points, for the direct introduction of load, or to whiffle-trees for the introduction of distributed loads. See Fig. 2. The fatigue spectrum for each FSD test was derived from the same load conditions and master event sequence (MES) used for design. The jack loads required to simulate each of the design fatigue conditions were defined and then applied to a vehicle level finite element model (FEM) of the test article. The results from each of these load cases were stored and then used in conjunction with the MES to produce stress spectra for a number of critical points throughout the test article. Detailed FCG analyses were conducted at the critical points, and exhaustive, comprehensive spectrum severity analyses were conducted throughout the airframe, in order to establish 1) the level of truncation required to reduce the number of load points in the test spectrum as required to meet schedule constraints, 2) the associated level of compensation (stress scaling) required to return the spectrum damage rates to their pre-truncation levels, and 3) the expected damage contributions due to maneuver and buffet loading, which had to be applied separately in the test. In order to accomplish the truncation objective, a multi-channel spectrum truncation capability was developed in which the stress vs. time histories for multiple reference control points were subjected simultaneously to a filtering algorithm which only removed turning points which satisfied the truncation criteria on all channels [15]. This process was repeated for a range of truncation levels until the target cycle count (based on the allowable test duration) had been achieved. The reference control point analyses were then re-run with the truncated spectra and a range of scale factors in order to determine the level of 'compensation' required to return each spectrum to its pre-truncation severity.

Figure 2. a) Conceptual view of F-35 full scale ground test article, actuators, and load pads, and b) photograph of CJ-1 test article and fixture.

These studies were conducted separately for maneuver spectra and buffet spectra due to the unique nature of each. The maneuver spectra represent the aerodynamic and inertial loading on the airframe during all ground and flight operations. They are designed to test the response of the complete airframe to a quasi-static loading environment and they are random in nature. The buffet spectra on the other hand, were developed to simulate the response of limited portions of the airframe (primarily outer wings and empennage) to the buffet environment. While buffeting is a dynamic phenomenon, the simulation of it in the F-35 FSD tests is being performed quasi-statically. The buffet spectra consist of blocks cycles whose amplitudes and cycle counts are derived from the response stress distributions (psd) for each buffet affected region of the aircraft and for one or more dominant buffeting modes. Buffet spectra generally consist of blocks of constant amplitude cycles with lower amplitude, so they can be applied at higher frequencies than maneuver spectra. Since it was not practical to run the maneuver and buffet tests with the same jack configurations and loading frequencies, the entire FSD test spectrum development process was performed separately for each [16].

The FSD test spectrum generation process was greatly facilitated through the use of the integrated DaDT tool suite described in [5]. This included everything from the multi-channel, targeted spectrum truncation at reference control points with representative geometric, loading and material parameters, to broad surveys of thousands of generic control points throughout the airframe used to compare the test spectra to the design spectra. Extensive test adequacy analyses were conducted to ensure that damage rates at the critical regions in the test article, based on the fatigue test spectrum, were comparable to those in the SDD configuration aircraft with design usage applied. In general, it was not possible to achieve a damage ratio of 1 throughout the entire airframe. This is an important consideration during the assessment of test anomalies: the significance of an anomaly may be mitigated (at least in part) if it occurs in a region of over-test.

It is important to note that all of the truncation studies performed during the spectrum generation process were analytical in nature. In order to gain confidence that the truncation and compensation levels prescribed by this process would indeed result in fatigue lives which were comparable to the un-truncated case, an extensive, coupon-level truncation verification test (TVT) program was conducted [17]. The TVT program included up to five spectrum types for each of the three variants, as well as three material types, and it addressed both FCI and FCG. The test program demonstrated that the calculated relative fatigue lives for un-truncated vs. truncated and compensated spectra, were accurate to within the normal scatter bands (factor of 2) expected for FCI and FCG analyses.

Initial Test Findings. According to the current structural design criteria, each F-35 FSD test must successfully complete two lifetimes fatigue cycling. As each test is conducted, displacement and strain measurements are taken to monitor load paths and detect damage within the aircraft structure. In addition, detailed inspections are conducted at regular intervals to identify and monitor damage initiation and growth. As of this writing, each of the three variants has completed one lifetime of testing (1LOT) and has started its second lifetime. As expected, there have been a number of findings on each test. This is consistent with legacy experience; in fact the rate of findings to date has been at or slightly below that for the comparable legacy fighter aircraft FSD tests [18]. This demonstrates the utility of the full scale ground test program: the purpose of these tests is to identify points in the structure that do not meet the durability requirements, and to demonstrate that the overall vehicle design is sound.

In most cases, the anomalies have been due to cracking at a geometric details or stress concentrations that were either not analyzed, or were incorrectly analyzed during SDD. These appear to be local, non-systemic 'escapes' in an otherwise sound design. None of the root cause analyses performed to date have indicated that any fundamental or widespread issues exist with any of the three variants. Examination of the test findings also reveals that the majority of the anomalies were the result of crack formation at surface stress concentrations (as opposed to cracking at fastener holes). This is likely a direct result of the increased use of unitized structure on the F-35 and points to the need for increased attention to the kinds of details that are formed (fillet radii, steps, compound Kts, etc.) when this design approach is used.

An example of one of the test findings which was fairly typical, is shown in Fig. 3. In this case a crack formed and grew along the radius between a vertical stiffener and a bulkhead cap. Investigation of this anomaly showed that there were three major contributors to the early cracking: 1) stress peaking at the cap to gusset radius due to curved beam bending effects, 2) stress peaking at the forward edge of the flange due to a significant forward-aft ('kick') loading introduced by a longeron at this location, and 3) significantly increased fatigue spectrum content, again due to the longeron kick load.

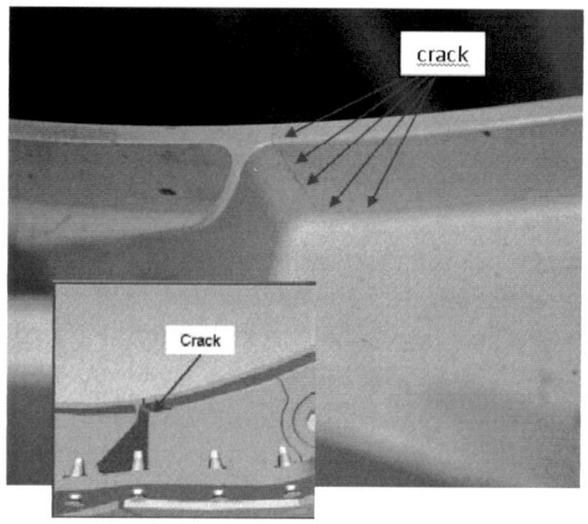

Figure 3. BH-1 FS472 bulkhead crack at 6750 EFH, views looking aft and up.

TEST CORRELATION ANALYSES

At the time of every major test finding, a root cause analysis (RCA) is performed in order to identify the cause(s) of the anomaly, as well as to ensure that the problem is sufficiently well understood to allow valid corrective actions to be taken. This includes both repairs to the test article and modifications and/or redesigns to production aircraft if required. A critical part of each RCA is an analytical simulation of the observed crack formation and growth behavior. The test findings and the simulations thereof provide the opportunity to evaluate every aspect of the F-35 DaDT approach. This includes everything from the characterization of fatigue related mechanical properties for affected materials, to the generation of fine grid finite element models for the definition of control point stress distributions, to the development of control point stress spectra based on one or more finite element component stresses from the vehicle level FEM of the test article. To a lesser extent, it also includes evaluation of the FCI and FCG analysis algorithms themselves, though most of the development/validation activity for the analysis tools was, by necessity, done during the building block test program when the various critical parts of each model could be studied in isolation and at the coupon level. It is very important to make the distinction between FCI and FCG analyses that are performed in order to demonstrate compliance with design requirements, and those that are performed in order to simulate a known cracking event. In the former, the FCI life or FCG life alone must meet or exceed the design life requirement. In the latter, since the durability tests involve the natural formation of cracks, the correlation analysis is based on the sum of the calculated FCI and FCG lives.

For each major test finding, both FCI and FCG life calculations have been made and the total compared against the observed data. An example of the simulation of the observed fatigue behavior for the BH-1 FS472 lower arch at BL0 is shown in Fig. 4. For the majority of the major findings that have occurred to date, with the application of any corrections/revisions identified during the respective RCAs, the computed total lives (FCI+FCG) have been in good agreement with the test findings. To date, there has been no use of 'test correlation factors' which are sometimes required to achieve agreement between analysis and test. A summary of the correlation of computed lives vs. test lives for all of the major findings on the three FSD tests is shown in Fig. 5.

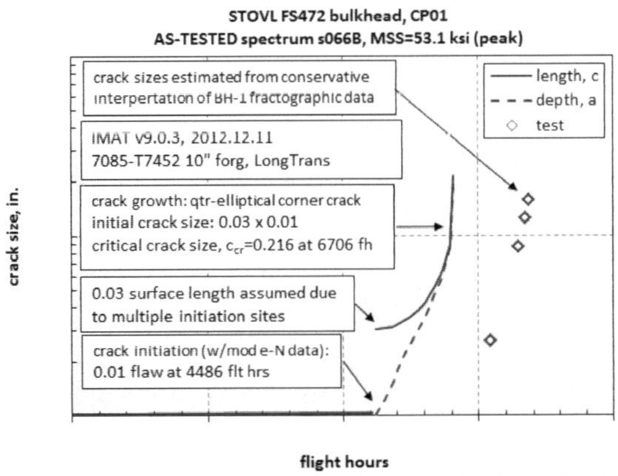

Figure 4. Example of simulation of test observed fatigue behavior using FCI and FCG analysis.

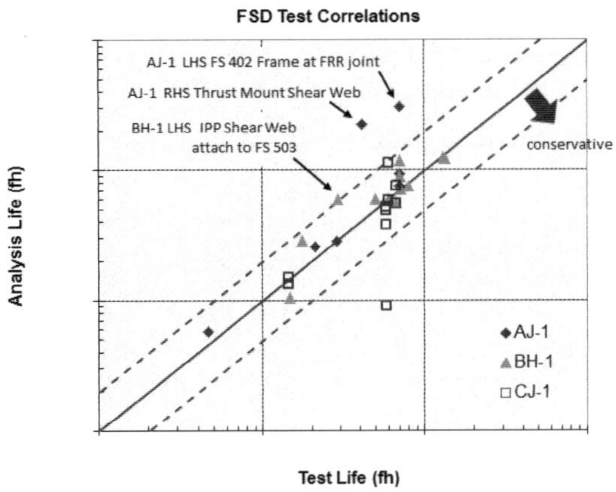

Figure 5. Summary of analysis vs. test correlation for F-35 FSD test findings.

SUMMARY

The F-35 Joint Strike Fighter is being designed and built in accordance with the durability and damage tolerance design philosophy prescribed in MIL-STD-1530C and JSSG-2006. With the support of a building block test program, the analytical tools required to support the design and manufacture of the airframe were developed and calibrated. These tools were integrated into a robust software system that permitted very comprehensive analyses of all three aircraft variants and their full-scale test articles. This system was used to develop the fatigue spectra used for each of the full scale durability tests, which are currently underway. Findings from the first lifetime of testing for each of the three variants have identified a number of structural issues which have been corrected, both on the respective test articles and for production aircraft. The rate of test findings to date has been consistent with that from comparable legacy programs. In short, F-35 experience has shown once again that the DaDT design process coupled with full scale durability testing is essential and effective for the production of aircraft which are both safe and economical to operate.

REFERENCES

[1] Joint Service Specification Guide, Aircraft Structures, JSSG-2006.

[2] Ball, D.L., Norwood, D.S. and TerMaath, S.C., "Joint Strike Fighter Airframe Durability and Damage Tolerance Certification," 47th AIAA/ASME/ASCE/AHS/ASC Structures, Structural Dynamics, and Materials Conference, May 2006.

[3] MIL-STD-1530C (USAF), Aircraft Structural Integrity Program (ASIP), Department of Defense Standard Practice, 1 November, 2005.

[4] Ball, D.L., "Examination of Durability and Damage Tolerance Design Criteria," USAF Aircraft Structural Integrity Program Conference, Nov. 2012.

[5] Limer, T.S., Ishee, D.M., Hamilton, J.L., Black, H.W., Ball, D.L., "An Advanced Software System for the Comprehensive Implementation of Durability and Damage Tolerance Requirements in the Design of New Aircraft Structure", Fatigue 2006 - The 9th International Fatigue Congress, Atlanta, GA, USA, 16 May, 2006.

[6] Porter, P. G. and Liu, A. F., A Rapid Method to Predict Fatigue Crack Initiation, NADC-8010-60, Naval Air Development Center, Warminster, PA, February 1983.

[7] Morrow, J., Martin, J.F., and Dowling, N.E., Local Stress-Strain Approach to Cumulative Fatigue Damage Analysis – Final Report, T.&A.M. Report No. 379, 1974.

[8] Neuber, H., "Theory of Stress Concentration for Shear Strain Prismatical Bodies with Arbitrary Nonlinear Stress-Strain Law", Journal of Applied Mechanics, Trans., ASME, Series E, December 1961.

[9] Nihei, M., Heuler, P., Boller, C. and Seeger, T., "Evaluation of Mean Stress Effect on Fatigue Life by Use of Damage Parameters," Intl. J. Fatigue, Vol. 8, 1986, p. 119.

[10] Gallagher, J.P., Giessler, F.J., Berens, A.P. and Engle, R.M. Jr., USAF Damage Tolerant Design Handbook, AFWAL-TR-82-3073, Air Force Wright Aeronautical Labs, Wright-Patterson AFB, OH, May 1984.

[11] Gallagher, J.P. and Hughes, T.F., Influence of Yield Strength on Overload Affected Fatigue Crack Growth Behavior in 4340 Steel, Air Force Flight Dynamics Laboratory, AFFDL-TR-74-27, July 1974.

[12] Christian, M.E., "Overview of the CTOL and STOVL Full Scale Static Tests on F-35 Lightning II Program," USAF Aircraft Structural Integrity Program Conference, Dec. 2009.

[13] Yates, J.B., "Durability Testing of the STOVL F-35 Lightning II," USAF Aircraft Structural Integrity Program Conference, Nov. 2011.

[14] Christian, M.E., "Overview of the Full Scale Durability Tests on F-35 Lightning II Program," USAF Aircraft Structural Integrity Program Conference, Dec. 2012.

[15] Ball, D.L., "Fatigue Stress Spectrum Generation Program, SPCGEN v2.3," FZM-9779, Lockheed Martin Aeronautics Co. 2009.

[16] Manders, C., "CTOL Spectrum Generation Report," 2ZGP01306, RevC., Lockheed Martin Aeronautics Co. 2010.

[17] Needler, S.D., "CO-53 Truncation Verification Test Program," Lockheed Martin Aeronautics Co., 2010.

[18] Yates, J.B., "Comparison of F-35 and Legacy FSD Test Finding Rates," Lockheed Martin Aeronautics Co., 2013.

Structural Integrity Assessment of Gripen NG Aircraft

Hans Ansell

SAAB AB
Linkoping, Sweden

Abstract: The Gripen weapon system was from the outset built to adapt to changing threats and operational requirements that armed forces face. Based on the proven Gripen C/D platform, the Gripen Next Generation (Gripen NG, also known as Gripen E/F) carries this heritage and continues to be a high-performance multi-role fighter with very low life cycle cost.

In this presentation, the accomplishment of aircraft structural integrity assessment task for the Gripen NG version is outlined. The interpretation of applied Mil-specifications i.e. MIL-STD-1530C/JSSG-2006 and applicable Structures Bulletins is described with focus on essential engineering tasks affecting the fatigue service life and damage tolerance assessment effort. Examples from the airframe development stage are given such as mission analysis and operational profile development, loads analysis, structural analysis, local stress and damage analysis. Major structural differences between C/D versions and NG version are described and the consequences regarding ground and flight test verification needs and efforts are outlined. The service life monitoring system for individual aircraft tracking, which is a heritage from the C/D version, is outlined and necessary adaption for application to the NG version is pointed out.

INTRODUCTION

Gripen NG (Next Generation) is the latest version of the Gripen family being developed presently and will be delivered to the Brazilian airforce and the Swedish armed forces in the 2020s. The Gripen weapon system was from the outset built to adapt to changing threats and operational requirements that armed forces face. Based on the proven Gripen C/D platform, the Gripen Next Generation carries this heritage and continues to be a high-performance multi-role fighter with very low life cycle cost a through a broad spectrum of enhancements making it prepared for a future that is difficult to predict.

The Gripen NG weapon system

Saab chose a balanced design solution for Gripen where multispectral reduction of the aircraft's visual, infra-red and radar signature reduces detection without compromising its performance, manoeuvrability and maintainability. It is designed for co-operation in large units, sharing real-time information using advanced data-links in order to create a complete situation awareness picture. The ability to acquire, process, share and display tactical information - such as positional data of hostile and friendly units, missile engagement zones and information from radar warning receivers - permits Gripen pilots to operate and interact effectively, whilst avoiding enemy threats. The Beyond Visual Range (BVR) air-to-air combat capability of Gripen is achieved by using high performance AESA radar and IRST (Infra-Red Search and Track) sensors combined with the fighter's inherent low radar signature. During an intercept Gripen can simultaneously engage multiple targets with the most advanced BVR active radar guided missiles. For close range air-to-air combat (Within Visual Range, WVR), Gripen with high off-bore sight missiles and Helmet Mounted Display (HMD) provides a capability to destroy highly manoeuvrable targets.

Gripen's high turn rate gives excellent performance in close-in combat. The fighter's full authority digital 'fly by wire' flight control system enables carefree handling in all areas of the flight envelope. This ensures that the pilot can concentrate on the mission and be confident that the fighter always will respond with maximum efficiency, performance and safety regardless of external stores configuration and aircraft weight.

The Gripen has an integrated defensive aids suite consisting of an integrated Electronic Warfare System and an Identification Friend or Foe Interrogator Transponder (IFF). The EW-system includes a Radar Warning Receiver (RWR), an internal jammer supporting active electronic counter measures and chaff and flare dispensers. The system provides a unique ability for each Gripen operator regarding Electronic Warfare System programming including sovereign control over threat and countermeasure libraries. Low visual, radar and infra-red signatures and a virtually smoke-free engine enhance survivability in the air.

Airframe design

A durable, light weight and damage tolerant structure has always been an essential design criterion for the Gripen. The aircraft is designed for manoeuvres between -3g and +9g, whilst having a service life of at least 8000 flying hours. This equates to more than 30 years operational life at normal operational flying rates. The airframe is qualified to meet world-wide environmental conditions such as high and low temperatures in combination with high humidity.

The Gripen airframe is designed as modules, see figure 1, in order to allow cost efficient production and to facilitate maintenance, structural repair, future changes and upgrades.

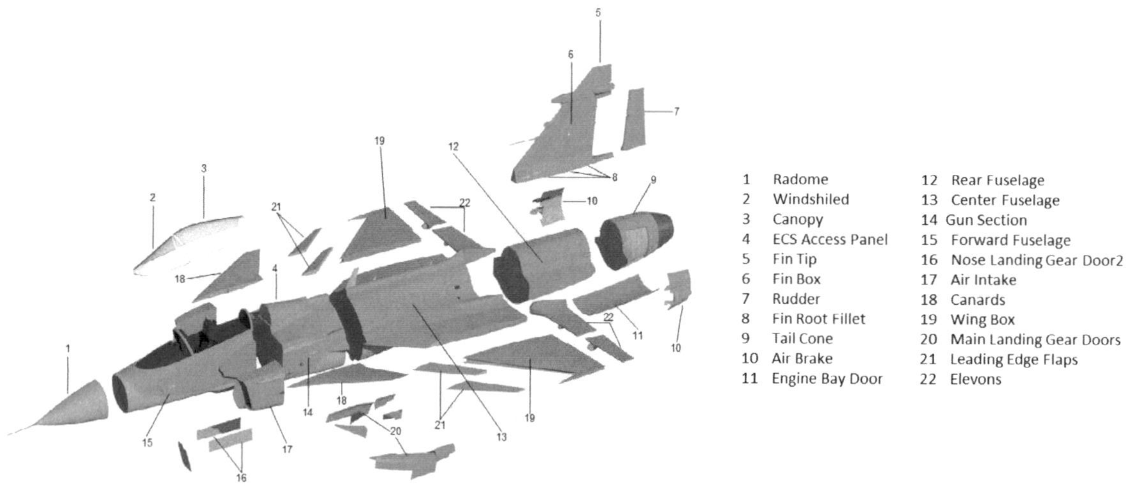

Figure 1. Gripen E modular construction.

Certain areas of the Gripen airframe have been designed as electrically shielded compartments with conductive, sealed walls and access panels to protect installations from electromagnetic interference. The composite material used in the airframe corresponds to approximately 20% of the airframe weight, see figure 2.

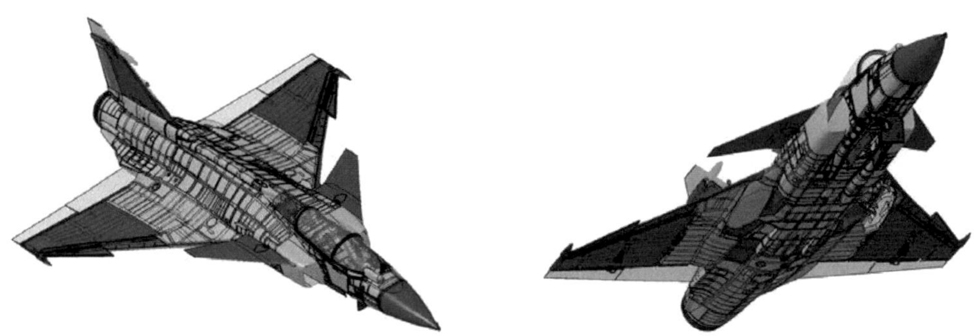

Figure 2. Gripen E use of composite materials.

Protective surface treatments are applied on metallic and composite airframe materials in order to protect the materials against corrosion and abrasive wear and to secure adhesion for adhesive, paint and sealant. High level of corrosion protection is also secured through careful selection of material and material combinations.

The Airframe is divided into different zones depending on environment exposure. Each zone has specified use of primer and top-coat paint, sealants and corrosion preventing compounds. The protective treatments give a corrosion protection level which is considered sufficient to give protection enough for an environment as of CSN ISO 9223, degree C4.

The Gripen NG main structural changes

The main structural differences between the legacy C/D versions and the NG version are:

- Engine GE F404 (RM12) has been upgraded to F414G-39E for increased thrust to maintain performance.
- The amount of internal fuel has been increased with nearly 40% for improved endurance.

- The main landing gear has been moved from the fuselage to the inner wing in order to accommodate the fuel increase in volumes within the fuselage and to make more room for system installations.
- The wing to fuselage connection concept has been changed.
- Furnished with two additional fuselage mounted pylons for external stores.
- Accommodation of an AESA radar in the radome area and with electronic counter measures and chaff and flare dispensers in various areas.
- Selected design changes for Design-for-Manufacturing purposes.
  - More complex machined part and less sheet metal design.
  - Modifications of build joints between fuselage sections
- Introduction of Aluminium-Lithium alloys maintain a low structural weight to design weight ratio despite extensively increased max aircraft weight 14,000 to 16,500 kg.
- Extended fuselage to optimize CoG.
- Increased span.

Figure 3. Major structural changes of Gripen E.

The decision to change the concept for the wing to fuselage connection was based on an analysis that it was possible to save 100-150 kg structure weight. It also increased the internal fuel volume with > 200 kg fuel, due to integration of fairings in the working skin. If the C/D wing connection had been kept, as it is on 39-Demo, the MLG connection has partly been integrated in the wing and partly in the center fuselage. This is a design solution not being preferred. The new MLG was already in the design due to another option to increase the internal fuel volume, to add supplementary store stations, and facilitated increased braking energy due to increased MTOW. The new MLG has the same foot-print as 39 C/D.

On the C/D versions is the wing designed as an interchangeable unit. The same is valid for all control surfaces and pylons connected to the wing. On Gripen NG the wing is not defined as an interchangeable unit. The connection is designed as a build joint with "fix" fasteners. The most important reasons for this decision are:

- Weight efficient design.
- The use for an interchangeable wing has shown to be of low value from an operational point of view (it has never been used as intended for A/B/C/D versions). In cases where it, after all, has been done, it has always been as a part of a major repair or update of the structure. The time and tooling constraint to change a wing from aircraft to aircraft is no longer considered essential.

This choice of design required some additional boundary conditions for the design work:

- There shall be a growth potential for routing of wiring, an option to be feasible later in the life of the aircraft with no need for removing or opening the wing.
- It shall be possible to remove a wing, which is damaged beyond repair with available methods, and replace it with a new wing. This repair may need personnel, tooling and jigs only available at depot level (customer or industry).

Some requirements from C/D are still valid for the new wing design, most important are:

- The wing shall have an inspection free life as a design goal.
- The wing structure shall be possible to be inspected from the outside (with boroscope type techniques).
- The top wing skin shall be removable for close inspection of internal structure, or to modify the design for added installation requirements or structural updates.
- All control surfaces and pylons, interfacing with the wing, shall be interchangeable with other wing units.
- Repair methods shall be available for the wing skins.

Changes have also been made with regard to methods and tools for defining structural design and structural analysis. The introduction of CATIA V5 design tools has led to a significant increase of three-dimensional models becoming available, and consequently, an increased number of solid models of stress analysis being created and used for sizing, primarily for fatigue and damage tolerance analyses.

Gripen NG is defined 100% as MBD, Model Based Definition (no drawings exist). This has introduced a lot of advantages for design and subsequent disciplines.

- It has increased the fit of structural items and minimized the need for shimming, trimming and tooling.
- Reduced lead time since all disciplines can access and use design data on line without hand overs between disciplines.
- Possibility to simulate assembly and maintenance at any time during design and before all design reviews to verify design.
- Support all disciplines that depends of geometry for simulation with data concurrent, for example; thermodynamics, RCS-analyses, fluid analysis (refuelling, venting, etc).
- Present design requirements for designers with requirement models which is overlaid on the design as zones containing rules for; corrosion protection, EMI-rules, particular risk, etc.
- Perform zonal analysis before each design review on the complete equipped airframe to verify; separations, redundancies, particular risks, etc.
- Analyse weight, CoG and inertia of a complete A/C with high accuracy end frequency.

STRUCTURAL INTEGRITY ASSESSMENT

General

The structural integrity assessment of Gripen E follows the same procedures and specifications as for the previous Gripen versions. Changes have been introduced when required such as specification development from MIL-STD-1530A and MIL-A-8860 series to MIL-STD-1530C and JSSG-2006.

A conceptual service life and damage tolerance assessment procedure was adopted from the beginning of the Gripen programme, shown in figure 4. The procedure consists of the various topics that constitute building blocks within the assessment philosophy for aircraft structures. An important feature of the model is that it possesses feedback from the involved tasks and complies with ASIP procedure as specified in MIL-STD-1530C.

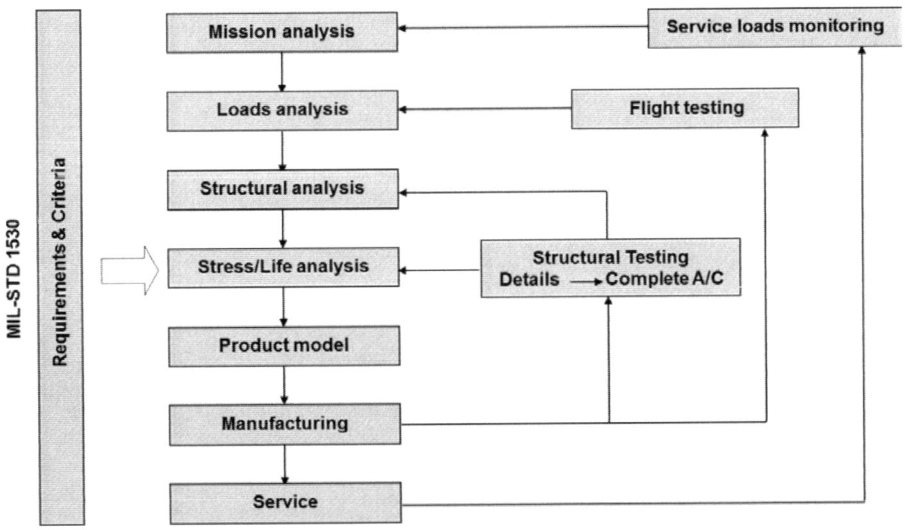

Figure 4. Service life and damage tolerance assessment procedure.

It is in this context essential to understand the design philosophy applied to Gripen in order to ensure service life and damage tolerance. All primary structural parts, of airframe, systems and systems installations are designed for a remotely low probability of fatigue crack initiation. A probability less than 10^{-4} at full service life has been applied in detail sizing i.e. what is normally required for pure safe-life (safety-by-retirement) type of design. In addition to this, damage tolerance has been imposed to the Gripen design in order to ensure the flight safety issue (safety-by-inspections). The inspection programme is entirely founded on the damage tolerance assessment of the aircraft. This assessment assumes conservatively that fatigue cracks are present at the most stressed spots in all primary structural parts from the very beginning of the life of the aircraft.

This "two-way" approach pave-the-way for a service life free from fatigue cracking although any such unlikely cracking are handled by the damage tolerance founded inspection programme. The statement is further strengthened by the excellent outcome of the extensive full-scale test programme of the Gripen aircraft, assemblies as well as complete airframes, comprising four lifetimes of testing. A "one-way" safe-life design approach would have been satisfied with this analysis and test evidence and allowed for a full service life without detailed inspections. It should however be kept in mind that such an approach would be less safe, not prepared for unforeseen damages, and would in addition not support future life extensions, if requested.

Safety-of-flight structure implies that structure of the airframe, systems, and systems installation whose mechanical failure could cause loss of the aircraft or aircrew. Safety of flight and other selected structural components of the aircraft are categorized as fracture critical items and be capable of maintaining adequate residual strength in the presence of material, manufacturing and processing defects and damage induced during normal usage and maintenance, until the damage is detected through periodic scheduled inspections.

The design goals for metal parts are as follows:

- All fracture critical metal parts are designed such that no part shall require scheduled inspections for cracks or fatigue damage before mid-life under the design operational profiles applied.
- Fracture critical parts that are difficult to inspect or replace are designed such that they shall not require scheduled inspections for cracks or fatigue damage before full-life under the design operational profiles applied.
- In addition to the assessment of damage tolerance, the assessment of fatigue service life of all structure, fracture critical as well as maintenance critical, are based on safe-life principles. The probability to have fatigue cracking, before full service life, shall be remotely low.

Operational profiles

The basis for all life design and test verification work is founded on the operational profile which shall reflect the expected future usage of the aircraft type. The operational profile of the Gripen aircraft is based on airforce outlook into future operations and expected mission types with associated armaments, tactics etc. as well as previous design basis and experience from service. The missions are simulated in flight simulators and flight parameter and other relevant configuration data are extracted with sufficient sample rate. Typical flight parameters are altitude, speed, angles of attack, roll and pitch rate, control surface deflections, weight and weight distributions etc. These parameters are included in the balancing of the external loads to equilibrium conditions.

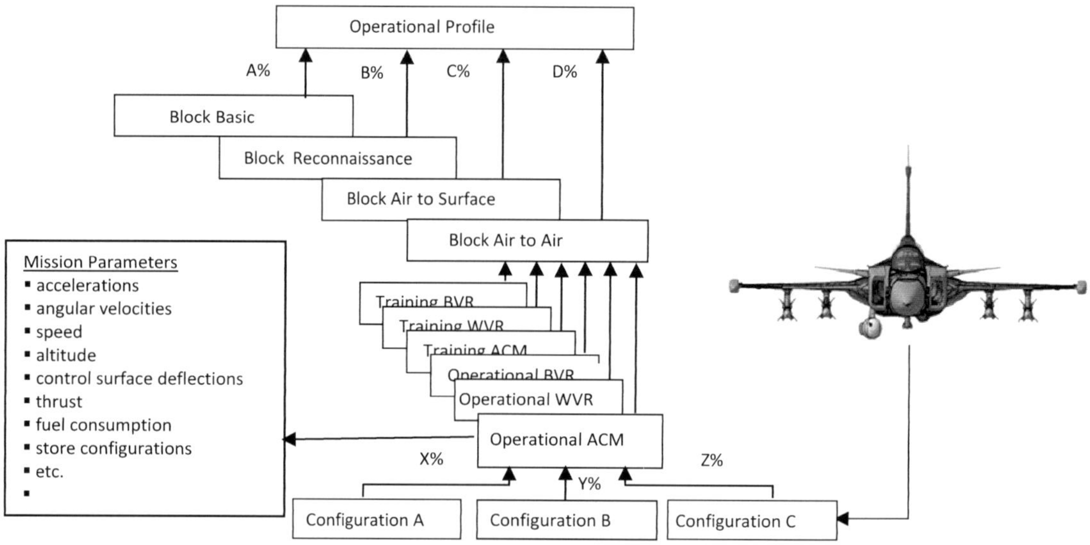

Figure 5. The hierarchic description of the operational profile.

The missions are defined by brief mission descriptions which have been the base for the mission mix data base (mission definitions done with simulated data). The missions included in the mission mix are built up by 'mission blocks' (segments/macros), which represents a certain mission phase such as; 'Take off', 'Climb', 'Navigation', 'Engagement', 'Descent', 'Landing' etc. These segments are simulated in the PMSIM, a simulator with a pilot in the loop. The frequency each mission have been assigned to is a judgment due to factors like; operational needs (pilot training needs etc.), general assumptions regarding a generic customer needs, and the resulting mean mission duration.

The operational profile is based on the assumption that the aircraft fleet is used for pilot training and in operational missions. The pilot training is normally done in the home country and the operational missions are performed in the home country (QRA), as deployed operations world-wide or in a war case. The flight intensity is normally higher during the deployed operations than during the "normal" service at the home base.

Two different operational profiles are applied simultaneously for life design purposes. One baseline design profile, defined for an aircraft service life of 8,000 hours, consists of 60 unique flight profiles/store configurations defined with durations from ~40 minutes up to 8 hours with a mean duration is 1.3 hours. The mission types consist of:

- General missions, 20% of the flight time, general category missions, such as; Ferry, AAR training, Basic Air Training etc.
- Training missions, 50% of the flight time, normal duty at MOB comprising Air-Air, Air-Surface and ISR mission types
- Operational missions, 30% of the flight time, QRA, deployed operations, homeland defense comprising Air-Air, Air-Surface, ISR and CAP mission types

One second operational profile defined by the Swedish armed forces with an alternative mix of Air-Air, Air-Surface and a different speed-altitude composition is used simultaneously for service life purposes. This operational profile is defined for an aircraft service life of 6,000 hours.

The purpose of using two operational profiles is to accomplish a robust design regarding the service life. The sequences are used for sizing, testing and monitoring purposes.

Loads analysis

Following the operational profiles, balanced load cases are derived using unit load cases for a number of unbalanced pressure distributions, acceleration distributions etc. Also other load cases representing internal pressures in fuel tanks, air intake, engine thrust etc., see figure 6, and temperature distributions are defined as unit load cases. Many load cases for e.g. attach points for weapon pylons, equipment, landing gears etc. are also included. The pressure distributions are obtained in wind tunnel tests supplemented with CFD analyses.

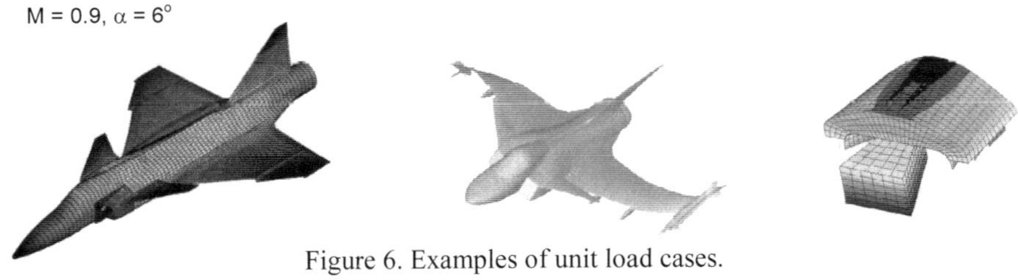

Figure 6. Examples of unit load cases.

The balanced flight load cases are obtained for every defined instant flight, landing and ground manoeuvre condition, with regard to the current configuration, defined in the operational profile in an optimization procedure under constraints governed by the flight control system and aeroelasticity.

The internal loads model

The unit external load cases are solved in the finite element model of the complete airframe. This load distribution model for Gripen NG follows the same principles as for previous Gripen versions but has been much more detailed with a higher element mesh density. Post processing including the derived equations for balancing the unit load cases together with a hierarchic description of the sequential load conditions as given by the missions in the operational profile results in local load sequences for load entities for any element of the FE model, see figure 7.

By use of an interactive computer code, the FEM database containing the solved unit load cases is combined with the global sequence database containing the loads and mission definitions. Any local load or stress sequence for arbitrary elements in the FEM model can then be generated and analysed using rain-flow count algorithms.

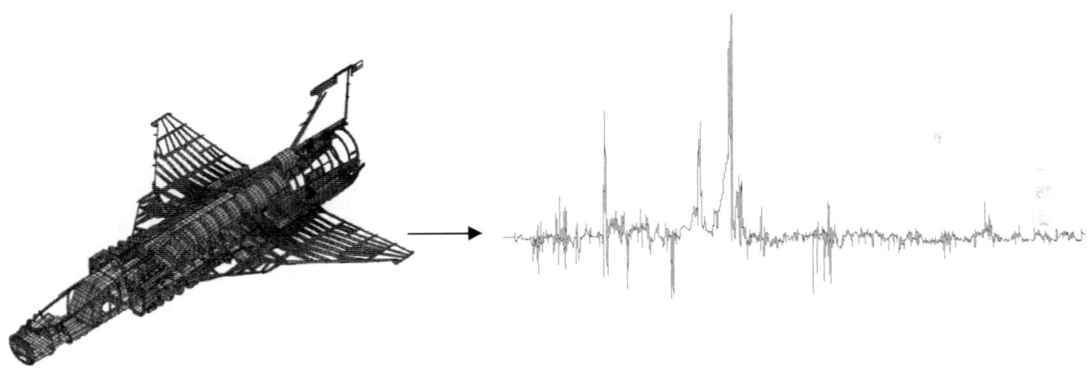

Figure 7. Local load sequences obtained from any element of the finite element model.

Application of requirements

The airframe is designed to sustain loads occurring in flight and ground conditions without permanent deformation at 1.15 x Limit Load (LL) and without failure at Ultimate Load (UL). LL is defined as a design load magnitude not expected to be exceeded during the aircraft service life. UL is defined as LL times applicable Safety Factors. In general, the Safety Factor is 1.5 or higher for certain cases where additional factors are applied.

Safety of flight and other selected structural components of the aircraft are categorized as fracture critical items and shall be capable of maintaining adequate residual strength in the presence of material, manufacturing and processing defects and damage induced during normal usage and maintenance until the damage is detected through periodic scheduled inspections. This requirement applies to metallic and non-metallic structures, including composites, with appropriate distinctions and variations.

The principles of fracture mechanics are the means for ensuring flight safety of fracture critical items. Exceptionally, for structure where such principles are impossible or impractical to apply, safe-life principles with dedicated factors of uncertainty shall apply. In addition to the assessment of damage tolerance, the assessment of fatigue service life of all structure, fracture critical as well as maintenance critical, shall be based on safe-life principles.

Fracture critical items have in general been categorized into one of two main categories:

- Slow crack growth – no growth, non inspectable or inspectable on depot level

Slow crack growth structure consists of those design concepts where flaws or defects are not allowed to attain the critical size required for unstable rapid crack propagation. Safety is assured through slow crack growth for specified periods of usage depending upon the degree of inspectability. The strength of slow crack growth structure with subcritical damage present shall not be degraded below 120% limit load during the period of unrepaired service usage.

Composite structure classified as safety-of-flight structure shall be designed for damage tolerance. The design usage must be identified and adequately defined such that the potentially damaging high load cases are properly represented. The structure shall be designed for no-growth of damages under both operational loads and high static loads. The damages to be considered are defects from manufacturing and service such as delaminations and barely visible impact damages (BVID). It shall be shown that the structure can tolerate these defects during its full design life

To obtain the desired confidence in the composite components the general criteria of reduced static strain based on overall notch factor are applied, i.e. design static strain for the laminate shall be reduced with a notch factor corresponding to a 6 mm open hole. Optionally, criteria based on allowable strains obtained from compression after impact (CAI) tests and fatigue after impact tests (FAI) are applied.

- Fail-safe

Fail-safe concepts where the required residual strength of the remaining intact structure shall be maintained for a period of unrepaired usage through the use of multiple load paths or damage arrest features after a failure or partial failure. The period of unrepaired usage necessary to achieve fail-safety must be long enough to ensure that the failure or partial failure will be detected visually and repaired prior to the failure of the remaining intact structure.

For mechanical systems and systems installations, the following categories are optional:

- High durability margin

For structure where damage tolerance principles are impossible or impractical to apply, safe-life principles under special conditions can be used. The aim with these principles is the same as with damage tolerance principles i.e. to lower the operational stress in order to deal with structural degradations due to the presence of defects, but with other means.

Structure of aircraft systems and system installations certified under this category shall have an average failure rate less than10^{-8}. This can be established by fatigue design to a local probability of failure of 10^{-5} within a typical full fighter lifetime. It is well established that the fatigue life and the operational stress spectra can be characterized by log-normal probability distributions. The stated local failure rate can, for metals, be achieved by combination of limit load spectra enlarged to represent 2 standard deviations above mean spectrum and a fatigue strength data reduced to represent at least 3.3 standard deviations below mean fatigue strength.

Supporting fatigue testing of components are required when relevant fatigue data are lacking e.g. for fatigue property affected materials as in welded of swage tubes, or when reliable stress analysis is too complex to accomplish e.g. gimbals or other system parts which perform a movement/function etc. A quantitative field data evaluation is recommended to justify the high durability margin method e.g. dominant failure modes shall be established. It shall be shown that established practices for design, analysis, inspection, process control, manufacturing, testing etc. result in parts that are durable and reliable in service. Applicable field data results shall be scaled to the current aircraft operational baseline. Design and material aspects which support damage tolerance but are not directly preserved by the classical fatigue methods shall still be considered e.g. materials with low fracture toughness shall be avoided.

- Leak before burst

Leak before burst is a concept where fracture mechanics is used to confirm that a through crack in a fluid container will leak fluid before the container bursts.

- Proof test

Proof testing is a concept where an overload test is performed on each production component and can demonstrate that the part does not contain flaws which otherwise will induce failure before a certain time of operation.

Inspection levels

In manufacturing quality control, the initial flaw sizes presume the inspection of 100 percent of all fracture critical regions of all structural components as required by the fracture control provisions. This inspection includes as a minimum a close visual inspection of all holes and cut-outs, and conventional ultrasonic, penetrant or magnetic particle inspection of the fracture critical regions. Where special non-destructive inspection procedures have demonstrated a detection capability better than indicated by the specified flaw sizes, and the resulting smaller assumed flaw sizes are used in the design of the structure, these special inspection procedures must be used in the aircraft manufacturing quality control. If development test data indicates that more severe flaw shapes than assumed are probable, worst case assumptions should prevail.

All scheduled in-service inspections for structural damages shall be possible to conduct at depot/base level. Structure is depot or base level inspectable if the nature and extent of damage will be detected utilizing one or more selected non-destructive inspection procedures. The inspection procedures may include NDI techniques such as penetrant, X-ray, ultrasonic, etc. Accessibility considerations may include removal of those components designed for removal.

Inspection intervals

The scheduled in-service inspection intervals shall be a multiple of 400 hrs. The target for first inspection shall not be less than mid-life of the aircraft. For structure difficult to inspect and costly to replace, the inspection interval shall be one full aircraft service life. This interval shall be established by larger margin than the normal slow crack growth factor of two for robustness reasons. This means that the predicted life shall be 1.5 x 2 times the aircraft service life using one design operational profile or, as for Gripen NG, 1.3 x 2 times since two operational profiles are used.

DESIGN ANALYSIS AND SIZING

The detail stress analysis and sizing methods and procedures for service life and damage tolerance have continuously been developed in the Gripen programme. Besides the developments of procedures for mission analysis (use of flight simulators), loads analysis (CFD and PSP in wind tunnel tests), structural design and analysis (3D CAD to 3D FEM) have also the detailed fatigue and damage tolerance analysis techniques been improved accordingly.

The fatigue and damage tolerance sizing of Gripen A/B versions in the 1980'ies were mainly based on analysis using handbook solutions of stress concentration and stress intensity factors. During the design of Gripen C/D versions in the 1990'ies the methods were almost the same although more analysis of hot spots and cracks in FE-models were done. The reasons were better computer capacity but also the increased need for such analyses due to the extensive transition from mainly built-up structure to high-speed machined integral structures as part of the parts-count reduction programme. Such design required more detailed models in order to better predict possible crack path developments in geometrical complex structural details.

The use of advanced 3D-CAD techniques for the Gripen NG version has made the conditions for detailed FEM much better by meshing of models of complex structural geometries more directly. Solid 3D models are solved in both NASTRAN and in ABAQUS FE-codes, see figure 8. Special functionality such as XFEM in ABAQUS use a modification of the conventional FE formulation. It is capable of modelling discontinuities, such as cracks, without having to adapt the element mesh to the crack topology. At the same time, the near-tip displacement and stress fields are well represented. Furthermore, the propagation of the crack can be modelled without re-meshing the domain. These characteristics give XFEM substantial advantages over other approaches such as adaptive re-meshing techniques.

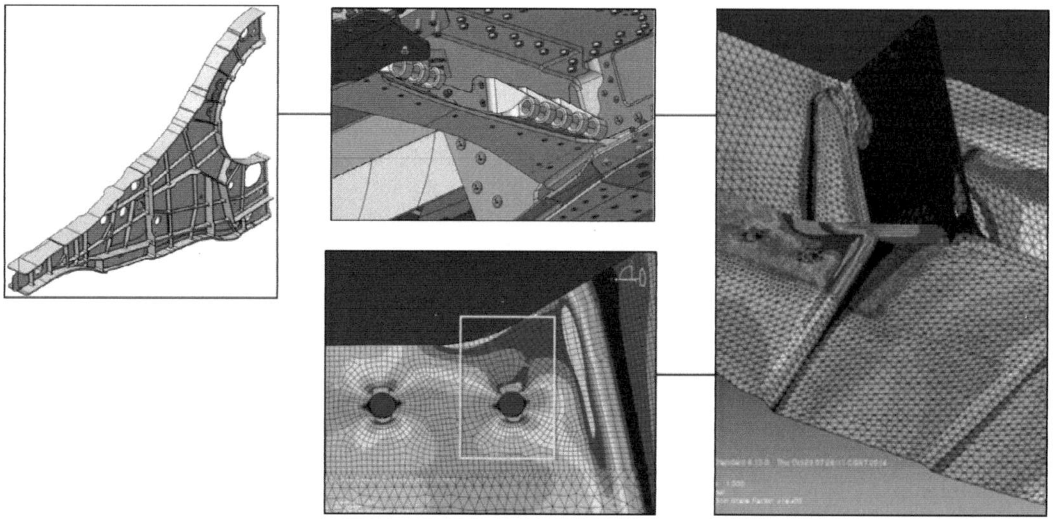

Figure 8. 3D Catia models to FE models for detailed fatigue and damage tolerance design.

Despite the potential for advanced analyses with this approach it is still in many engineering situations an advantage to quickly create a good enough model. Provided elementary cases are obviously the fastest but with a simple but good enough FEM programs can significantly increase the quality without costing too much time.

One such special duty application with advanced analysis capabilities for stress concentration and growing cracks have developed in collaboration with the University of Linkoping. The program can solve the stress intensity factors for both 2D and 3D cracks and also capable of predicting crack paths. The program is not compatible with Catia CAD systems, with its proc and cons, but has a very fast and efficient technique for creating geometries. The step from an elementary case to a good model of this system is in time very small but with a considerable increase in quality. In figure 9 (upper part) is a simplified model created and solved almost instantaneously of a complex structure but where essential features are included, features that is not easily approximated using compounding of elementary cases.

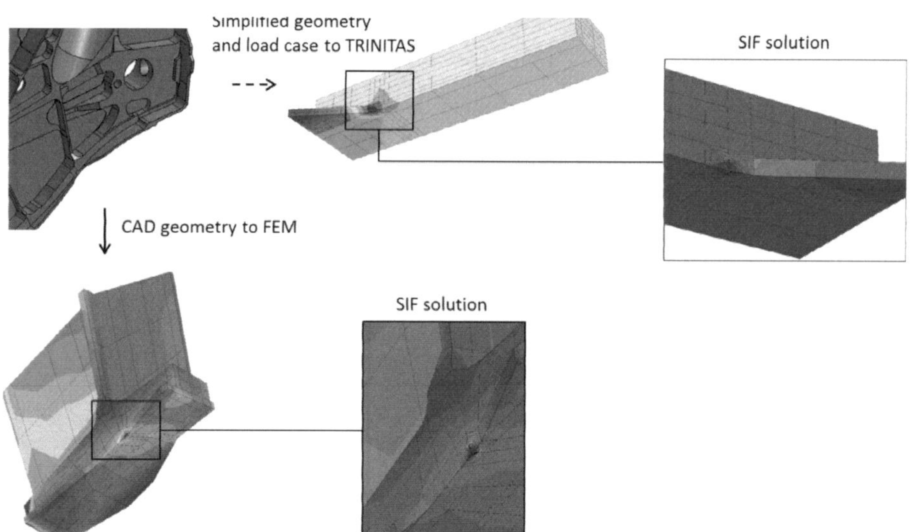

Figure 9. Simplification of crack growth analysis using easy-to-use and fast executed special duty program.

TEST VERIFICATION

Service life and damage tolerance capability needs to be verified by testing. These tests can either be based on similarities to previously conducted tests, if applicable, or by new tests when no previous test results are available. Full-scale testing is required when:

- Changed design principles in primary load paths
- Changed material in safety-of-flight critical parts
- Significant change of stress conditions in legacy primary load paths

When Gripen A/B was initially developed, the test verification task was huge since all of the above criteria applied. Especially the implementation of the damage tolerance task according to the MIL-A-83444 specification called for test verification. An extensive fatigue and damage tolerance test program of full-scale test assemblies was accomplished, see figure 10. Both airframe and systems parts, e.g. actuators in the flight control system, were included and a total of about 800 artificial defects, sizing between .05 and .25 inch, were introduced in the structural parts and in the correct structural surroundings. In addition conventional full-scale static and fatigue tests of both A and B versions were conducted covering testing beyond ultimate loads and 4 lifetimes respectively. More detailed description of this work can be found in [1] for airframe and [2] for mechanical system.

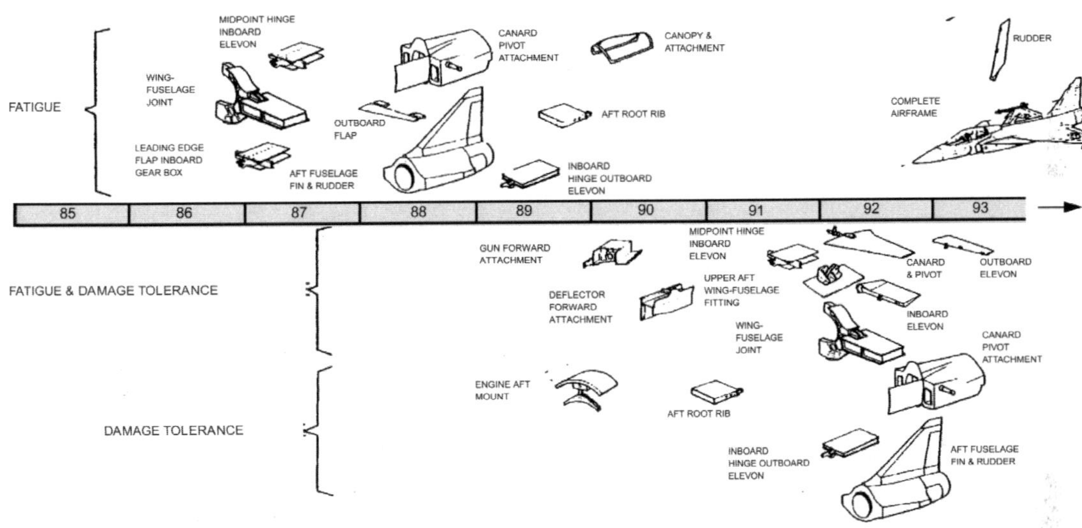

Figure 10. Test programme for fatigue and damage tolerance verification of Gripen A/B airframe.

When Gripen C/D was developed, no change of materials was done and no change of major load paths. Essential changes of local geometries were however done through the redesign to integral structures, extended service life (from 4,000 to 8,000 hours) and increased basic design mass (~10% compared to A/B versions) and some other load and structural changes, e.g. air-to-air refuelling probe, called for new full-scale fatigue tests, se figure 11. Damage tolerance tests were however not done in any large extent since the load path were the same although redesigned but the validation effort done on the fracture mechanic based damage tolerance methods during the A/B developments were deemed to be sufficient also for the C/D versions.

Figure 11. Full-scafe fatigue test of Gripen D version.

With the development of Gripen NG versions, structural changes were made that interrogated with several of the criteria for the need of test verification. Certification for airworthiness by full-scale testing required due to:

- New mid fuselage/wing design, fuselage joints, MLG attachments
- Change of classification of parts due to design-for-manufacturing purposes
- Use of a new material type (Al-Li AA2050 and AA2060)
- Increased basic design mass (~40% compared to C/D versions)
- New operational profiles.

These structural changes call for a more extensive test programme than what was needed for the C/D versions. A full-scale static test (including impact damaged composite parts) of the complete airframe is underway and a full-scale fatigue test of the complete airframe will follow thereafter and tested in the same rig for 4 lifetimes, see figure 12. In addition to the full-scale fatigue test, control surfaces (not included in the test airframe) will be tested in separate assemblies. These tests will be a combined fatigue and damage tolerance test by initially cycled for 2 lifetimes without artificial initial defect followed by 2 lifetimes with artificially manufactured defects installed.

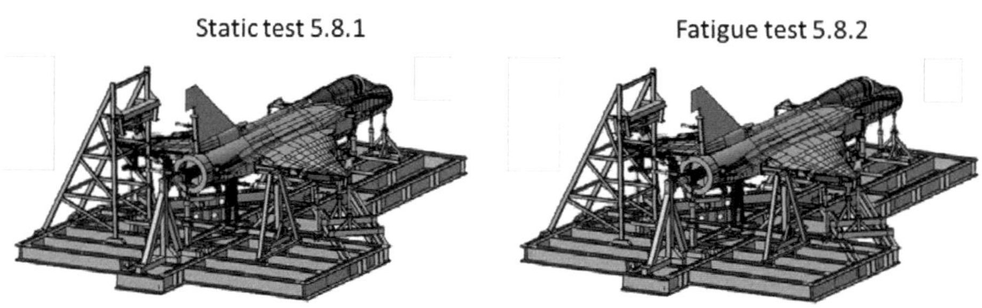

Figure 12. Full scale fatigue tests of Gripen E version.

Full-scale tests for compliance with damage tolerance requirements are also needed due to the structural changes in load paths and the upgraded classification of fuselage stringer joints and the design change of the wing to fuselage joint. The change from AA7050 alloy to AA2050 alloy in safety-of-flight critical parts also makes the damage tolerance test verification needed. The decision to use aluminium-lithium alloys in primary structural components was preceded by extensive validations of the fatigue and damage tolerance material properties. To assure necessary and sufficient confidence regarding damage tolerance, a significant test effort designed to challenge typical airframe crack scenarios with part through cracks (surface cracks in thickness steps, open and loaded holes etc.) was accomplished.

Figure 13 shows structural objects/assemblies which will be test verified for damage tolerance when having multiple artificial initial defects installed in critical sections.

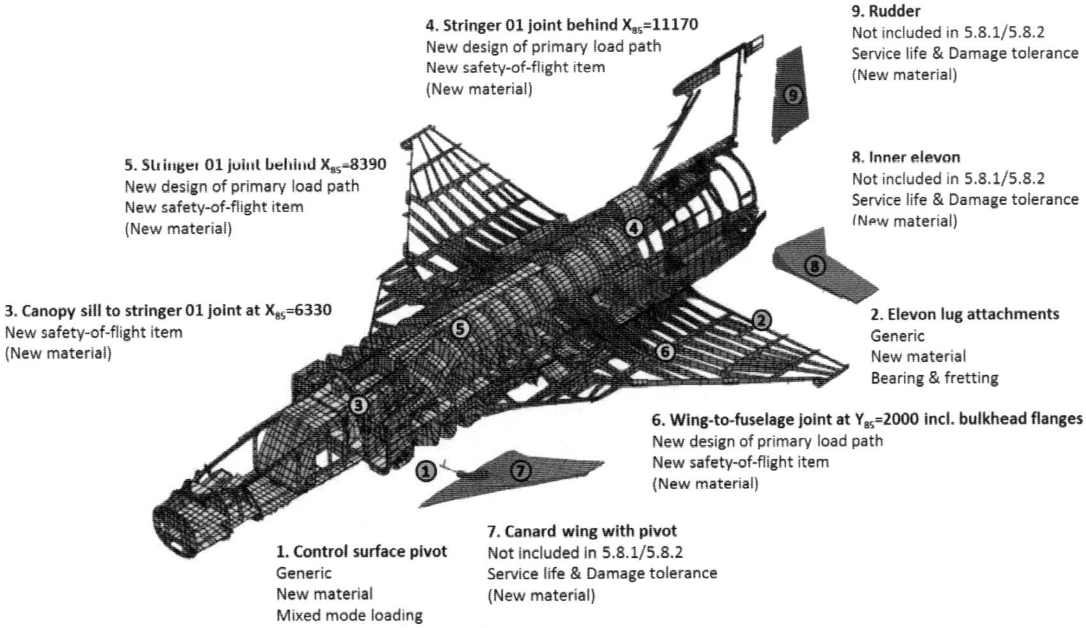

Figure 13. Full-scale damage tolerance tests of assemblies having artificial defects installed.

FLIGHT PARAMETER BASED LOADS MONITORING

All design and verification of durability and damage tolerance is based on predicted usage which is expected in the future. The synthetic operational profile is the best one available before the aircraft type is in operation. During service it is not unlikely that the usage conditions will change. In order to take this into account, each aircraft is equipped with a load monitor system and life management utilities.

The system is based on flight and configuration parameters and is processed on ground. The processing on ground enables the use of the same procedure as used in design of the aircraft. The procedures for external loads analysis and internal loads analysis (structural analysis) as schematically shown in figure 4 is used with the exception that the operational analysis is replaced by actual recorded flight and configuration parameters for each accomplished mission. From calculated local load sequences, range-pair-range cycles are identified, counted and analysed in terms of fatigue life consumption for all covered structure. A schematic picture of the system is shown in figure 14 and further descriptions can be found in [3].

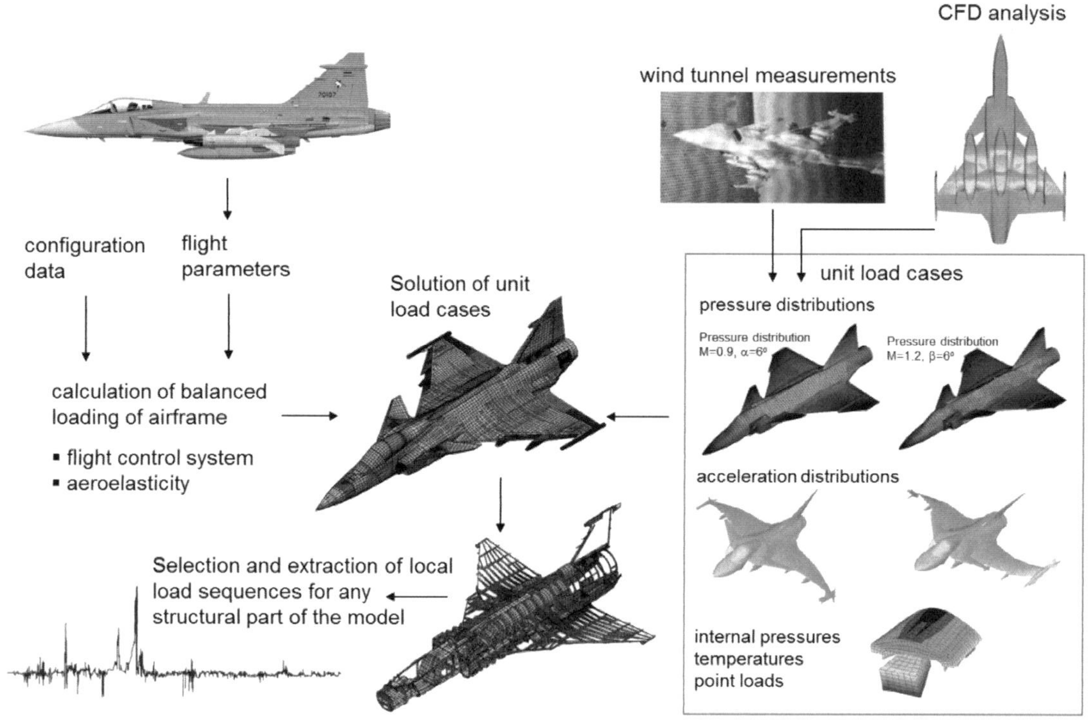

Figure 14. Schematic picture of the flight parameter based loads monitoring system.

In previous chapter, the procedure applied during the design process using the synthetic operational profile has been described. The only difference with the procedure used for loads monitoring is that the synthetic mission parameters are replaced by the actual recorded flight and configuration parameters.

The loads monitoring system as defined for operational use, has a number of predefined control points. These control points have been selected based on criticality i.e. safety-of-flight structural components and on other life limiting structure which may be of availability interest e.g. weapon pylons and weapon pylon attachments to the airframe. About 100 control points covering the airframe and pylons have been defined for the Gripen C/D versions. The control points represent both loads intersecting major structural parts i.e. principal load paths, as well as loads in structural members of the inner structure. The system will be inherited into the Gripen NG versions with necessary changes of control points.

The models for loads and fatigue life monitoring of the Gripen C/D airframe have been integrated into the Maintenance Ground Support System (MGSS). This product import true usage data for evaluation and has the competence to exchange data with the maintenance system. Aircraft data from different cycle types (Ground, APU, Hydraulic, Engine, Flight and Standard Cycles) are prepared and analysed by MGSS. Each cycle is evaluated and several operational parameters, related to the aircraft maintenance, are calculated. The MGSS fatigue function takes actions on the flight cycle data using the loads and damage models above discussed. After evaluation of cycles in the MGSS, a set of operational parameters is exported from MGSS to the maintenance system.

Fleet management

The export of parameters is done on a nearly daily basis, making the maintenance system up to date regarding aircraft maintenance data. The evaluated operational parameters are imported into the maintenance system. This system has access to aircraft configurations. Thereby values on operational parameters can be allocated to individual tracked equipment, e.g. fuselage, wing, canard and fin.

The MGSS fatigue function introduces possibilities to specify maintenance actions, e.g. crack inspection and discard, based on "monitored" deviations from design conditions. Deviations must be managed and accounted for on specified inspections intervals and fatigue life limits given by the maintenance publications (based on prescribed operational conditions). The maintenance plan describes the usage of maintenance parameters and the aircraft maintenance publications includes all the maintenance procedures that must be done on the equipment to make sure it will function correctly during operation and storage. It is described what is going to be maintained, when, where, how and by whom the procedure will be carried out and which maintenance procedure the maintenance includes.

Furthermore, MGSS and the maintenance system can exchange pairs of numbers at the predefined control points, fatigue consumed flight hours (FLH) versus direct recorded flight hours (FH), for equipment that MGSS keeps track of. The data can be visualized in MGSS by different kind of plots and the information can be used in service planning activities.

In depth analyses of true usage, evaluated by MGSS and exported on daily basis to the maintenance system, are available through a product named load matrix accumulator. The main task for this product is to store accumulated load matrices for all control points on individual tracked equipment. The accumulator import configuration data from the maintenance system and load matrices from the MGSS:

- Import of load matrices for every point of analysis per flight cycle, including:
  - Consumed damage
  - Minimum load
  - Maximum load
  - Range-pair-range matrix
- Import of configuration data for each flight cycle.

The export of configuration data is done on a regularly month basis. The data can be used in life extension/limitation issues and decisions on limitations specified by the maintenance publications, e.g. thresholds for crack inspection, crack inspection intervals and life limitations.

Dynamic loads events

Effects on structural loads by dynamic events are not possible to record using flight parameters due to insignificant sample rate and parameters not feasible to obtain. The solution under development is to define synthetic pre-defined load sequences for such events e.g. landing impacts, stores separation, arrest wire crossover, buffeting etc. These load sequences are obtained from measurements in instrumented test flights and from dynamic structural analyses. A number of such predefined sequences will be developed under specified conditions. Although the load sequences are not directly obtained from the flight under evaluation, it is included upon actual events recorded under associated conditions e.g. separation of a specific store from a specific pylon at a certain flight condition or a landing with a correct aircraft weight.

Substitution of lost flight data

Although the flight parameter recording is a priority, it cannot be guaranteed that data can become inaccessible. When such occasions occur, there are a number of predefined standard missions available for replacements. There are eight such missions defined, each with a predefined set of stores. When a conducted mission is lacking enough flight parameters, a standard mission is used instead in order to keep paste with the fatigue life consumption of the specific aircraft. The selection of standard mission is based on the code the missions have been assigned to in the mission planning system e.g. air-to-air, reconnaissance, control flight, ferry etc. The system allows for selection of stores.

REFERENCES

[1] Ansell H., Lilja T., Principles, procedures and outcome of the damage tolerance test verification of JAS39 Gripen Airframe. Aircraft Structural Integrity Program Conference (ASIP), San Antonio, Texas, USA, 1995.

[2] Ansell H., Fatigue and damage tolerance assessment of mechanical systems in a fighter aircraft – JAS39 Gripen. 18th Symposium of the International Committee on Aeronautical Fatigue (ICAF), Melbourne, Australia, 1995.

[3] Ansell H., Johansson T., A flight parametric method for loads and fatigue life monitoring of Gripen C/D aircraft. 27th Symposium of the International Committee on Aeronautical Fatigue (ICAF), Jerusalem, Israel, 2013.

Random vibrations fatigue analysis and tests

Etienne Deshaies, Jérémy Bernadou, Pierre Madelpech

DGA Aeronautical Systems, France

Abstract: DGA is in charge of qualification aspects and continued airworthiness for military aircrafts. In this context, DGA has to perform tests for the qualification of equipment in vibrations environment. But tests are expensive and can take a long time to do, relatively to operational issues. The objective of this study is not to replace all the tests by calculations – many data can only be obtained by test for now, damping for example – but to be able to anticipate a test in order to reduce hazard of non-predicted failures or to extend a structural qualification from a vibrations environment substantiated by test to another one slightly different. Moreover, some subjects are raised about crack initiations which cannot be explained by "classical" fatigue models, especially for rotorcrafts, because of the type of excitation: vibrations. So, with this study, DGA can also improve its knowledge about random vibration fatigue in order to explain and to bring solutions to technical facts relative to this particular subject. For this purpose, this study is split in two parts. The first one is a numerical assessment of the different models from literature to calculate damage and fatigue life for random vibrations environment. The comparison is done with "classical" damage calculations. Many power spectral densities are generated at random. For each one, many corresponding temporal signals are also generated. The damage is then calculated for each signal by "classical" method (Rainflow counting and Miner's rule are used). The average of all these calculations (for each power spectral density) is the reference. The damage is also calculated directly from the power spectral densities with the different models. A comparison of the results leads to assess their accuracy, scattering and conservatism. The aim of the second part of the study is to assess these models by confronting them to test results: some tests are made with a vibrating pot on two types of coupons. To be able to assess properly the models, the coupons have to be "wide band". Indeed, the standards for aircraft and rotorcraft, which are studied in the frame of this work, present generally random vibrations on a wide band and some typical frequencies excitations corresponding to rotor or blades speed. In fact, a wide band response for "simple" coupons is quite difficult to obtain, that is why they are designed to obtain a transfer function with two natural mechanical frequencies. The first design of coupon is found in literature and only has one critical point. The second one is "home-made" by DGA: the design is optimized to obtain two different critical points, two resonance frequencies and a fixed predicted test duration. This optimization and the random excitation definition are made thanks to finite elements modeling (FEM) of the coupons and damage calculations with the different previous models. Finally, the first coupons tests enable to adjust the complete model (FEM and damage calculation) and the second ones enable to compare the models and to assess the possibility to make this type of damage prediction.

INTRODUCTION

The DGA is in charge of qualification aspects and continued airworthiness for military aircrafts. In this context, DGA has to perform tests for the qualification of equipment in vibrations environment. But tests are expensive and can take a long time to do, relatively to operational issues. The objective of this study is not to replace all the tests by calculations – many data can only be obtained by test for now, damping for example – but to be able to anticipate a test or to extend a structural qualification. Anticipate a test is really important to reduce hazard of non-predicted failures. Indeed, the critical points which are found by calculations can especially be monitored and a fatigue life estimation of these parts gives some information about margins for the test. In case these ones are low, and for a test with several directions of excitation (the cross excitation of several directions can lead to overload a structure during a qualification test campaign), it is possible to organize the schedule of the different parts of the test: if damage appears, it is possible to analyze the signal and assess its overtested part to validate the qualification. Otherwise, when some vibration tests have already been done on a configuration of a structure, it could be interesting and possible to assess the impact of slight modifications of this structure by a calculation approach. After a first modeling of the initial configuration, which would be fitted with the test results, a second modeling of the modified configuration could be considered acceptable to validate the modifications.

For this purpose, and in order to improve our knowledge about random vibration fatigue, a study was launched, with two main parts. The first one is a numerical assessment of the different models found in literature to calculate damage and fatigue life for random vibrations environment. This part is only briefly exposed in the beginning of this article as it has already been presented in a previous ICAF Symposium [13]. Only the approach and the main results are reminded.

The aim of the second part of the study is to assess these models by confronting them to test results: some tests are made with a vibrating pot on two types of coupons. This part of the study, that is still running, is deeply described in this document. But first of all, a presentation is made about vibratory loading and fatigue damage assessment in the frequency domain.

PRESENTATION OF SPECTRAL FATIGUE

Vibratory loading and PSD function

Vibratory loadings are usually random and therefore, the signal $x(t)$ can only be described statistically. These time signals are generally considered as stationary, ergodic and Gaussian and can thus be statistically determined in the time domain univocally by their autocorrelation function $R_x(\tau)$ which is defined for a real process by:

$$R_x(\tau) = E[x(t)x(t-\tau)] \text{ where } E \text{ is the stochastic mean} \tag{1}$$

In the frequency domain, the Fourier transform of the autocorrelation function is often used to characterize the signal. Denoted $S_x(f)$, the one sided PSD usually used in mechanics can be expressed like that:

$$\begin{aligned} S_x(f) &= 2\int_{-\infty}^{\infty} R_x(\tau)e^{-2j\pi f\tau}d\tau \quad & f>0 \\ S_x(f) &= 0 & f \le 0 \end{aligned} \tag{2}$$

The PSD is usually characterized through its spectral moments of different orders, the nth moment being expressed as follows:

$$m_n = \int_0^{\infty} f^n S_x(f)df \tag{3}$$

Thanks to these spectral moments, the irregularity factor γ which is usually used to measure the bandwidth of the signal can be calculated:

$$\gamma = \sqrt{\frac{m_2^2}{m_0 m_4}} \tag{4}$$

An irregularity factor close to 1 corresponds to a narrow band process whereas a wide band process is characterized by an irregularity factor approaching 0. Another very important result from Rice [4] lies in the fact that for a zero-mean process, the average expected rate of occurrence of peaks $E[P]$ and of mean up-crossing $E[0^+]$ (i.e. zero crossing with positive slope) can be found from these spectral moments:

$$E[P] = \sqrt{\frac{m_4}{m_2}} \quad \text{and} \quad E[0] = \sqrt{\frac{m_2}{m_0}} \tag{5-6}$$

In the end, for a zero mean Gaussian process, the 0th moment of the PSD is equal to the well-known RMS value of the signal.

Methodology to assess fatigue damage in the frequency domain

The damage is directly estimated from the stress power spectral density function. Due to the randomness of the signal the damage is an average damage per unit of time expressed by the following formula:

$$E[D] = \int \frac{S^b}{C}.E[P].f_S(S).dS \tag{7}$$

In fact this formula is nothing else but the rewriting of the Miner's rule after having extracted different elementary cycles from the complex spectrum. Two terms can be distinguished in this expression:

- $E[P].f_S(S).dS$: $f_S(S)$ represents the probability density function of having extracted from the signal an elementary cycle with an amplitude S. The term $E[P]$ is the average number of peaks of the signal per unit of time. Once multiplied together with dS, it comes the average number of cycles per unit of time with amplitude comprised between S and $S+dS$ in the signal. In fact it corresponds to the results of a rainflow counting method.

- $\frac{S^b}{C}$ represents the damage of an S amplitude cycle when the Basquin model is used to approximate the Wöhler's curve.

Once integrated between 0 and $+\infty$, the average damage per unit of time of the signal is obtained.

The different studied models

In fact it can be said that one model corresponds to one expression of the probability density function $f_S(S)$.

The Narrow Band approach (NB). The first model found in literature has been proposed by Bendat [5] who assumed that the probability density function corresponds to a Rayleigh distribution. Some additional explanation about this model can be found in [1] but in this article we will just remind its expression:

$$f_s(S) = \frac{S}{4m_0} e^{\frac{-S^2}{8m_0}} \text{ with } m_0 \text{ the } 0^{\text{th}} \text{ moment of the PSD} \tag{8}$$

In the formula (7), it is common to replace $E[P]$ by $E[0^+]$ even if for a strictly narrow band process (i.e. $\gamma = 1$) they are equal to each other. Finally, once integrated in the formula, the following expression can be used:

$$E[D_{NB}] = E[0^+]\ C^{-1}\left(2\sqrt{2m_0}\right)^b \Gamma\left(1+\frac{b}{2}\right) \tag{9}$$

with the gamma function $\Gamma(x) = \int_0^{+\infty} t^{x-1} e^{-t} dt$

Models based on the narrow band approach. The NB approach is usually very conservative when applied to wide-band signals. Thus some authors proposed to modify the NB model through a correction factor leading to the following expression:

$$E[D_{WB}] = \lambda_{WB} \times E[D_{NB}] \tag{10}$$

Wirshing and Light (WL)

This is a purely empirical model [6] based on a large number of simulations and dependent on the irregularity factor and the slope of the Wöhler curve.

$$\lambda_{WL} = A(b) + [1 - A(b)] \times [1-\varepsilon]^{B(b)} \quad with \quad \begin{cases} A(b) = 0{,}926 - 0{,}033b \\ B(b) = 1{,}587b - 2{,}323 \\ \varepsilon = \sqrt{1-\gamma^2} \end{cases} \tag{11}$$

Ortiz and Chen (OC)

Proceeding with the same manner, Ortiz and Chen [7] developed the following correction factor:

$$\lambda_{OC} = \frac{\beta^b}{\gamma} \quad with \quad \beta = \sqrt{\frac{m_2 m_{2/b}}{m_2 m_{2+2/b}}} \tag{12}$$

Tovo and Benasciutti (TB)

The model developed by Tovo and Benasciutti [3] is a semi empirical factor based on Ryshlick results [8]. The NB correction factor is expressed as follows:

$$\lambda_{TB} = \beta + (1-\beta)\alpha_2^{b-1} \tag{13}$$

$$\text{with} \begin{cases} \beta = \dfrac{(\alpha_1 - \alpha_2)\left[A(1 + \alpha_1\alpha_2 - (\alpha_1 + \alpha_2))e^{B\alpha_2} + \alpha_1 - \alpha_2\right]}{(\alpha_2 - 1)^2} \\ \alpha_1 = \dfrac{m_1}{\sqrt{m_0 m_2}} \quad \alpha_2 = \dfrac{m_2}{\sqrt{m_0 m_4}} \\ A - 1.112 \qquad B - 2.11 \end{cases}$$

Direct models. The models which are presented hereafter are directly linked to the probability density function of rainflow extracted cycles.

Chaudhury and Dover (CD)

$$f_s(S) = \frac{1}{\sqrt{m_0}}\left[D_1 \exp\left(-\frac{Z^2}{2(1-\gamma^2)}\right) + D_2 ZQ \exp\left(-\frac{Z^2}{2}\right)\right] \tag{14}$$

$$\text{with} \begin{cases} D_1 = \dfrac{\sqrt{1-\gamma^2}}{\sqrt{2\pi}} \quad D_2 = \dfrac{\gamma}{2} \\ Q = 1 + erf\left(\dfrac{\gamma\ Z}{\sqrt{2(1-\gamma^2)}}\right) \end{cases}$$

and where *erf* is the error function expressed by:

$$erf(x) = \frac{2}{\sqrt{\pi}}\int_0^x e^{-t^2}\,dt = \frac{2}{\sqrt{\pi}}\sum_0^{+\infty}(-1)^n \frac{x^{2n+1}}{n!(2n+1)}.$$

Dirlick (DK)

Dirlick [2] proposed a model which is the sum of an exponential and two Rayleigh distributions:

$$f_s(S) = \frac{\dfrac{D_1}{Q}e^{\frac{-Z}{Q}} + \dfrac{D_2 Z}{R^2}e^{\frac{-Z^2}{2R^2}} + D_3 Z e^{\frac{-Z^2}{2}}}{2\sqrt{m_0}} \quad \text{with } Z = \frac{S}{2\sqrt{m_0}} \tag{15}$$

and several factors determined empirically

$$x_m = \frac{m_1}{m_2}\sqrt{\frac{m_2}{m_4}} \qquad D_1 = \frac{2(x_m - \alpha_2^2)}{1+\alpha_2^2} \qquad R = \frac{\alpha_2 - x_m - D_1^2}{1 - \alpha_2 - D_1 + D_1^2}$$

$$D_2 = \frac{1 - \alpha_2 - D_1 + D_1^2}{1-R} \qquad D_3 = 1 - D_1 - D_2 \qquad Q = \frac{1.25(\alpha_2 - D_3 - D_2 R)}{D_1}$$

Zhao and Backer (ZB)

The probability density function proposed by Zhao and Backer [11] is the sum of a Weibull and a Rayleigh distribution:

$$f_s(S) = \frac{w\alpha\beta Z^{\beta-1}e^{-\alpha Z^{\beta}} + (1-w)Ze^{\frac{-Z^2}{2}}}{2\sqrt{m_0}} \quad \text{with } Z = \frac{S}{2\sqrt{m_0}} \tag{16}$$

and several factors determined empirically

$$\alpha = 8 - 7\gamma \quad \beta = \begin{cases} 1{,}1 & if \quad \gamma < 0{,}9 \\ 1{,}1 + 9(\gamma - 0{,}9) & if \quad \gamma < 0{,}9 \end{cases} \qquad w = \frac{1-\gamma}{1 - \sqrt{\dfrac{2}{\pi}}\Gamma(1 + \dfrac{1}{\beta})\alpha^{-\frac{1}{\beta}}}$$

Single moment (SM)

This model proposed by Larsen and Lutes [12] holds its name from its dependency on only one spectral moment of the PSD. The integrated form of the formula can be expressed as follows:

$$E[D_{SM}] = C^{-1}\left(2\sqrt{2}\right)^{b}\left(\sqrt{m_{2/b}}\right)^{b}\Gamma\left(\frac{1+b}{2}\right) \tag{17}$$

NUMERICAL ASSESSMENT OF THE DIFFERENT MODELS

As it was explained in the introduction, this part of the study was already presented in a previous ICAF Symposium. That is why only the approach and the main results are briefly exposed here as a part of a complete study made by the DGA. The reader who is interested by this analysis is suggested to see the paper [13].

Comparison of methods

In order to compare the different models, extensive sets of simulations are performed. The basic principle of the method is given hereafter:

- More than 600 PSD of different types defined on the 0-100Hz band were used for the simulation. Special care is taken in order to use realistic PSD and it is chosen to work with multimodal PSD (see figure 1).

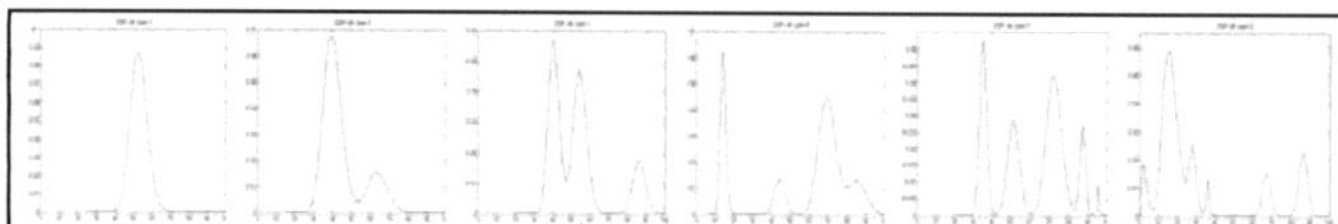

$$S(f) = \sum_{i=1}^{n} A_i \times \exp\left(-\frac{(f - f_{ci})^2}{B_i}\right) \quad with \quad n = 1..6$$

Figure 1. The different types of PSD used in this study.

- For each PSD, a set of time signals are generated thanks to the formula $X(t) = \sum_{k=0}^{N} \sqrt{2S_s(f_k)\Delta f}\cos(2\pi f_k t + \varphi_k)$ where N is the number of points chosen to draw the PSD, Δf the frequency step between two points ($f_k = f_{\min} + k\Delta f$) and φ_k a random phase angle defined independently for each frequency f_k. Then the fatigue life of each signal is calculated thanks to the rainflow counting method, a certain mean stress correction method and the Miner rule. An average fatigue life is calculated and defined as the reference fatigue life for the studied PSD.
- For each PSD the fatigue life is also directly calculated in the frequency domain by the different models presented previously.
- A comparison between both fatigue lives is made for each PSD.
- The following well known error indicator is calculated on the whole set of PSD in order to compare the different models: $IE = 100 \times \sqrt{\frac{1}{n_{tot}} \sum_{i=1}^{n_{tot}} \left(\log\left(\frac{T_{spec,i}}{T_{ref,i}}\right)\right)^2}$ where T_{spec} is the predicted fatigue life in the frequency domain and T_{ref} the predicted fatigue life in the time domain for the PSD n°i.

Studied parameters

The objective of the study is to determine the most important parameters on the error committed by the different models. Thus it is decided to study the influence of:

- the Basquin slope of the Wöhler curve. It is taken equal to 3, 5, 7, 9, 11 and 13 while the C factors is defined in order to have 1000 cycles to failure for an applied stress of 450 MPa, this value being defined as the tensile strength of the materials S_u.
- the RMS value expressed in term of percentage of S_u.
- the bandwidth of the signal in term of irregularity factor: thanks to well selected PSD, the irregularity factor covers the range between 0.1 and 1.0.

- the mean stress correction method: the Goodman and the Gerber models are compared to a calculation without any mean stress correction method.

Conclusions of the analytical part

The models to assess the fatigue life of structures subjected to vibratory loading are tested. In a first part, it is seen that the influence of the RMS value and of the used mean stress correction method is negligible. Then, the *b* parameter of the Basquin model and the irregularity factor, an indicator of the signal bandwidth, are identified as the most influent parameters on the error committed by the different models. Actually the studied models give very accurate fatigue life estimations for an irregularity factor above 0.9 but below, the results are more mitigated: a result between 0.5 et 1.5 times the reference fatigue life is often obtained, which could be sufficient depending on the application frame. Finally the different simulations show that the Dirlick model is the most accurate one even if the non conservatism of the approach implies to use a large safety factor when applied in the domain where the Wöhler curve is quite horizontal.

After this analytical assessment, it is interesting to compare the fatigue life predictions with some tests results. It concerns the second part of the study which is presented here below. According to the previous conclusions:

- the Dirlick model is the one used to calibrate tests (even if all the models give rather good results)
- the tests are imagined to have an irregularity factor below 0.9
- the parameters of Basquin are discussed below.

COMPARISON BETWEEN TESTS AND CALCULATIONS

Objectives of the tests and challenges for the coupons design

To complete the analytical approach, it is decided to continue the study with some vibration tests. This paragraph explains the objectives of these tests and the constraints on the coupons design to reach these objectives. The analytical study shows that the irregularity factor is very important: the thinner the bandwidth is, the better the results are. That is why, in order to properly assess the models, it is interesting to avoid a structure with a narrow band frequency response. To do that, it is needed to create coupons with more than one eigen mode. With simple coupons, it is not really easy to have many modes so it is chosen to create coupons with only two resonance frequencies.

In order to have a fatigue failure, it is needed to determine one or more "hot points" which are the weak points of the coupon. To avoid having a local narrow band response, these points have to be excited by both eigen modes of the structure, with equivalent intensities. It is important to note that the design of the coupon is a way to achieve these objectives but the selection of the excitation PSD can also modify the response values. The choice of the PSD is discussed below.

Finally, as it was written in the introduction, the global objective of the study is to anticipate a test or to extend a structural qualification, so it is interesting to evaluate the models to calculate the fatigue life but also to predict the localizations of critical points. That is why it is decided to create a coupon with two hot points, each excited by both resonance modes and with close fatigue lives to have a real challenge to determine which hot point will fail first.

Methodology of calculation

In the *Methodology to assess fatigue damage in the frequency domain* paragraph, it was presented the way to calculate damage and fatigue life from the local stress PSD of the considered critical point. The objective of this paragraph is to explain the way to obtain this stress PSD (called output PSD in this article), from the input PSD (acceleration PSD of excitation, which is the control of the vibrating pot). To do that, a Finite Element Modeling (FEM) is made to obtain the transfer function H(f) between acceleration of excitation and local stress. As levels of stress are low, it is indeed possible to consider the structure as linear.

	Input	System	Output	
Temporal domain	$e(t)$	$h(t)$	$s(t) = \int_0^t e(\tau)h(t-\tau)d\tau$	(18)
Spectral domain	$e(f)$	$h(f)$	$s(f) = H(f) \times e(f)$	(19)

$H(f)^2$ enables to link directly the input PSD and the output PSD.

$DSP_{input}(f)$	x	$H(f)^2$	=	$DSP_{output}(f)$
$\frac{g^2}{Hz}$		$\left[\frac{MPa}{g}\right]^2$		$\frac{MPa^2}{Hz}$

(20)

To obtain H(f), a sweep sine is done on the FEM with a 1g acceleration as input. The local VonMises stress function of frequency is directly the transfer function H(f). The bandwidth of the sweep sine is determined to have both natural frequencies of the structure: [10; 300] Hz.

The FEM is a 2D shell modeling made with Samcef software with a refined mesh next to critical points. For the boundary conditions, the area which is clamped on the vibrating pot is rigidly linked to a virtual point. This point is linked to a fixed point with a spring element: the z stiffness (in the acceleration direction) is very low (100 N/m) whereas the x and y stiffness are very high (200 000 N/m). For the excitation, all the "clamped" area (green area on the next figure) is loaded by a 1g acceleration (sweep sine).

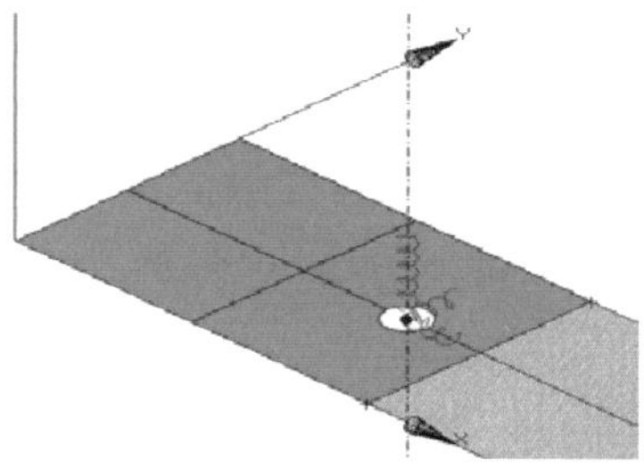

Figure 2. FEM and boundary conditions.

The main difficulty for the modeling is the choice of the damping value. The value of 4% which is a classical value for structure is initially considered. As the test results show it, this value is nevertheless largely overestimated for a simple coupon, without any fixation or assembly so without friction or energy dissipation. This value is anyway fitted with the results of test sweep sine. The values of the transfer function obtained by test for the resonance frequencies enable indeed to obtain the damping values, which can be different for each eigen modes.

Once the output DSP is obtained, the calculation of the fatigue life is done with an Excel file, by discretizing of signal and integration.

It is very important to remind that the material parameters shall be precisely known, especially the Basquin parameters b and C, to have accurate results (see the *Conclusions of the analytical part* paragraph).As no specific tests campaign to characterize the material used for coupons (which is 2024T3 aluminum alloy) is planned, those parameters are selected thanks to a bibliographic research. Many sources are found, with many different results, often obtained by classical fatigue test and not vibrating test [14], [15], [16] and [17]. Initially, the values from the reference [17] for Kt = 5 is selected.

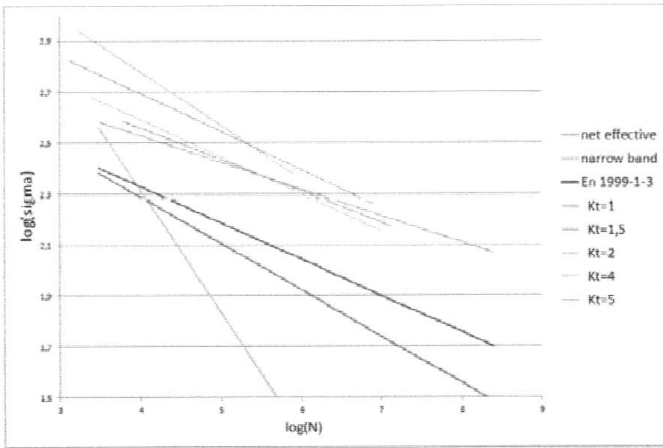

Figure 3. Different parameters for the Basquin model.

As it is just presented in the next paragraph, two types of coupons are manufactured. The first ones (type 1) enable to fit the Basquin parameters, using the durations of the test, then these parameters are used for the calculation of fatigue life for the type 2 coupons.

Design of coupons

Two types of coupons are tested, with three specimens for each type. Both types are described below.

Type 1: the design of this coupon is copied from reference [18], in which similar tests were made. In this reference [18], the material is different (7075 aluminum alloy) and so is the clamping area (the coupon is only fixed to the vibrating pot by a screw) but the coupon has two resonance frequencies and tests have already been done successfully with it.

It is chosen to use a clamping tool instead of a screw for two main reasons: a more reproducible assembly between each specimen and the suppression of non-linearity due to contact between the coupon and the vibrating pot caused by a direct fixation to the pot.

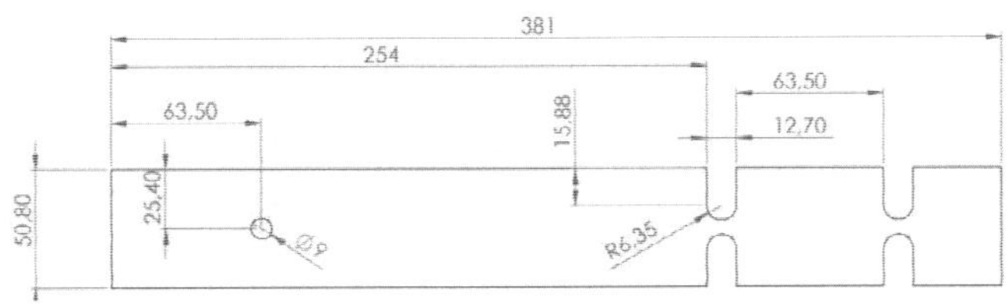

Figure 4. Design of type 1 coupon. Comment: the hole of 9 mm was not manufactured (it was used to fix the coupon with a screw in reference [18]).

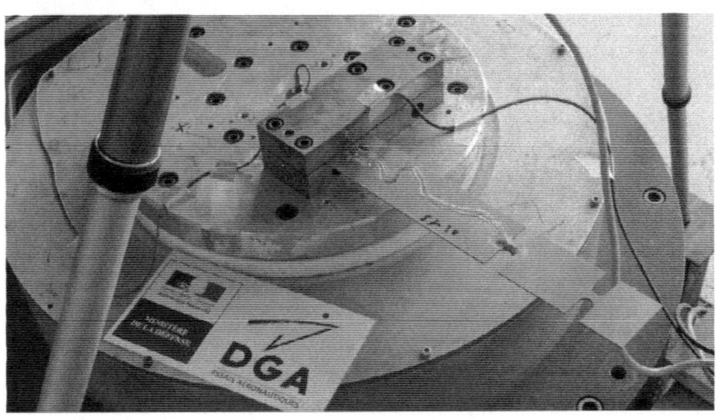

Figure 5. Type 1 coupon clamped to the vibrating pot.

There is only one hot point but two resonance frequencies thanks to the two small parts in the design.

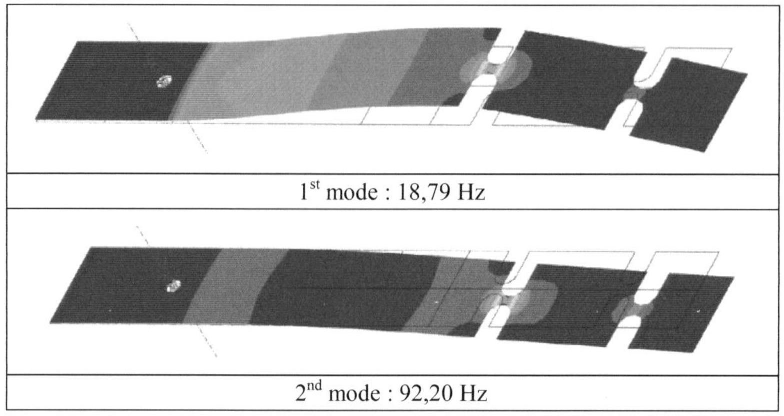

Figure 6. Modal analysis of type 1 coupon with FEM.

The transfer function is obtained as presented in the *Methodology of calculation* paragraph.

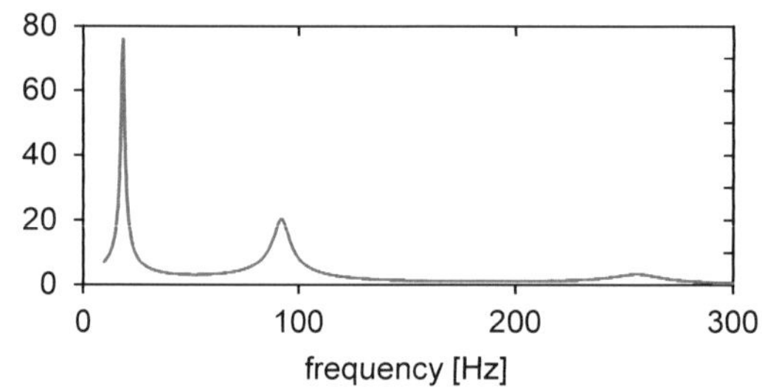

Figure 7. Transfer function of type 1 coupon.

The "shape" of the input PSD is also copied from reference [18] but the levels are modified as the material and the objective of fatigue life are different. A 30min fatigue life is aimed to have tests:

- long enough to be able to assess the scattering between each specimen
- not too long in case of a bad assessment of the fatigue life to be able to interrupt tests after some hours.

That is what happened during tests. As it is explained later in this paper, the levels of the next input PSD were thus modified several times, but always with the same shape.

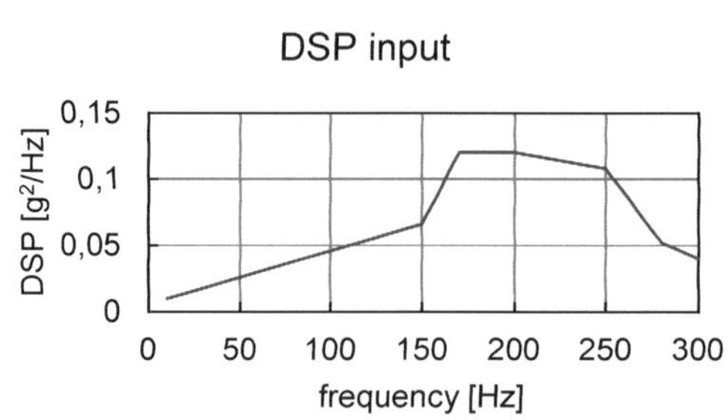

Figure 8. Initial input PSD for type 1 coupon.

It can be observed on the next table that the results with the different models are very close.

Table I. Fatigue life for type 1 coupon with different models.

Dirlik	0h 31m 3s
Benasciutti-Tovo	0h 32m 41s
Single moment	0h 34m 24s
Wirsching-Light	0h 33m 49s
Ortiz-Chen	0h 29m 17s
Zhao-Backer	0h 32m 21s
Chaudhury-Dover	0h 33m 57s
Steinberg	0h 24m 37s
Narrow-Band	0h 26m 1s

Type 2: the design of this second coupon is created especially for this study. The objective is to have two hot points and two resonance frequencies. The first critical point is the hole made in the coupon (zone B, in blue on the next figure). The second one is a double one and corresponds to the junction parts with the small fins (zone C, in red on the next figure).

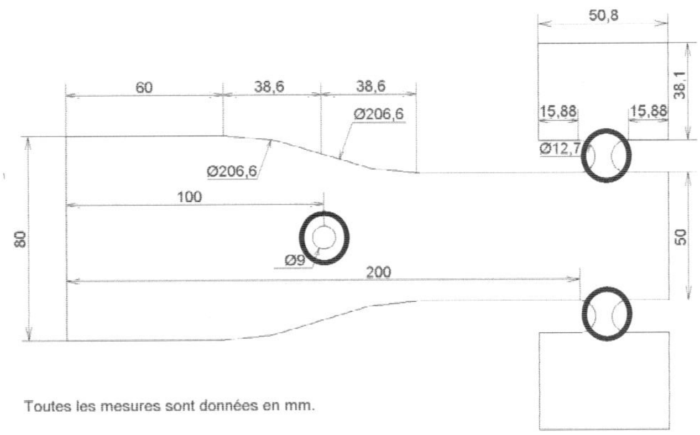

Figure 9. Design of type 2 coupon.

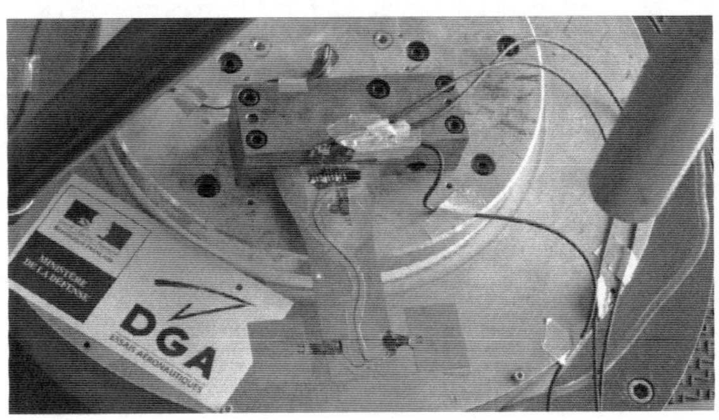

Figure 10. Type 2 coupon clamped to the vibrating pot.

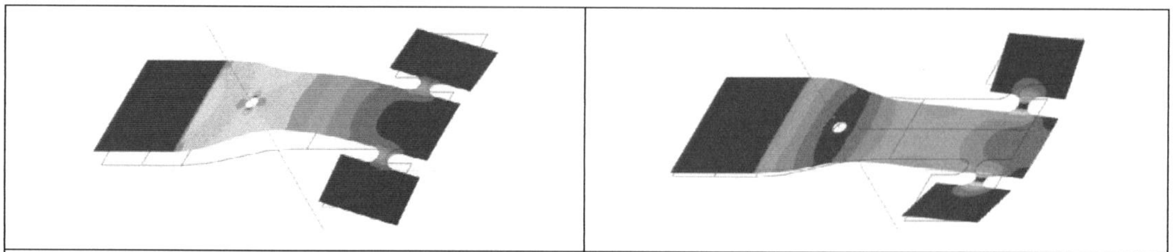

1st mode : 40,78 Hz	2nd mode : 251,87 Hz

Figure 11. Modal analysis of type 2 coupon with FEM.

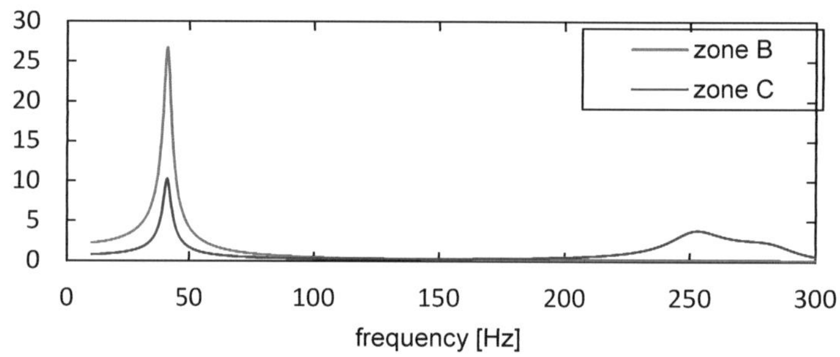

Figure 12. Transfer function of type 2 coupon for zones B and C.

The input PSD is chosen to have a fatigue life for each critical point about 30 min, with a small difference between each zone: it is foreseen that the zone B has damage before zone C. Like for type 1 coupons, it can be observed on the next table that the results with the different models are close, but with some inversion between both zones in the prediction of the first damaged area.

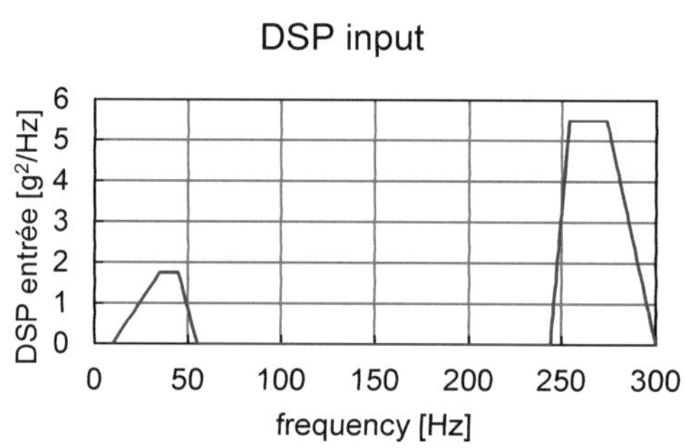

Figure 13. Initial input PSD for type 2 coupon.

Table II. Fatigue life for type 2 coupon with different models.

	Clamped zone	**Zone B**	**Zone C**
Dirlik	2h 35m 7s	**30m 25s**	**36m 48s**
Benasciutti-Tovo	2h 37m 56s	30m 7s	37m 51s
Single moment	2h 53m 33s	30m 5s	32m 24s
Wirsching-Light	2h 12m 39s	38m 32s	33m 3s
Ortiz-Chen	2h 28m 38s	22m 34s	32m 6s
Zhao-Backer	2h 31m 16s	34m 51s	27m 53s
Chaudhury-Dover	2h 0m 49s	35m 56s	26m 36s
Steinberg	1h 36m 34s	28m 3s	24m 13s
Narrow-Band	1h 42m 4s	29m 39s	25m 35s

Like for the input PSD of type 1 coupons, the levels of the previous PSD are modified before tests due to the fitting of damping and Basquin models, but not the shape.

Tests description

For instrumentation, an accelerometer, a laser vibrometer and some gauges are used.

The accelerometer, on the vibrating pot, is used for its control.

The gauges are set next to the critical points. They do not last because there are too many "cycles" during tests but they are useful to have information about stresses. With the transfer function obtained by sweep sine test at gauge location, it is possible to fit the damping of the coupon for the FEM.

The laser vibrometer gives information about displacement and not about stress. Stresses are however not needed to follow the fatigue endurance test: the knowledge of the resonance frequencies is enough. The vibrometer has the advantage to last during all the test.

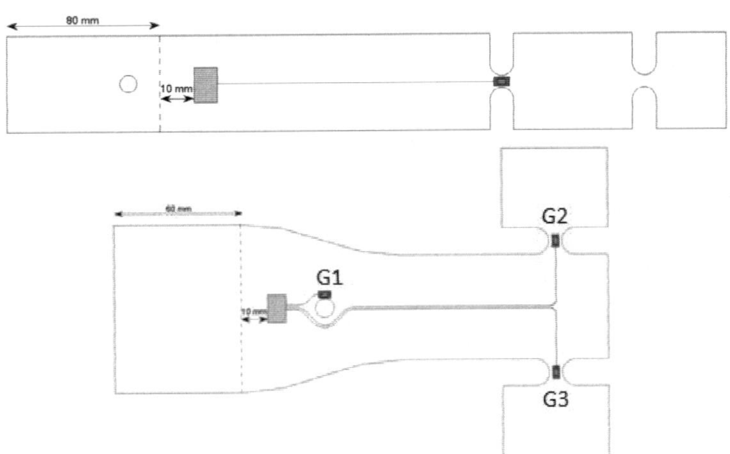

Figure 14. Type 1 and 2 coupons with gauges locations.

There are two tests phases:

- First, a sweep sine between 10 and 300 Hz is done in order to identify the test transfer function. The objective is to fit the damping and then to adjust the input PSD with the same objectives of fatigue life with this new damping.

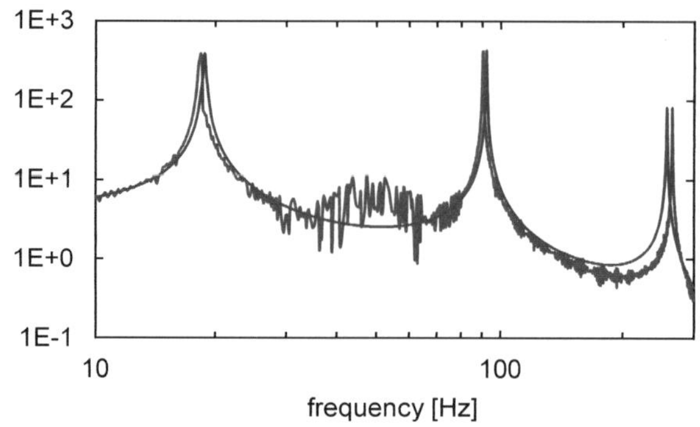

Figure 15. Transfer functions obtained by test (red) and FEM (blue) after fitting of damping for coupon #12 (type 1).

- Then, the input PSD is applied and the output PSD (for each gauge and vibrometer) are regularly recorded during the fatigue test. The detection of damage is done by the analysis of the resonance frequency value. Indeed, when a crack appears on a structure, its stiffness decreases and the eigen modes are modified: the resonance frequency values decrease also. The initial criterion which is chosen to detect these changes is a 2% decreasing. In fact, it is quite difficult to observe this gap of frequency during test: the end of the test is the failure of the coupon but it is possible to detect the gap by post treatment. The figures below show the vibrometer output PSD function of time, for the second type 1 coupon: both figures are detailed zones for PSD around the two first resonance frequencies. The first one (about 18 Hz) has no change but a decreasing of the frequency for the second one can be stated (red line, about 90 Hz).

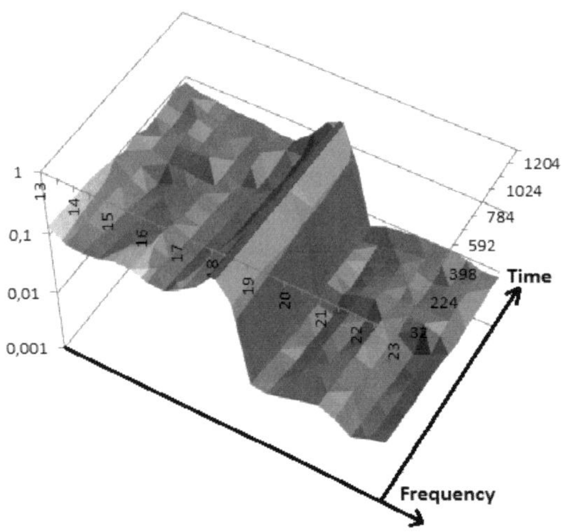

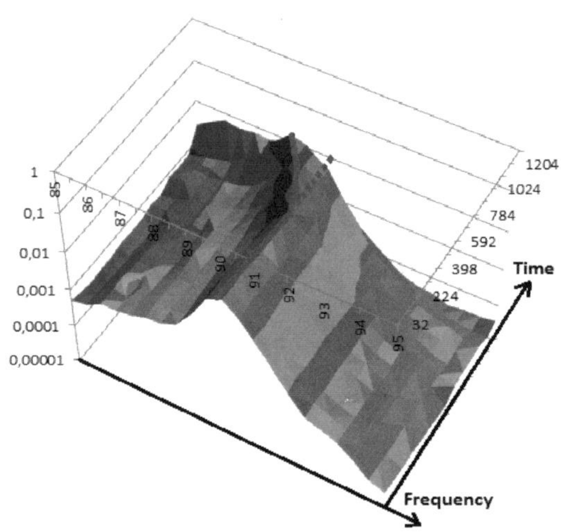

Figure 16 and Figure 17. Vibrometer output DSP function of time for coupon #12 (type 1). Details around the first resonance frequency at the left hand side and around the second one at the right hand side, with a time evolution.

Results

For the first type 1 coupon (#11), the damping values for each eigen modes are determined: 0.6% for the first resonance frequency, 0.2% for the second one and 0.1% for the last one. The initial input PSD (see *Design of coupon* paragraph) is adjusted with these new damping values to keep the objective of a fatigue life about 30 min.

This new input PSD is used but nothing appears after two hours of test so the level is increased step by step to obtain a failure. The results of this coupon are difficult to analyze because it was loaded by several different input PSD. This enables however to determine a level for the input PSD of the second (#12) and the third (#13) type 1 coupons. The fatigue lives which are obtained for coupons #12 and #13 are very close.

Table III. Fatigue lives for type 1 coupons.

Fatigue Life	
Coupons	Δ = 2%
11	-
12	19mln34
13	19min04

Because the damping value had already been fitted, the only parameters to explain the difference between tests results and calculation are the Basquin parameters. They are therefore fitted to obtain a fatigue life calculation for the type 1 coupon about 20min with:

- the damping values obtained by test (mean of the three coupons)
- the new input PSD which were used for the tests of #12 and #13 coupons.

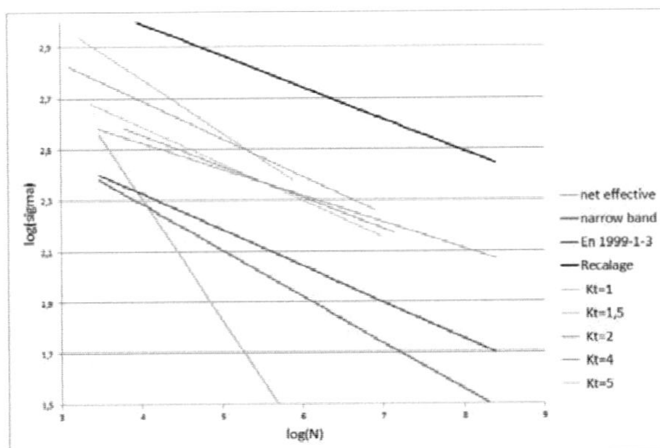

Figure 18. Different parameters for the Basquin model with the fitted parameters in black.

The black line of the previous figure corresponds to the fitted Basquin parameters ($b = 8$, $C = 9.10^{27}$) compared to the different bibliographic values (see *Methodology of calculation* paragraph). We currently have no explanation for these high values: are they the "true" values for this type of test or is it a compensation for a problem in modeling or models? A characterization tests campaign on the same material, with the same bandwidth, could be interesting to perform to provide the answer.

For the type 2 coupons (#21, #22 and #23), the levels of the initial input PSD (see *Design of coupon* paragraph) are adjusted to obtain a 30min fatigue life with:

- the new Basquin parameters
- the fitted damping.

The analysis of these results is not finished yet so only some comments are made about results.

On the transfer functions of type 2 coupons obtained by test (the one on the next figure, for gauge G2 of coupon #21, is representative of the others), it can be seen that:

- there is another resonance frequency around 100 Hz, but with a low value
- contrary to the transfer function obtained by modeling, the second (or third if the small eigen mode at 100 Hz is considered) resonance frequency is a double one. These two values probably correspond to each small fin which are probably not symmetrical (manufacturing, clamping, gauges positions…).

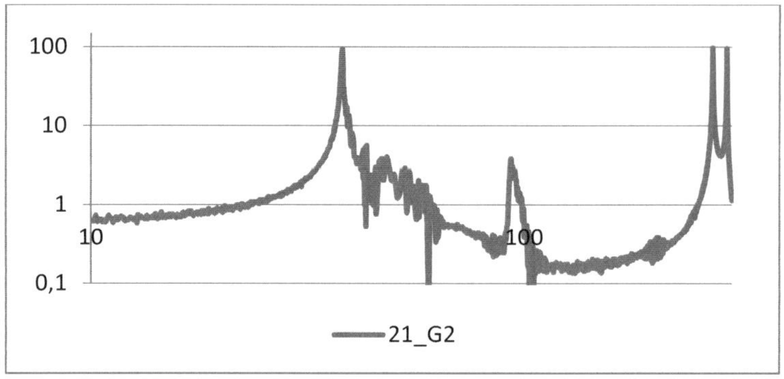

Figure 19. transfer function obtained by test for gauge #2 of coupon #21 (type 2).

It is more difficult to detect the damage than for type 1 coupons. That is why different criteria are used: a gap for the resonance frequencies of 1Hz, 1,5Hz, 2Hz or 2%.

Table IV. Fatigue lives for type 2 coupons. "?" means that data cannot be used. "-" means that the gap of frequency was not reached before the end of the test.

Fatigue Life					
Coupons	**Δ = 1Hz**	**Δ = 1,5Hz**	**Δ = 2Hz**	**Δ = 2%**	**Failure**
21 (fin)	?	47min04	?	57min34	58min04
22 (fin)	38min04	?	64min04	1h24min34	1h24min34
23 (fin)	?	50min16	60min01	-	1h38min50
23 (hole)	1h09min26	-	-	1h09min26	-

Except for the coupon #23, only the second resonance frequency (corresponding to zone C, the junction parts with the small fins, see *Design of coupons* paragraph) decreases during test, and the failure always happens for one fin. For coupon #23, after 1 h 09 min of test, a gap of 1 Hz (~2% for this frequency) appears for the first resonance frequency corresponding to zone B, next to the hole (see *Design of coupons* paragraph). A further investigation at the edge of the hole could confirm the initial hypothesis that both selected hot points are critical points with close fatigue lives if damage is stated in this area.

All data cannot be used so it is difficult to detect precisely the fatigue life but it seems to happen between 30 min and 1 hour according to the chosen criterion so the calculations are conservative and give results of the same order of magnitude.

A detailed analysis and investigations on the coupons have to be done. Moreover, it could be interesting to continue the study by removing the second fin (to obtain a symmetrical coupon) and then to carry on the fatigue test to assess the fatigue life of the hole (zone B) of the coupon. Obviously, a new modeling corresponding to the new geometry will be needed.

CONCLUSION

The models to assess the fatigue life of structures subjected to vibratory loading were tested. A first part of the study was analytical and the influent parameters were analyzed. The second part presented some tests to evaluate these models. For both simple structures which were studied, the different models gave close results and of the same order of magnitude as the test results. The damping for Finite Element Models was obtained by test: between 0.1% and 0.6%, which is very low according to the classical value of 4% for structure. The Basquin parameters, which are very important for fatigue life calculation, were also obtained by test. The values are very different from bibliography: a specific characterization test campaign could be interesting to precisely estimate these parameters. A detailed analysis and investigations on the coupons have to be done to confirm the initial hypothesis that both selected hot points are critical points with close fatigue lives. Moreover, it could be interesting to continue the study by removing the second fin (to obtain a symmetrical coupon) and then to carry on the fatigue test to assess the fatigue life of the hole (zone B) of the coupon. The next step will be to apply this methodology and these models to a real structure before a vibrating endurance test.

REFERENCES

[1] F. Sherratt, N. Bishop and T. Dirlik, (2004), *Predicting fatigue life from frequency domain data: current methods*, Part A.

[2] T. Dirlik, (1985), *Applications of computers in fatigue analysis*, Ph.D. Thesis, University of Warwick.

[3] D. Benasciutti, (2004), *Fatigue analysis of random loadings*, Ph.D. Thesis, University of Ferra.

[4] S.O. Rice, (1954), *Mathematical analysis of random noise, Papers on noise and stochastic processes*, Ed. N. Wax, Dover, New York.

[5] J.S. Bendat, (1964), *Probability functions for random responses*, NASA report.

[6] C. Lalanne, (1999), *Dommage par fatigue*, Ed. Hermès Science.

[7] P.H. Wirsching, T.L. Paez and K. Ortiz, (1995), *Random Vibrations, Dover Publications*, New York.

[8] M. Frendahl and I. Rychlik, (1993), *Rainflow analysis: Markov method*, International Journal of Fatigue, vol. 15, n. 4, pp. 265-272.

[9] G.K. Chaudhury and W.D. Dover, (1985), *Fatigue analysis of offshore platforms subject to sea wave loadings*, International Journal of Fatigue, vol. 7, pp. 13-19.

[10] J.C.P.T. Kam and W.D. Dover, (1988), *Fast fatigue assessment for offshore structures under random stress history*, Proc. of the Institution of Civil Engineers, Part 2, pp. 689-700.

[11] W. Zhao and M.J. Baker, (1990), *A new stress-range distribution model for fatigue analysis under wave loading*, Environmental forces on offshore structures and their prediction, vol. 26, pp. 271-291.

[12] C.E. Larsen and C.D. Lutes, (1991), *Predicting the fatigue life of offshore structures by the single-moment spectral method*, Probabilistic Engineering Mechanics, vol. 6, pp. 96-108.

[13] M. Fressinet, F. Fuchs, P. Madelpech (2011), *Fatigue life estimation of structures subjected to vibratory loading*, 26th ICAF Symposium.

[14] A. Lynn, D. DuQuesnay (2002), *Computer simulation of variable amplitude fatigue crack initiation behaviour using a new strain-based cumulative damage model*, International Journal of Fatigue 24, pp. 977-986, 2002.

[15] G. Brown, R. Ikegami (1970), *The Fatigue of Aluminium Alloys Subjected to Random Loading.*

[16] Y. Huang, A. Ringeval (2012), *Random Vibration Fatigue Analysis with LS-DYNA*, 12th International LS-DYNA Users Conference.

[17] Military Handbook MIL-HDBK-5H (1998), *Metallic materials and elements for aerospace vehicle structures*, Department of Defense, United States of America.

[18] V. K. Nagulapalli, A. Gupta, S. Fan (2007), *Estimation of Fatigue Life of Aluminum Beams subjected to Random Vibration*, Department of Mechanical Engineering, Northern Illinois University, DeKalb, IL.

Tackling the Characterization of H-60 Airframe In-Flight Dynamic Response and Fatigue Loading Using Small Scale High Performance Computing

John Vine, Luther Krake, Phil Jackson

Defence Science and Technology Organisation (DSTO), Australia

The Australian Defence Science and Technology Organisation (DSTO), in collaboration with the United States Navy (USN) Naval Air Systems Command (NAVAIR) and the Australian Defence Materiel Organisation (DMO), is embarking on a program which aims to demonstrate the technical feasibility of undertaking a Full Scale Fatigue Test (FSFT) on a helicopter airframe. A necessary precursor to conducting a FSFT on a helicopter is a detailed understanding of how the airframe responds to inflight loading. Helicopter airframe responses are induced by a complex combination of quasi-static manoeuvre loads and, primarily, rotor-induced vibratory loads. Vibratory loads result in oscillating airframe responses (i.e. vibrations) at the operating frequencies (and harmonics thereof) of rotating components, and can excite resonant responses at nearby natural frequencies of the airframe. Understanding these vibratory responses, how they vary across the airframe, how they vary by flight condition and, more importantly, their contribution and importance to material fatigue, are all necessary steps in defining the target airframe responses that a FSFT will need to simulate. To that end, DSTO has developed a novel methodology to characterize the measured in-flight dynamic response of a helicopter airframe and establish how the response varies with flight condition and airframe location. The methodology also takes advantage of modern graphics processing hardware, and concurrent programming and data aggregation techniques, to enable efficient processing and presentation of the large amount of flight data required by the activity. This paper describes the methodology and its application to the H-60 Black Hawk airframe, and includes some example cases to qualitatively demonstrate how well the characterisation correlates with that expected from the literature.

SOURCES OF IN-FLIGHT VIBRATIONS

Helicopter airframes experience high frequency vibratory loading as a result of their rotating systems. This vibratory loading produces high-cycle stress responses within the airframe which can contribute significantly to airframe material fatigue. Generating a thorough understanding of this vibratory loading is a vital first step in analysing and testing for material fatigue of helicopter airframes. A methodology has been developed to improve understanding of the vibratory loading of the Sikorsky H-60 family of helicopters by characterizing measured in-flight dynamic airframe response.

The Sikorsky H-60 Black Hawk is a four-bladed, twin-engine, medium-lift utility helicopter manufactured by Sikorsky Aircraft. It employs bifilar absorbers, in-cabin vibration absorbers and elastomeric bearings to reduce fuselage vibrations. Helicopter vibrations are generally characterised by dominant sinusoidal forced responses to periodic inputs produced by the major rotating components (e.g. main rotor, tail rotor, engines, etc.) superimposed on a broadband background (generally lower amplitude sinusoids and random vibrations from sources such as aerodynamic turbulence) [1]. Understanding helicopter vibrations requires understanding those periodic load inputs, specifically their frequencies of occurrence and potential to excite resonant responses.

Main Rotor Vibrations

The dominant vibratory response of a helicopter airframe is due to rotor hub forces and moments caused by the essentially unsteady aerodynamics of the main rotor. Theory [2, 3] suggests that for a four bladed helicopter (such as the H-60), blade root vertical shear (thrust), torque (lagwise moment), rotor drag and side forces (in-plane shears), pitch and roll moments (flapwise moments), and collective and cyclic control system loads (feathering moments) should all cancel each other at the hub, except at integer multiples of the blade passage frequency which are transmitted to the airframe. However, in reality the rotor hub is not a perfect filter of blade root forces and (due to variations in loading caused by compressibility and stall effects, and due to small aerodynamic (out-of-track) and inertia (out-of-balance) differences between blades) can transmit vibrations at frequencies other than multiples of the blade passage frequency. Although the H-60 fuselage will experience some 1/rev, 2/rev and 3/rev vibrations in flight, the dominant vibration will occur at integer multiples of the blade passage frequency (i.e. 4/rev, 8/rev, 12/rev, etc.). As the nominal main-rotor frequency of the H-60 is 4.3 Hz, these main rotor induced vibratory responses were expected to occur at 4.3 Hz, 8.6 Hz, 12.9 Hz, 17.2 Hz, 34.4 Hz, 51.6 Hz, etc.

Tail Rotor Vibrations

The tail rotor is another source of forced vibratory responses similar to, but generally of less severity than, the main rotor. Frequencies associated with tail rotor excitation are integer multiples of the tail rotor frequency, with multiples of the tail rotor blade passage frequency expected to be most significant. Forced vibratory responses caused by the tail rotor can be expected to be more severe in areas of close proximity to the tail rotor, such as the empennage and tail cone. As the nominal tail-rotor frequency of the H-60 is 19.83 Hz, tail-rotor vibrations were expected at 19.83 Hz, 39.67 Hz, 59.50 Hz, 79.33 Hz, etc.

Drive Train Vibrations

The H-60 has various drive-shafts, gear boxes, and other rotational machinery, which also produce vibrations in the airframe. The major components expected to produce vibrations within the aircraft and their nominal frequencies were: the twin-engines (348.3 Hz); the generator (196.8 Hz); the hydraulic pump module (119.8 Hz); the main gear box main module (95.8 Hz); the tail-cone drive shaft (68.6 Hz) and; the tail-pylon drive shaft (55.3 Hz). Vibrations associated with these sources are generally of a relatively low magnitude compared to vibrations caused by the main and tail rotors.

Other Aerodynamic Induced Vibrations

Additional sources of in-flight vibrations include aerodynamic pressure pulses imparted by the main rotor blades on the upper fuselage surfaces, and main rotor trailing blade-tip vortices that impact on the relatively flexible horizontal stabilizer and vertical tail pylon. Interaction between the rotor wake and the empennage is especially intense in hover, climb, and low-speed forward flight since in these flight regimes the wake is transported more directly onto exposed surfaces. Aerodynamic turbulence or vorticity generated by the airflow separating from some forward portion of the fuselage, can also impinge on the horizontal stabilizer and tail, inducing vibrations. This turbulence is generally broadband in nature and is often characterised by the resonant vibrations it may excite.

Modal Response

A key principle in helicopter design is to avoid the coincidence of structural modal (natural) frequencies with the frequencies of possible exciting forces, particularly the dominant ones associated with main rotor hub loads. To determine what modes are being excited as part of the overall airframe vibration response, it is first necessary to identify the inherent system dynamic properties (i.e. modal frequencies, mode shapes and damping) expected of the airframe structure.

Table V. UH-60A frequencies obtained from finite element analysis compared to modes obtain from Ground Vibration Test (GVT) data [4].

Mode No.	Mode Shape	Test Frequency FX	FY	FZ	Test (Avg)	NASTRAN	% Error of NASTRAN (n-t)/t	%Critical Damping FX	FY	FZ
1	1st Lateral Bending		5.58		5.58	5.34	-4%		1.03%	
2	1st Vertical Bending	6.76		6.82	6.79	6.36	-6%	0.98%		1.25%
3	Stabilator RB	10.59	10.12		10.36	9.43	-9%	4.00%	2.93%	
4	Stabilator RB	11.94			11.94	11.16	-7%	1.95%		
5	Xssn Pitch	12.32		12.56	12.44	11.00	-12%	1.36%		1.41%
6	Xssn Roll/ 2nd Lateral		13.81		13.81	12.74	-8%		2.69%	
7	2nd Vertical Bending	13.96		14.05	14.01	12.34	-12%	2.64%		2.15%
8	Cockpit/ Cabin Roll		15.41		15.41	14.48	-6%		2.84%	
9	2nd lateral					15.32				
10	Vertical					15.84				
11	Stabilator Yaw	16.14			16.14			3.09%		
12	3rd Vertical / Xssn Vert	19.32		19.40	19.36	18.07	-7%	2.14%		2.28%
13	Cabin Shear/ 3rd vert					19.52				
14	Nose Vert/ Long					20.00				
15	Cockpit Roll/ Lateral					20.65				
16	?					21.39				
17	?					22.71				
18	Cockpit/ Cabin Torsion					23.29				
19	Cabin Vertical					23.40				
20	Cabin Torsion					23.53				
21	?	23.16			23.16			5.94%		
22	Stabilator Vert Bend.	25.22		25.07	25.15	24.53		1.69%		1.90%
23		25.64			25.64			1.88%		
24		26.58		26.39	26.49	26.60		1.65%		2.12%
25		28.36			28.36	28.50		2.84%		
26		30.57	30.65		30.61			2.18%	2.27%	
27		32.07		32.76	32.42	32.09		2.21%		1.89%
28		33.78			33.78			2.09%		
29		34.52			34.52	34.48		3.21%		
30		37.10			37.10			3.82%		
31		38.67		39.66	39.17			3.34%		1.89%

Table I shows a comparison by [4] of measured frequencies from a modal test on a UH-60A helicopter and those predicted using a NASTRAN Finite Element Model (FEM). The average frequencies were assumed to be indicative of expected H-60 modal response frequencies, and were used for the purposes of this analysis as a means of differentiating between resonances and other forced vibrations in the in-flight responses of the airframe.

DATA

The primary source of airframe stress response and accelerometer data used within this paper is sourced from a joint United States Air Force (USAF)/Australian Defence Force (ADF) S-70A-9 Black Hawk Flight Strain Survey (FSS) conducted by Georgia Tech Research Institute (GTRI) and the Australian Aircraft Research Development Unit (ARDU) in 2000 [5-8].

The Black Hawk FSS dataset consists of in-flight measurements of 282 parameters including: 164 single active strain gauges; 33 rosette single active gauges (11 full rosettes); 30 flight state parameters; 20 bending bridges (16 full, 4 half); 10 axial bridges (8 full, 2 half); 4 full shear bridges; 18 accelerometers; and 3 channels of time measurements.

Parameters were measured at locations distributed over the cabin, main rotor pylon, tail cone, tail rotor pylon, horizontal stabilator and External Stores Support System (ESSS). Flight state parameters were recorded at a base sampling frequency of 52.1 Hz, with strain, bridge and accelerometer data recorded at integer multiples of 8 or 16 times the base frequency, depending on airframe location. The majority of data were recorded at 417 Hz, with the exception of data from near the tail of the aircraft, which were recorded at 833 Hz. These sampling rates were considered sufficient to capture the key response frequencies of the airframe in flight, as described previously.

The Black Hawk FSS consisted of a series of 75 flights, with each flight representing a particular aircraft configuration (i.e. combination of Gross Weight (GW), Centre of Gravity (CG), etc.). Various flight conditions (e.g. hover, turns, level flight at different airspeeds, etc.) were recorded during each flight, with each flight condition referred to as a run. There were typically 50 'runs' per flight, giving a total of 3,766 separate runs in the flight survey database.

METHODOLOGY

The methodology for characterising the in-flight dynamic response of the H-60 airframe can be summarised as follows:

1. *Data Integrity Checks:* Starting with a set of raw Black Hawk FSS data, a series of data integrity checks were performed to create a subset of 'clean' data suitable for analysis.
2. *Frequency Synchronisation:* To facilitate comparison across all flights and runs, the 'clean' data was resampled to reduce the effect of varying main-rotor speed (i.e. Rotations per Minute (RPM)).
3. *Spectral Density Estimation:* Power Spectral Density (PSD) estimates were computed for all airframe parameters within the 'clean' dataset.
4. *Classification of Energy by Source:* Energy within the PSDs was classified by source using the previously identified expected H-60 forced and modal response frequencies.
5. *Dynamic Response Output:* Data was aggregated and formatted so as to highlight variations in airframe in-flight dynamic responses by flight condition and airframe location based on excitation source.

Data Integrity Checks

Careful inspection of the raw Black Hawk FSS dataset revealed that it contained three distinct types of anomaly; namely: dropouts; spikes; and contiguity anomalies. These anomalies had to be identified, assessed and corrected to produce a 'clean' dataset as required for downstream dynamic analyses.

Dropouts occur whenever one or more sample points within a sequence of data contains a NULL or empty value. Spikes occur whenever one or more contiguous sample points vary in a way that is not consistent with the surrounding data; for example, a too rapid or disproportionately large change in magnitude. Contiguity anomalies are characterised by successive data samples that are not sufficiently close in time; that is, the time difference between successive samples is something other than the sampling period.

It is beyond the scope of this paper to decribe these anomaly detection techniques in detail; however, their application to the Black Hawk FSS data was necessary before any further analysis on the data could be conducted.

Frequency Synchronisation

As previously noted, helicopter vibrations are dominated by sinusoidal forced responses to periodic inputs produced by rotating components. The frequencies of these periodic inputs depend on engine RPM, which can fluctuate around its nominal design value. This fluctuation leads to a proportional fluctuation in all rotating components and hence the periodic inputs they produce. This RPM-induced variation complicates the process of comparing in-flight measured responses across multiple flights where engine RPM can be substantially different. Therefore, a correction was applied to account for this phenomenon.

The distribution of operating RPM as a percentage of nominal RPM for all data within the 'clean' FSS dataset is shown in Figure 1. For this activity a relatively straight forward two-step resampling process was used to correct this variation.

First, the elapsed time associated with each data sample was modified to the value it would have been had the operating RPM been at the nominal value. For example, for a period of operating RPM at 102%, the difference between successive timestamps was artifically reduced by 2%. Second, the data with modified elapsed time was resampled at the original sampling rate. For example, applying the technique to a 10 second manoeuvre operating at 102% of nominal RPM would cause all data recorded for that manoeuvre to be resampled at a rate that is in effect 2% faster, but whose time stamps reflect the original sampling frequency, resulting in a 10.2 second manoeuvre at 100% RPM.

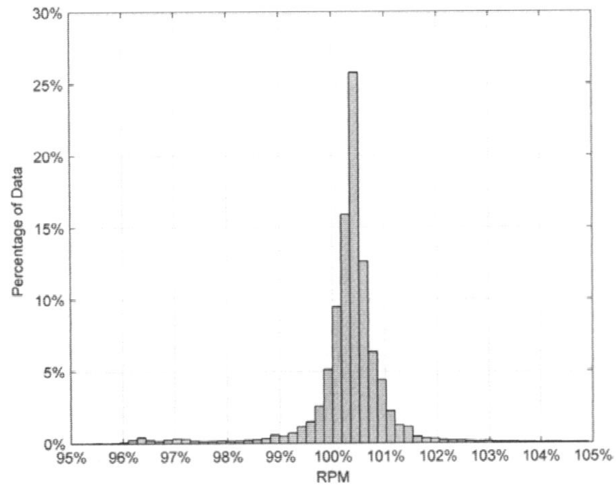

Figure 1. Histogram of recorded operating RPM as a percentage of nominal RPM for the FSS dataset, presented as a percentage of total data samples. Note: the average operating RPM is above 100%.

The major advantage of this resampling technique is that it makes it possible to directly compare the frequencies of in-flight responses for all flight/run combinations. Responses are easier to identify as they are better defined and located closer to the expected source frequencies. These effects are illustrated in Figure 2, which shows two overlaid PSD plots. The first plot (shown as dark/black) was generated from uncorrected FSS data. The second plot (shown as light/orange) was generated from the same subset of data corrected using the aforementioned resampling technique. As the figure shows, for responses that are a function of RPM (e.g. main-rotor and tail-rotor harmonics), resampling reduces RPM-induced smearing of the energy content and shifts the peaks closer to their expected source frequencies.

A disadvantage of this technique is that it will have the opposite effect on responses that are not a function of RPM, such as natural modes. The technique redistributes energy based on the observed distribution of RPM, and those responses that do not fluctuate with RPM will be smeared. Figure 3 shows an example of this occurring for the same data used in Figure 2 at a different frequency range. A well-defined, non-RPM driven response peak near 184 Hz is redistributed across a wide frequency range resulting in two less-defined peaks.

The extent that this technique redistributes energy is dependent on the frequency of responses. Those responses occurring at higher frequencies will see a larger shift for the same percentage change in RPM as responses at lower frequencies. As the majority of vibration within the FSS dataset is RPM driven, and that which is not is of relatively low frequency (e.g. natural modes < 40 Hz), the smearing did not have a major impact for this investigation.

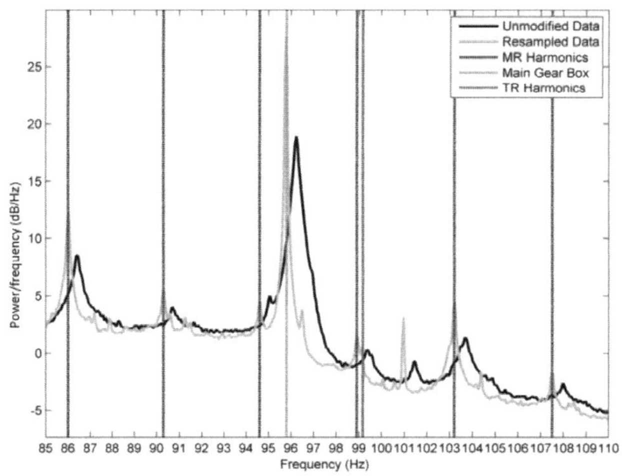

Figure 2. Comparison of PSD plots of resampled and non-resampled data demonstrating improved definition and alignment of response energy for RPM based sources.

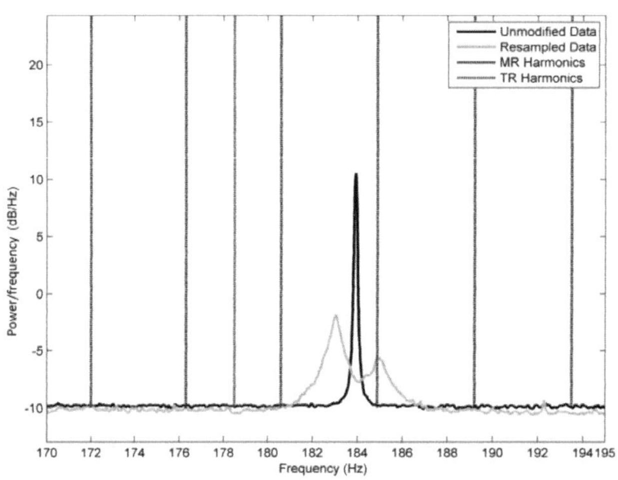

Figure 3. Comparison of PSD plots of resampled and non-resampled data demonstrating smearing and misaligning observed energy for non-RPM based sources.

Spectral Density Estimation

Spectral density estimates were computed using Welch's method [9] as a basis for characterising the in-flight dynamic response of the airframe. By comparing spectral density estimations for various gauge locations and flight/run combinations, we were able to assess how the in-flight dynamic response varied with flight condition and airframe location. Welch's method is an improvement on the standard periodogram spectrum estimation method in that it uses overlapping modified periodograms to reduce variance and spectral bias in the estimated PSD.

For the activities documented in this paper, Welch's method was varied in one aspect. Rather than using a time-based average of the overlapped segments, the median and 99th percentile values were used. Since helicopter vibrations are not stationary, this minimised the effect of outliers in the case of the median values, and captured 'worst case' transient events in the case of the 99th percentile values. A Hanning window with a 50% overlap was also used when calculating the periodograms, and the number of sample points used within the Fourier transforms corresponded to a final frequency resolution of approximately 0.2 Hz. The analysis was run on a HP Z800 12 Core Workstation, with the FFT based approach optimized for computation on a 2880 CUDA Core GTX Titan Black Graphics Processing Unit (GPU). Performing the analysis on the GPU significantly accelerated the computation of the four million PSDs required for this analysis.

Classifying Power by Source

Spectral density estimates were computed for all individual flight/run combinations for all data channels recorded in the FSS dataset. Given that the analysis of the entire FSS dataset yielded approximately four million PSD estimates, it was not practical for human analysts to manually analyse them all individually. Instead, an automated and scalable numerical method was developed to facilitate both detailed analysis (i.e. drilling down to the level of individual PSD estimates) and abstraction (i.e. collapsing multiple PSD estimates to produce aggregated views) depending on the requirements of the task at hand.

A PSD estimate consists of a series of discrete power values at evenly spaced frequency bins typically plotted as a continuous line. The square root of the PSD integral of a defined bandwidth corresponds to the Root Mean Square (RMS) power of that bandwidth. To facilitate analysis an algorithm was developed that compared each peak in a given PSD plot with the frequencies of the periodic inputs expected to generate responses, and automatically classified each peak by source. As each peak had been classified by frequency source, the total RMS energy associated with all main-rotor, tail-rotor, drive-train, or modal sources could be estimated by the square-rooted sum of the squared individual RMS values for that source, as shown in Figure 4.

The key benefit of this approach was that it allowed direct comparison of PSDs across flights, and increased the scalability of the approach. By classifying all energy within PSD estimates by source, data could be aggregated at varying levels of detail, allowing analysts to easily drill-down or roll-up results as required.

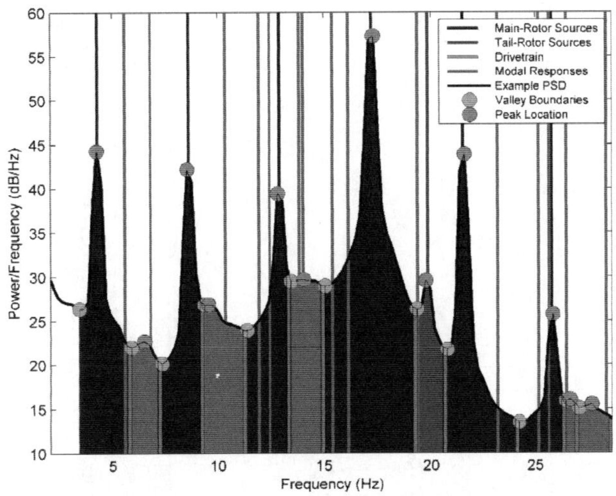

Figure 4. Example PSD displaying peak RMS coloured by frequency source.

Dynamic Response Output

The objective of this investigation was to characterise the in-flight dynamic response including its variation with flight condition and airframe location. This was achieved by generating a large number of Dynamic Response Signature (DRS) plots, which display the dynamic response data, including the aggregate data, for the instrumented locations on the airframe with sufficient clarity to aid understanding and decision making. These plots are intended to be rich in detail and at the same time easy to interrogate, aggregate and compare. An example of a DRS plot is shown in Figure 5.

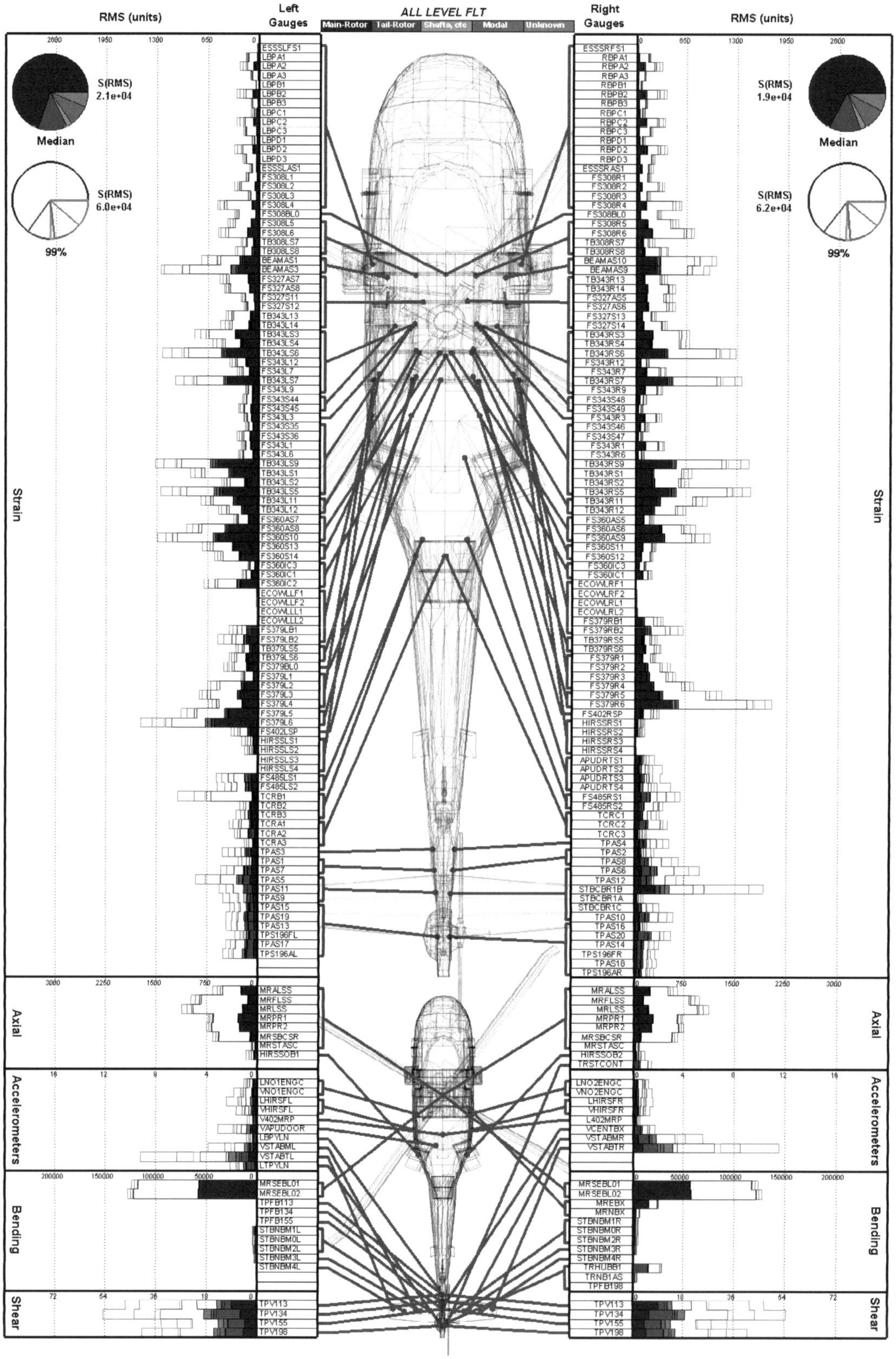

Figure 5. Example DRS plot highlighting key features.

For this activity a number of one-page DRS plots were generated, each containing the estimates of energy content aggregated by source (e.g. main rotor, tail rotor, modal sources, etc.), for all parameters for a specific manoeuvre sub-category. These energy contents are displayed as bar graphs of the median and 99th percentile values, and grouped based on parameter type (e.g. single active strain gauge, accelerometer, bending bridges, etc.). Overall summaries for all single active strain gauges are also provided in the form of pie charts. The energy estimates in the DRS plots have been coloured blue, red, green, magenta or grey to represent key expected vibration sources (i.e. main rotor, tail rotor, drive-train, modal and unknown, respectively).

EXAMPLE USES

The collection of DRS plots for all estimates calculated during this investigation represent an indicative H-60 airframe dynamic response characterisation. This section provides illustrative examples to highlight how these DRS plots can be aggregated and used to identify trends, characterise responses, and compare the FSS dataset with the literature.

Variation of Source Frequencies across Airframe Locations

Figure 6 shows a standard visual representation of the sources of vibration and their distribution for a typical helicopter. So an initial question we asked ourselves was how does our FSS data compare in terms of overall trends with respect to the sources and distribution of vibrations in the H-60 airframe?

To answer this question we generated Figure 7, which shows the median response of all FSS data across all manoeuvres and airframe locations grouped by source frequency. This involved aggregating the DRS plots for all manoeuvres, thereby summarising the energy associated with each source for each individual parameter. The sizes of the location markers in Figure 7 are proportional to the median energy response at that location, and the colours of the markers represent the relative contribution of each source. As in Figure 6, Figure 7 shows that the majority of the measured locations across the airframe are dominated by main rotor source frequencies, with a small region near the tail rotor dominated by tail rotor frequencies, and a region down the vertical tail showing a gradual shift from main to tail rotor frequency sources. The region of Figure 6 expected to be dominated by drivetrain sources was not reflected in Figure 7; however, this is because relatively few locations in that region were available in the FSS dataset and those that were, were located on structure not associated with drive train components. Further interrogation of DRS plots for those locations did show a higher presence of drive train sources than other areas; however, they were dwarfed in comparison to the accompanying main rotor source frequencies.

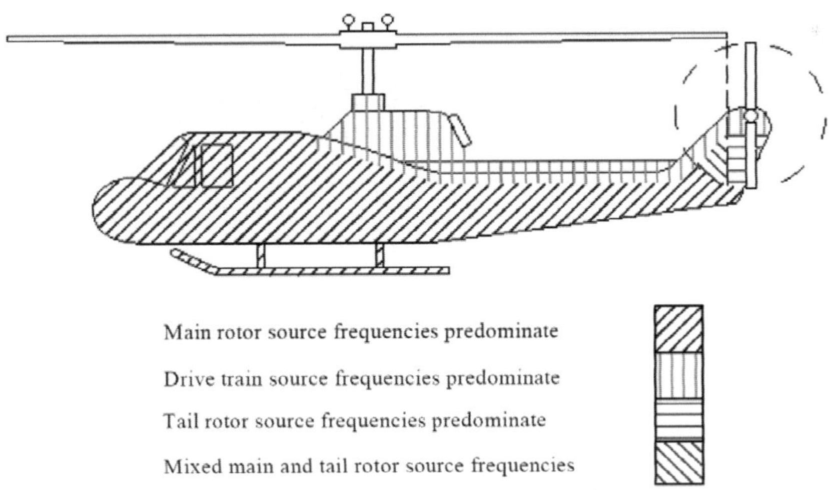

Figure 6. Helicopter vibration zones, adapted (coloured) from [1].

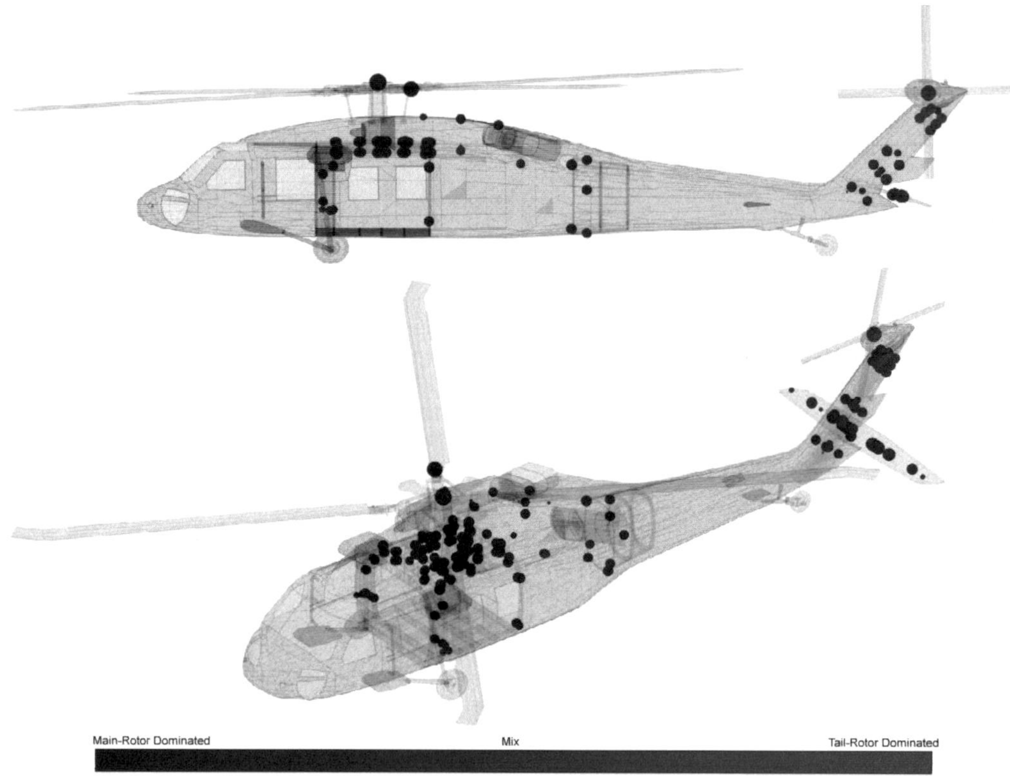

Figure 7. The median dynamic response of the FSS dataset for all manoeuvres across all airframe locations. Each marker is sized by the median energy response at that location and coloured according to the dominant contributing sources.

Variation in Vibration with Airspeed

Figure 8 shows typical variation in helicopter vibration with changing airspeed as presented in [3]. This plot is simplified and not specific to any particular helicopter type, but it typifies the level of detail commonly found in the literature.

Again, how does the FSS data compare? To conduct such a comparison, it was first necessary to generate a plot equivalent to Johnson's based on the FSS dataset. Since the FSS dataset did not contain many accelerometers, the total RMS energy was calculated from all single active strain gauges in the DRS plots to provide an estimation of vibration amplitude. This was done for all hover and level flight manoeuvre sub-categories, with manoeuvres sorted by increasing airspeed. Although this was a comparison of RMS vibratory stress versus forward speed rather than vibration amplitude and advance ratio, the similarities between the two were considered acceptable for the purpose of this comparison.

A visual comparison between Figure 8 and Figure 9 shows good agreement between the established trends and those observed in the FSS data. In both figures fuselage vibration amplitudes are typically lower in hover and increase with forward flight speed due to compressibility effects on the advancing blade. Increased vibration levels also occur in high thrust conditions due to the onset of retreating blade stall, and in transition (i.e. from hover to forward flight) and descent due to the effect of the main rotor wake on rotor blade loads.

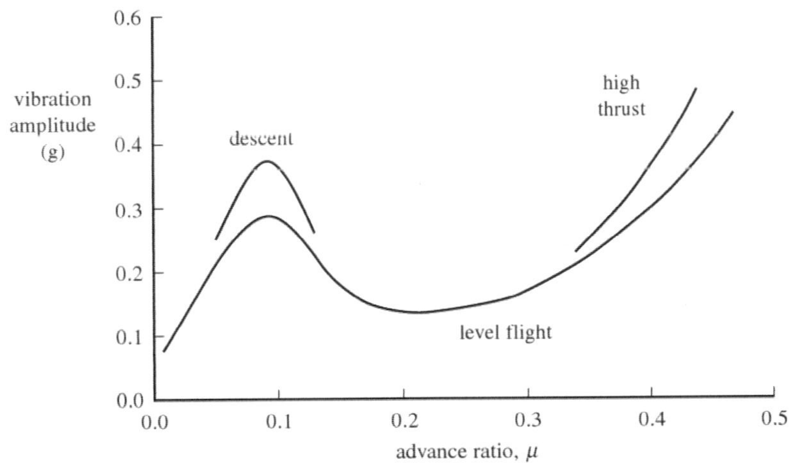

Figure 8. Typical variation in helicopter vibration with speed (Johnson (2013)).

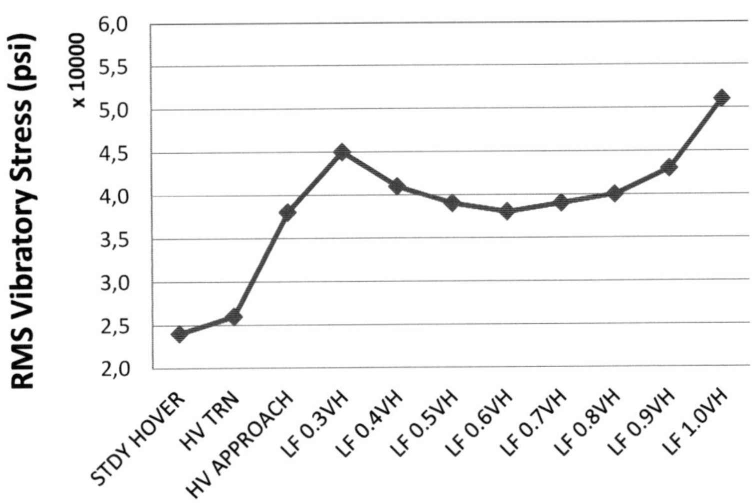

Figure 9. Variation in FSS dataset total vibratory stresses across the airframe, with manoeuvres of increasing airspeed.

CONCLUSION

DSTO has developed a methodology capable of characterising the in-flight dynamic response of the H-60 aiframe including its variation with flight condition and airframe location. Key features of this methodology include: techniques to assess the integrity of in-flight dynamic response data; techniques to resample response data to account for fluctuations in operating RPM; and techniques to classify vibration responses by excitation source. Specific computational and data analytic techniques were used to enable both the computation of four million PSD estimates in a timely manner, and the efficient aggregation and presentation of results. In following this methodology, key characteristics of vibration responses across all parameters for all sub-categories of manoeuvres were captured in the form of DRS plots.

REFERENCES

[1] Department of Defense (2008) *MIL-STD-810G, Department of Defense Test Method Standard for Environmental Engineering Considerations and Laboratory Tests*, United States Department of Defense.

[2] Bielawa, R. L. (2006), *Rotary Wing Structural Dynamics and Aeroelasticity Second Edition*, American Institute of Aeronautics and Astronautics (AIAA), Reston, VA, USA.

[3] Johnson, W. (2013), *Rotorcraft Aeromechanics*, Cambridge University Press, New York, NY, USA.

[4] Howland G. R., Durno, J. A., and Twomey, W. J. (1990), *Ground Shake Test of the UH-60A Helicopter Airframe and Comparison with NASTRAN Finite Element Model Predictions*, NASA CR-181993.

[5] Friend, D (2000), *Joint USAF-ADF S-70A-9 Flight Test Program, Equipment and Instrumentation Installation Report*, GTRI Project A-6048 Final Report, Georgia Tech Research Institute.

[6] Friend, D (2000), *Joint USAF-ADF S-70A-9 Flight Test Program, Equipment and Instrumentation Installation Report (Attachment 1 – Gage Photographs)*, GTRI Project A-6048 Draft Report, Georgia Tech Research Institute.

[7] Friend, D (2000), *Joint USAF-ADF S-70A-9 Flight Test Program, Equipment and Instrumentation Installation Report (Appendices)*, GTRI Project A-6048 Draft Report, Georgia Tech Research Institute.

[8] Aerospace, Transportation, & Advanced Systems Lab (2001), *Joint USAF-ADF S-70A-9 Flight Test Program, Summary Report with Appendices A thru D*, GTRI Project A-6196 Final Report, Georgia Tech Research Institute.

[9] Welch, P.D. (1967), *The Use of Fast Fourier Transform for the Estimation of Power Spectra: A Method Based on Time Averaging Over Short, Modified Periodograms*, IEEE Transactions on Audio Electroacoustics, AU-15, 70–73.

Fatigue Crack Growth Prediction Using PREFFAS Model

Bertrand Journet

Airbus Group Innovations, Suresnes, France

Abstract: Airbus Group has developed a model called PREFFAS for the purpose of carrying out damage tolerance analysis of metal structural parts. Since its first development in the early eighties within the former Aerospatiale organization, the model has been tested, implemented and extended to deal with various load spectra, materials or fracture mechanics features. It is the purpose of this paper to present some of the related case studies that led to develop additional schemes or to adapt the methodology according to the service application. The model is based on the crack closure concept to account for the fatigue crack growth behavior and for the load spectrum effects. Case studies dealing with residual stresses or with isothermal temperature conditions are presented. The case of residual stress is related to welding technology. A scheme dealing with the residual stress has been worked out so that the model needs not to be modified. With temperature and/or ageing effect, the methodology to identify the model parameters is simply adapted to the case. The concerned materials are aluminum alloys and steels. The conclusion recalls the features of the model together with the application cases and its limitations. The idea to try and apply the original scheme as much as possible in the reported cases has allowed not to increase the number of modelling parameters. These investigations have been undertaken by Airbus Group Innovations through public funded projects.

INTRODUCTION

Airbus Group Innovations has developed in the past a model to predict fatigue crack growth under spectrum loading, namely PREFFAS model, in the frame of linear elastic fracture mechanics. Effort to extend the model with new options has been continuous throughout the years. This model is used by Airbus Commercial Airplanes to determine the maintenance inspection intervals. This model was first developed in 1985 for cases of tensile loading on aluminum plates [1]. It was further implemented on thin aluminum sheets, steels and titanium plates. Later on improvements dealt with the occurrence of compressive loading on aluminum sheets. The model has been successfully evaluated on numerous types of flight spectra and related materials: inner wing, outer wing, pylon, fuselage of the Airbus aircraft family [2, 3, 4]. Recent evaluations of the model have been dealing with the presence of residual stresses due to welding and with the exposure to temperature and ageing effects during fatigue cycling. This is the purpose of the paper to report on these two recent investigations.

The basics of PREFFAS model is reminded together with the methodology to identify the model parameters. Then the investigated evaluation cases of the model are presented. The first case deals with a stiffened aluminum panel containing two welded stringers and a centre bay crack submitted to a fuselage load spectrum and to the built-in residual stresses due to welding. The second case concerns a cracked laboratory coupon made out of a stainless steel and submitted to a pylon load spectrum and a high temperature. The model has the capability to handle these two cases but in the first one, it needs to be worked out in order to treat the residual stresses. The investigations were carried out within the frame of public funded projects, DaToN and PREVISIA respectively.

PREFFAS MODEL

PREFFAS is a French acronym that stands for PREvision de la Fissuration en Fatigue AéroSpatiale, i.e. crack growth prediction in aerospace fatigue. A short presentation of the model is given. A more detailed description can be found in references [1, 2]. The model is based on plasticity induced clack closure to predict the crack growth rate under spectrum loading. Plasticity crack closure accounts for R ratio effect and retardation effect. The crack retardation analysis made in the model is based on a scheme to update the crack closure at each cycle taking into account the load history (preceding cycles) in the spectrum. The scheme is presented further down in this section.

Model parameters

The model depends on three parameters only: C_1, β and A. C_1 and β are the Paris' law coefficients of the material tested under R = 0.1 constant amplitude loading (CAL). Parameter A is one of Elber's constant accounting for crack closure [5]. The other needed Elber's parameter B is related to A through the relationship A + B = 1. These three independent parameters allow for determining the effective fatigue crack growth rate (FCGR) law of the alloy that is used to compute the crack growth at each cycle of the load spectrum in the scheme of the model together with the update of the crack closure. The basic equations highlighting the parameters are read as follows, when $R \geq 0$:

R0.1 FCGR law: $$\frac{da}{dN} = C_1(\Delta K)^{\beta} \quad (1)$$

stress intensity factor range: $$\Delta K = K_{max} - K_{min} = (1-R)K_{max}$$

load ratio R: $$R = \frac{K_{min}}{K_{max}}$$

crack closure (or opening) K: $$K_{op}$$

effective ΔK: $$\Delta K_{eff} = K_{max} - K_{op} \quad (2)$$

effective FCGR law: $$\frac{da}{dN} = C_{eff}(\Delta K_{eff})^{\beta} \quad (3)$$

defining U as: $$\Delta K_{eff} = U\Delta K \quad (4)$$

Elber's relation: $$U = A + BR$$

relation in PREFFAS: $$U = AR + B \quad (5)$$

(note that A in PREFFAS plays the role of B in Elber's)

assuming: $$A + B = 1$$

it is derived that: $$C_{eff} = \frac{C_1}{(0.1A + B)^{\beta}} \quad (6)$$

and $$K_{op} = K_{max}[1 - (AR + B)(1 - R)] \quad (7)$$

equations (6) and (7) are read with PREFFAS A and B. When $R < 0$, the PREFFAS model proposes a specific formulation, as presented further down in the extended options subsection.

Identification of the parameters

The three parameters are identified running two baseline tests [1] on laboratory specimens made out of the alloy of the structural part with the same orientation and thickness as those of the application, Figure 1.

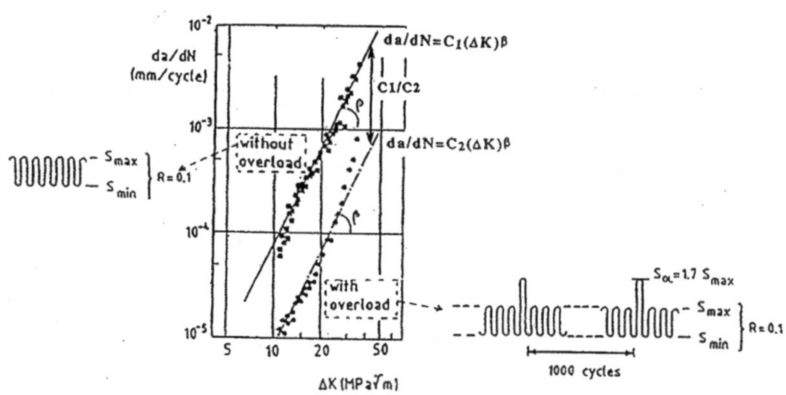

Figure 1. Identification of PREFFAS model parameters C_1, β and A on an aluminum 2124T351 alloy, with 6 mm thick CT specimens. Two FCGR tests: CAL R = 0.1 test and repeated spike overload test. β = 3.2, C1/C2 = 10, U = 0.42 R + 0.58 [2].

The two tests are characterized with Paris'laws:

- CAL R=0.1 test, with above Paris' law (1),
- Repeated 1.7 spike overload (OVL) test: 1 spike overload every 1000 R = 0.1 cycles, with Paris' law

$$\frac{da}{dN} = C_2 (\Delta K)^{\beta} \tag{8}$$

The retardation factor C_1/C_2 allows for the identifica ion of A. This is the ratio of average FCGRs without and with crack retardation, it is directly given by the test data of Figure 1. In order to derive A, C_1/C_2 is explicitely written using the effective Paris' law. The FCGR without retardation is that of one nominal R0.1 cycle: it is given by (3) with R = 0.1. The FCGR with crack retardation is the crack advance due one overload cycle + 1000 retarded R0.1 cycles, the whole sum then divided by 1001. It is calculated using (2), (3) and (7) with the relevant R ratios according to the retardation scheme as outlined in the next subsection. The computation yields:

$$\frac{C_1}{C_2} = \frac{1001\left[0.9(1-0.9A)\right]^{\beta}}{1000\left[1-\tau\left[1-\left(1-\frac{0.1}{\tau}\right)\left[1-A\left(1-\frac{0.1}{\tau}\right)\right]\right]\right]^{\beta} + \tau^{\beta}\left[\left(1-\frac{0.1}{\tau}\right)\left[1-A\left(1-\frac{0.1}{\tau}\right)\right]\right]^{\beta}} \tag{9}$$

with the τ of 1.7 and C1/C2 = 10, this yields A = 0.42. The property of having parallel Paris' lines is very important for the model because one value of parameter A can thus account for retardation effect over the whole range of Paris' regime. The model scheme for retardation computation relies on the assumption that the FCGR curves of the two tests are two parallel lines. In house investigations have shown that a spike overload ratio of 1.7 on thin sheets is too much because the retardation FCGR curve is no more linear. It is proposed to use an overload ratio of 1.4 on thin sheets [4]. However it has been checked in house [4] that test OVL test results with OVL ratios of 1.2 to 1.4 lead to the same value of parameter A, which strongly confirms the abilty of crack closure to characterize fatigue crack growth.

Spectrum analysis and crack opening calculation

Crack retardation is assumed not to vanish over one spectrum sequence (relatively short load spectrum). From experience, a load spectrum with about 200,000 cycles can be considered as a short one. A dedicated criterion is given later on.

The key point lies in the determination of the crack opening stress K_{op} at each cycle. It must take into account the load or stress history of the spectrum sequence and one must search for the most influencing (max-min) load path preceding the cycle, i.e. that with the highest crack closure. This key notion of (max-min) path has been originally introduced in the CORPUS model of the Dutch Aerospace National Lanboratory. The (max-min) values are local overload and underload, they are not necessarily consecutive in the load sequence. The crack opening of the cycle is then that given by this path since there is no crack retardation vanishing. It is computed applying relationship (7) to this load path, if the R ratio of this path is positive, or relation (15) if the ratio is negative, as detailed further down in the extended options subsection.

The PREFFAS scheme to search for the most influencing load path and to calculate the crack closure or opening $K_{op,i}$ at cycle i is thereafter shortly presented.

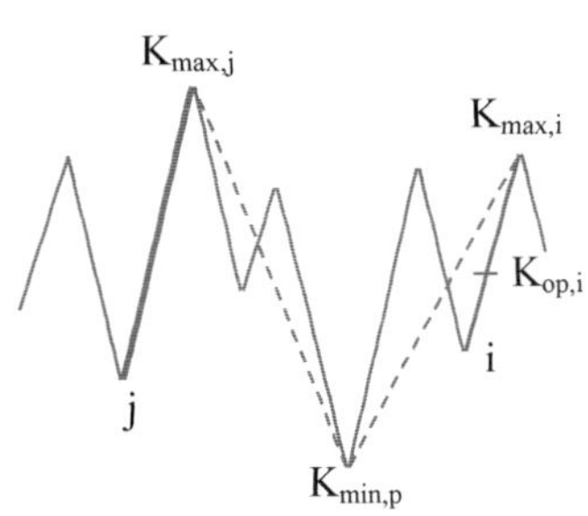

Figure 2. Oustanding overload - underload path $K_{max,j}$-$K_{min,p}$ preceding cycle i and between cycles j and i.

There are many preceding (max-min) load paths that may control the crack closure of cycle i. One of them will bear the strongest influence inducing the lagest retardation effect. This is the one to look for. The paths are pairs of outstanding overload and underload in the spectrum, like ($K_{max\ j}$ - $K_{min\ p}$) path as shown in Figure 2. There, $K_{max\ j}$ is the first K_{max} value preceding $K_{max\ i}$ and greater than $K_{max\ i}$ when going backwards from cycle i. $K_{min\ p}$ is the lowest K_{min} value occurring between $K_{max\ j}$ and $K_{max\ i}$. ($K_{max\ j}$ - $K_{min\ p}$) path is one load path that may control the closure of cycle i. There are others in the spectrum that may do so and they are to be found. Their max (K_{max}) values are located between j and the beginning of the spectrum. The associated min value (K_{min}) is between the max and cycle i. The search follows some pinciples:

- retardation due to a K_{max} level wipes out all preceding Kmax with lower levels,
- a Kmin level wipes out all preceding Kmin with higher levels until an outstanding preceding K_{max} is met.

Load path ($K_{max,j}$ - $K_{min,p}$) has a closure $K_{op,ij}$: it is calculated hereafter using (7) when R ratio o fthe path positive, otherwise applying (14) to get U direcly when R<0:

$$K_{\min,p} = \inf_{j+1\le k\le i} K_{\min,k}$$

$$R_{ij} = \frac{K_{\min,p}}{K_{\max,j}}$$

$$K_{op,ij} = K_{\max,j}\left[1-\left(AR_{ij}+B\right)\left(1-R_{ij}\right)\right] \quad (10)$$

Let $K_{op,i}$ denote the crack opening of cycle i. Therefore:

$$K_{op,i} = \sup_{1\le j\le i-1} K_{op,ij} \quad (11)$$

For next cycle, i+1, the same search scheme is applied. In order to gain time, the spectrum is analysed first in a way to sort out the outstanding (overload-underload) paths. This allows to define a series of outstanding KH history values ($KH_{max,r}$, $KH_{op,r}$, $KH_{min,r}$) with the following properties: $KH_{max,r}$ series is decreasing, $KH_{op,r}$ series is increasing, $KH_{min,r}$ series is increasing. The history series looks like as shown in Figure 3. It usually consists of about 10 terms. The crack opening stress of cycle i is then straightforward: it is that given by the closest preceding history load path (KH_{op3} for cycle i, in Figure 3).

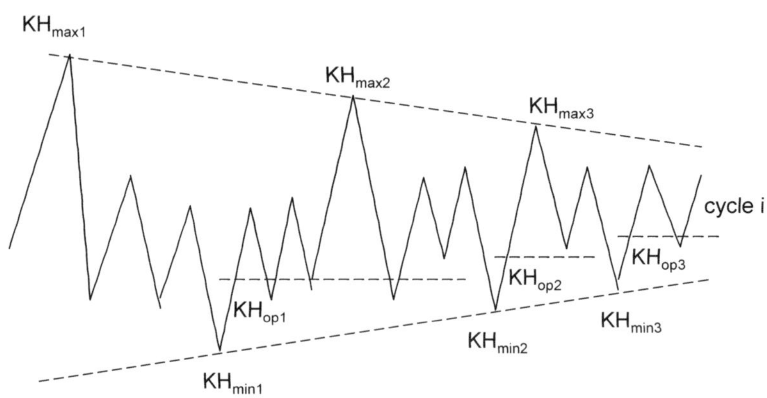

Figure 3. Series of KH history values.

Crack growth computation

Once the spectrum has been analysed in order to determine the crack opening stress at each cycle i in the spectrum, as explained above, crack growth at Δa_i cycle i is computed using (2) to (5) in case of positive R or (15) to (16) with negative R. The summation of the crack growth Δa_i is made over the spectrum, yielding the crack advance Δa for one spectrum sequence, with n denoting the number of cycles in the spectrum:

$$\Delta a = \sum_{i=1}^{n} \Delta a_i = C_{eff} \sum_{i=1}^{n} \left(\Delta K_{eff,i}\right)^{\beta}$$

The crack growth calculation at each cycle i also proceeds through a rainflow type summation when $K_{max,i}$ at cycle i is greater than the preceeding load history value $KH_{max,r}$. In the case when the last history value $KH_{max,r}$, before current cycle i, is exceeded by $K_{max,i}$ (Figure 4), the rainflow computation reads as follows :

$$\Delta a_i = C_{eff}\left(K_{\max,i} - K_{op,r-1}\right)^{\beta} + C_{eff}\left(KH_{\max,r} - KH_{op,r}\right)^{\beta} - C_{eff}\left(KH_{\max,r} - KH_{op,r-1}\right)^{\beta}$$

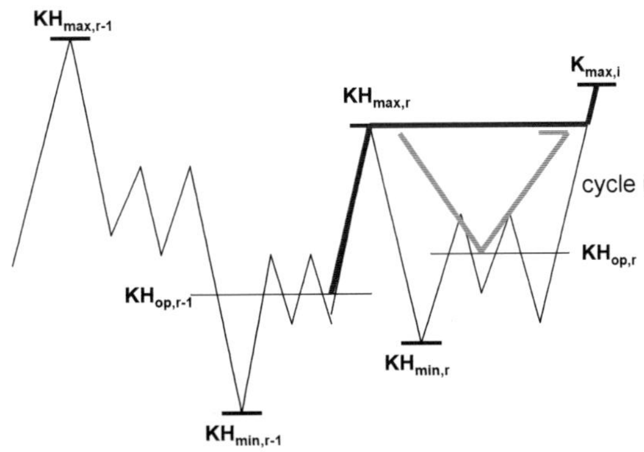

Figure 4. Rainflow crack growth summation when a history value is exceeded by current cycle i.

Generalized to cases when a range of load history values $KH_{max,p+1}$ to $KH_{max,q}$, before current cycle i, are exceeded by $K_{max,i}$, the rainflow computation yields:

$$\Delta a_i = C_{eff}\left(K_{\max,i} - K_{op,p}\right)^{\beta} + C_{eff} \sum_{r=p+1}^{q}\left(KH_{\max,r} - KH_{op,r}\right)^{\beta} - C_{eff} \sum_{r=p+1}^{q}\left(KH_{\max,r} - KH_{op,r-1}\right)^{\beta}$$

Assuming that crack length a does not change much over one spectrum sequence (short spectrum assumption) so that the change in K with a is negligible, one reads:

$$\Delta a = C_{eff}\left[f(a/W)\sqrt{\pi a}\right]^{\beta} \sum_{i=1}^{n}\left(\Delta S_{eff,i}\right)^{\beta}$$

$$\Delta a = C_{eff}\left[f(a/W)\sqrt{\pi a}\right]^{\beta} EF \tag{12}$$

where EF denotes the sequence efficiency: $EF = \sum_{i=1}^{n}\left(\Delta S_{eff,i}\right)^{\beta}$

Then an EQUI parameter is defined:

$$EQUI = \frac{EF}{\left(AR_{ref} + B\right)^{\beta}\left(S_{\max,ref} - S_{\min,ref}\right)^{\beta}} \tag{13}$$

where ($S_{min,ref}$, $S_{max,ref}$) is a CAL reference cycle, S denoting the applied stress or load.

Figure 5 depicts the meaning of EQUI: the crack advance due to one load spectrum sequence is the same as that given by EQUI reference cycles. In the example of Figure 5, with an inner wing spectrum, it is observed that the model is fairly well validated on laboratory compact tension (CT) specimens since the calculations are close to the experimental results. [2]. Figure 5 shows that ignoring the retardation effect would result in this case in a shorter crack growth time by a factor of 2. Therefore the aircraft would be inspected too much, twice as much as needed !

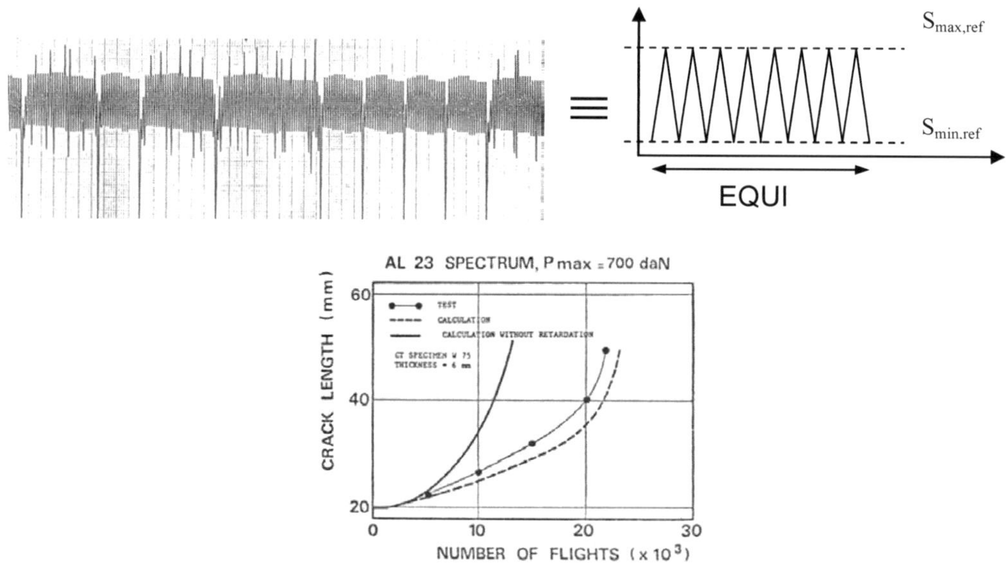

Figure 5. The spectrum analysis by PREFFAS yields an equivalent parameter EQUI which is used for crack growth computation. Validation of the model with an inner wing spectrum loading FCG test on a laboratory CT coupon [2].

The benefit in using EQUI is to easily compute the crack growth by just integrating the CAL FCGR equation with respect to ($S_{min,ref}$, $S_{max,ref}$) cycles. For a targeted crack growth range, i.e. [a_{start}, a_{final}], the obtained number of cycles devided by EQUI yields the number of flight spectra.

Criteria and assumptions

PREFFAS model is applicable under the following assumptions:

- Paris' law regime.
- FCGR tests results under CAL R = 0.1 and repeated spike overload every 1000 CAL R = 0.1 cycles yields parallel da/dN vs □□ K lines.
- Crack retardation is not vanishing: crack advance Δa per spectrum sequence should be small compared to the plastic zone produced by the highest overload K_{max} in the spectrum, i.e.:

$$\Delta a < \frac{1}{10}\left(\frac{K_{max}}{Y}\right)^2 \qquad (14)$$

Extended options

The model has been further extended to take into account two realistic situations: compressive loads and R effect with $A + B \neq 1$ [4].

A compressive load is going to reduce the retardation effect. This cannot be accounted for with the above formalism developed for positive R ratios. The idea is to take two limiting cases: when the compressive stress is equal to the yield stress, therefore the plasticity induced crack closure is wiped out and crack closure is equal to zero. When the compressive stress is just zero, one must find the situation with R=0. In between these limiting cases, U is linearly interpolated. The model proposes the following expression of U, Y and UTS being the yield and ultimate tensile strength of the material, S_{min} the compressive minimum stress of current fatigue cycle i and K_{max} the maximum value of K of fatigue cycle i:

$$U = B\left[1 + \alpha\frac{2|S_{min}|}{Y + UTS}\right] \qquad (15)$$

$$\alpha = \frac{1-B}{B}$$

$$\Delta K_{eff} = UK_{max} \tag{16}$$

The rationale directly deals with stresses and U. This formulation has been validated in house [4] on inner and upper wing flight spectra.

When $A + B \neq 1$, which is not a rare case, an additional test is run at high R ratio (e.g. 0.7). B is now an independent parameter. The three baseline FCGR tests therefore yield two equations with two unknowns A and B that can be calculated in the same spirit as (9). This option is now currently used. This leaves the model now with four parameters. This option has been validated in house [4].

Concluding remarks

PREFFAS model is based on the crack closure concept to predict fatigue crack growth under spectrum loading. It has been evaluated and validated on numerous aeronautical materials and flight spectra. The model has been extended to deal with compressive loads and more comprehensive R ratio effects. The model has four parameters. Yield and UTS of the material are also needed. Its simplicity makes it easy to use in industry. Three assumptions are key to apply the model: Paris' regime, parallel lines in the three baseline tests used to determine the model parameters, short spectra. The FCGR curves under constant R ratios are usually linear and parallel. The repeated spike overload curve is not always linear, especially on low gage coupons. Experience shows that by approximating a parallel straight line through it allows to get a good order of magnitude of the retardation effects.

Other situations are thought to be worth to extend the model to: presence of residual stresses, exposure to high temperature. In the first case stress intensity factors with residual stress need to be computed. A scheme to include the residual stresses for crack closure analysis is also to be set up. In the second case, viscous effects might show up. The proposed work presented thereafter is carried out with the spirit to keep the simplicity of the model as much as possible, trying not to increase the number of parameters.

PREFFAS AND RESIDUAL STRESSES

Residual stresses are known to exert an influence on the fatigue crack growth rates. They are of benefit when they are negative. Taking them into account by calculating more realistic inspection intervals may result in significant overhaul cost savings. However the analysis of the damage tolerant behavior of structural parts requires to calculate stress intensity factors with the residual stresses and to lay out a scheme in order to calculate the flight inspection intervals with the benefit of the compressive residual stresses. More precisely, FCG rates and crack retardation have to be computed taking into account the residual stresses. This is the object of the work presented herein. It has mainly been carried out within the frame of DaToN project [6]. It features the use of PREFFAS model.

The case under investigation deals with a stiffened panel with two laser beam welded stringers and with a centre crack in the middle bay. The welding process induces the build-up of compressive residual stresses in the mid bay. The study of the fatigue crack growth is limited to this mid bay. Stress intensity factors are calculated with the residual stresses. The effect of the residual stress on the FCG rates without and with retardation is considered as an R ratio effect. The equations of PREFFAS and its retardation scheme analysis are applied to compute predictions of the FCG rates under R = 0.1 and under flight spectrum loading. The comparison to the experimental FCG rates is discussed in order to validate the findings. On this ground, a prospective calculation with PREFFAS is made to compare the inspection intervals of stiffened panels manufactured with different technologies: integrally machined, laser beam and friction stir welded stringers.

Laser beam welded panel

The panel geometry is presented together with the residual stresses and the testing conditions.

The panel is referred to as 6056T6 LBW1. It has been manufactued by Airbus Group Innovations [7]. The material was supplied by Constellium who was partner in DaToN. The panel is made with a skin and two stringers, all out of 6056T6 alloy. The panel is 450 mm wide (skin width), the two stringers are 150 mm apart, Figure 6.

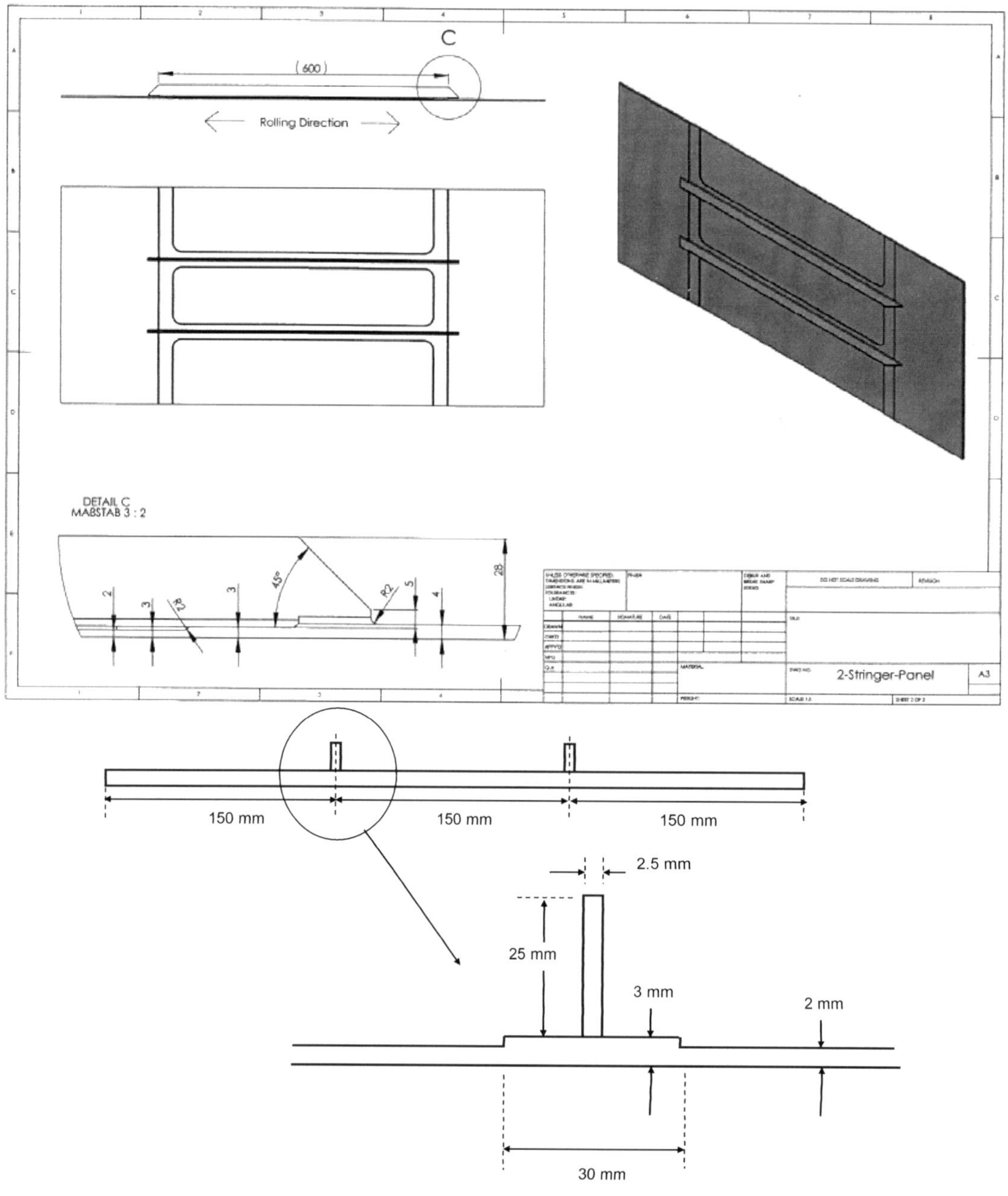

Figure 6. 6056 LBW1 panel geometry.

The skin was machined with three pockets out of an initial 5 mm thick 6056T6 sheet. The thickness of the final sheet is 2 mm in the pocketing areas and 3mm elsewhere in the testing area (outside the grip area and mainly under the stringers). The thickness is increased to 4mm in the grip area. The mid bay pocket is 120 mm wide. The skin thickness under the stringers is 3mm over a width of 30 mm. The L direction of the skin is parallel to the stringers. A 10 mm long central notch is machined in the mid bay in the L-T orientation.

After machining, the skins were distorted although the 5mm thick supplied sheets were flat. Two stringers were laser beam welded by Airbus Group Innovations: the weld seam is both in the sheet and stringers, Figure 7. The sheet was tightly flattened during welding but the initial distorsion could not be offset and the panel after welding was distorted as well. The panels did not undergo any post weld heat treatment.

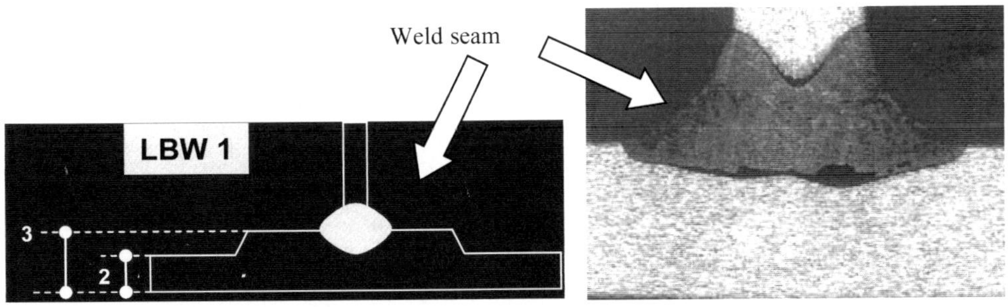

Figure 7. 6056 LBW1 panel: laser beam welding in sheet and stringer [7].

The residual stress in 6056T6 LBW1 panel was determined by Pisa University [8], who was partner in DaTon, applying a sectioning method. The panel was set up into the grips of a testing machine. Strain gauges were bonded on the panel over the width on both sides (flat and stringer side). Strain gauge response is recorded when a central cut is progressively extended. These measurements allow for calculation of the residual stress. Figure 8 shows the strain gauges location and the calculated residual stress profiles in the L direction of the panel. These profiles will be used as an input to the K calculations.

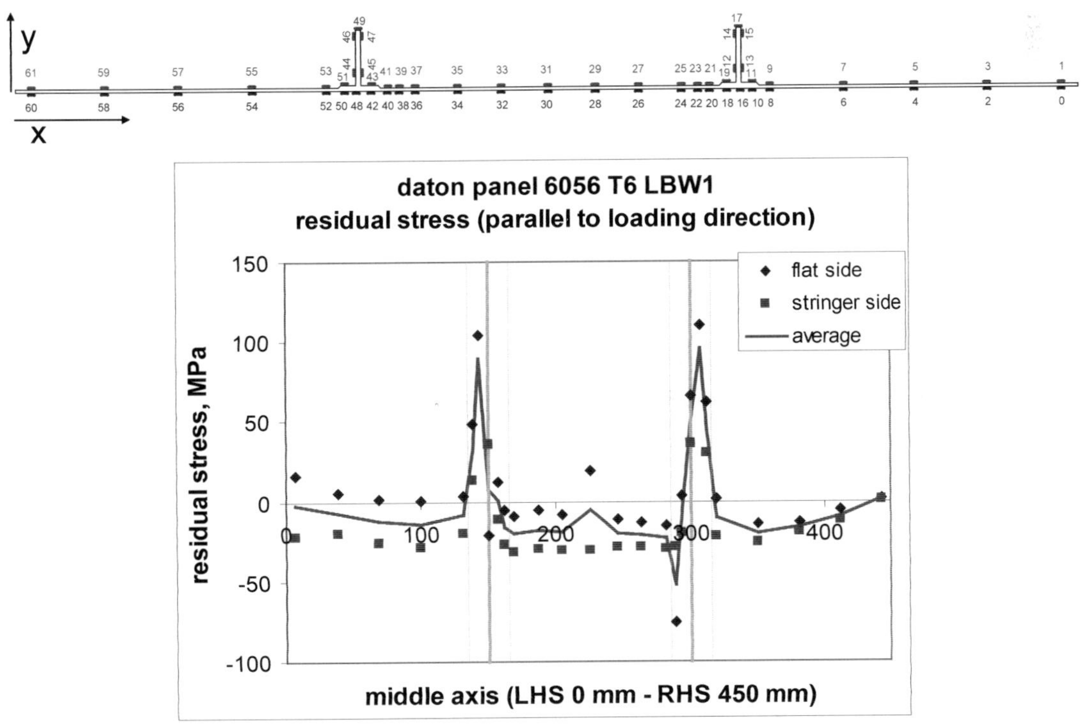

Figure 8. Residual stress determination by Pisa University on 6056T6 LBW1 panel applying the sectioning method. Schematics: strain gauges location on the panel at mid section. Graph: residual stress profiles.

The stress distribution is considered as fairly symmetrical with respect to the center line of the panel. This will help in simplifying the analysis through finite element analysis. The peak stress in the stringer, on the stringer side, could not be evaluated. This issue will be adressed later on when calculating K.

The panels are tested in fatigue under constant amplitude loading with R = 0.1, under repeated 1.4 pike overload every 1000 R0.1 cycles and under flight spectrum loading. The notch is precracked by fatigue with an R ratio of 0.1 over a few millimiters at the beginning of each fatigue crack growth tests. The load is applied in the grip area. Its direction is parallel to the stringers. Under CAL and spike overload, the value of the applied maximum load of the R0.1 fatigue cycles is 72 kN. The corresponding stress over the whole mid section cross area is 66.4 MPa. Under spectrum loading, the maximum applied load is 108 kN (99.5 MPa).

The base sheets of 6056 are tested as well on center crack tension (CCT) in the L-T direction under constant amplitude loading with R = 0.1, under repeated 1.4 pike overload every 1000 R0.1 cycles and under flight spectrum loading. The results will allow for the determination of PREFSA parameters. The sheets are 2.5 mm thin. Although the thickness of the pockets is 2 mm it is thought the FCR properties of 2 and 2.5 mm will be the same.

The flight spectrum is an Airbus wide body fuselage spectrum. The spectrum consists of 8000 flights totaling 192314 cycles, with a range of different severities and a typical overload ratio of 1.4. Figure 9 gives an idea of how the spectrum looks like by extraction of a sequence of cycles. The values are in % of the maximum stress.

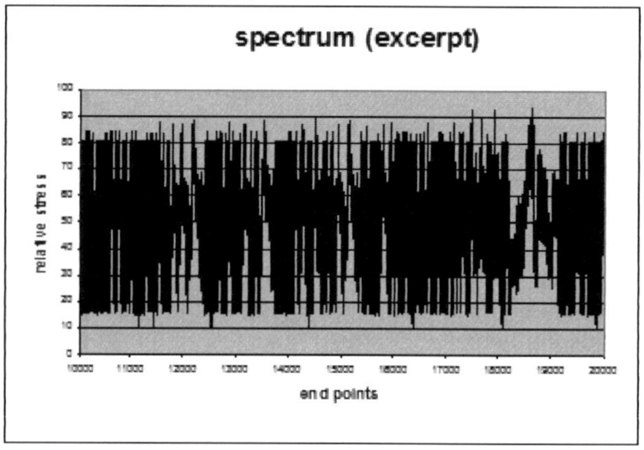

Figure 9. Except of the flight spectrum used for the tests in terms of relative stress (%) versus end points.

Stress intensity factors

The stress intensity factor is calculated with crack length using finite element analysis carried out on the panel geometry under the load of 72 kN. Stress intensity factor K is first calculated without the residual stress, then with. The residual stress profiles, determined by Pisa University through a sectioning method, were obtained setting up the panel into the grips of a testing machine, thus in a configuration actually representative to the starting one in any fatigue crack growth test run in this study. Therefore the calculations with the residual stress profile determined in the grip position makes sense. The distortion, as mentioned earmier, is not in this scheme a first order matter.

The finite element calculation of K is made in Abaqus, Figure 10. The calculation is linear elastic. Due to the presence of two symmetry planes, the geometrical model is reduced to one fourth of the panel. The mesh pattern is built up using quadratic rectangular shell elements (S8R5). There are five Gaussian points across the thickness.

Stress intesnity factor K is calculated with Abaqus Standard using the J integral method. A particular mesh pattern at the crack tip is thus built up. It consists of degenerated rectangular elements: the length of the side which is supposed to be at the crack tip is zero because it is reduced to one point (the three nodes of the side are merged). On the adjacent sides, the mid nodes are displayed to one fourth of the element (Barsoum type element) in order to match the stress singularity at the crack tip (Figure 10). The mesh pattern at the crack tip is thus a circular area with three rings. The length of the elements along the crack path is 0.5 mm. This size is reduced when the crack tip is in the vicinity of the thickness change which is located at 60 mm from the center of the panel. The idea is to avoid having the cricular area cross over the thickness transition line. For crack length a = 59 mm and a = 61 mm, the size of the three rings is 0.2/0.2/0.4 mm, starting from the crack tip. The J integral is evaluated on the three circular contours and are checked to be the same before giving K. The K value is an average result over the thickness.

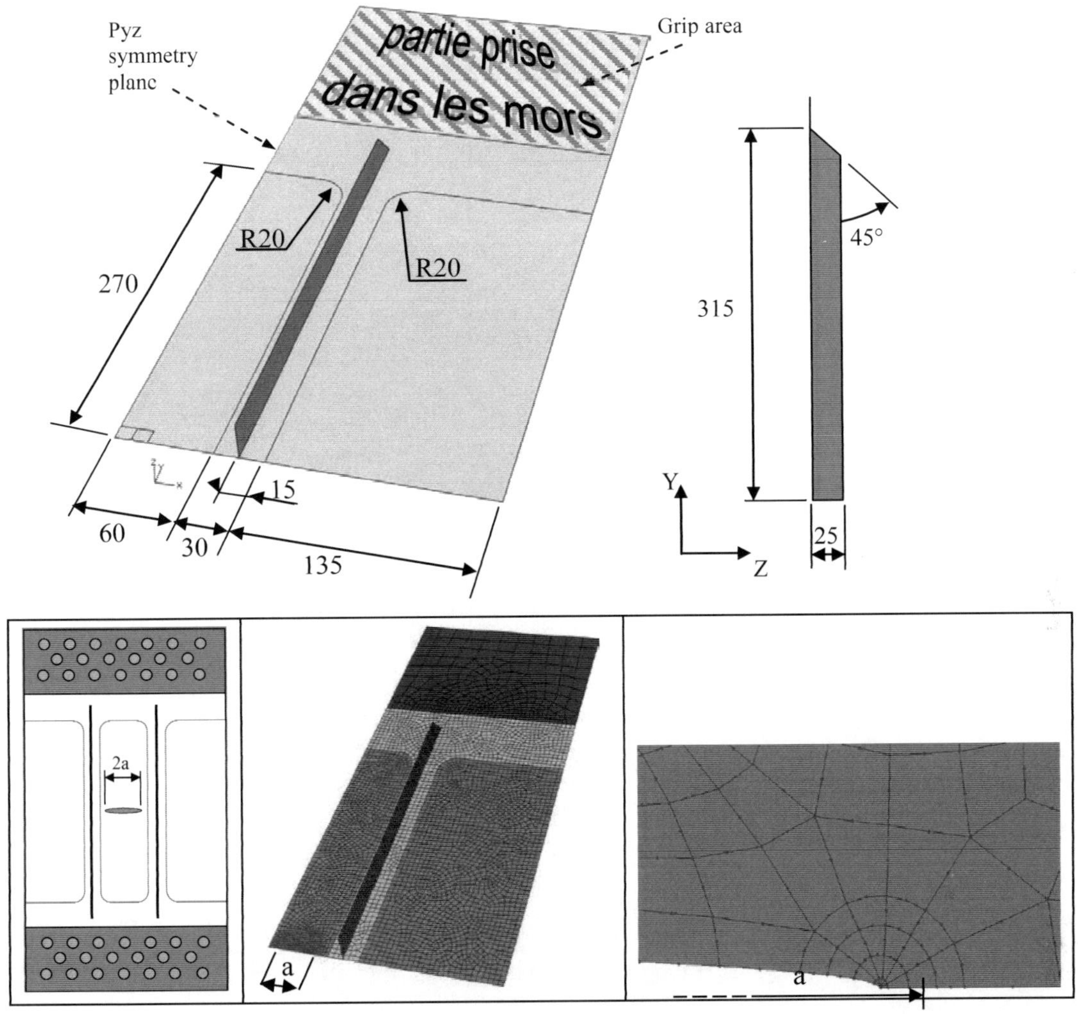

Figure 10. Geometrical model of 6056T6 LBW1 panel with shell elements and mesh. The load is applied on the grip area and the panel is free to bend.

Results of calculations without any residual stress is plotted in Figure 11. The stress profile in the skin (panel with no crack) due to the applied load of 72 kN is shown: it is slightly larger than 66.4 MPa due to secondary bending of the panel. The stress intensity factor with half crack length a, 10 mm < a < 75 mm, is also plotted: K increases from 13 $MPa\sqrt{m}$ to 28 $MPa\sqrt{m}$ before the change of skin thickness, then decreases down to 26 $MPa\sqrt{m}$ right after the overstep, then increases up to 30 $MPa\sqrt{m}$ just before the stringer web then slightly drops down when getting closer to the web.

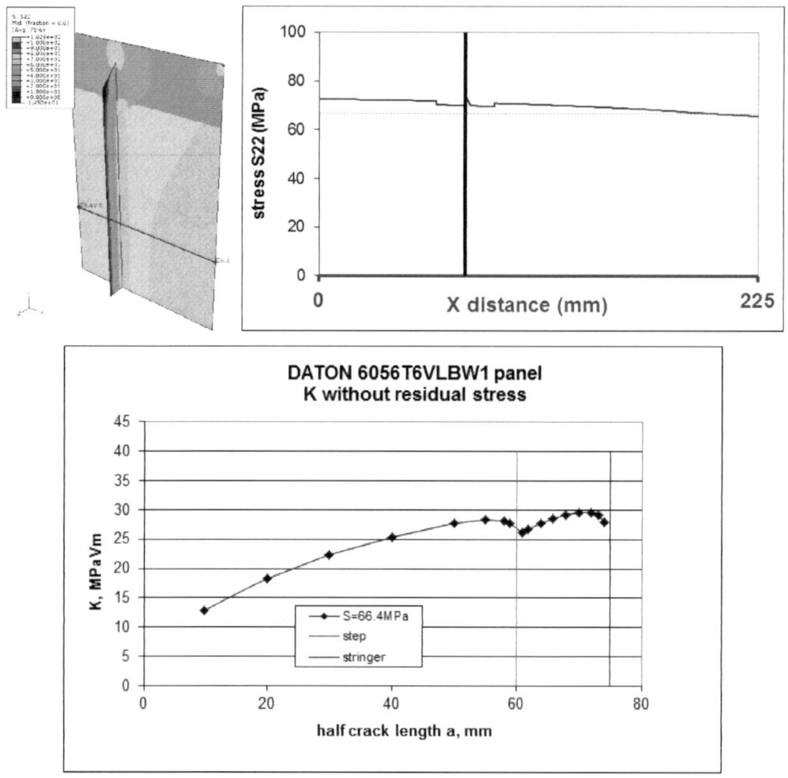

Figure 11. Finite element calculation on 6056T6 LBW1 panel geometry without any residual stress and under an applied load of 72 kN (S_{app} = 66.4 MPa). Top: stress distribution with no crack. Bottom: stress intensity factor in the mid bay area.

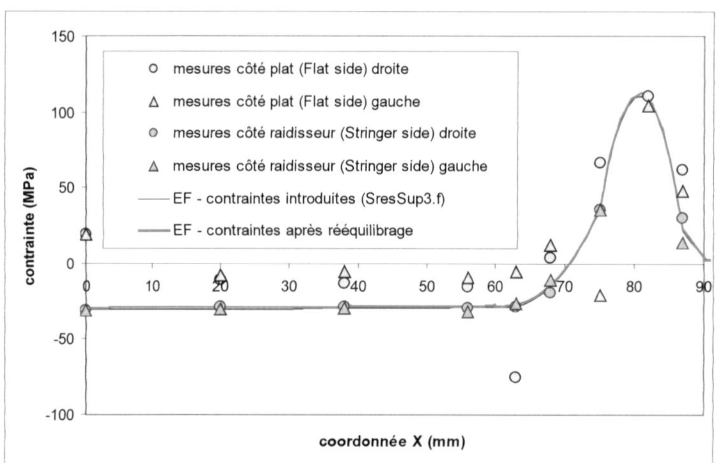

Figure 12. Residual stress profile in finite element analysis of 6056T6 LBW1 panel. Stringer side profile is introduced + peak flat side value in stringer area + slightly positive values beyond. No relaxation took place after re-equilibrium in the model. Stress values vs distance from symmetry plane Pyz.

The scheme concerning the introduction of the residual stress in the geometrical model using Abaqus finite element analysis code is as follows:

- STEP 0 or initial state, introduction of the residual stress profile in the mesh elements uniformly across the thickness.

- STEP 1, mechanical re-equilibrium of the state of stress (if the introduced residual stress field is already at equilibrium, this calculation step will not change the introduced residual stress values).

A subroutine in Abaqus allows to introduce a residual stress profile using polynomial fits. The stress values are then set at the Gaussian points in all the elements of the panel. The profile is also set uniform across the L direction. The values of residual stress determined on the stringer side are introduced but the peak in the stringer area is set to peak flat side value and, beyond the stringer, the values are set close to zero instead of negative. Figure 12 shows the profile at STEP 0 and at STEP 1 (profiel referred to as SresSup3.f). No stress relaxation takes place. The introduced profile was at equilibrium.

In order to calculate the stress intensity factors with the residual stress profile, the scheme using Abaqus is as follows:

- STEP 2, crack introduction in the meshing (by releasing boundary conditions).
- STEP 3, application of load on the panel.
- STEP5, calculation of K through the J-integral contour method.

All calculations are linear elastic with the mesh and elements presented above. K is the average value over the thickness. The obtained results with the load P = 72 kN (S_{app} = 66.4 MPa) are given in Figure 13. The reduction in K goes up to 10 $MPa\sqrt{m}$. The corresponding K_{res} values are also plotted: $K_{res} = K - K_{app}$, K_{app} is the applied K or K without residual stress. It is between -5 and -10 $MPa\sqrt{m}$.

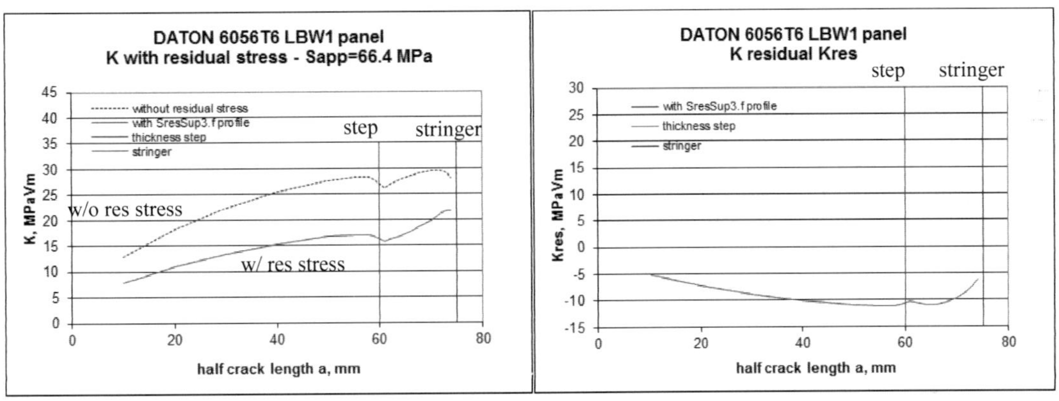

Figure 13. K with residual stress on 6056T6 LBW1 panel under applied stress of 66.4 MPa. Resulting residual K K_{res}.

PREFFAS parameters

The model parameters A, B, C1 and β are identified as previously explained from two baseline fatigue crack growth test results and using A + B = 1. The results, obtained on CCT specimens with a width W of 200 mm, are plotted in Figure 14. From R = 0.1 CAL tests results, the following is found: $C_1 = 1.83\ 10^{-7}$ and $\beta = 2.8$, with ΔK in $MPa\sqrt{m}$, da/dN in mm/cycle. A (and B) is determined from the gap between the two tests. When the two curves are parallel lines, the determination of A is unique. When this is not, several straight lines parallel to the CAL test may be drawn through the spike overload test results. Three lines seem to fit a global behavior. This lead to three C_1/C_2 retardation ratios: 2, 2.5 and 3. Plugging these in turns into eq. (9) leads to three possible values of A (and B), Table I. As it will be seen later on, the predictions of crack growth under spectrum loading with the three sets or parameters fall within 10% to each other.

Table I. PREFFAS parameters.

C_1/C_2	3	2.5	2
A	0.44	0.40	0.34
B	0.56	0.60	0.66

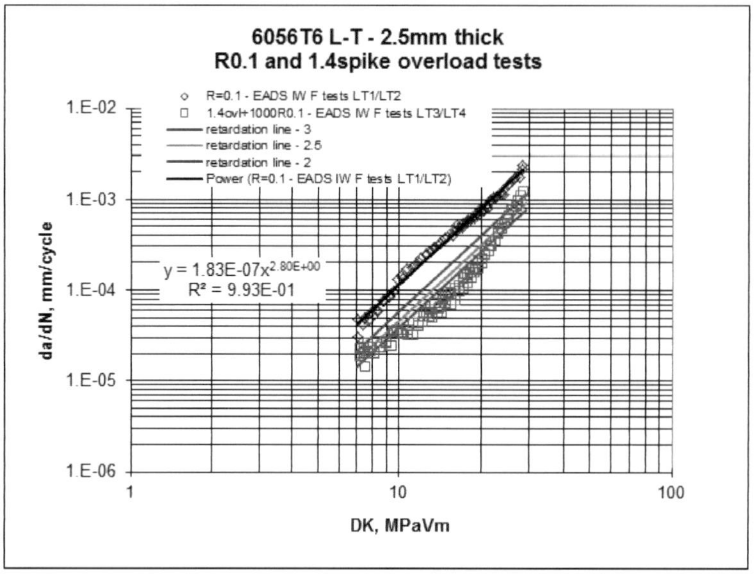

Figure 14. Fatigue crack growth on 6056T6 L-T alloy. CAL R = 0.1 test, repeated 1.4 spike overload test on CCT specimens.

Prediction of FCG on 6056T6 LB1 panel under CAL R = 0.1

Fatigue crack growth predictions on 6056T6 LBW1 panel are made for constant amplitude loading with R = 0.1 and $S_{app,max}$ = 66.4 MPa. A corresponding test was run on the panel.

The residual stress effect is interpreted as an R ratio effect, using the additive principle of K, PREFFAS formulas are used in order to extrapolate the baseline FCGR law at R = 0.1 to other R ratios and to compute the FCGRs:

$$K_{\max} = K_{app,\max} + K_{res}$$

$$K_{\min} = K_{app,\min} + K_{res}$$

$$\Delta K = \Delta K_{app}$$

$$R = \frac{K_{\min}}{K_{\max}} = \frac{K_{app,\min} + K_{res}}{K_{app,\max} + K_{res}}$$

Then:

- if R>0, $\frac{da}{dN} = C_{eff}(\Delta K_{eff})^{\beta} = C_{eff}(AR + B)^{\beta}(\Delta K_{app})^{\beta}$

- if R<0, $\frac{da}{dN} = C_{eff}\left[B\left(1 + \alpha\frac{2\left|S_{app,min} + S_{res}\right|}{Y + UTS}\right)\right]^{\beta}(K_{app,max} + K_{res})^{\beta}$

with PREFFAS parameters, namely $C_1 = 1.83\ 10^{-7}$, $\beta = 2.8$, A = 0.44, B = 0.56, Figure 15 plots the predictions together with the testing results. Prediction of crack growth without residual stress is also given. Term S_{res} is an equivalent remote residuel stress as detailed in the next subsection.

The prediction of the FCG rate with half crack length agrees fairly well with the experimental results. Some small differences occur in the vicinity of the skin step at half crack length of 60 mm and up to the stringer. In terms of half crack length a versus N, the predictions made with the two furthest sets of parameters agree within 10% with the experimental data. The predictions made without the residual stress is very conservative (factor of 4). This latter result highlights the significant benefit brought by the residual stress induced by the welding of the stringers and it must accounted for in teh damage tolerance analysis.

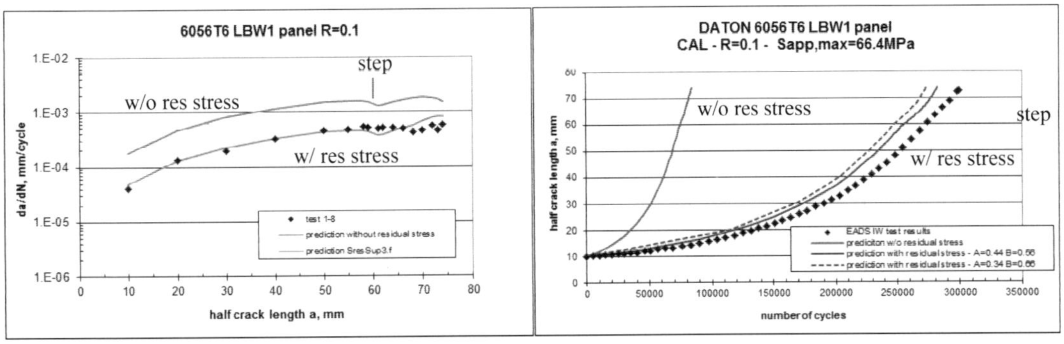

Figure 15. Prediction of fatigue crack growth on 6056T6 LBW1 panel under R = 0.1 CAL and applied S_{max} = 66.4 MPa.

Prediction of FCG on 6056T6 LB1 panel under flight spectrum loading

The predictions made on the CCT specimen are reported first. Then the panel is dealt with. In the latter case, it will be confined to the mid bay area.

The CCT W200 specimen was tested under spectrum loading with a maximum spectrum stress of 50 MPa. Figure 16 plots the predictions by PREFFAS, with the three sets of parameters. A fairly good agreement within 10% is obtained.

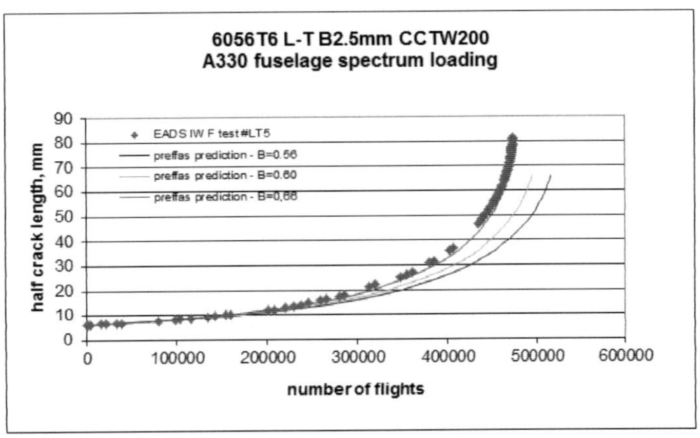

Figure 16. Fatigue crack growth on 6056T6 L-T alloy under spectrum loading. Test and predictions.

On 6056 LBW1 panel, the predictions under spectrum loading by PREFFAS will be made with two sets of parameters, those above giving the closest and the furthest agreements: B = 0.66 and B = 0.56 respectively.

In the crack retardation analysis of 6056T6 LBW1 panel, the residual stress effect is treated as an R ratio effect. The additive principle on K is applied: K(applied stress + residual stress) = K(applied stress) + K(residual stress). In order to use the model, the input (remote) stress spectrum is updated with the addition of an equivalent remote residual stress. The point is to determine this equivalent stress. On the panel without residual stress, K reads as:

$$K = S_{app}\sqrt{\pi a}\, f\left(\frac{a}{W}\right) \tag{17}$$

where S_{app} is the applied remote stress, a is half the crack length and W the panel width. The equivalent remote residual stress S_{res} is defined as:

$$K_{res} = S_{res}\sqrt{\pi a}\; g\left(\frac{a}{W}\right) \tag{18}$$

From the stress intensity factor results obtained on 6056T6 LBW1 panel above with S_{app} = 66.4 MPa, it is found that S_{res} = -26.1 MPa and functionals f and g have been determined as polynomials of a/W, with half crack length a in the range of crack length of [10 mm, 70 mm]. Therefore the value of -26.1 MPa is added to each value of the stresss spectrum. This modified spectrum is thus analysed by PREFFAS.

The 6056T6 LBW1 panel was tested under spectrum loading with a maximum spectrum stress of 99.5 MPa. In the PREFFAS analysis taking into account the residual stress, the maximum stress is thus 73.4 MPa. Consequently, the minimum stress is -15.4 MPa. The compression option of PREFFAS is thus used to make the predictions. Figure 17 plots the predictions by PREFFAS, with the two sets of parameters. Some marked difference is obtained: the predictions are conservative -23% and -46%, which is not as good as on the CCT specimens.

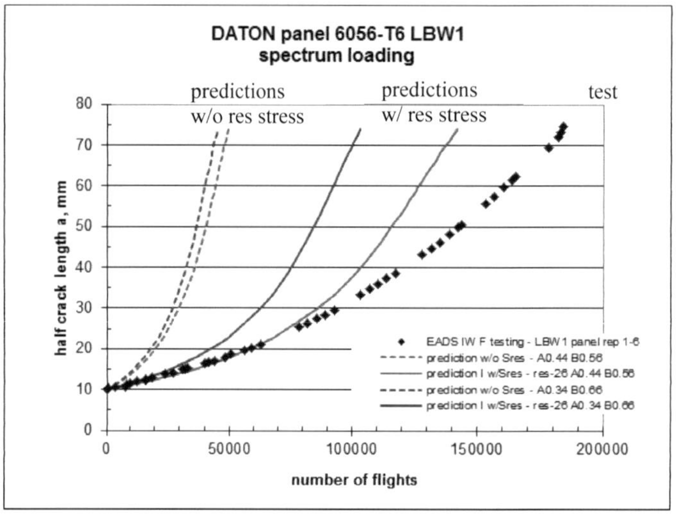

Figure 17. Fatigue crack growth on 6056T6 LBW1 panel under spectrum loading. Test and predictions.

Predictions without the residual stresses were also made. It shows evidence that the residual stress significantly contributes to increase the crack retardation, thus the crack propagation life.

Discussion

The predictions are somehow very conservative. This might hold to the assumptions of the model that are not fulfilled. Figure 14 shows that the retardation curve is not linear. But straight lines parallel to the R0.1 curve can go through in a satisfactory manner. It is confirmed by the prediction in Figure 16. When the crack length is around 20 mm, the crack advance per spectrum sequence is larger than $(1/10)(K_{max}/Y)^2$ but does not go beyond two to three times this value. In that case crack retardation might vanish a little bit although it is thought not to. Should retardation vanish, the prediction curve would be to the right hand side of the experimental one, which is not the case as observed in Figure 17. Therefore, there is another reason. It cannot be residual stress relaxation, because in that case the testing curve would be shifted to the left hand side of the prediction curves, which is not the case.

Looking at Figure 17, the prediction curves do start very well, right on the experimental curve with the same slopes. Then when the crack length is between 20 and 30 mm, the predicted behaviors depart from the tested one. Something is taking place at this crack length. Why do the slopes break from each other although they were in line. In order to try and understand what's going on, the tested behaviors of the panel under R = 0.1 and under the repeated 1.4 spike overload have been plotted together with those of the CCT specimen. Effective ΔK has been used to report on the behavior. On the CCT specimen, the effective ΔK is the applied effective ΔK, without any retardation effect, in the same spirit as of Figure 14 where ΔK is the applied one, no retardation is included. On the panel, the effective ΔK is also the applied effective ΔK without any retardation effect but it includes the residual stress. PREFFAS formulations and results of finite element calculations with residual stresses are used to compute the applied effective ΔK including the residual stresses.

Figure 18 plots the FCG rate results with the effective ΔK. The R0.1 test data of the CCT specimen and of the panel overlap. Those of the repeated 1.4 spike overload do not when the effective ΔK is greater than 8 $MPa\sqrt{m}$: the crack retardation on the panel is greater. For some reason, which is not clear, the retardation as on the CCT is not representative of that of the panel with the residual stresses.

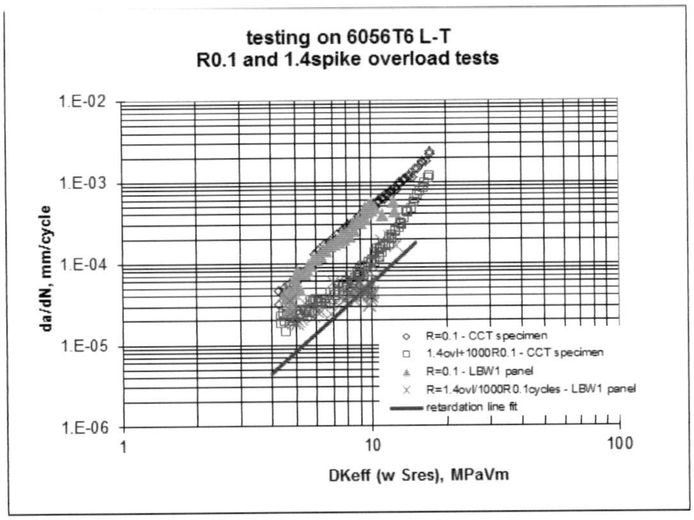

Figure 18. Fatigue crack growth on 6056T6 CCT specimen and LBW1 panel under R = 0.1 and repeated 1.4 spike overload.

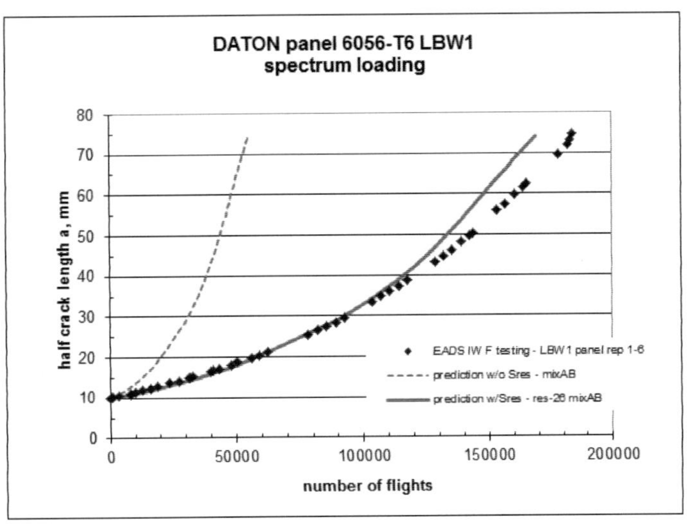

Figure 19. Fatigue crack growth on 6056T6 LBW1 panel under spectrum loading. Test and improved prediction.

The retardation being more significant for ΔK_{eff} greater than 8 $MPa\sqrt{m}$, new PREFFAS parameters were thus determined by drawing a straight line through the retardartion curve of the panel and parallel to the R0.1 line, Figure 18. This yields: A = 0.53, B = 0.47. Then a mixed prediction of the crack growth is made on the panel such as: when a < 20 mm, the prediction with previous parameters is made (A = 0.44), when a > 20 mm, the prediction with new parameters (A = 0.53) is made. These results are plotted in Figure 19.

The predictions have been significantly improved, being 8% conservative. When the crack is in the vicinity of the step, a around 60 mm, we know that the prediction of the crack growth under R = 0.1 is conservative.

Comparison of stringer technologies

Now that the model implementation is better off, it is used to compare stringer technologies: integrally machined (M), laser beam welded (LBW), friction stir welded (FSW). In terms of residual stresses, there is none for the integrally machined, there are in the welded ones. The friction stir welding brings about more compressive stresses as determines by Pisa University [8], Figure 20. The FSW panel is expected to be better in damage tolerance.

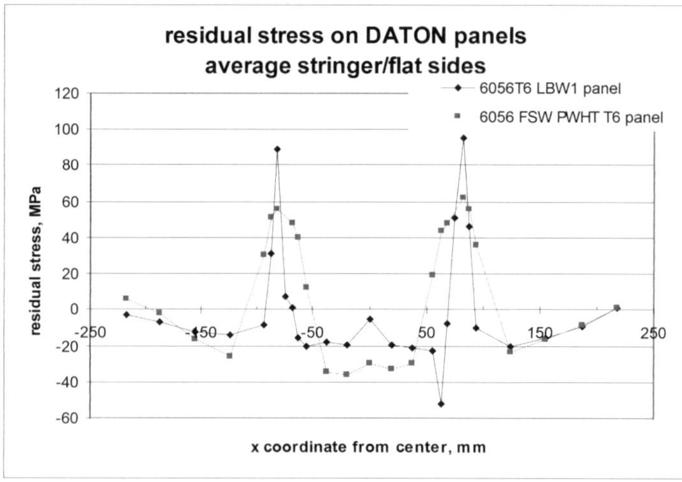

Figure 20. Residual stress determination by Pisa University on 6056T6 LBW1 and FSW panels applying the sectioning method. Average values between flat and stringer sides of the panel. Stress in loading direction.

The whole predictive methodology as developed for the LBW1 panel has been applied to the FSW panel. The three panels have the same geometry and are submitted to the same wide body fuselage spectrum. Fatigue crack growth predictions using PREFFAS model and taking into account the residual stresses have been made. The results are shown in Table II and Figure 21 in terms of relative inspection intervals. Taking a reference interval of 1 for the panel with machined stringers, the LBW panel is 200% better and the FSW panel 300% better. The FSW technology offers the best damage tolerance performance. Testing has confirmed the outstanding performance of the FSW panel.

Table II. Expected benefits in damage tolerance from welding panels over integral panels.

	LBW1 over M	FSW over M	FSW over LBW1
expected increase in inspection interval	+ 189%	+ 323%	+ 46%

M: machined stringers (no residual stress)

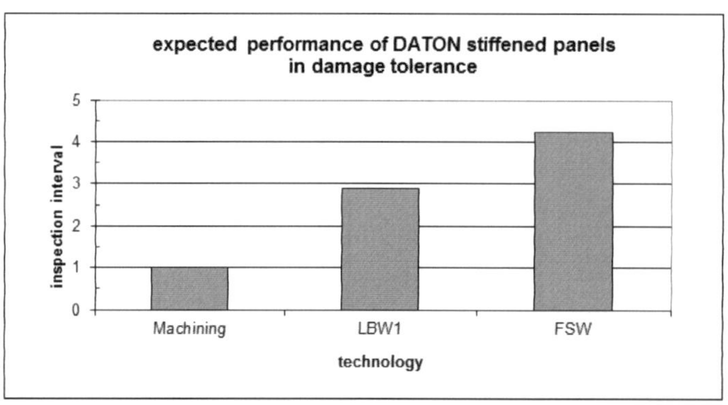

Figure 21. Expected damage tolerance performance of stiffened panels (DATON geometry and flight spectrum).

Conclusion

In order to apply the PREFFAS methodology to make fatigue crack growth prediction in presence of residual stresses under an applied wide body fuselage fatigue load spectrum, a scheme has been developed to process the spectrum with the residual stress in order to serve as an input to the PREFFAS analysis. The case study was a stiffened panel with two laser beam welded stringers and a centre crack in the mid bay. The crack growth of interest is confined to the mid bay where there are compressive residual stresses due to welding. From the knowledge of the residual stress profile in the panel, stress intensity factors have been calculated on the panel. The results have allowed to set up a scheme to process the spectrum with the residual stresses. In order to make predictions on the panels, the model parameters are determined on two baseline tests on coupons made out of the panel skin alloy. Then the model makes the retardation analysis on the processed spectrum and computes the crack growth. The application of the model led to fairly good predictions of crack growth under constant amplitude in the stiffened panel. The predictions of crack growth under a spectrum loading were fairly conservative. The reason holds to an unexplained enhanced retardation effect. The test results under repeated spike overload on the panel have allowed to update the model parameters and to propose an improved prediction. The simplicity of the methodology has been kept, dealing with two baseline tests on the skin sheets and the same number of modelling parameters. Despite the unexplained retardation, the model is able to give fairly good predictions. The methodology has been applied to evaluate the damage tolerance of other stiffened panels with machined or friction stir welded stringers. The last one by far offers the best performance thanks to its more compressive residual stresses.

PREFFAS AT TEMPERATURE

Temperature is known to exert an influence on the fatigue crack growth rates especially at low frequency where the effect may be detrimental. The reason holds to viscous deformation in the vicinity of the crack tip that may enhance crack growth and to microstructure ageing by phase transformation due to long exposure to temperature that may change the properties of the material. Since PREFFAS methodology proceeds with two baseline tests, it is proposed to evaluate the model running all the tests at the frequency of the application. Therefore, if any effect shows up during the service application under spectrum loading, it should be captured at the baseline tests level. This is the object of the work presented herein. It has been carried out within the frame of PREVISIA project [9]. It features the use of PREFFAS model.

The case under investigation deals with CT coupons made out of 15-5PH martensitic stainless steel. The service temperature is between 300 and 400 °C. Three cases are studied: unaged alloy tested at room temperature and at 300 °C, aged alloy tested at room temperature. At 300 °C, the test frequency is 2 Hz. The PREFFAS methodology is applied. Predictions of crack growth under flight spectrum loading are compared to the experimental results for the three cases.

Testing matrix

Four types of FCG tests have been run at room temperature on the non aged and aged steel, at 300 °C on the non aged: CAL R = 0.1, CAL R = 0.7, repeated 1.7 spike overload every 1000 R0.1 cycles, pylon spectrum loading. Table III lists the testing conditions. Compact Tension specimens were machined out of 15-5PH alloy which was supplied by Aubert & Duval. They were tested in the L-T orientation. They are 10 mm thick. Their width W is 75 mm.

Table III. Testing matrix.

Non aged	5000 h/375 °C aged
20 °C, 10-20 Hz frequency	20 °C, 10-20 Hz frequency
300 °C, 2 Hz frequency	

Baseline FCG behaviour

The baseline results obtained under CAL and repeated spike overload are plotted in Figure 22.

On the non aged tested at 20 °C, there is no R ratio effect. The spike overload test exhibits crack retardation with a factor of 5.

On the 5000 h/375 °C aged steel tested at 20 °C, the behaviour is the same as that of the non aged steel, except for the high crack growth rate regimes which starts earlier. The reason holds to a low fracture toughness due to ageing. Some slight R ratio effect is observed but seems to be very small in the Paris regimes.

On the non aged tested at 300 °C and 2 Hz, there is no R ratio effect. The FCG rates are 3 to 4 times larger than at 20 °C at R = 0.1.There is no retardation effect althought it shows up under a frequency of 20 Hz [10] with a factor of 5 or more.

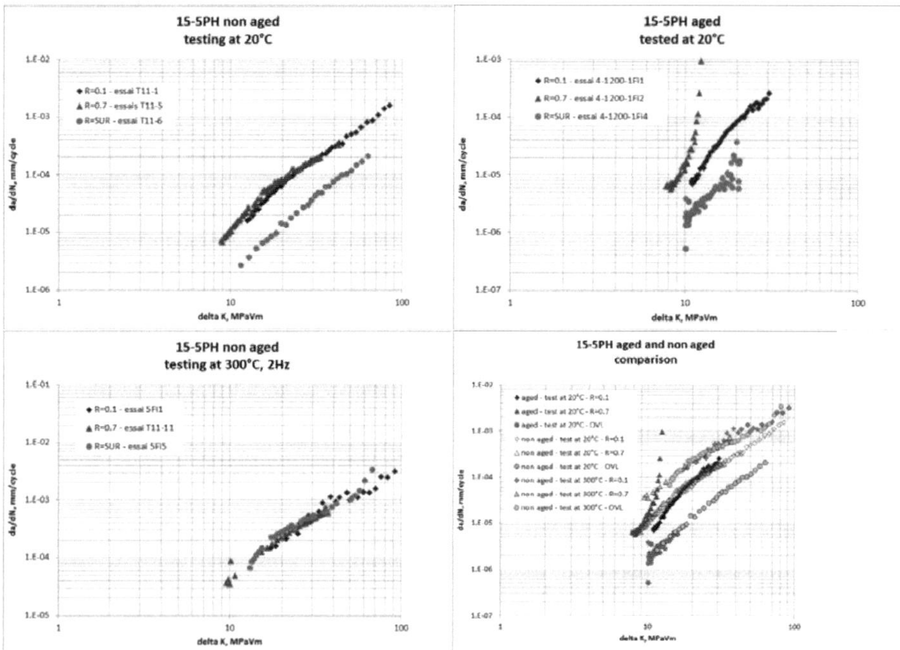

Figure 22. Baseline tests on 15-5PH steel: non aged tested at 20 °C (top left), 5000 h/375 °C aged tested at 20 °C (top right), non aged tested at 300 °C (bottom left), comparison (bottom right).

PREFFAS parameters

The option of PREFFAS with $A + B \neq 1$ is used. A and B are thus determined with the Paris's regime of the baseline tests, CAL R0.1 and R0.7 and the repeated spike overload. The identification scheme is plotted in Figure 23. Parallel Paris' lines can be drawn through the test data. For the aged steel tested at room temperature, the Paris' regime at R = 0.7 is very short and difficult to identify. Therefore an average behavior is considered which gives a slight R ratio effect. The PREFFAS parameters come out as: A = 0 and B = 0.59 on the non aged tested at 20 °C, A = 0.17 and B = 0.57 on the aged tested at 20 °C, A = 0. and B = 0.99 for the non aged tested at 300 °C.

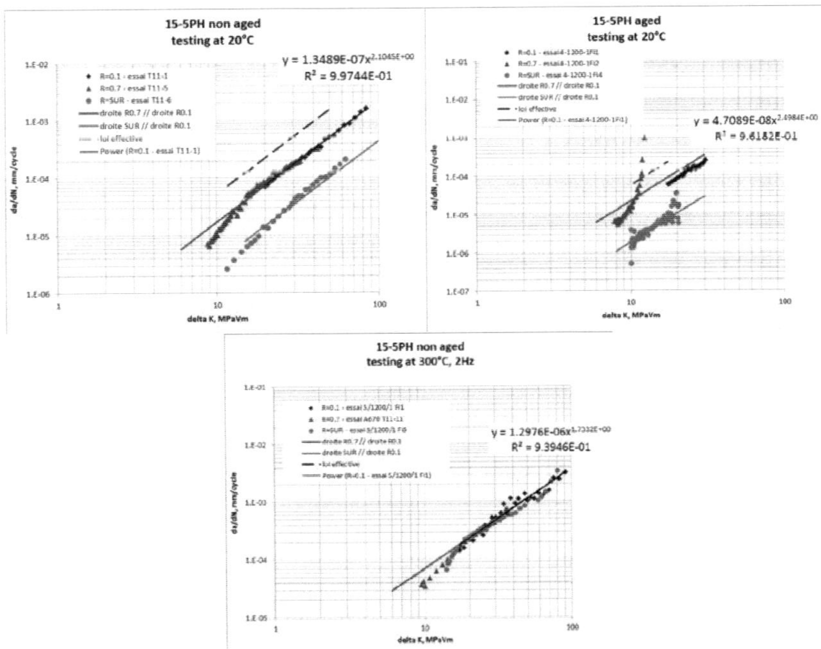

Figure 23. Determination of PREFFAS parameters on 15-5PH steel: non aged tested at 20 °C (top left), 5000 h/375 °C aged tested at 20 °C (top right), non aged tested at 300 °C (bottom).

FCG behaviour under flight spectrum loading and predictions

The flight spectrum is a pylon spectrum of an Airbus single aisle jet. The spectrum sequence consists of 4500 representative flights with different severities. It totals 104906 variable amplitude cycles. Figure 24 gives an idea of some of the flight cycles.

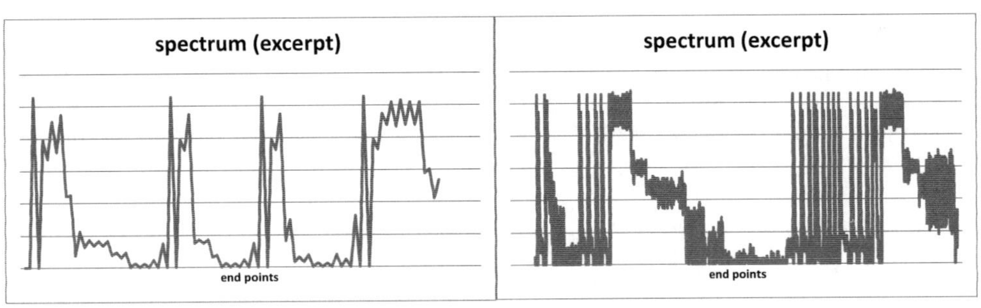

Figure 24. Except of the flight spectrum used for the tests in terms of relative stress versus end points.

In all three cases, the same load spectrum has been applied to the CT specimens. The PREFFAS predictions were made with the corresponding determined parameters. They are plotted in Figure 25 against the experimental data.

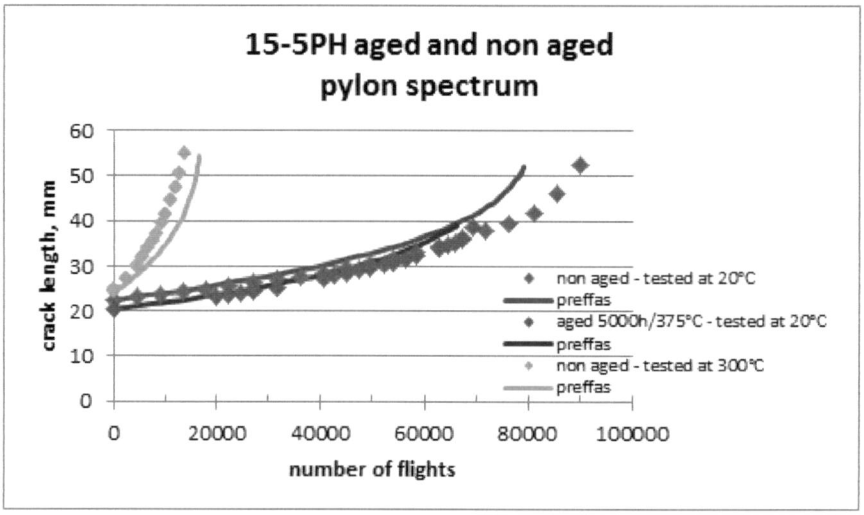

Figure 25. Fatigue crack growth under flight spectrum loading. Tests and predictions.

The FCG behaviors at room temperature of the aged and non aged steels under spectrum loading are very close, the aged one breaks a little bit earlier. This is consistent with the similar baseline FCG properties of both steels and a lower fracture etoughness of the aged one.In the non aged steel tested at 300 °C, the crack grows five to six times faster. This is consistent with baseline behaviors which are three to four times faster and with the absence of retardation.

The predictions are within 10% of the experimental results in terms of crack propagation life. They are conservative at room temperature, they are not at 300 °C. The conservatism is due to the low amplitude cycles which are overestimated by the Paris' law. However, looking at the spectrum, the damaging cycles are mainly the ground-air ground cycles which are numerous and which are in Paris' regime. The conservatism will thus be low. At 300 °C, the model predicts a slightly lower growth. Since the high crack growth regime takes place very soon, the Paris' law will underestimate the growth rate, with a small error yet when looking at the graphs.

Conclusion

Under baseline testing at room temperature, the FCG behavior of aged and non aged 15-5PH steel are similar: no R ratio effect, crack retardation, Paris' regime reduced for the aged steel. For the non aged tested at 300 °C and 2 Hz: no R ratio effect as well, higher crack growth rates, no crack retardation. Under flight spectrum loading under the same temperatures, the behaviors are consistent with the baseline behaviors. The predictions givent by PREFFAS agree fairly well with the experimental data. The methodology applied to tests run at 300 °C and 2 Hz have captures the temperature/ageing effect.

CONCLUSION

PREFFAS model computes fatigue crack growth using the crack closure concept. It features four parameters and a crack retardation analysis taking into account the load history. The model works in the Paris' regime, it assumes that crack retardation does not vanish and is in a stationary regime. The parameters are identified on two or three baseline fatigue crack growth tests. The object of the presented work was to evaluate the model in the presence of residual stresses or temperature. As far as residual stresses are concerned, the case study dealt with an aluminum alloy stiffened panels with built-in residual stresses due to the welded stringers and submitted to a fuselage load spectrum. The crack growth of concern was in the mid bay area where the residual stresses are compressive. A scheme to include the residual stresses to make the PREFFAS analysis of the flight load spectrum has been presented and applied, based on the additive principle of the stress intensity factor. The model gave fairly good predictions of fatigue crack growth. The benefit of the compressive residual stresses was highlighted as well. The case study dealing with temperature concerned a stainless steel laboratory coupon tested at 300 °C at 2 Hz under a pylon load spectrum. With temperature and low frequency, the fatigue crack growth rates are enhanced and the retardation effects disappear. The identification of the model parameters with the baseline tests run at 300 °C and 2 Hz have captured these behaviors. The predictions given by the model were quite satisfactory. The results of the whole work show that the model has the capability to handle these two situations without any development leading to increase the number of parameters. The model seems to show some robustness giving relevant orders of magnitude in the results. This aspect has to be more investigated.

ACKNOWLEGEMENTS

The financial support of the European Commission (EC) to public project DaTon and of the French National Research Agency (ANR) to public project PREVISIA are geatfully acknowledged. DaToN project was part of the FP6 program of the EC. The project ended in 2008. PREVISIA project is part of the MatetPro 2010 program of the ANR and is still running. The partnership is made of SIMaP Materials and Processing Science and Engineering department of the National Polytechnic Institute of Grenoble, CIRIMAT Materials Research and Engineering department of the National Polytechnic Institute of Toulouse, PPRIME Materials Physics and Mechanics department of ENSMA, LMT Mechanics and Technology laboratory of Ecole Normale Supérieure of Cachan, Aubert & Duval, Airbus Commercial Airplanes and Airbus Group Innovations (coordinator).

The work that has been presented in this paper is the result of collaborative works with people from different teams and organizations. First, the author would like to thank the contributions of Airbus Group Innovations colleagues to the results. Namely in DaToN: Fabrice Congurdeau for the finite element analysis, Delphine Alléhaux for the materials characterisation, Laëtitia Kirchner for the stringer welding, Christian Ithurralde and Frédéric Louis for the testing. In PREVISIA: Fernand Vinhas for the materials characterisation and potential drop technique, Christian Ithurralde, Antoine Decalom and Albert Perez for the testing and the electrical potential drop technique development and implementation. Overall, Gérald Martinez for developing testing and monitoring softwares, Gilles Thomas for supervising the machining and the fatigue crack growth testing and François Pradat, responsible of the testing platform. Second, thanks are also addressed to colleagues outside Airbus Group Innovations, whose contributions were in direct relation to the treated topics of the paper as inputs or comparison, namely at Pisa University, Claudia Polese and Luigi Lazzeri for their contribution to DaToN project in the residual stress determination on the welded panel which served as inputs to the herein pesented work, at PPRIME ENSMA, Loïc Dimithe, Mandana Arghazi and Gilber Hénaff for their contribution to PREVISIA project in the fatigue crack growth evaluation and discussion for comparison, at Airbus Commercial Airplanes, Zacharie Obadia for his contribution to PREVISIA project in the industrial service application, at Aubert & Duval, Sylvain Puech for his contribution to PREVISIA in the processing and supply of the steel that has been evaluated in fatigue crack growth.

REFERENCES

[1] Aliaga, D., Davy, A. and Schaff, H. (1985). In: *Durability and Damage Tolerance in Aircraft Design*, Proceedings of the 13th ICAF Symposium, Pisa.

[2] Héciak, B., Schaff, H., Aliaga, D., Davy, A.and Budillon, E.(1984). Proceedings of the ICAS Conference.

[3] Journet, B. (1993). In: *Task 4 Flight Simulation.* BRITE EURAM project 3250, BREU 128C, final overall task 4 report, n° DCR/M-60365/F-93.

[4] In-house Airbus Group reports on extension of PREFFAS on several grades of aluminum alloys, titanium allloys and steels, several spectrum types (fuselage, engine pylon, wing).

[5] Elber, W. (1971). *ASTM STP 486.*

[6] Journet, B. and Congourdeau, F (2008). In: *WP4 validation of models for fatigue and damage tolerance analyses of integrally stiffened panels*, FP6 program DaToN project, FP6--516053, report DaToN-WD-WP4-1.1/EADS-F.

[7] Alléhaux, D. and Kirchner, L.. DaToN (2007). In: *WP3 Manufacturing and Testing*, FP6 program DaToN project, FP6-516053, report DaToN-WD-WP3-x.1/EADS-F.

[8] Polese, C. and Lazzeri, L. (2007). In: *WP3 Manufacturing and Testing*, FP6 program DaToN project, FP6-516053, report DaToN-WD-WP3-x.1/Pisa.

[9] Journet, B. (2014). Presentation at progress meeting R6, PREVISIA project, ANR MatetPro 2010.

[10] Dimithe, L., Arzaghi, M. and Hénaff, G. (2013). Presentation at progress meeting R4, PREVISIA project, ANRMatet Pro 2010.

An Amphibian and the Concept of its Derivative Model for Fire-fighting Application

Masatoshi Tsujii, Hideyuki Mochida, Katsuo Tanaka, Atsuhiko Fujitani, Yushi Goda

ShinMaywa Industries, Ltd., Japan

Abstract: Following the lifting of the ban on aircraft development and manufacturing in Japan, ShinMaywa proceeded with the development of a new seaplane. With the UF-XS experimental plane, which was modified from the Grumman UF-1 provided by the US Navy, ShinMaywa pursued the studies of new technologies such as the hull design of high seaworthiness and the Boundary Layer Control (BLC) system, which were adopted in the PS-1 Anti-Submarine Flying Boat. The PS-1 then developed into US-1/US-1A Search and Rescue Amphibian, which had a landing gear, and made further progress to the latest US-2 Search and Rescue Amphibian, which is the world's only search and rescue amphibian that is capable of operating in rough sea. And now, the study to convert the US-2 for many applications is ongoing, including aerial firefighting.

Regarding corrosion due to seawater, which is peculiar to seaplanes, US-2 adopts some techniques for corrosion protection. Although it is very difficult to prevent corrosion perfectly, it is currently planned to monitor the corrosion status to develop a more effective protection method.

THE HISTORY OF SEAPLANES [1] [2]

In 1910, seven years later than the first flight of the Wright brothers, the world's first flight of seaplane was achieved by Henri Fabre in France. Flight by seaplanes had been tried a few years before the Fabre's success by some including Louis Bleriot. It shows that seaplanes had also received interest since the advent of aircraft, as with land-based aircraft. It can be supposed that it was much easier to choose a place to land and take off because seaplanes don't require runways. Also, people at the time might have had a feel like a sense of security for the fact that the surface for landing is water.

Jacques Schneider in France was also one who thought that seaplanes would have a brighter future than land-based aircraft since they could use water surface as runway. Schneider proposed an air race for seaplanes, which would be known as "the Schneider Trophy", aiming to emphasise the development of seaplanes and high-power engines. Although the contests were cancelled several times due to the First World War, it regularly took place since 1919 and speed record had been updated year after year. The seaplanes at the time were faster than land-based aircraft and easily smashed their records.

In the late 1920s, the development of new aircraft became active. The needs of aircraft as a new means of transport increased with the growing needs of flight from Europe and other countries to their colonies, such as South Africa, South-East Asia, and South Pacific Ocean region. At the time, however, there were no airports or even runways at the destination or way points, so the harbours for vessels and ships were the only reliable place. In addition, since the distance to the destination was very long, they had no choice but to make aircraft larger. Given such circumstances, seaplanes were the most suitable option. In 1929, a large flying boat "Do X" was experimentally developed by Dornier. Its all-up weight reached 52 tons and it was powered by twelve engines. Do X achieved a flight with approximately 160 passengers. In early 1930s, a double-hulled flying boat "S.55" developed by Savoia-Marchetti, a flying boat "Latécoère 521" by Pierre-Georges Latécoère, and a flying boat "P2Y-1" by Consolidated Aircraft achieved a formation flight from Italy to Brazil, transatlantic flight, and flight from San Francisco to Honolulu, respectively. It is said that DC-3, which was land-based aircraft equipped with a landing gear with shock absorber, was first adopted as a means of air transport in 1936. It is also said that Pan American World Airways began using Boeing 307 in 1940 after Laguardia Airport in New York opened. So before that, major airlines had been covered by seaplanes or flying boats.

From the late 1930s to the end of the Second World War, military seaplanes were experimentally used and adopted for many applications including reconnaissance, cargo, bombing strike, antisubmarine, and rescue, and seaplanes showed their ability to the fullest. However, with the end of the Second World War, the golden age of seaplanes came to an end. The Second World War strongly emphasised the development of aircraft including land-based aircraft, and the technology for land-based aircraft made significant progress. Long range and high speed transport became ordinary and the presence of seaplanes gradually decreased. Although some seaplanes that were mass-produced in the United States or England were modified to cargo aircraft or other applications after the end of the war, but most of them were replaced by land-based aircraft.

It is the truth that seaplanes are still active in particular field. Land-based aircraft has advantages in performance of flying speed, range, or carrying capacity over seaplanes, but requires large prepared runways for landing and take-off. Helicopters have the useful property that they are capable of hovering and require only small place for landing, but weak in terms of flying speed, range and carrying capacity. On the other hand, seaplanes are able to provide their performance to target that is out of range of helicopters, or places without prepared runways. Applications or missions that only seaplanes are capable of performing or performing more effectively and efficiently indeed exist, such as Search and Rescue at sea, transport to remote islands, disaster-relief or evacuation in case of large earthquake, or volcanic disaster on an isolated island, and aerial fire-fighting.

THE POST-WAR HISTORY OF AMPHIBIAN DEVELOPMENT BY ShinMaywa

Before the Second World War, even under war conditions, seaplanes could attempt to land on sea surface with wave height of 1.5 m at most. Even civil transoceanic seaplanes had only flown between harbors and they never performed landing on rough water surface at the open ocean. This low seaworthiness had limited operation of seaplanes. In response, improving seaworthiness with efforts in technology and landing bravely on rough water to enhance their operational capability, it is what seaplanes should be.

After the Second World War, Japan was forbidden from engaging aeronautical activities, such as production, repair and research and development of aircraft and aeronautical products, and even transport by aircraft. The period of no aeronautical industry continued for seven years. Kawanishi aircraft company was also prohibited from research and development and producing aircraft as other companies. Under such circumstances, Kawanishi aircraft company was re-established as "ShinMaywa Industries, Ltd." (hereafter referred to as "ShinMaywa") and was producing special-purpose trucks and other industrial equipment. After the restriction was removed in 1952, however, aeronautical industrial infrastructure in Japan was reconstructed with the licensed-production and development for Japanese defence aircraft. It also led ShinMaywa to the second stage of the challenge to aircraft business.

As described earlier, previous seaplanes were only able to perform limited operation because of their low seaworthiness. ShinMaywa started to address the technical challenge to improve seaworthiness of seaplanes. A primary issue for landing on rough water at the open sea is landing impact and splashing water. It may work to sharpen the angle of V-shaped hull bottom in order to reduce landing impact, but a flying boat having so sharp V-shaped hull bottom that it can absorb large landing impact would have a disadvantage in terms of reducing drag because draft of such flying boat would be very deep and height of aircraft would be larger. To solve this problem, it was proposed as an approach to make the speed for landing significantly low. Extremely low speed landing realises cushioning the impact of landing without drag increase.

A flying boat having such feature was developed on the assumption that it was capable of operating under the condition of wave height of three meters, which is Sea-State 5. The concept of the entire system was established as studies of element technologies progressed, but it was considered too much risk to develop actual aircraft in a single step. As a first step, integration technology experimental aircraft "UF-XS", which was modified from the Grumman UF-1 provided by the US Navy, was developed and used in experiments. Many studies and experiments for new element technologies such as long-thin hull, V-shaped hull bottom, grooved wave suppressor, large flap angle capability, Boundary Layer Control (BLC), and extremely low-speed flight were performed to improve seaworthiness of the flying boat. A new flying boat was developed based on these studies and experiments with UF-XS and subsequently named "PS-1".

In October of 1967, the first aircraft of PS-1 made its first flight. PS-1 showed low-speed take-off and landing at the open sea and landing on water with wave height of four meters. From July of 1968, technical evaluation test and operational test were performed. From 1971, operation of PS-1 was heavily reviewed and it was performed by Japan Self-Defense Forces and supported by involved private companies. This review, which continued for four years, provided the know-how for actual operations with many water take-off and landing experiences.

PS-1, which was capable of landing on water with wave height of three meters, had the highest seaworthiness that the world had ever seen, but it was not able to land on the ground. With keeping the high seaworthiness, US-1 was developed as amphibious aircraft, which was capable of operations on land and water. Outer shapes of fuselage of PS-1 and US-1 were almost the same, but US-1 had bulge overhanging on its fuselage to stow the landing gear. Since the shape of bulge affects aircraft seaworthiness, a wide variety of shapes for bulge were tested in towing tank in design phase. At first, US-1, which was powered by four engines, adopted an engine having approximately 3,000 horse-powers. However, the need for increasing the power arose for improving the performance of take-off from water, the engines were replaced by approximately 3,500 horse-powers engine. This US-1 with advanced engines was called US-1A.

As seen above, practical application of high seaworthiness technology was achieved by PS-1 and US-1/1A, which inherited PS-1's characteristic. It is no exaggeration to say that PS-1 and US-1/1A enhanced the operational capability of seaplanes with technological innovation.

THE FEATURES OF THE US-2 SEARCH AND RESCUE AMPHIBIAN

US-1A Search and Rescue amphibious aircraft is still in use. One reason is that it is the world's only aircraft having high seaworthiness and STOL capability with unique devices and systems, including wave suppressor and BLC system. On the other hand, demands on US-1A's performance were increasing through long experience of operations. Japan Maritime Self-Defense Force, which is user of ShinMaywa's amphibious aircraft, called for improved controllability for water landing and take-off, improvement of patient transport environment, and enhanced Search and Rescue capability. In 1991, ShinMaywa started the studies including those demands for subsequent aircraft, which would be successor of US-1A, and began design work in 1996.

Successor aircraft of US-1A, which would be named "US-2", was developed as major modified aircraft of US-1A. The latest technologies at the time were aggressively adopted with keeping the superior performance of PS-1/US-1, such as high seaworthiness. Taking the users' demands into consideration, the development was aimed at improving water take-off and landing performance, improving long-range flight performance, facilitating on-board first-aid treatment at cruise altitude for survivors rescued, improving stability and controllability at low-speed flight. To achieve these aims, some major modifications were implemented, including Replacement of the engines and propellers, Introduction of pressurised cabin, Changing the materials as a lightweight solution, Introduction of Fly-By-Wire into flight control system, and Introduction of an electronic integrated instrument system.

As US-2 was developed based on US-1A, it has similar shape and structure to US-1A and much the same dimensions. This chapter introduces the effects of the modifications.

Replacement of the engines and propellers

The engine power was substantially increased from 3,500 to 4,600 horse-powers by replacing with AE2100J. The previous propellers, which were three-blade metal propeller, were also replaced with Dowty R414, which is six-blade propeller made of composite material and a half feet shorter. These modifications improved the performance of long-range flight and speed and also shortened the time of running on water required for water take-off and landing. In addition, the shorter-blade propeller reduced the risk that the blades hit the water. As a whole, the performance of water take-off and landing and safety were considerably enhanced.

Introduction of pressurised cabin

US-1A has no pressurised system and its cruising altitude is limited to 10,000 feet. It means that US-1A's operational capability depends on weather conditions. Therefore, for example, even in case of emergency, such as disaster-relief operations or urgent patient transport, US-1A has to take a circuitous route if there is cumulonimbus cloud or a front on the route. US-2, which has a pressurised cabin, is capable of flying at up to 30,000 feet, which is high enough to make US-2's operations more flexible. In addition, since the introduction of pressurised cabin reduced the noise level in aircraft, the environment in aircraft, especially patient transport environment, was significantly improved.

Changing the materials as a lightweight solution

Special fuselage structure and cabin pressure control equipment and other equipment are required to introduce the pressurised cabin. These obviously lead to weight increase, which adversely affects flight and other performances. To solve this problem, changing the materials of some components was conducted, such as Main wing fuel tank with integral structure, which adopts an integral forming by machining, Changing the aluminum-alloy float to composite material, Changing the aluminum alloy wave suppressor to titanium alloy.

Introduction of Fly-By-Wire into flight control system

Fly-By-Wire control system, which was the latest at the time, was introduced into the flight control system to improve its function. US-2 has a triple Fly-By-Wire control system and another mechanical back-up control system to ensure redundancy. In addition, Stability and Control Augmentation System (SCAS) maintains its controllability and stability at low-speed flight. Moreover, Flight Path Control (FPC) performs accurate follow-up controls of flight path angle of water landing approach, and realises more stable water landing. Usually, water landing requires an adequate training and proficient skills, but these advances improved controllability and stability of water landing and contributes to reducing pilot workload.

Furthermore, the US-2 flight control system includes Engine Failure Compensation (EFC). One engine failure during take-off or landing would obviously put aircraft in a dangerous situation, including aircraft bank and loss of altitude. However, immediately after automatic detection of engine failure, EFC performs counter-control by integrated control of Fly-By-Wire control system and electronic fuel control system, and increasing engine power in order to maintain aircraft attitude and altitude. So, this function also enhanced US-2's safety and reliability.

Introduction of an electronic integrated instrument system

For instrument panel of the cockpit, previously-used analog instrument panel was replaced by an electronic integrated instrument, which consists of seven liquid crystal display units, two display processing units, and other equipment. The system indicates information with regard to flight, navigation, aircraft system, videos and others by integrating them into Electronic Flight Instrument System (EFIS) or Engine Indication and Crew Alerting System (EICAS). Comparing the previous-used analog type of US-1A and the modernised type of US-2, it is quite obvious that improvement of operability and visibility and reduction of pilot workload were achieved.

Success and failure of search and rescue operations by seaplanes at the open sea depend significantly on the capability of water take-off and landing. The US-2 is the world's only search and rescue amphibian that is capable of landing on water with waves up to three meters high. No other seaplanes have such performance. US-1A and US-2 have steadily built up a track record in search and rescue and patient transport, as of the end of February of 2015, they have accomplished 1,000 missions.

THE CONCEPT OF FIRE-FIGHTING APPLICATION

Wildfires occur around the world in dry season. Recently, major wildfires occur and cause serious damage, especially in Mediterranean region, United States, Australia, Russia, and other countries. In these countries, helicopters and fixed-wing aircraft have been used for aerial firefighting over many years. Seaplanes that are capable of flying to a fire site with high speed and repeating water scooping and water dropping within a short time are considered as an effective method and mainly used in Southern Europe and North America. Not only wildfires, aerial firefighting is also conducted against many kinds of fire, including tanker, train, airport, shopping centre, and other fires in large facilities.

Currently, the study to convert US-2 for the firefighting application is ongoing. Firefighting amphibious aircraft performs water scooping by running on water. After water scooping, aircraft takes off and drops water on target by opening doors provided at the hull bottom. Therefore, as a concept, the US-2 firefighting amphibian would have special features for aerial firefighting, including scoops for water intake, water tank, doors for water dropping, and another tank for extinguishing agent added to water to enhance firefighting efficiency.

There are mainly three points that can be listed as measures of extinguishing performance of firefighting aircraft including firefighting seaplanes.

Cumulative amount of water dropped per day

It indicates how much water in total the aircraft is able to drop on a target fire site within a day. Obviously, the amount would become larger for larger carrying capacity of water, higher flying speed, and shorter time of water intake. Turnaround time of water intake of Land-based aircraft is longer because they can load water from only an airport that has a water supply facility. Seaplanes with scoops have an advantage in turnaround time because seaplanes are able to load water from any nearby water source if it is large enough, such as rivers, lakes, and sea.

Water distribution pattern

"Water distribution pattern", which is measured in liter per square meter, is a density distribution of dropped water at ground level. The effective level of density distribution depends on the type of vegetation being burned, soil, and weather conditions. If the density is below a certain level, water dropping will not work. The density distribution is determined by the amount of water, opening space of the door, flow rate of released water, flying speed and altitude, and other factors. The effective area of water distribution pattern would become larger for larger amount of dropping water. Additionally, application pattern, which is the shape of distributed area, is determined by the combination of the flow rate, and the flying speed and altitude. For instance, in case of low altitude dropping, dropped water reaches the ground with high forward velocity before splitting into small pieces, and it would be a concentrative and destructive dropping. In case of high altitude dropping, dropped water is almost atomised before reaching the ground, and it would be a wide-area and safe dropping but ineffective for heavy fire.

Safety

Safety of aerial firefighting operations is also an important factor for firefighting aircraft, as well as fire extinguishing performance. In recent aerial firefighting operations, firefighting aircraft often performs water dropping at the lowest possible altitude in order to drop water on a target as accurate as possible. However, it makes the operations even more severe, and in fact many accidents have occurred so far. US-2 firefighting amphibian capable of extreme low-speed flight would be able to approach to a target at lower speed than other fix-wing aircraft. As a result of its capability, it can be expected that the US-2 firefighting amphibian would be able to achieve accurate water dropping from higher altitude than other aircraft and improved safety. For this concept, it would be a challenge to ensure effective water distribution pattern with water dropping from high altitude, but it is considered feasible with large water carrying capacity and low-speed water dropping capability.

A systematic experimental study of aerial water dropping is definitely needed to realise high efficiency and effective aerial fiefighting. As part of the study, collaborative studies were conducted between Japan Aircraft Development Cooperation (JADC), Japan Aerospace Exploration Agency (JAXA), and ShinMaywa from 2007 to 2012. Firstly, wind tunnel test and CFD analysis were conducted to establish basic data for calculating dropping water behavior. Secondly, flight simulation test was carried out to build appropriate operational procedures and reduce pilot workload. In addition, it is planed that the function, performance, and design of water tank for the US-2 firefighting amphibian will be evaluated with an actual-size mock-up in 2014 and 2015.

It is expected that these studies will lead to review of the water tank design, flying speed and altitude for water dropping, and establishment of a method for more effective aerial firefighting. We hope that it will be contributory to the realisation of the US-2 firefighting amphibian.

THE EVALUATION OF STRUCTURAL INTEGRITY OF AMPHIBIANS

The concept of structural integrity evaluation

Corrosion due to seawater can be cited as a cause for damage specific to amphibious aircraft. ShinMaywa has unique techniques for corrosion protection with long experience as an amphibious aircraft manufacturer. One of the techniques is unique coating system for corrosion prevention. In addition, ShinMaywa's amphibians have On-board Engine Washing System to prevent the main engines, APU (Auxiliary Power Unit), and BLC from salt accumulation and corrosion.

Corrosion Control -Chemical Means-

The unique coating system is applied to parts that are prone to corrosion, such as hull structure, fastener component of skin and so on. Standard coating system consists of Primer and Top coat on skin. (see LH of Figure 1), whereas the unique coating system consists of Primer, Protecting coating, Primer and Top coat on skin. (see RH of Figure 1) In addition, corrosion inhibitor is applied especially to inside of hull, hinge fitting and so on.

Corrosion Control -Mechanical Means-

As described above, ShinMaywa's amphibians have On-board Engine Washing System to prevent the main engines, APU (Auxiliary Power Unit), and BLC from salt accumulation and corrosion. The compressor blades of these are washed with spraying mixture IPA (isopropyl alcohol) and water. This system realises on-board washing of the engines on the sea or in-flight during operations. For in-flight washing, main engine to be washed is stopped and washed one by one. In addition, the washing system has a function that automatically informs appropriate washing timing by monitoring main engine power if salt accumulation rises beyond the level that can cause power reduction. The entire aircraft is also washed after operations if water landing is performed.

Influence of Corrosion for DT Analysis

In spite of the above, it is still very difficult to prevent corrosion perfectly. When some corrosion is detected and removed, structural member is reinforced because removal of corrosion leads to decrease in thickness of structural member. ShinMaywa considers that it is possible to develop an original analysis model that is suitable for amphibians by reflecting the influence of thickness decrease and reinforcement to Fatigue and Damage Tolerance analysis model.

Loss of thickness causes stress increase on the structural member. In case of corner crack, loss of thickness increases the crack growth rate in the thickness-direction and the crack reaches to the opposite end faster. In addition, if a part decreased in thickness is reinforced, it is required to change the model of stress intensity factor. For the reasons stated above, a mechanism is required to change the calculation condition intermittently with the calculation progress of DT analysis. (see Figure 2)

Since a load transfer varies with thickness decrease or reinforcement, stress change is not a linear change with a simple change in cross section area. Therefore, another FEM analysis may be required. Previously, the influence of thickness decrease due to corrosion and reinforcement has not been totally taken into account in Fatigue and DT analysis. However, some corrosion is found in every US-2 Progressive Aircraft Rework, which means scheduled heavy maintenance. Currently, it is planned to monitor the corrosion status and thickness decrease regarding in-service US-2. ShinMaywa intends to select structural critical points that should be monitored in preparation for the analysis model development.

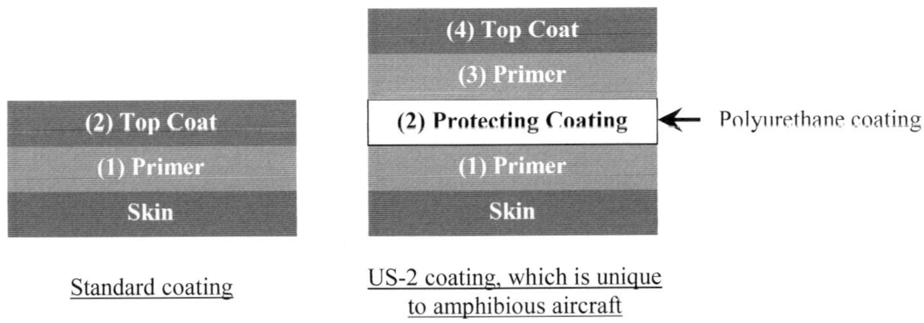

Figure 1. Unique Coating.

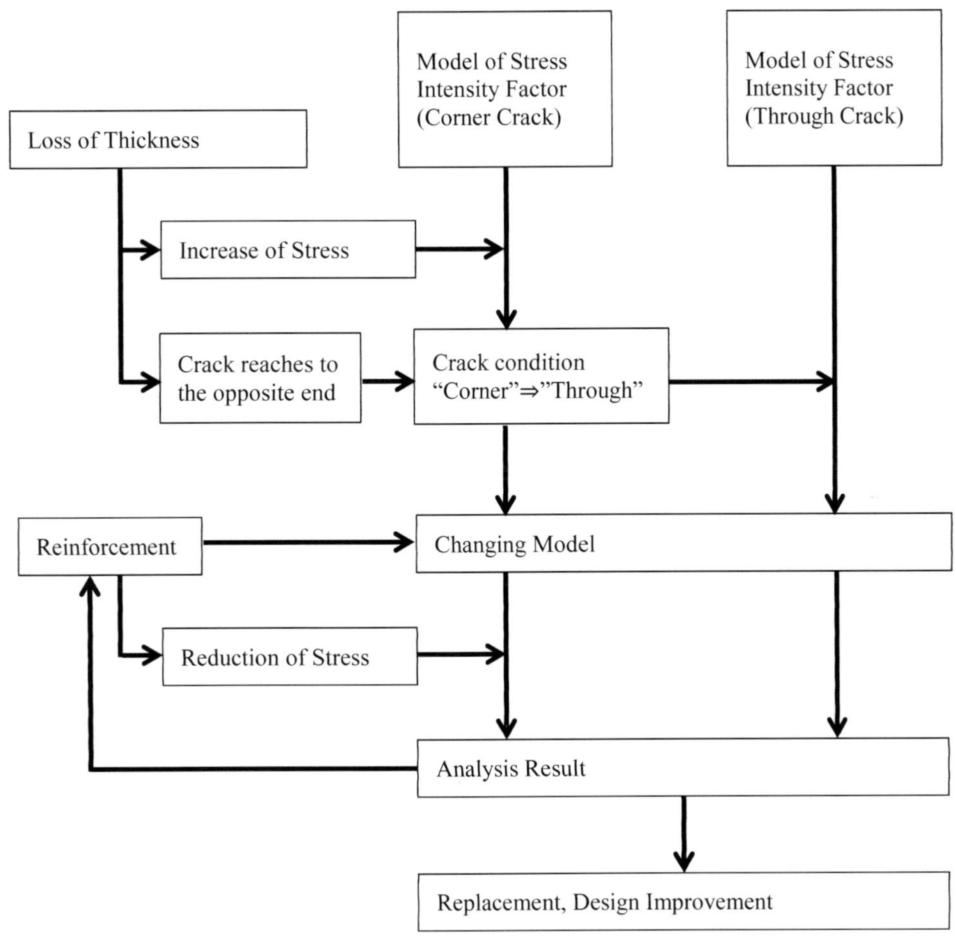

Figure 2. DT Analysis.

REFERENCES

[1] Inoue, K. (2003). “The history and technological prospects of seaplanes (the first volume)”, *Japan Association of Defense Industry*, vol. 673, pp. 1–16.

[2] Inoue, K. (2003). “The history and technological prospects of seaplanes (the second volume)”, *Japan Association of Defense Industry*, vol. 674, pp. 17–31.

Fleet monitoring

The A400M Usage Monitoring Function

Alexis Falga, Dr. Jean-Pascal Kleinermann, Dr. Alain Santgerma

Airbus SAS Operations

Abstract: The A400M is a versatile military transport aircraft, which is expected to cover very different types of usages during its service life. It has been designed, certified, and a maintenance programme has been established using a mix of different types of fatigue missions. Once the aircraft will be in service, a Usage Monitoring Function will be used as a mean among others to estimate the fatigue damage accumulated by each aircraft according to its actual usage.

This Usage Monitoring Function includes 3 complementary capabilities: Indirect Measurement Function is based on the recording of parameters by the Systems of the Aircraft, which are then used to rebuild loads and stress histories at typical structure locations and to estimate fatigue and crack growth damage. The Statistical Usage provides statistics on fatigue driving parameters derived from the actual usage of each individual aircraft. Direct Measurement Function, installed only on some aircraft, includes strain sensors allowing Operational Loads monitoring and cross-checking of estimated loads and stress histories.

This paper focuses on the method and algorithm used to estimate damage as part of Indirect Measurement Function. It also provides the validation results comparing estimated loads and stress histories with actual measurements Flight Tests and Full-Scale test.

The Usage Monitoring Function will be used in the frame of A400M Continued Airworthiness, as a mean among others to help ensuring structure integrity all along aircraft life thanks to inspection periodicities appropriate for the actual usage of each aircraft. It will also help the Air Forces manage their fleet cost-effectively by judicious use of operations.

INTRODUCTION

The A400M is a military transport aircraft adapted to the needs of the 21st century, as it is able to fly a wide range of missions, from pure military to humanitarian ones for many different customers. Due to this wide possible use, the maintenance program defined at the entry in service will have to be later adjusted to the actual utilisation of the aircraft.

The challenging part is to translate the actual usage into damage estimation on the different elements of the structure. Typically, this monitoring is performed using dedicated strain sensors installed on the structure, with the addition of data given by the crew or by systems recordings. The use of sensors has 2 major drawbacks which are the necessity to maintain them during the complete life of the aircraft and the need to recalibrate regularly the loads interpolation deduced from the recordings. It was thus decided to implement a new approach using only parameters recorded by the systems and to deduce from these data an estimation of the damage on the different structural zones. However, this approach must demonstrate a sufficient accuracy within the wide variety of aircraft utilisation. This function will be used to monitor the aircraft usage, helping the customer to decide for future A/C role allocation (tactical, logistic and training), in order to limit the accumulation of damage on particular locations.

This paper explains the process followed by Airbus to define this function and in particular the damage estimation algorithm.

BACKGROUND FOR A400M CERTIFICATION

The military transport aircrafts of the 21st century must answer the need for expanded operational capabilities both in purely military missions, such as deployment of Rapid Reaction Forces and in humanitarian interventions, as in the case of earthquake, flood or famine relief operations.

The A400M not only meets, but exceeds these needs in a number of key areas. It can perform missions which previously required two or more different types of aircraft, and which even then provided an imperfect solution. It has the greatest payload/range of any airlifter able to perform tactical flights, while also being able to fly long-range strategic sorties. It is the only tactical airlifter which can carry of the many kinds of large, outsize cargo currently used for military and civic / humanitarian missions. Designed from the outset to be a dual-role transport and tanker, the A400M provides a cost-effective way for air forces to acquire an aerial refuelling capability.

Figure 1. A400M Aircraft.

Different flight profiles have been selected and agreed with the Air Forces after review of their different operational needs. These are:

- Tactical multi-role: flight possibly including several different tactical manoeuvres among landing on semi-prepared runway, aerial delivery, contour flying, and combat off-loading
- Training: flight with max altitude under 5000 ft, and including touch-and-goes, and/or one engine out, and/or asymmetric approach
- Simple logistic: pure logistic flight without Air to Air Refuelling
- Refuelling logistic: pure logistic flight with Air to Air Refuelling

They are used for aircraft design, certification and initial maintenance program definition.

However, the real use of the aircraft will obviously diverge from these reference flights, be it in terms of weight, duration, range, payload, or profile. So, the customers required a monitoring of the real operational usage of the aircraft in order to catch any deviation compared to the reference usage. This will allow optimizing its utilisation and adequately ensure its continued airworthiness.

USAGE MONITORING FUNCTION PHILOSOPHY

Different methods exist for the type of monitoring function expected. Some are requesting additional sensors to be installed on the structure, whereas others are only using crew reporting or recordings. The next sections describe them with their pros and cons and explain the reasoning behind the retained one.

Choice of the method for usage monitoring function:

Method with additional sensors installed on structure. The use of sensors installed on the structure is the typical way of monitoring the differences between theoretical reference flights and real ones: the data recorded allow determining the loads on the equipped sections. The certification load model is used to determine the stresses on study points and to compare them to the ones obtained from the theoretical reference flights.

The main interest of this method is the accuracy of the measurements since the extraction of loads is direct. The main drawback is the installation and maintenance of the sensors. Indeed, they have to be installed with a dedicated system and associated wirings and have to be maintained and replaced if necessary during the complete life of the aircraft. In addition, a recalibration is necessary from time to time to ensure that the equations used for the determination of the loads are still valid.

Due to these drawbacks, this solution has not been retained for the complete fleet. However, this system has been implemented on a limited number of aircrafts. These aircraft will be used as Operational Loads Monitoring aircrafts and additionally will be used to help in the calibrating of the selected method not using sensors.

Methods without additional sensors. These methods are using data either given by the pilot through the crew logbook, or directly obtained from the systems recording, without the need to install additional sensors on the structure.

The simplest method is using basic parameters (weight, low-level flight, payload, flight duration, etc...) gathering information about the stages of the flight realized by the aircraft. From these parameters, it is then possible to determine on each pilot point which type of reference flight is closest to the one flown: this can lead to the selection of different flight profiles for different pilot points. Then, the damage from the selected reference flight profile is used for the pilot points. This method is quick and easy to implement, but also conservative as it always considers the worst case for each flight and for each pilot point, which may lead to unnecessary maintenance operations. In addition, it may be inaccurate when dealing with a new flight profile. These drawbacks were considered too important and this method was not retained for A400M.

An alternative method is an improvement on the previous one: it is based on the recording of a limited set of parameters describing the flight profile with few figures (e.g. flight time at low altitude, landing impact severity, number and level of pressure cycle, aircraft weight, centre of gravity position, etc…). These parameters must characterize the flown flight as a whole (i.e. no time histories), but may be different from one area to the other depending on the major fatigue contributors in this area. Then, a numerical model (i.e. interpolation function) is created that directly links aircraft usage, characterized by this limited set of recorded parameters, to resulting fatigue damage. This method will be less conservative than the previous one, but will require a high number of initial calculations in order to develop the numerical model. It will also be difficult to cover any new kind of flight profile as it may need parameters outside the range initially selected for the numerical model, leading to inaccuracy in the results. This method was also not retained, mainly because of its inability to take into account accurately enough any kind of flight profile.

The last and most complex method intends to rebuild local behaviour history on considered locations called pilot points. This is done through the recording of a high number of parameters and the creation of transfer functions from these parameters to loads or stresses at each time slice considered, through a simplified load model. A stress spectrum at each pilot point location is then built and the fatigue damage is calculated. This method is more complex to develop as it asks for a simplified load model to be created and validated, but it allows covering any kind of flight profile with the same level of accuracy.

This third method was thus selected for the Usage Monitoring Function (UMF) and will be applied by all the A400M customers to monitor each of their aircraft.

Pilot points selection:
40 pilot points (PP) are selected in order to cover all the structural components of the aircraft. These pilot points represent typical reference locations such as fuselage sections and main frames, wing root area and spars, junction of wing to fuselage, VTP and HTP sections and important features such as MLG fittings, pylon-to-wing attachments or VTP attachments.

Parameters recorded:
Hundreds of parameters are recorded continuously during the flight. From these parameters, 145 are selected for the UMF and characterize:

- Aircraft and fuel weights, repartition in fuel tanks
- Altitude
- Aircraft speed (on ground, in flight), Mach number
- Engine data: propeller pitch, rpm, torque, etc…
- Control surface position: angles of spoilers, ailerons, rudder, elevator
- Aircraft configuration: angles of slats, flaps
- Attitude, speed and acceleration at Centre of Gravity of aircraft along pitch, roll and yaw axes
- Air to air refuelling transfer: time of beginning and end, volume transferred
- Brakes pressure
- Cargo and paratrooper doors state (open/closed)

UMF Process:
The initial maintenance program defined for the A400M is based on fatigue and damage tolerance calculations performed on the complete structure, on particular points called study points. The calculations are done by considering the theoretical usage expected for the aircraft in terms of flight profiles, mission durations, disturbances statistics, etc...

The external loads applied on the structure are processed through a Load Model and ran through on a Finite Element Model to obtain the stresses applied on each study point at each time of the flight. The stress spectrum is then obtained by cumulating the stresses during each flight. Finally, the damage created during each flight is calculated.

The UMF process is similar in the principle, but with two main differences: the initial data are not coming from assumptions and statistical distribution, but from recorded parameters which represent the real flights performed. And the loads and stresses are estimated from the parameters using transfer functions.

Figure 2 gives the flowchart describing the UMF process.

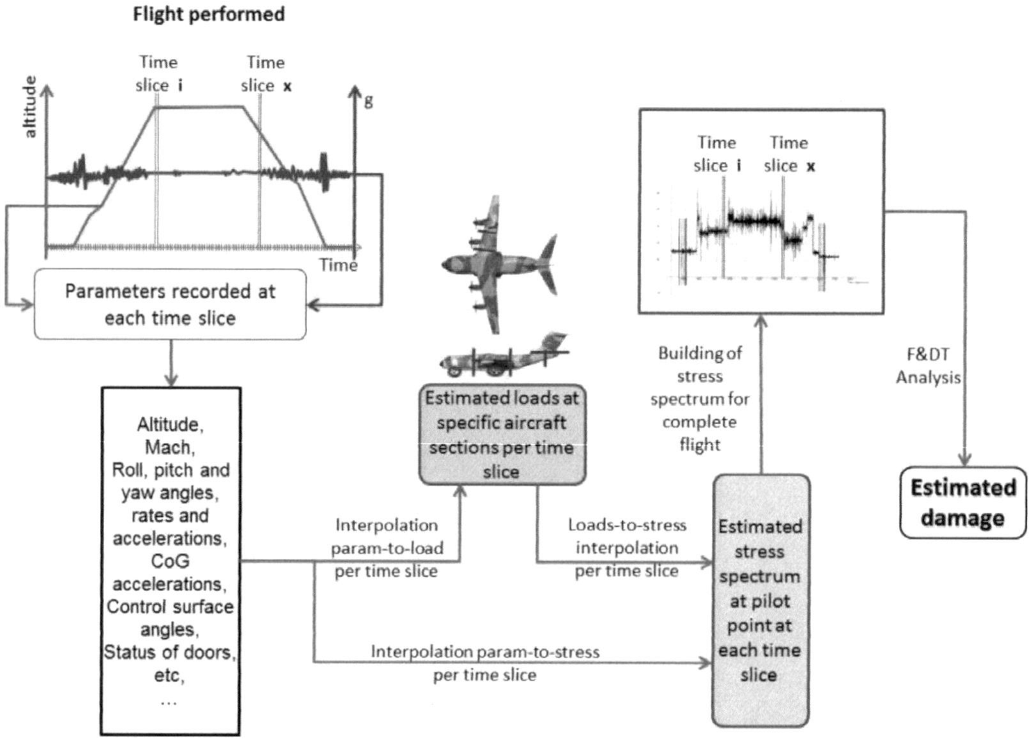

Figure 2. Usage Monitoring Function principle flowchart.

The 145 parameters are recorded continuously during the flight. For each time slice, an interpolation determines transfer functions from the parameters to loads on specific aircraft sections. From the loads, another interpolation based on a simplified load model allows determining the stress applied on a pilot point during the considered time slice. On some pilot points, the parameters can be directly linked per another transfer function to stresses. The fatigue stress spectrum for the studied pilot point is then the combination of all the stresses per time slices. A typical fatigue calculation is applied on the obtained spectrum to derive the fatigue damage created by the flight at this pilot point.

USAGE MONITORING FUNCTION DAMAGE ESTIMATION ALGORITHM

This section describes in details the various steps of the UMF process presented above.

Loads and stresses calculations:

The continuous recording of all parameters during the flight authorizes the determination of loads and stresses at each time slice.

The parameters can be used to directly interpolate stresses on some pilot points. However, on others, an intermediate step has to be performed. Indeed, accurate direct relationship cannot be established between parameters and stresses. It becomes then physically meaningful to estimate loads from parameters on specific aircraft sections, and then to derive stresses from these loads. These aircraft sections are independent from the pilot points but representative of the aircraft behaviour.

Loads components selection. A subset of reference loads components has been selected to represent the complete aircraft loading. The load stations are the ones instrumented on flight test aircrafts for load model validation, (fuselage, wing, VTP and HTP sections) plus some specific loads on main landing gear, nose landing gear and pylon to wing attachments as shown in Figure 3 below.

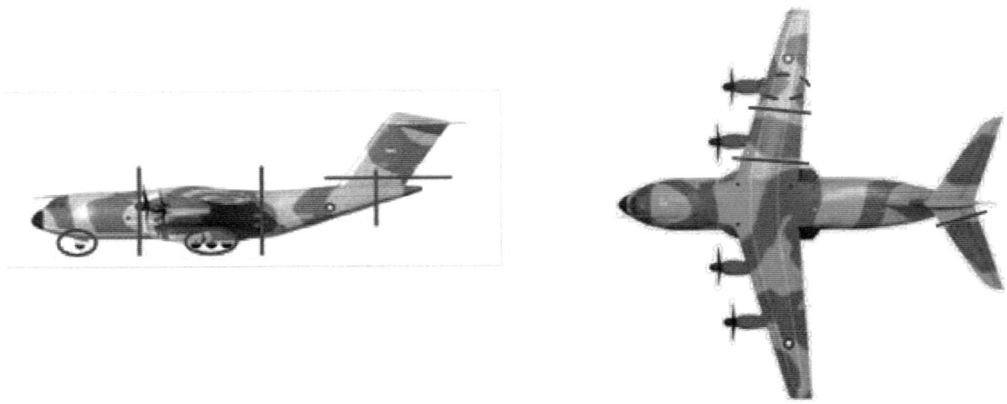

Figure 3. Locations of aircraft sections where loads are estimated by UMF.

It must be ensured that "param-to-load" interpolations provide histories for each of these loads.

Param-to-load transfer functions. The aircraft is subject to numerous phenomena during the flight representing the different actions applied on it. These phenomena include 1g loads for flight and ground, disturbances such as gusts and manoeuvers in flight and bumps, turns and braking on ground and specific stages such as landing impact, aerial delivery, etc… For each identified phenomenon, a "param-to-load" transfer function is developed to provide estimated instantaneous values for the load components. This interpolation function uses only some of the parameters available at each time slice.

Figure 4 gives an overview of the way to obtain the loads on a specific section for a time slice of the flight based on the parameters recorded and the param-to-load transfer functions.

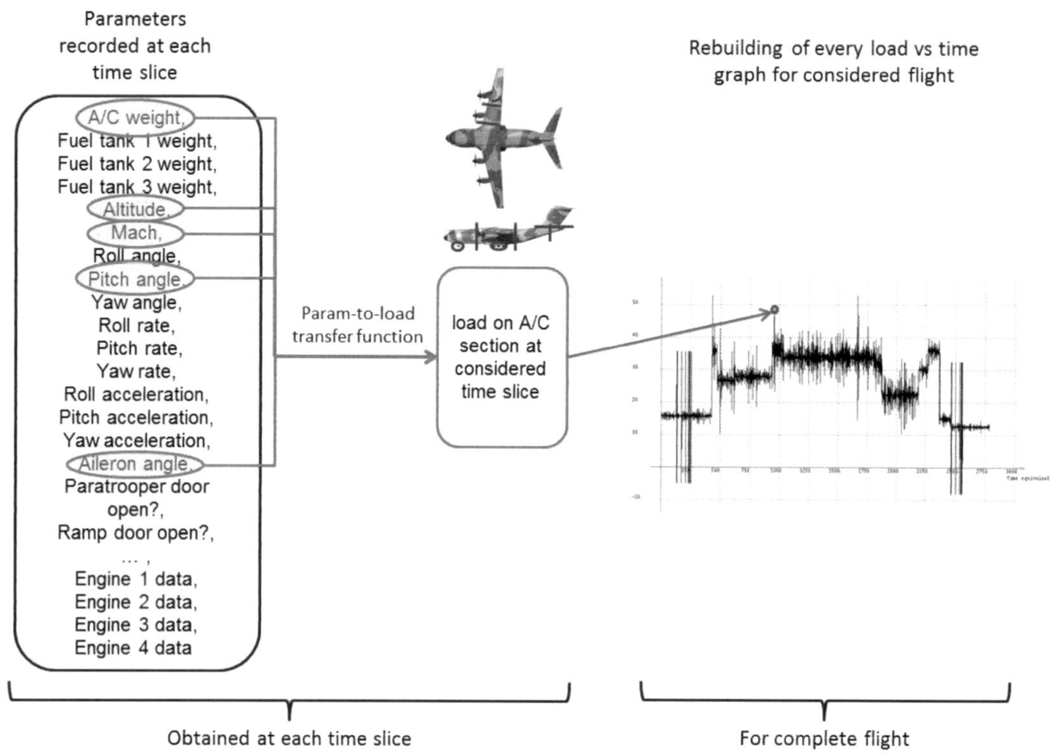

Figure 4. Principle of determination of loads per UMF.

Different transfer functions are used to determine the loads and stresses on the sections or pilot points depending on the considered phenomena.

Table I gives an overview of the principal phenomena and the type of method used to interpolate the corresponding loads and stresses.

Table I. interpolation methods used to obtain loads/stresses following the phenomena verified.

Interpolation method	Load denomination	Type of load
A	1g flight	1g
B	1g ground	
	Bumps on ground	Ground disturbance
	Slow turn	
	Rationale turn	
	MLG landing impact	
	NLG spring-back	
	Lateral gusts	Flight disturbance
	Vertical gusts	
	Manoeuvers	
	Other secondary phenomena	Disturbances

The fatigue damage is very sensitive to 1g flight loads, but these loads have a high and complex sensitivity to A/C configuration and usage. Method A consist thus in the use of an Artificial Neural Network (ANN) interpolation to determine 1g flight load to ensure good accuracy. This ANN is developed from additional calculations considering only equilibrated "1g" cases. One separated ANN is developed per load component. For each ANN, 6 to 8 input parameters are required depending on load component. Nevertheless, the main difficulty is to find ANN showing "smoothed" and meaningful shape. So, each selected ANN is a compromise between accuracy and physical meaning.

The accuracy of this method was demonstrated to be good as shown on Figure 5.

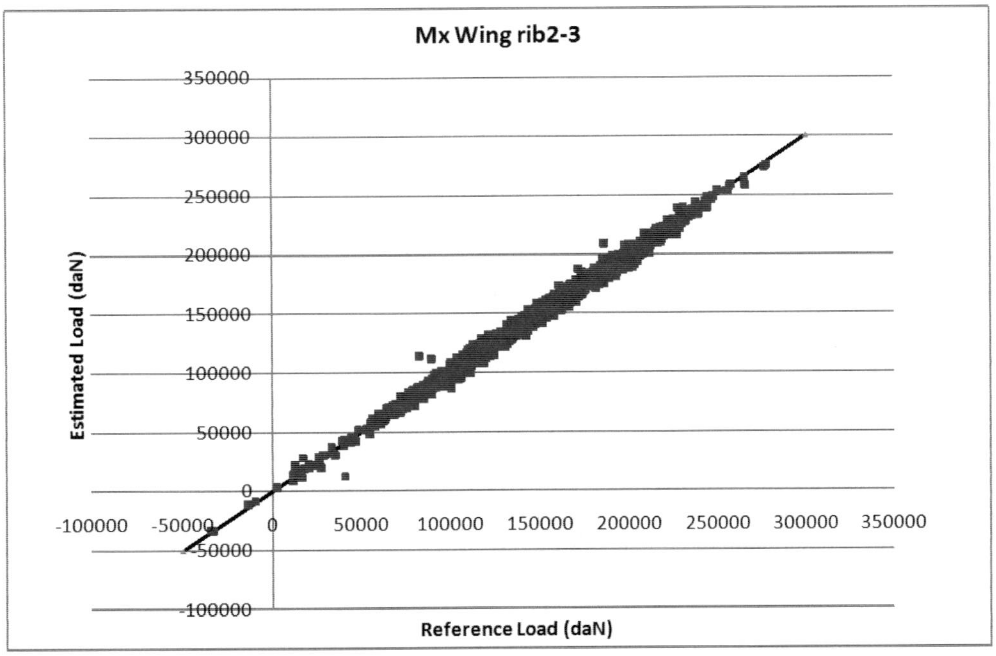

Figure 3. Illustrative example of estimated vs reference 1g flight load for the load cases used for interpolation development.

Method B is used when the phenomenon can be represented simply and accurately by a linear interpolation of some parameters to estimate the considered load some stages of the flight or during the complete flight. A best-fit interpolation allows the determination of the coefficients for each selected phenomena and load.

A typical equation resulting from method B estimation can be written as follow:

$$Load_x = a.Param_1 + b.Param_2 + c.Param_3 + \cdots + x.Param_x$$

Stresses calculations. The UMF damage estimation method following step consists in determining the stresses at each pilot point. This is done following "param-to-stress" interpolation for some pilot points and "Load-to-stress" interpolation when stresses are derived from loads on specific sections.

A linear interpolation is selected for "load-to-stress" function where estimated stress in each pilot point is given by a combination of a selected subset of the loads. A different interpolation is identified for each pilot point.

For "param-to-stress" interpolation, the same approach as for the "param-to-load" interpolation is used, meaning a linear interpolation with a limited number of global input parameters ($M_{A/C}$, M_{fuel}, CG,...). Best-fit interpolation is identified using available load cases.

Pressurization. Cabin pressurization Δp is measured on-board. For each pilot point sensitive to pressurization, a simple linear factor is identified providing local stress from measured pressurization. This stress is added to each peak and valley in the local stress sequences of the affected pilot point.

Thermal loading. The A400M structure is hybrid, mixing composite fibre elements with metallic ones. On the hybrid zones, the thermal effect has to be taken into account since it can represent a significant part of the stress. Different thermal missions with various on ground and in flight temperatures have been used to cover the possible use of the aircraft. The relation between the air temperature and the stress level established by theoretical analyses for design and certification is used to estimate the resulting thermal stress on the relevant pilot points. This thermal stress is also added to each peak and valley in the local stress sequences of the affected pilot point.

Damage calculation:

A typical method for damage calculation is applied to compute the estimated damage of the Flight Cycle in each pilot point from the estimated local stress sequences provided by previous steps. This F&DT method is identical to the one used for certification process and includes Rain-Flow counting, Miner's law, S/N data curves to obtain fatigue damage and crack growth laws and curves to estimate crack growth damage for each flight.

Estimated fatigue damage and crack growth sensitivity are then used to compute Fatigue Index (FI) and Crack Index (CI) in each pilot point.

Provided outputs:

The UMF process is implemented on a ground based tool. The data recorded in flight are downloaded on this tool and then processed per the damage estimation algorithm. The possible outputs are presented on next paragraphs.

Estimated loads & local stress sequences. UMF provides as an output the following intermediate results:

- Estimated sequences of each reference loads for the considered Flight Cycle
- Estimated sequence of local stresses in each pilot point for the considered Flight Cycle

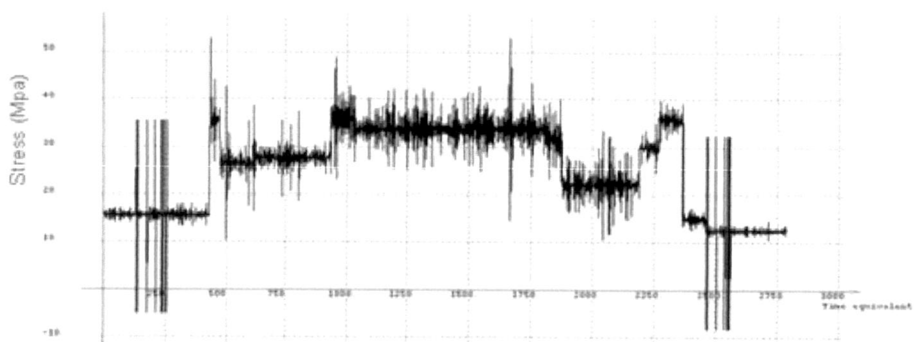

Figure 6. Estimated local stress sequence of "FCx" at pilot point "PPy".

Fatigue & Crack growth Indexes (FI & CI). UMF provides indexes as an image of fatigue and crack growth damage in each selected pilot point. These indexes are defined as the ratio between the damage of the Flight Cycle as estimated by UMF and reference damage of 1 Flight Cycle corresponding to reference usage (i.e. reference flight profile mix used for design).

For each Flight Cycle of each aircraft, UMF provides FI & CI in each pilot point:

$$FI_{\substack{FC\#i \\ PP\#j}}(\%) = 100.\frac{\text{Estimated fatigue damage of the FC\# i at PP\# j}}{\text{Fatigue damage of 1 reference usage FC at PP\# j}}$$

$$CI_{\substack{FC\#i \\ PP\#j}}(\%) = 100.\frac{\text{Estimated Crack Growth damage of the FC\# i at PP\# j}}{\text{Crack Crowth damage of 1 reference usage FC at PP\# j}}$$

Then, UMF is able to cumulate these indexes, Flight Cycle after Flight Cycle to provide a status of cumulated damage since entry in service (EIS). Similarly, Fatigue and Crack growth Indexes per aircrafts are defined as the ratio between cumulated damage estimated by UMF and cumulated damage corresponding to reference usage (i.e. reference flight profile mix used for design).

$$FI_{\substack{aircraft \\ PP\#j}}(\%) = \frac{\sum_{FC \text{ since } EIS} FI_{\substack{FC\#i \\ PP\#j}}}{\text{Nb of FC}} = 100.\frac{\text{Cumulated fatigue damage since EIS at PP\# j}}{\text{Cumulated fatigue damage since EIS with ref usage at PP\# j}}$$

$$CI_{\substack{aircraft\\PP\#j}}(\%) = \frac{\sum_{FC\,since\,EIS} CI_{\substack{FC\#i\\PP\#j}}}{\text{Nb of FC}} = 100.\frac{\text{Cumulated Crack Growth damage since EIS at PP\# j}}{\text{Cumulated Crack Growth damage since EIS with ref usage at PP\# j}}$$

The cumulative values can be represented as on Figure 7:

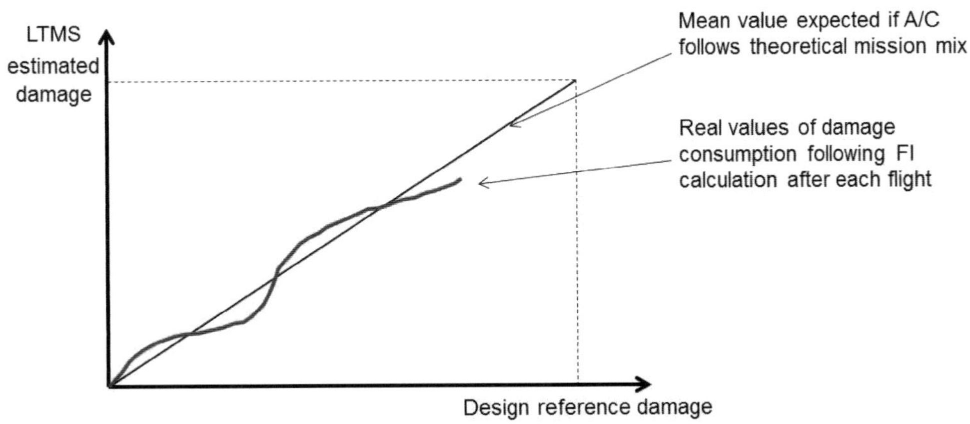

Figure 7. Cumulative damage consumed compared to expected damage.

Prediction tool. UMF also includes a prediction tool aimed to help the customer to decide for future A/C role allocation (tactical, logistic and training), in order to optimize the damage consumption on the complete structure.

Figure 6 gives an example of a possible use of this prediction tool. It represents the estimated damage on 2 pilot points for an aircraft having already performed 1500 flights. The damage estimated diverges from the reference assumed for design and certification. On the fuselage, the cumulated damage is below the reference value whereas on the wing, it is above the reference. This usually comes from the fact that the aircraft may have performed a significant number of tactical flights. Continuing operation of this particular aircraft in a logistic type of role would optimise his damage consumption all over the various airframe components.

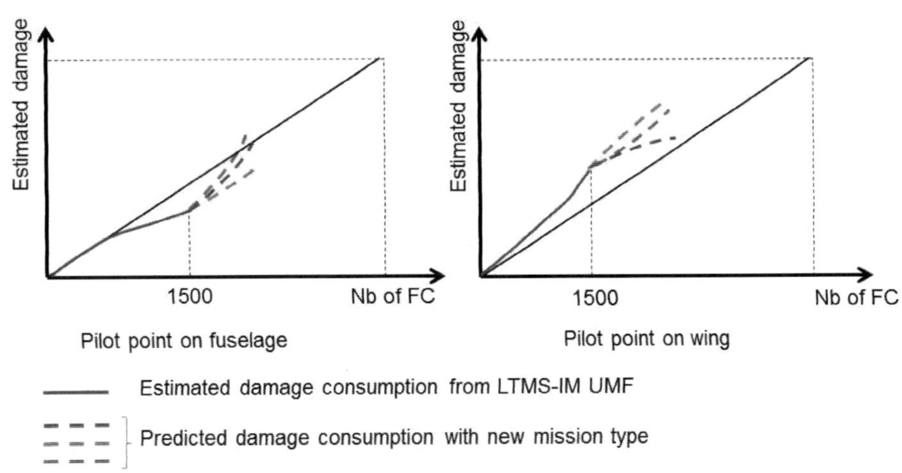

Figure 8. Use of prediction tool for selection of future flight profiles.

VALIDATION OF DAMAGE ESTIMATION ALGORITHM

Principle:
The principle used for the UMF has however to be validated by ensuring that the loads and stress calculated are in the range of the real loads and stresses. It is thus necessary to verify the validity of the different transfer functions.

This will be done by using the Flight Test (FT) on one hand and Full-Scale Static Test (FSST) on the other hand.

The Flight tests will allow the validation of the param-to-loads interpolation on the aircraft reference sections, whereas the FSST will validate the comparison of the loads to stress interpolations.

Flight Test (FT) comparison of measured and estimated load spectra:

Principle. The validation through FT consists in a comparison of the calculated and measured load sequences during a few complete flights, chosen for their similarity with the reference flight profiles.

The comparison is done as per Figure 9.The measured loads are obtained using the gauges located on the studied aircraft section. The calculated ones are obtained through the UMF param-to-load transfer function.

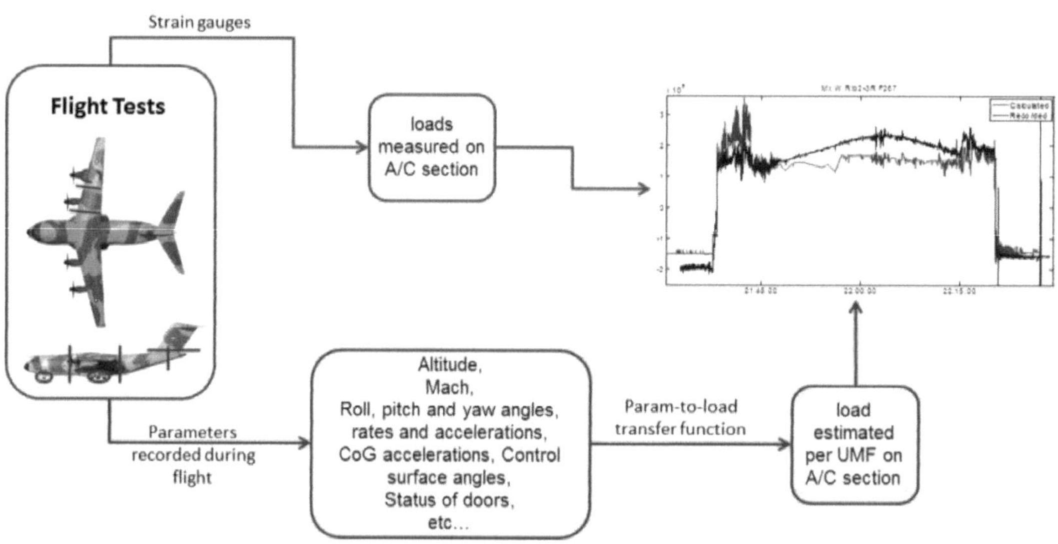

Figure 9. Principle of comparison of load spectra with flight test data.

Results and conclusion. Good correlation between measured and estimated load spectra is shown. The general shape and the main perturbations peaks are correctly captured. Some differences are evidenced but they can be tracked down to the effect of thermal loading, taken into account at a later stage in the process, or on the inaccuracy of the load measurements and of the UMF process for some secondary phenomena. These differences are not considered significant.

Full-Scale Static Test (FSST) validation of loads-to-stress transfer functions:

Principle. The validation consists in a comparison between "measured stress" and "calculated stress" for the pilot points on which strain gauges were installed for a restricted number of load cases.

The comparison is done as per Figure 10.

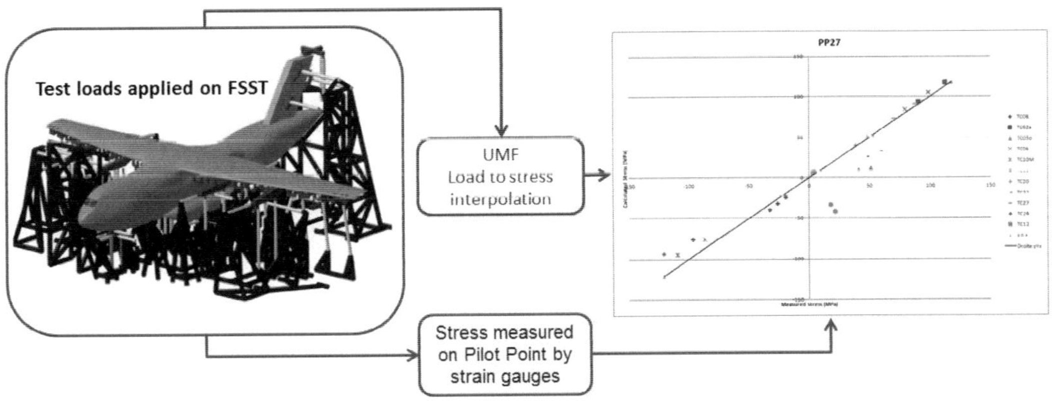

Figure 10. Stresses comparison between the ones measured on FSST and the ones obtained per UMF loads to stress interpolation.

Results and conclusion. Figure 11 gives an example of graph showing the comparison between measured and calculated stresses on a particular pilot point (PP7):

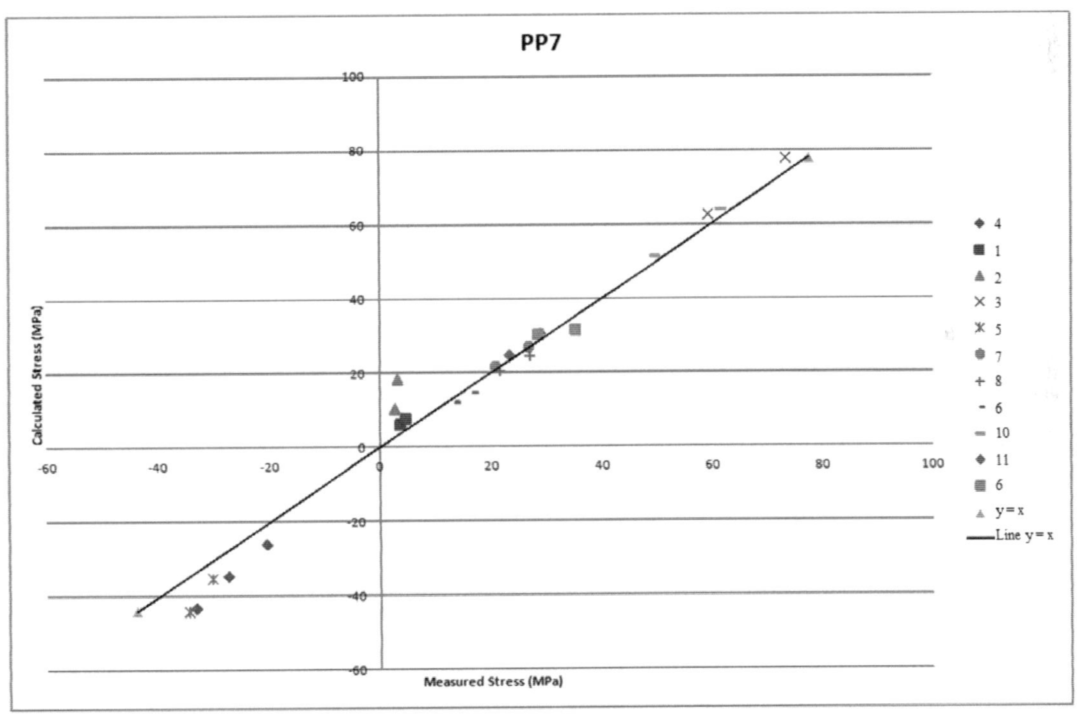

Figure 11. PP7 : Front fuselage skin longitudinal stress at frame 29 stringer 13.

This comparison is showing a good correlation between measured and estimated stresses, specifically for the most relevant load cases (wing bending, fuselage pressurization or bending, etc...).

Validation conclusion:

The UMF function's aim is to determine the criticity in terms of fatigue damage of one flight compared to another. The validation in two steps shown before is demonstrating a good level of correlation between measured and calculated loads and stresses. The level of accuracy obtained gives full confidence that:

- the relative damage cumulated between flights of different types will be well captured,
- the level of damage cumulated in the various structural parts will be properly predicted (areas under-damaged versus over-damaged).

The UMF process is thus validated.

CONCLUSION

The UMF has been developed following a requirement of the A400M customers. Its aim is to follow the real use of the aircraft in order to identify any deviation in F&DT damage due to its wide possible utilization. Different means were envisioned such as the installation of strain gauges on the aircraft or calculation methods of various complexities using data recorded by systems or crew. The strain sensors will be used only on a limited number of aircraft due to the complexity of the maintenance of such a system. The simple calculations methods using predefined damage values were discarded due their inaccuracy and their inability to cover any new flight profile that could be flown. The chosen method was the most complex calculation method, based on a simplified load model allowing the determination of loads or stresses on particular sections of the aircraft from data recorded by Systems during flight, the rebuilding of stress spectra on selected pilot points and the estimation of the F&DT damage created by the real flights compared to the expected reference damage used for certification. This method has been selected because of its accuracy and of its ability to determine damages created by any new flight profile not envisioned at the time of the certification.

The validation of the damage estimation algorithm at the heart of the UMF has been done through a comparison of loads or stresses measured during the Full-Scale Static test and Flight Tests compared to UMF estimated ones on the same locations. A good correlation has been evidenced, even if some differences have been seen between recorded and calculated spectra. These differences have been considered acceptable for the aim of the UMF process. Indeed, the expected result is not to determine the real absolute value of the damage created by the flight on pilot points, but to identify deviations of fatigue damage compared to the one expected for reference flight profiles and to predict the level of damage cumulated on the various structural parts (areas under-damaged versus over-damaged). The UMF process is thus validated.

The UMF will be a key element of A400M fleet management, without adding any additional weight or maintenance burden. The prediction tool included in the system can be used to anticipate on the future use of the aircrafts, and so to improve the dispatch of the aircraft in an appropriate squadron (dedicated to logistic, tactical or training) to limit the apparition of excess damage on specific locations.

The next step is to gather information on flights performed in real military environment since all data used for this study have been based on flight test aircraft. When the number of flights performed in service will become significant, it will be possible to refine the algorithm and to adjust the maintenance programme accordingly to better fit the actual aircraft utilisation.

The Use of Micro Electro-Mechanical Systems to characterise the Structural Usage of a Multi-Role Military Aircraft

Robin Kamiya, Michael J Duffield

QinetiQ, Farnborough, UK

Abstract: Airworthiness Policy in the UK Military Aviation Authority Regulatory Policy (MRP) mandates an Operational Loads Measurement (OLM) programme periodically throughout an aircraft's life. A full aircraft OLM programme can be highly complex as an airframe needs instrumenting with strain gauges and then the measurement system calibrated. Furthermore, from the beginning of such a programme, a clear understanding of how the multitude of recorded parametric and strain gauge data should be used is essential in order to realise the value of the expense and time put into such a programme. QinetiQ has worked with the Defence Science and Technology Laboratory (Dstl) to conduct a fleet wide Nz (vertical 'g) data capture exercise on the UK MoD fleet of multi-role Britten Norman Islander aircraft using a Modular Signal Recorder (MSR) incorporating micro electro-mechanical systems (MEMS) technology. The Islander fleet has been in service since the late 1980s and has a usage history defined in terms of Nz spectra obtained from traditional, electro-mechanical, counting accelerometers - otherwise known as fatigue meters. These meters were removed from the fleet approximately 10 years ago and since then, the aircraft operator has had to rely upon non-quantitative means of substantiating structural usage. Hence, the installation of the MSR device to capture Nz data for the Islander fleet provides tangible benefits as the resulting spectra can be compared to the past usage as corroborative evidence of how typical operational flying has evolved. Furthermore, due to the much greater resolution of the MSR data compared with the fatigue meter records, there is considerable scope for refining the associated fatigue analysis to assess the actual structural usage of individual aircraft from the perspective of historical, present and future flying.

INTRODUCTION

The UK Ministry of Defence (MoD) operates a small fleet of multi-role Britten Norman BN-2T Islander aircraft, Figure 1. The Islander does not have a permanently installed system for the continuous monitoring of structural usage, and has been operated to date without being subject to periodic programmes of Operational Loads Measurement (OLM); such programmes are a requirement of the UK Military Aviation Authority (MAA) Regulatory Article (RA) 5720. This deviation from MAA policy has to be supported by a formal deferment issued by the Operational Duty Holder (ODH) at 2* level as part of the air system safety management requirements for UK military aircraft.

Figure 1. UK MoD Islander Aircraft.

In order to provide evidence that could be used for the justification of the deferment of OLM, QinetiQ was tasked by the UK MoD to compare recorded Normal Acceleration (Nz) data from recent Islander flying with historical Nz data from the early 1990s onwards to identify any significant variations. To achieve this, the UK MoD funded a data collection exercise to record Nz data across the active Islander fleet by installation of a Modular Signal Recorder (MSR) [1] that uses Micro Electronic Mechanical Systems (MEMS) technology, Figure 2.

Micro-electromechanical systems (MEMS) is a technology that combines computing with tiny mechanical devices such as sensors, valves and gears, all embedded in semiconductor chips. MEMS devices generally range in size from 20 micrometres (20 millionths of a metre) to a millimetre (i.e. 0.02 to 1.0 mm), usually consisting of a microprocessor and components that interact with the surroundings i.e. a sensor. They will also have very low power consumption measured in μA when in active mode or tenths of μA when in standby mode[1].

Figure 2. MSR logger[2].

The installation of a MSR device on the fleet of Islanders is temporary, but also necessary, as the mechanical counting accelerometers previously used to gather the historical Nz data were removed from all the Islanders in 2005. The overall aim for QinetiQ was to analyse the recent Nz data and compare the results to an identical analysis conducted on the historical Nz data. The comparison would establish if current flying has changed since the counting accelerometer units were removed.

Additional detail is provided by calculating a 'nominal' fatigue damage from each Nz spectrum using design assumptions obtained from Britten Norman documentation. However it should be noted, that these assumptions were based on documentation for the Britten Norman Trislander aircraft and corresponding fatigue life calculations associated with the Trislander wing. The Trislander is a larger, three engine variant of the Islander with a wing that is different in detail to that of the Islander but for which a comprehensive fatigue test was completed during the 1970s for the purposes of civil certification. The results from that testing enable a consistent interpretation of the MSR-derived Nz spectra versus the historical data in terms of nominal fatigue damage rates. However, it is important to note that the nominal fatigue damage rates calculated in this way are for comparative purposes only in the context of this paper.

METHODOLOGY – DATA CAPTURE

The MSR device has an in-built tri-axial accelerometer and also the capability to record atmospheric pressure. The device has an internal battery for power, and a means of auto-starting the recording when specific conditions are met and reverting to stand-by mode when an alternate set of conditions are met.

By virtue of the MEMS technology, the MSR unit is very compact, measuring only 72mm x 39mm x 23mm and has a mass of 70 grams meaning its overall dimensions and mass are of secondary importance when choosing a location for installation. Furthermore, the battery is a rechargeable 800mAh lithium cell which conveys the advantage that a chosen mounting location is not dependent on having access to an external power source.

The MSR is supplied with the necessary software to programme the unit, and for Islander it was programmed to capture Nz and atmospheric pressure data at 50 samples per second. This chosen sample rate was considered consistent with capturing peak and trough Nz values for anticipated maximum gust frequencies of 7Hz [2]. Each MSR was statically calibrated at +1g, 0g and -1g before despatch to the Islander operators and rechecked upon return from the operators to ensure its operating performance remained within tolerance. All programming and testing of the MSR units used for the Islander project was performed by the UK Defence Science and Technology Laboratory (Dstl).

[1] http://en.wikipedia.org/wiki/Microelectromechanical_systems
[2] http://www.msr.ch/en/product/datalogger.php

Record mode was programmed to be triggered to start when Nz > +1.15g or Nz < +0.85g, while standby mode was triggered a pre-defined number of minutes after Nz had remained within a zone of Nz < +1.15g and Nz > +0.85g. Engine start was sufficient to provide Nz levels that triggered the start of recording. This combination of programming parameters ensured that the unit would be "awoken" prior to any structurally significant events and remain in record mode until after engine shut-down.

Each MSR was attached to a mounting plate, which itself was attached to the aircraft at a location as close as possible to the aircraft centre of gravity and which was known to be free of excessive vibration and would not be inadvertently knocked by the crew. A fully charged battery was capable of 75 hours data capture or up to two months in stand-by mode. When either of these criteria were approached the unit was removed and returned to Dstl for data download. A replacement MSR would be refitted immediately and set to standby mode prior to the next flight.

Acquisition of sufficient units to facilitate this strategy that allowed every aircraft to be subject to continuous data capture and the associated asset management was undertaken by the Dstl.

The Nz and pressure data were recorded to a 4GB Secure Digital (SD) memory card which Dstl accessed and downloaded upon each units return. The data from each flight was processed individually and the Nz composition of each flight was established. The processing conducted by Dstl is described in more detail in the coming paragraphs.

For the Islander, with an unpressurised cabin, the atmospheric pressure parameter can be used to provide an indicator of the sortie altitude profile. When plotted as a time history, the resulting profile provides an easily assimilated graphical illustration of the flight and a typical example is shown in Figure 3.

The upper chart displaying the sortie Nz time history shows green data points represent the aircraft on the ground prior to take-off, post landing and (if present) during roller landings, while the blue-coloured data is airborne content.

The lower chart displays the altitude profile derived from the atmospheric pressure recordings. The altitude has been estimated by relating the recorded pressure to the International Standard Atmosphere (ISA) and assuming the nominal pressure at ground level to be equal to 1013.25mb and an ambient temperature of 15°C. Obviously, it is unlikely that either ISA conditions prevailed on the day of any particular flight and the local nominal pressure seldom applies in the UK; it is for these reasons that the altitude plot can be used only as an indicator of the actual altitude variation through the sortie and why the ground phases of the sortie depicted in Figure 3 are shown at an altitude not equal to 0ft. This small ambiguity is of no consequence for the purposes of the current exercise and the altitude charts are of sufficient accuracy and resolution to provide a meaningful indication of the altitude variations at various stages of each sortie. This information is very useful for the wider assessment of sortie profiles but is outside the scope of the main subject of the work reported here and so is not considered further within this paper.

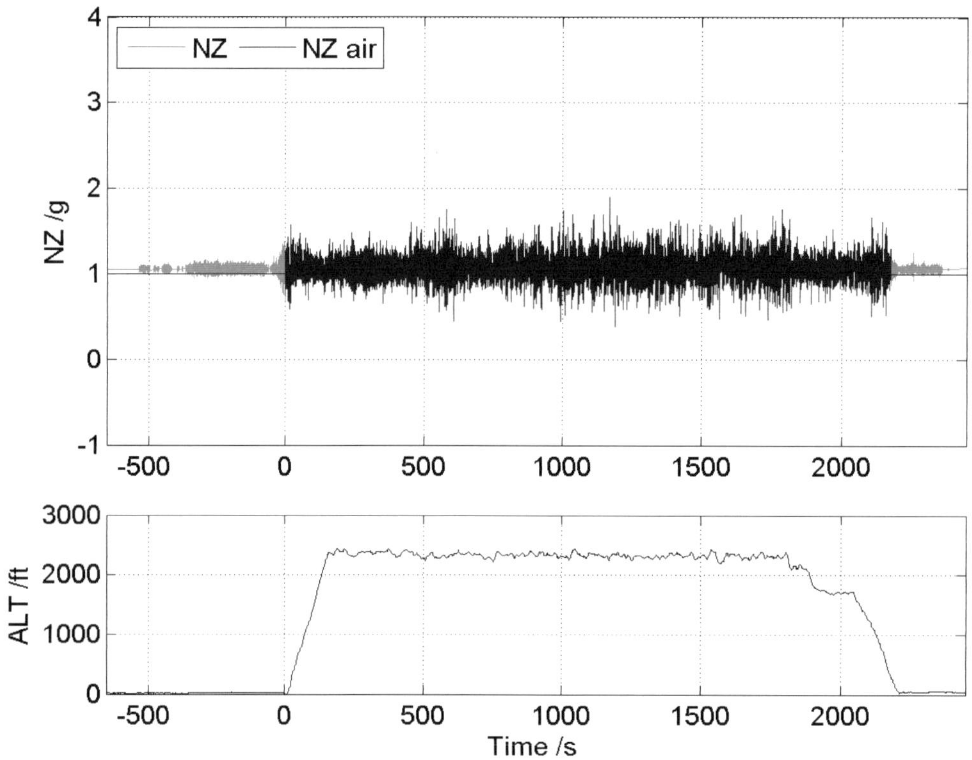

Figure 3. Typical Islander Nz and Altitude time history[3].

METHODOLOGY – DATA PROCESSING

The Nz sortie time history is processed using a Peak Trough counting algorithm, specifically, for this programme, Downing and Socie (ASTM E 1049-85). The algorithm needs 'g' crossing levels in which to count exceedences, and thus the levels chosen to count normal acceleration are based on those of the Meggit 11043/39 mechanical fatigue meter which was installed fleetwide on the UK MoD Islanders prior to 2005.

To analyse the Nz time history, the Peak Trough algorithm was used to count normal accelerations passing increasing levels of +1.3g, +1.5, +2.0g, +2.5 and +3.0g (hence, the peaks) and also normal acceleration passing decreasing levels of +0.7g, +0.5, +0.2g and +0.0g (the troughs). It will be noted that there are +0.5g gaps from +1.5g upwards representing the practical limitations of the mechanical fatigue meter; hence, when analysing the Nz time histories, it was decided to introduce intermediate counting levels of +1.8g, +2.2g and +2.7g. There were no additional intermediate levels considered necessary below +1.0g. The counting algorithm applied to the Nz data also required a set of release levels for each peak or tough exceedence. Again, these were based on the specification of the Meggit 11043/39 fatigue meter (typically half the range back to +1.0g), and the same assumption was applied to derive release levels for the new levels of +1.8g, +2.2g and +2.7g.

The analysis of the Nz time history was conducted for each flight using the Downing and Socie algorithm leading to cumulative exceedence results at each 'g' level in the form of the pseudo fatigue meter counts presented in Table for the example flight shown in Figure 3.

[3] Time history provided by Dstl

Table I. Exceedence results from MSR Flight data from Figure 3.

Pseudo Fatigue Meter Counts	Values
Window A (0.0g/0.5g)	0
Window B (0.2g/0.7g)	0
Window C (0.5g/0.8g)	5
Window D (0.7g/1.0g)	116
Window E (1.3g/1.2g)	549
Window F (1.5g/1.2g)	78
Window FF (1.8g/1.4g)	1
Window G (2.0g/1.5g)	0
Window GG (2.2g/1.6g)	0
Window H (2.5g/1.8g)	0
Window HH (2.7g/1.85g)	0
Window J (3.0g/2.0g)	0

Hence, for each and every flight recorded using the MSR, a pictorial representation of the Nz time history and altitude time history was available, as was the analysis of the Nz history in terms of Nz exceedences. The MSR data capture programme covered 12 months flying and 898 flights were recorded from 6 different Islander aircraft operated by the UK MoD. The distribution of this flying in terms of flights and flying hours is presented in Table II:

Table II. MSR Fleet Statistics by aircraft.

Aircraft	No of Flights	Flying Hours
F3	146	251:30
G5	160	232:15
G6	9	11:40
G8	103	124:40
H6	399	692:30
H7	75	145:40
	898	1448:15

The principal of the flight analysis conducted by QinetiQ was to group data from each aircraft into fleet and individual aircraft exceedence spectra and through manipulation derive a 'nominal' fatigue accrual per flight, both overall and by individual aircraft, where a flight represents a take-off, flight, followed by a braked landing where the aircraft is brought to a controlled stop. Hence from a flight mechanics perspective, there is a single Ground to Air to Ground (GAG) cycle, cycle(s) from encounters with gust and (perhaps) cycles from manoeuvre content. This is best described by using the results presented in Table I as an example; that flight was recorded by aircraft F3 in Table II.

The GAG cycle is a once per flight event, hence from the total g counts, remove 1 count from all levels greater than 1.0g. What is left is attributable to the combined Residual Gust and Manoeuvre (RGM) component. This can be seen in Table III:

Table III. Total, GAG and RGM Nz component from Figure 3.

	Total	GAG	RGM
Window B (0.2g/0.7g)	0	0	0
Window C (0.5g/0.8g)	5	0	5
Window D (0.7g/1.0g)	116	0	116
Window E (1.3g/1.0g)	549	1	548
Window F (1.5g/1.2g)	78	1	77
Window FF (1.8g/1.4g)	1	1	0
Window G (2.0g/1.5g)	0	0	0

This technique of removing the GAG related Nz data from each flight is applied to all 898 Nz flight records resulting in GAG and RGM exceedence totals for the 6 aircraft. GAG and RGM exceedence subtotals can also be derived for each aircraft.

The next manipulation is to normalise the data. For the purposes of this programme and to relate back to the available design information, it was chosen to normalise the data on an "exceedences per flight" basis, and thereby match the design data associated with results from the wing fatigue testing previously undertaken by the Islander aircraft manufacturer, Britten Norman. Hence the normalisation parameter was the total flights flown by each aircraft or the total number of flights, i.e. 898, as shown in Table III.

The normalised total Nz, GAG and RGM exceedence data for the combined fleet of 6 aircraft and 898 flights is presented in Table IV, Table V and Table VI, while Figure 4 is a graphical representation of the counts per flight data from each of the tables.

Table IV. Islander 2012/13, total Nz exceedences data for 'All Flying'.

Aircraft	Flts	0.0g	+0.2g	+0.5g	+0.7g	+1.3g	+1.5g	+1.8g	+2.0g	+2.2g	+2.5g	+2.7g	+3.0g
F3	146	0	4	215	4549	8513	611	20	1	0	0	0	0
H6	399	4	29	565	8819	17528	1649	135	53	28	9	3	1
H7	75	0	1	53	1063	2504	162	4	0	0	0	0	0
G5	166	2	13	520	9967	18720	2094	222	80	27	3	0	0
G6	9	0	1	33	608	2180	284	24	6	0	0	0	0
G8	103	3	9	260	4955	7107	776	99	33	11	0	0	0
Totals	898	9	57	1646	29961	56552	5576	504	173	66	12	3	1
Exceedences per flight		0.010	0.063	1.833	33.364	62.976	6.209	0.561	0.193	0.073	0.013	0.003	0.001

Table V. Islander 2012/13, total Nz exceedences data attributable to 'GAG'.

Aircraft	+1.3g	+1.5g	+1.8g	+2.0g	+2.2g	+2.5g	+2.7g	+3.0g
F3	145	88	10	1	0	0	0	0
H6	398	262	55	24	13	6	2	1
H7	74	34	3	0	0	0	0	0
G5	166	150	76	41	19	3	0	0
G6	9	9	6	4	0	0	0	0
G8	103	75	31	15	8	0	0	0
Totals	895	618	181	85	40	9	2	1
Exceedences per flight	0.997	0.688	0.202	0.095	0.045	0.010	0.002	0.001

Table VI. Islander 2012/13, total Nz exceedences data attributable to 'RGM'.

Aircraft	+1.3g	+1.5g	+1.8g	+2.0g	+2.2g	+2.5g	+2.7g	+3.0g
F3	8368	523	10	0	0	0	0	0
H6	17130	1387	80	29	15	3	1	0
H7	2430	128	1	0	0	0	0	0
G5	18554	1944	146	39	8	0	0	0
G6	2171	275	18	2	0	0	0	0
G8	7004	701	68	18	3	0	0	0
Totals	55657	4958	323	88	26	3	1	0
Exceedences per flight	61.979	5.52	0.360	0.098	0.029	0.003	0.001	0.00

The RGM counts at 0.0g to +0.7g are the same as presented in Table IV, hence are not repeated in Table VI.

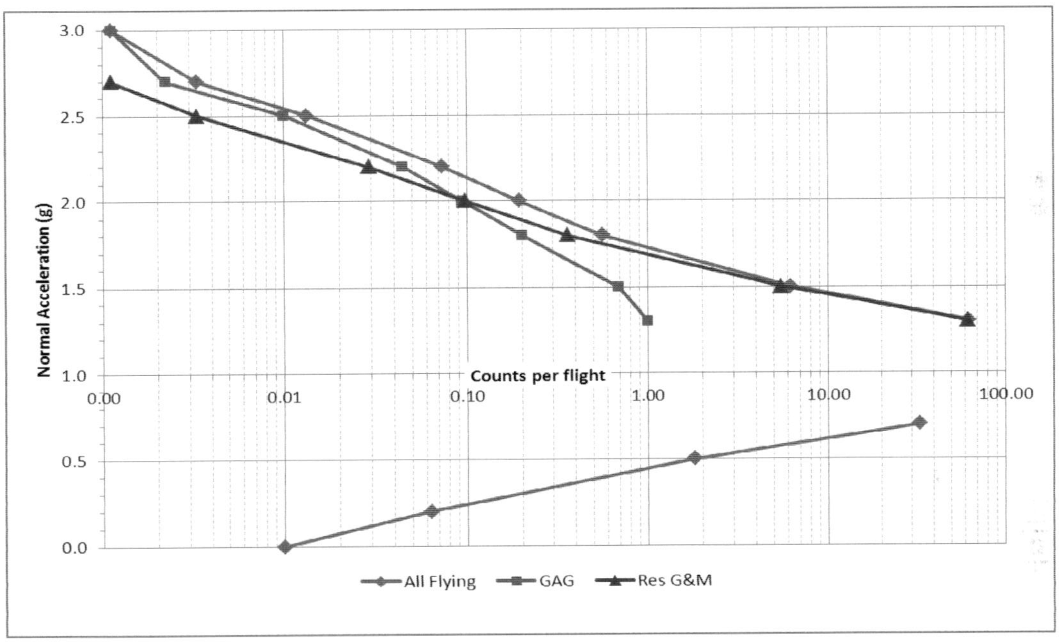

Figure 4. Islander exceedences (counts) per flight Nz spectra.

The Nz data are now in a form where they can be converted to stress values using information obtained from the Britten Norman design data. To calculate the nominal damage, the GAG and RGM spectra need to be converted from an Nz exceedences per flight spectrum (Figure 4) to a stress exceedence spectrum, such that the two components of flying, GAG and RGM, can be assessed using an appropriate stress-endurance (s-N) curve. This approach, which incorporates the Palmgren-Minor Cumulative Damage Rule, is considered in keeping with the design concepts, qualification testing and certification requirements of the Islander and the era when this simple, rugged aircraft was originally developed.

The Nz spectra presented in Figure 4, were converted to the equivalent stress spectra based on the data provided by Britten Norman, and this accounted for +1g level flight stress, stress per 'g' (gust or manoeuvre) and a stress for a nominal +0.5g ground bump.

GROUND STRESS ASSUMPTION

Britten Norman provided the contemporary technical documentation relating to the Islander/Defender/Trislander family of aircraft which was reviewed to identify the (numerical) design assumptions that are important to the calculation of fatigue lives based on aircraft service usage. These numerical assumptions were tested to ensure that fatigue life results declared by Britten Norman at the time the reports were written could be replicated by QinetiQ. One such value used in the BN reports is the magnitude of the assumed load factor associated with a 'ground bump', which is used in the calculation of the GAG component.

It had been assumed that a ground bump represented a load factor of +0.5g, and this value was used in the wing fatigue test loading spectrum for the Trislander aircraft, the Trislander being the only variant to have been subject of a dedicated fatigue test for the purposes of the original civil certification. Hence, with no better data available, this +0.5g ground bump assumption was employed to determine a ground stress for use in the fatigue life calculations associated with GAG component of flying.

However, with the MSR installed on the aircraft and recording continuously from engine start to engine shut-down, this ground bump assumption for Islander can be tested by review of the normal acceleration minima recorded during the ground taxi elements of each flight.

Data from the MSR

The ground minimum data from all 898 MSR flights have been used to assess if the assumed value of +0.5g is truly representative or should be revised. The outcome of the analysis can be summarised as follows:

868 flights or 95.8% of all flights show a ground minimum of +0.5g or above, and 30 flights have a ground minimum below +0.5g.

Of these 30, the lowest recorded minimum is +0.208g and their average is +0.419g.

20 of the 30 'flights' have a minimum of +0.5g > Min ≥ +0.4g, 8 'flights' have minima of +0.4g > Min ≥ +0.3g and the remaining 2 'flights' have minima of +0.239g and +0.208g.

The average of the minima from the 898 flights is +0.696g, and of the 868 flights above +0.5g, the average is +0.706g.

These measured data show that 95.8% of the 'flights' have minima that are at or above the current accepted assumption of +0.5g, and only 4.2% are below. The results from this review indicate that the current assumption of +0.5g is suitably conservative for 95.8% of the data sample. There is evidence to suggest that the ground bump assumption of +0.5g could be increased to a slightly higher (less severe) value. However, such a change would ignore the very small percentage of data that are non-conservative and outside the current ground bump assumption. As these data are in a significant minority, the current balance appears acceptable. The ground stress associated with +0.5g leads to a calculation of fatigue damage applicable to the GAG component which is likely to be conservative to an appropriate degree.

STRESS CONVERSION

Application of the assumed flight stress and stress per 'g' values to the Nz exceedences per flight data for the All Flying (Table IV), the GAG (Table V) and RGM (Table VI) components leads to the stress spectra presented in Table VII and plotted in Figure 5.

Table VII. Stress equivalent data from All Flying, GAG and RGM spectra, per flight.

	Exceedences per Flight			'g' and Stress	
Total 'g'	Total flying	GAG	RGM	Delta 'g'	Total Stress (psi)
3.0	0.001	0.001	0	2.0	17349
2.7	0.003	0.002	0.001	1.7	15669
2.5	0.013	0.010	0.003	1.5	14549
2.2	0.073	0.045	0.029	1.2	12869
2.0	0.193	0.095	0.098	1.0	11749
1.8	0.561	0.202	0.360	0.8	10629
1.5	6.209	0.688	5.521	0.5	8949
1.3	62.976	0.997	61.979	0.3	7829
0.7	33.364	-	33.364	-0.3	4469
0.5	1.833	-	1.833	-0.5	3349
0.2	0.063	-	0.063	-0.8	1669
0.0	0.010	-	0.010	-1.0	549

(where Delta 'g' = Total 'g'-1)

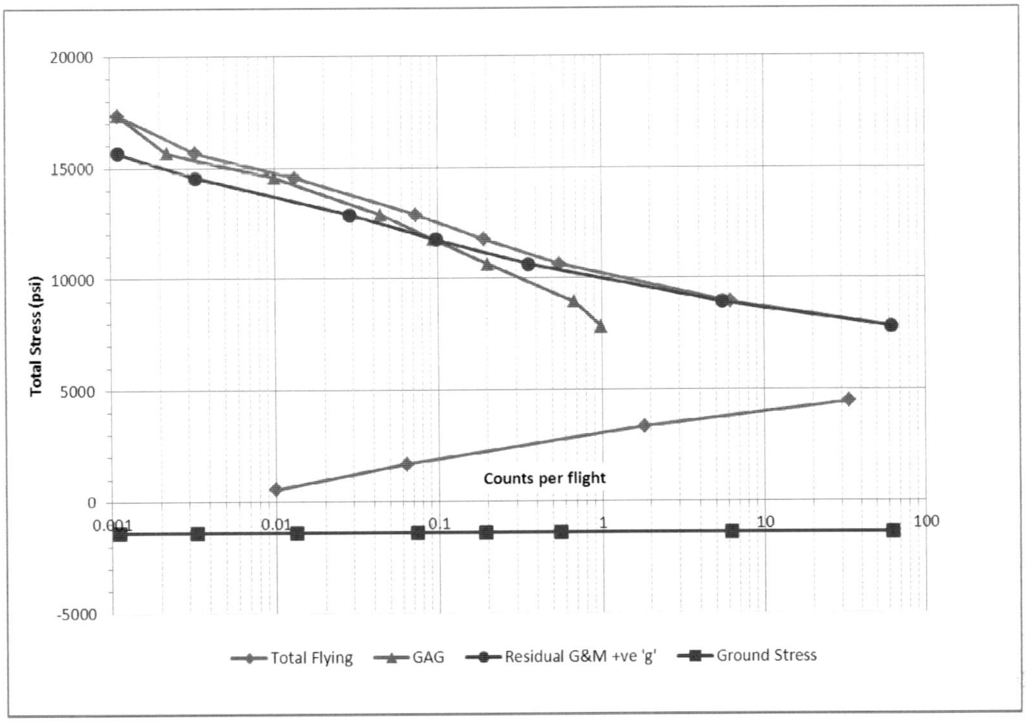

Figure 5. Stress Spectra, based on Table VII.

As the GAG and RGM spectra are identified, each can be converted to a block spectrum by taking the average stress between points; this process of averaging is conducted between all points, and thus produces a block stress spectrum for the GAG as shown in Figure 6. The block for the +3.0g data point - equivalent to a stress of 17349 lb/in^2 - continues back to the theoretical 0.0 exceedences point. Note that the minimum stress for each block is that for the nominal +0.5g taxi bump.

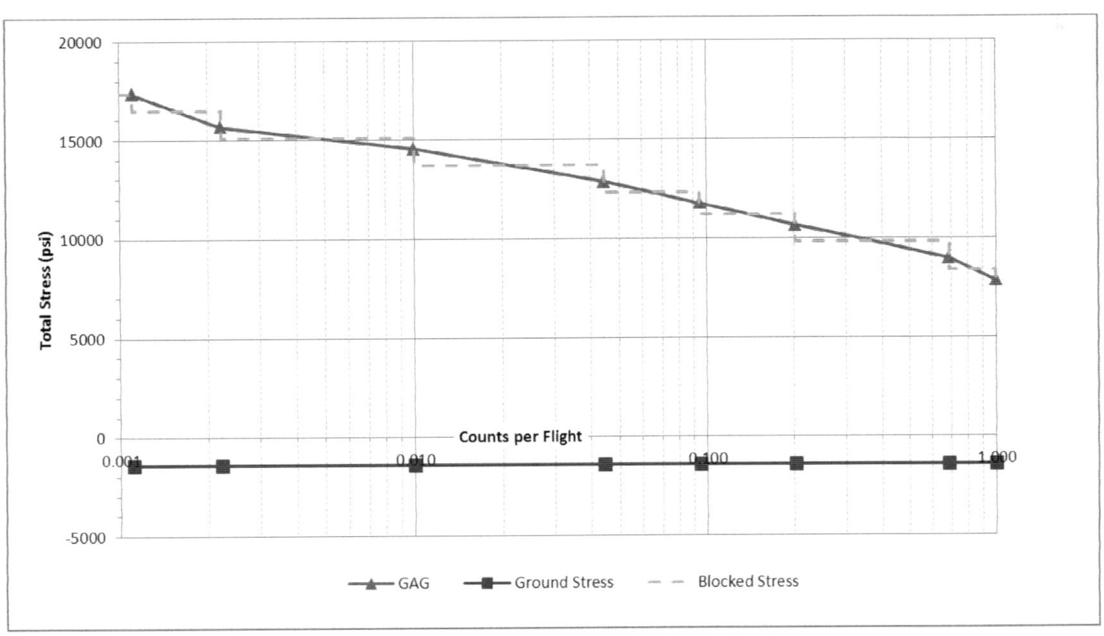

Figure 6. GAG block stress spectrum.

A similar process of 'blocking' is applied to the RGM spectrum; however, it is complicated by data points from the upper spectrum being mirrored onto the lower spectrum and data points on the lower spectrum to the upper spectrum. The stress values at these intermediate points on each spectrum are calculated by a simple log-linear interpolation - see Figure 7.

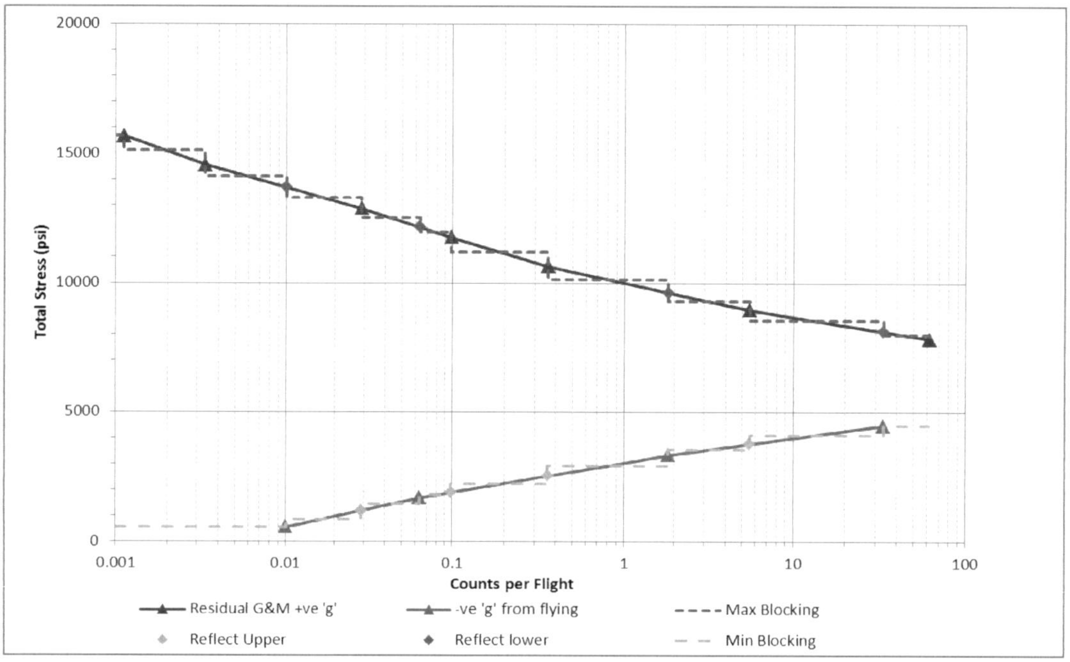

Figure 7. RGM block stress spectrum.

The block width (x-axis) representing the number of cycles per flight in that block and block height (y-axis) defining the stress range are now used to calculate the nominal fatigue damage from each block of the approximated GAG and RGM spectra. This is achieved in Table VIII below, which uses the stress values from Table VII. The stress values between points are averaged to define the upper and lower peak values for the typical stress cycle within each block; the difference between these values is halved to obtain the alternating stress, S_{alt}, from which a nominal endurance, N, is calculated using the formula below:

$$N = [(1500 / ((S_{alt}/1000) - 1.5)]^2 \quad (1)$$

Note: This is the formula used to describe the so-called Heywood's s-N curve and is the same as used by Britten Norman in the original fatigue calculations for the Islander/Trislander/Defender aircraft.

The number of cycles within each block, n, is achieved by subtraction of adjacent data points, and is then divided by N to calculate the fatigue damage for each block. In Figure there are 8 blocks and Figure there are 10 blocks. The fatigue damage calculations are summarised in Table VIII for the GAG and in Table IX for the RGM and are completed by applying the Palmgren-Miner cumulative damage rule to obtain a total damage estimate for each component.

Total fatigue damage (nominal) = 7.705616E-06 + 1.469721E-05 = 2.238482E-05 per flight

Table VIII. GAG block stress and damage calculations based on Figure 6.

Spectrum			Blocked spectrum							
Total 'g'	Stress	Counts	Total 'g'	Stress	Counts (n)	Stress range	Stress alt	N	n/N	% block damage
3.0	17349	0.001114	3.00	17349	0.001114	18744	9372	36311	3.06677E-08	0.4%
			2.85	16509	0.001113	17904	8952	40520	2.74826E-08	0.4%
2.7	15669	0.002227								
			2.60	15109	0.007795	16504	8252	49357	1.57934E-07	2.0%
2.5	14549	0.010022								
			2.35	13709	0.034521	15104	7552	61434	5.61919E-07	7.3%
2.2	12869	0.044543								
			2.10	12309	0.050111	13704	6852	78555	6.37913E-07	8.3%
2.0	11749	0.094655								
			1.90	11189	0.106904	12584	6292	97988	1.091E-06	14.2%
1.8	10629	0.201559								
			1.65	9789	0.486640	11184	5592	134379	3.62139E-06	47.0%
1.5	8949	0.688196								
			1.40	8389	0.308463	9784	4892	195562	1.57732E-06	20.5%
1.3	7829	0.996659								
Total Nominal damage (GAG)									7.705616E-06	

Table IX. RGM block stress and damage calculations based on Figure 7.

Spectrum			Blocked spectrum							
Total 'g'	Stress	Counts	Total 'g'	Stress	Counts (n)	Stress range	Stress alt	N	n/N	% block damage
2.7	15669	0.001114		15669	0.001114	15119	7560	61276	1.81733E-08	0.1%
			2.6	15109	0.002227	14559	7280	67357	3.30654E-08	0.2%
2.5	14549	0.003341								
			2.42	14121	0.006682	13572	6786	80525	8.29748E-08	0.6%
	13694	0.010022								
			2.27	13281	0.018931	12410	6205	101635	1.86263E-07	1.3%
2.2	12869	0.028953								
			2.14	12508	0.034521	11077	5539	137954	2.50236E-07	1.7%
	12148	0.063474								
			2.04	11948	0.034521	10170	5085	175043	1.97216E-07	1.3%
2.0	11749	0.097995								
			1.90	11189	0.261693	8978	4489	251858	1.03905E-06	7.1%
1.8	10629	0.359688								
			1.71	10128	1.473274	7186	3593	513737	2.86776E-06	19.5%
	9627	1.832962								
			1.56	9288	3.688196	5726	2863	1211088	3.04536E-06	20.7%
1.5	8949	5.521158								
			1.43	8532	27.842984	4410	2205	4524439	6.15391E-06	41.9%
	8116	33.364143								
			1.33	7972	28.614699	3503	1752	35537061	8.05207E-07	5.5%
1.3	7829	61.978842								
0.7	4469	33.364143								
			0.64	4122	27.842984					
	3775	5.521158								
			0.54	3562	3.688196					
0.5	3349	1.832962								
			0.43	2942	1.473274					
	2536	0.359688								
			0.30	2211	0.261693					
	1886	0.097996								
			0.22	1778	0.034521					
0.2	1669	0.063474								
			0.16	1431	0.034521					
	1193	0.028953								
			0.06	871	0.018931					
0.0	549	0.010022								
Total Nominal Damage (RGM)									1.467921E-05	

2012/13 FLYING DATA AND ANALYSIS

The analysis sequence presented as Table IV to Table IX, and Figure 4 to Figure 7 are for 898 flights from 6 different aircraft the UK MoD Islander fleet. The same process can and has been applied to individual aircraft data sets and groups of specific aircraft; the results of these analyses are shown in Table X.

Table X. Nominal fatigue damage analysis results from 2012/13 flying.

Ser.		Flts	Total hours	Mean duration (hrs)	GAG	RGM	RGM % of Total	Total damage per flight
1	Fleet	898	1448.24	1.61	7.706E-06	1.468E-05	65.6%	2.238E-05
2	G Ac Only	278	358.57	1.29	9.111E-06	2.511E-05	73.4%	3.422E-05
3	F3	146	251.50	1.72	6.679E-06	1.176E-05	63.8%	1.844E-05
4	G5	166	222.24	1.34	9.554E-06	2.785E-05	74.5%	3.740E-05
5	G6	9	11.67	1.30	9.830E-06	5.513E-05	84.9%	6.496E-05
6	G8	103	124.67	1.21	8.114E-06	1.937E-05	70.5%	2.749E-05
7	H6	399	692.50	1.74	7.367E-06	1.023E-05	58.1%	1.760E-05
8	H7	75	145.67	1.94	6.165E-06	5.855E-06	48.7%	1.202E-05
9	Fleet Old FMF Levels	898	1448.24	1.61	8.107E-06	1.592E-05	66.3%	2.403E-05

This table shows results from all flying grouped together (Ser 1), the 'G' aircraft only (Ser 2) and all individual aircraft analyses (Ser 3 to 8). Ser 9 is a further row of analysis results that are discussed later in the report covering results for the historic Nz data. The columns in the Table provide detail on the:

- total number flights
- total flying hours
- mean duration of the flights
- 'nominal' GAG damage of a flight
- 'nominal' RGM damage of flight
- 'nominal' Total damage of a flight

It should be noted that the MSR data span a short period of time of only 12 months versus the historical database with which the results are to be compared that spans a 12 year period. It is apparent from Table X that there is significant variation in the calculated damage rates and that, in particular, the 'G' aircraft appear to have a different typical sortie duration from the other aircraft, and that the balance between GAG and RGM damage rates is similarly different between the two groups of aircraft. This is likely to be due to the typical tasking of the various aircraft being in different specific roles but that aspect of the results presented in Table X is outside the scope of the assessment presented in this paper.

Summarising, the analysis of the MSR flight data has focused on calculating nominal fatigue damage rates from the GAG and RGM components of the Nz spectra based on assumed stress data. The GAG and RGM results can be compared with the nominal damage obtained from an identical analysis of the historical Islander flying data from 1992 to 2004 when all the aircraft flew with the Meggit 11039/43 mechanical counting fatigue meters and the Nz exceedence data were captured by manual recording of the window counts on the MoD Form 725 (F725) Flying Log after each flight. That comparison is presented in the following section of this paper.

HISTORICAL DATA

A review of historic Islander flying from F725 records identified a total of 3317 flights and the associated Nz data to allow a calculation of the historic fatigue damage per flight from the GAG and RGM components. The analysis process applied to these historical data is exactly the same as has been applied to the flying data collected from the MSR, the only difference is the Nz data are obtained from the Meggit 11039/43 mechanical counting fatigue meters at the previously mentioned 'g' levels of +0.0g +0.2g, +0.5, +0.7g, +1.3g, +1.5, +2.0g, +2.5 and +3.0g, i.e. there is no facility for generating data equivalent to the intermediate 'g' levels used in the analysis of the MSR Nz traces.

The results from the analysis of the 3317 historical flights are summarised in Table XI.

Table XI. Nominal fatigue damage analysis results from historical Islander flying.

Ser.		Flts	Total hours	Mean duration (hrs)	GAG	RGM	RGM % of Total	Total damage per flight
1	Fleet	3317	4139.36	1.25	8.971E-06	2.157E-05	70.6%	3.055E-05
	Individual A/c							
2	G4	637	732.07	1.15	8.946E-06	2.543E-05	74.0%	3.438E-05
3	G5	509	695.12	1.37	9.478E-06	1.652E-05	63.5%	2.600E-05
4	G6	590	717.14	1.22	8.894E-06	2.351E-05	72.6%	3.241E-05
5	G7	626	738.11	1.18	9.030E-06	2.514E-05	73.6%	3.417E-05
6	G8	505	618.66	1.23	8.409E-06	2.109E-05	71.5%	2.950E-05
7	G3	450	638.27	1.42	8.717E-06	1.588E-05	64.6%	2.460E-05

It is these results from historic flying that are compared with similar damage rates calculated from the more recent MSR data as presented earlier in this report at Table X. Comparisons between flying can now be drawn.

The data presented in Table XI are based on flying from only the 'G' aircraft. Consequently, Ser 1 from Table XI can be compared directly with Ser 2 from Table X. These two sets of results are very similar, while individual aircraft results also show broad similiarity, with the obvious exception of aircraft G6 in Table X, results of which are probably attributable to the small sample of MSR data from this aircraft.

Results from the analysis of historical flying, individual aircraft data
Single aircraft analyses have also been conducted to characterise the 'nominal' fatigue damage rates for the GAG and RGM components of flying. These individual aircraft results are shown as Ser 2 to 7 of Table XI.

Comparison of the Table X and Table XI shows average flight durations, and damage per flight for the GAG and RGM flying components appear to be very similar, with the exception of aircraft G6.

The obvious difference in the analysis of the MSR data versus that for the historical data is the additional 'g' levels at +1.8g, +2.2g and +2.7g, which are not present in the historical data. Thus the MSR results presented are not strictly a like for like comparison with the historical results, although they are considered to be the best balance between the complexity of the calculations and the accuracy of the results. Accordingly, an analysis of the MSR data using only the original fatigue meter levels is shown as Ser 9 in Table X to illustrate the refinement in the results that the additional 'g' levels has provided.

MSR data analyses at historical 'g' levels

The Islander MSR normal acceleration data analysed and supplied by Dstl has a greater degree of resolution in terms of level counting than that available from the mechanical fatigue meters previously fitted to the aircraft. As stated earlier, counting on the MSR data took place at additional intermediate levels of +1.8g, +2.2g and +2.7g in addition to the previous fatigue meter levels of +1.5g, +2.0g, +2.5g and +3.0g. By 'degrading' the MSR normal acceleration data to the original levels captured by the mechanical fatigue meter, an analysis of comparable detail to that of the historical Islander flying data can be done; Such a comparison has two benefits:

Firstly, a direct comparison of the MSR data to the historical data based on a common level in terms of data counting resolution is possible.

Secondly, the benefit of using higher resolution data is apparent in the improved precision from the fatigue calculations compared to a calculation using data counted at the same (Nz) levels recorded by the fatigue meter.

A summary of the nominal fatigue damagerates calculated on this basis is presented in Table XII, column 2. These can be compared directly to the values in column 3 based on the full counting spectrum results taken from the Ser 1 row of Table X. The table also includes the historical results representing the 3317 flight data set (from Ser 1 row of Table XI) for comparison.

Table XII. Nominal fatigue damage results, including results from degraded MSR data.

	MSR data 'degraded' to Historical 'g' levels	MSR data	Historic Damage per flight (3317)
GAG Damage	8.106782E-06	7.70562E-06	8.97137E-06
RGM Damage	1.592457E-05	1.46792E-05	2.15738E-05
Total Damage	2.403135E-05	2.23848E-05	3.05451E-05

The 'degraded' MSR results from column 2 of Table XII compared with column 3 demonstrate the improvement provided by the increased resolution at which the normal acceleration data has been counted. More, smaller stress blocks, rather than fewer, larger stress blocks, produce a better approximation of the area bounded within the upper and lower stress spectra and therefore a more precise estimate of the associated nominal fatigue damage rates. The difference between the totals from the 'degraded' and standard resolution damage rates is 5.2% on the GAG data and 8.5% on the RGM data.

Damage comparison MSR & Historic

This section covers a detailed interpretation of the MSR results versus the historical flying data, hence looking at the differences, and similarities, of the data presented in the 6th and 7th columns of Table XI and Table X. The data in Table XII show the comparison of the MSR fleet data for overall current flying is no more severe than the historic flying, but this is not the entire truth as there are underlying subtleties. The comparison of the 'G' aircraft MSR nominal damage rates is not a like for like comparison with historical damage rates due to the difference in resolution of the raw data used in the calculations. It has been shown earlier that the difference in damage calculations between the 'degraded' data and standard MSR data was 5.2% for GAG and 8.5% for RGM. Thus it is reasonable to assume that these factors should also be applied to the Ser 2 Table X results from the 'G' aircraft to make them directly comparable to the historical results which were for 'G' aircraft only.

Applying these factors to the Ser 2 results from Table X produces the values presented in row 1 of Table XIII. Historical results (Ser 1 Table XI) are also included for direct comparison in row 2, the percentage differences being included in the row 1 entries:

Table XIII. Comparison of current and historic UK MoD flying.

	GAG	RGM	Total
Factored MSR G Ac Only (278 flights)	9.58497E-06 (+6.84%)	2.72411E-05 (+26.27%)	3.68261E-05 (+20.56%)
Historical UK MoD (3317 flights)	8.97137E-06	2.15738E-05	3.05451E-05

These results now show the GAG component is slightly more severe (+6.8%) than the Table XI result, but the RGM data from the MSRdata are significantly more severe (+26.3%) than the comparable historical data. Why the current flying from the 'G' aircraft fleet is more severe than the past is not immediately obvious; however, one reason could be the MSR data sample of 278 flights is less than 10% of the historical data sample of 3317 flights. A second reason could be that detail aspects of the way the aircraft are flown have evolved during the 8 years or so between the two sets of data. It is these types of observations that demonstrate why regular reviews of usage data are appropriate in ensuring the structural usage of an aircraft fleet is properly understood and quantified. Further work is necessary to explore the significance of the differences between the results from the MSR exercise from 2012/13 and the historic records from 1992/2004.

NZ SPECTRA AND OPERATIONAL LOADS MEASUREMENT

Overview

The analysis of the exceedence spectra for historical and MSR Nz records of the Islander fleet provides a useful understanding of some key aspects of the way the aircraft are used. There is clear evidence of a difference in the MSR datasets of different groups of aircraft. These differences are evident in the tabulated data presented in Table and Table of this paper and result in substantial variations in the nominal fatigue damage rates. This is an important observation because it highlights how averaging results across a fleet of aircraft within which individual aircraft may be operated differently from others might produce misleading conclusions.

These findings may be considered in the context of the requirements of RA5720 which mandates the implementation of operational loads measurement (OLM) programmes as a key element of the structural integrity management philosophy for UK military aircraft. This requirement is considered in greater detail in the context of the UK MoD fleet of Islander aircraft in the following sub-sections of this paper.

OLM Objectives

As cited in Paragraph 103 of RA5720, comprehensive guidance relating to the objectives and conduct of an OLM programme are published in MASAAG Paper 109 [3] which includes the following definition:

OLM Definition: *Operational Loads Measurement is a structural usage substantiation activity involving the capture, analysis and reporting of representative, directly measured strain data or derived loads and associated flight parameters from a sample of suitably instrumented in-service aircraft within a fleet.*

It is important to note that, in this context, "structural usage substantiation" refers to the assessment of structural usage in service in relation to the assumptions and methods used in design and qualification. With this in mind, the objectives of an OLM programme can be summarised in terms of a number of typical aims that may be tailored to the needs of any particular project and which may vary according to the point in the service life cycle and/or the evolution of operational experience or intent; these aims can include but are not limited to the following purposes:

- to substantiate the fatigue spectra used in design and qualification and review fatigue clearances
- to identify local stresses in a structural feature
- to substantiate the monitoring systems (including identification of further monitoring requirements)
- to capture fatigue test spectra data
- to identify particularly damaging activity or manoeuvres
- to provide data to support investigations of structural issues or as life extension programmes
- to provide data for use in a review of the SOIU

The assessment of Nz spectra captured by the MSRs provides information that might contribute to several of these aims, including the 1st, 3rd, 5th, 6th and 7th purposes; however, none of these can be wholly satisfied by what has been done to date. The 2nd item would require direct measurements of strain at a feature and so is outside the scope of any investigation that relies solely on the gathering and analysis of Nz data whilst the 4th item is only relevant if there were plans to conduct fatigue testing of the Islander airframe; there are no such plans with respect to the UK MoD aircraft.

Furthermore, the extent to which these results could be considered as satisfying any of the cited aims, albeit partially, depends entirely upon what overall objectives for an OLM programme are considered appropriate for the Islander. Consequently, it is beneficial to consider the analysis results in terms of what can been learned about the structural usage of the airframe in UK military service and, hence, what further work might be necessary to satisfy the overall intent of RA5720 with respect to OLM.

Data Limitations

The MSR data acquisition is limited to a tri-axial accelerometer and atmospheric pressure. With the MSR unit being placed close to the aircraft centre-of-gravity, the accelerometer provides a basic measure of the overall loads affecting the aircraft both in flight and on the ground. The atmospheric pressure provides a useful measure of the flight profile in terms of altitude - which helps to confirm the overall nature of each flight and also aids in the identification of the transition to/from airborne flight and groundborne conditions. However, the tri-axial accelerometer has only been used to gather Nz data, i.e. vertical acceleration. The lateral and longitudinal acceleration data recorded by the MSR have not been used in any part of the analysis.

Scope of Analysis

The methodology applied in the fatigue analysis of both sets of data is optimised by the breaking down of the Nz spectra into its two principal components, i.e. the GAG and RGM spectra; this methodology applies only to the primary structure of the wing and, specifically, to the equivalent location of the limiting fatigue cracking observed on the full scale fatigue test of the Trislander aircraft as previously completed by Britten Norman during the 1970s. Due to the detail differences between the Islander and Trislander, this means that the fatigue damage calculations can only be considered as nominal indications of the relative severity of different data samples and they do not provide any measure at this stage of the likely fatigue life limit for the Islander aircraft.

The GAG spectra also include assumptions about the load increment used to define the extreme ground load condition for use in the fatigue calculations. As per the Trislander fatigue test, that increment is presumed to be +0.5g. An assessment of the MSR data has shown that this is a reasonably conservative estimate relative to over 95% of the 898 flights included in the MSR sample. Consequently, the default increment of +0.5g has been retained throughout the fatigue analysis; this simplifying assumption is unlikely to have any substantial effect on the results from the fatigue analysis in terms of the relative damage rates associated with each data set.

The peak load condition for the GAG is derived for each flight by identifying the highest Nz exceedence value from the F725 records for the historical data set and from the pseudo fatigue meter output of the MSR data processing as undertaken by Dstl. There is no way to identify whether the event that produced the peak load was due to gust or manoeuvre. Furthermore, it was assumed for the Trislander fatigue test that the RGM spectrum would be dominated by the gust content such that the associated stresses could be calculated based on the response of the aircraft to a nominal upgust; the equivalent stress-per-g appropriate to manoeuvring flight was not considered. Throughout the analysis of the Nz spectra considered in this report it has been assumed that the stress-per-g for manoeuvres is the same as for gusts but this is in the absence of any consideration of how the gust and manoeuvre content of the RGM compares with the Trislander assumptions. In other words, if the proportion of the Nz spectra due to manoeuvres is substantially greater than assumed for the Trislander, then the use of a single stress-per-g value may be insufficient to provide an accurate measure of the fatigue damage contribution of the RGM.

Exploitation of Results

The key finding from the analysis performed is that there are distinct fatigue damage rates for different groups of aircraft within the MSR data set. This is likely to be due to different operational usage patterns of aircraft with different primary roles and equipment fits. It is also suggested that such usage patterns have evolved between the period of coverage of the historical data set (1992 ~ 2004) and the latest 2012/13 sample. Neither of these observations should be unexpected but they do confirm the value of periodic reviews of usage data.

For instance, the evolution of usage patterns indicates that some of the differences apparent in the fatigue calculations may be due to the typical altitudes at which different groups of aircraft are typically flown. Insufficient analysis of all the available data has been performed to confirm that this is so but the results provide the necessary pointers towards where additional analysis could be directed in order to fully understand the reported findings.

Consequently, it may be concluded that the results so far can be regarded as an initial successful step in determining a thorough understanding of how the UK MoD Islander fleet is actually being used from a structural integrity assurance perspective. Such an understanding is necessary in the context of the overarching purpose of an OLM programme as expressed within MASAAG 109 [3], i.e. “a structural usage substantiation activity”.

Clearly, the information that may be extracted from the MSR data set is suitable for this purpose as it is possible to assess directly flight profiles in terms of altitude and time and thereby ascertain at what points during each flight the Nz spectra are significant. The linking of this information to other flight details recorded on F725 Flying Log data sheets will then allow the identification of particular types of flying where fatigue damage rates are highest.

Whilst such a “building block” approach to exploiting these results is feasible with the existing data set, it is likely that the sample sizes considered within this paper are too small to provide a comprehensive assessment of all variables that might affect the Nz spectra and, hence, the fatigue damage calculations. Additional periods of data acquisition, involving as many individual aircraft as possible and encompassing any obvious variations in operating conditions – such as the location of the main operating base, specific equipment fits which cause individual aircraft to be assigned to particular roles, etc. – are likely to be necessary.

As mentioned previously, the fatigue analysis methodology adopted for the interpretation of the Nz spectra for the Islander aircraft was adopted from an assessment of the Trislander wing fatigue test programme and the associated calculations. As such, the results provide only a measure of the relative fatigue damage rates for each group of aircraft within the historical and MSR data sets. A second aspect of the exploitation of findings will therefore be for the analysis results to be compared with the fatigue qualification basis for the Islander. This would entail such parameters as the stress levels at the critical location of the Islander wing in comparison with those of the Trislander being assessed and identifying any other aspects of the analysis process where the physical differences between the two aircraft types might demand revision of the assumptions used by QinetiQ for the purposes of the initial assessments summarised in this paper.

CONCLUSIONS

The UK MoD had no means to quantitatively compare current Islander flying to either the historical flying records or the fatigue performance baseline provided by the Trislander wing fatigue test programme; such comparisons are included in the requirements that must be satisified to comply with the objectives of an OLM programme. The fitting of a MSR to capture flying data in the form of Nz time histories, converted to a flight ‘g’ exceedence spectra followed by grouping of component parts of the resulting stress spectra (i.e. GAG and RGM) has permitted an effective quantitative analysis resolving the flying down to a single nominal fatigue damage value. The same analysis methodology has been applied to historical Nz data to provide directly comparable quantitative assessments of aircraft usage over two distinct time periods.

The analysis approach has used several assumptions all of which are justifiable and commensurate with the early stage life cycle of the analysis process and all of which can be incorporated for future analyses depending on the influence they hold within the fidelity of the overall fatigue calculations in accordance with the overall requirements of the structural integrity assurance framework applicable to UK military aircraft.

The results of comparing recent flying data and past flying data as a direct comparison on an aircraft-by-aircraft basis, sub-fleet basis and entire fleet basis quite clearly shows that significant value that can be extracted from the data recorded by an inexpensive and easy to use technology like the MSR as facilitated by advances in the associated MEMS technology

ACKNOWLEDGEMENT

The work presented in this paper was undertaken by the Structures Capability at QinetiQ. However thanks are due to the support of the UK MoD Islander Project Team for their permission to use this subject matter, to Dstl who have supplied QinetiQ the post flight data, and to B-N Aircraft without whose contributions the analysis would not have been achieveable.

REFERENCES

[1] MSR Electronics GmbH, http://www.msr.ch/en/index.php

[2] Reed Dr S C, (2012), Structural integrity usage monitoring capability development – Phase 2, pp 7, Physical Sciences, Dstl

[3] Reed Dr S C, Holford D M, (2007), Guidance for Aircraft Operational Load Measurement Programmes, Airiworthiness & Structural Integrity Group, QinetiQ

In-Flight Structural Health Monitoring System for Medium Altitude Long Endurance Unmanned Air vehicle

I. Kressel[1], J. Balter[1], N. Mashiach[2], I. Sovran[3], O. Shapira[4], N.Y. Shemesh[4], B. Glam[4], A. Dvorjetski[2], M. Tur[3]

[1]IAI Engineering Division, Ben Gurion International Airport 70100 Israel
[2]IAF UAVs Branch, Tel-Aviv Israel
[3]School of Electrical Engineering, Tel-Aviv University, Tel-Aviv Israel
[4]IAF Aircraft Structure Branch, Tel-Aviv Israel

Abstract: This work presents the design, qualification and flight service evaluation of an embedded Fiber Bragg Gratings (FBG) based, Health and Usage Monitoring System (HUMS) for the Israeli Air Force Medium Altitude Long Endurance (MALE) Unmanned Aerial Vehicle (UAV).A total of 54 FBG sensors were embedded on the wing and tail booms, enabling accurate tracking of both the vibrations signature and the actual loading conditions of these components. Reliable flight data, obtained during flight tests and normal UAV operational missions were recorded and analyzed both in the frequency and time domains so that normal structural behavior could be identified and tracked. Based on the data obtained, it is now possible to track the actual loads spectra of both the booms and wing during each flight, with emphasis on the landing impact vibration signatures. The sensing system has already gained over 1000 flight hours, including more than 300 logging hours. It is the purpose of this system to early detect and identify dangerous changes in the normal structural behavior of an individual UAV, prompting the required corrective action, thereby, paving the way to condition-based maintenance.

INTRODUCTION

The increasing value of modern Unmanned Aerial Vehicles (UAVs), together with their high usage under variety of missions, at extreme environmental conditions, demands continuous monitoring of their structural airworthiness over their life span. Reliability, safety and the economic implications of maintaining aging UAV fleet, subject to cyclic loading, corrosion, wear and material degradation are the prime concerns of both the operator and the UAV manufacturer due to their logistics implications.

In recent years, the concept of Health and Usage Monitoring System (HUMS) was proposed in order to assure safe and economic operation of UAVs. This concept is based on tracking the individual UAV by evaluating its real flight history and the impact of the latter on the structure, combined with real time identification of events that may eventually lead to undesired failure. HUMS has a unique benefit for UAVs since conventional maintenance, based on periodic inspections, requiring direct access to all of the UAVs critical structural components, is hindered by limited accessibility and the need for highly trained, i.e., costly technicians, not always available at remote sites. It should also be noted that UAVs are missing the human pilot, who practically acts as a full time highly sophisticated sensor, having the ability to identify unusual behaviors, caused by wear, fatigue, bird or hail impact, lightning strike etc. [1]. One of the purposes of HUMS is to fill this gap.

By applying the HUMS concept, maintenance action should be performed only when the need arises, reducing the UAV ownership cost. This concept also coincides with the recently introduced airworthiness requirements for UAVs, like the STANAG 4671 [1, 2] where, for example, the substantiation of structural bonded jointsmay be based on: "repeatable and reliable non-destructive inspection".

The concept of Structural Health Monitoring (SHM) which is included in HUMS, has been discussed for many years. Summary of early Health Monitoring literature 1996–2001 [4] indicates that a multidisciplinary effort should be directed to solve issues in sensors reliability, data processing for damage diagnostics and prognostics [5,6,7].

Among the available sensing technologies, the use of fiber-optic sensors to implement an online system appears quite attractive [8]. Due to their small diameter optical fibers can be easily embedded within composite materials, having currently ubiquitous presence in modern UAVs. Optical fibers are flexible, passive, quite tolerant to environmental conditions and insensitive to electromagnetic disturbances. Quite a few sensors can be multiplexed on a single fiber strand, and the overall technology, including the interrogators is already quite mature, and well-known to be highly reliable [4].

Structure Health Monitoring systems based on Surface bonding of fiber-optic sensors proved itself in lab scale and prototype demonstrations [9,10]. One of the first successful attempts to demonstrate an in-flight, real-time, fully embedded, airworthy SHM system was on the Indian Nishant UAV [11]. This Fiber Bragg Grating (FBG) based system successfully monitored the two tail booms during flight and traced the boom loading during the entire test flight.

This paper presents a complete airworthy HUMS system that was integrated in the Israeli Air Force Medium Altitude Long Endurance (MALE) Unmanned Aerial Vehicle. The purpose of this project is to evaluate such system benefits under tight military operational environment. This airworthy system is based on an array of FBG sensors, embedded in the two composite-made, tail booms, as well as in the central wing. The FBG sensor net was tailored to monitor critical locations in the wing and along the tail booms. The system was first tested and calibrated on ground in order to verify its ability to track both static and dynamic boom loadings. Structural characteristics like strain distribution under static loading and impact response were successfully monitored by the system. The high signal to noise ratio of the optical sensors enables tracking of booms vibration signature during all the UAV manoeuvers, including cruise, turns, steep ascents and descents and heavy landing. As a final and definitive proof of the concept, the system, including the FBG multi-channel interrogator and data storage device, was integrated in the UAV. The system entered service and reliable flight data were recorded by the sensors during a dedicated flight test, covering the full UAV flight envelope, and continued to be recorded during the normal operation of the aircraft. The on-board collected data are processed off-line on the ground and analyzed both in the frequency and time domains so that critical components such as the individual UAV wing and booms load spectra and vibration signatures can be identified, characterized and tracked. The sensors data were also integrated with the flight parameters on a single GUI, for visual and quantitative correlation of the sensors readings and flight manoeuvers. Based on the data obtained, it is now possible to track this UAV actual load spectrum and use its characteristics to make reliable predictions on the UAV structural integrity.

This paper discusses the system implementation on the UAV booms, including the sensor layout, system architecture and data processing.

THE IMPLEMENTED HUMS CONCEPT

A MALE UAV (Figure 1), designed and manufactured in Israel by Israel Aerospace Industries was selected by the Israeli Air Force as the test-bed for the evaluation of this fiber-optic based HUMS concept in operational real-field environment. Each of the two tail booms is a composite beam structure attachedto the wing in the front and at the back holds the empennage, comprising a horizontal tail (with elevator) and two vertical tailsand rudders.

Each boom is basically a cantilever beam with a relatively large mass at the back end. The main boom loading condition is vertical bending. In order to track these loading conditions, two fibers were embedded at the center of the boom ("Top" and "Bottom" in Figure 2). Eight FBG sensors were imprinted on each fiber. Thus, eight section of the boom are monitored by two FBGs; one on top and another on the bottom (Figure 2). For such a sensing net arrangement, the two fibers are only sensitive to vertical bending. Since the cross-section of the boom is not symmetrical with respect to the horizontal plane, vertical bending will introduce non-identical but opposite strains in corresponding top and bottom FBGs.

Figure 1. IAI's UAVs after takeoff. In the current project two fibers were embedded on each boom and wing, see Fig. 2 for details.

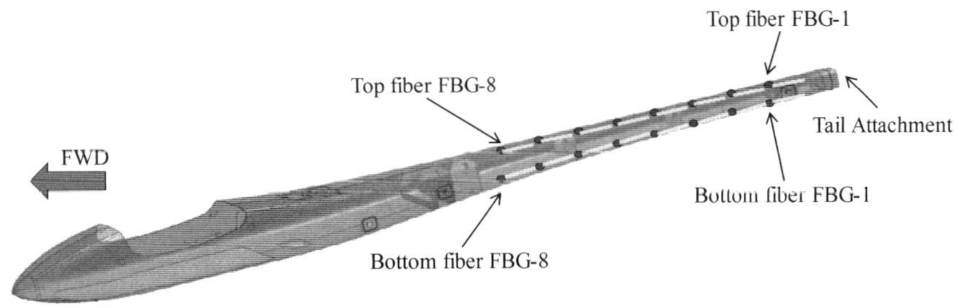

Figure 2. Boom general layout and the routing of the embedded optical fibers in the boom. 16 FBGs were located at points of interest (RHS boom shown. LHS is similar.).

The optical fibers were polyimide-coated to ensure good bonding to the composite structure during embedment. To successfully solve the ingress/egress issue, a special pre-cured composite protective patch, made of glassfabric was used for each pair of fibers at the egress point providing good shielding for the bare fiber as it leaves the structure and enters the protective (3 mm) sleeve strengthened by Kevlar fibers. The patch was located at the front of the boom, where it is attached to the wing. For accurate sensor placement, a pre-fabricated glass-fiber composite sensing mat, containing the polyimide-coated optical fiber, was fabricated. This sensing mat was then bonded to the UAV structure, minimizing the risk of optical fiber damage and FBG sensor reflection attenuation. The shielded fibers were internally routed inside the UAV from each boom and wing to the interrogation and logging unit located in the payload bay. FC/APC connectors were used to enable wing and booms disassembly. Similarly the wing fibers monitor the outer wing attachment and track the wing bending moment at the center.

A solid-state, high sampling rate (2.5kHz) FBG interrogation unit was used, capable of simultaneously tracking 4 fibers, having multiple FBGs on each. The interrogation included a data-logging system was placed in the UAV payload bay, and was powered by the UAV electrical system. While the data logger had significant processing power, all data analysis reported here was done offline. The interrogator containing data-logger unit has been successfully subjected to environmental testing fully compatible with the environmental requirements of the MALE UAV. The sensors data were also integrated with the flight parameters on a single GUI, Figure 3, for visual and quantitative correlation of the sensors readings and flight manoeuvers. Based on the data obtained, it is now possible to track this UAV actual load spectrum and its impact on UAV structural integrity.

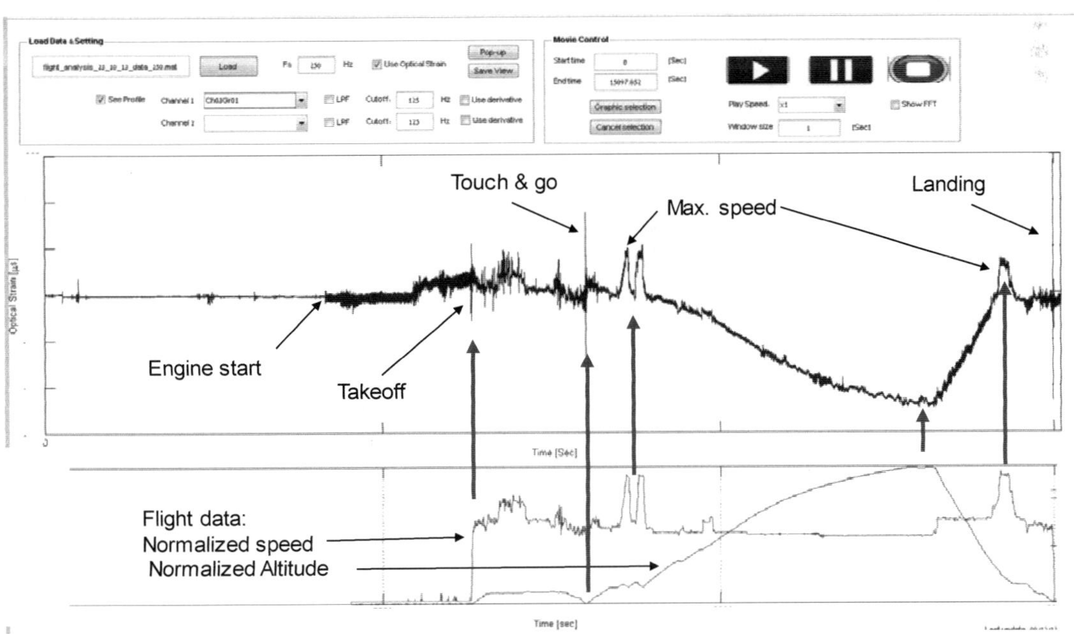

Figure 3. Flight data GUI; Typical Boom FBG readings during flight. Note the 'touch & go', maneuver and landing.

SYSTEM VALIDATION AND CALIBRATION

Static calibration tests were performed in order to correlate the embedded FBG readings with the boom loading under vertical bending load. Tail loadings were applied at the boom end near the tail, up to 30Kg. FBG readings were taken at 10Kg increments.

As explained earlier, the boom main loading condition is vertical bending. The bending moment distribution along the boom is complex and varies during UAV manoeuvers according to the loads induced by the wing and tail. It is extremely important to make sure the UAV do not exceed its designed limits during flight and automatic or manual controlled landing. Once the strain-to-bending conversion factors are evaluated by the above mentioned ground procedure, proper determination of the bending moment can be obtained from the measured strain at each FBG station. It should be noted that this strain to bending moment calibration, obtained by static test is also valid for dynamic loading.

Since the UAV flies at high altitudes, where the temperature is significantly lower than on ground, temperature compensation for the FBG data is required. In order to evaluate the sensitivity of the embedded sensors to temperature, a heating (or cooling) test was conducted. FBG records were taken, on ground, both at a hot day and a cold night, resulting in a 20°C difference. The derived overall sensors sensitivity at each FBG station, assuming the temperature is the same at each FBG pair location, was thus found to be:

$$\begin{bmatrix} \varepsilon_{top} \\ \varepsilon_{bottom} \end{bmatrix} = \begin{bmatrix} K_{1T} & K_{1M} \\ K_{2T} & K_{2M} \end{bmatrix} \begin{bmatrix} T \\ M \end{bmatrix} \tag{1}$$

where ε is the optical strain ($\Delta\lambda/\lambda$), T is the temperature change and M is the bending moment at a boom station, where strain is measured by pair of FBGs. The K matrix can be inverted making it possible to get both the temperature and bending moment for each pair of FBGs. Once the bending moment is known, using the ground calibration again, top and bottom mechanical strains can be calculated.

FLIGHT DATA EVALUATION

The first flight with the system fully operational was on October 2013. Typical FBG readings of boom mounted FBG (Top and bottom FBG-8, see Figure 2) is seen in Figure 4. The effect of engine start, takeoff, touch and go landing, acceleration to maximum speed and landing are clearly depicted by the FBG sensors. The effect of temperature is also seen as the UAV climbs to high altitude.

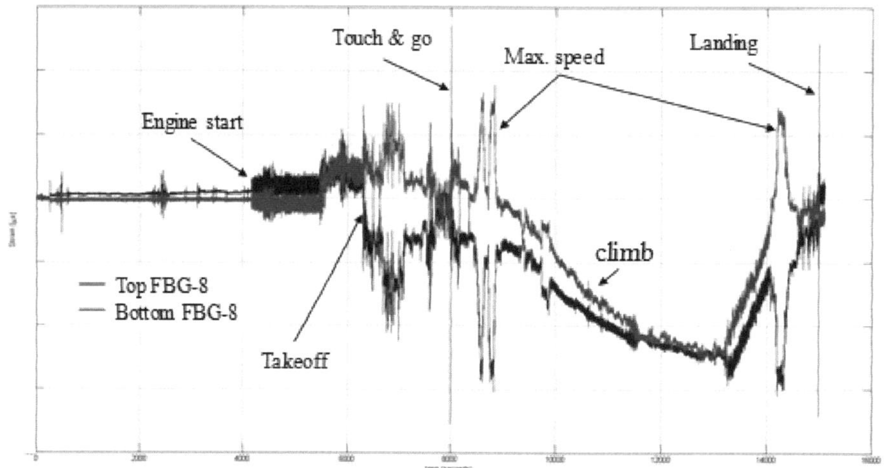

Figure 4. Typical FBG readings of boom mounted FBG (Top and bottom sensors).

By using Equation 1, the mechanical bending moment can be calculated and the individual mechanical strain at each sensor pair can be derived, as can be seen in Figure 5.

The high sampling rate of the interrogator enables tracking the dynamic behavior of the wing and booms. Since landing impact is the most critical boom loading it is of great interest to closely monitor the bending loads developed in the boom during the landing impact.

Typical boom bending moments derived from a specific landing are presented in Figure 6. The almost constant distribution of the bending moment along the boom indicates a complex loading condition, associated to its dynamic behavior.

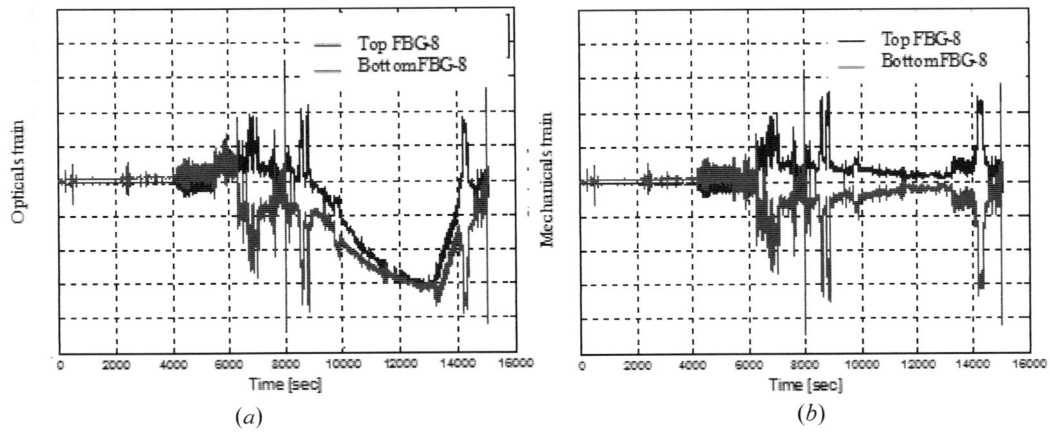

Figure 5. Boom FBG calibration. (a) Original FBG optical strain data. (b) Mechanical strain.

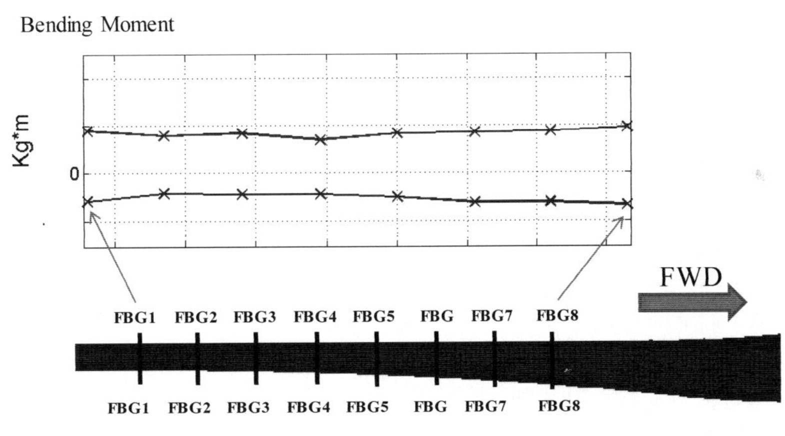

Figure 6. Boom bending moment envelope during landing.

CONCLUSION

This fiber-optic-based sensing concept has already accumulated hundreds of flight hours without any failure in "real world" service environment. The system is easy to operate by the squadron personnel and flight data unloading for ground evaluation after flight is fast. All major flight events were clearly detected by the FBG sensor net. The system is easy to operate and the visual GUI enables a quick and basic evaluation of flight data. It also provides a deeper analysis when required. Tracking the structural behavior over time can be used for Condition Based Maintenance (CBM), with the hope to eventually reduce maintenance cost and aircraft down-time.

REFERENCES

[1] M. Neubauer, G. Gunteher and K. Fullhas, Structural Design Aspects and Criteria for MilitaryUAV.RTO-MP-AVT 145 UAV Design Processesand Criteria 2007.

[2] "STANAG 4671 (Edition 1) – Unmanned Aerial Vehicles Systems Airworthiness Requirements(USAR)," North Atlantic Treaty Organization Standardization Agreements, 2009

[3] C.R. Farrar and H.Sohn, Structural Health Monitoring Using Statistical Pattern Recognition Techniques, Journal of Dynamic Systems, Measurement, and Control, Volume 123, Issue 4, 706, 2001

[4] A Review of StructuralHealth Monitoring Literature: 1996–2001, Los Alamos National Laboratory Report, LA-13976-MS, 2004

[5] Staszewski W, Boller C and Tomlinson G. eds. Health monitoring for aerospacestructures, Wiley, Chichester, UK, 2004.

[6] K. Worden, C. R. Farrar, G. Manson, G. Park, The fundamental axioms of structural health monitoring, Proc. R. Soc. A 2007 463, 1639-1664, 2007

[7] C.R. Farrar and K. Worden, An introduction to structural health monitoring, Phil. Trans. R. Soc. A 15 vol. 365 no. 1851 303-315, 2007

[8] Todd, Michael D., et al. "Bragg grating-based fibre optic sensors in structural health monitoring." Philosophical Transactions of the Royal Society A: Mathematical, Physical and Engineering Sciences 365.1851 (2007): 317-343.

[9] C.Y. Park, J.H. Kim, S.M. Jun, A structural health monitoring project for a composite unmanned aerial vehicle wing: overview and evaluation test, Structural Control and Health Monitoring, Vol 19 Issue 7, 2012

[10] Derkevorkian A; AlvarengaJ.a; Sami F. Masri ; Helen; W. Lance Richards. "Computational studies of a strain-based deformation shape prediction algorithm for control and monitoring applications."." SPIE Smart Structures and Materials+ Nondestructive Evaluation and Health Monitoring. International Society for Optics and Photonics, 2012.

[11] M. Tur, I. Kressel, Y.Botsev, A. Handelman, P. Gud's, S. Ronen, S. Cohen, ACR Pillai, M. Hari Prasad, GM Kamath, N. Gupta, A. Khatkhate, S. Gali. "Flight Data from an Airworthy Structural Health Monitoring System for an Unmanned Air Vehicle Using Integrally Embedded Fiber Optic Sensors." 8th International Workshop on Structural Health Monitoring. USA: DEStech Publications, Inc. Vol. 1. 2011.

Low-Cost Fixed-Wing Aircraft Usage Validation

Steve Reed

Technical Fellow, Defence Science and Technology Laboratory, UK MOD

Abstract: In addition to the well-known aircraft platforms, the United Kingdom Ministry of Defence (UK MOD) operates a wide range of smaller aircraft fleets, often undertaking specific military roles. These fleets range from traditional military aircraft, through military derivatives of civil aircraft to historic or vintage aircraft. The continuing air safety and duty-of-care responsibilities for these fleets are no less stringent than the traditional large military aircraft fleets, despite the often very limited budgets within which these aircraft fleets operate. However, for these fleets the traditional approach to mandatory regulatory requirements of an Operational Loads Measurement programme is often effectively unaffordable. Therefore, there was a clear need to develop more cost-effective solutions to allow comparison of in-service structural usage with design assumptions. Within this paper, the approach taken within the UK MOD to solving this problem is presented. In essence, two approaches, within a common framework have been developed. This framework, which includes data-capture hardware and data-analysis software, is termed the Aircraft Data Analysis and Monitoring (ADAM) system. For those aircraft with no existing data capture systems, advantage has been taken of commercially-available, low-cost sensor technologies to develop minimal interface aircraft instrumentation. For this part of the programme, the development of the commercial product, in conjunction with the equipment supplier, and the provision of the evidence necessary to use such devices in the military environment are detailed. This includes aspects such as Electromagnetic Compatibility (EMC) testing and shielding and thermal environment test programme. The design criteria for this approach also included method of improving traditional data capture rates, such as minimizing the burden upon the aircraft front-line operators and the methodologies adopted to achieve this requirement are discussed. For aircraft with viable existing data systems, approaches have been developed to extract and exploit data captured on these platforms and this aspect of the programme will be presented separately. Examples are provided from the range of air platforms that have benefited from the ADAM system. These range from the 1930s Swordfish historic torpedo bomber, operated by the Royal Navy Historic Flight, to some of the most recent acquisitions from within the MOD fleets. The examples also include methods of presenting the structural usage data in clear and concise frameworks to assist the decision making process for both Senior Commanders and the relevant Design Organisation.

INTRODUCTION

The history books are filled with examples of failures to understand the significance of aircraft in-service structural usage [1]. The United Kingdom Ministry of Defence (UK MOD), alongside a great many military aircraft operators, has learned this lesson the hard way, with loss of life and aircraft [2, 3]. Consequently, the UK MOD Military Aviation Authority (MAA) and its predecessors have developed a regulatory framework requiring the validation of in-service usage [4-6]. The technology is readily available to undertake such usage validation exercises [7]. However, these programmes can often be extremely costly, complex and very protracted. Moreover, a technical solution that is appropriate and justifiable for an agile combat aircraft may not be necessary for a civil-derivative aircraft with near-civil usage or for an historic aircraft understood to be operated in a relatively benign flight regime. Defence budgets are under unprecedented strain and many military aircraft fleets are now relatively small and are often derivatives of civil aircraft, pulled into military use for specific, often short-term roles. Therefore, there is a need to develop a rapid, cost-effective methodology for undertaking limited structural usage validation for a range of military aircraft fleets. This programme is aimed at addressing this requirement. This work is part of the Usage Validation Programme (UVP), within the Defence Science and Technology Laboratory (Dstl)-led UK MOD Understanding Ageing Aircraft Research and Development (UAAR&D) programme. The UVP is aimed at both fixed-wing and rotary-wing aircraft usage validation development. The rotary-wing programme development is reported by Perrett [8].

AIMS AND REQUIREMENTS

The clear definition of requirements is essential for every successful programme. Yet it is disheartening to read how often key programme failure causes are identified as poor definition of requirements, lack of understand of requirements, requirement growth or unrealistic or unaffordable requirements. These risks are equally applicable to usage validation programmes. Such approaches as, *"we will just capture all the data we can and work out what we are going to do with it later"* have been applied on occasion in the past and have proven to be unwise, costly and damaging, in terms of credibility with those tasked with capturing the data, and should be avoided at all costs.

Therefore, the first aspect addressed within this programme was setting realistic aims and requirements. The driving aim was adequate usage validation across a range of platforms for minimum cost. Initially, the development of requirements to meet this aim was done in generic terms as the intention was to cover a range of likely scenarios to identify what was likely to be achievable at various levels of complexity and cost. The starting point was considering what was likely to be needed for a minimum usage validation programme by looking at the typical information available for comparison. From the aircraft design perspective, this might include design usage spectra (DUS), fatigue test data, damage tolerance analysis, or known previous in-service data, where appropriate. This comparison data set is identified as design data, for ease of reference throughout the remainder of this paper. Having collated the design data, the key types of assumptions within that data were identified. For example, for many civil-derivative aircraft the airborne flight phase design assumptions are based upon relatively simple flight profiles. These might include a climb to cruise altitude, a cruise for a specified period, and a descent to land. Events are largely dominated by gust loading, with little or no manoeuvre content in the assumed spectrum, with the gust event frequency and magnitude based upon an assumed time at flight level and statistical gust data from a relevant published source alongside simple mass assumptions. From an in-service perspective, all UK MOD aircraft fleets have a Statement of Operating Intent and Usage (SOIU) document. This document, which is assembled by the relevant service authority, describes the current and intended usage of the fleet. The SOIU is provided to the Design Organisation (DO) for review and contains basic information on the type of flying undertaken in terms of sortie profile codes (SPC). Some aspects of the information in the SOIU can be validated from flight record information, such as aircraft mass at take-off and landing and flight time but this is limited without parametric data captured during flight.

Therefore, coming back to the simple example, it was clear that an initial validation of the military usage of such an aircraft, for the airborne flight phase, could be achieved with a relatively simple data set, including altitude (ALT), normal acceleration (NZ) and time, in conjunction with mass data from the flight records. Such an approach, although simplistic would allow a low-cost and rapid initial assessment to be made of the military usage. This approach would also allow an evidence-based assessment to be conducted to identify whether a more detailed, and hence more costly data capture programme was necessary.

Loads and usage

The simple example illustrated above is clearly aimed at validating usage rather the loads seen by the aircraft in-service. The distinction here is significant and has not always been fully understood. In this scenario there is no attempt being made to measure flight loads directly using loads-calibrated strain gauge systems, for example.

Minimising burden on the front line

The second key aim of the programme was to minimise the burden on the front line organisations. In this context these are typically Royal Air Force (RAF) or Royal Navy (RN) Squadrons or Army Regiments. Experience has shown, across a wide range of Operational Loads Measurement (OLM) Programmes [6], that data capture rates fall dramatically when a significant burden is placed on the front line organisations.

Existing flight data systems

As well as identifying requirements, a parallel exercise was conducted to identify the extent of flight data already available across target aircraft fleets. These fleets were those of civil origins and historic aircraft. The aim was to use existing data systems wherever possible. The logic behind this was simple. The cost of embodying existing systems had already been borne. Moreover, such systems, typically Flight Data Recorder (FDR), Accident Data Recorders (ADR) or Parametric Data Gathering (PDG) systems generally already had some infrastructure in place to ensure system serviceability and the ability to download and store data, although exploitation of the data was often very limited. Moreover, this had the potential to be invisible to the front line, if the existing data system download schedule was adequate for the UVP.

From this study it was identified that the BAE125, BAE146 CCMk2, Sentinel R1, E-3D Sentry, Avenger fleets and the Sea Hawk and Beaver historic aircraft were already fitted with flight data capture systems that, in principle, could be used to support a UVP. The BAE125 programme has been completed and programmes are either underway or planned for the remaining fleets. These will be reported in a later paper.

Minimum introduced system

A significant proportion of the target fleets were identified as having no existing flight data capture systems suitable for a UVP. These fleets included Islander, Defender, Shadow R1, King Air 200/200GT and the Lancaster, Spitfires, Hurricanes, Dakota and Chipmunks from the Battle of Britain Memorial Flight (BBMF) and the Swordfish from the

Royal Navy Historic Flight (RNHF). With the key aims having been identified as adequate validation for minimum cost and low burden on the front line, the minimum system requirements were identified as follows:

- Three-axis accelerometer (NZ primary sense , NX and NY for take-off and landing detection – also allows various mounting options)
- Sample rate for accelerometers minimum 50 sample per second – ideally user definable
- Accelerometer range >+/- 6 g
- Pressure sensor (0-1.5 Bar)
- Temperature range -40 deg C to +50 deg C
- Self contained power supply
- Capacity and power for at least one week's data capture when in service
- Small size (match box ideally)
- Small mass (< 100 g ideally)
- Robust
- Ease of mounting onto aircraft
- Low unit cost
- Likely manufacturer product production for several years
- Availability of test evidence

MINIMUM SYSTEMS SOLUTION – MODULAR SIGNAL RECORDER

Approach

A survey of commercially-available equipment was undertaken with the aim of identifying an off-the-shelf solution to meet the minimum system requirement. Several micro-electro-mechanical-systems (MEMS) based recording accelerometers were selected for initial testing. Of these, the Modular Signal Recorder (MSR), model MSR165®, manufactured by MSR Electronics GmbH [11] most closely matched the requirements.

Three-axis accelerometer

The MSR165 unit is fitted with a 3-axis Analog Devices ADXL345 micro-electro-mechanical systems (MEMS) accelerometer. It is designed to measure static accelerations (e.g. tilt) as well as dynamic accelerations (vibration or shock) and has a range of +/- 15 g, in each of the 3-axes. The analogue-to-digital conversion uses 13-bit resolution. Although the data range was greater than was necessary for the applications being considered within this programme (+/-6 g would probably be sufficient), the resolution of the unit, governed by the least significant bit (LSB) value of 4 mg (0.004 g), was still considered sufficient for this application. The MSR165 accelerometer has an accuracy specification of +/-0.15 g (at 25 deg C). This value was comparable with tolerances seen on fatigue load meters under test, reported by Sturgeon and Mitchell [12]. In addition, the accuracy specification represents a 1 % error at full scale and this magnitude of error is consistent with experimental evaluation of a range of commercial MEMS accelerometers tested by Acar and Shkel [13].

Data sample rate

The MSR165 units can be programmes to record accelerations at a range of sample rates from 25 samples per second to 1600 samples per second. For the applications reported here, the MSR was programmed to capture acceleration data at 50 samples per second initially and 100 samples per second for later applications, once the power consumption was better understood. This sample rate was considered consistent with capturing the peak and trough values for events up to around 7 Hz (anticipated maximum gust frequencies) [12]. The remaining data (typically pressure, temperature and relative humidity) were sampled once per second.

Operating temperature range

The operating temperature range for the MSR165 unit was specified as -20 deg C to +65 deg C. The hot-end limit was considered adequate for planned use. However, the cold-end limit was considered inadequate for some potential applications. Therefore, cold-soak testing to a DO178 Category C2 schedule [14] confirmed that a well-charged MSR165 performed as expected when subject to a -55 deg C environment for over 3 hours. The small size of the MSR (Figure 1) was attractive as would allow a great deal of flexibility as to where the unit could be located and thereby allowing the unit to be protected from extremes of temperature if necessary.

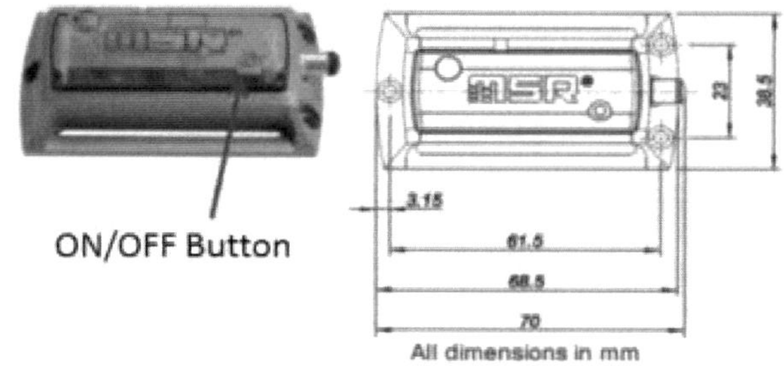

Figure 1. MSR165.

Time alignment

The MSR units are fitted with a real time clock (RTC), which is updated when the units are connected to a computer, with the MSR software installed. The RTC drift was tested, using several MSR units, over a two month period and was found to be less than two seconds per day. This was considered acceptable for these applications. However, further testing identified that the data streams from the accelerometer and the other sensors in the MSR (pressure, temperature and relative humidity) were not aligned using a single clock. This was significant when recording long flights where the misalignment could be several seconds and made identification and segregation of roller landings data very challenging. Therefore, MSR Electronics GmbH agreed to modify the workings within the MSR unit to use a common time stamp.

Power supply

The MSR165 is powered by an Advanced Electronics Energy AE471736P8H, 3.7 V, 900 mAh, re-chargeable lithium-polymer battery. The battery is a small button-type unit (similar to a watch battery) of approximately 5 mm diameter and with a mass of 5 g. The extensive test regime, including full battery abuse testing, for the 900 mAh battery, provided by MSR Electronics GmbH, was considered suitable to indicated fitness for purpose.

Protection

The MSR165, with SD card unit included has a mass of less than 70 g and is 39 mm x 23 mm x 72 mm. The basic MSR units have an ingress protection rating (IP) of 67. This rating indicates that the unit is dust tight and protected against heavy seas or powerful jets of water. With the SD card recorder included, the IP rating reduces to 60. This means that the unit was still considered protected from dust but not from moisture. In reality this has not proved to be a significant impediment in applications to date as the SD card access has been protected with appropriate adhesive tape (also to prevent any loose-article hazard should the SD card be ejected). The MSR165 units are compliant with European Union (EU) directives for Restriction on Hazardous Substances (RoHS) and Waste Electrical and Electronic Equipment (WEEE).

Robustness

A range of realistic tests were conducted to identify whether the MSR was suitably robust for the intended environment and possible accidental damage. These included rapid changes in atmospheric pressure, dropping from realistic heights onto concrete, to ensure no damage to the casing or potential loose articles. The MSR was found to be extremely robust. This was illustrated dramatically when the units were used for several non-aircraft programmes. One MSR suffered a direct hit from a 7.62 mm round. Although the unit was a little misshapen, it continued to function. Another unit spend several days in the sea and once rinsed out with clear water, again functioned as expected.

Programming

The MSR units were programmed to record acceleration data when normal acceleration (NZ) limits exceeded pre-defined trigger levels (typically NZ > 1.15 g or NZ < 0.90 g (or 0.80 or 0.85 g for tail-dragging aircraft)). In this mode the MSR is activated by holding in the start button (Figure 1) for 5 seconds and then fitting the unit to the aircraft. The NZ levels apparent during engine start are then sufficient to initiate recording, which stops after engine shut down. The remaining parameters (generally pressure, temperature and relative humidity) were captured at one sample per second continuously from activation of the MSR. This 'fit and forget' approach for a recording period (typically a month) significantly reduces the burden on operators and the triggering system increases the battery charge life.

Data recording

Data are recorded onto an internal SD card (4GB). Laboratory testing indicated that at typical sample rates for these programmes (accelerations sampled at 50 samples per second and pressure, temperature and relative humidity at one sample per second) a fully charged battery (approximately 4 V) would support continuous recording for approximately 11 days. Similar tests were conducted with a MSR unit which was tested using a more representative duty cycle (typically 3 hours recording per day). In this case the MSR recorded over 200 hours of data before the unit shut down.

Although no account was taken of the effect of temperature fluctuation on the battery efficiency, it was considered that for high utilisation aircraft limits of 75 flying hours and one month recording would be appropriate initially. For the low-utilisation aircraft, a 25 flying hour / one month replacement period was recommended. The shut-down protocol was also tested and it was confirmed that no data losses occurred when the charge value reached 3 V and the MSR shut down.

Electro-magnetic compatibility

MSR Electronics GmbH was requested to provide evidence of electro-magnetic compatibility (EMC) for the MSR units. The test evidence provided confirmed that the unit met the European standards for immunity for industrial, residential, commercial and light industrial applications. However, these standards do not cover the same range as the Def Stan 59-411 [15]. Although it was considered that the extremely low power consumption of the unit, should be sufficient to support the fit of the units in most applications, further EMC testing to Def Stan 59-411 was undertaken by QinetiQ Farnborough [16]. Testing confirmed that the MSR passed the relevant radiation emissions and static discharge tests but failed the radiated susceptibility test (E field) in the 2 GHz - 4 GHz range at 200 V/m. This was expected for a piece of commercial electronics not designed for an EMC hardened environment. The failure mode identified was that the MSR stopped recording data onto the SD card. Reprogramming of the unit was seen to reset the MSR with no apparent long-term effect on the performance. The MSR was retested inside a commercially available die-cast aluminium box (Figure 2) and sufficient protection was provided using this approach to pass the test. The Def Stan 59-411 regime is a particularly severe element of the EMC testing and this failure has no safety implications for any application; moreover no significant EMC issues have been reported in service to date. Hence, the approach taken has been if an EMC susceptibility issue is identified in service for any application, the MSR is mounted in a die-cast aluminium box.

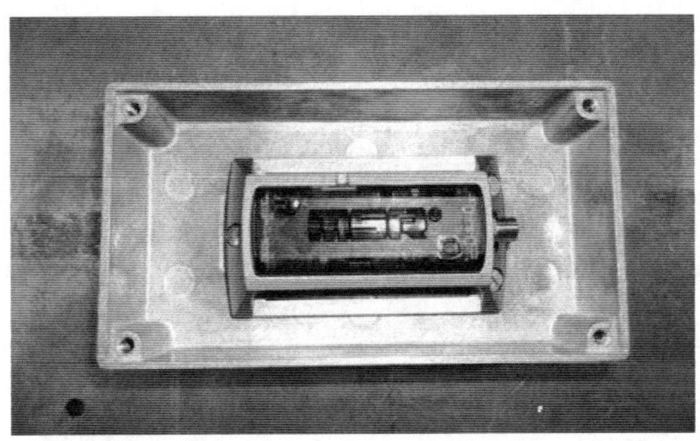

Figure 2. EMC testing in die-cast box (opened for illustration).

MSR in-service issues and vulnerabilities

Two in-service issues with the MSR were identified during these programmes. Firstly, despite face-to-face briefings and providing instruction to front-line operators, several programmes had initial issues with the MSR units not being turned on correctly. This result was no data being captured. As a result the programming method for the MSR was modified to remove any potential confusion. In the revised programming, the operator turns on the unit and shakes it. If the red LED flashes, the recording sequence has been started successfully.

Secondly, several occurrences of the SD card being ejected from the MSR were identified. No loose article hazard occurred as the SD card was retained in location by the insulating tape in both cases. Both cases were most likely handling issues with the MSR being held with a thumb on the SD Card. Revised instructions and warnings (Figure 3) were published in all instructions and this risk was included in all face-to-face briefings.

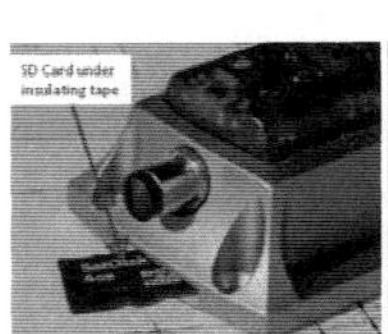

WARNING: IF PRESSURE IS APPLIED TO THE SD CARD LOCATED UNDERNEATH THE INSULATING TAPE BELOW THE CHARGING POINT OF THE MSR, THIS MAY CAUSE THE SD CARD TO DISENGAGE AND DATA RECORDING WILL CEASE. THIS MAY NOT BE VISIBLE TO THE USER. THEREFORE, HANDLING THE MSR UNIT IN THE VICINITY OF THE SD CARD ACCESS SHOULD BE AVOIDED.

Figure 3. Handling warning.

Flight trials

Flight trials were conducted during 2010 and early 2011, primarily on the Islander fleet, to confirm the performance of the MSR.

MSR calibration

From the outset of the programme static calibration have been undertaken for each MSR before dispatch and after receipt into service. All three axes of the accelerometer are tested at +1 g and -1 g with the MSR mounted in a calibration fixture. Also comparative checks are undertaken for the pressure, temperature and relative humidity sensors. The calibration data have been collated and monitored against tight limits over a 4-year period. To date, no MSR units have failed in service or have seen any significant deviation in calibration offset values.

Clearance approach

The approach taken in gaining clearance to fit the MSR to a range of fleets has varied depending upon the extant clearance approach for that fleet. For example, for fleets operated within a civilian authorisation procedure, generally the approach has been to develop a supplementary type certificate based clearance, with the appropriate authority. For aircraft governed by a traditional military authorisation system, a military modification approach has generally been applied.

AIRCRAFT DATA ANALYSIS AND MONITORING (ADAM) SYSTEM

Aim

The aim from the outset of this programme was to make as much of the Aircraft Data Analysis and Monitoring (ADAM) software as common as was possible. Common requirements were identified, based upon the likely technical requirements described earlier in this paper. Common modules were developed to meet these requirements. This approach did place a significant burden on software development initially, using a rigorous method of user requirements, system requirements and software requirements, with detailed verification and validation procedures, within a high-control software environment. However, the benefit of this approach have been realised in the longer term. As the programme has expanded each software release has been based heavily on pre-validated and pre-verified software modules. The generic ADAM architecture is illustrated in Figure 4. The code was written using the Mathworks Matlab® software environment. This was to ensure the longevity of the software and the ease of deployment.

Software design architecture

A brief explanation of the analysis package is provided in the following sections.

Analysis. The analysis program acts as the controller for the core code in the ADAM system, calling the relevant script and function files in sequential order. All modules in the software are subject to individual configuration control and each module called, including the analysis code, writes its respective configuration and date to the info structured array. The analysis code also reports the completion of the programme.

```
ADAM_analysis
    -   ADAM_analysis_log
    -   ADAM_directories
    -   ADAM_load_flight_data
    -   ADAM_flight_record_data
    -   ADAM_calculated_parameters
            o   ADAM_function_filter_lowPass
    -   ADAM_statistics
    -   ADAM_limits
    -   ADAM_anomaly_detector
    -   ADAM_turning_points_time
            o   ADAM_function_turning_points_time
    -   ADAM_rmp3_time
            o   ADAM_function_rmp3_time
    -   ADAM_fatigue_meter
            o   ADAM_function_fatigue_meter
    -   ADAM_plots
    -   ADAM_output_data
    -   ADAM_backup
```

Figure 4. Generic ADAM code architecture.

Analysis_log. The code analysis_log is used to open a text log of all events during the running of the programme, including any screen entries by the user. All calling scripts are identified in this file, along with all outputs to the analysis screen, including any error codes raised. The analysis log also contains details of the sortie date, number and the flight record data entered. At the end of the programme run, the analysis log is automatically saved in the sortie directory as the log.txt file.

Directories. The directories code is used to create automatically the directories within the sortie directory.

Load_flight_data. The load_flight_data program is used to extract the relevant data from the source file. Data are extracted from the msr.dat file into the Matlab binary format in preparation for further analysis. Full consistency checks are included within the extraction routine.

Flight_data. The flight_data code provides an interactive command-line data input facility for information captured from the flight record sheets (MOD Form 724/725 and F003 for civil aircraft) and provides initial visualisation of the data sets. The information required varies largely depending on what is recorded on the flight records and what information is required to be added to the data set. A typical example might include flight date, flight time, landings, sortie profile code and mass data. Each entry has an example for the user and full validation rules are applied. If an initial input is outside of the validation range for the entry, the user is provided with a warning and requested to re-enter the value. The program will accept the second value entered, irrespective of whether it falls within the validation range or not. This is to prevent the program terminating; however, the invalid entry is reported in the log.txt file. The data entered are written into the info structured array.

Calculated_parameters. The calculated_parameters program is used to apply corrections, unit changes, apply filters and offsets and to calculate any additional parameters from the captured flight data. All changes are documented in the info structured array. A low-pass finite impulse response filter, using a Hanning window was developed, based upon the method outlined by Weeks [17]. The filter is used as an anti-aliasing filter and to isolate different frequency content within the time history data. Most importantly, the filter works in a pass and return mode to ensure that the frequency content of the data remains aligned in the time history.

Statistics. The statistics program is used to generate statistical data for each of the ADAM data channels. These data are primarily generated for use by the anomaly_detector and for identifying and locating maximum values in data sets (e.g. NZ). All data are written to the info structure array.

Limits. The limits program is used to generate the limit values for the anomaly_detector program. The majority of the parameters have a maximum value, minimum value and a rate value. The maximum and minimum values are based upon knowledge of the measurand and the likely performance of the aircraft. The limit values are written to the info structure array. The rate value is the maximum difference expected between two adjacent data points in the time history for each applicable parameter. Again these values have been identified by considering the physics of the measurand and the sample rate of the data.

Anomaly_detector. The anomaly_detector program extracts the statistics for each channel from the info.statistics region of the structure array and compares the values with the data in the info.limits fields. Transgressions of these limits are reported and are written to the info.anomaly fields in the array.

Turning_points_time. The turning_points_time program and its associated function are used to extract turning points (local maxima and minima) from the normal acceleration time history files.

Rmp3_time. The rmp3_time program and its associated function are used to extract cycle files (described by maximum and minimum values from the turning point files. This program was coded from Downing and Socie's three-points algorithm [18], detailed in [19].

Fatigue_meter. The fatigue_meter program undertakes a software emulation of the counting mechanism in a fatigue meter accelerometer. This program can be amended to replicates the function of any appropriate fatigue meter. The architecture of this program is constant but the lock and release levels and the number or windows vary between platforms and hence the fatigue meter algorithm is subject to validation checks for each release. The counts in each window are written to the info.fatigue_meter field in the structure array. In addition, a plot is produced for each fatigue meter window, illustrating the lock and release points in the time history. An example is reproduced in Figure 5.

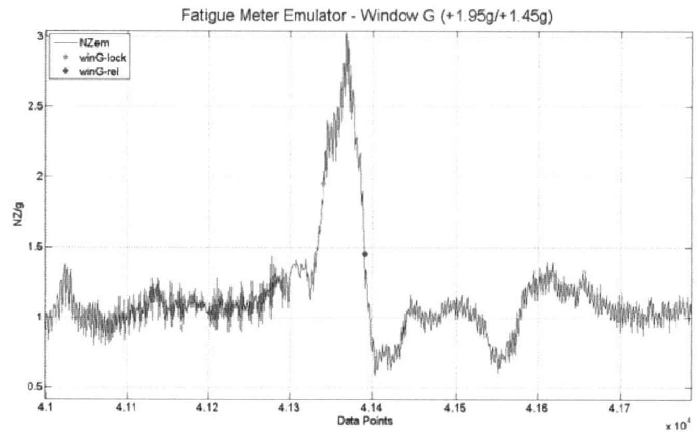

Figure 5. Fatigue meter emulation illustration.

Plots. The plots program is used to generate a range of plots of the data extracted and analysed within the software. The format is similar across the aircraft types but does vary with data sets.

Data_output. The data_output program is used to generate an output from fields in the info structure array into an MS Excel file, with accompanying titles.

Backup. The backup performs and automatic backup of the data files, with a handshake confirmation that the backup has been successful.

IN-SERVICE USAGE VALIDATION PROGRAMMES

Programmes conducted

To date, the MSR has been used to capture structural usage data from a range of aircraft types including: Islander and Defender, Shadow, Swordfish, Beaver, Lancaster, Spitfire, Hurricane, Dakota, Chipmunk and Hunter and further programmes are planned. Example time history outputs for an individual flight are illustrated in Figures 6 and 7. The example in Figure 6 is from a Hunter ground attack aircraft, whereas Figure 7 is more typical of the data seen in a civil-type aircraft fleet. Brief details of these programmes are provided in the following sections:

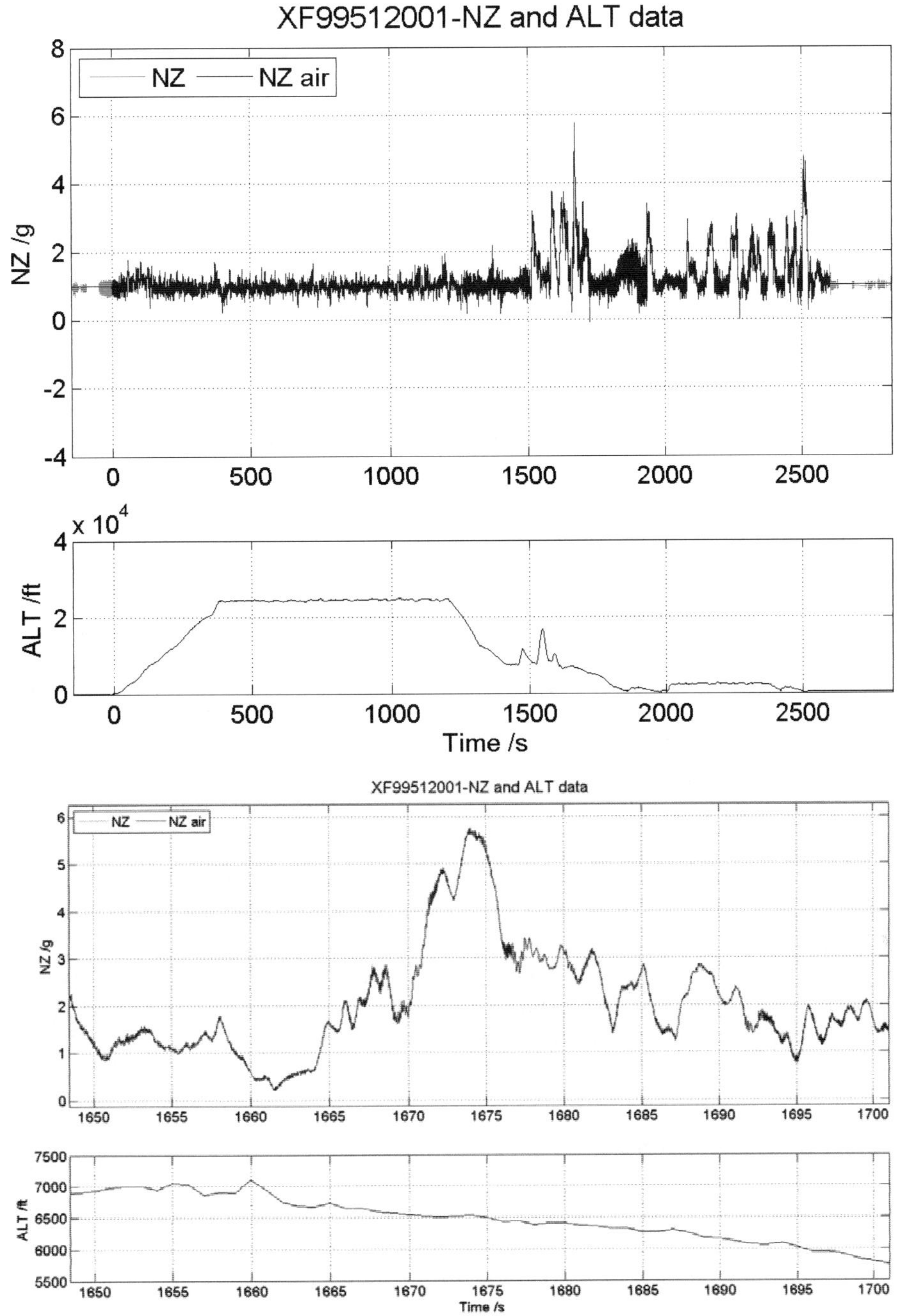

Figure 6. Example Hunter ground attack aircraft time history plot and zoom of peak NZ.

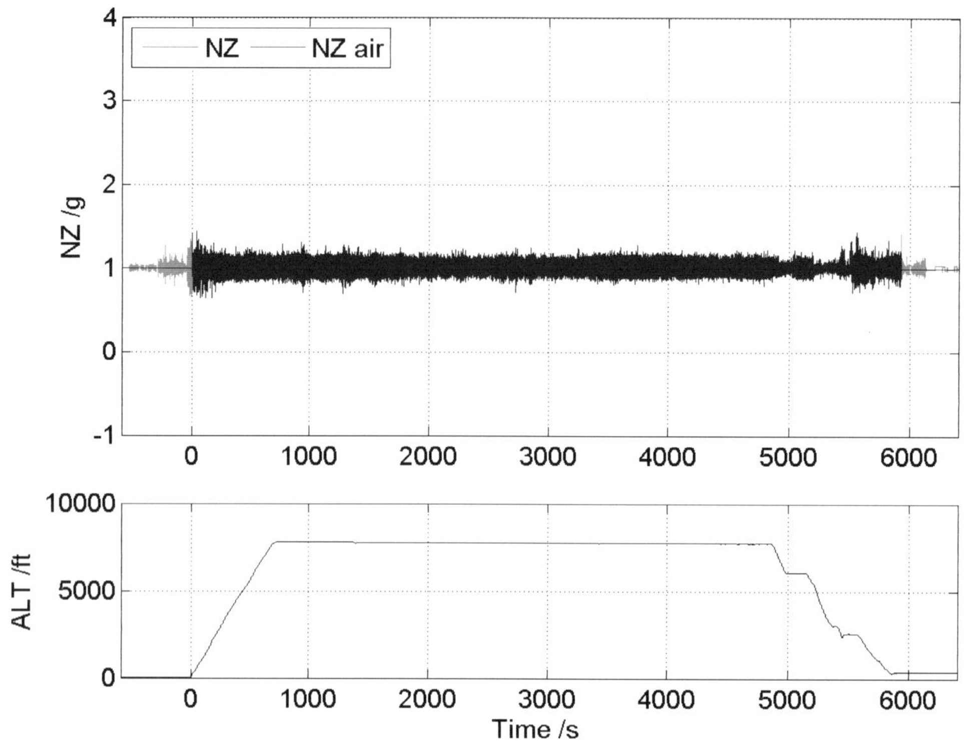

Figure 7. Typical time history plot for civil-type aircraft.

Islander and Defender
Data capture began for the Islander and Defender programme in later 2011. To date, over 4000 flying hours of data have been captured and analysed from across the fleet, in a collaborative programme with QinetiQ Ltd. This is reported in detail by Kamiya and Duffield [20].

Shadow
The initial data capture programme for the Intelligence, Surveillance, Target Acquisition and Reconnaissance (ISTAR) Shadow R1 fleet (Figure 8) was conducted during 2014 and over 1200 flying hours of data were captured. The primary output from the analysis programme was a NZ spectrum for each SPC and a combine spectrum for the fleet, alongside time-at-flight-level data. Analysed usage data were provided to the MOD Project Team in March 2014, in a format agreed with the Design Organisation for the aircraft type. The output data are currently been assessed by the Design Organisation against design usage assumptions. A further data capture programme is planned for the Shadow fleet in the next year.

Figure 8. Shadow R1 aircraft (MOD Copyright).

BBMF fleet
The initial data capture programme for the Lancaster aircraft on the BBMF was conducted in 2011. Approximately 100 flying hours were captured during the first year. The data were invaluable in understanding the effect of the gust

environment, superimposed on top of the benign manoeuvres flown by the Lancaster. The Lancaster was already fitted with a fatigue meter. However, the exceedance counts produced by the fatigue meter have no time base and therefore cannot be visualised in time. Therefore by presenting the NZ time history and by separating out the low-frequency (1 Hz) content from the signal likely to have been recorded by a fatigue meter (11 Hz) in the data, it was possible to separate out these effects.

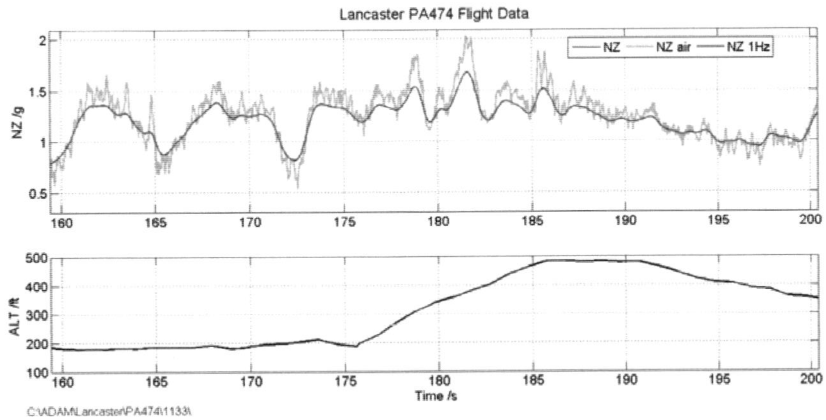

Figure 9. Lancaster data with separated frequency content.

Since the initial programme, MSR units have been fitted to all aircraft in the BBMF since 2013 and have already captured several hundreds of hours of data. These data will be used to baseline the usage of these aircraft in the BBMF.

Swordfish

For the Swordfish torpedo bomber (Figure 10), operated by the RNHF, the release-to-service (RTS) NZ limit for the aircraft was set to 2 g. The design of the aircraft was to at least 6 g and hence the usage was considered benign compared with design and original in-service usage. However, the absence of a monitoring system prevented any evidence of benign usage being presented. The MSR data, captured over two full seasons of flying (nearly 100 flying hours) was used to confirm benign operations. There were no counts recorded in the 2.5 g window of the fatigue meter emulator over the entire period.

Figure 10. Swordfish aircraft (MOD Copyright).

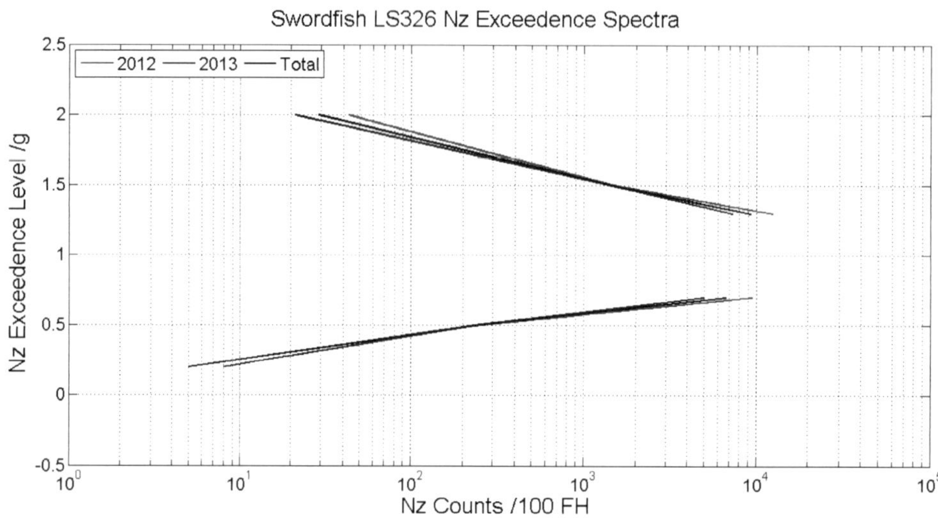

Figure 11. Swordfish NZ spectrum.

CONCLUSION

The aim of this programme was to develop a low-cost, low-operator-burden, structural usage validation system for aircraft without existing data capture facilities. To date, this programme, based upon development of the commercially available MSR data logger and the ADAM software, has proven successful in meeting this aim and further applications are planned.

REFERENCES

[1] Schutz, W., (1996), A History of Fatigue, *Engineering Fracture Mechanics*, Vol 54, No, Pp 263-300.

[2] Davies, T. C., (1991) Structural Integrity in the Royal Air Force, Keynote Address to Fatigue Management, AGARD Conference Proceedings CP 506, 72nd Meeting of the AGARD Structures and Materials Panel, Bath, UK, 29 April – 3 May 1991.

[3] Adams, P. D. and Cansdale, R., (1992), Analysis of UK Military Accidents 1970-1990, UK Defence Research Agency Technical Memorandum 1206.

[4] UK Ministry of Defence, (2007), Defence Standard 00-970 Section 3, Chapter 3, Issue 5.

[5] UK Ministry of Defence, (2010), Military Aviation Authority Regulatory Article 5720 – Management of Structural Integrity

[6] Reed, S. C. and Holford, D. M., (2007), Guidance for Aircraft Operational Loads Measurement Programmes, UK Military Aircraft Structures Advisory Group, MASAAG Paper 109.

[7] Reed, S. C., Holford, D. M., Gelder, M. S. and Roberts, S., (2001), Operational Loads Measurement of the Hawk Aircraft in RAF Service, In: *Proceedings of the 21st Symposium of the International Committee on Aeronautical Fatigue*, ISBN 2854285743, Rouchon, J. (Ed.), Toulouse, France.

[8] Perrett, B., (2015), Flight Condition Recognition and Independent Flight Severity Monitoring for the Management of Helicopter Fleets, In: *Proceedings of the 28st Symposium of the International Committee on Aeronautical Fatigue*, Siljander, A. (Ed.), Helsinki, Finland.

[9] Reed, S. C. and Egli, W., (2012) Measuring aircraft structural usage using the MSR165 data logger, *Elektronik Messen und Testen, April 2012.*

[10] Reed, S. C., (2011) Operational Data Recording, In: *Defence IQ Ageing Aircraft 2011 Conference Proceedings*, London.

[11] MSR165 – Data logger for shock and vibration, data sheet MSR Electronic GmbH, http://www.msr.ch

[12] Sturgeon, J. R. and Mitchell, C. G. B., (1968), R.A.E. Fatigue Meters and Counting Accelerometers, Royal Aircraft Establishment Technical Memorandum Structures 718, March 1968.

[13] Acar, C. and Shkel, A. M., (2003), Experimental evaluation and comparative analysis of commercial variable-capacitance MEMS accelerometers, *Journal of Micromechanics and Microengineering 13*, Pp 634–645 PII.

[14] RCTA/DO-160G (2010) Environmental Conditions and Test Procedures for Airborne Equipment, Chapter 4 (Temperature and Altitude) Category C2.

[15] UK Ministry of Defence (2007), Defence Standard 59-411 Electromagnetic Compatibility, Parts 1-5, Issue 1.

[16] QinetiQ Ltd, (2012), Certificate of Test, Def Stan 59-411, QientiQ/MS/AD/TC 1200010, Issue 1, 15 February 2012.

[17] Weeks, M., (2006) *Digital Signal Processing*, Infinity Science Press Electrical Engineering Series, ISBN: 0-9778582-0-0.

[18] Downing, S. D., and Socie, D. F., (1982) Simple Rainflow Counting Algorithms, In: *International Journal of Fatigue,* January 1982, Pp31-40.

[19] ASTM E 1049-85 (Reapproved 1997) Standard Practices for Cycle Counting in Fatigue Analysis - Simplified Method.

[20] Kamiya, R., and Duffield, M., (2015), The Use of Micro Electro-Mechanical Systems to characterise the Structural Usage of a Multi-Role Military Aircraft, In: *Proceedings of the 28st Symposium of the International Committee on Aeronautical Fatigue*, Siljander, A. (Ed.), Helsinki, Finland.

The Fatigue Damage Spectrum: Explained

Mr. Rob Plaskitt[1], Dr. Andrew Halfpenny[2]

[1]*UK Application Engineering, HBM-nCode, United Kingdom*
[2]*Chief Technologist, HBM-nCode, United Kingdom*

Abstract: The fatigue damage spectrum (FDS) is a very useful analysis technique to quantitatively assess the potential for a fatigue failure mode in components or systems operating in a mechanical vibration environment. The mathematical theory to calculate the FDS is presented in AECTP-240 Environmental Test Methods (Edition 4) Leaflet 2410, and is referenced by standards NATO STANAG 4370, UK DEF STAN 00-35 and USA MIL-STD-810G. However, using only the equations shown in Leaflet 2410, it is very difficult to reproduce the FDS (and MRS) examples included to illustrate the equations. To reproduce these examples requires the previous edition of the standard, and following the references back to the withdrawn French standard GAM-EG-13 and the Lalanne books from which it is based. This paper explains the FDS and demonstrates its calculation using the examples from Leaflet 2410.

INTRODUCTION

The fatigue damage spectrum (FDS) is a very useful analysis technique to quantitatively assess the potential for a fatigue failure mode in components or systems operating in a mechanical vibration environment.

This paper considers the environmental standards for mechanical vibration of defence materiel;

- UK National Defence Standard 00-35 Issue 4 (2006) [1], referred to as Def Stan 00-35
- International Defence Standard NATO STANAG 4370 (2009) [2], and specifically to AECTP-240.
- USA National Defence Standard Mil-Std-810G (2008) [3], referred to as Mil Std 810

In addition to these three current standards, the following withdrawn standards are also referenced;

- France National Defence Standard GAM-EG-13 (2002) [4], referred to as GAM-EG-13 (this France standard was the first to propose the use of FDS to quantify the fatigue damage potential of a mechanical vibration environment)
- International Defence Standard NATO STANAG 4370 (2006) [5], and specifically to Leaflet 2410/1.

When this paper references these standards, it gives the location within the standard because each standard is several hundred pages long.

The use of FDS is explicit within Def Stan 00-35 and AECTP-240, which define FDS as;

- Def Stan 00-35 [1], Chapter 1 Glossary: "The fatigue damage spectrum is a curve that represents the variation of damage, as defined by Miners Law, as a function of the natural frequency of the system with one degree of freedom, for a given damping factor."
- AECTP-240 [2], Leaflet 2410/1 Annex C: "The fatigue damage spectrum is a spectral representation of a fatigue damage index as a function of any system's natural frequency. This spectrum is computed directly from the power spectral density (PSD) function representing a test situation or a field environment, and provides a relative fatigue damage estimate based on acceleration level and exposure time.

The use of FDS is considered to be implicit within Mil Std 810 [3], Method 514.6 Vibration, Chapter 2 Tailoring Guidance. Though it does not refer to FDS by name, it uses the phrase 'fatigue damage potential across the frequency range of importance'. This is considered to be a means of defining FDS as a curve that represents the variation of damage across a frequency range without specifying that Miners Law must be used for cumulative fatigue damage and without specifying that the system must be represented by one degree of freedom with damping.

To understand the derivation and mathematics of FDS it is necessary to understand its origins in the maximum response spectrum (MRS). Again MRS is explicitly referred to in Def Stan 00-35 and AECTP-240, but is not in Mil Std 810. The MRS is defined exactly the same in Def Stan 00-35 [1], Chapter 1 Glossary and AECTP-240 [2], Leaflet 2410/1 Annex E as;

- When a vibration excitation is applied to a mechanical system with one degree of freedom, the maximum value of the response of this system for a deterministic signal, or the probability of a maximum value for a random signal, can be calculated. This value is called the 'maximum' or the 'extreme' value. The maximum response spectrum is the curve that represents variations of the 'maximum' response value as a function of the natural frequency of the system with one degree of freedom for a given damping factor.

This paper uses the simple examples included in AECTP-240 [2], Leaflet 2410/1 Annex E to demonstrate the calculation of MRS and FDS. It will be shown that some examples can be reproduced using only the information provided in the AECTP-240 annex, some are missing essential information, and some equations are beyond simple hand calculation verification. In these cases any missing information is explained and alternative approximate equations are shown.

BACKGROUND TO MRS AND FDS

The origins of MRS and FDS can be traced back to the shock response spectrum (SRS) proposed by Biot in the 1930's. Further development of Biot's approach was conducted by Lalanne and the French Ministry of Defense in preparation of the military design standard GAM EG-13 [4] in the 1980's. Lalanne [6] proposed an analogous fatigue damage spectrum (FDS), giving a relationship between potential fatigue damage and natural frequency.

Halfpenny and Walton [7] give a very good introduction and review of background theory for SRS and FDS calculated from measured accelerations and directly from a vibration test specification. They use the terms shock response spectrum (SRS) calculated by time domain convolution of a deterministic measured acceleration and extreme response spectrum (ERS) calculated by closed form equations using statistical probability methods from a vibration test specification.

The above definition of MRS from the Def Stan 00-35 and AECTP-240 standards does not make a distinction between whether the MRS has been calculated directly from a measured acceleration or from a vibration test specification. This definition of MRS states "the maximum value of the response of this system for a deterministic signal" (for example from a measured acceleration, a shock waveform or a sinusoidal vibration test) "or the probability of a maximum value for a random signal" (for example from a random vibration test).

However it has been calculated, the MRS represents a plot of the maximum response value (usually acceleration or displacement) vs. potential natural frequency of a one degree of freedom system with a given damping factor. Such maximum responses typically result from extreme shock events such as severe landing, impact, weapons discharge or nearby explosions. An MRS is used to determine the maximum response seen during a flight event or a vibration test. The safety margin of the test can be determined by comparing the test MRS with the flight MRS. It is insufficient to simply record the highest static acceleration level because this does not account for the frequency of the vibration. Dynamic systems are more sensitive to certain frequencies than others, the so called 'resonant' (or natural) frequencies.

The FDS is calculated in the same way as the MRS but rather than simply finding the maximum acceleration or displacement response, the filtered displacement response is Rainflow cycle counted and the 'potential' fatigue damage obtained using a stress-life fatigue approach using a Basquin power law relationship. The important assumption is that fatigue damage is proportional to excessive strain energy, and strain energy in a vibrating component is proportional to displacement rather than acceleration. Therefore the damaging effect of acceleration is seen to reduce with the power of the frequency, i.e. high frequencies become significantly less damaging than lower frequencies provided a natural frequency is not excited. This can be seen in the FDS examples presented later in this paper.

The calculation of MRS and FDS is easiest to understand from the perspective of a measured acceleration. Figure 1 shows a schematic of the analysis process from acceleration, through a single degree of freedom filtered displacement response whose peak amplitude gives the MRS value for that filter frequency, Rainflow cycle counting and fatigue analysis to give the FDS value for that filter frequency. Figure 2 shows a very similar schematic for the same process starting from an acceleration power spectral density (PSD) typically used to define a random vibration test.

Where the MRS represents the maximum response for a range of potential natural frequencies, the FDS represents the potential cumulative fatigue damage at each potential natural frequency. This difference is important because it means that MRS represents the extreme envelope of maximum responses independent of whether there are one or many occurrences, whilst the FDS represents the accumulation of many occurrences.

For example if a whole flight contains an extreme but short transient manoeuvre, the MRS of the whole flight will be identical to the MRS of short transient manoeuvre. (This assumes that the short transient manoeuvre contains the maximum responses through the MRS frequency range.) However the FDS of the whole flight will have higher magnitude than the FDS of the short transient manoeuvre because it includes a contribution from all other manoeuvres within the flight.

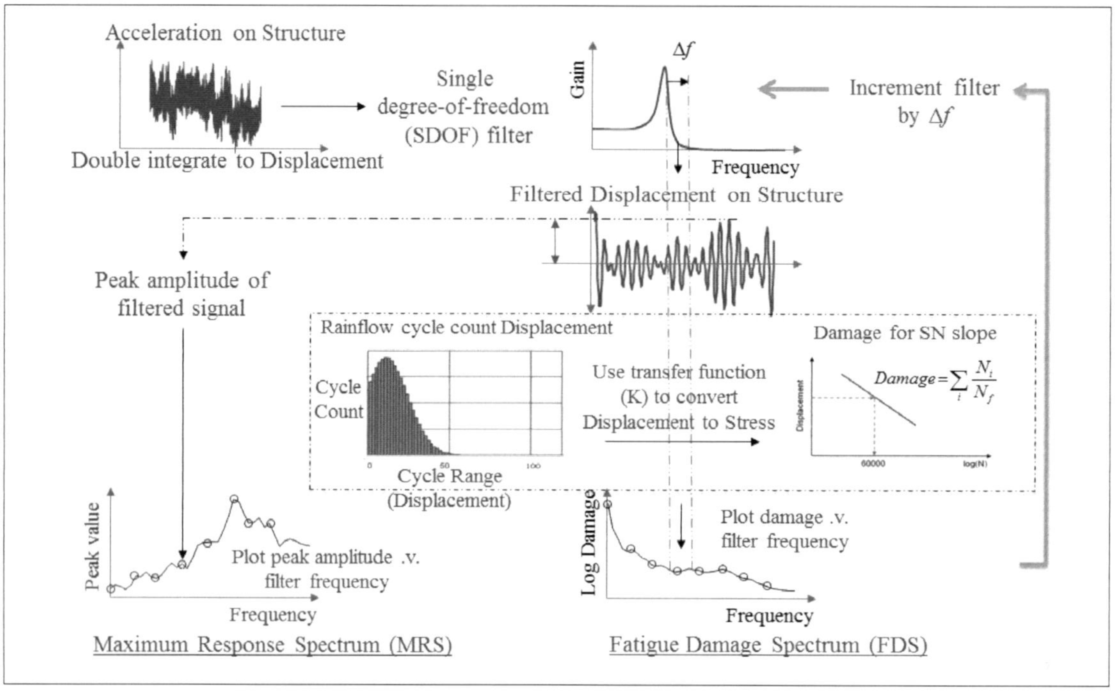

Figure 1. A schematic of the process to calculate MRS and FDS from a measured acceleration.

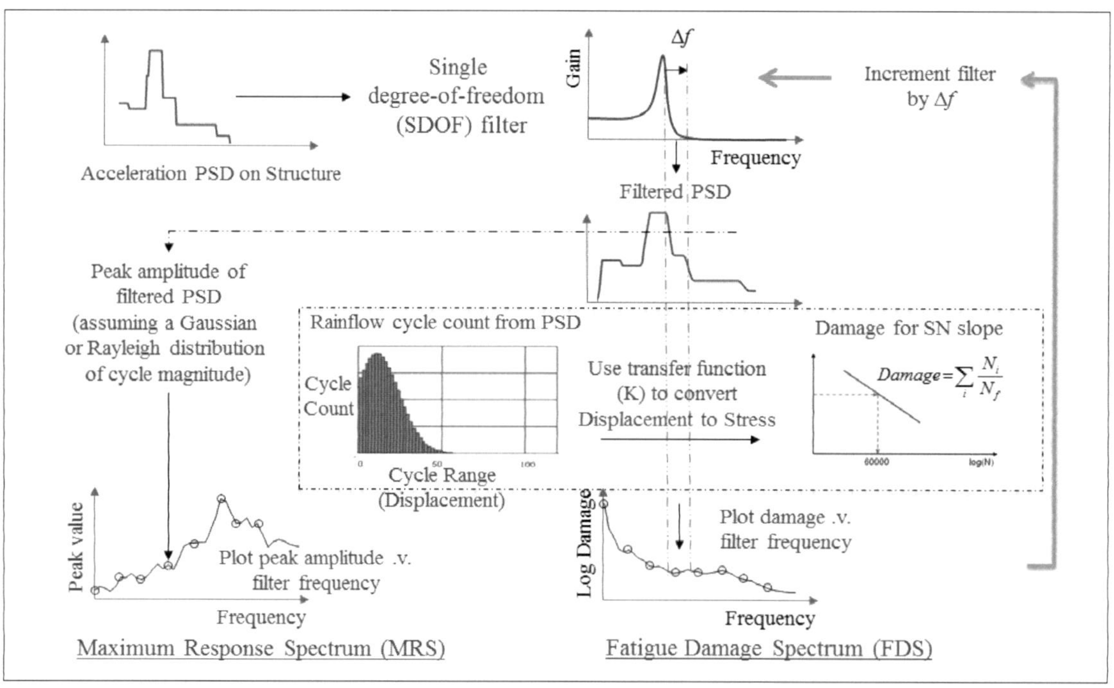

Figure 2. A schematic of the process to calculate MRS and FDS from an acceleration PSD.

VIBRATION TESTS

Vibration tests are performed for a number of purposes, for example;

- functional tests to demonstrate correct operation while exposed to worst case vibration
- endurance tests (usually accelerated) to produce the same fatigue damage as the expected service life
- resonance search tests to identify resonance frequencies

Vibration tests are usually performed by electro-dynamic or electro-hydraulic shakers, and may be single axis or multi-axis. Their vibration controllers generate the vibration environment required by the test specification.

Table I shows some examples of vibration environments. These can be broadly grouped into sinusoidal vibration (a and b), random vibration (c), mixed mode or composite tests (d, e, f and g), and transients (h). The vibration magnitudes for a test are determined by analysis of measured acceleration data and test tailoring, and by the purpose of the test.

All three standards provide default vibration magnitudes for various air/land/sea vehicles, their intended location within the vehicle (for example, near to engines or propellers) and for various vehicle operating environments.

Table I. Vibration environments.

(a) Sine dwell (alt. 'dwell test', 'vibration dwell', 'resonance dwell')	A sinusoidal vibration at a constant frequency, defined by its frequency (Hz), magnitude (acceleration: g, m/s^2, or displacement: m, mm, inDA [inches Double Amplitude]) for a test time duration (hours) or number of cycles.	Tests often include multiple dwell frequencies at known forcing frequencies (from propellers or rotor blades) or resonance frequencies. More common in older vibration test standards.
(b) Sine sweep (alt. 'cycling test', 'sinusoidal cycling')	A sinusoidal vibration where the frequency sweeps through a range, defined by its frequency sweep range (Hz), magnitude (see 'a') and sweep rate for a test time duration or number of sweeps.	In addition to its use in functional and endurance tests, sine sweeps are very commonly used with lower level magnitudes for resonance search tests, to identify resonance frequencies for sine dwell tests.
(c) Broadband random (alt. 'PSD', 'Random PSD'	A broadband random vibration, defined by the magnitude of a power spectral density (g^2/Hz) through a frequency range, for a test time duration.	Most modern vibration test standards define a random vibration environment, and are often supplemented by static or sweeping narrow bands and sine tones.
(d) Narrow band on random (alt. 'multiple narrow bands on broadband random')	A broadband random vibration background (defined as 'c'), with one or more narrow bands superimposed (defined as 'c') and test time duration.	Narrow bands are included where there are known fixed forcing frequencies, from propellers or rotor blades, usually operating at fixed rotational speeds.
(e) Sine on random (alt. 'multiple sine dwell on broadband random')	A broadband random vibration background (defined as 'c'), with one or more sine tones superimposed (defined as 'a') for a test time duration.	Sine tones are included where there are known fixed forcing frequencies, from propellers or rotor blades, usually operating at fixed rotational speeds.
(f) Swept narrow band on random (alt. 'multiple swept narrow bands on broadband random')	A broadband random vibration background (defined as 'c'), with one or more swept narrow bands superimposed (defined as 'c') and test time duration.	Sweeping narrow bands are included where the forcing frequencies are known to sweep through a range of frequencies, for example tracked land vehicles as their ground speed changes.
(g) Swept sine on random (alt. 'multiple swept sines on broadband random')	A broadband random vibration background (defined as 'c'), with one or more sine sweeps superimposed (defined as 'b') and test time duration.	Sweeping sine tones are included where the forcing frequencies are known to sweep through a range of frequencies, for example tracked land vehicles as their ground speed changes.
(h) Transient (alt. 'waveform replication', 'time history replication')	A transient test is defined as a waveform, or time history, of a specific magnitude at a specific time.	They preserve the time and frequency of the original waveform, and are commonly used for shock tests.

The following sections demonstrate the calculation of MRS and FDS using simple examples from the current and previous AECTP [2, 5] for sine dwell, sine sweep and broadband random vibration environments. They reproduce the equation and the vibration environment stated in the AECTP and an MRS curve digitised from the curve included in the AECTP.

Additional equations and comments are added as appropriate to help explain these MRS and FDS calculations. The additional equations are mostly taken from Halfpenny and Walton [7], and their distinction between "approximate" and "accurate" (advanced methods) are maintained. Some of the accurate equations cannot be calculated easily because they require numerical integration. For these cases an alternative, though approximate, closed form equation is provided that is calculable using a spreadsheet program. Specific frequency spot check values are provided to help understand the calculation or to explain differences where necessary.

The approximate equations used are based on several assumptions. The main assumptions are;

- the input PSD is broad-banded tending to white noise – (i.e. the PSD has a fairly flat profile and contains no significant peaks)
- the response is narrow-banded – this assumption is usually assured by having Q >= 10

These approximate equations are adequate for general comparative analyses provided the assumptions are valid. Lalanne [6] and Halfpenny and Walton [7] have further developed these approaches and present numerical solution algorithms offering much greater accuracy without the limitations inherent in these assumptions. They offer the ability to calculate MRS and FDS from a much wider range of vibration environments including the sine on random, swept sine on random, narrow band on random, swept narrow band on random and transients described in Table I. These complex vibration environments do not fulfil these assumptions and presentation of the algorithms to calculate MRS and FDS for these is beyond the scope of this paper. The reader is referred to Halfpenny [8] for more information.

Where an accurate equation is presented that requires numerical integration, the MRS or FDS curve has been calculated using the numerical solution algorithms within the specialist software, nCode GlyphWorks.

CALCULATION OF MAXIMUM RESPONSE SPECTRUM (MRS)

AECTP-240 [2] Leaflet 2410 Annex E "Theoretical Aspects of Maximum Response" presents the mathematics to calculate MRS and includes examples for sine dwell, sine sweep and broadband random vibration environments. These mathematics are similar to those presented in the previous AECTP release [5], and are based on those from GAM-EG-13 [4].

Def Stan 00-35 and Mil Std 810 do not include the mathematics to demonstrate the calculation of MRS.

Sine Dwell

The calculation of MRS for a sine dwell and its resulting MRS curve is shown in Figure 3. A sine dwell test is a constant amplitude sine wave defined by the dwell frequency, its peak acceleration magnitude and its number of cycles or its time duration. In addition to these, the MRS calculation requires either the damping factor Q (or the damping ratio ξ). The equations are simple to use and calculate the MRS for an assumed natural frequency f_0.

Inspection of the curves and spot check values in Figure 3 shows that the AECTP equation and Halfpenny and Walton accurate equation calculate MRS exactly the same. It is noted that the MRS curve shown for this example in the AECTP is different to the values calculated by these equations. The reasons for this difference are not known.

Sine Sweep

The calculation of MRS for a sine sweep and its resulting MRS curve are shown in Figure 4. A sine sweep is considered to be a series of sine dwells at all frequencies through the sweep range. This is shown by the inclusion of sine dwells at 300, 600, 900 and 1200 Hz (dashed black lines) in the plot. The same sine dwell equations are used; requiring the frequency sweep range (a series of dwell frequencies), the sweep acceleration levels (a series of acceleration levels at each dwell) and damping factor Q.

The same equations are used, so the calculated curves are similar as for the sine dwell. Again there are differences with the AECTP digitised curve at the lower (<300 Hz) and upper (>1200 Hz) frequencies.

AECTP-240 [2] Leaflet 2410/1 Annex E
"E2 Sinusoidal Excitation"

$$MRS = \frac{-\ddot{x}(m)}{\left\{\left[1-\left(\frac{f}{f_0}\right)^2\right]^2 + 4\xi^2\left(\frac{f}{f_0}\right)^2\right\}^{\frac{1}{2}}}$$

- Dwell frequency: $f = 500\ Hz$
- Damping factor: $Q = 5$

Accurate
Using Halfpenny and Walton [7 "Equation 1"]

$$Gain_{accel}(f_0) = \frac{1}{\sqrt{\left[1-\left(\frac{f}{f_0}\right)^2\right]^2 + \frac{1}{Q^2}\left(\frac{f}{f_0}\right)^2}}$$

$$MRS(f_0) = Gain_{accel}(f_0) * |\ddot{x}(m)|$$

In addition to the information given in AECTP-240 [2] shown above, the following are also required;

- Dwell acceleration: $\ddot{x}(m) = 1\ m/s^2$
- Damping ratio: $\xi = 1/(2Q)$
- Use the absolute value of the calculated MRS

Spot check MRS (m/s^2) values at frequencies;

Frequency (Hz)	AECTP Equation	Accurate Equation	AECTP Digitised
300	0.553	0.553	~1.3
500	5.000	5.000	~5.0
1000	1.322	1.322	~2.0

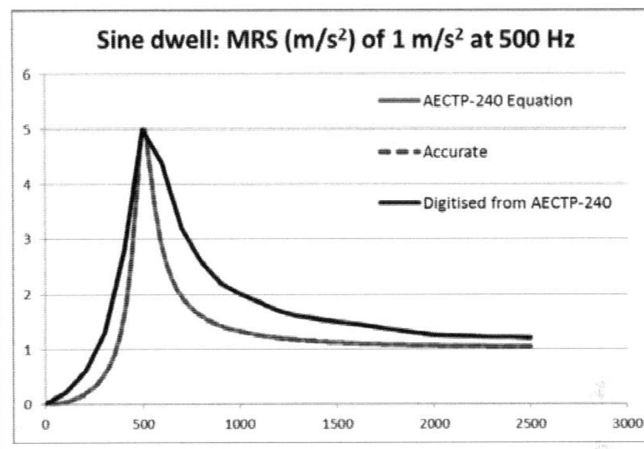

Figure 3. Calculation of MRS for a Sine Dwell Vibration Environment.

The MRS is calculated from the envelope of the corresponding sine dwell MRS for all frequencies in the sine sweep range, using the equations in Figure 1.

- Sweep range: $f = 300\ to\ 1200\ Hz$
- Damping factor: $Q = 5$
- Sweep acceleration: $\ddot{x} = 1\ m/s^2$

Spot check MRS (m/s^2) values at frequencies;

Frequency (Hz)	AECTP Equation	Accurate Equation	AECTP Digitised
200	0.778	0.778	~1.8
500	5.000	5.000	~5.0
2000	1.536	1.536	~2.2

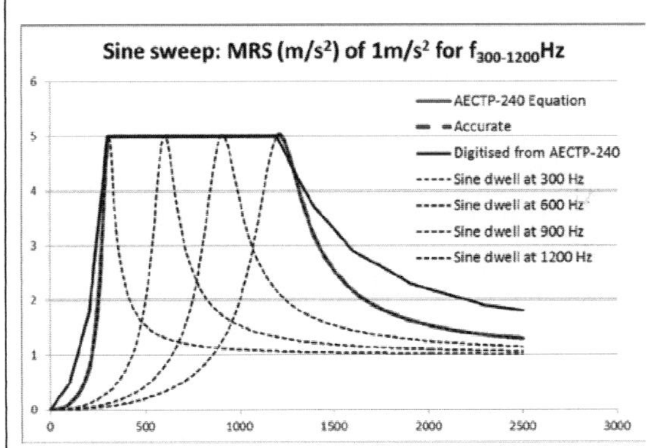

Figure 4. Calculation of MRS for a Sine Sweep Vibration Environment.

Broadband Random

The calculation of MRS for a broadband random vibration and its resulting MRS curve is shown in Figure 5. A broadband random vibration is defined by the duration and magnitude of a power spectral density (PSD) distribution; a series of breakpoints of frequency (Hz) and magnitude expressed as "acceleration power per Hz" (g^2/Hz, $(m/s^2)^2$/Hz, or similar). In addition to these, the MRS calculation requires either the damping factor Q (or the damping ratio ξ).

However Figure 5 shows that the information provided in this AECTP example for broadband random vibration is insufficient to demonstrate calculation of its MRS curve. The test duration is not provided, the 1200 Hz (from $f_{600\text{-}1200}$) is doubted (the plot suggests this drops at ~1000 Hz), and the input and MRS y-axis units "g^2/Hz" are doubted. Even if this information were provided clearly, it does not describe how to calculate z_{eff} used in the equation.

To determine the MRS of a broadband random vibration it is necessary to know the duration of the vibration. Whereas a sinusoidal vibration is deterministic and has a known maximum value, a random vibration has a probability of a maximum value within a given time period.

Inspection of the curves and spot check values in Figure 5 shows that the Halfpenny accurate equation calculated MRS very similarly to the AECTP digitised MRS curve. The Halfpenny and Walton approximate equation follows the general shape of the accurate curve, particularly within the frequency range of the broadband random vibration. Outside of the specified 100 to 1000 Hz frequency range this approximate equation will calculate MRS = 0 m/s^2, because $G(f_{0\text{-}100})=0$ and $G(f_{1000\text{-}1500})=0$. In practice when a broadband random vibration environment is reproduced by a vibration controller these levels will be close to zero, but will not be zero. A noise floor of 0.1 (m/s^2)^2/Hz is used in the approximate equation to account for this.

AECTP-240 [2] Leaflet 2410/1 Annex E
"E4 Random Vibration Excitation"

$$MRS = 4\pi_2 f_0^2 z_{eff}\sqrt{2\ln(n_0^+T)}$$

["π_2" is assumed to be a typing error for "π^2"]

- $G(f_{100-300}) = 0.5\ g^2/Hz$
- $G(f_{300-600}) = 1.0\ g^2/Hz$
- $G(f_{600-1200}) = 0.2\ g^2/Hz$
- Damping factor: $Q = 10$

The information given by AECTP for this example is shown above. However it is considered that there are further typing errors within this information. This is discussed and explained in the main text.

The PSD of acceleration and other information used in this example are;

- $G(f_{100-300}) = 5\ (m/s^2)^2/Hz$
- $G(f_{300-600}) = 10\ (m/s^2)^2/Hz$
- $G(f_{600-1000}) = 2\ (m/s^2)^2/Hz$
- Damping factor: $Q = 10$
- Duration: $T = 3600\ seconds$

Accurate

Halfpenny [8] presents the equation below, requiring numerical integration to calculate the MRS;

$$MRS(f_0) = \left\{ 2ln(f_0T)\int_0^{\infty} \frac{G(f)}{\left[1-\left(\frac{f}{f_0}\right)^2\right]^2 + \frac{1}{Q^2}\left(\frac{f}{f_0}\right)^2} df \right\}^{\frac{1}{2}}$$

Where: $G(f)$ is the PSD of acceleration at frequency f expressed in "$(m/s^2)^2$/Hz" and f_0 is the assumed natural frequency.

Approximate

Using Halfpenny and Walton [7, Equation 7]

$$MRS(f_0) \cong \sqrt{\pi \cdot f_0 \cdot Q \cdot G(f_0) \cdot \ln(f_0T)}$$

In addition this also uses a noise floor;

- $G(f_{0-100\ and\ 1000-1500}) = 0.1\ (m/s^2)^2/Hz$

MRS is not calculated using the AECTP equation because it does not explain how to use the "z_{eff}" term.

Spot check MRS (m/s^2) values at frequencies;

Frequency (Hz)	Approx. Equation	Accurate Equation	AECTP Digitised
200	651	664	~620
400	1335	1321	~1200
800	865	1075	~1000
1200	240	502	~550

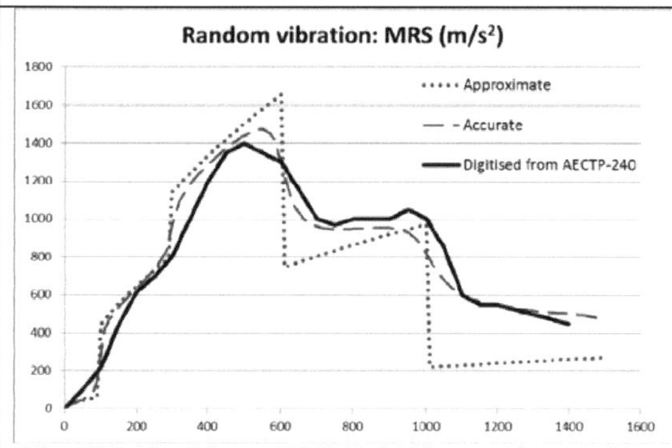

(The "Digitised from AECTP-240" curve is scaled by 10x, see the main text for the explanation.)

Figure 5. Calculation of MRS for a Broadband Random Vibration Environment.

It is considered that this AECTP example is derived from a very similar broadband random example included within GAM-EG-13 [4]. From inspection of the examples, the missing and corrected information is assumed as;

- the duration should be 1 hour *(In the GAM-EG-13 example)*
- the final breakpoint should be at 1000 Hz rather than 1200 Hz *(In the GAM-EG-13 example)*

- the input levels should be in units $(m/s^2)^2/Hz$
  (In the GAM-EG-13 example, $G(f_{100\text{-}300})$ is 5 $(m/s^2)^2/Hz$, becoming 0.5 g^2/Hz in the AECTP example. This is an "acceleration squared" unit, so the correct units conversion should be 0.05 g^2/Hz.)
- the output MRS curve plot should have y-axis units m/s^2
  (In the GAM-EG-13 example they are "m/s^2", but in the AECTP example they are "g^2/Hz", with magnitude values ~10x lower. The "/Hz" part of these units is considered to be a typing error, and the ~10x lower values are expected to be a similar "acceleration squared" units conversion error as above.)
- the "Digitised from AECTP-240" curve is scaled by 10x (g to m/s^2)
  (This is to correct for the "acceleration squared" units conversion error described above.)

CALCULATION OF FATIGUE DAMAGE SPECTRUM (FDS)

AECTP-240 [2] Leaflet 2410 Annex C "Combination of Spectra (Fatigue Damage Spectra)" presents the mathematics to calculate FDS from a power spectral density. The mathematics to calculate FDS for sine dwell and sine sweep vibration environments are presented in the previous AECTP release [5].

Again, Def Stan 00-35 and Mil Std 810 do not include the mathematics to demonstrate the calculation of FDS.

The fatigue term of the FDS is calculated by a stress-life approach using the Basquin power law relationship;

$$C = s^b . N_f \tag{1}$$

Where C is the Basquin coefficient (intercept of the stress-life curve with the y-axis), s is the stress range (twice the amplitude of the sinusoidal stress), b is the Basquin exponent (gradient of the stress-life curve in log space) and N_f is the number of sinusoidal cycles to failure.

An alternative Basquin equation is sometimes used where the Basquin coefficient C is replaced with A

$$A^b = s^b . N_f \tag{2}$$

The total fatigue damage D, for all stress cycles s, is the cumulative linear sum (according to Miner's rule);

$$Total\ Fatigue\ Damage: D = \sum \frac{1}{N_f} = \sum \frac{s^b}{C} = \sum \frac{s^b}{A^b} = \sum \frac{K^b \dot{x}^b}{C} \tag{3}$$

Where K is the spring stiffness, the transfer function from displacement x to stress s using $s = Kx$.

Sine Dwell

The calculation of FDS for a sine dwell and its resulting FDS curve is shown in Figure 6. The MRS calculation required the dwell frequency, peak acceleration magnitude and damping factor Q. In addition to these the FDS calculation requires the duration (or number of cycles), spring stiffness K, and Basquin exponent b and coefficient C. This equation is simple to use and calculates the FDS for an assumed natural frequency f_0.

Inspection of the curves and spot check values in Figure 6 shows that the AECTP equation and Halfpenny and Walton accurate equation calculate FDS exactly the same, and are similar to the AECTP digitised FDS curve.

Sine Sweep

The calculation of FDS for a sine sweep and its resulting FDS curve is shown in Figure 7. The MRS calculation required the frequency sweep range, sweep acceleration levels and damping factor Q. In addition to these the FDS requires the sweep rate and duration (or number of sweeps), spring stiffness K, and Basquin exponent b and coefficient C. The AECTP equation is not used to calculate FDS because it requires numerical integration and is beyond a simple spreadsheet calculation. The FDS curve shown for the accurate equation is calculated by the specialist software nCode GlyphWorks. The approximate equation is simple to use in a spreadsheet and calculates the FDS for an assumed natural frequency f_0.

Inspection of the curves and spot check values in Figure 7 shows that the Halfpenny and Walton accurate and approximate equations calculate FDS very similarly to the AECTP digitised FDS curve for natural frequencies between 100 and 500 Hz. It is noted that there are two upwards steps in the AECTP digitised FDS curve at ~100 and ~500 Hz. Such steps are unexpected for a constant acceleration logarithmic sine sweep between 10 and 500 Hz.

AECTP-200 [5] Leaflet 2410/1 Annex A "A2.2 Fixed Sine Excitation"

$$FDS = \frac{K^b}{C} f T \frac{\ddot{x}_m^b}{\omega_0^{2b}\left\{\left[1-\left(\frac{f}{f_0}\right)^2\right]^2+\frac{1}{Q^2}\left(\frac{f}{f_0}\right)^2\right\}^{\frac{b}{2}}}$$

- Dwell frequency: $f = 500\ Hz$
- Dwell acceleration: $\ddot{x}_m = 10\ m/s^2$
- Damping factor: $Q = 10$
- Stiffness: $K = 1\ N/m^3$
- Basquin exponent: $b = 8$
- Basquin coefficient: $C = 1\ N/m^2$

Accurate

Halfpenny [7] presents the equation below to calculate the FDS;

$$FDS(f_0) = \frac{K^b}{C}\cdot\left[\frac{|\ddot{x}|}{(2\pi f_0)^2}\right]^b \cdot \left[\frac{1}{\left\{\left[1-\left(\frac{f}{f_0}\right)^2\right]^2+\frac{1}{Q^2}\left(\frac{f}{f_0}\right)^2\right\}^{b/2}}\right].f.T$$

In addition to the information given in AECTP-200 [5] shown above, use;

- $\omega = 2\pi f$

Spot check FDS (damage) values at frequencies;

Frequency (Hz)	AECTP Equation	Accurate Equation	AECTP Digitised
300	6.86e-41	6.86e-41	~1e-40
500	2.00e-34	2.00e-34	~2e-34
1000	2.99e-46	2.99e-46	~2e-46

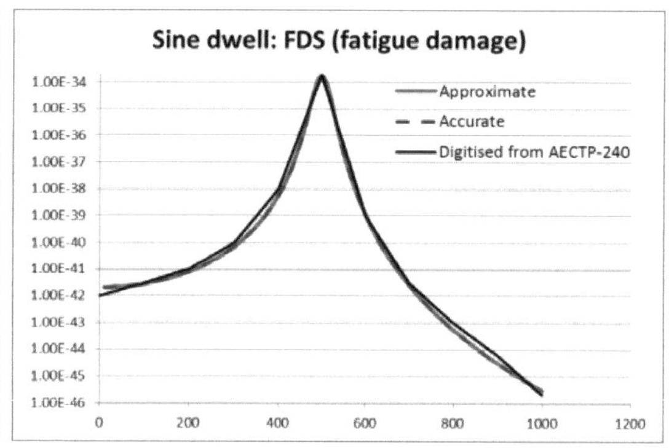

Figure 6. Calculation of FDS for a Sine Dwell Vibration Environment.

Broadband Random

The calculation of FDS for a broadband random and its resulting FDS curve is shown in Figure 8. The MRS calculation required the duration and magnitude of a power spectral density (PSD) distribution, and damping factor Q. In addition to these the FDS requires the spring stiffness K, and Basquin exponent b and coefficient C.

AECTP-240 [2] Leaflet 2410/1 Annex E "E4 Random Vibration Excitation" provides an equation to calculate FDS but no guidance to use the equation and no example are provided. This equation is not used in Figure 8.

The previous AECTP release [5], AECTP-200 [5] Leaflet 2410/1 Annex A "A2.4 Random Vibration Excitation" provides an equation to calculate FDS and an example to demonstrate it. This AECTP example states "600 to 10000 Hz" but by reference back to a similar example in GAM-EG-13 [4] the "10000" is considered a typing error and the frequency range "600 to 1000 Hz" is used. However there is no guidance to use the equation and again its solution requires numerical integration beyond the scope of a spreadsheet calculation.

Inspection of the curves and spot check values in Figure 8 shows that the Halfpenny and Walton accurate and approximate equations calculate FDS very similarly to the AECTP digitised FDS curve for natural frequencies within the frequency range of the broadband random. Outside of the specified 100 to 1000 Hz frequency range this approximate equation will calculate FDS = 0, because $G(f_{0\text{-}100})=0$ and because $G(f_{1000\text{-}1400})=0$. In practice when a broadband random vibration environment is reproduced by a vibration controller these levels will be close to zero, but will not be zero. A noise floor of 0.1 $(m/s^2)^2/Hz$ is used in the approximate equation to account for this.

AECTP-200 [5] Leaflet 2410/1 Annex A "A2.3 Swept Sine Excitation"

$$D = \frac{K^b}{C} \cdot f_0 \frac{T_1 \ddot{x}_m^b}{\left(4\pi^2 {f_0}^2\right)^b} \cdot \int_{h_1}^{h_2} f(t) \frac{dh}{\left\{[1-(h)^2]^2 + \frac{h^2}{Q^2}\right\}^{\frac{b}{2}}}$$

- Sweep acceleration: $\ddot{x}_m = 10\ m/s^2$
- Sweep range: $f = 10\ to\ 500\ Hz$ (logarithmic sweep rate)
- Duration: $T = 1800\ seconds$
- Basquin exponent: $b = 10$
- Damping factor: $Q = 10$
- Stiffness: $K = 1\ N/m^3$
- Basquin coefficient: $C = 1\ N/m^2$

FDS is not calculated using this AECTP equation because this requires numerical integration and is beyond a spreadsheet calculation.

Accurate

Halfpenny [8] presents the equation below, requiring numerical integration to calculate the FDS;

$$FDS(f_0) = \frac{K^b}{C} \cdot \frac{60 f_0}{\rho \ln(2)} \cdot \left[\frac{|\ddot{x}|}{(2\pi f_0)^2}\right]^b \cdot \int_{f_1}^{f_2} \frac{1}{\left\{\left[1-\left(\frac{f}{f_0}\right)^2\right]^2 + \frac{1}{Q^2}\left(\frac{f}{f_0}\right)^2\right\}^{b/2}} df$$

Where ρ is the logarithmic sweep rate in octaves per minute.

Approximate

It is necessary to extend the Halfpenny and Walton [7, Equation 12] to include K^b/C. They had omitted this for their comparative analyses where K=1 and C=1.

$$FDS(f_0) \cong \frac{K^b}{C} \cdot \frac{60 f_0}{\rho \ln(2)} \left[\frac{A(f_0)}{(2\pi f_0)^2}\right]^b \frac{\pi}{2} Q^{b-1} b^{\frac{-1}{\sqrt{\pi}}}$$

Where ρ logarithmic sweep rate in octaves per minute, and $A(f_0)$ is the acceleration amplitude in m/s^2 at frequency f_0 Hz.

In addition to the information given in AECTP-200 [5] shown above, use;

- Sweep rate: $\rho = 1\ octave\ per\ minute$

Spot check FDS (damage) values at frequencies;

Frequency (Hz)	Approx Equation	Accurate Equation	AECTP Digitised
50	2.11e-28	1.12e-27	~1e-30
200	7.69e-40	4.12e-39	~1e-39
400	1.47e-45	7.72e-45	~1e-45

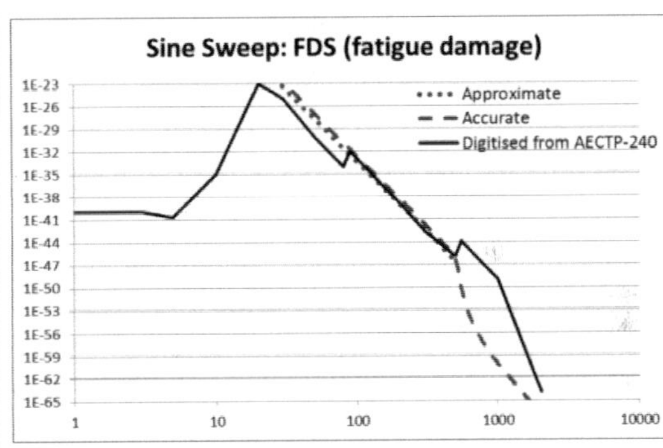

Figure 7. Calculation of FDS for a Sine Sweep Vibration Environment.

APPLICATIONS

Def Stan 00-35 [1] Chapter 2-19 Test M19 - General Time History Replication Test, lists FDS as one of a number of validation methods useful to quantitatively compare parameters associated with a time history waveform. It states that FDS allows comparison of the effects of the waveforms in terms of potential fatigue damage whilst retaining information relating to the frequency content of the waveforms. Though it notes that a limitation of FDS is that it requires assumptions relating to the fatigue properties of the materiel.

FDS should be used as a relative comparison method between two or more time history waveforms or vibration test environments. Used in this way the fatigue properties of the material are the same for each case and the relative similarity or difference of the time history waveforms or vibration test environments are revealed. For such relative analyses the only fatigue property required is the Basquin exponent 'b', the gradient of the stress-life curve in log space. The other primary assumption, and limitation, is the unknown damping factor Q. In practice this is usually defaulted to 10, and again for a relative analysis reveals the similarities and differences between the vibration environments.

AECTP-240 [2] Leaflet 2410/1 Annex E "E4 Random Vibration Excitation"

$$DP(f_n) = f_n T.\left(\frac{G(f_n)}{f_n.\zeta}\right)^{\frac{b}{2}}$$

The equation above is provided, but no guidance to use the equation and no example are provided.

AECTP-200 [5] Leaflet 2410/1 Annex A "A2.4 Random Vibration Excitation"

$$FDS(f_0) = \left(\frac{K}{A}\right)^b . n_p^+ T. \left(\sqrt{2}z_{rms}\right)^b . \Gamma\left(1 + \frac{b}{2}\right)$$

The breakpoints for the PSD of acceleration are;

- $G(f_{100-300}) = 5\ (m/s^2)^2/Hz$
- $G(f_{300-600}) = 10\ (m/s^2)^2/Hz$
- $G(f_{600-1000}) = 2\ (m/s^2)^2/Hz$
- Duration: $T = 1800\ seconds$
- Damping factor: $Q = 10$
- Basquin exponent: $b = 8$
- Stiffness: $K = 0.5e12\ N/m^3$
- Basquin coefficient: $A = 1e9\ N/m^2$

[The AECTP example states "600 to 10000 Hz" though is considered a typing error and "600 to 1000 Hz" is used.]

Accurate

Halfpenny [8] presents the equation below, requiring numerical integration to calculate the FDS;

$$FDS(f_0) = n_p.T.\frac{K^b}{C}.\int_0^\infty z^b.P(z)dz$$

Approximate

This is a modified form of the Halfpenny and Walton [7, Equation 12].

- The "K^b/C" term has been added (this was omitted from their comparative FDS analyses where K=1 and C=1)
- The "9.81^2" term has been removed (this vibration environment is defined in "$(m/s^2)^2/Hz$", rather than their "g^2/Hz" units.

$$FDS(f_0) \cong f_n.T.\frac{K^b}{C}.\left[\frac{Q.G(f_0)}{2(2\pi.f_0)^3}\right]^{b/2}\Gamma\left(1 + \frac{b}{2}\right)$$

- $G(f_{noise-floor}) = 0.1\ (m/s^2)^2/Hz$

This uses the Gamma function "$\Gamma(n)$". When "n" is an integer (if Basquin exponent b is an even number), the Gamma function reduces to "n-1 factorial"; $(n-1)!$

Spot check FDS (damage) values at frequencies;

Frequency (Hz)	Approx	Accurate	AECTP Digitised
200	4.25e-3	1.47e-3	1e-3
400	3.32e-5	9.86e-6	2e-5
800	2.59e-11	1.99e-11	1e-11

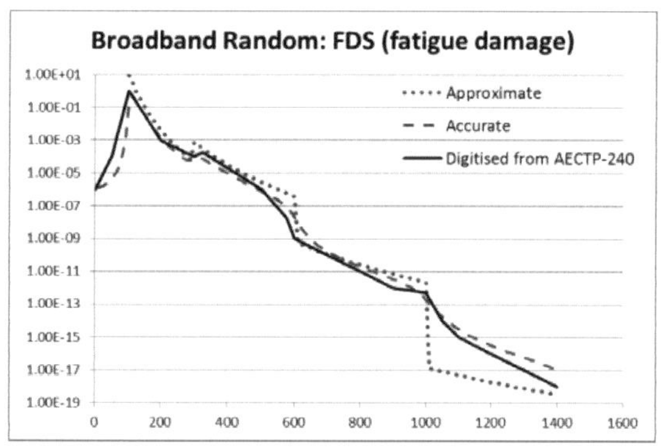

Figure 8. Calculation of FDS for a Broadband Random Vibration Environment.

This means that the FDS can be used to compare the potential cumulative fatigue damage for the duration of various vibration environments, such as within;

- Test Tailoring: Def Stan 00-35 [1], AECTP-240 [2] and Mil Std 810 [3] all recommend test tailoring of laboratory vibration test schedules. The vibration chapter of Mil Std 810 begins with the words "Tailoring is essential", and AECTP-240 [2] Leaflet 2410/1 Annex C describes the FDS as "an attractive technique in development of a laboratory vibration test schedule".
- Vibration Qualification by Read Across Evidence: In many cases existing vibration qualification test evidence is available for equipment to be used in a specific aircraft vibration environment or to an environment defined by one the Def Stan 00-35, AECTP-240 or Mil Std 810 standards. In such cases FDS comparison of existing test evidence with the manufacturer's vibration requirements may prove sufficient for the new aircraft enabling full type approval to be awarded without recourse to additional testing. This offers huge cost savings because vibration tests are very expensive on account of the direct testing costs and the cost of the test component which

is life expired at the end of the test. If the existing test evidence is insufficient for full type approval, it is often sufficient to justify experimental flight approval subject to a limited flight envelope or number of flight hours. For urgent operational requirements this may be sufficient to enable flights to begin whilst gaining time for full qualification testing to be completed or enable qualification through assessment of measurements during flight trials.

- Health Usage Monitoring: The FDS is an analysis method that can increase the capability of on-board health usage monitoring systems (HUMS). For many years aircraft have used fatigue meters and counting accelerometers to monitor their in-flight acceleration levels in order to estimate their structural life consumed during flight. This FDS approach is very suitable for assessment of vibration life consumed, with respect to the vibration qualification test evidence, for sensitive materiel (e.g. electronic equipment and missile structures, ref AECTP-200 [5] Leaflet 2410/1) and localised complex systems such as hydraulics, fuel systems, on-board oxygen generators, etc.

Halfpenny and Walton [7] present case studies using FDS from their work with helicopters for Test Tailoring and Vibration Qualification by Read Across Evidence.

Further work by the authors with Walton used all three of the above FDS applications during the successful integration of external fuel tanks to a helicopter. Existing vibration qualification test evidence was insufficient for full type approval, but was sufficient to justify a limited number of hours of experimental flight by read across evidence. The health of this system was monitored during experimental flight testing by comparing the cumulative FDS of accelerations measured at locations on the system with the FDS of the existing vibration qualification test evidence. The results of this monitoring and analysis showed that the cumulative FDS from experimental flight testing increased rapidly at certain frequencies and would exceed those from the qualification test evidence. Identification of this led the supplier to a successful redesign of their system to pass a new vibration qualification test tailored to this specific helicopter vibration environment. The qualification of the redesign was confirmed by further experimental flight monitoring and FDS analysis of the measured accelerations.

CONCLUSIONS

The purpose of this paper is to explain the fatigue damage spectrum. Simple sinusoidal and broadband random examples from the AECTP standard are used that can be easily reproduced by spreadsheet calculation. By demonstrating FDS for these simple vibration environments, though sometimes requiring approximate equations with their assumptions; the purpose is to give confidence in the use of their accurate equations for more complex vibration environments which cannot be so easily reproduced by spreadsheet calculation.

The authors would like to acknowledge Frédéric Kihm (HBM-nCode, France) for his help with the France national standard GAM EG-13 [4].

REFERENCES

[1] UK Ministry of Defence, Defence Standard 00-35 Issue 4 (2006). *Environmental Handbook for Defence Materiel: Part 3: Environmental Test Methods*

[2] NATO Standardisation Agency, STANAG 4370 (2009). *AECTP-240 (Edition 1) Mechanical Conditions,* (http://nso.nato.int/nso/nsdd/apdetails.html?APNo=1252)

[3] USA Department of Defense, Test Method Standard Mil-Std-810G (2008). *Environmental Engineering Considerations and Laboratory Tests*

[4] France Ministère de la Défense, Délégation Générale pour l'Armement (2002), *GAM EG-13 – Annexe Générale Mécanique*

[5] NATO Standardisation Agency, STANAG 4370 (2006). *AECTP-200 (Edition 3) Mechanical Conditions*

[6] Lalanne, C. (2002), *Mechanical Vibration and Shock, Volume 2 Mechanical Shock, Volume 4 Fatigue Damage, Volume 5 Specification Development*, Hermes Penton Ltd.

[7] Halfpenny, A. and Walton, T. (2010), *New Techniques for Vibration Qualification of Vibrating Components on Aircraft*, Proceedings of Aircraft Airworthiness & Sustainment 2010, (http://www.meetingdata.utcdayton.com/agenda/airworthiness/2010/agenda.htm)

[8] Halfpenny, A. (2008), *Accelerated Testing Theory and User Manual*, HBM-nCode GlyphWorks Product Documentation.

Review of Computational Methods for Payload Life Prediction

RJ Amos, A Whitelaw, MA Hussain

MBDA – UK Limited, Six Hills Way, Stevenage, Hertfordshire, SG1 2DA, United Kingdom

Abstract: This paper is concerned with computational methods for life prediction of payload structures using air-carry environmental data from a health and usage monitoring system (HUMS) mounted on an aircraft and monitored continuously throughout the air-carry life of the payload. Here, payload refers to any equipment such as munitions carried by the aircraft for the performance of a particular mission, and life is interpreted as referring to the remaining air-carry fatigue life of the payload structure and its structural components.

Literature searches have been carried out, but little has been found specifically on health and usage monitoring for predicting fatigue life of payloads. However, there is a much wider literature on monitoring for life prediction of structural components of aircraft, and particularly rotorcraft, which becomes directly applicable by simply regarding the payload as a structural component of an aircraft, but without the possibility of monitoring.

This paper, by reference to other people's work on life prediction of aircraft structures, described in the literature, develops the understanding of the problem and how it might be addressed for payload structures, and presents a review of computational methods previously used or proposed. Recommendations for the preferred tools and methods, catering for both metallic and composite structures, most likely to be applicable to this study are made.

UNDERSTANDING THE PROBLEM

Introduction

The current fatigue life prediction methodology for air-carried payloads is based on the demonstration of a ***safe fatigue life***, i.e. the maximum permitted number of hours of air-carry, for the payload design at a specific mounting position on a specific aircraft. The safe life is defined as the fatigue life of a statistically worst case payload structure under the ***design fatigue loads*** (the statistically worst case repetitive loads seen in a worst case flight) and the ***design fatigue spectrum*** (which defines the numbers of cycles and load levels seen in a worst case flight) for the aircraft. The structure must retain sufficient residual strength at the end of its safe fatigue life to meet all worst case design static load requirements. The safe fatigue life is demonstrated in qualification by analysis and/or testing.

Then, regardless of the actual severity of the loading environment seen by a particular payload structure (most flights will not involve continuous flying at the design extreme loading), it is withdrawn from service once its actual air-carry life reaches its safe fatigue life. This potentially results in many structural components being withdrawn from service long before they need to be. On the other hand predictions do not allow for statistically unlikely extreme conditions when the manoeuvres seen in air-carry are more severe that those assumed in design. In that case premature failure may occur.

This paper is concerned with the development of a methodology to more accurately predict the remaining allowable fatigue life of a payload structure mounted on an aircraft based on the outputs from a ***health and usage monitoring system (HUMS)*** mounted on the aircraft and monitored continuously throughout the air-carry life of the payload. The critical point is that the methodology will make use of monitoring systems already widely fitted and used on aircraft to predict the life for payloads, rather than modify and monitor the payload itself. In order to come up with the best possible solutions it has been assumed that there are no impenetrable boundaries between different manufacturers and design authorities (i.e. between aircraft and payload) and that information can be freely passed between.

Health and Usage Monitoring Systems

Health and Usage Monitoring Systems (HUMS) refer to equipment, techniques, and/or procedures applied to structures by which selected incipient failures or degradation, and/or selected aspects of service history can be determined. They are widely used as a predictive maintenance strategy in helicopters and fixed wing aircraft.

Health monitoring determines the current condition of the structure. A comparison is made with the initial condition to determine degradation and predict the likelihood of incipient failure. ***Usage monitoring*** determines what conditions the structure has seen and, by means of an appropriate algorithm, how far through its life it should be, but gives no information on the actual current condition or level of degradation. The current study is concerned with the use of devices already installed on an aircraft to give an improved estimate of the remaining fatigue life of a payload mounted on that aircraft. This eliminates the need to develop a separate system for the payload and any consequent cost, mass or space penalty for the payload. However only the usage rather than the health monitoring function of the HUMS can then be used.

The potential benefits of usage monitoring are shown in Figure 1, obtained from [1]. It is a graph of life consumption against actual time in air-carry. Here, life consumption is a measure of how much of the available fatigue life has been used up. The current situation, with no monitoring (other than duration of flight) carried out, is shown by the centre line marked ***predicted severity***. In this case, life consumed equals time spent in air-carry. Once a payload has reached its safe life, its life is all consumed and it is withdrawn from service. The service life of critical components can be extended if the actual usage, measured by usage monitoring, can be shown to be low compared to the usage assumed to obtain the safe life. This is shown by the lower line marked ***low severity***. This would happen when the manoeuvres seen in air-carry are more benign that those assumed in design. Life consumed is then less than time spent in air-carry. Usage monitoring can also provide a safety benefit when actual usage is more severe than predicted. This is shown by the upper line marked ***high severity***. Payloads which have seen severe usage can be withdrawn from service early thereby reducing the risk of failures in service. This would happen when the manoeuvres seen in air-carry are more severe that those assumed in design. Life consumed is then greater than time spent in air-carry.

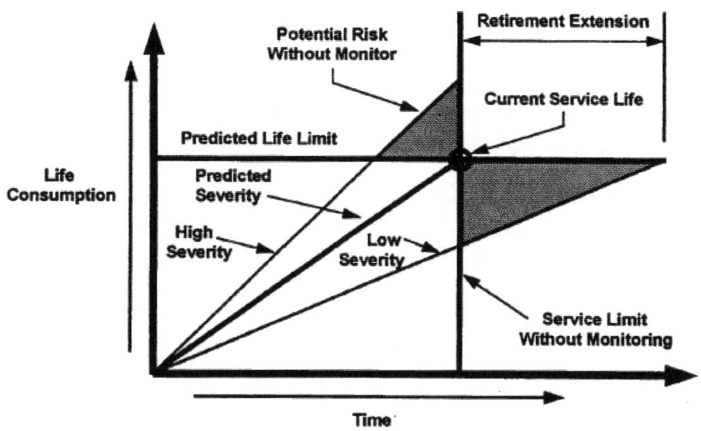

Figure 1. Potential Benefits of Usage Monitoring (from [1]).

Predicting Fatigue Life of a Payload

Loads are the governing factor in predicting fatigue lives. Without a knowledge of loads, fatigue prediction is impossible. In design/qualification the required loads for fatigue life prediction of an air-carried payload are defined by the design fatigue loads and the design fatigue spectrum, defined above, for the payload. These payload design loads and design spectrum are obtained from an assessment of the operational usage of, and consequent design loads applied to, the aircraft and its structural response to these applied external loads. Neither the aircraft nor the payloads are rigid structures and their structural responses to the applied external loads will affect the internal loads and stresses and hence fatigue life at different locations on the payloads and/or the aircraft. Using the payload design loads and design spectrum the fatigue life of the payload can be obtained either directly from structural fatigue tests or predicted from structural (finite element) analysis, which determines the internal load, stress and strain responses in the payload, followed by standard fatigue calculations such as a Miner's rule summation using an S-N curve for the material.

For fatigue life prediction during usage monitoring the responses at the fatigue critical locations on the payload will not be directly measured. In that case the required response parameters (the oscillatory loads, stresses and strains at the fatigue critical locations on the payload) must be estimated or predicted from the output data from the HUMS on the aircraft. Output data from the HUMS will be in the form of flight parameters such as airspeed, altitude, accelerations, stick position etc. Published guidelines for aircraft parameters and recording rates are referenced in [2]. This estimate of the required response parameters is achieved by means of an algorithm based on a mathematical model or relationship between the flight parameters monitored and the response parameters required for the fatigue life predictions. For components on an aircraft this mathematical relationship is generally derived using data fitting techniques from data directly gathered on an aircraft flight loads survey. The success of this process will depend on how well the monitored flight parameters define the actual usage and on the reliability of the data fitting techniques used.

HOW THE PROBLEM MIGHT BE ADDRESSED

Approaches to Estimating Fatigue Life Usage

From a general study of the literature there appear to be essentially five general approaches to the problem of estimating the data required to predict fatigue life

1. ***Regime Recognition (RR), also called Flight Condition Recognition (FCR);***
2. ***Flight Loads Synthesis (FLS);***
3. ***Direct Loads Measurement (DLM);***
4. ***Enhanced FLS, or FLS/DLM;***
5. ***Hours Based Life prediction.***

The first two approaches (RR and FLS) both involve an indirect determination (prediction) of the loads during flight at the fatigue critical components using measured flight data, and subsequent fatigue analysis to estimate the safe fatigue life that has been used up.

Regime Recognition (RR): A ***regime*** is a category of operation that an aircraft can be in at a given time. A regime can also be known as a flight condition or manoeuvre. A set of regimes covering all possible operational use must be defined. A worst case damage rate for each regime is estimated in advance for each fatigue critical component in terms of hours of safe fatigue life used up per hour of flight. The RR approach uses output data from the aircraft HUMS recorded during the flight together with a derived algorithm to determine the time spent in each regime during the flight as well as the regime sequence (i.e., flight profile). The known damage rate for each regime is applied to the recorded time durations to determine the accumulated damage in terms of total hours of safe fatigue life used up.

It is stated in [3] that *regime recognition models produce a conservative estimate of the fatigue damage experienced by critical dynamic components during each flight regime.* Using a regime recognition approach can only improve results if the fraction of time spent in the various regimes is different to the manufacturer's design spectrum. Regime recognition does not take into account the variability of loads experienced in any given regime. If the regime covers too wide a range of loads then this approach may clearly be conservative. It follows that the more detailed the regime definitions can be the less conservative the approach will be.

Flight Loads Synthesis (FLS): The FLS approach uses output data from the aircraft HUMS recorded during the flight together with a derived algorithm to predict the oscillatory loads and number of cycles seen at the fatigue critical locations on the payload during the flight. This may be either in the form of a time history or as a fatigue spectrum (load levels and number of cycles seen). If this is as a time history then a standard rainflow analysis could be carried out to give a fatigue spectrum. With a fatigue spectrum then standard fatigue calculations, as used in the current life prediction methodology, can be carried out to give an estimate of fatigue life used up.

Direct Loads Measurement (DLM): The DLM approach involves the direct measurement of the loads during flight at the fatigue critical components. This is followed by standard fatigue life calculations as for the FLS approach. The current study is concerned with the use of devices already installed on the aircraft rather than on the payload. Therefore it will be assumed here that sensors will not be mounted within the payload.

Enhanced Flight Loads Synthesis (FLS/DLM): One concern with the DLM approach would be the potentially large number of parameters and positions required to be measured to characterise the fatigue usage of a complex structure. In addition, by the nature of fatigue it may not be possible to make every measurement required; for example in tight radii where stress gradients are high. The enhanced FLS approach involves a combination of the FLS and DLM approaches whereby direct load measurements are made at a limited number of positions on the aircraft, but not necessarily at the fatigue critical locations, and the data is used to improve the loads synthesis at the fatigue critical components. For payload life prediction the obvious location for load measurement on the aircraft would be at, or close to, the interface positions to the payload.

Hours Based Life Prediction: This is simply the current method of monitoring the air-carry life in hours and directly comparing this against the safe fatigue life.

Derivation of the Algorithm

An algorithm to obtain the response parameters required at each applicable location on the aircraft and payload from the output data from the HUMS on the aircraft is required in each case.

For locations on the aircraft this algorithm is generally derived using data fitting or optimisation techniques from data directly gathered by flight testing (called a ***flight loads survey***) during aircraft qualification; see, for example, [3]: *Determine the strain in each critical dynamic component while experiencing the design flight regimes. This is accomplished by flight testing an instrumented aircraft to measure component strains,* and [2]: *The flight loads survey* *flight test manoeuvres are flown in accordance with a set of manoeuvre standards .... that are designed to cover the flight manoeuvres in the design usage spectrum. Both aircraft parameters and component loads generated for the flights are stored in the* ***flight loads survey database****, where the data for each scripted manoeuvre is stored with start and stop times for the manoeuvre.*

For locations on the payload this is not possible since the payload is not instrumented in the flight loads survey. However, if flight testing is conducted with an actual or representative payload installed, then responses at the interface positions to the payload on the aircraft may be obtained.

Regime Recognition: For locations on the aircraft the data stored in the flight loads survey database is used to generate the RR algorithm by developing the list of regimes based on the values of the flight parameters monitored by the aircraft HUMS, and matching these with loads data at each fatigue critical location. This last step will require a mathematical data fitting technique to obtain a best fit of component loads from the flight parameters. A regime will cover a defined range of values of each flight parameter and the worst case loads for each regime would be used for the damage rate calculation. The RR algorithm then applies standard fatigue calculations to determine the damage rate for each regime from the loads from the flight loads survey. From [4]: *actual regimes flown can be matched with flight loads data obtained in the aircraft flight test qualification program* and *since loads and damage rates have already been associated with each regime during the aircraft flight test qualification program, the ability to calculate incremental damage, manoeuvre-by-manoeuvre, and flight-by-flight, is readily available.*

The RR algorithm would then simply estimate the regime being seen and use the calculated damage rate for each regime applied to the recorded time duration to determine the accumulated damage in terms of total hours of safe fatigue life used up. If the aircraft HUMS provides regime recognition for the aircraft then the regime identified can be used directly for the fatigue critical locations on the payload with appropriately derived damage rates. However, in the case of a payload, where it is assumed that there will be no direct monitoring of loads during the aircraft flight loads survey, these damage rates can only be obtained by predicting the loads at the critical locations and then applying standard fatigue calculations to determine the accumulated damage.

Flight Loads Synthesis: For flight loads synthesis (FLS) the procedure is similar, but misses out the steps of identifying the regime and damage rates by making a prediction of the loads from the precise values of the flight parameters. For this a mathematical data fitting technique is required as in RR to define the algorithm between the loads required and the flight parameters monitored. The FLS algorithm would then estimate the loads at each fatigue critical location using the derived mathematical relationship and apply standard fatigue calculations to determine the accumulated damage. The loads would generally be in the form of time histories so that cycles could be counted.

For a payload both cases, RR and FLS, require the loads at the critical locations which could not be obtained during the aircraft flight loads survey. A general approach to the problem is indicated in [5]: *With flight parameter based systems, loads in the major load carrying members are calculated from flight parameters using regression techniques or neural network techniques. These loads in turn are related to stresses at critical locations via transfer functions. Ideally, provided that a sufficient number of primary load carrying structures are routinely monitored, stresses at all critical locations could be determined from strain surveys and/or finite element modelling, with a transfer function relating the monitored load to the critical location stresses. Therefore, a change in the critical location can be accommodated through the development of a new transfer function to the new critical location.* This proposes the algorithm relating the output data from the HUMS on the aircraft and the response parameters required at each fatigue critical location could consist of two factors which would be multiplied together to give the required relationship:

1. ***A first factor relating the output data from the HUMS to the loads at some intermediate locations.***
2. ***A second factor relating the loads at the intermediate locations to the required response parameters at the required locations on the payload.***

The first factor is obtained as previously described either from the results of a flight loads survey by using a mathematical data fitting or optimisation technique (RR or FLS), or by direct loads measurement (DLM). The second factor must be obtained as a transfer function either from test or finite element analysis. This transfer function relates the loads at the aircraft to payload interfaces to the required response parameters at the required critical locations on the payload. In the case where the transfer function is obtained by test then this would have to be a laboratory test of the component alone rather than a full scale flight test. Testing would require a lot of measured parameters and would probably be impractical. Alternatively, the use of finite element analysis allows the use of predictive methods where no test data are available or even possible. Predictive techniques also provide responses throughout the structure rather than at specific monitoring locations.

The load data at the intermediate locations, together with the output data from the HUMS, must be sufficient to adequately define the loads on a component so that it is possible to uniquely generate load, stress and strain distributions throughout that component. These intermediate locations must therefore be either the aircraft to payload interfaces or locations sufficiently near that the structural response of the aircraft has no significant effect. If the aircraft HUMS monitors loads at these locations then the first factor is obtained from these loads. If no such loads have been obtained during the aircraft flight loads survey then the first factor can only be obtained by finite element analysis.

Finite Element Analysis: The use of finite element analysis to completely derive the mathematical relationship between the output data from the HUMS and the loads and responses at locations on the aircraft, whether intermediate or fatigue critical, has not been an option discussed in any of the literature found in the literature searches. This may be due to a lack of confidence in the predicted results and the difficulties of convincing authorities of accepting such an approach where safety is concerned.

Historically, a detailed and accurate finite element analysis of a complete aircraft structure would have been considered a challenge due to the limitations of computer disk space, processor speed and software capabilities. However, with the rapid advances in computer hardware and software this approach is now a real possibility, and it presents real benefits in terms of reduced testing costs and increased flexibility. An example of this approach is the Simulated Loads Monitoring (enhanced FLS) method of [4]. Here, a few key sensors in critical locations are combined with an analytical model to provide an accurate picture of the loads throughout the rotor system. In a further example in [6] Technical Data Analysis Inc, partnered with Georgia Tech, developed a dynamic model of the UH-60 rotor system using DYMORE, a finite element-based tool for nonlinear flexible multi-body system modelling. Using data from a flight test program, the model was calibrated such that known blade loads input into the model would result in matching expected loads on the dynamic components throughout the rotor system.

Published Studies

An overview of health management issues and challenges for rotorcraft is presented in [7]. This includes a review of standard practices used for rotorcraft health management, lessons learned from fielded trials, and future challenges. It is stated that *rotorcraft health monitoring systems, while historically focussed on safety, can also provide economic benefits for rotorcraft operators,* and *such benefits were demonstrated on a flight trial program at Petroleum Helicopters in the form of reduced insurance cost through enhanced safety, lower operating costs, increased aircraft availability and improved operating efficiency.* Experience and lessons learned from 48 years of developing and implementing methods for life cycle management of aircraft structural integrity in the US are presented in [3].

Bell Helicopter Textron and Petroleum Helicopters for NASA: The first three and the fifth approaches were evaluated against each other in a feasibility study of a Rotorcraft HUMS carried out by Bell Helicopter Textron and Petroleum Helicopters for NASA. This evaluation is reported in [1, 8 and 9]. Instrumentation was mounted at the positions of interest to validate the predicted loads against measured data. In [8] it was stated that the predicted loads from the FCR (RR) method agreed very well with the measured loads. Figure 2 shows a typical comparison between the life usage predicted by FCR against that obtained from measured data and that predicted in advance (baseline) which is assumed to be the safe fatigue life.

For each critical position considered the measured data gave the longest life followed by FCR followed by the baseline with the shortest life. Some problems were encountered with the FLS method. The method was found to be sensitive to noise in the loads. The quality of the data and curve fits were not good and therefore led to errors. No consistent relationships between used lives were found between FLS and the other methods. Whilst it might be thought that the FLS method would give a more accurate prediction than the FCR method which makes more approximations the experience here appears to be that the FLS method is more difficult to get right.

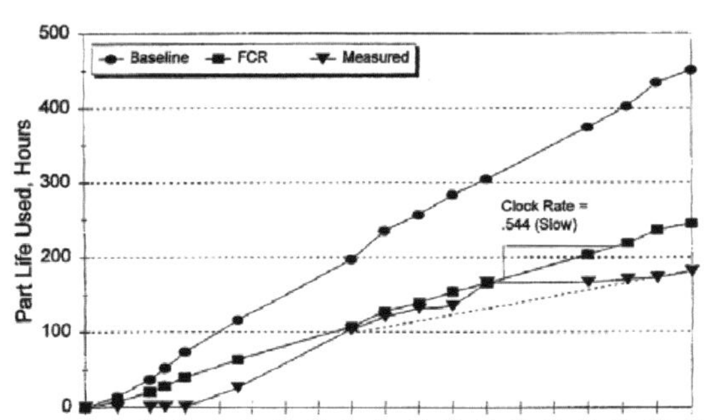

Figure 2. Typical Life Usage Comparison for FCR Method (from [8]).

Boeing Rotocraft: A more recent paper from Boeing Rotorcraft, [2], reports on the development and validation of a usage-based fatigue damage calculation using the Regime Recognition approach for selected life-limited dynamic components on the AH-64 Apache Helicopter. It is reported that *Damage Fraction Calculation based on actual usage for five dynamic components for 72 flight hours are calculated and compared against damage obtained for the design usage spectrum. Usage-based damage fraction for three components (main rotor pitch housing, lead-lag damper body and main drive shaft) were significantly lower when compared to the design spectrum based damage fraction. Two other components (tail rotor gearbox housing assembly and collective system) resulted in nearly the same level of damage for both approaches.*

Technical Data Analysis Inc: Technical Data Analysis Inc. in [4] present results from a similar study to that of [8] now comparing the first, fourth and fifth approaches. Results are shown in Figure 3. Here, Simulated Loads Modelling is an enhanced FLS method, HUMS-RR is RR or FCR and CRT (or Component Retirement Time) is the baseline safe fatigue life. Results are given for four individual aircraft trials. The FLS method used has been enhanced by a small number of direct load measurements to increase accuracy (i.e. a combined FLS/DLM approach). It can be seen that the enhanced FLS method gave significantly longer lives than the baseline or RR method, and that for one of the aircraft the RR method gave a shorter life than the baseline. This last result could happen if actual usage were close to the design spectrum rendering the RR method overly conservative.

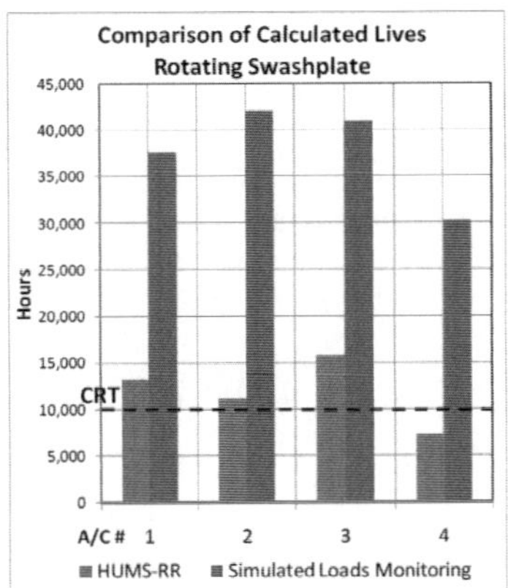

Figure 3. Comparison of Life Prediction Methods (from [4]).

An approach based on direct measurements (DLM or FLS/DLM) might be thought to be the most accurate and is recommended by various authors, see for example [5] which states that *The direct method of loads monitoring using strain gauges is the method advocated by the authors.* [5] draws from the extensive experience from Australian fatigue monitoring programs.

REVIEW OF COMPUTATIONAL METHODS

Introduction

It is critical for a HUMS device that the quality and reliability of the data fitting algorithms, and hence of the computational methods used within them, are high. A number of computational methods previously used or proposed in the literature to derive the algorithms relating the output data from the HUMS and the response parameters required at the appropriate locations have been reviewed. The methods can be split into three groups:

1. ***Methods for regime recognition (RR);***
2. ***Optimisation techniques for obtaining best data fits in order to synthesise loads (FLS);***
3. ***Methods for constructing transfer functions to uniquely define loads from measured or predicted data at other positions.***

Methods for Regime Recognition

During flight, the flight parameter and loads data obtained by the HUMS must be analysed through an algorithm to give the current regime together with time spent in that regime. [8] describes the RR algorithm simply as a *Deterministic computer program that checks basic aircraft parameter data against preprogramed "normal ranges" to establish flight conditions.* This sounds simple, but the reality is quite complicated. Regime recognition has been widely applied in HUMS systems for aircraft, and, in particular, rotorcraft. For example, [2] refers to the U.S. Army AH-64D Apache

Helicopter. However, generally very little detail is given in the literature on the methods used in the algorithms for identifying the regime from the flight parameters. [2] uses proprietary algorithms and software from Boeing Rotocraft. [10] states that *most existing RR programs are proprietary and thus the specific details and processes are usually not known.*

One report giving some detail is [11]. The objective of this research was to establish regime recognition models using ***classification algorithms***. Two earlier algorithms used by Goodrich Corporation were referred to: ***a logical test-based regime recognition model*** and ***a maximum likelihood estimator methodology***. The study looked at a number of classification algorithms and established two regime recognition models which gave the best results. These were compared with two other methods, ***neural networks and logical regression.*** The classification algorithms were found to give the better performance. Different ways to reduce the effect of noise in the data by sub-setting, transforming and filtering were also considered.

In [12] an algorithm is developed based on a ***Hidden Markov Model.*** The paper adopts the philosophy of regime recognition as basically a data mining problem, i.e., mining the measured parameter data and mapping them to a defined flight profile. The model is compared against two other methods, a logical test-based model and a model implemented as a hierarchical set of elliptical function neural networks. The Hidden Markov Model algorithms were found to give the better performance. Logical Test-Based Models are the simplest models. In these models the specific parameter cases are tested and if the test result is true for a regime, that regime is considered true. [12] states that *if the measured 'parameters were free of noise, logical tests would give accurate results.*

The basic idea of regime recognition is very simple, but due to noise in the measured data errors are introduced. The more complex methods listed all seek to minimise these noise errors. The overall impression is that regime recognition is widely used, conservative and reliable. A wide range of modelling techniques is available from very simple to highly complex mathematics, but algorithms are largely proprietary. The measure for the success of the RR algorithm used is the proportion of misclassifications to the correct regime. Several of the studies give numerical estimates of this. A summary of some of the results obtained is presented in Table I. Methods for Regime Recognition identified from the literature are listed as follows:

1. ***Logical test-based models;***
2. ***Maximum likelihood estimator methods;***
3. ***Classification algorithms;***
4. ***Neural networks;***
5. ***Logical regression;***
6. ***Hidden Markov model.***

Table I. Comparison of Classification Rates obtained using Different RR Algorithms.

Reference	Algorithm	Correct Classification Rate (%)	Remarks
2	Boeing Rotorcraft Proprietary	98	
11	Classification Algorithms	95	The models were not as powerful in classifying the low-speed regimes as in classifying high-speed regimes.
11	Neural Networks	55	
11	Logical Regression	55	
12	Logical Test-Based Model	90	
12	Neural Network	76	This model gave near perfect classification results for level flight regimes. However, the results for classification of all regime groups did not show a consistent effectiveness. For example, for level flight group the classification rate was 97.85% but it was 33.18% for turns group. Because of the low classification rates for some groups, the scheme gave an overall rate of 76.21%.

Optimisation Techniques for Flight Loads Synthesis

Flight loads synthesis (FLS) involves making a prediction of the required response parameters from the precise values of the flight parameters monitored by the HUMS. Where there are data from a flight loads survey this prediction is achieved by defining the relationship between the parameters required and the flight parameters monitored using data fitting or optimisation techniques on the data directly gathered during the flight testing.

Computational methods identified in [8] which have been used before in FLS algorithms are ***Holometrics, Neural Networks and Multiple Linear Regression***. The latter was selected for use in the feasibility study reported. The holometric method was not used as its output is a time history trace which was not required for the damage calculation method used in the study. Numbers of cycles were introduced by a different method. Essentially, the method was simplified so that one cycle corresponded to one rotor cycle. Neural network approaches were not used as it was believed that they lacked the maturity and tools that were available for multiple linear regression at the time of the study. Additionally, ***Genetic Algorithms*** were proposed in [3].

A review of methodologies for the estimation of structural component loads in helicopters is presented in [13]. In this review the parameters to be monitored for estimating component loads for helicopter component fatigue life estimation purposes fall into one of three broad categories: loads on fixed components (enhanced FLS), flight parameters (RR and FLS), or both loads on fixed components and flight parameters (enhanced FLS). Computational methods covered then include: Holometrics in the first category; Regression, Neural Networks, Flight Condition (or Regime) Recognition, and combined Regression and Neural Networks in the second category; and an extension of the Holometric method in the third category.

Multiple Linear Regression: Regression is the basic tool for optimisation problems. Some discussion on regression is included in [13]. This is the method used in [8].

Neural Networks: Neural networks techniques are widely used. Applications in regime recognition have already been discussed. Some discussion on neural networks for FLS is included in [13]. [14] describes the development of a neural network prototype to predict strain from data obtained from an F-14 flight test program. Results were evaluated by comparing the correlation coefficients between the predicted and measured strains. The correlation coefficient obtained by the neural network was 0.93 and by a regression equation was 0.94. Based on these preliminary results, the conclusion is made that *the neural network approach offers a viable alternative to standard regression analysis for predicting strains on airframes*. Several methods based on neural networks were also considered in [15] where it was concluded that *neural networks show promise for use in high-fidelity, real time modelling of rotorcraft vibration.*

Holometrics: A description of Holometrics is given in [16]. Some discussion is included in [13] which states that *In the application of the holometric technique to an Apache helicopter, excellent identification of dynamic system loads was achieved. The only exceptions to this excellent identification result were the identification of loads that were well below the endurance limit.*

Genetic or Evolutionary Algorithms: In the computer science field of artificial intelligence, a genetic algorithm (GA) is a search heuristic that mimics the process of natural evolution. This heuristic is routinely used to generate useful solutions to optimization and search problems. Genetic algorithms belong to the larger class of evolutionary algorithms, which generate solutions to optimization problems using techniques inspired by natural evolution, such as inheritance, mutation, selection, and crossover. [17] explores the use of evolutionary computational methods to help estimate helicopter dynamic loads.

The conclusions of [8], which used multiple linear regression, were that the method was not as reliable as regime recognition due to the effects of noise in the data. Noise is clearly a major problem in all the methods. Other literature appears to report good results with various FLS methods. Methods for flight loads synthesis identified from the literature are listed as follows:

1. ***Multiple linear regression***
2. ***Neural networks***
3. ***Holometrics***
4. ***Genetic or evolutionary algorithms***

Transfer Functions

The transfer function approach gives a unique relationship, called a transfer function, between the monitored parameters and the required response parameters. Transfer functions can be obtained either from test or from finite element analysis as discussed above. For a payload it is concluded that testing may be impractical and that finite element analysis would offer the best approach. In both cases the solution is unique i.e. no best fit is involved, but again there will be the problem of noise in the monitored parameters from the HUMS and errors in the predictions of the finite element analysis. Errors in the predictions of the finite element analysis might be addressed by sensitivity studies and worst case modelling. Noise in the monitored parameters might be addressed using one of the computational methods discussed above. For example, a regime recognition or optimisation approach could be used first to obtain a best fit to the monitored data. This best fit would then be used with the transfer function to derive the required response parameters. [18] presents an investigation of the effects that may have an influence on the development of a linear stress transfer function relating the stress in dynamic components to the stress in static components. Effects such as buckling, non-uniqueness, vibration, and solution procedure are considered. Two procedures for determining the stress transfer function are compared, and the effects of noise are included.

Inverse methods for constructing transfer functions are considered in [19]. These offer techniques for re-constructing load inputs from structural responses. They are relevant to the situation where there is direct monitoring of loads, stresses or strains on the payload itself. If sufficient measurements are made on the payload to reconstruct the loads then response data can be further predicted at any location on the payload.

RECOMMENDATIONS AND CONCLUDING REMARKS

1. The technique for deriving the algorithm relating the output data from the HUMS on the aircraft and the response parameters required at each fatigue critical location on the payload should consist of two factors which are multiplied together. The first factor relates the output data from the HUMS to the loads at defined intermediate locations at the aircraft to payload interfaces. The second factor relates the loads at the intermediate locations to the required response parameters at the required critical locations.
2. The first factor could be obtained by one of three methods:
    a. If the aircraft HUMS monitors the loads at the intermediate locations then the first factor can be directly obtained from these data. (DLM)
    b. If a representative payload was attached to the aircraft during the aircraft flight loads survey and loads have been monitored at, or near to, the intermediate locations then these loads can be used to derive an estimate of the first factor using a data fitting or optimisation technique. (RR or FLS)
    c. If no such loads have been obtained during the aircraft flight loads survey then the first factor can only be obtained by finite element analysis of the combined aircraft and payload structure.
3. The second factor would be obtained as a transfer function from a finite element analysis of the payload structure alone.
4. The use of a data fitting or optimisation technique is implicit in the second method for obtaining the first factor, but should also be applied to minimise noise in the monitored data in all the other methods.
5. The third method for obtaining the first factor is the most general, requiring least measured or test data, and the most novel. It is the preferred method recommended here. It does require finite element modelling of the aircraft structure and therefore close cooperation between aircraft and payload authorities.
6. Either a regime recognition or flight loads synthesis approach can be used. No recommended methods for data fitting or optimisation are given as the choice of method will depend on the form of the data, what parameters are available from the aircraft HUMS and ease of implementation.
7. Sensitivity studies and worst case modelling should be adopted in the finite element analysis to minimise the adverse effects of errors in the predictions.
8. The methods would cater for both metallic and composite structures for which material fatigue properties are known.

REFERENCES

[1] Romero R, Summers H, and Cronkhite J (1996), *Feasibility Study of a Rotorcraft Health and Usage Monitoring System (HUMS): Results of Operator's Evaluation.* NASA 198446, ARL-CR-289.

[2] Shanthakumaran P, Larchuk T, Christ R, Mittleider D and Hitchcock E (2010), *Usage Based Fatigue Damage Calculation for AH-64 Apache Dynamic Components*, Presented at The American Helicopter Society 66th Annual Forum, Phoenix, Arizona, May 11-13, 2010.

[3] White DJ (2011), *Using HUMS Data to Manage the Fatigue Life of Rotorcraft Critical Dynamic Components*, AIAC14 Fourteenth Australian International Aerospace Congress, Seventh DSTO International Conference on Health & Usage Monitoring (HUMS 2011).

[4] Algera D, et al (2010), *Rotary Wing Structural Life Tracking Architecture*, Presented at The American Helicopter Society 66th Annual Forum, Phoenix, Arizona, May 11-13, 2010.

[5] Molent L and Agius J. (2008), *Structural Health Monitoring of Agile Military Aircraft.* In *Encyclopedia of Structural Health Monitoring*, Edited by C. Boller, F Chang and Y. Fujino, John Wiley & Sons, Ltd. 2008.

[6] McColl C, et al (2010), *Comprehensive UH-60 Loads Model Validation*, Presented at The American Helicopter Society 66th Annual Forum, Phoenix, Arizona, May 11-13, 2010.

[7] Zakrajsek JJ, et al (2006), *Rotorcraft Health Management Issues and Challenges*, NASA/TM-2006-214022.

[8] Dickson B, Cronkhite J, Bielefeld S, Killian L, and Hayden R (1996), *Feasibility Study of a Rotorcraft Health and Usage Monitoring System: Usage and Structural Life Monitoring Evaluation*. NASA 198447, ARL-CR-290.

[9] Cronkhite J, Dickson B, Martin W and Collingwood G (1998), *Operational Evaluation of a Health and Usage Monitoring System (HUMS)*, NASA/CR-1998-207409, ARL-CR-420.

[10] Suresh Moon, Nam Phan and David Churchill (2011), *Manoeuvre Recognition Verification & Validation Using Visualization*, AIAC14 Fourteenth Australian International Aerospace Congress, Seventh DSTO International Conference on Health & Usage Monitoring (HUMS 2011).

[11] Dere AM (2006), *Flight Regime Recognition Analysis for the Army UH-60A IMDS Usage*, Master's Thesis, Naval Postgraduate School, USA.

[12] David He, Shenliang Wu and Eric Bechhoefer (2010), *A Regime Recognition Algorithm for Helicopter Usage Monitoring*, In *Aerospace Technologies Advancements*, edited by: Dr. Thawar T. Arif, ISBN 978-953-7619-96-1, pp. 492, January 2010, INTECH, Croatia.

[13] Polanco FG (1999), *Estimation of Structural Component Loads in Helicopters: A Review of Current Methodologies*, DSTO-TN-0239, Australia.

[14] Hoffman ME (1992), *A Neural Network Prototype for Predicting F-14B Strains at the B.L.10 Longeron*, NAWC Report No. NAWCADWA-92042-60.

[15] Kottapalli S (2000), *Neural-Network-Based Modeling of Rotorcraft Vibration for Real-Time Applications*, AIAA-2000-4305.

[16] George Fox Lang and Clifford T Gunsallus (1989), *Statistical Transduction ... an Introduction to Holometrics*, *Sound and Vibration*, April 1989.

[17] Valdes JJ, Cheung C and Wang W (2011), *Evolutionary Computation Methods for Helicopter Loads Estimation*, 2011 IEEE Congress on Evolutionary Computation (CEC), pp. 1589-1596, NRC Canada.

[18] Polanco FG (2000), *Development of a Stress Transfer Function for an Idealised Helicopter Structure*, DSTO-RR-0171, Australia.

[19] R. Vishwakarma, D. Turner, A. Lewis, Y. Chen and Y. Xu (2012), *The Use of Pseudo-Inverse Methods in Reconstructing Loads on a Payload Structure*, *International Journal of Modelling, Identification and Control*, Vol. 17, No. 3, 2012, p. 242-250.

Challenges and Proposed Methodologies for Structural Health and Usage Monitoring Systems (SHUMS) in the domain of Airborne Complex Weapons

C Lomax[1], A Whitelaw[1], J Raiker[2], A Groves[3]

[1]*MBDA (UK) Limited, Six Hills Way, Stevenage, Hertfordshire, SG1 2DA, United Kingdom*
[2]*MJVR Limited, Sunnyside, Porkellis, Helston, Cornwall, TR13 0LB, United Kingdom*
[3]*Dstl, Porton Down, Salisbury, Wiltshire, SP4 0JQ, United Kingdom*

Abstract: Structural Health and Usage Monitoring Systems (SHUMS) have existed within aerospace and civil engineering for many years, and although there are many parallelisms in the design, operating environment and support activities there are very few examples of their use in airborne Complex Weapons, i.e. guided missiles and bombs. The introduction of SHUMS to airborne Complex Weapons should bring numerous benefits, most notably increased safety and longer operational lives through known loading and reduced ownership costs.

This paper will discuss the unique challenges of developing an autonomous self-contained SHUMS for airborne Complex Weapons with reference to the techniques and strategies for recording and assessing the remaining air carriage life and ensuring the Structural Integrity of the installed Weapon System.

It will detail the difference between Operational Monitoring and Damage Detection approaches for SHUMS and the implications on the current design criteria and flight approval of airborne Complex Weapons

The relative merits of several analytical techniques including Regime Recognition, Direct Load Measurement and Indirect Load Reconstruction shall be considered, with a review of work which has developed, or is looking to develop, these techniques. A specific focus is drawn on the development of Indirect Load Reconstruction where various computational algorithms such as Inverse Methods and Pattern Recognition may be used to generate a Loads Reconstruction model to derive the real-time flight loads for use in a fatigue assessment of typical structural features.

The paper concludes with a development SHUMS proposal, utilising a combination of the various approaches and analysis techniques discussed within the report, and SHUMS technologies, such as sensors, that are available to meet the specific SHUMS requirements.

CONTEXT

The development of a Structural Health and Usage Monitoring System (SHUMS) for Complex Weapons discussed in this paper is carried out as part of the MAST STC Research Programme managed by Dstl under contract number 'DSTLX-1000086238'.

Structural Health and Usage Monitoring Systems (SHUMS) have existed within aerospace and civil engineering for many years and, although there are many parallelisms in the design, operating environment and support activities, there are very few examples of their use in airborne Complex Weapons, i.e. guided missiles and bombs.

Current airworthiness requirements for airborne Complex Weapons and the associated safety design criteria utilised by the UK is of a safe-life philosophy, which results in airborne Complex Weapons structural clearance being a deterministic process resulting in air carriage life being certified on a time based policy regardless of the actual individual asset utilisation and environment.

Cost conscious customers and competitive export markets are pushing airborne Complex Weapons suppliers towards longer service lives and limited, or no, through life maintenance. At the same time, as the performance of weapon platforms, such as military aircraft and unmanned air vehicles, is increasing there is a corresponding increase in the severity of the environment Complex Weapons are expected to operate in, which would typically lead to shorter safe structural lives using the traditional Safe-Life philosophy. This is in direct contrast to the increase in utilisation through peacekeeping operations; these are frequently conducted under benign conditions with only, if any, short periods of combat. SHUMS thus also offers a realistic solution to this dichotomy.

The introduction of SHUMS systems into airborne Complex Weapons will allow a fundamental change in the structural life clearance methodology to allow the customers to manage the structural life of individual assets. Whilst monitoring

SHUMS will require greater engagement of the customer in the short term as new asset management philosophies are introduced in the longer term reduced costs will emerge. Other benefits include;

i. Increase safety
ii. Extend operational life
iii. Reduce support costs
iv. Reduce downtime
v. Increase reliability
vi. Warranty protection and risk mitigation for availability contracting

It is also worthy of note that SHUMS is being considered as an integral part of the Integrated Vehicle Health Management (IVHM) philosophy which, for Complex Weapons also includes energetic materials and electronics. Knowledge gathered by a SHUMS should thus have wider application than just structural.

UNDERSTANDING THE CHALLENGES

The design challenges for Complex Weapons SHUMS are unusual. Missiles, by their very nature, are compact and tightly packaged, with critical structural features that interface with the Weapon Platform being subject to contact and fretting due to relative motion of adjacent components.

Throughout the majority of the service life of a Complex Weapon, particularly during extended periods in storage, there is neither internal nor external power available and it must be demonstrated that adding a new power source will not adversely affect the stringent safety requirements of a Complex Weapon.

Air carried Complex Weapons generally suffer a harsh loads environment. Wing tip carried stores can see accelerations in excess of 17g [1], broadband vibrations in excess of $8.9g_{rms}$ [1], and shocks that can be higher than 40g [1]. Thermal conditions can range from lower than minus 50°C to greater than 100°C. These and other environments all threaten the structural integrity of the Complex Weapon.

DESIGN CONSIDERATIONS

The most basic design considerations for SHUMS for Complex Weapons are to identify the areas of concern and the parameters that influence their behaviour. Engineering experience has shown that the typical structural features listed below are either life limiting, or are subject to more conservative analysis due to uncertainty in the expected operating environment.

i. Weapon Platform interfaces
ii. Body joints
iii. Unavoidable stress concentrations
iv. Bonded structures
v. Composite structures
vi. Specialist or delicate internal devices

Many of these features are designed using a Safe-Life philosophy and it is not within the scope of the SHUMS programme to change them or consider an Inspection Dependent or Damage Tolerant philosophy. This is significant as there are typically two fundamentally different monitoring approaches; Operational Monitoring and Damage Detection, and selecting one over the other can be related to the clearance philosophy used on the structure

Damage Detection, as the name suggests, uses sensors and analysis methods that identify the presence of structural defects and damage. More sophisticated systems could determine the location and magnitude of the damage and assess this against predetermined allowable values to advise the operator when to conduct maintenance or retire the structure from service. Operational Monitoring instead measures the environment the structure is exposed to and uses models to predict when structural failures are likely to occur. By managing the accrual of predicted damage it is possible for the operator to ensure that the structure is removed from service when the structure is realistically reaching it end of life, as opposed to an end of life based on hours flown which is likely not to reflect the state of the structure. Clearly the opposite is true for a missile that has experienced very severe loading thus increasing MOD missile asset safety as well.

Components which are designed with a Safe-Life philosophy, such as missile hangers which are very highly stressed and exploit very strong materials with an associated small critical defect size, have a very short period between detectable and critical damage limits; this means that it would become very difficult to safely manage the Structural Integrity of Complex Weapons using Damage Detection in isolation. For this reason it is that Safe-Life structures adopt an Operational Monitoring approach.

There are many threats to the Structural Integrity of an airborne Complex Weapon, but the priority of SHUMS will be on monitoring the structure for fatigue and overload during air carriage or transportation and handling. There are mitigating factors for corrosion damage already in place, such as coating and controlled storage conditions environments.

During the development programme of SHUMS for Complex Weapons many of the key stakeholders from the MOD, regulatory bodies and industry were invited to provide their input on how SHUMS should operate. This highlighted many practical factors which will influence the development of SHUMS.

It is unacceptable to significantly increase the maintenance required due to SHUMS, therefore the periodicity of data downloads and battery changes must be considered if SHUMS is to be implemented on an operational weapon. Ideally maintenance should only occur when the SHUMS systems requests human intervention and ideally less than current maintenance philosophies.

The security of SHUMS data must be protected; even the knowledge of the weapons location and stockpile size is often classified. The transmission of data wirelessly may impact the security of the weapons location; seemingly simple solutions to problems quickly become complicated.

The off-board storage, distribution and post-processing system, will also require handling of sensitive data and may need new, or revised, logistical support services to distribute and post-process the data in to usable information for the customer, user and industry. The level of detail presented should be tailored according to role and responsibility of the users. This may mean that users in-theatre have a concise Go/ No-Go statement while design authorities have access to flight by flight and fleet wide trends.

There is not an agreed UK certification process for the use of SHUMS on airborne Complex Weapons. Whilst this does not, and should not, prevent a SHUMS being designed and offered for evaluation it is a significant challenge to industry and the customer communities to agree a formal process. There are significant safety considerations to be addressed, not least of which are the reliability and accuracy of the SHUMS.

REVIEW OF CURRENT TECHNOLOGY

Structural monitoring has been a topic of much research in recent years, and the development of SHUMS for Complex Weapons included a review of 160 papers. Although no significant research was identified for Complex Weapons, there were numerous references to the implementation of structural health monitoring systems to comparable systems with development of algorithms, sensor and power technology that may be applied to provide a novel capability.

Algorithms

In the context of SHUMS for Complex Weapons, an algorithm is considered to be the means by which information from measured data is converted in to knowledge about the state of the structure. It is perceived that the most relevant algorithms may be Inverse Methods, Flight Loads Synthesis and Pattern Recognition.

The first two approaches are consistent with a Loads Reconstruction strategy, with a subsequent fatigue analysis to predict the remaining safe-life of the Complex Weapon. The third approach does not determine loads, but provides an output of a change is structural behaviour indicating possible structural defects.

Inverse Methods. In the field of mathematical modelling, two main frameworks exist to fully define the physical system being observed. If the physical system is fully defined and understood then modelling can be used to predict the outcome of measurements of the system with different inputs.

The first, and simplest, framework is known as the forward problem. In the forward problem, the model's parameters are fully defined and understood and these are used to generate data. An inverse problem, as one would imagine, is this inverse of this whereby data from a physical system is observed and used to infer the parameters characterising the system under investigation.

The application of Inverse Methods would capture missile response as measured accelerations and displacements that can be processed to determine the loads acting on the airframe and conduct a fatigue analysis to estimate the safe fatigue life that has been consumed. This approach has a major benefit that it is able to determine large amounts of useful data about the state of a structure using a small number of precise measurements [2].

Flight Loads Synthesis. This approach would involve an indirect determination of the loads during flight at the fatigue critical components using flight data from the Weapon Platform, together with a derived algorithm to synthesise the oscillatory loads and number of cycles seen at the fatigue critical locations on the missile and perform a subsequent fatigue analysis.

Pattern Recognition. An algorithm can be trained to monitor measured data, and attribute recognised patterns to a certain conditions using a statistical assessment. This may be applied in combination with a Regime Recognition strategy which categorises the rate at which structural damage is accrued with typical usage under specific types of operation. The alternative application for a Damage Detection methodology is in monitoring changes from a baselined response of a structure to identify when damage has occurred [3].

Sensors

A vast range of sensors are marketed for structural monitoring applications, with the most interesting being the development of Micro Electro-Mechanical Systems (MEMS) which are smaller, potentially cheaper and consume less power than their conventional equivalents.

Energy harvesting

Energy harvesting aims to capture energy from the area surrounding the Complex Weapon, either to supplement or replace the requirement for conventional batteries, which have a finite life. Energy harvesting can look to extract various forms of ambient energy including thermal, solar and electro-magnetic, however the most viable source may to mount piezo-electric generators on the missile airframe to utilise kinetic energy in the form of vibration caused by the very same cyclic loading which is expected to cause fatigue in the structure [4].

DEVELOPMENT OF A SHUMS PROPOSAL

The development and acceptance of SHUMS for Complex Weapons would need to take in to account the technical challenges and the requirements of the key stakeholders within the MOD, regulatory organisations and industry. The following proposal assumes that SHUMS shall not be retro-fitted to any individual Complex Weapon that have already been delivered into service, as it is deemed to be too difficult to accurately reflect the provenance and current status of the asset. SHUMS may however be retro-fitted to a new build Complex Weapons of an existing design.

SHUMS shall be capable of assessing the remaining life of multiple structural features, and providing an indication of structurally significant events, such as over-load, shocks, drops and impacts, which may need further actions before the asset can be returned to service. However any output from SHUMS shall be considered as advisory and may be disregarded by a user with appropriate authority, although a record of the decision shall be held by SHUMS. This should avoid disrupting operational use of the airborne Complex Weapon under unforeseen circumstances, or system faults.

The SHUMS solution shall consist of an on-board system to gather data and process information, and an off-board system which allows fleet-wide information to be stored and accessed by multiple users at different locations.

The on-board SHUMS system is expected to operate as an automated system which is able to monitor and manage its own data storage capacity and power supply as only a limited scheduled maintenance currently occurs to in-service Complex Weapons, and SHUMS should not dictate the periodicity of future maintenance. The indication that some further action must take place is to be integrated with the logistical activities, such as recording which Complex Weapons have been loaded and unloaded to the Weapon Platform.

The current proposal for SHUMS has selected an Operational Monitoring approach that shall use an Inverse Method for Loads Reconstruction with airframe response measured by accelerometers along the length of the structure. It is felt that this configuration could be most easily integrated with the structure as only several sensors will be required on a typical airborne Complex Weapons, while also providing information with sufficient accuracy to predict the accrual of fatigue damage and identify any static over-load.

Power shall be provided to the system by conventional battery technology, but to extend the period between battery maintenance energy shall be harvested from the motion of the missile using piezoelectric generators, as well as exploiting power from the Weapon Platform, when available. Battery replacement could be scheduled as part of routine maintenance and arranged to be no greater than current maintenance philosophies.

The on-board SHUMS architecture should include a maintenance interface module that can provide two-way communication with handheld Local Support Equipment (LSE). The LSE shall display sufficient data to allow a user to assess the status and suitability of individual assets to be used for the intended mission, and have sufficient data storage to download SHUMS data from the Complex Weapon, and upload software updates to the Complex Weapon. The Ground Station computer shall receive the information from the LSE, and can be used for more complex interrogation of the SHUMS data for an individual Complex Weapons, or analysis of fleet-wide information.

The function of the off-board SHUMS is to distribute information to stakeholders in a timely and efficient manner. The transfer of information has two stages; from Complex Weapon to local Ground Station computer, and from local Ground Station computer to other Ground Station computers via a centralised data storage device.

The introduction of SHUMS for Complex Weapons is now being considered as a component within the broader context of Integrated Vehicle Health Management (IVHM), this recognises that the life of a Complex Weapon may be limited by other features, most notably the energetic material in the rocket motors and the warhead. By having a consistent approach to the monitoring and management of these various features would allow similar sensor requirements to be shared, and minimising the amount of maintenance required to recovery information.

CONCLUDING REMARKS

This study was unable to identify any existing systems that provide the desired monitoring capabilities on airborne Complex Weapons, therefore is has been concluded that the proposed SHUMS would be introducing a novel capability to the market, and the resulting SHUMS capability shall be beneficial to both the customer and industry.

As SHUMS will be monitoring the loads on individual Complex Weapons it is recognised that the cleared life of a Safe-Life structure could be increased by at least 50% [5] if the in-service loading was equivalent to the design assumptions. Should the in-service loading be less severe, then the life may be extended further. By extending the service life of a Complex Weapon, the operating costs can be reduced

The ability to monitor and manage fatigue and static over-load which may threaten the structural integrity of individual airborne Complex Weapons also has a positive impact on safety, and can also provide industry with risk mitigation against warranty claims when the structure has been exposed to an environment beyond the agreed design requirements.

It is apparent that SHUMS will need to address certain technical challenges including airworthiness approval. However it appears that the technology building blocks exist, even if it hasn't been implemented in this manner before.

REFERENCES

[1] DEF STAN 00-35. (2006). *Environmental Handbook for Defence Materiel.*

[2] Ruderman G, (2010), *Health Management and Service Life for Air Force Missiles – Past and Future.* RTO-MP-AVT-176

[3] Noman A., Farah D. and Bagchi A. (2009). *Structural Health Monitoring Using Vibration-Based methods and Statistical Pattern Recognition Techniques.*

[4] Yuse K., Lallart M., Petit L., Richard C., Monnier T. and Guyomar D. (2010). *Self-powered structural health monitoring with nonlinear energy harvesting system.*

[5] DEF STAN 00-970. (2014). *Design and Airworthiness Requirements for Service Aircraft.* Leaflet 35

Flight Condition Recognition and Independent Flight Severity Monitoring for the Management of Helicopter Fleets

Brian H Perrett

HeliSAC Limited

Abstract: This paper describes Helicopter Flight Condition Recognition algorithms developed in parallel with a flight severity indicator used to provide independent assurance for the full identification of usage parameters. Helicopter usage has been identified in the context of the Design Usage Spectrum, used by the Designer to set life in service for critical components and structure. Additionally, usage is compiled with sufficient resolution to correlate with the loads data gathered during the life substantiation programme of the AgustaWestland AW159 Wildcat. The resulting usage analyses are designed to replace the expensive and intrusive Manual Data and Operational Data Recording programmes that are currently used in the UK to seek assurance that helicopters are being operated within their design envelopes. By using data from the Health and usage Monitoring systems of the Wildcat and by embedding the analysis functions in software having fleet management capability, it is proposed that this approach could provide continuous fleet-wide monitoring of sortie usage and flight severity.

INTRODUCTION

The work described in this paper has been sponsored as part of the UK, Dstl 'Understanding Ageing Aircraft' R&D programme, and evolved from the need to describe flight severity for the UK historic helicopter fleet to confirm that the aircraft were being operated within the design performance envelope. Whilst this early stage of the programme concentrated on describing the distribution of Normal accelerations over the range of angular pitch and roll displacements encountered in service, simple Flight Condition Recognition (FCR) algorithms were also developed to identify a limited range of usage parameters that could be related both to the Design Usage Spectrum and the manoeuvre activity. These algorithms used data from a low cost instrumentation suite providing only pitch, roll and yaw in addition to three-axis accelerations - all recorded on a common time base.

The programme emphasis changed when data containing a greater number of flight parameters (from a Eurocopter AS350 squirrel of the Defence Helicopter Flying School) became available. By changing the way in which the flight parameters were used to identify both steady state flight conditions and transitions, it became possible to increase the number of flight events that could be identified using a series of deterministic FCR algorithms. In addition, the manoeuvre severity algorithms based on pitch rates, roll rates and acceleration were developed to provide an independent measure of flight severity that could be compared in software with the flight profile and usage spectrum described by the FCR algorithms. These capabilities have been developed in parallel with features that provide elements of fleet management, allowing the review of individual tail histories, individual sortie histories and cumulative fleet histories.

The work is currently being expanded to provide an alternative to the Manual Data and Operational Data Recording (ODR) programmes that are a requirement of the UK Military airworthiness regulations. The ODR programme involves an extensive instrumentation fit and is required to provide information describing helicopter usage and usage severity (in the form of strain histories at a limited number of locations) for aircraft that must then be exposed to the full range of fleet usage. Typically this might involve the instrumentation of between three and five aircraft (including slip-ring systems), a data gathering programme lasting three to five years - during which the aircraft must be exposed to the full range of operational flying - followed by an intensive analysis programme. This is an expensive undertaking and very intrusive in terms of engineering support and system maintenance. To provide an alternative approach, the Fleet management capability described in this paper, developed using data from the AS350, has been adapted to operate with the Health and Usage Monitoring system (HUMs) outputs of the new AgustaWestland AW159 Wildcat fleet in order to provide a permanent fleet history that will obviate the need for both Manual Data Recording and the ODR Programme.

This paper describes the current Fleet management capability in terms of the FCR algorithms, the flight severity algorithms and the independent correlation of Flight conditions, flight events and flight severity metrics. It also describes how usage and usage severity will be related to damage by correlation with the original loads database and lifing procedures of the Design Authority.

THE FLIGHT CONDITION RECOGNITION ALGORITHMS

Flight Condition Recognition (FCR) has been of interest in the helicopter community for many years and various programmes of work have used information from control inputs and other aircraft sensors to describe usage histories [1].

Recent years have seen a surge in the production of affordable compact transducer systems providing a self-contained capability that can deliver the functionality of a Flight Data Recorder (FDR) [2]. In addition, many helicopters of the UK Military fleets contain Health and Usage Monitoring systems that use transducer outputs to provide a continuous record of key physical flight parameters. In this work, those flight parameters that describe the attitude of the aircraft and the rates and accelerations associated with changes in the aircraft attitude have been used to determine usage and usage severity.

Deterministic FCR algorithms have been developed to describe those features of flight that are used during structural engineering programmes to set the service lives of dynamic components and airframe structure through a knowledge of the loads generated during flight manoeuvres, transitions and steady-state flight conditions. The FCR outputs have been used to describe sortie profiles and the various relationships that are published in the Design Usage Spectrum.

Although early FCR relationships, using data from the Sioux helicopter of the historic fleet [3], used only pitch, roll and yaw to detect transitions, the greater range of flight parameters made available by access to Flight Data Recorder/HUMs outputs [4] have made it possible to 'drive' the algorithms in a more structured way and sensor inputs to the FCR algorithms now include the following key helicopter flight parameters:

- Ground speed (knots)
- Airspeed (knots)
- Vertical speed (feet/minute)
- Altitude (feet)
- Pitch-angle (degrees)
- Pitch-angle rate (degrees per second)
- Roll-angle (degrees)
- Roll-angle rate (degrees/second)
- Yaw-angle (degrees)
- Yaw-angle rate (degrees per second)
- Drift angle
- Weight-on-wheels (on/off)
- Rotor speed (% Nr)
- Vertical acceleration Nz (g) – manoeuvring frequency and high frequency
- Time – common time-base

As an example of the deterministic relationships used to identify flight conditions and manoeuvres, the following figure shows the identification of a transition to forward flight from low speed flight. The algorithm looks for a change in the ground speed of the helicopter to above 20 knots. The algorithm then scans the pitch channel from the same location on the time-base, in the direction of reducing time, to find a change in pitch to a nose-down attitude.

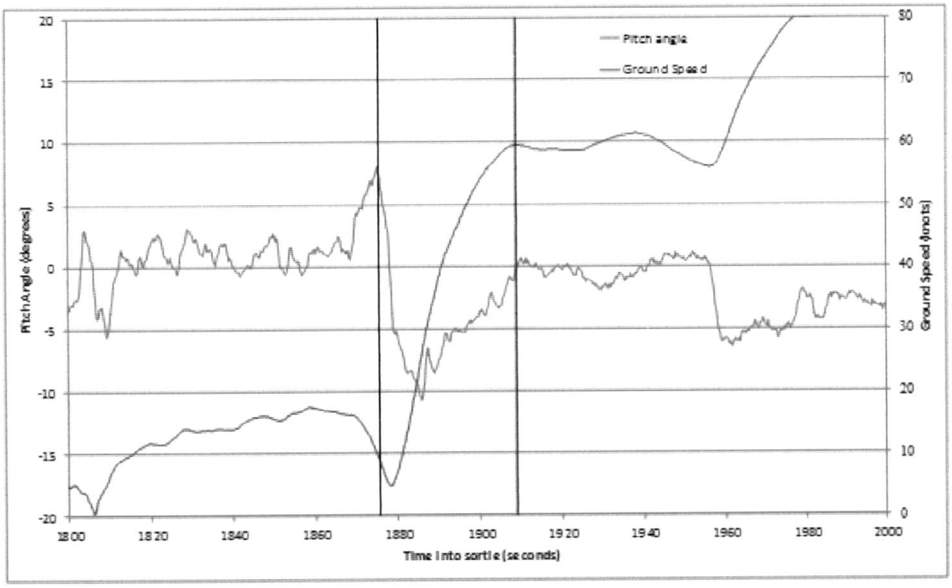

Figure 1. Change in pitch during a transition from low speed manoeuvring to forward flight.

Figure 1 above shows forward speed relative to the ground in red and pitch angle in blue. At the point that forward speed goes above 20 knots (against the scale on the right), the algorithm looks back in time for a change in pitch angle (value on the left scale) to 'nose down' – a change in aircraft attitude resulting from the pilot's input on the cyclic channel to redirect the main rotor thrust vector for forward motion. The algorithm then looks for the negative-going crossing of zero and the point where the nose-down transition is initiated, the point at which the most negative trough is reached (in this case at -10.75°) and the point at which the pitch 'disturbance' returns to zero following the event. The period between the two vertical black lines above is then recorded in a linear list of parameters showing the time at which each of these three events occurs and also the magnitude of the negative trough. In addition, a cumulative variable is incremented by the time in transition to forward flight and a counter incremented to record the frequency of occurrence of the transition. These are stored in a matrix for comparison with the Design Usage Spectrum. In this way, steady state conditions and the transitions between them are identified and stored as lists of discrete event 'types'. Although some events such as the quick-stop, Autorotation and wingover involve the review of a greater number of physical flight parameters and a more complex set of 'logical deciders', it has been possible to identify the flight conditions listed below using relatively uncomplicated relationships:

- Transition from hover to flight
- Turning transitions from hover to flight
- Transition from flight to hover
- Flares
- Quick-stops
- Auto-rotations
- Time in hover
- Spot turn spectrum
- Time in low-speed manoeuvring flight
- Time in forward flight
- Forward flight spectrum
- Rolling turn spectrum
- Wingover
- Altitude spectrum
- Altitude rate spectrum
- Quartering flight spectrum
- 'g' – exceedance spectrum

All flight conditions are recorded as unique entries in lists which hold the time of the event start, the time of the event end and a number of related 'deciders' (decelerations, accelerations, height lost, speed lost, etc) that help to define the event. Such 'deciders' would make it possible for instance to differentiate between a transition to hover, a flare or a quick-stop. These lists are later scanned in order to confirm the nature of the entry and to generate analyses that can be used to enhance fleet management. This issue is addressed later.

THE FLIGHT SEVERITY ALGORITHM

An additional software feature provides an independent measure of flight severity by recording rates and accelerations during the pitch and roll displacements associated with departures (upsets) from the current steady state of the aircraft. An independent pass of the pitch, roll and Normal acceleration (Nz) data is made, looking for excursions in pitch beyond ±5 degrees and in roll beyond 10 degrees. Where an excursion is found, then for the duration of the pitch change and/or the duration of the roll change, the rate of change of pitch and/or roll is calculated. Also, the Nz data are reviewed in order to extract statistics describing the behaviour of vertical accelerations during the 'upset'.

A Severity Index is then determined by simply summing the pitch and roll rates and the product of the square of the maximum normal acceleration reached during the upset :-

$$SI = \left\{\frac{(angular\ pitch\ range)}{(duration\ of\ pitch\ range)} + \frac{(angular\ roll\ range)}{(duration\ of\ roll\ range)}\right\} \times Nz^2$$

These values can then be plotted on the axes of a 3-dimensional matrix covering the full range of potential pitch and roll angle ranges as follows.

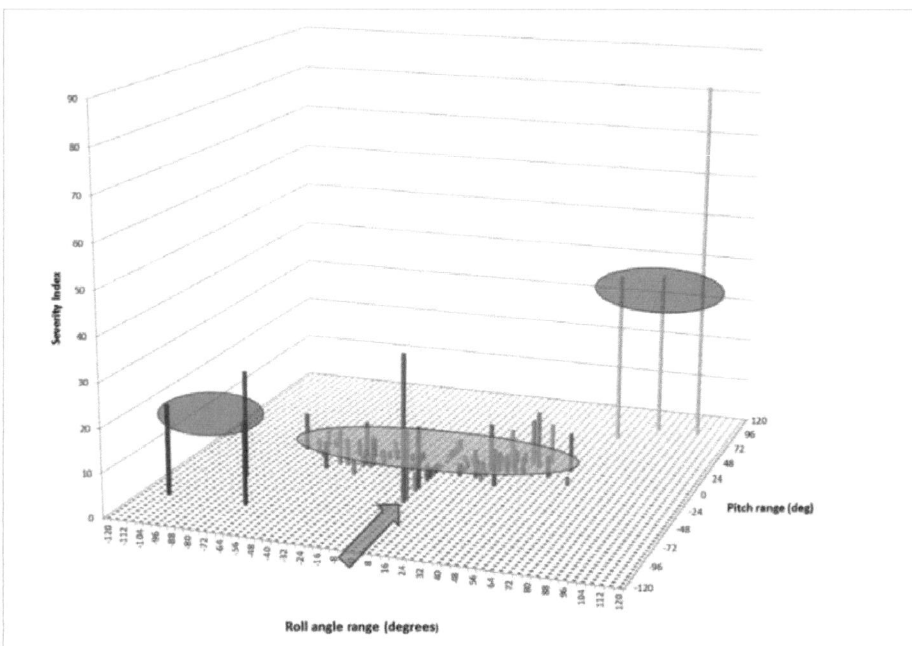

Figure 2. Severity Index distribution for a single display sortie of the AS350 Squirrel.

The severity index distribution, shown above in Figure 2, is the result of scanning data from a single display sortie of the AS350 squirrel of the UK Defence Helicopter Flying School [4]. The x-axis shows roll range (peak-trough), the y-axis shows pitch range (peak-trough) and the z-axis shows the cumulative severity factor at combinations of pitch and roll range. The broad blue arrow indicates the vector containing data having no roll content – only pitch. It can be seen that the bulk of the occurrences lie about the middle of the plot (shaded in orange) with the biggest spread occurring on the roll (x) axis. This would be expected, as in normal flying pitch displacements are relatively modest. The plot does show extreme occurrences in the top right and bottom left quadrants of the matrix (shaded in red) and these are due to three wingovers that formed part of the display training flight from which the data come. They can be considered as extreme by virtue of their position in the matrix and by their amplitudes on the z-axis scale. The three top-right excursions are due to positive going pitch combined with increasing roll and the three at the bottom left result from negative-going pitch and negative-going roll (return to the perpendicular). Although the wingover manoeuvres produce increased 'g', this did not exceed 2'g' during the manoeuvre.

The severity scan provides the quick visual review of the sortie (shown above) but also produces a linear list containing the severity contributors (pitch rate, roll rate and Normal accelerations) and the time of the beginning and end of each up-set/disturbance.

FLIGHT CONDITION RECOGNITION AND FLIGHT SEVERITY IN FLEET MANAGEMENT

Fleet management capabilities are required in order to ensure that the engineering Support Authorities and Design Authorities have an accurate and complete operational history for 'the Fleet' and for individual aircraft of the fleet. This needs to be recorded in terms that can be related to the consumption of service life for critical components and structure – as determined by the Designers life substantiation procedures.

On delivery of the aircraft the operator and support team are given various documents that describe how the aircraft are to be flown and operated as follows :-

Limiting conditions for flight	– published in the Aircrew Manual
Sortie Profiles	– published in the Statement of Operating Intent
Design Usage Spectrum	– published in the Statement of Operating Intent and Aircrew Manual

Currently, helicopters are operated by recording time in sortie, by recording written detail in form 724s and by repeating the MDRE and ODR referred to in the introduction. The MDRE is a manual data recording programme having the obvious limitations of human error and the ODR is a very expensive and intrusive programme that describes aircraft usage within a limited operational window [5]. Although ODR programmes should contain elements of both usage and usage severity, past applications of the ODR have lacked a full implementation of the usage monitoring requirement as disproportionate cost and effort is spent acquiring loads.

It is proposed here that a continuous programme – based on FCR and flight severity monitoring – would provide an improved and more complete history of individual and fleet usage. The challenge would be to provide usage information in a way that would give sufficient confidence to the Designer to be able relate flight histories to the procedures used to substantiate the strength and service life of critical dynamic components and structure. To a degree this is addressed in the processes described above; the flight severity 'pass' of the sortie data produces a linear list of 'upsets' which can then be compared with the linear list of flight conditions/events (Sortie Profile) that is produced by the FCR algorithm. Any entries in the 'severity list' that cannot be matched with an entry occupying the same time window in the sortie profile list (FCR output) would then represent an occurrence that might raise concern. This process is illustrated in the following figure.

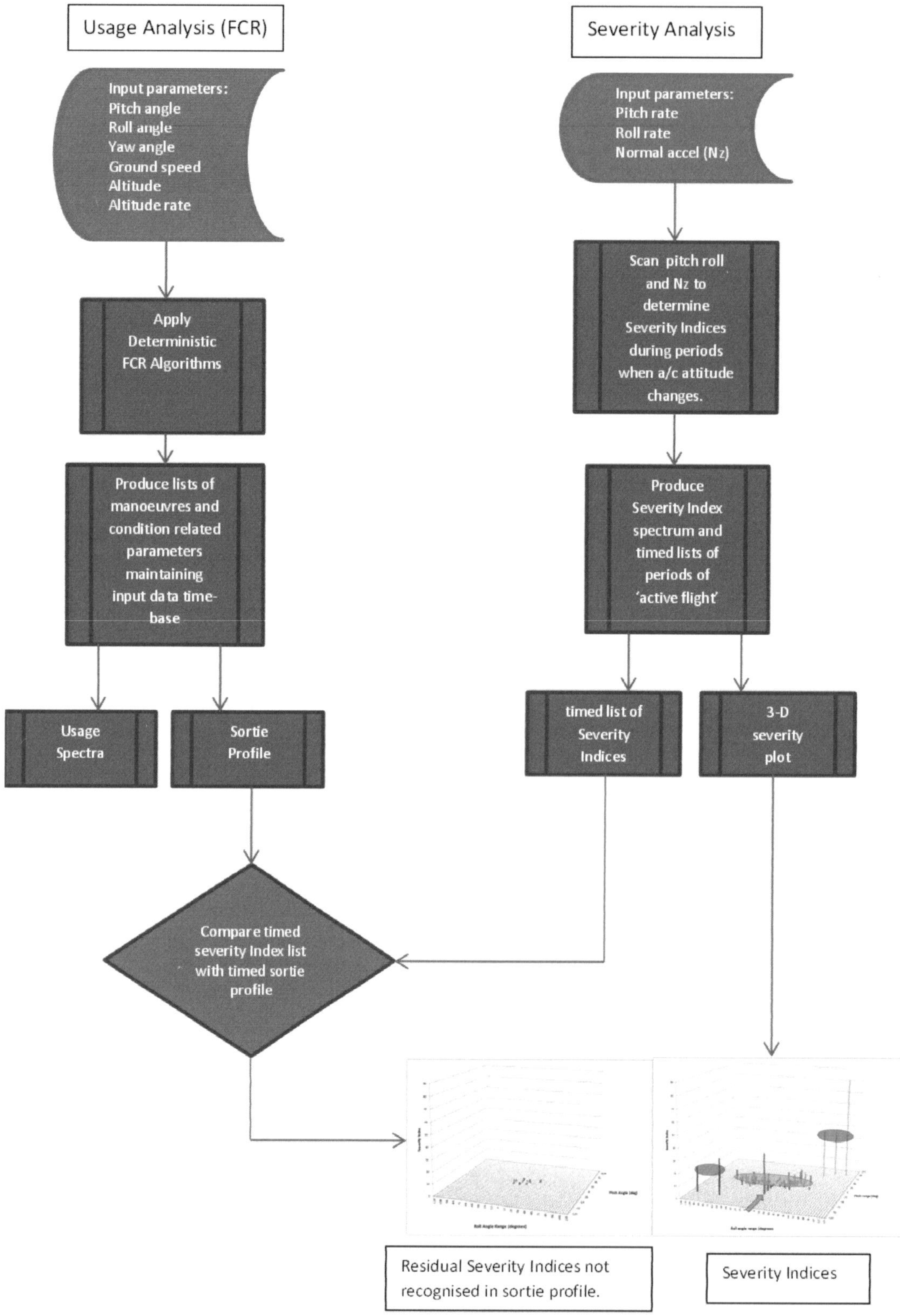

Figure 3. Flow chart for Flight Condition Recognition with Flight Severity 'sweep-up'.

The figure above shows the processes that have been applied to data from the AS350 Squirrel with the 3-D plot at the bottom right showing the total severity distribution for the sortie (as in figure 2 above) and adjacent 3-D plot on the left showing residual severity 'events' that have not been associated with flight conditions or manoeuvres from the sortie profile. It can be seen that, both in terms of their distribution and severity magnitudes, in this case these unaccounted activities are of no significance and therefore no concern. In fact they amount to only 4% of the 'activity' ascribed to recognised events in terms of their cumulative severity index and are clustered in the centre of the matrix. Apart from providing a means of comparing the severity of sorties, this 'sweep-up' procedure provides confidence that the FCR algorithms have captured all of the significant conditions and events. Furthermore, during the course of the scan, the severity data are combined with the sortie profile list (where they are coincident in time) in order to show the severity with which each manoeuvre or condition is encountered. The following table shows such an output from the analysis of the AS350 squirrel display practice flight.

Table I. List of damaging flight conditions from Squirrel display practice sortie.

	Manoeuvre	Rotor+start	Severity Index	Roll %	combined %	Pitch %
1	Wingover	1:1:51	92.35	0.00	100.00	0.00
2	Wingover	1:2:26	57.96	0.00	100.00	0.00
3	Wingover	1:2:53	53.46	0.00	100.00	0.00
4	Deceleration	1:2:19	36.91	0.00	0.00	100.00
5	Deceleration	1:2:49	32.42	0.00	100.00	0.00
6	Autorotation	0:58:58	26.38	0.00	0.00	100.00
7	Rolling turn	1:42:1	26.04	40.80	59.20	0.00
8	Rolling turn	0:56:7	24.21	6.58	93.42	0.00
9	Descending flight	1:12:5	18.73	47.78	52.22	0.00
10	Autorotation	1:24:21	16.72	0.00	0.00	100.00
11	Transition to flight	0:28:14	16.33	62.33	37.67	0.00
12	Rolling turn	0:43:23	15.75	0.00	96.16	3.84
13	Flare	1:12:49	13.86	71.09	28.91	0.00
14	Rolling turn	1:9:58	13.74	0.00	100.00	0.00
15	Rolling turn	0:39:22	13.34	0.00	100.00	0.00
16	Rolling turn	1:7:17	13.26	60.81	39.19	0.00
17	Rolling turn	1:14:31	12.18	79.56	20.44	0.00
18	Rolling turn	1:12:23	11.86	100.00	0.00	0.00
19	Rolling turn	0:55:25	11.44	0.00	100.00	0.00
20	Rolling turn	1:27:29	10.05	0.00	100.00	0.00
21	Transition to flight	1:17:38	9.91	44.85	55.15	0.00
22	Transition to hover	1:13:4	9.18	0.00	0.00	100.00
23	Rolling turn	1:14:1	8.96	0.00	100.00	0.00
24	Transition to hover	1:32:36	8.91	0.00	0.00	100.00
25	Rolling turn	1:34:12	8.78	31.68	68.32	0.00
26	Transition to hover	1:36:18	8.66	0.00	0.00	100.00
27	Rolling turn	1:26:36	8.06	53.21	46.79	0.00
28	Rolling turn	1:9:37	8.00	100.00	0.00	0.00
29	Rolling turn	0:36:28	7.88	60.71	39.29	0.00
30	Transition to hover	1:8:3	7.55	0.00	0.00	100.00
31	Rolling turn	1:23:45	7.53	0.00	100.00	0.00
32	Rolling turn	0:57:24	7.04	58.35	41.65	0.00
33	Rolling turn	1:17:54	7.00	100.00	0.00	0.00
34	Rolling turn	0:50:51	6.84	61.49	38.51	0.00
35	Rolling turn	1:4:5	6.58	32.65	67.35	0.00
36	Transition to flight	1:8:44	6.52	0.00	100.00	0.00
37	Rolling turn	1:4:57	6.39	37.57	62.43	0.00
38	Rolling turn	1:0:45	6.31	100.00	0.00	0.00
39	Descending flight	1:9:26	5.93	0.00	100.00	0.00
40	Rolling turn	1:11:57	5.93	0.00	100.00	0.00
41	Rolling turn	1:6:23	5.85	65.73	34.27	0.00

The 'screen dump' shown above lists conditions and transitions in order of decreasing 'severity' with the time at which they appear in the sortie shown in the third column. Alternative tables show the sortie flight conditions and events as they occur in time together with critical parameters such as climb rate, roll angle, deceleration rate, etc. Other features of the software allow the graphical display of various features from the sortie such as severity plots, normal acceleration distributions, g-exceedence plots, drift-angle plots (low speed and high speed), etc.

Compatibility with strength substantiation procedures

As discussed earlier, when an aircraft is delivered, its projected operational service is based on the published usage assumptions which the ODR programme is intended to revalidate – particularly with respect to the Design Usage Spectrum and loads in critical components and structure. Whilst being able to record the content of the DUS has obvious benefits in terms of fleet management, the DUS does not provide the full description of how usage relates to the 'partial damage sums' that determine the 'safe life' in service of critical components. The loads due to usage are related to that usage with a higher resolution through the Designers loads survey matrix and, although this information is less manageable than the DUS, it is perfectly possible to produce flight histories in this format using FCR algorithms having appropriate resolution.

To pursue this goal AgustaWestland have given their support to this programme by providing advice, access to their loads survey matrix and to associated test flight data for their new AW159 Wildcat helicopter. The following table shows the 34 unique conditions that comprise the Design Usage Spectrum published in the Statement of Operating Intent. For each of these conditions the published DUS lists the percentage time spent in each condition and the frequency of occurrence of transitions to the condition.

Table II. Design Usage Spectrum conditions for the Wildcat helicopter.

Forward flight :	0.4VNe	Banked flight:	0.4VNe	Low speed:	forwards
	0.6VNe		0.6VNe		rearwards
	0.8VNe		0.8VNe		sideways
	0.9VNe		0.9VNe		spot turns
	1.0VNe		1.0VNe		Hover
Manoeuvring:	Take-off		Autorotation	Ground:	Spin-up
	Transition from hover		2 Engines VNo		Rotor start
	Climb		Simulated failure		MPOG
	Descent		Train simulated failure		SMPOG
	Transition to hover		PPI		Taxi
	Landing		Total single engine		Shut-down
	G-A-G				

In contrast to the DUS, the loads survey matrix contains in excess of 220 unique conditions and transitions. For each of these, sequences of loads have been recorded that represent entry to the condition, the condition steady state and exit from the condition. Furthermore, these loads are acquired at a range of altitudes and for a range of weights and aircraft balance combinations. However, not all of these combinations will necessarily be used to calculate service life as worst case conditions may be used if they provide sufficient endurance.

The following two figures illustrate the difference between the DUS and the information in the loads survey matrix for the sub-spectra containing the banked turn distribution. The first figure shows the DUS values in blue and data from Wildcat sortie 'x' – as identified by the FCR algorithm – in red. The second figure below shows the same Wildcat data from sortie x, also identified by the FCR algorithm, but with a resolution compatible with the Loads survey matrix.

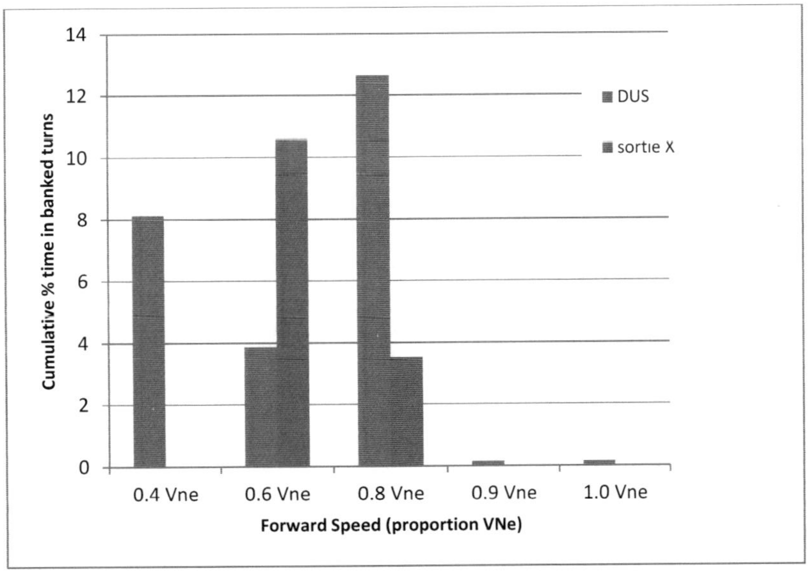

Figure 4. Banked turn sub-spectrum for the Wildcat DUS and FCR recordings for sortie 'x'.

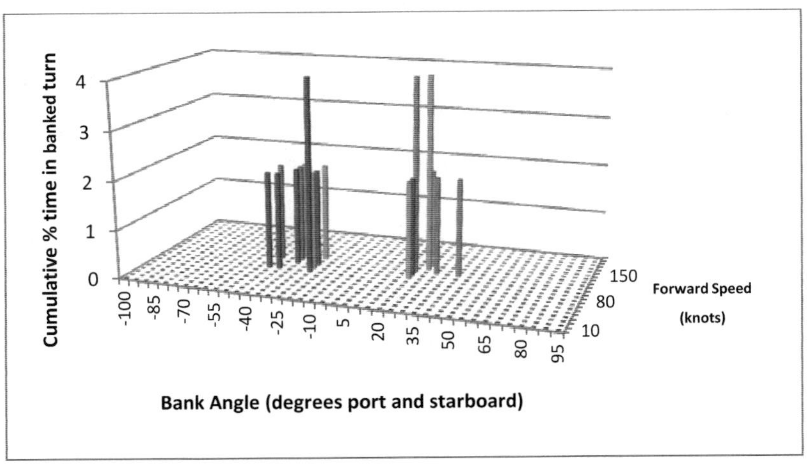

Figure 5. Banked turn recordings with loads matrix resolution for Wildcat sortie 'x'.

By providing analyses having the format and resolution of the loads matrix, it would be possible to apportion time to each individual entry in the matrix for each sortie and calculate the damage increment due to each element of usage. Summing all of the matrix inputs would produce an increment of life consumption for that sortie and provide a means of tracking cumulative service usage for each aircraft of the fleet. This of course would have to be applied to the damage sums for all critical dynamic component and structural features of the aircraft. The validity of such an approach would rely on the FCR algorithms having identified each flight condition such that its duration would encompass the significant loads attributed to the condition during the analysis of the loads survey data. At the time of producing this paper this has yet to be investigated and it will be at the discretion of the Designer as to whether/how such a programme of life tracking could be safely implemented.

Programme to ensure compatibility with the loads matrix

Flight condition recognition algorithms:

In order to validate the output of the FCR algorithms, AgustaWestland are providing HUM/FDR histories from the load survey flights. These can be used to confirm that the FCR algorithms are capable of repeatedly identifying conditions with sufficient resolution and certainty. Furthermore, it will be possible to compare the beginning and end of the events identified by the FCR algorithms with the loads associated with the condition in the loads matrix. If these can be shown to be compatible, then a 'FCR event' can be recorded as a fixed increment of life that is directly related to its duration.

It is for this reason that the beginning and end of the 'events' are associated with departures from the steady state condition of the aircraft as determined by changes in displacements, rates and accelerations.

The following figure provides an example of the scope of the information required in order to associate a unique set of loads for just the banked turn spectrum:

Figure 6. Information flow required in order to associate FCR events with unique load histories.

It can be seen that it will be necessary to resolve a significant volume of information in order to associate a feature of usage with a unique set of loads data. Although the first two spectra are relatively straight forward to resolve in software, the mass balance spectrum presents more of a challenge and a process separate to the FCR algorithms may be required to automate this part of the information flow. Alternatively, this information is available for each sortie in the aircraft form '724' flight documentation and fuel state is available in the HUMs data. If necessary it would not be difficult to include a fuel burn algorithm to model changes in sortie mass. Furthermore, even if mass/balance might not be resolved with the same resolution as the other parts of the process, in many cases it may be possible to associate 'worst case' mass/balance without incurring life consumption penalties.

Flight severity algorithm:

Although the flight severity algorithms themselves cannot be directly related to 'life consumption', it is possible that the relationships between angular displacements, rates and accelerations could be modelled – perhaps using a Flight Dynamic Simulator. While this could provide a means of relating life consumption to the vigour with which individual manoeuvres are flown, this is not a goal of the current programme which will concentrate on associating flight conditions with the loads measured during development flying and used during the strength substantiation programmes.

It is also appreciated that the current severity index may not be the most appropriate measure of flight severity for all systems of the aircraft. For instance, it would be more appropriate for the various drive shafts and gearboxes to use a measure of severity related to transmission torque. However, there is no reason why several flight severity indices should not be developed to better target the potential influence of 'flight severity' on specific systems of the aircraft.

DISCUSSION

Monitoring procedures such as the MDRE and ODR are introduced in order to maintain the Airworthiness of helicopter fleets by ensuring that current aircraft usage is reviewed within the framework of the fatigue life substantiation programme. A significant departure from the usage described in the Statement of Operating Intent could require that the lives of critical components be recalculated – at the discretion of the Designer. It has been the case in previous ODR programmes that usage analyses have been conducted rather 'piecemeal' when compared to the loads analysis. Furthermore, the strain installations for ODR programmes, whilst significant and costly, are not as comprehensive as those used to gather data for the substantiation programme. By continuously applying FCR to HUMs data, a more complete understanding of usage over a greater number of aircraft and for greater periods of operation should be achievable with the following enhancements to safety:

- Identification of potentially damaging manoeuvres
- Identification of damaging sortie types
- Identification of critical usage metrics post damage incident (wear, fatigue, overload)
- Identification of aircraft over exposed to damaging sorties 'fleets-within-fleets'
- Indication of life penalties in order to maintain safety

A further potential use for continuous usage monitoring is in the implementation of Condition Based Maintenance; this relies on the capability to characterise individual aircraft usage in terms of either an extension or a reduction in the published service replacement lives and inspection periods for critical design features. This aspiration would need to be implemented without increasing operational risk and a reliable knowledge of usage and flight severity for individual aircraft would be essential.

When the service lives of critical structural design features are substantiated, the Designer must adhere to a strict testing programme dependent upon conservative design factors to ensure adequate margins of safety. Whilst the substantiation methodology will be all-embracing, each critical feature will have a unique 'margin of safety' in service due to specific assumptions made in testing, during usage analysis and in the choice of engineering design curves (S-N curves). Furthermore, not all of the potential critical features of a component would necessarily be tested, as the first point of failure usually sets the service life. Thus, extending life in service could potentially reveal latent failures at other locations. For these reasons any decision that might result in changing a service life must be taken by the Designer with reference to the substantiation history of each critical design feature. It is also often the case that critical design features are threatened only by a small number of usage parameters – often the ground-air-ground cycle is responsible for setting the service life in a range of components. This would imply that much could be gained from having even a restricted knowledge of the 'in-service' usage spectrum. What is quite clear is that the level of reliability must be very high in order to make any judgement concerning service life and that the Design Authority is the only organisation that can endorse such judgements.

The determination of service life has been studied by Moon and Phan [6] using probabilistic analyses to examine the contribution to reliability of the various stages of the fatigue substantiation process. They concluded that 'six nines' reliability (failure rate of one in a million flying hours) can be maintained while increasing time in service where knowledge of the aircraft usage allows. However, they also found that the significance of load history was greater than that of usage history in determining service life, suggesting that having knowledge of loads in some form would be at least of equal importance.

In the work described in this paper, by associating usage - as determined by FCR - with the loads survey matrix, the conservative procedures of the substantiation programme are inherently retained and any subsequent increase or decrease in service life is applied within a regime that retains the same level of risk. Using this approach the flight severity algorithm is not used to imply loads but simply to increase confidence in the performance of the FCR algorithms and to indicate periods of particularly aggressive usage. However, by understanding how the properties of the severity metrics can be related to the loads measured during the documented flights of the load survey programme, it may well be possible to associate 'flight severity' with loads for individual aircraft of the fleet. An additional programme of work will be carried out to identify those parameters of the severity metrics that best correlate with load intensity.

CONCLUSIONS

Deterministic Flight Condition Recognition algorithms and flight severity algorithms have been developed and tested using data sets from three types of helicopter to provide fleet management functions. Recent collaboration with AgustaWestland provides access to a more complete range of data, engineering advice and the opportunity to correlate FCR outputs with loads data used in substantiating strength and service life for components and structure of Wildcat. In this way usage can be recorded within the same 'risk' management environment as for the strength substantiation programme.

FCR algorithms have been developed to successfully identify the following aspects of usage:

- Limited features of the Sioux usage spectrum
- All features of the AS350 Squirrel Design Usage Spectrum
- Conformity to the DUS for individual sorties, individual aircraft and the fleet
- Currently two thirds of the AW159 Wildcat Design Usage Spectrum has been identified
- Currently two thirds of the AW159 Wildcat Loads Survey Matrix has been identified

Those conditions that have not yet been identified for Wildcat will be detected using further channels of data from the HUM system.

A Flight Severity algorithm has been developed to provide the following features:

- Confirmation that the Sioux aircraft is flown by the UK Historic Flight within its Design Envelope
- Independent confirmation that FCR algorithms identify all periods of 'active' flight
- Association of a 'severity term' with individual flight conditions and manoeuvres
- Indication of sortie severity
- Cumulative flight severity indication for each aircraft of 'the fleet'

Further development is anticipated as follows:

Near term Goals

- Independent validation of the accuracy of FCR usage algorithms with the help of AgustaWestland
- Full correlation with the AW159 Wildcat Design Usage Spectrum
- Full correlation with the AW159 Wildcat loads survey matrix

Medium term goals

- Enhancement of the flight severity algorithms to introduce alternative 'system specific' metrics such as transmission torque.
- Correlation of 'flight severity' metrics with loads data from the load survey flying programme.
- Formulation of a mass/balance algorithm

REFERENCES

[1] Barndt,G. and Sarkar,S. Miller,C. (2006) 63rd In: *Manoeuvre Recognition Development and Verification for H-60 Structural Monitoring, AHS annual Forum, Virginia Beach.*

[2] Appareo Systems, In: *https://www.appareo.com/aviation/flight-data-monitoring. ALERTS (Aircraft Logging and Event Recording for Training and Safety)*

[3] Perrett, B. (2013). In: *The Development of Monitoring Technology to Enhance Structural Airworthiness in Ageing Helicopters, Report No: HS-013-A032*

[4] Perrett, B. (2014). In: *The Development of Flight Condition Recognition Software for Helicopter Fleet management, End-of-year Report Dstlx1-100083936*

[5] (2014), In: *Ministry of Defence, Military Aviation Authority Regulatory Publications. Regulatory Article 5720:* Structural Integrity Management – section 4, Validating Structural Integrity.

[6] Moon,S. and Phan,N. (2006) In: *Component Fatigue Life reliability with Usage Monitor,* 63rd *AHS annual Forum, Virginia Beach.*

Non-Destructive Inspection (NDI) development and implementation

Reliability assessment of different SHM and load monitoring systems under load and environmental conditions

Shashank Pant, David Backman, Tom Benak, Marc Genest

National Research Council Canada, Canada

Abstract: As part of an experimental program to assess and validate different Structural Health Monitoring (SHM) systems, the National Research Council Canada (NRC) has been developing several SHM test platforms ranging from a simple cantilever beam to a full-scale CF-18 fighter aircraft wing. In this ongoing study, a wingbox structure similar to the complexity of the inner structures found in the CF-18 outer wing, with fastened ribs, spars and skins was used to create representative in-service hidden fatigue crack of specific size and location. The hidden crack was grown using bending loads under room, hot/wet, and cold/dry environments to evaluate different SHM systems' abilities to detect such damage and to monitor its growth under simulated service conditions. Four different SHM techniques were evaluated, which include: Acoustic Emission, Acouto-Ultrasonics, fiber optic strain sensor, and a NRC developed CracK Indication Sensor (KIS). The ability of each sensor technology to survive in realistic environmental conditions and the effect of these environmental conditions on the sensor signal were assessed.

INTRODUCTION

As a part of a damage-tolerant design philosophy, aircraft components are inspected and as needed repairs are done at regular intervals based on calendar or flight hours. The inspections are performed using *routine* inspection, for example pre-flight visual examination or *detailed* inspection such as A, B, C, and D checks, with increase in inspection interval and in the amount of Non-Destructive Evaluation (NDE) required. NDE methods can be labour intensive, costly, and time-consuming. Also the aircraft has to be taken out of service in order to perform such inspections resulting in a loss of revenue or reduced operation capability. In order to improve upon the currently used time-based maintenance approach, Original Equipment Manufacturers (OEM), aircraft operators, and researchers are actively working on monitoring the health of aircraft structures using various Structural Health Monitoring (SHM) systems. These systems can be used to monitor, detect, and identify damage evolution in structures to confirm structural integrity throughout an aircraft lifecycle. Before an SHM system can be installed in an aircraft for monitoring and evaluation, it has to be certified by the civil or military aviation authority. The certification process may include safety assessment, verification, and validation. As a minimum, an SHM system should be able to detect damage exceeding a prescribed threshold set by the established Probability of Detection (PoD) with associated confidence level traditionally used in NDE techniques. PoD methodologies for NDE techniques are well established, which is not the case for SHM. The sources of errors in SHM can be different than NDE. For example, for SHM the major variability arises from sensor technologies, load, and environmental condition as compared to human factor and inspection process for NDE.

As part of an ongoing project to assess and validate different SHM systems, the National Research Council Canada (NRC) has been developing several SHM test platforms ranging in complexity from a simple cantilever beam to a full-scale CF-18 wing. In this study a wingbox structure similar to the complexity of the inner structures found in the CF-18 outer wing with fastened ribs, spars, and carbon-fiber composite skins was used to create representative, hidden, in-service fatigue damage of specific size and location in an aluminum (7075-T6 alloy) C-Channel, as shown in Figure 1.

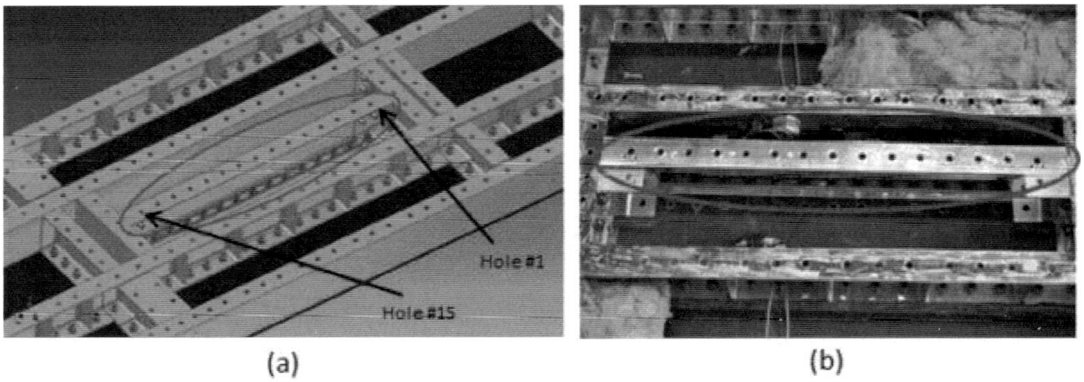

Figure 1. Wingbox structure showing the replaceable C-Channel (a) rendering and (b) actual.

The hidden crack in the C-Channel was grown using sinusoidal bending loads under room, hot/wet, and cold/dry conditions to evaluate different SHM systems' abilities to detect such damage and to monitor its growth under simulated service conditions. The aim was to study the effectiveness and reliability of different SHM sensors for damage detection under environmental conditions. Four different SHM systems were evaluated - an NRC developed CracK Indication Sensor (KIS), a Hybrid Acoustic-Emission / Acoustic-Ultrasonic (AE / AU), and a fiber optic strain sensors, which are briefly discussed in the next section.

STRUCTURAL HEALTH MONITORING SENSOR SYSTEMS

CracK Indication Sensor (KIS)

The CracK Indication Sensor (KIS) technology was developed at the National Research Council Canada (NRC) to monitor and detect crack growth in metallic structures (Figure 2a). The principal behind this technique is that a thin strip of conductive paint is applied to the surface to be monitored. A thin insulating base coat layer separates the strip of conductive paint from the metallic substrate. A 5 V electrical circuit is then connected to the KIS. When the crack propagates through the conductive paint strip, the sensor will break causing the circuit to open (zero voltage reading). The zero voltage reading activates a red LED light associated with the sensor in the interrogator (Figure 2d), notifying the operator about the presence of a crack.

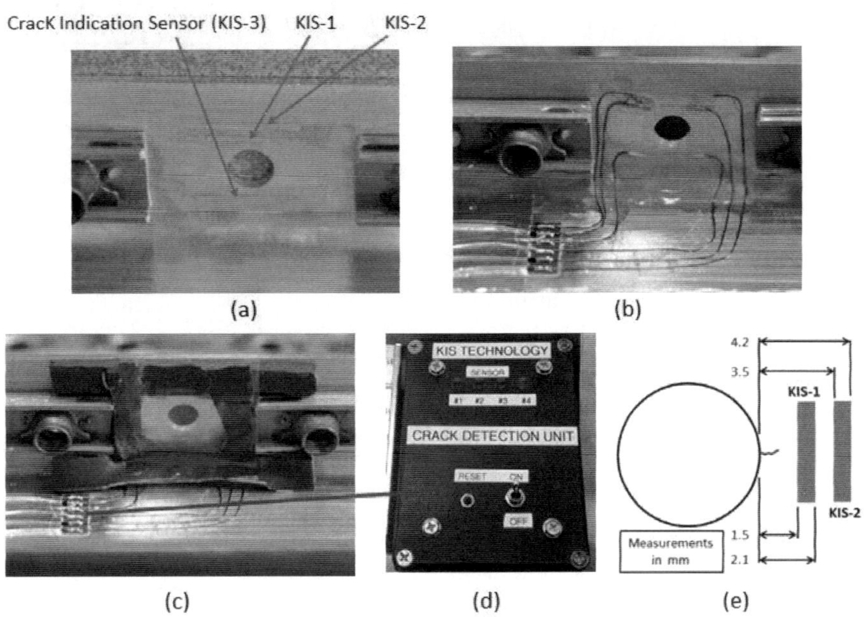

Figure 2. KIS system: (a) installation, (b) wiring, (c) insulation, (d) monitoring, and (e) location.

In this experiment, three KISs were used, two along the direction of the anticipated crack growth and one on the other side, as denoted by KIS-1/2/3 in Figure 2a. KIS-3 was used to check the performance of the KIS technology in the environmental loading conditions. Theoretically, KIS-3 should not trip because the anticipated crack growth was expected to be along KIS-1 and 2. Once the sensors were laid down, the ends of each sensor were wired, as shown in Figure 2b. The KISs were sealed using a double-sided tacky tape with a plastic bag on top for moisture insulation, as shown in Figure 2c. The KIS sensors were then connected to the interrogator for continuous monitoring, as shown in Figure 2d.

Hybrid Acoustic – Emission / Ultrasonic (AE / AU)

The hybrid Acoustic-Emission / Acoustic-Ultrasonic (AE / AU) system consisted of a separate AE and an AU system. However, both AE and AU system used the same permanently-bonded sensors; hence, the name hybrid. The sensors used were commercially available piezoelectric Lead Zirconate Titanate (PZT), which were bonded using AE-10 epoxy, as shown in Figure 3. AE system was used while the test was running; whereas AU system was used to gather data at the end of each test block. This way, the same sensors were used during and after the test to monitor crack growth.

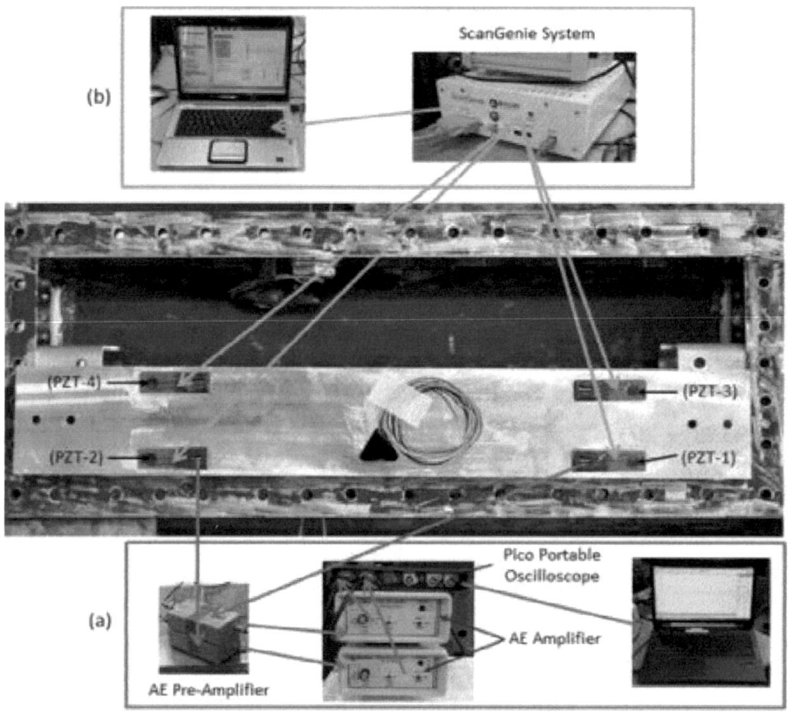

Figure 3. Hybrid SHM system (a) Acoustic-Emission and (b) Acoustic-Ultrasonic.

In the AE setup shown in Figure 3a, two sensors (PZT-1 and 2) were connected to Mistras 2/4/6 pre-amplifiers, which amplified the signals by 40 dB as well as filtered out the signals with a built-in bandpass filter of 100 kHz to 1200 kHz. The filtered signals were further amplified by 20 dB via Mistras A2EA amplifiers. The amplified signals were captured by the portable oscilloscope (Pico 5443) connected to the computer. Any signal with amplitude above 1 V was recorded by the oscilloscope at 2 MS/s and saved in the computer. The saved signals were post-processed using custom software and algorithm developed in Matlab.

In the AU setup, two sets of PZT sensors were used. The first set consisted of PZT-1 and 2, whereas PZT-3 and 4 were considered as the second set. This was done to cross examine the signal difference between the first and the second set (with and without a growing crack). The PZTs were connected to the Scangenie hardware, which was controlled by custom software developed in Matlab. Data was gathered using the Scangenie hardware at the end of each block at 24 MS/s by exciting the installed PZTs from 200 to 350 kHz at 50 kHz interval. Custom Matlab software was used to post-process the data gathered from the Scangenie hardware.

Strain Measuring System

For measuring strain in the C-Channel, a commercially available Distributed Fiber-Optic Sensing (DFS) technology based on Rayleigh backscattering was used. The DFS system uses light backscattered from the fiber to provide strain and temperature response over the entire length of a fiber sensor with an equivalent gauge length of 5 mm. The system used in this experiment is shown in Figure 4. It consists of a permanently bonded 10 m optical fiber connected to the interrogator, which was controlled by the provided software. The optical fiber was bonded to the C-Channel using moisture resistant AE-10 epoxy and the path chosen for the fiber installation was designed to allow for the detection of crack tip strains. Even though a 10 m fiber was used, only the last 1.75 m of the fiber was bonded on-to the C-Channel to measure the strain response along the web as well as in the flange.

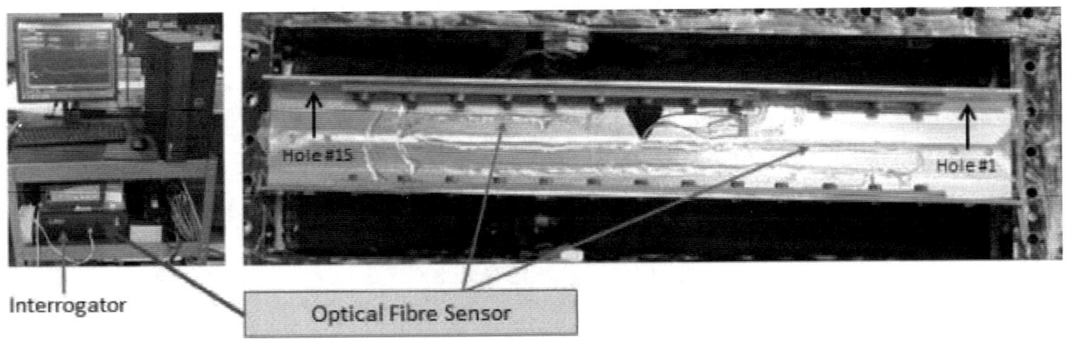

Figure 4. Optical fiber setup.

FATIGUE CRACK SEEDING IN THE REPLACEABLE C-CHANNEL

A fatigue crack of specific length and geometry was grown in the replaceable C-Channel before installing it in the wingbox. For the purpose of this test, an undersized hole of 3.2 mm diameter was machined in the C-Channel flange. A 0.25 mm thick jeweller's saw blade was used to introduce a saw-cut (i.e. starter crack) approximately 0.4 to 0.5 mm in length (Figure 5a) to act as a fatigue crack nucleation site. A crack was grown from the saw cut to a final crack length of 2.5 mm (Figure 5b). The initial hole was then re-drilled to a diameter of 6.4 mm. After drilling, a naturally grown, 1.0 mm fatigue crack remained in the C-Channel (Figure 5c) [1].

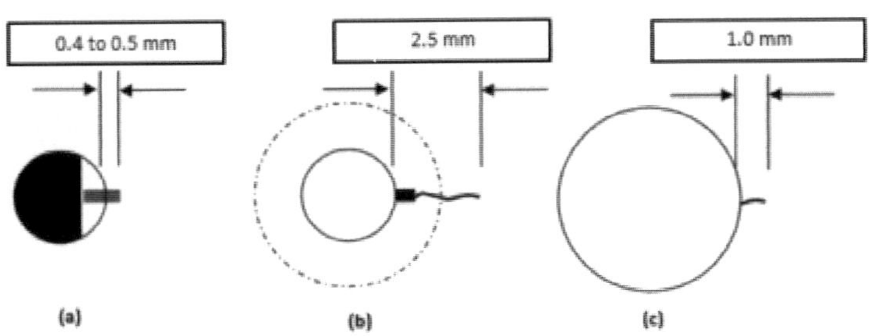

Figure 5. Fatigue crack growth in the C-Channel structure.

In order to grow a fatigue crack from the saw-cut, the C-Channel was installed in an MTS servo-hydraulic test frame and loaded in cyclical fatigue at a frequency of 2 Hz with a maximum loading of 11,121 N and an R ratio (ratio of minimum to maximum load) of 0.1. This portion of the testing was done at room temperature and ambient humidity. The crack was monitored during the fatigue cycle using a travelling microscope (with 15x magnification), a digital video camera, along with AE, AU, and KIS system. The fatigue crack growth portion of the test setup is shown in Figure 6.

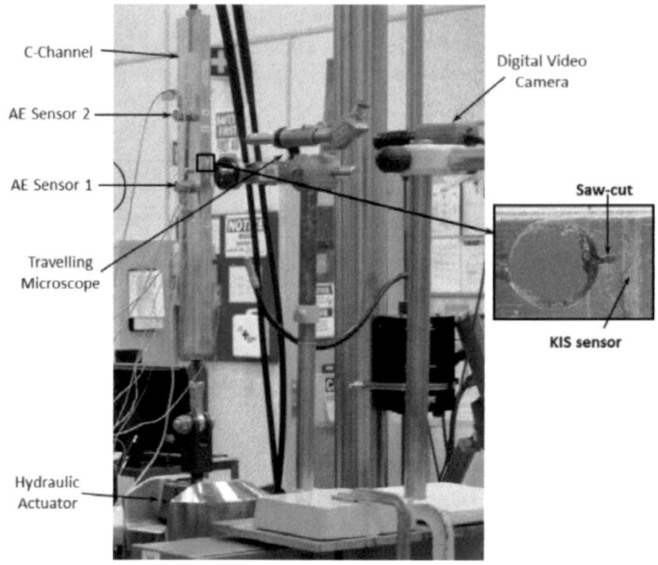

Figure 6. Fatigue crack test setup of C-Channel structure.

RESULTS AND DISCUSSION: SHM SENSORS AT ROOM TEMPERATURE

The process of growing the seeded crack at room temperature provided an opportunity to comment on the ability of various sensor systems to detect the growing crack under hot and cold temperatures. Since this portion of the test was performed at room temperature and in a laboratory environment, the ability to visually confirm the size of the crack provided an opportunity to correlate the recorded sensor signals to the measured crack length.

Visual Inspection

The images captured by the digital video camera were processed in Matlab. The crack length was tracked using image processing technique in Matlab, where the pixel distance of the crack was converted into real distance. The crack growth curve is shown in Figure 7.

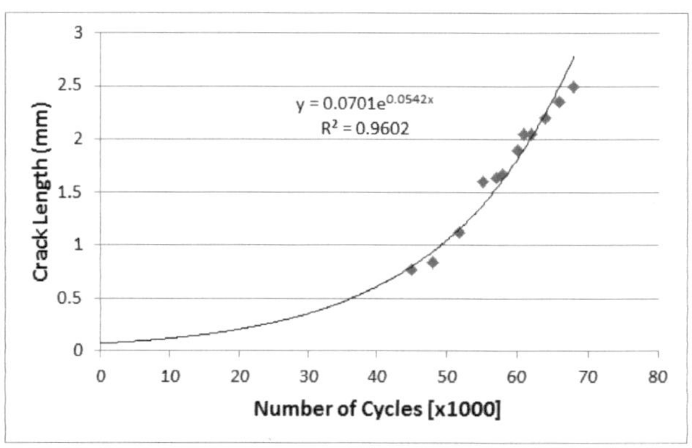

Figure 7: Fatigue crack growth.

CracK Indication Sensor (KIS)

An indication of the fatigue crack from the saw-cut in the machined hole was confirmed by the KIS Technology, when the crack length reached over 1.3 mm breaking the sensor at 51,727 cycles.

Acoustic Emission
At approximately 30,000 cycles into the fatigue test, the data from the AE system started to consistently suggest the presence of a crack, localized to the region of the pilot hole. These signals remained constant until 67,000 cycles when the fatigue cycling was terminated. A typical AE plot is shown in Figure 8.

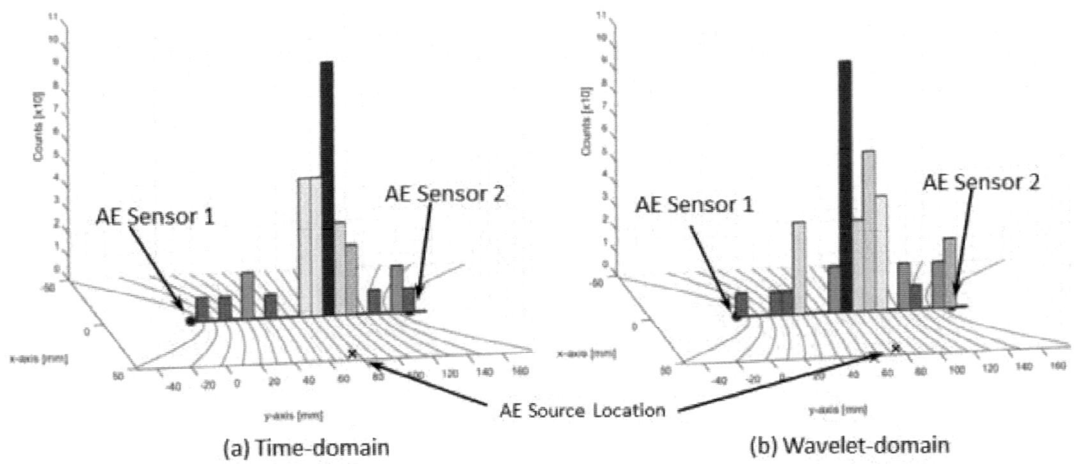

Figure 8. Typical AE plot; data processed in (a) Time-domain, (b) Wavelet-domain.

Acoustic Ultrasonic
For Acoustic-Ultrasonic (AU), the baseline data at 17,000 cycles were compared against the data acquired at the end of each 3,000 cycle block. At the end of each fatigue block, the AU system was excited from 50 kHz to 350 kHz at 50 kHz intervals. The outcome of the comparison was a correlation coefficient (measured between 0 and 1 providing information regarding how close the stress wave acquired at different cycles was to the baseline), an energy ratio (energy at different cycles divided by energy at baseline), and a difference in the maximum wave amplitude. Waves at 350 kHz were found to be most affected by the presence of the fatigue crack as compared to other frequencies. Energy ratio, amplitude difference, and correlation coefficient for 100 kHz and 350 kHz are shown in Figure 9, where path 0 and path 1 refer to wave propagating from Sensor 1 to Sensor 2 and vice-versa respectively. The indication of crack growth was found around 50,000 cycles with a crack length of approximately 1 mm.

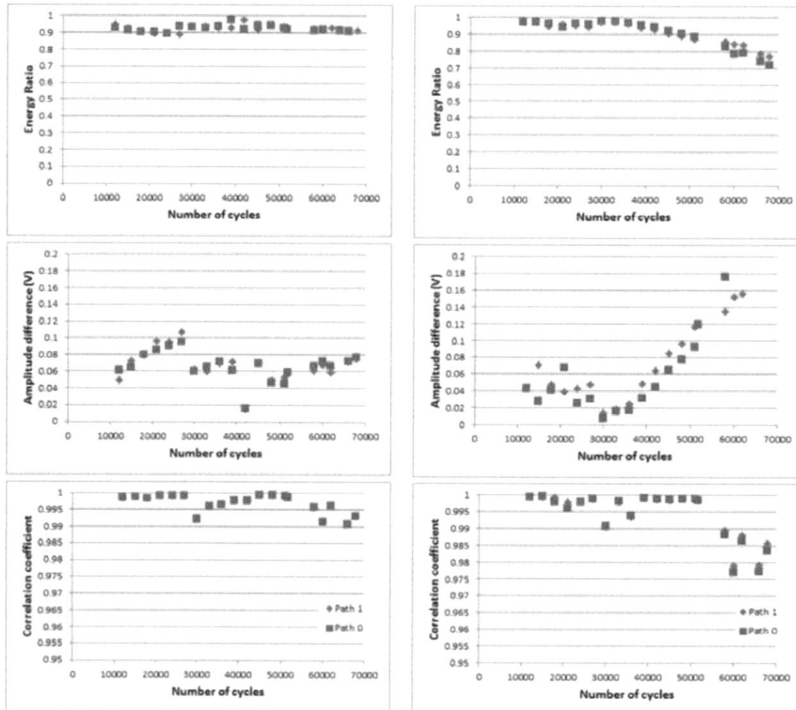

(a) 100 kHz (b) 350 kHz

Figure 9. AU comparison, (a) 100kHz vs. (b) 350kHz.

EXPERIMENTAL SETUP FOR ENVIRONMENTAL LOADING

The C-Channel with the naturally grown crack was installed in the wingbox. The wingbox was loaded as a cantilever beam, with one end fixed and a sinusoidal cyclic bending load applied at the other end via two servo-hydraulic actuators as shown in Figure 10. The actuators were tuned to run at a load level of 2,224 to 8,896 N (R=0.25) at a speed of 0.77 Hz. On top of fatigue loading, the internal temperature, pressure, and humidity levels were changed to simulate environmental conditions experienced by an in-service aircraft.

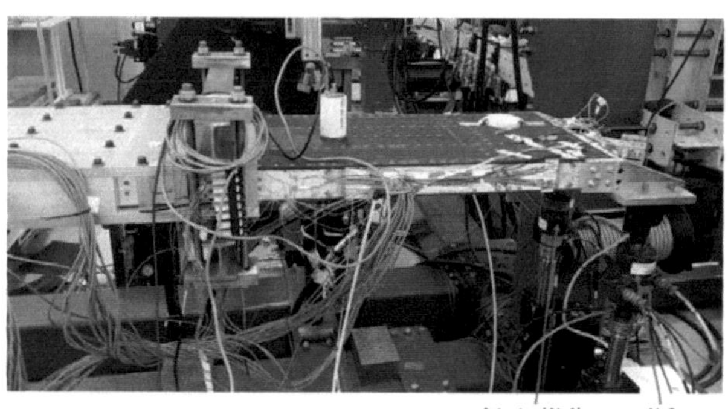

Figure 10. Wingbox loading setup.

The wingbox was loaded in blocks of 15,000 cycles under three loading conditions: room temperature/ambient humidity, hot/wet, and cold/dry conditions to assess the performance of the various SHM / strain measuring sensors described earlier. The cold/dry test was run at -60°C in addition to vacuum pressure of 61 kPa (approx. flying altitude of 7,000 m) and the applied fatigue loads. Liquid nitrogen was used to cool the test chamber within the wingbox. The test setup and the temperature distribution are shown in Figure 11 and Figure 12 respectively. In Figure 12, thermocouple

(TC-0) represents room temperature, whereas TC-5 was installed on the carbon-fiber skin. The rest of thermocouples were scattered within the gauge region.

Figure 11. Cold / dry loading.

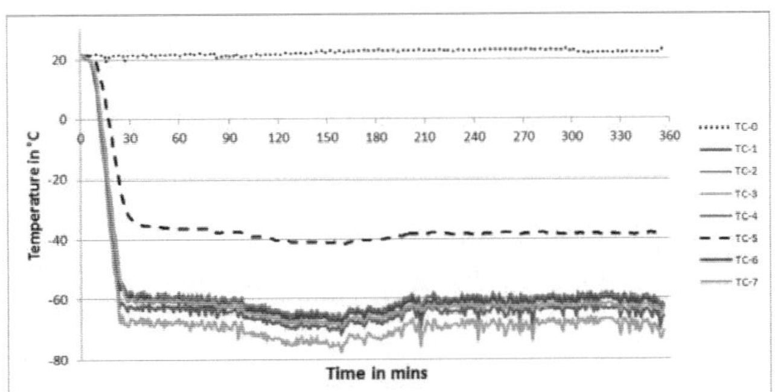

Figure 12. Cold / dry temperature distribution.

The hot/wet loading was run at a vacuum pressure of 88 kPa (approx. flying altitude of 13,700 m) in conjunction with 80% Relative Humidity (RH) at 50°C. The wingbox was heated with four-electric heaters installed within the chamber. Steam was discharged to change the relative humidity. The test setup and the temperature distribution are shown in Figure 13 and Figure 14 respectively.

Figure 13. Hot / wet loading.

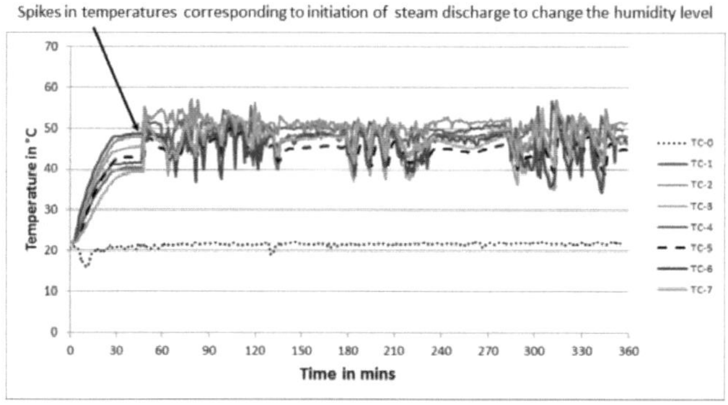

Figure 14. Hot / wet temperature distribution.

It may be noted from Figure 14, that the hot/wet temperature distribution over time is not as smooth as the cold/dry loading shown in Figure 12. This was because of the addition of steam (i.e humidity) within the gauge region which caused a change in the temperature profile as the two closed loop controllers (one for temperature and other for the humidity) worked to regain their set-points. Also the vacuum pressure was relatively low for cold/dry condition as compared to hot/wet because temperature has a direct effect on pressure for constant volume.

RESULTS AND DISCUSSION: SHM SENSORS UNDER ENVIRONMENTAL LOADING

The crack growth in the C-Channel was continuously monitored using Acoustic-Emission, CracK Indication Sensor, and fiber-optic strain measuring systems. At the beginning and the end of each block, the Acoustic-Ultrasonic method was used to gather the data. Also, at the end of each block 8,896 N force was applied to the wingbox and an iPlex Ultralite borescope from Olympus was used to take internal images of the hole while the crack was open. Although the initial plan was to measure crack lengths using the borescope images, the internal structure prevented the borescope from obtaining a sufficiently clear image to visualize the crack. The KIS sensor also did not trip during the testing suggesting the crack has not grown past 2.1 mm as verified by the post-test eddy current measurement. Results for the other SHM systems are provided below.

Altogether, the wingbox fatigue test ran for 135,000 cycles at which point PZT-1 and 2 failed to operate completely. At this point, the wingbox was removed from the hydraulic test frame and the upper carbon-fiber skin was opened for inspection. Upon inspection, PZT-1 and 2 were found to be disbonded. Post-test bolt hole eddy current inspection was conducted on hole numbers 1 to 15 in the C-channel (Figure 4), with no signs of cracking found except for the hole with the pre-existing crack (hole number 5). Post-test liquid penetrant and eddy current inspection measured the final crack length at approximately 2 mm.

Optical-fiber strain measuring system

The optical-fiber strain monitoring system failed during the initial temperature cycling. Post failure discussion with the manufacturer revealed that the failure may have originated at the termination point of the fiber. The fiber termination is used to reflect laser light back into the interrogator unit. The fiber itself has an advertised temperature rating of between -50°C to +200°C, but in discussion with the manufacturer it was revealed that the termination of the fiber is bonded on with epoxy and has reduced temperature range of between +5 to +35°C. Future testing may require a custom fiber optic sensor using high temperature epoxy at the termination point, or an effort will need to be made to locate the termination of the sensor outside of the temperature controlled zone.

Acoustic-Emission

The anti-symmetric (A_o) Lamb wave was found to be dominant in the C-Channel as established by the pencil-break tests. Therefore, the A_o dispersion curve (Figure 15) at room temperature, -60°C, and +50°C for C-Channel fabricated out of aluminium alloy 7075-T6 was used to calculate the A_o wave velocity at given frequency and temperature regime for the AE source localization algorithm. Material properties for calculating the dispersion curve are provided in Table I.

Table I. Material properties of 7075-T6 [2].

	Room Temperature	-60 °C	+50 °C
Elastic Stiffness (GPa)	71	73.8	69.6
Shear Stiffness (GPa)	26.9	27.8	26.2
Poisson's Ratio	0.33	0.33	0.33
Density (kg/m^3)	2796	2796	2796

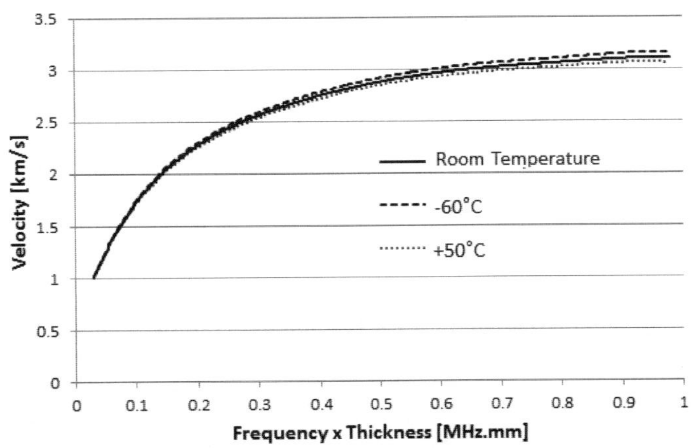

Figure 15. Theoretical anti-symmetric (A_o) Lamb wave dispersion curve for different temperatures.

Analysing the AE data, it was found that majority of the signals were between a narrow bandwidth of 50 to 150 kHz peaking at 110 kHz. Therefore, a digital Butterworth bandpass filter was applied to the entire AE signals for filtering out signals outside the 50 kHz to 150 kHz regime. The peak frequency of 110 kHz was used to find the A_o wave velocity using the dispersion data provided in Figure 15. At 110 kHz the percent difference between the A_o wave velocity at room temperature versus +50°C and -60°C was found to be 0.96% and 1.06% respectively, correspond to a shift in damage location by ±1.5 mm. A Morlet Wavelet was used to find the time of arrival for AE source localization. Figure 16 shows the results of the AE analysis, where the green circles denote the hole location (Figure 1a) and the X represent the existing crack location in the replaceable C-channel.

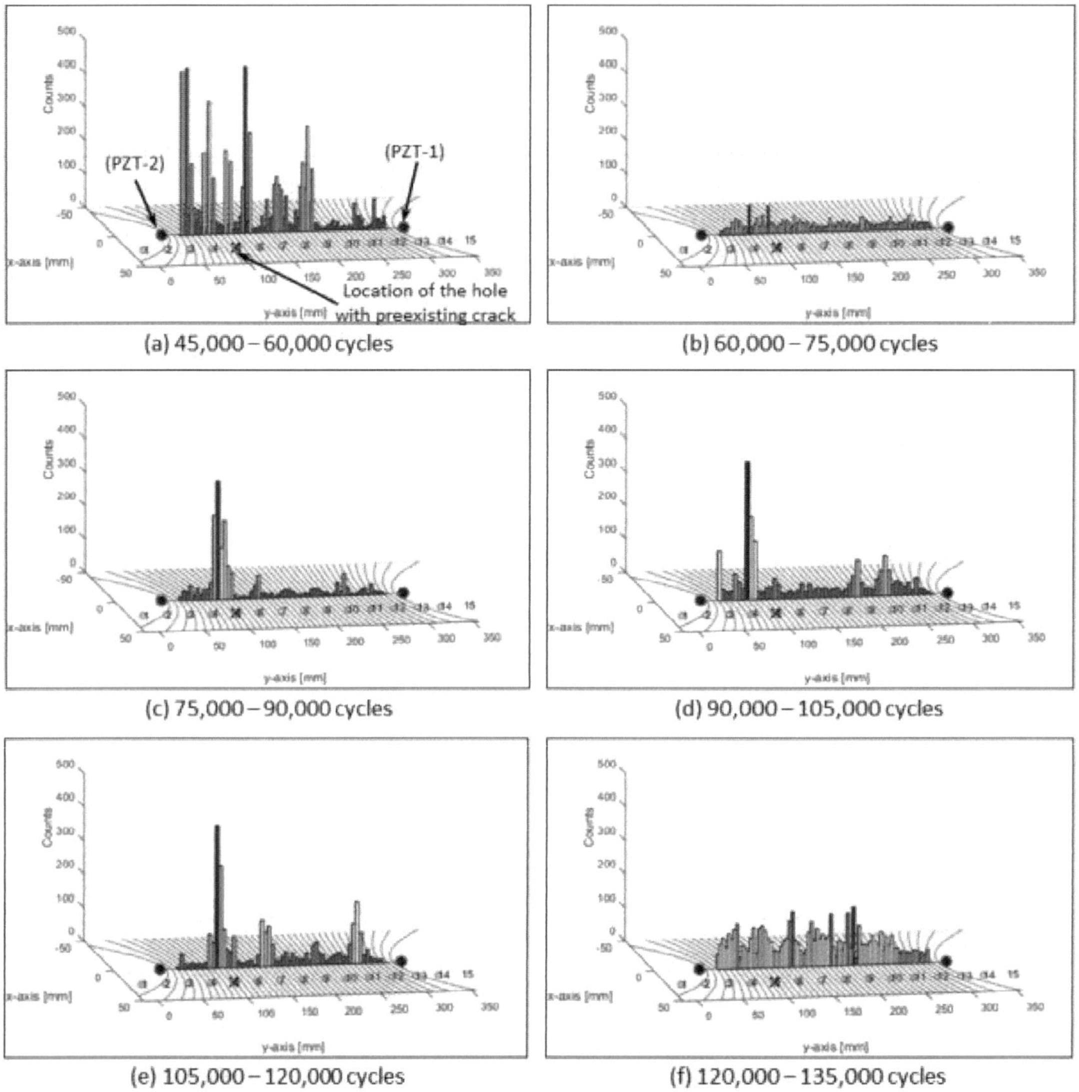

Figure 16. AE results at different fatigue cycles.

Between 0 to 45,000 cycles no significant hits within the sensor region were registered. From 45,000 to 75,000 cycles (Figure 16a and b), the AE analysis showed some hits within the sensor region; however, these hits were not localized to any consistent hole location. Also, it can be noticed that there is a significant reduction in AE hits between 45,00–60,000 cycles (Figure 16a) versus 60,000–75,000 cycles (Figure 16b). This was because starting at 60,000 cycles signal from the actuator was used as an additional trigger to discard any AE hits below 80% of maximum load.

It was interesting to note that between 75,000 and 120,000 cycles (Figure 16c to e), the AE analysis showed significant hits centered on the region of the pre-existing crack suggesting that the crack might be growing. No conclusive location information was provided after 120,000 cycles (Figure 16f) and subsequent analysis of the signals showed significantly lower signal amplitudes than in previous cycles suggesting possible degradation of either the epoxy or the sensor itself due to environmental and loading effects. The disbonding of the sensor itself was confirmed when the top carbon-fiber skin was removed from the wingbox, as shown in Figure 17. A summary of the AE analysis is provided in Table II.

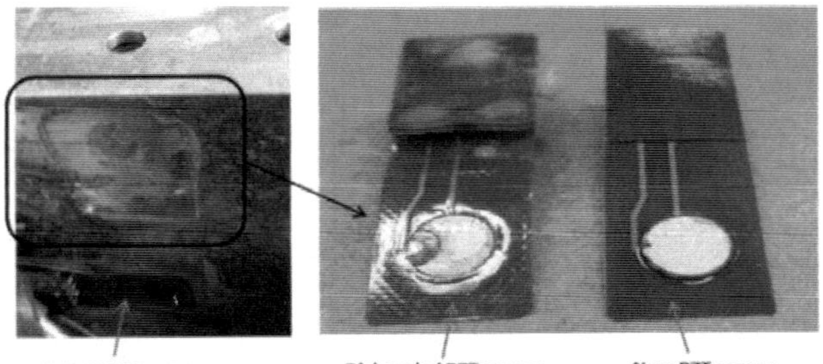

Figure 17. Disbonded PZT sensor (PZT-1).

Table II. Summary of the AE analysis.

Cycles in (1000s)	Temperature (°C)	Relative Humidity	Vacuum Pressure (kPa)	Crack Indication
0–15	Room (22°C)	Ambient	No Vacuum	No
15–30	-60	Ambient	61	No
30–45	Room	Ambient	No Vacuum	No
45–60	+50	80 %	88	No
60–75	-60	Ambient	61	No
75–90	+50	80%	88	Yes
90–105	-60	Ambient	61	Yes
105–120	+50	80%	88	Yes
120–135	Room	Ambient	No Vacuum	No

Acoustic-Ultrasonic

Both pairs of installed PZTs were excited with a five-cycle windowed sine-burst signal from 200 kHz to 350 kHz with 50 kHz interval. No conclusive evidence of crack growth was found using the data from PZT-1 and 2 as compared to the indications of crack growth found during the initial work, in which a starter crack was grown at room temperature in the same C-Channel. It was also found that PZT-1 and 2 were showing signs of degradation. The current hypothesis is that the epoxy bond was partially degraded but was still able to capture the passive acoustic emission signal; however, the PZTs were not able to fully excite the Lamb wave for AU analysis. Therefore, PZT-3 and PZT-4 were used for analysing the effect of loading and temperature. It was found that starting at 300 kHz and above, higher Lamb wave modes were present in the received signal which were absent during the initial C-Channel fatigue loading test in the MTS platform. This is due to added complexity of the installed C-Channel in the wingbox (top composite skin, nut plate, fastener holes, and fasteners) changing the boundary conditions as compared to just the C-Channel itself. The combined effects of temperature, humidity, and fatigue loading on the propagation characteristic of symmetric (S_o) Lamb wave were observed for different excitation frequencies for PZT-3 and PZT-4. Figure 18 shows the theoretical S_o Lamb wave dispersion curves of aluminium 7075-T6 at different temperatures, where the S_o wave slows down and speeds up at higher and lower temperatures respectively when compared to the room temperature.

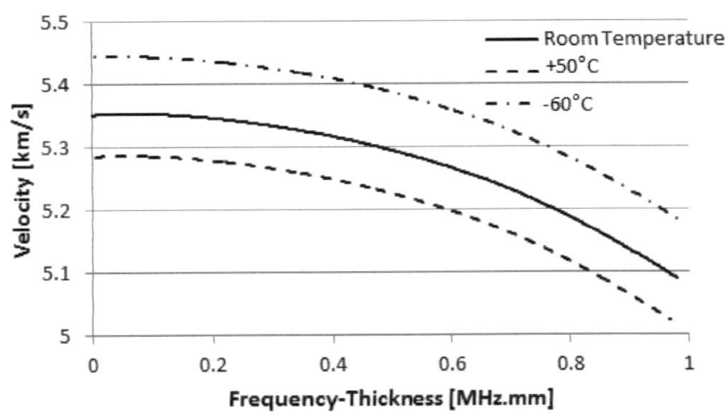

Figure 18. Theoretical symmetric Lamb wave (S_o) dispersion curves at different temperatures.

Comparing the analytical S_o wave to the experimental results shown in Figure 18 and Figure 19 respectively, it can be seen that the S_o wave slows down with an increase in temperature. It is to be noted that in Figure 19, the cold temperature reading was taken at 15,000 cycles, whereas the hot temperature was taken at 45,000 cycles. The effect of cycling can be seen in 350 kHz as an increase in amplitude but not in the other lower frequencies. This shows that in addition to the temperature, fatigue loading has a different effect at different excitation frequencies.

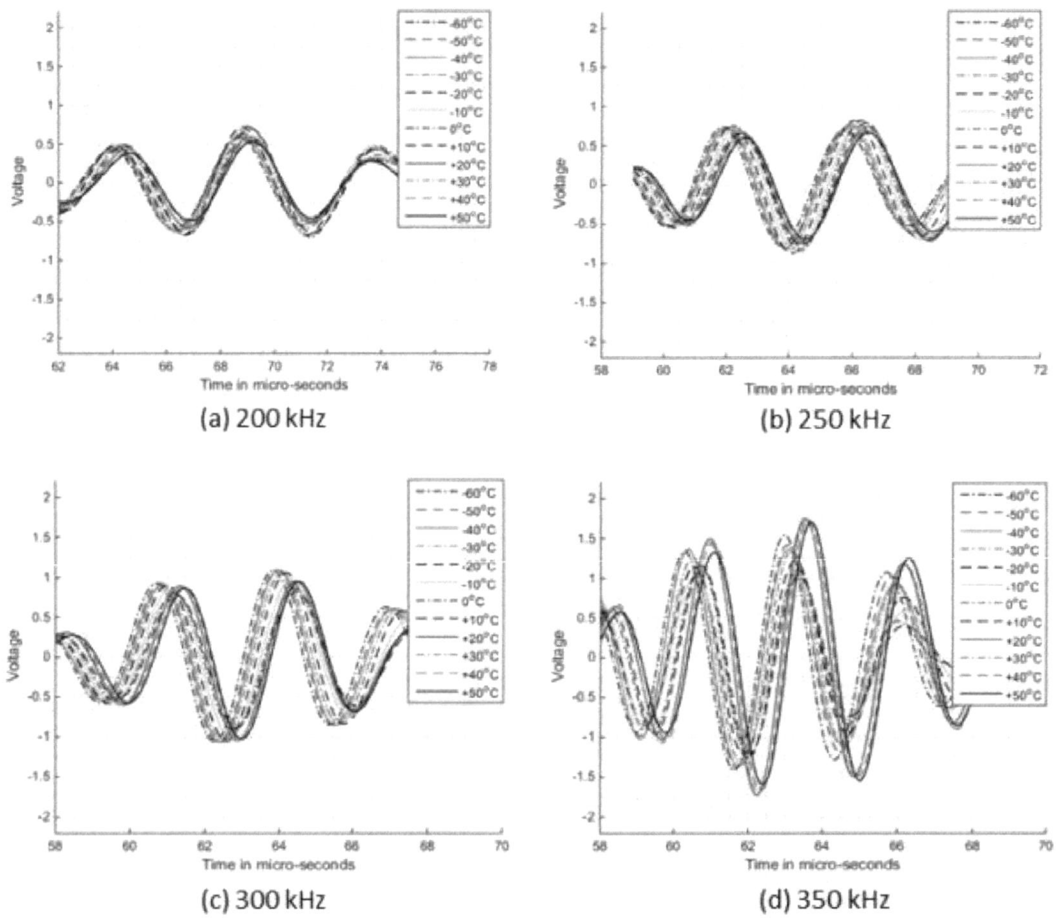

Figure 19. Effects of temperature on S_o Lamb wave velocity at different excitation frequencies.

The effect of cycling is provided in Figure 20, where the S_o data gathered at room temperature at the beginning of each block of cycles are plotted for different excitation frequencies.

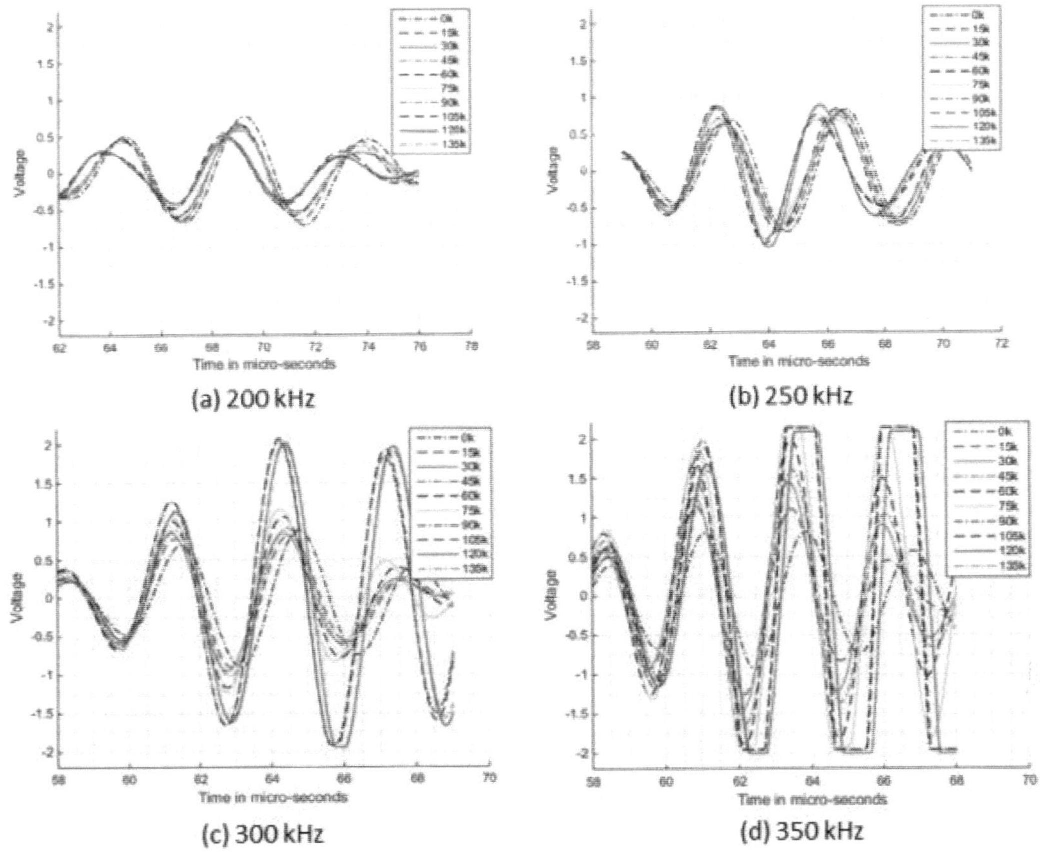

Figure 20. Effects of fatigue loading on S_0 Lamb wave velocity at different excitation frequencies.

From Figure 20, it can be seen that at 300 kHz and above, the amplitude of the S_0 wave is increasing with the number of cycles, the effect is opposite to what was expected in the presence of damage (Figure 9b for 350 kHz at room temperature), which will decrease the wave amplitude and energy ratio as some of the energy is dispersed around the damage region. The rate of increase in amplitude measured at different cycles is higher at 350 kHz when compared to other lower frequencies. It is to be noted that this is not a self-test (based on measuring electromechanical impedance at different excitation frequencies) where the disbonded PZTs show strong self-resonance at particular frequency when the sensor bond is degraded [3]. Upon final inspection PZT-3 and PZT-4 were found to be visually intact but might have been partially disbonded as compared to complete visual disbond of PZT-1 and 2 at the end of the test as mentioned earlier. Therefore, the effect of partial disbond on the propagation characteristic of Lamb wave at different frequencies needs to be taken into account for successful detection of damage using acoustic ultrasonic technique.

CONCLUSION

In this study a wingbox structure with fastened ribs, spars and carbon-fiber composite skins was used to assess the reliability of several SHM systems under fatigue and environmental loading. A representative in-service hidden fatigue crack was introduced in the replaceable C-Channel made out of aluminium 7075-T6 alloy. The hidden crack in the C-Channel was grown using sinusoidal bending loads under room (22°C , Ambient Humidity, No Vacuum), hot/wet (50°C, 80% Relative Humidity, Vacuum Pressure), and cold/dry (-60°C, Ambient Humidity, Vacuum Pressure) conditions.

The C-channel was installed in the wingbox and was fatigue cycled for a total of 135,000 cycles in blocks of 15,000 cycles. The crack growth in the C-Channel was monitored using borescope, Hybrid Acoustic-Emission / Acoustic-Ultrasonic (AE / AU), KIS, and fiber-optic strain measuring systems. Following are the observations made for this part of the experiment:

- Post-test confirmation by eddy current and liquid penetrant inspections showed that the existing crack had grown to 2 mm from its initial size of 1 mm. No signs of cracks were found on other holes except for the one with the existing crack.
- The optical-fiber strain measurement system completely failed to operate after the initial hot temperature cycling, likely as a result of temperature related disbonding of the epoxy bonded termination.
- The presence of a crack could not be confirmed using the acoustic ultrasonic technique under the effect of environmental and fatigue loads as the sensors close to the crack showed signs of degradation.
- The consistency of the PZT sensors appears to be an issue possibly due to partial and/or full degradation of the epoxy.
- As a result of the active excitation involved it would appear that bond strength is a more significant issue for acoustic ultrasonics compared to acoustic emission. In one case, sensors that worked passively (PZT-1 and 2) for AE were not able to perform properly when actively excited for AU applications.
- Acoustic emission was able to consistently detect and locate potential crack growth starting at 75,000 cycles to 120,000 cycles; however, the sensors failed to operate past 120,000 cycles due to complete disbonding.

The assessment results have shown conclusively that the SHM and load monitoring sensors that can function normally in room temperature may not be reliable under combined load and environmental conditions in a complex structure. Acoustic emission showed promising results during both room temperature and under environmental loading. It was found that effects of combined loading (load and temperature) on PZT sensor and epoxy at different excitation frequencies will need further investigation to reliably detect damage using acoustic ultrasonic techniques.

Several challenges are faced by researchers interested in implementing reliability assessment for SHM type sensors. For this current work, these difficulties were confounded by the addition of environmental conditions [temperature/humidity/pressure], as well as fatigue loading applied to a complex structure. Another issue is developing a reliability assessment methodology for different SHM systems under complex loading conditions such as the one in this work.

ACKNOWLEDGEMENTS

The authors would like to thank Defence Research and Development Canada (DRDC) for providing the funding as well as Michel Delannoy, Stephan Cloutier, and Mike Brothers for their technical support.

REFERENCES

[1] Yanishevsky, M. Martinez, M. Mandache, C. Khan, M. Fahr, A. Backman, D. (2010). The British Institute of Non-Destructive Testing, vol. 52, n. 12, p. 664.

[2] MIL-HDBK-5H. (1998). Military Handbook, p. 3-346 and 3-370.

[3] Giurgiutiu, V. (2002), SPIE's 10th Annual International Symposium on Smart Structures and Materials and 8th Annual International Symposium on NDE for Health Monitoring and Diagnostics, paper # 5056-17.

Probing the Future: Better F-16 Inspections Using Conformal Eddy Current Inspection Tools

Kimberli Jones[1], John Brausch[2], Ward Fong[3], Bryce Harris[1]

[1]*United States Air Force (F-16 ASIP), USA*
[2]*United States Air Force Research Laboratory, USA*
[3]*United States Air Force (Ogden ALC), USA*

Abstract: The Nondestructive Inspection (NDI) Capability Guidelines for United States Air Force (USAF) Aircraft Structures (EN-SB-08-012 Rev C) defines the recommended NDI flaw size capabilities for computing reinspection intervals for structures managed by the USAF Aircraft Structural Integrity Program (ASIP). USAF aircraft are expected to be in full compliance with this structures bulletin. The result of EN-SB-08-012 was an increase in the assumed flaw size detection capability for legacy eddy current pencil probe inspections, resulting in significantly shorter inspection intervals to ensure cracks do not propagate to critical size before the next inspection. Since EN-SB-08-012 was released in 2008, significant advances have been made in eddy current probe designs that reduce the influence of human factors while also improving detection sensitivity. This paper focuses on both the F-16 ASIP and the Ogden ALC NDI Program Office efforts to field these new conformal eddy current probe kits in order to be compliant with the latest USAF NDI structures bulletin.

Additionally, a case study in applying a conformal eddy current solution to inspect an upper wing carry through bulkhead radius will be detailed. The case study will review the validation and verification demonstrations and probability of detection (PoD) study. These efforts were not without continued challenges; lessons learned by the F-16 ASIP and NDI team will be presented for the benefit of other air vehicle programs.

INTRODUCTION

Damage tolerance is the prevailing method employed by the United States Air Force (USAF) to manage repair, modification, and inspection of fracture critical aircraft structures [1-2]. Damage tolerance is the ability of a structure to retain its required residual strength for a period of unrepaired usage after the structure has sustained specific levels of fatigue, corrosion, accidental, and/or discrete source damage. Within the ASIP process, damage tolerance requirements are applied to all safety-of-flight structures classified as fracture critical and fracture critical traceable.

Using this methodology, maintenance and inspection actions are scheduled such that damage does not grow beyond the size that poses an unacceptable risk to structural safety or can be economically repaired. Based on this approach, the initial inspection is required at half the time an initial flaw of size, a_o, (sometimes denoted as a_i) needs to grow to the critical flaw size, a_{CRIT}. The critical flaw size, $\mathbf{a}_{\mathbf{CRIT}}$, is the flaw size at which immediate and catastrophic failure of a structure can occur if the structure experiences design limit load, which requires that the inspections be implemented to ensure that cracks do not grow to a critical size before the defined NDI capability for the given inspection.

Recurring inspection intervals are established at half the time required for a crack to grow from the largest crack length that may be assumed to be missed during an inspection, a_{NDI} to the critical size a_{CRIT}. If no crack is found during an inspection, the crack size is analytically reset to a_{NDI} (i.e., the structure is assumed to have a crack of length a_{NDI} present) and the next inspection interval is calculated as before (at half the time required for a crack to grow from a_{NDI} to a_{CRIT}). If $\mathbf{a}_{\mathbf{NDI}}$ is not appropriately captured and a flaw larger than $\mathbf{a}_{\mathbf{NDI}}$ is missed, then it is possible that the structure could fail catastrophically before the next inspection comes due.

This methodology requires that the detection capability for these inspections be understood; however, after a series of large crack misses and subsequent field and depot probability of detection (PoD) studies, the USAF recognized that the inspection capability assumptions used for many aircraft programs were not realistic and did not consider human factors. To address this issue, Structures Bulletin EN-SB-08-012 was developed [3].

This Structures Bulletin, entitled "Nondestructive Inspection (NDI) Capability Guidelines for United States Air Force (USAF) Aircraft Structures", defines the recommended NDI flaw size capabilities (a_{NDI}) for computing reinspection intervals for structures managed by the USAF Aircraft Structural Integrity Program (ASIP). Deviation from these values must be supported by formal PoD experiments or approved by the Responsible NDI Level 3 and, if applicable, the aircraft specific Nondestructive Inspection Requirements Review Board (NDIRRB). Those deviations must be demonstrated through capability experiments guided by MIL-HDBK-1823A [4].

The initial release of the structures bulletin in October 2008 resulted in the reassessment of inspection intervals for the F-16 Program. The F-16 is managed using the damage tolerance approach discussed previously. The outcome of EN-SB-08-012 was an increase in the assumed flaw size detection capability for legacy eddy current pencil probe inspections, resulting in significantly shorter inspection intervals to ensure cracks do not propagate to critical size before the next inspection. Since EN-SB-08-012 was released, significant advances have been made in eddy current probe designs that reduce the influence of human factors while improving detection capability. Application of these probes to the F-16 ASIP will yield significant benefits by reducing inspection burden and reducing probability of structural failure. Presentations by F-16 ASIP personnel at the 2009 and 2012 ASIP Conferences (www.asipcon.com) [5-6] detailed the initial NDI related impacts and the benefits provided by implementing the improved eddy current conformal probe designs. In addition, a presentation from the 2012 Aircraft Airworthiness and Sustainment Conference (www.aasconference.com) [7] described the motivation and efforts to achieve compliance with the structures bulletin.

To discuss a few main points from the initial impacts of EN-SB-08-012, 14 control points generated 0 hour inspection intervals, with revision A of the bulletin reducing it to 10 control points. Zero hour inspection intervals are neither realistic nor possible in any maintenance realm. Inspection intervals are continuing to increase as the conformal probes are implemented within the depots and at the field level. Two control point examples of how these specialty probes help to provide better detection capability, and subsequently improved inspection intervals, are shown in Figures 1 and 2.

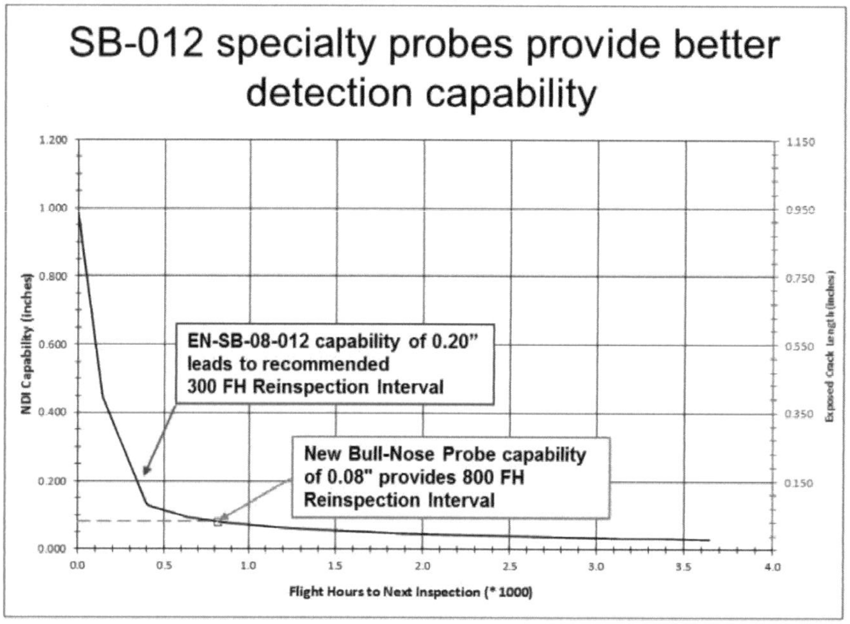

Figure 1. Control point example of specialty probe benefit to inspection intervals.

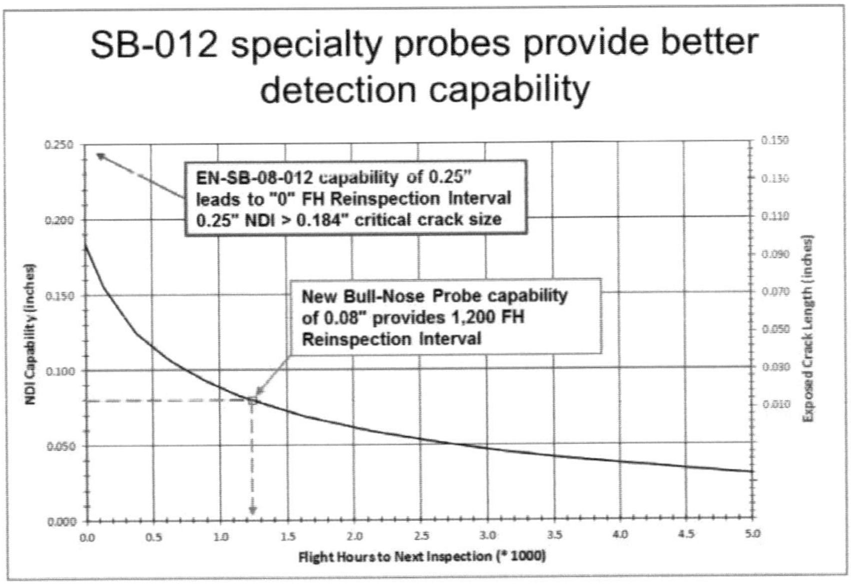

Figure 2. Control point example of bull nose probe benefit vs. legacy probe.

METHODOLOGY

The F-16 conformal eddy current probes better define the inspection area by actually conforming to structural geometries. This reduces inspection variability and uncertainty associated with standard probes, thereby decreasing detectable flaw sizes while increasing inspector confidence. The reduction of inspection intervals, along with significant time savings, can be realized using conformal probes and standardized inspections. It should be noted that the use of these probes is predicated on a detailed understanding of the location and direction of expected fatigue cracking as many of these probe designs are optimized to provide maximum detectability to cracks oriented relative to the scan direction.

The $a_{90/95}$ detection capability value for the bull-nose differential coil probe (Figure 3) is 0.080 inch surface length as listed in EN-SB-08-012, assuming aluminum structure. In contrast, the standard pencil probe using freehand scanning is listed at 0.250 inch surface length. For the inspection of an aluminum structure radius, usage of a conformal ribbon probe would yield an $a_{90/95}$ value of 0.200 inch surface length, while a standard freehand scanning pencil probe would have a capability of 0.500 inch surface length.

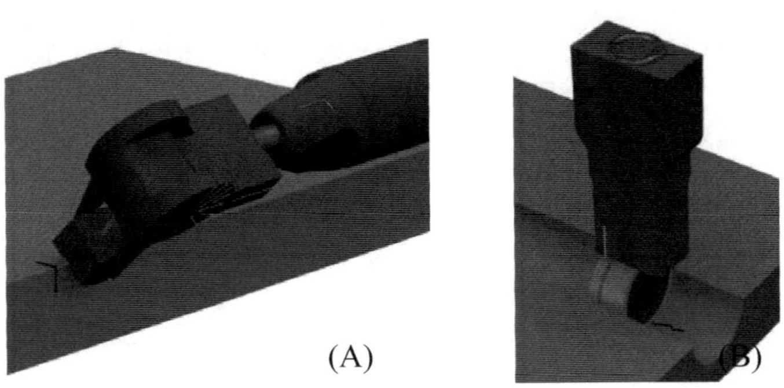

Figure 3. Conformal probe examples (bull nose (A) and conformal ribbon probe (B)).

In some cases, the detection capability of the generic conformal eddy current probe suite is not adequate for all control points. Furthermore, many areas on the F-16 have small critical crack sizes, are difficult to access, or have unique geometries that are not conducive to the use of the standard kit. Some areas can be managed using fail safety, per EN-SB-08-001, that can provide some relief from unrealistic inspection intervals. Relying upon fail safety can also be a good option when areas are inaccessible or it is not practical to inspect them regularly. An example of where fail safety is used on the F-16 is the fuel shelf joint bolt hole, shown in Figure 4. This area has been designated as a depot level inspection due to the difficulty in removing the bolt; field units may not have the capability to perform the labor intensive disassembly, including issues with complex fuel systems and pump removal. There is also a limitation on how many times hydraulic lines can be cut, which is required for this inspection, before complete replacement is required. Significant damage is often induced when the bolt is removed. A much better option is to manage this area using fail safety and visually inspect for large cracks or fuel leaks.

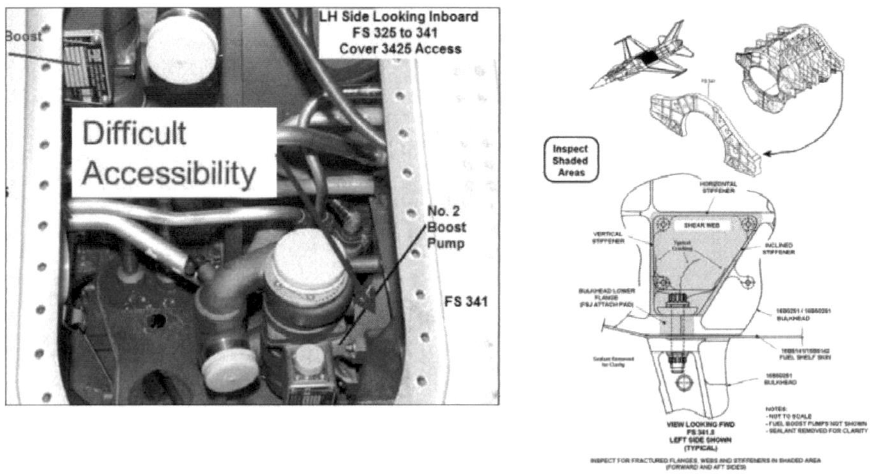

Figure 4.: Illustration of accessibility issues for fuel shelf joint bolt hole.

Case Study – Implementation of Tailored Conformal Probes

The F-16 airframe has developed cracking in the upper carry-through bulkheads, at the web-to-outboard end pad fillet radii where the lower wing attaches. Cracks nucleate adjacent to wing attach fastener holes (Figure 5). Because this area requires blended repairs and is difficult to access, it was considered a good candidate for applying a conformal eddy current solution to inspect the affected radii. Ideal access to the bulkhead radius area involves removal of wings, disassembly of some internal systems, and rigorous removal of thick fuel sealant. Unfortunately, maintenance induced damage often occurs during the sealant removal process. Ribbon radius probes (Figure 3B) were initially used for inspecting these radii; however, surface roughness, the proximity of the bolt hole and variability in blend repair geometry resulted in significant signal noise, false-positives and poor inspection confidence. Furthermore, a smaller detection capability value was required to extend the reinspection to a manageable interval than the ribbon radius coil could provide.

The F-16 System Program Office (SPO) teamed with the NDI Program Office at Hill AFB and the Air Force Research Laboratory, Materials and Manufacturing Directorate, Systems Support Division to develop and demonstrate an inspection solution to improve inspection confidence, reduce false positives and provide a goal inspection capability ($\mathbf{a_{90/95}}$) of 0.100 inch (length) x 0.050 inch (depth). An eddy current probe was developed in collaboration with Uniwest, featuring a pivoting head shaped to fit into both nominal radii and blend repaired radii. A driver-pickup coil design was selected to provide optimum sensitivity to cracks running along the axis of the radius while reducing sensitivity to part geometry changes typically found in a blend repair. This coil configuration also enabled the inspector to monitor coil liftoff during the inspection. Furthermore, the pivoting head and probe handle ensure coil contact with the part, providing tactile feedback enabling the inspector to confidently control the inspection scan and coverage without direct line-of-site to the inspection detail. An associated reference standard was also created, with notches in two different positions on the radius to establish sensitivity and insure radius coverage. Both the probe and the reference standard are shown in Figure 6.

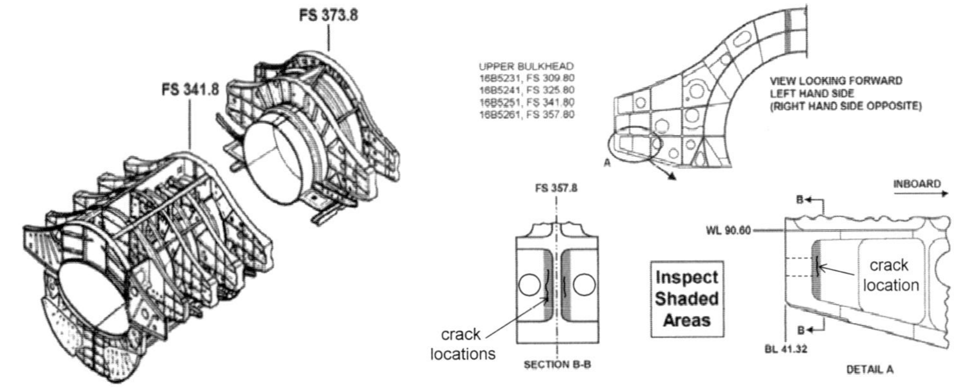

Figure 5. Upper carry-through Bulkheads, web-to-outboard end pad fillet radius at lower wing attach.

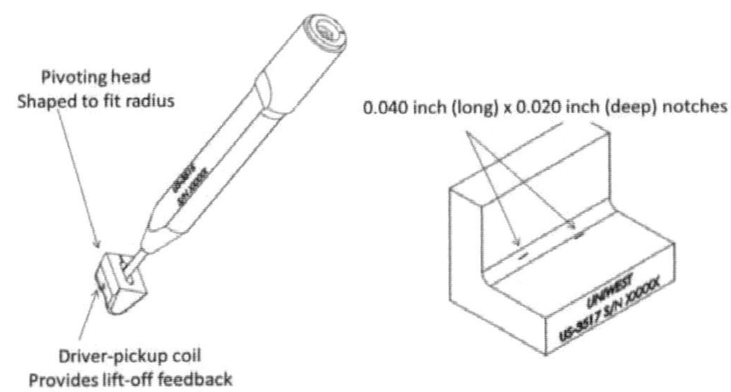

Figure 6. Radius probe and reference standard.

If cracking or damage is identified within a nominal radius, the damaged is removed by blending with a specially designed tool. If excessive coil liftoff signal was observed during the inspection resulting from inadequate blend taper, then the bulkhead would be rejected. Further blending was required to improve the horizontal taper and coil contact (Figure 7), with a repeated inspection following this step.

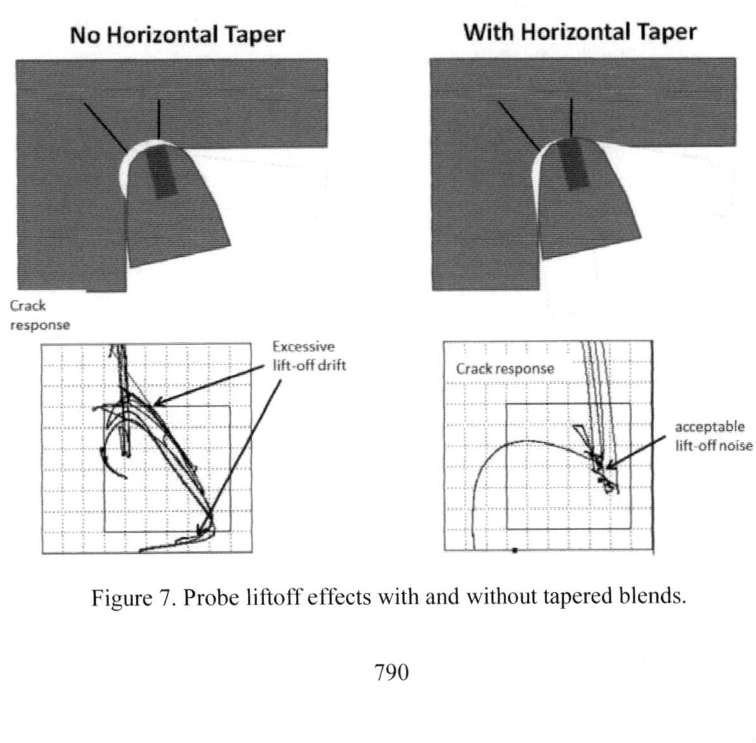

Figure 7. Probe liftoff effects with and without tapered blends.

The probe design and inspection procedure were validated using a simple mock-up built at Wright-Patterson AFB to simulate the inspection access and validate the inspection procedure and probe design shown in Figure 8(b). An initial PoD study was executed using this mock-up. This study involved the creation of 75 angle specimens with electric discharge machined (EDM) radial notch depths ranging from 0.010 inch to 0.100 inch with an aspect ratio of 2:1 (surface length-to-depth). Before notching, the specimens were machined to match the nominal bulkhead radius and then various depths of blend repairs were performed ranging from 0.030 inch to 0.060 inch in depth. Once blended, the EDM notches were placed in three different positions in the radius, such as in the center, 22.4 degrees from center, and at the tangency point as shown in Figure 9. A number of additional unflawed specimens were also manufactured and randomly introduced during the PoD experiment to represent unflawed bulkhead radii. These specimens also contained various blend depths.

Figure 8(a) shows the access area on the aircraft. Note that this inspection required removal of the bolt and thick fuel sealant to expose the radii surface. The access panel openings are not very large, with many structural and system obstacles limiting the way the probe can be manipulated. Within the mock-up, the manufactured specimens were placed within the bulkhead pocket (see Figure 8(c-d), 10), held into place with a foam block, thereby simulating the access and inspector probe manipulation limitations on an actual aircraft. The placement of hydraulic tubing was not simulated in this mock-up.

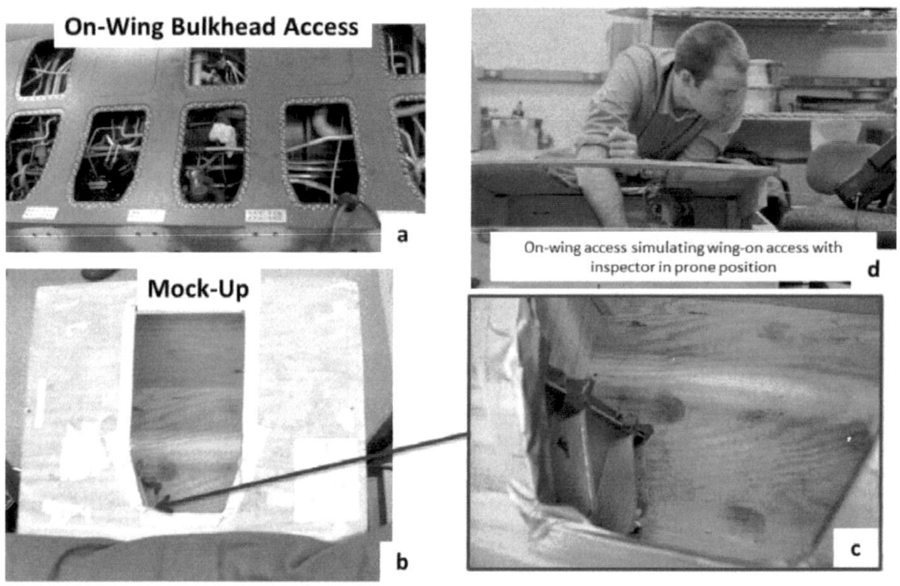

Figure 8. Inspection validation – laboratory mockup.

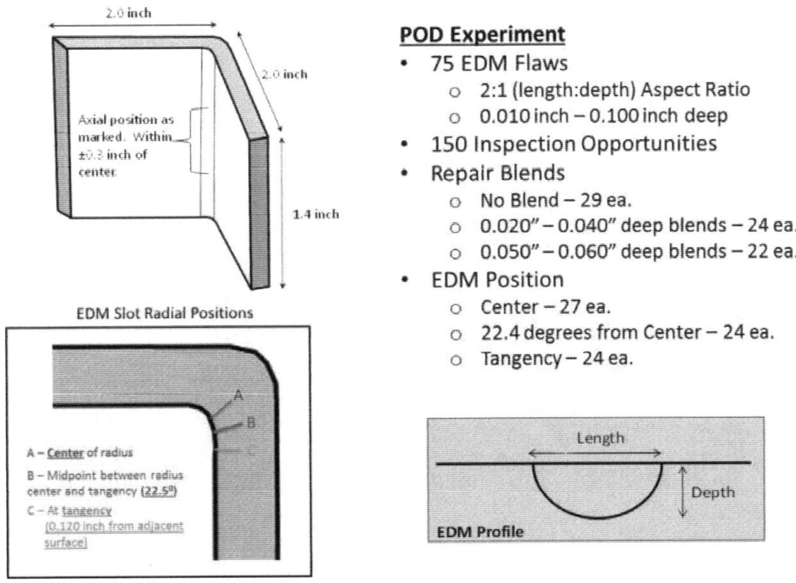

Figure 9. F-16 bulkhead PoD specimen diagram.

Figure 10. Inspection validation – specimen placement within lab mockup.

Once successfully validated, the inspection was verified via a PoD demonstration on the depot production line at Hill Air Force Base, Utah. Two eddy current Level 2 inspectors from the production shop performed the demonstration in accordance with a detailed written procedure. Although the inspectors were familiar with the inspection location, introductory training was provided to each inspector on usage of the procedure, new probe and scan technique before performing the PoD study. During the study, the PoD specimens were inserted into one of two different lower bulkhead pocket locations by a facilitator. This approach realistically replicated the on-aircraft inspection with the wing removed (see Figure 11) while evaluating the inspector performance with both left and right hand scanning. Each inspector required approximately six hours to complete the inspection task.

Figure 11. Inspection verification – on aircraft.

RESULTS

The raw inspection results for both the validation and verification demonstrations are presented in the eddy current amplitude versus flaw size plots of Figure 12. This figure also presents the same data after a -6dB correction was applied. Using the probe and inspection procedure, a comparison of eddy current response between the blended notches and cracks of similar size was also performed to ensure that the difference between the two flaw types was accounted for in the PoD analysis. The resulting comparison revealed very little difference in eddy current response between the two flaw types (Figure 13); however, for conservatism, a -6dB correction factor was applied to the raw inspection data to account for any unanticipated variance created by non-uniform shaped blends.

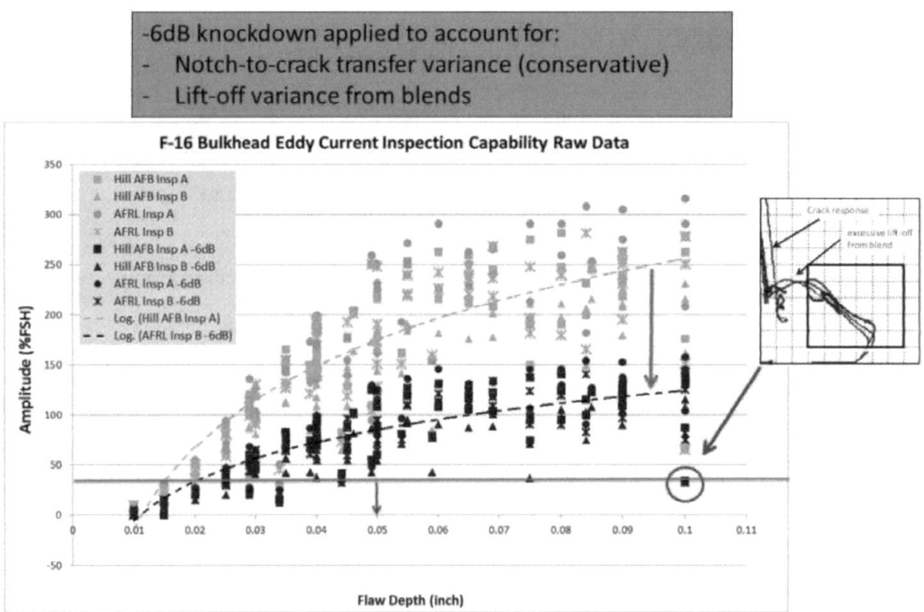

Figure 12. Verification/Validation study inspection results.

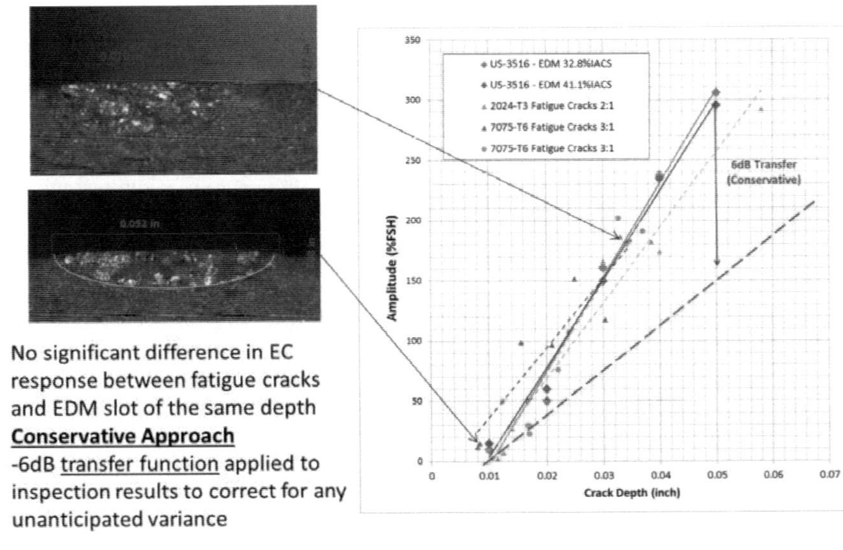

Figure 13. EDM notch-to-fatigue crack transfer function.

The PoD results for both the Hill and AFRL inspectors are presented in Figure 14. Part (a) shows the PoD curve, including a 95% confidence bound, for a rejection threshold of 30% full screen height (FSH) signal amplitude. Part (b) plots the data for EDM notches located at the outboard pad tangency described previously and shown in Figure 5 section cut B-B. Based on these results, an $a_{90/95}$ crack depth detection capability of 0.050 inch was established for this inspection. It should be noted that the inspection results generated no false positives.

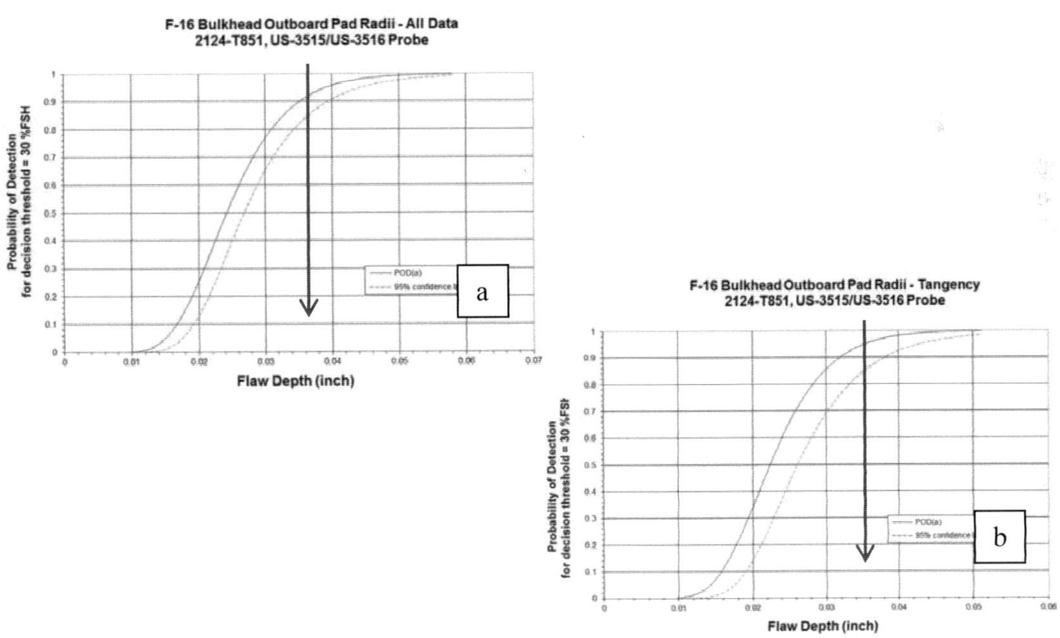

Figure 14. Inspection results – (a) all data, (b) tangency.

DISCUSSION

Inspection for small flaws is often made challenging by human factors variability and poor signal-to-noise performance introduced by unique and varying structural geometries and access constraints. Significant effort is often necessary to develop, validate and verify unique inspection solutions particularly where the potential for cracking and potential structural failure is high. Investment in these solutions can result in fewer intrusive inspections during the life of the aircraft and the capability to detect smaller repairable cracks. In this particular case, field inspections of the outboard radii will likely be needed in the future, but are not required immediately due to the significantly longer inspection intervals driven by the improvement in inspection capability. Based on the results of the demonstration, AFRL was able to provide PoD parameters to be used in probability of failure (PoF) risk calculations (see Table I). The use of PoF analysis has become significantly more important to managing the F-16 structural life as the airframe continues to age.

Table I. PoD parameters for PROF risk assessment.

a_{50}	0.030 inch (depth)
a_{90}	0.045 inch (depth)
$a_{90/95}$	0.050 inch (depth)

Due to the criticality and the unique challenges of the inspection, dedicated inspector training and qualification is justified. Specific training requirement and proficiency demonstrations have been implemented and are the main reason this inspection will remain a depot level task for the near term. Inspector qualification includes the requirement that each inspector perform a small-scale proficiency demonstration using a subset sample size of the full PoD sample set. This approach provides assurance that each inspector is sufficiently skilled to accomplish the inspection with high confidence. This small scale PoD is performed either on aircraft as shown in Figure 11 or on a sample bulkhead off-aircraft in a similar inspection position.

Efforts are ongoing to transition standard conformal eddy current probe kits to the field. Field level training is progressing and is anticipated to be completed by fall 2015 as initial probe kits enter the field. Implementing these probes in the field has not been a simple or fast process, but the results will be extremely beneficial for the fleet in terms of minimizing inspection impact as the fleet continues to age.

CONCLUSION AND RECOMMENDATIONS

As evidenced by the success of the bulkhead radius eddy current probe development and demonstration, specialty eddy current probes are viable solutions for difficult to inspect areas with unique geometries. It is imperative that such solutions be rigorously validated and verified via capability demonstrations particularly where the potential for cracking and structural failure is high.

Conformal probes designed for specific structural geometries and focused locations can be extremely beneficial when appropriately implemented (applicable geometry and crack orientation), and at the right time. In the case of the bulkhead radius probe development, the inspection was successfully demonstrated and transitioned providing the target capability of $a_{90/95}$ of 0.050 inch depth with high confidence yielding no false positives. Application of this solution in routine depot inspections has been very successful and addresses an extremely serious F-16 problem, extending unrealistic inspection intervals and mitigating unacceptable impact to aircraft availability.

The F-16 program continues to pursue transition of conformal eddy current probe designs to address multiple structural locations and improve inspection capability for damage tolerance-based inspections. Implementation of these solutions will help ensure aircraft safety as the F-16 fleet continues to age and the potential for cracking in many structural features increases.

ACKNOWLEDGEMENTS

Many people contributed to the success of this project including:

- Diane White (Hill AFB NDI Inspector)
- John Geary (Hill AFB NDI Inspector)
- Marsha Stermer (Hill AFB Supervisor Production NDI)
- Dan Laufersweiler (Universal Technology Corporation, AFRL/RXSA)
- Nick Heider (AFRL/RXSA)
- Dave Roberts (AFRL/RXSA)
- Matt Fowers (AFLCMC/WWMEX)
- Mike Shaw (AFLCMC/WWMEX)

- Steve Thompson (AFRL/RXSA)
- Mark Gehlen (Uniwest)
- Nick Jacobs (University of Dayton Research Institute, AFRL/RXSA)
- Adam Long (University of Dayton Research Institute, AFRL/RXSA)

We offer our sincere thanks to all who participated.

REFERENCES

[1] United States Air Force. (2005). *Aircraft Structural Integrity Program (ASIP)*, MIL-STD-1530C.

[2] Brausch, J., Butkus, L., Campbell, D., Mullis, T., and Paulk, M. (2008). *Processes and Best Practices for Nondestructive Inspection (NDI) of Safety of Flight Structures*, AFRL-RX-WP-TR-2008-4373, Air Force Research Laboratory, Materials and Manufacturing Directorate, Wright Patterson AFB, OH.

[3] Babish, C. and Reams, R. (2011). *Nondestructive Inspection Capability Guidelines for United States Air Force Aircraft Structures*, EN-SB-08-012.

[4] Department of Defense. (2009). *Nondestructive Evaluation System Reliability Assessment*, MIL-HDBK-1823A.

[5] Harris, B., Jeske, T., and Jones, K. (2009). *F-16 ASIP Impacts of NDI Capability Guidelines for USAF Structures*, In: Proceedings of the 2009 Aircraft Structural Integrity Program Conference, http://www.meetingdata.utcdayton.com/agenda/asip/2009/proceedings/presentations/P2791.pdf.

[6] Jones, K., Harris, B., and Killian, J. (2012). *Challenges and Lessons Learned from Conformal Eddy Current Probe Acquisition and Implementation*, In: Proceedings of the 2012 Aircraft Structural Integrity Program Conference, http://www.meetingdata.utcdayton.com/agenda/asip/2012/proceedings/presentations/P6273.pdf.

[7] Jones, K., Harris, B., Kennedy, K., Killian, J., and Lewis, B. (2012). *The F-16 Quest for NDI Capability Guidelines Compliance*, In: Proceedings of the 2012 Aircraft Airworthiness & Sustainment Conference, http://www.meetingdata.utcdayton.com/agenda/airworthiness/2012/proceedings/presentations/P5536.pdf.

Novel thermographic inspection method to detect the moisture in early stage of the water ingress and a procedure to remove the moisture from the composite structure

Eetta Saarimäki[1], Yrjö Laatikainen[2]

[1]VTT Technical Research Centre of Finland Ltd, Finland
[2]Patria Aviation Oy, Finland

Abstract: Penetrated water in the composite sandwich structures has caused problems in aircraft structures. Flight surfaces have been lost during the flights, because moisture corrodes the honeycomb and further reduces the strength of the adhesive. Water can also cause additional defects during the composite repairs, which have resulted because of the expansion of the moisture (in closed cavity), hence causing skin blow core phenomena during the curing cycle (heating) of the repair.

The number of penetrated water-induced failures is rising, therefore penetrated water in the composite structures operating in arctic conditions has been a research activity in Finland for several years.

The costs of thermography inspection + gentle drying are far less than the costs of repair planning + repair, not to mention the costs of new rudders. This thermographic inspection method has shown to be the only method that detects small amounts of penetrated water from large areas without removing aircraft composite parts from the aircraft – and the only method within the FINAF which can detect small amounts of water.

INTRODUCTION

Penetrated water in the composite sandwich structures has caused problems in aircraft structures. Flight surfaces have been lost during the flights, because moisture corrodes the honeycomb and further reduces the strength of the adhesive [1]. Water can also cause additional defects during the composite repairs, which have resulted because of the expansion of the moisture (in closed cavity), hence causing skin blow core phenomena during the curing cycle (heating) of the repair.

The number of penetrated water-induced failures is rising, therefore penetrated water in the composite structures operating in arctic conditions has been a research activity in Finland for several years.

DRYING AND ISPECTION PROCEDURE

VTT and Patria have worked recently in close co-operation to develop a method to detect moisture and remove it from the structure efficiently.

Procedure is divided to three phases:

1. X-ray inspection, which can indicate suspected water in one or multiple honeycombs.
2. Drying procedure (several hours)
3. Thermographic inspection to verify remaining water.

X-ray inspection

X-ray radiography is particularly useful for the detection of core defects in bonded sandwich honeycomb assemblies. The low density and thin composite skins usually provide easy penetration for the X-rays to image the honeycomb core materials. Core defects such as blown core, crushed core, condensed core, corroded or cut out core, and foaming adhesive voids can be detected by X-ray radiographic methods [2].

X-ray inspection can also detect assumed water ingress, if it fills most of one or several the honeycomb cells (Figure 1). The drawback is that there is no certainty, if the indication is from water or excess adhesive/resin from manufacturing process. If doubt of water ingress is observed during X-ray inspection, structure will go through special drying procedure.

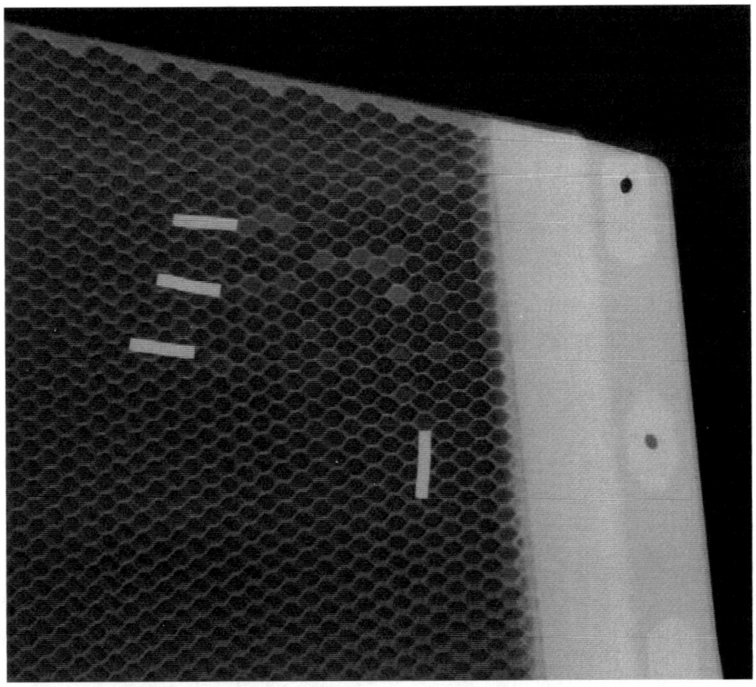

Figure 15. An example of the X-ray inspection. Assumed water ingress in multiple honeycombs. Picture by courtesy of Patria.

Drying procedure

A gentle procedure to remove the water is applied to the honeycomb composite structure, because it is essential not to cause skin blow core effect during the drying phase. Honeycomb composite structure is heated under a low vacuum to vaporize moisture from the structure (Figure 2). Heating takes place several hours.

If water is detected after thermographic inspection, drying procedure and thermographic inspection are repeated as long as structure is found to be dry.

Figure 16. Drying procedure, where the honeycomb composite structure is heated under a low vacuum to vaporize moisture from the structure. Picture by courtesy of Patria.

Thermographic inspection
Pulsed thermography is reported to indicate water ingress in the composite structure. However, water can´t be distinguished from other liquids easily or at all by this method [3]. VTT has developed a thermographic inspection routine exploiting phase transition of water for moisture detection in aircraft structures by cooling the structure under freezing conditions [4]. With skin thickness of 1 mm (CFRP), less than 0.05 g of water can be easily detected with this method.

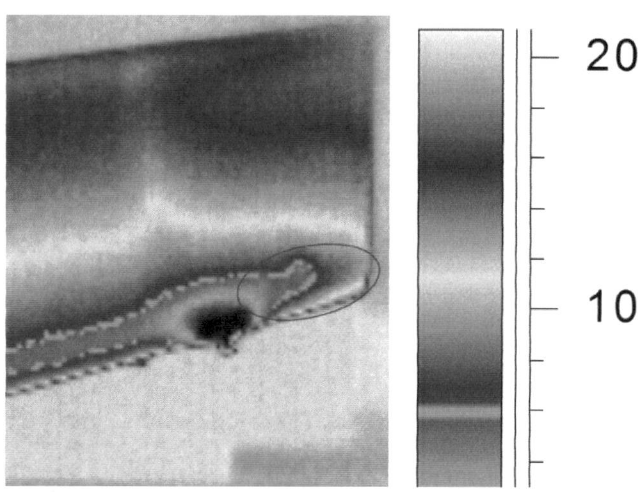

Figure 17. An example of the thermographic inspection based on the phase transition of water. Abnormal warming is observed within the circled area. Picture by courtesy of VTT.

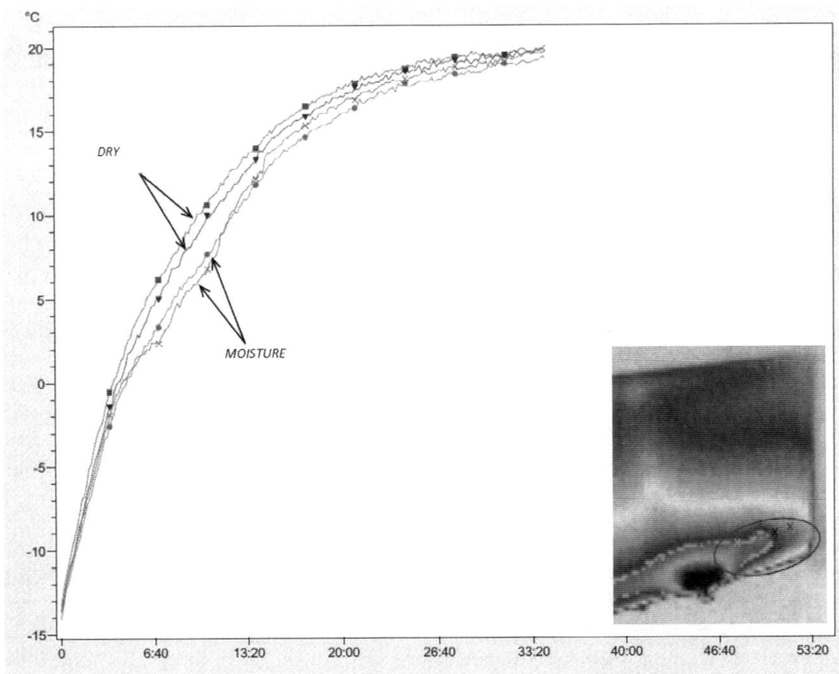

Figure 18. The measurement locations are displayed with symbols ”X” (right); each warm-up curve’s color is related to the same color ”X” (lower right within the circled area). Moisture ingress/penetrated water can be observed from the curves denoted with “MOISTURE”. Picture by courtesy of VTT.

Thermographic inspection based on the phase transition of water exploits the phase transition energy that is needed for the water ingress to be defrosted (melted). Water ingress indication is observed in specific phase transition temperature and insures the presence of water. An other advantage of this method is that no additional excitation source is needed for the tests. Method based on phase transition can be especially exploited during the long period of arctic weather conditions in Finland and other cold areas. Composite structures can be left outside in freezing conditions overnight and inspected when they have been brought in to warm conditions.

CONCLUSIONS

The costs of thermography inspection + gentle drying are far less than the costs of repair planning + repair, not to mention the costs of new rudders. This thermographic inspection method has shown to be the only method that detects small amounts of penetrated water from large areas without removing aircraft composite parts from the aircraft – and the only method within the FINAF which can detect small amounts of water.

BIBLIOGRAPHY:

[1] D. Chen, Z. Zeng, C. Zhang, X. Jin, and Z. Zhang, "Quantitative study of water ingress in pulsed thermography," Insight: Non-Destructive Testing & Condition Monitoring, vol. 55, no. 5, pp. 257-263, 2013

[2] A. Doyum and M. Dürer, "Defect characterization of composite honeycomb panels by non-destructive inspection methods," German Society for Non-Destructive Testing (DGZfP), 2002

[3] D. Chen, Z. Zeng, N. Tao, C. Zhang, and Z. Zhang, "Liquid ingress recognition in honeycomb structure by pulsed thermography," EPJ Applied Physics, vol. 62, no. 2 2013

[4] E. Saarimäki and P. Ylinen, "Development of thermographic inspection routine exploiting phase transition of water for moisture detection in aircraft structures," 2009, pp. 729911-1-729911-9.

Integrated Health Monitoring of the Composite Structures Based on PZT Sensors

Krzysztof Dragan, Michal Dziendzikowski, Artur Kurnyta, Andrzej Leski

Air Force Institute of Technology, Warsaw, Poland

Abstract: The composite structures monitoring is one of the most important branch of Structural Health Monitoring (SHM) technology based on the research made at FAA NDI Validation Center. Composites are vulnerable to random in nature impact damages, burdening the determined time interval between subsequent inspections with additional risk factor. One of the ideas for structural health monitoring systems built is based on the measuring of the mechanical properties of materials used for aircraft structural elements. Such approach is based on analysis of elastic waves propagation excited in the element by a network of PZT piezoelectric actuators. The state of a monitored structure is assessed based on chosen signal characteristics called the Damage Indices (DIs). In the article approach to built and selection of such damage indices will be presented. In order to test damage detection capabilities in the presented preliminary approach (based on the array of sensors), impact test causing Barely Visible Impact Damages (BVIDs) of CFRP (Carbon Fiber Reinforced Plastic) and GFRP (Glass Fibre Reinforced Plastics) specimens with assembled PZT sensor network were performed. Impacts with energies: 9J, 6J and 3J were carried out subsequently and after each of them a series of signals generated by the PZT network were collected. Different signal parameters, e.g. its frequency, gain and duration were tested to enhance the system performance. A model based on selected Damage Indices, allowing for BVIDs detection and localization was developed.

INTRODUCTION

The conventional nondestructive testing techniques are nowadays supposed to be complemented by systems of structure integrated sensors continuously monitoring its health. Application of such methods would definitely increase safety, especially when considering hardy accessible 'hot-spots', but also it could save up to 50% of necessary inspections time depending on the aircraft type. The composite structures monitoring is one of the most important branch of Structural Health Monitoring (SHM) technology based on the research made at FAA NDI Validation Center. Composites are vulnerable to random in nature impact damages, burdening the determined time interval between subsequent inspections with additional risk factor. Their wear-out process is to a high extent unknown, at least comparing to metallic structures which has another adverse effect on the safety. Furthermore even impacts of low energy can form an extensive network of cracks and delamination deployed under the surface of an element causing severe loss to their strength. Such damages can be invisible or barely visible on the boundaries (Barely Visible Impact Damage – BVID) hindering the use of low cost NDI methods, therefore a reliable SHM system allowing for automated detection of such damages is highly desired by the aerospace industry [1].

Figure 1. BVID in CFRP composite structure [2].

SIGNAL ANALYSIS METHODS

One of the ideas for structural health monitoring systems built is based on the measuring of the mechanical properties of materials used for aircraft structural elements. Such approach is based on analysis of small displacements propagation excited in the element by a network of PZT piezoelectric actuators. Solution for small deformation dynamics of the medium strongly depends on the boundary conditions, in particular the geometry of the object and its distortions caused by discontinuities and deformations. Structural damages can thus result in observable changes of the signal generated by the network sensors. The state of a monitored structure is assessed based on chosen signal characteristics called the Damage Indices (DIs) [3].

In some models the full information carried by obtained signals is employed via discretized integral transformation of the form:

$$\Phi(f_{gs}) = \int_T \varphi(t) f_{gs}(t) dt \quad , \tag{1}$$

where f_{gs} is the signal on a sensing path $g \to s$ determined by a generator g and a sensor s and φ a function designed to capture essential changes of f_{gs} due to a damage. The functional Φ are used to evaluate the structure by comparing its value for current and initial state of the structure. A main drawback of this approach is that such Damage Index is tailored to the specific problem considered. There are various random factors which may have influence on the signal at the chosen instants of time. Moreover also the changes of the relative geometry of the problem, i.e. the damage location with respect to sensors of a network, can significantly alter the proposed signal transformations causing the misclassification problem.

Therefore in the adopted approach the Damage Indices (DI's) carries marginal signal information content in order to make them more flexible in application. Denoting as f_{gs}^{env} the envelope of a signal generated by the transducer g and received by the sensor s and as $f_{gs,b}^{env}$ the envelope of the corresponding baseline, i.e. the reference signal obtained for the initial state of the structure, the proposed Damage Indices are given as follows:

$$DI_1(g,s) = 1 - \mathrm{cor}(f_{gs}^{env}, f_{gs,b}^{env}) \quad , \quad DI_2(g,s) = \left| \frac{\int (f_{gs}^{env} - f_{gs,b}^{env})^2 dt}{\int (f_{gs,b}^{env})^2 dt} \right| \tag{2}$$

where $\mathrm{cor}(f_{gs}^{env}, f_{gs,b}^{env})$ stands for the sample correlation of $f_{gs}^{env}, f_{gs,b}^{env}$. Both of the proposed DI's are correlated with the total energy received by a given sensor but also with its distribution in time during the measurement. In order to decrease further the dependence of DI's on the damage location the Averaged Damage Indices (ADI's) can be used [3]:

$$ADI_j := \frac{1}{n(n-1)} \sum_{g \to s} DI_j(g,s), \qquad j = 1,2 \tag{3}$$

where n is the number of transducers in a network measurement node and $g \to s$ denotes all of the network sensing paths. In this way the joint information from all of the sensing paths of the network is utilized in a single Damage Index. Averaged Damage Indices (ADI's) are better suited for damage size estimation and remain structure evaluation ability also in the case of improper functioning of several transducers of the network.

The Averaged Damage Indices are used for qualitatively indicating that a damage occurred – possibly with some kind of its severity assessment. Using DI's, even those carrying marginal signal information, it is also possible to estimate the damage location. The idea is that elastic wave interaction with a BVID damage is a local phenomena, thus any changes of the signal for sensing paths being far from a damage should be irrelevant, thus any signal transformation, e.g. DI's should not be influenced by a damage then. Thus having sensing paths visualized by a map $R_{gs}(x,y)$ (example on figure 2), one can consider a weighted sum

$$I(x,y) = \sum_{g \to s} DI(g,s) R_{gs}(x,y) \tag{4}$$

as a map allowing to localize the damage. Originally this algorithm was introduced in [4] and applied to PZT networks having homogeneously distributed sensing paths like shown on Figure 3(a). However for network geometries a special focal points can exist (Fig. 3(b)), which can heavily bias the damage location indicated by the map above. In this case a modification on the function (4) can be used:

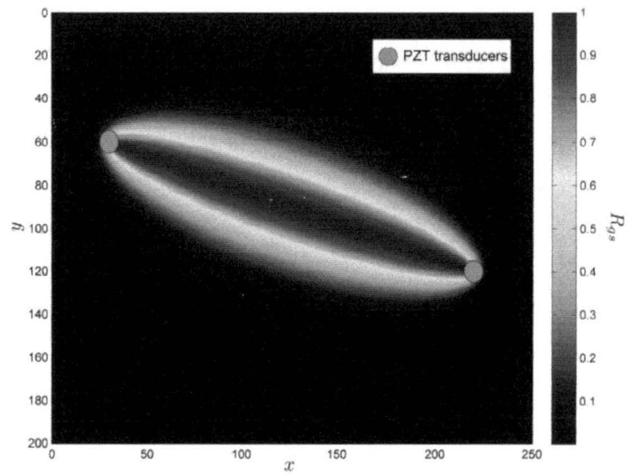

Figure 2. An example of a sensing path visualization, i.e. the map $R_{gs}(x, y)$.

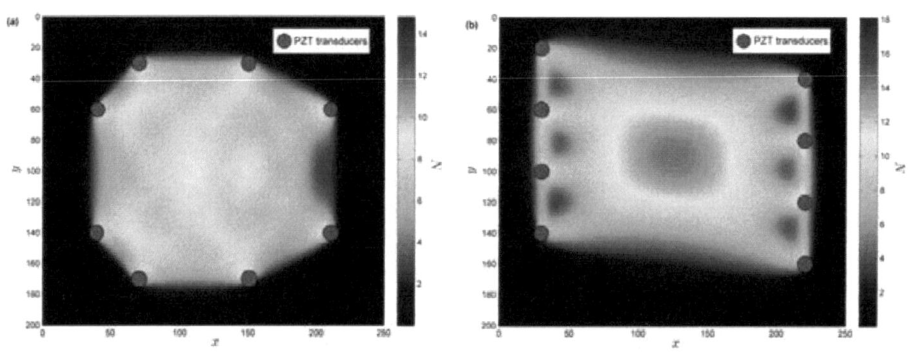

Figure 3. Sum of sensing paths function R_{gs} for different network geometries: (a) homogeneous geometry and (b) parallel geometry.

$$I(x,y) = \frac{1}{N(x,y)} \sum_{g \to s} DI(g,s) R_{gs}(x,y) \quad (5)$$

where

$$N(x,y) = \sum_{g \to s} R_{gs}(x,y) \quad (6)$$

IMPACT DAMAGE TESTS

In order to test BVID's damage detection capabilities of the presented approach, impact test of CFRP (Carbon Fiber Reinforced Plastic) specimens with assembled PZT sensor network were performed. Impacts with energies: 9J, 6J and 3J were carried out subsequently and after each of them a series of signals generated by the PZT network were collected. The following figures presents the subsequent impact localizations (Fig. 4–Fig. 7) as well as the change of the proposed Damage Indices (DI's) on all of the network sensing paths. The colors of data on the plot corresponds to the generator number, whereas their shape codes the information about the sensor.

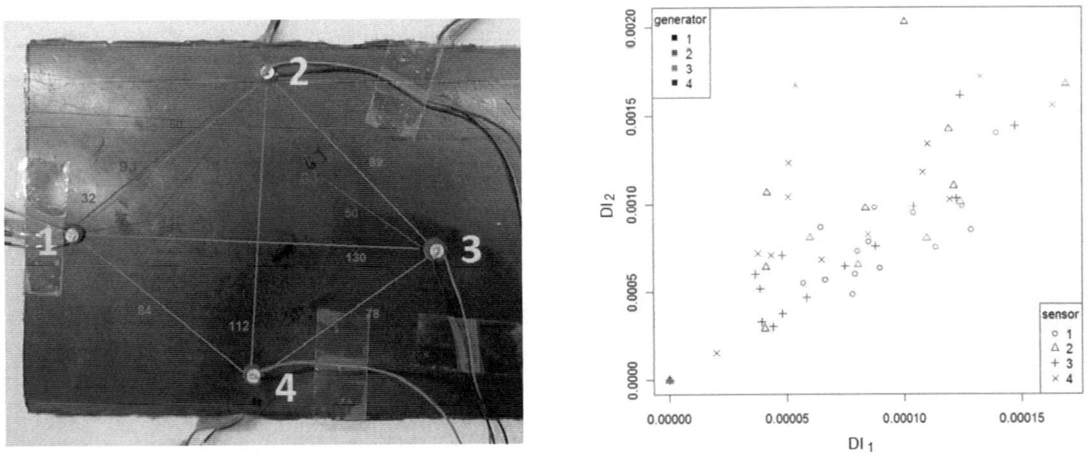

Figure 4. Damage Indices in the pristine state of the structure.

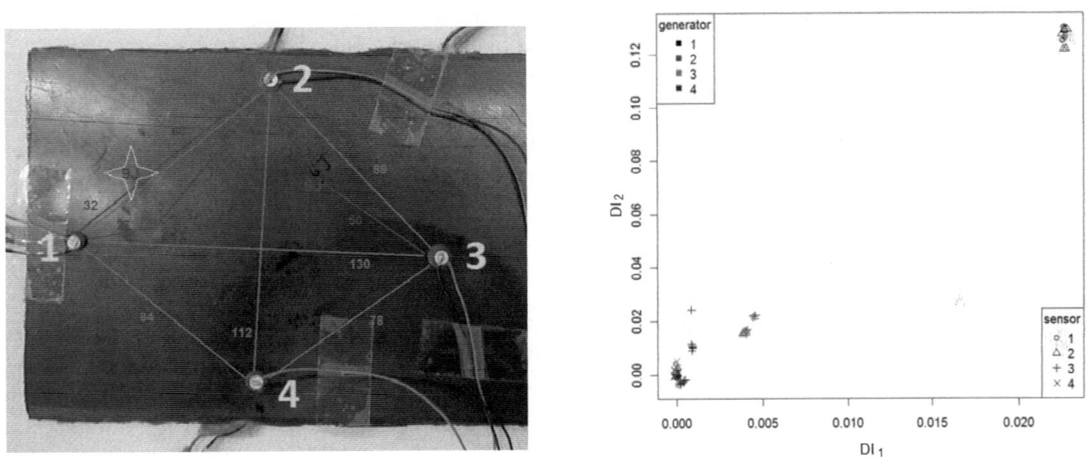

Figure 5. Damage Indices after the 9J impact.

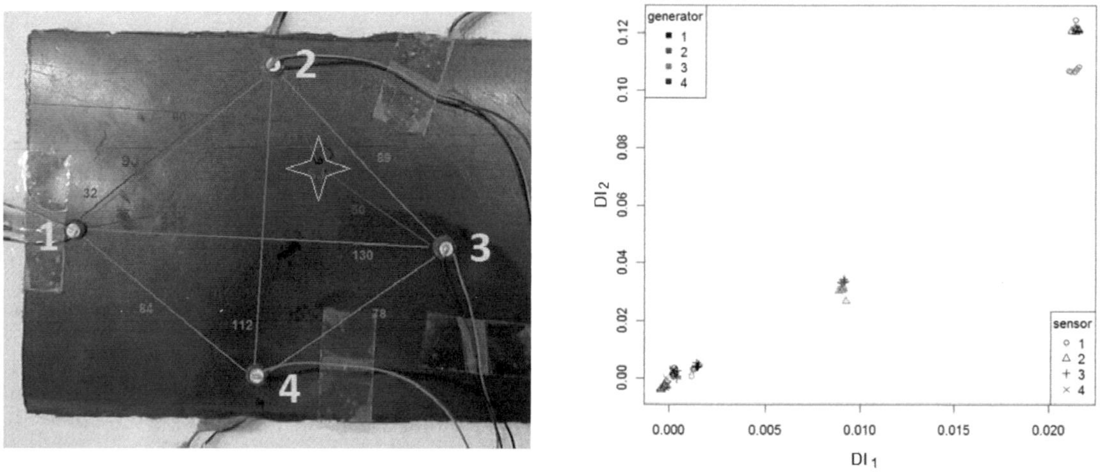

Figure 6. Damage Indices after the 6J impact.

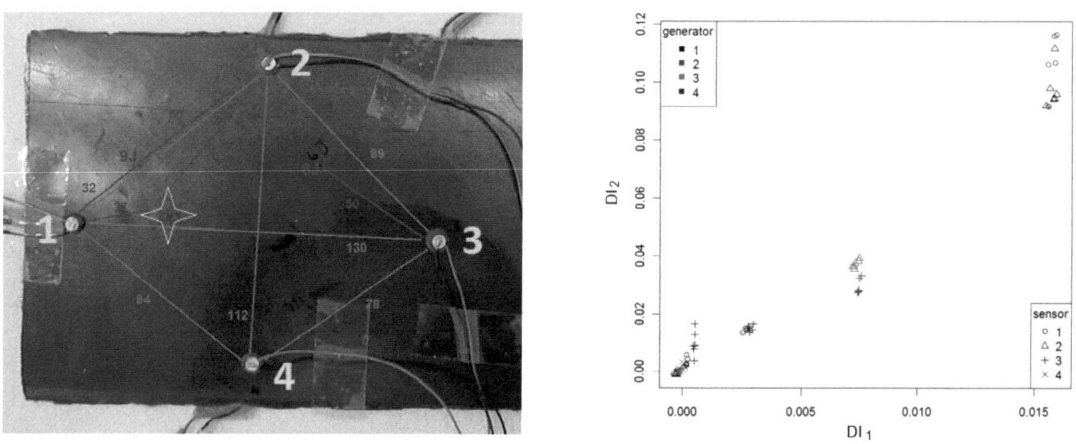

Figure 7. Damage Indices after 3J impact.

It is evident that the changes of Damage Indices are the effect of the BVID's. For each impact the change in the DI's occurs only for the sensing path in the proximity of the resultant damage. The value of this change depends on the impact energy and thus is correlated with the size of BVID damage. The plot (Fig. 8) shows the Averaged Damage Indices with respect to cumulative energy of the subsequent impacts. Both the data corresponding to 9J impact and the 6J impact are well separated from the base state of the structure as well as from each other. The impact of energy 3J results in a slight but significant ADI's separation which provide a hint about damage detection capabilities of the system.

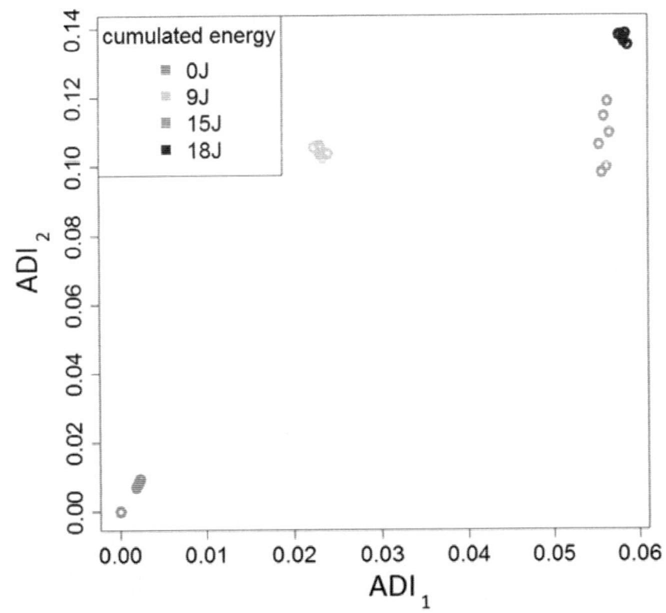

Figure 8. Averaged Damage Indices with respect to cumulative energy of impacts.

The network described above contains only 4 transducers thus and it is not enough to localize damage using the map *I* (Eq. 6). Below the map *I* obtained for a GFRP structure having two lines of embedded PZT transducers (Fig. 9) being hit by a low energy impact is presented. The damage is quite well localized (Fig. 10) along the line of the transducers since there is a lot of different sensing path – some being far from a damage and the other in its close proximity. In order to enhance the localization resolution along the perpendicular dimension additional transducers located between the two lines of sensors need to be introduced.

Figure 9. Averaged Damage Indices with respect to cumulative energy of impacts.

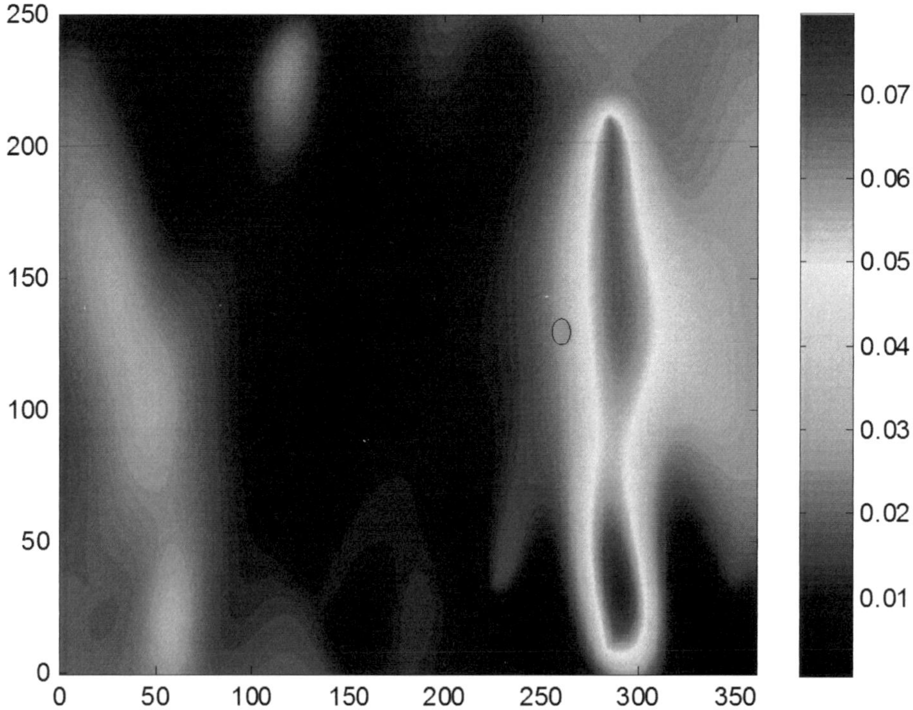

Figure 10. The map I for a GFRP structure with BVID damage (location of damage indicated).

SUMMARY

In the article an approach to structure monitoring based on PZT sensor network was presented. The proposed method was verified monitoring damages interacting with actuated waves in a different manner which confirms its robustness. The damage size resolution of the approach is restricted to indicate if sensing paths of PZT sensor network were intersected by an incident damage. This is due to limited information content carried by the Damage Indices used. In return they exhibit small sensitivity to signal decoherence in time which allows to apply the proposed DIs drift compensation formula decreasing significantly false positive indications, which besides damage detection capability is one of the crucial factors in SHM applications. The approach for a long term DIs drift compensation was validated using a data collected during a Full Scale Fatigue Test of an aircraft.

EXAMPLES:

[1] D. Roach (2012). Industry Survey of Structural Health Monitoring Technology and Usage. *Sandia National Laboratories Report.*

[2] L. Ilcewicz (2006). *CS&TA Federal Aviation Administration.*

[3] Z. Su and L. Ye (2009). Identification of Damage Using Lamb Waves. *Springer-Verlag.*

[4] X. Zhao, H. Gao, B. Ayhan, F. Yan, C. Kwan, J.L. Rose (2007). Active health monitoring of an aircraft wing with embedded piezoelectric sensor/actuator network: I. Defect detection, localization and growth monitoring. *Smart Materials and Structures,* vol. 16(4), p. 1208.

Sensor location studies for damage detection in Aerospace Structures using 3D Scanning Laser Vibrometry

R. Marks[1], A. Clarke[1], C. Featherston[1], C. Paget[2], R. Pullin[1]

[1]*Cardiff School of Engineering, Queen's Buildings, The Parade, Cardiff, United Kingdom. CF24 3AA*
[2]*Airbus UK Ltd, Filton, Bristol, United Kingdom. BS99 7AR*

Abstract: With the increasing complexity of aircraft structures and materials there is an essential need to continually monitor the structure for damage. This also drives the requirement to optimize the location of sensors for damage detection to ensure full damage detection coverage of the structure whilst minimizing the number of sensors required, hence reducing costs, weight and data processing requirements.

An experimental study using 3D scanning laser vibrometry was carried out to investigate Lamb wave interaction with a disbonded stiffener. A piezoelectric transducer was coupled to two different stiffened aluminium panels; one healthy and one with a 25.4mm long disbond. Three excitation frequencies were used; 100kHz, 250kHz and 300kHz, to investigate the effects of frequency on wave interaction with the disbond.

A computational study using local interaction simulation approach (LISA) was carried out for comparison with the experimental results. The objective of the simulation was to determine its viability as a tool for simulating wave interaction with disbonds for damage detection sensor network design. The results from the LISA model produced good results that qualitatively compared with the experimental results. This demonstrated that this simulation technique is well suited for performing sensor location studies for active sensor networks and has the potential to serve as a useful tool for damage detection sensor network designers.

INTRODUCTION

Acousto-ultrasonic induced Lamb waves are a long-established technique for detecting damage in structures. The principle involves exciting a piezoelectric transducer mounted to the structure's surface which induces a Lamb wave that is then detected by another transducer mounted at a different location on the structure. If damage occurs within the field between the two sensors, the signal propagation is altered resulting in a quantifiable difference in the signal received.

Adhesively bonded metallic joints have been used in the primary structure of aircraft for the past fifty years [1]. Bonding as opposed to mechanically fastening stiffeners to the aircraft's skin is advantageous as it can reduce the mass of the joints, distribute the stress loading more evenly and be more resistant to environmental effects which an aircraft experiences [2]. Adhesive bonding is also less labour intensive and more cost effective than using mechanical fasteners [3].

Though adhesives have been used in the secondary structure without arrestment fasteners they are rarely used in the primary structure (i.e. the structure in which the occurrence of failure would be catastrophic). Without arrestment fasteners there is a lack of confidence in the reliability of the joint [4]. From the experience of the Royal Australian Air Force (RAAF) 53% of defects detected on aircraft structures such as the F-111 were found to be bond failures [5]. Many cases of adhesive bond failure have been found to be the result of improper application of the adhesive or insufficient surface preparation [4], i.e. human error in the manufacture of the joint.

Though it has been proven that bonded joints are more resistant to fatigue loading as well as being able to withstand higher peak loading [5], the degradation of the adhesive layer over time is not fully understood [6]. This uncertainty in the condition of the bond throughout the aircraft's in-service life requires monitoring using non-destructive techniques to ensure airworthiness. If a sensor network could be installed to monitor damage and degradation of the bonded joint an "as required" inspection program could be utilised. However sensor networks need a high probability of detection whilst minimizing the weight penalty they add to the overall structure. Hence, understanding the optimal location of sensors for this monitoring is of the utmost importance.

The most straightforward method of implementing a structural health monitoring network onto an aircraft is during the design stage [7]. It has been suggested that by implementing SHM during design, weight savings of up to 15% could be achieved at component level [8]. In order for effective implementation, a validated method of successfully modelling damage conditions is vital in assisting sensor network designers with optimal sensor positioning.

LOCAL INTERACTION SIMULATION APPROACH (LISA)

There have been many advances in modelling Lamb wave interaction with damage in recent years. One of the most promising methods of modelling is the local interaction simulation approach (LISA). LISA is a finite difference method that was developed some twenty years ago [9-11]. Since then, there has been much research into assessing the accuracy, reliability and validity of the models produced particularly at boundaries and their interaction with features. These studies have demonstrated that LISA is a well suited method for the modelling of Lamb wave interaction [9, 12-16]. For further explanation on LISA the reader is referred to *Lee and Staszewski* [15].

Since the widespread advent of NVIDA® CUDA parallel computing architecture, LISA has come to the fore being able to process large simulations in minutes, allowing many scenarios to be simulated in a short period of time.

A LISA model was used in this study to investigate Lamb wave interaction with disbonds in stiffened metallic plates.

3D VIBROMETER EXPERIMENTAL SETUP

Two stiffened panels of dimensions 1250mm x 1250mm were constructed from 3mm thick 6082-T6 aluminium plate. An unequal angle stiffener made from 6082-T6 aluminium was bonded to each plate using commercially available Araldite® 420 structural adhesive. The film thickness of the adhesive was regulated using 0.1mm copper wire gauges to achieve the optimum shear strength [17].

On one of the panels an intentional disbond was induced at the centre of the stiffener of 25.4mm in length which covered the width of the stiffener. This was produced by installing PTFE tape over this region of the bond which was then removed once the adhesive had cured.

A square region of 145mm x 145mm in the centre of each panel, on the opposite face of the panel to the face to which the stiffener was bonded, was the designated area of investigation. This area of investigation was selected in an attempt to reduce reflections from the edges of the panels as shown in Figure 1.

The excitation site was located 312.5mm from the left edge of the plate; the midpoint between the plate edge and the stiffener. This ensured that there was sufficient distance for the Lamb modes to form and disperse before interacting with the stiffener whilst reducing the effects of edge reflections.

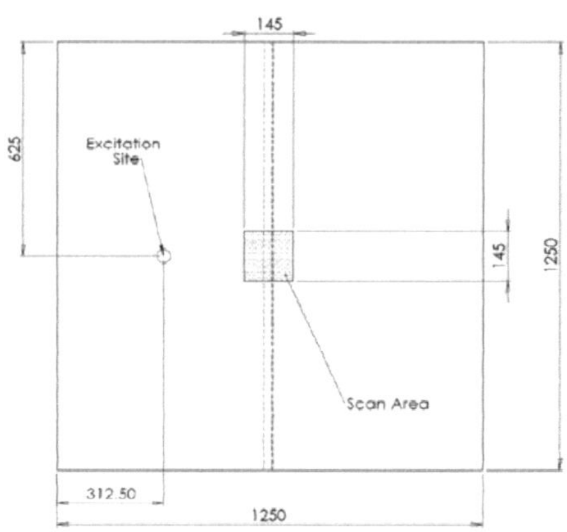

Figure 19. Dimensions of the Stiffened Panel (all dimensions in mm).

5329 vibrometer measurement points were set up within the investigation region which provided sufficient spatial resolution to construct a visual representation of the Lamb modes propagating in the panel.

A PANCOM Pico-Z (200kHz -500kHz) transducer was excited with a 10-cycle sine burst generated by Physical Acoustics WaveGen function generator software connected to Physical Acoustics µdisp/NB-8 hardware. The peak-to-peak amplitude of the excitation signal was 160 V. Three excitation frequencies of 100kHz, 250kHz and 300kHz were selected for this experiment. A 10 V peak-to-peak wave was also generated and used as a reference signal for triggering the acquisition of the vibrometer. A repetitive trigger rate of 20 Hz was used as this gave sufficient time for the induced wave energy to fully dissipate before taking the next measurement. Velocity amplitudes were investigated at the measurement points with two hundred measurements being recorded at each point. These measurements were then averaged in order to improve the signal to noise ratios.

The vibrometer used for this experiment was the Polytec PSV-500-3D-M. This vibrometer uses three laser heads to measure each scan point. The measurements recorded by each laser head can then be used to calculate, using trigonometry, both the in-plane and out-of-plane modes. By taking measurements from three heads it becomes less important for the laser heads to be perpendicular to the structure under test with a 3D scanning system when compared to a 1D system. The investigation region was coated with retro reflective glass spheres which improves the signal quality by increasing the amount of back scattered light. During testing each panel was supported on wooden supports at the edges of the plate. This allowed the plate to lie flat. A layer of 'bubble wrap' packaging material was placed between the wooden supports and the plate to acoustically uncouple it from its surroundings. A series of markers were laid within the laboratory and on the panels to ensure that each panel was positioned in the same place relative to the laser heads.

LISA MODEL

The LISA model was designed to simulate the Lamb wave interaction with stiffeners and disbonds in metallic panels which could be validated with the results from the 3D vibrometry.

Global cube edge length

One major drawback of LISA is that it is governed by a global cube edge length and hence the smallest component dimension drives the global cube edge length as opposed to finite element techniques where the mesh can easily be refined in regions of interest whilst adopting a coarser mesh elsewhere. From the panels used in the experimental study the smallest dimension was the adhesive film thickness of 0.1mm.

Using a 0.1mm global cube edge length generated a very large set of cube geometry data which was not possible to be processed by the CUDA graphics card. This was due to insufficient memory (2GB) on the graphics card. Therefore, the model was simplified to reduce the memory required.

Model setup

cuLISA3D version 0.8.5-x64 with a MATLAB interface was used for this simulation. The simulation was performed on a PC fitted with an NVIDIA GTX690 CUDA graphics card (2 Kepler GPUS with 256 cores running at 995MHz, 2GB of RAM).

The inner 625mm x 625mm stiffened aluminium plate was modelled with an ultrasonic actuator positioned on the edge to reduce the effects of edge reflections. The smaller geometry also reduced the computational costs and model processing time.

A global cube edge length of 0.5mm was used for the model. This introduced issues in the modelling of the adhesive film. Therefore, the effects of the adhesive film were assumed to be negligible. Due to the global cube edge length, a true 25.4mm stiffener width could not be modelled. This was rounded to the nearest cube integer of 25.5mm.

In the case of modelling the disbond, a section of cubes 25.5mm x 25.5mm x 0.5mm was removed from the underside of the stiffener to simulate the reduction in stiffness as shown in Figure 2.

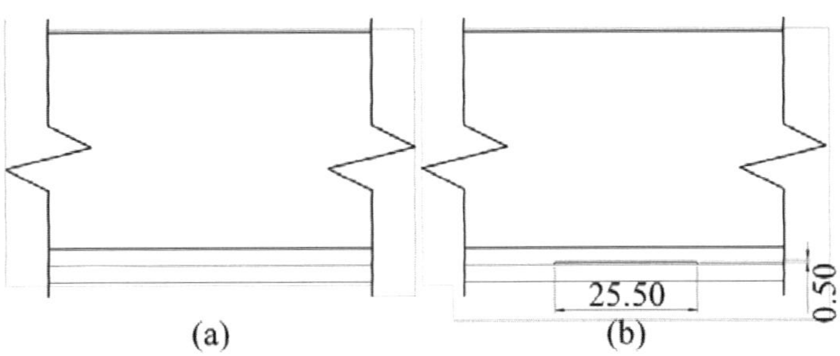

Figure 2. Side elevation of the LISA geometry for the healthy panel (a) and the disbonded panel (b).

A 10-cycle sine wave with a Hanning window was used for the excitation. As with the experimental setup, three frequencies of 100kHz, 250kHz and 300kHz were used for the excitation.

RESULTS AND DISCUSSION

As the 3D laser vibrometry technique measures velocities, these were integrated to calculate the displacements. 0μs was set when the wave first appeared in the area of interest on the model.

100kHz Healthy stiffened panel results

The LISA and 3D laser vibrometry displacement plots for the 100kHz Lamb wave interaction with the healthy stiffened panel are shown in Figure 3. The plots for the disbonded stiffened panel are shown in Figure 4.

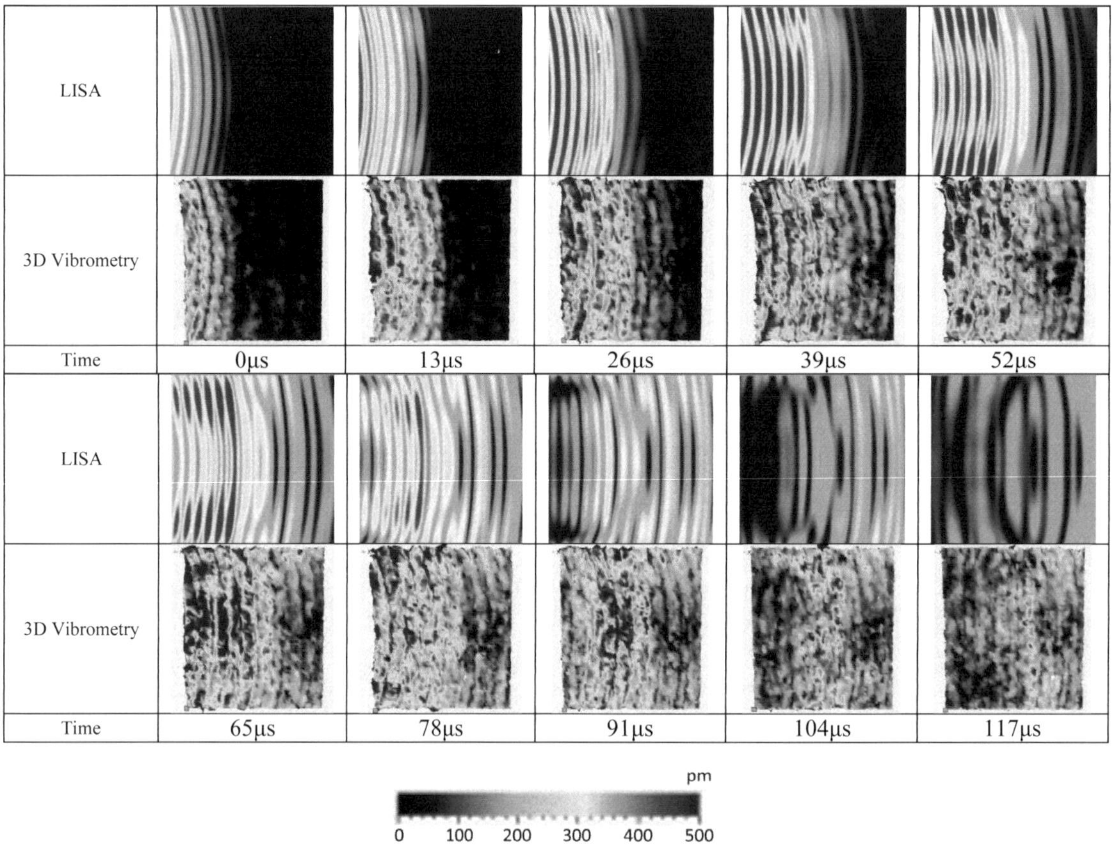

Figure 3. 100kHz healthy stiffened panel.

The results from the 3D vibrometry and the LISA model correspond well. In both sets of the results shown in Figure 3, the amplitude of the displacements are seen to significantly decrease as the Lamb wave passes through the stiffener. This can be attributed to the significant localised change in stiffness as well as a proportion of the wave energy being reflected back towards the source.

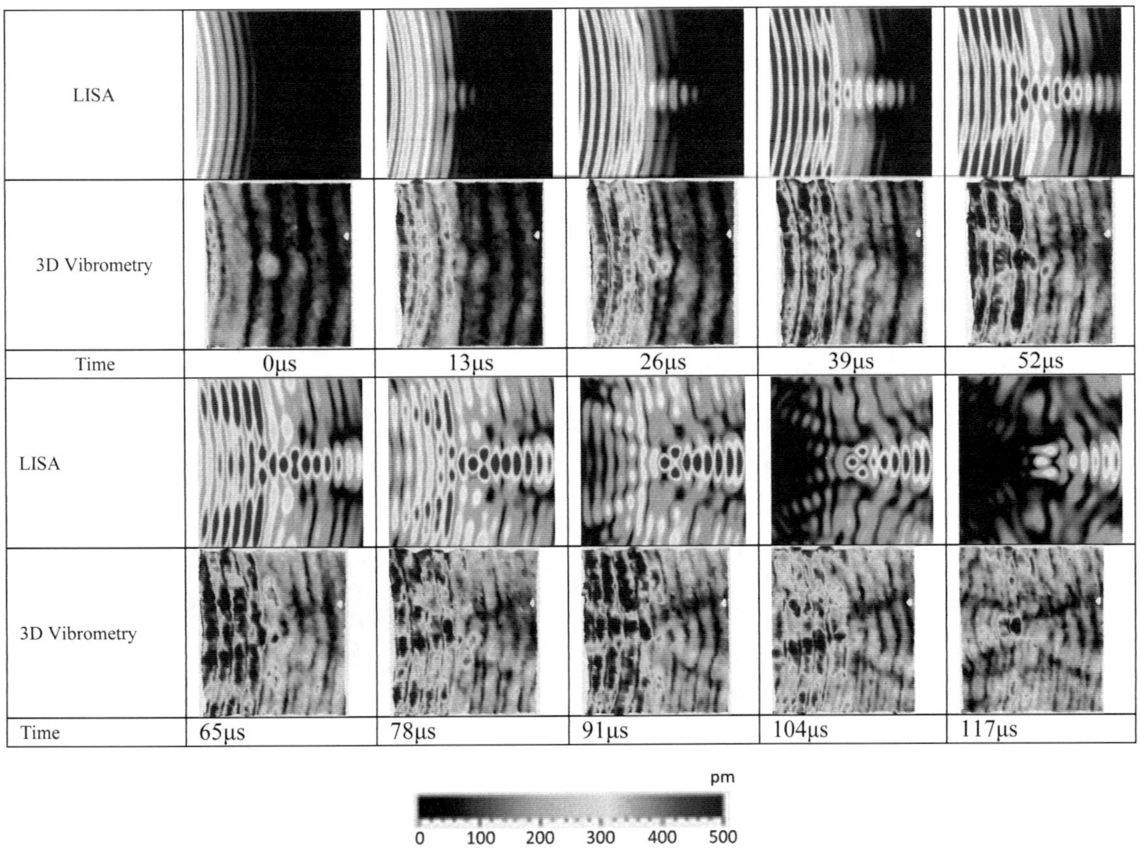

Figure 4. 100kHz disbonded stiffened panel.

Initially (0 - 26μs), the wave interaction with the disbond in both the results from LISA and the 3D vibrometry are very similar, as shown in Figure 4. It is possible to see from the 3D vibrometry that at 0μs the S_0 interaction with the disbond produced a significant increase in the amplitude whereas the results from the LISA model did not reflect this. However, the A_0 interaction with the stiffener showed a similar increase in amplitude when the two displacement plots at 26μs are compared.

As the wave continued to interact however, there was a difference in the results from the model and the 3D vibrometry. In the LISA model, there was a fringe pattern of high amplitude on the right hand side of the stiffener as the wave is transmitted through the disbond. The fringe pattern observed on the 3D vibrometry was of lower amplitude.

Considering only the fringe patterns however it can be seen that there were similarities in the conical fringes on both sides of the stiffener.

250kHz Displacement Plots

The LISA and 3D laser vibrometry displacement plots for the 250kHz Lamb wave interaction with the healthy stiffened panel are shown in Figure 5.

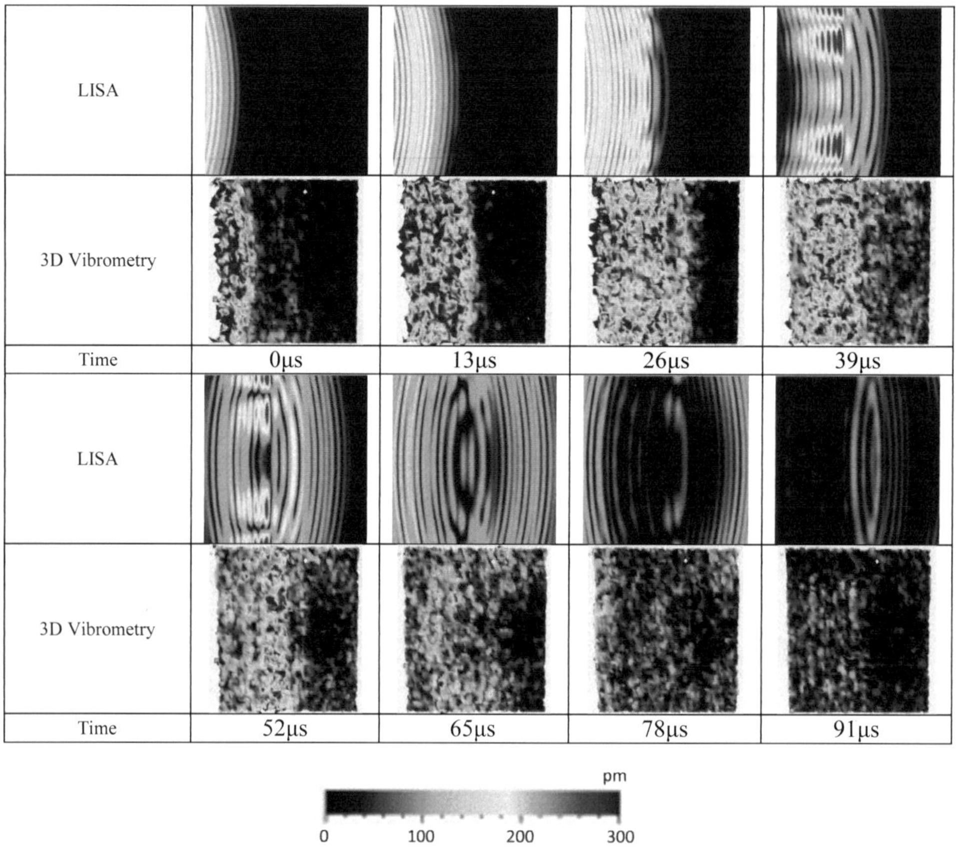

Figure 5. 250kHz healthy stiffened panel.

The first observation from the results in Figure 5 when making the comparison with the results from the 100kHz study was that the 250kHz wave front was travelling faster. This was due to the dispersive phenomena of Lamb Waves. LISA was derived from the fundamental wave equation for particle displacement which can be developed from Green's approach [14]. The dispersion phenomenon was observed in both the LISA model and the 3D vibrometry.

The high amplitude fringes that were observed in the LISA model as the incoming wave was guided along the length of the stiffener were present in the 3D vibrometry plots but to a lesser extent seen at 39μs and 52μs. This was due to the amplitudes of the waves reflected off of the stiffener being of significantly higher amplitude than those that were observed in the model. However, this wave energy rapidly attenuated producing a good similarity between the model and the experimental results.

The LISA and 3D laser vibrometry displacement plots for the 250kHz Lamb wave interaction with the disbonded stiffened panel are shown in Figure 6.

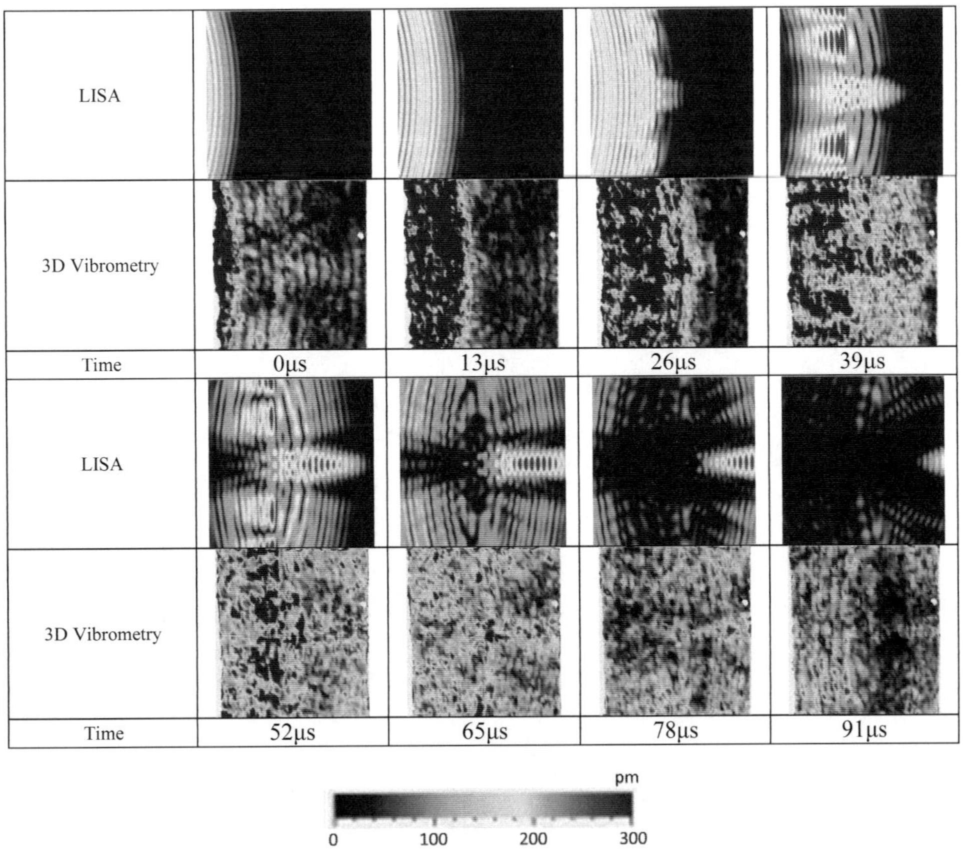

Figure 6. 250kHz disbonded stiffened panel.

The observations of the 250kHz LISA disbonded stiffened panel results shown in Figure 6 are qualitatively similar to those observed in the 3D vibrometry results. At 0μs the S_0 wave interaction with the disbond was apparent in both sets of results. However, the wave interaction in the LISA model displayed significantly smaller amplitudes than those observed in the 3D vibrometry results.

The interaction with the A_0 mode observed from 26μs onward shows great similarity. At 26μs there was high amplitude observed in the disbonded region in both the LISA model and the 3D vibrometry. As time progressed, there were high amplitudes of the Lamb wave displacement transmitted through the disbonded region. In the LISA model there was a more pronounced conical fringe pattern observed that was not as easily identifiable in the 3D vibrometry results.

300kHz Displacement Plots

The LISA and 3D laser vibrometry displacement plots for the 300kHz Lamb wave interaction with the healthy stiffened panel are shown in Figure 7.

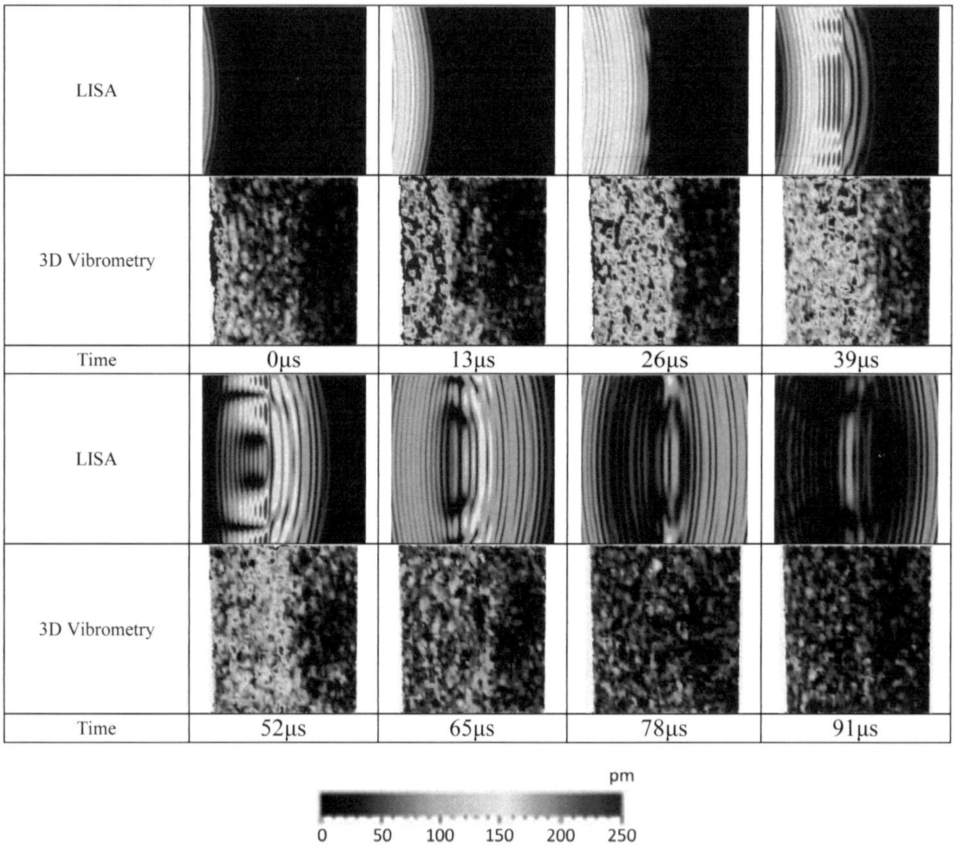

Figure 7. 300kHz healthy stiffened panel.

The results observed from the 300kHz studies once again demonstrated a reduction in wave amplitude as the wave interacted with the stiffener. There was presence of what appeared to be a mode conversion in LISA model as the A_0 mode interacted with stiffener shown by a longer wavelength. This was also present in the 250kHz study but was not observed in the results from the 3D vibrometry. Mode conversion in the presence of stiffeners has been treated theoretically in [18] which supports these findings.

It was also possible to observe once again similarities in fringe patterns as the wave was guided along the length of the stiffener at 52μs.

The LISA and 3D laser vibrometry displacement plots for the 300kHz Lamb wave interaction with the disbonded stiffened panel are shown in Figure 8.

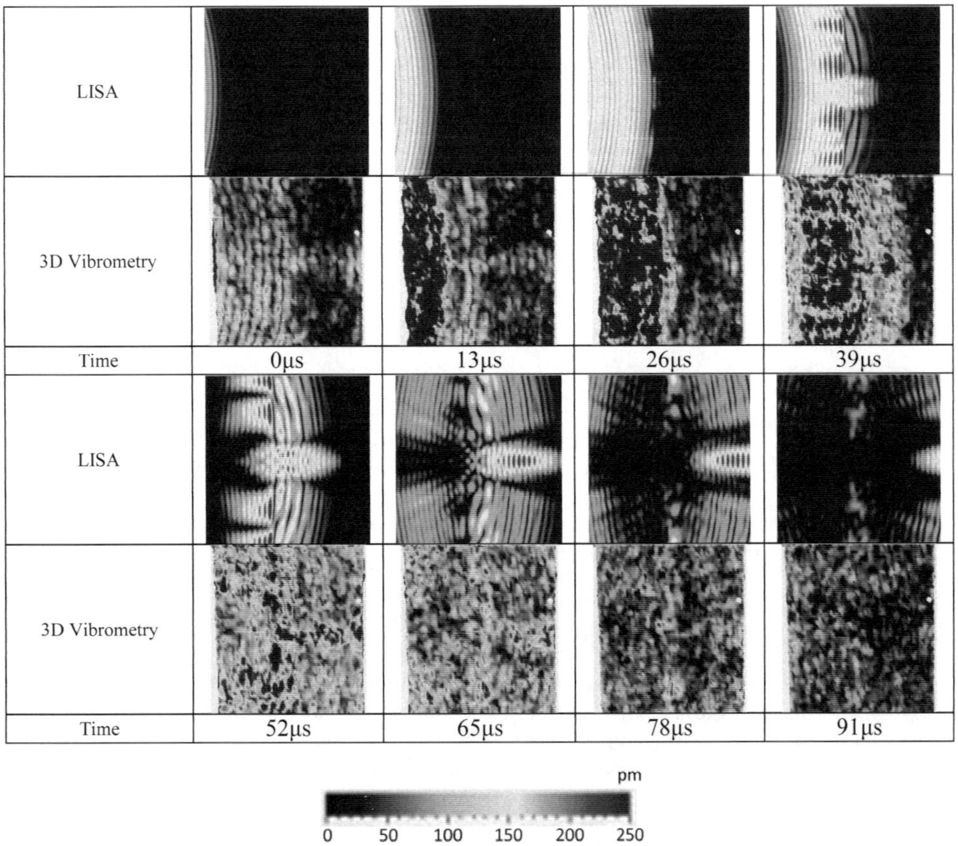

Figure 8. 300kHz disbonded stiffened panel.

As found in the 250kHz study there were similar wave interaction fringes observed between the LISA model and the 3D vibrometry. There was a similar interaction fringe with the S_0 mode observed at 0μs though the amplitude of the fringe observed in the 3D vibrometry results was significantly higher than those observed in the LISA model.

At 26μs the A_0 mode started to interact with the disbonded region of the stiffener. From this point onwards, it can be seen that the waves transmitted through the disbonded region exhibited higher amplitudes than the waves that that were transmitted through the healthy parts of the stiffener. It was observed from the 3D Vibrometry results however that the wave that was transmitted through the disbonded region appeared to attenuate more than the wave that was observed in the LISA model.

Comparative discussion

The results from the LISA simulation generally exhibited similar wave interaction to the results observed from the 3D vibrometry. There was however a reduction in the quality of the 250kHz and 300kHz plots compared to 100kHz plot. This could be attributed to the spatial resolution of the scan points. The number of scan points were kept consistent throughout the study in order obtain results that were comparable and to allow the study to be completed in a timely fashion. By increasing the number of scan points more detailed 3D vibrometry fringe plots could be achieved allowing a better comparison to be made between the experimental and simulated results.

There were also some differences in the amplitude of the displacements between the healthy and disbonded 250kHz and 300kHz results. This could be attributed to two separate panels being used which may have resulted in a difference in the acoustic coupling of the actuator.

The differences seen in the LISA results compared to the 3D vibrometry could be due to several factors. It is important to note that assumption of the adhesive film layer being negligible may have resulted in the differences seen. The adhesive film had different material properties to the aluminium. This would result in a change in acoustic impedance which would have influences on the wave interaction. The extent to which this is significant requires further study.

This may explain the higher reduction of the wave amplitude as the waves were transmitted through the stiffener in the LISA model.

During the manufacture, PTFE tape was used to induce the disbond. Though every effort was made to achieve a perfect disbond, it was possible that the disbond was not geometrically perfect as modelled in the LISA simulation. This could explain some of the differences in the interaction fringes particularly those observed in the 100kHz study.

The high amplitude of the waves transmitted through the disbond gives an indication that disbonds have a significant effect on wave interaction. The presence of high amplitude waves directly behind the disbond region and the surrounding low amplitude fringes compared to the low amplitude observed in the healthy bonds suggests that wave energy could be used to detect the presence of the damage.

A common observation that was made throughout all three frequency studies was the conical fringe pattern observed from the wave interaction with the disbond compared to the wave interaction with the healthy stiffener. By gaining a better understanding of this type of fringe pattern it a may be possible to design a sensor network that could not only detect the damage but also locate and quantify the size of the defect.

LISA has been shown to be a useful tool for performing studies investigating wave interaction with defects for sensor network design. LISA's ability to run on the CUDA parallel computing architecture has the advantage over traditional FEA methods by being able to run multiple simulations in a relatively short space of time. This has the exciting prospect of being able to model many different damage scenarios which could serve as an input to a sensor network optimisation tool.

The use of LISA also has cost and time advantages over experimental methods such as 3D vibrometry. Experimental methods require damage to be induced into structures in order for wave interaction studies to be performed. This typically requires the manufacture of many test specimens. However, it is important to validate models with experimental testing. It would be possible, with careful consideration given to experimental design, to perform a series of tests that would validate LISA's modelling of wave propagation and interaction with damage. It would then be possible to have a high confidence level in the validity of the model's results when simulating wave interaction with other damage scenarios.

The commercially available LISA software package does have drawbacks particularly regarding the modelling of composite materials at present. However, there have been studies conducted that have shown significant advances in adapting LISA to the modelling of composite materials [19] The ability to model composite materials will be of substantial benefit in the design of sensor networks for damage detection in advanced aerospace materials.

CONCLUSION

From the results presented it can be concluded that there was a good similarity of wave interaction between the 3D vibrometry and LISA model. This study has demonstrated that LISA is a very powerful tool in the simulation of wave interaction with stiffeners and disbonds and is representative of experimental observations.

It has been demonstrated that LISA is capable of modelling the wave interaction of both the S_0 and the A_0 modes which are the most common Lamb wave modes used in structural health monitoring of plate-like structures.

This study has only considered the wave interaction when the actuator is aligned with the disbond. In reality, this may not be the case. Therefore, it would be beneficial to investigate the effect of the actuator being at different relative angles and distances to the disbond. Other considerations could also be taken into account for defect size in order to work towards achieving an optimisation platform for sensor network design. However the potential of modelling tools such as LISA to provide information for the design of damage detection sensor networks has been clearly demonstrated.

ACKNOWLEDGEMENTS

The authors would like to the support of EPSRC DTG together with Airbus UK Ltd. for the funding of this project as well as Dr Mark Eaton and Dr Mathew Pearson for their advice during this project.

REFERENCES

[1] Higgins, A., *Adhesive bonding of aircraft structures.* International Journal of Adhesion and Adhesives, 2000. **20**(5): p. 367-376.

[2] Arenas, J.M., et al., *Considerations for the industrial application of structural adhesive joints in the aluminium-composite material bonding.* Composites Part B: Engineering, 2013. **44**(1): p. 417-423.

[3] Petrie, E.M., *Adhesives for the assembly of aircraft structures and components: Decades of performance improvement, with the new applications of the horizon.* Metal Finishing, 2008. **106**(2): p. 26-31.

[4]. Bossi, R. and M.J. Piehl, *BONDING PRIMARY AIRCRAFT STRUCTURE: THE ISSUES.* Manufacturing Engineering, 2011. **146**(3): p. 101-102,104-109.

[5] Davis, M. and D. Bond, *Principles and practices of adhesive bonded structural joints and repairs.* International journal of adhesion and adhesives, 1999. **19**(2): p. 91-105.

[6] N Quaegebeur, P.M., P Masson, M Castaings, *Methodology for optimal configuration in structural health monitoring of composite bonded joints.* Smart Materials and Structures, 2012. **21**(10): p. 11.

[7] Assler, H. and J. Telgkamp, *Design of aircraft structures under consideration of NDT*, in *WCNDT-World Conference of Non-Destructive Testing.* 2004: Montreal, Canada.

[8] Speckmann, H. and H. Roesner, *Structural Health Monitoring: A Contribution to the Intelligent Aircraft Structure*, in *ECNDT 2006.* 2006.

[9] Delsanto, P.P., et al., *Connection machine simulation of ultrasonic wave propagation in materials. II: The two-dimensional case.* Wave Motion, 1994. **20**(4): p. 295-314.

[10] Delsanto, P.P., R.S. Schechter, and R.B. Mignogna, *Connection machine simulation of ultrasonic wave propagation in materials III: The three-dimensional case.* Wave Motion, 1997. **26**(4): p. 329-339.

[11] Delsanto, P.P., et al., *Connection machine simulation of ultrasonic wave propagation in materials. I: the one-dimensional case.* Wave Motion, 1992. **16**(1): p. 65-80.

[12] Delsanto, P.P., et al., *Analysis of stability and convergence in FD simulations of the 1-D ultrasonic wave propagation.* Mathematical and Computer Modelling, 1997. **25**(6): p. 19-29.

[13] Ruffino, E. and P.P. Delsanto, *Problems of accuracy and reliability in 2-D LISA simulations.* Computers & Mathematics with Applications, 1999. **38**(5–6): p. 89-97.

[14] Lee, B.C. and W.J. Staszewski, *Sensor location studies for damage detection with Lamb waves.* Smart Materials and Structures, 2007. **16**(2): p. 399-408.

[15] Lee, B.C. and W.J. Staszewski, *Modelling of Lamb waves for damage detection in metallic structures: Part I. Wave propagation.* Smart Materials and Structures, 2003. **12**(5): p. 804.

[16] Lee, B.C. and W.J. Staszewski, *Modelling of Lamb waves for damage detection in metallic structures: Part II. Wave interactions with damage.* Smart Materials and Structures, 2003. **12**(5): p. 815.

[17] Huntsman, *Aerospace Adhesives: Araldite 420 A/B Two component epoxy adhesive.* 2000.

[18] Reusser, R.S., et al., *Reflection and transmission of guided ultrasonic plate waves by vertical stiffeners.* The Journal of the Acoustical Society of America, 2014. **136**(1): p. 170-182.

[19] Nadella, K. and C.S. Cesnik, *Local interaction simulation approach for modeling wave propagation in composite structures.* CEAS Aeronautical Journal, 2013. **4**(1): p. 35-48.

Long-term Durability of Adhesively Bonded Composite Joints Under Quasi-static and Fatigue Loading

Chun Li[1], Zhong Yi Wan[1], Gabriel LaPlante[2], Calvin Rans[3], Gang Li[1]

[1]*National Research Council Canada, Aerospace, Ottawa, Canada*
[2]*Department of Mechanical Engineering, Université de Moncton, Moncton, Canada*
[3]*Faculty of Aerospace Engineering, Delft University of Technology, The Netherlands*

Abstract: The 21st century marks an era where ultra-high specific strength and stiffness carbon-fibre reinforced composite materials are increasingly used for fabricating aircraft structures. It is well known that composite materials are susceptible to environmental degradation, which is of particular concern for adhesively bonded composite joints due to the lack of non-destructive methods for quantitative assessment of bond strength. Although fatigue is not considered to be an issue for composite material and structures, past experiences have indicated that a better understanding of fatigue problems of composites are important for risk analysis for structures that may be subjected to undetected incidental and environmental damages. Furthermore, it also offers the potential to develop tools for life extension programs and more efficient design of composites. This work studies fatigue and fracture behaviours of bonded composite joints, with a focus on the effect of long-term exposure to hot/wet environments on fatigue delamination/disbond growth under mode I and Mixed Mode I/II loading. The objective of this project is to build understanding of the behaviour of bonded composite joints under combined loading and environmental conditions.

The bonded joint specimens were manufactured and conditioned to an equilibrium state in several environments, including moisture infusion and immersion. The as-made and conditioned specimens were tested under quasi-static and fatigue loadings with load mixities (G_{II}/G) of 0, 0.2, 0.4, 0.6 and 1. Quasi-static and fatigue delamination/disbond growth rates were characterized for as-made and conditioned specimens for bonded composite joints. The conditioned specimens exhibited a significant reduction in fracture toughness, G_C, at up to 34%, demonstrating the significant effect of environmental ageing. Such effect was also apparent on fatigue disbond growth rates for the environmental aged composite joints.

INTRODUCTION

Although experience with composites has been mostly positive [[1]], there are concerns arising from their susceptibility to delamination and disbond. Delamination and disbond are among the major sources of part failure due to the fact that it shows little visual indication when the damage exceeds the threshold. Some of the sources of delamination and disbond include manufacturing defects, impact damage and environmental degradation [[2]-[4]]. A delamination and disbond may propagate at small loads, leading to catastrophic failure. An example of this is the F-18 composite structure where disbond is a major source of concern; the disbond of the control surfaces has led to reduced stiffness, strength and eventual in-flight failures [[4]].

The need for building the capability for life prediction of composite structures is important to the Royal Canadian Air Force because operational requirements are significantly expanding the role of military fleets, both fixed- and rotary-wing aircraft. This expansion has resulted in aircraft flying missions that are beyond their original design usage spectrum. Studies of long-term degradation of helicopter composite structures would help address growing concerns about continued airworthiness and structural integrity of these structures.

The conventional material strength approach is defined by applied stress and material failure strength. In this traditional approach, a design is considered sound when the anticipated applied stress is less than the material failure strength. This approach may be adequate for brittle fracture by imposing a safety factor, combined with minimal tensile elongation requirements on the material, but is not adequate for cases where flaws or damage propagate in a progressive manner, leading to failure of the structure. In contrast, fracture mechanics approach employs three variables: applied stress, flaw size and fracture toughness. Within this energy-based approach, failure occurs when the energy available for damage growth exceeds the resistance of the material. Strain energy release rate (G), one form of fracture toughness, is defined to be the loss of potential energy, dU, in the test specimen per unit of specimen width for an infinitesimal increase in delamination length, da, for a disbond growing under a constant displacement (1).

$$G = -\frac{1}{b}\frac{dU}{da} \tag{1}$$

where b is the width of the specimen. Analogous to the power relationship commonly used to describe the dependency of crack growth rate in metals on the stress intensity factor, K or ΔK, the fatigue delamination growth rate, *da/dN*, of a fibre-reinforced composite for a given *R*-ratio can be captured by a power relationship of the maximum strain energy release rate, G_{max}, as shown in Equation (2) [[5],[6]].

$$\frac{da}{dN} = C(G_{max})^n \tag{2}$$

where, *C* and *n* are material parameters. This is analogous to the Paris relationship for stress intensity factors.

This study aims to characterize and model quasi-static and fatigue behaviour of bonded composite specimens under Mixed Mode I/II bending (MMB) that are subjected to varying mode mixities and environmental conditions. By convention, Mode I is a peeling motion while Mode II is an in-plane shearing motion. Each mode has a strain energy release rate associated with it, namely G_I and G_{II}, with mode mixity defined as:

$$\frac{G_{II}}{G} \tag{3}$$

$$G = G_I + G_{II} \tag{4}$$

When G exceeds the G_C for the same mode mixity, the onset of delamination/disbond occurs. In the case of an existing delamination or disbond, the damage will propagate. Please note that the energy that is required for the onset of damage may not be the same as that for propagation. Thus the value of G_C for delamination/disbond onset could be different from that for propagation.

EXPERIMENTAL

Specimen Preparation, Design and Fabrication

The test specimens were manufactured as per ASTM D 6671M-06 [[7]], which was designed for fracture toughness testing of unidirectional composite laminates with a single-phase polymer matrix under quasi-static loading. The mixed-mode bending (MMB) specimens were made with two 13-ply unidirectional continuous carbon fibre-reinforced composite material, Cytec CYCOM 5276-1, bonded using two plies of 3M AF163-2K 0.06 psf adhesive. A 0.013 mm (0.0005″) thick Teflon insert was placed at the one end of the specimen between the two plies of adhesive to introduce a pre-crack. Modifications in specimen configuration and result analysis were made to expand its application to the bonded joint configuration. The configuration of the specimen is shown in Figure 1, with final dimensions of 140 mm (5.5″) length, 21.5 mm (0.844″) width (*b*). The average of the specimen thickness (2*h*) is 3.8 mm (0.15″).

Aluminium piano hinges were then adhesively bonded to the end of the specimen, locating the loading point to be 25.4 mm (1″) from the initial disbond (a_0). Hysol EA9360 paste adhesive was used for hinge bonding. The adhesive was cured at 80°C for 2 hours. The manufacturing procedures for laminate lay-up, surface preparation, adhesive curing, and coupon machining are described in the manufacture report [[8]].

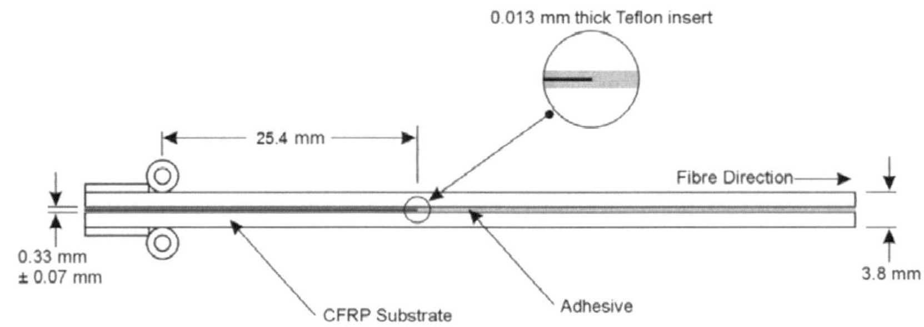

Figure 1. Schematic of the MMB Specimen.

Experimental Setup

All experiments were conducted in laboratory conditions, at a temperature and humidity level of 22±5°C and 50±20%, respectively. All the tests were conduct under displacement control at a frequency between 0.5 to 2.5 Hz. The test utilized a MTS 858 Tabletop hydraulic load frame, using a 250N or a 500N load cell with calibrations traceable to National Standards and load accuracy of 1% of reading. The data acquisition station used TestStar II Station Manager v3.4B to record the load and crosshead displacement during testing. Prior to testing, inspection and metrology were performed on each specimen. To aid in the detection of delamination, a thin layer of white lacquer paint was sprayed on the side of the specimen along with a paper scale. Below is figure of the MMB fixture used in testing; where c is the lever length and L is the half-span of the fixture base.

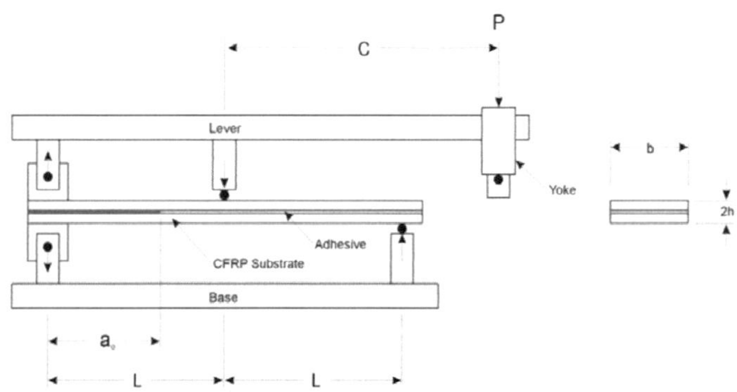

Figure 2. Schematic and image of the Mixed Mode I/II bending fixture.

The specimens were placed into the fixture via the piano hinges. The load and displacement were zeroed prior to commencement of each test. Once the specimen was installed and properly aligned, a loading ramp according to ASTM D-6671 [[7]] was applied. Once the displacement reached the target maximum value, fatigue loading was applied to the specimen at a set frequency. Time, cycle count, load and displacement data were collected by the load frame control software.

A reduced set of information was also captured by the Labview control software, along with images of disbond tip location at the peak displacements. An in-house automated delamination/ disbond monitoring system (ADMS) [[9]]was developed to take images at specific intervals during the test to obtain the actual delamination length without interrupting the test.

Environmental Conditioning and Testing

To study the degradation effects of environmental conditioning on fracture toughness, a portion of the manufactured specimens were exposed to designated conditions. Three different conditions were devised for this purpose and these are summarized in Table I.

Table I. Summary of Environmental Conditions.

Environmental Conditons	Description
RTA	Room Temperature Ambient; at a temperature and humidity level of 22±5 °C and 50% RH±20%
IVWN	Intermediate temperature at 70 °C, Vapour at 95% relative humidity; Water; No freeze-thaw
HISN	High temperature at 80 °C Immersion in Salt water at 5% Salinity; No freeze-thaw

The specimens were conditioned in the environments list in Table I until they reached an equilibrium state, where additional conditioning over a 2-week period led to no additional weight gain. All the MMB tests of RTA and conditioned specimens were performed at RTA conditions. Care was taken to minimize the exposure time of IVWN and HISN specimens to the RTA test conditions.

To demonstrate the effect of environments on bond performance under Mixed Mode I-II loading, a range of load mixities of 0.2, 0.4, and 0.6 were investigated. The onset mode mixity was fixed by changing the lever length, *c* (see Figure 2). In the course of a fatigue test, it was found that the actual mode mixity deviated slightly, which was likely due to the change in crack length, *a*, as it grew during the test. It can be noted, however, that the deviation in mode mixity was small and on the order of 10% or less.

Data Collection and Analysis

In the quasi-static tests, the Mixed Mode I/II critical strain energy release rate of bonded joints, G_C, was first measured in tests with a displacement rate of 0.5 mm/min, following the procedure described in ASTM D6671[[7]]. A slower loading rate of 0.25 mm/min was applied to conditioned specimens or specimens of higher load mixity because of the observed unstable delamination growth in those specimens. The test frame collected the load and displacement data while the ADMS system took pictures of the specimen to track the delamination growth. This allowed for the calculation of the strain energy release rate as a function of delamination length. The formulae used for Mixed Mode I/II strain energy release rate are given in Equations (6)-(8).

In the fatigue tests, the maximum strain energy release rate, G_{max}, was measured as per ASTM D6671 [[7]] at the peak displacement of the sinusoidal motion. The displacement ratio, R_D, for a fatigue propagation test is defined as:

$$R_D = \frac{\delta_{min}}{\delta_{max}} \tag{5}$$

where δ_{min} is the minimum cyclic displacement.

Due to the combination of loading modes involved, the displacement ratio is not equal to the stress ratio. In the fatigue testing portion of this paper, a constant ratio R_D of 0.1 was applied to all specimens. Multiple tests were conducted with varying values of δ_{max} for the same load mixity, in order to capture the whole life of Mixed Mode I/II specimens under high to low stresses. Fatigue delamination/disbond growth was conducted beyond the stable delamination/disbond growth region. Tests were conducted at higher displacements and higher values of G_{max}/G_c (>1) to investigate the unstable growth region, while tests at lower displacements and lower G_{max}/G_c were performed to determine whether there was an observable threshold hypothesized by Chen et al. [[10]], below which no fatigue delamination/disbond growth would occur. At the conclusion of both the quasi-static and fatigue tests, the values of the critical strain energy release rate are calculated as follows [[7]]:

$$G_{Ic} = \frac{12[P(3c-L)+P_g(3c_g-L)]^2}{16b^2h^3L^2E_{1f}}(a+\chi h)^2 \tag{6}$$

$$G_{IIc} = \frac{9[P(c+L)+\ P_g(c_g+L)]^2}{16b^2h^3L^2E_{1f}}(a+0.42\chi h)^2 \tag{7}$$

$$G_c = G_{Ic} + G_{IIc} \tag{8}$$

where,
P = Load;
b = Specimen width;
a = Disbond length (distance from hinge point to disbond tip);
h = Half thickness of the specimen;
L = Half-span of the MMB test fixture;
c = Lever length of the MMB test fixture;
E_{1f} = Specimen bending modulus, calculated based on a formula given in [[7]];
χ = Crack length correction parameter, a constant based on Young's and shear moduli;
P_g = Weight of the lever and attach apparatus of the MMB fixture;
c_g = Lever length to centre of gravity of the MMB fixture lever.

RESULTS AND DISCUSSIONS

Determination of Onset and Propagation G_C

Mode Mixity Effect on G_C. The critical energy release rate, G_c, represents the threshold of resistance to delamination/disbond growth under Mixed Mode I/II quasi-static loading. The values of damage onset and damage propagation strain energy release rates were generated in a quasi-static test, where pre-cracking was performed following ASTM D5528 [[11]] to remove the effect of the Teflon insert. A follow-on failure mode analysis showed that most specimens failed cohesively for lower mode mixities (the failure occurred in the adhesive layer) according to ASTM D 5573-99 [[12]], a mixed cohesion/substrate failure was observed for higher mode mixities.

The critical energy release rates at damage onset and during damage propagation were calculated using Eqns (6)-(8) based on the load, disbond length, as well as the specimen compliance. Figure 3 shows typical strain energy release rates versus delamination length of quasi-static tests for various mode mixities. The values for pure Mode I (with mode mixity of 0) were generated on the same specimens and tested following ASTM D5528 double cantilever tests [[9]]. It was found in Figure 3 that the critical strain energy release rates at onset, $G_{c,onset}$, as well as those during propagation, $G_{c,prop}$, were highly dependent on the load mixity. The average values for $G_{c,prop}$ was calculated for each specimens and the summary of critical strain energy release values at different load mixities G_{II}/G are given in Table II. Figure 4 and Figure 5 show the power law equations assumed for the trend in G_c versus mode mixity for damage onset and propagation, respectively.

It can be seen in Table II that G_c values generally increase with each increment in load mixity, G_{II}/G, due to the increasing percentage of Mode II shear loading, which confers a greater resistance to disbond than Mode I peeling. Figure 3 also showS that G_c values for the same specimen may vary significant as damage propagates. At a lower load mixity where damage propagation is highly determined by Mode I peel loading, $G_{c,prop}$ was found to be fairly constant with the increase of delamination/disbond length. $G_{c,\ prop}$ values for mode mixity G_{II}/G of 0 and 0.2 are relatively reliable. However, when loading was dominated by mode II shear, such damage propagation became unstable, and $G_{c,prop}$ tends to vary greatly with the damage size. In this study, the average G_c values were calculated based on data points for MMB tests that are deems to be valid [[7]]. As a result, there a high level of uncertainty associated with those average values for mode mixities G_{II}/G of 0.4, 0.6 and 0.8, posing a challenge for model development of fatigue delamination of bonded joints under Mixed Mode I/II loading.

Although the measured G_c values for these adhesively bonded joints are relatively consistent within a batch, an apparent batch-to-batch difference in the average G_c value was observed. All specimens were made and tested following the same protocols and the cohesive failure mode was observed for all specimens, but G_c values of composite joints bonded with AF163-2K were still shown to vary. For example, G_c values in mode mixity of 0.4 are in the range of 4380-4840 J/m^2. Therefore, it is highly recommended that quasi-static tests be performed on each batch of the specimens, whenever possible.

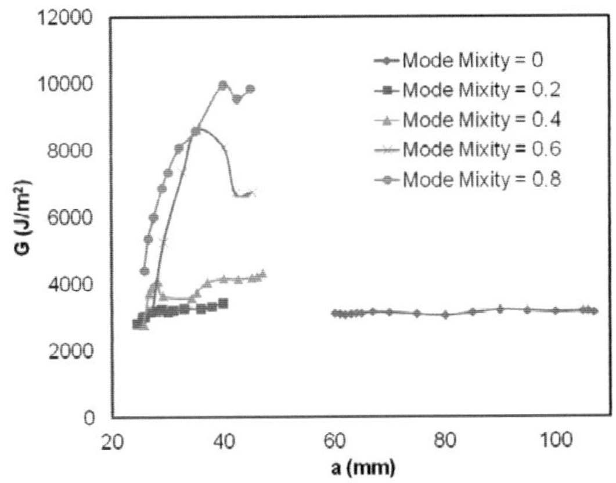

Figure 3. Strain energy release rate of bonded composites joints versus disbond length.

Table II. Average values of G at different mode mixities G_{II}/G.

G_{II}/G	$G_{Ic,prop}$ (J/m^2)	$G_{IIc,prop}$ (J/m^2)	$G_{c,prop}$ (J/m^2)	$G_{c,onset}$ (J/m^2)
0	2900	0	2900	2906
0.2	2780	695	3475	2931
0.4	2709	1927	4636	3458
0.6	2340	3992	6332	3663
0.8	1331	5598	6929	3218
1	0	N/A	N/A	5435

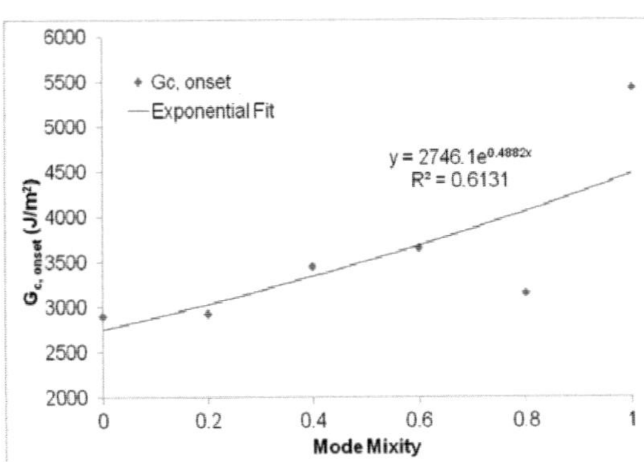

Figure 4. Graph of $G_{c,\ onset}$ versus mode mixity.

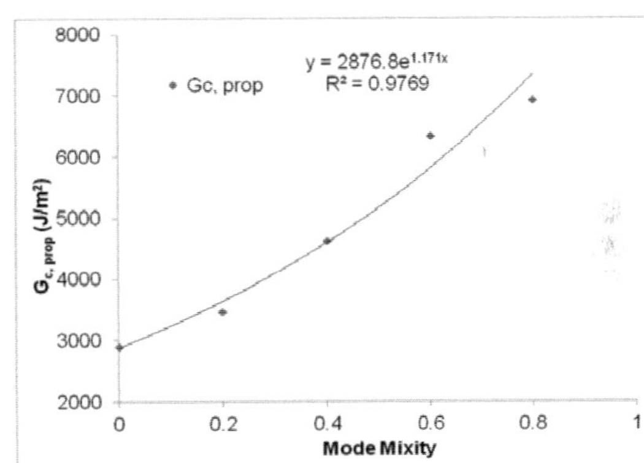

Figure 5. Graph of $G_{c,\ prop}$ versus mode mixity.

Environmental Effect on G_C. The critical energy release rates, G_c, for the conditioned specimens were measured in quasi-static tests for a range of load mixities of 0, 0.2, 0.4, as shown in Table III. Environmentally conditioned specimens were also tested under MMB loading of higher load mixity such as 0.6, but no data was captured due to catastrophic failure of the specimens under these test conditions.

It can be seen that long-term environmental exposure led to a significant reduction in fracture toughness of the adhesive bonded composite joints. For Mode I, the fracture toughness of the bonded joints was reduced by as much as 34%, from 2900 J/m^2 for an as-made to 1917 J/m^2 for a bonded joint immersed in salt water at 80 ^{o}C (or HISN). Such reduction in fracture toughness can also be seen on bonded joints under MMB loading of varying load mixity.

Table III. Average values of G_c for Adhesively Bonded Joints Exposed to Different Environments.

G_{II}/G	$G_{c,prop}$ (J/m^2)		
	RTA	*IVHN*	*HISN*
0	2900	2356	1917
0.2	2780	1654	2146
0.4	2709	2298	2489

Fatigue Behaviour of Adhesively Bonded Composite Joints

From the study by Poursartip [[13]], it is postulated that G_R, the delamination resistance, is the strain energy release rate that needs to be overcome for the delamination to grow. In order to normalize the G_{max} values obtained in the fatigue test, G_R is to be determined to find G_{max}/G_R. This paper proposes that G_R is best represented by $G_{c,prop}$, the critical strain energy release rate during the propagation of the disbond, when $G_{c,prop}$ is different from $G_{c,onset}$. This is based on the assumption that $G_{c,prop}$ takes into account any changes in delamination/disbond resistance as damage propagates and such changes are consistent for the specimens of the same conditions. All fatigue tests were conducted at 0.5 to 2.5 Hz under displacement control, at a maximum constant displacement value. As delamination/disbond propagated, the growth rate reduced along with the drop in G_{max}/G_c. The test terminated when reached the value of G_{max}/G_c reached 0.2 or lower.

Effect of Mode Mixity.

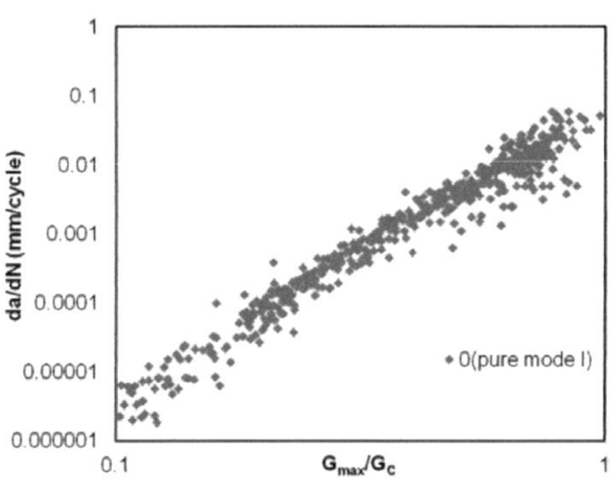

(a) mode mixity *GII/G* of 0 (pure Mode I)

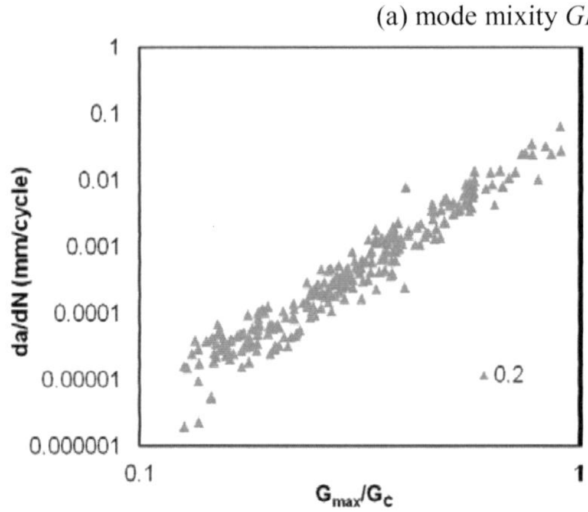

(b) GII/G of 0.2

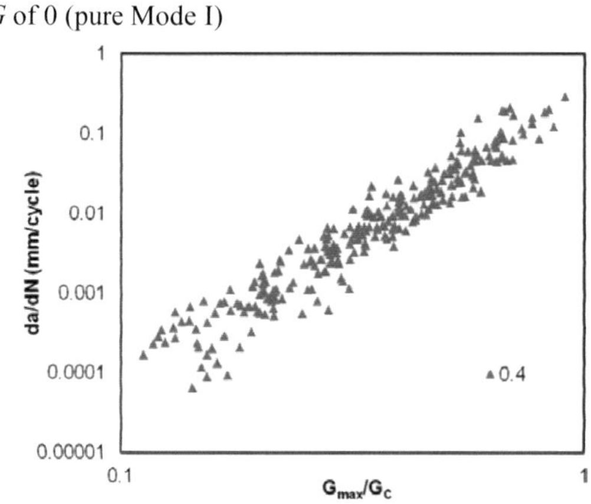

(c) GII/G of 0.4

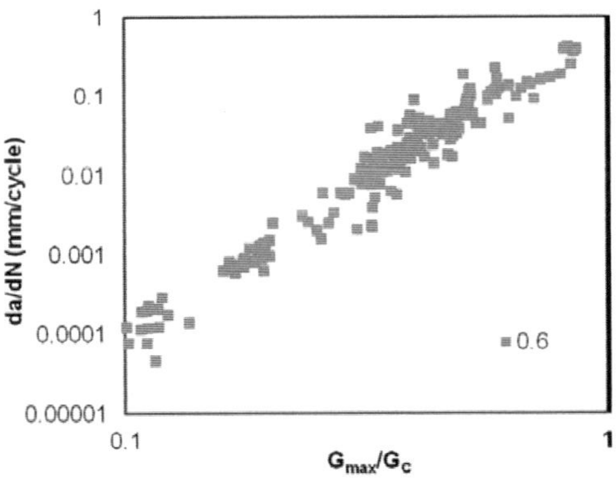

(d) GII/G of 0.6

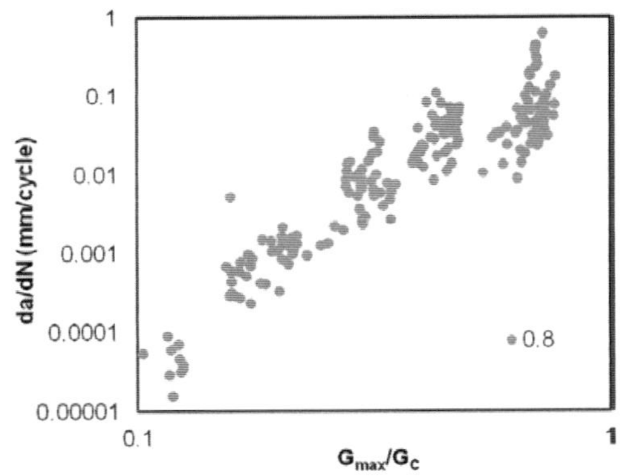

€ *GII/G* of 0.8

Figure 6 shows the delamination/disbond growth rates of an as-made bonded composite joint as a function of strain energy release rate under a range of mode mixities. Each curve consists of data from test results from five specimens. It is clear that a consistent power relationship can be found for each of the mode mixities. The test results showed that there was a clear and observable steady growth region, where the disbond growth of the adhesively bonded composite joints can be characterized by a simple power law relationship covered up to $\boldsymbol{G_{max}} = \boldsymbol{G_c}$ for all the studied mode mixities.

A comparison of the effect of mode mixities on fatigue growth rate is plotted in Figure 7. A least square fit was conducted to fit the data in each graph using a power law relationship, in the form of $da/dN = b(G_{max}/G_c)^n$. A summary of the fitting coefficients is presented in Table IV. Fitting coefficients of $\mathrm{da/dN} = \mathrm{b(G_{max}/G_c)^n}$Table IV. It can be seen that the parameter, n, an indicator of dependency of the growth rate on the load, is similar among all the studied mode mixities. However, the constant *B* which is subjected to total critical fracture toughness, G_c, for each mode mixity varies. This constant B is almost identical for mode mixity of 0, or pure Mode I tested under DCB, and that of 0.2. Given that G_c for mode mixity of 0 and 0.2 was relatively reliable to measure as compared to those of higher mode mixities (see Figure 3). For the normalized fracture toughness G_{max}/G_c, it is hypothesized that the curves may collapse on top of itself, meaning that a simple power equation may be used to describe fatigue delamination of adhesively bonded composites joints with consideration of mode mixity. Further studies are recommended to investigate test and data analysis protocols that help generate a reliable measurement of G_c for MMB tests to evaluate whether such hypothesis holds true.

Table IV. Fitting coefficients of $da/dN = b(G_{max}/G_c)^n$.

Mode mixity	*b*	*n*	R^2
0	0.0628	4.1836	0.9551
0.2	0.0604	4.2059	0.9363
0.4	0.3199	3.632	0.9197
0.6	0.8741	3.9791	0.951
0.8	0.401	3.6946	0.8994

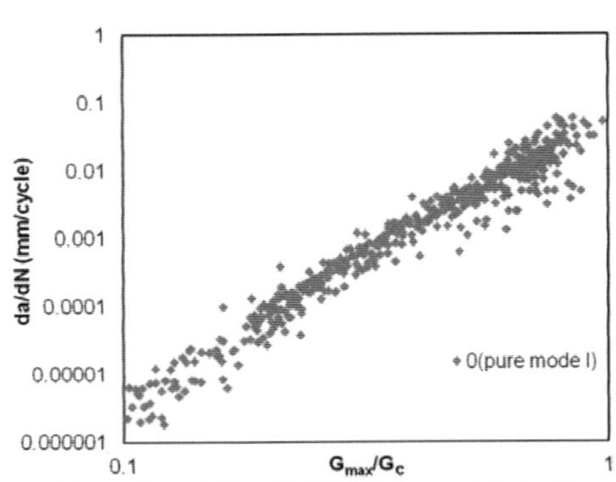

(a) mode mixity G_{II}/G of 0 (pure Mode I)

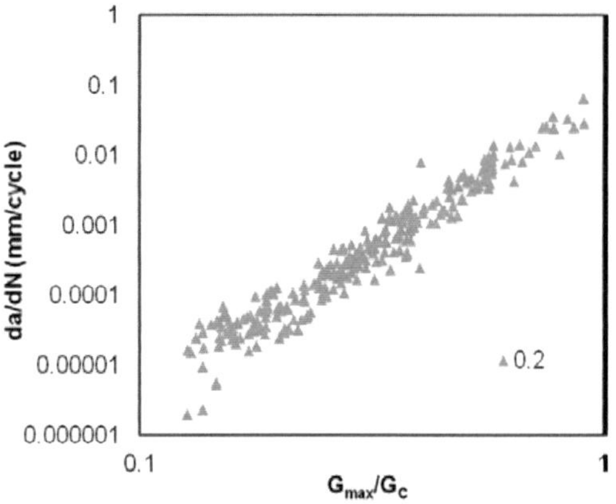

(b) G_{II}/G of 0.2

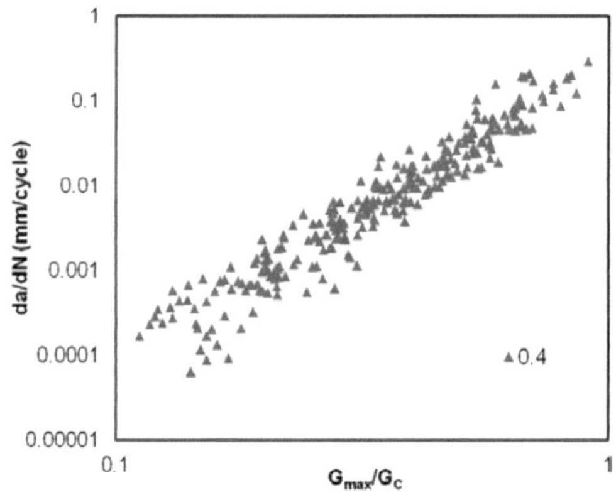

(c) G_{II}/G of 0.4

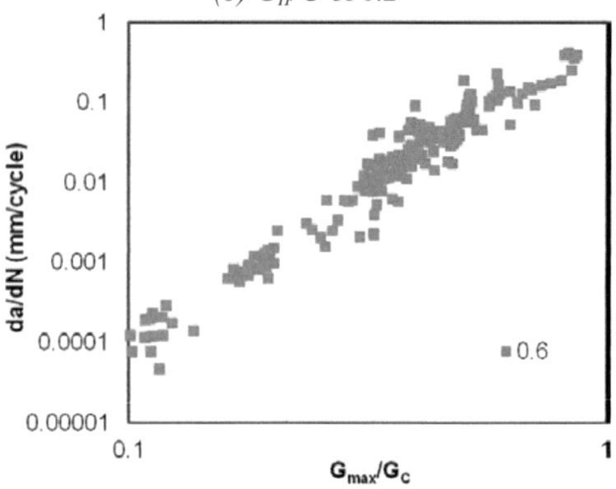

(d) G_{II}/G of 0.6

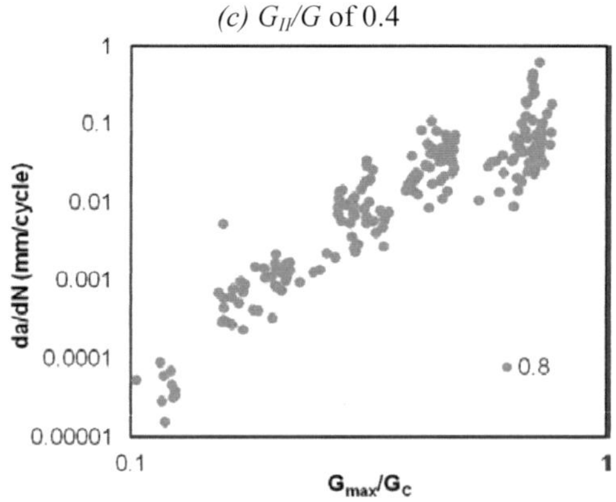

€ G_{II}/G of 0.8

Figure 6. Fatigue delamination/disbond growth rate curves under Mixed Mode I/II of mode mixities of (a) 0, (b) 0.2, (c) 0.4, (d) 0.6 and € 0.8 on the log-log scale.

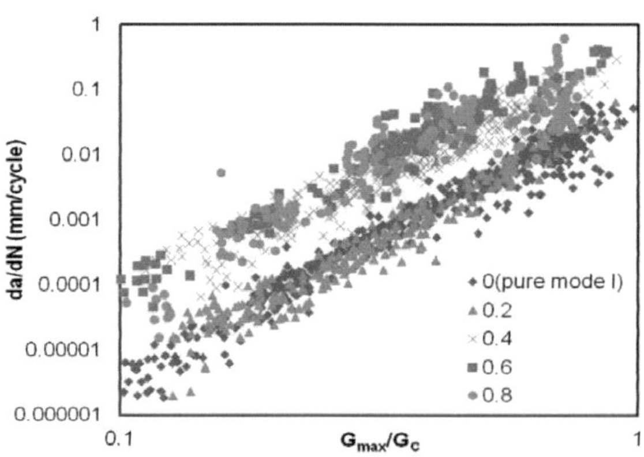

Figure 7. Comparison of fatigue delamination growth of bonded composite joints under Mixed Mode I/II loading on the log-log scale.

Hygrothermal Effects on Mixed Mode I/II Fatigue Delamination/ Disbond Growth. Bonded composites joints were subjected to three environments, RTA, IVWN and HISN until the specimens reached the equilibrium state prior to mechanical testing. $\boldsymbol{G_c}$ of these specimens was first obtained via quasi-static tests. It was found that the conditioned specimens exhibited significant effect of environmental ageing on fracture toughness. Such effect was also apparent on fatigue disbond growth rates for the environmental aged joints, where typically a faster rate was observed as compared to the behaviour of the as-made ones. Fatigue growth under Mixed Mode I/II with mixities of 0.2, 0.4 and 0.6 were studied. However, the conditioned bonded composite specimens failed abruptly when subjected to a load mixity of 0.6, therefore no reliable data was available for comparison.

Error! Reference source not found. shows a comparison of fatigue delamination/disbond growth of RTA, IVWN and HISN specimens under Mixed Mode I/II fatigue loading with a mode mixity of 0.2. It appears that greater disbond propagation rate was found for the conditioned specimens when they are subjected to higher strain energy release rate levels. As the energy release rate reduces, however, the delamination/disbond behaviour among all three conditions tends to converge. However, due to high scatter of delamination/disbond growth rate data for the conditioned specimens such trend needs to be read carefully.

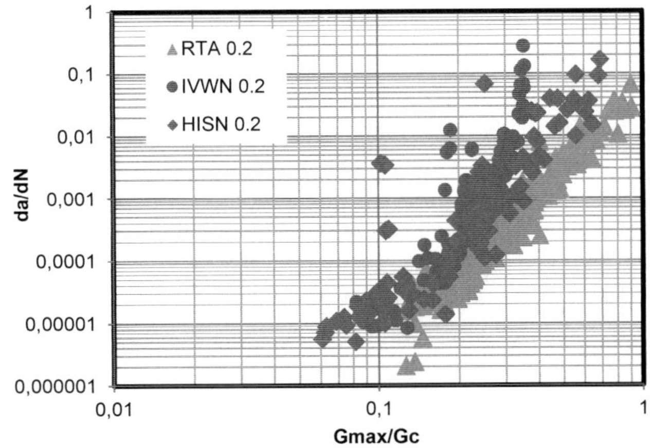

Figure 8. Delamination/disbond fatigue growth of conditioned bonded composite joints subjected to Mixed Mode I/II loading on log-log scale (mode mixity of 0.2).

At the higher mode mixity of 0.4 (see Figure 9), the effect of environmental conditioning on fatigue dibond growth was less prominent on the log-log scale for the range of strain energy release rates studied. Fatigue disbond growth of the HISN and IVWN specimens are in the same range given the level of scatter in the data. They both showed a greater dependency of disbond growth rate on the loading than the as-made specimens. This trend is consistent with that of the case of mode mixity of 0.2.

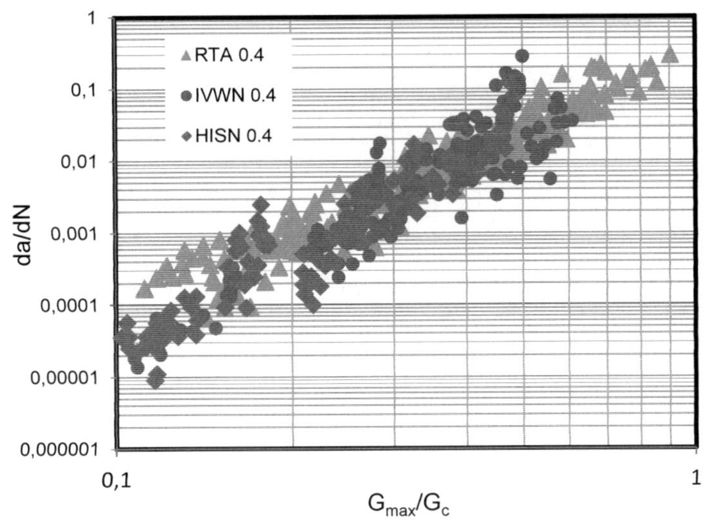

Figure 9. Delamination/disbond fatigue growth of conditioned bonded composite joints subjected to Mixed Mode I/II loading on log-log scale (mode mixity of 0.4).

CONCLUSIONS

The effect of environmental ageing on fatigue delamination/disbond growth of bonded composite joints under Mixed Mode I/II loading was investigated experimentally. The test results showed that the disbond growth of the adhesively bonded composite joints, as-made or conditioned, for the studied mode mixities of 0, 0.2, 0.4, 0.6 and 0.8 can be characterized by a simple power law relationship. The conditioned specimens exhibited a significant reduction in fracture toughness G_C at up to 34%, demonstrating the significant effect of environmental ageing. Such effect was also apparent on fatigue disbond growth rates for the environmental aged composite joints. The study also revealed unstable nature of delamination/disbond growth when the bonded composite joints were subjected to shear dominated loading, posing a challenge in developing fatigue models that take into account mode mixities and environmental ageing.

REFERENCES

[1] Tomblin J. and L. Salah, (2009), Aging of Composite Aircraft Structures Teardown of a Decomissioned B737-200 Horizontal Stabilizer, FAA Joint Advanced Materials and Structures (JAMS) Fifth Annual Technical Review Meeting July 21-22.

[2] Shafizadeh J. E, J. C. Seferis, E. F. Chesmar, and R. Geyer, (1999), Evaluation of the In-service Performance Behaviour Honeycomb Composite Sandwich Structures, Journal of Materials Engineering and Performance, 8(6): 661-668.

[3] Kumar P. and B. Rai, (1993), Delaminations of Barely Visible Impact Damage in CFRP Laminates, Composite Structures 23, 313-318.

[4] Li C., J. Teuwen and V. Lefebvre, (2006) Investigation of Moisture Ingress and Migration Mechanisms of an Aircraft Rudder Composites Sandwich Structure, ICCM-16, Kyoto, Japan.

[5] Wilkins D.J., J.R. Eisenmann, R.A. Camin, W.S. Margolis, R.A. Benson, (1982), Characterizing Delamination Growth in Graphite-Epoxy. In: Damage in Composite Materials, ASTN STP 775, p. 168–183.

[6] O'Brien T.K., (1982) Characterization of Delamination Onset and Growth in a Composite Laminate. In: Damage in composite materials, ASTM STP 775, p. 140–67.

[7] Standard Test Method for Mixed Mode I – Mode II Interlaminar Fracture Toughness of Unidirectional Fiber Reinforced Polymer Matrix Composites, ASTM International, D 6671M-06, 2007.

[8] Teng T., Hamilton J., Li. C, (2013) Fatigue and Fracture of Adhesively Bonded Composite Joints under Mode I Loading, Part 1: Speciment Manufacturing LM-SMPL-2012-0051

[9] Li C., Teng T., Li G., Yanishevsky M. (2013) Fatigue Disbond Growth of Environmentally Aged/Degraded Adhesively Bonded Composite Joints under Mode I Loading, International Conference on Composite Materials, 19th International Conference on Composite Materials

[10] Chen H., Shivakumar K., Abali F. (2005) A Comparison of Total Fatigue Life Models for Composite Laminates, Fatigue & Fracture of Engineering Materials & Structures Volume 29, Issue 1

[11] Standard Test Method for Mode I Interlaminar Fracture Toughness of Unidirectional Fiber-Reinforced Polymer Matrix Composites, ASTM D5528-01, 2007.

[12] Standard Practice for Classifying Failure Modes in Fiber-Reinforced-Plastic (FRP) Joints. ASTM D5573-99(2012). In: Annual Book of ASTM Standards.

[13] Poursartip A. (1987) The Characterization of Edge Delamination Growth in Laminates Under Fatigue Loading. Toughened Composites, ASTM STP 937, Norman J. Johnston, Ed., American Society for Testing and Materials, Philadelphia, p. 222-241

Design of health and usage monitoring systems based on optical fibers for composite wing spars

Alessandro Airoldi[1], Paolo Bettini[1], Paolo Belotti[1], Theodoros Loutas[2], Christos Koimtzoglou[3], Giuseppe Sala[1]

[1]*Dept. of Aerospace Science and Technology, Politecnico di Milano, Italy*
[2]*Dept. of Mechanical Engineering and Aeronautics, University of Patras, Greece*
[3]*Integrated Aerospace Science Corporation (INASCO), Athens, Greece*

Abstract: Health and usage monitoring systems may have a key role in the definition of new strategies for the design and maintenance of modern aerospace composite structures. Design of composite components is notoriously affected by damage tolerance considerations, since accidents occurring during production processes, as well as in-service exogenous events may induce barely visible damages. At first, this paper describes the most viable general techniques for integrating optical fibre sensors into composite materials. Then the design of a specific strain sensing systems to be integrated into composite spars of a wing-box is presented, aimed to exploit the load and damage monitoring features offered by such type of sensors. Finally, the capability of detecting delaminations inside the caps of a composite spar through OF-based monitoring systems is discussed. Numerical techniques are adopted for studying the influence exerted by delaminations on the strain fields in a composite element. As a benchmark, a C-shaped composite spar is chosen, characterized by different thicknesses and stacking sequences.

INTRODUCTION

The design of composites aircraft structures cannot help considering damage tolerance issues. Nowadays, additional uncertainties may arise, since innovative composite and hybrid structural architectures behave in more complex ways than traditional metallic structures subjected to thermo-mechanical loading conditions typical of aeronautical service. Besides, damage not only affects the behavior of damaged part, but possibly modifies load paths and local stress distributions in other primary components, sometimes unpredictably. Therefore, the application of damage tolerance strategies has to face several uncertainties, which eventually could prevent the complete exploitation of composites advantages, at the same time increasing design and maintenance costs. Optical fiber-based monitoring systems can provide a substantial aid for addressing such issues. Through Fiber Bragg gratings (FBG) carried by optical fibers, local values of strain, stress, temperature, moisture and chemicals concentration can be measured, being immune from electro-magnetic jamming. Their weight and invasivity is low compared to traditional strain gages, including all the auxiliary systems required for sensor interrogation [1]. Moreover, optical fiber-based measuring networks can be either bonded *to* the surface or embedded *into* host material, so receiving protection from external agents. Anyway, to implement an efficient and reliable HUMS (heath and usage monitoring system), two main issues should be tackled at first, i.e. the assessment of fibers integration techniques and the prediction of system measuring capability and sensitivity.

In the following, viable technologies for integrating optical fibers into composite materials are described, and then the design of a specific strain sensing systems is presented, to be installed in the composite spars of a wing-box. From the technological standpoint, an appealing solution consists in the adoption of pre-cured ribbons (Quick-Packs), which are produced separately by embedding the optical fibers between two thin and compliant glass-epoxy laminates [2,3]. These ribbons can be easily co-bonded to the surfaces of laminates, thus providing a simple way of applying an OF-based measurement system to composites parts. In addition, such a solution allows avoiding direct exposure of fibers to external agents and offers a straightforward procedure to repair or replace parts of the network. A second viable method of integration is represented by surface co-curing of pre-cured ribbons during manufacturing process of composite parts. Finally, a third integration strategy is based on Quick-Packs embedment into composite laminates during production phases. This provides an even more accurate monitoring of strains during damage propagation [4]. Moreover, such an early integration allows to investigate the build-up of residual thermo-elastic stresses, often responsible for damage onset and growth (*from-the-cradle-to-the-grave* process monitoring). In the following, the outcomes of technological and experimental investigations are presented, pros- and cons- of different solutions are highlighted and best practices for the design of health and usage monitoring systems are pointed out.

Once assessed the issues relevant to technological integration of fibers and sensors, the possibility of detecting damages in composite laminates through OF-based monitoring systems should be investigated. In particular, the delaminations occurring inside the caps of a composite wing spar are taken into account here below. Numerical techniques are applied to reproduce the effects of delamination on the strain fields in a composite element [5]; more specifically, a C-shaped composite spar, with different thicknesses and lamination sequences is considered. The results of numerical analysis show that the chance of detecting damage actually depends on the structural role played by interlaminar layers when the spar is loaded in shear, as well as on the degree of proximity between measuring points and damages themselves. In particular, it appears that the *signature* of damage on longitudinal strains increases as the stiffness mismatch between web and cap (depending on the respective thicknesses and stacking sequences) increases. These evidences have to be exploited to design a system for spar caps damage monitoring based on embedded sensors, which could detect the anomalies in the longitudinal strain field by applying different damage indices. The performance of such system are numerically evaluated and discussed considering a series of damage scenarios and design hypotheses for the positioning of OFs carrying the strain sensors.

FO INTEGRATION

The integration of fibre optic sensors with composite structures can be accomplished according to different techniques: they can be simply bonded to the external surface (co-bonding), co-cured to the laminate or embedded inside the laminate itself during the manufacturing process. In many case, owing to their small dimensions and intrinsic brittleness, direct handling of bare fibres is not conceivable; besides, the exiguity of bonding surface cannot guarantee adequate load-transfer capability. To overcome these drawbacks, a special device should always be used, consisting in a thin, narrow and compliant GFRP laminate embedding optical fibres, called Quick-Pack (QP) (Fig. 1). Such a ribbon, besides protecting and insulating OF, allows a perfect adhesion to curved surfaces and does not prevent the adoption of high-temperature processes, being irrespective of possible low thermal stability of OF coating.

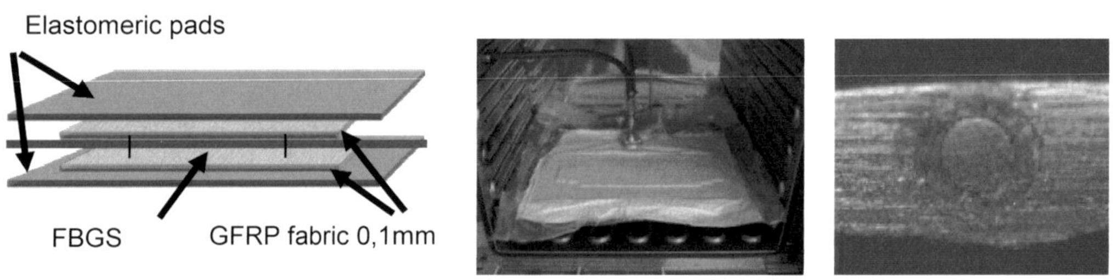

(a) (b) (c)

Figure 1. Quick-Pack production.

QPs can be directly co-bonded to the outer surface of composite laminates according to standard processes, by means of low-temperature film or paste adhesives, cured in oven + vacuum bag or electric heaters + pressure pads.

Possible critical issues refer to surface preparation, OFs protection, QPs positioning and adhesive amount optimization. With reference to the set-up shown in Fig. 2, the process check-list should include:

- accurate definition of QPs position on laminate surface;
- protection laminate regions to be left unaffected by surface preparation;
- surface roughing (wire netting) and cleaning (MEK);
- positioning elastomeric layers having the same thickness of the laminate;
- positioning thin (0.2mm) and flexible CFRP caul plates acting as QPs template and fibres protection;
- spreading a thin and homogeneous layer of two-components adhesive on QPs surface;
- securing QPs to laminate; caul plates act as template, meanwhile protecting the fibres;
- additional vacuum bag consumable layers: perforated films, bleeder, elastomeric layer, vacuum bag.

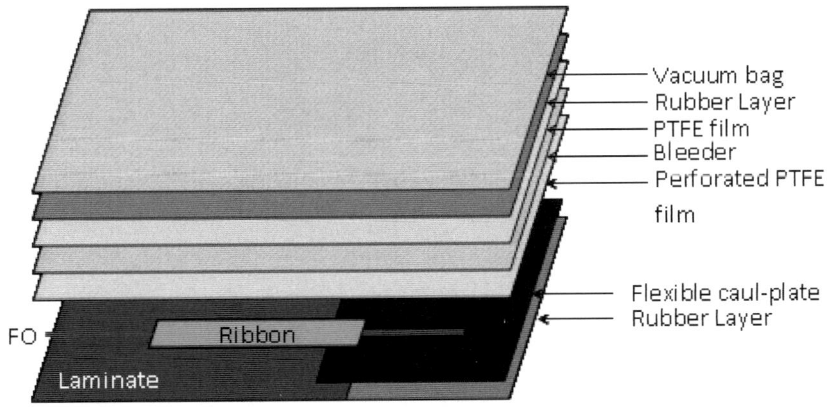

Figure 2. Co-bonding set-up.

Co-curing technique can adopted to get a better integration between sensors and host material. Adhesion of QPs to underlying laminate is granted by uncured resin in excess, no needing additional adhesive. As a consequence, all materials required for producing QPs, as well as vacuum bag consumables, must be compatible with high temperatures deserved to cure composite structural components. Such a process improves mechanical properties of adhesive bonding and its strain-transfer capability, at the expense of increased process complexity and of decreased QPs replacement easiness. Further difficulties may arise, owing to the need to protect OFs during curing cycle, as well as to possible sinking of QPs and PTFE tubing into pre-pregs, creating unacceptable dents in final laminates.

Referring to Fig. 3, the main phases of the procedure consist of:

- pre-pregs lamination over the mould;
- accurate definition of QPs position on peel-ply;
- peel-ply puncturing to allow PTFE tubing crossing;
- positioning of QPs + peel-ply + PTFE tubing assembly on the laminate;
- insertion of two thin (0.2mm) and flexible caul plates between peel-ply and OFs, to avoid OFs failure and QPs misalignments;
- addition of vacuum bag consumable layers: perforated films, bleeder, elastomeric layer, vacuum bag.

Finally, a temperature-pressure-vacuum curing cycle suited to resin system is applied (Fig. 4)

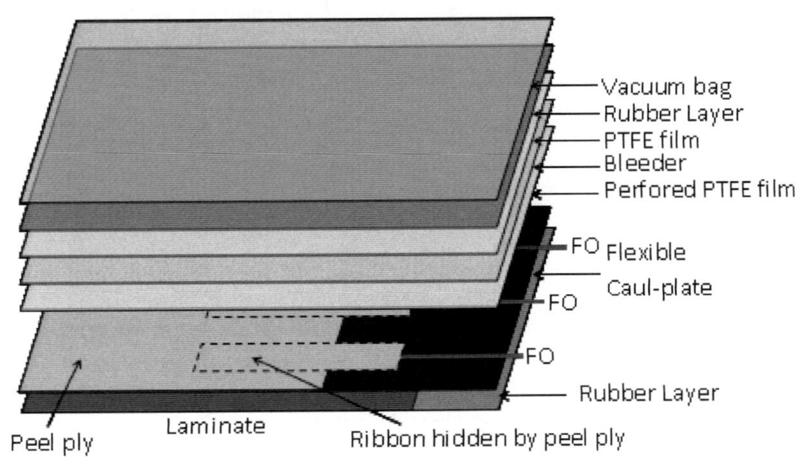

Figure 3. Co-curing set-up.

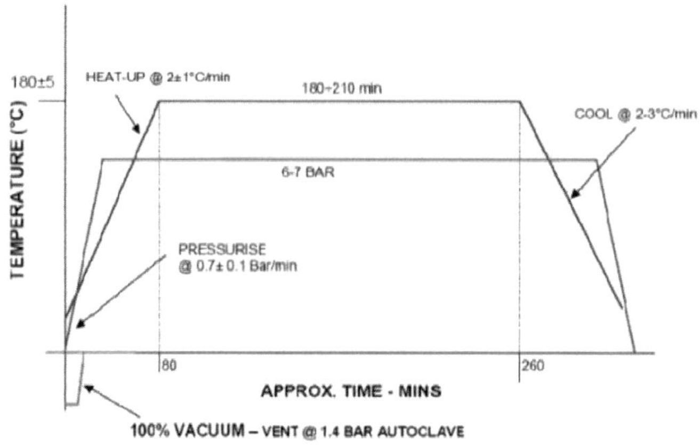

Figure 4. Autoclave co-curing cycle.

Alternatively to co-bonding and co-curing techniques, OFs can be integrated into composite laminates by direct embedment during lamination phase, leading to peculiar advantages, like protection from environmental effects and capability to monitor internal hot-spots; besides, aerodynamics is not affected. On the other hand, concerns about inspectability, repairability, replaceability and maintainability may arise. Moreover, OFs embedment has to address other critical issues, like preservation of OFs coating integrity during curing process and avoidance of cross-section distortions, which may affect measurement accuracy. Both issues are influenced by lamination sequence of host material. Indeed, owing to high values of pressure and to OFs bending sensitivity, the presence of off-axis fibres may induce the distortion of sensors spectra (Fig. 5) as well as the permanent deformation or even the failure of coating (Fig. 6).

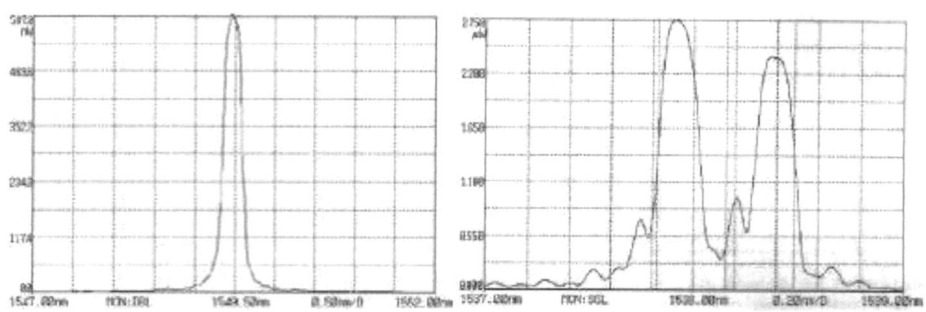

Figure 5. Effects of cross-section deformation on FBG's spectrum: un-deformed (a) and deformed (b).

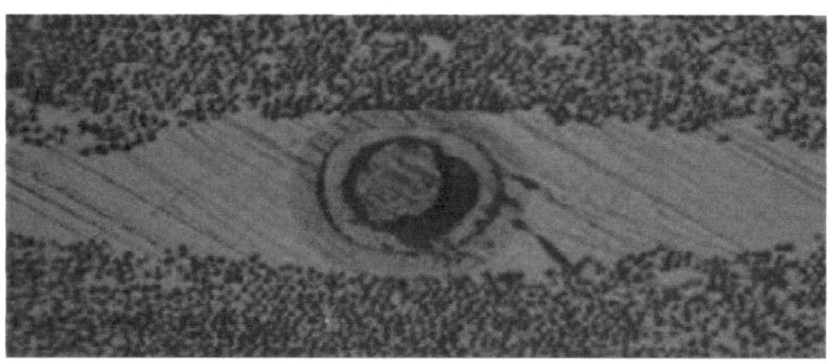

Figure 6. Deformed and failed poly-acrylate coating.

To circumvent both drawbacks, high-temperature (poly-imide) coatings or QPs (Figs.7a-b) should be adopted.

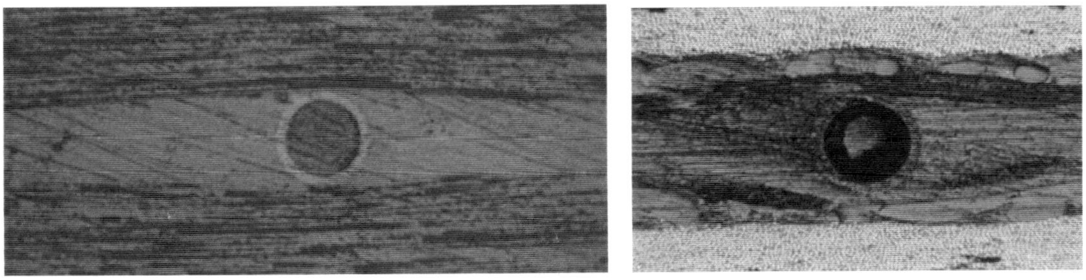

Figure 7. Un-affected poly-imide coating (a); embedment through QP (b).

As a final suggestion, when possible, OFs should always be embedded on-axis inside a UD reinforcements layer.

LOAD IDENTIFICATION

In the most straightforward application, FBGs can be used to get local values of strain, averaged along gratings' length. However, the inherent characteristics of OF-sensor networks, i.e. simplicity, lightness and immunity from EM jamming, make conceivable the integration of a relatively large number of sensors into a structural component. Such networks involve moderate weight increase and allow reliable interrogation in flight conditions at adequate frequencies, so to make possible in-service reconstruction of the strain field evolution affecting structural components. The spatial resolution and the degree of structural integration of the sensors necessary to extract meaningful information depends on the scope of strain sensing. If identification of internal local damages in a composite structures is required, a dense network of embedded sensors is needed to detect damage signature, as it will be shown in the following paragraph. On the other hand, few sensors are sufficient to measure the approximate resultant of external forces applied to a structure, while a relatively thicker network of sensors should be integrated if a more detailed evaluation of internal loads, as well as of loads transmitted by the surrounding components, is pursued. Such level of application becomes particularly interesting for innovative composite and hybrid aircraft structural architectures, since a detailed knowledge of actual loading acting on main structural parts provides significant information to define and update maintenance operations. Continuous monitoring of structural response allows us to detect load paths modification due to aging of materials, damaging of redundant components and introduction of new flight configurations.

In aeronautical components, loading conditions can result in a pretty complex combination of external loads and forces transferred by the surrounding structure. In spite of this, the knowledge of basic structural behaviour of stressed skin constructions allows us to define a simplified, but acceptable parameterised set of load components.

The complete strain field in a structural component can be reconstructed starting from local strains (inverse problem) through a numerical technique reported in Appendix I. Such method relies on the availability of a finite element model of the structure and is based on a least-square approximation. Other methods to solve these types of inverse problems can be found in [6,7]. As a numerical example, let's assume a seven-bays C-shaped carbon/epoxy wing spar, having a uniform lamination sequence, with about 48% of reinforcement fibres in +/-45° direction, 34% in 0° direction and 14% in 90°, being 0° direction coincident with the longitudinal axis of the spar, subjected to a generic load conditions and equipped with a virtual FBGs network. Each bay can be instrumented with 18 FBGs, bonded to the external surface of the spar and able to measure longitudinal strains. By varying number and position of active sensors, 2^{18} possible configurations exist, which can be coded in a codifying vector (Fig. 8), where positions hosting an active sensors = 1 and void positions = 0.

The generic parameterized load condition (Fig. 9) consists of 17 concentrated forces and 12 distributed load systems. Concentrated forces include longitudinal and horizontal resultants as well as bending moments. Distributed loads are applied as nodal force systems and include 8 longitudinal loads, divided into upper and lower components, and 4 systems representing torsional moments. The spar is clamped at the root.

To evaluate the performance of the load identification system, an identical configuration of n_b active sensors per each bay is assumed, according to the vector reported in Fig. 8. The strains to be identified are obtained through a virtual test model subjected to a generic combination of parameterized loads. The load components in the force vectors are chosen to represent a realistic flight load condition (cruise). Internal wing forces are used to estimate vertical forces, bending and torsional moments, while horizontal loads are separately added. The output of such virtual test model is used to surrogate the strain vector to be acquired by monitoring system [ε*].

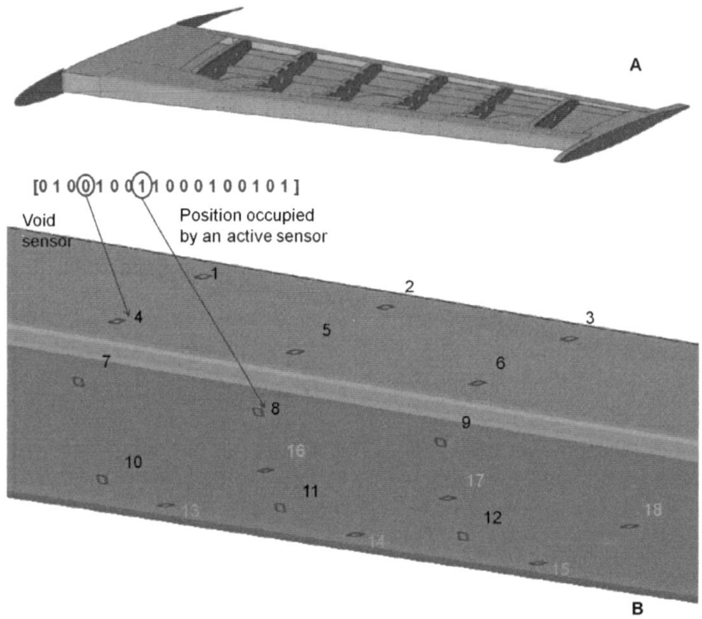

Figure 8. Active sensors positioning and related codifying vector (typical).

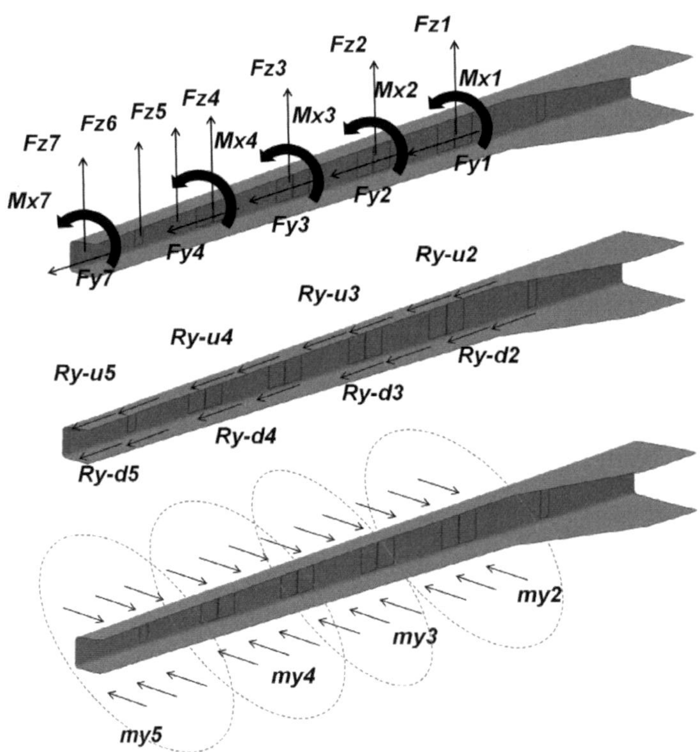

Figure 9. Generic parameterized load condition.

Once fixed the number of active sensors, a Monte-Carlo approach can be followed, by randomly varying codifying vector. For each configuration, root mean square error (RMSE) and maximum error (MAXERR) can be evaluated by comparing load components *identified* by the algorithm and *actual* load components.

Assuming no discrepancies between nominal model and model used for virtual test, an indication of system performances is reported in Fig. 10, where % RMSE index is given for the best 50 configuration at n_b = 5,6,8, considering 1000 possible combinations for each n_b value. Load components are accurately identified: RMSE < 0.5% for n_b = 5. The index presents minimum value which progressively decreases as n_b increases.

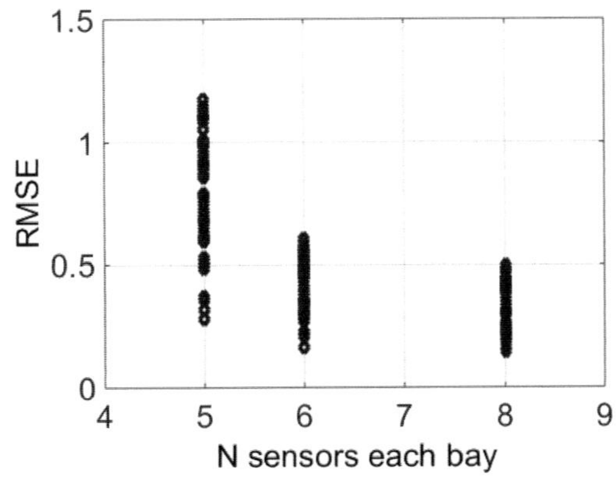

Figure 10. Nominal spar: system performances - RMSE index.

However, it is more realistic to consider a few differences between nominal and virtual models. Having this in mind, the virtual test spar model is divided in 18 sectors, featuring thickness alteration and plies mis-orientation. Gaussian distributions are adopted to attribute a stacking sequence to each sector (3.6% covariance in thickness distribution and 2° in plies orientation referring to nominal values). The performances of monitoring system in such a realistic condition are presented in Fig. 11, where RMSE and MAXERR indices are provided (best 10 configurations and 10000 combinations for each n_b value).

Although the number of configuration is not adequate to identify optimal performances, the results point out well-defined trends. The number of configurations with acceptable performances is very low compared with the total number of configurations considered. This depends on the importance of proper sensors positioning, as well as on the sensitivity to specific load conditions. The number of sensors holds a paramount importance for improving the efficiency of the system. Minima values of RMSE and MAXERR reduce respectively from 25% at n_b = 5 to 13% at n_b = 8 and from 88% at n_b = 5, to 34% at n_b = 8. The latter values are quite high, but refer to small force components, which marginally affect the strain field.

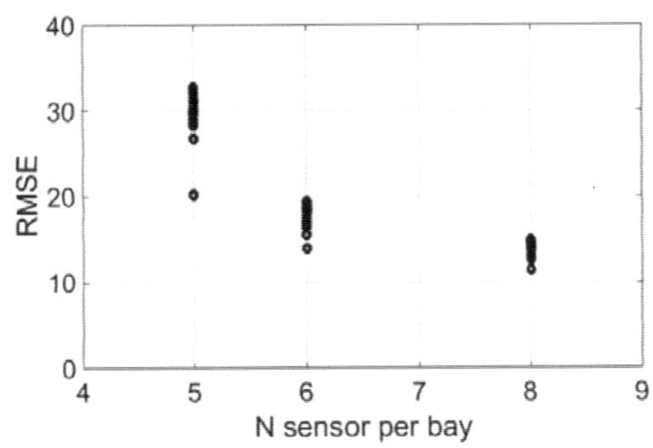

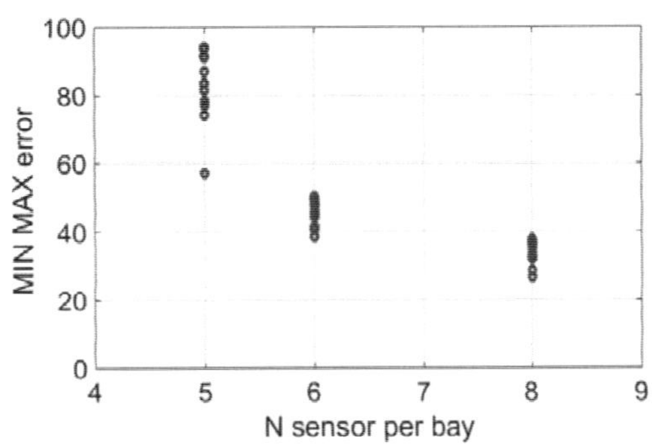

Figure 11. Realistic spar: system performances - RMSE and MAXERR indices.

DAMAGE DETECTION

In the previous paragraph, the capability of a FBGs network to identify with sufficient accuracy the complex load conditions acting on aeronautical components has been assessed, provided an adequate number of sensors is used. Now the issue of *damage detection* should be tackled. It is well known that, under structural loading, the presence of a damage induces a variation in the local strain distribution, due to the variation of load path. As a consequence, the damage itself can be detected by comparing such altered strain distribution with the strain field affecting the same un-damaged structure. The main drawback of such a procedure resides in the intrinsically *local* nature of damage-induced strain anomalies, which makes damages only detectable in vicinity of strain sensors. As a matter of fact, this does not represent a real hindrance, since structural *hot-spots* can be preliminarly pointed out through FEM analysis. On the other hand, while numerical predictions are generally accurate and straighforward, experimental results may be affected by noise, loading/constrainig imperfection and production tolerances. So direct comparison may result impracticable. An alternative approach makes use of statistical methods, according to which the outcomes of different sensors at different times belong to a single statistical population, provided the measurements are normalized for filtering the effects of load time-variation.

However, the strain field in many structural components is expected to present a smoothly varying distributions in the zones far from the application of concentrated loads. Such aspect can be exploited to detect anomalies that indicate a local alteration with respect to a smooth nominal solution. An example of such approach can be found in [8,9], where an algorithm was devised to identify failures in bonded junction. The method is reported in Appendix II. Comparison between the local strain signal and a set of strain acquisition in the same region is required to develop this type of damage detection strategy. The apparent advantage is the elimination of a reference base-line and of a reference load condition, since anomalies are detected with respect to an instantaneous strain distribution, which can be originated by any load condition.

To investigate and validate the method, again a C-shaped carbon/epoxy wing spar is assumed and modelled as test-bed, but added with two flanges on top and bottom surfaces (see Fig. 12).

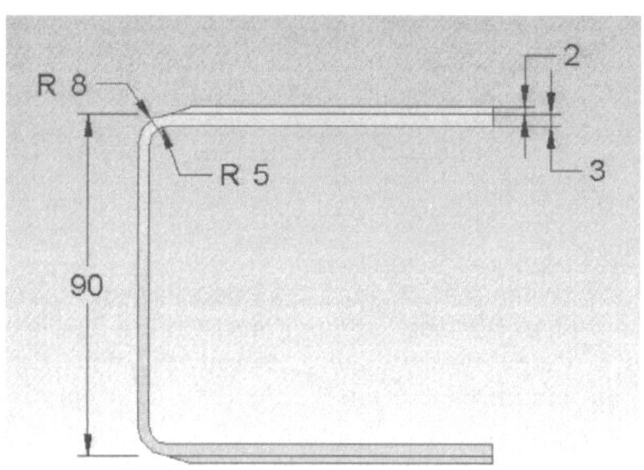

(a)

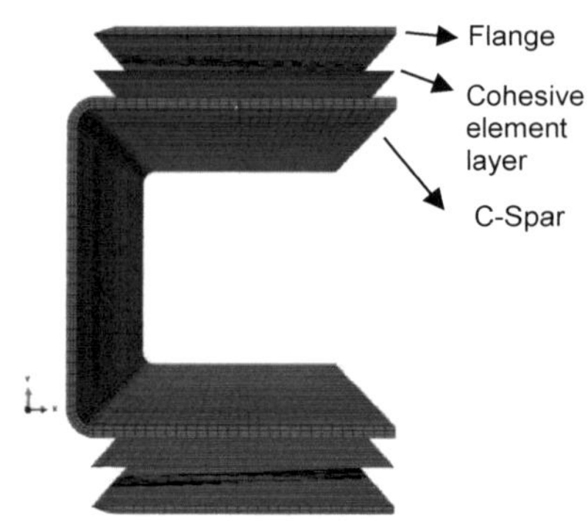

(b)

Figure 12. Flanged C-shaped spar considered for damage detection.

The C-shape spar consists of 15 plies; its stacking sequence is made of 63% $[45°]_{fabric}$, 25% $[0°]_{unidirectional}$ and 12% $[90°]_{unidirectional}$. Each flange consists of 10 plies, whose lamination includes 58% $[0°]_{unidirectional}$, 21% $[0°]_{fabric}$ and 21% $[45°]_{fabric}$. Spar and flanges are modelled by using a solid laminated elements and joined together through a layer of cohesive elements, inside which delaminations can be modelled by attributing no stiffness. The component is subjected to a three-point-bending-like loading condition. To check different damage scenarios, different sizes (Fig. 13b) of the same delamination (Fig. 13a) are considered.

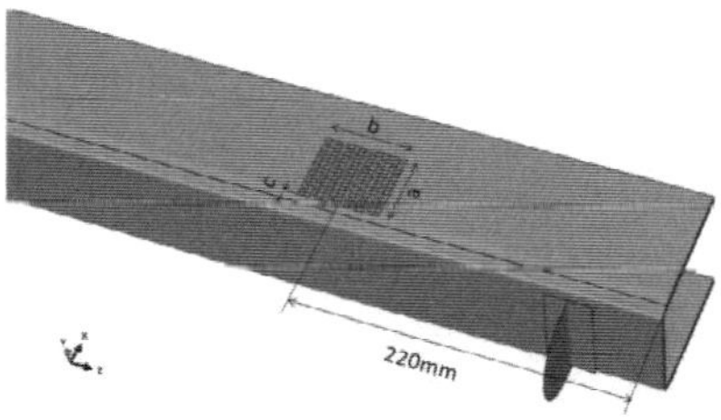

(a)

Damage configuration	a (mm)	b (mm)	c (mm)	Area (mm^2)
#C1	48	56	0	2688
#C2	48	72	0	3456
#C3	48	56	6	2688
#C4	48	72	6	3456
#C5	69	72	6	4968
#C6	51	72	36	3672

(b)

Figure 13. Delamination position and size.

Strains are evaluated along paths (which correspond to the optical fibres carrying FBGs), whose traces are shown in Fig. 14. The model of the spar was subject to a three-point bending load condition and typical strain-span curves, for every combination of damage size and measuring path can be obtained, showing different degree of sensitivity. The examples reported in (Fig. 15) show that the alteration of the strain distribution along external paths, like paths 1 and 9 (see Fig. 14), is characterized by a moderate amplitude distributed along an appreciable span length. However, alteration is very sensitive to the position of the path with respect to the damage location, since path 1 shows very limited variation with respect to the undamaged conditions.

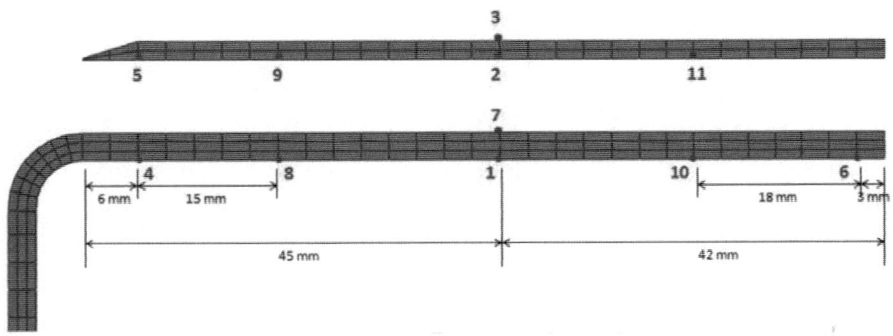

Figure 14. Delamination position and size.

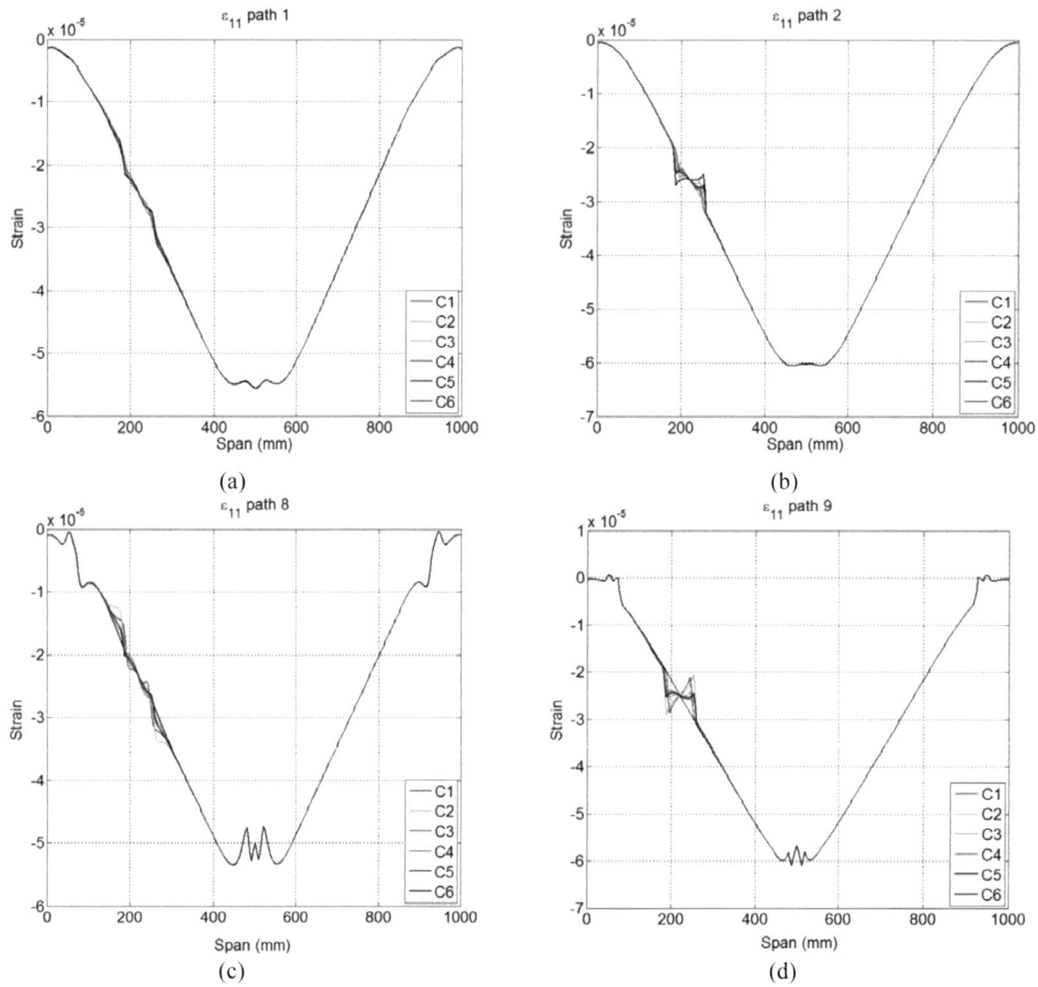

Figure 15. Strain distribution along different paths for different damage scenarios.

The alteration along internal paths located in the flange, two plies above the layer of cohesive elements, are characterised by higher amplitudes, though they are extended over a lower span. Moreover, internal paths appear less sensitive to damage localization.

Once completed such a screening procedure, a damages/paths the configuration shown in Fig. 16 is defined to assess the damage detection capability of different methods.

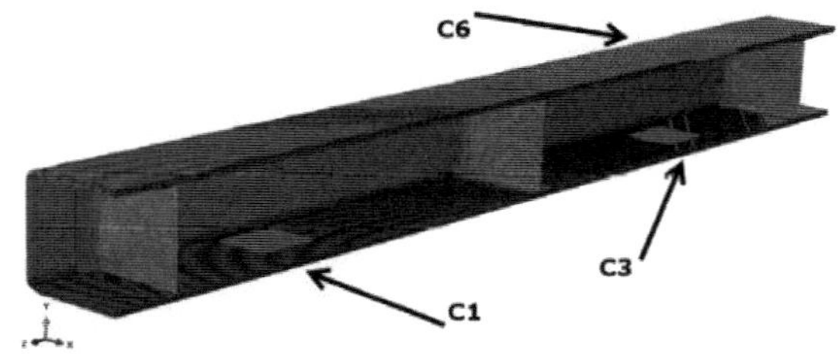

(a)

Damage configuration	x (mm)	z (mm)
#C1	48	56
#C3	48	72
#C6	48	56

(b)

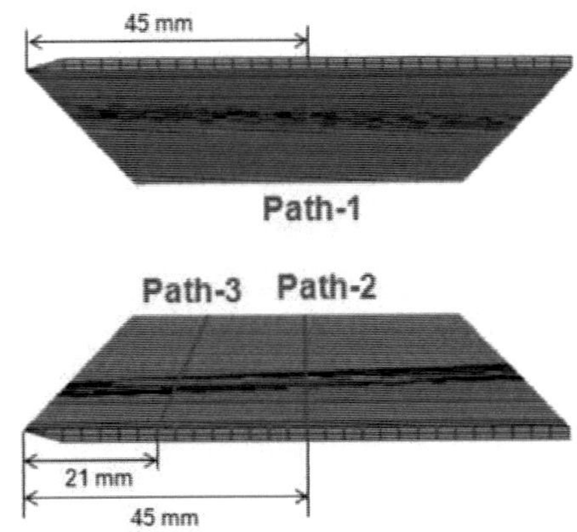

(c)

Figure 16. Final configuration of damage scenarios and damage detection path.

Finally, the influence exerted by stacking sequence on damage detectability has to be evaluated. Comparing the four laminations reported Fig. 17 it appears that the more spar and flanges laminations differ, the more easily damage can be detected (see typical curves shown in Fig. 17).

	Lay-up #1	Lay-up #2	Lay-up #3	Lay-up #4
C-Spar				
FB 45°	50%	100%	58%	41.6%
UD 0°	25%	-	25%	41.6%
UD 90°	25%	-	17%	16.8%
Flange				
FB 45°	50%	-	25%	15%
UD 0°	25%	100%	50%	12.5%
UD 90°	25%	-	-	62.5%
FB 0°	-	-	25%	-

(a)

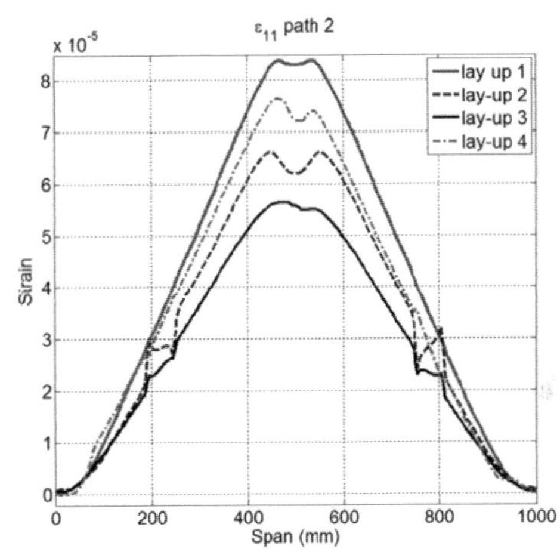

(b)

Figure 17. Sensitivity of damage detectability to different lay-ups.

In conclusion, global information about strain distribution can be processed through statistical techniques for providing synthetic damage indexes. For this purpose, five points are extrapolated in the damaged area along measuring path #1, mimicking real FBGs: three ahead and two across damage itself, then residual strain indexes are evaluated.

Adopting the criterion reported in Appendix I, through the formula:

$$R_{x_i} = \frac{2\varepsilon_{x_i}}{\left(\varepsilon_{x_i} + \varepsilon_{x_{i+1}}\right)\Delta x} - \frac{1}{\Delta x} \tag{1}$$

the results reported in Fig. 18-a are obtained for damage type C6, along the path #1. The x-axis refers to the abscissa along the span of the second bay in spar element. Results are inconsistent since the basic hypotheses of this method have been violated.

Viceversa, if strain residuals are computed as the difference between measured value and least-squares-approximated value, the formula:

$$S_{x_i} = \frac{\varepsilon_{x_i} - \sum_{n=0}^{N} k_n x_i^n}{\varepsilon_{x_i}} \tag{2}$$

supplies the results shown in Fig. 18-b, where the first three points (ahead damage) are characterized by almost null residuals, while the other two points (across damage) show residuals whose values are far away from zero.

Finally, the adoption of a further approach, according to which damage indices depend on the variation of the gradient of strain distribution along the path, as shown by the formula:

$$G_{x_i} = \frac{\left(\varepsilon_{x_{i+1}} - \varepsilon_{x_i}\right) - \frac{d}{dx}\left(\sum_{n=0}^{N} k_n x_i^n\right)}{\frac{d}{dx}\left(\sum_{n=0}^{N} k_n x_i^n\right)\Delta x} \tag{3}$$

allows to anticipate the effect of damage, which can be pointed out by the third strain residual already (see Fig. 18-c). By all means, the two latter techniques prove to be efficient and able to supply consistent results.

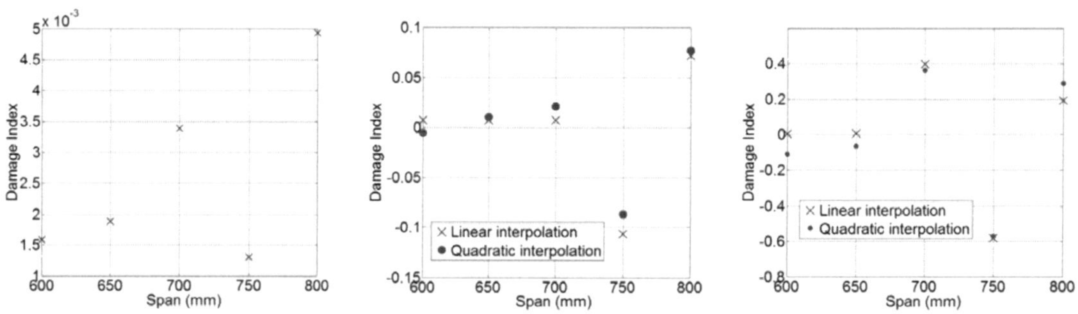

(a) (b) (c)

Figure 18. Damage indices for damage C6 along path #2: normalized strains (a), strain interpolation (b), strain gradients (c).

Since the three-point-bending-like loading condition can be considered a sound test-bed, but not truly representative of the real forces system acting of the wing, a second loading condition is taken into account, where the spar acts a cantilever beam, built-in at the root and loaded by vertical forces and bending moments at span mid-point and free edge.

In such a condition, typical strain span distribution, as measured along path #2 for damage C3, is shown in Fig. 19-a, while corresponding damage indices computed as the difference betweeen measured and least-squares-approximated gradients of strain distributions are reported in Fig. 19-b. Again, it appears that damage can be pointed out at a sufficient degree of accuracy, provided a suitable measuring path is preliminary chosen.

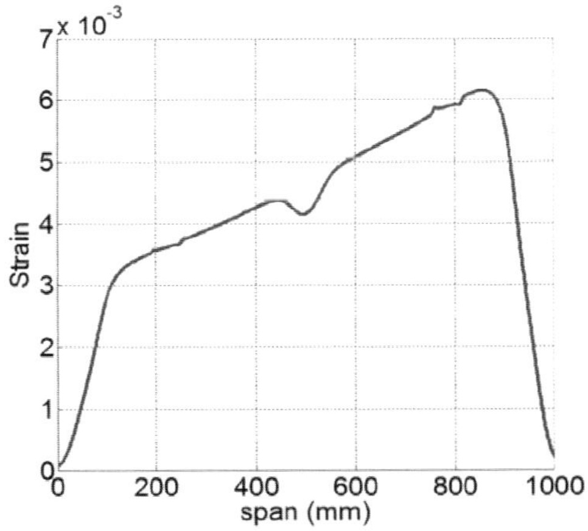

(a)

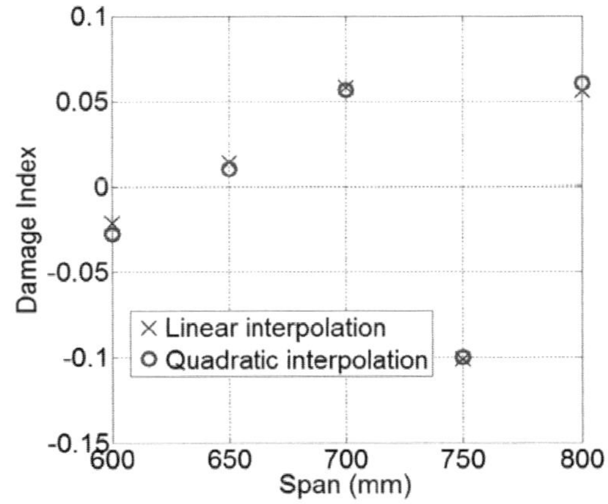

(b)

Figure 19. Strain distribution along path #2 for the cantilever beam load conditions (a) and damage index based on strain gradients (b).

CONCLUSIONS

The very basic issue of integrating optical fibre sensors into host composite materials for setting-up an health and usage monitoring system can be accomplished according to three different technological approaches, each of them characterized by pros and cons. Co-bonding technique is the simplest, even though criticalities can arise, due to surface preparation, OFs environmental protection, QPs positioning, adhesive optimization and structure-to-sensors strain-transfer capability. In turn, co-bonding technique offers the best performance in terms of sensors inspectability, removability/replaceability, repairability and maintainability. Co-curing process guarantees better integration between sensors and underlying structure, improves mechanical properties of adhesive bonding and increases strain-transfer capability; in addition, sensors can be better protected from exogenous/environmental effects. On the other hand, QPs and vacuum bag consumables must be compatible with high temperatures required to cure composite structural component; besides, OFs should be protected during curing cycle and sinking of QPs and PTFE tubing into pre-pregs should be avoided, not to create unacceptable dents in final laminates. In conclusion, process complexity is increased and QPs removability/replaceability easiness is reduced. The third approach, consisting in direct embedding fibre optic sensors into host material during lamination phase, leads to several advantages, i.e. protection from environmental effects, capability to accomplish health monitoring as well as process monitoring, preservation of surface aerodynamics. On the other hand, severe concerns about inspectability, repairability, replaceability and maintainability may arise. Finally, OFs embedment technique should address issues regarding OFs coating degradation due to high curing temperatures, as well as OFs cross-section deformation, owing to high pressures in presence of off-axis reinforcing layers.

As far as load identification issue is concerned, the outcomes of this study substantiate the possibility of identifying to a good extent the complex load conditions acting in an aeronautical component, provided an adequate number of sensors is used. Hence, lightness, reliability and easiness of installation of the sensors network play a fundamental role to determine the feasibility of such type of monitoring systems. If uncertainties are present, the system should be designed considering realistic load conditions and exploring as many configurations as possible to identify the best trade-off between number of sensors and robustness of identification. On the other hand, as regards damage detection capability, a network of FBGs sensors proves to be able to identify typical damages (i.e. delaminations) affecting composite laminates. By the way, sensitivity of monitoring system and soundness of results are remarkably affected by laminate stacking sequence, damage-to-measuring path mutual distance and, above all, by statistical techniques chosen for evaluating strain residuals, i.e. synthetic damage indexes. Here above, two approaches demonstrated their efficiency. Finally a last concern, still open, has to be pointed out, i.e. the comparison between minimum damage size to be detected (dictated by damage tolerance considerations) and sensitivity threshold of such systems (dependent on weight, cost and reliability constraints).

ACKNOWLEDGEMENTS

This research activity has received funding from the European Union's Seventh Framework Program for research, technological development and demonstration under grant agreement N° 284562.

The experimental activity of this research has been performed at A.dvanced MA.terials LA.boratory (AMALA) of Politecnico di Milano.

REFERENCES

[1] Zhou G, Sim LM. (2002). *Damage detection and assessment in fiber reinforced composite structures with embedded fibre optic sensors-review.* Smart Mater Struct, 11(6), 925.

[2] Bettini P., Sala G. (2007). In: *Preliminary Assessment of Helicopter Rotor Blades Fatigue Endurance Through Embedded F.O. Sensors*, Proceeding of the 24th ICAF Symposium, May 16–18 (Naples), 2007

[3] Gaudenzi P., Olivier M., Sala G., Sciacovelli D., Wheelan M., Bettini P., Nosenzo G., Tralli A. (2003). In: *Development of an Active Composite with Embedded Piezoelectric Sensors and Actuators for Structure Actuation and Control*, Proceedings of the 54th International Astronautical, IAC-03-I.4.03, Bremen

[4] Baldi A., Airoldi A., Bettini P., Fanteria D., Lazzeri L., Mariani U., Sala G., Scapinello F. (2013). In: *Methods for Design and Health Monitoring of Damage Tolerant Helicopter Structures in Fiber-Reinforced Polymer Composites*, Understanding Failure Mechanisms of Composites for Sustaining and Enhancing Military Systems Structures, p. 1-14 Riga, Latvia

[5] Airoldi A., Sala G., Bettini P., Baldi A. (2013). *An Efficient Approach for Modeling Interlaminar Damage in Composite Laminates with Explicit Finite Element Codes*, Journal of Reinforced Plastics and Composites, 32(15), 1075-1091.

[6] Tessler A., Spangler J.L., *A least square variational method for full-field reconstruction of elastic deformation in shear-deformable plates and shell*, Computer Methods in applied mechanics and engineering, 194; 2005;327-329

[7] Shkkarayev S., Krashantisa R., Tessler A., *An inverse interpolation method utilizing in-flight strain measurements, for determining loads and structural response of aerospace vehicles*, Third International Workshop on Structural Health Monitoring, Stanford, CA, 2001.

[8] Li H.C.H., Herszeberg I., Davis C.E., Mouritz A.P., Galea S.C., *Health Monitoring of marine composite structural joints using fibre optic sensors*, Composite Structures 75; 2006; 321-327

[9] Herszeberg I. , Mouritz A.P., *Automated characterization of structural disbonds by statistical examination of bond-line strain distribution*, Structural Health Monitoring 5;2006;83-94

APPENDIX I

The reconstruction of the whole strain field in a structural element on the basis of local strains is an inverse problem that can be addressed by methods based on least square approximation. Assuming a linear behaviour, the problem can be significantly simplified if the loading conditions applied to the element can be approximated by a set of m parameterized loads, acting at known location in the structures, which are represented by a vector $\{F\}$. Each term of the parameterised load set induces, when applied to the structural elements, a strain field. Considering a set of n local strains, $\{\varepsilon^*\}$, acquired at given position by the sensor network, the application of a single term F_i of the load vector $\{F\}$ lead to a set of strain values that can be expressed as in Eq. a, under the assumption of linear response.

$$\{\varepsilon^*\}_i = \{\alpha\}_i F_i \tag{a}$$

where $\{\alpha\}_i$ is a vector of influence coefficients, which provides the local strains for a unit value of the load term F_i. The superimposition of all the parameterised loads leads to a vector of strain measures that is expressed by defining a rectangular matrix of influence coefficient, $[\alpha]$, whose columns are the vectors $\{\alpha\}_i$, as indicated in Eq. b.

$$\{\varepsilon^*\} = \sum_{i\_1}^{m} \{\varepsilon^*\}_i = [\{\alpha\}_1 \quad \dots \quad \{\alpha\}_i \qquad \{\alpha\}_n]\{F\} = [\alpha]\{F\} \tag{b}$$

If the matrix $[\alpha]$ is known and the vector of strains $\{\varepsilon^*\}$ is given, the vector of the unknown loads $\{F\}$ can be evaluated by minimizing the squared norm of the distance between the experimentally acquired strains and the strains computed by using $[\alpha]$, as expressed in Eq. c.

$$\left\|\{\varepsilon^*\} - [\alpha]\{F\}\right\|^2 = \sum_{j=1}^{m}\left(\varepsilon_j^* - \sum_{i=1}^{n}\alpha_{ij}F_i\right) \tag{c}$$

The minimisation of the norm in Eq. c lead to a least square problem with the solution expressed in Eq. d.

$$\{F\} = \left([\alpha]^T[\alpha]\right)^{-1}[\alpha]^T\{\varepsilon^*\} \tag{d}$$

Hence, the system of parameterized loads can be identified if the coefficient of influence $\{\alpha\}_i$ are known. Such vectors can be evaluated by experiments or by developing a finite element model of the structural component and by separately applying the F_i loads, on the basis of Eq. a.

APPENDIX II

Let's assume that the axial strain distribution measured by sensors at every time step could be expressed through an *m*-order Taylor's series as follows:

$$\varepsilon_x(x) = \sum_{n=0}^{m} k_n x^n \tag{e}$$

Where k_0 represents *0th*-order strain (due to constant axial load), k_1 represents *1st*-order strain (due to mid-span concentrated force-induced bending), k_2 represents *2nd*-order strain (due to distributed load-induced bending) and so on. Such an hypothesis holds provided the strain field is homogenous and without severe singularities, so to be *x*-derivable up to *m*-order. A normalization method neglecting the effect of unknown loading conditions needs to consider the strain values in every point being normalized through strain increments between present and following point. The expression:

$$\varepsilon_{x_i}^{N} = \frac{\varepsilon_{x_i}}{\int_{x_i}^{x_{i+1}} \varepsilon_x(x)dx} \tag{f}$$

through Newton-Raphson's rule, can be re-written as:

$$\varepsilon_{x_i}^{N} = \frac{2\varepsilon_{x_i}}{\left(\varepsilon_{x_i} + \varepsilon_{x_{i+1}}\right)\Delta x} \tag{g}$$

then, by means of finite differences formulation, it becomes:

$$\frac{1}{\varepsilon_{x_i}^{N}} = \frac{\Delta x}{2}\left(2 + \frac{\varepsilon'_{x_i}\Delta x}{\varepsilon_{x_i}}\right) \tag{h}$$

finally, considering Taylor's series, we get:

$$\frac{1}{\varepsilon_{x_i}^{N}} = \frac{\Delta x}{2}\left(2 + \frac{\sum_{n=1}^{m} n k_n x_i^{n-1}}{\sum_{n=0}^{m} k_n x_i^{n}}\Delta x\right) \tag{i}$$

Normalized index very much depends on sensors spacing Δx and, to a lesser extent, on Taylor's coefficients. If all coefficients of the series but k_0 nullify, strain index only depends on Δx. The latter is assumed as reference value, corresponding to pure constant axial load. So doing, a strain residual function can be defined, by subtracting reference values from measured values:

$$R_{x_i} = \varepsilon_{x_i}^{N} - \varepsilon_{x_{ref}}^{N} = \frac{2\varepsilon_{x_i}}{\left(\varepsilon_{x_i} + \varepsilon_{x_{i+1}}\right)\Delta x} - \frac{1}{\Delta x} \tag{l}$$

Such a normalization technique and related strain residuals representation allow to get data no longer dependent on loads type and intensity. In case of unaffected structure, one should expect to find strain residual values very small or approaching to zero. On the contrary, within a damaged area, strain residuals values should remarkably increase. Besides, the smooth behaviour should be disrupted, owing to strain singularity. In particular, strain residual should be small if:

1. the coefficient k_0 is larger than the coefficients of higher order;
2. the function is never null inside the area considered;
3. the order of the Taylor's series is low.

Conditions #1–2 can be satisfied thanks to a *smart* sensors positioning; condition #3 is naturally satisfied since the effects of highest-order strains (> 2) have got small or negligible values compared with the ones of lowest order. Finally, it shall be mentioned that not every kind of damage can be pointed out, but only those which remarkably modify the strain field.

Improved Acoustic Emission Damage Source Location During Fatigue Testing of Complex Structures

Matthew Pearson, Mark Eaton, Carol Featherston, Rhys Pullin, Karen Holford

Cardiff School of Engineering, Cardiff University UK

Abstract: Structural health monitoring (SHM) allows the continuous monitoring of a structure's deterioration by the use of permanently embedded or surface mounted sensors. One such non-destructive technique that could be used to detect and locate damage is Acoustic Emission (AE). AE is a passive technique that can allow global monitoring of large scale structures for many different applications. One advantage of AE is the ability to locate damage sources. The traditional technique of Time of Arrival (TOA) is sufficient to locate AE damage sources in simple metallic plates, however complications arise when trying to locate in complex geometries and materials. Delta-T mapping was developed to overcome these problems, however errors still occurred due to the arrival time calculation. This paper outlines a further improved Delta-T mapping technique which uses the Akaike Information Criteria (AIC) to determine the true sensor arrival time. All three techniques (TOA, Delta-T and AIC-Delta-T) were used to locate a fatigue crack in a complex aluminium specimen and a comparison was drawn between the techniques. The AIC Delta-T mapping technique showed a significant improvement over Delta-T and TOA.

INTRODUCTION

Structural health monitoring (SHM) systems allow engineering components to be continuously monitored by permanently mounted or embedded sensors within a structure. These systems would reduce maintenance costs, improve safety and lead to lighter more optimised structures. One such Non-Destructive Testing (NDT) technique that could be used in this way is Acoustic Emission (AE). AE is a passive approach, essentially listening to the ultrasonic stress waves generated from initiation and propagation of damage in a structure. One of the primary advantages of AE is the ability to locate damage sources as they occur using a Time of Arrival (TOA) [1] technique. This techniques works well in simple flat plates made from homogenous materials, however in complex structures and anisotropic materials significant errors can occur. The reasons for this are discussed in full by Eaton et al. [2] but are fundamentally linked to an accurate arrival time determination and the variation in propagation velocity due to geometry and material properties. An initial cause for error is that TOA assumes a straight propagation path between source and sensor, in complex structures such as those found in the aerospace industry, this is rarely the case. Furthermore TOA assumes a single uniform propagation velocity to calculate a source location, again in components where thickness changes occur or in advanced materials such as composites this is incorrect. Finally and importantly for TOA and Delta-T the arrival time of a signal at a sensor is determined using a user defined threshold crossing; where the signal crossing the threshold defines the arrival time. However this can introduce significant errors as in many examples where low energy events occur the actual start of the waveform goes undetected. All these issues can lead to large errors occurring in the estimated source location.

There are many techniques which have been used to improve arrival time estimation and these include wavelet transforms [3], wavelet decomposition [4] and statistical techniques [5]. These techniques have shown improvements over the conventional threshold crossing techniques. There has also been vast improvements in location algorithms over the past number of years that account for variation in velocity with propagating angle. Ciampa and Meo [6] used a specific layout of 6 sensors in 3 closely spaced pairs to solve a set of non-linear equation to resolve impact events in composite plates with accuracies of 3mm. However this techniques still assumed a constant propagating velocity on a particular propagation angle. In terms of complex structures mapping techniques can be used, Baxter et al [7] developed Delta-T mapping which used artificial AE sources to generate maps of difference in arrival time between sensor pairs. These maps were used as training data to then compare with experimental data to accurately determine the source location. This paper presents an improved Delta-T mapping technique by incorporating Akaike Information Criteria (AIC) [8] which is a statistical approach which has been shown to accurately determine the actual arrival of the wave. This technique has been tested on a complex specimen under fatigue to accurately locate damage and was shown to provide improvements in location accuracy over the TOA and conventional Delta-T mapping technique.

DELTA-T MAPPING

The traditional Delta-T mapping technique was developed by Baxter el al [7], a summary of the technique is given below:

1. Determine an area of interest on a structure, this could be an area of complex geometry or an area of known higher stress concentrations.
2. Construct a grid on the area of interest. The resolutions of grid will affect the accuracy of the estimated source locations. The orientation of the grid with respect to the sensors is not important and is not important to know the sensor positions, as the grid is used as the reference in the technique.
3. Conduct artificial (H-N source [9]) source events at each node position within the grid to record the arrival time of the AE event at each sensor. At least five sources are used at each nodal position in order to provide an average result and remove erroneous data. Often it is not possible to have data from all node positions however the data can be interpolated from surrounding nodes.
4. Determine the difference in arrival time (Delta-T) maps. Calculate the Delta-T for each node position for each sensor pairing (fours sensors results in six sensor pairings). Knowing the average Delta-T for each node alongside the Cartesian co-ordinates generates delta-t maps for each sensor pairing.
5. Determine the Delta-T for test data events for each sensor pairing and compares these with the training maps. This identifies a contour of Delta-T where a possible source location could occur. Overlaying these identified contours from all sensor pairing indentifies a crossing point which is the calculated source location.

This technique has successfully been used to determine AE source locations in both metallic [7] and composite structures [10]. However a significant disadvantage of this technique is that the time of arrival is still determined on a threshold crossing technique. This can introduce erroneous data in the technique due to triggering from different parts of the AE wave due to either attenuation or dispersion of the AE signal. The AIC technique can be used to overcome these errors and is able to determine the actual arrival time of the AE wave. The AIC technique replaced the threshold crossing technique for arrival time estimation and was incorporated into the training and location parts of the Delta-T programme and resulted with the AIC Delta-T mapping technique. The form the AIC used in the programme was developed by Maeda [8] and is described by Equation 1.

$$AIC(t) = t\log_{10}\big(var(x[1;t])\big) + (T - t - 1)\log_{10}\big(var(x[t;T])\big)) \tag{1}$$

The function splits the signal into two parts, from time 0 to time "t" and that from "t" until the end of the signal "T". Equation 1 describes the similarity in variance between the two parts for every time "t" throughout the whole signal. When "t" corresponds to the onset of the signal, the minimum similarity is seen between the high variance noise prior to signal onset and the low variance after the signal onset. Kurz et al [11] and Sedlak et al [12] successfully applied the AIC technique to AE signals. Figure 1 shows a comparison of the threshold crossing technique with a 40dB threshold and the AIC techniques. For the high amplitude events there is very little difference in the arrival time determined by the threshold and the AIC techniques meaning little error is introduced into the Delta-T mapping technique. For example an H-N source can be described as a large amplitude AE event. However when you consider a low amplitude events, such as a small fatigue crack, the actual AE signal onset goes undetected by the threshold triggering technique but the AIC technique is still accurately able to determine the correct onset of the signal. Due to the wave speed in metallic structure being in the order of 5000m/s this difference in between the actual signal onset and the one determined by the threshold crossing significant errors are introduced into the Delta-T mapping technique. Nonetheless the ability of the AIC technique to accurately determine the actual arrival time for low amplitude events means that these errors are eliminated from the AIC Delta-T mapping technique.

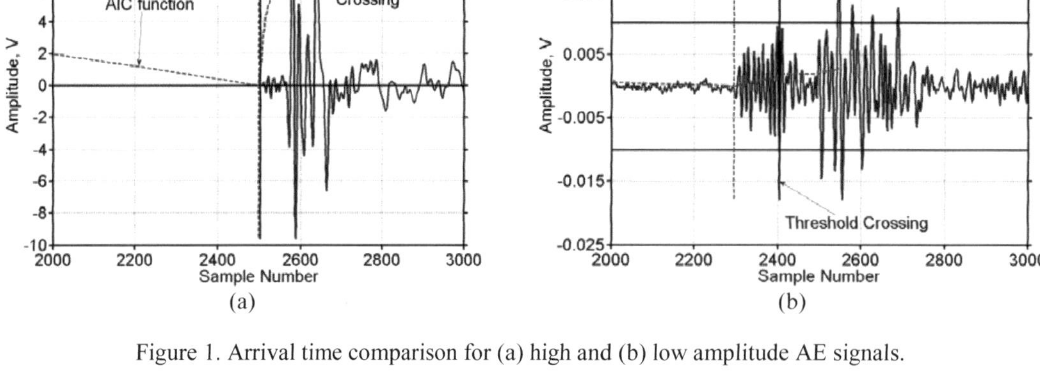

Figure 1. Arrival time comparison for (a) high and (b) low amplitude AE signals.

EXPERIMENTAL PROCEDURE

A complex geometry specimens was manufactured from aerospace grade aluminium (2024-T3) (Figure 2). The overall specimen dimension was 370 x 200mm with a thickness of 3.18mm. The specimen had a series of different diameter holes machined to create a complex geometry, this was undertaken in order to assess the performance of TOA, Delta-T and AIC Delta-T mapping techniques. The area of interest identified on the specimen was 200 x 160mm and had a grid resolution of 10mm. Four Mistras Group Limited (MGL) Nano 30 sensors, which have an operating frequency of 125-750kHz, were coupled to the specimen using silicon RTV (Loctite 595). The sensors were connected to pre-amplifiers with a gain of 40dBA. An MGL PCI-2 system was used to record AE data for the Delta-T mapping techniques and to record data from the fatigue test. A 40dB threshold and 2MHz sample rate was used throughout. Before testing commenced five H-N sources were used at each node position within the area of interest to generate artificial AE sources in the specimen for the Delta-T mapping techniques. To establish the effectiveness of the Delta-T training data a series of five H-N sources at six random node positions within the grid were recorded. 5mm steel clamping plates were used to distribute the load into the specimens and a 20mm loading pin was used to attach the specimen to the hydraulic load machine. The specimen was cyclically loaded under tension from 0.25kN to a maximum load of 24kN at a rate of 1Hz. During the fatigue test AE data was continuously collected.

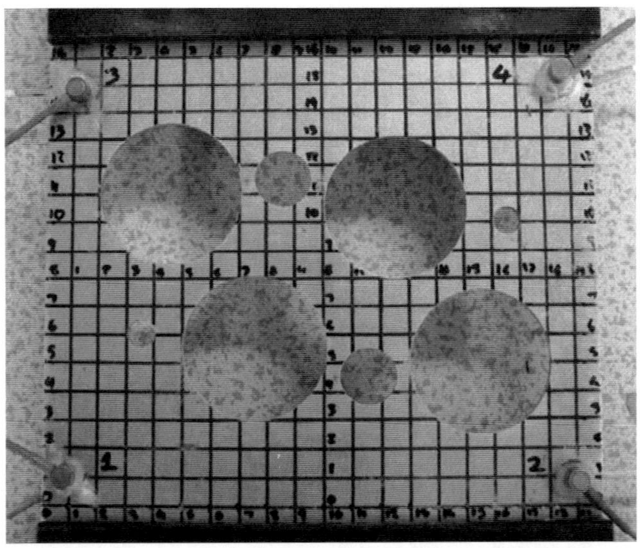

Figure 2. 2024-T3 complex geometry specimen.

EXPERIMENTAL RESULTS AND DISUSSION

H-N Source Investigation

Figure 3 shows the calculated AE source locations from a series of H-N sources at six random node positions for the TOA, Delta-T and AIC Delta-T techniques. The figure shows that the Delta-T and AIC Delta-T techniques were able to accurately locate H-N sources from all 6 positions, whilst the TOA could only estimate the actual; location for four of the six locations. This demonstrated the robustness of the Delta-T techniques in accurately determining source locations in complex geometry specimens. Overall both Delta-T techniques are able to resolve locations to a higher level of accuracy than that of the TOA techniques. Position 4 was the only location where the TOA technique had a higher level of accuracy than the other two techniques but all techniques are of high accuracy.

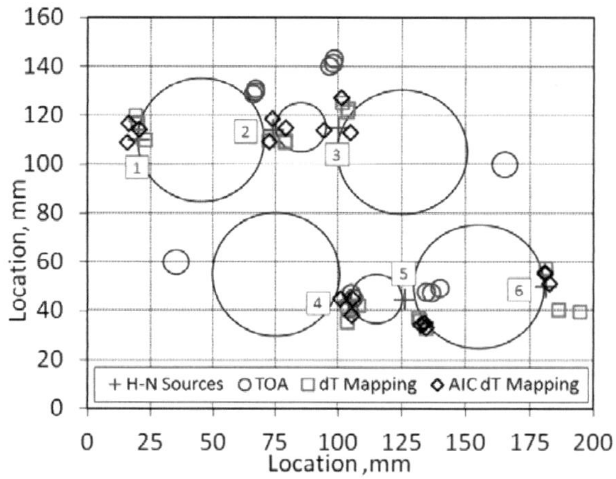

Figure 3. Calculated source locations from H-N sources.

Figure 4 shows the Euclidean distance error, the difference between the actual H-N locations and the calculated source locations, for the TOA, Delta-T and AIC Delta-T techniques. Overall the TOA technique performs poorly for the majority of the positions, whilst the AIC Delta-T technique shows an improvement over the traditional Delta-T technique in five of the six positions. The reasons for the similarity between the two Delta-T techniques are due to the use of the H-N source. The H-N sources creates a very large amplitude AE event in the structure which means there will be very little difference in the arrival time calculated by the threshold crossing and the AIC technique as shown in Figure 1.

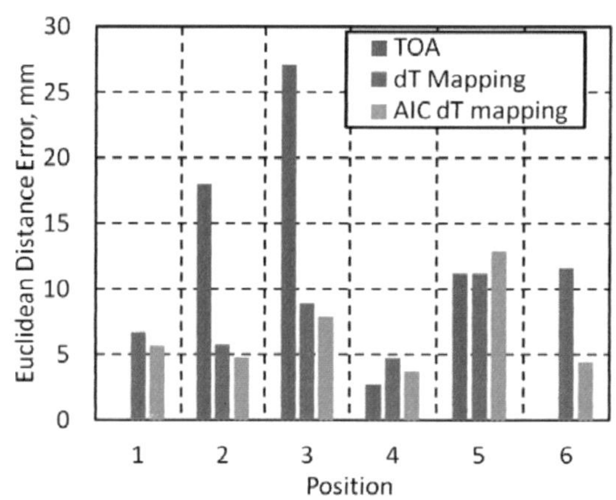

Figure 4. Euclidean distance errors for the calculated locations of the H-N sources.

Complex Geometry Fatigue Investigation

Figure 5 shows the specimen failure regions after the specimen had been subjected 96,000 fatigue cycles. From visual observation of the failure regions it was evident a small fatigue crack initiated at the right hand side thin section (Figure 4 (a) – solid arrow). This propagated a small distance before leading to a rapid failure, this then caused rapid failure of the of left hand side thin section (Figure 4 (b) – dashed arrow). No evidence was found of a fatigue crack through visible observation at the last inspection prior to failure (76,000 cycles); suggesting that the damage initiated, propagated and caused failure within 20,000 cycles.

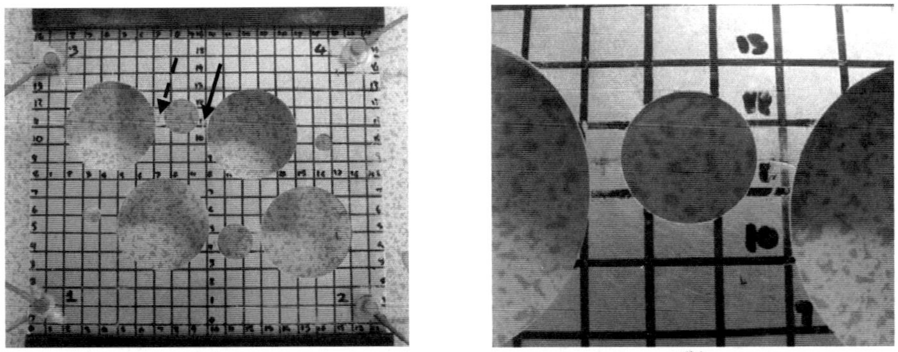

(a) (b)

Figure 5. Damage locations for the complex geometry specimen.

Figure 6 shows all the calculated AE source locations from the fatigue investigations in terms of a spatial binning surface plot. The figure shows the total number of events which were located in a 5 x 5mm spatial bin for the TOA, Delta-T and AIC Delta-T techniques respectively. The TOA results show located events all across the specimen with significant cluster events on the top half of the specimen, from these location results alone it would be very difficult to determine the location of the damage present in the structure. There are three distinct clusters located approximately 20, 42 and 40mm away from the fatigue crack. The results from Delta-T mapping are encouraging with most of the located events situated towards the top central region of the specimen. The highest number of events is clustered approximately 35mm from the fatigue crack. The results from the AIC Delta-T mapping show one region which has a high number of located events in the fatigue crack area with most of the events in the a cluster 20mm away from failure region. This highlights not only the advantage of the mapping technique combined with an accurate time of arrival estimation. Overall it would be difficult to determine damage is present in the TOA results due to the large amount of scattering in the data, however the AIC Delta-T mapping techniques shows an improvement of 15mm in terms of the highest number of events in a cluster when compared with the traditional Delta-T mapping.

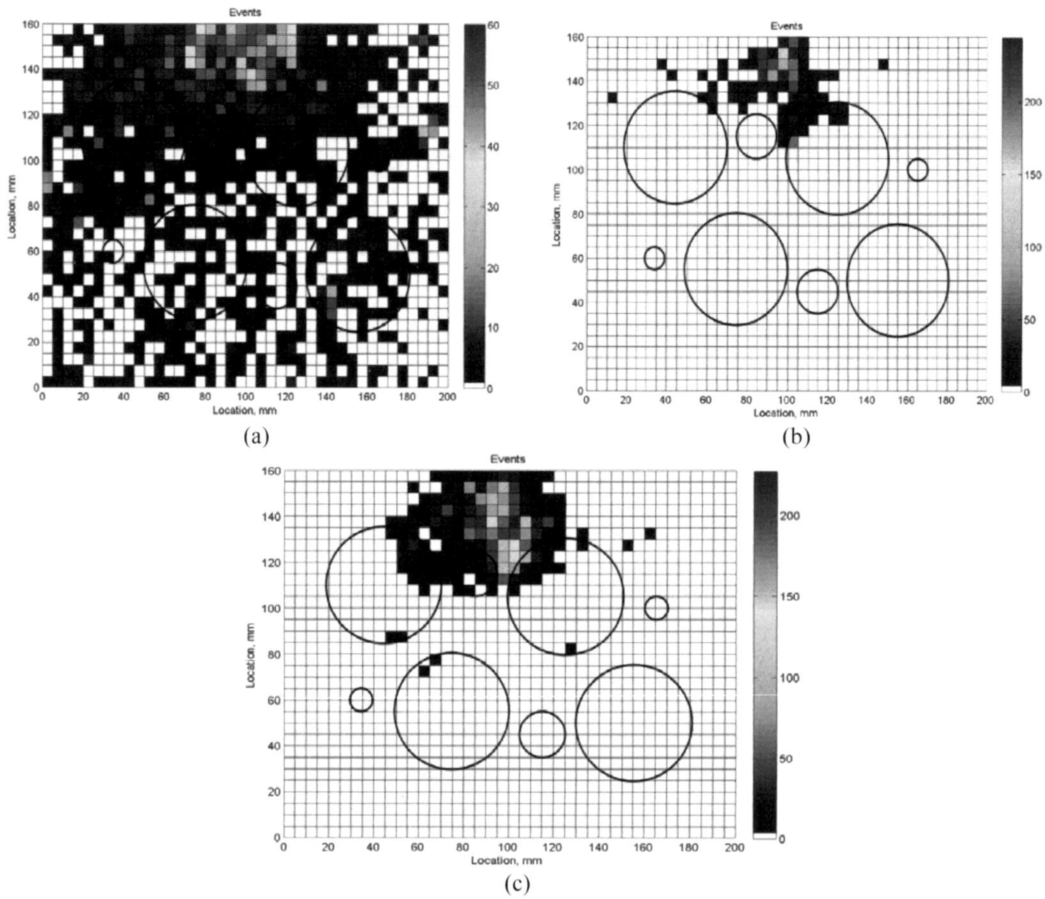

Figure 6. Spatial binning in terms of the number located events from the fatigue investigations for the a) TOA b) delta-t c) AIC delta-t techniques

Figure 7 shows the same spatial binning approach as used in Figure 6, however instead of the number of events in a spatial bin, the analysis is conducted for the summation of the energy of the AE events within that particular spatial bin. Unfortunately due to the way the original Delta-T mapping algorithms works it is not possible to obtain the energy of the located AE events as well as the location. This is an improvement which has been incorporated in the AIC Delta-T algorithm. The results for TOA shows a region of considerable higher energy located 20mm from the actual failure region. The results from the AIC Delta-T mapping however shows a region of higher energy located in the precise location of the fatigue crack and resulted in an error of 3mm between the actual failure area and the higher energy region. Both Figure 6 and Figure 7 do not show many located events on the right hand thin section, this is thought to be due to the nature of the failure, which for this section was thought to be very rapid due to no visible evidence of a fatigue crack. This might be the reasons for the lack of locations with any of the techniques in this region. Another reason could be the failure of left hand section was enough to cause significant differences between the actual arrival times in the structure and those of the training maps.

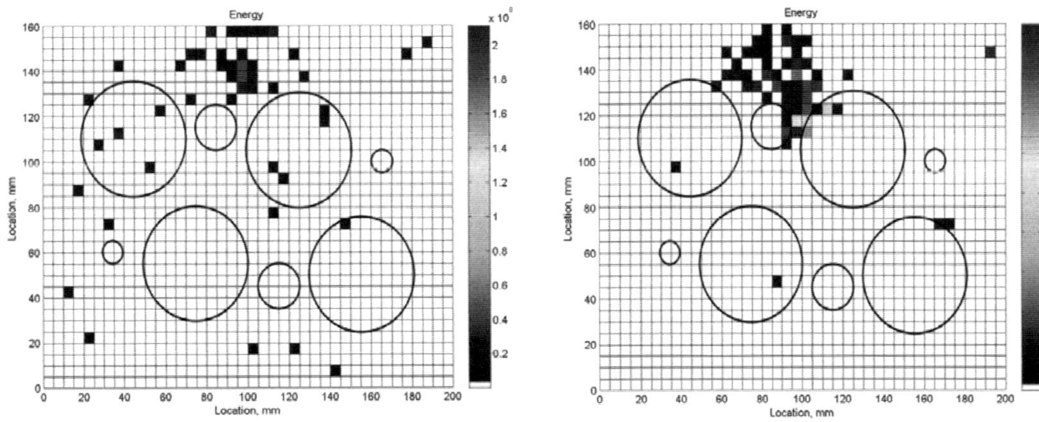

(a) (b)

Figure 7. Spatial binning in terms of the energy of the located events from the fatigue investigations for the a) TOA b) delta-t c) AIC delta-t techniques.

CONCLUSIONS

The combination of the Delta-T mapping techniques combined with an AIC picker to accurately determine the actually first arrival of the waveform showed significant improvements over the conventional TOA and Delta-T mapping techniques. The location of H-N sources showed an average improvement in error of 8.2 and 1.6mm over the TOA and Delta-T mapping techniques respectively, resulting in an improvement of 125% and 25% respectively. Using event spatial binning the AIC delta-t technique showed an average improvement of 75% over the traditional Delta-T technique. It was also not possible to confidently detect the damage using the TOA technique and demonstrated the effectiveness of the mapping approaches. Alternatively using an energy spatial binning technique the AIC Delta-T mapping technique located the fatigue crack within 3mm of the actual location this was a significant improvement over the TOA technique. Improved source location, in SHM systems, can lead to greater sensor spacings being used to achieve a similar accuracy to that of the TOA technique. This means less sensors are required to monitor the same area which in reality would significantly reduce the costs and weight for sensors and cabling which is a currently a limiting factor of installing an SHM system.

REFERENCES

[1] Miller, R.K., Hill, E V.K. and Moore, P.O. (2005). *Non destructive Testing Handbook,: Volume 6, Acoustic Emission Testing,* ASNT.

[2] Eaton, M.J., Pullin R. and Holford, K.M. (2012) *Towards Improved Damage Location Using Acoustic Emission,* Proc. IMECHe, Part C, vol. 226 n. 9 pp. 2141-2153.

[3] Hamstad, M.A., O'Gallagher, A and Gary, J. (2002). A Wavelet Transform Applied to Acoustic Emission Signals: Part 2: Source Location. Journal of AE, vol. 20, pp. 62-82.

[4] Ding, Y.,Reuben, R.L. and Steel, J.A. (2004). A new Method for Waveform Analysis for Estimating AE Wave Arrival Times using Wavelet Decomposition. NDT & E International, vol. 37, n. 4, pp. 279-290.

[5] Lokajicek, T and Kilma. K. (2006). A First Arrival Identification System of Acoustic Emission Signals by Means of a High-Order Statistics Approach. Meas. Sci. and Tech., vol. 17, n. 9, pp. 2461-2466.

[6] Ciampa, F. and Meo, M. (2010). A New Algorithm for Acoustic Emission Localization and Flexural Group Velocity Determination in Anisotropic Structure. Composites: Part A, vol. 41, n. 12, pp. 1777-1786.

[7] Baxter, M.G., Pullin, R., Holford, K.M. and Evans, S.L. (2007). Delta T Source Location for Acoustic Emission. MSSP, vol. 21, n.3, pp. 1512-1520.

[8] Maeda, N. (1985). A Method for Reading and Checking Phase Times in Auto-Processing System of Seismic Wave Data. Zisin = Jishin, vol. 38, n. 3, pp. 365-379.

[9] Hsu, N.N. and Breckenbridge, F.R. (1981). Characterisation and Calibration of Acoustic emission Sensors. Materials Evaluation, vol. 39, n. 1, pp. 60-68.

[10] Eaton, M.J., Pullin, R. and Holford, K.M. (2012). Acoustic Emission Source Location in Composite Materials using Delta T Mapping. Composites: Part A, vol. 43, n. 6, pp. 856-863.

[11] Kurz, J.H., Grosse, C.U., Reinhardt, H.W. (2005). Strategies for Reliable Automatic Onset Time Picking of Acoustic emissions and of Ultrasound Signals in Concrete. Ultrasonics, vol. 7, pp. 538-546.

[12] Sedlak, P., Hirose, Y., Khan, S.A., Enoki, M and Sikula, J. (2009). *New Automatic Localization Technique of Acoustic Emissions Signal in Thin Metal Plates. Ultrasonics, vol. 49, n. 2, pp. 254-262.*

Full Scale Fatigue Tests and Management of aging fleets

A350 XWB EW Test - Combined static and fatigue testing of composite aircraft structures within one test set up

Fin Schorr[1], Michael Stodt[1], Timo Plate[2]

[1]*IABG mbH, Germany*
[2]*AIRBUS Operations GmbH, Germany*

Abstract: In 2009 IABG was contracted by AIRBUS Operations to design, erect and perform the A350 XWB EW test. The structure of the AIRBUS A350 XWB combines usage of metallic and composite materials in the primary structure of the airframe and so fatigue testing of both kinds of material was specified. As part of the full scale tests AIRBUS decided to perform a dedicated test on the left hand wing structure with combination of static and fatigue testing as well as functional testing within one test set up, the so called A350 XWB EW test, with main focus on the composite material. IABG was responsible for the provision of the overall test set-up to load the specimen including all mechanical and hydraulic components as well the equipment for control and monitoring and data acquisition.

This paper presents the special requirements to the overall test set up to perform static and fatigue testing as well as functional tests within one dedicated test set up. Furthermore this paper gives an overview about state-of-the-art full scale testing currently being performed by IABG.

INTRODUCTION

For the final qualification and type certification of a new aircraft type a huge amount of test activities is necessary to prove static and fatigue behaviour of the primary structure of the airframe and as well demonstrate functionality of individual sub-systems like high lift systems or landing gears as part of the overall aircraft. To fulfil all the requirements defined by the aviation authorities a test plan has to be defined by the manufacturer of a new aircraft type how to demonstrate compliance with the individual requirements. AIRBUS decided to perform an additional full scale test in comparison to former qualification and type certification processes for the AIRBUS aircraft family. This test was called EW and comprised the left hand (LH) wing structure of the A350 XWB. The main focus of the EW test, see Figure 1, was to prove static and fatigue behaviour of the composite material of the primary structure of the wing made of more than 50% reinforced plastics (mainly CFRP). In addition functional testing of high lift systems was a major task to be performed within this test.

The combination of static and fatigue testing as well as functional testing within one dedicated test set up generated a complex and extensive spectrum of requirements for the design of the test set up. All these requirements had to be respected in every single step of the design phase to be able to fulfil the dedicated requirements of the individual test phases without too much modification work to be time and cost efficient.

REQUIREMENTS FOR THE TEST SET UP

The overall requirements for the A350 XWB EW contained requirements from each of the single test campaigns for static, fatigue and functional testing. The static test campaign requirements were mainly defined by the requirement to perform tests up to ultimate load. This requirement defined the maximum loads and strokes of the hydraulic actuators and different positions of the hydraulic actuators to ensure perpendicular load introduction into the specimen during ultimate load level to reduce side loads.

The fatigue test campaign required a very high test speed to perform about 4,000 flights per week with an average number of 40 load cases per flight and deflections during fatigue testing up to limit load due to the testing according to a dedicated CFRP loading spectrum which was described during 27th ICAF Symposium (for details see Ref. [1]. As the loading program contained a lot of different loading conditions and corresponding deflections between limit load up- and down-bend and different torque conditions the hydraulic actuators moved in a wide range during the fatigue testing. In addition good access conditions to the specimen were required to perform regular inspections within short time frames.

The functional testing contained mainly a test of the movement of the high lift system under 1g deflection conditions and a torque test of the high lift system drive shafts under limit load up-bend conditions. To perform the 1g deflection the specimen had to be loaded by the hydraulic actuators and blocked in deflected position. As the high lift systems were moved during the test a certain clearance to the surrounding area of the high lift systems was necessary for the movement. For the torque test of the high lift system drive shafts the installation of the drive shafts and in parallel a load introduction at the leading edge was mandatory.

Figure 1. A350 XWB EW – Ultimate Load Up-Bend Test.

MECHANICAL TEST SET UP

The mechanical test setup included the restraint system, the load introduction, the loading rigs and the inspection rig. The purpose of the mechanical test set up was mainly to spread loads generated by the hydraulic actuators and connect to the airframe, to react loads generated by the hydraulic actuators into the strong floor and to gain access to the specimen.

A very important factor for the sizing of the mechanical test set up was to grant that the mechanical test set up did endure the whole test duration although even the mechanical test set up is kind of static and fatigue tested. The factor of 1.5 between loads up to ultimate load during static testing and up to limit load during fatigue testing was a main influence for the sizing of the mechanical test set up for the A350 XWB EW test. In former full scale tests with fatigue and static testing up to limit load or static testing up to ultimate load explicit guidelines for sizing of steel work structures were used whereas for the sizing of this mechanical test set up it was necessary to combine individual technical guidelines and sizing methods incl. non-linear FE analysis to demonstrate endurance of the mechanical test set up over the whole test duration. This was a new step in sizing of a mechanical test set up for a full scale test.

Restraint System

In deviation to former full scale tests performed by IABG the A350 XWB EW specimen i.e. the LH wing structure was not supported and restraint by rods to allow load measurement in discrete directions but was restraint by a clamped solution. Therefore the specimen was attached by the serial root joint to a dummy part called dummy centre wing box (DCWB). The DCWB presented the serial root joint components of the centre wing box section attaching to dummy parts to direct loads into the steel structure of a wing root clamp assembly (WRCA). The WRCA was connected to a steel frame loading rig to react the loads to the strong floor. To ensure representative stiffness of the restraint system DCWB and WRCA were designed by AIRBUS as close as possible to the properties of the serial centre wing box section.

Load Introduction System

The load introduction system included several whiffle trees to distribute the load of the hydraulic actuators to the individual load introduction points at the specimen and dummy structures for the high lift systems, the engine pylon, the main landing gear (MLG) and the wingtip.

Wing Box Load Introduction. The loading of the wing box of the LH specimen structure was realised by 18 hydraulic actuators located above the wing at the inner half of the wing and below the wing at the outer half of the wing. The arrangement of the hydraulic actuators ensured that all kind of vertical loading of the specimen like bending and torque loading could be realised.

Figure 2. Wingbox Whiffle Trees.

The loads of the individual hydraulic actuators were distributed by whiffle trees, see Figure 2, which connected up to 26 load introduction points split over up to 4 ribs at the specimen. The load introduction was mainly carried out by bolted load fittings on the upper wing cover. Due to the rigid connection the whiffle trees were constructed as a statically determinate framework to avoid constraining forces. This was reached by a special bearing and axis arrangement within the whiffle trees. In deviation to wing whiffle trees in former full scale tests occurring side loads due to changing angle of attack of hydraulic actuators during test performance were only introduced into the wing at dedicated positions. To reduce these side loads to a minimum a main focus in the design process of the whiffle trees was to position the hydraulic actuator head ends as near as possible to the specimen structure. Due to the big deflections of the specimen during static test campaigns under ultimate load the analysis of the clearance within the whiffle trees in deflected positions was done by CAD deflection checks. In the past statically under-determinate whiffle trees were used for areas of big deflections in static tests but due to cost and time efficiency it was decided to use only one set of whiffle trees for static and fatigue testing.

The whiffle trees had to be sized to resist the high loads up to ultimate load but it was also necessary to perform weight optimisation as the requirement regarding high test speed for the fatigue test campaigns required the reduction of weight of moved masses. Therefore the whiffle trees were sized by using FEM models and state of the art technical guidelines and design principles.

Each of the 244 load fittings that connected the whiffle trees to the specimen had an individual design corresponding to the hole pattern of the rib feet below. Therefore a general design was developed which was adapted to the individual load fittings by parametric construction. Using this design method IABG was able to handle changes of the hole pattern of the rib feet in the design process very easily. This was mandatory due to the concurrent engineering of the aircraft and the test set up which means that the design of the aircraft was not finalised at the stage where the design of the test set up was performed.

High Lift Load Introduction. The aerodynamic loads on the high lift systems of a serial aircraft were introduced into the specimen via steel dummy structures which were connected to the serial interfaces for the high lift systems in the leading and trailing edge area. These dummies were then directly loaded or connected via whiffle trees which then were loaded by hydraulic actuators.

The test of the high lift drive shaft system at limit load up-bend deflection required the possibility of an installed drive shaft system and at the same time installed dummies at the leading edge. Therefore the design of the dummies had to respect the space allocation of the drive shaft and geometry had to be adapted accordingly.

MLG Direct Loading. In former full scale tests performed for AIRBUS the landing gears were replaced at least partly by steel dummy structures which transferred the load generated by the hydraulic actuators in the area of the wheels of a serial aircraft into the attachments of the landing gears at the fuselage and the wings. For the A350 XWB EW test it was required by AIRBUS to directly load the MLG attachments at the wing structure by hydraulic actuators. This requirement arose from the fact that the mounting of the MLG was in deviation to former AIRBUS aircrafts overdetermined which means that the stiffness of the MLG itself has an influence on the load distribution to the individual MLG attachments at the wing. Therefore IABG developed a special direct loadıng solution, see Figure 3, to load the individual attachments at the wing which react loads gained by the MLG.

The individual MLG attachments at the wing specimen structure were loaded with up to four hydraulic jacks per attachment to simulate loads and moments created by the MLG at a serial aircraft (touch-down, taxi bump, breaking and ground handling loads). To transfer the load from hydraulic actuators to the attachments at the specimen special load introduction fittings were designed and sized which were directly connected to the serial airframe attachments. The number of hydraulic actuators and load introduction fittings in a relative small area increased the complexity to realize the MLG direct loading solution. As part of the MLG direct loading solution partly special hydraulic actuators had to be used to be able to allocate the overall load introduction in that area.

Figure 3. MLG Direct Loading.

Loading Rigs

The loading rigs mainly mounted the hydraulic actuators and reacted their generated loads in to the test hangar´s strong floor. Therefore complex steel frame work structures were used which were designed to the individual positions of the hydraulic actuators.

For time efficiency between the different static and fatigue testing campaigns and to avoid too much modification work in between these phases IABG decided to design the loading rigs in a way to cover all the different setting positions of the individual hydraulic actuators for the different test campaigns, see Figure 4.

As part of this design a complex requirement regarding load combinations and angles of attack of loads had to be considered as the hydraulic actuators moved with the deflection of the specimen. To decrease the complexity the first intention was to design a single loading rig for each hydraulic actuator and each setting position but this was not possible as the setting positions where too close to each other and loading rigs needed to have some space to react the loads into the strong floor while considering the load allowable of the strong floor. Therefore loading rigs with up to 12 different setting positions for hydraulic actuators had to be designed and sized incl. the usage of a big number of permutations of load variations. Meeting this requirement was only possible by usage of modern FE calculation tools and exceeding known calculation efforts for loading rigs.

The close distance between the setting positions of hydraulic actuators sometimes let no other possibility than using small adapter constructions to be able to modify the hydraulic actuator mounting position in a short time frame.

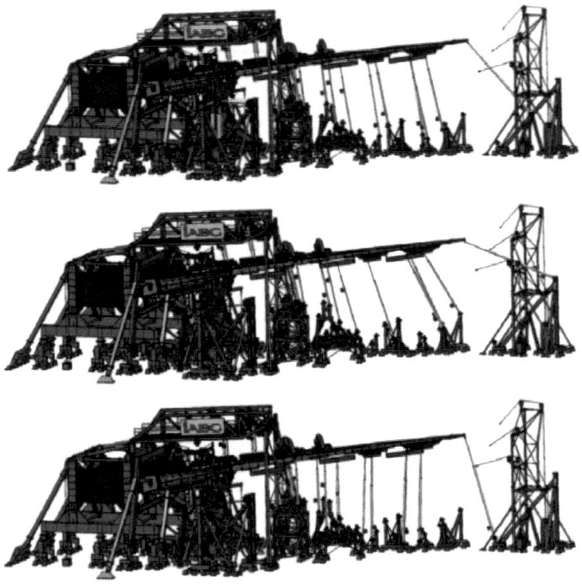

Figure 4. Different Positions for Hydraulic Actuators (Specimen in Jig-Shape).

Inspection Rig

The purpose of the inspection rig was to provide suitable access conditions for inspections, repairs and modifications at the specimen. Therefore the main platform of the inspection rig was installed below the wing and a second platform was installed to provide access to the leading and trailing edge as well as the MLG area. As hydraulic actuators loaded the wing partly from below it was necessary to provide cut-outs for the hydraulic actuators within the platforms of the inspection rig. These cut-outs were optimized to the requirements regarding hydraulic actuator movement within the fatigue test campaign as the cut-outs had to be large enough to allow hydraulic actuator movement but as small as possible not to disturb access conditions to the specimen. For the ultimate load tests within the static test campaigns the positions of the hydraulic actuators were changed and so the cut-outs had to be adapted to these positions.

To allow the modifications mandatory for the different test campaigns IABG decided to install an inspection rig made out of standard scaffolding. The modular character of a standard scaffolding system allowed especially the modifications of cut-outs for hydraulic actuators by changing the geometry of the scaffolding in the dedicated areas. These modifications were possible in a short time period and did not require any machine work on the inspection rig. In addition a standard scaffolding system had the advantage to re-install it for different uses at a later time.

HYDRAULIC TEST SET UP

Different requirements regarding loads, strokes and operation speed of the hydraulic actuators existed in the test due to the combination of static and fatigue testing. For the static test phases up to ultimate load requirements regarding loads and strokes were higher compared to the requirement for the fatigue test phases whereas the requirement for high speed operating hydraulic actuators was defined by the fatigue test phases.

Due to cost and time efficiency IABG decided to use a common set of hydraulic actuators covering all the requirements from the static and fatigue test phases as otherwise hydraulic actuators would have to be exchanged between static and fatigue test phases. To avoid oversized hydraulic actuators which result in an increased consumption of pressurized hydraulic oil and so energy IABG used different hydraulic pressure values for the performance of the static and fatigue test phases. In the fatigue mode a hydraulic pressure of 280 bar was used to generate loads by the hydraulic actuators. This is the normal hydraulic operation pressure in the IABG performed fatigue tests and so it was possible to use existing components for hydraulic infrastructure and for the control of the hydraulic actuators. For application of loads during static phases the hydraulic pressure was increased to 315 bar to allow generation of higher loads. Due to the lower requirements regarding operation speed and the corresponding lower flow velocity there was no influence on the above mentioned existing components.

CONTROL AND DATA ACQUISITION SYSTEM

The requirement for the fatigue testing campaign included the simulation of 4,000 flights per week with an average number of 40 load cases per flight and still meeting the high accuracy requirements known from former full scale tests. The achievement of this requirement was only possible by using and optimising IABG own developed control and monitoring system incl. simulation of the fatigue test performance. Therefore a kind of a virtual test set up was used which contains on one side a GFEM model of the specimen and on the other side a simplified FE model of the IABG test set up, see Figure 5. With the usage of multi-body simulation tools the complete fatigue test performance was simulated and control parameters for the individual controllers of the overall system and as well the hydraulic actuators were pre-defined. Already having these very accurate pre-defined control parameters determined only smaller adjustments at the controllers were necessary to meet the AIRBUS requirements regarding test speed and test accuracy.

Figure 5. Multi-Body simulation model.

For the data acquisition a system was used which was able to measure and record more than 3,000 channels with up to 100 Hz sample rate. This high sample rate was necessary as a possible failure of the specimen during static test campaigns up to ultimate load or an emergency shut-down during static or fatigue testing campaigns had to be recorded very exactly and detailed.

To reduce the dead weight of cables for the more than 3,000 measurement channels a CANbus system was used which enabled IABG to save more than 50% of cables.

CONCLUSION

All in all the combination of static, fatigue and functional testing within one dedicated test set up raised a complex set of different requirements which required compromises to meet all the requirements.

The provision of a fitted test set up to perform the A350 XWB EW test in time and quality required the usage of newest design principles and sizing methods and requested a lot more customised solutions then known from former tests.

By using these newest state-of-the-art design methods combined with the substantial knowledge within IABG about full scale testing for more than 40 years it was possible to provide relevant input for first flight clearance and type certification for the new AIRBUS A350 XWB EW.

REFERENCES

[1] Ratier, L. (2013). In: *How to Test Hybrid Aircraft in Fatigue*, Proceedings of the 27th ICAF Symposium, vol. I, pp. 153-162, Brot, A. (Ed.), Paragon Israel – Dan Knassim Ltd. Jerusalem, Israel

Advanced life assessment of the lead crack configuration of the RNLAF F-16 wing damage enhancement test

Frank Grooteman, Marcel Bos

National Aerospace Laboratory (NLR)

Abstract: In order to verify the current estimates for the service life and maintenance requirements of the RNLAF and FACh F-16 Block 15 wings, a durability test has been conducted at NLR on the left hand side wing of a decommissioned F-16 Block 15 aircraft of the RNLAF that had accumulated 4,200 operational flight hours. This durability test aimed to grow in-service cracks of sub-detectable size to a size where they could readily be detected. The main objective of the test was to determine if the ex-service wing contained damage not accounted for in the durability test programme of the 1970s by General Dynamics or in the analyses of Lockheed Martin. Other objectives were to generate data that can be used for an assessment of the current maintenance programme and to establish the most critical locations and the most likely fail scenario. Upon completion of the test the wing had been subjected to a load spectrum that was equivalent to a total of 18,200 RNLAF flight hours. During the test a number of fatigue cracks had initiated and grown, some of them to a significant size. The wing was able to sustain the three limit load cases that were applied after the durability test, however.

A lifing framework was developed at NLR to enhance the life prediction for complex geometries and/or complex loads by improving the accuracy of the applied stress intensity factor solution, the computation of which can be very time consuming. Due to the increase in computer capacity, the better integration of computer aided design and finite element tools, and new numerical algorithms, the computation of a stress intensity factor solution today can be done much more efficiently. A framework was developed with which a crack growth life analysis for a complex geometry and load can be performed much more efficiently, automating the different steps as much as possible, especially the time consuming parts. The framework provides a coupling of an (integrated) CAD tool to define parameterised crack geometries, a finite element tool to compute the stress intensity factor solution and an advanced crack growth tool to compute the life of the structure. The framework was successfully applied to the complex lead crack configuration that occurred in the full scale fatigue test on an F-16 Block 15 wing of the RNLAF, yielding an accurate damage tolerance life estimate.

INTRODUCTION

The Defence Materiel Organisation (DMO) of the Netherlands has tasked the National Aerospace Laboratory NLR to conduct a so-called damage enhancement test on the LHS wing of a decommissioned F-16 Block 15 of the Royal Netherlands Air Force (RNLAF). The wing had a known history and had accumulated 4,200 in-service flight hours. The results of the test programme are relevant for both the RNLAF and the Chilean Air Force (FACh), who shared the programme results with the RNLAF on a government to government basis.

The damage enhancement test essentially was a durability test (and will be referred to as such in the remainder of the paper) that aimed to grow in-service cracks of sub-detectable size to a size where they may readily be detected. This significantly enhanced the teardown inspection programme that was conducted in parallel on the RHS wing of the same aircraft. The main objective of the test was to determine if the tested ex-service wing contained damage not accounted for in the early durability test programme that was performed in the late 1970s by General Dynamics or in the current durability and damage tolerance analysis (DADTA) of Lockheed Martin (LM). Other objectives were to generate data (i.e. crack growth curves for the locations that are most critical) that can be used for an assessment of the current aircraft structural integrity program (ASIP) inspection programme and to establish the most likely fail scenario, incl. an estimate of the associated technical end of life time.

Adamage tolerance analysis was performed for the lead crack configuration of the damage enhancement test. The result of such a damage tolerance life prediction is in particular sensitive for the applied loads and stress intensity factor (SIF) solution. Loads are measured extensively nowadays often even on individual aircraft, as in the case of the RNLAF. In the DADTA use is made of handbook SIF solutions. An overestimation of the SIF by 25% can yield a factor 2 reduction in life, yielding an extensive reduction in operational life. Hence, most significant improvements can be obtained in the applied stress intensity factor solution. For the F-16 lead crack growth analysis therefore accurate SIF solutions were applied. For this an advanced lifing framework was applied that has been developed at NLR to enhance the crack growth life prediction for complex structures and complex loads. The main objective was to automate the different steps as much as possible, especially the time consuming parts such as the stress intensity factor solution (SIF) computation.

F-16 WING DAMAGE ENHACEMENT TEST

The durability test covered 18,200 flight hours (incl. the 4,200 in-service hours of the tested wing), which is somewhat more than the minimum of two nominal lifetimes of 2 x 8,000 = 16,000 flight hours that is specified by MIL-STD-1530C [1]. In order not to miss any potentially critical cracks because of insufficient loading conditions, the test setup was fairly complex and involved the use of 23 hydraulic load actuators and the application of a representative fatigue load spectrum that was based on the 2004 L/ESS spectrum as used by LM in the development of the current fleet structural maintenance plan (FSMP) for the RNLAF. The added advantage was that the observed crack growth rates in the tested wing were representative of those in service. This enabled the translation of the test results to service conditions and provided an experimental validation of the theoretical crack growth curves provided by LM in the DADTA. It is emphasized that the results are not meant for certification purposes, however. The available budget and time precluded the presence of the leading edge flap (LEF) and flaperon (the calculated interface loads have been applied, however) and the pressurization of the fuel tank. The durability test involved a simplification by considering the wing only configuration; the design of a representative wing root support has been dealt with using calculations with the F-16 Block 15 coarse grid Finite Element Model (cgFEM) of LM that is available at the NLR.

The test has been carried out under room temperature ambient conditions. No artificial damages have been applied. Only naturally existing damages, caused by in-service fatigue loading, corrosion, etch pits, tool marks, etc. were considered.

No formal residual strength test has been conducted at the conclusion of the test; three different limit load cases have been applied, however.

Prior to the installation of the test article in the test rig, all wing-specific inspections as called out in the Fleet Structural Maintenance Plan have been performed where possible.

Figure 1 provides an impression of the test setup. The static and fatigue loads on the wing box were introduced by a total of 23 force controlled hydraulic actuators. With these actuators, forces were applied that were representative of the aerodynamic loads, the inertia loads (incl. those from the stores), the LEF loads at the four rotary actuators and the brake, and the flaperon loads at the three flaperon hinges. The simulated aerodynamic loads were introduced at relevant wing stations with two hydraulic actuators per station, to enable simultaneous introduction of wing bending loads and torsion (also called 'pitch') loads. The general layout and the numbering are indicated in Figure 2.

Figure 1. Impression of the F-16 Block 15 wing durability test setup. The yellow frame is the measurement frame to which the six displacement transducers on the wing upper surface were connected.

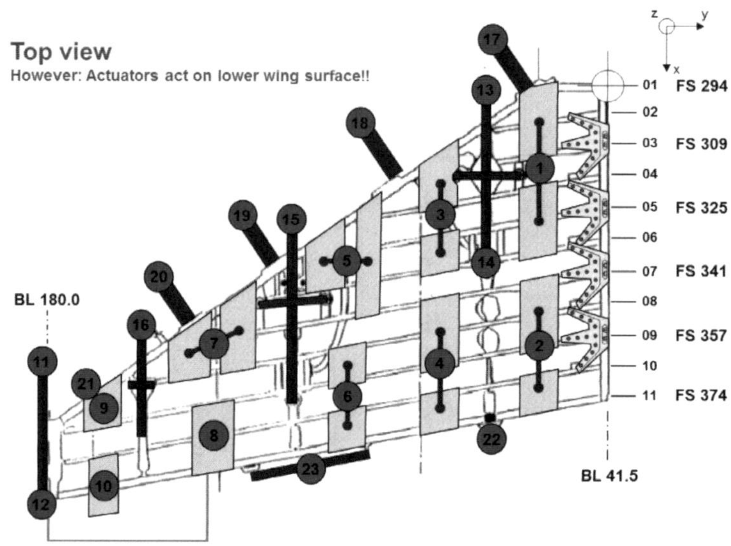

Figure 2. Numbering and position of the hydraulic actuators; the coordinates are in the F-16 reference frame. The red dots indicate the actuator positions, the grey areas represent the load pads and the blue lines represent the dummy store pylons, dummy LEFRAs and whiffletrees.

The following groups of actuators were distinguished:

- Actuators 1 to/incl. 10 were used to introduce the simulated aerodynamic loads and wing inertia loads on the lower surface of the wing. They were connected to the wing by means of bonded load pads, which consisted of rubber pads that were sandwiched between the wing skin and steel brackets. The brackets were interconnected by means of whiffletrees, to ensure a uniform distribution of the actuator loads.
- Actuators 11 to/incl. 16 were used to introduce the simulated aerodynamic and inertia loads from the stores and launchers under the wing. They were connected to the wing by means of steel dummy brackets. The bracket-wing interfaces were representative of the real launcher hardware.
- Actuators 17 to/incl. 21 were used to introduce the simulated aerodynamic and induced bending loads from the LEF. Actuators 17 to/incl. 20 were connected to the wing by means of steel dummy brackets with a wing interface that, in terms of stiffness, was representative of the rotary actuators. Actuator 21 was directly connected to the wing part of the LEF hinge bracket.
- Actuators 22 and 23 were used to introduce the simulated aerodynamic and induced bending loads from the flaperon, using a sliding pin and a dummy bracket respectively.

The actuator loads were measured with calibrated load cells that were placed between the actuators and the wing. The load trains (i.e. all elements between the wing brackets and the test frame, including the actuators and load cells) included spiral washers to ensure proper fixation.

The test article was installed in a closed self-contained steel test frame which had a steel floor on which the hydraulic actuators were mounted. The test article was connected to four steel H-beams that were vertically mounted in the test frame in such a way that they were statically determinate. These Wing Attachment Beams (WABs) simulated the center fuselage carry-through bulkheads of the F-16. It has been verified by means of FEA that this setup was such that the distribution of the wing interface loads over the simulated bulkheads was sufficiently similar to that in the real aircraft. The upper side of each of the vertical beams was connected to the test frame by means of an adjustable link. Each link was instrumented with a calibrated strain gauge bridge. The outputs of the four bridges have been monitored during the durability test to verify the proper distribution of the transferred bending moments over the four simulated bulkheads during the test. The lower flanges of the vertical beams were connected to the test frame through instrumented pins in spherical plain bearings that measured the loads in y- and z-direction. These loads were monitored as well. The shear ties at the front and rear spars of the wing were fixed to the test rig by means of vertical steel straps and the upper sides of the forward and aft WABs were fixated to the test rig by means of rod ends. The reaction loads in these straps and rod ends were continuously monitored with load cells.

The instrumentation on the test article consisted of strain gauges and displacement transducers. The strain gauges were placed such that all relevant load paths were covered.

Prior to the start of the durability test, a series of static commissioning load cases have been applied to verify the correct functioning of the test setup. In addition, strain surveys have been performed regularly during the durability test programme to monitor the quality of the test and to detect a possible change of load paths due to the growth of fatigue cracks. The same load cases have been used for the commissioning test and the strain surveys. The load cases were selected from the set of 211 Limit Load (LL) cases that LM has provided with the F-16 Block 15 coarse grid FEM. They were scaled down to 50–80% of the LL values, depending on the load case.

The flight-by-flight load spectrum consisted of a series of load lines or condition codes, where each condition code represented a set of actuator loads that simulate a point in the sky. The spectrum has been developed and validated using the cgFEM of the F-16 Block 15 (which is sufficiently detailed as to resolve the relevant load paths), the associated set of 211 static limit load cases, the NLR developed tool AELEV-MIL (for the calculation of the operational flight loads [2]) and the NLR developed spectrum generation software CLASS [3]. The spectrum represented a block of 500 simulated flight hours and 412 flights that was repeatedly applied up to the end of the durability test. The 500-flight hours spectrum contained 117,732 load lines, which is in line with the spectrum used by LM in the F-16 Block 50 full-scale durability test [4]. The process to generate the load spectrum is outlined in Figure 3. The input to the process was formed by the set of measured flight parameters for the 2004 L/ESS batch of 412 flights (500 flight hours) that has been used in the development of the current FSMP for the RNLAF fleet.

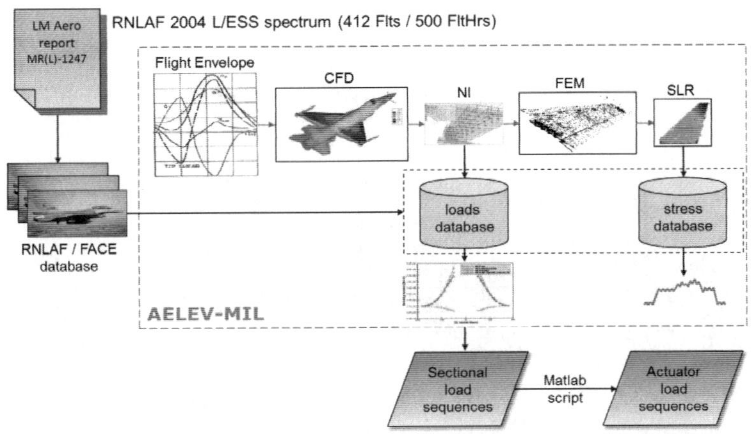

Figure 3. Process to generate the actuator load spectrum used in the test.

The actuator load spectrum included three different so-called marker load blocks to improve the readability of the fatigue crack surfaces; this greatly enhanced the post-test Quantitative Fractography (QF) efforts, see for instance Figure 4. It was verified that the addition of the marker loads does not significantly affect the severity of the load spectrum in terms of fatigue life or crack growth life (and that the small influence is conservative).

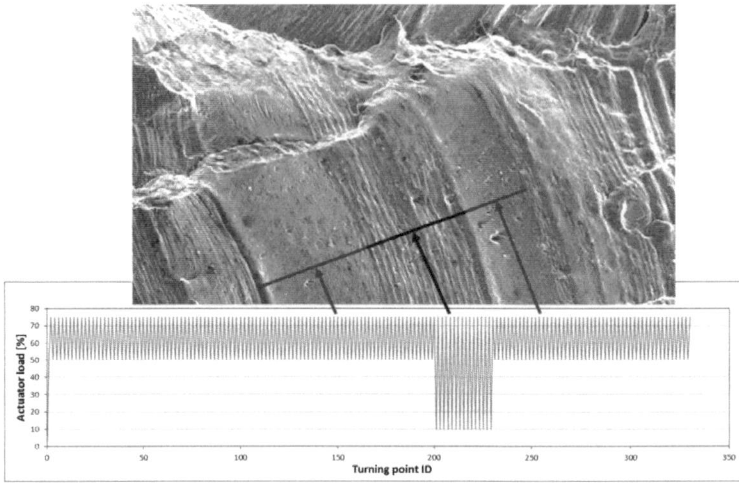

Figure 4. Definition of marker load sequences and corresponding fracture surface pattern.

Figure 5 shows the maximum deflections of the wing tip in upward and downward direction. Measured deflections at actuator #12 (dummy tip launcher aft position) were +435 mm up and -255 mm down, with a total range of 690 mm.

Figure 5. Composite photographs that show the wing deflections during the applied limit load cases "-3g down bending" and "+9g up bending".

During the test a number of fatigue cracks have initiated and grown, some of them to a significant size. The wing was able to sustain the three limit load cases that were applied after the durability test, however.

The lead crack configuration that occurred in the test consisted of a crack in the lower skin starting at the rectangular cutout at Butt-Line (BL) 71, depicted in Figure 6, growing to the spar #6 fasteners and finally to the pylon hardpoint hole, and exactly matched the assumed crack configuration in the DADTA for this location, depicted in Figure 11. During this period, other relevant cracks started at nearby locations in the rectangular cutout, pylon hardpoint hole and in the underlying spar #6, Figure 7, interacting with this crack.

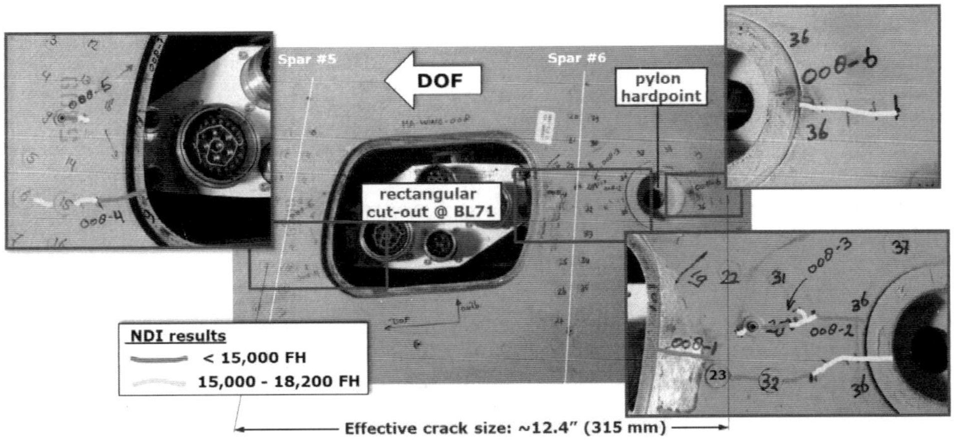

Figure 6. Widespread Fatigue Damage in the area of ASIP CP MA-WING-008: "Lower wing skin, pylon Rectangular Cutout at BL 71, Cutout and Fastener Holes". A total of ten cracks were present after completion of the durability test. The cracks have been highlighted with red (initial phase) and yellow (final phase).

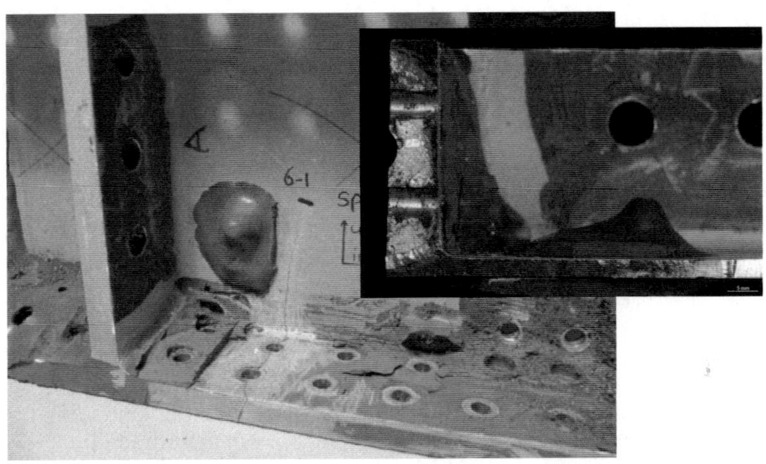

Figure 7. F-16 Block 15 Wing spar #6 crack.

After removal of the test article from the test rig, the upper wing skin was removed at Leeuwarden air force base. Upon return of the wing and upper skin at NLR in Marknesse an extensive detailed visual inspection (DVI) and non-destructive inspection (NDI) was performed and a number of additional cracks were found. The largest of the detected cracks were opened for fractographic examination. Summaries of the most significant findings are provided in Figure 8.

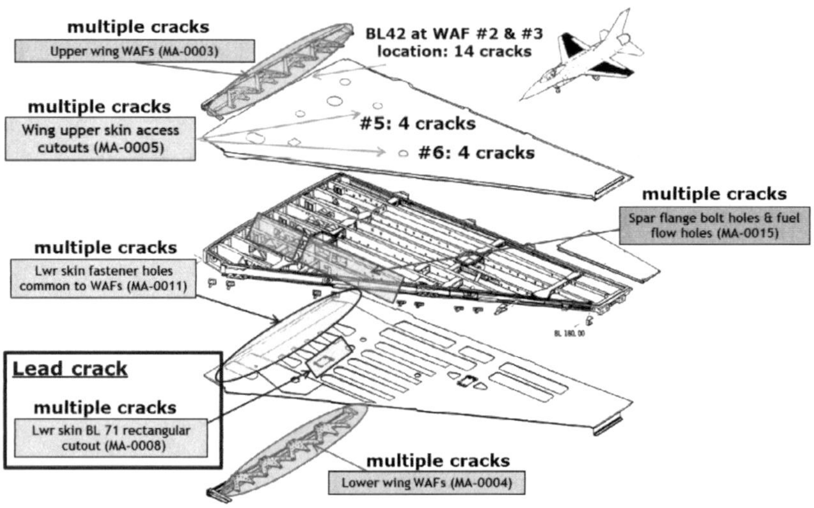

Figure 8. Summary of the most significant fatigue damages detected during inspection after the durability test.

The loads and strains that have been measured during the static tests (i.e. strain surveys and final LL test) and the durability test have been used in a correlation study to verify the following points:

6. the validity of the available F-16 Block 15 cgFEM;
7. the consistency and quality of the applied loads in the durability test;
8. the validity and relevance of the durability test itself, by comparison of the local loading conditions at fatigue critical locations in the wing as installed in the test rig to those in the wing as installed on a real F-16 Block 15 aircraft.

The results confirmed each of the above three points.

LIFING FRAMEWORK

Fatigue is one of the main design drivers for aircraft structures and airworthiness regulations require proof that aircraft can be operated safely. This implies that critical components must be replaced or repaired before safety is compromised. The philosophies underlying the approaches for guaranteeing safety are called Safe-Life and Damage Tolerance. The Damage Tolerance philosophy recognises that damage can occur and develop during the service life of a component. Also, it assumes that cracks or flaws can be present in pristine structures. Safety is obtained from this approach by the requirements that either (1) any damage be detected by routine inspection before it results in a dangerous reduction of the static strength (inspectable components), or (2) initial damage shall not grow to a dangerous size during the service life (non-inspectable components). For Damage Tolerance analysis to be successful it must be possible to predict crack growth during the time until the next inspection or until the design service life is reached.

The DADTA of Lockheed Martin for the F-16 is based on a crack growth model in which the life in cycles or flight hours is computed to grow a crack from an initial flaw up to a critical crack size causing fracture of the structural part or functional impairment (e.g. fuel leakage). The crack growth model uses the stress field in front of the crack tip. It can be shown that this stress field is of the following form, in polar coordinates, see for example [5]:

$$\sigma(r,\theta) = \frac{K_I}{\sqrt{2\pi r}} f^I(\theta) + \frac{K_{II}}{\sqrt{2\pi r}} f^{II}(\theta) + \frac{K_{III}}{\sqrt{2\pi r}} f^{III}(\theta) + HOT \tag{1}$$

where r is the radius denoting the distance from the crack tip and θ the angle with respect to the crack surface. Three different crack opening modes (tension, in-plane shear and transverse shear) can be distinguished with corresponding stress intensity factors (SIF) *K*. The tensile mode-I is by far the most important. SIFs for the same mode may be added (superposition), but not for different modes. For homogeneous, isotropic materials the following relation exists for combined mixed-mode SIF:

$$G = (1 - \nu^2)\left(\frac{K_I^2 + K_{II}^2}{E}\right) + (1 + \nu)\frac{K_{III}^2}{E} \tag{2}$$

in which G is the energy release rate and represents the energy per unit new crack area due to an infinitesimal crack extension. The mode I SIF solution is the most important since the crack in general will quickly change direction perpendicular to the load direction, becoming a mode I case only.

The SIF solution forms the basis of all linear elastic crack growth analyses. More in general, the SIF can be written as:

$$K_I = \beta(a)S\sqrt{\pi a} \tag{3}$$

where the function $\beta(a)$ is a factor correcting for the difference in geometry with respect to the infinite plate and is called the normalised SIF, i.e. $\beta(a) = K_I/K_0$. K_0 is a normalisation SIF which for a remote stress is often selected equal to the infinite plate solution $K_0 = S\sqrt{\pi a}$.

For simple geometries (mostly flat plates with straight cracks) and loading conditions (mostly uniform stress, bending stress or point loads) many (semi)-analytical solutions for mode-I have been generated and collected in handbooks such as [6-8].

Besides simple geometries and loading conditions, most handbook solutions are for mode-I solutions under load control. There exist only a very limited number of mode II and mode III solutions and also only a few so-called displacement controlled solutions, were the applied load is a remote displacement instead of a remote stress.

To apply these handbook solutions to real structures, it has to be simplified in geometry and loading. By means of combining different solutions (superposition), somewhat more complicated geometries and loading conditions can be approximated. A well-known approach is the compounding method [9].

The crack growth life is determined by computing the crack growth during a load cycle from the minimum to the maximum load, causing a change in the SIF, i.e. $\Delta K = K_{max}-K_{min}$. The most simple, but often applied, crack growth equation is the Paris equation:

$$\frac{da}{dN} = C\Delta K^m \tag{4}$$

Other more advanced equations exist, such as the Forman and NASGRO equation. The parameter m in general has a value around 3.0, indicating that *an overestimation of the SIF by 25% can result in a factor 2 reduction in life*, yielding an extensive reduction in operational life. Equation (4) indicates that such an error can be easily introduced by the use of an erroneous normalised SIF solution β and/or by an error in loads. Loads are measured extensively nowadays often even on individual aircraft like in the case of the RNLAF F-16s. Hence, most significant improvements in the life prediction can be obtained in SIF values.

A National Technology Project (NTP) has been performed with the objective to enhance the life prediction for complex geometries and/or complex loads by improving the accuracy of the applied normalised SIF solution. This can be done with the aid of numerical methods like the finite element method (FEM) and the boundary element method (BEM). In this way, more accurate SIF solutions can be obtained for complex geometries and loads. FEM is more versatile (BEM for instance is less suitable for anisotropic materials and non-linear elastic-plastic problems) and widely available and was therefore selected as the preferred numerical method. The calculation of a SIF solution can be very time consuming. Due to the increase in computer capacity, the better integration of computer aided design (CAD)/FEM and new numerical algorithms, the computation of a SIF solution today can be done much more efficient. The calculation of the crack growth life using a SIF solution computed with FEM requires the following steps:

1. Generation of the baseline finite element model, often already available
2. Compute the SIF solution for several crack sizes
    a. Insert the new crack length(s)
    b. Regenerate the mesh, e.g. focused mesh at crack tip(s)
    c. Compute the stress/strain distribution
    d. Post-processing of the results to derive the SIF
3. Importing the SIF solution(s) in a crack growth tool
4. Perform the life analysis

The objective was to automate each of these steps as much as possible, especially the time consuming parts, and couple the required tools. This resulted in a framework with which a life analysis can be performed much more efficiently.

To obtain a much more automatic generation of SIFs the FEM model has to be parameterised, i.e. the crack geometry (and even better also the part geometry) has to be defined in terms of modifiable parameters, allowing automatic changes to the crack geometry, i.e. to simulate crack growth. The right way to do so is by means of coupling the FEM tool with a CAD tool. The combination CATIA and Abaqus was examined for this purpose, both from the same software vendor. A so-called associative interface is offered with which modifications made to the CAD model are automatically transferred to the Abaqus finite element model. However, this only has very limited functionality and is by far not (yet) a viable option for the following main reasons:

- Materials are imported but are not assigned to parts, which has to be done manually
- Only parts are imported but no assembly, which still has to be created in Abaqus
- Boundary conditions and loads are not imported
- There exist no equivalent type for a partition used to model the crack or facilitate the generation of the finite element mesh
- Cracks cannot be modelled in CATIA and transferred to Abaqus
- Modifications made in the CATIA model are not always correctly exported to the Abaqus model and might cause unexpected errors

Nevertheless, the CATIA CAD model can still be used to import an existing geometry of the part(s), serving as a good basis for the creation of the baseline/reference model. Abaqus comes with an internal CAD engine as well that is tightly linked with the finite element model and as such much more suitable. Although limited, it still offers a versatile toolset to construct complicated models. Hence, the integrated CAD engine was applied instead of a CATIA CAD model.

The crack growth life of a structural detail is determined with a crack growth program. Various crack growth programs exist of which the most well-known is NASGRO [10]. A more recent NLR in-house developed crack growth program is Cragro++ [11], written in the object oriented programming language C++, which provides for much better data management, encapsulation, extendibility and maintainability. Existing crack growth tools can perform crack growth analyses for various crack growth laws and models on various simple crack geometries. The available geometries are the most important handbook solutions. To enable more accurate life prediction for complex geometries and/or loads, the SIF solution needs to be determined by FEM (or BEM) and imported in the crack growth tool. The latter can be done in two ways:

1. Calculate the normalised SIF $\beta(a)$ for various increasing crack sizes and import this table of SIF values in the crack growth tool. For intermediate crack sizes the normalised SIF solution is determined by means of interpolation.
2. For more complicated problems a complete table cannot be generated in advance and a SIF solution has to be generated on-the-fly for the crack size(s) at hand. This, for instance, is the case when the crack path is unknown in advance. Another example is crack bifurcation, where a single crack is split in two or more cracks growing in separate directions, which for example occurs in integrally machined stiffened parts, e.g. Chinook frames. A third example is a two- or even three-dimensional crack, such as a corner or surface crack, where the crack front changes during growth, e.g. a crack in a turbine blade. For this, it is necessary to couple the crack growth tool with a CAD/FEM tool. The crack growth program then becomes the driver and requests the CAD/FEM tool to provide a SIF solution for specific crack geometries.

For the above mentioned reasons, a framework, schematised in Figure 9, was built to enable a more accurate highly automated life prediction for complex geometries and/or loads. It consists of the crack growth program Cragro++ coupled with the finite element program Abaqus.

Cragro++ was selected for two important reasons. First, it could be extended with an FEM interface that passes the current crack geometry and reads the resulting SIF solution, for which access to the source code is required. Second, it has no limitations in the number of cracks and load cases, important for realistic cases like the lead crack of the F-16 wing test. Third, the specific Lockheed Martin crack growth model is implemented for life analyses for the F-16, other models can be easily added.

Abaqus was selected, because it comes with an integrated CAD environment that enables parameterised models, which is a prerequisite for automatic SIF computation. Second, it has extensive capabilities for SIF computation with both conventional crack tip elements and the more recently implemented XFEM approach. Third, it has an extensive application programming interface (API) that provides easy access to the FE model for modifications (e.g. new crack size) as well as the result database to extract and process the (SIF) results.

The conventional crack tip element approach gives the most accurate SIF solution. The XFEM approach is not as mesh independent as sometimes suggested, but still requires a sufficiently fine structured mesh around the crack tip that correctly captures the stress field. Nevertheless, for complex geometries and/or crack shapes XFEM remains a very interesting approach, especially when a number of issues with the applied Abaqus 6.12 XFEM implementation are solved in a future version.

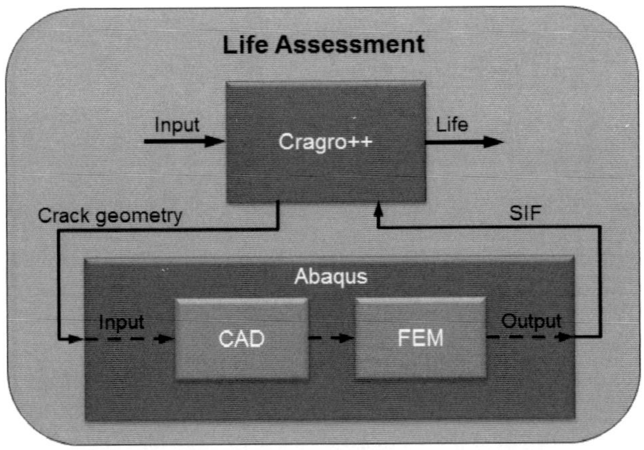

Figure 9. Life assessment framework for complex geometries and loads.

To facilitate an interface with a CAD/FEM tool enabling automatic computation of the SIF solution for a given crack size, a special crack geometry model was implemented in Cragro++. The objective was to provide a very generic interface applicable to as broad a range of problems as possible. Hence, it allows the definition of an unlimited number of cracks, for all three crack opening modes, as well as the definition of constants and variables (expressions). The constants and variables, for instance, allow the definition of any normalisation stress intensity factor K_0, but also the definition of new variables that can be exported to the CAD/FEM tool (here Abaqus), for instance to change mesh properties or dimensions. The latter, for example, can be applied to alter the focused mesh dimensions or to automatically generate a set of SIF solutions for varying geometrical details such as for instance different length over width ratios.

To verify the functioning of the framework it was applied to a variety of available one- and two-dimensional handbook solutions that include different geometries and loads. An example is presented next.

Central slant crack in a rectangular sheet loaded by a parabolic tensile stress

Here a crack in a rectangular sheet is under an angle with respect to the load, being a parabolic tensile stress distribution, as depicted in Figure 10. Since the crack is under an angle it yields a combined mode I and mode II stress intensity factor solution. The handbook solution [7] is depicted in the Figure as black dots. The crack size is kept constant and the effect of the height to width ratio is varied, demonstrating the automatic change in geometrical properties besides the crack geometry, requiring a parameterised geometry as well.

The computed normalised SIF solution using 2D and 3D crack tip elements and XFEM are depicted. The 2D crack tip element solution coincides with the handbook solution that assumes a plane stress condition. The 3D crack tip solution is somewhat higher for the mode I SIF, but the mode II solution shows a larger deviation caused by the more realistic non plane stress condition. The mode I XFEM solution is somewhat higher than the crack tip element solution, but the mode II solution lies closer to the plane stress handbook solution. Apart from small differences, all three methods thus show comparable results for this case.

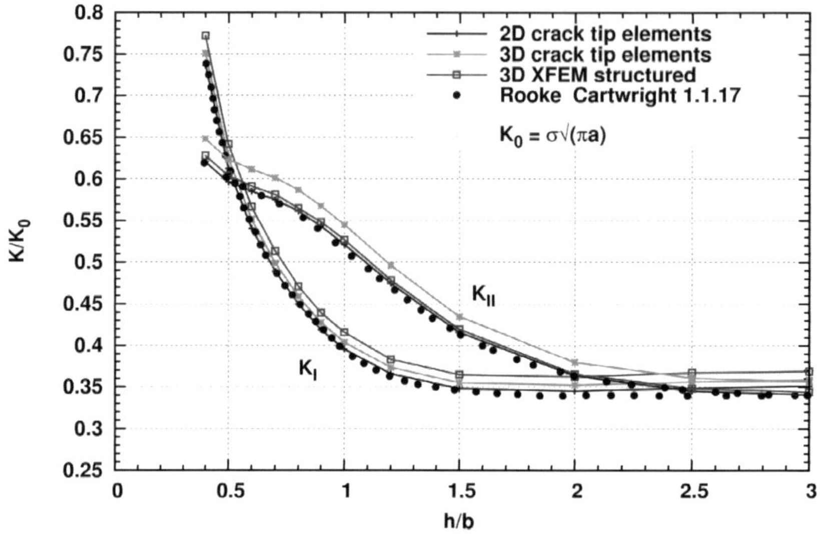

Figure 10. Computed normalised SIF for a central slant crack in a rectangular sheet loaded by a parabolic varying tensile stress.

F-16 WING LEAD CRACK GROWTH ANALYSIS

An advanced crack growth analysis was performed for the lead crack configuration that occurred in the test in the lower skin starting at the rectangular cutout at butt-line 71, depicted in Figure 6, and compared with the wing test results and the damage tolerance life analysis performed by LM for this location. The objective was to obtain a good estimate of the remaining life of the wing structure and to determine possible necessary actions. A second objective was to determine the life prediction capability of such an analysis and demonstrate the highly automated framework presented in the previous section.

Mode I and mode II Stress Intensity Factor solutions were computed for each of the 23 unit actuators applied in the F-16 Block 15 wing test using a coarse grid finite element model obtained from LM that is available at NLR. For this, the model was significantly refined in the region of the lead crack configuration around the rectangular cutout. The analysis resulted in accurate SIF solutions instead of the simple handbook solutions normally applied, taking into account the correct wing structure, the true test loads and the realistic crack configuration including interaction effects.

These 23 Mode I and II SIF solutions together with the load sequence applied in the wing test for all 23 actuators were used in a crack growth analysis, using a special in-house developed crack growth tool Cragro++ [11], to predict the damage tolerance life and inspection interval of the F-16 wing for this most critical location.

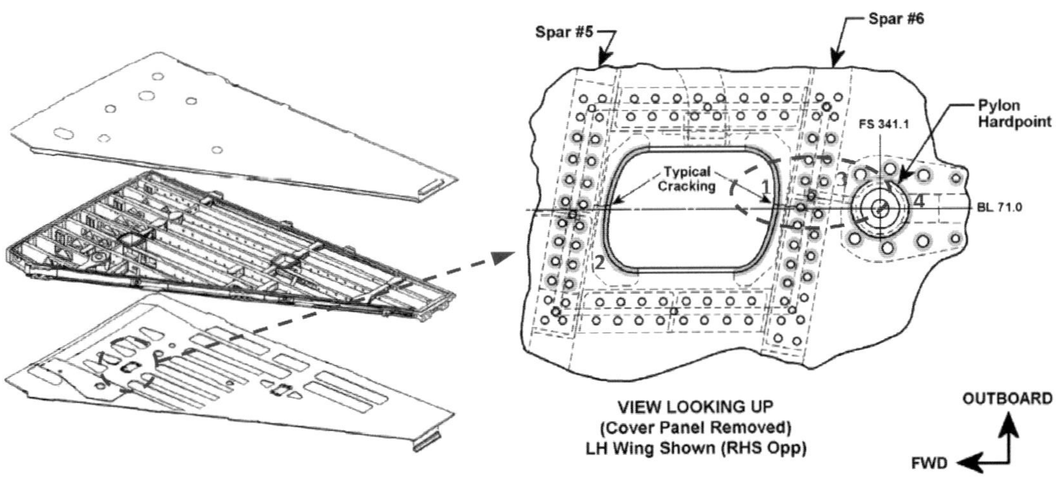

Figure 11. Schematised lead crack location, from DADTA LM.

The computation of the SIF solution for the lead crack of the F-16 Block 15 wing consists of a number of steps. The lead crack configuration that occurred in the test, depicted in Figure 6, consisted of the lead crack in the lower skin starting at the rectangular cutout (location 1 in Figure 11) at Butt-Line (BL) 71 growing to the spar number 6 and finally to the pylon hardpoint hole. During this period, other relevant cracks started at nearby locations in the rectangular cutout (location 2), pylon hardpoint hole (location 3 and 4) and in the underlying spar, depicted in Figure 11, interacting with this crack. The spar crack runs under the fastener, one row away from the lead crack.

During the test non-destructive inspections (NDI) were performed at regular intervals both visually as well as with eddy current. Furthermore, marker load sequences were applied after every block of 500 flight hour (FH) leaving a characteristic striation pattern (marker bands) on the fracture surface that allows the crack growth to be tracked after the test by means of fractography. Figure 12 depicts the NDI (open symbols) and fractography data (closed symbols) that was obtained for the lead crack configuration. NDI data was only available after the crack was detected. Fractography data was only available when the marker bands were visible on the fracture surface.

Only one marker band was found for the spar crack together with the final crack length, from which the most likely crack growth line was constructed (black line), also supported by additional crack growth analyses. The crack sizes are plotted on a logarithmic scale resulting in more or less straight crack growth curves indicating exponential crack growth, apart from the NDO data close to breakthrough of the lead crack. Exponential crack growth has been frequently observed, by for instance DSTO and also NLR, especially for fighter aircraft spectra, see for instance [12].

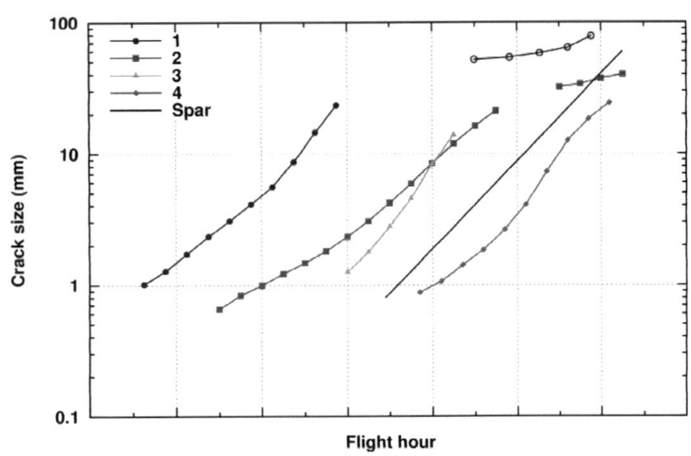

Figure 12. F-16 Block 15 wing test lead crack configuration NDI (open symbols) and fractography data (closed symbols).

As stated before, a small error in the SIF solution yields a much larger error in the computed life. Therefore the objective here was to compute an accurate SIF solution for the lead crack by means of finite element analyses. The left-hand-side wing part of the coarse grid finite element model of the F-16 Block 15 was used as a starting point. Besides the SIF computation, the wing model was applied in the definition of the wing interface in the test and the actuator loads. For each of the 23 actuators the displacement and stress/strain field for the wing in the test frame was computed for a unit load. Only part of the wing model was applied in the SIF computations, consisting of a region around the lead crack configuration. The corresponding sub-model is depicted in Figure 13. The applied load on the edges is a displacement load since the sub-model is constraint by the surrounding wing structure. The edges of the sub-model were selected sufficiently away from the crack region, such that the displacement loads at the edges were not influenced by the crack development. For each of the 23 unit actuator loads the corresponding displacement boundary conditions on the sub-model edges were determined from the computed displacement fields for the whole wing model. For some actuators, partly acting in the area of the sub-model, additional forces were active as well. An example of the boundary conditions and internal forces is shown for actuator 1 in Figure 13.

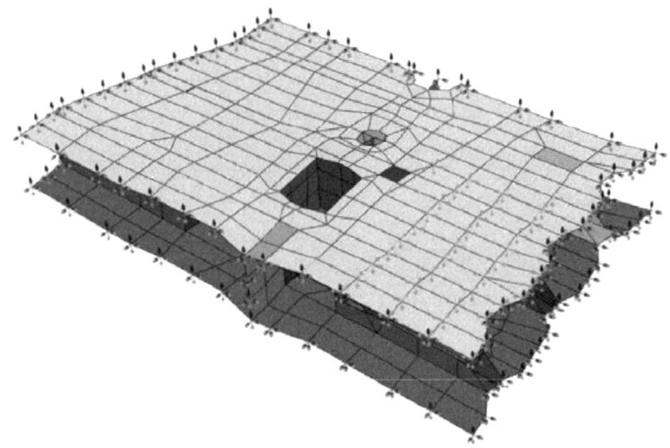

Figure 13. Example cgFEM sub-model of lead crack region including displacement and actuator loads.

The coarse grid finite element model sufficiently represents the stiffness of the wing structure, but cannot represent the local stress/strains field, which is required for an accurate SIF computation. Hence, the model was refined around the crack locations, for the skin and spar, yielding an accurate model of the true geometry of the rectangular cutout together with a refined mesh. The thickness and offset values of the coarse grid model were hereby preserved. Apart from much more elements, 6900 instead of 37 elements for the skin and 2607 instead of 10 for the spar, the mesh is also build from quadratic instead of linear shell elements further improving the accuracy. The quadratic mesh is also required for the SIF computation using crack tip elements to accurately represent the stress field around the crack tip.

Figure 14 shows the combined refined and coarse mesh of the final complete sub-model. Special tie-constraints were applied to connect the refined meshes with the coarse grid model. For the skin this consisted of a tie-constrain to connect the outer edges, tie constraints to connect the skin to the underlying spars at the areas were the fasteners are located and tie-constraints to connect the skin and spars at the fasteners were cracks were located. These last constraints were updated once the crack tip arrives at the fastener hole. The tie-constraint for the corresponding fastener hole then was released to simulate a non-load transferring fastener.

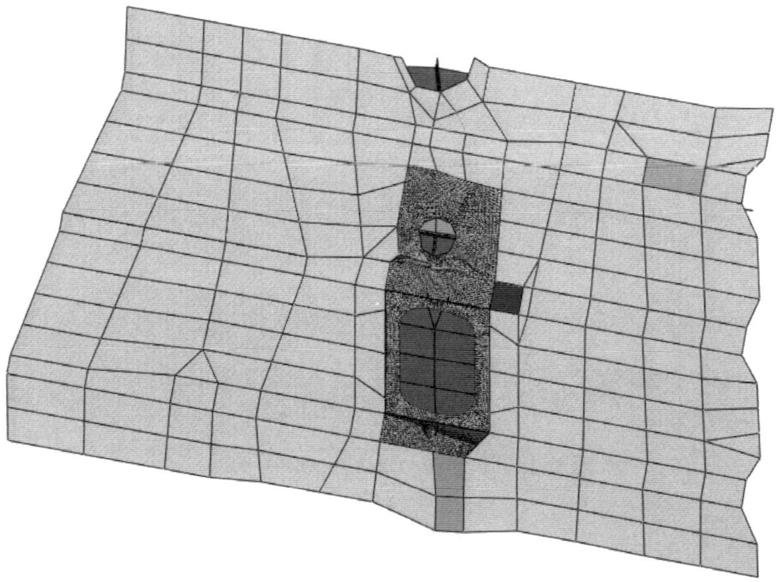

Figure 14. F-16 wing sub-model with refined mesh in the lead crack region.

The refined sub-model was applied in the SIF solution computation with the lifing framework. From the crack growth data obtained in the wing test, as depicted in Figure 12, various crack configurations were determined for different stadia of the lead crack. The crack configurations hereby consisted of the lead crack size that grows from the rectangular cutout to the pylon hard point and the corresponding crack sizes of the other three skin cracks and the crack in the spar. For this the exponential behaviour of the crack growth curves was applied allowing a realistic determination of the various crack sizes for a given lead crack size. The maximum crack advance was limited to 2 mm to get a detailed SIF solution. At start, the SIF solution for the lead crack is not influenced by the other cracks, however, for larger crack sizes it does. The crack paths for the different cracks consist of straight lines coinciding as much as possible with the crack paths observed in the wing test. Around each crack tip a focused-mesh was created that consist of a ring of crack-tip elements, surrounded by a number of additional rings of rectangular elements for accurate computation of the SIF solution.

Fastener holes were not modelled, because of the interference fit of the fasteners in the real structure causing a beneficial plastic deformation of the skin material around the fastener hole. The hole can therefore be seen as a filled hole. When the crack arrives at a fastener hole, the tie-constraint connecting the skin and underlying spar is released assuming no more load transfer is possible by the fastener. This is the case for the lead crack, the second crack in the left lower corner of the rectangular cutout and the crack in the spar. For the spar crack it is assumed that the fasteners keep on carrying load until the flange is fully cracked.

Because the fastener holes are not modelled, the crack is grown across the fastener hole, while in reality the crack will need time to reinitiate at the other side where it will restart as a corner crack. Hence, the computed SIF solution is somewhat conservative at these locations.

Since the finite element model was built from shell elements, all cracks are through cracks. However, the SIF solution is corrected for the fact that the lead crack starts as a corner crack by using the ratio of a known SIF solution for a corner and through crack. The crack becomes a through crack when the size equals the thickness of the skin.

Figure 15 depicts the computed normalised mode I and II SIF solution for the skin cracks for actuator 7, the colours refer to the crack locations as indicated by the insert. Each dot in the plot represents a separate finite element analysis, requiring: automatic crack advance, regeneration of the (focused) mesh, solution of the problem, computation of the J-integral and extracting the results. The overall behaviour shows a decreasing SIF solution due to loss of stiffness of the structure and load transfer to the spar. Crack interaction effects and release of fasteners when the crack arrives at a fastener hole causes a slight increase in the SIF solution.

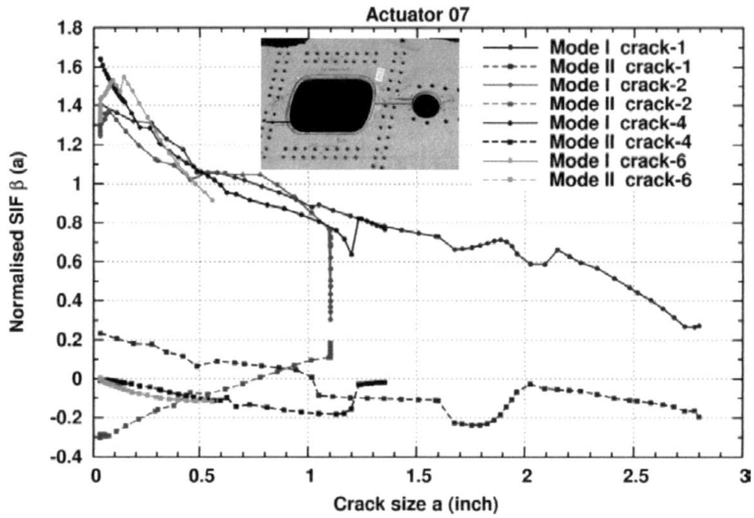

Figure 15. Normalised mode I and II SIF solution for skin lead crack configuration, actuator 7.

Figure 16 and 17 depict the mode I, respectively, mode II SIF solutions for all 23 actuators, showing a more or less similar behaviour. All SIF solutions are normalised with $K_0=\sigma\sqrt{\pi a}$ with a unit actuator load. The difference in SIF level is due to the location of the actuator. The upper curves correspond to the actuators around the tip of the wing (see also Figure 2) and the bottom curves to the actuators near the wing root for which a unit load causes much less stress in the region of the rectangular cutout. Both mode I and mode II normalised SIF solutions are used in the crack growth life analysis discussed next.

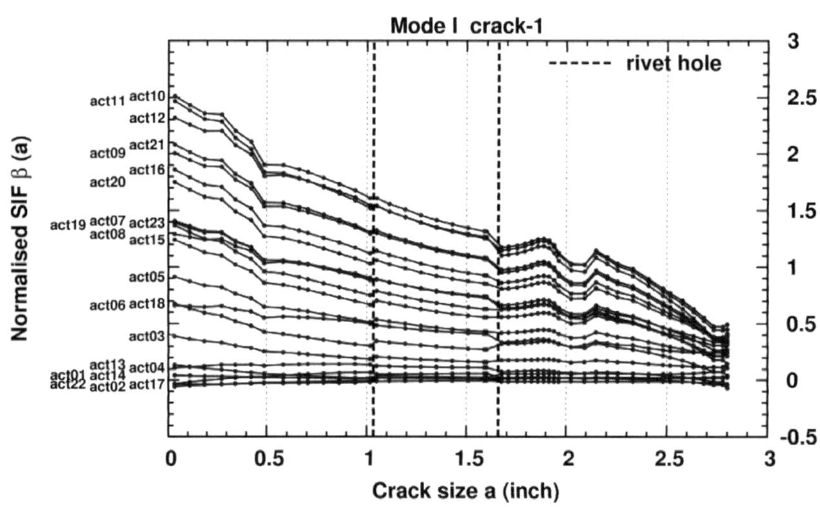

Figure 16. Normalised mode I SIF solution for skin lead crack for all actuators.

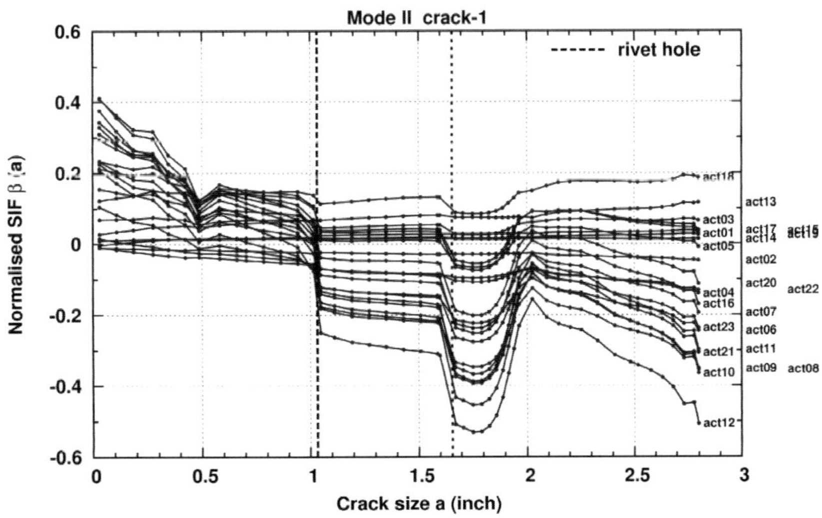

Figure 17. Normalised mode II SIF solution for skin lead crack for all actuators.

The RNLAF baseline load spectrum as applied by LM in the DADTA was used to derive the actuator loads of the wing test. The wing test results therefore can be directly compared with the LM DADTA result and also with NLR analysis result. The actuator spectra as applied in the test were used in the crack growth life analysis, however, these spectra contain intermediate points and intermediate load cycles (range pairs) which both needs to be removed. Hence, the spectrum was filtered to remove the intermediate points and rainflow counted to remove the intermediate cycles. Normal rainflow counting destroys the order of the cycles in the stress spectrum, which affects the history and therefore the retardation effects. Hence, an indexed rainflow filter scheme was applied that preserves the location of the peaks in the spectrum.

The same crack growth model as used by LM in their DADTA was applied, using the same material properties. The main difference with respect to the LM analysis is the applied SIF solutions, which are the finite element based solutions presented in the previous section instead of stress corrected handbook solutions as applied by LM. Also the stress spectrum is directly derived from the actuator load spectrum as applied in the test, which was similar as the stress spectrum applied by LM since both are derived from the same baseline load spectrum. This was checked afterwards.

The results of the crack growth life analysis are presented in Figure 18 (in blue) together with the wing test results (in black) and LM crack growth results (in green). The red line denotes the critical crack size which is defined as functional impairment due to fuel leakage. The following observation can be made:

- From the wing test a damage tolerance life, starting with an initial crack size of 0.05 inch up to functional impairment, of 4100 FH is obtained.
- The LM crack growth result predicts the crack growth life up to the first fastener hole reasonably well, although predicting a somewhat faster crack growth. The crack growth from the first fastener to the second fastener is much slower than observed in the test causing a much longer (un-conservative) damage tolerance life of 10,500 FH up to functional impairment. The much slower crack growth cannot be explained by the neglection of the spar crack, having a limited effect on the crack growth up to the second fastener as can be observed from the NLR analysis results, but most likely is caused by the accuracy of the applied SIF solution and the neglection of the cracked spar.
- The NLR analyses predicts a damage tolerance crack growth life up to functional impairment that matches well with the corresponding life found in the test (without any tweaking), although the shape of the crack growth curve differs. The shape of both LM and NLR crack growth curves do not match the crack growth curve coming from the wing test. The reason for this is that the exponential crack growth as observed in the wing test cannot be represented well by conventional crack growth models. The remaining crack growth life after functional impairment, not of interest for the damage tolerance life prediction, consists of growth to the second rivet hole and up to the pylon hard point. The predicted life is small, whereas in the test a longer life was observed represented by the NDI data, although this data unfortunately could not be fully correlated with the fractography data and may possibly shift somewhat to the left. The most likely explanation for the larger test life is that a large part of the load is transferred to the remainder wing structure not taken in to account in the presented SIF analyses.

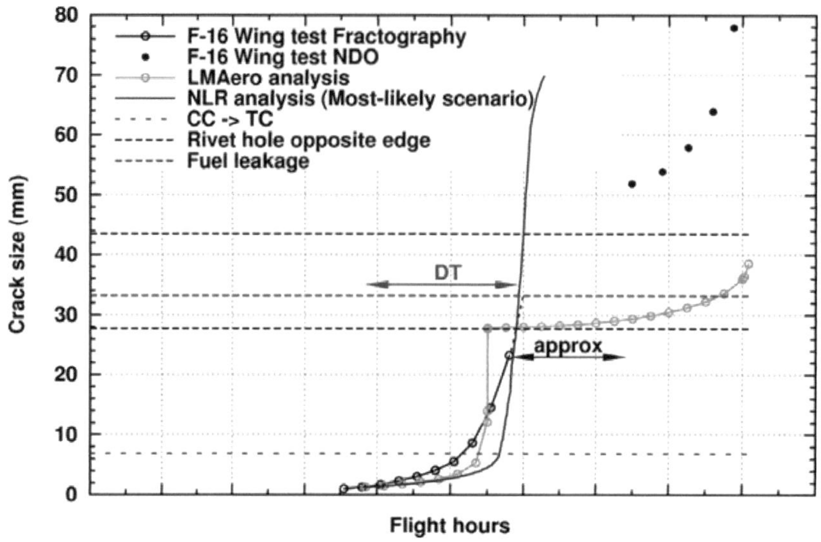

Figure 18. Crack growth life for the lower wing skin rectangular cutout lead crack.

CONCLUSIONS

In order to verify the current estimates for the service life and maintenance requirements of the RNLAF and FACh F-16 Block 15 wings, a durability test has been conducted at NLR on the left hand side wing of a decommissioned F-16 Block 15 aircraft of the RNLAF that had accumulated 4,200 operational flight hours. This durability test aimed to grow in-service cracks of sub-detectable size to a size where they could readily be detected. The main objective of the test was to determine if the ex-service wing contained damage not accounted for in the durability test programme of the 1970s by General Dynamics or in the analyses of Lockheed Martin. Other objectives were to generate data that can be used for an assessment of the current maintenance programme and to establish the most critical locations and the most likely fail scenario.

Upon completion of the damage enhancement test the wing had been subjected to a load spectrum that was equivalent to a total of 18,200 RNLAF flight hours (incl. the 4,200 service hours). During the test a number of fatigue cracks had initiated and grown, some of them to a significant size. The wing was able to sustain the three limit load cases that were applied after the durability test, however.

A lifing framework was developed at NLR to enhance the life prediction for complex geometries and/or complex loads by improving the accuracy of the applied stress intensity factor (SIF) solution. This can be done with the aid of the finite element method. The disadvantage is that the calculation of such a solution can be very time consuming. Due to the increase in computer capacity, the better integration of computer aided design and finite element tools, and new numerical algorithms, the computation of a SIF solutions today can be done much more efficiently. A framework was developed with which a crack growth life analysis for a complex geometry and load can be performed much more efficiently, automating the different steps as much as possible, especially the time consuming parts. The framework provides a coupling of an (integrated) CAD tool to define parameterised crack geometries, a finite element tool to compute the SIF solution and an advanced crack growth tool to compute the life of the structure.

The framework was verified by several one-dimensional and two-dimensional crack geometries. The SIF solution was automatically computed and compared against handbook solutions. The framework can also compute the SIF solution on-the-fly in a crack growth analysis, calling a finite element program from inside the crack growth program when a new solution is required, for instance enabling the SIF computation in case of a change in crack shape or crack bifurcation such as in an integrally stiffened panel.

The framework was successfully applied to predict the damage tolerance life for the complex lead crack configuration that occurred in the full scale fatigue test on an F-16 Block 15 wing of the RNLAF.

ACKNOWLEDGEMENT

The research presented was funded by a shared effort of the Royal Netherlands Air Force and the Chilean Air Force.

REFERENCES

[1] Aircraft Structural Integrity Program (ASIP), *MIL-STD-1530C* (USAF), Department of Defense Standard Practice, 1 November 2005.

[2] Muijden, J. van, Prananta, B.B. and Veul, R.P.G., Efficient aeroelastic simulation in a parametric procedure for fatigue analysis, National Aerospace Laboratory *NLR-TP-2008-587*, April 2009.

[3] Houwink, R., Veul, R.P.G. and Hoeve, H.J. ten, Computer aided sequencing of loads and stresses for fatigue analysis and testing, National Aerospace Laboratory *NLR-TP-2001-287*, 2001.

[4] Welch, K., The F-16 Block 50 Full Scale Durability Test, *Lockheed Martin presentation at ASIP conference in San Antonio, TX*, December 2011.

[5] Schijve, J., Fatigue of structures and materials, Springer, second edition, 2009.

[6] Murakami, Y., et al., Stress intensity factors handbook, Vol. 1 and 2, Pergamon press, 1988.

[7] Rooke, D.P. and Cartwright, D.J., Compendium of Stress Intensity Factors, 1976.

[8] Tada, H., Paris, P.C. and Irwin, G.R., The stress analysis of cracks handbook, ASME press, Third edition, 2000.

[9] Rooke, D.P., Compounding stress intensity factors, the Parthenon press, 1986.

[10] NASGRO, Fracture Mechanics and Fatigue Crack Growth Analysis Software, *http://www.nasgro.com/*.

[11] Grooteman, F.P., Crack growth analysis tool CRAGRO++ - User Manual, National Aerospace Laboratory *NLR-TR-2014-279*, 2014.

[12] Molent, L., Barter, S.A. and Wanhill, R.J.H., The Lead Crack Fatigue Lifing Framework, *DSTO-RR-0353*, 2010.

Practical Experience of Neural Network Based Fatigue Life Monitoring

Jarkko Tikka, Tuomo Salonen

Patria Aviation, Finland

Abstract: Patria Aviation uses a neural network based fatigue life monitoring system for Finnish Air Force F-18 fleet since 2007. After being in operational use for several years, practical experiences of analysis performance and usefulness are now reviewed.

The fatigue life results cover each aircraft, being one tool for Aircraft Structural Integrity Program's individual aircraft tracking task. Based on in-service performance assessment, modelling error is below 20% in fatigue life expenditure, when comparing results to direct strain gauge measurements.

The neural network analysis uses as input flight parameter data, which is recorded by each aircraft, and flight. This data is used to model strain history at critical locations using neural networks. Based on the modelled strain history, fatigue life consumption for selected details can be calculated.

Results of vertical tail attachment area, which experiences a high amount of buffet, indicate that neural networks can model the fatigue life consumption at reasonable accuracy. These results are used to select only a subset of aircraft to high-cost non-destructive inspections. Similar use of results is feasible to fuselage longeron and engine door former, where dynamic loading content is smaller. One detail, where neural network modelling capability fails, is horizontal tail attachment area. In this detail available inputs are insufficient to model structural response.

Based on our experience it can be stated that neural network based analysis brings, at low cost, valuable data to aircraft and location specific fatigue life tracking and Aircraft Structural Integrity Program.

INTRODUCTION

Finnish Air Force (FINAF) maintains a fleet of 62 F-18 Hornets that is planned to provide the main fighter capability over the next decade. The fleet is passing the halfway of the planned service life and it is time to implement preventative modifications, design structural repairs and fleet inspection procedures. It is important to keep up the high availability till the end of the service life of the aircraft.

The FINAF Hornet's ASIP (Aircraft Structural Integrity Program) is based on MIL-STD-1530C. The ASIP has been developed to ensure safe and economical operation of the Hornet fleet up to the planned withdrawal date. An essential part of the ASIP is the requirement to gather Individual Aircraft Tracking (IAT) data. The tracking target is to have 100% coverage of the fleet with a data capture rate over 90%. The purpose of the tracking is to ensure that no aircraft are flown outside the tested and analyzed usage spectrum, and to enable the capability to adjust and allocate inspection and modification needs for all critical structures based on individual usage history.

The FINAF Hornet's ASIP includes two IAT systems: the original system SAFE (Structural Appraisal of Fatigue Effects) developed by Boeing, and the flight parameter based fatigue life analysis called also the NN (Neural Network) based analysis. The flight parameter based fatigue life monitoring system was developed for the Finnish Air Force F-18 fleet, and it has been in operational use since 2007. The system uses neural networks to model strain histories at selected structural locations as a function of recorded flight parameters, allowing flight and location specific fatigue life analysis. State of the analysis at that time has been described in the ICAF 2007 papers [1, 2]. The NN based analysis is a partial fulfilment of ASIP's individual aircraft tracking task. Its appeal is enhanced by capability to estimate fatigue damage retrospectively, for whole fleet and for all flights flown, thus, being able to cover also of late discovered critical locations. More data has been gathered since 2007, and experience of practical use has increased, promoting to look at results from practical point of use.

MONITORING METHOD

The neural network (NN) based fatigue life analysis shown in Figure 1 uses as input flight parameter data, gathered by each fleet aircraft and from each flight [1]. NN's are used to model strain at selected locations, based on the input data. NN is trained with a set of pre-processed flight parameter (input data) and strain gauge data (desired target) from two extensively instrumented Hornet Operational Loads Measurement (HOLM) aircraft. The modelled strain history is used as an input to fatigue life analysis, which in F-18 case is usually ε-N method. The output of the monitoring method is a history independent Fatigue Index (FI) for all monitored locations and flights for each individual aircraft in the fleet.

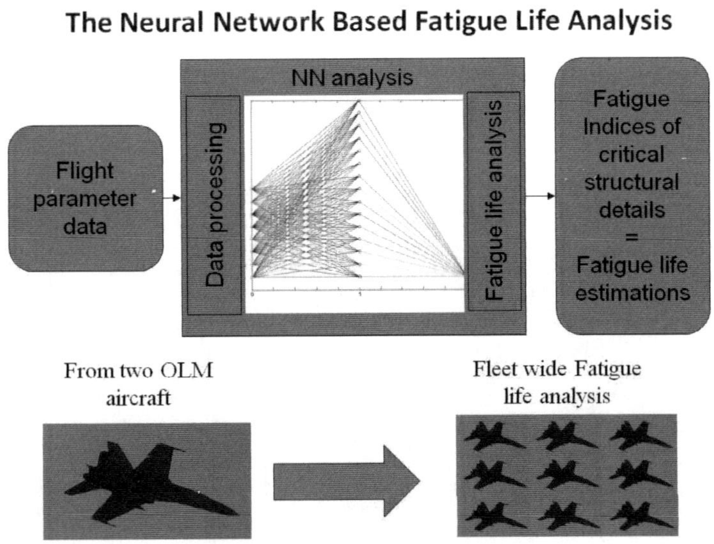

Figure 1. Data flow of the applied monitoring method.

The main benefits of the NN based system compared to traditional strain gauge based IAT are:

- New structural location can be implemented by instrumenting only one HOLM aircraft
- The whole flight history can be analyzed afterwards when the flight parameter data is available
- Maintenance of the system is cost-efficient because no fleet-wide modification is required

Dynamic fatigue content

The analysis environment models dynamic strain history using low-pass filtered steady state condition into which a statistically modelled vibration content is added. The dynamic signal is modelled with method developed by Tampere university of Technology [2]. The results have been very useful in estimating dynamic fatigue content on fleet-wide and tailnumber specific scope. In F-18 case the importance of dynamic damage estimation is increased due to limited or missing dynamic loading in fatigue tests.

Improvements/upgrades during operational usage

Since the original analysis environment introduction in 2007, some improvements have been done and the amount of produced information has multiplied. The number of monitored locations is increased from 10 to 17. Furthermore, the fatigue damage (Safe-life) estimates are calculated for multiple features (e.g 1-3 critical fastener holes in the same structure) for the each location. In this paper the features are called sub-locations.

The core of the analysis environment has, however, remained unchanged, and no major needs for change have been identified.

Current monitoring capability

One of main reasons for NN based analysis development were limitations of the original IAT system. There is an acute need of information about structural integrity of several structural locations. The current NN analysis system includes 17 actively monitored locations (Figure 2), which are handled with 15 virtual strain gauges, and a capability to calculate fatigue indexes for 3 sub-locations (fastener hole, geometric location etc.) per location. Therefore the amount of result data is substantial and comprehensive.

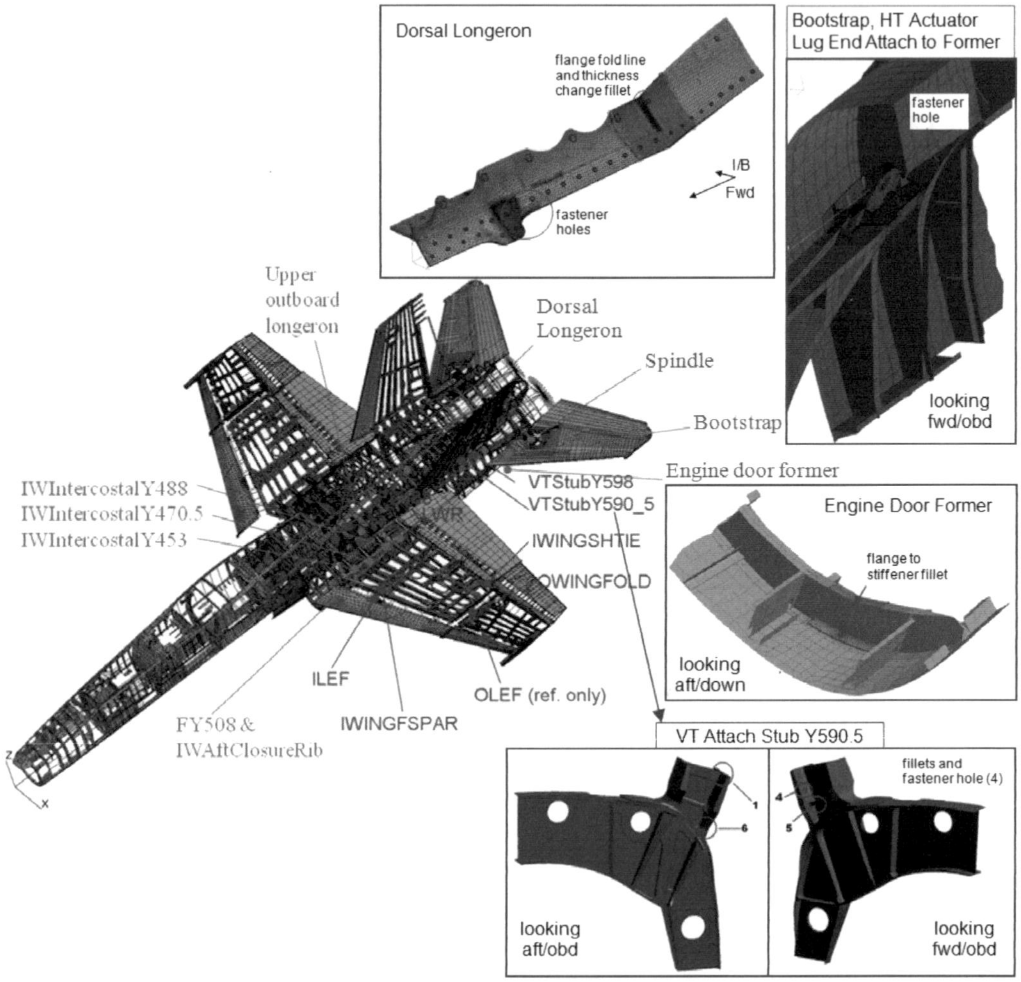

Figure 2. Monitored structural locations, and illustrations of four locations discussed in this paper.

Factors of the analysis accuracy

Analysis from flight parameters to cumulative FI is a multistage process and therefore the overall accuracy is a sum of errors in individual stages. The main sources of error in each stage are:

- Stage: Modelling strain in a strain gauge location (virtual strain gauge)
    - Flight parameter data quality
        - Availability of significant parameters
        - Recording frequency and data quality
    - Neural network training result
        - Training data set coverage
        - NN modelling and generalization capability
        - Strain gauge data quality
    - Modelling vibration components
        - Performance of vibration model
        - NN modelling capability
- Stage: Transfer function from strain gauge location to critical point
    - Distance between strain gauge locations and critical point
    - Non-linearity of transfer function
- Stage: Crack initiation analysis (ε-N method)
    - Material data quality
    - Stress level

For most monitored details the quality of transfer function from strain gauge location to critical point becomes the limiting factor for results accuracy. The availability of significant input parameters at reasonable frequency, like control surface deflections, can be seen as a second error source.

Validation

Compared to the ICAF 2007 presentation and to the set of 22 HOLM flights, a larger quantity of operational flight data has been used for model training and analysis verification. In 2009 an extensive and representative set of 100 flights was selected for NN training purposes. Additional independent set of 200 flights was reserved for verification. [3]

A good and representative indicator of the results' quality is Fatigue Index (FI), because evaluation of modelled strain history is complicated. Cumulative FI is also the result of the analysis. The reference value is the FI calculated from two HOLM aircraft strain gauge measurements. The performance requirement for the analysis was defined in [1] as: A scatter factor of two for individual flight and 1.5 for a set of 25 flights.

Reliable operation of the structural monitoring process is ensured with an in-service maintenance plan. It consists of an in-service assessment procedure including periodic revalidation against recent HOLM data. The assessment has been conducted twice since the original training phase. Based on the assessments, the structural monitoring quality has remained unchanged over the years and no NN re-training or analysis improvements were required on the grounds of the results. Based on in-service performance assessments better accuracy than ± 20% (measured in fatigue life expenditure) was achieved typically with data of two HOLM aircraft. All results stay inside a scatter factor of 1.5 with the exception of one location. [4][5]

An illustrative example of the in-service maintenance assessment is shown in graph form below (Figure 3). The data is from engine door former which is discussed also later in this paper. Points in the figure indicate an individual flight. On the left are shown scatter graphs where NN based FI is shown versus strain gauge based FI. On the right are cumulative damage curves of the both data.

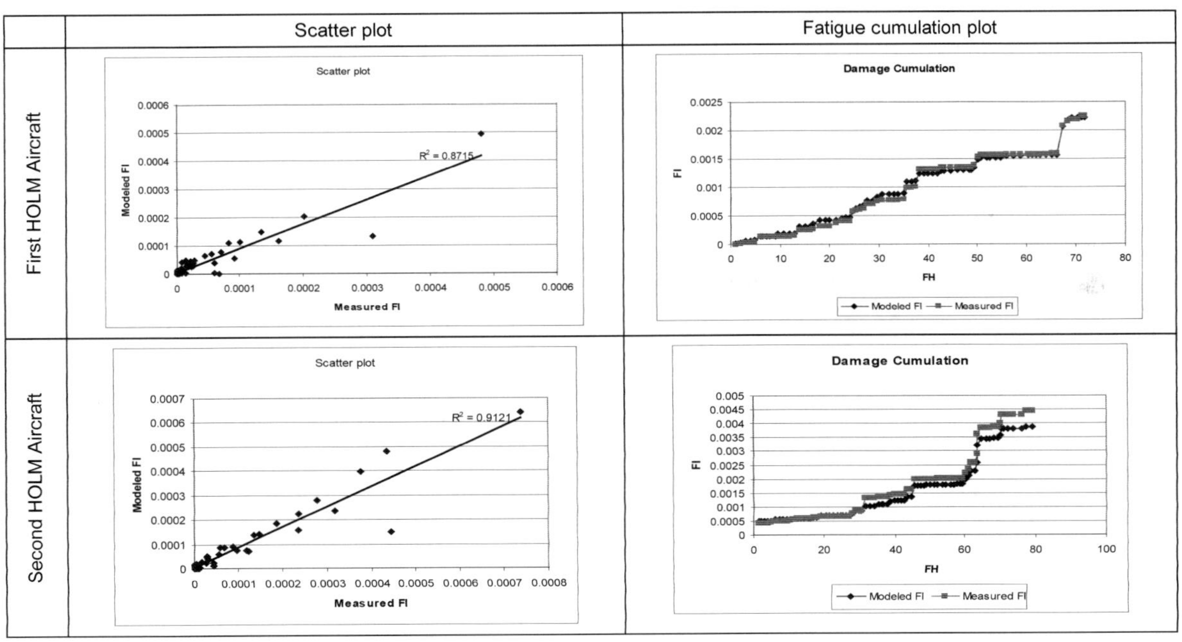

Figure 3. Scatter and cumulative damage graphs from in-service maintenance assessment. [5]

Cost of work

The cost of work required for the ongoing analysis has been minimal compared to fleet wide dedicated strain gauge instrumentation and non-destructive inspections. During the period 2007 to 2015 maintenance and execution of NN analysis has required about a half of one engineer's workload. This supports fleet wide flight parameter based monitoring as a cost effective solution

PRACTICAL USE OF RESULTS

Observations from the long-term monitoring will be discussed using 4 structural locations, which are selected as representative for more generalized conclusions. The locations are briefly introduced here and in Figure 2.

As of year 2000, complete flight history of the whole fleet has been analyzed. That is almost full service life of the FINAF's F-18s except of early flights from 1995 to 1999. Those flights are planned to be analyzed in the near future as they require more manual work due to the changes of aircraft software configuration. Every aircraft records flight parameter data and, thus, the fleet wide analysis meets the 100% coverage requirements of ASIP.

On the other hand, 90% data capture rate is not reached, but typically 50–80% of flight hours of an aircraft have been analyzed. The highest achievable percentage is about 90%, because there is about 10% difference between reported flight hours and recorded data hours. The rest of the missing flight data are explained by absence of flights before year 2000, being covered soon, and data capturing shortages. SAFE's data capturing rate is about 90%, but it is not achievable for the NN based analysis because of requirement for much larger set of flight parameters.

Vertical Tail Stub

The vertical Tail Stub Y590.5 is a part of fuselage in F-18's aft outboard former and it is one attachment point for the vertical stabilizer (Figure 2). This example demonstrates the importance of dynamic fatigue effects, which originate from high angle of attack manoeuvres. The OEM fatigue monitoring system has limited capability to monitor fatigue consumption in such flight conditions.

FINAF usage trends are demanding, and the structure requires inspections near the end of the service life of the aircraft. It would be labor intensive and expensive to inspect all the critical locations over the whole fleet and hence there is call for more efficient approach. Current plan is to inspect approximately a dozen most stressed aircraft in the first phase, and make a decision about expanding inspections on the grounds of that. For the vertical tail stub there are two methods for arranging aircraft in order by fatigue life expenditure:

- SAFE's T*-time which is a sum of time spent in certain angle of attack versus dynamic pressure bins. The method has been developed by the international F-18 user community (FISIF) for assessing the fatigue of F-18 Vertical Tail.
- Vertical Tail Stub FI from the NN based analysis

Based on Figure 4 it is evident that SAFE's Left Wing Root Fatigue Life Expenditure (LWR FLE) has a poor correlation with T*-time, thus, LWR FLE is not a suitable parameter for tail area monitoring. Alternatively NN analysis based FI correlates better with T*-time as can be seen on the right Figure. The outlying aircraft in right Figure is used extensively for flight test purposes (fleet leader), and likely indicates that NN can capture maneuvering loads which are not detected by T*-time. On the grounds of the result, both parameters are suitable for scheduling inspections, and much better than usually applied approach of flight hour based periods.

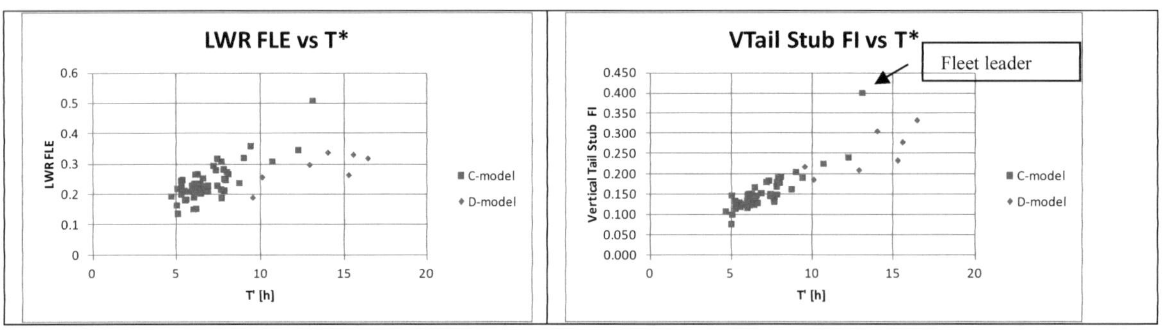

Figure 4. Comparison of Vertical tail stub fatigue parameters.

In Figure 5 the whole fleet is shown in NN based FI order. The red rectangle indicates good candidates for vertical tail area structural inspections, highlighting the possibility for fatigue life consumption based NDI. The graph also indicates that fatigue damage cumulation is not equal within the fleet, although the number of flight hours is high and relatively even.

Traditional way to schedule structural inspections is to start inspecting aircraft after certain number of flight hours. In Figure 6 if shown tail number specific FI distribution (fleet average scaled to 1) when flight hours of each aircraft are fixed to 2000. This could be a realistic time to start inspecting the structure. The graph shows that there is a significantly wide FI distribution and that savings could be achieved by starting inspections based on fatigue life consumption.

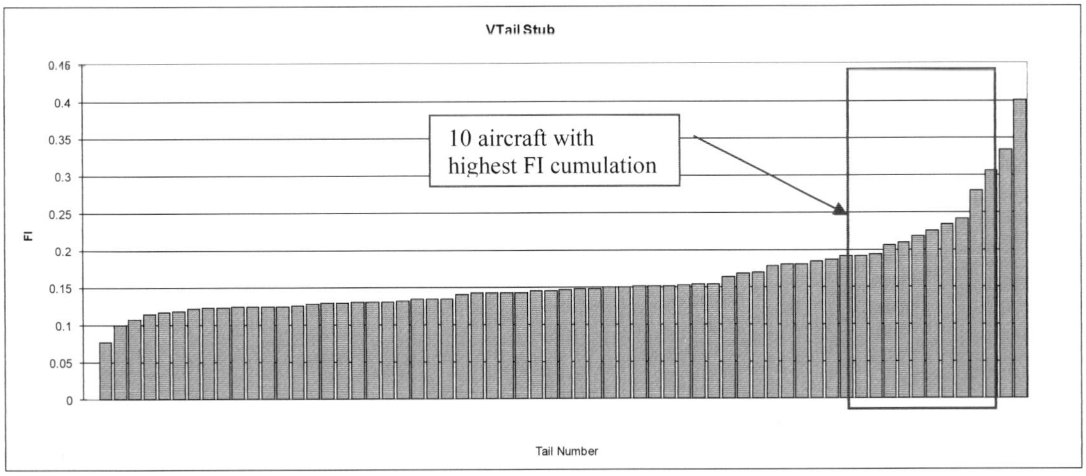

Figure 5. The fleet in Vertical Tail Stub FI order.

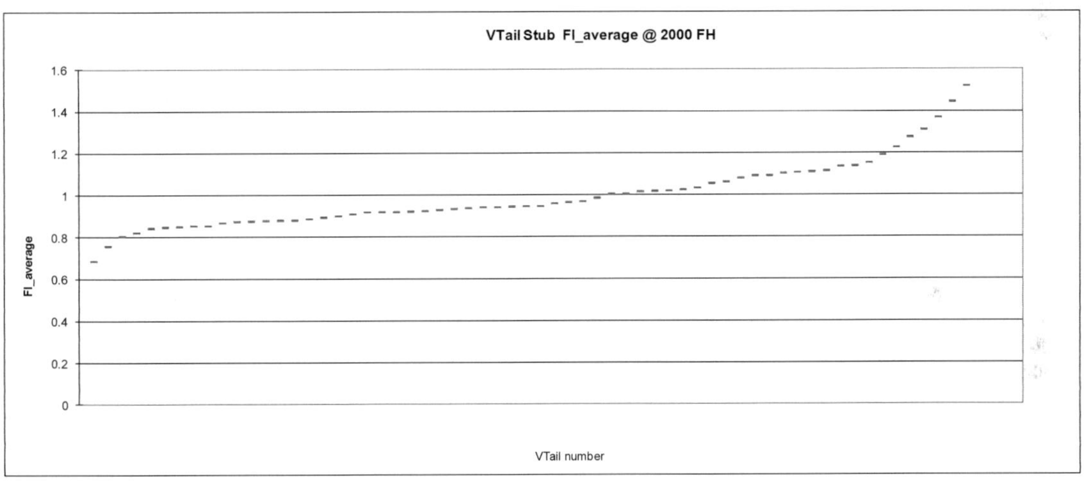

Figure 6. Tail number specific FI of Vertical Tail Stub when the fleet average FI is scaled to 1 and flight hours of each aircraft is scaled to 2000 FH.

Fuselage Longeron

This structure is the upper inboard longeron in the center fuselage (called dorsal longeron in Figure 2). It is a good example of a fatigue critical structure, where it is uncertain if the original aircraft manufacturer's fatigue monitoring system works on the fatigue problem. Currently open issues are, what is the share of dynamic fatigue and should the whole fleet be repaired? The LWR FLE does not measure dynamic fatigue well, and the loads of the longeron likely depend also on horizontal tail loads, for which the original system doesn't have reliable monitoring. The NN based analysis is in this case a useful tool for the problem.

The correlation between LWR FLE and the NN based FI (Figure 7) indicates that the LWR FLE is a moderate parameter for fuselage longeron fatigue. Nevertheless, it could be recommend to use the NN based FI as the base of fleet inspection scheduling (Figure 8). The accuracy of NN modeled strain response in this case is good compared to measured data, promoting that also absolute FI value should be accurate. But the most fatigue critical sub-location is a fastener hole, for which the transfer function produced by finite element analysis is challenging in general, and the accuracy of absolute FI result in figure 8 is questionable.

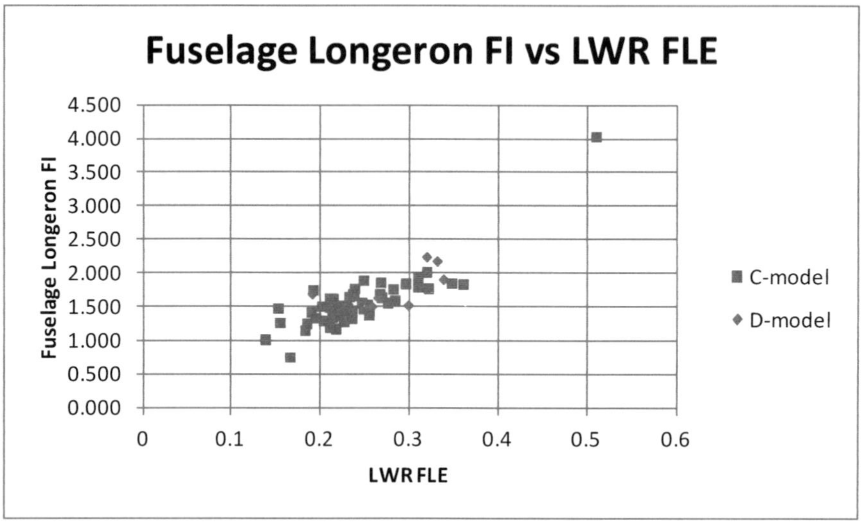

Figure 7. Comparison of Fuselage Longeron and LWR FLE parameters.

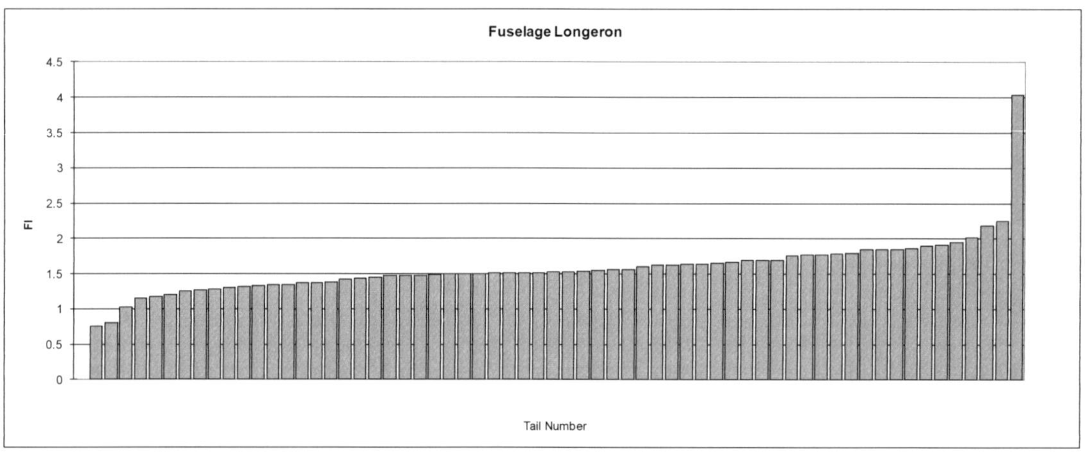

Figure 8. The fleet in Fuselage Longeron FI order.

Engine Door Former

The structure in question is a main former in the engine door (Figure 2). It is loaded by fuselage bending, shear and torsion, caused by horizontal and vertical tail loads, which contain quite heavy buffet loads. It is a fatigue critical structure, but also easily accessible, and a doubler modification has been developed for the most critical area. It can be replaced with a reasonable amount of work.

For this kind of structure the NN based analysis results could be used for inspection and modification scheduling. According to results in Figure 9, the fatigue life expenditure of the structure is not estimated satisfactorily enough by the OEM fatigue monitoring method. If the former is taken into replacement program, the NN based analysis is a good tool to assess replacement need and schedule for individual aircraft, because there is significant dispersion of fatigue life expenditure among the fleet (Figure 10).

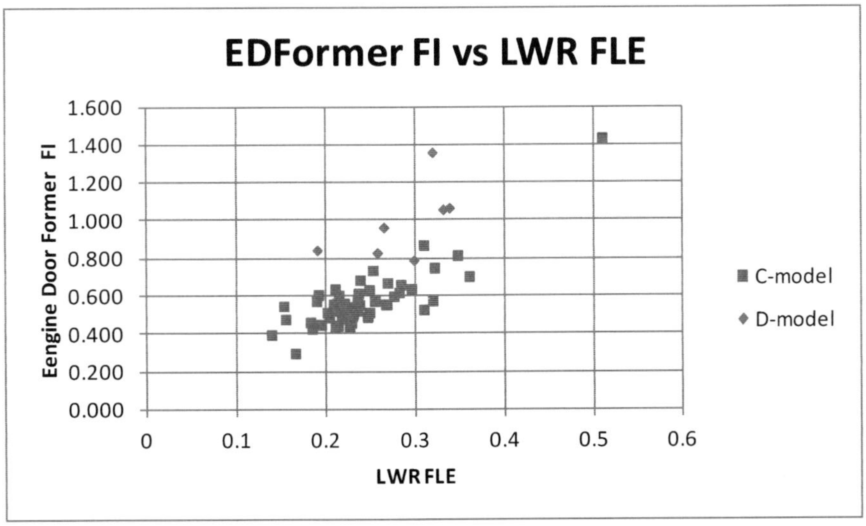

Figure 9. LWR FLE versus FI of the engine door former.

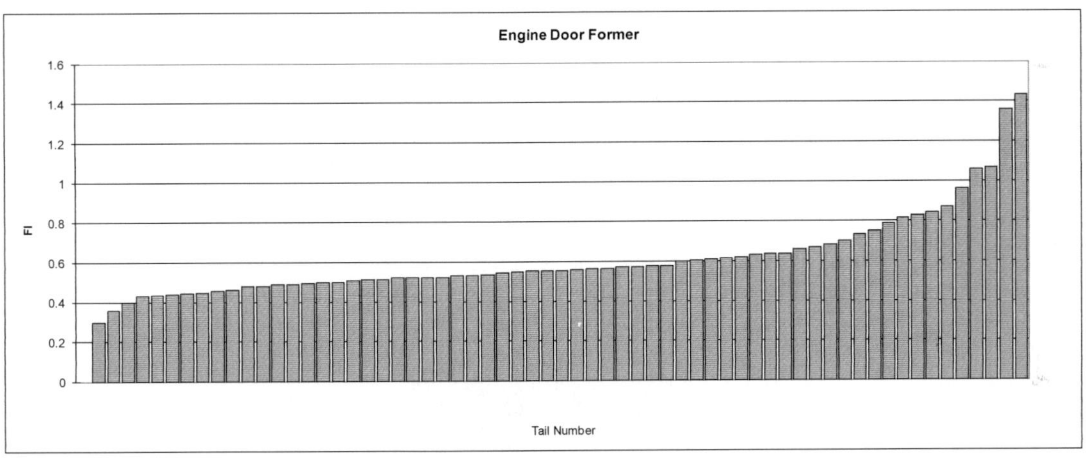

Figure 10. The fleet in Engine Door Former FI order.

Bootstrap

Bootstrap is a part of horizontal tail support structure which forms a stiff box structure between horizontal tail supports on both sides of the aircraft. Bootstrap contains an attachment lug for horizontal stabilator actuator and the critical area is the forward flange attachment to a fuselage former (Figure 2). Bootstrap is loaded by horizontal stabilator actuator loads, which interact with wing control surfaces (e.g. trailing edge flap), and also (through the horizontal tail support structure) by overall fuselage loads from tails of both sides. Therefore the loading of bootstrap is complex. The majority of its fatigue is caused by dynamic damage. It is discussed as an example of the structure difficult to model by the NN based method.

According to the in-service performance assessments and analysis capability assessment comprising NN training evaluation, the bootstrap fatigue life modelling accuracy is degraded. The deviation compared to FI calculated from strain gauge measurements is significant using NN training flight set and independent validation flight sets (Figure 11). There are two major factors that degrade the analysis accuracy. The structure is mainly loaded by horizontal stabilator, and the stabilator deflection data is not recorded in a standard aircraft. Therefore an important NN input parameter is missing, which has a significant effect to the NN analysis. In addition, the transfer function between global strain and critical point is, to a large extent, unreliable.

The root of the error could be found when studying the most damaging flight in the right lower graph. In Figure 12 is shown a short period of modelled and measured (HOLM strain gauge) strain in the flight. There is a fast change from banked turn to opposite side banked turn at transonic speed where the maneuver is executed with differential horizontal stabilator deflections. The NN miscalculates strain on the event because of the lack of necessary input parameters. That event in Figure explains the majority of the error in modeled damage of the flight. Even so, the long term fatigue accumulation accuracy is moderate.

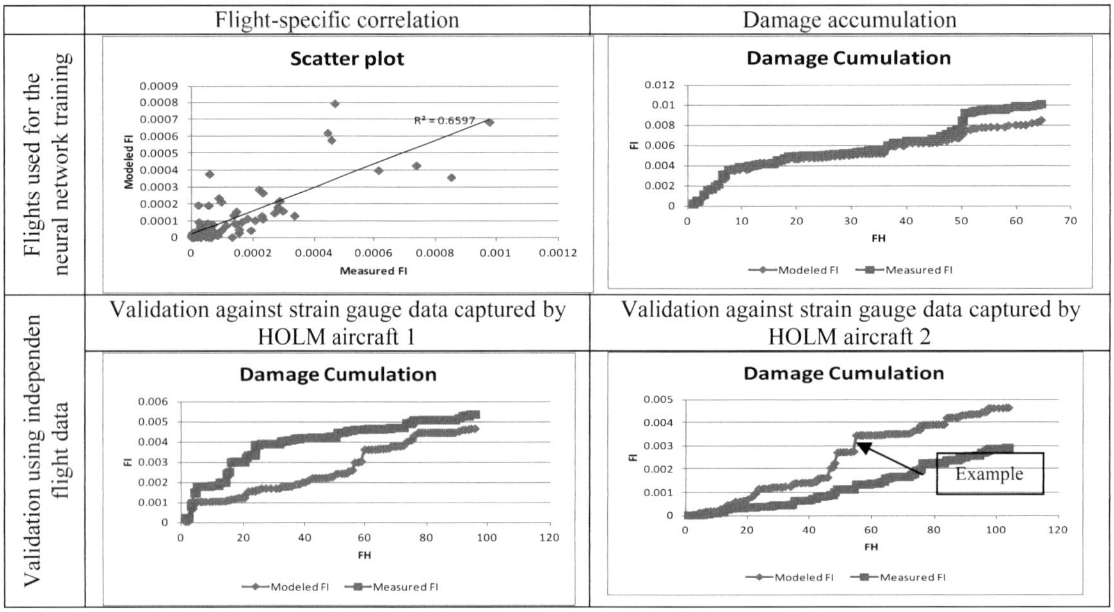

Figure 11. Bootstrap modelling capability assessment.

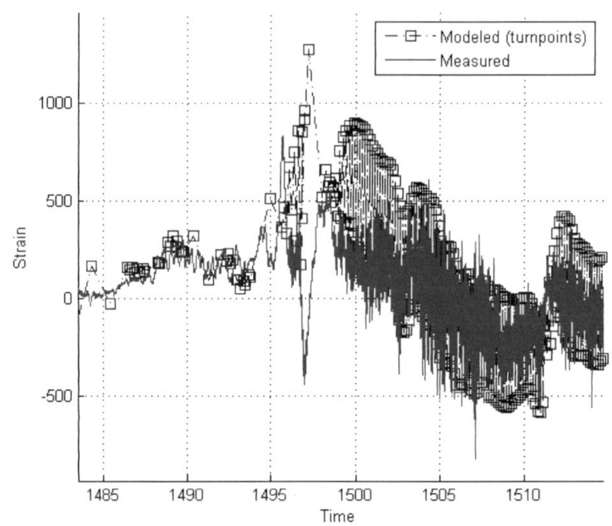

Figure 12. Modelling error of Bootstrap strain.

CONCLUSION

The NN based analysis is shown to be a good tool to produce usage spectra for fatigue critical structures of individual aircraft. Nevertheless, the prediction of aircraft structures fatigue life using flight parameters as input is challenging in the case of agile fighter aircraft. The most limiting phase in the NN analysis chain is the same as in traditional strain gauge based system - the transfer function from strain gauge location to critical detail. It is also important to monitor the analysis accuracy by performing in-service performance assessments at regular intervals.

The extent of the analysis has been increased since the implementation of the NN based analysis in 2007. Currently 17 structural locations are included in the analysis encompassing center fuselage, wing and tail areas of the aircraft, providing important information to the FINAF Hornet's ASIP (Aircraft Structural Integrity Program). The Hornet's primary fatigue life criteria is calculated for wing root by OEM's system. Supplementing that criteria, the NN based analysis provides FLE estimates for several critical areas/structures of each aircraft. The results of NN based system enable adjusting structural inspections and repairs individually.

As the FINAF F-18 fleet goes through structural repairs and modifications, the fatigue life analysis parameters may change. For example, some structural part could be replaced, or its fatigue life expenditure is removed (removed initial cracks), or slowed down (reduced stress level). Future development is to define individual fatigue analysis parameters for each aircraft which take into account life improvement factors of the repairs.

ACKNOWLEDGEMENTS

The authors acknowledge the Finnish Air Force for possibility to realize this work. The support of the Technical Research Center of Finland is highly appreciated.

REFERENCES

[1] Tikka, J., Salonen, T. (2007). Parameter Based Fatigue Life Analysis of F18 Aircraft. Durability and Damage Tolerance of Aircraft Structures: Metals vs. Composites. Proceedings of the 24th Symposium of the International Committee on Aeronautical Fatigue (ICAF), Naples, Italy, 16-18 May 2007.Volume I (lecture papers; L. Lazzeri & A. Salvetti, Eds.) pp 412-426.

[2] Jylhä, J & all. (2007). Modelling Structural Vibrations for Automated Aircraft Fatigue Monitoring. Durability and Damage Tolerance of Aircraft Structures: Metals vs. Composites. Proceedings of the 24th Symposium of the International Committee on Aeronautical Fatigue (ICAF), Naples, Italy, 16-18 May 2007.Volume II (poster papers; L. Lazzeri & A. Salvetti, Eds.) pp 1031-1042.

[3] Malmi, S. (2012). Parameter based fatigue life analysis – Training procedures and analysis capability 2012, HN-S-0062 Rev. C, Patria, Finland, (confidential).

[4] Salonen, T. (2011). Parameter based fatigue life analysis - In-service performance assessment 2011, HN-S-0070, Patria, Finland, (confidential).

[5] Malmi, S. (2012). Parameter based fatigue life analysis - In-service performance assessment 2012, HN-L-1601, Patria, Finland, (confidential).

Full-Scale Fatigue Testing of Two T-38 Wings

David Wieland[1], Jon Cutshall[1], Michael Blinn[2]

[1] *Southwest Research Institute*
[2] *AFLCMC/WLDEJ United States Air Force*

Abstract: SwRI is performing two T-38 -29 wing full-scale fatigue tests with the overarching goal of determining the wing test demonstrated life for two unique usages and to determine the number of wings that will need to be purchased over the next 10 to 15 years. Results of the wing fatigue testing and analyses will provide a baseline for continuing development of field service replacement criteria for the -29 wing to include both Introduction to Fighter Fundamentals (IFF) and Specialized Undergraduate Pilot Training (SUPT) usages. This paper will discuss the full-scale fatigue testing of two T-38 -29 wings.

INTRODUCTION

This paper describes the test set-up for the full-scale fatigue testing of the T-38 -29 wing, testing of the wings and current status of the program. The Sabreliner Team (consisting of Sabreliner Corporation, Southwest Research Institute® (SwRI®), and NexOne) is supporting the Center for Aircraft Structural Life Extension (CAStLE) at the United Stated Air Force (USAF) Academy and the T-38 Aircraft Structural Integrity Program (ASIP) office in performing the T-38 -29 wing full-scale fatigue test, analysis, and reporting efforts as described in the USAF Statement of Objectives.

The overarching goal of this full-scale wing testing is to determine the economic life of the T-38 -29 wings. Results of the wing teardown and analysis will provide a baseline for continuing development of field service replacement criteria for the -29 wing to include both Introduction to Fighter Fundamentals (IFF) and Specialized Undergraduate Pilot Training (SUPT) usages. This test program provides for the full-scale fatigue testing of two T-38 -29 wings, teardown of said wings, damage tolerance analysis update and an economic life analysis of the -29 wing.

Background

The T-38 initially entered USAF service in the early 1960s. How the plane has been flown and the purpose for which the aircraft is utilized has changed dramatically over the years. AETC flies the severe IFF and the moderate SUPT missions. Euro-NATO Joint Jet Pilot Training (ENJJPT) also flies their own mission, similar to AETC's SUPT. Other commands and services fly an assortment of missions (e.g., Air Combat Command (ACC), Navy, NASA, etc.).

As the missions and usages change, a reassessment of the aircraft structural evaluation is required. The reassessment requires full-scale structural testing and teardown activities to provide a complete understanding of how the fleet usages affect the -29 wing economic life. Test and teardown information are used for two key items when managing the T-38 structure: 1) prediction (how the fleet will sustain damage in the future); and 2) prevention (understanding the current state of damage to the T-38 airframe). Ultimately, the scheduling of structural modification programs, predicting wing requirements and parts procurement, accuracy of tech order inspections, and a number of other sustainment activities (such as robust repairs) are all driven by test and teardown information.

It is anticipated the T-38 -29 wing will be in service until at least 2029 with the possibility of remaining in service well past this date. This requires the -29 wing to be sustained and supported for at least the next 15 years. Accurate, updated engineering information is vital to the viability of the T-38 fleet.

The last full-scale wing test of the -29 wing was performed by SwRI in the early 1990s for a now dated version of the ENJJPT usage. The test usage is no longer being flown by the T-38 and was a much less severe usage than is currently being flown by the T-38. The full-scale -29 wing test information is potentially no longer valid and is out-dated (nearly 20 years old). In order for the T-38 -29 wing to be supportable through at least 2029, the full-scale test data and resulting teardown data and analyses need to be updated.

Test Item Description. Two T-38 -29 Wings removed from service were utilized in the test program. Figure 1 shows a drawing of a T-38 wing. One wing is tested to the SUPT usage and the other to the IFF usage. Each wing is attached to a section of fuselage retrieved from the Aerospace Maintenance and Regeneration Group (AMARG). The fuselage section is loaded and constrained so that the proper boundary conditions are applied to the wing attachment fittings. The test item also contains the instrumentation as described in this paper. All fatigue testing was accomplished on a complete wing, including the wing tips, leading edge and control surface. Excluded are the landing gear and mechanical systems.

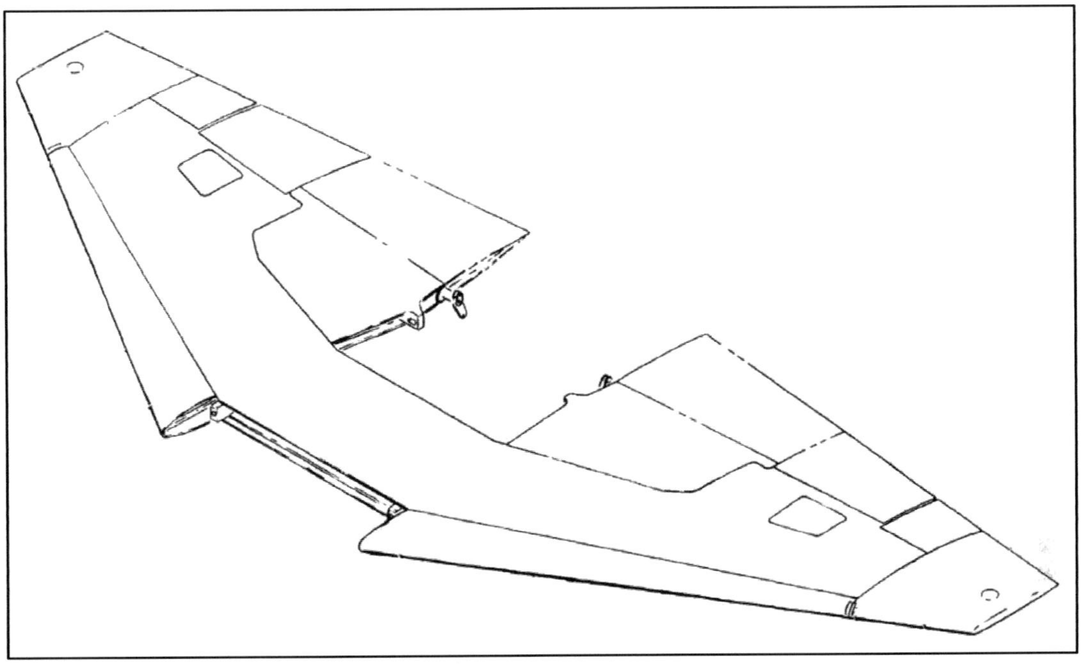

Figure 1. Wing test article.

Purpose of Test. The fatigue test goal was to apply an estimated 5.8 blocks (1000 test hours/block) of simulated flight loads of the severe IFF and 23.25 blocks of the moderate SUPT usage. Since the wings were removed from service the IFF test wing already had approximately 2.2 equivalent blocks of prior usage and the SUPT wing had 0.7 blocks of prior usage. The goal was to accumulate, through testing, approximately two lifetimes for each usage. Fatigue testing to be performed to support the determination of the T-38 -29 wing economic life. The test also could allow for identification of new fatigue critical locations and allow for testing of various repair options.

FATIGUE TEST LOADS

The load conditions selected for full-scale fatigue testing are identical to the loads used in the T-38 Durability and Damage Tolerance Analysis (DADTA) report. Nine basic load conditions are represented for the T-38. The loads are developed for three normal load factors for each representative load case. From these 27 unique load cases, linear interpolation can be used to determine the loads at any given normal load factor for a given load case. The fatigue test loads are based upon the same mission types, mission mix and usage as analysed in the DADTA report. The same procedure used to develop the DADTA loading spectra are used in this test program. The resulting loading sequences are not truncated.

Test Spectrum Development

Two distinct usages will be applied to the wings during the fatigue test. One wing will be tested using the IFF usage while the second wing will be tested to the SUPT usage. The IFF usage was generated based on measurements collected at Moody AFB over the period 2005 to 2007 using flight data recorders installed on T-38C aircraft by Northrop Grumman (NG). The SUPT usage was collected from T-38A aircraft flown at Laughlin AFB using flight data recorders installed and monitored by SwRI. The data collection was during 2004 through the first part of 2005. Figure 2 shows a comparison of the normal load factor (Nz) spectra for each usage per test block.

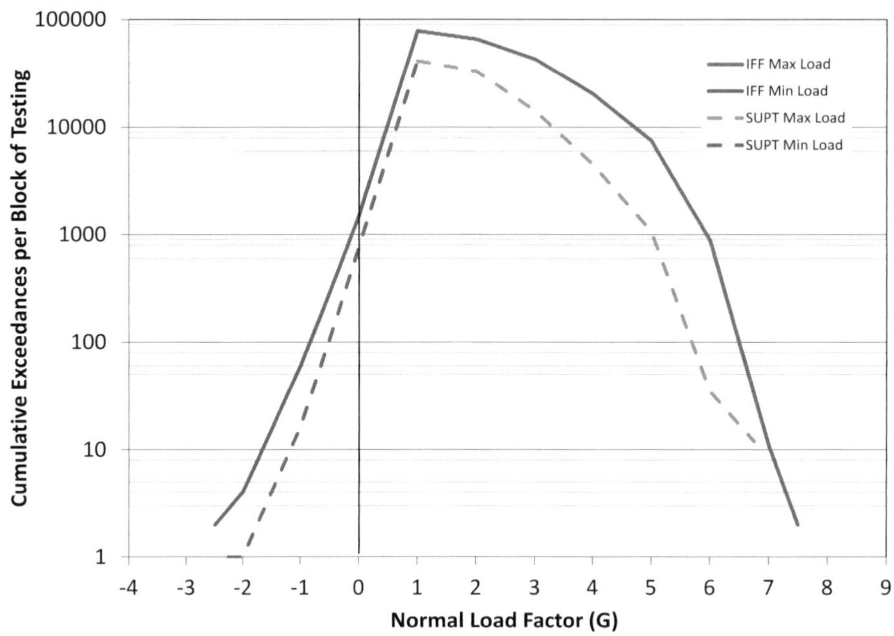

Figure 2. Nz spectra comparison.

Maneuver Spectra Development

The flight loads recorded data from the instrumented aircraft were processed using the AirStats computer program. This program processes flight data and assigns the critical events in the flight data to specific representative flight conditions (RFC). These conditions have been defined for the T-38 aircraft for specific points in the aircraft flight envelope (in terms of altitude and Mach number) along with a gross weight.

Since the loads to be applied to the wings will represent the T-38C configuration, the T-38A flight data collected for SUPT was processed as if it were T-38C data. The T-38A maneuver (i.e., the Mach number and altitude) are assumed to be the same as those of a T-38C aircraft. The gross weight data for the recorded T-38A flights were adjusted to be compatible with the T-38C aircraft. The difference in representative gross weights for the T-38C versus the T-38A aircraft is approximately 2.3%.

AirStats develops a maneuver spectra database for the given usages. From that database, a matrix of discrete Nz occurrences by representative flight condition per test block was produced. This matrix for each usage was required to produce the sequence of events to be applied during the test. This matrix contains Nz occurrences for each mission and phase (climb, cruise, primary, descent, and pattern), flight condition, gross weight ratio (GwR), average gross weight (GW), and Nz by range. It should be noted that these are the same matrices used to generate stress sequences for the T-38 DADTA.

Load Event Sequence Development

The Aircraft Stress Sequence (ACStrSeq) software program was used to create a sequence of load events to be applied during the fatigue test. The program uses the matrices output by AirStats as described in the previous sections to create a flight-by-flight ordering of events for load application. The primary purpose of the ACStrSeq program is to create a sequence of stresses for use in crack growth analysis for chosen fatigue critical locations. A special version of the ACStrSeq program was used to develop this load event sequence. This version outputs each event in the flight sequence as a triplet. For example, if an event in the flight is 4.25g, then a 1.0g event precedes and follows the 4.25g event at the same flight condition.

Culling the Load Event Sequence File

Normally the ACStrSeq program performs the culling of the stress triplets before creating the stress sequence. Since it is only dealing with a single FCL it is relatively simple to peak pick the FCL stress sequence. This culling becomes more complex for a full-scale test article that has many FCLs. There are numerous situations where two 1.0g events are next to each other. For the fatigue tests, elimination of at least one of these 1.0g events would result in reducing the number of test points by about one third. In some cases, the significant load event on one side might be a positive Nz value while on the other side it might be a negative value. In that situation, it might be possible to eliminate both 1.0g events.

However, just the fact that two consecutive events might be 1.0g does not mean that the effect to the structure will be the same. In fact, often times the 1.0g events are for different flight conditions and at different gross weight ratios. Stress computations using these flight parameters will likely result in two different stress values. Also, the wing lower surface has numerous FCLs and elimination of any event must account for the effects to all FCLs.

SwRI performed a systematic culling process of consecutive 1.0g events to remove unneeded 1.0g events from the test spectrums. For the IFF usage, the unculled number of events per test block was 238,968 which were reduced to 157,506 events after culling. Similarly for the SUPT usage, the number of events dropped from 124,803 to 83,008 after culling. The total number of events after culling is roughly two-thirds of the pre-culled number.

FATIGUE TEST SET-UP

The fatigue test set-up for the T-38 -29 wing provides the required mechanical equipment, load control, and load distributions to fulfil the requirements for airframe structural testing. The approach taken to develop the necessary hardware and load control software to accomplish the required structural testing is discussed below. Test item safety during the conduct of the test is also discussed.

Airframe Structural Test Set-up

The general approach for applying loads to the wing is to apply jack loads that distribute the load through whiffletrees glued to pads on the upper wing skin. The pads are glued along the 15%, 44%, and 66% spars. In the area of the forward fuselage, jack loads are applied to whiffletrees that attach to external load straps in order to apply discrete reaction loads to the fuselage. In order to facilitate inspections of the lower wing skin and simplify the test frame, the test articles are inverted for cyclic load application. This set-up allows for easy access to the lower wing skin in order to accomplish periodic inspections, and it allows the test frame to be shorter. All testing was accomplished with a fuselage acting as a restraint fixture. An MTS Aero ST loads controller was used to control load application and achieve load equilibrium. The following sections discuss details of the load fixtures, the load control system, the approach used to generate sets of discrete loads, the distribution of load jacks used to load the airframe, and the structural layout of the load reaction frame.

Load Application Design

Whiffletrees are used to apply discrete loads to the T-38 wing and fuselage. The whiffletree transfers loads into the upper wing skin via bonded pads along spars. The whiffletree arrangement for the left wing is shown in Figure 3. The whiffletree is designed in such a way so that both tension and compression jack loads can be applied to the same pads. This design removes the necessity of having pads glued to the lower wing skin. The whiffletree is also designed such that only vertical shear load is applied to the pads.

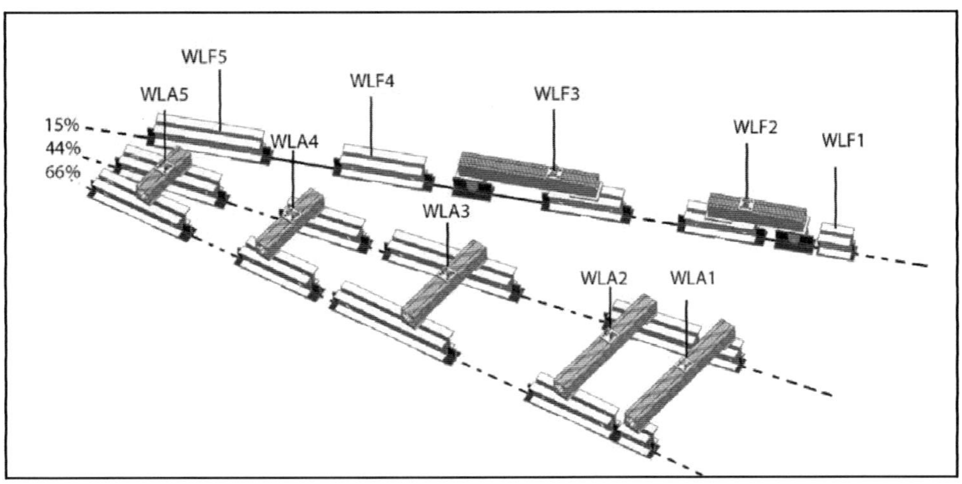

Figure 3. T-38 whiffletree arrangements for the left wing.

Distribution of Load Jacks
24 MTS model 243 (or equivalent) load jacks are used to load the airframe. These load jacks are equipped with load cells; have actuation ranging in stroke from 4 inches to 40 inches, and have load capabilities ranging from 1,500 lbs to 15,000 lbs in tension. All load cells were dual bridge models allowing for A/B compare (load cell comparison) to ensure accurate and reliable load applications. Twenty load jacks are used to provide symmetric loads to the wing. To control the shear load in the fuselage, four jacks are used.

Each side of a wing is fitted with nine whiffletrees and one point load. There were a total of ten load jacks per wing side. Four vertical load jacks were on the fuselage. Where required, jacks are mounted with jack stands to position them at the proper height. Figure 4 shows the whiffletree and jacks installed on the SUPT right wing.

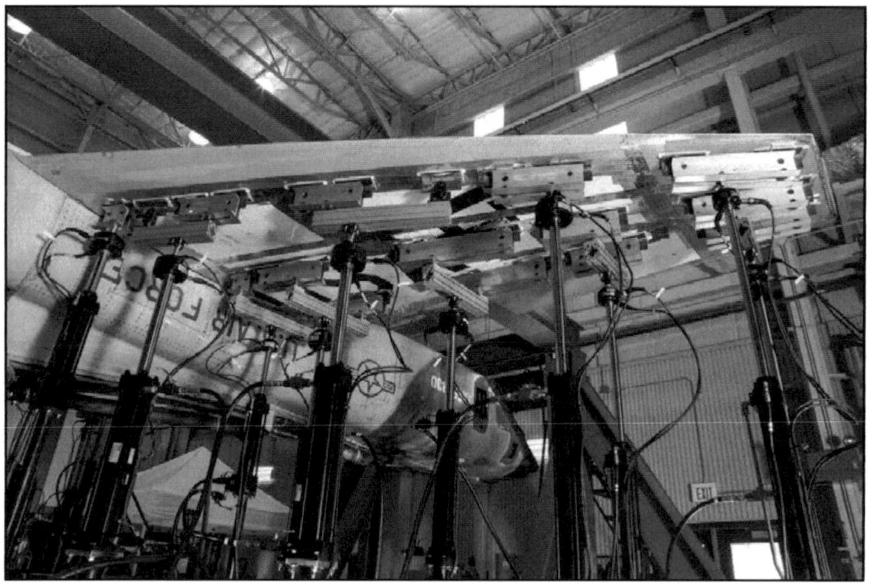

Figure 4. Whiffletree and jacks as installed.

This distribution of load jacks and corresponding whiffletree is capable of very accurately matching the target shear, moment and torque distributions. Figure 5 demonstrates how well the test applied wing bending moments match the target moments.

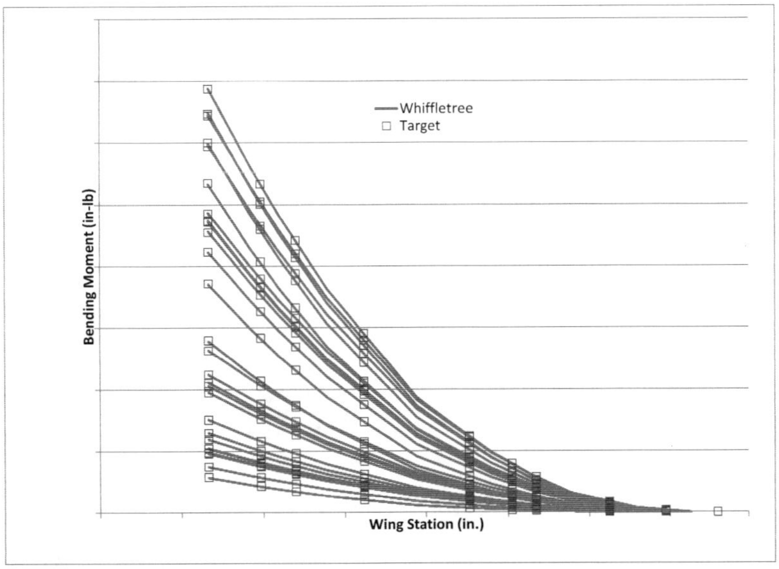

Figure 5. Whiffletree wing bending moment vs. target representative load cases.

Fuselage Constraint Fixture

All testing was accomplished with a fuselage acting as a restraining fixture. This fixture ensured the proper boundary conditions at the wing attachment fittings. The fuselage was constrained at Fuselage Station (FS) 284 and canted FS 505.075. The constraint at canted FS 505.075 was attached to the removable aft fuselage section, and it was then mounted to the test frame to prevent translation (or rotation) in the local roll and yaw axes. It was free to rotate in the local pitch axis. The FS 284 constraint was attached at the upper and lower longeron splice fittings. It was fixed in the lateral direction and cannot rotate about the local roll and yaw axes. Links in the forward attachment allowed the FS 284 bulkhead to translate fore-aft and to rotate about the local pitch axis. The links also constrained the vertical deflection at FS 284.

Load Frame

Two load frames designed to provide rigid support for application of the various load distributions were constructed in the Hi-Bay of Building 249 at Southwest Research Institute, San Antonio, Texas. Each load frame was designed to allow no more than 0.15 inches of deflection at the wing tip jacks under 1.5 times the maximum loads applied to the wing. Structural members were predominantly W12x79 beams with smaller auxiliary members for stiffeners made of W9x31 beams. The general arrangement of the load frame, with the airframe installed, is shown in Figure 6. The top of the fixture and the beam that the fuselage constraint fixtures are attached to are easily removable. This design feature allows for the easy removal of the wing and or fuselage if required. Figure 6 shows a fuselage being installed in one of the test frames.

Figure 6. Fuselage in the test frame.

Test Item Safety

Test item safety is a major concern with the force capacity that exists with the proposed electro-hydraulic test equipment. Two systems work jointly to provide the necessary measures and ensure that airframe overloads do not occur and that the system is gradually brought to a gentle rest state during loss of pressure or signal abort. The Aero ST controller monitors the feedback from dual load cells on each of the load jacks. The A/B compare feature ensures that the two load cell readings are within specified limits. This feature allows for detection of one bridge of the load cell going bad and pausing the test so the load cell may be repaired or replaced, ensuring accurate load application. The AeroPro software allows monitoring of all load and instrumentation channels for exceeding pre-set limits on all data channels. Pre-set lower and upper limits were specified on all data channels, and a number of controller predetermined actions were assigned when the limits are exceeded. Predetermined actions available are disable, indicate, interlock, program interlock, program stop, program hold, and station power off. Hydraulic safety components in the system are the Series 293 Hydraulic Service Manifold and the Series 297 Overload Protection Modules. The service manifold is a pressure and flow control regulation device that is capable of controlling pressure to test stations independent of the hydraulic power unit pressure. The overload protection modules were equipped with load limiter valves. The modules were capable of accurately limiting tension and compression output forces of the actuator to which they are coupled. They were used to limit the maximum force capable of being produced at any test article reaction point. Load abort valves were also used on each of the units in order to allow controlled bleed down of the load jack for a graceful shutdown when loss of pressure occurs.

INSTRUMENTATION

The instrumentation employed during the conduct of the fatigue test was grouped into three categories. There are a primary load control set of instrumentation, a set of structural strain gauges, and a set of deflection transducers. All of these sensors were monitored and recorded during the conduct of the tests. The AeroPro software and the Aero ST test control system were used for performance of the fatigue test. The AeroPro software provides coordination between the Aero ST test control and a data acquisition and test monitoring system. The Aero ST control system includes the Aero ST controller and a PC with AeroPro software working together as a tightly integrated system. A 100 Mb Ethernet connection linked the PC and the controller, providing fast, reliable communications and data transfer. Closed loop control was completely managed by the digital controller that contains signal conditioning and data acquisition modules. The Aero ST software provided test monitoring at remote stations and display of test data. The local area network provided a high-speed link to the Agilent Technologies data acquisition subsystem.

Primary Load Control

Each of the 24 load jacks were equipped with dual bridge MTS Model 661 axial load cells or similar equivalents to accurately indicate the applied load and provide reliable backup if problems with one of the load cell internal gauges develops. The Aero ST provided closed loop control. The Aero ST outputs a load control signal and the load cell provides input back to Aero ST on the applied load.

Strain Gage Instrumentation

Two basic types of strain gauges are used: Axial and Rosette. As the name implies, axial strain gages were used to measure uniaxial strain (e.g. along spar flanges). Rosette strain gages are three axis (axial) gages with two orthogonal axes and one at 45 degrees. These gages were generally used on panels or spar webs where biaxial stresses are anticipated (e.g., wing spar webs, skin panels, etc.). All Rosette and axial strain gages used were manufactured by Vishay Micro-Measurements, Inc.

The SUPT wing has 64 strain gages installed and the IFF wing was more heavily instrumented with 136 gages. The SUPT gages are a subset of the IFF gages. During testing the applied load, measured strains and deflections were recoded for each load event. All gages on the left wing have an odd number. Subsets of the left gages were repeated on the right side of the wing.

Strain gage locations were determined by Northrop Grumman coordinated with their finite element model (FEM) mesh. These locations allowed for easy comparison to the FEM for overall model verification and accuracy. Northrop Grumman provided Mylar templates with the exact position desired for the gages. The mylars were aligned on the wing and the strain gage locations were transferred to the wing.

Aircraft strain and displacement responses were acquired with a 1024 channel Agilent Epsilon Data Acquisition System. The strain conditioning and data acquisition system operated on a standard VXI platform. The system supports full, 1/2, and 1/4 bridge strain gages, bridge excitation, amplification and low pass filtering on each channel with self-calibration (< 5 minutes for 1,000 channels). Control of the Agilent DAC, from channel set-up, sampling rate, calibration, filtering, etc., was within the AeroPro software.

Deflection Transducers

Each wing was instrumented with twenty deflection transducers for recording the wing displacements. Fourteen deflection transducers were on the left wing, and six on the right wing. Figure 7 depicts the locations of the deflection transducers.

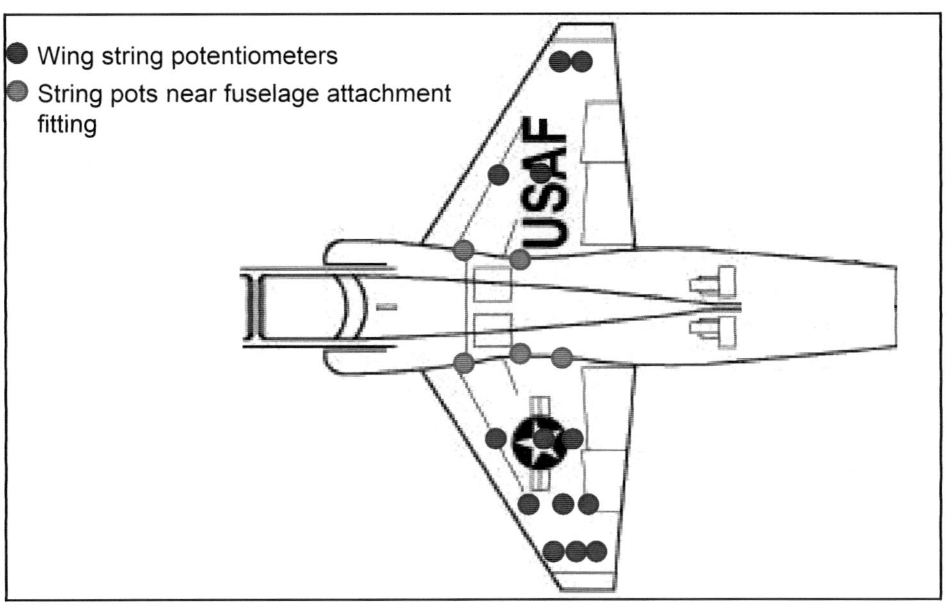

Figure 7. Deflection transducer locations.

TESTING

Once both tests were set-up and configured, both wings were tested to the target number of test blocks or failure - whichever occurred first. The basic tests were performed in the following sequence: a baseline strain survey was performed, followed by application of a fatigue load sequence blocks. After each fatigue block a strain survey was repeated.

Strain Surveys

A strain surveys was performed before fatigue loading was applied to the wing. The strain survey was repeated after completion of each block of simulated flight hours. The strain survey load condition file and the strain survey profile files are loaded into the AeroPro software before beginning the strain survey. The strain survey load profile is summarized as follows;

1. Apply the tare loads equivalent to weight of the wing and loading systems.
2. When this load has stabilized, zero all strain gages and deflection pots, record readings (some channels may not be exactly zero) as a baseline from which all calculated stresses and deflections will be taken.
3. Apply 80% of maximum flight condition 1 loads in progressive 10% increments and then decrease to zero load in 10% increments. Record data for each load step.
4. Repeat for each of the additional 8 flight conditions.
5. Apply 50% of maximum wing down bending for flight condition 1 load in progressive 10% increments and then reduce to zero load in 10% load increments. Record data for each load step.
6. Repeat for flight condition 2 and 6.

During and after the strain survey, all transducer data (strain gages, pots, and load cells) channels were reviewed for anomalies. Any anomalies were address before performing the fatigue test. Depending on the anomaly the action taken varied. The most common anomalies were bad strain gages or deflection transducers.

Fatigue Testing

After the first strain survey, the wing fatigue testing commenced. The fatigue testing was accomplished in simulated flight hour blocks. The fatigue load condition file and the fatigue spectrum profile files will be loaded into the AeroPro software for either the SUPT or IFF usage. The fatigue test will be paused or stopped as required to perform the inspections in accordance with the inspection requirement outlined below.

The test spectrum for the first 100 load events in the test sequence file was run, recording all transducer data at each event. The data was reviewed upon completion and no anomalies were found. The fatigue testing was then continued for the remainder of the first block stopping at intermediate flight-hour points for inspections, as needed. The transducer data at each event was recorded, as well as inspection results at proper intervals. At the completion of a block the data was downloaded and archived for further processing and review. The spectrum load blocks were repeated as necessary, to achieve the structural life goals for each wing usage. Figure 8 shows both wings being tested and Figure 9 shows a close up of one wing with motion blur showing the wing deformation.

Figure 8. Both -29 wings being tested.

Figure 9. Motion blur showing wing deformation.

INSPECTIONS

A series of inspections were performed to document the condition of the wing specimens over the duration of the testing. These are outlined in the following descriptions.

Nondestructive Inspection Methods

Several NDI methods are available at SwRI for inspection of both aluminum and steel components. These include:

- Visual
- Eddy-current (both surface and bolt-hole)
- Dye-penetrant
- Ultrasonic wave
- X-ray
- Magnetic particle (applicable to magnetic steels only)

These methods were used for this program as needed and called out for in the T-38 technical orders that cover inspection timing (when and where) and NDI procedures (how).

Initial Inspection

Since both wings had prior service usage before the fatigue testing began, a detailed receiving inspection was performed. A baseline inspection of the lower wing skin fastener holes and radii was accomplished. An x-ray inspection of the substructure was also performed. Particular attention was paid to those areas that have been designated as fatigue critical locations or have a history of in-service cracking. The SUPT did not have any detectable pre-existing damage. The IFF wing had multiple pre-existing cracks that were repaired before the fatigue testing started.

Inspection During Testing

Inspection intervals and methods for the wings are specified in the T-38 tech orders. Currently published and newer (in development) requirements were used to determine where and how often to inspect the wing structure. In addition, SwRI personnel assigned to the test program visually inspected the exterior of the wings daily for any anomalies. They also monitored recorded strain gage data for significant changes that might indicate a needed inspection of an area or component. If cracks or crack indications were noted, inspection intervals of that area or component may have been reduced accordingly and or the cracked component may have been repaired.

RESULTS TO DATE

Both the IFF and SUPT wings have developed fatigue cracking during the cyclic test. The SUPT wing was the first wing to begin testing. At the time of this paper it has experienced over 14 blocks of simulated flight loading. Numerous cracks in the lower wing skin fastener holes have nucleated, the majority of which have been repaired by simply oversizing the fasteners. A few holes have required an installation of a bushing and so far only one hole has required the installation of a doubler. Both the left and right side 66% spars cracked at known crack prone locations and typical repairs were installed.

The most interesting cracking to date on the SUPT wing has been along the 15% spar at the main landing gear rib. Cracking was discovered between (and not into) cold worked fastener holes on both the left and right side of the wing. They appear to be surface cracks emanating from the residual tension stress field of the cold worked holes. Multiple surface cracks were found some of which nearly ran between (into) fastener holes. A bonded boron composite repair was designed and installed on both the left and right side of the SUPT wing. Figure 10 shows a typical crack on the left and the bonded repair on the right. The bonded repair has experienced over 5 blocks of testing with no signs of debonding or degradation.

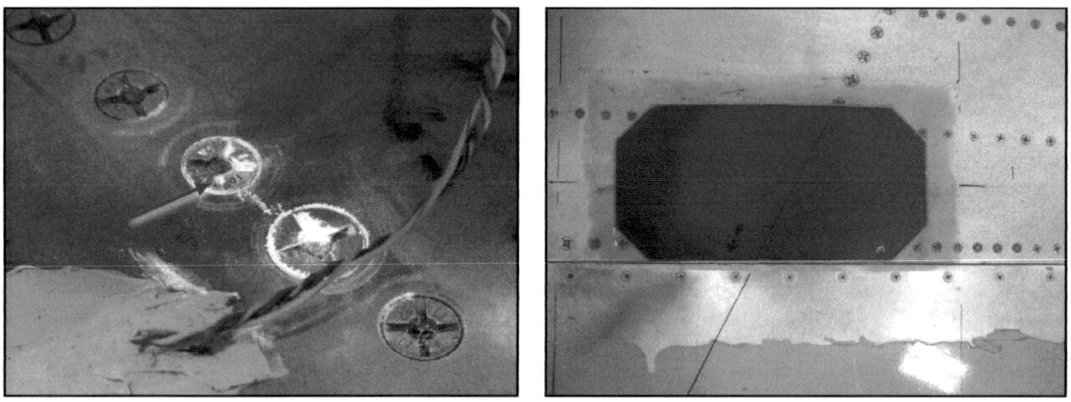

Figure 10. Typical cracks on left and a bonded (non-typical) repair on right.

The IFF test was delayed due to availability of the wing. The IFF experienced approximately 2.75 blocks of simulated flight loading before it could no longer meet load requirements, due to a large crack on the right wing. It had experience numerous cracks in lower wing skin fastener holes. The majority of which were repaired using a simple oversizing of the fasteners/holes. A few holes have required installation of a bushing and so far only one hole has required the installation of a doubler. Both the left and right side 66% spars have cracked at known crack prone locations and the typical repair was installed. The main landing gear bay skin radius cracked and was repaired by milling it out in accordance with the standard repair process.

A smaller crack on the left wing was discovered at a similar location as the large crack on the right wing. The IFF wing test was reconfigured to only apply loads to the left wing in order to obtain additional crack grow rate data on cracks discovered on the left side. An additional 0.35 blocks of spectrum was applied to the left wing before completing the test. The IFF wing is currently undergoing a detailed teardown evaluation.

FUTURE ACTIVITIES

The primary effort in the short term will be to complete the SUPT wing test while also disassembling the IFF wing for a thorough teardown inspection and metallurgical evaluation. Every fastener hole in the main wing box will be inspected multiple times as part of the teardown. A baseline inspection with fasteners installed using Automated Fastener Hole Inspection System (AFIS) and Rotoscan was performed for the IFF wing which will be followed by bolt hole eddy current inspection and florescent penetrant inspection after the fasteners are removed. Each indication will be recorded, broken open and crack size measured if present. A significant number of cracks will have metallurgical evaluations performed on them. The test finding will be used to update the T-38 -29 wing durability and damage tolerance analysis and an economic life analysis of the -29 wing.

SUMMARY

This paper summarizes the progress made to date on fatigue testing of two T-38 -29 wings. It discusses the background and identifies the need for performing two fatigue test of the same model wing. A description of the test set, loading, whiffle tree and instrumentation is provided. The test finding to date are outlined and future activities are discussed.

Integrating Engineers Into the PDM Enterprise

Dave Gaddis

L-3 Communications Mission Integration Division

Abstract: With multiple competing requirements of increased operational tempos, diminishing fleet sizes and decreasing recapitalization funding, increasing airworthiness requirements and aging fleets, and increased emphasis on aircraft availability, military Depot operators are being required to discover, invent, and implement innovative ways of improving aircraft throughput for Programmed Depot Maintenance (PDM). While many have focused on improving and increasing support to the mechanics on the floor, L-3 Communications has placed additional focus on proactive development of a new breed: Sustainment Engineers, and their integration into daily PDM and heavy maintenance execution.

"Sustainment Engineering" (SE) as used in this context denotes a proactive, cross-functional extrapolation of the typical "find it, fix it" Liaison Engineering function combined with both a continuing airworthiness approach and an overarching emphasis on aircraft availability improvement. Sustainment Engineers are engaged not only in day-to-day engineering liaison duties, but are also fully integrated in the total life cycle sustainment and continuing relevance of their weapon system. From the General Expectations section of the Sustainment Engineering Competence Model:

Sustainment Engineers are expected to exhibit a high degree of initiative, motivation, interest, and enthusiasm in their work, taking a proactive, entrepreneurial approach to their assignments. Each Sustainment Engineer is expected to develop the knowledge, skills, and abilities required to perform at his current level. Sustainers are also expected to take a larger view of their contributions, and to support team as well as individual goals.

BACKGROUND AND INTRODUCTION

In aircraft serial production and manufacturing environments (such as aircraft factories), Engineering Liaison traditionally provides answers and clarifications to engineering design questions from Manufacturing, and remedies to discrepancies encountered in the manufacturing process. In the former role, providing answers and clarifications to engineering design questions, Liaison engineers work with Design Engineering, Production, Planning, Purchasing, and other functions to correct drawing errors, resolve producibility concerns, and provide alternate materials types and sources. The resulting engineering changes and drawing revisions are generally reviewed with the same rigor as original design engineering (Stress Analysis, Materials & Processes, Manufacturing, et c.), as they normally will be incorporated into design drawings to maintain configuration control in continued production.

In the latter role, providing remedies to discrepancies encountered in the manufacturing process, Liaison engineers normally must attain authorization or certification under a Material Review Board (MRB, or some equivalent review authority), prior to giving disposition for manufacturing discrepancies. As the MRB Liaison engineer in this instance is working in a serial production environment, engineering dispositions may have been previously developed, either as a collection of existing repair dispositions, or as a set of generic repairs developed in anticipation of discrepancies that may be encountered during manufacturing, or some combination of these methods. These repairs are generally pre-approved by the design authority for application in specific circumstances, with some limited additional approval required within a range of parameters. Further expansion of these limited repair applications generally requires complete repair approval from original design authority (again, Stress, Materials & Processes, Manufacturing, et c.). Any anticipated problems would have been resolved in design, planning, or other up-front process, rendering both of Liaison's former and latter roles as described inherently reactive in nature.

In an aircraft maintenance and modification environment (such as an aircraft Depot or overhaul facility), Engineering Liaison still works principally in a reactive mode, providing engineering disposition to discrepancies encountered during Programmed Depot Maintenance (PDM), C-/D-check inspections, or other Maintenance, Repair, & Overhaul (MRO) activities. The discrepancies identified during this type work typically are not manufacturing errors, but are discrepancies between what "is" and what "should be", comparing existing on-aircraft conditions to as-manufactured or as-designed conditions.

PDM Liaison engineers don't necessarily interface directly with Design Engineering, as the aircraft Original Equipment Manufacturer (OEM) many times is no longer associated with the Depot aircraft in question; also, new parts are not being manufactured in appreciable quantities to warrant a dedicated manufacturing Liaison role, and certification under MRB (or some equivalent review authority), becomes a necessity for Depot Liaison engineers. MRB or equivalent certification for Depot Liaison engineers invokes a level of engineering authority that transcends manufacturing Liaison

MRB authority, as the discrepancies and non-conformances encountered are seldom predictable or repeatable, requiring creation of new engineering for each situation.

One definition of "repair" indicates that an engineering disposition will reduce but not eliminate a non-conformance, and that that repair is intended for a single application only. In Depot Liaison engineering execution, repairs are commonplace—dispositions for non-conformances and other discrepancies are created and released on a daily basis. Potentially confusing to the uninitiated, the simple execution of a repair generally creates a new set of non-conformances—replacements or acceptable substitutes are created to meet original form/fit/function requirements, but do not match the original, as-manufactured condition. In essence, the Depot Liaison engineer is creating a new set of non-conformances to remediate the original non-conformances.

Typical Depot organizations use their Liaison engineers as "firefighters", to resolve maintenance and repair engineering issues as quickly as possible, as they arise. These engineers are not routinely integrated into the program execution processes, such as purchasing, scheduling, or planning, but rather will semi-autonomously assess damage, develop a repair concept, perform stress analysis as required to ensure airworthiness of the repair, qualitatively evaluate any damage tolerance impacts of the repair, and then design the repair, defining the parts and materials needed, and release the engineering requirements for the repair. If parts and materials are not readily available for repair work, "the system" will work to order, receive, process, fabricate, and install the repair parts, while the engineer has moved on to the next problem.

The engineering repair process works relatively independent of other sustainment functions (logistics, planning, integrity programs, usage/maintenance tracking), relying on after-the-fact reporting mechanisms to ensure the discrepancy is correctly documented and coded for tracking, and the final repair is eventually incorporated into aircraft records and usage tracking schemes. The "system" collects these metrics over time, so that programs can conduct trend analyses and make decisions for modification and maintenance actions. Allowed to work unabated, the process will eventually yield results for systematic weapon system sustainment.

"SUSTAINMENT ENGINEER" CONCEPT

Multiple competing requirements of aircraft availability improvement, mission operational tempos, and increasing airworthiness requirements compound the sustainment challenge. In resolving these multiple competing requirements, the realization occurred that fleet sustainment activities coalesce in one place, in a regular pulse, like a heartbeat—Depot. And one group of people regularly experiences the real and immediate outcomes of fleet sustainment processes—Depot Liaison Engineers. If your Damage Tolerance program or Corrosion Prevention and Control Program is (or isn't) working, Depot Liaison Engineers see the results. If your sustainment logistics system is (or isn't) effective, they see those results, too, in parts shortages and workarounds. Schedule pressures resulting from procurement funding strategy; more frequent and invasive inspections due to increased age and usage; major structural replacements due to design material choices and/or aging fleets—the Depot Engineer is there to see the results. But maintaining aging fleets while simultaneously improving availability seems to be diametrically opposing goals, resulting in pressures to perform "minimum but adequate" maintenance and repairs. Time and budget pressures can cause the Depot engineer to concentrate on addressing the problems at hand, and the outcomes of those potentially valuable observations may not make timely feedback to the sustainment system.

This type and amount of information flowing into one functional organization, coupled with a broad-spectrum approach to aircraft availability improvement, yielded an evolution in thinking to leverage the collective expertise of this group of engineers by expanding inherent Depot engineering capabilities to provide proactive output. Rather than these Depot engineers functioning as a reactive force, a perpetual bottleneck irritant, the functional realization was to train the engineers on

- Damage Tolerance requirements and best practices,
- individual aircraft missions, usage, and tracking (and why it's important),
- airworthiness and certification requirements,
- predictive logistics (based on engineering processes), parts design, fabrication, and procurement, and
- Depot schedule as it relates to availability and mission accomplishment,

resulting not only in an understanding of life cycle sustainment as a product, but taking ownership of it—let the "firefighters" help prevent the fires before they get started.

As a consequence, routine individual engineering dispositions would no longer be single, stand-alone engineering incidents, but would include total life cycle considerations such as Procurement, Program Management, Scheduling, PDM Management, Fleet Management, and continuing airworthiness, among many others. Repair dispositions would no longer be limited to "What's best for the airplane today?", but would transform to "What is the best solution to simultaneously enhance life cycle sustainment and to increase aircraft availability for the next 20 years (or more)?" This proactive, broad-spectrum approach is the conceptual basis for creating the Sustainment Engineer.

WHAT DOES IT TAKE TO BE A SUSTAINMENT ENGINEER?

The key characteristics of a Sustainment Engineer are creativity, innovation, and an entrepreneurial spirit—navigating complexity while managing risk—resulting not only in reduced Depot Maintenance throughput times, but also improved airworthiness compliance and increased aircraft availability to the end user. Sustainers must be well-versed in

- ✓ Metallurgy (Aluminum, Steel, Titanium, Magnesium, et c.)
- ✓ Stress Analysis, Damage Tolerance, ASIP/"WFD Rule"
- ✓ Machine Processes (Milling, Grinding, Boring, et c.)
- ✓ Fabrication Processes (Metals, Composites, Fabrics, et c.)
- ✓ Aircraft Assembly Processes (Tooling, Shoring, Fasteners, et c.)
- ✓ Non-Destructive Inspection techniques (Ultrasonic, Eddy Current, X-ray, et c.)
- ✓ Finishing Processes (Paint, Plating, Anodizing, Corrosion Prevention, et c.)
- ✓ Human Factors (Safety, Producibility, Skill Mix, et c.)
- ✓ Aerodynamics
- ✓ Aircraft Engines
- ✓ Aircraft Systems, MECSIP
- ✓ Aircraft Landing Gear
- ✓ Programmatic impacts of dispositions
- ✓ Logistics implications of dispositions
- ✓ Financial impacts of dispositions
- ✓ Schedule impacts of dispositions
- ✓ Life cycle impacts of dispositions
- ✓ **Every Sustainer has a mentor**

Every Sustainer is a specialist at being a generalist.
Generalists have broad exposure and experience across a variety of situations, which allows strong confidence in general decision-making, with the possibility of errors in decisions requiring esoteric or specialized knowledge. Conversely, specialists have deep exposure and experience in a well-defined class of situations, resulting in a lower decision error rate on an individual basis, but with limitations on scope of endeavor.

"Specialists at being Generalists" indicates a deliberate balance of these conflicting concepts, such that the individual contributor gains enough experience and exposure to understand his system well enough to recognize when it is functioning properly, and also develops the insight to *recognize and take or recommend preemptive corrective action* when it is not.

SUSTAINMENT ENGINEERING ORGANIZATIONAL IMPLEMENTATION

Sustainment Engineering implementation is by necessity a multi-dimensional matrix organization, a complex adaptive system even within its own shell. As described in this paper and illustrated in Figure 1, SE comprises four tiers of execution, each interdependent and receiving external input, building and feeding back upon the others, creating a synergy that ensures the whole is greater than the sum of its parts. Sustainment Engineering consists of

- PDM/Mod Support—Day-to-day execution of aircraft PDM, maintenance, and repairs, and activities in support of Modification[4] installations
- Fleet Support—Engineering, procurement, fabrication, and deployment of PDM-critical assets, assemblies, and installations; also includes development of specialized tooling
- Fleet Management—Monitoring and managing Integrity Program requirements; management, assessment, and evaluation of PDM/inspection results; and long-term fleet life cycle sustainment planning
- Field Support—Immediate, real-time support of fielded assets, and feedback of field concerns into fleet sustainment planning

[4] The experiences described in this paper are based on applications to highly-modified mission aircraft. As used here, the term "modification" addresses mission system installations, as opposed to fleet sustainment modifications applied to aging aircraft.

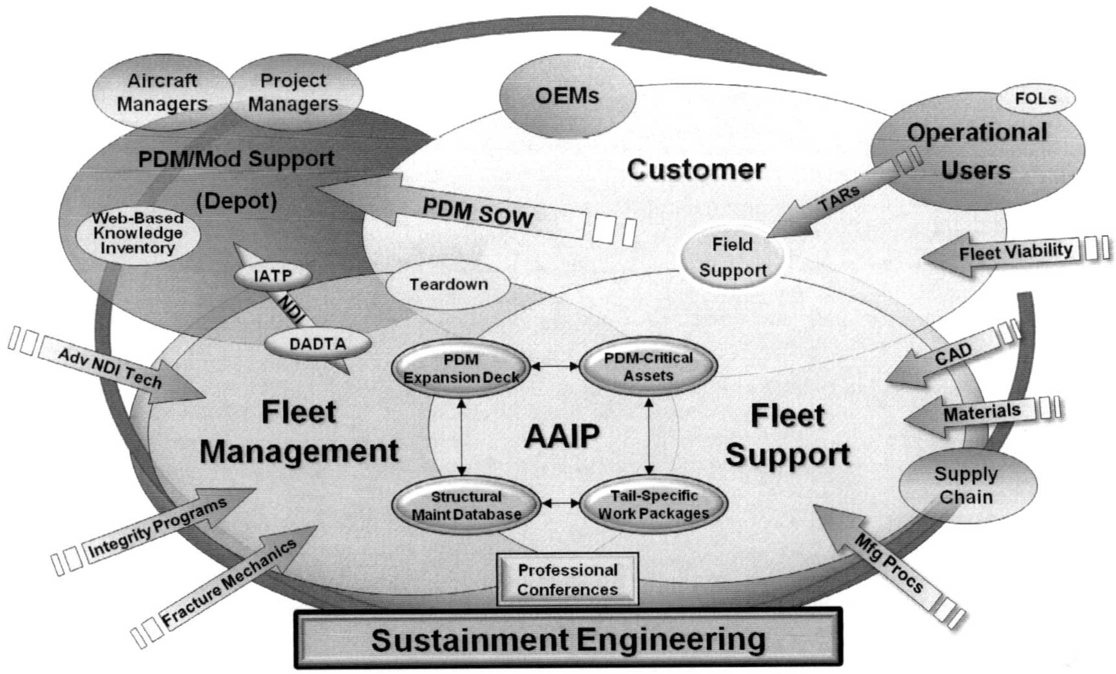

Figure 1. Sustainment Engineering Organizational Implementation.

In many sustainment endeavors, these four functions would be performed by different groups, under varied funding accounts and profiles, possibly across multiple commands, services, or other administrative constructs. However, the SE organization described here integrates all four functions into one coherent organization, allowing immediate and comprehensive visibility across the life cycle sustainment spectrum.

PDM/Mod Support

PDM/Mod support consumes the majority of Sustainment Engineering resources, supporting scheduled inspections and maintenance required for fleet sustainment. This function has the quotidian task of shepherding aircraft through the PDM and Modification processes, and includes development of repair dispositions in support of Over & Above (O&A) repair and maintenance activity.

Of particular interest is SE's involvement in PDM inspections: Sustainers are fully integrated into the Arrival Assessment (AA) process (a comprehensive and exhaustive set of inspections based on and in addition to the PDM Work Specification), meeting daily with Aircraft Managers and Project Managers to monitor findings and help anticipate repair requirements. Early identification of potential major structural repairs/replacements (MSRs) is paramount in developing PDM schedule and materiel requirements, as any injection of O&A maintenance activity can have adverse schedule implications. Current goal is to identify 75% of all PDM O&A maintenance requirements within the first few weeks of aircraft input[5].

For each aircraft in PDM/Mod, Sustainment Engineering assigns a Point of Contact (POC), a single-face-to-the-customer representative, responsible for all SE activities on his or her aircraft. The SE POC meets daily with his Aircraft Manager and Project Manager to prioritize work items in SE's queue, ensuring timely engineering dispositions for maximum Production efficiency. The SE POC is also responsible for maximizing concurrency of multiple major structural repairs or replacements, assisting in developing PDM/MSR and Mod work sequencing interactions, and ensuring aircraft and worker safety by implementing local personnel and/or weight restrictions during MSR execution.

Sustainment Engineering is the principal Engineering Material Review Board authority, leading and setting requirements in the non-conformance disposition process. SE coordinates engineering disposition through the required subject matter experts (Stress, Airworthiness, Customer), and, as such, is in the position of monitoring and overseeing airworthiness.

[5] For the aircraft described here, mission system modification efforts are generally so extensive as to dwarf PDM inspection and maintenance requirements. As a result, SE workload includes significant effort to ensure timely basic aircraft structure repairs/replacements to minimize disruption of mission system modification installations.

Fleet Support

The Fleet Support function develops engineering in support of procurement and fabrication of PDM-critical assets, including detail parts, assemblies, and installations required for major structural repairs and replacements. The engineering effort comprises

- Close interaction with PDM/Mod Support to identify immediate and potential material shortfalls
- Computer-Aided Design (CAD) modeling and associated drawings for repair part fabrication
- Generating concise, comprehensive fabrication, assembly, and installation instructions for MSRs
- Defend against obsolescence by working with Customer, Materials & Processes Engineering, Procurement, and Project Management to discover, develop, and bring on line new vendors, suppliers, and processes, and to then obtain materials and parts from those suppliers and vendors
- Close interaction with Fleet Management to ensure supportability for upcoming fleet sustainment requirements

In addition to engineering oversight of offsite vendors and suppliers, Fleet Support Sustainers are also best-positioned to serve as SE POCs for onsite shop fabrication and assembly activities.

Fleet Management

Fleet Management (FM) is responsible for monitoring and tracking fleet aircraft usage and health, and from those results developing and implementing engineering processes in compliance with Integrity Program requirements (MIL-STD-1530 for ASIP, MIL-STD-1798 for MECSIP…). The Fleet Manager then translates those requirements into inspections and airworthiness actions, and disseminates them in executable form across Sustainment Engineering and the PDM Enterprise. The Fleet Manager is uniquely trained and positioned not only to develop and implement the engineering processes required, but to also advise and assist in applying technologies to physically validate and verify compliance with requirements. In execution, the tasks include:

- ASIP Tasks IV & V Compliance
  - Durability and Damage Tolerance Analysis (DADTA), Force Structural Maintenance Plan (FSMP) compliance, and Individual Aircraft Tracking Program (IATP) compliance, and their associated working groups
  - Structural Maintenance Database (SMD) development and application
  - Corrosion Prevention and Control Plan (CPCP)/Advisory Board (CPAB) implementation
  - Monitor and improve usage data capture rates and effectiveness
  - Testing & Teardown program support
- MECSIP Compliance
  - Classification and characterization of systems, components
  - Safety/hazard analyses and evaluations, assessment of fleet sustainment impact
- Technology deployments
  - NDI technology deployments for enhanced/improved inspection capabilities
  - Structural technology deployments for enhanced fatigue/corrosion performance
- Government/Industry best practices
  - Product Improvement Working Groups
  - Fleet Sustainment Roadmap development
  - Fleet viability assessments
- Development of executable airworthiness tasks
  - Development of inspections, validation/verification, and reporting activities
  - Recommendations for fleet repairs, sustainment modifications, or component replacements
  - Cross-functional implementation (SE, QA, NDI, Production, Customer…)

As a clear demonstration of the synergy resulting from integrating engineering functions, Fleet Management leverages the requirements of aircraft integrity programs to proactively manage airworthiness of assigned assets, and at the same time drive aircraft availability improvements. Through the IATP, FM has developed a usage projection method which is then applied against FSMP inspection requirements. These results are relayed to the Fleet Support function, who then ensure materials and equipment are available for a particular aircraft's inspection requirements at next PDM. Since both functions are under the same organizational umbrella, this process is seamless and immediate.

Field Support

Field Support is the 24/7/365, quick-reaction engineering support necessary to maintain airworthiness and availability of mission-capable assets. A request for support takes precedence over all other activity, and is generally reserved for the most experienced Sustainers.

Field issues generally jeopardize mission capability; however, to preclude airworthiness concerns and reduce non-mission capable time, these field issues can also inform Fleet Management, Fleet Support, and PDM Support requirements of preventative repairs and inspections that can be added to the PDM workflow. For example, minor damage requiring repair to secondary structure might not ordinarily cancel a mission; however, if the secondary structure in question is the wing leading edge, virtually any damage would likely ground the aircraft till repaired. One solution might be to implement a fleet-wide inspection during PDM, and to develop a repair kit that would be available for installation on aircraft during PDM. A more aggressive approach would be to procure kits and implement installation across the fleet (resulting in a fleet sustainment modification), with attendant modification certification requirements met.

Interdependence and Interrelationship
Interdependence is described pictorially in Figure 2.

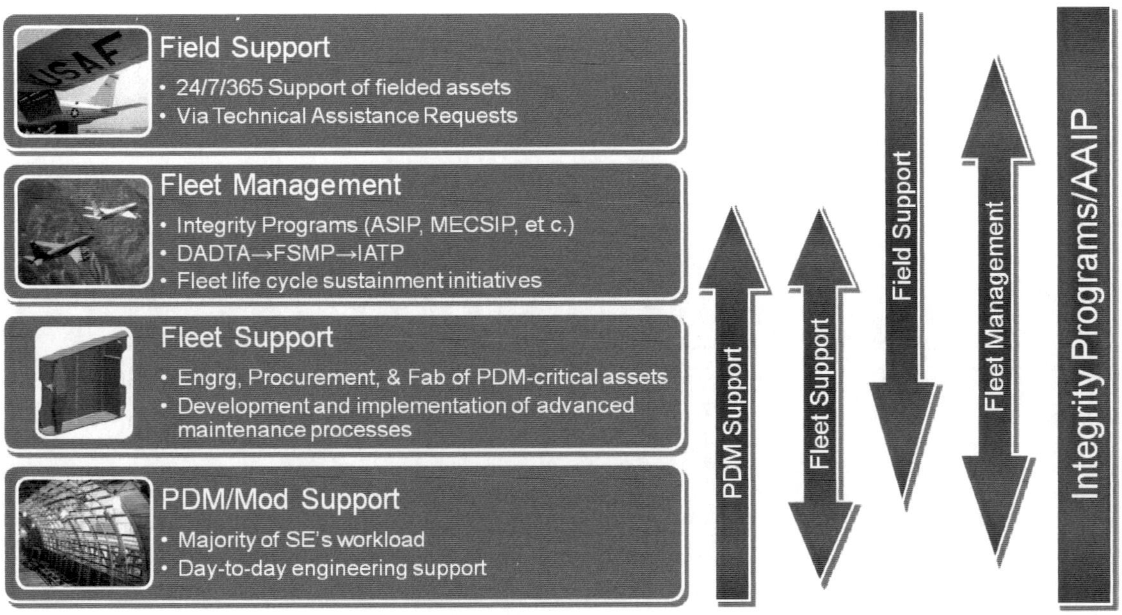

Figure 2. Interdependence and Interrelationship.

From this graphic description, each individual function exerts influence across at least two domains outside its own: PDM/Mod Support informs Fleet Support of peculiar upcoming PDM asset needs, and can further inform Fleet Management of the need for additional inspections across the fleet; Field Support concerns can in turn drive maintenance and inspection requirements all the way to the basic PDM SOW. Integrated Fleet Management, based on Integrity Programs and driven by fleet availability improvement needs[6], serves as the cohesive force, ensuring all facets are considered in life cycle sustainment planning and execution.

SUSTAINMENT ENGINEERING PROCESS

In the perpetual quest for mission-capable aircraft, availability can be viewed as a driving factor across the sustainment spectrum. In the integrated construct of Sustainment Engineering, the contributions of a successful AAIP cannot be overstated. Capabilities and processes developed for AAIP are directly applicable to Fleet Management execution, and Fleet Management is correspondingly able to leverage Integrity Program requirements to drive availability improvement.

Database development and analysis capabilities obtained during AAIP implementation are cross-decked to Fleet Management data, yielding weekly or monthly aircraft usage tracking that is compared with damage tolerance information and inspection requirements across the Fleet, both to preclude inspection overflights and to help ensure availability of assets for PDM/Integrity Program inspection and repair needs. As aircraft approach PDM input, Field Support information (TARs, Maintenance Bulletins, TCTOs, et c.) is also reviewed for any inspections or repairs needed as a result of field actions.

[6] In the context of this paper, "availability" is not just an indicator of "iron on the ramp" (aircraft not at Depot), but is also taken as an indicator of "iron in the air".

Once these PDM asset requirements are identified, Fleet Support works with Procurement to ensure asset availability. Computer Aided Design (CAD) capabilities are leveraged to create detailed engineering instructions for repair/replacement instructions, or to create specialized parts for peculiar repairs or replacements.

Once the aircraft arrives for PDM, PDM/Mod Support provides daily Engineering Liaison leadership to shepherd the aircraft through the PDM/Mod process, and to feed inspection and repair results back to Fleet Management for input to the Structures Maintenance Database and other Integrity Program tracking.

The Sustainment Engineering process is graphically represented in Figure 3.

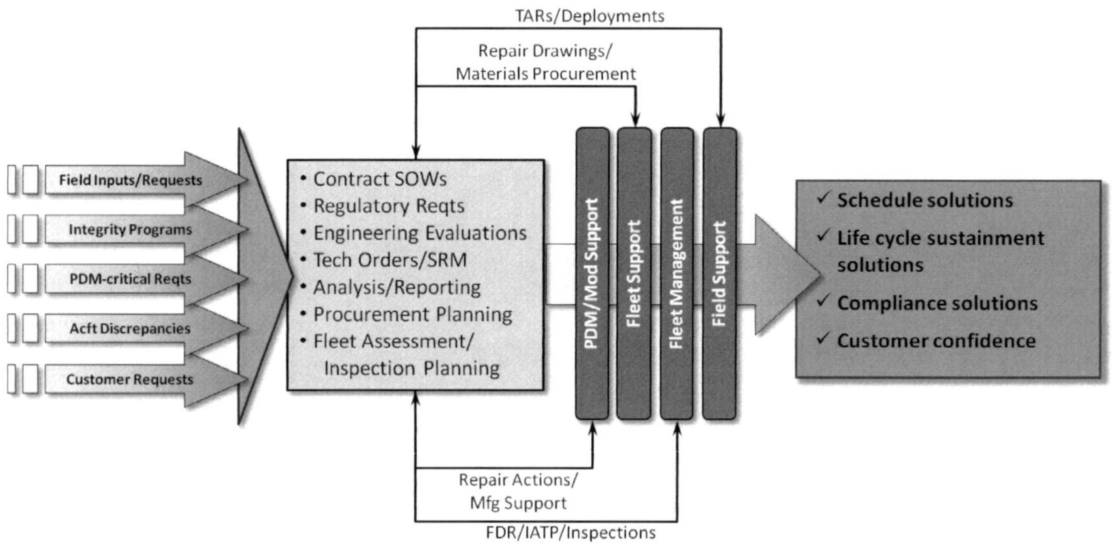

Figure 3. Sustainment Engineering Process.

Sustainment Engineers are materially contributing members of all PDM/Mod Integrated Product Teams (IPTs), providing engineering leadership at all levels of the PDM Enterprise. PDM Work Specifications are reviewed during development and again at release to determine upcoming requirements. Inspections of Modification areas not covered by the basic Work Spec are added to the PDM Expansion Deck (PDMXD), where specific inspections are developed and implemented early to preclude "late finds" requiring repair or replacement and jeopardizing schedule.

To improve aircraft availability, PDMXD inspections are also developed as a result of those "late finds" when they do occur. As an example, one aircraft had a very late corrosion discovery (on the flight line, after PDM), in an area not inspected by the basic PDM Work Specification. Repair action was initiated—Fleet Support developed CAD models and drawings for the required repair parts, worked with Procurement to get the parts built and brought in-house, and PDM/Mod Support provided engineering instructions to Production for the repair installation. Concurrently, Fleet Management developed a detailed inspection for another aircraft on the flight line, resulting in the need for the same repair.

"Two points make a trend"

Fleet Support immediately contacted Procurement to increase the repair parts order to a quantity of five, Fleet Management implemented inspections on aircraft in the hangar undergoing PDM, and PDM/Mod Support provided dispositions for hangar aircraft repairs. Within one week, all onsite PDM aircraft were inspected. The inspection was added to the PDMXD for all aircraft at PDM input, parts procured in advance, and the entire fleet has now received a one-time inspection and repairs for this discrepancy.

RESULTS

Sustainment Engineering implementation began in 2007, under the auspices of US Air Force's Aircraft Availability Improvement Program (AAIP). The AAIP was developed in response to increasing numbers of aircraft staying longer at Depot for maintenance and repairs due to age and increased operational tempo (usage), resulting in fewer aircraft available to users for mission accomplishment. One goal of the AAIP was to improve fleet availability by 20% by 2011, by improving Depot operations and reducing throughput times.

The aircraft considered in this paper were similarly affected by age and increased usage, with inordinately long Depot durations. Implementation of the Sustainment Engineering concept, as one component of a comprehensive set of process and materiel improvements, has resulted in dramatic reductions of aircraft Depot durations—up to 44% reductions for some aircraft types—achieving and surpassing the goal of 20% increase in aircraft availability, by 2011. In fact, although the goal was reached and surpassed, ongoing process improvements continue to reduce Depot throughput times even further.

WHAT DOES IT MEAN TO BE A SUSTAINMENT ENGINEER?

The majority of Sustainment Engineering's quotidian workload derives from aircraft discrepancies discovered during Arrival Assessment Inspections, conducted on each aircraft at PDM induction. However, for the weapon system described here, PDM is only a portion of the overall PDM/Modification effort, and, in most cases, aircraft discrepancies must be remedied before mission system modifications can be installed. In this regard, SE is largely a reactionary force, which is not the optimum condition to support reduced Depot times. To overcome (or at least minimize) reaction times, and to be a fully effective 24/7/365 organization, Sustainers must exhibit strong initiative and a creative, innovative, entrepreneurial spirit. They must develop and maintain strong professional relationships with customers (internal and external), support daily IPT meetings, and actively engage with Production on work prioritization. But the primary ingredients to a Sustainer's success are his or her reputation and relationships—both must be developed, cultivated, and protected.

Customer service is core to SE's purpose, and Sustainers are expected to be fully engaged, energetic, enthusiastic, and ready to help at a moment's notice. Sustainers are located in the hangars with their aircraft, for immediate access both to the aircraft and by the customers they support. When notified of a discrepancy or problem on the aircraft, the Sustainer's initial reaction is to get up from the desk, go to the airplane, and look at the problem. To that end, each Sustainer is issued his own flashlight, mirror, and scale. There is no substitute for eyes on the situation.

Between modification and maintenance activities, it is not uncommon to have four, five, or more major structural repair, replacement, and/or modification efforts in work at the same time on a given aircraft. Sustainers develop work sequencing and concurrency plans based on tech data restrictions, maximizing available work for Production at all times. Sustainment Engineering POCs work directly with Aircraft Managers, supervisors, and mechanics to determine the best course of action for each discrepancy (Figure 4).

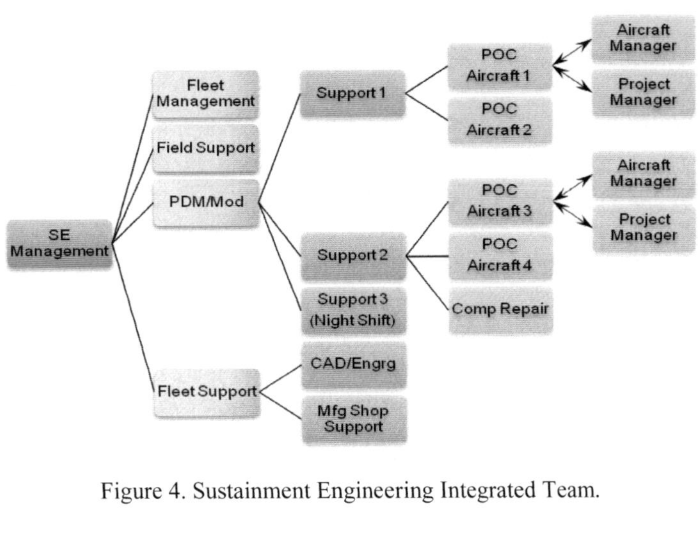

Figure 4. Sustainment Engineering Integrated Team.

Three times weekly, SE POCs meet to discuss aircraft issues, manpower needs, and internal workload apportionment and prioritization. Fleet Support and Fleet Management also attend the POC meetings, to gather and disseminate fleet sustainment information. Field Support activity is also discussed, to pass field findings to the PDM/Mod Support group through the POCs.

Another facet of a Sustainer's makeup is an institutionalized understanding and acceptance of creative standardization—although every disposition is unique, SEs are creating, using, and reusing standardized engineering dispositions, and storing those dispositions in a way that is searchable and traceable. Sustainers have created an internal website using SharePoint to collect, store, and disseminate information in an electronic "handbook", complete with Google Search functionality so that thousands of documents can be rapidly searched across the entire website.

To reduce errors in implementation, each engineering disposition receives a peer review before release to ensure it is complete, accurate, and correct for its intended application. Depending on the type of disposition and experience and certification level of its creator, a disposition may receive a third or fourth "set of eyes" before release. The goal is to reduce PDM cycle time by putting out the right information first time, every time.

Finally, and likely most critical to success in this endeavor, every Sustainer has a mentor. Although it is important to gain the knowledge and skills required to perform their assigned tasks, Sustainment Engineers must be able to navigate a complex programmatic landscape. Concerns such as cost, schedule, material availability, mission capability, funding profiles, facilities, skill mix, tooling—all must be continually balanced to achieve maximal result. Ultimately, each Sustainer will have an understanding of how the entire PDM Enterprise works, and of his place in it and contributions to its success.

THE UMBRELLA

Obviously, Sustainment Engineers do not work in an organizational vacuum, nor can they autonomously create engineering solutions without a business structure to support them. Sustainment Engineering's primary function is to provide engineering solutions to problems encountered on the basic airframe, which, in this case, is heavily modified for mission equipment installation. Aircraft structures and systems must be compatible with modification designs before installations can begin; so, by definition, almost any problem requiring Sustainment Engineering intervention will likely interfere with mission system modifications in some fashion, and must be resolved as quickly as possible to minimize mod schedule delay. As a result, Sustainers are routinely in a position to receive "critique" as to the appropriateness of their engineering dispositions and levels of support, and evaluations of the timeliness thereof.

Therefore, Sustainment Engineering management personnel must provide top-cover protection to their people ("the umbrella") by running programmatic interference on their behalf with Production, Project Office, Customer, and other Engineering and support organizations: Sustainers must be allowed to perform unbiased engineering evaluations and provide disposition while balancing pressures from multiple competing customer requirements such as schedule, cost, and materiel availability. Just as a physical umbrella keeps rain off its bearer, so too must the leadership umbrella minimize the programmatic effluents reaching their intended recipients.

Other concepts covered under the Umbrella are leadership, training and development, and planning. Sustainment Engineers need to know they are making contributions on a daily basis, their contributions are important, and *why* these things are so. If a Sustainer (or almost any other professional) knows the *why* of his job, he is able to find ways to improve performance and efficiency.

The basic building blocks for this development are

- **Technical/Operations**—how to do your job/how Engineering and industry works
- **Leadership/Professionalism**—how to execute with aplomb and get others to do the same
- **Administrative/Programmatic**—tracking/reporting, training, management, funding…

With such a critical set of contributions in their bailiwick, Sustainers need to know they are being *led*, and not just *managed.* It is incumbent upon Management to provide active leadership, to set the vision; let the engineers know there is a plan, that the plan is being followed, and that they are a consequential part of the plan. This type of leadership goes beyond bulletin boards, all-hands meetings, multimedia presentations, and mission statements; it is an active engagement with the people in the organization, challenging engineers to stretch capabilities and grow—even if that growing means preparing them well enough that other organizations will want to steal them.

Training is important to increasing knowledge and skills, but development is where critical Sustainment Engineering abilities are engendered. Training provides concrete information in a low-abstraction environment, to increase factual knowledge of technical subject matter and to facilitate maturity and expansion of existing skills. Some of the items included in "training" are

- Technical process specs, component and material specs
- Drafting manuals, drawing requirements, engineering processes
- Tech data, Tech Orders, Structural Repair Manuals

- Company/Customer directives and processes, security requirements
- Skills—analysis, CAD, technical writing

Development occurs after training, when sufficient knowledge and skills are acquired and the basic Sustainment Engineering process becomes fairly ingrained. At this point, Sustainers will start to speak a shared language of aircraft engineering so that mentors and trainees can begin to communicate in shorthand, symbolic terminology based on common experience, introducing increasing conceptual abstraction—discussing ideas rather than facts. Development centers around abilities—compounding the knowledge and skills from training into useable products such as comprehensive engineering dispositions and instructions, drawings, reports, et c. Once Sustainers reach this level, their abilities appear to grow exponentially.

Development also must take the form of professional and career enhancement for Sustainers. Although there is much to learn and accomplish in the sustainment enterprise, bright, intelligent, creative engineers must continually be challenged to learn and grow. Concentration on increasing technical prowess will consume most engineers for the first four to six years, when they are learning the "what" and "how" of their jobs; for development beyond the execution phase, engineers must know and understand the "why" of their jobs. Once a young Sustainer reaches that stage, learning and development become multiplicative rather than additive.

Professionalism and leadership are also critical to the Sustainer's development, as well as critical thinking and decision-making abilities. These are difficult subjects to "teach", and mentoring is critical during this time. Above all, the most valuable development needs for Sustainers are reputation and relationships—only the Sustainers themselves can develop these. And they must be protected, because without these two elements, the Sustainment Engineer will be just another voice in the wind to the customers he is assigned to serve.

As a final note, Dwight Eisenhower said "Plans are useless, but planning is indispensable". Sustainment Management is always in a state of planning—planning ways to improve engineering dispositions and processes, planning methods to increase contributions and improve programmatic execution, and planning ways to further develop their people. As abilities increase across the Sustainment enterprise, younger Sustainers can begin making material contributions to planning life cycle sustainment activities, furthering their own professional development and career opportunities, and completing the development cycle.

CONCLUSION

Engineers can make many contributions to help ensure program success, improve availability, or decrease Depot turn times. Regardless of the reason for doing it or the specific organizational construct used, integrating engineers into the PDM enterprise can have tremendous benefits. Every application is unique and no single solution will answer all needs; it is therefore one of the most difficult and at the same time one of the most rewarding tasks to undertake.

This document consists of general capabilities and concept information that does not contain controlled technical data as defined within the International Traffic and Arms Regulations (ITAR) Part 120.10 or Export Administration Regulations (EAR) Part 734.7-11

Reducing the maintenance impact of an additional vibration spectrum added into a classical fatigue spectrum

Sarah Payros[1], Dort Daandels[1], Nikolaus Ohrloff[1], Anna Campmany Roselló[2]

[1]*Airbus Operations GmbH*
[2]*ELAN-Ausy GmbH*

Abstract: This abstract deals with the addition of a vibration spectrum into an existing fatigue spectrum. It provides an approach how to minimise the analytical damage caused by additional load cycles using time correlated data of multiple loads components to create a real stress time history (STH). To be able to use this approach it is necessary to implement it early in the process, which means starting with the load database generation. The stress calculations need then to be adjusted to process the load data together with an update of the F&DT software. Finally improved test concepts with high frequency cylinders ensure that the structural tests can be performed in an acceptable time with a similar complex spectrum as in the calculations. In line with CS 25.571 [1], the additional loads have been considered in the spectrum in both analyses and test while at the same time the impact on the maintenance program is minimised.

INTRODUCTION

During some phases of a flight, some parts in the aircraft may endure locally a level of vibration loading, which needs to be considered in the fatigue and damage tolerance assessment. In case of a relative high steady loading, such as internal pressure in fuselage or lift on wing structure, the vibration loading may seem non-significant. However, if the classical loading is close to the level of the vibration loading, the latter may become significant. Depending on the amount of load cycles in it, the additional vibration event can become as well dominant in the fatigue spectrum, even if the amplitude of the vibration loads is considerably lower than the steady loads. The numerous repetitions of small cycles can indeed contribute significantly to the overall damage. On top of that, the crack growth can also be sensitive to small cycles. Overall an additional vibration spectrum has an impact on the fatigue life and can contribute to a life reduction, which has to be estimated and taken into account in the aircraft maintenance program.

Some of the vibrations have harmonic behaviour. They can be added as a block of additional loads using occurrence (frequency*time) against amplitude. Typical examples are incremental loads such as ground manoeuvres. If vibration loads become more irregular or random, then square blocks cannot represent the vibration realistically anymore. Non-harmonic vibration spectra can be better sorted and summarized accumulatively as exceedance vs occurrence in so-called exceedance diagrams. They are then added as a triangular block to the spectrum or on top of a steady load case, similar to gust loads. An alternative way to represent the spectrum of such event is to use power spectra density. This is particularly relevant for random events which are environmentally driven or frequency driven. Whether power spectra density or exceedance diagram are used, this implies that the random non-harmonic vibrations are known and can be statistically represented based on fleet survey and previous or current in-service experience. Many examples are described in [2] and in [3].

The next step in complexity of spectrum generation is considering chaotic loading. A chaotic loading is here defined as a load set with loads in various directions which are not correlating with each other. These loads have different frequencies and amplitudes. An example of a vibration event for two load components A and B is shown in Figure 1.

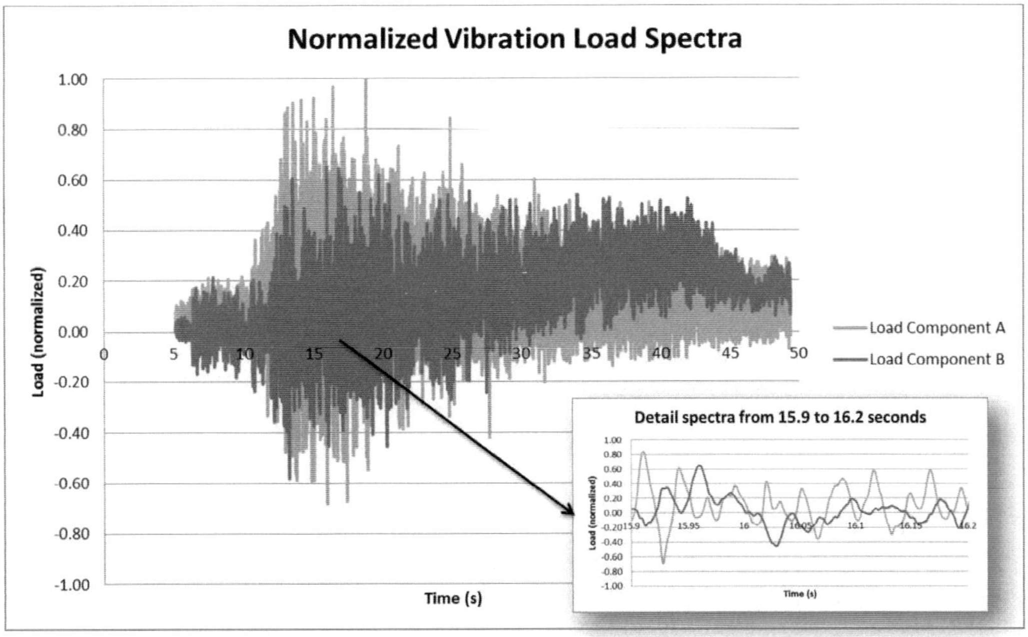

Figure 10. Example of a chaotic loading time slice for two component parts A and B.

For each of the load components the accumulative occurrence of exceedance diagram can still be generated individually, similar as described previously. The application of these diagrams to generate a multi-axial load spectrum is however not straight forward. Pragmatically the maxima of several components can be added together at each number of occurrences. This is shown in the diagram in Figure 2 and referred as Case 1.

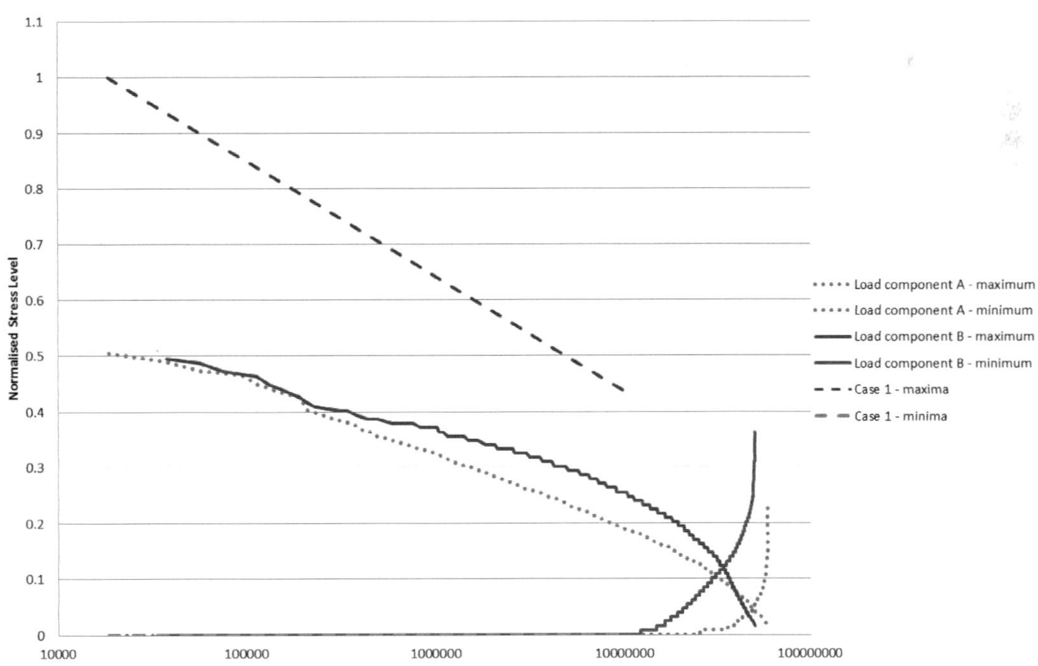

Figure 11. Case 1: multiple loaded stress spectrum created by addition of individual exceedance diagrams.

Loads from the given exceedance diagrams were measured during flight test campaigns and post-processed using different load generation techniques to get a load spectrum representative of a full flight. More details are given later. It is assumed that the stress at one critical location α comes from two loads in two different directions A and B. The proportion of each load contribution to the stress at location α is known. From each single load spectrum at A and B, a stress exceedance diagram is generated (in green and blue in Figure 2). To get the final exceedance diagram at location α, each stress level from A and B at each number of occurrences has been added together.

The resulting curves in red for maxima and in orange for minima are also shown in Figure 2 and defined as Case 1. However if looking a bit more detailed into load time history, the maxima peaks from each component might not appear at the same time. This can be seen in the zoom of the vibration time slice shown in Figure 1. Using the approach of Case 1, the big cycles from each component will be still added together. This will lead to a very conservative spectrum and consequently to unnecessary maintenance inspections. Another extreme solution is to consider the loads components as separate events, decouple them and add them one after the other. This time the stress spectrum might become un-conservative, which is not acceptable.

The truth is somewhere in between these two extremes. Therefore an alternative approach for adding a vibration spectrum is developed and is presented in the following section.

FATIGUE SPECTRUM – A PRAGMATICAL ALTERNATIVE TO BLOCK SPECTRUM

For each load introduction point and load directions, load data sets from a vibration event can be obtained from flight test measurements. Flight test campaigns are typically part of load process in the aircraft industry and allow validating loads while consolidating the analytical models. During those flight test campaigns, several locations and loading directions are monitored. More than thousands of data points can be measured at a single location for 1 second, which generates a huge amount of data for analysis.

As explained in the introduction the difficulty with chaotic loads is to determine a realistic load set to be used for fatigue calculations with theoretical load models or statistical analysis only. In order to get a representative dynamic stress time history, it is then necessary to combine these methods with an appropriate flight test campaign, which will include an appropriate number of flight tests in relevant conditions. This flight test campaign will support the selection and determination of a load spectrum representative of a daily flight.

Once loads as load time history are available for each relevant component, the next step is to build the spectra for those multiple loaded locations. Instead of adding the loads from exceedance diagrams as described in previous section (Case 1), loads from each component are added at each time step as measured in the flight test. From different measured loads spectra, a single real-time stress time history is thus created for each location of interest. Using the same example than Case 1, load components for a critical location α are first added time–correlated. That means that at real measured time (i), load component A (i) is added to component B (i) considering the appropriate contribution of each load component as shown in Eqn. 1:

$$\sigma_i(\alpha) = c_A \times F(A)_i + c_B \times F(B)_i \tag{1}$$

With:

$\sigma_i(\alpha)$ the stress at location α at a time step i
c_A and c_B being the load to stress factors representative of the contribution of respectively load components A and B
i the time step as measured during flight test
F(A) and F(B) the measured loads from flight test

The resulting vibration stress time history is then added directly to the existing classical spectrum at the appropriate phase of the flight, where it occurs.

The new resulting exceedance diagram resulting from this approach is generated and shown in Figure 3 as Case 2. For comparison Case 1 is also included in the diagram.

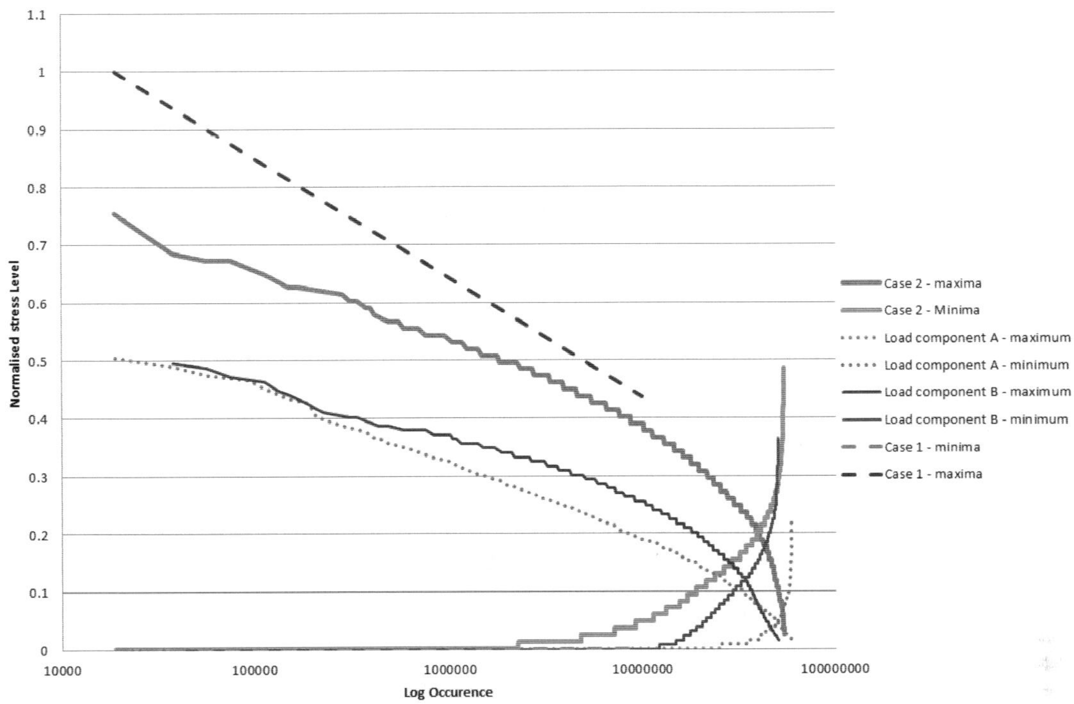

Figure 12. Case 2: Exceedance diagram with loads combination added time-correlated compared to Case 1.

It can be seen that the new spectrum resulting from Case 2 is less severe than for Case 1. For multiple-loaded component parts, this conservatism from Case 1 approach can be then minimised by considering the load measurements based on time of occurrence and not on the number of exceedances.

To be able to process the extended spectrum, the F&DT tool data capacity is improved to handle it in an acceptable run time per calculation point, as this latter includes a huge amount of data point due to the additional real stress time history.

By considering the stress time history from the vibration event, all relevant peaks and their sequence can be considered at the appropriate part of the flight and processed using conventional fatigue methods. Additionally for parts driven by crack growth, using the real stress time history for a specific vibration event can be an advantage, as it allows using the benefit of crack retardation effect. Overall this alternative approach enables to be as close as possible to the real vibrational loading experimented by the aircraft and which cannot be simplified by conventional means due to its chaotic behaviour.

TESTING APPROACH – DYNAMIC VS STANDARD

Test evidence is also used to support analysis as stated in [1]. Representative testing can be conducted at different levels – from the very coupon material and single component test to the full-scale fatigue test. In aircraft industry this means furthermore showing compliance for a full life of the plane. The typical fatigue flight needs consequently to be tested and repeated as long as necessary to validate fatigue results during a full life of the aircraft in service.

Since the spectrum for one flight can contain thousands of data points, the classical way of fatigue testing using standard hydraulic jacks would become too time-consuming within the industry frame. Applied frequency is indeed typically 1 Hz with such cylinders. Testing a single flight with more than 50000 points at this frequency would cost approximately 14 hours, which is even longer than a real flight. This would lead consequently to an unrealistic overall testing time during many years, way beyond the entry into service.

For a component loaded with one significant load, the full fatigue spectrum including steady loads and vibration events can be reduced, even if multiple locations for fatigue calculations are identified. This structural part might indeed contain different fatigue sensitive design features such as lugs, bolts or radii, where different stress concentrations and stress intensity factors will lead to different fatigue or crack propagation results. Still the reduction can be done by removing the not significant cycles based on fatigue damage rather than on stress criteria, similar like in [4], considering all locations as necessary. Additionally for parts to be justified with crack propagation, the representativeness of the reduced spectrum based on fatigue damage is iteratively checked at relevant locations using crack growth analysis.

However higher up in the aircraft test pyramid, the reduction does not converge anymore for all locations and is only applied for the quasi static part of the spectrum. The reason of the non-convergence is that in structural tests of (sub-) assemblies, various locations are significant and some are driven by different combinations of load components. Additionally, in case of a chaotic behaviour of the loads, it is not possible to reduce the spectrum based on stress cycles and corresponding damage and still keep the reduced spectrum representative for multiple locations.

Consequently, it is decided to enhance the structural tests by using high frequency hydraulic cylinders. This type of hydraulic jacks allows to reproduce dynamic flight test loads as-is with a high frequency. To test a full flight, a combination of standard fatigue spectrum for steady loads and dynamic spectrum for the vibration part can be applied. During test load validation, additional calculation checks of test spectrum are performed based on damage comparison. Loads are enhanced or tuned correspondingly using load enhancement factor or fitting the dynamic behaviour to the one of the spectrum, in order to be representative and on the conservative side. Good accuracy between desired loads and load response of test can be obtained.

By testing this way the test spectrum remains representative of the vibration loading with similar complexity as used in the fatigue calculations. Furthermore the test time of approximately 2 minutes per flight stays at an acceptable level.

CONCLUSION

The addition of new significant loads into an existing fatigue spectrum will reduce the maintenance intervals and consequently increase the maintenance tasks for the airlines. In case of adding chaotic vibration loads in multiple directions, a simplification of the load spectrum using addition of exceedance diagrams will lead to an overestimation of local stress cycles. A solution to avoid the overestimation is to directly consider stress time history of the vibration part and include it in the classical spectrum. This stress time history can be obtained either from flight tests measurements as described in this paper or from other applicable methods such as dynamic models. The creation of the local stress spectrum is then based on the time of occurrence of each contributing load components or directions. By considering this chaotic vibration loading as it is during the whole fatigue and damage tolerance assessment, this over-conservatism in stresses can be then reduced.

Consequently the impact on maintenance program of this additional chaotic loading in spectrum is also minimised. Nowadays computer performance is capable of processing the necessary amount of data needed for this approach. Furthermore the improved test concepts with high frequency cylinders make sure the structural testing can be performed with a similar complex spectrum as in the calculations and within an acceptable timescale. This enables to justify the calculation approach and be as close as possible to the real loading conditions.

REFERENCES

[1] Certification Specification for Large Aeroplanes, EASA CS 25.571 / FAR 25.571

[2] Broek, D. (1988). *The practical use of fracture mechanics*, Kluwer Academic Publishers, Dordrecht, Netherlands

[3] Schijve, J. (2009). *Fatigue of structures and materials*, 2nd Edition, Springer Science+Business Media, B.V.

[4] Dosman, S. (2013). In: *Joint RAF and RAAF C-130J Full Scale Wing Fatigue Test Load Development and Test Running*, 27th ICAF Symposium, Jerusalem, Israel.

Delamination analysis of a trailing edge flap

Jarno Jokinen*, Markus Wallin, Olli Saarela

Aalto University, Finland,
**corresponding author*

Abstract: A hinge of the trailing edge flap (TEF) was modified during the midlife upgrade program of the Finnish Air Force F-18 fleet. Ultrasonic testing after the hinge removal revealed delaminations in the composite web of the flap. In this work, we analysed two flaps using Virtual Crack Closure Technique (VCCT) to evaluate energy release rates (ERR) at edges of the delaminations. A fracture criterion was further used to evaluate criticality of the values. The ERR values and computed fracture criterion values were relatively small in one flap. In the other flap delamination was more critical.

INTRODUCTION

Composite structures have a layer like structure that may encounter interlaminar damages such as delaminations. Delaminations may be a result of manufacturing defects or faults during overhaul process, or they may be caused e.g. by impacts during the service of a structure.

Modifications for increasing lifetime of F-18 were performed during the midlife upgrade program of the Finnish Air Force F-18 fleet. These included modifications of the trailing edge flap (TEF). During the work, aluminium hinges were replaced with titanium hinges. An ultrasound inspection performed for the spars after the hinge removal revealed delaminations around the fastener holes in the spar web. Thus, it was necessary to evaluate criticality of the delaminations to determine whether the flaps are still airworthy without repair.

Several fracture mechanics based analysis methods are available for analysing criticality of delaminations. One of the methods is Virtual Crack Closure Technique (VCCT). VCCT is a finite element based method for computing energy release rate values at the crack tip for three fracture modes. The computation is based on nodal displacements and forces at the crack tip. The criticality of a delamination can be evaluated by comparing computed energy release rate values to the corresponding critical values. Critical energy release rates are measured experimentally for the applied material and interface.

We applied VCCT in the analysis of two delaminated F-18 TEF spars. We had two targets for the analyses. Firstly, our interest was to study applicability of VCCT for the case. Secondly, we wanted to evaluate criticality of the delaminations.

MATERIALS AND METHODS

Materials

The TEF spar is made of carbon/epoxy laminate AS4/3501-6. It consists of 15 plies with the stacking sequence of $[45/-45/0/45/-45/45/-45/90]_{SO}$. The thickness of the spar is close to two millimetres.

Fracture mechanics properties of the material are required for defining the criticality of a delamination. Hintikka [1] performed experimental testing which provided critical energy release rates G_{IC}=72 J/m^2 and G_{IIC}=779 J/m^2 for the current material. G_{IIIC} was not measured. Typically it is assumed to be equal with G_{IIC}. Hintikka [1] also provided mixed-mode results (mode I and II) for the current material. The Power law curve fit was defined for the data by Hakuri [2]. The Power law is a mixed-mode criterion for evaluating criticality of a delamination. A delamination is expected to grow, when the criterion gets the value of one or a higher value. With the current material and test data the Power law takes the form

$$f = (G_I/72)^{0.75} + (G_{II}/779)^{0.69}. \tag{1}$$

It should be noted that the failure criterion includes only the modes I and II. Experimental results and the Power law fit are shown in Figure 1.

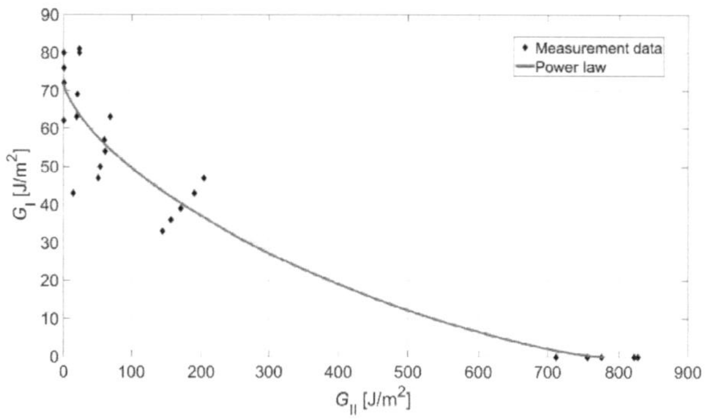

Figure 1. Experimental results and Power law fit for AS4/3501-6.

Delaminations

The trailing edge flap without a hinge is shown in Figure 2. The position of the hinge is seen as the grey area in the spar with six rivet holes.

Figure 2. Trailing edge flap composite spar with hinge rivet holes.

We studied two trailing edge flaps in our analysis. Delaminations of the flap spars are shown in Figure 3. Flap A included only one delamination that located around the rivet hole c. Based on ultrasonic tests, the diameter of the delamination was estimated to be 15 mm. The delamination was estimated to exist in the depth of 1.7 mm measured from the front side of the spar. This states that the delamination locates between the 13th and 14th ply (2 mm/15≈0.132 mm). The delamination is thus close to the backside of the spar.

The second flap under study (flap B) had two delaminations. The smaller delamination B1 in Figure 3 has initiated from the rivet hole (b). The delamination is circular with the diameter around 5 mm. The larger delamination B2 is around two rivet holes (e and f). It continues to the junction of the web and flange. The width of the delamination is around 50 mm and the height close to 30 mm. The delaminations B1 and B2 and their dimensions are further illustrated in Figure 4. Delamination B was estimated to exist in depth of 1.5 mm which states location between the 11th and 12th ply.

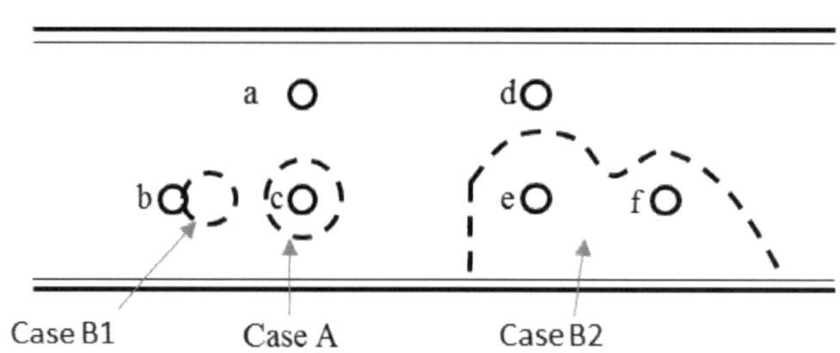

Figure 3. Delaminations being studied.

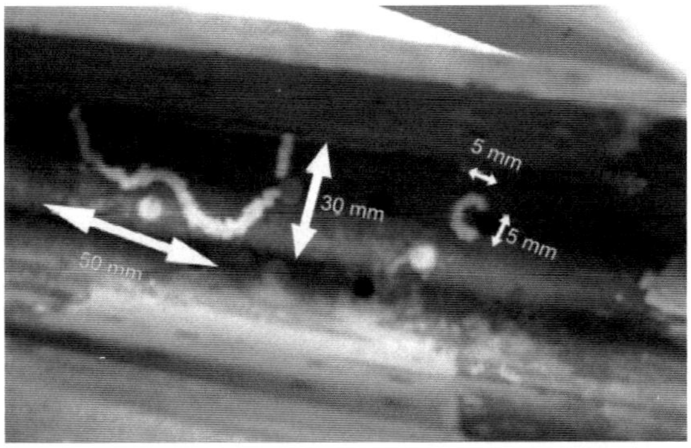

Figure 4. Delaminations B1 and B2. [3]

Virtual crack closure technique (VCCT)

Virtual crack closure technique (VCCT) is a linear elastic fracture analysis method. It is originally based on Irwin's crack closure integral which was later modified by Rybicki and Kanninen [4] and further developed by several researchers. A wide review of the method is presented by Krueger [5]. A crack tip modelled with shell elements is shown in Figure 5. Energy release rate equations for VCCT can be presented in the form

$$
\begin{aligned}
G_I &= -\frac{1}{2b\Delta a} Z_{Li}(w_{Ll} - w_{Ll^*}) \\
G_{II} &= -\frac{1}{2b\Delta a} X_{Li}(u_{Ll} - u_{Ll^*}) \\
G_{III} &= -\frac{1}{2b\Delta a} Y_{Li}(v_{Ll} - v_{Ll^*})
\end{aligned}
\quad (2)
$$

where Δa is the element length, b is the element width, w, u, v are the displacements and Z, X, Y are the forces in the z, x and y coordinate axis directions [5].

We applied VCCT by studying only crack onset, i.e., by computing energy release rate values at the nodal points of the delamination edges.

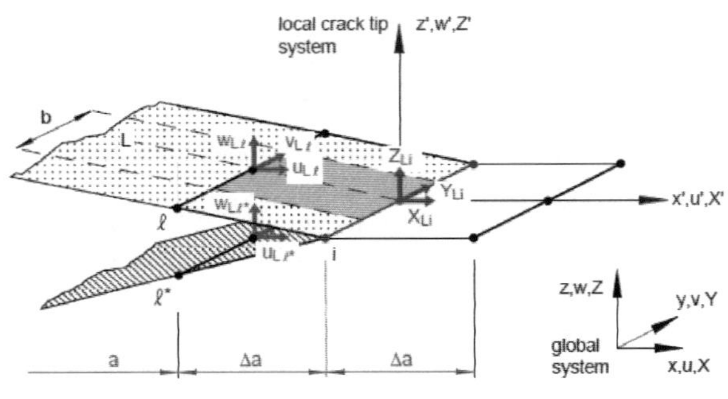

Figure 5. VCCT quantities at the crack tip when utilizing shell elements. [5]

FINITE ELEMENT ANALYSIS

Analysis approach

We received a submodel of the trailing edge flap from Patria Aviation. The submodel contained load information representing the most critical flight condition. The submodel is presented in Figure 6 where the spar is highlighted. We modified the submodel for our analysis by replacing the original web of the spar with two geometry duplicates of the web. This allowed us to model delaminations in between the web duplicates. Material properties of the two webs were defined based on sublaminate properties. These properties were dependent on the studied interface where the delamination was assumed to exist. The modelled webs were positioned equally with the removed web. This created a slight geometrical error because none of the delaminations was at the symmetry plane of the web. We utilized tie constraint restricting both translation and rotation degrees of freedom for attaching the sublaminates to the original submodel.

Delaminations were created using partition of the created geometry. Partitions were also applied for improving mesh generation. We tried to utilize rectangular and regular sized elements at the crack tip. Since we did not analyse delamination growth, the mesh around the delamination edges was merely for transferring loads. Material properties of web sublaminates were defined using the ESAComp software from which bending stiffness matrices were achieved.

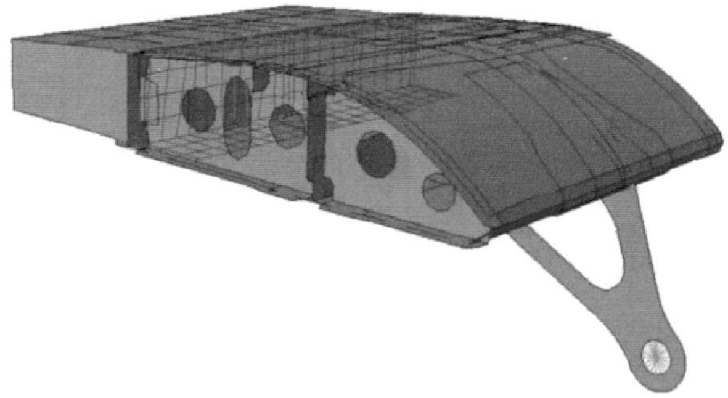

Figure 6. The submodel of the trailing edge flap where spar is highlighted.

In the delaminated area we had three parts that were defined using shell elements: two sublaminates of the web and the hinge. VCCT definition was modelled in between the sublaminates. The thicker sublaminate was defined as a master surface in all analyses. One interest in our analysis was to study how to model the joint between the spar web and the hinge. We studied possible options in a preliminary analysis using a tie constraint in between the hinge and the spar and by utilizing connector elements [6]. When using connector elements, we further applied different techniques for defining how translations and rotations are transferred between the parts. The four techniques studied in our preliminary analysis were:

1. Tie constraint in between the whole surface between hinge and the web (without connector elements)
2. A connector element in between the hinge and the web, translation restricted
3. A connector element in between the hinge and the web, translation and rotation restricted
4. A connector element in between the hinge and the web and in between the sublaminates, translation restricted.

The option 2) was noted to provide most conservative ERR values and was selected for final analyses [6].

We evaluated ERR values at the edges of delaminations. In the delamination case B2, the edge reaching the junction of the web and flange was excluded due to the limitations of the applied model. The results are presented in the following chapter in terms of maximum ERR values existing at delamination edges. In addition, values provided by the mixed-mode criterion with the maximum ERR values are given. It should be noted that the latter values are conservative because maximum ERR values for different modes do not occur at the same nodal point.

Interface effect

Delaminations in the trailing edge flap were detected with ultrasonic testing. The method did not provide exact information on the position of the delaminations in the thickness direction. Therefore we computed ERR values for delaminations by assuming them to exist in different interfaces. The finite element model was created utilizing shell elements. This made it easy to study the interface effect since in shell elements the thickness is provided for the software with one value. With this procedure the position of the spar is also changed but it does not affect much analysis results because ply thickness is relatively low.

Initial imperfection effect

In a delaminated area the surrounding plies are not perfectly in contact with each other, i.e., some distance exists between the plies as illustrated in Figure 7. We also studied how such initial imperfection affects ERR values. We modelled the imperfection using partition in an area inside the delamination. In case A the imperfection was assumed to be circular in shape. The diameter of the initial imperfection was close to 5 mm. In the delamination case B1 the distance between the initial imperfection edge and the delamination edge was around 1 mm. In the delamination case B2 the area of initial imperfection did not completely follow the delamination edge. In the vertical direction the imperfection extended from lower edges of the rivet holes e and f (see Figure 3) to the distance of 4 mm from delamination edge closest to the flange. In the horizontal direction the imperfection extended approximately to the distance of 4 mm from the delamination edges. The delamination is illustrated in Figure 8.

The magnitude of the enforced displacement was in all studied cases 0.1 mm. The imperfection was realized with an enforced displacement while the delamination edge displacements were restrained. The loading of flight condition was then applied to the deformed delamination geometry.

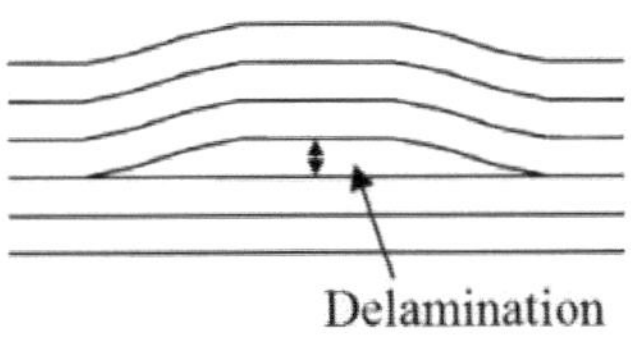

Figure 7. Characterization of initial imperfection.

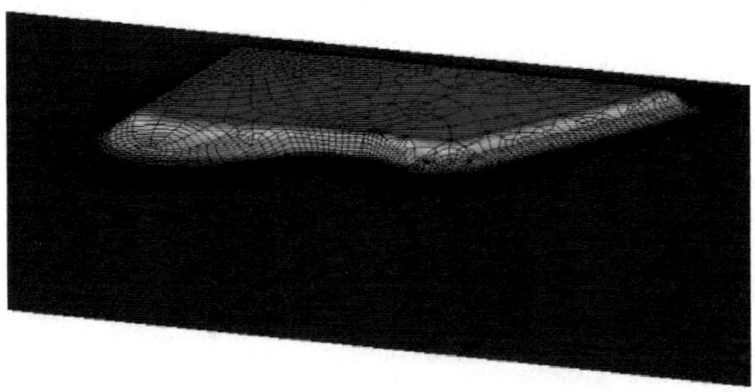

Figure 8. Exaggerated initial imperfection in delamination B2.

RESULTS

Rivet forces

We defined rivet forces using connector-elements before studying delaminations. The rivets are identified as shown in Figure 3. The connector-element forces using this convention are shown in Table I in the form of out-of-plane loads and resultant in-plane loads. The out-of-plane loads are seen to be low when compared to the in-plane loads. This suggests that loading at delamination edges should be mode II and III dominated. The delamination A exists around the rivet c with the lowest in-plane load. The delamination B1 is close to the rivet b which is more loaded than the rivet c. The rivet f achieves the highest in-plane and out-of-plane loads. This rivet is inside the delamination B2. The other rivet inside this delamination, rivet e, is also relatively highly loaded.

Table I. Connector forces.

Rivet	In-plane [N]	Out-of-plane [N]
a	387.6	9.4
b	814.0	0.0
c	270.3	-10.6
d	838.4	14.4
e	771.4	-19.5
f	2420.1	-251.0

Interface effect

The interface effect was studied by assuming the delamination A to exist in the interface 11/12, 12/13 or 13/14. Based on the results of ultrasonic testing, the interface 13/14 is the most probable location of the delamination. Table II presents the interface effect on ERR values. It can be seen that the ERR values are low in all cases. Mode I values are almost negligible. The results further reveal that mode I values decrease and mode II values increase when the delamination is moved closer to the middle surface of the spar. This result is understandable since shear stresses increase towards the middle surface. The mode III values do not have a clear trend. Fracture criterion values are low in all studied cases suggesting that the delamination is not critical.

Table II. Interface effect on ERR in delamination A.

Interface	G_I [J/m^2]	G_{II} [J/m^2]	G_{III} [J/m^2]	f
11/12	1.36E-03	0.40	0.23	0.006
12/13	1.71E-03	0.33	0.14	0.005
13/14 ref.	2.25E-03	0.23	0.14	0.004

The delamination B2 results are shown in Table III. It can be seen that fracture criterion values are much higher than in case A. This could already be assumed based on the rivet force study and size of the delamination. All ERR values are relatively high indicating that the delamination B2 could be critical. The original modelling was performed by assuming the delamination to exist in the interface 11/12 but we also studied the cases in which the delamination was assumed to exist in the neighbouring interfaces. Table III presents that all ERR values increase towards the middle surface (interface 10/11 vs. interface 11/12). In the other direction, mode I and III values decrease and mode II value slightly increases (interface 12/13 vs. interface 11/12).

Table III. Interface effect on ERR in delamination B2.

Interface	G_I [J/m^2]	G_{II} [J/m^2]	G_{III} [J/m^2]	f
10/11	24.82	61.66	70.40	0.624
11/12 ref.	17.58	46.92	53.69	0.491
12/13	8.41	49.40	35.17	0.349

Results of the delamination case B1 are presented in Table IV. It can be seen that fracture criterion values are smaller when compared to the delamination case B2 but higher when compared to the delamination case A. The latter result is in line with the magnitude of the connector forces (b vs. c). Case B1 is also clearly mode II dominated. Mode II ERR values are seen to increase and mode III values to decrease towards the middle surface.

Table IV. Joint modelling effect on ERR in delamination B1.

Interface	G_I [J/m^2]	G_{II} [J/m^2]	G_{III} [J/m^2]	f
10/11	0.56	13.01	2.49	0.086
11/12 ref.	0.09	11.72	4.26	0.062
12/13	0.63	8.98	5.04	0.075

Initial imperfection effect

The effect of initial imperfection on ERR values at edges of the delamination A is given in Table V. It is seen that an initial imperfection increases mode II and III values while the mode I value decreases. The value of failure criterion increases but is still relatively small.

The imperfection effect on ERR values at the edges of the delaminations B1 and B2 was studied separately. The results are given in Table VI and Table VII. In the delamination B2 the imperfection increases mode II ERR values considerably. The effect on mode I and III values is not as high. In the delamination B1 the effect of imperfection on the mode II ERR value is even more dramatic. The imperfection increases the mode II ERR value to the same level as in the delamination B2.

Table V. Initial imperfection effect on ERR in delamination case A.

Imperfection [mm]	G_I [J/m^2]	G_{II} [J/m^2]	G_{III} [J/m^2]	f
-	2.25e-3	0.23	0.14	0.004
0.1	9.76e-4	0.93	1.77	0.01

Table VI. Initial imperfection effect on ERR in delamination B2.

Imperfection [mm]	G_I [J/m^2]	G_{II} [J/m^2]	G_{III} [J/m^2]	f
-	17.58	46.92	53.69	0.491
0.1	17.64	162.08	76.55	0.687

Table VII. Initial imperfection effect on ERR in delamination case B1.

Imperfection [mm]	G_I [J/m^2]	G_{II} [J/m^2]	G_{III} [J/m^2]	f
-	0.09	11.72	4.26	0.062
0.1	0.09	157.23	17.69	0.338

CONCLUSION

We studied applicability of the VCCT method in the computation of ERR values at edges of delaminations detected in the trailing edge flap of F-18. We found that modifying an existing structure submodel for the VCCT analysis is relatively easy. The modification requires changes mainly in geometry because an interface for the delamination needs to be modelled. Numerically the utilization of the method for delamination onset study is effective and straightforward.

We studied two trailing edge flaps. The delamination A in one flap was circular and around a rivet hole. Based on our analysis, the ERR values at the edge of this delamination are small. This was expected since the load transferred by the rivet was computed to be small. The values were not sensitive to uncertainties covered by the analysis. The ERR values provided by VCCT and the computed fracture criterion values suggest that the delamination is not critical.

Another flap including the delaminations B1 and B2 was assumed to be more critical. The small delamination B1 was around a rivet hole, similarly to the delamination A. However, the rivet hole and its loading were different. VCCT analysis results indicated that ERR values at the edge of the delamination were considerably higher than in the delamination A. The main interest was in the larger delamination B2 which had propagated to the junction of the web and flange. The VCCT analysis provided high ERR values. The fracture criterion value based on maximum ERR values was also high. We recommend that this flap should be studied more extensively, e.g. with delamination propagation analyses. A more detailed model should also be used since we could not analyse with our simplified model the delamination edge at the junction of the web and flange.

REFERENCES

[1] Hintikka, P. (2009) Determination of interlaminar fracture toughness of composite laminates, Master's thesis, Aalto University (in Finnish)

[2] Hakuri, O. (2015) Mixed-mode failure criterions for delamination analysis, Bachelor's thesis, Aalto University (in Finnish)

[3] Crack inspection document 893/08, Patria Aircraft (in Finnish)

[4] Rybicki, E. F. and Kanninen M. F. (1977) *Eng. Fract. Mech.*, vol. 9, pp. 931-938

[5] Krueger, R. (2002) The Virtual Crack Closure Technique: History, Approach and Applications, NASA/ CR-2002-211628

[6] Jokinen, J. (2012) Delamination modelling of trailing edge flap spar, Laboratory report KRT-T366, Department of Applied Mechanics, Aalto University (in Finnish)

Full Scale Wing Test – Test Health Monitoring

Stephen Dosman, Naumaan Hamid

Marshall Aerospace and Defence Group

Abstract: Full scale wing fatigue tests can involve dozens of actuators and hundreds of strain gauges. Ensuring any such test is operating properly will require a sophisticated monitoring system that will need to look at more than just actuator accuracy; as well as a monitoring system it will also require a comprehensive maintenance programme over the years that the test may run.

Details of the test health monitoring carried out on the RAF/RAAF C-130J Full Scale Wing Fatigue Test are provided along with details of the rig maintenance schedule.

Lessons learned from test included a better understanding of the importance of detailed trend monitoring in order to identify issues early and show that rectification was effective, an understanding of the difficulties of identifying dynamic running effects, an understanding of where dither can improve servo-valve performance in the short term, an understanding that monitoring of strain-gauge output creep can provide early warning of failures, and also when actuator tremble my occur.

INTRODUCTION

So you've set up a full scale wing test, you have your load spectrum, and you have a rig that can apply the spectrum within your stroke and load limits – are you ready to go? Not quite. Test 'health' is a key concern when carrying out a test programme over several years, and this can cover such things as ensuring that you are applying loads consistently as well as looking after the response of your equipment and the impact the equipment has on test fidelity as it ages.

The RAF/RAAF C-130J Full Scale Wing Fatigue Test (WFT) currently under way at Marshall Aerospace and Defence Group's Cambridge UK facility is a complex system that must run for many years at a high accuracy level; There is a proportional-integral-derivative (PID) control system including airbag loading, 40 hydraulic loading actuators, an applied loading spectrum derived directly from on-aircraft measurements rather than constructed from idealised wing states, hundreds of strain gauge measurements to monitor, and a customer requirement to identify and deal with developing damage via trend monitoring as well as via normal maintenance inspections. The test was contracted to meet tight levels of applied loading Accuracy, Repeatability and Dynamic Overshoot, and a means to monitor, assess, and confirm that these requirements continue to be met is a requirement of the contract. For more details on how the loading system was developed see reference [1].

To meet this challenge a comprehensive Trend Monitoring Process was developed in close collaboration with DSTO that would confirm metrics achieved against these requirements for every one of over 750,000 Load Lines (LL) applied during every 3,000-Equivalent-Flight-Hour test Superblock. This ongoing process needed to confirm that the applied loads were as intended, identify incipient damage developing based on strain response, and aid in the diagnosis of any misbehaving equipment.

How the trend monitoring process works, component ageing and impact on test health, and lessons learned will all be discussed. These results are of interest for any organization attempting to carry out full scale fatigue testing on an aircraft structure.

DISCUSSION

Criteria for fatigue test health can be difficult to define: one could define a 'load fidelity' target for example, but that wouldn't tell you the impact on the fidelity of test results. On the other hand one could define a 'results fidelity' target based on fatigue performance or on crack growth rate performance – the results will be different depending on the metric chosen. In the case of the WFT this dilemma is simplified in that there is a Def Stan 00-970 (reference [2]) Acceptable Means of Compliance as design input. These requirements and how they are met and reported to the Authority (The RAF and RAAF) will be discussed, followed by details on the trend monitoring processes used, a discussion of component ageing and suggested maintenance periods, and finally details of lessons learned on this programme.

Requirements

For the WFT the primary test fidelity and health requirements stemmed from Def Stan 00-970, reference [2], Leaflet 37, which highlights the high sensitivity of fatigue test fidelity to the loading accuracy and repeatability. In summary the loads should be:

- Accurate – accuracy should be within 3% of maximum demanded load for each load line
- Repeatable – load lines in a test block should be within 2% of the particular applied load or 0.5% of the maximum demanded load, which ever the greater, as each block is repeated
- Free from dynamic effects such that accuracy or repeatability are achieved throughout the structure

Figure 1 illustrates the repeatability requirements – essentially repeatability tolerance goes up with increasing load, but even for small loads there is a minimum limit of 0.5% of the maximum applied load. It should be noted here that the Leaflets define an acceptable means of compliance (AMC), but not the only means. Other means of demonstrating adequate test fidelity can be considered if necessary.

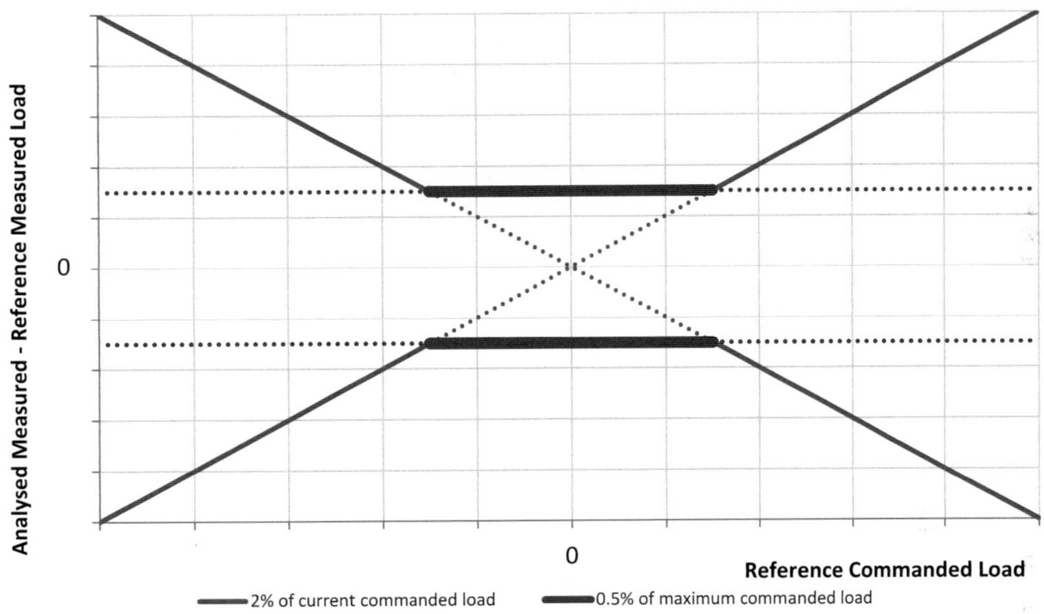

Figure 1. Def Stan 00-970 – Acceptable Means of Compliance, Repeatability Limits.

Whilst the details of the accuracy and repeatability requirements can take some work to develop[7], it is the dynamic effects requirement for which it can be difficult to establish a suitable measurement metric. Essentially the 'a priori' assumption used when developing the loads for the test is that the test article is in equilibrium for every applied Load Line; however in reality when a Load Line is applied to the test article it is in motion (accelerating/decelerating), and the assumption that Total Force equals Mass times Accelerations equals Zero is therefore not strictly correct. Furthermore if the frequency of the wing fundamental vibration modes couples with the load application rate, then loading away from the actuators may not reflect that which is intended.

How these requirements are demonstrated to be met is discussed in the next section.

It should be noted that if one assures individual actuator accuracy and repeatability to the levels given in the Def-Stan, then the effect on the wing internal loads for a multi-actuator test will always be significantly higher accuracy/repeatability than for the individual actuators. Put another way: unless there is a systemic error in the loads, then accuracy and repeatability errors on individual actuators will tend to average out. This means that compliance with the Def Stan suggested limits will generally lead to an extremely high fidelity on a multiple actuator test.

[7] For example when considering accuracy does one refer to the maximum load or the maximum absolute load? When considering repeatability does one compare load lines back to a reference block at the beginning of test, or against the most recent load line? Etc.

Trend Monitoring Process

The test was contracted to meet exacting Def Stan 00-970 requirements on actuator absolute accuracy, actuator repeatability from test block to test block, and on actuator dynamic effects; the customer had specified that this monitoring process would need to be actively reviewed and checked and formally reported for every Load Line applied to the test article.

In order to achieve these monitoring goals software and methods were developed that generated automated reports with summary plots on prime metrics going into a regularly issued report; behind these reports soft copy annexes contained thousands of html generated plots covering the detailed response of each actuator and strain gauge that could be queried if a metric summary indicated a possible issue. Also generated is a strain gauge follow-on report where any strain gauges with anomalous behaviour would be assessed and classified in order to identify possible damage development. The follow-on report was also used to confirm the health of the strain gauges and their ability to function as expected to capture valid data and provide a secondary data source for evaluating actuator performance, if needed.

These high standards may not be the industry norm, but putting together tools to report metrics against every load line in the test allowed Marshall to provide objective evidence of how well the test was meeting the requirements, allowed any signs of deterioration in the performance to be quickly identified, and allowed the benefits of treatments to correct deterioration to be objectively measured. For example deteriorating servo-valves could be identified through trend monitoring checks, and steps to improve performance utilised the trend monitoring tools to confirm whether or not the steps taken had succeeded.

Some examples of some of the checks that are carried out are provided in Figures 2 through 4 below. The checks are reported every 3,000 Test Equivalent Hours[8] with the body of the report consisting largely of summarising charts as that shown in Figures 2 and 4. Soft copy annexes to the report contain more detailed charts as Figure 3 to allow sufficient granularity[9] if a particular problem is being investigated and more detail is required.

[8] One 'Superblock' of 6 test blocks is approximately 3,000 Test Equivalent Hours.

[9] For the WFT this amounts to over 2200 plots/tables per report provided in soft-copy html format.

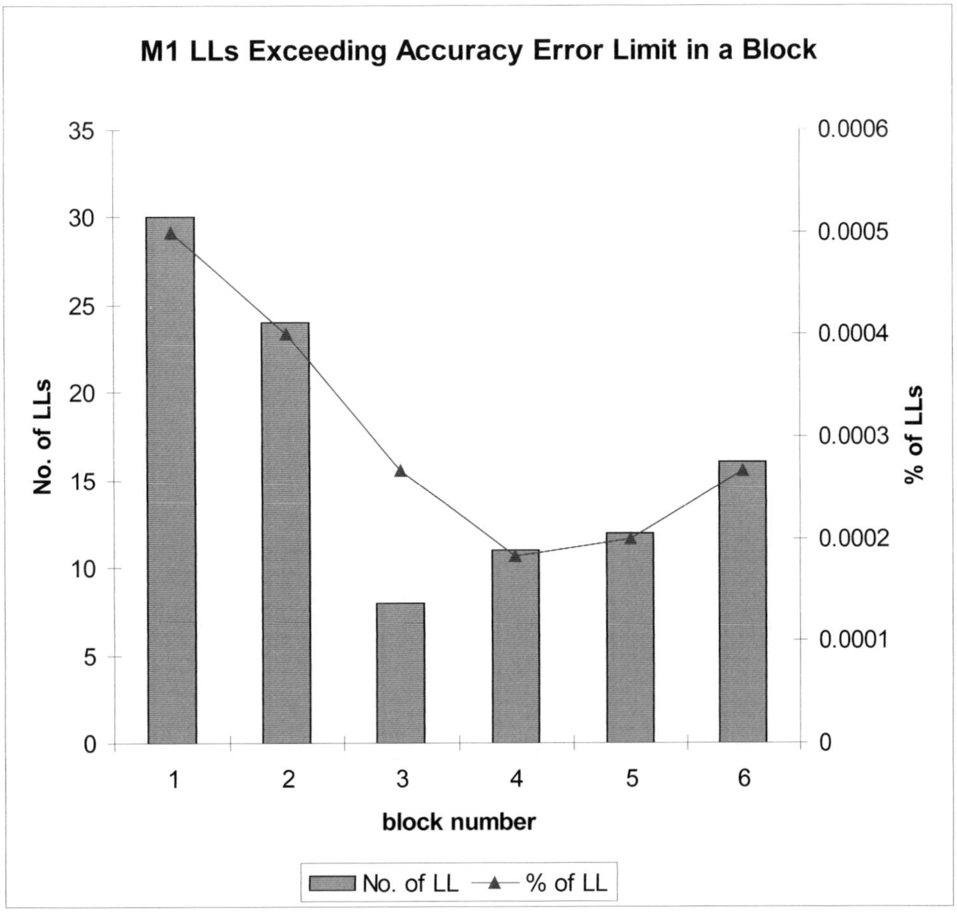

Figure 2. Example Trend Monitoring Report Plot – Load Lines exceeding Accuracy Limit – Sample Data.

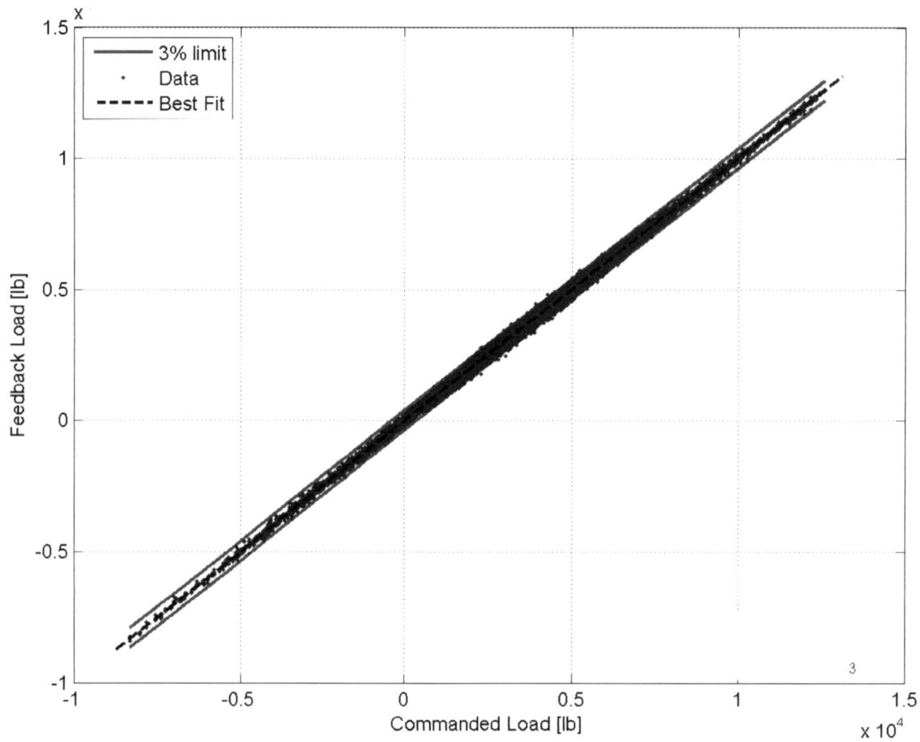

Figure 3. Example Trend Monitoring Report Plot – Load Lines command vs feedback – Sample Data.

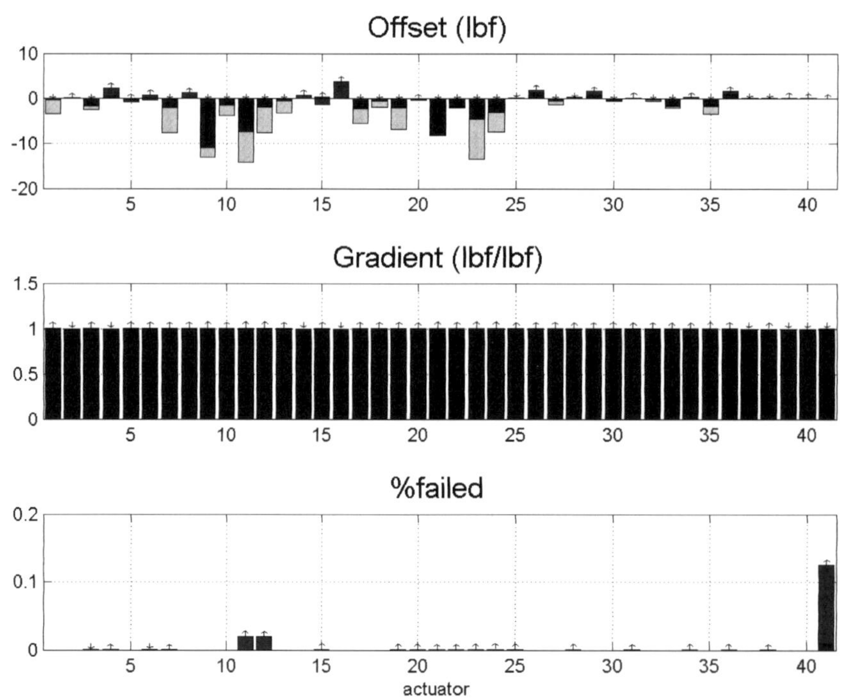

Figure 4. Example Trend Monitoring Report Plot – Command vs Feedback crossplot summaries – Sample Data.

One aspect of health monitoring that is difficult to put a specific metric on, but which has been found to be very helpful, is the use of strain gauge monitoring in critical areas to identify incipient damage prior to failure. It was thought at the beginning of the test that although identifying growing cracks from strain gauge data prior to failure might be possible, it was not something that would prove practical; however after failures on the test article where a subsequent review of the data clearly showed signs of significant shifts in strain gauge response prior to failure, it is now a key activity to review gauge results for signs of shifts that may indicate early signs of cracking. An example plot is shown in Figure 5 below.

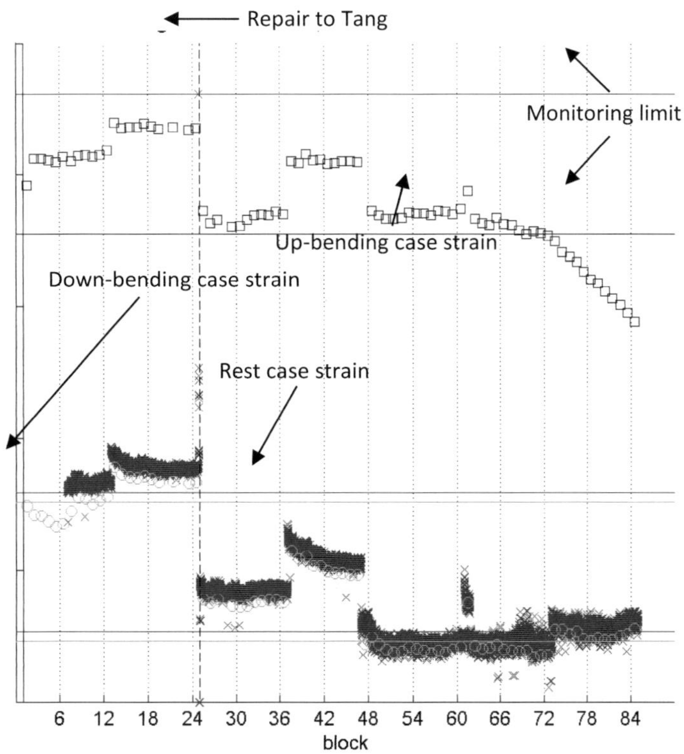

Figure 5. Strain Trending Data, micro-strain vs test block.

Figure 5 shows an engine truss-mount tang strain gauge result for three load lines repeated over many test blocks. It can be seen that over time the strain reading under a wing up-bending case started to drift beyond warning limits, whilst the others were essentially unchanged. This could be symptomatic of the strain gauge being in the vicinity of a crack which would open in up-bending but closed in down-bending. Here the drop in strain could be due to load dropping in the member due to the increased flexibility compared to the surrounding structure. In this case it was found that when inspected at Block 84 there was a 1/2" crack present in the tang.

The benefit to the test of this strain gauge drift monitoring is that inspections for suspect areas can be identified early and directed to areas of possible cracking; although gauge drift may not mean cracking is present.

On the WFT all strain gauges are monitored this way, and are regularly reviewed[10]. If a gauge exceeds a monitoring limit, then the strain response is investigated, and if the evidence warrants a directed inspection carried out to look for signs of cracking or other degradation of the test article. The strain-gauge follow-on reports attempt to characterise why a particular gauge exceeds a particular limit, and will recommend inspections or gauge rectification if warranted.

Overshoot monitor. In order to ensure that the loads applied to the test are as intended then their dynamic effects must be understood. The actuator forces are developed on the assumption that the test article is static; however the test is always in motion, and therefore the test article acceleration is non-zero and between the actuators vibration effects may lead to unintended loads developing. As long as the wing primary bending modes are not close in frequency to the

[10] Albeit that there is a short-list of gauges in more critical locations that is reviewed at a higher frequency in order to identify problems earlier if possible.

applied loading rate and as long as the acceleration as a load line is relatively small, then these dynamic effects are also small.

In order to demonstrate that the dynamic effects are small, Marshall developed a method to compare strain-survey strain data from low-speed running with that measured at full running rate; however it was found that this method only measured the natural variability in loading seen from block to block on the test. Therefore a new method was developed – this was called the Strain Phase Monitoring (SPM) method. This new method considered the Force equals Mass times Acceleration (F=ma) effects separately, and concentrated on showing that for strain gauges on the wing that the peak strain occurred at the point when each load line was achieved. I.e. that there was no phase lag and no overshoot (see Figure 6 below). For the F=ma question it was shown that for the peak accelerations seen on the test article that the equivalent force error was negligible.

The overshoot is measured in terms of micro-strain for each OLM load bridge on the test article (These are strain gauges that are 'bridged' to generate a mV output that can be correlated to an internal load value (bending moment, shear, torsion, etc); however each bridge also generates a micro-strain value[11]). An overshoot limit of 10-micro-strain for all bridges has been accepted by the Authority based on the recommendations of the Defence Science and Technology Organisation (DSTO).

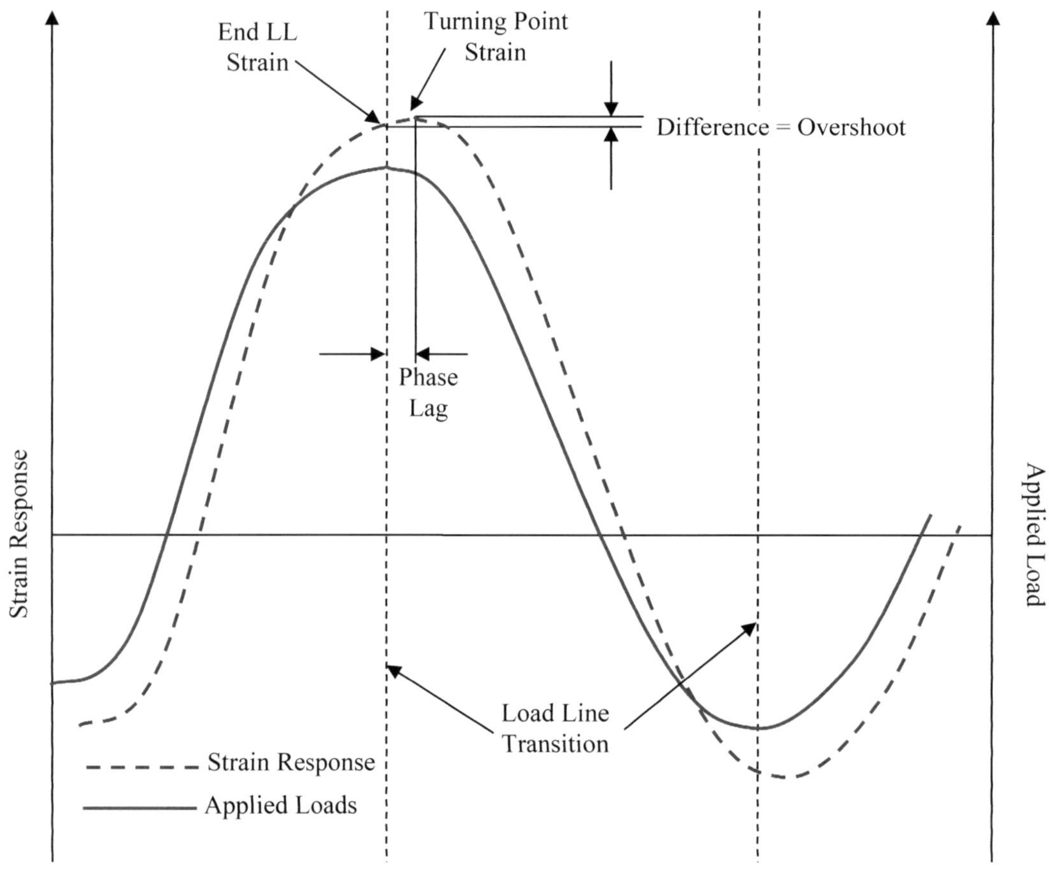

Figure 6. Strain Phase Monitor example method.

No sign of any dynamic overshoot that exceeds 10-micro-strain has been found on the test article; all investigations of load lines that flag against the 10-micro-strain limit have been ascribed to issues not associated with overshoot.

Component Ageing and Suggested Maintenance

[11] The relationship between measured micro-strain output and actual strain will vary depending on the load bridge type. E.g. For a full bridge measuring bending strain the measured output will be four times the actual bending strain (reference [4] section 8.4.2).

On a multi-year full scale test programme the ageing of rig components and their maintenance becomes a significant issue to manage. A list of primary areas of concern is provided below in Table I along with suggested maintenance periods suitable for a full scale wing test

Table I. Suggested maintenance to address component ageing in full scale fatigue test

	Component	Suggested Maintenance	Comments
1	Hydraulics	Filters & Pumps (power packs), yearly Oil cleanliness check, 6 to 12-monthly	Oil contamination can strongly affect servo valve operation and thereby actuator accuracy and repeatability
2	Servo-valves	Normally replaced or overhauled 'on condition' (Trend monitoring repeatability performance is used to indentify poorly performing servo-valves)	Subject to wear and tear over time. On WFT the servo-valves are returned to the manufacturer for overhaul or replacement based on trend monitoring
3	Actuators	Weekly 'walk around' check for leaks Close visual inspection every quarter. If signs of pitting, scoring, or heavy leakage then return to manufacturer for re-chroming or seal replacement	Actuators with low displacement and high frequency can be expected to exhibit greater wear and tear. Actuator rod wear can lead to loss of seal effectiveness thus causing weeping/leakage Seal replacement can be a large undertaking
4	Hydraulic pipelines and hoses	Replace rubber hoses every 5 years, or earlier if usage or inspections dictate	Hard pipes not generally an issue Hoses are rubber and can suffer from internal degradation that causes oil contamination. Excessive pulsing due to heavy/rapid actuator use can exacerbate this
5	Load cells and pressure transducers	Full recalibration against an 'appropriate standard' every 2 years maximum (yearly calibrations may be necessary and is sometimes considered best practice) Shunt calibration of system every year	Calibration of the load cells and the whole signal chain (via shunt calibration) is a key part of maintaining the effectiveness of testing equipment. The calibration ensures that actual loads are the same as what is recorded as applied.
6	Mechanical test rig joints	Greasing at an appropriate interval depending on the type of joint and bearing design. On the WFT this was often at every quarter Replace if wear and tear too high (Inspect for excessive play, and in some cases checking tolerances), inspect at an appropriate interval and at natural breaks in the test programme. On the WFT this was often at every quarter.	E.g. trunnion bearings, gimbals, etc. Excessive play can lead to backlash, and interfacing problems in joints. These issues in turn can lead to load repeatability issues.

Notes:

- Maintenance & calibration of support system is crucial. Software upgrades are only recommended if there is a defined and present need for the upgrade
- Trend data is the prime source to identify any potential issues that are not immediately visible in walk-around inspections
- General maintenance and upkeep of test equipment is easier and more effective to increase longevity than fault finding after the fact

Lessons Learned

Of the many things that have been learned on this extensive test programme there are several associated with test health that are pertinent here. Overarching this it became clear that a proper trend monitoring system provides valuable information on the running of the fatigue test, flags issues before they occur, and allows the impact of rectification to be objectively measured. In detail the lessons learned include the realisation of the link between servo-valve health and test repeatability, the understanding that use of command signal dither can, in some circumstances, temporarily improve a deteriorating servo-valve's performance, and the understanding that with complex multi-actuator control systems that minor load fluctuation instability can be unavoidable and that strain surveying at extremely slow running speeds can generate worse results than simply running at 'normal' speed.

Servo Valves. Servo valves along with load cells provide the control in a hydraulic actuator system. The valves control the flow of high pressure hydraulic fluid in order to apply actuator load, and within a PID feedback control system along with the feedback from the load cell will keep the actuator load at the desired level. To operate optimally a servo valve should be rated appropriately for the oil flow expected in the application (i.e. litres/minute), which may require some judgment of the trade-off between sizing to accommodate the peak flow rate or ensuring good performance at the nominal flow rate. A full description of how servo valves work is beyond the scope of this paper, but a good primer can be found at reference [3].

Because the valves are in a feedback loop it can be difficult to identify faults or deteriorating performance, and on the WFT it was found that the only reliable way to assess possible servo-valve condition was through monitoring for changes in performance over a large number of load lines.

The units used on the WFT are MOOG G761/761 series Flow Control Servo Valves.

Repeatability here is measured for each actuator and each load line in a block against a reference standard using the metrics shown in Figure 1 above. In this case each load line is assessed against the Figure 1 chart and the difference in load against the reference block divided by the repeatability limit at that load level is plotted (i.e. (load – reference block load)/(Def-Stan limit at that reference block load)). It should be noted that the plots show many hundreds of thousands of load lines, so a small number beyond a particular limit may indicate only a 10th of a percent of the total number of load lines or less that have minor repeatability exceedences.

Well performing actuator/servo-valve. In Figures 7 and 8 below repeatability is displayed for a well performing servo-valve and for one which has deteriorated.

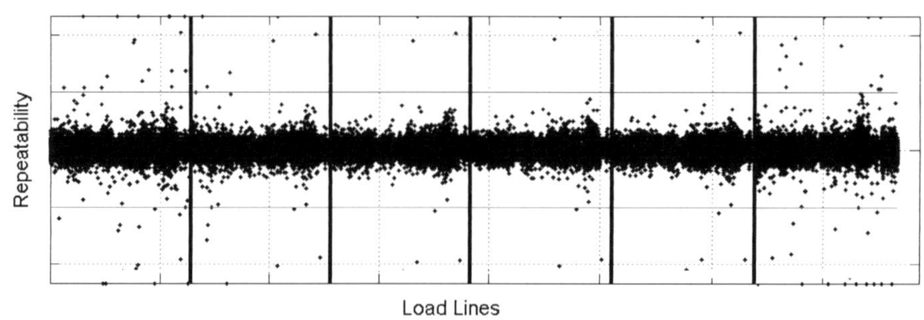

Figure 7. Well performing servo-valve.

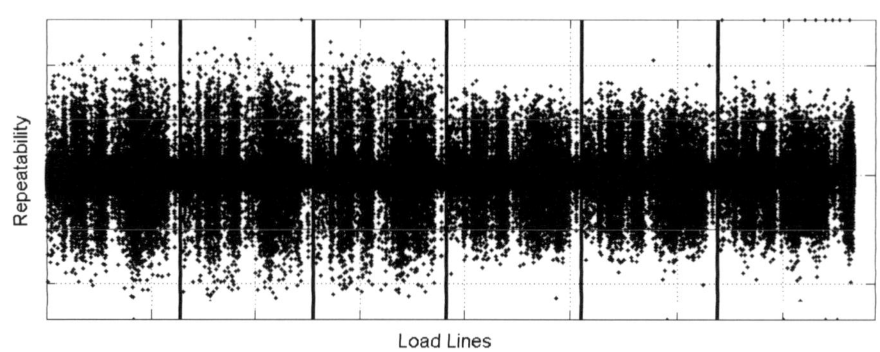

Figure 8. Poorly performing servo-valve.

Impact of Dither. Whilst the application of dither is well known in the industry as a means to improve a poorly behaving test, it is unusual to be able to show the impact of dither on what would otherwise be a fairly nebulous quality, that being the repeatability of the loading. In figures 9 and 10 below we show the impact on repeatability of applying dither to the drive current on a servo-valve that had begun to deteriorate[12]. Applying dither keeps the servo-valve spool moving, and in some cases significantly improving repeatability performance on the test article. The performance increase is sufficient to bring the actuator into normal limits for the vast majority of load lines. Dither has a downside, however, as it can affect overall servo-valve life. Therefore on the WFT it was used to address specific problems, and where possible poorly performing servo-valves were replaced as soon as was practicable rather than relying on dither in the long term.

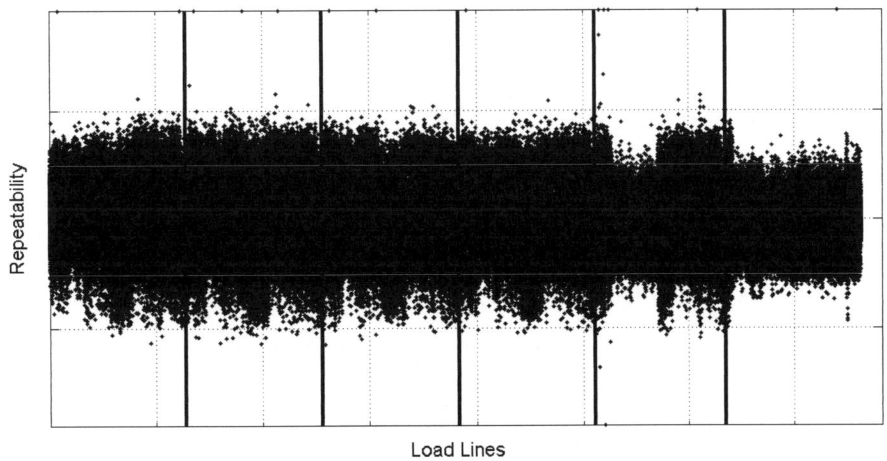

Figure 9. Without Dither.

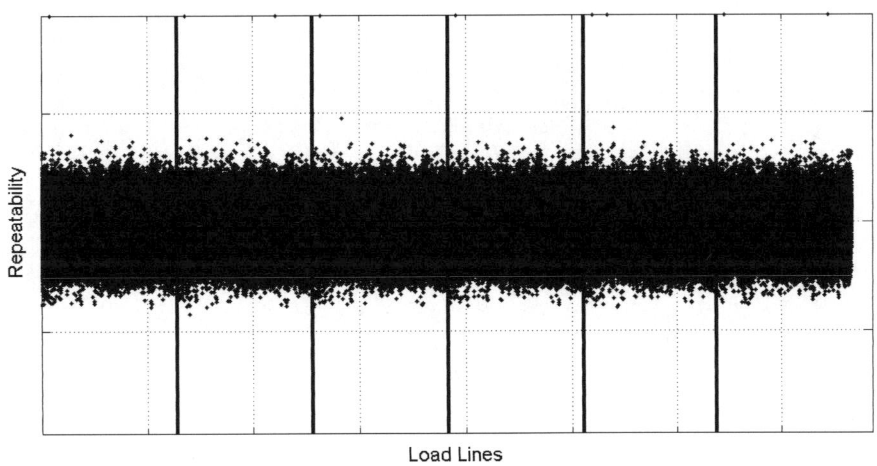

Figure 10. After application of Dither.

Where servo valve replacement did not help. In Figure 11 and 12 below we show how in isolated cases changing out a servo-valve did not improve repeatability performance. Here something else was causing the problem, so a further investigation would be warranted. If it has been found that the servo-valve is not at fault then Marshall will notify the Authority of the issue, and a process to resolution (investigation, reporting, etc) will be agreed and documented.

[12] Dither is an oscillatory signal applied to the servo-valve drive signal current. It is designed to only excite the first or pilot stage of the servovalve, and is normally at a frequency of 100Hz or higher, and so will not normally translate through the servovalve into the actuator or the test article. The intent here is to keep the spool moving so that 'sticktion' effects do not affect performance. The dither signal does not affect the command signal so a 1,000 lbf command will remain 1,000 lbf regardless of whether dither is used. Generally on the WFT dither has been applied, where necessary, at 2-3% of the absolute value of the ±40mA current used (e.g. 2% dither would be ±0.08mA oscillation on the drive signal).

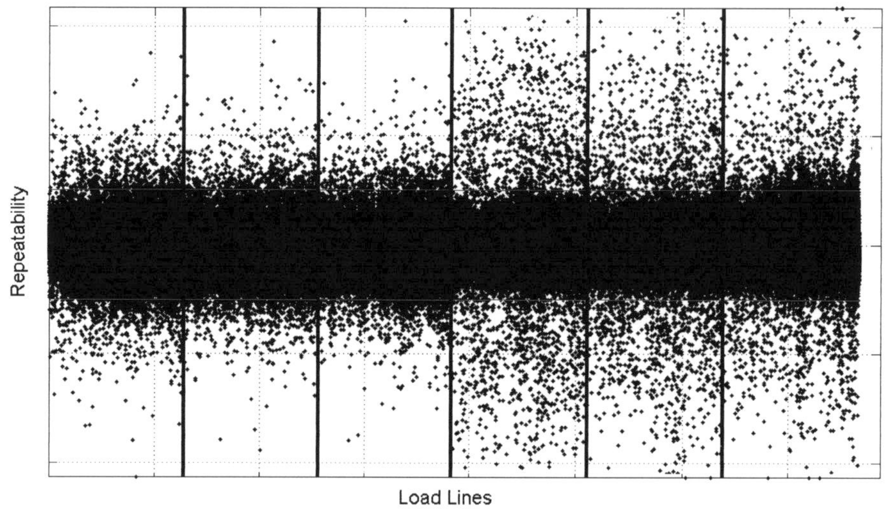

Figure 11. Before servo-valve replacement.

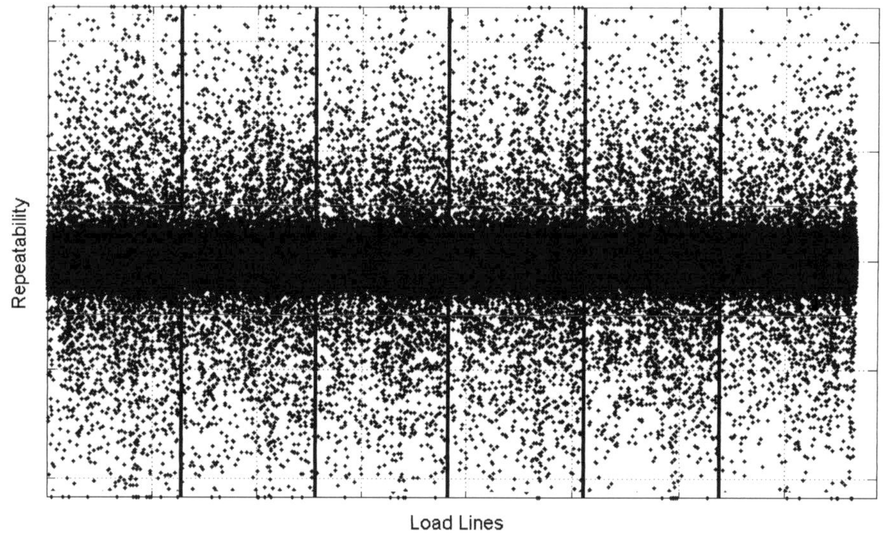

Figure 12. After servo-valve replacement.

Slow speed running and tremble. In Figure 13 below one can observe the effect of undesirable oscillations, tremble, that are set up when an actuator attempts to hold a load, and see their impact on the measured strains on the test article. The activity around the 45 second mark in this example primarily involved tremble frequency content at 3.1 and 4.8 Hz. The tremble tends to be uncorrelated across actuators, so for much of the structure the effect is 'averaged' out by the effect of multiple actuators contributing to wing internal loads. The tremble can be tuned out to some degree with changes to the PID controller parameters, but with a large number of actuators and a highly variable load spectrum it can be extremely difficult to fully tune out all undesired tremble. In the case of Figure 13 this represented the worst case seen prior to final tuning, and the final tuned state (not shown here) was considered to reflect acceptable fidelity.

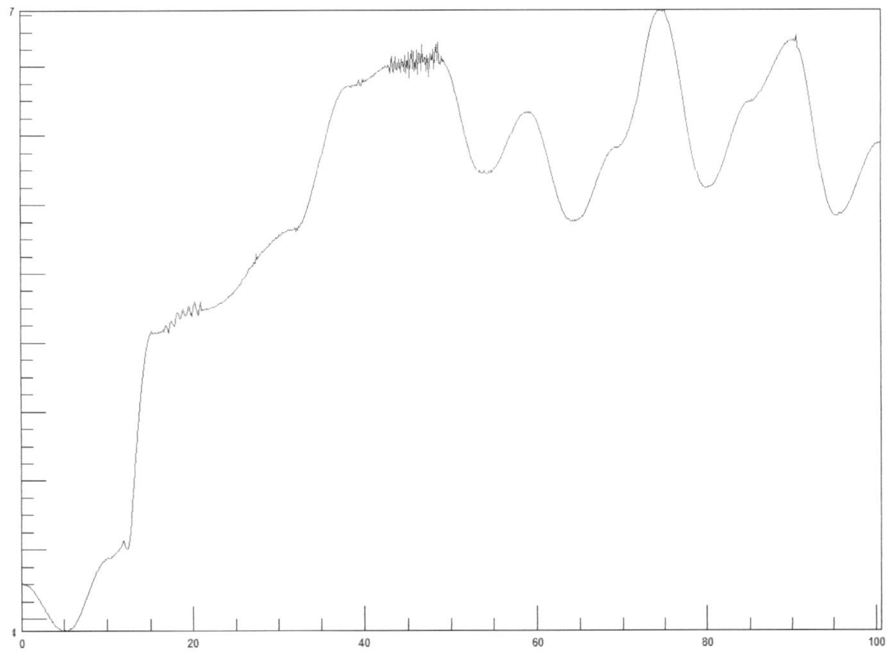

Figure 13. Oscillations in measured strain during actuator holds (45-sec mark), worst case gauge, pre-tuning.

The possible presence of tremble in data should be considered if strain surveys are being taken, and steps should be taken to ensure that any slow speed running as part of a strain survey does not make the situation worse by generating undesired tremble in the strain results.

It should be noted that whilst tremble is occasionally seen on full scale tests, in the experience of the engineers involved it had not previously been seen on loads holds and at slow speed (as in Figure 13 where it can be seen to occur when the commanded load is relatively static). This may indicate that the occurrence of tremble is specific to particular circumstances on an individual rig.

CONCLUSIONS

Ensuring test health is maintained on a multi-year multi-actuator full scale wing fatigue test is not trivial. Measuring and understanding any trends takes dedicated tools to be developed, and a comprehensive maintenance schedule to be established.

Lesson learned included:

- That a comprehensive test monitoring programme allowed issues to be identified early and allowed the impact of rectifications to be objectively measured
- Monitoring dynamic effects is particularly difficult to carry out, and care should be taken to ensure that any monitor is measuring actual dynamics and not another source of variability.
- It was found that some components, such as servo-valves, need to be assessed using trend monitoring tools, and that in the short term use of dither can in some circumstances improve performance. That said sometimes valve replacement is necessary, and in some cases the problem will be due to other causes.
- Monitoring of strain-gauge drift over time could provide early warnings of failure, although in some cases the cause of the drift may never be found.
- Actuator tremble became an issue when running at slower speeds than what the test was tuned to operate at, and steps to avoid tremble during strain surveys needed to be taken.

ACKNOWLEDGEMENTS

A complicated program of this type inevitably owes a great deal to many contributors spread across many organisations. The author would particularly like to acknowledge the contribution and efforts of the members of:

- the RAF C-130J Project Team,
- the members of the RAAF Directorate General of Technical Airworthiness, who were instrumental in the development of the WFT Trend Monitoring process,
- the QinetiQ Airworthiness and Structural Integrity Group,
- members of the Loads, Structural Mechanics and Transport Structures groups at DSTO, and
- members of Marshall Aerospace and their sub contractors

who have made this programme a practical reality.

REFERENCES

[1] Dosman, S. (2013). In: *Joint RAF and RAAF C-130J Full Scale Wing Fatigue Test Loads Development and Test Running*, Proceedings of the 27th ICAF Symposium, vol. I, pp. 163-180, Brot, A. (Ed.), Paragon Israel – Dan Knassim Ltd., Geneva.

[2] UK Ministry of Defence, (11 July, 2014), "*Design and Airworthiness Requirements for Service Aircraft Part 1: Fixed Wing*", Def Stan 00-970, Issue 13, Section 3.

[3] McCormick, C. Star Hydraulics. (2012). "*How It Works – Using Servo Valves*", Engineering Integrity Society, Vol. 32, p. 18-21.

[4] Hoffmann, K, (1989) "*An Introduction to Measurements using Strain Gauges*", Hottinger Baldwin Messtechnik GMbH, Darmstadt.

Digital twins

Fatigue Analysis of a Virtual Airframe Structure

Chris Wallbrink, Michael Opie, Xiaobo Yu

Defence Science and Technology Organisation, Australia

Abstract: Fatigue analysis of a virtual airframe structure is a relatively new concept and understanding the limitations of this modelling technique is important for interpreting the results of this method. Presented in this paper is a detailed fatigue analysis of a virtual airframe structure where the results have been compared to real damage observed during the full scale fatigue test conducted on the same airframe. The fatigue analysis conducted used an elemental based approach to develop a fatigue index for each element in the model. The fatigue index map of the structure was qualitatively compared to known cracked locations. The results of the investigation indicate that the present elemental based fatigue analysis methodology was able to identify many locations known to crack during the full scale fatigue test. However, it was found that absolute fatigue life estimates were difficult to achieve and highly sensitive to the mesh used in the analysis. A number of issues are raised and discussed in this paper relating to achieving the appropriate model fidelity. Approaches to improving analysis of geometric details using a combined p-element and h-element analysis and the use of a critical distance methodology is also presented and discussed.

INTRODUCTION

The conduct of a full scale fatigue test (FSFT) is an important step in the certification of an aircraft as it is used to identify the unforeseen fatigue related issues that could compromise safety of flight. It would be hard to ever argue to replace such a test with a virtual counterpart due to the inherent unknowns and discrepancies with what is expected and what is reality. However, with continual advances in computation capability an opportunity exists to supplement management strategies with additional information through improved identification of locations that are potentially fatigue critical. This paper will discuss the use of such a methodology to supplement the fleet management strategy for the PC-9/A fleet in the Royal Australian Air Force.

Industry motivation to use Finite Element Analysis (FEA) in structural assessments of complex structures for the purposes of development, manufacturing and maintenance to deliver increased financial and engineering benefits is clear [1]. Fatigue analysis of a virtual airframe structure is an appealing concept in terms of cost and time and can greatly aid in targeted engineering effort to address fatigue related issues. The method has been used within the aerospace and automotive industries [2-4] but as yet there are limited publications on the topic in the literature. Patria [3] uses FEA to construct local stress histories for a Hawk jet trainer aircraft which are then used in a fatigue analysis at chosen locations. Ball [4] has presented a methodology used on the F-35 that also constructs the spectrum on each element using FEA and then combining this stress history with a strain-life methodology to determine fatigue life estimates. Many leading finite element software packages now incorporate fatigue analysis tools such as the MSC.Nastran tool Embedded Fatigue analysis [5] and the Abaqus tool FE.Safe [6]. The methodology is not complex though relies heavily on computational power to linearly combine various load histories on each element in a finite element model and then conduct a strain life analysis on each element. But the question remains, how accurate is this method? The issues underlying the answer are somewhat complicated. This paper attempts to examine the issues associated with accuracy with comparisons to results from a full scale fatigue test of a PC-9/A trainer aircraft conducted and completed at DSTO in 2000.

The present investigation adds to the knowledge base underpinning efforts to develop the digital twin concept [7] of fleet management proposed by the Air Force Research Laboratory (AFRL). That is the idea of completely modelling, in the virtual world, each individual aircraft to such detail as to match the real world behaviour of its real world counterpart or twin. The idea itself is a grand scientific challenge but offers significant potential benefits in monitoring the effects of changing fleet management profiles and usage to safely and economically manage each individual aircraft through its operational life. While the idea is grand in nature, can we step toward this goal while obtaining benefits to fleet management?

BACKGROUND

Buffet was not originally characterised in the original loads monitoring program that contributed to the development of the load spectrum used during the FSFT of the PC-9/A. A follow-on operational loads monitoring program was undertaken to measure edge of envelope flight and better characterize the flight dynamics of the aircraft including buffet events. To contribute to a reassessment of the aft fuselage, a single detailed finite element model was developed for both static and dynamic analysis. The purpose of unifying the models was to aid in mapping the mode shapes derived from the dynamic analysis onto the static analysis from which element stress/strain responses can be easily combined for a subsequent fatigue analysis. This combined model will be hence forth be referred to as the "unified" finite element model (UFEM). The development of this model has provided the unique opportunity to conduct fatigue analysis of a virtual airframe structure using an elemental based fatigue analysis (EBFA) method. The results of which can be assessed against observed fatigue crack development on the same airframe structure during the FSFT. This paper focusses on some of the issues and limitations of the technique to identify locations and predict when crack development will occur.

Historically, finite element analysis of aircraft structures has been performed in a two-step approach. The first step is to construct a global coarse grid model (often referred to as a loads model) for the purpose of generating appropriate boundary conditions for the second step which develops a local refined finite element model. The local refined model or sub model is then used to estimate local stresses within the structure. The present work introduces local refinement to the global finite element mesh for the aft section of the PC-9/A, such that local load paths and stress concentrations critical to fatigue analysis are reasonably captured. This eliminates the inaccuracies introduced by the two step approach. Further, it simplifies superposition (via mode shapes) of dynamic load induced stress/strain with those introduced by static loads to study the damage developed by both dynamic and static load cases. The unified model for both static and dynamic analysis developed by the Original Equipment Manufacturer (OEM) with consultation from the DSTO contained 890,000 degrees of freedom. The model was used in various configurations including: ground loads, ground vibration, edge of envelope and full scale fatigue test configuration. The configuration discussed here is the FSFT configuration and is presented in Figure 1. A linear finite element analysis was conducted by applying the same FSFT spectrum via simulated form boards and whiffletree attachments.

The EBFA is conducted for every element in the UFEM to estimate a fatigue index. This index is the estimated fatigue life of that individual element in the finite element model. The idea of conducting fatigue analysis on every element in the model is to identify all fatigue critical areas or "hot spots"[13]. The motivation behind the analysis was the realisation that detecting an area of peak stress or stress hot spot may not necessarily indicate a fatigue critical location. The original EBFA was conducted by the OEM using a program call StructLife (recently rebranded as DVL Life [8]) under contract to the Australian Defence Force. A validation step in the process was to compare the EBFA results from the UFEM to the observations on the original full scale fatigue test conducted and completed at the DSTO in 2000. To further evaluate the methodology, DSTO has developed an in-house capability to replicate the algorithm present in StructLife and other similar programs. The results presented in this paper were built on the original work conducted by the OEM to further assess the capability and limitations of the elemental based fatigue analysis methodology.

[13] "hot spot" refers to the most likely area where local stresses are at a maximum or the estimated fatigue life is at a minimum.

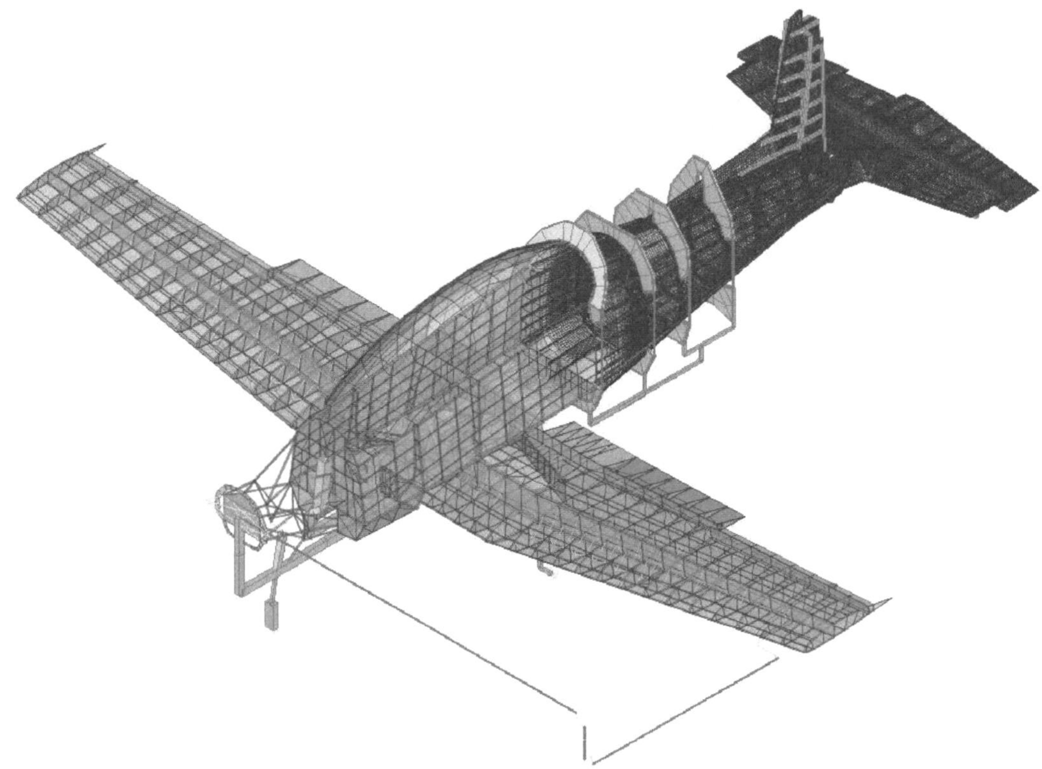

Figure 1. Original refined finite element model for the PC-9/A in FSFT configuration.

The FSFT load spectrum that was applied through the 24 hydraulic jacks during the FSFT was applied to the finite element model [9]. The spectrum block applied contained 93,492 load lines and contains one -3g event and one +7g event. The spectrum represents 250 flight hours that comprises a wide variety of manoeuvres including take-off and landings, touch and go's, spins, etc. Linear supposition of loads from each jack was used to construct a load history for every element in the model. Thus, every element in the model was analysed using the strain-life approach to determine a relative fatigue index to rank locations. For simplicity a signed von-Mises approach was used to develop the local load history at each element centre. A number of control points were chosen in the analysis to serve as metrics on which to assess the methodology. These control points correspond to locations where fatigue cracking was observed in the FSFT and/or fleet.

FATIGUE HOT SPOT OR STRESS HOT SPOT?

A stress hot spot in a structure refers to a location which has the highest peak stress under all possible load events. Meanwhile, the fatigue hot spot in a structure refers to a location which accumulates the most fatigue damage under a given load spectrum. In many cases the fatigue hot spot overlaps with the stress hot spot. Thus, in fatigue analysis, the stress hot spot is often chosen as the critical location for fatigue. However, fatigue hot spots can differ from stress hot spots in the following cases (but, not limited to):

- Where multi-axial loading is present
- Where the dominant loading direction changes or fluctuates
- Where notch size effects on fatigue are present
- Where residual stresses are present (sources include manufacturing, build up stresses, contained plasticity),

The possibility that multi-axial loads can produce a fatigue hot spot at a separate location to a stress hot spot is important in ensuring safety against fatigue. This is particularly poignant given that this consideration is commonly ignored with traditional linear elastic approaches. Other effects such as residual stresses are harder to model and are ignored in this study but it should be noted that their contribution could be significant. Again this highlights the point that the analysis presented here can elucidate potential failure locations but cannot replace physical testing.

The Appendix presents a simple example of a cantilever beam demonstrating that fatigue hot spots can differ from stress hot spots. Real structures can be subjected to multiple forcing functions which may or may not be related but will produce complex local load histories at structural details. It is not a stretch of the imagination to see the similarity between the cantilever beam example and the tail of an aircraft. The tail of an aircraft is a perfect example of a complex structure that is subjected to multiple load inputs which may or may not be related through various manoeuvres. Hence, local stress histories at structural details are hard to determine without direct measurement or a highly accurate loads model. A key to improved fatigue analysis is an accurate understanding of the local stress history at structural details. Simply, the better the local stress history is understood the better the fatigue prediction.

FATIGUE ANALYSIS USING THE UNIFIED FINITE ELEMENT MODEL (UFEM)

As part of the program of work to identify fatigue critical locations in the PC-9/A aft fuselage, input from both DSTO and the OEM contributed to the development of a refined NASTRAN finite element model representative of the PC-9/A test article in the FSFT shown in Figure 1. The model contained approximately 285,000 elements and 24 load jack inputs with simulated form boards and whiffletree connections. Given that the main structural elements are manufactured from thin sheet aluminium, the model mostly consisted of shell elements. Linear analysis was performed on the model without consideration of nonlinear effects such as buckling, contact, or prying. The refinements to the finite element mesh were concentrated toward the aft end of the fuselage. The original fatigue analysis of the UFEM was conducted with the analysis code StructLife. However, the EBFA results presented in this paper are based on a DSTO developed approach where the strain life analysis is performed with the well-known fatigue analysis code FAMS [10]. In this study, DSTO has further refined the mesh at specific locations to assess the EBFA methodology. These further refinements are needed to achieve a better representation of the local geometric features and understand how these affect the resulting fatigue index. The following subsections look more closely at specific components and assess the performance of an EBFA in identifying fatigue critical locations. A further discussion will follow considering the quantitative aspects of this analysis methodology.

Forward Fin Connecting Strap

The forward fin connecting strap is arguably the most fatigue sensitive component in the aft section of the PC-9/A airframe. The connecting strap connects the forward section of the tail fin to frame 9 in the aft fuselage. A point to note in the following discussion is that the forward fin connecting strap in the FSFT was designed as a bent structural element. Prior to completion of the FSFT, cracking had already been detected in the forward fin connecting strap within 2000 flight hours on a number of RAAF aircraft. The location of the forward fin connecting strap and a picture of the failed component from the teardown at the conclusion of the FSFT is shown in Figure 2.

In the original refined UFEM model the forward fin connecting strap contained 2,181 elements, but in this study to further resolve the analysis the component was remeshed with 6,530 elements with side length of approximately 1 mm shown in Figure 3a. The uniform mesh size was used in response to the uncertainty relating to the locations that may be fatigue critical. A discussion relating to the mesh will follow in a later section. Referring to the component that failed in the FSFT in Figure 3c, post failure analysis reveals that rivet holes labelled 1 and 3 cracked first. The EBFA of the component is shown in Figure 3b and indicates that both rivet 1 and 3 are the most critical (i.e. have the shortest fatigue lives). The fatigue analysis does not factor how the load is redistributed once a crack begins to grow. The test article failed from a crack that emanated from rivet 5 and travelled back along the bend line to the edge of the strap. It is interesting to note that the fatigue analysis seems to indicate that the bend line in the forward fin connecting strap is also a highly critical feature for fatigue. This seems plausible when considering that fatigue cracks growing from rivets 5 and 7 travelled down to the bend line rather than to neighbouring cracks at rivets 1 and 3. For this component the EBFA appears to give a good qualitative indication of areas that are fatigue critical.

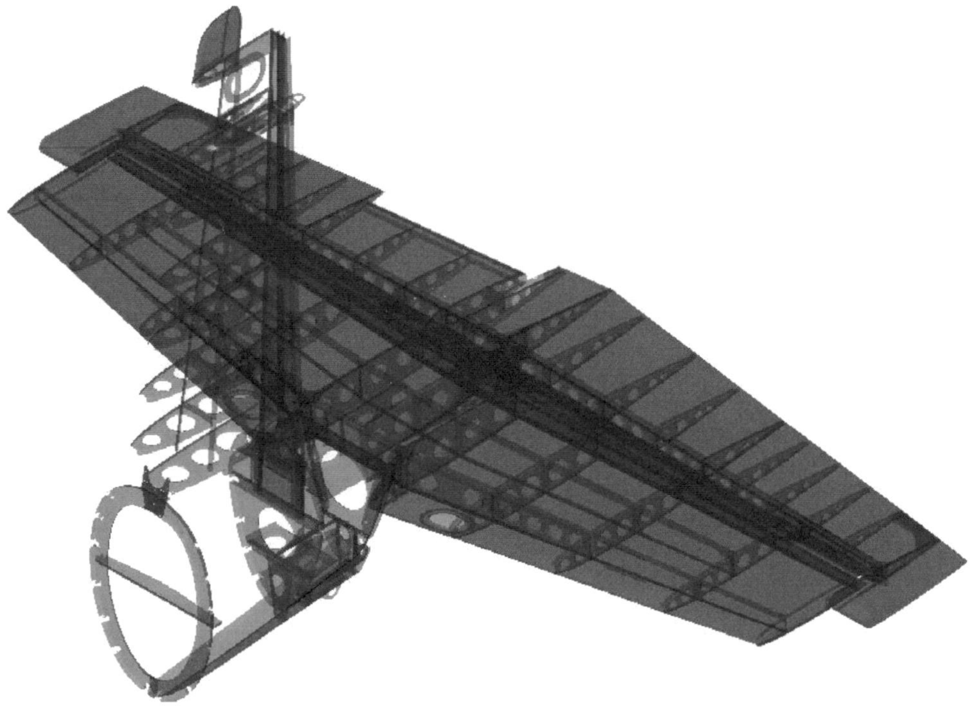

Figure 2. The location of the forward fin connecting strap in the aft section of the PC-9/A and a picture [11] of the failed component at teardown after FSFT.

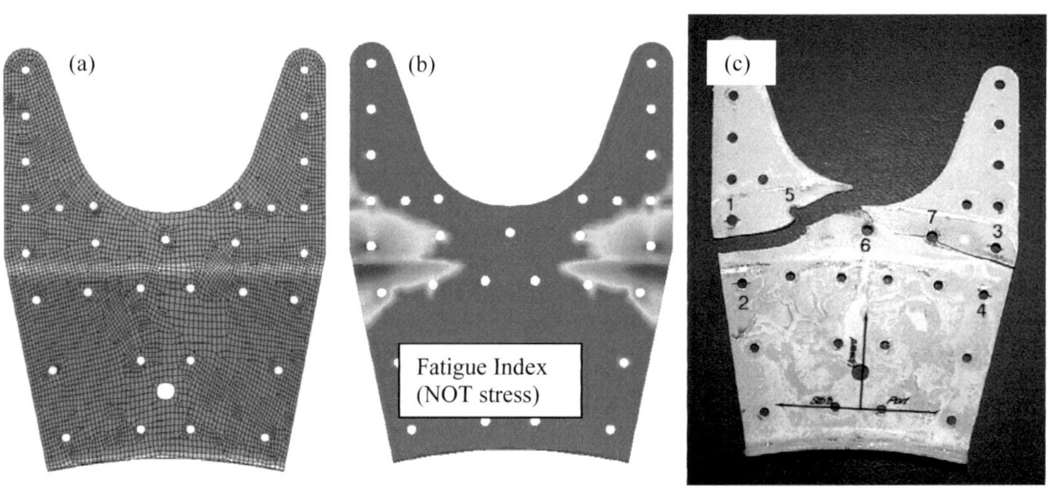

Figure 3: a) Finite element mesh b) Elemental based fatigue analysis of the forward fin connecting strap and c) the failed forward fin connecting strap at teardown [11].

Top Hat Section

The top hat section connects directly to frame 10 and is highlighted in green in Figure 4. During the FSFT, cracking was detected at 33,000 flight hours and those cracks were located at the rivet holes labelled A, B and C in Figure 4c. Rivet hole A contained 3 cracks ranging between 1.5 mm to 2 mm and rivet hole B had seven crack ranging from 1mm to 4 mm. Finally, rivet hole C had two cracks ranging in size of 2 mm and 3.5 mm long. At 33,000 flight hours, the top hat section was replaced before recommencement of the test. At the conclusion of the FSFT (100,000 flight hours), the subsequent teardown revealed several large cracks shown in Figure 5. Significant cracking had occurred with cracks ranging up to 71 mm in length. Figure 5c shows a large jagged hole produced by the coalescence of two large cracks.

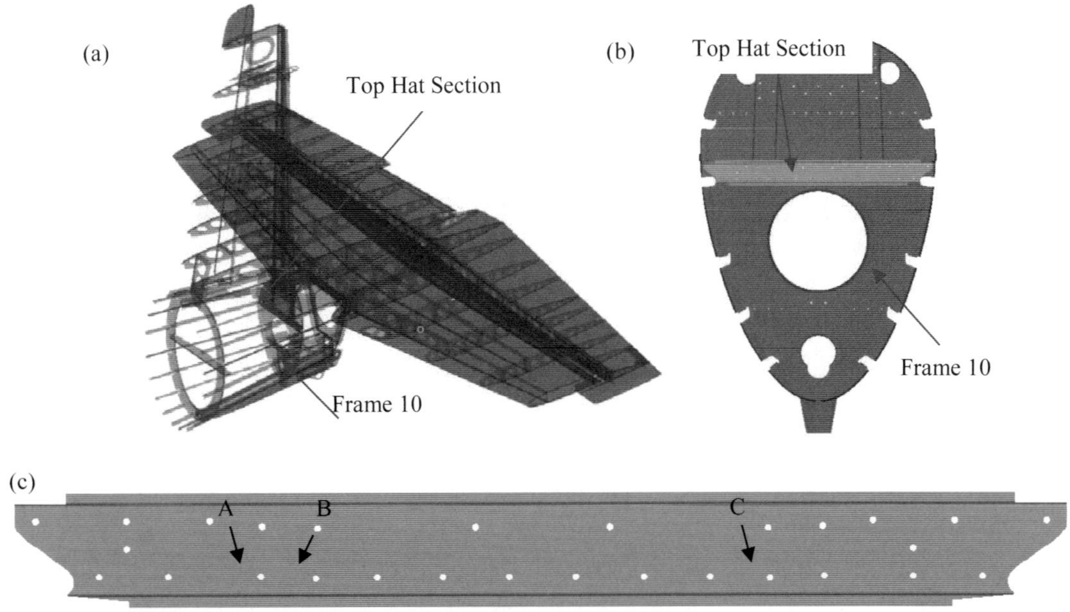

Figure 4. a) Aft tail section highlighting the top hat section in green b) Frame 10 with the top hat section highlighted in green, c) Top hat section with locations of cracks found at 33,000 hours.

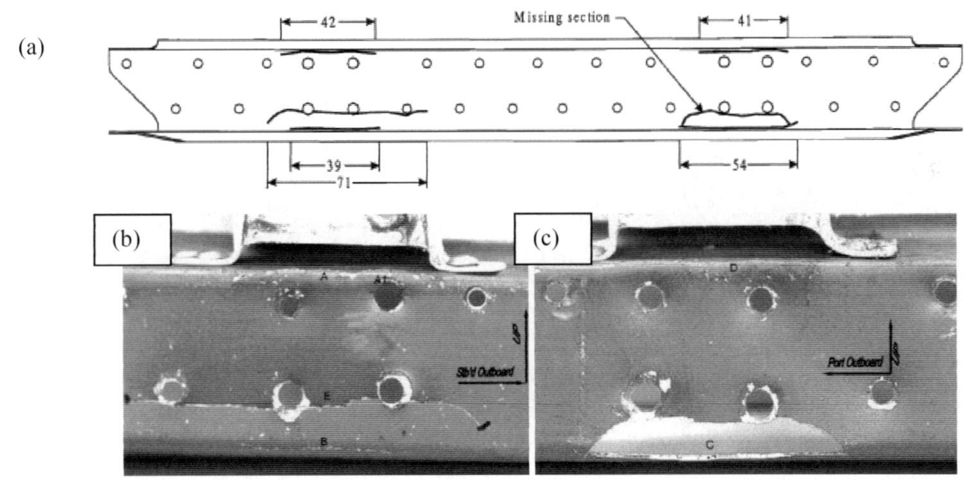

Figure 5. a) Schematic drawing indicating the locations of cracks detected at the conclusion of the FSFT, b) photo of cracks on the starboard side of the top hat section [12] and c) photo of cracks on the port side of the top hat section [12].

The original refined model of the top hat section contained 518 elements where rivets were modelled by merged nodes. The model has been rebuilt in the present analysis shown in Figure 6a with the rivets and rivet holes explicitly modeled by interpolation constraint elements (RBE3) spanning open holes. The new mesh for the top hat section contains 26,589 elements with a side length of approximately 1 mm. As is observed in Figure 6b, the analysis in this case has produced a strong correlation to the observed cracking in the FSFT. Approximately 6 hydraulic jacks contribute significantly to fatigue at this location and they include loading to both the horizontal stabilisers and vertical tail.

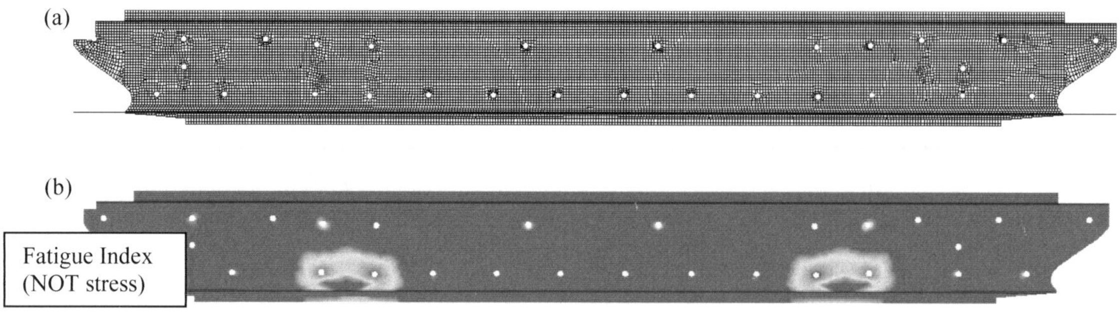

Figure 6. a) Mesh used in the analysis and b) Fatigue index combining top and bottom surfaces for the top hat section.

Frame 10

Frame 10 is located to the rear of the aft fuselage and the main vertical stabiliser connects directly to this component. Early cracking was detected at around 7,500 fight hours during the FSFT at the stringer cut outs. The stringer cut outs are shown on the edge of frame 10 in Figure 7b. Later in the testing, at around 40,000 flight hours, a 3 mm crack (labeled 3 in Figure 8c) was detected around the rudder cable service hole on the port side. At 65,000 flight hours, this crack had propagated to 10mm with an additional 3mm crack found (labeled 4 in Figure 8c). At the conlcussion of the test (100,000 flight hours), significant cracking had occurred with the largest crack about 35 mm long (labelled 1 in Figure 8c) [13].

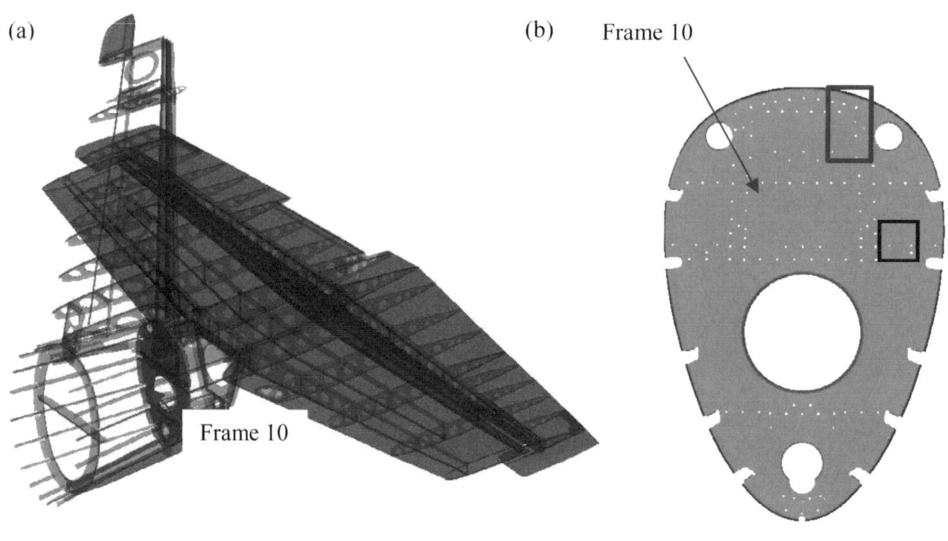

Figure 7. a) Frame 10 is highlighted green in the aft section b) frame 10 with the port side rudder cable service hole highlighted in red.

Again, this component was further refined relative to the original refined UFEM. The new frame 10 model contains 157,000 elements as opposed to the 4,613 elements in the original model. A section of the mesh is shown in Figure 8a. Rivet holes were modelled in the main plane of frame 10 but the rivets along the flange bordering the part were modelled by merged nodes. The results of the EBFA are presented in Figure 8b versus the final cracking observed at 100,000 flight hours in Figure 8c. In this case the analysis seems to have identified the location of most probable cracking but also the preferential direction.

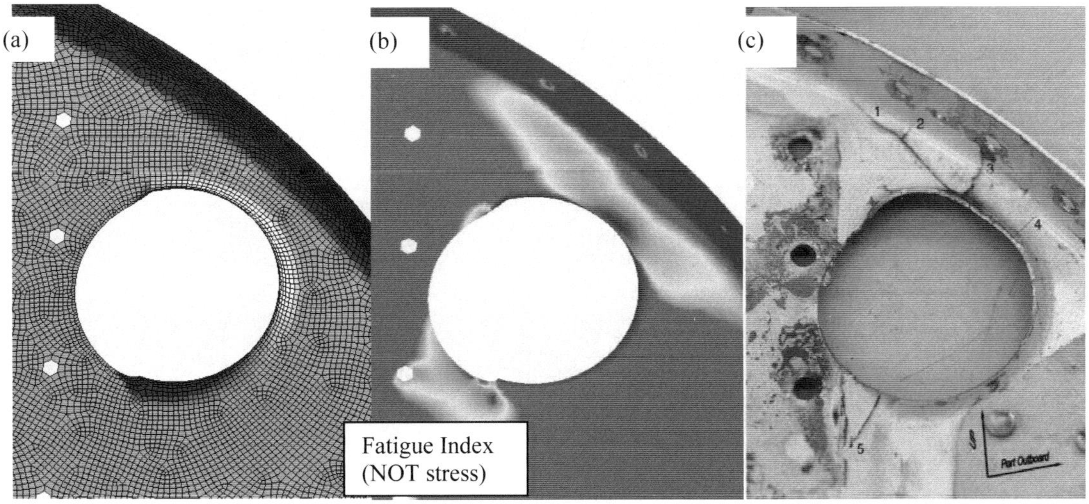

Figure 8. a) Mesh used in analysis b) Elemental based fatigue analysis c) Crack detected at the conclusion of the test in frame 10 [13].

Top fuselage skin between frame 9 ad frame 10

The top fuselage skin between frame 9 and frame 10 experienced significant cracking during the FSFT [11]. The top skin of interest in the present investigation is highlighted green in Figure 9. A 12 mm crack was first detected at 3,500 flight hours at the unstiffened service hole forward of frame 10 and also at the slotted hole on the top fuselage skin at frame 9. After subsequent repairs and further cracking a doubler repair was designed and integrated at 33,000 flight hours. At this time the top fuselage skin was replaced. A point to note is that all cracking from the service holes occurred at 45 degrees to the main fuselage. Throughout testing, extensive cracking was also observed along the rivet line connecting the tail skin to the main fuselage skin, an example of which is shown later in Figure 12a.

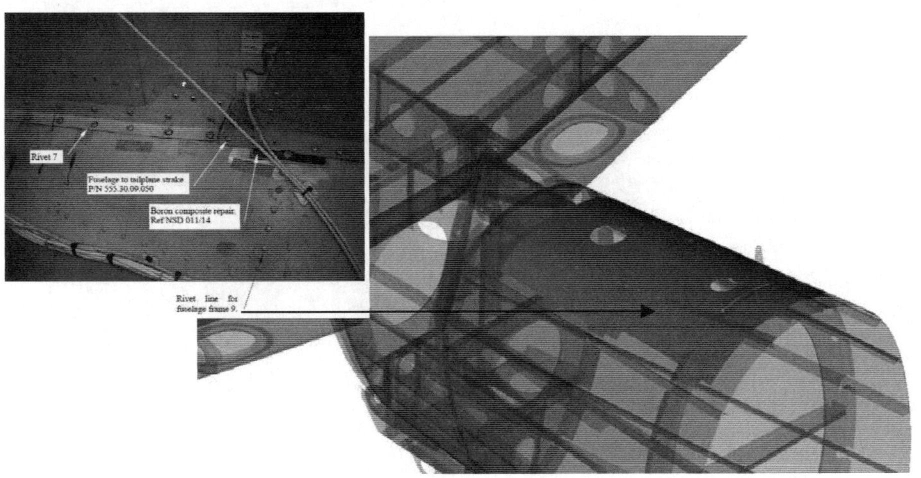

Figure 9. Top fuselage skin between frame 9 and frame 10, (photo ref [12]).

At teardown (100,000 flight hours), a number of cracks were detected around the service holes forward of frame 10 and aft of frame 9. The cracks around the service holes are labelled 1-4 in Figure 10. Cracks were again detected at the ends of the slotted hole at frame 9 labelled 5 and 6 in Figure 10. Also cracks were detected along the rivet line that connects the skin of the tail to the main fuselage. These cracks are labelled 7 and 8 in Figure 10. Aditional rivet holes are observed around the service holes and around the slotted hole. These correspond to additional doubler repairs made to the structure that were not present at the beginning of the test. The UFEM does not contain these doubler repairs.

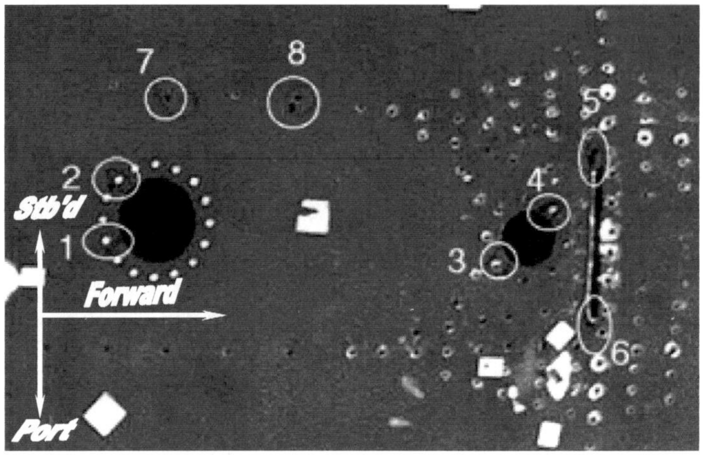

Figure 10. The underside of the top fuselage skin between frame 9 and frame 10 at teardown with cracks indicated by numbers 1-8 [11].

Two models (with and without explicitely modeled rivet holes) are presented in Figure 11. Figure 11a and Figure 11b show the original mesh and results for rivets modelled with merged nodes. Figure 11c and Figure 11d show the refined mesh and results for rivet holes modelled explicitely. Here it is evident that where rivets were modelled with merged nodes, the refinement of the model is not high enough to identify potential cracking around some of the rivet locations. However, it is interesting to note that there were a few rivet locations that gave an indication that they might be fatigue sensitive. When the rivet holes were explicitly modelled (Figure 11b) it is clear that the rivet line that connects the vertical tail skin to main fuselage is fatigue sensitive. It also gives a qualitative indication as to which areas along the rivet line are more fatigue sensitive.

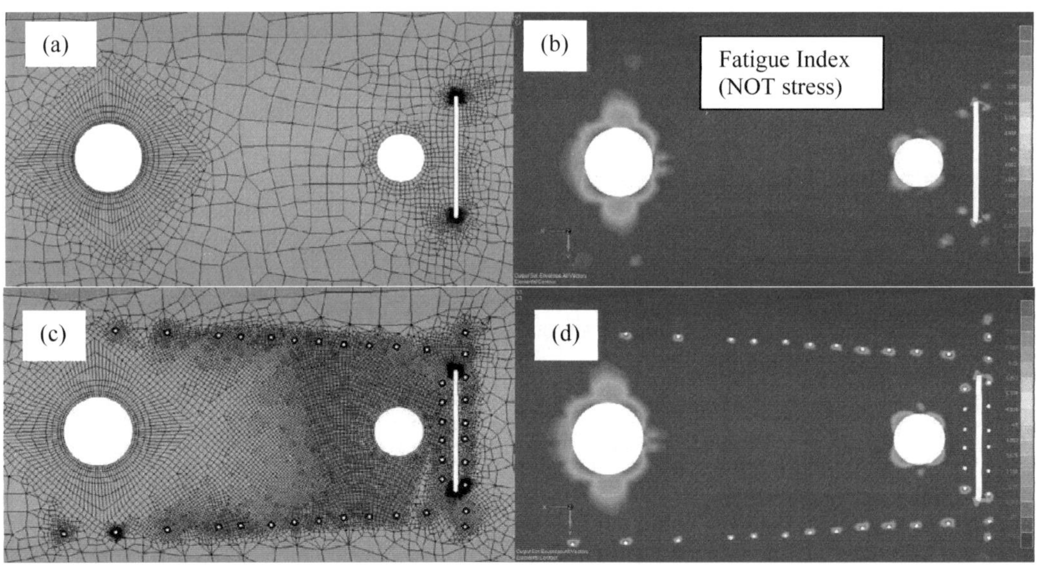

Figure 11. Models of top skin between frame 10 and frame 9 showing a) a mesh without rivet holes explicitly modelled b) fatigue index results without rivet holes explicitly modelled c) a mesh with rivet holes explicitly modelled and d) fatigue index results with rivet holes explicitly modelled.

In Figure 12a, the cracking along the rivet line is shown and compared to the analysis in Figure 12b. Here in a qualitative sense the methodology seems to correlate well. It is however interesting to note that the rivets around the end of the slotted hole also indicate a high propensity for fatigue cracking, but to the authors knowledge that has never been observed in the fleet or the FSFT. This could be the result of inadequate modelling of bypass stresses as the skin at this point is sandwiched between two other components above and below.

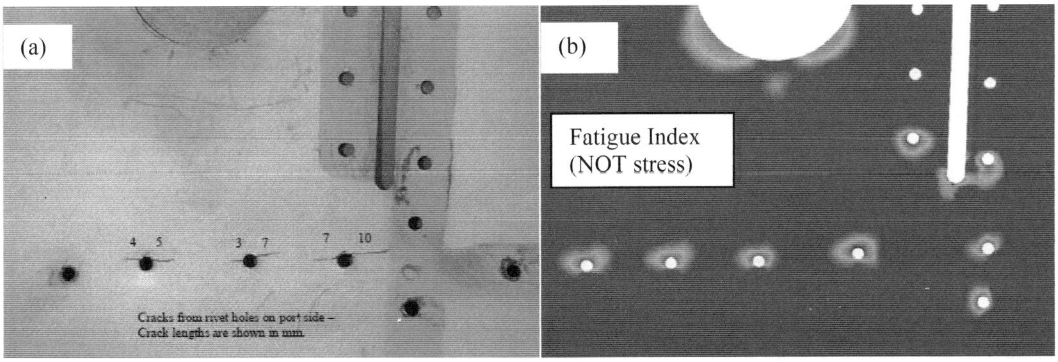

Figure 12. a) Extensive cracking along the rivet line connecting the vertical tail skin to the aft fuselage top skin at 33,000 hours on the FSFT [14] b) distributed fatigue model showing locations of high potential for fatigue.

It was clear that in a number of locations, the finite element model did not exactly reflect the actual geometry. An example is shown in Figure 12 where the rivet locations differ to that on the actual FSFT test article. To what degree these variations affect the analysis is uncertain, but it is worth noting for future application of this methodology. This is another example of how design expectation can differ from as-built reality.

Rivets were predominantly modelled with merged nodes, but in regions considered to be of structural significance they were modelled with interpolation constraint elements (RBE3) connected with a spring elements (CBUSH). This was considered to be a better representation of the rivets but is still an approximation to the real rivet connections. On closer inspection of Figure 5b, it is evident that the crack either initiated or passed around the head of the rivet as the hole itself remains intact with a small ligament of material left behind. Difficulties in modelling rivet connections do contribute to uncertainty and is an area for further development in conjunction with virtual fatigue analysis techniques.

It is inherently very difficult to achieve a high level of correlation between a computational model and the real structure. In a previous study [14], the strains predicted using a coarse finite element model, which was available prior to the development of the UFEM, were compared to the strains measured during the FSFT at two airframe locations. One such comparison relates to the bending strain at the root of the vertical tail and is reproduced here as Figure 13. It is clear that while the linearity is good, the FE predicted strains were systematically higher – approximately 30% higher than the measured strains. There are many causes that may contribute to this disparity, including fleet variations, inaccuracy in both finite element analysis and the measurement and mismatch of strain sampling locations at a region of high strain gradient. As will be discussed in following sections, the fatigue model is highly sensitive to stress and stress-gradients, and errors in the predicted stress can significantly be amplified in subsequent fatigue life predictions. There are certainly challenges in quantifying the uncertainty with the present methodology and further work will be required to understand the implications. While uncertainties exist in both modelling and reality an interesting question to ask is "will a second airframe behave in the same way"?

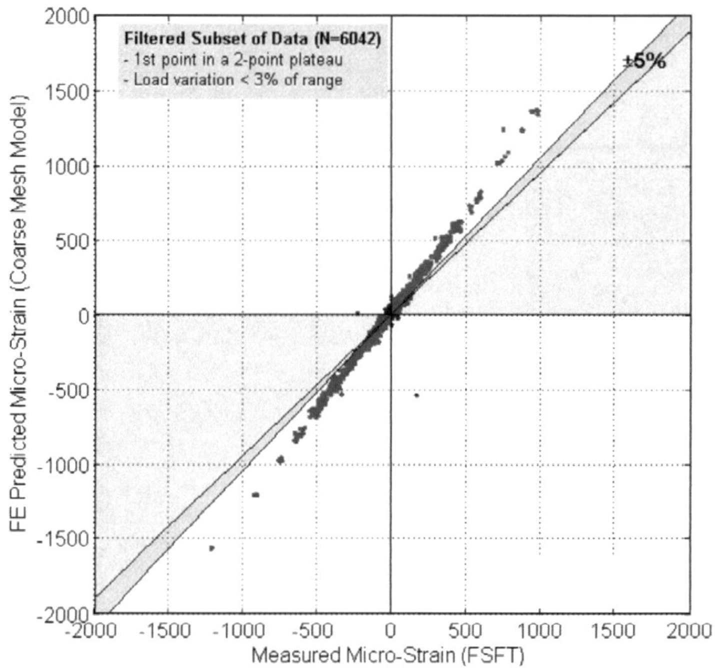

Figure 13. Strains predicted on the vertical tail versus the recorded strain gauge readings from the full scale fatigue test [15].

FACTORS AFFECTING QUANTITATIVE ASSESSMENT

The previous section reporting on fatigue analysis results has concentrated on qualitative assessments of the EBFA model as compared to the observed cracking on the FSFT. Quantitative assessments were not discussed as it became evident through this process that obtaining total life predictions was inherently very difficult and highly sensitive to the mesh density applied to the model and local stress gradients. In estimating the fatigue life of a structure using this strain life method an assumption is made that the fatigue behaviour of a small volume of material anywhere on a structure can be directly related to the fatigue behaviour of the smooth test specimens. This assumption ignores "notch effects" which introduces a size dependency to the fatigue life of a structure. Commonly in estimating the fatigue life of a structure using the strain life method at chosen areas of peak stress corrections for the local geometric details are used. At these local geometric details or stress concentrations the notch effect is commonly accounted for by the use of notch factors, the most common of which is that produced by Neuber [16]. In the present elemental or nodal based fatigue analysis the strain life method is applied with disregard to the local geometry. As such no direct correction is applied to account for notch effects. Thus direct use of the fatigue index as a measure of fatigue could be erroneous. This paper discusses aspects of this behaviour with respect to interpreting the fatigue index with some suggested methods but does not extend the discussion to an in-depth analysis. Two factors affecting quantitative assessment were examined: mesh convergence (in respect to stress and predicted fatigue life) and the sensitivity of the fatigue life prediction to stress gradients.

Mesh Convergence

To investigate mesh convergence in terms of both stress and fatigue life, a thin 2024-T351 aluminium alloy plate with centre hole subject to a remote tension was considered. The ½ width to hole-radius ratio was 10:1 with a plate height (along the loading direction) 1.5 times the width. To estimate the fatigue life, a representative PC-9/A spectrum was chosen, which was constructed based on one of the strain gauge measurements recorded during the FSFT [17]. Loading was adjusted such that the peak stress in the spectrum, applied remotely, was 200 MPa. This ensured that notch plasticity would occur during application of the spectrum. The fatigue life material properties used in the present analysis were sourced from Ref [18]. Five quarter models presented in Figure 14 were created using linear elements; each successive model increased the mesh density by a factor of 4 over the area from the previous. To assess errors in these models, a sixth model was created using second order isoparametric elements with a mesh with 16 times as many elements than the finest of the five models. The results based on the sixth model were used here as references to assess mesh convergence.

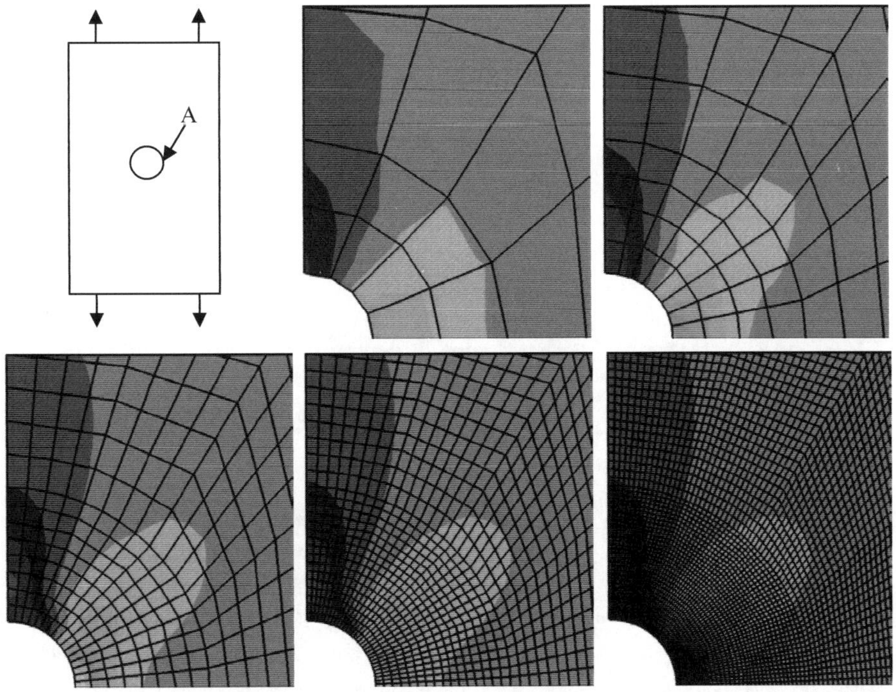

Figure 14. Open hole finite element models used to study mesh convergence.

The von Mises stress was calculated at the vertex at the edge of the hole at location “A” in Figure 14 and at the centroid of the element closest to “A”. The von Mises stress in these elements was compared directly with the vertex von Mises stress evaluated in the high fidelity model at location “A” in the plate. Using the stress result from the high fidelity model as reference, the percentage error in both stress and life for each of the models in Figure 14 is presented in Figure 15. Note that the presented errors in fatigue life are only related to those caused by the errors in stress calculation. It does not account for errors inherent in the spectrum, material modelling, input material data, etc. It is well known that errors in stress can have a large influence on predicted fatigue life and for this example the errors in fatigue life ranged from 5 to 15 times larger than errors in stress.

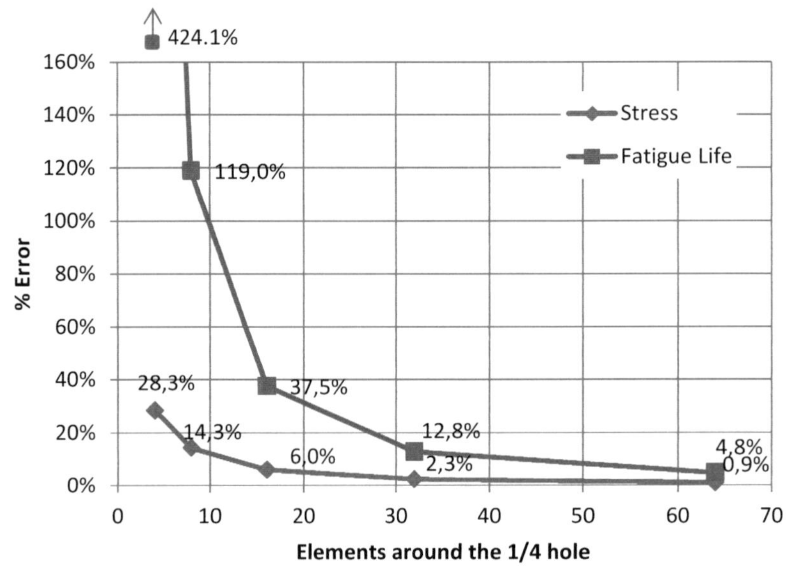

Figure 15. Percentage error in stress and fatigue life evaluated at the edge of the hole. Results obtained on the vertex at the edge of the hole for both the high fidelity model and the models used in the convergence study.

Now if we compare the stress evaluated at the centroid of the element closest to the peak stress and compare that to the peak stress evaluated in the high fidelity model we see (in Figure 16) that the errors have substantially increased. These errors presented in Figure 16 are associated with both non-convergence in stress and also the location of the centroid of the element. Here it is evident that the location of the centroid relative to the edge of the hole introduces much larger errors than does the errors in stress alone. This is an important consideration when interpreting results from an elemental based approach.

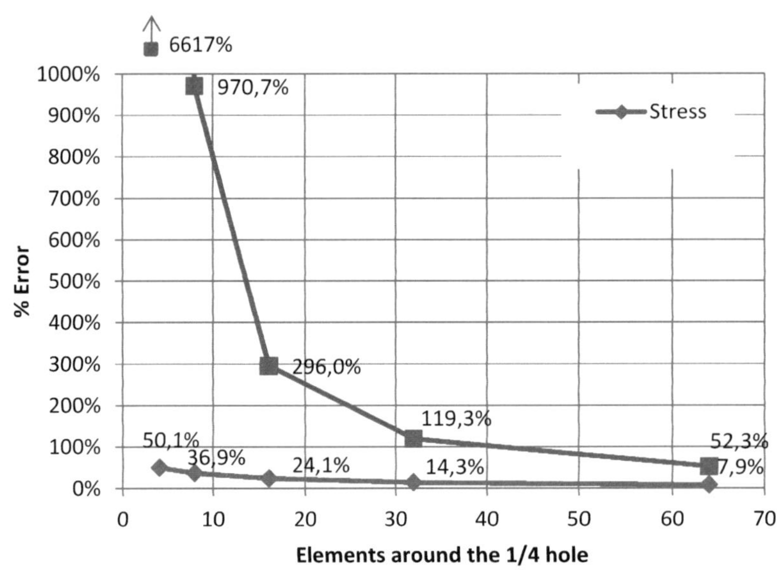

Figure 16. Percentage error in stress and fatigue life evaluated at the edge of the hole. Results obtained at the centroid of the element closest to the edge of the hole (location A) for each of the models in the convergence study and in combination with the result at the vertex at the edge of the hole for the high fidelity model.

The results here indicate (and not surprisingly) that convergence in fatigue life is much harder to achieve than convergence in stress. Therefore much higher mesh densities are required to achieve similar levels of convergence to that in stress. In the present analysis (Figure 15), 20 elements around the ¼ hole were required to achieve 5% error in stress, but approximately 64 elements around the ¼ hole were required to achieve the same level of convergence in fatigue life. Furthermore, the results in Figure 16 indicate that the location of stress estimation in combination with stress gradients can have a more significant effect on estimating the fatigue life.

Sensitivity to stress gradient

Now consider an example of a hole in an infinite plate in combination with an arbitrarily selected PC-9/A spectrum to calculate the fatigue life. The theoretical solution for stresses in an infinitely large plate subjected to uni-axial loading is well known and eliminates any errors associated with stress. The percentage differences of fatigue lives calculated at various points away from the hole relative to the fatigue life evaluated at the edge of the hole are presented in Figure 17. From this figure it is quite clear that fatigue life estimates can change significantly even over very small distances. For the finest mesh considered in Figure 14, the smallest element at the notch root has a side length of 0.03 which has been normalised with the radius of the hole. In other words using an element with a centroid positioned at a distance 3% of the radius of the hole produced a fatigue life 54.8% greater than would be expected at the edge of the hole. Note: the use of the terminology "percentage difference" in Figure 17 is deliberate as this example assumes no notch effect exists.

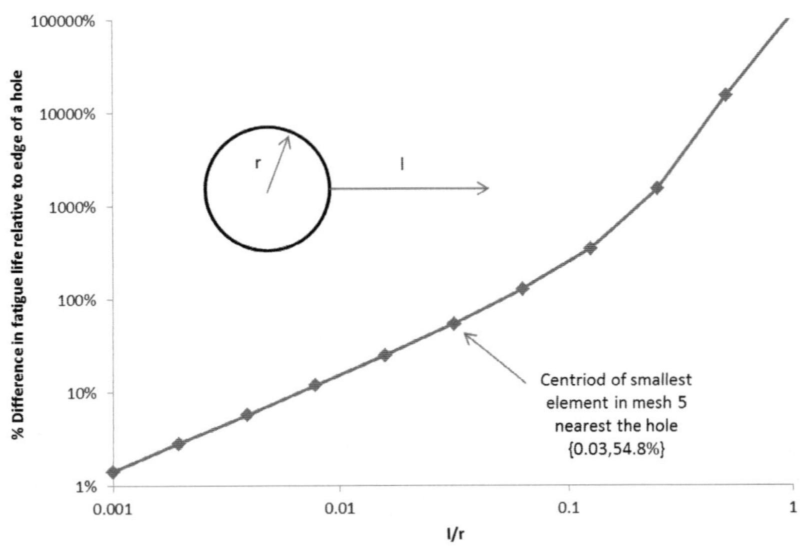

Figure 17. The percentage difference between fatigue life evaluated at the edge of the hole to locations away from the edge of the hole. Stresses were evaluated with the theoretical solution of a hole in an infinite plate.

In using an EBFA or even a nodal-based fatigue analysis, there is a tendency by the user to pick the element or node with the shortest fatigue life, usually near a stress concentrator in a fatigue assessment. It is well known that using the fatigue life calculated at the peak stress is unrealistic as it does not account for the stress decay. When small cracks nucleate at the notch root they quickly grow into areas of lower stress and hence their fatigue life tends to be longer than cracks nucleating from more benign geometric features with similar loading. As mentioned earlier the calculated fatigue life at notches tends to be shorter than that expected. It is not only the peak stress on the surface of the notch root that influences fatigue, but the average stress in a small volume of material at the notch root. With knowledge of the fatigue notch factor it could be possible to engineer the size of the elements or position the elements such that the fatigue index is evaluated at a location with a local to remote stress ratio equivalent to the fatigue notch factor.

Although not implemented in the present analysis, the theory of critical distances could be one way to interpret the results from an EBFA [19]. In using this approach, one way to account for the stress gradient could be to evaluate the fatigue life at a critical point just behind the notch. This critical distance might be used to determine the appropriate mesh size to use in the analysis. The line method of critical distances, although more involved, is another alternative.

IMPROVED STRUCTURAL DETAIL MODELLING

A difficulty that arises in the previous analysis is the required mesh density to achieve a convergent solution in life and the extreme sensitivity to stress gradients. Both issues highlight the need for accurate stress estimates and accurate representation of the notch feature. A possible resolution to this issue could be the use of p-elements. While simplifying the modelling process and producing highly accurate stress estimates, the formulation containing local geometry provides a possible mechanism to correct fatigue life estimates at geometric details.

To compare the use of the p-element formulation with the original refined PC-9/A FSFT UFEM, a cut-out in frame 10 was considered. This analysis considers one loading scenario to produce stress contour plots. The cut-out is highlighted in Figure 7b in blue. The original loads model used 3 elements to describe the boundary of the cut-out. Frame 10 was refined with in the present analysis and defines the boundary with 50 elements. The elements in the original model were converted to p-elements and the boundary of the notch specified in the element formulation by defining only 3 additional points between the existing nodes. The dark lines in Figure 18a and Figure 18b show the original mesh and the contours in Figure 18b are generated here using the p-element analysis. A more refined h-element mesh is used in Figure 18c for comparison. Although this comparison is qualitative, it is apparent that the p-element analysis using the same course grid mesh that was used in Figure 18a produces a result very similar to the more refined mesh in Figure 18c. This location has experienced one crack in the fleet at this location and was located at approximately the 11 o'clock position in the images in Figure 18. It is interesting to note that in both the refined h-element analysis and p-element analysis the peak stress has moved to the 11 o'clock position from the 1 o'clock positon in the original loads model. The model took 1 minute 23 seconds to run with only h-elements and 1 minute 42 seconds with 12 p-elements specified around the notch run at a p-order of 9. The addition of p-elements increased the run time by 22%, but compared to the further refined model (Figure 18c), the analysis was significantly faster and arguably more accurate.

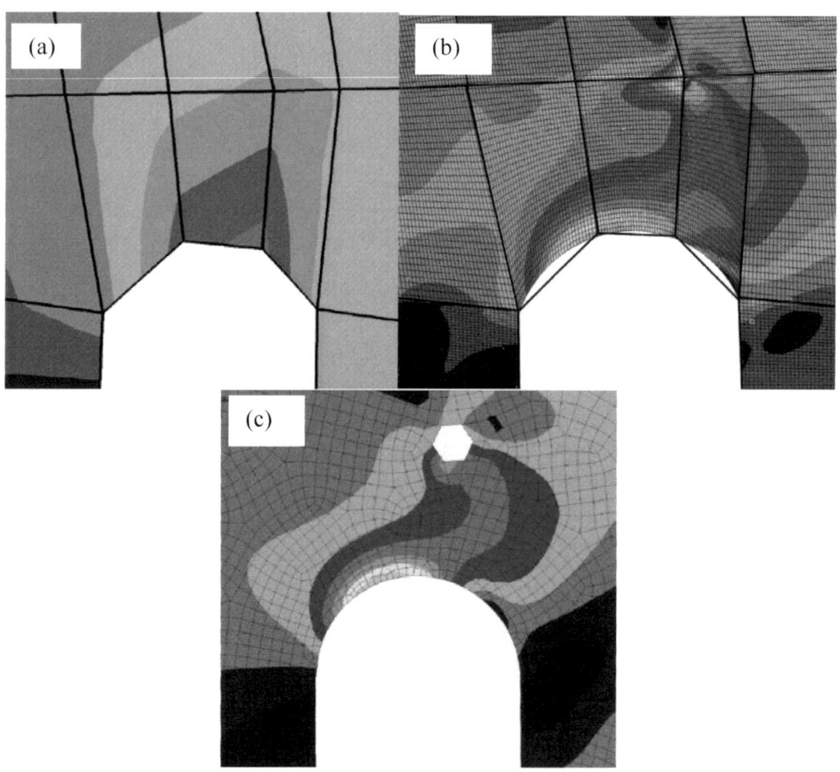

Figure 18. Cut-out in frame 10 with von Mises stress contours resulting from jack loading on the vertical tail. a) Original mesh b) p-element analysis with dark lines showing the mesh used in the analysis c) refined h-element analysis.

CONCLUSION

A fatigue analysis of a virtual airframe structure of a PC-9/A was conducted. The fatigue analysis involved the use of an elemental based fatigue approach to determine a fatigue index on each element. Selected examples of the EBFA are discussed and have been presented in comparison to observation from the full scale fatigue test. The EBFA appeared in many cases to qualitatively provide a good indication of the location of fatigue cracking on the full scale fatigue test article. It is important to note that the modelling did not consider non-linear effects such as buckling, contact and prying. This investigation has shown that both geometric notch effects and mesh density affect interpretation of the fatigue index. The predicted fatigue lives showed a high sensitivity to the mesh density, especially around stress concentrating geometric features. Significant caution is advised when interpreting fatigue life predictions directly from this methodology without further consideration to the local geometry. Specific attention must be placed on the effect of notches on fatigue life estimates. Results from the EBFA should be considered in conjunction with a method such as "critical distances" to appropriately make judgements on relative fatigue sensitivity. A local p-element formulation capturing the geometry features in the element formulation is proposed. This approach alleviates significant burden on mesh refinement and potentially provides a mechanism to adjust fatigue life estimates based on local geometry. Inherent errors in the modelling process make it also difficult to compare or rank the criticality of locations over large distances. Further work is required to refine the modelling process accounting for among other things, non-linear interactions, variations in components, build up stresses, and build quality, all with a view to improve model correlation.

ACKNOWLEDGMENTS

The authors wish to acknowledge the contributions of their DSTO colleagues that both worked on and supported the PC-9/A project. Further, we would like to acknowledge the efforts of the OEM, specifically noting the contribution of Domenico Quaranta.

REFERENCES

[19] Mahe, M., Imbert, J.-F. and Brand, C. (2005) Aircraft Virtual Structural Testing: Challenges for Structure Analysis.

[20] Firat, M., et al. (2009) *Engineering Failure Analysis* vol. 16, n. 5, p. 1533-1541.

[21] Tikka, J. (2002). In: *ICAS,* Toronto, Canada

[22] Ball, D. L. (2014). In: *11th International Fatigue Congress* Melbourne, Australia: 5 March

[23] *MSC Nastran 2013.1 Nastran Embedded Fatligue.* (2013) USA, MSC.Software Corporation.

[24] *Fe.Safe.* (2014). Dassault Systems.

[25] Tuegel, E., et al. (2011) *International Journal of Aerospace Engineering.*

[26] *Dvl Life.* (2014). DVL Fatigue.

[27] Anderson, I. and Green, P. (1999) *PC9/A Fatigue Test: Summary of Loads Applied to the Test Article.* DSTO-TN-0226, Melbourne, DSTO.

[28] Naval Air Warfare Center (1995) *Fatigue Analysis of Metallic Structures (FAMS) - A Computer Program to Calculate Fatigue Damage by Local Stress/Strain Approach.*

[29] Bates, P. (2001) *A Compendium of Abridged Notices of Structural Damage Developed During the PC9/A Fatigue Test – Part 1 of 6.* DSTO-TN-0338, DSTO.

[30] Bates, P. (2001) *A Compendium of Abridged Notices of Structural Damage Developed During the Pilatus PC9/A Fatigue Test – Part 4 of 6.* DSTO-TN-0338, DSTO.

[31] Bates, P. (2001) *A Compendium of Abridged Notices of Structural Damage Developed During the PC9/A Fatigue Test – Part 5 of 6.* DSTO-TN-0338, DSTO.

[32] Anderson, I. (2000) *PC9/A Fatigue Test, Summary of Structural Damage.* DSTO-TN-0269, DSTO.

[33] Yu, X., Kaye, R. and Opie, M. (2013). In: *15th Australian International Aerospace Congress (AIAC15),* Melbourne, Australia: 25-28 February 2013

[34] Neuber, H. (1937) *Kerbspannungslehre.* Berlin, Springer.

[35] Gregory, A. (2014) *Coupon Fatigue Testing of Thin Sheet Aluminium Alloy 2024-T3 Using the Direct Current Potential Drop Method.* DSTO-TR-2962, Melbourne, DSTO.

[36] Quaranta, D. (2014) *Pearla Phase 1 - Analysis with FSFT Loads.* TM-09-000137, Pilatus.

[37] Taylor, D. (2008) *Engineering Fracture Mechanics* vol. 75, n. 7, p. 1696-1705.

APPENDIX: A DISCUSSION ON MULTI-AXIAL LOADING AND THE POSSIBILITY OF IDENTIFYING A FATIGUE CRITICAL LOCATION SEPARATE TO THE LOCATION OF PEAK STRESS

In simple linear elastic structures, where a single load path exists, there is a high probability to identify the location where the component is expected to fail as it will usually coincide with the highest stress concentration. However, if more than one forcing function exists or even if the direction of only one input forcing function changes then the position of the fatigue critical location is less clear. The simple explanation to why this is lies within the driving mechanisms of fatigue. Although peak stresses in a spectrum can be very damaging, fatigue is typically driven by the changes in applied loading, the number of cycles, and the sequence in which they are applied. A single peak load event could introduce large damage to a structure but the equivalent damage could be introduced by many small load cycles with a lower peak stress. Thus, for a fatigue critical location to occur separately from a location known to produce the peak stress, two distinctly different local load spectrums must exit.

It can easily be visualised by considering a square cross section cantilever beam as shown in Figure 19. The cantilever beam is subjected to two independent forcing functions at right angles to one another (Spectrum A and Spectrum B). Consider the locations A and B which are located on the centre faces on the cantilever beam at equal distance from the fixed end in Figure 19. Location A is positioned at the neutral axis for any of the loads applied by spectrum B and similarly Location B is positioned at the neutral axis of loads applied by spectrum A. The result, as demonstrated by this simple example, is the existence of two independent local load spectrums in the one structure.

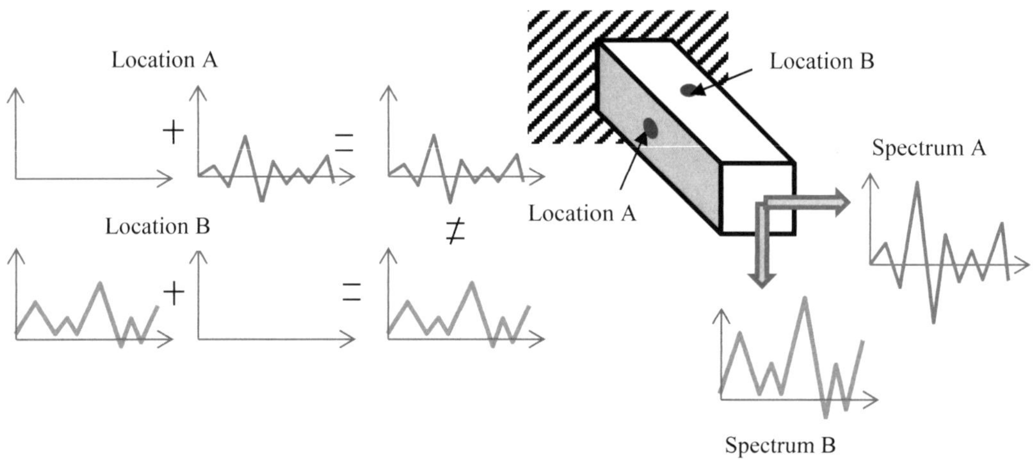

Figure 19. Cantilever beam with two forcing functions showing different local spectra at Location A and B.

A hypothetical example

At least from an analytical point of view, can we demonstrate a situation where the peak stress hot spot in the structure differs to the fatigue critical location? Consider the cantilever beam in Figure 20. Again we apply two spectrums: spectrum B contains a load 1.2 times greater than next largest load in both spectrums. Spectrum B is a transport type spectrum where the majority of the loads are small relative to the occasional gust load. Spectrum A represents a fighter type spectrum containing a lot of large load cycles. The spectrums are both tensile but are applied 180 degrees out of phase.

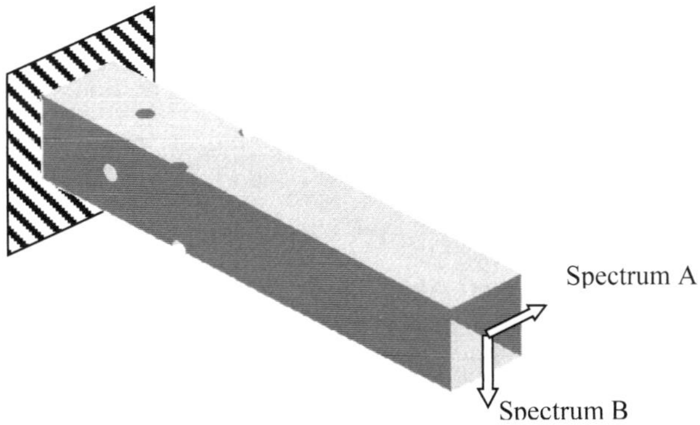

Figure 20. Cantilever beam with holes and edge cut outs.

A finite element model was created and a fatigue analysis was conducted on all elements in the model by constructing local spectrums on each element due to the applied spectrum loads A and B simultaneously. Firstly, if we consider just the peak load in Spectrum B the stress hot spot appears around the hole on the top surface, see Figure 21a. If we consider the peak load in spectrum A then the hot spot is located on the side around the hole (see Figure 21b), but has a magnitude 20% less than that produced by spectrum B. If we consider the spectrums in combination and extract results for the peak load on each spectrum generated for each element then we identify a new hot spot around the notch further down from the fixed end. This stress hot spot is clearly shown in Figure 21c. Finally conducting a fatigue analysis for each spectrum generated on all the elements, a map of fatigue lives can be created. These fatigue lives displayed as a fatigue index (NOT stress) are presented in Figure 21d. The fatigue index indicates that the fatigue critical hot spot is located around the hole on the side of the cantilever beam. Although when combining spectrums A and B together produced a hot spot in stress at the edge notch (Figure 21c) the analysis shows that a fatigue crack would more likely form at the hole on the side which will have loading dominated by Spectrum A. As the spectrums were tensile and applied 180 degrees out of phase the size of the load cycles were reduced around the edge notch relative to the hole. Also even though Spectrum B had the largest peak load 1.2 time that of spectrum A the rest of the spectrum was benign relative to Spectrum B. This is a simple example but one engineered to demonstrate the possibility that fatigue critical areas can differ from areas that may periodically experience a higher local stress.

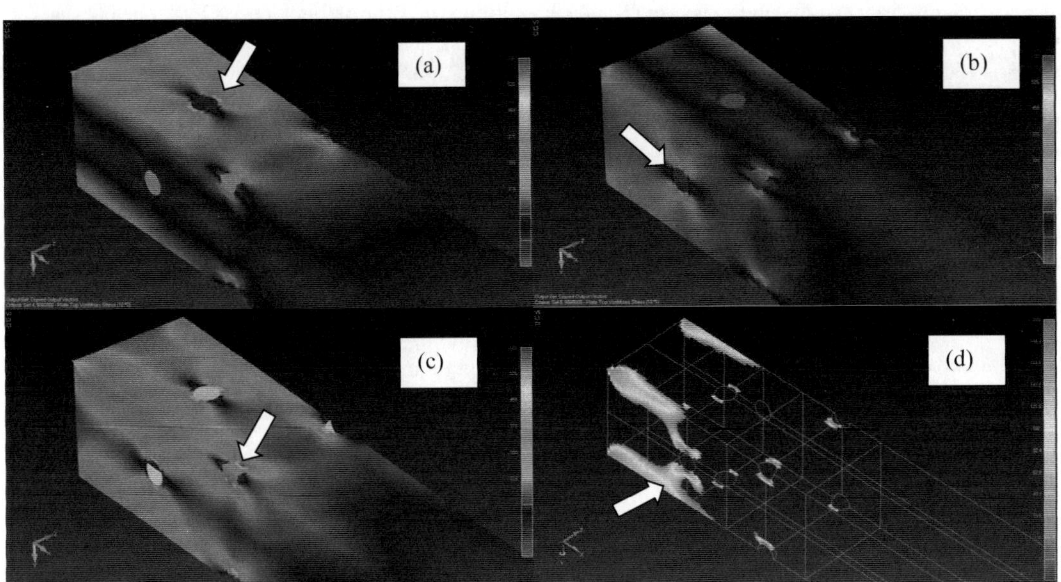

Figure 21. a) Von Mises stress due to peak load in Spectrum B, b) Von Mises stress due to peak load in Spectrum A, c) Peak load in each element due to the combined spectrums and d) Fatigue index (not stress) where red indicates areas of shortest life.

Towards the Virtual Fatigue Test: Hardware-in-the-Loop integrated Fatigue Test Simulation (HiLiFTS)

Albert K. Wong, Phil Jackson

Defence Science and Technology Organisation, Australia

Abstract: The costs and times associated with Full Scale Fatigue Tests (FSFTs) are enormous, but to date, they have proved to be the only reliable means for verification and validation of aircraft durability and safety and are thus mandated by aviation regulators for any new aircraft type. At the same time, however, analyses or models have been playing an increasing role in the qualification/certification process, and the question beckons whether the FSFT can one day be dispensed with. In this paper, rationale for both sides of the argument is presented, but concludes that the important question is not so much whether the FSFT can be done away with, but rather how FSFT can be performed more efficiently? To this end, a new concept is proposed that is a hybrid of the conventional FSFT and the complete digital model. Borrowed from the control systems engineering field, the hardware-in-the-loop simulation (HiLS) approach is adapted to provide the best of both worlds – the realism of actual testing for the critical components, together with the flexibility and cost-effectiveness of modelling. Whilst this approach has been made possible by a unique DSTO-developed technology that allows fatigue *hotspots* to be identified, its eventual success will also require some other technical problems to be solved. However, whilst these hurdles may be challenging, it is shown that they are not insurmountable and some suggestions are presented on how these may be tackled.

INTRODUCTION

Full Scale Fatigue Tests form the cornerstone in the qualification and certification of new aircraft types in both the civil and military sectors. These tests consume enormous resources in time, effort and costs, but have thus far proved to be essential in providing independent verification and validation of the design and manufacture of aircraft in terms of safety and durability. But with the rising complexity and costs of physical testing, together with the ever increasing computational capabilities, great focus has been given towards increasing reliance on analysis and models. The NATO ATO Report titled "Qualification By Analysis" [1] wastes no time in setting the rationale for turning towards analyses by its Executive Summary's opening sentence: "The time and cost associated with designing, developing and qualifying new military air vehicles is prohibitive…"; true to its title, the report then goes on to advocate the use of analysis for the qualification of all manners of aircraft systems and sub-systems to control both costs and time to market. The economics or benefits of qualification by analysis are alluring and as models become increasingly credible, terms such as "Virtual Tests" or more specifically, "Virtual Fatigue Tests" have emerged in the literature a little over a decade ago ([2], [3]). These early "tests" amount to nothing more than detailed models of materials or components which can produce results that compared well to physical tests under certain conditions. More recently, the concept of the virtual or digital aircraft assumes even greater prominence, where the aim is for high fidelity simulations and models of the entire aircraft and its environments to play a major role in the management of the aircraft over its entire life cycle. These are being pursued by European programs such as VIVACE[14], MAAXIMUS[15], CRESCENDO[16], and the US led programs on the Digital Twin[17], all of which have generated much excitement and interest in the aerospace world. Considering only the qualification/certification aspect, the scopes and expectations of these programs span a broad spectrum - from using simulations to simply improve efficiency of design, to forming the certification basis, as exemplified by the following:

- "to improve the efficiency of aircraft systems design, integration and certification by using simulation widely." (Homsi [6])
- "… to reduce need for repeat physical testing" (Coleman [7])
- "… will form a basis for certification of vehicles by simulation…" (Glaessgen & Stargel [5])

---

[14] Value Improvement through a Virtual Aeronautical Collaborative Enterprise (http://ec.europa.eu/research/transport/projects/items/vivace_en.htm)

[15] More Affordable Aircraft through eXtended, Integrated and Mature nUmerical Sizing (http://www.maaximus.eu/)

[16] Collaborative and Robust Engineering using Simulation Capability Enabling Next Design Optimisation (http://www.crescendo-fp7.eu/)

[17] See USAF Broad Agency Announcement Airframe Solicitation Number: BAA-13-03-RQKP "Digital Twin Spiral 1" (https://www.fbo.gov/index?s=opportunity&mode=form&id=0b10f8d15837d4ad47ca81da9e97cfcd&_tab=core&_cview=1); see also [4], [5].

Figure 1 shows a schematic that traces the evolution of the growing roles played by models in the context of qualification/certification of aircraft. The qualitative scale shown may be considered representative of a variety of metrics such as epoch, levels of difficulty, degree of utility or pay-off. In the beginning, tests were conducted to validate models, and as the latter became more powerful, they have been used increasingly to supplement or even replace certain full scale tests. Whilst early models could hardly be considered as virtual tests, more recent approaches whereby detailed models of the test article complete with the loading rig may be so considered (see, for example, [8]-[10]). The fidelities of these models are such that certain tests (eg., some static load or stability tests) can be used to confidently predict the results of actual tests, indicating that qualification by such means may be valid. Airbus cites that in the automotive industry, virtual crash testing is "now commonplace" [8]. Ultimately, the most challenging goal and one which would yield the greatest payoff would be the replacement of FSFTs by models or virtual tests.

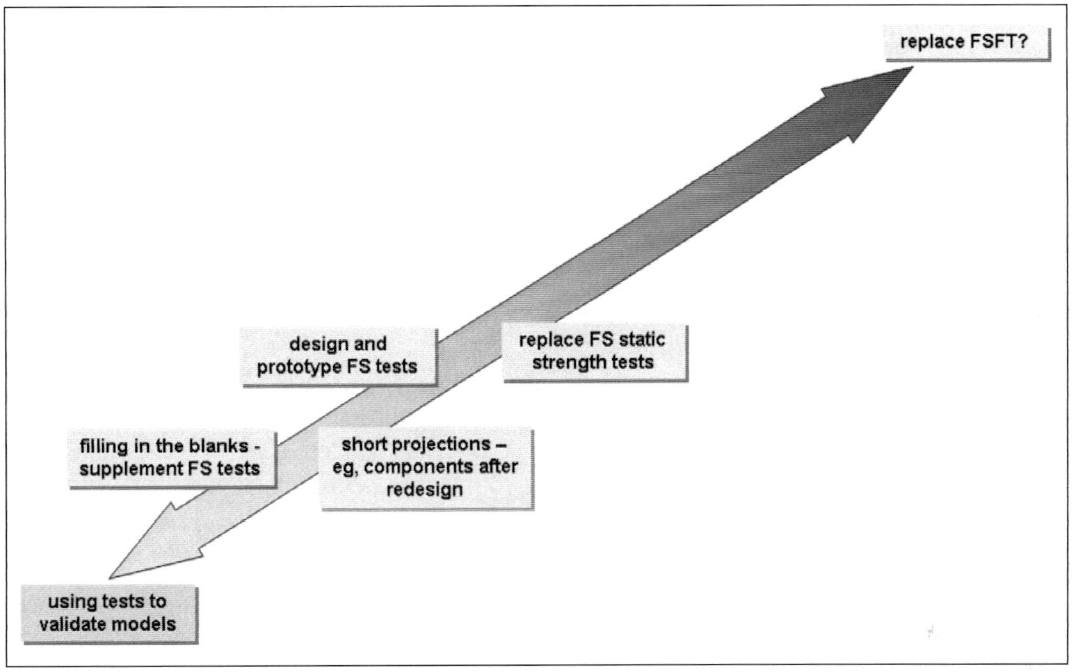

Figure 1. Evolution of the Full Scale Virtual Structural Test.

WILL VIRTUAL TESTS EVER REPLACE FSFTs?

With the growing role of models in the certification process, one obvious question is whether the ultimate goal of replacing FSFTs by virtual tests will eventually be met? In an effort to address this, let's consider the following arguments that may support one case versus the other:

The Case For:

- Moore's Law: the trend depicting the exponential increase in computing memory and power is continuing unabated, providing hope that computational power will be sufficient to meet whatever is required.
- Increasing model fidelity: structural models are getting increasingly sophisticated and realistic, incorporating non-linear, high-order elements, multi-disciplinary/multi-physics, and multi-scale implementations.
- Enormous pay-offs: FSFTs consume extraordinary resources and time to conduct, and their elimination, if possible, would be of significant impact and thus serve as a great incentive for driving progress towards such a goal.
- Visionary programs: significant programs such as VIVACE, MAAXIMUS, CRESCENDO, Digital Twin, are ambitious but are backed by and/or involving well resourced and well credentialed organisations with impeccable track records of achievements.

The Case Against:

- Models lack physics: whilst models for predicting displacement, deformation, strains, and even stresses could arguably be based on physics[18], the prediction of fatigue crack growth and, in particular, crack initiation are invariably phenomenological. As such, these models lack robustness and their limits of validity may not be known.
- Models are only as good as the assumptions made: assumptions of linearity, uniformity, reduced dimensionality can all lead to under-representation of reality. It is customary (and often necessary) to use coarse finite element meshes in regions that are thought to be *safe,* but every now and then, this supposition becomes a fatal flaw. If a certain model neglects the details of fasteners, then it is unlikely that it can reveal issues that are associated with them. It is also difficult to imagine how models can reveal the effects of manufacturing defects, build-qualities or fit-up stresses if these were not assumed in the first place. Essentially, models are ineffective for finding the *unknown unknowns*.
- Catch 22: There is no doubt that models are increasing in their fidelity and are being increasingly used with greater confidence. This has the tendency to drive down conservatisms in the design in pursuance of performance and economics. As the margins of errors are thinned out, such optimal designs are arguably more prone to surprises or to the *unknown unknowns*. Ironically then, the increasing powers of models only increase the importance of independent verification and validation (IV&V) through actual tests.
- No IV&V: Suppose that one day, a model is so good that it can essentially perform a virtual fatigue test for the purpose of qualification/certification. If such a model exists, then why wouldn't one just use it for design? That would make perfect sense, except that using the same model for design and IV&V would be fundamentally flawed, unless we accept that IV&V is *not* required.

Having set out both sides of the argument, the reader is invited to draw one's own conclusion to the original question. The reason for such irresolution is that arriving at the correct answer is not the most important. As the great American novelist and Nobel laureate Ernest Hemingway famously quoted "It is good to have an end to journey towards, but it is the journey that matters, in the end", what is important is for the pursuance of that ultimate goal to continue so that important technological developments will continue to emerge to benefit the field.

TAKING STOCK

Nevertheless, it's instructive to gauge how close (or far) we've come towards that goal after over 60 years of concerted effort on aeronautical fatigue research. In the same aforementioned NATO RTO report [1] in which it actively advocates the use of analyses for all manners of design, development and qualification of new military air vehicles, it concedes that

"... Full Scale Fatigue Tests (FSFTs) continue to produce premature fatigue cracking, demonstrating the current inadequacies of the fatigue analysis process. Until a time is reached when FSFTs do not reveal significant premature failures or cracking, designers and regulators alike will not have enough confidence in analysis methods to allow deletion of full scale testing."

This is aptly illustrated by the current System Development and Demonstration Program for the 5th Generation fighter, the F-35 Lightning II, otherwise known as the Joint Strike Fighter (JSF). To certify the structural integrity of the airframe, three separate FSFTs[19] were instigated – one for each of the three variants of the jet. For the Short Takeover and Vertical Landing (STOVL) variant, severe cracking of the bulkhead occurred in approx. 1500 simulated flight hours – less than just 10% of the scheduled 16,000 hr target (see Gertler [11]). This failure caused over 12 months of test program delay while the failed component was being reanalysed and reengineered. Soon after recommencement of testing, another major structural component (this time on the Carrier variant, but with implications for the others) suffered an unanticipated catastrophic failure, only some 15% into the scheduled program, causing another lengthy delay (see Capaccio [12]). These premature failures were exacerbated in this particular program because of the high degree of concurrency in production and tests, compelling the JSF's Executive Program Officer to concede such level of concurrency to be a "miscalculation" and called for a slow down in production (see Whittle [13]). Since that candid admission, the STOVL test once again broke catastrophically at one of the bulkheads at approx. 9,480 simulated flight hours (still 6,520 hrs short of the target) causing yet another lengthy stoppage in the test program (see Capaccio [14]).

[18] This is at least true in the linear regime where the basic laws of equilibrium and compatibility are enforced. In the nonlinear or plastic regime, additional rules such as those for hardening or hysteresis are introduced to match observations under certain conditions rather than necessarily meeting any rigorous physical requirements, and such models are therefore also deemed partially phenomenological.

[19] The F-35 Program refers to these tests as Full Scale Durability Tests (FSDT).

MINIMISING SURPRISES IN FSFTs

In many ways, the FSFTs in the JSF program did exactly what these tests were designed to do - they identified design faults that will now prevent those from being present in the sky. Even though the costs of the tests and, in particular, the failures, are extremely high, the costs of not doing them are unequivocally even higher. The question to tackle is not so much 'how can tests be done away with?' but 'how can these tests be conducted more effectively?' In the same NATO report, it identified that "the most practical and efficient way of reducing test costs is to ensure that all mandatory tests are successful and performed only once. Finding deficiencies early in the development program is thus a key in reducing program costs."

In the preceding ICAF symposium, Wong [15] presented the DSTO-developed microbolometric TSA (Thermoelastic Stress Analysis) technology, MiTE[20], and showed that it has proven to be effective for identifying stress *hotspots* (and therefore fatigue critical locations) during fatigue tests and argued that such technology should become a standard tool for any FSFT. Subsequently, DSTO has released a freeware version of the software in late 2013 that allows low cost TSA systems to be assembled by anyone (see http://www.dsto.defence.gov.au/mite/). To date, downloads of the MiTE software have been registered from all corners of the world and are on the rise, pointing to a rapidly growing user-base for this technology. For this to be a practical tool for seeking out FSFT *hotspots* systematically, however, it needs to acquire some form of automated and omnipresence capability. This requirement has led to the initiation of the TSA Robot (TSAR) demonstrator program in DSTO that aims to illustrate the possibility of pervasive and persistent monitoring of FSFTs using TSA (see Wong *et al* [16]). The demonstrator, designated as TSAR Mk 1, is based on the integration of the MiTE TSA technology with a 6 degree-of-freedom robotic arm, together with a video camera for 3-D reconstruction using a *Structures from Motion* algorithm (eg., [17]). The objective is to perform pervasive and persistent monitoring of a large section of a FSFT such as that for the F/A-18 Centre Barrel [18]. Figure 2 shows the setup of this demonstrator.

This task will not be trivial as a number of challenges still need to be tackled. Amongst these are the relatively large motion effects (particularly at the extremities of the test article where there will be significant displacements under load), and the rendering of the 3-D reconstructed virtual net with TSA data in an automated and seamless manner. Whilst these may be challenging, they are not by any means impregnable.

(a) (b)

Figure 2. DSTO's TSAR Mk 1 Program; a) General setup showing the incorporation of a 6DOF robotic arm with MiTE system at the F/A-18 Centre Barrel test rig; b) close up of the camera systems used, viz., the FLIR A35 microbolometer for TSA and a digital camera for 3-D reconstruction.

Once the basic capabilities of an automated TSA system to perform systematic wide-area monitoring has been demonstrated, the program will be expanded to pursue the grand vision of pervasive and persistent monitoring of FSFTs for complete airframes. By then, microbolometers will be of such small size and suitable sensitivities that they can be incorporated into small and/or flexible robots which can gain access to almost all parts of an aircraft. These robots will have great degrees of autonomy and work as teams to scour all over the airframe, charting out the geometries and stress states at every location, beaming the data wireless to a "Cloud" where a virtual wireframe of the aircraft is rendered with the TSA data. Any *hotspots* are then automatically flagged, and the analyst is alerted to inspect these locations via a virtual reality environment. Figure 3 shows a schematic of such a vision.

[20] MiTE is a DSTO developed technology which stands for Microbolometer Thermoelastic Evaluation.

HARDWARE-in-the-LOOP SIMULATION (HiLS)

Once the TSAR concept comes to fruition, its ability to identify potential failure locations early will greatly benefit FSFTs in allowing failures and weak spots to be managed effectively and efficiently. Not only will it prevent surprise catastrophic failures that may lead to collateral damage to the test article, but will allow fault propagations to be managed or controlled, and design fixes to be initiated and implemented well before actual failures occur. The realisation of such a goal will undoubtedly be one of the most important capabilities in controlling the cost and schedule of any new aircraft test program.

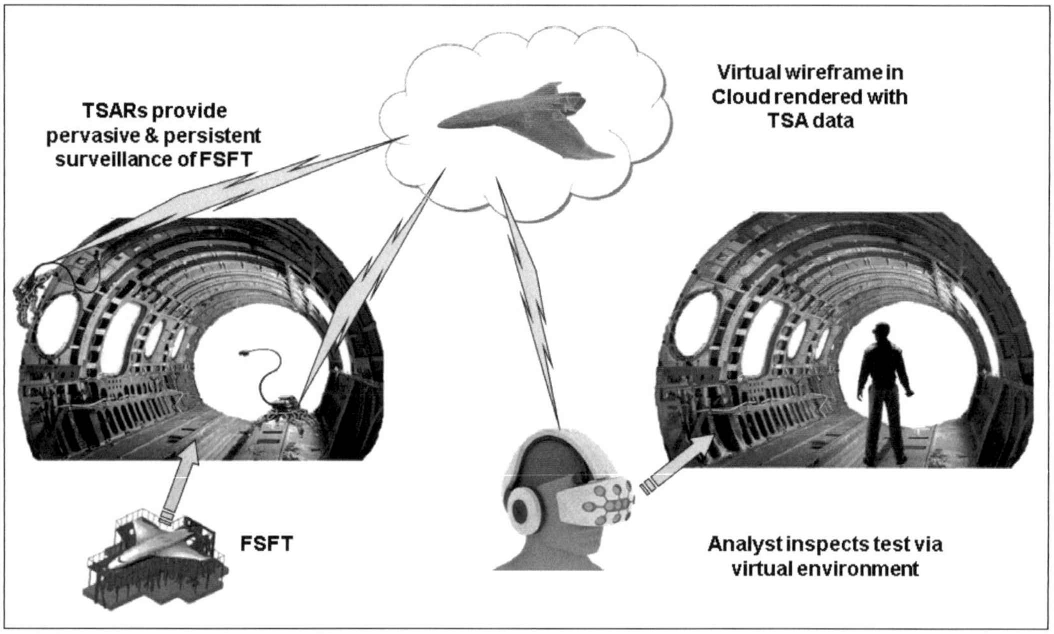

Figure 3. Visionary schematic of the roles played by future TSARs in FSFTs.

There is, however, an important corollary from such a realisation that can potentially lead to an even greater impact – one that can potentially revolutionise how future certification tests may be performed. If it transpires that future TSARs can indeed successfully survey and locate all potential failure locations[21], it then follows that after such an initial survey, only those locations need be tested provided of course the same loading conditions can be replicated. This suggestion appears to be quite radical at first, but in a way, this approach in the form of major component or sub-component tests is already being taken and accepted under certain circumstances. A good example of this is the F/A-18 Centre-Barrel test program being conducted by DSTO where it has been ascertained that the life of the aircraft, under a certain prescribed configuration, role and environment, is determined solely by the life of that component (see Molent *et al* [18]). The idea being explored in this paper is taking this approach to the extreme, viz., to consider whether sub-elements of some critical sub-components can be tested to determine the life of an airframe? The concept is akin to a technique known as Hardware-in-the-Loop Simulation (HiLS) that has been well established and utilised in the field of control systems engineering. Interestingly, HiLS has roots in the aviation industry, associated with the development of flight control systems some four decades ago (see eg., [19]), and has since become a standard technique for developing and testing critical sub-systems for a wide range of engineering products, ranging from automotive anti-lock brakes, traffic signals, unmanned aerial systems, to guided weapons. At its most fundamental level, any complex system that are responsive to external inputs can be depicted by the schematic shown in Fig. 4 where the overall system may be divided into the part(s) of interest (referred here as the "Critical Unit(s) of Interest", CUI), and everything else ("Rest of the System") (see [20]). In a HiLS, the latter is replaced by a computer model/simulation, and only the CUI is retained as the real part (ie., the hardware) in the system. Provided that the simulated part of the process is done with some accuracy (ie., the model resembles reality), then it is easy to see how HiLS can be an extremely attractive alternative compared to working with the entirely real system. Advantages listed in [20] include significantly decreased development time and costs, as well as increasing safety and reducing risks to property and test article.

---

[21] This would predicate on the advancement of TSA sensors that are so small and pervasive that they can either gain access to or be embedded in, virtually all locations. The promising trend towards this direction has been reported in Rajic & Galea [21].

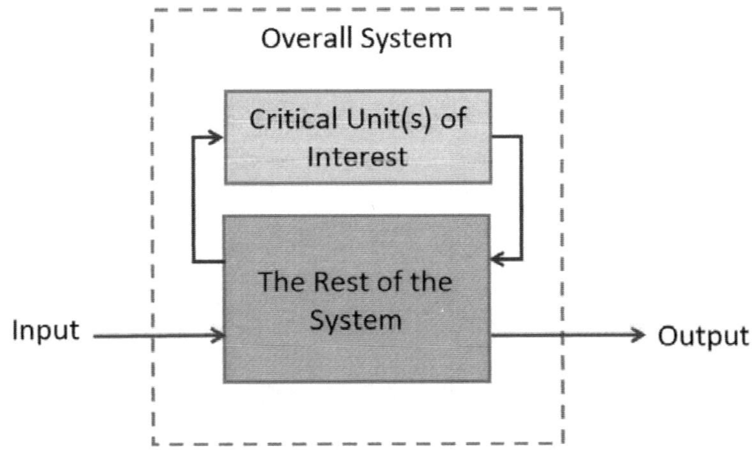

Figure 4. Decomposition of a complex dynamic system to form basis for a HiLS.

HARDWARE-in-the-LOOP integrated FATIGUE TEST SIMULATION (HiLiFTS)

Although the HiLS technique is aimed towards product development so that the hardware is often the product of concern (usually some type of electronic control unit), other hardware such as sensors, actuators, and even complete craft such as an unmanned aerial vehicle (see Gans *et al* [22]) can be included to improve the simulation itself. This latter goal is in fact our objective, viz., to find a means to improve our fatigue modelling capabilities by embedding real elements of the system into the simulation. For this, the same template as that shown in Fig. 4 can be applied as follows:

1. Let the "Overall System" be our fatigue test article in which the "Input" can represent the loading configuration and spectra, and the "Output" can be the outcome of the fatigue test, say the life of the test article.
2. Let the "Critical Unit(s) of Interest" be the critical location(s) that determine/s the life of the test article. As it is known that the modelling or life-prediction for such locations is difficult, we will opt for such part(s) to remain real.
3. Let the "Rest of the System" be the remainder of the test article. It is known that away from the critical areas, reliable models can be established. Furthermore, St. Venant's Principle informs us that exact form of loading distributions become less important as we move away from the regions of interest as long as the gross load and moments are represented.

The concept of embedding real material into a model can be visualised with the help of Fig. 5. In a structural simulation such as a finite element (FE) model, it is customary (and necessary) to refine the mesh in the region of concern (eg., see Fig 5a, near the 10 o'clock position around the modelled access hole). In a HiLS, we could imagine this region to be replaced by a piece of the actual component instead (see Fig. 5b).

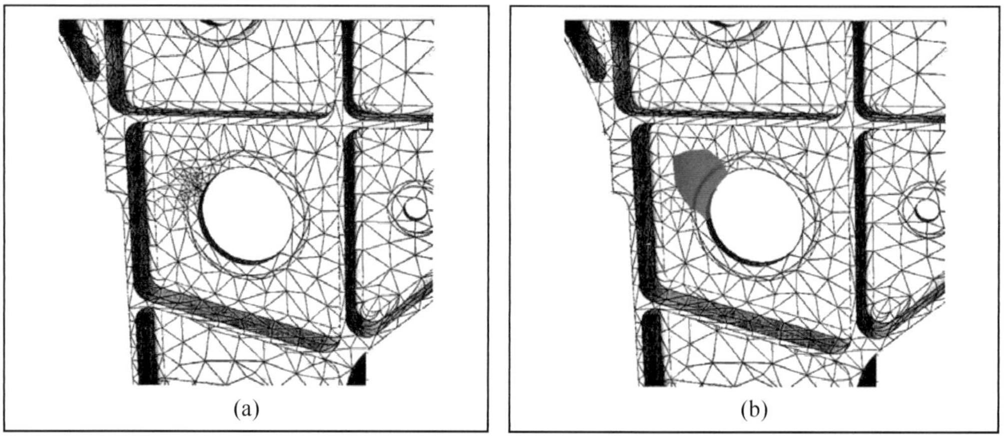

Figure 5. Conceptual illustration of HiLS implementation in a structural mechanics context: (a) an FE model of a structure with a refined mesh in the region of concern, (b) the same FE model, but with the region of concern replaced by a piece of the actual structure.

Whilst this concept may be easily pictured, its implementation is significantly more difficult as the application of complex localised loads to a small piece of component will not be easy, if at all possible. One conceivable implementation of such a sub-element test is depicted in Fig. 6 where complex load boundary conditions may be applied via a distributed set of electromechanical actuators. For simplicity, a circular boundary has been shown for the illustration; however, in the real application, it would be logical to use a contour of principal stresses to eliminate or minimise any requirement for shear transfer. To connect this sub-element test to the FE model, the HiLS template (as shown in Fig. 4) can be applied in which data are passed between these two sub-systems as depicted in Fig. 7.

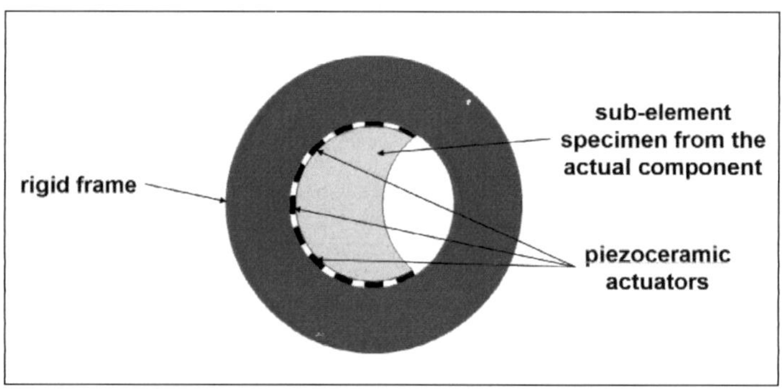

Figure 6. Concept of sub-element test rig showing an array of electromechanical actuators to apply known boundary loads (as determined by modelling).

Although the foregoing description has been for a specimen test, this approach is scalable and can be applied for the entire FSFT. Figure 8 shows a schematic on the implementation of this concept: it commences with a FSFT of the complete airframe, but right from the beginning, teams of TSARs will be deployed to map the entire airframe's stress distributions under test as described in the preceding section and in Fig. 3. This will perhaps take a few weeks to complete, where all critical locations[22] will be identified. Thereafter, the FSFT will no longer be needed and the airframe can be used for some other purpose, perhaps even returned back to the production line for finishing into a functioning aircraft. In the meantime, a global FE model will be run and adjusted or tuned to achieve localised stresses at the critical locations that match those found by TSA. Once this is achieved, sub-element tests such as that described

[22] This will be true if fatigue critical locations can be inextricably linked to stress *hotspots* – perhaps not necessarily ones displaying the maximum stress, but *hotspots* nonetheless (ie., spots with at least high tensile stresses or stress gradients). It is reasonable to assume that fatigue cracks will not grow in regions of benign stresses.

earlier (also Fig. 7) can be performed and continued until fatigue lives at these locations have been determined. A schematic of this concept is shown in Fig. 8.

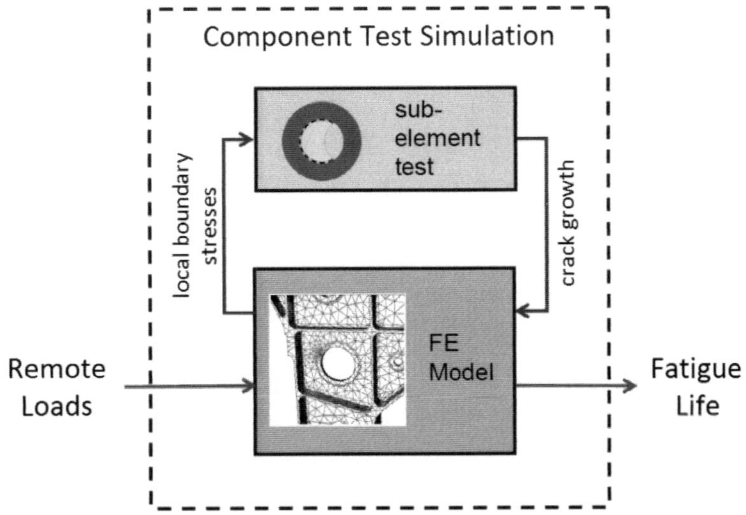

Figure 7. Schematic implementation of HiLS for fatigue testing to form a Hardware-in-the- Loop integrated Fatigue Test Simulation (HiLiFTS).

Of course, this vision has been idealised and its realisation is predicated on achievement of several challenging goals, viz., successful development of the future TSAR[23] concept; the ability to gain access to almost every important part of the airframe; and the ability to perform sub-element tests faithfully. Each of these goals is indeed not easily achievable, but at the same time, doesn't appear to be impossible. What is clear is that if the concept of HiLiFTS is achieved, then significant benefits will be reaped and new possibilities presented. These include:

1. FSFTs no longer need to be tested to destruction. In fact, it is envisaged that the test article only needs to be tested for a relatively short time, and perhaps even at non-damaging loads as TSA sensitivities continue to improve.
2. Because of the above, it will be possible to perform such FSFTs on multiple airframes, thus allowing evaluation of such things as build-quality and airframe-to-airframe scatter.
3. Sub-element tests can be performed at much higher rates, say 10-20 Hz or higher compared to less than 0.5 Hz typical of FSFTs. Such significant (20-40 fold) increase in test speed will undoubtedly translate to much shorter development and test times as experienced already by other industries that have embraced HiLS.
4. The ability to perform sub-element tests as envisioned will be of great significance on its own. Because of the small size, these can be performed anywhere, perhaps on desktops and be anywhere in the world. This opens up the following possibilities:
   a. tests can be widely distributed and be done at much lower costs.
   b. many more sub-element tests can be performed, thus allowing the evaluation of scatter (even if only a single airframe has been put through the FSFT).
   c. If unanticipated fatigue damage is found on in-service fleets (eg., when service usage turned out to be different to that of the test spectra), high speed sub-element tests can allow new analyses to be done and for the tests to quickly catch up to the actual fleet thus offering an unprecedented capability.

[23] Although the present HiLiFTS concept is built around the TSA technology, there is no reason why this can't be replaced by any other full-field experimental technique that can reliably and comprehensively map out the stress state of the structure in a FSFT.

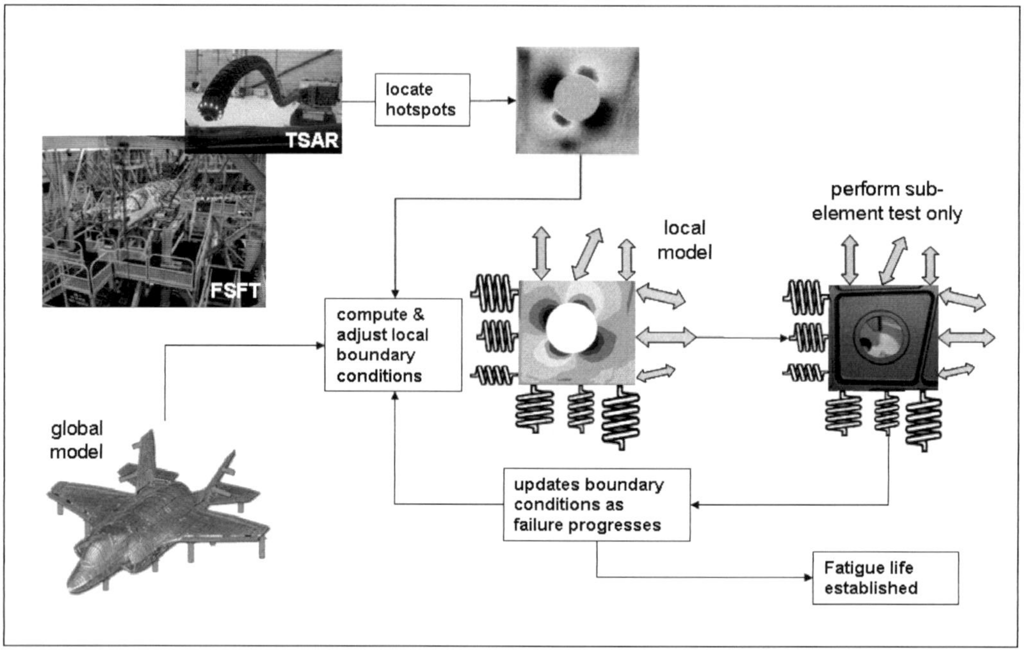

Figure 8. Schematic of the implementation of HiLiFTS to a full airframe FSFT.

SUMMARY

The burden of proof by aircraft manufacturers on the structural integrity of any of their new aircraft types is indeed heavy, culminating in the need to perform FSFT on the airframe as mandated by the regulators. As material systems and designs have become increasingly exotic and complex, so too have certification tests become more specialised and complicated. In an effort to control costs and schedule, engineers and researchers alike have turned towards analyses and modelling, perhaps hoping that the FSFT will one day be superseded by the virtual fatigue test. Although it is true that certain tests, such as some static or stability tests, can already be confidently replaced by models, the prospect of achieving a truly reliable virtual fatigue test within the foreseeable future is unrealistic. The main reasons for this are several, starting with the fact that virtually all fatigue models to date are phenomenological and therefore lack the necessary physics-base on which robust predictive capabilities can be built. Other reasons include the sensitivity of the fatigue process on a raft of factors such as stress state, surface conditions, and build-quality. Last but not least, models are ineffective in finding *unknown unknowns* – assumptions made early in the development of the models such as linearity, reduced dimensionality, mesh size, restraints, representation of fasteners and joints, and build-up strains effectively predetermine what features or criticalities can and cannot be discovered. As a result, early unanticipated failures in present FSFTs are still being experienced.

So, if continuing to depend on FSFTs for certification in the future is deemed to be less than palatable on the one hand, and the hope of relying on the fully digital aircraft is unrealistic on the other, it would appear logical to look for a solution in between. The proposal presented in this paper amounts to exactly this in that it combines the best of both worlds by retaining the realism of testing on the most critical parts whilst incorporating the modelling of the bulk of the system that can be modelled with confidence. Borrowing from the field of control systems engineering, the concept of hardware-in-the-loop simulation (HiLS) is adapted for this purpose. The benefits of HiLS in those industries have been undisputed and the technique has become standard practice for product development and certification. Whilst the adaptation of this technology to determine airframe durability has yet to be proven as there are yet some practical problems to be solved, the prospect of significant impact should stimulate and encourage the pursuance of this challenge.

ACKNOWLEDGMENTS

The author wishes to acknowledge the sound advice, counsel and input made by Messrs Philip Jackson, Loris Molent, and Len Meadows. Comments and suggestions made on the manuscript by Mr John Vine and Dr Nik Rajic were also greatly appreciated.

REFERENCES

[1] NATO RTO (2009) "Qualification by analysis", *NATO-RTO Report RTO-TR-AVT-092 AC/323(AVT-092)TP/229.*

[2] Tikka, J. (2002) "Fatigue life evaluation of critical locations in aircraft structures using virtual fatigue test", *Proc 23rd ICAS, 8-13 Sept, Toronto, Canada.*

[3] Farahmand, B. (2004) "Virtual testing versus physical testing for material characterisation", *Proc 45th AIAA/ASME/ASCE/AHS/ASC Struct, Struc Dynamics & Mat Conf, Apr 19-22, Palm Springs, California.*

[4] Tuegel, E.J., Ingraffea, A.R., Eason, T.G., Spottswood, S.M. (2011) "Reengineering aircraft structural life prediction using a Digital Twin", *Int J. Aerospace Eng'g,* v2011, n154798, DOI:10.1155/2011/154798.

[5] Glaessgen, E.H., Stargel, D.H. (2012) "The Digital Twin paradigm for future NASA and U.S. Air Force vehicles", *53rd Struct, Struct Dyn & Mat Conf; Special Session on Digital Twin, http://ntrs.nasa.gov/archive/nasa/casi.ntrs.nasa.gov/20120008178.pdf* (accessed 8 Apr 2014).

[6] Homsi, P. (2007) " VIVACE – Value Improvement through a Virtual Aeronautical Collaborative Enterprise", *http://www.transport-research.info/Upload/Documents/201210/20121011_124433_10190_ Technical_leaflet_final.pdf* (accessed 8 Apr 2014).

[7] Coleman, P. (2013) "CRESCENDO - Collaborative and Robust Engineering using Simulation Capability Enabling Next Design Optimisation" *Final Publishable Summary, http://cordis.europa.eu/ docs/results/234344/final1-crescendo-d017-final-report-20130430-r1.pdf* (accessed 8 Apr 2014).

[8] Mahé, M., Imbert, J. F., and Brand, C., "Aircraft virtual structural testing,", *Proc. 7th Nat Colloq Stress Analysis*, invited lecture, v1, Hermes ISBN 2-7462-1139-4, Giens France (2005).

[9] Tabaste, O., (2010) "A route toward virtual certification of aircraft", *Proc 27th ICAS, 19-24 Sept, Nice, France.*

[10] Ostergaard, M.G., Ibbotson, A.R. (2011) "Virtual testing of aircraft structures", *CEAS Aero J,* v1, pp 83-103.

[11] Gertler J. (2011) "F-35 Joint Strike Fighter (JSF) Program: background and issues for Congress". *Congressional Research Service Rept, Apr 26.*

[12] Capaccio A. (2011) "Lockheed F-35 Fighter has 'Design Flaw' in wing part, Pentagon tester says" *Bloomberg News, 2 Sep.,* http://www.bloomberg.com/news/2011-09-01/lockheed-martin-f-35-wing-part-has-design-flaw-tester-says.html (accessed 1 May 2012).

[13] Whittle R. (2011) "JSF's build and test was 'Miscalculation,' Adm. Venlet says; production must slow" *AOL Defense, 1 Dec.*

[14] Capaccio A. (2014) "Lockheed F-35 for Marines delayed as test exposes cracks" *Bloomberg News*, 21 Feb., http://www.bloomberg.com/news/2014-02-21/lockheed-f-35-for-marines-delayed-as-test-exposes-cracks.html (accessed 15 June 2014).

[15] Wong, A.K. (2013) "Making the invisible visible: taking a look at stresses in aircraft fatigue analysis and testing" *Proc ICAF Symposium, 5-7 June, Israel.*

[16] Wong, A.K., Rajic, N., Nguyen, Q. (2014) "Seeing stresses through the thermoelastic lens - a retrospective and prospective from an Australian viewpoint" *Int J. Expt Mech. (STRAIN),* Article first published online: 21 NOV 2014 | DOI: 10.1111/str.12116, also accessible via *http://onlinelibrary.wiley.com/doi/10.1111/str.12116/epdf.*

[17] Snavely, N. (2010) "Bundler: Structure from motion (SFM) for unordered image collections" Available online: http://www.cs.cornell.edu/~snavely/bundler/ (accessed on 1 May 2014).

[18] Molent, L., Dixon, B., Barter, S., White, P., Mills, A., Maxfield, K., Swanton, G., Main, B. (2009) "Enhanced teardown of ex-service F/A-18A/B/C/D centre fuselages" *Proc ICAF Symposium, 27-29 May, Rotterdam.*

[19] Ayari, L. (2008) "Hybrid testing & simulation - the next step in verification of mechanical requirements in the aerospace industry" *Chpt 20 in Hybrid Simulation,* Eds Saouma, V.E., Sivaselvan, M.V., ISBN/ISSN 9780415465687; 041546580, 9780203892947.

[20] Schlager, M. (2008) "Hardware-in-the-loop simulation" *Pub. VDM Verlag Dr, M. Aktiengesellschaft & Co*, Germany.

[21] Rajic, N., Galea, S. (2014) "Thermoelastic stress analysis and structural health monitoring: an emerging nexus." in *Structural Health Monitoring* DOI: 10.1177/1475921714548936.

[22] Gans, N.R., Dixon, W.E., Lind, R., Kurdila, A. (2009) "A hardware in the loop simulation platform for vision-based control of umanned air vehicles" *Mechatronics*, v19, pp 1043–1056.

Probabilistic modelling and risk analysis of structural integrity

Application of Quantitative Risk Analysis in Support of Continuing Airworthiness Management

Min Liao, Guillaume Renaud, Yan Bombardier

Aerospace Portfolio, National Research Council Canada (NRC), 1200 Montreal Road, Ottawa, Canada K1A 0R6

Abstract: This paper presents some of the best practices and lessons learned in the application of quantitative risk analysis methods and tools to support the Royal Canadian Air Force (RCAF) Aircraft Structural Integrity Program (ASIP) and continued airworthiness management. First, this paper presents a brief introduction of the RCAF Record of Airworthiness Risk Management (RARM) process in support of the continuing airworthiness management. An overview of the RCAF airframe lifing policy is also presented along with the applicability of different risk analysis methods to the Safe-Life and Damage-Tolerant designed structures. Second, the damage tolerance and risk analysis methods and tools developed by NRC are presented for traditional fatigue reliability analysis and advanced risk analysis based on probabilistic fracture mechanics. The fatigue reliability analysis includes a practical Weibull/Lognormal analysis approach that uses in-service data with censored data. The advanced risk analysis methods and tools include the NRC in-house tools, ProDTA and CanGROW, which were developed to calculate the probability of failure (POF) for airframes containing multi-site fatigue damage (MSD) and multi-element damage (MED). At last, some risk analysis examples were presented for the RCAF CC-130 and CP-140 aircraft by applying the NRC's methods and tools. These examples include a Safe-Life based risk analysis, as well as a Damage-Tolerant based risk analysis considering widespread fatigue damage (WFD). Data preparation and technical challenges for properly accounting for initial quality, usage scatter, repair quality, and inspection capability are discussed, along with the practices developed to increase the confidence of the quantitative risk analysis. In conclusion, this paper shows that the risk-based management process, specifically the RCAF RARM, has served as a powerful tool to support the continuing airworthiness management to maintain or increase availability, while reducing operational costs. The methods and tools developed by NRC have provided significant support to the quantitative risk analyses for various RCAF aircraft fleets.

INTRODUCTION

Through the Technical Airworthiness Authority (TAA), the Royal Canadian Air Force (RCAF) is responsible for the regulation of the technical airworthiness aspects of design, manufacturing, maintenance, and material support of aeronautical products, as well as the determination of the airworthiness acceptability of those products prior to operational service. The technical airworthiness program consists of three primary elements: initial airworthiness, continuing airworthiness, and disposal. The processes incorporated into the technical airworthiness program include the Technical Airworthiness Risk Management, which is not only a qualitative decision-making tool but also a quantitative tool used to assist individuals in making informed risk management decisions. The key document in the process of managing airworthiness risk is identified as the Record of Airworthiness Risk Management (RARM) [1]. Since 2003, the RCAF has started adopting the RARM process to manage Technical Airworthiness for all its aircraft.

The RARM process was designed to cover the airworthiness for all aircraft systems, i.e. hydraulics, structural, mechanical, avionics, etc. For the aircraft structures, the RARM is being implemented in conjunction with the Aircraft Structural Integrity Process (ASIP) program when compliance with the basis of certification cannot be demonstrated due to lack of data. The potential hazards on structures include the failures that can cause injuries or death to personnel, damages to or loss of the aircraft, or reduction of mission readiness or availability. Since the majority of the structural failures are due to fatigue fracture under both cyclic loading and environment-related aging, it is more difficult to conduct a quantitative risk analysis for aircraft structures due to the complexity of the fatigue and corrosion processes and to the scarcity of data on catastrophic structural failures. Given that different design philosophies, such as Safe-Life and Damage-Tolerant, have been used for designing aircraft structures and determining their life limits, different methods are needed to determine the airworthiness risk, i.e. the single flight hour probability of failure (SFHPOF), in a quantitative risk analysis for aircraft structures.

To support the RCAF ASIP and RARM process for maintenance and service life assessment, the National Research Council Canada (NRC) has been developing practical statistical and probabilistic methods and tools for Safe-Life and Damage-Tolerant based quantitative risk analyses (QRA). Through numerous projects tasked by the RCAF, NRC has conducted various quantitative risk analyses for different air fleets, such as the CC-130 and the CP-140. This paper intends to present methods developed by NRC with some of the best practices and lessons learned in the application of these methods and tools to support the RCAF ASIP and RARM processes.

RISK-BASED CONTINUING AIRWORTHINESS MANAGEMENT

Record of airworthiness risk management (RARM)

The RCAF RARM process includes five steps for risk management [1]: hazard identification, risk assessment, risk control, RARM approval, and risk tracking. The key airworthiness risk index, as defined in Figure 1, is measured by a product combing both Hazard Severity and Hazard Probability. The risk index definitions are aligned with other military and civil standards including FAA/JAR 25-1309, AC 25-1309, SAE ARP 4761, and MIL-STD-882. Detailed definitions for the Hazard Severity, i.e. Catastrophic (A), Hazardous (Severe Major) (B), Major (C), Minor (D), and Negligible (E), are documented in [1]. Detailed definitions for qualitative Hazard Probability (Probability of Occurrence), i.e. Frequent (Level 1), Probable (Level 2), Remote (Level 3), Extremely Remote (Level 4), and Extremely Improbable (Level 5), are also documented in [1]. The corresponding quantitative hazard probability thresholds (HPT) per flight hour, such as 10^{-3}, 10^{-5}, 10^{-7}, or 10^{-9}, are presented in Figure 2. Note that the HPT defined in Figure 2 also depends on the type of RCAF aircraft platforms, such that an aircraft carrying passenger will require lower HPT than an unmanned air vehicle.

The RARM process has affected all RCAF fleets (DND-AD-2007-01), covering all the airworthiness and safety related aircraft systems including hydraulics, structural, mechanical, avionics, etc. [2]. Today, the RARM has become the single most critical decision-making tool for the RCAF air fleet management [2]. When there are sufficient data available, a quantitative risk assessment can be performed to substantiate the assignment of a risk index number. When a qualitative risk assessment indicates a high or medium risk, a detailed quantitative RA is often requested to calculate the hazard probability, i.e. the SFHPOF, to gain additional confidence in the decision-making. In practice, the RCAF's first intention is always to conduct a *quantitative* risk assessment; when there are not enough data, a *qualitative* risk assessment is performed supported by engineering judgement.

Some specific features in aircraft structural risk analysis are,

- Compared to other aircraft systems, there are much fewer structural catastrophic failure data, so simple data-driven reliability or empirical statistical models cannot produce high-confidence structural risk assessment.
- Since the majority of structural failures are due to fatigue fracture under both cyclic loading and environment-related aging, compared to other systems, it is more difficult to conduct a quantitative risk analysis for aircraft structures, due to the complexity of the fatigue and corrosion processes, which are both time-related deterioration processes with large intrinsic and extrinsic variability.
- Given the fact that all aircraft structures have in-service life limits and that different design methods, such as Safe-Life and Damage-Tolerant, have been used for designing aircraft structures and for determining these limits, different technical methods have been used to determine the hazard probability or the SFHPOF.
- The SFHPOF threshold in Figure 2 is a constant value (e.g. 10^{-8} per flight hour) for a certain type of aircraft and is independent of the aircraft lifetime and fleet size. Maintenance, modification, or replacement are needed to ensure this constant SFHPOF threshold (or reliability level, defined as 1.0-SFHPOF), otherwise a different SFHPOF would be reviewed/revisited for longer lifetime aircraft and larger size of fleet.

Hazard		Probability	Category				
		Severity	A	B	C	D	E
			Catastrophic	Hazardous (Severe Major)	Major	Minor	Negligible
LEVEL	1	Frequent	A1 Extremely High	B1 Extremely High	C1 Medium	D1 Low	E1
	2	Probable	A2 Extremely High	B2 High	C2 Low	D2	E2
	3	Remote	A3 High	B3 Medium	C3	D3	E3
	4	Extremely Remote	A4 Medium	B4	C4	D4	E4
	5	Extremely Improbable	A5	B5	C5	D5	E5

Risk Threshold

NOTE

Risks below the Risk Threshold are considered to be within the Acceptable Level of Safety for all approved missions and flying operations.

Figure 1. RCAF airworthiness risk indices [1].

Hazard Probability Level	Hazard Probability Thresholds (Per Flight Hour)			
	DND Passenger Carrying Aircraft *(Derived from FAR 25/29 Civil Designs)*	Military Aircraft		
		Military Aircraft	Military Aircraft - *Ejection Seat Equipped*	Unmanned Aerial Vehicles (UAVs) * *Above 150Kg TOW*
Frequent	Greater than 1×10^{-3}	Greater than 1×10^{-3}	Greater than 1×10^{-2}	Greater than 1×10^{-2}
Reasonably Probable	Less than 1×10^{-3}	Less than 1×10^{-3}	Less than 1×10^{-3}	Less than 1×10^{-2}
Remote	Less than 1×10^{-5}	Less than 1×10^{-5}	Less than 1×10^{-4}	Less than 1×10^{-3}
Extremely Remote	Less than 1×10^{-7}	Less than 1×10^{-6}	Less than 1×10^{-5}	Less than 1×10^{-5}
Extremely Improbable	Less than 1×10^{-9}	Less than 1×10^{-8}	Less than 1×10^{-7}	Less than 1×10^{-6}

Figure 2. Quantitative hazard probability [1].

Review of the RCAF aircraft structural lifing policy

The RCAF aircraft fleets have been certified using different airworthiness standards. A summary of the RCAF aircraft structural lifing policies, including original OEM and current RCAF methods, is presented in Table I [4]. Table I indicates that two major lifing methods, i.e. Safe-Life and Damage-Tolerant, are used across the RCAF fleets, either in a separate or combined manner. Both military and commercial airworthiness specifications are considered when deciding the current RCAF lifing methods, such as DEF-STAN 00-970, MIL-STD-1530, FAR 23, FAR 25, and JAR 29. However, from previous experience [6], it is difficult to quantify and compare the level of "safety" used by the different airworthiness standards. This could raise challenges when using the same RARM risk index to manage different fleets which are designed using different lifing methods. Other challenges have been experienced by RCAF during the aircraft structural integrity program (ASIP) management, such as,

Table I. Summary of RCAF aircraft structural lifing policies, original and current.

No.	Aircraft (fleet age in 2012)	Service Usage (fleet size)	Manufacturer (when acquired)	Original Lifing Policy	Current Lifing Policy
1	CC115 Buffalo (55 years)	SAR (6)	de Havilland (1967)	Safe Life	Safe Life, Damage Tolerance
2	CC130 E&H Hercules (21 to 52 years)	Transport, Tanker, SAR (23)	Lockheed Martin (1960 to 1991)	Safe Life	Damage Tolerance
3	CC130J Hercules (2 years)	Transport, Tanker, SAR (17)	Lockheed Martin (2010)	Damage Tolerance	Damage Tolerance
4	CC138 Twin Otter (42 years)	Transport, SAR (4)	de Havilland (1970)	Residual Strength, Fail Safe	Safe Life, Damage Tolerance
5	CC144 Challenger (28 to 33 years)	VIP Transport (6)	Bombardier (1980 to 1985)	Damage Tolerance	Damage Tolerance
6	CC150 Polaris (A310) (20 years)	Transport, Cargo, Tanker (5)	Airbus (1992 to 1993)	Damage Tolerance	Damage Tolerance
7	CF188 Hornet (24 to 30 years)	Fighter (77) (modernized)	McDonnel Douglas / Boeing (1982 to 1988)	Safe Life (US Navy Mil Std 1530)	Safe Life, Damage Tolerance (safety-by-inspection)
8	CP140 Aurora (32 years)	Patrol (18)	Lockheed Martin (1980)	Residual Strength, Fail Safe	Safe-Life, Damage Tolerance
9	CC177 Globalmaster III (5 years)	Transport (4)	McDonnel Douglas / Boeing (2007)	Damage Tolerance	Damage Tolerance
10	CT114 Tutor (Snowbirds) (50 years)	Training, Aerobatic (24)	Canadair (1962)	Safe Life	Safe Life, Damage Tolerance
11	CT142 Dash 8 (22 to 25 years)	Navigation Training (4)	de Havilland (1987 to 1990)	Damage Tolerance (FAR 25)	Damage Tolerance
12	CT155 Hawk (leased) (12 years)	Fighter Trainer (16)	BAE Systems (2000)	Safe Life	Safe Life
13	CT156 Harvard II (leased) (13 years)	Pilot Training (25)	Raytheon (1999)	Damage Tolerance	Damage Tolerance (FAR 23)
14	CH-139 Jet Ranger (31 years)	Training (13)	Bell Helicopter Textron (1981)	Safe Life	Safe Life (under civilian contract)
15	CH124 Sea King (43 to 49 years)	SAR, Patrol (27)	Sikorski (1963 to 1969)	Safe Life	Safe Life
16	CH146 Griffon (15 to 18 years)	Utility, Transport, Combat Support	Bell (1995 to 1997)	Safe Life (FAR 29)	Safe Life
17	CH149 Cormorant (9 to 11 years)	SAR (14)	AgustaWestland (2001 to 2003)	Safe Life, Flaw Tolerant (JAR 29)	Safe Life, Flaw Tolerant
18	CH148 Cyclone	SAR, Patrol (28)	Sikorski (under certification)	Safe Life, Damage Tolerance	Safe Life, Damage Tolerance

- For some aircraft, the current RCAF lifing methods are different from the OEM ones, the current lifing methods maintain safety well, not necessarily cost-effective
- The actual RCAF aircraft usage/loads and in-service damage can be very different from the OEM design conditions, including those used during full scale certification tests.
- The RCAF lacks some data/information such as the load models and spectra, and is sometimes faced with incomplete component tracking (components interchanged between aircraft but the swap was not recorded) and component history cards.

- The RCAF lacks some data/information on new materials (composite and new metals, etc.), on new manufacturing processes (friction stir and laser welding, etc.), and fatigue enhancement technologies (cold work/interference fits, peening etc.).
- Complex damage, such as MSD (multi-site fatigue damage), MED (multi-element fatigue damage) or WFD (widespread fatigue damage) and Canadian environment related damage modes that can be present in the various materials were not originally taken into account by the OEM, for most RCAF aircraft.

Given these challenges related to "lack of data/info.", "unknowns", and "differences", a risk-based approach, like the RARM process, is adopted by RCAF to manage the *impact* of uncertainties. For Safe-Life designed structures, Weibull and Lognormal statistical analysis, also called conventional fatigue reliability analysis, is often used to determine fatigue life distributions and the associated scatter factors for different probability and confidence levels. For typical Damage-Tolerant designed structures [7], probabilistic fracture mechanics is used to simulate fatigue crack growth, calculate life distributions, and then calculate the SFHPOF of critical locations subjected to random loading. In this paper, this is also called modern or advanced risk analysis [8].

QUANTITATIVE RISK ANALYSIS METHODS AND TOOLS

Conventional fatigue reliability analysis using Weibull and Lognormal distributions

Typically, fatigue lifetime data for metallic materials can be well fitted by a Weibull or a Lognormal distribution. A classic Lognormal analysis, including scatter factor determination, can be found in [9] (which is also used by DEF-STAN 00-970). This paper presents some Weibull analysis processes and equations based on NRC's best practices on risk analysis [3]. The cumulative distribution function (CDF), $F(t)$, and probabilistic density function (PDF), $f(t)$, of a Weibull can be expressed as,

$$F(t) = 1 - \exp\left(-\left(\frac{t}{\alpha}\right)^{\beta}\right) \tag{9}$$

$$f(t) = \frac{\beta}{\alpha}\left(\frac{t}{\alpha}\right)^{\beta-1} \exp\left(-\left(\frac{t}{\alpha}\right)^{\beta}\right) \quad t \geq 0; \alpha, \beta > 0 \tag{10}$$

where α is the scale parameter and β is the shape parameter. The hazard function or hazard rate (HR), $h(t)$, for the Weibull function can be obtained in an exact closed-form, as,

$$\mathrm{HR} = h(t) = \frac{f(t)}{1 - F(t)} = \frac{\beta}{\alpha}\left(\frac{t}{\alpha}\right)^{\beta-1} \tag{11}$$

The hazard function (or hazard rate (HR), failure rate, risk of failure) specifies the instantaneous failure rate of death or failure at time t, given that the individual item survives up to t [10]. From its mathematical form, HR is a PDF normalized by 1 - $F(t)$, thus it is sometimes called a "conditional failure rate" (but not a conditional probability) since the denominator 1 - $F(t)$ (i.e. the population survivors) converts the expression into a conditional rate, given survival past time t [11]. From Eq. (3), it is shown that,

- For a Weibull distribution, when $\beta > 1.0$ (or < 1.0), the Weibull HR will monotonically increase (or decrease) with the time t.
- In general, a HR ($h(t)$) is dependent on both $f(t)$ and 1- $F(t)$, so it can be increasing or decreasing with the time t. It is only when the decrease of the denominator 1- $F(t)$ is faster than the decrease of $f(t)$, then a HR ($h(t)$) will keep increasing with the time t.

In a small time increment (piecewise), a CDF can often be well fitted by an Exponential distribution (which is the special case of the Weibull distribution with β = 1.0), such as,

$$F(t) = 1 - \exp(-t/\alpha) = 1 - \exp(-\lambda t), \lambda = 1/\alpha \tag{12}$$

$$f(t) = \lambda \exp(-\lambda t) \tag{13}$$

Eq. (4) shows that there is a log-linear relationship between (1-$F(t)$) and t. Note that in a real problem, a location parameter γ may be needed to best fit the data; in such case, t is simply replaced by t - γ. The hazard function of the Exponential distribution is the constant λ, which can be used for a piecewise fitted probability function in a small time increment, as,

$$h_{Exp}(t) = \lambda = \ln\left[\frac{1-F(t_1)}{1-F(t_2)}\right] \Big/ (t_2 - t_1) \tag{14}$$

In a structural risk analysis, when $F(t)$ is calculated as a function of time t using numerical integration or Monte Carlo simulation, the HR in a ***small*** time increment can be approximately calculated by Eq. (14). The smaller the time increment, the more accurate Eq. (14) is.

For aircraft structures or coupons made of aluminium alloys, a shape parameter of $\beta = 4.0$ is usually used for fatigue life Weibull distributions. Assuming a scale parameter of 8000 flight hours (characteristic fatigue life), the relationship of the Weibull CDF (Eq.(1)), PDF (Eq.(2)), exact HR (Eq.(11)), and an approximated HR (Eq. (6)) are presented in Figure 3 in which both the x axis and the y axis are in logarithmic scale. It is shown that, for the studied case, the Weibull PDF (Eq. (10)), exact HR (Eq.(11)), and approximated HR (Eq.(14)) are almost identical before the probability reaches 1.0E-4. The exact Weibull HR keeps increasing as the function of time t, while the approximated HR stops increasing when both $F(t_1)$ and $F(t_2)$ reach 1.0 at high lifetime (t > 20,000 FH). In other words, when the fatigue life distribution CDF is available, the HR can be calculated from the exact HR Eq.(11) and approximated HR Eq. (14), or the PDF Eq. (10) when a probability is very small and the time increment is small (how small may depend on the real problem).

If using a Lognormal distribution to fit the same lifetime data, which was fitted by the Weibull distribution ($\beta = 4$, $\alpha =$ 8000), by maintaining the same median and standard deviation (SD) values, the closest Lognormal distribution can be found to have a mean of log10(t) of 3.8633 and a SD of *log*10(t) of 0.14 (or a mean of *ln*(t) of 8.90 and a SD of the *ln*(t) of 0.14**ln*(10) = 0.32). For this Lognormal distribution, the relationship between the CDF, the PDF, and the HR are presented in Figure 4. Also this figure shows that the HR can be calculated from the exact HR, an approximated HR, or the PDF when a probability is very small (in this case it is 1.0E-5, other cases depend on the distribution parameters). As both $F(t_1)$ and $F(t_2)$ reach to 1.0 at a high lifetime (t > 90,000 FH), the numerical error in the approximated HR calculation (Eq. (6)) is increasing.

By comparing Figure 3 with Figure 4, it can be seen that the Weibull HR is very different from the Lognormal HR. In particular, the Weibull HR keeps increasing while the Lognormal HR starts decreasing after a certain time (~120,000 FH). Physically, a Weibull distribution is suitable to describe a fatigue process as it is based on the weakest link model [10]. This case shows again that one should be cautious when using a Lognormal distribution to fit fatigue life data. Although the Lognormal distribution may fit the data in the normal/middle probability region well, the decreasing HR is not physically suitable for modelling a fatigue process which has irreversible damage accumulation/crack growth process.

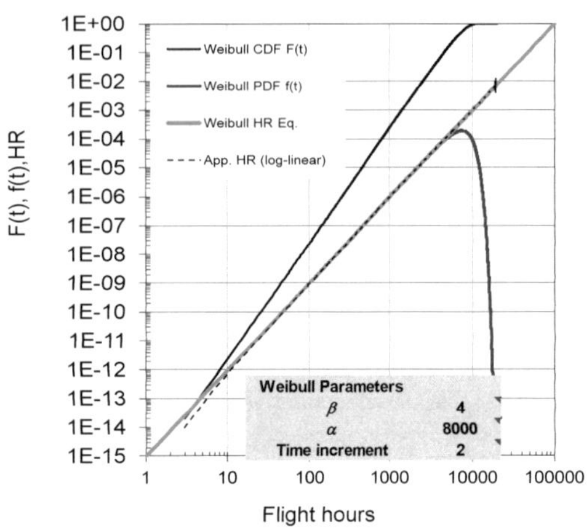

Figure 3. Weibull CDF, PDF, and Hazard Function for a typical aircraft fatigue lifetime.

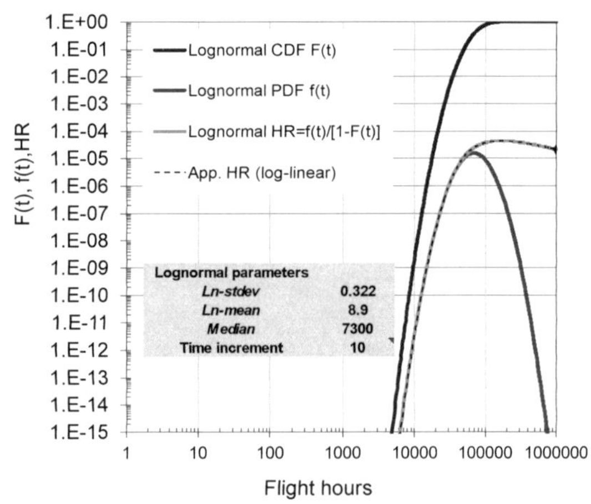

Figure 4. Lognormal CDF, PDF, and Hazard Function for a typical aircraft fatigue lifetime.

Modern risk analysis for SFHPOF calculations

The conventional fatigue reliability analysis based on lifetime data has a fundamental issue as "*it did not account for the fatigue characteristics of the individual materials in the structure*", consequently "*the same scatter factor independent of the structural material or the stress spectra*" was used [12]. Today, it has been increasingly realized that there are considerable differences between the Weibull shape parameters depending on the material and spectra. Also, the Damage-Tolerant designed structures allow fatigue crack growth below a critical length, which is not considered in the conventional fatigue reliability analysis. For risk-based airworthiness management, such as the RCAF RARM, it is required to calculate a SFHPOF (instantaneous risk) to permit flight based on what is conceived as an acceptable risk for the next or future flights.

Note all the equations below are presented for calculating single flight POF (SFPOF), they can also be used to calculate the SFHPOF by just replacing a per-flight maximum stress distribution with a per-flight-hour maximum stress distribution.

Lincoln and Berens SFPOF calculation: In 1985, Lincoln published a fracture mechanics based method to calculate the SFPOF [8] by assuming that at a given time the crack size distribution function is independent of the stress density function for a particular control point. The following equation, presented in [13], is used to represent the Lincoln method for the SFPOF calculation at the i-th flight (Lincoln's original formula did not explicitly show all variables. PROF V1.0 [13] presented them using the same SFPOF definition):

$$POF(i) = P\left[\sigma > \frac{K_c}{\alpha(a_i)}\right] = \int_0^\infty f_i(a_i) \int_0^\infty g(K_c)\left[1 - H\left(\frac{K_c}{\alpha(a_i)}\right)\right] dK_c \, da_i \quad (7)$$

where,

$POF(i)$: POF at the i-th flight
σ: applied stress
K_c: fracture toughness
a_i: crack size at the i-th flight
$\alpha(a_i)$: stress intensity at a location divided by the applied stress, or $\beta(a)\sqrt{\pi a}$
$f_i(a_i)$: probabilistic density function (PDF) of the crack size a at the i-th flight
$g(K_c)$: probabilistic density function (PDF) of the fracture toughness K_c
$H(\sigma)$: distribution of the peak/maximum stress per single flight.

In 1991, Berens used the Lincoln SFPOF definition in the USAF tool PROF (Probability of Fracture) V1.0 [13], and then modified Eq. (7) in PROF V2.0 [14] by using a residual strength (RS) curve, $\sigma_{RS}(a)$, as a function of crack size,

$$POF(i) = P[\sigma > \sigma_{RS}(a)] = \int_0^\infty f_i(a_i)[1 - H(\sigma_{RS}(a_i))] \, da_i \quad (8)$$

In practice, the RS function, $\sigma_{RS}(a)$, can be determined by taking the minimum stress from multiple failure criteria, including fracture toughness, net section yielding, and plastic zone linkup.

In PROF, the influence of crack inspection and repair on $POF(t)$ is taken into account by renewing the crack size distribution, $f_{a\ after}(a,t)$, based on a probability of detection (POD) from a non-destructive inspection (NDI) method and a repaired crack size distribution, $f_{RCSD}(a)$, as follows,

$$f_{a,after}(a,t) = \int_0^\infty POD(a) f_{a,before}(a,t) da \cdot f_{RCSD}(a) + [1 - POD(a)] f_{a,before}(a,t) \quad (9)$$

The repaired crack size distribution represents the repair quality and uncertainties. An initial crack size distribution can be used as $f_{RCSD}(a)$ when the repaired location is assumed to be restored to the original as-manufactured condition.

In 2005, USAF PROF V3.0 [15] slightly revised Eq. (8), as an indirect way to calculate a hazard rate ($h(t)$), which actually calculates the single flight hour POF for the K_C and RS failure criteria [3].

Freudenthal and Shinozuka SFPOF calculation: It was recently re-discovered [3][16][17] that Freudenthal and Shinozuka [18] had developed comprehensive reliability methods in 1966 to calculate the hazard rate based POF considering all non-prior failure events. The Freudenthal equation is:

$$POF(i) = h(t_i) = \frac{f(t_i)}{1-F(t_i)} \quad (10)$$

where

$$f(t_i) = \int_0^\infty g(K_c)\left\{\int_0^\infty f_0(a_0)\left[\prod_{j=1}^{i-1} H\left(\frac{K_c}{\alpha[a(a_0,t_j)]}\right)\right] * \left[1 - H\left(\frac{K_c}{\alpha[a(a_0,t_i)]}\right)\right] da_0\right\} dK_c \quad (11)$$

$$F(t_i) = \int_0^\infty g(K_c)\left\{\int_0^\infty f_0(a_0)\left[1 - \prod_{j=1}^{i} H\left(\frac{K_c}{\alpha[a(a_0,t_j)]}\right)\right] da_0\right\} dK_c \quad (12)$$

Notes:

- For a very small probability (< 10^{-7}), it was shown that $POF(i) = h(t_i) \cong f(t_i)$ [3].
- When the product of the probability of non-prior failure events $\prod_{j=1}^{i-1} H\left(\frac{K_c}{\alpha(a(a_0,t_j))}\right) = 1.0$, the Freudenthal equation, Eq. (11), is essentially the same as the Lincoln equation, Eq. (7). This means that the Lincoln Eq. (7) is a probability density function, which can be very close to a hazard rate (HR) as shown in Figure 3 with Figure 4 for small probability. Although this product could be very close to 1.0 for high reliability problems or at the early stage of an aircraft usage, mathematically it should always be less than 1.0. Consequently, the Lincoln equation, Eq. (7), should always give higher (more conservative) POF results than the Freudenthal equation, Eq. (10). Also, due to this product, the computation time for the Freudenthal equations is significantly increased.
- For residual strength (RS) failure criterion, the Eqs (11-12) can be modified by replacing the item $H\left(\frac{K_c}{\alpha[a(a_0,t_i)]}\right)$ with $H[\sigma_{RS}(a)]$, where $\sigma_{RS}(a)$ is the residual strength curve as a function of the crack size:

$$f(t_i) = \int_0^{\infty} f_0(a_0) \cdot \prod_{j=1}^{i-1} H(\sigma_{RS}[a(a_0,t_j)]) \cdot [1 - H(\sigma_{RS}[a(a_0,t_i)])] da_0 \quad (13)$$

$$F(t_i) = \int_0^{\infty} f_0(a_0)[1 - \prod_{j=1}^{i} H(\sigma_{RS}[a(a_0,t_j)])]da_0 \quad (14)$$

It should be noted that these RS criteria based Eqs. has only one dimension of integration, such that the numerical difference between including or not the probability of non-prior failure, i.e. $\prod_{j=1}^{i-1} H(\sigma_{RS}[a(a_0,t_j)])$, should be less since it has only single integration. This is why, in some cases such as the USAF risk analysis example published in [3], the POF difference between the Lincoln and Freudenthal equations can reach two orders of magnitude. In other cases, especially when using the residual strength failure criterion, there is virtually no difference until the POF becomes very high (e.g. 10^{-4} to 10^{-5}) [3].

In 2012, NRC conducted a detailed review and comparison of different SFPOF calculation methods along with several numerical examples [3], it was verified that the Lincoln equations did give more conservative (higher) SFPOF than the Freudenthal equations, while the difference was reduced when a deterministic residual strength failure criterion was applied. The review supported the need to formally define a standard SFPOF in a controlled document for aircraft structural risk analysis, to that end, more benchmark examples may be needed. In the examples presented in the following Sections 4 and 5, the Lincoln equation was used by NRC for the SFPOF or SFHPOF calculation for RCAF aircraft structural risk analyses.

NRC advanced risk analysis methods and tools: Since 1999, NRC has been developing structural risk analysis methods and tools in collaboration with Defence Research and Development Canada (DRDC). In the beginning, NRC evaluated and used USAF PROF [13] for risk analysis [19]. Because of the limitations of PROF for some applications, especially related to corrosion risk assessment, NRC developed an in-house tool, ProDTA (Probabilistic Damage Tolerance Analysis), for structural risk analysis by taking into account both conventional fatigue damage and age-related environmental damage (i.e. corrosion) [20][21]. This tool was created to support the Holistic Structural Integrity Process (HOLSIP) framework [22], that NRC is developing in collaboration with other organizations with the ultimate goal to evolve it into a new paradigm for aircraft design and sustainment.

Figure 5 presents the major inputs and supporting tools for ProDTA, which are grouped into fatigue and corrosion related inputs. Without the corrosion effect, ProDTA uses similar methodologies to those of PROF V2.0. However, ProDTA uses different numerical techniques/algorithms or a Monte Carlo simulation to calculate the POF. More importantly, ProDTA has the flexibility to use various statistical models for different inputs, depending on the available fleet data. For corrosion or MSD risk assessment, ProDTA uses a Monte Carlo Simulation coupled with numerical integration to calculate the POF [20][21].

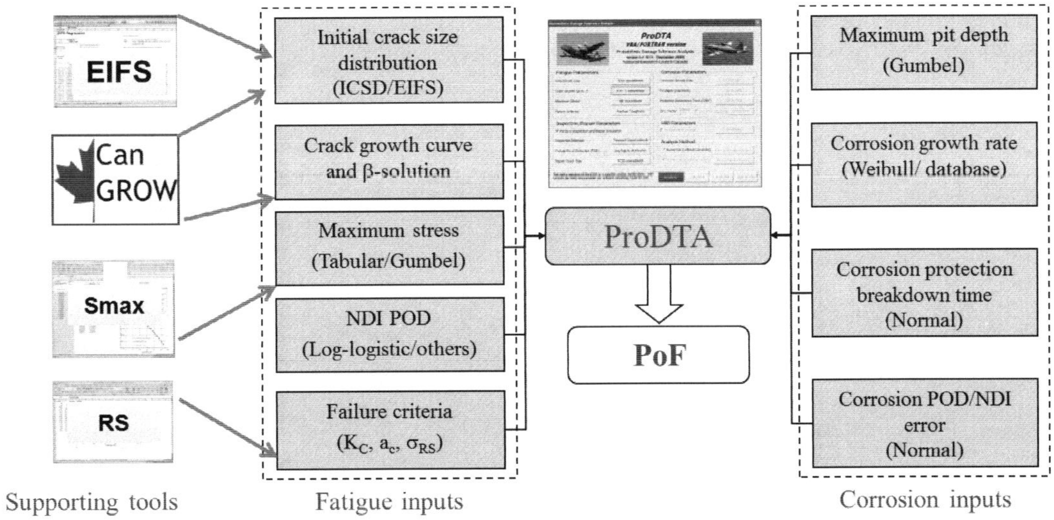

Figure 5. Main inputs for NRC in-house tool ProDTA.

In the past years, ProDTA has been improved with different subroutines to calculate SFHPOF using Lincoln and Freudenthal equations. In terms of failure criteria, ProDTA was designed to use three criteria: critical crack size, fracture toughness, and/or residual strength. These criteria can be used separately or jointly in a POF calculation.

In a practical risk analysis, most of the effort is spent on preparing various inputs and repeating the SFHPOF calculations including sensitivity analyses. To help prepare the inputs for a risk analysis, NRC also developed a number of Excel tools [23], including,

- In-service damage analysis tool to determine the equivalent initial flaw size distribution (EIFSD) or the initial crack size distribution (ICSD) [23][24].
- Maximum stress worksheet to fit stress exceedance data with a Gumbel distribution or Tabular data [23].

For advanced risk analysis with multi-site fatigue damage (MSD) and multi-element damage (MED) [25][26], more comprehensive tools were developed at NRC to provide inputs to ProDTA, including,

- A library of stress intensity factors, geometrical correction factors (β-factors), for MSD cracks [25][27].
- Generic FE-based β-factor tool for complex structures with MSD/MED [28].
- MSD/MED residual strength analysis tool to calculate residual strength for MSD/MED scenarios [29].
- An in-house crack growth analysis program, CanGROW, specifically developed for MSD crack growth [30].

These NRC developed methods and tools have been used for a number of risk analyses for different RCAF aircraft structures, including build-up structures containing MSD and MED. Some examples are presented in the following sections.

EXAMPLES ON CONVENTIONAL FATIGUE RELIABILITY ANALYSIS

CF-188 Hornet

The F/A-18 was originally certified by the OEM to 6000 spectrum flight hours (SFH), based on the Safe-Life method and a USN design spectrum. In 1990s, the RCAF conducted a review of various airworthiness standards [6], and then issued the first major revision of the CF-188 lifing policy in 2002. The current CF-188 lifing policy combines Safe-Life and Damage-Tolerant methods, revises scatter factors (SF) using the Bullen Equations [9] including censored data, and incorporates a residual strength (RS) concept to avoid having to repair/address all cracked locations.

Despite some updates to the lifing policy and allowance for unrepaired cracks (RS criteria), the RCAF lifing policy remains relatively stringent and required a large number of modifications/repairs to demonstrate exact compliance. The requirements are relaxed when risk based methodologies are put in place to accept cases that are not exactly compliant but are judged to pose acceptable risk levels. As shown in Figure 6, the CF-188 risk-based method in ASIP is based on the Safe-Life (i.e. "crack initiation" (CI) stage, defined as the time to a crack length of 0.01") and Damage-Tolerant (i.e. crack growth (CG)) approaches [31]. For some cases, the CI or CI+CG lives (after applying the revised SFs) can meet the full-life target so no maintenance action is needed. But for many cases, the full-life target cannot be met, and then the airworthiness risk assessment is followed with logistic risk assessment, which may lead to a maintenance action or no action required. In the risk assessment, a certain level of risk can be accepted in specific circumstances.

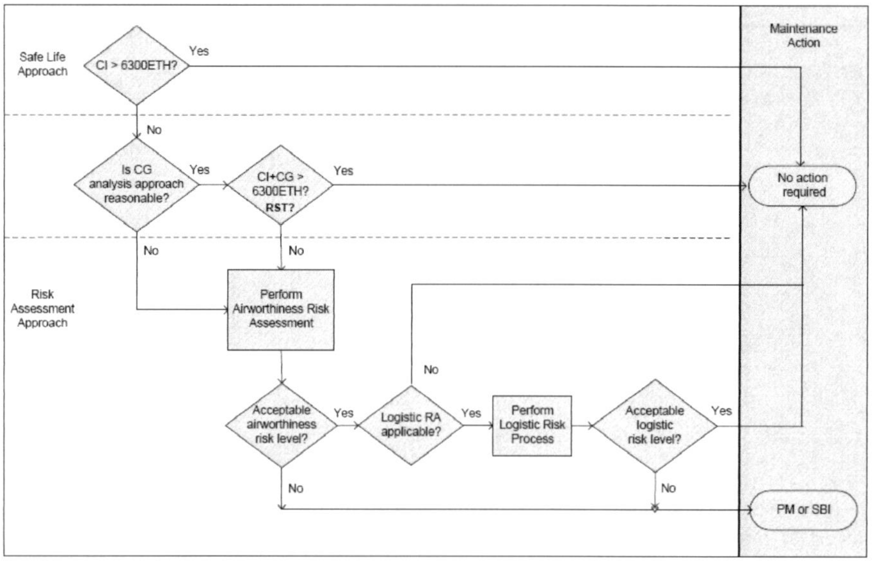

ETH: Equivalent test hour
CI: Crack initiation
CG: Crack growth
RST: Residual strength test
RA: Risk assessment
PM: Planned maintenance
SBI: Safety by inspection

Figure 6. CF-188 Lifing Policy highlights - risk-based methodology in ASIP.

The safety level (or acceptable risk level) of the CF-188 is largely based on the DEF-STAN 00-970, and defined as total (cumulative) probability of failure (TPOF) of 0.001 over the lifetime of the aircraft. In practice, the TPOF is calculated from a Weibull or Lognornal distribution whose statistical characteristics (location, mean parameter etc.) are calculated from the crack initiation (CI) or/and crack growth (CG) methods, along with shape factor/standard deviation determined from the CF-188 lifing policy [6]. Therefore, the hazard probability (TPOF) used in the CF-188 risk analysis is different from the SFHPOF used in the current RARM risk index. A recent risk analysis on the CF-188 V-tail stubs was presented in [32].

Recently, the change to a risk mitigation strategy has led to increased requirements of NDI to detect cracks for some critical locations [31]. The requirement of NDI technologies and reliability would be defined and assessed through a risk assessment. It is deemed that the current CF-188 risk-based methodologies are deterministic, and there is an intention to pursue with probabilistic methods, especially for risk assessment after inspection and repair. Given that inspection and repair can introduce damage, a Damage-Tolerant method is usually used in the airworthiness assessment. As such, fracture mechanics based risk assessment would be suitable to such airworthiness risk analysis, which can include the NDI reliability (e.g. POD) and quantify the effect of a certain NDI POD on the risk index. A NRC risk analysis on upper outboard longeron was previously presented in [33].

CC-130E Hercules

In 2002, a catastrophic failure of a US Forest Service C-130A centre wing in Walker, California, has prompted a review of the RCAF centre wing fatigue usage on the CC-130E aircraft [34][35]. The review included the CC-130E "Safe Life" and "Service Life" limit or "Limit of Validity" (LOV) for damage tolerance considering the presence of WFD. In that review:

- "Safe Life" was defined by the OEM (Lockheed Martin (LM)) Wing Durability Test (WDT) test life divided by a scatter factor of 4 (originally used by the OEM for the C-130). This corresponds to the lifetime to local cracking and before a significant probability (e.g. 10^{-6}) of WFD exists.
- "Service Life" of the C-130 centre wing was initially defined by OEM as a "2% probability of generalized cracking" (uneconomical to repair) [36]. The 'Limit of Validity' for WFD, based on a FAA recommendation [37], was defined as the WDT test life divided by a scatter factor of 2. The LOV lifetime was considered as a "Service Life" limit by DND for the CC-130E fleet [34].

The RCAF review and OEM analysis did not describe how the safety factors were obtained. Thus NRC carried out a Lognormal analysis and found that, assuming a standard deviation of 0.1296 (used in DEF STAN00-970 for Log10 life, in Table 1, Part1, Section 3, Leaflet 35, issue 2, Dec. 1999), these safety factors can be calculated for the stated probabilities. If a different standard deviation is used, the test life scatter factor to determine the LOV service life at 2% (probability/percentile) will be different from 2.

In 2002, L-3 Communications Spar Aerospace Limited was tasked by DND to perform an investigation on the correlation between the CC-130E centre wing in-service cracks and the OEM C-130 WDT cracks at eight locations. Based on that, DND conducted a direct comparison that indicated that the "Safe Life" and the "Service Life" of the CC-130E centre wing would be significantly less than that of the original estimate from LM [34][35]. This comparison raised great concern about the risk of catastrophic failure of a CC-130E aircraft. Meanwhile, it was realized that there were some limitations in the direct comparison, including not accounting for clean and non-inspected aircraft (censored data), different crack sizes between in-service and WDT, and load changes in WDT.

In October 2003, the Operational Research Division (ORD) of DND was then tasked to conduct comparative statistical analyses of the in-service and WDT fatigue cracks in the C-130 centre wing [38]. The censored/null-finding data were included in the Lognormal analysis using the Maximum Likelihood (ML) approach. In March 2004, NRC was requested by DND to review and validate the 'soundness' of the statistical/mathematical methods used in the ORD analysis. For that task, NRC evaluated various goodness-of-fit tests to verify the Lognormal distributions fitting to the CC-130 data and developed an in-house FORTRAN algorithm to verify the ORD analysis [39]. In addition, NRC provided the first batch of confidence bands (±95%) life results using the simplest large-sample method and conducted a Weibull analysis using the same dataset for comparison. Furthermore NRC verified the results from the in-house algorithm using the commercial software SuperSMITH, which also provided extra confidence on the level of accuracy [39].

Based on the ORD statistical analysis and the NRC verification, a DND risk assessment (RARM-CC-130-2004-TH2-037) determined a service life limit of 50,000 EBH-CF1980 (equivalent baseline hours, based on the CF 1980 usage) for the CC-130E centre wing, and the primary means of risk mitigation was inspection per OEM service bulletin (SB82-790, centre wing lower surface generalized cracking & WFD inspection requirements starting from ~40,000 EBH, and repeat very 10,000 EBH). Note that the hazard probability used in the CC-130E risk assessment was based on the 2% probability of an aircraft having reached the WDT test life (without scatter factor), which is a cumulative POF, not a hazard probability per flight hour (SFHPOF) [34][40].

In 2010, the OEM updated the CC-130 Durability and Damage Tolerance Assessment (DaDTA) , which states that the service life of 45,000 EBH-OEM (based on the USAF baseline usage, not the same as the Canadian Forces (CF) 1980 usage) is applicable for the CF CC-130B/E/H centre wing, based on component generalized cracking data and a multi-site damage (MSD) risk analysis. This updated service life could have greatly affected the availability and maintenance/replacement costs of the CC-130 fleets, especially of the remaining nine CC-130E aircraft. The OEM MSD based risk analysis determined the SFPOF for the C-130 centre wing, which used the fracture mechanics based risk analysis method.

Again, NRC was tasked by DND to repeat the previous ORD and NRC statistical analysis, and then continue the analysis considering structural modification CF-519 (SB82-481) conducted for the CC-130 in 1980's [41]. One of the results is presented in Figure 7. Along with the results for other scenarios and locations, the updated NRC analysis showed again that the original 50,000 EBH-CF1980 is acceptable for CC-130E fleet given that the SB82-790 or an equivalent action was done, which supported the CC-130E RARM process (RARM-CC-130-2004-TH2-037 V5) and assured the airworthiness and availability of the nine CC-130E aircraft. In addition, NRC was tasked by DND to calculate the SFHPOF for MSD/MED scenarios for verifying the OEM POF analysis and then calculate the CC-130 SFHPOF using the RCAF in-service damage and usage data.

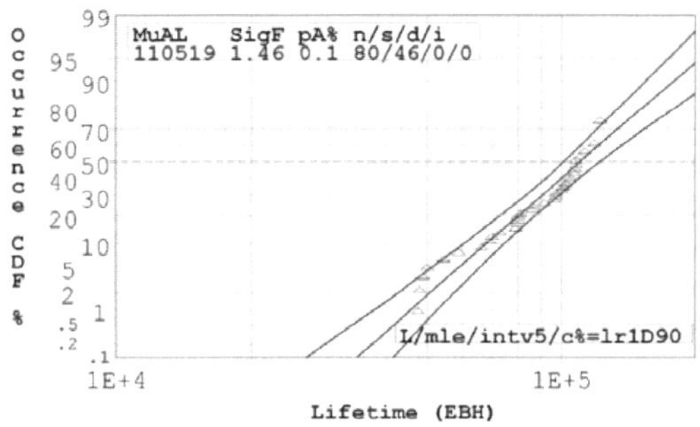

B%	95% Lower	Lifetime (EBH)	95% Upper	Hazad Rate (HR)
0.01	19721	27037	33982	3.85E-08
0.1	26469	34303	41442	2.59E-07
1	37747	45807	52869	1.55E-06
2	42793	50788	57739	2.57E-06
10	60602	68033	74695	7.57E-06
20	73192	80363	87444	1.15E-05
50	101360	110519	122765	1.91E-05

a) Lognormal analysis

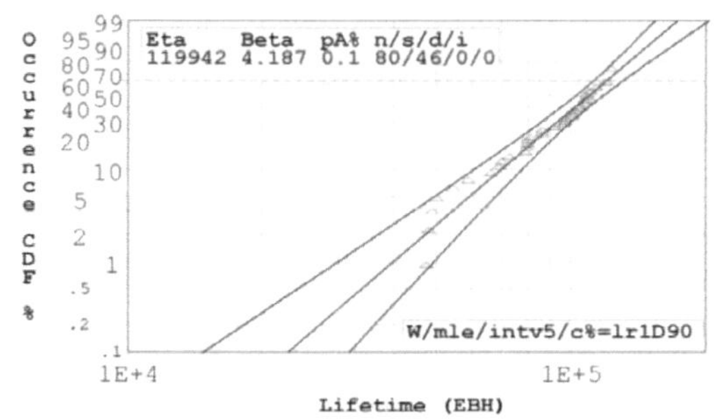

B%	95% Lower	Lifetime (EBH)	95% Upper	Hazad Rate (HR)
0.01	7306	13292	20487	3.15E-08
0.1	14801	23040	31765	1.82E-07
1	29967	39976	49370	1.05E-06
2	37063	47231	56463	1.79E-06
10	60953	70072	77944	6.30E-06
20	75916	83827	90895	1.11E-05
50	102996	109889	118486	2.64E-05

b) Weibull analysis

Figure 7. An NRC statistical analysis example for CC-130E fleet (CORA Scheme 3/DOR Scheme 1+CG to WDT DI, w/o CW-1, 2006 CORA data+2010 HS correction).

EXAMPLES ON ADVANCED RISK ANALYSIS

CC-130E/H MSD risk analysis

Since 2007, NRC has been developing in-house MSD/MED based risk analysis methods and tools (ProDTA and CanGROW) through DND-NRC collaborative projects. The developed methods/tools are used in the following risk analysis examples for MSD/MED scenarios. MSD and MED were found in service at the lower surface panels of the centre wings of the CC-130 aircraft as shown in Figure 8 [35][42]. Using the NRC developed tools; the input data for the risk analysis were prepared as follows [43].

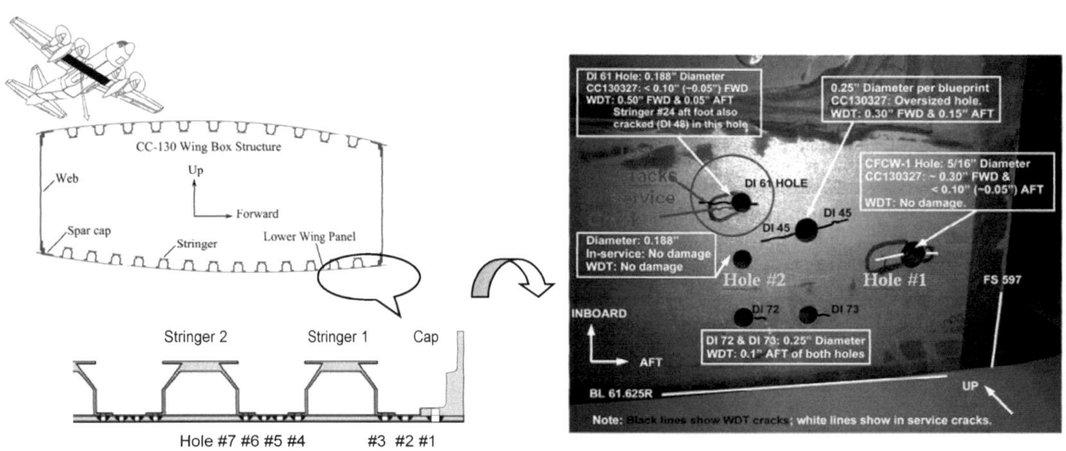

Figure 8. CC-130 centre wing lower surface panel fatigue critical location (CW-1) [35][42].

Initial crack size distribution (ICSD/EIFSD): The RCAF CC-130 in-service crack data were used to determine the EIFSD, which involves the following steps and assumptions:

1. Determine the EBH for all in-service cracks from different aircraft. Since each aircraft has its own flight hours (FH) based on individual mission and usage, the EBH is obtained using a severity factor between the individual usage and a baseline usage (Reference OEM usage). Therefore, the severity factor is an important factor, which is separately evaluated through a re-baselining RARM task for CC-130 (RARM-CC-130-2010-008). The main reason for the re-baselining is because a more accurate load model has superseded an old load model, and a new DaDTA in 2010 was completed by the OEM, which led to modified severity factors. The EBH based on the OEM reference usage is hereby called EBH-OEM.
2. Determine in-service crack size for CW-1 location (Figure 8 photo). In total, 34 in-service crack data were collected from the CW-1 fastener hole #1 ("true CW-1 scenario") from the CC-130 fleet. The RCAF inspection reports recorded the actual crack lengths from both sides of the hole. It should be noted that sometimes, due to poor inspection reporting, crack data reports combined the lengths of double-sided cracks, possibly including the hole diameter, without specifying the length on each side of the hole. When combined cracks are used as a single-sided crack, the calculated EIFS could be much larger, which could lead to a significantly higher POF. In addition, for the hole sides without a crack finding using the bolt hole eddy current inspection technique, it was assumed that a below threshold crack of 0.030 inch existed, which was treated as a censored datum to determine the EIFSD.
3. Use the NRC crack growth tool (CanGROW) to regress the in-service crack data and determine the EIFS values. The EIFS are then best-fitted by a Weibull or Lognormal distribution using a goodness-of-fit test. It is noted that some EIFS values were very small so near threshold short/small da/dN-ΔK curve is needed in the regression. Such short/small crack growth data are not normally available, or determined by a questionable long crack growth testing. In addition, LEFM may not be applicable to the short/small crack growth modelling. As such, the regressed EIFS cannot be related to the physical features of crack nucleation, and reliable short/small crack growth data should be reviewed and used whenever available.

Figure 9 presents the EIFS results, as a function of the probability of exceedance, for both the RCAF in-service data and the Reference OEM (Ref-OEM) data for CW-1 hole #1, in which the latter has more data points but contains double-sided cracks combined into a single crack. It is shown that the EIFS values are very different between the RCAF and Ref-OEM scenarios. Later, it was found that the difference was mainly due to the EBH used for all the cracks which did not consider a modification (described as CF-519 in the next section) for CW-1 location.

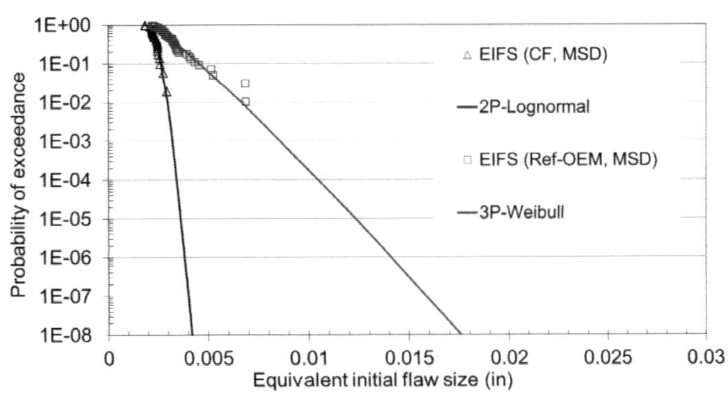

Figure 9. Equivalent initial flaw size (EIFS) distributions for CW-1 hole #1.

Crack growth curve: Using the NRC developed β-library available in CanGROW and the NRC generic FE-beta tools, the total β-factor was determined for the CW-1 location for the MSD scenario, as shown in Figure 10 a). The MSD crack growth curve was obtained using CanGROW, as shown in Figure 10 b). As a comparison, the crack growth curve for the Standard Crack (SC) scenario (used in a typical DTA) is also presented in Figure 10 b), which shows a longer life than the MSD scenario. Based on the same EIFSD, random EIFS values were generated for the 17 holes (2 cracks per hole) and a tail sampling based Monte Carlo simulation was conducted in CanGROW to determine the lead crack size distributions, which were validate with in-service findings. The tail sampling technique with 10^5 trials, from the 0.9-1.0 probability region, was actually used to save computation time and was verified with a full distribution region (0.0-1.0) sampling with 10^6 trials.

Maximum stress distribution and residual strength: In this case study, a table look-up format fitted the stress data better than a Gumbel distribution, so the table look-up format was used in ProDTA. Figure 11 a) presents the maximum stress distributions for several RCAF CC-130 usages, their sensitivities are also examined later on the SFHPOF results.

Residual strength: Using the analysis methods described in [25] as well as the total β-factor for the CW-1 location, the RS curve was calculated for the MSD scenarios, as presented in Figure 11 b). It is shown that the difference between the SC and MSD RS curve, as a function of the lead crack, is marginal in this case. This is due to the fact that for this location the MSD lead crack mode is the same as that of the SC, although the former grows faster due to secondary cracks.

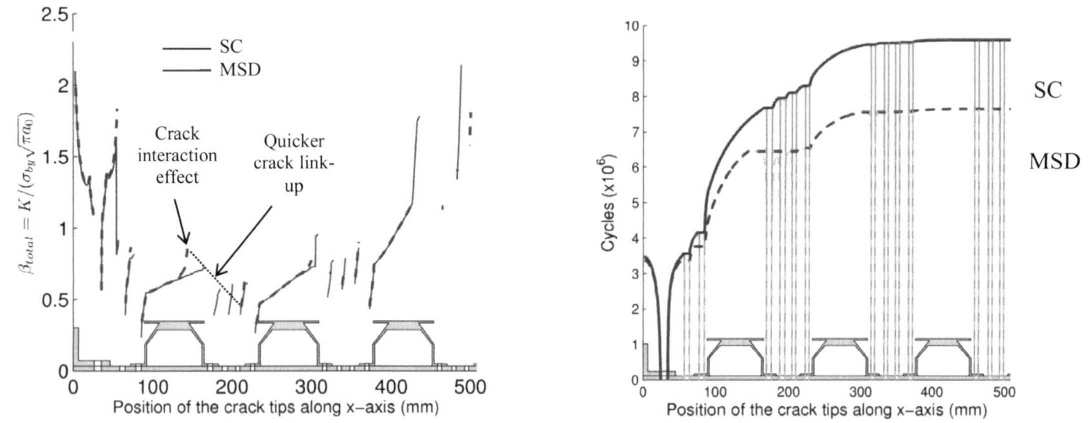

a) β-solutions of the lead cracks for the standard crack (SC) and multiple site fatigue damage (MSD) scenarios

b) Crack growth curves for SC and MSD scenarios (lead crack)

Figure 10. NRC MSD analysis using an in-house tool CanGROW.

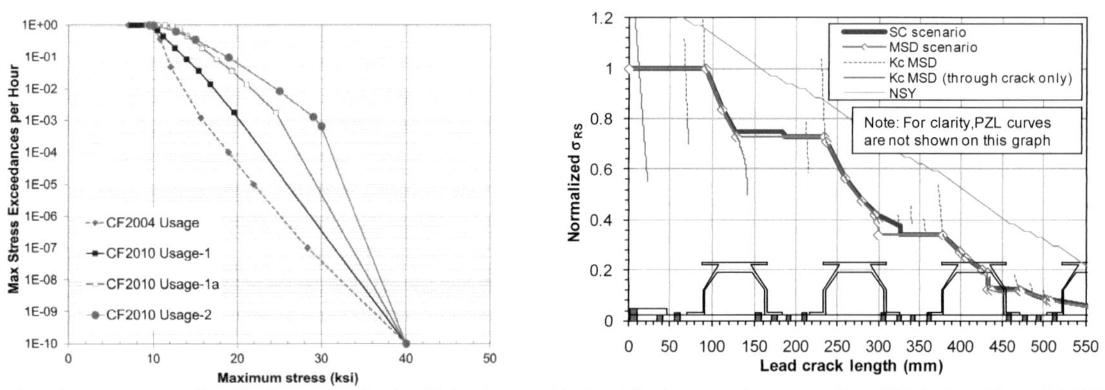

a) Maximum stress distribution per single flight hour for the CC-130 CW-1 location.

b) Residual strength curves for CW-1 for SC and MSD scenarios.

Figure 11. Maximum stress distribution and residual strength curves for CW-1.

POF results: The NRC in-house tool ProDTA was used to calculate SFHPOF. For the CW-1 POF study, the residual strength failure criterion, Eq. (8), was used. This criterion accounts for the failure probability that the crack length is smaller than a certain size, but the stress level encountered during the single flight hour is greater than σ_{RS} (a), dominated by crack size a. Since previous works [23][25][42][50] have shown that a POF of a MSD scenario is much higher than a SC scenario, this paper only presents the MSD POF analysis.

Based on the DaDTA, the CW-1 hole #1 is treated as the most critical hole. It has a larger diameter than the other holes and carries more loading from the spar cap (Figure 8). This roughly agrees with the CC-130 in-service findings. The other holes, which are smaller and carry different loading, should have different EIFSD. However, there are not enough in-service cracks found in the holes #2, #3, and others, thus an assumption was made to use the hole #1 EIFSD for all other holes on the CW lower surface panel. This approach would lead to a conservative risk analysis. The limitation of the EIFS method needs further investigation for MSD/MED problem and recent NRC progress is presented in [29].

The Monte Carlo based POF study was conducted for the following two scenarios [43]:

1. Worst MSD scenario: In each Monte Carlo simulation, one random EIFS value was generated from the EIFSD and used for all 34 cracks at the 17 holes. This scenario implies that the 17 holes are as critical as hole #1 and that the largest EIFS occurred at all 17 holes simultaneously, which would give the shortest life. The MSD scenario is illustrated in Figure 8.
2. Random MSD scenario: In each Monte Carlo run, 34 EIFS values were randomly generated for the 34 cracks at the 17 holes. This scenario implies that the 17 holes are as critical as hole #1, but that the EIFS randomly occurs at different locations, which should be a more reasonable scenario given the in-service cracks were found at holes #1 - #3 which also dominated the failure cracking modes.

Considering all the sensitivity and conservatism of the inputs used in the POF studies, the highlighted POF results are provided in Figure 12. These results include the RCAF Worst-MSD and Random-MSD cases. These relatively conservative POF results show that the Ref-OEM service life of CW-1 is about 43,500 EBH at SFHPOF = 10^{-7} (note that this life would be longer if the null-findings/censored data were used). The RCAF CW-1 service life is about 48,500 EBH at SFHPOF = 10^{-8} for the Worst-MSD scenario, and 54,000 EBH at SFHPOF = 10^{-8} for the Random-MSD scenario.

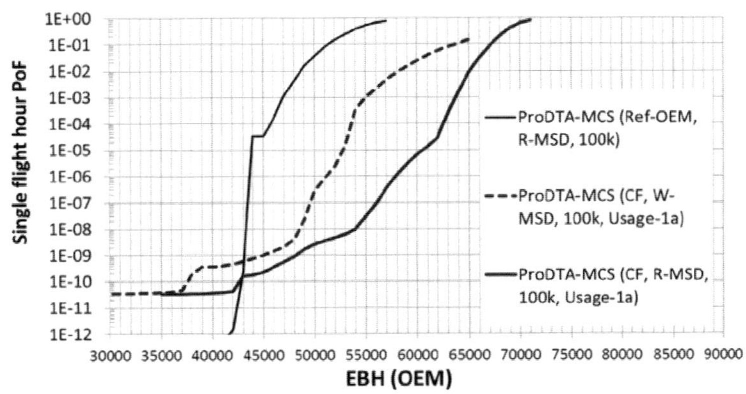

Figure 12. The highlighted POF rsults for RCAF MSD cases.

Figure 13 presents one of the sensitivity studies on the effect of different usage on the SFHPOF, which is not available from the conventional fatigue reliability analysis (Weibull/Lognormal), but can provide very useful information for risk management decision-making. Along with other sensitivity studies, the NRC SFHPOF results were used in the RARM process (RARM-CC-130-2011-003) to support the service life determination of the CC-130E fleet.

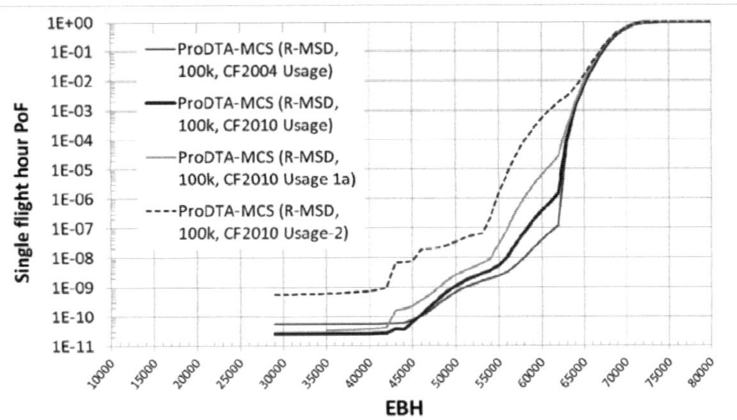

Figure 13. Sensitivity study on the effect of different usages (in Figure 11 a)) on SFPOF.

Effect of a structural modification/repair including inspection: After completing the CW-1 SFHPOF, NRC was informed that a structural modification (CF-519 or SB82-481) had actually been conducted on the CW-1 location of the CC-130 aircraft. The CF-519 modification (CF-519 mod) was implemented on the CC-130 fleet in the mid-1980s, at approximately 13,000 EBH on average, except for two aircraft for which it was implemented during production. The CF-519 mod involved a bolt hole eddy current inspection, oversizing the hole, and replacing four Taper-Loc (TL) fasteners with Hi-Tigue (or Hi-Lok/Tigue, HLT) fasteners at CW-1 [44]. As such, all the CC-130 CW-1 in-service cracks were actually nucleated from the HLT holes. Although it was stated that the CF-519 would improve the fatigue performance against the TL fasteners, the same DaDTA methods were used in the OEM analysis. The effects of the CF-519 mod on the risk analysis are:

1. Considering the CF-519 mod, the EIFS values should be larger since the in-service cracks should be regressed back to about 13,000 EBH, not to time zero (as shown in Figure 9). Since the same DaDTA method (including β-factors, bearing bypass stress ratio, stress level, etc.) was used by the OEM for both TL and HLT fasteners DTA analyses and almost no retardation effect was seen in the crack growth analysis, the EIFSD for the TL and HLT fasteners should be identical. This was verified through another NRC task [45], which found that the RCAF EIFSD is very close to that of the OEM EIFSD for the TL fasteners when the effect of the CF-519 modification is considered and double-sided cracks are combined. If the hole oversize effect is considered, they are even closer to each other [44]. It should be noted that, before the CF-519 mod was considered, the RCAF and OEM EIFSD were significantly different (see Figure 9), which was originally thought to be mainly due to different sample sizes of the in-service crack data.
2. Considering only the EIFS effect, the SFHPOF curves presented in Figure 12 was found to remain approximately the same (if each aircraft received the CF-519 at the exactly 13,000 EBH then the SFPOF would be exactly the same), but the time axis (EBH) should be reduced by about 13,000 EBH to represent the updated SFPOF after the CF-519 mod [46].

By considering only the inspection effect, an OEM analysis has shown that an inspection with a 75% POI (probability of inspection) could reduce the SFHPOF by about 10 to 100 times for different usages [47]. In a qualitative risk assessment (RARM-CC-130-2011-003), the SB82-790 is deemed to reduce risk by one level based on the risk index in Figure 1, e.g., reduce the SFHPOF by 100 times in this case. Considering all the effects from the CF-519 modification, including the change in hole size, the clamping force, the interference fit, residual stresses, and possibly repair/inspection induced damages, would likely result in additional differences in crack growth curves and EIFSD, which could affect the SFHPOF results of CW-1. As such, different crack growth curves should be used in the SFHPOF study after the CF-519 modification. Further investigation is needed to quantify the effect of inspection or SB82-790 on the SFHPOF using an updated crack growth curve and different POI, such as 95-100% POI, which is acceptable for the RCAF CC-130.

CC-130 CW risk analysis comparisons: Sections 4.2 presents the risk analysis of the CC-130E CW based on the conventional fatigue reliability associated with the scatter factor approach, while Section 5.1 presents the risk analysis for CC-130E/H using a DaDTA based SFPOF study. In comparison, it is shown that,

1. The risk analysis presented in Section 4.2 used eight CW critical locations, including CW-1 that seemed to have the most effect on the risk level of CC-130 CW. The CC-130 fleet has received the CF-519 modification for CW-1, but not for other CW locations.
2. The risk analysis presented in Section 5.1 calculated the SFHPOF for CW-1 only, which assumed that the CW-1 failure would lead to a catastrophic failure of the aircraft. Another SFHPOF study [48] was also conducted for location CW-14, which showed earlier cracking in CC-130. However, the failure of CW-14 would not cause a catastrophic failure of the CW/aircraft, as seen in the WDT and an NRC analysis [49].
3. Given that different methods and assumptions are used in the above risk analyses, they likely did not produce the same lifetime statistical distribution (at least, a careful comparison has not been done yet for the lifetime distribution of the same location). The RCAF decided to use different RARMs to manage the airworthiness risks of the CC-130E and H fleets. The E model airworthiness risk (RARM-CC-130-2004-TH20037) was managed based on the statistical analysis, first conducted by ORD [38], verified and updated by NRC analyses [39] and [41], and then supported with additional NRC SFPOF analysis. The H model airworthiness risk (RARM-CC-130-2011-003) was managed by adopting the OEM recommendation which was based on both conventional fatigue reliability and SFPOF analysis to determine the service life limit of C-130.

In 2005, the OEM fatigue analysis [50] established 45,000 EBH-OEM as the upper bound Economic Service Life, based on a Scatter Factor of 2 on the WDT test life. It appears that the WFD average lifetime is considered to be 90k EBH-OEM, at which time the WDT CW testing loads were reduced. According to the FAA WFD process description, the WFD average lifetime represents when, without intervention, 50% of the airplanes in a fleet fall below the minimum strength requirements. The WDT lifetime (90,000 EBH-OEM) is also considered as the economic lifetime after which it is deemed not economic to inspect and repair the exponentially increasing number of cracks in the structure.

In 2006, the OEM POF study [47] established the SFPOF results and showed that risk mitigation by inspection is possible, but a high uncertainty exists in the POD and POI due to the large number of fastener holes requiring inspection. For CW-1 only, however, NRC's SFPOF study showed that lower risks may be achievable using the RCAF in-service findings considering the CF-519 effects, and potentially higher POI for NDI POD but this detailed POF analysis has not been conducted yet. From the experience gained on the CC-130H risk analysis, the OEM also included the SFPOF analysis in the DaDTA section of aircraft structural integrity management plan (ASIMP) for the new aircraft (C-130J).

CP-140 risk analysis with limited in-service damage

Following some wing lower forward spar cap inspection results from the United States Navy (USN) P-3 fleet in 2007, the CF launched the RARM process and an initial *qualitative* risk assessment indicated that the CP-140 Aurora fleet (the Canadian version of the P-3) was under a *high* risk when reaching the targeted operational lifetime (24,500 hours). As this risk analysis was solely based on the USN P-3 findings and no CF inspections, the CF initiated a Canadian Special Inspection (CSI) line to inspect the affected wing areas in order to re-assess the risk level. Some wing structures were first removed from three CP-140 aircraft and sent to the Quality Engineering Test Establishment (QETE) for detailed inspections. In the meantime, NRC was requested by the CF to carry out some *quantitative* risk analysis to calculate the POFs of some critical locations in the CP-140 wing, including a location at the front spar cap aft flange at wing station (WS) 130, also referred to as Location 3 in this paper. The typical geometry and cracking paths of the critical locations between WS65 and WS167, including Location 3, are shown in Figure 14.

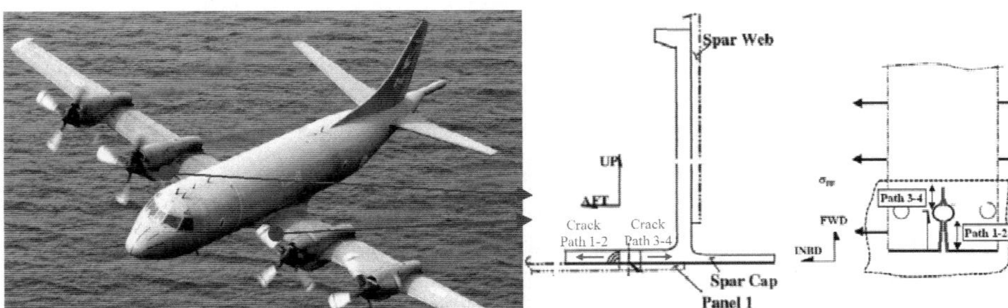

Figure 14. Typical geometry and multi-phase cracking paths for critical locations within WS65-167, including Location 3 - front spar cap aft flange at WS130 (not to scale).

Initial crack size distribution: For the spar cap between WS65 and WS167, the full scale fatigue test showed about 20 cracks and the USN P-3 showed over 20 cracks only between WS90 and WS140. However, from the first three CF CP-140 aircraft inspected, only one "crack indication" was reported by QETE at Location 3 and adjacent holes within WS90-140 from a total of 396 holes inspected in three aircraft. For a quick conservative risk analysis, this "indication" was assumed as a crack whose size was 0.030", which represents p*=1/396 or the 0.253^{th} percentile in a crack size distribution. This crack, along with the USN P-3 in-service crack findings, was used to determine an EIFSD as follows:For the CF aircraft, it was assumed that the time to crack size (TTCS) distribution followed a Lognormal distribution that had the same standard deviation as the USN TTCS. Since the USN in-service findings were obtained at different times, the TTCS data were regressed to a common crack size of 0.03" using a master crack growth curve. Two standard deviations of the natural logarithmic TTCS data (log-TTCS) were calculated as 0.103, before TTCS regression, and 0.133 after TTCS regression. In total 28 cracks from the USN P-3 were used in the analysis, and no null-findings (censored data) were used. Excluding the null-findings would make the POF results more conservative.

1. Using the percentile of one crack finding (p*= 0.25%), the mean of the log-TTCS distribution $\mu_{\log-TTCS}$, for the CF aircraft was calculated as,

$$\mu_{\log-TTCS} = \ln(T^*) - \Phi^{-1}[p^*] \times \sigma_{\log-TTCS} \tag{15}$$

where T^* is the inspection time for the CF aircraft (22,162 hours), $\Phi^{-1}[p]$ is the inverse function of a standard Normal distribution, and $\sigma_{\log-TTCS}$ is the standard deviation of the log-TTCS distribution.

2. Using a master crack growth curve, the CF TTCS distribution was regressed to time zero to determine an EIFS distribution, as shown in Figure 15. In this approach, a certain percentile p^* of crack was regressed to an EIFS but with $1-p^*$ percentile. Two EIFSDs were determined using $\sigma_{\log-TTCS}$ = 0.103 (before TTCS regression), and

$\sigma_{\log-TTCS} = 0.133$ (after TTCS regression), as presented in Figure 16. If more CF in-service cracks were available, a maximum likelihood method could be used to estimate the parameters of the TTCS distribution, with and without censored data (null-findings) [41].

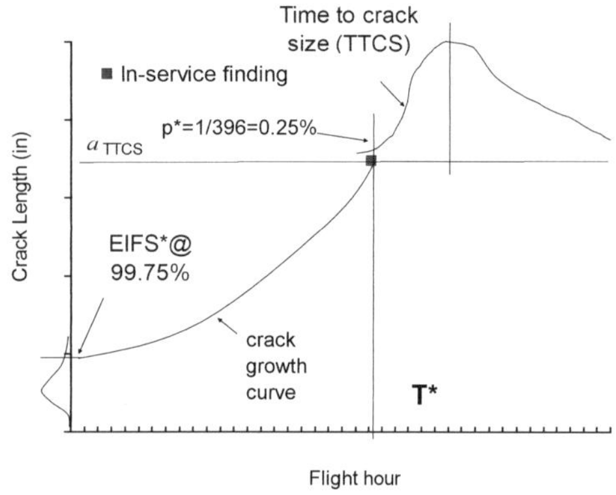

Figure 15. Schematic of the regression of EIFSD from TTCS.

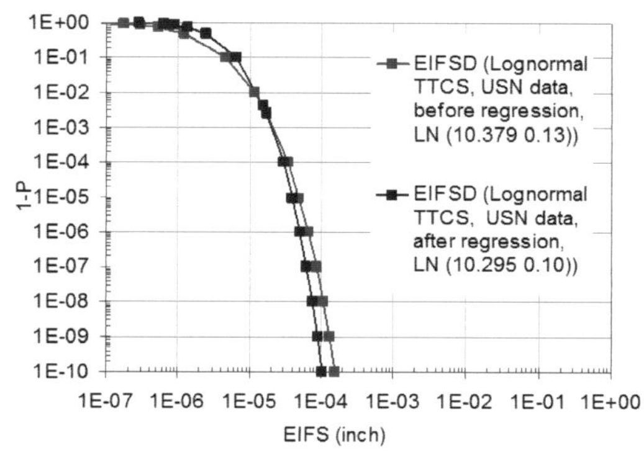

Figure 16. EIFSDs determined from Lognormal TTCS distributions before and after regression.

Crack growth curve: A phase-by-phase crack growth analysis was carried out by IMP Aerospace and the results are presented in Figure 17. The analysis was performed using FASTRAN, and the crack growth analysis included crack paths 1-2 and crack paths 3-4 (Figure 14). The hole diameter was simply added to crack path 3-4 after the crack path 1-2 had failed. A spectrum representing 15,000 hours was generated using the Service Life Assessment Program (SLAP) software: Database Interface/Spectra Sequencing Tool (DBI/SST) for the FASTRAN analysis. As required by the NRC risk analysis, the IMP crack growth analysis started from an initial crack size of 0.002 inch.

As mentioned before, the crack growth curve for crack path 1-2 was used to regress the only crack finding (0.03" at 22,162 hours) to the EIFS. The regressed EIFS of 1.736×10^{-5} inch (or 0.04 µm) was found to be much smaller than previously observed material intrinsic discontinuities (crack-nucleating particles or pores) or manufacturing discontinuities (scratches, marks). This implies that the crack growth analysis could be conservative, especially in the small/short crack regime.

It should be noted that using the more accurate stress spectra generated from the strains recorded from the CP-140 Structural Data Recording Set (SDRS), the updated crack growth analyses predicted much longer fatigue lives for many CP-140 locations. Recently, NRC also developed the stress intensity factor solutions for simulating simultaneous crack growth for paths 1-2 and 3-4 by accounting for the interaction of the two radial cracks [51]. This approach was shown to provide more accurate crack growth analysis than the phase-by-phase crack growth analysis. For conservative POF study, this paper still used the original IMP crack growth curve based on the DBI/SST tool.

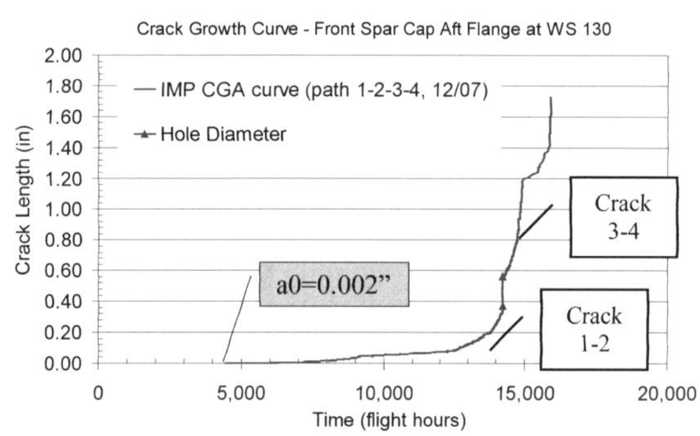

Figure 17. Crack growth curve, combined from crack path 1-2 and 3-4.

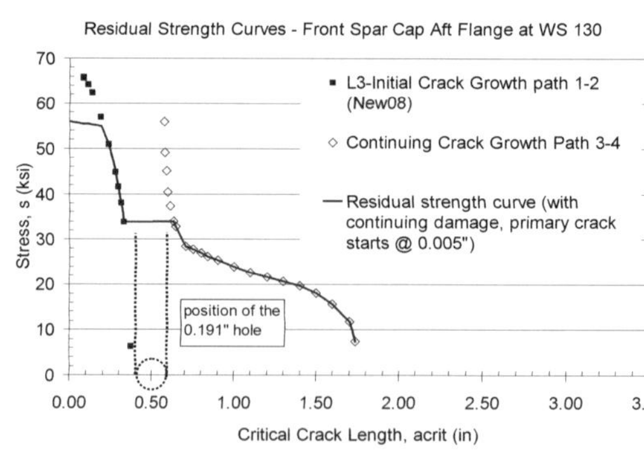

Figure 18. Residual strength data and curve.

Stress exceedance data and maximum stress distribution: The stress exceedance data, as shown in Figure 19, was generated by the P-3 SLAP DBI/SST software for the 15,000-hour spectrum generated for this location. It was assumed that cracked structures fail under a maximum (peak tension) stress during a cyclic loading. Therefore, only the maximum stress data was used to determine a maximum stress (σ_{max}) distribution, i.e. the probability of exceeding a certain maximum stress value, per flight hour. Figure 20 presents the σ_{max} data per flight hour, which was converted using the Berens approach [13]. A cutoff stress of 30 ksi at a probability of exceedance of 10^{-10} was added based on engineering judgment of the CF, IMP and NRC. These σ_{max} data can be directly used by the NRC tool ProDTA, or fitted with a Gumbel distribution (Type-I extreme value distribution of maxima) as,

$$H(\sigma_{\max}) = Exp[-Exp(-(\sigma_{\max} - B)/A)] \tag{16}$$

where A and B are scale and location parameters of the Gumbel distribution. Based on the Berens' approach, the last five actual σ_{max} data points (except the last cutoff stress point) were used for the Gumbel fitting, resulting in A = 1.60, B = 14.69 ksi. Figure 20 shows that the Gumbel distribution fitted the actual data very well, and it would give a higher (conservative) probability of exceeding than the actual data point when σ_{max} > 27ksi. Thus the Gumbel distribution, which was also used for the POF calculation, was expected to result in conservative (higher) POF results.

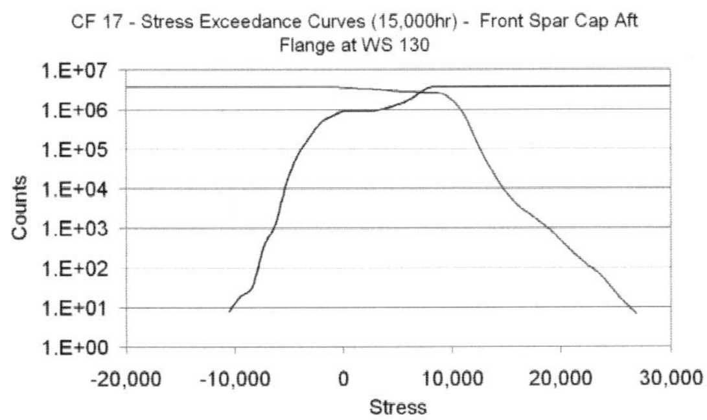

Figure 19. Original stress exceedance curves for 15,000 flight hours.

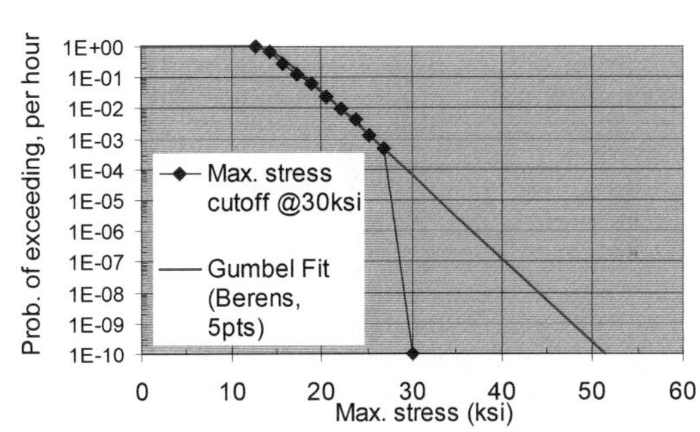

Figure 20. Probability exceeding maximum stress (per flight hour) and Gumbel distribution fit.

Residual strength: The failure scenario simulated in this work is a consecutive process with the first failure of the crack path 1-2 (reaches to the edge of the flange) and then the failure of the crack path 3-4. At first, the residual strength (RS) data, shown as data points in Figure 18, were calculated by IMP separately for crack paths 1-2 and 3-4, based on fracture toughness (Kc) criterion. The Kc value used was 51.5 ksi$\sqrt{in}$, which corresponds to the lower bound of the Kc distribution, which has an average of 71 ksi$\sqrt{in}$ for the 7075-T6 spar cap with a thickness of approximately 0.1 inch. Assuming a 10% coefficient of variation (COV) for a Normal distribution of Kc, there would be less than 3-in-1000 Kc values which would be lower than the 51.5 ksi$\sqrt{in}$ lower bound (about 3 standard deviations below the average).

Next, NRC combined the two RS curves together to represent the consecutive crack growth scenarios of crack paths 1-2 and 3-4, and considering the continuing damage growth of crack path 3-4. The continuing damage size was approximately calculated by growing an initial crack of 0.005" to the time when the crack path 1-2 reached the edge of the flange (edge failure). As shown in Figure 18, the combined RS curve (line) is much lower than the starting RS strength of the crack path 3-4, which is expected to result in conservative (higher) POF results.

POF results: Using the NRC in-house tool ProDTA, the SFHPOF of Location 3 was calculated using the conservative Lincoln Eq. (2). The SFHPOF (for crack path 1-2-3-4) are presented in Figure 21 for two EIFSDs based on 1) $\sigma_{\log-TTCS}$ = 0.103 (before TTCS regression) and 2) $\sigma_{\log-TTCS}$ = 0.133 (after TTCS regression). It is shown that the EIFSD based on $\sigma_{\log-TTCS}$ = 0.133 gave higher/ conservative POF results.

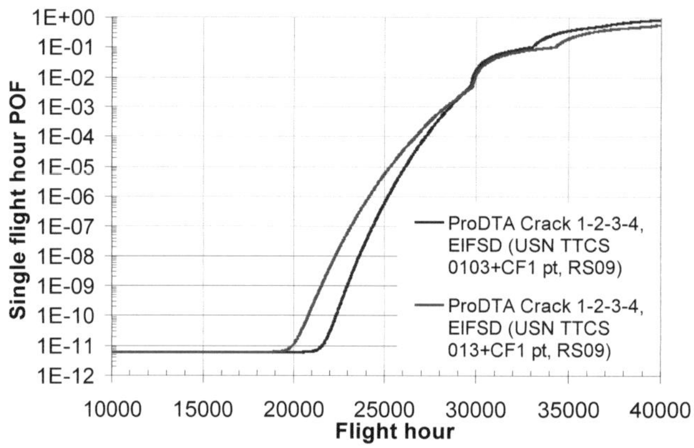

Figure 21. Single flight hour POF results for Location 3, using the two EIFSD derived from two Lognormal TTCS distributions before ("USN TTCS 0103") and after ("USN TTCS 013") regression.

According to the RARM risk index matrix (Figure 1 and Figure 2), the NRC POF results indicated a *low* risk index at the inspection time of 22,162 hours (SFPOF = 1.2×10^{-9} to $6.5\times10^{-8} < 10^{-7}$), and a *medium* risk index at the targeted operational lifetime of 24,500 hours (SFPOF = 1.5×10^{-7} to $2.1\times10^{-6} < 10^{-5}$) when *assuming* a 'Military aircraft-Hazard Probability' and a 'Hazard Severity-Category B (Hazardous)' for the analyzed Location 3 without future inspection/repair. Note that the P-3 full scale fatigue test showed that the wing did not fail catastrophically even with much larger crack size (3~4 inches) or severed spar cap.

It should be noted that this NRC POF study covered only one wing location; but it still provided additional support for the RCAF to downgrade the risk level of the wing structures to medium, which was also based on the inspection and assessment performed by QETE and IMP. This case study demonstrated a conservative POF study using limited in-service damage, which may be applied for other critical locations. Of course, for an entire wing or aircraft risk assessment, other damage findings, such as cracks on other wing locations, corrosion, hole elongation, and mechanical damage, would have to be taken into account. The follow-on inspection and repair, as well as the potential mechanical damages induced during these actions should also be taken into account in a risk assessment.

DISCUSSIONS

Relationship between conventional fatigue reliability and modern risk analysis

Historically the fatigue scatter has been taken into account by using a scatter factor associated with the Safe-Life approach. Some scatter factors are documented in standards such as DEF-STAN 00-970 (leaflet 35), which is based on a statistical analysis using the Bullen Equation for Case iii (mean known, standard deviation unknown), which assumes that the fatigue life follows a lognormal distribution with a known standard deviation of 0.1296. However, the scatter factor approach does not account for the fatigue characteristics of the individual materials in the structure, which means that the same scatter factor is used, independently of the structural material or the stress spectra [12].

It is been increasingly realized that there are considerable differences between the Lognormal standard deviation or Weibull shape parameters depending on the material and stress spectra. The scatter factor approach could lead to very conservative results and yet the probability of maximum stress is not accounted for. Therefore, a probabilistic fracture mechanics based risk analysis method was proposed for Damage-Tolerant structures to determine the POF/HR as a function of flight hours. The SFHPOF is also required by the risk assessment in the RCAF RARM process, but sometimes a cumulative POF is also used in the RARM for Safe-Life structures.

One can also calculate a scatter factor and a HR using a statistical distribution. Figure 22 shows the correlation between the scatter factor and the SFHPOF based on a representative lifetime distribution. This distribution may be established using the time to a common crack size (TTCS), which may be determined from the in-service crack data. For the same location, under the same fatigue failure model, using the same distribution, the scatter factor approach and the HR/SFPOF approach should produce the same risk level, so that a common safety level should exist for different airworthiness standards.

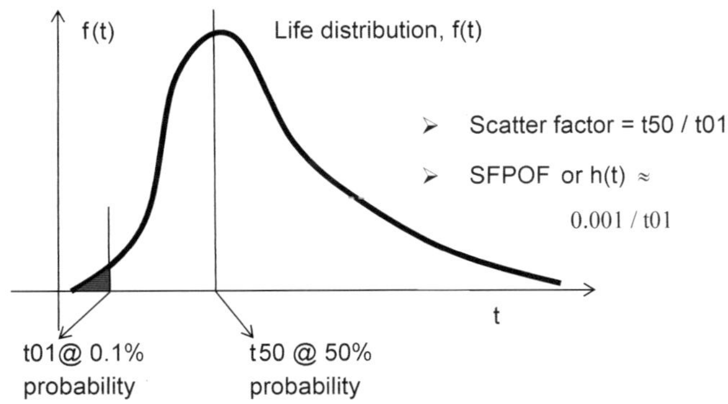

Figure 22. Correlation between Scatter factor and HR/SFPOF for the same lifetime.

In summary, Table II gives a brief comparison between conventional fatigue reliability analysis and fracture mechanics risk analysis.

Table II. Brief comparison between conventional fatigue reliability and modern risk analysis.

Conventional fatigue reliability analysis using the scatter factor (SF) approach	**Modern risk analysis using the Damage-Tolerant approach**
Using SF to cover fatigue life scatter caused by material, load, environment etc. Statistical basis: 1) Fatigue life distribution 2) Load distribution 3) Total probability of failure (TPOF) (e.g. 0.001 in DEF-STAN 00-970) More empirical, qualitative. Only a few SFs for all materials/loads, but in practice it is still comfortable/ defensible/confident to use the SFs from Standards/Handbooks with less intention to review/challenge the physics and statistics. Application: for multiple locations with more failure data, whole aircraft, whole fleet.	Using fracture mechanics, probabilistic modeling for material, load, environment, inspection and repair etc. Statistical basis: 1) Fatigue life distribution 2) Load distribution 3) Total probability of failure (TPOF) 4) Instantaneous risk/single flight POF (SFPOF) or SFHPOF 5) NDI reliability (POD) More physical, quantitative, sensitivity study advantage, but less comfortable/defensible/confident to use the SFPOF in practice (which is still varying due to recent Handbooks/Standards/case studies being published). Application: SFPOF based risk management for daily operation, for single location, and individual aircraft. TPOF for multiple materials, locations, and fleet.

Future challenges on quantitative risk analysis

Considering various uncertainties, including human factors involved in NDI/POD and handling damages in inspection and repair, it is still challenging to carry out a realistic quantitative risk analysis to simulate the effect of inspection and repair. In addition, often cold working and interference fitting are used for metallic structural repair, the lack of validated analytical models for quantifying engineering residual stress effects of cold working and interference fitting increases the complexity of the risk analyses.

In the past years, many quantitative risk analyses have been carried out especially for military aircraft structures. Most of these analyses are for the most critical locations, e.g. Category A (Catastrophic) locations defined in Figure 1. Global risk analysis methodologies with case studies are still rare and challenging for the determination of a combined risk for a group of components, including less critical locations (Categories B to E defined in Figure 1).

Current risk analyses are mainly for fatigue and fracture failures in metallic structures. It is still challenging to do a quantitative risk analysis for other failure mechanisms like corrosion (stress corrosion cracking, intergranular corrosion, corrosion pitting etc.), delamination, and impact damage in composites mainly due to the lack of physics-based models.

CONCLUDING REMARKS

A brief review has shown that the RCAF uses different airworthiness standards in aircraft structural lifing policy and some lifing policy are different from the original OEM. Various challenges have been experienced during the ASIP, which are related to various uncertainties induced by the lack of information, different usages, new material/manufacturing, and complex damage modes. A risk-based approach, the RCAF RARM process, has been established for all RCAF aircraft to quantify and manage these uncertainties in support of the technical airworthiness management. This approach consists of assessing the risk of failure for the next/future flights. This approach is proactive and has great potential to save cost while maintaining safety and availability. Also, this paper has shown that the RARM has a potential to use the same airworthiness standard to manage both Safe-Life and Damage-Tolerant aircraft, but in practice this is not achieved yet. The Safe-Life based risk analysis uses a cumulative POF to manage the risk; while the Damage-Tolerant based risk analysis uses the SFHPOF to manage the risk.

To support the RCAF ASIP and RARM processes for maintenance and service life assessment, NRC has been developing various statistical and probabilistic methods and tools for Safe-Life and Damage-Tolerant based quantitative risk analyses, including advanced damage tolerance analysis for complex damage modes like MSD and MED. Through the presented risk analysis examples, it was shown that the methods and tools developed by NRC have provided significant support to the quantitative risk analyses for various RCAF aircraft fleets. More importantly, the examples also showed how to prepare different inputs, and how these inputs affected the risk analysis for Safe-Life and Damage-Tolerant aircraft. With more data and experience being accumulated, a quantitative structural risk analysis can be performed faster and easier; similar to an extended DaDTA performed by the OEMs for new aircraft. For the RCAF, the technical airworthiness authority is held by the RCAF even under the in-service support contract framework (ISSCF). Therefore, it is expected that, the lessons learned here can not only increase the confidence of quantitative risk analysis to manage continuing airworthiness, but also can support RCAF as being a smart buyer to assure initial airworthiness for the RCAF air fleets.

ACKNOWLEDGEMENT

This work was performed with financial support from DRDC (Defence Research and Development Canada) and NRC (National Research Council Canada) through projects, "*Integrated Structural Life Assessment for RCAF Air Fleets*" and "*Aircraft Structural Life Assessment and Prognosis*". NRC risk analysis examples have been supported by various tasks from the Directorate of Technical Airworthiness and Engineering Support (DTAES) of the Department of National Defence (DND).

REFERENCES

[1] Department of National Defence of Canada (July 2007) Technical Airworthiness Manual (TAM), Document No. C-05-005-001/AG-001, 530 p.

[2] Department of National Defence of Canada (August 2007) Review of Open Records of Airworthiness Risk Management, Airworthiness Directive DND-AD-2007-01, 5 p.

[3] Liao, M., Comparison of Different Single Flight Probability of Failure Calculations for Aircraft Structural Risk Analysis, The 2012 Aircraft Airworthiness & Sustainment Conference (AA&S2012), Baltimore USA (http://www.meetingdata.utcdayton.com/agenda/airworthiness/2012/proceedings/techpapers/TP5435.pdf)

[4] Liao, M., Renaud, G., Bombardier, Y., and Bellinger, N., Review of Fatigue Life Assessment Methods for Aircraft Structures, LTR-SMPL-2013-0057, March 2013.

[5] Capt. Breton, M., Development of CF-18 Lifing Policy – New Fatigue Management Requirements and Integration of Full Scale Fatigue Test Requirements, CASI Conference, Montreal, 2003.

[6] APPENDIX 9, TO ANNEX A, TO CF-188 CMATC, DATED SEP 01, CF-188 STRUCTURAL INTEGRITY STANDARD FOR CONTINUED AIRWORTHINESS, V1 – DATED 6 MAR 2002.

[7] Gallagher, J. P., Giessler, F. J., Berens A. P., and Engle, R. M. Jr., USAF Damage Tolerant Design Handbook: Guidelines for the Analysis and Design of Damage Tolerant Aircraft Structures, Wright-Patterson AFB, Ohio, Air Force Wright Aeronautical Labs, AFWAL-TR-82-3073, 835 p., May 1984.

[8] Lincoln, J.W., Risk assessment of an aging military aircraft, J. of Aircraft, 22/8, 1985, pp 687-691.

[9] Bullen, N.I., A Note on Test Factors, Aeronautical Research Council, Reports No. 3166, 1956.

[10] Lawless, J.F., "Statistical Models and Methods for Lifetime Data", 1982, by John Wiley &Sons, Inc.

[11] NIST/SEMATECH e-Handbook of Statistical Methods, http://www.itl.nist.gov/div898 /handbook/, National Institute of Standards and Technology (NIST, USA).

[12] Lincoln, J. W., Risk assessment of aging aircraft, RTO AVT Lecture series on "Aging Aircraft Fleets: Structural and Other Subsystem Aspects", Sofia, Bulgaria, Nov. 2000 and published in RTO EN-015.

[13] Berens, A.P., Hovey, P.W., and Skinn, D.A., Risk analysis for aging aircraft, Vol.1 – Analysis, WL-TR-91-3066, Air Force Research Laboratory, Wright-Patterson Air Force Base, Ohio, 1991.

[14] Hovey, P.W., Berens, A.P., Loomis, J., Update of the probability of fracture (PROF) computer program for aging aircraft risk analysis, AFRL-VA-WP-TR-1999-3030, 1999.

[15] Miedlar, P.C., Berens, A.P., Hovey, P.W., Boehnlein, T.R., Loomis, J.S., PRoF v3 PRobability Of Fracture, Aging Aircraft Risk Analysis Update, UDR-TR-2005-00240, 2005.

[16] Email From: Tuegel, Eric J USAF, Sent: Friday, April 29, 2011 11:59 AM, To: Babish, Charles USAF; joegallagher; joseph.cardinal; rpbell3; Tombrussat1; Jim Rudd, Cc: tony.y.torng; Liao, Min; laura domyancic; hal.burnside; David Rusk; Peter Hovey USAFA, Subject: Continuing Saga of SFPOF.

[17] Brussat, T., When "what we always do" won't solve the problem, The 2013 USAF Aircraft Structural Integrity Program (ASIP) Conference, Bonita Springs Florida, USA, Dec. 2013.

[18] Freudenthal, A.M., Garrelts, J.M., and Shinozuka, M., The Analysis of Structural Safety, Journal of the Structural Division, Proc. of the American Society of Civil Engineers, Vol. 92, ST 1, Feb. 1966, pp. 267-325.

[19] Liao, M. and Xiong, Y., PRISM -- A Computer Code for Risk Assessment of Aging Aircraft Structures, LTR-ST-2248, National Research Council Canada (NRC), 2000.

[20] Liao, M., Forsyth, D.S., and Bellinger, N.C., Risk Assessment of Corrosion Maintenance Actions in Aircraft Structures, Proceedings of the 2003 Aircraft Structural Integrity Program Conference, Savannah, USA, 2003.

[21] Liao, M., Bellinger, N.C, Forsyth, D.S., and Komorowski, J.P., A new probabilistic damage tolerance analysis tool and its application for corrosion risk assessment, Proceeding of the 2005 International Committee on Aeronautical Fatigue Sym. (ICAF 2005), Hamburg, Germany, EMAS Publishing, pp. 241-252, 2005.

[22] Komorowski, J., New tools for aircraft maintenance, Aircraft Engineering and Aerospace Technology, Vol. 75, pp. 453-460, 2003.

[23] Liao, M., Bombardier, Y., and Renaud, G., Quantitative risk assessment for CC-130 centre wing, Phase II Final Report, NRC Internal Report, LTR-SMPL-2009-0098, 2009.

[24] Renaud, G. and Liao, M., CC-130 In-service damage assessment methodology and analysis Tool, LTR-SMPL-2013-0036, 2013.

[25] Liao, M., Bombardier, Y., Renaud, G., Bellinger, N.C. and Cheung, T., Development of advanced risk assessment methodologies for aircraft structures containing MSD/MED, Proceedings of the 2009 International Committee on Aeronautical Fatigue Sym. (ICAF2009), Rotterdam, Netherlands, June 2009.

[26] Liao, M., Renaud, G., Bombardier, Y., and Bellinger, N., Probabilistic risk analysis for aircraft structural life assessment with limited in-service damage findings, The 28th Congress of the International Council of the Aeronautical Sciences (ICAS 2012), 23-28 September 2012, Brisbane, Australia.

[27] Bombardier, Y., Liao, M., and Renaud, G., CGCC130MSD: A crack growth analysis program for panels with multiple-site fatigue damages (Version 0.1.0), LTR-SMPL-2009-0193, 2009.

[28] Renaud, G., Liao, M., and Bombardier, Y., Development of an efficient methodology and tool to determine stress intensity correction factors for complex aircraft structures, Proceedings of the 2011 International Committee on Aeronautical Fatigue Sym. (ICAF2011), Montréal, Canada, 2011.

[29] Renaud, G., Liao, M., and Bombardier, Y., Monte carlo based quantitative risk assessment for multiple site damage in aircraft structures, The 2014 Aircraft Airworthiness & Sustainment Conference (AA&S2012), Baltimore, Maryland, USA, 2014.

[30] Bombardier, Y., Liao, M., and Renaud, G, CanGROW for windows - User Guide (Version 1.1), LTR-SMPL-2013-0148, 2013.

[31] Capt. Morency, R., Current ASIP programs and issues in the Canadian Forces, HOLSIP2012, Utah, 2012.

[32] Beauvais, Y., Benefits of using a risk process in ASIP – the CF-18 experience, Proceedings of the 2011 International Committee on Aeronautical Fatigue Symposium (ICAF2011), Montréal, Canada, 2011.

[33] Liao, M and Xiong, Y., Risk assessment of aging aircraft structures using PRISM and PROF, Canadian Aeronautics and Space Institute (CASI) Journal, Vol. 46, No. 4, December 2000, pp. 191-203.

[34] Johnson, J., CC-130 centre wing fatigue risk assessment, presentation at NRC, the Directorate of Technical Airworthiness (DTA) of the Department of National Defence (DND), April 2004.

[35] Caron, Y., CC-130 lifing considerations, Canadian Aeronautics and Space Institute 2005 Conference on Aerospace Technology and Innovation, Toronto, 2005.

[36] Bateman, G., Wing service life assessment methodology, 2004 Hercules Operation Conference (HOC), Atlanta, USA, 2004.

[37] FAA, Aging Airplane Program: Widespread Fatigue Damage; Final Rule, 14 CFR Parts 25, 26, 121, and 129 [Docket No. FAA–2006–24281; Amendment Nos. 25–132, 26–5, 121–351, 129–48] RIN 2120–AI05, Nov.15th, 2010.

[38] Bourdon, S. and Guillouzic S., A statistical analysis of fatigue accumulation in CC-130 (Hercules) centre wings, ORD Project Report PR 2004/02, Directorate of Operation Research (DOR), Department of National Defence Canada, 2004.

[39] Liao, M., Review of a statistical analysis for fatigue accumulation of CC130 centre wing, LTR-SMPL-2004-0147, National Research Council of Canada, 2004.

[40] Black, D. and Johnson, J., Managing risk – the Canadian Forces experience, Hercules Operators Conference (HOC), Atlanta, USA, 2008.

[41] Liao, M., and Renaud, G., Statistical analysis for assessing the CC-130 center wing service life limit, LTR-SMPL-2011-0023, 2011.

[42] Liao, M., Bombardier, Y., Renaud, G., and Bellinger, N.C., Advanced damage tolerance and risk analysis methodologies and tools for aircraft structures containing multiple-site and multiple-element fatigue damages, Proceedings of the Institution of Mechanical Engineers, Part G: Journal of Aerospace Engineering, November 2012 226: pp. 1412-1423.

[43] Liao, M., Bombardier, Y., and Renaud, G., Updated quantitative risk analysis for the CC-130 center wing structures: Part 1 CW-1 location, LTR-SMPL-2010-0222, 2010.

[44] Liao, M., Renaud, G., Bombardier, Y. and Yanishevsky, M., CC-130 in-service damage analysis: review and tool development - final report, LTR-SMPL-2013-0039, 2013.

[45] Renaud, G. and Liao, M., CC-130 In-service damage assessment methodology and analysis tool, LTR-SMPL-2013-0036, 2013.

[46] Renaud, G., Liao, M., and Bombardier, Y., Effect of a structural modification on the life estimation and probability of failure of the CC-130 center wing, The 2015 Aircraft Airworthiness & Sustainment Conference (AA&S2015), Baltimore, Maryland, USA, 2015.

[47] Christiansen, P., Bateman, J., and Navarrete, A., C-130 center wing MSD/MED risk analysis, Proceedings of the 2006 Aircraft Structural Integrity Program Conference, San Antonio, USA, 2006.

[48] Liao, M., Bombardier, Y., and Renaud, G., Updated quantitative risk analysis for the CC-130 center wing Structure: Part 2 CW-14B Location, LTR-SMPL-2010-0232, 2010.

[49] Li, G., Renaud, G., and Liao, M., Analysis of load profile at the CC-130 CW-1 critical location using global finite element modeling, LTR-SMPL-2012-0100, 2012.

[50] Bateman, G. R. and Christiansen, P., C-130 Centre Wing Fatigue Cracking - A risk management approach, 2005 USAF Aircraft Structural Integrity Program Conference, 2005.

[51] Bombardier, Y., Liao, M., and Renaud, G., Modeling of continuing damage for damage tolerance analysis, Proceedings of the 2011 International Committee on Aeronautical Fatigue Sym. (ICAF2011), Montréal, 2011.

Probabilistic approach to the inspection interval determination in a multi-site damage susceptible structure

J. Segurado, P. Caffyn, J. García, B. de Nicolás, J. I. Armijo, D. Iñesta, E. Mirón

AIRBUS DEFENCE and SPACE Fatigue & Fracture Mechanics Department

Abstract: This paper presents the results of advanced Probabilistic Damage Tolerance Analysis and the methodology and tools developed to predict the crack growth and the inspection interval of potentially critical locations susceptible of generating Widespread Fatigue Damage (WFD).

A computational procedure, using the principles of fracture mechanics and a compounding β-solution method, has been developed and applied with success by AIRBUS Defence and Space (ADS) during the last decades. A new β-solution to predict the crack growth of collinear cracks interacting between them in panels subjected to multi-site damage (MSD) has been included in the last evolution of this software called MULTIGRIETA. The comparison of this new solution with test results is presented in this paper.

The traditional deterministic approach does not take into account explicitly the uncertainty sources (it does, however, include their effects via the Scatter Factor) involved in a damage tolerance assessment. Therefore, an alternative probabilistic method based on the structural risk analysis theory and carried out with Monte-Carlo simulations has been developed in this work and its results have been compared with the outcome of the deterministic analysis to demonstrate its capability.

MULTIGRIETA software has been incorporated to a wider scope software project, written in MATLAB, called MARATHON that has been programmed ad-hoc to simultaneously grow multiple cracks in stiffened panels and to facilitate the probabilistic analysis of the case study in this paper. Although this methodology can be easily extended to a general case, the case study in this paper is a riveted stiffened panel prone to generate MSD.

INTRODUCTION

As the aeronautical community knows, the likelihood of the occurrence of fatigue damage in an aircraft's structure increases with the aircraft's usage. The design process generally establishes a design service goal (DSG) during which only local cracking is expected. If the DSG is extended, in what have been called aging aircraft, there is a risk that the structure develops several fatigue cracks in the same structural element details. These cracks may interact, and they can have an adverse effect on the strength of the structure before they become detectable. This phenomenon, known as Widespread Fatigue Damage, was identified after the Aloha accident in 1988 as one of the major concerns for the continued airworthiness of ageing aircraft structures. Therefore, predicting the crack growth and the inspection needs of potentially critical locations, in order to ensure a minimum specified reliability, is a critical requirement that the methods for local cracking calculation cannot properly quantify.

The work in this paper is focused on a stiffened panel that is susceptible to generate Widespread Fatigue Damage. The analysis is focused on the calculation of the inspection interval obtained from the WFD assessment that could be established in these structures to complement the maintenance program obtained with the traditional Damage Tolerance assessment. The traditional deterministic approach to calculate the crack growth and inspection interval uses Scatter Factors (SF) to take into account the uncertainty sources involved in the engineering problem, but it is not able to quantify the over conservatism introduced by these SF in the results. Therefore, an alternative probabilistic method based on the structural risk analysis theory has been applied in this work and its results have been compared to the outcome of the deterministic analysis.

One of the essential elements in probabilistic based design is to identify and quantify the sources of uncertainty. This probabilistic analysis requires a careful identification of the variables involved in the process of crack growth, characterization of these variables with the most appropriate probability distribution functions and the reliability analysis calculations, which will be performed using Monte Carlo experiments. A comparison between the probabilistic and deterministic results will also be included in order to evaluate the differences between them.

The crack propagation analysis in this paper incorporates probability distributions of equivalent initial flaw sizes (EIFS) obtained from ADS's tests correlated to a Weibull and a Lognormal distribution, fracture toughness, probability of detection (both from literature review) and crack propagation curves (using ADS's average data, and variances from published data).

The work is divided in the following main parts:

1. Validation of the crack growth algorithm used to perform crack growth analysis in panels with collinear cracks interacting between them. This is done to prove the accuracy of this program, in which the rest of the work is supported and uses test results as a reference.
2. Deterministic analysis of the case study. Determination of the inspection interval for different initial crack scenarios to comply with the requirements of the WFD rules.
3. Stochastic analysis by means of Monte-Carlo simulations. Determination of the inspection interval of the same case study for a specified reliability level goal. A sensitivity analysis of the probabilistic variables considered in the problem is carried out together with a comparison between the deterministic and the probabilistic results.

These objectives are accomplished with the aid of the program MULTIGRIETA, combined with other software. This software, based on the linear elastic fracture mechanics theory, has been incorporated to a new program called MARATHON that has been programmed ad-hoc in MATLAB to simultaneously grow multiple cracks in stiffened panels so that it is possible to perform a complete crack growth and a residual strength analysis of the panel. This automation has helped to perform the probabilistic analysis of the case study in this paper.

VALIDATION OF THE CRACK GROWTH ALGORITHM

AIRBUS Defence and Space (ADS) has developed for decades its own numerical algorithms to predict the fatigue life and the crack growth in metals. In the computer program GRIETA, the linear fracture mechanics theory and the principle of the superposition method is used to obtain a solution for the stress intensity factor of complex geometric configurations by superimposing a set of appropriate individual β-solutions, usually associated with simple configurations of cracks and component boundaries with known solutions. The resultant stress intensity factor (ΔK) with the presence of all simple configurations and boundaries is then expressed in terms of the formula:

$$\Delta K = \Delta\sigma \cdot \sqrt{\pi \cdot a} \cdot \prod \beta_i \qquad (1)$$

where βi corresponds to the stress intensity factor associated to the individual configuration i. The fatigue crack growth is then predicted by a power relation between the crack growth rate (da/dN) and the stress intensity factor range (ΔK) by means of da/dN curves property of ADS.

Recently, an evolution of the program GRIETA called MULTIGRIETA, has been developed by ADS to predict the crack growth of collinear cracks interacting between them in panels prone to develop multi-site damage. The evolution includes new β-solutions for collinear cracks in an infinite plate [3]. The program MULTIGRIETA is the core of the program MARATHON developed to perform the deterministic and probabilistic analysis presented in the following chapters.

This chapter includes a correlation between MULTIGRIETA results and a series of published tests results [1, 2]. A complete validation of the program will need a long and complex process that is out of the scope of this work, but the comparison made in the following sections shows how MULTIGRIETA, and in particular the combination of the new β-solutions for collinear cracks with the rest of the β-solutions extensively proven by ADS, correlates with the reality to a great extent. This indicates that a complete validation is worth to finally implement this methodology in the engineering process.

The prediction method has been verified using the results of a series of fatigue crack growth tests [1, 2] performed on six 2024T3 (L) alloy sheet specimens with a collinear row of open holes and artificially introduced through cracks at the three central holes with three different geometrical configurations.

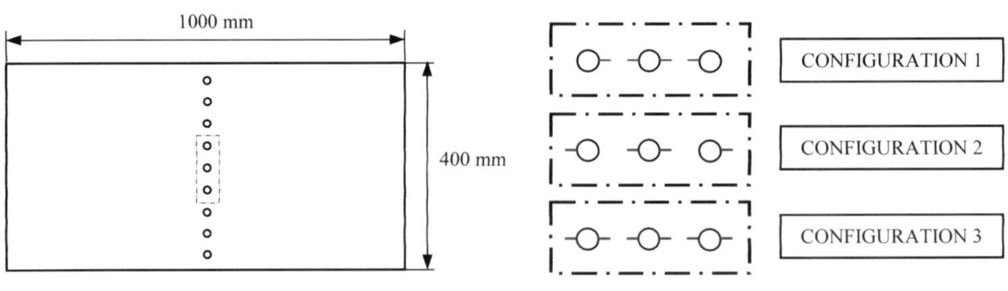

Figure 1. Test specimens configuration.

Table I. Initial crack lengths.

Specimen	Conf.	Left hole from the central hole (mm)		Central hole (mm)		Right hole from the central hole (mm)	
		Tip A	Tip C	Tip A	Tip C	Tip A	Tip C
H1	1	-	2.225	1.2	1.323	2.292	-
H2	1	-	2.223	1.038	0.915	1.618	-
H3	2	2.123	-	0.934	1.383	-	2.121
H4	3	1.063	0.861	0.951	1.101	0.927	0.971
H5	3	1.146	1.25	1.168	1.238	1.29	1.153
H6	2	1.973	-	1.315	1.171	-	1.738

The specimens were loaded with constant amplitude (CA) cyclic loads of $\Delta\sigma = 70$ MPa and $R = 0$ except specimen H4 that was loaded with CA cycles of $\Delta\sigma = 65$ MPa and $R = 0$.

The stress intensity factor is calculated with the following combination of β-factors:

- β-factor for a single or double through crack emanating from a hole subjected to axial and bearing load.
- β-factor for the finite width sheet correction for cracks emanating from holes.
- β-factor for an eccentric through crack in a finite width sheet.

In addition, the algorithm considers automatically another β-factor that takes into account the effect of the interaction between collinear cracks when necessary. The fatigue crack growth is predicted by means of da/dN curves for Al2024 property of ADS.

Table II. Spectrum definition.

Specimen No.	**Marker loads**	**No. of cycles of CA blocks**	**No. of cycles of marker blocks**	**Δσ (MPa)**	**Configuration**
H1	Yes	10000	10000	70	1
H2				70	1
H3				70	2
H4				65	3
H5				70	3
H6		5000		70	2*

*Hole pitch of 20 mm instead of 40 mm.

The following table shows a comparison between tests and MULTIGRIETA results for the number of cycles from initial crack size to the first link-up (first joint of two collinear cracks).

Table III. Cycles until 1st link-up.

Specimen	**Configuration**	**Tests (No. of cycles)**	**MULTIGRIETA (No. of cycles)**	**Error (%)**
H1	1	131520	130286	0.9
H2	1	138000	134701	2.4
H3	2	134827	142173	5.2
H4	3	185800	142197	30.7
H5	3	107600	110685	2.8
H6	2	85000	75800	12.1

Specimens H1, H2, H3 and H5. The theoretical approximation for these specimens correlates well with the test measurements, especially in specimen H1 where the error is lower than 1%. Results of specimen H1 are shown as an example. The tests' and MULTIGRIETA's results are plotted in the same graph to compare the crack growth of all the crack tips in every hole.

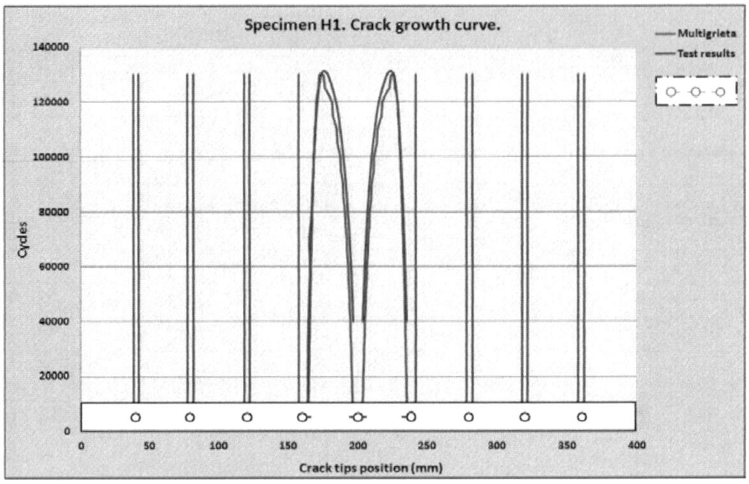

Figure 2. Crack growth in Specimen H1 (configuration 1).

Specimen H4 (configuration 3). The difference in specimen H4 between the test's and predicted crack growth is related to the shape of the theoretical da/dN and not to the values of ΔK which are almost the same in both cases. This difference is even bigger because the applied stresses are lower than in other specimens ($\Delta\sigma$ = 65 MPa instead of 70 MPa). This makes that, at early stages (with small crack lengths that usually are difficult to simulate) the values of ΔK are also small (~200 MPa$\sqrt{\text{mm}}$). These values correspond to the left part of the curve da/dN, where the differences in crack growth rate between the test derived da/dN and the curve used in the algorithm of MULTIGRIETA are the biggest ones. The growth ratio with the test da/dN is about 10^{-6} mm/cycle, while it is about 10^{-5} mm/cycle with the theoretical da/dN. As a consequence, the correlation is poor in terms of absolute crack growth life.

However, the aim of the algorithm is to obtain an inspection interval and this interval is calculated with the relative crack growth life from detectable to critical crack length. Typical detectable lengths are not usually lower than 3 mm with the typical Non Destructive Tests used in the aeronautical industry, therefore the actual interest must be the crack growth stages from about 3 mm onwards. Comparing the crack growths in this range of crack lengths, the correlation between the theoretical and the test curves is much better. The new error at the first link-up point is shown in Table IV.

Specimen H6 (configuration 2). The effect observed in specimen H4 appears again in H6 although less marked, and the error is much lower after the adjustment. But another effect is observed in this specimen. The specimen H6 has 19 holes with a separation of 20 mm between them instead of 40 mm as in the rest of specimens. This causes this specimen to clearly show the 'growth stop' effect since the crack reaches one hole edge until it appears at the other hole edge. This is due to the reduction of the stress intensity factor at the crack tip because of the hole and, the longer is the crack, the lower is its importance in the global crack growth.

The program MULTIGRIETA does not have any β-solution available to simulate this effect. It replaces the crack and the hole by a crack with a total length equals to the sum of the crack length and the hole diameter. This is the reason why the progress of the crack in the following holes is faster than in reality, which is conservative from design point of view.

Table IV. Cycles from cracks of 2.8 mm until 1st link-up.

Specimen	Configuration	Tests (No. of cycles)	MULTIGRIETA (No. of cycles)	Error (%)
H4 (adjusted)	3	95500	99197	3.87
H6 (adjusted)	2	55000	53448	2.9

Conclusions. The program MULTIGRIETA has shown a good correlation between the tests crack growths and the simulated crack growths in a MSD scenario. Only one test (specimen H4) has shown a bigger difference in terms of total cycles from the initial crack length to the first link up. It has been observed that the differences come from the crack growth rates predicted by the theoretical da/dN for small ΔK values at the early stages of the crack growth. However, a good correlation has been found when considering the relative crack growth life from the typical detectable lengths onwards, that is, the part of the crack growth of interest to establish the inspection interval. Another non correlation is observed in specimen H6, where the algorithm is not able to simulate the 'growth stop' effect when a crack reaches a hole and instead it uses a conservative approach.

In conclusion, a good agreement has been found between the test results and the predictions, and in those cases where the adjustment is not so good, the program obtains conservative results. Further investigations should be done with other scenarios and other materials to perform a complete validation of the program. Meanwhile, it has been used to develop a methodology to calculate automatically the crack growth and inspection interval in stiffened panels with collinear holes and cracks.

DETERMINISTIC ANALYSIS OF THE CASE STUDY

Overview on WFD

After the Aloha Airlines Boeing 737 accident in 1988, the industry specialists (airworthiness authorities, operators and manufacturers) made a great effort to identify the underlying mechanisms of the WFD phenomenon in ageing aircraft. The FAA sponsored activities on WFD in ageing aircraft through the Airworthiness Assurance Working Group (AAWG) whose main tasks were to develop and evaluate methods suitable for WFD assessment.

The FAA and industry groups concluded that the best approach against the phenomenon of WFD is to preclude it and the result of the investigations was the issuance of a set of rules in 2010 applicable to transport aircraft. FAR §25.571 amendment 132 for new Type Certificates (TC) after January 2011 and FAR §26.21 & FAR §26.23 for all aircraft models with MTOW ≥ 75000 lb, with existing TC, with pending TC before January 2011, or derivatives with reversion to earlier amendment.

Compliance with these new rules requires the establishment of a Limit of Validity (LOV) of the engineering data by the manufacturers, this means that WFD must be demonstrated to be unlikely to occur during the service life of the aircraft up to the LOV. The assessment for WFD must be done in addition to fatigue and damage tolerance evaluations already performed under FAR/CS §25.571 and airlines will have to implement, if necessary, possible additional tasks into their maintenance programs as a result of the WFD evaluation.

The process used to perform a WFD assessment of a structure must be based on analysis, tests, evaluation of service experience and teardown inspections. AC 25.571-1D and AC 120-104 provide guidance for compliance with the new requirements, including evaluation of WFD and establishing a LOV of the engineering data that supports the structural maintenance program. Safarian [6] made a compilation of the requirements for the WFD assessment and together with the Airbus Policy on WFD [7] have been followed to perform the evaluation of the case study in this work.

Following this guidance the WFD evaluation has five main steps:

1. **Identification of a candidate for LOV.** Every LOV can be valid only if it can be demonstrated that the model of aircraft is free of WFD until that LOV. This candidate depends on high level technical considerations as the age or the service experience of the fleet, analysis and tests results, and economical considerations.
2. **Identification of WFD susceptible structures** including the list of 16 examples given in AC 120-104 or any other specific structures of the aircraft including evidences of tests/service.
3. **Evaluation of WFD** following the next steps:
   - **Determine of the "WFD average behaviour" (N_{WFD}).** N_{WFD} is the point in time at which, without intervention, it is expected that 50% of the fleet develops WFD in a particular structure. For this, the full scale fatigue test (FSFT) is the most accurate method. Likewise the service experience and a specific analytical evaluation may be used in addition to the FSFT.
   - **Establish of the Inspection Start Point (ISP) and the Structural Modification Point (SMP).** The ISP and SMP are determined through a statistical analysis of crack initiation based on fatigue testing, teardown, or service experience of similar structural details. It is assumed that the ISP and SMP are equivalent to lower bounds with a specific probability in the statistical distribution of cracking events. Alternatively, in a deterministic analysis, the ISP and SMP may be established by applying appropriate scatter factors to the N_{WFD}. If the inspections are effective, they must be implemented from the ISP up to the SMP during what the AAWG [8] called the Monitoring Period. If MSD inspections are not effective, then only the SMP must be set.
   - **Establish of the inspection interval (I_{WFD}).** Whenever inspections are demonstrated to be effective, an interval should be chosen to provide a sufficient number of inspections during the Monitoring Period so that there is a high confidence that no MSD condition will reach the final cracking scenario without detection.

Conservative scenarios should be assumed to set the inspection interval unless other assumptions can be consistently supported by test and service experience. The work in this paper is focused on the establishment of this I_{WFD}.

4. **Establish additional maintenance actions** derived from the WFD evaluation in locations where the maintenance program derived from the traditional Fatigue and Damage Tolerance evaluation is not enough to safeguard the structure against WFD.
5. **Establish of the LOV**. Once all the structures from step 2 have been evaluated following step 3 points, a LOV is established maintaining, lowering or increasing the candidate identified in step 1. Again, technical and economical considerations must be evaluated to take a decision.

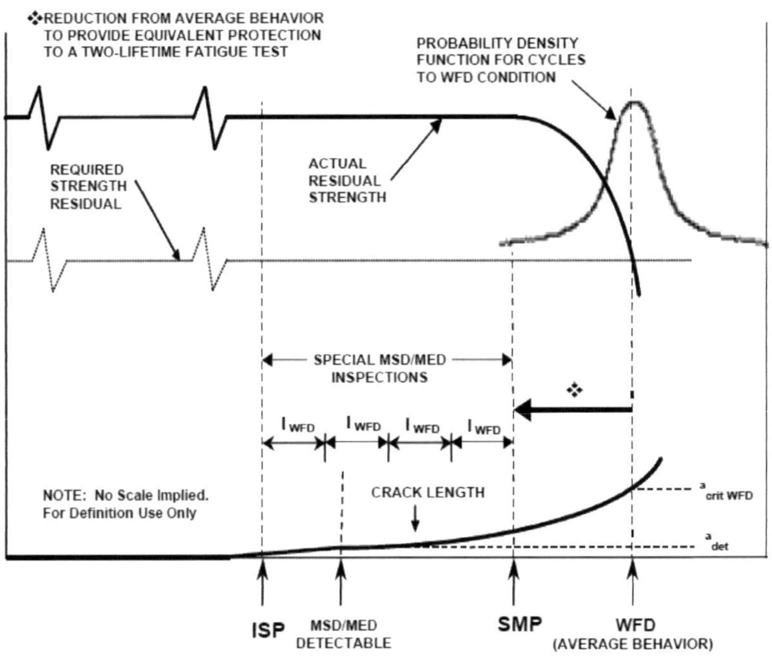

Figure 3. Determination of the WFD critical points in structures able to be inspected.

Case study

The analysed region is the wing lower skin of a light transport turboprop. The section of the analysis corresponds to the first row of the riveted joint between the wing lower skin and an external doubler. Eleven rivet holes will be considered at the area around stringers 2 and 3 because this is the most vulnerable zone according to the results of the wing-box stress analysis and the in-service experience. The material of the wing skin is Al 2024-T42. The fatigue stress spectrum used for the analysis has been calculated for a typical transport mission mix of Short Range, Medium Range, Long Range and Training flight profiles.

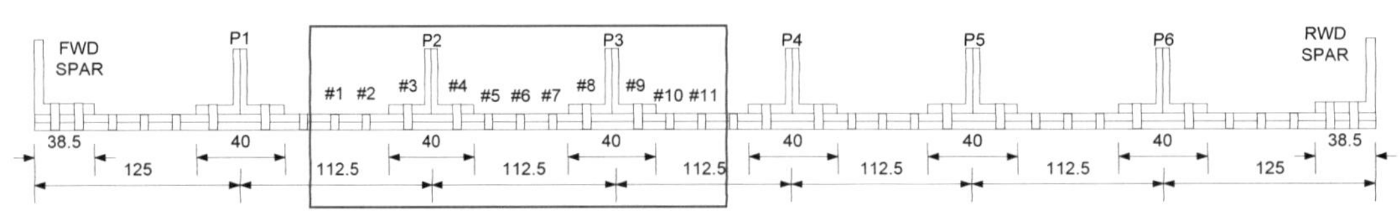

Figure 4. Section A-A. Section of analysis.

This case study has been evaluated analytically under the WFD requirements and the following values have been obtained for the critical points N_{WFD}, ISP, SMP. Airbus policy [7] has been used in the calculations:

$LOV_{candidate} = DSG = 30000$ flights

$N_F = 290000$ flights
$N_{WFD} = N_F * 0.9^P / k_0 = 290000 \cdot 0.9^5 / 1.5 = 445600 \cdot 0.39366 = 114000$ flights
$ISP = N_{WFD}/3 = 38000$ flights
$SMP = N_{WFD}/2 = 57000$ flights
SMP ISP = 19000 flights.

According to these results, the Monitoring Period (SMP - ISP) seems to be long enough to establish effective inspections, therefore, three different scenarios are considered to calculate the I_{WFD}. The selection of these scenarios as acceptable scenarios to comply with the requirements of the WFD rule has been an agreement between the manufacturers and the authorities (FAA). In particular, the experience of Airbus in exploring various means of analysis to determine the interval [11], and comparing several configurations, including correlation with test/in-service examples [12] make possible the use of the following scenarios in the work in this paper.

1. Multiple site damage scenarios:
   - Cracks in three adjacent holes with cracks of the same size at both sides of the hole. The rest of the holes without cracks. The initial crack length is the detectable crack length. A Scatter Factor of 3.0 is used.

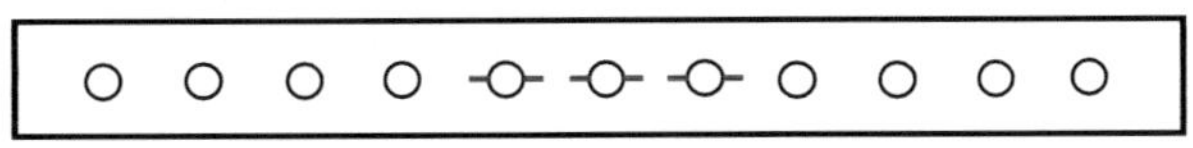

Figure 5. Deterministic scenario 1a.

   - Same size cracks in all the locations. Cracks of 0.125 mm and 1.25 mm have been tried. The probability of occurrence is extremely remote. A Scatter Factor of 3.0 is used.

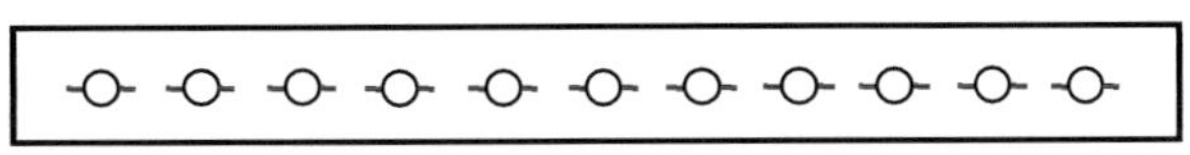

Figure 6. Deterministic scenarios 1b & 1c.

2. Standard scenario of fatigue and damage tolerance:
   - One "Rogue Flaw" of 1.25 mm and "Quality Flaws" of 0.125 mm.
   - A Scatter Factor of 4.0 is used. A Scatter Factor of 3.0 is used only if the minimum crack length of the Quality Flaws (when the length of the Rogue Flaw reaches its detectable crack length) is above 2/3 of the detectable crack size. This condition shall be fulfilled for at least 2 holes on each side of the hole containing the primary crack.

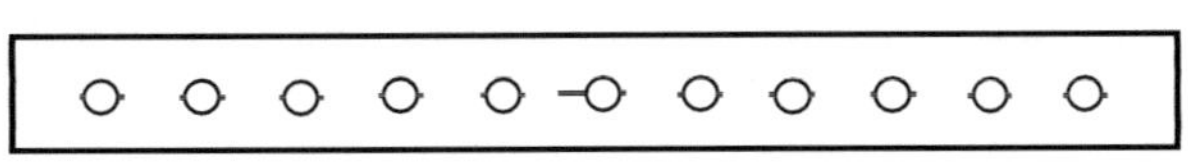

Figure 7. Deterministic scenario 2.

For the calculation it is necessary to know the type of inspection that will be carried out in the structure, its associated detectable length and the length of the hidden zones of the structure for each crack tip. Typically the crack size of the Probability of Detection (POD) curve at 90% probability of detection with 95% confidence is used to define and rate the capability of an inspection technique. In this case study a HFEC (High Frecuency Eddy Currents) method has been chosen with the crack dimensions defined in the following figure, which depend on the crack tip inspected.

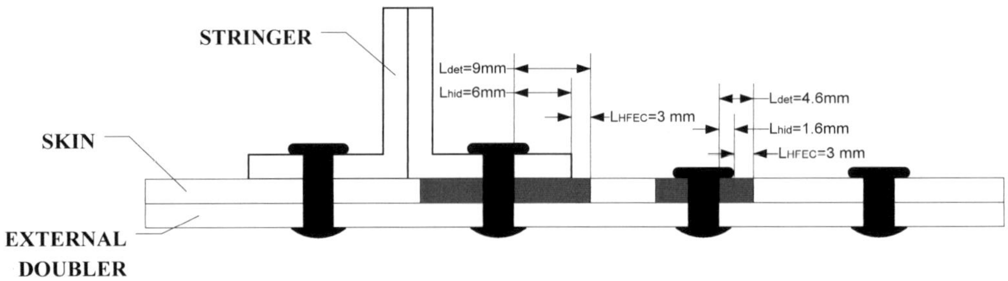

Figure 8. Detectable crack lengths.

Calculation of the I_{WFD}

As mentioned previously, MARATHON has been programmed to simultaneously and automatically grow multiple cracks in stiffened panels, so that it is possible to perform a complete crack growth and a residual strength analysis of the panel. This program integrates MULTIGRIETA Program (MG), which contains the algorithm that, calculates the crack growth. MARATHON can work in two different modes, deterministic and probabilistic, this last one to perform Monte-Carlo simulations. This section describes the global process developed to determine automatically the I_{WFD} using deterministic methods. Figure 9 shows the flowchart of the process followed by MARATHON that is roughly described in the next steps:

1. A subroutine called GuiData stores the initial data introduced through the graphic user interface of the program. The data contains information related to:
   - **Geometry**: panel dimensions, holes dimensions and positions, stringers properties
   - **Material Properties**: da/dN, Kc, K1c, R-curve, Fty
   - **Fatigue Spectrum**
   - **Limit Load or Stress**
   - **Initial Cracks**: dimensions and locations
   - **Inspection Data** detectable length $(L_{det})_i$ according to the chosen inspection method and hidden length $(L_{hid})_i$ for each crack tip i
   - **Retardation Model, Scatter Factor (SF) and other execution control data**

2. A subroutine called MG_Bridge recovers the input data stored by GuiData and prepares the input of the first run of MG. In the computer program MG, the linear fracture mechanics theory and the principle of the superposition method is used to obtain a solution for the stress intensity factor of complex geometric configurations by superimposing a set of appropriate individual β-solutions, usually associated with simple configurations of cracks and component boundaries with known solutions. A β-factor library has been developed by ADS for use in their crack growth algorithms and it has been incorporated to MG program as well. Some β-factors were directly obtained from the literature and others were obtained by the principle of superposition using various published β-factors. Depending on the geometry of the problem and the type of cracks, different beta factors must be used. For instance, β-factors used in the case study of this work are:

   - β-factor for a single or double corner crack emanating from a hole subjected to axial and bearing load.
   - β-factor for a single or double through crack emanating from a hole subjected to axial and bearing load.
   - β-factor for the finite width sheet correction for cracks emanating from holes.
   - β-factor for an eccentric through crack in a finite width sheet.
   - β-factor for an edge crack in a finite width sheet.
   - Tabular β-factors used to store the effect of the stringers on every crack tip. These factors are calculated with program REMACHADO, property of ADS, in each step before running MG. The β-factor of the crack tip growing towards the stringer is shown to be significantly reduced by the presence of the stringer that plays the role of a crack stopper due to the load redistribution produced during the crack growth, while the crack tip growing away from the stringer is marginally affected. In case the load redistribution is big enough to exceed the stringer strength, the current step of MG is stopped and a new set of β-factors is calculated with the new configuration of stringer broken and the corresponding load redistribution.
   - β-factor considered automatically by MG to take into account the effect of the interaction between collinear cracks.

Once the geometry of the step has been defined, and with the input data stored in the GuiData, the MG_Bridge subroutine apply the suitable combination of these factors to calculate the total ΔK and prepares the input of the first run of MG.

3. MG performs the crack growth of the first step. The crack growth is predicted by the power relation between the crack growth rate (da/dN) and the stress intensity factor range (ΔK) by means of da/dN curves property of ADS. The program calculates cycle-by-cycle the crack growth of all the cracks defined in the input data by entering the da/dN curves with the values of ΔK. The output is a crack length vs. number of cycles/flights discrete curve per crack tip.
4. The outputs of MG are read and stored by the Results Interpreter in a matrix of results. The Results Interpreter is an algorithm that was developed to identify and update the crack information at the end of the previous step. This includes extracting the location and size of every crack, determining if the crack is a corner or through crack, determining if the crack is an edge, a hole or an eccentric crack, determining the type of interaction with adjacent structural elements (edge, hole, other cracks or stiffeners) and merging the cracks if applicable. With this information the algorithm updates the information stored in the GuiData for the following step.
5. The MG_Bridge subroutine recovers the updated GuiData and prepares the next step of the crack growth to be run with MG, so that the whole process of crack growth can be calculated automatically step by step. The loop continues until the panel is completely cracked.
6. When MG finishes the last step all the steps are joined together in a unique matrix with the subroutine All Results and then Marathon Program proceeds to perform the Residual Strength Analysis. At this point, the critical crack (a_{cr}) and the corresponding number of flights (N_{cr}) are calculated automatically with three criteria: Net Section Yield (NSY), Rapid Crack propagation (RCP) and the R-Curve. The program calculates and stores the results of the three criteria and identifies the most critical pair of a_{cr}, N_{cr} among them.
7. Finally, with the inspection data saved by GuiData, a new subroutine calculates the detectable crack length of every crack tip $(a_{det})_i$ as the sum of $(L_{det})_i$ plus $(L_{hid})_i$. Then the program looks in all the outputs for the number of flights $(N_{det})_i$ at which $(a_{det})_i$ is reached, and identifies the crack tip whose $(N_{det})_i$ is minimum, that is, the first detectable crack tip. Finally, the subroutine calculates the inspection interval as $I_{WFD} = (N_{cr} - N_{det}) / SF$ for the three abovementioned criteria and the minimum among them as well.

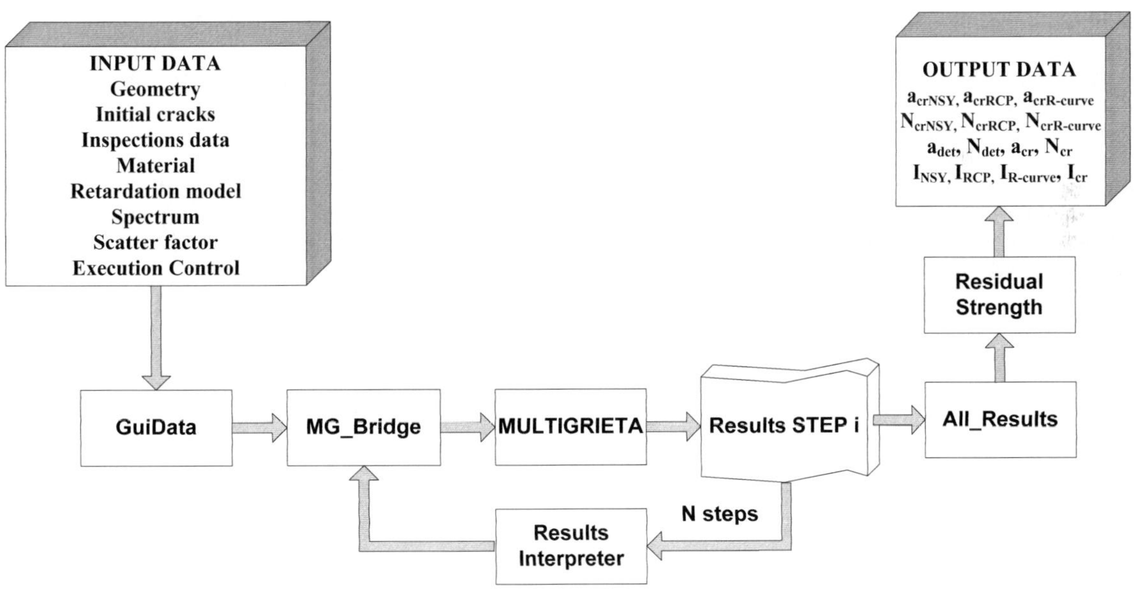

Figure 9. Flowchart of MARATHON program. Deterministic module.

Results and conclusions

The following results have been achieved using Marathon Program in its deterministic mode.

Table V. Deterministic results for multi-site damage scenarios.

SCENARIO	Description	a_{det} (mm)	a_{cr} (mm)	N_{det} (flights)	N_{cr} (flights)	N_{cr}-N_{det} (flights)	Residual Strength Criterion	Scatter Factor	I_{WFD} (flights)
1a	Cracks of 4.6 mm in Holes 5, 6 & 7. (SF = 3)	4.6	350	501	22235	21734	NSY	3	7245
1b	Cracks of 0.125 x 0.125 mm in all Holes (SF = 3)	4.6	350	22433	31335	8902	NSY	3	2967
1c	Cracks of 1.250 x 1.250 mm in all Holes	4.6	350	6114	14199	8085	NSY	3	2695
2(1)	Crack of 1.25 mm in Hole 3C. Cracks of 0.125x0.125 mm in rest	9	350	15902	27267	11365	NSY	4	2841

(1) The analysis has been carried out locating the rogue flaw at each crack tip A and C of each hole from hole 1 thru 11. Therefore, for instance, the most critical scenario is number 3C, that is, the rogue flaw of 1.25 mm is located in Hole 3, Tip C.

As it was expected, the scenarios with the lower values of N_{cr} - N_{det} correspond to the ones with the same size cracks in all the locations. This is due to the big interaction among all the collinear cracks. The more cracks and longer, the more interaction exists and the quicker is the crack growth. At the same time, when the first of these cracks becomes detectable, the length of the remaining cracks is so big that the uncracked section of the panel have been reduced considerably, and therefore the residual strength capability is, at that moment, very small. As a result, the interval from detectable to critical crack is short and thus the inspection intervals calculated with these scenarios are typically small. As the probability of occurrence of this type of scenario is extremely remote, the results of this analysis have not been considered.

On the other hand, the less critical of the analyzed scenarios is 1a, the one with cracks in three adjacent holes. This can be explained in this case because the cracks of the 3 holes quickly join together in a long crack and the effect of interaction disappears, growing from that moment on as a single crack.

Finally, with respect to the standard scenario number 2, the most critical one produces a result for the inspection interval calculated with a SF = 4 that is not very different to the scenario 1c with a SF = 3. That is to say that, in case this scenarios are used with the typical scatter factor of 4, the conservatism assumed is of the same order of the one assumed with the unlikely scenario 1b or 1c as mentioned before. If this conservatism is assumed, the manufacturer might be introducing an extra cost in the maintenance program of the airplane.

However, part of the reason for these low intervals has to do with the hidden crack lengths in this location. The inspection must be made from the inside of the wing box because the external doubler covers the area to be inspected. Inside the wing-box, the hidden length of holes 3, 4, 8 and 9 is bigger because they are covered by a stringer and the crack is not visible until it appears under its foot, resulting in lower inspection intervals. A way to avoid this could be the use of an alternative inspection method that is able to detect the cracks under the strap, for example Angular Ultrasounds.

Using the standard scenario number 2 has the advantage for the manufacturer that it can use the same scenario for both, the standard damage tolerance assessment and the WFD assessment with the appropriate Scatter Factors in each case. For all these reasons the results of this scenario are chosen to set the deterministic I_{WFD}.

STOCHASTIC ANALYSIS

The stochastic method

The deterministic approach has been found very useful and adequate to keep a high level of safety for operational aircraft since the adoption of the damage tolerance assessment requirements in mid 70s and it can be easily applied to the design process of new aircraft structure. However, the stochastic nature of fatigue crack growth continues to be a well-identified problem. The uncertainty of material properties, loading, manufacturing quality or inspection capabilities often causes difficulties in providing an appropriate and economical structural maintenance plan to maintain reliability. This variation is especially problematic in older aircraft where, in addition, the higher probability to develop WFD must be considered. This leads to the need of considering more critical scenarios and higher scatter factors that, in the end, turns into lower inspection intervals and higher maintenance costs.

This makes the probabilistic approach a potential additional tool to assist and determine more economical and effective inspection plans whilst maintaining a high level of safety. This method replaces scatter factors by probability density distributions of the material properties, initial crack length, detectable crack length and any other variable affecting the damage tolerance assessment.

According to the classification established by R. Rackwitz [13] the stochastic method followed in this work consists of 5 simplified steps:

1. **Identification of the problem:**
   - *Modes of failure:* crack propagation and residual strength (damage tolerance, for repetitive inspection). *Target reliability:* The target reliability is interpreted here as the complement of the probability of failure accepted for the structure. Reference [26] chapter 7 gathers the conclusions of the investigations of several researchers and the content of several airworthiness standards regarding the acceptable level of fatigue failure in aircraft structures. According to this, a usual acceptable level of fatigue failure that produces the loss of the whole aircraft is 10^{-7} per flight in civil aircraft and 10^{-6} in military aircraft. The 10^{-3} cumulative probability of failure of an aircraft (DSG in the order of 10^4) specified in DEFSTAN-970 [14] produces similar instantaneous risk levels as the 10^{-7} proposed by these authors. But, what should be an acceptable safety level for each location when only the total probability of failure of the whole airplane is specified. Lincoln suggested that the aircraft should be zoned in approximately 10 critical zones, each one with a probability of failure of 10^{-8} (10^{-1} x 10^{-7}). Thus it seems a reasonable guide to set an acceptable probability of failure for any critical location as around 10^{-8} for civil aircraft and 10^{-7} for military aircraft. Therefore an acceptable cumulative probability of failure for one critical location could be 10^{-4} for civil aircraft and 10^{-3} for military aircraft. This means a fail probability of 0.1% and a safe probability of 99.9% for one critical location.

2. **Choice of model for problem representation.**
   - *Failure criteria:* For crack propagation failure is produced when $a = a_{crit}$. (Minimum among NSY, RCP & R-Curve criterion)
   - *Analytical model:* The same model as used in the corresponding deterministic analysis. In this case, Marathon/MULTIGRIETA program is used based on linear elastic fracture mechanics, using da/dN vs. stress intensity factor curves and modified generalized Willenborg model for crack retardation effects.

3. **Uncertainty model:** The uncertainties considered in this job are:
   - *Material properties and residual strength (da/dN, K_c, K_{1c}, F_{ty})*
   - *Manufacturing quality (manufacturing tolerances, Equivalent Initial Flaw Size)*
   - *Detection crack size (POD curve)*

   The variables related to the load and stress spectrum have not been considered due to the absence of real monitored data that could have been transformed in a probabilistic variable to consider in the process. Therefore only a partial comparison with the deterministic scatter factor can be made with the present analysis. Further details of the probability distributions are given in the following sections.

4. **Limit state function g(X):** The limit-state function shall be defined in accordance with the chosen failure criterion. It shall be positive when the structural component is safe with respect to the failure mode.

   $$g(X) = I_{WFD}(X) - I_{WFD\ aim} \qquad (2)$$

   where $I_{WFD}(X)$ is the probabilistic inspection interval, defined as the crack growth period from detectable to critical size, obtained for the vector of random variables X and $I_{WFD\ aim}$ is the inspection interval for the target reliability.

5. **Model updating:** Once the probabilistic model has been established with all the previous information presented, an updating process has to be performed to assure its efficiency. This updating would be divided in two aspects, monitoring and calibration. The model updating is not in the scope of this work.
   - Monitoring: Monitoring refers to the verification that all the assumed probability distributions are adequate, through fabrication quality controls, inspections, in-service usage monitoring, etc.
   - Calibration: Calibration refers to the verification that the overall probabilistic approach used gives the adequate level of safety or reliability. The probabilistic model should be calibrated using a reference structure design for which adequate reliability has been demonstrated, in order to assure that the most relevant variables and distribution functions have been taken into account.

The general formulation for computing the probability of failure is:

$$P_f = \int_{g(x)\leq 0} f(x)dx \qquad (3)$$

The reliability method used to perform the probabilistic analysis within this work is direct Monte-Carlo simulation. Monte Carlo simulation consists of drawing samples of the basic variables X according to their probabilistic distributions and then feeding them into the performance function, therefore, equation (3) turns into the following expression when it is implemented using direct Monte Carlo simulation:

$$P_f = \frac{1}{N}\sum_{i=1}^{N} I[g(X_i)] \quad (4)$$

where:

N: number of simulations
Xi: is the i-th simulation of the vector of random variables
g(X)]: is the limit state function
I[g(X)]: = 0 if g(X) > 0 (safe)
= 1 if g(X) ≤ 0 (failure).

Probabilistic variables (X)

One of the essential elements in probabilistic analysis is to identify and quantify the sources of uncertainty. This section shows the variables considered in this simulation and the probability distribution functions they have been approximated to.

- ***Equivalent initial flaw size (EIFS):*** The concept of EIFS was introduced after the works of Rudd and Gray [15] in 1977. The EIFS is an equivalent parameter that substitutes the poorly understood small cracks growth by means of a simple crack propagation event. ADS performed its own investigation [16] using fatigue test results for specific structural components. In the company research, an estimation of the initial flaw distribution was obtained based on back extrapolation techniques and in house crack growth models. From this distribution it was possible to obtain an EIFS model for crack propagation analysis represented by both Weibull and Log-Normal statistical distributions.

  In the case study in this paper, this statistical distribution of EIFS has been used independently for each tip A and C of each one of the 11 holes of the crack growth scenario. Therefore, a matrix of initial lengths of Nx22-dimension has been generated, where N is the number of cases to run in the Monte-Carlo simulation and 22 is the total number of tips A and C supposed in the problem. It has been considered as additional hypothesis that the tips B and D have the same initial crack lengths than tips A and C respectively assuming the common practice in the industry of supposing quarter circular shaped initial cracks. This methodology allows us to reflect the quality of the manufacturing process for the item evaluated.

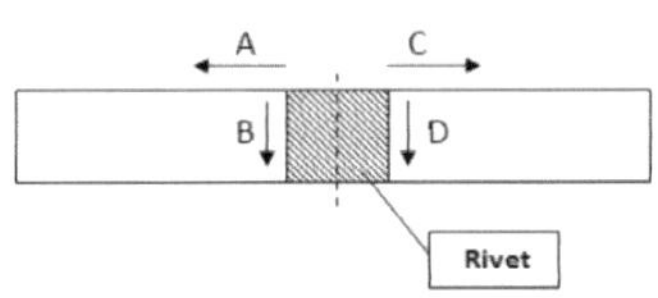

Figure 10. Crack tips identification.

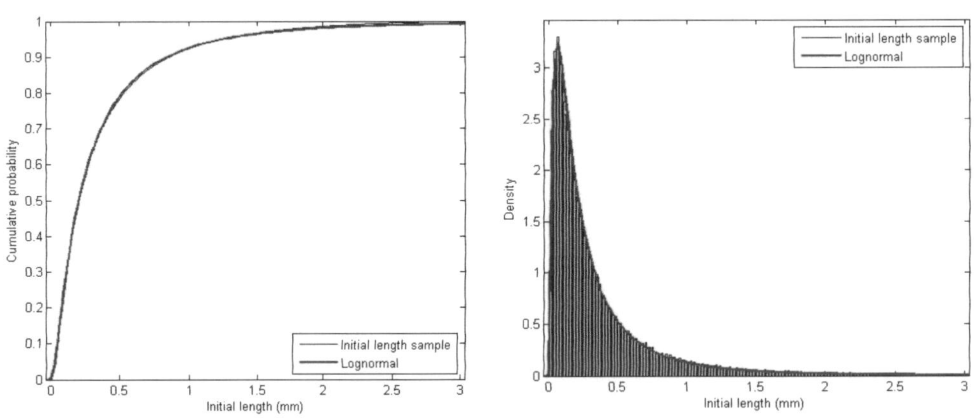

Figure 11. Initial length (mm) cumulative density function (CDF) & probability density (PDF).

- ***Inspection detection:*** Successful life management of the aging airplanes depends at a great extent on the ability to quantify non-destructive test (NDT) methods capability and reliability. A frequently used quantitative measure of the capability of a NDT method is the Probability of Detection (POD).

  The POD curve, which is a cumulative density function (CDF), is used to calculate a probability density function (PDF) for the crack lengths that inspectors are able to detect in a particular scenario and under particular inspection conditions. Therefore, the POD and the corresponding PDF is particular to a NDT method but also to all the parameters that affect the application of that NDT method to a particular scenario. Typically, the crack size 90/95 (cumulative probability of 90% with a 95% of confidence) is used to define and rate the capability of an inspection technique and is used as the nominal value of detectable crack length in a deterministic analysis.

  The NDT method chosen to inspect the case study location is High Frequency Eddy Currents (HFEC). Considering that HFEC is one of the NDT methods whose capabilities are best known, a POD curve obtained from [17] for a typical riveted lap joint has been used in this work. According to [18] and [19], HFEC POD curves can be approximated by a lognormal cumulative distribution function. Therefore, the previous POD curve has been approximated by the lognormal CDF and the corresponding PDF shown in Figure 12.

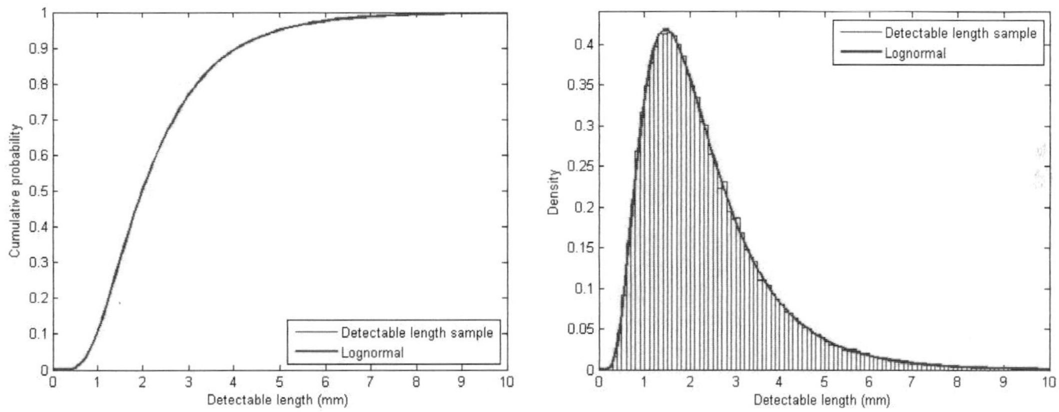

Figure 12. Detectable length (mm). POD curve and probability density function.

- ***da/dN curves:*** The following method to develop a probabilistic distribution of da/dN curves has been used in absence of something relevant in the literature. The starting point is the tabular da/dN for the Al 2024-T42 property of ADS. The area where the variability of the slope is almost depreciable is adjusted by the Paris equation that best approaches this zone and the two constants, C and n, of the equation are calculated:

$$\frac{da}{dN} = C \cdot \Delta K^n \tag{5}$$

Several authors, as for instance J.N.Yang [26], have proposed lognormal models for the crack growth. According to [22], C and n can be approximated by lognormal distributions with a Coefficient of Variation of $CoV_C = 0.44$ and $CoV_n = 0.06$ respectively and a correlation parameter of $\rho = -0.99$. Therefore it is possible to obtain a distribution of Paris curves with the above distribution of parameters C and n. Finally, one curve per curve of the distribution of Paris' curves is created, parallel to the original curve of ADS database, so that it is shifted upwards or downwards the same distance that the Paris curve is separated from the mean of the Paris curves distribution.

The following figure shows a sample of 1000 curves of the distribution obtained starting from the da/dN curve of Al2024T42 for R = -0.4.

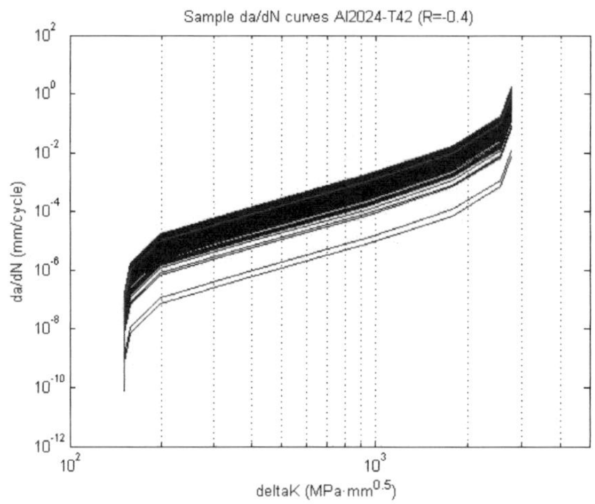

Figure 13. Variations of the da/dN curves parallel to the original tabular curve.

- ***Yield strength (Fty):*** The damage tolerance analysis within this work is based on the elastic linear fracture mechanics theory, therefore, it is considered that the material does not reach its Yield Strength. The values of F_{ty} are used to calculate the critical crack length by means of the Net Section Yield method. According to [20] the value of F_{ty} is defined using values A (F_{ty} for a cumulative probability of 99% with a 95% of confidence) or B (F_{ty} for a cumulative probability of 90% with a 95% of confidence).

  The same reference [20] and other references in the literature suggest that F_{ty} follows a lognormal distribution. Therefore it is possible to obtain a unique lognormal PDF that complies $P[F_{ty} \geq A] = 0.99$ and $P[F_{ty} \geq B] = 0.90$. Figure 14 shows the PDF and the CDF of the variable F_{ty} used in this work.

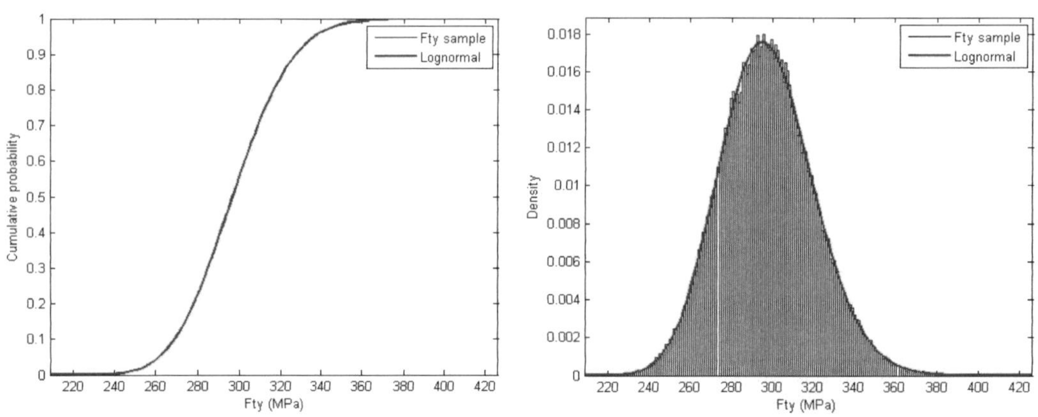

Figure 14. Fty (MPa) cumulative density function (CDF) & probability density (PDF).

- ***Kc and K_{IC}:*** The fracture toughness for the aperture mode I is used for the case study under analysis. Normal distributions with mean values equal to the deterministic values of K_C and K_{IC} (property of ADS) and a standard deviation of 10% of the mean are proposed according to [21]. However, K_C and K_{IC} cannot be varied independently because they are part of the definition of the same material property. In absence of a better approximation the variation of K_{IC} and K_C was made following a unique normal standard distribution with mean = 1 and std = 0.1 (10% of mean). Then, the distributions of K_{IC} and K_C were obtained by multiplying the normal standard by the deterministic mean of each parameter.

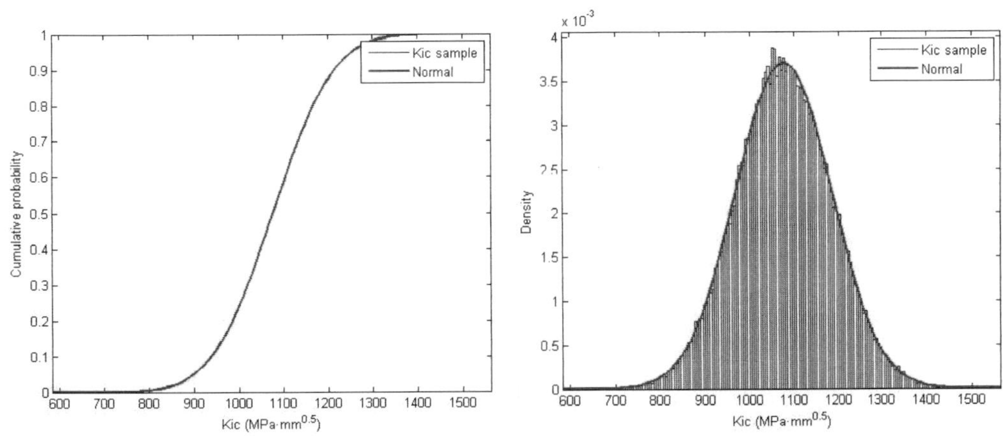

Figure 15. K_{IC} (MPa√mm) cumulative density function (CDF) & probability density (PDF).

- ***Manufacturing quality (manufacturing tolerances):*** As a result of the manufacturing process it is possible that the dimensions of the structure vary around its nominal values according to the tolerances defined in the design phase affecting the damage tolerance capability of the analysed structure. The thickness of the skin panel is considered the probabilistic variable that can affect the damage tolerance to a greater extent in this case study. Supposing that the thickness can vary the same amount around its nominal value, it is assumed that the thickness follows a standard normal distribution with mean = 1 and standard deviation calculated with the drawing tolerances.

  The thickness variation is taken into account in the crack growth algorithm through the geometrical parameter β_1. This is a factor applied to the stress spectrum that will be equals to:

$$\beta_1 = A_{gross}(t)/A_{gross\ nominal} \tag{6}$$

where A_{gross} (t) is the gross section calculated with the value of thickness (t) obtained in each case from the probability density function and $A_{gross\ nominal}$ is the gross section calculated with the nominal thickness.

The effect of the thickness variability is considered as well during the Residual Strength analysis performed with the net section yield criterion. Again, in the Net Section Yield (NSY) criterion, it is necessary to calculate the A_{gross} (t) for each case as a function of the real thickness:

$$\sigma_{NSY} = \frac{P_{lim}}{A_{uncracked}} = \frac{\sigma_{lim} \cdot A_{gross}}{A_{uncracked}} = F_{ty} \rightarrow A_{uncracked} = \frac{\sigma_{lim} \cdot A_{gross}}{F_{ty}} \tag{7}$$

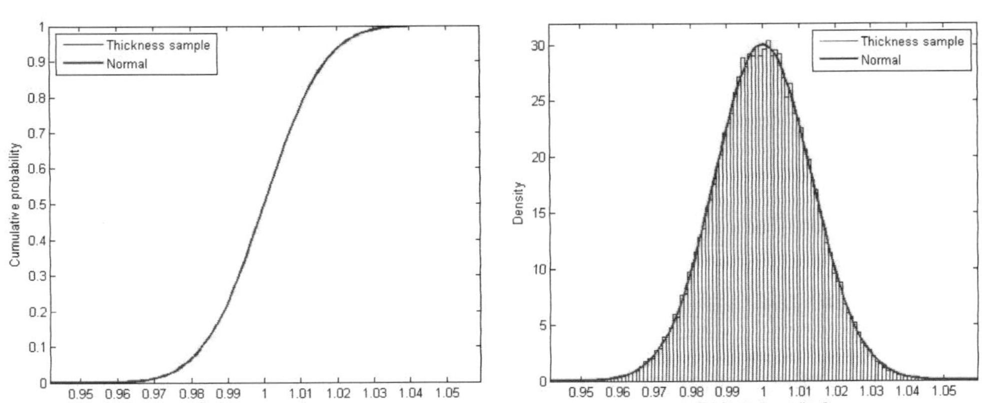

Figure 16. Thickness (normalized) cumulative density function (CDF) & probability density (PDF).

Calculation of the WFD Inspection Interval ($I_{WFD}(X)$)

Figure 17 shows the flowchart of the process followed by the probabilistic module of the program Marathon to determine automatically the $I_{WFD}(X)$ using Monte-Carlo simulations in the case study described in previous sections.

Marathon Program generates a random vector of each probabilistic variable according to the probability density function defined previously and whose parameters are introduced in the visual interface. The dimension of each vector is equals to the number of cases selected for the run in the simulation (N). Once defined the random vectors, Marathon Program gets one value of each variable vector and prepares the input to perform a deterministic run as explained in Figure 9. This process is repeated until N deterministic runs are made, one per value of the vectors of N-dimension of the probabilistic variables. The result of these N runs is a vector with N values of inspection intervals $I_{WFD}(X)$ that can be adjusted by a probabilistic distribution. The following diagram helps to understand better the process.

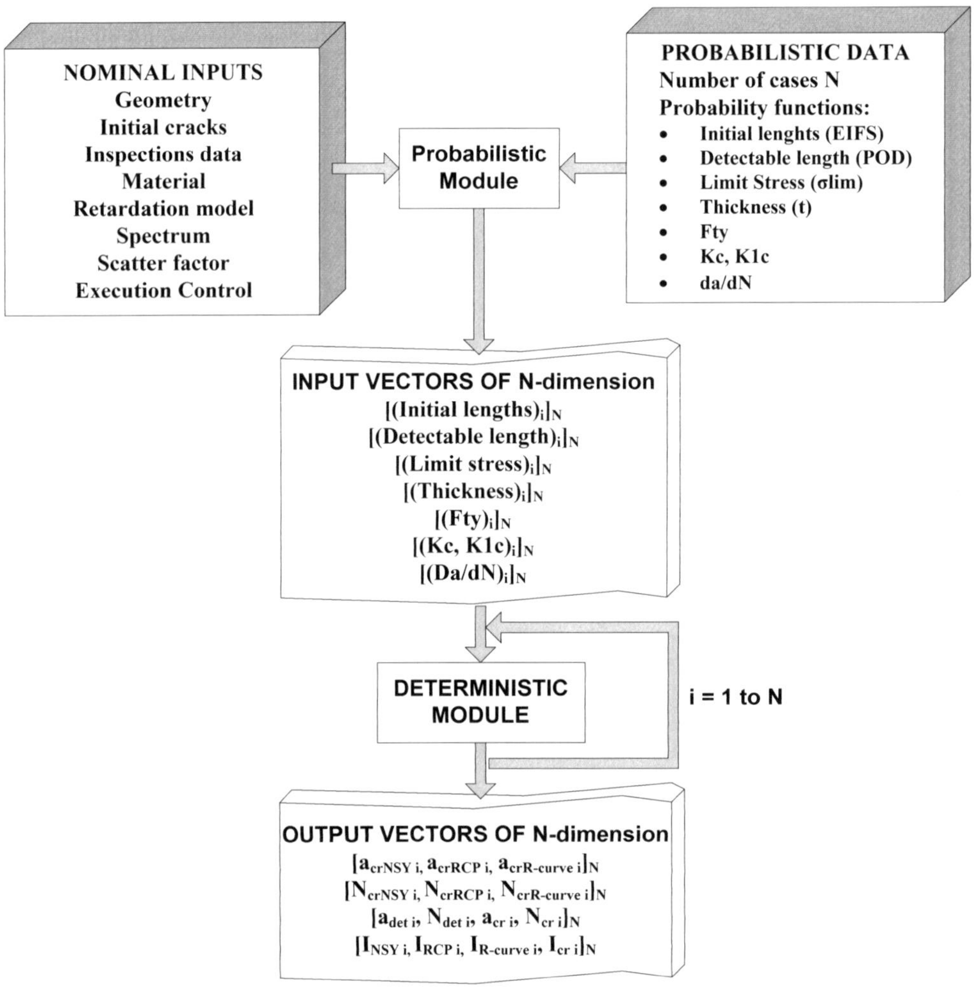

Figure 17. Flowchart of MARATHON program. Probabilistic module.

Results

The vector of inspection intervals $I_{WFD}(X)$ obtained with the Monte-Carlo simulation can be represented by a histogram of occurrences (see Figure 18 for a 45000 cases sample size). Figure 19 shows as well the Burr function that best approximates the $I_{WFD}(X)$ distribution. The Burr distribution is a continuous probability density function and is one of the called generalized log-logistic distributions.

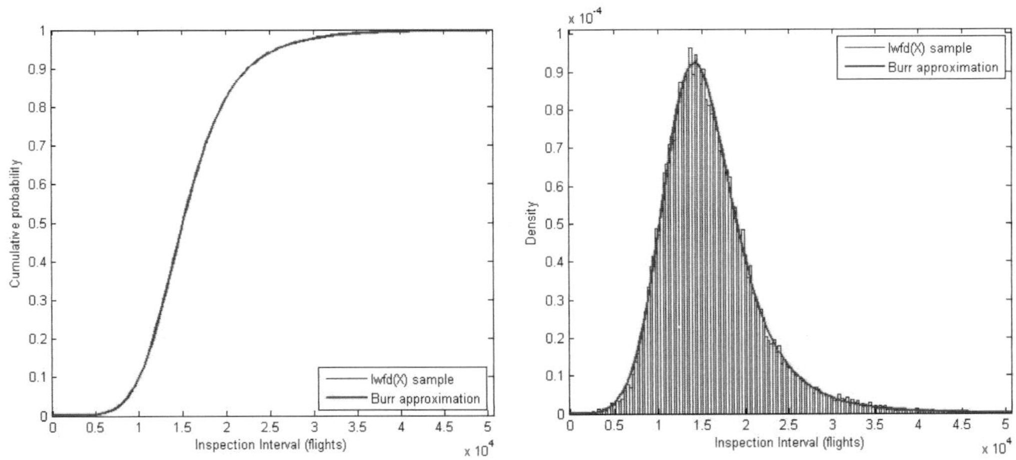

Figure 19. Minimum inspection interval. Probability density function and cumulative density function.

The approximation with a continuous distribution allows the analyst to calculate the $I_{WFD\ aim}$ for different safe cumulative probabilities. Table VI shows the results for the case study approximated with the Burr distribution.

Table VI. Stochastic analysis. Inspection interval.

Monte-Carlo simulation (45000 cases)	
Safe probability	$I_{WFD\ aim}$ (flights)
50%	15071
90%	10011
95%	8705
99%	6390
99.90%	4141
99.99%	2692

One of the most important drawbacks of Monte Carlo techniques is the amount of computer time needed for calculations due to the large sample sizes needed to get accurate results, especially in this case where a complex computer code involves a long calculation per simulation cycle. An estimation of the accuracy on the failure probability obtained with a direct Monte Carlo simulation is given in [25].

The coefficient of variation of the failure probability (p_f), denoted δ_{pf}, measures the level of accuracy of the failure probability p_f. The necessary number of samples N to achieve a prescribed coefficient of variation for a specific target failure probability can be calculated as:

$$N = \frac{1}{\delta_{pf}^2}\left(\frac{1-p_f}{p_f}\right) \qquad (8)$$

A coefficient of variation around 2-5% is usually considered acceptable. Higher values may draw the accuracy of p_f into question. For example, for a failure probability of 10^{-3} considered for the case study (safe probability of 99.9%) it is expected that around 400000 samples are needed to achieve a 5% coefficient of variation. Running such an amount of cases would take around 15 days with the current computational capabilities what is affordable but still expensive for a design office. More efficient Monte-Carlo methods as, for instance, Importance Sampling, have been developed to get better accuracy on the failure probability than direct Monte-Carlo simulation for the same sample size. Exploring an optimization algorithm based on one of these methods is a task for future works.

For the moment, several direct Monte-Carlo simulations have been carried out for different sample sizes so that it has been possible to gain accuracy in the results. The following table shows the I_{WFD} obtained for a 99.9% of safe probability for runs with different sample sizes and the related δ_{pf}. The table shows that, despite the relatively small samples with high values of δ_{pf}, the approximation obtained for I_{WFDaim} does not vary significantly and therefore, the result obtained with these samples can be considered acceptable and reliable.

Table VII. Stochastic analysis. Accuracy of the I_{WFD}.

Monte-Carlo simulation (99.9% safe probability)		
Sample Size (N)	$I_{WFD\ aim}$ (flights)	CoV (δ_{pf})
10000	4140	0.316
15000	4192	0.258
20000	4115	0.223
30000	4126	0.182
45000	4141	0.149

Sensitivity analysis

The analysis of uncertainty importance involves measuring the degree to which each random variable and its distribution parameters contribute to the uncertainty in the output limit-state function. A sensitivity analysis of the parameters affecting the crack growth is performed for the case study. The sensitivity is calculated by running Monte Carlo simulations with only one random variable at a time. That means that all the variables are set to their deterministic values except one. Results are plotted the following pie chart.

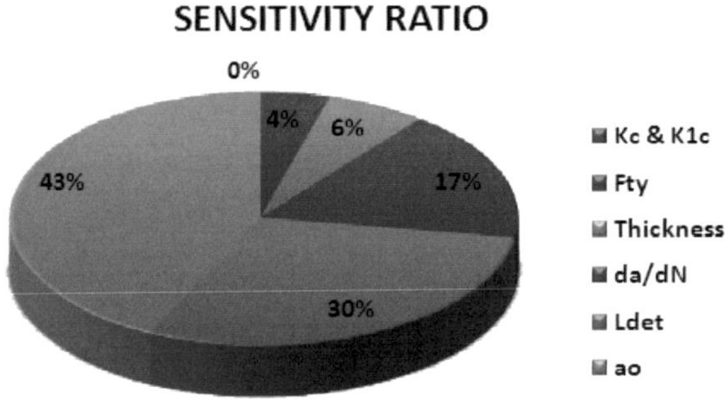

Figure 20. Crack growth sensitivity analysis (standard deviation).

Two critical parameters in any fracture analysis, deterministic or stochastic, are the size, location and distribution of initial cracks, and the ability of NDT techniques to detect a crack smaller than a certain size. As the sensitivity analysis pie chart shows, the initial crack length is the most influent parameter (43%) and the detectable crack length is the second most influent (30%).

The next most influent parameter in this case study is the da/dN curves. The absence of in house probabilistic da/dN parameters for the da/dN curves' variation has been addressed using a probabilistic model from the literature. A calibration should be done now to confirm if this model represents the actual material behaviour properly and to confirm the influence of the da/dN obtained in the sensitivity analysis. However this monitoring is out of the scope of this work.

Note that variables K_C and K_{IC} have a negligible influence in the analysis. This can be explained because the material of the skin panel of the case study has high fracture toughness and even for the lower bounds of the distributions K_C and K_{IC} are very high and the fracture criterion has been always the less critical through the whole analysis of this case study due to, precisely, the high values of fracture toughness.

Comparison between deterministic and probabilistic approach

At the beginning of this section it was said that deterministic fracture analysis uses relatively large scatter factors to account for the many inherent sources of uncertainty of the problem. Stochastic analysis methods extend the accepted deterministic methods by allowing to, explicitly, account for these uncertainties by treating them as random variables. The use of a stochastic approach can be beneficial in obtaining a quantifiable reliability.

During the deterministic analysis it was said that the typical Scatter Factor used in crack propagation analysis is 3 for the standard scenario. However, bigger Scatter Factors are usually required in the analysis of Multi-Site damage (MSD) due to the uncertainties, especially in the crack detection and in the crack growth itself due to the high interaction between cracks. That is why the Scatter Factor used in the standard scenario 2 was 4.

The uncertainties considered in the stochastic analysis in this case study come from the variability of parameters that affect the crack growth such as the material properties, the manufacturing quality data and the detection parameters. However, the variable related to the load spectrum has not been considered due to the absence of real monitored data, even though it has been demonstrated to be one of the most influencing parameters of the crack growth as shows the work of reference [23]. Therefore only a partial comparison with the deterministic scatter factor can be made with the present analysis.

According to [14], "*the inspectable life must be divided by a factor of 1.5 when loads are unmonitored*". This factor was previously applied too by Abelkis for the fatigue life cumulative probability. The internal work of reference [23] suggests scatters factors even higher due to the spectrum uncertainties (SF = 1.87). This means that a typical Scatter Factor of at least 1.5 might be added to the stochastic results to take into account the uncertainties of the spectrum and to make a more realistic comparison with deterministic results.

Table VIII shows the comparison between the I_{WFD} obtained with the deterministic analysis of scenario 2 (SF = 4) and the $I_{WFD\ aim}$ obtained with the stochastic analysis for different safe cumulative probabilities. Note that the column to be compared with is the one of $I_{WFD\ aim}/1.5$ due to the reason explained above.

Table VIII. Comparison between stochastic and deterministic results.

Monte-Carlo simulation (45000 cases)			SCENARIO 2 (SF=4)
Safe probability (s)	$I_{WFD\ aim}$ (flights)	$I_{WFD\ aim}/1.5$ (flights)	I_{WFD} (flights)
50%	15071	10047	2841
90%	10011	6674	
95%	8705	5803	
99%	6390	4260	
99.90%	4141	2761	
99.99%	2692	1795	

According to the reliability level goal that was set at the beginning of this project (99.9%) the inspection interval to be applied by the operator according to this simulation would be of 4141 flights between inspections divided by the scatter factor coming from the loads spectrum and not considered in the stochastic analysis, that is:

$$I_{WFD} = 4141/1.5 = 2761 \text{ flights}$$

Therefore, the run with a sample of 45000 cases is slightly more conservative than the deterministic approach for the same reliability goal.

CONCLUSIONS

This paper presents the methodology and tools developed to predict the crack growth and the inspection interval of potentially critical locations prone to generate WFD. It presents a comparison between a deterministic analysis and a probabilistic approach based on direct Monte-Carlo simulations. Exploring an optimization algorithm based on more efficient Monte-Carlo methods as, for instance, Importance Sampling, in order to make better use of the available resources, is a task for future works.

Anyway, the work in this paper has shown that the output of stochastic analysis is particularly useful for fleet management. The output given in terms of probability of failure provide operators with information regarding the expected chance of failure of their fleet within the next hours or flights of operation, and hence allows better management of the use of the aircraft. Once the model of all the variables were calibrated, it would be possible to get more reliable inspection intervals. This would serve to reduce maintenance costs by reducing the failure region of the distribution function obtained for the inspection interval or, in other words, by reducing the standard deviation of this distribution.

The standard deviation of this function is influenced by the standard deviation of the parameters considered in the probabilistic analysis, and this influence is expected to be proportional to their sensitivity ratio. Therefore, if the standard deviation of the probabilistic variables involved were reduced, especially in the more influent according to the sensitivity ratio, the inspection interval could be increased maintaining the same reliability level.

Reducing the standard deviation means an increment in the quality of the probabilistic variables, for example, a reduction in the manufacturing tolerances would mean a reduction in the thickness dispersion; an investment in a manufacturing quality program to reduce the number and sizes of the manufacturing defects would reduce the EIFS distribution standard deviation; the development of better equipment for HFEC inspections or the development of an inspection training program for operators could reduce the dispersion of the detectable crack length, etc.

However, it is a decision of the manufacturer if it is worth to perform an investment in all these areas to reduce the maintenance costs of their airplanes and be more competitive. The amount of that investment requires a compromise solution as usual in the industry and a stochastic analysis including cost evaluation variables would help to make a decision.

ACKNOWLEDGEMENTS

We would like to express our gratitude to D. Fernandez, head of the Methods and Tools Department, who has developed MULTIGRIETA and has offered invaluable assistance and support. We would like also to convey thanks to F. Sanchez for providing the hardware. And special thanks also to all the Fatigue and Damage Tolerance Department members for sharing their knowledge and invaluable assistance.

REFERENCES

[1] Pártl, O. and Schijve, J. (January 1992). *Multiple-site-damage in 2024-t3 alloy sheet.* Report, LR-660. Faculty of Aerospace Engineering. Delft University of Technology.

[2] Pártl, O. (August 1991). *Compilation of results associated to the problems of multiple-site-damage in open-hole sheet.* Document B2-91-04.Faculty of Aerospace Engineering. Delft University of Technology.

[3] D. P. Rooke, D. J. Cartwright, Compendium of Stress Intensity Factors, Her Majesty's Stationery Office, ISBN 0117713368, 1976, 330 p.

[4] FAR 25 § 25.571 Amdt 25-132. (15th November 2010). Damage Tolerance and Fatigue Evaluation of Structure.

[5] FAR 26 Subpart C Amdt 26-5. (15th November 2010). Aging Airplane Safety – Widespread Fatigue Damage.

[6] Safarian P. (ICAF June 2013) *Widespread Fatigue Damage Evaluation Method.*

[7] Turrel N., Vicario M. and Santgerma A. (28th November 2013) *Policy on Widespread Fatigue Damage.* Technical Report X022RP1124258. Property of Airbus.

[8] McGuire, J. and Foucault, J. (11th March 1999). *Recommendations for Regulatory Action to Prevent Widespread Fatigue Damage in the Commercial Airplane Fleet.* A report of the Airworthiness Assurance Working Group.

[9] AC 25.571-1D, (13th January 2011) *Damage Tolerance and Fatigue Evaluation of Structure.* US Department of transportation. Federal Aviation Administration.

[10] AC 120-104 (10th January 2011) *Establishing and implementing limit of validity to prevent widespread fatigue damage.* US Department of transportation. Federal Aviation Administration.

[11] Santgerma A., Vicario M., Lahorgue PY. and Turrel N., (3rd August 2007). *Approach to assess repairs susceptible to WFD.* Report X022RP0502393. Property of Airbus.

[12] Santgerma A., Vicario M., Lahorgue PY. and Turrel N., (21st September 2012). *AIRBUS WFD actions - Rationale for the Inspection Interval Methodology.* Report X022PR1247207. Property of Airbus.

[13] Rackwitz, R. (July 1976). *Practical Probabilistic Approach to Design, First Order Reliability Concepts for Design Codes.* Comite European du Beton, Bulletin D'Information, No. 112.

[14] (December 1999) *Leaflet 36. Fatigue Inspection based substantiation.* DEF STAN 00-970 Part 1 Section 3. Military Aviation Authority, Ministry of Defence, UK Government.

[15] Rudd, J.L. and Gray, T.D., (March 21-23, 1977) *Quantification of Fastener Hole Quality.* Procedures of AIAA/ASME 18th Structures, Structural Dynamics and Materials Conference, San Diego, California.

[16] D. Fernandez (2002), *Equivalent flaws size determination from fatigue test results,* NT-T-ADP-02005 Issue B. Airbus Military internal document.

[17] Rummel, W.D. and Matzkanin, G. A. (May 1996). *Non destructive Evaluation (NDE) Capabilities Data Book,* NTIAC.

[18] M. Tuttle (November 2005). *Overview of the Centre of Excellence on Advanced Materials in Transport Aircraft Structures (AMTAS).* SAMPE Fall Technical Conference.

[19] DOT/FAA/AR-96/65 (September 1996). *Visual Inspection Research Project Report on Benchmark Inspections*. U.S Department of Administration.

[20] Metallic Materials Properties Development and Standardization (MMPDS-05), FAA (2010), *Metallic Materials design data acceptable to Government procuring or certification agencies,* Battelle Memorial Institute.

[21] Ledbetter, H.M. (January 1982). *Physical properties data compilations relevant to energy storage*. Fracture and Deformation Division. Boulder, Colorado.

[22] Socie, D. *Probabilistic statistical simulations technical background.* https://www.efatigue.com/

[23] Iñesta D., Mirón E., Armijo J.I., Cabrejas J., Boluda G. (ICAF 2013). *Fatigue and damage tolerance probabilistic approach for fleets of medium and light aircraft under military transport usage.* Airbus Military Fatigue & Fracture Mechanics Department.

[24] Abelkis P., Bivins T. and Swift T. (1972) *Aircraft fatigue and Fail-Safe Analysis and Design.* Mc Donnell Douglas Corp. Structural Mechanics Subdivision.

[25] Haukaas T. *http://inrisk.sites.olt.ubc.ca/files/2012/11/Sampling3.pdf.* University of British Columbia.

[26] J. N. Yang, W, S. D. Manning, H. Hsi. (September 1985) *Stochastic Crack Propagation with applications to Durability and Damage Tolerance Analyses*. Report AFWAL-TR-85-3062. Flight Dynamics Laboratory. Air Force Wright Aeronautical Laboratories. Air Force Systems Command. Wright-Patterson Air Force Base, Ohio 45433.

Crack Growth Variability and Its Effect on Risk Analysis of Fracture Prediction

Ribelito F. Torregosa, Weiping Hu

Aerospace Division, Defence Science and Technology Organisation
506 Lorimer Street, Fishermans Bend, Victoria, Australia, 3207

Abstract: The probability distribution of crack lengths at a fatigue-critical location on an airframe is a key input for the evaluation of its risk of failure by fracture. At any given time instance, this distribution depends on the initial surface condition of the structural component, the crack growth behaviour of the material and the applied load history. This paper reports the results of a study that aims to quantify the scatter on the crack growth rates of 7075-T7351 aluminium alloy subjected to spectrum loading and to assess its effect on the prediction of probability of fracture. The test results of 85 specimens are presented and show that the time at which fracture occurred varied significantly. The variability in crack growth curves was investigated using both the log-normal and normal distribution models. The crack size distributions at five time instances were analysed and the coefficient of variation of crack sizes for different initial crack sizes was analysed to derive a relationship between the coefficient of variation and load history. The probabilistic risk analysis of fracture for different confidence levels of crack growth curves is presented based on the coefficient of variation. The outcome of this study shows that the results of risk analysis strongly depends on the confidence level of the crack growth curves used, and this needs to be taken into consideration when a master crack growth curve method is used.

INTRODUCTION

The probability distribution of crack lengths at a critical location on an airframe is a key input for the evaluation of risk of failure by fracture. At any given time during the flight history, this distribution depends on the initial surface condition of the structural component, the crack growth behaviour of the material and the applied load history. MIL-STD 1530C [1] allows the use of a single master crack growth curve to derive the subsequent crack length distributions from the initial crack length distribution, without explicitly taking into consideration the well-recognized scatter in crack growth rates due to uncertainties in material properties. Virkler et. al. [2] studied the scatter in crack growth rates of 2024-T3 aluminium alloy using centre-crack specimens subjected to constant amplitude loading. The results clearly show significant scatter in crack growth rates and different models were examined to represent the scatter. Other researchers also studied the problem under different loading conditions for different materials, e.g., [3] and [4]. Our previous work [5, 6] indicates that the distribution of initial flaw size and the crack growth law plays a major role in the results of risk analysis of fracture. When crack growth curves are variable, the assumption of a master crack growth curve for the risk analysis of fracture may underestimate the risk prediction, if the master crack growth curve only refers to the mean crack growth curve. Thus, the need to consider the variability of crack growth curve is necessary for accurate prediction of the risk of failure. In order to determine the variability of the crack growth curve to be used in the risk analysis of fracture, the Defence Science and Technology Organisation conducted a test of 85 middle-tension specimens of 7075-T7351 aluminium alloy. A variable amplitude spectrum loading block, consisting of 19032 turning points and equivalent to 100 flight hours, was used in the test.

Partial results of the testing and analysis were reported in reference [7] which showed a significant scatter of the crack growth curve that are attributable to material properties. It was observed that the standard deviation of the crack sizes increases with the increase of crack size. Since fracture occurs at higher load cycles, it was concluded that the use of a master crack growth curve, based on the mean crack growth, in risk analysis of fracture may underestimate the risk of failure prediction due to increased variability as indicated by high standard deviation of crack sizes at higher load cycles.

This paper investigates the effect of crack growth variability in the context of probabilistic risk analysis of fracture of airframe structures. To statistically quantify the crack size variability, the probability distribution of crack size corresponding to the number of load cycles was analysed. Two distribution models are used to model the distribution of crack sizes; these are the normal distribution and the log-normal distribution. The coefficient of variation (δ), which is the ratio of the standard deviation to the mean crack size, is determined for different number of load cycles. To illustrate how crack variability affects the results of the probabilistic risk analysis (PRA) of failure, the problem that was presented in reference [8] is re-analysed using the crack growth variation models developed here. Risk values were obtained for different confidence levels and compared to those reported in [8].

TEST SPECIFICATION AND RESULTS

Test specimen

The coupon test specimen configuration is shown in Figure 1. It is manufactured from 6.35 mm thick 7075-T7351 aluminium bare plate. Each specimen has a 2.0 mm hole in the centre with a 0.5 mm slot on each side of the hole. A sharp pre-crack was generated by applying a constant amplitude load with a maximum stress equalling to 70% of the peak spectrum load and a stress ratio of R = 0.1, until the pre-crack reached a length of about 2.0 mm on each side of the notch. During pre-cracking, load shedding was applied by reducing the load from 70% to 60%, in order to reduce the plasticity effects at the crack tip. The pre-crack was measured from the edge of the slot.

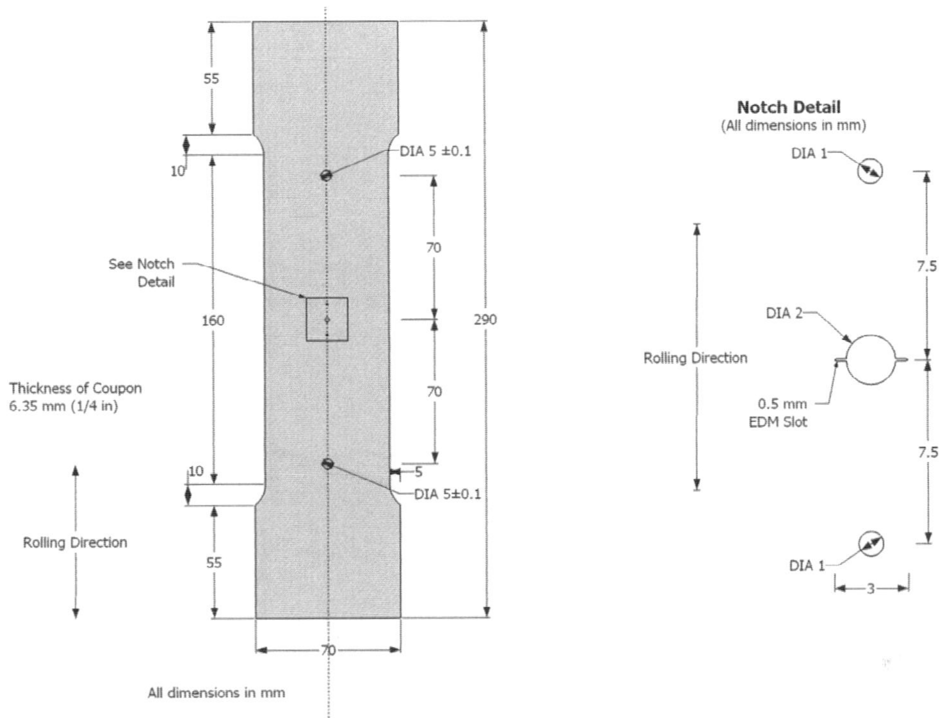

Figure 1. Coupon test specimen geometry.

Spectrum loading and crack length measurement

The Mini FALSTAFF spectrum consisting of 19032 turning points (i.e., 9516 cycles) which is called one load block was used in the specimen testing. The Mini FALSTAFF spectrum load block includes a marker band spectrum composed of 1020 turning points at the end of the load spectra to make it easier to recognise the crack growth at the end of each load block in case a quantitative fractography has to be conducted. The crack size was measured and obtained using the direct current potential drop (DCPD) technique. The crack length was measured at an interval of 1000 cycles. The test continued until the specimen failed by fracture.

Coupon test results

The actual crack growth curves obtained from the test results using DCPD technique showed some inconsistencies in the measured crack lengths. As shown in Figure 2a, at some points within the curves notably when the crack size is small (see inset of Figure 2a), the test results did not show monotonically increasing crack lengths which is not realistic. One possible reason for this is that the measurement interval of 1000 cycles is too short and that the actual difference of the crack sizes within the 1000 cycle interval is too small compared to the error in measurement. Another possibility is that measurement took place at the moment crack closure was significant. For this reason, a smoothing procedure was applied to all the crack growth curves. The equation used for the smoothing is given by the equation below[9].

$$S_j = \frac{Y_{j-3}+3Y_{j-2}+6Y_{j-1}+7Y_j+6Y_{j+1}+3Y_{j+2}+Y_{j+3}}{27} \quad (1)$$

In Eqn. (1), S_j is the jth point in the smoothed curve, Y_j the jth point in the original curve. After applying the smoothing procedure, the crack growth curves of the 85 specimens tested are shown in Figure 2b. It can be observed that the crack growth curves did not start from identical crack size. It varies from 3.38 mm to 4.111 mm. This is because the crack measurement started after the pre-crack loading procedure, which gave non-uniform crack sizes at the end of the procedure. It is emphasized here that although the testing procedure stated that the pre-crack loading is to be conducted until the crack size reaches 2.0 mm, in the actual test an exact 2.0 mm crack size is not always possible. Therefore, variability of initial crack sizes of the resulting crack growth curves is observed. To investigate the variability of crack sizes, caused by uncertainties in material properties, with respect to the number of cycles, it was necessary that cracks start from identical sizes. For this reason, one initial crack size was selected before analysing the variability. Since the crack size of 4.111 mm is the largest initial crack size of all crack growth curves results, the variability investigation started from 4.111 mm. It is emphasized here that conducting extrapolation to approximate the initial crack sizes was ruled out due to the potentially large errors resulting from the procedure. In each crack growth curve, the zero cycles is assumed to occur at 4.111 mm and the crack sizes less than 4.111 mm are not included in the crack growth curve plotting. This methodology is illustrated in Figure 3. In the figure, a1 refers to the artificial initial size of all crack growth curves, which in the case of this paper is 4.111 mm. The modified crack growth curves with artificial initial crack size of 4.111 mm are shown in Figure 4. Using this figure, statistical analysis was conducted to determine the crack growth variability.

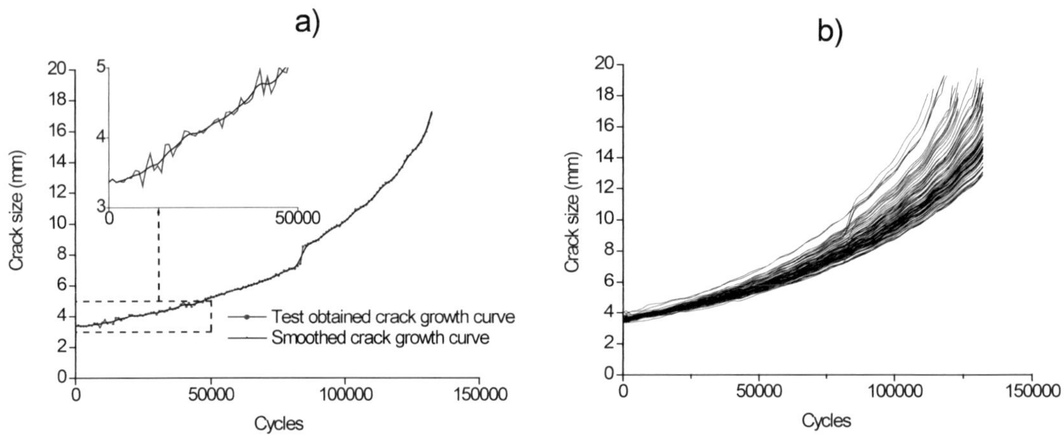

Figure 2. a) Comparison between measured crack growth curve and smoothed crack growth curve from one of the test specimens, b) Smoothed crack growth curves of 85 test specimens.

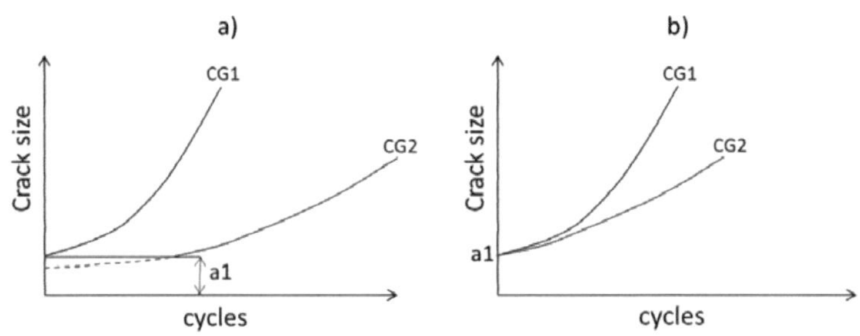

Figure 3. a) Typical crack growth curve variability observed from testing, b) Comparing the crack growth curve variability based on a single point of origin.

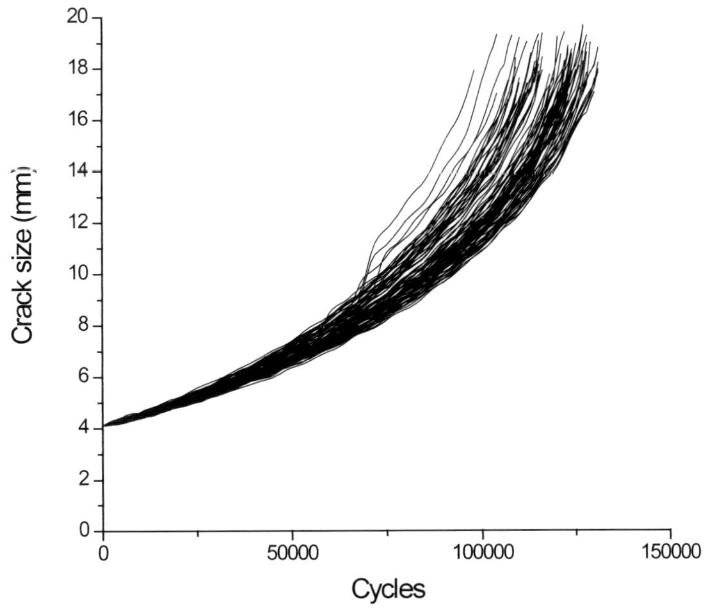

Figure 4. Crack growth curves starting from 4.111 mm.

STATISTICAL ANALYSIS OF CRACK SIZES CORRESPONDING TO NUMBER OF LOAD CYCLES

Statistical analysis of the crack size was conducted at every 2000 cycles after the crack reached 4.111 mm. In the analysis, two types of distribution models were used. These are the normal distribution and the log-normal distribution models. For each distribution model, crack growth curves corresponding to different confidence levels (CL) are shown in Figure 5. Figure 5a to 5d show the variability of crack sizes for confidence level of 50% (i.e., the mean), 95%, 99% and 99.999% CLs. It can be observed that crack growth curves using normal distribution and log-normal distribution models are nearly identical except at 99.999 CL where the latter gives higher prediction of the crack sizes with increasing load history. Clearly, log-normal distribution model gives a more conservative estimate of the two. Another observation from the figure is that crack growth curves corresponding to a given CL showed some fluctuations especially at higher load cycles and is more prominent at higher confidence levels. The fluctuation of the crack growth sizes at higher load cycles is due to the inconsistent number of data used in the statistical analysis at higher fatigue lives. Test specimens failed at different cycles, thus, the number of crack size data decreases at higher fatigue lives.

Statistical analysis of the variability of crack sizes corresponding to the number of load cycles was conducted at five load cycle points. These are at 30000, 45000, 60000, 75000, and 89000 cycles. The initial idea was to compare the variability at every 15000 cycle interval starting from 30000 cycles. However, it was noted that at 90000, one specimen had failed, thus at fifth cycle of interest (i.e., at 90000 cycles) the number of data is less than the other four. Since inconsistent data will lead to unreliable statistics of the crack growth variability, it was imperative that the number of data must be maintained at all points of comparison. Therefore, it was decided that the fifth cycle point of comparison is at 89000 cycles where failure was yet to occur in any specimen.

Variability of crack size corresponding to initial crack size

The statistical analysis shown in Figure 5 was based on the curves shown in Figure 4 which assumed an initial crack size of 4.111 mm. The assumption of 4.111 mm as the initial crack size was based on the observation that the largest initial crack size of test result is 4.111 mm. By using this value, both the number of data and the load histories that can be used in the analysis were maximised without extrapolation of test data. However, it is emphasized here that initial crack size for statistical analysis is arbitrary thus the use of artificial initial crack size is acceptable. Statistical analysis was then conducted for different initial crack sizes corresponding to 4.111 mm, 4.510 mm, 5.084 mm, and 6.029 mm. These four values were chosen because one of the crack growth curves obtained from testing has these four values thereby minimizing the number of interpolation for statistical purposes. It should be noted that when crack growth curves observed from test results were adjusted to start from a specified initial size (see Figure 4), the cycle data point at each individual curve also changed. Thus to get crack size data at the specified cycles of interest (i.e., 30000, 45000, 60000, 75000 and 89000), interpolation is necessary.

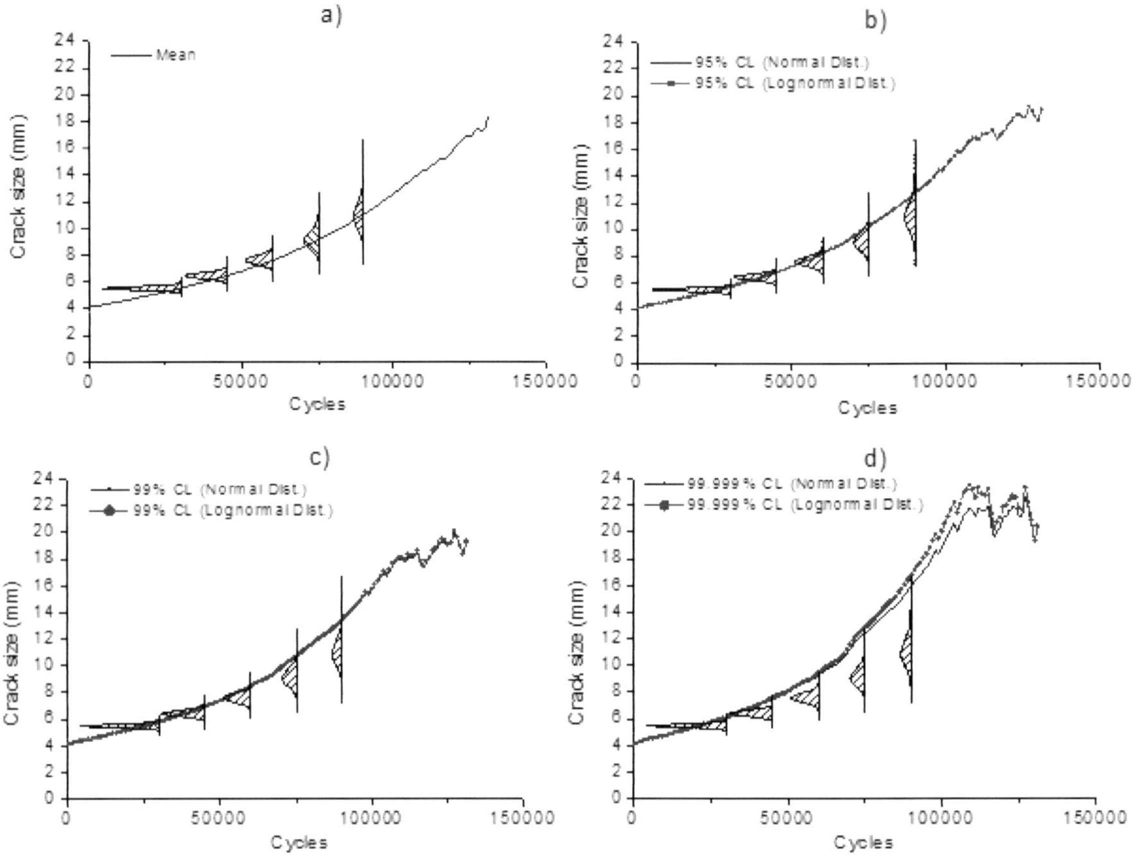

Figure 5. Variability of crack size with respect to the number of cycles, showing the crack growth curves corresponding to a confidence level (CL), a) 50% CL b) 95% CL c) 99% CL and d) 99.999% CL.

The result of the statistical analysis is given in Table I. In the table, the three statistical parameters are shown. These are the mean $\overline{a}$, the standard deviation σ_a, and the coefficient of variation δ, which is the ratio $\sigma_a/\overline{a}$. The first column gives the number of load cycles after the crack size reach a_0. In the table, for a_0 values of 4.111 mm and 4.510 mm, the statistical parameters corresponding to 30000, 45000, 60000, 75000, and 89000 are tabulated. However, for a_0 values of 5.084 mm, the statistical parameters at 89000 cycles and at 6.029 mm the statistical parameters at 75000 and 89000 cycles are not calculated. This is because when a_0 = 5.084 mm, at 89000 cycles after, there were already specimen failures that occurred thereby reducing the number of crack size data. Similarly, when a_0 = 6.029 mm, there were already failures at 75000 and 89000 cycles. As explained earlier, the authors only conducted statistical analysis where there are no specimen failures in order to maintain the same number of data. Figure 6 illustrates the evolution of the variability of crack sizes based on a specified initial crack size. It can be observed that the coefficient of variation is more correlated to the number of cycles after the initial crack size rather than to the crack size itself. As shown in Table I, at 30,000 cycles, δ is equal to 0.023 for initial crack sizes 4.111 mm, 4.510 mm and 5.084 mm respectively. At a_0 = 6.029 mm which is significantly larger than the first three a_0's, the coefficient of variation is 0.034 which is slightly higher. At 45000 cycles, δ's of the first three are still closer. At higher load cycles however, it appears that the values of δ's began to diverge, wherein higher initial sizes showing higher variability. These show that variability of crack size (i.e., δ) is highly correlated with the number of flight hours and also correlated with the size of the crack.

Table I. Variability of crack size from a specified crack size, a_0 (number of data = 85).

Cycles after a_0	a_0 = 4.111 mm			a_0 = 4.510 mm			a_0 = 5.084 mm			a_0 = 6.029 mm		
	$\bar{a}$	σ_a	δ	$\bar{a}$	σ_a	δ	$\bar{a}$	σ_a	δ	$\bar{a}$	σ_a	δ
30000	5.5	0.126	0.023	6.1	0.141	0.023	7.0	0.161	0.023	8.4	0.286	0.034
45000	6.4	0.228	0.035	7.2	0.248	0.035	8.3	0.337	0.041	10.1	0.587	0.058
60000	7.6	0.356	0.047	8.6	0.459	0.054	9.9	0.669	0.067	12.3	0.910	0.074
75000	9.1	0.688	0.075	10.3	0.815	0.079	12.1	1.019	0.085			
89000	10.9	1.000	0.092	12.4	1.252	0.101						

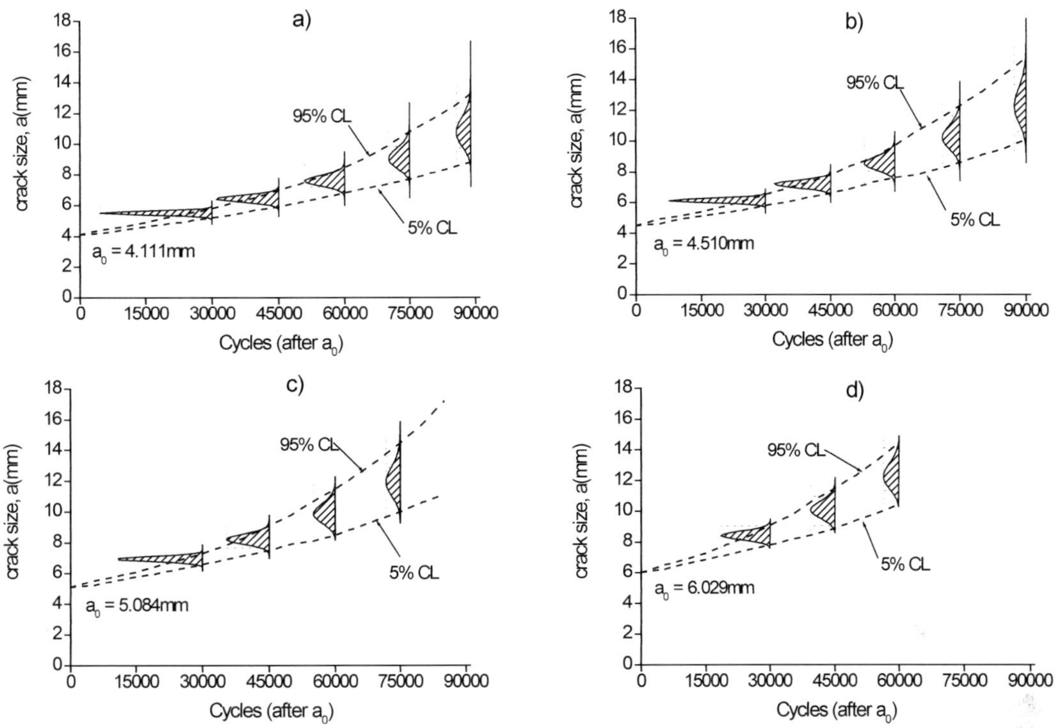

Figure 6. Crack growth variability from a specified initial crack size, a_0, of a) 4.111 mm b) 4.510 c) 5.084 mm and d) 6.029 mm.

Variability of crack size corresponding to the number of load cycles

The main purpose of determining the crack growth variability is to apply the results to the probabilistic risk analysis of fracture of airframes. The prevailing procedure of conducting probabilistic risk analysis [10] is to use a master crack growth curve which is a deterministic assumption of the crack size evolution with respect to the number of flight hours. As shown in this paper, even for identical test conditions, crack growth curves have shown variability. Therefore, by identifying the coefficient of variation of the crack size corresponding to load history, a crack growth curve for a given confidence level can be generated. Thus, probability of failure (PoF) prediction will give varying risk curves with each curve associated to a specified confidence level. This paper called this procedure a Second Order Probabilistic Risk Analysis of Failure (SOPRAF).

It was observed that crack size variability is highly correlated with the number of cycles of applied load. The coefficient of variation δ of the crack size was then analysed with respect to number of load cycles. The load spectrum used in the specimen testing was mini-FALSTAFF fighter spectrum with each load spectrum block (i.e., 19032 turning points or 9516 cycles) equivalent to 100 flight hours. Using this conversion, the coefficient of variation of crack sizes with respect to flight hours for the four initial crack sizes analysed are shown in Figure 7. The figure shows the coefficient of variations of crack sizes corresponding to different initial crack sizes (i.e., 4.111 mm, 4.510 mm, 5.084 mm, and 6.029 mm). Regression analysis was then conducted to get the equation of the mean curve. The mean curve is given by the equation

$$\delta = 0.00002t^{1.2212}, \quad (2)$$

where t is the number of flight hours in which the initial crack size a_0 has been subject to cyclic load.

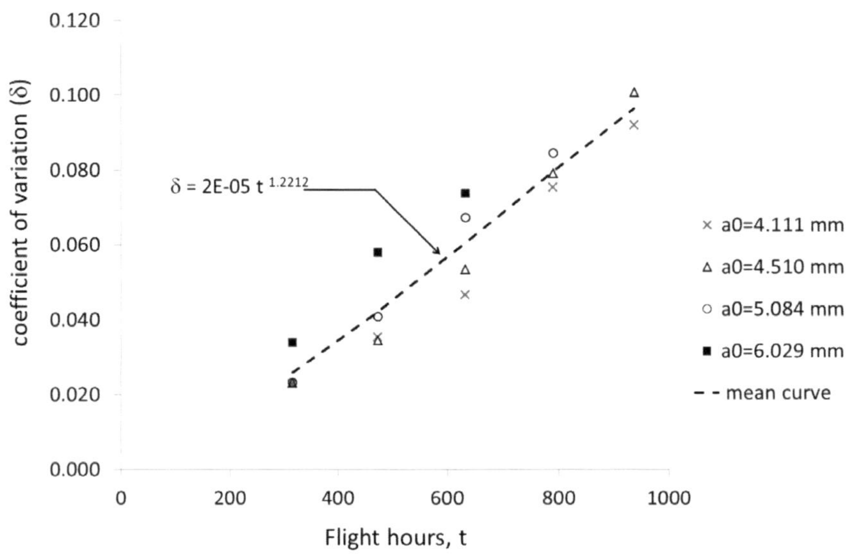

Figure 7. Coefficient of variation of crack sizes with increasing number of flight hours for various initial crack size.

APPLICATION OF THE VARIABILITY OF CRACK GRWOTH CURVES TO THE PROBABILISTIC RISK ANLAYSIS OF FRACTURE OF AIFRAMES

In conventional probabilistic risk analysis of fracture of aircraft structures [10], crack growth is assumed to be deterministic, hence a master crack growth curve is used to derive the crack size distribution at a given time, from the initial crack size distribution. This method works well if an appropriate master curve is used but the confidence level of the risk analysis prediction cannot be indicated. As shown earlier in this paper, crack growth curves vary even for identical test conditions and that variability becomes more prominent with increasing load history. To illustrate the effect of the variable crack growth curves, this paper presents a probabilistic risk analysis of fracture with additional analysis indicator, the confidence level.

Reference [8], presented a probabilistic risk analysis of fracture for F/A-18. Since the load spectra used in the testing was mini-FALSTAFF fighter spectra, the severity of stresses is more appropriate for fighter aircraft. Thus, this paper presents a risk analysis of fracture identical to the problem in reference [8] but uses a crack growth curve with a corresponding confidence level. The probability of failure (PoF) was calculated using the equation [11],

$$PoF(t) = \int_0^{\infty} f_a(a_t)\left[1 - \int_0^{S_{cr}(a)} f_s(s)ds\right] da, \quad (3)$$

where $f_a(a_t)$ is the probability density function of crack size, and the subscript t indicates the dependence of crack size on flight hours, t. $f_s(s)$ is the peak stress probability density function for a single flight. At $t = 0$, $f_a(a_0)$ refers to the equivalent initial flaw size (EIFS) distribution [12]. The peak stress distribution is given by the equation [8],

$$f(s) = 0.207e^{-0.207(s-262.2)-e^{-0.207(s-262.2)}}, \quad (4)$$

which is a Gumbel probability density function. The critical stress, S_{cr} is calculated using the equation,

$$S_{cr} = \frac{K}{\beta\sqrt{\pi a}} \quad (5)$$

where the stress intensity factor, K is assumed to be a normal distribution with mean value of 30 MPa $m^{1/2}$ and standard deviation, $\sigma_k = 2$ MPa $m^{1/2}$ given by the equation,

$$f_K(K) = \frac{1}{\sqrt{2\pi}\sigma_k} e^{\frac{-(K-m)^2}{2\sigma_k^2}} \quad (6)$$

n this analysis just the mean value of 30 MPa m^1/2 was used. The EIFS distribution is represented by a log-normal distribution published in [8] with mean $\mu = 0.0125$ mm and standard deviation of the natural logarithm of initial crack sizes σ_0 is given by $\sigma_a = 0.665$. The equation is given by,

$$f(a_0) = \frac{1}{\sqrt{2\pi}\sigma_a\sigma_0} e^{\frac{-(ln(a_0)-\mu)^2}{2\sigma_a^2}} \quad (7)$$

The master crack growth curve is obtained using the equation [8]

$$a(t) = a_0 e^{0.0003t} \quad (8)$$

where t is the flight hours and $a(t)$ is the crack size after t flight hours given the initial size at time zero is equal to a_0.

Second order probability of failure analysis

Using Eqns. (3) to (8), a second order probabilistic risk analysis of fracture was conducted. It was assumed that the crack growth curve obtained using Eqn. (8) is the mean crack size which corresponds to a 50% confidence level. Since the mean value of the crack size at a specific flight hours can be obtained from the crack size distribution, and δ of crack size can be obtained using Eqn. (2), the crack size corresponding to a given confidence level can be calculated. The analysis used the DSTO developed FracRisk program [13]. Three confidence levels were considered in the analysis. These are the 90%, 95% and 99% CLs and shown in Figure 8.

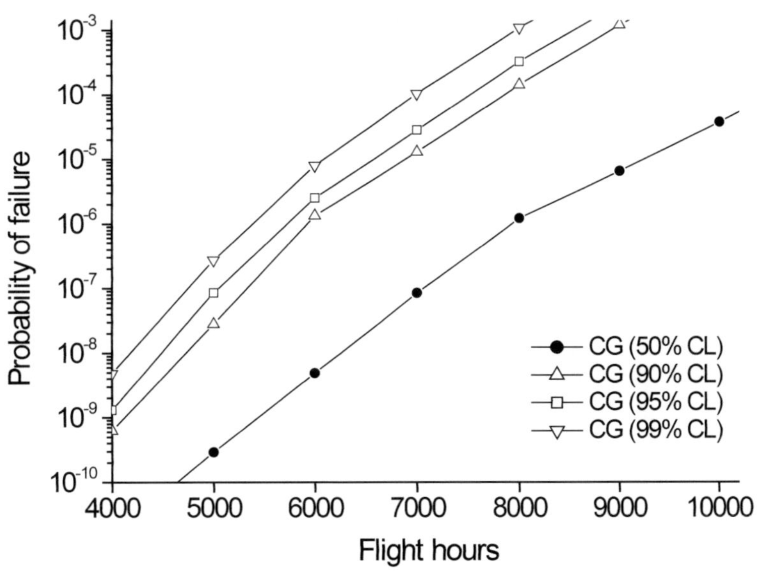

Figure 8. Probability of failure with increasing number of flight hours for each corresponding confidence level.

Discussion of results

Figure 8 shows that specifying confidence levels of a crack growth curve in the risk analysis will result in a significant difference in the predicted risk values. It is observed that increasing the confidence from 90% to 99% increases the risk prediction by approximately a magnitude. It is emphasized here that using deterministic crack growth curve without specifying its accuracy may only be construed as the mean curve from the perspective of the probabilistic nature of the crack sizes. Based on the logical interpretation by the authors, since there was no indication on the accuracy of the crack growth curve given in Eqn. (8), the most logical assumption that can be made is that it is the mean curve. As a consequence, given the observed coefficient of variation from the test results, increasing the confidence levels of the risk predictions of fracture resulted in a significant increase in risk values. The varying risk values corresponding to a confidence level shows that probability of failure itself is variable. Additional parameters, such as the confidence level is a good measure to indicate the accuracy of risk prediction. Furthermore, this shows that the prevailing procedure of assuming a deterministic crack growth is highly dependent on the accuracy of the master crack growth curve.

CONCLUSION AND RECOMMENDATION

This paper presents the results of a test program involving 85 aluminium 7075-T7351 test specimens under variable amplitude spectrum loading. The test results showed that under identical test conditions, crack growth curves still exhibit variability, which is attributable to uncertainties in material properties. The coefficient of variation was analysed and a second order probabilistic risk analysis of fracture was conducted for different confidence levels. The following conclusion may be drawn:

1. Variation of crack sizes is highly correlated with number of load cycles (i.e., flight hours) and, to a lesser degree, correlated with the initial crack size.
2. Risk analysis of fracture prediction can vary significantly depending on the confidence level of the crack growth curve used.
3. Second order probabilistic risk analyses of fracture may be performed for different confidence levels when crack growth variability data is available.

The load spectrum used in the laboratory experiment was the Mini FALSTAFF spectrum which is for fighter aircraft. Since severity of loads is different for each aircraft platform, it is desirable to conduct specimen test using different load spectra such as the load spectra for transport aircraft or other platforms. In this paper, the equation used to demonstrate crack growth variability is defined only as a function of load history. In the future a more accurate equation which includes the initial crack size in the equation will be dealt with.

Acknowledgement
he authors wish to thank Dr. Colin Pickthall of DSTO for his comments during the review of this paper.

REFERENCES

1. US Department of Defence, *MIL-STD-1530C, Aircraft Structural Integirty Program (ASIP)*. 2005.

2. Virkler, D.A., B.M. Hillberry and P.K. Geoel, *The statistical nature of fatigue crack propagation.* Journal of Engineering Materials and Technology, 1978. **101**(2): p. 148-153.

3. Wu, W.F. and C.C. Ni, *Statistical aspects of some fatigue crack growth data.* Engineering Fracture Mechanics, 2007. **74**: p. 2952-2963.

4. Kim, J.-K. and D.-S. Shim, *The variation in fatigue crack growth due to the thickness effect.* International Journal of Fatigue, 2000. **22**(7): p. 611-618.

5. Torregosa, R.F. and W. Hu, *Probabilistic risk analysis of fracture of aircraft structures using a Bayesian approach to update the distribution of the equivalent initial flaw sizes.* Fatigue & Fracture of Engineering Materials & Structures, 2013.

6. Torregosa, R.F. and W. Hu,*C-130H Centre Wing Lower Surface Panel Number 3 Probabilistic Risk Analysis*,Defence Science and Technology Organisation,DSTO-RR-0371,2011.

7. Hu, W., B. Krieg and R. Torregosa, *Material Scatter in Crack Growth Rates of 7075-T7351 under Spectrum Loading*, in *Eleventh International Fatigue Congress*. 2014: Melbourne, Australia.

8. Macheret, Y. and J. Teichman. *Improved estimation of aircraft probability of failure.* in *IEEE Aerospace Conference*. 2008. Montana, USA.

9. The University of Maryland at College Park. *A Pragmatic Introduction to Signal Processing with applications in scientific measurement.* [cited 2015 21 January 2015]; Available from: http://terpconnect.umd.edu/~toh/spectrum/Smoothing.html.

10. Hovey, P.W., A.P. Berens and J.S. Loomis,*Update of the Probability of Fracture (PROF) Computer Program for Aging Aircraft Risk Analysis*,1998.

11. Liao, M., Y. Bombardier and G. Renaud,*Quantitative Risk Assessment for the CC-130 Centre Wing Structure (Phase II)*,National Research Council Canada,2009.

12. Rudd, J., J. Yang, S. Manning and B. Yee. *Probabilistic Fracture Mechanics Analysis Method for Structural Durability*. in *55th Meeting AGARD Structures and Materials Panel*. 1982. Toronto, Canada.

13. Torregosa, R. and W. Hu, *A Methodology for the Analysis of the Probability of Failure by Fast Fracture of Aircraft Structural Components*, in *8th Structural Integrity and Fracture*. 2013: Melbourne, Australia.

Corrosion and environmental control (Management of aging fleets)

Standardized Methodology for the Investigation of Corrosion Fatigue Nucleation and Propagation

Sarah E. Galyon Dorman, Saravanan Arunachalam, Justin W. Rausch, Matthew J. Hammond, Scott A., Fawaz

SAFE Inc.

Abstract: The United States Department of Defense (DoD) estimated that the annual cost of corrosion, the unintended material degradation due to interaction with the environment, to weapon systems and infrastructure in 2010 exceeded $21 billion, and that the number was likely to continue to rise. Corrosion affects military readiness by taking critical weapon systems out of action, due to the degradation of equipment. Unfortunately, as the warfighters demand more from their systems, corrosion prevention and control is frequently traded during the acquisition cycle for weapon system performance. As a result, the DoD remains entrenched in a find-and-fix corrosion management philosophy which is expensive and unsustainable. Better standardized laboratory procedures are need to help the DoD develop (1) a fundamental understanding of corrosion damage, (2) material performance data relevant to corrosion damage, (3) prediction methodologies to help mitigate the effects of corrosion nucleated damage and (4) to develop an understanding of how corrosion preventative coatings can slow mechanical damage.

INTRODUCTION

As the United States Air Force (USAF) and the Department of Defense (DoD) extend the life of current weapon systems, corrosion damage becomes a larger concern to the structural integrity community. Corrosion damage can cause a reduction in static strength and stability capability and can include situations in which corrosion damage may act as a stress raiser and initiate fatigue cracks. At the same time the DoD has moved to phase out the use of chromates as a corrosion inhibitor due to the environmental and personnel risks [1]. Chromates are used in a variety of corrosion prevention coatings including conversion coatings and primers. While it has been documented that high levels of chromate and a chromate replacement inhibitor, molybdate, added to a bulk solution can inhibit fatigue crack propagation, it has not been shown that these inhibitors leaching from a coating can do the same [2-5]. However, if chromate does provide protection that slows corrosion fatigue crack propagation, current systems have unaccounted for corrosion fatigue protection. If that protection is not documented, then the protection will be lost in replacement chromate-free systems. At the same time, chromate is considered the benchmark for all other corrosion inhibitors to meet, so an understanding of the protection current chromate coatings provides to corrosion fatigue is needed.

Most laboratory testing methods which examine the effect of environment on mechanical damage uses samples which initiate fatigue cracks from sharp notch defects while fully immersed in an accelerating environmental (sodium chloride solution or humid air for example). Other testing has shown effects of commercial chromate coating systems on corrosion fatigue damage under acidified salt spray [6]. While these laboratory tests can provide useful relative information they can have limited applicability to service environments. During teardown analysis of several aircraft, corrosion damage was noted to initiate fatigue cracks making understanding and predicting the interaction between service environment and mechanical damage of critical importance [7]. However there are no current standardized test methodologies which account for this interaction between corrosion and mechanical damage.

In an effort to address these limitations in understanding of corrosion fatigue damage and inhibition, a large corrosion fatigue program has been undertaken to examine the effects of corrosion inhibitors on environmentally assisted fatigue as well as to expand the accepted testing methodologies to include a standardized method for looking at the pit to crack transition under environment. Under this effort research was completed examining the effect of polymer corrosion prevention coatings on environmental fatigue damage and development of a standardized test method for examining the corrosion pit and fatigue crack transition. To develop the new test protocol the following steps were required (1) document the effect of coatings using current corrosion fatigue samples, (2) validation the new sample design and procedures and (3) develop new K-solutions for the pit to crack sample. Each of these sections will be covered in detail.

EXPERIMENTAL PROCEDURE

Inhibition of Corrosion Fatigue

To determine the effect chromate or the chromate replacement inhibitor molybdate, has on corrosion fatigue damage in DoD-relevant materials, a single edge notch (SEN) specimen, shown in Figure 1, was used for all fatigue testing. The samples were made from a peak-aged, legacy age-hardenable Al-Zn-Mg-Cu aluminum alloy and temper (7075-T651). The sample was fatigue tested using a computer-controlled servo-hydraulic test frame and the crack growth rate was measured using a direct current potential drop (dcPD) system. The test load was controlled to provide a constant ΔK = 6 MPa√m with a stress ratio (R) of 0.65, the loading frequency was varied from 0.02 to 20 Hz.

Leaching Effects. All of the prior corrosion fatigue work with aluminum alloys was completed using inhibitors added to a bulk NaCl solution or deliquesced onto the surface of a sample rather than the migration of inhibitors from polymer matrix coatings [2-5]. The inhibitors used for most of these fatigue studies were high rather than low solubility inhibitors, the latter of which are typically used in corrosion prevention military aviation coatings [2-5, 8, 9]. This difference may be of importance as the leaching of inhibitors is purposely controlled in organic coatings via salt solubility in order to make the life of the coating acceptable. That said, this inhibitor leaching would be the source of any inhibitor that could affect fatigue crack propagation. All of these variables make understanding the leaching rate of known coatings and inhibitors critical to designing appropriate fatigue test criteria. An understanding of the chromate leaching rate from coatings and how chloride concentration and pH and could change the amount of chromate present in a solution was needed. To answer these questions a literature review of chromate leaching rates, inhibitor pigment solubility effects and leaching studies was performed.

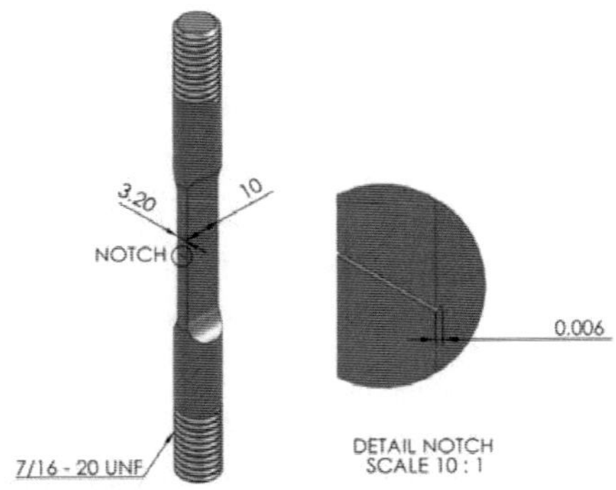

Figure 1. Single edge notch (SEN) specimens were made of 7xxx series aluminum alloy. All dimensions in millimeters.

For current commercial coatings the loading amounts of inhibitors can vary, but are designed such that the coating is not depleted of inhibitor during a maintenance cycle, typically 6 to 8 years [8-11]. Organic coatings with inhibitors are generally understood to readily absorb water or other liquid environments through inherent defects in the coating resulting in interconnected pores which allows inhibitor pigments within the coating to dissolve and dissociate in electrolyte within the coating itself [8-10]. Typically pigments with low water solubility are selected for use in coatings so that (a) osmotic pressure-induced blisters are avoided, and (b) the inhibitor remains in the coating over long times rather than rapidly dissolving out when first in contact with corrosive environments [8,9]. Because of the low solubility, only low concentrations of inhibitors would be expected in the surrounding solution before precipitation of solids occurs. For the previously completed bulk solution corrosion fatigue testing (sample fully immersed), Na_2CrO_4 (solubility in water: 87.6 g/100 mL) and Na_2MoO_4 (65.0g/100 mL) were used as inhibitors; these salts have a much higher water solubility than coating pigments ($SrCrO_4$: 0.096g/100 mL, $CaMoO_4$: 0.0011g/100 mL) [8, 9, 12]. Table I gives the properties desired for inhibitor pigments for loading into a polymer coating [8, 9].

For the bulk solution fatigue testing previously completed, Na_2CrO_4 and Na_2MoO_4 were typically used, which are salts with much higher water solubility than typical in an inhibitor pigment loaded into a coating [8,9]. Table II shows the solubility of typical coating pigments, $SrCrO_4$ and $CaMoO_4$, in water compared to those of Na_2CrO_4 and Na_2MoO_4. It should be noted that all other physical properties of Na_2CrO_4 and Na_2MoO_4 are within the acceptable limits for coating use other than solubility [8]. The high water solubility can cause blistering of the polymer and cause the inhibitor to leach from the polymer too quickly which also make them undesirable for use in coatings [8, 9].

Table I. Property requirements of inhibitor pigments [8].

Inhibitor Pigment Parameter	Desired Value
Solubility in Water	< 2g/100 mL
pH of Saturated Solution	7-9.5
Specific Gravity	1.5-5
Particle Size Distribution	2-6 μm
Solubility in Organic Medium	Practically Insoluble
Vapor Pressure at 20 C	<1/10,000 mmHg
Melting Point	>100 C

Table II. Solubility in water for typical chromate and molybdate pigments compared to Na_2CrO_4 and Na_2MoO_4 [12]. It should be noted sodium chromate is within acceptable ranges for all of the other parameters in Table I.

Inhibitor Pigment	Solubility in Water
$SrCrO_4$	0.096 g/100 mL
Na_2CrO_4	87.6 g/100 mL
$CaMoO_4$	0.0011 g/100 mL
Na_2MoO_4	65.0 g/100 mL

To examine leaching effects, four coatings containing different loading amounts of $SrCrO_4$ (either 12 or 17 weight percent) and epoxy bases were developed by Luna Innovations [13]. The leaching studies were completed by exposing the free-standing films in 100 mL of DI water for 2 and 4 days. The surface area in contact with water for each film was 29 cm^2. The leached liquid samples were then examined using UV-Vis Spectroscopy using 0.001299 M and 0.00033 M CrO_4^{2-} standards serially diluted from a 1000 ppm CrO_4^{2-} (0.0065 M) purchased standard solution (K_2CrO_4 was used for the standards) to determine the amount of CrO_4^{2-} present. The leaching data from the free films were extrapolated from the standards by the WinUV software on the Varian Cary Series spectroscope.

The results of the leaching from the $SrCrO_4$ films appear in Table III in terms of mg CrO_4^{2-}/cm^2 of coating. Table IV gives the leaching data converted into mol/L of CrO_4^{2-} to allow comparison with published results. Based on the leaching results, Film 1 was selected for fatigue crack growth testing given the high leaching amount and relatively high remaining CrO_4^{2-} remaining in the film, both desired for fatigue testing. Based on the original inhibitor loading amounts Film 3 had already leached the majority of the inhibitor making it undesirable for long immersion fatigue tests.

Table III. Leaching results for $SrCrO_4$ primer films with 29 cm^2 surface area leached into 100 mL DI water. Original leaching data obtained by UV-Vis Spectroscopy.

Film Number	Concentration (mg/cm^2) 2 Days	Concentration (mg/cm^2) 4 Days
1	0.310	0.414
2	0.241	0.310
3	0.483	0.552
4	0.310	0.379

Table IV. Leaching results for $SrCrO_4$ primer films with 29 cm^2 surface area leached into 100 mL DI water converted into millimol/L (mM) CrO_4^{2-}. Original leaching data obtained by UV-Vis Spectroscopy.

Film Number	Concentration (mM) 2 Days	Concentration (mM) 4 Days
1	0.4	0.6
2	0.3	0.4
3	0.7	0.8
4	0.4	0.5

Table V shows the expected leaching concentrations of CrO_4^{2-} based on the film leaching results for the different testing geometries used in the program. Figure 2 shows the single edge notch (SEN) sample and three different test configurations. The Bulk Testing label refers to a standard test cell filled with 500 mL of solution with a SEN sample primed on all four sides (Coating Surface Area: 7.92 cm^2). The Restricted Volume refers to a very small cell containing approximately 0.1 mL of solution covering a 2 cm high portion of the flat on the SEN sample (Coating Surface Area: 5.28 cm^2) which seeks to mimic areas such as lap joints and other occluded regions of aircraft. The Coating Surface Area to Volume ratio for the reduced volume cell was determined using aircraft components at USAFA. Thin Film refers to salt deliquesced to form a 100 µm thickness film ($8x10^{-5}$ L) onto the four flat surfaces of the SEN sample (Coating Surface Area: 7.92 cm^2). The Thin Film environment is used to mimic how atmospheric corrosion is understood to occur on aircraft structure rather than a large pool of liquid around the sample.

Table V. Expected leaching results for $SrCrO_4$ primer films converted to the test three geometries. All concentrations are in terms of mol/L (M) CrO_4^{2-}. Original leaching data obtained by UV-Vis spectroscopy.
*above the $SrCrO_4$ solubility limit of 0.0047 M CrO_4^{2-} [12].

Film Number	Expected Concentration (M) Bulk Solution (500mL) 4 Days	Expected Concentration (M) Restricted Volume (0.1 mL) 4 Days	Expected Concentration (M) Thin Film (0.08 mL) 4 Days
1	3.21 x 10^{-5}	0.021*	0.201*
2	2.41 x 10^{-5}	0.016*	0.151*
3	4.28 x 10^{-5}	0.029*	0.268*
4	2.95 x 10^{-5}	0.020*	0.184*

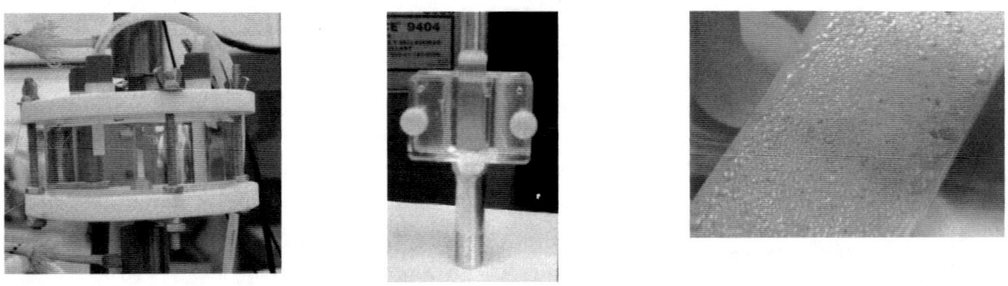

Bulk Solution Test Cell (500 mL)

Reduced Volume Test Cell (0.1 mL)

Atmospheric (Thin Film) Exposure[3]

Figure 2. Sample Volumes used for Concentration Calculations.

The leach rates from the model films were compared to that expected from published CrO_4^{-2} leaching in Table VI and Table VII [9, 10, 14, 15]. To make the comparison, the published leaching data were converted to molarity using the geometries and volumes of the Luna film leaching experiments and the testing geometries reviewed above. The Luna films were able to leach about 6-20 times more chromate into solution than the commercial epoxy primer. This result suggests that the Luna film is likely more porous than the commercial primer as the loaded salt was the same ($SrCrO_4$). Porosity and how the pores are connected in a polymer greatly affects leaching of inhibitors, so porosity differences could account for the observations [8,9]. The porosity is related to the pigment volume concentration present in a coating and the resin used in the polymer [9]. Often these parameters are not reported. However pore size, crosslink density binders and other polymer matrix information that would be helpful to understanding the mechanism behind the porosity effect were not provided [8, 9].

Table VI. Chromate leaching data from a commercial epoxy primer in different chloride solutions [9]. Original leaching data obtained by capillary ion analysis spectroscopy.
*above the $SrCrO_4$ solubility limit of 0.0047 M CrO_4^{2-} [12].

Chloride Concentration	**Expected Concentration (M) per Film Dimensions (0.1 L) 225 Hours**	**Expected Concentration (M) Bulk Solution (500 mL) 225 Hours**	**Expected Concentration (M) Restricted Volume (0.1 mL) 225 Hours**	**Expected Concentration (M) Thin Film (0.08 mL) 225 Hours**
0.1 M	8.5 x 10^{-5}	4.7 x 10^{-6}	0.003	0.029*
0.01 M	7.0 x 10^{-5}	3.8 x 10^{-6}	0.003	0.024*
0.001 M	8.5 x 10^{-5}	4.7 x 10^{-6}	0.003	0.029*
0.0001 M	10.0 x 10^{-5}	5.4 x 10^{-6}	0.004	0.034*

Table VII. Leaching data from a $SrCrO_4$ commercial epoxy primer. Original data obtained by atomic absorption spectroscopy [8].
*above the $SrCrO_4$ solubility limit of 0.0047 M CrO_4^{2-} [12].

Solution	**Expected Concentration (M) per Luna Film Dimensions (0.1 L) 50 Days**	**Expected Concentration (M) Bulk Solution (500 mL) 50 Days**	**Expected Concentration (M) Restricted Volume (0.1 mL) 50 Days**	**Expected Concentration (M) Thin Film (0.08 mL) 50 Days**
H_2O	5.1 x 10^{-5}	2.8 x 10^{-6}	0.002	0.012*
0.85 M NaCl	8.7 x 10^{-5}	4.8 x 10^{-6}	0.003	0.020*

Table VIII. Chromate leaching data from a commercial epoxy primer in 0.85 M NaCl adjusted to different pH levels [14]. Original leaching data obtained by inductively coupled plasma atomic emission spectroscopy (ICP-AES).
*above the $SrCrO_4$ solubility limit of 0.0047 M CrO_4^{2-} [12].

pH adjusted 0.85 M NaCl	**Expected Concentration (M) per Luna Film Dimensions (0.1 L) 60 Days**	**Expected Concentration (M) Bulk Solution (500 mL) 60 Days**	**Expected Concentration (M) Restricted Volume (0.1 mL) 60 Days**	**Expected Concentration (M) Thin Film (0.08 mL) 60 Days**
1	3.6 x 10^{-4}	1.9 x 10^{-5}	0.013*	0.121*
3	3.6 x 10^{-4}	1.9 x 10^{-5}	0.013*	0.121*
5	3.6 x 10^{-4}	1.9 x 10^{-5}	0.013*	0.121*
7	3.6 x 10^{-4}	1.9 x 10^{-5}	0.013*	0.121*

Table IX. Chromate leaching data from a commercial epoxy primer in 0.85 M NaCl adjusted to different pH levels [14]. Original leaching data obtained by inductively coupled plasma atomic emission spectroscopy (ICP-AES). *above the $SrCrO_4$ solubility limit of 0.0047 M CrO_4^{2-} [12].

pH adjusted 0.85 M NaCl	**Expected Concentration (M) per Luna Film Dimensions (0.1 L) 60 Days**	**Expected Concentration (M) Bulk Solution (500 mL) 60 Days**	**Expected Concentration (M) Restricted Volume (0.1 mL) 60 Days**	**Expected Concentration (M) Thin Film (0.08 mL) 60 Days**
1	8.5×10^{-4}	4.7×10^{-5}	0.031*	0.291*
3	4.1×10^{-4}	2.3×10^{-5}	0.015*	0.141*
5	3.1×10^{-4}	1.7×10^{-5}	0.011*	0.107*
7	3.6×10^{-4}	1.9×10^{-5}	0.013*	0.121*

Interestingly, in the published data on the leaching of four commercial epoxy primers in Tables VI-IX studied varied by about 16 times between the high and low leaching coatings [8, 9, 14, 15]. The loading amount of chromate for all of the coatings can vary, however as the porosity of the film can also greatly affect the leaching rates there may be some difference in the polymer filler between the two primers [8, 9]. More research is needed into how the polymer matrix changes the leaching rates.

From the data in Table VI, there is no significant difference in leaching rates with the variation of chloride content, which are all relatively low concentrations (0.1 M or less NaCl). In Table VII however the higher chloride content (0.85 M) makes a large difference, nearly doubling the amount of CrO_4^{2-} leached from the coating. This result suggests there could be a threshold of chloride content over which the chloride is detrimental to the polymer matrix allowing for more inhibitor release. Because the baseline leaching rates for the two commercial primers are different it is difficult say what the overall effect of chloride has, but some publications state that chloride content affects leaching because chloride is detrimental to the polymer matrix [8, 9].

In the primers shown in Table VIII and Table IX, the change in pH does not cause large changes in the amount of chromate leached. There is some suggestion in Table IX that more acidic environments cause more leaching for that coating, but the change in the amount of chromate leached is about the same as the scatter with NaCl concentration changes, which is less that the leaching amount changes from one epoxy primer to the next.

The University of Southern Mississippi (USM) developed three $CaMoO_4$ containing primers [16]. Each primer had a different loading amount of MoO_4^{2-} (20.8%, 28.7%, 36.9% weight percent). USM coated and scribed AA 7075 panels (Surface Area: 6.45 cm^2) prior to completing leaching experiments in 100 mL DI water. The leaching results appear in Table X along with the concentrations expected for the three test conditions. Table X shows that the higher the loading amount of MoO_4^{2-}, the lower the leaching amount. This result is highly unexpected, and the mechanism behind this result is currently not understood. USM originally thought they might be exceeding the solubility limit between the $CaMoO_4$ and solution, but that is not the case as the leaching values are always below the solubility limit of 0.05 mM [12]. Also, as the inductively coupled plasma mass spectrometry (ICP-MS) is capable of detecting 1 part per billion, it is unlikely they are having trouble measuring the molybdate leaching from the coating. The explanation for these results remains to be determined.

Table X. Leaching results for the $CaMoO_4$ coatings developed by USM. Original leaching data produced by inductively coupled plasma mass spectrometry (ICP-MS) [16].
*above solubility $CaMoO_4$ limit of 0.055mM [12].

Loading Amount	**Leaching Concentration (100 mL) 4 Days**	**Expected Concentration (M) Bulk Solution (500 mL) 4 Days**	**Expected Concentration (M) Restricted Volume 4 Days**	**Expected Concentration (M) Thin Film (0.08 mL) 4 Days**
20.8% MolyWhite	1.8×10^{-6}	4.3×10^{-7}	2.9×10^{-4}*	0.003*
28.7% MolyWhite	4.1×10^{-7}	1.0×10^{-7}	6.7×10^{-5}*	6.0×10^{-4}*
36.9% MolyWhite	4.7×10^{-7}	1.1×10^{-7}	7.7×10^{-5}	7.0×10^{-4}*

Based on the leaching results presented, the plan was to move forward with the test matrix in Table XI for all corrosion fatigue inhibition testing. This testing was to determine the effect of the corrosion inhibitors chromate and molybdate in amounts that could be leached from at a coating on fatigue crack growth rates. Another theory to be tested was the low solubility inhibitors could inhibit fatigue crack propagation by forming solids within the crack and causing crack closure.

Table XI. Test matrix to determine the effect of chromate and molybdate in leached concentrations.

Inhibitor	**Inhibitor Form**	**Environment**	**Amount (concentration in solution or wt % in coating)**
Na_2CrO_4 /Na_2MoO_4	Salt	Bulk Solution NaCl	<0.5/0.002mM
$SrCrO_4$ /$CaMoO_4$	Salt	Bulk Solution NaCl	<0.5/0.002mM
$SrCrO_4$ /$CaMoO_4$	Salt	Bulk Solution NaCl	>4.7/0.05mM
$SrCrO_4$ /$CaMoO_4$	Primer	Reduced Volume NaCl (< 0.1 mL)	17%/20.8%
$SrCrO_4$ /$CaMoO_4$	Primer	Atmospheric NaCl	17%/20.8%
$SrCrO_4$ /$CaMoO_4$	Primer	Bulk Solution NaCl	17%/20.8%

Corrosion to Fatigue Damage Standardization

To develop the methodology for examining the corrosion pit to fatigue crack transition a sample with a center hole was designed. A preferential corrosion pit is placed at the corner of the hole to act as a crack initiation location. The morphology is carefully controlled so that the surface of the pit is rough to allow for crack initiation to better replicate service corrosion damage. Figure 3 shows the pit to crack fatigue sample design.

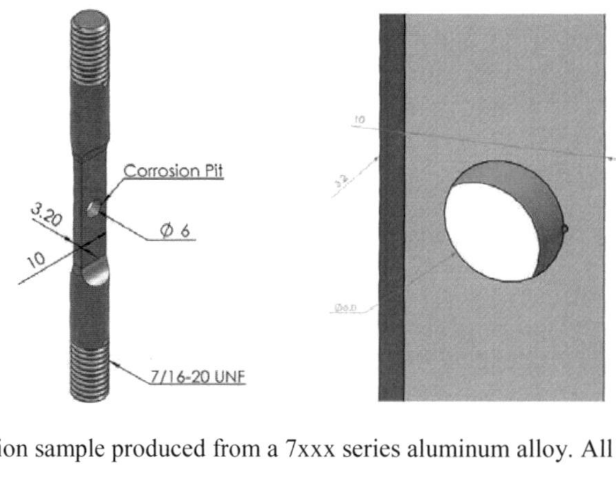

Figure 3. Pit to crack transition sample produced from a 7xxx series aluminum alloy. All dimensions are in millimeters.

To place the pit onto the sample, the entire sample other than the pit location was masked with tape or an electroplating lacquer. The sample was then placed into an agitated acidic solution and a small current, calculated from Faradaic mass loss, applied for a specific length of time [17, 18]. Once the pit was placed onto the sample platinum potential lead wires for use with the dcPD method to measure crack growth were attached by spot welding. The leads are placed equidistant on either side of the pit, with a spacing approximately 800 μm apart. The bare portion of the probe wires away from the weld joints are coated around with a liquid insulation tape so that the potential probes are free from any electrical contact with the alloy specimen [17].

Force controlled fatigue tests at an R of 0.65 were carried out at 10 Hz on a servo-hydraulic test frame until the crack achieved 1.5 mm of planar growth. During fatigue testing a 10-4-6 marker band spectrum, 8,170 cycles, was applied to validate the dcPD crack measurements, schematic Figure 4. Then the sample was pulled to failure.

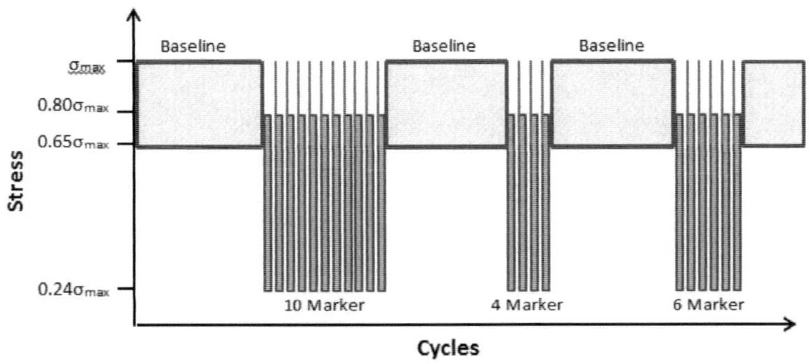

Figure 4. Marker band spectrum, 8,170 total cycles, 10-4-6 used to map crack shape development.

Post-test scanning electron microscopy (SEM) was used to analyze the fractured surface of the fatigued specimens. The nucleating feature on the pit, crack shape evolution from the marker bands and the final crack dimensions (planar c, and hole bore, a) of the specimen were all documented. Further, roughness of the electrochemically developed pit was studied using the Zygo laser interferometer as well as Confocal Laser Scanning Microscope (CLSM). A list of all specimen dimensions, pit dimensions, probe spacing, and stress conditions tested are given in Table XII.

Table XII. All specimen dimensions used for standardized test method validation.

Stress, σ_{max}, MPa	Specimen ID	Specimen Width, mm	Web Width, mm	Specimen Thickness, mm	Pit Dimensions a, mm	Pit Dimensions c, mm	a/c	Wire diameter, inches	Wire Spacing, mm	P_{max}, kN
70	S70-1	9.84	1.92	3.13	0.0984	0.100	0.987	0.003	0.512	2.2
	S70-2	9.76	1.80	2.96	0.084	0.117	0.718	0.005	0.520	2.0
	S70-3	9.98	2.00	3.13	0.121	0.091	1.330	0.003	0.458	2.2
124	S124-1	10.02	2.01	3.15	0.116	0.096	1.208	0.005	0.925	4.0
	S124-2	10.22	2.11	3.15	0.111	0.112	0.991	0.005	1.025	4.0
	S124-3	10.22	2.11	3.15	0.207	0.136	1.552	0.005	0.831	4.0
180	S180-1	10.04	2.11	3.11	0.113	0.134	0.843	0.005	0.688	5.8
	S180-2	10.06	2.08	3.12	0.084	0.109	0.771	0.005	0.838	5.7
	S180-3	10.14	2.04	3.20	0.113	0.123	0.919	0.005	0.806	5.8

K-solution development

The efforts to develop the procedures for both the inhibitor testing and pit-to-crack transition sample highlighted limitations within the existing K-solution space. An effort was undertaken to refine both for use with better models and predictions. In the case of the SEN sample, the new K-solution was needed to produce crack growth rate results that account for the shift in load line as a result of the free rotation at the pin and clevis connection.

The quantification of the efficacy of corrosion inhibitors in the pit-to-crack transition is paramount to structural integrity and damage tolerance analyses (DTA) process. One way in which the impact of inhibitors is assessed is by determining the effective changes in crack growth rates (da/dN) as a function of the Stress Intensity Factor (SIF) range (ΔK) and stress ratio (R) for very small cracks. Therefore, accurate SIFs must be known for the proposed laboratory test specimen under a wide range of crack shape scenarios. Many damage tolerance analysis codes use the readily available Newman-Raju (N-R) semi-empirical SIF corrections for finite width plates, which are singled valued and universally applied to all locations along the crack front [19-22]. Investigations into the validity of this application indicated that these correction procedures produce stress intensity values +/- 30% from the new solutions for relatively narrow plates (W/D < 6). Furthermore, the crack depth to length ratio and depth to thickness ratio can significantly influence the applicability of historical solutions and finite width corrections. The analytical investigation seeks to determine the stress intensity solutions for crack geometries outside the existing valid solution space and to capture representative crack growth behaviour of single quarter-elliptical corner cracks emanating from a centrally located hole in finite width plates under various loading conditions (remote tension, bending, and pin loading), see Figure 5.

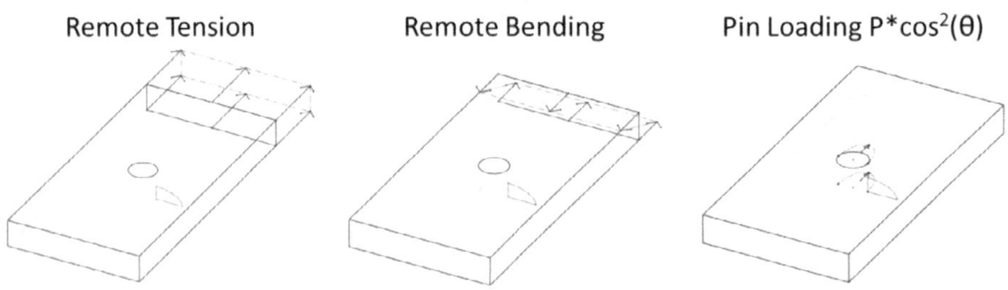

Figure 5. Load cases investigated.

Stress intensity values are generated for a range of through-thickness crack depth, a, to crack length in the transverse direction, c, ratios (or crack aspect ratios) of $0.1 \leq a/c \leq 10$, and crack depth to plate thickness, t, ratios of $0.1 \leq a/t \leq 0.99$ for the test specimen plate geometry.

Solution Process. Well-structured, fully hexahedral, finite element (FE) cracked half-plate models, similar to the one seen in Figure 6, are automatically generated and interrogated for mesh quality. Many models are used to generate a sufficiently high-fidelity solution space as to minimize interpolation error in SIF extraction at intermediate crack geometries.

The FE models are generated by a multi-step process aimed to reduce the total number of degrees-of-freedom (DoF), which is a major driver in the time to solution. Maintaining a sufficiently dense mesh to minimize the discretization and numerical error inherent in FE analyses and ensure the applicability of the virtual crack closure technique (VCCT) is also a major priority [23]. Two of the main drivers of the mesh's element size along the various crack fronts are: 1) the proximity of the crack front to free surfaces; and 2) the curvature of the crack tip. All mesh density data is defined through several MATLAB subroutines which create the input files for mesh and boundary condition application in the pre-processor, TrueGrid [24, 25]. Many other typical "rules-of-thumb" checks are performed to ensure mesh quality prior to solution submittal (e.g. element aspect ratio, skew, and Jacobian checks).

After the mesh generation the mesh geometry and boundary condition files are uploaded onto one of several DoD high performance computing (HPC) centers for execution. The FE analysis solvers ZIP3D [26] and STRIPE (developed by The Swedish Defense Research Agency) have been used in this effort.

After successful execution of the solver, the results files are interrogated in order to extract the needed crack front nodal forces and crack wake nodal displacements used in the determination of the strain energy release rate, G, and subsequently the SIF, as presented in, along the entire crack front [23].

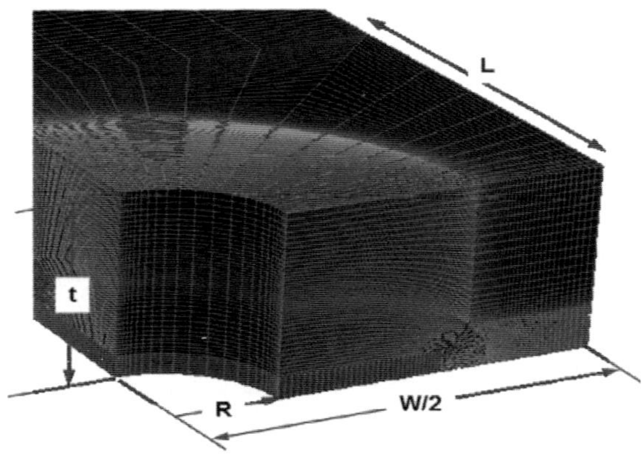

Figure 6. One quadrant of a plate FE model showing the crack plane.

RESULTS

Inhibition of Corrosion Fatigue

Testing was completed using inhibitor pigments that would be found in commercial molybdate coatings, calcium molybdate ($CaMoO_4$). All testing was completed at a constant ΔK = 6 MPa$\sqrt{}$m, R = 0.65 in a bulk (500 mL) 0.06 M NaCl solution. The results for this test are shown in Figure 7 The starting concentration of $CaMoO_4$ was 0.002 mM (yellow star) which is below the solubility limit, the concentration was then increased to 0.01 mM $CaMoO_4$ (orange star), for these concentrations the solubility limit would be expected to have been exceeded in the crack. The crack growth rates for the two concentrations of $CaMoO_4$ are in line with the pure 0.06 M NaCl showing no inhibition with the low concentrations of calcium molybdate added to a bulk solution and no inhibition due to crack closure. All other data in Figure 7 are from other published work [3].

Testing has also been completed to validate the methodology for holding the relative humidity constant for the atmospheric salt tests. The test procedure requires the deposition of salt on the surface of the sample; the salt layer is allowed to dry and then is rehydrated and held at a constant relative humidity. Precise control of the chamber relative humidity is needed to ensure the concentration of salt on the surface of the sample remains constant during the test. Baseline testing has been completed using glycerin baths from ASTM D5032, in these tests the relative humidity is held constant without the application of the salt layer [27]. Figure 8 shows the results of these baseline tests, which are in line with the full immersion crack growth rates [3].

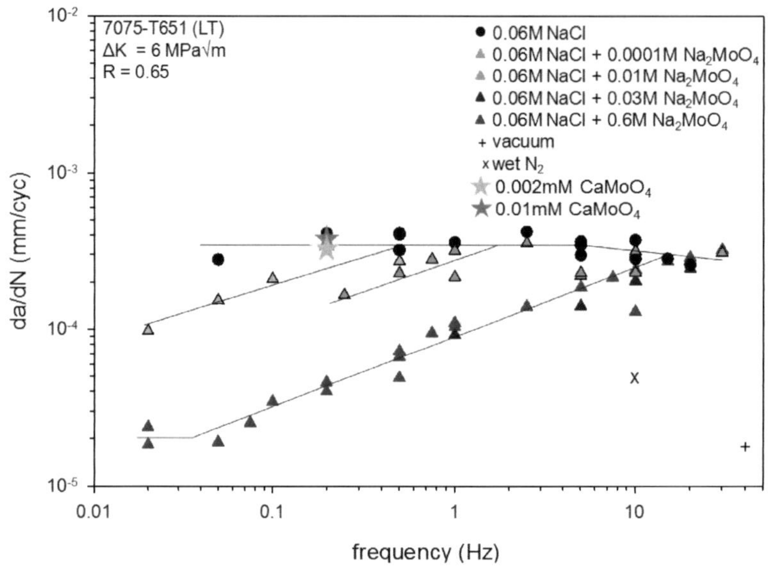

Figure 7. Measured crack growth rate curves compared from coating relevant calcium molybdate to published data for bare specimens [3].

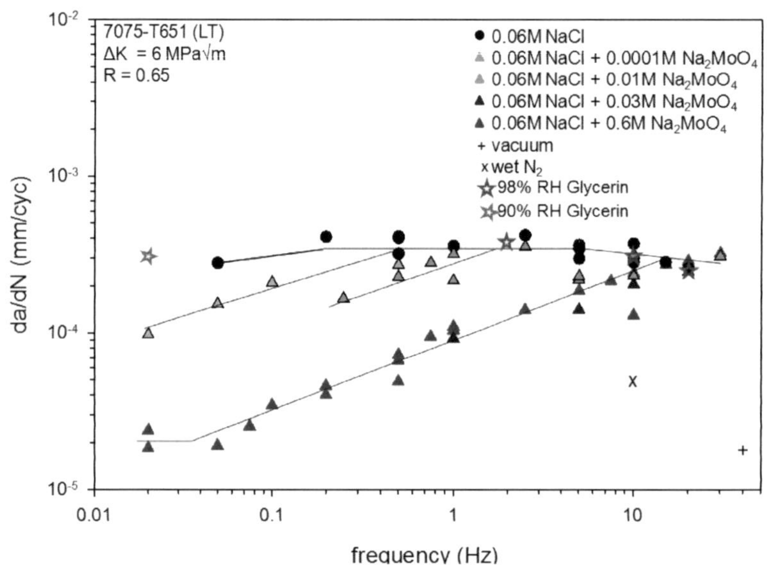

Figure 8. Measured crack growth rates for samples held at 98% relative humidity as compared to full immersion testing and the corrosion fatigue inhibitor sodium molybdate.

Pit to Fatigue Damage Standardization

The dcPD data was validated by comparing the post-test measured starter notch (corrosion pit) and final crack lengths as well as marker band dimensions to that obtained from experimental crack dimensions (from normalized dcPD voltage values vs crack dimensions, "*a*" and "*c*"). Experimental crack dimensions "*a*" and "*c*" were calculated using the Roe-Coffin potential solution [28]. The Roe-Coffin equation relates the potential measured between the spot weld wires to that of the crack length [28]. Calculated crack dimensions i.e. "*a*" and "*c*" were corrected with respect to the initial flaw dimensions (i.e. corrosion pit) and the final crack front dimension using the Eqn 1.

$$a_{corrected} = a_{init} + (a_{calculated} - a_{init})\frac{(a_{final-measured} - a_{init})}{(a_{final-calculated} - a_{init})} \tag{1}$$

Where, a_{init} and $a_{final\text{-}measured}$ refer to the dimensions measured from the corrosion pit and the final crack dimension, while $a_{calculated}$ and $a_{final\text{-}calculated}$ refer to the instantaneous and final crack dimensions obtained from the calculation procedure described above. The same procedure was carried out for all crack lengths in the *c*-direction.

Once the crack dimensions were corrected, crack growth rates for the experimental data in both directions were calculated using the incremental polynomial method outlined in the ASTM E647 [29]. The stress intensity factors K_a and K_c were calculated using Eqn 2 and Eqn 3.

$$K_a = \sigma\sqrt{(\pi a)}\beta_a(\frac{a}{c},\frac{a}{t},\frac{r}{t},\frac{c}{W}) \quad (2)$$

$$K_c = \sigma\sqrt{(\pi c)}\beta_c(\frac{a}{c},\frac{a}{t},\frac{r}{t},\frac{c}{W}) \quad (3)$$

Where, K_a, K_c are the stress intensity factors, β_a, β_c are the geometric correction factors, σ is the remote applied stress, *a*, *c* are the crack dimensions, depth and length, respectively, and *t*, *r* and *W* are geometric inputs. The β_a, β_c values were determined from AFGROW simulations using the periphery crack as the initial crack dimensions. The term periphery crack relates to the observation of the crack front of the first marker band that extends from specimen surface to hole bore, i.e. from side "*c*" to "*a*". The corner crack solutions developed by Fawaz and Andersson were used for the crack growth rate calculations [30]. The experimental crack growth rates thus calculated from this study were then compared with that of DARPA structural integrity prognosis system (SIPS) database for AA7075-T651 at *R* of 0.65.

The *da/dN vs. ΔK* curve developed under the SIPS program were generated using C(T) specimens and constant amplitude loading. The crack growth rates (*da/dN* and *dc/dN*) in the present investigation were generated using the variable amplitude marker band spectrum (variable *R*). In order to compare the present crack growth rates with those of the SIPS data, the growth rates were corrected to account for load sequence effects due to the marker band loading. The correction was carried out by adopting the plasticity induced crack closure concept and simple Elber equation [31]. Elber postulated that the residual plastic tensile deformation left in the wake of the crack extension creates compressive residual stresses thus reducing crack opening and resulting in crack closure during unloading. He proposed that the crack tip stress intensity factor range, *ΔK*, would be modified by crack closure by a factor *U*, shown in Eqn 4.

$$U = (\sigma_{max} - \sigma_{op})/\Delta\sigma = \Delta\sigma_{eff}/\Delta\sigma \quad (4)$$

Where, σ_{op} is the lowest stress at which the crack is completely open, σ_{max} is the maximum remote stress applied to the specimen and $\Delta\sigma = \sigma_{max}*(1-R)$. The Elber equation for *U* is given by $U = 0.5 + 0.4R$, where *R* is the stress ratio [31].

A decrease or reduction in the crack growth rate during the test is represented in Eqn 5 [32].

$$\frac{\left(\frac{da}{dN}\right)_{marker}}{\left(\frac{da}{dN}\right)_{baseline}} = \left[\frac{(\Delta\sigma_{eff})_{marker}}{(\Delta\sigma_{eff})_{baseline}}\right]^m \quad (5)$$

In essence, the ratio of crack growth rates during marker cycles to those of the baseline cycles is the factor by which the marker band cycles are replaced in the calculation with those of the baseline cycles. For example; if the ratio is 0.10, then every block of 100 marker band load cycles are replaced by 10 baseline cycles. The crack growth rates in both directions (*da/dN* and *dc/dN*) for all the stress condition were corrected using the above mentioned procedure and then compared with those of the SIPS data.

Fatigue data analysis. Table XIII gives a summary of the fatigue test analysis for all stress conditions tested. Figure 9(a-c) shows the normalized dcPD (V/V_0) signal for the 70, 124 and 180 MPa specimens tested at *R* of 0.65, 10Hz. In Figure 9(a) a little scatter in dcPD values is observed around 700,000 cycles which is due to stopping and starting the test. However, from the post-test fracture surface analysis, all plots can be divided into 3 stages based on the crack evolution characteristics as shown in Figure 10. Stage I corresponds to early disturbances observed in the dcPD signal that most likely corresponds to crack nucleation. Stage II corresponds to growth of the corrosion pit nucleated crack to a full periphery crack, where, the dcPD signal shows a steady rise or transient behavior. Stage III corresponds to growth of the periphery crack. A steady rise in the dcPD voltage is indicative of the crack becoming a fully formed (continuous crack front from the plate surface to hole bore) fatigue crack.

Table XIII. Summary of fatigue analysis for all stress conditions tested.

Specimen ID	Norm dcPD at Nucleation	Cycles to Nucleation	Periphery Crack Dimension		Norm dcPD at Periphery Crack	Cycles to Periphery Crack	Total Cycles	%Life for Nucleation	%Life for Periphery
			a, mm	c, mm					
S70-1	1.00036	728,658	0.107	0.161	1.00366	782,480	906,080	80	86
S70-2	1.00025	634,921	0.245	0.134	1.02480	810,072	930,582	68	87
S70-3	1.00009	1,881,617	0.126	0.196	1.00413	1,943,417	2,097,917	90	93
S124-1	1.00004	9,406	0.259	0.213	1.01832	31,709	60,399	15	52
S124-2	1.00002	8,608	0.249	0.119	1.00190	25,413	50,523	17	50
S124-3	1.00023	14,521	0.445	0.276	1.01850	27,539	48,348	30	56
S180-1	1.00018	225	0.369	0.279	1.00610	7,725	13,101	1.7	58
S180-2	1.00013	370	0.253	0.121	1.00520	7,210	16,012	2.3	45
S180-3	1.00034	138	0.473	0.638	1.01520	10,815	18,216	0.8	59

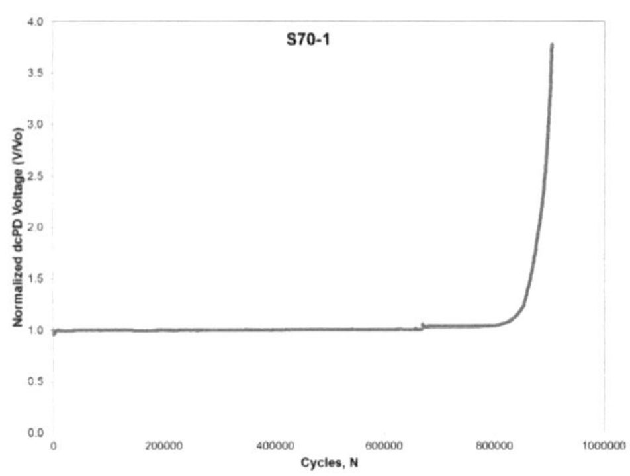

(a)

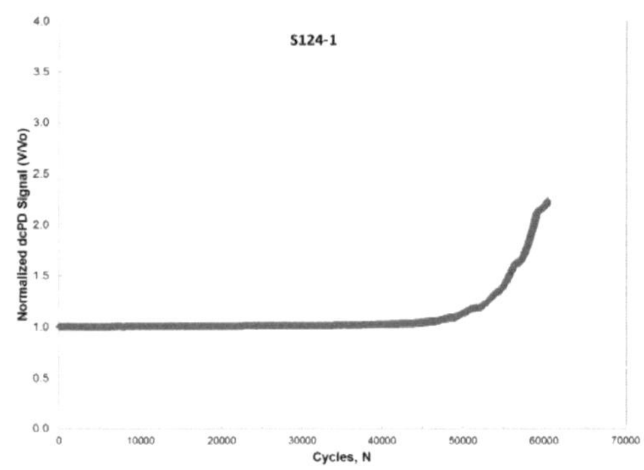

(b)

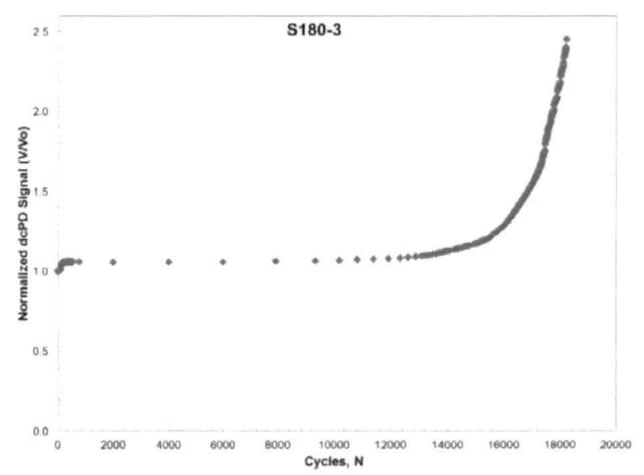

(c)

Figure 9. Normalized direct current potential drop (dcPD) voltage signal as function of cycles for (a) S70-1, (b) S124-1 and (c) S180-3 specimens.

For the low stress condition, 70 MPa, the average life for crack nucleation is about 80% of the total life, and, another 10% of the total life is required for the nucleated crack to grow into a periphery crack. In other words, about 90% of the total life is spent on growth of the mechanically small crack to a periphery crack. On the other hand, for the high stress condition (180 MPa), the crack nucleation starts as early as less than 1% of the total life, and for all practical purposes it can be considered that crack nucleation occurs at the beginning of the test.

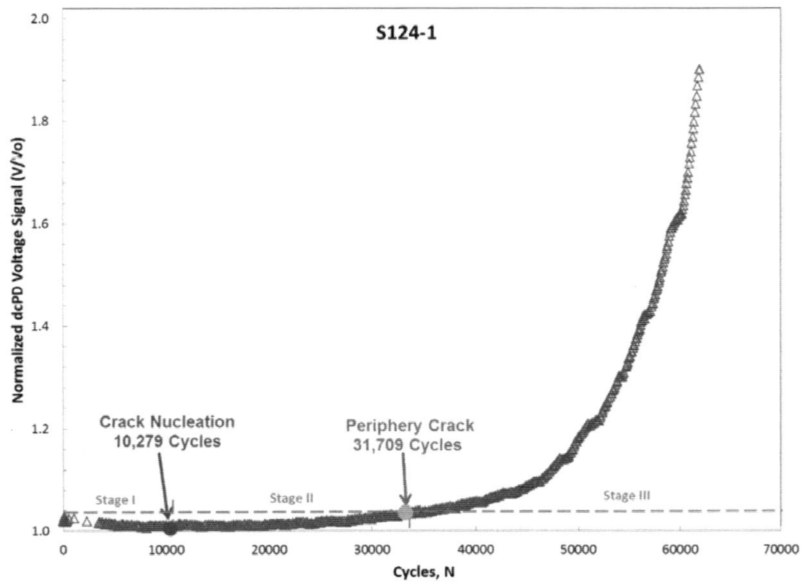

Figure 10. Identification of periphery crack formation from the raw dcPD voltage and 3 stages of crack formation for the S124-1 specimen at R of 0.65, 10 Hz frequency.

From this study, it is difficult to draw a single threshold normalized dcPD signal value for crack nucleation or periphery crack formation for all stress conditions. This can be understood based on the effect of stress on crack nucleation and growth. In the case of the low stress condition, the crack will nucleate from the high stress intensity location (micro stress concentration) and this depends purely on the corrosion pit profile and geometry. Similarly, for a high stress condition, the crack will nucleate at all possible locations. Thus for these two extreme stress conditions, crack nucleation will have some scatter reflected in the normalized dcPD signal. Irrespective of the initial pit dimensions, stress level and stress ratio, crack nucleation occurs before V/Vo is 1.00036 and the periphery crack is already formed at $V/Vo > 1.02$. This work also suggests that possibly the crack nucleation for all stress conditions can be determined by increasing the sensitivity of dcPD signal by reducing the probe spacing.

In general, the dcPD calculated crack growth in both the "*a*" and "*c*" directions as a function of cycles shows a good correlation to those measured from the marker band measurements for all stress conditions tested. A representative plot of the calculated crack growth and measured are shown in Figure 11(a-c) for the specimens tested at 70, 124 and 180 MPa, $R = 0.65$ at 10 Hz. It should be noted here that for the 70 MPa stress level, a minimum of 40 marker bands were traced back from the outermost crack front. A detailed crack front profile measured for the 70 MPa specimen S70-1 is shown in Figure 12.

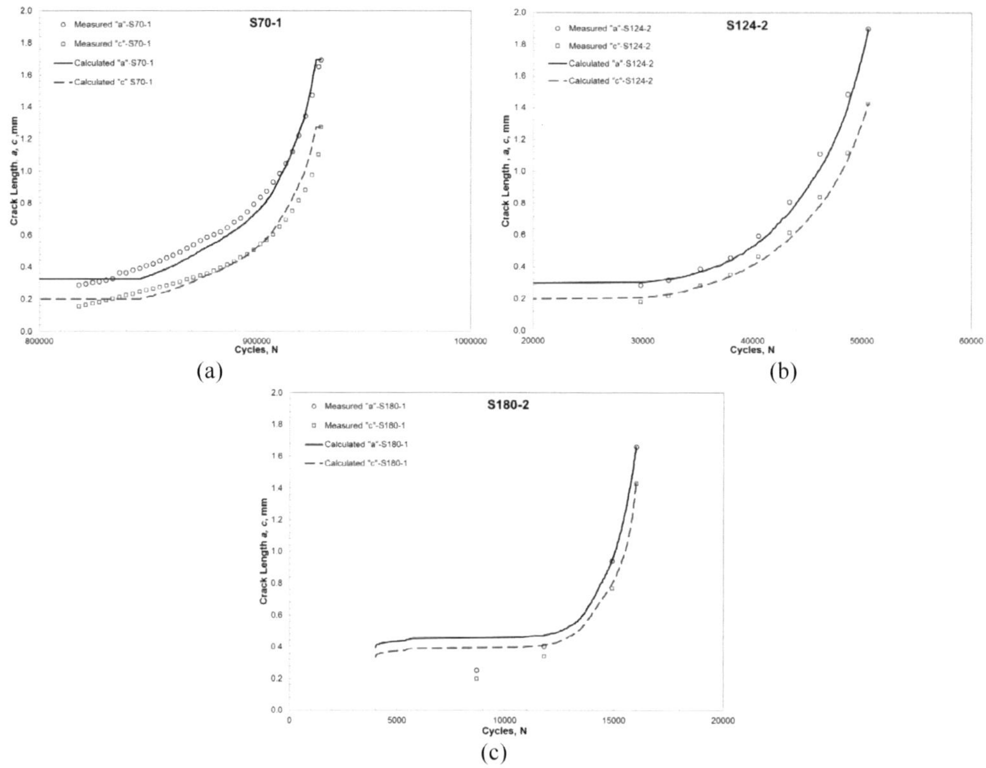

Figure 11. Calculated and measured crack dimensions for (a) S70-1, (b) S124-2 and (c) S180-2 specimens.

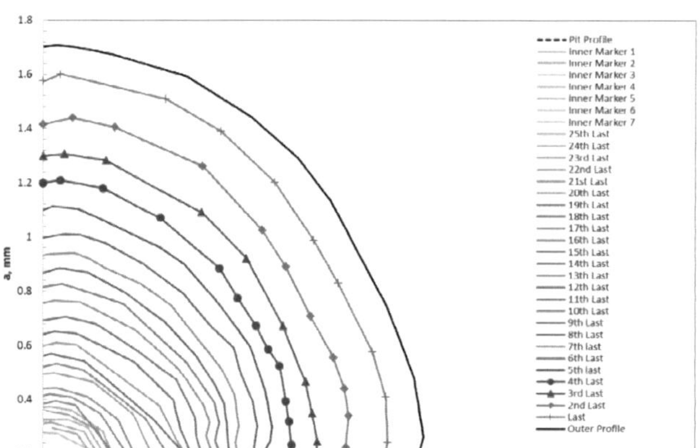

Figure 12. Crack front profile measured from marker band analysis for the specimen S70-1 tested 70 MPa, R of 0.65 and at 10 Hz.

Crack growth rates were calculated from the measured crack dimensions using the incremental polynomial method; ΔK_a and ΔK_c were calculated from the measured crack dimensions, corresponding remote stress, and β_a, β_c values from the Fawaz/Andersson solutions implemented in AFGROW. The ΔK stress intensity values in both directions ranges from 1.6 to 12 MPa√m for all the stress condition tested. The stress intensity range 1 to 4.0 MPa√m corresponds to the near threshold region for AA7075 [32-34]. Calculated crack growth rates in both directions ("a" and "c") from this present study are comparable and in good agreement with SIPS data over the stress intensity ranges tested. Corrected crack growth rates using the Elber equation shows a net influence of slightly shifting up, indicating a marginally higher growth rates as shown in Figure 13 (a) and (b) for the "c" direction. As expected, the calculated as well as corrected growth rates correspond well with SIPS data. Crack growth rate discrepancies near the threshold regime can be understood from the fact the determination of crack growth rates near threshold ΔK is largely affected by several factors including the type of test. For example, Wanhill reported large scatter and higher crack growth rates near the threshold for the data collected from small cracks in AA2024-T3 sheet samples compared to rates acquired from large cracks [35]. Similarly, in the present study the test was carried out in a force controlled mode with ΔK continuously increasing. Furthermore, variation in the crack nucleation location along the corrosion pit will result in variation in the crack growth rates for small cracks. This variation is understood qualitatively but is not the focus of the present work.

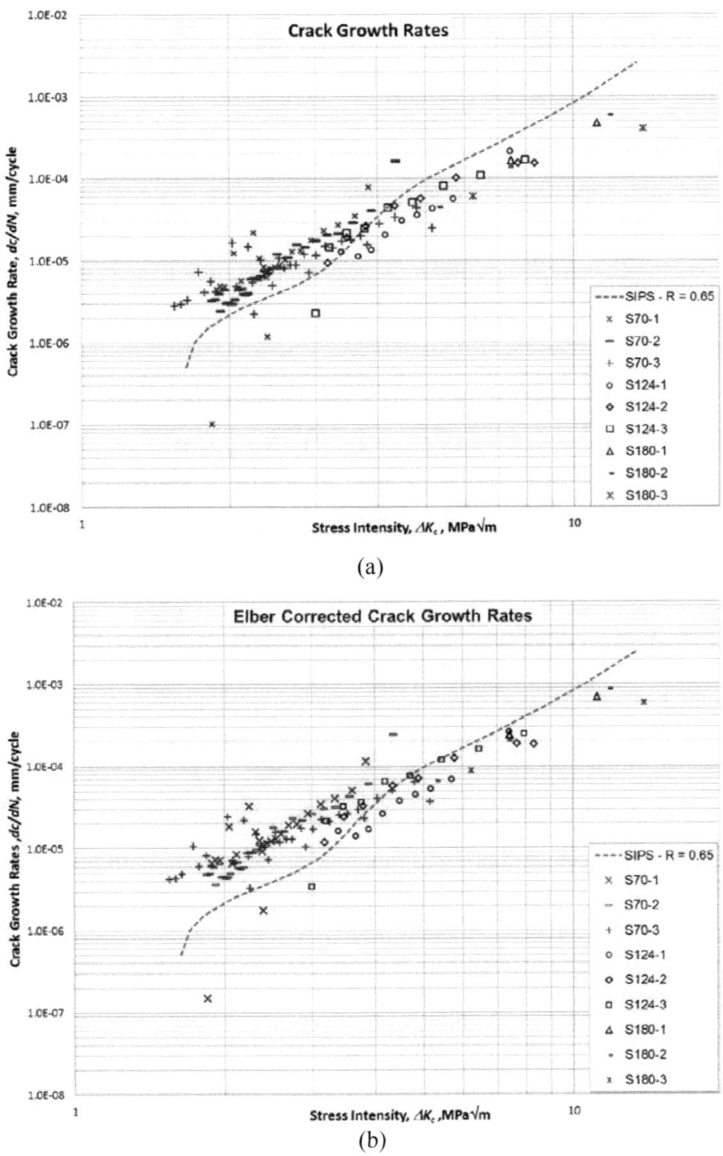

Figure 13. Comparison of crack growth rates calculated along the "c" direction with SIPS Data (a) without correction and (b) with correction using the Elber equation.

Fractography studies. Post-test fractographic analyses were carried out on all the fractured samples to identify: the corrosion pit geometry, crack nucleation feature and location, and crack front evolution using the marker bands. The corrosion pit geometry was primarily used for correcting normalizing the dcPD measured crack lengths. Figure 14 shows an overview of the post-test fracture surface of the S70-1 sample with the corrosion pit profile, fatigue region and post-test fracture region. Similarly, Figure 15 shows an example of fatigue marker bands signature of the marker bands used to study the crack growth and crack profile evolution.

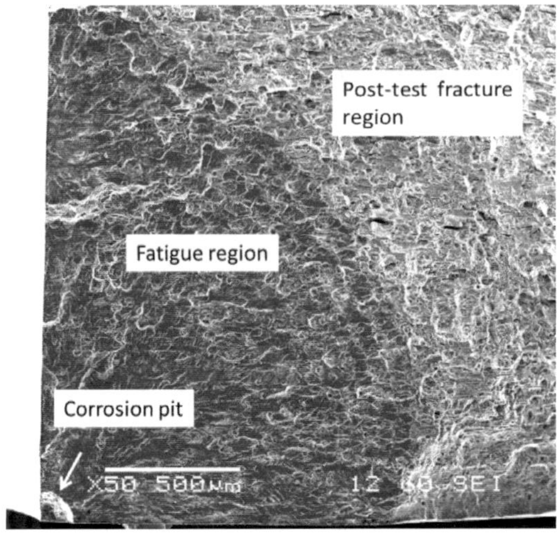

Figure 14. SEM micrograph shows overview of the fracture surface with location of the corrosion pit and fatigue region after post-test facture.

Figure 15. SEM image shows an example of marker bands used to study the crack profile and crack growth (S70-1 specimen R = 0.65, 10 Hz).

In general, the crack nucleation region for all the samples tested for all stress conditions was at or near the "*a*" location, as expected since the macro stress concentration is higher at that location for corner crack geometry. Figure 16 shows the SEM image of the crack nucleation site of S180-1 along with the first marker band close to the nucleation site. However, there are very few instances where the fatigue crack nucleated away from the "*a*" location as shown in Figure 17 for the specimen tested at 124 MPa maximum stress level. The site for crack nucleation from a corrosion pit is a complex phenomenon and depends on several factors. Unlike in the case of a mechanical flaw, where, the crack nucleation site solely depends on the maximum stress concentration region, nucleation from a corrosion pit is considered to be driven by four major factors: site geometry, microstructure-scale plasticity and two forms of environmental interactions [36]. From the fracture mechanics point of view, local stress and plastic strain concentration induced by macro and micro topography of the corrosion pit influence the crack nucleation and corrosion to crack transition [36, 37-40]. Thus, it is difficult to predict the nucleation site purely based on the pit geometry or pit aspect ratio. Microscopic features and size of the intrusion or extrusion on the pit surface and also known as "micro-pit" and "jut-in" along with macro pit profile dictate the crack nucleation site. Figure 17 shows an example of "micro-pit" where the crack nucleates on the pit profile.

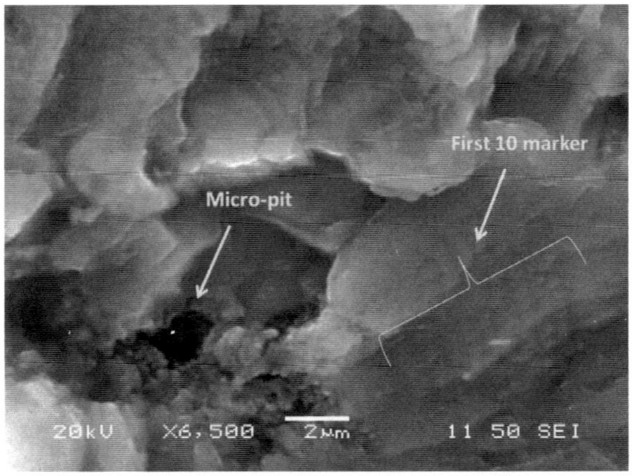

Figure 16. SEM images of fractured specimens of the tests carried out at 180 MPa and R = 0.65.

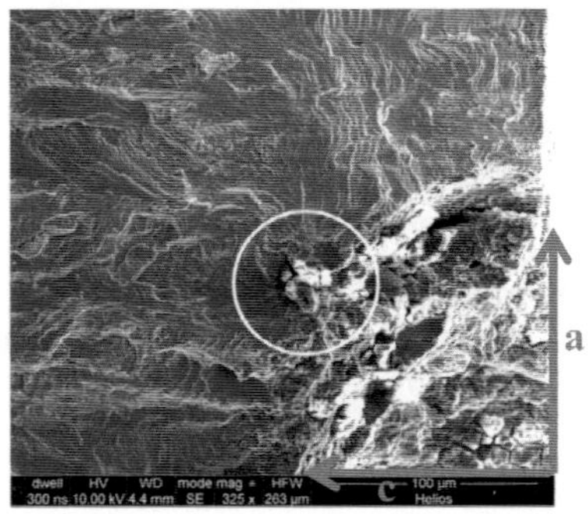

Figure 17. SEM images of fractured specimen and fatigue crack initiation location of a 124 MPa sample.

K-solution Validation

Multiple models were generated at the onset of the program to determine the mesh density parameters needed to ensure acceptable accuracy in the calculated SIFs. Several different model results are presented as a function of the normalized parametric angle, ϕ, as defined in Figure 18. These models are all loaded in uniform tension with the geometry values shown within Figure 19, and three (3) different crack aspect ratios are investigated. Each crack aspect ratio is plotted with three (3) levels of mesh density along the crack front.

Note that all of the mesh density cases lay nearly perfectly on top of one another for each crack aspect ratio. The only significant differences occur at the crack vertices, where the crack front intersects the free surfaces. It appears that the coarsest mesh still adequately captures the variation in SIF along the crack front and completely contains the boundary layer effect within the vertex elements at the free surfaces, where the SIF is, by definition, zero.

Comparisons to available industry results were also performed for several cases. The results are shown in Figure 20 for several double symmetrical corner crack (DSCC) cases. (Note that these cases where chosen to reduce the number of DoF and to avoid use of the Shah correction factor typically employed to extract single corner crack SIFs from double corner crack solutions [41].) Results of the N-R equations [22], the AFGROW Advanced solutions [19], which employ the Fawaz-Andersson [42] solutions along with an additional width correction factor in addition to the N-R correction, and the current investigations' solutions are presented in Figure 20.

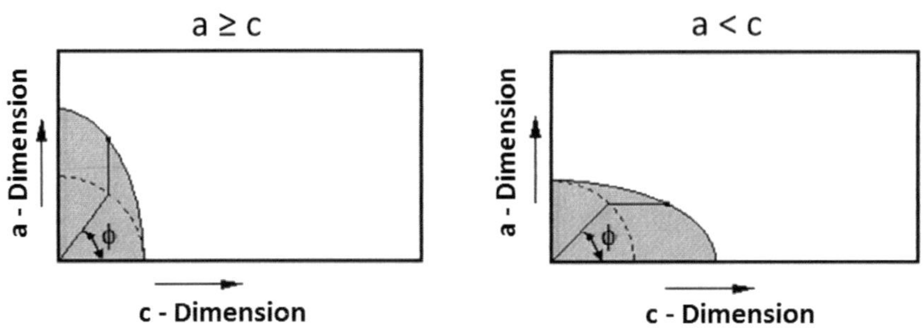

Figure 18. Parametric angle definition.

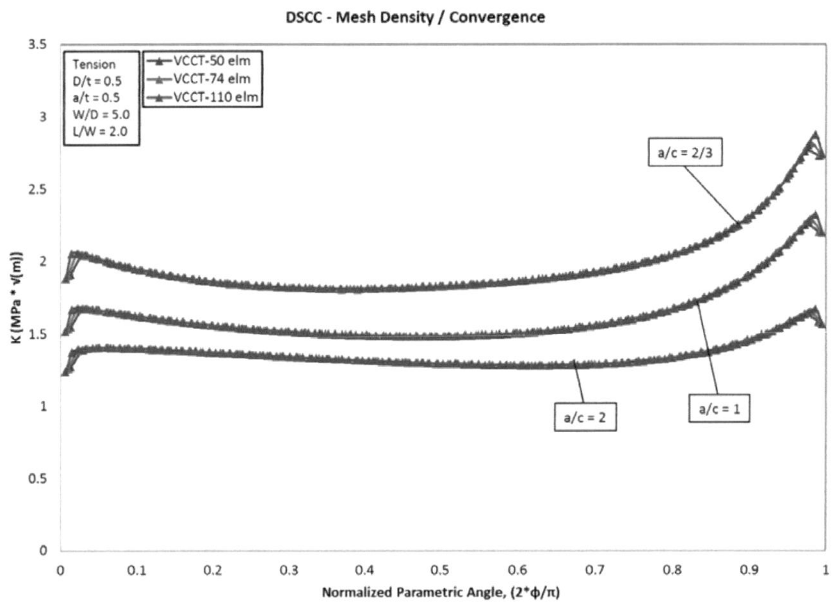

Figure 19. Mesh density determination results.

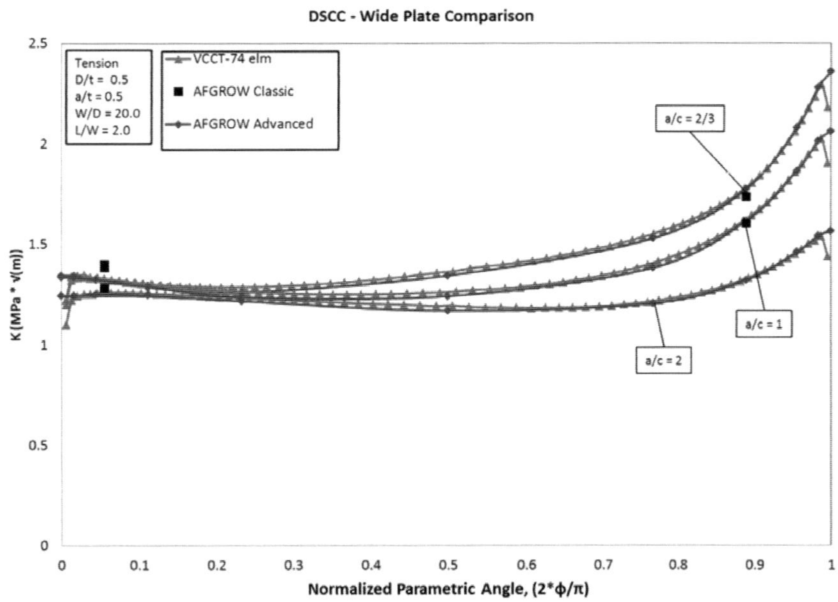

Figure 20. Comparison of SIF values for several DSCC cases.

As can be seen in Figure 20, there is excellent agreement between the AFGROW Advanced solutions and the current work for the given geometry. However, there is a larger difference in the N-R solutions when compared to the others for these likely crack scenarios, especially near the crack vertices. It should be noted that the vertex behavior displayed in Figure 20 for the AFGROW Advanced solutions is more pronounced in the raw data sets than the data displayed here, which are curve fits used in the DTA simulations [43].

Because there are hundreds of different crack geometry models investigated in this effort, only a small snapshot of the analysis results are presented here. The most dramatic differences between the results of the present investigation and those available in industry are expected to exist at the extreme crack geometries (e.g. very large or small crack aspect ratios, significant portions of the crack plane occupied by a crack, crack tip(s) near the free surface(s)). Therefore, in order to gain further confidence in the new numerical results, several comparisons are made with existing industry solutions. Note that only AFGROW Advanced solutions are presented here, as much of the crack geometry investigated is well outside of the reported solution space of the N-R equations.

Figure 21 shows several cracking scenarios with an aspect ratio of a/c = 2.0. As can be seen from the figure, the relative difference is not uniform between the two solution methods for varying crack sizes. The relative difference between the two solutions is also a function of the position along the crack front. This variation in absolute and relative differences between the two solutions indicates that a simple functional form to correct the N-R equations may not exist.

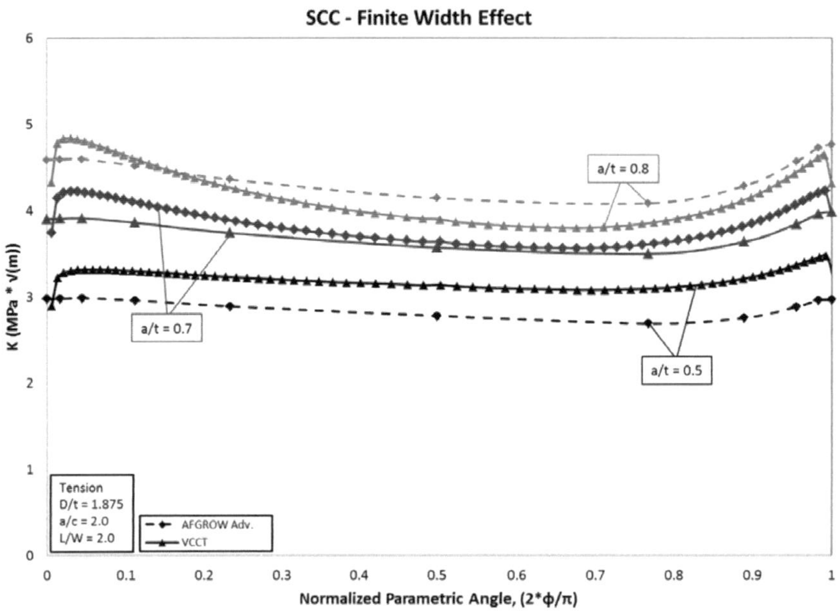

Figure 21. Results comparisons with a/c = 1.0.

CONCLUSIONS

Low solubility corrosion inhibition pigments ($CaMoO_4$) have not shown the ability to reduced fatigue crack growth rates in bulk 0.06 M NaCl solutions whether through passive film formation or crack closure. The effect of chromate on small scale fatigue damage can be characterized for the different primer types. The analysis of surface area-to-volume for leaching suggests that a thin salt film may be more applicable than full immersion testing for aircraft applications, particularly lap joints. New coating designs that might apply high levels of inhibitors at areas of damage also should be explored. A methodology for the examination of the corrosion pit to fatigue crack development has been standardized using the dcPD method for crack growth measurement. Going forward in aggressive environments marker bands will not be needed based on the understanding of how the dcPD voltage signal changes relate to crack initiation and propagation. Using these results and the standardized method testing will continue examining the effect of environment and inhibitors on the transition from a corrosion pit to a fatigue crack. The stress intensity solution space has been greatly expanded to provide valid K-solutions for the samples being used in this program. The expansion of the solution space will allow for better modeling and prediction of crack growth rates based on environmental conditions.

REFERENCES

[1] Undersecretary of Defense, John, J. Young, Jr., (2009) "Limiting the Use of Hexavalent Chromium," *Memorandum for Secretaries of the Military Departments.*

[2] Gasem, Z. and Gangloff, R.P., (2001) "Rate-Limiting Processes in Environmental Fatigue Crack Propagation in 7000-series Aluminum Alloys", in *Chemisty and Electrochemistry of Corrosion and Stress Corrosion Cracking*, p. 501-521. R.H. Jones, Editor. TMS-AIME: Warrendale, PA.

[3] Warner, J.S. (2010) "The Inhibition of Environmental Fatigue Crack Propagation in Age-Hardenable Aluminum Alloys." *PhD Dissertation,* University of Virginia, Charlottesville, VA.

[4] Warner, J.S., Kim, S. and Gangloff, R.P., (2009) *International Journal of Fatigue* vol. **31**. Pp. 1952-1965.

[5] Lui, X.F., Huang, S.J. and Gu, H.C. (2003) *Corrosion Science* vol. **45** (9) pp.1921-1938.

[6] Wanhill, R.J.H., De Luccia, J.J. and Russo, M.T. (1989) *The Fatigue in Aircraft Corrosion Testing (FACT) Programme* AGARD Report 713 (AGARD-R-713).

[7] Shoales, G.A., Walters, M.R., Fawaz, S.A. (2009) "Compilation of Damage Findings from Multiple Recent Teardown Analysis Programs," *Proceedings of the 2009 International Conference on Aeronautical Fatigue (ICAF) Symosium*, Rotterdam, The Netherlands. pp. 187-208.

[8] Sinko, J. (2001) *Progress in Organic Coatings*, vol. **42.** pp. 267-282.

[9] Sinko, J. (2009) "Pigment grade corrosion inhibitors: a review of chemistry and relevant concepts". *DoD Corrosion Conference*, Washington D.C.

[10] Petry, L. and Dante, J.F. (1999), *Evaluation Report No. 99-71*, AFRL/MLSA.

[11] Office of the Inspector General Department of Defense, (1996) "Air Force Aircraft Painting and Corrosion Control," *Report Number 96-062*.

[12] Haynes, W.M. Ed, (2011) *The Handbook of Chemistry and Physics*, 92nd Ed.

[13] Luna Innovations, Charlottesville, VA., USA, www.lunainc.com.

[14] Scholes, F.H. et al, (2006) *Progress in Organic Coatings*, p. 23-32, vol **56**.

[15] Funke, W., (1983) "Towards Environmentally acceptable corrosion protection by organic coatings", *Journal of Coating Technology,* p.705, vol **55**.

[16] University of Southern Mississippi, Hattiesburg, MS. http://www.usm.edu/polymer.

[17] Divakar M. (2013) USAFA-TR-2013-04, Pitting Protocol, Center for Aircraft Structural Life Extension (CAStLE), USAF Academy, CO 80840.

[18] Jones, D. A. (1996) Principles and Prevention of Corrosion. 2nd edn. , p. 572 Prentice Hall, Upper Saddle River, NJ, xvi.

[19] AFGROW, LexTech, Inc., http://www.afgrow.net.

[20] Cracks2000, University of Dayton Research Institute (UDRI), http://cracks.udri.udayton.edu.

[21] NASGRO, Southwest Research Institute (SwRI), http:// www.nasgro.swri.org.

[22] Newman Jr., J.C. and Raju, I.S. (1986). In: *Computational Methods in the Mechanics of Fracture*, pp. 312-334, Atluri, S.N. (Ed.), Elsevier Science Publishers B.V., Amsterdam.

[23] Shivakumar, K.N., Tan, P.W. and Newman Jr., J.C. (1988). *Int. J. Fract.*, vol. 36., n. 3, pp. R43-R50.

[24] MATLAB, The MathWorks, Inc., http://www.mathworks.com/products/matlab.

[25] TrueGrid, XYZ Scientific Applications, Inc., http://truegrid.com.

[26] Shivakumar, K.N. and Newman Jr., J.C. (1990). NASA Technical Memorandum 102753.

[27] ASTM D5032 (2011) Standard Practice for Maintaining Constant Relative Humidity by Means of Aqueous Glycerin Solutions West Conshohocken, PA, ASTM.

[28] Gangloff RP. (1981) *Fat Eng. Mater. Struct.* vol 4. pp. 15-33.

[29] ASTM E647 (2000) Standard Test Method for Measurement of Fatigue Crack Growth Rates. West Conshohocken, PA, ASTM.

[30] Fawaz S.A. and Andersson B. (2004) pp. 1235-54 *Eng. Fract. Mech.* 2004, vol. 71.

[31] Fawaz SA. (1997) Fatigue Crack Growth in Riveted Joints. Delft University, The Netherlands.

[32] Alcoa Aluminum alloy 7075 Data Sheet https://www.**alcoa**.com/mill_products/ *.../**alloy**7050tech**sheet**rev.pdf, Dec 13, 2014.*

[33] Metallic Materials Properties Development and Standardization (MMPDS) DOT/FAA/AR-MMPDS-01, (2003) Scientific Report, Office of Aviation Research, Washington, D.C. 20591. pp. 3-414.

[34] Petit. J, Zeghloul. A, (1986) "On the effect of Environment on Short Crack Growth Behavior and Threshold, The Behavior of Short Fatigue Cracks," Ed. Miller. KJ, de los Rios. ER *Mechanical Engineering Publications*, London pp.163-177.

[35] Wanhill R.J.H, (1991) "Durability analysis using short and long fatigue crack growth data, Aircraft Damage Assessment and Repair." The Institution of Engineering, Barton, Australia, *Fatigue of structural materials*, 2nd edition, by Jaap Schijve p.221.

[36] Burns, J.T. Larsen, J.M. and R.P. Gangloff, (2011) *Fatigue Fract. Engng. Mater. Struct.,* vol 34. pp. 745-773.

[37] Gao Y.X., Yi, J. Z., Lee, P.D. and Lindley, T.C. (2004) *Fatigue Fract Eng Mater Struct* vol 27 pp. 559-570.

[38] As, S. K., Skallerud, B. and Tveiten, B. W. (2008) *Int J Fatigue* 30, p. 2200-2209.

[39] Rusk, D. T. and Hoppe,W. (2009) *Int J Fatigue* 31, p. 1454–1463.

[40] Rusk, D. T., Hoppe, W., Braisted, W. and Powar, N. (2009) *Int J Fatigue* 31, p. 1464–1475.

[41] Shah, R.C., (1976). In: *American Society for Testing and Materials STP-590*, pp. 429-459.

[42] Fawaz, S.A., and Andersson, B. (2004). *Eng. Fract. Mech.*, vol. 71, n. 9-10, pp. 1235-1254.

[43] Fawaz, S.A., (2011). Personal correspondence.

Simulation of corrosion damage, stress concentration, and fatigue crack growth

Sharon Mellings, John Baynham, Andres Peratta

CM BEASY, Ashurst Lodge, Ashurst, Southampton UK sharon@beasy.com

Abstract: Airframe structures regularly operate in environments that allow high levels of corrosion damage, and this damage leads to stress concentrations within the structure and potential development of cracks. Even when only a thin film of electrolyte is present on the structure, this can still lead to an electrical field that causes surface damage.

Computation of this electrical field can be used to identify areas in the airframe structure that are most susceptible to corrosion damage and which, after possible fatigue crack initiation, may lead to structural failure. Corrosion simulation can be used to take account of the properties of the electrolyte as well as the structural materials, to determine the rate of material loss from the structure. Having removed material from the surface (corresponding to corrosion occurring over a given exposure time) the stress concentrations can be evaluated and, if required, cracks can be initiated in each potential problem area, to identify vulnerability to fatigue failure.

Different corrosion damage could arise when different electrolytes are present, for example airframes operating at or near the sea will be subjected to salt water spray, and the electrolyte properties may vary with operational temperatures. Electrolytes produced by condensation, overflows and spillage may be of significant depth inside the aircraft, and may not be pure water. External surfaces may be subject to de-icing chemicals.

All-metal airframes are vulnerable to general corrosion, and depending on construction, crevice corrosion and pitting may also occur. Composite airframes which combine aluminium alloy and carbon fibre are at greater risk of galvanic corrosion because of the significant difference between the positions of aluminium and carbon in the electrochemical series. These properties, as well as the effects of surface coatings, all need to be included in the corrosion simulation. Simulation of the galvanic effects leading to corrosion has been successfully used for many years to assist design of cathodic protection systems for marine vessels and offshore structures, and has more recently been applied to aircraft structures.

This paper will firstly use simulation to investigate the influence of parameters, including electrolyte thickness and conductivity, on the rate of corrosion for a galvanic cell caused by a metallic sample in contact with a more noble material. The two materials will be exposed to differing environmental conditions, including immersion in a deep layer of electrolyte, exposure to thick layers of electrolyte at junctions between the different materials, and to thinner layers of electrolyte which might be caused by condensation. The paper will reach conclusions regarding the type of environment that is likely to produce higher penetration rates for the assumed geometry and materials and where these higher rates are likely to occur.

The paper will determine the damage caused by corrosion mass loss over a given operational time, and investigate the effect of such geometry change on the electrical fields in the electrolyte and the resulting rate of corrosion.

The geometry change caused by corrosion mass loss will be used to perform stress analysis of the same structure, to determine the stress concentration in the component at the corresponding time in the life of the aircraft. It will then be assumed that cracks initiate, and the subsequent crack growth will be simulated. This crack growth will take into account the corrosion damage and will inherently include local stress concentration due to the damaged surface. In the crack growth simulation, the full crack path and direction will be determined along with the fatigue life.

This paper describes a methodology that can be used by design engineers, to identify potential problems in an airframe structure during initial design, giving scope for design changes to reduce potential failures and in-service repair costs. This methodology identifies areas of the structure that have the greatest risk of damage - which may not be obvious without combined corrosion and fracture simulation; and so provides more informed targeting of locations where "what if" fracture mechanics should be applied.

INTRODUCTION

Airframe structures regularly operate in environments that allow high levels of corrosion damage, and this damage leads to stress concentrations within the structure and potential development of cracks. Even a thin film of electrolyte on the structure can lead to an electrical field that causes surface damage.

Computation of this electrical field can be used to identify areas in the airframe structure that are most susceptible to corrosion damage and which, after possible fatigue crack initiation, may lead to structural failure. Corrosion simulation can be used to take account of the properties of the electrolyte as well as the structural materials, to determine the rate of material loss from the structure. Having removed material from the surface (corresponding to corrosion occurring over a given exposure time) the stress concentrations can be evaluated and, if required, cracks can be initiated in each potential problem area, to identify vulnerability to fatigue failure. Different corrosion damage could arise when different electrolytes are present, for example airframes operating at or near the sea will be subjected to salt water spray, and the electrolyte properties may vary with operational temperatures. Electrolytes produced by condensation, overflows and spillage may be of significant depth inside the aircraft, and may not be pure water. External surfaces may be subject to de-icing chemicals.

"All-metal" airframes are vulnerable to general corrosion, and depending on construction, crevice corrosion and pitting may also occur. Composite airframes which combine aluminium alloy and carbon fibre are at greater risk of galvanic corrosion because of the significant difference between the positions of aluminium and carbon in the electrochemical series. These properties, as well as the effects of surface coatings, all need to be included in the corrosion simulation.

Simulation of the galvanic effects leading to corrosion has been successfully used for many years to assist design of cathodic protection systems for marine vessels and offshore structures, and has more recently been applied to aircraft structures.

The surface damage resulting from corrosion creates stress concentrations in the structure which will increase the chances of cracks initiating in the part. This damage will also affect the crack growth rate and direction.

This paper is focused on the modelling of a test specimen exposed to a corrosive environment and subject to cyclic fatigue load, which can lead to the initiation and propagation of cracks.

This paper is in two parts. Firstly, the galvanic corrosion modelling of a test specimen exposed to thin and deep layers of electrolyte is presented and discussed. Secondly, the corrosion damaged specimen, after initiation of cracks, is subjected to a fatigue load, with corresponding crack propagation presented.

CORROSION MODELLING

Introduction

The BEASY software has been used to predict the rate of corrosion on a test specimen consisting of a stretch part in contact with two clamps as shown in Figure 1, where the stretch part is aluminium AA2024 and the clamps are carbon fibre reinforced polymer (CFRP).

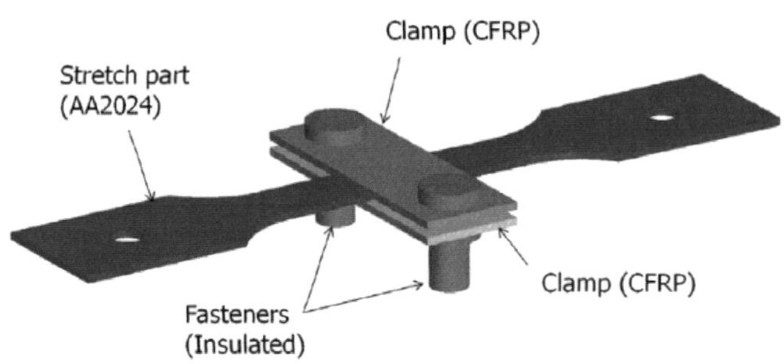

Figure 1. Test specimen used for corrosion modelling.

The stretch part is considered to be perfectly connected to both clamps with zero electrical resistance. In addition, the internal electrical resistance of CFRP and AA2024 are considered negligible. Because of the dissimilar electrochemical nature of the materials involved, a corrosion cell between the aluminium and the CFRP clamps will be established whenever the specimen is exposed to an electrolyte. CFRP is more noble than AA2024; therefore the stretch part will corrode rather than the clamps. In other words, the stretch part will behave anodically, the CFRP clamps cathodically, and an ionic current driven by the different polarisation properties of the coupled materials, will flow through the electrolyte. The corrosion damage occurs on the anodic region (AA2024) and the distribution of normal current density flowing from the anodic surface to the electrolyte becomes a direct measurement of the corrosion damage, as prescribed by equation (1) based on Faraday's law:

$$CR = B_1 \frac{j_n}{\delta} EW, \tag{1}$$

where:
CR = Corrosion rate, expressed in mm/year.
j_n = Corrosion current density, expressed in A/m^2
δ = Mass density, in kg/m^3
EW = Equivalent weight of the corroding sample
B_1 = Constant involving Faraday's constant and unit conversion factors such that the result appears in mm per year. The value of B_1 is 327.2 mm kg/(A m yr).

The specific mass loss rate (S_R) is related to j_n by the expression $S_R = B_2 j_n EW$ where B_2 is a constant involving Faraday's constant and unit conversion factors, K2 = 895.3 mg/(A day), so that S_R is expressed in mg/(day m^2).

Similarly, the mass loss rate for a component can be expressed as:

$$M_R = B_2 \times I \times EW, \tag{2}$$

where I is the total anodic current for the component, and M_R is expressed in mg/day

The equivalent weight for an alloy can be calculated as follows:

$$EW = \frac{1}{\sum_i \frac{n_i \phi_i}{W_i}}, \tag{3}$$

where:
Φi = mass fraction of the i-th element in the alloy
Wi = atomic weight of the i-th element in the alloy expressed in grams per mol
ni = the number of electrons involved in the oxidation process of the i-th element in the alloy.

Using the above expression, the equivalent weight for AA2024 is calculated to be 9.38.

The form of the electric field in the electrolyte is affected by the geometry of the polarising surface, the polarisation curves, the geometry of the electrolyte and electrical properties of the electrolyte.

The following sections provide some theoretical background, describe main features of the test specimen under study, and report the modelling results of corrosion damage obtained under the following exposure conditions:

- Specimen fully immersed in deep electrolyte
- Specimen covered by a thin layer of electrolyte.

Theoretical background

Simulation involves solution of the 3D Poisson equation for the electrolyte potential with non-linear boundary conditions imposed by the polarisation curves on the active electrodes. The physical and mathematical background for the corrosion modelling in deep electrolyte can be followed from references [1], [2] and [3] and references therein.

In the case of thin film electrolyte modelling, the basic formulation is similar to the deep electrolyte case, except that the electrolyte is treated as a two dimensional object warped in three-dimensional space to adopt to the geometry of the specimen. The electrolyte is regarded as a thin sheet that covers the specimen. Referring to Figure 2, we make the following assumptions:

- the electrolyte thickness (w) and conductivity (σ) are functions of the two-dimensional coordinates X and Y;
- for the model to be regarded as "thin" the condition $w \ll L$ must be satisfied;
- the electric field ($\bar{E}$) in the bulk of the electrolyte is mostly oriented in the X-Y direction, parallel to the underlying surface (i.e. E_Z is negligible compared to E_x and E_y).

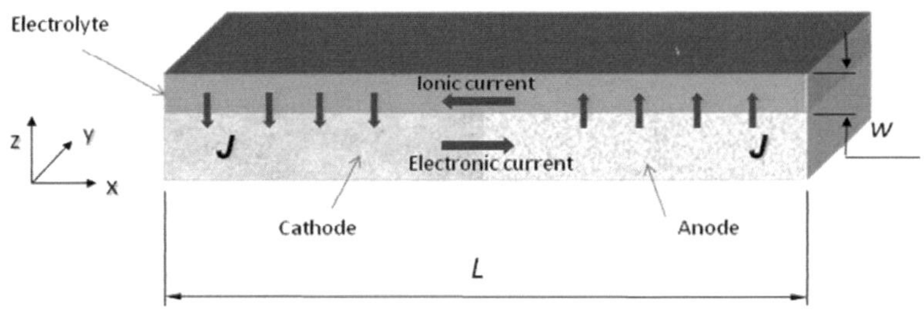

Figure 2. Schematic showing current flow between cathodic and anodic regions.

Under the above mentioned assumptions, the equation for solution is:

$$w\nabla_{2D} \cdot \left(-\sigma\nabla_{2D}u_e\right) = -j_n \qquad (4)$$

Where ∇_{2D} represents the two dimensional "gradient" operator acting on x and y coordinates, u_e is the electrolyte potential, w is the electrolyte thickness, and j_n is the current density coming from the active electrodes, or zero if the surface underneath the thin film electrolyte is perfectly coated and acts as an insulator. The difference between potential in the electrolyte above the metal surface and potential of the metal surface is referred to as "polarisation potential", and the polarisation curve dictates the way in which j_n varies with $u_e - u_m$ (where u_m is the metal potential). This is a non-linear polarization curve of the electrode and the electrolyte, which can be experimentally determined. Further details on the theoretical background for thin film modelling and the corresponding theoretical validation can be found in reference [4] and [5] and references therein.

Modelling Assumptions

Figure 3 shows the specimen dimensions. The stretch part is 2 mm thick, and the fasteners are considered to be perfectly insulated.

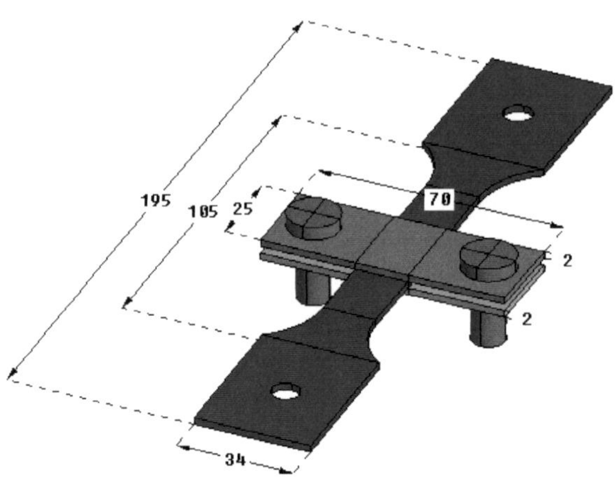

Figure 3. Specimen dimensions (in millimetres).

Unless specified otherwise, in all cases the electrolyte is a saline solution NaCl 0.3 M with conductivity 0.5 S/m. The polarisation properties of CFRP (cathodic branch) and AA2024 (anodic branch) used in this study are shown in Figure 4.

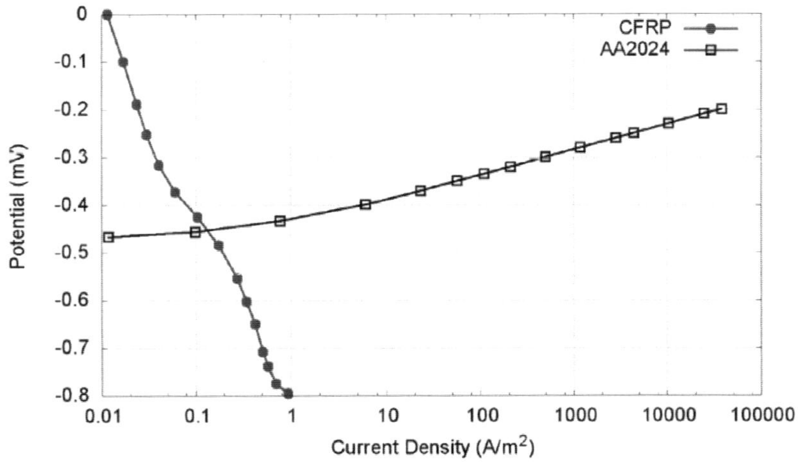

Figure 4. Polarisation curves.

Unless otherwise indicated, potentials are in Volts relative to a saturated Ag/AgCl reference electrode; currents are expressed in Amperes, and current density in Amperes per square metre.

Case 1: Deep electrolyte

In this case, the specimen immersed at the centre of a 'large' (2 m x 2 m x 2 m) tank of electrolyte. The tank surfaces are perfectly insulating. Because the dimensions of the tank are big compared to the specimen, the solution obtained is representative of a sample submerged in an infinitely large tank (i.e. the results do not change with increased tank size.).

The distribution of polarisation potential on the cathodic and anodic regions is shown in Figure 5. The polarisation potential ranges approximately from about -0.438 V to -0.460 V (22 mV difference).

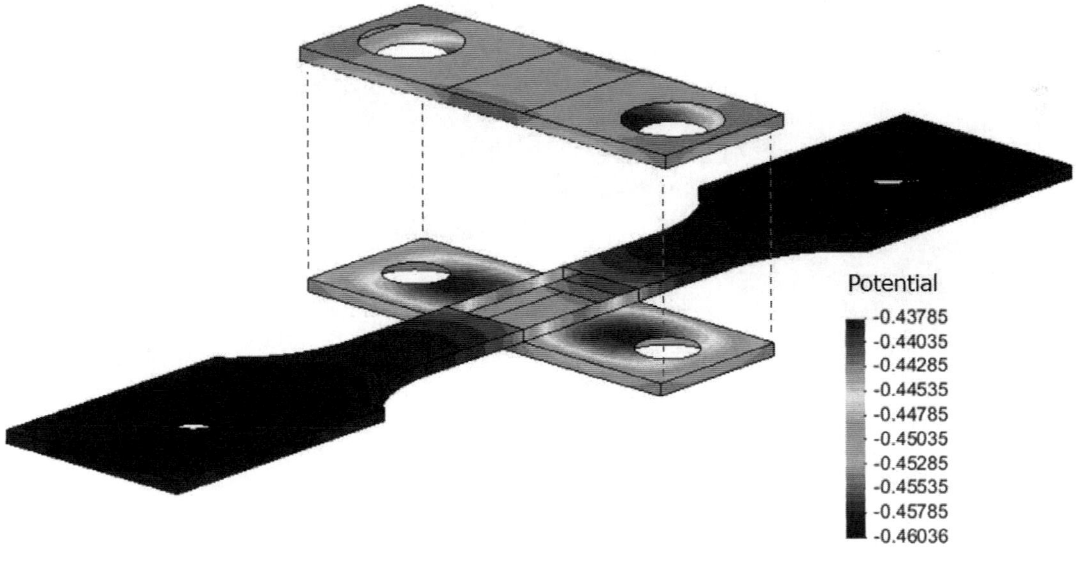

Figure 5. Contours of polarisation potential (u) in Volts on anodic and cathodic surfaces.

Figure 6 shows the distribution of normal current density on the stretch part and bottom clamp. The stretch part behaves entirely anodically and the current density ranges from approximately 0.07 A/m^2 to 0.532 A/m^2. The highest value of anodic current density occurs on the side of the stretch part between the two clamps, while the minimum occurs on the most distant areas from the clamps. The cathodic current density (on the clamp) is comparatively more uniformly distributed and averages to approximately 0.12 A/m^2.

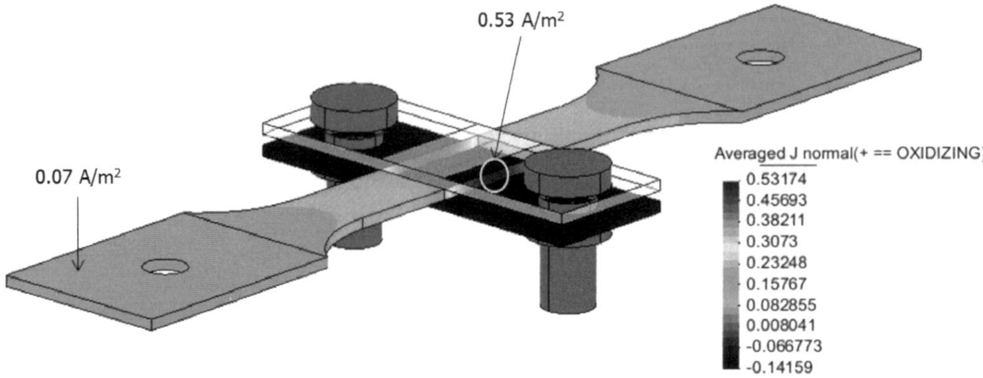

Figure 6. Contours of normal current density (j_n) in A/m^2 on anodic and cathodic surfaces (positive current density indicates anodic behaviour).

The current flowing between the stretch part and the clamps is about 0.81 mA, which corresponds to 6.8 mg/day mass loss rate.

Case 2: Thin film electrolyte

Figure 7 shows the distribution of polarisation potential caused by exposure to a uniform electrolyte film of thickness 100 μm covering the clamps and stretch part. The polarisation potential ranges from -0.157 V to -0.468 V (311 mV difference).

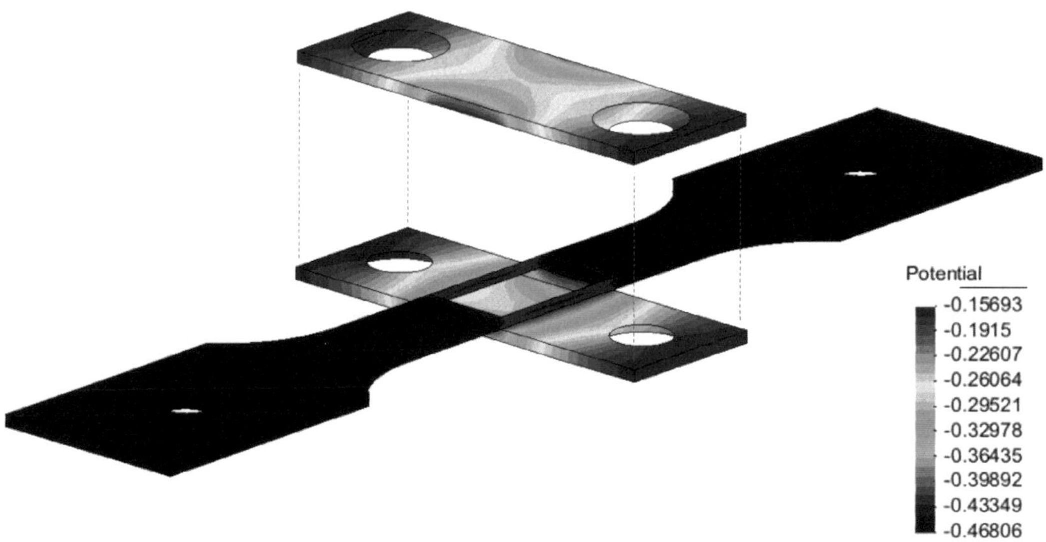

Figure 7. Contours of polarisation potential (u) in Volts on the surfaces of the specimen exposed to thin film.

Figure 8 shows the corresponding distribution of normal current density using two different scales, one for the cathodic currents on the clamps (with negative values), and the other for the stretch part (with positive values). The highest anodic current density (approximately 3.4 A/m^2) occurs at the edges of the stretch part in contact with the clamps, and it rapidly decays to nearly zero in all regions of the aluminium more distant than 20mm from the clamps.

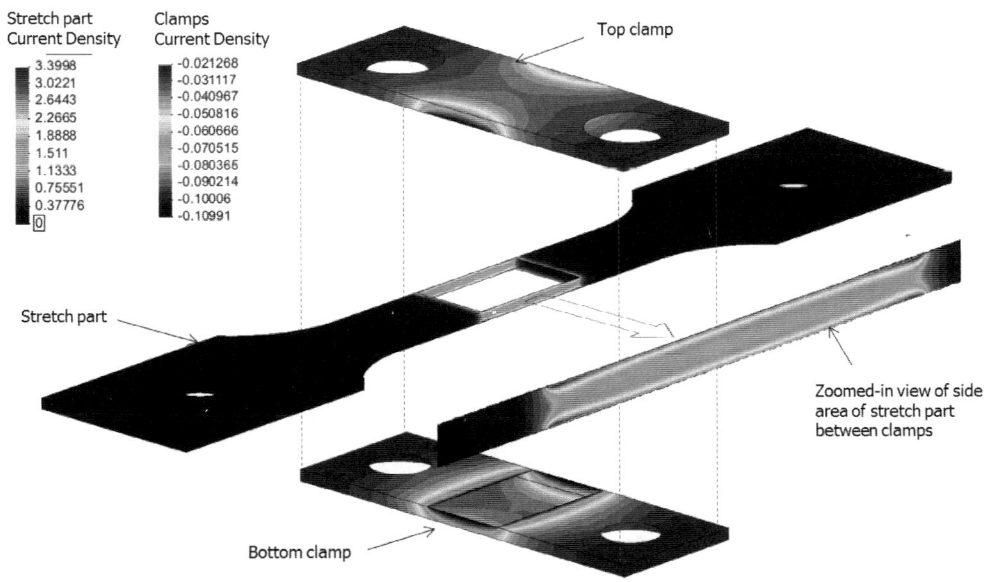

Figure 8. Contours of normal current density (j_n) in A/m^2 on the surfaces of the specimen exposed to thin film electrolyte (Electrolyte thickness = 100 μm).

Figure 9 shows the distribution of corrosion rate in the region of the stretch part with highest damage, and vectors showing the electric field.

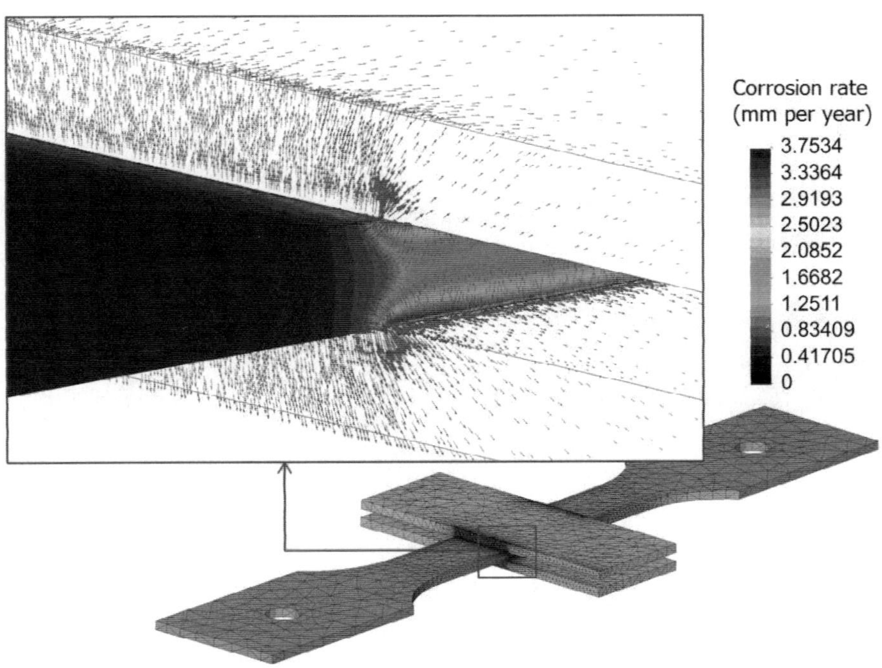

Figure 9. Showing contours of corrosion rate expressed in millimetres per year zoomed-in on the area of the stretch part with highest damage. The arrows represent electric field in the electrolyte.

The current flowing between the stretch part and the clamps when exposed to a 100 μm thick electrolyte is about 0.2 mA, which corresponds to 1.73 mg/day mass loss rate.

Figure 10 shows how the mass loss rate changes with the electrolyte thickness. The results were obtained by performing simulations with electrolyte thickness ranging from 100 μm to 1000 μm.

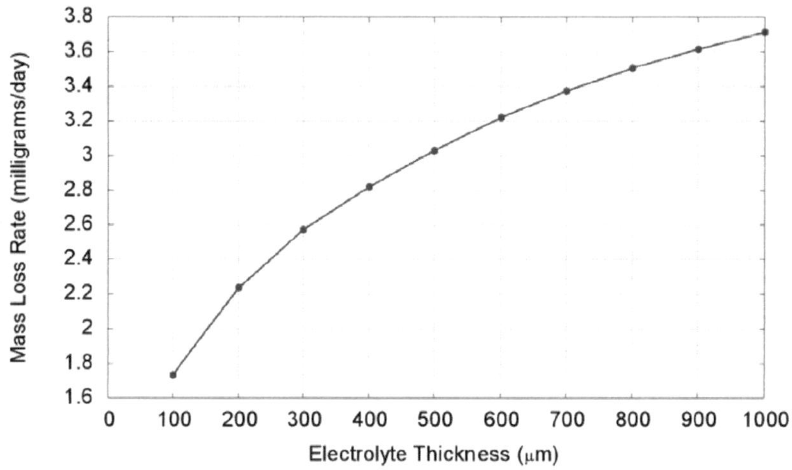

Figure 10. Variation of total mass loss in the stretch part as a function of electrolyte thickness.

The mass loss due to anodic dissolution implies material removal from the stretch part. Without considering any change of properties as mass is lost, the geometry of the specimen damaged by corrosion will eventually take the form shown in Figure 11. The mesh representing the damaged specimen shown in the figure has been obtained by iteratively modifying the position of each mesh point, with a translation proportional to the corrosion penetration rate and directed along the normal to the sample surface, then re-solving for new corrosion rates.

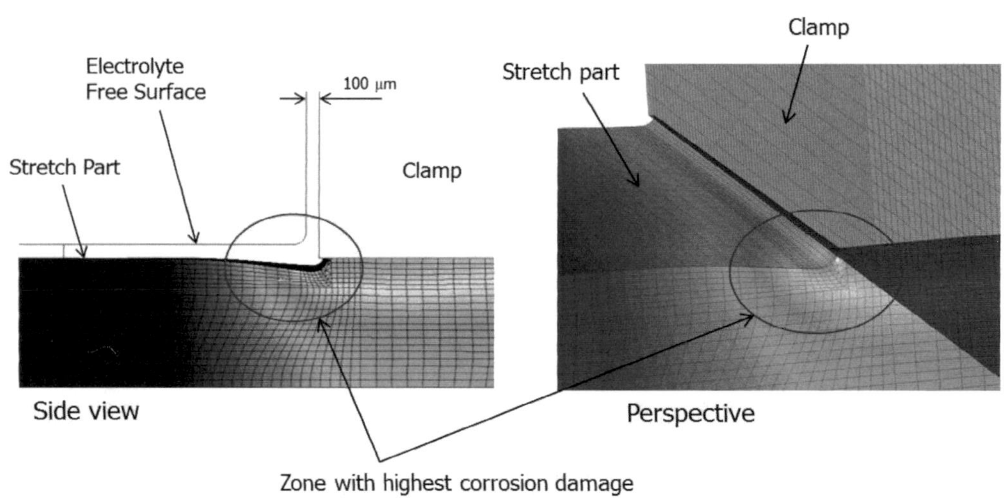

Figure 11. Showing geometry changes in the specimen due to corrosion damage.

STRESS CONCENTRATIONS AND CRACK GROWTH RESULTING FROM CORROSION DAMAGE

As discussed earlier, corrosion results in loss of material from an affected structure. This leads to localised stress concentrations, which can accelerate development and growth of cracks and fatigue failure of the part. Using the corrosion simulation results, a prediction of material loss after a given time has been determined and used to modify the geometry of the part prior to computation of the stress concentrations.

By identifying the areas of highest stress concentration, cracks were initiated in the component using an automatic crack modelling tool. Crack growth simulation was then performed, growing these cracks automatically using fatigue crack growth, under a defined cyclic loading.

Since the geometry includes corrosion damage in the "un-cracked" model, the crack growth simulation inherently computes the growth including the corrosion damage. This incorporates stress concentrations due to the material loss directly within the crack growth simulation, rather than requiring any numerical correction in the computed fatigue results.

Stress concentration due to the corrosion damage

The corrosion simulation showed that the greatest susceptibility to corrosion damage is at the interface between the carbon fibre and the aluminium. The crack growth shown here assumes that there is a line of damage across the part, with a damage depth of 0.15bmm.

Since the surface damage is a surface feature, the use of boundary element modelling for the stress analysis of the damaged part is ideal. In this example, a line of material has been removed from the stretch part near the intersection with the carbon fibre tie beam (as shown in Figure 12); however it would be possible to simulate any damage that is predicted by the corrosion simulation.

Figure 12. Damaged stretch component.

The stress has been computed in the stretch part by applying a uniform load through the axis of the stretch component with a nominal 1 N/mm^2 load applied to one of the bolt holes and a clamp restraint applied to the other bolt hole. In addition normal restraint was applied to the central section where the carbon fibre tie-bar will restrict out of plane deformation of the stretch part. The load applied to the part is 50.2 N.

Resulting stresses are shown in Figure 13 and Figure 14

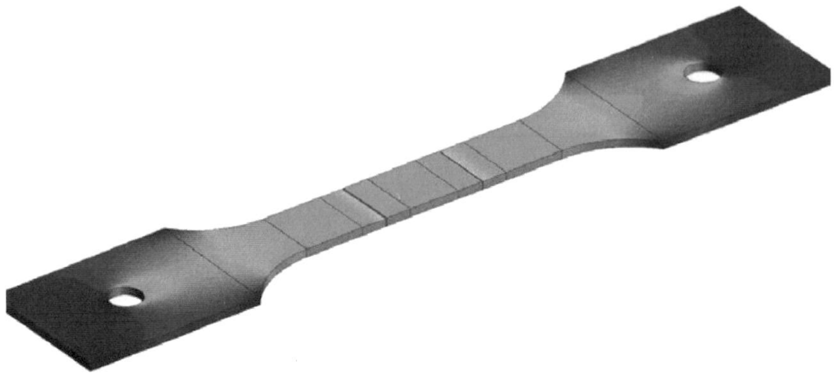

Figure 13. Stresses in damaged stretch component.

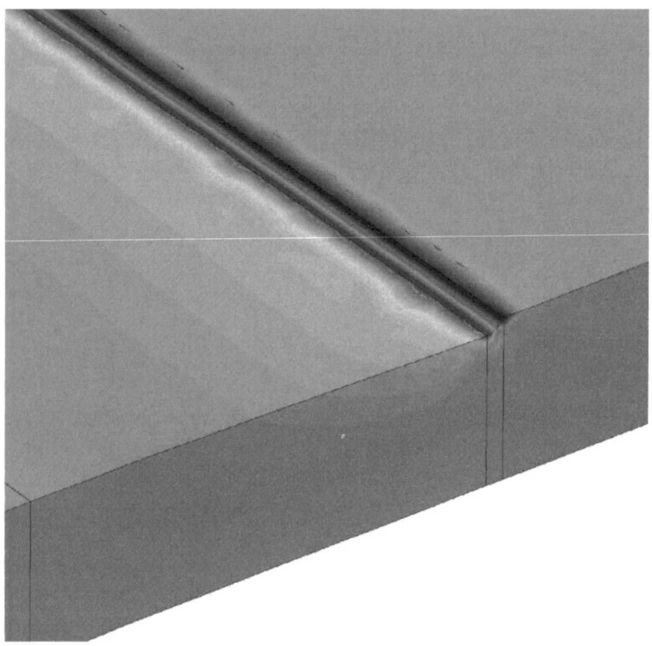

Figure 14. Localised stresses in damaged stretch component.

The results show stress concentration factor of 2.52 caused by the corrosion damage to the surface of the part. The peak stress occurs along a line at the bottom of the corrosion damage, and this indicates a likely location for an initial crack.

Crack analysis with a single edge crack

Whilst cracks could initiate anywhere, in this example, the stress concentration results in a localised area where there is increased chance of crack initiation. The highest stress value occurred at the bottom of the corrosion damage, a short distance from the corner of the part. In addition to increasing the chance of cracks initiating in the part, the damaged shape of the structure may also affect the crack growth rate and direction. Therefore it is preferable to simulate the crack growth in the damaged part.

The damaged model as shown in Figure 12 has been used for crack growth simulation, using an initial surface edge crack, of radius 0.1 mm, initiated at the point where the highest stress concentration has been computed in the "damaged" model.

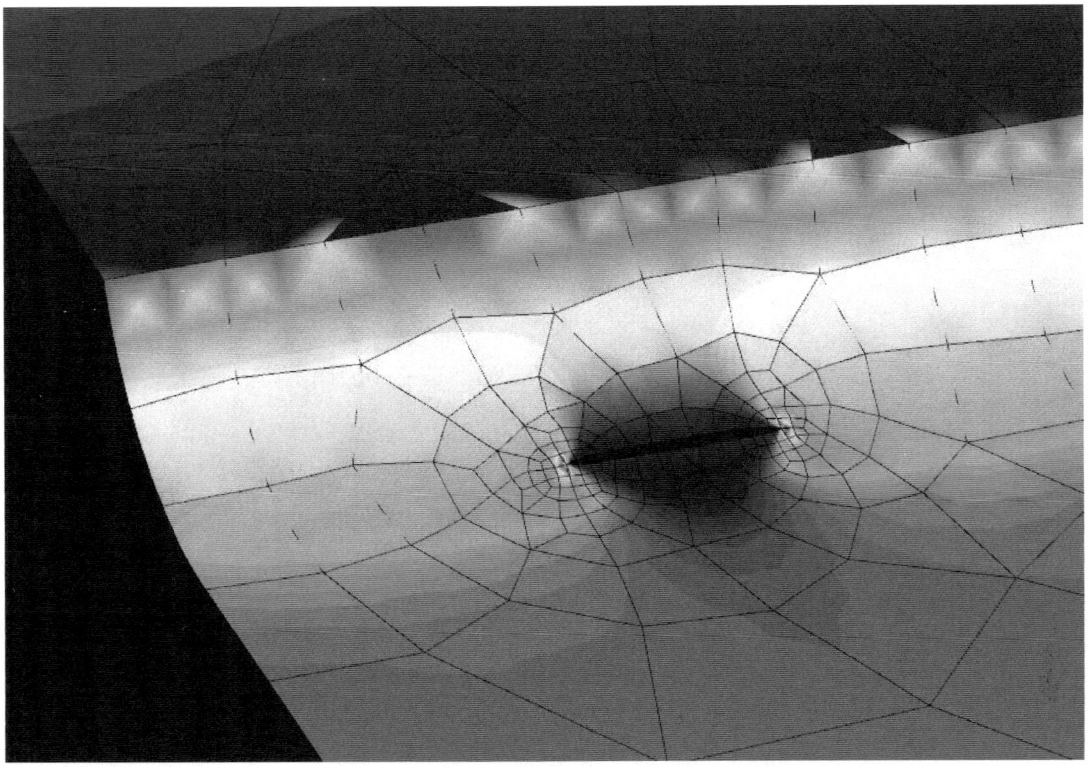

Figure 15. Stresses around the initial edge crack.

The stresses are computed using the Dual Boundary Element Methods [6-9] and the stresses around the crack are shown in Figure 15, with the resultant SIF values, for the applied uniaxial load, shown in Figure 16.

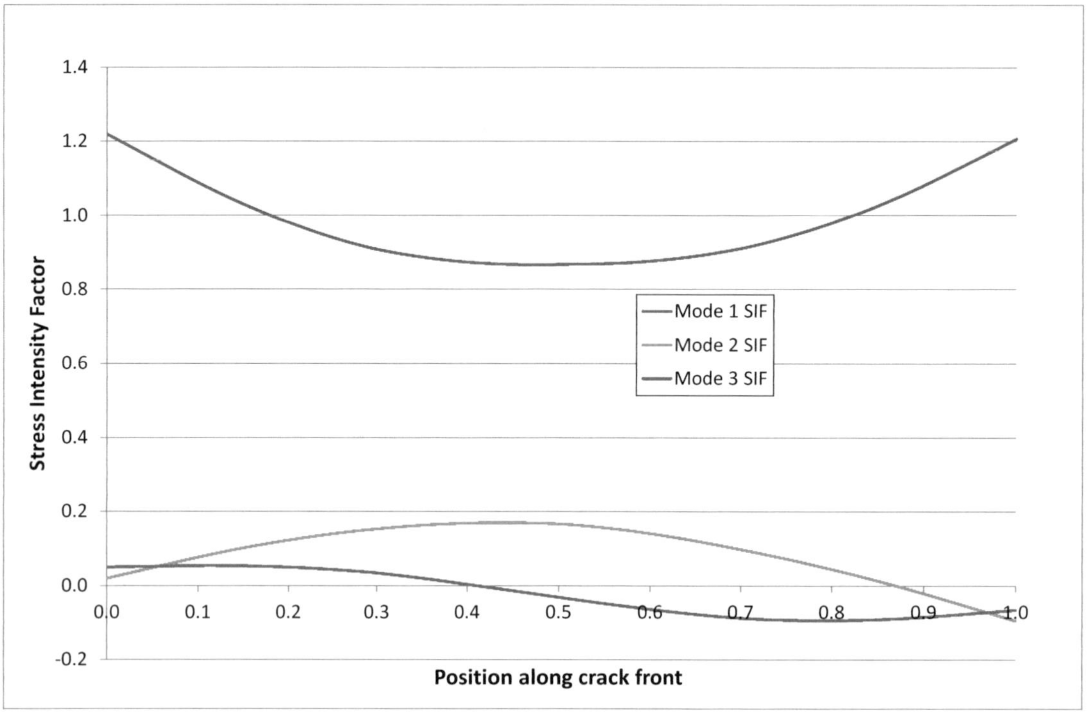

Figure 16. SIF values along the crack front.

Fatigue crack growth of a single edge crack

The stress intensity factors obtained, using an applied load of around 50 N, are not sufficiently high to cause the crack to start growing. However the SIF values have been computed using a nominal crack loading and different load scenarios can be investigated. This could include combined multi-axial loading or complex loading histories.

In this example, we assumed that the crack is subject to a fatigue load cycling between 0 and 7530 N. This gives a load factor of 150 and, since the stretch part has been analysed using a linear analysis, this allows the SIF values to be computed simply by scaling up the results from the "nominal" load. If the simulation required multiple load cases, this can be achieved by combining the SIF values from each of the load cases.

Using this load regime in conjunction with a set of NASGRO 3.0 fatigue properties for an aluminium material, the crack has been grown automatically, re-computing the SIF values periodically during the crack growth.

After a number of growth increments, the edge crack reaches the corner of the structure, after which the crack growth automatically continues as a corner crack.

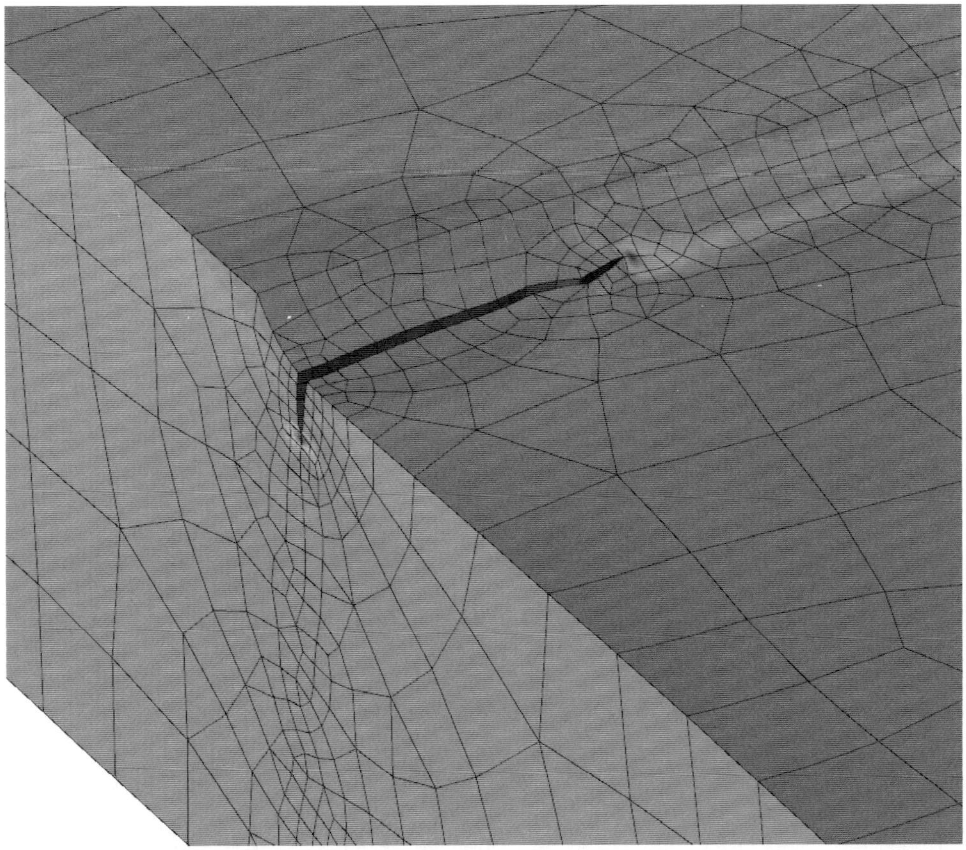

Figure 17. Stress plot after crack grows around the corner.

The crack growth simulation proceeds, with the stresses and SIFs re-computed at every stage. After a number of additional growth steps the crack grows through the thickness of the stretch part and transitions to a through crack as shown in Figure 18. The life of the crack is shown in Figure 19.

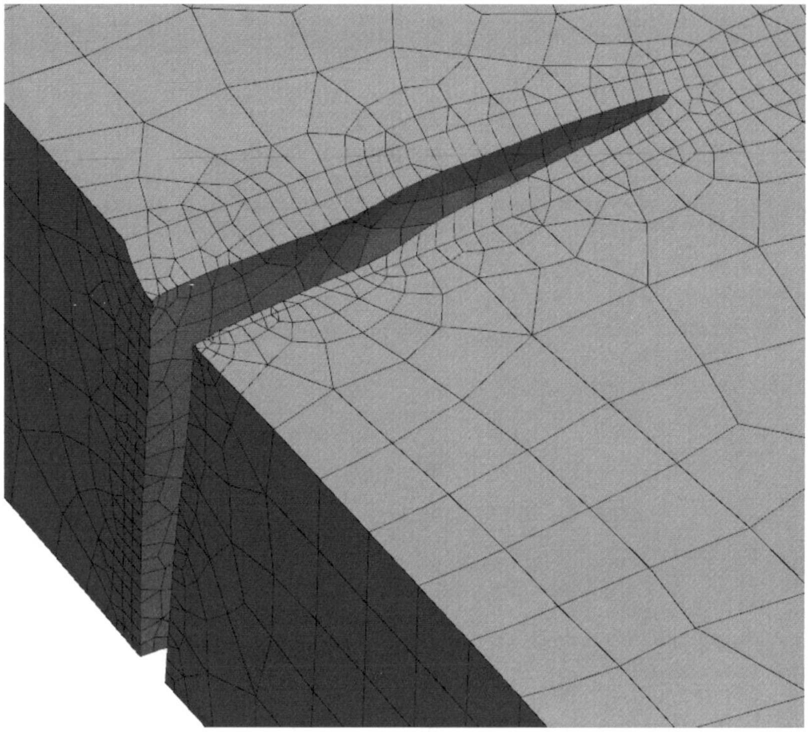

Figure 18. Crack transitions to a through crack.

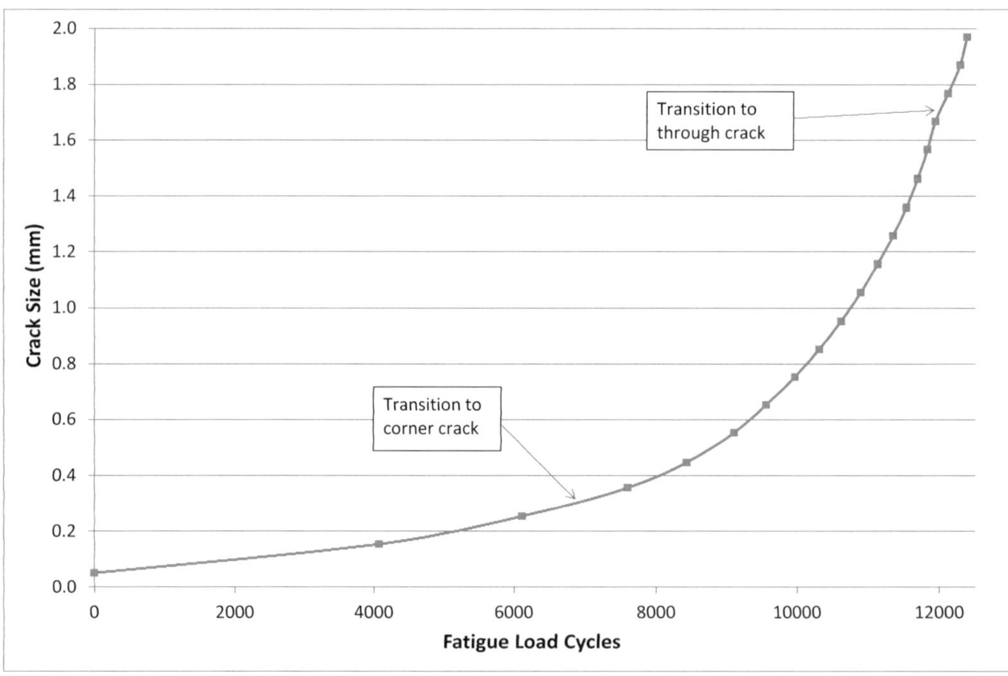

Figure 19. Fatigue results for the growth of the single crack.

Stress intensity factor calculations for multiple cracks

The fatigue crack growth simulation shown above assumed that a single surface crack had initiated, in the damaged area of the part where the stresses are highest. However, the stress concentrations, shown in Figure 14, indicate there is a line of high stress values across the part in the damaged region. Cracks are more likely to initiate along this a line than elsewhere in the part, so it is possible for multiple cracks to initiate in this region.

In this second example of fatigue simulation, two cracks have been initiated along the line where the stress concentration is the highest, with both of the cracks close to the corner of the part. The first crack is, as in the previous simulation, a thumbnail edge crack, with radius 0.1 mm and located 0.45 mm from the corner, whilst the second crack is a corner crack, also with a radius of 0.1 mm.

In this simulation both cracks are added to the part and a new mesh is created to incorporate both cracks (as shown in Figure 20), before the stresses are re-computed. This process results in inherent modelling of any load interaction between the two cracks, with no numerical correction required. Figure 21 shows the model stress intensity factors computed from this model.

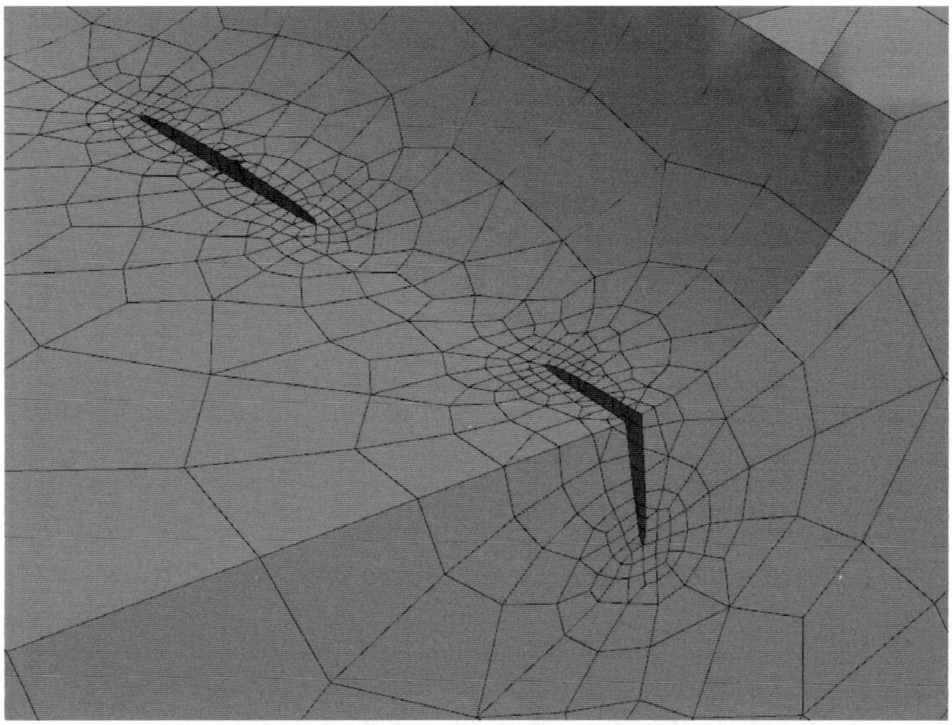

Figure 20. Two cracks modelled in the stretch part.

Comparison of the SIF results on the edge crack in this example (Figure 21) with those given earlier for the single edge crack (Figure 16) shows that the presence of the additional crack has raised the computed SIF values by around 10% but, for these initial cracks, there does not appear to be any significant distortion of the shape of the graph of K_I for crack 1 in Figure 21.

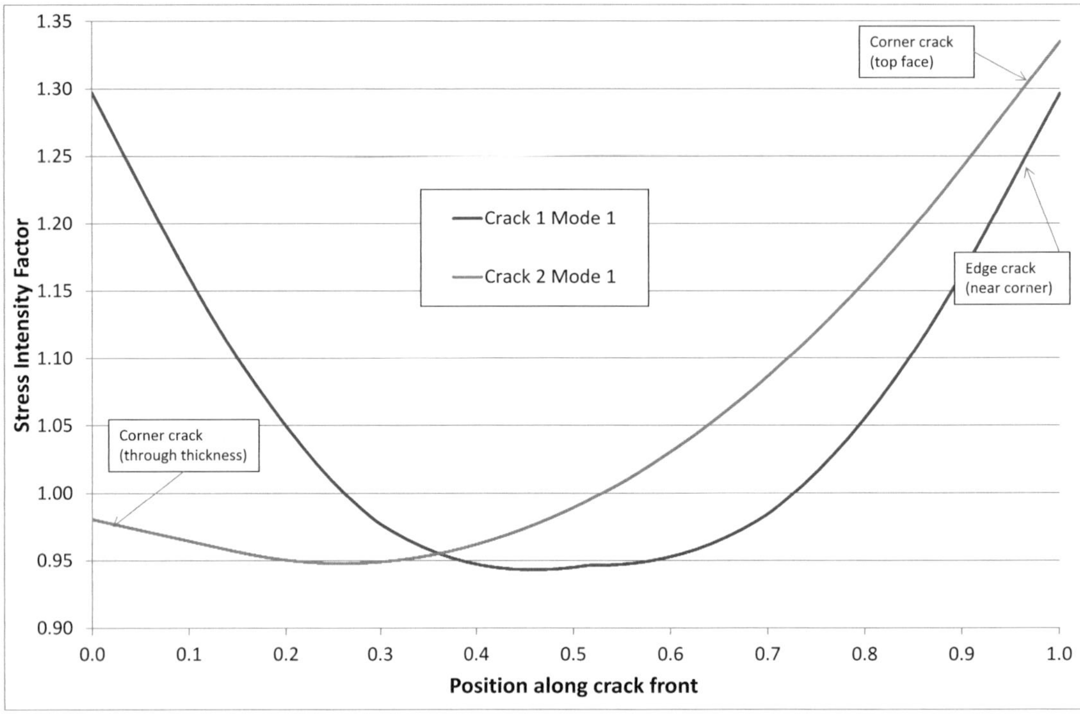

Figure 21. Mode 1 stress intensity factor results for the two cracks.

The simulation was continued using automatic growth of these two cracks; remeshing and recomputing the stress field at each growth increment. After a number of growth steps the cracks appear to interfere, as shown in Figures 22 and 23. The SIF plots in Figure 23 show greatly increased SIF at the nearby ends of the two cracks.

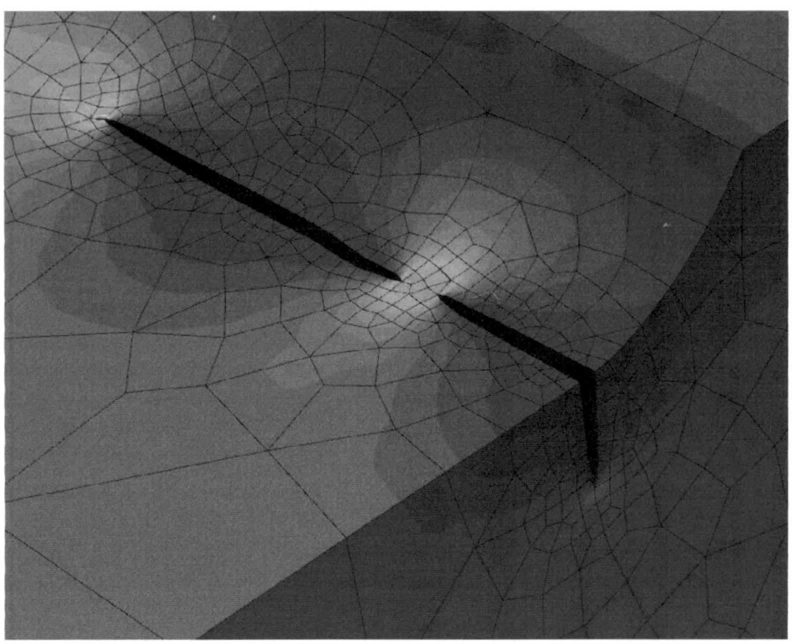

Figure 22. Stress plots after a number of crack growth steps.

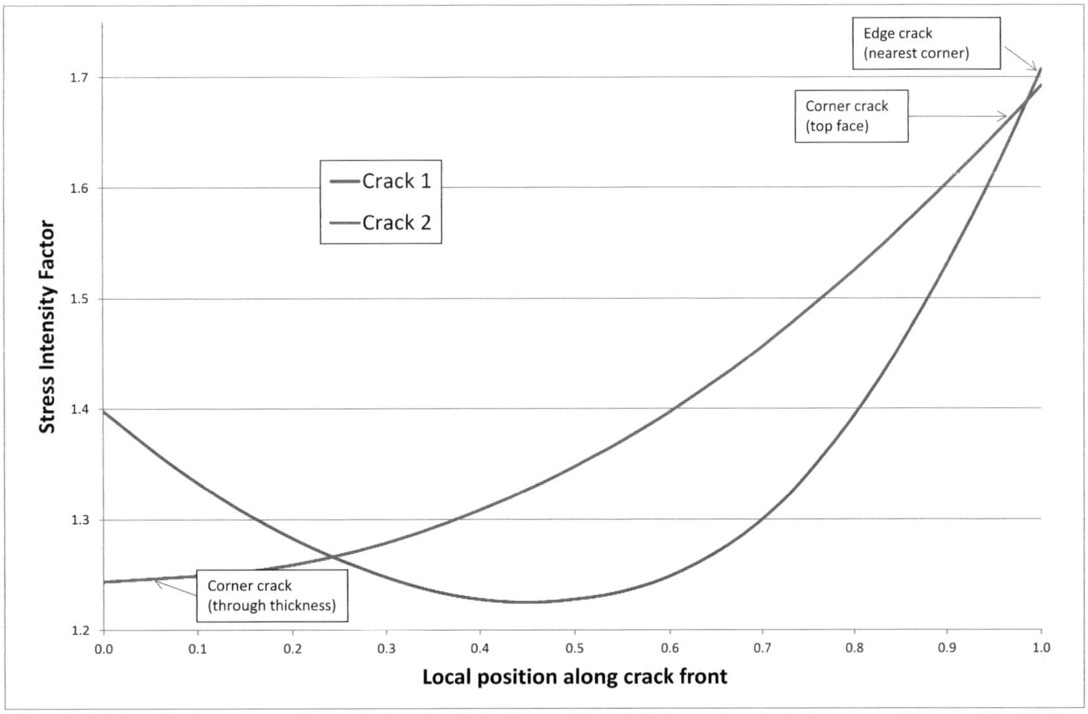

Figure 23. Mode 1 stress intensity factor results for two cracks after a number of crack growth steps.

The simulation could be continued by creating a coalesced crack. The coalesced crack would then grow in a similar way to the previous example, growing through the thickness to transition into a through crack. Such further growth simulation has not been presented here, but would continue automatically.

SUMMARY

The damage caused by galvanic corrosion and subsequent mechanical loading of a specimen consisting of two dissimilar materials exposed to different types of electrolyte has been quantified using a computational modelling approach based on the BEASY software.

The corrosion simulation can predict mass loss rate, corrosion penetration rate, current density, electric fields and polarisation potentials on the specimen exposed to different types of electrolytes. With fatigue loading of the corrosion the damaged part, the future simulation can grow an initial crack through to failure, determining the fatigue life from any given time.

An example of an aluminium stretch part in between two CFRP clamps has been solved for different environmental conditions.

The corrosion damage resulting from anodic dissolution induced a geometry change in the aluminium, which affected the initial distribution of mechanical stress and subsequent crack propagation.

Although the example shown assumed no charges at polarisation property with mass removal, it is possible that loss of surface films could lead to locally changed properties, and accelerated corrosion such as pitting.

The example shown is for specific materials (CFRP and AA2024) and specific geometry and electrolyte. Nevertheless, the results support the following general observations:

- The galvanic effects observed with exposure to a thin film of electrolyte are much higher than in the case of deep electrolyte.
- The distribution of anodic current density (and therefore corrosion damage) becomes significantly more localised as the thickness of the electrolyte decreases. In general, when the sample is exposed to thin films, the

corrosion damage tends to accumulate at the edges near the junctions between dissimilar materials, whereas for bulk electrolytes, the damage becomes more uniformly distributed, extending to farther regions.

- Assuming no changes of the concentration of oxygen in the electrolyte, or of the polarisation curves of the materials involved, the overall mass loss rate in the case of fully immersed samples is always bigger than in the case of exposure to a thin film. This is because the electrical resistance through by the electrolyte is increased when the electrolyte thickness decreases.
- Although it may depend on details of geometry, it is clear that the penetration rate can be much higher with a thin film of electrolyte than with deep electrolyte. The overall mass loss rate, obtained with a thin film of electrolyte, tends as the film thickness increases to the value obtained for the sample in a large tank of electrolyte. Figure 24 shows how the mass loss rate calculated for increasing values of electrolyte thickness using case 2 conditions tend to the limiting value found in case 1.

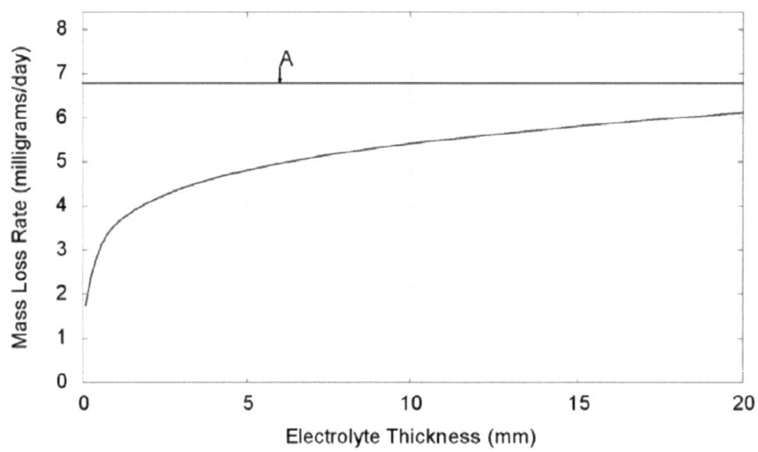

Figure 24. Showing simulation results for mass loss rate as a function of electrolyte thickness. Line (A), indicating mass loss rate = 6.8 mg/day, corresponds to the result obtained in case 1 for the specimen immersed in a large pool of electrolyte.

REFERENCES

[1] Adey, R. and Niku, S. (1988). In: *Galvanic Corrosion*, ASTM Committee G-1 on Corrosion of Metals, Hack, H. (Ed.), ASTM International.

[2] Roberge, P. (2008). "Corrosion Engineering. Principles and Practice". McGraw-Hill.

[3] Peratta A, Hack, T., Adey, R., Baynham J., Lohner, H. (2009). "*Galvanic Corrosion Modelling for Aircraft Environments*". Eurocorr, Nice.

[4] Peratta A. and Adey R (2013). "*Modeling Galvanic Corrosion in Multi-Material Aircraft Structures*", NACE Corrosion.

[5] Palani S, Hack T., Peratta A., Adey R., Baynham J and Lohner H. (2010). "*Validation Of A Galvanic Corrosion Model For AA2024 And CFRP with Localised Coating Damage*". Eurocorr.

[6] Portela, A.; Aliabadi M.H; Rooke, D.P., "The Dual Boundary Element Method: Efficient Implementation for Cracked Problems", International Journal for Numerical Methods in Engineering, Vol. 32, pp 1269-1287, (1992).

[7] Mi Y; Aliabadi M.H., "Three-dimensional crack growth simulation using 6EM", Computers & Structures, Vol. 52, No. 5, pp 871-878, (1994).

[8] Mellings, S., Baynham, J., Adey, R.A., "Advances in crack growth modelling of 3D Aircraft Structures", International Committee on Aeronautical Fatigue, Rotterdam, Netherlands, May 2009.

[9] Rigby R.; Aliabadi M.H; "Decomposition of the mixed-mode J-integral—revisited"; International Journal of Solids and Structures, Volume 35, Issue 17, June 1998, Pages 2073-2099.

The Decoupling of Corrosion and Fatigue for Aircraft Service Life Management

Loris Molent[1], Simon Barter[1], Russell Wanhill[2]

[1]Aerospace Division, Defence Science and Technology Organisation, 506 Lorimer Street, Fishermans Bend, Vic., 3207, Australia.

[2] National Aerospace Laboratory NLR, Amsterdam, the Netherlands.

Abstract: Modern military aircraft have highly optimised metallic structures in order to achieve their performance objectives. These structures therefore tend to be uniformly highly stressed and susceptible to fatigue cracking from small discontinuities. Aluminium alloys are the most widely used structural metals, and they are susceptible to corrosion as well as fatigue. There is thus the potential for interaction between both these forms of degradation, leading to reduced service lives and/or costly repair or modification programmes. The fatigue and environmental degradation management of aircraft is generally well understood, as may be seen from typical Aircraft Structural Integrity Management Plans (ASIMPs). However, corrosion repair involves little or no remedial action beyond *find and fix*. This policy can cope with straightforward go/no-go assessments based on original equipment manufacturer's clean-up limits on static strength, but there are no operator-approved tools for assessing the structural significance of *in situ* corrosion on the fatigue life. A factor contributing to this problem is uncertainty about the interactions between corrosion and fatigue, since corrosion is time-dependent while fatigue depends on the cyclic load history.

Much aircraft corrosion can be classified as pitting corrosion. This is a form of localised metal attack and is known to cause fatigue cracking. Hence pitting corrosion is specifically addressed in the present paper, which summarises a body of work that demonstrates that corrosion is predominately a ground-based phenomenon, whilst fatigue mainly occurs in flight. In other words, these two degradation mechanisms generally *do not* interact. This observation is significant and in contrast to conventional thinking, and implies that laboratory generated and realistically simulated fatigue data are applicable to in-service aircraft in the presence of pitting caused by corrosion. Thus, given a suitable metric to characterise the significance of the pits (or more accurately the crack-like nature of the pits) a fatigue analysis can be conducted to determine the remaining life of a pit-corroded structure. In this work we use and discuss the Equivalent Pre-crack Size (EPS) metric.

INTRODUCTION

Background

Corrosion is an important source of fatigue cracking in aluminium alloy aircraft structures, causing fatigue failures in all types of aircraft and a variety of components [1]-[9]. Corrosion pitting of propellers and rotor blades is particularly dangerous [2][3] for the following reasons: (i) only very small defects can be tolerated before the high frequency of the loading will begin rapid fatigue crack growth (FCG), (ii) the time to failure is correspondingly short, and (iii) such failures usually - but not always [6] make the aircraft uncontrollable. These types of rapid fatigue failure are mentioned here particularly because they are incontrovertible instances of 'decoupled' corrosion and fatigue. In other words, corrosion pitting occurred before the fatigue damage could propagate, and once FCG began there was no further contribution by corrosion. An example of fatigue cracking growing from pitting is given in Figure 1.

Figure 1. Fatigue crack growth from a corrosion pit on a 2014-T651 (peak aged) aluminium alloy forged propeller blade [2]. The blade surface had not been regularly cleaned, allowing corrosion to occur.

Opinions about other types of failures involving aluminium alloy corrosion and fatigue are less clear-cut: sometimes it has been assumed that corrosion continues to play a significant role during FCG, and sometimes not. These contrasting assumptions have influenced the choice of laboratory test conditions to support the development of life-prediction models incorporating both corrosion and fatigue [10][11]. Over the last 20 years there has been considerable effort to develop such models. These have become increasingly sophisticated, culminating in an internationally-supported life management methodology known as the Holistic Structural Integrity Process (HOLSIP) [11].

No matter how sophisticated the models, it is essential that they take actual case histories into account. HOLSIP does this, but other models, most notably those investigating corrosion and fatigue of the industry standard alloy 2024-T3 [10]-[14], have not. Experiments and modelling for this alloy were intended to be applicable to corrosion and fatigue problems in transport aircraft pressure cabins, specifically the longitudinal lap splices. However, for these problems the basic assumption that fatigue initiation would be due to corrosion pitting is incorrect; this was shown by investigating lap splices taken from service aircraft [15]. Also, it appeared that there was either a corrosion problem or a fatigue problem, i.e. these problems could be decoupled. Similarly, although the Aloha Airlines accident involved corrosion and fatigue; the pressure cabin failed because of an initial corrosion problem (skin splice and tear strap disbonding, *not corrosion pitting*) followed by a fatigue problem [16].

Present work

The possibility and actuality of decoupling corrosion and fatigue when considering life prediction is important for aircraft service life management. Firstly, there is the basic problem that corrosion is time-dependent, while fatigue depends on the cyclic load history. This means that inspection and maintenance schedules for corrosion and fatigue are largely incompatible. Secondly, corrosion repair involves little or no remedial action beyond *find and fix*. This policy can cope with straightforward go/no-go assessment based on OEM clean-up limits on static strength, but there are no operator-approved tools for assessing the structural significance of *in situ* corrosion on the fatigue life. A factor contributing to this problem is uncertainty about the interactions between corrosion and fatigue in the service environment, a situation that was recognised more than 40 years ago [17].

In these respects it would be advantageous to be able to specify maintenance actions based on the understanding that corrosion and fatigue are decoupled. In particular, it might be possible to delay the removal/repair of corrosion until a more opportune and scheduled time if it is found during active service, with the delay based on corrosion-unaffected FCG life assessment from the 'initial' observed corrosion damage.

In the present paper service cases for several aircraft and aircraft types are reported that show that pitting is the principal form of corrosion that can lead to fatigue, and that this is predominantly a ground-based phenomenon, while the fatigue that grows from these pits mainly occurs in flight. Consequently, these two degradation mechanisms generally *do not* interact and are effectively decoupled. The paper then goes on to show that a suitable metric, the Equivalent Pre-crack Size (EPS) can characterise the pitting and enable FCG analyses of the remaining lives of corroded components. Finally, we propose a framework for implementing the method into the service life management schedules.

THE DECOUPLING OF CORROSION AND FATIGUE

Case studies

As mentioned above, Wanhill and Koolloos [15] showed that there was no primary association between fatigue and corrosion in 2024-T3 Alclad aluminium alloy lap splices from several civil transport aircraft. This is supported by the study carried out by Trathen [18], who surveyed the corrosion rates in internal bays of several types of Royal Australian Air Force (RAAF) aircraft using corrosion monitors. These aircraft types included the F-111, Seahawk, AP3C and B707. This study was the first to report that there appeared to be little corrosion activity occurring during flight. For instance, for the F-111 over a period of 4536 hours (189 days and 64 flights) there were 105 monitored corrosion events[24] of which 93 were detected while the aircraft was on the ground; 10 extended into a flight; while only 2 that began on the ground continued during and after a flight.

Barter and Molent [19]-[21] subsequently found that fatigue cracks in 7050-T7451 aluminium alloy components in F/A-18 aircraft showed fracture surface corrosion only when there had been extended periods of downtime (inactivity). This is supported by the work of Burns *et al.*, who state that aircraft fatigue cracking occurs in cold and dry in-flight environments [22] where corrosion activity is largely suspended[25]. Jones and Hoeppner also showed that in some cases fatigue initiated preferentially from material discontinuities despite the presence of corrosion pits and a corrosive environment [23]. Finally Crawford *et al.* showed that FCG rate was not affected by the presence of corrosion pits [24].

Table 1 presents several cases, including those mentioned above, where fatigue had grown from corrosion damage and where a corrosive environment existed during service while these cracks were growing. In the main these case studies involved the use of quantitative fractography (QF) measurements and observations of the fracture surfaces to support an assessment of the FCG rate under the experienced conditions. Generally, and particularly for aluminium alloys, the case studies 1-8 show that the presence of a corrosive environment did not have a marked effect on FCG. Note also that for case 6 the corrosion environment enabled some stress corrosion cracking (SCC) that most probably occurred during long downtimes (aircraft on the ground), and that subsequent FCG was retarded because the stress corrosion cracks were perpendicular to the fatigue cracking[26].

Notwithstanding the results of case studies 1-8, Table 1 includes two cases, 9 and 10, where there *was* some interaction between corrosion and FCG: either apparently faster FCG (case 9) or extension of fatigue cracking by SCC (case 10). As noted in Table 1, these cases were due to unusual circumstances:

- For case 9 there was a particularly aggressive environment (salt-containing water trapped in a crevice from which it could not escape during flight) and the presence of tensile residual stresses in the hinge lug (unexpected because the lug wall underwent net section yielding during cold expansion of the hole in which cracking later occurred).
- For case 10 the material (ultra-high strength steel) was very susceptible to SCC and also contained high tensile residual stresses from local yielding of the pivot fitting during proof testing of the airframe.

These two cases are included here as caveats: the methodology proposed later in this paper is applicable specifically to pitting corrosion, and may not be helpful for cases where there are other possible effects due to corrosion.

Figures 2-5 give examples of the corrosion and fatigue associations for cases 2, 3, 6 and 7.

[24] Corrosion event = significant corrosion induced current on a monitor's detector indicating the presence of condensed water.

[25] This may not apply to aircraft operating for extended periods at low levels in hot and salty environments, or to structural locations in which such environments accumulate.

[26] This is a general observation when SCC is associated with fatigue in heavily wrought aluminium alloys, since SCC nearly always occurs in the short-transverse plane, which is perpendicular to both the long-transverse and transverse-longitudinal planes that are most likely to be fatigue cracking planes.

Table I. Case studies involving corrosion and fatigue.

Case	Aircraft / Component	Material	Loading	Fatigue Initiation	Comments	Ref.
1	F-28, BAC 1-11, 727-100 lap splices	2024-T3 Alclad	Service pressurisation cycles	Faying surfaces near or at rivet hole corners	***No*** primary association between fatigue and corrosion	[15]
2	F/A-18 A-D vertical tail attachment stub frame	7050-T7452	Service (including significant aero-buffet)	Etch pits	Fracture surface showed corrosion areas corresponding to significant downtime periods. ***No*** evidence that FCG rate was affected by corrosion (see Figure 2).	[19]
3	F/A-18 A-D wing carry-through bulkhead	7050-T7451	Service and mini-FALSTAFF	Etch pit	In-service portion of crack showed exposure to corrosive environment. ***No*** evidence that FCG rate was affected by corrosion. (see Figure 7)	[20]
4	F/A-18 A wing carry- through bulkhead	7050-T7451	Mini-FALSTAFF	Corrosion pit	Pitting occurred before laboratory testing. ***No*** evidence that FCG rate was affected by corrosion (see Figure 3).	[21]
5	F/A-18 A-D MLG wheel hub	2014-T6	Service	Corrosion pit	In-service exposure: fatigue initiated from pitting. ***No*** evidence that FCG rate was affected by corrosion, despite harsh environment.	[25]
6	C130 Hercules FS 497 bow beam fitting	7075-T6	Service	Corrosion pits	In-service exposure produced pits, and also stress corrosion cracks perpendicular to fatigue cracking. Stress corrosion cracks apparently retarded fatigue crack growth (see Figure 4).	[26]
7	Iroquois main rotor yoke	4340 steel	Service	Corrosion pit	In-service exposure. Once fatigue cracking had started, it did not appear to have been influenced by corrosion (although corrosion had attacked the fatigue fracture surface (see Figure 5).	[27]
8	Macchi centre section lower spars (several)	4340 steel	Service/test	Corrosion pits	In-service exposure produced the initial pits ***and*** also FCG. Subsequent laboratory testing showed ***No*** evidence that the corrosive conditions had had any effect on the FCG rate.	[28]
9	F/A-18 A-D trailing edge flap hinge lug	7050-T73652	Service	Corrosion pit	In-service exposure. Early FCG appeared affected by corrosive conditions, as seen from brittle-looking progression markings. High tensile residual stresses and a water-entrapping crevice probably influenced this.	[29]
10	F-111 wing pivot fitting (several cases)	D6ac steel	Service	Corrosion pit	In-service exposure caused surface pitting that transitioned to FCG. A long downtime resulted in crack extension by stress corrosion cracking.	[30]

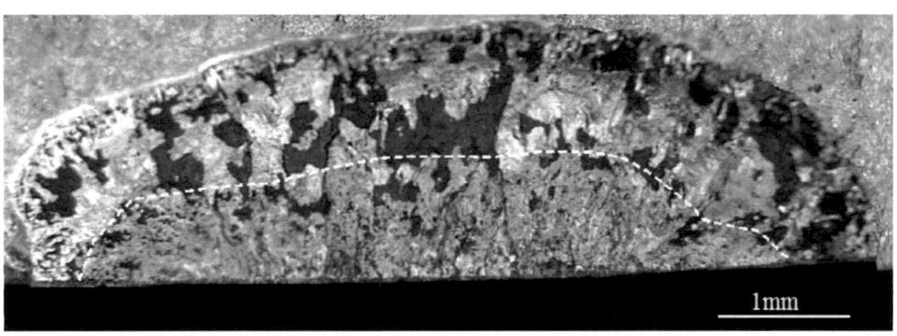

Figure 2.One of the cracks in a service-exposed stub frame from an F/A-18A aircraft (case study 2 in Table I) prior to cleaning. The white dashed line indicates the approximate boundary of light corrosion attack.

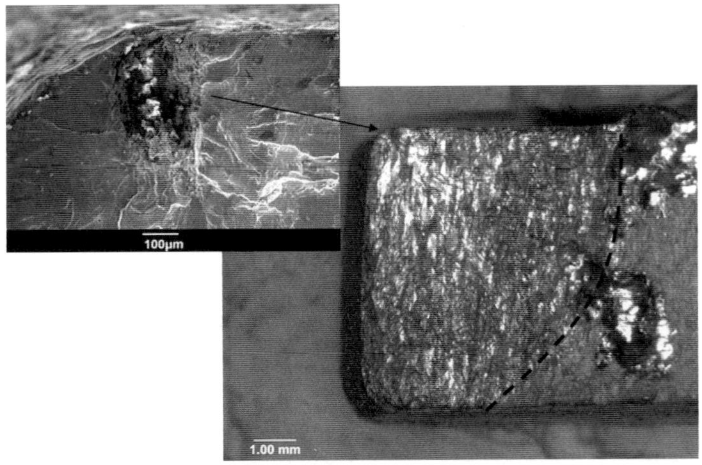

Figure 3. A fatigue crack in an F/A-18A-D wing carry-through bulkhead (case study 4 in Table 1). The black dashed line gives the approximate final depth of the fatigue crack. The large pit from which the fatigue cracking initiated is shown in the insert to the upper left of the main view. N.B: the shiny large areas beyond the fatigue crack in the main view were caused by post-failure contact between the fracture halves.

Figure 4. One of the cracks in a C130 bow beam fitting, showing SCC perpendicular to the fatigue cracking (case study 6 in Table 1). The fatigue crack front has clearly been retarded by the stress corrosion cracks.

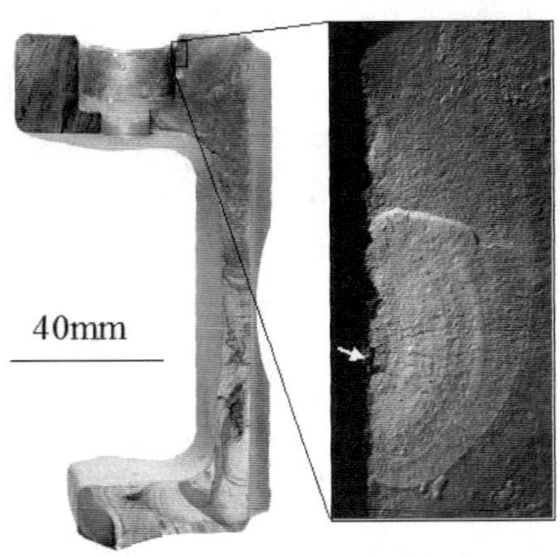

Figure 5. Fracture of a helicopter main rotor yoke showing fatigue cracking starting from corrosion pitting (white arrow): case study 7 in Table 1. Although the fatigue crack surface was corroded there was no evidence that corrosion had accelerated the cracking [27].

Coupon tests

A test programme was carried out at the Defence Science and Technology Organisation to further investigate the effects of corrosion pitting and ground-based environments on FCG. Several hourglass aluminium alloy 7050-T7451 coupons were corrosion pitted in 3.5% sodium chloride solution, washed and dried, and subjected to block-spectrum representing a fighter wing root bending moment sequence. Twenty blocks of spectrum loading, equivalent to about 300 flying hours, were applied at a nominal frequency of 10 Hz in laboratory air. The coupons were then removed, wetted and placed in an environmental chamber at 37 °C with a relative humidity of 98% for one week. Then the coupons were fatigue loaded again for five blocks, removed from the test machine and again placed in the environmental chamber for one week. This (five block + 1 week exposure) schedule was repeated until coupon failure by fatigue, whereupon the crack growth was measured by QF and the fatigue fracture surface examined for evidence of corrosion influences.

The results are illustrated in Figure 6. Each fatigue crack grew from a corrosion pit, and each exposure period discoloured the fatigue fracture surface, as may be seen in Figure 6B. This type of discoloration can also be seen in Figure 7. This shows a service crack (the discoloured region bounded by the dashed black line) that was subsequently extended by fatigue without intermittent exposure (case 3 in Table 1).

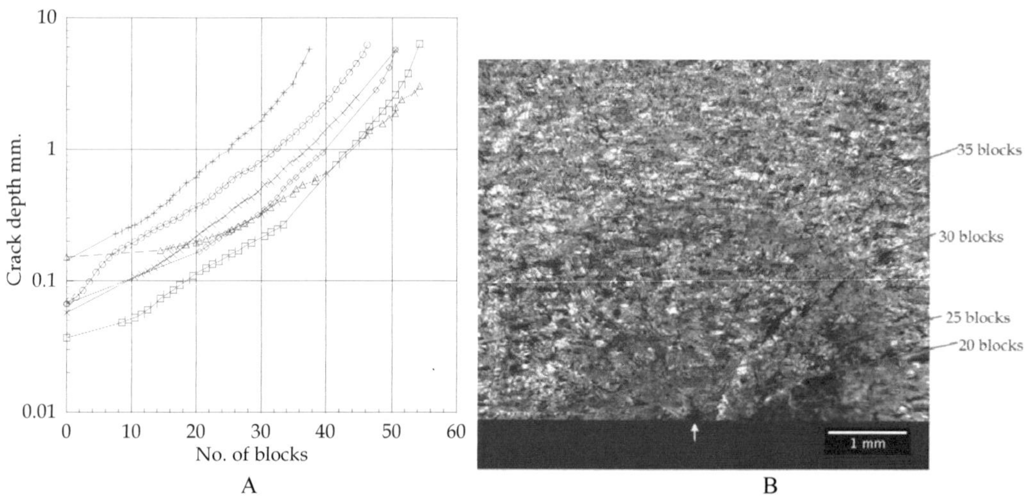

Figure 6. (A) The FCG curves for all of the coupon cracks plotted on log crack depth vs. linear life scales, starting from the depth of the fatigue-initiating discontinuities which were pitting in each case. (B) The fracture surface of coupon LM417 after the intermittent exposures, and indications of the crack depths after each exposure. Note that the darker discoloration of the fracture surface back towards the fatigue origin is the result of repeated intermittent exposures, i.e. with each exposure the corrosion on the fracture surface increased.

Figure 7. The fatigue origin region of an F/A18 A-D bulkhead crack that was discoloured due to service exposure during its growth (case 3 in Table 1). Fatigue crack growth in laboratory air began at the boundary marked by the dashed black line [20] and the fracture surface is clean and corrosion-free.

The FCG curves in Figure 6 were plotted from the depth of the corrosion pit from which each crack grew. In Figure 8A the curves are re-plotted from initial depths that have been corrected for their equivalent crack-like depths (EPS), see the next main section of this paper). In Figure 8B these curves are compared with FCG curves (green) obtained from a different coupon study of the same material, subjected to the same spectrum loading with the same peak net section stress, but without any exposure to a corrosive environment. Comparing these results shows that the intermittent environmental exposures had no appreciable effect on the FCG rates in the tests.

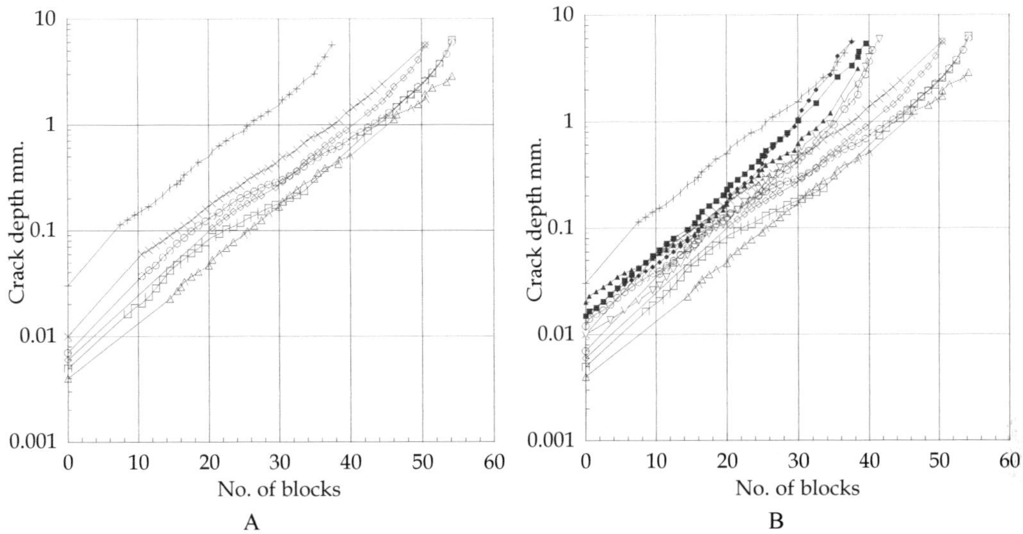

Figure 8. (A) FCG curves for the cracks initiating from corrosion pits after correcting for the effectiveness of the pits as crack-like initiators. (B) shows the FCG curves from (A) compared with crack growth curves from a different coupon study of the same material subjected to the same spectrum loading with the same peak net section stress, but without any exposure to a corrosive environment.

Summary

From the case studies and the coupon test programme it has been found that:

1. For case studies 1-8 there was no evidence that exposure to the in-service environment resulted in enhanced FCG rates. This was confirmed by laboratory testing in air: in other words, fatigue testing in laboratory air[27], under spectrum loading representative of the service load history, will result in similar FCG rates.
2. Prolonged periods of downtime-induced corrosion were evident as small bands of discoloration near the fatigue crack tips. However, the periodic presence of these bands did not affect the overall crack growth.
3. Case studies 9 and 10 were exceptional (and useful caveats) in that the FCG rates *were* influenced by the environmental conditions and also residual stresses. However, in detail these cases were very different from each other.

From these points we may conclude that for high strength aerospace aluminium alloys it will often be possible to use in-service FCG data or laboratory spectrum FCG data interchangeably and complementarily to predict fatigue crack growth with intermediate exposures to a corrosive environment.

LEAD CRACKS AND THE EPS CONCEPT

Subject to a few caveats [32], lead fatigue cracks (i.e. the ones leading to failure of a component or coupon) have the following *general* characteristics:

1. Crack growth from an initial discontinuity commences essentially from the time the aircraft first experiences in-service loading: a regular crack front forms and stabilises quickly.
2. The cracks grow in an approximately exponential manner up the point where quasi-static failure commences.

[27] Note humidity significantly influences the FCG rate [31].

3. For some discontinuity types (e.g. mechanical damage, small pits due to surface chemical etching) the equivalent crack-like depth (or EPS) correlates strongly with the discontinuity depth [33].
4. For some aluminium alloys a typical (mean) EPS for these discontinuities is 10μm [33].

The EPS is defined as the back-projection to time zero from the QF derived crack growth curve (see [32][33]), as shown in Figure 8A and 8B. In these figures it is seen that the EPS is always smaller than the depths of the corrosion pits. This is because corrosion pits are voluminous [34][36] and do not have sharp profiles like fatigue cracks, and also tend to induce multiple initiations on different planes that can interfere with one another or may be poorly located for optimum growth.

CORROSION PIT STRUCTURAL ASSESSMENT METHODOLOGY

In order to consider the impact of pitting corrosion on the structural integrity of an airframe, its structural certification basis must be understood. For the following discussion a RAAF reference Structural Certification Basis (SCB) will be considered, namely the UK Defence Standard 970 [37][28]. The salient points from this standard are:

1. Corrosion is not considered with respect to structural integrity, only good design practice;
2. The cumulative probability of failure (CPOF) for airframe structural fatigue must be limited to 1/1000 during the life of type[29]. This is principally achieved by following the SCB guidelines during design, and conducting a representative full-scale fatigue test (FSFT). The demonstrated FSFT-simulated flight hours until reaching a residual strength criterion are then divided by a scatter factor (SF) to achieve the required CPOF. (Here the FSFT is assumed to be representative of the mean quality and build standard of the fleet, and the applied load spectrum is assumed to be representative of the mean aircraft utilisation);
3. For primary structure it is required that any crack in the FSFT does not reach the critical size corresponding to 1.2 Design Limit Load (DLL) within a lifetime of the spectrum loading times the SF; and
4. Fatigue driving loads are monitored (i.e. tracked) in service and periodically reviewed.

The lead crack framework can be used to graphically illustrate these certification requirements, see Figure 9. For an optimally designed airframe (i.e. one that just achieves the design requirement at the end of the FSFT) the lead (i.e. largest) fatigue crack will be just below the critical crack size (a_{crit}), which is typically about 10 mm [30] for highly stressed structures. Also, since the test article is assumed to represent the mean quality of the fleet, it can be assumed that this lead crack had initiated from an aircraft production-quality defect or discontinuity typically represented by an EPS of 0.01mm deep [32][33].

For the purposes of illustration it is assumed (i) that the required service life is 6000 flight hours for a typical SF = 3.0; and (ii) the exponential crack growth labelled 'Max certification basis' in Figure 9 is the fastest possible (in the ambient FSFT environment) for an airframe that has achieved SCB certification. The other curves 'Max pit day one' and 'Max pit at 4000 hours' will be discussed in the following subsection 'Crack growth from corrosion pits'.

---

[28] The USAF damage tolerance approach [38] could also have been considered. However, the present method has major differences (see [32]) including (i) use of a measured initial discontinuity size or EPS rather than an assumed initial flaw size, and (ii) use of a specified (standard) residual strength criterion.

[29] The SCB only considered (or assumed) two critical items (e.g. left and right wing) for a total probability of failure of 1/1000.

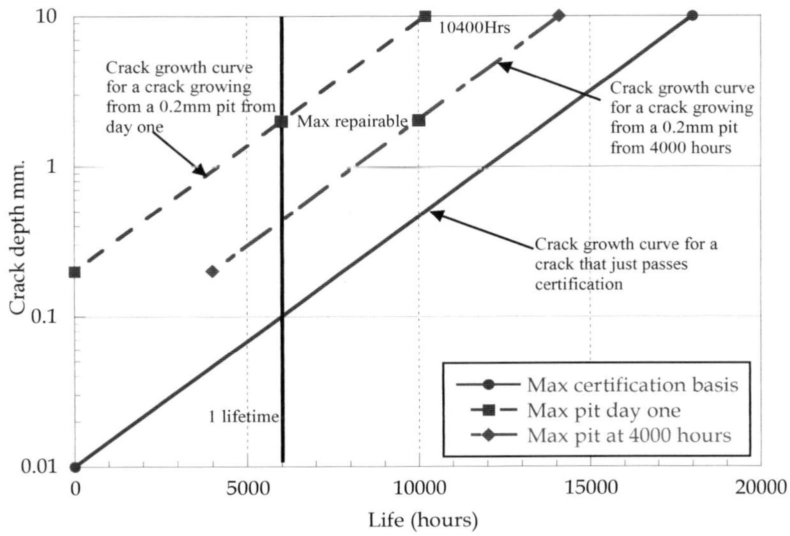

Figure 9. Schematic combination diagram for fatigue crack growth from a production discontinuity and a corrosion pit (see text above) in an aluminium alloy component from a tactical aircraft. For fatigue crack growth from a corrosion pit it is assumed that the fatigue-initiating pit is 0.2 mm deep, and that the crack grows at the maximum growth rate, i.e. the pit is fully fatigue-like in its effect on subsequent growth. Also shown is the case when the corrosion pit initiates a fatigue crack after 4000 hours. An assumed maximum repairable crack depth of about 2 mm is also shown.

Crack growth from corrosion pits

Unlike cracks growing from initial material discontinuities, corrosion is likely to occur only after some time in service, namely once the corrosion protection scheme has at least partially degraded. Whether or not a fatigue crack will grow from a corrosion pit will depend on the local stress levels and microstructural texture: not all pits will grow a crack [23]. Since the time to initiation of corrosion pitting cannot be predicted, the following rationale and assessment considers only the case of pitting corrosion that has been detected during routine inspections.

If left undetected or unrepaired, corrosion in a certified airframe will impact structural integrity only if a corrosion-initiated fatigue crack reaches the critical size before a crack growing from an initial production discontinuity does so. Also it is important to note, as discussed earlier, that corrosion is likely to occur when the aircraft is on the ground. Thus the growth of the fatigue crack itself (under in-flight loading) will not be significantly affected by a corrosion-inducing environment[30]. In other words, the FCG data generated or observed in ambient environments will be appropriately conservative for the assessment of this crack growth. If there is a concern about corrosion interaction, the application of corrosion inhibiting compounds (CICs) should alleviate this, as mentioned in the following main section of this paper.

It is proposed here that a corrosion pit will eventually reach a particular depth capable of starting FCG. For this discussion a corrosion pit depth of 200 μm for typical aircraft aluminium alloys is assumed (see [33][31] [39][40]), noting that the pit's EPS would be considerably smaller. The pit is assumed to be located at the highest stressed location in the airframe, i.e. a fatigue crack initiating from the pit will grow at the maximum growth rate.

From Figure 9 it can be seen that if corrosion to the fatigue-initiating depth were already present when the aircraft entered service (Max pit day one), then a corrosion-initiated fatigue crack would 'win the race' to a_{crit}. However, the *a priori* presence of such a corrosion pit is most improbable. It is almost certain that - with standard corrosion protection systems, corrosion pitting to the fatigue-initiating depth would take several thousand hours. As an example, Figure 9 shows fatigue cracking beginning from a corrosion pit after 4000 hours (Max pit at 4000 hours). This fatigue crack would not reach a_{crit} until well beyond the service life.

On the other hand, instead of leaving a corrosion-initiated fatigue crack, or cracks, to grow unhindered, a logistically-motivated repair option should be considered. If we assume, for example, that a 2mm deep crack is repairable, this still

[30] Note: By this we mean that in-service FCG will occur under environmental conditions similar to those for standard laboratory testing in ambient air. In both cases there is a significant environmental effect compared to FCG *in vacuo* (this is very well documented in the fatigue literature, e.g. [35]).

[31] Note: the number of corrosion pit EPS values in [33] is not considered statistically significant.

gives a fatigue crack growth interval of 6000 hours ('Max repairable' in Figure 9). This is a most useful interval, offering several opportunities for carrying out scheduled repairs.

THE MANAGEMENT OF PITTING CORROSION

The assessment methodology described in the previous section of this paper (also see [34][36]) currently applies only to pitting corrosion and ensuring the structural *safety*. The methodology is designed such that the removal/repair of pitting corrosion can be delayed until a more opportune *and* scheduled time[32]. This may be the next scheduled maintenance, when any pitting corrosion identified in the interval between schedules would be repaired in a planned fashion. The management framework is as follows:

1. Walk-around or other scheduled inspections are typically those when corrosion damage is found. These inspections would continue.
2. If corrosion were found, its type and severity[33] (i.e. extent or coverage[34]) would be determined. It would also be inspected for associated fatigue cracks. Any detected fatigue cracks would be managed using established methods[35]. The specific non-destructive testing (NDT) method used, and hence the NDT detectability threshold, should be governed by the local geometry. The presence of pitting corrosion should not affect the sensitivity of a method like eddy current.
3. If the corrosion *is* pitting, and minor, then the damaged area would be treated periodically[36] with CICs and left unrepaired until the next scheduled maintenance. Application of a CIC is intended to prevent further spread of the corrosion or the activation of other corrosive modes of damage, thereby limiting any potential increase in the difficulty of repair.
4. During the remainder of the unrepaired interval the corrosion should ideally be re-inspected to check for further spreading or for (detectable) fatigue cracking. The demands on detectability will determine the NDI technique to be used.
5. Ideally, any corrosion damage should be mapped by location to facilitate its removal during the next scheduled maintenance.
6. At the next appropriate maintenance the corrosion would be removed: this is normally done by carefully blending-out. Assuming that this repair can be done within the allowable thickness limits (as it should be when the assessment methodology is correctly used), the corrosion protection system would be restored and the aircraft returned to service.

It is worth noting here that in an area of a Seahawk airframe where corrosion had been previously found, a corrosion monitor recorded no corrosion events after the area had been sprayed with CICs [18]. This indicates the effectiveness of such a maintenance action at preventing further corrosion, and also that this is a method of ensuring that corrosion will not affect any fatigue cracking that may occur.

CONCLUSIONS

Pitting corrosion is known to be a major maintenance driver for aircraft production aluminium alloys. This paper summarises several examples of corrosion-induced pitting or fatigue cracks exposed to a potentially aggressive environment. These examples support the concept that fatigue crack growth (which occurs primarily in the sky) is not significantly affected by environmental exposure, which occurs mainly on the ground.

A safety-based methodology (or framework) is presented to justify leaving unrepaired aluminium alloy pitting corrosion for appropriate intervals. The method relies on the fatigue metrics of pitting corrosion, namely the effective pit depth for fatigue crack initiation as well as the lead crack concept. By delaying the repair of pitting corrosion until the next scheduled maintenance (rather than 'find and fix'), it is considered that this methodology will save considerable resources and improve aircraft availability. Looking further, the development of an analytical tool capable of accurately assessing pitting corrosion effects on the durability of a structure would be a major advancement in aircraft structural integrity management.

---

[32] Alternatively if corrosion is found on an aircraft during depot maintenance, the method can be used to justify the need for an immediate full-fleet inspection and allow time to plan fleet action without compromising fleet safety.

[33] The definitions from [41] for light, moderate and severe corrosion are useful here.

[34] Minor severity here relates to the potential for static failure modes related to the spread of corrosion and the depth and amount of material to be removed in a subsequent repair.

[35] Note that the probability of multi-site or widespread cracking effects is considered very remote in view of the short intervals that it is expected that the corrosion would remain in service (i.e. next convenient maintenance period) once it had been detected, if the methodology suggested here were to be followed.

[36] Defined by the CIC type and environment where the aircraft is based.

REFERENCES

[1] Cohen, B., Corrosion fatigue in the aerospace industry, Corrosion Fatigue: Chemistry, Mechanics and Microstructure, (Editors O.F. Devereux, A.J. McEvily and R.W. Staehle), 65-83, National Association of Corrosion Engineers, Houston, Texas, USA, 1972.

[2] Gestel, G.F.J.A. van, Investigation into the cause of failure of a McCauley propeller blade, National Aerospace Laboratory NLR Technical Report NLR TR 77086 C, Amsterdam, the Netherlands, July 1977.

[3] Wanhill, R.J.H., Graaf, E.A.B. de and Delil, A.A.M., Significance of a rotor blade failure for fleet operation, inspection, maintenance, design and certification, Paper Nr. 38, Fifth European Rotorcraft and Powered Lift Aircraft Forum, Amsterdam, the Netherlands, September 1979.

[4] Forsyth, P.J.E., The fatigue performance of service aircraft and the relevance of laboratory data, Journal of the Society of Environmental Engineers, Vol. 19, 3-10, March 1980.

[5] Campbell, G.S. and Lahey, R., A survey of serious aircraft accidents involving fatigue fracture, International Journal of Fatigue, Vol. 6, 25-30, 1984.

[6] Kolkman, H.J., On the cause of failure of a propeller blade of Air Tractor PH-CPR, National Aerospace Laboratory NLR Technical Report NLR TR 87086 C, Amsterdam, the Netherlands, July 1987.

[7] Latham, M.C., Harlow, D.G. and Wei, R.P., Nature and distribution of corrosion fatigue damage in the wingskin fastener holes of a Boeing 707, ICAF ' 2001, Design for Durability in the Digital Age, (Editor J. Rouchon), Vol I, 469-483, Cépaduès-Éditions, Toulouse, France, 2002.

[8] Hoeppner, D.W., A review of corrosion fatigue and corrosion/fatigue considerations in aircraft structural design, ICAF 2003, Fatigue of Aeronautical Structures as an Engineering Challenge, (Editor M. Guillaume), Vol. I, 425-438, EMAS Publishing, Sheffield, UK, 2004.

[9] Barter, S.A., Sharp, P.K., Holden, G. and Clark, G., Initiation and early growth of fatigue cracks in an aerospace aluminium alloy, Fatigue and Fracture of Engineering Materials and Structures, Vol. 25, 111-125, 2002.

[10] Wei, R.P. and Harlow, D.G., A mechanistically based probability approach for predicting corrosion and corrosion fatigue life, ICAF 1993, Durability and Structural Integrity of Airframes, (Editor A.F. Blom), 347-366, Engineering Materials Advisory Services, Warley, UK, 1993.

[11] Bellinger, N.C. and Liao, M., Corrosion and fatigue modeling of aircraft structures, Chapter 8 in: Corrosion Control in the Aerospace Industry, (Editor S. Benavides), 172-191, Woodhead Publishing Limited, Cambridge, UK, 2009.

[12] Chen, G.S., Gao, M., Harlow, D.G. and Wei, R.P., Corrosion and corrosion fatigue of airframe aluminum alloys, FAA/NASA International Symposium on Advanced Structural Integrity Methods for Airframe Durability and Damage Tolerance, NASA Conference Publication NASA-CP-3274, (Editor C.E. Harris), Part 1, 157-173, NASA Langley Research Center, Hampton, Virginia, USA, 1994.

[13] Chen, G.S., Wan, K. C., Gao, M., Wei, R.P. and Flournoy, T.H., Transition from pitting to fatigue crack growth − modeling of corrosion fatigue crack nucleation in a 2024-T3 aluminum alloy, Materials Science and Engineering A, Vol. A219, 126-132, 1996.

[14] Walde, K. van der and Hillberry, B.M., Characterization of pitting damage and prediction of remaining fatigue life, International Journal of Fatigue, Vol. 30, 106-118, 2008.

[15] Wanhill, R.J.H. and Koolloos, M.F.J., Fatigue and corrosion in aircraft pressure cabin lap splices, International Journal of Fatigue, Vol. 23, S337-S347, 2001.

[16] Aircraft Accident Report, Aloha Airlines, Flight 243, Boeing 737-200, N73711, Near Maui, Hawaii, April 28, 1988, NTSB Report No. NTSB/AAR-89/03, National Transportation Safety Board, Washington, D.C., USA, 1989.

[17] Anderson, W.E., Fatigue of aircraft structures, International Metallurgical Reviews, Vol. 17, 240-263, 1972.

[18] Trathen, P., Corrosion monitoring systems on military aircraft, Proceedings of the 18th International Conference on Corrosion, Perth, Australia, 20-24 November, 2011.

[19] Barter, S.A. and Molent, L., Investigation of an in-service crack subjected to aerodynamic buffet and manoeuvre loads and exposed to a corrosive environment, Proceedings of the 28th International Congress of the Aeronautical Sciences, Brisbane, Australia, 23-28 September, 2012.

[20] Barter, S.A. and Molent, L., Service fatigue cracking in an aircraft bulkhead exposed to a corrosive environment, Engineering Failure Analysis, Vol. 34, 181–188, 2013.

[21] Barter, S.A. and Molent, L., Fatigue cracking from a corrosion pit in an aircraft bulkhead, Engineering Failure Analysis, Vol. 39, 155-163, 2014.

[22] Burns, J.T., Gangloff, R.P. and Bush, R.W., Effect of environment on corrosion induced fatigue crack formation and early propagation in aluminum alloy 7075-T651, Proceedings of the Department of Defense Corrosion Conference, La Quinta California, USA, 31 July–5 August, 2011.

[23] Jones K. and Hoeppner D.W., The interaction between pitting corrosion, grain boundaries, and constituent particles during corrosion fatigue of 7075-T6 aluminum alloy, International Journal of Fatigue, Vol. 31, 686–692, 2009.

[24] Crawford B.R., Loader C., Hay D.G., Urbani C., Stonham A.J. and Spence S.H. initiation and growth of fatigue cracks from pits in pre-corroded 7010-T7651, proc. Structural Integrity and Failure - September 2006.

[25] Smith, C. and Donnelly, D., Case studies of fatigue failures in defence aircraft components, Advanced Materials Research, Vols. 891-892, 81-86, 2014.

[26] Athiniotis, N., Examination of cracked bow beam fitting FS 497 from C130 Hercules A97-004, Department of Defence, Aeronautical and Maritime Research Laboratory, Letter report M37/95/NA, 1995.

[27] Lamb, S., Goldsmith, N. and Bishop, B., Interim report, Iroquois main rotor yoke investigation, Department of Defence, Aeronautical Research Laboratory, Letter report M57/85/SRL/NTG/BCB, 1985.

[28] Athiniotis, N., Barter, S. A., Blaricum, T. van, Clark, G. and Sammut, L., RAAF Macchi MB326H centre section lower boom fatigue testing, Technical Report DSTO-TR-0328, AR-009-681, Department of Defence, Defence Science and Technology Organisation, Aeronautical and Maritime Research Laboratory, May 1996.

[29] Barter, S.A. Sharp, P.K. and Clark, G., The failure of an F/A-18 trailing edge flap hinge, Engineering Failure Analysis, Vol. 1/4; 255-266, 1994.

[30] Cox, A.F., Goldsmith, N. T., Bland, L. M., Glanvill, D. W. and Patching, C. A., Preliminary investigation into failure of F-111C A8-129 wing pivot fitting in cold proof load test, Defence Science and Technology Organisation, Aeronautical Research Laboratory, ARL-MAT-Tech-Memo-383, ARL-STRUC-Tech-Memo-360, 1983.

[31] Bray G.H., Bucci R. J. and Brazill R. L., Lessons Neglected: Effects of Moist Air on Fatigue and Fatigue Crack Growth in Aluminum Alloys, Materials Science Forum 2000; 331-337:1413-1426.

[32] Molent, L., Barter, S.A. and Wanhill, R.J.H., The lead crack fatigue lifing framework, International Journal of Fatigue, Vol. 33, 323–331, 2011.

[33] Molent, L., A review of equivalent pre-crack sizes in aluminium alloy 7050-T7451, Fatigue and Fracture of Engineering Materials and Structures, Vol. 37, 1055-1074, 2014.

[34] Loader C, Crawford BR, Sharp K and Molent L. A proposal for the management of detected pitting corrosion for the F/A-18 Hornet, Defence Science and Technology Organisation, DSTO-TR-2976, 2014.

[35] Wanhill RJH. Fractography of fatigue crack propagation in 2024–T3 and 7075–T6 aluminum alloys in air and vacuum. Metall Trans A 1975;6A:1587–96.

[36] Molent, L., Managing airframe fatigue from corrosion pits – a proposal, Engineering Fracture Mechanics, 2015; 137: 12–25.

[37] Design and Airworthiness Requirements for Service Aircraft Volume 1 – Aeroplanes. Defence Standard 00-970, Issue 1, December 1983.

[38] Aircraft Structural Integrity Management Plan (ASIP), USAF, MIL-STD-1530C, November 2005.

[39] Smith, S.W., Newman, J.A. and Piascik, R.S., Simulation of fatigue crack initiation at corrosion pits with EDM notches, NASA/TM-2003-212166. March 2003.

[40] Burns, J.T., Larsen, J.M. and Gangloff, R.P., Driving forces for localized corrosion-to-fatigue crack transition in Al–Zn–Mg–Cu, Fatigue and Fracture of Engineering Materials and Structures, Vol. 34, 745-773, 2011.

[41] Wallace, W., Hoeppner, D.W. and Kandachar, P., AGARD Corrosion Handbook Vol 1: Aircraft Corrosion: Causes and Case Histories, AGARD-AG-278, Advisory Group for Aerospace Research and Development, Neuilly-sur-Seine, France, 1978.

Fatigue Crack Growth from Environmentally Induced Damage in 7075 Alloy

Michael Lo[1], Rhys Jones[1], Daren Peng[1], Adam Bowler[2], Dustin Edwards[3]

[1]*Department of Mechanical and Aerospace Engineering, Monash University, Wellington Rd, Clayton, Vic 3800, Australia*
[2]*Maritime Systems Project Office, RAAF Base Edinburgh, Edinburgh, SA 5111, Australia*
[3]*Airbus Australasia Pacific, PO Box 49, RAAF Edinburgh, SA 5111, Australia*

Abstract: Ageing aircraft fleets commonly experience environmentally induced degradation. This requires an accurate assessment of fatigue performance including any influence of corrosion damage. Since the majority of the life of a fatigue crack is spent in propagation of small cracks, the predictive capabilities of crack growth laws must ideally capture small and intermediate crack behaviour as well as long crack growth for fatigue life assessment. Here, studies were conducted on corroded coupons including "dome nut hole" coupons with intergranular corrosion (IGC) representative of a critical location in the RAAF AP-3C Orion aircraft. The examination of cracks grown from corrosion with quantitative fractography revealed that fatigue cracks can nucleate from corrosion sites and are strongly affected by the morphologies and geometries of the corrosion damages. Further, the locations of the crack origins and the properties of the local microstructure can have a significant impact on the initial crack propagation. The presence of an intergranular corrosion (IGC) region at the dome nut hole can influence the fatigue crack paths such that multiple cracks originating from corrosion sites can be segregated by the IGC. Beyond the depth of the IGC, coexisting cracks coalesce and propagate as a single main crack. The experimental crack growth measurements and predictive techniques investigated herein confirms that a Hartman-Schijve variant of the NASGRO equation can accurately capture small, intermediate and long crack growth, and therefore allow for the total fatigue life to be assessed.

INTRODUCTION

Environmentally incited degradation is a prevalent issue which can be detrimental to the integrity and fatigue life of aircraft structures. Aviation operators employ a damage tolerant approach, namely a Safety by Inspection method, in managing the occurrence of corrosion [1-6]. In many instances the detection of corrosion can lead to unscheduled servicing of an aircraft to remove the corrosion damage and therefore lead to increased costs, unplanned for maintenance and unavailability of the aircraft [1-6]. Given these drivers, there is an ongoing necessity to develop effective methods for both detecting and assessing the impacts of corrosion damage. To this end, the current research addresses the effect of environmentally assisted degradation on fatigue crack nucleation and propagation in aircraft structures.

In this paper, two forms of localised corrosion damage of 7075-T6 and 7075-T651 are examined, namely, corrosion pitting and intergranular corrosion (IGC). A particular interest of the work is the effect that IGC has on a "dome nut" fastener hole detail representative of the structural geometry that is found at a critical location of the RAAF AP-3C Orion maritime patrol aircraft. Some instances of intergranular corrosion of up to 5 mm in depth have been recorded and IGC is a widespread problem on the AP-3C Orion fleet [7]. The P-3C Service Life Assessment Program (SLAP) was developed and one early focus of this work was to establish tools for assessing/predicting crack growth in critical areas of the aircraft [8-10]. In line with this program, for extended operational life beyond the designed safe life, it is necessary to account for the impact of the IGC at flight critical locations by the introduction of inspections.

The maritime operational environment of AP-3C Orion is a significant contributor to the corrosion observed in structural locations [11]. Corrosion monitoring of the AP-3C fleet revealed that corrosion appears to be largely a ground and post-flight activity [11, 12]. This reported data supported the proposition that because of cold and dry environment, corrosion does not occur inflight [13]. A number of other independent findings in Refs [2-4, 14-16] suggest further evidence that fatigue and corrosion are predominantly decoupled, although there is also some evidence that they cannot be completely decoupled for all ageing aircrafts [17]. Nevertheless, for the purpose of the results and the current discussion presented here, the studies of fatigue and corrosion processes are decoupled and investigated as separate activities.

Corrosion damage is detrimental to the strength and fatigue life of structural components. If not mitigated, corrosions can cause severe structural issues in aircrafts [2-7, 18]. Corrosion degradation has been proven to initiate fatigue cracks and drastically reduce fatigue life in high strength aluminium alloys [2-5, 19]. Corrosion pits often nucleate multiple small fatigue cracks that simultaneously grow and coalesce to form larger cracks [2-5, 19]. Parameters such as the size and depth of the corrosion pits are generally not a complete indicator of the significance of the impact of corrosion damage on structural fatigue performance [20]. Several critical factors including specific positions of crack origins relative to irregularly formed corrosion pits and the properties of the local microstructure in the region of stress concentrating discontinuities, described by the fatigue threshold parameter, can also have a substantial influence on crack propagation [19, 21-24].

Although significant intergranular corrosion has been identified on AP-3C Orion aircrafts, specifically at fastener joints and fatigue critical areas [7, 25], the effects of the IGC on fatigue crack life and structural integrity has not been adequately quantified. Recently theoretical models revealed that, to some extent, intergranular corrosion on its own has minimal effect on the stress state around fastener holes subjected to representative operational flight loads [24-26]. In a related experimental study, a four week intergranular corrosion protocol was established by DSTO (Defence Science and Technology Organisation) in order to produce IGC at dome nut fastener holes (DNH), see Figure 6. 7075-T651 DNH coupons were filled and refilled weekly with a candidate solution consisting of 0.6 M sodium chloride, 0.1 M sodium bicarbonate, 0.03 M sodium nitrate, 0.03 M sodium sulphate [7, 25, 27]. To prevent the corrosion activity continuing, i.e. during fatigue testing, the coupons were heated in a furnace at 50 °C for a further eight weeks at the conclusion of the four week protocol.

A Hartman-Schijve variant of NASGRO Fatigue Crack Growth Formulation

The methodology by which small fatigue crack is measured is dependent on the quantities of interest, and can be defined as either microstructurally small, chemically small, small relative to the plastic crack tip field, or physically small [28, 29]. The majority of the life of a fatigue crack is consumed in propagation of small cracks (up to about 1 mm), consequently the understanding of small crack growth behaviour is crucial for fatigue life assessment [29]. When assessing the service life of an aircraft, the ability to capture small, intermediate and long crack growth is critical. Any such analysis should use a single formulation which should be accurate for small cracks, the transition from small to long, and capture the long crack growth behaviour.

In the following research the interaction between fatigue crack growth and pitting and intergranular corrosion is assessed and characterised by an empirical crack growth formulation. It has been documented in literature that the relationship between fatigue crack growth rate da/dN and stress intensity factor ΔK of 7075-T6 and 7075-T651 test coupons can be accurately represented by the Hartman-Schijve (HS) variant of the NASGRO formulation, which is given by

$$da/dN = D\left(\frac{\Delta K - \Delta K_{thr}}{\sqrt{1-K_{max}/A}}\right)^2 \qquad (1)$$

where $D = 1.86\times10^{-9}$, and $A = 47$ MPa m$^{1/2}$ [19, 21-24].

It is known from Refs [21-24] that this formulation (1) can capture the variation in the crack growth rates from small to long cracks by allowing for appropriate changes in the fatigue threshold parameter ΔK_{thr}. Typically, small cracks exhibit lower thresholds stress intensity levels than long cracks [21-24, 30]. Small crack growth also exhibit minimal plasticity induced crack closure (PICC) behaviour [21, 22, 28, 29]. The variant of the HS formulation (1) evaluated in the present crack growth predictive analyses has not accounted for PICC, however for analysis where PICC is significant, the crack-closure based fatigue crack growth formulation should be used as described in Refs [21, 22].

This paper will reveal that the variability in the small and intermediate crack growth rates seen in the 7075-T6 and 7075-T651 fatigue test coupons examined here can be accurately predicted using Equation (1) allowing for the correct fatigue threshold values. The effect of corrosion degradations such as pitting and intergranular corrosion on fatigue crack growth life and associated difficulties in assessment of structural integrity will also be discussed.

LOCALISED PITTING CORROSION

Alloys such as 7075 have a layer of passivating oxide film on their surface which aids in protecting the material against corrosion. Pitting corrosion can take place at small localised areas with high concentration of anions, e.g. chloride, or low concentration of oxygen, and generally where inconsistencies in the surface are present such as grain boundaries, inclusions or other crystal discontinuities [31-33]. Thus, while the majority of the metal surface is protected by the oxide film, pitting corrosion attacks tend to occur at any regional disparities in the oxide layer through which corrodents can reach the metal surface, this in turn results in degradation of the surrounding oxide film, and corrosion of underlying alloy [31-33].

In order to investigate the effect of prior corrosion damage on fatigue crack growth, fatigue test coupons in the form of dog-bone coupons with the nominal thickness of 4.60 mm were machined from 7075-T6 extruded sheet. The test coupons were pre-corroded with 0.606 M sodium chloride droplet for exposure of 168 hours on the surface of the L-T plane on the test section area, see Figure 1. These test coupons were then fatigue tested at remote peak stress of 250 MPa using a cyclic loading sequence designed to mark the surface of any cracks produced. The marker loading sequence involved two alternating load ratios $R = 0.1$ and $R = 0.8$. The marks produced on the fatigue crack surfaces were used for failure analysis and crack growth measurement via quantitative fractography [34-35], see Figure 2. The experimental results of four test coupons are tabulated on Table I.

The corrosion pits produced here led to fatigue crack nucleation and propagation. Quantitative fractography of the fracture surfaces revealed that in each coupon fatigue crack nucleation occurred about the substantial pitting damages, see Figures 3 and 4. The corrosion pits appeared as small cavities which had a variety of profiles, and were mostly irregularly shaped. In general, some forms of intergranular corrosion are present, particularly in the deeper regions of the pitting sites. Fatigue cracking in the test coupons grew from the surface in the vicinity of the localised pitting damage due to higher local stress concentration, or K_t, in the material under fatigue loading, consistent with the findings in Refs [2-5, 13, 34, 35].

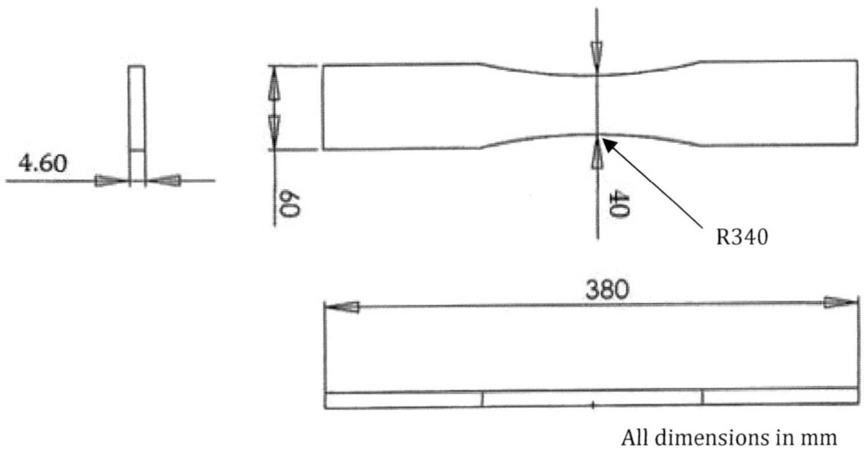

Figure 1. Specification of the 7075-T6 pitting corrosion fatigue test coupon.

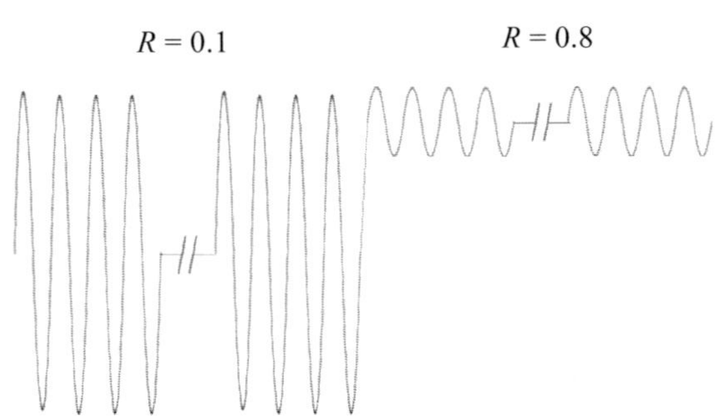

Figure 2. Illustration of the loading block where the R=0.8 loads produce a marker band in a single application of the block used in the fatigue tests.

Table I. Summary of stresses, loading conditions and failure mechanisms of pitting corrosion coupons.

Coupon ID	Remote peak stress (MPa)	Number of cycles at respective *R* ratios in one loading block		Fatigue life in number of blocks to failure	Failure mechanism
		***R* = 0.1**	***R* = 0.8**		
DBS-2-9	250	3000	12000	23	Cracking from corrosion pit with crack-like shape of about 20 μm in depth; pit depth closely matches the EPS of 18 μm
DBS-2-15	250	3000	15000	37	Cracking from corrosion pit with irregular shape and IG attack. Pit depth of about 50 μm was similar to the EPS of 57 μm
DBS-2-16	250	3000	15000	38	Cracking from an origin in the region between two relatively large corrosion pits. Initial flaw depth does not match the EPS of 57 μm
DBS-2-17	250	3000	15000	25	Cracking from non-crack-like shallow and wide pit of about 62 μm in depth. Pit depth does not match EPS of 17 μm

The optical micrographs shown in Figures 3 and 4 indicate that in some occurrences, the small initial fatigue cracks traversed around corrosion damage in order to propagate and grow, for instance, in DBS-2-15 and DBS-2-16 on Figures 4 (a) and (b) respectively. In the DBS-2-16 coupon on Figure 4(b), an origin of the main fatigue crack is located in a region on the surface of the material adjacent to a corrosion pit. This region is a small but sharply shaped ligament where there is a higher local stress concentration. The fatigue crack path is strongly impacted by the jarring features of the pitting and the resultant small crack growth is uneven and tortuous. Thus, these corrosion pits are usually non-ideal crack starters due to their rough and irregular geometries as discussed further below.

Fatigue crack growth rates from corrosion pits may depend on whether the pit is "crack-like" [2-5, 20]. As described in Refs [2-5], a pit which is crack-like should have a single crack initiation origin, a stable crack front without local interference, as well as an optimal position and location of the crack origin relative to the microstructural orientations and local stresses. For non-crack-like pits, the interference of small cracks with origins from corrosion pits is possible and may reduce overall crack growth rates [2-5, 20]. A rough corrosion pit or multiple pitting damages can often generate multiple cracks which interfere with one another, resulting in "ill-defined crack fronts" [2-5, 20]. The initial cracks can also interact considerably with irregular shaped corrosion pits as seen in these test coupons. As shown on Figure 4 (a) and (b), for both DBS-2-15 and DBS-2-16 the fatigue lives are longer and the fatigue crack propagation rates are lower than if the pits were to have relatively more semi-elliptical profiles such as those observed in DBS-2-9 and DBS-2-17 in Figure 3 (a) and (b) respectively. Further, small crack growth is inherently affected by local microstructural properties and sub-granular features [2-5, 20, 30]. The orientations and properties of crystal lattices structures and slip bands can result in early crack growth that is highly inconstant [36-37].

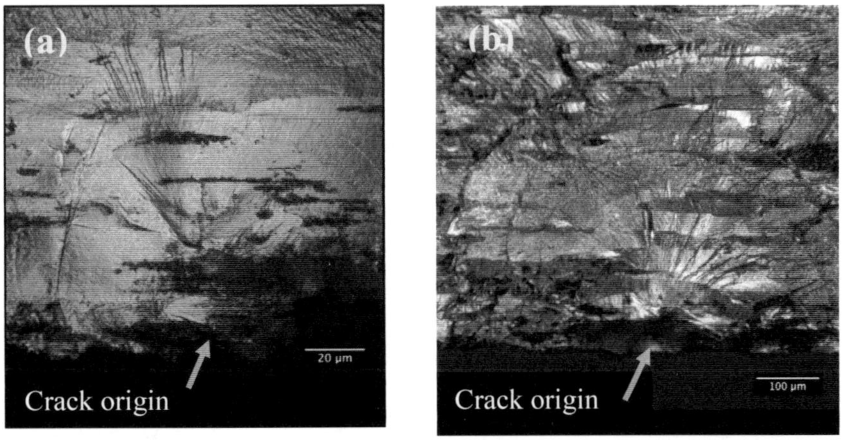

Figure 3. (a) Optical micrograph of DBS-2-9 illustrating cracking from a corrosion pit with depth of approximately 20 µm. (b) Optical micrograph of DBS-2-17 displaying fatigue crack growth from shallow and wide pit about 62 µm in depth. Note the bands on the fatigue surface from the marker band loading.

The ability of the modified NASGRO formulation, i.e. Equation (1), to replicate the measured crack growth histories is evident in Figure 5. This technique required considerations of appropriate initial crack size and fatigue threshold parameter ΔK_{thr} to account for the varying crack growth behaviours. The Equivalent Pre-crack Size (EPS) approach was used here where the physical length of the crack was approximated by an initial crack length based on fatigue crack growth formulation (1) [2-5, 20, 34, 35]. For instance in the DBS-2-9 coupon, fatigue cracking initiated from corrosion pit with a semi-elliptical crack-like shape of approximately 20 µm in depth and the EPS of 18 µm closely matched this depth. However, there are generally discrepancies between EPS and measured corrosion pit depth. For the DBS-2-17 coupon, the fatigue cracking originated from a non-crack-like shallow and wide pit of about 62 µm in depth, and the EPS of 17 µm does not match this pit depth. As a consequence, the correlation between EPS and corrosion pit depth size is usually difficult to determine since the cracks tend to grow slowly about the more complex pits discussed above [2-5, 20].

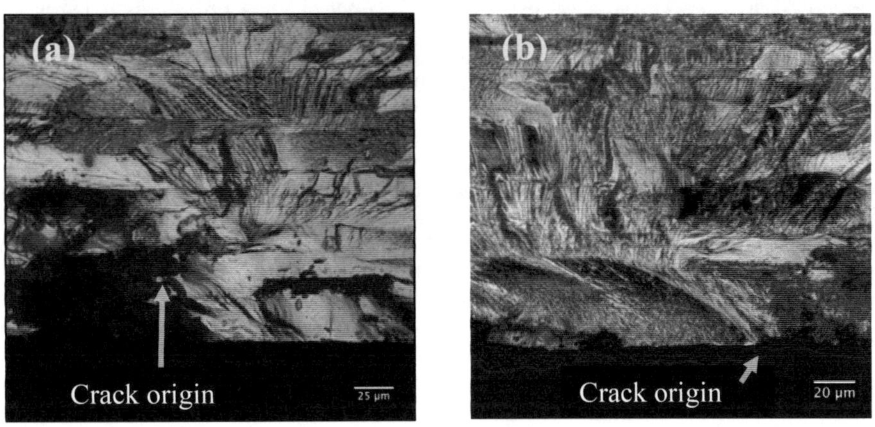

Figure 4. (a) Optical micrograph of DBS-2-15 illustrating cracking from a corrosion pit with irregular shape and IG attack. The pit depth was about 50 µm. (b) DBS-2-16 displaying fatigue crack growth from an origin in the region between two relatively large corrosion pits.

For the current model, in order to allow for the numerous origins and multiple crack paths associated with where fatigue cracks at corrosion pits nucleate, the resultant variability in crack growth was quantified by the appropriate initial crack size (EPS) and the fatigue threshold parameter used in the formulation (1) [19, 21-24]. Figure 5 shows that the formulation (1) can closely capture the nature of the varying crack growth lives for small cracks through their transition to long cracks. Although some information about the pit depth or the geometry of the pit is known, they may not completely describe the behaviour of the fatigue crack growth. Particularly for small to intermediate crack depths, where the precise interactions between the fatigue cracks, pitting corrosion and material microstructures are complex. This is an area that can be further studied.

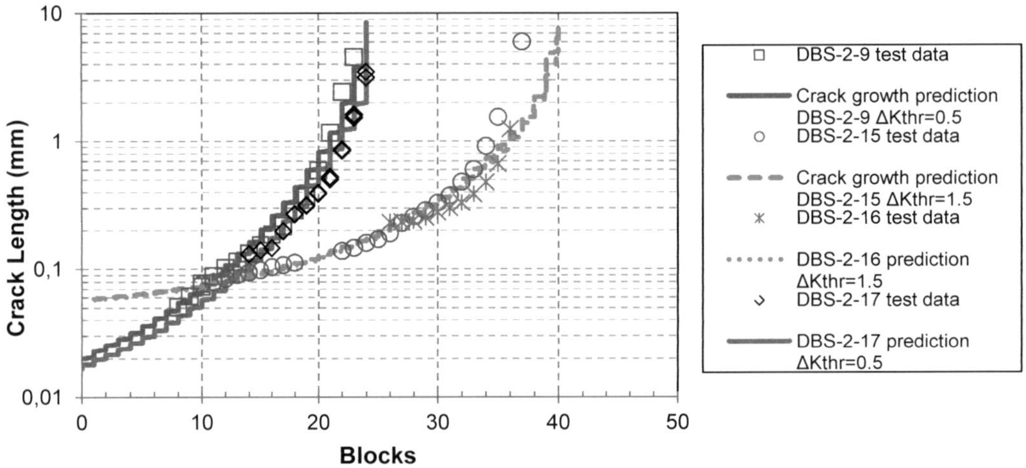

Figure 5. Measured fatigue crack growth and life prediction curves of pitting corrosion test coupons.

INTERGRANULAR CORROSION AT FASTENER HOLE

In aircraft structures, Intergranular Corrosion (IGC) is a localised microstructural phenomenon where preferential corrosion occurs at the vicinity of grain boundaries. Cold work such as extrusion or rolling that produces anisotropic forms out of high strength aluminium alloys will result in long, flat grains in the microstructure that are prone to intergranular and exfoliation corrosion in the peak aged tempers. IGC can be instigated by corrodents such as aqueous chlorides that depoleris the regions either at the grain boundaries (2XXX aluminium alloys) or the regions adjacent to the grain boundaries (7xxx aluminium alloys) [6, 33]. In the case of the 7XXX alloys, dissolution progresses at the grain boundaries in a precipitate free region adjacent to coarse grain boundary precipitates [6]. The precipitates contain elements such as copper and zinc and are formed during heat treatment by denuding the adjacent material of these elements, thereby producing a galvanic couple between the more anodic precipitate free aluminium and more cathodic coarse grain boundary precipitates [6, 32, 33]. Residual stresses which are attributable to stress corrosion cracking (SCC) can add to the rate of attack so that heavily worked 7075-T6 and 7075-T651 (peak aged) aluminium alloys are particularly vulnerable to exfoliation and intergranular attack [33].

The 7075-T651 DNH coupons tested here were machined from RAAF AP-3C Orion wing panels, and the schematic is illustrated on Figure 6. Note however that the average thickness of these DNH coupons were approximately 1.40 mm rather than the nominal thickness of 2.032 mm due mainly to manufacturing difficulties. The DNH coupons were corroded according to the DSTO IGC protocol as discussed previously [7, 25, 27]. The fatigue tests of pre-corroded DNH coupons were conducted with the marker band loading sequence with two alternating *R* (load) ratios for identification of crack growth with quantitative fractography, see Figure 2. Table II outlines the parameters and results of six fatigue tests. Here the marker bands at small crack sizes of below approximately 100 μm were not clearly distinguishable under quantitative fractography, since there was evidence of surface roughness in the initial cracks, as well as interference of multiple small cracks on some fracture surfaces of these test coupons.

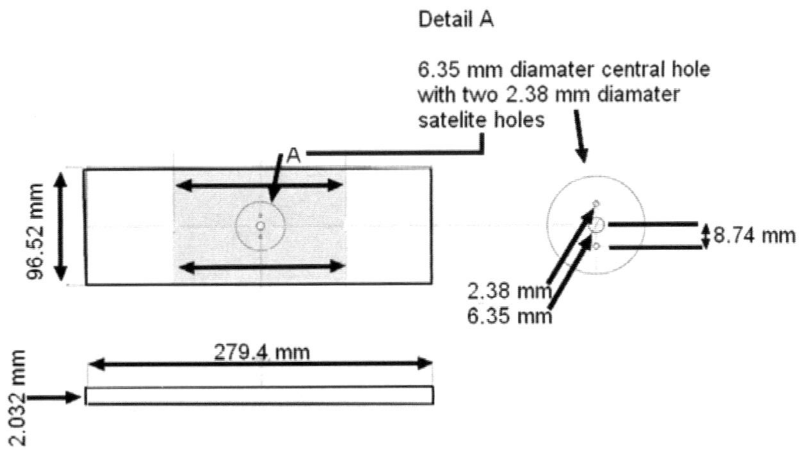

Figure 6. Schematic of the 7075-T651 dome nut hole (DNH) coupon used in analysis of the interaction between IGC and fatigue crack growth.

Figure 7 depicts Scanning Electron Microscope (SEM) fractographs of fatigue cracking at an inner ligament of the pre-corroded dome nut hole coupon DNHS-1-1 tested at a remote peak stress of 150 MPa. A large transgranular corrosion pit and a number of layers of intergranular corrosion were observed at the bore of the central hole and are evident in the figures. The roughly formed corrosion pit is likely to have initiated multiple out-of-plane fatigue cracks. The small cracks near the bore of the hole could be further influenced by an intergranular corrosion of about 0.5 mm in depth as shown on the figure. Under fatigue cracking, an IGC region can act as a weak free boundary providing partitioning between two cracking ligaments on either side of the IGC, described as "pinning" in [25]. In DNHS-1-1, the fatigue cracks in both ligaments grew simultaneously and eventually coalesced to form a single crack away from the depth of the IGC. The fatigue crack transitioned to fast fracture as the magnitude of the crack depth approaches the through-section thickness.

Table II. Summary of stresses, loading conditions and failure mechanisms of pre-corroded dome nut hole coupons.

Coupon ID	**Remote peak stress (MPa)**	**Number of cycles at respective R ratios in one loading block**		**Fatigue life in number of blocks to failure**	**Failure mechanism of first inner ligament to show fatigue cracking**
		$R = 0.1$	$R = 0.8$		
DNHS-1-1	150	3000	15000	6 (116173 cycles)	Cracking from corrosion damage at the bore of the central hole with IGC depth of ~0.5 mm
DNHS-1-6	150	3000	15000	8 (145,327 cycles)	Cracking from surface corrosion damage near the central hole
DNHS-1-7	150	3000	15000	5 (92,932 cycles)	Cracking from corrosion damage at the corner of the central hole
DNHS-1-2	80	3000	15000	33* (594000 cycles)	Cracking from surface corrosion damage near the central hole
DNHS-1-3	80	3000	15000	44* (795000 cycles)	Cracking from surface corrosion damage near the central hole
DNHS-1-4	80	3000	15000	90 (1655736 cycles)	Cracking from corrosion damage at the bore of the central hole with IGC depth of ~0.4mmm

*number of blocks until failure at first inner ligament to show fatigue cracking only.

Figure 7. Secondary and backscatter SEM fractographs (a) and (b) respectively showing fatigue cracking in DNHS-1-1. Fatigue cracks initiated from out of plane locations around pitting and intergranular corrosion at the bore of the central hole.

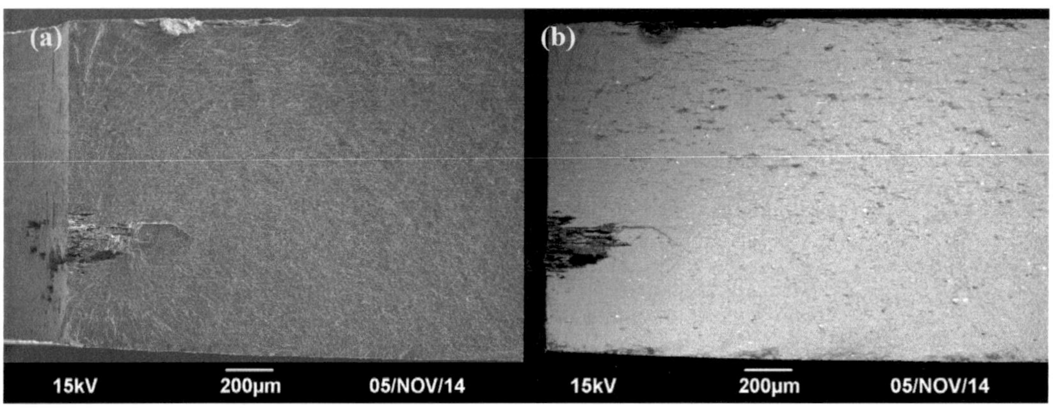

Figure 8. Secondary and backscatter SEM fractographs (a) and (b) respectively showing fatigue cracks in DNHS-1-4 initiating from a corrosion pit at the bore of the central hole.

A similar phenomenon as above was also present in the DNHS-1-4 coupon as shown on Figure 8. The corrosion damage in the DNHS-1-4 coupon consisted of a relatively large pitting and a co-located IGC at the centre of the bore of the hole. The corrosion damage created multiple initiation sites for fatigue cracks. These small fatigue cracks propagate concurrently and gradually coalesced to form larger cracks. Evidence of the IGC that was about 0.4mm deep from the bore of the hole at the base a corrosion pit is revealed on the backscatter SEM fractograph on Figure 8 (b). Consistent with DNHS-1-1, the presence of an IGC partitioned the paths of multiple fatigue cracks up to the depth of the IGC. Beyond that depth, the slightly off-plane cracks coalesced and propagated as a single main crack. As expected in these results, the IGC and the fatigue cracks primarily occurred in orthogonal geometrical planes such that the IGC is located in the L-T plane whereas fatigue crack growth is situated in the T-S plane, similar to the findings in Ref [25].

The fatigue crack surfaces of DNHS-1-2 and DNHS-1-3 are shown in Figures 9 (a) and (b) respectively. The dominant fatigue cracks resulted from the growth of small cracks from two distinct origins in DNHS-1-2 and three distinct origins in DNHS-1-3. In both of the coupons, the small cracks nucleated from corrosion pitting damage located on the surface and grew with little interaction with the corrosion and minimal interference between the multiple small cracks. The propagation behaviours of these fatigue cracks were similar to corner cracks at the bore of the holes. The DNHS-1-4 coupon ran for a significantly longer number of cycles than DNHS-1-2 and DNHS-1-3, see Table II. Although the large corrosion pit in DNHS-1-4 acted as stress concentrator and initiated fatigue cracks, it appears that it was a less efficient crack starter than those of the other two coupons. The interference between multiple small initiating cracks was evident as indicated by the rough and tortuous initial crack surface as well as the coarsely formed crack fronts, as has been seen elsewhere [2-5, 34, 35].

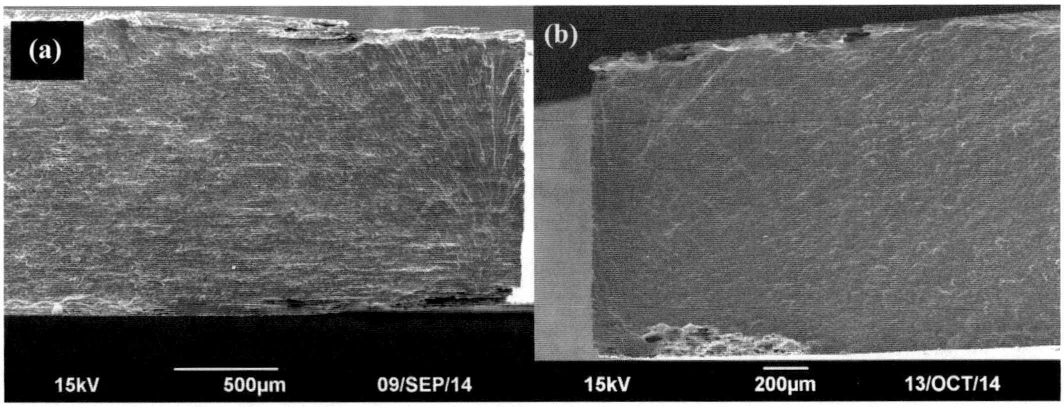

Figure 9. Secondary SEM fractographs showing fatigue cracks initiating from corrosion damage on the surface near the central holes in dome nut hole coupons (a) DNHS-1-2 and (b) DNHS-1-3.

In these results, the corrosion pits created larger discontinuities than the IGC and these damages were located closer to the bore of the fastener holes. Thus pitting corrosion has proved to be the dominant form of stress concentrator and the main source of crack initiation in these tests. While IGC can also initiate fatigue cracks, in contrast, the presence of a typical intergranular corrosion region is found to generally have a relatively minor effect on the stress states [24-26]. However, IGC is a potential fatigue crack initiating defect and can have a range of influences on the crack growth behaviour [25]. In cases where the fatigue cracks "cut" through thin layers of IGC, the IGC can have minimal effect on fatigue crack growth [25]. Conversely, the segregation of fatigue cracks by IGC in coupons DNHS-1-1 and DNHS-1-4 can cause unusual crack growth paths and can affect the rate of fatigue crack propagation, in conjunction with the multiple and irregularly formed cracks that originated from the corrosion pits as seen here.

Figure 10 presents a plot of the measured and computed crack length histories using Equation (1) with the same values of D and A used to represent large crack growth in Ref [24], viz: $D = 1.86\times10^{-9}$, and $A = 47$ MPa m$^{1/2}$. The DNHS-1-2 and DNHS-1-3 coupons were modelled as corner cracks at the bore of the central hole while the DNHS-1-4 coupon was modelled as a semi-elliptical crack at the centre of the bore of the central hole. This analysis further demonstrates that the correlation between fatigue crack growth rate da/dN and stress intensity factor ΔK can be accurately described by the Hartman-Schijve (HS) variant of the NASGRO formulation (1) and that the scatter in crack growth rates can be captured by allowing for the associated initial crack size (EPS) and the fatigue threshold ΔK_{thr} parameter.

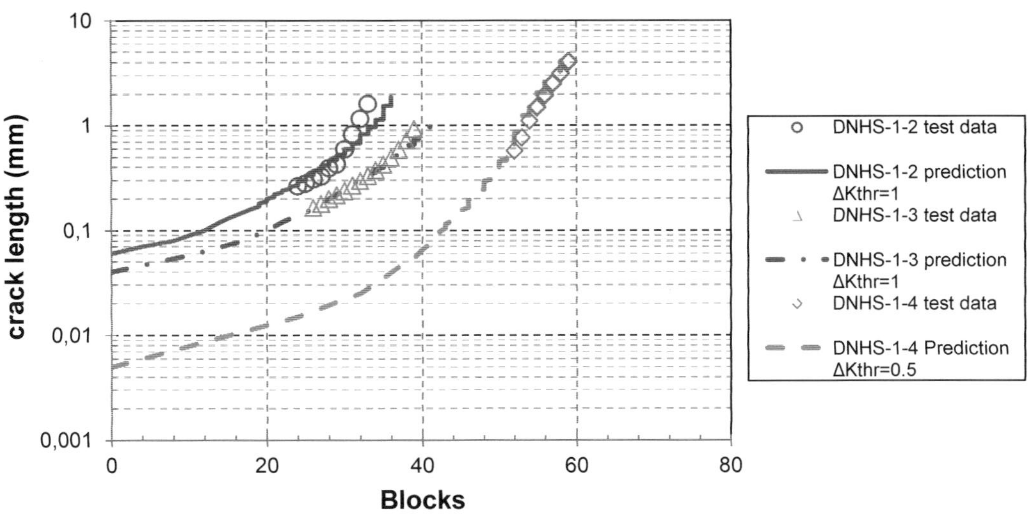

Figure 10. Fatigue crack growth and life prediction of corroded dome nut hole test coupons.

For an associated EPS, the proposed fatigue threshold compatibly accounts for the non-homogenous local microstructural grain orientations as well as corrosion pits which are usually non-ideal crack starters [21-24]. As shown on Figure 10, the physical depth of corrosion damage and initial crack depths in the analytical prediction (EPS) generally do not match, and as discussed above, the physical size of the discontinuity is not the only feature that can strongly dictate fatigue crack growth behaviour. Therefore, for the purpose of predicting the entire scope of the aircraft life and crack growth history, the formulation (1) accounts for the varying EPS and the fatigue threshold parameter of fatigue crack growth at the corrosion sites. The experimental results prove that Equation (1) can predict from small to long crack growth and the total fatigue life. In circumstances where the life of an aircraft is largely dependent on environmentally initiated fatigue crack growth, the present fatigue failure analysis technique requires a definitive understanding of the impact of the diverse morphologies of pitting and intergranular corrosions.

CONCLUSIONS

The research presented here is focused on the effect of environmental degradation on aircraft structural integrity. Under conditions where the surface protection systems have failed and corrosion exists in the airframe, fatigue cracks can initiate in the vicinity of the localised corrosion damage. The origins of fatigue cracks are typically located at material or corrosion features that when fatigue loaded produce high stress concentrations. In such cases, the occurrences of multiple crack origins around irregularly formed corrosion pitting are typically observed. Fatigue crack growth rates from corrosion pits can depend significantly on whether the morphologies of the sites are optimal as crack starters. The interferences of small cracks growing around the pits have the potential to reduce overall crack growth rates while the cracks are small. Further, to some extent, the presence of IGC that results in disconnected ligaments in the corroded area can produce deep penetrations which act as weak free boundaries segregating the propagation paths of multiple fatigue cracks. The multiple fatigue cracks typically coalesce beyond the depth of intergranular corrosion. For these reasons and in concurrence with factors including features in the local granular microstructures, the growth rates of small to intermediate cracks can fluctuate drastically.

When assessing the fatigue life of a flight critical structural component, the associated analysis tools for predicting the fatigue crack growth should be appropriate for the full scope of small to long cracks. In the present class of problems this means that furthering the understanding of impact of the different morphologies of environmentally produced degradation is required. This study has shown that the Hartman-Schijve variant of the NASGRO equation, i.e. Equation (1), can reasonably capture the entire gamut of the crack growth history as it grows from a small to a long crack and eventually to final failure. As such it is expected that this approach would contribute to improvements in techniques to manage structural integrity where the life is determined by the growth of a fatigue crack that initiates from corrosion pitting damage as well as corrosion damage which also has co-located intergranular degradation.

ACKNOWLEDGEMENT

This work was performed under funding from RAAF MPSPO and Airbus Group Australia Pacific. The authors would like to thank Dr Simon Barter, Mr Chris Loader, Mr David Gouldie, Ms Maria Salagaras and Mr Andrew Walliker of DSTO, and Prof. Raman Singh of Monash University for their help and advice.

REFERENCES

[1] Shekhter, A., Loader, C., Hu, W. and Crawford, B. (2007). DSTO-TR-2080, Defence Science and Technology Organisation.

[2] Molent, L. (2013). In: Proceeedings 15th Australian International Aerospace Congress (AIAC15), Melbourne Convention Centre, Australia.

[3] Barter, S. A. and Molent, L. (2013). *Eng. Fail. Anal.*, vol. 34, pp. 181–188.

[4] Barter, S.A., Molent, L. (2014), *Eng. Fail. Anal.* Vol. 39, pp. 155-163, ISSN 1350-6307.

[5] Molent, L. (2014). *Eng. Fract. Mech.* ISSN 0013-7944.

[6] Sudholz, A., Knight, S., and A. Trueman, A. (2011), In: 18th International Corrosion Congress, Paper 224.

[7] Salagaras, M., Wythe, A. & Trathen, P. (2012). In: ACA Corrosion & Prevention 2012, Paper 120, The Australasian Corrosion Association Incorporated, Melbourne, Victoria, Australia.

[8] Iyyer, N. S., Kwon, Y. S, and Phan, N. (2003). In: 2003 International Committee on Aeronautic Fatigue, Lucerne, Switzerland.

[9] Iyyer, N.S., Sarkar S., Merrill R., Phan N. (2007). *Int. J. of Fatigue.*, vol. 29, pp. 1584–1607.

[10] Veul, R. and Ubels, L. (2003). NLR, NLR-CR-2003-488, p. 163, Netherlands.

[11] Trathen, P. (2011). In: Proceedings 18th International Conference on Corrosion, Perth, Australia.

[12] Trathen, P.N. (2009), In: Materials Forum Vol 33, pp. 450-456. Galea, S., Chiu, W.K. and Mita, A. (Eds.), Institute of Materials Engineering Australasia Ltd.

[13] Burns, J. T., Gangloff, R. P. and Bush, R. W. (2011) In: Proc. DoD Corrosion Conference, La Quinta Ca.

[14] Du, M. L., Chiang, F. P., Kagwade, S. V. and Clayton, C. R. (1998). *Int. J. Fatigue*, vol. 20, pp. 743–748.

[15] Wei, R.P. and Harlow, D.G. (2006). DOT/FAA/AR-95/76, U.S. Department of Transportation Federal Aviation Administration.

[16] Barter, S. A. and Molent, L. (2012). In: Proceedings 28th International Congress of the Aeronautical Sciences, Brisbane.

[17] Baldwin, J. D. (2002), Air Force Office of Scientific Research (AFOSR).

[18] Hoeppner, D.W., Grimes, L., Hoeppner, A., Ledesma, J., Mills, T., and Shah, A. (1995), In: Proceedings of the 18th Symposium on the International Committee of Aeronautical Fatigue, (ICAF '95), Grandage J.M. and Jost, G.S. (Eds.), vol. 1 of Estimation, Enhancement and Control of Aircraft Fatigue Performance, pp. 87–106, EMAS, Melbourne, Australia.

[19] Jones R, Molent, L., and Barter, S. (2013). *Int. J. of Fatigue.* vol. 55, pp. 178-182.

[20] Molent, L. (2014). *Fatigue Fract. Engng. Mater. Struct.*, vol. 37, pp. 1055–1074.

[21] Jones R. (2014). *Fatigue Fract. Engng. Mater. Struct.*, vol. 37, n. 5, pp. 463–483.

[22] Jones R. and Tamboli D. (2013). *Eng. Fail. Anal*, vol. 29, pp.149-166.

[23] Jones, R., Molent, L., and Walker K. (2012). *Int. J. Fatigue*, vol. 40, pp 43-50.

[24] Jones R., Lo M., Peng D., Bowler A., Dorman M., Janardhana M., Iyyer NS. (2013). *Meccanica*, vol. 50, n. 2, pp. 517-532.

[25] Loader, C., Goudie, D., Salagaras, M., Underwood, J. and Walliker, A. (2014). DSTO-TR-3048, Defence Science and Technology Organisation.

[26] Harrison, T.J., Crawford, B.R., Clark, G., and Brandt, M. (2014). *Adv. Mater. Research,* vols. 891-892, pp. 242-247.

[27] Braun R. (2000). *Mater. Corrosion*, vol. 51, n. 9, pp. 607-615.

[28] Blom, A. F. (1996). In: Proceedings of the Second Engineering Foundation International Conference/Workshop on Small Fatigue Cracks, Metallurgical Society Inc, pp. 623-638.

[29] Suresh, S. and Ritchie, R.O. (1996). *Int. Met. Rev.* Vol. 29, n. 6, pp.445-475.

[30] Murakami, Y. and Endo, M. (1986). In: EGF Pub. 1. pp. 275-293. Miller K.J. and de los Rios E.R. (Eds.), Mechanical Engineering Publication, London.

[31] Foley, R.T. (1986). *Corrosion*, vol. 428, n. 5, pp. 277–288.

[32] Hoeppner, D.W. and Arriscorreta, C.A. (2012). *Int. J. Aero. Eng.,* Article ID 191879.

[33] Jones, D. (1992). *Principles and Prevention of Corrosion*, Macmillan Publishing, New York, NY, USA.

[34] Barter, S.A. (2011). *Int. J. Struct. Integrity*, vol. 2, n. 3, pp. 243–263.

[35] Barter, S. (2007). PhD Thesis, Monash University, Clayton Victoria.

[36] Kamp, N., Gao, N., Starink, M.J., & Sinclair, I. (2007). *Int. J. Fatigue*, vol. 29, n. 5, pp. 869-878, ISSN 0142-1123.

[37] Proudhon, H. and Buffière, J-Y. (2003). In: Proceedings of Fatigue Crack Paths (FCP 2003), pp. 520.

AUTHOR INDEX

AUTHOR INDEX

AUTHOR INDEX

VTT Information Service
PO Box 1000
FI-02044 VTT
Finland

ISBN 978-1-5108-5774-2

CURRAN ASSOCIATES INC.
proceedings.com